2012 7th IEEE International Conference on Nano/Micro Engineered and Molecular Systems

(NEMS 2012)

Kyoto, Japan
5-8 March 2012

IEEE Catalog Number: CFP12NME-PRT
ISBN: 978-1-4673-1122-9

Copyright © 2012 by the Institute of Electrical and Electronic Engineers, Inc
All Rights Reserved

Copyright and Reprint Permissions: Abstracting is permitted with credit to the source. Libraries are permitted to photocopy beyond the limit of U.S. copyright law for private use of patrons those articles in this volume that carry a code at the bottom of the first page, provided the per-copy fee indicated in the code is paid through Copyright Clearance Center, 222 Rosewood Drive, Danvers, MA 01923.

For other copying, reprint or republication permission, write to IEEE Copyrights Manager, IEEE Service Center, 445 Hoes Lane, Piscataway, NJ 08854. All rights reserved.

***This publication is a representation of what appears in the IEEE Digital Libraries. Some format issues inherent in the e-media version may also appear in this print version.

IEEE Catalog Number: CFP12NME-PRT
ISBN 13:   978-1-4673-1122-9

Additional Copies of This Publication Are Available From:

Curran Associates, Inc
57 Morehouse Lane
Red Hook, NY 12571 USA
Phone: (845) 758-0400
Fax: (845) 758-2633
E-mail: curran@proceedings.com
Web: www.proceedings.com

2012 7th IEEE International Conference on Nano/Micro Engineered and Molecular Systems (IEEE-NEMS 2012)

Table of Contents

Portable E-Nose Based on Polymer/CNT Sensor Array for Protein-Based Detection ················ 1
P. Lorwongtragool[1,2], T. Seesaard[1], C. Tongta[3], T. Kerdcharoen[4]
[1]Mahidol University, THAILAND
[2]Rajamangala University of Technology Suvarnabhumi, THAILAND
[3]Thaksin University, THAILAND
[4]National Nanotechnology Center, Thailand

Multilayer Graphene Growth by a Metal-Catalyzed Crystallization of Diamond-Like Carbon ············ 7
P. Peng, D. Xie, Y. Yang, C. Zhou, H. Tian, T. Feng, X. Li, T. Ren, H. Zhu
Tsinghua University, CHINA

Chemical Vapor Deposition of Nanocrystalline Graphene Directly on Arbitrary High-Temperature Insulating Substrates ················ 11
J. Sun[1], N. Lindvall[1], M. T. Cole[2], K. B. K. Teo[3], A. Yurgens[1]
[1]Chalmers University of Technology, SWEDEN
[2]University of Cambridge, UNITED KINGDOM
[3]AIXTRON Nanoinstruments Ltd., UNITED KINGDOM

Characterization of Strain Fields in Graphene Films ················ 15
R. Dewanto, C. Dale, Z. Hu, N. Keegan, B. Gallacher, J. Hedley
Newcastle University, UNITED KINGDOM

Transfer-Free Fabrication of Suspended Graphene Grown by Chemical Vapor Deposition ·············· 19
N. Lindvall, J. Sun, A. Yurgens
Chalmers University of Technology, SWEDEN

Photovoltaic Response of N-Doped Graphene-Based Photodetector ················ 23
W. Wang[1,2], T. Li[1], Y. Wang[1]
[1]Shanghai Institute of Microsystem and Information Technology,
  Chinese Academy of Sciences, CHINA
[2]Graduate School of Chinese Academy of Science, CHINA

Development of Roll Press UV Imprint Process for Replication of Micro Lens Arrays on the Large and Thin Quartz Substrate ················ 27
L. Li, K. Ishii, Y. Tsutsui, S. Shoji, J. Mizuno
Waseda University, JAPAN

Fabrication Metal Roller Mold with Sub-Micrometer Feature Size Based On Contact Printing Photolithography ················ 31
K.-F. Huang, S.-W. Tsai, Y.-C. Lee
National Cheng Kung University, TAIWAN

Fabrication of Metal Embedded Photo-mask for Sub-Micrometer Scaled Photolithography and Patterning Sapphire Substrate ················ 36
J.-N. Yan, Y.-C. Lee
National Cheng Kung University, TAIWAN

Metal Contact Printing Photolithography for Fabricating Sub-Micrometer Patterned Sapphire Substrates in Light-Emitting Diodes ·········· 40
Y.-T. Hsieh, Y.-C. Lee
National Cheng Kung University, TAIWAN

Nano-Scale Mechanical Relays Fabricated by Nanoimprint Technology ·········· 45
Y. J. Chang, D. Y. Liu, C. L. Kuo
National Yunlin University of Science and Technology, TAIWAN

A Facile Nanowire Fabrication Approach Based on Edge Lithography ·········· 49
Y. Liu, W. Wang, H. A. Zhang, W. Wu, Z. Li
Peking University, CHINA

Hybrid Mask Lithography for Fabrication of Micro-Pattern with Nano-Pillars ·········· 53
S. Sakuma[1], M. Sugita[1], F. Arai[1,2]
[1]Nagoya University, JAPAN
[2]Seoul National University, KOREA

Adaptable Chip-Level Microfluidic Packaging for a Micro-Scale Gas Chromatograph ·········· 57
N. Ward[1], X. Mu[1], G. Serrano[2], E. Covington[2], C. Kurdak[2], E. T. Zellers[2], A. J. Mason[1], W. Li[1]
[1]Michigan State University, USA
[2]University of Michigan, USA

Detecting Vapor Traces of Explosives using a Self-Assembled Mono Layer on a Surface-Modified MEMS Capacitor and CMOS Electronics ·········· 61
D. Strle[1], B. Štefane[1], I. Muševič[2]
[1]University of Ljubljana, SLOVENIA
[2]Jožef Stefan Institute, SLOVENIA

A High Heating Efficiency Two-Beam Microhotplate for Catalytic Gas Sensors ·········· 65
L. Xu[1,2], T. Li[1], X. Gao[1], Y. Wang[1]
[1]Shanghai Institute of Microsystem and Information Technology,
  Chinese Academy of Sciences, CHINA
[2]Graduate School of Chinese Academy of Science, CHINA

Vibration Mode Localization in Coupled Beam-Shaped Resonator Array ·········· 69
K. Chatani[1], D. F. Wang[1], T. Ikehara[2], R. Maeda[2]
[1]Ibaraki University, JAPAN
[2]National Institute of Advanced Industrial Science and Technology, JAPAN

Piezoelectrically Actuated Circular Diaphragm Resonator Mass Sensors ·········· 73
N. Keegan, J. Hedley, Z. X. Hu, J. A. Spoors, W. Waugh, B. J. Gallacher, C. J. McNeil
Newcastle University, UNITED KINGDOM

A Carbon Nanotube Network Conjugated by Semiconductor Nanoparticles with Defined Nanometer-Scaled Gaps ·········· 78
S. Kumagai[1], N. Okamoto[2], M. Kobayashi[2,3], I. Yamashita[2]
[1]Toyota Technological Institute, JAPAN
[2]Nara Institute of Science and Technology, JAPAN
[3]The Cancer Institute of Japanese Foundation for Cancer Research, JAPAN

Closed-Looped Nano Stimulation Microscope for Living Cell Membrane ·········· 82
T. Hoshino[1], K. Morishima[2]
[1]Tokyo University of Agriculture and Technolog, JAPAN
[2]Osaka University, JAPAN

Pyramidal Nanowire Tip for Atomic Force Microscopy and Thermal Imaging ⋯⋯⋯⋯ 86
N. Burouni[1], E. Sarajlic[2], M. Siekman[1], L. Abelmann[1], N. Tas[1]
[1]University of Twente, NETHERLANDS
[2]Smarttip, NETHERLANDS

A Pyrex Nanochannel Device Fabricated by AFM Nanolithography ⋯⋯⋯⋯⋯⋯ 90
O. Hibbert, T. Busch, S. Tung
University of Arkansas, USA

Effect of Nonlinear Vibration on Double Region of Synchronized Frequency Responses in Mechanically Coupled Beam-Shaped Oscillator System ⋯⋯⋯⋯⋯⋯ 95
T. Itoh[1], D. F. Wang[1], T. Ikehara[2], M. Nakajima[1], R. Maeda[2]
[1]Ibaraki University, JAPAN
[2]National Institute of Advanced Industrial Science and Technology, JAPAN

Characterization of a Multi-Layered MEMS Pressure Sensor Using Piezoresistive Silicon Nanowire within Large Measurable Strain Range ⋯⋯⋯⋯⋯⋯ 99
L. Lou[1], S. Zhang[1], W.-T. Park[2], L. Lim[2], D.-L. Kwong[2], C. Lee[1]
[1]National University of Singapore, SINGAPORE
[2]Institute of Microelectronics, Singapore, SINGAPORE

A Novel Stretchable CMUT Array Using Liquid-Metal Electrodes on a PDMS Substrate ⋯⋯⋯ 104
X. Shi[1], C.-H. Cheng[1], J. Peng[2]
[1]The Hong Kong Polytechnic University, HONG KONG
[2]Shen Zhen University, HONG KONG

Miniature Pyranometer with Asteroid Shape Thermopile ⋯⋯⋯⋯⋯⋯ 108
J. Zhang, Z. Wu, Z. Zhao, X. Guan
Institute of Electronics, Chinese Academy of Sciences, CHINA

Design and Fabrication of Thin-Film Aluminum Microheater and Nickel Temperature Sensor ⋯⋯⋯ 112
R. Phatthanakun[1], P. Deekla[2], W. Pummara[2], C. Sriphung[1], C. Pantong[2], N. Chomnawang[2]
[1]Synchrotron Light Research Institute, THAILAND
[2]Suranaree University of Technology, THAILAND

Micro Device Array Design and Fabrication in Monolithic MEMS SoC ⋯⋯⋯⋯⋯⋯ 116
J.-H. Wen, W. Fang
National Tsing Hua University, TAIWAN

Integration the Back-Side Inclined Exposure Technology to Fabricate the 45º k-Type Prism with Nanometer Roughness ⋯⋯⋯⋯⋯⋯ 120
K.-Y. Hung, Y.-W. Tsai, C.-F. Lee, Y.-H. Chu
Ming-Chi University of Technology, TAIWAN

Study on Piezoelectric Properties of Near-Field Electrospinning PVDF/MWCNT Nano-Fiber ⋯⋯⋯ 125
Z. Y. Ou[1], Z. H. Liu[1], C. T. Pan[1], L. W. Lin[2], Y. J. Chen[1], H. W. Lai[1]
[1]National Sun Yat-Sen University, TAIWAN
[2]University of California, Berkeley, USA

Local Ablation by Micro-Electric Knife ⋯⋯⋯⋯⋯⋯ 129
Y. Yamanishi, H. Kuriki, S. Sakuma, K. Onda, F. Arai
Nagoya University, JAPAN

In-Vitro Transgenic Mice Liver Tissue Culture Via Hydrodynamic Flow Perfusion Bioreactor ········· 133
C.-W. Wu[1], S. Sivashankar[1], S. Valagerahally Puttaswamy[1], H.-L. Lin[2], K.-W. Chang[1], C.-T. Yeh[2], C.-H. Liu[1]
[1]National Tsing Hua University, TAIWAN
[2]Chang Gung Memorial Hospital, TAIWAN

High Quality Factor Parylene-Based Intraocular Pressure Sensor ·········· 137
J. C.-H. Lin[1], Y. Zhao[1], P.-J. Chen[2], Y.-C. Tai[1]
[1]California Institute of Technology, USA
[2]Robert Bosch LLC, USA

Light-Addressable Electrochemical Micropatterning of Cell-Encapsulated Alginate Hydrogels for Cell-Based Microarray ·········· 141
S.-H. Huang[1], H.-T. Chu[1], C.-W. Wu[2], Y.-Y. Chuang[2]
[1]National Taiwan Ocean University   , TAIWAN
[2]Ming Chuan University, TAIWAN

A Free-Standing and Flexible Parylene PCR Device ·········· 146
P. Satsanarukkit, H. Lo, Y.-C. Tai
California Institute of Technology, USA

The Effect of Cytochalasin D on F-Actin Behavior of Single-Cell Electroendocytosis using Multi-Chamber Micro Cell Chip ·········· 150
R. Lin, D. C. Chang, Y.-K. Lee
Hong Kong University of Science and Technology, HONG KONG

Cryogenic Frozen Device for Hepatocyte Culture and Responses ·········· 154
L.-Y. Ke[1], Y.-S. Chen[1], J. Liu[2], C.-H. Liu[1]
[1]National Tsing Hua University, TAIWAN
[2]Tsinghua University, CHINA

3D Microstructure Integrated Bioreactor System for Transgenic Mice Thick Liver Tissue Culture ·········· 158
S. Sivashankar[1], S. Valagerahally Puttaswamy[1], H.-L. Lin[2], S.-M. Yang[3], H.-P. Chen[1], C.-T. Yeh[2], C.-H. Liu[1]
[1]National Tsing Hua University, TAIWAN
[2]Chang Gung Memorial Hospital, TAIWAN
[3]National Chiao Tung University, TAIWAN

Experimental Investigation of Bulk Response of Cells on Optoelectronic Dielectrophoresis Chip ·········· 162
S. Valagerahally Puttaswamy[1], S.-M. Yang[2], S. Sivashankar[1], K.-W. Chang[1], L. Hsu[2], C.-H. Liu[1]
[1]National Tsing Hua University, TAIWAN
[2]National Chiao Tung University, TAIWAN

PEGDA-Based Photocrosslinking Platform for Real Time Cell Trapping ·········· 166
L.-Y. Ke[1], Z.-K. Kuo[2], Y.-S. Chen[1], H.-W. Tseng[2], C.-H. Liu[1]
[1]National Tsing Hua University, TAIWAN
[2]Industrial Technology Research Institute, TAIWAN

Unified Theory to Evaluate the Effect of Concentration Difference and Peclet Number on Electroosmotic Mobility Error of Micro Electroosmotic Flow ·········· 170
W. Wang, Y.-K. Lee
Hong Kong University of Science and Technology, HONG KONG

Blood Vessels by Fractal Gelatin ⋯⋯⋯⋯⋯⋯⋯⋯⋯⋯⋯⋯⋯⋯⋯⋯⋯ 175
  L.-J. Yang, B.-H. Chen
  Tamkang University, TAIWAN

Design, Simulation, and Verification of Fluidic Light-Guide Chips with Various Geometries of Micro Polymer Channels ⋯⋯⋯⋯⋯⋯⋯⋯⋯⋯⋯⋯⋯⋯ 179
  G.-W. Huang, T.-Y. Hung, C.-T. Chen
  National Kaohsiung University of Applied Sciences, TAIWAN

Separation of Dendritic and T Cells Using Electrowetting and Dielectrophoresis ⋯⋯⋯⋯⋯ 183
  C.-A. Chen[1], C.-H. Chen[1], A. M. Ghaemmaghami[2], S.-K. Fan[1]
  [1]National Chiao Tung University, TAIWAN
  [2]University of Nottingham, UNITED KINGDOM

Development of Microbead-Based Affinity Biosensor by Insulator-Based Dielectrophoresis ⋯⋯⋯⋯ 187
  T.-M. Chuo, W. Hsu, S.-K. Fan
  National Chiao Tung University, TAIWAN

Difference Proportional Cell Contact Platform for 3D Hepatocyte Culture ⋯⋯⋯⋯⋯⋯⋯ 191
  Y.-S. Chen[1], Z.-K. Kuo[2], L.-Y. Ke[1], C.-W. Lin[1], H.-W. Tseng[2], C.-H. Liu[1]
  [1]National Tsing Hua University, TAIWAN
  [2]Industrial Technology Research Institute, TAIWAN

Inducing Self-Rotation of Melan-a Cells by ODEP ⋯⋯⋯⋯⋯⋯⋯⋯⋯⋯⋯⋯⋯⋯⋯ 195
  L.-H. Chau[1], M. Ouyang[1], W. Liang[2], G.-B. Lee[3], W. J. Li[1,2,4], W. K. Liu[1]
  [1]The Chinese University of Hong Kong, HONG KONG
  [2]Shenyang Institute of Automation, Chinese Achademy of Sciences, CHINA
  [3]National Tsing Hua University, TAIWAN
  [4]City University of Hong Kong, HONG KONG

Microfluidic System for Rapid Detection of Influenza Infection by Utilizing Magnetic $MnFe_2O_4$ Nanoparticle-Based Immunoassay ⋯⋯⋯⋯⋯⋯⋯⋯⋯⋯⋯ 200
  L.-Y. Hung[1], F.-Y. Cheng[2], C.-C. Huang[2], Y.-C. Tsai[2], C.-S. Yeh[2], H.-Y. Lei[2], G.-B. Lee[1]
  [1]National Tsing Hua University, TAIWAN
  [2]National Cheng Kung University, TAIWAN

An Optical Diagnostic System using Isothermal Amplification Technique for *Phalaenopsis Orchids* ⋯⋯⋯⋯⋯⋯⋯⋯⋯⋯⋯⋯⋯⋯⋯⋯⋯⋯⋯⋯⋯⋯⋯⋯ 204
  W.-H. Chang[1], S.-Y. Yang[1], C.-H. Wang[1], P.-C. Li[2], F.-J. Jan[2], T.-Y. Chen[3], G.-B. Lee[1]
  [1]National Tsing Hua University, TAIWAN
  [2]National Chung Hsing University, TAIWAN
  [3]National Cheng Kung University, TAIWAN

Automated Immunoassay System based on the Colorimetric Detection ⋯⋯⋯⋯⋯⋯⋯ 208
  K. F. Lei
  Chang Gung University, TAIWAN

Three-Dimensional Lab-on-a-CD with Enzyme-Linked Immunosorbent Assay ⋯⋯⋯⋯⋯ 213
  M. Ishizawa[1], T. Azeta[1], H. Nose[1], Y. Ukita[2], Y. Utsumi[1]
  [1]University of Hyogo, JAPAN
  [2]Japan Advanced Institute of Science and Technology, JAPAN

3D Biomimetic Chip Integrated with Microvascular System for Studying the Liver Specific Functions ⋯⋯⋯⋯⋯⋯⋯⋯⋯⋯⋯⋯⋯⋯⋯⋯⋯⋯⋯⋯⋯⋯⋯ 218
  K.-W. Chang, C.-T. Lee, P. Tushar Harishchandra, H.-P. Chen, T.-R. Yueh,
  S. Valagerahally Puttaswamy, S. Sivashankar, C.-H. Liu
  National Tsing Hua University, TAIWAN

Plasma-Treated Switchable Wettability of Parylene-C Surface · · · · · · 222
X.-P. Bi, N. L. Ward, B. P. Crum, W. Li
Michigan State University, USA

In Situ Heating to Improve Adhesion for Parylene-on-Parylene Deposition · · · · · · 226
D. Kang, J. H.-C. Chang, J. Y.-H. Kim, Y.-C. Tai
California Institute of Technology, USA

Engineering a Biomimetic Villus Array for In Vitro 3-Dimensional Culture of Intestinal Epithelial Cells · · · · · · 230
Y. Chen, W. Yang, Y. Huang, C.-C. Fu, Y. Fu, S. Tang
National Tsing Hua University, TAIWAN

Reduction of AC Resistance in MEMS Intraocular Foil Coils Using Microfabricated Planar Litz Structure · · · · · · 234
Y. Zhao, M. Nandra, C.-C. Yu, Y.-C. Tai
California Institute of Technology, USA

Design and Fabrication of Flexible Parylene-Based Inductors with Electroplated NiFe Magnetic Core for Wireless Power Transmission System · · · · · · 238
X. Sun[1], Y. Zheng[1], Z. Li[1], M. Yu[1], Q. Yuan[1], X. Li[2], H. A. Zhang[1]
[1]Peking University, CHINA
[2]Beijing Jiaotong University, CHINA

Design of a Small Wankel Engine · · · · · · 243
L.-J. Yang, T.-H. Wang
Tamkang University, TAIWAN

Fabrication of High-Aspect-Ratio PZT Structure by Nanocomposite Sol-Gel Method for Laterally-Driven Piezoelectric MEMS Switch · · · · · · 247
N. Wang, S. Yoshida, M. Kumano, Y. Kawai, M. Esashi
Tohoku University, JAPAN

Silicon-Polymer Electro-Thermal Bimorph Actuators with SiC Bottom-Layer for Large Out-of-Plane Motion and Improved Power Efficiency · · · · · · 253
M. Aarts, J. Wei, P. M. Sarro
Delft University of Technology, NETHERLANDS

Study of a Novel Bi-Stable and Easy Integrated MEMS ETBS · · · · · · 257
Y. Zhao, W. Lou, D. Li
Beijing Institute of Technology, CHINA

Rapid Thermal Plasma Deposition of Transparent Nanocrystalline ZnO Thin Films and the Effects of Annealing · · · · · · 261
K. S. Teh, J. Pedersen, H. Esposito
San Francisco State University, USA

Various Carbon Composite Pyrolyzed Polymers and Their Electrical Characterization · · · · · · 266
A. Akazawa, K. Okamoto, A. Syunori, S. Konishi
Ritsumeikan University, JAPAN

Reliability Prediction of 3C-SiC Cantilever Beams using Dynamic Raman Spectroscopy · · · · · · 270
R. Dewanto[1], T. Chen[2], R. Cheung[3], Z. Hu[1], B. Gallacher[1], J. Hedley[1]
[1]Newcastle University, UNITED KINGDOM
[2]The University of Edinburgh, UNITED KINGDOM
[3]Edinburgh University, UNITED KINGDOM

Characterization of Wafer-Level XeF$_2$ Gas-Phase Isotropic Etching for MEMS Processing ············ 274

D. Xu[1], B. Xiong[1], G. Wu[1,2], Y. Ma[1], Y. Wang[1], E. Jing[1,3]

[1]Shanghai Institute of Microsystem and Information Technology,
  Chinese Academy of Sciences, CHINA
[2]Graduate School of Chinese Academy of Science, CHINA
[3]CSMC Technologies Corporation, CHINA

Precise Width Control of Single Crystalline Silicon Nano-Wall Structure Based on Wet Etching Process on (111) Wafer ······································· 278

X. Yu, Q. Jin, T. Li, Y. Wang

Shanghai Institute of Microsystem and Information Technology,
  Chinese Academy of Sciences, CHINA

Conductive Micro Silver Wires via Aerosol Deposition ································· 282

B. Xu, W. Lv, X. Wang, T. Lei, G. Zheng, Y. Zhao, D. Sun

Xiamen University, CHINA

Dry Mechanical Liftoff Technology for Metallization on Parylene-C using SU-8 ·························· 286

J. H.-C. Chang, D. Kang, Y.-C. Tai

California Institute of Technology, USA

Fabrication of Nanogap Electrode Using Electromigration Method During Metal Deposition ··········· 290

T. Ohata[1], Y. Naitoh[2], M. Horikawa[2], D. F. Wang[1], R. Maeda[2]

[1]Ibaraki University, JAPAN
[2]National Institute of Advanced Industrial Science and Technology, JAPAN

Pseudo-Spin Filter in Metallic Single-Walled Carbon Nanotubes ························· 294

D. Bercioux[1], L. Mayrhofer[2]

[1]University of Freiburg, GERMANY
[2]Fraunhofer IWM, GERMANY

Top-Down vs. Bottom-Up Coarse-Graining of Graphene and CNTs for Nanodevice Simulation ··· 298

D. Kauzlarić[1], O. Liba[2], Y. Hanein[2], P. Español[3], A. Greiner[1], S. Succi[1,4], J. G. Korvink[1]

[1]University of Freiburg, GERMANY
[2]Tel-Aviv University, ISRAEL
[3]Universidad Nacional de Educación a Distancia, SPAIN
[4]Italian National Research Council, ITALY

Electrowetting Droplets Investigated with Smoothed Particle Hydrodynamics and Moving Least Squares ··· 304

D. Weiß[1], A. Greiner[1], J. Lienemann[2], J. G. Korvink[1]

[1]University of Freiburg, GERMANY
[2]Schmidt & Partner Engineering AG, SWITZERLAND

Processing of 3D Multilevel SU-8 Fluidic Networks Assisted by PerMX Dry-Photoresist Lamination ······································· 308

R. C. Meier, V. Badilita, U. Wallrabe, J. G. Korvink

University of Freiburg, GERMANY

Conductive and Transparent Gel Microstructures Fabricated by Inkjet Printing of Ionic Liquid Based Fluids ······································· 312

U. Löffelmann, D. Mager, J. G. Korvink

University of Freiburg, GERMANY

Nanomechanical DNA Origami Devices as Versatile Molecular Sensors ··············· 318
A. Kuzuya[1], T. Yamazaki[2], K. Yasuda[2], Y. Sakai[2], Y. Yamanaka[1], Y. Xu[2], Y. Aiba[2],
Y. Ohya[1], M. Komiyama[2]
[1]Kansai University, JAPAN
[2]The University of Tokyo, JAPAN

**Gold-Coated Polystyrene Bead Array and the Investigation of Their Plasmon Coupling
Abilities** ·············· 322
H.-Y. Hsieh[1], T.-W. Huang[1], C.-S. Yang[1,2], P.-C. Wang[1], J.-L. Xiao[3,4], C.-H. Lee[3,4], F.-G. Tseng[1,3]
[1]National Tsing Hua University, TAIWAN
[2]National Health Research Institutes, TAIWAN
[3]Academia Sinica, TAIWAN
[4]National Yang-Ming University, TAIWAN

**Development of Microfabricated Phononic Crystal Resonators Based on Two-Dimensional
Silicon Slab** ·············· 326
N. Wang[1], F.-L. Hsiao[2], M. Palaniapan[1], C. Lee[1]
[1]National University of Singapore, SINGAPORE
[2]National Changhua University of Education, TAIWAN

**Electric Field Design of Metallic Sub-Wavelength Hole Arrays for Optical Permittivity
Sensing** ·············· 331
E. Maeda, T. Matsuki, I. Yamada, J.-J. Delaunay
The University of Tokyo, JAPAN

**A Stable Frequency References Platform Based on Packaged Microsphere-Taper Coupling
System** ·············· 335
J.-J. Xiong[1], Y.-Z. Yan[1], C.-L. Zou[2], F.-W. Sun[2], S.-B. Yan[1], C.-Y. Xue[1], J. Liu[1], W.-D. Zhang[1]
[1]North University of China, CHINA
[2]University of Science and Technology of China, CHINA

**Arrayed Metallic Micro/Nano Particles for Localized Surface Plasmon Resonance Based on
Metal Contact Transfer Lithography** ·············· 339
H.-Y. Chung, C.-Y. Wu, C.-H. Chen, Y.-C. Lee
National Cheng Kung University, TAIWAN

Ligand-Exchange of TOPO-Capped CdSe Quantum Dots with Quinuclidines ·············· 343
J. M. Obliosca[1,2], F.-G. Tseng[1], C.-M. Huang[3], L.-W. Lo[3], P.-C. Wang[1]
[1]National Tsing Hua University, TAIWAN
[2]Academia Sinica, TAIWAN
[3]National Health Research Institutes, TAIWAN

**A Study of Nano-Structured Manganese Dioxides and Their Composites as Electrode
Materials for Micro Supercapacitors** ·············· 347
S. Li[1,2], X. Wang[1,2], C. Shen[1,2], J. Wang[1,2], F. Kang[1,2]
[1]Tsinghua National Laboratory for Information Science and Technology, CHINA
[2]Tsinghua University, CHINA

CoNiMnP-AAO Hard Magnetic Nanocomposite Film for MEMS Applications ·············· 351
T.-Y. Chao, J.-R. Lin, Y.-T. Cheng
National Chiao Tung University, TAIWAN

**An Adhesion Strength Measurement Method for Particle Transfer and Assembly in Dry
Environment** ·············· 355
S.-F. Wang, Y.-T. Lin, Y.-B. Lin, K. Wang
National Changhua University of Education, TAIWAN

Automatic Somatic Cell Operating Process for Nuclear Transplantation ········ 359
Y. Wang, X. Zhao, Q. Zhao, M. Sun, G. Lu
Nankai University, CHINA

High Speed Enucleation of Oocyte Using Magnetically Actuated Microrobot on a Chip ········ 364
M. Hagiwara, A. Ichikawa, T. Kawahara, F. Arai
Nagoya University, JAPAN

Catalytic Nano-Mobile Robot with Finely Designed Geometry ········ 368
J. Bao, M. Nakajima, Z. Yang, M. Kojima, T. Fukuda
Nagoya University, JAPAN

Design and Fabrication of Diffractive Phase Element for Minimizing the Focusing Spot Size beyond Diffraction Limit ········ 372
N. Atthi[1], S. Boonruang[1], W. Mohammed[2], W. Jeamsaksiri[1], C. Hruanun[1], A. Poyai[1]
[1]National Electronics and Computer Technology Center, THAILAND
[2]Bangkok University, THAILAND

Enhancing Light Output of GaN-Based Light-Emitting Diodes With Nanoparticle-Assembled On-Top Layers ········ 376
C. Zheng[1], L. Sun[1], X. Chen[1], Y. Shen[2], P. Mao[1], M. Han[1]
[1]Nanjing University, CHINA
[2]Shandong University, CHINA

The Chromatic Dispersion Module with Large Chromatic Focal Shift ········ 380
M. C. Wei, K. Y. Hung, Y. J. Chuang, S. H. Huang
Ming Chi University of Technology, TAIWAN

A Multi-View Reflective Three-Dimensional Display ········ 384
J.-F. Chuang, K. Wang
National Changhua University of Education, TAIWAN

Use Bionic Microlens Array and CMOS Image Sensor for Three-Dimensional Motion Detection ········ 388
C.-Y. Liu, J.-F. Chuang, T.-C. Yu, K. Wang
National Changhua University of Education, TAIWAN

A High Sensitivity and Low-Cost Polycarbonate(PC)-Based Biosensor ········ 392
Y.-S. Chen, G.-J. Wang
National Chung Hsing University, TAIWAN

An Electro-Enzymatic Flexible Molecular Lactate Sensor ········ 398
N. Thomas, I. Lähdesmäki, B. A. Parviz
University of Washington, USA

Spontaneous Motion of a Water Droplet on Hydrophilic and Curvature Gradient Conical-Shaped Surfaces ········ 403
Y. C. Chuang[1], H. Y. Hsieh[1], Q. Zheng[2], F.-G. Tseng[1,3]
[1]National Tsing Hua University, TAIWAN
[2]Tsinghua University, CHINA
[3]Academia Sinica, TAIWAN

Electric Manipulations of Hydrogel on a Digital Microfluidic Platform ········ 407
M.-Y. Chiang, S.-K. Fan
National Chiao Tung University, TAIWAN

Microbubble and Microplasma Manipulations for Gas Analyses ·············· 411
Y.-T. Shen, L.-P. Tsai, S.-K. Fan
National Chiao Tung University, TAIWAN

Three-Dimensional Digital Microfluidics and Applications ·············· 415
G. Wang[1], D. Teng[1], S.-K. Fan[2]
[1]University of Saskatchewan, CANADA
[2]National Chiao Tung University, TAIWAN

Specific Design and Implementation of a Piezoelectric Droplet Actuator for Evaporative Cooling of Free Space ·············· 419
H.-Y. Wang, C. Huang, C.-T. Chen
National Kaohsiung University of Applied Sciences, TAIWAN

Micro-Droplet Formation with Non-Newtonian Solutions in Microfluidic T-Junctions with Different Inlet Angles ·············· 423
Z. Gu, J.-L. Liow
University of New South Wales at the
Australian Defence Force Academy, AUSTRALIA

Investigation of Electrical Properties of DNA-Attached Carbon Nano-Particles for Biological-Applications ·············· 429
M. Ouyang[1], W. J. Li[2], K. W. Wong[3], W. K. Liu[1]
[1]The Chinese University of Hong Kong, HONG KONG
[2]City University of Hong Kong, HONG KONG
[3]Chengdu Green Energy and Green Manufacturing Technology R&D Center, CHINA

Electrical Performance of Micro-Assembled Beads under Different Temperatures and Loadings ·············· 433
Y.-L. Tzeng, K. Wang
National Changhua University of Education, TAIWAN

Vertical Deposition of Nanospheres on the Open Sidewalls of Silicon Pillars ·············· 437
Y. F. Wang, Y. Tian, K. J. Feng, C. Li, D. D. She, W. G. Wu
Peking University, CHINA

Nanoscale Laser Writing of Indium-Tin-Oxide Nanowires ·············· 441
M. Afshar, D. Feili, H. Voellm, M. Straub, K. Koenig, H. Seidel
Saarland University, GERMANY

Fabrication and Characters of Squama-Shape Micro/Nano Multi-Scale Structures ·············· 445
X.-S. Zhang, F.-Y. Zhu, H. A. Zhang
Peking University, CHINA

A High Efficient POM Micro-Methanol Reformer ·············· 449
H.-S. Wang, K.-Y. Huang, H.-C. Peng, Y.-J. Huang, F.-G. Tseng
National Tsing Hua University, TAIWAN

Proton Exchange Membranes Based on Aryl Epoxy Resin for Fuel Cells Operated at Elevated Temperatures ·············· 453
T.-Y. Lee[1], T.-C. Ho[1], C.-J. Chang[2], P.-C. Wang[1], F.-G. Tseng[1,3]
[1]National Tsing Hua University, TAIWAN
[2]Industrial Technology Research Institute, TAIWAN
[3]Academia Sinica, TAIWAN

Material Nonlinearity Limits on a Lamé-Mode Single Crystal Bulk Resonator ·········· 457
H. Zhu, C. Tu, J. E.-Y. Lee
City University of Hong Kong, HONG KONG

Evidence on the Impact of T-Shaped Tether Variations on Q factor of Bulk-Mode Square-Plate Resonators ·········· 463
Y. Xu, J. E.-Y. Lee
City University of Hong Kong, HONG KONG

Analysis of Air Damping in Micromachined Resonators ·········· 469
G. Wu[1,2], D. Xu[1], B. Xiong[1], Y. Ma[1], Y. Wang[1], E. Jing[1,3]
[1]Shanghai Institute of Microsystem and Information Technology,
  Chinese Academy of Sciences, CHINA
[2]Graduate School of Chinese Academy of Science, CHINA
[3]CSMC Technologies Corporation, CHINA

Benchmarking the Passive Differential Input Technique to Shielded GSG Probes ·········· 473
Y. Xu, H. Zhu, J. E.-Y. Lee
City University of Hong Kong, HONG KONG

Study on Thermoelastic Dissipation in Bulk Mode Resonators with Etch Holes ·········· 478
C. Tu, J. E.-Y. Lee
City University of Hong Kong, HONG KONG

A Piezoresistive Normal and Shear Force Sensor Using Liquid Metal Alloy as Gauge Material ·········· 483
X. Shi, C.-H. Cheng, C. Chao, L. Wang, Y. Zheng
The Hong Kong Polytechnic University, HONG KONG

Development of A Novel Force Sensor System Built with an Industrial Multilayer Ceramic Capacitor (MLCC) ·········· 487
K.-R. Lin[1], C.-H. Chiang[2], C.-H. Chang[1], C.-H. Lin[2]
[1]National Cheng Kung University, TAIWAN
[2]National Sun Yat-Sen University, TAIWAN

Implementation of a Subwavelength Bragg Reflector for Terahertz Applications ·········· 491
V. Singal, S. Smaili, Y. Massoud
University of Alabama at Birmingham, USA

On Sustaining Robustness of Molecular Pathway Circuits of the HSR Network of *E. coli* under Spatial Configuration ·········· 496
J.-Q. Liu[1], T. Yamanishi[2], H. Nishimura[3], S. Nobukawa[3], H. Umehara[1]
[1]National Institute of Information and Communications Technology, JAPAN
[2]Fukui University of Technology, JAPAN
[3]University of Hyogo, JAPAN

Numerical Simulation of CZTS Thin Film Solar Cell ·········· 502
W. Zhao, W. Zhou, X. Miao
Huazhong University of Science and Technology, CHINA

On the Design of Subwavelength Waveguiding Structures for Terahertz Applications ·········· 506
V. Singal, S. Smaili, Y. Massoud
University of Alabama at Birmingham, USA

Polypyrrole (PPy) Nanowire Arrays Entrapped with Glucose Oxidase Biosensor for Glucose Detection ·········· 511
G. Q. Xu, J. Lv, Z. X. Zheng, Y. C. Wu
Hefei University of Technology, CHINA

The Electromigration Investigation of Cu-Ni Nanocomposites · · · · · · 515
Y. C. Chen, C. W. Chu, T.-Y. Chao, Y. T. Cheng, C. Chen
National Chiao Tung University, TAIWAN

A SWNTs Thin Film Solar Microcell Prepared by Simple Solution-Evaporation Method · · · · · · 519
C. C. Chen, Y. Y. Chang, J. Zhang
Peking University, CHINA

Investigation of Particle Dispersion and Deposition in a Channel with Elliptic Obstruction using Lattice Boltzmann Method · · · · · · 523
A. Tehrani, A. Moosavi
Sharif University of Technology, IRAN

A Study of Hydrogen Peroxide Microfluidic Fuel Cells · · · · · · 529
J. C. Shyu[1], C. L. Huang[1], T. S. Sheu[2], H. Ay[1], J. W. Huang[2]
[1]National Kaohsiung University of Applied Sciences, TAIWAN
[2]R. O. C. Military Academy, TAIWAN

Microfluidic Circulatory System for the Raise of Liver Urea Assay · · · · · · 533
Y.-S. Chen, T.-H. Dai, L.-Y. Ke, C.-H. Liu
National Tsing Hua University, TAIWAN

Preparation and Physicochemical Study of Liposomes Containing Nicotinamide · · · · · · 537
N. Langlah, S. Pinsuwan, T. Amnuaikit
Prince of Songkla University, THAILAND

Fabrication of Controllable Profile Microlens Array by Nanoimprinting Process · · · · · · 542
M. C. Cheng, L. K. Chen, C. K. Sung
National Tsing Hua University, TAIWAN

Analysis on 3-Dimensional Spatial Electric Field of AFM Based Anodic Oxidation · · · · · · 547
Z. Liu, N. Jiao, Z. Wang, Z. Dong
Shenyang Institute of Automation, Chinese Academy of Sciences, CHINA

Scanning Electron Beam Induced Deposition for Conductive Tip Modification · · · · · · 553
P. L. Chen[1], J. Su[1], M. H. Shiao[1], M. N. Chang[2], C. H. Lee[3], C. W. Liu[3]
[1]National Applied Research Laboratories, TAIWAN
[2]National Chung Hsing University, TAIWAN
[3]National Taiwan University, TAIWAN

In situ Study of Thermal Deformation of Metal Resistive Heater on Silicon Nitride Membrane by Digital Holographic Microscopy · · · · · · 557
Y. W. Lai, J. E.-Y. Lee
City University of Hong Kong, HONG KONG

Comparison of Glass Etching Properties Between HCl and HNO$_3$ Solution · · · · · · 562
W. Tao, W. Lv, Z. Zhan, W. Zuo, X. Qiu, L. Wang, D. Sun
Xiamen University, CHINA

Surface Analysis and Process Optimization of Black Silicon · · · · · · 567
F.-Y. Zhu[1,2], Q.-L. Di[1], X.-J. Zeng[1], X.-S. Zhang[1], X. Zhao[2], H.-X. Zhang[1]
[1]Peking University, CHINA
[2]Nankai University, CHINA

Application of Nonlinear Driving in Frequency Matching of Tunneling Gyroscope · · · · · · 571
L. Wang, X. Du, Y. Su, Z. Zhan, W. Zuo, D. Sun
Xiamen University, CHINA

Integrated Flexible Micro Pressure, Temperature and Flow Sensors for Use in PEMFC ·············· 575
C.-Y. Lee, T. Yang, Y.-M. Lee, T.-H. Chien, Y.-T. Cheng
Yuan Ze University, TAIWAN

Sensitivity Enhancement in SGOI Nanowire Biosensor Fabricated by Top Surface Passivation ··············· 579
K.-M. Chang[1,2], C.-F. Chen[1], C.-H. Lai[3], C.-T. Hsieh[1], C.-N. Wu[1], Y.-B. Wang[1], C.-H. Liu[1]
[1]National Chiao Tung University, TAIWAN
[2]I-Shou University, TAIWAN
[3]Chung Hua University, TAIWAN

A Silicon-on-Glass Z-axis Accelerometer with Vertical Sensing Comb Capacitors ·············· 583
J. Wang, Z. Yang, G. Yan
Peking University, CHINA

The Vertical MSM Diamond X-Ray Detector ·············· 587
S. Cheirsirikul[1], S. Jesen[1], C. Hruanun[2]
[1]King Mongkut's Institute of Technology Ladkrabang, THAILAND
[2]National Electronics and Computer Center, THAILAND

Effect of Geometrical Design of Support on Frequency Shift and Energy Loss of Piezoelectric Ring Resonator Applicable to Liquid Circumstance ·············· 591
T. Sagawa[1], D. F. Wang[1], J. Lu[2], R. Maeda[2]
[1]Ibaraki University, JAPAN
[2]National Institute of Advanced Industrial Science and Technology, JAPAN

Effective Force Generation for ECLIA Composed of Si Bone Structure and Conductive Polymer Flexible Slider ·············· 595
T. A. Nguyen, S. Konishi
Ritsumeikan University, JAPAN

The Study of Forward and Reverse Schottky Junction for Dual Magnetodiode ·············· 599
T. Phetchakul[1], W. Luanatikomkul[1], W. Yamwong[2], A. Poyai[2]
[1]King Mongkut's Institute of Technology Ladkrabang, THAILAND
[2]National Electronics and Computer Technology Center, THAILAND

Electrode Design Optimization of a CMOS Fringing-Field Capacitive Sensor ·············· 603
Y.-T. Li, Y.-L. Tzeng, C.-M. Chao, K. Wang
National Changhua University of Education, TAIWAN

Dimensions and Capillary Effects of Microfluidic Channel for Blood Plasma Separation ·············· 607
Y.-H. Zhan, J.-N. Kuo
National Formosa University, TAIWAN

Fabrication and Analysis of Integrated MEMS Pyramidal Horn Antenna for Terahertz Applications ·············· 611
C. Li[1], L. Guo[2], W. G. Wu[1], X. S. Tang[2], F. Y. Huang[2]
[1]Peking University, CHINA
[2]Southeast University, CHINA

Characterization of Super-Harmonic Effect Using Piezoelectric Film Cantilever with A Proof Mass in the Point ·············· 615
H. Ishinabe[1], T. Kobayashi[2], D. F. Wang[1], T. Itoh[2], R. Maeda[2]
[1]Ibaraki University, JAPAN
[2]National Institute of Advanced Industrial Science and Technology, JAPAN

High-Q Maintenance of Microcavity by using a Sealed and Packaged Structure ························ 619
S.-B. Yan, Y.-Z. Yan, Y.-G. Zhang, L. Wang, C.-Y. Xue, J. Liu, W.-D. Zhang, J.-J. Xiong
North University of China, CHINA

On the Effect of Width of Metallic Armchair Graphene Nanoribbons in Plasmonic Waveguide Applications ··· 623
S. Smaili, V. Singal, Y. Massoud
University of Alabama at Birmingham, USA

The Facile Transferral of Graphene onto Interdigitated Electrodes for Sensing Applications ··········· 627
C. Dale, S. Rana, R. H. Page, J. Hedley, N. Keegan
Newcastle University, UNITED KINGDOM

Excitation of Mechanical Oscillations in Double-carbon-nanotube System by Terahertz Radiation ·· 631
V. Semenenko[1], V. Leiman[1], A. Arsenin[1], Y. Stebunov[1], V. Ryzhii[2]
[1]Moscow Institute of Physics and Technology, RUSSIA
[2]University of Aizu, JAPAN

Capillary Kinetics of Water in Hydrophilic Microscope Coverslip Nanochannels ···················· 636
J.-N. Kuo, Y.-K. Lin
National Formosa University, TAIWAN

A Visualization Study of Venting Gas via Hydrophobic Nanoporous Membrane ···················· 640
J.-C. Shyu[1], S.-M. Dai[1], K.-S. Yang[2], C.-C. Wang[3]
[1]National Kaohsiung University of Applied Sciences, TAIWAN
[2]Industrial Technology and Research Institute, TAIWAN
[3]National Chiao Tung University, TAIWAN

Switching Characteristic Model and Biochemical Application Analysis for Electrolyte-Oxide-Semiconductor Structure Diodes ································ 644
G. C. Sun[1], X. Y. Ma[1], A. S. Tang[2], Y. F. Chen[1], W. G. Wu[1]
[1]Peking University, CHINA
[2]Massachusetts Institute of Technology, USA

A 3D Micro-Channel Cooling System Embedded in LTCC Packaging Substrate ···················· 649
S. Jia[1], M. Miao[2,3], R. Fang[1,2], S. Guo[1,2], D. Hu[1,2], Y. Jin[1,2]
[1]Information Engineering Institute of Peking University, CHINA
[2]Peking University, CHINA
[3]Beijing Information Science & Technology University, CHINA

A Flexible Evaporation Micropump with Precision Flow Rate Control for Micro-Fluidic Systems ··· 653
K.-Y. Chen, K.-E. Chen, K. Wang
[1]National Changhua University of Education, TAIWAN

Nanofluidic Device with Self-Assembled Nafion Membrane Utilizing Capillary Valve ···················· 657
S. Wang, H. Yu, W. Wang, Z. Li
Peking University, CHINA

Effect of Coating Organic Film on Dropwise Condensation in Microgrooves with Nanostructure Surface ··· 661
T. S. Sheu[1], J. C. Shyu[2], J. W. Hsiao[1], Y. C. Pan[1]
[1]R. O. C. Military Academy, TAIWAN
[2]National Kaohsiung University of Applied Sciences, TAIWAN

Novel Core Etching Technique on Synthesized Gold Nanoparticles for Colorimetric Detection of Dopamine Biosample ·········· 665
H.-C. Lee, T.-H. Chen, W.-L. Tseng, C.-H. Lin
National Sun Yat-sen University, TAIWAN

Fast Self-Resonant Startup Procedure for Digital MEMS Gyroscope System ·········· 669
F. Ge, D. Liu, L. Lin, Z. Yang, G. Yan
Peking University, CHINA

The Influence of Experimental Parameters on the Assembly of SWNTs by AC Dielectrophoresis ·········· 673
Z. Wang, F. Yu, W. Li, J. Zhang
Peking University, CHINA

The Manufacture of Micropillars with High Depth-to-Width Ratio, and the Comparison between Two Typical Materials ·········· 677
Z. Wang[1], X. Qin[2]
[1]Singapore-MIT Alliance for Research and Technology Center, SINGAPORE
[2]Tsinghua University, CHINA

Surface-Modified Diamond Embedded in Nickel Matrix Composite for Intrinsic Polishing Application ·········· 681
C.-J. Shih, W.-C. Lin, C.-S. Lin, Y.-N. Pan
National Taiwan University, TAIWAN

Fabrication and Performance Optimization of the Microplasma Reactor with Composite Dielectrics ·········· 685
Z. Yuan, L. Wen, L. Cheng, J. Ma, J. Chu
University of Science and Technology of China, CHINA

Fabrication of Deep Lateral Single-Crystal-Silicon Blaze Micro-Grating by Inductively-Coupled-Plasma Reactive Ion Etch ·········· 689
Y.-H. Lin[1,2], C. J. Weng[2], C. Y. Su[2], W. Hsu[1]
[1]National Chiao Tung University, TAIWAN
[2]National Applied Research Laboratories, TAIWAN

A Study of Tin Oxide Thin Tilm Gas Sensors with High Oxygen Vacancies ·········· 693
C. Lin[1], D. Zhang[1], X. Liu[2]
[1]Peking University, CHINA
[2]BOE Technology Group Co., Ltd., CHINA

***In situ* Monitoring of Temperature using Flexible Micro Temperature Sensors inside Polymer Lithium-Ion Battery** ·········· 698
C.-Y. Lee, S.-J. Lee, Y.-M. Lee, M.-S. Tang, P.-C. Chen, Y.-M. Chang
Yuan Ze University, TAIWAN

Design and Simulation of Fully-Symmetrical Resonant Pressure Sensor ·········· 702
Y. Jiang, X. Du, Z. Zhan, B. Xu, W. Lv, L. Wang, D. Sun
Xiamen University, CHINA

Effect of Oxidation on SGOI Nanowire Biosensor Fabrication Using Ge Condensation ·········· 708
K.-M. Chang[1,2], C.-F. Chen[1], C.-H. Lai[3], C.-N. Wu[1], C.-T. Hsieh[1], Y.-B. Wang[1], C.-H. Liu[1]
[1]National Chiao Tung University, TAIWAN
[2]I-Shou University, TAIWAN
[3]Chung Hua University, TAIWAN

A Capacitive Readout Circuit with DC Sensing Method for Micromachined Gyroscopes ················ 712
K. Zhou, L. Sun, F. Ge, Z. Yang, G. Yan
Peking University, CHINA

The Optimal Vibrational Shear Stress for Bovine Endothelial Cell Proliferation ················· 716
C.-W. Li[1], J.-L. Chen[1], C.-C. Wu[2], G.-J. Wang[1]
[1]National Chung Hsing University, TAIWAN
[2]National Cheng Kung University, TAIWAN

An Optimized Fabrication of High Yield CMOS-Compatible Silicon Carbide Capacitive Pressure Sensors ················· 721
B. Meng, W. Tang, Z. R. Wang, H.-X. Zhang
Peking University, CHINA

Study of Thin Film Adhesion Properties of Multi-Layer Flexible Electronics Composites ············· 725
C. C. Li[1], Z. H. Liu[1], C. T. Pan[1], J. K. Tseng[2], H. L. Huang[2], S. W. Mao[2], S. C. Shen[3], S. J. Chang[4]
[1]National Sun Yat-sen University, TAIWAN
[2]R.O.C Military Academy, TAIWAN
[3]National Cheng Kung University, TAIWAN
[4]National Yunlin University of Science and Technology, TAIWAN

Using a Canny-Edge-Detection Based Method to Characterize In-Plane Micro-Actuators ············· 729
C.-Y. Cheng, Y.-B. Lin, K. Wang
National Changhua University of Education, TAIWAN

Thermal Switch and Variable Capacitance Designed for Micro Electrostatic Converter by using CMOS MEMS Process ················· 733
J.-C. Chiou[1], L.-C. Chou[1], Y.-L. Lai[2], S.-C. Huang[1]
[1]National Chiao Tung University, TAIWAN
[2]National Chip Implementation Center, TAIWAN

A Research of the Bandwidth of a Mode-Matching MEMS Vibratory Gyroscope ·················· 738
C. He, Q. Zhao, J. Cui, Z. Yang, G. Yan
Peking University, CHINA

Graphene Nanoribbon Based AM Demodulator of Terahertz Radiation ················· 742
Y. Stebunov[1], A. Arsenin[1], V. Leiman[1], V. Semenenko[1], V. Ryzhii[2]
[1]Moscow Institute of Physics and Technology, RUSSIA
[2]University of Aizu, JAPAN

Atomic Layer Deposited Protective Coatings for Integrated MEMS Flow Sensor ················· 747
D. Li[1], A. Abdulagatov[2], F. Yang[1], D. C. Zhang[1]
[1]Peking University, CHINA
[2]University of Colorado, USA

Additional Papers

Nems Based Tools for Nanoscience and Atomic Clocks 751
N.F. de Rooij, S. Gautsch, T. Akiyama, F. Loizeau, G. Mileti, Y. Petremand,
U. Staufer, R. Straessle, G. Yoshikawa

Beyond Watson and Crick: Programming DNA Self-assembly for Nanofabrication 753
P. Rothemund

Recent Advances on Nano-materials for Advanced Packaging Applications 755
C.P. Wong

Printed Carbon Nanotube Devices and Their Applications 756
Z. Cui, J. Zhao, T. Zhang

Sensing and Noise Characteristics of Si-Nanowire Ion-Sensitive-Field-Effect-Transistors 758
for Future Biosensor Applications
J.S. Lee

Atomic Resolution Nanofabrication and Dynamic Characterization 759
L. Sun

A Nanopatterning by Phase Change Nanolithography 761
X.S. Miao, B.J. Zeng, Z. Li, W.L. Zhou

Oral Sessions   Tuesday, March 6

IEEE-NEMS 2012 Program Schedule
Monday, March 5, 2012
Welcome Reception: Cafe Restaurant Camphora

16:30 – 18:30   Welcome Reception

Tuesday, March 6, 2012
Plenary Session: Centennial Hall

9:15 - 9:45   Opening Ceremony

9:45 - 12:00   T1G   Plenary Lectures

Session Chairs: **Osamu Tabata,**   Kyoto University, JAPAN

Toshiyuki Tsuchiya,   Kyoto University, JAPAN

Qing-An Huang,   Southeast University, CHINA

T1G-1   **NEMS Based Tools for Nanoscience and Atomic Clocks**
N. F. de Rooij[1], S. Gautsch[1], T. Akiyama[1], F. Loizeau[1], G. Mileti[2], Y. Pétremand[1], U. Staufer[3], R. Straessle[1], G. Yoshikawa[4]
[1]Ecole Polytechnique Fédérale de Lausanne, SWITZERLAND
[2]Université de Neuchâtel, SWITZERLAND
[3]Technical University of Delft, NETHERLANDS
[4]National Institute for Materials Science (NIMS), JAPAN

T1G-2   **Beyond Watson and Crick: Programming DNA Self-Assembly for Nanofabrication**
P. W. K. Rothemund
California Institute of Technology, USA

T1G-3   **Recent Advances on Nano-Materials for Advanced Packaging Applications**
C. P. Wong
The Chinese University of Hong Kong, HONG KONG

ORAL SESSIONS
ROOM A (International Conference Hall I)

14:30 - 16:00   T3A   Invite I/Carbon Nanotubes
Session Chair: **Alice H. X. Zhang,**   Peking University, CHINA

T3A-1   (invite) **Printed Carbon Nanotube Devices and Their Applications**
Z. Cui, J. Zhao, T. Zhang
Suzhou Institute of Nanotech, Chinese Academy of Science, CHINA

T3A-2   **Portable E-Nose Based on Polymer/CNT Sensor Array for Protein-Based Detection**
P. Lorwongtragool[1], T. Seesaard[1], C. Tongta[2], T. Kerdcharoen[1]
[1]Mahidol University, THAILAND
[2]Thaksin University, THAILAND

T3A-3   **Carbon Nanotubes Nanoarray in Anodized Alumina Nanopores for Glucose Biosensing**
A. Wisitsoraat, D. Phokharatkul, C. Karuwan, P. Sritongkham, A. Tuantranont
National Electronics and Computer Technology Center, THAILAND

T3A-4   **Investigations on Passivation Layers for Carbon Nanotube Transistors for Sensor**

Oral Sessions   Tuesday, March 6

Applications
K. Chikkadi, M. Politou, E. Cagin, O. Kurapova, M. Döbeli, C. Hierold
ETH Zürich, SWITZERLAND

T3A-5   **Study on Carbon-Based Electrode for Air-Cathode Microbial Fuel Cell**
Y.-C. Huang, H.-Y. Tsai
National Tsing Hua University, TAIWAN

16:30 - 18:00   T4A   Graphenes

Session Chair: **Fan-Gang Tseng,**   National Tsing Hua University, TAIWAN

T4A-1   **Multilayer Graphene Growth by a Metal-Catalyzed Crystallization of Diamond-Like Carbon**
P. Peng, D. Xie, Y. Yang, C. Zhou, H. Tian, T. Feng, X. Li, T. Ren, H. Zhu
Tsinghua University, CHINA

T4A-2   **Chemical Vapor Deposition of Nanocrystalline Graphene Directly on Arbitrary High-Temperature Insulating Substrates**
J. Sun[1], N. Lindvall[1], M. T. Cole[2], K. Teo[3], A. Yurgens[1]
[1]Chalmers University of Technology, SWEDEN
[2]University of Cambridge, UNITED KINGDOM
[3]AIXTRON Nanoinstruments Ltd., UNITED KINGDOM

T4A-3   **Characterization of Strain Fields in Graphene Films**
R. Dewanto, C. Dale, Z. Hu, N. Keegan, B. Gallacher, J. Hedley
Newcastle University, UNITED KINGDOM

T4A-4   **Transfer-Free Fabrication of Suspended Graphene Grown by Chemical Vapor Deposition**
N. Lindvall, J. Sun, A. Yurgens
Chalmers University of Technology, SWEDEN

T4A-5   **Photovoltaic Response of N-Doped Graphene-Based Photodetector**
W. Wang, T. Li, Y. Wang
Shanghai Institute of Microsystem and Information Technology, CHINA

T4A-6   **Actuators Based on Graphene/Graphene Oxide Papers**
H. Bi, X. Xie, K. Yin, S. Wan, L. Sun
Southeast University, CHINA

ROOM B (International Conference Hall III)

14:30 - 16:00   T3B   Nanoimprint

Session Chair: **Francesc Perez-Murano,**

Insitituto de Microelectronica de Barcelona, SPAIN

T3B-1   **A Novel Multi-Step Programmable Thermal Nanoimprint Lithography**
Y.-J. Chang, J.-W. Hsu
National Yunlin University of Science and Technology, TAIWAN

T3B-2   **Development of Roll Press UV Imprint Process for Replication of Micro Lens Arrays on the Large and Thin Quartz Substrate**
L. Li, K. Ishii, Y. Tsutsui, S. Shoji, J. Mizuno
Waseda University, JAPAN

T3B-3   **Nanoimprinting of Sub-Wavelength Structure by Using Cone-Shape Anodic Alumina Oxide**

Oral Sessions   Tuesday, March 6

Template for the Enhancement of Solar Cell Energy Conversion Efficiency
C.-H. Chuang[1], F.-F. Chuang[1], S.-W. Tsai[1], Y.-M. Shen[2], C.-P. Chen[3], S.-C. Wan[1]
[1]Southern Taiwan University, TAIWAN
[2]National Cheng Kung University, TAIWAN
[3]Metal Industries Research & Development Centre, TAIWAN

T3B-4   **Fabrication Metal Roller Mold with Sub-Micrometer Feature Size Based on Contact Printing Photolithography**
K.-F. Huang, S.-W. Tsai, Y.-C. Lee
National Cheng Kung University, TAIWAN

T3B-5   **Fabrication of Metal Embedded Photo-Mask for Sub-Micrometer Scaled Photolithography and Patterning Sapphire Substrate**
J.-N. Yan, Y.-C. Lee
National Cheng Kung University, TAIWAN

T3B-6   **Metal Contact Printing Photolithography for Fabricating Sub-Micrometer Patterned Sapphire Substrates in Light-Emitting Diodes**
Y.-T. Hsieh, Y.-C. Lee
National Cheng Kung University, TAIWAN

16:30 - 18:00   T4B   Lithography
     Session Chair: **Pasqualina M. Sarro,**
                                   Delft University of Technology, THE NETHERLANDS

T4B-1   **Nano-Scale Mechanical Relays Fabricated by Nanoimprint Technology**
Y.-J. Chang, D. Y. Liu, C. L. Kuo
National Yunlin University of Science and Technology, TAIWAN

T4B-2   **A Facile Nanowire Fabrication Approach Based on Edge Lithography**
Y. Liu, W. Wang, H. A. Zhang, W. Wu, Z. Li
Peking University, CHINA

T4B-3   **Hybrid Mask Lithography for Fabrication of Micro-Pattern with Nano-Pillars**
S. Sakuma, M. Sugita, F. Arai
Nagoya University, JAPAN

T4B-4L   **In-situ Actuated Gap Reduction and Clogging-Free Apertures for Quasi-Dynamic Stencil Lithography**
S. Xie, V. Savu, J. Brugger
Ecole Polytechnique Fédérale de Lausanne, SWITZERLAND

T4B-5L   (withdrawn)

T4B-6L   **Roll-to-Roll Microcontact Printing with Roller Stamp Fabricated by Optical Soft Lithography**
S. Makino, J. Park, N. Takama, B. Kim
The University of Tokyo, JAPAN

ROOM C (Meeting Room IV)

Oral Sessions   Tuesday, March 6

14:30 - 16:00   T3C   Chemical & Molecular Sensing I
                  Session Chair: **Yi-Kuen Lee**,
                              Hong Kong University of Science and Technology, HONG KONG

T3C-1   **Adaptable Chip-Level Microfluidic Packaging for a Micro-Scale Gas Chromatograph**
          N. Ward[1], X. Mu[1], G. Serrano[2], E. Covington[2], C. Kurdak[2], E. T. Zellers[2], A. J. Mason[1], W. Li[1]
          [1]Michigan State University, USA
          [2]University of Michigan, USA

T3C-2   **Detecting Vapor Traces of Explosives Using a Self-Assembled Mono Layer on a Surface-Modified MEMS Capacitor and CMOS Electronics**
          D. Strle[1], B. Štefane[1], I. Mušević[2]
          [1]University of Ljubljana, SLOVENIA
          [2]Jožef Stefan Institute, SLOVENIA

T3C-3   **A High Heating Efficiency Two-Beam Microhotplate for Catalytic Gas Sensors**
          L. Xu, T. Li, X. Gao, Y. Wang
          Shanghai Institute of Microsystem and Information Technology, CHINA

T3C-4   **Vibration Mode Localization in Coupled Beam-Shaped Resonator Array**
          K. Chatani[1], D. F. Wang[1], T. Ikehara[2], R. Maeda[2]
          [1]Ibaraki University, JAPAN
          [2]National Institute of Advanced Industrial Science and Technology (AIST), JAPAN

T3C-5   **Piezoelectrically Actuated Circular Diaphragm Resonator Mass Sensors**
          N. Keegan, J. Hedley, Z. Hu, J. A. Spoors, W. Waugh, B. Gallacher, C. J. McNeil
          Newcastle University, UNITED KINGDOM

T3C-6   **Clay Nanocomposite Hydrogels Applied to MEMS-Based Chemical Microsensors**
          V. Schulz, B. Ferse, A. Grosse, K.-F. Arndt, G. Gerlach
          Technische Universitaet Dresden, GERMANY

16:30 - 18:00   T4C   OS: Emergence in Chemistry for Integrated Nano System
                  Session Chair: **Keisuke Morishima**,   Osaka University, JAPAN

T4C-1   **Electron Transport and Photoresponce of Self-Assembled Molecular Layer of Ru Complex with Phosphonic Acids on ITO**
          T. Ishida[1], K. Terada[1], H. Nakamura[1], Y. Asai[1], T. Sumi[1,2], K. Kanaizuka[2,3], M. Haga[2]
          [1]NRI-AIST, JAPAN
          [2]Chuo University, JAPAN
          [3]Yamagata University, JAPAN

T4C-2   **Noise-Induced Stochastic Enhancement for a Device Based on Redox-Active Huge Molecule and DNA Nanonetwork**
          T. Matsumoto, Y. Hirano, Y. Segawa, T. Kawai
          Osaka University, JAPAN

T4C-3   **Pattern Size Dependence of Reorientation of Photoinduced Liquid Crystalline Polymer by Thermal Nanoimprinting**
          M. Okada, M. Kurita, M. Kondo, Y. Haruyama, N. Kawatsuki, S. Matsui
          University of Hyogo, JAPAN

T4C-4   **Magnetoresistance of Uni-Molecular Junctions with Ferromagnetic Electrodes**

Oral Sessions   Tuesday, March 6

M. Noguchi, R. Yamada, H. Tada
Osaka University, JAPAN

T4C-5   **A Carbon Nanotube Network Conjugated by Semiconductor Nanoparticles with Defined Nanometer-Scaled Gaps**
S. Kumagai[1], N. Okamoto[2], M. Kobayashi[2,3], I. Yamashita[2]
[1]Toyota Technological Institute, JAPAN
[2]Nara Institute of Science and Technology, JAPAN
[3]Japanese Foundation for Cancer Research, JAPAN

T4C-6   **Closed-Looped Nano Stimulation Microscope for Living Cell Membrane**
T. Hoshino[1], K. Morishima[2]
[1]Tokyo University of Agriculture and Technolog, JAPAN
[2]Osaka University, JAPAN

ROOM D (Meeting Room III)

14:30 - 16:00   T3D   Probe & Cantilever Technology and Application
        Session Chair: **Yoko Yamanishi,**   Nagoya University, JAPAN

T3D-1   **Pyramidal Nanowire Tip for Atomic Force Microscopy and Thermal Imaging**
N. Burouni[1], E. Sarajlic[2], M. Siekman[1], L. Abelmann[1], N. Tas[1]
[1]University of Twente, NETHERLANDS
[2]Smarttip, NETHERLANDS

T3D-2   (withdrawn)

T3D-3   **AFM Imaging Experiment Using Cantilever Probe Integrated with Microplasma Reactor**
L. Cheng, L. Wen, Z. Yuan, D. Niu, L. He, J. Chu
University of Science and Technology of China, CHINA

T3D-4   **A Pyrex Nanochannel Device Fabricated by AFM Nanolithography**
O. Hibbert, T. Busch, S. Tung
University of Arkansas, USA

T3D-5   **Effect of Nonlinear Vibration on Double Region of Synchronized Frequency Responses in Mechanically Coupled Beam-Shaped Oscillator System**
T. Itoh[1], D. F. Wang[1], T. Ikehara[2], M. Nakajima[1], R. Maeda[2]
[1]Ibaraki University, JAPAN
[2]National Institute of Advanced Industrial Science and Technology (AIST), JAPAN

T3D-6L   **Patterning Property of a Novel Anti-Wear Probe for SPM LAO Lithography**
Y. F. Li[1], Y. Tomizawa[1], A. Koga[1], G. Hashiguchi[2], M. Sugiyama[3], H. Fujita[3]
[1]BEANS Project, JAPAN
[2]Shizuoka University, JAPAN
[3]The University of Tokyo, JAPAN

16:30 - 18:00   T4D   Physical Sensors
        Session Chair: **Daoheng Sun,**   Xiamen University, CHINA

Oral Sessions   Tuesday, March 6

T4D-1   **Flexible and Biocompatible Pressure Sensor Sheet for a Treatment Patch of Myelomeningocele**
          K. Kuwana, K. Masamune, T. Dohi
          The University of Tokyo, JAPAN

T4D-2   **Characterization of a Multi-Layered MEMS Pressure Sensor Using Piezoresistive Silicon Nanowire within Large Measurable Strain Range**
          L. Lou[1], S. Zhang[1], W.-T. Park[2], L. Lim[2], D.-L. Kwong[2], C. Lee[1]
          [1]National University of Singapore, SINGAPORE
          [2]Institute of Microelectronics, Singapore, SINGAPORE

T4D-3   **Analysis, Modeling and Verification of Air-Venting Effect on Frequency Response of a Capacitive MEMS Microphone**
          C.-T. Chen
          National Kaohsiung University of Applied Sciences, TAIWAN

T4D-4   **A Novel Stretchable CMUT Array Using Liquid-Metal Electrodes on a PDMS Substrate**
          X. Shi[1], C.-H. Cheng[1], J. Peng[2]
          [1]The Hong Kong Polytechnic University, HONG KONG
          [2]Shen Zhen University, CHINA

T4D-5   **Miniature Pyranometer with Asteroid Shape Thermopile**
          J. Zhang, Z. Wu, Z. Zhao, X. Guan
          Institute of Electronics, Chinese Academy of Sciences, CHINA

T4D-6   **Design and Fabrication of Thin-Film Aluminum Microheater and Nickel Temperature Sensor**
          R. Phatthanakun[1], P. Deekla[2], W. Pummara[2], C. Sriphung[1], C. Pantong[2], N. Chomnawang[2]
          [1]Synchrotron Light Research Institute, THAILAND
          [2]Suranaree University of Technology, THAILAND

Oral Sessions   Wednesday, March 7

Wednesday, March 7, 2012
ORAL SESSIONS
ROOM A (International Conference Hall I)

9:00 - 10:15   W1A   Invite II/Award Finalist Presentations I
          Session Chair: **Steve Tung**,   University of Arkansas, USA

W1A-1   (invite) **Sensing and Noise Characteristics of Si-Nanowire Ion-Sensitive-Field-Effect-Transistors for Future Biosensor Applications**
        J.-S. Lee
        Pohang University of Science and Technology, KOREA

W1A-2   **Micro Device Array Design and Fabrication in Monolithic MEMS SoC**
        J.-H. Wen, W. Fang
        National Tsing Hua University, TAIWAN

W1A-3   **Integration the Back-Side Inclined Exposure Technology to Fabricate the 45degree k-Type Prism with Nanometer Roughness**
        K.-Y. Hung, Y.-W. Tsai, C.-F. Lee, Y.-H. Chu
        Ming Chi University of Technology, TAIWAN

10:45 - 12:00   W2A   Award Finalist Presentation II
          Session Chair: **Qing-An Huang**,   Southeast University, CHINA

W2A-1   **Study on Piezoelectric Properties of Near-Field Electrospinning PVDF/MWCNT Nano-Fiber**
        Z. Y. Ou[1], Z. H. Liu[1], C. T. Pan[1], L. W. Lin[2], Y. J. Chen[1], H. W. Lai[1]
        [1]National Sun Yat-Sen University, TAIWAN
        [2]University of California, Berkeley, USA

W2A-2   **Local Ablation by Micro-Electric Knife**
        Y. Yamanishi, H. Kuriki, S. Sakuma, K. Onda, F. Arai
        Nagoya University, JAPAN

W2A-3   **In-Vitro Transgenic Mice Liver Tissue Culture Via Hydrodynamic Flow Perfusion Bioreactor**
        C.-W. Wu[1], S. Sivashankar[1], S. Valagerahally Puttaswamy[1], H.-L. Lin[2], K.-W. Chang[1], C.-T. Yeh[2], C.-H. Liu[1]
        [1]National Tsing Hua University, TAIWAN
        [2]Chang Gung Memorial Hospital, TAIWAN

W2A-4   **High Quality Factor Parylene-Based Intraocular Pressure Sensor**
        J. C.-H. Lin[1], Y. Zhao[1], P.-J. Chen[2], Y.-C. Tai[1]
        [1]California Institute of Technology, USA
        [2]Robert Bosch LLC, USA

Oral Sessions   Wednesday, March 7

14:30 - 16:00   W4A   Award Finalist Presentation III
                Session Chair: **Qing-An Huang**,   Southeast University, CHINA

W4A-1   **Light-Addressable Electrochemical Micropatterning of Cell-Encapsulated Alginate Hydrogels for Cell-Based Microarray**
           S.-H. Huang[1], H.-T. Chu[1], C.-W. Wu[2], Y.-Y. Chuang[2]
           [1]National Taiwan Ocean University , TAIWAN
           [2]Ming Chuan University, TAIWAN

W4A-2   **A Free-Standing and Flexible Parylene PCR Device**
           P. Satsanarukkit, H. Lo, Y.-C. Tai
           California Institute of Technology, USA

W4A-3   **Cell Culture on MEMS Materials in Micro-Environment Limited by a Physical Condition**
           M. Inoue[1], A. Okonigi[2], K. Terao[1], H. Takao[1], F. Shimokawa[1], F. Oohira[1], H. Kotera[2], T. Suzuki[1]
           [1]Kagawa University, JAPAN
           [2]Kyoto University, JAPAN

W4A-4   **The Effect of Cytochalasin D on F-Actin Behavior of Single-Cell Electroendocytosis Using Multi-Chamber Micro Cell Chip**
           R. Lin, D. C. Chang, Y.-K. Lee
           Hong Kong University of Science and Technology, HONG KONG

W4A-5   **Cryogenic Frozen Device for Hepatocyte Culture and Responses**
           L.-Y. Ke[1], Y.-S. Chen[1], J. Liu[2], C.-H. Liu[1]
           [1]National Tsing Hua University, TAIWAN
           [2]Tsinghua *University*, CHINA

16:30 - 18:00   W5A   Cell Engineering
                Session Chair: **Wen J. Li**,   City Universiyt of Hong Kong, HONG KONG

W5A-1   **Rapid Millions Cell-Assembly Chip for the Formation of High-Density Monolayer Cells Array**
           T.-J. Chen, Y.-C. Chang, F.-G. Tseng
           National Tsing Hua University, TAIWAN

W5A-2   **Proliferation Rate Regulation of Primary Chondrocyte Cells by Applying a Cycling Mechanical Stimulation on Flexible Micro Cells Well**
           T.-Y. Lin, T.-J. Chen, F.-G. Tseng
           National Tsing Hua University, TAIWAN

W5A-3   **3D Microstructure Integrated Bioreactor System for Transgenic Mice Thick Liver Tissue Culture**
           S. Sivashankar[1], S. Valagerahally Puttaswamy[1], H.-L. Lin[2], S.-M. Yang[3], H.-P. Chen[1], C.-T. Yeh[2], C.-H. Liu[1]
           [1]National Tsing Hua University, TAIWAN
           [2]Chang Gung Memorial Hospital, TAIWAN
           [3]National Chiao Tung University, TAIWAN

Oral Sessions   Wednesday, March 7

W5A-4   **Experimental Investigation of Bulk Response of Cells on Optoelectronic Dielectrophoresis Chip**
S. Valagerahally Puttaswamy[1], S.-M. Yang[2], S. Sivashankar[1], K.-W. Chang[1], L. Hsu[2], C.-H. Liu[1]
[1]National Tsing Hua University, TAIWAN
[2]National Chiao Tung University, TAIWAN

W5A-5   **PEGDA-Based Photocrosslinking Platform for Real Time Cell Trapping**
L.-Y. Ke[1], Z.-K. Kuo[2], Y.-S. Chen[1], H.-W. Tseng[2], C.-H. Liu[1]
[1]National Tsing Hua University, TAIWAN
[2]Industrial Technology Research Institute, TAIWAN

W5A-6   **Gene Transfection Enhancement of Electroporation Microchip by Combining High and Low Electric Fields**
Y.-C. Chung[1], W.-J. Liao[1], Y.-T. Huang[2], C.-Y. Wu[1]
[1]Ming Chi University of Technology, TAIWAN
[2]Fu Jen Catholic University, TAIWAN

ROOM B (International Conference Hall III)

9:00 - 10:15   W1B   Invite III/Membranes
Session Chair: **Yu-Cheng Lin**,   National Cheng Kung University, TAIWAN

W1B-1   (invite) **Atomic Resolution Nanofabrication and Dynamic Characterization**
L. Sun
Southeast University, CHINA

W1B-2   **A High Performance Nano Desalination by the Manipulation of EDL among AAO Nanochannels**
Y.-S. Huang, C.-J. Chang, W.-C. Chang, Y.-L. Chueh, F.-G. Tseng
National Tsing Hua University, TAIWAN

W1B-3   **Development of Conductive Nanoporous Polymer Membrane for Selective Transport of Charged Biomolecules**
P.-R. Chen[1], S.-H. Huang[2], Y.-J. Chuang[1]
[1]Ming Chuan University, TAIWAN
[2]National Taiwan Ocean University , TAIWAN

W1B-4   **A Magnetic-Driven Membrane Made of Photosensitive Nanocomposite without Alignment Process**
T. Nakahara[1], Y. Hosokawa[1], K. Terao[1], H. Takao[1], F. Shimokawa[1], F. Oohira[1], H. Miyagawa[1], T. Namazu[2], H. Kotera[3], T. Suzuki[1]
[1]Kagawa University, JAPAN
[2]University of Hyogo, JAPAN
[3]Kyoto University, JAPAN

Oral Sessions   Wednesday, March 7

10:45 - 12:00   W2B   Microchannels and Fluidics
                Session Chair: **Gwo-Bin Lee**,   National Tsing Hua University, TAIWAN

W2B-1   **Unified Theory to Evaluate the Effect of Concentration Difference and Peclet Number on Electroosmotic Mobility Error of Micro Electroosmotic Flow**
         W. Wang, Y.-K. Lee
         Hong Kong University of Science and Technology, HONG KONG

W2B-2   **Fabrication of Sub-Spot-Size Microstructure of Microfluidic Channels Using CO_2 Laser Processing with Metal Film Protection**
         C. K. Chung, S. L. Lin, T. K. Tan, K. Z. Tu
         National Cheng Kung University, TAIWAN

W2B-3   **Blood Vessels by Fractal Gelatin**
         L.-J. Yang, B.-H. Chen
         Tamkang University, TAIWAN

W2B-4   **Controlled Orientation of Zeolites Based on Viscosity Segregation for Functional Optofluidic Systems**
         H. S. Khoo[1], M. Otter[2], L. De Cola[2], A. D. Griffiths[1]
         [1]Université de Strasbourg, FRANCE
         [2]Westfälische Wilhelms-Universität Münster, FRANCE

W2B-5   **Design, Simulation, and Verification of Fluidic Light-Guide Chips with Various Geometries of Micro Polymer Channels**
         G.-W. Huang, T.-Y. Hung, C.-T. Chen
         National Kaohsiung University of Applied Sciences, TAIWAN

14:30 - 16:00   W4B   Dielectrophoresis
                Session Chair: **Beomjoon Kim**,   the University of Tokyo, JAPAN

W4B-1   **Separation of Dendritic and T Cells Using Electrowetting and Dielectrophoresis**
         C.-A. Chen[1], C.-H. Chen[1], A. M. Ghaemmaghami[2], S.-K. Fan[1]
         [1]National Chiao Tung University, TAIWAN
         [2]University of Nottingham, UNITED KINGDOM

W4B-2   **Development of Microbead-Based Affinity Biosensor by Insulator-Based Dielectrophoresis**
         T.-M. Chuo, W. Hsu, S.-K. Fan
         National Chiao Tung University, TAIWAN

W4B-3   **Difference Proportional Cell Contact Platform for 3D Hepatocyte Culture**
         Y.-S. Chen[1], Z.-K. Kuo[2], L.-Y. Ke[1], C.-W. Lin[1], H.-W. Tseng[2], C.-H. Liu[1]
         [1]National Tsing Hua University, TAIWAN
         [2]Industrial Technology Research Institute, TAIWAN

W4B-4   **Enhancement of Fluorescent Intensity by Using DEP Manipulations of Polyaniline-Coated Al_2O_3 Nanoparticles for Immunosensing**
         C.-H. Chuang[1], H.-P. Wu[1], Y.-W. Huang[2], C.-H. Chen[1], C.-P. Jen[2]
         [1]Southern Taiwan University, TAIWAN
         [2]National Chung Cheng University, TAIWAN

Oral Sessions   Wednesday, March 7

W4B-5   **Inducing Self-Rotation of Melan-a Cells by ODEP**
         L.-H. Chau[1], M. Ouyang[1], W. Liang[2], G.-B. Lee[3], W. J. Li[1,2,4], W. K. Liu[1]
         [1]The Chinese University of Hong Kong, HONG KONG
         [2]Shenyang Institute of Automation, Chinese Achademy of Sciences, CHINA
         [3]National Tsing Hua University, TAIWAN
         [4]City University of Hong Kong, HONG KONG

16:30 - 18:00   **W5B**   Microfluidic Devices
                 Session Chair: **Lung-Jieh Yang,**   Tamkang University, TAIWAN

W5B-1   **Microfluidic System for Rapid Detection of Influenza Infection by Utilizing Magnetic
         MnFe$_2$O$_4$ Nanoparticle-Based Immunoassay**
         L.-Y. Hung[1], F.-Y. Cheng[2], C.-C. Huang[2], Y.-C. Tsai[2], C.-S. Yeh[2], H.-Y. Lei[2], G.-B. Lee[1]
         [1]National Tsing Hua University, TAIWAN
         [2]National Cheng Kung University, TAIWAN

W5B-2   **An Optical Diagnostic System Using Isothermal Amplification Technique for *Phalaenopsis*
         *Orchids***
         W.-H. Chang[1], S.-Y. Yang[1], C.-H. Wang[1], P.-C. Li[2], F.-J. Jan[2], T.-Y. Chen[3], G.-B. Lee[1]
         [1]National Tsing Hua University, TAIWAN
         [2]National Chung Hsing University, TAIWAN
         [3]National Cheng Kung University, TAIWAN

W5B-3   **Automated Immunoassay System Based on the Colorimetric Detection**
         K. F. Lei
         Chang Gung University, TAIWAN

W5B-4   **Three-Dimensional Lab-on-a-CD with Enzyme-Linked Immunosorbent Assay**
         M. Ishizawa[1], T. Azeta[1], H. Nose[1], Y. Ukita[2], Y. Utsumi[1]
         [1]University of Hyogo, JAPAN
         [2]Japan Advanced Institute of Science and Technology, JAPAN

W5B-5   **Multiplex Immunoassay on a Power-Free Microchip for Point-of-Care Testing**
         K. Hosokawa, H. Okada, M. Maeda
         RIKEN, JAPAN

W5B-6   **3D Biomimetic Chip Integrated with Microvascular System for Studying the Liver Specific
         Functions**
         *K.-W. Chang, C.-T. Lee, P. Tushar Harishchandra, H.-P. Chen, T.-R. Yueh, S. Valagerahally
         Puttaswamy, S. Sivashankar, C.-H. Liu*
         National Tsing Hua University, TAIWAN

ROOM C (Meeting Room IV)

9:00 - 10:15   **W1C**   Fabrication I
                Session Chair: **Yuichi Utsumi,**   University of Hyogo, JAPAN

W1C-1   **Plasma-Treated Switchable Wettability of Parylene-C Surface**
         X.-P. Bi, N. L. Ward, B. P. Crum, W. Li
         Michigan State University, USA

Oral Sessions   Wednesday, March 7

W1C-2   **In Situ Heating to Improve Adhesion for Parylene-on-Parylene Deposition**
         D. Kang, J. H.-C. Chang, J. Y.-H. Kim, Y.-C. Tai
         California Institute of Technology, USA

W1C-3   **Engineering a Biomimetic Villus Array for In Vitro 3-Dimensional Culture of Intestinal Epithelial Cells**
         Y. Chen, W. Yang, Y. Huang, C.-C. Fu, Y. Fu, S. Tang
         National Tsing Hua University, TAIWAN

W1C-4   **Reduction of AC Resistance in MEMS Intraocular Foil Coils Using Microfabricated Planar Litz Structure**
         Y. Zhao, M. Nandra, C.-C. Yu, Y.-C. Tai
         California Institute of Technology, USA

W1C-5   **Design and Fabrication of Flexible Parylene-Based Inductors with Electroplated NiFe Magnetic Core for Wireless Power Transmission System**
         X. Sun[1], Y. Zheng[1], Z. Li[1], M. Yu[1], Q. Yuan[1], X. Li[2], H. A. Zhang[1]
         [1]Peking University, CHINA
         [2]Beijing Jiaotong University, CHINA

10:45 - 12:00   W2C   Actuators
         Session Chair: **Satoshi Konishi,**   Ritsumeikan University, JAPAN

W2C-1   **Design of a Small Wankel Engine**
         L.-J. Yang, T.-H. Wang
         Tamkang University, TAIWAN

W2C-2   **Fabrication of High-Aspect-Ratio PZT Structure by Nanocomposite Sol-Gel Method for Laterally-Driven Piezoelectric MEMS Switch**
         N. Wang, S. Yoshida, M. Kumano, Y. Kawai, M. Esashi
         Tohoku University, JAPAN

W2C-3   **Silicon-Polymer Electro-Thermal Bimorph Actuators with SiC Bottom-Layer for Large Out-of-Plane Motion and Improved Power Efficiency**
         M. Aarts, J. Wei, P. M. Sarro
         Delft University of Technology, NETHERLANDS

W2C-4   **Study of a Novel Bi-Stable and Easy Integrated MEMS ETBS**
         Y. Zhao, W. Lou, D. Li
         Beijing Institute of Technology, CHINA

W2C-5L   **Tensile Testing of SWCNT Using Thermal Actuator Clamped with Electrolessly Deposited Gold Layer**
         T. Kataoka, Y. Hirai, K. Sugano, T. Tsuchiya, O. Tabata
         Kyoto University, JAPAN

Oral Sessions   Wednesday, March 7

14:30 - 16:00   W4C   Materials
               Session Chair: **Jung-Sik Kim**,   University of Seoul, KOREA

W4C-1   **Carbonaceous Magnetic Nanocapsules for Drug Delivery**
        M. A. Zeeshan[1], S. Pané[1], E. Pellicer[2], S. Schürle[1], J. Sort[2,3], M. D. Baró[2], B. J. Nelson[1]
        [1]ETH Zürich, SWITZERLAND
        [2]Universitat Autònoma de Barcelona, SPAIN
        [3]Institucio Catalana de Recerca i Estudis Avancats, SPAIN

W4C-2   **Rapid Thermal Plasma Deposition of Transparent Nanocrystalline ZnO Thin Films and the Effects of Annealing**
        K. S. Teh, J. Pedersen, H. Esposito
        San Francisco State University, USA

W4C-3   **Various Carbon Composite Pyrolyzed Polymers and Their Electrical Characterization**
        A. Akazawa, K. Okamoto, A. Syunori, S. Konishi
        Ritsumeikan University, JAPAN

W4C-4   **Reliability Prediction of 3C-SiC Cantilever Beams Using Dynamic Raman Spectroscopy**
        R. Dewanto[1], T. Chen[2], R. Cheung[2], Z. Hu[1], B. Gallacher[1], J. Hedley[1]
        [1]Newcastle University, UNITED KINGDOM
        [2]The University of Edinburgh, UNITED KINGDOM

W4C-5   **Effect of Spin-Orbit Coupling on Piezoresistivity in p-Doped Semiconductor Bulks and Nanofilms on the Basis of First-Principles Theory**
        K. Nakamura
        Kyoto University, JAPAN

W4C-6L **Thermal Radiation from Non-Rigid Nanoparticle Surrounded by Medium**
        N. N. Sharma
        Birla Institute of Technology & Science, INDIA

16:30 - 18:00   W5C   Fabrication II
               Session Chair: **Takaaki Suzuki**,   Kagawa University, JAPAN

W5C-1   **Characterization of Wafer-Level XeF$_2$ Gas-Phase Isotropic Etching for MEMS Processing**
        D. Xu, B. Xiong, G. Wu, Y. Ma, Y. Wang, E. Jing
        Shanghai Institute of Microsystem and Information Technology, CHINA

W5C-2   **Precise Width Control of Single Crystalline Silicon Nano-Wall Structure Based on Wet Etching Process on (111) Wafer**
        X. Yu, Q. Jin, T. Li, Y. Wang
        Shanghai Institute of Microsystem and Information Technology, CHINA

W5C-3   **Conductive Micro Silver Wires via Aerosol Deposition**
        B. Xu, W. Lv, X. Wang, T. Lei, G. Zheng, Y. Zhao, D. Sun
        Xiamen University, CHINA

W5C-4   **Dry Mechanical Liftoff Technology for Metallization on Parylene-C Using SU-8**
        J. H.-C. Chang, D. Kang, Y.-C. Tai
        California Institute of Technology, USA

Oral Sessions   Wednesday, March 7

W5C-5   **Fabrication of Nanogap Electrode Using Electromigration Method During Metal Deposition**
         T. Ohata[1], Y. Naitoh[2], M. Horikawa[2], D. F. Wang[1], R. Maeda[2]
         [1]Ibaraki University, JAPAN
         [2]National Institute of Advanced Industrial Science and Technology (AIST), JAPAN

W5C-6L   **Fabrication and Operation of Nanomechanical Structures Defined by FIB Ion Beam Implantation**
         J. Llobet[1], X. Borrisé[2], M. Gerbolés[1], G. Rius[3], F. Perez-Murano[1]
         [1]Insitituto de Microelectronica de Barcelona, SPAIN
         [2]Institut Catala de Nanotecnologia, SPAIN
         [3]Toyota Technological Institute, JAPAN

ROOM D (Meeting Room III)

9:00 - 10:15   W1D   FRIAS Special Session
                    Session Chair: **Andreas Greiner,**   University of Freiburg, GERMANY

W1D-1   **Pseudo-Spin Filter in Metallic Single-Walled Carbon Nanotubes**
         D. Bercioux[1], L. Mayrhofer[2]
         [1]University of Freiburg, GERMANY
         [2]Fraunhofer IWM, GERMANY

W1D-2   **Top-Down vs. Bottom-Up Coarse-Graining of Graphene and CNTs for Nanodevice Simulation**
         D. Kauzlarić[1], O. Liba[2], Y. Hanein[2], P. Español[3], A. Greiner[1], S. Succi[4], J. G. Korvink[1]
         [1]University of Freiburg, GERMANY
         [2]Tel-Aviv University, ISRAEL
         [3]National Distance Education University, USA
         [4]Italian National Research Council, ITALY

W1D-3   **Electrowetting Droplets Investigated with Smoothed Particle Hydrodynamics and Moving Least Squares**
         D. Weiß[1], A. Greiner[1], J. Lienemann[2], J. G. Korvink[1]
         [1]University of Freiburg, GERMANY
         [2]Schmidt&Partner Engineering AG, GERMANY

W1D-4   **Processing of 3D Multilevel SU-8 Fluidic Networks Assisted by PerMX Dry-Photoresist Lamination**
         R. C. Meier, V. Badilita, U. Wallrabe, J. G. Korvink
         University of Freiburg, GERMANY

W1D-5   **Conductive and Transparent Gel Microstructures Fabricated by Inkjet Printing of Ionic Liquid Based Fluids**
         U. Löffelmann, D. Mager, J. G. Korvink
         University of Freiburg, GERMANY

Oral Sessions   Wednesday, March 7

10:45 - 12:00   W2D   DNA origami
            Session Chair: **Xiao-Hong Wang,**   Tsinghua University, CHINA

W2D-1 **AuNPs Conjugate DNA Origami Nanotubes for Nanophotonic Application**
S. Z. Kiss[1], D. S. Hautzinger[1], O. Tabata[2], J. G. Korvink[1]
[1]University of Freiburg, GERMANY
[2]Kyoto University, JAPAN

W2D-2 **Nanomechanical DNA Origami Devices as Versatile Molecular Sensors**
A. Kuzuya[1], T. Yamazaki[2], K. Yasuda[2], Y. Sakai[2], Y. Yamanaka[1], Y. Xu[2], Y. Aiba[2], Y. Ohya[1], M. Komiyama[2]
[1]Kansai University, JAPAN
[2]The University of Tokyo, JAPAN

W2D-3L **Manifesting a 2D Layer of DNA Origami Tiles Using Base-Pair Shape Recognition**
I. C. Robertson, Y. Yanagida, S. Oda
Tokyo Institute of Technology, JAPAN

W2D-4L **Temperature Dependency of DNA Origami Self-Assembly Rate**
C. Huang, T. Akishiba, N. Tamura, Y. Hirai, K. Sugano, T. Tsuchiya, O. Tabata
Kyoto University, JAPAN

14:30 - 16:00   W4D   Photonic Devices
            Session Chair: **Wengang Wu,**   Peking University, CHINA

W4D-1 **Gold-Coated Polystyrene Bead Array and the Investigation of Their Plasmon Coupling Abilities**
H.-Y. Hsieh[1], T.-W. Huang[1], C.-S. Yang[2], P.-C. Wang[1], J.-L. Xiao[3], C.-H. Lee[3], F.-G. Tseng[1]
[1]National Tsing Hua University, TAIWAN
[2]National Health Research Institutes, TAIWAN
[3]National Yang-Ming University, TAIWAN

W4D-2 **Photothermal and Photoacoustic Converters Using Local Plasmon Resonator**
M. Suzuki, K. Namura, K. Nakajima, K. Kimura
Kyoto University, JAPAN

W4D-3 **Development of Microfabricated Phononic Crystal Resonators Based on Two-Dimensional Silicon Slab**
N. Wang[1], F.-L. Hsiao[2], M. Palaniapan[1], C. Lee[1]
[1]National University of Singapore, SINGAPORE
[2]National Changhua University of Education, TAIWAN

W4D-4 **Electric Field Design of Metallic Sub-Wavelength Hole Arrays for Optical Permittivity Sensing**
E. Maeda, T. Matsuki, I. Yamada, J.-J. Delaunay
The University of Tokyo, JAPAN

W4D-5 **A Stable Frequency References Platform Based on Packaged Microsphere-Taper Coupling System**
J.-J. Xiong[1], Y.-Z. Yan[1], C.-L. Zou[2], F.-W. Sun[2], S.-B. Yan[1], C.-Y. Xue[1], J. Liu[1], W.-D. Zhang[1]
[1]North University of China, CHINA
[2]University of Science and Technology of China, CHINA

Oral Sessions

W4D-6L **Confocal Distance Sensor with Varifocal Liquid Lens**
K. Noda, B.-K. Nguyen, Y. Takei, T. Takahata, K. Matsumoto, I. Shimoyama
The University of Tokyo, JAPAN

16:30 - 18:00   W5D   SERS and Quantum Optics
Session Chair: **Motofumi Suzuki,**   Kyoto University, JAPAN

W5D-1 **Non-Labeled Qualitative and Quantitative Analysis of Hepatitis Antigen Using SERS-Active Nano-Cavity Array**
C.-K. Yao, J.-D. Liao, C.-W. Chang, J.-R. Lin
National Cheng Kung University, TAIWAN

W5D-2 **A Nanostars Structure of Highly Efficient Surface Enhanced Raman Scattering Active Gold Nanoparticles Array**
T.-F. Kuo, T.-Y. Liu, T.-Y. Li, R.-G. Wu, F.-G. Tseng
National Tsing Hua University, TAIWAN

W5D-3 **Gold Nanorod Arrays for Near Infrared Optofluidic Device**
T. Fukuoka[1], M. Yoshida[1], R. Takahashi[1], M. Suzuki[2], Y. Utsumi[1]
[1]University of Hyogo, JAPAN
[2]Kyoto University, JAPAN

W5D-4 **Arrayed Metallic Micro/Nano Particles for Localized Surface Plasmon Resonance Based on Metal Contact Transfer Lithography**
H. Y. Chung, C. Y. Wu, C. H. Chen, Y. C. Lee
National Cheng Kung University, TAIWAN

W5D-5 **Ligand-Exchange of TOPO-Capped CdSe Quantum Dots with Quinuclidines**
J. M. Obliosca[1], F.-G. Tseng[1], C.-M. Huang[2], L.-W. Lo[2], P.-C. Wang[1]
[1]National Tsing Hua University, TAIWAN
[2]National Health Research Institutes, TAIWAN

W5D-6 **Crystalline Structure Dependent Photoluminescence of ZnO Nanosheet-Covered Carbon Fibers**
Y.-H. Pai
National Dong Hwa University, TAIWAN

Oral Sessions   Thursday, March 8

Thursday, March 8, 2012
ORAL SESSIONS

ROOM A (International Conference Hall I)

9:00 - 10:15   Th1A   Invite IV/Materials
Session Chair: **Wibool Piyawattanametha,**

National Electronics and Computer Technology Center, THAILAND

Th1A-1   (invite) **Nanopatterning by Phase Change Nanolithography**
X. Miao, B. J. Zeng, Z. Li, W. L. Zhou
Huazhong University of Science and Technology, CHINA

Th1A-2   **Exploration of Crystal Surface Dependent Nanopiezotronic Properties in the Obliquely Aligned InN Nanorod Array**
N.-J. Ku, J.-H. Huang, C.-H. Wang, H.-C. Fang, C.-P. Liu
National Cheng Kung University, TAIWAN

Th1A-3   **A Study of Nano-Structured Manganese Dioxides and Their Composites as Electrode Materials for Micro Supercapacitors**
S. Li, X. Wang, C. Shen, J. Wang, F. Kang
Tsinghua University, CHINA

Th1A-4   **CoNiMnP-AAO Hard Magnetic Nanocomposite Film for MEMS Applications**
T.-Y. Chao, J.-R. Lin, Y. T. Cheng
National Chiao Tung University, TAIWAN

10:45 - 12:00   Th2A   Robotics & Assembly
Session Chair: **Kazuo Hosokawa,**   RIKEN, JAPAN

Th2A-1   **An Adhesion Strength Measurement Method for Particle Transfer and Assembly in Dry Environment**
S.-F. Wang, Y.-T. Lin, Y.-B. Lin, K. Wang
National Changhua University of Education, TAIWAN

Th2A-2   **Automatic Somatic Cell Operating Process for Nuclear Transplantation**
Y. Wang, X. Zhao, Q. Zhao, M. Sun, G. Lu
NanKai University, CHINA

Th2A-3   **High Speed Enucleation of Oocyte Using Magnetically Actuated Microrobot on a Chip**
M. Hagiwara, A. Ichikawa, T. Kawahara, F. Arai
Nagoya University, JAPAN

Th2A-4   **Catalytic Nano-Mobile Robot with Finely Designed Geometry**
J. Bao, M. Nakajima, Z. Yang, M. Kojima, T. Fukuda
Nagoya University, JAPAN

Th2A-5   **Magnetic Manipulation of Liposomal Microstructures in Plant-Based Blood Vessel Phantoms**
S. Schuerle[1], S. Pané[1], E. Pellicer[2], J. Sort[2,3], M. D. Baró[2], B. J. Nelson[1]
[1]ETH Zürich, SWITZERLAND
[2]Universitat Autònoma de Bracelona, SPAIN
[3]Institucio Catalana de Recerca i Estudis Avancats, SPAIN

Oral Sessions   Thursday, March 8

13:00 - 14:30   Th3A   Optical System
              Session Chair: **Toshiyuki Tsuchiya**,   Kyoto University, JAPAN

Th3A-1   (withdrawn)

Th3A-2   **Design and Fabrication of Diffractive Phase Element for Minimizing the Focusing Spot Size beyond Diffraction Limit**
N. Atthi[1], S. Boonruang[1], W. Mohammed[2], W. Jeamsaksiri[1], C. Hruanun[1], A. Poyai[1]
[1]National Electronics and Computer Technology Center, THAILAND
[2]Bangkok University, THAILAND

Th3A-3   **Enhancing Light Output of GaN-Based Light-Emitting Diodes with Nanoparticle-Assembled On-Top Layers**
C. Zheng[1], L. Sun[1], X. Chen[1], Y. Shen[2], P. Mao[1], M. Han[1]
[1]Nanjing University, CHINA
[2]Shandong University, CHINA

Th3A-4   **The Chromatic Dispersion Module with Large Chromatic Focal Shift**
M.-C. Wei, K.-Y. Hung, Y.-J. Chuang, S.-H. Huang
Ming Chi University of Technology, TAIWAN

Th3A-5   **A Multi-View Reflective Three-Dimensional Display**
J.-F. Chuang, K. Wang
National Changhua University of Education, TAIWAN

Th3A-6   **Use Bionic Microlens Array and CMOS Image Sensor for Three-Dimensional Motion Detection**
C.-Y. Liu, J.-F. Chuang, T.-C. Yu, K. Wang
National Changhua University of Education, TAIWAN

ROOM B (International Conference Hall III)

9:00 - 10:15   Th1B   Biosensors
              Session Chair: **Ken-Ichiro Kamei**,   Kyoto University, JAPAN

Th1B-1   **Integration of Solid-State Sensor and Microfluidic Chip for Glucose, Urea and Creatinine Measurement**
Y.-H. Lin, S.-H. Wang, C.-P. Chu, M.-H. Wu, T.-M. Pan
Chang Gung University, TAIWAN

Th1B-2   **Fluorescent Hydrogel Fiber for Highly-Accurate Glucose Monitoring**
M. Takahashi[1,3], Y. J. Heo[1,2], T. Kawanishi[1,3], T. Okitsu[1,2], S. Takeuchi[1,2]
[1]BEANS Project, JAPAN
[2]The University of Tokyo, JAPAN
[3]Terumo Co., JAPAN

Th1B-3   **A High Sensitivity and Low-Cost Polycarbonate (PC)-Based Biosensor**
Y.-S. Chen, G.-J. Wang
National Chung Hsing University, TAIWAN

Oral Sessions   Thursday, March 8

Th1B-4   An Electro-Enzymatic Flexible Molecular Lactate Sensor
N. Thomas, I. Lähdesmäki, B. A. Parviz
University of Washington, USA

Th1B-5L (withdrawn)

10:45 - 12:00   Th2B   Droplet and Bubble Manipulation
         Session Chair: **Yu-Ting Cheng**,   National Chiao Tung University, TAIWAN

Th2B-1   Spontaneous Motion of a Water Droplet on Hydrophilic and Curvature Gradient Conical-Shaped Surfaces
Y. C. Chuang[1], H. Y. Hsieh[1], Q. Zheng[2], F.-G. Tseng[1]
[1]National Tsing Hua University, TAIWAN
[2]Tsinghua University, CHINA

Th2B-2   Electric Manipulations of Hydrogel on a Digital Microfluidic Platform
M.-Y. Chiang, S.-K. Fan
National Chiao Tung University, TAIWAN

Th2B-3   Microbubble and Microplasma Manipulations for Gas Analyses
Y.-T. Shen, L.-P. Tsai, S.-K. Fan
National Chiao Tung University, TAIWAN

Th2B-4   Sub-Nanoliter Color-Resist Droplets Inkjet-Printed on a Commercial Black-Matrix Glass
C.-T. Chen, C.-T. Chuang
National Kaohsiung University of Applied Sciences, TAIWAN

Th2B-5   Three-Dimensional Digital Microfluidics and Applications
G. Wang[1], D. Teng[1], S.-K. Fan[2]
[1]University of Saskatchewan, CANADA
[2]National Chiao Tung University, TAIWAN

13:00 - 14:30   Th3B   Droplet Technologies
         Session Chair: **Gou Jen Wang**,   National Chung Hsing University, TAIWAN

Th3B-1   Specific Design and Implementation of a Piezoelectric Droplet Actuator for Evaporative Cooling of Free Space
H.-Y. Wang, C. Huang, C.-T. Chen
National Kaohsiung University of Applied Sciences, TAIWAN

Th3B-2   Using Developed Microfluidic Chip for Producing the Droplets with Different Concentrations
C.-H. Yeh, Y.-C. Chen, Y.-C. Lin
National Cheng Kung University, TAIWAN

Th3B-3   Micro-Droplet Formation with Non-Newtonian Solutions in Microfluidic T-Junctions with Different Inlet Angles
Z. Gu, J.-L. Liow
University of New South Wales at the Australian Defence Force Academy, AUSTRALIA

Oral Sessions   Thursday, March 8

Th3B-4 **Investigation of Electrical Properties of DNA-Attached Carbon Nano-Particles for Biological-Applications**
M. Ouyang[1], W. J. Li[2], K. W. Wong[3], W. K. Liu[1]
[1]The *Chinese* University of Hong Kong, HONG KONG
[2]City *University* of Hong Kong, HONG KONG
[3]Chengdu *Green* Energy and Green Manufacturing Technology R&D Center, CHINA

Th3B-5L **Evaluation of Negative Photoresists on Phenotypes of Human Indiced Pluripotent Stem Cells (hiPSCs)**
K. Kamei, Y. Hirai, Y. Makino, L. Liu, Q. Yuan, M. Yoshioka, Y. Chen, O. Tabata
Kyoto University, JAPAN

ROOM C (Meeting Room IV)

9:00 - 10:15   Th1C   Particle Assembly
             Session Chair: **Koji Sugano**,   Kyoto University, JAPAN

Th1C-1 **Electrical Performance of Micro-Assembled Beads under Different Temperatures and Loadings**
Y.-L. Tzeng, K. Wang
National Changhua University of Education, TAIWAN

Th1C-2 **Local Particle Assembly Using a Microfluidic Setup**
M. J. K. Klein[1], T. Tamulevicius[2], M. Manning[3], U. Drechsler[1], C. Kümin[1], T. Visegrady[1], H. Wolf[1]
[1]IBM Research GmbH, SWITZERLAND
[2]Kaunas University, LITHUANIA
[3]Union College, USA

Th1C-3 **Vertical Deposition of Nanospheres on the Open Sidewalls of Silicon Pillars**
Y. F. Wang, Y. Tian, K. J. Feng, C. Li, D. D. She, W. G. Wu
Peking University, CHINA

Th1C-4 **Hydrodynamic Trap for Directed Self-Assembly of MEMS**
M. R. Gullo, L. Jacot-Descombes, J. Brugger
Ecole Polytechnique Fédérale de Lausanne, SWITZERLAND

10:45 - 12:00   Th2C   Nanostructures
             Session Chair: **Litao Sun**,   Southeast University, CHINA

Th2C-1 **Nanoscale Laser Writing of Indium-Tin-Oxide Nanowires**
M. Afshar, D. Feili, H. Voellm, M. Straub, K. Koenig, H. Seidel
Saarland University, GERMANY

Th2C-2 **Controllable Diameter of Naturally Polymer-Coated Gold Nanowires and Their Biocompability**
H.-H. Hsieh, C.-W. Chou
China Medical University, TAIWAN

Th2C-3 **Mechanical Property Characterizations of Complex Shaped Helical Nanowires**
G. Hwang[1], L. Couraud[1], R. Braive[1], I. Robert-Philip[1], I. Sagnes[1], S. Bouchoule[1], L. Yu[2]
[1]Laboratoire de Photonique et de Nanostructures, FRANCE
[2]Laboratoire de Photonique des Interfaces et Couches Minces, FRANCE

Oral Sessions   Thursday, March 8

Th2C-4   **Facile Controlled Preparation of Natural Polysaccharide-Capped Gold Nanostructures**
          C.-W. Chou, S.-Y. Hung
          China Medical University, TAIWAN

Th2C-5   **Fabrication and Characters of Squama-Shape Micro/Nano Multi-Scale Structures**
          X.-S. Zhang, F.-Y. Zhu, H.-X. Zhang
          Peking University, CHINA

13:00 - 14:30   Th3C   Energy
                 Session Chair: **Dong Fang Wang,**   Ibaraki University, JAPAN

Th3C-1   **Development of a Silicon-Based Suspending Microthermoelectric Generator with Series Array Structure Using Surface Micromachining Technology**
          G.-M. Chen, I.-Y. Huang, T.-Y. Wu
          National Sun Yat-sen University, TAIWAN

Th3C-2   **A High Efficient POM Micro-Methanol Reformer**
          H.-S. Wang, K.-Y. Huang, H.-C. Peng, Y.-J. Huang, F.-G. Tseng
          National Tsing Hua University, TAIWAN

Th3C-3   **Proton Exchange Membranes Based on Aryl Epoxy Resin for Fuel Cells Operated at Elevated Temperatures**
          T.-Y. Lee[1], T.-C. Ho[1], C.-J. Chang[2], P.-C. Wang[1], F.-G. Tseng[1]
          [1]National Tsing Hua University, TAIWAN
          [2]Microsystem Technology Center, TAIWAN

Th3C-4L   **Oscillating Type Piezoelectric DC Current Sensor Integrated with a Micro Magnet**
          Y. Suzuki[1], D. F. Wang[1], T. Kobayashi[2], K. Isagawa[1], T. Itoh[2], R. Maeda[2]
          [1]Ibaraki University, JAPAN
          [2]National Institute of Advanced Industrial Science and Technology (AIST), JAPAN

ROOM D (Meeting Room III)

9:00 - 10:15   Th1D   Bulk Resonators
                Session Chair: **Sang-Seok Lee,**   Tottori University, JAPAN

Th1D-1   **Material Nonlinearity Limits on a Lamé-Mode Single Crystal Bulk Resonator**
          H. Zhu, C. Tu, J. E.-Y. Lee
          City University of Hong Kong, HONG KONG

Th1D-2   **Evidence on the Impact of T-Shaped Tether Variations on Q Factor of Bulk-Mode Square-Plate Resonators**
          Y. Xu, J. E.-Y. Lee
          City University of Hong Kong, HONG KONG

Th1D-3   **Analysis of Air Damping in Micromachined Resonators**
          G. Wu, D. Xu, B. Xiong, Y. Ma, Y. Wang, E. Jing
          Shanghai Institute of Microsystem and Information Technology, CHINA

Th1D-4   **Benchmarking the Passive Differential Input Technique to Shielded GSG Probes**
          Y. Xu, H. Zhu, J. E.-Y. Lee
          City University of Hong Kong, HONG KONG

Oral Sessions   Thursday, March 8

Th1D-5   **Study on Thermoelastic Dissipation in Bulk Mode Resonators with Etch Holes**
         C. Tu, J. E.-Y. Lee
         City University of Hong Kong, HONG KONG

10:45 - 12:00   **Th2D**   Physical Sensing
                 Session Chair: **Dong Fang Wang**,   Ibaraki University, JAPAN

Th2D-1   **A Piezoresistive Normal and Shear Force Sensor Using Liquid Metal Alloy as Gauge Material**
         X. Shi, C.-H. Cheng, C. Chao, L. Wang, Y. Zheng
         The Hong Kong Polytechnic University, HONG KONG

Th2D-2   **Development of A Novel Force Sensor System Built with an Industrial Multilayer Ceramic Capacitor (MLCC)**
         K.-R. Lin[1], C.-H. Chiang[2], C.-H. Chang[1], C.-H. Lin[2]
         [1]National Cheng Kung University, TAIWAN
         [2]National Sun Yat-Sen University, TAIWAN

Th2D-3   **High Spatial Resolution 2D Cell Traction Force Measurement by Light Scattering from Nanopillar Array**
         L.-M. Liu, F.-G. Tseng, C.-C. Chen
         National Tsing Hua University, TAIWAN

Th2D-4L (withdrawn)

13:00 - 14:30   **Th3D**   Chemical & Molecular Sensing II
                 Session Chair: **Pen Cheng Wang**,   National Tsing Hua University, TAIWAN

Th3D-1   **Implementation of a Subwavelength Bragg Reflector for Terahertz Applications**
         V. Singal, S. Smaili, Y. Massoud
         University of Alabama at Birmingham, USA

Th3D-2   **On Sustaining Robustness of Molecular Pathway Circuits of the HSR Network of *E. coli* under Spatial Configuration**
         J.-Q. Liu[1], T. Yamanishi[2], H. Nishimura[3], S. Nobukawa[3], H. Umehara[1]
         [1]National Institute of Information and Communications Technology, JAPAN
         [2]Fukui University of Technology, JAPAN
         [3]University of Hyogo, JAPAN

Th3D-3L **Temperature Effects on Calixarene Capped Silver Nanoparticle Sensing of Nucleotides**
         Y. Tauran[1], R. Ueno[1], A. W. Coleman[2], B. Kim[1]
         [1]The University of Tokyo, JAPAN
         [2]University of Lyon, FRANCE

Th3D-4L **Scale Effect on Electrochemical Impedance of Nanoelectrode**
         A. Inaba, K. Matsumoto, I. Shimoyama
         The University of Tokyo, JAPAN

Th3D-5L **Fabrication of Nanogap Electrodes by Gold Nanorod Growth on Substrate**
         S. Nishino, Y. Takenaka, Y. Hirai, K. Sugano, T. Tsuchiya, O. Tabata
         Kyoto University, JAPAN

Poster Session   Tuesday, March 6

POSTER SESSIONS

Tuesday, March 6, 2012

13:00 – 14:30   T2P   Poster Session I

T2P-1   **Numerical Simulation of CZTS Thin Film Solar Cell**
W. Zhao, W. Zhou, X. Miao
Huazhong University of Science and Technology, CHINA

T2P-2   **Enhanced Raman Signal of Graphene between a Gold Layer and Gold Nanoparticles**
H.-Y. Hsieh[1], Y.-H. Lee[2], L.-J. Li[2], P.-C. Wang[1], F.-G. Tseng[1]
[1]National Tsing Hua University, TAIWAN
[2]Academia Sinica, TAIWAN

T2P-3   **On the Design of Subwavelength Waveguiding Structures for Terahertz Applications**
V. Singal, S. Smaili, Y. Massoud
University of Alabama at Birmingham, USA

T2P-4   **PPy Nanowires Array Entrapped with Glucose Oxidase for Glucose Detection**
G. Q. Xu, J. Lv, Z. X. Zheng, Y. C. Wu
Hefei University of Technology, CHINA

T2P-5   **Direct Electrochemistry of Cholesterol Oxidase on Multi-Wall Carbon Nanotubes and its Application for Cholesterol Determination**
S. Pakapongpan[1], P. Sritongkham[2], A. Tuantranont[1]
[1]National Electronics and Computer Technology Center, THAILAND
[2]Mahidol University, THAILAND

T2P-6   **The Electromigration Investigation of Cu-Ni Nanocomposites**
Y. C. Chen, C. W. Chu, T.-Y. Chao, Y. T. Cheng, C. Chen
National Chiao Tung University, TAIWAN

T2P-7   **Fabrication of Ternary Metal Sulfide Nanowires Through Solid State Reaction**
C.-K. Wu, Y.-C. Li, C.-P. Liu
National Cheng Kung University, TAIWAN

T2P-8   (withdrawn)

T2P-9   (withdrawn)

T2P-10   **A SWNTs Thin Film Solar Microcell Prepared by Simple Solution-Evaporation Method**
C. C. Chen, Y. Y. Chang, J. Zhang
Peking University, CHINA

T2P-11   **High Efficient Heat Removal By Droplet Impinging on Nano Structured Silicon Surface with Straight CNTs Array**
C.-J. Chen, C. Lin, C. C. Chieng, F.-G. Tseng
National Tsing Hua University, TAIWAN

T2P-12   **Fabrication of Hydrogel-Based Antibody Microarray for Immunoassays**
C.-T. Huang[1], C.-H. Chuang[2], C.-P. Jen[1]
[1]National Chung Cheng University, TAIWAN
[2]Southern Taiwan University, TAIWAN

Poster Session   Tuesday, March 6

T2P-13   **Dielectrophoretic Preconcentration for Cells Utilizing Dual-Planar Electrodes in Stepping Electric Fields**
          C.-Y. Hsieh, C.-P. Jen
          National Chung Cheng University, TAIWAN

T2P-14   **Using Cross-Flow Filtration Chip to Collect Plasma from Whole Blood**
          C.-H. Yeh[1], C.-W. Hung[1], C.-H. Wu[2], Y.-C. Lin[1]
          [1]National Cheng Kung University, TAIWAN
          [2]Ritek Corporation, TAIWAN

T2P-15   **Investigation of Particle Dispersion and Deposition in a Channel with Elliptic Obstruction Using Lattice Boltzmann Method**
          A. Tehrani, A. Moosavi
          Sharif University of Technology, IRAN

T2P-16   **A Study of Hydrogen Peroxide Microfluidic Fuel Cells**
          J.-C. Shyu[1], C.-L. Huang[1], T.-S. Sheu[2], H. Ay[1], J.-W. Huang[2]
          [1]National Kaohsiung University of Applied Sciences, TAIWAN
          [2]R. O. C. Military Academy, TAIWAN

T2P-17   **Microfluidics Circulatory System for the Raise of Liver Urea Assay**
          Y.-S. Chen, T.-H. Dai, L.-Y. Ke, C.-H. Liu
          National Tsing Hua University, TAIWAN

T2P-18   **Droplet Evaporation Based Ring Type for DNA Separation**
          C.-S. Yu[1], Y.-C. Ou[1], C.-C. Yang[1], M.-Y. Lin[1], F.-G. Tseng[2]
          [1]National Applied Research Laboratories , TAIWAN
          [2]National Tsing Hua University, TAIWAN

T2P-19   **A Smart Microfluid Device for Electrodeposition in a Single Droplet**
          C.-T. Lin[1], C.-S. Yu[1], C.-C. Yang[1], Y.-C. Ou[1], F.-G. Tseng[2], J.-S. Kao[1], M.-H. Shiao[1]
          [1]National Applied Research Laboratories, TAIWAN
          [2]National Tsing Hua University, TAIWAN

T2P-20   **Diffuser-Type Micro Fluidic System for High-Throughput Sperm Sorting Based on Sperm's Moving Against Flow Behavior**
          P.-C. Chen[1], R.-G. Wu[1], L.-C. Pan[2], F.-G. Tseng[1]
          [1]National Tsing Hua University, TAIWAN
          [2]Taipei Medical University, TAIWAN

T2P-21   **Real-Time Droplet-Based Polymerase Chain Reaction Detection System by Convection Flow**
          C.-S. Yu[1], C.-C. Yang[1], Y.-C. Ou[1], J.-S. Kao[1], F.-G. Tseng[2]
          [1]National Applied Research Laboratories, TAIWAN
          [2]National Tsing Hua University, TAIWAN

T2P-22   (withdrawn)

T2P-23   **Dose-Dependent Inverse Relationship of Gold Nanoparticles Concentration and the NIH-3T3 Fibroblast Cell Viability and Proliferation**
          C. Danladkaew, A. Sereemaspun
          Chulalongkorn University, THAILAND

T2P-24   **Preparation and Physicochemical Study of Liposomes Containing Nicotinamide**
          N. Langlah, S. Pinsuwan, T. Amnuaikit
          Prince of Songkla University, THAILAND

Poster Session   Tuesday, March 6

T2P-25   **A Novel Fe$_3$O$_4$-Gold Chitosan-Polyethylene Glycols Hydrogel Beads Prepared by Photochemical Green Method and Their Catalytic Application**
T.-H. Yeh, C.-W. Chou
China Medical University, TAIWAN

T2P-26   **Fabrication of Controllable Profile Microlens Array by Nanoimprinting Process**
M. C. Cheng, L. K. Chen, C. K. Sung
National Tsing Hua University, TAIWAN

T2P-27   **Reversible Creation of Nanostructures between Identical or Different Species of Materials**
J. H. Park[1], H. I. Jang[1,2], J. Y. Park[3], D. E. Lee[4], S. W. Jeon[3], C. W. Ahn[1], K. S. Yoo[2]
[1]Korea National NanoFab Center, KOREA
[2]University of Seoul, KOREA
[3]Korea Advanced Institute of Science and Technology, KOREA
[4]Seoul National University, KOREA

T2P-28   **Analysis on 3-Dimensional Spatial Electric Field of AFM Based Anodic Oxidation**
Z. L. Liu[1], N. D. Jiao[1], Z. D. Wang[2], Z. L. Dong[1]
[1]Shenyang Institute of Automation Chinese Academy of Sciences, CHINA
[2]Chiba Institute of Technology, JAPAN

T2P-29   **Scanning Electron Beam Induced Deposition for Conductive Tip Modification**
P.-L. Chen[1], J. Su[1], M.-H. Shiao[1], M.-N. Chang[2], C.-H. Lee[3], C. W. Liu[3]
[1]Instrument Technology Research Center, TAIWAN
[2]National Chung Hsing University, TAIWAN
[3]National Taiwan University, TAIWAN

T2P-30   **Application of Surface Modification and Photo-Etching of Polytetrafluoroethylene for LIGA**
H. Kido[1], T. Kuroki[2], M. Okubo[2], Y. Utsumi[1]
[1]University of Hyogo, JAPAN
[2]Osaka Prefecture University, JAPAN

T2P-31   **In situ Study of Thermal Deformation of Metal Resistive Heater on Silicon Nitride Membrane by Digital Holographic Microscopy**
Y. W. Lai, J. E.-Y. Lee
City University of Hong Kong, HONG KONG

T2P-32   **Comparison of Glass Etching Properties Between HCl and HNO$_3$ Solution**
W. Tao, W. Lv, Z. Zhan, W. Zuo, X. Qiu, L. Wang, D. Sun
Xiamen University, CHINA

T2P-33   **The Study on Deep X-Ray Lithography to Fabricate SU-8 Hard Mask of Burnishing Head Patterns**
C. Maneekat[1], K. Siangchaew[2], R. Phatthanakun[3], K. Leksakul[1]
[1]Chiang Mai University, THAILAND
[2]Western Digital (Thailand) Company Limited, THAILAND
[3]Synchrotron Light Research Institute, THAILAND

T2P-34   **Surface Analysis and Process Optimization of Black Silicon**
F.-Y. Zhu[1,2], Q.-L. Di[1], X.-J. Zeng[1], X.-S. Zhang[1], X. Zhao[2], H.-X. Zhang[1]
[1]Peking University, CHINA
[2]NanKai University, CHINA

Poster Session   Tuesday, March 6

T2P-35   **Electrochemical Ethanol Sensor Based on NiFe Alloyed Thin Film Prepared by Co-Sputtering**
          W. Srichaisiriwech[1], A. Wisitsoraat[2], D. Phokharatkul[2], A. Tuantranont[2], T. Kerdcharoen[3]
          [1]Mahidol University, THAILAND
          [2]National Electronics and Computer Technology Center, THAILAND
          [3]National Nanotechnology Center, THAILAND

T2P-36   **Application of Nonlinear Driving in Frequency Matching of Tunneling Gyroscope**
          L. Wang, X. Du, Y. Su, Z. Zhan, W. Zuo, D. Sun
          Xiamen University, CHINA

T2P-37   **Integrated Flexible Micro Pressure, Temperature and Flow Sensors for Use in PEMFC**
          C.-Y. Lee, T. Yang, Y.-M. Lee, T.-H. Chien, Y.-T. Cheng
          Yuan Ze University, TAIWAN

T2P-38   **Performance of an AC-EO Micromixer with FAPPES in Mixing the Nano- and Micro-Scale Bio-Particles**
          J.-L. Chen[1], W.-H. Shih[2], W.-H. Hsieh[1]
          [1]National Chung Cheng University, TAIWAN
          [2]Metal Industries Research & Development Centrer, TAIWAN

T2P-39   **Sensitivity Enhancement in SGOI Nanowire Biosensor Fabricated by Top Surface Passivation**
          K.-M. Chang[1], C.-F. Chen[1], C.-H. Lai[2], C.-T. Hsieh[1], C.-N. Wu[1], Y.-B. Wang[1], C.-H. Liu[1]
          [1]National Chiao Tung University, TAIWAN
          [2]Chung Hua University, TAIWAN

T2P-40   **A Silicon-on-Glass Z-Axis Accelerometer with Vertical Sensing Comb Capacitors**
          J. Wang, Z. Yang, G. Yan
          Peking University , CHINA

T2P-41   **The Vertical MSM Diamond X-Ray Detector**
          S. Cheirsirikul[1], S. Jesen[1], C. Hruanun[2]
          [1]King Mongkut's Institute of Technology Ladkrabang, THAILAND
          [2]National Electronics and Computer Center, THAILAND

T2P-42   **Effect of Geometrical Design of Support on Frequency Shift and Energy Loss of Piezoelectric Ring Resonator Applicable to Liquid Circumstance**
          T. Sagawa[1], D. F. Wang[1], J. Lu[2], R. Maeda[2]
          [1]Ibaraki University, JAPAN
          [2]National Institute of Advanced Industrial Science and Technology (AIST), JAPAN

T2P-43   **Effective Force Generation for ECLIA Composed of Si Bone Structure and Conductive Polymer Flexible Slider**
          T. A. Nguyen, S. Konishi
          Ritsumeikan University, JAPAN

T2P-44   **The Study of Forward and Reverse Schottky Junction for Dual Magnetodiode**
          T. Phetchakul[1], W. Luanatikomkul[1], W. Yamwong[2], A. Poyai[2]
          [1]King Mongkut's Institute of Technology Ladkrabang, THAILAND
          [2]National Electronics and Computer Technology Center, THAILAND

T2P-45   **Electrode Design Optimization of a CMOS Fringing-Field Capacitive Sensor**
          Y.-T. Li, Y.-L. Tzeng, C.-M. Chao, K. Wang
          National Changhua University of Education, TAIWAN

T2P-46   **Dimension and Capillary Effects of Microfluidic Channel for Blood Plasma Separation**
          Y.-H. Zhan, J.-N. Kuo
          National Formosa University, TAIWAN

Poster Session

Tuesday, March 6

T2P-47   **Fabrication and Analysis of Integrated MEMS Pyramidal Horn Antenna for Terahertz Applications**
C. Li[1], L. Guo[2], W. G. Wu[1], X. S. Tang[2], F. Y. Huang[2]
[1]Peking University, CHINA
[2]Southeast University, CHINA

T2P-48   **Characterization of Super-Harmonic Effect Using Piezoelectric Film Cantilever with a Proof Mass in the Point**
H. Ishinabe[1], T. Kobayashi[2], D. F. Wang[1], T. Itoh[2], R. Maeda[2]
[1]Ibaraki University, JAPAN
[2]National Institute of Advanced Industrial Science and Technology (AIST), JAPAN

T2P-49L   **The Optical Properties of Ga_2S_3 Nanowires**
Y.-W. Cheng, Y.-C. Li, C.-P. Liu
National Cheng Kung University, TAIWAN

T2P-50L   **Precisely Controlled Micro Droplet Merging Device Using Horizontal Pneumatic Valves**
M. Igaki, D. H. Yoon, T. Sekiguchi, S. Shoji
Waseda University, JAPAN

T2P-51L   **Near-Field Electrospinning for Preparation of Piezoelectric Microfibers Based Cantilever**
X. Li[1], D. F. Wang[1], R. Maeda[2]
[1]Ibaraki University, JAPAN
[2]National Institute of Advanced Industrial Science and Technology (AIST), JAPAN

T2P-52L   **Particle Sensor Using an Ultra-Thin Piezo Resistive Cantilever**
H. Takahashi, T. Kan, K. Matsumoto, I. Shimoyama
The University of Tokyo, JAPAN

Wednesday, March 7, 2012

13:00 – 14:30   W3P   Poster Session II

W3P-1   **High-Q Maintenance of Microcavity by Using a Sealed and Packaged Structure**
S.-B. Yan, Y.-Z. Yan, Y.-G. Zhang, L. Wang, C.-Y. Xue, J. Liu, W.-D. Zhang, J.-J. Xiong
North University of China, CHINA

W3P-2   **On the Effect of Width of Metallic Armchair Graphene Nanoribbons in Plasmonic Waveguide Applications**
S. Smaili, V. Singal, Y. Massoud
University of Alabama at Birmingham, USA

W3P-3   **Growth Mechanism of Tellurium Nanotubes and Their Room-Temperature CO Sensing Properties**
Y.-C. Her, S.-L. Huang
National Chung Hsing University, TAIWAN

W3P-4   **Disposable Inkjet-Printed Graphene-Based Electrochemical Sensor on Paper-Based Devices**
C. Karuwan, C. Sriprachuabwong, A. Wisitsoraat, D. Phokharatkul, A. Tuantranont
National Electronics and Computer Technology Center, THAILAND

W3P-5   **The Facile Transferral of Graphene onto Interdigitated Electrodes for Sensing Applications**
C. Dale, S. Rana, R. H. Page, J. Hedley, N. Keegan
Newcastle University, UNITED KINGDOM

W3P-6   **Fabrication and Hydrophilic Property of Titanium Dioxide Thin Film Using Sol-Gel Method and CO_2 Laser Irradiation**
C. K. Chung, S. L. Lin, K. P. Chuang, K. Y. Shie, K. Z. Tu
National Cheng Kung University, TAIWAN

W3P-7   **Enhanced Biocompatibility and Catalytic Activity of Size Controlled Gold Nanoparticles by Photochemical Green Synthesis Method**
X.-R. Liu, C.-W. Chou
China Medical University, TAIWAN

W3P-8   (withdrawn)

W3P-9   (withdrawn)

W3P-10   **Excitation of Mechanical Oscillations in Double-Carbon-Nanotube System by Terahertz Radiation**
V. Semenenko[1], V. Leiman[1], A. Arsenin[1], Y. Stebunov[1], V. Ryzhii[2]
[1]Moscow Institute of Physics and Technology, RUSSIA
[2]University of Aizu, JAPAN

W3P-11   **Rapid and Sensitive Detection of Shrimp Taura Syndrome Virus by Loop-Mediated Isothermal Amplification Combined with Quartz Crystal Microbalance**
W. Kiatpathomchai, T. Kaewphinit, J. Phromjai, W. Jaroenram, S. Santiwatanakul, K. Chansiri, A. Tuantranont
National Center for Genetic Engineering and Biotechnology, THAILAND

W3P-12   **Level Set Simulation of Droplet Formation in a T-Shaped Microchannel**
Y. Yan, D. Guo, S. Z. Wen
Tsing Hua University, CHINA

Poster Session   Wednesday, March 7

W3P-13 Protein Preconcentration Utilizing Nanogaps Formed by Junction Gap Breakdown
C.-C. Kuo, C.-P. Jen
National Chung Cheng University, TAIWAN

W3P-14 Capillary Kinetics of Water in Hydrophilic Microscope Coverslip Nanochannels
J.-N. Kuo, Y.-K. Lin
National Formosa University, TAIWAN

W3P-15 A Visualization Study of Venting Gas via Hydrophobic Nanoporous Membrane
J.-C. Shyu[1], S.-M. Dai[1], K.-S. Yang[2], C.-C. Wang[3]
[1]National Kaohsiung University of Applied Sciences, TAIWAN
[2]Industrial Technology and Research Institute, TAIWAN
[3]National Chiao Tung University, TAIWAN

W3P-16 Switching Characteristic Model and Biochemical Application Analysis for Electrolyte-Oxide-Semiconductor Structure Diodes
G. C. Sun[1], X. Y. Ma[1], A. S. Tang[2], Y. F. Chen[1], W. G. Wu[1]
[1]Peking University, CHINA
[2]Massachusetts Institute of Technology, USA

W3P-17 A 3D Micro-Channel Cooling System Embedded in LTCC Packaging Substrate
S. Jia[1], M. Miao[2], R. Fang[1], S. Guo[1], D. Hu[1], Y. Jin[1]
[1]Peking University, CHINA
[2]Beijing Information Science & Technology University, CHINA

W3P-18 A Flexible Evaporation Micropump with Precision Flow Rate Control for Micro-Fluidic Systems
K.-Y. Chen, K.-E. Chen, K. Wang
National Changhua University of Education, TAIWAN

W3P-19 Nanofluidic Device with Self-Assembled Nafion Membrane Utilizing Capillary Valve
S. Wang, H. Yu, W. Wang, Z. Li
Peking University, CHINA

W3P-20 Effect of Coating Organic Film on Dropwise Condensation in Microgrooves with Nanostructured Surface
T.-S. Sheu[1], J.-C. Shyu[2], J.-W. Hsiao[1], Y.-C. Pan[1]
[1]R. O. C. Military Academy, TAIWAN
[2]National Kaohsiung University of Applied Sciences, TAIWAN

W3P-21 A Remark on the Crucial Point with Respect to the Robustness of the Heat Shock Response Pathway of *E. coli* under the Variance of the Noise Scale
J.-Q. Liu[1], T. Yamanishi[2], H. Nishimura[3], S. Nobukawa[3], H. Umehara[1]
[1]National Institute of Information and Communications Technology, JAPAN
[2]Fukui University of Technology, JAPAN
[3]University of Hyogo, JAPAN

W3P-22 Novel Core Etching Technique on Synthesized Gold Nanoparticles for Colorimetric Detection of Dopamine Biosample
H.-C. Lee, T.-H. Chen, W.-L. Tseng, C.-H. Lin
National Sun Yat-sen University, TAIWAN

W3P-23 Fast Self-Resonant Startup Procedure for Digital MEMS Gyroscope System
F. Ge, D. Liu, L. Lin, Z. Yang, G. Yan
Peking University, CHINA

Poster Session   Wednesday, March 7

W3P-24   **Green Synthesis of Magnetic Core-Shell Fe$_3$O$_4$-Au Nanoparticles**
         C.-H. Lin, C.-W. Chou, C.-L. Yang
         China Medical University, TAIWAN

W3P-25   **The Influence of Experimental Parameters on the Assembly of SWNTs by AC Dielectrophoresis**
         Z. Wang, F. Yu, W. Li, J. Zhang
         Peking University, CHINA

W3P-26   **Chemical Vapor Deposited Graphene Layers on Sputtered Stainless Steel Thin Film for Electrochemical Sensing**
         D. Phokharatkul, A. Wisitsoraat, C. Sriprachuabwong, T. Pogfay, A. Tuantranont
         National Electronics and Computer Technology Center, THAILAND

W3P-27   **The Manufacture of Micropillars with High Depth-to-Width Ratio, and the Comparison between Two Typical Materials**
         Z. Wang[1], X. Qin[2]
         [1]Singapore-MIT Alliance for Research and Technology Center, CHINA
         [2]Tsinghua University, CHINA

W3P-28   **Surface-Modified Diamond Embedded in Nickel Matrix Composite for Intrinsic Polishing Application**
         C.-J. Shih, W.-C. Lin, C.-S. Lin, Y.-N. Pan
         National Taiwan University, TAIWAN

W3P-29   **Fabrication and Performance Optimization of the Microplasma Reactor with Composite Dielectrics**
         Z. Yuan, L. Wen, L. Cheng, J. Ma, J. Chu
         University of Science and Technology of China, CHINA

W3P-30   **The InGaN/GaN Mutiple Quantum Wells Nanopillars Light Emitting Doides Fabricated by Focused Ion Beam**
         C.-H. Wang[1], Y.-W. Huang[1], C. H. Tu[1], S. E. Wu[2], C.-P. Liu[1]
         [1]National Cheng Kung University, TAIWAN
         [2]Genesis Photonics Inc., TAIWAN

W3P-31   **Fabrication of Deep Lateral Single-Crystal-Silicon Blaze Micro-Grating by Inductively-Coupled-Plasma Riactive Ion Etch**
         Y.-H. Lin[1], C. J. Weng[2], C. Y. Su[2], W. Hsu[1]
         [1]National Chiao Tung University, TAIWAN
         [2]National Applied Research Laboratories, TAIWAN

W3P-32   **Surface Morphology Dependent Depolarization Properties of Nano-Porous Alumina**
         C.-W. Tseng[1], Y.-H. Pai[1], G.-R. Lin[2]
         [1]National Dong Hwa University, TAIWAN
         [2]National Taiwan University, TAIWAN

W3P-33   **Improvement of Coating Uniformity for Thick Photoresist Using a Partial Spray Coat**
         M. Akamatsu, K. Terao, H. Takao, F. Shimokawa, F. Oohira, T. Suzuki
         Kagawa University, JAPAN

W3P-34   (withdrawn)

W3P-35   **A Study of Tin Oxide Thin Tilm Gas Sensors with High Oxygen Vacancies**
         C. Lin[1], D. Zhang[1], X. Liu[2]
         [1]Peking University, CHINA
         [2]BOE Technology Group Co., Ltd., CHINA

Poster Session   Wednesday, March 7

W3P-36 *In situ* **Monitoring of Temperature Using Flexible Micro Temperature Sensors Inside Polymer Lithium-Ion Battery**
C.-Y. Lee, S.-J. Lee, Y.-M. Lee, M.-S. Tang, P.-C. Chen, Y.-M. Chang
Yuan Ze University, TAIWAN

W3P-37 **Design and Simulation of Fully-Symmetrical Resonant Pressure Sensor**
Y. Jiang, X. Du, Z. Zhan, B. Xu, W. Lv, L. Wang, D. Sun
Xiamen University, CHINA

W3P-38 **Effect of Oxidation on SGOI Nanowire Biosensor Fabrication Using Ge Condensation**
K.-M. Chang[1], C.-F. Chen[1], C.-H. Lai[2], C.-N. Wu[1], C.-T. Hsieh[1], Y.-B. Wang[1], C.-H. Liu[1]
[1]National Chiao Tung University, TAIWAN
[2]Chung Hua University, TAIWAN

W3P-39 **A Capacitive Readout Circuit with DC Sensing Method for Micromachined Gyroscopes**
K. Zhou, L. Sun, F. Ge, Z. Yang, G. Yan
Peking University, CHINA

W3P-40 **The Optimal Vibrational Shear Stress for Bovine Endothelial Cell Proliferation**
C.-W. Li[1], J.-L. Chen[1], C.-C. Wu[2], G.-J. Wang[1]
[1]National Chung Hsing University, TAIWAN
[2]National Cheng Kung University, TAIWAN

W3P-41 **An Optimized Fabrication of High Yield CMOS-Compatible Silicon Carbide Capacitive Pressure Sensors**
B. Meng, W. Tang, Z. R. Wang, H. X. Zhang
Peking University, CHINA

W3P-42 (withdrawn)

W3P-43 **Study of Thin Film Adhesion Properties of Multi-Layer Flexible Electronics Composites**
C. C. Li[1], Z. H. Liu[1], C. T. Pan[1], J. K. Tseng[2], H. L. Huang[2], S. W. Mao[2], S.-C. Shen[3], S. J. Chang[4]
[1]National Sun Yat-Sen University, TAIWAN
[2]R.O.C Military Academy, TAIWAN
[3]National Cheng Kung University, TAIWAN
[4]National Yunlin University of Science and Technology, TAIWAN

W3P-44 **Using a Canny-Edge-Detection Based Method to Characterize In-Plane Micro-Actuators**
C.-Y. Cheng, Y.-B. Lin, K. Wang
National Changhua University of Education, TAIWAN

W3P-45 **Thermal Switch and Variable Capacitance Designed for Micro Electrostatic Converter by Using CMOS MEMS Process**
J.-C. Chiou[1], L.-C. Chou[1], Y.-L. Lai[2], S.-C. Huang[1]
[1]National Chiao Tung University, TAIWAN
[2]National Chip Implementation Center, TAIWAN

W3P-46 **A Research of the Bandwidth of a Mode-Matching MEMS Vibratory Gyroscope**
C. He, Q. Zhao, J. Cui, Z. Yang, G. Yan
Peking University, CHINA

W3P-47 **Graphene Nanoribbon Based AM Demodulator of Terahertz Radiation**
Y. Stebunov[1], A. Arsenin[1], V. Leiman[1], V. Semenenko[1], V. Ryzhii[2]
[1]Moscow Institute of Physics and Technology, RUSSIA
[2]University of Aizu, JAPAN

Poster Session   Wednesday, March 7

W3P-48   **Atomic Layer Deposited Protective Coatings for Integrated MEMS Flow Sensor**
          D. Li[1], A. Abdulagatov[2], F. Yang[1], D. C. Zhang[1]
          [1]Peking University, CHINA
          [2]University of Colorado, USA

W3P-49L   **Modeling and Elastic Property Simulation of Epoxy-Based Negative Photoresist Using Coarse-Grained Molecular Dynamics**
          H. Yagyu[1], Y. Hirai[2], A. Uesugi[2], Y. Makino[2], K. Sugano[2], T. Tsuchiya[2], O. Tabata[2]
          [1]Mitsuboshi Belting Ltd., JAPAN
          [2]Kyoto University, JAPAN

W3P-50L   **Toward Chemical Sensing on Micro-Sized Interfaces Using a Micro-Fluidic Resonator**
          C. Pigot, A. Hibara
          The University of Tokyo, JAPAN

W3P-52L   **50nm-Thick Piezo-Resistive Cantilever**
          T. Usami, A. Nakai, K. Matsumoto, I. Shimoyama
          The University of Tokyo, JAPAN

W3P-53L   **A MEMS-Based Micro Sensor for the Detection of Formaldehyde Gas**
          B.-J. Kim, J.-H. Yoon, J.-S. Kim
          University of Seoul, KOREA

Welcome Message

Welcome to Kyoto, Japan, and the 7th IEEE International Conference on Nano/Micro Engineered and Molecular Systems (NEMS). This is the first IEEE-NEMS Conference held in Japan and the second IEEE-NEMS Conference held at non-native countries of Chinese. Originally, this IEEE-NEMS Conference series started in 2006 in Zhuhai, China. Since then, having Asia as its arena, IEEE-NEMS has been taking the lead in exploring the new engineering field to realize new functional systems by fusing nanotechnology and Micro Electromechanical Systems (MEMS).

The technical program for this year's conference consists of 3 plenary lectures, 4 invited lectures, 181 contributed oral presentations and 104 contributed poster presentations. The contributed presentations were selected from a total of 337 abstracts including 32 Late-News submitted from 21 countries. The Technical Program Committee (TPC), with 58 members in total, and made up with representatives from three regional divisions, including Asia, America and Europe, volunteered their time to a paper selection process by sharing their technical expertise and insight. In order to facilitate detailed review of the submitted abstracts, each abstract was peer reviewed by three committee members. The oral presentations are arranged as four parallel sessions. The posters are arrangements to be on display throughout the conference at the three main rooms to facilitate interaction and discussion between authors and participants. All presentations at the conference are included in the Technical Digest and the Conference Proceedings CD.

As usual, three types of awards, including the Best Conference Paper Award, the Best Student Paper Award, and the CM HO Best Paper Award in Micro/Nano Fluidics, in honor of Prof. C.-M. Ho of the University of California, Los Angeles, are offered. I believe these awards will greatly encourage researchers in this field, and which will eventually lead to innovative achievements.

Finally, I would like to express my sincere gratitude to everyone who contributed to IEEE-NEMS 2012 Conference. First of all, I would express my great appreciation to all the authors of all the submitted contributions, whose highest quality serves as the solid foundations for the success of this IEEE-NEMS Conference series. I am greatly indebted to the Technical Program Co-Chairs, Prof. Toshiyuki Tsuchiya and Prof. Qing-An Huang, together with three Regional Program Co-Chairs, Prof. Satoshi Konishi, Prof. Steve Tung, and Prof. Bradley Nelson, for their impressively great effort on paper selection and grading process to formulate the excellent conference program of IEEE-NEMS 2012. I am also grateful to the International Advisory Committee and Steering Committee for generously sharing their experiences. Furthermore, I am thankful to all the supporters and exhibitors for their contributions and also an acknowledgement is due to the IEEE Nanotechnology Council for their continuous support of the IEEE-NEMS Conference series. At last but not least, dedicated work for the conference by the staff members of Nano/Micro System Engineering Laboratory, Graduate School of Engineering, Kyoto University and MV Destination

Management Limited were indispensable, and I have no words to express my appreciation to them.

The number seven represents completeness and represents rest because God rested on the seventh day after his work of creation was completed. However, the organization of IEEE-NEMS Conference doesnot rest, but rather, continues and endeavors to make the 7th year as the new starting point for further progress. I hope all of you will enjoy the 7th IEEE-NEMS Conference, which is offered with historical surroundings and atmosphere of Kyoto.

Sincerely,

Osamu Tabata
Kyoto University
General Chair

IEEE-NEMS 2012 General Chair
Osamu Tabata .... Kyoto University, JAPAN

Organizing Committee Co-Chairs
Fumihito Arai .... Nagoya University, JAPAN
Yu-Cheng Lin .... National Cheng Kung University, TAIWAN

Technical Program Committee
Co-Chairs
Toshiyuki Tsuchiya .... Kyoto University, JAPAN
Qing-An Huang .... Southeast University, CHINA
Asian/Oceania Regional Chair
Satoshi Konishi .... Ritsumeikan University, JAPAN
European/African Regional Chair
Bradley Nelson .... ETH Zürich, SWITZERLAND
America's Regional Chair
Steve Tung .... University of Arkansas, USA
Asian/Oceania Members
Fumihito Arai .... Nagoya University, JAPAN
Pei-Zen Chang .... National Taiwan University, TAIWAN
Shuo Hung Chang .... National Taiwan University, TAIWAN
Yu-Ting Cheng .... National Chiao Tung University, TAIWAN
Dong-il Cho... Seoul National University, KOREA
Kunkjin Chun .... Seoul National University, KOREA
Kazuo Hosokawa.. RIKEN, JAPAN
I-Ming Hsing .... HKUST, HONG KONG
Kyung Chun Kim .... Pusan National University, KOREA
Yi-Kuen Lee .... HKUST, HONG KONG
Xinxin Li.. SIMIT-CAS, CHINA
Yung-Chun Lee .... National Cheng Kung University, TAIWAN
Yu-Cheng Lin .... National Cheng Kung University, TAIWAN
Cheng-Hsien Liu .... National Tsing Hua University, TAIWAN
Ai Qun Liu .... Nanyang Technological University, SINGAPORE
Ming Liu .... IMECAS, CHINA
Shoji Maruo.. Yokohama National University, JAPAN
Yuji Miyahara .... Tokyo Medical and Dental University, JAPAN
Keisuke Morishima .... Osaka University, JAPAN
Tianling Ren .... Tsinghua University, CHINA
Shuichi Shoji .... Waseda University, JAPAN

Kahp-Yang Suh	Seoul National University, KOREA
Daoheng Sun	Xiamen University, CHINA
Litao Sun	Southeast University, CHINA
Motofumi Suzuki	Kyoto University, JAPAN
Shoji Takeuchi	The University of Tokyo, JAPAN
Takaaki Suzuki	Kagawa University, JAPAN
Ping Wang	Zhejiang University, CHINA
Xiao-Hong Wang	Tsinghua University, CHINA
Wengang Wu	Peking University, CHINA
Lung-Jieh Yang	Tamkang University, TAIWAN
Weizheng Yuan	Northwestern Polytechnical University, CHINA

European/African Members

Manish K. Tiwari	ETH Zürich, SWITZERLAND
Guus Rijnders	University of Twente, THE NETHERLANDS
Franck Chollet	FEMTO-ST, FRANCE
Francesc Perez-Murano	IMB-CNM/CSIC, SPAIN
Christoph Stampfer	RWTH Aachen University, GERMANY
Maria Lucia Curri	CNR-IPCF, ITALY
Liviu Nicu	LAAS-CNRS, FRANCE
Zachary J. Davis	Danish Technological Institute, DENMARK
Heiko Wolf	IBM Research Laboratory, SWITZERLAND

American Members

Dean Ho	Northwestern University, USA
Tony Huang	Pennsylvania State University, USA
Hongrui Jiang	University of Wisconsin-Madison, USA
Tingrui Pan	UC Davis, USA
Bill Tang	UC Irvine, USA
Pak Kin Wong	University of Arizona, USA
Xiaojing Zhang	University of Texas at Austin, USA
Jeff Wang	Johns Hopkins University, USA
Qiao Lin	Columbia University, USA
Leidong Mao	University of Georgia, USA
Samuel Sia	Columbia University, USA
Gymama Slaughter	University of Maryland, USA

International Advisor Committee
Chair

Tzyh-Jong Tarn .... Washington University, USA

Members

Chih-Ming Ho .... UCLA, USA

Masayoshi Esashi .... Tohoku University, JAPAN

Meyya Meyyappan .... NASA AMES, USA

Nicolas F. de Rooij .... EPFL, SWITZERLAND

Toshio Fukuda .... Nagoya University, JAPAN

IEEE NEMS Conference Steering Committee
Chair

Ning Xi .... Michigan State University, USA

Members

Alice H. X. Zhang .... Peking University, CHINA

Daoheng Sun .... Xiamen University, CHINA

Gwo-Bin Vincent Lee .... National Tsing Hua University, TAIWAN

Osamu Tabata .... Kyoto University, JAPAN

Wen J. Li .... City University of Hong Kong, HONG KONG

Yu-Chong Tai .... California Institute of Technology, USA

Awards Committee
Chair
Nicolas F. de Rooij ....EPFL, SWITZERLAND
Members
Toshio Fukuda..Nagoya University, JAPAN
Chih-Ming Ho .... UCLA, USA
Tzyh-Jong Tarn .... Washington University, USA
Alice H. X. Zhung .... Peking University, CHINA

Publications Committee
Chair
Zhidong Wang .... Chiba University, JAPAN
Co-Chair
Wen J. Li .... City University of Hong Kong, HONG KONG

Local Organizers
Local Arrangements
Yoshikazu Hirai .... Kyoto University, JAPAN
Exhibitions Committee
Koji Sugano... Kyoto University, JAPAN
Conference Management
Miri Okino .... Kyoto University, JAPAN
Momo Watanabe .... Kyoto University, JAPAN
Sachiko Matoba .... Kyoto University, JAPAN
Clemson Lo .... MV Destination Management, HONG KONG

Conference Sponsors

 IEEE

 IEEE Nanotechnology Council

Conference Supporters

 Chinese International NEMS Society (CINS)

 Freiburg Institute for Advanced Studies (FRIAS)

 City University of Hong Kong

 Kyoto-Advanced Nanotechnology Network

 Integrated Nanotechnology Foundation Research Project (Kyoto Univ.)

Financial Sponsors
- The Murata Science Foundation
- The Kyoto University Foundation
- Office of Naval Research Global
- Asian Office of Aerospace Research and Development (AOARD)
- Kyoto Convention Bureau
- Interface Focus (Royal Society Publishing)
- Nanotechnology (IOP)

Special Thanks
Dr. Chong Ong of the Office of Naval Research Global passed away a few months ago. Dr. Ong has been an avid supporter of the IEEE-NEMS Conference series from the beginning. We thank him for his support and pray that he will rest in peace. We will miss him greatly.

Conference at a Glance

Monday, March 5				
16:30-18:30	Welcome Reception (Cafe Restaurant Camphora)			

Tuesday, March 6				
9:15-9:45	Opening Ceremony and Welcome Address (Centennial Hall)			
9:45-12:00	T1G Plenary Lectures			
12:00-13:00	Lunch			
13:00-14:30	T2P Poster Session (Int. Conf. Hall I, II, III)			
	Room A Int. Conf. Hall I	Room B Int. Conf. Hall III	Room C Meeting Room IV	Room D Meeting Room III
14:30-16:00	T3A Invite I / Carbon nanotubes	T3B Nanoimprint	T3C Chemical & Molecular Sensing I	T3D Probe & Cantilever Technology and Application
16:00-16:30	Coffee Break and Poster Inspection			
16:30-18:00	T4A Graphenes	T4B Lithography	T4C OS: Emergence in Chemistry for Integrated Nano System	T4D Physical Sensors

Wednesday, March 7				
9:00-10:15	W1A Invite II /Award Finalist Presentations I	W1B Invite III / Membranes	W1C Fabrication I	W1D FRIAS Special Session
10:15-10:45	Coffee Break			
10:45-12:00	W2A Award Finalist Presentation II	W2B Microchannels and Fluidics	W2C Actuators	W2D DNA origami
12:00-13:00	Lunch			
13:00-14:30	W3P Poster Session (Int. Conf. Hall I, II, III)			
14:30-16:00	W4A Award Finalist Presentation III	W4B Dielectrophoresis	W4C Materials	W4D Photonic Devices
16:00-16:30	Coffee Break and Poster Inspection			
16:30-18:00	W5A Cell Engineering	W5B Microfluidic Devices	W5C Fabrication II	W5D SERS and Quantum Optics
19:00-21:30	Conference Banquet (Kyoto Royal Hotel & SPA)			
9:00-10:15	Th1A Invite IV /Materials	Th1B Biosensors	Th1C Particle Assembly	Th1D Bulk Resonators
10:15-10:45	Coffee Break			
10:45-12:00	Th2A Robotics & Assembly	Th2B Droplet and Bubble Manipulation	Th2C Nanostructures	Th2D Physical Sensing
12:00-13:00	Lunch			
13:00-14:30	Th3A Optical System	Th3B Droplet Technologies	Th3C Energy	Th3D Chemical & Molecular Sensing II

Portable E-Nose Based on Polymer/CNT Sensor Array for Protein-Based Detection

Panida Lorwongtragool[1,2], Thara Seesaard[1], Chadarpon Tongta[3], Teerakiat Kerdcharoen[4*]

[1]Materials Science and Engineering Programme, Faculty of Science, Mahidol University, Bangkok, 10400 Thailand
[2]Faculty of Science and Technology, Rajamangala University of Technology Suvarnabhumi, Nonthaburi, 11000 Thailand
[3]Department of Chemistry, Faculty of Science, Thaksin University, Phatthalung, 93110 Thailand
*[4]NANOTEC Center of Excellence at Mahidol University, National Nanotechnology Center, Bangkok, 10400 Thailand

Abstract— **Portable electronic nose (e-nose) consisting of polymer/carbon nanotube (CNT) sensor array was developed to detect protein-based foods. Gas sensors were fabricated by spin-coating functionalized CNT/polymer nanocomposite materials onto interdigitated electrodes. The sensors were tested with various types of volatile compounds such as ammonia, amine compounds, acetic acid, water and organic solvents in the range of ppm level. It was found that most sensors yield strong signals to ammonia, amine compounds and acetic acid as well, while they present quite low response to organic solvents and water. To understand the relation of the interaction of amine species related to the sensor response, we have performed molecular modelling based on the density functional theory (DFT) on one polymer structure providing the best response to volatile ammonia. Based on the principal component analysis (PCA), this portable e-nose was successfully applied for the classification of the seafood releasing different amount of amine compounds.**

Keywords— *Electronic nose; Amine sensor; Seafood quality assessment; Polymer/CNT sensor*

I. INTRODUCTION

In recent decades there are significantly increasing interest in the applications of e-nose for qualitative analysis of odors. This technology is non-destructive, rapid, reliable and inexpensive [1-2] as compared to the conventional techniques such as GC/MS [3] and most importantly the measurement can be performed at the point of use. For these reasons, numerous publications have been devoted to the development and improvement of the sensing materials, device system as well as methods of data analysis. The performance of e-nose relies on the combination of gas sensor array. It has been recommended to use broadly-cross reactive arrays of chemical sensors for generating a unique pattern for each odorant [4].

For the applications in fish and seafood industry, the research works on amine sensors are focused on improving sensitivity and selectivity as well as decreasing the manufacturing cost. Amines are well-known as common compounds released from bio-degradation process of protein-based foods, as usually referred to Total Volatile Basic Nitrogen (TVB-N) [5-6]. Several research groups developing in the amine sensors have contributed on the electrochemical transducer [7-8], optical transducer [9-10] and gravimetric transducer [11-12] to employ in the e-nose system.

For instance, Rakow et al. [13], Tang et al. [14] and Sutarlie and Yang [15] were successful on the colorimetric responses to amine compounds in range of ppm. However, the sensors based on chemically responsive dyes were also reported about non-reproducibility of the colour response limited with very low concentration of volatile amine [14]. Quartz crystal microbalance (QCM) sensors have also been widely used for the detection of amines [16]. Although the QCM sensor can respond to the compounds and can be reusable after flushing with reference gas or hot air, it requires for a complex fabrication process and expensive equipments [17-18].

In this work, we have extended the development of chemiresistive sensor based on polymer/CNT nanocomposite for the detection of the amine compounds as proposed in the previous work [19-20], by developing more sensors and integrating the sensor array into a handheld e-nose system. The composite materials have been demonstrated for the selectivity and sensitivity to be improved depending on the polymers. The designed e-nose system takes into account the energy consumption and the cost as well as the weight.

The sensing mechanism of the non-conductive polymer added with conductive filler can be simply understood by a fractional volume increase of the polymer matrix [21-22]. The kind of interaction between the analyte gas and the sensing material highly determines the sensitivity, selectivity and reversibility of the gas sensors. For example, the weak interaction of binding molecules on sensing surface will result in low sensing response but better reversibility in turn. In other words, the good response can be achieved in case of the strong binding but the sensor may not recover to the original states [23]. In this paper, the underlying principles of gas adsorption on the polymer have been explained by quantum chemistry calculations based on the density functional theory.

In the real application, the e-nose based on eight elements of polymer/CNT sensors was employed to classify three kinds of seafood such as dried fish, dried squid and dried shrimp, which normally generate different level of total volatile amines.

II. EXPERIMENT

A. Sensor Fabrication

Fabrication method of the sensor based on CNT/polymer was given in the details as presented in the previous work [19].

978-1-4673-1122-9/12 $31.00 © 2012 IEEE

TABLE I. THE CHEMICAL COMPONENTS OF SENSING MATERIALS

Sensor ID	Polymer	Carbon Nanotube	Solvent
S1	Polyvinyl chloride (PVC)	SWNT-COOH	THF
S2	Cumene terminated polystyrene-co-maleic anhydride (Cumene-PSMA)	SWNT-COOH	Acetone
S3	Poly(styrene-co-maleic acid) partial isobutyl/ methyl mixed ester (PSE)	SWNT-COOH	Acetone
S4	Polyvinylpyrrolidon (PVP)	SWNT-COOH	Ethanol
S5	Polyvinyl chloride (PVC)	SWNT-OH	THF
S6	Cumene terminated polystyrene-co-maleic anhydride (Cumene-PSMA)	SWNT-OH	Acetone
S7	Poly(styrene-co-maleic acid) partial isobutyl/ methyl mixed ester (PSE)	SWNT-OH	Acetone
S8	Polyvinylpyrrolidon (PVP)	SWNT-OH	Ethanol

The chemical compounds used as sensing materials are summarised in Table I.

The polymers were varied as 4 types to introduce the difference of the physical and chemical properties for generating the specific patterns of activation across the sensor array, according to the principle of e-nose [24]. For the same reason, two types of the functionalized single-walled carbon nanotubes were used to disperse in each of the polymer matrix. The carboxylic-functionalized single-walled carbon nanotubes (SWNT-COOH) and the hydroxyl functionalized single-walled carbon nanotubes (SWNT-OH) were purchased from Cheap Tube Inc containing 90wt% carbon with 1-2 nm in diameter and 0.5–2.0 μm in length. The degree of functionalization on SWNT is 2.73% and 3.96% for SWNT-COOH and SWNT-OH, respectively, according to the supplier.

Briefly, each polymer was dissolved in the proper solvent to obtain the polymer solutions. CNT powder was loaded as 20 wt% of polymer matrix to obtain a high sensor response according to the recipe as proposed by Piromjitpong, et al [20]. Spin –coating method was used to form the sensing film onto the interdigitated gold electrode by setting the spinning rate around 1500-2000 rpm for 30 s. After film formation, it was heated at 150 °C for 3 h to obtain a stable sensor.

B. Electrical response to the analyte gases

Eight fabricated sensors (S1-S8) were tested to the single volatile compounds: dimethylamine, dipropylamine, pyridine, and ammonium hydroxide, acetic acid, tetrahydrofuran, ethyl alcohol, acetone and water by varying the concentration in a closed chamber as 50, 200, 500 and 1000 ppm. The measurement circuit based on voltage divider method was used to obtain the resistances of each sensor. The data were acquired through a USB DAQ device (NI USB-6008). The resistances are recorded with 2 periods: 2 min to obtain baseline of reference resistance without injecting any volatile

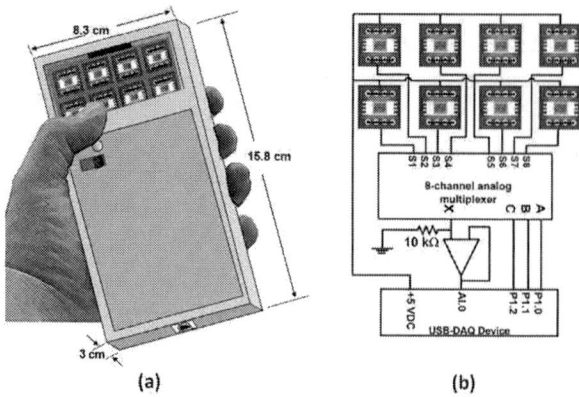

Fig. 1(a) Handheld e-nose and (b) schematic of measurement circuit based on voltage divider method

and 7 min to obtain the responsive resistance with injecting the volatile into the chamber. The last minute was represented as a steady-state condition of the response.

The sensor exposed to the analyte gas was determined from the fractional method [24-25] to obtain a dimensionless response and provide a normalized response signal. The average of baseline resistance in the first period (R_o) is subtracted and then divided from the average of the signals in the steady-state condition (R_s).

$$\% \text{Sensor response} = \frac{R_s - R_o}{R_o} \times 100 \qquad (1)$$

C. Handheld E-Nose System

The handheld e-nose was designed to support key requirements such as lightweight, low power consumption, easy to use, low cost and low maintenance.

Fig.1 (a) shows the picture of the handheld e-nose and Fig. 1(b) shows a measurement circuit inside the case. The system consists of three main parts: (i) air flow unit, (ii) sensing unit and (iii) data acquisition unit [26-28]. All parts are packed in a small plastic case with a dimension of 8.3 cm x15.8 cm x 8cm. The total weight is only 170 g. A small electric fan is used for sucking the odors into the sensing unit. The S1-S8 sensors were placed in the prepared sockets of the sensing unit. For the data acquisition unit, a simple measurement circuit based on the voltage divider method was used and the data were acquired through a USB DAQ device as shown in Fig. 1(b).

D. Seafood Discrimination

In this work, we have illustrated the real-world application by classification of dried seafood which normally generate different amount of amine and ammonia compounds. Dried squid, dried fish and dried shrimp were obtained from a supermarket in Thailand. The closed packages indicating the same manufacturing date were used as the analyte samples. Measurement of each sample odor was performed by the handheld e-nose. 20 g of each sample was kept in a closed glass bottle at room temperature for 10 min to generate the odor into the headspace. The handheld e-nose was used to

978-1-4673-1122-9/12 $31.00 © 2012 IEEE

measure the fresh air to obtain the reference baseline for 2 min. After that, the sample bottle was opened and the handheld e-nose was directly pointed at the top of the bottle for 5 min.

With help of the electric fan inside the e-nose, the sensor signal can quickly reach a steady state. The recorded data were determined in term of the percent sensor response as shown in Eq. (1). Principal component analysis (PCA) was employed to discriminate the states of various protein-containing foods, using the advantage of non-specific chemical interactions in the sensor array.

III. MOLECULAR MODEL

To understand the interaction of the polymer and amine compounds, the PSE presenting the best response to amine and ammonia [19], as compared to other polymers according to the experimental results, was chosen to study by the first principles calculation. Thus, we shall consider only at the active site (carboxyl group of the PSE molecule) which can be adsorbed by base molecules such as ammonia and amine compounds via hydrogen bond interactions [29-32]. For the PSE structure, only the main part of poly(styrene-co-maleic acid) was considered by neglecting the partial part of the isobutyl/methyl mixed ester.

Full geometry optimization of the individual PSE and analyte molecules was carried out using density functional theory (DFT) [33] with Becke's three-parameter functional with the gradient-corrected correlation of Lee, Yang and Parr (B3LYP) [34] along with the 6-31+G(d,p) basis set.

The diffuse function was recommended to be included in the basis set to improve the long-range interaction and provide reliable prediction of the properties of hydrogen-bonded complexes [30]. Conformation and total energy of the complex molecule between the PSE and the base molecule was fully relaxed with the same calculation level. The adsorption energy (ΔE_{ad}) is calculated by

$$\Delta E_{ad} = E_{PSE@Base-molecule} - E_{PSE} - E_{Base-molecule} \qquad (2)$$

Where $E_{PSE@Base-molecule}$ is the total energy of the complex molecule between the PSE and the base molecule. E_{PSE} is the total energy of isolated PSE molecule and $E_{Base-molecule}$ is the total energy of isolated the base-molecule, i.e. ammonia, dimethylamine, pyridine and dipropylamine. Beside the adsorption energy, the intermolecular distance between the nitrogen heteromolecule and the donating proton, $r(H\cdots N)$ and the elongated distance of the covalent bond holding the proton, $r(O-H)$ were also considered.

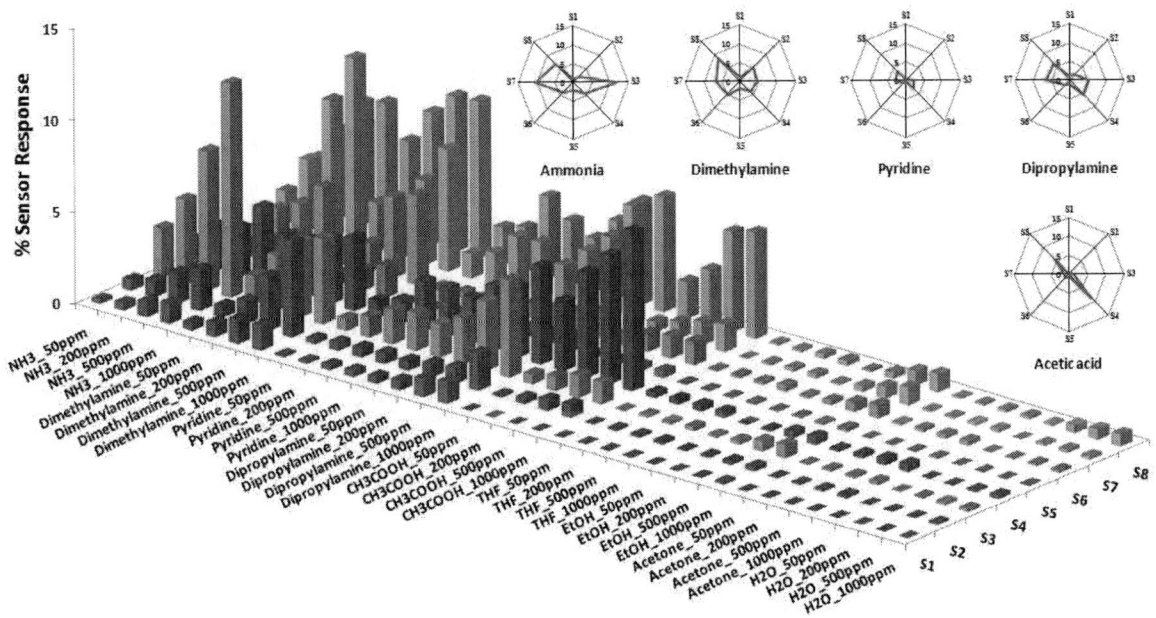

Fig. 2. Percent sensor response of 8 sensor elements (S1-S8) tested with 50 ppm-1000 ppm of ammonia(NH₃), dimethylamine, pyridine, dipropylamine, acetic acid (CH3COOH), tetrahydrofuran (THF),ethanol (EtOH), acetone and water (H2O) and radar plot of the percent sensor response when tested with 1000 ppm of acid-base compounds

IV. RESULTS AND DISCUSSION

Fifty to one thousand parts per million concentrations of analyte gases such as ammonium hydroxide, dimethylamine, pyridine, dipropylamine, acetic acid, tetrahydrofuran, ethanol, acetone and water were used to test the S1-S8 sensors at room temperature. The sensor resistance increases when the analyte gas was injected into the chamber. It is mainly related to the increased volume of the polymer matrix due to the swelling effect [21-22]; however the process of the charge transfer between CNT and the binding gas [35] is also possible [19, 36].

Fig. 2 shows percentage change of the resistance for all sensors as determined by Eq. (1).

According to the experimental results as shown in Fig.2, all sensors show chemoselectivity in the range of the acid-base detection. The S1-S8 sensors can respond to ammonia and amine compounds as well as acetic acid whereas these sensors show low responses to organic solvents such as THF, ethanol acetone and water. The increasing of resistance with rising volatile concentration can be modelled by the Plateau–Bretano–Stevens law [37-38] in accordance with the previous work [19]. The insets in Fig.3 demonstrate the radar plot of the percent sensor responses when the sensors were tested with 1000 ppm of ammonia, amines and acetic acid. The sensor responses show differences in their patterns when exposed with the different volatiles.

Differences of the functional groups existing in the sensing materials can generate various patterns [39]. In this work, we found that varying of polymers has influence to the change of the patterns rather than varying the functionalized CNT. Therefore, the results can imply that the response of the sensor based on CNT/polymer composite is caused by the swelling effect rather than the charge transfer process.

According to the experimental evidence in Fig. 2, we have focused on the PSE polymer (S3 and S7) which presents specific response to basic volatiles such as ammonia and amines, rather than the PVP polymer (S4 and S8) that provide good response to both acid and base molecules. Molecular modelling can give the information to understand the underlying principles of the sensors [40]. Moreover, the computer modelling approaches are possible to be used for obtaining affinity and selective sensors [41].

A carboxyl site on the PSE was chosen as an important target for forming interaction with the base molecule. In addition, the presence of carboxyl group allows the nitrogen heteroatom of ammonia or amines to form hydrogen bond [29-31]. Based on molecular calculation, the strength of intermolecular interaction can be indicated by the adsorption energy, the intermolecular distance $r(H\cdots N)$ and the elongated distance $r(O-H)$ [29]. Table II summarizes the results of the first principles calculation of the PSE/base-molecule complex, pK_b values from literatures and the percent responses of S3 and S7 from the experimental observation whereas the optimized geometries of complex forming are shown in Fig. 3. The calculated results show the strong correlation of three values, $r(O-H)$, $r(H\cdots N)$ and ΔE_{ad}. It was found that the strength of ammonia or amines interaction on the carboxyl group of the PSE can be ranked as dipropylamine > dimethylamine > ammonia > pyridine. Moreover these calculated values also correspond to pK_b values as presented in the literatures.

In most cases, the sensor responses follow the same trend as the above-mentioned ranking of acid-base interactions, except for ammonia. Although dipropylamine and dimethylamine can form stronger interaction with PSE than ammonia, we thus found that ammonia with high mobility can easily diffuse into the sensing material resulting in superiority of the sensor response among other base molecules. As the results, the sensor responses of the PSE to the base-molecules is in the order as ammonia > dipropylamine > dimethylamine > pyridine [19].

TABLE II. THE BOND DISTANCE R(O-H), R(H···N) AND ADSORPTION ENERGY (ΔE_{AD}) OF VOLATILE MOLECULE ON PSE POLYMER SHOWN IN FIG. 2 CALCULATED BY B3LYP/ 6-31+G(D,P)

Volatile Molecule	r(O-H) $\overset{o}{A}$	r(H···V) $\overset{o}{A}$	ΔE_{ad} kcal/mol	pK_b	% Sensor response S3/S7
Ammonia	1.010	1.711	-17.09	4.74[42]	11.7/10.2
Dimethylamine	1.020	1.679	-17.46	3.29[43]	3.9/6.7
Pyridine	1.000	1.754	-15.26	8.75[42]	2.0/3.0
Dipropylamine	1.030	1.660	-18.60	3.09[44]	5.3/6.5

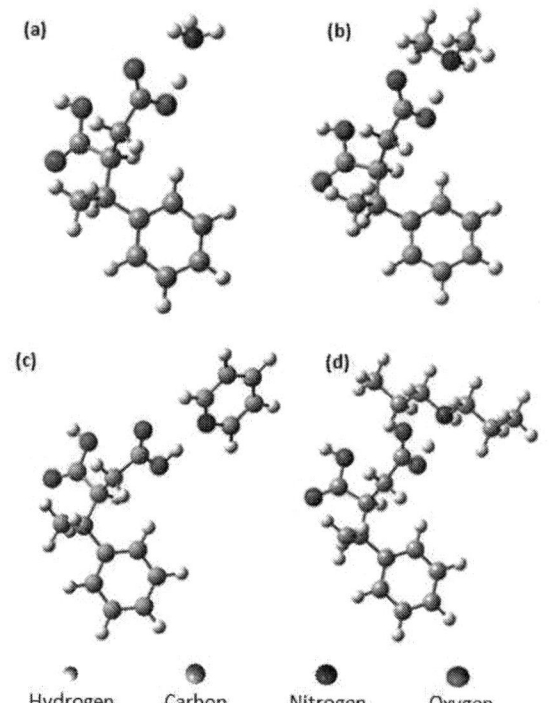

Fig. 3. Optimized conformation of (a) ammonia, (b) dimethylamine, (c) pyridine and (d) dipropylamine interacting on a carboxyl group of PSE polymer by B3LYP/6-31+G(d,p)

To compare the differences of the functionalized CNT in the same polymer matrix, it was observed that most sensors produced from the SWNT-OH exhibit significantly larger sensor response than the sensor produced from the SWNT-COOH. This can be attributed to the degree of functionalization of SWNT that the hydroxyl function was presented on the SWNT 31% more than the carboxyl function.

For the real application, we have used all sensors (S1-S8) installed in the sensing unit of the handheld e-nose. Decomposition of nitrogen-containing compounds such as amino acid in protein based food can form ammonia and amine compounds [45-46] due to microbial spoilage, enzymatic spoilage, chemical spoilage and physical spoilage [47].

According the experimental observation of the static testing, these sensors have been demonstrated that they can respond to ammonia, amines or acid and can generate the specific patterns as shown by the radar plots in Fig. 2.

In Fig. 4, we illustrate the application of the fabricated sensors to classify the differences of three dried seafood, i.e. dried fish, dried squid and dried shrimp. We applied PCA to the data set extracted from the sensor responses after exposed with the headspace of each samples. The PCA plot can discriminate three samples. Three clusters of PCA plot indicate that there are differences of volatiles generated from the different dried seafood.

V. CONCLUSIONS

The sensor array based on polymer/CNT nanocomposite has been developed to generate the specific patterns by using differences of the polymer matrix and functionalized CNT. The fabricated sensors have been demonstrated to show good responses to specific volatiles such as ammonia, amines and acid whereas they present very low response to organic solvents. The first principles calculation based on B3LYP/6-31+G(d,p) approach could provide the understanding involving the interaction of the base molecule with the PSE polymer. It was found that the ammonia or amines can form hydrogen bond with the carboxyl group on the PSE molecule with adsorption energy ~15.26-18.60 kcal/mol corresponding to the values of pK_b of the binding molecules. Based on the theoretical calculation and the experimental sensor testing, the mechanism underlying the sensor response were concluded to be the outcome of the counterbalance between basicity and mobility of the volatiles. In the last part, the handheld e-nose containing these eight sensor elements can be used to discriminate the different dried seafood.

ACKNOWLEDGMENT

This work was supported by Mahidol University and the National Science and Technology Agency. A research career development grant from the Thailand Research Fund to TK and a CHE-Ph.D.-SW-NEU. scholarship from the Commission of Higher Education to PL are acknowledged.

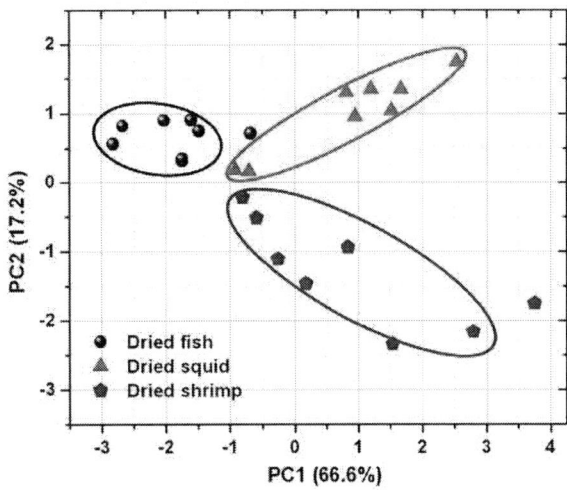

Fig. 4. 2D PCA plot in PC1 and PC2 components for classification of three kinds of seafood: dried fish, dried squid and dried shrimp

REFERENCES

[1] A. Pacquit, K. T. Lau, H. McLaughlin, J. Frisby, B. Quilty, D. Diamond, "Development of a volatile amine sensor for the monitoring of fish spoilage," *Talanta*, 69, p. 515–520, 2006.

[2] J.K. Heising, M. Dekker, P.V. Bartels, M.A.J.S. van Boekel, "A non-destructive ammonium detection method as indicator for freshness for packed fish: Application on cod," *Journal of Food Engineering*, in press.

[3] A. D. Wilson, M. Baietto, "Applications and Advances in Electronic-Nose Technologies," *Sensors*, , 9, 5099-5148, 2009.

[4] B. J. Doleman, N. S. Lewis, "Comparison of odor detection thresholds and odor discriminablities of a conducting polymer composite electronic nose versus mammalian olfaction," *Sensors and Actuators B*, 72, p. 41-50, 2001.

[5] A. K. Anderson, "Biogenic and volatile amine-related qualities of three popular fish species sold at Kuwait fish markets" *Food Chemistry*, 107, p.761–767, 2008.

[6] C. R.-Capillas, C. M. Gillyon, W. F.A. Horner, "Determination of volatile basic nitrogen and trimethylamine nitrogen in fish sauce by flow injection analysis," *Eur Food Res Technol*, 210, p.434–436, 2000.

[7] N. Guernion, R.J. Ewen, K. Pihlainen, N.M. Ratcliffe, G.C. Teare, "The fabrication and characterisation of a highly sensitive polypyrrole sensor and its electrical responses to amines of differing basicity at high humidities," *Synthetic Metals, Volume,* 126, P. 301-310, 2002.

[8] C. Zamani, O. Casals, T. Andreu, J. R.Morante, A. R.-Rodriguez, "Detection of amines with chromium-doped WO3 mesoporous material," *Sensors and Actuators B*, 140, p. 557–562, 2009.

[9] A. Gräfe, K. Haupt, G. J. Mohr, "Optical sensor materials for the detection of amines in organic solvents," *Analytica Chimica Acta*, 565, p. 42–47, 2006.

[10] G.J. Mohr, U.-W. Grummt, "Photochemistry of the amine-sensor dye 4-N,N-dioctylamino-4'-trifluoroacetylazobenzene," *Journal of Photochemistry and Photobiology A: Chemistry*, 163, p.341–345, 2004.

[11] K. Aoki, L. C. Brousseau and T. E. Mallouk, "Metal phosphonate-based quartz crystal microbalance sensors for amines and ammonia," *Sensors and Actuators B*, 13-14, p.703-704, 1993.

[12] H.C. Hao, K.T. Tang, P.H. Ku, J.S. Chao, C.H. Li, C.M. Yang, D.J. Yao, "Development of a portable electronic nose based on chemical surface acoustic wave array with multiplexed oscillator and readout electronics," *Sensors and Actuators B*, 146, p. 545–553, 2010.

[13] N. A. Rakow, A. Sen, M.C. Janzen, J. B. Ponder, K. S. Suslick, "Molecular Recognition and Discrimination of Amines with a Colorimetric Array," *Angew. Chem. Int. Ed.* 44, p. 4528 –4532, 2005.

[14] Z. Tang, J.Yang, J.n Yu, B.Cui, " A Colorimetric Sensor for Qualitative Discrimination and Quantitative Detection of Volatile Amines," *Sensors*, 10, 6463-6476, 2010.

[15] L. Sutarlie, K.-L. Yang, "Colorimetric responses of transparent polymers doped with metal phthalocyanine for detecting vaporous amines," *Sensors and Actuators B*, 134, p. 1000–1004. 2008.

[16] M.M. Ayad, N. L. Torad, "Quartz crystal microbalance sensor for detection of aliphatic amines vapours," *Sensors and Actuators B*, 147, p. 481–487, 2010.

[17] H.T. Nagle, R. Gutierrez-Osuna, S.S. Schiffman, "The how and why of electronic noses," *IEEE Spectrum*, 35 (9), p. 22-34, 1998.

[18] A. J. Lilienthal, A. Loutfi, T. Duckett, "Airborne Chemical Sensing with Mobile Robots," *Sensors*, 6, p. 1616 – 1678, 2006.

[19] P. Lorwongtragool, A. Wisitsoraat, T. Kerdcharoen, "An Electronic Nose for Amine Detection Based on Polymer/SWNT-COOH Nanocomposite," *J. Nanosci. Nanotechnology*, 11, 2011, inpress.

[20] T. Piromjitpong, P. Lorwongtragool, P. Piromjitpong, T. Kerdcharoen, "The Development in an Effective of Ammonia-Odor Sensor Based on PSE-Polymer/SWNT Nanocomposite," in *Proceedings of Chiang Mai International Conference on Biomaterials & Applications (CMICBA 2011)*, Chiang Mai, Thailand, August 2011.

[21] N. K. Kang, T. S. Jun, D.-D. La, J. H. Oh, Y. W.Cho, Y. S. Kim, "Evaluation of the limit-of-detection capability of carbon black-polymer composite sensors for volatile breath biomarkers," *Sensors and Actuators B*, 147, p.55–60, 2010.

[22] S. Maldonado, E. G.-Berríos, M. D. Woodka, B. S. Brunschwig, N.S. Lewis, "Detection of organic vapors and NH3(g) using thin-film carbon black–metallophthalocyanine composite chemiresistors," *Sensors and Actuators B*, 134, p.521–531, 2008.

[23] D. James, S. M. Scott, Z. Ali, W.T. O'Hare, "Chemical Sensors for Electronic Nose Systems," *Microchim. Acta*, 149, p.1–17, 2005.

[24] T.C. Pearce, S.S. Schiffman, H.T. Nagle, and J.W. Gardner, *Handbook of Machine Olfaction; Electronic Nose Technology*, Wiley-VCH, 2002.

[25] K. Arshak, E. Moore, G.M. Lyons, J. Harris and S. Clifford, "A review of gas sensors employed in electronic nose applications," *Sensor Review*, 24 (2), pp. 181–198, 2004

[26] C. Wongchoosuk, S. Choopun, A. Tuantranont and T. Kerdcharoen, "Au-doped Zinc Oxide Nanostructure Sensors for Detection and Discrimination of Volatile Organic Compounds," *Materials Research Innovation*, Vol. 13, pp. 185-188, 2009.

[27] C. Wongchoosuk, A. Wisitsoraat, A. Tuantranont and T. Kerdcharoen, "Portable Electronic Nose Based on Carbon Nanotube-Sn_{O2} Gas Sensors: Feature Extraction Techniques and Its Application for Detection of Methanol Contamination in Whiskeys," *Sensors and Actuators B: Chemical*, Vol. 147, pp. 392-399, 2010.

[28] P. Lorwongtragool, C. Wongchoosuk, T. Kerdcharoen, "Portable artificial nose system for assessing air quality in swine buildings," *Proceedings of ECTI-CON2010 7th International Conference of Electrical Engineering/Electronics, Computer, Telecommunications and Information Technology Association*, pp. 532 - 535, ISBN 978-1-4244-5606-2, Chaing Mai, THAILAND, 19-21, 2010.

[29] L. Tao, J. Han, F.-M.Tao, "Correlations and Predictions of Carboxylic Acid pKa Values Using Intermolecular Structure and Properties of Hydrogen-Bonded Complexes," *J. Phys. Chem. A, 112*, 775-782, 2008.

[30] D. Singh, P. K. Bhattacharyya, J. B. Baruah, "Structural Studies on Solvates of Cyclic Imide Tethered Carboxylic Acids with Pyridine and Quinoline," *Crystal Growth & Design*, 10, p.348-359, 2010.

[31] V. Dimitrova, S. Ilieva, B.Galabov, "Electrostatic potential at nuclei as a reactivity index in hydrogen bond formation. Complexes of ammonia with C–H, N–H and O–H proton donor molecules," *Journal of Molecular Structure (Theochem)*, 637 , p.73–80, 2003.

[32] R. S. Miller, D. Y. Curtin, I. C. Paul, "Reactions of Molecular Crystals with Gases. I. Reactions of Solid Aromatic Carboxylic Acids and Related Compounds with Ammonia and Amines," *J. Amer. Chem. Soc.*96, p.6340-6349, 1974.

[33] A. D. Becke, "Density-functional thermochemistry. III. The role of exact exchange," *J. Chern. Phys.* 98 (7), p.5648-5652, 1993.

[34] P. J. Stephens, F. J. Devlin, C. F. Chabalowski, M. J. Frisch, "Ab Initio Calculation of Vibrational Absorption and Circular Dichroism Spectra Using Density Functional Force Fields," *J. Phys. Chem.*, 98 (45), p. 11623–11627, 1994.

[35] J. A. Robinson, E. S. Snow, S.C. Bădescu,T. L. Reinecke, F. K. Perkins, "Role of Defects in Single-Walled Carbon Nanotube Chemical Sensors," *Nano Lett.*, 6, p. 1747-1751, 2006.

[36] B. Safadi, R. Andrews, E.A. Grulke, Multiwalled carbon nanotube polymer composites: synthesis and characterization of thin films, *J. Appl. Polym. Sci.* 84 (2002) 2660–2669.

[37] A. Szczurek, M. Maciejewska, "Relationship between odour intensity assessed by human assessor and TGS sensor array response, " *Sensors and Actuators B*, 106, p. 13–19, 2005.

[38] V. V. Kamadia, Y. Yoon, M. W. Schilling, D. L. Marshall, Relationships between Odorant Concentration and Aroma Intensity,"*J. Food Sci.* 71, p. S193-S197, 2006.

[39] S. Ampuero, J.O. Bosset, "The electronic nose applied to dairy products: a review," *Sensors and Actuators B*, 94 p.1–12, 2003.

[40] H. Chang, .J. D.Lee, S. M.Lee, Y. H. Lee, "Adsorption of NH3 and NO2 molecules on carbon nanotubes," *Appl. Phys. Lett.*, 79, p.3863-3865, 2001.

[41] M. A. Ryan, A. V. Shevade,C. J. Taylor, M. L. Homer, M. Blanco, J. R. Stetter, *Computational Methods for Sensor Material Selection*, Springer; 1st ed., 2009.

[42] T. Morimoto, J.Imai, M.Nagao, "Infrared Spectra of n-Butylamine Adsorbed on Silica-Alumina," *The Journalof Physical Chemistry*, 78, p.704-708, 1974.

[43] K. Kandori, N. Ohkoshi, A. Yasukawa, T. Ishikawa, "Morphology control and texture of hematite particles by dimethylformaminde in forced hydrolysis reaction," *J. Mater. Res.*, 13, p. 1698-1706, 1998.

[44] D. H. Feldman, J. C. Cavagnol, "Estimation of 1-Butanol and Ethyl Cellosolve in Waste Streams with Sudan III Reagent," *Anal. Chem.*, 28 (11), p. 1746–1748, 1956.

[45] M. Y.Ali, M. I. Sharif, R. K. Adhikari, O. Faruque, "Post mortem variation in Total Volatile Base Nitrogen and Trimethylamine Nitrogen between Galda (*Macrobrachium rosenbergii*) and Bagda (*Penaeus monodon*)," *Univ. j. zool. Rajshahi. Univ.* 28, p.7-10, 2010.

[46] P. Howgate, "A Critical Review of Total Volatile Bases and Trimethylamine as Indices of Freshness of Fish. Part 1.Determination," *EJEAFChe*, 9 (1), p.29-57, 2010.

[47] C. Wongchoosuk, P.Lorwongtragool, T. Kerdcharoen. *Malodor Detection Based on Electronic Nose, Air Quality Monitoring, Assessment and Management*, Nicolás A. Mazzeo (Ed.), ISBN: 978-953-307-317-0, InTech, 2011. Available from: http:// www. Intechopen. com/articles/show/title/malodor-detection-based-on-electronic-nose.

978-1-4673-1122-9/12 $31.00 © 2012 IEEE

Multilayer Graphene Growth by a Metal-catalyzed Crystallization of Diamond-like Carbon

Pinggang Peng[1], *IEEE Student Member*, Dan Xie[1,*], Yi Yang[1], Changjian Zhou[1], *IEEE Student Member*, He Tian[1], Tingting Feng[1], Xiao Li[2], Tianling Ren[1,*], *IEEE Senior Member*, Hongwei Zhu[2,3]

[1]Tsinghua National Laboratory for Information Science and Technology (TNList) & Institute of Microelectronics, Tsinghua University, Beijing, China
[2]Department of Mechanical Engineering, Tsinghua University, Beijing, China
[3]Center for Nano and Micro Mechanics (CNMM), Tsinghua University, Beijing, China
*Contact Author: Dan Xie (Email: xiedan@mail.tsinghua.edu.cn), Tianling Ren (Email: RenTL@tsinghua.edu.cn)

Abstract—**Graphene has attracted significant attention due to its excellent electrical, optical, mechanical and thermal properties. In the paper, we report a simple and efficient growth method of multilayer graphene using a metal-catalyzed crystallization of diamond-like carbon (DLC) by thermal annealing without any extraneous carbon sources. The Ni/DLC/Si multilayered structure was fabricated on silicon substrate at room temperature (RT) and then was thermally annealed. Multilayer graphene was formed on the nickel surface after cooling down to RT by the analysis of Raman spectra. The quality of the multilayer graphene is comparable to chemical vapor deposition methods (CVD). This method of synthesizing graphene sheets with DLC films provides an important way toward the integration of DLC-based electronic devices with graphene.**

Keywords- diamond-like carbon (DLC); graphene; metal-catalyzed crystallization; Raman spectra

I. INTRODUCTION

Graphene has attracted significant attention in recent years due to its unique structure and fascinating electrical, optical, mechanical, and thermal properties [1-4]. Graphene films with different thickness such as single-layer, few-layer and multilayer have been prepared by different techniques, such as exfoliation [1, 5], chemical vapor deposition (CVD) [6], epitaxial growth from SiC and Ru single crystals [7, 8] or other chemical methods [9]. Although many efforts have been made to produce high-quality and large area graphene, controllable surface growth is still difficult, which determines the properties and large-scale fabrication of graphene electronics. Among these methods, CVD technique is useful and available for the preparation of large-area graphene. But the accurate control of graphene layers is also difficult due to its unacceptable uniformity and sensitivity to various parameters. It is reported that the graphene can be produced via metal-catalyzed crystallization of amorphous carbon [10]. The graphene layer can be controlled by the thickness of the amorphous carbon film. Furthermore, it does not need any extraneous carbon sources, presenting a great way toward massive production of controllable graphene layers. Diamond-like carbon (DLC) is a kind of amorphous carbon and has become increasingly interesting in electronic applications for the feasibility of large area depositing and high performances

as the substrate of carbon-based electronic devices. It is reported that the graphene transistors fabricated on diamond-like carbon [11] have better properties than those on SiO$_2$ substrate because of a less impurity and phonon scattering between DLC film and graphene layer. Therefore, the method of synthesizing graphene sheets from DLC film provides an important way toward the application for the integration graphene with DLC-based electronic devices.

In this paper, a simple and efficient method was proposed to produces multilayer graphene films via metal-catalyzed crystallization of DLC film by thermal annealing process. The properties of the DLC film and as-produced multilayer graphene were investigated by the Raman spectroscopy analysis. The effect of the temperature on the growth of multilayer graphene was also studied in detail.

II. EXPERIMENTS DETAILS

N-type, (100) orientation silicon wafers were prepared as the substrates. The DLC layer of 100nm thick was deposited on the substrates at room temperature (RT) by the filtered cathodic vacuum arc (FCVA) technique under a pulse bias with duty ratio of 20% with the incident ion energy of 100 eV in a vacuum chamber in which the base pressure is lower than 5×10^{-3} Pa at RT. Then the nickel metal thin film (200 nm) with a diameter of 1200μm was deposited by dc sputtering at RT with an in situ metal shadow mask. The fabricated Ni/DLC/Si multilayered structure is shown in the left panel of Fig.1. Then the structure was thermally annealed at 650°C, 800°C and 900°C, respectively, for 10min using a tube furnace under a normal pressure with the argon flow of 340 mL/min. The nucleation and growth of multilayer graphene on the nickel surfaces occur after the DLC film has been dissolved. Multilayer graphene of Ni-catalyzed crystallization was formed on the metal surface after cooling down to RT with the cooling rate of 15°C/min. Fig.1 shows the whole process of the preparation of multilayer graphene. The surface morphology of deposited DLC film was determined by atomic force microscope (AFM). The photographs of the fabricated multilayered structure before and after thermal annealing were observed by optical microscope. The properties of the

978-1-4673-1122-9/12 $31.00 © 2012 IEEE

produced multilayer graphene were determined by Raman spectroscopy with a Renishaw R-1000 system using wavelength of 514.5 nm.

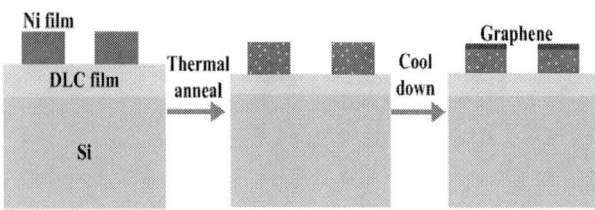

Fig. 1. The process schematics for the Ni-catalyzed crystallization of DLC film to multilayer graphene by thermal annealing and cooling in a tube furnace.

III. RESULTS AND DISCUSSION

Fig. 2 (a) shows the AFM image of the deposited DLC film on Si wafers substrates with the root mean square (RMS) of the roughness 0.52 nm and it shows uniformly distributed particles over the measured range of 5μm × 5μm. Fig. 2 (b) shows the atomic bonding features of the DLC films defined by the Raman spectra. The DLC film before annealing process shows a broad peak over 1400–1700 cm^{-1} range and consists of D peak (centered at 1392 cm^{-1}) and G peak (centered at 1552 cm^{-1}) from the deconvolution of the Raman spectra. The intensity ratio of I_D/I_G of the DLC film is 0.598, from which it can be concluded that the DLC film is composed of 15% insulating bonds and 85% conducting bonds [12].

Fig. 3 (a) and (b) show the optical microscope photographs of the fabricated structure before and after thermal annealing, which exhibits great difference on the surface morphology of the structure. The nickel film is uniform before thermal annealing as shown in Fig. 3 (a), while the film turns to be non-uniform after annealing at 800°C and some black spots and light spots were formed as shown in Fig. 3 (b). From the analysis of Raman spectra, it can be concluded that graphene sheets were formed in the light spot area. The black spots formed in the surface of the nickel are the amorphous carbon films and will be discussed later in detail.

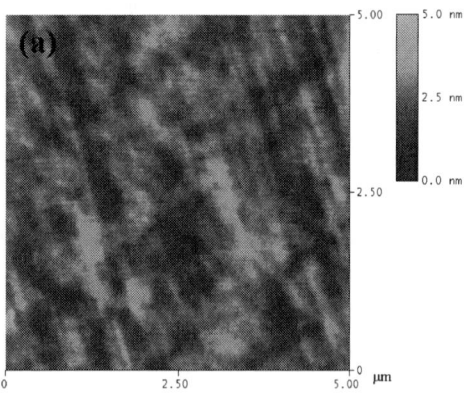

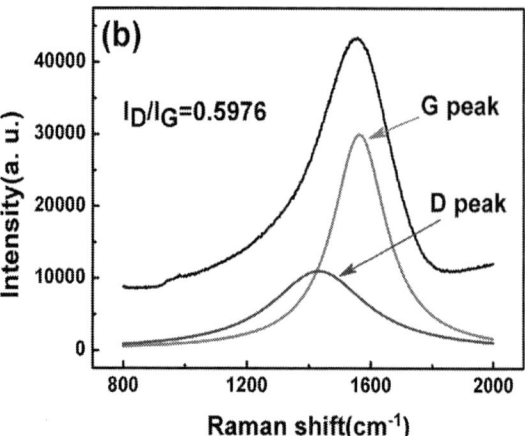

Fig. 2. (a) AFM image of the deposited DLC film, the RMS of the roughness of the films deposited on Si is 0.52 nm over the measured range of 5μm×5μm. (b) Raman spectra of the deconvoluted data of the deposited DLC film before thermal annealing process.

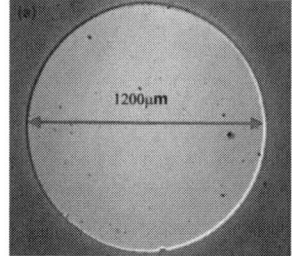
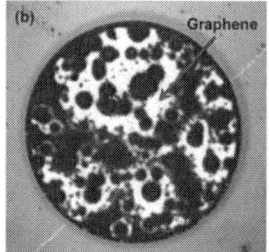

Fig. 3. Optical microscopic photographs of the fabricated Ni/DLC/Si multilayered structure (a) before and (b) after thermal annealing.

Fig. 4 (a) and (b) show the Raman spectra of the produced amorphous carbon and graphene layer on nickel after Ni-catalyzed crystallization by thermal annealing at 800°C. From Fig. 4 (a), it can be concluded that carbon atoms in DLC film can move to the surface under certain thermal condition. The formed amorphous carbon film consists of D peak and G peak as shown in the figure. The inset shows the area of the amorphous carbon that represents one of the black spots in Fig. 3 (b). The composition of the formed amorphous carbon on the surface of the nickel is quite different with the DLC film deposited on silicon by comparing the Raman spectra shown in Fig. 2 (b) and Fig. 4 (a). In contrast, Fig. 4(b) exhibits the Raman spectra of the produced graphene layer which consists of D peak (centered at 1357 cm^{-1}), G peak (centered at 1583 cm^{-1}), and 2D peak (centered at 2713 cm^{-1}). The presence of the 2D peak in the Raman spectra indicates the formation of the multilayer graphene. The inset in the figure shows that the produced multilayer graphene is uniform. The quality of the produced multilayer graphene is similar to the reported result that the graphene was grown by CVD on Ni film [13]. Future

978-1-4673-1122-9/12 $31.00 © 2012 IEEE

investigations will be processed to determine the exact properties of the produced multilayer graphene with other characterization techniques, especially the accurate area of the graphene when the nickel is melted.

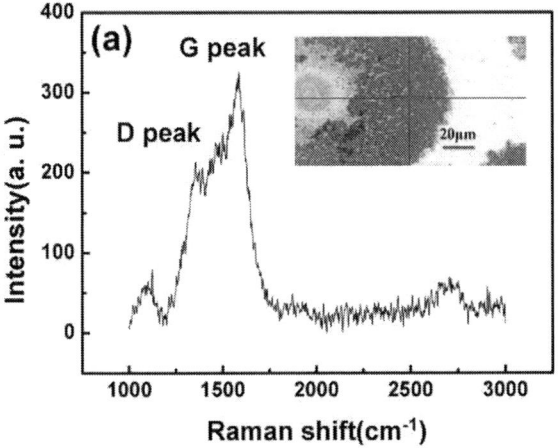

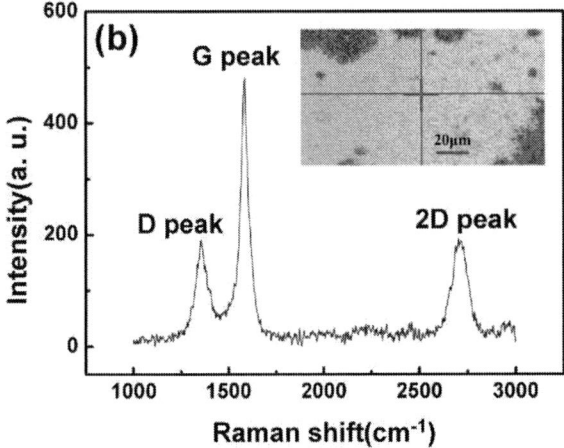

Fig. 4. Raman spectra of (a) the produced amorphous carbon and (b) multilayer graphene on nickel after Ni-catalyzed crystallization by thermal annealing at 800°C. Insets show representative optical images after annealing.

The annealing temperature is one of the vital factors that affect the metal-catalyzed crystallization process and the quality of the graphene layer. Fig. 5 shows the Raman spectra of the produced graphene layer on nickel after annealing at 650°C, 800°C and 900°C, respectively. There is no 2D peak formed when the annealing temperature is 650°C, exhibiting that graphene will not be produced when the temperature is below 650°C although the amorphous carbons can move to the nickel surface. It can be observed that the 2D peak appears when the annealing temperatures increase to 800°C and 900°C. The amplitude of the 2D peak at 900°C is higher than the peak at 800°C, suggesting that the optimal temperature for the graphene growth is about 900°C.

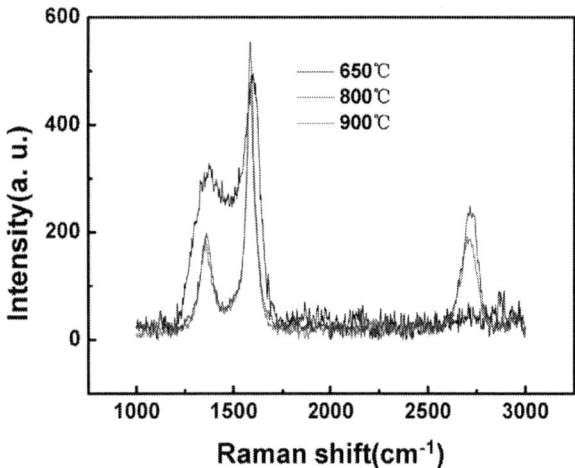

Fig. 5. Raman spectra of the produced graphene layer on nickel after annealing at 650°C, 800°C and 900°C, respectively.

The growth process of multilayer graphene involves a series of steps. First, carbon atoms in DLC film are dissolved in nickel film when the temperature increases to about 650°C. Then the nucleation and growth of graphene on the nickel film surface take place during the cooling process. The nickel crystals act as diffusion channels to transport carbon atoms in DLC film to each location at the surface. When the annealing temperature is about 650°C, only the amorphous carbons can move to the nickel surface. The graphene layer will be formed when the annealing temperature is higher than 800°C. Higher annealing temperature is propitious to the nucleation and growth of graphene. So the produced multilayer graphene exhibits good quality under higher annealing temperature. It is also proved that the number of produced graphene layers can be determined by the thickness of the deposited carbon layer [10]. So the graphene layer can be controlled by the thickness of the DLC film. The DLC film can easily be deposited uniformly on metal layers with large area by FCVA method. After heating, the metal could be removed by etching or other methods and the free-standing graphene layers can be maintained in preselected positions. So the technique of transforming DLC film to graphene may has many applications in graphene-based electronic devices and systems. Furthermore, it does not need any extraneous carbon sources, so this method can presents an advantageous way toward massive production of controllable graphene layers as compared to a CVD process during which the uniformity is strongly affected by the substrate, CVD process and other parameters.

IV. CONCLUSIONS

In summary, a simple and efficient method was proposed to produce multilayer graphene via metal-catalyzed crystallization of DLC by thermal annealing process of the Ni/DLC/Si multilayered structure at different temperature. Multilayer graphene was formed on the nickel surface after cooling down to RT. The annealing temperature is a vital factor that affects formation of the graphene layer. The optimal temperature for the graphene growth is about 900°C. The nickel crystals act as diffusion channels to transport carbon atoms in DLC film to each location at the nickel surface. This method presents an advantageous way toward massive production of controllable graphene layers. This method of synthesizing graphene sheets with DLC films also provides an important way toward the integration of DLC-based electronic devices with graphene.

ACKNOWLEDGMENT

This work is supported by National Natural Science Foundation (61025021, 60936002, 51072089, 50972067, 61011130296, 61020106006), National Key Project of Science and Technology (2009ZX02023-001-3, 2011ZX02403-002) of China and Beijing NSF (3111002).

REFERENCES

[1] K.S. Novoselov, A.K. Geim, S.V. Morozov, D. Jiang, Y. Zhang, S.V. Dubonos, et al, "Electric field effect in atomically thin carbon films," *Science*, vol. 306, pp. 666-669, 2004.

[2] A. K. Geim, K. S. Novoselov, "The rise of graphene," *Nat. Mater.*, vol. 6, pp. 183-191, 2007.

[3] M. I. Katsnelson, "Graphene: carbon in two dimensions," *Mater. Today*, vol. 10, pp. 20-27, 2007.

[4] R. R. Nair, P. Blake, A. N. Grigorenko, K. S. Novoselov, T. J. Booth, T. Stauber, et al, "Fine structure constant defines visual transparency of graphene," *Science*, vol. 320, pp. 1308, 2008.

[5] K.S. Novoselov, D. Jiang, F. Schedin, T.J. Booth, V.V. Khotkevich, S.V. Morozov, et al, "Two-dimensional atomic crystals," *Proc. Natl. Acad. Sci*, vol. 102, pp. 10451-10453, 2005.

[6] X. Li, W. Cai, J. An, Kim S, J. Nah, D. Yang, et al, "Large-area synthesis of high-quality and uniform graphene films on copper foils," *Science*, vol. 324, pp. 1312-1314, 2009.

[7] P.W. Sutter, J.I. Flege, E.A. Sutter, "Epitaxial graphene on ruthenium," *Nat. Mater.*, vol. 7, pp. 406-411, 2008.

[8] K.V. Emtsev, A. Bostwick, K. Horn, J. Jobst, G.L. Kellogg, L. Ley, et al, "Towards wafer-size graphene layers by atmospheric pressure graphitization of silicon carbide," *Nat. Mater.*, vol. 8, pp. 203-207, 2009.

[9] S. Park, R. S. Ruo, "Chemical methods for the production of graphenes," *Nat. Nanotechnolgy*, vol. 4, pp. 217-224, 2009.

[10] M. Zheng, T. Kuniharu, H. Benjamin, F. Hui, X.B. Zhang, F. Nicola, et al, "Metal-catalyzed crystallization of amorphous carbon to graphene," *Appl. Phys. Lett*, vol. 96, 063110, 2010.

[11] Y.Q. Wu, Y.M. Lin, A.A. Bol, K.A. Jenkins, F.N. Xia, D.B. Farmer, et al, "High-frequency, scaled graphene transistors on diamond-like carbon," *Nature*, vol. 472, pp. 74-78, 2011.

[12] B.K. Tay, X. Shi, H.S. Tan, H.S. Yang, Z. Sun, "Raman studies of tetrahedral amorphous carbon films deposited by filtered cathodic vacuum arc," *Surf. Coat. Technol.*, vol. 105, pp. 155-158, 1998.

[13] Q. Yu, J. Lian, S. Siriponglert, H. Li, et al, "Graphene segregated on Ni surfaces and transferred to insulators," *Appl. Phys. Lett*, vol. 93, 113103, 2008.

Chemical Vapor Deposition of Nanocrystalline Graphene Directly on Arbitrary High-temperature Insulating Substrates

Jie Sun[*1], Niclas Lindvall[1], Matthew T. Cole[2], Kenneth B. K. Teo[3], *Member, IEEE,* and August Yurgens[1]

[1]Department of Microtechnology and Nanoscience, Chalmers University of Technology, Göteborg 41296, Sweden
[2]Department of Engineering, University of Cambridge, Cambridge CB30FA, United Kingdom
[3]AIXTRON Nanoinstruments Ltd., Cambridge CB244FQ, United Kingdom

Abstract—**Large area uniform nanocrystalline graphene is grown by chemical vapor deposition on arbitrary insulating substrates that can survive ~1000 °C. The as-synthesized graphene is nanocrystalline with a domain size in the order of ~10 nm. The material possesses a transparency and conductivity similar to standard graphene fabricated by exfoliation or catalysis. A noncatalytic mechanism is proposed to explain the experimental phenomena. The developed technique is scalable and reproducible, compatible with the existing semiconductor technology, and thus can be very useful in nanoelectronic applications such as transparent electronics, nanoelectromechanical systems, as well as molecular electronics.**

Keywords—**Graphene, chemical vapor deposition, insulator, nanoelectronics**

I. INTRODUCTION

Graphene is a novel nanoelectronic material which has recently been intensively studied. The intrinsic mobility is extremely high. Together with its atomic-scale thickness, this material has been suggested as a potential candidate for post-silicon electronics. Nevertheless, it does not have a bandgap, leading to a small on-off ratio in transistors. Also, the experimentally demonstrated mobility falls far below the theoretically predicted value. Thus, graphene based computer processors are unlikely to be realized very soon.

This notwithstanding, there are several fields in nanoelectronics where graphene can play a key role in future: transparent electronics, nanoelectromechanical systems (NEMS), molecular electronics, etc. With the recent advances in chemical vapor deposition (CVD), large area graphene can be fabricated massively [1]. This technology is cost-effective and scalable. It is the most promising route towards industrial applications. However, metal catalysts used in the CVD need to be etched away for the transfer of graphene onto foreign insulating substrates. There is an urgent need to develop a catalyst-free method to directly synthesize graphene on dielectrics.

For some time, the noncatalytic graphene CVD in large area has been considered impossible, or at least very difficult. But recently, for the first time, we have demonstrated the uniform graphene on insulators such as Si_3N_4 and HfO_2 by metal-free CVD [2]. We also predict that this type of graphene can be grown on virtually any substrate that withstands ~1000 °C [2]. A few other groups also achieved similar results [3]. Nevertheless, the mechanism is not understood and the experiments are largely experience based. In this paper, we propose a noncatalytic mechanism of graphene growth based on the pyrolysis of hydrocarbon. More detailed description will be published elsewhere [4]. We also give examples of such a graphene on SiO_2, Al_2O_3 and GaN substrates and discuss its optical, electrical, and mechanical properties.

II. GROWTH MODEL

Herein, we propose a model to explain the noncatalytic

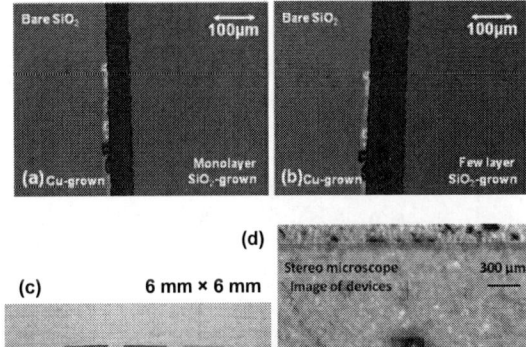

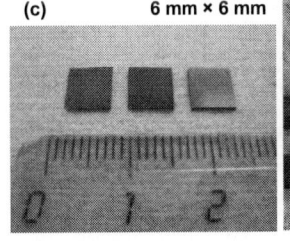

Fig. 1. (a) and (b) Color comparison of the Cu-catalyzed graphene (transferred to 300 nm SiO_2/Si) and the graphene grown directly on the same type of substrate. (c) Photo of (from left to right) bare SiO_2/Si substrate and SiO_2/Si with deposited monolayer graphene and thick graphite (~70 nm). (d) Stereo microscope image of Hall-bar devices fabricated in the multilayer graphene deposited on GaN/sapphire wafer.

* Corresponding author. Email: jiesu@chalmers.se.

978-1-4673-1122-9/12 $31.00 © 2012 IEEE

graphene growth. It is known that hydrocarbons decompose without the need of any catalyst. For example, in industry, methane pyrolysis is widely used to synthesize carbon black in large quantities [5]. This process is important, because it is a simple method for C and H_2 production without generating any CO_2. However, it has so far been over looked in terms of graphene growth. Note that carbon black is randomly stacked nanoscale graphene flakes [6]. Therefore, if a substrate is flat and high-temperature stable, the carbon black can be turned into textured graphene thin films. Due to the fact that metal catalysts are absent, high carbon precursor concentration and long deposition time are also needed to form graphene thin films.

After the hydrocarbon is decomposed, the carbon atoms readily form airborne nanographene flakes. Larger flakes have enough adhesion energy to the substrate, whereas smaller ones will have to leave the substrate at ~1000 °C. The growth is not self-limited, which means that the thickness of the graphene is controllable from submonolayer to thick graphite. In the hot-wall quartz-tube CVD reactor used in this work, the middle region is heated to ~1000 °C while the two ends (gas inlet and outlet) of the quartz reactor are at temperatures less than 600 °C. No carbon deposition is found at the gas inlet part of the quartz tube due to the absence of effective pyrolysis of methane. In the middle zone, graphene growth occurs on various substrates such as SiO_2, Al_2O_3, MgO, SiC, Si, etc. Our detailed growth conditions are described elsewhere [2, 4]. At the gas outlet part of the quartz tube, carbon black deposition occurs. Figs. 1(a) and (b) show optical images of nominal monolayer and few layer graphene grown directly on 300 nm SiO_2/Si, where the left section in each picture is a transferred Cu-grown graphene on the same type of substrates for comparision. It is known that Cu-catalyzed graphene is a monolayer (within certain parameter windows) because of the self-limiting growth mechanism [1, 7]. Based on the color comparison [8], the SiO_2-grown graphene shown in in Fig. 1(a) is much likely also a single layer. Fig. 1(c) shows the photo of (left to right) bare 300 nm SiO_2/Si substrate, monolayer graphene and thick graphite (~70 nm) on SiO_2/Si, respectively. In this figure, the films keep their metallic luster even for hundreds of layers. If, however, the temperature is not high enough, or the wafer is rough, carbon black films with a dull black color are then produced, which can hardly be useful in electronics. Fig. 1(d) demonstrates the photo of a finished Hall-bar device fabricated on GaN(0001)/sapphire substrate at 950 °C. For the first time, large-area graphitic thin films are grown directly on GaN for potential applications in optoelectronics.

We also grow graphene on double-side polished 5 mm × 5 mm quartz and sapphire substrates. Fig. 2 shows a photo of the samples after the CVD. In each row, three nominal monolayer graphene samples grown in different runs are placed to the right. Graphene which is unintentionally deposited at the bottoms of samples has been removed by oxygen plasma etching. These samples are very similar, indicating high

5 mm × 5 mm

Fig. 2. Photo of 5 mm × 5 mm graphene samples on quartz and sapphire. In each row, the left is the bare substrate. Three nominal monolayer graphene samples are placed in the right showing high process reproducibility. The text behind is clearly visible due to the high transparency.

reproducibility in the CVD process. The text on the paper below the samples is clearly seen, showing their high transparency. The left-most sample is the bare quartz (sapphire) for comparison. The edges of the quartz samples are rough after dicing, resulting in carbon black deposits after the CVD (notice the black edge of the quartz samples in Fig. 2). In contrast, the edges of sapphire substrates have been polished after cutting, and therefore carbon black is absent. This fact suggests the importance of substrate flatness in the deposition of nanocrystalline graphene. At rough places, there are numerous kinks, pits and steps, etc. Smaller or misaligned (inclined or vertical) graphene flakes are easier to be adsorbed onto these sites forming carbon black. These defects on the substrate would not be important if the thin film were grown from individual carbon atoms. Therefore, it provides evidence that the major mechanism of the deposition is indeed based on graphene nanoflakes. In our recipe, we completely remove the carbon precursor gas immediately after the graphene growth preventing flake formation during cooling down. This growth mechanism can be used to explain nonocrystalline graphene formation in common CVD systems with relatively large quantities of precursor material. Nevertheless, in molecular beam epitaxy [9, 10] with typically much smaller carbon-atom fluxes and much longer deposition times, the growth mode is most likely based on single C atoms. Further study is needed to understand the mechanism thoroughly in these cases.

Here, due to the above-mentioned mechanism, the graphene is fundamentally nanocrystalline. Therefore, one can only speak about average number of layers. The exact number of layers is only meaningful locally. For example, the middle sample in Fig. 1(c) primarily consists of monolayer flakes; but that does not exclude the coexistence of multilayer flakes at some other spots. Besides, flakes do not have enough mobility to freely diffuse along the surface to rearrange themselves at 1000 °C, and thus at the grain boundaries some overlap may happen, resulting in a slightly rougher film (roughness ~1 nm) compared with single-crystalline graphene (resting on thermally oxidized Si).

III. OPTICAL AND ELCTRICAL PROPERTIES

After the graphene coating, the transmittances of quartz and sapphire are reduced by only 2-3% measured by transmission spectroscopy [4], which is similar to the case of exfoliated or metal-catalyzed monolayer graphene. Both on quartz and sapphire, G and 2D peaks are clearly detected in the Raman spectra [4], confirming that the deposited thin films have sp^2 hybridized graphitic structures. The laser used in the Raman measurement has a beam spot of ~10 μm in size, which covers numerous crystallites in the graphene thin film. These grain boundaries naturally lead to a relatively high D peak [4] (D stands for disorder).

Hall-bar devices are fabricated by photolithography on these graphene samples on quartz, sapphire, GaN and SiO_2/Si. At room temperature, the sheet resistance R_s of the quartz and sapphire based graphene is 2.9 and 13 kΩ/□, respectively. These values are again comparable to that of intrinsic graphene produced by other means. Interestingly, on GaN, we have found that the deposited graphene has quasi-ohmic contacts to the underlying GaN, which is very promising towards solving the "current crowding" problem in GaN based lasers, etc. Also, the GaN is strongly doped after the CVD. We ascribe the observed effects to the high-temperature process during the CVD. Currently, we are investigating this and the results will be published elsewhere. For the nanocrystalline

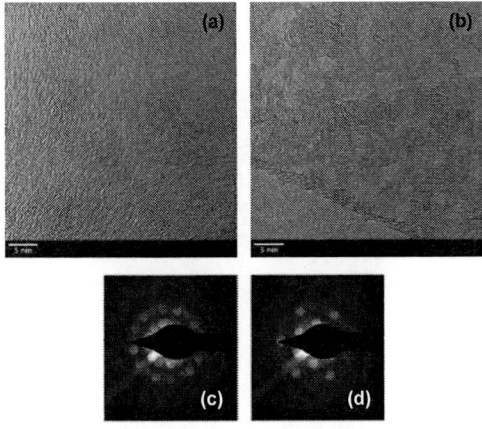

Fig. 3. (a) and (b) High-resolution TEM images of the SiO_2/Si-grown graphene transferred to Cu grids. After very long time observation, holes are burnt in the thin film, as are visible in (b). Scale bar: 5 nm. (c) and (d) Converged beam electron diffraction pattern obtained from the graphene thin films. (c) Mixed signals from two or more domains. (d) Typical monolayer graphene diffraction pattern is clearly observed reproducibly everywhere in the membrane, as long as the beam spot is ~10 nm or smaller..

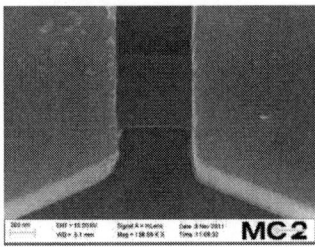

Fig. 4. SEM image of a typical suspended-channel graphene device fabricated from the SiO_2/Si-grown graphene. Scale bar: 200 nm.

graphene grown on 300 nm SiO_2/Si, it is possible to apply gate voltages using the doped silicon substrate as a back gate. By analyzing the field effect at room temperature, we obtain a hole mobility of ~40 cm^2/Vs, which is further confirmed by Hall-effect measurements.

IV. SUSPENDED GRAPHENE

Figs. 3(a) and (b) show two high-resolution transmission electron microscopy (TEM) images of the graphene (transferred to Cu grids with holey carbon films) grown directly on SiO_2/Si. In Fig. 3(a), the graphene membrane is rather uniform. Converged beam electron diffraction pattern clearly shows a monolayer crystal lattice feature (Fig. 3(d)). As far as we know, the hexagonal diffraction pattern of nanocrystalline graphene has not been reported before. Such a pattern is a direct evidence of the graphene nature of this type of thin films. Fig. 3(c) is a diffraction pattern obtained when the focused beam spot is somewhat larger than ~10 nm. Signals from two or multiple randomly oriented domains destroy the hexagonal pattern. By varying the beam size, we can thereby obtain an indirect estimate of the domain size. Although the domains are only ~10 nm large, the graphene thin films on the whole have very good mechanical properties and can be suspended.

We have successfully fabricated arrays of suspended-channel two-terminal devices from the graphene directly grown on 300 nm SiO_2/Si. The silicon dioxide beneath the graphene channel is locally removed by wet etching in buffered HF acid. Fig. 4 is a SEM image of a finished device. The graphene membrane is very clean without any holes or buckles. The success rate in device fabrication is nearly 100%. Suspended graphene can be used as mechanical resonators for sensitive mass sensors [11], as well as for other NEMS. Details about this type of devices can be found in another paper of this conference proceeding [12].

V. CONCLUSION

In summary, we have proposed a noncatalytic mechanism for nanocrystalline graphene CVD. The model is used to explain the formation of graphene without any metallic catalysts directly on arbitrary insulating substrates that can survive high temperature of ~1000 °C. Raman spectroscopy and TEM observation reveal the graphene nature of the synthesized thin films. The produced thin films are optically transparent, electrically conducting, and have excellent mechanical properties, resembling standard single crystal graphene. The graphene is scalable, uniform and cost-efficient. Even though the carrier mobility is much lower compared with the graphene grown by catalysis, the developed method is very promising in future industrial applications in fields such as transparent electronics, NEMS, as well as molecular electronics. Further improvements of the deposition- and fabrication techniques are underway and will be published in the future.

ACKNOWLEDGMENT

The authors thank Dr. T. Booth for the help in TEM. Drs. Å. Haglund, T. Ive and Mr. S. A. Ahamd have contributed in the graphene deposition on GaN. This work gets support from the Swedish Research Council and the Swedish Foundation for Strategic Research.

REFERENCES

[1] X. Li, W. Cai, J. An, et al., "Large-area synthesis of high.quality and uniform graphene films on copper foils," *Science* vol. 324, pp. 1312-1314, 2009.

[2] J. Sun, N. Lindvall, M. T. Cole, et al., "Large-area uniform graphene-like thin films grown by chemical vapor deposition directly on silicon nitride," *Appl. Phys. Lett.* vol. 98, p. 252107, 2011.

[3] J. Chen, Y. Wen, Y. Guo, et al., "Oxygen-aided synthesis of polycrystalline graphene on silicon dioxide substrates," *J. Am. Chem. Soc.* vol. 133, pp. 17548-17551, 2011.

[4] J. Sun, M. T. Cole, N. Lindvall, et al., "Noncatalytic chemical vapor deposition of graphene on high-temperature substrates for transparent electrodes," To be published.

[5] A. W. Weimer, J. Dahl, J. Tamburini, et al., "Thermal dissociation of methane using a solar coupled aerosol flow reactor," *Proceedings of the 2001 US DOE hydrogen program review* NREL/CP-570-30535, 2011.

[6] J. Biscoe and B. E. Warren, "An X-ray study of carbon black," *J. Appl. Phys.* vol. 13, pp. 364-371, 1942.

[7] J. Sun, N. Lindvall, M. T. Cole, et al., "Low partial pressure chemical vapor deposition of graphene on coper," *IEEE Trans. Nanotechnol.* In press, doi: 10.1109/TNANO.2011.2160729.

[8] Y.-F. Chen, D. Liu, Z.-G. Wang, et al., "Rapid determination of the thickness of graphene using the ratio of color difference," *J. Phys. Chem. C* vol. 115, pp. 6690-6693, 2011.

[9] S.-K. Jerng, D. S. Yu, J. H. Lee, et al., "Graphitic carbon growth on crystalline and amorphous oxide substrates using molecular beam epitaxy," *Nanoscale Research Lett.* vol. 6, p. 565, 2011.

[10] G. Lippert, J. Dabrowski, M. C. Lemme, et al., "Direct graphene growth on insulator," *Phys. Status Solidi B* vol. 248, pp. 2619–2622, 2011.

[11] R. A. Barton, J. Parpia, and H. G. Craighead, "Fabrication and performance of graphene nanoelectromechanical systems," *J. Vac. Sci. Technol. B* vol. 29, p. 050801, 2011.

[12] N. Lindvall, J. Sun, and A. Yurgens, "Transfer-free fabrication of suspended graphene grown by chemical vapor deposition," *This Conference Proceeding (IEEE NEMS 2012).*

Characterization of Strain Fields in Graphene Films

Raden Dewanto[*], Carl Dale, Zhongxu Hu, Neil Keegan, Barry Gallacher, John Hedley

[*]Mechanical and Systems Engineering, Newcastle University, Newcastle upon Tyne, UK

r.s.dewanto@ncl.ac.uk

Abstract—**This paper reports on the Raman shifts corresponding to strain induced in graphene films. A direct correlation is demonstrated between the shifts in the D, G and 2D peaks of graphene compared to the characteristic 521cm^{-1} peak of the underlying silicon substrate. The approach is shown to be suitable for characterizing the graphene Raman spectrum under load conditions.**

Keywords-strained graphene, Raman spectroscopy

I. INTRODUCTION

In the ever expanding area of sensor development towards faster, cheaper, more accurate and reliable sensor systems there has been an expanding interest in the exploitation of nanomaterials, in particular, the employment of carbon nanotubes as transduction elements. Recently the emergence of the nanomaterial graphene [1], which is effectively an unrolled nanotube, has become an attractive alternative material with significant potential advantages for exploitation in the sensor arena. Due to the atomic thickness planar nature

of graphene, graphene is predicted to revolutionize performance in nanoelectronics and NEMS. Sensors based on its unique electrical/mechanical properties have the potential to be extremely sensitive. Part of any new technology is the characterisation of the material properties and in terms of both electrical and mechanical properties, it is important to understand graphene under strained conditions.

Our group have recently reported mapping of strain fields in vibrating silicon cantilevers utilizing Raman spectroscopy. In this work, the 521cm^{-1} characteristic LO_z silicon peak is shown to shift by 5.2×10^{-4} cm^{-1}/volumetric μstrain, measurements were done using both analysis of the broadening of the Raman peak [2] and a direct measure of the shift using a strobed probe laser [3]. Results were is good agreement with modeling. Studies into the properties of graphene under tensile and compressive forces are only just starting to appear. Both Mohiuddin et al [4] and Frank et al [5] have investigated peak shifts due to tensile uniaxial loading. This work aims to characterize the Raman shifts of graphene produced through

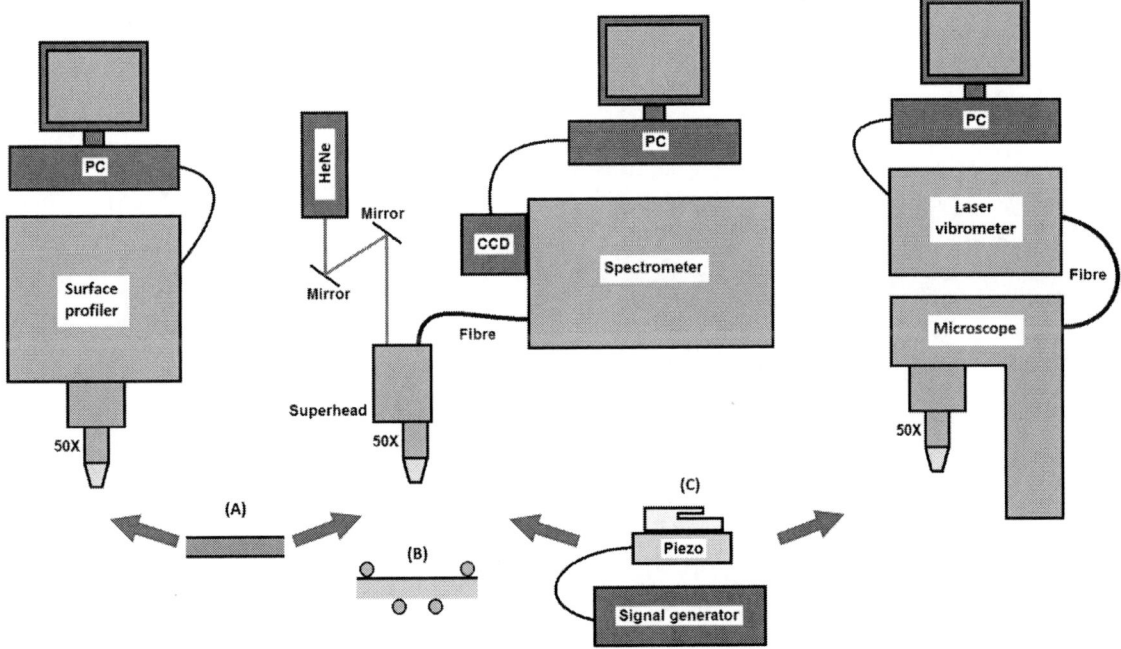

Fig. 1. Schematic of the experimental setup. (A) Curvature and Raman measurements of epitaxially grown graphene on copper foil. (B) Comparison of induced strain in graphene and silicon substrate when loaded via four point bending. (C) Dynamically induced strain loading of graphene and characterization with laser vibrometry and Raman spectroscopy.

complex strain fields in graphene films giving calibration data for further studies in this field. Graphene films are characterized by 3 Raman peaks, the D, G and 2D peaks located at 1320cm⁻¹, 1600 cm⁻¹ and 2650cm⁻¹ respectively. Reported here is the development of 3 techniques to perform this characterization on each of these peaks. Fig. 1 gives an overview of the experimental approaches used in this work.

II. METHODOLOGIES

A. Epitaxially grown graphene from copper foil

Various methods are being researched into fabrication of good quality graphene films, the 3 most notable approaches are mechanical exfoliation of graphene from graphite [1], epitaxially grown graphene on copper foil [6] and heating of silicon carbide to 1800^0C in vacuum [7]. The epitaxially grown is known to give good single layer coverage and therefore acts as a good starting point for investigation.

For this experiment, commercially available graphene produced by chemical vapor deposition (CVD) on a 20 μm thick copper sheet was purchased from Graphene Laboratories. The copper foil was shaped to a range of curvatures and a surface profile and Raman spectrum taken for each of these induced strains. Although the Raman spectrum showed appreciable shifts with measured values ranging from 1320.7cm⁻¹ to 1325.9cm⁻¹, 1587.9cm⁻¹ to 1590.2cm⁻¹ and 2638.8cm⁻¹ to 2646.0cm⁻¹ for the D, G and 2D peaks respectively, it proved impossible to track specific locations between transitions of the sample from the surface profiler to the Raman microscope. The surface profile shown in Fig. 2 indicates the variation of curvature over the copper foil and therefore this method was deemed unsuitable in providing high precision measurements.

B. Graphene on silicon wafers

It is possible to transfer graphene to other substrates. The disadvantage of this approach is a poorer quality graphene

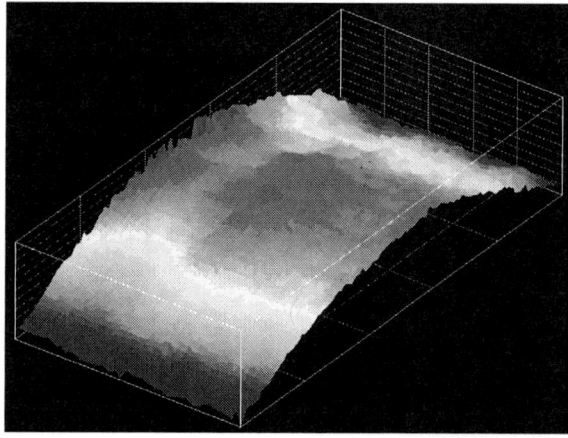

Fig. 2. Surface profile image of epitaxially grown graphene on copper foil.

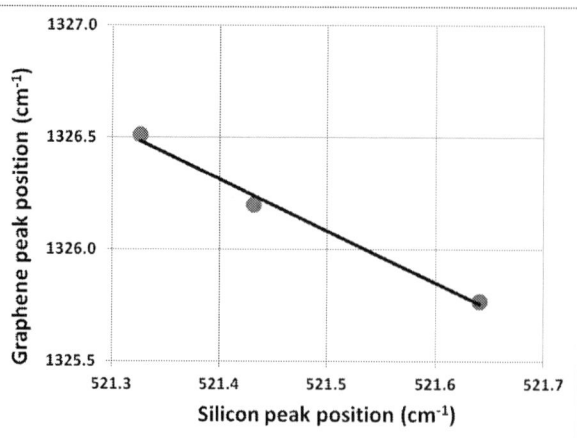

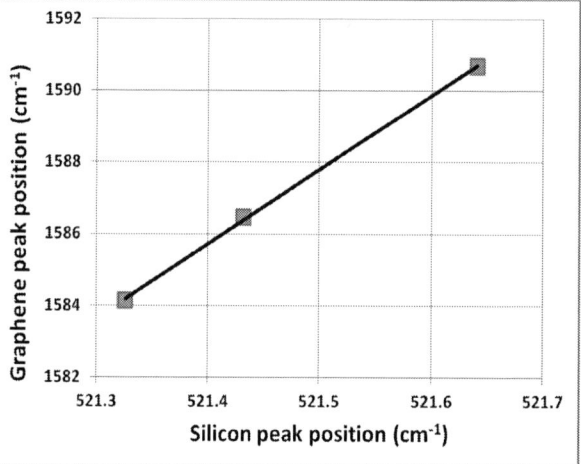

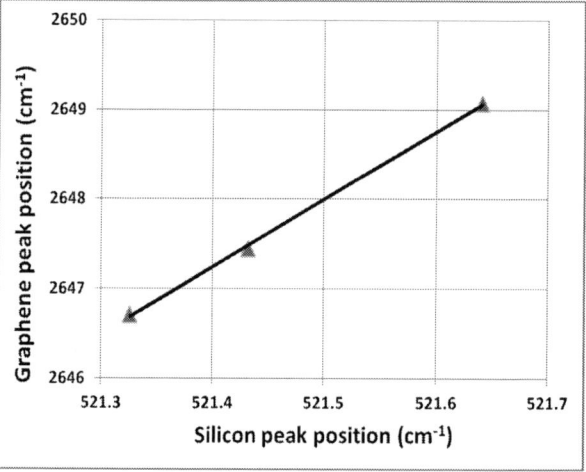

Fig. 3. Peak shifts in the D, G and 2D spectra of graphene compared with peak shifts in the supporting silicon substrate. The corresponding gradients of the line of best fit are -2.32, 20.82 and 7.55 respectively.

film. In principle, a similar experiment may be performed where measurement of curvature of the silicon surface implies strain in the graphene film. However one major advantage of using silicon as the substrate is that the silicon lattice may be used as the reference for curvature. The Raman shift due to strain in silicon is well characterised and therefore a measure of the substrate silicon Raman peak position along with the graphene Raman measurements gives us a well defined measure of the strain present at the silicon surface.

Again, commercially available graphene on silicon/silicon oxide wafers were purchased from Graphene Laboratories. The 300nm silicon oxide layer facilitates visualizing the graphene layer. The samples were divided into 10mm×2mm pieces and place in a micro 4 point bending apparatus. Raman spectra were taken of the silicon and graphene at various loadings, results of which are shown in Fig. 3.

C. Graphene on microcantilevers

The principle behind the work was to attach graphene films onto vibrating microcantilevers and from the dynamically induced strain fields, characterize the Raman shifts. An scanning electron microscope (SEM) image of the devices used is shown in Fig. 4.

1) Device fabrication

Devices were fabricated in the clean room facilities in INEX at Newcastle University from SOI wafers. Firstly, an oxide layer was desposited onto the 500μm of handle silicon and a photolithography step used to define the back etch regions. The exposed oxide was removed with a plasma etch. Photolithography was then performed on the 15μm device layer to define the microcantilever geometries and a deep reactive ion etch (DRIE) down to the 2μm buried oxide layer performed. To protect the devices during the remaining processing, this device layer was covered with a protective layer and bonded to a support wafer. A DRIE was then used on the back of the wafer to remove the masked handle silicon with the now exposed buried oxide layer being plasma etched. Finally the support wafer was removed with application of a solvent. Graphene is most easily visualised when viewed on a 300nm SiO_2 surface. To grow this layer, a dry thermal oxidation was performed by placing dies in a furnace, ramping up to 1150^0C and maintaining this temperature for 3½ hours.

2) Graphene transferral

Once fabricated the microcantilever devices were cleaned using piranha solution (70% H_2SO_4: 30% H_2O_2) to remove all possible organic contamination, producing a substrate suitable for graphene placement. Several methods are available for the transference of graphene onto a surface. The devices underwent a range of transfer methods to deposit graphene on the microcantilevers. The mechanical exfoliation of graphene from Kish graphite on to the devices was performed [8] although it was deemed not suitable for this application as it produces flakes in random locations. Also it can be damaging to the microcantilever structures when the tape is peeled from the device. An alternative method was also investigated by

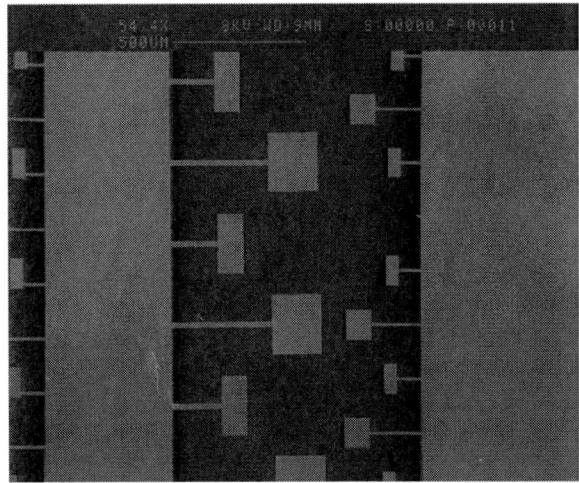

Fig. 4. *SEM image of the fabricated silicon microcantilevers.*

using commercially available monolayer graphene flakes in a liquid dispersion to transfer graphene. However this method was not effective to measure strain in graphene as the average lateral flakes size was 0.5μm that made it difficult to achieve a suitable Raman signal with the laser spot size. The method that was most effective to transfer graphene on the devices was by using the thermal release (TR) tape transfer method [5], a schematic of which is shown in Fig.5.

The graphene was placed onto the microcantilevers by the transferral of epitaxially grown graphene from copper foil substrates. The graphene produced on copper sheet from Graphene Laboratories was again used here. The copper sheet was stuck down firmly onto thermal release tape and placed into a copper etch solution of 40 % Fe_3Cl_2 for 60 minutes. This ensured that all the copper was removed by chemical etching and a layer of graphene remained present on the TR tape. The TR tape was rinsed carefully with water 3 times and left to dry at room temperature for 20 minutes. The microcantilever dies were placed and stuck onto the TR tape

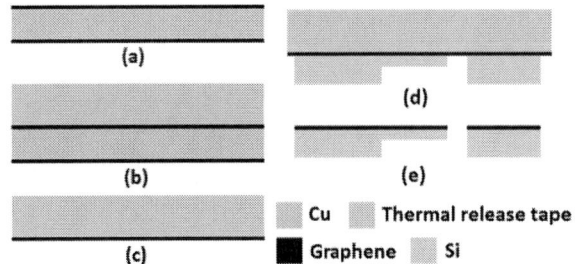

Fig. 5. *Schematic of the process flow for graphene transfer onto the silicon microcantilevers. (a) Cu foil with epitaxial grown graphene. (b) Application of thermal release tape. (c) Etching of Cu. (d) Transfer of microcantilevers onto the graphene / tape. (e) Removal of tape by heating to 95 °C for 15 minutes.*

978-1-4673-1122-9/12 $31.00 © 2012 IEEE

carefully so that the graphene came into contact with the structures. The TR tape with the dies stuck down was placed into an oven for 15 minutes at 95 °C so that the tape lost the adhesive properties leaving graphene deposited on the microcantilevers.

3) Raman measurement

The surface was first viewed under a microscope and visually it appeared that a good graphene layer was present. However the Raman analysis identified only the D and G peaks, the 2D peak was not evident. Work is continuing on this preparation method to obtain more consistent Raman spectra from the transfer process.

III. DISCUSSION

Of the 3 methodologies undertaken, comparison of measurement of the Raman peak shifts in strained graphene with the corresponding silicon peak shifts shows the most repeatable approach for characterization. For this experiment, the sample was placed under tension, corresponding to a lowering of silicon Raman peak position. Both the G and 2D peak shifted towards lower wavenumbers, in agreement with that reported by Mohiuddin [4], whilst the D peak is shifted to higher values. The slopes in Fig. 3 (i.e. relative shift in graphene peak compared to the silicon peak) for the D, G and 2D peaks are -2.32, 20.82 and 7.55 respectively.

IV. CONCLUSIONS

The work has examined 3 approaches to measuring Raman shifts in strained graphene. Results are in good agreement with previously published literature showing an appreciable shift for the G and 2D peak of graphene to lower wavenumbers under the application of tensile strain whilst the D peak is shifted towards higher wavenumbers. Work is ongoing to further quantify these shifts and examine the material properties of graphene under such conditions.

ACKNOWLEDGMENT

R.S.D. acknowledges the Directorate General of Higher Education, Ministry of National Education, Indonesia for the funded studentship. J.H. thanks EPSRC for the feasibility research grant on graphene sensors EP/I015930/1 and the previous research grant EP/C015045/1 thereby enabling the Raman spectroscopy aspects of this work to be performed. The authors wish to thank Barry Dunne (INEX, Newcastle University) and Oana Bretcanu for fabrication of the devices and Tracey Davey (EM Research Services, Newcastle University) for the SEM imaging.

REFERENCES

[1] K.S. Novoselov et al, "Electric Field Effect in Atomically Thin Carbon Films," *Science*, vol. 306, pp. 666-669, 2004.

[2] J. Hedley, Z. Hu, I. Arce-Garcia, and B.J. Gallacher, "Mode shape and failure analysis of high frequency MEMS/NEMS using Raman spectroscopy," *3rd IEEE International Conference on Nano/Micro Engineered and Molecular Systems, NEMS*, pp. 842-846, January 2008.

[3] Z.X. Hu, J. Hedley, B.J. Gallacher, and I. Arce-Garcia, "Dynamic characterization of MEMS using Raman spectroscopy," *Journal of Micromechanics and Microengineering*, vol. 18, 095019, 2008.

[4] T.M.G. Mohiuddin et al, "Uniaxial strain in graphene by Raman spectroscopy: G peak splitting, Grüneisen parameters, and sample orientation," *Physical Review B*, vol. 79, 205433, 2009.

[5] O. Frank et al, "Raman 2D-band splitting in graphene: Theory and experiment," *ACS Nano*, vol. 5, pp. 2231-2239, 2011.

[6] X. Li and W. Cai, "Large-Area Synthesis of High-Quality and Uniform Graphene Films on Copper Foils," *Science*, vol. 324, pp.1312-1314, 2009.

[7] Z. G. Cambaz, G. Yushin, S. Osswald, V. Mochalin, and Y. Gogotsi, "Noncatalytic synthesis of carbon nanotubes, graphene and graphite on SiC," *Carbon*, vol. 46, pp. 841-849, 2008.

[8] S. Bae et al, "Roll-to-roll production of 30-inch graphene films for transparent electrodes," *Nat. Nano.*, vol. 5, pp. 574-578, 2010.

Transfer-free Fabrication of Suspended Graphene Grown by Chemical Vapor Deposition

N. Lindvall[*], J. Sun and A. Yurgens

Department of Microtechnology and Nanoscience, Chalmers University of Technology, Gothenburg, Sweden
[*]Corresponding author. E-mail: lindvaln@gmail.com

Abstract—**Graphene, a true two-dimensional material with extraordinary mechanical- and electronic properties, is thought to be ideal for nanoelectromechanical systems (NEMS), like mass- and force sensors. Here, we present two different ways to fabricate suspended graphene for the intended use in future NEMS applications. The fabrication schemes do not require transfer of graphene from a catalyst where the graphene is grown on to another supporting substrate. The transfer is a source of several issues causing irreproducibility in large-scale production of graphene devices. We obtain suspended graphene membranes by locally removing the copper thin film on top of which the graphene is catalytically grown. The membranes are uniform and exhibit mechanical properties similar to those of exfoliated graphene. Also, suspended graphene beams with electrical interconnects are fabricated from non-catalytically grown graphene on SiO_2. Both approaches represent the first steps towards transfer-free fabrication of suspended graphene for NEMS applications.**

Keywords-graphene; suspended;chemical vapor deposition; transfer-free

I. INTRODUCTION

Graphene, a one atom thick hexagonal lattice of sp^2-hybridized carbon atoms, has attracted tremendous interest during the past years due to its unique properties [1-3]. Charge carriers in graphene can be tuned from electrons to holes and exhibit very high carrier mobility. They have zero rest mass and for low energy excitations they behave relativistically, making graphene a platform for Dirac physics experiments. Graphene also inhibits exceptional optical- [4], thermal- [5] and mechanical properties [6].

Many applications for graphene have been proposed, including transparent electrodes [7], high-frequency transistors [8, 9] and various sensors [10, 11]. More specifically, having extremely low mass density, high Young's modulus (1 TPa), surviving strain up to 25% and true two-dimensionality, it is ideal for NEMS applications [12]. These include mass- and force sensing. The fact that it combines ultimately small thickness and remarkable mechanical properties with high conductivity and electric field effect makes it a unique NEMS material.

Several experiments have already been realized using graphene in NEMS applications. Graphene resonators made from exfoliated graphite have been investigated using both optical- [13] and electrical [14, 15] measurements. These

devices are impressive for proof of concept experiments, but the exfoliation fabrication technique is not scalable and not applicable in industrial applications. Recently, great improvements in the fabrication of large-scale graphene have been made, primarily using chemical vapor deposition (CVD) on Ni or Cu catalysts [7, 16-20]. This has been utilized also for NEMS applications where arrays of CVD-grown graphene resonators have been realized [21].

However, the fabrication of such devices includes the transfer of graphene from the metal catalyst to a dielectric substrate. This process is associated with unavoidable issues with holes, wrinkles and metal residues. Hence, there is need for reliable large-scale fabrication of graphene NEMS systems without the transfer of graphene.

In this report we demonstrate the transfer-free fabrication of suspended graphene intended for use in NEMS applications. Two different approaches are explored. In the first method, we make suspended graphene membranes by locally collapsing the supporting copper thin film on top of which the graphene is catalytically grown. We present the membrane material characterization including scanning electron microscopy (SEM) and atomic force microscopy (AFM). Second, by removing SiO_2 from underneath graphene, we fabricate suspended graphene beams grown directly on SiO_2/Si substrates [22-25]. We show electrical- and SEM characterization of such devices.

II. GRAPHENE GROWTH ON CU THIN FILMS

Catalytic CVD of graphene on both bulk copper foils and thin films is the standard technique today for producing large-area, high quality graphene. The process is self-limiting, resulting in primarily monolayer graphene. To make devices on insulating substrates, the copper has to be etched away and the graphene transferred to the desired substrate. In this work we propose a new method for producing suspended graphene membranes. Previously, growth of graphene on the bottom side of copper with poor adhesion to SiO_2 has been achieved by carbon diffusion through the grain boundaries of copper [26]. When copper is removed, graphene remains on the insulating substrate. However, the uniformity of graphene is not good and it is not suspended. Suspended graphene forms a web over disintegrated nickel thin films [27] but it is not uniformly monolayer graphene and it does not form well-defined geometries.

In our fabrication method, we use silicon chips with 300 nm thermal oxide and lift-off e-beam lithography to pattern 200 nm copper thin films. The chips are then put into a cold-wall low-pressure CVD reactor equipped with a graphitic resistive heater. The samples are heated up to 800 °C in 1000 sccm Ar / 20 sccm H$_2$ atmosphere and annealed for 5 minutes. Then, 10 sccm of acetylene is introduced into the chamber initiating the growth of graphene. The growth continues for 10 minutes. At this high temperature, the copper partly melts and the thin films disintegrate, leaving parts of the sample covered and some parts uncovered. Finally, the samples are cooled down to room temperature in argon and hydrogen atmosphere.

At many places where the copper thin film disintegrates, it leaves graphene fragmented lying down on the substrate. However, at other places the graphene stays suspended over the SiO$_2$ substrate supported by the copper around it. A typical SEM micrograph is shown in Fig. 1. This particular copper feature was pre-patterned to a 10 μm square. As the metal collapses forming a ring-like shape, the graphene in the middle becomes suspended. Even without patterning, similar structures are easily formed. However, it becomes more controllable with pre-patterned geometries.

In Fig. 2, a SEM micrograph with the sample tilted at a high angle is shown where the suspended graphene is seen more clearly. It is possible to induce damage to the graphene by focused high electron current irradiation in the SEM, making it even more clear that the graphene is suspended.

To further characterize the material morphology and mechanical properties, we perform AFM measurements using low force tapping mode scanning. A three-dimensional representation of the height profile is shown in Fig. 3. The graphene is attached close to the top of the copper ring, suspended approximately 200 nm above the SiO$_2$ surface.

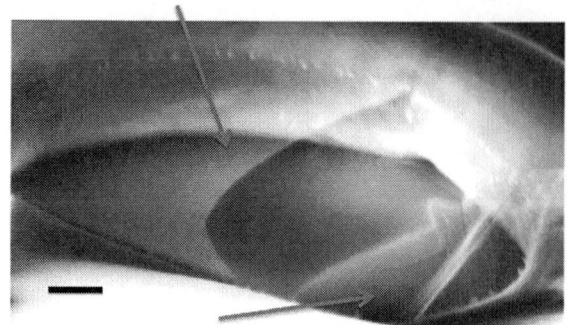

Fig. 2. Tilted SEM micrograph of a deliberately damaged membrane. Damage can be induced to the graphene by extensive high intensity electron irradiation. The two arrows point at the areas where graphene is still suspended. The scale-bar is 200 nm.

It is possible to obtain quantitative mechanical properties from AFM nano-indentation experiments. After proper calibration of the AFM, the tip is pushed down in the middle of the suspended graphene membrane and the corresponding force is recorded. This way, we obtain force-displacement data. Such force-displacement curves we fit by [6]:

$$F = \sigma_0^{2D}(\pi a)(\delta/a) + E^{2D}(q^3 a)(\delta/a)^3 \qquad (1)$$

Where F is the force applied to the membrane, a is the membrane diameter, δ is the deflection and $q = 1.02$ is a dimensionless constant. The pretension σ_0^{2D} and elastic stiffness E^{2D} are the fitting parameters. The force-displacement curves that we obtain are reproducible and follow theory well. Surprisingly, all of them show a non-linear feature at moderate applied force, which remains unexplained and will be subject of future work.

Typical values of the elastic stiffness and pre-tension are in

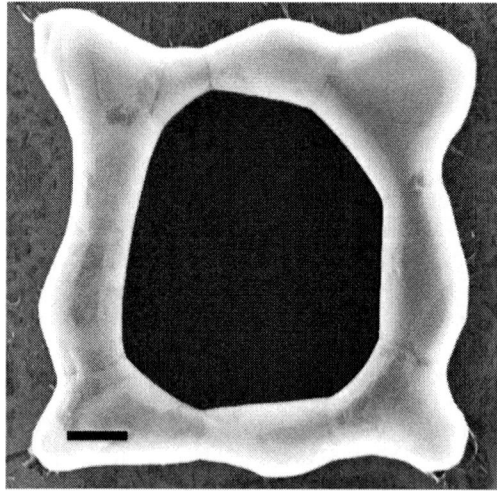

Fig. 1. SEM micrograph of a suspended graphene membrane. The bright part is disintegrated Cu. Inside, the black area is the suspended graphene membrane. Outside the Cu, graphene has fallen down on the substrate, or is only partly suspended. The scale-bar is 1 μm.

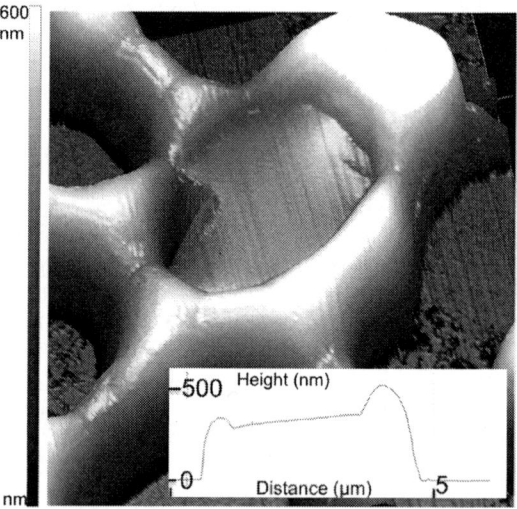

Fig. 3. 3D AFM height representation of a suspended membrane, reveling the morphology of the sample. The vertical scale is 600 nm and the later size is 7x7 μm^2. (inset) AFM line-scan showing that the graphene is clearly suspended, approximately 300 nm over the SiO^2.

the order of 350 N/m and 0.3 N/m, respectively. This corresponds to effective Young's modulus of ~1.0 TPa and effective pre-stress of slightly less than 1.0 GPa. In Fig. 4, the force-displacement data for an approximately circular membrane with 1.9 µm diameter is shown. At around 20 nm of deflection the non-linear feature is clearly seen. The mechanical properties are similar to those of high-quality exfoliated graphene, making it interesting for NEMS applications. Even though pre-patterning of copper gives some reproducibility of the formation of graphene membranes, there is still need for improvement of the fabrication process to obtain good process control.

III.   NON-CATALYTIC PROCESS

Recently, it was realized that graphene can grow non-catalytically on virtually any substrate, including dielectric ones like SiO_2 and Si_3N_4 [22, 25]. Since the graphene is already on a dielectric, there is no need to transfer it. Non-catalytically grown graphene exhibits smaller grain size and worse electronic properties than catalytically grown graphene. Its mechanical properties are yet to be determined.

Nano-crystalline graphene is grown directly on silicon substrates with 300 nm thermal oxide. The substrate is loaded into a low-pressure CVD reactor. It is heated to 1000 °C in 1000 sccm Ar / 20 sccm H_2 atmosphere. After annealing for 3 minutes, acetylene (20 sccm) is introduced to start the graphene growth. The growth time can be varied to achieve different film thicknesses and properties. A growth during 15-20 minutes gives films with optical properties similar to that of uniform monolayer graphene. In Fig. 5, transmission electron microscopy (TEM) diffraction pattern and high-resolution images of the film after transfer to TEM grids are shown. When focusing the e-beam to ~10 nm, the monolayer graphene diffraction signature is seen. Reference [25] provides more details on the material growth and characterization of this material.

A two-step e-beam lithography process is used to pattern the graphene and Au/Cr (150 nm / 3 nm) electrodes. Two-terminal resistance measurements reveal a sheet resistance of around 20 kΩ. By applying a gate voltage to the conducting silicon substrate, the charge-carrier density in the graphene can be tuned. In Fig. 6, the sheet resistance is shown as a function of the gate voltage, revealing a variation of ~20 %.

To suspend the graphene, the device is put in a buffered oxide etchant for 3 minutes to remove ~250 nm of SiO_2. The metal leads are used as etching mask. The etching is almost immediate underneath the graphene. Finally, the sample is critically point dried. In Fig. 6 (b), a SEM micrograph of the final device is shown at high viewing angle. The graphene membrane is straight and uniform. After suspension, the resistance increases, typically by ~50 %.

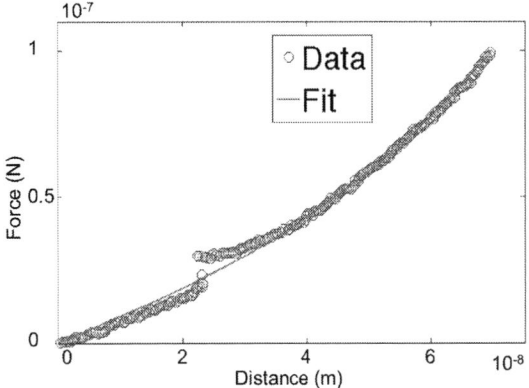

Fig. 4.   Nano-indentation measurement of a 1.9 µm diameter circular membrane. Blue dots represent membrane deflection data (force-vs-displacement) when the tip is pushed down in the middle of the membrane. The red solid line is an elastic response model fit. The overall fit gives an elastic stiffness of 373 (N/m) and pre-tension of 0.29 (N/m), similar to that of exfoliated graphene. However, the membrane is not perfectly circular and the fit is not perfect giving rise to some uncertainty. Also, the response shows a reproducible, so far not explained behavior around 20 nN force.

IV.   SUMMARY AND OUTLOOK

We fabricate suspended graphene systems for the intended use in future NEMS by two different methods. Both do not need transfer of graphene from metal catalysts. Suspended graphene membranes can be realized by copper thin film disintegration during CVD. By pre-patterning the copper film, some control of the membrane shape is achieved. The membranes are uniform and mechanically strong, exhibiting a Young's modulus similar to that of exfoliated graphene. Future work includes the integration of this growth process into the real device fabrication followed by electrical characterization.

We also fabricate suspended graphene beams from non-catalytic CVD-grown graphene. The beams are uniform, conducting and mechanically rigid. However, the electronic properties of the material need to be improved.

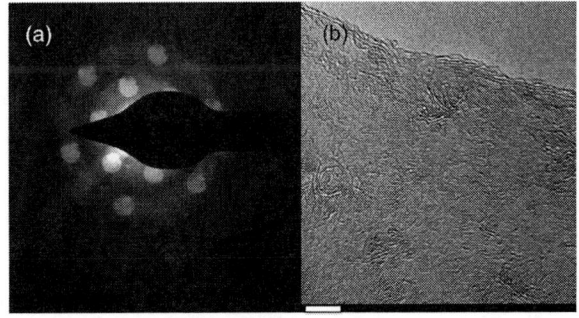

Fig. 5. (a) Electron diffraction pattern obtained by focusing the beam to a spot size of ~10 nm. It shows the signal of monolayer graphene. (b) High-resolution image of the graphene film. The scale-bar is 5 nm.

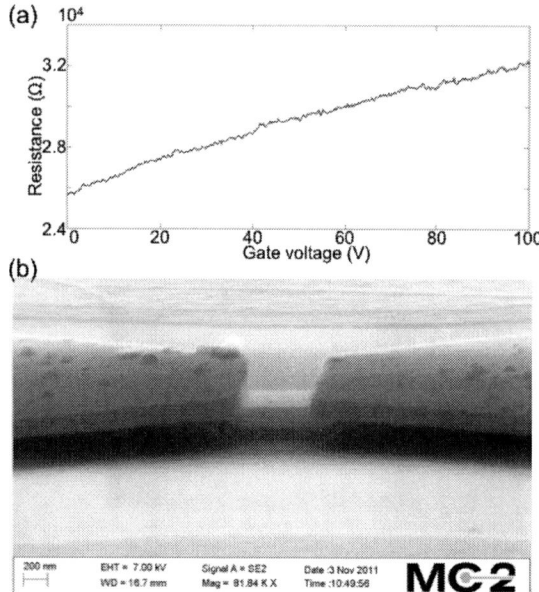

Fig. 6. (a) Electrical characterization of graphene grown on SiO₂ before suspension. Resistance is plotted as function of applied gate voltage to the substrate. The resistance can be modulated ~20 % by applying +100 V to the substrate. (b) SEM micrograph of the final device. The graphene is suspended between the two gold electrodes. The micrograph is taken at a high tilt angle. The scale-bar is 200 nm.

Both approaches represent the first steps towards transfer-free fabrication of graphene NEMS.

ACKNOWLEDGMENT

The authors thank professor T. Booth for the help in TEM. Financial support from the Swedish Research Council and the Swedish Foundation for Strategic Research is greatly appreciated. The cleanroom processing was performed on the equipment funded by the Knut and Alice Wallenberg Foundation.

REFERENCES

[1] K. S. Novoselov *et al.*, "Electric field effect in atomically thin carbon films," Science vol. 306, pp 666-669, 2004.

[2] K. S. Novoselov *et al.*, "Two-dimensional atomic crystals," PNAS vol. 102, pp 10451-10453, 2005.

[3] A. K. Geim, and K. S. Novoselov, "The rise of graphene," Nature Mater vol. 6, pp 183-191, 2007.

[4] F. Bonaccorso, Z. Sun, T. Hasan, and A. C. Ferrari, "Graphene photonics and optoelectronics," Nature Photon vol. 4, pp 611-622, 2010.

[5] A. A. Balandin *et al.*, "Superior thermal conductivity of single-layer graphene," Nano Lett vol. 8, pp 902-907, 2008.

[6] J. Hone, C. Lee, X. D. Wei, and J. W. Kysar, "Measurement of the elastic properties and intrinsic strength of monolayer graphene," Science vol. 321, pp 385-388, 2008.

[7] S. Bae *et al.*, "Roll-to-roll production of 30-inch graphene films for transparent electrodes," Nature Nanotech vol. 5, pp 574-578, 2010.

[8] Y. Q. Wu *et al.*, "High-frequency, scaled graphene transistors on diamond-like carbon," Nature vol. 472, pp 74-78, 2011.

[9] J. Svensson *et al.*, "Carbon Nanotube Field Effect Transistors with Suspended Graphene Gates," Nano Lett vol. 11, pp 3569-3575, 2011.

[10] A. K. Geim, "Graphene: Status and Prospects," Science vol. 324, pp 1530-1534, 2009.

[11] M. Tarasov, N. Lindvall, L. Kuzmin, and A. Yurgens, "Family of graphene-based superconducting devices," JETP Lett vol. 94, pp 329-332, 2011.

[12] C. Lee, X. D. Wei, J. W. Kysar, and J. Hone, "Measurement of the elastic properties and intrinsic strength of monolayer graphene," Science vol. 321, pp 385-388, 2008.

[13] J. S. Bunch *et al.*, "Electromechanical resonators from graphene sheets," Science vol. 315, pp 490-493, 2007.

[14] C. Y. Chen *et al.*, "Performance of monolayer graphene nanomechanical resonators with electrical readout," Nature Nanotech vol. 4, pp 861-867, 2009.

[15] Y. H. Xu *et al.*, "Radio frequency electrical transduction of graphene mechanical resonators," Appl Phys Lett vol. 97, 2010.

[16] A. N. Obraztsov, E. A. Obraztsova, A. V. Tyurnina, and A. A. Zolotukhin, "Chemical vapor deposition of thin graphite films of nanometer thickness," Carbon vol. 45, pp 2017-2021, 2007.

[17] X. S. Li *et al.*, "Large-Area Synthesis of High-Quality and Uniform Graphene Films on Copper Foils," Science vol. 324, pp 1312-1314, 2009.

[18] M. P. Levendorf, C. S. Ruiz-Vargas, S. Garg, and J. Park, "Transfer-Free Batch Fabrication of Single Layer Graphene Transistors," Nano Lett vol. 9, pp 4479-4483, 2009.

[19] Y. Lee *et al.*, "Wafer-Scale Synthesis and Transfer of Graphene Films," Nano Lett vol. 10, pp 490-493, 2010.

[20] J. Sun *et al.*, "Low Partial Pressure Chemical Vapor Deposition of Graphene on Copper," IEEE T Nanotechnol, In press.

[21] A. M. van der Zande *et al.*, "Large-Scale Arrays of Single-Layer Graphene Resonators," Nano Lett vol. 10, pp 4869-4873, 2010.

[22] J. Sun, N. Lindvall, M. T. Cole, K. B. K. Teo, and A. Yurgens, "Large-area uniform graphene-like thin films grown by chemical vapor deposition directly on silicon nitride," Appl Phys Lett vol. 98, p. 252107, 2011.

[23] S. K. Jerng *et al.*, "Nanocrystalline Graphite Growth on Sapphire by Carbon Molecular Beam Epitaxy," J Phys Chem C vol. 115, pp 4491-4494, 2011.

[24] J. Y. Chen *et al.*, "Oxygen-Aided Synthesis of Polycrystalline Graphene on Silicon Dioxide Substrates," J Am Chem Soc vol. 133, pp 17548-17551, 2011.

[25] J. Sun, M. Cole, N. Lindvall, K. Teo, and A. Yurgens, "Noncatalytic chemical vapor deposition of graphene on high-temperature substrates for transparent electrodes," To be published.

[26] C. Y. Su *et al.*, "Direct Formation of Wafer Scale Graphene Thin Layers on Insulating Substrates by Chemical Vapor Deposition," Nano Lett vol. 11, pp 3612-3616, 2011.

[27] Y. H. Lee, and J. H. Lee, "Catalyst patterned growth of interconnecting graphene layer on SiO(2)/Si substrate for integrated devices," Appl Phys Lett vol. 95, 2009.

Photovoltaic Response of N-doped Graphene-based Photodetector

Wenrong Wang[1,2], Tie Li[1,*], Yuelin Wang[1]

[1,*]State Key Laboratories of Transducer Technology, Science and Technology on Microsystem Laboratory, Shanghai Institute of Microsystem and Information Technology, CAS, PR China

tli@mail.sim.ac.cn

[2]Graduate School of Chinese Academy of Science, PR China

In this paper, photovoltaic response has been found in an N-doped graphene based photodetector with metal-graphene-metal structure. Raman and XPS spectra indicate that nitrogen atoms are doped into the graphene after synthesized in a CVD system by introducing CH_4 and NH_3. With the semiconducting behavior of I-V curves, N-doping graphene exhibits an opened band-gap, which generates photocurrent under light excitation.

Keywords-CVD; graphene; N-doped; photovoltaic

I. INTRODUCTION

Graphene is a two-dimensional material of carbon atoms packed in honeycomb lattice. It has recently attracted significant attention since its discovery by Geim in 2004 [1], primarily due to its extraordinary electronic properties [2].

Graphene-based photodetectors are promising new devices for optoelectronic applications. It has been proposed that graphene nanoribbons [3] or bi-layer [4] can open the band-gap of graphene and be used as photodetectors, but these kinds of graphene material should be carefully prepared when precisely control the lattice edge or number of the sheets in complex processes. In 2009, Liu et al. has found that the band-gap of graphene can also be formed by introducing nitrogen atoms as dopant when grown by chemical vapor deposition (CVD) method [5]. Here, we explored N-doping to open graphene's bandgap and made a simple metal-graphene-metal structure of N-doped graphene-based photodetector with brief process and new mechanism, in which Schottky barriers were the main contribution to photovoltaic effect.

II. EXPERIMENT

A. N-doped Graphene Synthesis

The N-doped graphene was synthesized on 99.9% copper foils with the mixed gas of CH_4, NH_3, Ar and H_2 at 1000°C by ourselves-made CVD system shown in Fig. 1, which can be found at Transducers' 2011 [6]. In this work, we used the gas flow of CH_4:NH_3:Ar:H_2 =10:10:1000:100sccm for 5 min to get few-layer N-doped graphene.

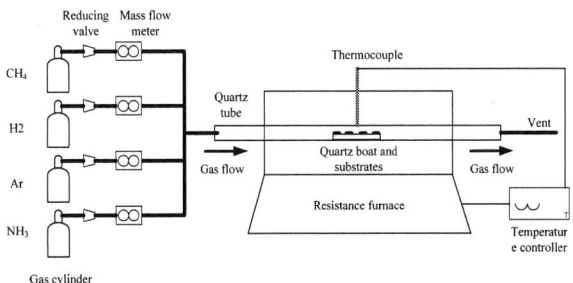

Fig. 1. Schematic diagram of the thermal CVD system.

B. Graphene transfer and Device fabrication

The transfer of graphene to insulating substrate was carried out after the wet-etching of the underlying Cu foil by the solution of $FeCl_3$. The process are as follows: firstly, a supported material PMMA is coated on the Cu/Graphene surface; secondly etch away Cu substrate by $FeCl_3$ solution; thirdly put the free-standing PMMA/graphene membrane on the target substrate; finally dissolve the PMMA with acetone to yield a graphene film on the desired substrate. In this way, the transferred graphene films can be attached to almost any material, such as semiconductor, glass, metal, plastic and so on. Here, we used the Si with 300nm thick thermally grown SiO_2 layer as the substrate.

Graphene is particular sensitive to surface contaminations, especially resist residues in the lithography process, which would modify the electronic structure of graphene. In order to prevent further chemical contaminations, we used hard silicon mask to fabricate electrodes, which is a lithography free process shown in Fig. 2 (a). The hard silicon mask was made by Micro-Electro-Mechanical Systems (MEMS) technology. First, the substrate with graphene was covered with hard silicon mask; then aluminum was deposited as electrodes by thermal evaporation. Finally, the silicon mask was lifted up and moved away. The schematic view and photo of the device can be found in Fig. 2 (b) and (c).

978-1-4673-1122-9/12 $31.00 © 2012 IEEE

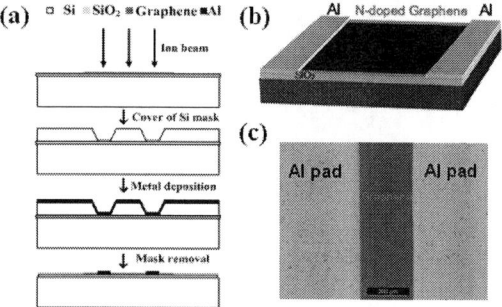

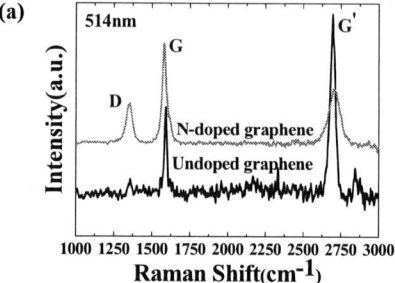

Fig. 2. (a) The lithography free fabrication process of the device, in order to maintain the properties of graphene. (b) Bird's-eye view of a schematic device configuration. (c) Optical photo of the N-doped graphene-based device.

C. Photovoltaic Response Test

Testing of the device was carried out in a high vacuum environment (10^{-7} Torr) with a cryogenic probe system. The testing system is schematically shown in Fig. 3. The device was mounted on the sample holder of the cryogenic system. An IR lamp was used as light source, which has a peak emission around 1.2μm. The distance between the lamp and the devices was set to be 50cm. The light intensity was estimated to be 0.32mW/mm^2. The electrical properties and the photoresponses of the devices were measured using a Keithley 4200 semiconductor parameter analyzer.

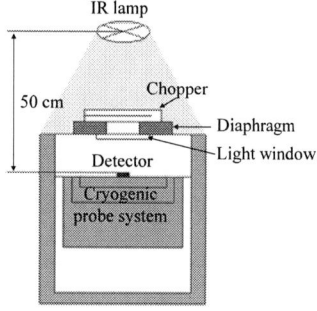

Fig. 3. Schematic diagram of the testing system.

D. Characterization of graphene

For graphene characterization, optical microscope, transmission electron microscope (TEM, Philips, 160KV), X-ray photoelectron spectroscopy (XPS, Kratos AXIS Ultra DLD) and Raman spectrum (Renishaw, with Ar$^+$ laser excitation at 514nm) were used to confirm the graphene film grown on the copper foils and obtain information about the quality and the number of the graphene.

III. RESULTS AND DISCUSSIONS

A. N-doped Graphene Synthesis

Characterizations of N-doped graphene are shown in Fig. 4. Raman spectra is the most effective way to confirm the quality

of graphene, which provides a quick and facile structural and characterization of the produced material [7]. Fig. 4(a) shows the Raman spectra of N-doped graphene compared with undoped graphene. The broader, lower G' band in Raman spectrum implies that it is few-layer graphene [8], which is also confirmed directly by the HREM in Fig. 2(c). The shifted G band (from 1590 to 1580 cm^{-1}) as well as the strong D band (1351.6cm^{-1}) indicates the defects in the as-grown graphitic lattice, which with the N-1s peak at 401.5 in the XPS spectra, just suggests that the doping N atoms in the graphene.

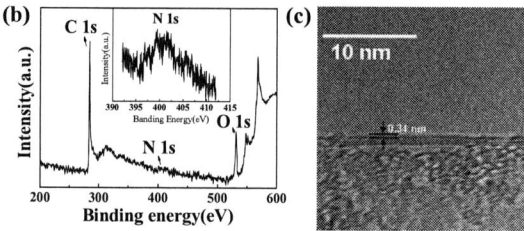

Fig. 4: (a) The Raman spectra of N-doped graphene, the black line corresponds to the un-doped graphene. (b) XPS spectra of N-doped graphene. The inset shows N-1s spectrum of the N-doped graphene. (c) HRTEM image of the N-doped graphene with few-layer.

B. Photovoltaic Response

The IV characteristic of N-doped graphene based device is shown in figure 4. We can see that the IV curves are nonlinear, implying the semiconducting property of N-doped graphene, which means the band-gap of graphene was successfully opened by N dopant.

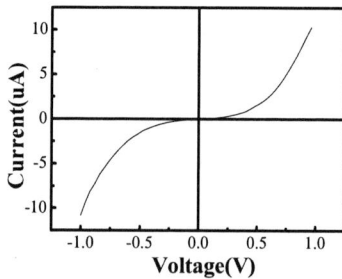

Fig. 5. IV characteristic of the device, which indicates semiconductor behavior.

Fig. 6 shows the small voltage IV curves of the device in dark and under illumination. From the figure, we can see clearly that the IV curve shifted downwards, indicating photocurrent generated under illumination without applied voltage, which showed photovoltaic effect under illumination.

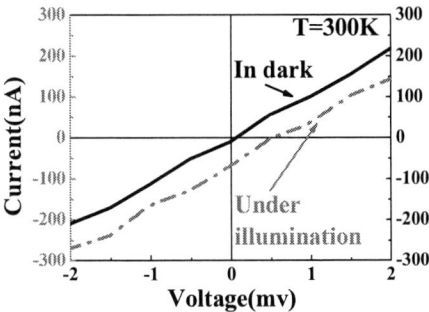

Fig. 6. *S*mall voltage IV curves of the device in dark and under illumination. IV curve shifted downwards under illumination.

The behavior of the device under illumination can be interpreted by the energy diagram shown in Fig. 7. Schottky barriers were formed at the contact regions between metal electrodes and N-doped graphene. In dark, the amount of electrons injected from graphene to aluminum equals with the amount of electrons injected from aluminum to graphene per unit time for a Shottky barrier. Under illumination, the electron balance at the Schottky barrier is broken; more electrons are injected from graphene to aluminum, inducing electric current at the barrier. At different locations, the graphene synthesized may have different structures leading to different band structures, so the barrier heights at the junctions are different. Asymmetry in barrier height leads to the asymmetry of photocurrent generated at two barriers, resulting in the photovoltaic effect.

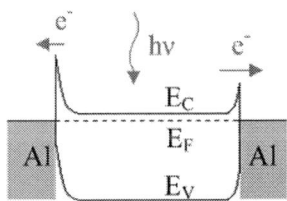

Fig. 7. The energy band diagram of the device. Due to the differences in N-doped graphene energy band and contact conditions, the Schottky barriers formed at the contacts between metal electrodes and N-doped graphene might be different.

Further experiments on the responses of the device at different temperatures were also investigated. We can see from Fig. 8(a) that the photocurrent of the device was found to decrease as the temperature decreased from room temperature to 4.1K, which suggests the hot electron emission mode in this N-doped graphene device. In the hot electron emission mode of Schottky barrier, carriers with energy larger than the barrier height moving towards the barrier would travel through. In dark at a certain temperature, the current from metal to semiconductor counterbalances that from semiconductor to metal. Under illumination, part of the photoexcited carriers also travel through the barrier, resulting in the observed photocurrent.

In this photodetector, 60nA photocurrent was generated under 0.32mW/mm^2 IR lamp illumination at room temperature, which is much higher than SWCNT-based photodetector under the same conditions of design, manufacture, and measurement. Fig. 8(b) shows the typical responses of the devices at different temperature.

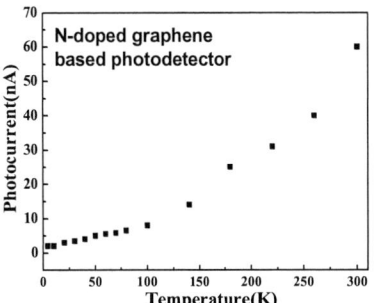

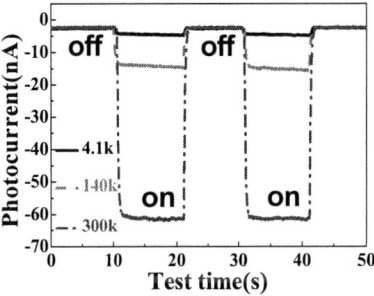

Fig. 8. (a) Photovoltaic response of N-doped graphene-based (black dots) photodetectors at different temperatures; (b) The responses of N-doped graphene device at 4.1K, 140K and 300K, which show obvious response under illumination.

IV. CONCLISION

In conclusion, N-doped graphene was synthesized by CVD method and the photodetector based on N-doped graphene was fabricated, which is a novel photodetector based on graphene. The photovoltaic effect mechanism of the device was studied in this work, in which the asymmetrical Schottky barriers formed at the contacts between metal electrodes and N-doped graphene were considered as the main reason caused photovoltaic effect. Under $0.32 mW/mm^2$ IR lamp illumination, 60nA photocurrent was generated. The photovoltaic response at different temperatures was investigated and the photovoltaic effect was found to increase as the temperature rose.

ACKNOWLEDGMENT

This work was supported by the Important National Science and Technology Specific Projects (No. 2011ZX02707), the National Program on Key Basic Research Project (No. 2011CB309501), the Foundation for Innovative Research Groups of the National Natural Science Foundation of China (No. 61021064), the National Natural Science Foundation of China (No. 60936001 and No. 60876037) and the Chinese Academy of Sciences Special Grant for Postgraduate Research, Innovation and Practice.

REFERENCES

[1] K. S. Novoselov, A. K. Geim, S. V. Morozov, et al., "Electric filed effect in atomically thin carbon films," *Science*, vol. 306, pp. 666-669, 2004.

[2] A. K. Geim and K. S. Novoselov, "The rise of graphene," *Nature material*, vol. 6, pp. 183-191, 2007.

[3] V. Ryzhii, V. Mitin, et al., "Device model for graphene nanoribbon phototransistor," *Applied Physics Express*, vol. 1, p. 063002, 2008.

[4] T. Mueller, F. Xia and P. Avouris, "Graphene photodetector for high-speed optical communications," *Nature photonics*, vol. 4, pp. 297-301, 2010.

[5] D. Wei, Y. Liu, et al., "Synthesis of N-doped graphene by chemical vapor deposition and its electrical properties," *Nano Letters*, vol. 9, pp. 1752-1758, 2009.

[6] W. Wang, Y. Zhou, et al., "Magnetoresistance behaviors of undoped and N-doped graphene grown by CVD method," *Transducers'11*, Beijing China, 2011.

[7] A. C. Ferrari, J. C. Meyer, et al., "Raman spectrum of graphene and graphene layers," *Physical Review Letters*, vol. 97, pp.187401, 2006.

[8] Z. Ni, Y. Wang, et al., "Raman spectroscopy and imaging of graphene," *Nano Res.*, vol. 1, pp. 273-291, 2008.

Development of roll press UV imprint process for replication of micro lens array on the large and thin quartz substrate

L.Li[*], K.Ishii, Y.Tsutsui, S.Shoji, J.Mizuno

* Major in Nano-Science and Nano-Engineering, Waseda University, JAPAN,
li-ly@asagi.waseda.jp

Abstract— **In this study, micro lens arrays (MLAs) fabrication process was carried out using flexible replica mold and an original roll press UV imprint system. We successfully demonstrated MLAs fabrication process with high throughput. The lens patterns of 30 × 100 mm2 in area replicated by two photo-curable resins were formed on the quartz substrate of 0.15 mm in thickness. This fabrication process is applicable for MLAs fabrication of next generation optical devices.**

Keywords-micro lens arrays; quartz substrate; UV imprint; roll press UV imprint system; flexible replica mold

I. INTRODUCTION

Micro lens arrays (MLAs) are one of the important optical devices that are widely used in the area of optical communication and optoelectronics. Many researches have been carried out on MLAs including the fabrication techniques and also their potential applications for the past 1 decade. MLAs are a series of miniaturized lenses that are arranged in certain forms, normally rectangular or hexagonal in shape. They have been applied commonly in the optical and lighting devices in recent years, such as in optical fibers, image systems and illuminations [1,2].

Now, MLAs are requested to fabricate on large and thin quartz substrates with high throughput and low cost for next generation optical devices as 3-dimention flat panel display, widely light-emitting diode, and so on. We suggested MLAs fabrication process using replica mold and UV imprint method which is widely employed as an effective process of micro/nano fabrication [3-14]. However, during replication on large and thin quartz substrates using conventional UV imprint, non-conformal contact often causes the reduction of replication uniformity and the damage for quartz substrates, due to local flatness distortion of the mold [15].

For solving this problem, we employed UV roller imprinting method. The method has been developed as a high throughput micro/nano fabrication for mass production [16-18]. This technique has much attention in micro/nano fabrication technology for flexible optical element, electric paper, liquid crystal display and personal computer [19-21]. UV roller imprinting is the most cost-effective fabrication method.

In this study, an original UV roller imprinting, called "roll press UV imprint process" and prototype of MLAs fabrication are described.

II. DETAIL OF SYSTEM

Fig. 1 shows a photograph of the developed roll press UV imprint system, and Table.1 shows its performances. Experiments described below were carried out by using this system.

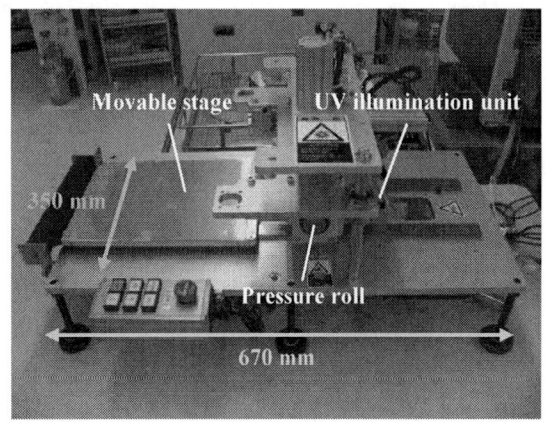

Fig. 1. Photograph of the roll press UV imprint system.

TABLE I.   PERFORMANCES OF THE ROLL PRESS UV IMPRINT SYSTEM

UV wavelength (nm)	375
UV intensity (mW/cm2)	100
Stage velocity (m/min)	6-125
Pressure (MPa)	0-4.8
Stage size (mm2)	200 × 270 mm2
System size (mm2)	350 × 670 mm2
Material of roll	Urethane rubber

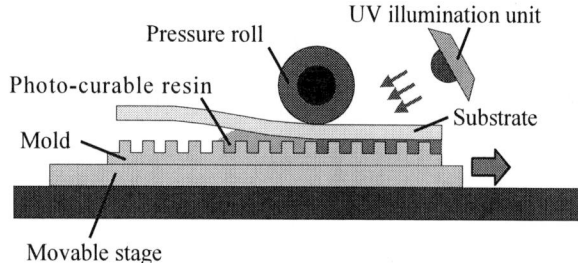

Fig. 2. Schematic diagram of the roll press UV imprint system. The mold and the substrate can change positions each other, if the top one has light permeability.

Fig. 2 shows a schematic drawing of the developed system. This imprint system can prevent mold from non-conformal contacting, due to the equipped pressure roll applying pressure on the mold or substrate. It can achieve large area pattern replication (max: 200 × 270 mm²) with high throughput. The process time for one cycle of replication is ranged from 20 to 30 sec. In addition, it can employ a wide variety of molds, flexible molds as well as standard hard molds. These are remarkable advanteges of the research level tool.

III. FABRICATION PROCESS OF MLAS

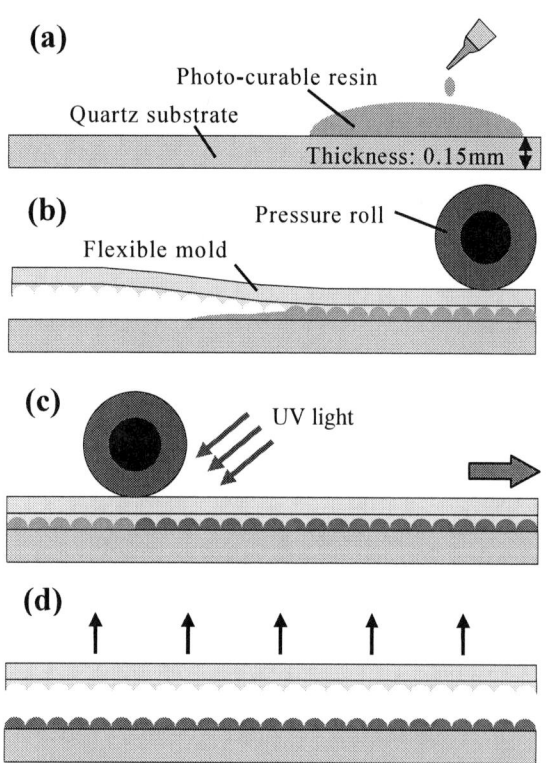

Fig. 3. Schematic diagram of the roll press UV imprint process. (a) Dispensing the photo-curable resin onto the quartz substrate, (b) Pressing the flexible mold on the resin, (c) movable stage moving to the terminal point and irradiating UV light and (d) releasing the flexible mold from the resin.

Fig. 3 shows a fabrication process flow of the roll press UV imprint process, and details are described below. First, a photo-curable resin was dispensed onto the quartz substrate (largeness: 30 × 100 mm², thickness: 0.15 mm), and a flexible replica mold was placed on the resin. We used two photo-curable resins "LGR-1001 and PAK-02". Table. 2 shows comparisons of LGR-1001 and PAK-02. LGR-1001 (from Matsunami glass Ind., Ltd.) has a high refractive-index of 1.62 for application to optical devices. And, PAK-02 (from from Toyo Gosei Co., Ltd.) has a low viscosity of 9 mPa·sec being suitable for the UV roller imprinting method. Fig. 4 (a) shows a flexible replica mold (from Soken Chemical & Engineering Co., Ltd.) placed on the resin. Fig. 4 (b) shows a SEM image of micro pattern to replicate MLAs on the flexible mold, whose size of 7.8 μm diameter was replicated.

TABLE II.   COMPARISONS OF TWO PHOTO-CURABLE RESINS

	LGR-1001	PAK-02
Refractive index	1.62	1.51
Viscosity (mPa·sec)	2400	9
Purpose of development	For optical devices	For roll UV imprint method

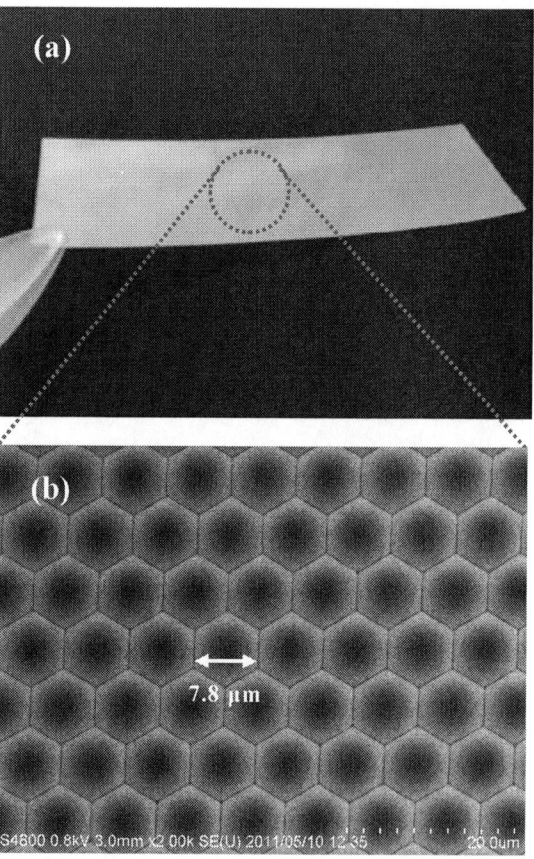

Fig. 4. Images of the flexible replica mold to replicate MLAs. (a) Photograph and (b) SEM images.

Next, uniform pressure of 4.0 MPa was applied on the flexible mold by the pressure roll. By keeping load pressure, the movable stage moved to the terminal point with velocity of 10 m/min. Then, UV light was exposed for curing the photo-curable resin with UV intensity of 100 mW/cm² at wavelength of 375 nm. Finally, the substrate with the photo-curable resin was released from the flexible mold. Thus, MLAs were replicated on the quartz substrate. This fabrication process can archive high throughput replication, and the process time for one cycle is ranged from 20 to 30 sec.

IV. RESULTS AND DISCUSSION

The SEM images of the MLAs replicated by LGR-1001 and PAK-02 are shown in Fig. 5 and Fig. 6. The lens patterns replicated by LGR-1001 have 7.8 μm diameter and 3.70 μm height, and that replicated by PAK-02 have 7.74 μm diameter and 3.68 μm height. In comparison with the pattern size of the flexible mold shown in Fig. 4, MLAs were successfully formed on the quartz substrate with high throughput.

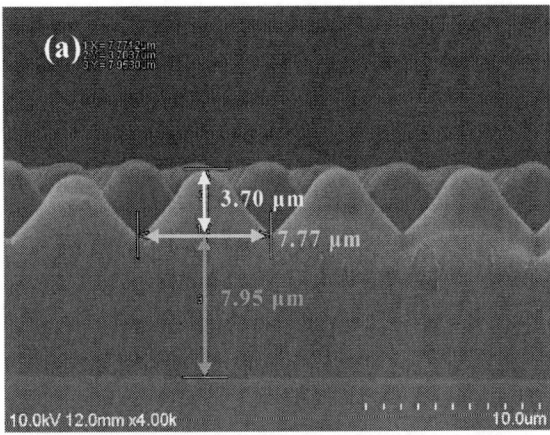

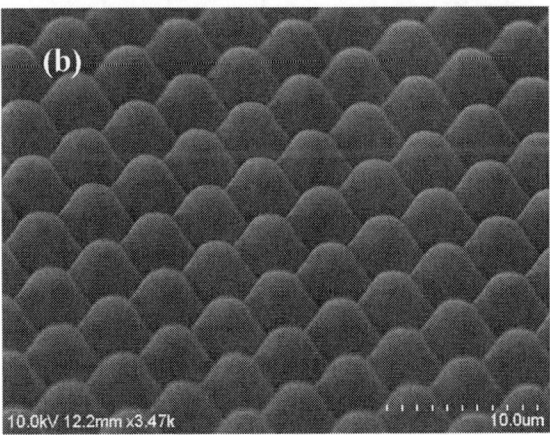

Fig. 5. Images of MLAs replicated by LGR-1001.
(a)Cross-sectional and (b) skew images.

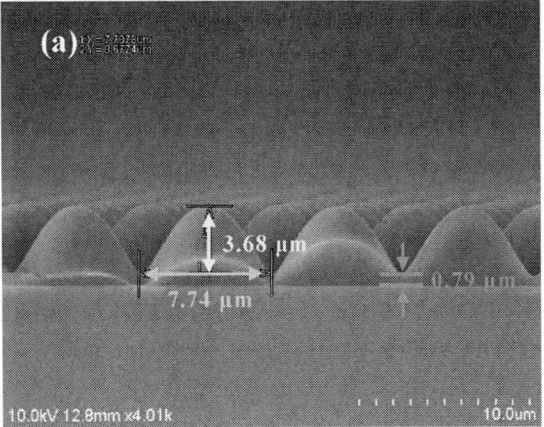

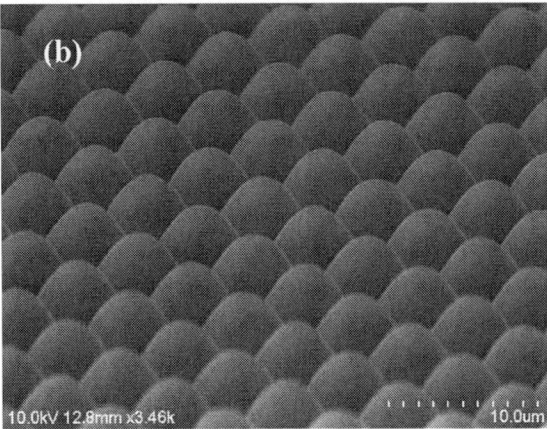

Fig. 6. Images of MLAs replicated by PAK-02.
(a)Cross-sectional and (b) skew images.

Comparing MLAs replicated by LGR-1001 and PAK-02, there are two different issues. First, micro lenses of LGR-1001 have concave slope, and that of PAK-02 have convex slope. And the other is that a layer thickness of LGR-1001 is much thicker than that of PAK-02. These are caused by difference of viscosity as shown in Table. 2. From results of MLAs replication by using LGR-1001 and PAK-02, a formability of PAK-02 on the roll UV imprint process is better than that of LGR-1001 due to a low viscosity. But, high refractive-index of LGR-1001 is suitable for some optical devices such as 3-dimention flat panel display. So, we suggest choosing appropriate photo-curable resins as the device demands.

V. CONCLUSION

This paper describes a development of roll press UV imprint process for replication of MLAs on the large and thin quarts substrate with high throughput. Replication tests were carried out by using the flexible replica mold and two photo-curable resins. The lens patterns of 30 × 100 mm² in area replicated by two photo-curable resins were formed on the quartz substrate of 0.15 mm in thickness. Since, the process can prevent non-conformal contacting, it is useful for large area replication of micro/nano structures on thin quartz substrates.

In future work, we will improve the roll press UV imprint process to achieve more precise and efficient for micro/nano fabrications. In addition, we will evaluate further characteristics of the roll press UV imprint for wider application fields.

ACKNOWLEDGMENT

This work is partly supported by Japan Ministry of Education, Culture, Sports Science & Technology Grant-in-Aid for Scientific Basic Research (S) No. 23226010 and the Grant-in-Aid for Specially Promoted Research "Establishment of Electrochemical Device Engineering from the Ministry of Education, Culture, Sports, Science and Technology (MEXT), Japan. The authors thank for Toyo Gosei Co., Ltd, Soken Chemical & Engineering Co., Ltd and Nanotechnology Support Project of Waseda University for their technical advices.

REFERENCES

[1] C.S. Lim, M.H. Hong, A.S. Kumar, M. Rahman, and X.D. Liu, "Fabrication of concave micro lens array using laser patterning and isotropic etching," *Int. J. Mach. Tool. Manufact.*, vol. 46, pp. 552-558, 2006.

[2] K.H. Liu, M.F. Chen, C.T. Pan, M.Y. Chang, and W.Y. Huang, "Fabrication of various dimensions of high fill-factor micro-lens arrays for OLED package," *Sens. Actuator A*, vol. 159, pp. 126–134, 2010

[3] S.Y. Chou, P.R. Krauss, and P.J. Renstrom, "Nanoimprint lithography," *J. Vac. Sci. Technol.*, vol. B14, pp. 4129-4133, 1996.

[4] H. Tan, A. Gilbertson, and S.Y. Chou, "Roller nanoimprint lithography," *J. Vac. Sci. Technol.* vol. B16, pp. 3926-3928, 1998.

[5] B. Heidari, I. Maximov, and L. Montelius, "Nanoimprint lithography at the 6 in. wafer scale," *J. Vac. Sci. Technol.*, vol. B18, pp. 3557-3560, 2000.

[6] C.M.S. Torres, S. Zankovych, J. Seekamp, A.P. Kam, C.C. Cedeno, T. Hoffmann, J. Ahopelto, F. Reuther, K. Pfeiffer, G. Bleidiessel, G. Gruetzner, M.V. Maximov, and B. Heidari, "Nanoimprint lithography: an alternative nanofabrication approach," *Mater. Sci. Eng.*, vol. C23, pp. 23-31, 2003.

[7] C. Guo, L. Feng, J. Zhai, G. Wang, Y. Song, L. Jiang, and D. Zhu, "Large-area fabrication of a nanostructure-induced hydrophobic surface

from a hydrophilic polymer," *Chem. Phys. Chem.*, vol. 5, pp. 750-753, 2004.

[8] F. Lazzarino, C. Gourgon, P. Schiavone, and C. Perret, "Mold deformation in nanoimprint lithography," *J. Vac. Sci. Technol.*, vol. B22, pp. 3318-3322, 2004.

[9] L.J. Guo, "Recent progress in nanoimprint technology and its applications," *J. Phys. D: Appl. Phys.*, vol. 37, pp. R123-R141, 2004.

[10] B.D. Gates, Q. Xu, M. Stewart, D. Ryan, C.G. Willson, and G.M. Whitesides, "New approaches to nanofabrication:molding, printing, and other techniques," *Chem. Rev.*, vol. 105, pp. 1171–1196, 2005.

[11] C.Y. Chang, S.Y. Yang, and J.L. Sheh, "A roller embossing process for rapid fabrication of microlens arrays on glass substrates," *Microsyst. Technol.*, vol. 12, pp. 754-759, 2006.

[12] M. Nakajima, T. Yoshikawa, K. Sogo, and Y. Hirai, "Fabrication of multi-layered nano-channels by reversal imprint lithography," *Microelectron. Eng.*, vol. 83, pp. 876-879, 2006.

[13] S.W. Youn, M. Takahashi, H. Goto, and R. Maeda, "Fabrication of micro-mold for glass embossing using focused ion beam, femto-second laser, eximer laser and dicing techniques," *J. Mater. Process. Technol.*, vol. 187-188, pp. 326-330, 2007.

[14] J. Mizuno, L. Li, Y. Kawaguchi, T. Kentaro, H. Shinohara, and S. Shoji, "Anti-Sticking Cuing of Fluorinated Polymers for Improvement of Mold Releasability," *J. Photopolym. Sci Technol.*, vol. 24, pp. 89-93, 2011.

[15] S.W. Youn, M. Ogiwara, H. Goto, M. Takahashi, and R. Maeda, "Prototype development of a roller imprint system and its application to large area polymer replication for a microstructured optical device," *J. Mater. Process. Technol.*, vol. 202, pp. 76-85, 2008.

[16] J. Taniguchi, M. Komuro, S. Inoue, N. Kimura, Y. Tokano, H. Hiroshima, and S. Matsui, "Preparation of diamond mold using electron beamlithography for application to nanoimprint lithography," *Jpn. J. Appl.Phys.*, vol. 39, pp. 7070-7074, December 2000.

[17] N. Unno, J. Taniguchi, M. Shizuno, and K. Ishikawa, "Fabrication of thenanoimprint mold using inorganic electron beam resist with post exposure bake," *J. Vac. Sci.Technol*, vol. B26, pp. 2390-2393, 2008.

[18] T. Shibazaki, H. Shinohara, T. Hirasawa, N. Sakai, J. Taniguchi, J. Mizuno, and S. Shoji, "Desktop Type Equipment of Thermal-assisted UV Roller Imprinting," *J. Photopolym. Sci. Technol.*, vol. 22, pp. 727-730, 2009.

[19] C. Stuart, and Y. Chen, "Roll in and roll out: A path to high-throughput nanoimprint lithography," *ACS Nano.*, vol. 3, pp. 2062-2064, August 2009.

[20] L.P. Yeo, S.H. Ng, Z. Wang, Z. Wang, and N.F.de Rooij, "Microfabrication of polymeric devices using hot roller embossing," *Microelectron. Eng.*, vol. 86, pp. 933-936, April 2009.

[21] T.C. Huang, J.T. Wu, S.Y. Yang, P.H. Huang, and S.H. Chang, "Direct fabrication of microstructures on metal roller using stepped rotating lithography and electroless nickel plating," *Microelectron. Eng.*, vol. 86, pp. 615-618, April 2009.

Fabrication Metal Roller Mold with Sub-Micrometer Feature Size Based on Contact Printing Photolithography

Kuo-Feng Huang, Sung-Wen Tsai, Yung-Chun Lee[*]

[*]Department of Mechanical Engineering, National Cheng Kung University, Taiwan

Corresponding author: yunglee@mail.ncku.edu.tw

Abstract—**This paper presents an innovative approach for directly forming surface micro-structures at sub-micrometer scale on the cylindrical surface of a metal roller. This roller can then serve as a roller mold in roller imprinting processes for large-area micro/nano-fabrication. The key element in this roller mold fabrication mold processes is a new type of contact-printing method with can transfer a patterned metal film from a planar mold to a thin photo-resist layer coated on the roller's cylindrical surface. The pattern definition capability can easily reach sub-micrometer or nanometer over a large patterning area. Subsequent ultraviolet light exposure patterned surface micro-structures with sub-micrometer–scaled feature sizes can be directly formed on the roller's surface by either metal etching or metal deposition.**

Keywords- Roller Mold, Roller Imprinting, Contact Printing Lithography

I. INTRODUCTION

Roller imprinting is one of the most important methods for large-area, high-throughput, and low-cost fabrication of micro/nano-structures. It can be integrated into a roll-to-roll (R2R) type of manufacturing process for continuous replication of polymeric micro/nano-structures. There are two types of roller imprinting: hot-embossing and UV-curing, and both are well developed and implemented for mass-production of micro/nano-structures [1] with a wide range of applications in micro-optics, electro-optics, flexible electronics, organic photo-electronics, flat panel displays, back-light units, e-papers,... etc.

The key issue in roller imprinting technology is the fabrication of roller molds which have certain desired patterns of micro/nano-structures on their cylindrical surfaces. Direct machining using diamond-tip cutters and a precision turning machine on a metal roller's cylindrical surface is commonly seen in industries. Although this method can produce robust roller molds with longer lifetime of roller imprinting, it is limited to micro-structures with less complicated patterns and relatively large feature sizes above few tens of micrometers.

To achieve smaller feature sizes and more complex patterns, the most commonly used method in preparing roller molds is thin mold wrapping [2-10]. Although the thin mold wrapping methods have been widely used in preparing roller molds for many years, there are several serious problems not resolved

yet. First of all, to firmly attach a planar mold onto the cylindrical surface of a roller is not a trivial job, particularly when precise alignments of the mold with respect to itself or to the roller are required. Mold sliding is another problem commonly encountered because of the excessive mechanical contact forces or mismatched thermal expansion between thin mold and roller when operating at a higher temperature. Finally, there is always a discontinuity, both laterally and vertically, on the surface of a thin-mold-wrapped roller mold. Besides thin-film-wrapping methods, the curved surface lithography which applies standard photolithography to a cylindrical curved surface instead of a planar one, is a promising way to fabricated roller molds with complicated patterns.

In this work, we have developed and applied a new type of contact printing photolithography for directly fabricating roller molds with feature sizes in the sub-micrometer or even nanometer scale. Figure 1 schematically depicts the concept of this new roller mold fabrication process. First of all, a modified dip-coating method is developed which can uniformly coated a thin PR layer of desired layer thickness on the roller surface. Secondly, a pre-fabricated planar mold is first coated with a thin metal film and then brought into contact with the PR-coated roller. By applying proper contact pressure and thermal heating, the patterned metal film defined by the convex or extruded surface of the planar mold can be transferred to the PR layer coated on the roller. Thirdly, following standard photolithography, the PR layer is then exposed to ultraviolet (UV) light while using the transferred metal film as an adhered photo-mask. Subsequent PR developing process can remove those UV-exposed PR and form a layer of patterned PR micro-structures on the roller's surface. Finally, the roller mold can be completed in several ways including: (a). direct chemical etching on the metal roller using the PR micro-structures as an etching mask, (b). electroplating a layer of metallic micro-structures from the roller surface and then strip off the PR layer, and (c). evaporating a metal film on the roller's surface and then light-off the PR layer.

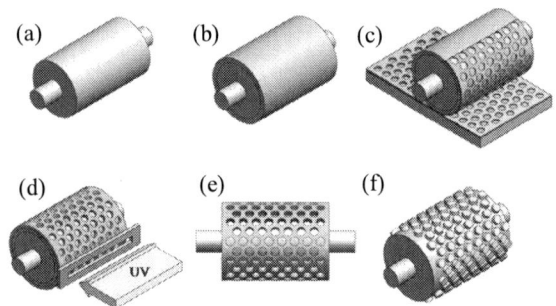

Fig. 1. Fabrication processes of a roller mold with sub-micrometer feature size using metal-film contact printing photolithography.

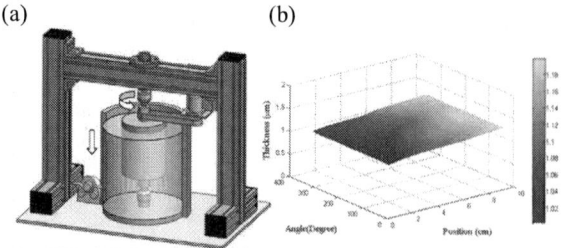

Fig. 2. (a) A schematic of a new dip-coating method for photo-resist coating on a cylinder, (b) an experimental result of the measured PR film thickness along both longitudinal and circumferential directions of a roller.

II. FABRICATION OF METAL ROLLER MOLD

A. Coating a thin photo-resist layer on the surface of a roller

In this work, a modified dip-coating method for coating a PR film on rollers is proposed and designed as shown in Fig. 2(a). The idea is to vertically hold the roller with one end attached to a frame and the other end on the bottom pate of a circular tank. The roller can self-spin freely along its axis at a given angular velocity through a belt and a servo-controlled motor. The circular tank is first filled with diluted PR so that the roller is completely immersed. A hose is connected to the bottom part of the circular tank and a peristaltic pump can slowly drain the PR away from the tubular tank to a reservoir, and therefore gradually lower down the surface of PR solution. By simultaneously spinning the roller and bringing down the surface of PR solution, it achieves the same effects as in conventional dip-coating or spinning-and-pulling methods but with the easiness of handling heavy and bulky rollers and the precise controls on roller's spinning speed and the moving down velocity of the PR liquid. As being described in the literatures, the coated PR layer thickness and its uniformity are determined by several important parameters such as the viscosity of PR solution, the roller spinning speed, the moving down velocity of PR solution,... etc.

Figure 2(b) shows one experimental result of our new type of PR coating method. The thickness of a coated PR layer is measured along the circumferential direction ($0°$ to $360°$) as well as the longitudinal direction (0 to 100 mm) of the stainless steel roller which is 50 mm in diameter and 150 mm in length. The PR used here is a positive-tone AZ4620 (ECHO chemical Corp., Tokyo, Japan) diluted with its solvent to a viscosity of 69 cps. From Fig. 2(b), it shows that the coated PR layer thickness is uniformly controlled in between 1.0 to 1.1 μm over a 100 mm long cylindrical area of the roller.

B. Contact printing for transferring a patterned metal film

After coating a PR layer on a roller's cylindrical surface, the next step is to transfer a patterned metal film from a planar mold to the PR layer as schematically depicted in Fig. 1(c). To ensure the pattern transfer to be completely successful over large patterning area, a Poly-dimethylsiloxane (PDMS) soft mold is adopted because its compliance can readily form an intimate and conformal contact with the steel roller. Another reason for using a PDMS mold is because it can be easily and repeatedly replicated from a mother mold through a molding process.

Using a molding process, PDMS (184, Sylgard, Dow Corning, Michigan, USA) molds are obtained from the negative replication of the silicon mold. The features on the PDMS molds are now arrayed holes with a number of different diameters. The PDMS molds are then placed in an e-beam thermal evaporator (VT1-10CE, ULVAC, Kanagawa, Japan) and a 70 nm thick gold (Au) film is deposited on its surface. Figure 3 is a photo of the Au-film-coated planar PDMS mold with micro-hole arrays. The diameters of micro-hole in the six sub-area are 2.0, 1.2, 0.8, 0.6, 0.4, and 0.3 μm, respectively, and the size is 60x60 mm^2.

The mechanism of contact printing for transferring a patterned metal film from a planar PDMS mold to a PR layer coated on a roller's surface is schematically depicted in Fig. 4(a). An experimental setup, as shown in the photo of Fig. 4(b), is designed and constructed to perform the contact printing process. Both ends of the roller are first mounted on a frame through roller bearings so that the roller can spin freely with minimized rotation resistance. The PDMS mold is place on top of a glass plate which is guided by a servo-controlled linear stage for horizontal movement. Beneath the glass plate sit an IR lamp which can focus its light onto the roller surface and heat it up. The heating temperature is controlled by adjusting the input power to the IR Lamp. The roller is pressed against the PDMS mold through a vertical loading mechanism and a load cell, and the contact force between the PDMS mold and the roller is controllable. When the PDMS mold is horizontally translated, the roller will rotate accordingly through the contact force exerted by the PDMS mold. After the contact printing process, the Au-coated PDMS mold is

978-1-4673-1122-9/12 $31.00 © 2012 IEEE

examined. It is observed that, over the 6x6 cm^2 surface area of the PDMS mold, 95% of the patterned metal film is completely transferred to the roller.

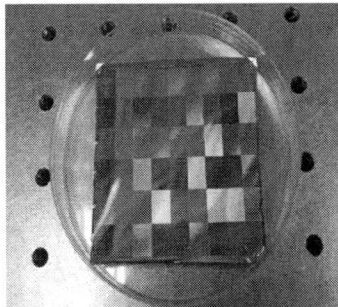

Fig. 3. Schematics of a photo of a 6x6 cm2 Au-coated PDMS mold containing hole-arrays of different hole-diameters of 2, 1.2, 0.8, 0.6, 0.4, and 0.3 μm

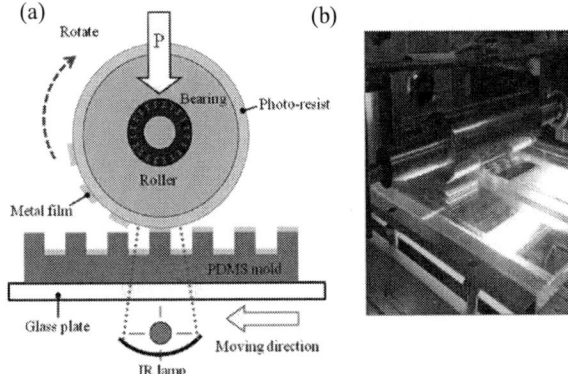

Fig. 4. An experimental contact printing system for transferring patterned metal films from a PDMS mold to a PR layer coated on roller's surface; (a) a schematic drawing and (b) a photo of the system.

C. Step-and-rotated UV exposure and pattern-transfer

After transferring a patterned metal film onto the PR layer, it is then followed by standard photolithography to complete the PR micro-structures on the roller's cylindrical surface. As shown schematically in Fig. 5, the UV exposure is carried out through a slot opening to ensure good resolution in pattern definition. After soft baking of the PR by 9 min at 90 ℃, the roller is mounted on a rotation mechanism and a thin metal plate with a 2 mm wide and 10 cm long slot opening is placed closely to the roller. Through the slot opening, a 6" inch parallel UV light source (AG1000-6N, M&R Nano Technology Co., TaoYuan, Taiwan) is used to expose the roller which is in constant spinning. The UV dose amount is controlled by the UV light intensity, the exposure time, and the slot opening width. When proper UV dosage is achieved, the roller is hard baked for 18 min at 90 ℃, and then immersed in the PR developer. Finally, the PR micro-structures are formed.

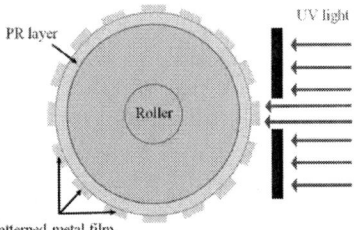

Fig. 5. UV exposure of PR layer using an open slot and a 6" parallel UV light source; the patterned metal film is acting as a photo-mask for the PR layer.

D. Fabrication of Roller Mold

As examples, Fig. 6(a) and 6(b) are two optical microscope images of the PR hole-arrayed microstructures with a hole-diameter of 1.2 μm and 0.6 μm, respectively, formed on the roller's surface. A 20 nm thick chromium (Cr) film and a 200 nm thick nickel (Ni) film are subsequently deposited on the cylindrical surface of the roller using the same e-beam thermal evaporator as described before. A specially designed mechanism is installed inside the chamber of the e-beam evaporator to hold the roller in position and to continuously rotate and spin the roller during the thermal evaporation of Cr and Ni films. After thermal evaporation, the roller mold is immersed in an aceton solution in an ultrasonic cleaner to lift-off the PR layer and the metal films on top of the PR layer.

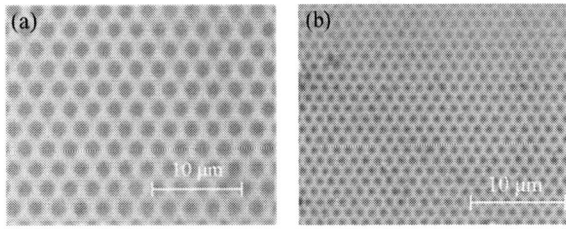

Fig. 6. Optical microscope images of the developed hole-arrayed PR micro-structures on roller's surface after UV exposure and developing; the holes have a diameter of (a) 1.2 μm and (b) 0.6 μm

III. EXPERIMENTAL RESULT

A finished roller mold is shown in the photo of Fig. 7. Details of its surface microstructures in different areas with different features sizes are examined by both optical microscope and 3D confocal microscope. For example, Fig. 8(a) and 8(b) show the optical image and the 3D surface profile, respectively, of the pillar-arrayed surface micro-structures with a diameter of 1.2 μm. Figure 8(c) displays the cross-section profile in details. It shows that the lateral dimensions well match their counterparts designed in the original silicon mold, and the height match with the deposited 20 nm Cr and 200 nm Ni layers. To demonstrate the micro-structures with smaller feature size, Fig. 9(a), 9(b), and 9(c) show the optical image, the 3D micro-structure profile, and the

978-1-4673-1122-9/12 $31.00 © 2012 IEEE

cross-section profile, respectively, of the fabricated arrays of micro-pillars which have a diameter of 0.6 µm. Again, it demonstrates the metallic surface micro-structures are successfully established on the roller's cylindrical surface. As for the patterned areas which have the smallest feature sizes of 0.4 and 0.3 µm, the micro-structures are not formed on the roller's surface. This failure is because the PR layer (1.0~1.1 µm in thickness) is too thick in comparison with the hole-diameter (0.3 or 0.4 µm) so that the hole-arrayed PR micro-structures are not completely developed through the UV exposure and PR developing processes. Apparently, it is necessary to make the coated PR layer even thinner before one can successfully obtained further sub-micrometer or even nano-meter scaled surface features on the roller surface.

Fig. 7. A photo of the completed roller mold after metal film evaporation and PR lift-off.

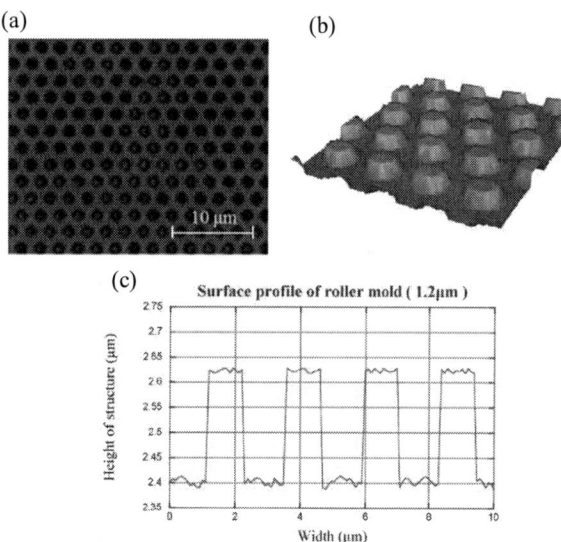

Fig. 8. Arrayed micro-pillars with a diameter of 1.2 µm on the surface of roller mold, (a) a 2D image taken by an optical microscope, (b) and (c) the 3D profile and cross-section profiles, respectively, measured by an confocal microscope

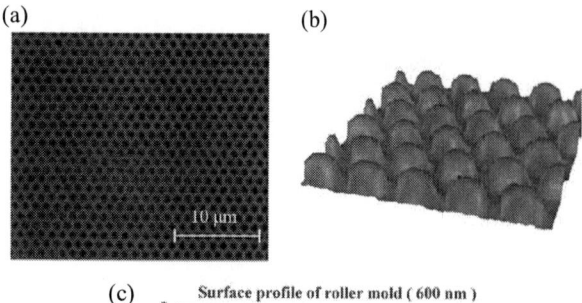

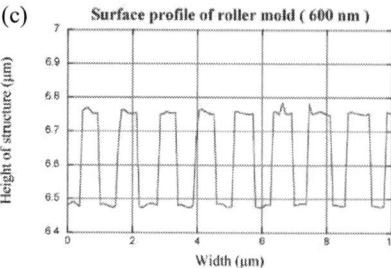

Fig. 9. Arrayed micro-pillars with a diameter of 0.6 µm on the surface of roller mold, (a) a 2D image taken by an optical microscope, (b) and (c) the 3D profile and cross-section profiles, respectively, measured by an confocal microscope

IV. CONCLUSION

In this paper, we demonstrate experimentally a new way of directly fabricating sub-micrometer scaled surface structures on the surface of a metallic roller. A new type of dip-coating method is proposed and experimentally verified for coating a thin and uniform PR layer on a roller. This new design is particularly usefully in handling larger and heavier roller molds and in the mean time maintaining precise controls on the working parameters of dip-coating process. Based on this PR coating method, a new type of contact printing method is proposed and experimentally tested. The idea is based on the non-sticky characteristics of PDMS materials so that patterned metal films on top of a PDMS mold can be easily transferred to a PR layer coated on a roller's surface at an elevated temperature. This continuous sheet-to-roller type of metal film transfer also relies on properly chosen thermal and mechanical conditions as well as carefully designed mechanism. Finally, through UV exposure and PR developing, a layer of PR micro-structures is formed and roller mold fabrication is completed using either subtractive etching process or additive metal deposition processes. In this work, the smallest feature size we achieve is 0.6 µm and the patterned are size is 6x6 cm². It is possible to pursue smaller feature sizes, even down to nanometers, if a thinner PR layer was achieved by further adjusting the viscosity of PR and related dip-coating parameters. Of course, the feature size and the patterned area size are also determined by the original silicon mold.

ACKNOWLEDGMENT

This work is supported by the National Science Council (NSC) of Taiwan through projects NSC 098-2811-M-006-012 and NSC 100-2120-M-006 -007 -CC1.

REFERENCES

[1] Lan, S., Lee, H., Ni, J., Lee, S., and Lee, M., "Survey on Roller-type Nanoimprint Lithography (RNIL) process," *Int. Conf. Smart Manufacturing Application (ICSMA)*, 2008, pp. 371-376.

[2] Makela, T., Haatainen, T., Majander, P., Ahopelto, J., and Lambertini, V., "Continuous double-sided roll-to-roll imprinting of polymer film," *Japn. J. Appl. Phys.* Vol. 47 No. 6, 2008, pp. 5142-5144.

[3] Chang, C.-Y., Yang, S.-Y., and Sheh, J.-L., "A roller embossing process for rapid fabrication of microlens arrays on glass substrates," *Microsystem Technologies* Vol. 12, 2006, pp. 754-759.

[4] Ting, C.-J., Chang, F.-Y., Chen, C.-F., and Chou, C. P., "Fabrication of an antireflective polymer optical film with subwavelength structures using a roll-to-roll micro-replication process," *J. Micromech. Microeng.* Vol. 18, No. 7, 075001, 2008

[5] Marques, C., Desta, Y. M., Rogers, J., Murphy, M. C., and Kelly, K., "Fabrication of High-Aspect-Ratio Microstructures on Planar and Nonplanar Surfaces Using a Modified LIGA Process," *J. Microelectromech. Syst.* Vol. 6 No. 4, 1997, pp. 329–336.

[6] Ahn, S. H. and Guo, L. J., "High-speed roll-to-roll nanoimprint lithography on flexible plastic substrates," *Advanced Materials* Vol. 20 No. 11, 2008, pp. 2044-2049.

[7] Liu, S.-J., and Chang, Y.-C., "A novel soft-mold roller embossing method for the rapid fabrication of micro-blocks onto glass substrates," *J. Micromech. Microeng.* Vol. 17, No. 1, 2007, pp. 172-179.

[8] Yang, S.-Y., Cheng, F.-S., Xu, S.-W., Huang, P.-H., and Huang, T.-C., "Fabrication of microlens arrays using UV micro-stamping with soft roller and gas-pressurized platform," *Microelect. Eng.* Vol. 85, No. 3, 2008, pp. 603-609.

[9] Chang, C.-Y., Yang, S.-Y., and Chu, M.-H., "Rapid fabrication of ultraviolet-cured polymer microlens arrays by soft roller stamping process," *Microelect. Eng.* Vol. 84, No. 2, 2007, pp. 355-361.

[10] Park, S.Y., Choi, K. B.; Kim, G. H., and Lee, J. J, "Nanoscale patterning with the double-layered soft cylindrical stamps by means of UV-nanoimprint lithography," *Microelect. Eng.* Vol. 86, 2009, pp. 604-607.

Fabrication of metal embedded photo-mask for sub-micrometer scaled photolithography and patterning sapphire substrate

Jhih-Nan Yan, Yung-Chun Lee [*]
[*] Department of Mechanical Engineering, National Cheng Kung University, Tainan, Taiwan
yunglee@mail.ncku.edu.tw

Abstract—This paper describes the fabrication processes of a new type of metal-embedded photo-mask, which will be used in standard photolithography and for fabricating patterned sapphire substrates. This new metal-embedded photo-mask is prepared by metal contact printing lithography and therefore can easily achieve smaller feature size around or below 1 μm. Besides its easiness in fabricating and obtaining smaller line-width, this new metal-embedded photo-mask differs from a conventional Cr/glass photo-mask in that the metallic patterns are embedded and therefore are not in contact with photo-resist during UV exposure. This unique feature can minimize the damage to photo-mask in use and prolong its lifetime. In this work, this metal-embedded photo-mask is experimentally prepared and applied to photolithographic patterning of PR microstructures on sapphire substrates, which are important in LED industries.

Keywords- metal-embedded photo-mask; patterned sapphire substrates

I. INTRODUCTION

The needs of more efficient and brighter GaN-based light emitting diodes (LEDs) are enormously increasing, such as traffic signal, full color display, and backlight in liquid crystal display [1]. There have been many research works discussing on how to obtain these LEDs with better performance. It is well known that high density threading dislocations are inherent in the epitaxially grown GaN layers on a sapphire substrate due to the lager difference in lattice constants of crystal structures and thermal expansion coefficient between the epitaxial layer and sapphire substrate. Therefore, how to further reduce the dislocation density is an important issue for fabricating high-performance LED. Many methods to grow GaN layers such as epitaxial lateral overgrowth (ELOG) and patterned sapphire substrate (PSS) have been proposed in order to reduce the dislocation densities and thus to improve optical emission property. Using ELOG technique has shown to reduce the dislocation density significantly [2-3], but the two-step growth procedure is time consuming. To improve these problems, the single growth technique by PSS technique was growth interruption-free means to achieve a reduced dislocation density. Furthermore, the measurement of the LEDs grown on different PSSs with different feature sizes was found that the luminous intensity was increasing with decreasing feature size [4-5]. There are several different methods for fabricating PSSs such as photolithography [6-8], nanosphere lithography [9-10], wet-etching process [11-12]

and nano-imprint lithography [13]. However, the most commonly used method is by photolithography and dry etching process. The conventional contact photo-mask has the metal patterns deposited on the transparent glass or quartz surface, which can lead to photo-mask damage because metal patterns in contact with photo-resist (PR) during ultraviolet (UV) exposure. The contact exposure process of conventional photo-mask is schematically shown in Fig. 1(a). First, the conventional quartz mask is in contact with the PR layer and has a uniform pressure to submissive to contact of photo-mask on the substrate, and then PR layer is exposed to collimate UV light. When removing the photo-mask from the substrate, the metal film may come off from photo-mask surface or metal pattern may be contaminated by the photo-resist, and the lift-time of the photo-mask may be cut shorter.

In this work, we propose an innovative and easily implemented method for the manufacture of a metal-embedded photo-mask. It is developed based on our previously work "contact-transferred and mask-embedded lithography" (CMEL) [14]. This new metal-embedded photo-mask differs from a conventional Cr/glass photo-mask in that the metallic patterns are embedded and therefore are not in contact with photo-resist during UV exposure. This unique feature can minimize the damage to photo-mask in use and prolong its lifetime, as shown in Fig. 1(b). In particular, the procedures and equipements needed for manufacturing these new type of metal-embedded photo-mask masks are simple and inexpensive, and therefore can be widely applied in industries.

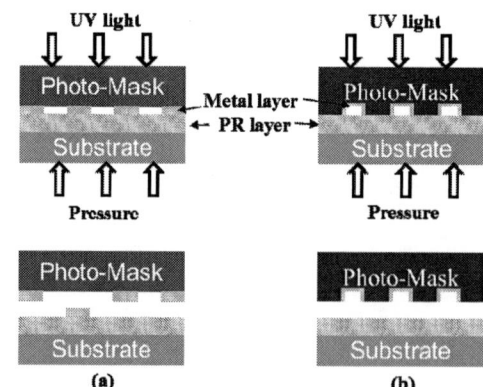

Fig. 1. Schematic diagram of the exposure procedures of (a) conventional contact photo-mask and (b) metal-embedded photo-mask.

978-1-4673-1122-9/12 $31.00 © 2012 IEEE

II. Experiments

Figure 2 shows a schematic diagram of the fabrication procedures of a metal-embedded quartz photo-mask, and the application of this metal-embedded photo-mask for photolithography on patterning sapphire substrate. First of all, a pre-polymer of polydimethlysiloxane (PDMS) mold with some micro/nano-scaled feature cast and cured against the hexagonal dot-arrayed patterns on silicon mold. The silicon mold which has pre-fabricated hexagonal arrayed pillars features with diameter 1μm and a center-to-center distance of 2 μm. A 50 nm thick Au is then deposited on the PDMS mold by electron-beam evaporator (VT1-10CE, ULVAC, Kanagawa, Japan) at a rate of 0.6 Å/s. For the quartz substrate, we used a positive resist (AZ1500, 20cps, ECHO chemical Corp., Japan) mixed with thinner of propylene glycol monomethyl ether acetate (PGMEA, ECHO chemical Corp., Japan), which allocated by weight percentage of 2:1, and it is spin-coated on top of the quartz substrate with a thickness of around 900 nm at a spin-coating rate of 4000 rpm, then soft bake at 100 ℃ for 60 second, as shown in Fig. 2(a). Figure 2(b) shows a nano-imprinting technology called "contact-transfer and mask-embedded lithography" (CMEL) which would be applied to fabricate micron or sub-micron patterns on quartz substrate surface. The mechanism for CMEL has been described in detail previously [14]. Following standard photolithography, the PR layer is then exposed to UV light (365nm, AG100-6N, M&R Nano Technology, Taiwan) while using the transferred metal film as an adhered photo-mask and the exposure time is 1.4 second. Subsequently PR developing process can remove those UV-exposed PR by developer (AZ 300MIF) and form a layer of patterned PR structures on the quartz substrate surface, then hard bake at 110 ℃ for 90 second, as shown in Fig. 2(c).The transferred metal pattern and PR layer are then serving as an etching mask for subsequent etching the underlying quartz by reactive ion etching (RIE, NE-550EX, ULVAC, Japan) using the patterned Au film and PR layer as an etching mask. The power of RIE is set to 200W, the pressure of the reaction chamber is 70 mTorr and the flow rate of gas CF_4 and SF_6 are controlled at 50 and 40 sccm, respectively, as shown in Fig. 2(d). The 50 nm thick Cr and 20 nm thick Au are then deposited sequentially on the quartz substrate surface by electron-beam evaporator at a rate of 1 and 0.6 Å/s, respectively, as shown in Fig. 2(e). Finally, we lifted off the photoresist and metal using acetone, the metal-embedded photo-mask is fabricated, as shown in Fig. 2(f).

For the sapphire substrate, we used a positive resist (AZ1500) mixed with thinner (propylene glycol monomethyl ether acetate, PGMEA), which allocated by weight percentage of 5:1. The diluted PR is spin-coated on top of the sapphire substrate with a thickness of around 1.3 μm at a spin-coating rate of 3000 rpm, then soft baked at 100 ℃ for 60 second. After the metal-embedded photo-mask and the sapphire substrate are prepared, the metal-embedded photo-mask is placed on top of the sapphire substrate and a loading force is applied to be submissive to contact of metal-embedded photo-mask on the sapphire substrate. Following standard photolithography, the PR layer is then exposed to collimate UV light passing through the metal-embedded photo-mask of the non-metals region and the exposure time is 4 second, as shown in Fig. 2(g). Finally, PR developing process can remove those UV-exposed PR and form a layer of patterned PR structures on the sapphire's surface, then hard bake at 110 ℃ for 90 second.

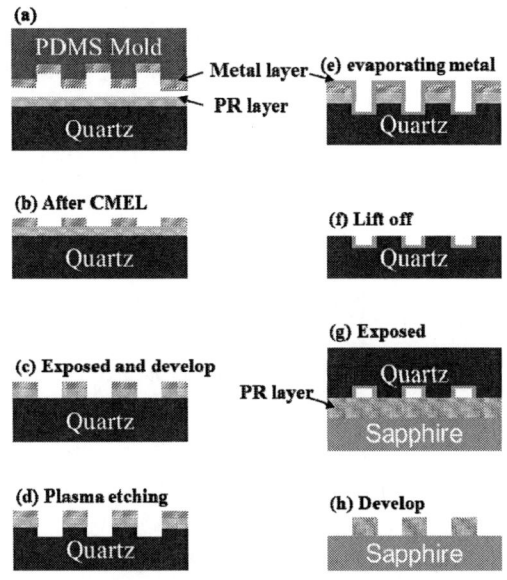

Fig. 2. (a) to (f) A schematic diagram of the fabrication procedures of a metal-embedded quartz photo-mask, (g) and (h) the application of this metal-embedded photo-mask on patterning sapphire substrate.

III. Results and Discussion

After the contact imprinting process, the patterned 50 nm thick Au film is embedded into the PR layer on top of the quartz substrate surface, as shown in Fig. 3. As seen from figure 3(a), we can complete imprinting pattern metal films on 2 inch or more than quartz substrate by CMEL techniques. Figure 3(b) shows the Au films has arrayed of hexagonally close-packed holes pattern on the 2 inch PR/ quartz substrate (Fig. 3a).

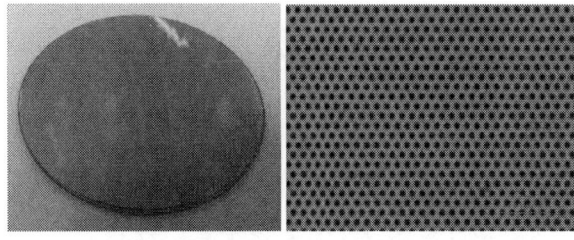

Fig. 3. After contact imprinting, the patterned gold films are embedded into the PR layer on top of the quartz surface: (a) 2 inch complete pattern gold films, (b) Au films has arrays of hexagonally close-packed holes pattern on the 2 inch PR/ quartz substrate (Fig. 3a).

978-1-4673-1122-9/12 $31.00 © 2012 IEEE

Figure 4 shows the metal-embedded photo-mask after lifted off by using acetone with the arrays of hexagonally close-packed holes feature. Experiments have been carried out for fabricating a metal-embedded photo-mask on a 2 inch quartz wafer. Figures 5(a) and (b) are the optical microscope images of the metal-embedded photo-mask by optical microscope positive light source and optical microscope back light source, respectively. As shown in Figure 5, the mask has arrays of hexagonally close-packed holes with a diameter of 1 μm and a pitch of 2 μm. Figure 5(b) shows, metal-embedded photo-mask has well light transmittance (non-metal region) and shading (metal region). Therefore, we can employ UV light through the metal-embedded photo-mask of the green color regions (Fig. 5b) to expose PR layer. Subsequently PR developing process can remove those UV-exposed PR and form a layer of patterned PR structures on the sapphire substrate surface.

Figure 6(a) and 6(b) are the SEM images from top and cross-section view, respectively, of the fabricated photo-mask. The metal-embedded photo-mask has arrays of hexagonally close-packed holes with a diameter of 1.2 μm and a 2 μm center-to-center pitch, as shown in Fig. 6(a). One can see that the bottom surface and side-wall of the holes are covered by a 0.3 m Cr/Au metal film, as shown in Fig. 6(b). We have also applied this metal-embedded photo-mask for patterning PR microstructures on top of a sapphire substrate. The experimental result is shown in Fig. 7 and Fig. 8. Hexagonally array of PR micro-pillars with a diameter around 1 μm and a height of 1.3 μm are successfully created on a 2 inch sapphire substrate. A PSS for LEDs can be readily obtained by applying ICP etching on this PR-patterned sapphire substrate.

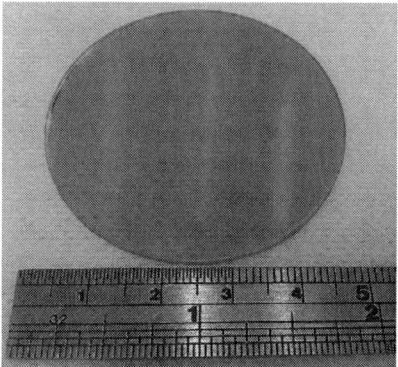

Fig. 4. The image shows the metal-embedded photo-mask after lifted off by using acetone with the arrays of hexagonally close-packed holes feature.

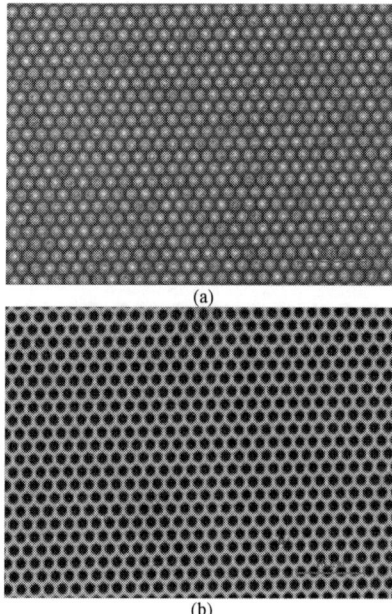

(a)

(b)

Fig. 5. The optical microscope images with (a) positive light source and (b) back light source of the metal-embedded photo-mask.

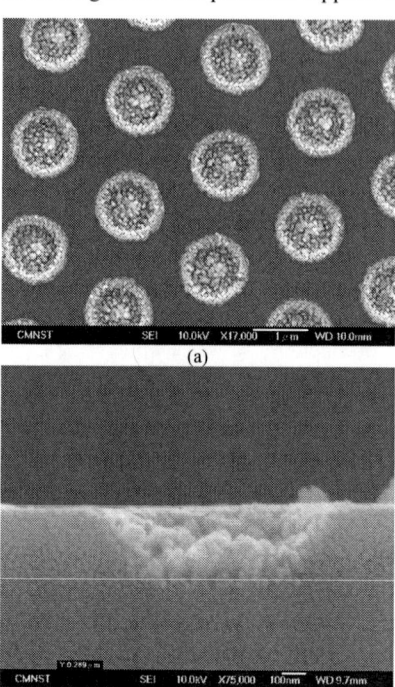

(a)

(b)

Fig. 6. SEM images of (a) the top view of a metal-embedded quartz photo-mask which contains hexagonal arrays of metal-deposited micro-holes with a diameter of 1.2 μm and a period of 2 μm, and (b) the cross-section view of the metal-deposited micro-holes.

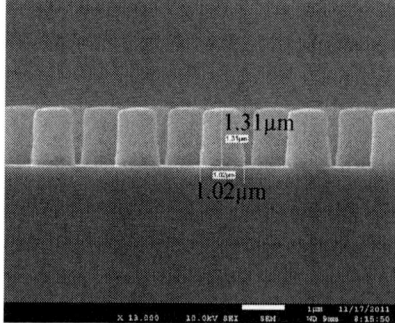

Fig. 7. A cross-section view SEM image of the developed PR micro-structures on sapphire substrate using the metal-embedded quartz photo-mask; the hexagonal array of close-packed micro-pillars has a diameter around 1 μm and a height of 1.3 μm.

Fig. 8. A SEM image of the developed PR micro-structures on sapphire substrate using the metal-embedded quartz photo-mask.

IV. CONCLUSION

In this paper, we demonstrate experimentally the fabrication processes of a new type of metal-embedded photo-mask, which will be used in standard photolithography and for fabricating patterned sapphire substrates. The metal-embedded photo-mask is ease of implementation without using sophisticated and expensive facilities. Although this paper only used the arrays of hexagonally close-packed holes structure on metal-embedded photo-mask, it can also carry out on several samples with linear grating and dot-arrayed features on metal-embedded photo-mask. Metal-embedded photo-mask used the metal to embed within quartz mask, this unique feature can minimize the damage to photo-mask in use and prolong its lifetime, and furthermore the photo-mask has a smooth surface to contact onto polymer surface while in exposure process. Experiments have been carried out for fabricating a metal-embedded photo-mask on a 2 inch quartz wafer, which has arrays of hexagonally close-packed holes with a diameter of 1.2 μm and a pitch of 2 μm.

We have also applied this metal-embedded photo-mask for patterning PR microstructures on top of a sapphire substrate. It can successfully fabricate the hexagonal array of PR micro-pillars with a diameter around 1 μm and a height of 1.3 μm on a 2 inch sapphire substrate. A PSS for LEDs can be readily obtained by applying inductively coupled plasma (ICP) etching on this PR-patterned sapphire substrate.

ACKNOWLEDGMENT

This works is supported by the National Science Council (NSC) of Taiwan through projects NSC 098-2811-M-006-012 and NSC 100-2120-M-006-007-CC1.

REFERENCES

[1] E. F. Suchubert, J. K. Kim, "Solid-state light sources getting smart," Science 308, no. 5726, 2005,pp. 1274-1278.

[2] S. Nakamura, M. Senoh, S. Nagahama, N. Iwassa, T. Yamada, T. Matsushita, Y. Sugimoto, H. Kiyoku, "Room-temperature continuous-wave operation of InGaN multi-quantum-well-structure laser diodes with a long lifetime," Appl. Phys. Lett. 70, 868, 1997.

[3] C. I Chang, Y. L. Lai, C. P. Liu, R. C. Wang, "The influence of mask area ratio on GaN regrowth by epitaxial lateral overgrowth," J. Phys. Chem. Sol. vol. 69, pp. 420-424, 2008.

[4] Y. K. Su, J. J. Chen, C. L. Lin, S. M. Chen, W. L. Li, C. C. Kao, "Pattern-size dependence of characteristics of nitride-based LEDs grown on patterned sapphire substrates," J. Cryst. Growth, 311, 2973-2976, 2009.

[5] Y. T. Hsieh, C. H. Chen, Y. C. Lee, X. F. Zeng, S. C. Shei, H. Y. Lin, "Pattern-size dependence of characteristics of nitride-based LEDs grown on patterned sapphire substrates," 5th IEEE International Conference on Nano/Micro Engineered and Molecular Systems, 915-918,2010.

[6] D. S. Wuu, W. K. Wang, W. C. Shih, R. H. Horng, C. E. Lin, J. S. Fang, "Enhanced output power of near-ultraviolet InGan-Gan leds grown on pattern sapphire substrate," IEEE Photonics Technol. Lett. vol. 17, no. 2, pp. 288-290, 2005.

[7] T. S. Oh, S. H. Kim, T. K. Kim, Y. S. Lee, H. Jeong, G. M. Yang, E.-K. Suh, "Gan-based light-emitting diodes on micro-lens patterned sapphire substrate," Jpn. J. Appl. Phys., vol. 47 ,No. 7, pp. 5333-5336, 2008.

[8] J. H. Cheng, Y. S. Wu, W. C. Liao, B. W. Lin, "Improved crystal quality and performance of Gan-based light-emitting diodes by descreasing the slanted angle of patterned sapphire," Appl. Phys. Lett.96, 051109, 2010.

[9] H. W. Huang, H. C. Kuo, J. T. Chu, C. F. Lai, C. C. Kao, T. C. Lu, S. C. Wang, R. J. Tsai, C. C. Yu, C. F. Lin, "Nitride-based LEDs with nano-scale textured sidewalls using natural lithography," Nanotechnology 17, 2998-3001, 2006.

[10] T. S. Kim, S. M. Kim, Y. H. Jang, G. Y. Jung, "Increase of light extraction from GaN based light emitting diodes incorporating patterned structure by colloidal lithography," Appl. Phys. Lett., 91, 171114, 2007.

[11] J. wang, L. W. Guo, H. Q. Jia, Z. G. Xing, Y. Wang, J. F. Yan, N. S. Yu, H. Chen, J. M. Zhou, "Investigation of characteristics of laterally overgrown GaN on striped sapphire substrates patterned by wet chemical etching," J. Crystal Growth, 290, 398-404, 2006.

[12] Y. J. Lee, J. M. Hwang, T. C. Hsu, M. H. Hsieh, M. J. Jou, B. J. Lee, T. C. Lu, H. C. Kuo, S. C. Wang, "Enhancing the output power of Gan-based LEDs grown on wet-etched patterned sapphire substrates," IEEE photon. Technol. Lett., vol. 18, pp.1152-1154, 2006.

[13] H. W. Huang, C. H. Lin, J. K. Huang, K. Y. Lee, C. F. Lin, C. C. Yu, J. Y. Tsai, R. Hsueh, H. C. Kuo, S. C. Wang, "Investigation of GaN-based light emitting diodes with nano-hole patterned sapphire substrate (NHPSS) by nano-imprint lithography," Mater. Sci. Eng. B, 164, 76-79, 2009.

[14] Y. C. Yee, C. Y. Chiu, "Investigation of GaN-based light emitting diodes with nano-hole patterned sapphire substrate (NHPSS) by nano-imprint lithography," Micro-/ nano-lithography based on the contact transfer of thin film and mask embedded etching," J. Micromech. Microeng., 18, 075013, 2008.

Metal Contact Printing Photolithography for Fabricating Sub-Micrometer Patterned Sapphire Substrates in Light-Emitting Diodes

Yi-Ta Hsieh[1] and Yung-Chun Lee[2]

[1]Institute of Nanotechnology and Microsystems Engineering, NCKU, Tainan, Taiwan
[2]Department of Mechanical Engineering, NCKU, Tainan, Taiwan
Q28991075@mail.ncku.edu.tw

Abstract—**This paper reports a novel process which is combine the contact metal transfer method and traditional photolithography process for fabricate nano-scale pattern sapphire substrate (NPSS) used in high brightness light emitting diodes (LEDs). The novel process can directly transfer a metal pattern onto the PR layer which above the sapphire substrate, the transferred metal pattern can as a perfect photo-mask for subsequent photolithography process. In this work, the high aspect ratio PR structures with the aspect ratio of 5 and line width of 500 nm are created by this novel process. Furthermore, the PR structure can as a etching mask for inductively coupled plasma (ICP) etching on the sapphire substrate. During the ICP etching, we successfully to obtain the NPSS with a perfect cone shape. Experiments have been demonstrate the feasibility of using this new approach for obtaining sub-micrometer surface structures on the complete surface area of a 2 inch and 4 inch sapphire substrates.**

Keywords-LEDs;NPSS; metal contact printing

I. INTRODUCTION

Gallium nitride (GaN) based light-emitting diodes has widely used in solid state lighting, laser diode and backlight source for displays. However, there still have some problem need to overcome for the high brightness application. Due to the large lattice mismatch and different thermal expansion coefficient between epitaxial GaN and sapphire substrates, GaN layers grown on sapphire substrates exhibit high dislocation densities [1]. Another major issue is the low light extraction efficiency, the high refractive index of GaN (n=2.5) cause the internal light difficulty escape into air from GaN layer , the light will surrounded by total internal reflection and absorbed by the GaN material [2-3]. Improving light extraction efficiency and epitaxial quality has become a significant researches. In recent years, GaN based LEDs grown on patterned sapphire substrates (PSS) have attracted great of researchers to studies. It is know that the PSS not only can reduce the dislocation density by the epitaxial lateral overgrowth technique [4-6], but also can enhance light extraction efficiency by the textual sapphire surface. PSS substrates used in LED have been demonstrated a high efficiency way for promoted the efficiency of LED [7-9]. Sapphire substrates can be patterned with several methods,

such as wet etching [10-11], dry etching [12-13] and nano-imprint lithography [14-15]. However, the most commonly used one in LED industries is by photolithography and plasma dry etching. Furthermore, there are more and more researches indicating that using nano-scale pattern sapphire substrate (NPSS) in LED can obtain higher light extraction efficiency than PSS [16-17]. However, when the PSS feature size reaches sub-micrometer scale or smaller, optical diffraction problem occurs on the traditional photolithography method and it will lead to poor exposure resolution. Moreover, the sapphire wafer has a natural warpage about 5 to 10 μm and 10 to 20 μm for 2 inch and 4 inch sapphire wafer respectively, the curved surface will caused the worse yield rate and the uniformity in the photolithography process, and it will become more serious when the structures scale down to sub-micrometer. Thus, fabricating NPSS by photolithography method is still challenging and expensive.

In this work, we propose an innovative and easily implemented method for fabricating PSSs with sub-micrometer feature sizes and with 3D cone shape. Based on our previous work [14], the proposed method is combining standard photolithography and metal contact printing technique. The metal contact printing method can transfer a metal film to a photoresist layer deposited on a sapphire substrate surface, and the transferred metal film then will act as a photo-mask for subsequent UV exposure. The exposure principle is similar to contact mold photolithography [18]. However, since the metal contact printing process is transferring a metal-film to the photo-resist (PR) layer as a perfect contact photo-mask, it is easier to obtain PR patterns with high resolution and high aspect ratio on sapphire substrate through photolithography process. The PR structures will act as an etching mask for subsequent ICP etching process.

II. EXPERIMENT

Figure 1 shows the schematic diagram of metal contact printing method and photolithography process for fabricate NPSS. This process start from the PDMS stamp preparation which is reproducing from silicon master mold. A silicon mold is first prepared with standard photolithography method and dry etching process commonly used in semiconductor

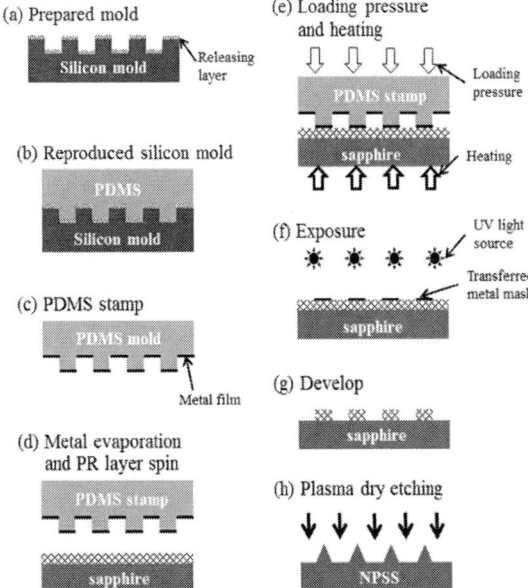

Fig. 1. A schematic diagram for the procedures of metal contact printing lithography and photolithography process for fabricating sub-micrometer surface structures on a sapphire substrate.

industries, the pattern is arranged for the hexagonally packed dot-array, and the spacing between adjacent dots is equal to the diameter of dot and with two kind of diameter 1 μm and 500 nm, respectively. The silicon mold with pre-fabricated surface structures is first coated with an anti-adhesion layer by the thermal deposition method [19], as shown in Fig. 1(a). In order to diminish the damage of the expensive silicon mold and to promote the transfer process yield rate, the silicon mold will be reproduced by the flexible material of PDMS (sylgard 184, Dow Corning, USA) which is cheaper and easy to use. The PDMS is directly drop casted on the silicon mold surface and cured at 70 °C for 3 hours. After cooling down to the room temperature, the PDMS stamp with surface relief feature is than separated from the silicon mold. After the fabrication of the PDMS stamp, a 50 nm thick Au is deposited sequentially on the PDMS stamp by E-Beam evaporation system with the rate of 0.6 Å/s, the Au film is just weakly attached to the PDMS stamp surface since the low surface energy of PDMS [20], the process is depicted in Fig. 1(b) and 1(c). As for the sapphire substrate, an intermediate photosensitive polymer is spin-coated for metal bonding and subsequence pattern transfer by photolithography method, as shown in Fig. 1(d). In this work, we used the PR of AZ1500 (AZ1500, AZ Electronic Materials, Japan) as the intermediate photosensitive polymer, the PR will be diluted by PGMEA in two type of concentration for different thickness application, the weight percent of the PR and PGMEA are 8:1 and 2:1, respectively. The PR thickness as a function of spin speed is

measured and displayed in Fig. 2. In the metal transfer process, the metal pattern on the stamp will directly transfer to the PR layer surface which above the sapphire substrate, the transfer way is by the difference in surface adhesion properties between the PDMS-metal and the metal-PR layer interfaces. The intimate contact between the PDMS stamp and sapphire substrate is the important issue for higher yield rate. It is also known that heating can enhance the adhesion ability between metal pattern and PR layer. Thus a uniform pressure of 2 MPa and temperature of 95 °C for 2 min will be applied for the process, the diagram is shown in Fig. 1(e).

After the imprinting process, the sapphire wafer can be successfully and entirely patterned with the 50 nm thick Au film from the PDMS mold. The metal patterns can as a perfect photo-mask for subsequence UV expose in the PR layer which is not protect by the metal mask. the I-line is then employed to expose the Au-patterned sample with the energy of about 22 mW/cm^2. Development is carried out with AZ 300MIF developer, the wafer is immersed and then slightly shake in the bath to enhance the developer reaction. The exposure and development procedure is shown in the Fig. 1(f) and 1(g). After the photolithography process, the metal mask will clean by the metal etching solution and the PR structures with micro/nano scale feature size are formed on the sapphire substrate. Before dry etching process. a post bake is applied to the sample with 100 °C and 10 min for obtained more strongly PR structure. Finally, the sapphire substrate with PR pattern is sent into the ICP system (NE-550EX, ULVAC, Japan) for dry etching. The PR structure can as a good etching mask for etching the unprotect sapphire. During ICP etching process, it can obtain a cone ship NPSS through control the etching parameter, such as RF bias power, etching gas, BCl$_3$ flow rate and chamber pressure [12]. The etch profiles are examined by a field emission scanning microscope (SEM).

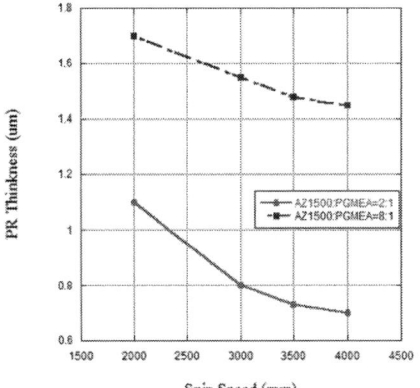

Fig. 2. PR thickness as a function of spin speed.

III. RESULT AND DISCUSSION

For the metal transfer process, the accurate control on the imprinting pressure is important since excessive pressure can

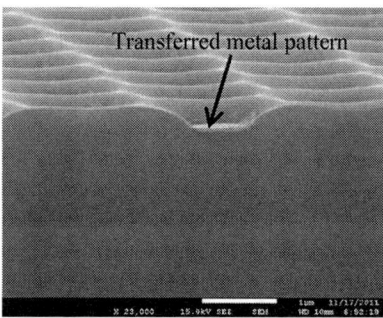

Fig. 3. A dot-array pattern of 50 nm thick Au film transferred to an 1.4 μm thick PR layer on top of a sapphire substrate, the metal pattern is perfect contact on the PR surface.

Fig. 4. SEM images for the high aspect ratio linear grating structures with the aspect ratio of 5 and line width of 500 nm.

cause the PR to flow into the concave area of the PDMS mold and make unwanted metal pattern transfer. Figure 3 show the SEM cross-section view of the metal pattern mask on PR layer surface with the diameter of 1 μm. As shown, the metal pattern is embedded into the PR layer by metal contact printing lithography. It is known that the gap between photomask and PR layer will seriously affect the exposure resolution in photolithography process, the traditional way to solve this condition is applied a pressure to force the mask and PR intimate contact. However, the sapphire substrate has a natural warpage surface, for this reason, the tradition way can not totally solve the contact problem on the PSS fabrication. However, the metal transfer process can directly obtain a photo-mask on the PR layer surface since the mask is perfect contact on the PR layer, the perfect contact mask can serve as a perfect photo-mask for subsequent exposure process. After exposure and development, it can obtain a high resolution and high aspect ratio PR structure onto the sapphire substrate. Figure 4 show the SEM image of a linear grating PR structure after exposure and development with 2 second and 1 min, respectively, the line width of 500 nm and the aspect ratio about 5.

In this work, we apply this novel process for fabricate the PSS, the pattern is designed in a hexagon dot array with two different diameter, Figure 5(a) and 5(b) show the PDMS mold with diameter of 1 μm and 500 nm, respectively. After the metal contact printing for pattern transfer and exposure to the PR, the SEM image for the obtained PR pillar shape structures

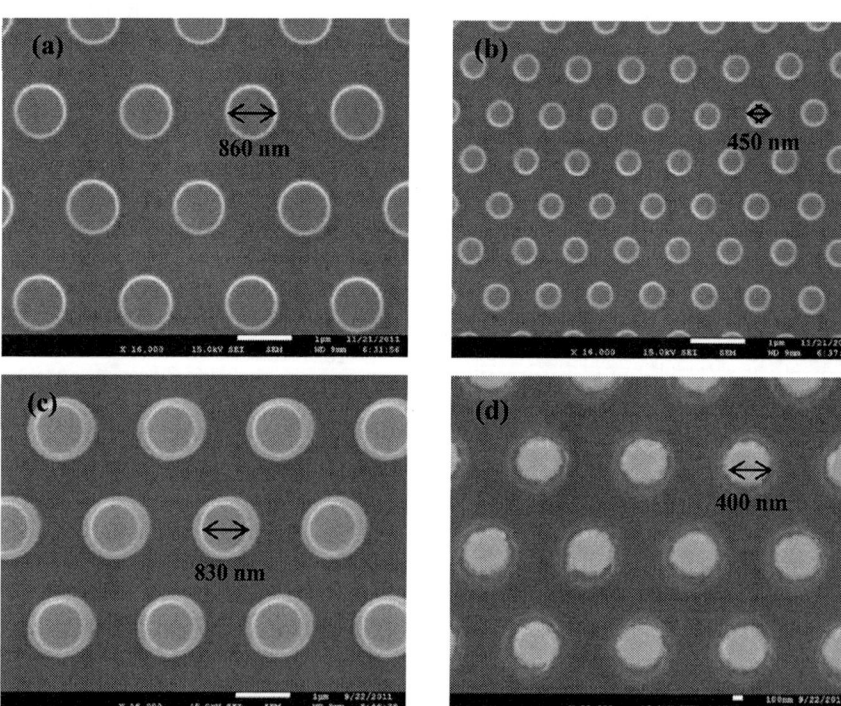

Fig. 5. SEM images for the hexagonally packed arrayed pillar structures; (a) and (b) are the original PDMS mold with 1 μm and 500 nm diameters, respectively, and (c) and (d) are the PR structures after photolithography process with 1 μm and 500 nm diameters, respectively.

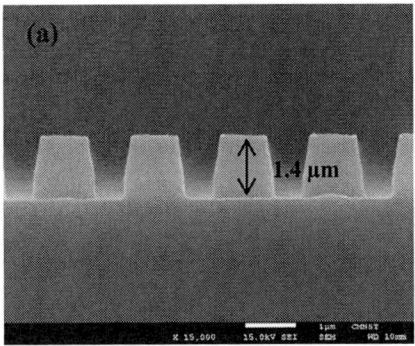 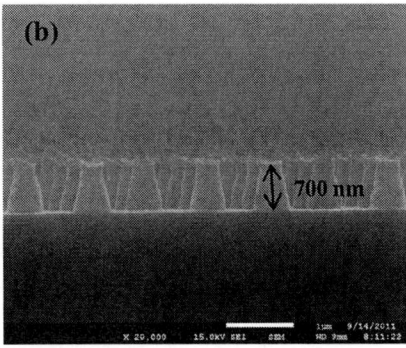

Fig. 6. SEM cross-section view of the two different PR structures with the structure height of (a) 1.4 μm and (b) 700 nm for the diameter of 1 μm and 500 nm, respectively.

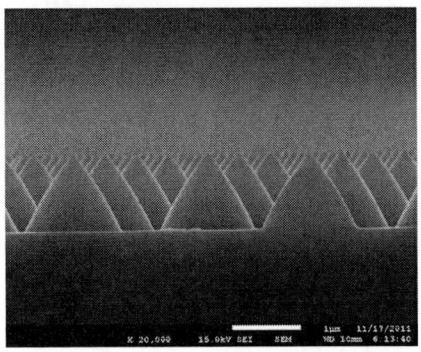 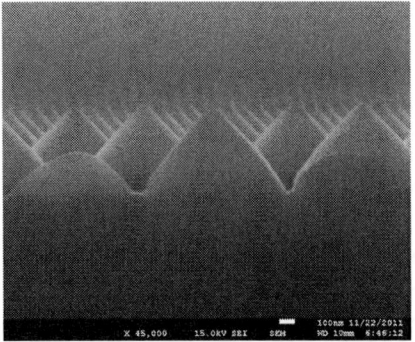

Fig. 7. The pattern sapphire substrates with cone shape structure after ICP etching, there are two type of PSS with diameter/spacing/height being (a) 1.4 μm/0.6 μm/1.1 μm and (b) 720 nm/280 nm/400 nm.

are shown in Fig. 5(c) and 5(d) with diameter of 1 μm and 500 nm, and the exposure time are 1.5 sec and 0.8 sec, respectively. There is only five percent diameter to lose between the mold and the PR structures since the light diffraction during the exposure process. Figure 6(a) and 6(b) show the SEM cross section image of the PR structures, as show, the PR structure height are 1.4 μm and 800 nm with diameter of 1 μm and 500 nm, respectively. The aspect ratio is about 1.4 for both of the 1 μm and 500 nm structures, it is height enough to resist the consumption for the subsequent dry etching process. During the etching process, ICP power, RF bias power, BCl3 flow rate and chamber pressure are kept at 900 W, 150W, 40 sccm and 5 mtorr, respectively. Figure 7(a) and 7(b) shows the morphologies of the two type of sample after ICP etching, the etching time are 20 min and 12 min, respectively. As show, both of two samples are form a perfect cone shape structure after ICP etching process, two type of PSS are prepared with diameter/spacing/height being 1.4 μm /0.6 μm /1.1 μm and 0.72 μm/0.28 μm/0.4. μm. Figure 8 show the 2 inch and 4 inch pattern sapphire wafers which is fabricated by metal contact process and photolithography method.

IV. CONCLUSION

In this study, the cone shape PSS with sub-micro meter scale by a novel process which combining the metal contact printing method and the tradition photolithography process is

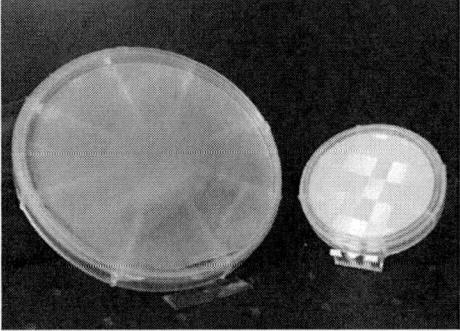

Fig. 8. A photo of 2 inch and 4 inch pattern sapphire substrate.

demonstrated. The metal pattern mask will directly transfer from PDMS stamp to a thick PR layer surface above the sapphire substrate. For the successfully transfer of the metal pattern, a uniform pressure of 2 MPa and a proper temperature of 90 °C are needed. Experiments result have been carried out to demonstrate the transferred metal mask can as s perfect photo-mask for subsequent photolithography process to create high aspect ratio PR structure, a structure with aspect ratio of 5 is achieved at the line width of 500 nm. The hexagonally dot array PR structures with two different diameter of 1 μm and 500 nm are prepared on the sapphire substrate by this novel process, PR structure can as a good etching mask in subsequent ICP etching for fabricate cone shape PSS. We have successfully patterned the full wafer area of the 2 inch and 4 inch sapphire wafer with micrometer and sub-micrometer feature sizes. The propose method is inherit all the merits of both metal contact printing lithography and the tradition photolithography, such as easily to obtain a high yield rate patterning with large area and sub-micro meter scale in only one step imprinting process, having a high exposure resolution and reliability by a commercial photolithography process. In conclusion, the novel process presented here is simplicity and easiness for a great quantity to produce PSS with sub-micro meter scale, we believe it has great potential for commercialize in the nearly future.

ACKNOWLEDGMENT

This work is supported by the National Science Council (NSC) of Taiwan through projects NSC 098-2811-M-006-012 and NSC 100-2120-M-006 -007 -CC1.

REFERENCES

[1]. W. K Wang, D. S. Wuu, S.H. Lin, S. H. Huang, K. S. Wen and R. H. Horng, J. Phys. Chem. Solids, 69 (2008) 714-718.

[2]. X. H. Huang, J. P. Liu, Y. Y. Fan, J. J. Kong H. Yang and H. B. Wang, IEEE Photon. Technol. Lett., 23 (2011) 944-946.

[3]. J. K. Sheu, C. J. Pan, G. C. Chiu, C. H. Kuo, L. W. Wu, C. H. Chen, S. J. Chang and Y. J. Su, IEEE Photon. Lett., 14 (2002) 450-452.

[4]. X. H. Huang, J. P. Liu, J. J. Kong, H. Yang and H. B. Wang, OPTIC EXPRESS, 19 (2011) 949-955.

[5]. M. T. Wang, K. Y. Liao and Y. L. Li, IEEE Photon. Technol. Lett., 23 (2011) 965-967.

[6]. Ishibashi, I. Kidoguchi, G. Sugahara and Y. Ban, J. Cryst. Growth, 221 (2000) 338-344.

[7]. S. Kissinger, S. M. Jeong, S. H. Yun, S. J. Lee, D. W. Kim, I. H. Lee and C. R. Lee, Solid-State Electron., 54 (2010) 509-515.

[8]. H. Y. Shin, S. K. Kwon, Y. I. Chang, M. J. Cho and K. H. Park, J. Cryst. Growth 311 (2009) 4167-4170.

[9]. Y. J. Lee, T. C. Lu, H. C. Kuo and S. C. Wang, J. Display Technol., 3 (2007) 118-125.

[10]. R. M. Lin, Y. C. Lu, S. F. Yu, Y. S. Wu, C. H. Chiang, W. C. Hsu and S. J. Chang, J. Electrochem. Soc., 156 (2009) 874-876.

[11]. D. S. Wuu, W. K. Wang, K. S. Wen, S. C. Huang , S. H. Lin, R. H. Horng, Y. S. Yu and M. H. Pan, J. Electrochem. Soc., 153 (2006) 765-770.

[12]. W. F. Yang, Q. Z. Zhang, M. G. Wang and Y. Xia, Sci. China Tech. Sci., 54 (2011) 2232-2236.

[13]. Y. J. Sung, H. S. Kim, Y. H. Lee, J. W. Lee, S.H. Chae, Y. J. Park and G. Y. Yeom, Mat. Sci. Eng., 82 (2001) 50-52.

[14]. Y. T. Hsieh and Y. C. Lee, J. Micromeh. Mocroeng., 21 (2011) 015001.

[15]. H.W Huang, C. H. Lin, J. K. Huang, K. Y. Lee, C. F. Lin, C. C. Yu, J. Y. Tsai, R. Hsueh, H. C. Kuo and S. C. Wang, Mat. Sci. Eng., 164 (2009) 76-79.

[16]. Y. K. Su, J. J. Chen, C. L. Lin, S. M. Chen, W. L. Li and C. C. Koa, J. Cryst. Growth, 311 (2009) 2973-2976.

[17]. J. Lee, D. H. Kim, J. Kim and H. Jeon, Curr. Appl. Phys., 9 (2009) 633-635.

[18]. H. Miyajima, M. Mehregany, J. Microelectromech. Syst., 4 (1995), pp. 220-228.

[19]. D. Suh, J. Rhee and H. H. Lee, Nanotechnology, 15 (2004) 1103-1107.

[20]. S. H. Hur and D. Y. Khang, Appl. Phys. Lett., 58 (2004) 5730-5732.

Nano-scale Mechanical Relays Fabricated by Nanoimprint Technology

Y. J. Chang[*], D. Y. Liu, and C. L. Kuo

[*]Department of Mechanical Engineering, National Yunlin University of Science and Technology, TAIWAN
changy@yuntech.edu.tw

Abstract—**In this paper, we first demonstrate nanoscale mechanical relays fabricated by a nanoimprint technology, called Contact-Transfer and Mask-Embedded Lithography (CMEL). With this technology, the nanoscale metallic source electrode can be easily fabricated in one step at a low cost. We successfully fabricated the three-terminal nanorelays with various lengths. The nanorelays are demonstrated by measuring the I-V curve of each device. The measured pull-in voltages are compared with the simulation results.**

Keywords-NEMS; nanoimprint; nanorelays

I. INTRODUCTION

In the past several decades, silicon-based microelectronics, i.e. integrated circuits (ICs), have achieved unprecedented improvements in device density and performance, as Moore's Law predicts [1]. With high density, the minimum feature size has reached a sub-100 nm regime. The power consumption of ICs increases enormously due to significant power leakage when the solid state structures enter a sub-100 nm scale [2].

Nanoelectromechanical systems are considered to be an effective solution to the leakage problem and nanoscale mechanical relays could be a new generation of transistors [3]. Due to their excellent mechanical and electrical properties, two main nanomaterials are employed to construct the nanoscale mechanical relays: carbon nanotubes [4][5] and nanowires [6]. The main challenges involved in using these nanomaterials are the uniformity of material properties and fabrication processes. Material properties and dimensions of nanomaterials vary between nanotubes and nanowires, even in the same fabrication batch. The fabrication process, however, is cumbersome [4]. The growth of nanotubes on a predefined location is a challenge as well. An alternate approach is the "top-down" method. Nanolithography, such as e-beam lithography or focus ion beam lithography, can be utilized [7][8]. The throughput, however, is low and these methods are not suitable for mass production. Therefore, nanofabrication technology for mass production is very important for commercialization. In this study, we utilize a novel nanoimprint technology, called Contact-transfer and Mask Embedded Lithography (CMEL) to fabricate TiN-based nanorelays [9].

II. DESIGN AND FABRICATION PROCESS

The structure of a nanoscale mechanical relay is shown in Fig. 1. It is a clamped-clamped type, three-terminal relay, consisting of one source, one drain, and two gate electrodes. It is in a vertical configuration, i.e. the source electrode is suspended over the drain and gate electrodes. The gate electrodes are placed symmetrically on both sides of the drain electrode. The mechanical nanorelay is actuated by the electrostatic force generated by applying voltage to these two gate electrodes simultaneously while the source electrode is grounded. The relay is a device normally in an "off" state. The relay is "on" once the suspending source electrode is bent by the electrostatic force and contacts the drain electrode.

The fabrication process is divided into two parts: mold and bottom substrate fabrications. In the mold fabrication, we first deposit a 300 nm thick silicon dioxide layer on silicon by a wet thermal process. We then pattern the oxide layer by photolithography using a stepper and wet etching process to create the submicron features. The silicon substrate is then patterned by ICP-RIE. The etching depth of silicon substrate is 300 nm. The oxide layer is completely removed by wet etching afterwards. The silicon pattern defines the shape of the source electrode when applying CMEL. We then dice the wafer into 2 × 2 cm² pieces and coat the molds with an anti-stiction material: 1H, 1H, 2H, 2H-perfluorooctyltrichlorosilane, using a vapor deposition method. Finally, a 50 nm thick Au layer is deposited by e-beam evaporation to complete the mold fabrication.

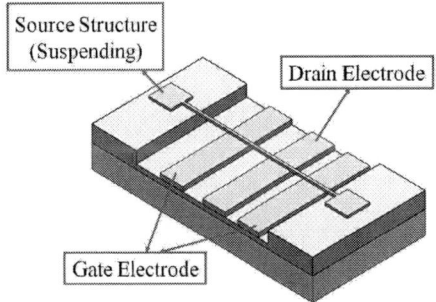

Fig. 1 Structure of a nanoscale mechanical relay

The bottom substrate fabrication process is more complicated than the mold fabrication. The bottom substrate fabrication process is shown in Fig. 2. A 100 nm thick oxide layer is first deposited on silicon wafers. Then a 300 nm thick polysilicon layer is deposited and patterned to define the area for the drain and gate electrodes. The polysilicon layer is completely removed to expose the oxide layer in the patterned area. We then deposit and pattern a 100 nm thick TiN layer to form the drain and gate electrodes in the patterned area defined previously. The width of the electrodes varies with the size of the area. After fabricating the drain and gate electrodes, we deposit a thick oxide layer and perform CMP to obtain a flat surface and expose the polysilicon layer. A 50 nm thick TiN layer is deposited on the substrate as the source electrode material. Finally, the CMEL process is applied to complete the device fabrication. The CMEL process is shown in Fig. 3. A layer of A4 950 PMMA is spin-coated on top of the substrate. By applying air pressure of 2kgf/cm^2, the Au layer on the mold contacts the PMMA layer. Meanwhile, both the substrate and the mold are heated to 95℃ for 6 minutes. After cooling down, the mold and substrate are separated and the Au pattern is transferred to the PMMA layer. The Au on the PMMA layer is used as the mask in RIE to define the source electrode. Two RIE steps are applied: one is with oxygen to pattern the PMMA layer while the other is with SF$_6$ to pattern the top TiN layer. After lifting off the Au layer with acetone, the oxide layer under the top TiN layer is removed using BOE (Buffer Oxide Etch). Finally, a CO$_2$ critical dryer is applied to prevent the source electrode from stiction and the devices are completed.

In this process, the length of the source electrode is defined by the spacing between the polysilicon patterns on the bottom substrate. We have successfully fabricated the electrode with various lengths from 4 μm to 20 μm. The width of the source electrode is around 600 nm.

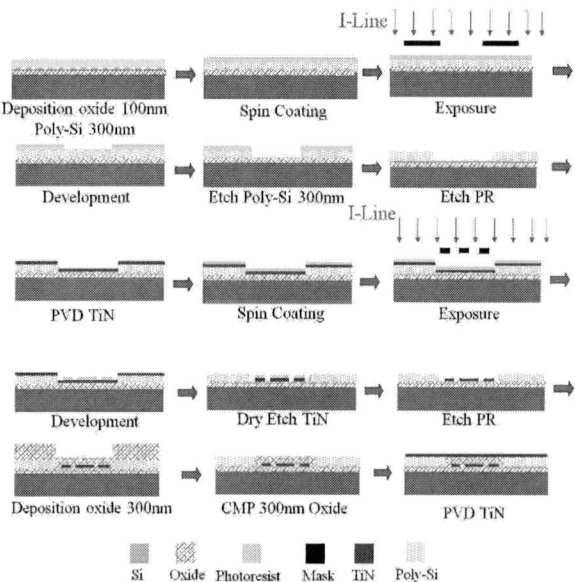

Fig. 2 Fabrication process of bottom substrate

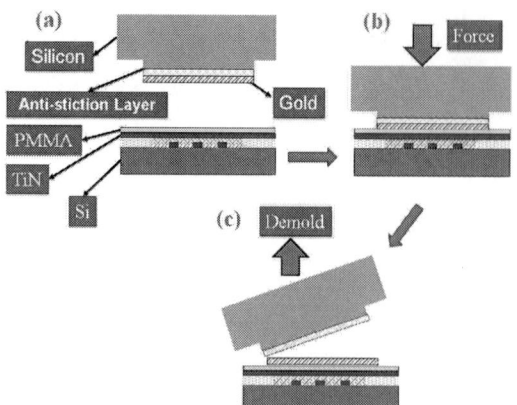

Fig. 3 CMEL Process

III. SIMULATIONS

Simulation of pull-in voltage at different lengths is performed by COMSOL^{TM} Multiphysics before measurements are taken. The geometries of devices in the simulation are listed in Table 1. These geometries match with the fabricated devices. The simulation is performed in two-dimension. Young's Modulus of TiN is set to 600 GPa and the structures have no residue stress. The thickness of the structure is only 50 nm and is divided into 8 elements. The aspect ratio of mesh element is kept within 2:1 for excellent meshing quality. The suspended source electrode is the only moving object. Both ends of the source electrode are set as fixed while the rest of the source electrode is set free. The gap between the source electrode and the bottom electrodes, i.e. drain electrode and gate electrodes, is set as air. The gap is 100 nm in the simulation. One of the simulation results is shown in Fig. 4. The color shows the displacement in y direction, i.e. normal to the source electrode. The simulation results of the pull-in voltages are listed in Table 1 and shown in Fig. 5.

Table 1: List of Geometry of Simulated Devices

Length of S (μm)	Width of D (nm)	Width of G (nm)	Spacing b/w D & G (nm)	Pull-in voltage (Volt)
4	800	600	500	89
5	800	600	500	47
8	800	600	500	28
10	2000	3000	500	16.2
20	2000	3000	500	6.6
40	2000	3000	500	1.4

S: Source; D: Drain; G: Gate; b/w: between

IV. RESULTS AND DISCUSSIONS

We have successfully fabricated devices with different lengths. An SEM picture of a 5 μm long device is shown in Fig. 6. The width of all the devices is 633 nm. The variation in length is due to the misalignment between the mold and substrate. The variation depends on the length and is controlled within 0.7 μm.

The pull-in voltage of the devices is characterized by measuring the I-V curves. A probe station with a semiconductor parameter analyzer is utilized. The actuation voltage is applied to two gate electrodes simultaneously while the source electrode is grounded. The increment of the actuation voltage is 0.1 V. At the same time, a small voltage of 50 mV is applied to the drain electrode. Once the pull-in voltage is reached, a sudden current change can be detected and an ohm contact between the source and drain electrodes is formed.

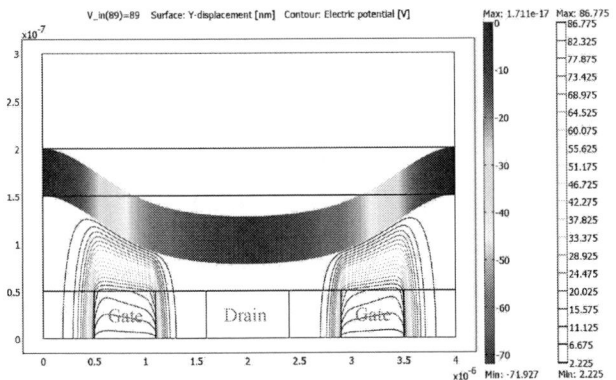

Fig. 4 Pull-in voltage simulation of a 4μm long relay.

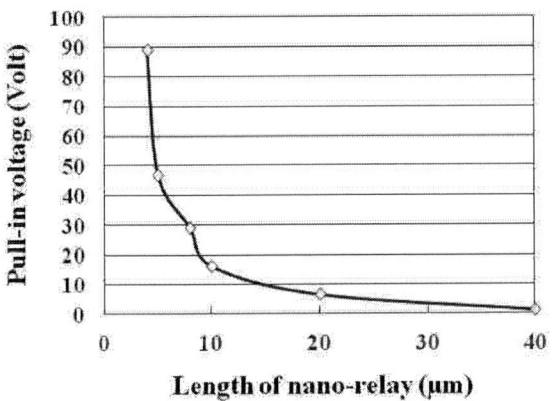

Fig. 5 Simulation results of pull-in voltage with different lengths of nano mechanical relays.

The measurement results of nanorelays with different lengths are shown in Fig. 7 to 10. The designed lengths of the devices in each figure are 4 μm, 5 μm, 8 μm, and 20 μm, respectively. A sudden drain current (i.e. IDS) jumps from a nano-ampere to micro-ampere and the pull-in phenomenon is determined. The measured pull-in voltages are 16 volts, 12 volts, 4 volts, and 1.75 volts, respectively. After pull-in occurs, an ohm contact can be easily seen from the linear relationship

in the I-V curve. The contact resistance of devices which are 4 μm and 5 μm in length is 30 kΩ. The contact resistances of 8 μm and 20 μm long devices are 6.67 kΩ and 2.3 kΩ, respectively. The difference between the contact resistances of the devices is mainly due to the contact area between the source and drain electrode. In our design, the longer suspending structure corresponds to the wider drain electrode.

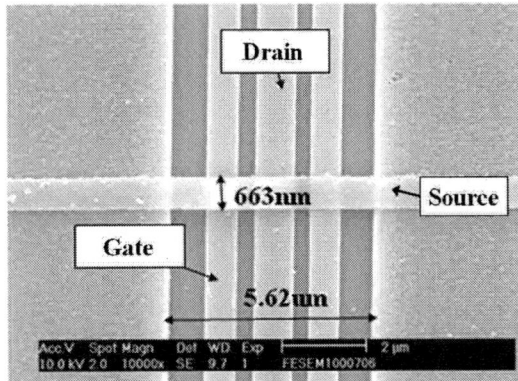

Fig. 6 SEM picture of a fabricated 5μm long relay.

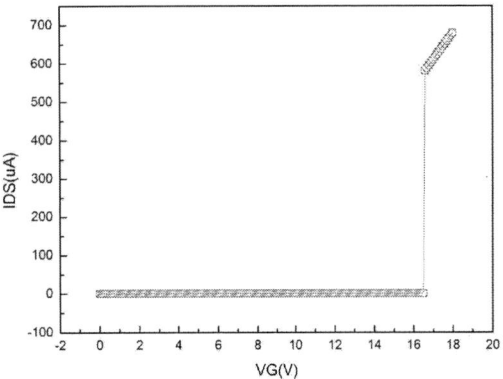

Fig. 7 I-V curve of a fabricated 4μm long relay.

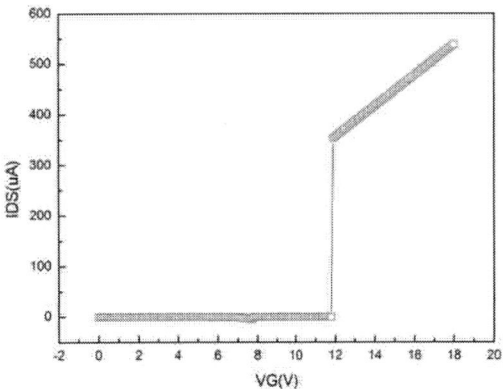

Fig. 8 I-V curve of a fabricated 5μm long relay.

Comparison with simulation results is shown in Fig. 11. It is clearly seen that the measured pull-in voltages are all smaller than the simulated ones. Three factors are considered to be the causes:

1. The BOE used in the process could slightly etch the TiN. The etch rate is around 2.5 nm/min in BOE [10]. A thinner structure results in a smaller pull-in voltage.
2. The air gap is determined by the CMP process and the variation range is not controllable.
3. Due to the thin TiN structure, bending of the structure could happen and result in smaller air gaps. The pull-in voltage is, therefore, smaller than the simulated results.

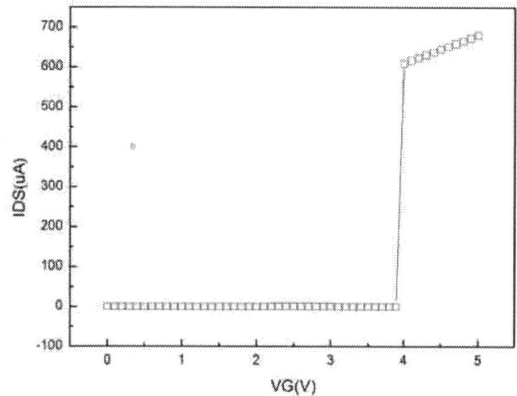

Fig. 9 I-V curve of a fabricated 8μm long relay.

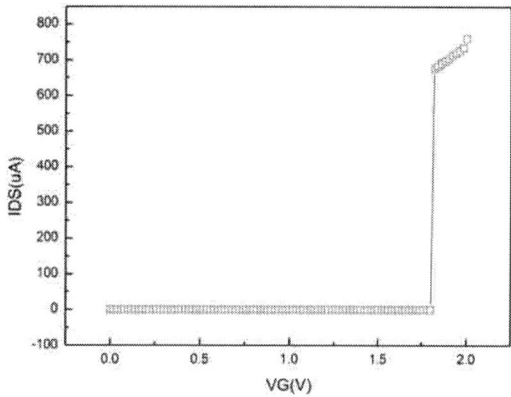

Fig. 10 I-V curve of a fabricated 20μm long relay.

V.   CONCLUSIONS

In conclusion, we have successfully applied nanoimprint technology to fabricate nanoscale TiN-based mechanical relays. The relays are demonstrated by measuring their I-V curves. The pull-in voltages range from 16 volts to 1.75 volts depending on their lengths. An ohm contact is formed between the source and drain electrodes and it shows a successful relay operation. The simulated pull-in voltage and the measured

results are compared. The measured pull-in voltages are all smaller than the simulated ones.

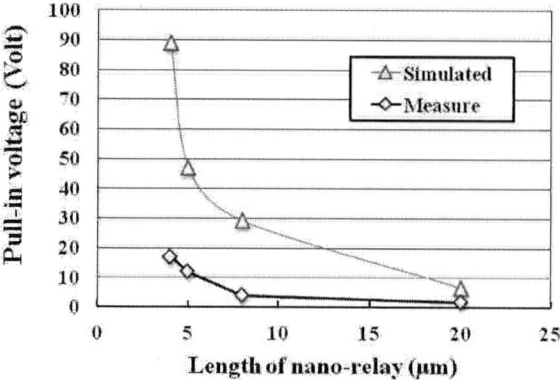

Fig. 11 Comparison of pull-in voltage between the simulated and measured results.

ACKNOWLEDGMENT

The authors would like to thank the National Science Council (NSC98-2218-E-224-009-) for their financial support and the National Nano-Device Laboratory for supplying equipment, as well as the Center for Micro/Nano Science and Technology at National Cheng Kung University, Taiwan.

REFERENCES

[1] I.T. R. F. Semiconductors Emerging research devices, Technical report, International Technology Roadmap for Semiconductors, 2005

[2] K. Roy, S. Mukhopadhyay, and H. Mahmoodi-Meimand, "Leakage current mechanisms and leakage reduction techniques in deep-submicrometer cmos circuits," *Proc. of the IEEE*, vol. 91, pp.305-327, 2003.

[3] V. Pott, H. Kam, R. Nathanael, J. Jeon, E. Alon, and T. J. K. Liu, "Mechanical computing redux: relays for integrated circuit applications," *Proc. of the IEEE*, vol. 98, pp.2076-2094, 2010.

[4] J. E. Jang, S. N. Cha, Y. Choi, G. A. J. Amaratunga, D. J. Kang, D. G. Hasko, J. E. Jung, and K. M. Kim, "Nanoelectromechanical switches with vertically aligned carbon nanotubes," *Appl. Phys. Lett.*, vol. 87, No. 16, 113102, 2005.

[5] S. W. Lee, "Carbon-nanotube based nano electromechanical switches," *J. of the Korean Phys. Soc.*, vol.55, pp. 957-961, 2009.

[6] K. J. Ziegler, D. M. Lyons, J. D. Holmes, d. Erts, B. Polyakov, H. Olin, K. Svensson, and E. Olsson, "Bistable nanoelectromechanical devices," *Appl. Phys. Lett.*, vol. 84, No. 20, pp.4074-4076, 2004.

[7] W. W. Jang, "NEMS switch with 30 nm thick beam and 20 nm thick air-gap for high density non-volatile memory applications," *Solid-State Electronics*, vol. 52 , pp. 1578–1583, 2008.

[8] W. W. Jang, "Fabrication and characterization of a nanoelectromechanical switch with 15-nm thick suspension air gap, '*Appl. Phys. Lett.*, vol. 92 , 103110, 2008.

[9] Y. C. Lee, C.Y. Chui," Micro-/nano- lithography on the contact transfer of thin film and mask embedded etching," *J. of Micromech. and Microeng.*, vol. 18, 075013, 2008.

[10] K. R. William, K. Kupta, M. Wasilik, "Etch rates for micromachining processing – Part II," *J. of Microelectromechanical System*, Vol. 12, No. 6, 2003, pp.761-778

978-1-4673-1122-9/12 $31.00 © 2012 IEEE

A Facile Nanowire fabrication Approach Based on Edge Lithography

Yaoping Liu*, Wei Wang, Haixia Alice Zhang, Wengang Wu and Zhihong Li

National Key Laboratory of Science and Technology on Micro/Nano Fabrication, Institute of
Microelectronics, Peking University, Beijing, 100871, China

w.wang@pku.edu.cn

Abstract—This paper developed a facile nanofabrication approach at a 4-inch wafer level based on edge lithography defined hundreds nanometer-sized features. Aluminum nanowires with width around 600 nm were prepared and the wafer-level width non-uniformity was less than 8.45%. The aluminum nanowires functioned as masks in the deep reaction ion etching for vertically-stacked silicon nanowires and silicon nanoridge preparation. The obtained nanoridges acted as the nano-master in the molding of a PDMS micro/nano integrated channels. The proposed approach has the potential of nanofabricating with mass-production at effective cost, thereby holds promising future in MEMS (Microelectromechanical system) compatible micro/nano integration and nanobiotechnology applications.

Keywords-nanofabrication; wafer-level; edge lithography; nanowires; nanoridge

I. Introduction

Over years, nanostructures have attracted many attentions because of their promising potential in electronic, optical, chemical and biological applications. There are several methods used in the fabrication of these nanostructures, which are commonly characterized as "top-down" and "bottom-up". The bottom-up approach mainly utilizes chemical synthesis while the top-down one generally relies on various nanometer resolution lithography techniques to make nanopatterns. Resolution is a key issue for nanofabrication. Advanced lithography techniques have been developed to facilitate nanowire preparations by introducing new light sources, such as E-beam, ion beam and X-ray, or nanoscale imprinting strategy (nanoimprint). Electron beam lithography (EBL) has shown sub-10 nm fabrication ability in patterning ultra-fine nanostructures [1, 2], but huge time and cost occupation restrict its usage in mass production. The other emerging technologies, such as the focused ion beam (FIB), suffer the similar issues. More seriously, it is difficult for both EBL and FIB to get a wafer-level process considering their limited throughput capacity. X-ray lithography has also shown the ability to realize 20 nm dimensions and even below [3], but improvements in the mask material and resist systems are still required for high throughput fabrication. Nanoimprint lithography [4, 5, 6] is free from the dependence of wavelength, but high-accuracy alignment needs further improvement. So an effective nanofabrication approach with mass production and low cost is in urgent demand in many nanostructure-related fields, especially in research groups which do not have the advanced lithography facilities.

This work proposed a facile nanowire fabrication approach at a 4-inch wafer level. Edge lithography [7, 8, 9], which combines standard photolithography, controllable wet etching, physical vapor deposition of aluminum and dry advanced metal etching in this work, defined hundreds nanometer-sized features. And all the included techniques are compatible with traditional microelectromechanical system (MEMS). The so-prepared aluminum nanowires functioned as etching mask for the following vertically-stacked silicon nanowires and nanoridge preparation. Geometrical size and non-uniformity on a 4-inch wafer of the so-prepared nanowires and nanoridges were characterized preliminarily.

II. Experiments

A. Aluminum nanowire fabrication

Fig. 1 (a)-(d) schematically illustrated the present aluminum nanowire fabrication process. At first, LOL-2000 beneath the photoresist (RZJ-304, Suzhou Ruihong Electronics Chemicals. LTD, Suzhou, China) was overdeveloped and formed undercutting, as shown in Fig. 1 (a). 80 nm thick aluminum layer was then sputtered onto the wafer. During the sputter, the aluminum atom dropped to the substrate in almost every direction, so some aluminum atoms would be transported into the previously formed undercuts as illustrated in Fig. 1 (b). Then, an advanced metal etching (AME) process etched the entire exposed aluminum off, shown in Fig. 1 (c). After removing photoresist along with the LOL-2000 in a PRS-3000 bath @60 °C for 5 min, aluminum residues, which located at the photoresist pattern edges, formed the targeted nanowires, as shown in Fig.1 (d). The so-prepared aluminum nanowires served as a good mask in the following silicon etching.

B. Vertically-stacked Silicon nanowire and silicon nanoridge fabrication

As exhibited in Fig. 1(e), after a 160-second DRIE (Deep reactive ion etching), vertically-stacked silicon nanowires could be obtained if the scallop with a critical size at the both

978-1-4673-1122-9/12 $31.00 © 2012 IEEE

sides of nanowire would just right to overlap, when the etching/passivation cycle in DRIE was carefully set.

300 nm silicon was etched in ASE (Advanced Silicon Etching) Shown in Fig. 1(f), and aluminum layer was removed with standard aluminum enchant (H_3PO_4: HAc: HNO_3=16:2:1@39 ℃), then the silicon nanoridges could be achieved.

C. Micro/nano integrated channels fabrication

After preparation of wafer-level silicon nanoridges, second photolithography was conducted to define micro-scale structures of photoresist, which smoothly connected with the obtained nanoridges to form micro/nano integration on the wafer. The micro/nano integrated structure functioned as the master for PDMS (Polydimethylsiloxane) chip with integrated micro/nanochannel. In the PDMS molding process, PDMS pre-polymer was prepared by mixing the base and the curing agent with a weight ratio of 10:1 (Sylgard 184，Dow Corning Corp., USA). The mixture was degassed for 20 min in a vacuum desiccator. Then the pre-polymer was poured onto the silicon master and cured in an oven at 70℃ for 60 min. After released from the mold, the PDMS replica could be bonded with a flat PDMS sheet by an O_2 plasma treatment of 10s to get a micro/nanofluidic integrated PDMS chip.

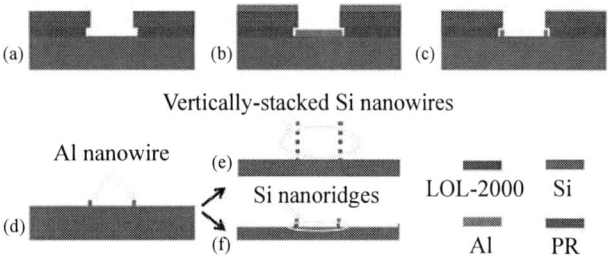

Fig. 1. Schematical illustration of aluminum nanowire fabrication. (a) Formation of undercutting at LOL-2000 layer by overdeveloping. (b) Deposition of aluminum by sputterring. (c) ASE of entire exposed aluminum. (d) Removal of photoresist along with LOL-2000 in a PRS-3000 bath. (e) DRIE of silicon with prepared aluminum nanowire as the etching mask. (f) 300nm depth of ASE with prepared aluminum nanowire as mask.

III. RESULTS AND DISCUSSIONS

Part Ⅰ: In the present experiments, the aluminum layer was 80 nm thick, and the aluminum nanowire width was controlled by the sputtering process as the undercutting is usually large enough, as indicated in Fig. 2 (a). The width of the prepared aluminum nanowire varied from 515nm to 631nm with the non-uniformity less than 8.45% on a 4-inch wafer and 10.12% between processes. Using the aluminum nanowires prepared by the edge lithography as the nanopatterns transferring mask in following silicon etching process, vertically-stacked silicon nanowires and nanoridges were successfully obtained.

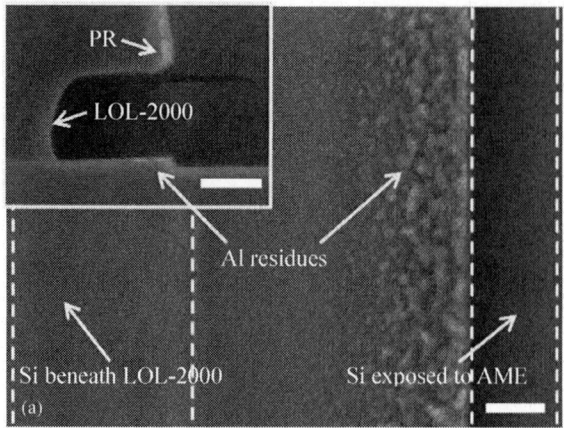

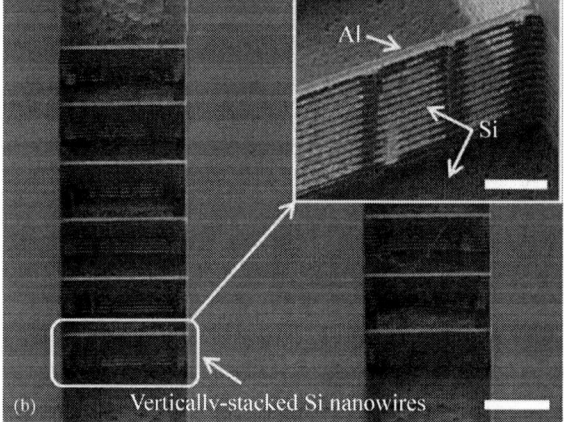

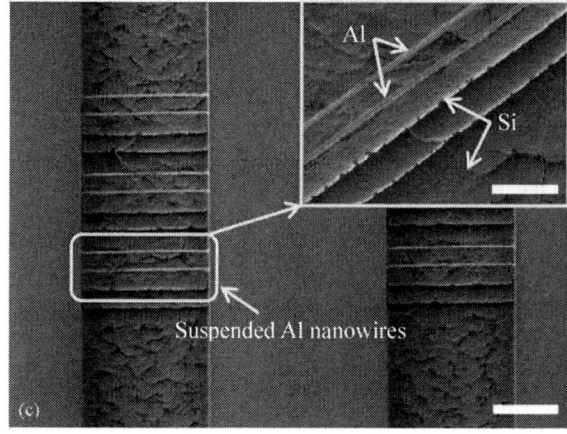

Fig. 2. (a), SEM photos of aluminum nanowire and the inserted one is the sectional view of undercutting and aluminum residues. (b), SEM photos of the vertically-stacked silicon nanowires after the DRIE. (c), SEM photo of the suspended aluminum nanowires after DRIE with silicon being overetched. The bars are 250nm in (a), 7μm in (b) and (c) and the bars for the inserted ones are 400nm in (a), 3μm in (b) and (c).

978-1-4673-1122-9/12 $31.00 © 2012 IEEE

Fig. 2 (b) showed the SEM photos of vertically-stacked silicon nanowire generated after a 160-second DRIE. In some cases, if the etching/passivation cycle was not well controlled, the scallop size would be smaller or bigger than the critical one, both sides of silicon beneath aluminum nanowire would not come across or be etched completely with suspended Al nanowires left only, shown in Fig. 2 (c). Etching/passivation cycle needs further optimization to achieve a robust and high-quality vertically-stacked silicon nanowires preparation. After DRIE, the aluminum nanowire width could be decreased to 364±30 nm, and the non-uniformity was less than 14.54% on a 4-inch wafer and 8.12% between wafers.

Part II: Fig. 3 showed the typical scanning electron microscopy (SEM) photos of the silicon nanoridges with a corresponding 4-inch wafer as the background. The non-uniformity of the silicon nanoridges width in the whole wafer was 458±70.1nm and the non-uniformity was less than 6.28% on a 4-inth wafer and 6.11% between wafers. Fig. 4 (a) and (b) showed the SEM photos of the silicon nanoridges integrating with micro-scale photoresist on the wafer and integrated micro/nanochannels on the molded PDMS replica. The inserted ones are the magnification of interconnection of micro/nano structures.

Width of nanostructures and non-uniformity on a 4-inch wafer and between wafers were shown in the following table.

TABLE I.   WIDTH AND NON-UNIFORMITY

	After removal of PR/LOL-2000	After 150-second DRIE	After 300-nm ASE
Width of nanostructure	576±38.9nm	364±31.3nm	458±70.1nm
Non-uniformity on a wafer	8.45%	14.54%	6.28%
Non-uniformity between wafers	10.12%	8.12%	6.11%

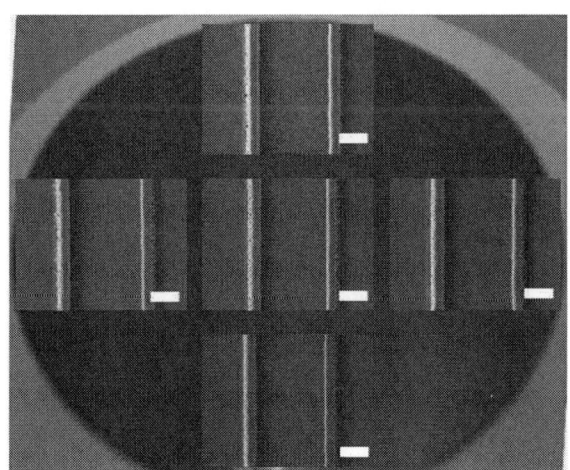

Fig. 3. Typical SEM photos (top-view) of the aluminum nanowires after the DRIE with corresponding 4-inch wafer as the background. The scale bars are 1 μm.

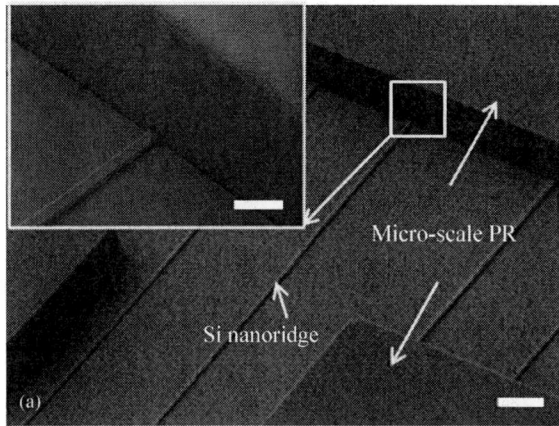

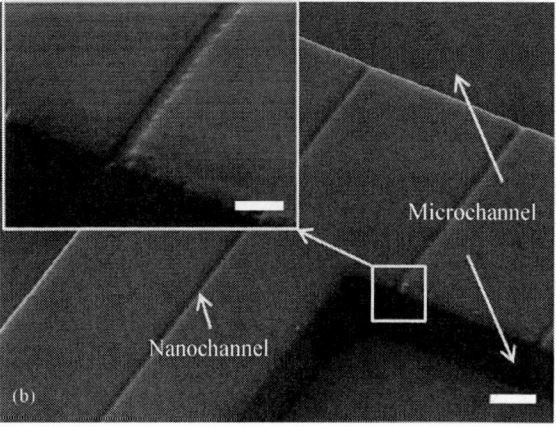

Fig. 4. (a), SEM photos of silicon nano----s integrating with micro-scale photoresist on the wafer. (b), SEM photos of the micro/nano nanochannels on the molded PDMS replica. The inserted ones are the magnification of interconnection of micro/nano structures. The scale bars are 5 μm in (a) and (b) and the scale bars for the inserted ones are 2μm in (a) and (b).

IV.   CONCLUSIONS

In summary, a wafer-level nanowire fabrication was developed based on edge lithography defined undercutting. The obtained aluminum nanowires could function as the nanopattern transfer mask in the preparation of vertically-stacked silicon nanowires and silicon nanoridges. Being a mass-producible and cost-effective manufacturing approach of versatile nanostructures, the present nanofabrication technique holds promising future in nanobiotechnology applications. The technique made integration of microchannels and nanochannels with smooth interconnection being possible in a single fabrication step, which will find wide applications in micro/nanofluidic researches. A high yield and good reproducible nanowires can be an option for the nano-biosensors and nano-actuators applied in biomolecular detection, DNA quantification and other bio-manipulations.

ACKNOWLEDGMENT

This work was financially supported by the Major State Basic Research Development Program (973 Program) (Grant No. 2009CB320300) and the National Natural Science Foundation of China (Grant No: 91023045).

REFERENCES

[1] A. N. Broers, J. M. E. Harper and W. W. Molzen, "250-Å linewidths with PMMA electron resist", Appl. Phys. Lett., Vol.33, pp.392-394, 1978.

[2] J. G. Bai, C. L. Chang, J. H. Chung and K. H. Lee, "Shadow edge lithography for nanoscale patterning and manufacturing", Nanotechnology, Vol.18, 405307, 2007.

[3] G. Simon, A. M. Haghiri-Gosnet, J. Bourneix, D. Decanini, Y. Chen, F. Rousseaux, H. Launois and B. Vidal. J. Vac, "Sub-20 nm x-ray nanolithography using conventional mask technologies on monochromatized synchrotron radiation", Sci. Technol. B, Vol.15, pp.2489-2494, 1997.

[4] T. Jun, T. Yuji, M. Iwao, K. Masanori, H. Hiroshi, "Diamond nanoimprint lithography", Nanotechnology, Vol.13, pp.592-596, 2002.

[5] S. Y. Chou, P. R. Krauss, and P. J. Renstrom, "Imprint of sub-25 nm vias and trenches in polymers", Appl. Phys. Lett., Vol.67, pp.3114-3117, 1995.

[6] S. Y. Chou, P. R. Krauss, and P. J. Renstrom, "Nanoimprint lithography", J. Vac. Sci. Technol. B, Vol. 14, pp.4129-4133, 1996.

[7] J. C. Love, K.E. Paul, G. M. Whitesides, "Fabrication of Nanometer-Scale Features by Controlled Isotropic Wet Chemical Etching", Adv. Mater., Vol. 13, pp.604-607, 2001.

[8] B. D. Gates, Q. B. Xu, M. Stewart, D. Ryan, C. G. Willson, G. M. Whitesides, "New Approaches to Nanofabrication: Molding, Printing, and Other Techniques", Chem. Rev., Vol.105, pp.1171-1196, 2005.

[9] Y. Li, Q. Xie, W. Wang, M. Zheng, H. Zhang, Y. Lei, H. Zhang, W. Wu, Z. Li, "Parylene C-on-photoresist (POP): a low temperature spacer scheme for polymer/metal nanowire fabrication", J. Micromech. Microeng., Vol.21, 067001, 2011.

978-1-4673-1122-9/12 $31.00 © 2012 IEEE

Hybrid Mask Lithography
for Fabrication of Micro-pattern with Nano-pillars

Shinya Sakuma[*], Masakuni Sugita[*], and Fumihito Arai[*,**]
[*]Department of Micro-Nano Systems Engineering, Nagoya University, JAPAN
[**]Soul National University, KOREA
sakuma@biorobotics.mech.nagoya-u.ac.jp

Abstract— **This paper presents the fabrication process of nano-pillar with micro-pattern simultaneously. We proposed the hybrid mask of micro-pattern photo mask and nano-particle mask to obtain the simple fabrication of nano-pillar with micro-pattern. The features of this process are summarized as follows. (1) It is possible to fabricate arbitrary 2D pattern using photolithography. (2) We can get nano-pillar whose size was diffraction-limited by using nano-particle mask. (3) We can control the density of the nano-pillar by changing the weight ratio of nano-particle. (4) We can control the height of the pillar as same the micro-pattern which is the original surface of the substrate.**

Keywords-component; hybrid exposure; nano-pillar; non-wall microchannel;

I. INTRODUCTION

In the field of micro-nano engineering, it is important to control the surface wettability such as the hydrophobic / hydrophilic surface [1]-[5]. The wettability is characterized by the contact angle which is defined as the angle between a liquid droplet and a surface. Nanostructures on the surface contribute to enhance the contact angle in two different ways. The two representative methods to model contact angle are wenzel model and cassie-baxter model. In the wenzel model [1], the nanostructure on the surface enhances the wettability of the surface. In the cassie-baxter model [2], the liquid contact the surface at the top of the nanostructure and the surface become the more hydrophobic than only flat surface. Several different physical and chemical patterning approaches have been employed for fabrication of nanostructure on the surface. Especially, physical modification provide more efficient surface than that of the only by the chemical modification because nano-structure create more surface area [5]. On the other hand, in the biomedical filed, wettability contributs not only to control the surface tension of the liquid in a microfluidic chip but also to restricted flow direction [6]-[13]. The wettability is essentially important to build functional microfluidic surface on a substrate for complex biomedical operations.

Previously, several methods commonly used to fabrication of nanostructure such as black silicon produced by the deep reactive ion etching (DRIE), photolithography, electron-beam lithography, and selective growth of carbon nanotubes. The important features of the fabrication method are the controllability of the pillar density and the size of patterning

area. Generally speaking black silicon which is needle like structure was not preferable condition of DRIE because it is difficult to fabricate the flat surface [14]. On the other hand, the sharp structure contributes to obtain the hydrophobic surface [15]. However, it is difficult to control of the height and the density of the nanostructure because this fabrication process depends on the condition of the etching gas and deposition gas. Nanostructure fabrication based on the photolithography technique [16]-[17] can pattern large area. However it is difficult to control of the density of the nanostructure because the density depends on the distance of the aligned nanospheres. Electron-beam lithography [18] is technique possible to fabricated the nanostructure with the order of under diffraction-limited of the photolithography. In addition this technique is able to control of the density o the nanostructure because this process based on the scanning method. However the technique based on the scanning method, it takes long time to pattern large area. Selective growth of carbon nanotubes [19] is used to fabricate the nanostructure. This technique is possible to fabricate the nanostructure in large area because this method based on a kind of stencil mask process. However this stencil mask process needs the precise alignment between the substrate and the mask.

In this paper, we proposed the hybrid mask of micro-pattern photo mask and nano-particle mask to obtain the simple fabrication of the nano-pillar with micro-pattern simultaneously. The concept of the hybrid mask lithography is shown in Figure 1. Details fabrication process of hybrid mask lithography describes in section2. Briefly, the hybrid mask lithography was utilized the composite of positive photoresist and nano-particle was exposed by using the two different masks of the photo mask and nano-particle mask. Section 3, we describe the basic experiments.

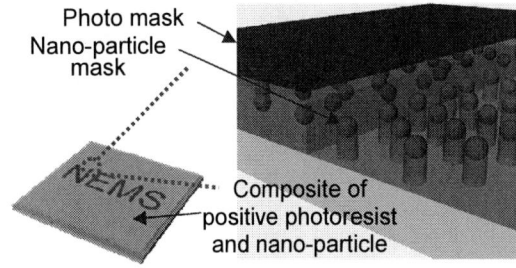

Fig.1 Concept of hybrid mask exposure

Further, we evaluate the effectiveness of the proposed approach to fabricate the nanostructure with microstructure. Finally, the conclusions and our future plan are shown in section 4.

II. Hybrid Mask Lithography

A. Concept of hybrid mask lithography

Fabrication method of nano-pillar with microchannel is based on a simple two step procedure of the standard photolithography process and the DRIE process. We proposed the hybrid mask lithography. In the photolithography process, the composite of positive photoresist and nano-particle is exposed by using the two different masks of the photo mask and nano-particle mask. The photo mask was adapted to fabricate micro-channel pattern and the nano-particle mask was adapted to fabricate nano-pillar pattern. By using these two masks, we can get the multi-scale etching mask of DRIE which are the microchannel pattern as well as nano-pillar pattern simultaneously. Our lithography method allows the fabrication of nano-pillar enclosed micro-pattern with diffraction-limited accuracy using only standard microfabrication processes.

The features of this process are summarized as follows. (1) It is possible to fabricate arbitrary 2D pattern using photolithography. (2) We can get nano-pillar at the range of diffraction-limited by using nano-particle mask. (3) We can control the density of the nano-pillar by changing the weight ratio of nano-particle. (4) We can control the height of the pillar as same the micro-pattern which is the original surface of the substrate.

B. Fabrication process of nano-pillar enclosed micro-pattern

Figure 2 shows process flow of nano-pillar with micro-pattern fabrication. AT first, the composite of OFPR and Fe nano-particle (Tateyama machine co., Ltd. TEC201101-NPC), Fe particle : 1.0 wt.%, diameter = 2.3 nm) was spin coated (figure 2 (a)). OFPR (Tokyo Ohka Co., Ltd.) is a kind of positive photoresist. After spincoating, the composite was exposed using photo mask (figure 2 (b)). In this process, nano-particle was also worked as the photo mask to fabricate the nano-pillar pattern and we could get the nano-pillar enclosed micro-pattern simultaneously. After that, Si layer was etched by using DRIE process (Etching gas : SF_6, Deposition gas : C_4F_8) (figure 2 (c)). Finally, the composite was removed. Eventually, we can get nano-pillar with micro-pattern of Si after the composite was removed (figure 2 (d)).

As an example of the fabrication of the micro-pattern with nano-pillar, we fabricated the round shape micro-pattern (diameter = 150 µm). SEM images of the fabricated structure are shown in Figure 3 (a). We have succeeded in fabricating multi-scale structure which are the micro-channel as well as nano-pillar simultaneously. In addition, we have also succeeded in the fabrication of the uniform height nano-pillar and micro-pattern (figure 3 (a)). The height and the diameter of nano-pillar was evaluated. The measured height by SEM image was about 6.2 µm and the diameter was about 0.35 µm.

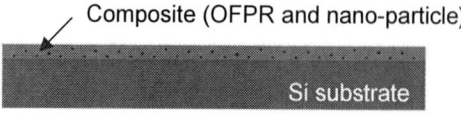

(a) Spin coating of composite

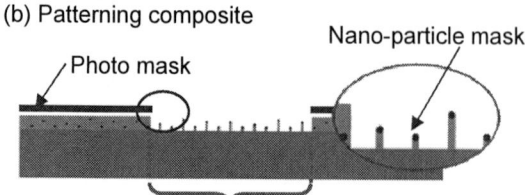

(b) Patterning composite

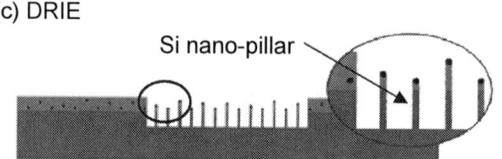

(c) DRIE

(d) Removal of composite

Fig.2 Si nano-pillar fabrication process flow

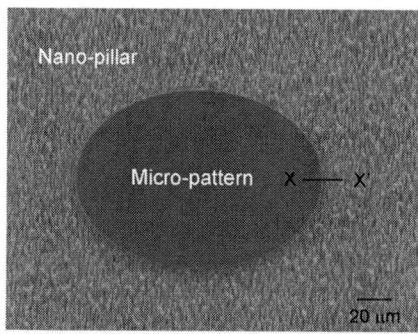

(a) Fabricated micro-pattern (ϕ =150 µm)

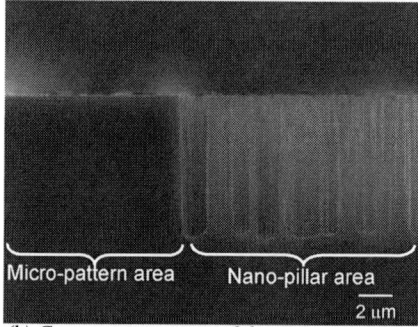

(b) Cross-section view of the micro-pattern along the line X-X'

Fig.3 SEM images of fabricated micropattern with nano-pillar by using the hybrid mask lithography

978-1-4673-1122-9/12 $31.00 © 2012 IEEE   54

III. BASIC EXPERIMENTS

The density of nano-pillar as a function of weight ratio of nano-particle was evaluated. Figure 4 shows the SEM images of the nano-pillar. The density was measured by the SEM image. Figure 5 shows the density of the nano-pillar as a function of weight ratio of nano-particle. Horizontal axis shows the weight ratio of nano-particle and vertical axis shows the density of the nano-pillar. It was confirmed that the density of the magnetite can be controlled simply by changing the weight ratio of the nano-particle.

Figure 6 shows that the contact angle of DI water droplet (2 μl) as a function of the weight ratio of nano-particle. The contact angle was measured by the CCD image. Horizontal axis shows the weight ratio of nano-particle and vertical axis shows the contact of between the original surface and the droplet. It was confirmed that the contact angle also can be controlled simply by changing the weight ratio of the magnetite. Therefore, the hydrophobic surface can be also achieved by changing the weight ration of magnetite.

Hence, it is likely to say that the contact angle was controlled by the density of the nano-pillar

Finally, we applied the hybrid mask technique to fabricate the nano-pillar enclosed the micro-channel. The nanostructure on the surface enhances the wettability. It is possible to control liquid motions by applying the surface modification. As an example of the application, SiO_2 was sputtered on the nano-pillar as a surface modification to obtain the hydrophilic surface. Figure 7 shows the process flow of the modification. This modification process was simple lift-off process. First, the OFPR was spincoated and exposed on the surface which was fabricated in figure 2 (figure 2 (a)). After spincoating, the SiO_2 was sputtered (figure 2 (b)). In this process, we could fabricated the nano-particle which was coated SiO_2. Finally, the OFPR layer was removed (figure 2 (c)). Figure 8 shows the fabricated nano-pillar enclosed micro-pattern. We succeeded in transportation of micro-particle by using the modification of nano-pillar (Figure 7).

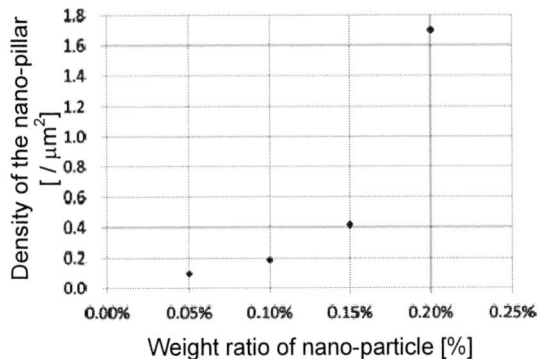

Fig.4 Measured density of spin coated magnetite as a function of weight ratio of nano-particle

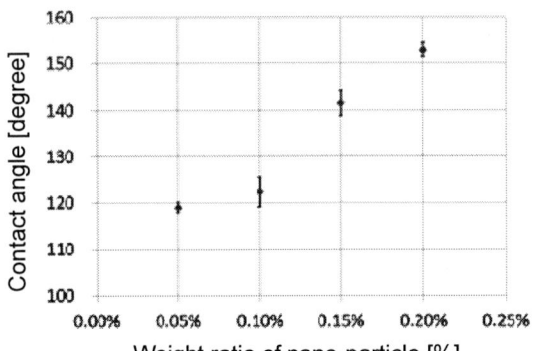

Fig.6 Measured contact angle of water droplet as a function of weight ratio of magnetite

(a) Spin coating of OFPR

(b) Sputtering of SiO_2

(c) Removal of OFPR and SiO_2

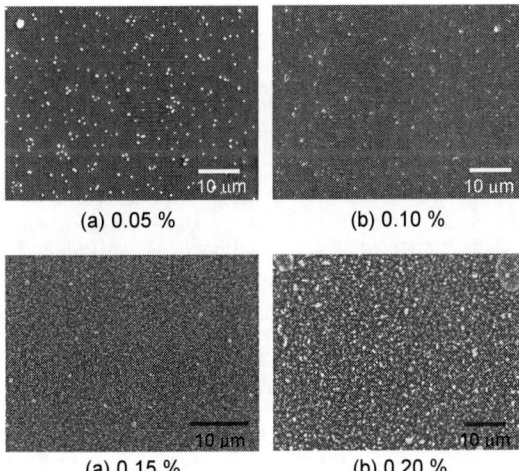

Fig.5 SEM images of the nano-pillar on the surface as a function of weight ratio of nano-particle

Fig.7 Process flow of the surface modification

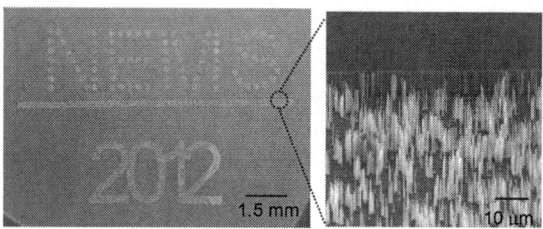

Fig.8 SEM images of fabricated micro-channel

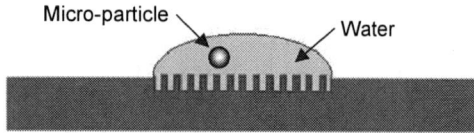

(a) Side view of the micro-channel with nano-pillar

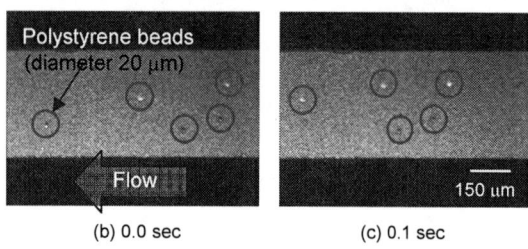

(b) 0.0 sec   (c) 0.1 sec

Fig.9 Confirmation of production of
micro-channle due to hydrophobic surface

IV. CONCLUSIONS AND DISCUTIONS

We proposed the hybrid mask of micro-pattern photo mask and nano-particle mask to obtain the nano-pillar with micro-pattern simultaneously. This fabrication method is based on a simple two step procedure of the standard photolithography process and the DRIE process. In the photolithography process, the composite of OFPR (positive photoresist) and Fe nano-particle was exposed by using the two different masks of the photo mask and nano-particle mask. By using these two masks, we can get the multi-scale etching mask of DRIE which are the microchannel pattern as well as nano-pillar pattern simultaneously. In this simple two step process, we could get nano-pillar enclosed the microchannel-pattern easily.

The density of nano-pillar was evaluated as a function of weight ratio of nano-particle. It was confirmed that the density of the magnetite can be controlled simply by changing the weight ratio of the magnetite. The contact angle of droplet was evaluated as a function of the weight ratio of magnetite. It was confirmed that the hydrophobic surface can be also achieved by changing the weight ration of magnetite.

Finally, we applied the hybrid mask technique to fabricate the nano-pillar enclosed micro-channel. To obtain the hydrophilic surface, SiO_2 was sputtered on the nano-pillar. We succeeded in transportation of micro-particle by using the modification of nano-pillar .

The proposed method will be one of the promising methods to fabricate the functional microchannel based on the surface structure for biomedical application because this fabrication method is based on a standard photolithography process.

ACKNOWLEDGMENT

This work is partially supported by Scientific Research from Ministry of Education, Culture, Sports, Science and Technology (23106002) and the Nagoya University Global COE program for Education and Research of Micro-Nano Mechatronics.

REFERENCES

[1] R. N. Wenzel, J. Phys. Colloid Chem. 1949, 53, 1466.

[2] A. B. D. Cassie, S. Baxter, Trans. Faraday Soc. 1944, 40, 546.

[3] D. Quere, A. Lafuma, "Slippy and sticky microtextured solids", Bico, J. Nanotechnology, *14*, 1109-1112. 2003

[4] Z. Yoshimitsu, A. Nakajima, T. Watanabe, K. Hashimoto," Effects of Surface Structure on the Hydrophobicity and Sliding Behavior of Water Droplets", Langmuir, 18, 5818-5822, 2002

[5] E. Stratakis, et.al., "Biomimetic micro/nanostructured functional surfaces for microfluidic and tissue engineering applications", Biomicrofluidics, 5, 013411, 2011

[6] B. Zhao, J. S. Moore, and D. J. Beebe, "Surface-directed liquid flow inside microchannels", Science 291, 1023, 2001

[7] G. M. Walker, D. J. Beebe, "A passive pumping method for microfluidic devices", Lab Chip, 2, 131–134, 2002

[8] P. Lam, K.J. Wynne, G.E. Wnek, "Surface-tension-confined microfluidics", Langmuir 18, 948-951, 2002

[9] J. Melin, W. Wijingaart, G. Stemme, "Behaviour and design considerations for continuous flow closed-open-closed liquid microchannels", Lab Chip, 5, 682–686, 2005

[10] S. Bouaidat, O. Hansen, H. Bruus, C. Berendsen, N. K. Bau-Madsen, P. Thomsen, A. Wolff and J. Jonsmann, "Surface-directed capillary system; theory, experiments and applications", Lab Chip 5, 827–836, 2005

[11] M. J. Swickrath, S. D. Burns, G.E. Wnek, "Modulating passive micromixing in 2-d microfluidic devices via discontinuities in surface energy", Sens Actuators B 140, 656–662, 2009

[12] Watanabe M, "Surface-directed channels filled with organic solvents", Lab Chip, 9, 1143–1146, 2009

[13] L. Hong and T. Pan, "Surface microfluidics fabricated by photopatternable superhydrophobic nanocomposite", Microfluid. Nanofluid, 10, 991–997, 2011

[14] H. Jansen, M. Boer, R. Legtenberg and M. Elwenspoek, "The black silicon method: a universal method for determining the parameter setting of a fluorine-based reactive ion etcher in deep silicon trench etching with profile control", J. Micromech. Microeng. 5. 115-120, 1995

[15] C. H. Choi and C. J. Kim, "Large slip of aqueous liquid flow over a nanoengineered superhydrophobic surface", PRL 96, 066001, 2006

[16] J.Y. Shiu, C.W. Kuo, P. Chen, and C.Y. Mou," Fabrication of tunable superhydrophobic surfaces by nanosphere lithography", Chem. Mater. 16, 561, 2004

[17] C.C. Yu, Y.D. Yao, S. C. Chou, and Y. Liu," Magnetic nanostructures fabricated through nanosphere lithography", Trans. Magn. Soc. Japan, 5, 9-12, 2005

[18] E. Martines, K. Seunarine, H. Morgan, N. Gadegaard, C. D. W. Wilkinson, and M. O. Riehle, "Superhydrophobicity and superhydrophilicity of regular nanopatterns ", Nano Lett. 5, 2097, 2005

[19] K. K. S. Lau, J. Bico, K. B. K. Teo, M. Chhowalla, G. A. J. Amaratunga, W. I. Milne, G. H. McKinley, and K. K.Gleason, "Superhydrophobic carbon nanotube forests", Nano Lett. 3, 1701 2003

Adaptable Chip-Level Microfluidic Packaging for a Micro-Scale Gas Chromatograph

Nathan Ward[1], Xiaoyi Mu[1], Gustavo Serrano[2], Elizabeth Covington[3], Cagliyan Kurdak[3], Edward T. Zellers[2], Andrew J. Mason[1], Wen Li[1]

[1]Department of Electrical and Computer Engineering, Michigan State University, East Lansing, MI 48824, USA
[2]Department of Environmental Health Sciences, University of Michigan, Ann Arbor, MI 48109, USA
[3]Department of Physics, University of Michigan, Ann Arbor, MI 48109, USA
E-mail: wenli@egr.msu.edu

Abstract— **In this paper, we present a robust and adaptable technique to integrate microfluidics with an on-chip thiolate-monolayer-protected gold nanoparticle coated chemiresistor-array for vapor analyte detection in a micro-scale gas chromatograph (µGC). The process involves mounting a sensing chip and capillary tubes within a silicon "extension carrier" (EC), capping the chemiresistor-array with a glass lid, and sealing the microfluidic package with non-sorbent epoxy. The stability and efficacy of the integrated detector cell is elucidated by consistent chip responses induced by the diffusion of vapor analytes though the detector cell.**

Keywords-microfluidics; chip-level packaging; µGC; chemiresistor array

I. INTRODUCTION AND MOTIVATIONS

Integrated microfluidics and lab-on-a-chip systems have been used in a wide variety of applications such as biological/chemical detection, drug delivery, and clinical diagnosis, to just name a few [1-3]. Particularly, micro-gas-chromatography (µGC) permits fast detection, classification and quantification of gas/vapor mixtures for applications in environmental monitoring, military surveillance, and healthcare diagnostics [4, 5]. Integration of pre-fabricated electronics and sensors with microfluidic package can significantly enhance system miniaturization, reduce sample volume, and minimize environmental interferences for low-cost, efficient, stable detection of vapor analytes in µGC. However, current existing packaging solutions for microfluidic systems mainly rely on the molding of inexpensive polymers such as PDMS. Such a packaging methodology is not suitable for some µGC packaging due to the sorption of target analytes by PDMS [6], which can greatly compromise the accuracy and sensitivity of the vapor detection. Beyond advances in the creation of microfluidic features, research into generalized packaging technology for the integration of microfluidics with pre-fabricated circuitry still has a lot of ground to cover. In many cases, designing individualized packaging systems for specific devices can be a large component of a project's time and cost. Therefore, there is an increasing need for the development of robust and adaptable packaging strategies, which will ideally facilitate easier integration between microfluidic devices and other processing stages of the system at large.

In this paper, a chip-level microfluidic packaging scheme which enables alignment-free integration of microfluidic components and a foundry-fabricated CMOS sensing chip is presented. Although this packaging can be used with a variety of chip layouts, it has been developed primarily for use with a particular µGC sensing chip featuring a thiolate-monolayer-protected gold nanoparticle (MPN) coated chemiresistor-array and on-chip drift compensation circuitry [7]. During the µGC detection, vapor analytes pass through a pre-concentrator and are separated by a microcolumn. The sample then enters the detector cell through an inlet capillary tube and reaches the chemiresistor array (Fig 1). Our custom-designed detector cell is a key component that connects the fabricated microfluidic channel and chemiresistor array to the integrated system via input and output capillary tubes that have an inner diameter of 150 µm and an outer diameter of 350 µm.

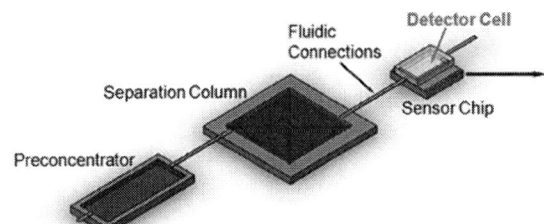

Fig. 1. Visual representation of complete system. Gas is collected in preconcentrator and then separated by diffusion rates in separation column.

In the design of the integrated microfluidic detector cell, a central motivation is to achieve an optimal environment for the miniaturized gas detection chips. Factors such as turbulence and vapor sorption must be minimized to achieve a rapid, accurate reading of the vapor analytes, however ease of fabrication and assembly are factors that must also be considered. To accomplish these goals, we have proposed a post-fabrication integration approach, as illustrated in Fig. 2. In this approach, a modular glass lid and silicon "extension carrier" (EC) are constructed separately and then assembled with the sensor-readout chip and interconnection capillaries using non-sorbent epoxy. Specifically, glass lid dimensions will be determined based on data produced from finite element simulations in COMSOL and Coventorware, to permit laminar

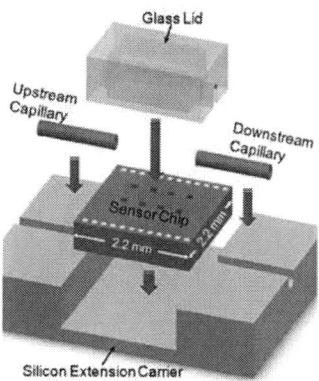

Fig. 2. Schematic view of total device assembly. Components of packaging placed sequentially.

flow for use in this integrated system. This silicon EC serves the purpose of securing the gas detection chip within it and allowing the fine alignment of chip to interface with the microfluidic channel and input/output capillary tubes.

II. DESIGN AND FABRICATION

A. Extension Carrier

In order to provide a stable platform for the gas detection chip, microfluidic channel, and input/output capillary tubes, a silicon EC has been designed and fabricated to secure these various components in place (Fig. 3). In this work, ECs are fabricated via deep reactive ion etching with a photoresist mask. Key design features of the EC which allow these various components to be mounted include an anisotropically etched inner cavity and trenches. Each of these features increase the tolerance for error in the packaging assembly step, particularly error caused by the process of hand-assembly and variance in chip topology. One such feature is the extended margins of the chip cavity. The dimensions of the chip used in this system are 2.2 mm x 2.2 mm, however, the cavity of the EC has an extended width of 900 μm.

This feature allows the chip to be shifted during the assembly step, increasing the compatibility of the packaging for variations in chip layout, particularly for change in location of the sensing array on the chip. Additionally, this feature allows the EC to mount a variety of other chip dimensions which have widths greater than 2.2 mm. Furthermore, the EC can be diced horizontally along the

center to host chips with lengths larger than 2.2 mm, by simply placing the halved segments of the EC at each end of the sensor chip. Operations like this do not introduce leaks to the packaging and provide the same stability as un-diced EC's.

Another feature of the EC which aids in the error-prone process of hand assembly is the inclusion of a "stop" in the trenches which hold the capillaries. These capillary trenches are designed to serve as a location for the capillary tubes to be mounted within the ends of the microfluidic channel, adjacent to the mounted gas sensing chip, however without this additional stop, the capillary tubes would be prone to receding and advancing towards the chip. This movement could cause the capillaries to advance onto the chip itself perhaps damaging the sensing array or fabricated electronics. Additionally, the capillary tubes could be placed too far away from the chip during assembly, introducing turbulence or posing a hazard for leaks. By including a tapered stop, the capillary tubes will not advance too close to the chip and will consistently remain at the same point in the trench for each packaged device. Additionally, although the stop will prevent the capillary tubes from advancing towards the chip, there is still sufficient clearance for gasses to pass through the device, unimpeded.

Glass lid fabrication involves wet etching of a soda-lime substrate in buffered HF with an amorphous silicon mask to create vapor chambers with a controllable surface roughness of less than 30 nm, as shown in Fig 4. The primary purpose of the lid is to provide a microfluidic channel over the chip connected to the input and output capillaries. The lid consists of a wet etched channel for gases to pass through with a wider microfluidic chamber designed to distribute gases over the whole of the chemiresistor array. The channel is meant to match the outer diameter of the input and output capillary tubes (350 μm) and allow for easy sealing via non-sorbent epoxy. Conceptually, the sealing of the interface of the lid presents moderate challenge. Not only must the lid seal over the capillaries it must also contact the lid and the extension carrier and provide a seal over these elements as well. Due to sheer number of components which must be sealed by the lid, leaks do present a potential hazard. However, through certain assembly procedures discussed later in this paper, like the mindful placement of non-sorbent epoxy, risk of creating leaks at the boundary of components is reduced. Beyond the issue of device sealing, other considerations must be taken into account in the design of the lid.

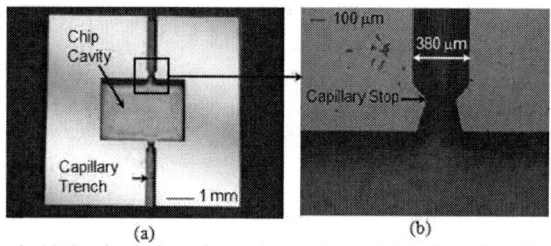

(a)   (b)

Fig. 3. (a) Overhead view of extension carrier, central cavity and capillary trenches are shown. (b) Capillary stop shown in detail, stop tapered to 100 μm to prevent impedance of gas flow.

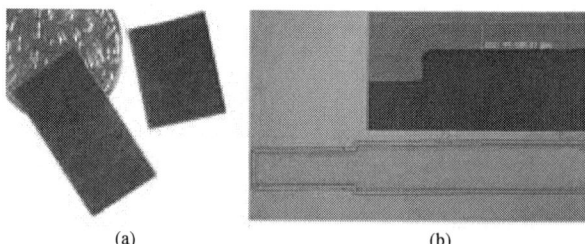

(a)   (b)

Fig. 4. (a) Array of lids with different dimensions shown with etchant protection layer still intact. (b) Closer view of lid shows undercut due to isotropic etching process.

One constraint in the design of the lid, mentioned previously, is chemical inertness. To avoid the potential problem of chemical sorption caused by polymers such as PDMS, soda lime glass was chosen as the material substrate for the lid. Glass is not perfectly immune to negative interactions with the target analytes [8], but it significantly reduces the diffusion of vapor through the material itself and chemical sorption. Furthermore, glass allows for easy alignment and diagnosis of poor component placement because of its transparency.

Beyond merely providing a robust microfluidic environment for the gas sensing chip, the glass lid must be adaptable to differences in chip layout to be compatible with multiple chip generations and designs. One design feature that aims to provide this adaptability is variation in the dimensions of the microfluidic channel with inner chamber lengths and widths varying from 1.2-2.2 mm and 500-600 μm respectively. These dimensions allow specifically for changes in the scale of the on-chip region containing the chemiresistors while the position of this region is compensated for by the EC, as mentioned above.

III. SYSTEM ASSMEMBLY

Assembly of the total packaged device follows the diagram in Fig. 2. First, a layer of non-sorbent epoxy is applied to the central cavity of the EC. The gas sensing chip is then aligned in the central cavity such that the chemiresistor array lies in line with the EC's trenches which will eventually mount the capillary tubes. At this point, the packaging is placed and secured within a header which will connect the chip to external readout circuitry. Next, the chip is wirebonded to the header. Wirebonding is performed at this time because the lid often interferes with the machinery which performs the wirebonding. After this step is complete, capillary tubes are placed within the EC's trenches with epoxy applied around the perimeter of these tubes to prevent leaks. Once the epoxy on the capillary tubes has been cured, more of the same adhesive is applied to the bottom of the glass lid. Finally, the glass lid is placed over the current packaged device, which will seal over the chip, EC, and capillaries, forming a microfluidic channel going from the input capillary tube, over the surface of the chip, and then through the output capillary tube. Fig. 5 shows an example of the assembled detector cell, which is wirebonded on a 40-pin DIP carrier.

Fig. 5. Assembled device mounted and wirebonded on a 40-pin DIP carrier.

IV. TESTING AND RESULTS

Beyond merely testing the chip response to various target analytes, the viability of the packaging must be evaluated in other ways. One key aspect of the packaging which could potentially affect the performance of the gas sensing chip is the existence of leaks. To test for this, helium was passed through the capillaries and the packaging was probed with a helium leak detector (Fig 6). If leaks were detected at a certain point along the packaging, additional adhesive was applied to seal the source of the leak. Once the packaging was confirmed to be leak free, testing of the gas sensing chip commenced.

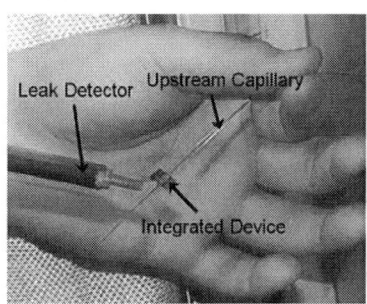

Fig. 6. Method of testing leaks. Leak detector probe placed along the packaging while helium was passed through the device.

The function of the integrated detector cell was validated using a custom-designed setup as shown in Fig. 7. In this study, the sensing responses were taken from a 4×2 chemiresistor-array fabricated on the silicon dioxide surface of the readout chip. The 75x75 μm² active area of each chemiresistor comprised 62 pairs of interdigitated gold electrodes, with electrode width and spacing of 300 nm. The interdigitated electrodes were made by electron-beam lithography and liftoff procedure of Ti/Au [9]. The electrodes were coated with a solution of Au nanoparticles with n-octanethiol ligands (C8) suspended in toluene, which was synthesized with the Rowe method [10]. Nanoparticles coating was done using a Microfab Jetlab 4 micro-dispensing system, which permits precise control of film placements.

During the experiments, the chemiresistor response was measured against various concentrations of different vapors, consisting of acetone, ethanol, n-heptane, and toluene. Samples of these vapors (0.25 mL) were injected via a gas sampling loop through the capillary column of a bench-scale GC (Agilent 7890A) at a flow rate of 1.2 mL/min. The

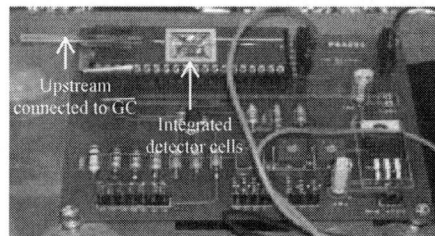

Fig. 7. Packaged sensor mounted on the test PCB for vapor sensing measurements.

978-1-4673-1122-9/12 $31.00 © 2012 IEEE

transient voltage changes of the chemiresistor (i.e., peaks) yielded a chromatogram that was logged via data capture software run on a laptop computer.

The chemiresistor response varies with the physical characteristics and concentration of the vapor in question. In addition, each of the vapors has a characteristic retention time allowing distinct characterization of a wide variety of chemical compounds. Fig. 8 gives the responses from a representative chemiresistor upon exposure to vapor samples over 5 replicates. It can be seen that the values of peak area and full width at half maximum (FWHM) were highly reproducible. Acetone, for instance, which gave the largest peak, maintained an average peak area of 0.89 +/- 0.08 V and an average FWHM of 1.18 +/- 0.07 sec (Table 1).

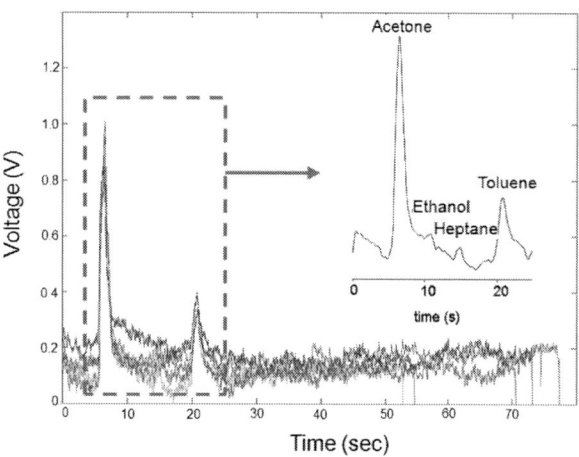

Fig. 8. Chromatograms recorded from a single chemiresistor sensor over 5 replicates. Measurements were taken with a flow rate of 1.2 mL/min, a temperature of 30 °C, and vapor concentrations ranging from 200 to 500 ppm.

TABLE I.   SENSOR RESPONSES TO ACETONE (5 REPLICATES)

Trial number	1	2	3	4	5	Avg	SD	RSD %
Retention time (s)	6.47	6.69	6.57	6.59	6.58	6.58	0.08	1.2
FWHM (s)	1.12	1.16	1.30	1.14	1.20	1.18	0.07	6.0
Peak area	0.78	0.83	0.93	0.95	0.94	0.89	0.08	8.6
Peak height (V)	0.61	0.64	0.83	0.86	0.82	0.75	0.12	15

V. CONCLUSION

This package has two central components: the glass lid and extension carrier, which were integrated manually. This package has several attributes which contributed to the ease of the assembly step, including a widened inner cavity, a stop to position the capillary tubes, and variation in lid dimensions. Upon the sealing of these components, leak detection tests and chip response verified the quality of the packaging. Measurements verified the stability and suitability of this platform for GC detection applications. This inexpensive and robust packaging strategy could be applied to a plethora of different gas detection platforms. Although hand assembly

was used to prove the concept, automated schemes can be envisioned. Future improvement is expected by use of parylene-based sealing of microfluidics to reduce the need for epoxy.

VI. ACKNOWLEDGEMENT

This work was supported by the Science and Technology Directorate of the U.S. Department of Homeland Security (06-G-024). The authors would like to acknowledge Steven Sostrom and Brenden Casey for their help with wirebonding these devices, and the staff at the Lurie Nanofabrication facility at the University of Michigan. Additionally, the staff at the Michigan State University College of Electrical and Computer Engineering, including Roxanne Peacock, Brian Wright, and Gregg Mulder for their technical assistance.

REFERENCES

[1] D.J. Beebe, G.A. Mensing, and G.M. Walker, "Physics and applications of microfluidics in biology," *Annual Review of Biomedical Engineering,* vol. 4, pp. 261-286, 2002.

[2] C.H. Ahn, C. Jin-Woo, G. Beaucage, J.H. Nevin, L. Jeong-Bong, A. Puntambekar, and J.Y. Lee, "Disposable smart lab on a chip for point-of-care clinical diagnostics," *Proceedings of the IEEE,* vol. 92, pp. 154-173, 2004.

[3] R. Lo, P.Y. Li, S. Saati, R.N. Agrawal, M.S. Humayun, and E. Meng, "A passive MEMS drug delivery pump for treatment of ocular diseases," *Biomedical Microdevices,* vol. 11, pp. 959-970, 2009.

[4] P.R. Lewis, R.P. Manginell, D.R. Adkins, R.J. Kottenstette, D.R. Wheeler, S.S. Sokolowski, D.E. Trudell, J.E. Bymes, M. Okandan, J.M. Bauer, R.G. Manley, and G.C. Frye-Mason, "Recent advancements in the gas-phase µChemLab," *IEEE Sensors J.* vol. 6, pp. 784-95, 2006.

[5] S.K. Kim, H. Chang, and E.T. Zellers, "Microfabricated gas chromatograph for the selective determination of trichloroethylene vapor at sub-parts-per-billion concentrations in complex mixtures," *Analytical Chemistry,* vol. 83, pp. 7198-7206, 2011.

[6] M.W. Toepke and D.J. Beebe, "PDMS absorption of small molecules and consequences in microfluidic applications", *Lab Chip,* vol. 6, pp. 1484-1486, 2006.

[7] X. Mu, D. Rairigh, and A.J. Mason, "125ppm resolution and 120dB dynamic range nanoparticle chemiresistor array readout circuit," *IEEE International Symposium on Circuits and Systems (ISCAS),* pp. 2213-2216, 15-18 May 2011.

[8] R. Mukhopadhyay, "When microfluidic devices go bad," *Anal Chem.,* vol. 77, pp. 429A–432A, 2005.

[9] F.I. Bohrer, E. Covington, Ç. Kurdak, and E.T. Zellers, "Characterization of dense arrays of chemiresistor vapor sensors with submicrometer features and patterned nanoparticle interface layers," *Anal. Chem.,* vol. 83, pp. 3687-3695, 2011.

[10] M.P. Rowe, K.E. Plass, K. Kim, Ç. Kurdak, E.T. Zellers, and A.J. Matzger, "Single-phase synthesis of functinalized gold nanoparticles," *Chem. Matter,* vol. 24, pp. 3513-3517, 2004.

Detecting Vapor Traces of Explosives Using a Self-Assembled Mono Layer on a Surface-Modified MEMS Capacitor and CMOS Electronics

Drago Strle[*], Bogdan Štefane, Igor Muševič

[*]Department of Electrical Engineering, University of Ljubljana, Slovenia
drago.strle@fe.uni-lj.si

Abstract—A miniature detection system, which is able to detect and selectively recognize different vapour traces of explosives, is presented in this article. It is based on surface-functionalised MEMS capacitive sensors and an extremely low noise integrated circuit. The instrument is sensitive and selective, consumes a minimum amount of energy, is very small and cheap to produce in large quantities, and is insensitive to mechanical influences. It is possible to detect 3ppt of TNT in the atmosphere (3 TNT molecules in 10^{+12} molecules of air) at 25 °C in 1 Hz bandwidth using very small volume (few mm³) and only approx 20 mA current from a 5 V supply voltage.

Keywords: detection of vapor traces, surface modification, capacitance measurements

I. INTRODUCTION

Detecting vapor traces of explosives in the atmosphere is a potentially powerful method to reveal the presence of explosives and land mines. The principle of detection method is based on the fact that any explosive device will constantly emit rather small, but detectable number of different molecules, constituting the explosive. Numerous detection systems on the market are capable of detecting explosive devices [1]. However, their common limitations are: rather large size; high power consumption; unreliable detection with false alarms; insufficient sensitivity and chemical selectivity, or hypersensitivity to mechanical influences; and very high price.

The chemo-mechanical sensor with optical or electrical detection is currently the most promising and the most popular method for the detection of vapor traces of explosive. The method is based on measuring the deflection of the sensing micro-cantilever, which is caused by adsorption of target molecules on the chemically modified surface of the cantilever [1]. This deflection is measured very precisely using either quadrant photodiode detection, or measurement of change of the capacitance or resistance of the cantilever. Optical cantilever position measurement is difficult because the apparatus is bulky. The measurements of the cantilever bending are very sensitive to environmental influences. Therefore, we consider such measurements not practical. In this work, we measured the change of the capacitance of a miniature functionalized MEMS capacitor with fixed COMB fingers in order to avoid the problems described above. Because of the adsorption of explosive molecules on the functionalized surfaces of MEMS capacitors, the dielectric constant of the capacitor is changed, which can be measured using an extremely sensitive electronic measurement system.

In section II, the principles of operation are described, followed by section III where a short explanation of COMB sensor design and implementation are given. In section IV, the principles of low noise signal processing are explained together with an estimation of the noise properties. The simulation results of the measurement channel are also presented. In section V, the response of the sensor to the mixture of nitrogen and TNT vapors are given. Sensitivity is calculated based on measured results and compared to the estimates.

II. PRINCIPLES OF OPERATION

The central part of the detection measurement system is a surface-modified, differential capacitive sensor with electrical equivalent circuit shown in Fig. 1. It is composed of two COMB capacitors, each covered with a thin layer of silicon dioxide: C_p is chemically modified, while C_n is not. They are very close to each other; on average, equal numbers of target molecules are present in the space between the plates of each capacitor. There are also other molecules in the air around the sensors, but only target molecules adsorb preferentially for a short time to the surface of the modified capacitor C_p [2]. The relative dielectric constant of the left capacitor changes until the surface-adsorbed molecules desorbs. Using appropriate sensing signals and low noise electronics, one can measure the difference of the two capacitors. If the measurements are performed with extremely low noise, a detection of small number of adsorbed target molecules in the air is possible.

III. SENSOR DESIGN AND FABRICATION

Fig. 2 shows an SEM micrograph of the implemented COMB sensor using a modified 1um MEMS process. Poly silicon COMB fingers are 1um apart and 2.5um high; on the bottom, they are attached to the thick layer of silicon dioxide and covered with approximately 10nm of silicon dioxide.

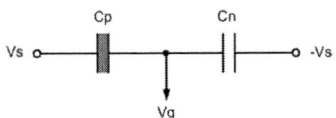

Fig. 1. Detection principle circuit diagram; Cp is functionalized.

One capacitor consists of 51 fingers with a length of approx. 350um. They are interconnected using Al metal lines forming the capacitor with approximately 0.5pF capacitance. At the beginning, both capacitors are equal; the modification layer on one capacitor is later removed using selective laser erosion. After processing, the capacitors do not have exactly the same capacitance. A 3σ statistical variation of matching accuracy can reach up to 5%; therefore, the initial difference might be as big as 25fF. During production only those differential sensors are selected that have initial difference smaller than 5fF so, that the measurement channel is not saturated at high gain.

The surface used for molecular sensing is a SiO_2 surface covered with a monolayer of trialcoxyalkylamino and trialcoxyarylamino silanes (Fig. 3). The modification of the surface was carried out by dipping the sensors into a diluted solution of the corresponding silane in organic solvent at room temperature. It was found that for successful modification different procedures had to be applied regarding the utilized silane. Solvents with different polarity were used to assure its solubility [3]. Finally, modified surfaces were investigated by XPS-spectroscopy. For the modification using APS organosilane, 10 mL of 3mM dimethylformamide solution of the corresponding silane in dry flat-bottomed glassware under an argon atmosphere was prepared. The substrates with the oxygen plasmapre-treated SiO_2 surface were immersed in the corresponding solution. The self-assembly process was carried out for a period of 5 hrs. at room temperature. Upon completion, the modified MEMS sensors were removed from the solution and rinsed several times with dimethylformamide and methanol to remove any organic residue. Finally, the sensors were thoroughly dried under the argon stream. Modification of the sensors with APhS and UPS organosilanes was nearly identical; with a difference, that dry toluene was used as a solvent. Additionally, XPS measurements confirmed the presence of an amino group on the SiO_2 sensor surface, thus, conforming successful modification with organosilanes [4].

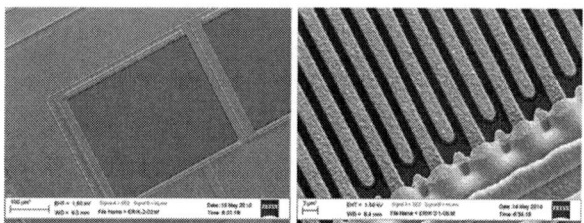

Fig. 2. SEM micrograph of a COMB sensor

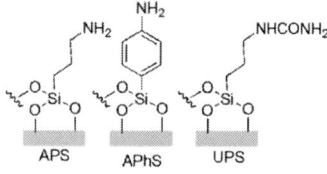

Fig. 3. Chemical modification of SiO_2 surface

IV. ELECTRONIC SYSTEM DESIGN

Fig. 4 shows a simplified block diagram of the signal processing electronics implemented in a 0.35um CMOS process. The whole system architecture is a kind of lock-in amplifier and consists of low noise analog front-end electronics and the DSP. Analog front-end electronic is implemented on the ASIC, while the DSP is implemented on the FPGA. One channel can process signals from one differential sensor; one of the COMB capacitors is modified using an appropriate chemical modification layer. Two measurement channels are integrated on one chip; therefore, two differently modified differential sensors may be connected in one chip. If sensors are modified differently, they show different sensitivity to the target molecules in the air and different response to the other molecules. In this way using differently modified sensors, the selectivity could be further improved. For now, only four differently modified sensors are produced, each of them connected to a different measurement channel. The difference in charges from modified capacitors C_p and corresponding unmodified capacitor C_n are translated to the voltage using low noise charge amplifier and two sensing voltages V_s with opposing phases. To reduce the influence of flicker noise of the whole measurement channel, sensing signal frequencies are selected well above the flicker noise corner of the measurement channel that is in the range of several $100\ kHz$ for modern CMOS process [5]. Driving sensor capacitances with differential square wave with adjustable frequency and amplitude results in quasi-square wave signal at the output of the charge amplifier with amplitude $V_{cho} \cong V_s \left(\Delta C / C_f \right)$, which is proportional to the difference of sensor capacitances $\Delta C = C_p - C_n$, amplitude of sensing signal V_s and inversely proportional to the feedback capacitor C_f of the charge amplifier. The signal V_{cho} is amplified and the resultant signal is mixed with signal with frequency f_m. The spectrum after passive mixer is a reach composition of sums and differences of all fundamental spectral components and their odd harmonics. The important part is the difference of the first harmonics frequencies: $f_o = f_s - f_m$. The information of the capacitance difference is hidden in the amplitude of corresponding component, which is proportional to sensing signal amplitude and capacitance difference. The low frequency components are amplified while HF components are attenuated.

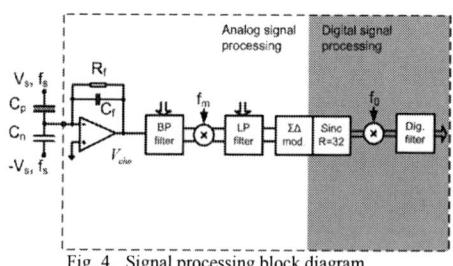

Fig. 4. Signal processing block diagram

The signal is quantized using an 18 bits switched capacitor Σ-Δ modulator and a 3rd order digital decimation filter with an oversampling ratio R=32. The higher harmonics of the mixing process fall outside the band of interest in the band of noise shaped quantization noise, which is attenuated by the digital decimation filter implemented in the DSP. All further signal processing is digital and consists of digital mixers that translate the signal from f_o down-to DC; the remaining higher spectral components are attenuated by a digital low-pass filter. The DSP output rate is approx. 12 samples per second with word-length of 32 bits. The results are transferred to the PC via USB interface, where further averaging is performed and the signals are presented on the screen.

Sensing signals are generated on chip and connected to the sensor capacitors. They provide a possibility to transfer information from slowly varying sensor capacitances to the trans-impedance charge amplifier at high frequency that is well above corner frequency of a flicker noise. The whole measurement system looks like a modified lock-in amplifier [6], using double mixing architecture to sense at high frequency and perform amplification above 1\f noise corner of the measurement channel. The frequencies of all signals are derived from one master clock with $f_{clk} = 25\ MHz$ using programmable counters, where all frequencies are adjusted in such a way that they are coherent. The selection of high frequencies for f_s, makes it possible to sense slowly changing capacitance difference at a relatively high frequency, while preserving all fluctuations of the sensors capacitances even those at DC. The "DC" signal-processing happens only in the DSP, and thus all DC drift and 1/f noise problems of the analogue electronic modules are removed. Sensing signal frequencies are selected in such a way that the ratio f_{dec1}/f_o is an even integer. In this way, the coherence of all signals in the measurement system is maintained in a very efficient way. The decisive parameter regarding detection sensitivity of a measurement system is input referred noise of complete electronics, which is mainly determined by the noise performances of the charge amplifier V_{ndop}, the parasitic capacitance of the ASIC, the capacitances $C_{parASIC} + C_{par}$ and noise properties of the gain stage following the charge amplifier. The signal-to-noise ratio at the output of the charge amplifier is defined as (1):

$$(S/N) \cong 20 \cdot \log_{10} \left(V_{Cho\_rms}/V_{ndCho} \right) \qquad (1)$$

$$V_{cho}(s) \cong V_s \cdot \left(\Delta C/C_f \right) \cdot H_{CHA}(s) \qquad (2)$$

Signal amplitude at the output of the charge amplifier is estimated with (2), where $|H_{CHA}(s)| \cong 1$. If sensing signal frequencies are above the pole frequency of the charge amplifier, than total noise power density at the output of charge amplifier is mainly determined by the charge amplifier noise V_{ndop} multiplied by factor $\left(1+\sum C_{vg}/C_f\right)$ where C_{vg} is the sum of all capacitors connected to the virtual ground of the charge amplifier and C_f is its feedback capacitor. Usually,

noise contributions of all other sources in the measurement channel are selected in such a way that they together contribute approximately the same noise power as a major noise contributor (charge amplifier). The estimate of detectable capacitance difference of the sensor is (3) for charge amplifier having $V_{ndop}(f_s) \leq 15 nV/\sqrt{Hz}$, $V_s = 4V$ and $C_f = 2pF$.

$$\delta C \cong \left(4/V_s\right)V_{ndop}\left(C_f + \sum C_{vg}\right) = 0.06\left[aF/\sqrt{Hz}\right] \qquad (3)$$

Factor 4 in (3) takes into considerations the reduction of S/N ratio caused by analogue and digital mixers; each contributes approx. $\sqrt{2}$ if designed properly [7].

Assuming that adsorbed molecules do not change dielectric constant of the modified capacitors significantly, the minimum detectable distance change between fingers of the COMB capacitors due to adsorption can be estimated with (4); the number is well below the thickness of one layer of adsorbed target molecules that is estimated to be approx 0.5nm for the TNT [7].

$$d_m \cong \left(d_0 \cdot \delta C/(C_0 + \delta C)\right) = 120\left[fm\right] \qquad (4)$$

The electronic measurement system was first modeled on a high hierarchical level using Matlab. The concept, sensitivity, selectivity, and functionality of the sensors were checked. Most important non-ideal effects of the measurement channel were taken into considerations (nonlinearities, thermal noise, 1\f noise, kT/C noise, spread of elements, offset voltages, etc.). A bit-true model of a complete DSP part was used. In this way, simulation results on a system level are very close to the simulation results on a circuit level and to the measurements of real circuit. Measurements confirm the validity of the models. Fig. 5 shows the spectrum at the output of the Σ-Δ modulator for the capacitance difference of 5.1 fF. Detection levels from the simulation results are very close to the estimates (3). The amplitude of the main component at ~3 kHz is proportional to the capacitance difference. The sensitivity is thus better than $0.05\ aF/\sqrt{Hz}$. Higher frequency remains of the first mixing process are attenuated in the digital decimation filter implemented in the FPGA.

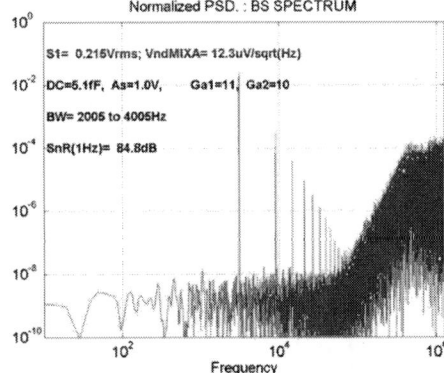

Fig. 5. Spectrum at the output of the analog front-end. Spectral line at 3 kHz is proportional to ΔC=5.1fF. Signal-to-noise ratio (SnR) is calculated for the main spectral line and 1Hz bandwidth.

V. MEASUREMENTS

For laboratory measurements we built a TNT vapor generator. The demonstrator is built as a System in Package (SiP) composed of the ASIC with analog front-end having two analog channels and 2 differently modified sensors in one package. The dimension of the SiP is approx. 18x6x2 mm. The PCB of the demonstrator with open package, which contains two differential sensors and the ASIC is shown on the right of Fig. 6. In reality, the package is covered (cover shown on the left). The mixtures of N_2 gas with vapor pressure TNT molecules is delivered via one hole through a Teflon tube and expelled through another hole. The gas is pumped to the sensor using miniature piezo-electric pumps. For laboratory experiments, the gas input is switched between dry N_2 gas and N_2 contaminated with vapor pressure target molecules. At room temperature, the vapor pressure of TNT is estimated to $5 \cdot 10^{-6}$ Torr, which means that the density of target molecules relative to the N_2 molecules is in a range of $X_{targ} = 10^{-9}$ [8].

Fig. 7 shows the response of one channel to the gas switched between N_2 and N_2 contaminated with TNT at vapor pressure in equal proportions, therefore, the estimated number of TNT molecules in the mixture is $0.5 \cdot X_{targ}$. From the difference between two readings ΔN, standard deviation of the result σ and bandwidth BW one can estimate the normalized sensitivity of the detector in 1Hz and thus the detection level using (5).

$$S_{TNT} \cong \left(0.5 \cdot X_{targ} \cdot \sigma / \left(\Delta N_{TNT} \sqrt{BW} \right) \right) = 3.5 \cdot 10^{-12} / \sqrt{Hz} \quad (5)$$

The lower level of the response on the y axis of Fig. 7 is proportional to the intrinsic difference between capacitors, while the change due to adsorbed molecules is proportional to ΔN. Long measurement times are a consequence of slow gas flow through piezzo-electric pumps and small diameter of the tubes used.

The sensitivity of our sensor is still 2 to 3 orders of magnitude worse than the sensitivity of a dog's nose. Currently, the selectivity measurements are going on. The first results show that differently modified sensor show different responses to target molecules and different responses to other molecules, which can be used to improve the selectivity.

VI. CONCLUSIONS

Detecting vapor traces of explosives in the atmosphere is a potentially powerful method to reveal the presence of the explosives. In this article, a miniature detection sensor system

Fig. 6. SiP of two differential sensors and the ASIC on a PCB

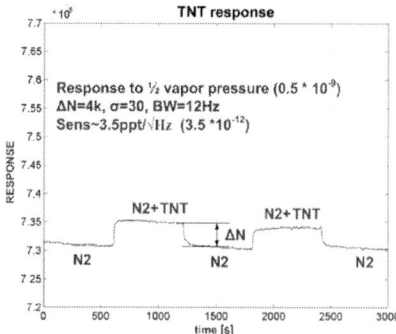

Fig. 7. Measured response to switching between N_2 and N_2+TNT. The concentration of TNT molecules is 0.5 molecule of TNT in 10^{+9} molecules of nitrogen.

has been presented. It is sensitive, selective, consumes a minimum amount of energy, is very small and cheap to produce in large quantities, and is insensitive to mechanical influences. It is based on extremely sensitive measurements of modified differential COMB capacitor difference using low noise analogue signal processing and fixed point DSP. Such an instrument could be massively deployed for detection of hazardous molecules in the air. In laboratory environment, a detection level of 3ppt of TNT, (i.e., 3 molecules of TNT in 10^{+12} molecules of N_2) has been achieved. Measurements in the real environment show very promising results.

Future work will be directed towards improving the sensitivity and the selectivity by changing the sensor geometry and size and by reducing further the analog channel noise.

ACKNOWLEDGMENT

This work was supported partly by Slovenian centre of excellence NAMASTE, and partly by Ministry of Defence of the Republic of Slovenia.

REFERENCES

[1] Anja Boisen, Søren Dohn, Stephan Sylvest Keller, Silvan Schmid and Maria Tenje, "Cantilever-like Micromechanical Sensors," Rep. Prog. Phys., vol. 74, Feb. 2011.

[2] Z. Peiji, W. Dwight, "Electrostatic Characteristics of Tether Atoms in Connecting Organic Molecules to the Surface of Silicon," Applied Physical Letters, vol. 91, nu. 6, Aug. 2007.

[3] J.C. Love, L.A. Estroff, J.K. Kriebel, R.G. nuzzo, G.M. Whitesides, "Self-Assembled Monolayers of Thiolates on Metals as a Form of Nano-Technology," Chem. Rev., 2005, 105, 1103-1169.

[4] G. Ashkenasy, D. Cahen, R. Cohen, A. Shanzer, A. Vilan, "Molecular Engineering of Semiconductor Surfaces and Devices," Acc. Chem. Res. Vol. 35, No. 2, Feb 2002.

[5] D. Šiprak, M. Tiebout, N. Yanolla, P. Baumgartner, C. Fienga, "Noise Reduction in CMOS Circuits Through Switched Gate and Forward Substrate Bias, " IEEE JSSC, vol. 44, No. 7, July 2009.

[6] M. Tavakoli, R. Sarpeshkar, "An Offset Cancellation Low Noise Lock in Architecture for Capacitive Sensing," IEEE JSSC, vol. 38, No. 2, Feb. 2003.

[7] S. Chehrazi, A. Mirzaei, A.A. Abidi, "Noise in Current Commutating Passive FET Mixers," IEEE Trans. On CAS. I. Vol. 57, No. 2, Feb. 2010.

[8] J.Chen, P. Xu and X.Li, "Self-assembling Siloxane Bilayer Directly on SiO2 Surface of Micro-Cantilevers for Long-Term Highly Repeatable Sensing to Trace Explosives," IOP Science Nanotechnology, vol. 21, Jun. 2010.

A High Heating Efficiency Two-Beam Microhotplate for Catalytic Gas Sensors

Lei Xu[1,2], Tie Li[1*], Xiuli Gao[1], Yuelin Wang[1]

[1*]State Key Laboratory of Transducer Technology, Science and Technology on Micro-system Laboratory, Shanghai Institute of Microsystem and Information Technology, Chinese Academy of Sciences, Shanghai, China

tli@mail.sim.ac.cn

[2]Graduate School of Chinese Academy of Sciences, Beijing, China

Abstract—**This paper presents a suspended-membrane-type microhotplate with high heating efficiency for catalytic gas sensors. This microhotplate has only two supporting beams through which heat loss via conduction can be effectively reduced. By isolating heat in a rectangular active membrane, high heating efficiency can be achieved. Power per active area (PPAA) of the microhotplate is only about 30% in comparison with current microhotplates. Based on the two-beam microhotplate, a catalytic gas sensor was fabricated by sol-gel process introducing Pd-Pt as the catalytic material. Test results indicate that this catalytic gas is sensitive to methane with a high sensitivity. And the sensor output has a fairly linear relation to methane concentration.**

Keywords-microhotplates; heating efficiency; power per active area; catalytic gas sensors

I. INTRODUCTION

Due to the low operation power, low manufacturing cost, fast response, and high compatibility with microelectronic processing, microhotplates have been widely researched in recent years in a variety of fields including environmental monitoring, military applications, and industrial processes. There are two typical types of microhotplates. One is the closed-membrane-type microhotplate which using a whole closed dielectric membrane to support the heating resistor [1-3]. Most of these microhotplates are fabricated by anisotropic silicon etching from back-side of the substrate, which is time-consuming and with low utilization of wafer. Another type of microhotplate is suspended-membrane-type microhotplate whose active area is supported by several dielectric beams [4-6]. This kind of microhotplate has lower power, smaller size, and faster response than the closed-membrane-type microhotplate.

Nowadays, microhotplates have been widely used in gas sensors, particularly in semiconductor gas sensors and catalytic gas sensors. Instead of using traditional platinum coil with high power consumption, the microhotplate-based catalytic gas sensors usually consist of a catalytic surface coated on a hotplate with Pt resistor that heats the catalyst to a sufficiently high temperature, at which any flammable gas molecules present can burn and release combustion heat [7]. Gas concentrations can be detected by monitoring resistance change of the Pt resistor resulting from temperature increases produced by combustion heat.

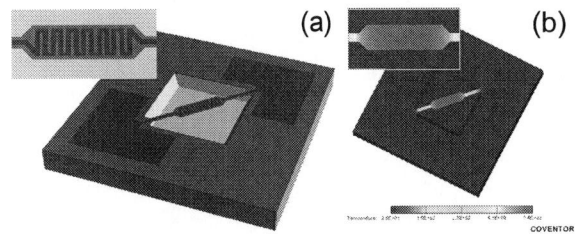

Fig. 1. (a) 3D model of the microhotplate, (b) Temperature distribution obtained by electro-thermal analysis.

Since the suspended-membrane-type microhotplates have better thermal isolation than the closed-membrane-type microhotplates, catalytic gas sensors based on the suspended-membrane-type microhotplates usually have higher sensitivity, because of the higher utilization of combustion heat.

In order to improve the heating efficiency of suspended-membrane-type microhotplates and the performance of catalytic gas sensors, we report a two-beam microhotplate which can reduce thermal loss via conduction and in turn improve the sensitivity of catalytic gas sensors.

II. MICROHOTPLATE

A. Design

When electric current flow through the heating resistor, electric energy transforms into thermal energy in the active area. Heat can be lost via three ways: through the supporting beams by conduction, through air by convection, and through radiation. Among the three ways, most of the heat is lost through the beams. To achieve high heating efficiency, heat should be well isolated. Most of current reported suspended-membrane-type microhotplates have four supporting beams [4-6] which still cause much heat loss by conduction. According to the traditional catalytic gas sensors using Pt coil as the heater, two supporting beams are enough for the microhotplates when applied in catalytic gas sensors. So the structure of present suspended-membrane-type microhotplate could be more simple if applied in catalytic gas sensors.

978-1-4673-1122-9/12 $31.00 © 2012 IEEE

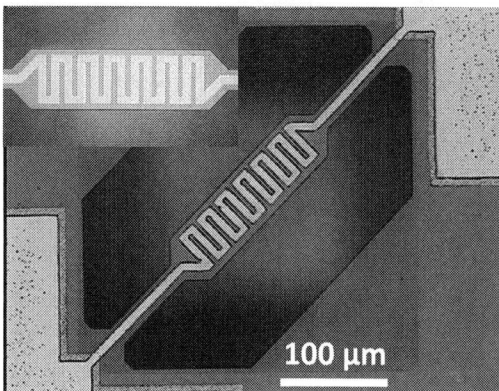

Fig. 2. Optical photo of the microhotplate.

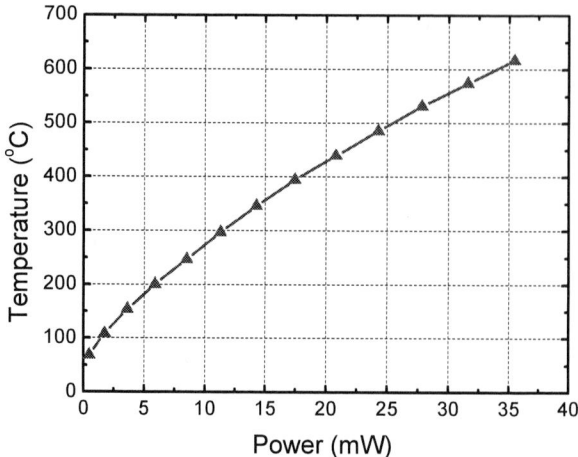

Fig. 3. Temperature versus power of the microhotplate.

The structure of our two-beam microhotplate is shown in Fig. 1 (a). A 70 μm × 210 μm rectangular microhotplate with Pt heating resistor on it is supported by two 25 μm × 150 μm slender beams only. For thermal isolation, a silicon pit under the suspension structure was formed by wet-chemical anisotropic etching. Electro-thermal analysis results obtained by FEM software Coventor shown in Fig. 1 (b) indicate that heat can be well isolated in the rectangular membrane with a good thermal homogeneity.

B. Fabrication

The microhotplate was fabricated by MEMS technology. The fabrication process is as follows.

(a) It started with a double side polished N-type <100> oriented silicon wafer with a layer of SiO_2 (200nm) thermally grown at 1100°C.

(b) Then a layer of low stress SiN_x (600nm) was deposited on each side of the silicon substrate by low pressure chemical vapor deposition (LPCVD) at 800°C.

(c) The Pt/Ti electrodes and bonding pads were patterned by lift-off process.

(d) Then the SiO_2/SiN_x membrane was selectively etched by RIE etching to form front etching windows.

(e) At last the membrane was released in a solution of TMAH (25 wt. %) at 80°C.

Fig. 2 shows the optical photo of the microhotplate.

C. Characteristics

Electro-thermal characteristics of the micorhotplate has been tested by applying constant voltages. Temperature versus power of the microhotplate is shown in Fig. 3. We can see that the microhotplate can be heated up to 400°C with a low power of 18mW. Temperature of the microhotplate was calculated by equation (1) which is widely used of estimating the temperature of microhotplate in gas sensing application.

$$T=(R-R_o)/(\alpha R_o)+25 \qquad (1)$$

where α is the temperature coefficient of resistance of Pt, R is the measured resistance, R_o is the original resistance at room temperature, and T is the temperature in °C.

Table I shows the comparison between the two-beam microhotplate and another three microhotplates with four supporting beams. Active area of the two-beam microhotplate

TABLE I.   COMPARISON BETWEEN THE TWO-BEAM MICROHOTPLATE AND ANOTHER THREE MICROHOTPLATES WITH DIFFERENT STRUCTURES.

Microhotplates	Rectangular active area with two beams	Square active area with four beams [8]	Circular active area with four beams [9]	Square active area with four beams [10]
Structure				
Active area (μm²)	17070	2500	4780	10000
Power @ 400°C (mW)	18	9	14	12
PPAA (mW/μm²)	1.05×10^{-3}	3.60×10^{-3}	2.93×10^{-3}	1.20×10^{-3}

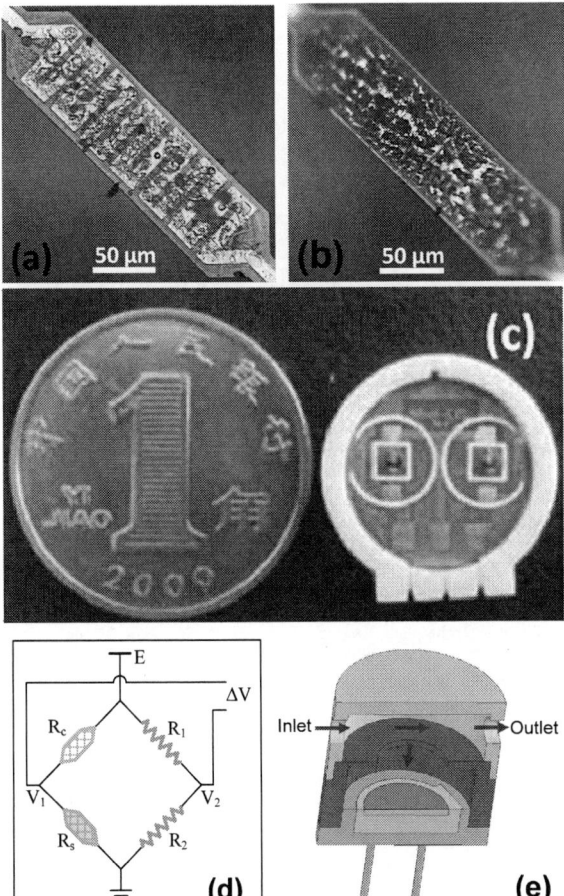

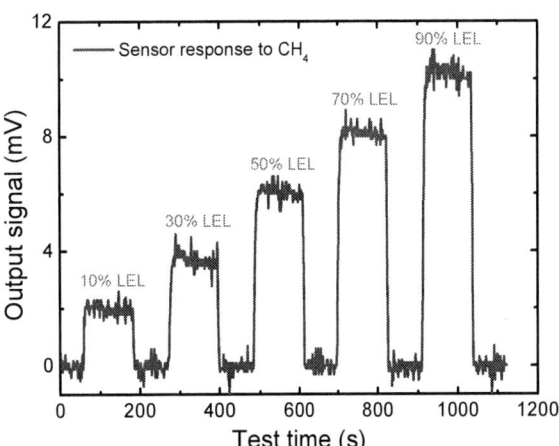

Fig. 5. Sensor response to methane.

Fig. 4. (a) Reference element, (b) Sensitive element, (c) PCB with two elements, (d) Bridge circuits for test: Rc, resistance of the reference element, Rs, resistance of the sensitive element, R_1,R_2, fixed resistors, $\Delta V=V_1-V_2$, (e) Testing chamber.

is larger than other three microhotplates, so it needs more power to achieve the same temperature. Power per active area of our microhotplate is about 30% of that in those microhotplates [8-9], which demonstrates that the two-beam microhotplate features higher heating efficiency.

Supported by four beams with very high length-to-width ratio, this four-beam microhotplate [10] also achieved high heating efficiency. But the structure might be too weak for the application in catalytic gas sensors. While our two-beam microhotplate with thermal strength well released has achieved high mechanical strength [11].

III. CATALYTIC GAS SENSOR

A. Sensor Fabrication and Package

By introducing γ-Al$_2$O$_3$ supported catalytic metal (15wt.% Pd-Pt) as sensing material and γ-Al$_2$O$_3$ as compensation material, palladium and platinum chloric acid mixed alumina aerogel and alumina aerogel were spin coated separately on microhotplates. After electrical heating at 650°C for 5min in air, the alumina aerogel was sintered into mesoporous γ-Al$_2$O$_3$ support, while high activity palladium and platinum particles were also formed. Fig. 4 (a) and Fig. 4 (b) shows the reference element and the sensitive element.

Then a sensitive element and a compensation element were electrically bonded to a circular PCB which was embedded in a plastic holder. The package of the sensor is shown in Fig. 4c.

A Whitestone bridge circuit shown in Fig. 4 (c) was used to get output signals from the catalytic combustion gas sensor. The bridge circuit was comprised of a sensitive element, a reference element, and two fixed resistors. The sensor was operated with a constant voltage mode. Output signal was defined as the voltage difference: $\Delta V=V_1-V_2$.

B. Sensor Response to Methane

Gas detection was carried out in a small chamber of a volume of 2 cm^3 with a gas inlet and outlet, shown in Fig. 4 (e). In the test, we used methane as the target gas. Because methane is a typical combustible gas which is widely used as the primary calibration gas for a catalytic gas sensor. In addition, methane gas is a very common gas and is often encountered in many applications.

Fig. 5 shows the original output signal of sensor response to methane with different concentrations. Response time of the catalytic gas sensor is less than 10 s. The sensitivity to 50% LEL CH$_4$ is 2.4 mV/ % CH$_4$ which is higher than that of our previous catalytic gas sensor [12] based on a four-beam microhotplate tested under the same operation temperature.

Linear fit of the sensor output to methane with different concentration is shown in Fig. 6. Output signal has a fairly linear relationship to methane concentration. And the sensor has a 2 mV output voltage with a good signal to noise ratio

978-1-4673-1122-9/12 $31.00 © 2012 IEEE

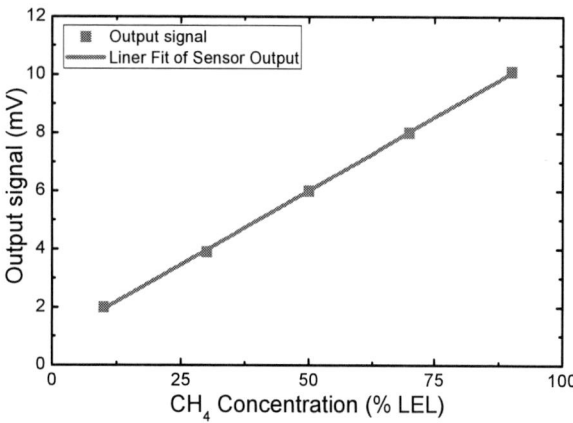

Fig. 6. Linear fit of sensor output to methane concentration.

(SNR) to 10% LEL CH4 which is the alarm level of industrial devices.

IV. CONCLUSIONS

A two-beam microhotplate with high heating efficiency was fabricated by MEMS technology. Power per active area of the microhotplate is only about 30% in comparison with current high performance microhotplates. Based on the two-beam microhotplate, a catalytic gas sensor was developed by sol-gel process introducing Pd-Pt as the catalytic material. Test results indicate that the sensors achieved high sensitivity, fast response time, and good SNR. The sensor was sensitive to 10% LEL CH4 with fairly good output voltage in relation to methane concentrations.

ACKNOWLEDGMENT

This work was supported in part by the National Basic Research Program of China under Grant No. 2011CB309501, by the National Basic Research Program of China under Grant No. 2012CB934102, by the National Hi-tech Research and Development Program of China under Grant No. SS2012AA040402, by the Fund for Creative Research of NSFC under Grant No. 61021064, by the Key project of NSFC under Grant No.60936001, 91123037, and by the National Natural Science Foundation of China under Grant No. 60876037.

REFERENCES

[1] M. Baroncini, P. Placidi, G.C. Cardinali, and A. Scorzoni, "Thermal characterization of a microheater for micromachined gas sensors," *Sens. Actuator A*, vol. 115, no. 1, pp.8–14, September 2004.

[2] P.K. Guha, S.Z. Ali, C.C.C. Lee, F. Udrea, W.I. Milne, T. Iwaki, J.A. Covington, and J.W. Gardner, "Novel design and characterisation of SOI CMOS micro-hotplates for high temperature gas sensors," *Sens. Actuator B*, vol. 127, no. 1, pp. 260–266, October 2007.

[3] C. Lu, K. Liao, F Udrea, J A Covington, and J W Gardner, "Multi-field simulations and characterization of CMOS-MEMS high-temperature smart gas sensors based on SOI technology," *J. Micromech. Microeng.*, vol. 18, pp. 075010 1-11, May 2008.

[4] G. Yan, Z. Tang, P. C.H Chan, J. K.O Sin, I M. Hsing, and Y. Wang, "An experimental study on high-temperature metallization for micro-hotplate-based integrated gas sensors," *Sens. Actuator B*, vol. 86, no. 1, pp. 1–11, August 2002.

[5] B. Guo, A. Bermak, P. C.H Chan, and G. Yan, "A monolithic integrated 4×4 tin oxide gas sensor array with on-chip multiplexing and differential readout circuits," *Solid-State Electron.*, vol. 51, no. 1, pp. 69–76, January 2007.

[6] B. Raman, D. C. Meier, J. K. Evju, and S. Semancik, "Designing and optimizing microsensor arrays for recognizing chemical hazards in complex environments," *Sens. Actuator B*, vol. 137, no. 2, pp. 617–629, April 2009.

[7] P T Moseley, "Solid state gas sensors," *Meas. Sci. Technol.* Vol. 8, pp. 223-237, 1997.

[8] K. Sadek, and W. Moussa, "Investigating the effect of deposition variation on the performance sensitivity of low-power gas sensors," *Sens. Actuator B*, vol. 107, no. 2, pp. 497-508, June 2005.

[9] I. Elmi, S. Zampolli, E. Cozzani, F. Mancarella, and G. C Cardinali, "Development of ultra-low-power consumption MOX sensors with ppb-level VOC detection capabilities for emerging applications," *Sens. Actuator B*, vol. 135, no. 1, pp. 342-351, December 2008.

[10] J. Lee, C. M. Spadaccini, E. V. Mukerjee, and W. P. King, "Differential Scanning Calorimeter Based on Suspended Membrane Single Crystal Silicon Microhotplate," *J. Microelectromech. Syst.*, vol. 17, no. 6, pp. 1513-1525, December 2008.

[11] L. Xu, T. Li, X. Gao, and Y. Wang, "Development of a Reliable Micro-Hotplate With Low Power Consumption," *IEEE Sensors J.*, vol. 11, no. 4, pp. 913-919, April 2011.

[12] L. Xu, T. Li, X. Gao, Y. Wang, R. Zheng, L. Xie, and L. Lee, "A Low Power Catalytic Combustion Gas Sensor Based on a Suspended Membrane Microhotplate," *IEEE NEMS 2011*, pp. 92-95, Kaohsiung, Taiwan, February 2011.

Vibration Mode Localization
in Coupled Beam-shaped Resonator Array

Keisuke Chatani [1], Dong F. Wang [1,*], Tsuyoshi Ikehara [2], and Ryutaro Maeda [2]

[1] Micro Engineering & Micro Systems Laboratory, Ibaraki University (College of Eng.), Hitachi, Ibaraki 316-8511, JAPAN
[2] Research Center for Ubiquitous MEMS and Micro Engineering (UMEMSME), AIST, Tsukuba, Ibaraki 305-8564, JAPAN
([*] Tel: +81-294-38-5024; Fax: +81-294-38-5047; E-mail: dfwang@mx.ibaraki.ac.jp)

Abstract— The use of vibration mode localization in arrays of mechanically-coupled, nearly identical beam-shaped resonators has been studied for ultra-sensitive mass detection and analyte identification. This study seeks to enhance the amount of amplitude change due to vibration mode localization with a beam shaped 3-resonator array. The preliminarily results have been discussed from view point of vibration characteristic by comparing experimental results with analytical ones, without a small mass perturbation.

Keywords- Vibration mode localization ; Eigenstate shift (amplitude change); Coupling overhang; Mechanically-coupled resonator array; Ultrasensitive mass detection; Analyte identification

I. INTRODUCTION

Micro resonator sensor [1-5] is used to detect external stimulus, i.e., force, mass, molecular as well as atomic adsorption et al. The amount of relative change in resonant frequency or corresponding amplitude, or any other related eigenstate changes is measured to detect external stimulus. If the amount of changes can be enhanced by certain amplification mechanism, ultimate sensing can be achieved, and small stimulus can be detected. In this study, vibration mode localization is used to enhance the amount of amplitude change for magnification.

II. VIBRATION MODE LOCALIZATION

Vibration mode localization [6-15] is one kind of vibration phenomena, usually happened in resonator array system. For example, there are two objects (named A and B) which are connected to each other by spring. If the A is vibrated by any force, the B also starts vibrate due to vibration of the A, or in the state where both are vibrating. If the vibration condition of the A is changed by any force, the vibration condition of the B is also changed. This kind of phenomenon is believed to be related to vibration mode localization.

Some expected advantages of mode localized sensing can be listed below. Firstly, times or orders of magnitude in parametric sensitivity of micromechanical mass detection compared to the conventional frequency-shift approach can be obtained. Secondly, such sensors can offer the important advantages to intrinsic common mode rejection that renders it less susceptible to false-positive readings that frequency-shift

based sensors. Thirdly, both the ultra sensitive detection and analyte identification of small perturbation can be achieved at same time with a single coupled resonator array.

Mode localization in coupled 2-resonator array was studied [6] and indicated that the relative change in eigenstates due to the added mass could be two orders of magnitude greater than the relative change in resonance frequency.

Recently, the effect of different geometrical designs of the coupling overhang [16] on the relative change (%) of amplitude shifts was theoretically studied using a coupled 15-resonator array for each vibration mode before and after a small mass perturbation of 10 pg. However, using a 5-resonator array without a small mass perturbation [16], the vibration characteristics were complicated, and the analytical results were also found to be not clearly corresponding to the experimental ones.

In this report, a mechanically-coupled 3-resonator array was thus selected and prepared for vibration localization evaluation, because 3 is the large integer next to 2. It is expected that the amount of amplitude change of coupled 3-resonator would be higher than that of frequency sift of single-resonator, and the analytical results would clearly corresponding to the experimental ones.

III. PHYSICS OF THE AMPLITUDE ENHANCEMENT

A. Vibration localization in coupled two-beam array

A schematic and a discretized model of two identical beam-shaped resonators coupled by an overhang are shown in Fig. 1(a) and 1(b), respectively. Each resonator is modeled as a undamped simple harmonic oscillator, while the effect of the overhang coupling is modeled as spring connecting the two oscillators.

Considering first the case of two initially identical resonators, the eigenvalue governing the undamped free oscillations of the system can be written as follows [6]:

$$\begin{bmatrix} 1 + Kc/K & -Kc/K \\ -Kc/K & (1+Kc/K)/(1+\delta) \end{bmatrix} u = \lambda u \quad (1)$$

where K_1 $(=K)$, M_1 $(=M)$ and k_2 $(=K)$, M_2 $(=M)$ are, respectively the bending stiffness and suspended mass of the two resonators, while δ represents the ratio of the effect mass (ΔM) being detected to the single resonator mass (M). Kc is the stiffness of the overhang coupling the two resonators.

After analyzing the two conditions of $\delta = 0$ and $\delta \neq 0$, it can be seen that the relative change in normalized eigenstate is given by

$$\frac{\left| u_i - u_i^0 \right|}{\left| u_i^0 \right|} = \left(\frac{1}{4} + \frac{1}{4Kc/K} \right) \delta , \quad i = 1,2 \qquad (2)$$

while the relative change in the eigenvalue or resonance frequency of a single resonator is given by

$$\frac{\lambda - \lambda_0}{\lambda_0} = \frac{-\delta}{2} \qquad (3)$$

Noted that the perturbed eigenstates U_i, i=1, 2 start becoming localized in the sense that in each eigenstate one resonator oscillates more than the other.

Equation (2), which defines the sensed quantity in this sensing paradigm, suggests that simply by decreasing the scaled coupling between the two resonators Kc/K, the relative changes in eigenstates can be made orders of magnitude greater than the relative change in eigenvalue of a single resonator.

(a)

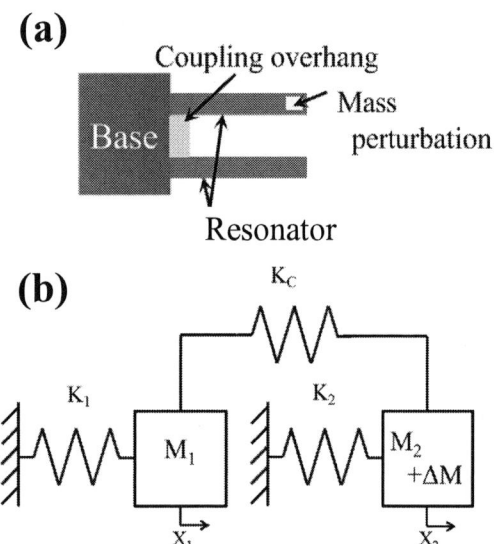

(b)

Fig. 1. (a) Schematic of the coupled 2-resonator array with a mass perturbation placed at the end of one beam-shaped resonator, and (b) simplified model of the coupled 2-resonator array.

B. Vibration localization in coupled n-beam array

A discretized model of identical resonators coupled by overhangs in a large array is shown in Fig. 2. Considering a perfect array of identical spring-mass resonators with each resonator connected to its neighbor by a coupling spring, the sensitivity to mass added of the eigenstates of the coupled array can then be estimated as follows [7]:

$$\widetilde{M}^{-1} \widetilde{K} u = \omega^2 u \qquad (4)$$

where u is a normalized eigenstate ($|u_i| = 1$) of the system representing the tip amplitudes of each resonator of the array at the corresponding eigenfrequency, ω is an eighenfrequency of the system, and M and K are the mass and stiffness matrices of the system, respectively, given by the following Equations (5) and (6), respectively.

$$\widetilde{M} = \begin{bmatrix} M_1 & 0 & \cdots & 0 \\ 0 & M_2 & \cdots & 0 \\ \vdots & \vdots & \ddots & \vdots \\ 0 & 0 & \cdots & M_n + \Delta M \end{bmatrix} \qquad (5)$$

$$\widetilde{K} = \begin{bmatrix} K_1 + K_c & -K_c & \cdots & 0 \\ -K_c & K_2 + K_c & \cdots & 0 \\ \vdots & \vdots & \ddots & \vdots \\ 0 & 0 & -K_c & K_n + K_c \end{bmatrix} \qquad (6)$$

where M_i corresponds to beam-shaped resonator, and $M_i + \Delta M$ are corresponds to the resonator with mass perturbation, and K_i corresponds to the stiffness of each resonators. Solving Equation (4) when $\Delta M = 0$ and $M_i = M$, $K_i = K$ yields n eigenstates of the initially perfectly ordered system, while the primary mode consists of all resonators vibrating in phase with identical amplitude.

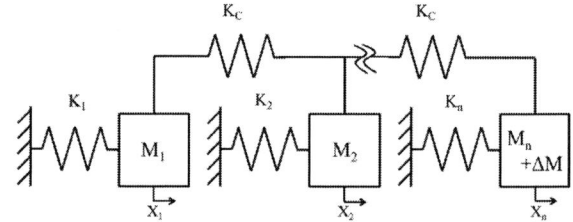

Fig. 2. Schematic of the coupled n-resonator array, where K_i (=K), and k_i (=K) are, respectively the bending stiffness and suspended mass of the identical resonators, while Kc is the stiffness of the overhang coupling the resonators.

IV. EFFECT OF COUPLING OVERHANG

Fig. 3 shows the model of coupled 3-resonator array using on analysis. The model consists of three resonators and four coupling overhangs, and each resonator is connected by two coupling overhangs. This study expects that vibration mode localization happens in this array due to coupling overhangs.

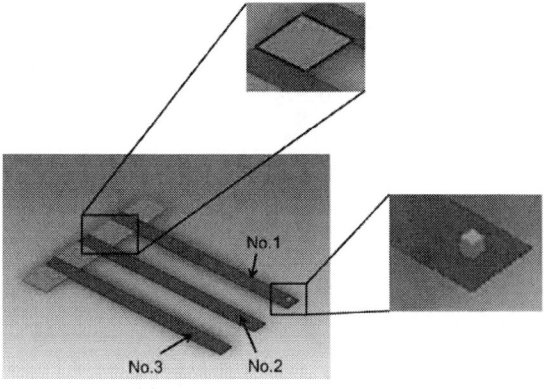

A coupled 3-resonator array

Fig. 3. The model of coupled 3-resonator array in analysis.

The analysis results of array without coupling overhangs and with them are shown in Fig. 4 and Fig. 5. If take off couplings from the model in Fig. 3, resonators are just away from neighbor one by the size of coupled overhang on monitor, but in analysis, they are far away from neighbor one. Therefore, combine the base with each array to be away from neighbor one by the size of it in analysis. From the result of without coupling overhang, there is only one resonance point. Therefore, all resonators in this array vibrate individually and vibration mode localization doesn't happen in this case. On the other hand, in the array with coupling overhang, there are three resonance points, and amplitudes of No.1 and No.3 are different from that of No.2 due to resonators interacting under vibration. It is means that vibration mode localization happens due to coupling overhang in this array.

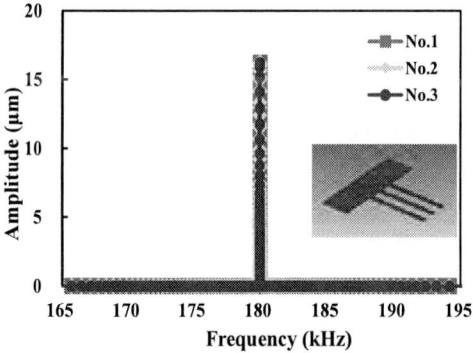

Fig. 4. The result of uncoupled 3-resonator array in analysis.

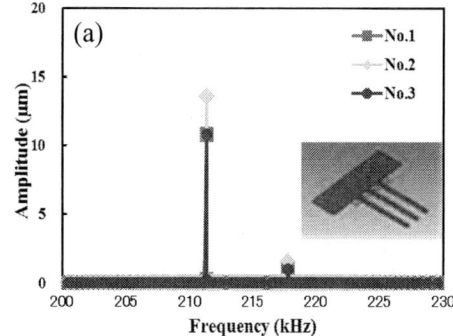

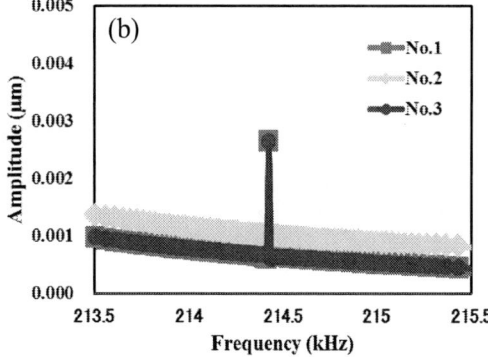

Fig. 5. (a) The result of coupled 3-resonator array in analysis and (b) magnification around 214.5 kHz point.

V. MICRO FABRICATION OF 3-RESONATOR ARRAY

For fabricating the above mechanically-coupled beam-shaped resonator arrays, an SOI (silicon on insulator) wafer with a 2.5-μm-thick top silicon layer, 300-nm-thick SiO_2 layer, and 400-μm-thick silicon substrate was used as a starting material, as shown in Fig. 6.

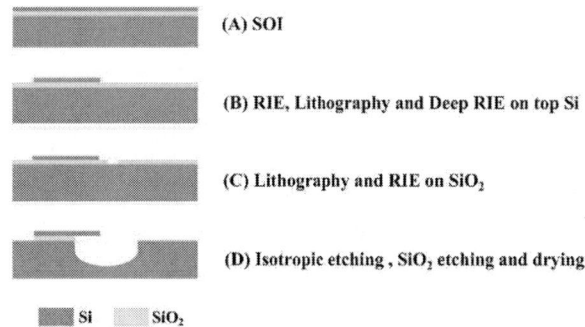

Fig. 6. Typical process chart for the coupled bema-shaped resonator array system.

The topside silicon was first thinned to 500 nm by reactive ion etching (RIE) using SF6, and then patterned by lithography and etched using a deep reactive ion etching (deep RIE) to form the mechanically-coupled resonator pattern. The substrate silicon was isotropically etched by RIE through the etching window of insulating SiO2. The coupled resonator structures were released by wet etching of SiO2 in HF and following supercritical point drying. Several oscillator systems are fabricated with micromechanically-coupled overhangs. The coupled 3-resonator array which used our current study is made by this micro fabrication. And the coupled 3-resonator array shown Fig. 7 is made from array by breaking both side resonators of the coupled 5-resonator array.

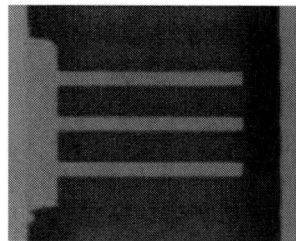

Fig.7. Typical micrograph of the micro-fabricated coupled 3-resonator arrays.

VI. EXPERIMENTAL RESULTS

Fig. 8 shows the experimental result of the coupled 3-resonator array. It is noted that 3 resonance points can be observed for all 3 resonators, which are similar to those of analytical results. However, the measured resonance frequencies are shifted from those of analysis result. It is because that resonators and couplings size changes due to fabrication error. On the other hand, the relation of amplitude of each resonator at each resonance point is almost same for both analysis and experiment. At each resonance point, the amplitude value of No.1 and No.3 is almost the same. At lowest resonance point and highest resonance point, the amplitude value of No.2 is higher than that of other resonators. And at middle resonance point, the amplitude value of No.2 is smaller than that of other resonators. And then, the amplitude value of each resonator at each resonance points is high in order of lowest resonance point, highest resonance point, and middle resonance point.

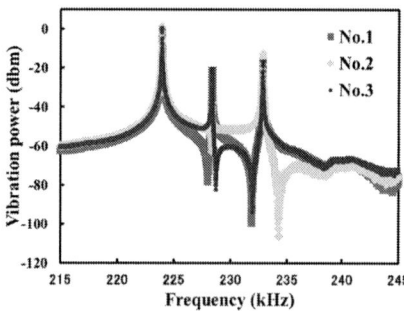

Fig.8. Experimental result of coupled 3-resonator array as shown in Fig.7.

Furthermore, vibration phase of each resonator is also same at each resonance point by comparing experiment with analysis. In light of the above comparison, the vibration characteristics of analysis and experiment are almost the same.

VII. CONCLUSIONS

It is verified from analytical result that vibration mode localization happens due to coupling overhang in the resonator array. And it is expected that the number of resonance point is the same with the number of coupled resonator. Therefore, in this report, there are 3-resonance points in the coupled 3-resonator array.

When comparing experimental results with analytical ones, the main vibration characteristics including the number of resonance points, the high order of each resonator at each resonance point, as well as the vibration phase at each resonance point, are found to be almost the same and corresponding to each other.

ACKNOWLEDGEMMENTS

Part of this work was supported by MEMS Inter University Network and performed in Research Center for Ubiquitous MEMS & Micro Engineering (UMEMSME) of National Institute of Advanced Industrial Science & Technology (AIST).

REFERENCES

[1] K. L. Ekinci, X. M. H. Huang, and M. L. Roukes, Appl. Phys. Lett. 84, 4469, 2004.
[2] T. Ono, X. Li, H. Miyashita, and M. Esashi, Rev. Sci. Instrum. 74, 1240, 2003.
[3] Z. Davis and A. Boisen, Appl. Phys. Lett. 87, 013102, 2005.
[4] H. Sone, Y. Fujinuma, and S. Hosaka, Jpn. J. Appl. Phys. Part 1 43, 3648, 2004.
[5] B. Ilic, D. Czaplewski, M. Zalalutdinov, and H. G. Craighead, J. Vac. Sci. Technol. B 19, 2825, 2001.
[6] M. Spletzer, A. Raman, A.Q. Wu, X. Xu, and R. Reifenberger, Appl. Phys. Lett. 88, 254102, 2006.
[7] M. Spletzer, A. Raman, H. Sumali and J.P. Sullivan, Appl. Phys. Lett., 92, 114102, 2008.
[8] P. W. Anderson. Phys. Rev. 109, 1492, 1958.
[9] C. Pierre, D. M. Tang, and E. H. Dowell, AJAAJ. 25, 1249, 1987.
[10] O. O. Bendiksen, AJAAJ. 25, 1492, 1987.
[11] M. Sato, B. E. Hubbard, A. J. Sievers, B. Ilic, D. A. Czaplewski, and H. G. Craighead, Phys. Rev. Lett. 90, 044102, 2003.
[12] E. Buks and M. L. Roukes, J. Microelectromech. Syst. 11, 802, 2002.
[13] M. Napoli, W. H. Zhang, K. Turner, and B. Bamieh, J.Microelectromech. Syst. 14, 295, 2005.
[14] L. Nicu and C. Bergaud. J. Micromech. Microeng. 14, 727, 2004.
[15] A. Qazi, D. Nonis, A. Pozzato, M. Tormen, M. Lazzarino, S. Carrato, and G. Scoles, Appl. Phys. Lett. 90, 173118, 2007.
[16] Keisuke Chatani, Dong F. Wang, Tsuyoshi Ikehara, and Ryutaro Maeda, Amplitude Enhancement Using Vibration Mode Localization with A Single Micro-mechanically Coupled Beam-shaped Resonator Array, Proc. of Symposium on Design, Test, Integration, & Packaging of MEMS/MOEMS (DTIP 2011), pp. 339-343.

Piezoelectrically Actuated Circular Diaphragm Resonator Mass Sensors

N. Keegan[*], J. Hedley, Z.X. Hu, J.A. Spoors, W. Waugh, B.J. Gallacher, C.J. McNeil

[*]*Institute of Cellular Medicine, Medical School, Newcastle University, UK.*
[*]*Neil.Keegan@newcastle.ac.uk*

Abstract - **This work reports a piezoelectrically driven and sensed Circular Diaphragm Resonator (CDR) mass sensor. The work is a development of an electrostatically-activated version and aims to simplify the microfabrication process and signal recovery electronics. A range of device geometries were fabricated and both optical and electrical testing performed to assess performance. Electrical sensing, using a charge amplifier, achieved a signal to noise ratio of 10:1 at 6 MHz and a preliminary sensitivity of 55 fg Hz^{-1}. The devices are nano-enabled as biosensors using a high resolution bio-molecule patterning technique and preliminary results are introduced in this regard.**

KEYWORDS – Microelectromechanical systems (MEMS); Biosensor; PZT

I. INTRODUCTION

The ability to diagnose diseases, such as cancer, at its early stages has a profound impact on successful treatment. This driver is encouraging the scientific community to realise clinical diagnostics that have increased sensitivity and the potential to be used in remote locations. The desired effect is often achieved through fluidic, sensor or instrument miniaturisation [1]. This style of technology could also aid in the rapid screening of nosocomial and community-acquired infections, which are a major cause of morbidity and mortality in the developed world. While comparatively rapid screening systems have been developed (primarily PCR-based), they are relatively expensive. Moreover, the results are not usually available, at best, for 5 to 6 hours or, more usually, until the following day in a real clinical laboratory setting. The ultimate aim of this research is to create nano-enabled microelectromechanical systems (MEMS) capable of detecting biological entities quickly and at extreme sensitivities, while remaining compatible with complementary miniaturisation technologies, for lab on a chip scenarios.

The MEMS under development is termed a circular diaphragm resonator (CDR), mass sensor. The physical principles of modal degeneracy, which underpin this sensor, have already been discussed in detail [2]. Since the early concept the CDR design has shown promise in an empirical setting, demonstrating excellent environmental stability e.g. temperature sensitivity of only 2 Hz °C^{-1} and biological detection [3] The environmental stability aids in the ultimate

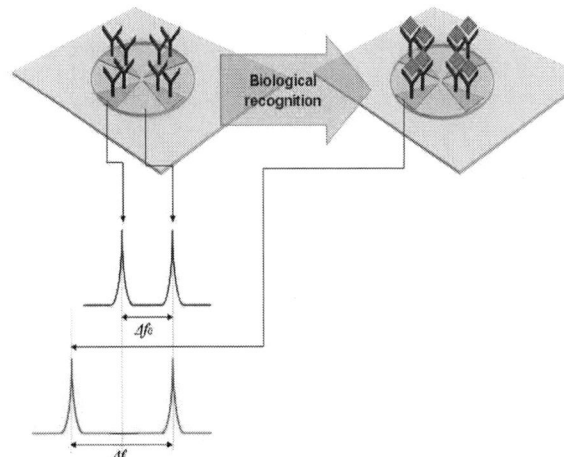

Fig. 1. *The schematic depicts the historical CDR working principle. In this illustration the biological immobilization is set-up to disrupt the cos2Θ (sense), while leaving the sin2Θ as a blank reference. In the self compensating device, the frequency split tracks the added mass independently of environmental effects. The current work uses the same general principle, but the mode of degeneracy tracked demands an alternative configuration for biological immobilization and device geometry.*

sensitivity of the CDR, which has the potential to reach 100 pg cm^{-2} Hz^{-1}, for reported designs [2]. The mass sensitivity of previously reported quartz crystal microbalance devices can be 100 fold less sensitive [4]. However, the previous designs proved to be challenging from a fabrication perspective, producing variable yields on numerous fabrication runs. In addition, measurements were mainly performed in a laboratory setting using a laser Doppler vibrometer, as the signal processing solution for resonant frequencies [3]. An electronic solution has recently been developed (manuscript in preparation), but the required proximity of the signal amplifiers to the device would drive up the cost of a commercial product. The preliminary results for the electronic signal recovery are described in [5]. The current work moves away from capacitive, electrostatic activation and describes a piezoelectrically driven device, which simplifies the recovery electronics. In addition, the design and fabrication processes have been modified to allow high resolution biological patterning of the surface for the ultimate aim of analytical detection. Fig. 1 shows a schematic of the CDR working principle.

II. DESIGN AND FABRICATION

Two major developments were incorporated in this new design. Firstly the geometry of the new piezoelectric sense electrodes onto the silicon diaphragm resonator was critical to maximize drive and signal recovery. Work by Burdess et al on active control of vibration demonstrated that the placement of a piezoelectric actuator on a structure is equivalent to introducing a bending moment in the structure along the perimeter of this actuator [6]. Therefore in order to determine optimum placement for the piezoelectric electrodes in this sensor, a line integral was performed to determine the total moment produced for a given geometry, the two design parameters being the inner and outer radii of the electrode. The calculation, results of which are shown in Fig. 2, indicate that maximum actuation occurs at inner and outer positions of 21% and 77% of the total radius respectively (A). This positions the electrode over the antinode of the mode, providing maximum sensitivity to mass addition. A secondary maximum occurs at 77% to 100% of the total radius (B). This geometry is only 59% effective compared with (A), but it does allow for the electrodes and mass addition sectors to be independently addressable. Both geometries were fabricated for testing.

A second important design criterion was the inclusion of electrically addressable immobilization sectors. Previous designs utilized thiol-gold chemistry for immobilization however this was directed towards passive (electrically unconnected) sectors of the sensor. This new design variation gave the option for a potential to be set at each immobilization site, allowing electrochemical cleaning of the immobilization sites.

Devices were fabricated at Tronics Microsystems, France and involved a four mask process flow, see Fig. 3. The first step was to create the circular diaphragm resonator. An oxide was thermally grown on the handle silicon and a lithography step defined the cavity region which would form the diaphragm structure. Next a silicon-on-insulator wafer was bonded onto this patterned oxide surface and its handle silicon etched. A thermal oxidation then produced an electrical insulation layer. With the cavity formed, the second stage was to produce the actuation and sensing electrodes. A PZT layer was deposited onto a Pt base layer and the PZT layer etched to allow electrical contact to this Pt layer. An oxide layer was deposited onto the PZT followed by lithography/etch step to remove the oxide for subsequent electrode deposition. Finally a gold deposition, lithography and etch produced the upper electrodes and bond pads. This gold layer would also be used for the immobilization. SEMs of fabricated devices are shown in Fig. 4. For polarization, the devices were placed in an oven at 120°C and a voltage, applied across the electrodes, ramped from 0V to 50V over a 1 minute period. This voltage was held for 15 minutes and then ramped back to 0V, over a 1 minute period.

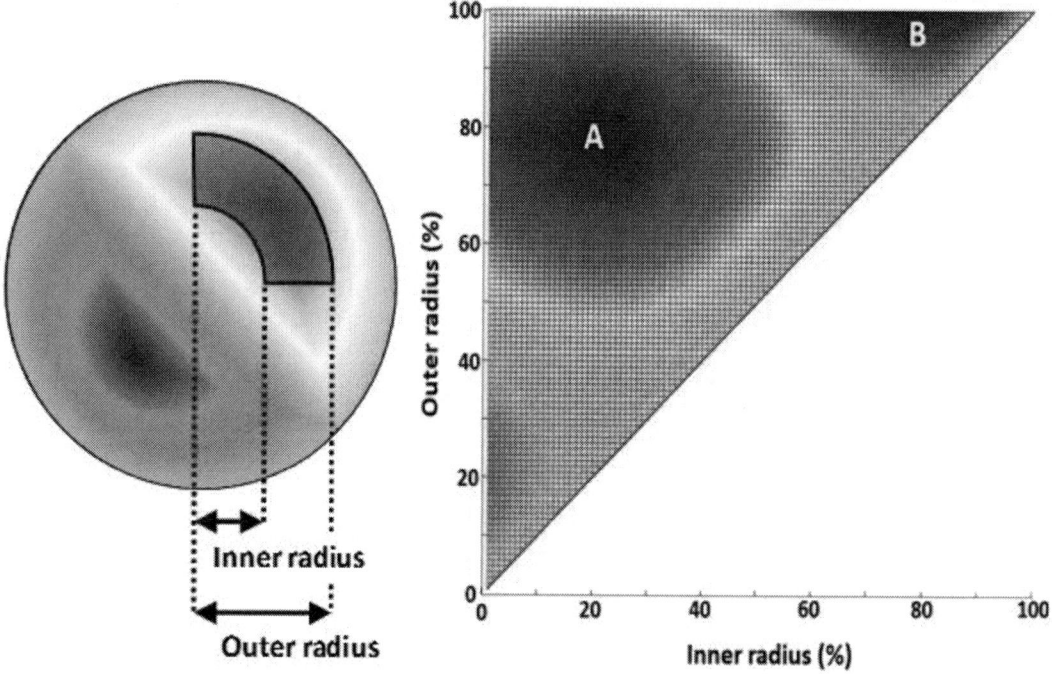

Fig. 2: Optimum electrode geometry is determined by calculating the moment produced by the PZT actuation electrode as a function of electrode inner and outer radii. The highest point (A) corresponds to maximum positive moment and occurs when the electrode is placed towards the centre of the diaphragm; the lowest point (B) corresponds to maximum negative moment and occurs when the electrode is placed towards the diaphragm circumference.

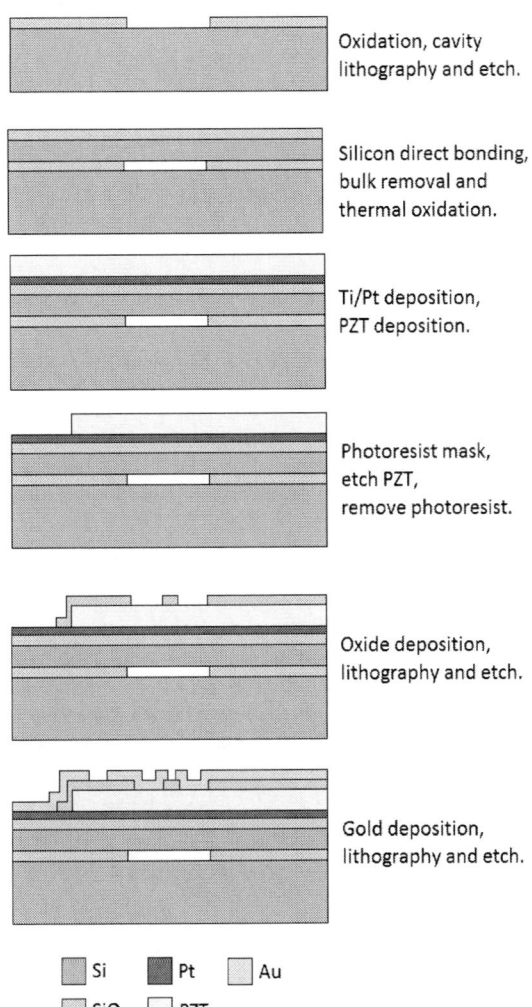

Oxidation, cavity lithography and etch.

Silicon direct bonding, bulk removal and thermal oxidation.

Ti/Pt deposition, PZT deposition.

Photoresist mask, etch PZT, remove photoresist.

Oxide deposition, lithography and etch.

Gold deposition, lithography and etch.

▨ Si   ■ Pt   ▨ Au
▨ SiO₂   □ PZT

Fig. 3: Schematic of the process flow for sensor fabrication.

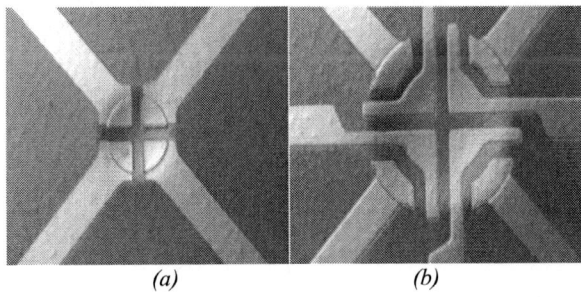

(a)   (b)

Fig. 4: SEM of the two diaphragm designs, (a) corresponds to electrode geometry (A) in Fig. 2, (b) corresponds to electrode geometry (B) in Fig. 2.

III.   EXPERIMENTAL METHODS

A.   Device characterization:

The mechanical properties of a PZT thin film near to the resonant frequency can be expressed by means of an equivalent circuit, such as the extended Butterworth-Van Dyke model [7]. When the PZT thin film is deposited on a mechanical structure, the heavily loaded PZT film experiences multiple resonances of the mechanical structure. The interaction between the PZT film and the structure changes the mechanical boundary conditions of the PZT and therefore the electrical parameters of the equivalent circuit. A simple impedance spectrum test was used to detect the mechanical resonances of the diaphragm. Full electrical vibration characterization of the device was conducted both in vacuum and in air. The diaphragm was driven into resonance electrically via the PZT thin film, and the vibration detected from charges generated within the sensing PZT layer by a differential charge amplifier. The results were compared with the diaphragm vibration as measured by a laser vibrometer.

B.   Electroplating CDR devices:

An AutoLab potentiostat (Metrohm) was set-up in galvanostatic mode to provide a constant current. The current and time were varied depending on the desirable deposition depth and the area of the deposition electrode [8]. A gold electroplating solution ECF62 (Metalor Technologies) was placed between a positive electrode and the negative deposition electrode (CDR interface), allowing defined quantities of gold to be deposited which is subsequently assessed using a Zygo surface profiler.

C.   Electrochemical cleaning and high resolution patterning:

CDR devices underwent an initial cleaning procedure using a UV-ozone system (NovaScan Technologies). The devices were placed on the stage and sealed within the chamber, which was purged for 1 minute in pure oxygen, followed by 5 minutes UV exposure and 30 minute incubation. Cyclic Voltammetry (CV) of the CDR electrodes was conducted using the following parameters; sweep from -150 mV to +600 mV at a scan rate of 100 mV s^{-1}, assessing the oxidation and reduction of 1mM potassium ferricyanide (100 mM phosphate buffer, pH7). If the initial characterization demonstrated good electrochemistry the device diaphragms were submersed in 0.5 mM HS-C$_{11}$-PEG$_3$-OH (Prochimia) blocking thiol solution (ethanol), for 24 hours. In a final step the surfaces were rinsed excessively in ethanol and dried under a stream of nitrogen. The CV of potassium ferricyanide was re-assessed at the diaphragm electrode surfaces as described above. Confirmation of dense monolayers on all electrodes was followed by site-selective cleaning of the separately addressable electrodes. The AutoLab was used in potentiostatic mode and -1.4 V applied to the requisite electrodes (sense) for 2 minutes, in 100 mM sodium phosphate pH 10. The diaphragm electrodes were rinsed in phosphate buffer and a repeat CV of 1mM potassium ferricyanide was

978-1-4673-1122-9/12 $31.00 © 2012 IEEE   75

recorded. Finally, an 80:20 ratio of HS-C_{11}-PEG$_3$-OH and HS-C_{11}-PEG$_6$-COOH in ethanol was incubated over the diaphragm surface for 24 hours, total thiol concentration 0.5mM. The CV of potassium ferricyanide was assessed at the diaphragm electrode interface as described above.

IV. RESULTS AND DISCUSSION

A. Device Characterization:

Testing of the fabricated devices confirmed the resonance at 6MHz which was to be used in the mass sensing applications. A detailed frequency response, both in vacuum and at atmospheric pressure, showed a slightly non-degenerate mode of vibration with a frequency split of about 8 KHz due to fabrication tolerances, as shown in Fig. 5. This non-degeneracy also leads to mode misalignment and coupling. The quality factor in vacuum was measured to be 750.

Device sensitivity, for a 120 µm diameter diaphragm, was designed at 38fg Hz^{-1} for perfect modal alignment. The operational principle of the circular diaphragm mass sensor requires the mode antinodes to align with the centre of the drive electrode. One way of mode realignment is to deliberately add mass to a pair of opposite electrodes; initial testing indicated this to be an effective approach. To calibrate mass sensitivity, the addition of the electroplated gold onto sense regions, coupled with the optical profiler depth measurement provided fixed mass addition to be determined. Initial measurements have demonstrated a mass sensitivity of 55 fg Hz^{-1}, which is in good agreement with the theory.

B. Site Selective Patterning:

Biomolecule immobilization was also a critical consideration in the fabrication design process. Both blocking (reference) and recognition (sense) molecules need to be site-specifically immobilized at the transducer interface. This was achieved using electrically addressable regions as shown schematically in Fig. 6.

The CDR device was initially immersed in a PEG thiol OH blocking solution, modifying the electrode surfaces with a dense layer, which is resistant to non specific binding of biological material. In the next step selective cleaning/regeneration was achieved by applying a -1.4 V potential, to the sense electrodes, but not the reference electrodes. The reference electrodes maintained a dense monolayer of blocking material, while the sense electrodes were clean and available to undergo a subsequent immobilization procedure. A PEG thiol layer, containing functional groups for subsequent recognition biomolecule coupling was added to the sense electrodes. This whole procedure was monitored using CV and the oxidation/reduction of potassium ferricyanide at the gold electrode surface (situated at the CDR diaphragm interface). The oxidation and reduction peaks can be monitored at clean/regenerated surfaces, but the electrochemistry was effectively blocked after immobilization procedures. The data depicted in Fig. 7 acts as confirmation that selective blocking and recognition layers can be patterned at the CDR interface. In future work, the base layers will be used to site-specifically couple recognition molecules, which will confirm that the biological immobilization and recognition can be tracked as a function of CDR frequency split.

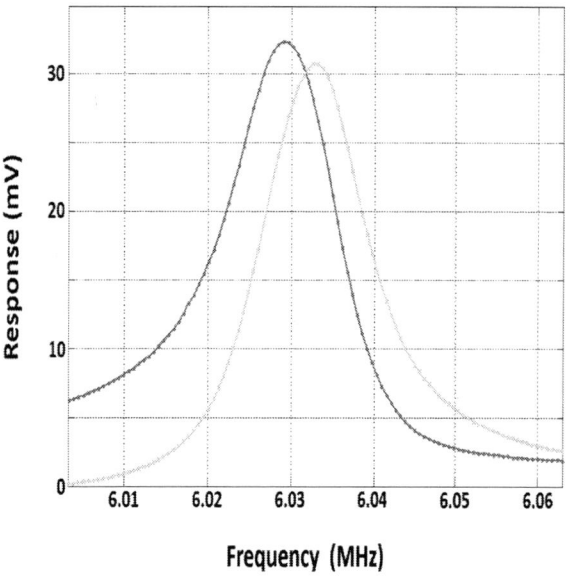

Fig. 5: The electrically recovered signal from each of the operational modes.

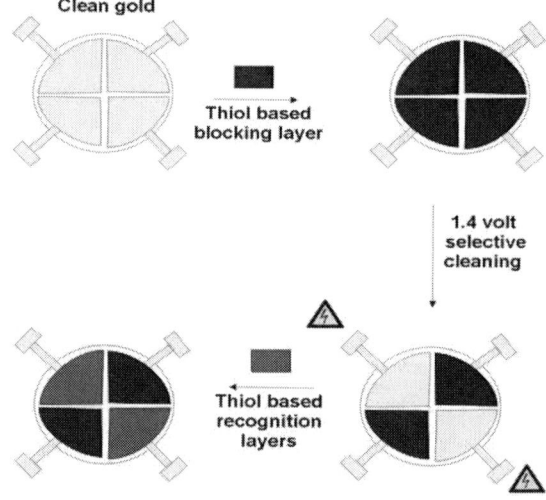

Fig. 6: The schematic depicts a selective cleaning process, which was used to achieve immobilization at specific regions of the diaphragm, via electrodes. The sense sectors are shown in red and the reference sectors in blue

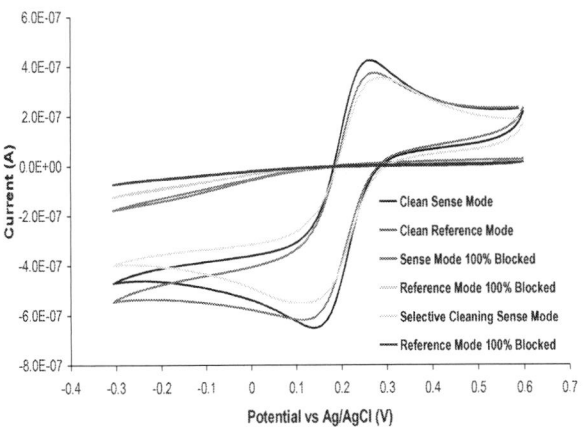

Fig. 7: Cyclic Voltammetry demonstrating selective immobilization of blocking and recognition layers on the sense and reference electrodes of CDR interfaces. The electrochemical behaviour of potassium ferricyanide on clean gold surfaces demonstrated a typical voltammogram. After blocking / protein resistant PEG thiol layers were immobilised, the potassium ferricyanide can't access the electrodes and the oxidation/reduction peaks are lost. After the sense sectors are cleaned (-1.4 V for 2 min) the electrochemical activity of potassium ferricyanide is detected again, but the reference electrodes remain blocked. In a final step, which is omitted from this figure, the freshly cleaned sense sectors are available for recognition layer functionalization.

V. CONCLUSIONS

A piezoelectrically driven and sensed CDR mass sensor has been successfully fabricated. The fabrication procedure described is relatively simple, in comparison to the historical device fabrication [2,3,5]. The signal recovery electronics of the current device is also considerably simplified when compared to an electrostatically actuated device [5]. In this instance electrical sensing was achieved, using a charge amplifier, with a signal to noise ratio of 10:1 at 6 MHz. A preliminary sensitivity of 55 fg Hz^{-1} was achieved using a gold electroplating calibration measurement on a 120 μm diameter device. The patterning procedure, which was built into the fabrication concept, was also demonstrated to be viable. This procedure will be used in future work to better define the utility of the next generation CDRs in terms of biosensing.

ACKNOWLEDGMENT

The authors would like to thank the EPSRC for providing financial support (AptaMEMS-ID, EP/G061394/1) and F.X. Boillot/J. Collet of Tronics Microsystems, France for MEMS fabrication.

REFERENCES

[1] L.J. Kircka, "Miniaiturization of analytical systems," Clin. Chem., vol. 44, pp. 2008-2014, 1998.

[2] A.K. Ismail, J.S. Burdess, A.J. Harris, C.J. McNeil, J. Hedley, S-C. Chang and G. Suarez "The principle of a circular diaphragm mass sensor" J. Micromech. Microeng. Vol. 16, pp 1487–1493, (2006).

[3] A.K. Ismail, J.S. Burdess, A.J. Harris, G. Suarez, N. Keegan, J.A. Spoors, S-C. Chang, C.J. McNeil and J Hedley "The fabrication, characterisation and testing of a circular diaphragm mass sensor" J. Micromech. Microeng. Vol. 18, (2008)

[4] D. M. Ward and D. A. Buttry "In situ interfacial mass detection with piezoelectric transducers". Science, Vol 249, 1000–7 (1990)

[5] P. Ortiz, N. Keegan, J. Spoors, R. Burnett, J. Hedley, A. Harris, J. Burdess, T. Velten, M. Biehl, W. Haberer, M. Solomon, A. Campitelli and C. McNeil. "Development of a biosensor cartridge integrating active microfluidics, MEMS sensor technology and detection electronics". MicroTAS, Groningen, (2010).

[6] J. S. Burdess and J. N. Fawcett. "Experimental evaluation of a piezoelectric actuator for the control of vibration in a cantilever beam". Proc. Inst. Mech. Eng., Vol 206 , pp. 99-106 (1992).

[7] J. Kim, B. L. Grisso, J. K. Kim, D. S. Ha, and D. J. Inman. "Electrical modeling of piezoelectric ceramics for cnalysis and evaluation of sensory systems". IEEE Sensors Applications Symposium, Atlanta, (2008).

[8] T. Fujita, S. Nakamichi, S. Ioku, K. Maenaka, Y. Takayama. "Selective and direct gold electroplating on silicon surface for MEMS applications". IEEE, MEMS, Istanbul, Turkey (2006).

A Carbon Nanotube Network Conjugated by Semiconductor Nanoparticles with Defined Nanometer-Scaled Gaps

[1]Shinya Kumagai, [2]Naofumi Okamoto, [2,3]Mime Kobayashi, [2]Ichiro Yamashita[*]

[1]Department of Advanced Science and Technology, Toyota Technological Institute, Nagoya, JAPAN
[*2]Department of Materials Science, Nara Institute of Science and Technology, Nara, JAPAN
[3]The Cancer Institute of the Japanese Foundation for Cancer Research, Tokyo, JAPAN
ichiro@ms.naist.jp

Abstract—**Carbon nanotubes (CNTs) have been attracting broad interest in many research fields including nano-electronics and micro-/nano-electromechanical systems because of their unique physical and electronic properties. Utilizing a biological material, a novel nanostructure of carbon nanotube conjugated by semiconductor nanoparticles were synthesized intended for electronics devices with unprecedented properties. A peptide aptamer that has an affinity for carbonaceous materials was fused to a cage-shaped protein, LiDps, to be displayed on its outer surface. Semiconductor nanoparticle was synthesized within a cavity of the engineered protein, NHBP-LiDps. By the affinity of the peptide aptamer, the NHBP-LiDps proteins accommodating semiconductor nanoparticles were adhered to surround the CNTs with nanometer-scaled gap defined by the thickness of the protein-cage. The CNTs conjugated by the NHBP-LiDps were adsorbed onto a pair of electrodes, making CNT network with dispersed nanoparticles. We report our attempt to characterize the electronic property of the structure.**

Keywords-carbon nanotube, nanoparticle, cage-shaped protein, peptide aptamer, Coulomb blockade

I. INTRODUCTION

Carbon nanotubes (CNTs) have been attracting broad interest in many research fields including nano-electronics and micro-/nano-electromechanical systems because of their unique physical and electronic properties [1,2]. Intriguing nano-electronic properties may emerge from a CNT network conjugated by metallic or semiconductor nanoparticles (NPs) (Fig.1). The metal or semiconductor NPs can trap and release electrons to contribute for the novel electronic characteristics. Novel memory effects may emerge as a result of multiple electron pathways formed in the complex. There are several method to decorate CNTs with the NPs [3-5]. In the previous report, utilizing a peptide aptamer with an affinity for carbonaceous materials [NHBP-1 (DYFSSPYYEQLF)] [6], a structure in which single-walled CNTs are surrounded by cage-shaped protein, *Listeria innocua* Dps (LiDps), was fabricated (Fig.2) [7]. Compared to the chemical approaches, which could damage the surface of CNTs and degrade electronic property of the CNTs, utilization of peptide aptamer and protein cages has a number of advantages. Cavity of the

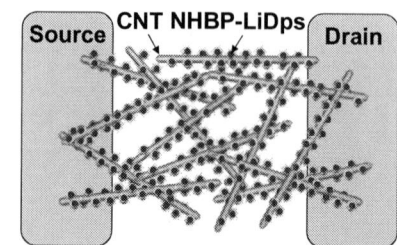

Fig. 1: Schematic drawing of a CNT network conjugated by nanoparticles.

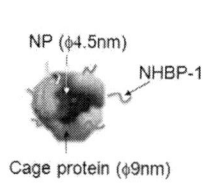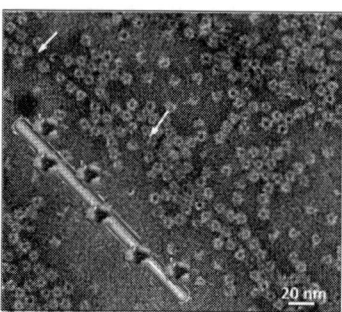

Fig. 2: A TEM image of CNTs surrounded by LiDps proteins with Co oxide NP Cores. Stained with 3% phosphotungstic acid. The LiDps proteins adhere to CNTs by NHBP-1 peptide aptamer displayed on the outer surface of LiDps protein. The cavity of LiDps functions as a spatially restricted chemical reaction chamber for NP synthesis [8].

LiDps can be used as a spatially restricted chemical reaction chamber [8]. Various metal- or semiconductor- materials can be synthesized inside the cavity to form NPs. The NPs synthesized inside proteins can function as charge storage nodes [9,10]. The LiDps proteins with NP cores adhere to the surface of CNTs. The NPs are separated from the SWNTs by the nanometer-scaled gaps that are defined by the thickness of the protein cage. The gaps can function as tunneling gaps during electron transportation [11]. In the present study, we report our attempt to characterize the electronic property of the CNT network conjugated by semiconductor NPs.

978-1-4673-1122-9/12 $31.00 © 2012 IEEE

II. EXPERIMENTAL

A. Protein Preparation

An engineered protein that has the NHBP-1 peptide fused to the N-terminus of the LiDps (NHBP-LiDps) protein was expressed in *E. coli* [7]. Size of the LiDps is ϕ9.5 nm. The inner cavity size is about ϕ4 nm. The cells were lysed and the supernatant was subjected to a thermal denaturation at 75°C for 20 min. After centrifugation at 6,000 *g* for 10 min, the supernatant was subjected to salt dialysis with a final concentration of 0.5 M NaCl. After centrifugation at 6,000 *g* for 10 min, the precipitate was suspended in 50 mM Tris-HCl (pH 8.0). The salt dialysis was repeated 2 times, and protein purity was confirmed with sodium dodecyl sulfate polyacrylamide gel electrophoresis (SDS-PAGE) and transmission electron microscopy (TEM: JEM-2200FS, JEOL) analysis.

B. Synthesis of Co Oxide NPs within NHBP-LiDps Proteins

The NHBP-LiDps protein (0.1 mg mL-1) was dissolved in 100 mM HEPES-NaOH (pH 8.2) buffer and incubated in the presence of 0.5 mM $(NH_4)_2Co(SO_4)_2$ and 1 mM H_2O_2 at 50°C for 15 min. The reaction solution was centrifuged at 30,000 rpm on a 45Ti rotor (Beckman Coulter), and the supernatant was applied to a filtration column (NMWC: 30K, Amicon Ultra) and centrifuged at 3,000 rpm at 4°C, then washed by adding H_2O. The protein was further purified through a Sephacryl S-300 in XK 26/100 column (GE Healthcare), and applied on top of a layer of 15%, 25% and 50% sucrose. After centrifugation at 30,000 rpm on an SW32Ti rotor (Beckman Coulter), fractions with high 400 nm/280nm absorbance ratio were collected. TEM and energy dispersive X-ray spectroscopy (EDS: JED-2200 analyzer, JEOL) confirmed the existence of Co oxide cores inside the proteins

C. CNTs Conjugated by NHBP-LiDps Proteins

To construct a nanostructure in which the cage-shaped protein surround single-walled carbon nanotubes (SWNTs), 15 mg of SWNTs (Aldrich 519308) were mixed in 50 mL of water in a glass vial. Next, the protein was mixed with the SWNTs in H_2O in a final concentration of 0.3 mg mL-1. The final concentration of SWNTs is roughly 0.2 mg mL-1. Ultrasonication (Sonifer 250; Branson Ultrasonic Corp., Danbury, CT, USA) was applied for 5 min (1 sec ON/ 3 sec OFF cycles, total of 20 min) and the solution was centrifuged at 15,000 rpm for 10 min at 4°C. The supernatant was put on a Cu TEM grid and stained with 3% phosphotungstic acid (PTA). TEM observation confirmed that the NHBP-LiDps proteins were aligned along SWNTs.

D. Elerctronic Characteristics Measurements of CNT Networks with Nanoparticles

Pairs of Au/Cr electrodes were fabricated by lift-off process. Electron beam (EB) lithography resist (ZEP-520A, Zeon) was spin-coated on a Si substrate with SiO_2 layer. Thickness of the

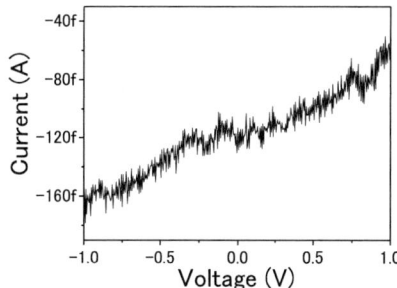

Fig. 3: Background noise level of current-voltage measurement.

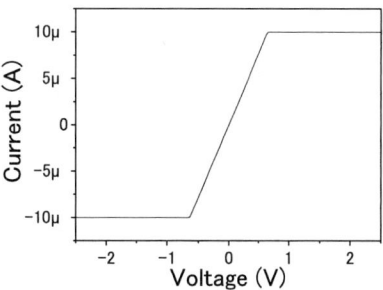

Fig. 4: Current-voltage characteristics of CNTs. Compliance current level was set to 10μA.

EB resist was 100 nm. Electrode structures, which were 500 nm apart, were patterned in the resist film by EB lithography (CABL-8200 TFT, CRESTEC). Cr and Au were evaporated onto the substrate with the patterned resist. Thickness of Cr and Au layers were 5 nm and 20 nm, respectively. The substrate was immersed in dimethylacetamide (40°C, 10 min) and sonicated to lift-off the deposited Au/Cr films. After the lift-off process, pairs of Au/Cr electrodes that have 500-nm gaps were obtained.

Before the adsorption of CNTs conjugated by NHBP-LiDps proteins, surface of the Au/Cr electrodes were cleaned by UV/Ozone treatment at 110°C with an oxygen gas flow rate of 0.5 slm for 10 min. (UV-1, SAMCO). Solution of CNTs conjugated by NHBP-LiDps was dropped onto the Au/Cr electrodes, dried at ambient conditions, and rinsed with pure water. The sample was baked at 110°C for 5 min to remove any residual liquids. Current-voltage characteristics were measured at ambient conditions using metal probes connected to a semiconductor parameter analyzer (4156C, Agilent). The structures of networks of CNTs with NPs on each electrode pairs were confirmed with scanning electron microscope (SEM: JSM-7400, JEOL, SU-6600, Hitachi) after the electric measurements.

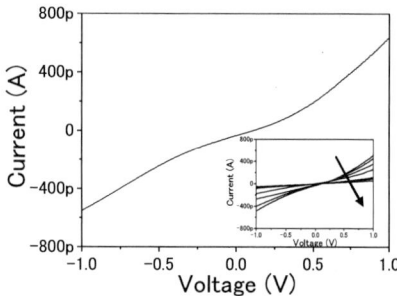

Fig. 5: Current-voltage characteristics of the dried buffer. Repeated measurements resulted in decreased currents as shown in the inset.

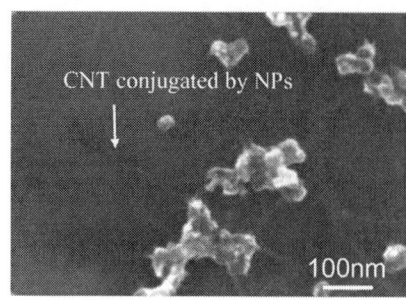

Fig. 7: A SEM image of the network of CNTs conjugated by Co NPs. Protein shell cannot be observed with SEM, and the accommodated Co oxide NPs were observed as white dots. The network was constructed on the Si substrate with oxidized surface.

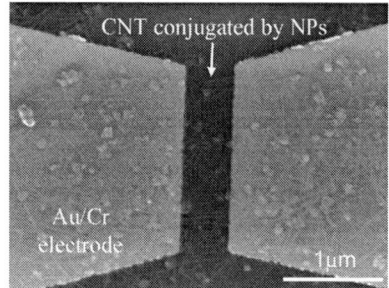

Fig.6: A SEM image of the CNT network with NPs.

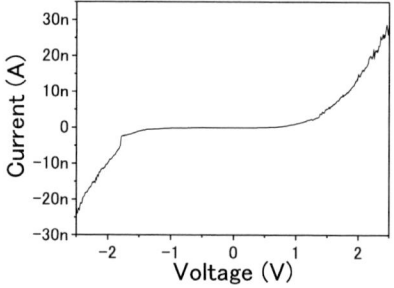

Fig. 8: Current-voltage characteristics of the CNT network with semiconductor NPs.

III. RESULTS AND DISCUSSION

Before the current-voltage measurement of the CNTs conjugated by NHBP-LiDps proteins, we assessed the background current of the Au/Cr electrode device. Bare electrodes showed background current of a few hundreds of femtoamperes in the range of ± one volt as shown in Fig. 3.

Current-voltage characteristics of CNTs were also measured. CNT-dispersed water was dropped onto Au/Cr nanogap electrodes and dried. The current-voltage curve showed Ohmic characteristics as shown in Fig. 4. With increasing bias voltage, current reached the compliance level of the measurement setup (10 μA).

Although buffer is effective to prevent the proteins from denaturation, it may affect the current-voltage characteristics. Therefore, the influence of the residual buffer components on electrical characteristics was also investigated. A buffer solution without CNT nor protein was dropped onto a pair of Au/Cr electrodes and dried. Current-voltage characteristic of the dried buffer was in the order of several hundreds of picoamperes (Fig. 5). Repeated measurements resulted in decreased current, which indicates degradation of the remaining buffer components.

To eliminate the influence of the buffer and other salts, buffer exchange of the solution of CNTs conjugated by NHBP-LiDps proteins was performed using centrifugal device with an ultrafiltration membrane. CNT networks with NPs are fabricated as shown in Fig. 6. The CNT networks are observed between the Au/Cr electrode pairs. Magnified image of the CNT network is shown in Fig. 7. The structures that NHBP-LiDps proteins surround CNTs are maintained. The NHBP-LiDps proteins have sufficient adhesive force to keep the structure during the rinse process. The presence of NHBP-LiDps proteins is expected to avoid direct contact between CNTs.

The current-voltage characteristics of the buffer-free CNT network with semiconductor NPs were measured (Fig.8). At low bias voltage, current did not flow. With increasing bias voltage, the current begins to flow. The order of the current was a few tens of nanoamperes, which is larger than the current level observed for the dried buffer. The current suppression at low bias voltage is considered to be the effect of Coulomb blockade. The current suppression may have been caused by the Co oxide NPs separated by the nanometer-scaled gaps from the CNTs. Tunneling junctions between CNTs and some of the NPs surrounding them may have been formed. Further electronic characterization is necessary to reveal the true nature of the CNT-protein complex.

IV. CONCLUSION

CNTs conjugated by semiconductor (Co oxide) NPs were synthesized using NHBP-LiDps proteins. The protein cage of NHBP-LiDps kept nanometer-scaled gap between CNTs and inner NPs. CNT networks with Co oxide NPs was constructed between Au/Cr electrode pairs. Direct contacts between CNTs are avoided by the adhered NHBP-LiDps proteins that surround CNTs. In the electronic characteristics measurements, it was revealed that residual buffer components show current-voltage characteristics on the order of several hundreds of picoampere. The electric currents were decreased by repeating the measurement. Electronic characteristics of CNTs conjugated by NHBP-LiDps (Co oxide core) showed Coulomb blockade effect at low bias voltage.

ACKNOWLEDGMENT

We thank M. Yamane, S. Fujita and Drs. B. Zheng and M. Uenuma for technical assistance. We gratefully acknowledge the support of this work by Human Frontier Science Program (RGP61/2007), Grant-in-Aid for Scientific Research on Innovative Areas "Emergence in Chemistry" (No. 2311171400) from JSPS, CREST from Japan Science and Technology Agency, and "Nanotechnology Network Japan Project" at Toyota Technological Institute in Chubu Area from the Ministry of Education, Culture, Sports, Science and Technology (MEXT), Japan.

REFERENCES

[1] G. A. Steele, G. Gotz, and L. P. Kouwenhoven, "Tunable few-electron double quantum dots and Klein tunnelling in ultraclean carbon nanotubes", Nat. Nanotechnol. vol. 4 pp. 363-367, 2009.

[2] V. Sazonova, Y. Yaish, H. Üstünel, D. Roundy, T. A. Arias, and P. L. McEuen, "A tunable carbon nanotube electromechanical oscillator", Nature, vol.431, pp.284-287, 2004.

[3] B. Kim, and W. M. Sigmund, "Functionalized Multiwall Carbon Nanotube/Gold Nanoparticle Composites", vol. 20, pp. 8239-8242, 2004.

[4] H. Ozawa, X. Yi, T. Fujigaya, Y. Niidome, T. Asano, and N. Nakashima, "Supramolecular Hybrid of Gold Nanoparticles and Semiconducting Single-Walled Carbon Nanotubes Wrapped by a Porphyrin Fluorene Copolymer", J. Am. Chem. Soc., vol. 133, pp. 14771-14777, 2011.

[5] N. A. Kumar, A. Bund, B. G. Cho, "Novel amino-acid-based polymer/multi-walled carbon nanotube bio-nanocomposites: highly water dispersible carbon nanotubes decorated with gold nanoparticles", Nanotechnology, vol 20, 225608, 2009.

[6] D. Kase, J. L. Kulp, M. Yudasaka, J. S. Evans, S. Iijima, K. Shiba, Langmuir, "Affinity Selection of Peptide Phage Libraries against Single-Wall Carbon Nanohorns Identifies a Peptide Aptamer with Conformational Variability" vol. 20, pp. 8939-8941, 2004.

[7] M. Kobayashi, S. Kumagai, B. Zheng, Y. Uraoka, T. Douglas, I. Yamashita, "A water-soluble carbon nanotube network conjugated by nanoparticles with defined nanometre gaps" Chem. Commun, vol. 47, pp. 3475-3477, 2011.

[8] I. Yamashita, K. Iwahori, S. Kumagai, "Ferritin in the field of nanodevice" Biochim. Biophys. Act. vol. 1800, pp.846-857, 2010.

[9] A. Mirura, T. Hikono, T. Matsumura, H. Yano, T. Hatayama, Y. Uraoka, T. Fuyuki, S. Yoshii, and I. Yamashita, "" Floating Nanodot Gate Memory Devices Based on Biomineralized Inorganic Nanodot Arrayas a Storage Node", Jpn. J. Appl. Phys. vol. 45, pp. L1-L3, 2006.

[10] A. Miura, R. Tanaka, Y. Uraoka, N. Matsukawa, I. Yamashita, and T. Fuyuki, "The characterization of a single discrete bionanodot for memory device applications", vol. 20, 125702, 2009

[11] S. Kumagai, S. Yoshii, N. Matsukawa, K. Nishio, R. Tsukamoto, and I. Yamashita, "Self-aligned placement of biologically synthesized Coulomb island within nanogap electrodes for single electron transistor", Appl. Phys. Lett. vol. 94, 083103, 2009.

Closed-looped Nano Stimulation Microscope for Living Cell Membrane

Takayuki Hoshino [*], Keisuke. Morishima

[*]Bio-Application and Systems Engineering, Tokyo University of Agriculture and Technology, Tokyo, Japan

t_hoshi@cc.tuat.ac.jp

Abstract— Electron-beam could stimulate a living cell membrane through a 100-nm-thick SiN nanomembrane. We designed a co-axial dual microscope with an electron-beam lithography system and a fluorescence microscope to investigate bio-molecular dynamic system of living cell membrane. This microscope had a closed-looped controlling system using the fluorescence live cell imaging and the electron beam induced stimulation. The live cell imaging of the fluorescence microscope provides the chemical identity of the target molecule and the response to a following stimulation on the cell membrane, and the electron-beam can induce electro-chemical stimulation in hundreds nanometer resolution. Scanning of the electron-beam could provide high-speed, precise, and large scale stimulation on the target molecules. We attempted that the combination with this stimulation system and fluorescence microscope would provide the closed-loop control of the bio-molecule system on cell membrane. We proposed here the concept of the virtual molecule environmental display to study the functional changes on the target behavior due to our closed-loop control system on this co-axial microscope.

Keywords-virtual molecular display; electron-beam; cell membrane; molecular dyanamics

I. INTRODUCTION

Cellular response to mechanical and chemical stimulations is a result of changes of protein reactions and membrane traffic in a cell. For example, changes in membrane protein dynamics based on its fluidity affect cancer metastasis. The diffusion constant of a specific protein on a cell membrane was found to be related to cancer migration into healthy tissue.[1] Therefore the analysis of molecular level dynamics can provide new approaches to understand how medical treatment affects their mechanisms.

The protein-scale dynamic analysis to understand the live cell system is generally carried out using nanoprobing techniques in a liquid environment. Generally, a cell membrane consists of multiple floating proteins which show high velocity random walk movements (Brownian motion) on the surface of the lipid bilayer[2]. Ideally, nanoprobes, therefore, need to be designed with high resolution both spatially and temporally, and also to have a precise selectivity for the biological molecule of interest. In previous studies, therefore, nanoprobing and manipulation of the biological materials were done using a silicon nano needle[3] and high speed scanning[4] by an atomic force microscope, a magnetic bead with a labeling of an antibody,[5] a scanning femtosecond laser for cell ablation,[6], [7] two-photon

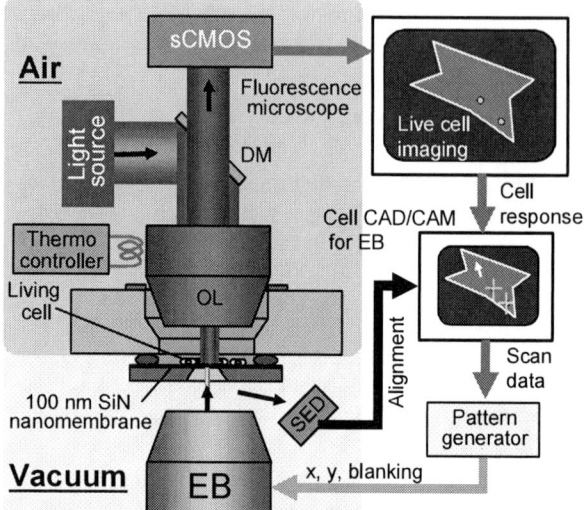

Fig. 1. Co-axial microscope with closed-looped control system. The electron-beam was mounted on the counter position of a fluorescence microscope. 100-nm thick SiN nanomembrane placed between the electron-beam and fluorescence microscope to separate an atmospheric environment and a vacuum environment. The nanomembrane was also worked as a cell culture surface and an energy transfer window of electron beam energy. The secondary electron microscope image was aligned to the fluorescence live cell image to generate the stimulation pattern on the CAD/CAM system. High speed closed-loop feedback via the CAD/CAM system was contribute virtual molecular dynamic control on real live cell membrane.

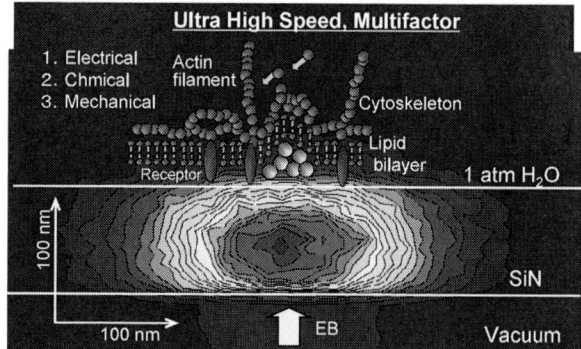

Fig. 2. Schematics of the electron-beam induced reaction on the cell membrane. The energy distribution through the SiN nanomembrane was about several tens to 100 nm in plane, and 10 nm in depth. This energy would generate electrical, chemical, and mechanical changes to the molecular system on cell membrane. Electron beam was 2.5 keV.

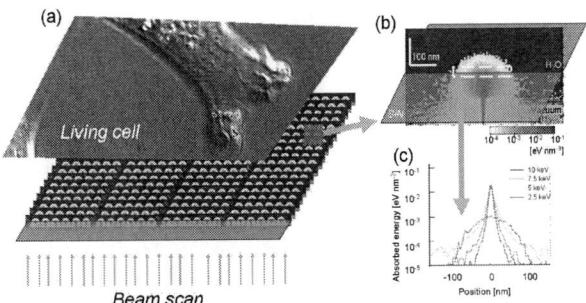

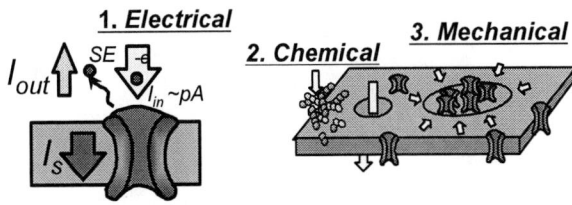

Fig. 3. Schematics of virtual molecule environmental display for large scale molecular control system on living cell. (a)The scanning electron-beam induced two-dimensional chemical reaction field on the live cell membrane with molecular resolution. (b) The energy distribution of single spot irradiation of 2.5 keV EB into the H2O through the 100-nm-thick SiN nanomembrane. (c) The absorbed energy for the surface reaction between the SiN nanomembrane and H2O. [meV nm-3] was similar energy to hydrogen bonding energy.

Fig. 4. Electron-beam induced reaction on cell membrane. The scanning pattern and the absorbed energy could control the stimulation mode as electrical, chemical, and mechanical.

uncaging of glutamate[8] and GABA,[9] and micro fluidic device.[10], [11] Each probing device could reach a single molecule and sub-cellular scale to stimulate the cellular functional signal pathway. Although conventional probing techniques have been successfully realized with enough spatial resolution to carry out nano mechanisms on biomolecular systems, the ideal goal is particularly challenging because, that bio-molecule responded dynamically for a high spatio-temporal period to accomplish a multi-molecular interaction in a lipid bilayer on living cell.

This paper reports an electron beam induced nano stimulation system on fluorescent microscope to modulate molecular systems in a live cell (figure 1). Our method provided nano space energy transfer in liquid water using focused electron beam. And this performed both nano deposition and ablation on a living cell in nanoscale resolution. This paper will show nanostructures fabrication using this electron beam (EB) induced molecular modification for the evaluation of cell response to energy flow. Our process has nanoscale resolution because of using focused electron beam. We proposed our method for live cell processible nanofabrication to analyze protein scale dynamic reaction on cell system. And We proposed here the concept of the virtual molecule environmental display to study the functional changes on the target behavior due to molecular dynamic control using our closed-loop feedback system on this co-axial microscope.

II. ELECTRON-BEAM INDUCED NANOPROCESSING ON LIVING CELL

A. Energy transmission through SiN nanomembrane

Highly accelerated Electron-beam is transmitted through a micro scale and light mass material as transmitted electron microscope. Low energy electro-beam also transmitted through a nanometer thick material. This phenomenon was

applied to observe a wet sample in atmospheric environment through a hundreds nanometers thick nano membrane in a wet capsule [12–17]. We focused the nanometer size energy transfer into the liquid environment for the virtual nano manipulation probe. . The 2.5 keV electron beam was transmitted the electron kinetic energy the opposit side of the 100-nm-thick nanomembrane (Fig.2). The transmitted energy was distributed several tens to hundreds nanometers in a plane and tens nanometers in depth. This reactive area due to electron transmitted energy was distributed in the similar dimension of the molecular system on the cell membrane.

B. Chemical reaction

Electro beam induces a surface chemical reaction to deposit precursor materials due to an emission of secondary electron form the surface [2]. This is a result of equivalent energy of secondary electron to redox reaction. This electron beam induced chemical reaction could be applied for a modification of the cell membrane traffic as nano stimulation probes. We chose the precursor material 3,4-ethylenedioxythiophene (EDOT). When a low energy electron beam was irradiated through thin nanomembrane into the culturing chamber which was containing with EDOT solution. To utilize the electron beam for in situ fabrication probe, the cultivation medium and electron optics was separated with a nanomembrane as shown in figure 1. The irradiated electron beam was penetrated through the nanomembrane into the culturing chamber (figure 2). This process has three types of mode for nanofabrication, electrical, chemical, and mechanical on cell component. We could choose the mode of the process by selecting the situation. Higher acceleration voltage could induce ablation mode for the process (figure 3 (c)).

C. Virtual molecule environmental display

Figure 3 (a) shows a virtual molecule environmental display for a large scale molecular control system on living cells. The scanning electron-beam induced two-dimensional chemical

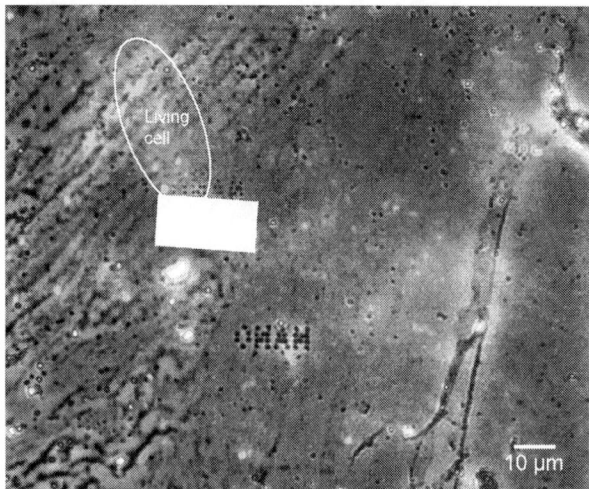

Fig. 5 Nano patterning on living Hep G2 cell using electron beam-induced chemical reaction on the SiN nanomembrane. Electron beam-induce chemical solution deposition on live Hep G2 cell. Merged image of differential interference contrast (DIC) image of the cell and the fluorescent cell image. The 'NANO' word pattern (mirror writing, gray) was deposited on the living cell (fluorescent Calcein-AM, green). Cells were stained with both live/dead indicating fluorescent dye Calcein-AM (live cell) and PI (dead cell).

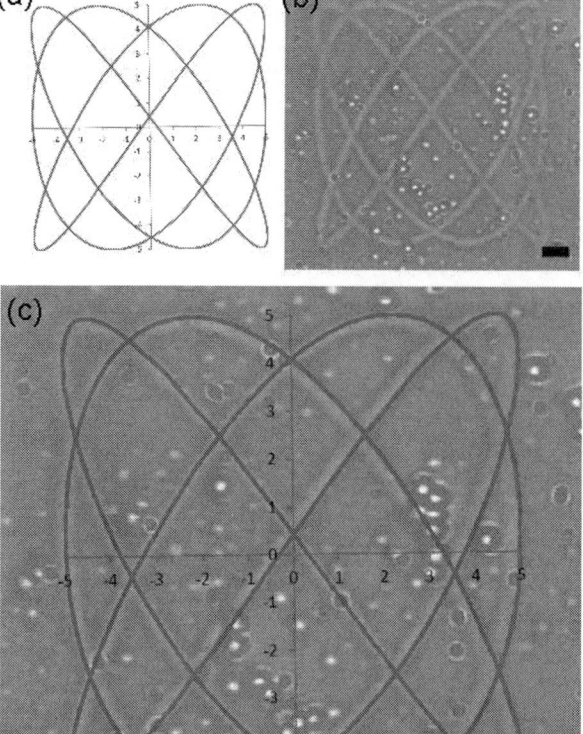

Fig. 6. On-site observation of nano pattern of PEDOT on SiN nanomembrane. (a) Two-dimensional CAD model for the nano patterning. (b) Nano deposition pattern. (c) Merged image of CAD model and deposition result. Particles on the image were 1 μm polystyrene beads. Scale bar = 5 μm.

reaction field on the live cell membrane with molecular resolution. Figure 3 (b) shows the energy distribution of single spot irradiation of 2.5 keV EB into H_2O through the 100-nm-thick SiN nanomembrane. The absorbed energy for the surface reaction on the interface between the SiN nanomembrane and H_2O. The energy range was similar to hydrogen bonding energy on bio-molecule. The energy range of the adsorption enegy could be chose by acceleration voltage of the electro-beam. So, we attempted that the stimulation could be controlled by the scanning pattern and absorbed energy using kinetic energy and electrostatic/dielectric force from the electron-beam. (Figure 4)

The chemical reaction was already confirmed on the living cell. [12] We allied electro-beam induced deposition of EDOT on Hep G2 cell. Electrolytic polymerization of PEDOT was induced by electron-beam energy through the 100-nm-thick SiN nanomembrane. This nano deposition was fabricated on the intact cell membrane. Figure 5 showed nano depositions on the nanomembrane. Spot beam could fabricate nano dot about 310 nm in diameter. The nano deposition pattern would mechanically stimulate the cell membrane and mechano receptor.

III. TWO-DIMENSIONAL PATTERNING

A. Materials and methods

10 mM EDOT was diluted in conventional cell culture medium D-MEM (Dulbecco's Modified Eagle Medium) with 10% fetal bovine serum. 5 keV electron beam (Apco, MINI-ECO) was irradiated into the medium though the 100-nm-thick SiN nanomembrane (NORCADA). The medium was warmed to 36-37 degree C using a lens heater (Tech-alpha).

The scanner for the electron-beam was connected to the two-channel signal generator. Figure 5 (a) was input signal to the scanner (Lissajous figure: 5 Vp-p, 15 and 20 Hz, 5 degree for phase delay).

B. Results and disucussion

The deposition pattern was observed in real-time using the bright field microscope. The deposition pattern was precisely fabricated as figure 5 (b, c). The accuracy of the patterning was < 1 μm.

The deposition pattern was monomer polymerization in nano size in the cultivation medium during the optical observation. Therefore, the deposition pattern would be applied for the mechanical stimulation of mechanosensitive channel. Generally, the mechanosensitive channel contribute the mechanical transduction for tactile sensing, the cell differentiation and growth of myotube, and blood vessel stress. These important function of the mechanosensitive channel was distributed on the cell membrane to detect the stress of the cell membrane [18–20]. Therefore, our electron-beam induced stimulation would be possible to be apply the mechanosensitive channel on the cell membrane.

,

IV. CONCLUSION

The electron beam induced oxidation-reduction reaction in cell culturing medium was confirmed. This in situ nano patterning and cellular nano stimulation would be utilized for real time analysis of cell membrane traffic and molecular function o intracellular protein systems. The electron beam has, in theoretically, less than nanometer scale wave length, and the electron range in water was also nano scale. Therefore this process could be applied to molecular level analysis for micro scale system, for example single mechanosensitive channel spine and cell migration.

ACKNOWLEDGMENT

This work was supported in part by the Industrial Research Program of NEDO, Grants-in-Aid for Scientific Research from the Japanese Ministry of Education, Culture, Sports, Science and Technology (Nos. 23680052, 20860031, 21676002, 21111503, and 21225007).

REFERENCES

[1] K. Gonda, T. M. Watanabe, N. Ohuchi, and H. Higuchi, "In Vivo Nano-imaging of Membrane Dynamics in Metastatic Tumor Cells Using Quantum Dots," *Journal of Biological Chemistry*, vol. 285, no. 4, pp. 2750 -2757, Jan. 2010.

[2] A. Kusumi, Y. Sako, and M. Yamamoto, "Confined lateral diffusion of membrane receptors as studied by single particle tracking (nanovid microscopy). Effects of calcium-induced differentiation in cultured epithelial cells," *Biophysical Journal*, vol. 65, no. 5, pp. 2021-2040, Nov. 1993.

[3] I. Obataya, C. Nakamura, S. Han, N. Nakamura, and J. Miyake, "Nanoscale operation of a living cell using an atomic force microscope with a nanoneedle," *Nano letters*, vol. 5, no. 1, pp. 27-30, Jan. 2005.

[4] T. Ando, N. Kodera, E. Takai, D. Maruyama, K. Saito, and A. Toda, "A high-speed atomic force microscope for studying biological macromolecules," *Proceedings of the National Academy of Sciences of the United States of America*, vol. 98, no. 22, pp. 12468 -12472, Oct. 2001.

[5] B. D. Matthews et al., "Mechanical properties of individual focal adhesions probed with a magnetic microneedle," *Biochemical and Biophysical Research Communications*, vol. 313, no. 3, pp. 758-764, Jan. 2004.

[6] C. Hosokawa et al., "Femtosecond laser modification of living neuronal network," *Applied Physics A*, vol. 93, no. 1, pp. 57-63, Jun. 2008.

[7] C. Hosokawa, S. N. Kudoh, A. Kiyohara, and T. Taguchi, "Resynchronization in neuronal network divided by femtosecond laser processing," *NeuroReport*, vol. 19, no. 7, pp. 771-775, May 2008.

[8] M. Matsuzaki, G. C. Ellis-Davies, T. Nemoto, Y. Miyashita, M. Iino, and H. Kasai, "Dendritic spine geometry is critical for AMPA receptor expression in hippocampal CA1 pyramidal neurons," *Nature Neuroscience*, vol. 4, no. 11, pp. 1086-1092, Nov. 2001.

[9] S. Kantevari, M. Matsuzaki, Y. Kanemoto, H. Kasai, and G. C. R. Ellis-Davies, "Two-color, two-photon uncaging of glutamate and GABA," *Nat Meth*, vol. 7, no. 2, pp. 123-125, Feb. 2010.

[10] S. Takayama, E. Ostuni, P. LeDuc, K. Naruse, D. E. Ingber, and G. M. Whitesides, "Laminar flows: Subcellular positioning of small molecules," *Nature*, vol. 411, no. 6841, p. 1016, Jun. 2001.

[11] H. Wu, A. Wheeler, and R. N. Zare, "Chemical cytometry on a picoliter-scale integrated microfluidic chip," *Proceedings of the National Academy of Sciences of the United States of America*, vol. 101, no. 35, pp. 12809-12813, Aug. 2004.

[12] T. Hoshino and K. Morishima, "Electron-beam direct processing on living cell membrane," *Applied Physics Letters*, vol. 99, no. 17, p. 174102, 2011.

[13] T. Hoshino and K. Morishima, "Electron-beam induced in situ spatiotemporal nanofabrication toward intracellular nanorobotics," in *Proc. The 14th International Conference on Miniaturized Systems for Chemistry and Life Science (μTAS 2010), W63A*.

[14] H. Nishiyama et al., "Atmospheric scanning electron microscope observes cells and tissues in open medium through silicon nitride film," *Journal of Structural Biology*, vol. 169, p. 438, 2010.

[15] W. Inami, K. Nakajima, A. Miyakawa, and Y. Kawata, "Electron beam excitation assisted opticalmicroscope with ultra-high resolution," *Optics Express*, vol. 18, no. 12, pp. 12897-12902, Jun. 2010.

[16] E. U. Donev and J. T. Hastings, "Electron-Beam-Induced Deposition of Platinum from a Liquid Precursor," *Nano Letters*, vol. 9, no. 7, pp. 2715-2718, Jul. 2009.

[17] E. U. Donev and J. T. Hastings, "Liquid-precursor electron-beam-induced deposition of Pt nanostructures: dose, proximity, resolution," *Nanotechnology*, vol. 20, no. 50, p. 505302, Dec. 2009.

[18] Z. Qi, S. Chi, X. Su, K. Naruse, and M. Sokabe, "Activation of a mechanosensitive BK channel by membrane stress created with amphipaths," *Molecular Membrane Biology*, vol. 22, no. 6, pp. 519-527, Jan. 2005.

[19] K. Naruse and M. Sokabe, "Involvement of stretch-activated ion channels in Ca2+ mobilization to mechanical stretch in endothelial cells," *American Journal of Physiology - Cell Physiology*, vol. 264, no. 4, p. C1037 -C1044, Apr. 1993.

[20] S. Ito et al., "A Novel Ca2+ Influx Pathway Activated by Mechanical Stretch in Human Airway Smooth Muscle Cells," *Am. J. Respir. Cell Mol. Biol.*, vol. 38, no. 4, pp. 407-413, Apr. 2008.

Pyramidal Nanowire Tip for Atomic Force Microscopy and Thermal Imaging

Narges Burouni[*], Edin Sarajlic, Martin Siekman, Leon Abelmann and Niels Tas

[*]Transducer Science and Technology, Department of Electrical Engineering, Mathematics and Computer Science, University of Twente, The Netherlands
[*]MESA+ Research Institute, University of Twente, The Netherlands
n.burouni@utwente.nl

Abstract—We present a novel 3D nanowire pyramid as scanning microscopy probe for thermal imaging and atomic force microscopy. This probe is fabricated by standard micromachining and conventional optical contact lithography. The probe features an AFM-type cantilever with a sharp pyramidal tip composed of four freestanding silicon nitride nanowires with a diameter of 60 nm. The nanowires, which are made of silicon nitride coated by metal, form an electrical cross junction at the apex of the tip, addressable through the electrodes integrated on the cantilever. The cross junction on the tip apex can be utilized to produce heat and detect local temperature changes. Electrical and thermal properties of the probe were experimentally determined. The temperature changes in the nanowires due to Joule heating can be sensed by measuring the resistance of the nanowires. We employed the scanning probe in an atomic force microscope.

Keywords- Pyramidal Nanowire; Thermal Imaging; Atomic Force Microscopy; Corner lithography

I. INTRODUCTION

Atomic force microscopy and Scanning thermal microscopy (SThM) [1-3] are widely applied techniques for the study of nanoscale phenomena. At the heart of the SThM technique is a modified scanning probe, which has a sharp tip with a nanowire cross junction integrated at its apex. Such a probe can be realized by crafting a AFM cantilever using direct deposition of platinum by focused electron beam [1]. However, this fabrication method is rather impractical and time consuming. Another approach is based on micromachining and multiple level direct-write electron beam lithography [2-4]. The high cost of an E-beam system and the serial nature of its writing process make this method both expensive and unsuitable for high-volume manufacturing. We present a scanning microscopy probe for AFM and thermal imaging fabricated by standard micromachining and conventional optical contact lithography.

Thermal probes can be used in two ways, either with temperature feedback, also called active mode, or without temperature feedback, called passive mode. The passive mode is the constant current mode; a small constant current is applied to the probe, in this way the probes works as a thermometer.

The active mode is a constant temperature mode; the temperature feedback keeps the probe at a constant elevated temperature by heating the thermal element. The way this is done depends on the method used. For the resistive wire this is achieved by resistive joule heating. When a probe comes near a sample, heat flows between the tip and the sample, depending on the difference in temperature. The power delivered to the tip to keep it at constant temperature is a measure for the local temperature and/or thermal conductivity of the sample [5,6].

A Scanning Thermal Microscope (SThM) can be used in Atomic Force Microscopes (AFM) and Scanning Tunneling Microscopes (STM). In this way both topography and thermal properties can be investigated simultaneously. The advantage of using an AFM is that insulators well as conductors can be examined.

We apply a recently discovered nano-fabrication technique [7-9] to develop sophisticated probes with electronic functionality for scanning probe microscopy. This method, called corner lithography, can be used to realize three-dimensional wireframes with nanometer diameters, using simple micrometer scale optical lithography. This method is inexpensive and can be applied on a wafer scale, is perfectly suited for small scale production and is compatible with conventional micromachining.

The probe, shown in Figure 1, features an AFM-type cantilever with a pyramidal tip composed of four freestanding silicon nitride nanowires. The nanowires, which are made of silicon nitride coated by metal, form an electrical cross junction at the apex of the tip, addressable through the

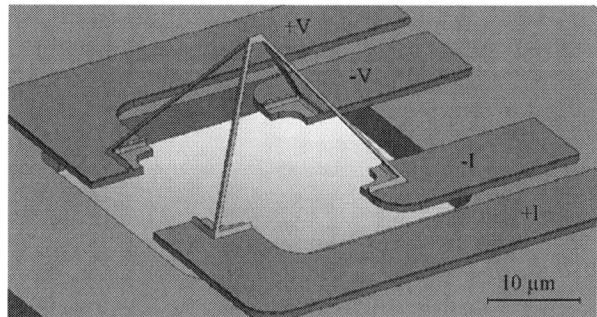

Fig. 1. Schematic illustration of a silicon nitride wireframe tip, coated with conducting layer to enable thermal imaging.

electrodes integrated on the cantilever. The cross junction on the tip apex can be utilized to produce heat and detect local temperature changes and perform AFM and thermal imaging by scanning the probe tip over a surface.

II. FABRICATION

A. Corner Lithography

Corner lithography method results a well size-controlled fabrication procedure for well-defined nanometer scale structures with exact position and spatial arrangement fully determined by the template. We employed this technique to define uniform nanowires. Corner lithography is based on the material that is left in sharp concave corners after conformal deposition and isotropic etching (figure 2). Controlling the size of remaining material (l) which is depends to the angle of the corner (α) and initial thickness (t), is related on the etching time with respect to the removal layer during etching (r).

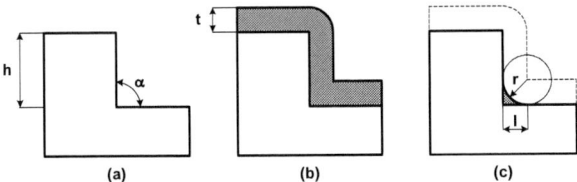

Fig. 2. Principle of corner lithography: exploiting conformal deposition and isotropic etching.

B. Pyramidal Nanowire Tip

We have successfully fabricated a first probe prototype with a nanowire tip composed of approximately 60 nm width and 11 μm long silicon nitride wires metalized by 6 nm Ti and 35 nm Au layers as shown in figure 3[12]. To fabricate nanowires with approximately 60 nm width, first a 192 nm low stress silicon nitride was deposited. Then, the layer was overetched for 10% of initial thickness to have 60nm width as result (figure 4). Consequently, the radius of the tip (r) will be 211nm at the end.

The corner lithography technique leads to highly uniform wires. Figure 5 shows the standard deviation of the silicon nitride layer as a function of etching time. This standard deviation results in an uniformity of the over etch factor of 1.1±0.04.

To be able to use this nanowire pyramid in applications such as SThM and SHPM a conductive layer has to be applied on the pyramid and the leads. For this a sputtering process with a titanium bonding layer and a gold layer is used. In this way a resistive nanowire tip is created. The conduction through the metal layers on the cantilever and the nanowires is mainly determined by the gold layer, as it has a thickness in the order of 35 nm or more, and the thickness of the titanium layer is less than 10 nm, and the resistivity of titanium is approximately 20 times higher than the resistivity of gold.

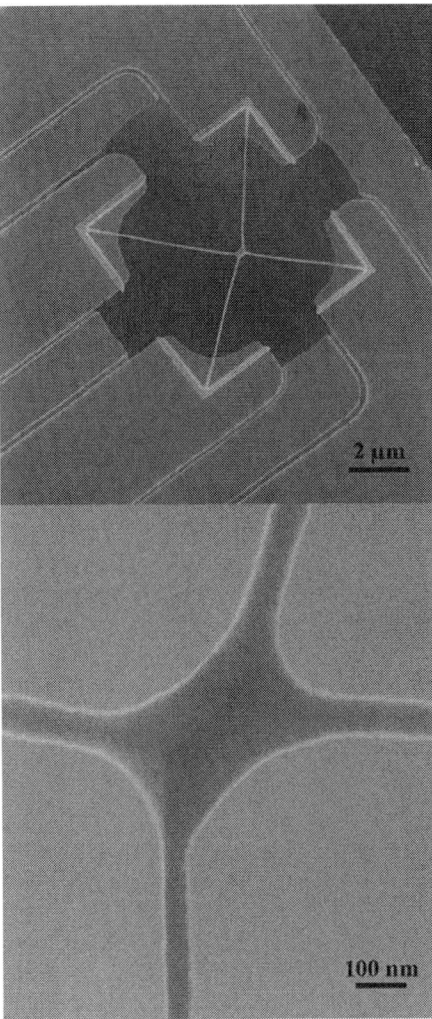

Fig. 3. HRSEM images of the fabricated probe. Top: overview of the wireframe and electrodes. Bottom: tip apex. The wire width is approximately 60 nm. The wires are coated with a Ti(6 nm)/Au(35 nm) layer.

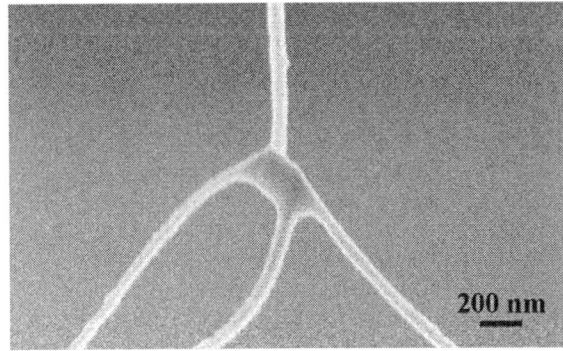

Fig. 4. Side view of the tip apex. The inner radius is around 211nm.

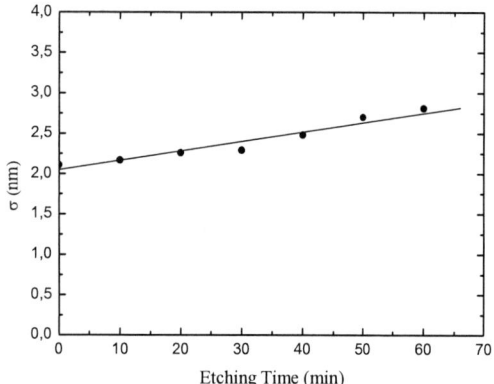

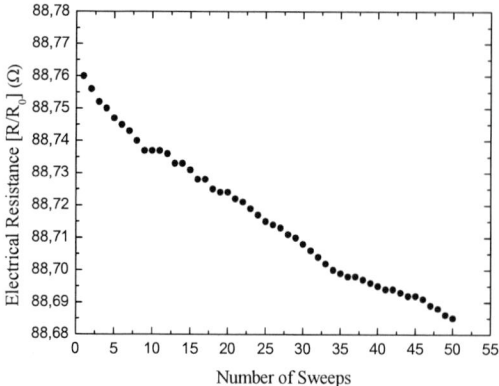

Fig. 5. Standard deviation of the silicon nitride layer during etching by 50% HF in wafer scale which is related to the initial thickness of the nitride ,etching time and number of the measurements in each point.

Fig. 6. Electrical resistance of the nanowires when sweeping the current from 0 up to 0.5 mA.

III. RESULTS

A. Electrical and Thermal Properties

When a constant current runs through the nanowires, the entire wire heats up. The uniform heat power delivered to the nanowires is V^2/RL , where V is the voltage applied to one pair of the nanowires and R is the resistance and L is the length of the nanowire. The temperature distribution reaches a steady-state after heating in fraction of second. This gives an indication that steady-state heat transfer can assumed.

The electrical resistance of the nanowires forming the pyramidal tip is a function of temperature. In order to demonstrate the temperature dependence, we have resistively heated the tip by passing a DC current through one pair of nanowires. For each measurement point (V and R), the current is kept constant until the resistance stabilizes.

The change in electrical conductance to other nanowire pair induced by the heating of the tip was measured using lock-in techniques (figure 1). The resistance was around 89Ω which fits well with 119 Ω as theoretical expected resistance. By sweeping the current, the resistance decreases and can be described as a function of the number of sweeps as shown in figure 6. Sweeping the current improves the electric conductance through the probe. After a current sweep the wire resistance slightly decreases (see figure 6). A possible explanation is that the current heats the bottlenecks in the gold layer, which have the highest resistance, enough to cause the gold layer to become somewhat more mobile and in that way reorder the gold atoms in and around those points. This decreases the resistance of those points, resulting in a decrease of the overall resistance.

The temperature changes in the nanowires due to Joule heating can be sensed by measuring the resistance of the nanowires. It is expected that temperature changes of a surface can be sensed in the same way. The current through the nanowires can be expected to be limited by a maximum current density because this is the most limiting factor of the

lifetime of the wire, and not the temperature. Electro migration will destroy the wire before Joule heating does [10]. In figure 6, as expected, the electrical resistance increases with increasing the heating current from 11 to 18mA. At a current of 18 mA, we estimate the maximum temperature to be [390K], based on the temperature coefficient of thin film gold (0.0017 K^{-1} [13]). Above 18 mA we suspect that electromigration occurs.

B. Atomic Force Microscopy

We employed the scanning probe in an atomic force microscope. Figure 8 shows a contact AFM scan of a magnetic hard disk taken with a sharp pyramidal wireframe probe. The resolution is higher than the tip outer dimensions, most likely since we scan with only one corner. The probe was scanned many times on the surface of the sample without damaging either the sample or the tip.

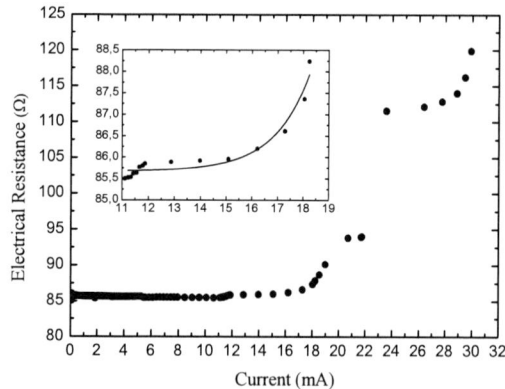

Fig. 7. Electrical resistance of the wire as a function of current. In each measured point, current is kept constant until resistance stabilizes.

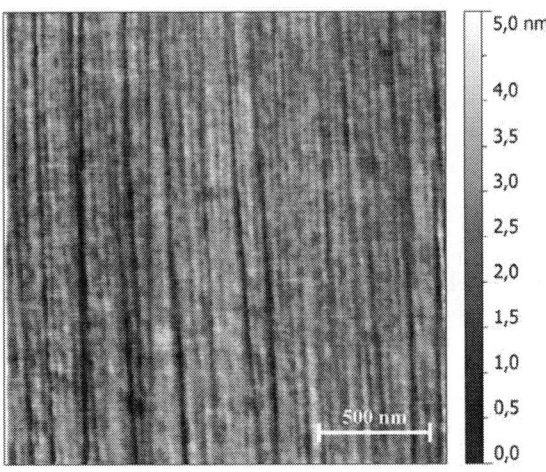

5,0 nm
4,0
3,5
3,0
2,5
2,0
1,5
1,0
0,5
0,0

500 nm

Fig. 8. AFM scan of a magnetic hard disk surface, using a sharp wireframe probe.

IV. CONCLUSION

We present a novel wireframe probe for atomic force and scanning thermal microscopy, based on corner lithography. The batch fabrication process results in silicon nitride wires with approximately 60 nm width and a standard deviation as low as 2-3 nm. These nanowires are coated with a 41 nm Cr/Au layer.

The wires can be heated by means of an electrical current. From the increase in resistance with current, we estimated the maximum temperature to be 390 K at a current of 18 mA. When sweeping the current several times from 0 to 0.5 mA, we observed a non-reversible reduction in the resistance of about 15 ppm per sweep. The probes work well in tapping mode AFM, and we observed a resolution of 200 nm when scanning the surface of a magnetic hard disk.

These results give us confidence that these exiting new probes can be successfully applied in scanning thermal microscopy, to either measure the thermal conductance or temperature of surfaces.

ACKNOWLEDGMENT

This work was partially done within the "FunTips" project and funded by the Dutch Technology Foundation (STW).

REFERENCES

[1] K. Edinger et al. 2001 J. Vac. Sci. Technol. B 19 (6), pp. 2856-2860.

[2] H. Zhou et al. 1998 J. Vac. Sci. Technol. B 16 (1), pp. 54-58.

[3] G. Mills et al. 1998 Applied Physics Letters 72 (22), pp. 2900-2902.

[4] B.K. Chong et al. 2001 J. Vac. Sci. Technol. A 19 (4), pp. 1769-1772.

[5] Gmelin et al. Sub-micrometer thermal physics - An overview on SThM techniques. Thermochimica Acta. 1997, Vol. 310, 1-2.

[6] A. Hammiche et al. Meas. Sci. Technol. 1996, Vol. 7.

[7] N. Burouni et al. Proceeding of the 6td IEEE int. Conf. on Nano/Micro Engineered and Molecular Systems, Kaohsiung, Taiwan, 2011.

[8] E. Berenschot et al. Proceeding of the 3rd IEEE int. Conf. on Nano/Micro Engineered and Molecular Systems, Sanya, China, 2008.

[9] E. Sarajlic et al. The 13th International Conference on Solid-State Sensors, Actuators and Microsystems, Seoul, Korea, 2005.

[10] Pierce et al. P.G. Electromigration: a review. Microelectron. Reliab. 1997, Vol. 37, 7.

[11] M. Aguilar et al. Surf. Scie. 1998, Vol. 409.

[12] E. Sarajlic et al., Proceeding of the 23rd IEEE International Conference on Micro Electro Mechanical Systemes, MEMS 2010, Wanchai, Hong Kong, 24-28 Jan 2010.

[13] X. Zhang et al. Int. J. Thermophys. 2007, Vol. 28, 1.

A Pyrex Nanochannel Device Fabricated by AFM Nanolithography

Orain Hibbert[1], Taylor Busch[1*], Steve Tung[2], *Member IEEE*

[1]Micoelectronics-Photonics Graduate Program, University of Arkansas, Fayetteville, AR 72701, USA

[2]Department of Mechanical Engineering, University of Arkansas, Fayetteville, AR 72701, USA

*Contact author: tbusch@uark.edu

Abstract—**A Pyrex nanochannel device was fabricated using AFM nanolithography in conjunction with MEMS-based microfabrication. The fabrication process began with the patterning of microchannels on a Pyrex substrate using photolithography and wet etching. AFM nanolithography was then performed to realize a nanochannel between the micro reservoirs. A diamond-coated AFM probe with a relatively large force constant served as the cutting tool. A detailed study was conducted to determine the relationship between the AFM input parameters and the resultant nanochannel dimensions. A linear trend was obtained between the AFM scratch force and the nanochannel depth. A similar trend was also observed between the number of repeated scratches and the channel depth. Minimal material pileup on the nanochannel sidewall was observed during the scratch process. Flow patency in the capped-off nanochannel was demonstrated through fluorescence microscopy.**

Keywords – Atomic force microscopy (AFM); Nanofluidics; Nanolithography; Nanochannel; DNA sequencing

I. INTRODUCTION

Nanofluidics is an increasingly important field in nanotechnology, as nanofluidic-related publications have doubled every two years for the last decade. Recent publications have suggested the possibility of using nanofluidic devices such as nanopores to achieve rapid chip-based DNA sequencing. In this approach, single-stranded DNAs are translocated through a nanoscale hole and sequencing is accomplished by measuring the corresponding blockage current [1]. At present, it is difficult to achieve single-base resolution using nanopores due to the high translocation speed as a result of the externally imposed electric field [2]. Using a magnetic field that opposes this electric field can decrease the translocation speed, allowing for a more accurate DNA reading [3]. Alternatively, it has been suggested that a nanochannel with embedded sensors can detect single DNA bases and ultimately sequence DNA [4].

Successful nanochannels have been realized through different processes, including surface nanomachining and nanoimprint lithography [5-7]. Surface nanomachining can consistently produce 1-D nanochannels where the channel depth is in the nanometer scale. Nanoimprint lithography can be used to realize 2-D nanochannels where both the channel depth and width are in the nanometer scale. Atomic force microscopy (AFM) is a new technique for fabricating 2-D nanochannels. Recently, AFM-based nanolithography has been demonstrated to realize nanochannels in various substrates such as silicon [8-9]. Other substrates such as Pyrex which is more suitable for biomedical applications can also be used for AFM nanochannel fabrication.

The present paper describes the development of a glass nanochannel device designed to characterize nanoscale biomolecules such as DNA. The nanochannel is realized by applying AFM nanolithography to a patterned Pyrex substrate. The nanochannel is capped off by anodic bonding and preliminary flow tests are conducted using fluorescence microscopy that demonstrates successful fluid flow through the nanochannel.

II. FABRICATION OF NANOCHANNEL SYSTEM

A. Design and fabrication of a Pyrex chip

Fig. 1 shows the schematic of the design of the Pyrex nanochannel device. The design width of the microchannels varied, as well as the distance between the two microchannels. Microchannel widths ranged from $40 – 250$ µm and the area located between the microchannels ranged from $20 – 100$ µm. Each microreservoir had a radius of 2.50 mm. The fabrication process began with the patterning of microchannels on a Pyrex 7740 – 500 µm thick – wafer through photolithography. The channels were etched at an 8 µm depth in a BOE:HCl:H$_2$O wet isotropic etch at an etch rate of 1 µm/min. The BOE etchant undercuts the photoresist, resulting in sloped channel walls and a smaller nanochannel region that the original design. Typically, the nanochannel region is approximately decreased by the design gap value minus two times the etch depth. After the wafer is patterned, etched, and diced, it is ready for AFM nanolithography to realize a nanochannel.

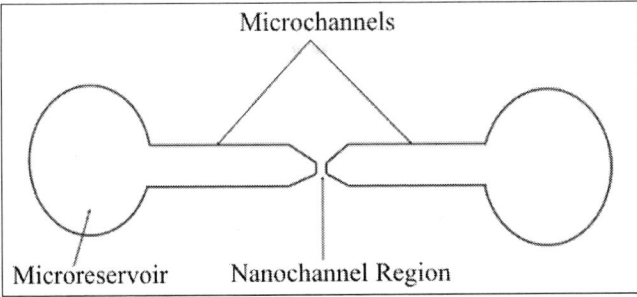

Fig. 1. Design schematic of the completely fabricated Pyrex device

B. AFM-based nanolithography

AFM nanolithography is used to scratch into the Pyrex substrate as shown in Fig. 2. The tip is forced down into the substrate along a straight path to realize a nanochannel.

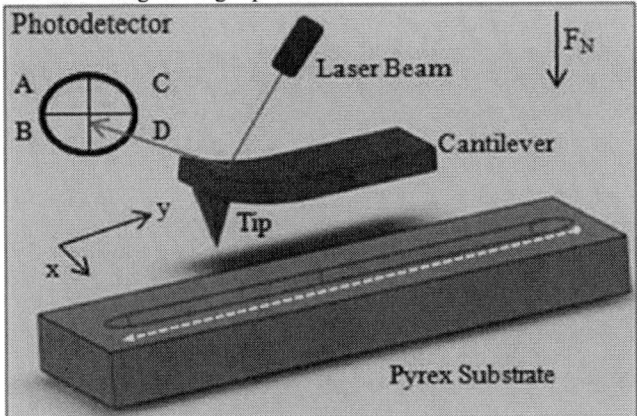

Fig. 2. Basic layout of the AFM nanolithography process

An Agilent 5500 Scanning Probe Microscope was used to mechanically scratch out a nanochannel between the upstream and downstream microchannels. The AFM probe was a diamond-coated Tap 190 DLC (Innovative Solutions Ltd.) with a relatively large force constant of 48 N/m and a tip radius of less than 15 nm [10]. Before scratching was performed, several AFM calibration experiments were carried out to obtain the scratching characteristics of the AFM tip on the Pyrex substrate.

AFM nanolithography is operated in contact mode, and there is a large normal force applied by the AFM tip on the substrate to be able to realize a nanochannel. Nanochannel dimensions varied based on input parameters such as the force setpoint (volts), number of scratches, and the tip speed. The force setpoint is a voltage input with a force output. Equation (1) shows how to convert from input voltage to output force F_N and is based on Hooke's Law:

$$F_N = (k)(D_s)(F_s), \qquad (1)$$

where k represents the force constant of the AFM cantilever (μN/nm), D_s is the deflection sensitivity (nm/V), and F_s is the force setpoint (V).

Fig. 3 shows that the AFM scratch force is linearly proportional to the nanochannel depth. Input force setpoint voltages used for AFM calibration were 3.5 V, 5.5 V, 7.5 V, and 10 V. From (1), the respective output force applied to the substrate was 9.06 μN, 14.24 μN, 19.42 μN, and 25.89 μN. For each trial, a constant tip speed of 1 μm/s was used. The linear trend is in agreement with previous work with Si nanochannels [11]. However, previous papers indicate a logarithmic trend for lower force setpoint values [12]. The logarithmic trend is dominant only when the force setpoint results in a pressure on the surface close to the yield strength of the substrate material. For higher force setpoint values, a linear trend is dominant.

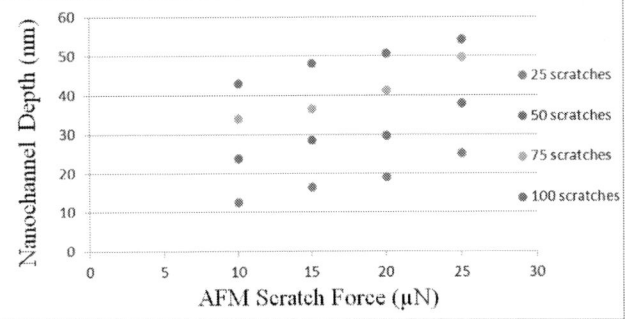

Fig. 3. AFM nanolithography: nanochannel depth versus scratch force for different scratch numbers.

Since Pyrex is a relatively 'hard' material, multiple scratches are usually required to achieve the desired nanochannel depth. Fig. 4 shows the cross-sectional topography of three nanochannels realized on a flat Pyrex surface with different number of scratches. It can be seen that both the channel depth and width increases as the number of scratches increases. The width increases due to the 'drift' of the AFM probe on each cycle. Although the AFM is operating with a feedback loop, the tip does not always cut from the same starting point, resulting in a wider channel with increasing number of cuts. Minimal material pileup on the nanochannel sidewall is observed; this is critical to achieving a tight seal around the channel through anodic bonding. A tight anodic bonding seal will allow for the device to be tested for nanochannel flow.

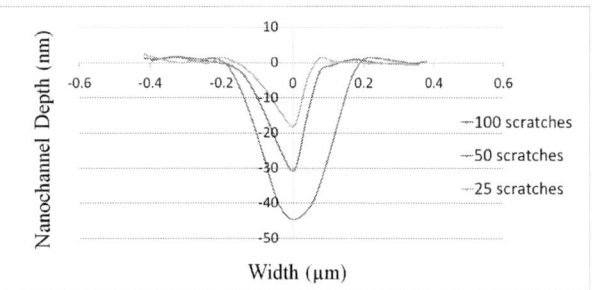

Fig. 4. AFM nanolithography: nanochannel dimensions versus number of repeated scratches.

To realize the nanochannel in the patterned Pyrex chip, AFM nanolithography must be applied to a non-flat substrate surface as demonstrated in Fig. 5. The nanochannel region is the flat area in the middle of the figure. The microchannels in the bottom left and top right areas, each having a sloped edge. The AFM probe is required to climb over the contour of the microchannel walls to scratch a nanochannel from the upstream microchannel to the downstream.

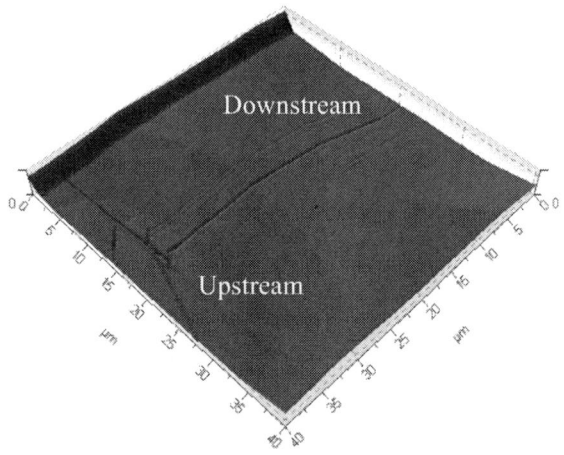

Fig. 5. Three-dimensional image of nanochannel realized by AFM nanolithography.

Fig. 6 shows the comparison between a successful and unsuccessful scratch. Some scratches only span halfway across the two microchannels, resulting in a failed scratch. In this case, fluid will not flow from the upstream to the downstream. Since the tip is not scratching on a flat surface, it can be rather difficult to obtain a successful, continuous scratch.

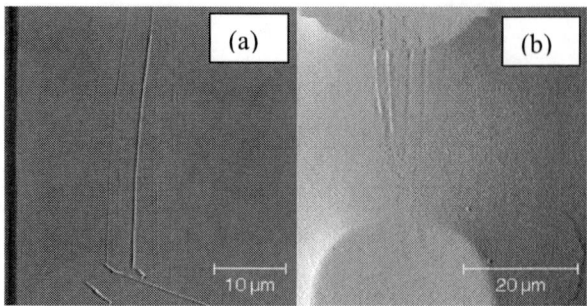

Fig. 6. Two-dimensional AFM images of (a) successful scratch and (b) failed scratch.

AFM nanolithography imposes a tremendous amount of wear and tear on the probe tip, and a worn-out tip can result in an unsuccessful scratch. Worn-out tips are the main cause of scratches that are undefined and do not span the entire distance between microchannels. The average lifetime of a tip is approximately 300 – 350 scratches. After this point, the tip will tend to realize non-continuous nanochannels and produce more pileup than normal.

Scanning electron microscopy (SEM) can be used to compare new tips versus damaged tips, as shown in Fig. 7. The AFM probe in Figure 6b has clearly succeeded its lifetime, as indicated by the fractured tip from the SEM image.

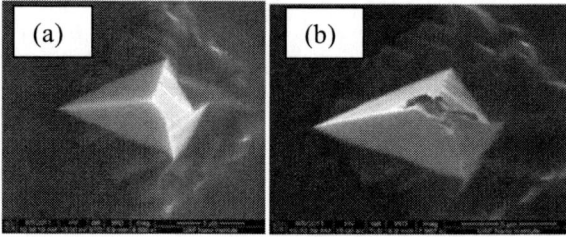

Fig. 7. SEM micrographs of Tap 190DLC AFM probes. (a) pristine tip. (b) worn-out tip.

C. Anodic Bonding

Access holes were drilled in the scratched Pyrex chip and it was capped off by a silicon chip through anodic bonding. For proper bonding, a temperature of 450° C and a DC voltage of 900 V were applied. A typical bond takes between 10 – 15 minutes. The dashed area in Fig. 8 represents the permanent, transparent layer of SiO_2 that is formed during the bonding process that seals off the nanochannel.

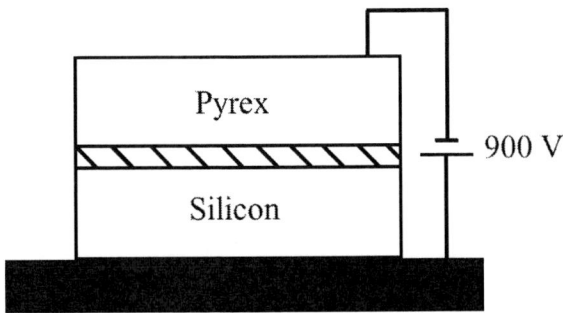

Fig. 8. Anodic bonding set up sealing off the nanochannel with a robust SiO₂ layer at a temperature of 450° C.

Fig. 9 shows a schematic of the completed nanochannel device. The inlet and outlet connectors are glued on directly above the two thru holes drilled in the Pyrex. The nanochannel region can be inspected via optical microscopy through the transparent Pyrex layer and is ready for flow testing.

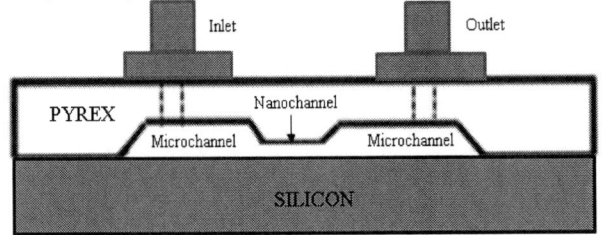

Fig. 9. Schematic of Pyrex nanochannel device.

III. FLOW TESTS

A. Wetting Steps

Prior to flow testing, several wetting steps were mandatory to ensure proper fluid flow. The channels were sequentially treated with acetone, methanol, isopropyl alcohol, and DI water. The device was placed in a vacuum desiccator and each wetting fluid was added to the device by a syringe and pulled through the channel due to the high pressure gradient. Each step lasted 30 minutes and increased the overall wettability of the channels. Flow patency in the capped-off Pyrex nanochannel device was verified through two different methods.

B. Fluorescein Isothiocyanate

Fig. 10 demonstrates the passage of fluorescein isothiocyanate (FITC) solution initiated by simultaneously imposing high pressure at the inlet and vacuum at the outlet of the nanochannel. FITC was introduced by a connecting a syringe to the inlet. The syringe exerted a constant pressure of FITC to the device, as the lid contacted the top of the syringe with a constant force. The outlet was under vacuum, resulting in the flow of FITC from the upstream to the downstream through the nanochannel.

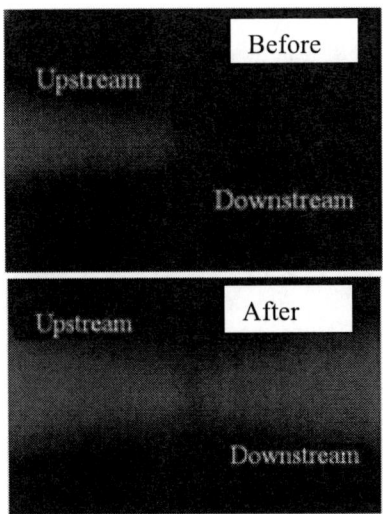

Fig. 10. Passage of fluorescein isothiocyanate (FITC) solution through the nanochannel.

C. Nanobeads

The second method of flow patency is through nanobead translocation. Fig. 11 demonstrates the translocation of FluoSpheres® (20-nm in diameter), negatively charged nanobeads through the nanochannel. The inlet is filled with the nanobeads and the outlet with Phosphate Buffer Saline (PBS). Electrodes are inserted in both the inlet and outlet connectors and hooked up to a DC voltage supply. A +10 V_{DC} bias is applied to the downstream reservoir. The positive voltage pulled the nanobeads through the nanochannel to the downstream, verifying the flow patency of the device.

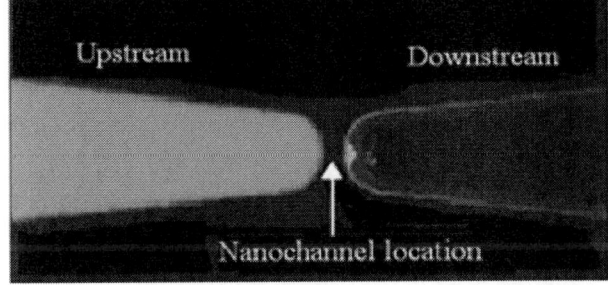

Fig. 11. Translocation of 20-nm negatively-charged nanobeads (FluoSpheres, Invitrogen Corp.) through the nanochannel.

IV. CONCLUSIONS

The present paper describes the design, fabrication, and testing results of a nanochannel device fabricated by the integration of MEMS microfabrication and AFM nanolithography. Detailed studies were carried out to characterize the relationship between AFM control parameters and the resulting nanochannel dimensions. For flow testing, a microchannel pattern was etched into a Pyrex substrate at an 8 μm depth. A nanochannel with a depth of 45 nm and width of

978-1-4673-1122-9/12 $31.00 © 2012 IEEE

approximately 175 nm that bridges the microchannels was then realized via AFM nanolithography. Anodic bonding was used to cap off the Pyrex chip with a matching silicon chip. Flow patency of the device was verified through two separate tests. First, FITC was pumped through the nanochannel by the combination of a syringe and a vacuum desiccator. Second, translocation of negatively charged nanobeads through the nanochannel demonstrated the potential of the nanochannel to characterize biomolecules.

ACKNOWLEDGEMENT

The current project is partially supported by the NSF ECCS division (ECCS-1137948) and the NSF GK-12 program.

REFERENCES

[1] J. A. Schloss, "How to get genomes at one ten-thousandth the cost," Nature Biotechnology, Vol. 26, pp. 1113-1116, 2008.

[2] V. Tabar-Cossa, D. Trivedi, M. Wiggin, et al., "Noise analysis and reduction in solid-state nanopore," Nanotechnology, vol. 18, 305505, 2007.

[3] Hongbo Peng, Xinsheng Sean Ling "Reverse DNA Translocation Through a Solid-state Nanopore by Magnetic Tweezers," Nanotechnology, vol. 20, 185101, 2009.

[4] S. K. Min, W. Y. Kim, Y. Cho, K. S. Kim, "Fast DNA sequencing with a graphene-based nanochannel device," Nature Nanotechnology, 6, 162-165, 2011.

[5] M. B. Stern, M. W. Geis, J. E. Curtin. "Nanochannel fabrication for chemical sensors," J. Vac. Sci. Technol. B, vol. 15, pp. 2887-2891, 1997.

[6] H. Cao, Z. N. Yu, J. Wang, et al., "Fabrication of 10 nm enclosed nanofluidic channels," Appl. Phys. Lett., vol. 81, pp. 174-176, 2002.

[7] L. J. Guo, X. Cheng, C. F. Chou, "Fabrication of size-controllable nanofluidic channels by nanoimprinting and its application for DNA stretching," Nano Lett., vol. 4, pp. 69-73, 2004.

[8] J. C. Rosa, M. Wendel, H. Lorenz, et al., "Direct patterning of surface quantum wells with an atomic force microscope," Appl. Phys. Lett., vol. 73, pp. 2684-2686, 1998.

[9] J. Regul, U. F. Keyser, M. Paesler, et al., "Fabrication of quantum point contacts by engraving GaAs/AlGaAs heterostructures with a diamond tip," Appl. Phys. Lett., vol. 81, pp. 2023-2025, 2002.

[10] Budget Sensors AFM Probes. http://www.budgetsensors.com/?gclid=CJL675-h8aoCFSlgTAodWyVjOA. Date accessed: August 21, 2011.

[11] Z. Q. Wang, N. D. Jiao, S. Tung, Z. L. Dong, "Atomic Force Microscopy-Based Repeated Machining Theory for Nanochannels on Silicon Oxide Surfaces," Applied Surface Science, 257, 3627-3631, 2011.

[12] Ampere A. Tseng, "A comparison study of scratch and wear properties using atomic force microscopy," Applied Surface Science, 256, 4246–4252, 2010.

Effect of Nonlinear Vibration on Double Region of Synchronized Frequency Responses in Mechanically Coupled Beam-Shaped Oscillator System

Takumi Itoh [1], Dong F. Wang [1,*], Tsuyoshi Ikehara [2], Mamoru Nakajima [1], and Ryutaro Maeda [2]

[1] Micro Engineering & Micro Systems Laboratory, Ibaraki University (College of Eng.), Hitachi, Ibaraki 316-8511, JAPAN
[2] Research Center for Ubiquitous MEMS and Micro Engineering (UMEMSME), AIST, Tsukuba, Ibaraki 305-8564, JAPAN
(*Tel: +81-294-38-5024; Fax: +81-294-38-5047; E-mail: dfwang@mx.ibaraki.ac.jp)

Abstract-- **This paper reports that the effect of nonlinear vibration on double region of synchronized frequency responses in a mechanically coupled beam-shaped oscillator, so as to further expand the applicable limits for practical mass perturbation. The resonant frequency was multiplied via synchronization by introducing a coupling overhang, and the double region was magnified from 30 Hz to over 400 Hz by increasing the driving voltage (induced power). This magnification is believed to be related to nonlinear characteristic of coupled oscillator, and the relation between nonlinearity and synchronization should be clarified to achieve a design principle of mechanically coupled mass sensor for various ultimate sensing applications.**

Keywords- Synchronization, Nonlinear vibration, Miniaturization; Coupling overhang, Beam-shaped oscillator system

I. INTRODUCTION

The sensitivity and resolution of an independent cantilever (self-supporting) depend on various factors, i.e., total effective resonant frequency, linear and nonlinear spring constant, quality factor, and noise level at a given circumstance [1-2].

Synchronization of coupled oscillator systems has been studied for over three centuries, which is a common phenomenon in both nature and human physiology. Christian Huygens, who remarked that two slightly out-of-step pendulum-like clocks become synchronized after they are attached to a same thin wooden board [3].

Synchronized oscillation of a mechanically coupled oscillator is represented not only the resonant frequency ratio 1:1 but also the resonant frequency ratio 1:2, 1:3 in case of high degree vibration [4-5]. As schematically shown in Fig. 1, the resonant frequency of the long cantilever is ω_1, while the short one is ω_2, respectively. if the resonant frequencies of two geometrically designed cantilevers obey the relation of $\omega_2 \approx n\omega_1$ (n is integer), they will synchronize with each other and lock at $\omega_2 = n\omega_1$ when they are coupled through a coupling overhang. Even if the resonant frequency ω_1 slightly shifts to ω_1', the synchronized frequency ω_2 also shifts slightly to $\omega_2' = \omega_1'$. This means that the signal can be enhanced by a factor of n if mechanically coupled system is constructed by two independent oscillators. This might be applied to further improve the detectable limit (resolution) of cantilever-based resonators for various sensing applications. [6-10]

Generally, nonlinearity, existed in all vibration systems, is believed to be attributed to synchronized oscillation, but factors related to strengthening nonlinearity have not yet been well elucidated [11]. However, geometrical miniaturization (length, width, and thickness etc.), spring constant of coupling, and displacement (vibration amplitude) are important factors influencing the nonlinearity so as to influencing the synchronized responses in a coupled system.

In our past studies, a synchronized oscillation with improved phase noise was reported at MEMS 2010 [12] using a mechanically coupled oscillation system, consisting of two singly-clamped beam-shaped cantilevers. A reformed mechanically coupled oscillation system was further designed and frequency enhancement was reported at Transducers 2011 [13] that the frequency response can be doubled when the detecting cantilever (higher frequency ω_2) is synchronized with the sensing one sensing one (lower frequency ω_1).

Based on a simplified coupled two-beam oscillator system, this paper focuses on the effect of vibration amplitude on synchronized responses (double region) through changing driving voltage.

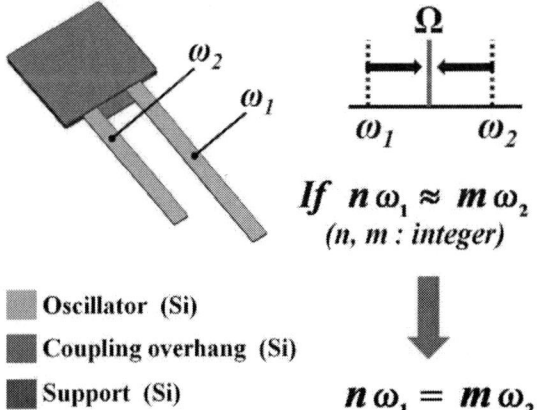

Fig. 1. Schematic figure for mutual synchronization, and the definition of oscillator, support and coupling overhang.

II. STRUCTURAL DESIGN, MICRO-FABRICATION AND MEASUMENT SET-UP

In light of foregoing studies [12-13], a mechanically coupled oscillator comprising of two cantilevers and coupling overhang was selected. The simple geometric structure of the oscillator is believed to be suitable for clarifying the factors related to strengthening nonlinearity and their functions in synchronized oscillation.

$$\omega_0 = \frac{0.162t}{l^2}\sqrt{\frac{E}{\rho}} \qquad (1)$$

Equation (1) is often used as a resonant frequency of theoretical beam-shaped cantilevers. E is the Young's modulus of oscillator, ρ is degree of density and t, l are the thickness, and length [14-15], respectively. The vibration modes have been simulated using a commercial Coventor Ware^{TM} software to guarantee the two resonant frequencies to mutually obey the relation of $\omega_2 \approx 2\omega_1$, which are expected to be coupled through a coupling overhangs. Two cantilevers, with theoretical resonant frequencies of 300.18 kHz and 150.11 kHz respectively, have been therefore fixed for this work. As a result, the resonant frequency ratio became $1.9997 \approx 2.000$.

For fabricating the above mechanically coupled beam-shaped oscillator system, an SOI (silicon on insulator) wafer with a 2.5-μm-thick top silicon layer, 300-nm-thick SiO_2 layer, and 400-μm-thick silicon substrate was used as a starting material. The topside silicon was first thinned to 600 nm by reactive ion etching (RIE) using SF_6, and then patterned by lithography and etched using a deep reactive ion etching (deep RIE) to form the mechanically-coupled cantilever pattern. The substrate silicon was isotropically-etched by RIE through the etching window of insulating SiO_2. The coupled cantilever structures were released by wet etching of SiO_2 in HF and following supercritical point drying. Several oscillator systems are fabricated with a micro-mechanically coupled overhang.

As typically shown in Fig. 2, the "mechanically coupled beam-shaped oscillator" system consists of the longer one (67.6 μm long ×5.00 μm wide × 600 nm thick), functioned as sensing part and referred to as oscillator 1, and the other (48.1 μm long ×5.00 μm wide × 600 nm thick), functioned as detecting part and referred to as oscillator 2, coupling overhang (20 μm long ×5.00 μm wide × 600 nm thick), has been geometrically designed, respectively.

All measurements were performed in a vacuum of ~7.5 Pa at room temperature, and velocity range of laser Doppler vibrometer fixes at 1000 mm/sec/V. The sample was mounted on a piezoelectric ceramic plate, which can be vibrated by applying an AC voltage using a signal generator. The resonant frequency and effect of linear and nonlinear vibration of oscillator 1 and oscillator 2 are measured using a network analyzer. Moreover, double region of synchronized frequency responses is measured using a spectrum analyzer, as shown in Fig. 3.

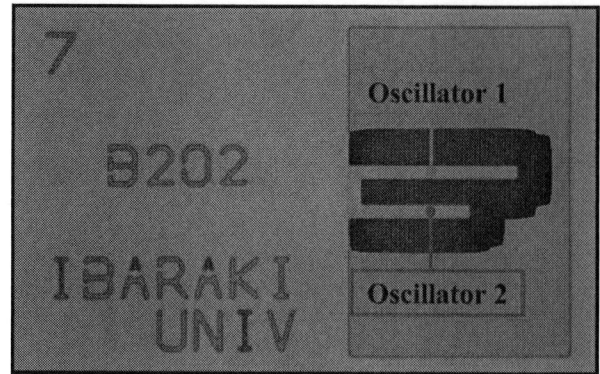

Fig. 2. A typical micrograph of the mechanically coupled beam-shaped oscillation system.

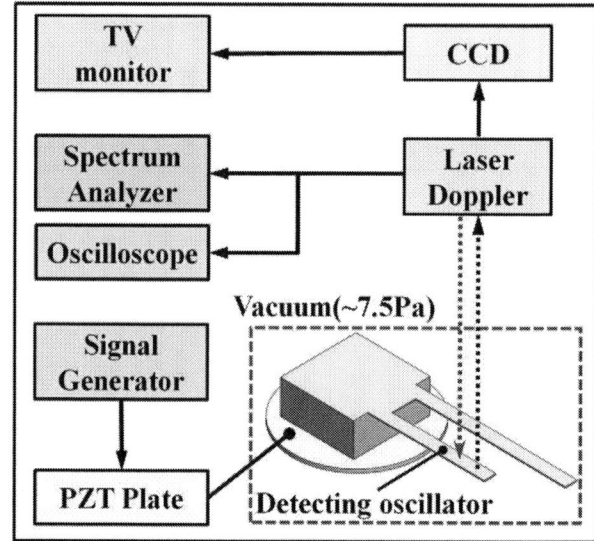

Fig. 3. Experimental set-up to measure for the double region of synchronized frequency response by spectrum analyzer.

III. MEASURING EXPERIMENT

A. Resonant frequency

As shown in Fig.4, resonant frequencies of the mechanically coupled beam-shaped oscillator at span of frequency 5 kHz from network analyzer (50 Hz/step with average of 10 scans). The resonant frequency of oscillator 1 is 177.38 kHz and its Q-factor is 3502 as shown in Fig. 4(a). The resonant frequency of oscillator 2 is 351.66 kHz and its Q-factor is 5104 as shown in Fig. 4(b). The resonant frequency increased approximately 18% from the value of structural design. This might be due to thickness change as well as unevenness. The resonant frequency ratio of those oscillators is 1.983, which decreased approximately 0.8% from the value of structural design. However, the empirically measured frequency ratio is approximately obeying the relation of $\omega_2 \approx 2\omega_1$ as shown in Fig. 1.

978-1-4673-1122-9/12 $31.00 © 2012 IEEE

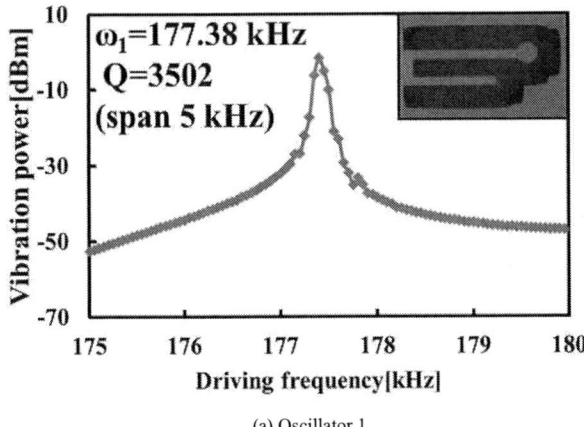

(a) Oscillator 1

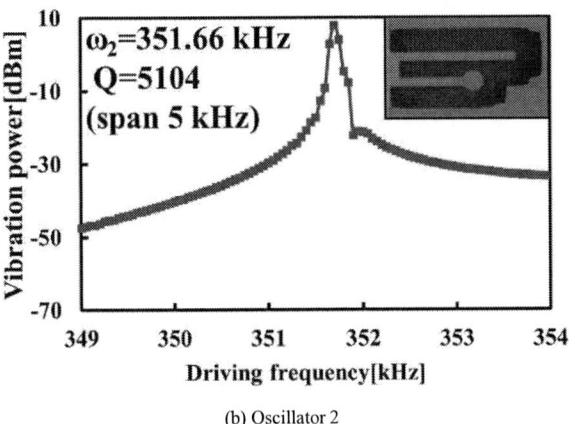

(b) Oscillator 2

Fig. 4. Resonant frequencies of the mechanically coupled beam-shaped oscillator at span of frequency 5 kHz from network analyzer, where (a): Oscillator 1; and (b): Oscillator 2.

B. Effect of driving AC voltage on resonant frequency

The nonlinear responses of the oscillator with higher resonant frequency (ω_2) in primary vibration (mode 1) were shown in Fig. 5. Oscillator 2 is measured by network analyzer. The X-axis is the driving frequency with 2.5 Hz/step, and the Y-axis is vibration power (amplitude). The resonant frequency was measured as 351.64, 351.67 and 351.69 kHz when the AC voltage was applied at 1.0, 3.0 and 5.0 Vpp. In 3.0, and 5.0 Vpp, un-symmetric behaviors can be observed at the resonant point, which is believed to be attributed to an obvious nonlinear vibration, caused by increasing the driving AC voltage. Applying an increased AC voltage to a piezoelectric ceramic plate means that vibration power becomes large. This larger vibration power is believed to attribute to increasing amplitude (displacement), and to lead to strengthening the nonlinearity.

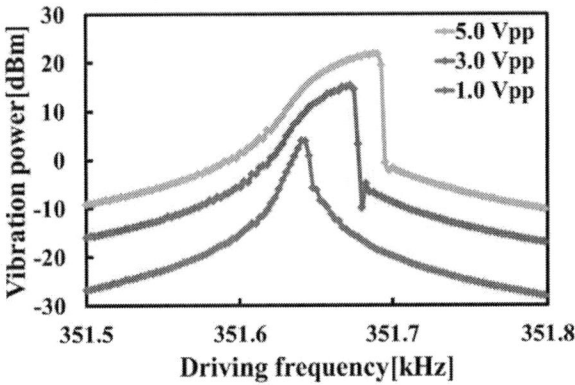

Fig. 5. Linear and nonlinear response of the oscillator with higher resonant frequency in mode 1 from network analyzer.

C. Demonstration of synchronized oscillation

Double synchronization was demonstrated by spectrum analyzer (375 Hz/step with average of 10 scans) when the oscillator with lower resonant frequency ω_1 was driven by an AC voltage of 5.0 Vpp at its own resonant frequency from a signal generator and the mechanically coupled oscillator with higher resonant frequency ω_2 was self-oscillated. A typical example of demonstration of synchronized oscillation was shown in Fig. 6. 177.38 kHz (ω_1) was approximately doubled to 354.76 kHz ($2\omega_1$) from lower frequency oscillator (oscillator 1) to higher frequency oscillator (oscillator 2) by double synchronization through a coupling overhang. The result less than 5.0 Vpp was shown in the figure, since the signal of synchronized oscillation is weaker under a lower driving AC voltage. Two noise spikes are also observed around 300 kHz of the driving frequency.

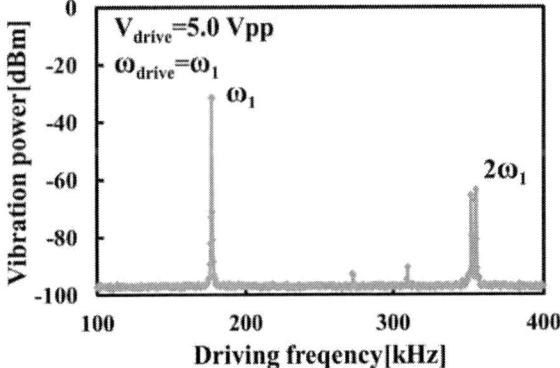

Fig. 6. Demonstration of synchronized oscillation (double synchronization) Span from 100 to 400 kHz.

D. Magnification of synchronized oscillation

AS shown in Fig.7, from the plot, a plateau, with a frequency ratio of 2.000 corresponding to a double region was observed by spectrum analyzer (10 Hz/step with average of 10 scans) and this double region can be magnified from 30 Hz to over 400 Hz when the driving AC voltage was increased from 1.0 to 5.0 Vpp with the driving frequency around the lower resonant frequency ω_1. Usually, the diagram which shows the relation between double region of synchronized frequency responses and external force is called "Arnold's tongue", and the synchronized region is often represented by Arnold's tongue, which is the region of frequency entrainment in the parameter space.

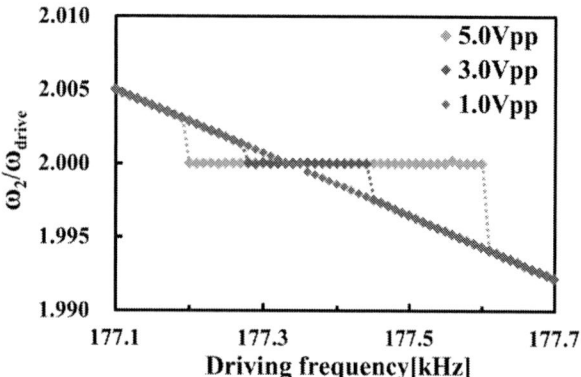

Fig. 7. Magnification of the double region (plateau) by increasing the driving Span of driving AC voltage from 1.0 to 5.0 Vpp on spectrum analyzer.

SUMMARY

Based on a micro-fabricated "mechanically coupled beam-shaped oscillator" system, the resonant frequency was multiplied via synchronization by introducing a coupling overhang, and the double region was magnified from 30 Hz to over 400 Hz by increasing the driving voltage (induced power). This magnification is believed to be related to nonlinear characteristic of the coupled oscillator system, but factors related to strengthening nonlinearity should be further studied in detail.

ACKNOWLEDGMENTS

Part of this work was supported by MEMS Inter University Network and performed in the Ubiquitous MEMS & Micro Engineering Research Center (UMEMSME) of National Institute of Advanced Industrial Science & Technology (AIST).

REFERENCES

[1] T. R. Albrecht, P. Grütter, D. Horne, and D. Rugar, J. Appl. Phys. 69, 668 (1991).

[2] A. N. Cleland and M. L. Roukes, J. Appl. Phys. 92, Noise processes in nanomechanical resonators, pp. 2758 - 2769 (2002).

[3] A. Pikovsky, M. Rosenblum, and J. Kurths,"Synchronization: A Universal Concept in NonlinearSciences", Cambridge University Press, Cambridge, 2001.M. Zalalutdinov, K. L. Aubin, M. Pandey, A. T.

[4] M. Zalalutdinov, K. L. Aubin, M. Pandey, A. T.Zehnder, R. H. Rand, H. G. Craighead, and J. M.Parpia, Frequency entrainment for micromechanical oscillator, Appl. Phys. Lett. 83, 16, 3281 (2003).

[5] S. Shim, M. Imboden, P. Mohanty, Synchronized Oscillation in Coupled Nanomechanical Oscillators, Science, 316, 6 (2007).

[6] E. A. Thundat, E. A. Wachter, S. L. Sharp, and R. J. Warmack, Micromechanical sensors for chemical and physical measurements, Appl. Phys. Lett. 66, pp. 3662 - 3667 (1995).

[7] H. J. Mamin and D. Rugar, Sub-attonewton force detection at millikelvin temperatures, Appl. Phys. Lett. 79, pp. 3358-3360 (2001).

[8] T. Ono, D.F. Wang and M. Esashi, Mass sensing with resonating ultrathin double beams, IEEE Sensors 2003, pp.825-829.

[9] T. Ono and M. Esashi, Magnetic force and optical force sensing with ultrathin silicon resonator, Rev. Sci. Instrum. 74, pp.5141-5146 (2003).

[10] K. L. Ekinci, X. M. H. Huang, and M. L. Roukes, Ultrasensitive nanoelectromechanical mass detection, Appl.Phys.Lett. 84, pp. 4469-4471 (2004).

[11] Y. Jiang, T. Ono and M. Esashi, Modeling and experrimental analysis on the nonliearily of single crystal silicon cantilevered microstructures, IEEJ Transactions on Sensors and Micromachines, pp.195-196.

[12] D.F. Wang, J. Feng, T. Ono, M. Esashi and X. Ye, Micromechanically-coupled resonated system for synchronized oscillation with improved phase noise , IEEE MEMS 2010, pp. 703-706 (2010).

[13] M. Nakajima, D.F. Wang, T. Ikehara and R. Maeda, Synchronized oscillation in micro mechanically coupled opposite C-shaped cantilever-based oscillator system, IEEE Transducers 2011, pp. 1492-1495 (2011).

[14] N.V. Lavrik, M.J. Sepaniak, P.G. Datskos, Cantilever transducers as a platform for chemical and biological sensors, Rev. Sci. Instr. 75, 2229-2253 (2004)

[15] N. Kianoush, T. Prateek and B. Oliver, Geometrical optimization of resonant cantilever sensors, Tranducers 2007, pp. 10-14.

Characterization of a Multi-layered MEMS Pressure Sensor Using Piezoresistive Silicon Nanowire within Large Measurable Strain Range

Liang Lou, Songsong Zhang, Woo-Tae Park, Member, IEEE, Lishiah Lim, Dim-Lee Kwong, Fellow, IEEE, and Chengkuo Lee[*], Member, IEEE

[*]Department of Electrical & Computer Engineering, National University of Singapore, Singapore 117576
elelc@nus.edu.sg

Abstract— **Multilayered pressure sensors using piezoresistive silicon nanowires (SiNWs) are characterized using center displacement loading approach. The silicon nanowire (SiNW) is embedded in a multilayered diaphragm structure comprising of silicon nitride and silicon oxide. By leveraging the high fracture stress and intrinsic tensile stress of silicon nitride layer to produce a flat diaphragm, we can create compressive strain to the SiNW as large as 1.7% without damaging the diaphragm. The equivalent pressure to break the diaphragm is derived as high as above 500 psi. The sensitivity at low pressure application region (<45 psi) is derived as around 0.25% psi^{-1}. The relationship between SiNW resistance change and applied strain is measured and investigated with 2 μm and 5 μm SiNWs for both scientific and practical points of view. This approach also demonstrates the validity to reveal the SiNW properties under large strain and the exploration provides a good reference for future SiNW based MEMS sensor design.**

Keywords- pressure sensor; silicon nanowire

I. INTRODUCTION

Piezoresistive transduction, as one of the earliest demonstrations of mechanisms suitable for microdevices, has been widely used in microelectromechanical system (MEMS) sensors since the first report by Smith in 1954 [1,2]. Micromachined piezoresistive sensors are the most widely used microsensors in industry today, partially due to the relatively straightforward interface circuitry and the ease of process integration. Other than the well-known automotive applications for pressure sensors including engine manifold monitoring, tire pressure monitoring, and both oil and brake fluid pressures [3-7], pressure is one of the most important physical parameters for various biomedical applications [8-11], for example, to measure intrauterine pressure during birth, to monitor the inlet and outlet pressures of blood in kidney dialysis, to monitor the pressure in the cardiovascular system, to measure and control the vacuum level used to remove fluid from the eye for eye surgery, etc. One of the earliest research efforts in biomedical applications is a pressure sensor developed by Samaun et al. for biomedical instrumentation applications including cardiovascular catheterization [12]. A 50-μm-thick silicon substrate was used to fabricate a single-crystal silicon diaphragm with 1.2 mm in diameter and 5 μm

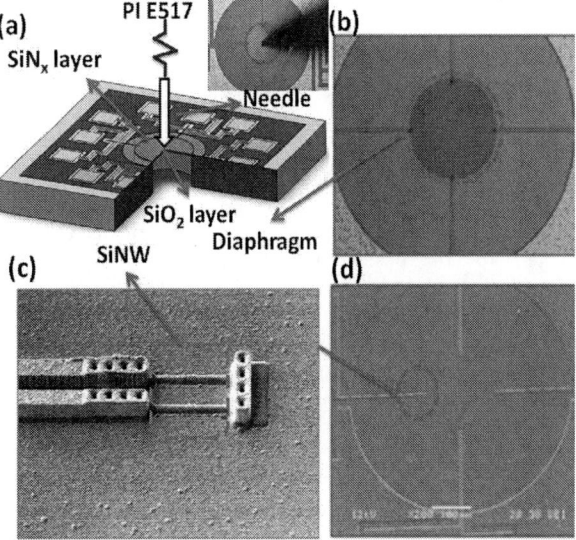

Fig.1. (a) Schematic drawing of the pressure sensor (b) OM photo of the fabricated device (c) SiNWs after metal deposition and patterning (d) SEM photo of the fabricated device.

in thickness, where the diaphragm with four integrated piezoresistors made by a diffusion process was released the anisotropic wet etching technique. They also developed a technique to precisely define the thickness of the membrane within 1 μm thickness, while the size of diaphragm could be reduced to as small as 0.8 mm. Another milestone in pressure sensor development was the merging of a sensor and its interface circuitry on a monolithically fabricated die. This was first implemented by Borky and Wise in 1980 with their development of a micromachined piezoresistive pressure sensor integrated into a triple-diffused bipolar circuit process [13]. By leveraging the advanced semiconductor process technology, NEMS based biosensors using SiNW have been reported as promising DNA and protein sensors [14, 15]. Due to the large piezoresistive effect of SiNWs [16-18], using SiNWs as the piezoresistive sensing element in the pressure-deformable diaphragm operated in the tensile strain region has

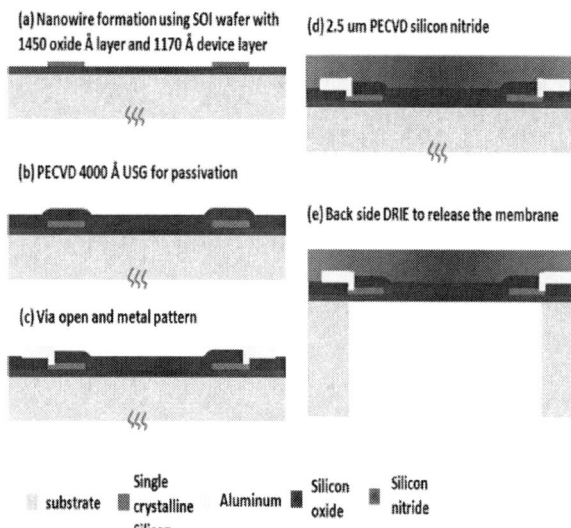

(a) Nanowire formation using SOI wafer with 1450 oxide Å layer and 1170 Å device layer

(b) PECVD 4000 Å USG for passivation

(c) Via open and metal pattern

(d) 2.5 um PECVD silicon nitride

(e) Back side DRIE to release the membrane

substrate　Single crystalline Silicon　Aluminum　Silicon oxide　Silicon nitride

Fig.2 Fabrication process flow of the SiNW and the diaphragm structure.

II. DESIGN AND FABRICATION

A. Pressure Sensor Design

The schematic drawing of the pressure sensor is illustrated in Fig.1(a), while the SEM photo of a microfabricated multi-layered diaphragm with diameter of 200 μm and the optical microscope (OM) photo of a whole device chip are shown in Fig.1(b) and (c). Inset shows a 5 μm SiNW after metal patterning. The multi-layered diaphragm comprises the SiNx layer and the SiO2 layer. The pressure sensor chip shown in Fig.1(c) is in square shape with dimension of 2 mm by 2 mm. The yellow color refers to the SiNx. The SiNWs are located along <110> direction at the edge of the diaphragm for maximum strain extraction.

B. Fabrication Process

The process flow to fabricate the diaphragm starts with a Si (100) wafer as shown in Figure.2. After photolithography, the width of photoresist pattern with respect to the nanowires is about 160 nm. By using plasma comes from feeding gas of He/O2 + N2, where the ratio of He/O2 is 70/30, the width of photoresist pattern is reduced to 110 nm. The He/O2 is deployed to oxidize the photoresist and the N2 is used to smoothen the surface of the photoresist. Thus the photoresist trimming process can achieve the critical dimension to around 110 nm. After DRIE for patterning Si NWs, thermal oxidation is conducted to further shrink down the dimension of SiNWs such that the derived cross section of a SiNW is around 90 nm by 90 nm. Then p-type implantation using BF2+ is performed with a dosage of 1E14 ion/cm2, and followed by annealing for activation. Next, an extra SiO2 layer of 4000 Å is deposited for passivation. After via open and metal patterning, a 2.5 μm

been reported [19]. A major research efforts have been devoted to the characterization of SiNWs mechanical properties based on miniaturized or microelectromechanical systems (MEMS) based testing platform [16,20-22]. Generally, these characterization approaches are kinds of method using the bulky test platform to conduct the four points bending for uniform stress [20] or utilizing specially designed MEMS based test platforms [21,22]. However, up to date only piezoresistive property of SiNWs of large tensile strain has been reported, while the measured data of compressive strain is limited to 0.06% [18]. Four-point-bending bulky test platform suffers the fracture issue under large compressive strain, and MEMS device based testing platform usually measures a SiNW by artificially assembling the SiNW to the MEMS device or with synthesized SiNW, in which such SiNWs are restricted in a small compressive strain range due to buckling issues. In practical applications, the SiNWs are usually embedded in the thin films to form a sensor structure which is used not only in tensile state but in compressive state as well. As such investigation of SiNW behavior within a wide range of compressive strain is indispensible [20]. In the recent report by S.I. Kozlovskiy et al, the SiNW under large compressive stress (~1 GPa) is computationally investigated for SiNW oriented along <100>, <110> and <111> directions respectively. Here we report a MEMS pressure sensor comprising SiNW piezoresistors characterized in the large compressive strain range. By leveraging the large fracture stress of silicon nitride, the measured resistance data versus compressive strain up to 1.7% is recorded for <110> SiNWs embedded in the multilayered pressure sensor, and the pressure sensing range of the sensor is derived as high as above 500 psi.

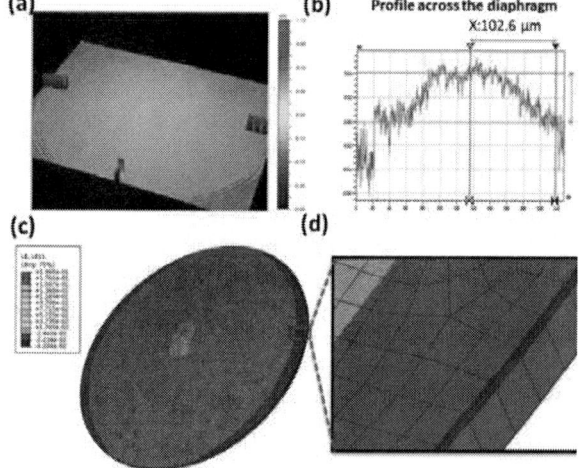

Fig.3. (a) 3-demensional topography of fabricated diaphragm structure (b) Curve showing surface roughness across the wafer (c) ABAQUS simulation of the longitudinal strain in the SiNW (d) Zoom-in picture showing the meshing and multi-layered structure at the membrane edge of SiNW area.

978-1-4673-1122-9/12 $31.00 © 2012 IEEE

silicon nitride film is deposited to compensate the compressive stress in the SiO2 layer. Finally, deep reactive ion etch (DRIE) is conducted to release the diaphragm structure.

III. MODELING AND CHARACTERIZATION

A. Characterization Set-up and Modeling

A white light interferometer (Vecco NT3300) system is used to record the surface profile of the microfabricated diaphragm as shown in Fig. 3(a). Fig. 3(b) shows the measured surface roughness across the membrane. This data gives almost zero deflection, indicating that the SiNW is basically at a normal state without pre-stress. In comparison with the previous work as shown in [19], the diaphragm is improved to eliminate the initial buckling profile, avoiding making the diaphragm potentially impractical for real situation in the long run. To investigate the strain distribution across the wafer, especially at the SiNW area under a center point displacement load, finite element analysis (FEA) software ABAQUS is used in a non-linear analysis mode. Three layer structure model is built comprising of 1450 Å BOX layer, 4000 Å PECVD oxide and 2.5 μm SiN layer as show in Fig (c) and (d). The SiNW is located between the two oxide layers at the diaphragm edge. The longitudinal strain of the SiNW is extracted and averaged from the elements at the membrane edge as the displacement load is given from 0 μm to 20 μm. The simulation results shows that the strain of the diaphragm edge is in a fairly good linear relationship against the center displacement (Fig.4 inset).

To experimentally characterize the pressure sensor with SiNWs embedded in the MEMS devices to a large strain range, a specially fabricated tungsten needle is attached to a manipulator controlled by a PZT system using E-517 Digital Piezo Controller. The diameter of the fabricated needle tip is around 300 nm, which can be reasonably viewed as point in comparison with the 200 μm diaphragm in diameter. The needle is deployed to push the diaphragm center as illustrated by the white arrow mark in Fig 1 (a), thus to deform the membrane and transmit the strain to the SiNW. Meanwhile, the electrical measurement of the SiNW resistance is conducted using the semiconductor parameter analyzer system (Agilent 4156C). The experiment is conducted on a probe station platform under a microscope. The tip is carefully aligned to be positioned directly on top of the center of the membrane, and then moves in perpendicular to the membrane with 1 μm displacement in each step. This approach takes advantage of the high fracture stress of silicon nitride and the PZT-based precise displacement control. Such a set-up can exert quite high strain to the SiNW without demanding high pressure which is necessary to be applied in the bulge test [23-25].

B. Results and Discussions

Fig.4 shows the measured resistance change against tip displacement for the sensors with 2 μm and 5 μm SiNWs. It shows that the SiNW resistance keeps constant initially until the on-set point at about 2.0 μm, indicating that the strain introduced by the tip is transmitted onto the SiNW. The resistance keeps dropping down to 17.7% for the 5 μm SiNW and to 15.8% for the 2 μm SiNW respectively as more displacement is continuously applied to tip, while the membrane breaks when tip displacement reaches at 22 μm and 14 μm, i.e., diaphragm center displacement 20 μm and 12 μm accordingly. This data also exhibits a larger resistance change in comparison with previous reported data for bulky silicon under a compressive state, where such data are usually less than 8%. The enlarged resistance range further proves the effectiveness of this testing approach. Moreover, the two curves show good linear behavior up to center displacement of about 5 μm in Fig. 4. The 2 μm SiNW curve gives slightl deeper slope than the one of 5 μm SiNW as indicated by the linear fitting line. This is due to that the 5 μm SiNW sensor senses wider span of longitudinal strain across its length than the one of 2 μm SiNW; in other words, the 5 μm SiNW has a lower average compressive strain than the 2 μm SiNW under the same diaphragm center displacement.

To derive the equivalent pressure with the displacement loading that provides identical strain at the SiNW area, the loading condition in the Finite-Element-Method (FEM) model is changed from the displacement at the center of the diaphragm to a uniformly applied differential pressure over the diaphragm. The strain at the SiNW area under pressure application is extracted to compare with that under a certain amount of displacement loading. These two pieces of strain data are iteratively converged together by adjusting the pressure loading onto the diaphragm. For both of the measured pressure sensors, the equivalent pressure is 20 psi when the

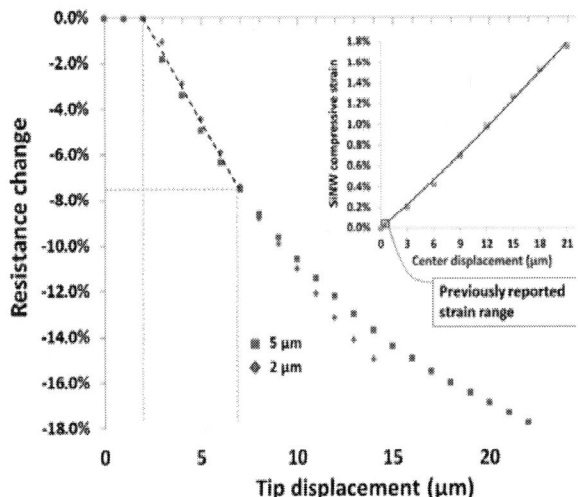

Fig.4 The SiNW resistance change against tip displacement, inset shows the SiNW strain against the center displacement.

diaphragm center displacement is 3 μm. Thus, the sensitivity of the pressure sensor is derived as around 0.25% psi^{-1} ; In contrast, the equivalent pressure for the 20 μm center displacement is 600psi, indicating a sensitivity of 0.03 %. The sensitivity of the pressure sensor drops in a nonlinear manner as the pressure increases. Such relationship is mainly attributed to the nonlinear relationship between the diaphragm strain against the applied pressure [26]. Based on the plate theory, the diaphragm displacement in the perpendicular direction and the diaphragm strain is considered to be in a linear relationship against the applied pressure when its central deflection is smaller than the diaphragm thickness. As introduced above, the thickness of the pressure sensor is 3 μm, thus, the linear region of pressure application is derived approximately as within 45 psi. Furthermore, according to the displacement measured in the above fracture experiments, the equivalent pressure to break the diaphragm is derived to be above 500 psi for pressure sensor with multilayered diaphragm of 1450 Å BOX / 4000 Å oxide layer / 2.5 μm SiNx, indicating very high mechanical strength of the diaphragm that is able to survive even under very high pressure without damage.

Besides, there is nonlinear region observed for both cases when the center displacement is larger than 5 μm till the fracture point. The resistance change decreases as the strain increases in this region, in which it indicates the gauge factor of the SiNW drops as the strain increases. Such nonlinear behavior has been investigated firstly by Katuhisa Suzuki et al. [27], and stress decoupling of the degenerate valence band into two bands of parlate and oblate ellipsoidal energy surface is proposed to explore the origin of the piezoresistance of p-silicon diffused layers. Our result is a further evidence to show the non-linear behavior of the piezoresistance of p-type silicon in an extended compressive strain region. Besides, the SiNW is reported with giant factor by He and Yang [17]. The giant factor appeared in the compressive range is lower than 0.06%, which corresponds to the red box shown in the inset of Fig.4. The evolution of the SiNW behavior under extended region of compressive strain could be of great value towards revealing the origin of the giant piezoresistive effect. From scientific point of view, the p-type silicon piezoresistive mechanism is still not fully understood [28], and it requires more experimental effort. Additionally, in the practical applications, this result indicates that calibration is required to offset the non-linearity when such sensors using the SiNW under large compressive strain.

IV. CONCLUSION

In this paper, we reported the experimental data of a multilayered pressure sensor using <110> direction SiNW as sensing element. The equivalent pressure to break the diaphragm is derived as high as above 500 psi. The sensitivity at low pressure application region (<45 psi) is derived as

around 0.25% psi^{-1}. We also successfully achieved measurement in large compressive strain, i.e., 1.7%, with resistance change up to 17.7% by leveraging the intrinsic tensile stress of silicon nitride to produce a flat multi-layered diaphragm comprising of silicon nitride, silicon oxide and embedded SiNWs, and by forming a strong membrane attributed to the high fracture stress of silicon nitride. Our results also revealed the SiNW properties in the large compressive strain region where it has not been reported until now. It fills the missing link between actual behavior of SiNWs in sensor configuration and preliminary data of suspended SiNWs measured at bulky testing platform.

ACKNOWLEDGMENT

This work was supported by grants from Academic Research Committee (ARC) Fund MOE2009-T2-2-011 (R-263000598112) at the National University of Singapore, and A*STAR, SERC under Grant Nos. 0921480070, 1021650084, 1021010022 and 1021520013. Liang Lou would like to thank the PhD research scholarship received from Electrical and Computer Eng. Dept of National University of Singapore.

REFERENCES

[1] C. S. Smith, "Piezoresistance effect in germanium and silicon", *Phys. Rev.*vol. 94, pp. 42–49, 1954.

[2] Peake E R, Zias A R, and Egan J V, "Solid-state digital pressure transducer IEEE Tran. Electron Devices" vol.16 870–876,1969

[3] Barlian A A, Park W-T, Mallon J R, Rastegar Jr A J and Pruitt B L, " Review: Semiconductor Piezoresistance for Microsystems" Proc. of The IEEE , vol. 97, pp. 513-552, 2009.

[4] Fleming W J, "Overview of automotive sensors", IEEE Sensors J.vol.1 pp.296-308, 2001

[5] Esashi M, Sugiyama S, Ikeda K, Wang Y, and Miyashita H,"Vacuum-Sealed Silicon Micromachined Pressure Sensors", Proc. of The IEEE, vol. 86, pp.1627-1639,1998.

[6] Eddy D and Sparks D, "Applications of MEMS technology in automotive sensors and actuators", Proc. of The IEEE,vol. 86,pp.1747-1755, 1998.

[7] Eaton W P and Smith J H , " Micromachined pressure sensors: review and recent developments" Smart Mater. Struct.vol. 6, pp530–9, 1997

[8] Marco S, Samitier J, Ruiz O, Morante J R and Steve J E "High performance piezoresistive pressure sensors for biomedical applications using very thin structured membranes", Meas. Sci. Technol.vol.7, pp.1195–203, 1996

[9] Katuri K C, Asrani S, and Ramasubramanian M K, "Intraocular pressure monitoring sensors", IEEE Sensors J, vol. 8, pp.12-19, 2008

[10] Peng C, Ko W H, Young D J, "Wireless Batteryless Implantable Blood Pressure Monitoring Microsystem for Small Laboratory Animals" IEEE Sensors J, vol. 10, pp, 243-254, 2010

[11] Chatzandroulis S, Tsoukalas D and Neukomm.P.A, "A miniature pressure system with a capacitive sensor and a passive telemetry link for use in implantable applications", IEEE J Microelectromech. Syst. vol. 9 pp. 18-23, 2000.

[12] Samaun, Wise.K.D and Angell.J.B, " IC piezoresistive pressure sensor for biomedical instrumentation", IEEE Trans. Biomedical Eng. BME-20 pp.101-109, 1973

[13] Borky.J and Wise.K.D, "Integrated signal conditioning for silicon pressure sensors IEEE Trans. Electron Devices ED-26 pp. 1906–1910, 1979.

[14] T. Toriyama, Y. Tanimoto, S. Sugiyama, "Characteristics of silicon nano wire as piezoresistor for nano electro mechanical systems", The 14th IEEE International Conference on Micro Electro Mechanical Systems, pp. 305 - 308, 2001

[15] P.R.Nair, M.A.Alam, "Design Considerations of Silicon Nanowire Biosensors", Electron Devices, IEEE Transactions, Vol.54 , pp.3400 – 3408, 2007

[16] Neuzil.P, C.C.Wong, J.Reboud, "Electrically controlled giant piezoresistance in silicon nanowires", Nano Letters., Vol.10, pp.1248-1252, 2010.

[17] R. He, P. Yang, "Giant piezoresistance effect in silicon nanowires", Nature Nanotechnology, Vol. 1, pp 42 – 46, 2006.

[18] K. Reck, J. Richter, O. Hansen, and E. V. Thomsen "Piezoresistive effect in top-down fabricated silicon nanowires", Proc. IEEE MEMS, pp. 717-720, 2008.

[19] B.Soon, P. Neuzil, C. Wong, J. Reboud, H. Feng, C. Lee, "Ultrasensitive nanowire pressure sensor makes its debut", Procedia Eng, Vol.5, pp. 1127-1130,2010.

[20] Eivind. Lund and Terje G. Finstada , "Design and construction of a four-point bending based set-up for measurement of piezoresistance in semiconductors" Review of Scientific Instruments, Vol 75, pp. 4960-4966, 2004

[21] A. Lugstein, M. Steinmair, A. Steiger et al., "Anomalous Piezoresistance Effect in Ultrastrained Silicon Nanowires," Nano Letters, Vol. 10, no. 8, pp. 3204-3208, Aug, 2010

[22] J. M. Chen, and N. C. MacDonald, "Measuring the nonlinearity of silicon piezoresistance by tensile loading of a submicron diameter fiber using a microinstrument," Review of Scientific Instruments, Vol. 75, no. 1, pp. 276-278, Jan, 2004.

[23] Hatty.V., Kahn. H, Heuer.A.H., Gen. Electr., Cleveland. OH , " Fracture Toughness, Fracture Strength, and Stress Corrosion Cracking of Silicon Dioxide Thin Films", J.Microelectromech. Syst.,Vol.17 ,pp. 943 – 947, 2008.

[24] H. Huang, X.Z. Hu, Y. Liu, M. Bush, K. Winchester, C. Musca, J. Dell, L. Faraone, "Characterization of mechanical properties of silicon nitride thin films for MEMS devices by nanoindentation", J. Mater. Sci. Technol , Vol. 21,pp. 13-16, 2005.

[25] L. J. Chen, C. L. Hsin, W. J. Mai et al., "Elastic Properties and Buckling of Silicon Nanowires," Advanced Materials, Vol. 20, no. 20, pp. 3919-+, Oct 17, 2008.

[26] Eaton W P, Bitsie F, Smith J H and Plummer D W 1999 A new analytical solution for diaphragm deflection and its application to a surface-micromachined pressure sensor *Proc. Intern. Conf. on Modeling and Simulation of Microsystems, MSM 99*, San Juan, Puerto Rico, USA

[27] Katuhisa Suzuki, Hiroshi Hasegawa and Yozo Kanda, "Origin of the Linear and Nonlinear Piezoresistance Effects in p-Type Silicon", Jpn. J. Appl. Phys. Vol. 23, pp. L871-L874, 1984.

[28] K. Matsuda, ''Strain-dependent hole masses and piezoresistive properties of silicon,'' IWCE-10 10th International Workshop on Computational Electronics, pp. 173–174, 2004.

A Novel Stretchable CMUT Array Using Liquid-Metal Electrodes on a PDMS Substrate

Xiaomei Shi[1], Ching-Hsiang Cheng[1*], Jue Peng[2]

[1*] Department of Industrial and Systems Engineering, the Hong Kong Polytechnic University, Hong Kong

[2] Department of Biomedical Engineering, School of Medicine, Shen Zhen University, China

mfcheng@inet.polyu.edu.hk

Abstract—This paper introduces a new method for fabricating a stretchable capacitive micromachined ultrasonic transducer (CMUT) array with liquid-metal (Ga-In-Sn) electrodes on a polydimethylsiloxane (PDMS) substrate. A stretchable CMUT array can make it fully comply with a 3D curved surface for biomedical applications. The transducer membrane and cavity are fabricated separately using PDMS and bonded together by using O_2 plasma, which allows us to form the concave bottom electrodes on top of the reflowed photoresist. By using concave bottom electrodes, the effective capacitance will be increased by reducing the gap distance on the membrane edge, especially when pulled in by a DC bias. This can increase the device sensitivity, fill factor and output pressure. The device is designed to operate in the range of 100 kHz to 500 kHz for low frequency applications. The preliminary experimental results show a resonant frequency at around 200 kHz by using an impedance analyzer.

Keywords- capacitive micromachined ultrasonic transducer; polydimethylsiloxane (PDMS) substrate; liquid metal; concave bottom electrode

I. INTRODUCTION

Capacitive micromachined ultrasonic transducers (CMUTs) employ the fabrication technology of standard integrated circuits, which have recently been applied to the field of intracardiac medical imaging. The microfabrication has made possible to fabricate silicon-based electrostatic transducers competing in performance with the piezoelectric transducers. Since 90s, people start to explore more into each development stage of CMUTs including design, modeling, analysis, fabrication, characterization, testing and packaging. Several effects have been carried out to implement such transducer for medical application. The basic structure of a CMUT consists of a moveable thin membranes and a conductive silicon substrate separated by a vacuum cavity. The membrane is coated with metal on top to create a parallel plate capacitor. The CMUT can transmit and receive ultrasound by vibrating its membrane like a drum. DC bias is applied to bring the membrane closer to the bottom electrode for increasing its sensitivity.

The first CMUT [1] was fabricated on silicon by using surface micromachining techniques, such as thin film deposition, photolithography, and thin film etching. CMUTs on a silicon substrate have been reported from other groups with different fabrication techniques, including surface and bulk micromachining using different membrane, insulator, and substrate materials.

A CMUT with reverse top-to-bottom fabrication has been reported [2], which has the benefit of having the backing layer deposited right on the backside of the membrane to reduce the ringing effect. In recent year, we have reported a flexible CMUT array with concave bottom electrodes to increase effective capacitance [3]. Fabrication of flexible CMUT arrays based on trench refilling with PDMS has been reported [4]. It has preserved the advantage of silicon-based transducers with flexibility provided by the soft PDMS refills. A flexible polymer-based CMUT has been reported, which is also known as "Sonic Paper" [5]. Since solid metal electrodes and a rigid polymer substrate are used, it is only possible to flex the device array but not able to stretch it. Liquid metal alloy has been used to make stretchable antenna [6].

In this paper, we introduce a novel stretchable capacitive micromachined ultrasonic transducer array using liquid-metal electrodes on a PDMS substrate. This process is very much simplified the traditional fabrication steps. Furthermore, we propose to use the room temperature liquid metal alloy as electrodes for CMUT arrays. By using new materials we believe there is a great potential to improve the flexibility and durability without reduced the sensitivity. The details of the fabrication process are described in the third section followed by the experimental results obtained from CMUTs that are fabricated by the new method and new material.

II. MOTIVATION

There are a several advantages associated with fabrication of capacitive micromachined ultrasonic transducers (CMUTs) by using liquid-metal electrodes on a PDMS substrate as compared to the traditional one:

A. Vacuum Cavity Formation

. In the traditional case, the fabrication of vacuum cavity [8] starts with a silicon wafer with a metal sacrificial layer. The details of the fabrication have been introduced in several papers. Because of the high stresses involved with the metal thin film depositions. It is very difficult to release any size membranes, and very unpractical to deposit any thickness sacrificial layers.

In the novel design, we propose to reverse the fabrication steps to have the membrane on the bottom and then build the cavity and bottom electrode on the top. This allows us to form the concave bottom electrode on top of the reflowed photoresist in convex spherical shape using over-plating technique, which cannot be achieved by the conventional bottom-up surface-micromachining technique. After removal of sacrificial photoresist to form the CMUT cavity, the membrane is released by single-side wet etching to remove the backside silicon substrate. Since device side remains dry, there is no membrane stiction problem caused by drying inside the cavity, which can realize a thinner and bigger membrane. By using the concave bottom electrode to increase the effective area of the membrane, it can increase the effective capacitance to improve the fill factor, output pressure, bandwidth, and sensitivity of the transducer.

B. Membrane Material

The membrane is usually made of silicon nitride in the traditional CMUT fabrication, which is deposited using a Lower Pressure Chemical Vapor Deposition (LPCVD). Silicon nitride films deposited in this way turn out with high residual stress. The resulting devices are limited in working under small forces. It means these CMUTs can get broken when a large force is applied. And the silicon substrates usually are too thick and lack of flexibility to fit into a curved interface. The resulting devices are limited in working under small forces. It means these CMUTs can get broken when a large force is applied. And the silicon substrates usually are too thick and lack of flexibility to fit into a curved interface.

C. Liquid Metal Electrodes

One other limitation the traditional CMUT design brings is using a thin solid metal film as the electrodes for CMUT arrays.

In novel stretchable CMUTs design, the membrane is made by using liquid-metal electrodes on a PDMS substrate, which has better mechanical properties. Therefore, the PDMS substrate and membrane improve the reliability and predictability as well as the performance of the device. As shown in Figure 1, the CMUTs can be stretched or bended to certain extent. Compared with the existing curved or flexible CMUT, our device has higher flexibility and stretchability because of the liquid metal electrodes is used.

These advantages brought by the new design are more pronounced in the application of interest, where the intended frequency of operation requires the fabrication of CMUTs with very large membranes. In the following sections we will describe the realization of this device.

III. FABRICATION

A. Material—liquid Metal

Liquid metal alloy means alloy of several very low melting point metal at room temperature. As a replacement for toxic

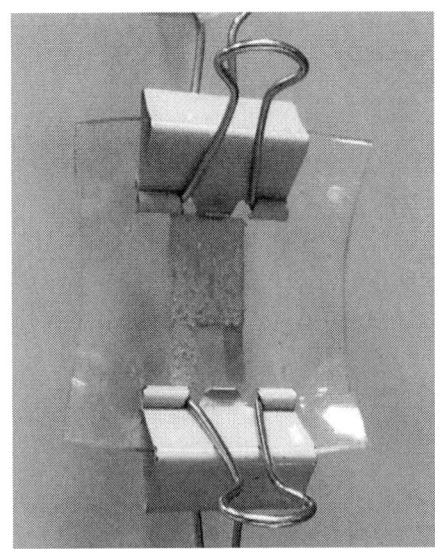

Fig. 1. Flexible 2-D CMUT array stretched by a pair of clamps

mercury, liquid metal alloy is used in various applications. These liquid metal alloys have a high degree of thermal conductivity far superior to ordinary nonmetallic liquids. Another advantage of these liquid metals is lower toxic, lower vapor pressure, high densities and electrical conductivities. Liquid metal alloy will wet and adhere to most metallic surface. It forms a thin looking oxide layer on the surface of metallic.

In this design, we choose a kind of liquid metal called Coollaboratory Liquid Pro. Alloy of the metal components are gallium, indium, rhodium, silver, zinc and stannous. It has high performance hear conducting and good electrical conducting that consists of 100% liquid metal alloy. It is liquid at room temperature (freezing temperature 8 ℃), but it is absolutely nontoxic and has a high moistening ability for several materials. It is neither contains non-metallic additives nor solid particles. Due to these properties, Coollaboratory Liquid Pro surpasses the best high performance heat conducting pastes by a multiple.

B. CMUT Fabrication

Fabrication of a stretchable capacitive micromachined ultrasonic transducer (CMUT) array using liquid-metal electrodes on a PDMS substrate is a five-step process as shown in Fig.1. The process starts with a glass substrate as a carrier wafer. Photoresist AZ5214 is coated by spin coater to get a thickness of around 1 μm and patterned to define the active area of the CMUT cell. A thermal reflow process is carried at 150°C for 30 minutes to melt the patterned photoresist to form spherical profile by surface tension (Figure 2(1)).

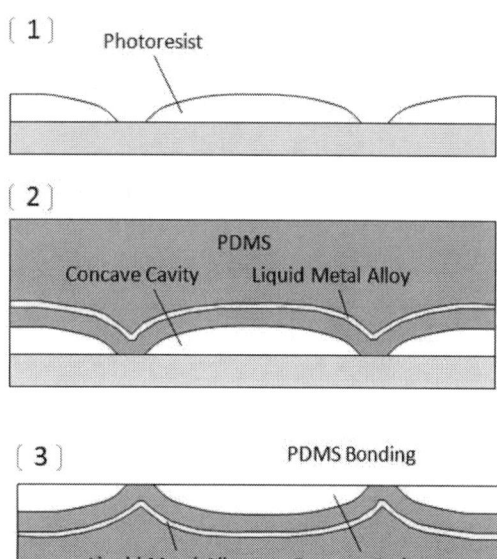

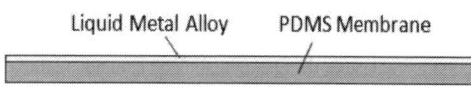

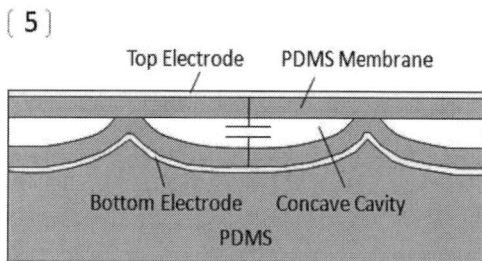

Fig. 2. Microfabrication process of the CMUT: (1) Patterning of the photoresist to form the sacrificial layer for cavity areas. Thermal reflow at 150°C to have the photoresist become convex shape. (2) Covering with a thin layer of PDMS printing with liquid-metal electrodes and passivating with another PDMS layer. (3) Removing the glass substrate for the cavity layer. (4) Coating of PDMS and printing liquid metal pad for the membrane layer. (5) Bonding two layers of PDMS together by O2 plasma.

A thin layer of PDMS film was made by coating the Dow Corning Sylgard 184 PDMS pre-polymer and its curing agent mixture on the glass substrate then cured at elevated temperature of 70°C in a convection oven. A layer of liquid metal pad was printed as electrode (Figure 2(2)).

After last step, another thin layer of PDMS was needed by coating for protecting the metal electrode and the bond pads then cured at room temperature (Figure 2(3)). A thin layer of

PDMS was spun coated on another glass substrate (Figure 2(4)). A layer of liquid metal pad was printed and covered by a thin layer of PDMS then cured at room temperature. Two layers of PDMS were bonded together by O_2 plasma less than 2 minutes (Figure 2(5)).

This project is to fabricate a stretchable CMUT array by using liquid metal alloy as electrodes to prevent metal breakage when a tensile force is applied.

IV. RESULTS

We successfully fabricated several CMUT arrays. One example is shown in Fig. 3; the array has 1 element; each element consists of 2500 membranes connected in parallel. Each element was arranged on a square area of 10mm×10mm

Fig. 3. The photo of the fabricated Capacitive Micromachined Ultrasonic Transducers. Each element was arranged on a square area of 10mm×10mm with 2500 cells inside.

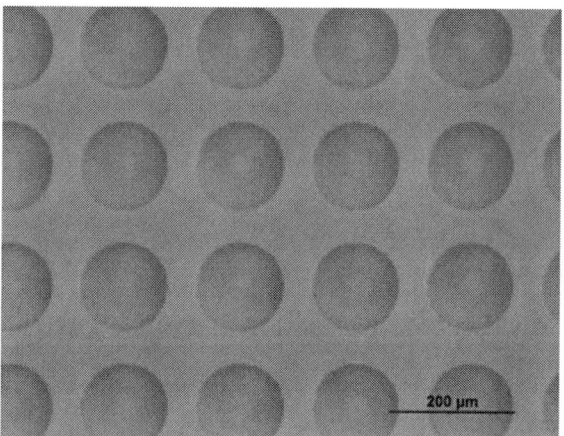

Fig. 4. Photo of Concave Cavity.

with 2500 cells inside. Most of the device area is covered by PDMS; used the room temperature liquid metal alloy as electrodes [Fig. 3]. Figure 4 shows the surface of concave electrodes of PDMS-CMUT. To further investigate the structure, we cleaved one cell and imaged the part of PDMS bonded concave cavity SEM image by using a scanning electron microscope (SEM). In Figure 5, the fracture of PDMS

978-1-4673-1122-9/12 $31.00 © 2012 IEEE

shows river like veins typically observed in the cutting of polymer materials. It indicates that the thickness of concave cavity is about 10~12 μm. Fabrication of the concave bottom electrode was done by thermal reflow of photoresist.

The electrical input impedance of the device was measured at room temperature using an impedance analyzer (Agilent Technologies, Model 4294A), and a variable DC bias voltage connected via a bias-T.

The result for the CMUT shown in figure 6 clearly indicates that the input impedance has a peak value at its mechanical

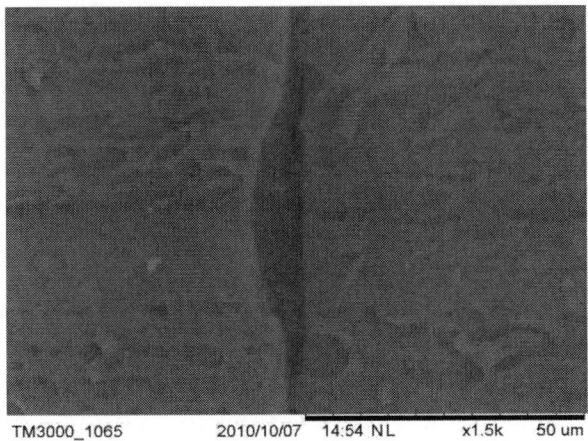

Fig. 5. A SEM image of PDMS bonded concave cavity. The diameter and thickness for each membrane were 160μm and 10μm respectively

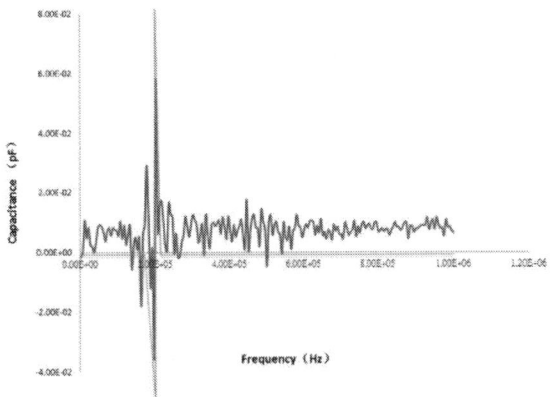

Fig. 6. The measured relationships of impedance vs. frequency for the CMUT array. The resonant frequency of the CMUT is around 200 kHz.

resonant frequency. In general, our devices show similar characteristics to other CMUTs fabricated with liquid-metal electrodes on a PDMS substrate or thin solid metal film electrodes silicon based. The resonant frequency was at 200 KHz in air. Without applying a DC bias, the peak of signal is relatively small compared with the background noise. Fabrication of the concave bottom electrode was done by thermal reflow of photoresist.

ACKNOWLEDGMENT

The work described in this paper was partially supported by a grant from the RGC General Research Fund (RGC project reference number PolyU 513208), The Hong Kong Polytechnic University, Hong Kong; partially supported by a grant from the Innovation and Technology Fund (ITF project reference number ITS/024/09), The Hong Kong Polytechnic University, Hong Kong; and partially supported by a grant from the PolyU Internal Funding, The Hong Kong Polytechnic University, Hong Kong (Project Account Code A-PE1G)

REFERENCES

[1] M. I. Haller and B. T. Khuri-Yakub, "A surface micromachined electrostatic ultrasonic air transducer," in Proc. IEEE Ultrason. Symp., 1994, pp. 1241-1244.

[2] G. Caliano*, A. Caronti, A. Savoia, C. Longo, M. Pappalardo, E. Cianci, and V. Foglietti, "Capacitive Micromachined Ultrasonic Transducer (cMUT) Made by a Novel "Reverse Fabrication Process"," 2005 IEEE International Ultrasonics Symposium, 18-21 Sept. 2005, Vol. 1, pp. 479-482.

[3] Ching-Hsiang Cheng, Chen Chao, Xiaomei Shi, Wallace Leung, "A Flexible Capacitive Micromachined Ultrasonic Transducer (CMUT) Array with Increased Effective Capacitance from Concave Bottom Electrodes for Ultrasonic Imaging Applications, " Proceedings of 2009 IEEE International Ultrasonics Symposium (IUS), 20-23 September 2009, Rome, Italy, pp. 996 – 999.

[4] X. Zhuang, D.S. Lin, Ö. Oralkan and B. T. Khuri-Yakub, "Fabrication of Flexible Transducer Arrays with Through-Wafer Electrical Interconnects Based on Trench Refilling with PDMS," Journal of Microelectromechanical Systems, Vol. 17, No. 2, May 2008, pp. 446-452.

[5] Ming-Wei Chang, Hsu-Cheng Deng, Da-Chen Pang, Mu-Yue Chen, "A Novel Method for Fabricating Sonic Paper, "2007. IEEE International Ultrasonics Symposium, 28-31 October 2007, New York, NY, pp. 527 – 530.

[6] Shi Cheng, Anders Rydberg, Klas Hjort and Zhigang Wua, "Liquid Metal Stretchable Unbalanced Loop Antenna," Applied Physics Letters, Vol. 94, Issue 14, April 2009, pp. 144103 - 144103-3.

[7] Caliano, G.; Carotenuto, R.; Caronti, A.; Pappalardo, M., "cMUT echographic probes: design and fabrication process, "Ultrasonics Symposium, 2002. Proceedings. 2002 IEEE , pp. 1067 - 1070 vol.2.

[8] Huang, Y.; Ergun, A.S.; Haeggstrom, E.; Khuri-Yakub, B.T., "Fabrication of Capacitive Micromachined Ultrasonic Transducers (CMUTs) using wafer bonding technology for low frequency (10 kHz-150 kHz) sonar applications"OCEANS '02 MTS/IEEE, 2002 , Page(s): 2322 - 2327 vol.4

Miniature pyranometer with asteroid shape thermopile

Jiangang Zhang*, Zhengwei Wu, Zhan Zhao, Xin Guan

*State key lab of transducer technology, Institute of electronics, Chinese academy of sciences, Beijing, China
jgzhanghli@yahoo.com.cn

Abstract— We have designed and fabricated a miniature pyranometer with a novel asteroid shape thermopile. Comparing with existing pyranometer, the size of our miniature pyranometer has been reduced to 8mm×4mm×0.5mm, and much metal material has been saved in our pyranometer. Test results of this pyranometer show: (a). the output of this pyranometer is proportional to input illuminance. (b). the resolution of this pyranometer with a pre-amplifier is $0.191\,mV/10^5$Lux. (c). As the maximum illuminance of sunshine is about 100,000Lux, the sensitivity of this fabricated pyranometer is receptible. The further test and improvement such as the reliability test of the sensor and the optimum design of material and structure of the sensor will be performed in the next work in the future.

Keyword—miniature pyranometer, asteroid shape thermopile, solar radiation

I. INTRODUCTION

Measurements of solar radiation are essential to both atmospheric science and renewable energy system design and research. Pyranometer is the instrument that measures the solar radiation flux density [1]. So far, almost all the pyranometers are thermopile type formed by a number of twisted metal wires. Such thermopile normally has hot and cold junctions with equal number. Hot junction is also called measuring junction, and cold junction is also called reference junction. The temperature difference between the measuring and reference junctions produces a voltage that is proportional to the solar radiation. Recently, since the renewable energy system technology and wireless sense network technology have been developed rapidly, a large quantity of pyranometer with small size was highly needed. Depending on microelectromechanical systems (MEMS) technology, the size of a device can be reduced much more, and a number of the material such as the metal of thermopile can be saved. Based on MEMS technology, we have fabricated a miniature pyranometer with novel designs on thermopile and the structure of device. The resent test result of the fabricated miniature pyranometer show the output of the sensor is proportional to input illuminance, and the sensitivity of this fabricated pyranometer is receptible. The further test and improvement such as the reliability test of the sensor and the optimum design of material and structure of the sensor will be performed in the next work in the future.

II. DESIGN OF DEVICE

The core principle of our miniature pyranometer is the mechanism of thermopile. Normally, in order to ensure a thermopile has a continuous output, hot spot and cold spot are necessary. The hot junction of thermopile should be located at the hot spot, and the cold junction of thermopile should be located at the cold spot. The stable difference of the temperature between hot spot and cold spot ensure the continuity of the output of the thermopile. Fig.1 presents a schematic illustration of our design of the micro-electro-mechanical system (MEMS) based miniature pyranometer. As shown in Fig.1, this device has three parts mainly: substrate, silicon membrane and thermopile layer.

A. The two functions of silicon membrane

In this device, both the substrate and 20μm thickness silicon membrane are silicon. The silicon membrane of this device has two functions. One is the support of the thermopile unit, the other one is the thermal insulative structure. There are two factors to ensure that the heat loss through silicon membrane is little. The first factor is the relative lower thermal conductivity of 20μm thickness silicon membrane compared with bulk silicon material. In solid state physics, to a certain kind of material, the thin film has a relative low thermal conductivity than bulk material [2]. The second factor is the very little contact area between the silicon membrane and substrate, due to the fact that the thickness of silicon membrane is very thin.

978-1-4673-1122-9/12 $31.00 © 2012 IEEE

approximately equal. The whole thermopile of our pyranometer forms an asteroid shape shown in Fig.2.

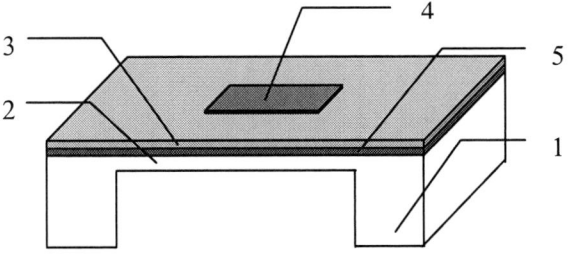

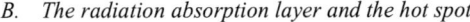

Fig.1. Schematic of miniature pyranometer:
1. silicon substrate; 2. silicon membrane;
3. thermopile layer; 4. radiation absorption layer;
5. electric insulative layer.

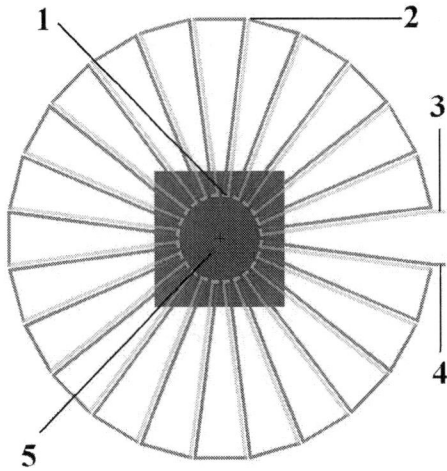

Fig.2. Schematic of thermopile layer :
1. hot junction; 2. cold junction; 3. NiCr electrode;
4. Ni electrode; 5. radiation absorption layer

B. The radiation absorption layer and the hot spot

As shown in Fig.1, the radiation absorption layer was located at the central area of the silicon membrane. Because the film of Si_3N_4 is superior radiation absorption material [3], we chose Si_3N_4 film fabricated by plasma-enhanced chemical vapor deposition (PECVD) technology as the radiation absorption layer with thickness of $0.5\,\mu$ m. Due to the thermal insulative function of the silicon membrane, most of the heat absorbed by the radiation absorption layer was insulated by the silicon membrane. Since the substrate is the bulk material of Si, we can regard the substrate of the device as a heat sink. Therefore, when the solar radiation irradiates the device continually, there is a continuous difference of temperature between the radiation absorption layer and the substrate of device. This difference of temperature depends on the solar radiation flux density. So, we chose the central area of silicon membrane as the hot spot of the thermopile of our pyranometer, and chose the substrate of the device as the cold spot of the thermopile of our pyranometer. Thus, the output of thermopile is proportional to the solar radiation flux density.

C. Asteroid shape thermopile

We chose Ni and NiCr to make the junctions of the thermopile in our pyranometer. As shown in Fig.2, the hot junction of this thermopile beneath the radiation absorption layer, is located at the central area of the silicon membrane. The cold junction of this thermopile above the electric insulative layer, is located at the edge of thermopile layer. Because the electric insulative layer (Si_3N_4 film with thickness of $0.2\,\mu$ m) is very thin comparing with the silicon substrate (thickness of $350\,\mu$ m), the temperature of the edge of thermopile layer and the silicon substrate of the device are

III. FABRICATION OF DEVICE

There are two steps in the fabrication of this miniature pyranometer mainly. Figure.3 showed the Schematic diagram of the fabrication process of this device.

As shown in Fig.3 (a), the first step is the formation of the thermopile layer and radiation absorption layer. The thermopile unit is formed by sputtering technology. Then the radiation absorption layer (Si_3N_4 film) is deposited above the thermopile layer by plasma-enhanced chemical vapor deposition (PECVD) technology.

As shown in Fig.3 (b), the second step is the formation of the silicon membrane. Deep reactive ion etching (DRIE) was used in another side of the wafer to form this silicon membrane.

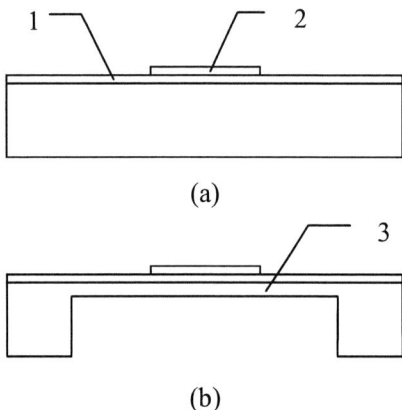

Fig.3. The fabrication process of miniature pyranometer:
(a): the formation of the thermopile layer and radiation absorption layer: 1.thermopile layer, 2. radiation absorption layer;
(b): the formation of silicon membrane: 3.silicon membrane

Fig.4. The photography of fabricated miniature pyranometer

IV. TEST AND RESULT

Fig.4. shows a fabricated miniature pyranometer. The size of this device is 8mm×4mm×0.5mm. The width of the beam of the thermopile in this fabricated pyranometer is 30 μ m. The thickness of the radiation absorption layer of this pyranometer is 0.5 μ m. The thickness of the silicon membrane is 20 μ m. A pre-amplifier is used to amplify the output of this fabricated pyranometer. Fig.5. shows the performance of this fabricated pyranometer. The test results show: (a). the output of this pyranometer is proportional to input illuminance. (b). the resolution of this pyranometer with a pre-amplifier is 0.191mV/10^5Lux. (c). As the maximum illuminance of sunshine is about 100,000Lux, the sensitivity of this fabricated pyranometer is receptible. The further test and improvement such as the reliability test of the sensor and the optimum design of material and structure of the sensor will perform in the next work in the future.

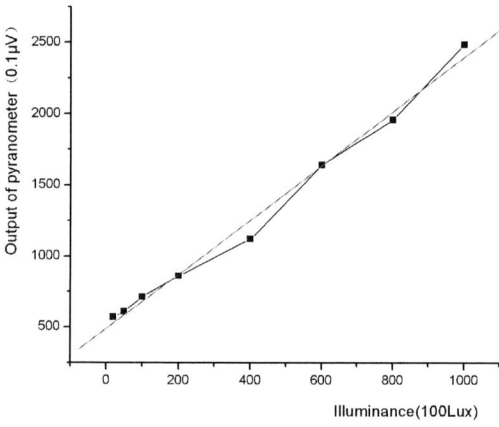

Fig.5. Test result of fabricated miniature pyranometer

V. CONCLUSION

We have designed and fabricated a miniature pyranometer with a novel asteroid shape thermopile. Comparing with existing pyranometer, the size of our miniature pyranometer has been reduced to $8mm \times 4mm \times 0.5mm$, and much metal material has been saved in our pyranometer. Test results of this pyranometer show: (a). the output of this pyranometer is proportional to input illuminance. (b). the resolution of this pyranometer with a pre-amplifier is $0.191mV/10^5$Lux. (c). As the maximum illuminance of sunshine is about 100,000Lux, the sensitivity of this fabricated pyranometer is receptible. The further test and improvement such as the reliability test of the sensor and the optimum design of material and structure of the sensor will be performed in the next work in the future.

ACKNOWLEDGMENT

The authors appreciate both the support from commonwealth R&D special program of China on meteorology under grant GYHY201106040 and the fusion collaboration from State key lab of transducer technology, Institute of electronics, Chinese academy of sciences.

REFERENCES

[1] A. Lester, D.R. Myers, A method for improving global pyranometer measurements by modeling responsivity functions, *Solar Energy*, vol 80, pp:322 – 331,2006.

[2] Tien C-L, M.A. Gerner, *Microscale Energy Transport Series in Chemical and Mechanical Engineering*, Taylor & Francis, 1998.

[3] Wood R A, Uncooled thermal imaging with monolithic silicon focal arrays. *SPIE,* vol 2020, pp:33-37, 1993.

Design and Fabrication of Thin-Film Aluminum Microheater and Nickel Temperature Sensor

R. Phatthanakun[*], P. Deekla[**], W. Pummara[**], C. Sriphung[*], C. Pantong[**], N. Chomnawang[**]

[*]Synchrotron Light Research Institute (Public Organization), Thailand
rungrueang@slri.or.th

[**]School of Electrical Engineering, Suranaree University of Technology, Thailand
sut.mems@gmail.com

Abstract— **This paper presents development of a thin film aluminum microheater and a nickel temperature sensor for low temperature applications by using Micro-Electro-Mechanical Systems. Both of them are fabricated onto a glass substrate and protected by thin PDMS membrane. The microheater is energized to find sensor characteristic. As linearity and accuracy of nickel sensor give a wide temperature range, its electrical resistance variations are calibrated directly by temperatures of energized microheater. Variations of resistance signal are transformed and fed back to control temperature of microheater in PI closed-loop feedback control. K_p and K_i values are adjusted to obtain the optimal time response. Experimental testing of controlled temperature ranges from 40°C to 140°C is presented for their integration in stability system.**

Keywords-microheater; sensor; MEMS; PI controller

I. Introduction

Microheater and temperature sensor have been widely used in many applications such as micro explosive boiling, micropump, pressure sensor, flow rate sensor and microvalve [1-3]. Advance in Micro-Electro-Mechanical Systems (MEMS) made it possible for thin-film microheater to be applied for smart devices because of their extensive applications. Generally, materials which provide wide temperature range, accuracy, and stability are selected to fabricate as microheater such as platinum and gold. In low temperature system, however, aluminum has been chosen to work as actuators instead because it offers good properties with lower price. To control these microheaters in integrated systems, thin-film sensors have been used as resistance temperature detectors (RTDs) which give a linear positive change in resistance with respect to temperature. Platinum still supports this requirement with expensive cost and operates from -200°C to +800°C. Like as many applications in low temperature system, nickel can be chosen to be sensor to operate from -100°C to +260°C with low cost and the best sensitivity.

In this paper, a thin-film aluminum microheater and nickel temperature sensor are developed by MEMS technology for actuator and sensor applications. They are constructed on a glass substrate because of its low thermal conductivity and high electrical resistance. Thin film PDMS membrane is then coated to protect them from environmental hazards. Nickel resistance is calibrated directly by temperature of microheater to estimate its characteristic using FLIR infrared camera. These resistance signals are transformed and fed back to control microheater temperatures through PI closed-loop feedback control. The variations of the microheater temperature with time and a linear function of metal sensor are investigated. Controlled temperature ranges from 40°C to 140°C are displayed to show the capability in application of low temperature electronic device.

II. Design And Fabrication

Aluminum and nickel are selected to use as the metal of microheater and temperature sensor, respectively, because of its good linearity, low cost, and chemical stability. The pattern of microheater is designed to encompass sensor as shown in Fig. 1, resulting in fast response of RTDs material.

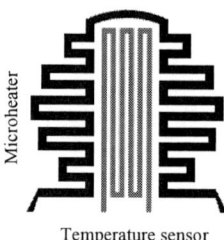

Fig. 1. Schematic of microheater and temperature sensor.

By applying a voltage (V) across the two end of a resister with a resistance (R), the heating power (P) of a microheater can be generated. Equation (1) is a heating power calculation which implies that the smaller resistance results in the larger heating power with the fast temperature rise.

$$P = V^2 / R \qquad (1)$$

A resistance of thin-film microheater appeared in (1) can be calculated by (2).

$$R = \frac{\rho L}{wt} \qquad (2)$$

where ρ is the resistivity of material, L is the length, w and t are the width and the thickness of thin-film, respectively.

978-1-4673-1122-9/12 $31.00 © 2012 IEEE

For a RTDs thin-film sensor, the relationship between the resistance and temperature (> 0°C) can be calculated over its operating temperature range by (3) [4].

$$R(t) = R_0\left(1 + At + Bt^2\right) \qquad (3)$$

where $R(t)$ is resistance at temperature t, R_0 is base resistance at 0°C, A and B are constant coefficient of equation resulting from experimental results.

To achieve a large heating power with small voltage source, the microheater is designed to obtain a small resistance about 100 Ω, while nickel temperature sensor is considered a resistance around 1000 Ω. The characteristics of them are displayed in Table I. After fabrication, their resistances (R_{mea}) are measured and compared with the required resistances (R_{req}) which are calculated by (2).

TABLE I.   CHARACTERISTICS OF MICROHEATER AND SENSOR

Materials	L (μm)	w (μm)	t (Å)	P (Ω·m)	R_{req} (Ω)	R_{mea} (Ω)	Err (%)
Al	8703	50	450	2.6×10^{-8}	100	94.5	5.5
Ni	7030	30	163.8	6.9×10^{-8}	1000	1195	19.5

Microheater and sensor are fabricated on a microscope glass slide by using UV lithography and lift-off process as illustrated in Fig. 2. The glass slide is first deposited with aluminum thin film using thermal evaporation and coated by AZ 1512 photoresist to create protective layer by UV exposure as shown in Fig. 2(a) and (b), respectively. Aluminum microheater is left on the substrate by etching in HF solution, followed by photoresist removal as shown in Fig. 2(c). The layer of AZ photoresist is coated again to cover all patterns and exposed to create the sensor mold as shown in Fig. 2(d) and (e), respectively. Nickel is filled into the resist mold by evaporation as shown in Fig. 2(f) and acetone is used to lift-off photoresist as shown in Fig. 2(g). Finally, thin-film PDMS is spin-coated to perform as a protective layer as shown in Fig. 2(h). A complete microheater and temperature sensor are displayed in Fig. 3.

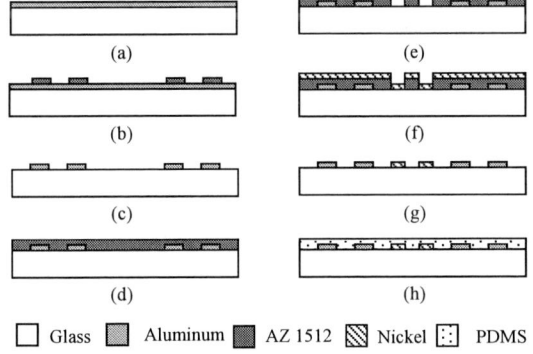

Fig. 2. Fabrication sequence of microheater and temperature sensor.

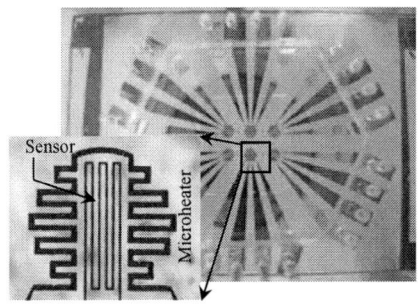

Fig. 3. Aluminum microheater with nickel temperature sensor on a glass substrate covered by thin PDMS membrane.

III.   ELECTRICAL CHARACTERIZATION OF SENSOR

The DC voltage source is applied across the two ends of the microheater. Its temperature is detected to find the temperature relationship as a function of the applied power presented in Fig. 4. The temperature is measured by a FLIR infrared camera as shown in Fig. 5.

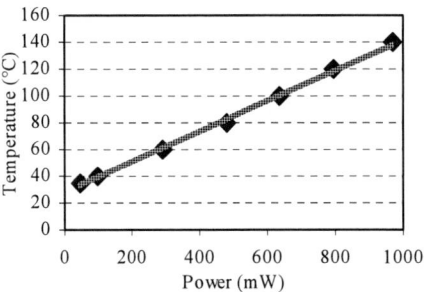

Fig. 4. Temperature of microheater as a function of the applied power.

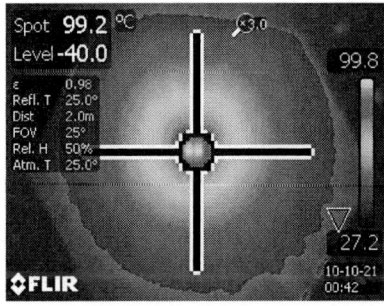

Fig. 5. Temperature distribution measured by a FLIR infrared camera.

Nickel sensor is characterized as a function of the resistance-temperature (R-T) curve as shown in Fig. 6. Its resistance is checked by applying of DC voltage to microheater and adjusted its temperature. The R-T result is then fitted by a second order regression curve and arranged to form the relationship of R-T as in (4). The 0°C resistance of nickel sensor is 1133.2 Ω with constant coefficients of A and B are 1, 2.091×10^{-3} and 2.294×10^{-6}, respectively.

$$R(t) = 1133.2(1 + 2.091 \times 10^{-3} t + 2.294 \times 10^{-6} t^2) \qquad (4)$$

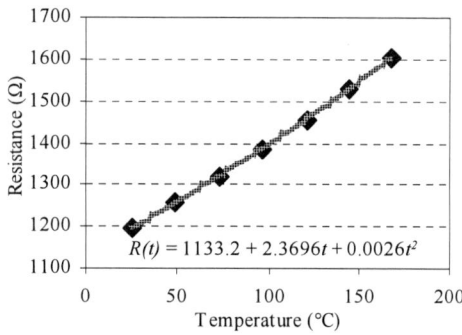

Fig. 6. Resistance of RTD nickel sensor as a function of temperature.

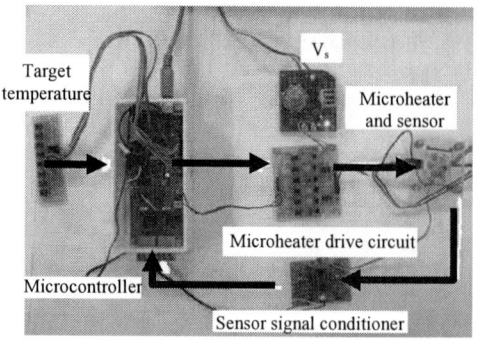

Fig. 9. The completed temperature control kit for microheater.

IV. TEMPERATURE CONTROL

To control the process or the parameter (temperature, flow, or speed), the principle of control is similar. Signals of input and output have to be appropriate to the application. The simplest and cheapest control system is closed-loop feedback control with On/Off function. However, controlled temperatures using this method have output oscillations as shown in Fig. 7. Although it provides fast responses, the stability systems that require high accuracy temperature output cannot utilizes this method.

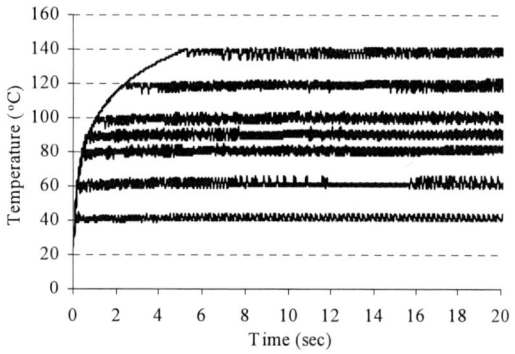

Fig. 7. Time responses of closed-loop feedback control with On/Off function.

Another of control function which can be used to avoid oscillation output problem is PI function. Microheater is controlled by PI closed-loop feedback control which observes real-time temperatures via the sensor [5]. The block diagram is illustrated in Fig. 8 and a picture of the completed temperature control kit with PI controller is illustrated in Fig. 9.

Figure 10 shows a comparison of time response between On/Off and PI closed-loop feed back control. The On/Off control offers a rise time of 1.5 sec faster than PI control that requires 7.6 sec. Temperature oscillation of On/Off control, however, is about 8°C while the PI control gives the slow temperature rise when approaching set-point. The On/Off control energizes the microheater at maximum energy to increase the temperature when the output lower than the set-point, and stops current flow when the output higher than the set-point, resulting in temperature oscillation in dead band. Meanwhile, variation of PI control is more smoothly because it calculates an error value as the difference between a target temperature set-point and a measured temperature from sensor.

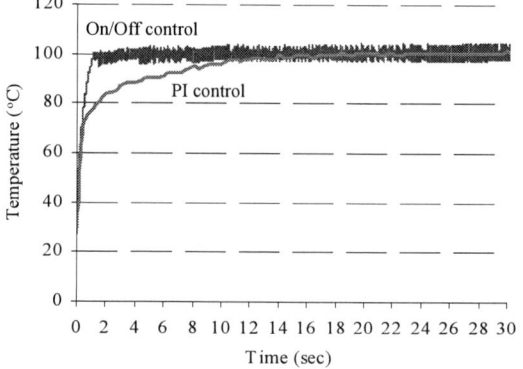

Fig. 10. A comparison between On/Off and PI function at 100°C.

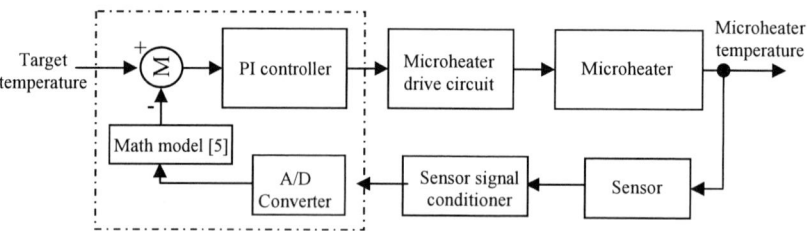

Fig. 8. Block diagram of PI closed-loop feedback control for microheater.

V. EXPERIMENTAL RESULTS

A PI controller which has been widely used in control system is selected to be a control loop feedback mechanism. It calculates an error value as the difference between a target temperature set-point and a measured temperature from sensor. K_p and K_i values are adjusted by considering overshoot and rise time of step response. The K_p value is first adjusted to get rid of overshoot with minimum rise time (t_r). Then, the K_i value is tuned to obtain the minimum steady state error (ess) of 6 degree temperatures for 40 60 80 100 120 and 140°C, respectively. The rise times of each temperature which are defined as the time to go from 10% to 90% of its final value are observed and presented in Fig. 11. Furthermore, the temperature variations between 40°C to 120°C are varied to estimate the performance of the temperature control kit as shown in Fig. 12. It can be observed that the temperature quickly responded when the system was actuating to high temperature. In the other hand, the responses of cool down states do not provide suitable variations, especially a change of temperature from 60°C to 40°C. To solve these problems, the PI controller has to be adjusted for both step-up and step-down procedures to achieve the optimum values.

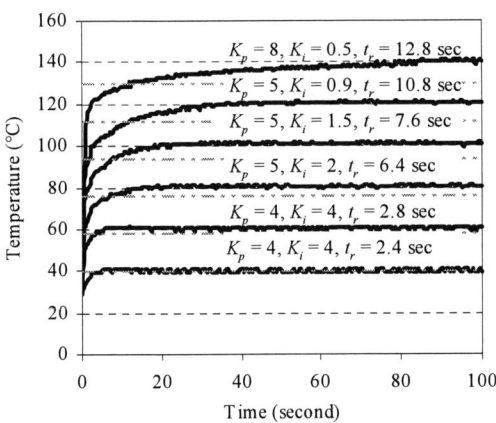

Fig. 11. Step response of microhrater by adjusting PI controller.

VI. CONCLUSIONS

Thin-film aluminum microheater and nickel temperature sensor were designed and fabricated by MEMS technology. Aluminum which is low cost material and linear response was selected to be microheater. Nickel which provides the best sensitivity and linear resistance was chosen to act as temperature sensor, and placed inside microheater pattern to attain fast temperature variation. The characteristics of microheater and sensor were characterized and applied to the PI closed-loop feedback temperature control kit. The PI controller was adjusted to achieve the optimum time response of temperature between 40°C to 140°C. The impressive results can obtain for step up, but the PI values have to be adjusted for step down. However, the thin-film microheater which can be controlled its temperature is

suitable to apply in many low temperature electronic device applications.

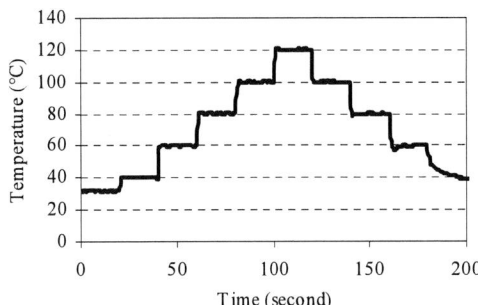

Fig. 12. The temperature variation based on 5 target temperatures.

ACKNOWLEDGMENT

We are grateful to the Accelerator Technology Division and Beamline 6a (DXL) of the Synchrotron Light Research Institute (Public Organization) for the help in the FLIR infrared camera and the processing of the devices.

REFERENCES

[1] K.L. Zhang, S.K. Chou, and S.S. Ang, "Fabrication, modeling and testing of a thin film Au/Ti microheater," *International Journal of Thermal Sciences*. vol. 46(6), pp 580-588, June 2006.

[2] M.A. Gajda, and H. Ahmed, "Application of thermal silicon sensors on membranes," *Sensor and Actuators A* vol. 49 (1-2), pp. 1-9, 1995.

[3] J. Puigcorbe, D. Vogel, B. Michel, A. Vila, I. Gracia, C. Cane, and J.R. Morante, "Thermal and mechanical analysis of micromachined gas sensors," *Journal of Micromechanics and Microengineering* vol. 13 (5), pp. 548-556, 2003.

[4] E.J.P. Santos, and I.B. Vasconcelos, "RTD-based smart temperature sensor: Process development and circuit design," *26th International Conference on Microelectronics*, pp. 333-336. May 2008.

[5] R. Phattanakun, P. Deekla, W. Pummara, C. Pantong and N. Chomnawang. "Fabrication and Control of Thin-Film Aluminum Microheater and Nickel Temperature Sensor," *Proceedings of The 8th international conference on Electrical Engineering/ Electronics, Computer, Telecommunications and Information Technology*, pp. 14-17, 2011.

Micro Device Array Design and Fabrication in Monolithic MEMS SoC

Jung-Hung Wen[1], *Member, IEEE* and Weileun Fang[1,2]

[1]Institute of NanoEngineering and MicroSystems, National Tsing Hua University, Hsinchu, TAIWAN
[2]Power Mechanical Engineering, National Tsing Hua University, Hsinchu, TAIWAN

Abstract—This study presents novel micro-device-array designs and fabrications potentially for accelerometer, gyro or even bio-sensor applications based on an advanced 0.35µm 4 metal bipolar-CMOS-DMOS (BCD) technology. Structures built by pure metal single layer have better mechanical properties and multi-electrodes design provide more actuating and sensing possibilities for both out-of-plane and in-plane motions and signal fine-tunes. MEMS devices and integrated circuits can be fabricated monolithically. SEMs show the delicate array matrices by post-process micromachining and finished die with circuit area protected by passivation. No warpages are observed by interferometer in our devices, and linear out-of-plane movements are being electrically driven. At last, frequency responses of 14.22 kHz and 2.71 kHz are also characterized for device mode shapes.

Keywords- Bipolar/CMOS/DMOS; MEMS

I. INTRODUCTION

Mature CMOS technology gives promising platform to produce challenging MEMS devices [1]. Nevertheless, conventional CMOS-MEMS devices are built by complex layers of metals and oxides with post DRIE [2,3] and facing unpreventable temperature-coefficient-mismatching structure curling [4]. Fundamental ideas of our design are to fabricate single material structure which can multiply its sensing signals by the scalable matrix or actuate/sense signals pixel-by-pixel with an advanced 0.35µm 4 metal bipolar-CMOS-DMOS (BCD) technology [5].

New MEMS- or NEMS-related system on chip (SoC) designs, such as RF-MEMS, have now focus more on mature silicon-or GaAs-based integrated circuits (IC) processing technologies. They provide high speed and high frequency operation, high quality factor and low loss performance, adaptive and high density components. Today's silicon-based IC technologies are proved to be a promising technology to integrate kinds of miniature mechanical parts as system on chips. The single chip feasibility offers possibilities on portable electronics, wearable communication, and also high-definition video applications. Examples of successful products adopting MEMS SoCs are Analog Devices Inc.'s ADXRS and Texas Instrument's DMD. Nevertheless, these technologies are fabricated by their in-house patented manufacturing, which using intermediate process (*i*MEMS) or post process with add-on layers. Numerous techniques are also tried hard in order to build moving parts in regular IC process sequences, with their signal processing circuits on the same die, to gain the benefits of high-yield, high-volume, low-interference, and low-cost. These methods include using pre-CMOS, intra-CMOS, and post-CMOS processes, with surface or bulk micromachining. Although unique, these processing steps might degrade yield, cost, and make those difficult in implementation. Here, we propose designs for monolithic MEMS SoCs using the 0.35 µm BCD technology major for mixed-mode IC applications.

This research proved advanced IC process with simple BOE (buffered oxide etching) for inter-metal dielectrics etching could release conducting moving structures, without attacking passivation-protected circuit part. We also demonstrate array-type in-plane and out-of-plane motion structures, which resonant frequency could wide-ranged between sub-kHz to several MHz. This proves the feasibility of BCD technology to fit various MEMS applications. Thus, the monolithic processing electronics, sensing and actuating function with novel quadrant electrodes design could be easily modified. From this, we present novel promising designs over conventional methods.

II. PROCESS AND FABRICATION

A. Process Technology

Mixed-mode technologies have potentials for MEMS SoC (system on a chip) applications. The 0.35 µm Bipolar/CMOS/DMOS (BCD) 4 metal process with 3 µm top metal (see Fig. 1(a)) has tightly controlled fine design rules. It had been proven to be able to fabricate uniformly delicate array-type microstructures monolithically on a single chip. With BJT's precise and linear feature, CMOS for logic functions, this technology provides DMOS (double-diffused MOS) for high driving capability additionally. In this aspect, it could actuate devices up to 60 volts and is thus to be suitable for some MEMS applications.

B. Post Micromachining

Thick top metal option offers flexibilities in proof mass design, prevents stress-induced warpage, and therefore could be used as a structure layer. After standard IC processing and our in-house (post-process) surface micromachining by using 6:1 BOE etching and IPA treatment for anti-stiction, the top thick metal micro-structure could be released. This would not affect passivation-protected circuit part spaced 40 µm away as Fig. 1(b) shown, and interconnection underneath by timing

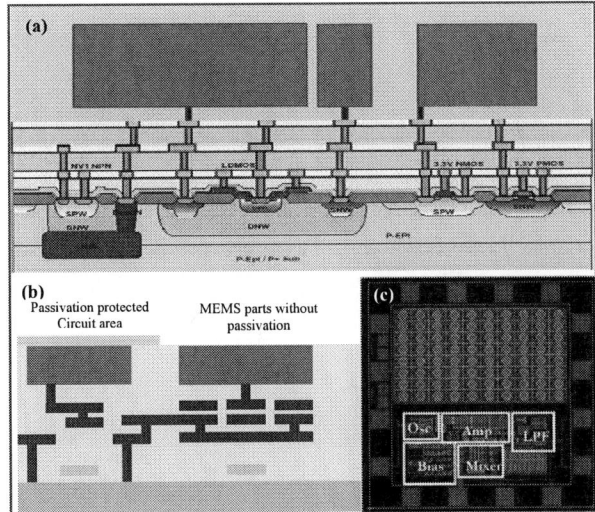

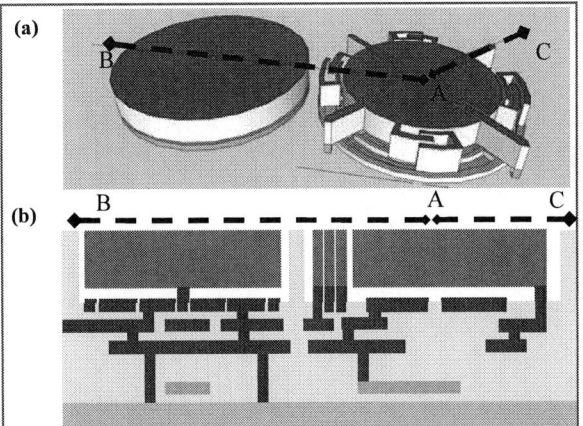

Fig. 1: (a) Cross-section view of the 0.35μm 4M BCD technology, (b) monolithically combined with circuits, and (c) MEMS SoC layout.

Fig. 2: Design concept of micro devices. 3D and cross-section of the structures in (a), (b).

controlled etching stop, the MEMS and micro-electronic devices are then being fabricated monolithically. Since the structured layer is the top-metal, underneath area could be used entirely for more complex signal processing function needed, which saves die cost effectively. Furthermore, flexibilities on altering conduction provide ways for fast application-oriented design, without suffering generic performances such as revealed in the Fig. 1(c)'s final MEMS SoC layout that can fulfill conditioning functions blocks as perform.

III. DESIGN CONCEPTS

In Fig. 2(a), we choose 3μm Al top metal as the proof mass, and springs could either use via for out-of-plane design or same-top-metal for in-plane structure (see detailed cross-section view in Fig. 2(b)). As Fig. 3 described, interconnection layer underneath metal 4 can be flexibly design patterns for electrodes of electro-static actuating or capacitive sensing, DC-tuning or calibration, and also for moving parts landing stoppers or posts. This gives the possibilities and flexibilities for the device designs on the resonator, accelerometer, gyro and versatile applications.

Table 1 summaries the key design parameters of the devices in our MEMS SoC. Proof mass for type X10 and X20 are in 25μm and 50μm diameter, respectively, and both are 3μm thickness. Spring rigidness is decided by its geometries, and we can notice that the symmetrical axial-length ratio of X10's mass and spring is as huge as 62.5.

IV. RESULTS AND DISCUSSIONS

Case studies will be introduced here (type X10 and X20 as examples) to show the feasibility of the present 0.35 μm BCD 1P4M process.

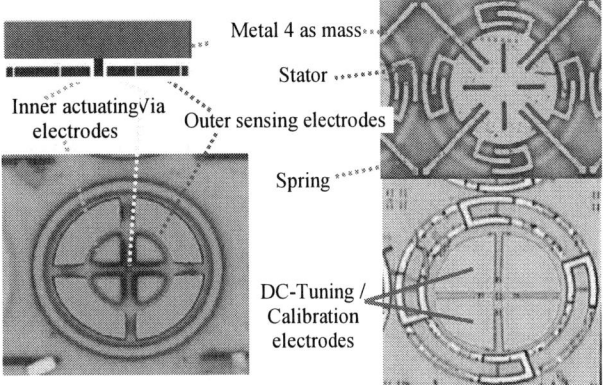

Fig. 3: Detailed descriptions of our designs by OM

Table 1: Key design parameter of the devices in the MEMS SoC.

Key Design Parameters	Structure Type	
	X10	X20
Proof Mass Geometry (μm)	Φ=25 t=3	Φ=50, t=3
Spring Geometry (μm)	W/L=0.4/0.4 H=0.9	W/L=1/45 N=4
Effective Electrode Area Ratio	42%(sen.)/14%(act.)	62%(DC-tune)
Array Matrix (Row/Column)	15/25	4/6
Potential Applications	Accelerometer, Sensor Array, Mirror Array,	Gyro, Mirror Array

After the inter-metal dielectrics being etched by the mixture solution of ammonium fluoride and acetic acid, different types of our novel MEMS array designs are demonstrated in Fig. 4(a) and 4(b) for type X10 (mushroom-like) and X20 (gyro-like), respectively. 500nm gap under the top circle plate is the travel distance of X10's out-of-plate vibration (Fig. 4(c)). 4 crossover beams out-stretched the mass in Fig. 4(d) are springs of X20's, and the 4-comb-pairs are to present its capacitive sensing signal.

978-1-4673-1122-9/12 $31.00 © 2012 IEEE   117

Uniform tiny structured pixel-array gains the most benefits of the IC high yield rate. Small area for each pixel (12.5 μm radius in the case shown) guarantees tightly arranged pitch matrix on every chip. By thick metal structuring devices using BCD technology, not only in-plane motion devices are feasible, following the post micromachining process, we can also adopt structures for out-of-plane torsion application. These are 25μm diameter disk array posted at the center by a via3-plug of 0.5 μm squared-width. Fig. 4(c) is a closer-up view of X10's moving parts, showing inter-metal (metal 3 and metal 4) dielectric (oxide) layer to be totally cleaned. Gap height between metals is 500 nm, which provides a sufficient displacement range for some MEMS applications, such as inertial sensors or resonators.

In type X20's case, novel quadrant electrodes are being designed for programmable actuation and sensing processing, potentially, under those movable plates. With parallel arrangement of the array-type disks, signal being sensed could be multiplied and therefore highly leverage up its sensitivity with the quadrant electrode functions described earlier. Inner ring of the quadrant-type electrodes can be used for driving top plate, which could be programmed to set different specifications of sensing/sensitivity ranges, or pre-conditioned to calibrate or fine-tuning plate tilt, or individually corrected their phase-shifts in high speed operation applications. Outer rings have larger overlap areas with upper structured vibration disk, which could sensibly read out capacitance difference in quadrant electrodes, respectively. This sensing mechanism could effectively compute any 2-dimensional directed-force source by simple algorithms. Landing stopper ring is used as a limiter for disk vibrating protection (and hence higher reliability.) Similar as the other design case, flexible mass and spring designs could result in different frequency responses and sensitivities for wide-range applications, e.g. strong via gives one of their resonant frequency as high as several MHz.

The sensing elements are matrix-positioned to multiply the small single variation. It has the merit for inertial sensor application if one parallel connects all the pixels; or for optical mirror application if one controls each pixel with added addressing function in this example. Smaller array pixel design could also gain the benefit of semiconductor's high yield rates for fewer particle damage issue concerns. This case is also with highly potential to be combined with on-chip sensing and processing electronics as to fulfill system-level requirements.

SEM graphs also reveal springs connected to their outer frame (type X20) with 4 diagonal ones which are fixed at end posts in Fig. 4(b). Those pictures show that array-type devices are well-processed and prettily dimensioned as the design rules we drawn. This gives us great advantages to estimate device characteristics at the very beginning we design, includes mass, spring, and even damping factor (from the slots that we open for etching holes.) The potential function of this structure and metal rings saw from Fig. 4(b) and 4(d) underneath top metal are as explained earlier, we could find the underlying structured electrodes are smartly designed for DC-tuning, offset-calibration/compensation-pads, electrostatic driving electrodes, capacitive sensing electrodes, and landing stoppers. Detailed zoomed-in of type X20 is pictured in Fig. 4(b). The spring in this case adopts thick metal-4 to constrain in-plane

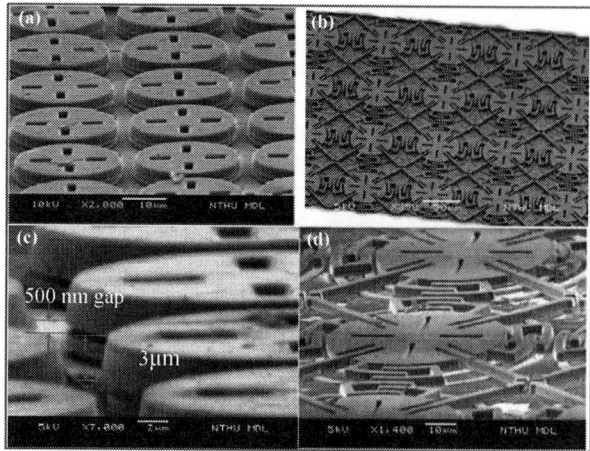

Fig. 4: SEMs of the novel micro arrays. (a) Type X10, (b) type X20, (c) and (d) are released structures of X10 and X20, respectively.

rotation, and outer frame is posted by via-3. Metal-3 could also be flexibly used for signal conditioning.

Within the 1000*1000μm^2 whole chip area (including processing circuit part), we could have small array-pixel-pitch of 45*45 in one of our designed cases in Fig. 5, with its SEM and OM photos shown.

Wide area range array-type plate structures without any stress-induced warpage or etch-induced surface roughness characteristics are easily obtained run by run, proved by optical interferometer measurement. The curvature radius of 37.73 mm shows pretty flat surface (compare to 12.5μm disk radius), which are nearly identical for every structure cell in the disk matrix. Therefore, this process offers reliable results, again indicates that standardized IC technologies have great potentials to fulfill monolithic integrated MEMS SoCs. The single metal layer structure of type X10 also proves its good performance within the wide temperature range (30 to 80 degree Celsius) as no warpages are observed by the optical interferometer (Fig. 6(a)). Measurement by applying different DC-biased voltages on one of the driving pad underneath its top plate, X10 shows linear displacement below 40 volts without snap-down as in Fig. 6(b). Under these situations (i.e., small disk area, 1/4 small actuating electrode, and rigid spring), out-of-plane torsion with several nm/volt linearly displacement is observed as depicted. When the driving voltage is higher

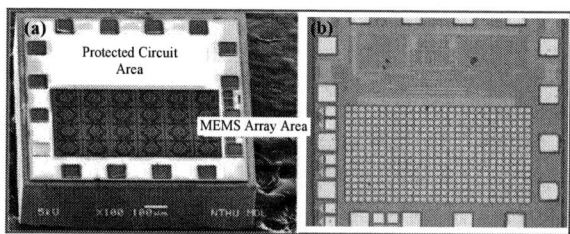

Fig. 5: SEM of finished die (a) and its OM that reveals circuits integrated (b).

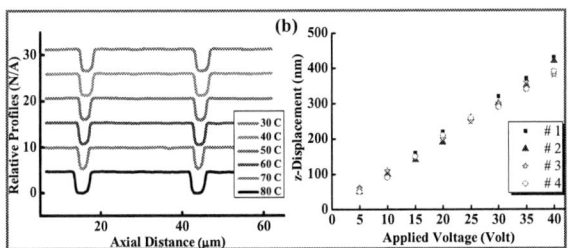

Fig. 6: No temperature induced warpages observed in WYKO scan of type X10 (a) and its z-displacements by apply voltages to single driving electrode (b).

than 40V, the disks become unstable and might snap-down to strike the landing pads in 500 nm gap height. At least within the 200 nm linear route of disk vibration, the structure could be used by high frequency switching or small displacement torsion applications with highly reliable operation, such as micro-mirror and accelerometer, respectively. Extending underneath actuating and sensing electrodes area, and modifying mass/spring geometries, the driving voltage could be lower down and sensing sensitivity could be leverage up.

Frequency responses are also characterized by laser Doppler velocimetry (LDV) and MEMS motion analysis (MMA) for X10 and X20 mode shapes, respectively. One out-of-plane motion peak happens at 14.22 kHz (Fig. 7(a)) and Fig. 7(b) indicates the in-plane fundamental motion sat on 2.71 kHz. The resonant frequencies and mode shapes have been predicted by finite element analysis, different design geometries would fall into hundreds kHz for those small (tens micron scale) proof mass. The rigid spring and tiny proof mass in one of the design gives lower 1st mode frequency, to high frequency of 193 kHz in the other one, with its center mass almost identical in each other. Other higher frequency responses (275 kHz and 505 kHz are indicated) of different design structure are also observed by the same measurements. Wider FWHM (full width half magnitude) and noisy shapes are mainly due to small mechanical signals, compare to un-shield testing background. Nevertheless, those measured data coincidence with simulation ones, which proves non-composite material (in this case, pure single metal of aluminum) structure gives better result prediction than composite ones (i.e. metal-oxide-metal back-end multi-layer composite structures).

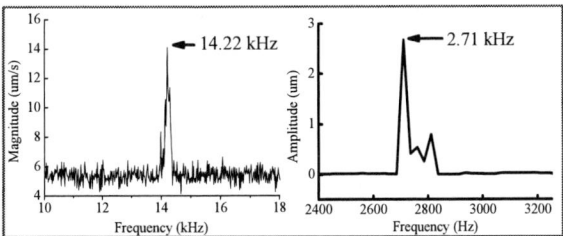

Fig. 7: LDV spectrum of X10 (a) and MMA frequency response of X20 (b) imply their fundamental motions.

V. CONCLUSIONS

This study reports the feasibility of 0.35μm 1P4M BCD technology with thick top aluminum metal layer (3μm) for monolithic MEMS SoC applications. Demonstrated array-type micro-structures designs are listed in Table 1 with some key parameters and their potential applicable areas. The DMOS in this BCD process is designed to handle operation range between 20 and 60 volts, therefore it is suitable for feedback force control for device driving conditioning to fulfill monolithic MEMS SoC applications. Nevertheless, the original feasibility-test-chip layout, with sensing electronic circuit blocks on the same die. These function blocks include oscillation clock which provide a reference modulation frequency, basic band-gap reference for biasing, low-noise amplifier (LNA) to read the capacitive signal out, mixing frequency to demodulate carrier with mechanical vibration frequencies, and then filters high band signals (Detailed circuit design and result will not be discussed herein.) Interconnections are run by 1st and 2nd metal layers, which could multiply sense out structured array signals or selectively pixel-by-pixel with addressing functions. Circuit blocks are protected by conventional passivation layer (Si_3N_4/SiO_2) but MEMS device areas are not – which intentionally being attacked by BOE etching away inter-metal-dielectric (SiO_2) to release top metal structure layer freely without post-processing related issues were found.

The array matrix can multiply signal received, which is again suitable for featured sensor applications such as accelerometer and gyros. Respectively addressing each device or triggering each divided electrode underneath the structured pattern with pixel-to-pixel signal process circuits can fulfill more intelligent MEMS SoC. This study also combines those MEMS devices with sensing circuit blocks here for future monolithic uses.

Although further characterizations have to be facilitated, the reliable post processing results observed show their credible performance and easy adoption. More dedicate structures could build directly on more functional circuit blocks to promisingly fulfill high-end MEMS SoC needs.

REFERENCES

[1] H. Luo, Z. Xu, H. Lakdawala, L. R. Carley, and G. K. Fedder, "A copper CMOSMEMS Z-axis gyroscope," Fifteenth IEEE International Conference on Micro Electro Mechanical Systems, MEMS2002, Las Vegas, NV, 2002, pp. 631-634.

[2] M.-H. Tsai, C.-M. Sun, Y.-C. Liu, C. Wang, and W. Fang, *IEEE TRANSDUCERS 2009*, pp. 672-675.

[3] H. Xie, L. Erdmann, X. Zhu, K. J. Gabriel, and G. K. Fedder, "Post-CMOS processing for high-aspect-ratio integrated silicon microstructures," *Journal of Microelectromechanical Systems*, vol. 11, 2002, pp. 93-101.

[4] H. Luo, "Integrated multiple device CMOS-MEMS IMU systems and RF-MEMS applications," Ph.D. Thesis, Department of Electrical and Computer Engineering, Carnegie-Mellon University, Pittsburg, 2002.

[5] BD350BA Design Manual, DongbuAnam Semiconductor, BD350BA-A-DMB (2.0), 2006.

Integration the Back-Side Inclined Exposure Technology to Fabricate the 45° k-type Prism with Nanometer Roughness

Kuo-Yung Hung[*], Yi-Wei Tsai, Chun-Fu Lee, Yi-Hao Chu

[*]Institute of Mechanical and Electrical Engineering, Ming-Chi University of Technology, Taiwan.

kuoyung@mail.mcut.edu.tw.

Abstract—This paper describes the design for a special k-type microprism structure for application in the lateral-type blu-ray semiconductor laser of an optical pickup head system. This design solves the current frontal type blu-ray semiconductor laser problem, and thus reduces the size of the optical pickup head. This study combines front- and back-inclined exposure technology to develop a k-type microprism structure. Thick film negative photoresist are used in this study as the main structural material. For obtaining the optimal structural surface roughness, the effect of solvent loss percentage of the polymer material was controlled. Besides, the bottom half of the k-type was generated through backside exposure to solve problems such as diffraction phenomena due to uneven photoresist surface during front-side inclined exposure. This design also avoids the problem of refractive index matching by omitting the step of filling the gap between the mask and the photo resistor with glycerol. For improving the roughness problem of the front-side inclined exposure, backside inclined exposure is implemented when fabricating a 45 ° polymer micro mirror. The use of front-side exposure for making the top half of the k-type solves the problem of undesirable surface roughness caused by insufficient light penetration during front-side inclined exposure. This paper utilizes front- and back-side inclined exposure technology to fabricate a 45 ° k-type polymer micro mirror with 15.2 and 12.4 nm (400 μm X 400 μm) roughness, respectively. The roughness level could meet the standards (λ/10, λ = 405 nm) of blue ray specifications. This type of micro prism can be used as a key component in Pico-projector, Interferometer, bio detection systems, data storage systems, and linear encoder optical systems. This novel technology also has the characteristics of high throughput and wafer-level assembly.

keywords- polymer; k-type prism; backside inclined exposure

INTRODUCTION

Generally, the micro-prism was fabricated by machining or injection molding methods. This is time consuming and expensive, especially for Pentagonal, Rochon, and Porro prisms. Develeing a new technology for fabricating the micro prism with high throughput, low cost, and high surface quality is a new challenge for MEMS technology. Recently, much research has been published on the wide range of applications of the SU-8 thick-film photoresist, and its use in fabricating many high aspect-ratio micro-optical structures [1-12].

C. H. Lee et al. (2007) [1] described the use of an ultra-thick SU-8 process (1 mm), analyzed the sidewall surface roughness after vertical exposure, and developed and compared the process with the silicon deep RIE (reactive ion etching) process. This study also used AFM to inspect an area of approximately 70 μm² and found that the surface roughness of the SU-8 was approximately five times less than that of silicon after deep RIE etching. The author had proven the polymer material could get the better roughness than silicon etching. However, many optical applications require 45 ° inclined mirrors or prisms. W. J. Kang [2] compared the effects of different gaps, exposure doses, and substrate reflectivity on the structure and surface roughness; the best roughness obtained was approximately 50 to 60 nm, and the experimental thickness was approximately 70 μm. The author found that sidewall roughness increases with exposure dose or substrate reflectivity, showing that interface reflection can have a large impact on roughness. Thus, the research purpose of this paper was to present back-side (conformal mask) inclined exposure technology to resolve the above problems.

K. Y. Hung et al. (2004) [3] proposed, for the first time, that inclined exposure technology could be used to fabricate thick-film micro-optical elements, and successfully used such elements in integrated optical pickup heads. The advantage of this approach is that it can avoid the alignment errors that may occur during the manual assembly of ordinary optical pickup heads, and it can also result in smaller pickup heads. Hung and colleagues [3] also used AFM to determine that the surface roughness of a small area of 25 μm² was less than 7 nm. It is apparent that researchers [4-11] have used many different exposure techniques to improve the roughness of thick-film SU-8 structures and meet the needs of relevant applications. Previous studies have indicated that the h-line (405 nm) wavelengths are better when manufacturing high aspect-ratio SU-8 structures [4]. By reducing the gap and the interface reflection of the substrate, it may be possible to reduce the sidewall roughness during perpendicular exposure.

Fresnel or near-field diffraction in thick positive and negative resist for microstructures resulting from a small gap in contact or proximity printing has been previously investigated [4-6]. For resolving the above problems, Kim et al. [12] also used UV-LIGA-like processing, and manufactured a circular micro pinhole array structure of approximately 33.6 to 101 um through backside vertical exposure. Mert and Mutlu [13] used SU-8 as the gate dielectric layer, and produced a polymer thin-film transistor structure with the back-side and front-side exposure technique. The polymer thin-film transistor structure was applied in an optical display to simplify the number of masks required in the manufacturing process. Song et al. [14] produced a micro prism structure of different aperture through the back-side exposure technique, and explored the effect of Fraunhofer

978-1-4673-1122-9/12 $31.00 © 2012 IEEE

diffraction on micro lens structures at different exposure doses.

However, very few researchers have attempted to fabricate new optical applications and improve the roughness of three-dimensional inclined structures; even though such three-dimensional Porro prism or beam-splitter structures are extremely important in optical and biomedical applications. Consequently, this paper sought to investigate the fabrication method and surface roughness of the k-type prism structures (~1.4 mm) on an SU-8 3035 photoresist. An optical interferometer was used to measure large-scale (~millimeter-scale) surface roughness, and a beam profiler (Ophir Optronics, Inc., Beamstar FX-50) was applied to measure the light power of the splitter.

STRUCTURAL DESIGN AND APPLICATION PRINCIPLES OF THE K-TYPE PRISM

As the demands for high-definition video and audio increase, home audio-visual equipment and portable electronic devices both develop towards the trend of high-definition. This trend facilitates the development of high-density storage technology, and the blue ray reading system is one of the most popular products in the current market.

Although numerous micro optical pickup head designs and the manufacturing processes of individual components are able to achieve the optical standard of red or blue light, each component must be assembled and aligned. Assembling and aligning micro components is a big challenge, and human labor is required for the process; therefore, labor costs cannot be reduced. In addition, blue ray semiconductor lasers of the high power frontal type have not yet been mass produced, making minimization of the system more difficult. Generally, the fewer the components in optical reading heads, the better; for the use of few components reduces (1) the weight of the reading head and effectively reduces the average reading time; (2) the cost for materials; and (3) the assembly and alignment time during mass production, and thereby reduce the production cost. Above mentions are the three key design points of this paper.

Fig. 1 is a schematic diagram of the k-type prism component for the application system this paper wishes to develop. The optic theory of the system is that a lateral type blue ray semiconductor laser emits a beam, and this beam horizontally strikes a first reflective micro prism. The laser beam, reflected from the first micro prism, vertically strikes a second reflective micro prism. Next, this laser beam is reflected from the second micro prism and horizontally strikes a third reflective micro prism. The laser beam is reflected from the third micro prism, passes through a holographic optical element, reaches a focusing objective lens, and focuses on the surface of a disc. The laser beam is then reflected from the disc and forms a reading beam retrace. The reading beam passes through the focusing objective lens, and is diffracted by holographic optical elements to form a first-order or multi-order diffracted beam. The diffracted beam is then reflected by the second and third micro prisms, enters the light

receiving area of the photo detector, generates a reading signal, and finally finishes the reading and writing processes of the micro optical pickup head.

Fig. 2 is a schematic diagram of the light path of three prisms simulated by OSLO software. Using the reflection mirror structure of two 45 ° k-types changes the light path by 180 °.

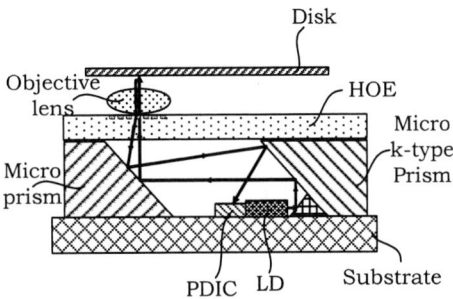

Fig 1. Schematic diagram of the k-type prism component for the micro pickup head application system

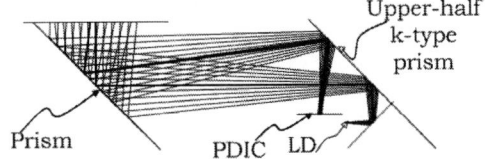

Fig 2. Schematic diagram of the light path of three prisms simulated by OSLO software.

With the design of this optical system, the right half of the optical components is a k-type prism. The bottom-half of the k-type prism is closer to the laser light source and is therefore smaller in size; it is approximately one-third of the upper-half of the k-type prism. This design significantly simplifies the component manufacturing and assembly complexity. It achieves the purpose of micro optical pickup head integration. The aim of this study is to develop the concept of k-type prism integration and wafer level using micro-processing integration technology. Therefore, the key components of this study are the fabrication method of the k-type prism and the surface roughness.

In this paper, the optimal soft-bake times 14.5 hrs (for 1.4 mm thickness) were used (softbake at 105 °C and PEB at 125 °C for one hour) for controlling the solvent percentage of the polymer [15]. The calculation of the solvent percentage of the polymer is as follows: (Spin photoresist weight- Photoresist weight after softbake) / (Spin photoresist weight x (1-solid contents percentage of the photoresist)).

FABRICATION PROCESS

Using the incline exposure process parameters [5-6], this paper presents a process for optimizing the roughness of the k-type prism by front-side (E-beam mask) and back-side (conformal mask) inclined exposure. Fig. 3 describes the fabrication process of the k-type prism in detail. The inclined exposure technology used in this paper was a previously developed technology and exposure system [5-6]. A system schematic was available for reference [5-6]. The UV exposure system used in this paper employed an optical filter mounted parallel to a light outlet. This optical filter (OMEGA OPTICAL, PL 360LP) removed the portion of the light emitted by the mercury-vapor lamp with a wavelength of less than 365 nm, but admitted 405 nm and 436 nm UV light. The long wavelength light effectively penetrates UV stabilized or pigmented films. This filter could partially eliminated diffraction on the top of the SU-8 structure.

The detailed description of the manufacturing procedure shown in Fig. 3 is as below.

- After cleaning the glass wafer, a layer of Chromium (Cr) was sputtered to make a masking layer, and was spin coated with an AZ-5214 photoresist to make a protective layer. The mask structure required for the backside exposure manufacturing process by exposure and development was defined **(due to the near-field or Fresnel diffraction of the gap between the mask and photoresist, secondary surface reflection of the substrate)**, and the Wafer was etched with Cr-7. Finally, the photoresist was removed to obtain the conformal mask (Fig. 3a to d).

- After the patterning of the Cr layer, a layer of Teflon was applied to the outer surface of the glass wafer. Because Teflon has a significantly lower surface energy (it is hydrophobic), applying it around the wafer prevents the photoresist from producing overflow due to high temperatures during soft-baking. Therefore, photoresist SU-8 3035 can be controlled in the glass wafer. Teflon cures after heating at 180 °C for 3 min (fig. 3d).

- SU-8 was spin coated on the glass wafer (Fig. 3e) by dripping approximately 10.5 g of SU-8 3035 on the wafer while it was hand spun, accompanied with spin coater spinning at 160 rpm for 30 s. Using this method once, a 1.4 mm thick coat of SU-8 can be applied to the glass wafer. Before soft-baking, the SU-8 photoresist was inspected for smoothness due to its effect on diffraction during exposure and inclined surface roughness. The purpose of soft-baking is to dry the solvent in the photoresist, because excessive solvent will cause caving or distortion to the surface during exposure. Therefore, time and temperature during soft-baking must be controlled precisely, with gradually increasing temperatures. The soft-baking temperature was gradually increased; starting at room temperature, the temperature was increased by 10 °C every 10 min. Once the temperature reached 105 °C, the k-type prism was soft-baked for 14.5 hrs.

After back- and front-side inclined exposure (Fig. 3f-g), post-exposure baking (125 °C, one hr), and development, the designed micro k-type prism structures were obtained (Fig. 3h).

Because the SU-8 layer is thicker at the center of the chips than at the edges, the gap between the photoresist and the mask is smaller at the center than at the edges; diffraction is therefore smaller at the center. In short, because the gap between the photoresist and mask is uneven, this paper focuses on investigating the effect of process parameters on sidewall structure. For the sake of consistency, measurements were therefore taken only at the center of chips; edge samples were not examined.

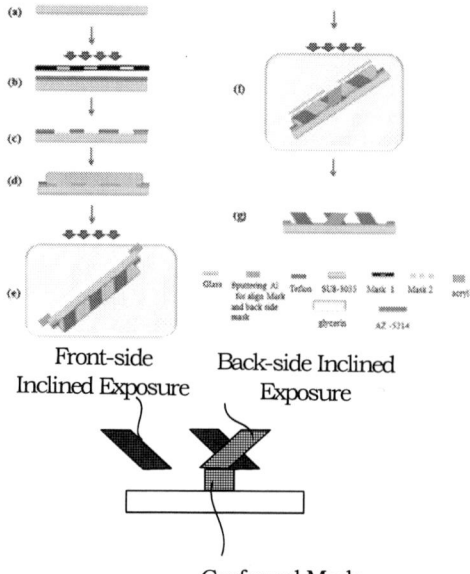

Fig 3. The fabrication process of the k-type prism in details.

EXPERIMENT RESULTS

Fig. 4 shows the pictures of the fabricated micro k-type prisms by optical microscope (Fig. 4a), and surface roughness results by a white light interferometer. The optimum roughness through backside exposure is about 12.4 nm (range ~ 400 X 400 um) (Fig. 4b). The optimum roughness through frontside exposure is about 15.2 nm (range ~ 400 X 400 um) (Fig. 4c). The smallest surface roughness achieved by frontside exposure in experiments using SU-8 3035 with a thickness of 1.4 mm was about 36 nm when the softbake time was 14.5 hrs (area: 730 X 730 μm) (fig. 4d). In this paper, the softbake temperature was 105℃, and post-exposure-bake (PEB) was 125℃ for one hr. The relationships of the solvent contents in the polymer material with structure surface roughness are important for optical application. The solvent removal percentage of the polymer was about 86.5 %; meaning that only about 13.5 % of the solvent contents remained, compared to the original. These are the optimum parameters applied to the structures.

Table 1 shows a comparison of the roughness of micro structures fabricated across different studies. These results have proved that utilizing the back-side inclined exposure technology could resolve the diffraction problem resulting from the gap between the mask and the uneven thick-film photoresist, because the roughness through backside inclined exposure is lower than that of frontside exposure. And if the pattern size is the same, this technology could also save the mask. Therefore, this technology can be used to develop the inclined polymer micro prism structure with blue-ray surface roughness.

The reflection of the fabricated k-type prism is tested (Fig. 5a) and Fig. 5b shows the two laser spots are reflected from k-type prism successfully.

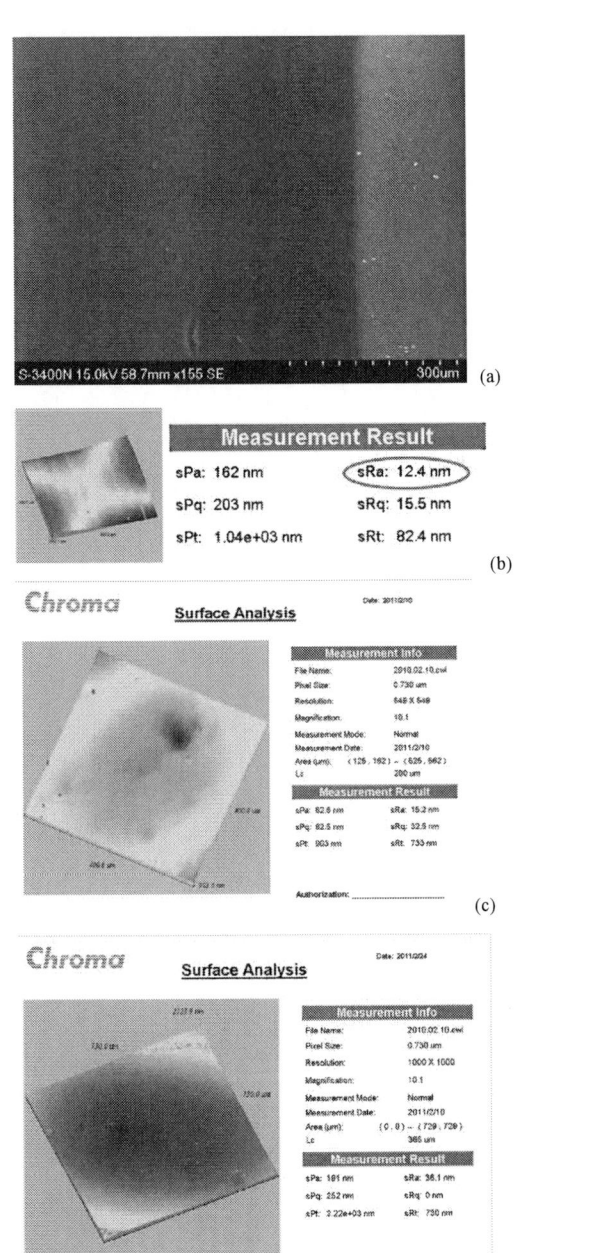

Fig 4. (a) The pictures of the fabricated micro k-type prisms by optical microscope, and (b) The optimum surface roughness through backside exposure by a white light interferometer. (c) The optimum roughness through frontside exposure (area: 400 X 400 um). (d) The optimum roughness through frontside exposure (area: 730 X 730 μm).

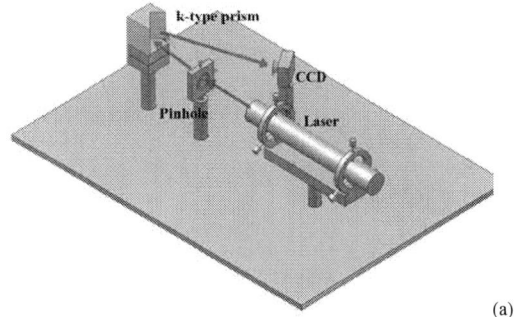

(a)

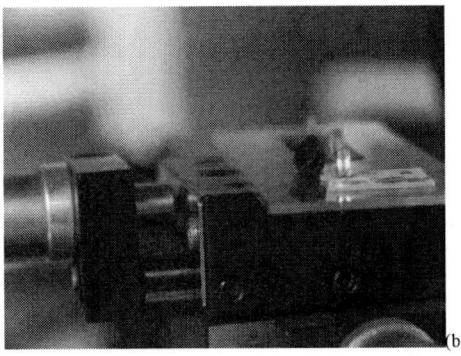

(b)

Figure 5. (a) The test mechanism for k-type prism. (b) The picture shows two laser spots are reflected from k-type prism successfully.

Table 1. Comparison of the roughness of micro structures fabricated across different studies.

Paper	Thickness	Measure area	Exposure mode	Measured roughness
[Rabe 2007]	E,90μm	149 X 66 μm	Vertical	30 nm
[Kang J,2006]	W70 μm	Hasn't mentioned	Vertical, partial Incline	50-60 nm
[Lee C 2007]	H,50-900μm	70 X 70 μm	Vertical	38.28 nm
[Yang 2005] [4]	R,1150μm	Hasn't measured	Vertical	Hasn't measured
This paper	500 μm	400 X um	40045° Incline (Backside inclined exposure)	12.4 nm
This paper	1400 μm	400 X um	40045° Incline (frontside inclined exposure)	15.2 nm

CONCLUSIONS

This paper experimentally verified that special (front- and backside) inclined exposure technology could be used to fabricate a polymer k-type prism, and therefore offers an alternative to conventional mechanical processing. This paper successfully used front- and backside inclined exposure technology to fabricate a wafer-level k-type micro prism with 45 ° optical grade surface (~1.4 mm thick). The smallest surface roughness achieved in the experiments using SU-8 3035 material with a thickness of 1.4 mm was less than 12 nm (400μm X 400μm area) through backside inclined exposure technology. Backside inclined exposure was able improve the roughness problem (due to the near-field or Fresnel diffraction of the gap between mask and photoresist, secondary surface reflection of the substrate) of the front-side inclined exposure when fabricating a 45 ° polymer micro prism. The reflection of the fabricated k-type prism is also tested successfully. This type of micro prism can be used as a key component in Pico-projector, Interferometer, bio detection systems, data storage systems, and linear encoder optical systems. This novel technology also has the characteristics of high throughput and wafer-level assembly.

ACKNOWLEDGMENT

We would like to thanks the National Science Council of Taiwan support the project budget, under the grant number NSC 100-2221-E-131 -024.

REFERENCES

[1] C. H. Lee, K. Jiang, G. J. Davies, "Sidewall roughness characterization and comparison between silicon and SU-8 microcomponents," *Journal of Materials Characterization,* vol. 58, pp. 603-609, 2007.

[2] W. J. Kang, E. Rabe, S. Kopetz and A. Neyer, "Novel exposure methods based on reflection and refraction effects in the field of SU-8 lithography," *Journal of Micromechanics and Microengineering,* vol. 16, pp. 821–831, 2006.

[3] K. Y. Hung, H. T. Hu, and F. G. Tseng, "Application 3D Glycerol-Compensated Inclined-Exposure Technology to Integrated Optical Pick-Up Head," *Journal of Micromechanics and Microengineering,* vol. 14, pp. 975-983, 2004.

[4] R. Yang and W. J. Wang, "A numerical and experimental study on gap compensation and wavelength selection in UV-lithography of ultra-high aspect ratio SU-8 microstructures," *Sensors and Actuators B,* vol. 110, pp. 279–288, 2005.

[5] K. Y. Hung and J. C. Liao, "The Application of Fresnel Equations and Anti-Reflection Technology to Improve Inclined Exposure Interface Reflection and Develop a Key Component Needed for Blu-ray DVD--Micro-Mirrors," *Journal of Micromechanics and Microengineering,* vol. 18, no. 7, 075022, 2008.

[6] K. Y. Hung, T. H. Liang, "Application of Inclined-Exposure and Thick Film Process for High Aspect-Ratio Micro Structures on Polymer Optic Devices," M*icrosystem Technologies-Micro-and Nanosystems-Information Storage and Processing Systems,* vol. 14, no. 9-11, pp. 1217-1222, 2008.

[7] Y. K. Yoon, J. H. Park, M. G. Allen, "Multidirectional UV Lithography for Complex 3-D MEMS Structures," *Journal of microelectromechanical system,* vol. 15, no. 5, pp. 1121-1130, 2006.

[8] K. D. Vora, B. Y. Shew, E. C. Harvey, J. P. Hayes, and A. G. Peele, "Sidewall slopes of SU-8 HARMST using deep X-ray lithography," *Journal of Micromechanics and Microengineering,* vol. 18, 2008.

[9] K. D. Vora, B. Lochel, E. C. Harvey, J. P. Hayes and A. G. Peele, "AFM-measured surface roughness of SU-8 structures produced by deep x-ray Lithography," *Journal of Micromechanics and Microengineering,* vol. 16, pp. 1975-1983, 2006.

[10] E. Rabe, S. Kopetz and A. Neyer, "The generation of mould patterns for multimode optical waveguide components by direct laser writing of SU-8 at 364nm," *Journal of Micromechanics and Microengineering,* vol. 17, pp. 1664-1670, 2007.

[11] E. F. Reznikova, J. Mohr, H. Hein, "Deep photo-lithography characterization of SU-8 resist layers," *Microsystem Technologies,* vol. 11, pp. 282–291, 2005.

[12] K. Kim, D. S. Park, H. M. Lu, W. Che, K. Kim, J. B. Lee and C. H. Ahn, "A tapered hollow metallic microneedle array using backside exposure of SU-8," *Journal of Micromechanics and Microengineering,* pp. 14, vol. 597–603, 2004.

[13] O. Mert, S. Mutlu, "Self-Aligned Polymer Thin Film Transistors Fabricated Using Backside Exposure," *The 5th International Thin Film Transistor Conference(ITC 2009)* Paris, France, pp. 102-05, 2009.

[14] I. H. Song, K. N. Kang, Y. Y. Jin, D. S. W. Park and P. K. Ajmera, "Microlens array fabrication by backside exposure using Fraunhofer diffraction," *Microsystem Technologies,* vol. 14, no. 9-11, pp. 1285-1290, 2008.

[15] K. Y. Hung, Y. C. Chen, C. F. Lee, S. H. Huang, Y. J. Chuang, "A Novel Fabrication Method and Optical Test of the Micro Beam-splitter," 9th International Workshop on High Aspect Ratio Micro Structure Technology (HARMST 2011) Hsinchu, Taiwan, pp. 143-144, June 2011.

Study on Piezoelectric Properties of Near-field Electrospinning PVDF/MWCNT Nano-fiber

Z.Y. Ou[1], Z.H. Liu[1], C.T. Pan[1], L.W. Lin[2], Y.J. Chen[1], H.W. Lai[1]

[1] Department of Mechanical and Electro-Mechanical Engineering, and Center for Nanoscience & Nanotechnology,
National Science Council Core Facilities Laboratory for Nano-Science and Nano-Technology in Kaohsiung-Pingtung
area, National Sun Yat-Sen University, Kaohsiung 804, Taiwan
[2]Department of Mechanical Engineering and Berkeley Sensor and Actuator Center, University of California, Berkeley,
California 94720
panct@faculty.nsysu.edu.tw

Abstract—In this study, near-field electrospinning (NFES) was used to fabricate PVDF (Polyvinylidene fluoride) piezoelectric nano-fibers mixed with additional multiwalled-carbon nanotubes (MWCNT). Both mechanical strength and piezoelectric characteristics of a single nano-fiber were discussed. NFES technology can be used to fabricate PVDF piezoelectric fibers with an excellent piezoelectric property. By adjusting velocity of a x-y stage, DC voltage, and the distance between the needle and collector, the morphology and polarization intensity of piezoelectric fiber can be controlled. In addition, the optimal parameters of PVDF solution such as weight percentage of PVDF powder and MWCNT were also discussed. From the observation of XRD (X-ray diffraction), it reveals a high diffraction peak at $2\theta=20.8°$ of piezoelectric crystal β-phase structure with PVDF/MWCNT spherical composite structures in fibers. Actuation property of fixed-fixed single PVDF fiber structure was tested using DC voltage supply, and the fiber has significant deflection in the experiment. The vertical deflection can be observed and compared with model solution.

*Keywords- Near-field electrospinning; Piezoelectric fiber;
MWCNT ; Micro-transducer; Spherical composite structures*

I. INTRODUCTION

PVDF is a potential piezoelectric polymer and attractive in energy conversion applications between micro-electric-mechanical device, electromechanical actuators and energy harvester because of its high flexibility, biocompatibility and inexpensive. [1]. PVDF actuators based on thin film were widely studied [2-4]. There were few studied in the single PVDF fiber [5]. A direct-write electrospinning technique by means of near-field electrospinning (NFES) [6-7] was developed to produce the controllable fiber deposition of various materials. NFES was used to cause PVDF/MWCNT fibers in situ electrical poling and transform non-polar α phase in the crystalline into polar β phase. The modified MWCNT was mixed in the PVDF fibers as spherical reinforcement of composite to enhance the fiber's tensile strength and piezoelectric properties. In this study, controllability of electrospinning PVDF/MWCNT fibers using NEFS has been demonstrated. The manufacturing process and material

property of the composite fibers were also discussed to analyze its crystalline structures and surface morphology. From XRD observation, PVDF fiber with 0.03 wt% MWCNT reveals a high diffraction peak at $2\theta=20.8°$ (β-phase structure). To predict the complicated piezoelectric responses of fixed-fixed single PVDF fiber structure under a high electric field, ANSYS FEA was used to investigate the coupled field behavior. Actuation property of PVDF/MWCNT fibers was tested under different DC voltage supply

II. EXPERIMENT

A. Near-field Electrospinning process

The equipment includes a needle, high voltage power supply, collector (silicon wafer or glass), X-Y motion stage (controlled by X-Y stage controller via computer), and needle holder (see Fig. 1). Inner diameters of needles are ranging from 0.15 mm-0.3 mm. The anode of high voltage power supply was connected to the needle, and the silicon wafer was grounded as a collector. A high potential between needle and collector was formed by supplying high voltage 600-1200 V. In addition, the collector was installed on X-Y stage with motion speed of 20-100 mm/sec and travel distance of 50 mm, respectively. The gap between needle and collector is adjustable about 0.5-2 mm. The route of collector mounted on X-Y stage can be controlled by programmed path. The applied electrical field generated sufficient electrostatic forces to deform the polymer meniscus into a conical shape (Taylor cone) and then induced a polymer jet from Taylor cone. The droplet overcame the surface tension of the solution under a high potential. Then the extremely fine PVDF fiber in sequence was spun out on the collector. In the preparation of PVDF solution, dimethylsulfoxide (DMSO) was used as solvent for PVDF powder (Mw=534000), with acetone and fluorosufactant (ZONYL®UR) to improve the evaporation rate and reduce the surface tension of PVDF solution, respectively. MWCNT with proper percentage was dispersed in DMSO, and PVDF was dispersed in acetone by stirring at the same time. The MWCNT-DMSO solution was put on the ultrasonic oscillator to break possible agglomerates. Finally,

978-1-4673-1122-9/12 $31.00 © 2012 IEEE

two co-solvents of DMSO/MWCNT and acetone/PVDF were mixed, and stirred until the PVDF/MWCNT solution was homogeneously formed.

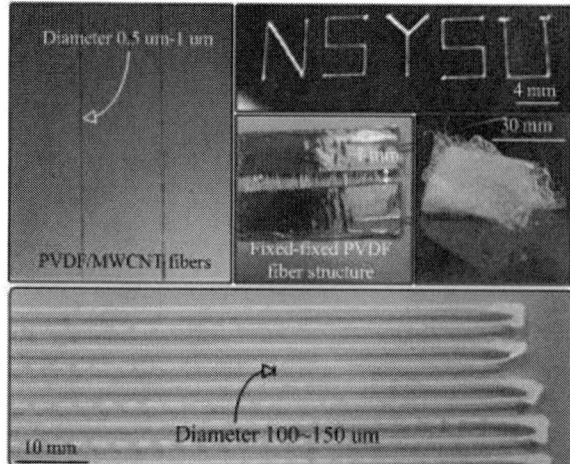

Fig. 1: Near-field electrospinning equipment and direct-write PVDF fiber process with in situ poling

B. Analysis of single PVDF fiber with FEA

Fig. 2 shows that static analysis to calculate the center displacement, a positive electric field (input voltage of 1000 V) was applied on the left side of the fixed-fixed fiber structure. The fiber diameter and length are 15 μm and 1 mm, respectively. The simulation result shows the downward center displacement was 5.04 μm. There are two significantly deformation such as necking phenomenon at fixed end and the internal compression strain in the z-axis. It reveals that a maximum internal strain appeared in the central part of PVDF fiber that induced by the repulsive force along the fiber axis and the moment near the fixed end. Since PVDF fiber has negative piezoelectric strain constant (d_{33}=-46×10^{-12} m/V [8]), the positive electric field caused compression strain which leaded to the fiber buckling.

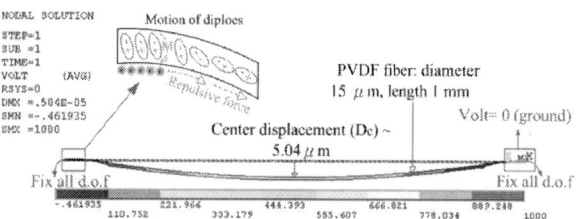

Fig. 2: Simulation results illustrate the deformation of the PVDF fiber

III. RESULTS AND DISCUSSION

A. Optimal parameters of NFES PVDF/MWCNT composite fibers

With using the needle (internal diameter of 0.25 mm), when the droplet overcame the surface tension of the solution, PVDF fiber was spun from the Talyor cone tip with a high voltage about 600-1200 V (the gap between needle and collector of 1 mm and high-voltage electric field of 7×10^5 V/m) To demonstrate NFES controllability using a programmable path X-Y stage, a variety of orderly patterns were deposited onto collector. The PVDF fibers of diameters (0.5-150 μm) and length (1mm-70mm) were spun on collector. The as-spun

diameter of PVDF fibers with different patterns are shown in Fig. 3.

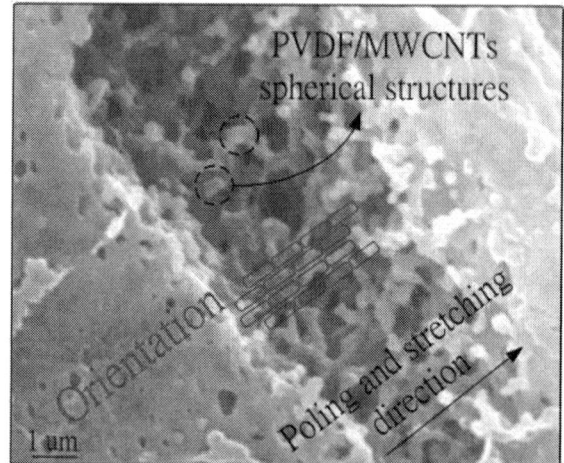

Fig. 3: The diameter of PVDF fibers and the other patterns

B. SEM observations of PVDF/MWCNT composite fibers

Fig. 4 shows SEM observation of PVDF/MWCNTs crystal structures, which were aligned in sequence and MWCNT aggregated in PVDF resin to form the spherical composite structures. Fig. 5 (a) shows that there were tiny pores on the surface when the fibers with MWCNT content of 0.03 wt% were dried. In addition, when the heating temperature was too high (set up over 50 ℃) that resulted in acetone evaporated quickly, and the PVDF solution had not enough time to reflow in the voids. After MWCNT content in PVDF solution exceeded 0.03 wt%, PVDF fibers with non-uniform chain structures on the surface would be suggested that excessive MWCNT agglomerating phenomenon and an unsteady solution flow rate in initially electrospinning process could take place as shown in Fig. 5 (b)

Fig. 4: SEM observation of PVDF/MWCNTs composite fiber crystal structures

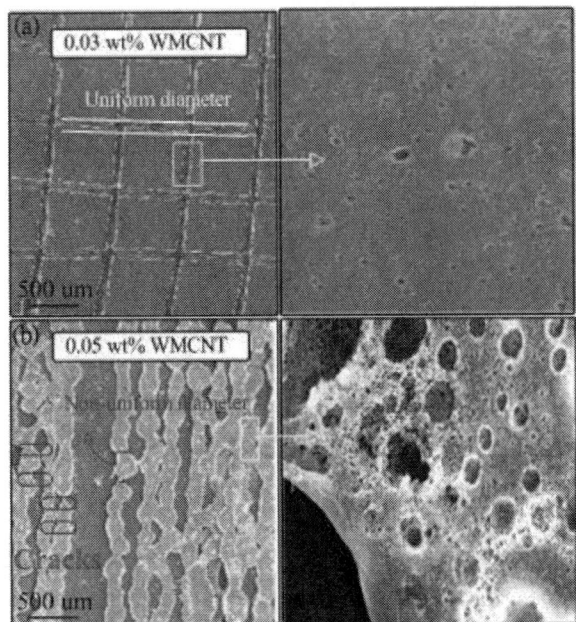

Fig. 5: SEM observations of PVDF/MWCNT composite fibers containing: (a) 0.03 wt% MWCNT and (b) 0.05 wt% MWCNT

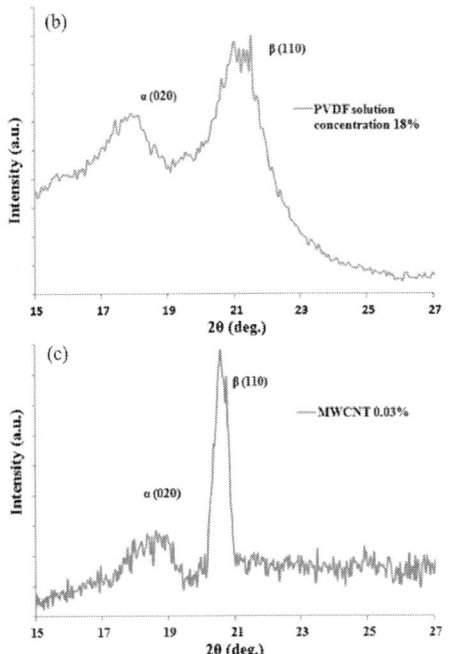

Fig. 6: XRD patterns of the PVDF fibers concerning (a) voltage of 1200V, (b) PVDF concentration of 18% and (c) MWCNT concentration of 0.03%

C. XRD analysis of PVDF fiber

In order to verify the in situ electric poling and mechanical stretching effects during NFES process, XRD (X-ray diffraction) observation of PVDF composite fibers was characterized. When a higher voltage of 1200 V applied in NFES process, β-crystalline structure in PVDF fiber was obviously increased as shown in Fig. 6 (a). Since a higher electric field resulted in completely arranging polarity direction of PVDF fiber and could assist growing-crystalline structure (diffraction peak at 2θ = 20.6°) in forming piezoelectric PVDF fibers. After increasing the concentration of PVDF solution up 18%, apparently the intensity of β-phase didn't change as shown in Fig. 6 (b). The intensity of β-phase increased significantly compared with the PVDF fiber without adding MWCNT as shown in Fig. 6 (c). The results indicated that there were a little different of β-phase intensity when MWCNT concentration was enhanced to 0.03 wt%. It means that even much more MWCNT added in PVDF fibers resulted in piezoelectric properties without enhancing obviously.

D. Observation of PVDF/MWCNT fiber actuation

After PVDF/MWCNT fiber transferred onto copper foil electrodes, silver glue was coated on the fixed end of the fiber to mount it on electrodes and decrease the effect of electrostatic effect. Then high voltage was respectively supplied to the electrodes for the purpose of easily analyzing and observing the fiber actuation of the suspended and the fixed-fixed fiber structures. Based on suspended piezoelectric PVDF/MWCNT fiber under 1 kV (see Fig. 7), the result reveals that a greater moment (+M) near the fixed end and the tensile strain to cause upward center displacement. Fig. 8 shows the observation of central displacement of polarization fiber under different positive electric fields (+E). When fixed-fixed piezoelectric PVDF/MWCNT fiber was applied different DC voltages (1, 1.2 and 1.5 kV), the out-of-plane displacements at the central part were measured.

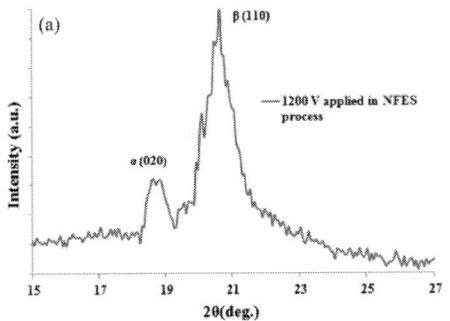

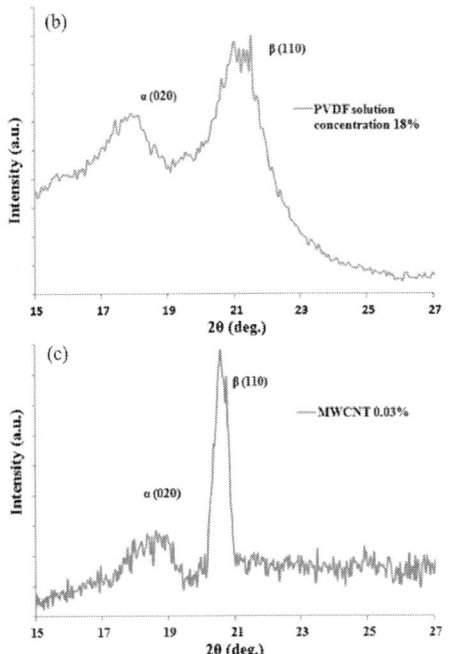

Fig. 7: Deformation of suspended PVDF/MWCNT fiber under electric field.

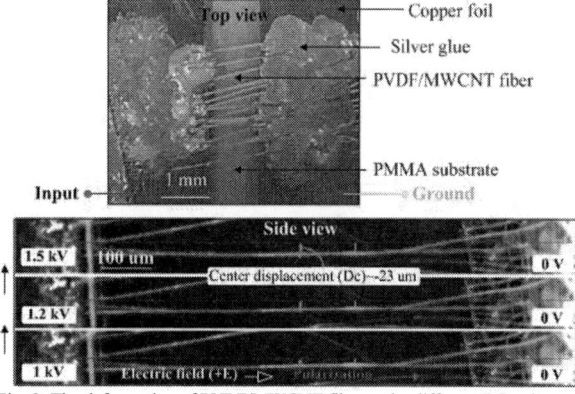

Fig. 8: The deformation of PVDF/MWCNT fiber under different DC voltages.

The center displacement of polarization fiber, non-polarization fiber as a function of voltage under a positive electric field (+E) compared with model solution as shown in Fig. 9 This results agreed with the previous theoretical prediction on piezoelectric responses. When a lower voltage was applied, electrostatic attraction force (F) was not enough to form an obvious displacement. Therefore, a conspicuous displacement can be induced after applied a higher voltage because of strong negative piezoelectric effect. The total center deformation of the buckled fiber is equal to the combination of slightly electrostatic downward deformation and strongly piezoelectric downward deformation generated by repulsive force along the fiber axis and a moment near the fixed end.

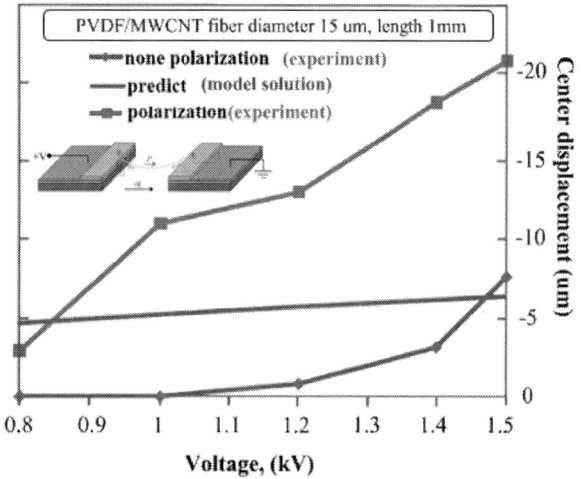

Fig. 9: The center displacement as a function of voltage among the polarization fiber, non-polarization fiber and model solution

IV. CONCLUSIONS

In this study, NFES was proposed to fabricate PVDF/MWCNT nano-fibers. By adjusting velocity of x-y stage, DC voltage, and the gap between needle and collector, the morphology and polarization intensity of piezoelectric fiber can be controlled. The optimal parameters of PVDF solution such as weight percentage of PVDF powder and MWCNT were discussed. In addition, the modified MWCNT was formed in PVDF fibers that can increase the crystalline of β phase for enhancing piezoelectric properties. From XRD observation of PVDF fiber with 0.03 wt% MWCNT, it reveals a high diffraction peak at 2θ=20.8° of piezoelectric crystal β-phase structure. The simulation of piezoelectric PVDF fiber was described by using commercial software ANSYS FEA with coupled field analysis to determine its center displacement. Actuation property of the composite fiber was tested under different DC voltage. The relationship of center displacements as a function of voltage among the polarization fiber, non-polarization fiber and model solution based on fixed-fixed single fiber structure was realized. The results show that the PVDF/MWCNT composite fiber has downward center displacement of 23 μm and upward center displacement of 16 μm under a high electric field.

ACKNOWLEDGMENT

The authors would like to thank the National Science Council of Taiwan for its financial support under grant NSC-100-2628-E-110-006-MY3. We also sincerely thank Prof. L.W. Lin, for their generosity in letting us share their resources and experimental equipment for conducting our experiments.

REFERENCES

[1] X. Chen, S.Y. Xu, N. Yao, "1.6 V Nanogenerator for Mechanical Energy Harvesting Using PZT Nanofibers," *Nano Letters*, Vol. 10, No. 6, pp. 2133-2137, 2010.

[2] G. Holmes-Siedle, P. D. Wilson, and A. P. Verrall, "PVDF: An Electronically-active Polymer for Industry," *Materials & Design*, Vol. 4, pp. 910-918, 1984.

[3] T. Sato, H. Ishikawa, and O. Ikeda, "MultilayeredDeformable Mirror using PVDF films," *Applied Optics*,vol. Vol. 21, No. 20, pp. 3664-3668, 1982.

[4] J.M. Ha, H.O. Limb and N.J. Jo, "Actuation Behaviorof CP Actuator Based on Polypyrrole and PVDF," Advanced Materials Research Vol. 29, pp. 363-366, 2007.

[5] C. Chang, Y.K. Fuh, and L.W. Lin, "A Direct-Write Piezoelectric PVDF Nanogenerator," *15th International Conference on Solid-State Sensors*, Actuators and Microsystems, Denver, pp. 1485-1488. 2009.

[6] D. Sun, C. Chang, S. Li, L. Lin, "Near-field electrospinning, " *Nanoletters* 6 839–842,2006.

[7] C. Chang, K. Limkrailassiri, L. Lin, "Continuous near-field electrospinning for large area deposition of orderly nanofiber patterns,"*Appl. Phys. Lett.* 93 123111, 2008.

[8] J. Pu, X.J. Yan, Y.D. Jiang, C. Chang, Liwei Lin1, "Piezoelectric Actuation of a Direct Write Electrospun PVDF Fiber," Micro Electro Mechanical Systems (MEMS), 2010 *IEEE* 23rd International Conference on , 2010.

Local Ablation by Micro-electric Knife

Yoko Yamanishi, Hiroki Kuriki[*], Shinya Sakuma, Kazuhisa Onda, Fumihito Arai
[*]Department of Micro-nano Systems Engineering, Nagoya University, JAPAN
kuriki@biorobotics.mech.nagoya-u.ac.jp

Abstract— **We have firstly succeeded in enucleation of oocyte by using micro-electric knife. The discharged output power and conductive area were controlled to adapt the cellular-order ablation under water and under atmospheric pressure environment. The local area ablation was obtained by the glass shell insulation around the copper wire with silver paste, and which prevent the bubble generation when it was discharged under the water. This fabricated sharp electric knife has a designed small space at the end of the knife between the wire and the glass tip, and which contribute to stabilize the electric discharge. The width of the ablation region was successfully reduced to more than a half compare to the conventional wire electrode, by using the robust glass shell insulation and a single micro-scale bubble injected from inside of the electrode. Moreover, the nucleus dyed by Hoechst was also removed successfully with limited damage region which is the difficult task for manual operation with glass capillary. This low cost micro-electric knife can fabricate any objective material under various environments and it has a possibility to contribute to a new fabrication method in micro-nano bioengineering field.**

Keywords-component; formatting; micro-ablation; electric knife, cell surgery

I. Introduction

The electric knife is one of indispensable surgical devices and it was used in wide surgical operation, however this technique has not been remarkably improved since it has invented several decades ago. For example, thermal damage of the current radiofrequency electric knife, which is one of the most non-intrusive type electric knives, is more than several hundred micrometers. On the other hand, a laser device, which was invented about 70 years ago, has been developed remarkably, and many researches reports the interaction between the cells and laser to fulfill the cell-order ablation [1-2]. Recently, the Palanker et al [2008] reported the micro-order ablation electric knife by a 12 μm metal blade with SiO_2 insulation and a non-intrusive cut was produced in a fresh porcine cornea in vitro [3-6]. For the present study, high-accuracy and non-intrusive micro-electric knife was proposed targeting cell surgery.

Manipulation and processing technology of micro-organisms and cells have been one of the important subjects to be developed with the development of the micro-nano processing technology and gene technology and neurology recently [7-10]. However they require a special environment such as in the medium (water) and under atmospheric pressure and which limited the device to be operated. For example, electron beam machining (EB)[11] or focused ion beam machining (FIB)[12], which are good at nano-scale fabrication,

cannot be used in the atmospheric environment, and water content sample cannot be processed. Moreover such devices tend to be huge size and high in cost and it takes long time to process a wide area.

For the present study, we propose a micro-electric knife as a novel top-down processing technology to fabricate wide range materials under atmospheric pressure and even in water environment. In this paper we confirm the enucleation of oocyte using micro-electric knife which is one of the complicated process in the cloning technique to proof non-intrusive processing.

II. Enucleation of Oocyte

A. Conventional manual enucleation of oocyte

Enucleation of oocytes is an important step in the cloning process. At present, manual manipulators with glass capillaries are used to remove a nucleus as micromanipulations under microscope as shown in Figure 1 [13]. However, the conventional manual manipulation has drawbacks of contamination, a low success rate, and low repeatability; therefore, complicated cell manipulation can be carried out only by skilled technicians. One of the essential drawbacks of the manual operation is the undesired large removed region when a nucleus is removed (Figure 1(4)). This is unavoidable for manual enucleation because a polar body is used as a mark to remove the invisible nucleus which believe to be located near the polar body when the oocyte in the MII stage. Therefore operator has to remove periphery region of the polar body and which reduce the production rate eventually. It is ideal to remove only the nucleus region to obtain high quality and homogeneity oocytes which increase the production rate. Thus non-intrusive ablation method is required to remove the nucleus itself.

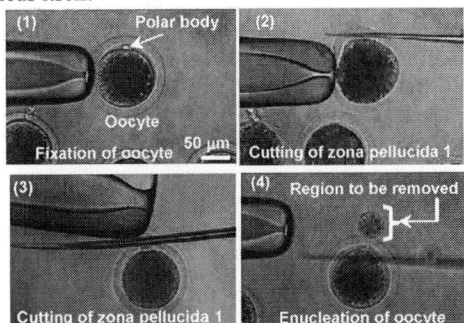

Fig. 1. Manual enucleation using a glass capillary. [13]

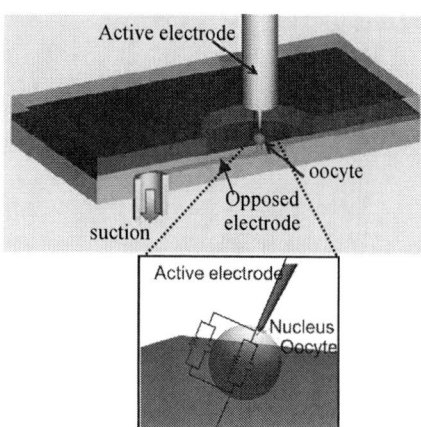

Fig. 2. Concept view of the cell surgery system.

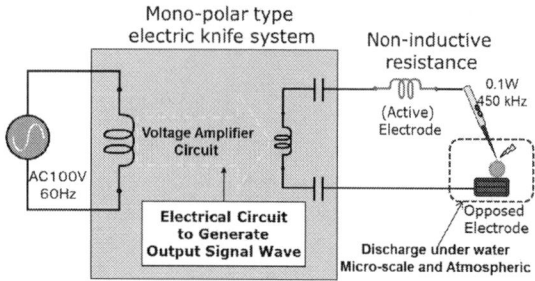

Fig. 3. Overview of electrical circuit of electric knife system.

B. Concept of micro-electric knife system for enucleation

The concept view of the micro-electric knife is shown in Figure 2. Generally micro-electric knife ablate or coagulate the tissue by a thermal input with a short time which is generated by a high frequency current. There are two types of the electrode setting. One of the settings is called the mono-polar type electrode which composed of the active electrode and opposed electrode, and the size of the opposed electrode is relatively larger than the active electrode to collect the discharged current efficiently. The other setting is called bipolar type electrode whose active and opposed electrode is similar small size and entire shape of the electrodes is like a "tweezers". For the current study a mono-polar type electrode was employed for the robust positioning and fixing of cells. A single hole in the opposed electrode in Figure 2 fix the cell position so that the active electrode can easy to access the target region of the cell. It is important to note that the cell which is the current ablation target has a resistance itself and their impedance changes with the progress of cutting of oocyte. Hence, it is important to monitor the impedance of cell during the ablation to provide appropriate current input with high speed feedback for the accurate cutting. Figure 3 shows the overview of the electrical circuit. Non-inductive resistance (10.8 kΩ) was installed to the commercial electric knife power supply (Hyfrecator2000, Kobayashi Medical) to adapt the

output power for the cellular-scale cutting. For the current system, the sampling ratio was set to 450 kHz and 3500 times feedback was carried out to adjust the output power to the impedance of cell which is changing while the fabrication. A foot switch of the electric knife was replaced by the digital input & output board (CSI-360116, Interface Co.) to reduce the width of the input pulse to less than 0.1 msec.

III.   LOCAL ABLATION BY MICROELECTRODE

A.   Ablation test using a knife with imperfect insulation

Initial ablation test of oocyte was carried out in the medium and under atmospheric environment. Figure 4 shows the ablation of the oocyte by a sharpest commercially available coated electric knife (714S tungsten needle, Kobayashi Medical) which was designed to local ablation for the plastic surgery. When the current was applied, it was observed that the generation of the undesired large bubble around the knife blade which prevents the observation for ablation. Also the ablations through the bubble accelerate the ablation to damage the oocyte in the region of more than 100 μm. Besides, it was also observed that the protein residue out of oocyte stuck on the surface of knife, and which prevents longer time usage of knife. It can be observed that the generation of bubble wrapped the protein to accelerate the contamination of the electric knife. Therefore it is required to perfect insulation of the knife to limit the conductive region to obtain non-intrusive ablation.

Power discharge under medium (TCM 199) and under atmospheric pressure

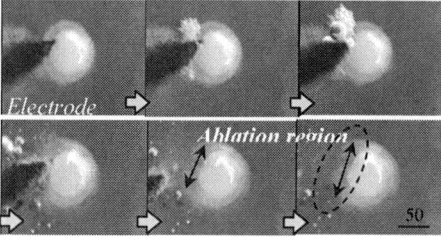

Fig. 4. Ablation of large region with imperfect insulated Electric knife (0.1W).

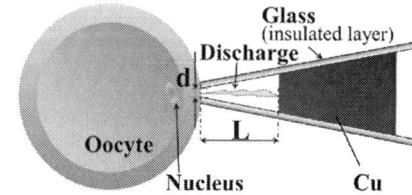

Fig. 5. Concept view of the fabricated electric knife

B.   Fabrication of the electric knife with perfect insulation

After several preliminary experiments, it was confirmed that the coating is not sufficient method to insulate electrode

978-1-4673-1122-9/12 $31.00 © 2012 IEEE

perfectly. It is ideal to have a glass insulated shell protecting the electrode by preventing unnecessary water invasions. The concept view of the ideal electric knife was shown in Figure 5. To produce such an electrode, the insulated layer and a wire were fabricated simultaneously to produce a knife for the perfect insulation. We used the glass capillary tube (50 μL, Drummund Scientific, Co, Ltd.,) and copper wire whose diameter of 30 μm and silver paste. The glass puller (P-1000IVF, Sutter Instrument, Co, Ltd.,) was used to pull the glass capillary. Figure 6 shows the fabrication process of the glass capillary. The copper wire was passed through the glass tube and then it was set to the glass puller. After applying the programmed thermal input, the tube and a wire can be disconnected. After that, the silver paste was stuffed to the end of the tube to connect to the power supply of the electric knife. Figure 6 also shows a photo of the fabricated micro-electric knife.

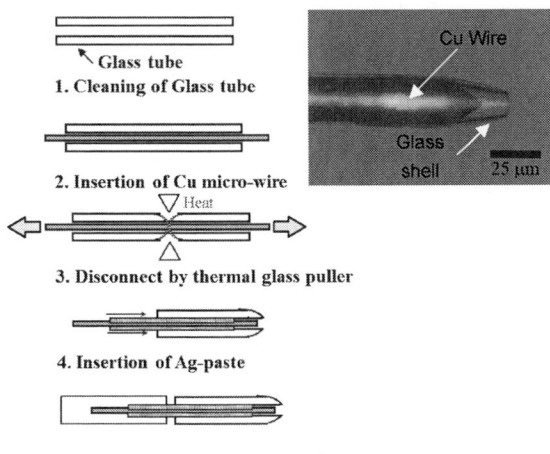

Fig. 6. Concept view of the fabricated electric knife and a photo of fabricated electric knife

C. Enucleation of the oocyte by micro-electric knife

After the evaluation of electrode, the electrode was fabricated with smallest diameter of d and largest distance of L as possible within the current fabrication technique. Inner diameter of d was about 15 μm and the L was about 20 μm. Figure 7 shows the sequence photos of the enucleation of oocyte by using the fabricated micro-electric knife. The oocyte was located in the medium (TCM 199) under atmospheric pressure environment. The fabricated micro-electrode was moved to touch the surface of the oocyte. Then the electric discharge was started with DIO control by triggering the power supply. The nucleus was dyed by the Hoechst and the nucleus was illuminated by the mercury lamp with blue light. The experiment was carried out under the mixture of bright field light and mercury lamp light. The region of the zona pellucida which is the transparent shell of the oocyte tends to be hard physical property than that of the egg cell inside. The

micro-electric knife was successfully make a small hole on the zona pellucida first without damaging the inside of the cell (Figure 7 ②). Then the small discharge by the electric knife makes a small blow around the nucleus and then only a part of the bluish illuminated nucleus start to be moved (Figure 7 ③). And finally the bluish nucleus was successfully blown out of the cell by additional several pulses without any undesired part of the oocyte. There is no comparison between the bluish removed part and that of the conventional method shown in the Figure 1(4). The ablation region was successfully reduced to less than 50 μm as shown in Figure 7④, which is less than at least a half of damage shown in Figure 1. No contamination was observed to the surface of the electric knife even after the longer usage.

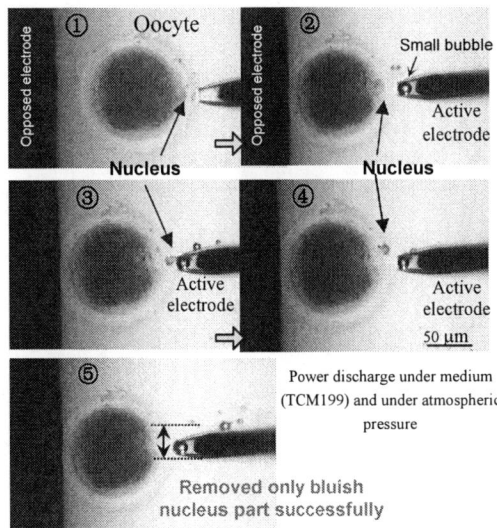

Fig. 7. Ablation of large region with imperfect insulated Electric knife (0.1W)

D. Non-intrusive fabrication

To fabricate smaller size scale object, it is required to obtain less-intrusive fabrication. The length L and d shown in Figure 5, which makes a space between the end of the electrode and the knife have an important role to control the resolution of the fabrication. It is important to evaluate controlled by the pre-programmed thermal input of the glass puller. During the fabrication of the insulated electrode, it was observed that the copper wire was disconnected faster than the glass capillary due to a difference thermal conductivity. It was also observed that the edge of the copper wire is slightly melted by the heat to fit the inner diameter of the glass tube. One of the important parameter to produce L and d is the speed of the puller which is magnetically controlled by the applied current to the electromagnet. Figures 8 and 9 show that the relationship between the length L, d and the applied current to the puller. As you can see in the Figure 8, the diameter d at the end of the knife decreased with increase the applied current (pulling speed) of the glass puller, however the decrease of the diameter has a certain limitation due to the viscosity of the glass. The distance L increase with increase of the current input (pulling seed) and there is a certain limitation due to the viscosity of the glass. Hence it is important to use the minimum value of d and maximum value of L to obtain non-intrusive fabrication.

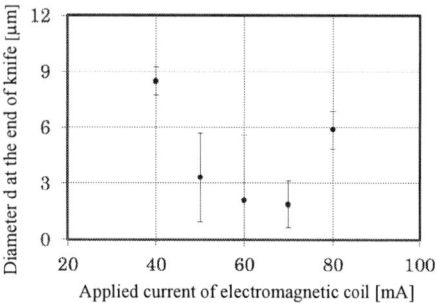

Fig. 8. Diameter d at the end of the knife as a function of the applied current for the electromagnetic coil of the glass puller.

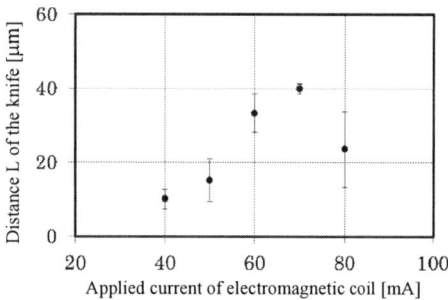

Fig. 9. Distance L at the end of the knife as a function of the applied current for the electromagnetic coil of the glass puller.

Ongoing research tries to find out the physical mechanism to explain less-intrusive fabrication during the ablation. It is likely related to the fluidic phenomenon with high speed. The magnified view in Figure 10 confirmed that a single high speed bubble generated inside the knife provide less-intrusive fabrication effect. The minimum damage of the fabrication in the zona pellucida is observed less than 5 μm.

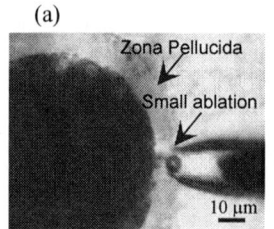 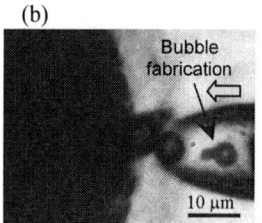

Fig. 10. Magnified view of electric knife and limited damaged region of the zona pellucida.

IV. CONCLUSIONS

We have firstly succeeded in enucleation of oocyte by using micro-electric knife in the medium and under atmospheric environment. It should be noted that a far lower-intrusive ablation has been achieved using the micro-electric knife compare to the conventional enucleation method without any contamination to the knife surface. The nucleus dyed by Hoechst was also removed successfully without undesired region which is the difficult task for manual operation with glass capillary. The width of the ablation region was successfully reduced to up to 5 μm by using a single micro-bubble trapped inside the electric knife. This low cost micro-electric knife has a possibility to extend to fabricate any objective material under various environments and contribute to a new top-down fabrication method in the micro-nano bioengineering field.

ACKNOWLEDGMENT

This work was supported in part by JST PRESTO program.

REFERENCES

[1] Y. Hosokawa, J. Takabayashi, C. Shukunan, Y. Kai and H. Masuhara, "Cell Manipulation by a shock wave induced by femto-second laser", Laser research, Vol.32, p.94-98, (2004).

[2] G. Jones, D. Cram, B. Song, G. Kokkali, "Novel strategy with potential to identify developmentally competent IVF blastocysts", Human Reproduction, pp. 1–12, (2008).

[3] D. Palanker, H. Nomoto, P. Huie, A. Vankov and D. Chang, "Anterior capsulotomy with a pulsed-electron avalanche knife", Journal of cataract refract surgery, Vol.38, p.128-132, (2010).

[4] D. Palanker, A. Vankov and P. Jayaraman, "On mechanisms of interaction in electrosurgery", New journal of physics, Vol.10, p.1-15, (2008).

[5] S. Loh, G. Carlson, E. Chang, E. Huang, D. Palanker and G. Gurtner,"Comparative healing of surgical incisions created by the PEAK Plasmablade, conventional electrosurgery, and a scalpel", Plastic and reconstructive surgery, Vol.124 (6), p.1849-1859, (2009).

[6] D. Palanker, J. Miller, M. Marmor, S. Sanislo, P. Huie and M. Blumenkranz, "Pulsed electron avalanche knife (PEAK) for intraocular surgery", Investigative ophthalmology and visual science, Vol.42 (11), p.2673-2678, (2001).

[7] M. Hagiwara, T. Kawahara, Y. Yamanishi, T. Masuda, L. Feng and F. Arai, Lab on a Chip, 11, (2011), pp. 2049-2054.

[8] N. Inomata, T. Mizunuma, Y. Yamanishi and F. Arai, J. Microelectromech. Syst, 20, (2011), pp. 383-388.

[9] F. Zeng,C. B. Rohde and M. F. Yanik, "Sub-cellular precision on-chip small-animal immobilization, multi-photon imaging and femtosecond-laser manipulation", Lab on a chip, Vol. 8, p.653-656, (2008).

[10] J. Teramoto, Y. Yamanishi, E.S. Magdy, A. Hasegawa, A. Kori, M. Nakajima, F. Arai, T. Fukuda and A. Ishihama, "Single-bacterial cell assay of promoter activity and regulation", Genes to Cells, Vol.15, pp.1111-1122, (2010).

[11] A.A. Tseng, K. Chen, C.D. Chen, and K. J. Ma, "Electron Beam Lithography in Nanoscale Fabrication:Recent Development", IEEE Transaction on electronics packing manufacturing, Vol. 26 (2), p.141-149, (2003).

[12] H. Hosokawa, K. Shimojima, Y. Chino, Y. Yamada, C.E. Wen, M. Mabuchi, "Fabrication of nanoscale Ti honeycombs by focused ion beam", Material and Science and Engineering A344, p.365-367, (2003).

Shinya Watanabe (edit.), "Manual for the enucleation of bovine oocyte and evaluation of quality of embryo", No.9, National Institute of Livestock and grassland science Technical report, ISSN 1347-2712, p.5, (2011).

In-Vitro Transgenic Mice Liver Tissue Culture Via hydrodynamic Flow Perfusion Bioreactor

Chen-Wei Wu[1], Shilpa Sivashankar[1*], Srinivasu Valagerahally Puttaswamy[1], Hui-Ling Lin[2], Kuo-Wei Chang[1], Chau-Ting Yeh[2] and Cheng- Hsien Liu[1]

[1]Department of Power Mechanical Engineering, National Tsing Hua University, Hsinchu, Taiwan, R.O.C
[2]Liver Research Unit, Chang Gung Memorial Hospital, Linkou, Taiwan, R.O.C
*megha_shilpa2001@yahoo.com

Abstract—we report a Poly methyl methacrylate (PMMA) bioreactor in which hepatitis B virus (HBV) infected transgenic mice liver tissue has been cultured for long term with the continuous flow of culture medium, surrounded by mesothelial cells to provide 3-dimensional culturing conditions with adequate nutrient exchange. PDMS membrane was fabricated and a layer of mesothelial cells were cultured on it. Further this membrane was deformed and positioned on the microchannel by suction through the holes at the bottom of the channel. Liver tissue of transgenic mice was introduced on PDMS membrane. The experimental results proved that the proposed bioreactor portrays the in-vivo conditions better, with substantial improvement of liver-specific functions such as sufficient antigen expression, better structural integrity when compared to conventional static culture method. The in-vitro culture period of sliced liver tissue in our bioreactor was extended for 9 days to facilitate drug screening applications.

Keywords- Bioreactor; HydrodynamicFlow; Tissue Culture; Transgenic Mice

I. INTRODUCTION

Long-term in vitro cultures of liver slices are needed to investigate metabolism and accumulation of toxic chemicals in liver tissues. Presently the tissue slices can only be cultured by supplying the nutrients through their surface. Therefore, a continuous process of liver tissue degradation cannot be avoided, as the core of the tissue cannot be fed with the fresh nutrients. In clinics, attempts have been made to culture liver tissues outside human body through conventional static culture method [1]. Some researchers tried to use dynamic methods to culture liver tissues such as, rocker, shaker, roller and stirrer incubation systems [2, 3] but, none of these methods could maintain liver specific functions beyond 24 hours of culture duration [4].

Flow perfusion bioreactor has been developed by Bancroft et al in which the culture medium is forced through the internal porous network of the scaffold [5]. The uni-axial and bi-axial rotational schemes are studied and compared, based on a vessel rotating speed to enhance the function of the bioreactor [6]. Support structures are also fabricated from poly (glycolic acid) [PGA] fibers reinforced with a second polymer to create and maintain a potential space for engineering a large tissue [7]. However, none of these bioreactors is proved efficient in delivering the nutrients to the core of the tissue in practical. In Recent time, novel intra-tissue perfusion system, for culturing thick liver tissue has been used in which culture medium is transported through hollow micro needles to reach core of liver tissues [8]. However, the cell viability was assesed only for three days, which is not suitable for long-term drug testing applications. In addition, mechanical piercing would damage the cells around micro needles.

In the proposed paper, we describe the design and function of our bioreactor suitable for long-term tissue culture consisting of interconnected wells. Gas permeable Polydimethylsiloxane (PDMS) membrane cultured with a layer of mesothelial cells is incorporated to provide favorable environment to the liver tissue in the bioreactor. The primary function of mesothelial cells in vivo is to act as a protective barrier against physical damage and invading organisms and a frictionless interface for the free movement of apposing organs and tissues [9].

Continuous perfusion of culture medium is achieved by using syringe pump with suitable flow rate to minimize the shear stress effect. The holes are drilled at the bottom of the chip to allow O_2 and CO_2 exchange which is vital for cell survival. Via this extracorporeal liver tissue vitalizing system, in-vitro culture period of sliced liver tissues of 1.5mm thickness could be extended up to nine days to ensure long-term culture. Conventional static culture method is used as control group for comparison.

II. EXPERIMENTAL

A. Design and simulation

The shape of the biorector well and the flow channel is designed in such a way that the culture medium is forced to flow through the core the liver tissue to deliver nutrients. In order to achive this the liver tissue is held at the neck of the nozzle shaped flow channel as the size of the tissue is more than the size of the flow channel. Meanwhile the continuous flow of fresh medium helps in removal of toxins. The simulation results reveal that the angle of the nozzle shaped channel has an influence on the flow velocity. The

978-1-4673-1122-9/12 $31.00 © 2012 IEEE

flow velocity is for 84° nozzle angle and is best suited for delivering nutrients to the core of the tissueand is represented in Fig. 1. The arrow heads in the figure represents the direction of medium flowing from inlet to outlet.

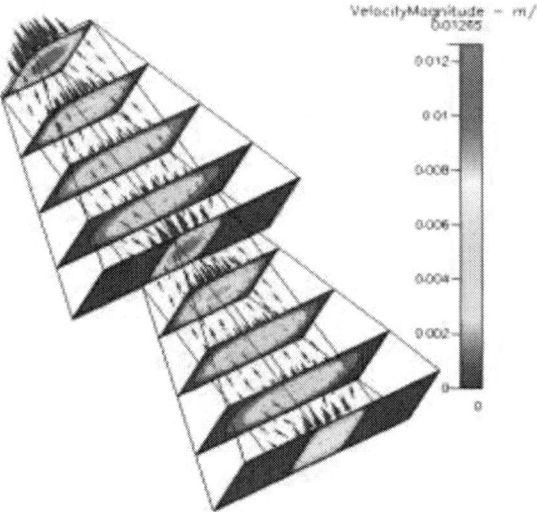

Fig. 1. The simulation result of the flow channel. The flow velocity will be maximum at the neck of the channel where tissue is held.

B. Fabrication of bio-reactorchip

Poly methyl methacrylate (PMMA) engraved with the nozzle-shaped well array by using engraving machine (Roland EGX-400). The single unit of the bioreactor chip comprises of three parts; nozzle shaped well substrate, PDMS membrane and a top cover as illustrated in Fig. 2. PDMS is a polymeric organosilicon material compound, which has been widely used in engineering and biomedical field. To fabricate PDMS membrane, PDMS prepolymers were mixed with curing reagent with a 10:1 mass ratio and degassed in a vacuum chamber to remove the air bubbles. The prepolymers of PDMS were left at the room temperature for two hours before spin coating to increase the viscosity. PDMS was poured carefully on a cleaned glass wafer, followed by spin coating at 2500 rpm for 30 seconds resulting in 500 µm thick PDMS membrane. After degassing in a vacuum chamber again, PDMS was cured by heating-up in an oven. The PDMS layer was peeled off from the glass wafer. Further, a layer of mesothelial cells is cultured on PDMS membrane for four days to let the cells adhere onto the surface of PDMS membrane and obtain optimum cell density.

C. Liver tissue slices preparation

Liver tissues of transgenic mice have been utilized throughout the experiments. The mice were implanted with a hepatitis B virus (HBV) genome organization, which is called as HBsAg gene. Once these mice were confirmed infected, liver tissues of these mice start to secrete HBsAg antigen. Therefore, by detecting the concentration of HBsAg antigen from the infected mice we can assess the viability trend of liver tissue. We take biopsy from the transgenic mice after they are implanted with HBsAg gene. The thickness of liver tissue used throughout the experiments is 1.5 mm. Thickness can be measured directly during slicing by using an automatic gauge provided with the tissue slicer. Tissue chopper machine (McIlwain TC752) is used to get tissue slices of smaller and uniform thickness. So obtained liver tissues slices are placed into bioreactor well and into conventional static culture dish.

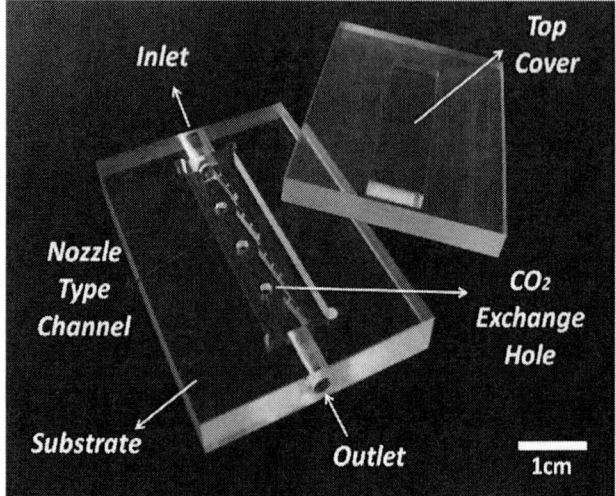

Fig. 2. The hydrodynamic-flow microbioreactor chip with a nozzle-shaped microwell culturing array.

D. Working principle of nozzle-shaped of bio-reactor chip

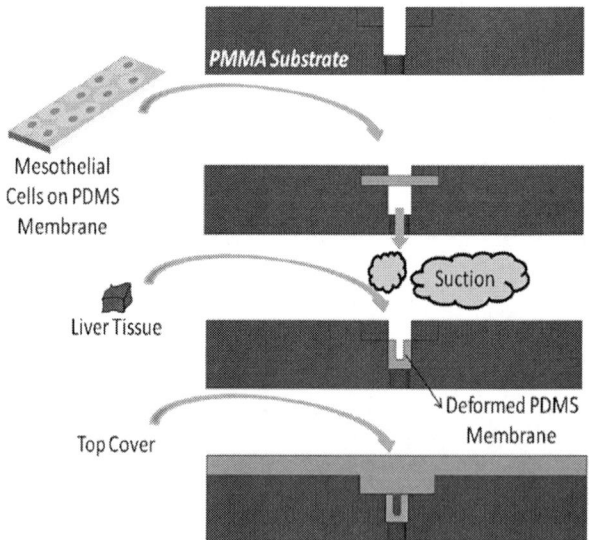

Fig. 3: Side view of our proposed hydrodynamic-flow microbioreactor. (a) PMMA bioreactor with the nozzle-shaped microchannel (b) PDMS membrane preliminarily cultured with mesothelial cells (c) Positioning of PDMS membrane by suction. (d) The liver tissue is inserted and the chip is covered by another PMMA cover.

The side view of nozzle-shaped bioreactor and its working mechanism is illustrated Fig. 3. After, mesothelial cells were cultured on PDMS membrane, it was placed on the chip and then by suction through the holes provided at the bottom of the chip, the PDMS membrane is deformed to fit into the flow channel. The deformation of PDMS membrane increases the contact area between the liver tissue and the mesothelial cells providing 3-D environment favorable for long-term tissue growth. After inserting the liver tissue into the bioreactor well, the top cover is placed to avoid contamination. Further, fresh medium is supplied by using a syringe pump (SP230IW, WPI, Florida, USA) with a flow rate of 80µl/min

III.   RESULTS AND DISCUSSION

A. HBsAg antigen expression

The concentration of antigen during the control group and hydrodynamic culture culture is plotted in Fig. 4. This plot shows decreasing trendline with steeper slope indicating low concentration of HBsAg as the time progress. This is an indication of tissue degradation due to lack of nutrients, inadquate O_2 and CO_2 exchange. The antigen concentration expression of liver tissue when cultured in bioreactor chip is enumerated. From the figure it is evident that the treandline is decreasing but with a gentle slope. This implies that the liver tissue was undergoing apoptosis at a very slow pace emphasizing the suitability of bioreactor chip for long term culture.

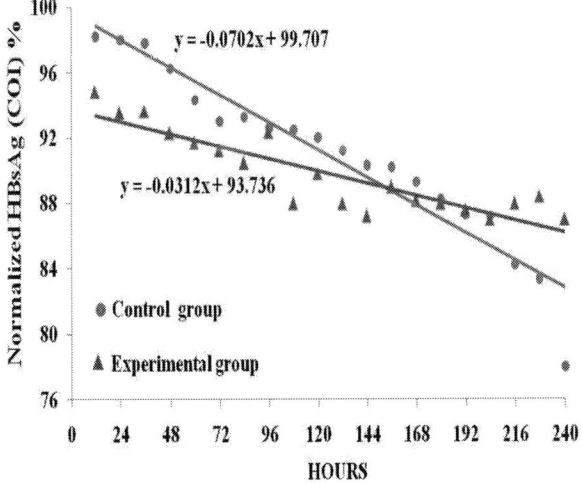

Fig. 4. HBsAg antigen expression. Circular entities' representing Steep declined trendline shows HBsAg antigen concentration in the static culture control. Triangular entities represent gentle declined trend line with a better HBsAg expression for the hydrodynamic-flow bioreactor culture.

B.   Hematoxylin & eosin

Tissue slices were isolated on the first, fifth and ninth day from bioreactor chip and control group and the viability was examined using H&E stain. The architecture of the liver tissue was intact on the first day indicating good morphology of the tissue in both the system [Fig. 5(a, b)]. On fifth day, structural dissociation was initiated in bioreactor group and structural dissociation was observed in control group [Fig. 5(c, d)]. At the end of ninth day structural integrity of liver tissue was maintained with mild interstitial spaces in hydrodynamic flow bioreactor chip where more cells were stained blue. On the other hand, severe structural dissociation was noticed in static

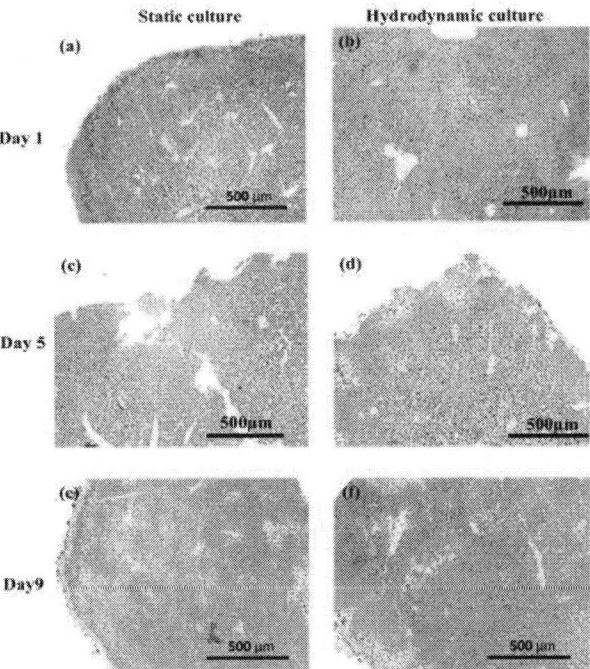

culture method with few cells stained blue [Fig. 5(e, f)].

Fig. 5. H&E stain. (a&b) intact structural integrity(c) Low density stained nucleus (blue spots) (d) slight structural disintegration observed (e) severe structural disassociation seen (f) High density of stained nucleus which implies more number of viable cells at the core with intact architecture of liver tissue.

The images were taken on first, third, fifth, seventh and ninth day. (Images taken on first, fifth and ninth day is shown in Fig. 5(a-f)). The image was divided into three regions such that each region at least inculcates 200cells. The average and the standard deviation were plotted for these three regions and

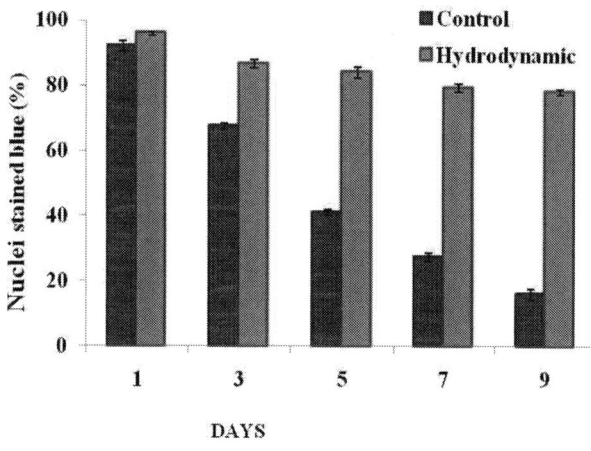

DAYS

the percentages of viable cells were counted and are as plotted in graph represented in Fig. 6.

Fig. 6. Percentages of viable cells when stained with H&E.

Both H&E stain and HBsAg expression have confirmed enhanced viability, better structural integreity and liver specific function upto nine days when compared to control group. The enhanced viability and intact architecture of the tissue is mainly due to the presence of mesothelial cells that acts as extra cellular matrix (ECM) providing required nutrients to the liver tissue. Apart from being a non-adhesive layer for liver tissue, the mesothelial cells will synthesize ECM molecules. It plays an important role in local fibrin deposition and clearance within serosal cavities [9]. Mesothelial cells participate in initiating and resolving serosal inflammation, repair by secreting various pro-, anti- and immune modulatory mediators, and release mediators in response to injury that initiate cell proliferation, migration and ECM synthesis. All the above mentioned facts of mesothelial cells contribute to the long-term survival of liver in our bioreactor chip. Secondly, though the thickness of the liver tissue is more than 1 mm, the nutrients are supplied to core of the tissue to over come the mass transfer problem. This is achieved by holding the tissue at the neck of the nozzle shaped flow channel causing the medium to perfuse into the core of the tissue. The simulation result support this fact with maximum flow velocity of 19,660 μm/s for a nozzle angle of 84˚. The toxins and the waste are removed from the liver tissue by continuous supply of fresh culture medium.

Many studies report that oxygen supply is one of the most important nutrient limiting tissue growth. meanwhile, it is often the limiting nutrient in successful tissue growth in vitro. The reason for this arises from the difficulty of bringing sufficient amounts of oxygen to the surface of the cells mainly because of the poor solubility of oxygen in culture media. The oxygen solubility in a typical culture medium is limited to 0.2mmol O_2/l when atmospheric oxygen is used, twice its solubility in pure water. To

facilitate transfer of oxygen,the holes are drilled at the bottom substrate of the bioreactor chip. The oxygen pass through these holes and diffuse through the PDMS membrane to oxiganate mesothelial cells and liver tissue.

IV. CONCLUSIONS

From the experimental results it was proved that the proposed bioreactor portrays the in-vivo conditions better, which apparently explains this substantial improvement of liver-specific functions such as sufficient antigen expression, the better structural integrity when compared to conventional static culture method. Via this hydrodynamic-flow microbioreactor chip, the in-vitro culture period of sliced liver tissue was extended up to nine days to facilitate drug screening application.

ACKNOWLEDGMENT

This research was financially supported by National Science Council under grant NSC 98-2120-M-007-003. The authors thank the semiconductor research center and National Nano Device Laboratory for the access.

REFERENCES

[1] P. Olinga, K. Groen, I. H. Hof, R. De Kanter, H. J. Koster, W. R. Leeman, A.A.J.J.L. Rutten, K. V. Twillert, and G.M.M. Groothuis, "Comparison of Five Incubation Systems for Rat Liver Slices Using Functional and Viability Parameters", *Journal of Pharmacological and Toxicological Methods*, vol. 38, pp. 59-69, 1997.

[2] R. D. Kanter, H. J. Koster, "Cryopreservation of Rat and Monkey Liver Slices", *Alternatives to laboratory animals Y.*, vol. 23, pp. 653-665, 1995.

[3] P. Olinga , M. T. Merema , D. K. F. Meijer, M. J. H. Slooff , G. M. M. Groothuis , "Human Liver Slices Express The Same Lidocaine Biotransformation Rate as Isolated Human Hepatocytes," *Netherlands Alternatives to Animal Experiments Platform.*, Bilthoven, vol. 21, pp. 466-469, 1993.

[4] W. R. Leeman, I. A. van de Gevel, A. A. J. J. L. Rutten , "Cytotoxicity of Retinoic Acid, Menadione and Aflatoxin B, in Rat Liver Slices Using Netwell Inserts as a New Culture System, " *Toxic. in Vitro*, vol. 9, pp. 91-298, 1995.

[5] G.N. Bancroft, V.I. Sikavitsas, A.G. Mikos, "Design of a Flow Perfusion Bioreactor System for Bone Tissue-engineering Applications," *Tissue Engineering*, vol. 9, 549, 2003.

[6] H. Singh, S.H. Teoh, H.T. Low, and D.W. Hutmacher, , "Flow Modelling within a Scaffold under The Influence of Uniaxial and Bi-axial Bioreactor Rotation," *J. Biotechno*, vol. 119, 181, 2005.

[7] P. Eiselt, B.S. Kim, B. Chacko, B. Isenberg, M.C. Peters, K.G. Greene, W.D. Roland, A.B. Loebsack, K.J.L. Burg, C. Culberson, C.R. Halberstadt, W.D. Holder and D.J. Mooney, "Development of Technologies Aiding Large-tissue Engineering," *Biotechnol. Prog*, vol. 14, 134, 1999.

[8] Y. M. Khong, J. Zhang, S. Zhou, C.Cheung, K. Doberstein, V. Samper, H. Yu, "Novel Intra-Tissue Perfusion System for Culturing Thick Liver Tissue," *Tissue Engineering*, vol. 13, pp. 9, 2007.

[9] S. E. Mutsaers, "Mesothelial Cells: Their Structure, Function And Role In Serosal Repair," *Respirology* 7, pp. 171–191, 2002

High Quality Factor Parylene-Based Intraocular Pressure Sensor

Jeffrey Chun-Hui Lin[1,*], Yu Zhao[1], Po-Jui Chen[2], Yu-Chong Tai[1]

[1]Electrical Engineering, California Institute of Technology, Pasadena, CA, USA
[2]Research and Technology Center, Robert Bosch LLC, Palo Alto, CA, USA
*Corresponding author's email: linch@mems.caltech.edu

Abstract—A new concept of the intraocular pressure (IOP) sensor design and its implantation approach are presented in this paper. A parylene-based sensing part with about 30 μm in thickness was fabricated, and then integrated with an implantation tube attached to sensor's backside pressure access hole. During the implantation, only the implantation tube was implanted into the anterior chamber to fulfill minimally invasive implantation. The IOP sensor membrane is thin and flexible so that it can attach to the cornea. Because the sensing area was exposed outside to the air all the time, the quality factor can be kept at 27-30 to maintain the sensing distance during the whole testing pressure range. The sensitivity is obtained as high as 542 ppm/mmHg while the responsivity is about 205 kHz/mmHg, which is suitable for biomedical applications.

Keywords-Intraocular implant, Intraocular pressure, pressure sensor, Glaucoma, Parylene

I. INTRODUCTION

People have been working on intraocular pressure (IOP) sensors measuring human's eye pressure (or IOP) for years to prevent and treat eye pressure-related diseases. For example, glaucoma is an eye disease caused by elevated eye pressure in patients' eyes. Without proper treatment, the elevated IOP would damage patients' optic nerves in the backside of the eye, and causes the blindness in the end. Statistics show that glaucoma is the second leading cause of blindness in the world according to World Health Organization [1]. Because there could be no symptoms of pain and human's eye tend to compensate a small peripheral vision loss, glaucoma patients in early stage usually don't know they are developing glaucoma and turns out to treat the disease late. It is shown that only half of the glaucoma patients in the U.S are aware of their having glaucoma.Thus an early stage diagnostic becomes important to find and treat glaucoma.

To diagnose the glaucoma, people usually measure a person's IOP to determine if the person is having glaucoma or not. Currently there are several ways to measure IOP. One of the most common ways is puff air onto or directly touch person's eye to calculate the current IOP by measuring the bouncing force of the eye ball. The drawbacks of these approaches are not just only the fact that some of them are contact methods, but also there exists some limitations of it. For example, as it is reported that the IOP can actually fluctuate depending on patients' daily lives [2], it is necessary to monitor patients' IOP 24 hours to understand eye pressure progress.

In order to fulfill the goal of monitoring IOP automatically, continuously and wirelessly, in the past we have presented a flexible parylene-based IOP sensor[3]. The sensor has one inductor and one capacitor combined in series to fulfill a passive LC-tank resonance circuit. The IOP sensor was implanted into the anterior chamber and anchored on the iris. The resonant frequency shift was registered by an external reader coil through a wireless inductive coupling link, as shown in Fig.3. In Fig.4-(a), when the sensor's surrounding pressure increases, the capacitance increases due to the sensing plate's deforming concavely, causing the resonant frequency shifts to the lower range. However, with the high loss tangent of the eye fluid in the anterior chamber and the blockage of the cornea, the quality factor measured was seriously degraded and the sensing distance reduced [3-5]. To overcome this problem, the new IOP sensor structure presented here has a very similar sensing part design, but the new implantation approach leaves the sensing part exposed to the air keeping its high quality factor, as shown in Fig.2. The sensing part is still composed of one sensing inductor and one sensing capacitor, as shown in Fig.1-(a). A pressure access hole connecting the sensing capacitor was created during the fabrication. An implantation tube, which inner diameter is larger than the pressure access hole was mounted onto the back of the device to cover the pressure access hole (Fig.1-(b)). The biocompatible epoxy was applied to seal the gap between the implantation tube and the sensing part to make it airtight and ensure the biocompatibility.

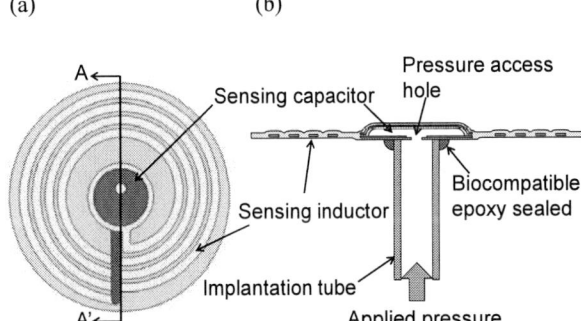

Fig.1. Thenew IOP sensor design: (a) Top view of the sensing part; (b) AA' cross-section view of the IOP sensor.

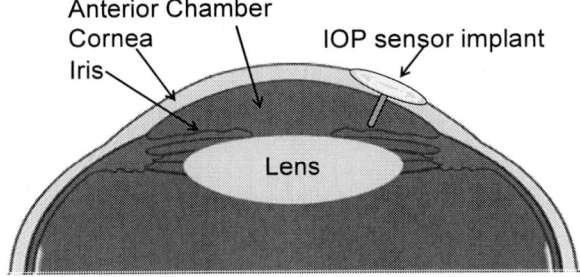

978-1-4673-1122-9/12 $31.00 © 2012 IEEE

Fig.2: The concept of minimally invasive implantation. The IOP sensor was implanted through the cornea and the flexible parylene-based sensing part is lying on the cornea.

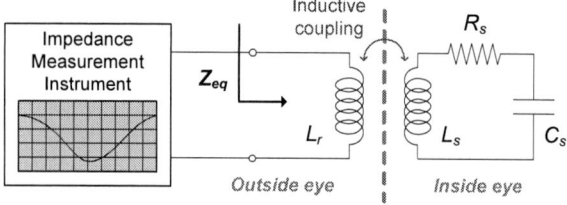

Fig.3. The concept of wireless inductive coupling link: The frequency shift is registered through an external oil reader.

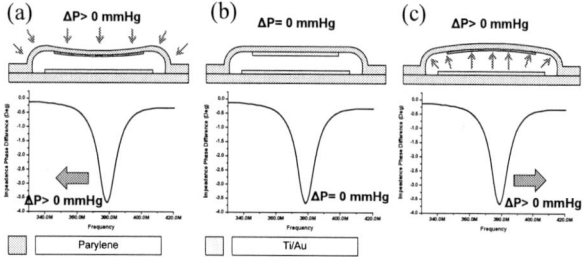

Fig.4: Resonant frequency shift corresponds to the applied pressure. (a) Frequency decreases as the capacitance increases. (b) Nofrequency shift when no pressure difference exists. (c) Frequency increases as the capacitance decreases.

II. SENSING SCHEME AND THE DEVICE DESIGN

A. Sensing scheme

The concept of wireless sensing scheme is shown in Fig.3. The right RLC circuit represents the implanted IOP sensor and the resonant frequency can be expressed as [4]

$$f_s = \frac{1}{2\pi}\sqrt{\frac{1}{L_s C_s} - \frac{R_s^2}{L_s^2}} \cong \frac{1}{2\pi\sqrt{L_s C_s}} \; if \; R_s^2 \ll \frac{L_s}{C_s}, \qquad (1)$$

where R_s, L_s, and C_s represent the sensor's resistance, inductance and capacitance, respectively. The equivalent impedance viewed from the external coil reader and apparatus is derived as[5-7]

$$z_{eq} = j2\pi f L_r \left[1 + k^2 \frac{\left(\frac{f}{f_s}\right)^2}{1 - \left(\frac{f}{f_s}\right)^2 + \frac{1}{Q_s}j\frac{f}{f_s}}\right], \qquad (2)$$

where f is the excitation frequency, k is the coupling coefficient of the inductive link depending on the physical geometries of the sensor [5-9]. Q_s is the quality factor of the sensor at the resonance and can be represented as

$$Q_s = \frac{1}{R_s\sqrt{L_s C_s}}. \qquad (3)$$

When the sensor is excited at the resonant frequency, Z_{eq}, (2), becomes

$$Z_{eq} = j2\pi f_s L_r (1 + jk^2 Q_s), \qquad (4)$$

and its phase dip magnitude can be approximated as

$$\Delta\phi \cong \tan^{-1}(k^2 Q_s). \qquad (5)$$

When the capacitance of the IOP sensor changes, it can be shown from (3)-(5) that the impedance phase dip shifts to either lower or higherfrequency and can be detected by a network analyzer.

B. Electrical and mechanical design of the device

The electrical design of the IOP sensor can be designed by the well-developed equations as follows[3, 10, 11]. The inductance of the spinal coil can be represented as:

$$L_s \cong \frac{\mu_0 n^2 d_{avg} c_1}{2}\left[ln\left(\frac{c_2}{F}\right) + c_3 F + c_4 F^2\right], \qquad (6)$$

where n is the number of turns of the inductor, d_{avg} is the averaged diameter of the coil windings, $F=(d_{out}-d_{in})/(d_{out}+d_{in})$ is the fill factor of the coil windings, and c_1-c_4 are constant coefficients determined by the winding geometry. The coil inductor inherently comes with a resistance which can be calculated as:

$$R_s = \frac{\rho l}{w\delta\left(1 - e^{-h/\delta}\right)}, \qquad (7)$$

where ρ is the electrical resistivity of the metal, w and h are the metal line width and height, respectively. δ is the frequency-dependent metal skin depth which can be written as:

$$\delta = \sqrt{\frac{\rho}{\pi f \mu}}, \qquad (8)$$

where μ is the magnetic permeability of the metal. The capacitance of the IOP sensor is given by

$$C_s = C_{s,g} + C_{s,p}, \qquad (9)$$

where $C_{s,g}$ is the capacitance of the parallel metal plates capacitor at the center of the IOP sensor and $C_{s,p}$ is the parasitic/stray capacitance introduced by other components in the entire device.

To have a detectable impedance phase dip shifts, a deformable circular metal plate was designed at the center of the IOP sensor. Once the plate deforms either downward or upward, the capacitance of the parallel metal-plates diaphragm capacitor changes and the impedance phase dip shifts can be registered. The deformation of the metal plate corresponding to the pressure difference can be predicted as[12]:

$$w(r) = \frac{\Delta P a^4}{64D}\left[1 - \left(\frac{r}{a}\right)^2\right]^2, \qquad (10)$$

where Δp is the pressure difference, r is the radius calculated from the center of the plate, a is the diaphragm radius and D is the flexural rigidity of the diaphragm.

In our new sensor implantation approach, the sensing metal plate deforms convexly with higher surrounding pressure transmitted to the metal diaphragm capacitor chamber through the implantation tube. It is shown that according to (1), this higher eye pressure causes the capacitance to reduce and thus the resonant frequency shift to the higher range, as described in Fig.4.-(c).

III. DEVICE FABRICATION AND CHARACTERIZATION

A. Device fabrication

The fabrication procedure is shown in Fig.5. The sensing part was made of parylene-3 μm gold-parylene sandwich structure. The first layer of parylene was first deposited and the pressure access hole with 180μm in diameter was opened by oxygen plasma. A 3 μm Ti/Au was deposited on top of the first layer parylene and patterned. The distance between 2 capacitor metal plates was designed as 10 μm which was fulfilled by spin coating a 10 μm photoresist and patterned. The second layer of parylene was deposited to cover and protect the 3 μm Ti/Au, followed by a 0.5 μm Ti/Au and then the third parylene layer with 7 μm in thickness. The final sensing part was released from the substrate by soaking in the acetone. The final completed sensing part is shown in Fig.6-(a).

After the sensing part fabrication was done, an implantation tube was attached onto the backside of the sensing part, as shown in Fig.6-(b). The inner diameter of the implantation tube was chosen as 320 μm to fully cover the pressure access hole. The outer diameter of the implantation tube is 450 μm. The implantation tube was manually mounted onto the sensing part. A precision XYZ stage was used to control the position of implantation tube, maneuvered to be concentric with the pressure access hole, as shown in Fig.1-(b). The implantation tube and sensing part were glued together by putting few drops of biocompatible epoxy.

B. Device Characterization

The completed IOP sensor was then integrated to a bigger testing capillary tube and sealed by photoresist, as shown in Fig.6-(c). The inner diameter of the testing tube was chosen as 500 μm to accommodate the implantation tube. The complete sensor with testing capillary tube assembly was stayed overnight to dry the photoresist.

The device characterization setup is shown in Fig.7. The whole assembly was mounted onto a pressure characterization setup. During the characterization, a HP 4195A Network/Spectrum analyzer was hooked up with a 1.5-mm-diameter hand-wound coil serving as the reader coil. The characterization signal was accessed via a data acquisition system and then analyzed in personal computer. The qualified IOP sensor was released by soaking the whole assembly in the acetone to remove the photoresist. The final complete IOP sensor is shown in Fig.6-(d) and is ready for the next *in vivo/ ex vivo* test.

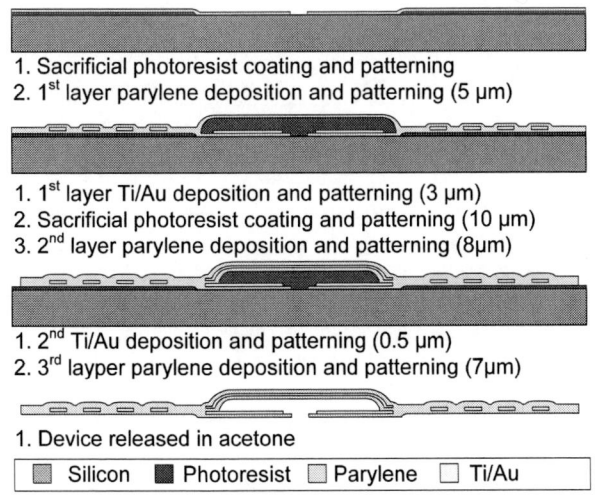

Fig.5. Fabrication procedures of the IOP sensor's sensing part.

1. Sacrificial photoresist coating and patterning
2. 1st layer parylene deposition and patterning (5 μm)

1. 1st layer Ti/Au deposition and patterning (3 μm)
2. Sacrificial photoresist coating and patterning (10 μm)
3. 2nd layer parylene deposition and patterning (8μm)

1. 2nd Ti/Au deposition and patterning (0.5 μm)
2. 3rd layper parylene deposition and patterning (7μm)

1. Device released in acetone

▨ Silicon   ■ Photoresist   ▢ Parylene   ▢ Ti/Au

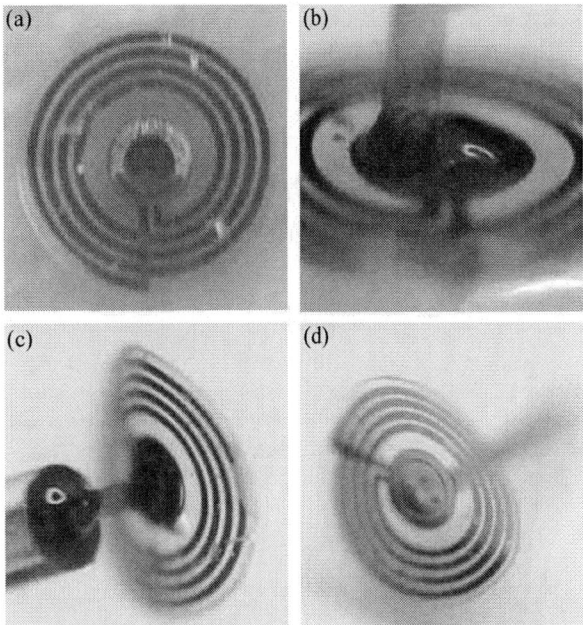

Fig.6. Fabrication and assembling results: (a) Complete sensing part; (b) Implantation tube attached onto the backside of the sensing part concentric with the pressure access hole; (c) IOP sensor mounted to the testing tube by photoresist. (d) Final complete IOP sensor.

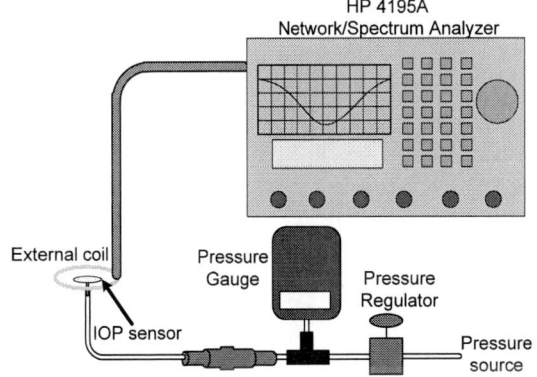

Fig.7. IOP sensor characterization setup: a 1.5-mm-diameter hand-wound coil served as the reader coil and a HP4195A network/spectrum analyzer was used to register the frequency shift of the phase dip.

IV. CHARACTERIZATION RESULTS AND DISCUSSION

The bench top characterization results are shown in Fig.8. The results showed that the resonant frequency is 379 MHz when the applied pressure difference is 0 mmHg. When the applied pressure difference increases, the resonant frequency shifts to the right because the metal plate deforms convexly as expected. The IOP sensor's electrical parameters are obtained and shown in Table 1. As the sensing part can always be maintained exposed to the air, the problem of quality factor drop caused by the lossy medium is solved in our new IOP sensor implantation approach. Therefore, the sensing distance can always be maintained as far as 2.5 cm, which was originally designed for the raw sensing part. The 2.5 cm

978-1-4673-1122-9/12 $31.00 © 2012 IEEE   139

sensing distance can fulfill the concept of glass reader paradigm to accomplish the autonomous, continuous, and wireless IOP monitoring.

The result of sensitivity analysis is shown Fig.9. The sensitivity of the IOP sensor is defined as: [4]

$$\text{IOP sensor sensitivity} = \left| \frac{\partial R}{\partial(\Delta P)} \right|_{\Delta P = 0}, \quad (11)$$

where R is the frequency ratio defined as:

$$R = \frac{f_{min}}{f_{min(\Delta P = 0)}} \quad (12)$$

The sensitivity of the IOP sensor is obtained as 542 ppm/mmHg, corresponding to the responsivity as 205 kHz/mmHg. With a proper designed high resolution external coil reader, the IOP sensor can resolve the pressure difference < 1 mmHg, which is suitable for glaucoma diagnostics.

V. CONCLUSION

We have successfully demonstrated the feasibility of the new concept and design of IOP sensor implant and its implantation approach. In this new implant, the IOP sensor can be implanted with sensing part exposed to the air, which maintains the high quality factor. This benefit makes the glass reader paradigm in reality, achieving autonomous, continuous and wireless IOP monitoring. The characterized IOP sensor is ready for use, and the *ex vivo* test is scheduled in the near future to verify the biological feasibility.

VI. ACKNOWLEDGMENT

This work is supported by the Biomimetic MicroElectronic Systems (BMES) Center under the grant number H31068. The authors would like to thank Trevor Roper's help in terms of sample preparation, machines' maintenance, and instrument's installation.

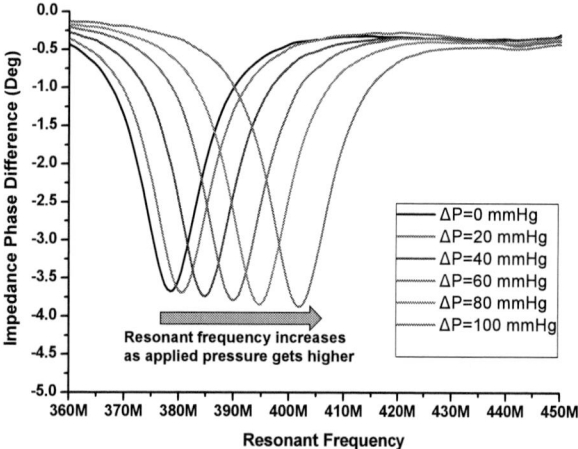

Fig.8. Characterization results.

Table 1. Measured IOP sensor's electrical parameters.

Pressure (mmHg)	0	20	40	60	80	100
Frequency (MHz)	379	381	385	390	395	402
Q Factor	27	27	28	30	28	29
Sensitivity	542 ppm/mmHg					
Responsivity	205 kHz/mmHg					

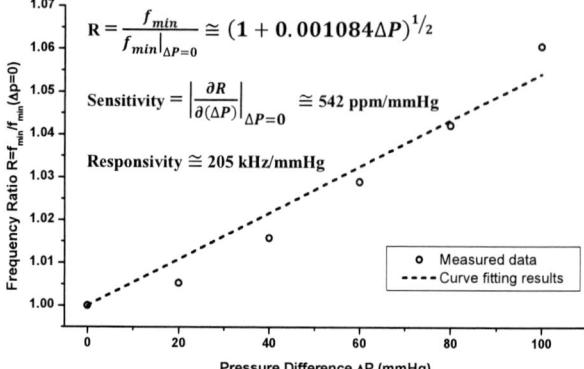

Fig.9. Sensitivity analysis of the IOP sensor.

VII. REFERENCES

[1] Glaucoma Research Foundation. (Nov. 30). [Online]. Available: http://www.glaucoma.org/

[2] E. Hughes, P. Spry, and J. Diamond, "24-hour monitoring of intraocular pressure in glaucoma management: A retrospective review," *Journal of Glaucoma*, vol. 12, pp. 232-236, Jun 2003.

[3] P. J. Chen, S. Saati, R. Varma, M. S. Humayun, and Y. C. Tai, "Wireless Intraocular Pressure Sensing Using Microfabricated Minimally Invasive Flexible-Coiled LC Sensor Implant," *Journal of Microelectromechanical Systems*, vol. 19, pp. 721-734, Aug 2010.

[4] P.-J. Chen, "Implantable Wireless Intraocular Pressure Sensors," Ph.D. dissertation, Electrical Engineering, California Institute of Technology, Pasadena, CA, 2008.

[5] M. A. Fonseca, M. G. Allen, J. Kroh, and J. White, "Flexible wireless passive pressure sensors for biomedical applications," in *12th Solid-State Sensors, Actuators, and Microsystems Workshop*, Hilton Head Island, SC, 2006, pp. 37-42.

[6] M. A. Fonseca, J. M. English, M. von Arx, and M. G. Allen, "Wireless micromachined ceramic pressure sensor for high-temperature applications," *Journal of Microelectromechanical Systems*, vol. 11, pp. 337-343, Aug 2002.

[7] J. Shih, J. Xie, and Y. C. Tai, "Surface micromachined and integrated capacitive sensors for microfluidic applications," *Boston Transducers'03: Digest of Technical Papers, Vols 1 and 2*, pp. 388-391, 2003.

[8] A. Baldi, W. Choi, and B. Ziaie, "A self-resonant frequency-modulated micromachined passive pressure transensor," *Ieee Sensors Journal*, vol. 3, pp. 728-733, Dec 2003.

[9] A. DeHennis and K. D. Wise, "A double-sided single-chip wireless pressure sensor," *Fifteenth Ieee International Conference on Micro Electro Mechanical Systems, Technical Digest*, pp. 252-255, 2002.

[10] O. Akar, T. Akin, and K. Najafi, "A wireless batch sealed absolute capacitive pressure sensor," *Sensors and Actuators a-Physical*, vol. 95, pp. 29-38, Dec 15 2001.

[11] T. H. Lee, *The design of CMOS radio-frequency integrated circuits*, 2nd ed. Cambridge, UK ; New York: Cambridge University Press, 2004.

[12] S. Timoshenko and S. Woinowsky-Krieger, *Theory of plates and shells*, 2d ed. New York,: McGraw-Hill, 1959.

Light-addressable Electrochemical Micropatterning of Cell-encapsulated Alginate Hydrogels for Cell-based Microarray

Shih-Hao Huang[*], Hsiao-Tzu Chu, Chih-Wei Wu, Yun-Yu Chuang

[*] Department of Mechanical and Mechatronic Engineering, National Taiwan Ocean Universit, TAIWAN

shihhao@mail.ntou.edu.tw

Abstract—We proposed a light-addressable electrolytic system to perform a micropatterning of calcium alginate hydrogels using a digital micromirror device (DMD). In this system, a patterned light illumination projected to a photoconductive substrate served as a photo-anode to electrolytically produce protons ions, which can lead to a decreased pH gradient. The low pH generated at the anode can locally release calcium ions from insoluble calcium carbonate $(CaCO_3)$ to produce gelation of the calcium alginate. By controlling illumination patterns on the DMD, the micropatterning of calcium alginate hydrogels with different shapes and sizes as well as multiplexed micropatterning were performed. The concentration effects of the alginate and $CaCO_3$ solution on the dimensional resolution of alginate hydrogel formation were experimentally examined. An array of 3×3 cell-encapsulated alginate hydrogels was successfully demonstrated through light-addressable electrochemical micropatterning. Our proposed method provides a programmable method for the spatiotemporally controllable assembly of cell populations for cell-based microarray.

Keywords-electrodeposition; digital micromirror device; calcium alginate; cellular microarray

I. INTRODUCTION

Recently, cellular microarrays have been proven to be a powerful experimental tool for biological studies and drug discovery [1]. Thus, a variety of novel methods have been developed to address and cultivate cells in array and microfluidic formats [2]. To assemble cells at specific addresses with a controllable pattern, Payne et al. [3] reported the electroaddressing of calcium alginate hydrogels with the ability to entrap viable cell populations within the electrodeposited films. However, achievable gel patterns such as locations, shapes and dimensions are completely subject to pre-defined configurations of microelectrodes. Changing the pattern requires the re-design and re-fabrication of the photo-masks, microelectrodes and sometimes the chip structures themselves. This inflexibility of the microelectrodes has hampered the feasibility in biological applications of achieving dynamical and multiplexed micropatterning of cell-encapsulated alginate hydrogels with different cell types on the same device. Moreover, electrodeposition was typically performed by immersing glass slides with pre-defined electrodes into a centimeter-scale electrolytic container, which

resulted in a low current density (3 A m^{-2}) during electrolysis. The low current density leads to a longer time (5 min) required for the formation of the calcium alginate hydrogels. Here, we propose an alternative approach to produce cell-encapsulated alginate hydrogels by utilizing a light-addressable electrolytic system to perform a micropatterning of calcium alginate hydrogels [4].

II. MATERIALS AND METHODS

A. Design Concepts

Figure 1 shows a light-addressable electrolytic system used to perform an electrodeposition of calcium alginate hydrogels using a DMD. A photoconductive substrate, which consists of 0.2 μm of heavily doped hydrogenated amorphous silicon (n^{+} a-Si:H) and 1 μm of undoped a-Si:H on a 700-μm indium tin oxide (ITO) glass substrate, serves as a light-addressable electrode. A microchamber either 0.5 mm or 1 mm in height was fabricated of polydimethylsiloxane (PDMS) and then bonded to the photoconductive substrate. The deposition solution, which contains soluble sodium alginate (80-120 mPa · S, Sigma-Aldrich) plus insoluble calcium carbonate $(CaCO_3)$ nanoparticles (70 nm, Specialty Minerals, UK), was introduced into the microchamber. We used $CaCO_3$ nanoparticles (70 nm in diameter) as the calcium complex to drastically increase the sedimentation time to over 10 minutes and to ensure a homogeneous dispersion within the microchamber, which is important for the production of calcium alginate hydrogels with high shape fidelity and small sizes of less than 100 μm. An anodic voltage was applied to the photoconductive substrate via the ITO layer, and metallic or platinum wire was inserted into the microchamber to serve as the cathode. When a DC voltage is applied concurrently, light illumination on a photoconductive substrate generates a conducting point that acts as a photo-anode to electrolytically produce protons, which can lead to a decreased pH gradient. The low pH generated at the anode can locally release calcium ions (Ca^{2+}) from insoluble $CaCO_3$ nanoparticles (Eq. 1) and cause gelation of the calcium alginate through sol-gel transition (Eq. 2).

$$2H^+ + CaCO_3 \rightarrow Ca^{2+} + H_2O + CO_2 \qquad (1)$$

$$Ca^{2+} + 2Na^+ Alg^- \rightarrow Ca^{2+}(Alg^-)_2 + 2Na^+ \qquad (2)$$

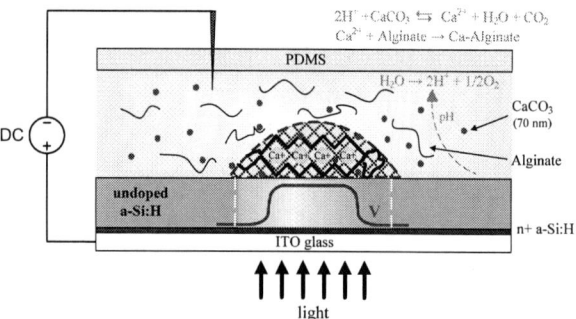

Fig. 1. Schematic of a light-addressable electrolytic system to perform a micropatterning of calcium alginate hydrogels

B. Experimental setup

Figure 2 shows an experimental setup for a light-addressable electrolytic system for performing electrodeposition of calcium alginate hydrogels. We modified the commercial DMD projector by simply removing the original lamp and the color filter wheel. The original projection lens was replaced with a commercial projection lens with a suitable focal length and adjustable apertures. The structured light patterns of the DMD were controlled by a computer. The continuous light illuminated the DMD uniformly through the built-in condensing lens (L1) within the DMD projector and then spatially projected it onto the photoconductive substrate through a projection lens (L2), a focus lens (L3), a 50/50 dichroic mirror (DM), and an objective lens with 10X magnification. An anodic voltage was applied to the photoconductive substrate via a DC power source, and metallic wire served as the cathode. We can control the illumination pattern on the DMD, which enables the performance of an electrodeposition of calcium alginate hydrogels with different shapes and sizes and multiplexed micropatterning.

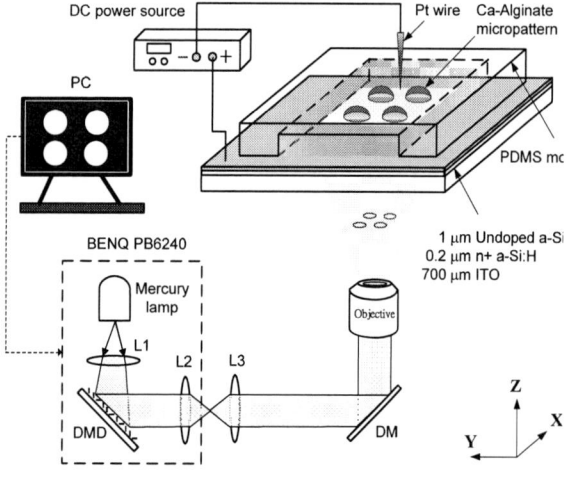

Fig. 2. Schematic of experimental setup for a light-addressable electrolytic system using DMD

III. EXPERIMENTAL RESULTS AND DISCUSSION

To demonstrate the feasibility of the light-addressable electrolytic system for performing the electrodeposition of calcium alginate hydrogels, calcium alginate hydrogels of different shapes and sizes and with multiplexed micropatterning were micropatterned, as shown in Fig. 3. In these experiments, the insoluble $CaCO_3$ powder (70-nm particles; 0.5 wt%) was blended into a sodium alginate (1.0 wt%) solution and sonicated for 10 min. FITC fluorescent dye (Sigma-Aldrich) was also blended into the deposition solution so that it was easy to observe. Then, the deposition solution was introduced into the microchamber. A DC voltage was applied between the photoconductive substrate and metallic wire for 15 s to achieve a current density of 180 A m^{-2} (typical voltage was about 10 V). A high current density with a short reaction time was used to prevent destruction of the photoconductive layer. The illumination patterns on the DMD can be either triangular or square shapes with a characteristic length of D_0=1.5 mm or circular shapes with different diameters ranging from 100 μm to 1000 μm, as shown in Fig. 3 (a). The shaped light pattern was then projected onto the photoconductive substrate to electrolytically produce protons for the electrodeposition of calcium alginate hydrogels. After 15 s, we turned off the light pattern and DC power supply and immediately introduced deionized (DI) water into the microchamber to flush away the deposition solution. The calcium alginate hydrogels deposited on the photoconductive layer were imaged using a fluorescence microscope (Olympus IX-71) connected to a digital camera (WAT-221s, Watec) to examine the shape fidelity between the alginate hydrogels produced by the process and the illuminated DMD. As shown in Fig. 3 (a), the produced calcium alginate hydrogels showed a high shape fidelity to the illumination patterns. These calcium alginate hydrogels contained several trapped bubbles due to the generation of CO_2 during electrolysis, as indicated in Eq. 1. To demonstrate the dimension limitation of the produced alginate hydrogels for our proposed method, we projected an illumination pattern of circular shapes with different diameters ranging from 100 μm to 1000 μm that was passed through an objective lens with 10X magnification. As shown in Fig. 3 (a), we can successfully produce 100-μm-sized alginate hydrogels on the photoconductive substrate after flushing with DI water. However, the alginate hydrogels smaller than 100 μm were flushed away from the photoconductive substrate due to poor adhesion between the alginate hydrogels and the photoconductive substrate. This problem might be overcome by treating the surface of the photoconductive substrate with poly-L-lysine (PLL). The positively charged PLL can promote the attachment of the negatively charged polysaccharide constituent of alginate upon gelation. We also demonstrated the capability for multiplexed micropatterning of calcium alginate hydrogels by performing light-addressable electrodepositions sequentially, as shown in Fig. 3(b). Two different deposition solution colors were electrodeposited using two different illumination light patterns. We first performed an electrodeposition of alginate hydrogels with a green color using an illumination pattern with four circles, and then we repeated the electrodeposition to form alginate hydrogels with a blue color adjacent to the former using an illumination pattern with five circles.

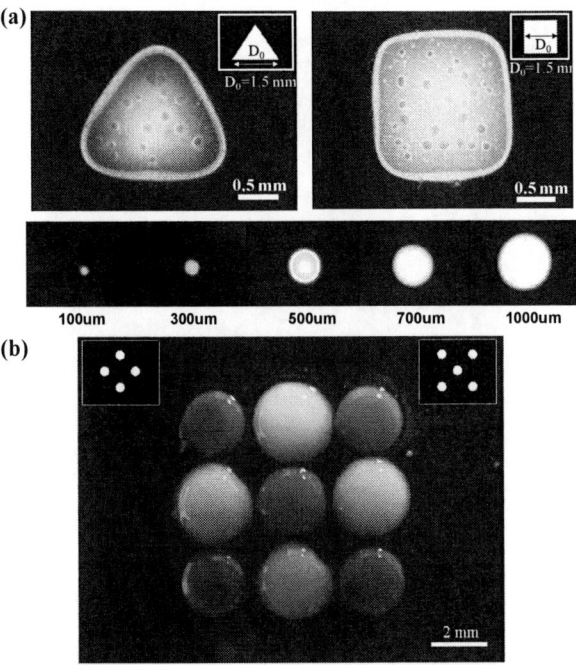

Fig. 3. Images of calcium alginate hydrogels with (a) different shapes and sizes as well as (b) multiplexed micropatterning

To characterize the light-addressable electrodeposition of alginate hydrogels, we examined the effect of the concentration of the alginate and $CaCO_3$ solutions on the dimensional resolution of calcium alginate formation, as shown in Fig. 4. The ratio of D/D_0 is defined as the dimensional resolution, where Do and D denote an illumination pattern of circular shapes with $D_0 = 600$ μm in diameter and the corresponding diameter (D) of the calcium alginate hydrogel produced by the illumination pattern, respectively. Figure 4 (a) shows the effect of the illumination time of the light pattern on the dimensional resolution (D/D_0) for the deposition solution (1% alginate; 0.5% $CaCO_3$) with applied voltages of 10 V and 20 V. The data indicate the average value of the five experiments, and the error bar shows the standard deviation at the same operating conditions. The ratio of D/Do increases with increasing illumination time for applied voltages of both 10 V and 20 V, whereas for a fixed illumination time the D/Do for an applied voltage of 20 V is larger than that for an applied voltage of 10 V. These results indicate that an increase in the illumination time or in the applied voltage for electrolysis results in an increase in the calcium ions released from the insoluble $CaCO_3$ nanoparticles. The calcium ions diffuse and then react with the alginate

solution to produce alginate hydrogels of larger sizes (D). These results demonstrate that alginate hydrogels of varying sizes can be controllably electrodeposited by changing either the illumination time or the applied voltages.

Figure 4 (b) shows the effect of the concentration of an alginate solution with 0.5% $CaCO_3$ on the dimensional resolution (D/D_0) for a deposition solution with an applied voltage of 10 V or 20 V and an illumination time of 15 s. The ratio of D/Do decreases as the concentration of the alginate solution increases for applied voltages of both 10 V and 20 V. We attribute this trend to the viscosity effect; the increase in the alginate concentration leads to an increase in the viscosity of the deposition solution, which hinders the diffusion of the produced Ca^{2+} ions within the illumination area during formation of alginate hydrogels. As the alginate concentration increases, the value of D/Do approaches 1.0, indicating high shape fidelity between the illumination pattern and the produced alginate hydrogels. However, for a concentration of the alginate solution higher than 4 wt%, the deposition solution is too viscous for the $CaCO_3$ nanoparticles to be homogeneously dispersed, which can significantly influence the formation of alginate hydrogels. Figure 4 (c) shows the effect of the concentration of $CaCO_3$ nanoparticles dispersed within a 1 wt% alginate solution on the dimensional resolution (D/D0) for a deposition solution with applied voltages of 10 V and 20 V and an illumination time of 15 s. The results show that the D/Do increases as the concentration of $CaCO_3$ nanoparticles increases for applied voltages of both 10 and 20 V. Increasing the concentration of $CaCO_3$ nanoparticles can result in more Ca^{2+} ions being released from the insoluble $CaCO_3$ nanoparticles during electrolysis. The calcium ions diffuse and then react with the alginate solution to produce larger-sized alginate hydrogels. However, as the concentration of $CaCO_3$ nanoparticles increases past 1 wt%, the D/D_0 gradually increases to a constant value, such as $D/D_0 = 1.6$ and 2.0 for applied voltages of 10 V and 20 V, respectively. This might be due to the fact that the low pH generated at the photo-anode, which was operated at a fixed applied voltage and illumination time, can release only a certain amount of Ca^{2+} ions from the insoluble calcium carbonate ($CaCO_3$). Increasing the concentration of $CaCO_3$ nanoparticles over a critical value cannot further increase the number of Ca^{2+} ions released from the insoluble $CaCO_3$ nanoparticles to cause gelation of the alginate hydrogel.

Finally, we demonstrated the potential for light-addressable electrodeposition of cell-encapsulated alginate hydrogels for 3D cellular microarrays as shown in Fig. 5. Baby hamster kidney-21 fibroblast cells (BHK-21) were homogeneously suspended in deposition solutions of either 1 wt% or 2 wt% sodium alginate solution with $CaCO_3$ concentrations of 0.5, 1 and 2 wt%. The concentrations of 1% and 2% for the alginate solution were chosen for this experiment because they are optimal concentrations with suitably low viscosities for facilitating the homogeneous dispersion of BHK-21 cells and $CaCO_3$ nanoparticles within the deposition solutions. Figure 5 shows a 3×3 array of cell-encapsulated alginate hydrogels produced through light-addressable electrodeposition. Cell viability was determined using a live/dead assay (Invitrogen, CA) containing calcein AM (live cells, green) and ethidium homodimer (dead cells, red). Cell-encapsulated alginate hydrogels were incubated for 1 h after electrodeposition, after which they were stained by incubation with the live/dead assay agents for 10 min to allow the stain to diffuse into the hydrogels. The stain was removed by washing the hydrogels with culture media before the hydrogels were imaged. The average cell viability percentage was calculated by counting the number of pixels in the green (living cells) channel from three different experiments.

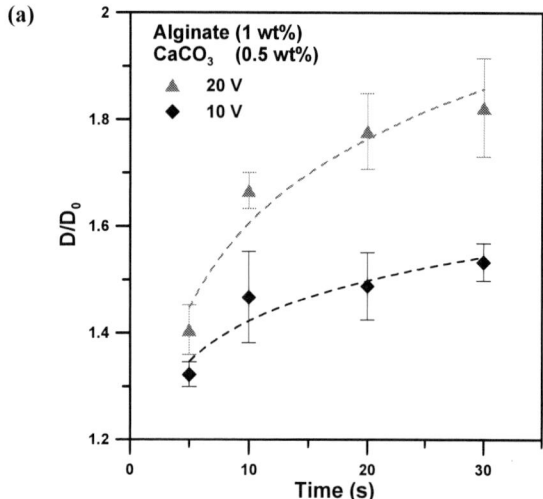

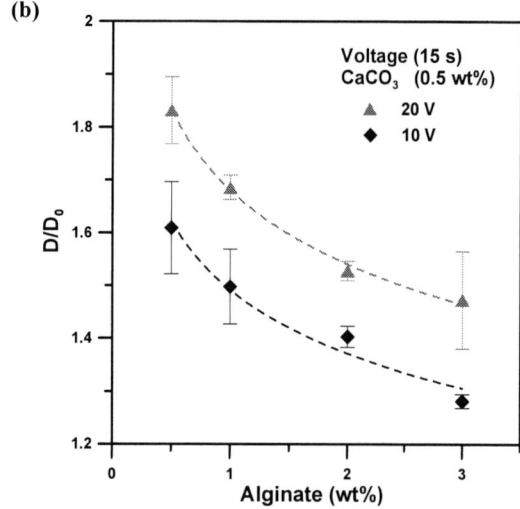

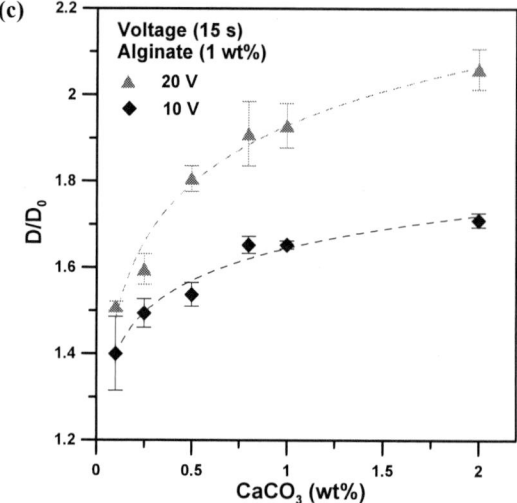

Fig. 4. (a) The effect of the illumination time of the light pattern on the dimensional resolution (D/D_0) for a deposition solution (1% alginate; 0.5% $CaCO_3$) with applied voltages of 10 V and 20 V. The effect of (b) the concentrations of alginate solutions with 0.5% $CaCO_3$ and (c) the concentrations of $CaCO_3$ nanoparticles dispersed within 1 wt% alginate solutions on the dimensional resolution (D/D_0) for a deposition solution with applied voltages of 10 V and 20 V, for an illumination time of 15 s. (Do and D denote the illumination pattern of circular shapes with D_0= 600 μm in diameter and the corresponding diameter (D) of the calcium alginate hydrogels produced by the illumination pattern).

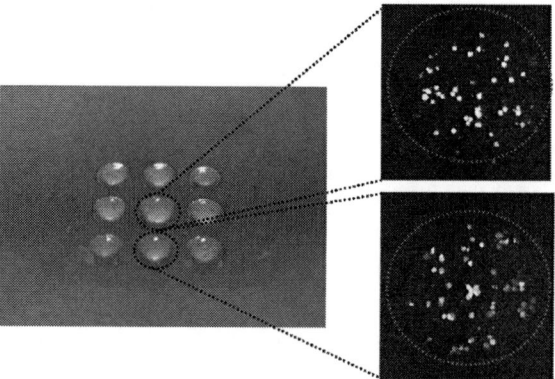

Fig. 5. Images of an array of 3×3 of cell-encapsulated alginate hydrogels

IV. CONCLUSION

In this study, we proposed a novel electrolytic system for performing an electrodeposition of calcium alginate hydrogels at desired locations on a photoconductive electrode substrate. Light illumination spatially patterned by the DMD produces a virtual photo-anode (i.e., a pH decrease) that triggers a localized release of Ca^{2+} from insoluble $CaCO_3$ nanoparticles to cause gelation of the calcium alginate. Flexible addressability, or patternability of the photo-electrodes, would allow the generation of desired alginate hydrogel patterns at a specific address, providing a great advantage over the electroaddressing of calcium alginate hydrogels using conventional metal microelectrodes. We demonstrate the ability to perform micropatterning of calcium alginate hydrogels with different shapes and sizes and with multiplexed micropatterning. A 3×3 array of cell-encapsulated alginate hydrogels was also successfully created without destroying viability through light-addressable electrodeposition. We also examine the effects of varying the concentration of the alginate and $CaCO_3$ solutions on the dimensional resolution of alginate hydrogel formation and the cell viability of cell-encapsulated alginate hydrogels after light-addressable electrodeposition. Our proposed electrolytic system does not need metal microelectrodes and provides both a reagentless method and the flexible electroaddressing of cell-encapsulated alginate hydrogels into cellular microarrays inside a microfluidic chamber or channel. We anticipate that this simple, rapid, and flexible method for light-addressable electrodeposition of cell populations into cellular microarrays could have a wide range of applications in drug discovery, toxicology, stem cell research, and potentially therapy.

ACKNOWLEDGMENT

This work was partially supported by the National Science Council, Taiwan, through grant NSC 99-2627-B-019-003.

REFERENCES

[1] T. G. Fernandes, M. M. Diogo, D. S. Clark, J. S. Dordick, and J. M. Cabral, "High-throughput cellular microarray platforms: applications in drug discovery, toxicology and stem cell research," *Trends Biotechnol.*, vol. 27(6), 2009, pp.234-240.

[2] W. H. Tan and S. Takeuchi, "Monodisperse Alginate Hydrogel Microbeads for Cell Encapsulation," *Adv. Mater.* Vol 19, 2007, pp. 2696-2701

[3] X. W. Shi, C. Y. Tsao, X. Y. , Y. Liu, P. Dykstra, G. W. Rubloff, R. Ghodssi, W. E. Bentley, and G. F. Payne, "Electroaddressing of Cell Populations by Co-Deposition with Calcium Alginate Hydrogels," *Adv. Funct. Mater.*, Vol 19, 2009, pp.2047-2052.

[4] S. H. Huang, H. J. Hsueh, and Y. L. Jiang, "Light-addressable electrodeposition of cell-encapsulated alginate hydrogels for a cellular microarray using a digital micromirror device," *BioMicrofluidics*, Vol. 5, 2011, 034109

A Free-standing and Flexible Parylene PCR Device

Penvipha Satsanarukkit[*], Hsiwen Lo, and Yu Chong Tai
[*]Department of Electrical Engineering, California Institute of Technology, USA
penvipha@mems.caltech.edu

Abstract—**We present a free-standing and flexible parylene-channel PCR device that allows a simple sealing mechanism to prevent bubbling caused by PCR temperature cycling. This sealing mechanism saves complicated, time-consuming in-channel valve fabrication processes and achieves excellent sealing performance. It is proven experimentally that this new PCR device plus its unique packaging is capable of locking PCR solution in the PCR chamber at 95°C for at least 5 hours without bubbling and leaking. As a result, we have successfully obtained the on chip PCR amplification without the bubble and leaking problem**

Keywords-PCR; parylene-C; microfluidics; bubbling and leaking problems

I. INTRODUCTION

After the discovery of PCR technique by Karis Mullis in 1986 [1], the PCR technique has become a widely-used, powerful technology for various molecular biological applications such as disease diagnosis and detection, forensic, cell analysis, etc.

On-chip PCR technology is particularly useful for cases with limited DNAs such as in single-cell analysis. On-chip technology is capable of fast and easy cell loading and precise cell alignment, two critical steps in single-cell analysis. The free-standing on-chip heater and sensor reduces the system's thermal mass and increases the heating and cooling rates. There are many works on miniature PCR including the use of silicon, silicon dioxide, glass, metal and polymers i.e., PDMS [2, 4, 5, 9]. Our group has a focus on using parylene for on-chip PCR applications because parylene is a superior biocompatible and chemically inert material. Our previous work [3] proved the feasibility of parylene technology for on-chip PCR. However, there are two issues, which are also the common issues for all other on-chip PCR devices, still remain to be improved. The first issue is the complexity of the device fabrication, and the second issue is the bubbling inside the PCR chamber usually caused by the high denaturation temperature step, evaporation or generated during sample loading [2, 4, 5, 6, 7, 8]. The bubbles create the insulating area leading to the nonuniform heat distribution across the PCR chamber. The small bubbles could grow up very quickly during higher temperature operation and push PCR reagents away from the heating zone, and affect the PCR efficiency. This work then presents a new free-standing parylene-channel on-chip PCR device that improves over both issues.

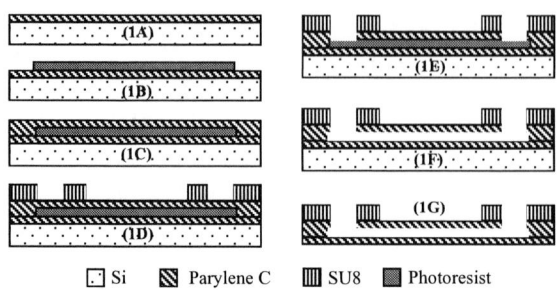

Fig. 1. Micro-PCR channel fabrication process using photoresist as a sacrificial layer. (A) Deposit 10-micron parylene C on a silicon substrate; (B) Spin and pattern sacrificial photoresist; (C) Deposit 10-micron parylene C; (D) Spin and pattern 100-micron SU-8 50; (E) Open the ports using laser ablation; (F) Release sacrificial photoresist; and (G) Release device from the silicon substrate.

☐ Si   ▧ Parylene C   ▥ SU8   ▨ Photoresist

II. DEVICE FABRICATION

This paper described a new totally free-standing and flexible parylene PCR device, which is detached from a rigid substrate so no more micromachining of the substrate is needed. As shown in Fig.1, the free-standing microchannels have two layers of 10-micron thick parylene and, at the inlet and outlet ports, a SU-8 50 (MicroChem Corp.) protection layer. Device fabrication starts with depositing 10-micron parylene C on a Si wafer. Then we deposited and patterned around 36-micron sacrificial AZ 9260 photoresist to form the microchannel. Before depositing another 10-micron parylene C, the wafer was hard-baked at 120°C for 8 hours to completely remove the solvent. Then, SU-8 50 was spin-coated and patterned to form a protection layer at the inlet and outlet ports. Next, the inlet and outlet ports were opened via laser ablation. The sacrificial layers were then released in room temperature isopropyl alcohol and acetone for around one week. The releasing time could be shortened to 2.5 days when increasing temperature of IPA and acetone to 40°C. After releasing the photoresist, the microchannel film was peeled off the Si substrate. To improve the adhesion between the two parylene layers, the microchannel film was annealed in vacuum at 200°C for 2 days. The fabricated microchannel film contains two microchannels, each of which is 100 micron wide.

The free-standing microheater/sensor was fabricated by depositing 20nm Ti/200nm Pt on 10-micron parylene C using ebeam and liftoff process. The metal film was peeled out from the Si substrate before operation.

978-1-4673-1122-9/12 $31.00 © 2012 IEEE

III. EXPERIMENTS AND RESULTS

A. Sealing Test

In this work, our device uses a simple but effective sealing mechanism to prevent bubbling for our parylene PCR microchannels by clamping the microchannels with two hard material strips and tightening with a silicone rubber sheet in between. This clamping mechanism is possible because of the flexibility of the parylene device and it saves complicated, time-consuming in-channel valve fabrication processes and achieves satisfactory sealing performance. Moreover, our simple sealing mechanism provides almost zero dead volumes, is compatible with existing PCR protocols, is dispensable and is reusable without contamination by harnessing the pinhole-free parylene characteristic.

For the sealing testing, first, the diluted food color was loaded into the microchannels. The microchannel film was heated with a digital hotplate (Dataplate Cole Parmer). Film temperature sensor was placed on an aluminum chuck and glued by thermal grease (Omegatherm 201) for good thermal contact. The temperature sensor was calibrated with a thermocouple and temperature tags (Wahl). The sealed microchannels were heated to 100°C for three hours. No bubble and leaking occurred.

Then, the channel was injected with PCR solution using the protocol stated in [3]. The chip was heated and taken the infrared images at 1X magnification using the Infrascope (EDO Corporation) which has standard Peltier and precise temperature controller (Fig. 2). The radiance images which depend on the material were checked every 5 minutes to observe if any air bubble happening. Fig. 3 shows the infrared and temperature images at 40 and 300 minutes of heating at 95°C. The images show the uniform of temperature distribution at 95°C with the PCR solution filled in the microchannels.

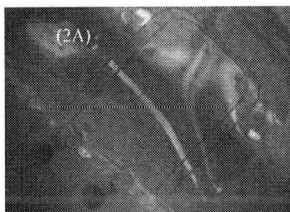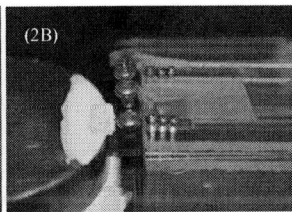

Fig. 2. The free-standing parylene-C micro-PCR channel. Each micro-PCR channel film contains two 100-micron wide channels. (A) after diluted food color loading; (B) after PCR solution loading and clamping and; (C) IR image setup.

The radiance images also clearly show the difference between the air spot close to the channel and filled PCR solution in the channel. At 95°C, the microchannels could hold the PCR solution for at least five hours without generating any bubble and leaking.

B. Device Characterization

Before performing the on-chip PCR, the free-standing 20nm Ti/200nm Pt microheater/sensor on parylene C film was characterized.

The temperature coefficient of resistance (TCR) of the microsensor was characterized using the oven (Delta Design 2300) and calculated using the protocol stated in [3]. The resistances at the temperature in between 35°C to 95°C, covering the PCR operational range, were measured. After the chip was put into the oven and waited 3 times of the heating time constant, low input voltage was applied and the output current was read from Universal Source (HP 3245A). The resistance was calculated from (1) and plotted in Fig. 4. The TCR of 1.6E-3 /°C is achieved.

$$\frac{R(T) - R(T_0)}{R(T_0)} = \alpha(T - T_0) \tag{1}$$

where T_0 is the reference temperature, and α is the temperature coefficient of resistance.

The on-chip heating uniformity was checked under the infrared microscope (Fig. 5). The free-standing microheater and microchannel were glued together using thermal grease. The complete micro-PCR device was supplied with the input power both in transient and equilibrium modes. The integrated device showed good uniform heating area in the targeted fluorescence detecting zone.

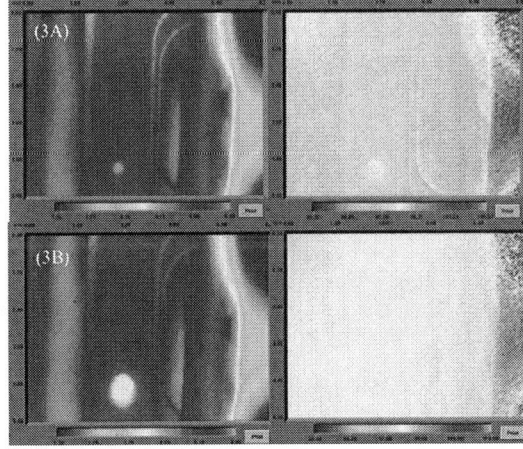

Fig. 3. Radiance and temperature images of microchannels filled with PCR solution heating at 95°C. The radiance images which depend on the material were recorded every 5 minutes to observe if any air bubble happening. (A) after 40 minutes; and (B) after 300 minutes. The images show the uniformity of temperature distribution at 95°C with the PCR solution filled in the channels. The air bubble spot on the left hand side of the channels in the radiance images helped to identify if there was any leaking.

C. On-chip PCR experiment

The PCR mixture consists of 1X Quanta PerfeCTa™ MultiPlex qPCR SuperMix (Quanta Biosciences); 200 nM of forward primer 5'-TGGAGAGGCTATTCGGCTATGACTG-3'; 200 nM of reverse primer 5'-ATACTTTCTCGGCAGGA-GCAAGGTG-3'; 200 nM of probe 5'-FAM-TAGCAGCCA-GTCCCTTCCCGCTTCAGTGA-BHQ-3'(Integrated DNA Technologies), designed by Arbel D. Tadmor; 1X ROX; and high copy plasmid bearing the ColE1 origin of replication and the kanamycin resistance gene pZS25O1+11-YFP, kindly provided by Hernan G. Garcia. The amplicon fragment is 294 base pairs.

Before loading PCR solution, the parylene channels were cleaned with DNA decontamination solution (Ambion) and rinsed with DEPC- treated and sterile filtered water (Sigma Aldrich). The fluids were manually injected using a micro syringe (Hamilton) with the PDMS and machined acrylic gasket. The PCR solution was well-mixed and centrifuged until no bubble observed. The first channel was then loaded with the PCR solution and the second channel was filled with the ROX reference passive dye working as a control. Approximately 2 nl of the PCR mixture with about 100 molecules of starting template was targeted at the 10X magnification fluorescence detecting zone. The microchannels were sealed by the same clamping technique.

On-chip PCR experiments were performed using protocols as described in [3]. The free-standing microheater and microchannel were glued together using thermal grease (Fig. 6). The PCR thermal cycling started with 95°C for 3 minutes, followed by 37 cycles of 95°C for 15 seconds (denaturation) and 60°C for 90 seconds (annealing/extension) respectively. The free-standing Ti/Pt resistor was used as both the heater and temperature sensor. The voltage was supplied from HP 3245A Universal Source to the chip and the current was measured with a precision Agilent 34401A multimeter. The LabView PID feedback control program was used to control the PCR thermal cycling.

The fluorescence images were taken under the fluorescence microscope at the end of each 37 thermal cycles with 10X magnification using a Nikon Eclipse E800 fluorescence microscope. The obtained fluorescence intensity from PCR channel was normalized against the ROX channel and parylene background. Fig. 7 shows the amplification fluorescence intensity. There was no bubbling and leaking during the on-chip PCR testing. We successfully obtained a PCR amplification curve comparable to that of a conventional real time PCR machine.

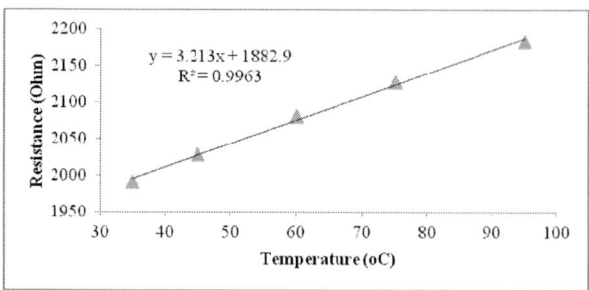

Fig. 4. 20nm Ti/200nm Pt microsensor on 10-micron parylene C characteristics. The TCR is 1.6E-3 /°C.

IV. CONCLUSION

Our free-standing and flexible parylene PCR device enables a simple clamping packaging to prevent bubbling during temperature cycling, which is then capable of locking PCR solution at 95°C for at least 5 hours. As a result, we have successfully obtained PCR amplification curves without any bubbling and leaking problems. The successful amplification from the starting of about 100 molecules of templates in approximate 2 nL volume is a promising step to further develop the on-chip single/rare cell analysis in a single chip.

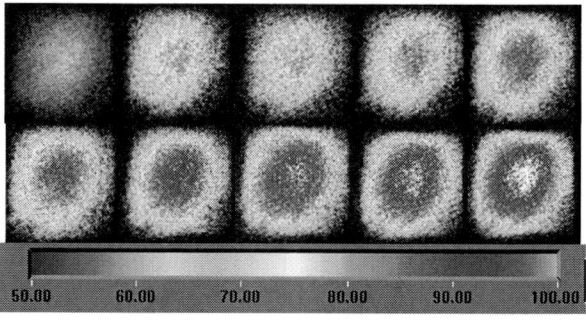

Fig. 5. Selected transient infrared thermal 1X magnification image characterization for the complete micro-PCR device with the input power 21 to 30 mW respectively. The integrated device showed good uniform heating area in the fluorescence detecting zone.

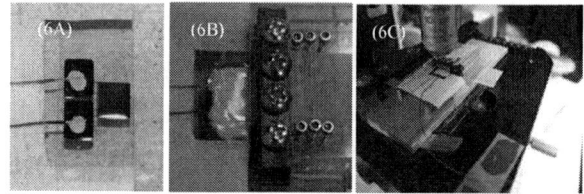

Fig. 6. (A) Free-standing 20nm Ti/200nm Pt microheater; (B) The heater and channel were glued together using thermal grease; (C) Detection of fluorescence intensity under the fluorescence microscope with 10X magnification.

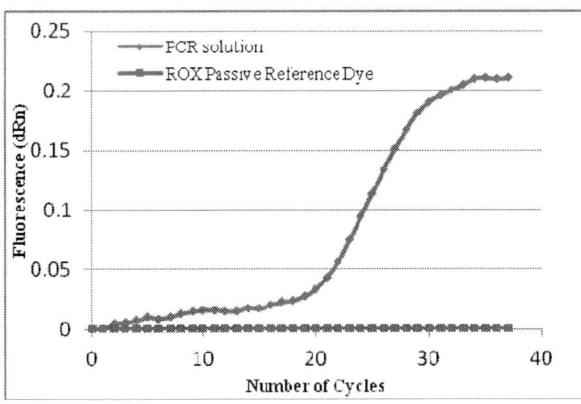

Fig. 7. The micro-PCR amplification fluorescence as a function of the number of cycles with the starting of about 100 molecules of templates in approximate 2 nL volume in the fluorescence detecting zone. The PCR curve were plot against the ROX passive reference dye injected into the parallel second channel with 150 micron apart from the PCR channel.

ACKNOWLEDGMENT

This work was supported by The Royal Thai Government scholarship and The Boeing Company. The authors appreciate all generous help and precious advice from the members of Caltech Micromachining Laboratory.

REFERENCES

[1] A. J. Mello, "DNA amplification: does 'small' really mean 'efficient'?," Lab on a chip, 1 (2001), pp. 24N-29N.

[2] C. Zhang and D. Xing, "Miniaturized PCR chips for nucleic acid amplification and analysis: latest advances and future trends," Nucleic Acids Research, 35, 13 (2007), pp. 4223-4237.

[3] P. Satsanarukkit, H. Lo, Q. Quach, and Y.C. Tai "On-chip PCR with free-standing parylene channel," *Proceedings of the 14th in the series of Hilton Head Workshops on the science and technology of solid-state sensors, actuators, and microsystems (Hilton Head 2010)*, pp. 439-442.

[4] H.B. Liu et al., "Micro air bubble formation and its control during polymerase chain reaction (PCR) in polydimethylsiloxane (PDMS) microreactors," *J. Micromech. Microeng.* **17** (2007) pp.2055–2064.

[5] Y.S. Shin et al., "PDMS-based micro PCR chip with Parylene coating," J. Micromech. Microeng. **13** (2003) pp.768–774.

[6] Z.Q. Niu, W.Y. Chen, S.Y. Shao, X.Y. Jia and W.P. Zhang, "DNA amplification on a PDMS–glass hybrid microchip," J. Micromech. Microeng. **16** (2006) pp.425–433.

[7] N.C. Cady, S. Stelick, M.V. Kunnavakkam, and C.A. Batt, "Real-time PCR detection of *Listeria monocytogenes* using an integrated microfluidics platform," Sensors and Actuators B 107 (2005) pp.332–341.

[8] T. Nakayama et al., "Circumventing air bubbles in microfluidic systems and quantitative continuous-flow PCR applications," Anal Bioanal Chem (2006) 386: pp.1327–1333.

[9] L.J. Kricka and P. Wilding, "Microchip PCR," Anal Bioanal Chem (2003) 377 : pp.820–825.

The Effect of Cytochalasin D on F-Actin Behavior of Single-Cell Electroendocytosis Using Multi-Chamber Micro Cell Chip

Ran Lin[1,2], Donald. C. Chang[3], Yi-Kuen Lee[2,4*]

[1]Bioengineering Graduate Program, HKUST, [2*]Dept of Mechanical Engineering, HKUST,
[3]Division of Life Science, HKUST, [4]Joint KAUST-HKUST Micro/Nanofluidics Laboratory, HKUST
Clear Water Bay, Kowloon, Hong Kong
Tel: +852 2358-8663, Fax: +852 2358-1543, E-mail: meyklee@ust.hk

Abstract—Electroendocytosis (EED) is a pulsed-electric-field (PEF) induced endocytosis, facilitating cells uptake molecules through nanometer-sized EED vesicles. We herein investigate the effect of a chemical inhibitor, Cytochalasin D (CD) on the actin-filaments (F-Actin) behavior of single-cell EED. The CD concentration (C_{CD}) can control the depolymerization of F-actin. A multi-chamber micro cell chip was fabricated to study the EED under different conditions. Large-scale single-cell data demonstrated EED highly depends on both electric field and C_{CD}.

Keywords—*Electroendocytosis; Micro Cell Chip; Actin filament; Cytochalasin D*

I. INTRODUCTION

Electroendocytosis (EED), i.e. electric-field-induced endocytosis-like process, is a technique to facilitate molecules delivery to cells using a pulsed electric field train (PEF). PEF enhances the cell plasma membrane invagination and fission via nanometer-sized endocytotic vesicles and herein benefits exterior molecules uptake. Taking advantages of high cell viability and long-term persistence, EED can be a promising technique for molecular delivery [1].

Cytoskeleton is system of intracellular filaments crucial for cell shape, division, and function in all three domains of life [2,3]. Typically, actin filament (F-Actin) is one of the most important types of cytoskeleton in the eukaryotic cells. F-Actin allows a cell to move, maintains its shape, and works in different cellular events. F-Actin is required for cellular endocytosis and the transfer of cargoes within the endocytotic system [4, 5], as shown in Fig. 1 (a). The formation of membrane extensions and endocytosis pits involves the remodeling of F-Actin and their corresponding protein regulators. In addition, the molecular motors that move on tracks of F-Actin are responsible in mediating the movement of endocytosis vesicles and the trafficking between compartments. Although EED has been shown to depend on F-Actin [5], to date, a systematic study of F-Actin behavior of single-cell EED under different conditions was not available.

To address this issue, we fabricated a multi-chamber micro cell chip using MEMS technology to overcome the limitations of conventional EED experiments. In this work, we applied cytochalasin D (CD), a chemical inhibitor, to study F-Actin dependent EED mechanisms at the single-cell level.

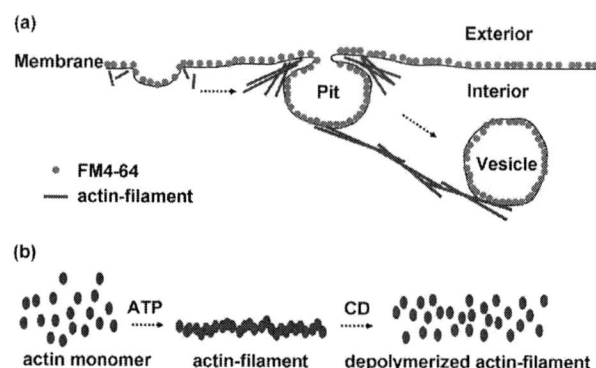

Fig. 1. (a) F-Actin regulates endocytosis process, including membrane invagination as endocytosis pit and intracellular vesicle trafficking, (b) ATP helps actin monomer polymerize as F-Actin, which is depolymerized by cytochalasin D (CD).

II. EXPERIMENTAL

A. Multi-Chamber Micro Cell Chip

In order to increase EED experimental efficiency in different biochemical environments, we proposed a novel multi-chamber design of our micro system. As illustrated in Fig. 2, a micro chip was designed and fabricated with 24 different electric inputs, 4 separated cell culture chambers and corresponding independent temperature sensors. The EED experiments of HeLa cells are available to carry out in parallels in different biochemical environments (*e.g.* different molecular probes, chemical inhibitors with distinct concentration, etc.), together with multiple electric inputs and temperature control.

In addition, this design can be simply further extended to a much higher density of units, which is able to replace conventional setup (*e.g.* multi-well plate + external electrodes). The integration of microelectrode array and separated temperature sensors will greatly benefit the investigation of different electrobiology and temperature-related phenomenon, which are difficult to achieve using conventional multi-well setup. The fabrication process of micro chip was reported elsewhere in details [6].

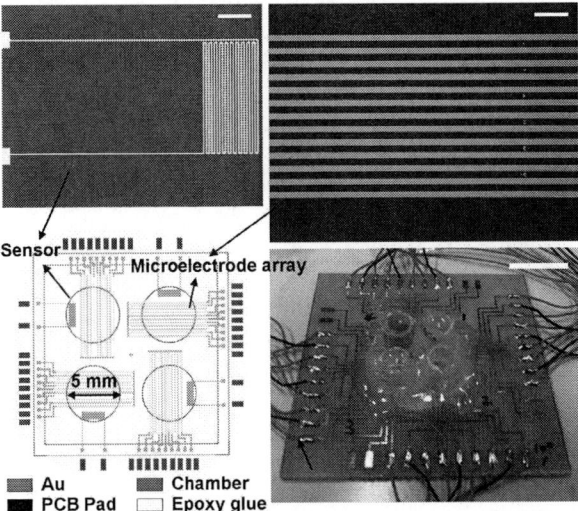

Fig. 2. (a) optical micrographs of the temperature sensor, (b) optical micrographs of the microelectrode array (scale bar: 500 μm), (c) a schematic of micro cell chip, and (d) photograph of a packed micro chip (scale bar: 10 mm).

B. Experimental Setup

A programmable arbitrary waveform generator (HP 33120A, Agilent Technologies, Inc., Santa Clara, CA, USA) was employed to produce an adjustable train of unipolar rectangular electric pulses as shown in Fig. 4. The parameters of applied rectangular pulse trains were as follows: electric field strength (E) = 0-200 V cm^{-1}; pulse duration (t_D) = 1.0 ms; pulse repetition frequency (f) = 100 Hz; and, total electric treatment time (t_T) = 180 s.

An epi-fluorescence microscope (Eclipse TE2000-U, Nikon Corp., Japan) with a filter (C-FL Epi-Fl filter Block N G-2A, Nikon Corp., Japan) was used to observe EED uptake by HeLa cells on micro chips at single-cell level. A cooled digital CCD camera (SPOT™ RT-SE18 Monochrome, Diagnostic Instruments Inc., USA) mounted on the microscope and the control software package (SPOT Imaging Software Advanced Version, Diagnostic Instruments Inc., USA) were used to record 8-bit fluorescent micrographs with spatial resolution of 1360 × 1024 pixels. Digital image processing software (AlphaDigiDoc, Alpha Innotech Corp., USA) was used to quantify each single-cell EED efficiency and EED vesicles distribution based on fluorescence micrographs.

HeLa cells (CCL-2™, ATCC, VA, USA) were cultured as a monolayer in a 60 mm petri dish containing Minimum Essential Medium (MEM) (Invitrogen Inc., USA) supplemented with 10% fetal bovine serum (ATCC, VA, USA) and 1% Streptomycin/Penicillin (GIBCO, Invitrogen Inc., USA), at 37 °C in a 5% CO$_2$ atmosphere. The cells were harvested at the log phase of growth by 0.25% trypsin/EDTA

(GIBCO, Invitrogen Inc., USA) from the petri dish, and then were resuspended in the culture medium at a concentration of 2×10^5 cells mL^{-1}. The cell suspension of 100 μL was put on the micro chips, and allowed to grow for 10 hr prior to the experiment (37 °C, 5% CO$_2$).

FM4-64 (Invitrogen Inc., USA; molecular weight = 607.51 Da; positively charged with Z = +2), a lipophilic styryl fluorescence dye for lipid labeling, was used to characterize membrane internalization and EED vesicles formation and trafficking. In this experiment, FM4-64 was dissolved in the culture medium at a concentration of 8 μM. The optical spectrum of FM4-64 is excitation/emission maximum at 515/670 nm.

C. Actin filament and Cytochalasin D

Actin filaments (F-Actin) are flexible polymers made of actin molecules. They can bind with various actin binding proteins that enable the filaments to serve a variety of functions in cells. In different pathways of endocytosis, the polymerization of actin meshwork is widely involved in the occurring of membrane ruffles, formation of endocytosis pits and intracellular trafficking of endocytosis vesicle structures.

As shown in Fig. 1(b), the polymerization of F-Actin is through adding of actin monomers at current actin filaments ends with the help of adenosine triphosphate (ATP). This process can be disrupted by cytochalasin D, a cell permeable and potent inhibitor of actin polymerization. As a result of the inhibition of actin polymerization, cytochalasin D can further modify cellular morphology, inhibit cell division, and even cause cells to undergo apoptosis.

III. RESULTS AND DISCUSSIONS

A. Cytochalsin D Effect on Cell Viability

Because of the cytotoxicity of cytochalasin D (CD), we firstly quantify the cell viability treated by cytochalasin D. HeLa cells were incubated in culture medium with 0-50 μg/mL CD for 3 hours, followed by a propidium iodide viability assay [1]. After 30 min of CD treatment, HeLa cells were observed to gradually round up owing to F-Actin depolymerization. However, CD at C_{CD} = 0-50 μg/mL has negligible cytotoxicity, and cell viability was kept at larger than 99%. In addition, the cellular viability in treated group keeps comparable to control group without CD inhibition.

B. F-Actin Dependent EED Efficiency

Before FM4-64 EED experiments, we pre-incubated the HeLa cells on a micro chip in culture medium supplemented with 0-50 μg/mL CD for 60 min in the incubator (37 °C, 5% CO$_2$). Then we removed CD and incubated the cells with 8 μM FM4-64 for 2 min. FM4-64 stain medium was replaced by exposure medium [1] and PEF was applied (E = 100 or 200 V cm^{-1}, t_D = 1 ms, t_T = 180 s, f = 100 Hz). PBS rinse was followed after 5 min post-exposure incubation and cells were then observed and imaged in the culture medium. The EED efficiency $I_{avg,FM}$ of HeLa cells in both CD-treated and CD-free groups were quantitatively determined.

As illustrated in Fig. 3, FM4-64 characterized EED efficiency increases with pulsed electric field (PEF) from 0 to 200 V/cm in both control cells and CD treated cells. When we compare corresponding micrographs, it is also obvious that CD is able to effectively inhibit both endocytosis and EED efficiency at any electric field strength.

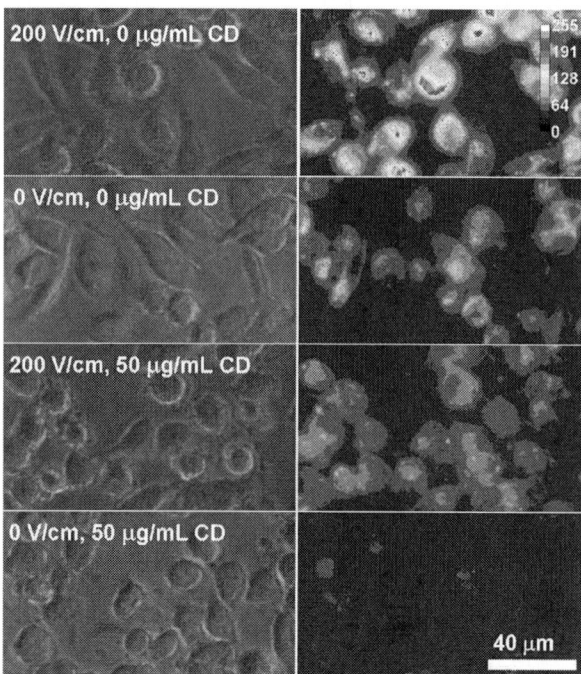

Fig. 3. Bright field (left) and pseudocolor fluorescent (right) micrographs. PEF is able to increase endocytosis efficiency with or without CD. CD reduces both endocytosis and EED efficiency and makes cell round up.

More quantitative results are provided in Fig. 4. We can see the normalized average fluorescent intensity of FM4-64 characterized HeLa cells ($I_{avg,FM}$) as a function of CD concentration (0-50 μg/mL) at different applied average electric field strengths (E = 0-200V/cm). It is clear that both endocytosis and EED efficiency decline dramatically with CD concentration increase from 0 to 20 μg/mL, and gradually reach a steady low status from 35 μg/mL. However, at any CD concentration level, PEF is capable to augment FM4-64 uptake efficiency significantly. As an example, even the meshwork of F-Actin was partially depolymerized by CD at a concentration of 10 μg/mL, PEF still facilitate EED with higher efficiency than intrinsic endocytosis without CD.

Both endocytosis and EED mechanism depend on the function of F-Actin, which undergoes partial depolymerization by CD at different concentration levels. However, PEF stimulation is able to compensate the F-Actin deficiency to

some extent and increase the endocytosis efficiency even the meshwork of F-Actin was partially depolymerized.

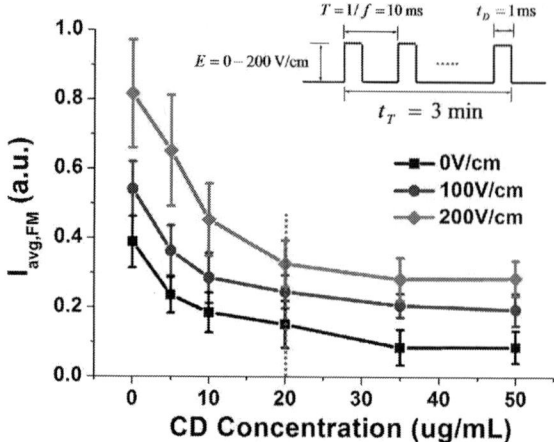

Fig. 4. $I_{avg,FM}$ as a function of CD concentration at different applied electric field strengths (E).

C. F-Actin Dependent Intracellular Trafficking

Detailed information involving intracellular trafficking of endocytosis/EED vesicles is provided from the distribution of endocytosed FM4-64 molecule within single-cell at 30 min after endocytosis and EED. As shown in Fig. 5 (a), CD depolymerizes F-Actin and significantly inhibits the endocytosis of HeLa cell, uptaking only a few FM4-64 molecules in the sparse fluorescent vesicles. The pulsed electric field at 200 V/cm is able to improve cell endocytosis with the deficiency of F-Actin, as much higher fluorescent intensity shown in Fig. 5 (b). However, we found the fluorescence distribution is uniform in the cellular cytoplasm at even 30 min after EED. As a comparison to the same EED treatment but without CD inhibition in Fig. 5 (c), the undisturbed polymerization of actin filament meshwork benefits not only the EED uptake, but also the subsequent intracellular trafficking of endocytosis vesicular structures. We can see that endocytosed FM4-64 in Fig. 5 (c) mainly undergoes endosome translocation and might highly aggregate at the location close to Golgi apparatus and endoplasmic reticulum.

Fig. 6 shows the local fluorescent intensity of FM4-64 ($I_{avg,FM}$) as a function of the radial coordinate from the nucleus under different PEF and CD. PEF compensates the depolymerization of F-Actin and therefore, FM4-64 molecules distribute uniformly in the cytoplasm (7-13 μm from cell nucleus). Without CD inhibition, the uptaken FM4-64 molecules through EED normally aggregate near Golgi apparatus (6 μm from nucleus) when F-Actin is intact.

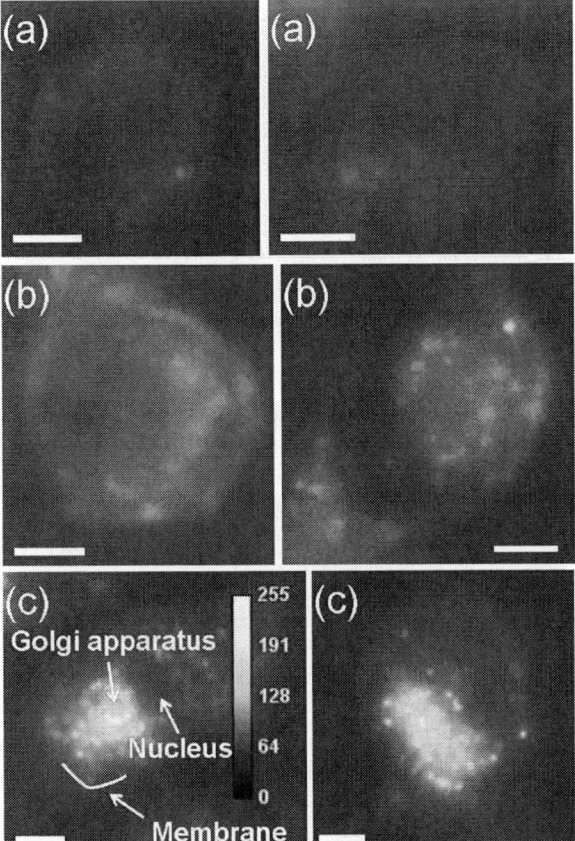

Fig. 5. Pseudocolor micrographs of intracellular FM4-64 distribution at (a) 0 V/cm PEF, 50 µg/mL CD, (b) 200V/cm PEF, 50 µg/mL CD, and (c) 200V/cm PEF, 0 µg/mL CD (all scale bar:5µm)

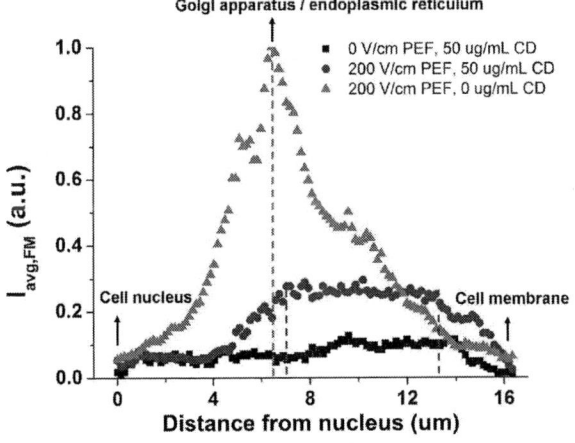

Fig. 6. Fluorescence intensity of FM4-64, $I_{avg,FM}$, of EED of a cell as a function of radial coordinate in different conditions.

These experimental results suggest that although PEF can facilitate the uptake of exterior material through endocytosis pathway even at F-Actin inhibition to some extent, however, the intracellular trafficking of EED is similar to intrinsic endocytosis, which might still follow the existing trafficking mechanism and be highly dependent on the F-Actin.

IV. CONCLUSIONS

In this work, the effect of Cytochalasin D (CD) on F-Actin behavior of single-cell EED was investigated. A multi-chamber micro cell chip was proposed, which benefits the high throughput micro EED experiments in different biochemical environments and multiple electric inputs. When the CD concentration increases from 0 to 50 µg/mL, the EED efficiency can be reduced by a factor of ~2.5 and subsequently intracellular trafficking is blocked. The experimental results also suggested that pulsed electric field (PEF) can facilitate endocytosis efficiency even at F-Actin depolymerization to some extent. However, vesicular organelles in EED still follow the same intracellular trafficking as intrinsic endocytosis, which is highly dependent on the integrity of F-Actin meshwork.

EED mechanism is highly dependent on the cellular cytoskeleton. The dependence of F-Actin in EED mainly lies in the vesicle formation and intracellular trafficking, both of which are alike to intrinsic endocytosis process. Electric field makes limited influences on intracellular trafficking, however, is able to compensate the partial deficiency of F-Actin, facilitating endocytosis efficiency in different biochemical environments possibly owing to enhanced cell membrane invagination.

ACKNOWLEDGMENT

The research was sponsored by a grant from Hong Kong Research Grants Council (Grant Ref. No. 615907) and by an award from the King Abdullah University of Science and Technology (KAUST Award No. SA-C0040/UK-C0016).

REFERENCES

[1] R. Lin, D. C. Chang, Y.-K. Lee, "Single-cell electroendocytosis on a micro chip using in situ fluorescence microscopy," *Biomedical Microdevices*, vol. 13, pp. 1063-1073, Nov 2011

[2] B. Alberts, *et al.*, *Essential Cell Biology*: Garland Science/Taylor & Francis Group, 2003.

[3] B. Wickstead and K. Gull, "The evolution of the cytoskeleton," J. Cell Biol., vol.194(4), pp. 513-525, Aug 2011.

[4] T. Soldati and M. Schliwa, "Powering membrane traffic in endocytosis and recycling," *Nat Rev Mol Cell Biol*, vol. 7, pp. 897-908, Dec 2006.

[5] G. Apodaca, "Endocytic traffic in polarized epithelial cells: role of the actin and microtubule cytoskeleton," *Traffic*, vol. 2, pp. 149-59, Mar 2001.

[6] M. Glogauer, W. Lee and C. A. G. McCulloc, "Induced Endocytosis in Human Fibroblasts by Electrical Fields," *Experimental Cell Research*, vol. 208, pp. 232-240, Sep 1993.

[7] R. Lin, D. C. Chang, Y.-K. Lee, "Study of temperature effect on single-cell fluid-phase endocytosis using micro cell chips and thermoelectric devices," *Proc. MicoTAS2010*, pp. 962-964, Nov. 2010.

Cryogenic Frozen Device for Hepatocyte Culture and Responses

Ling-Yi Ke[*1], Yu-Shih Chen[2], Jing Liu[3] and Cheng-Hsien Liu[1,2]

[1]1 Department of Power Mechanical Engineering, National Tsing Hua University, Hsinchu,
[2] Institute of NanoEngineering and MicroSystems, National Tsing Hua University, Hsinchu, Taiwan.
[3] Department of Biomedical Engineering , Tsinghua University, Beijing, China.
* lingyi0412@gmail.com

Abstract—In clinical medicine, freezing treatment has been applied to patients for years. However, the combination of the micro/nano biochip (or Lab Chip) techniques and cryogenic frozen techniques were seldom proposed in the past. This paper reports a cryogenic frozen apparatus for the study of microfluidic liver tissue mimic hepatic cords responses on three dimension hepatocyte culture chip. We designed the micro-cylinder about the height of 85 micrometer for loading hepatocyte to form three dimension tissue-mimic structures. The liver is organized into lobules which take the shape of polygonal prisms. Each lobule is typically hexagonal in cross section and is centered on the central vein. The bulk of the lobule consists of hepatocytes which are arranged into hepatic cords that are separated by sinusoid space. In liver function testing, the albumin secretion was affected 12% by freezing the hepatocyte ten minutes.

Keywords- Cryogenic Frozen, hepatocyte culture, liver funsction

I. INTRODUCTION

In mid-2009, researchers from Professor G. Whitesides' group at Harvard University reported a microfluidic instrument that produced drops of supercooled water suspended in a moving stream of liquid fluorocarbon. This is the first time for human being to observe the ice nucleation and the defrosting process of the small water drop in details. A metastable liquid such as supercooled water can exist because its transition to the thermodynamically stable phase must start with the formation of a microscopic amount of the stable phase within the bulk metastable phase[1]. Using a whole-chip cooling chamber, researchers' controlled the ambient temperature surrounding a microfluidic chip and induce cooling and freezing inside the channels [2]. For 3D tissue-mimic chip, microfluidic design and assembly of the linear concentration gradient generator and multiplexed cell culture chip to construct the 3D HepaTox Chip [3]. A major concern involved in a cryosurgery is to maximize the freezing efficiency in killing target focus while minimize the irreversible damage to the surrounding normal tissues [4]. Further, a three dimension lobule-mimetic chip is observed depend on the hydrogel-based geometry microstructure and the novel cell manipulation method [5].

Jing Liu had used selective freezing of target biological tissues in his group. They proposed a flexible method to control the size and shape of the iceball by injecting solutions

with specific thermal properties into the target tissues, to enhance freezing damage to the diseased tissues while preserving the normal tissues from injury [6]. The freezing technology performs increased control of the local disease in the area (i.e., reduced local recurrence of disease). And also can be used to enhance the natural destructive mechanisms of freezing within tissue [7].

In our group, we performed explore the studies of liver tissue recovering via the integration of liver lab chip techniques and the cryogenic frozen techniques. Under temperature control and frozen/defrosting control which had quite close relationship between the fever and the hepatocyte activity.

II. MATERIALS AND METHODS

A. Design concept

In order to understand the freezing step, understanding about the physical state of water in heterogeneous systems is needed, such as our cryogenic frozen device. In particular, below the freezing point of bulk water, two classes of water can be observed. One class is water molecules present far from any surface and macromolecule. These water molecules form regions of bulk ice. The other class, termed non-freezing water, interfacial water is constituted of water molecules that retain molecular mobility even below the temperature at which

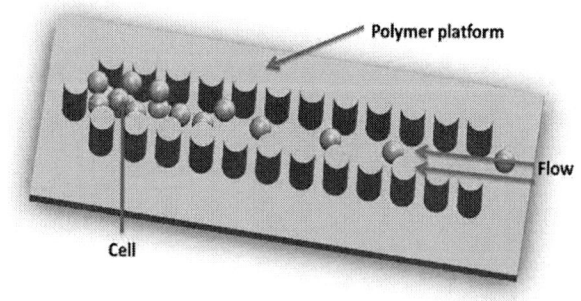

Figure 1: Schematic diagram of microcylinders about the height of 85 micrometer for loading hepatocyte to form three dimension tissue-mimic hepatic cords. This Schematic diagram is the unit of the 3D hepatocyte culture chip

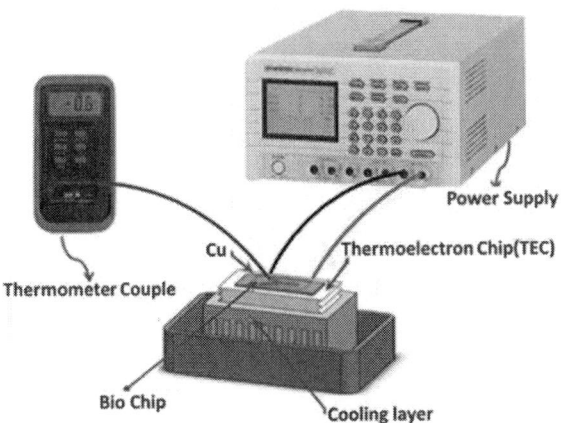

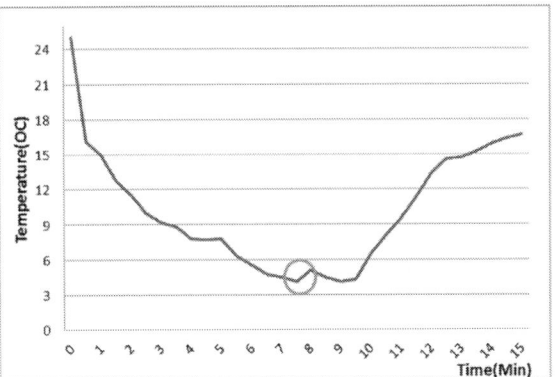

Figure 2: Schematic diagram of a cryogenic frozen apparatus. The 3D hepatocyte culture chip was frozen by thermoelectron chip (TEC) which was controlled by power supply and was measured by thermometer couple.

Figure 3: The freezing process to the 3D hepatocyte culture chip. We changed the temperature from 25℃ to the 4℃. Finally, the temperature rose to the room temperature. Blue circle shows the thermomter pase transistion.

regions of bulk ice have formed.

In figure 1, this schematic diagram is microcylinders about the height of 85 micrometer for loading hepatocyte to form three dimension tissue-mimic hepatic cords. This Schematic diagram is the unit of the three dimentional hepatocyte culture chip. We used microfabrication process has been used for making three dimentional hepatocyte culture chip platform. We have using microfluidic to load hepatocyte cells to form three dimension tissue-mimic hepatic cords. This integrated assembly allowed control of the direction of flow of medium across the cultured cells in this platform.

By the way, we used this method to form the hepatic cords.

Therefore, these microcylinders were arranged U-shaped to mimic the hepatic cords in vivo. Each lobule consists of hepatocytes which are arranged into hepatic cords that are separated by sinusoid space. However, we designed three dimentional hepatocyte U-shaped to be an array. It was difference to the radiate hepatic cords in vivo.

B. Cryogenic Frozen Device

In figure 2, schematic diagram of a cryogenic frozen apparatus. The 3D hepatocyte culture chip was frozen by thermoelectron chip (TEC) which was controlled by power supply and was measured by thermometer couple. We used

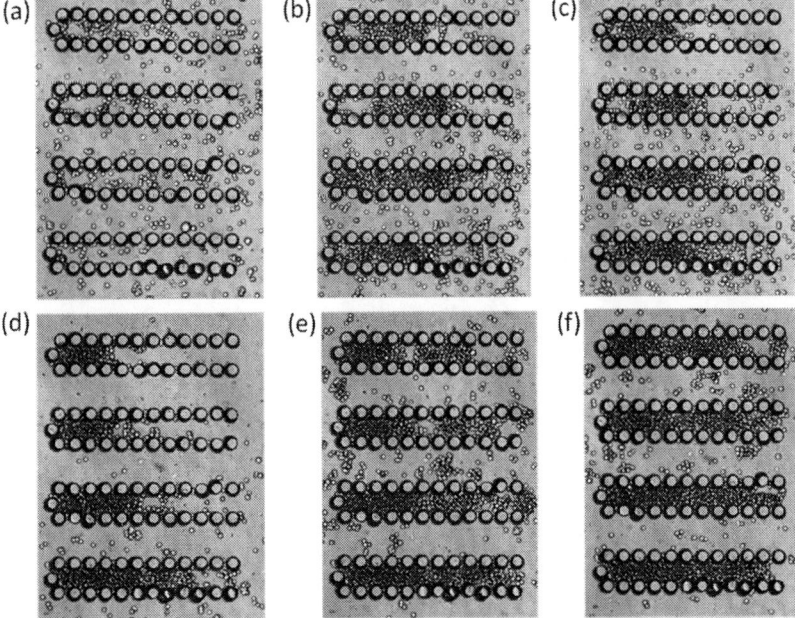

Figure 4: Hepatocyte loading into the microcylinders. In flow velocity is 1-10ul/min, the hepatocyte with the medium was flowed by the syringe pump.

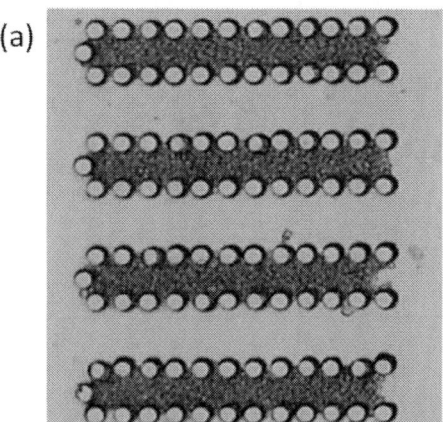

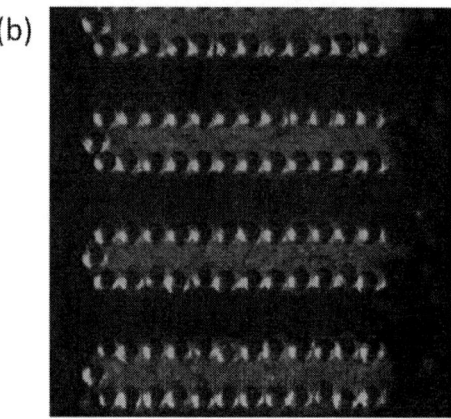

Figure 5: Image of the loading hepatocyte by microcylinders in the 3D hepatocyte culture chip to mimic hepatic cords in vitro. (a) The microscope photo of hepatic cords consists of hepatocyte cells. (b) Fluorescent photo of the hepatocyte cells used CMFDA dye(green), which culture to achieve continuous flow.

microfabrication process to set up microfluidic chip and loading hepatocyte cells to form three dimension tissue-mimic hepatic cords. In this cryogenic frozen procedure comprise three steps. Frist put on bio chip at thermoelectron chip (TEC). Second, we used power supply to control voltage supplied. The final step was using thermometer couple to gauging changes temperature.

III. RESULTS AND DISCUSSION

The freezing process to the three hepatocyte culture chip was shown in figure 3. Under temperature control and frozen/defrosting control, large-area artificial lobules were selectively frozen at target location on chip. This process is continued in order to achieve effective frozen/defrosting that cooling layer. Hence, we changed the temperature from 25℃ to the 4℃. Finally, the temperature rose to the room temperature. Blue circle shows the thermomter pase

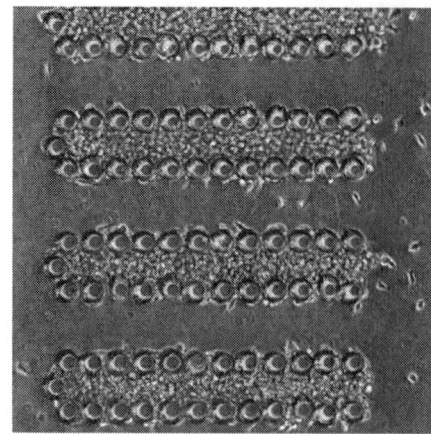

Figure 6: Image of the loading hepatocyte by microcylinders in the 3D hepatocyte culture chip to mimic hepatic cords in vitro, which culture by the medium on the third day.

transistion. Because we control the temperature, the cell activity was changed.

We took advantage of the cryogenic frozen techniques developed performs hepatocyte culture chip. In our design, the important unit of the microcylinders in three dimension hepatocyte culture chip. Figure 4 shows hepatocyte loading into the microcylinders. In flow velocity is 1-10ul/min, the hepatocyte with the medium was flowed by the pump. This integrated assembly allowed control of the direction of flow of medium across the cultured cells in this platform. In figure 5, these images were the loading hepatocyte by microcylinders in the 3D hepatocyte culture chip to mimic hepatic cords in vitro. Figure 5(a) shows the microscope photo of hepatic cords consists of hepatocyte cells. Figure 5(b) Fluorescent photo of the hepatocyte cells used CMFDA dye (green), which culture to achieve continuous flow. Figure 6 shows image of the

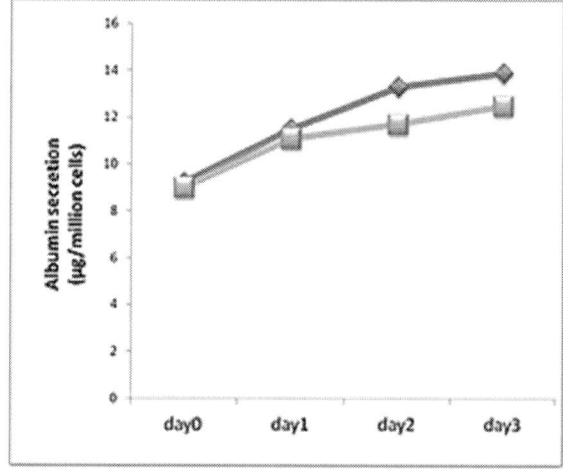

Figure7: In the same conditions, the deep green line shows the control setup and the light green line shows the 3D hepatocyte culture chip was frozen by thermoelectron chip (TEC). The albumin secretion was affected 12% by freezing the hepatocyte ten minutes.

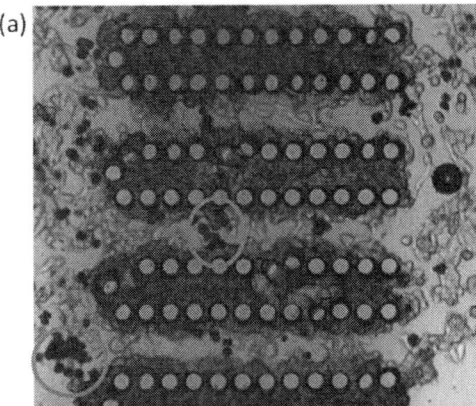

(a)

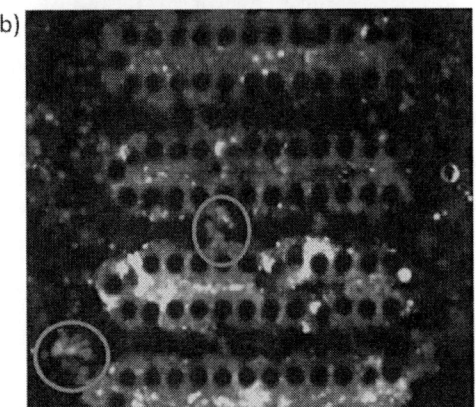

(b)

Figure 8. In cryogenic frozen case, those photos are the loading hepatocyte by microcylinders in the 3D hepatocyte culture chip on the third day.

loading hepatocyte by microcylinders in the 3D hepatocyte culture chip to mimic hepatic cords in vitro, which culture by the medium on three days.

Figure7 shows in the same conditions, the deep green line shows the control setup and the light green line shows the 3D hepatocyte culture chip was frozen by thermoelectron chip (TEC). The albumin secretion was affected 12% by freezing the hepatocyte ten minutes. Base on the contrast, we showed two photos of the cryogenic frozen case in long-term culture on the third day as shown in figure 8. By the way, we can see some cells were separated by the cryogenic frozen situation.

IV. CONCLUSION

We successfully to temperature control and frozen control which has quite close relationship between the temperature and the hepatocyte activity. We changed the temperature from 25℃ to the 4℃. The liver is organized into lobules which take the shape of polygonal prisms. Each lobule is typically hexagonal in cross section and is centered on the central vein.

Finally, the albumin secretion was affected 12% by freezing the hepatocyte ten minutes.

ACKNOWLEDGMENT

This work was supported partially by National Tsing Hua University of R.O.C. under the grant 100N2752E1.

REFERENCES

[1] C. A. Stan, et al., "A microfluidic apparatus for the study of ice nucleation in supercooled water drops ", Lab on a chip, vol. 9, 2009, pp2293-2305.

[2] A. E. Sgro and d. T. Chiu, "Droplet freezing, docking, and the exchange of immiscible phase and surfactant around frozen droplets", Lab on a chip, vol. 10, 2010, 1873-1877.

[3] Y. C. Toh," A microfluidic 3D hepatocyte chip for drug toxicity testing", Lab on a chip, Vol. 9, 2009, 2026-2035.

[4] Rabin, Y, and Shitzer, A, "Numerical solution of the multidimensional freezing problem during cryosurgery," ASME Journal of Biomech Eng., 120, 1998,pp. 32-37.

[5] Y. S. Chen, L. Y. Ke, C. H. Liu, "3D lobule-mimetic chip via positive dielectrophoresis force with sinusoidal spacing poly (ethylene glycol)-diacrylate microwalls", in *Digest Tech. Papers Transducers'11 Conference*, Beijing, June 5-9, 2011, pp. 1837-1840.

[6] Yu T.-H. Yu, Liu J.*, and Zhou Y.-X., "Selective freezing of target biological tissues through injection of solutions with specific thermal properties", Cryobiology, vol.50, 2005.pp.174-182,

[7] Raghav Goel et.al, "Adjuvant Approaches to Enhance Cryosurgery", Journal of Biomechanical Engineering, 131, 074003-1-11, 7, 2009

3D Microstructure Integrated Bioreactor System for Transgenic Mice Thick Liver Tissue Culture

Shilpa Sivashankar [1*], Srinivasu Valagerahally Puttaswamy[1], Hui-Ling Lin[2], Shih-Mo Yang[3], Hung-Po Chen[1], Chau-Ting Yeh[2] and Cheng-Hsien Liu[1]

[1]Department of Power Mechanical Engineering, National Tsing Hua University, Hsinchu, Taiwan, R.O.C
[2]Liver Research Unit, Chang Gung Memorial Hospital, Linkou, Taiwan, R.O.C
[3]Department of Electrophysics, National Chiao-Tung University, Taiwan, R.O.C.
*megha_shilpa2001@yahoo.com

Abstract—**A bioreactor system with 3-diemential (3D) Poly ethylene glycol diacrylate (PEGDA) structures in the culture-well to provide 3D culturing conditions is proposed here. The culture medium diffused through the tissue cause of the nozzle shaped flow channel at the outlet of each bioreactor and tissue being held between PEGDA microstructures. This helps in delivering the required nutrients to core of the tissue, eliminating the mass transfer problem. Presence of PEGDA based microstructures resulted in better circulation of medium around the liver tissue and mesothelial cells providing 3D microenvironment and to retain intact architecture. The flow rate was 20 µl/min and the flow direction is changed every 12hrs for effective removal of waste products and toxins. Tissues slices of 2mm thick have been cultured for 12 days with 76% viable cells.**

Keywords-**Microstructures; Bioreactor; Tissue Culture; Microfluidic Flow**

I. INTRODUCTION

It is extremely difficult maintaining tissue in vitro for longer duration that has been removed from the body and devoid of their normal in vivo vascular sources of nutrients and gas exchange. Microfluidic platforms that epitomize the physiological cellular microenvironment of the hepatic capillary bed, with perfusion flow that allows development of physiological oxygen gradients, may be helpful in understanding liver toxicity, disease, inflammation, and drug metabolism [1]. Variety of bioreactor technologies have been authenticated for the culture of engineered tissues, in a number of long-term tissue engineering studies, but failed to restore in vivo functions due to failure in delivering nutrients into the inner core of the tissue, resulting in cell death. Mass transfer limitation also affects the accuracy of drug transport studies because the drug penetration rate contributes to the drug uptake rate. Efforts have been made to improve mass transfer of in vitro liver slice culture, ranging from multiwell static culture to dynamic culture using rocker, roller, rotational culture methods and perfusion [2]. So far, none can maintain tissue slices of thickness beyond 200–300 mm for more than 24hrs due to mass transfer limitations. Novel intra-tissue perfusion system, for culturing thick liver tissue has been used in which culture medium is transported through hollow micro needles to reach core of liver tissues [3]. However, the cell viability was assed only for three days, which is not suitable for long-term drug testing applications. In addition, mechanical piercing would damage the cells around micro needles. In recent time, perfused multiwell plate has been used for 3D liver tissue culture by Domansky. K et al. [4]. However in this work, liver tissue has been cultured only for seven days with the culture medium being recirculated within the scaffold. Hydrogels, due to their relative biocompatibility, tissue-like water contents and tissue-like elasticity enable them to be a prime candidate for many tissue engineering applications [5]. Moreover, it can be eliminated from the body via liver and kidney and forms non toxic metabolite which makes it more suitable for tissue engineering applications. In the proposed paper, PEG-DA-based hydrogel microstructure pallets were fabricated on glass covers to provide 3D microenvironment within the bioreactor.

In the proposed paper, we describe the design and function of bioreactor suitable for long-term tissue culture consisting of interconnected bioreactor wells. Ease of use is achieved by designing the bioreactors as an array of PMMA microwells, enclosed with glass covers to provide closed microfluidic chamber. Microstructures of PEGDA fabricated on glass covers were cultured with a layer of mesothelial cells on its surface to provide favorable 3-D environment to the liver tissue in the bioreactor [6]. Via this liver tissue vitalizing system, in-vitro culture period of sliced liver tissues of 2 mm thickness could be extended up to twelve days to ensure long-term culture. Conventional static culture method is used as control group for comparison.

II. EXPERIMENTAL

A. Materials

The PEG-DA with a molecular weight of 575 was purchased from Sigma to be used as a precursor. 2-hydroxy-4'-(2-hydroxyethoxy)-2-methyl-1- propiophenone (Sigma-Aldrich) was used as a photo initiator. pentaerythritol tetraacrylate, PETA (MW 352.34, Sigma), phosphate-buffered saline, PBS (Applichem GmbH, Germany), RPMI 1640 supplemented with L-glutamine and containing 15% fetal calf serum (FCS), supplemented with insulin (Gibco), 2-mercaptoethanol (Sigma), 20 mM 4-(2-hydroxyethyl)-1-piperazineethanesulfonic acid (HEPES,Sigma), Hydrocortisone 400 [mu]g/l (Sigma) and antibiotics (benzylpenicillin 120 mg/l, gentamicin 4 mg/l, Amphotericin 2.5 mg/l; Sigma), Dulbecco's Modified Eagle Medium,DMEM (12800, Sigma).

B. Preparation of photo polymerizationsolution to fabricate microsructures

To generate PEG hydrogels, a solution containing poly(ethylene glycol)-diacrylate polymer, PEG-DA, (MW 575, Sigma) and pentaerythritol tetraacrylate, PETA (MW 352.34, Sigma) in PBS (pH 7.4) to yield final concentrations of 60% (w/w) was prepared prior to experiments in order to allow the PEG-DA to adequately dissolve into solution. Immediately prior to UV photo polymerization, photoinitiator solution was added to the prepolymer solution at 0.5 wt%.

C. Design and Simulation

There are eight square shaped bioreactor wells fluidically connected with each other with a cross sectional area of 2.5mm^2 with connecting microchannels of 300μm width. When the medium flow through the bioreactor well, the tissue will be held at the neck and the medium is forced to diffuse through the tissue. A commercial finite element software CFD-ACE+ (CFDRC, Huntsville, AL) was used to simulate the flow field. The CFD-ACE+ simulation software has wide range of application and has been used by our group to simulate microfluidic flow pattern and electric field distribution. The simulation of flow pattern for the medium flow is indicated by arrows in the Fig.1. It is evident from the Fig.1 that, the flow velocity is uniform in all the bioreactor wells. The flow direction was alternatively change for effective removal of toxins and other waste products.

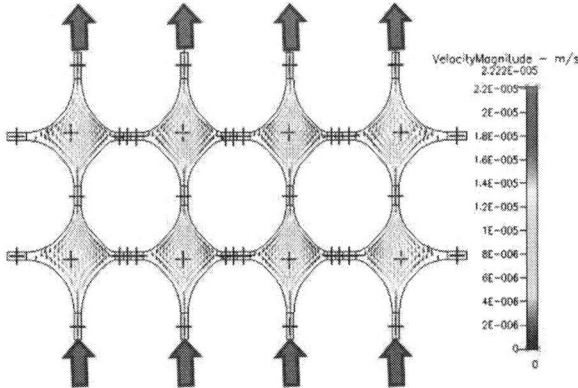

Fig.1. CFD-ACE+ simulation result elucidating fluid flow pattern through the bioreactor well.

D. Fabrication of bioreactor chip

The main body of the bioreactor was fabricated from Poly methyl methacrylate (PMMA) engraved with the square-shaped well array by using engraving machine (Roland EGX-400). The single unit of the bioreactor chip comprises of three parts; square-shaped well array, top glass cover and bottom glass cover as depicted in Fig. 2.

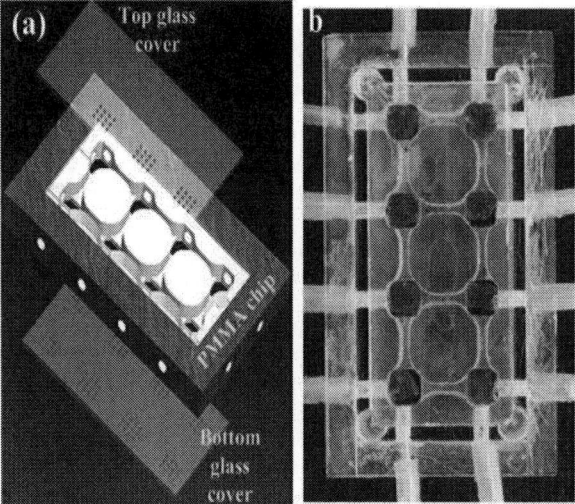

Fig. 2. (a) The PMMA microbioreactor chip with top and bottom glass covers fabricated with PEGDA structures. (b) Top view of the fabricated PMMA bioreactor.

After, mesothelial cells were cultured on bottom glass cover, it will be placed in position and sealed with PDMS to prevent any leakage and the liver tissue units are introduced into the bioreactor wells. Further, the top cover glass is accommodated to provide closed Microfluidic chamber and to avoid contamination. Finally the bioreactor chip is connected to the external fluidic system via silicone tubes. Further, fresh medium is supplied by using a peristaltic pump (Mp-1000, Eyelampe, Tokyo Rikakikai Co.Ltd) with a flow rate of 20 μl/min.

E. Fabrication of PEG-DA microstructures

Both the glass covers are fabricated with PEG-DA microstructures with the following procedure. A uniform layer of the prepared PEG-DA photo polymerization solution was smeared onto the glass. It was then exposed through a mask less photolithography machine (SF-100xpress, Intelligent Micro Patterning, LLC). Regions of PEG-DA exposed to UV underwent free-radical polymerization and became cross-linked, while unexposed regions were flushed out. The pallet array was rinsed with DI water and incubated in water for 5 minutes to dissolve remaining non-polymerized PEGDA. Fabrication method is well explained and the dimensions of each PEGDA microstructure are shown in Fig. 3.

978-1-4673-1122-9/12 $31.00 © 2012 IEEE

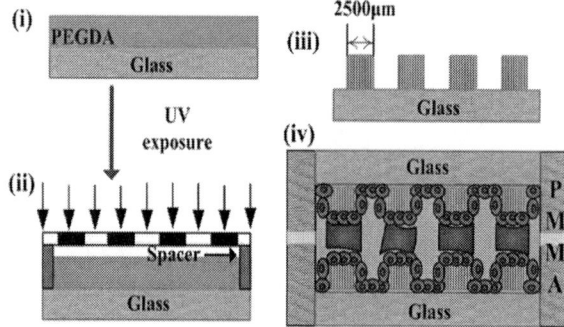

Fig. 3. Fabrication of 3D PEGDA structures. (i) PEGDA smeared onto surface-modified glass slide (ii) translucent photo mask with spacers in ben and exposed to UV light to obtain desired structures (iii) desired pattern is obtained (iv) cross sectional side view of bioreactor. Mesothelial cells are cultured on PEGDA prisms with tissue being sandwiched between them providing 3D environment.

III. RESULTS AND DISSUSION

A. HBsAg antigen expression

The biopsies from the transgenic mice after they are implanted with HBsAg gene were utilized for determining the concentration of HBsAg during twelve days of culture.

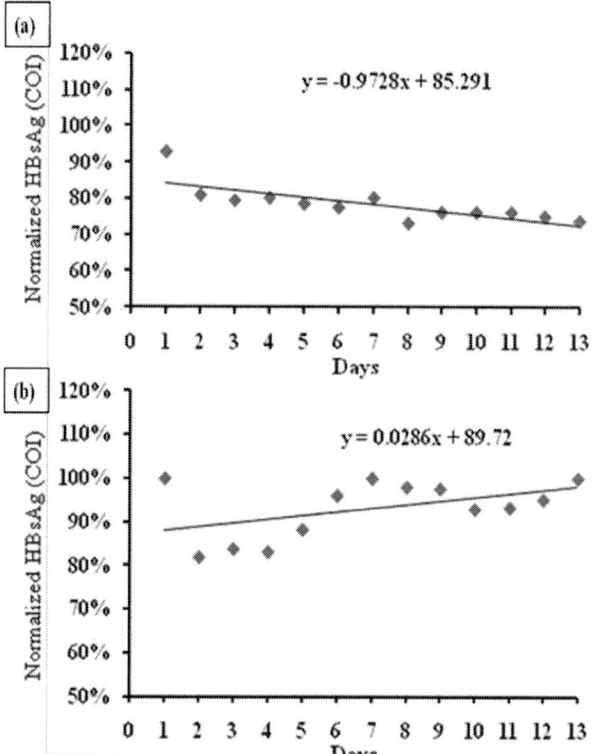

Fig. 4. HBsAg antigen expression. (a) Steep decline in HBsAg antigen concentration in the static culture (b) Gentle increased trend line slope with a better HBsAg expression in our bioreactor chip.

The supernatant medium is collected every 24 hours and assayed for HBsAg concentration. The trend line is plotted with the horizontal coordinate representing duration of tissue culture and vertical coordinate representing normalized concentration of HBsAg antigen. The antigen expression during the control group culture is plotted Fig. 4(a). This plot shows decreasing trend line with steeper slope. The decreasing trend line slope of HBsAg expression in the static culture system indicates the tissue degradation due to lack of nutrients and inadequate O2 and CO2 exchange. On the other hand the trend line is increasing for experimental group as represented in Fig. 4(b). It implies that the liver tissue was undergoing apoptosis at a very slow pace emphasizing the suitability of bioreactor chip for long term culture.

B. TUNEL assay

TUNEL is an established method for detecting DNA fragments and is a hallmark characteristic of apoptosis. This assay is based on incorporation of biotinylated nucleotides conjugated to bromodeoxyuridine (BrdU) at the 3' OH ends of the DNA fragments that form during apoptosis. This detection system utilizes a biotin conjugated anti-BrdU antibody and streptavidin-horseradish peroxidase. The cell viability was measured using ImageJ software (National Institute of Health, USA) by measuring cell population. The viability was nearly 95±1.5% in both the systems and not much difference was observed on first day (Fig. 5(a, b)) Next, cell viability was about 68±2.1%, 84±2.4%, on sixthth day in static culture and biorector system respectively (Fig. 5(c, d)). At the end of twelfth day, only a few cells survived with a vaiability of 15% in control group (Fig. 5(e)). On the other hand, the viablity of cells was measured to be 70% on twelfth day (Fig. 5(f)) in our bioreactor chip which is significant improvement when compared to control group. The result vindicates that more cells were viable at the core of the tissue supporting the fact that the nutrients were delivered even to the core of the tissue.

The tissue slices were cultured for up to 12 days. Tissues were isolated for assay on first, third, fifth, sixth, seventh, ninth and twelfth day. TUNEL assay was employed to know the distribution of live and apoptosis cells. After assaying the samples images were taken and were divided into three regions such that it inculcates 200 cells per region. Based on the number of cells viable (blue) and apoptosis cell (green), the viable cells were quantified and the graph has been plotted as represented in Fig. 6.

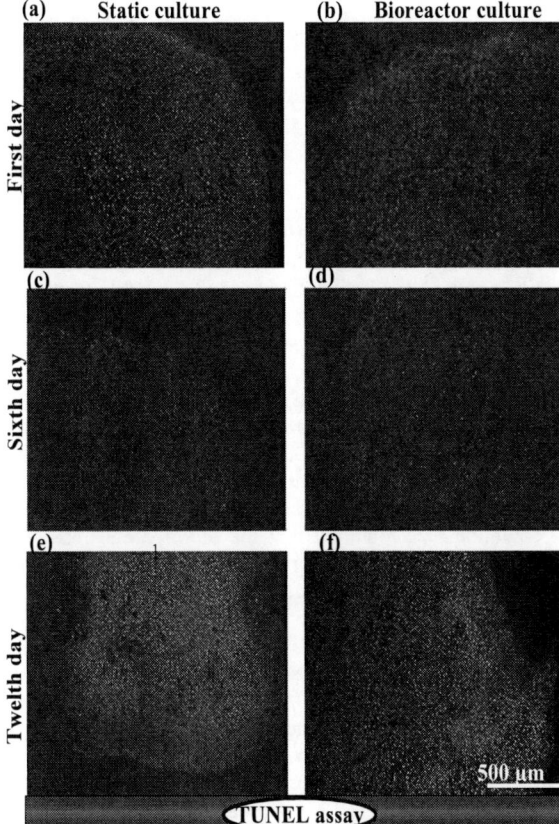

(a) Static culture **(b)** Bioreactor culture

First day

Sixth day

Twelth day

500 μm

TUNEL assay

Fig. 5: TUNEL assay indicating live (blue) and apoptosis (green) cells. (a,b) substantial viability observed in both the systems on the first day. (c) Viability reduced to 55% (d) viability was maintained at 82% (e) Apoptosis cells at the core indicating deterioration of liver tissue (f) considerable amount of live cells at the core indicating enhanced viability.

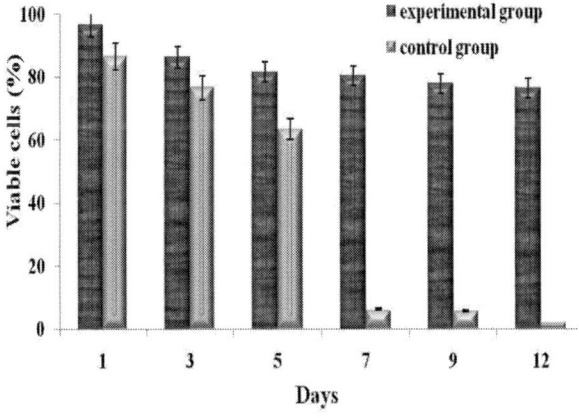

Fig. 6. The percentage of cells viable in control group and experimental group as measured by Image J software.

Many studies report that oxygen supply is one of the most important nutrients limiting tissue growth. Meanwhile, it is often the limiting nutrient in successful tissue growth in vitro. The reason for this arises from the difficulty of bringing sufficient amounts of oxygen to the surface of the cells mainly because of the poor solubility of oxygen in culture media [7]. To facilitate transfer of sufficient oxygen, the bioreactors were connected by gas-permeable silicon tubing to an external peristaltic pump that continuously supplies fresh cell culture medium between a reservoir and the micro-bioreactor. In addition the thin PDMS sealing provided at the top and bottom cover allow O_2 and CO_2 diffusion to oxiganate mesothelial cells and liver tissue.

IV. CONCLUSIONS

This bioreactor portrays the in vivo condition better, with enhanced viability and antigen expression when compared with the static group. This bioreactor has enabled us to culture tissue of more than 2mm size upto 12 days. The proposed bioreactor is an accomplishable and a good model system for in vitro studies of effects that drugs and therapeutic molecules have on liver-specific functions when cultured for a long period.

ACKNOWLEDGMENTS

This research was financially supported by National Science Council under grant NSC 98-2120-M-007-003. The authors thank the Semiconductor Research Center and National Nano Device Laboratory for the access.

REFERENCES

[1] W. R. Leeman, I. A. van de Gevel, A. A. J. J. L. Rutten , "Cytotoxicity of Retinoic Acid, Menadione and Aflatoxin B, in Rat Liver Slices Using Netwell Inserts as a New Culture System, " *Toxic. in Vitro*, vol. 9, pp.91-298, 1995.

[2] P. Olinga, K. Groen, I. H. Hof, R. De Kanter, H. J. Koster, W. R. Leeman, A.A.J.J.L. Rutten, K. V. Twillert, and G.M.M. Groothuis, "Comparison of Five Incubation Systems for Rat Liver Slices Using Functional and Viability Parameters", *Journal of Pharmacological and Toxicological Methods*, vol. 38, pp.59-69, 1997.

[3] Y. M. Khong, J. Zhang, S. Zhou, C.Cheung, K. Doberstein, V. Samper, H. Yu, "Novel Intra-Tissue Perfusion System for Culturing Thick Liver Tissue," *Tissue Engineering*, vol. 13, pp. 9, 2007.

[4] Domansky K, et al. Perfused multiwell plate for 3D liver tissue engineering. Lab on a Chip, vol 10,pp.51-58, 2010.

[5] Bryant SJ & Anseth KS (2001) The effects of scaffold thickness on tissue engineered cartilage in photocrosslinked poly(ethylene oxide) hydrogels. *Biomaterials* 22(6):619-626.

[6] S. E. Mutsaers, "Mesothelial Cells: Their Structure, Function And Role In Serosal Repair," *Respirology* 7, pp. 171–191, 2002.

[7] P. Eiselt, B.S. Kim, B. Chacko, B. Isenberg, M.C. Peters, K.G. Greene, W.D. Roland, A.B. Loebsack, K.J.L. Burg, C. Culberson, C.R. Halberstadt, W.D. Holder and D.J. Mooney, "Development of Technologies Aiding Large-tissue Engineering," *Biotechnol. Prog*, vol. 14, 134, 1998.

Experimental Investigation of Bulk Response of Cells on Optoelectronic Dielectrophoresis Chip

Srinivasu Valagerahally Puttaswamy[1*], Shih-Mo Yang[2], Shilpa Sivashankar[1], Kuo-Wei Chang[1], Long Hsu[2] and Cheng-Hsien Liu[1]

[1*]Microsystems and Control Laboratory, National Tsing-Hua University, Hsinchu, Taiwan, R.O.C
[2]Department of Electrophysics, National Chiao-Tung University, Taiwan, R.O.C.
*sri_shilu2007@yahoo.co.in

Abstract—The mechanism of the particle interactions is still inconclusive, even though many researchers have tried to figure out the particle-particle interactions under the influence of electric field. In this paper we report an experimental investigation of electrostatic attraction and repulsion of endothelial cells in an image-driven light induced dielectrophoresis (DEP) apparatus. The optoelectronic tweezers (OET) device consists of a planar structure having one or more portions which are photoconductive to convert incoming light to a change in the electric field pattern. When the cells are manipulated in OET chip at a frequency near the threshold of electrolysis, cells exhibits numerous interesting phenomena such as spinning, collision, rotation and pearl chain effect in three-dimensional (3D) space. All of these cell behaviors at such critical frequencies on this OET chip are reported.

Keywords-Dielectrophoresis, optoelectronic tweezers, cell behavior, pearl chain effect

I. INTRODUCTION

The expedition of the use of electric fields for measuring the cell property has a long history. The ability to analyze particles in the micro and nano-scale stage has favored many research fields. Biologists have conventionally gone into cell behavior by observing the bulk response of group of cells during their manipulation for biological and chemical applications. Biological cell manipulation can be achieved by employing variety of forces such as, mechanical, magnetic, fluidic, optical and electrokinetic forces. Wide varieties of devices such as micro grippers [1] and atomic-force microscope tips [2] have been used to control these forces. Non-contact methods are promoted by the biologists to prevent contamination and damages to bio-samples. Magnetic force is also considered suitable however; it requires attachment of particles to magnetic beads. The application of microfluidic device as another non-contact method is limited as it is tough to target a specific single cell even if they are suitable for sorting composite cell populations.

Opto-electronic tweezers is a prominent technology for bio-particle manipulation using dynamic optical images. The manipulations of both single and multiple cells based on novel optoelectronic tweezers (OET) with the operation AC frequency have been first reported and successfully demonstrated by Chiou et al. [3]. OET manipulation has been used for manipulating variety of microparticles, such as polystyrene beads, E. coli bacteria, red and white blood cells, HeLa cells, Jurkat cells, yeast cells, and protozoa. Moreover, OET is also capable of manipulating semiconducting and metallic nanowires [4]. OET has number of advantages over conventional optical tweezers, in particular the ability to perform operations in parallel and over a large area without damaging living cells. Unlike electrode-based DEP, OET is capable of trapping a specific single particle from a larger population. We highlight that the reported cell manipulation technique with low AC operation frequency (less than 100 kHz) is different from the previously reported operation modes such as positive OET, negative OET [5] and electrorotation [6]. On the whole there has been relatively little publication about the cell at this critical frequency of threshold until recently.

The mechanism of the particle interactions is still inconclusive, even though many researchers have tried to figure out the particle-particle interactions under the influence of electric field [7]. Meanwhile, some researchers have theoretically investigated the relative motion between two spherical particles in electrophoresis, based on the broadened model of dipole-dipole interaction mechanism [8]. On the other hand, Kadaksham et al. have reported about the dynamic behavior of electrorheological suspensions subjected to non-uniform electric field [9]. They have also considered the DEP force which acts on the particles under non-uniform electric field as well as the electrostatic particle-particle interactions which acts even when the electric field is uniform. In the proposed paper, we utilized the OET device to experimentally investigate the electrostatic interactions between cells. By using the OET device, we could manipulate cells to demonstrate the cell-cell interactions and observe the different cell behaviors as well. We could observe both attractive and repulsive cell motions using the proposed experimental setup.

II. EXPERIMENTAL

A. Fabrication of the device and System setup

The schematic of the system setup is shown in Fig. 1. By using the plasma-enhanced chemical-vapor-deposition (PECVD) method in a foundry, a photoconductive electrode is deposited with 50 nm n^+ a-Si:H and 1μm intrinsic a-Si:H in sequence on a transparent indium tin oxide (ITO) electrode. The a-Si:H at the OET chip edge is removed by KOH etching to expose the bottom ITO layer to establish contact with

978-1-4673-1122-9/12 $31.00 © 2012 IEEE

electric wire. The cells were re-suspended in DEP-manipulating medium which are sandwiched between an upper transparent ITO electrode glass and lower photoconductive electrode substrate with a spacer of 150μm thickness. The programmable light pattern is illuminated from a projector, through 10 X optical lenses and focused on the OET chip.

In the dark areas which are not illuminated, the undoped a-Si:H layer has higher electrical impedance than the liquid medium and the majority of the applied AC voltage drops across the photoconductive layer. When the light illuminated on the specific area of the chip, the photoconductivity magnitude of the undoped a-Si:H layer increases by several orders. The virtual electrode having programmable pattern, formed on the a-Si:H layer was used for DEP manipulation. The process of cell motion and manipulation process is observed and analyzed by the 40X objective lens and the charge-coupled device (CCD).

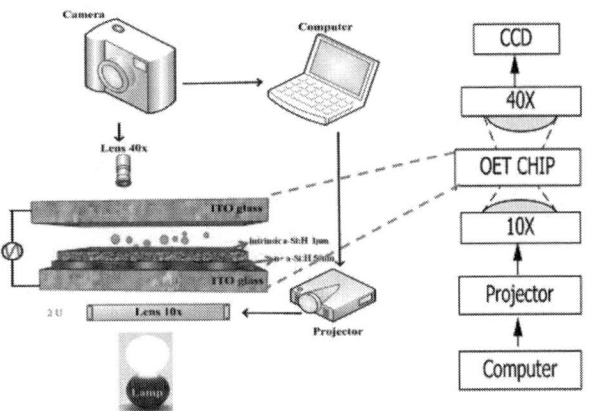

Fig. 1. The optical system setup and the cross-sectional view of the OET chip

A commercial finite element software CFD-ACE+ (CFDRC, Huntsville, AL) was used to simulate the electric field. The CFD-ACE+ simulation software has wide range of application and has been used by our group to simulate electric field distribution and microfluidic flow pattern. The simulation of electrorotation effect, which rotates beads/cells, is indicated by arrows in the Fig.2.

B. Cell preparation

The DEP phenomenon occurs when high conductivity cells which have numerous ions inside the membrane are exposed to an environment of low conductivity. Therefore, the cells for OET experiment are re-suspended in the low conductivity Dulbecco's modified Eagle medium (DMEM) medium composing of sucrose and dextrose with fairly increased 10% of calcium chloride with 1% of DMEM salt concentration. Three types of cells such as, HUVECs (human umbilical vein endothelial cells), HepG2 (Human hepatocellular liver

carcinoma cell line) and HEK293T (human embryonic kidney) are used during the experiment. However, there were no much differences observed in their behaviors.

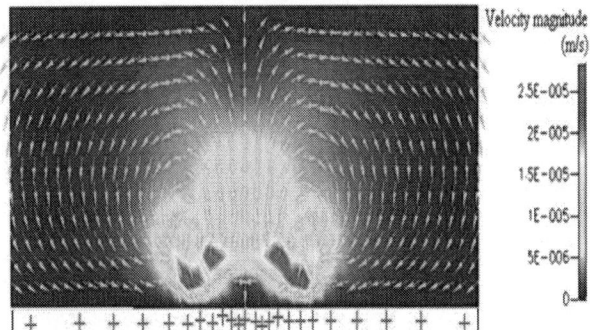

Fig. 2. CFD-ACE+ simulation result elucidating electro rotation effect

III. RESULTS AND DISSUSION

A. Spinning of cells

The fact that the small particles can be made to spin when placed in rotating electric field has been reported by many research groups [10].During spinning process, rotational axis is invariably perpendicular to enforced electric field and this modus operandi has been used as a non-invasive tool to oversee the physiological state of cells. During electrokinetic manipulation when the cells come close to each other, inter mutual interactions between them formulate a rotating field component resulting in electric torque on each cell. When the cells are neither parallel nor perpendicular alignment with the linearly polarized electric field display no concurring rotation. However, the cells tend to rotate when they are in any other alignment if the electric field frequency is within certain narrow bands. Electrorotation is the principle tool used for the dielectric characterizations. The primitive cell properties like, the membrane capacitance, conductance, as well to observe changes in these properties when cells are administered to various treatments, can be assessed by analyzing the rotational spectra

The model of transient processes in double cell configuration has been established and the behavior could be observed on the OET chip in our experiment. When the frequency of external electric field was in the range of 25to50 kHz, the cell spun itself driven by the induced static charges on the cell membrane. This phenomenon is more apparent when two or more cells are alongside each other and collide each other at the same time. The series of pictures were captured from the video as represented in Fig. 3 to reveal the behavior of cells during different time intervals. The experimental results proved the influences of size of the cell on their rotational speed. The angular velocity of smaller cell is higher compared with the larger ones during OET manipulation.

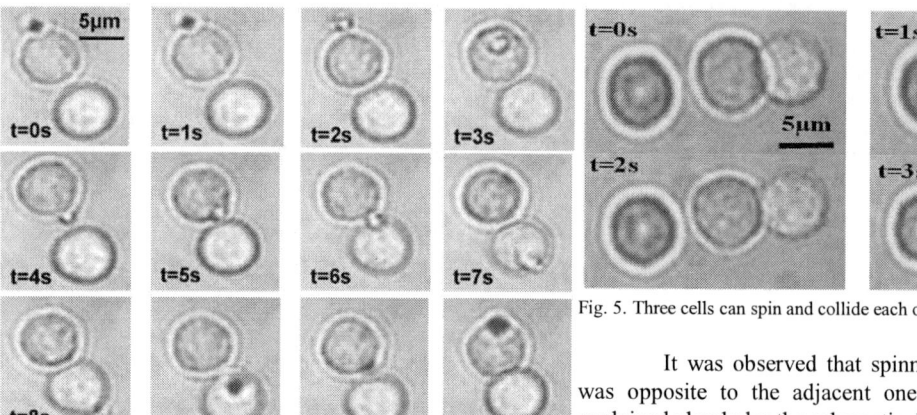

Fig. 3. The particle attracted on cell membrane by surface static charge could move from one cell to the other with the cell spin on the contrary direction at 40 kHz frequency

B. Rotation of Cell Wheel

The phenomenon of cell wheel rotation was observed when the applied AC frequency is increased to either 60 kHz or 70 kHz at which the density of induced static charges on the cell membrane increases significantly. High-density static charges results in stronger OET force causing number of cells to assemble and rotate together as a wheel as represented in Fig. 4. Initially the cells are spread and in a due course of time they gather together to form a wheel structure. It was observed that the more number of cells can be grouped to form a wheel structure and causing them to rotate at different applied voltage and frequency.

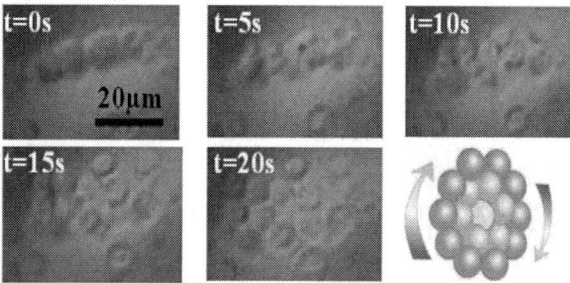

Fig. 4. About 17 cells, assemble and rotate as a disk, are collapsed and spread when the applied electric field was turned off.

C. Collisions of cells

In biomedical research cell-cell electro fusion is considered as an important approach. This modus operandi has wide range of applications such as, production of monoclonal, transfer of membrane components and production of hybrid cells. The high electro fusion efficiency can be obtained by achieving close physical contact between cells. Fig. 5. Represents the spinning and collision phenomenon of three adjacent cells during OET manipulation.

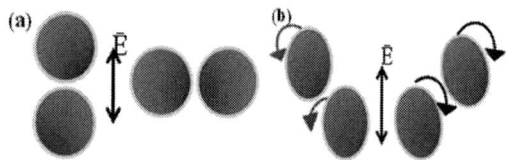

Fig. 5. Three cells can spin and collide each other at 40 kHz frequency.

It was observed that spinning direction of each cell was opposite to the adjacent one and the phenomenon is explained clearly by the schematic diagram in Fig. 6. If one of three cells leaves themselves, it would be more difficult for individual cell to spin, indicating that the external electric field through the neighbor cells would have comprehensive influence during the rotational process. The results show that the angular velocity of cell is highly pertinent to the applied frequency and voltage of the external electric field as evident from Fig.7.

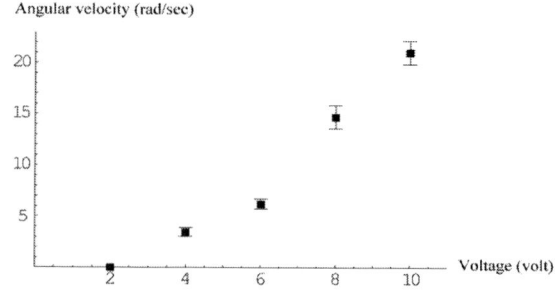

Fig. 6. (a) Particles in parallel or perpendicular alignment do not rotate (b) misaligned particles tend to rotate in the directions shown due to mutual interactions.

Fig.7. The relationship between peak-to- peak value of supplied voltage and the angular velocity of the rotating cell measured from experiments.

D. Cell 3D Pearl Chain Effect

Muth in 1927 has made detailed observation of pearl chain phenomenon using insoluble fat particles suspended in water. The same phenomena can also be observed when the biological cells are subjected to non-uniform electric filed suspended in media. The pearl-chain effect, a unique phenomenon of dielectrophoresis has been demonstrated by using fixed electrodes in two-dimensional plane [11].

According to DEP theory, the stronger electric field will be at the opposite poles of the cell along the external electric field and more static charges are accumulated at these two position. When a large number of cells are subjected to DEP, one cell would connect to neighbor cell and continue to form a pearl chain in three dimensional spaces. Due to dipole interactions the two cells repels each other when aligned end to end and attract each other in side by side orientation. The attractive force between the cells induces pearl chain formation.

In our OET chip, this phenomenon occurs at a frequency of 100 kHz. The process of cell connecting with the neighboring one during vertical DEP manipulation is illustrated in Fig. 8. The cells are classified as two groups and are marked by circle and square. Because the depth of the field is fixed, the clear cell images which are marked with square is the pearl chain connected from the bottom to the top and the blurred ones enclosed by circle are hung at the top ITO glass. The above phenomenon was observed during the interval of 21 to 27 seconds.

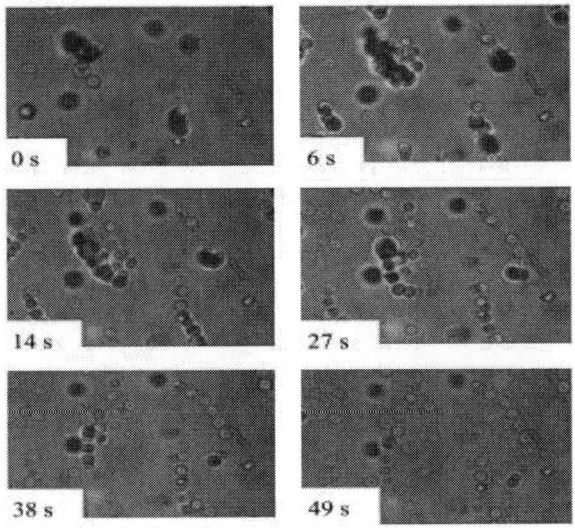

Figure 8: The process of cell pearl chain.

There are seventeen cells connected in this pearl chain and the schematic diagram is shown in Fig. 9. The gap between upper ITO glass and lower a-Si:H substrate was 160 μm based on the number and size of the connected cells.

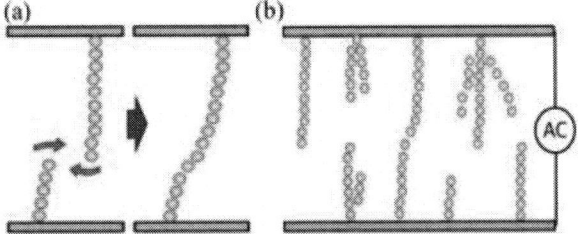

Figure 9: (a) the process of two pearl chains connecting together. (b) The cell in 3D forming different sorts of pearl chain shapes

However the pearl chains established by the external electric field would collapse when the applied field is turned off.

IV. CONCLUSIONS

In the proposed study, we applied the OET device to investigate the electrokinetic interactions between the suspended cells. It was evident that the OET, which is a novel manipulation technology using the light-induced electrokinetic mechanisms, can be a useful tool to study cell-cell interactions. Meanwhile, we have prosperously analyzed the phenomenon such as spinning, collision, rotation and pearl chain effect of cells.

ACKNOWLEDGMENTS

This research was financially supported by National Science Council under grant NSC 98-2120-M-007-003. The authors thank the Semiconductor Research Center and National Nano Device Laboratory for the access.

REFERENCES

[1] N. Chronis, L. P. Lee, "Electrothermally activated SU-8 microgripper for single cell manipulation in solution," Journal of Microelectromechanical Systems, vol.14, pp. 857-863, 2005

[2] P. Avouris, T. Hertel, R. Martel, T. Schmidt, H. R. Shea, R. E. Walkup, "Carbon nanotubes: nanomechanics, manipulation, and electronic devices," Applied Surface Science, vol. 141, pp. 201-209, 1999.

[3] P.Y. Chiou, "Cell addressing and trapping using novel optoelectronic tweezers," IEEE 17th Annual International Conference on Micro Electro Mechanical Systems (MEMS), pp. 21-24, 2004.

[4] K.Y. Kim, Recent Optical and Photonic Technologies, INTECH, Croatia, pp. 450, January 2010.

[5] P.Y. Chiou, "Massively parallel manipulation of single cells and microparticles using optical images," *Nature*, pp. 436, 2005.

[6] T.B. Jones, "Basic theory of dielectrophoresis and electrorotation," *IEEE Engineering in Medicine and Biology Magazine*, 22, 6, pp. 33-42, 2003.

[7] L. D. Reed, F. A. Morrison, Jr. "Hydrodynamic interactions in electrophoresis", J. Colloid Interface Sci. vol.54, pp.117–133, . 1976,

[8] O. Shramko, V. Shilov, T. Simonova, "Polarization interaction of disperse particles with not thin Debye atmosphere", Colloids Surf., A vol.140,pp.385–393, 1998, .

[9] J. Kadaksham, P. Singh, N. Aubry," Dynamics of Electrorheological Suspensions Subjected to Spatially Nonuniform Electric Fields", J. Fluids Eng. Vol.126, pp.170–179, 2004,

[10] X.B. Wang, Y. Huang, P.R.C. Gascoyne, F.F. Becker, R. H"olzel, R. Pethig, "Changes in friend murine erythroleukemia cell membranes during induced differentiation determined by electrorotation," Biochim.Biophys. Acta, 1193, pp. 330–44,1994.

[11] J. Kadaksham, P. Singh, N. Aubry, "Dynamics of Electrorheological Suspensions Subjected to Spatially Nonuniform Electric Fields," Vol. 126, pp.170-170, 2004

PEGDA-based Photocrosslinking Platform for Real Time Cell Trapping

Ling-Yi Ke[1,*], Zong-Keng Kuo[2], Yu-Shih Chen[3], Hsiang-Wen Tseng[2] and Cheng-Hsien Liu[1,3]

[1] Department of Power Mechanical Engineering, National Tsing Hua University, Hsinchu, Taiwan, R.O.C
[2] Biomedical Engineering Research Lab, Industrial Technology Research Institute, Hsinchu, Taiwan, R.O.C
[3] Institute of NanoEngineering and MicroSystems, National Tsing Hua University , Hsinchu, Taiwan, R.O.C
*lingyi0412@gmail.com

Abstract—**Poly (ethylene glycol)-diacrylate (PEG-DA) is a commonly used material in tissue engineering. It is cross-linking via UV-initiated photopolymerization to form the PEG-DA microstructures. This paper reports the fabrication of capillary PEG-DA architectures on photopolymerizable hydrogel characteristic. In the UV-initiated photopolymerization, oxygen can damage the photopolymerization reaction that to from microinterstices. In PDMS-based chip, a little oxygen in PDMS material is stored in PEG-DA material. By the way, we designed PEGDA-based V-shape microbarriers to trapped HMEC-1 cell for another biotechnology. For V-shaped microbarriers, we used the solution of the PEG-DA had been exposed to form V-shaped barriers by mask pattern. Cell trapping array platform was used for understanding fundamental cell studies which are cell-cell interactions and cell responses, such as hydrodynamic trapping system.**

Keywords-hydrogel, PEG-DA, photopolymerization, cell trapping, cell-cell interaction

I. INTRODUCTION

In recent study, the PEG hydrogels were changed an increasingly popular material for tissue engineering because of their ability to absorb water, mechanical properties and biocompatibility. Using photopolymerizable PEG hydrogels, a multilayer photopatterning platform were fabricated for embedding cells in hydrogels [1]. In our group, we had developed a double trapped single cell chip which was designed to the movable poly (-ethylene glycol) diacrylate (PEG-DA) microstructures, the novel geometry double-railed micro-fillister by positive dielectrophoresis (DEP) and microfluidic force [2]. Using photopolymerizable PEG hydrogels, a multilayer photopatterning platform was fabricated for embedding cells in hydrogels [3-4]. PEG-DA hydrogels with varying polymer which in studies of mechanisms of diseases such as cancer and atherosclerosis, in which changes in tissue stiffness may inform cell behavior. Produced three-dimensional hepatic tissues by photopatterning of poly (ethylene glycol) (PEG) hydrogels containing cells. The researchers incorporated cell-adhesive peptides, representing specific ECM proteins, in the hydrogels to support hepatocyte survival [5]. From the same research group, Underhill et al. studied the function of liver cells embedded in PEG hydrogels and found. And the albumin secretion of mouse embryonic liver cells was significantly influenced by the peptide sequence incorporated in the PEG hydrogel [6].

Further, A three dimension lobule-mimetic trapping chip is observed depend on the hydrogel-based geometry microstructure and the novel cell manipulation method [7]. By the way, we had developed a concentric-stellate-tip electrodes pattern to trap hepatic and endothelial cells for biomimetic heterogeneous [8].

In this paper, we demonstrate V-shaped microbarriers for cell trapping real time. V-shaped microbarriers were formed by UV-initiated photopolymerization. In V-shaped microbarriers array, the purposes of this research are to develop the engineered cell trapping regulated not only by cell mass culture but also by drug testing.

II. METHODS

A. Photopolymerization

In recent biotechnology, researcher had demonstrated a potopolymerizable biomaterials had been developed and could be used to photoencapsulate cells in peptide-derivatized PEG-DA networks. In our design, we shows schematic diagram of the photopolymerization process in figure 1. In figure 1(a) shows the unit mask photo pattern of the V-shaped microbarrier. The exposure of cross section views shows oxygen damage the photopolymerization reaction that to from microinterstices by UV light. The photopolymerization has

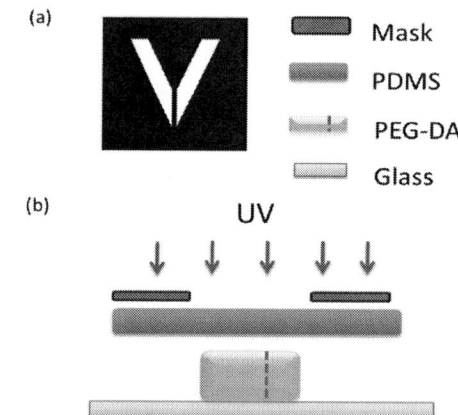

Figure 1: Schematic diagram is the photopolymerization process. (a)Mask photo pattern. (b)The PEG-DA solution was formed by UV light. Oxygen can damage the photopolymerization reaction that to from microinterstices.

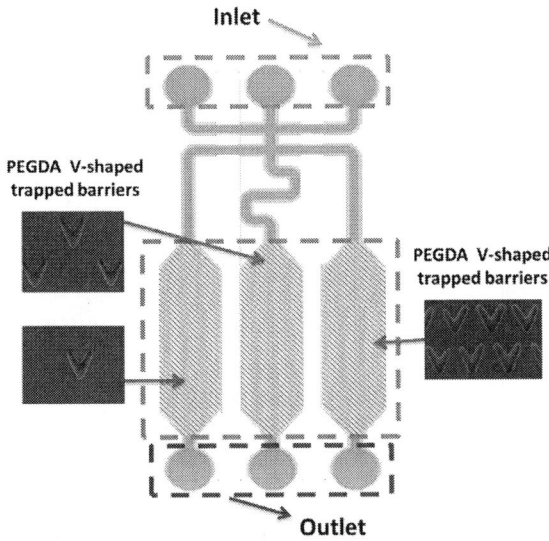

Figure 2: Schematic diagram is the top view of the PEGDA-based Photocrosslinking Platform. It was consisted of hydrogel V-shaped barrier chambers. For V-shaped barriers, we used the solution of the PEG-DA had been exposed to form V-shaped barriers array by mask pattern.

been used to form PEG-DA microstructures and to form trapping platform. At this fabrication, the 365 nm ultraviolet (UV) light was continuously radiated on maskless exposure system.

B. Chip design

An ideal material for using as a structure would be biocompatibility, be adhered cell by collagen, and capable of supporting long-term cell culture. In this study, we designed PEG-DA-based photocrosslinking platform. It was consisted of hydrogel V-shaped microbarriers chamber. In three V-shaped microbarriers chambers, we were used UV-initiated photo-polymerization to form PEG-DA microstructure to achieve trapping features.

We show schematic diagram which is the top view of the PEGDA-based Photocrosslinking Platform as shown in figure 2. There are three chambers in this platform. We designed difference proportional number about V-shaped microbarriers in these three chambers for real time cell trapping. Firstly, with regard to flow in microfluidic chambers are continuous. Secondly, cells were trapped by difference proportional at the same time in this microfluidic function.

C. PEG-DA V-shaped microbarriers

We were used UV-initiated photopolymerization to form hydrogel microstructure to achieve trapping features. When PEG-DA was exposed by maskless UV exposure, we observed the size of the microinterstices is 2.4 micrometer smaller than the size of the cells. For V-shaped barriers, we used the solution of the PEG-DA had been exposed to form V-shaped

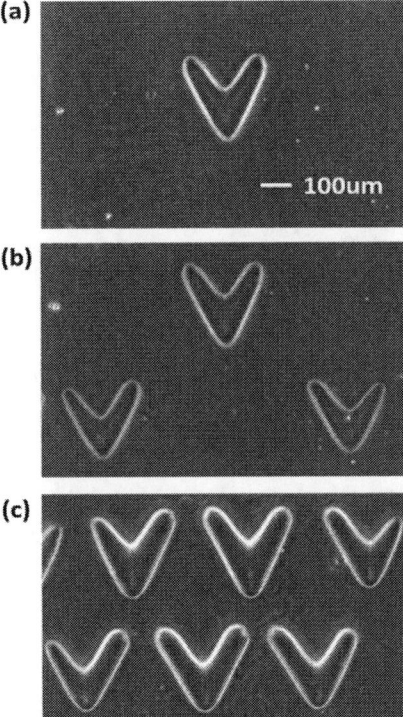

Figure 3: The microscope photo shows difference quantity in the same area. (a)Two V-shaped barriers in 2.56mm × 2.88mm. (b)Six V-shaped barriers in the same area. (c) Twelve V-shaped barriers for trapping. The scale bar is shown 100um.

barriers by mask pattern. The microscope photo shows difference quantity in the same area (2.56mm ×2.88mm) in figure (3a~3c). The scale bar is shown 100um. PEG-DA V-shaped barriers after exposing and swelling architectures photo of the V-shaped barriers. We calculated PEG-DA V-shaped microbarriers assay to support different cell count trapping.

III. RESULTS AND DISCUSSION

We demonstrated the ability to perform cell trapping by PEG-DA V-shaped microbarriers assays. Using PEG-DA hydrogels, we tailored the solution and architecture of the hydrogels to support the further cell trapping and signal transduction. In this research, we had developed the microstructures by conventional PEGDA-based process for producing biocompatibility microbarriers. In the UV-initiated photopolymerization, oxygen can damage the photo-polymerization reaction that to from microinterstices. Hence, there is need of PEG-DA to pay an important factor in microfluidic force which is used for cell sorting and trapping with the microinterstices of the PEG-DA V-shaped micro-barriers. The photo shows the cells successfully into the PEG-

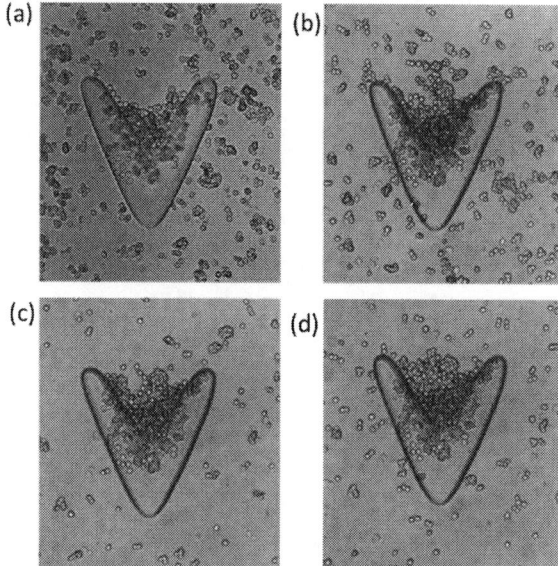

Figure 4: The continuous diagram shows the cells into the PEG-DA microbarriers. The cell trapping was shown the cell number in the different time at the same area. The flow direction in these photos was from the upper to the down.

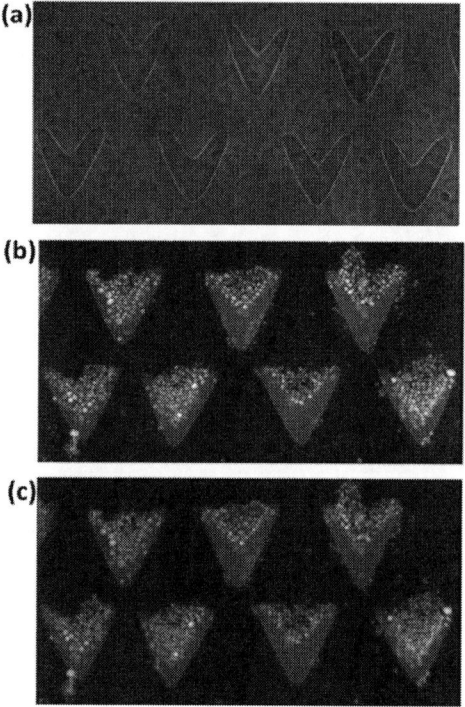

Figure 5: The microscope photo shows difference quantity in the same area. (a) PEG-DA V-shaped microbarriers after exposing and swelling architectures photo of the V-shaped barriers. (b) These photo shows barriers trapping HMEC-1cells at the same time. (c) The photo show fluorescence of PEG-DA microbarriers used FITC dye (green) and cells used CMFDA dye (green).

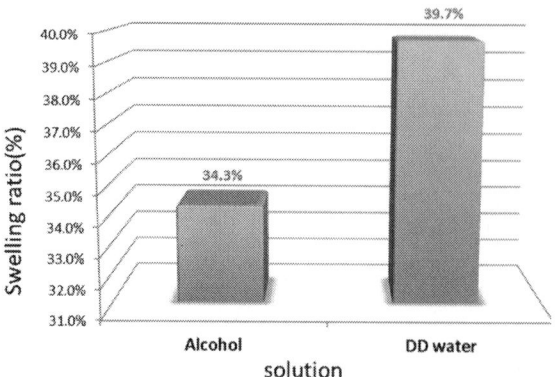

Figure 6: The photo shows the dependence of swelling ratio on the solution of PEG-DA microbarriers. The swelling of the solution affected the PEG-DA of the microbarriers.

DA microbarriers as shown in figure 4. The continuous diagram shows the cells into the PEG-DA microbarriers. The cell trapping was shown the cell number in the different time at the same area. The flow direction in these photos was from the upper to the down. We can see a series continuous diagram to represent the effect of PEG-DA microbarriers. This integrated assembly allowed control of the direction of flow of medium across the cultured cells.

In this paper, a simple PEG-DA microbarriers fabrication via UV photolithography has been proposed which can be integrated with microstructures for microfluidic systems. Figure 5(a) shows the photos of the PEG-DA V-shaped barriers after exposing and swelling architectures of the V-shaped barriers. We could successful to control the size of PEG-DA microbarriers. The photo shows barriers trapping cells at the same time in Figure 5(b). Figure 5(c) shows the fluorescence photo of HMEC-1cells were loaded by PEG-DA microbarriers with by CMFDA dye (green).

We were exposure PEGDA to guide the microstructures through different solutions. In figure 6, this photo shows the dependence of swelling ratio on the washing solution of PEG-DA microbarriers. There were many factors that influenced the swelling ratio of PEG-DA microbarriers (e.g. temperature, time, solution, etc.). The value of DD water was significantly greater than alcohol. The swelling ratio of DD water was higher than alcohol which values is about 5%.

$$Swelling\ ratio(\%) = \frac{W_s - W_d}{W_d} \times 100$$

W_s : weight of swollen PEG-DA

W_d : weight of exposed PEG-DA

About the different number microbarriers of the V-shaped chambers, we could achieve trapping different cells count in the continuous microfluidic at the same chamber. By our V-shaped microbarriers design in figure 7, we calculated the quantity of cells in barriers to remove by trypsin. Trypsin is used to cleave proteins bonding the cultured cells. In three V-shaped barriers, we set 36, 108 and 216 microbarriers. We respective got 12248, 36724 and 73443

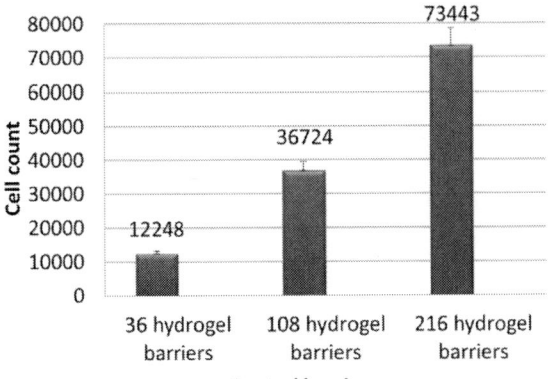

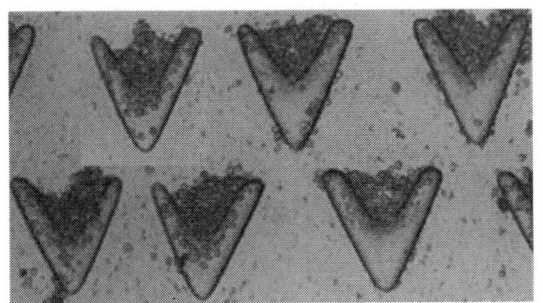

Figure 7: By our V-shaped microbarriers design, we calculated the quantity of cells in barriers to remove by trypsin. Trypsin is used to cleave proteins bonding the cultured cells. In three V-shaped barriers, we set 36, 108 and 216 microbarriers. We respective got 12248, 36724 and 73443 cells. And then one V-shaped barrier about has 340 cells.

Figure 8: Experiment photos of seven V-shaped barriers trapping cell culture on the third day.

cells. Finally, we calculated a barrier with an average of 340 cells.

Regarding the long-term for V-shaped microbarriers trapping cells for the catalysts, our results show that cell culture on the third day as shown in figure 8. Among these cell viability and culture factors, the most important one is the trapping structures which signal-communicated, bio-compatibility, their ability to absorb water performs microfluidic in tissue engineering.

IV.   CONCLUSION

Presently, we successfully had developed cells were loaded by PEG-DA-based V-shape microbarriers via microinterstices which were based on special photopolymerizable hydrogel characteristic for the enhancement of the liver function. The cell viability and culture in the PEG-DA-based photo-crosslinking platform that the result can prove to PEG-DA microbarriers is a novel method for tissue engineering.

ACKNOWLEDGMENT

This work was supported partially by Cheng-Hsien Liu of Department of Power Mechanical Engineering, National Tsing Hua University of R.O.C. under the grant NSC 99-2l20-M-007-00 1. We also would like to express our appreciation to Zong-Keng Kuo and Hsiang-Wen Tseng of Biomedical Engineering Research Lab, Industrial Technology Research Institute.

REFERENCES

[1]   L. G. Griffith, M. A. Swartz, "Capturing complex 3d tissue physiology in vitro", *Nature Reviews Molecular Cell Biology*, vol. 7, pp221-224, 2006.

[2]   L. Y. Ke, Y.S. Chen, Z.K. Kuo, C. H. Liu, "A double trapped single cell contact and interaction system via movable poly (ethylene glycol) diacrylate (peg-da) microstructure for immune analysis ", *Transducer 2011*, pp.302-305.

[3]   V. L. Tsang, A. A. Chen, L. M. Cho, K. D. Jadin, R. L. Sah,   S. DeLong, J. L. West, and S. N. Bhatia, "Fabrication   of 3D   hepatic   tissues   by   additive photopatterning of cellular hydrogels", *The FASEB Journal*, Vol. 21, pp.790-801, 2007.

[4]   Jennifer Elisseeff, "Hydrogels : Structure starts to gel", *Nature Materials*, 7, PP271-273, 2008.

[5]   Baker, M., "A living system on a chip". *Nature*, 2011. 471(7340): p. 661-665.

[6]   Tsang, V.L., et al., "Fabrication of 3D hepatic tissues by additive photopatterning of cellular hydrogels", *Faseb Journal*, 2007. 21(3): p. 790-801.

[7]   Y. S. Chen, L. Y. Ke, C. H. Liu, "3D lobule-mimetic chip via positive dielectrophoresis   force with sinusoidal spacing poly (ethylene glycol)-diacrylate microwalls", in Digest Tech. *Papers Transducers'11 Conference*, 2011, pp. 1837-1840.

[8]   C. T. Ho, R. Z. Lin, W. Y. Chang, H. Y. Chang, C. H. Liu, "Rapid heterogeneous liver-cell on-chip patterning   via   the enhanced field-induced dielectrophoresis traps", *Lab on a Chip*, vol.   6, pp.724-734, 2006.

Unified Theory to Evaluate the Effect of Concentration Difference and Peclet number on Electroosmotic Mobility Error of Micro Electroosmotic Flow

Wentao Wang[1] and Yi-Kuen Lee[1,2*]

[1*]Dept of Mechanical Engineering, HKUST, Kowloon, Hong Kong
[2]Joint KAUST-HKUST Micro/Nanofluidics Laboratory, HKUST, Kowloon, Hong Kong
Tel:+852 2358-8663, Fax:+852 2358-1543, E-mail: meyklee@ust.hk

Abstract—**Both theoretical analysis and nonlinear 2D numerical simulations are used to study the concentration difference and Peclet number effect on the measurement error of electroosmotic mobility in microchannels. We propose a compact analytical model for this error as a function of normalized concentration difference and Peclet number in micro electroosmotic flow. The analytical predictions of the errors are consistent with the numerical simulations.**

Keywords—*EOF, DNA, mobility, Peclet number, Current-monitor method.*

I. INTRODUCTION

Electroosmotic flow (EOF), as one of the important methods to electrically manipulate fluids in micro/nanofluidics, has been used in microchip electrophoresis for biochemical analysis [1], electroosmotic flow-based micro pumps [1] and nanoporous pumps [2-4], EOF-based micro-mixer with patterned surface charge [5-8] et al . Surface properties are the key features for controlling the electroosmotic flow [9]. Therefore, characterization of surface properties in micro/nano fluidics is one of important research areas for the applications of micro/nano fluidics. The accurate determination of electroosmotic mobilities, μ_{EO}, of biofluids, such as DNA or protein mixtures in microchannels is also useful for quality control of microfluidics. Current-monitoring method (CMM) has been widely used to measure average μ_{EO} ($=L/Et_0$) in capillary electrophoresis and microfluidic chips [10], where E is the electric filed, L the effective channel and t_0 the elapsed time to completely replace the low-concentration (c_L) solution by the high-concentration (c_H) counterpart.

To have a deeper understanding of CMM's replacement process, Ren *et al.* proposed two models to predict the interface position in a capillary [11]: one model assumes the presence of sharp interface and the other assumes a mixing zone between the two solution. Wang *et al.* also developed a similar model of the current response in rectangular microchannels [12] by solving the governing equations for flow and concentration profile. The electric current versus time was predicted based on the concentration profiles. However, these models can only be applied to the low concentration difference $\Delta c/c$ ($<<1$) cases. When the concentration difference Δc increases, $\Delta\mu_{EO}$ in the both regions would build up the pressure gradient (Δp) along the capillary as shown in Fig. 1(b) [12]. Mampallil *et al.* developed an implicit 1D model to predict the current response by assuming the pressure difference Δp is constant [13]. However, during the solution replacement process, Δp keeps changing and μ_{EO} is also correlated to the ionic concentration [14]. The velocity profile may change to the parabolic shape from the plug shape due to the pressure difference. As shown in the Fig. 1, the pressure difference at the interface would generate dispersion which would increase the effective diffusion constant [15]. In order to investigate the convection and diffusion effect on the measurement results, we conducted both theoretical analysis and numerical study to investigate the effects of the concentration difference Δc, molecular diffusion and convection at the fluid interface by considering coupling of the flow, concentration and electric field.

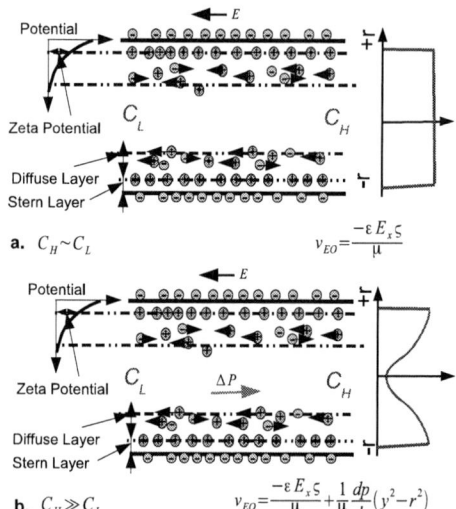

Fig. 1. Two extreme cases of electroosmotic flow in microchannel: (a) low concentration difference, (b) high concentration difference; the induced pressure gradient along the channel change the velocity profile from plug to parabolic. $\zeta \equiv$ zeta potential, $\varepsilon \equiv$ permittivity.

II. DEVICE AND EXPERIMENTAL RESULTS

A straight microfluidic chip with one simple microchannel (100 μm × 5 μm × 1.5 mm) was fabricated on a 4-inch Si wafer with MEMS technology at HKUST. The fabrication process was similar to our previous work [16]. A high-voltage power supply (PS310, SRS, Inc., USA) was connected to a microchip. A 20 kΩ resistor connected in series with the microchip was used to monitor the electric current $I(t)$ and recorded by a data acquisition system (Agilent 34970A, Agilent Technologies, USA).

Experimental setup for the current-monitoring method has been used for measurement of electroosmotic mobility μ_{EO} in the fabricated microchannel. The microchannel of the device and its one reservoir was firstly filled with 5X TBE buffer (Sigma-Aldrich, Inc., USA), at the other reservoir, a diluted TBE buffer (4X TBE buffer diluted by DI water) was added into the reservoir. Immediately, a voltage of 60 V was applied through two platinum electrodes immersed in the two reservoirs to drive the diluted solution flow through the microchannel by electroosmotic flow. The voltage response on the 20 kΩ resistor was recorded. During the solution replacement process, the current passing through the microchannel kept decreasing and become steady when the whole microchannel was completely replaced by the diluted solution. After the experiments finished, the microchannel was flushed with Isopropyl alcohol for 5 minutes, then flushed with DI water for 15 minutes and gently blew dry with compressed air. The experiments were repeated for different kind applied voltages and another simple electrolyte NaCl (18 mM NaCl replaced 20 mM).

The comparison between the experimental results and our numerical simulations is shown in the Fig. 4. For the low concentration difference ($\bar{c}$ = 0.053), the current response shows linear behavior, while for the high concentration difference ($\bar{c}$ = 0.263) the current as function of time shows strongly nonlinear behavior.

III. Modeling and Simulation

For the CMM, we are interested in estimating the effective diffusion length in the streamwise direction [17], which determine the measurement accuracy of electroosmotic mobility. The diffusion and convection in the streamwise direction are major factor to determine the effective diffusion length l_{Diff}. For the given time interval t, the present molecular diffusion yields a contribution of Dt to the square of the effective diffusion length along the streamwise direction. For the convection flow, if the normalized concentration difference ratio $\bar{c}$ is small (<0.05), the velocity distribution along the streamwise direction can be considered as plug flow. Then the contribution of convection flow to the effective diffusion length can be neglected. The electric current would be linearly proportional to time as shown in Fig. 4.

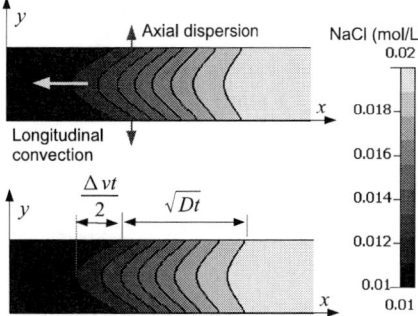

Fig. 2. Prediction of the μ_{EO} measurement error in the analytical model. Consider the axial dispersion due to diffusion and longitudinal convection due to combined effect of electroosmotic flow and pressure gradient.

However, as $\bar{c}$ increases, the velocity distribution along the streamwise direction is parabolic. Then the convection flow would influence the effective diffusion length l_{Diff}. For simplicity, the square of the effective diffusion length can be written as the sum of the contribution of flow convection and molecular diffusion as shown in Fig. 2:

$$l^2_{Diff} = Dt + (\frac{\Delta vt}{2})^2 \qquad (1)$$

where D is the diffusion constant, t is the time, Δv is velocity difference from two difference concentration region:

$$\Delta v = v_H - v_L \qquad (2)$$

where v_H and v_L are the velocity in the high and low concentration region, respectively.

At the micron scale and typical ionic concentrations, the electroosmotic velocity can be simplified as follows [13]:

$$v = \mu_{EOF}E = -\frac{\sigma^*}{\eta\kappa}E \qquad (3)$$

where η is the viscosity of the fluid in the microchannel, κ^{-1} is the Debye length and E is the electric field strength, σ^* is the surface charge density. For the symmetric monovalent electrolyte, the Debye length is inversely proportional to $c^{0.5}$. Then Equation (3) can be rewritten as:

$$v = \mu_{EOF}E \propto \frac{1}{\eta c^{0.5}}E \qquad (4)$$

Assume that the current passing through the microchannel is $I(t)$, and the electrical conductivity of the solution is λ. Then the electrical field strength E is equal to $I(t)\lambda$.

The electrical conductivity of the solution can be calculated using Kohlraush's law [18].

$$\lambda = c(L_m^0 - bc^{0.5}) \qquad (5)$$

where the constants of L_m^0 and b depend on the electrolyte. For NaCl, L_m^0 = 131.8±0.59 [18] and b = 43.3±1.64 [18] for the electrical conductivity unit is Sm^{-1} and concentration unit is mol. Then Eqn. (4) can be rewritten as:

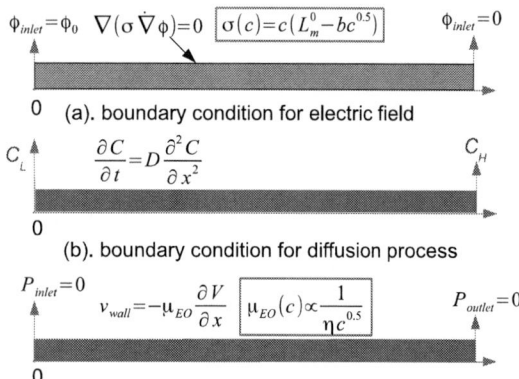

(a). boundary condition for electric field

(b). boundary condition for diffusion process

(c). boundary condition for N-S equation

Fig. 3. Nonlinear 2D electromechanical coupled model for electroosmotic flow in microchannel with large concentration difference between the two ends. The coupling relationships between the solution conductivity and electroosmotic mobility as function of electrolyte concentration are shown in red box.

$$v = \mu_{EOF}E \propto \frac{1}{\eta}c^{0.5}(L_m^0 - bc^{0.5}) \tag{6}$$

So the normalized velocity difference is:

$$\frac{\Delta v}{v} = \frac{1 - \lambda_1 c_H^{0.5}/\lambda_2 c_L^{0.5}}{1 + \lambda_1 c_H^{0.5}/\lambda_2 c_L^{0.5}} \tag{7}$$

where λ_1 and λ_2 are the electrical conductivity of solution with concentration c_H and c_L respectively, v is the average velocity.

We define the normalized concentration difference $\bar{c}$ as:

$$\bar{c} = |c_L - c_H|/(c_L + c_H) \tag{8}$$

From the analysis of the dispersion of fluid interface as illustrated in Fig. 2, we propose a unified analytical model to predict $\Delta\mu_{EO}$, normalized to the reference mobility $\mu_{EO,0}$, i.e., $\bar{\varepsilon}$, as a function of Peclet number (Pe $= L\mu_{EO}E/D$) and normalized Δc:

$$\bar{\varepsilon} = \frac{\Delta\mu_{EO}}{\mu_{EO,0}}(L, E, D, \bar{c}) = f(Pe, \bar{c}) = \sqrt{\frac{1}{Pe} + \frac{\Delta v^2}{4v^2}} \tag{9}$$

For small $\bar{c}$ (< 0.2), with the help of asymptotic expansion, Eqn. (9) can be reduced to the following expression:

$$\bar{\varepsilon} = \sqrt{\frac{1}{Pe} + a_1\bar{c}^2 + O(\bar{c})^4} \tag{10}$$

where a_1 is defined as:

$$a_1 = \frac{0.0625(4bc^{0.5} - 3L_m^0)^2}{(bc^{0.5} - L_m^0)^2} \tag{11}$$

for the NaCl at the concentration of c = 19mM, the constant a_1= 0.257413 based on the electrical conductivity properties of NaCl electrolyte in Ref. [18].

However, Eqn. (10) is only valid for low concentration $\bar{c}$. For large concentration $\bar{c}$, we have to construct a 2D nonlinear model using a commercial CFD code (CFD-ACE+, CFD-ESI, USA) to determine the flow, diffusion and electric fields. The corresponding governing equations are summarized as follows:

$$\rho\frac{\partial \vec{V}}{\partial t} = -\nabla P + \eta\nabla^2\vec{V}$$
$$\nabla \cdot \vec{V} = 0 \tag{12}$$

$$\nabla \cdot [\lambda\nabla\phi + \varepsilon\frac{\partial\nabla\phi}{\partial t}] = 0 \tag{13}$$

$$\frac{\partial c}{\partial t} = D\nabla^2 c \tag{14}$$

where $\vec{V}$ is a 2D velocity vector, P is the pressure, ρ is the fluid density, ϕ is the electrical potential, ε is the electrical permittivity and c is concentration.

The flow and electrical fields are coupled via electroosmotic flow; the diffusion and electrical field are coupled through electrical conductivity expressed in Eqn. (5). The corresponding boundary condition and initial condition are shown in Fig. 3.

From Fig. 4, the current response shows highly nonlinear behavior for high concentration $\bar{c}$ and linear behavior for low concentration $\bar{c}$. In order to obtain accurate measurement results by current-monitoring method, it is important to get a clear turning point (to determine the elapsed time t_0). However, due to molecular diffusion at the interface between two different concentration regions and the dispersion in the streamwise direction, the turning point becomes unclear as shown in Figs. 4 and 5. The maximum error induced by the diffusion and dispersion can be estimated by comparing simple 1D model with the 2D numerical simulation results.

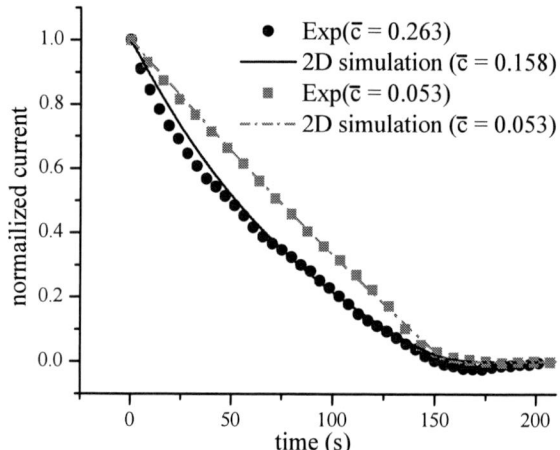

Fig. 4. Comparison of nonlinear current response for different normalized concentration difference $\bar{c}$ shows that 2D simulation is valid for both low and high $\bar{c}$.

For the simple 1D model, we assume that the electric field and electroosmotic mobility are constant and the diffusion time in the spanwise direction is much smaller compared in the streamwise direction. Therefore, the concentration in the spanwise direction is constant. The governing equation can be simplified into 1D transition diffusion process and the analytical results can be expressed as:

$$c(t,x) = C_L - (C_H - C_L) \cdot erfc(\frac{x - vt}{2\sqrt{2Dt}}) \qquad (15)$$

where x is the streamwise coordinate along the microchannel and $erfc(x)$ is the complementary error function defined as:

$$erfc(x) = \frac{2}{\sqrt{\pi}} \int_x^\infty e^{-t^2} dt$$

From Eqns. (5) and (15), we can calculate the total electric resistance of the microchannel $R(t)$

$$R(t) = \int_0^L \frac{1}{\lambda(c)} dx \qquad (16)$$

where L is the length of the microchannel.

The current response is: $I(t) = V/R(t)$, where V is the applied voltage.

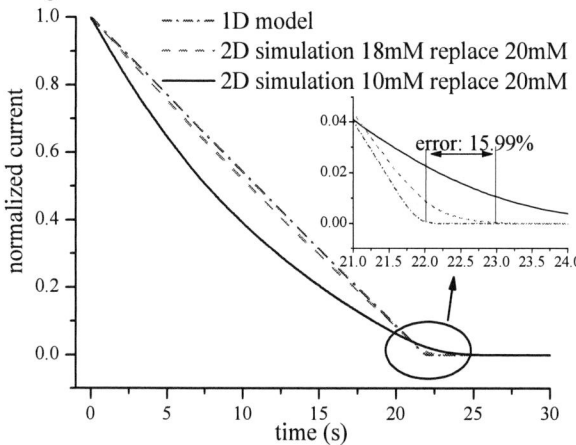

Fig. 5. Comparison of nonlinear current response for different cases shows that 1D model is only valid for low Δc, Inset shows the error compared with the idea elapsed time required to finish the solution replacement.

Obviously, the current response is highly nonlinear for large $\overline{c}$ as shown in Fig 4. Comparing the current responses with selected concentrations $\overline{c}$, the 2D numerical model can predict the cases with much higher $\overline{c}$ (Fig. 4). The normalized error $\overline{\varepsilon}$ at different $\overline{c}$ can be determined from Eqns. (11) and 2D simulations (Fig. 5). For typical Pe (=463) used in the previous measurements [17], the error $\overline{\varepsilon}$ for $\overline{c} = (0.05, 0.33)$ are 5% and 15.99%, respectively.

In order to validate our unified analytical model, a serial of simulations with different $\overline{c}$ was conducted and compared with the results in equation as shown in Fig. 6. Obviously, the theoretical predictions are consistent with the 2D simulations. As the normalized concentration difference ratio $\overline{c}$ increases, the difference between 2D simulation and our model increase. However, for $\overline{c}$ in the range of 0~0.2, the estimated maximum error from our unified model is acceptable using the 2D simulation results as reference. On the other side, the commonly used normalized concentration difference $\overline{c}$ in CMM falls into this range.

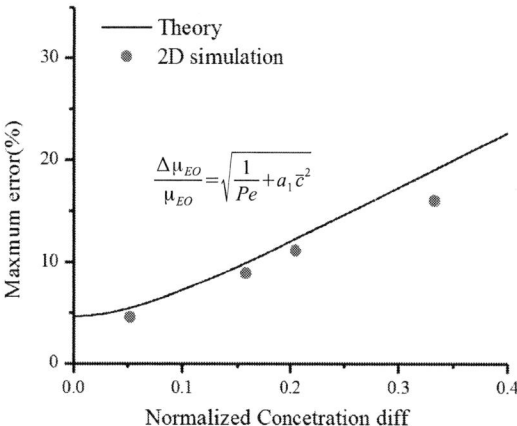

Fig. 6. Comparison of analytical prediction and 2D simulation for $\overline{\varepsilon}$ as a function of $\overline{c}$ (Pe = 463, [17]).

Considering the practical application of our research work, the unified compact model can be used as a guideline to determine the experimental condition for electroosmotic mobility measurement by using CMM with specific acceptable measurement error. To understand the combined effects of diffusion and convection on the maximum measurement error $\overline{\varepsilon}$, a contour plot of $\overline{\varepsilon}$ as a function of Pe and $\overline{c}$ was shown in Fig. 7. This plot can be used to pinpoint the practical region for accurate electroosmotic mobility measurement μ_{EO}.

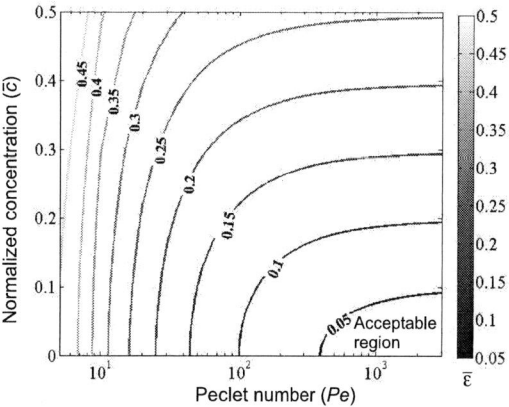

Fig. 7. Contour plot of the maximum measurement error of electroosmotic mobility $\overline{\varepsilon}$ as a function of Peclet number and normalized concentration difference.

IV. CONCLUSION

In conclusion, we propose a unified analytical model to evaluate the μ_{EO} measurement error due to the concentration difference and Peclet effect. With the 2D nonlinear numerical simulations, our analytical model with the normalized parameters is also useful for complicated biofluids in microfluidics.

ACKNOWLEDGMENTS

This research was partially supported by a Hong Kong RGC grant (No. 616106) and by an award from the King Abdullah

University of Science and Technology (KAUST Award No. SA-C0040/UK-C0016).

REFERENCES

[1] X. Y. Wang, *et al.*, "Electroosmotic pumps and their applications in microfluidic systems," *Microfluidics and Nanofluidics*, vol. 6, pp. 145-162, Feb 2009.

[2] Y. Ai, *et al.*, "A low-voltage nano-porous electroosmotic pump," *Journal of Colloid and Interface Science*, vol. 350, pp. 465-470, Oct 15 2010.

[3] Y. Berrouche, *et al.*, "Optimization of high flow rate nanoporous electroosmotic pump," *Journal of Fluids Engineering-Transactions of the Asme*, vol. 130, Aug 2008.

[4] A. Brask, *et al.*, "Long-term stable electroosmotic pump with ion exchange membranes," *Lab on a Chip*, vol. 5, pp. 730-738, 2005.

[5] L. M. Lee, *et al.*, "In-plane vortex flow in microchannels generated by electroosmosis with patterned surface charge," *Journal of Micromechanics and Microengineering*, vol. 16, pp. 17-26, Jan 2006.

[6] L. W. Hau, "Electrokinetically-driven liquid flows in microchannels using surface-chemistry technology," Thesis (Ph D), Hong Kong University of Science and Technology, 2005., 2005.

[7] I. Glasgow, *et al.*, "Electroosmotic mixing in microchannels," *Lab on a Chip*, vol. 4, pp. 558-562, 2004.

[8] M. Jain and K. Nandakumar, "Novel index for micromixing characterization and comparative analysis," *Biomicrofluidics*, vol. 4, Sep 2010.

[9] P. K. Wong, *et al.*, "Electrokinetics in micro devices for biotechnology applications," *IEEE/ASME Transactions on Mechatronics*, vol. 9, pp. 366-376, Jun 2004.

[10] X. Huang, *et al.*, "Current-monitoring method for measuring the electroosmotic flow rate in capillary zone electrophoresis," *Analytical Chemistry*, vol. 60, pp. 1837-1838, 1988.

[11] L. Q. Ren, *et al.*, "Experimental and theoretical study of the displacement process between two electrolyte solutions in a microchannel," *Journal of Colloid and Interface Science*, vol. 257, pp. 85-92, Jan 1 2003.

[12] C. Wang, *et al.*, "Characterization of electroosmotic flow in rectangular microchannels," *International Journal of Heat and Mass Transfer*, vol. 50, pp. 3115-3121, Jul 2007.

[13] D. Mampallil, *et al.*, "A simple method to determine the surface charge in microfluidic channels," *Electrophoresis*, vol. 31, pp. 563-569, Jan 2010.

[14] R. Sadr, *et al.*, "An experimental study of electro-osmotic flow in rectangular microchannels," *Journal of Fluid Mechanics*, vol. 506, pp. 357-367, May 10 2004.

[15] R. Aris, "On the Dispersion of a Solute in a Fluid Flowing through a Tube," *Proceedings of the Royal Society of London Series A*, vol. 235, pp. 67-77, 1956.

[16] Y. C. Chan, *et al.*, "Design and fabrication of an integrated microsystem for microcapillary electrophoresis," *Journal of Micromechanics and Microengineering*, vol. 13, pp. 914-921, Nov 2003.

[17] W. Wang and Y. K. Lee, "Optimum Peclet Numbers for Accurate Measurement of Electroosmotic Mobility of Complex DNA Buffers in Micro/Nanofluidics," in *MicroTAS 2010*, Groningen, The Netherlands, 2010, p. 1766.

[18] R. A. Robinson and R. H. Stokes, *Electrolyte solutions : the measurement and interpretation of conductance, chemical potential, and diffusion in solutions of simple electrolytes*, 2d ed. London: Butterworths, 1959.

Blood Vessels by Fractal Gelatin

Lung-Jieh Yang[*], Bo-Hong Chen

[*]Department of Mechanical and Electromechanical Engineering, Tamkang University, TAIWAN

Ljyang@mail.tku.edu.tw

Abstract—**In this study the authors developed a novel gelatin patterning technique applied to making blood vessel with fractal dendrite configuration. Such a chaotic tree-like pattern has been obtained through precipitating among the gelatin matrix which is spun coating on a glass substrate at room temperature. The weight percentage of the original gelatin solution is over-saturated. As the temperature decrease, the gelatin crystallizes and forms a natural fractal pattern in the thin film. The process parameters are changed and the hydraulic diameters of the fractal patterns are verified from 0.1 μm to 23 μm. Finally the authors used the biocompatible PDMS making blood vessel like capillary channels by the de-molding process of the soft lithography. The filling test of the PDMS fractal channel chip is done accordingly.**

Keywords-gelatin; blood vessels; fractal; PDMS

I. INTRODUCTION

Artificial blood vessels are important to the implant surgery subject to many irrecoverable damages of human organs. The material for the blood vessels should meet the requirement of good bio compatibility and long life time. Any polymer material extracted from the animal tissues or natural creatures seems more suitable to implant applications other engineering materials. One choice of them is gelatin which poses 18 kinds of amino acid and can be shaped into different geometries or patterns in the previous studies correlated with the weak microstructure strengthening and the selective stem cell culture [1-2]. In these prior arts, the gelatin micro patterns are very regular according to the designers' mask layouts hence far from the bio mimicking manner, e. g., the fractal patterns discussed in this paper. In other words, if we regard the gelatin as a material for making the artificial blood vessels to mimic the fractal-like capillary blood vessels, developing fractal gelatin patterns seems the first difficulty needed to be overcome.

Fractal patterns are commonly found during the re-crystallization process of over-saturated salty solution [3]. In NEMS-2011, the authors have ever chosen the gelatin aqueous solution dissolved with excess amount of a photo-sensitizer agent, potassium dichromate ($K_2Cr_2O_7$) [1] to occasionally generate the tree-like, fractal micro structures. One example is shown as Fig. 1 [4-5]. The classical weight percentage of the gelatin aqueous solution for generating the fractal gelatin is 10-20 wt%, with the matching amount of $K_2Cr_2O_7$ as 5-15 wt%. In our prior art [4-5] the confinement effect mentioned by R. Feynman was investigated to influence the repetitive characteristics of the gelatin fractal patterns and its application to the design of micro mixers. In this paper, however, the authors would like to use gelatin fractal pattern to fabricate

Fig. 1. The gelatin fractal patterns in the $K_2Cr_2O_7$-gelatin collagen matrix [4].

some chaotic capillary channels for liquid transportation and even more as the promising artificial blood vessels in the future.

II. FABRICATION AND EXPERIMENT

A. How to Make the Fractal Gelatin

The gelatin solution needs gentle heating up to 50-60°C with proper mixing and bubble removing. Spin coating is accessed to spread the gelatin film on a glass substrate. The dispensing time of the gelatin gel on the substrate should be controlled as short as possible to avoid gelatin solidification before the uniform spin coating. The re-crystallization happens no sooner than the completion of spin coating, and proceeds for proper duration corresponding to different concentrations of gelatin solution.

A variety of fractal gelatin patterns corresponding to different process parameters are observed in this work. These process parameters are the spin-coating (500, 1000, 1500, 2000 rpm), the weight percentages of gelatin (10, 15, 20 wt%), and the weight percentages of $K_2Cr_2O_7$ (5, 10, 15 wt%).

B. Process Parameters of the Fractal Gelatin

In this work, the authors specifically evaluated the hydraulic diameters of the gelatin fractal patterns as the characteristic length. The hydraulic diameter D_h of the fractal gelatin according to different process parameters are shown Fig. 2. Each datum value of Fig. 2 is obtained by averaging at least 5 measured points from one gelatin sample. The error bar of each datum is also evaluated as well. The cross section of a fractal gelatin tree-like branch bumped and embedded in the

978-1-4673-1122-9/12 $31.00 © 2012 IEEE

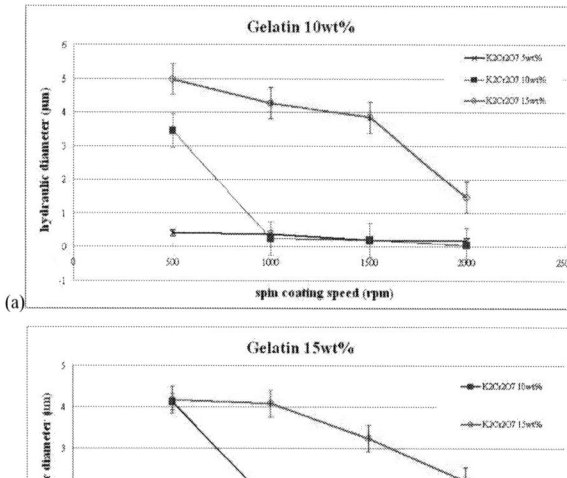

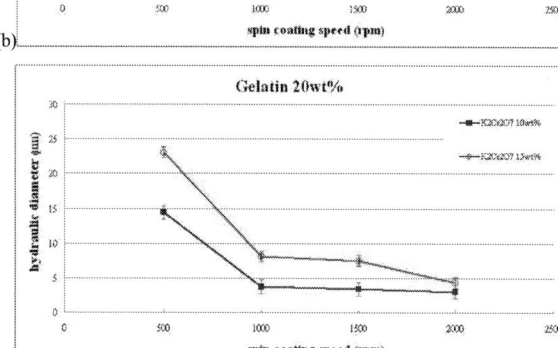

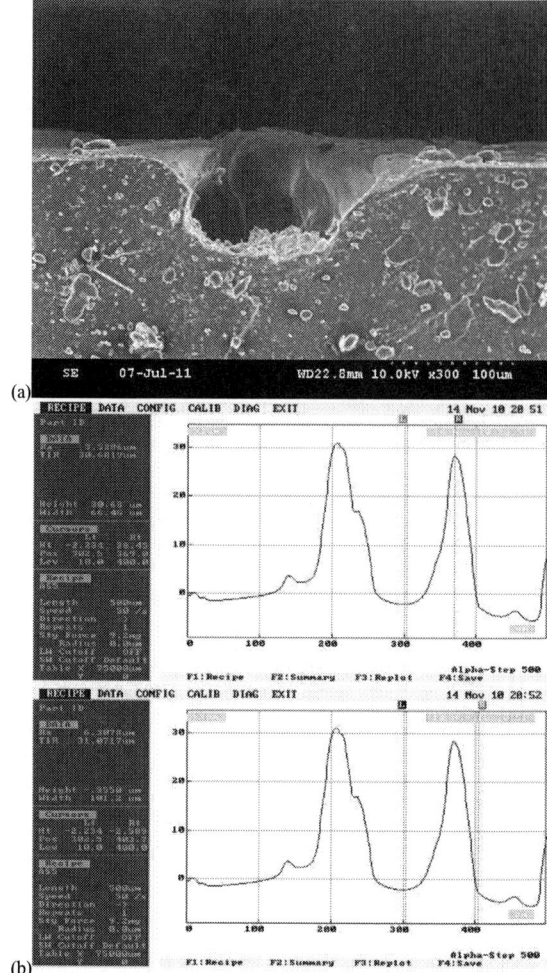

Fig. 2. Hydraulic diameters of the fractal gelatin versus spin coating speed with respect to different dichromate weight percentages: (a) in the 10 wt% gelatin; (b) in the 15 wt% gelatin; (c) in the 20 wt% gelatin.

Fig. 3. The gelatin fractal patterns scanned by: (a) SEM; (b) surface profiler. The SEM of (a) is taken from a slice specimen cut from a PDMS block molded by the fractal gelatin. In (b) the height b is 30.68 μm and the width $2a$ is 101.2 μm.

gelatin film is regarded as a semi-ellipse herein. Therefore all the values of D_h are calculated by the elliptic cross section formulas as follows.

$$D_h = 4A_c \, / \, P \tag{1}$$

$$P \fallingdotseq \pi \, [3(a+b) - \sqrt{(3a+b)(a+3b)} \,] \tag{2}$$

$$A_c = \pi ab \tag{3}$$

where A_c and P denote the cross section area and the perimeter of the flow channel under investigation. The values of a and b in (2) and (3) denote the long semi-axis and the short semi-axis respectively of the ellipse. They are quantitatively determined from the scanned curve of the fractal gelatin patterns by SEM or the surface profiler (alpha-step 500). One example is shown in Fig. 3.

From Fig. 2, the fractal gelatin dimension or the hydraulic diameter globally decreases with the increasing coating speed and the lower wt% concentration of gelatin as well as dichromate salt. In Fig. 2(a), the finest hydraulic diameter of

fractal gelatin branch is 0.1 μm subject to 10 wt% dichromate in 10 wt% gelatin solution spun-coating at 2000 rpm.

Except the case of 10 wt% gelatin solution in Fig. 2(a), there's no fractal gelatin appeared with the dichromate wt% less than 5%. Additionally, the fractal gelatin dimension changing trend subject to lower wt% behaves more unpredictable in Fig. 2(a-b). Their hydraulic diameters also don't match with the general blood vessels. Therefore, the thicker case of 20 wt% gelatin solution in Fig. 2(c) is finally chosen as the candidate for the fractal pattern transfer to PDMS in the next section.

C. Fractal Patterns Transfer to PDMS

Artificial blood vessels with chaotic dendrite tree shape mimicking the portal veins of livers were proposed by Ref. [6]. So far the authors could not find a proper method to fabricate the circular blood vessels directly by the gelatin material. An easier approach is to regard the fractal gelatin as a mother template, and to use the PDMS de-molding technique to transfer the bumped fractal patterns into the concave fractal hollow channels in a PDMS block. The drawback of this method is that the cross section of the fractal channel is only semi-circular. But the elastic PDMS with large deformation capability could somewhat compromise this shortcoming. The PDMS process is shown in Fig. 4. The fabricated PDMS flow chip with fractal channels is shown in Fig. 5. The process parameters for the fractal gelatin are 20 wt% gelatin aqueous solutions with 15 wt% $K_2Cr_2O_7$ subjected to 500 rpm spin coating speed. According to the process data in Fig. 2(c), the hydraulic diameter of this fractal channel is estimated to be 23 μm.

D. Water Filling Test

The fabricated PDMS flow chip is performed with water filling to confirm the mechanical strength before the bio-compatible verification. The volumetric flow rate of the syringe pump is set as 5-40 mL/min. The authors observed the width change of the fractal channels under the optical microscope and summarized them in Table 1. The experiment setup and the OM pictures are shown in Fig. 6. As the flow rate is up to 30 mL/min, the width change of 7% is observed. The corresponding pressure drop across the fractal channel is calculated as 6.24 MPa by the Hagen-Poiseuille equation [7].

$$Q = uA_c = \frac{\Delta P \cdot \pi \cdot D_h^4}{128\mu L} \qquad (4)$$

where Q, u, μ denote the volumetric flow rate, flow speed, and viscosity of the fluid. L is the channel length of the flow channel. $\triangle P$ is the pressure drop across the channel length L (=3.5 mm from Fig. 5.)

This pressure drop $\triangle P$=6.24 MPa is 4.46 times of the Young's modulus of PDMS, 1.4 MPa. Therefore the apparent channel deformation is reasonable. The flow speed of 59.3 m/s is far faster than the test speed of 2 mm/s for red blood cells [8], and even tremendously faster than the flow speed of 10 μm/s for tumor cells in capillaries [6]. Therefore the PDMS fractal blood vessels are robust enough to be used in the general biomedical flow filling.

III. CONCLUSION

In conclusion, the PDMS artificial blood vessel molded from fractal gelatin patterns has no damage issue subjected to MPa pressure loading so far. The characteristic hydraulic diameter of the chaotic micro channel ranges from 0.1 μm to 23 μm in this work. The potential applications of this fractal channel network include artificial blood vessels, cell culture attachment investigation, and chaotic micro mixers.

ACKNOWLEDGMENT

The authors would like to thank the financial and travel support from the National Science Council of Taiwan with the project no. of NSC-98-2221-E-032-025-MY3.

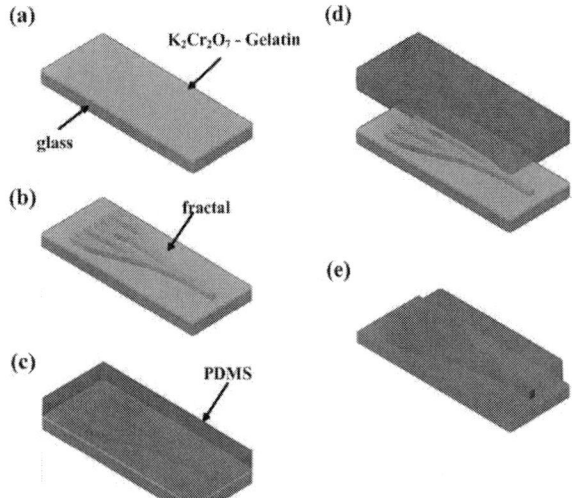

Fig. 4. Process flow of the PDMS fractal channels animating blood vessels. (a) Spin-coating the gelatin film with oversaturated potassium dichromate; (b) Fractal gelatin precipitation and UV exposure; (c) Dispensing PDMS; (d) PDMS de-molding; (e) A glass substrate bonded with the PDMS slide with fractal channel pattern.

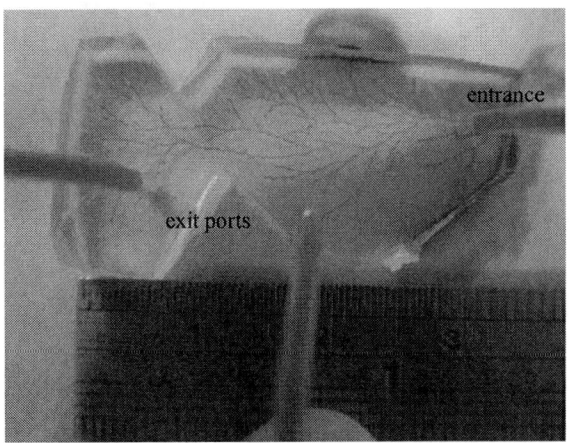

Fig. 5. A filling test of the PDMS fractal channel chip.

TABLE I. The testing results of the filling experiment on the PDMS chip.

Volumetric flow rate Q (mL/min)	Flow speed u (m/s)	Pressure difference $\triangle P$ (MPa)	Width change of fractal channels (L=0.035m; μ=1.003mPa · s)
5	10.6	1.19	None (width =100μm)
10	21.1	2.38	None (width =100μm)
15	31.7	3.57	None (width =100μm)
20	42.3	4.76	None (width =100μm)
25	52.9	5.95	None (width =100μm)
30	59.3	6.24	7% larger (width =107μm)
35	64.4	6.30	15% larger (width =115μm)
40	70.5	6.61	20% larger (width =120μm)

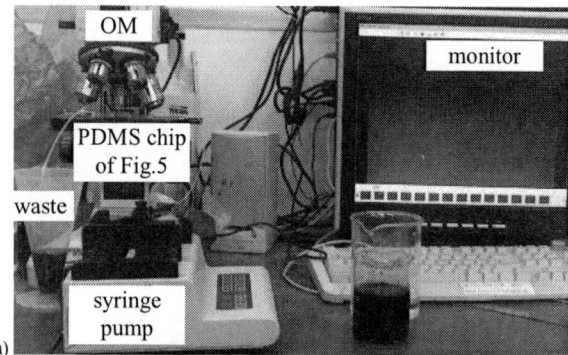

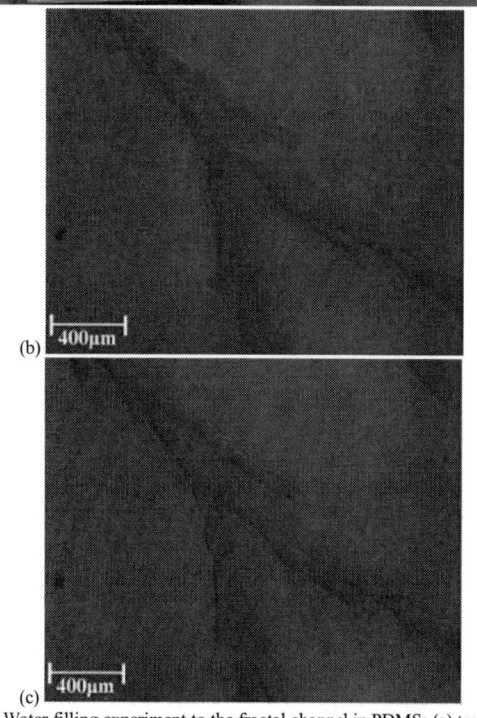

Fig. 6. Water filling experiment to the fractal channel in PDMS: (a) test setup; (b) channel without water filling; (c) channel filled with 35 mL/min water and the width changed with 15%.

REFERENCES

[1] L.J. Yang, W.Z. Lin, T.Y. Yao, and Y.C. Tai, "Photo-patternable gelatin as protection layers in surface micromachinings," *Sensors and Actuators A*, vol. 103(1-2) pp. 284-290, 2003.

[2] L.J. Yang and Y.C. Ou, "The micro patterning of glutaraldehyde(GA)-crosslinked gelatin and its application to cell-culture," *Lab on a Chip*, vol. 5, no. 9, pp. 979-984, 2005.

[3] A. Mersmann, *Crystallization Technology Handbook*, 1st ed., Marcel Dekker, New York, Chapter 1, pp. 48-56, 1995.

[4] L.J. Yang, C.T. Lee, P.H. Chen, and C.W. Hsu, "Confined fractal patterns in gelatin," The 6th IEEE International Conference on Nano/Micro Engineered and Molecular Systems (NEMS-2011), art. no. 6017292, pp. 49-52, 2011.

[5] A.F. Kao, L.J. Yang, F.W. Yeh, "Fractal grooves applied to passive micro-mixers," The 6th IEEE International Conference on Nano/Micro Engineered and Molecular Systems (NEMS-2011), art. no. 6017315, pp. 142-145, 2011.

[6] L.J. Yang, C.W. Hsu, and Y.C. Ou, "The minimum time estimation for initiating tumor-cell attachment," Tech. Dig. of Transducers-2011, Beijing, June 5-9, W3P.104, 2011.

[7] M.J. Madou, *Fundamentals of Microfabrication- The Science and Miniaturization*, 2nd ed., CRC Press, Boca Raton, p. 565, 2002.

[8] T. Secomb, R. Skalak, N. Ozkaya, and J.F. Gross, "Flow of axisymmetric red blood cells in narrow capillaries," *Journal of Fluid Mechanics*, vol. 163, pp. 405–423, 1986.

Design, Simulation, and Verification of Fluidic Light-Guide Chips with Various Geometries of Micro Polymer Channels

Guo-Wei Huang, Tzu-Yi Hung, and Chin-Tai Chen[*]

[*]Department of Mechanical Engineering, National Kaohsiung University of Applied Sciences, TAIWAN, ROC
chintai@kuas.edu.tw

Abstract—We present for the first time that fluidic and optical behavior of liquid within the micro channels has been modeled, simulated and experimented for the various geometries with constant depth of 160 μm. These channels are all rectangular in cross section made of transparent PDMS (Polydimethylsiloxane) through a molding process from a master of a SU-8 patterned wafer (4 inches in diameter). Each flexible channel varies in two-dimensional (2D) outline including rectilinear, curved and rectangular shapes. An optical camera and a spectrum analyzer were equipped to capture and analyze the dynamic evolutions of optofluidic flows supplied by a liquid injector. The light guide by the total reflection was proved feasible for many liquid solutions.

Keywords- MEMS; light-guide chips; microfluidics; optofluidics

I. INTRODUCTION

Growing interest in the field of optofluidics has resulted in a continuous thrust toward the development of more compact and robust lab-on-a-chip (LOC) devices. In recent years, many researchers have paid attention to this novel LOC technology. Psaltis *et al* clearly describe, in Nature [1], the significance of optofluidic microsystems with the fluidic and optical interplay. This kind of technology involves the "fluid-solid", "purely fluidic", and "colloidal suspensions" in the three different major fields. Cho *et al* have realized such an important application in waveguide using PDMS material [2]. Spin-coating the Teflon film within the PDMS channel was applied with the principle of total internal reflection (TIR) of light. We further investigated, in the paper, the optofluidic phenomena extended to the varying shapes in 2D outlines microfabricated with an array of micro channels (see Fig. 1). As reported by Nguyen [3], the fluidic media can also exhibit the more remarkable variation (e.g., lens shaping) and functions (e.g., biological information). C. Yang *et al* have reported a hand-held corona system used as an alternative for PDMS-PDMS bonding processes [4]. T. Zhang *et al* have reported a new simple microfluidic valve, which utilizes titanium dioxide (TiO_2) nanoparticles tunable surface wettability between super hydrophilic and super hydrophobic to switch the valve on/off [5]. However, the influence of fluidic paths on the interplay of optical lights is significant but unclear in those applications. Therefore, in the study, we aimed to perform simulations and experiments of optofluidic

flows through the channels that could quantitatively provide the data of the rate used as the fluidic analysis and design [6]. The optical measurement and configuration for the optofluidic system introduced here would be complementary to those of the other studied systems.

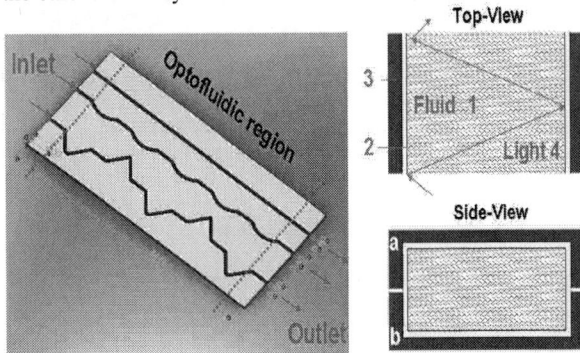

Fig. 1. Schematic configuration of an optofluidic chip designed for transporting lights(4) and fluids (1), in which the inner sidewalls (3) of micro closely bonded (a and b) channels coated with thin polymer films (2).

II. DESIGN AND FABRICATION

First, a polished silicon wafer was spin-coated with the thick photoresist (SU-8) at 500 rpm for 120s and 3000 rpm for 60s yielding an 80μm-thick layer, and was further patterned by a specific mask to form the fluidic channels. Next, the PDMS (Polydimethylsiloxane) layer was then formed by molding the SU-8-coated wafer as a master (see Fig. 2). With the two PDMS chips bonded together, one dyed fluid was filled into inner channels for flowing experiments as shown in Fig. 3, compared to the simulated results by customer software as demonstrated in Fig. 4.

Note that our experiments were carried out by filling the channels with blue-dyed fluid (see Fig. 3), in order to observe and analyze the corresponding fluid velocity, whileas the simulations were performed using the CFD software (Flow-3D). Next, the three different types of fluid flows were generated for the optofluidic region (~2 cm) and time (~1s) within the channels by using the appropriate UV curing glue. Hence, the entered lights can be created with total reflection at the interface between the SU-8-coated layer and fluid, thus

978-1-4673-1122-9/12 $31.00 © 2012 IEEE

simultaneously guiding the flows and light rays through the channels together. After testing the long-wavelength lights from optical fibers effects on the chips, we found that the lights can be transported through channel filled with glue, in which the associated spectral intensity was decreasing over time with the average spectral intensity of the outlet in the range of -50 dBm to -68 dBm at the peak wavelength of 1550nm. More importantly, the largest intensity (*i.e.*, -50 dBm) was generated by the fluidic flow with the smallest resistance ((a): rectilinear shape), as the smallest intensity (*i.e.*, -68 dBm) associated with the largest-resistance fluid ((c): rectangular shape). The optofluidic experiments were detailed in the following.

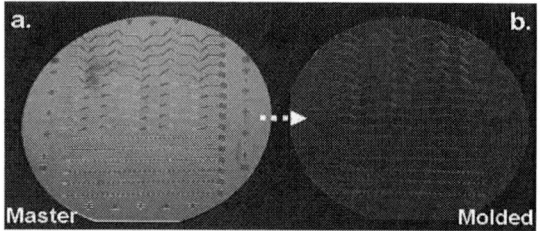

Fig. 2. Fabrication process of a micro optofluidic chip with (a) the master of SU-8 patterned on a silicon wafer (b) the PDMS directly poured, molded, and peeled off the master.

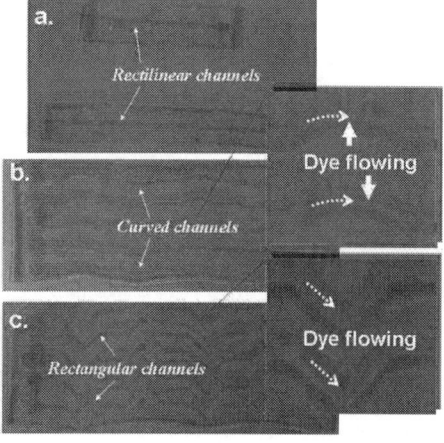

Fig. 3. Three different types of channel geometries implemented and studied in the chip showing the dyed-fluid flowing inside, including (a) rectilinear (b) curved (c) rectangular shapes, respectively.

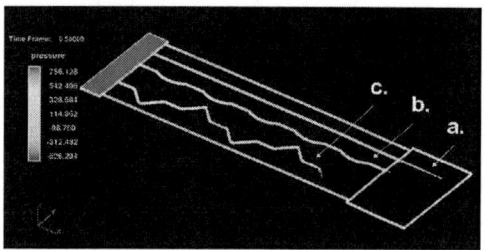

Fig. 4. Simulation of the fluid flow through the three micro channels encountered with different resistances for (a) rectilinear (b) curved (c) rectangular geometries, respectively.

III. RESULTS AND DISCUSSION

Because the lights in the study were intended to be transported by total internal reflection within the PDMS channels, which had a refractive index of n=1.5, we varied this refractive index inside the channel surfaces by coating four different materials including SU-8 (n=1.58), aluminum (n=1.7), SiO₂ (n=1.45) and Teflon (n=1.3). To establish a measurement system, we manually connected the optofluidic chips fabricated above to the light source and optical sensor spectrometer for capturing light signals in intensity, as shown in Fig. 5. And we also verified this design of TIR by the Snell's law, as described below in (1):

$$n_1 \sin\theta_1 = n_2 \sin\theta_2 \qquad (1)$$

Where n_1 is the refractive index of medium 1, θ_1 is the reflective angle of the medium 1, n_2 is the refractive index of medium 2 and θ_2 is the reflective angle of the medium 2.

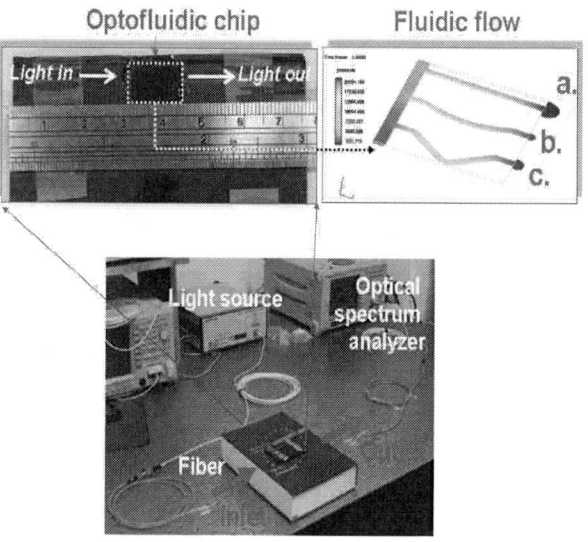

Fig. 5. Measurement apparatus equipped with the light source, the optical fibers for the entrance and exit of lights into the chip, the optical spectrum analyzer for indication of the outlet-light intensity.

With the four different materials coated within PDMS micro-channels we used a multi-mode optical fiber with a core diameter of 20 μm for the measurement, although some alignment errors led to slight loss of signal intensity. The measurement results showed that among four types of channel geometries the rectilinear one of 1cm in length yielded a peak value of SiO₂ of -55 dBm at most, as compared to other three signals with -62 dBm, as seen in Fig. 6. The curves in Fig. 6 are denser than those on both sides. When passing through a rectangular flow channel, the signal on the end side was almost the noise signal. As the result, the light can not be smoothly passed if the channel is in rectangular shape. But there were still signals on the chip with Teflon coated. The light intensity was -66 dBm on the peak of 1550 nm wavelength, as seen in Fig. 7.

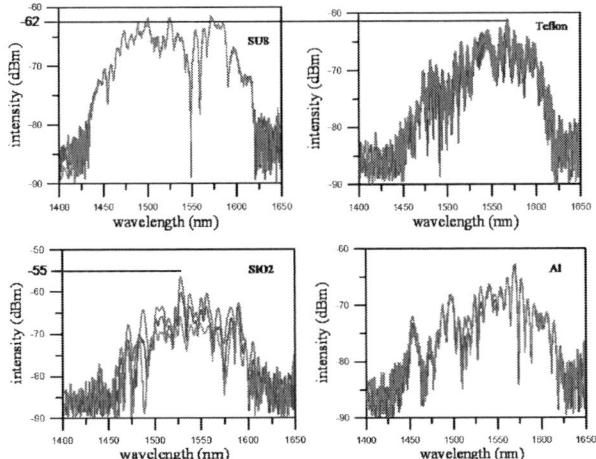

Fig. 6. Light signals were measured at the distance of 1cm between the chips and the detector.

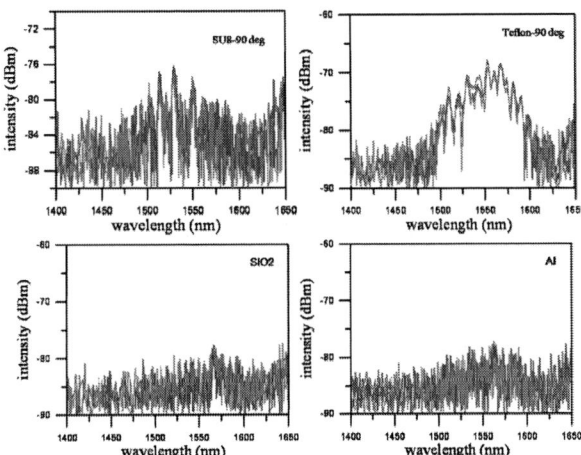

Fig. 7. Light signals were obtained for the chips coated with different materials

Furthermore, we filled a liquid polymeric material into the Teflon coated channels to achieve light transmission via the fluid within the micro-channels. As seen in Fig. 8, the outlets of the channels received the output of lights. There were three different types of the flow channels on the chip: (a) rectilinear (b) curved (c) rectangular. The peak intensity in a wavelength of 1550 nm was found to be -50 dBm, whereas the rectilinear and rectangular channels yielded the maximum and minimum respectively.

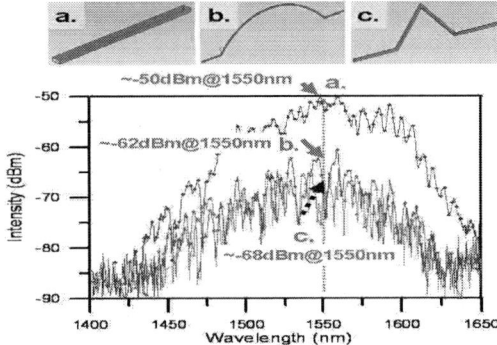

Fig. 8. Spectral intensities of outlet lights from the fabricated optofluidic chip for three different channels filled with polymer liquid flow, indicating the intensity decrease with (a) -50 dBm (b) -62 dBm (c) -68 dBm at 1550 nm

According to the experimental data, the highest light intensity (-55 dBm) was obtained when the PDMS chip was coated with SiO_2. The light intensity decreased with the change of the channel geometry. And when the channel coated with Teflon was full of the photosensitive liquid, the fluid passing through the light guide was found to be enhanced very clearly. In this case, we can enhance the light delivery by coating the low index of refraction materials on the channel that is full of the high index of refraction into the chip.

Therefore, we redesigned and concentrated onto fabricating the PDMS light-guide chip (n=1.3) with the channel coated Teflon that was full of polymer materials. The laser light can be guided from the left to right side through the channel. Microfluidic channels with the cross-sectional size of 200 μm × 160 μm were fabricated by PDMS and had two reservoirs. The PDMS was formed by molding the SU-8 coated wafer as a master and heated at 100 °C for one hour. The fabricated chip was eventually obtained as shown in Fig. 9.

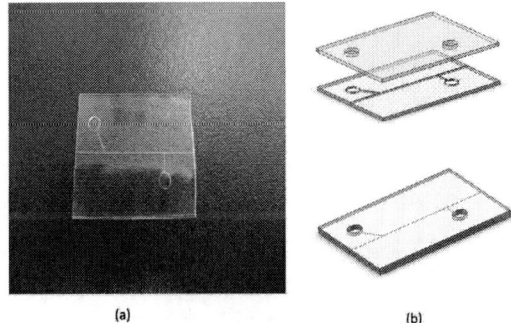

Fig. 9. (a) The fabricated PDMS light-guide chip (b) Schematic of two PDMS layers bonded to form the microchannels.

Then we used the corona triggered bonding method for microfluidic system fabrication. The corona treater was a model BD-20AC with a custom power-line filter, pure chased from Electro-Technic Products Inc [4]. This device has three parts of electrodes: a straight spring, a 2-inch disk, and a 2.5 inch wire. When the treater is powered, it can generate a high voltage potential across the electrodes at the tip of the unit, thus ionizing the air to create the localized corona discharge. Fig. 10 shows the treating operation of the corona treater.

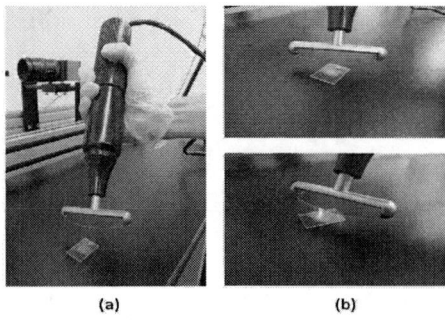

(a)   **(b)**

Fig. 10. (a) Corona treater experiment device. The bonding process can be performed at room temperature (b) the working of the corona treater. It was powered to generate a high electrical voltage across the electrodes.

This treating process can be operated at room temperature and normal atmospheric pressure. Teflon AF was needed for the lower refractive index (n) of the microfluidic channels due to surface modification [2]. When they were filled, one manual syringe was applied to remove excess Teflon AF from the channels. Then the Teflon-coated PDMS chip was heated up to 65°C for 5 min to form a smooth Teflon layer. Next, the polymer materials (n>1.3) were injected to the channel to generate a total internal reflection (TIR) of the light as seen in Fig. 11. The polymer materials were mixed with water with a weight ratio of 10:1. The laser light was injected into the channel from the left end and guided to the right end by the fluid flow microfluidic channel as seen in Fig. 12.

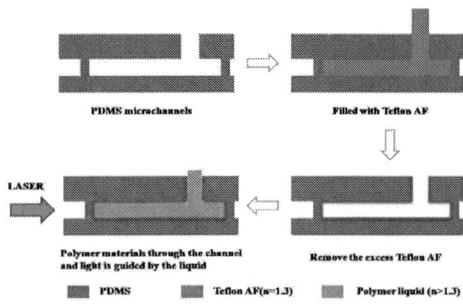

Fig. 11. Fabrication process for Teflon AF-coated core waveguides.

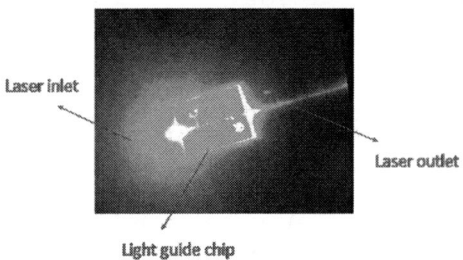

Fig. 12. The green laser can be guided from the left side to the right side through the fluidic chip.

IV. CONCLUSION

According to the present results in the study, the waveguide by the total reflection of lights was proved quite feasible for many liquid solutions. Further, we expected that the guided lights can be largely manipulated through fluidic flow simply by some micro actuations such as instant switching and pumping after the flow being filled, although experiments are not shown here. It implicitly represents that the optofluidic motion might dominate the transportation of lights added and consequently control the fundamental interaction of various fluids and lights, *e.g.* in optofluidic sensors and integrated circuits [7, 8].

V. ACKNOWLEDGMENT

The authors thank research grants by the National Science Council (NSC) under NSC-100-2221-E-151-042, Taiwan, ROC. The Research Center for MEMS and Precision Machines in National Kaohsiung University of Applied Sciences (KUAS), Taiwan, ROC, is also appreciated for the access of the surface processing and analysis equipments.

VI. REFERENCES

[1] D. Psaltis, S. R. Quake, and C. Yang, "Developing optofluidic technology through the fusion of microfluidics and optics," *Nature*, vol. 442, pp.381-386, 2006.

[2] S. H. Cho, J. Godin, and Y. H Lo, "Optofluidic waveguides in Teflon AF-coated PDMS Microfluidic Channels," *IEEE Photonics Technol. Lett.*, vol. 21, pp.1057-1059, 2009.

[3] N. T. Nguyen, "Micro-optofluidic lenses: a review," *Biomicrofluidics*, vol. 4, 031501, 2010.

[4] C. Yang, W. Wang, and Z. Li, "Optimization of corona-triggered PDMS-PDMS bonding method," *Proc. of the 4th IEEE international conference on Nano/Micro Engineered and Molecular Systems*, 2009.

[5] T. Zhang , M. Zhang, and T. Cui, "Microfluidic valves based on TiO$_2$ coating with tunable surface wettability between super hydrophilic and super hydrophobic," *Proc. of Transducers'11*, pp. 306-309, Beijing, China, June, 5-9, 2011.

[6] G. W. Huang, T. Y. Hung, K. Z. Tu, and C. T. Chen, "Geometric design, simulation and experiments for optofluidic chips," *The 15th Nano and Microsystem Technology Conference*, Taipei, Taiwan, Sep. 6-7, 2011 (Chinese).

[7] S. Mandal and D. Erickson, "Nanoscale optofluidic sensor arrays," *Opt. Express*, vol. 16, pp.1623-1631, 2008.

[8] S. H. Cho, J. Godin, C. H. Chen, F.S. Tsai, and Y.H. Lo, "Microfluidic photonic integrated circuits," *Proc. Optoelectron Mater. Devices*, vol. 7135, 2008.

978-1-4673-1122-9/12 $31.00 © 2012 IEEE

Separation of Dendritic and T Cells Using Electrowetting and Dielectrophoresis

Chang-An Chen[1], Chiun-Hsun Chen[1], Amir M. Ghaemmaghami[2], and Shih-Kang Fan[1,3*]

[1]Department of Mechanical Engineering, National Chiao Tung University, Hsinchu, Taiwan
[2]Allergy Research Group, School of Molecular Medical Sciences and Respiratory Biomedical Research Unit, University of Nottingham, Nottingham, United Kingdom
[3]Department of Materials Science and Engineering, National Chiao Tung University, Hsinchu, Taiwan
skfan@mail.nctu.edu.tw

Abstract—**The research of immune cells is fundamental to many biological studies. Dendritic cells have the ability to induce a primary immune response in resting naive T cells. The aim of this work is to separate the activated T cells from dendritic cells on a digital microfluidic device where droplets are driven by electrowetting-on-dielectric (EWOD). The cells suspended in EWOD-driven droplets are separated and concentrated using a high frequency electric signal which generates non-uniform electric fields and dielectrophoresis (DEP) forces exerting on dendritic cells and T cells. Separation and concentration of dendritic cells and T cells are demonstrated in a droplet using EWOD and DEP.**

KEYWORDS-dendritic cell;T cell; electrowetting-on-dielectric; dielectrophoresis

I. INTRODUCTION

Immune system plays an important role in human disease protection. In the immune system, T cell-mediated adaptive immune is triggered by antigen-presenting cells (APCs), such as dendritic cells. Hence the interaction between dendritic cells (DCs) and T cells is essential to understanding the start of immune responses and the mechanism of capturing foreign antigens [1][2]. In order to investigate the interaction between DCs and T cells, several biomedical tools were employed, including ELISA and flow cytometry [3]. In addition, investigators used some labels (antibody modified-magnetic bead) to capture cells for cell separation and concentration [4]. However, the reagents are usually costly, while the activity of the bead-captured cells may degenerate.

Here we report a label-free cell separation technique based on dielectrophoresis (DEP). DEP has been demonstrated in many biotechnological applications, for example, in the separation of leukocytes from human blood [5], the separation of viable and nonviable yeast cells [6], and the separation of latex particles and viruses [7]. Based on our previous studies, DEP and electrowetting-on-dielectric (EWOD) can be integrated on a single device for cross-scale manipulations of liquid droplets on mm scale and cells on μm scale [8].

This study presents a device for the separation of DCs and T cells, where EWOD pumps droplets and DEP captures cells. The proposed device uses square electrodes to move the droplets and strip electrodes to concentrate the cells. After operations, the cells can be separated and precisely located on the different sides of a droplet.

II. PRINCIPLE

EWOD and DEP are the two basic mechanisms to separation of the DCs and T cells.

A. EWOD (electrowetting-on-dielectric)

Figure 1 shows the droplet pumped in a parallel-plate device by EWOD. When applying an AC signal on one of the patterned driving electrodes (i.e., the right one in Fig. 1) on the bottom plate and the un-patterned electrode on the top plate, contact angle of droplet is decreased by EWOD. Because of the difference of the contact angles between the two sides of the droplet (Fig. 1(b)), the droplet moves toward the energized electrode. The patterned driving electrodes on the bottom plate are covered with a dielectric layer and a hydrophobic layer. The un-patterned electrode on the top plate is coated with a hydrophobic layer. The EWOD driving force in the parallel-plate device is described as:

$$F_{EWOD} = \frac{\varepsilon_0 \varepsilon_{dielectric} w}{2t} V_{dielectric}^2 \qquad (1)$$

where ε_0 is the permittivity of vacuum, $\varepsilon_{dielectric}$ is the relative permittivity of the dielectric layer, w is the electrode width, t is the thickness of the dielectric layer, and V is the applied voltage across the across the dielectric layer.

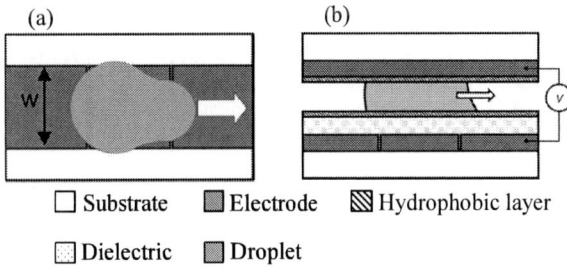

□ Substrate ■ Electrode ▨ Hydrophobic layer
▥ Dielectric ▨ Droplet

Fig. 1. Illustration of a parallel plate EWOD device to pump droplets between the plates with an applied AC voltage. (a) Top view. (b) Cross section.

B. DEP (dielectrophoresis)

A suspended particle in a medium with different dielectric characteristics will become electrically polarized in an electric field. The direction of the dipole moment induced in the suspended particles varies with the relative polarizability between the particles and the ambient medium. A non-uniform electric field will interact with the induced dipole and the subsequent asymmetric action of the electrical force on the particle. The time-averaged dielectrophoresis force, exerted on the particle suspended in a dielectric fluid medium and exposed to a non-uniform AC electric field can be described as [9]-[11]:

$$F_{DEP} = 2\pi a^3 \varepsilon_m \, \mathrm{Re} \left\{ \frac{\varepsilon_p^* - \varepsilon_m^*}{\varepsilon_p^* + 2\varepsilon_m^*} \right\} \nabla E^2 \qquad (2)$$

where a is the particle radius, ε_m is the permittivity of the suspension medium, E is the electric field, ε_p^* and ε_m^* are the complex permittivities of the manipulated particles and the suspension medium, respectively, which are frequency dependent and can be expressed. As a result, the electric field at high frequencies is almost uniform in a droplet as indicated by the electric field lines in (Fig. 2(a)). To create a non-uniform electric field inside a droplet in a dielectric-coated parallel plate device, shrunk electrodes (Fig. 2(b)) are as important as high frequency signals.

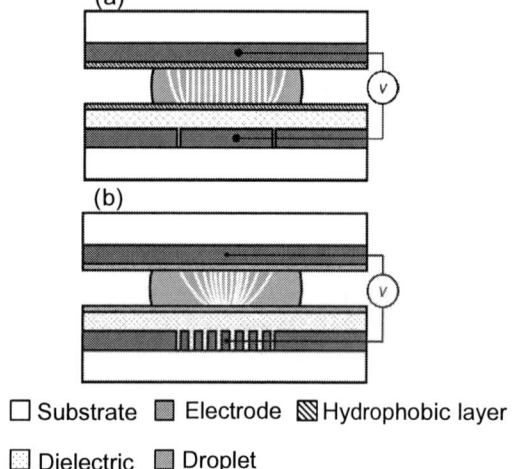

Fig. 2. Electric field lines displaying the uniformity of the electric field in a droplet when applying a high frequency signal on the designed electrodes. (a) Square EWOD driving electrodes generate a nearly uniform electric field. (b) Shuck electrodes establish a non-uniform electric field that is necessary for DEP actuations.

III. DESIGN AND FABRICATION

A. Design

Fig. 3 shows the parallel palate device containing square and strip electrodes capable of driving droplets by EWOD force and separating two kinds of suspended cells by DEP forces.

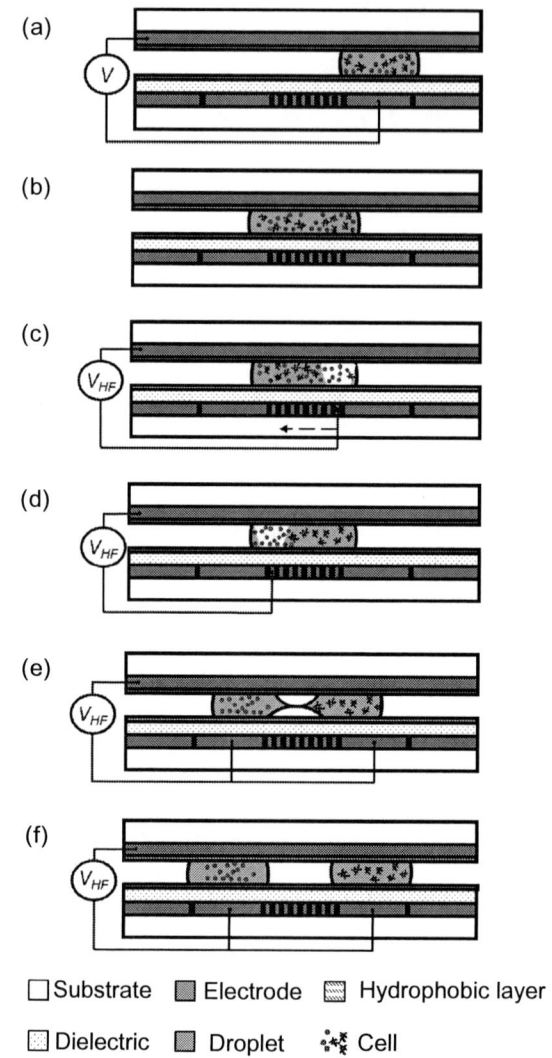

Fig. 3. Configuration and cell culture procedure of a parallel palate device containing square and strip electrodes for droplets and suspended cells by EWOD and DEP. (a) The droplet contain DC cells and T cells was pumped by EWOD force. (b) The droplet was pumped on the strip electrodes. (c)-(d) By applying a high frequency signal on one of the strip electrodes from right to left, the non-uniform electric field was generated to separate the cells. (e)-(f) After concentrating the cells, the droplet was separated into two individual droplets by EWOD force.

B. Fabrication

The devices are fabricated on glass substrates for ideal electrodes and for easier observation of the motion of biological cells. The photolithographic technology is used to define the pattern of electrode layout and the processes are described as below. The bottom driving electrodes were then covered with 2 μm-thick SU-8 as a dielectric layer and coated with a 55 nm-thick as a hydrophobic material Teflon layer. The top glass plate consisted of an un-patterned blank ITO layer covered by Teflon.

IV. EXPERIMENT SETUP

The fabricated devices were placed on a microscope (OLYMPUS Inverted microscope, IX-71) for testing. The applied voltage was produced by a function generator (Agilent, Dual channel arbitrary/function generator, 33522A), an amplifier (A.A.Lab Systems, A303), and a digital signal input/output controller (National Instruments, USB-6251), control circuits, and a personal computer.

Prior to the experiments, the DCs and T cells were detached from a 24-well plate by trypsin-EDTA. Subsequently, the DCs and T cells solution was centrifuged and washed for 3 times with PBS. After washing, the DCs and T cells mixed solution was centrifuged and the supernatant was removed by pipetting. The final DCs and T cells pellets were resuspended in suspension medium which was prepared by adding 3% PBS in a 280 mM isotonic sucrose solution, resulting in a conductivity of 480 μS·cm^{-1}. Cell solution droplets were pumped by EWOD force and cells were captured by DEP force.

V. RESULT

A. Concentrate T Cells

An isotonic sucrose solution droplet (~200 nl) containing 8.6 × 10^2 T cells was generated from its reservoir as shown in Fig. 4(a)-(b). The isotonic sucrose solution droplet was pump to strip electrodes by EWOD (60 V$_{RMS}$, 1 kHz AC signal) as shown in (Fig. 4(c)-(d)). After T cells were scatter on isotonic sucrose solution droplet, the T cells were concentrated from the right side to the left side by positive DEP (30 V$_{RMS}$, 3.2 MHz AC signal) as shown in Fig. 4(e)-(f). Two sub-droplets of different cell concentrations were formed by applying a low frequency signal (60 V$_{RMS}$, 1 kHz AC signal) to the two adjacent square electrodes as shown in Fig. 4(g)-(h).

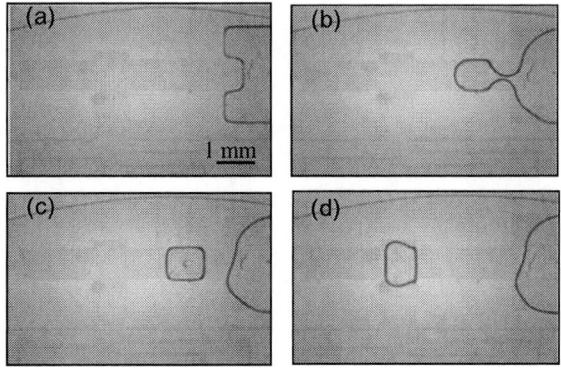

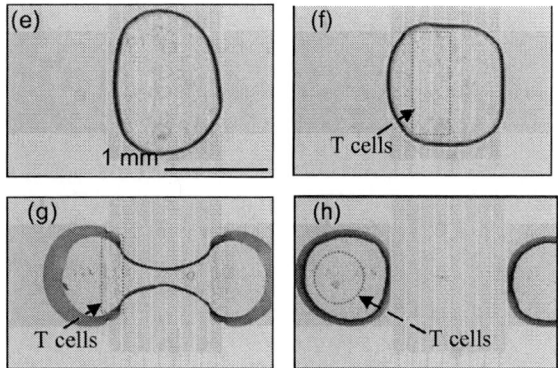

Fig. 4. (a) T cells evenly distributed (b) Create a droplet (c)-(d) Moved droplet to strip electrodes (e)-(f) The cell concentrated from right to left (g)-(h) Droplet cut into two droplets with different concentration.

B. Repel DCs

An isotonic sucrose solution droplet (~200 nl) containing 8.6 × 10^2 DCs was generated from its reservoir. The isotonic sucrose solution droplet was pump to strip electrodes by EWOD (60 V$_{RMS}$, 1 kHz AC signal). After DCs were scatter on isotonic sucrose solution droplet, the DCs were repelled from right to left by negative DEP (30 V$_{RMS}$, 3.2 MHz AC signal) as shown in Fig. 5(a)-(f).

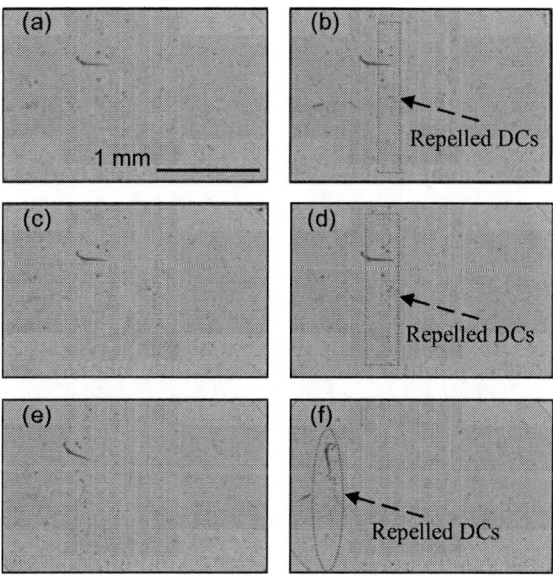

Fig. 5. (a)-(f) The DCs were concentrated from right side to left side by negative DEP.

C. Separate DCs and T Cells.

An isotonic sucrose solution droplet (~200 nl) containing 8.6×10^2 DCs and T cells was generated from its reservoir as shown in Fig. 6. The isotonic sucrose solution droplet was pump to strip electrodes by EWOD (60 V_{RMS}, 1 kHz AC signal). After DCs and T cells were scattered on isotonic sucrose solution droplet, the T cells were attracted from right to left by positive DEP, while the DCs were repelled from right to left by negative DEP (Fig. 6(a)-(e)). After sweeping the applied signal (30 V_{RMS}, 3.2 MHz AC signal) from left to right, the T cells were attracted from left to right as shown in Fig. 6(e)-(f). In our device, the volume of created droplet was quantitative, so the concentration of cells was determined. In the future we can create sub-droplets containing different kinds of cells. By moving the droplet and polarizing cells, cells can be separated and concentrated.

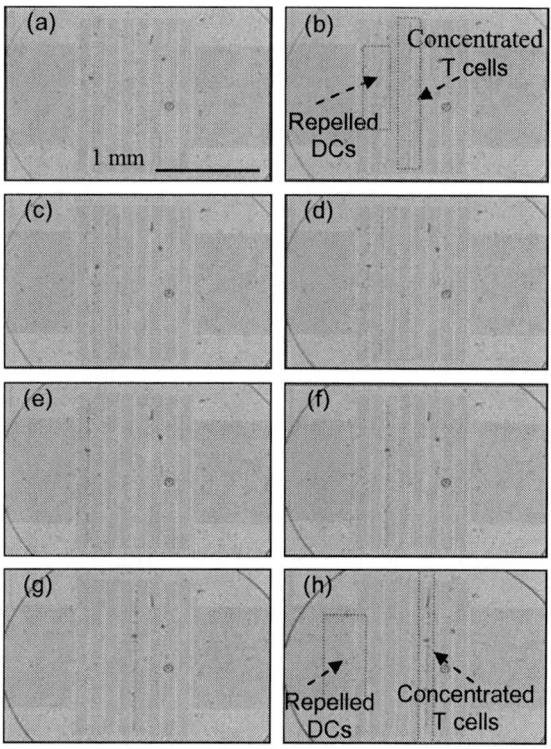

Fig. 6. (a)-(d) The T cells concentrated on the strip electrode from right to left by positive DEP and the DCs repelled from right to left by negative DEP (e)-(h) The T cells concentrated from left to right by positive DEP DEP.

VI. CONCLUSION

In this study, we successfully demonstrated that the DCs and T cells were separated by DEP force in the isotonic droplet driven by EWOD force. After concentration, the mother droplet was cut into two individual droplets. However, the DCs and T cells were not completely separated, which were resulted from different DEP forces. The alive or dead DCs, electrode design and conductivity of buffer will influence the effect of DEP force on cells. In the future, we will modify the conditions, including all alive cells in droplet, proper electrode design, and appropriate conductivity of solution. In addition, immune response of DCs and T cells will be achieved in our system.

ACKNOWLEDGMENT

This work was financially supported by the National Science Council of Taiwan under grant NSC-100-2120-M-009-009 and made use of shared facilities provided.

REFERENCES

[1] M.F. Lipscomb, and B.J Masten, "Dendritic cells: Immune regulators in health and disease," *Physiol. Rev.,* vol. 82, pp. 97-130, 2002.

[2] F. Sallusto, and A. Lanzavecchia, "The instructive role of dendritic cells on T-cell responses," *Arthritis. Res. Ther.,* vol. 4, pp. 127-132, 2002.

[3] T. Jung, U. Schauera, C. Heusserb, C. Neumannc, and C. Rieger, "Detection of intracellular cytokines by flow cytometry," *J. Immunological Methods.,* vol. 159, pp. 197-207, 1993.

[4] Martin A. M. Gijs, "Magnetic bead handling on-chip: new opportunities for analytical applications", *Microfluid. Nanofluid.,* vol. 1, pp. 22-40, 2004.

[5] X.-B. Wang, J. Yang, Y. Huang, J. Vykoukal, F.F. Becker, and P.R.C. Gascoyne, "Cell separation by dielectrophoretic field-flow-fractionation", *Anal. Chem.,* vol. 72, 832-839, 2000.

[6] G.-H. Mark, M.-S. Talary, and R. Pethig, "Separation of viable and non-viable yeast using dielectrophoresis", *J. Biotechnol,* vol. 32, pp. 29-37, 1994.

[7] T. Schnelle, T. Muller, G. Gradl, S.G. Shirley, and G. Fuhr, "Dielectrophoretic manipulation of suspended submicron particles", *Electrophoresis,* vol. 21, pp. 66–73, 2000.

[8] S.-K. Fan, P.-W. Huang, T.-T. Wang, and Y.-H. Peng, "Cross-scale electric manipulations of cells and droplets by frequency-modulated dielectrophoresis and electrowetting", *Lab Chip,* vol. 8, pp. 1325-1331, 2008.

[9] T. Schnelle, T. Muller, G. Gradl, S.G. Shirley, and G. Fuhr, "Dielectrophoretic manipulation of suspended submicron particles," *Electrophoresis,* vol. 21, pp. 66–73, 2000.

[10] H. Morgan and N. G. Green, AC Electrokinetics: colloids and nanoparticles, Research Studies Press Ltd., Baldock, 2003.

[11] T. B. Jones, Electromechanics of Particles, Cambridge University Press, Cambridge, 1995.

Development of Microbead-based Affinity Biosensor by Insulator-Based Dielectrophoresis

Tsung-Min Chuo[1], Wensyang Hsu[1], Shih-Kang Fan[1,2]

[1]Department of Mechanical Engineering, National Chiao Tung University, TAIWAN
[2]Department of Materials Science and Engineering, National Chiao Tung University, TAIWAN
skfan@mail.nctu.edu.tw

Abstract—**This research describes a high sensitivity microfluidic bead-based immunosensor based on the principle of insulator-based dielectrophoresis (iDEP). An insulator film with small holes between two electrodes creates a nonuniform electric field. By applying appropriate voltage and frequency, the fluorescent beads are concentrated to lower electric field regions due to the difference of dielectric properties. This concentrating step enhances the fluorescence intensity of analytes and decreases the detection limit of immunosenser. In this research, the fluorescence dye is conjugated with streptavidin which has high affinity to biotin. We use biotin-labeled polystyrene beads to bind with streptavidin, therefore, we can further detect fluorescent streptavidin conjugates by a fluorescence microscope. The biotin-labeled polystyrene beads perform not only various chemical characteristics by labeling different functional groups but also offer an increased surface area for antibodies or antigens to immobilize on. Finally, we fabricate a microfluidic bead-based immunosensor with high sensitivity (1 pg/ml), short analysis time (~10 minutes), few sample consumption (~0.5 μl) and without physical microchannel.**

Keywords-component; immunosensor; iDEP; fluorescence

I. INTRODUCTION

Immunoassays are biochemical analysis methods based on the high selectivity between antibody and antigen; they normally measure the presence or concentration of a specific substance in solutions or mediums that frequently contain a complex mixture of substances. Moreover, immunoassays are among the most sensitive and specific analytical methods that are routinely used in a clinical laboratory and other biological research applications.

In recent years, a new technique that uses microbeads as a solid support in immunoassays has become usual. There are several advantages in the use of microbeads. First, the microbeads' surface to volume ratio is greater than that of a microtiter plate commonly used in conventional immunoassays. For example, 1 g of microbeads with a diameter of 0.1μm has a total surface area of about 60 m^2 [1]. The large surface area provides a large interface and the reaction field between samples and reagents. The sensitivity of immunoassays would be increased as a result of the higher efficiency of the immunoreactions between the immobilized antibody and the antigen present in a continuous flow. In addition, the reaction rate may be increased because of the greater reaction field.

Second, the immunoassays which use microbeads as a solid support can be easily integrated into a microfluidic chip. The samples and reagents that used in immunoassays can be easily transported in a fluidic system by a syringe pump or another way. Third, there are various available surface modifications for microbeads. DNA, RNA, antibodies, antigens and a vast number of other biological molecules can be easily fixed on the surface of microbeads. Moreover, transportation and analysis in a fluidic system is easy [2].

Furthermore, the dynamic condition that utilizing both diffusion and convectional forces to deliver or mix samples with reagents in microfluidic system. In contrast, conventional immunoassay on a microtiter plate, likes enzyme-linked immunosorbent assay (ELISA), is a static condition that merely depends on diffusion of the molecules for interaction and binding.

Microfluidic technology is widely used in immunoassays available to improve the analytical characteristic performances, such as short analysis time, high reliability and high detecting sensitivity, easy handling and low consumption of reagents [3]. However, a retention method is necessary for trapping or fixing microbeads in microfluidic system in order to avoid the microbeads washing away in the microfluidic system. For example, microbeads can be trapped by arrayed microstructures [4]-[6], Kitamori and coworkers fabricated a dam structure for retaining polystyrene microparticles in a glass-based microchannel [6]. Magnetic beads are also used for immunoprotein support and separation, since these beads can be easily manipulated in the channel by applying a magnetic field [7]-[9]. Dielectrophoresis [10][11], and electrostatic forces [12][13] are another way to be a retention method.

Sensitivity means the lowest concentration or the smallest amount of analytes that can be detected above the baseline, which is perhaps the most widely touted measures of an assay since it is easy quantified. Compared to conventional immunoassays, those relying on fluorescence detection, are known to be highly sensitive [14]. They have the potential that can replace the traditional ELISA technique if the fluorescence signal arising from fluorophores bound with analytes can be effectively reinforced.

In this research, we can enhance the fluorescence intensity in a simple way instead of complex chemical operations. The fluorescence intensity can be increased by concentrating beads. We have developed a fluorescent bead-based immunoassay using iDEP adapted from existing ELISA. This microfluidic bead-based immunosensor has high sensitivity, short analysis time, few sample consumption and without any microchannel.

II. THEORY

A. Dielectrophoresis

Dielectrophoresis (DEP) is an electrokinetic phenomenon which can drive particles by using electrodes instead of moving actuators. A dielectric particle suspending in a solution would be affected by a force caused by the interaction between the spatially inhomogeneous electrical fields causing polarization. The DEP force has been widely used to manipulate, transport, separate and sort different types of particles.

DEP can be classified into two types: positive DEP (p-DEP) and negative DEP (n-DEP). Particles are attracted to the region of a stronger electric field with the p-DEP force because their permittivity is greater than that of the solution. In contrast, particles are attracted to the region of a weaker electric field with the n-DEP force because their permittivity is smaller than that of the solution. In addition, p-DEP [15][16] and n-DEP [17][18] have been used to manipulate particles and biological cells with microelectrode systems.

The DEP force, F_{DEP}, on a suspended spherical particle in a solution is given by

$$F_{DEP} = 2\pi\varepsilon_m a^3 \operatorname{Re}(f_{CM})\nabla E^2_{RMS}, \qquad (1)$$

where a is the particle radius [m], ε_m is the permittivity of the suspension solution [F/m], E_{RMS} is the root-mean-square electric field [V/m], ∇ is the del vector operator, and $\operatorname{Re}(f_{CM})$ is the real part of the Clausius-Mossotti factor, given by

$$f_{CM} = \frac{\widetilde{\varepsilon}_p - \widetilde{\varepsilon}_m}{\widetilde{\varepsilon}_p + 2\widetilde{\varepsilon}_m}, \qquad (2)$$

where $\widetilde{\varepsilon}_p$ is the complex permittivity of the particle, and $\widetilde{\varepsilon}$ is given by

$$\widetilde{\varepsilon}_m = \varepsilon_m - \frac{\sigma_m}{\omega}j,$$
$$\widetilde{\varepsilon}_p = \varepsilon_p - \frac{\sigma_p}{\omega}j, \qquad (3)$$

where σ is the conductivity [S/m], ε is the permittivity, ω is the angular frequency and j equals $\sqrt{-1}$.

B. Insulator-based dielectrophoresis

Insulator-based (electrodeless) dielectrophoresis (iDEP) is a technology to produce the nonuniform electrical field by insulators for driving DEP. Hence iDEP would avoid the problems caused by electrodes.

In this research, iDEP is used to collect fluorescent beads on a specific device which includes two electrodes with a patterned insulator film in between as shown in Fig. 1. When voltage is applied on the device, the charged fluorescent beads are gathered at the region of weaker electric field on the insulator film, so that the beads collection can be completed by the n-DEP force. The fluorescence signal will be enhanced by concentrating fluorescent beads. The method can increase fluorescence intensity and sensitivity. Our device is based on the iDEP technique to detect a limited amount of streptavidin.

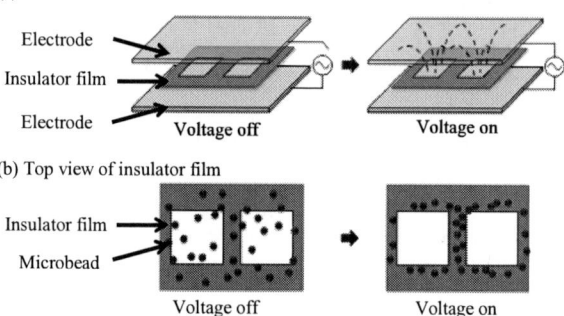

Fig. 1. Principle behind the assay methodology, combining the insulator film and the n-DEP-based manipulation techniques. (a) Using insulator film to form nonuniform electric field. (b) Beads concentrated on the gap of a weaker electric field.

III. EXPERIMENT

A. Fabrication

Indium tin oxide (ITO) is one of the most widely used transparent conducting oxides because of its electrical conductivity and optical transparency, as well as the ease of depositing and pattering. The device was fabricated on glass substrate with deposited ITO thin film. The dielectric layer of 2 μm Parylene C was deposited on ITO thin film by chmical vaporization deposition (CVD). The hydrophobic layer of Teflon is spin-coated onto the dielectric layer for increasing the contact angle of droplets.

The insulator film was fabricated by dry film photoresist (PerMX3020, DuPont). The pattern on the PerMX dry film is shown in Fig. 2. The pattern consists 25 squares arranged in a 5×5 array. The length of the squares is 35 μm and the interval between the squares is 60 μm. To fabricate it, a PerMX dry film was baked and is exposed by UV light (350-400 nm). Afterward, the PerMX dry film was developed by PerMX developer and rinsed by IPA.

978-1-4673-1122-9/12 $31.00 © 2012 IEEE

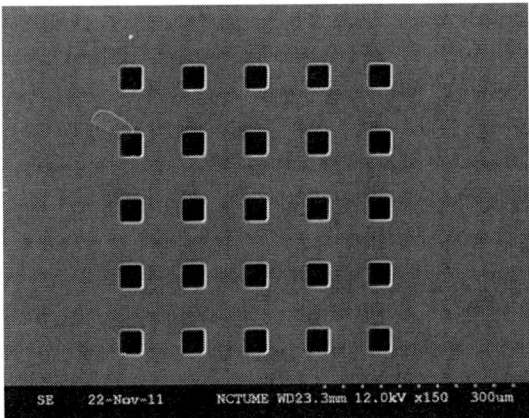

Fig. 2. SEM photo of the patterned PerMX dry film.

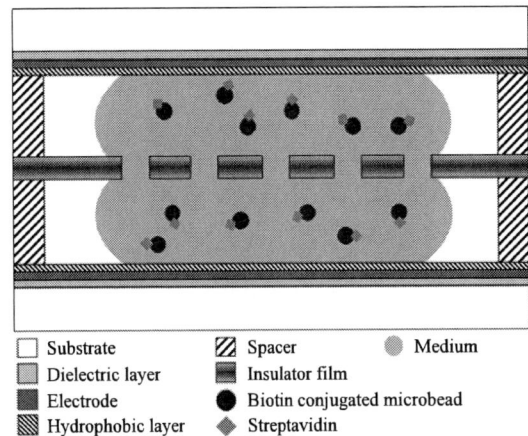

Substrate	Spacer	Medium
Dielectric layer	Insulator film	
Electrode	Biotin conjugated microbead	
Hydrophobic layer	Streptavidin	

Fig. 3. The cross section of experiment construction.

B. Reagents and Materials

Biotin covalently coupled to Fluoresbrite® YG fluorescent microbeads (2μm diameter) was purchased from Polysciences, Inc. (U.S.A.). The biotin-labeled polystyrene beads were dispensed into a microcentrifuge tube and centrifuged at 10000 ×G for 5 minutes. The supernatant in the tube was then rded. The beads are resuspended in PBS/BSA binding buffer. These steps will be repeated three times to wash the microsbeads. After the last washing, the beads can be resuspended to any volume, however higher concentrations usually work better (at least 5×10^8 particles/ml).

Rhodamine B labeled streptavidin purchased from Invitrogen (U.S.A.) was incubated with the processed beads for 30 minutes at 4C to ensure that the reaction between two species was sufficient. The tube was centrifuged for 5 minutes and the supernatant in the tube was discarded. The beads are resuspended in PBS/BSA binding buffer. After repeating three times, the beads were ready for use in the following experiment.

C. Experiment Setting

Double side tapes were pasted between the ITO glass plate and the insulator film as spacers. A 0.5 μl droplet was created by a pipette. Cross section of the experiment setting is shown in Fig. 3. AC voltage was applied to the sandwich constructer from a function generator. Finally, wires are connected between the chip and the control circuits. The condition of concentrated beads can be observed under a fluorescence microscope.

AC signal is produced by a function generator and passed through an amplifier to the electrodes. The electrodes were connected to a PC via an I/O card (DAQCard-USB6251). The signal is controlled with LabVIEW software (National Instruments). Detection processes was visualized by a fluorescence microscopy (IX71, Olympus) equipped with a CCD camera. The fluorescence images were analyzed by image analysis software (Image Pro Plus).

IV. RESULTS AND DISCUSSION

A. Concentrate Beads

By applying voltage, the beads would be concentrated of the gap between two square holes of the insulator film by the n-DEP force. In this experiment, two different fluorescence dyes are labeled on biotin-labeled polystyrene beads and streptavidin separately. Therefore, we would observe the condition of biotin-labeled polystyrene beads and streptavidin simultaneous by using different optical filter, respectively. The results are shown in Fig. 4 and the operation time was 15 minutes.

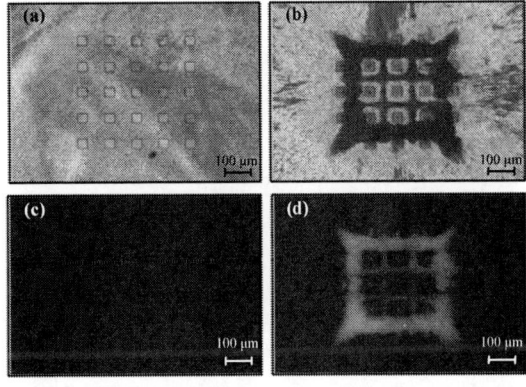

Fig. 4. The photos of concentrated beads on insulator film. (a) Before apply voltage (bright field). (b) Apply voltage to concentrate beads after 15 minutes (bright field). (c) Before apply voltage (fluorescence field). (d) Apply voltage to concentrate beads and observe Rhodamine B labeled streptavidin after 15 minutes (fluorescence field).

B. Operation Time

The fluorescence intensity of concentrated beads approaches to a limit after a long time. It is important to determine how long we need to concentrate beads. For quantitative evaluation of concentrated beads, the value of fluorescence intensity was calibrated to be 0 for the initial state and 1 for the final state. We applied a 140 V_{RMS} voltage at 100 kHz and analyzed the captured photos caught during operation. The results are shown in Fig. 5. From the results, the fluorescence intensity becomes stable about 10 minutes. The correlation coefficient for these 15 samples is ~0.92, which seems to be rather high for an immunoassay.

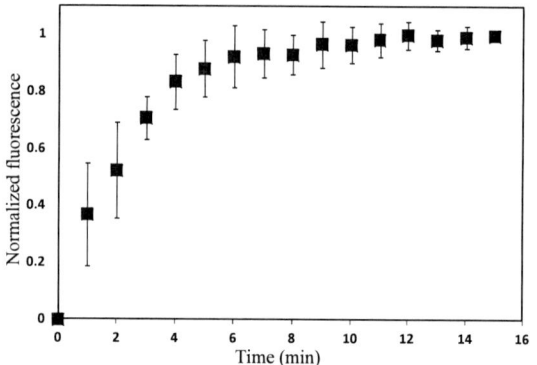

Fig. 5. The measured intensity of concentrated fluorescent beads plotted along with time.

C. Results

The fluorescence intensity increased while the concentration of fluorescent dye increased. In this research, we used fluorescent beads and Rhodamine B labeled streptavidin. For quantitative evaluation of the relation between intensity and the concentration of Rhodamine B labeled streptavidin. First, we choose a value of fluorescent beads' fluorescence intensity. Then, the value of streptavidin's fluorescence intensity divides by this number. Finally, we can get a concentration-fluorescence intensity graph and is summarized in Fig. 6. Concentrations of the Rhodamine B labeled streptavidin ranges from 1 pg/ml to 1 ng/μl and the determination limit is 1 pg/ml.

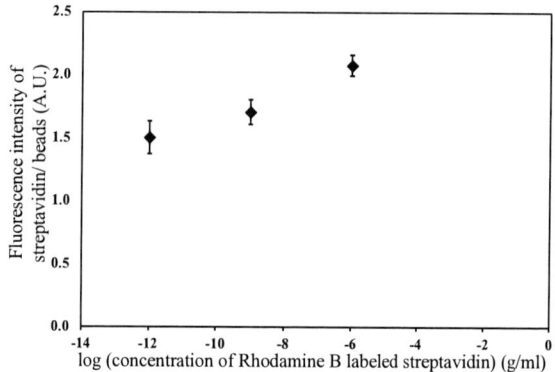

Fig. 6. The calibration curve of Rhodamine B labeled streptavidin.

V. CONCLUSION

In this paper, we increase the sensitivity of a bead-based immunosensor by concentrating beads instead of complex chemical operations. The concentration step by iDEP is finished less than 10 minutes and increases the fluorescence

under 140 V_{RMS} at 100 kHz. We anticipate that this mechanism may be useful in Pico-scale for clinical diagnosis or other biological research applications.

ACKNOWLEDGMENT

This work is supported by the National Science Council of Taiwan under grant NSC-100-2120-M-009-009.

REFERENCES

[1] E. Verpoorte, "Beads and chips: new recipes for analysis," *Lab Chip,* vol. 3, pp.60N-68N, 2003.

[2] C.J. Lim, and Y. Zhang, "Bead-based microfluidic immunoassays: The next generation," *Biosens. Bioelectron.,* vol. 22, pp. 1197-1204, 2007.

[3] A. Bange, H.B. Halsall, and W.R. Heiniman, "Microfluidic immunosensor systems," *Biosens. Bioelectron.,* vol. 20, pp. 2488-2503, 2005

[4] H. Andersson, W.v.d. Wijngaart, P. Enoksson, and G. Stemme, "Micromachined flow-through filter-chamber for chemical reactions on beads," *Sens. Actuators, B,* vol. 67, pp. 203-208,2000.

[5] K. Sato, M. Tokeshi, T. Odake, H. Kimura, T. Ooi, M. Nakao, and T. Kitamori, "Integration of an immunosorbentassay system: analysis of secretory human immunoglobulin A on polystyrene beads in a microchip," *Anal. Chem.,* vol. 72, pp. 1144-1147, 2000.

[6] K. Sato, M. Yamanaka, H. Takahashi, M. Tokeshi, H. Kimura, and T. Kitamori, "Microchip-based immunoassay system with branching multichannels for simultaneous determination of interferon-γ," *Electrophoresis,* vol. 23, pp. 734-739, 2002.

[7] J.W. Choi, K.W. Oh, J.H. Thomas, W.R. Heineman,H.B. Halsall, J.H. Nevin, A.J. Helmicki, H.T. Henderson, and C.H. Ahn, "An integrated microfluidic biochemical detection system with magnetic bead-based sampling and analysis capabilities," *Lab Chip,* vol. 2, pp. 27-30, 2002.

[8] C.A. Wijayawardhana, S. Purushothama, M.A. Cousino, H.B. Halsall, and W.R. Heineman, "Rotating disk electrode amperometric detection for a bead-based immunoassay," *J. Electroanal. Chem.,* vol. 468, pp. 2-8, 1999.

[9] E. Zacco, M.I. Pividori, S. Alegret, R. Galve, and M.P. Marco, "Electrochemical magnetoimmunosensing strategy for the detection of pesticides residues," *Anal. Chem.,* vol. 78, pp. 1780-1786, 2006.

[10] T. Yasukawa, M. Suzuki, T. Sekiya, Shiku H, and Matsue T, "Flow sandwich-type immunoassay in microfluidic devices based on negative dielectrophoresis," *Biosens. Bioelectron.,* vol. 22, pp. 2730-2736, 2007.

[11] D. Holmes, J.K. She, P.L. Roach, and H. Morgan, "Bead-based immunoassays using a micro-chip flow cytometer," *Lab Chip,* vol. 7, pp. 1048-1056, 2007.

[12] V. Sivagnanam, B. Song, C. Vandevyver, and M.A.M. Gijs, "On-Chip Immunoassay Using Electrostatic Assembly of Streptavidin-Coated Bead Micropatterns," *Anal. Chem.,* vol. 81, pp.6509-6515, 2009.

[13] V. Sivagnanam, A. Bouhmad, F. Lacharme, C. Vandevyver, and M.A.M. Gijs, "Sandwich immunoassay on a microfluidic chip using patterns of electrostatically self-assembled streptavidin-coated beads," *Microelectron. Eng.,* vol. 86, pp.1404-1406, 2009.

[14] R.S. Yalow, and S.A. Berson, "Immunoassay of endogenous plasma insulin in man", *J.Clin. Invest.,* vol. 39, 1157-1175, 1960.

[15] D.S. Gray, J.L. Tan, J. Voldman, and C.S. Chen, "Dielectrophoretic registration of living cells to a microelectrode array," *Biosens. Bioelectron.,* vol. 19, pp. 1765-1774, 2004.

[16] S. Masuda, M. Washizu, and T. Nanbu, "Novel method of cell fusion in field constriction area in fluid integration circuit," *IEEE Trans. Ind. Appl.,* vol. 25, pp. 732-737, 1989.

[17] M. Suzuki, T. Yasukawa, T. Mase, D. Oyamatsu, H. Shiku, and T. Matsue, "Dielectrophoretic Micropatterning with Microparticle Monolayers Covalently Linked to Glass Surfaces," *Langmuir,* vol. 20, pp. 11005-11011, 2004.

[18] T. Matsue, N. Matsumoto, and I. Uchida, "Rapid micropatterning of living cells by repulsive dielectrophoretic force," *Electrochim. Acta,* vol. 42, pp. 3251-3256, 1997.

Difference Proportional Cell Contact Platform for 3D Hepatocyte Culture

Yu-Shih Chen [1, *], Zong-Keng Kuo [2], Ling-Yi Ke [3], Chiou-Wen Lin [3], Hsiang-Wen Tseng [2] and Cheng-Hsien Liu [1,3]

[1] Institute of NanoEngineering and MicroSystems, National Tsing Hua University , [2] Biomedical Engineering Research Lab,
Industrial Technology Research Institute, [3] Department of Power Mechanical Engineering, National Tsing Hua University,
Hsinchu, Taiwan, R.O.C
* ericys2004@gmail.com

Abstract—**this paper reports a difference proportional cell contact platform for three dimension hepatocyte culture. We designed the cell wires about the width of 25 micrometer by dielectrophoresis for difference contact area. The platform was consisted of two dimensional cell contact chambers and three dimensional liver tissue chambers by dielectrophoresis pattern method and microfluidic channel design. For cell contact chambers, we used dielectrophoresis pattern method to make difference contact area for difference proportional cell communication. For liver tissue chambers, we used the UV-initiated solution of the poly (ethylene glycol)-diacrylate (PEG-DA) had been exposed to form microwells to pattern three dimension lobule-mimetic structure liver tissue in vitro. HepG2 cells were loading with three dimensional PEG-DA-dased microstructures for artificial lobule pattern. Finally, we showed liver functional albumin testing. The albumin secretion of liver pattern chambers with HMEC-1 and C2C12 co-culture communication was 34% higher than the albumin secretion of liver chip without cells communication.**

Keywords- Microfluidics Circulatory System; cell pattern; dielectrophoresis

I. INTRODUCTION

The ability to pattern 3D tissue-mimic pattern is a rapidly developing technique in biotechnology research, especially for the re-establishment of cell pattern and culture which are essential to functional tissue engineering. In recent study, the researcher developed microfluidic technology for the study of the hepatocytes. For example, an artificial liver sinusoid with a silicon-based endothelial-like barrier capable of transporting mass is reported [1]. Microfluidic methods have been developed to biomimic hepatic cords, as 3D culture systems for investigating 3D cell biology [2]. By the way, a 3D cell perfusion-culture in microfluidic system microenvironment is used in 3D cell-matrix interaction via a microfluidic channel based system [3]. For drug toxicity testing, hepatocytes are trapped by some micropillars as a microfluidic 3D hepatocyte chip [4]. Further, a perfusion-based device is supported to observe the expression of MRP2 protein for hepatocyte transport function assay [5]. In our group, we had developed a concentric-stellate-tip electrodes pattern to trap hepatic and endothelial cells for biomimetic heterogeneous patterning [6]. Further, A three dimension lobule-mimetic chip is observed depend on the hydrogel-based geometry microstructure and the novel cell manipulation method [7]. For cell-cell contact

we demonstrated a single cell trapping chip by DEP force with the movable hydrogel-based microstructure [8].

This paper reports a 2D cell contact difference pattern ration by positive dielectrophoresis(DEP) for 3D microfluidic hepatocyte culture in cell-to-cell communication array chip. Endothelium cells were patterned to form cell wires on the surface of the cell contact chambers. Hepatocyte cells was loading with 3D poly (ethylene glycol)-diacrylate (PEG-DA) microstructures for artificial lobule pattern and functional albumin testing. The liver is organized into lobules which take the shape of polygonal prisms.

II. METHID

A. Difference proportional cell contact platform

The cell sorting and tissue-mimetic is developing technique in soft lithography and microfluidic biotechnology. In liver study, liver is organized into lobules which take the shape of polygonal prisms. Each lobule is typically hexagonal in cross section and is centered on the central vein.

We demonstrated a difference proportional cell contact platform for three dimension hepatocyte culture as shown in

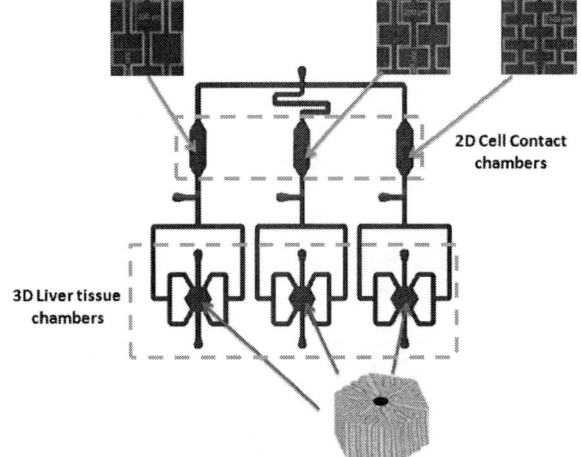

Fig. 1. Schematic diagram is a difference proportional cell contact platform for three dimension hepatocyte culture. It was consisted of two dimensional cell contact chambers in the upper of the orange dotted line rectangle and three dimensional liver tissue chambers in the below of the orange dotted line rectangle. We designed difference proportional cell communication real time.

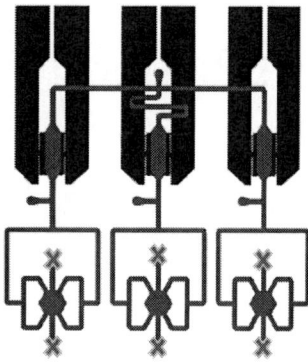

Fig. 2. Schematic diagram is the operation principle of cell contact chamber. Uniformly distributed HMEC-1 cells were loaded into the chamber from one inlet to the three chambers. The HMEC-1 cells were attracted and positioned via the DEP patterning electrodes.

Side View

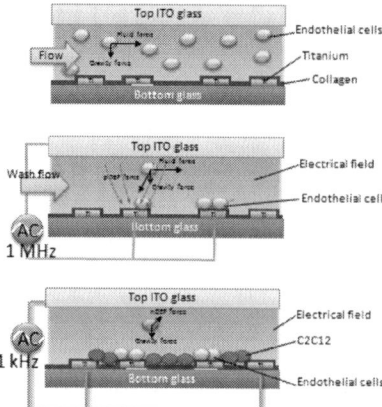

Fig. 4. Schematic diagram is operation principles of cell contact patterning. (a) Uniformly distributed HMEC-1 cells are loaded into the chamber. (b)The HMEC-1 cells are attracted and positioned via the DEP patterning electrodes. (c) The C2C12 cells are repelled and patterned via negative DEP force.

figure 1. It consisted of two dimensional cell contact chambers and three dimensional liver tissue chambers. These chanbers were controlled by dielectrophoresis pattern method and microfluidic channel design. For cell contact chambers, we used dielectrophoresis method to pattern the cell wires by HMEC-1 cells about the width of 25 micrometer. For liver tissue chambers, we used UV light to expose the solution of the PEG-DA to form microwells for patterning three dimensional lobule-mimetic structure liver tissues in vitro. The purposes of this research are to develop the engineered liver tissue regulated not only by cell signal-communicated but also by liver pattern mimic platform.

B. Two dimensional pattern

We had a novel idea which demonstrated cell wires by difference electrode pattern to make difference proportional cell contact for cell communication in figure 2. Six outlets about liver tissue chambers were sealed. The black area was the pattern of the electrode by the photolighography process for the dielectrophoresis pattern method. Uniformly

distributed HMEC-1 cells were loaded into the chamber from one inlet to the three chambers. Base on three kinds of the pattern of the electrode, the HMEC-1cells were attracted to focus three kinds of the size of the cell wires. In figure 3, we show the black area is electrode pattern. The white area is the cell wire area about the width of the 25 um. We designed three kings difference contact area which was depended on difference rectangles. For example, The rectangle area is 100umx300um in figure 3(a). In figure 3(b), The rectangle area is 200umx300um. The largest rectangle area is 300umx 300um in figure 3(c). And the orange lines were showed the position of the AC electric wire.

We can see the operation principles of cell contact patterning in figure 4. Uniformly distributed HMEC-1 cells are loaded into the chamber in step 1. The HMEC-1 cells are attracted and positioned via the DEP patterning electrodes in step 2. The C2C12 cells are repelled and patterned via

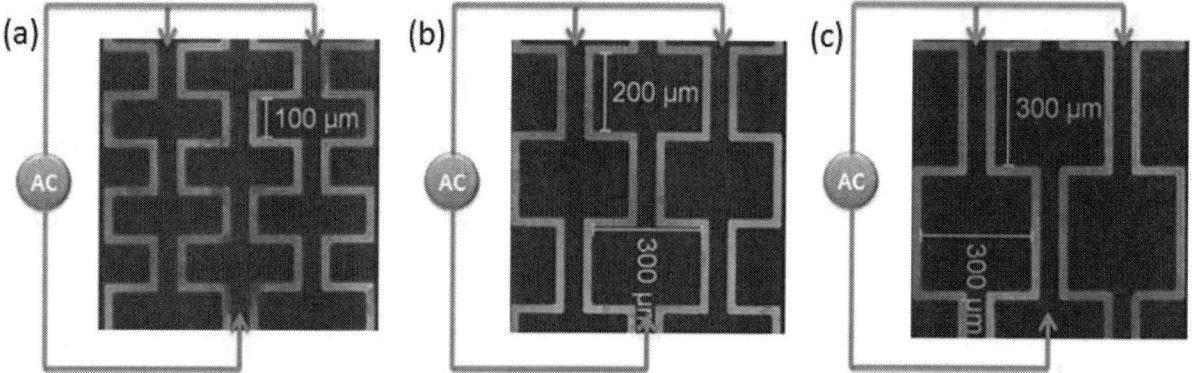

Fig. 3. We showed a novel idea which demonstrated the difference electrode pattern to make difference proportional cell contact for cell communication. (a) The rectangle area is100um×300um. (b) The rectangle area is 200um×300um area. (c) The rectangle area is 300um×300um area. The black area in figure 2 is electrode pattern. The white area is the cell wire area (the 25 um width).

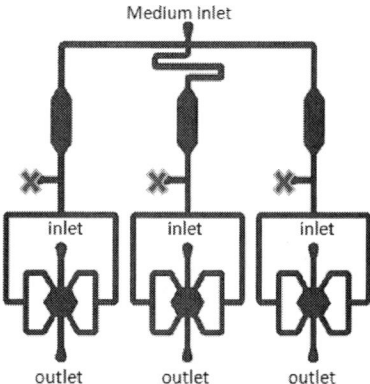

Fig. 5. Schematic diagram is the operation principle of liver tissue chamber. Uniformly distributed HepG2 cells were loaded into the chamber from three inlets to the three outlats. The HepG2 cells were trapped by the PEG-DA-based microwalls.

negative DEP force in step 3.

C. Three dimensional pattern

We showed the operation principle of liver tissue chamber. Four outlets about cell contact chambers were sealed. Uniformly distributed HepG2 cells were loaded into the chamber from three inlets to the three outlats. The HepG2 cells were trapped by the PEG-DA-based microwalls. In these three liver tissue chambers, we designed culture micro environment in the same situation.

III. RESULTS AND DISCUSSIO

Figure 6 shows on-chip cell contact patterning result of HMEC-1 cells and C2C12 cells in case of the rectangle area is 300um×300um. Firstly, the HMEC-1 cells were attracted and positioned via the positive DEP force on the surface of the cell contact chambers in figure 6(a). We can see the fluorescent photon of patterned artificial lobule in figure 6(b). And then, C2C12 cells were patterned on the electrode via negative DEP manipulation. We also can see the fluorescent photon of the two kinds of the cells pattern in figure 6(c).

On the other hand, HepG2 cells were patterned in liver tissue chambers and fluorescent photon of patterned artificial lobule in figure 7. We used the solution of the PEG-DA had been exposed to form microwells to pattern three dimension lobule-mimetic structure liver tissue in vitro.

In liver albumin experiment result in figure 8, the albumin secretion of liver pattern chamber with HMEC-1 and C2C12 co-culture communication (condition setup) was 34% higher than the albumin secretion of liver chip without cells communication (control setup). Finally, we successful not only develop difference proportional cell to cell contact patterning but also nicely the 3D hepatic lobule.

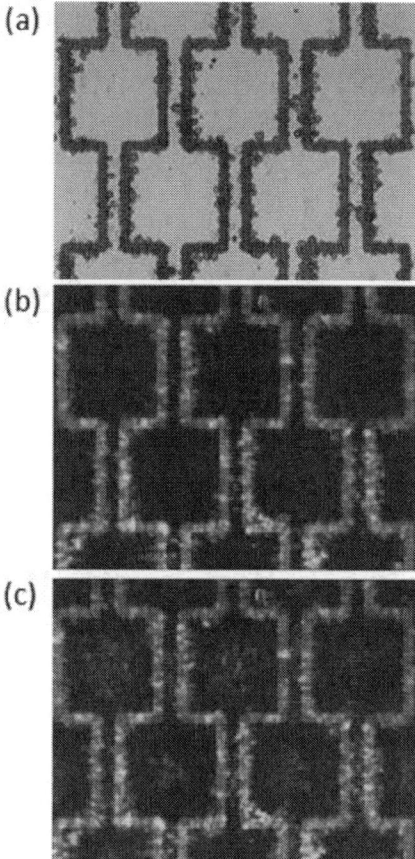

Fig. 6. On-chip cell contact patterning result of HMEC-1 (green) and C2C12 (red) cells in case of the rectangle area is 300um×300um.

Fig. 7. HepG2 cells were patterned in liver tissue chambers and fluorescent photon of patterned artificial lobule. We used the solution of the PEG-DA had been exposed to form microwells to pattern three dimension lobule-mimetic structure liver tissue in vitro.

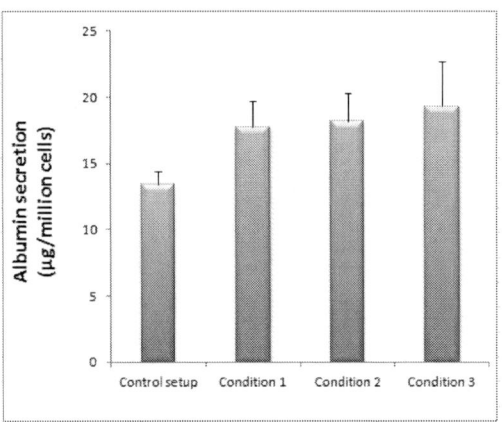

Fig. 8. In liver albumin experiment result, the albumin secretion of liver pattern chamber with HMEC-1 and C2C12 co-culture communication (condition setup) was 34% higher than the albumin secretion of liver chip without cells communication (control setup).

IV. CONCLUSION

We demonstrated a difference proportional cell contact platform for three dimension hepatocyte culture. The platform was consisted of two dimensional cell contact chambers and three dimensional liver tissue chambers. Those were patterned or trapped by dielectrophoresis pattern method and microfluidic channel design. For cell contact chambers, we used dielectrophoresis pattern method to make difference contact area for difference proportional cell communication. For liver tissue chambers, we used the UV light to expose the solution of the PEG-DA to form microwells for patternning three dimension lobule-mimetic structure liver tissues in vitro. HepG2 cells were loading with three dimensional PEG-DA-dased microstructures for artificial lobule pattern.

Finally, we showed liver functional albumin testing. The albumin secretion of liver pattern chambers with HMEC-1 and C2C12 co-culture communication was 34% higher than the albumin secretion of liver chip without cells communication.

ACKNOWLEDGMENT

This work was supported partially by National Science Council of R.O.C. under the grant NSC99-2120-M-007-00 1. We also would like to express our appreciation to Zong-Keng Kuo and Hsiang-Wen Tseng of Biomedical Engineering Research Lab, Industrial Technology Research Institute.

REFERENCES

[1] P. J. Lee, P. J. Hung, L, P. Lee, "An artificial liver sinusoid with a microfluidic endothelial-like barrier for primary hepatocyte culture" *Biotech.nology and Bioengineering*, vol. 97, no.5,2007,pp. 1340-1346.

[2] D. A. Bruzewicz, A. P. McGuigan, G. M. Whitesides, "Fabrication of a modular tissue construct in a microfluidic chip", *Lab on Chip*, vol. 8, 2008, pp.663-671.

[3] Y. C. Toh, C. Zhang, J. Zhang, Y. M. Khong, S. Chang, V. D. Samper, D. V. Noort, D. W. Hutmacher, H. Yu, "A novel 3D mammalian cell perfusion culture system in microfluidic channels", *Lab on Chip*, vol. 7, 2007, pp.302-309.

[4] Y. C. Toh, T. C. Lim, D. Tai, G. Xiao, D. V. Noort, H. Yu, "A microfluidic 3D hepatocyte chip for drug toxicity testing", *Lab on Chip*, vol. 9, 2009, pp.2026-2035.

[5] V. N. Goral, Y. C. Hsieh, O. N. Petzold, J. S. Clark, P. K. Yuen, R. A. Faris, "Perfusion-based microfluidic device for three-dimensional dynamic primary human hepatocyte cell culture in the absence of biological or synthetic matrices or coagulants", *Lab on Chip*, vol. 10, 2010, pp.3380-3386.

[6] C. T. Ho, R. Z. Lin, W. Y. Chang, H. Y. Chang, C. H. Liu, "Rapid heterogeneous liver-cell on-chip patterning via the enhanced field-induced dielectrophoresis trap", *Lab on Chip*, vol. 6, 2006, pp.724-734.

[7] Y. S. Chen, L. Y. Ke, C. H. Liu, "3D lobule-mimetic chip via positive dielectrophoresis force with sinusoidal spacing poly (ethylene glycol)-diacrylate microwalls", in *Digest Tech. Papers Transducers '11 Conference*, Beijing, June 5-9, 2011, pp. 1837-1840.

[8] L. Y. Ke, Y. S. Chen, C. H. Liu, "A double trapped single cell contact and interaction system via movable poly (ethylene glycol) diacrylate (peg-da) microstructure for immune analysis", in *Digest Tech. Papers Transducers '11 Conference*, Beijing, June 5-9, 2011, pp. 302-305.

Inducing Self-Rotation of Melan-a Cells by ODEP

Long-Ho Chau[1], Mengxing Ouyang[1], Wenfeng Liang[2], Gwo-Bin Lee[3], Wen J. Li[1,2,4,*], and W. K. Liu[5,#]

[1]Centre for Micro and Nano Systems, The Chinese University of Hong Kong, Hong Kong
[2]State Key Laboratory of Robotics, Shenyang Institute of Automation, Chinese Academy of Sciences, China
[3]Department of Power Mechanical Engineering, National Tsing Hua University, Hsinchu, Taiwan
[4]Department of Mechanical and Biomedical Engineering, City University of Hong Kong
[5]Department of Anatomy, Faculty of Medicine, The Chinese University of Hong Kong, Hong Kong

Contact Authors: *wenjli@cityu.edu.hk and #ken-liu@ana.cuhk.edu.hk

Abstract—**This paper presents our discovery that self-rotation of Melan-a pigment cells can be induced by applying appropriate optical dielectrophoretic (ODEP) parameters. Under optically induced DEP force, which is generated by specific optical electrode patterns and with a band-width of AC bias frequencies, Melan-a cells can be trapped or repelled away from the optical electrodes. In addition, the self-rotation motion of the Melan-a cells was observed. In particular, the applied frequency and voltage dominate the overall cell motion. Hence, the cells can be manipulated on an electrode-free surface. This project studies the rotation and translation motions of the cells in an ODEP system. We speculate that the unbalanced distribution of melanin inside the pigment cells causes this self-rotation phenomenon. Therefore, both pigment cells (Melan-a) and non-pigment cells (Raw 264.7) have been tested in the ODEP experiments to compare their motion behavior. Potential applications for this novel observation are to use the self-rotation phenomenon to elucidate the pigment cells' physical property, and separate the pigment and non-pigment cells.**

Keywords-ODEP; Melan-a; Raw 264.7; Electro-kinetics; Microfluidics

I. Introduction

The traditional micro-scale cells or particles manipulation adopts electroosmosis, electrophoresis, dielectrophoresis, mechanical obstacles and magnetic force. However, those methods require complex fabrication or preparation processes and often limit cell motions. For example, DEP cell trapping method requires some fixed geometric electrodes on the substrate so that they can guide the cells to flow in a particular direction when there is a DEP force [1]. Recently, scientists have shown an increased interest in using the optical dielectrophoresis (ODEP) technique for cells manipulation [2].

Using ODEP in cells manipulation can solve the above problems. First, the system can be built easily, since it only consists of a smooth photoconductive layer, a conductive chip, a microscopic station, a power supply and a light source. It

*For information related to ODEP system, contact Wen J. Li, who is a professor at the City University of Hong Kong and also an adjunct professor at The Chinese University of Hong Kong. #For information related to cell culturing and viability, contact Ken W. K. Liu, who is a professor at The Chinese University of Hong Kong. This project is partially supported by a startup fund from the City University of Hong Kong (W. J. Li) and a Direct Grant from The Chinese University of Hong Kong (K. W. K. Liu).

avoids the sometime complex micro-electrode and micro-channel fabrication procedures. Second, the cells can be moved in any direction (on the plane of the chip) in the micro-chamber. The position of the cells is controlled by the optical electrode under the localized DEP force. Hence, cells trapping and cells separation can be performed flexibly in the ODEP system.

Since the cells are suspended in the medium, they appear to be a sphere-like shape under the microscope. Hence, the time averaged DEP force acting on a suspended cell [3] is expressed as

$$F_{DEP} = \pi \varepsilon_m R^3 \, \text{Re}[K(\omega)] \nabla |E^2| \tag{1}$$

where ε_m is the dielectric permittivity of the medium, R is the radius of the particle, $K(\omega)$ is the Clausius-Mossotti (CM) factor, ω is the applied angular frequency across the medium, and E is the electric field. The CM factor is expressed as:

$$K(\omega) = \frac{\varepsilon_p^* - \varepsilon_m^*}{\varepsilon_p^* + 2\varepsilon_m^*} \tag{2}$$

where $\varepsilon_i^* = \varepsilon_i - j\sigma_i/\omega$, and $i = p \, or \, m$, denotes the particles and the medium, respectively, and σ is the conductivity. The direction of the DEP force is determined by the sign of the real part of the CM factor. If $\text{Re}[K(\omega)] > 0$, a positive DEP force is generated. The particles tend to move towards a strong electric field region. If $\text{Re}[K(\omega)] < 0$, a negative DEP force is generated, and hence the particles are pushed towards a weak electric field region.

An animal cell consists of a nucleus, cytoplasm, and other organelles and is surrounded by a membrane. It can be considered as a single-shell model instead of a homogeneous sphere, as shown in Fig. 1. Assume that the transconductance g_m can be negligible, the complex dielectric permittivity of the medium ε_m^* is given by [4]:

$$\varepsilon_m^* = c_m R \left[\frac{j\omega\tau_c + 1}{j\omega(\tau_c + \tau_m) + 1} \right] \tag{3}$$

978-1-4673-1122-9/12 $31.00 © 2012 IEEE

where $\tau_s = c_s R / \sigma_c$ and $\tau_s = c_s R / \sigma_c$. Substituting Equation (3) into (2), CM factor $K(\omega)$ becomes

$$K(\omega) = -\left[\frac{\omega^2(\tau_s \tau_m + \tau_c \tau_m') + j\omega(\tau_m' - \tau_m - \tau_s) - 1}{\omega^2(2\tau_s \tau_m + \tau_c \tau_m') - j\omega(\tau_m' + \tau_m + 2\tau_s) - 2} \right] \quad (4)$$

where $\tau_m' = c_s R / \sigma_m$ and $\tau_m = \varepsilon_m / \sigma_m$.

In literatures, several groups have reported that the particular cells could rotate in a two-phase-shifted DEP E-field, and in a travelling wave DEP due to electrorotation [5]. The real part of Equation (4) determines the direction of DEP force, while the imaginary part of Equation (4) determines the electrorotation of the particles in the rotating E-Field. DEP states that the action of an externally applied E-field on a polarizable particle results in the formation of an induced dipole moment. When a dipole sits in a uniform E-field, each charge on the dipole is parallel to the field, hence, it experiences a torque. If the field vector changes in direction, the induced dipole moment vector must realign itself with the E-field vector, causing the particle rotation. The torque of the particle in a rotating E-field [3] is expressed as,

$$\Gamma_{ROT} = -4\pi\varepsilon_m R^3 \operatorname{Im}[K(\omega)]|\nabla|E|^2 \quad (5)$$

Electrorotation occurs if the E-field has a spatially dependent phase, otherwise $\operatorname{Im}[K(\omega)] = 0$. Dielectrophoresis and electrorotation are common tools to measure the dielectric properties of biological cells [5]. The data obtained can be used as identification of cells, or forming a cell library. Scientists and engineers often use these electrical parameters for cells manipulation, such as cells separation and transportation using DEP or ODEP.

Our group has performed the self-induced rotation of Melan-a cells in an ODEP system. The cells could rotate during stationary and translational motion in a non-uniform E-Field. Noticeably, the generated AC E-field in OET does not rotate to cause electro-rotation, hence, zero torque is generated according to Equation (5).

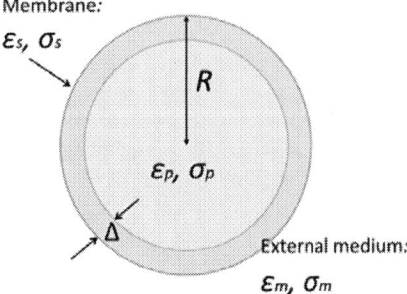

Fig. 1. The single-shell model of a spherical particle with radius R, permittivity ε_p and conductivity σ_p, which is covered by a uniform layer of thickness $\Delta \ll R$, permittivity ε_s and conductivity $\sigma_s \ll \sigma_p$, and is surrounded by a solution of permittivity ε_m and conductivity σ_m.

II. EXPERIMENTAL SETUP

A. Optoelectronic tweezers (OET)

The OET device is composed of a sandwich structure illustrated in Fig. 2. The top layer is a glass coated with a conductive and transparent Indium Tin Oxide (ITO) thin film. The bottom layer is an ITO glass coated with a photoconductive material, and a 0.3μm hydrogenated amorphous silicon (a-Si:H) thin film. The middle layer is a spacer with a thickness of 100μm used to contain samples.

The fabrication of the glass-ITO-a-Si:H structure was described by Gwo-Bin Lee et al. [6]. It was further processed by etching a part of the a-Si:H for electrical connection. An area of 5mm x 8mm a-Si:H was patterned by standard photolithography and dry-etching, using Oxford Plasma Lab 80 Plasma Etching System with 2% oxygen, 12.5% CF$_4$ gas, 30mTorr etching chamber, and 6-minute plasma exposure. Finally, the chip was rinsed and cleaned by acetone and DI water, and was blown dry by N$_2$ gas.

B. The ODEP System

The ODEP system is shown in Fig. 3. The OET chip was placed on a 2D stage which is integrated into an optical microscope (Nikon ECLIPSE TE2000-U). Light was projected from a commercial projector (DELL 1510X) and then passed through the ITO glass and liquid medium, and finally was focused on the a-Si:H surface. Light image patterns were controlled by a computer. The cells' motions were recorded by a high speed camera (PCO 1200S) through 40x objective lens. The top and bottom ITO glass chips were connected to an AC signal generator and CRO. In each set of the experiment, a drop of cell solution was injected into the OET device and there was no net fluid flow in the micro-chamber. The applied frequencies for cells manipulation were from 10kHz to 300kHz, and the applied voltages were from 6V to 20V peak to peak.

C. Cells Preparation

The melanomas from mouse, Melan-a pigment cells and Raw 264.7 (Mouse leukemic monocyte macrophage cell line) non-pigment cells were processed and suspended in 0.2M sucrose in DI water. The measured conductivity of the sucrose is 0.37mSm-1.

D. Rotation Measurement

The cells' motions were captured by a high speed camera and saved as a movie in avi format, 25 frames/sec. The frame images in bmp were extracted by a software called IrfanView. The velocity and the angular speed were obtained by importing the bmp images into the MATLAB program.

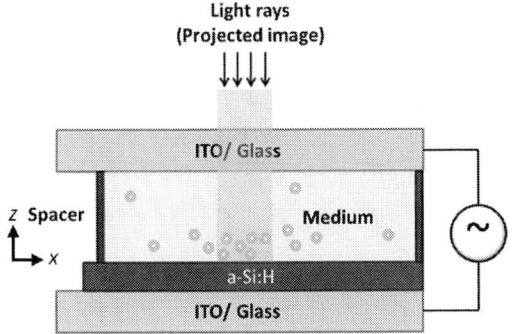

Fig. 2. An illustration of the ODEP system used by our group to manipulate biological cells. The patterned optical image is focused by a condenser lens and projected onto the hydrogenated amorphous silicon surface.

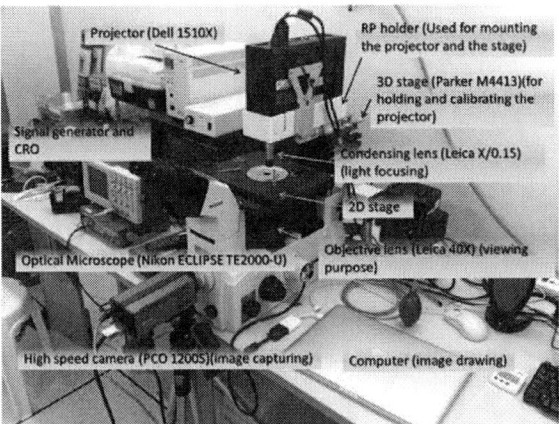

Fig. 3. The actual ODEP system setup for the cells manipulation used in our experiments.

III. CELLS ROTATION EXPERIMENTS

A. Cells Behavior Under ODEP System

1) Melan-a pigment cells

Melan-a cell is a type of melanomas which are malignant tumors of melanocytes. Melanocytes are melanin-producing cells located in the bottom layer (the stratum basale) of the skin's epidermis. Melanin is commonly found in most organisms. It determines their color of skin, hair, and iris of the eye. The Melan-a cell's size varies from 50µm to 100µm, and appears as a sphere-like shape in the medium, although its membrane is not smooth.

Frequencies from 25kHz to 300kHz have been used for testing. Results are listed in Table I. The Melan-a cells were at rest initially where there was no light and no AC voltage applied.

When the light spot was projected on the a-Si:H and the AC voltage with peak-to-peak value of 20V, 25k to 300kHz was switched on, the cells which were outside the spot started to move and accelerate towards the spot due to a positive DEP

force. At the same time, they also rotated in the direction towards the spot with the axis of rotation along x-y plane (please refer to Fig. 2). Then, some cells moved into the spot and then stopped, while other cells stayed stationary at 10-15µm near the spot, and kept rotating. If the spot moved to other positions, the cells near the spot would follow the moving spot, and again, they would not move into the spot, rather, they rotated and kept a distance of 10 to 15µm from the spot. This DEP force field also affects the cells which were far from the spot. We observed that the cells which were 200µm away from the spot rotated but remained stationary once the DEP field was applied.

On the contrary, the cells already attracted to the spot did not rotate. They would burst if the applied frequency was less than 25kHz in 20Vpp.

Fig. 4 shows the cells rotation near the spot. The rotation speed of cells is around 90 rpm (i.e., 1 cycle takes 0.68sec). The rotation direction of Cells A and B was towards the spot, while Cell C was away from the spot. However, all three cells have three behaviors in common: (i) they initially moved towards the spot by a positive DEP and the spinning rotation was the same; (ii) they would not enter the spot; and (iii) they stayed and self-rotated near the spot.

The simulation plot of the distribution of the magnitude and direction of the electric field at the applied voltage of 20Vpp at 100kHz, is showed in Fig. 5. The arrows indicate the direction and the magnitude of the DEP force, with the longer arrow showing the larger DEP force. The cells experienced a positive DEP force and moved from the dark-field region to the light spot. At around 10µm to 20µm near the spot in Fig. 5, the force is not pointing towards the spot, rather, it repels.

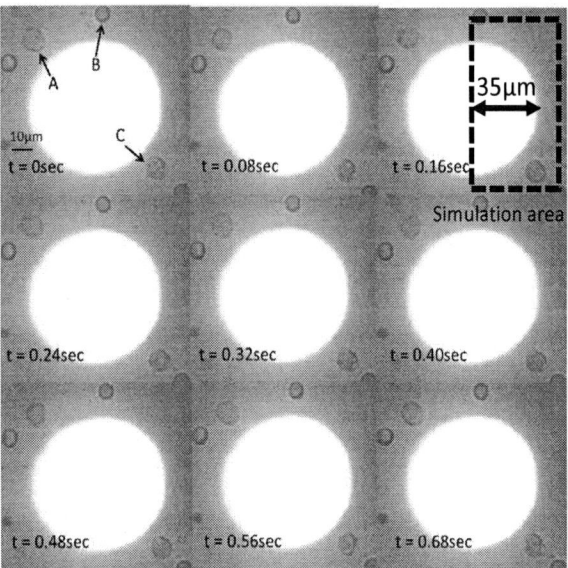

Fig. 4. Time lapse for the Melan-a cells rotation with the applied frequency 100kHz at 20Vpp recorded by a high speed camera. The rotation speed of the cells is 90rpm.

TABLE I. THE RESPONSE OF MELAN-A CELLS AND RAW 264.7 CELLS TOWARDS DEP FIELD

Applied Frequency (Hz) in 20Vpp		< 25k	25k-50k	50k-100k	100k-300k	> 300k
DEP Force	Melan-a cells	Positive DEP	Positive DEP	Positive DEP	Positive DEP	No Response
		All cells were strongly attracted to the image ,and then burst.	80% cells were attracted to the image. Other cells stayed stationary near the image.	Some cells attracted to the image and some cells stayed stationary near the image.	70% cells moved towards the image and then stayed stationary near the image.	No movement of the cells was observed.
	Raw 264.7 cells	No Response	No Response	Weak negative DEP	Negative DEP	No Response
		No movement of the cells was observed.	No movement of the cells was observed.	Cells started to repel from the image slowly.	Cells moved away from the image at the speed of 2μm/s at 150kHz.	No movement of the cells was observed.

By considering the weight and density of Melan-a cells, this simulation result can explain that they would not tend to move inside the spot, and they kept 10μm to 15μm from the spot, although they experienced a positive DEP force with the applied frequency 100kHz.

2) Raw 264.7 non-pigment cells

The Raw 264.7 is a cell line of Abelson murine leukemia virus-induced tumor from the mouse leukaemic monocyte macrophage. This Raw 264.7 macrophage cells do not contain pigments. The cells were measured to be around 10μm in the OET device. They experienced a negative DEP force in a range of 50kHz to 150kHz with the applied voltage 20Vpp. The cells which were in both bright-field region and dark-field region did not rotate in all applied frequencies, from few kHz to 3MHz. Fig. 6 shows the Raw 264.7 cell was expelled away from the light pattern at a speed of 2μm/s. During its translational motion, no rotation was observed.

B. Investigation of Induced Self-rotation

Cell rotation can be identified by tracking the change of the cells' structure, such as the shape of the organelles inside the cells and the shape of the cells. Since the cells (both Melan-a and Raw 264.7 cells) are not perfect spherical in the medium, we can easily observe the rotation behavior of the cells under the microscope by tracking the change of the cells' geometry. The rotation phenomenon was only observed for the Melan-a cells, the pigmented skin cells in the specific range of frequency and voltage. Melan-a cells only rotate in the dark-field region no matter they are stationary or moving towards the projected image listed in Table II. In addition, the Raw 264.7 cells did not rotate.

Self-rotation of the Melan-a was observed when they were experiencing a sufficient positive DEP force. The minimum voltage applied to cause self-rotation in our configuration is 3V and the frequency applied is less than 300kHz. The larger the voltage applied, the faster the cells spun. In terms of frequency, the cells started rotation when the frequency is less than 300kHz. The cells were observed self-rotation 10 μm near the spot. However, there are no rotating fields in that region, as shown in Fig. 5. We highly suspect that it is related

to the uneven mass distribution of the cells caused by the unbalanced distribution of melanin. If the centre of gravity (C.G.) of the pigment cells is not at the geometric centre, it is possible that the DEP force creates a torque on the cells. Hence, this physical property of the Melan-a cells causes them to undergo self-rotation. Since, the direction of rotation depends on the AC E-field direction, the induced moment of the cells mainly comes from the DEP force as stated in Equation (1).

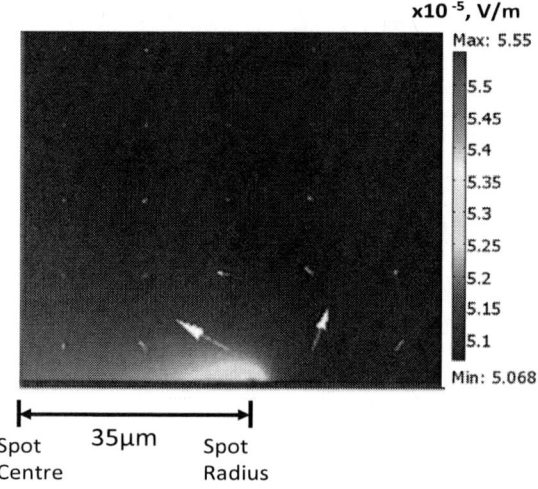

Fig. 5. The distribution of the magnitude and direction of an electric field. The arrows represent the direction of the positive DEP force and the surface plot indicates the distribution of the electric field.

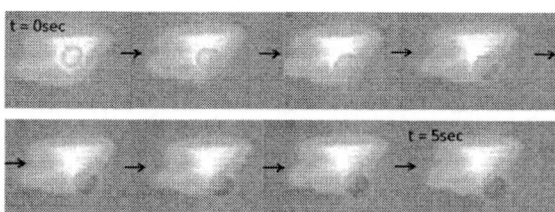

Fig. 6. Time lapse for the Raw 264.7 cells undergone a negative force motion with the applied frequency 150kHz at 10Vpp recorded by a high speed camera.

TABLE II. Rotation behavior of Melan-a and Raw 264.7 cells.

Location / Motion		At Light Spot	At Dark-field Region	Stationary	Moving
Rotation	Melan-a cells	No rotation was observed.	Cells rotated and its direction of rotation was pointing to the light spot from few kHz to 300kHz, 20Vpp.	Cells only rotated in dark-field region only.	Cells moved and rotated at the same time.
	Raw 264.7 cells	No rotation was observed.	No rotation was observed.	No rotation was observed.	No rotation was observed.

IV. CONCLUSION

Cell manipulation using ODEP technology was conducted using OET device and the results are discuss in this paper. The motion study of two cells, Melan-a pigment cells and Raw 264.7 non-pigment cells had been performed in the experiment. The self-rotation of Melan-a cells (pigmented cells) in the dark-field region were observed. The Melan-a cells experience a positive DEP force only from 25kHz to 300kHz. However, Raw 264.7 cells (non-pigment cells) did not rotate at all throughout the experiment. Results show that the frequency and voltage applied do affect the rotation speed. It is speculated that the uneven distribution of the melanin inside the cells causes the induced moment. Results of this self-rotation phenomenon of the pigment cells imply the possibility of using the rotating speed to predict the physical properties of the cells. In addition, it may also be used as biomarkers for cell discrimination in the future.

ACKNOWLEDGMENT

The author would like to thank Ms. Florence W. K. Cheung of School of Biomedical Sciences, The Chinese University of Hong Kong, for her help in culturing and preparing the cells for our experiments. Prof. Wen J. Li would like to thank The Chinese University of Hong Kong for its continual support of his graduate students at the University.

REFERENCES

[1] Unyoung Kim, Chih-Wen Shu, Karen Y. Dane, Patrick S. Daugherty, Jean Y. J. Wang, and H. T. Soh, "Selection of mammalian cells based on their cell-cycle phase using dielectrophoresis," *PNAS*, Vol. 104 (52), pp. 20708-20712, 2007

[2] Yen-Heng Lin, Wang-Ying Lin, Gwo-Bin Lee, "Image-driven cell manipulation," *IEEE Nanotechnology Magazine*, Vol. 3, No. 3, pp. 6-11, 2009.

[3] Hywel Morgan and Nicolas G Green, "AC Electrokinetics: colloids and nanoparticles," Research Studies Press Ltd., 2003.

[4] Minglin Li, Yanli Qu, Zaili Dong, Yuechao Wang, and Wen J. Li, "Limitations of Au Particle Nano-Assembly using Dielectrophoretic Force ——A Parametric Experimental and Theoretical Study," *IEEE Transactions on Nanotechnology* (Letter), March 2008.

[5] Jan Gimsa, Piotr Marszalek, Ulrike Loewe, and Tian Y. Tsong, "Dielectrophoresis and electrorotation of neurospora slime and murine myeloma cells," *Biophys. J. Biophysical Society*, Volume 60, pp. 749 – 760, October 1991.

[6] Yen-Heng Lin and Gwo-Bin Lee, "An integrated cell counting and continuous cell lysis device using optically induced electric field," *Sensors and Actuators B: Chemical*, Vol. 145, Issue 2, pp. 854-860, 2010.

Microfluidic System for Rapid Detection of Influenza Infection by Utilizing Magnetic MnFe$_2$O$_4$ Nanoparticle-based Immunoassay

Lien-Yu Hung[1], Fong-Yu Cheng[2], Chih-Chia Huang[2], Yi-Che Tsai[3], Chen-Sheng Yeh[2], Huan-Yao Lei[3] and Gwo-Bin Lee[1*]

[1]Department of Power Mechanical Engineering, National Tsing Hua University, Hsinchu 300, Taiwan
[2]Department of Chemistry, [3]Department of Microbiology and Immunology, National Cheng Kung University, Tainan 701, Taiwan

Abstract— **In this study, new magnetic** manganese ferrite **(MnFe$_2$O$_4$) nanoparticles with a size around 100 nanometer (nm) in diameter were used to improve the performance of an immunoassay for detection of influenza infection. A new microfluidic system was developed to implement the detection process. In order to apply these new nanoparticles for influenza detection, the design of the micromixer was modified to reduce the dead volume. Furthermore, the operating condition for the magnet was optimized such that magnetic MnFe$_2$O$_4$ nanoparticles can be collected in 120 seconds. The optical signals showed this nanoparticle-based immunoassay could successfully distinguish the virus sample from the negative control sample. This developed microfluidic system can automatically perform the entire process involved in the immunoassay, including virus purification and detection, and therefore may provide a promising platform for fast diagnosis of the infectious diseases.**

Keywords-nanoparticles; microfluidic system; influenza; virus

I. INTRODUCTION

Emerging, infectious diseases caused by viruses such as the seasonal and novel influenza have been a serious concern recently. This kind of viral infection can cause serious outbreaks and rise up to be the worldwide pandemic. For example, Hong Kong flu happened in 1968 and Bird flu happened in 2004 have caused significant deaths. Furthermore, "swine flu" outbreaks on June, 2009 from Mexico [1], have caused severe worry about "other flu" emerging around the world. Therefore, the ability to rapidly diagnose the presence of the influenza virus is important to provide immediate and proper clinical treatment.

There are several diagnostic procedures to detect influenza viral infections. For instance, an immunofluorescence (IF) assay [2], an enzyme-linked immunosorbent assay (ELISA) [3] and molecular diagnosis using real-time polymerase chain reaction (PCR) [4] have been demonstrated in the literature. Furthermore, assays utilizing three-dimensional (3D) microspheres and specific antibodies (or nucleotide probes) have been reported to significantly increase the signals [5]. Recently, various bead-based immunoassays incorporated with microfluidic systems have also been demonstrated [6]. For example, the current research group has reported an integrated microfluidic system for rapid detection of purified influenza viral particles in 15 minutes [7]. However, previous studies used magnetic beads with a size of about 4.5 micrometer (μm), which is much larger than influenza virus

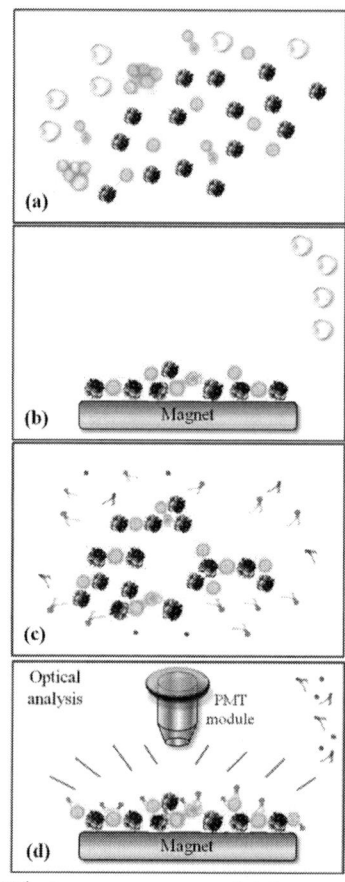

Fig. 1: A schematic diagram of the immunoassay used in this study: (a), Incubation of magnetic nanoparticles surface-coated with anti-influenza NP mAbs and influenza virus particles; (b), Purification of nanoparticle-virus complexes and washing out non-specific interferents; (c), Incubation of nanoparticle-virus complexes and direct-conjugated PE mAbs; (d), Purification of nanoparticle-virus-PE Abs complexes, washing out non-binding Abs and detection of optical signal.

(~100 nm in diameter). This size difference may affect the detection limit, especially, as it be applied to clinical samples.

978-1-4673-1122-9/12 $31.00 © 2012 IEEE

Nanoparticles have been used for biological applications because of their tiny size distribution, long-term stability and bio-compatibility [8]. Moreover, nanoparticles with magnetic characteristics may have advantages including multiplexing, decreased analysis time, large surface-to-volume ratio, selectively controllable and low background noise [9]. Recently, superparamagnetic nanoparticles have been used to capture nm-size particles, such as virions [10]. High capture rate, high throughput and extra high sample concentration than the traditional assay have been demonstrated. However, no attempt has been made to implement a nanoparticle-based immunoassay on a microfluidic system for rapid influenza detection. In this study, an integrated microfluidic system utilizing magnetic $MnFe_2O_4$ nanoparticles for influenza detection with FIA was therefore demonstrated. With this approach, the detection limit of the diagnosis is expected to be improved due the large surface to volume ratio caused by nanoparticles.

II. EXPERIMENTAL

A. Working Principle

A schematic illustration for a simplified diagram of the detection assay used in this study is shown in Fig. 1. The magnetic nanoparticles were surface-coated with anti-influenza-A-nucleoprotein (α-A-NP) monoclonal antibodies (mAbs) and loaded in the incubation chamber. Then they were incubated with the purified viral particle samples for 5 minutes. Then purification of nanoparticle-virus complexes was realized by washing out non-specific interferents when an external magnetic field was applied. After washing, the direct-conjugated R-phycoerythrin (PE) developing Abs was then transported into the detection chamber to incubate with nanoparticle-virus complexes for another 5 minutes. Then non-binding interferents were washed away and purification of the nanoparticle-virus-developing Ab complexes was performed. Finally, the optical detection module was used to detect the optical signal of purified nanoparticle-virus-developing Abs complexes. The specific α-influenza-A-NP mAbs were obtained from the Microbiology and Immunology Laboratory of NCKU. Furthermore, they were directly conjugated with PE fluorescent dye on the developing antibody.

B. Chip Design and Fabrication

Figure 2(a) shows a schematic illustration of the integrated microfluidic system composed of micropumps, microvalves and micromixers. The shape and size of the detection and incubation chambers were specially designed for magnetic nanoparticles. The original incubation chambers contained microfluidic side-channels which may adhere nanoparticles along the edges with the dead-volume regions. Therefore, the micromixer was modified into a circular shape to reduce magnetic nanoparticles lose and allow nanoparticles surface-coated with Abs to perform immunological diagnosis.

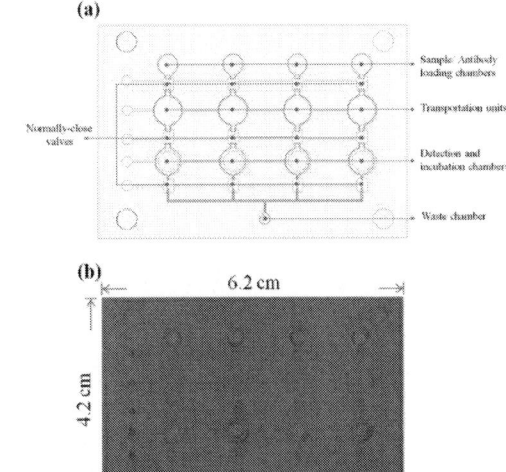

Fig. 2: (a), A photograph of the developed microfluidic chip. The dimensions of the chip were measured to be 6.2 cm x 4.2 cm. (b), Schematic illustration of the microfluidic chip for performing the immunoassay using magnetic $MnFe_2O_4$ nanoparticles.

A photograph of the developed microfluidic chip is shown in Fig. 2(b). The dimensions of the chip were measured to be 6.2 cm x 4.2 cm. In additionally, the chip device consisted of two PDMS layers including a thick PDMS structure with air chambers, a thin-film PDMS membrane as the fluidic channel layer and one glass substrate as the bottom layer.

C. Nanoparticles Production

A series of transmission electron microscope (TEM) images of magnetic $MnFe_2O_4$ nanoparticles are shown in Fig. 3. These nanoparticles were produced through serial chemical heating generation using Fe $(acac)_3$, Mn $(ac)_2$, oleic acid and trioctylamine as starting materials [11]. The synthesized nanoparticle has an average size of 98 ± 19.5 nm.

III. RESULTS AND DISCUSSION

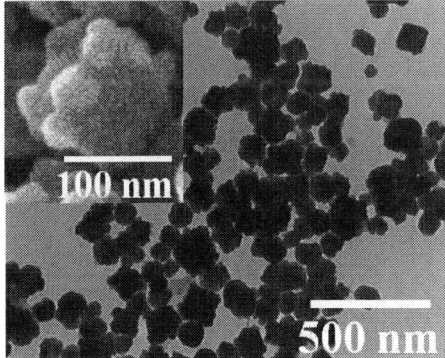

Fig. 3: A series of TEM images for magnetic $MnFe_2O_4$ nanoparticles.

A. Activity Test of the mAbs

In order to prove the activity of specific mAbs after binding onto these magnetic MnFe2O4 nanoparticles, influenza-infected cell line (Madin-Darby canine kidney cells, MDCK) and magnetic nanoparticles bound with specific α-NP mAbs

Light Field   **Dark Field**

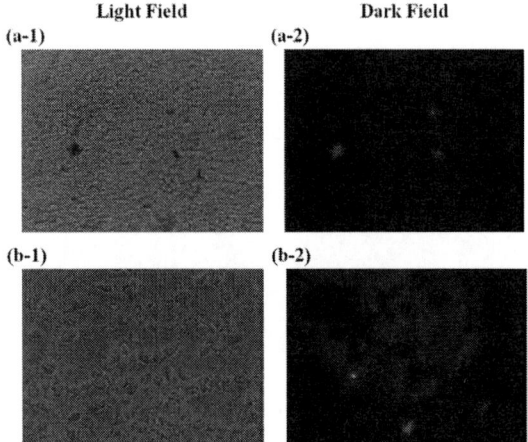

Fig. 4: A series of microscopic images for testing the specific mAbs activity by utilizing MDCK cell line. (a-1) and (a-2): the MDCK cells were not infected by influenza and therefore no fluorescent signals were detected under dark field. (b-1) and (b-2): the MDCK cells were infected by influenza and PE fluorescent signals were detected in the dark field.

which were modified with PE fluorescent dye were used. Note that Figs. 4(a) and 4(b) show the cells without and with virus infection, respectively. As shown in Fig. 4(a-1) and (a-2), the cells were not infected by influenza virus, and therefore the magnetic nanoparticles bound with α-NP mAbs-PE did not capture the cells. No fluorescent signals were detected. On the other hand, as shown in Fig. 4(b-1) and (b-2), the cells have been infected by influenza virus, and therefore the magnetic nanoparticles bound with α-NP mAbs-PE captured the infected cells. It is clearly seen that fluorescent signals can be detected.

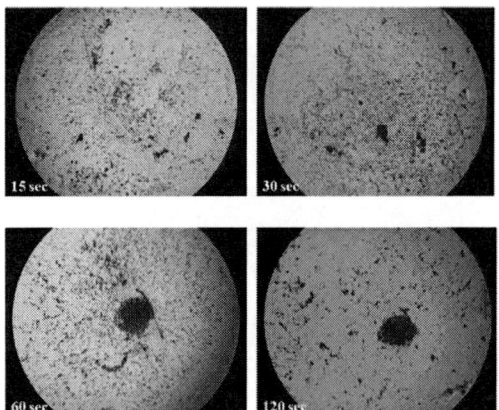

Fig. 5: A series of microscopic images for magnetic MnFe₂O₄ nanoparticles at different periods of collection time.

B. Nanoparticles Manupulation and Antibody Coating

The manipulation of magnetic nanoparticles was then tested by applying an external magnet. Figure 5 shows a series of images at different time periods for the collection of magnetic MnFe₂O₄ nanoparticles. Even though the size of the nanopartice is relatively small, experimental data showed that it can be collected within 120 seconds.

Different amounts of α-NP mAbs were then used to conjugate magnetic MnFe₂O₄ nanoparticles to explore the optimal antibody coating conditions. Briefly, 4 and 8 μg of primary mouse α-influenza A-NP mAb in phosphate buffer saline (1×PBS) solution with a total volume of 200 μL (Fe^{3+} 600 ppm) were shaken for 18 hr at 4 ℃, followed by purifying and re-suspending the mAb-coated magnetic nanoparticles in 100 μL of PBS. Next, a blocking process was then performed at 4 ℃ for another 18 hr by incubating the mAb-coated magnetic nanoparticles with the blocking solution (1×PBS with 1% (w/v) bovine serum albumin (BSA) to prevent non-specific binding in later immunological processing. Finally, the mAb-coated magnetic nanoparticles were stored at 4 ℃ prior to the FIA processes.

After conjugating the 4 and 8 μg α-NP mAbs with magnetic MnFe₂O₄ nanoparticles, PBS buffer and influenza virus (8 Hau) were used to test the coating efficiency. Figure 6 shows a series of optical images indicating that the magnetic nanoparticles complexes with influenza A virus and 1×PBS buffer can be successfully distinguished. Figure 6(a) shows

Light Field   **Dark Field**

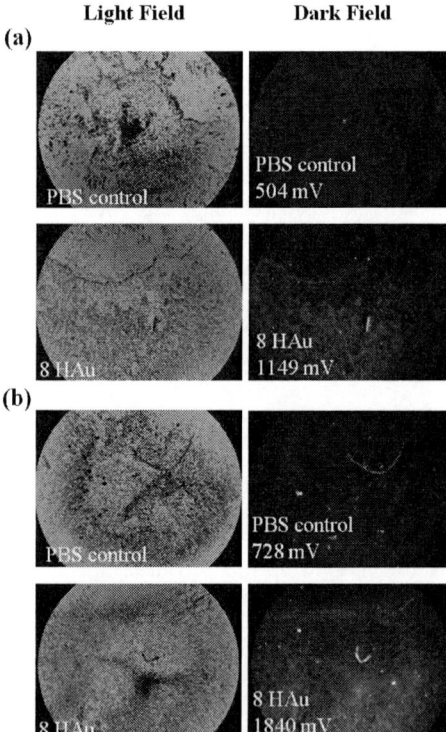

Fig. 6: A series of optical images for magnetic MnFe₂O₄ nanoparticle complexes with PBS buffer and influenza A virus.

978-1-4673-1122-9/12 $31.00 © 2012 IEEE

that the positive samples using nanoparticles (coated with 4 μg α-NP mAbs) have signals higher than PBS buffer (1149 to 504 mV). Figure 6(b) shows that the optical signal of nanoparticles (coated 8 μg α-NP mAbs) is also stronger than PBS buffer (1840 to 728 mV). The results show that although the nanoparticles with 8 μg α-NP mAbs have higher signals than nanoparticles with 4 μg α-NP mAbs (1840 to 1149 mV), it comes with a higher background signal (728 to 504 mV). However, the signal to noise ratio of 4-μg α-NP mAb nanoparticles is close to the 8-μg case (around 2.5:2.3). In order to save antibodies, the nanoparticles coated with 4 μg α-NP mAb were used for testing of different viral concentrations.

Figure 7(a) shows the optical signals for the detection of influenza virus in the microfluidic system with different viral concentrations. Figure 7(b) is the optical images from different viral concentrations of the proposed microfluidic system. It indicates that the proposed system can successfully detect the influenza infection in 20 minutes.

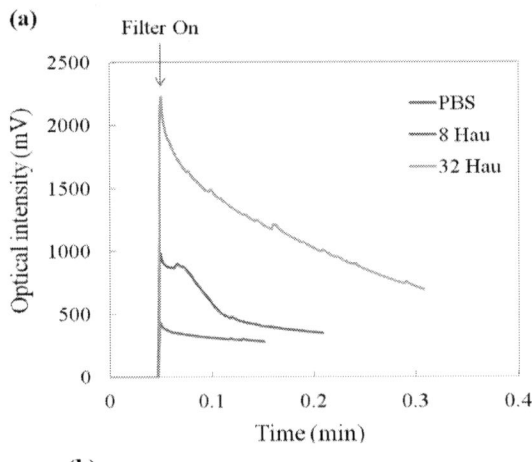

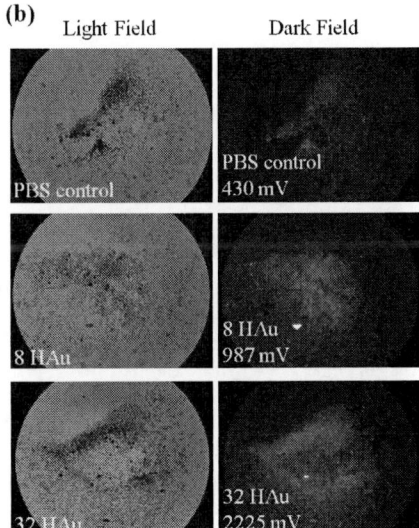

Fig. 7: *(a) Optical signals for the detection of influenza virus HA in the microfluidic system. (b) Sensitivity test of the proposed microfluidic system.*

III. CONCLUSION

In this study, an integrated microfluidic system utilizing magnetic $MnFe_2O_4$ nanoparticles for influenza detection with FIA has been demonstrated successfully. With this approach, the detection limit of the diagnosis is expected to be improved due the large surface-to-volume ratio caused by nanoparticles. This microfluidic system can automatically perform an immunoassay for influenza virus detection. It may be promising for fast diagnosis of influenza infection.

ACKNOWLEDGEMENTS

The authors want to thank the National Science Council in Taiwan for financial support (NSC 100-2120-M-007-014).

REFERENCES

[1] G. Chowell, Bertozzi, S.M., Colchero, M.A., Lopez-Gatell, H., Alpuche-Aranda, C., Her- nandez, M., Miller, M.A., "Severe Respiratory Disease Concurrent with the Circulation of H1N1 Influenza," New Engl. J. Med., 2009, vol. 361, pp. 674–679.

[2] T. Shibata, Tanaka, T., Shimizu, K., Hayakawa, S., Kuroda, K.,. "Immunofluorescence imaging of the influenza virus M1 protein is dependent on the fixation method," *J. Virol. Meth.,* 2009, vol. 156, pp. 162–165.

[3] G. Sala, Cordioli, P., Moreno-Martin, A., Tollis, M., Brocchi, E., Piccirillo, A., Lavazza, A., "ELISA test for the detection of influenza H7 antibodies in avian sera," *Avian Dis.,* 2003, vol. 47, pp. 1057–1059.

[4] G. Boivin, Côté, S., Déry, P., De Serres, G., Bergeron, M.G., "Multiplex Real-Time PCR Assay for Detection of Influenza and Human Respiratory Syncytial Viruses," *J. Clin. Microbiol.* 2004, vol. 42, pp. 45–51.

[5] C. Cao, Dhumpa, R., Bang, D.D., Ghavifekr, Z., Høgberg, J., Wolff, A., "Detection of avian influenza virus by fluorescent DNA barcode-based immunoassay with sensitivity comparable to PCR," *Analyst.,* 2010, vol. 135, pp. 337–342.

[6] K. Y. Lien, W. C. Lee, H. Y. Lei, G. B. Lee, "Integrated reverse transcription polymerase chain reaction systems for virus detection," *Biosen. Bioelectron.,* 2007, vol. 22, pp. 1739–1748.

[7] K. Y. Lien, L. Y. Hung, T. B. Huang, Y. C. Tsai, H. Y. Lei, G. B. Lee, "Rapid detection of influenza A virus infection utilizing the immunomagnetic bead-based microfluidic system," *Biosen. Bioelectron.,* 2011, vol. 26, pp.3900-3907.

[8] F. Y. H. Lin, M. Sabri, J. Alirezaie, D. Li, and P. M. Sherman, "Development of a Nanoparticle-Labeled Microfluidic Immunoassay for Detection of Pathogenic Microorganisms," *Clin Diagn Lab Immunol.* 2005, vol. 12, pp. 418–425.

[9] J. S. Beveridge, J. R. Stephens, and M. E. Williams, "The Use of Magnetic Nanoparticles in Analytical Chemistry," *Annu. Rev. Anal. Chem.,* 2011, vol. 4, pp. 251-273

[10] G. D. Chen, C. J. Alberts, W. Rodriguez, M. Toner, "Concentration and purification of human immunodeficiency virus type 1 virions by microfluidic separation of superparamagnetic nanoparticles," *Anal. Chem.,* 2010, vol. 82,pp. 723-728.

[11] C. C. Huang, C. N. Chang and C. S. Yeh, "A thermolysis approach to simultaneously achieve crystal phase- and shape-control of ternary M–Fe–O metal oxide nanoparticles," *Nanoscale,* 2011, vol. 3, pp. 4254-4260.

Corresponding author: Dr. Gwo-Bin "Vincent" Lee, address: No. 101, Section 2, Kuang-Fu Road, Hsinchu, Taiwan 30013; tel: +886-3-5715131 ext. 33765; fax: + 886-3-5722840; e-mail: gwobin@pme.nthu.edu.tw

An optical diagnostic system using isothermal amplification technique for *Phalaenopsis orchids*

Wen-Hsin Chang[1], Sung-Yi Yang[1], Chih-Hung Wang[1], Ping-Chen Li[2], Fuh-Jyh Jan[2], Tzong-Yueh Chen[3] and Gwo-Bin Lee[1]*

[1]*Department of Power Mechanical Engineering, National Tsing Hua University, Hsinchu, Taiwan*
[2]*Department of Plant Pathology, National Chung Hsing University, Taichung, Taiwan*
[3]*Institute of Biotechnology, National Cheng Kung University, Tainan, Taiwan*

Abstract—**Early detection of pathogens is important for surveillance and control of infectious diseases among *Phalaenopsis* orchids. Therefore, the current study presents an integrated microfluidic system for rapid and automatic detection of *Phalaenopsis* orchid viruses. The entire process, including pathogen-specific ribonucleic acid (RNA) purification, nucleic acid amplification using reverse transcription loop-mediated-isothermal-amplification (RT-LAMP) and optical detection by measuring turbidity change, can be automatically performed on a single chip within 65 minutes. This is the first time that an integrated microfluidic system for the detection of *Phalaenopsis* orchid viruses has been demonstrated. The specificity and sensitivity of the system were also explored in this study to validate its performance. One of the most prevalent orchid viruses, *Cymbidium mosaic virus* (CymMV), was used in current study to demonstrate the capabilities of the developed system.**

Keywords- loop-mediated-isothermal-amplification (LAMP); microfluidics; Phalaenopsis orchid

I. INTRODUCTION

Phalaenopsis orchids industry is one of the most promising economic activities in Taiwan. However, pathogen infection seriously impairs the quality of *Phalaenopsis* orchids and decreases their export value in consequence. Therefore, pathogen detection is crucial in elevating international competitiveness of *Phalaenopsis* orchids industry. There are several techniques that could carry out *Phalaenopsis* orchids pathogen detection. For example, histology, immuno-interaction [1] or polymerase chain reaction [2] has been demonstrated in the literature. However, these approaches have some intrinsic disadvantages. They either consume long time or require sophisticated techniques and tools, which are not suitable for on-site detection.

Loop-mediated-isothermal-amplification (LAMP) and reverse transcription loop-mediated-isothermal-amplification (RT-LAMP) have been shown to be a promising method with high sensitivity, high specificity for DNA (deoxyribonucleic acid) or RNA amplification in various kind of organisms, including bacteria and viruses [3,4]. LAMP and RT-LAMP are isothermal nucleotide acid amplification methods which are widely used in pathogen detection because it possesses some advantages over others such as high sensitivity, short time and single working temperature. Moreover, the result could be observed by naked eye because of the turbidity changes if there are LAMP products. Nevertheless, only few studies used this approach to detect plant RNA viruses [5,6].

Recently, a microfluidic chip was reported to detect the LAMP product by turbidity change [7]. However, DNA has to be extracted manually before it is subjected to a microfluidic assay. Another integrated microfluidic system including sample pre-treatment and LAMP reaction was reported by our research group to detect fish virus by isolating pathogen specific RNA with nucleotide probes first and then performing LAMP [8]. However, on-chip detection scheme has not yet been integrated. Therefore, this study endeavors to develop an integrated microfluidic RT-LAMP system that allows rapid detection of *Phalaenopsis* orchid infected with pathogens in an automatic format. The entire process including RNA purification, RT-LAMP and optical detection can be performed such that it can be performed within a shorter period of time. The developed system may be promising for rapid detection of plant pathogens.

II. MATERIAL AND METHODS

A. Experimental procedure

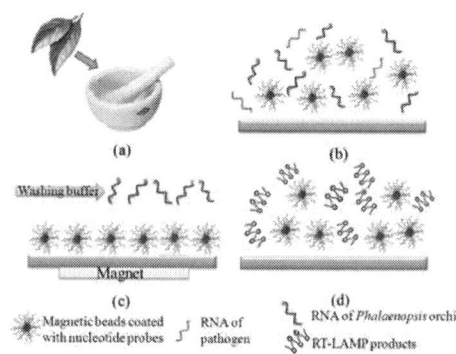

Fig. 1: Schematic illustration of the experimental procedure for sample pretreatment, RNA purification and RT-LAMP performed in the microfluidic system. (a) The leaves of Phalaenopsis orchids were first grinded with orchid lysis buffer. (b) Thermo-denature and hybridization. (c) Pathogen-specific RNA purification. (d) RT-LAMP of extracted RNA.

The leaves of Phalaenopsis orchids were first grinded with orchid lysis buffer. The total RNA sample from *Phalaenopsis* orchid leaves was then subjected to thermo-denature at 95°C for 5 minutes. Magnetic beads coating with pathogen specific probes was employed to carry out the sample pre-treatment step. After hybridization at 63°C for 15 minutes, the RNA of pathogens complementary to pathogen-specific probes was collected by applying an external magnetic field. Then, the washing step and RT-LAMP reagent transportation were controlled by a home-made system composed of electromagnetic valves (EMVs) and a control circuit automatically. Finally, the absorbance change was detected by using an optical detection device (Fig. 1).

B. Chip design

A microfluidic chip was used to carry out the above experiment procedure in current study. Figure 2(a) shows a schematic diagram of the integrated microfluidic RT-LAMP microfluidic chip. Red lines delimited the region for air control while black lines described liquid channels. This chip contained a normally-closed microvalve, a micropump unit and a waste unit which, three RT-LAMP reagent chambers, one reaction chamber, a positive control chamber, a negative control chamber, a wash buffer chamber. The exploded view of the integrated microfluidic RT-LAMP system was shown in Fig. 2(b), which consisted of two layers of polydimethylsiloxane (PDMS) (one layer for air chambers and one for liquid channels) and one layer of glass (used as a bottom layer for sealing the microchannels and chambers).

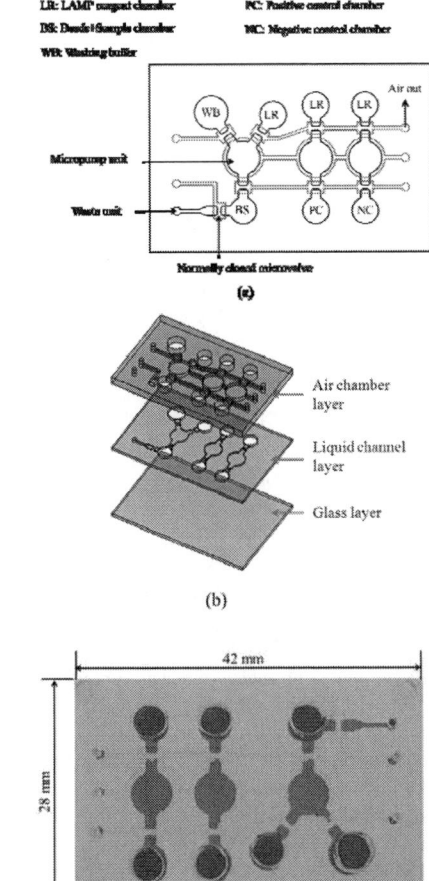

Fig. 2:

Schematic diagram, (b) the exploded view and (c) a photograph of the integrated microfluidic RT-LAMP system.

The microfluidic chip was fabricated by using a typical PDMS replication process. First, two PMMA master molds which have inverse microstructures of air chamber layer and liquid channel layer were formed by using a computer-numerical-control (CNC) machining process. Then the PDMS were casted to form two membrane layers. Finally, these two PDMS membrane layers and the glass layer were bonded together by using O₂ plasma treatment to form the microfluidic chip. Figure 2(c) is a photograph of the microfluidic chip. The dimensions of the microfluidic chip were measured to be 28 mm in width, 42 mm in length and 5 mm in depth.

In prior to the detection process, the wash buffer and RT-LAMP reagents were pre-loaded into the wash buffer chamber and the three RT-LAMP reagent chambers. RNA samples from *Phalaenopsis* orchid leaves were mixed with magnetic beads coated with pathogen-specific nucleotide probes and loaded into reaction chamber. Then, the temperature of the chip was heated up to 95 °C for 5 minutes for RNA thermo-denature to avoid any unwanted secondary structure and then decreased to 63 °C for 15 minutes for hybridization between the released RNA and the pathogen-specific nucleotide probes. After using a magnet to collect magnetic beads, wash buffer was transported into the reaction chamber twice for performing the washing step. Finally, the LAMP reagents were transported into the reaction chamber, the positive control chamber and the negative control chamber when the temperature of the chip was heated up to 63 °C for the LAMP reaction. Note the entire procedure could be performed programmably by using a computer and the home-made control system.

C. The microfluidic RT-LAMP detection system

In order to perform above-mentioned procedure automatically, an integrated microfluidic system was developed. A schematic illustration of the microfluidic RT-LAMP detection system was shown in Fig. 3, including a temperature control module to regulate temperature, an optical detection module to detect absorbance signals, an air compressor to provide the required suction force, several EMVs for PDMS membrane motion control and a personal computer.

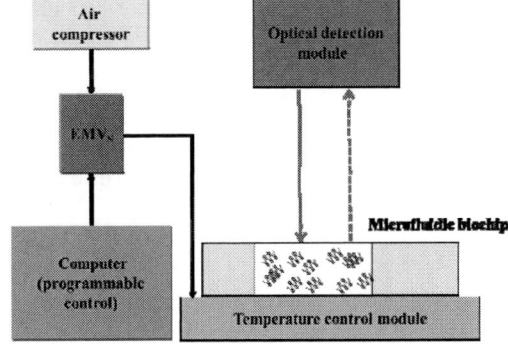

Fig. 3: A schematic illustration of the microfluidic RT-LAMP detection system including a temperature control module, an optical detection system, an air compressor, electromagnetic valves and a personal computer.

III. RESULTS AND DISCUSSION

CymMV is one of the most prevalent *Phalaenopsis* orchid viruses which often impairs flower color, size and growth and usually causes huge economical loss. The diagnosis of CymCV is crucial for quality and quantity control of *Phalaenopsis* orchids [6]. Therefore, CymMV has been chosen to be the target of this study. The optimal operating temperature for the microfluidic RT-LAMP system have been optimized by testing the several important parameters including the reaction temperature for the one-step RT-LAMP process and the reaction temperature for RNA hybridization. Figure 4 indicated that RT-LAMP can be successfully performed under hybridization temperatures of 57~65°C. Previous test showed 63°C was the optimal temperature for LAMP reaction and absorbance has significant change under 63°C (Table 1) Therefore, 63 °C was chosen as the optimal operating temperature for this integrated microfluidic system. Note that the slab-gel electropherograms shown here are to demonstrate the capability of the RT-LAMP module. The final detection of the virus was performed by using an optical detection system.

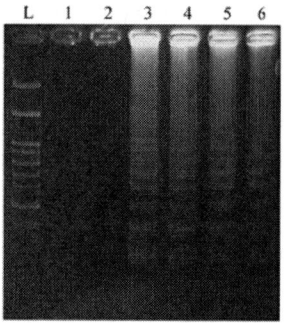

Fig. 4: Optimization of hybridization temperature for the probe-conjugated magnetic beads. L: 100-bp ladders, 1: ddH₂O; 2: RNA from healthy Phalaenopsis orchids; 3-6: RT-LAMP products using the probes pre-hybridized with RNA extracted from CymMV infected Phalaenopsis orchids at 67°C, 63°C, 60°C, and 57°C for 15 min, respectively.

RNA from three kinds of samples, including healthy *Phalaenopsis* orchids, *Odontoglossum ringspot virus* infected *Phalaenopsis* orchids and *Capsicum chlorosis virus* infected *Phalaenopsis* orchids were used to investigate the specificity of the developed assay to validate its performance. Experimental results showed that the developed assay had high specificity for CymCV detection (Fig. 5). The other two viruses cannot not be detected by using this assay.

The sensitivity test was also carried out. 10-fold serial dilutions were performed on a 250-ng total RNA extracted from CymMV infected *Phalaenopsis* orchids. The results showed that a concentration as low as 250 pg total RNA can be successfully detected by using the developed microfluidic system (Fig. 6). Moreover, LAMP products' signals on the slab gel from the developed microfluidic system were brighter than that from the conventional system, indicating the microscale RT-LAMP system has a superior performance.

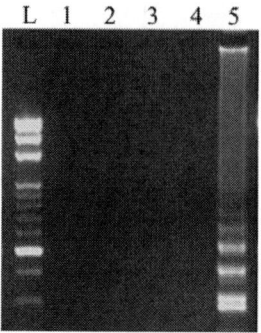

Fig. 5: Specificity test of CymMV LAMP primers. L: 100-bp DNA ladders; 1: DEPC-H₂O; 2: RNA from healthy Phalaenopsis orchids; 3: RNA from Odontoglossum ringspot virus infected Phalaenopsis orchids; 4: RNA from Capsicum chlorosis virus infected Phalaenopsis orchids; 5: RT-LAMP products from CymMV infected Phalaenopsis orchids.

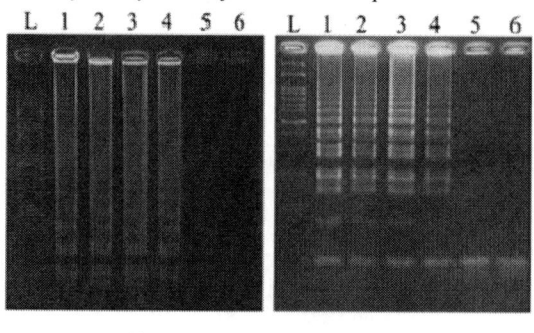

Fig. 6: Sensitivity test of CymMV LAMP primers. RT-LAMP products of (a) conventional system and (b) microfluidic system. L: 100-bp DNA ladders; Lanes 1-5: 10-fold serial dilution of 250-ng (5 μl) total RNA from CymCV infected Phalaenopsis orchids. 6: dd-H₂O as negative control.

In order to explore the optimal condition for optical density measurement, optical density under different wavelengths were first measured. Table 2 lists the optical density at different wavelengths when using different RNA concentrations. It is clearly seen that a light with 450 nm is optimal for the detection. Note that the optical density has about two-fold difference between 2.5-ng total RNA and negative control (ddH₂O).

The optical density under a light with 450 nm was then measured to compare its performance between the conventional and the microfluidic system. Table 3 indicated that the optical density of RT-LAMP carried on the microfluidic system is much higher than that carried on a conventional PCR machine. Moreover, the optical density difference is greater than that of the conventional system. The optical density has four-fold difference between 2.5-ng total RNA and negative control (ddH$_2$O). In summary, the proposed microfluidic system might be a promising tool for *Phalaenopsis* orchid pathogen detection.

Table 1 Optical temperature and optical density (OD) under different wavelengths.

Sample type	Sample concentration (ng/µl)	Temperature (°C)	Optical density (600 nm)	Optical density (450 nm)
Healthy	50	60	-0.007	-0.009
Infected	50	60	-0.005	-0.009
Healthy	50	63	0.01	0.014
Infected	50	63	0.059	0.088

Table 2 Optical density of different concentrations and different wavelengths. Negative control: ddH$_2$O.

	OD 450 (conventional)	OD 450 (microfluidic)
50 ng/µl (5µl, total RNA)	0.024	0.904
10^{-1} dilution	0.026	0.476
10^{-2} dilution	0.021	0.547
10^{-3} dilution	0.010	0.162
10^{-4} dilution	0.009	0.105
Negative control	0.010	0.136

Table 3 Comparison of optical density detected by a light with 450 nm between conventional and microfluidics system. Negative control: ddH$_2$O.

	OD 450	OD 600	OD 650
50 ng/µl (5µl, total RNA)	0.024	0.016	0.009
10^{-1} dilution	0.026	0.015	0.008
10^{-2} dilution	0.021	0.015	0.003
10^{-3} dilution	0.010	0.008	0.002
10^{-4} dilution	0.009	0.006	0.002
Negative control	0.010	0.010	0.001

IV. CONCLUSION

A new integrated microfludic system was designed and fabricated for carrying out the entire process for *Phalaenopsis* orchid pathogen detection. The entire process including sample pre-treatment, RT-LAMP process and on-chip optical detection can be performed in a short period of time (65 minutes) automatically. The sensitivity and specificity of the assay have been verified by using virus-infected *Phalaenopsis* orchid samples. Therefore, the proposed microfluidic system is promising for rapid isolation and detection of plant pathogens. Not only is the developed system useful for the *Phalaenopsis* orchids quality surveillance but it is also crucial for quarantine of *Phalaenopsis* orchids.

ACKNOWLEDGMENTS

The authors would like to thank Council of Agriculture in Taiwan for financial support (100AS-5.3.1-ST-aL).

REFERENCES

[1] A.J.C. Eun, S.M. Wong, " Detection of Cymbidium mosaic potexvirus and Odontoglossum ringspot tobamovirus using immuno-capillary zone electrophoresis." *Phytopathology*, vol. 89, pp. 522−528, 1999.

[2] M.L. Seoh, S.M. Wong, L. Zhang , "Simultaneous TD/RT-PCR detection of cymbidium mosaic potexvirus and odontoglossum ringspot tobamovirus with a single pair of primer", *J. Virol. Methods*,vol. 72, pp. 197−204, 1998.

[3] C. H. Wang, K.Y. Lien, J.J. Wu, G.B. Lee, "A magnetic bead-based assay for the rapid detection of methicillin-resistant *Staphylococcus aureus* by using a microfluidic system with integrated loopmediated isothermal amplification" , *Lab. Chip.*, vol. 173, pp. 43-48, 2011.

[4] C.H. Huang, G.H. Lai, M.S. Lee, W.H. Lin, Y.Y. Lien, S.C. Hsueh, J.Y. Kao, W.T. Chang, T.C. Lu, W.N. Lin, H.J. Chen, M.S. Lee,"Development and evaluation of a loop-mediated isothermal amplification assay for rapid detection of chicken anaemia virus", *J. Appl. Microbiol.* vol. 108, pp. 917−924, 2010.

[5] A. Varga, D. James, "Use of reverse transcription loop-mediated isothermal amplification for the detection of Plum pox virus", *J. Virol. Methods,* vol. 138, pp.184−190, 2006.

[6] M.S. Lee, M.J. Yang, Y.C. Hseu, G.H. Lai, W.T. Chang, Y.H. Hsu, M.K.Lin, "One-step reverse transcription loop-mediated isothermal amplification assay for rapid detection of *Cymbidium mosaic virus*", *J. Virol. Methods*,vol. 173, pp. 43 - 48, 2011.

[7] X. Fang, Y. Liu, J. Kong, X. Jiang, "Loop-mediated isothermal amplification integrated on microfluidic chips for point-of-care quantitative detection of pathogens", *Anal. Chem.*, vol. 82, pp.3002–3006, 2011.

[8] C.H. Wang, K.Y. Lien, T.Y. Wang, T.Y. Chen, G.B. Lee, "An integrated microfluidic loop-mediated-isothermal-amplification system for rapid sample pre-treatment and detection of viruses", *Biosens. Bioelectron.,* vol. 26, pp. 2045–2052, 2011.

CONTACT

*Dr. Gwo-Bin "Vincent" Lee, e-mail address: gwobin@pme.nthu.edu.tw; address: No. 101, Section 2, Kuang-Fu Road, Hsinchu, Taiwan 30013; Tel: +886-3-5715131 Ext.33765; Fax: + 886-3-5722840

Automated Immunoassay System based on the Colorimetric Detection

Kin Fong Lei

Graduate Institute of Medical Mechatronics, Department of Mechanical Engineering,
Chang Gung University, Taoyuan, Taiwan
Email: kflei@mail.cgu.edu.tw; Tel: +886-03-2118800 ext. 5345

Abstract—**In this work, an automated immunoassay system is developed based on colorimetric detection for the application of clinical diagnostics. In this system, most of the biological processes of sandwich immunoassay can be performed automatically. The immunoassay result can be represented by color intensity and observed by a regular camera or even naked eye. The biological reactions were performed in a microfluidic chip and the result of sandwich immunoassay could be observed directly. An example of the sandwich immunoassay has been demonstrated using this automated system. Anti-human IgG, human IgG, and anti-human IgG-Biotin conjugates were selected as the primary antibody, target antigen, and secondary antibody, respectively. The target antigen concentration ranging from 0.5 to 160 ng/ml can be detected quantitatively and the detection limit was 0.5 ng/ml. Since the detection method is simplicity, low cost, and miniaturization; therefore, the proposed system has the potential to be developed to an automated disease screening system.**

Keywords: Microfluidics; Sandwich immunoassay; Clinical diagnostics; Gold nanoparticles.

I. INTRODUCTION

In the statistical report from Department of Health in Taiwan [1], the 1st killer in the past 28 years was the cancer related diseases. A person was died from cancer every 13 minutes and 10 seconds in the average. Early treatment is an effective way to reduce the death rate. However, there is no significant symptom in the early stage. In order to provide the prevention of the disease, regular tumor marker screening is desired to identify the early stage of cancer. Currently, tumor marker screening is based on the technology called immunoassay, which is commonly used in clinical laboratories [2, 3].

Immunoassay is commonly used in many clinical, pharmaceutical, and scientific research laboratories. The principle of sandwich immunoassay is to use two antibodies to determine the concentration of the target antigen. Briefly, primary antibody which is highly specific for the target antigen is immobilized on a sold surface. Target antigen and secondary antibody are then added respectively. The secondary antibody binds the target antigen to a different site than the primary antibody. Consequently, the target antigen is sandwiched between two antibodies. When the concentration of the target antigen increases, the amount of the secondary antibody increases which leads to a higher measured response. This bio-analytical technique is capable to detect the specific antigen through the binding of the corresponding antibody. Based on different labels attaching on the secondary antibody, immunoassay can be classified as enzyme immunoassay [4-6], radioimmunoassay [7], magnetic immunoassay [8], fluorescent immunoassay [9], metalloimmunoassay [10], etc. However, immunoassay performing in conventional laboratory is laborious and requires various equipment and long analysis time. And also, most of the clinical immunoassay systems are based on fluorescent detection method. It has the advantage of high sensitivity, but requires dedicated optical equipment which has the limitation of miniaturization. Therefore, immunoassay is difficult to be developed to the point-of-care equipment for clinics.

Recently, microfluidic technology has been applied to the clinical applications because of its ability of miniaturization. From early single channel devices [11] to current complex analysis system [12], microfluidics has been rapidly developed. Immunoassay has also been implemented on the microfluidic chip [13-14]. In order to develop a miniaturize detection method, metallic nanoparticles have been applied for the labeling materials in immunoassay. A label-free visual immunoassay on a glass slide has been demonstrated [15]. Antibody was first immobilized on a glass slide and silver nanoparticles were electrostatically adsorbed on the glass slide under particular conditions. A white light-emitting diode (LED) torch was employed to illuminate the glass slide. Then, the presence of the antibody can be observed by the naked eye. Also, colorimetric immunoassay using antibody-gold nanoparticle conjugate and silver enhancement has been reported [16]. A regular flatbed scanner was utilized for the optical scanning and measuring the immuno-reaction signal on the glass slide. Moreover, electrical detection of immunoassay using silver enhanced gold nanoparticles has been presented [17]. In situ assembly of colloidal particles was across the micro-electrodes. The proteins were captured by the colloidal particles and the proteins were indicated by the silver enhanced gold nanoparticles. The concentration of the proteins can be determined by the conductivity changes across the micro-electrodes.

In this work, an automated immunoassay system was developed and sandwich immunoassay was demonstrated

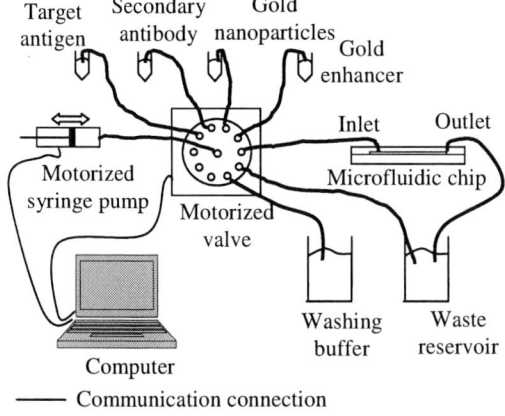

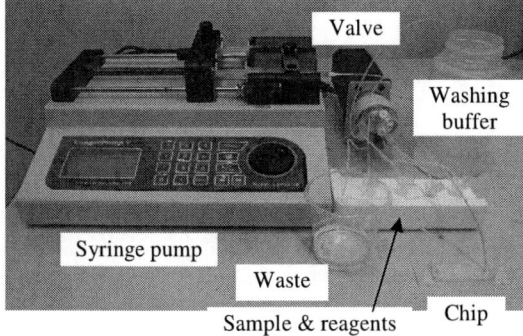

Figure 1. Setup of the automated immunoassay system.

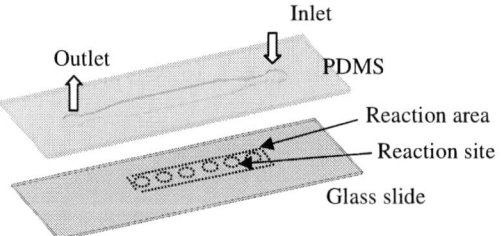

Figure 2. Illustration of the microfluidic chip. The dash area is the reaction area where is encapsulated by the PDMS chamber and the primary antibodies are immobilized on the reaction sites.

using the system. The immunoassay result was detected by colorimetric method [18]; that is, the concentration of the target antigen can be represented by visible color intensity and measured by a regular camera or naked eye. Therefore, the detection can be performed without bulky equipment and is suitable for portable applications. The system consisted of a motorized syringe pump, a motorized 10-position valve, and a microfluidic chip. Most of the immunoassay processes, including sample and reagent loading, washing, and signal development, could be handled automatically. The

immunoassay and the colorimetric detection were performed on the microfluidic chip. The entire process was around 1 hour. This automated immunoassay system is potentially developed to a portable system for clinical disease screening equipment.

II. MATERIALS AND METHODS

A. Chemicals and reagents

Anti-human IgG, human IgG, and bovine serum albumin (BSA) were purchased from Sigma, USA. Anti-human IgG-biotin conjugate was purchased from Millipore, USA. 3-aminopropyltriethoxysilane (APTES) and glutaraldehyde (GA) were purchased from Pierce, USA. Nanogold-streptavidin conjugate and gold enhancement solution were purchased from Nanoprobes, USA. Buffer used in this study was phosphate-buffered saline (PBS; 50mM phosphate, 150mM NaCl, 10mM EDTA, pH 7.6). Distilled water was used throughout the experiments.

B. Design of the automated colorimetric immunoassay system

The system is designed to be automated for most of the immunoassay processes including sample and reagent loading, washing, and signal detection. The system consists of a motorized syringe pump (Baoding Longer Precision Pump CO., Ltd.), a motorized 10-position valve (Valco Instruments Co. Inc.), and a microfluidic chip. The experimental setup is shown in Figure 1. The syringe pump provides the driving force for the fluidic manipulation. The valve is to select the fluidic path as shown in Figure 1. The syringe pump and valve are connected to a computer through RS-232 interface. Computer software is written to control the actuation sequence of the pump and valve. Sample and reagents can be pumped to the microfluidic chip sequentially. Moreover, the incubation time of the whole process can be controlled precisely. The microfluidic chip provides a biological reaction and detection environment for the sandwich immunoassay. The chip is designed to be disposable in order to eliminate the cross-contamination. Colorimetric detection method is used and the result of the immunoassay can be observed in the microfluidic chip. The microfluidic chip is a glass slide covered by a poly-dimethyl siloxane (PDMS) microfluidic channel, as shown in Figure 2. The channel is 40 mm in length and 5 mm in width and works as a closed volume chamber for the biological reaction and detection. 6 detection sites on the glass slide are assigned along the channel.

C. Surface modification of the microfluidic chip

Since the immunoassay is performed on the surface of the glass slide (reaction site), surface modification is necessary for the immobilization of the primary antibody, i.e., anti-human IgG. The procedure of the protein immobilization on glass surface followed the reported method [19]. Briefly, a regular glass slide was first cleaned in a 70:30 (v/v) mixture

978-1-4673-1122-9/12 $31.00 © 2012 IEEE   209

of H$_2$SO$_4$ and H$_2$O$_2$ for 30 minutes at room temperature. Then, it was washed several times in distilled water and dried in oven. The glass slide was immersed in 5% (v/v) APTES in acetone for 30 minutes at room temperature. Hence, it was rinsed thoroughly in acetone and distilled water, dried in nitrogen flow, and baked in oven at 80°C for 2 hours. This process could form a reactive amine on the glass surface. Then, 2 µl 2.5% (v/v) GA in PBS was pipetted to the detection sites for 2 hour incubation at room temperature. The GA reacted with amino group on the glass surface of the detection sites and worked as a cross-linker. After carefully rinsing with distilled water, the APTES-GA treated areas were ready to be covalently immobilized by the protein. 1 µl protein solution of the primary antibody of a certain concentration was pipetted to the surface of the detection sites. The primary antibody was immobilized covalently after the incubation of 1 hour at 37°C or overnight at 4°C. Then, 1 hour blocking treatment (1% BSA in PBS) at room temperature was performed to block the active site of the un-reacted GA. Finally, the PDMS microfluidic channel was attached to the glass slide with the alignment of the detection sites along the channel.

D. Colorimetric detection of sandwich immunoassay

The colorimetric detection of the sandwich immunoassay was based on gold nanoparticle indication and gold enhancement amplification. The schematic illustration of the protocol is shown in Figure 3. The optimized conditions of the colorimetric detection can be referred to our previous work [18]. The concentration of the target antigen can be represented by visible color intensity and the results can be observed by a regular camera or naked eye. Gold nanoparticle was utilized to indicate the secondary antibody through the biotin-streptavidin linkage. Since the gold nanoparticle was too tiny to be observed, gold enhancement process was performed for enlarging the gold nanoparticle physically. The enhanced gold particles became visible and increased the color intensity gradually with time. Therefore, the results of the sandwich immunoassay can be observed without sophisticated laboratory equipment.

The primary antibody, target antigen, and the secondary antibody were anti-human IgG, human IgG, and anti-human IgG-biotin conjugate, respectively. The primary antibody was first immobilized on the detection sites during the preparation process of the microfluidic chip. Then, the target antigen and secondary antibody were added to the detection sites respectively. Each antibody-antigen binding process required incubation at room temperature. In between each binding process, the detection sites were thoroughly washed by washing buffer (PBS with 0.05% Tween-20). Therefore, sandwich-like complex was formed on the detection sites. The colorimetric detection was carried out by tagging the detection antibody to the gold nanoparticles. Nanogold-streptavidin conjugates suspended in the buffer in a certain dilution containing 1% BSA were added to the detection sites and incubated at room temperature. Gold nanoparticles were bound to the secondary antibodies via biotin-streptavidin

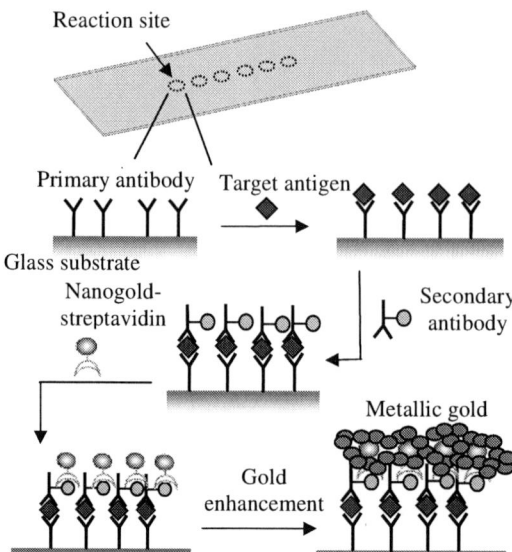

Figure 3. Illustration of the protocol of colorimetric sandwich immunoassay.

linkage. Theoretically, the more secondary antibodies locate on the detection sites, the more gold nanoparticles can bind to the secondary antibody. Because the gold nanoparticles are 1.4 nm in diameter, they are too tiny to generate observable visual signal. Gold enhancement, i.e., gold ions suspended in solution were catalytically deposited onto the gold nanoparticles as metallic gold, was performed to enlarge and fuse the gold nanoparticles. These gold precipitations became visible and darkened the color gradually with time. After completion of the process, the detection sites were rinsed by water and dried. The results, i.e., color intensity, were captured by a regular camera with an appropriate illumination. They were analyzed by commercial software and converted to 8-bit gray level values for representation.

III. RESULTS AND DISCUSSION

After the preparation of the microfluidic chip, its inlet and outlet were connected and loaded on the system. Then, the target antigen, secondary antibody, nanogold solution and gold enhancement solution were installed into the system. The system followed the preset procedure in the computer software and ran automatically. The whole process was around 1 hour. First, the target antigen, i.e., sample, was pumped to the microfluidic chip for 15 minutes. After washing, the secondary antibody was pumped to the chip for another 15 minutes. Therefore, the sandwich-like complex formed on the detection sites along the channel. After washing, nanogold solution was pumped to the chip for 15 minutes. Gold nanoparticles bound to the detection antibody via biotin-streptavidin linkage. The amount of the bound gold nanoparticles depended on the amount of the target antigen.

978-1-4673-1122-9/12 $31.00 © 2012 IEEE

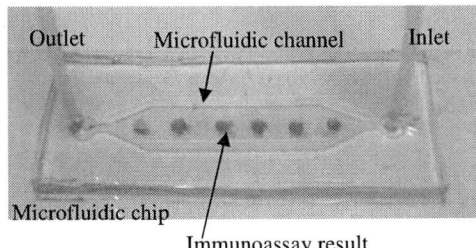

Figure 4. Photo of the microfluidic chip. The target antigen was present and the immunoassay result could be observed.

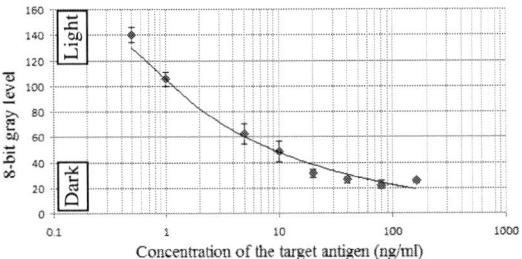

Figure 5. Results of the colorimetric sandwich immunoassay performed by the automated system. The concentration of the target antigen was 0.5, 1, 5, 10, 20, 40, 80, and 160 ng/ml. The color intensity was represented by 9-bit gray-level values.

After washing, gold enhancement process was performed for 10 minutes and the detection sites became visible when the target antigen was present. The colorimetric signal can be captured by a regular camera or observed by the naked eye with an appropriate illumination, shown in Figure 4. In Figure 4, the color intensities of the 6 detection sites were the same because the same concentration of the target antigen was applied to all detection sites simultaneously. Furthermore, more experiments were repeated by applying different concentrations of the target antigen. The colorimetric signals of different concentrations of the target antigen could be captured by running several experiments. Hence, the mapping between the color intensities and the different concentrations of the target antigen could be figured out. Quantification of the color intensities represented by 8-bit gray level values as a function of increasing concentrations of target antigen was plotted in Figure 5. The target antigen concentration ranging from 0.5 to 160 ng/ml can be detected quantitatively and the detection limit was 0.5 ng/ml.

IV. CONCLUSION.

An automated colorimetric immunoassay system has been developed for the application of clinical diagnostics. The system can perform sandwich immunoassay around 1 hour automatically. Compared to the conventional laboratory process, this system has the advantage of fast processing time, automation, and small in size. An example of the sandwich immunoassay has been demonstrated using this automated system. Anti-human IgG, human IgG, and anti-human IgG-Biotin conjugates were selected as the primary antibody, target antigen, and secondary antibody, respectively. The target antigen concentration ranging from 0.5 to 160 ng/ml can be detected quantitatively and the detection limit was 0.5 ng/ml. Since the detection method is simplicity, low cost, and miniaturization; therefore, the proposed system has the potential to be developed to clinical diagnostic equipment.

ACKNOWLEDGEMENT

The author would like to thank for the financial support from National Science Council, Taiwan (Project number: NSC99-2218-E-182-008 and NSC100-2221-E-182-022)

REFERENCES

[1] Cancer registry annual report in Taiwan area, 2008. Department of Health, Executive Yuan, Republic of China.
[2] Foekens JA, Portengen H, van Putten WLJ, Peters HA, Krijnen HLJM, Alexieva-Figusch J, Klijn JGM (1989) Prognostic value of estrogen and progesterone receptors measured by enzyme immunoassays in human breast tumor cytosols. Cancer Res 49:5823-5828.
[3] Cesaro-Tadic S, Dernick G, Juncker D, Buurman G, Kropshofer H, Michel B, Fattinger C, Delamarche E (2004) High-sensitivity miniaturized immunoassays for tumor necrosis factor a using microfluidic systems. Lab Chip 4:563-569.
[4] Yakovleva J, Davidsson R, Lobanova A, Bengtsson M, Eremin S, Laurell T, Emneus J (2002) Microfluidic enzyme immunoassay using silicon microchip with immobilized antibodies and chemiluminescence detection. Anal Chem 74:2994-3004.
[5] Kurita R, Yokota Y, Sato Y, Mizutani F, Niwa O (2006) On-chip enzyme immunoassay of a cardiac marker using a microfluidic device combined with a portable surface plasmon resonance system. Anal Chem 78:5525-5531.
[6] Arenkov P, Kukhtin A, Gemmell A, Voloshchuk S, Chupeeva V, Mirzabekov A (2000) Protein microchips: use for immunoassay and enzymatic reactions. Anal Biochem 278:123-131.
[7] Ma Z, Gingerich RL, Santiago JV, Klein S, Smith CH, Landt M (1996) Radioimmunoassay of leptin in human plasma. Clinical Chem 42:942-946.
[8] Chemla YR, Grossman HL, Poon Y, McDermott R, Stevens R, Alper MD, Clarke J (2000) Ultrasensitive magnetic biosensor for homogeneous immunoassay. Proc National Academy of Sciences 97:14268-14272.
[9] Li S, Floriano PN, Christodoulides N, Fozdar DY, Shao D, Ali MF, Dharshan P, Mohanty S, Neikirk D, McDevitt JT, Chen S (2005) Disposable polydimethylsiloxane/silicon hybrid chips for protein detection. Biosens Bioelectron 21:574-580.
[10] Fan A, Lau C, Lu J (2005) Magnetic bead-based chemiluminescent metal immunoassay with a colloidal gold label. Anal Chem 77:3238-3242.
[11] Harrison DJ, Fluri K, Seiler K, Fan Z, Effenhauser CS, Manz A (1993) Micromachining a miniaturized capillary electrophoresis-based chemical analysis system on a chip. Science 261:895-897.
[12] Melin J, Quake SR (2007) Microfluidic large-scale integration: the evolution of design rules for biological automation. Annu Rev Biophys Biomol Struct 36:213-231.
[13] Sato K, Yamanaka M, Takahashi H, Tokeshi M, Kimura H, Kitamori T (2002) Microchip-based immunoassay system with branching multichannels for simultaneous determination of interferon-γ. Electrophoresis 23:734-739.
[14] Cho JH, Han SM, Paek EH, Cho IH, Paek SH (2006) Plastic ELISA-on-a-chip based on sequential cross-flow chromatography. Anal Chem 78:793-800.
[15] Ling J, Li YF, Huang CZ (2008) A label-free visual immunoassay on solid support with silver nanoparticles as plasmon resonance scattering indicator. Anal Biochem 383:168-173.

[16] Yeh CH, Hung CY, Chang TC, Lin HP, Lin YC (2009) An immunoassay using antibody-gold nanoparticle conjugate, silver enhancement and flat bed scanner. Microfluid Nanofluid 6:85-91.

[17] Velev OD, Kaler EW (1999) In situ assembly of colloidal particles into miniaturized biosensors. Langmuir 15:3693-3698.

[18] Lei KF, Butt YKC (2010) Colorimetric immunoassay chip based on gold nanoparticles and gold enhancement. Microfluid Nanofluid 8:131-137.

[19] Williams RA, Blanch HW (2004) Covalent immobilization of protein monolayers for biosensor applications. Biosens Bioelectron 9:159-167.

Three-Dimensional Lab-on-a-CD with Enzyme-Linked Immunosorbent Assay

M. Ishizawa[1], T. Azeta[1], H. Nose[1], Y. Ukita[2], and Y. Utsumi[1]

[1] Laboratory of Advanced Science and Technology for Industry, University of Hyogo, Japan
[2] Japan Advanced Institute of Science and Technology, Japan

Abstract—**This paper reports a new type of lab-on-a-CD device with three-dimensional microchannel networks and vertical capillary bundle structure. The device consist of the multiple lab-on-a-CD chips with conventional planer microchannels and vertical microchannels which constructed on the disks made of poly-dimethylsiloxane (PDMS) and poly-methylmethacrilrate (PMMA). The PMMA layer with vertical channel were fabricated by deep X-ray lithography using synchrotron radiation. We designed and fabricated the chips for aim of taking enzyme-linked immunosorbent assay(ELISA). Three-dimensional liquid transportation through vertical capillary bundle structure is suggested by computational fluid dynamics and it is successfully demonstrated chips. Detection of immunogroblin G(IgG) from mouse is successfully demonstrated in the three-dimensional devices and it is notable that the result suggests quite high-sensitive implying detection limit down to few ng/mL by primitive result.**

Keywords: X-ray Lithography, Synchrotron Radiation, ELISA, Endocrine Disrupter,PCB

I. INTRODUCTION

The miniaturized analysis chip such as microchip electrophoresis have been well developed. However the microchemical systems with whole bio-chemical process, including sample pre-treatment, have not been developed, while it requires much efforts and time over separation and detection. The various chemical operations must be realized on one platform to realize totally automated bio-process. The concept of centrifugal microfluidics is promising to realize fully automated microfluidic systems without external syringe

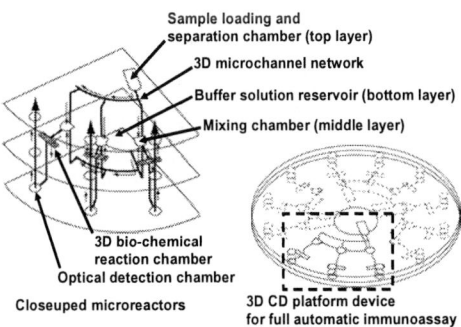

Fig 1. Schematic diagram of new CD-like microfluidic platform with three-dimensional

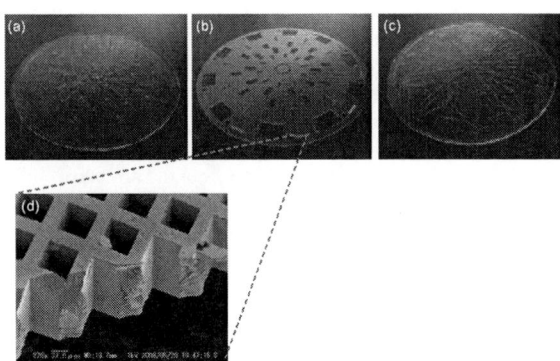

Fig 2. photographs of fabricated devices
(a) top layer with planer microchannel made by PDMS
(b) middle layer with vertical microchannel made by PMMA
(c) bottom layer with planer microchannel made by PDMS
(d) scanning electron microscope image of vertical vapillary bundle structure

pumping systems by using single motor control.[1][2][3] The parallel unit operation in integrated multiple microreactors, which integrated along radial direction on a rotational disk, is realized by simply spinning the disk-like chip. Moreover, centrifugal microfluidics can simultaneously carry out detection of each units. In this way, centrifugal microfluidics enables automation of chemical operations for overall assay protocol. The other benefits of the automation are realization of highly precise and reliable operations by reducing human errors or mistake. However, there is string restriction to the centrifugal microfluidic due to limited area for integration of multiple function and multiple reactor units and different modules. We propose a new concept of centrifugal microfluidic with three-dimensional (3D) microchannel interconnection, which enables fast, high-sensitive and high-throughput analysis. The concept of centrifugal microfluidic with 3D microchannel interconnection contributes to the large-scale integration of various modules and reactor systems by stacking layers of CD-like disks with different functions. For example, by stacking layers of centrifugal microfluidic can be compactly integrated with sample loading, 3D microchannel network, buffer solution reservoir, mixing chamber, 3D bio-chemical reaction chamber and detection chamber. Each multiple reservoirs and chamber with different functions are integrated and connected to each other through a 3D microchannel network. We were able to increase the density of the integrated microreactor units by utilizing three-dimensionally stacked layers. The flexibility of the structural design, which arises from the 3D microstructure, improves the performance of various modules. For example, our device has 3D scaffolds for the immobilization of antibody onto the surface [3], and the stacked structure of the 3D detection reservoir results in the increased of signal-to-noise ratio. The 3D scaffolds for the immobilization of antibodies is the microfillter with capillary bundle structure to obtain large surface area for high sensitive detection.

We performed an ELISA using the microfillter with capillary bundle structure, and we succeeded high sensitive detection of nonylphenol by competitive ELISA.[4][5][6] We expect that by using the 3D centrifugal microfluidic involving capillary bundle structure as a reaction chamber realize fast reaction, high-sensitive detection, high-throughput analysis and high efficiency mixing. This paper presents novel concept of centrifugal microfluidic device with 3D design and the demonstrate of ELISA by using capillary bundle structure.

II. *DESIGN AND FABRICATE*

The fig.1 shows the concept of centrifugal microfluidic with 3D microchannel interconnection. Stack concept offers various benefits to the centrifugal microfluidics device that we discribed in a foregoing paragraph. The 3D centrifugal microfluidic is stacking structure that laminates two planer channel disk made of PDMS and vertical channel disks made of PMMA. The CD-like PDMS disks with planer microchannel is fabricated by conventional soft lithography.[7] The CD-like and CD-like PMMA disk with vertical microchannel is fabricated by x-ray lithography.[8][9][10]The fabricated CD-like PMMA disk were formed by using large-area x-ray lithography system BL-2, which constructed in the NewSUBARU synchrotron radiation facility of University of Hyogo.[11][12] x-ray photoresist PMMA of thickness 200 μm was exposed to x-ray and the exposed sheet was developed by immersing into the GG developer to develop complete through-hole. To construct three-dimensional micro channel networks these disks are aligned and bonded together. The fig.2 shows photographs of fabricated devices and scanning electron microscope image of vertical vapillary bundle structure. The device consists of three-stacked layers with 8 assay units for parallel assay of 8 samples.

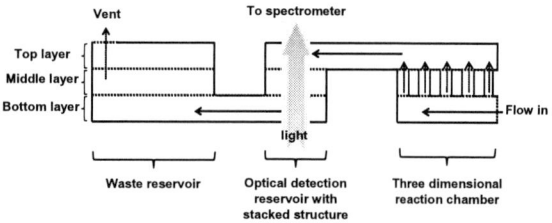

Fig 3 Cross sectional view of three-dimensional microchannel interconnection

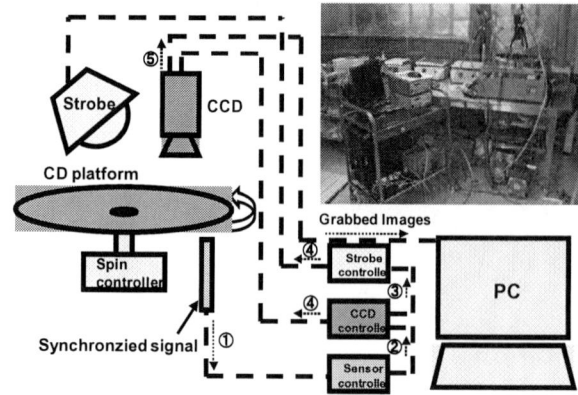

Fig 4. The schematic illustration of strobe scope system

Each unit consists of four liquid loading reservoirs and biochemical chamber with capillary bundle structure and optical detection chamber. All layers are aligned and stacked together by using self-adhesion of CD-like PDMS disks. The thickness of the reservoirs and chambers set to adjust the suitable sample, reagent volumes and optical path length of the reaction products for UV absorption and fluorescence such as figure 3.[13] The high aspect ratio structure of liquid loading reservoir is effective to integrate assay units by reducing planer area of reservoirs.

III. *EVALUATION OF LIQUID TRANSPORTATION*

We reported that we carried out computational fluid dynamics (CFD) using "FLUENT" software to investigate the feasibility of valving and liquid transportation by centrifugal force. And we succeeded to control vertical liquid transport using centrifugal force. (Kondo et al. 2009) As a next step, we evaluated overall transport liquid involving vertical liquid transport to perform ELISA. To demonstrate the automatic sequencing in 3D centrifugal microfluidic, we loaded stained

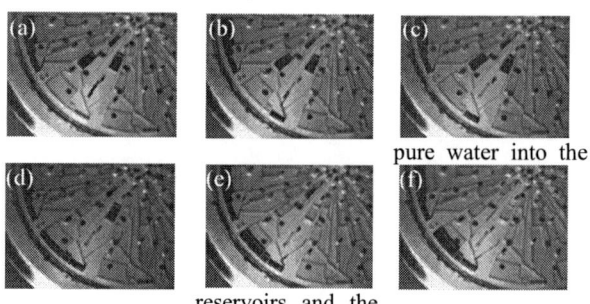

pure water into the reservoirs and the

Fig 5. images of flow sequencing

(a) Initial stage of spinning. (b)The 1st reservoir release the liquid at 650rpm (c)The liquid held on the vertical capillary transport through vertically and the whole liquid injected into the reaction reservoir. (d)The 2nd reservoir releases the liquid. (e) The 3rd reservoir releases the liquid. (f) The 4th reservoir releases the liquid.

978-1-4673-1122-9/12 $31.00 © 2012 IEEE   214

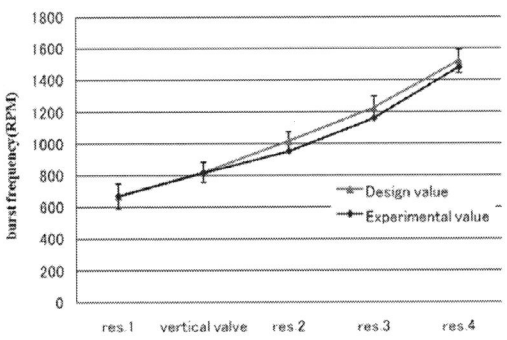

Fig 6. measured burst frequency curve

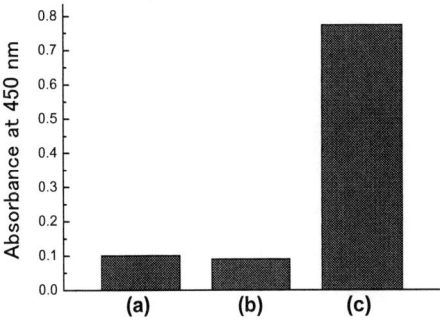

Fig. 7. Comparison of three different reaction conditions. (a) with 1st antibody, and without analyte (Mouse IgG) (b) without 1st antibody, and with analyte (10μg/ml of Mouse IgG) (c) with 1st antibody, and with analyte(10μg/ml fo Mouse IgG)

device is spun for the sequencing. The snapshot images are obtained by strobe scope

system to observe and measure the burst frequency of each chamber. The schematic illustration of strobe scope system is shown in Fig. 4. The optical sensor will generate the trigger signals for the synchronized control of CCD and strobe. The taken images are automatically grabbed with time information into PC and burst frequency was calculated by checking holding state and the time. Fig. 5 show the obtained images of automatic flow sequencing in 3D centrifugal microfluidic and measured burst frequencies are shown in fig 6. Figure 5(a) is Initial stage of spinning. All liquids are contained in reservoirs and the disk start to accelerate from 0rpm. The 1st reservoir release the liquid at 650rpm, however released liquid will be held on the surface of vertical capillary (fig. 5(b)). The liquid held on the vertical capillary transport through vertically and the whole liquid injected into the reaction reservoir at 800rpm (fig. 5(c)). The 2nd reservoir releases the liquid at 950rpm (fig. 5(d)). The 3rd reservoir releases the liquid at 1150rpm (fig. 5(e)) and the 4th reservoir releases the liquid at 1450rpm (fig. 5(f)). The four step flow sequencing involving vertical liquid flow around the reaction reservoir was successfully performed. The reproducibility is indeed good between 8 individual measurements. The result suggests the proposed 3D centrifugal microfluidcs is available to automated ELISA analysis.

IV. Application To ELISA

A. Protocol

To demonstrate the ELISA was carried out by means of standard sandwich ELISA in fabricated device. To estimate the possibility of ELISA in our device, we carried out detection of immunogroblin G(IgG) from mouse by standard protocol of sandwich ELISA. Capillary bundle structure in reaction chamber is locally immobilized with antibody to the analyte (mouse IgG) by physical adsorption and whole part of PMMA and PDMS disks are blocked by bovine serum albumin (BSA). Then these disks are aligned and stacked together by utilizing self adhesion of PDMS. To estimate the possibility of ELISA in our 3D centrifugal microfluidic, we carried out standard protocol of sandwich ELISA. The detection of the sample is carried out by using optical fiber spectroscopy system by probing three-dimensionally stacked

detection reservoirs on the device. The precise experimental procedure is described as follows. The 3 μl of sample solutions with varied concentration of analyte were pre-loaded in the reservoir and sample solution transported to reaction chamber by spun disk. The incubation was carried out for 10 min at room temperature for 1st antigen-antibody reaction. Then the stacked disk was disassembled and washing step was manually carried out by immersing the whole disks in the washing buffer for five times. Then disks were assembled again and 3 μl of 2nd antibody solution was injected. Incubation for 2nd antigen-antibody reaction was carried out by same manner. Next step is again washing step. To estimate the efficiency of on-chip washing, we carried out two different experiments with off-chip washing and on-chip washing. Because washing of enzyme conjugated 2nd antibody is the most critical step to reduce background, on-chip washing is tested by washing 2nd antibody. 10 μl of washing regents were preloaded in the reservoirs to carryout automatic washing of enzyme conjugate followed by injection of substrate.Then 10 μl of TMBZ (substrate) was loaded in the 3rd reservoir, while one of the reservoirs was loaded with acetic acid buffer as blank for spectroscopy, and injected into the reaction chamber by spinning and incubated for 5 min to develop reaction. After the incubation, the 3 μl of stop regent was loaded into the 4th

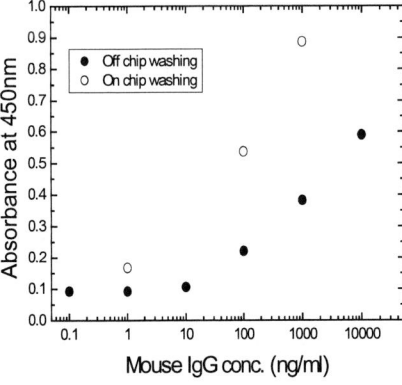

Fig 8. Obtained calibration curve by using three-dimensional lab-on-CD.

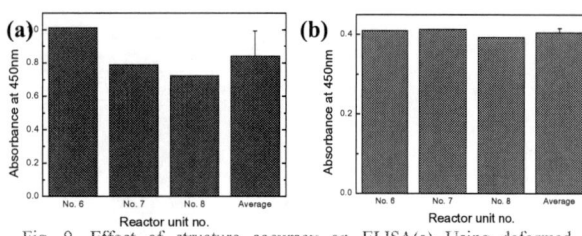

Fig. 9. Effect of structure accuracy on ELISA(a) Using deformed structure(b) Using improved structure

reservoir and injected into the reaction chamber by spinning the disk. And the regent was mixed with colorized substrate. The mixture was transferred into the detection reservoir with volume of 2.8 µl which located the next of the reaction chamber. The measurement of the absorbance spectroscopy was carried out by using optical fiber spectroscopy system with approximately 30 msec integration time (HR4000, Ocean Optics) by probing detection reservoirs of disk. The absorbance spectrum was obtained ranging from 200nm to 1200nm. And absorbance at 450nm and 650nm was read to plot the calibration curves. To evaluate the state of mixing and stopping of the enzyme reactions, the absorbance at 650nm was monitored every time. (14)

B. Immobilization

To estimate the immobilization and specificity of antibody, we compared three different reaction conditions in the device. As shown in fig.7, the absence of 1st antibody or absence of analyte result in low level signal, however if the immobilized 1st antibody and analyte present together, it exhibit higher signal. This result suggest successful immobilization of antibody with surficient activity and specificity on the PMMA capillary bundle structure. It is also notable that we succeeded in on-chip detection by means of absorption spectroscopy in three-dimensionally stacked detection reservoir with optical path of 700µm.

C. Wash

We demonstrated and compared the results of ELISA by on-chip and off-chip washing step. As shown in fig. 8, we successfully detected reasonable dose response of the signal reflecting the concentration of analyte. The corresponding assays to the each plot in the fig. 8 were simultaneously carried. The estimated detection limit is quite sensitive less than several ng/ml and oval reaction time is approximately 25 min. By comparing the results of off-chip washing with on-chip washing, it is very encouraging that the result of on-chip washing also exhibit clear dose response and plot of 1ng/ml analyte is quite close to the back ground level of off-chip washing.

D. Analysis Deviation

We found out the strong effect of micromachining accuracy of the bundle-like capillary structure onto the reproducibility of the assay. The fig.9 shows simultaneously taken results of the three tests with same condition by two structures of CD. As shown in fig. 9, the reproducibility of the signal intensity is quite improved structure with higher accuracy(error below the 5%).

E. Analysis Of PCB

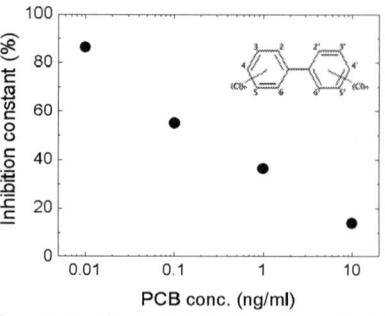

Fig. 10: Fluorescence microscope image of trapping vertical capillary

We demonstrated competitive ELISA and fig. 10 shows the result of the ELISA. The measurements corresponding to each plot are simultaneously carried out on same chip of three-dimensionally stacked detection reservoirs by means of absorption spectroscopy. As shown in fig.10, the decreasing signal with analyte concentration clearly suggests successful detection of PCB. Estimated detection limit is lessthan 0.01ng/ml, and the sensitivity is comparable to the commercial kit.

V. CONCLUSION

We proposed and fabricated a new type centrifugal microfluidic device with 3D microchannel network to perform for wide biochemical operation such as ELISA. Our proposed device has many advantages as follows. It can expand the application of centrifugal microfluidic by realizing higher integration of multiple microreactors and various functional components by using three-dimensional structure. Three-dimensional liquid transportation can be realized by transfer of liquid between stacked layers of disks. ELISA achieved on this 3D platform is widely application to food testing, environmental analysis, clinical diagnosis, and point of care testing. We demonstrated vertical liquid transport passing through the high aspect ratio capillary bundle structure by controlling spinning speed of device and centrifugal force as a pumping force. we confirmed capillary bundle structure could be employed as vertical microvalve for 3D fluidic systems using centrifugal force as a pumping method. And we succeeded four step flow sequencing involving vertical liquid flow around the reaction reservoir.

To evaluate our proposed device performance, we performed ELISA by using 3D centrifugal microfluidic. We succeeded to obtain calibration curve of Mouse IgG and detection limit was quite sensitive less than several ng/mL and oval reaction time was approximately 25min. The results indicate that we succeeded to develop the analytical device for rapid, high sensitive, high-throughput diagnosis systems by the 3D centrifugal microfluidics. The reproducibility of the assay is strongly affected by quality of structure. We successfully detected PCB in the device with detection limit down to 0.01ng/ml.

References

(1) Siyi Lai, Shengnian Wang, Jun Luo, L. James Lee, Shang-Tian Yang, and Mark J. Madou (2004) Design of a Compact Disk-like Microfluidic Platform for Enzyme-Linked Immunosorbent Assay. *Analytical chemistry* 76:1832-1837

(2) Nobuo Honda, Ulrika Lindberg, Per Andersson, Stephan Hoffmann, and Hiroyuki Takei (2005) Simultaneous Multiple Immunoassays in a Compact Disk-shaped Microfluidic Device Baced on Centrifugal Force. *Clinical Chemistry* 51:1955-1961

(3) Gregor Welte, Sascha Lutz, Berit Cleven, Hero Brahms, Claudia Gärtner, Günter Roth, Daniel Mark, Roland Zengerle, and Felix von Stetten (2010) MICROFLUIDIC LAB-ON-A-CHIP SYSTEM WITH INTEGRATED SAMPLE PREPARATION FOR PROCESSING IMMUNOASSAYS. *Micro TAS 2010* :818-820

(4) Y Utsumi, T Asano, Y Ukita, (2006) Proposal of a new microreactor for vertical chemical operation. Jpn J Appl Phys 45:2606-2611

(5) Y Ukita, T Asano, K Fujiwara, T Yokoyama, K Matsui, M Takeo, S Negoro, T Saiki, Y Utsumi (2006) Novel characteristics of multifunctional fluid filters fabricated by high-energy synchrotron radiation lithography. Jpn J Appl Phys 45:7203-7208

(6) Y Ukita, T Asano, Y Utsumi (2007) Fluid filter fabricated by deep X-ray lithography for micro fluidics. Microsyst Tevhnol 13:349-435

(7) David C. Duffy, J. Cooper McDonald, Oliver J. A. Schueller, and George M. Whitesides (1998) Rapid Prototyping of Microfluidic Systems in Poly(dimethylsiloxane). *Analytical chemistry* 70:4974-4984

(8) Yuichi Utsumi, Toshifumi Asano, Yoshiaki Ukita, Katsuhiro Matsui, Masahiro Takeo, and Seiji Negoro (2006) Proposal of a new microreactor for vertical chemical operation. *Journal of vacuum science and technology B* 24:2606-2611

(9) Yoshiaki Ukita, Toshifumi Asano, Kuniyo Fujiwara, Takuya Yokoyama, Katsuhiro Matsui, Masahiro Takeo, Seiji Negoro, Tsunemasa Saiki, and Yuichi Utsumi (2006) Novel Characteristics of Multifunctional Fluid Filters Fabricated by High-Energy Synchrotron RadiationLithography. *Japanese journal of applied physics* 45:7203-7208

(10) Yoshiaki Ukita, Toshifumi Asano, Kuniyo Fujiwara, Katsuhiro Matsui, Masahiro Takeo, Seiji Negoro, Tomohiko Kanie, Makoto Katayama, and Yuichi Utsumi (2008) Application of vertical microreactor stack with polystylene microbeads to immunoassay. *Sensors and actuators A* 145-146:449-455

(11) Yuichi Utsumi, Takefumi Kishimoto, Tadashi Hattori, and Hirotsugu Hara (2005) Large-Area X-ray Lithography System for LIGA Process Operating in Wide Energy Range of Synchrotron Radiation. *Japanese journal of applied physics* 44:5500-5504

(12) Yuichi Utsumi, and Takefumi Kishimoto (2005) Large area and wide dimension range x-ray lithography for lithographite, galvanoformung, and abformung process using energy variable synchrotron radiation. *Journal of vacuum science and technologies B* 23:2903-2909

(13) Katsuhiro Matsui, Isao Kawaji, Yuichi Utsumi, Yoshiaki Ukita, Toshifumi Asano, Masahiro Takeo, Dai-ichiro Kato, and Seiji Negoro (2007) Enzyme-linked Immunosorbent Assay for Nonylphenol Using Antibody-Bound Microfluid Filters in Vertical Fluidic Operation. *Journal of Bioscience and Bioengineering* 104:347-357

(14) P. David Josephy, Thomas Eling, and Ronald P. Mason (1982) The horseradish peroxidase-catalyzed oxidation of 3,5,3',5'-tetramethylbenzidine. Free radical and charge-transfer complex intermediates. *The journal of biological chemistry* 257:3669-3675

3D Biomimetic chip Integrated with Microvascular system for studying the Liver specific functions

Kuo-Wei Chang[1], Chia-Tung Lee[1], Punde Tushar Harishchandra[2], Hung-Po Chen[1], Ting-Ru Yueh[1], Srinivasu Valagerahally Puttaswamy[1], Shilpa Sivashankar[1], Cheng-Hsien Liu[1]

[1] Department of Power Mechanical Engineering, National Tsing-Hua University, Taiwan, R.O.C.
[2] Institute of NanoEngineering and MicroSystem, National Tsing-Hua University, Taiwan, R.O.C.
takeshivivi@hotmail.com

Abstract—**We describe the design and fabrication of 3D chip that enables the cell co-culture under continuous flow perfusion and repeated in situ observation by light microscopy. Thin polydimethylsiloxane (PDMS) membrane was etched to create an array of through-holes, and sandwiched between two PDMS channels to form an upper chamber and lower chamber. In the upper chamber hepatocytes (HepG2) and fibroblasts (3T3) were co-cultured to mimic the liver whereas in the lower structure endothelial cells (HMEC-1) were cultured to mimic the vascular system at the cellular scale for providing nutrients to the in vitro liver mimicked in the upper channel.**

Keywords-3D chip; Thin PDMS membrane; mimetic liver

I. INTRODUCTION

The liver is one of the most complicated organs of human body. It has over 500 kinds of important functions such as, secreting albumin, urea producing, bile, carbohydrate storing, supersession and also the first line of defense in human body [1, 2].
We find a lot of obstacles in developing and repairing of the organs. To rebuild an organ is extremely difficult in medical science and therefore we try to reconstruct a liver tissue in vitro in order to carry on the research (drug screening) into liver that function easier.

The relation between the liver and blood circulation is close. There is a big amount of crack between the cells -cell interaction in this microenvironment and many phenomenon are involved in this micro circulation. Liver is an organ with many microvascular systems present within in order to synthesize and metabolize various proteins. The blood capillaries present will have minute hole through which the liver in macrocosm receives proteins. It's crucial to synthesize plasma protein and metabolizes the product in combination with blood plasma protein.
They are two main function of liver.
1. The nutriment transport to the interstitial fluid through the capillary,
2. The uptake from the blood through the cell membrane.
Besides, the waste products will be transported through the cell membrane to the interstitial fluid, entering the endothelial cell which is located at the interface between the blood and the vessel wall, next to the plasma to accomplish the metabolism is as illustrated in Fig. 1.
We integrate above-mentioned phenomena as the foundation we plan to mimic the capillary by using micro-

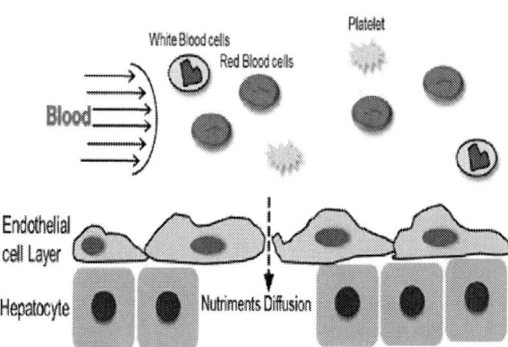

Fig. 1. The nutriment transport to the interstitial fluid through the capillary, then uptake from the blood through the cell membrane. Besides, the waste product transports through the cell membrane to the interstitial fluid, then going to the endothelial cell which is located at the interface between the blood and the vessel wall, next to the plasma to accomplish the metabolism.

fluid channel technology that is applied to rebuild a liver tissue in vivo.

II. MOTIVATION

In many areas of biomedical field, research is aimed at mimicking the in vivo organs of human body for studying its structure and function [3]. While mimicking this constructs it is very important to nourish the individual cells by providing the essential nutrients, oxygen and growth factors which in case of liver is achieved by the sinusoid. Micro-fluidic technology [4] offers great advantages over the conventional methods towards mimicking the cellular microenvironment [5] which makes it possible to reconstitute the organ level function on chip [6]. We developed a 3D bio-mimetic system which not only mimics the function of sinusoid but also mimics the liver and thus can serve as an efficient platform for studying the liver specific functions.

The designed chip is divided into two parts. One is liver tissue culture structure on the upper layer; this structure is a square space providing cell growing area. The other layer is a structure to imitate the intersection of blood capillary. The most important design is a porous film which is imitative structure that helps a substance diffuse via the membrane to the other side. Whilst in other words it is acting like the vessel wall or cell membrane in liver tissue. The diameter of the hole on the porous membrane is about 13 μm. It's too small to let the cells pass through but it lets the nutrient and waste product. The design enhances the cell culture on this membrane integrated liver tissue in vitro.

III. FABRICATION

The chip fabrication process is illustrated in Fig 2(a-d). SU-8(SU-8 2050, MicroChem Company, Newton, Massachusetts state) is a thick negative resist, and is spun coat on the wafer. It forms a pattern after exposure and development using developer RER600. The structure is represented in Fig. (c).

Next, we mix elastomer base and curing agent of PDMS (184 Sylgard, dishes of Kang Ning Company, Mydland, Michigan state) in ratio 10:1 and stir well to avoid of formation of air bubble in the mixture.

And then degas the PDMS mixture to jell and as a result cause discharging of bubbles inside the PDMS. Pour the PDMS mixture jells down on the wafer which is designed and patterned by lithography as a mold. Put them all in the oven immediately for baking by 90 degree for an hour to be hardened. the fabrication procedure is well explained in Fig. 3(a-h).

We obtain the upper layer and underside layer by the way of the process step3(a) to step 3(d) eventually .

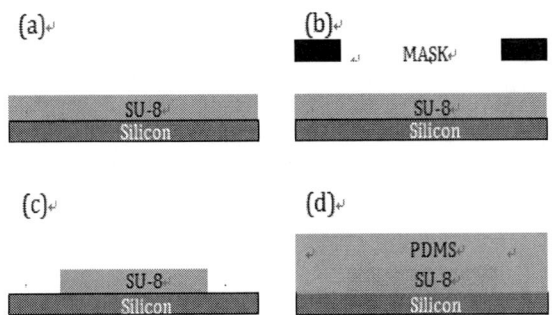

Fig. 2. The chip fabrication process.

The middle membrane is the core of the whole chip. The whole procedure is shown in Fig. 3, the preceding process step(a) to step(d) are the same as the general process of SU-8 fabrication .

First of all, define the columnar structure position on the wafer; the columnar structure is 40μm high with the base area of 6μm in radius.

Afterward we apply PDMS to the wafer especially the region of structure. It is represented in Fig. 3(d). most of PDMS not only gets distributed in the whole structure but also, sludges aside the column and some stick on the top of the column due to high viscosity of PDMS. Therefore, the PDMS will cover on the structure and wafer completely. It makes us an obstacle to mold a porous PDMS. We bond the upper layer and bottom layer together to turn it into a closed microchannel. We use tetra-butylammonium fluoride (TBAF) (75 wt% in H2O, Sigma-Aldrich, St Louis, MO) to etch extra PDMS in order to obtain a porous PDMS membrane. The membrane is

then peeled off and bonded to PDMS channel to obtain the complete chip

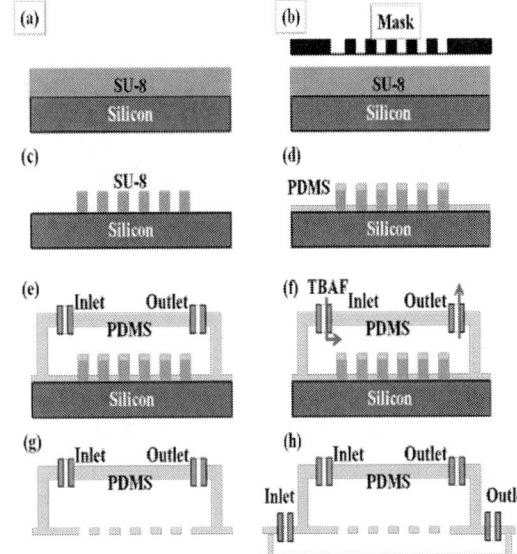

Fig. 3 Microfabrication process of the biomimetic chip.
(a) SU-8 negative photoresist is spin coated on Si wafer. (b)~(c) By photolithography process cylindrical structures of SU-8, 13~15 μm in diameter are defined on the Si wafer. (d) Thin layer of PDMS is spin coated on the Si-wafer. (e) PDMS structure is bonded to the thin layer to create the upper channel of the chip. (f) TBAF is injected in the upper channel for etching the PDMS in order to create a porous PDMS membrane. (g)~(h) The membrane is then peeled off and bonded to PDMS channel to create the complete chip.

Schematic cross-section of the overall chip is illustrated in figure 4, the upper space of liver tissue culture, the middle layer of porous film, and the lower is the imitation of microvascular structure.

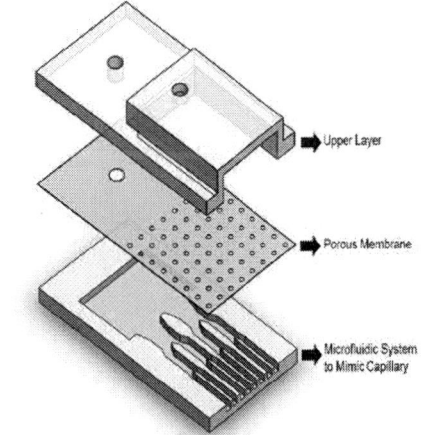

Fig.4 Porous PDMS membrane is sandwiched between two PDMS channels.

Chip is divided into two structure with one structure on top and other at bottom as shown in Fig. 5, the porous film is shown in Fig. 6, the size of holes was almost about 13~14μm. And the bottom microchannel is shown in Fig. 7, the maximum size of structure is 120μm, and the minimum size of structure is 30μm. The channel is descending to mimic the concept of artery in microvascular system.

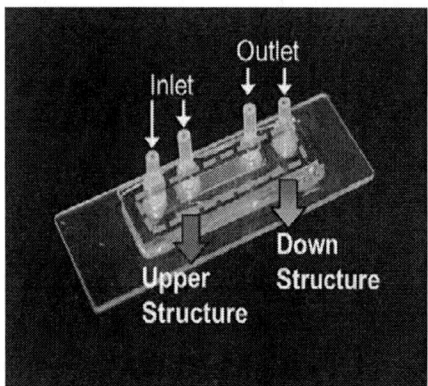

Fig.5 Porous PDMS membrane is sandwiched between two PDMS channels.

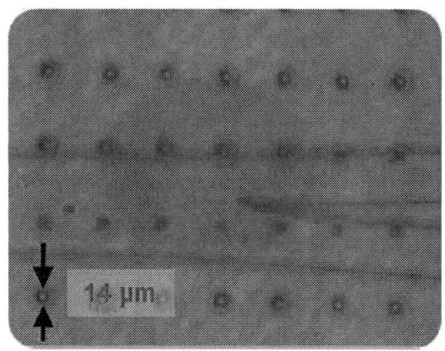

Fig.6 The porous membrane serves as a large intercellular cleft of capillary wall.

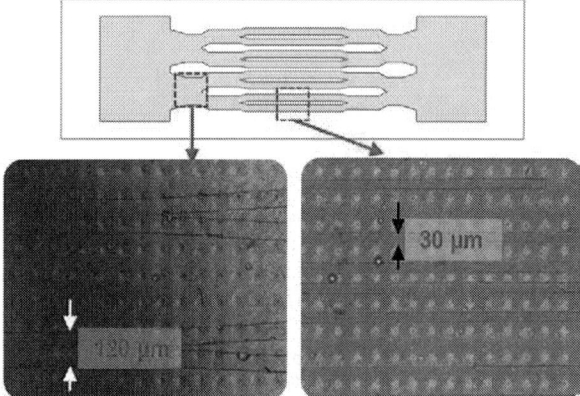

Fig.7 The channel is descending to mimic the concept of artery to microvascular.

IV. SIMULATION

To check the phenomenon of diffusion in the chip, we run the simulation of this chip. We used to simulate the circumstances of fluid by CFD-ACE+ (CFD Research Corporation, AL, USA). Figure 8(a)~(j) show the cross-section of chip. The top structure(up rectangle) is to culture liver cells and bottom structure(down rectangle) is for mimicking real blood vessel. In this simulation, we draw ten holes to represent the porous membrane to simulate the phenomenon of diffusion. The velocity of bottom structure is about 1(cm/s) to mimic the real velocity of blood vessel. Besides, we block the top channel to make sure fresh nutrients can diffuse to upper structure. The simulation results were obtained for the above mentioned conditions.

By this simulation, fresh nutrient (pink color) is diffusing to upper structure gradually shown in Figure 8(a)~(b). Finally, fresh nutrient is full of whole channel as shown in Figure 8 (c)~(j). The simulation shows the concept works in this chip. In other words, the liver tissue can get fresh nutrient by diffusion from bottom channel in this chip.

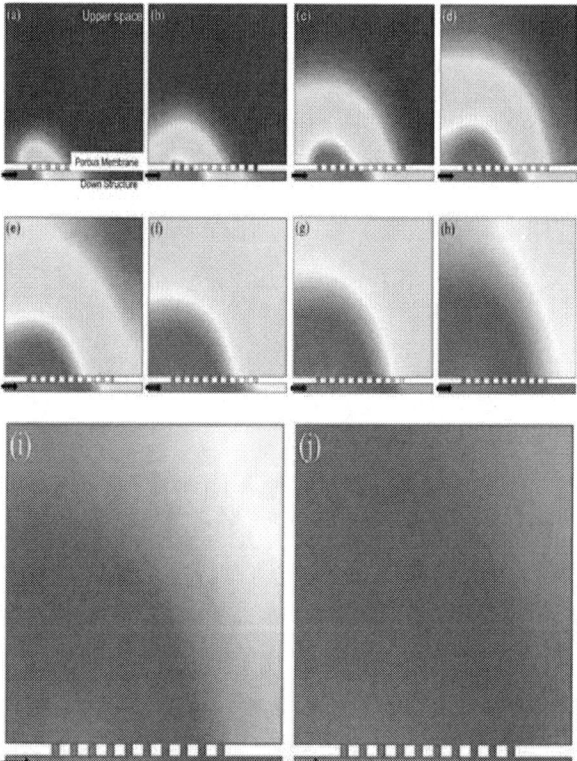

Fig.8 To check the phenomenon of diffusion in the chip, we run the simulation of this chip. We used to simulate the circumstances of fluid by CFD-ACE+ (CFD Research Corporation, AL, USA).

V. TESTING EXPERIMENT

For mimicking the in vivo liver and blood vessel, we culture HepG2 (ATCC; HB8065) and 3T3(ATCC; CCL163) in upper channel. Generally, 3T3 cell can help the growth of many kinds of cell, including HepG2. we hope to culture the HepG2 better in this case. In bottom channel, we culture HMEC-1(SV40-transformed human microvascular endothelial cell line. Derived by Ades) to mimic the blood vessel. As illustrated in Fig. 9 and Fig. 10, the green Fluorescence accomplishes the presence of HMEC-1 and red Fluorescence accomplishes the presence of HepG2. By this 3D structure we hope, that we can mimic the real liver and culture the liver tissue in vitro.

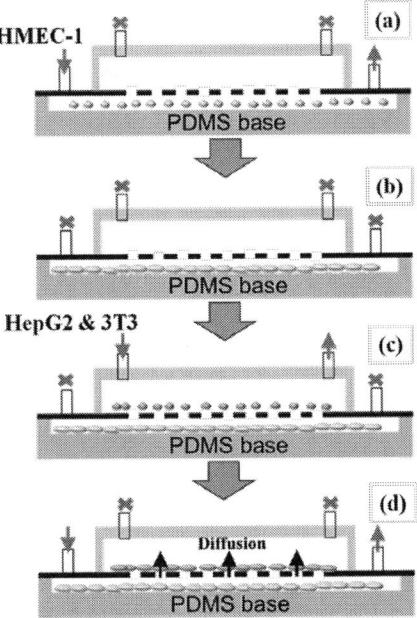

Fig.9 Schematic illustration of the experimental process. (a)~(b)HMEC-1 cells are injected in the bottom channel and cultured. (c)~(d)HepG2 and 3T3 cells are injected in the top channel and cultured in continuous flow condition

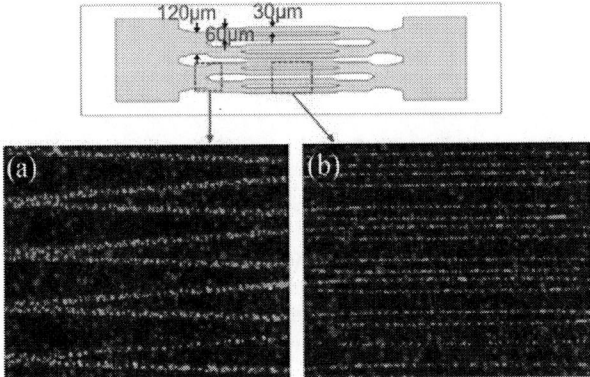

Fig.10 Schematic diagram of lower channel and fluorescence images. Green fluorescence shows that HMEC-1 cells are cultured in the lower channel to exhibit a sinusoid like structure for mimicking the microvascular system. HepG2 and 3T3 cells exhibit red fluorescence.

VI. CONCLUSION

We were successful in mimicking the nutrient transport function of microvascular system in the liver. We believe that this cell co-culture device with continuous perfusion system will serve as a platform for studying hepatocyte derived functions and analyze the effect of hepatotoxic chemicals and drugs.

ACKNOWLEDGMENT

This work is supported by National Science Council of R.O.C under the grant NSC 98-2120-M-007-003.

REFERENCES

[1] Y. Nahmias, R.E. Schwartz, Verfaillie, C.M. & Odde, D.J. "Laser-Guided Direct Writing for Three-Dimensional Tissue Engineering.," Biotechnol Bioeng. 92, pp.129-136, 2005.

[2] R. Taub "Liver regeneration: from myth to mechanism.," Nature Reviews Molecular Cell Biology 5, pp.836-847, 2004.

[3] A. D. v. d. Meer, A. A. Poot, M. H. G. Duits, J. Feijen, I. Vermes, *Journal of Biomedicine and Biotechnology*, pp. 1-10, (2009).

[4] G. Whitesides, E. Ostuni, S. Takayama, X. Jiang, D. Ingber, *Annu Rev Biomed Eng*, 3, pp.335 – 373, (2001).

[5] A. Khademhosseini, R. Langer, J. Borenstein, J. P. Vacanti, *Proceedings of the National Academy of Sciences of the United States of America*, 103, pp. 2480-2487, 2006.

[6] D. Huh, B. Matthews, A. Mammoto, M. Montoya-Zavala, H. Y. Hsin, D. Ingber, *Science*, 328, pp. 1662-1668, 2010.

Plasma-treated Switchable Wettability of Parylene-C Surface

Xiao-Peng Bi, Nathan L. Ward, Brian P. Crum, Wen Li

Department of Electrical and Computer Engineering, Michigan State University, USA

bixiaope@msu.edu

Abstract—The wetting behavior of biomaterials is of great importance for the issues such as biofouling control and biocompatibility improvement. Therefore, tailoring of their wettability is particularly useful and has been attracting a lot of interests. This paper focuses on the modification of surface wettability on the parylene-C film, which is exclusively used as a coating material for insulating implantable biomedical devices. The oxygen (O_2) and sulfur hexafluoride (SF_6) plasma were applied to treat parylene-C samples successively. Super hydrophilic (~0°) and super hydrophobic (~160°) surfaces were achieved under very low plasma power. The Atomic force microscopy (AFM) and X-ray photoelectron spectroscopy (XPS) results strongly suggest that the surface roughness and the incorporation of oxygen or fluorine bonds on the surface are the two main factors accounting for the significant change of wettability. Moreover, the parylene-C surface has been proved to be switchable between super hydrophilicity and hydrophobicity by applying only an additional short O_2 or SF_6 plasma treatment, which greatly benefits the use of parylene-C in a wide range of biomedical applications.

Keywords-parylene-C; super hydrophobic; super hydrophilic; plasma treatment

I. INTRODUCTION

At present, polymeric biomaterials have been employed everywhere in the field of medicine and biotechnology. Two major disciplines are tissue engineering and drug delivery systems [1, 2]. Each application owns its specific requirements which further demand materials with unique physical, chemical, mechanical and biological properties. Among them, the surface wettability is of great importance in influencing biological interactions of cells/tissues with materials [3-6]. For instance, it has been reported that cells tend to adhere and spread on hydrophilic surfaces [4, 5], while certain proteins like bovine serum albumin (BSA) adsorb more onto hydrophobic surfaces [4-6]. In addition, the hydrophobic surface is a favorable choice in minimizing the biofouling [7, 8]. Therefore, surface modification of wettability on biomaterials is often in need to match various applications.

As one of the most commonly used polymeric coating materials in biomedical micro- and nano-devices [9, 10], parylene-C has superb combination of biocompatibility, flexibility, mechanical strength and optical transparency. A uniform thickness of parylene-C film with conformal morphology can be prepared by vacuum chemical vapor deposition on the specific coating machine. Although parylene-C has been widely recognized and employed, little attention has been paid to its surface modification [11].

During the past years, many efforts have been devoted to the development of surface modifications on polymeric materials in order to achieve super hydrophobic surfaces (with a contact angle larger than 150°). Materials with low surface energies, such as Teflon [12] and highly porous polypropylene [13], have been used as the topmost coatings to improve the contact angle [14, 15]. On the other hand, such coatings greatly limit the possibility of successive patterning. Two well-established models proposed by Wenzel [16] and Cassie [17] suggest a strong effect of surface roughness on the wetting behavior. Inspired by this concept, researchers have developed a variety of ordered structures such as nanotube arrays [18], micropillar patterns [19] and the dual-scale micro-/nano-structured surfaces [11] to enhance the surface hydrophobicity. However, most of them require multi-step processes and are thus complicated and expensive.

In this study, we present a simple, time-efficient and pattern-free method to obtain the superior switchable wettability on parylene-C by low-power oxygen (O_2) and sulfur hexafluoride (SF_6) plasma treatment, eliminating the need for complex hierarchical dual-scale micro/nanostructure deign [19]. A super hydrophobic parylene-C surface with the contact angle up to 160.5° has been achieved. The wetting switchability has been demonstrated through an additional short plasma treatment of O_2 and SF_6 alternately. The effects of surface morphology and surface chemistry on the wettability have also been investigated.

II. MATERIALS AND METHODS

4-μm thick parylene-C films were prepared on the 10 mm × 10 mm silicon substrates by vacuum chemical vapor deposition. After that, surface treatments were performed in a reactive ion etching (RIE) system using O_2 and/or SF_6 with a constant flow rate of 100 sccm and a low power of 100 W. The O_2 plasma treatment was first carried out for various time (1 min ~ 10 min), followed by 1 min SF_6 plasma treatment. For the wetting switchability experiments, alternately additional 1 min plasma treatment of O_2 and SF_6 was run.

Water contact angles measurements were conducted at room temperature to characterize the surface wettability, using a contact angle analyzer. All the measurements were done immediately after the plasma treatments. 1 μL droplet of

978-1-4673-1122-9/12 $31.00 © 2012 IEEE

deionized water was applied and placed onto the parylene-C surface with a syringe. The droplet image was recorded by a video camera and subsequently solved to obtain the contact angle. The surface morphology was characterized by atomic force microscopy (AFM) in terms of the root mean square (rms) surface roughness. Scan area of 3 μm × 3 μm were randomly selected on the substrates. X-ray photoelectron spectroscopy (XPS) was finally performed to examine the change of surface chemistry for plasma-modified parylene-C surfaces.

III. RESULTS AND DISCUSSION

A. Contact Angle Measurements

The measured water contact angles on parylene-C surfaces were summarized in Table I. Starting from 90.2° prior to the treatment, the contact angle dropped gradually with O_2 treatment time, reaching nearly 0° after 10 min treatment. The contact angle evolution for 0, 2 min, 5 min and 10 min of O_2 processing time is presented in Fig. 1.

As shown in Table I, the following 1 min treatment of SF_6 plasma resulted in a significant change on contact angles. A completely reversed wetting behavior of parylene-C was observed. The surface repellency to water was enhanced as the previous O_2 plasma treatment time increased and finally super hydrophobicity on parylene-C was achieved with the contact angle up to 160.5°, corresponding to the recipe of 10 min O_2 and 1 min SF_6 plasma treatments.

These results agree well with the previous findings that O_2

plasma can effectively improve the hydrophilicity of polymer surfaces [14], while SF_6 plasma leads to a more hydrophobic surface [11]. More importantly, the superior contact angle was received through only two simple steps of low-power and time-efficient treatments without any pattern fabrication, which is post-fabrication compatible and favorable for applications.

B. Surface Morphology

The plasma-induced changes of surface morphology were examined quantitatively by AFM, as shown in Fig. 2 and Fig. 3. It can be seen that the parylene-C surface was continually roughened upon exposure to O_2 plasma. The untreated parylene-C displays a typical cotton-like surface, with rms roughness of 6.127 nm. During the O_2 plasma treatment, the columnar-like structures were formed on the surface and aggregated with time. The rms roughness was increased by ~5 nm per minute on average. On the other hand, the SF_6 plasma

TABLE I. CONTACT ANGLES OF PLASMA TREATED PARYLENE-C FILMS

No.	O_2 plasma		SF_6 plasma		Contact angle
	Power	Time	Power	Time	
01	/	/	/	/	90.2°
02	100W	2min	/	/	23.9°
03	100W	5min	/	/	10.2°
04	100W	10min	/	/	~0°
05	/	/	100W	1min	120.5°
06	100W	2min	100W	1min	133.4°
07	100W	5min	100W	1min	142.1°
08	100W	10min	100W	1min	160.5°

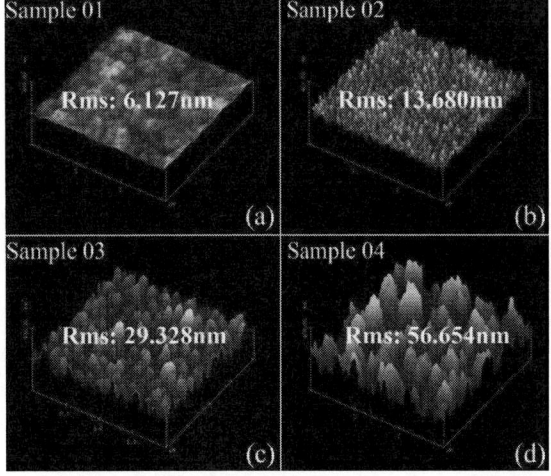

Fig. 2. Surface roughness of plasma-treated parylene-C films measured by AFM. Images (a)-(d) correspond to samples 01-04 in Table I, respectively.

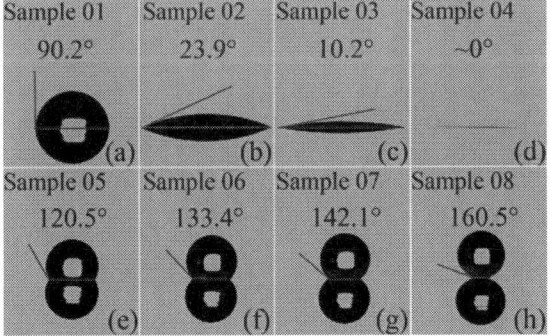

Fig. 1. Images of water contact angle measurements on parylene-C surfaces with various plasma treatments. Images (a)-(h) correspond to samples 01-08 in Table I, respectively.

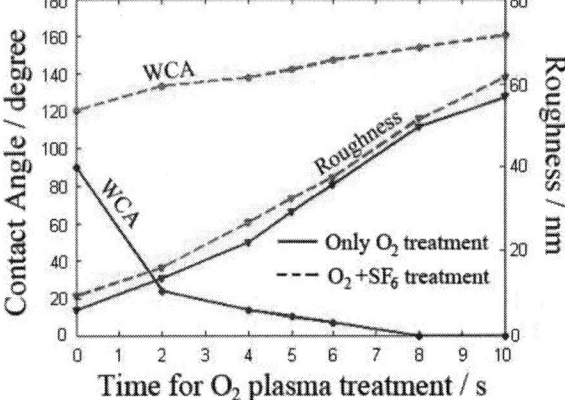

Fig. 3. Water contact angles and surface roughness on the parylene-C films as a function of O_2 plasma treatment time.

only caused ~2 nm increase of roughness in 1 min treatment, which indicates the SF_6 plasma influenced the surface morphology very slightly.

The results of contact angles and rms roughness for parylene-C samples were plotted in Fig. 3 as a function of O_2 processing time. It can be seen that the surface with increased roughness is always accompanied with an enhanced "intrinsic" wetting behavior (i.e. more hydrophilic after O_2 plasma while more hydrophobic after SF_6 plasma). In other words, the surface roughness plays a role like polarization in the contact angle variation. It also implied that some other factors besides surface roughness, would also affects the surface wettability.

C. Surface Chemistry

The surface chemistry was analyzed using XPS spectra, as shown in Fig. 4. Table II summarized the atomic percentage.

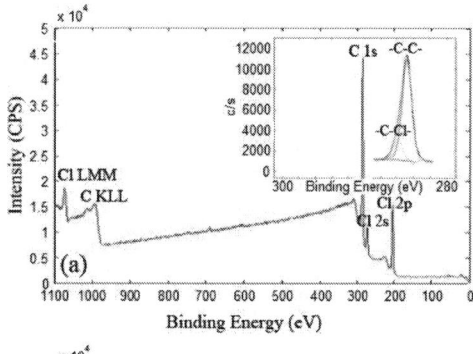

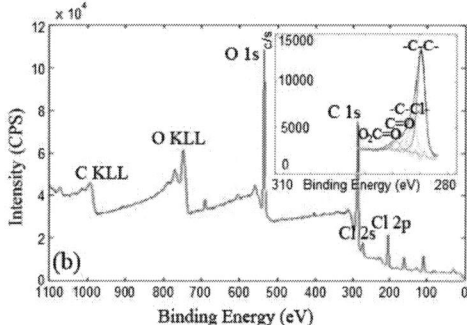

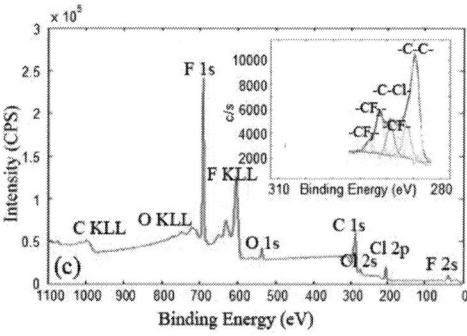

Fig. 4. XPS spectra of: (a) an untreated parylene-C film, (b) a parylene-C film treated with 100W, 10min O_2 plasma and (c) a parylene-C film treated with

TABLE II.   ATOMIC PERCENTAGE OF PARYLENE-C SURFACES

Recipe of plasma treatment	C	O	F	Cl	S
Untreated	88.0	0.5	0	11.5	0
10 min O_2	61.8	33.4	0	4.8	0
10 min O_2 and1 min SF_6	49.9	4.6	41.8	3.3	0.4

Compared to untreated parylene-C films, significant increases in oxygen and fluorine atomic concentrations were observed on the plasma-treated surfaces, due to the formation of new functional groups formed through the chemical interaction between active species from the plasma and surface atoms. For the surface treated with only O_2 plasma, the deconvolution of carbon 1s spectrum reveals the presence of two new peaks at 287.8 eV and 289.3 eV, which can be attributed to the carbon atom in free carbonyl group (C=O) bond and carbonate group (O_2C=O), respectively. The forming of these functional groups increased the polarity of parylene-C surface and thus the surface free energy was raised, resulting in the improved wettability. Similarly, new peaks representing the carbon atom in $-C-F-$, $-CF_2-$ and $-CF_3-$ have appeared in the carbon 1s spectrum corresponding to the parylene-C film treated by both O_2 and SF_6 plasma. Due to the chemical inertness, the fluorinated groups help lower the surface energy and thus contribute to the enhanced hydrophobicity.

These results suggest that the surface wettability can be effectively altered as a combinatory result of both physical roughening and chemical modification through the plasma treatment. The new functional groups formed on the surface influence the surface energy and consequently affect the general wetting property (hydrophilic or hydrophobic), while the surface roughness further magnifies the enhancement of the wetting behavior (affinity or repellency).

D. Switchablity

The switchability between super hydrophilicity and super hydrophobicity on parylene-C films was investigated through an alternately additional 1min O_2 or SF_6 plasma treatment. The contact angle variation is shown in Fig. 5, which indicates the

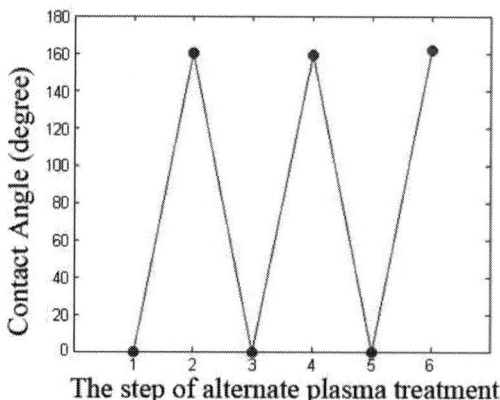

Fig. 5. The switchability of wetting behavior on the parylene-C surface, which is controlled by 1 min O_2 or SF_6 plasma treatment alternately. The step 1, 3and 5 corresponds to O_2 plasma, while 2, 4 and 6 corresponds to SF_6.

wettability can be switched flexibly through a brief and simple treatment. It provides an effective way to control parylene-C's wettability for a wide range of applications.

IV. CONCLUSION

Both super hydrophilic and super hydrophobic parylene-C films were achieved through O_2 and SF_6 plasma treatment. The process is low-powered, time-efficient and pattern-free. The change of surface wettability can be attributed to a combinatory effect of new oxygen or fluorine-containing functional groups and surface roughness, evidenced by XPS and AFM. The switchability between super hydrophilicity and super hydrophobicity has also been demonstrated.

ACKNOWLEDGMENT

This work is supported in part by the Electrical, Communications and Cyber Systems Division of the National Science Foundation under the Award Number ECCS-1055269. The authors would also like to thank Dr. Baokang Bi from the Department of Physics at Michigan State University for his assistance with fabrication.

REFERENCES

[1] A. Khademhosseini, R. Langer, J. Borenstein, and J.P. Vacanti, "Microscale Technologies for Tissue Engineering and Biology," *Proc. Natl. Acad. Sci. U.S.A.,* vol.103, pp. 2480-2487, February 2006.

[2] L.S. Nair and C.T. Laurencin, "Polymers as Biomaterials for Tissue Engineering and Controlled Drug Delivery," *Adv. Biochem. Engin./Biotechnol.,* vol. 102, pp. 47-90, 2006.

[3] D.C. Miller, A. Thapa, K. M. Haberstroh, and T.J. Webster, "Endothelial and Vascular Smooth Muscle Cell Function on Poly(lactic-co-glycolic acid) with Nano-structured Surface Features," *Biomaterials,* vol. 25, pp. 53-61, January 2004.

[4] J.N. Lee, X. Jiang, D. Ryan, and G.M. Whitesides, "Compatibility of Mammalian Cells on Surfaces of Poly(dimethylsiloxane)," *Langmuir,* vol. 20, pp. 11684-11691, December 2004.

[5] T.Y. Chang, V.G. Yadav, S. De Leo, A.Mohedas, B. Rajalingam, et al., "Cell and Protein Compatibility of Parylene-C Surfaces," *Langmuir,* vol. 23, pp. 11718-11725, November 2007.

[6] E.M. Harnett, J. Alderman, and T. Wood, "The Surface Energy of Various Biomaterials Coated with Adhesion Molecules Used in Cell Culture," *Colloid Surface B,* vol. 55, pp. 90-97, March 2007.

[7] S. Krishnan, C.J. Weinman, and C.K. Ober, "Advances in Polymers for Anti-biofouling Surfaces," *J. Mater. Chem.,* vol.18, pp. 3405-3413, 2008.

[8] I. Banerjee, R.C. Pangule, and R.S. Kane, "Antifouling Coatings: Recent Developments in the Design of Surfaces that Prevent Fouling by Proteins, Bacteria, and Marine Organisms," *Adv. Mater.,* vol. 23, pp.690-718, February 2011.

[9] D.C. Rodger, W. Li, H. Ameri, A. Ray, J.D. Weiland, et al., "Flexible Parylene-based Microelectrode Technology for Intraocular Retinal Prostheses," *Proc.1st IEEE NEMS,* pp. 743-746, January 2006.

[10] D.C. Rodger, A.J. Fong, W. Li, H. Ameri, A.K. Ahuja, et al., "Flexible Parylene-based Multielectrode Array Technology for High-density Neural Stimulation and Recording," *Sensor Actuat. B-Chem.,* vol. 132, pp. 449-460, January 2008.

[11] B. Lu, J. Lin, Z. Liu, Y. Lee, and Y. Tai, "Highly Flexible, Transparent and Patternable Parylene-C Superhydrophobic Films with High and Low Adhesion," *Proc. 24th IEEE MEMS,* pp. 1143-1146, January 2011.

[12] P. van der Wal and U. Steiner, "Super-hydrophobic Surfaces Made from Teflon," *Soft Matter,* vol. 3, pp. 426-429, 2007.

[13] H.Y. Erbil, A.L. Demirel, Y. Avci, and O. Mert, "Transformation of a Simple Plastic into a Superhydrophobic Surface," *Science,* vol. 299, pp. 1377-1380, February 2003.

[14] K. Tsougen, N. Vourdas, A. Tserepi, E. Gogolides, and C. Cardinaud, "Mechanisms of Oxygen Plasma Nanotexturing of Organic Polymer Surfaces: from Stable Super Hydrophilic to Super Hydrophobic Surfaces," *Langmuir,* vol. 25, pp.11748-11759, October 2009.

[15] L.M. Lacroix, M. Lejeune, L. Ceriotti, M. Kormunda, T. Meziani, et al., "Tuneable Rough Surfaces: a New Approach for Elaboration of Superhydrophobic Films," *Surf. Sci.,* vol. 592, pp. 182-188, November 2005.

[16] R.N. Wenzel, "Resistance of Solid Surfaces to Wetting by Water," *Ind. Eng. Chem.,* vol. 28, pp. 988-994, 1936.

[17] A.B.D. Cassie and S.Baxter, "Wettability of Porous Surfaces," *Trans. Faraday Soc.,* vol. 40, pp. 546-550, 1944.

[18] S.J. Zhu, Y.F. Li, J.H. Zhang, C.L. Lu, X. Dai, et al., "Biomimetic Polyimide Nanotube Arrays With Slippery or Sticky Superhydrophobicity," *J. Colloid Interf. Sci.,* vol.344, pp. 541-546, April 2010.

[19] J. Yeo, M.J. Choi, and D.S. Kim, "Robust Hydrophobic Surfaces with Various Micropillar Arrays," *J. Micromech. Microeng.,* vol. 20, 025028, February 2010.

In Situ Heating to Improve Adhesion for Parylene-on-Parylene Deposition

Dongyang Kang[*], Jay Han-Chieh Chang, Justin Young-Hyun Kim, Yu-Chong Tai
[*]Department of Electrical Engineering, California Institute of Technology, USA
dkang@caltech.edu

Abstract—A new technique of using "in-situ heating" to enhance adhesion for parylene-on-parylene deposition is reported in this paper. This method is compared with existing physical or chemical adhesion-enhancing methods and the results show clear advantages of this new technique. The physics is believed to be that the mobility of deposition-involved molecules (including the substrate parylene polymer chains and adsorbed monomers during deposition) is enhanced when deposition temperature rises, especially above the glass transition temperature of the substrate parylene. Each sample is patterned and then soaked in 0.9% saline at 90°C, and the undercut between two parylene layers due to the attack from saline during the soaking test could be observed. The undercut rate is measured to quantify the adhesion strength.

Keywords: in-situ heating; PA-C substrate; adhesion; undercut

I. INTRODUCTION

Parylene C (PA-C) has been widely recognized as a superb biocompatible coating material for implantable biomedical devices [1]. The parylene-metal-parylene structure is often needed for metallization, which then requires reliable parylene-on-parylene adhesion [2, 3].

Several adhesion-promoting techniques such as chemical surface cleaning and physical plasma have been investigated to enhance the adhesion between two PA-C layers [4]. They create either a clean hydrophobic surface that parylene molecules prefer to bond with or a roughened and porous surface with the surface area in direct contact with parylene molecules increased. Although these techniques have proven effective in promoting adhesion, there is still room for improvement since the PA-C interface discontinuity problem has not been completely resolved (Fig. 1, taken from [4]).

In this paper, we employ a new idea in search of better adhesion by increasing the deposition temperature during top layer PA-C deposition. Glass transition temperature marks the onset temperature for polymers with enhanced polymer mobility. Our hypothesis is that enhanced polymer mobility during deposition will improve interface polymer mixing and, therefore, adhesion. The glass transition temperature of PA-C has been reported to be in the range from 35°C to 80°C, depending on the history of PA-C [5].

This in-situ heating technique is compared with buffer HF surface cleaning, propylene carbonate (P.C.) surface cleaning, and O_2 plasma surface treatment. Test samples are soaked in 0.9% saline solution that simulates the body fluid environment since the parylene-on-parylene deposition is mainly used to construct the protection layer for the medical implant. The

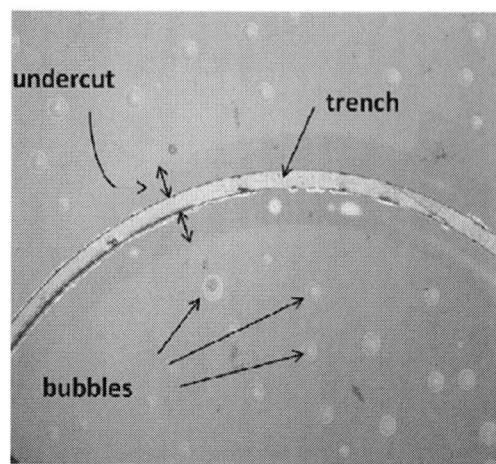

Fig. 1. Samples are soaked in 0.9% NaCl solution, and the trenches in double-layer parylene samples are deep enough for saline molecules to attack the interface of two PA-C layers. The undercut and bubbles after soaking are observed.

soaking test undercut rate is used to quantify the adhesion strength.

II. DEVICE DESIGN AND FABRICATION

One real picture of our experimental setup for the in-situ

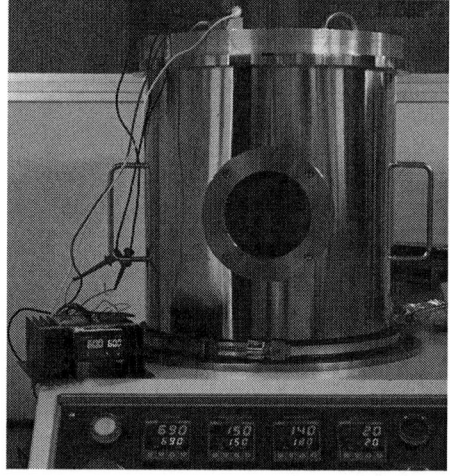

Fig. 2. The experimental setup of PA-C in-situ heating deposition. It shows PA-C is being deposited at 60°C.

heating deposition is shown in Fig. 2, in which a temperature controller is connected to a commercial parylene coating equipment, and PA-C in-situ 60°C deposition is underway. A closed loop temperature process control is integrated to the conventional parylene room temperature deposition equipment through electrical feedthroughs, and a diagram of this scheme is drawn as Fig. 3. The thermocouple adhering to the surface of the PA-C substrate measures its temperature and sends a signal to the temperature controller placed outside the vacuum chamber wall. The temperature controller compares the received signal with the temperature setting point to determine whether the heater should be switched off or not. Since the PID control mechanism is utilized by the temperature controller used in our experiment, the temperature fluctuation around the setting point during deposition turns out to be small. The purpose of the aluminum plate being sandwiched between the silicon wafer and the heater is to ensure the silicon wafer is uniformly warmed up until the temperature setting point is reached and the in-situ equilibrium temperature distribution on the PA-C substrate is even, because the parylene deposition rate is very sensitive to the substrate temperature [6].

The starting material is a slice of 500-μm-thick single-sided polished silicon wafer. And this slice of prime wafer is treated with buffer HF to make strong enough the adhesion between the silicon wafer and the PA-C substrate, followed by 5 μm PA-C substrate deposition. The twelve different ways that double-layer PA-C samples are prepared for the posterior adhesion test are illustrated in Fig. 4, and those ways differ from each other in PA-C substrate and parylene-to-parylene interface treatment, and deposition temperature. To study adhesion promotion techniques, it is important to make test samples fabricated under as many conditions that are encountered frequently during real device fabrication as possible. The as-deposit PA-C substrate experiences no treatment, while the "processed" PA-C substrate undergoes photoresist spinning, soft-baking, lithography, hard-baking, and photoresist stripping, which compose a process flow close to that for real device fabrication. Interface treatment is only

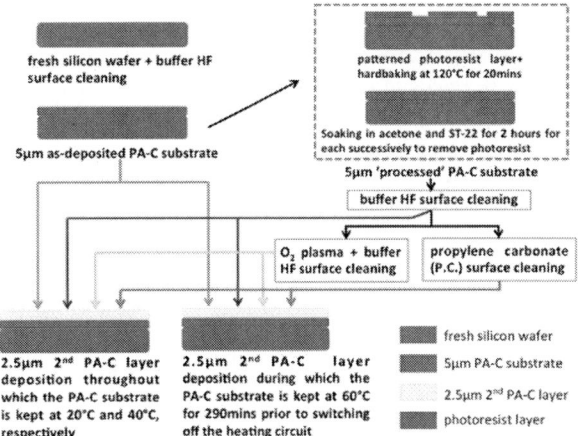

Fig. 4. Process steps for fabricating double-layer PA-C samples in twelve different ways to study which way leads to the strongest adhesion between two PA-C layers.

applied to the "processed" PA-C substrate, including buffer HF dipping, propylene carbonate (P.C.) surface cleaning, and O_2 plasma roughening. Buffer HF treatment is to dip samples for 30 seconds in the buffer HF solution. P.C. treatment is to spin-coat liquid P.C. onto the desired sample surface and wait until the chemical is completely spun off of the sample. The recipe for O_2 plasma treatment is 200 Watt, 200 mtorr, and 1 minute in a plasma etcher. 20°C (room temperature), 40°C and 60°C deposition are performed following each unique combination of PA-C substrate and interface treatment. It is noteworthy that 60°C deposition is performed in the way that at first keep PA-C substrate at 60°C for the first 290 minutes of deposition, then switch off the heater, and wait until the deposition ends. PA-C substrate would cool down naturally after the heater is turned off, and PA-C deposited subsequently is labeled orange in Fig. 5. The possibility that a new interface between parylene deposited at 60°C and that from the heater being switched off through the end of the deposition (labeled as interface 2 in Fig. 5) is produced cannot be ruled out. Therefore, a sample to test whether this unwanted interface exists or not is made for the soaking test.

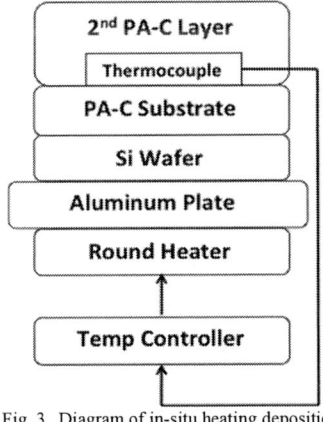

Fig. 3. Diagram of in-situ heating deposition.

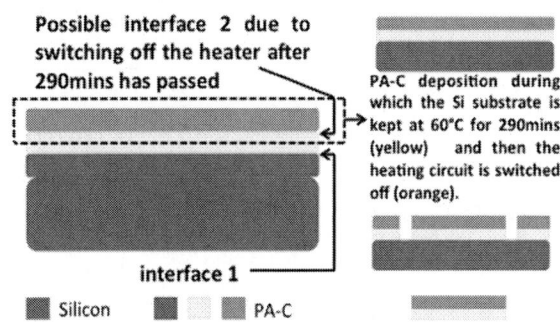

Fig. 5. Possible interface 2 generated during PA-C in-situ 60°C deposition and sample fabrication to test whether it exists or not.

Each sample (excluding that for testing interface 2) is pre-patterned to form a 250μm-wide trench with the depth greater than the thickness of the second PA-C layer and smaller than the total thickness of PA-C on silicon wafer (Fig. 6). Then all the samples are ready for the soaking test.

III. EXPERIMENTS AND RESULTS

The deposition rate when PA-C substrate temperature is set at 20°C (room temperature), 40°C, 60°C and 80°C, respectively, is experimentally determined at a pressure of 10mTorr. Fig. 7(a) shows the curve of the PA-C deposition rate versus PA-C substrate temperature, which is then used to estimate the required deposition time and PA-C dimer weight for the target thickness of the second PA-C layer that is 2.5μm in our design (Fig. 7(b)). Especially for the PA-C in-situ 60°C deposition, when the heater is switched off after the first 290 minutes, the deposition is still underway, and the temperature of the PA-C substrate as a function of time is recorded (Fig. 8). The time constant of this cooling curve is calculated as 15.9 ± 0.5 minutes.

The thickness non-uniformity of PA-C deposited by in-situ heating technique is also investigated. As in Fig. 9, for the

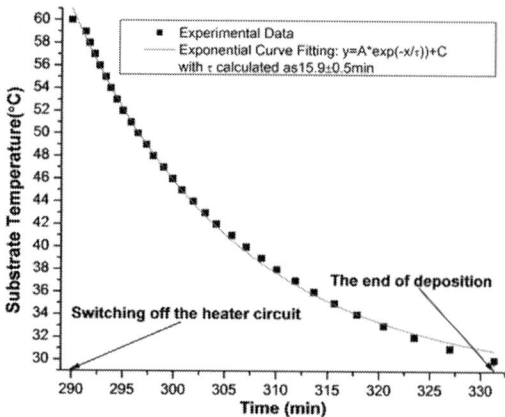

Fig. 8. Naturally cooling curve for PA-C substrate after the heater is switched off for PA-C in-situ 60°C deposition.

PA-C film deposited at 60°C, the horizontal axis represents the position of a point x along a straight line on the wafer surface through the center (x=0) of the wafer, and the vertical axis is the relative thickness defined as the ratio of the PA-C thickness t at a point x to the maximum PA-C thickness t_{max}. The maximum thickness is found around the center. The diameter of the wafer is about 100mm, and the curve in Fig. 9 implies that a circle of the diameter 62mm concentric with the wafer covers the region of PA-C film with the thickness not less than 90 percent of the maximum thickness.

Soaking test is done by soaking samples with 250μm-wide trenches in 0.9% saline solution at 90°C for over 30 days. The undercut between two PA-C layers due to the attack from saline during the soaking test could be observed (Fig. 10), and the undercut rate is calculated to quantify the adhesion strength. Finally, the undercut rate for the sample made with as-deposit PA-C substrate and room temperature deposition is 0.125μm/day. The ratio of undercut rate for each sample to

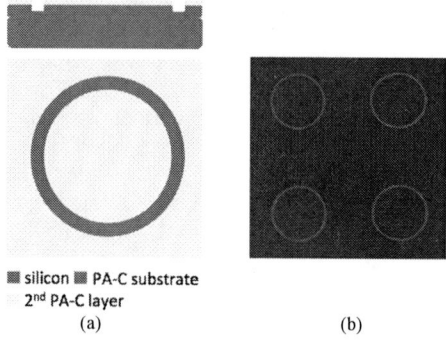

silicon ■ PA-C substrate
2nd PA-C layer
(a)   (b)

Fig. 6. (a) Patterning a double-layer PA-C sample for the soaking test. (b) The photo of a real sample.

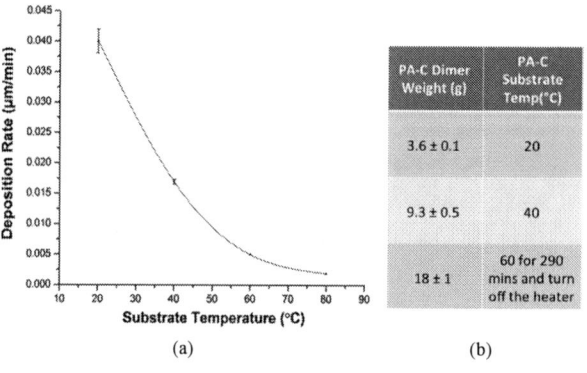

(a)   (b)

Fig. 7. (a) Experimentally determined PA-C deposition rate vs. PA-C substrate temperature at a pressure of 10 mTorr. (b) PA-C dimer weight vs. PA-C substrate temperature in order to achieve ~2.5μm 2nd PA-C layer.

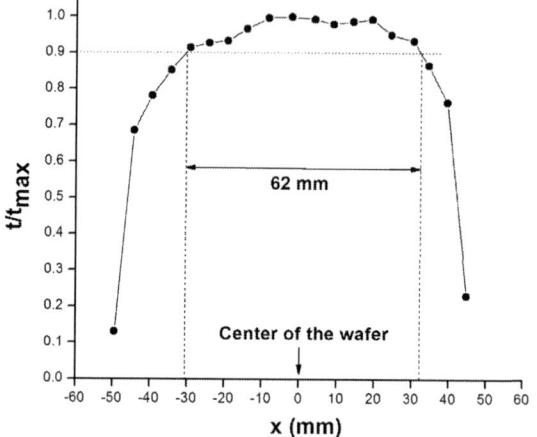

Fig. 9. Relative thickness of PA-C deposited at 60°C as a function of the position along a straight line through the center of the wafer.

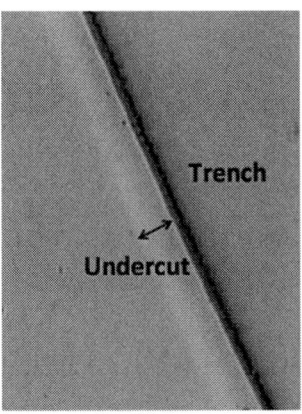

Fig. 10. Undercut between two PA-C layers caused by the attack from saline during soaking test.

TABLE I. UNDERCUT RATE FOR EACH DOUBLE-LAYER PA-C SAMPLE

Sample treatment		Substrate Temp (°C)	Relative undercut rate
As-deposited PA-C substrate		20	1
		40	< 0.60
		60	< 0.48
'Processed' PA-C substrate	HF	20	~1.10
		40	< 0.64
		60	< 0.55
	HF+P.C.	20	~1.15
		40	< 0.73
		60	< 0.68
	HF+O$_2$ plasma+HF	20	~1.42
		40	~1.35
		60	~1.23
Sample for testing interface 2		60	0

0.125μm/day is listed in Table. 1. The sample for testing interface 2 shows no undercut, meaning turning off the heater during deposition would not disrupt the continuity of the substance being deposited. All the samples show better adhesion as deposition temperature is increased. And each sample from 60°C in-situ heating deposition shows almost or even more than twice as low undercut rate as its room temperature deposition counterpart except for the O$_2$-plasma-treated samples. It is likely that the glass transition temperature of O$_2$-plasma-treated PA-C samples is increased, resulting from the oxidation and baking effects of O$_2$ plasma. Finally, 60°C in-situ heating deposition combined with HF or P.C. cleaning provides the best adhesion enhancement.

The underlying physics is likely that during parylene deposition, heating the PA-C substrate beyond its glass transition temperature should then increase parylene polymer's mobility (and even free volume, including the number of surface pores and gaps) to the degree that the number of parylene monomers diffusing into the bulk of the PA-C substrate is increased notably. The following polymerization of the monomers inside the PA-C substrate and further mixing/entanglement would greatly help anchor the second parylene layer onto the substrate.

IV. CONCLUSION

A new technique of using "in-situ heating" to enhance adhesion for parylene-on-parylene deposition is reported in this paper. This method is compared with existing physical or chemical adhesion-enhancing methods. Each double-layer PA-C sample is pre-patterned before being soaked in 0.9% saline at 90°C for over 30 days, and the undercut between two PA-C layers in each sample due to the attack from saline during the soaking test could be observed. The undercut rate is measured to quantify the adhesion strength. The results show that samples from 60°C in-situ heating deposition have almost or even more than twice as low undercut rate as those from room temperature deposition except for the O$_2$ plasma treated samples, likely due to their increased glass transition temperature resulting from the oxidation and baking effects of O$_2$ plasma. 60°C in-situ heating deposition combined with HF or propylene carbonate cleaning provides the best adhesion enhancement. The physics is believed to be that the mobility of deposition-involved molecules (including the substrate parylene polymers and adsorbed monomers during deposition) is enhanced when deposition temperature rises, especially above the glass transition temperature of the substrate.

ACKNOWLEDGMENT

This work is supported by Biomimetic MicroElectronic Systems (BMES). The authors would like to thank Mr. Trevor Roper and other members of the Caltech Micromachining Laboratory for their generous help and useful advice.

REFERENCES

[1] C. Hassler, R. P. von Metzen, P. Ruther, and T. Stieglitz, "Characterization of parylene C as an encapsulation material for implanted neural prostheses," *J Biomed Mater Res B Appl Biomater,* vol. 93, pp. 266-74, Apr 2010.

[2] J. H. Chang, R. Huang, and Y. C. Tai, "High density 256-channel chip integration with flexible parylene pocket," in *Solid-State Sensors, Actuators and Microsystems Conference (TRANSDUCERS), 2011 16th International,* 2011, pp. 378-381.

[3] R. Huang and Y. C. Tai, "Parylene-Pocket Chip Integration," in *Micro Electro Mechanical Systems, 2009. MEMS 2009. IEEE 22nd International Conference on,* 2009, pp. 749-752.

[4] J. H. Chang, B. Lu, and Y. C. Tai, "Adhesion-enhancing surface treatments for parylene deposition," in *Solid-State Sensors, Actuators and Microsystems Conference (TRANSDUCERS), 2011 16th International,* 2011, pp. 390-393.

[5] J. J. Senkevich and S. B. Desu, "Morphology of poly(chloro-p-xylylene) CVD thin films," *Polymer,* vol. 40, pp. 5751-5759, 1999.

[6] E. M. Charlson, E. J. Charlson, and R. Sabeti, "Temperature selective deposition of parylene-C," *IEEE Transactions on Biomedical Engineering,* vol. 39, pp. 202-206, Feb 1992.

Engineering a Biomimetic Villus Array for In Vitro 3-Dimensional Culture of Intestinal Epithelial Cells

Y. Chen, W. Yang, Y. Huang, C. Fu*
Institute of Nano Engineering and Microsystems,
Department of Power Mechanical Engineering,
National Tsing Hua University
Hsinchu,Taiwan
*Email: ccfu@mx.nthu.edu.tw

Ya. Fu, S. Tang
Department of Chemical Engineering
National Tsing-Hua University
Hsinchu, Taiwan

Abstract—**Small intestinal villi are projective microstructures from the mucosa that provide a large surface area for digestion and absorption. On the mucosa, intestinal epithelial cells undergo terminal differentiation in space -- along the crypt-villus axis -- until they slough off into the lumen. Despite this unique physiological feature, to date in vitro cultivation of the intestinal epithelial cells is routinely done at the planar tissue-culture surface. In this research, we fabricated a projective, 3-dimensional (3-D) tissue-culture environment to provide a physiologically relevant condition for establishing the enterocyte cell culture in vitro. We used the mouse small intestinal epithelium as the model and applied a microfabrication process, UV-LIGA, to generate an array of microneedles with a similar projective structure and size (height: 400 µm, base: 135 µm in diameter) as those of the duodenal villi. In addition, we shaped the LIGA-derived poly (lactic acid) microneedles by acetone/ethanol erosion to create a smooth tip structure for the engraftment of human Caco-2 enterocytes. The engineered villus array had a total surface area of 4.81 cm^2 per sq.cm of planar surface, which led to a 2.48-fold increase in the cell number of enterocytes on the 3-D construct relative to that on the planar control surface. Our work presents an initial step toward constituting a physiological gut in vitro by using an engineering approach for large-scale preparation of the biomimetic small intestine.**

I. INTRODUCTION

Small intestinal villi are microscopic, finger-like projections from the mucosa that range from several hundred micrometers to nearly one millimeter in length, depending on the species and location of the intestine [1]. The unique projections increase the absorptive area in space, which leads to efficient food digestion and nutrient uptake. The digested molecules contact the projective surfaces and pass through the epithelial boundary for absorption. In addition to nutrient absorption, the stereo villus structure also allows gut mucosa to be constantly renewed by terminal differentiation of the epithelial cells in 3 to 5 days along the crypt-villus axis until they slough off into the gut lumen [2,3].

Despite the distinctive feature of the villus microstructure, to date in vitro cultivation of the intestinal epithelial cells for the absorption screening, such as the drug permeability assay

[4-7], or for the gut tissue regeneration [8-10] has not been studied on the physiologically relevant 3-dimensional (3-D) scaffold. Instead, planar surfaces are routinely used to simplify the cell-culture conditions [11-13]. However, the simplified planar environment does not have the correct volume-to-surface ratio for analysis of the absorption kinetics, nor the geometric structure and cues for terminal differentiation.

In this research, we aimed to create a biomimetic 3-D microenvironment for in vitro culture of the intestinal epithelial cells. We applied a microfabrication process, UV-LIGA (derived from the German words lithographie, galvanoformung, and abformung, meaning lithography, electroplating, and molding, respectively) [14-19], to generate an array of microneedles with a similar projective structure and size as those of the mouse duodenal villi.

In the process, we applied ultraviolet light (UV) lithography to generate the micro-projective structures from the SU-8 photoresist. In the subsequent plating and molding steps, we replaced SU-8 substrate with the degradable, adhesion-friendly poly(lactic acid), PLA, for cell seeding and proliferation of the intestinal epithelial cells. Results on design, production, and modification of the biomimetic villus array as well as the engraftment of the human Caco-2 enterocytes onto the 3-D construct will be presented and discussed in this report.

II. EXPERIMENTAL

Fabrication of the SU-8 microneedle array via UV lithography

SU-8 photoresist (MicroChem Corp., Newton, MA, USA) was used in the UV lithography to fabricate the microneedle array following our previous studies [16-18]. Briefly, Figure 2A shows the process of the backside exposure. SU-8 was first attached to a 4-inch glass via a spinning process of 500 revolutions per minute (rpm) for 30 seconds and then left on the surface for 30 minutes for stress relaxation. Afterward, the SU-8 layer was baked at 90°C for 3 hours to evaporate the solvent. Polymerization of SU-8 was initiated by UV radiation (365 nm) with irradiance at 900 mJ/cm^2 through a mask with an array of 50-µm aperture openings (125-µm interval).

Afterward, SU-8 was baked at 95°C for 1 hour. The crosslinked SU-8 array was then developed in propylene glycol methylether acetate for 1 hour and subsequently rinsed with isopropyl alcohol before air drying.

Formation of the PDMS mold

We used the fabricated SU-8 microstructure as the template to cast the master polydimethylsiloxane (PDMS) mold (Figure 2B, in transition to the fabrication of the PLA villus array, Figure 2C). Before the molding process, we first deposited a Ti/Au (30 nm/100 nm) layer on the SU-8 surface by sputtering, which helped separate the SU-8 and PDMS boundaries after the PDMS mold formed on the SU-8 surface. We then poured the PDMS prepolymer (with 1/10 of the curing agent) to the SU-8 master and reduced the air pressure to 70 cm-Hg for 30 minutes to remove the potential bubbles in the solution. Afterward, the PDMS-SU-8 structure was placed onto a hot plate at 65°C for 1 hour to generate a PDMS mold, which was ~5 mm in thickness after peeling it off from the SU-8 structure.

Fabrication of the PLA biomimetic villus array

Figure 2C shows the process of using the PDMS mold as the template for the PLA array casting. First, we applied hot embossing machine HEX-01 from Jenoptik Mikrotechnik AG with 2 kg-force at 200°C for 10 minutes to fill in the PDMS mold with PLA. We then cooled the system down to room temperature for solidification. Afterward, we immersed the PLA array in the acetone/ethanol (volume ration = 1/2) solvent for 450 seconds, shaping the flat PLA tips into the curved, villus-like tips through erosion (Figure 2D) by dissolving PLA into the organic solvent.

Cell culture and imaging

Human Caco-2 cells (ATCC, cell line number HTB-37, Manassas, VA, USA) were grown in Dulbecco's modified Eagle's medium (DMEM) supplemented with 10% fetal bovine serum, 1% non-essential amino acids, 100 U/ml penicillin, and 100 µg/ml streptomycin. Cell cultures were maintained at 37°C in a 5% CO_2 and 95% air humidified atmosphere [20,21]. To improve cellular adhesion, the planar surface and the 3-D structure of PLA were coated with Poly-L-Lysine (Sigma, St. Louis, MO, USA) by incubating the surfaces with the solution for 3 hours. Afterward, 1.5×10^5 of the Caco-2 cells were seeded onto the planar surface (1 cm x 1 cm) or the villus array (3,249 villi above a 1 x 1 cm^2 unit square) for 2 weeks before additional analysis. Cell counting was performed by using the hemocytometer to estimate the number of cells on the surfaces after they were detached by the trypsin solution. In parallel, the Caco-2 engrafted PLA planar surface and 3-D construct were washed with phosphate-buffered saline (PBS) and then fixed with 4% paraformaldehyde for 20 minutes. Afterward, samples were dehydrated in 50%, 70%, 90%, and 100% ethanol solutions for 15 minutes twice at each concentration. Samples were then vacuum-dried and coated with gold by ion sputtering prior to being examined by the scanning electron microscopy (SEM, Hitachi S-3000N, Hitachi Ltd., Tokyo, Japan).

Confocal imaging of the mouse intestinal specimen was performed with a Zeiss LSM 710 confocal microscope (Carl Zeiss, Jena, Germany) equipped with an objective of 25 x LD "Plan-Apochromat" glycerine immersion lenses following our previous approach [22]. Briefly, the 488-nm argon laser was used to induce autofluorescence from the intestine to outline the tissue structure. The detector bandwidth was set at 493-740 nm to collect the signals from autofluorescence. The micrographs were recorded with 1,024 × 1,024 pixels of the X/Y plane and the increment of the Z-axis optical section was 2.5 µm. The acquired confocal image stack was projected by the Avizo 6.2 image reconstruction software (VSG, Burlington, MA, USA) using the "Voltex" module for 3-D visualization [23].

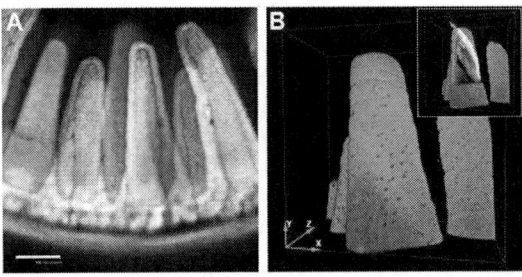

Figure 1. (A) Confocal micrograph of mouse duodenal villi. The duodenal microstructure was revealed by the tissue's autofluorescence [22]. (B) 3-D projection of the villus tips. In the inset, a cuboid of the autofluorescent signals was digitally subtracted to reveal the epithelial layer of the tissue. Dimensions of the scanned volume: 326 µm (X) × 326 µm (Y) × 285 µm (Z, the focal depth).

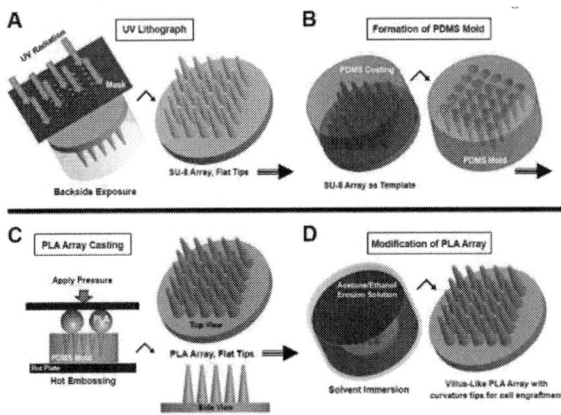

Figure 2. Schematic of UV-LIGA fabrication and erosive micro-modification of the PLA biomimetic villus array. Four steps were involved: (A) UV lithograph, (B) formation of the PDMS mold, (C) PLA array casting, and (D) erosive modification of the PLA array.

III. RESULTS AND DISCUSSION

Structural similarity and differences between the intestinal villi and LIGA-fabricated microneedles

Figures 1A-C compare the features of the mouse duodenal villi and the LIGA-fabricated microneedle array that we generated previously [16,17]. The similarity of the two projective microstructures and their repetitive patterns motivated us to further engineer the microneedle array to become a biomimetic villus array. However, despite their resemblance, two modest differences -- the body geometry (the height and base diameter) and the tip curvature -- were seen between the microneedles and villi. We therefore devised a process of integrating UV lithograph, molding, casting, and erosive modification to change the geometric dimensions of the projective microstructures. The individual steps of the process are summarized in Figure 2 and are presented and discussed in the following sections.

Fabrication of the villus-like SU-8 microstructures via UV lithography

We first adjusted the lithographic parameters used to produce the microneedles to create the microstructures with similar heights and widths as those of the mouse duodenal villi. Figure 2A is an illustration of the UV lithographic backside exposure. In this step, the SU-8 photoresist was crosslinked by UV radiation to create the projective structures. Specifically, a circular aperture on the UV mask was used to guide the light into the photoresist for reaction. In addition, we created a 1-mm gap by placing the transparent glass between the UV mask and the photoresist to induce the near-field Fresnel diffraction while the UV waves propagated in the photoresist [17, 24].

Figure 3 shows the lithographic SU-8 microstructures with various heights, widths, and tip shapes (n = 5). They were derived by increasing the aperture's diameter on the UV mask for exposure. Specifically, we observed that the height and size of the microstructure increased in association with the aperture's diameter until the height reached a plateau at 420 μm, where the crosslinking stopped due to the attenuation of the UV intensity along its path.

In the meantime, we observed a continuous increase in the base diameter (or width) of the structure in association with the aperture's diameter. In addition, the sharp tips of the microstructures gradually disappeared due to the increased size of the aperture. Among the structures, we chose the one with a 420-μm height and 130-μm width in dimensions (derived from the 50-μm-diameter aperture) as the villus candidate for large-scale array synthesis.

Formation of the PLA villus array with a smooth tip curvature

Because the hydrophobic SU-8 surface is not suitable for cell engraftment, we next used the projective SU-8 microstructure as the template to create a female poly-dimethyl-siloxane (PDMS) mold for casting the poly(lactic acid) (PLA) microstructures. Figures 4A and B are the gross view and close-up micrographs of the PLA microstructure array, respectively. Specifically, a density of 3,249 microstructures/cm^2 was generated using the hot embossing process (Figure 2C). It should be noted that the casting process can be repeatedly performed using the same PDMS mold, providing a straightforward method for large-scale preparation of the artificial villi.

The intrinsic limitation of this LIGA-based microfabrication was the apparent flat tips with an abrupt change of angles between the top and side surfaces (Figure 4B). The sharp boundary at the needle tip created a barrier to cell migration during the epithelial engraftment. Therefore, we employed a post-LIGA structural modification to smooth the PLA tip curvature.

This modification step (Figure 2D) was performed by taking advantage of the PLA's solubility in the organic solvent. We devised an erosion process by immersing the PLA array in a mixture of acetone and ethanol (1/2 volume ratio) to modify the microneedles' tip structure. Figures 4C and D are the gross view and the close-up micrographs of the modified PLA array, respectively. In the erosion process, the sharp areas of the PLA structure dissolved faster because they had more contact with the organic solvent in comparison to that of the smooth surfaces. The kinetic differences of their surface erosion led to a reformation of the microstructure, particularly at the peak area. Figure 4D shows a smooth peak curvature 450 seconds after the acetone/ethanol immersion, changed from a flat-top structure shown in Figure 4B.

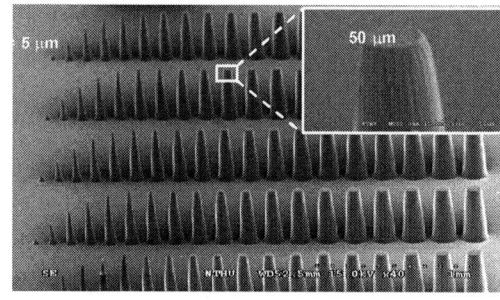

Figure 3. SEM micrograph of the UV lithograph-fabricated SU-8 microstructures. The incremental changes of the heights, widths, and tip shapes were derived by increasing the aperture's diameter on the exposure mask to adjust the UV radiation. The enlarged micrograph indicates a flat tip surface when the aperture diameter was at 50 μm. Similar flat tips were seen in other needles when the exposure diameters were larger than 40 μm. Bar = 50 μm.

Engraftment of the intestinal epithelial cells on the PLA villus structure

The lining of the digestive tract is covered by a layer of epithelial cells for digestion/absorption as well as protection against microorganisms from the environment. We mimicked the lining of gut mucosa by seeding human Caco-2 enterocytes onto the 3-D PLA structure (4.81 cm^2 per sq. cm of planar surface) for cellular adhesion and proliferation. In parallel, we used the planar PLA surface of the same 1 cm2 base area as the control to estimate the increase in cell population after we created the villus microstructures on top.

Figure 5A shows the typical morphology of the Caco-2 enterocytes on the planar surface. In comparison, Figures 5B and C show the side and top views of the Caco-2-engrafted 3-D PLA structure, where the enterocytes adhered and proliferated on the surface as successfully as those on the planar PLA sheet. A similar villus tip structure can be seen in

Figures 5D and E, which compares the close-up micrographs of the mouse duodenal villi and the engrafted PLA tips.

We next quantified and compared the numbers of enterocytes engrafted on the planar surface and the 3-D structure. Figure 5F shows that the engineered villus construct increased the surface area to 4.81 cm^2 from 1 cm^2 of the planar control, leading to a 2.48-fold increase in cell numbers and a 48% decrease in cell density on the 3-D construct. It should be noted that the decrease in cell density could be attributed to the geometric influence on the cellular migration and proliferation, which allowed each cell to occupy a larger area on average in comparison to the cells on the planar surface.

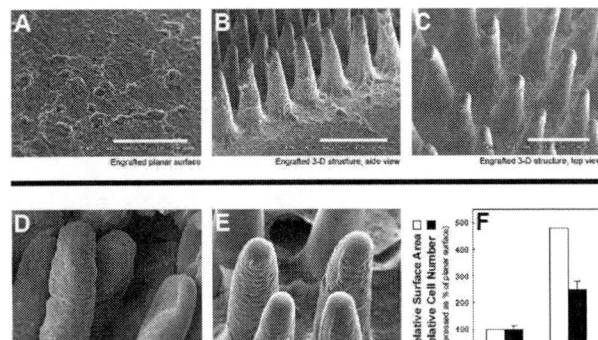

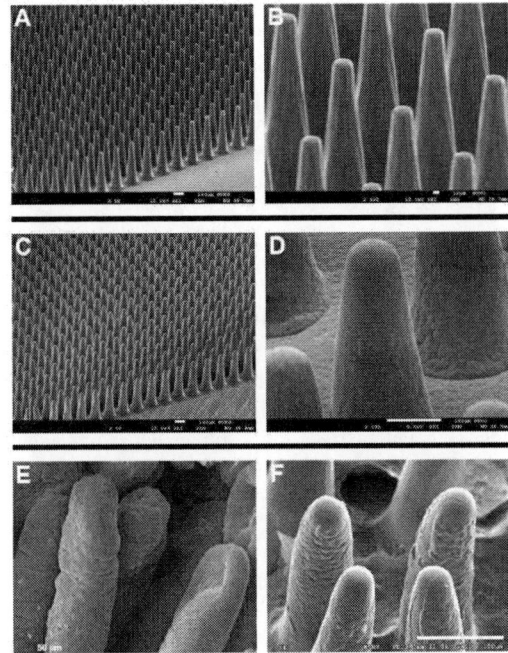

Figure 4. (A and B) Gross view and enlarged SEM micrographs of the PLA microstructure array derived from the PDMS template. (C and D) Gross view and enlarged SEM micrographs of the PLA biomimetic villus array. The flat tips in panels A and B were shaped by the acetone/ethanol immersion to create smooth tips for cell engraftment.

IV. CONCLUSIONS

In this report, we present an LIGA-based fabrication process to generate a biomimetic villus array for *in vitro* 3-D culture of the intestinal epithelial cells. The integration of UV lithography, molding, hot embossing, and solvent erosion enabled us to design and synthesize a tissue-culture PLA villus array with smooth tip curvature. The enterocyte-engrafted 3-D PLA villi allowed a higher cell population. We thus provided a biomimetic 3-D environment for *in vitro* culture of the intestinal epithelial cells. Future work will be aimed at using the *in vitro* epithelial device in absorption screening, absorption kinetic study, host-microbial study, and as a platform for intestinal tissue regeneration.

Figure 5. (A-C) Electron micrographs of the Caco-2 enterocyte-engrafted PLA planar surface and 3-D structure. Similar cellular morphologies were seen in both conditions. (D and E) Comparison of the villus tips of the mouse duodenal villi and the Caco-2-engrafted PLA 3-D structure. The engineered Caco-2-PLA villus tips have a similar curvature relative to those of the mouse duodenum. (F) Comparison of the surface areas and cell numbers on the PLA planar surface and 3-D structure. Bars indicate standard deviation (n = 6).

V. ACKNOWLEDGMENTS

This work was supported in part by grants from the National Science Council (NSC 99-2221-E-007-094 to S.-C. Tang and NSC-94-2627-M-007-004, NSC-95-2627-M-007-004, NSC-96-2627-M-007-004, NSC-97-2627-M-007-005, NSC-98-2627-M-007-004. to C.-C. Fu), and the 5-year Research Program in NTHU, Taiwan. We thank BRC in NTHU for technical support in confocal imaging.

REFERENCES

[1] Y. Y. Fu, C. W. Lin, G. Enikolopov, E. Sibley, A. S. Chiang and S. C. Tang, Gastroenterology, 2009, 137, 453-465.

[2] J. I. Gordon and M. L. Hermiston, Curr Opin Cell Biol, 1994, 6, 795-803.

[3] L. G. van der Flier and H. Clevers, Annu Rev Physiol, 2009, 71, 241-260.

[4] B. Press and D. Di Grandi, Curr Drug Metab, 2008, 9, 893-900.

[5] H. Sun, E. C. Chow, S. Liu, Y. Du and K. S. Pang, Expert Opin Drug Metab Toxicol, 2008, 4, 395-411.

[6] R. L. Oostendorp, J. H. Beijnen and J. H. M. Schellens, Cancer Treat Rev, 2009, 35, 137-147.

[7] K. C. Cheng, C. Li and A. S. Uss, Expert Opin Drug Met, 2008, 4, 581-590.

[8] Javaid-Ur-Rehman and T. Waseem, Surg Today, 2008, 38, 484-486.

[9] A. Gupta, A. Dixit, K. M. Sales, M. C. Winslet and A. M. Seifalian, Biomacromolecules, 2006, 7, 2701-2709.

[10] Y. Nakase, A. Hagiwara, T. Nakamura, S. Kin, S. Nakashima, T. Yoshikawa, K. Fukuda, Y. Kuriu, K. Miyagawa, C. Sakakura, E. Otsuji, Y. Shimizu, Y. Ikada and H. Yamagishi, Tissue Eng, 2006, 12, 403-412.

[11] P. M. van Midwoud, M. T. Merema, E. Verpoorte and G. M. M. Groothuis, Lab Chip, 2010, 10, 2778-2786.

[12] J. Kim, M. Hegde and A. Jayaraman, Lab Chip, 2010, 10, 43-50.

[13] H. Kimura, T. Yamamoto, H. Sakai, Y. Sakai and T. Fujii, Lab Chip, 2008, 8, 741-746.

[14] J. Zhang, M. B. Chan-Park and S. R. Conner, Lab Chip, 2004, 4, 646-653.

Reduction of AC Resistance in MEMS Intraocular Foil Coils Using Microfabricated Planar Litz Structure

Y. Zhao, M. Nandra, C. Yu, Y. Tai[*]

[*]Department of Electrical Engineering, California Institute of Technology, USA
yzhao@mems.caltech.edu

Abstract—The next-generation implantable high-power prosthetic devices, such as retinal prostheses, will be powered by inductively coupled coils of high efficiency. However, the use of high-frequency power carriers almost always induces significant AC resistance due to skin and proximity effects, which limits the overall coupling efficiency. Inspired by the widely used round Litz wire, planar coils with planar Litz structures were investigated on the printed circuit board (PCB) and showed the reduced AC resistances in some specific frequency ranges. Herein, various planar Litz structures were introduced into previous MEMS foil coils, which have been proved to be superior over the conventional planar counterparts. The experimental results provided by the aforementioned prototypes showed a reduction of AC resistance up to 13.3% compared to a solid foil coil. A microfabrication process, which combines a single layer metal deposition and patterning with a post folding step, was developed to construct the planar Litz structure.

Keywords-planar Litz structure; MEMS foil coil

I. INTRODUCTION

For most of the implantable devices, for example, retinal prostheses, an internal battery is not an ideal candidate for the power supply, due to its limited lifetime, large size and possibility of leakage[1]. Wireless power transfer is often required in order to reduce the risk of infection and patient discomfort, which can result from transcutaneous wires penetrating the skin. And wireless solutions are also more robust and minimally invasive[2].

Inductive coupling is the most common choice for wireless power and data transfer in biomedical telemetry systems [3, 4]. An inductive coupling system normally consists of an external primary coil and an implantable secondary coil. Inductive coupling would be ideal if it has reasonable power transfer efficiency, which is related to the square of the coupling coefficient and the respective quality factors of the primary and secondary coils. Further, the extreme size and mass constraints of the implantable secondary coil place an upper limit on its quality factor. Therefore, the design of the coil has to overcome these two limitations to deliver high performance.

In general, it is advantageous to operate the inductive coupling link at higher frequency to realize better power transmission efficiency. However, at high frequencies, severe AC losses are induced in the coils because of high frequency impedance effects such as skin and proximity effects [5], resulting in reduction of the quality factor at the operating frequency. Traditionally, Litz wires have been used to reduce the losses in conductors at high frequencies.

Inspired by the round Litz wire, planar Litz structure was realized on a printed circuit board (PCB) and showed reduced AC resistance in some specific frequency ranges [6]. However, there are optimal operating frequency ranges for both the round Litz wire and planar Litz structures on PCB. For example, the optimal operating frequency of the planar Litz structure realized on PCB is below 1 MHz and the optimal frequency range for 48 AWG Litz wire is about 3 MHz. Without more sophisticated machinery, the dimension feature cannot be decreased and achieve higher optimal frequency range. In order to make use of higher frequency carrier, micro fabrication is chosen to realize the planar Litz structure.

Previously it has been shown, that with the same amount of metal deposited, MEMS foil coil has superior performance (lower AC resistance) compared to those of its planar and wire-wound counterparts for intraocular implantable applications [7]. A typical MEMS intraocular foil coil has a high aspect ratio, with turn thickness of 2.5 μm and width of 1 mm. Thus, at high frequencies, the current tends to approach the edges in the width direction, while the non-uniform current distribution in the thickness direction is negligible. In other words, the high frequency effects are evident only in the width direction, thus the foil coil can be treated as one dimensional. The current distribution in the width direction is assumed to be identical from turn to turn.

In order to further improve the quality factor of the MEMS foil coil, planar Litz structures and a post fabrication folding technique were introduced into the coil design and explored in this paper. First, COMSOL simulations were done to compare the current distribution and AC resistances of the Litz-designed foil coil and standard foil coil. After that, a series of prototypes with different design parameters of dividing the solid gold foil into different number of strands were also investigated. Last, experimental results verified the Litz coil design, showing a reduced AC resistance compared to a solid foil coil. Different Litz design manifested different optimal operating frequency ranges.

978-1-4673-1122-9/12 $31.00 © 2012 IEEE

II. MICROFABRICATED PLANAR LITZ STRUCTURE CONCEPT

A. Design Principle

The winding AC resistances depend on both the operating frequency and the winding configuration [6]. At high frequencies, severe skin and proximity effects are readily encountered, causing a drastic increase in AC resistances and therefore reduction of the quality factor at the operating frequency.

Skin effect describes the tendency of high frequency alternating current to distribute itself on the surface of the conductor. Proximity effect is associated with the magnetic fields of two conductors in close proximity. The alternating magnetic field of one conductor induces eddy current in the adjacent one, causing overall current to flow in loops or concentrate in an uneven manner. Both effects cause a non-uniform current distribution throughout the cross section of the conductor.

Conventionally, round Litz wires are used to mitigate the high frequency losses caused by skin and proximity effects.

Litz wires are made of several insulated strands interwoven

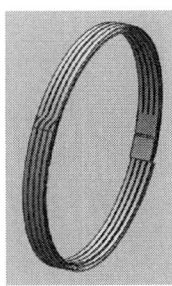

Fig. 2: A 3D illustration of the microfabricated Litz foil coil with 4 lines per fold.

and twisted together. In the current flowing direction, every single strand samples all the possible positions in the cross-sectional area, which results in a more uniform current distribution and therefore reduced AC resistances.

To realize the planar Litz structure in the foil coil, each fold is divided into insulated strands with its width maintaining an equivalent DC resistance as the solid foil (Figure 2). At least two metal layers were required to achieve the rotations of the insulated strands in the width direction. Instead of multi-layer metal deposition, a shifting technique and a post folding technique were utilized during and after the microfabrication processes.

The traces of the last fold have to be removed except for one to give space for traces shifting. Then, the insulated

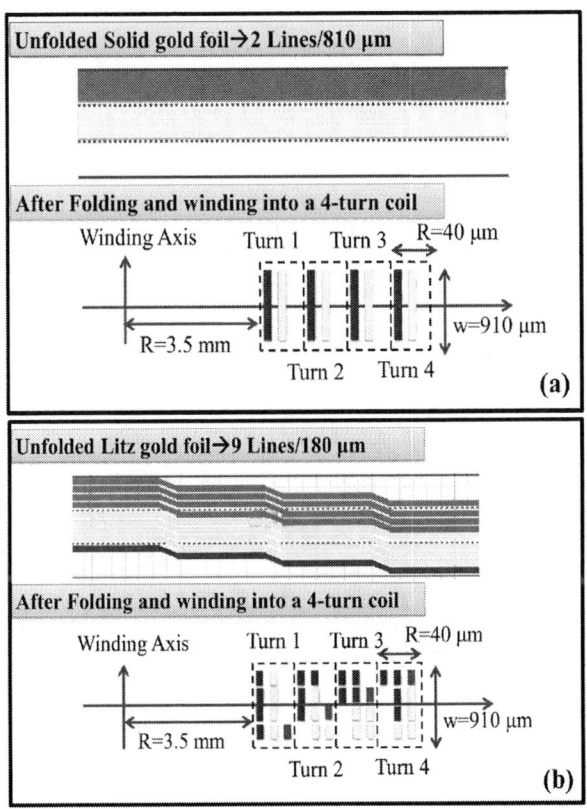

Fig. 1: The planar Litz structure design realized on a single layer gold foil. (a) Non-Litz foil design. (b)Litz foil design. The 9 insulted strands are shifted respectively to appear at different positions in the width direction. After folding and winding, the highlighted red strand moved from one edge to the other in the width direction, which results in a more uniform current distribution as shown in the simulated current density surface plot.

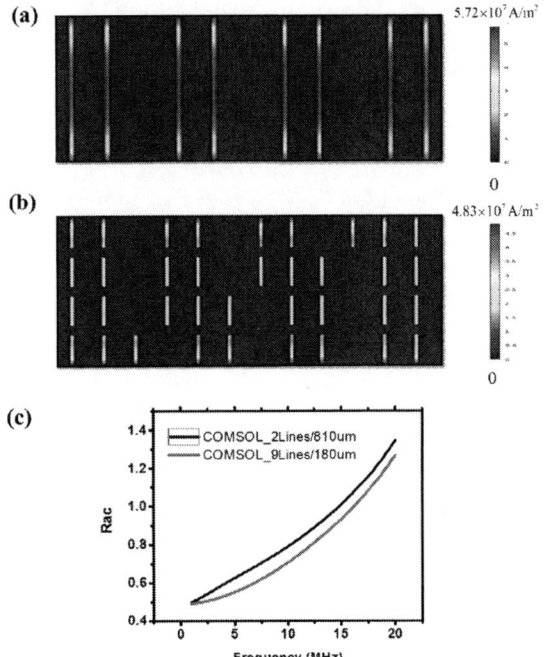

Fig. 3: (a) Simulated current distribution for non-Litz foil coil (b) Simulated current distribution for 9lines/180 μm Litz foil coil. (c) Simulated AC resistances vs. frequencies. 9 strands/180 μm showed constantly smaller AC resistance than the 2 strands 810 μm one.

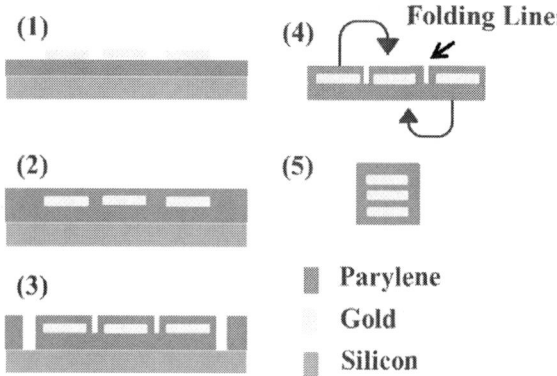

Folding Lines

■ Parylene
■ Gold
■ Silicon

Fig. 4: The microfabrication process: (1). 5 μm bottom parylene and 2.5 μm gold thermal deposition and patterning with wet etching. (2). 5 μm top parylene deposition. (3) Electrode opening, folding strands and foil outlines defining by O₂ plasma etching. (4). Foil release from wafer and folding. (5). Folded foil with three folds.

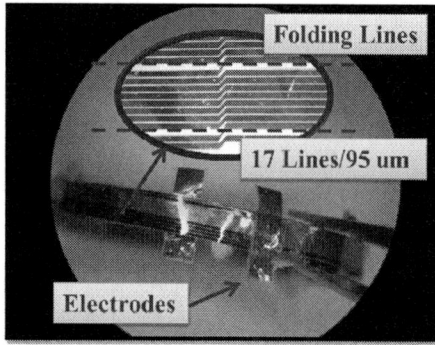

Fig. 5: Microfabricated MEMS foil coil with 17 strands/95 μm planar Litz structure. The coil is mounted to a measurement PCB and then connected to a measurement impedance analyzer HP 4194a for a 4-point measurement.

strands were shifted (patterned) respectively to appear at different positions in the width direction. If there were n lines per fold, then there should be at least n turns to realize the full shifting rotation of the strands (1 shift/turn).

After the microfabrication process, the coil is then folded into 3 folds aiming to increase the equivalent cross-sectional area and for shifting purposes. It is then wound into a 4-turn coil with a 3.5mm radius. As in Figure 1b, the highlighted red strand was moved from the bottom edge to the top edge in the width direction as well as the other eight strands, which is expected to exhibit a more uniform current distribution at high frequencies.

B. Definitions for a MEMS Litz foil coil

Some definitions and parameters of the MEMS Litz foil coil are established below to facilitate a better understanding of the Litz coil design.

Number of strands per fold: Each fold of the solid gold foil is divided into number of strands for shifting purposes.

TABLE I. *DIFFERENT PLANAR LITZ STRUCTURES REALIZED ON THE SIGNAL LAYER GOLD FOIL.*

Total No. of strands/Width of each strand	No. of strands Per fold	Separated width Between two strands	Total width Of metal	Total width of fold
2 strands/810 μm	1	0	1620μm	910μm
5 strands/325 μm	2	240	1625μm	910μm
9 strands/180 μm	4	60	1620μm	910μm
17 strands/95 μm	8	20	1615μm	910μm

Total number of strands/width of each strand: Similar to common Litz wire, the Litz foil coils are characterized as total number of strands/width of each strand, for example, 17 strands/95 μm.

Separated width between two strands: Each strand in one fold is separated by parylene C insulation.

Total width of metal: The total amount of metal is tried at best to be kept constant for a fair comparison of AC resistance.

Total width of fold: The total width of one fold includes the width of metal and two layers of parylene. It is kept constant.

C. COMSOL Simulation

In order to verify the Litz foil coil design, a 2D-symmetric COMSOL simulation was done to compare the current distribution of a standard foil coil and Litz foil coil in the width direction. A prototype of 9 strands/180 μm Litz foil coil was simulated along with the solid foil coil. From the COMSOL simulation of the above two designs, the 9 strands/180 μm Litz foil coil has shown a more uniform current distribution (Fig. 3b) and constantly smaller AC resistances (Fig. 3c) compared to the non-Litz one(Fig. 3a).

III. EXPERIMENTAL VERIFICATION

A. Microfabrication Process

The microfabrication and post-fabrication processes of the MEMS Litz foil coil are illustrated in Fig. 4. Using the silicon wafer as the carrier substrate, a 5-μm-thick Parylene C was first deposited on an HMDS-coated silicon wafer. A Cr/Au metal layer was thermally deposited and patterned using lift-off process. Chromium was involved for adhesion promotion purpose between parylene and gold layers. For the metal lift-off procedure, AZ nLOF 2035 was used. The new negative photoresist is developed for lifting metal layers of several micron thickness and undercut profiles were preferred. After another 5 μm top parylene C deposition, the electrode contacts were exposed. Then, parylene patterned etching was carried out to form the folding line and the foil outline. After the foil released and folded, the 3-fold foil was wound into a coil.

By dividing the solid gold foil into different number of strands, a series of prototypes, with detailed design parameters listed in Table. 1, were microfabricated following the described protocol. One of the fabricated coils is shown in Fig. 5, with 17 strands altogether and 95 μm in width for each strand.

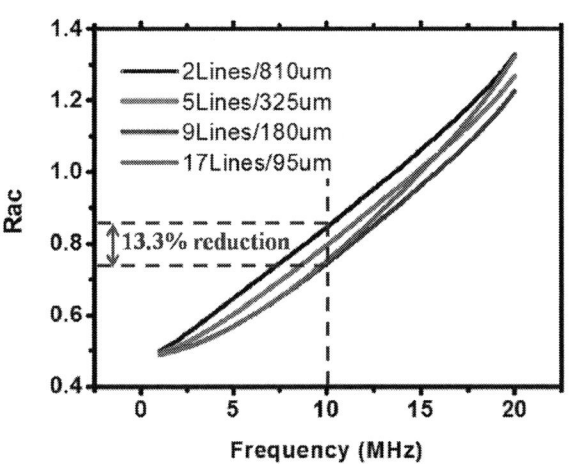

Fig. 6: Measured AC resistances vs. frequencies for 4 different foil coils. At 10 MHz, the 17 strands/95 μm design showed a 13.3% reduced AC resistance compared with the 2 strands/810 μm one.

B. Impedance Analyzer Characterization

The performance of the Litz foil coil needs to be verified experimentally. For AC resistance characterization, the coil was mounted to a measurement PCB and its four electrodes were connected to impedance analyzer HP 4194A for a 4-point measurement. The measured AC resistances are compared in Fig. 6. The foil coil with the 17 strands/95μm planar Litz design has shown a 13.3% reduced AC resistance compared to the non-Litz foil coil. Different Litz designs manifested different optimal operating frequency ranges.

IV. CONCLUSION

A MEMS intraocular foil coil with microfabricated planar Litz structure was investigated in this paper, aiming to reduce the AC resistance at high frequencies. A shifting technique was employed during the microfabrication process and a post-folding technique was utilized to achieve rotation of strands in the width direction, in which the non-uniform current distribution resulted in high frequency losses. COMSOL simulations indicated that the Litz foil coil is effective in achieving a more uniform current distribution compared to the standard foil coil. Experimental prototypes with different number of insulated strands were also constructed. The measurement results showed a reduction of AC resistance up to 13.3% compared to a solid foil coil with a 17 strands/95μm planar Litz design. The planar Litz structure realized by folding instead of multiple layer process is low-cost and effective in reducing the high frequency effect in an appropriate frequency range. Therefore, it is promising to further reduce the AC loss of the implant power coil.

ACKNOWLEDGMENTS

The funding is provided by the NSF BMES ERC Center under Award Number H31068. The authors would like to sincerely thank Mr. Trevor Roper and other members of the Caltech Micromachining laboratory for their assistance with fabrication.

REFERENCES

[1] Hmida, G.B., *Design of wireless power and data transmission circuits for implantable biomicrosystem.* Biotechnology, 2007. **6**(2): p. 153.

[2] Uei-Ming, J. and M. Ghovanloo, *Design and Optimization of Printed Spiral Coils for Efficient Transcutaneous Inductive Power Transmission.* Biomedical Circuits and Systems, IEEE Transactions on, 2007. **1**(3): p. 193-202.

[3] Weiland, J.D., W. Liu, and M.S. Humayun, *Retinal prosthesis.* Annu Rev Biomed Eng, 2005. **7**: p. 361-401.

[4] Wang, G., et al., *A dual band wireless power and data telemetry for retinal prosthesis.* Conf Proc IEEE Eng Med Biol Soc, 2006. **1**: p. 4392-5.

[5] Guoxing, W., et al., *Design and analysis of an adaptive transcutaneous power telemetry for biomedical implants.* Circuits and Systems I: Regular Papers, IEEE Transactions on, 2005. **52**(10): p. 2109-2117.

[6] Shen, W., et al., *Reduction of high-frequency conduction losses using a planar Litz structure.* Power Electronics, IEEE Transactions on, 2005. **20**(2): p. 261-267.

[7] Zhao, Y., M.S. Nandra, and Y.C. Tai. *A MEMS intraocular origami coil.* in *Solid-State Sensors, Actuators and Microsystems Conference (TRANSDUCERS), 2011 16th International.* 2011.

978-1-4673-1122-9/12 $31.00 © 2012 IEEE

Design and Fabrication of Flexible Parylene-based Inductors with Electroplated NiFe Magnetic Core for Wireless Power Transmission System

Xuming Sun[1], *Student Member, IEEE*, Yang Zheng[1], Zhongliang Li[1], Miao Yu[1],
Quan Yuan[1], *Student Member, IEEE*, Xiuhan Li[2], *Member, IEEE*, Haixia Zhang[1*], *Senior Member, IEEE*

[1]National Key Laboratory of Science and Technology on Micro/Nano Fabrication, Institute of Microelectronics, Peking University, Beijing 100871, China

[2]School of Electronic and Information Engineering, Beijing Jiaotong University, Beijing 100044, China

zhang-alice@pku.edu.cn

Abstract—Integrated flexible parylene-based MEMS inductors were successfully designed, fabricated and analyzed. By elctrodeposition technique, $Ni_{80}Fe_{20}$ Permalloy with high permeability was acquired as the magnetic core to increase the inductance and quality factor of inductors. Using parylene as substrates, these inductors became more flexible and biocompatible. For small scale inductors, by HFSS simulation, a maximum quality factor of 51.52 was achieved at 3.10GHz. The inductance of inductors with magnetic core was enhanced by 3.3% than those with air core. As the width of the core increased from 25μm to 100μm, the inductance was also increased by 11.3% maximally. For large scale inductors fabricated, the tested maximum quality factor was 110 at 6MHz. Based on the IC compatible fabrication process, these inductors could be used in wireless power transmission system for implantable medical devices.

Keywords- inductor; NiFe; soft magnets; electrodeposition

I. INTRODUCTION

Nowadays, the emergency of MEMS implantable devices is really a milestone in the development of medicine [1]. These devices can be implanted within the human body to perform a variety of therapeutic, prosthetic, and diagnostic functions [2]. For example, in [3], a wireless in-vivo communication system based on MEMS biosensors could realize interactive sensing of pH in stomach, internal body temperature, and so on.

However, the problems in seeking a proper power source greatly limit the applications of these MEMS devices [4]. Although the traditional batteries have been improved tremendously, they are still not suitable for in-vivo devices because of their limited lifespan, relatively large volume, and bad biocompatibility. Compared with other power sources, due to its long lifespan and small volume, wireless power transmission system with coupling of inductors has been one of the most promising solutions [5, 6]. Ref. [5] reported a battery-free compact (3.1×1.5×0.3mm) neurostimulator which used an inductively coupled wireless power delivery system to realize the power supply. In [6], the authors also reported a wireless power transmission link targeting Micro-Ball Endoscopy applications. Due to the long lifetime of wireless power supply, surgery is not required any more to replace batteries, which will greatly reduce the health risk of the patients.

However, nowadays, to realize high performance, most of the wireless systems use handmade coil inductors with a big volume which are difficult to be integrated with IC fabrication [6]. Compared with the handmade coils, MEMS inductors have considerable advantages especially for implantable devices, such as small volume and good biocompatibility. But the poor properties of integrated inductors have been a critical factor limiting the overall performance of this system [5, 7]. For instance, in [5], using MEMS technology, the device was more than factor of 50 times smaller than traditional device, but due to the small inductance value and low quality factor of MEMS inductors, the energy transmission distance was only 1mm which was too small for many applications. Therefore, in recent years, how to improve MEMS inductors is becoming a hotspot. Numerous improvements have been realized such as 3D high quality factor inductors [8], high aspect ratio inductors with an automatic wire bonder [9]. But as the enhancement of the inductor performance, the complexity of fabrication also increased.

Therefore, in this paper, we suggested a simple fabrication method based on electrodeposition technique to integrate MEMS planar coils on parylene substrate with $Ni_{80}Fe_{20}$ soft magnetic core. The high permeability of Permalloy can increase the inductance and quality factor of inductors. Meanwhile, the parylene substrate could provide good flexibility and biocompatibility for MEMS inductors, which is very important for implantable devices.

II. DESIGN AND HFSS SIMULATION

A. Inductor Design

In this study, to enhance the inductance and quality factor, MEMS planar inductors with $Ni_{80}Fe_{20}$ soft magnetic core ware designed and fabricated. The structure of inductors was illustrated in Fig. 1, which mainly consists of copper coils, NiFe magnetic core and parylene substrate.

978-1-4673-1122-9/12 $31.00 © 2012 IEEE

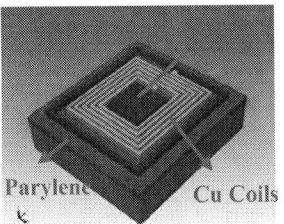

Fig. 1. 3D schematic of MEMS planar inductors with magnetic core

Copper is one of the common materials for coils. We fabricated copper coils by electrodeposition technique.

Use of soft magnetic core with high permeability in the integrated inductor can significantly increase the inductance by the relative permeability of magnetic material used [10]. There are many kinds of soft magnetic materials such as NiFe, CoNbZr, FeHfN and so on [11]. In our study, we chose NiFe as the soft magnetic material. Compared with other materials, NiFe has a relatively high permeability which can lead to large inductance enhancement as well as low resistivity which can reduce the additional energy loss. Furthermore, NiFe can be fabricated by eletrodeposition technique, which has many advantages such as simple setup, low cost, precisely controlled operation and good compatibility with IC fabrication.

Parylene layer was selected to be the substrate because of its good biocompatibility and flexibility. According to [12], parylene can well protect the silicon chip and its lifetime is more than 60 years at body temperature. Therefore, parylene is an excellent structural and packaging material especially for biomedical applications. In this paper, we realized the metal electrodeposition on flexible parylene layer.

Meanwhile, for the design of patterns, as illustrated in Fig. 2, we not only made comparison between inductors with and without magnetic core, but also considered the geometry effects of inductors. For instance, several inductors with different magnetic area were designed to do HFSS simulation and fabrication.

B. HFSS simulation

In order to help our design and analysis, the performance of inductors was simulated by HFSS as shown in Fig. 3. We not only focused on the effects of soft magnetic core but also compared inductors with different magnetic core area, as listed in Table Ⅰ.

For inductor I1 with NiFe soft magnetic core, Fig. 4 exhibited its inductance (L) and quality factor (Q). The maximum quality factor was 51.52 at 3.10GHz. The typical inductance was 3.2nH at 6GHz. It could be observed that the resonant frequency was about 3GHz.

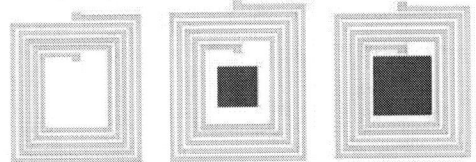

(a) Without core (b) With magnetic core (c) With large area core
Fig. 2. Design of different patterns for comparison

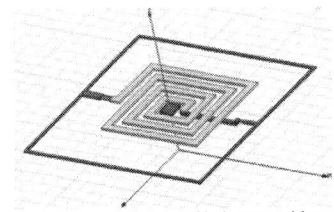

Fig. 3. HFSS simulation project view of inductors with magnetic core

TABLE I. DIFFERENT INDUCTORS FOR HFSS SIMULATION

No.	Coil Dimension	Core Dimension
I1	Line width: 50µm; Gap: 50µm; Thickness: 15µm	100µm×100µm×15µm
I2		No magnetic core
I3		50µm×50µm×15µm
I4		25µm×25µm×15µm

Fig. 5 showed the distribution of magnetic field. The main amounts of magnetic flux were confined within the area near the magnetic core, which could increase the coupling inductance of the inductor. Fig. 6 compared inductor I1 and I2. From this figure, it could be seen that due to the integration of soft magnetic core instead of air core, the inductance was enhanced by 3.3%, which showed the effect of NiFe core.

When the effect of core area was taken into consideration, we made a comparison with I1, I3 and I4 in Fig. 7.

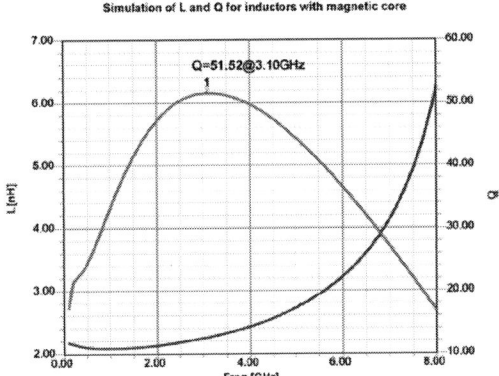

Fig. 4. Inductance (L: Blue) and quality factor (Q: Red) of the planar inductor (I1) with magnetic core

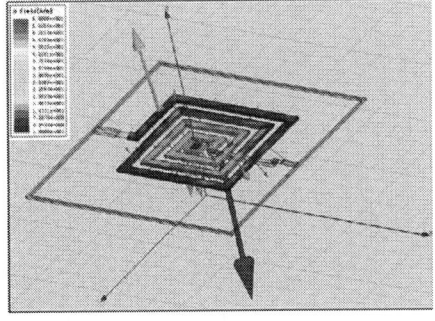

Fig. 5. Simulation results of magnetic field distribution of planar inductor (I1) with magnetic core

978-1-4673-1122-9/12 $31.00 © 2012 IEEE

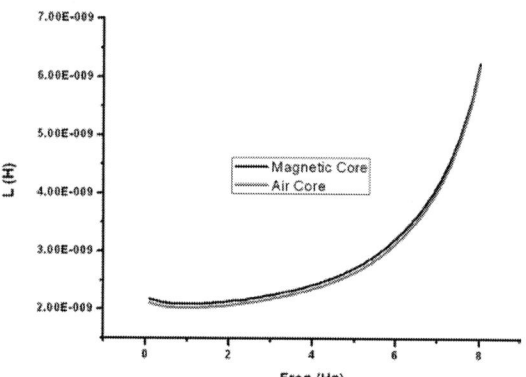

Fig. 6. Comparison of inductance between inductors with magnetic core (I1) and air core (I2)

For the inductance, as the width of the core increased from 25μm to 100μm, it was also increased by 11.3% maximally. This enhancement was caused by the increase of magnetic flux generated by magnetic core in the center. Since the volume of magnet increased, more magnetic fluxes would be generated. Therefore, the effect of magnetic core on inductance enhancement got improved. But for the quality factor, it was decreased as the increase of magnetic core area. It can be explained by the increase of the resistance loss in magnetic core. Therefore, there was a tradeoff between inductance and quality factor.

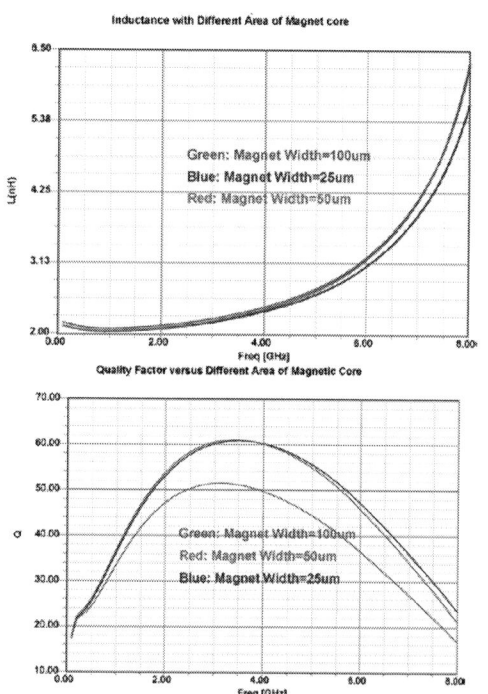

Fig. 7. Inductance and quality factor versus different magnetic core area

III. FABRICATION AND ELECTRODEPOSITION

A. Fabrication Process

In this study, a relatively simple CMOS compatible fabrication process was designed based on copper and NiFe electrodeposition technique. Fig. 8 showed the fabrication process flow.

(a) 10μm parylene layer was deposited as the substrate to provide good flexibility for the inductor. The Titanium (15nm)/Copper (150nm) layer was sputtered on as the seed layer for electrodeposition. Titanium can increase the adhesion between parylene and copper.

(b) Thick photoresist AZ4620 was spin-coated on the wafer to build patterned mold. Then, the copper coils were electroplated on.

(c) After copper electrodeposition, thin photoresist was spin-coated on the wafer to realize a flat surface. Then, the pattern for magnetic core was built by the 2nd photolithography.

(d) By electrodeposition, we got the NiFe magnetic core.

(e) Both the photoresist and seed layer were removed. At last, the wafer was diced into small samples for testing.

During the process above, step (c) was a little difficult. Since there were two layers of photoresist requiring exposure, a good control of exposure time was required, and there should be no photoresist remaining in the central area. Otherwise, because it is not conductive, the area covered by remaining photoresist can not be electroplated on, which will greatly affect the quality of magnetic core.

B. NiFe Electrodeposition Technique

Obviously, in the above process, the NiFe electrodeposition is the most important step. Based on our previous study on electrodeposition of magnets [13], the electroplating setup consists of a glass container, a pulse power source and electroplating bath.

Table II listed the main compositions of NiFe electroplating bath. The bath mainly contains nickel sulphate salts, nickel chloride salts, iron sulphate salts and other additives such as saccharin and boric acid. The saccharin can decrease the residual stress and improve the surface morphology of magnets. The boric acid can prove a constant pH environment for electrodeposition.

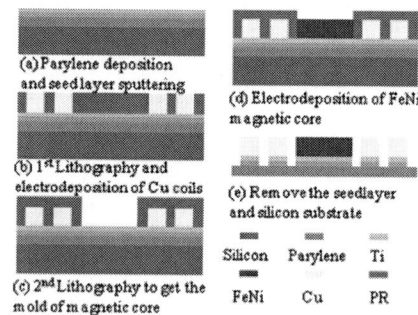

Fig. 8. Schematic illustration of the fabrication process flow

TABLE II. MAJOR COMPOSITIONS OF NIFE ELECTROLYTE

Composition	Grams/Liter	Composition	Grams/Liter
$NiSO_4 \cdot 6H_2O$	187	$FeSO_4 \cdot 7H_2O$	18
$NiCl_2 \cdot 6H_2O$	50	NaCl	20
H_3BO_3	40	$NaC_{12}H_{25}SO_4$	0.2
Saccharin	3.5	H_3PO_3	5
Current Density=1ASD			

During the electrodeposition, the nickel foil was used as the anode to maintain the constant nickel cation concentration in the electrolyte. Meanwhile, a bubble agitator was essential for this system to ensure uniform ion transportation in the bath so that the surface morphology of the film would be much improved. Furthermore, the current density is another key parameter. The speed of the electrodeposition will ascend by increasing the current density. But the quality will fall off due to many micro cracks and high surface roughness. Therefore, by many experiments, the current density was set to1ASD.

IV. CHARACTERIZATION AND DISCUSSION

In this section, we will focus on the properties of NiFe soft magnets as well as the performance of planar inductors with magnetic core.

A. Magnetic Properties of NiFe Soft Magnets

The magnetic properties of electroplated soft magnets were tested by Alternating Gradient Magnetometer (AGM), MicroMag 2900.

Fig. 9 showed the demagnetization loops of the NiFe soft magnets, which can provide the important parameters such as remanence, coercivity and the maximum energy density. From this figure, it could be observed at first that the coercivity of NiFe magnet was very small, which verified its soft magnetic property. Meanwhile, the slope of the loops showed the high permeability of the magnetic core. This high permeability can significantly increase the performance of inductors.

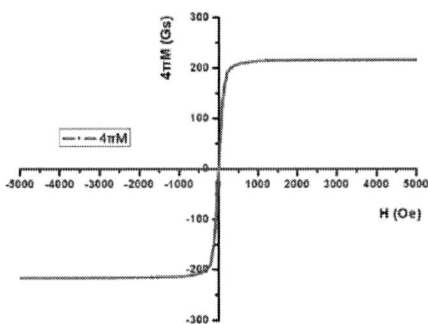

Fig. 9. The demagnetization loops of NiFe soft magnetic core

B. Characterization of Planar Inductor

Fig. 10 showed the photos of fabricated planar inductors. From the microscope photo (a), it can be seen that both the NiFe magnetic core and the copper coils were well arranged as the layout design. The surface of magnetic core was relatively smooth with several big lumps caused by nonuniform current

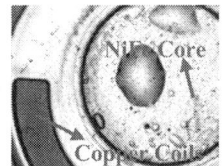 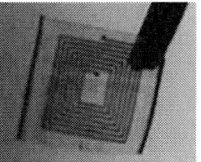

(a) Optical microscope photo   (b) Camera photo

Fig. 10. Fabrication results of MEMS planar inductors

Fig. 11. Good flexibility of MEMS planar inductor fabricated on parylene

distribution. From the photo (b), it could be observed that the inductor was well fabricated on transparent parylene layer. Meanwhile, Fig. 11 exhibited its good flexibility which will be helpful for implant applications.

Till now, we have tested a 3-turn planar inductor with copper coils of 500μm width and 100μm gap as well as 800×800μm area magnetic core. Fig. 12 is the measurement results of quality factor and inductance. The coil acquired highest quality factor in 6MHz. This frequency is quite low because of the large scale of the inductor. The typical inductance was 2nH to 7nH in frequency range of 60MHz to 70MHz.

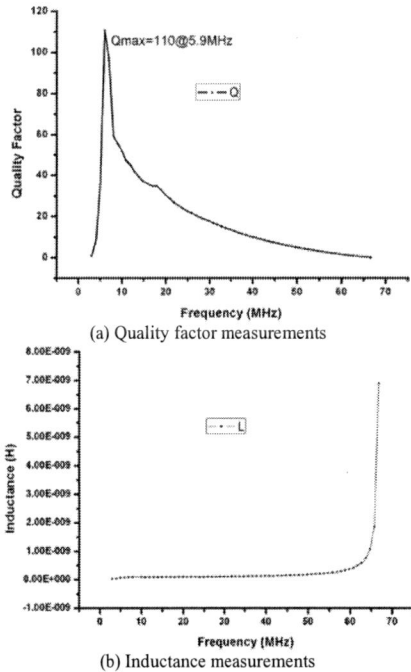

(a) Quality factor measurements

(b) Inductance measurements

Fig. 12. Quality factor and inductance tests of planar inductor

C. Further Research

The performance of the inductor could satisfy several applications in wireless power transfer system for implantable devices. However, we need to go further to test and compare the performance of inductors with different geometry factors. Meanwhile, the parylene package for this inductor is also required in the future work. Now, the more detailed testing is in progress.

V. CONCLUSIONS

Nowadays, lots of researches focus on the wireless power transmission system for implantable MEMS devices. But how to improve the performance of MEMS inductors in this system is a crucial issue limiting the application of this new technology.

In this paper, a new planar MEMS inductor with $Ni_{80}Fe_{20}$ soft magnetic core was successfully designed, simulated and fabricated. By design of electroplating setup and electrolyte, NiFe soft magnetic core with high permeability was integrated with the copper coils, which can increase the inductance and quality factor. Meanwhile, parylene substrate can provide good flexibility and biocompatibility for the inductors. By HFSS simulation, for small scale inductors, a maximum quality factor of 51.52 was achieved at 3.10GHz. The inductance of inductors with magnetic core was enhanced by 3.3% than those with air core. With the consideration of geometry factors, as the increase of magnetic core area, the inductance was also increased by 11.3% maximally. For large scale inductors fabricated, the tested maximum quality factor was 110 at 6MHz. Based on the full IC compatible fabrication process, these inductors could be used in wireless power transmission system for implantable medical devices.

ACKNOWLEDGMENT

This work was supported by Fund of National Natural Science Foundation of China (No. 61176103), National Natural Science Foundation of China (Grant No. 91023045) and National Key Laboratory of Nano/Micro Fabrication Technology (9140C7901080902).

REFERENCES

[1] A. Oki, Y. Takamura, T. Fukasawa, H. Ogawa, Y. Ito, T. Ichiki, et al., "Study on elemental technologies for creation of healthcare chip fabricated on polyethylen trephthalate plate," *IEICE Trans. on Electron.*, vol. E84-C, no. 12, pp. 1801-1806, Dec. 2001.

[2] Xiaoyu Liu, Fei Zhang, Steven A. Hackworth, Robert J. Sclabassi, and Mingui Sun, "Modeling and simulation of a thin film power transfer cell for medical devices and implants," *Circuits and Systems, 2009. ISCAS 2009. IEEE International Symposium on*, pp. 3086 – 3089, May 2009.

[3] T. Yamada, T. Uezono; H. Sugawara, K. Okada; K. Masu, A. Oki, et al., "Battery-less wireless communication system through human body for in-vivo healthcare chip," *Digest of Papers, 2004 Topical Meeting on Silicon Monolithic Integrated Circuits in RF Systems*, pp. 322-325, Sept. 2004.

[4] D. Ferber, "Re-engineering your body," *Reader Digest special 1000th issue*, pp. 98-113, August 2005.

[5] Sung-Hoon Cho, Lawrence Cauller, Will Rosellini, and Jeong-Bong Lee, "A MEMS-Based Fully-Integrated Wireless Neurostimulator", *Proceedings of 2010 IEEE 23rd International Conference on Micro Electro Mechanical Systems (MEMS)*, pp. 300-303, Jan. 2010.

[6] Tianjia Sun, Xiang Xie, Guolin Li, Yingke Gu, Yangdong Deng, Ziqiang Wang, et al., "An Asymmetric Resonant Coupling Wireless Power Transmission Link for Micro-Ball Endoscopy", *32nd Annual International Conference of the IEEE EMBS Buenos Aires*, pp. 6531-6534, Sept. 2010.

[7] S. Kima, K. Zoschke, M. Klein, D. Black, K. Buschick, M. Toepper, et al., "Switchable polymer-based thin film coils as a power module for wireless neural interfaces", *Sensors and Actuators A*, vol. 136, pp. 467-474, 2007.

[8] Dohi T, Kuwana K, Matsumoto K and Shimoyama I, "A standing micro coil for a high resolution MRI", *Proc. IEEE Transducers*, pp. 1313-1316, 2007.

[9] K. Kratt, V Badilita, T Burger, J G Korvink and UWallrabe, "A fully MEMS-compatible process for 3D high aspect ratio micro coils obtained with an automatic wire bonder", *J. Micromech. Microeng.* Vol. 20, pp. 1-11, Jan. 2010.

[10] Dok Won Lee, Kyu-Pyoung Hwang, and Shan X. Wang, "Fabrication and Analysis of High-Performance Integrated Solenoid Inductor with Magnetic Core", *IEEE TRANSACTIONS ON MAGNETICS*, vol. 44, NO. 11, November 2008.

[11] Liangliang Li, Dok Won Lee, Kyu-Pyung Hwang, Yongki Min, Toru Hizume, Masato Tanaka, et al., "Small-Resistance and High-Quality-Factor Magnetic Integrated Inductors on PCB," *IEEE TRANSACTIONS ON ADVANCED PACKAGING*, vol. 32, NO. 4, pp. 780-787, November 2009.

[12] Wen Li, Damien C. Rodger, Ellis Meng, James D. Weiland, Mark S. Humayun, and Yu-Chong Tai, "Wafer-Level Parylene Packaging With Integrated RF Electronics for Wireless Retinal Prostheses," *JOURNAL OF MICROELECTROMECHANICAL SYSTEMS*, vol. 19, NO. 4, pp. 735-742, AUGUST 2010.

[13] Xuming Sun, Quan Yuan, Dong-Ming Fang, Haixia Zhang, "Electrodeposition and Characterization of CoNiMnP-based Permanent Magnetic Film for MEMS Applications", *IEEE-NEMS 2011 Proceedings*, Feb. 20-23, 2011.

Design of a Small Wankel Engine

Lung-Jieh Yang[*], Tsan-Hsiang Wang

[*]Department of Mechanical and Electromechanical Engineering, Tamkang University, TAIWAN
Ljyang@mail.tku.edu.tw

Abstract—**This work presents a novel design of an ultra-small Wankel engine. With a device size of mm range and required power of mW, the rotation speed is theoretically calculated up to thousands of rpm. The PDMS MEMS process has been employed to make the Wankel engine planar and tiny. How to selecting the proper materials heterogeneously in the design stage of this engine is demonstrated herein.**

Keywords-component; Wankel engine; energy harvest

I. INTRODUCTION

Finding clean renewable energies has become a popular research topic in recent years. Using Stirling engine in harvesting the waste heat is one of the corresponding studies. Stirling engines are thermally driven by the mechanism of temperature difference and include the configurations of translational and rotary types. The rotary type Stirling engine invented by Felix Wankel in 1959 was also known as the Wankel engine. Wankel engines have several advantages over the traditional combustion piston engines in the past, for example, higher efficiency, lower vibration, and light weight [1-2].

In this work, MEMS machining method is used to fabricate an ultra-small Wankel engine for energy harvesting. Before the design of device manufacture, a theoretical discussion of the rotor dimension on the rotating speed of the Wankel engine is firstly done. Then a thermal resistance analysis is also addressed for selecting the proper materials and the corresponding MEMS processes. Silicon, glass, SU-8 resist, and PDMS are all considered as the device materials. The different working fluids inside the small Wankel engines are also proposed.

II. DEVICE DESIGN AND FABRICATION

A. Configuration of the Wankel Engine

The authors would like to manufacture an ultra-small Wankel engine by MEMS process here. Owing to the excuse that the structure of translational Stirling engine is complicated but the MEMS process is planar, so the rotary or the Wankel engine configuration is selected. However, the dimension and the materials for the engine should be assigned as well.

The housing inner wall of the Wankel engine has a mathematical form known as a trochoid or an epitrochoid shape [3-5]. The rotor is an eccentric rotor. The contour of the inner wall is shown in Fig. 1. The trajectory is depicted as the (x,y) coordinate in (1) and (2).

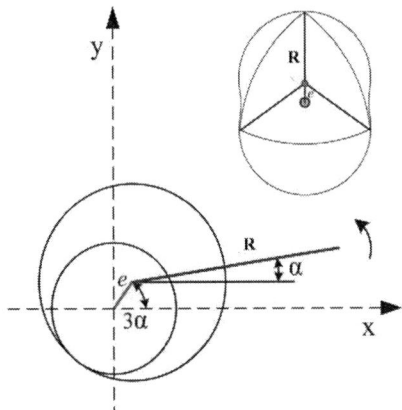

Fig. 1. The epitrochoid contour of Wankel engine.

$$x = R\cos(\alpha) - e\cos(3\alpha). \qquad (1)$$
$$y = R\sin(\alpha) - e\sin(3\alpha). \qquad (2)$$

where the angle α can be regarded as follows assuming the constant rotating speed ω,

$$\alpha = \omega t. \qquad (3)$$

The symbols R is the longest distance from the center circle to the periphery; e is the radius of the central circle.

Basically, the relative contour of Fig. 1 depends on e and R simultaneously. Herein, the authors fixed the ratio (e/R) as 0.162, and discussed the proper value of R on the rotating speed of the rotor in the following.

B. Rotation Speed of the Rotor

Ideally consider a circle disc with a radius of r_0 shown in Fig. 2, and let it rotate with a speed of ω above a supporting substrate with a gap of d. The viscous drag torque T could be taken as the integral around the disc area by the assumption of Newtonian fluid stress on the substrate as follows.

$$
\begin{aligned}
T &= \int_A (stress)\ (area)\ (force\_arm) \\
&= \int_A \mu\left(\frac{du}{dz}\right)_{substrate} dA \cdot r \\
&= \int_0^{r_0} \mu \frac{r\omega}{d} 2\pi r^2 dr \\
&= \frac{\pi\mu\omega r_0^{\,4}}{2d}.
\end{aligned}
\qquad (4)
$$

The power needed to overcome the drag torque T under the rotation speed ω is shown below.

978-1-4673-1122-9/12 $31.00 © 2012 IEEE

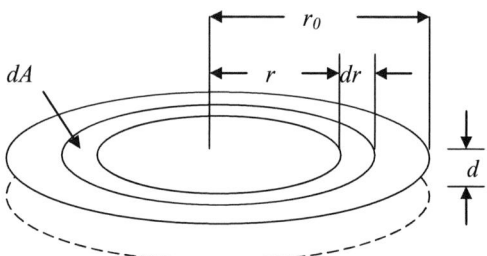

Fig. 2. The ideal disc model of the Wankel engine.

$$P = T\omega = \frac{\pi\mu\omega^2 r_0^4}{2d}. \tag{5}$$

As the authors applied the above power formula to the case of the Wankel engine, the rotating tip is not only rotating but also wobbling according to (1) and (2). So the averaged instantaneous tangential velocity $r\omega$ in (4) should be replaced with $\overline{r\omega}$ as follows.

$$\overline{r\omega} = r\omega\sqrt{1 + (3e/R)^2}. \tag{6}$$

and r^2 in (4) should be replaced with $\overline{r^2}$.

$$\overline{r^2} = r^2\left[1 + (e/R)^2\right] \tag{7}$$

Finally, the modified power and the rotation speed are in the following.

$$\overline{P} = \frac{\pi\mu\omega^2 R^4}{2d}\sqrt{1 + (3e/R)^2}\left[1 + (e/R)^2\right] \tag{8}$$

$$\overline{\omega} = \sqrt{\frac{2d\overline{P}}{\pi\mu R^4\sqrt{1 + (3e/R)^2}\left[1 + (e/R)^2\right]}}. \tag{9}$$

From (8) and (9), the required power P would be as small as possible when the rotor dimension R approaches to zero; the rotating speed ω would also approach to infinity when the rotor dimension is as small as possible.

As a viable device for harvesting any infinitesimal waste heat P, the trend of making the rotor R smaller is beneficial to the success of the Wankel engine herein.

C. Thermal Resistance Analysis

The proper choice of the heat conduction path in the Wankel engine is considered in this section. Then the required power P discussed in the previous section will be firstly estimated by this conduction heat transfer rate.

Here, the authors would like to compare the thermal conductivities of different internal working fluids in the central chamber of the engine which may be made of silicon and the PDMS substrates, respectively. These two kinds of substrate materials are of the most popular favor in MEMS area. The one-dimension, steady state, thermal conductivity equation (Fourier law) is shown as below.

$$q = kA\frac{\Delta T}{\Delta X}. \tag{10}$$

where q is heat transfer rate; k is thermal conductivity; A is cross section area; ΔT is temperature difference; ΔX is the assumed heat path length.

As people know well, the thermal resistance concept is beneficial to the first prediction of the proper heat path of heat conduction design problems. The thermal resistance R_T is defined as the temperature difference per unit heat transfer rate.

$$R_T = \frac{\Delta T}{q}. \tag{11}$$

Fourier law (10) and the definition (11) combine as below.

$$R_T = \frac{\Delta X}{kA}. \tag{12}$$

For the Wankel engine herein, there are two possible pathways for the thermal conduction heat to pass through. As shown in Fig. 3, the pathway (1) is along the periphery solid wall. The other pathway (2) is directly through the working fluid. For the sake of providing the required thermal power for rotating the engine rotor, the pathway (2) is of course the proper one.

Assume that ΔX_1 and ΔX_2 are the same; $A_1 : A_2 = 1 : 10$, and the rotor matter is negligible. If the pathway (2) is the favor one, the proper thermal resistance of $R_{T,2}$ should be much smaller than $R_{T,1}$ or expressed as the following inequality.

$$10 >> \left(\frac{k_1}{k_2}\right). \tag{13}$$

If material (1) or the chamber engine is made of silicon and $k_1 = 157$ W/mK, the thermal conductivity k_2 of the proper working fluid inside the engine chamber should be much greater than 15.7 W/mK. Only some liquid metals meet this requirement. It will encounter difficulties as packaging liquid metals into the engine chamber and the working temperature must be high for melting the metals inside the chamber.

On the other hand, if the engine chamber is made of PDMS or glass, and $k_1 = 0.16 \sim 1.38$ W/mK. Then the thermal conductivity k_2 of the proper working fluid inside the engine chamber should be much greater than $0.016 \sim 0.138$ W/mK. Water and even mercury meet this requirement. The corresponding thermal conductivities of the interested

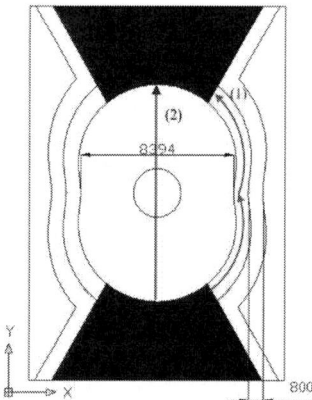

Fig. 3. Schematic diagrams of the small Wankel engine (unit: μm).

materials discussed above are listed in TABLE I [6].

Fig. 4 conclusively shows a design of the Wankel engine with the side chamber and bottom material of PDMS and with the cap material of glass. The central rotor is made of SU-8 resist. The working fluid inside the chamber is supposed to be mercury.

Three dimensions of the small Wankel engines have been designed. The values of R are assigned as 4990, 2495, and 1248 µm, respectively. In addition, there is an insulation groove has been designed to avoid the heat leak from the chamber in Fig. 4.

By the rotating speed ω_1 prediction of (9) and the conduction heat transfer rate (P is assumed to be converted from 40% of the heat conduction) of (10), the theoretical output data of the three engines with different rotor R are listed in TABLE II. The much faster ω_2 and ω_3 denote the cases that the gap between the rotor and the substrate is filled with water or air, respectively.

D. Fabrication Process

The soft lithography is selected as the fabrication technique for the small Wankel engine. A silicon mother mold is initially patterned by ICP etching and transfers to a PDMS chamber. This PDMS chamber block is bonded with a cover glass [7]. The whole processes are shown in Figs. 5-6.

The working fluid water or mercury should be filled into the chamber after the PDMS bonding. A filling hole is left in advance for the filling and the afterward sealing is done by epoxy gluing and parylene conformal coating [7]. One finished device filled with water is shown in Fig. 7. The functions of the Cr-Au electrodes made inside the chamber are two-folded. One function is as the thermal sensors for on-site monitoring the temperature field in the engine. The other function is as the joule heating source (or the thermal actuators) for adding so enough power into the engine as to push the rotor rotating.

E. Consideration of Testing Setup

The authors have applied the ultrasonic cleaner to shaking the rotor so as not to stick to the bottom surface of the PDMS chamber.

The rotor is detached from the substrate and of a little bit rotating around the center with several RPM. So far how to set up an ultrasonic shaking for the engine rotor is one concern for the successful experiment testing.

The temperature sensing function of the embedded electrodes have also been verified to monitor the temperature

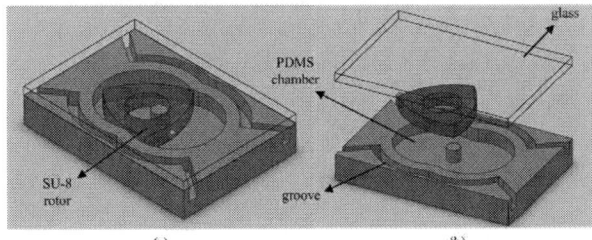

(a)   (b)

Fig. 4. Three dimensional diagram of the small Wankel engine.

TABLE I.   THERMAL CONDUCTIVITIES OF MATERIALS [6]

	Silicon	PDMS	Glass	Hg	Air	Water
Thermal conductivity (W/mK)	157	0.16	1.38	8.34	0.024~0.026	0.33~0.68

TABLE II.   PERFORMANCE DATA OF THE WANKEL ENGINE

R (µm)	4990	2495	1248
ΔT (K)	70 (20→90°C)		
P (mW)	84.4×40% (working fluid: mercury)		
μ (N/s m²)	1.4×10⁻² (gap fluid: mercury)		
ω_1 (RPM)	1,920	7,720	30,700
μ (N/s m²)	5.1×10⁻⁴ (gap fluid: water)		
ω_2 (RPM)	10,000	40,400	16,100
μ (N/s m²)	2×10⁻⁵ (gap fluid: air)		
ω_3 (RPM)	50,800	20,400	81,400
d=18.75 µm; ΔX=11607 µm; A=8394×200 µm²; k=8.34 W/mK			

change inside the chamber from the room temperature to the water boiling state. The Cr-Au metal is still fine as the resistive thermal detecting (RTD) material herein. However, how to combine the ceramic heating source with the fabricated engine to do the thermal testing is not easy since the thermal conductivity of PDMS wall is very small. The authors have tried to partially merge the fabricated PDMS small engine in the ultrasonic pool, and add heat directly into the hot side of the engine. Even though the temperature difference is up to 70 degree C, the engine rotor is only slightly shaking without apparent rotating. The authors re-calculated the output data of TABLE II by (9) and (10) with the working fluid changed from mercury to water. It's found that the theoretical rotating speed is around 2500 rpm subject to P=2 mW. Since the water electrolysis from the Cr-Au electrodes during heating would generate non-condensed bubbles to retard the rotation of the engine. Using water as the working fluid is then not feasible herein. The experiment of packaging the testing mercury as the working fluid is still under developing. Moreover, a cooling end packaged with a heat pipe is revealed in Fig. 8 for promoting the condensation of this small Wankel engine.

III. CONCLUSION

By the analysis of rotating speed and the thermal resistance of the possible designs of the ultra small Wankel engine in this study, the mm-size energy harvesting device is supposed to have thousands of RPM rotation speed subjected to mW power input. The fabrication process and the testing setup are also considered. Ultrasonic shaking is so far necessary to the testing for effectively levitating the center rotor without sticking issue. The authors also tried water as the working fluid inside the engine chamber and did not gain the successful result. The left choice of selecting mercury as the working fluid in PDMS engine chamber is still under way. The necessary packaging technique with conformal parylene is

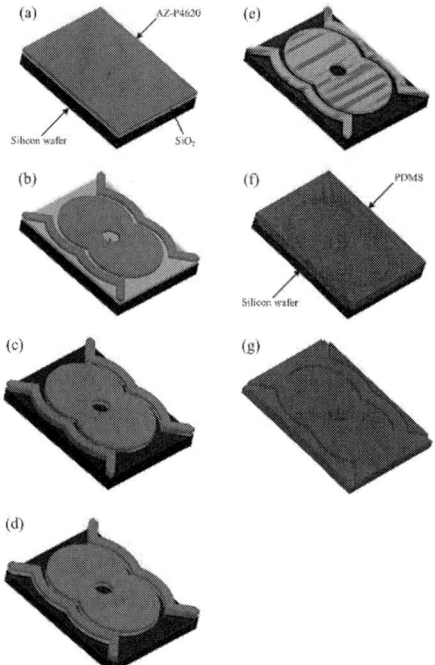

Fig. 5. Fabrication process for the PDMS chamber: (a) resist coating; (b) UV exposure and developing; (c) oxide opening; (d) ICP etching; (e) resist stripping; (f) dispensing PDMS; (g) PDMS demolding.

looked forward to avoiding mercury leakage. The Cr-Au electrodes designed as the thermal sensors and evaporator, a heat pipe installed at the cold end for condensation, are both hoped to augment the functionality of this small Wankel engine as well in the future.

Acknowledgment

The authors would like to thank the financial and travel support from the National Science Council of Taiwan with the project no. of NSC-98-2221-E-032-025-MY3. The technical helps from Prof. Shung-Wen Kang of Tamkang University and Dr. Yu-Cheng Ou of National Applied Research Laboratories are also highly acknowledged.

References

[1] P. Jin, K. Jiang and N. Sun, "Microfabrication of ultra-thick SU-8 photoresist for micro engines," Proceeding of SPIE, vol. no. 4979, pp. 105-110, 2003.

[2] C.H. Lee, K.C. Jiang, P. Jin, and P.D. Prewett, "Design and fabrication of a micro Wankel engine using MEMS technology," Microelectronic Engineering, vol. 73-74, pp. 529-534, 2004.

[3] R.F. Ansdale, The Wankel RC engine: design and performance, 1st edition, South Brunswick, New Jersey, 1969.

[4] D.E. Cole, "The Wankel engine," Scientific American, vol. 227, no. 2, pp. 14-23, 1972.

[5] K.C. Weston, "Computer simulation of a Wankel rotary engine analysis and graphics," Proceeding of the Conference of the Society for Computer Simulation, pp. 213-216, 1986.

[6] F. Kreith and M.S. Bohn, Principles of Heat Transfer, sixth edition, Thomson, .

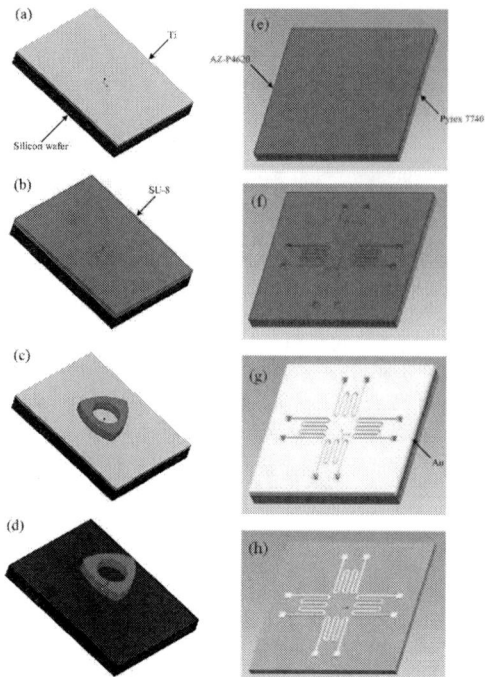

Fig. 6. Fabrication process for the rotor: (a) evaporating Ti; (b) SU-8 coating; (c) UV exposure and developing; (d) detaching SU-8 rotors from the substrate; Process for the electrodes: (e) resist coating; (f) UV exposure and developing; (g) evaporating the electrode metal film; (h) lift off process.

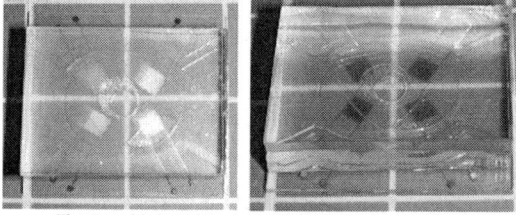

Fig. 7. A fabricated small Wankel engine (R=4990μm).

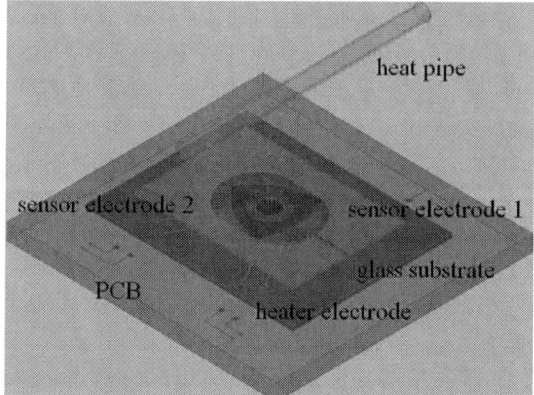

Fig. 8. The modified package of the small Wankel engine.

[7] L.J. Yang and T.Y. Lin, "A PDMS-based thermo-pneumatic micropump with Parylene inner walls," Microelectronic Engineering, vol. 88, pp. 1894-1897, 2011.

Fabrication of High-aspect-ratio PZT Structure by Nanocomposite Sol-gel Method for Laterally-driven Piezoelectric MEMS Switch

Nan WANG[*], Shinya YOSHIDA, Masafumi KUMANO, Yusuke KAWAI and Masayoshi ESASHI

[*]Graduate School of Engineering, Tohoku University, JAPAN

wang.nan@mems.mech.tohoku.ac.jp

Abstract—**in this study, we have proposed a novel laterally driven piezoelectric MEMS (MicroElectro Mechanical Systems) switch using a high-aspect-ratio (AR) PZT (Pb[Zr$_x$Ti$_{1-x}$]O$_3$) structure. Then, the fabrication process of the PZT structure based on PZT filling process in a deep Si trench is developed to realize a laterally driven PZT microactuator. At first, the process of the Si trench with a thin Al$_2$O$_3$ layer as a Pb-diffusion barrier layer and with a Pt film as the electrode for the actuator is successfully developed. Then, it is demonstrated that the dense and crack-free PZT structure with high-AR can be fabricated by the PZT filling process in the Si trench by nanocomposite sol-gel method. In addition, it is speculated from the X-ray diffraction pattern that the composite PZT has pure perovskite phase and piezoelectric property. The remnant polarization (P_r) and the coercive field (E_c) of a nanocomposite PZT thick film measured 11 μ cm^2 and 71.2 kV/cm, respectively. As the result of these experiments, it is demonstrated that the fabrication process of the high-AR PZT structure has the potential to realize the laterally driven piezoelectric MEMS switch.**

Keywords-Piezoelectric MEMS switch; Nanocomposite PZT; High-aspect-ratio structure

I. INTRODUCTION

So far, various kinds of electrical and optical switches based on Micro-Electromechanical Systems (MEMS) actuators have been developed. Recently, MEMS switch has been expected to be employed as a high-performance switch for high-frequency circuit. Compared with conventional semiconductor switches such as field effect transistor (FET) and PIN diode, MEMS switch has the advantage to achieve not only high isolation and low insertion loss, but also low power consumption [1]. In particular, since a piezoelectric actuator has excellent responsiveness and conversion efficiency from electrical energy to mechanical energy, various types of piezoelectric MEMS switch have been developed in recent years [2]. PZT (Pb[Zr$_x$Ti$_{1-x}$]O$_3$) is a widely-developed material for the piezoelectric actuators because of its superior piezoelectric property. Thus, PZT MEMS switches have been fabricated as one of the strongest candidates until now. For example, unimorph-type PZT MEMS switches with the shape of cantilever and fixed-fixed beam were fabricated [3]. However, the initial bending problem of the actuators was easily caused by the residual stress of the PZT film due to the difference of thermal expansion between PZT and the substrate. It is difficult to control the residual stress completely. Recently, a multilayer-stacked PZT switch was developed to achieve low operation voltage and generation of a large switching force [4].

Also in this case, the residual stress problem is considered to be more obvious.

In this study, we have proposed a laterally driven PZT MEMS switch in order to solve this problem. The conception of the PZT MEMS switch is described in Fig. 1. The PZT bimorph actuators with high-aspect-ratio (AR), which have Si elastic plates in the centers, are connected at the ends of the beams to generate a large actuating force in horizontal direction [5]. This switch has the symmetrical structure. Thus, it is expected that the initial bending problem can be reduced in principle. In addition, if the high-AR actuator can be fabricated, this problem can be reduced further due to the large stiffness in vertical direction. In order to achieve the PZT MEMS switch, it is necessary to develop the fabrication process of the high-AR PZT structure with side-wall electrodes and investigate the piezoelectric property of the PZT structure.

Therefore, as the preliminary study, we developed the fabrication process of the single-beam bimorph piezoelectric actuator with high-AR PZT structure fabricated by nanocomposite sol-gel method. Then, we investigated the potential ability of this method to fabricate the bimorph PZT actuator.

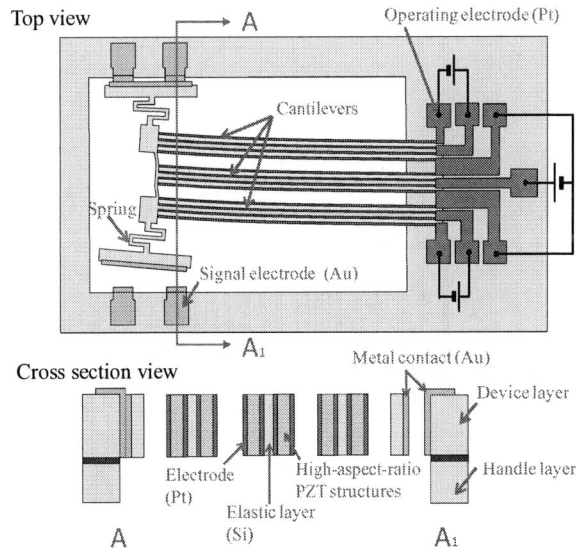

Fig. 1. Conception of the laterally driven piezoelectric MEMS switch

II. DEVICE STRUCTURE AND FABRICATION PROCESS OF LATERALLY-DRIVEN PZT ACTUATOR

Fig. 2 shows the fabrication process for a single-beam bimorph PZT actuator utilizing nanocomposite sol-gel method. The each process step in detail is shown as follows.

(1) A high-AR Si trench is fabricated by reactive ion etching (RIE) on a SOI wafer.

(2) An Al_2O_3 thin film is deposited by Atomic Layer Deposition (ALD) as a Pb diffusion barrier layer. Then, a conformal Pt film is deposited by electroless plating.

(3) Photoresist is patterned to fabricate the electrode pads for the actuator.

(4) A Pt film on the top and bottom sides of the trench is etched by Fast Atom Beam (FAB) etching to form the side-wall electrode and the electrode pads for application of an operation voltage.

(5) Photoresist is patterned for partial etching of the side-wall Pt electrodes.

(6) The side-wall Pt is partially etched by wet etching to prevent the shortage.

(7) PZT is filled in the trench as a template by nanocomposite sol-gel method. Then, the surface PZT is etched by wet etching. The nanocomposite sol-gel method can provide a dense, crack-free PZT thick film [6] [7]. Thus, this method is also expected to be available for fabricating the high-AR PZT structures.

(8) The beam structure of the bimorph actuator is fabricated by etching of the Si layer and Al_2O_3 film.

(9) The beam structure is released by etching Si and SiO_2 from the back side.

This paper mainly reports on the development of fabrication process of the silicon trench with Pt film on the side-wall surface for the actuator electrode and on the nanocomposite PZT filling process into deep Si trenches.

III. FABLICATION OF THE SILICON TRENCH WITH PT FILM ON THE SIDE WALL SURFACE FOR THE ACTUATOR ELECTRODE

A. Conformal deposition of Al_2O_3 and Pt films on Si trench

In the ALD process of Al_2O_3, Trimethyl aluminum and H_2O were employed as the precursors. N_2 was used as the carrier and purge gas. The Al_2O_3 film was deposited on a deep Si trench fabricated in a SOI wafer by repeating the ALD process cycle of 2000 times at 250 °C. Then, the Al_2O_3 film was heated at 900 °C for 3 hours in N_2 atmosphere to raise the chemical resistivity [8] [9]. After that, The Al_2O_3 film surface was treated by hydrofluoric acid (12.5 %) at room temperature to raise the surface roughness. This acid treatment leads to improvement of the adhesion between the Al_2O_3 film and a Pt film prepared by electroless plating. Then, a Pt catalyst (Pt activator, Tanaka-Kikinzoku Inc.) was coated on the Al_2O_3 film by spin-coating. After two-step drying for 30 min at 500 °C, the sample was dipped into an electroless plating solution (LECTROLESS Pt100, Tanaka-Kikinzoku Inc.) at 60 °C for 10 min, which resulted in deposition of a Pt film.

Fig. 3 shows the cross section view of the trench with the AR of 10 after the ALD and electroless plating processes. It can be seen that Al_2O_3 and Pt thin films could be conformally deposited on the Si trench structures. Peeling off between the Pt and Al_2O_3 films was not observed. The thicknesses of the Al_2O_3 and the Pt films were measured approximately 180 nm and 50nm.

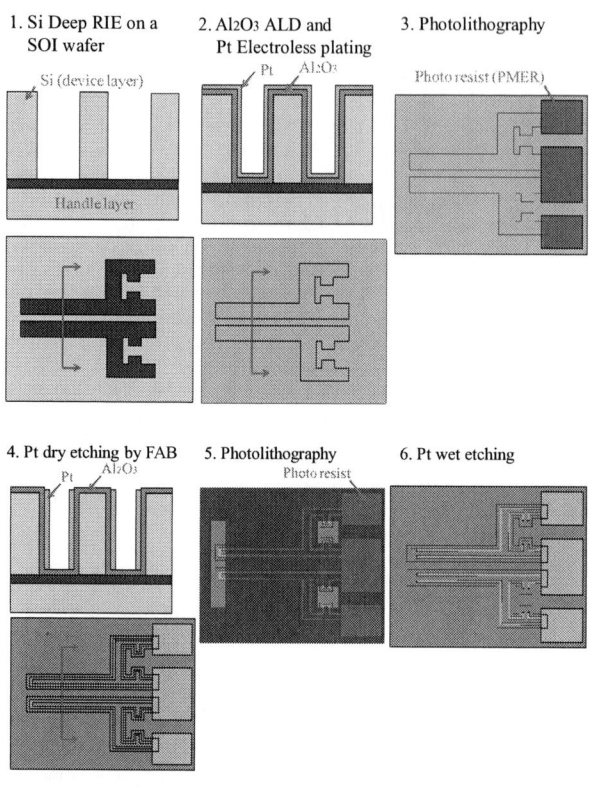

Fig. 2. Fabrication process of a single-beam bimorph PZT actuator utilizing nanocomposite sol-gel method.

B. Fabrication of the electrode pad and selective remove of the top- and bottom-side Pt films by FAB

Then, we developed the process steps of (3) and (4) in Fig. 2. In general, it is not so easy to conduct photolithography on three-dimensional structure such as a deep trench because the conformal resist coating is difficult. In particular, the resist film break at the corner on the trench is often caused when a lowly viscous resist is coated by spin-coating. The defective coating of the resist film causes bad connection between the electrode pads and the side-wall Pt electrodes which are fabricated by FAB etch process shown in Fig. 2(4). Thus, a thick photoresist (PMER P-LA 900 PM, Tokyo Ohka Kogyo co., ltd.) was employed in this process. The substrate surface is expected to be completely covered with the thick resist by filling the deep trench with the resist.

Fig. 4(a) shows the result of the resist patterning on the substrate which had the deep trench with the AR of 4. The resist film was successfully patterned. It can be also seen in Fig. 4(b) that the Si trench was filled and completely covered with the resist film. This result means that the connection part between the side-wall electrodes and the electrode pads can be protected well from the next Pt etching process by FAB.

Then, we investigated the effectivity of the FAB etching process to selectively etching the top-side and bottom-side Pt films of the high-AR trench without etching of the side-wall Pt film. In FAB process, chlorine gas was used as the etching gas, and the acceleration voltage was set 3.0 kV.

Fig. 5 shows the result of the etching process to the trench with the AR of 10. The selective etching was successfully conducted. Only the side-wall film could be remained. The Pt electrode pads were also successfully patterned, as show in Fig. 6. It is observed that the Pt electrode pads could be connected to the side-wall Pt films. Therefore, it could be demonstrated that the above-mentioned deposition methods, lithographic technique and the FAB etching process can fabricate the deep Si trench with the side-wall Pt electrode and the electrode pads for the high-AR PZT actuator.

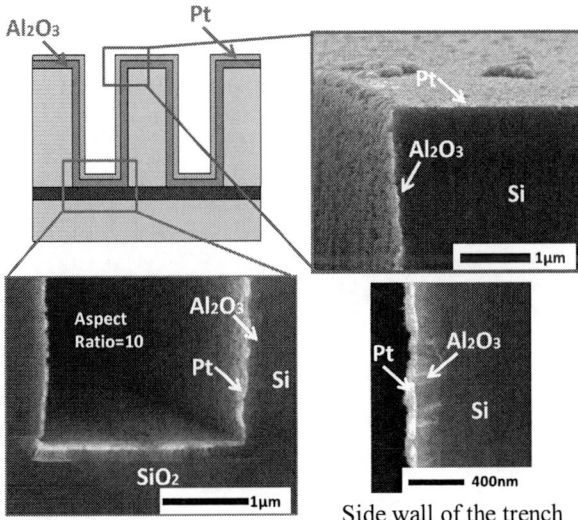

Fig. 3. Cross section views of the trench after deposition of the Al_2O_3 and Pt films.

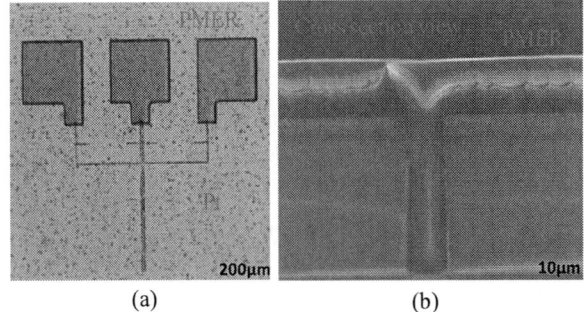

Fig. 4. (a) Result of the photolithography on the substrate with the deep Si trench. (b) Cross section view of the trench with AR of 4 after the resist coating.

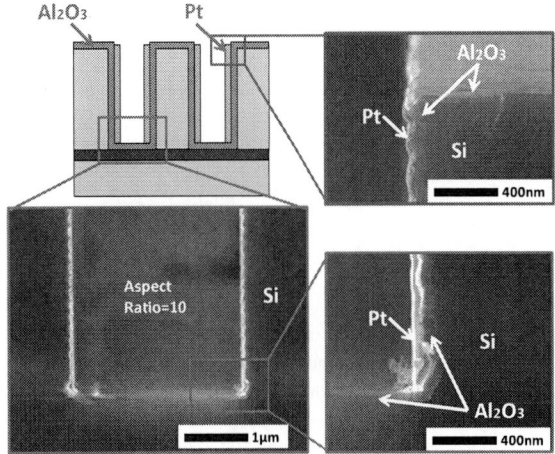

Fig. 5. Cross section views of the side wall Pt film on the deep trench after the FAB etching process.

Fig. 6. Top views of the Pt electrode pads fabricated by FAB etching.

IV. FABRICATION OF THE HIGH-ASPECT-RATIO PZT STRUCTURES BY NANOCOMPOSITE SOL-GEL METHOD

A. Development of filling process of the Si trench with PZT by nanocomposite sol-gel method

The detailed sequence of the fabrication process of high-AR PZT structures by nanocomposite method are shown in Fig. 7 and as follows.

(1) A nanocomposite slurry was prepared by mixing a PZT sol-gel solution of 100 ml (PZT-YM9, YOUTEC Inc., 20

wt% of PZT) and PZT nanopowders of 5 g (HIZIRCO AC750, HAYASHI Chemical Inc., average powder size: 450nm)[10]. Then, this mixture was stirred under ultrasonic agitation for 30 min.

(2) Appropriate amount of the PZT nanocomposite solution was dropped on the substrate with the deep Si trench.

(3) Vacuum treatment of the sample substrate was conducted for 1 min in order to accelerate the filling of the PZT nanocomposite solution.

(4) Spin-coating at 2,000 rpm for 20 sec was carried out for uniform coating of the solution.

(5) Vacuum treatment of the sample was carried out again.

(6) The PZT layer was dried by a two-step heating process. In first drying process to remove the solvent, the substrate was heated at 150 °C for 3 min on a hotplate. In second drying process for decomposition of the organic compounds, the substrate was heated at 400 °C for 3 min. As the result, an amorphous PZT layer was formed on the substrate.

(7) The amorphous PZT layer was crystallized at 680 °C by rapid thermal annealing (RTA) for 10 min.

(8) The Si trench was filled with PZT by repeating the steps from (2) to (7) for several times. As the result, a high-AR PZT structure was formed in the trench.

In this experiment, the PZT filling process was carried out using several-size trenches with various ARs and depths in order to investigate the possibility of this method to form high-AR PZT structure. The filling process by the conventional PZT sol-gel solution without the nanopowder was also carried out as the comparison experiment.

Fig. 8 (a-1), (a-2), (a-3) show the results of the PZT filling process on the trenches with the depth of 20 μm by nanocomposite sol-gel method. As the images shows, dense and crack-free PZT structures were successfully formed in the trenches with the AR of 2 and 4. Meanwhile, in the case of the AR of 10, the clogging of the PZT film was observed at the top of the trenches, as shown in Fig. 8 (a-3). Even in the filled trench, large pores were observed. Fig. 8 (b) shows the results of the filling process on the trench with the AR of 4 by the PZT sol-gel solution without the nanopowders. The large cracks were generated in the PZT structure due to the shrinkage in the annealing process, which resulted in failure of complete PZT filling of the trench. This experimental result means that the nanocomposite processing can prevent from the crack generation and enhance to form the dense PZT structure in the trench. It was attributed to decrease of the shrinkage volume of the PZT film in the annealing process due to the presence of a large amount of PZT nanopowders. It was also attributed to the reaction between the hydroxyl groups on the nanopowders' surface and the polymer macrocluster species in the sol-gel solution. [11] It is speculated that the strong bond was formed between the powders and sol-gel matrix, which can fabricate the dense PZT structure without large cracks. Fig. 8 (c-1) and (c-2) show the result of the filling process with the 50-μm-deep trench. PZT filling process was also succeeded into the trench with the AR of 4, and failed with the AR of 6.

Therefore, it could be demonstrated that the nanocomposite PZT filling process in the Si trench can provide dense PZT structures with high AR. The result in this study demonstrated that the PZT structure with AR of 4 at least can be fabricated when 20 and 50-μm-deep trenches are used as the template. Optimization of the filling process and the size selection of the nanopowder are expected to fabricate the higher AR PZT structure.

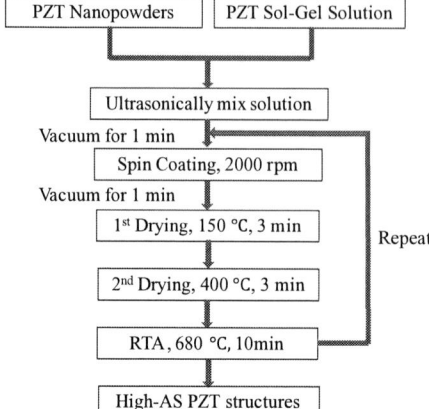

Fig. 7. Flow chart of the fabrication process for the high-aspect-ratio PZT structure.

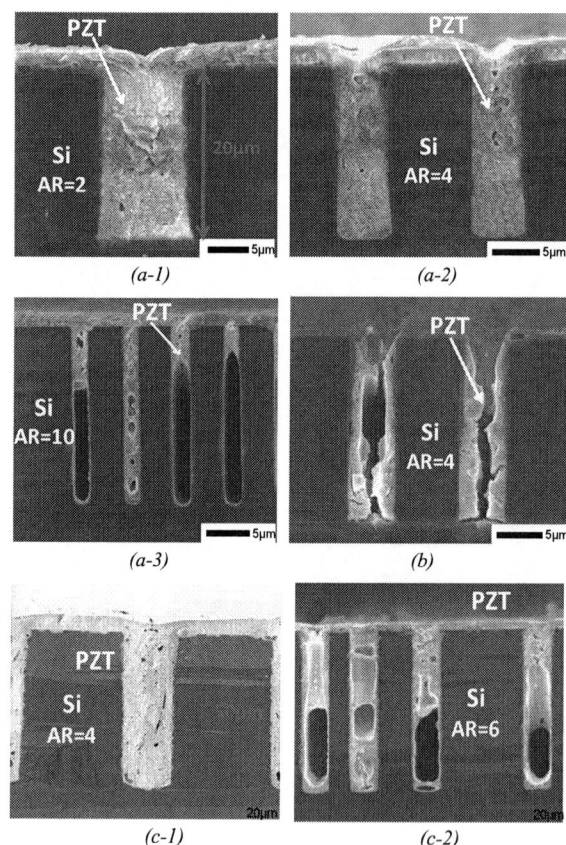

Fig. 8. Cross section view of the nanocomposite PZT filled in the Si templates

B. Orientation measurement of the nanocomposite PZT film

It is difficult to directly measure the orientation of the PZT structure sandwiched with the side-wall Pt film in the trench by X-ray diffraction (XRD). Thus, in order to speculate it, the orientation of the nanocomposite PZT film on the Pt film via electroless plating on a flat substrate was investigated.

In this experiment, we prepared and analyzed 4 kinds of samples, which are listed as follows, for the comparison study to investigate the orientation dependency of the composite PZT film on the orientation of the under layer. Si substrates were used as the flat substrate.

Sample (1): an 8-μm-thick nanocomposite PZT film on electroless-plated Pt/ALD-deposited Al$_2$O$_3$ films. The Pt and Al$_2$O$_3$ films were prepared in the same manner shown in chapter III.

Sample (2): an 8-μm-thick nanocomposite PZT film on sputtered Ti and Pt films with the thicknesses of 10 nm and 50 nm, respectively. The Ti layer was used as an adhesion layer. The sputtering was conducted at room temperature.

Sample (3): a 1-μm-thick PZT film prepared using a conventional sol-gel solution on the electroless-plated Pt/ALD-deposited Al$_2$O$_3$ films.

Sample (4): a 1-μm-thick PZT film prepared using a conventional sol-gel solution on the sputtered Pt/Ti films.

Fig. 9 shows the XRD pattern of these PZT films. Sol-gel PZT can form two crystallization phases: a pyrochlore phase and a perovskite phase. The pyrochlore phase does not exhibit the desired ferroelectric properties, which is exhibited in the perovskite phase [12]. The XRD pattern of the nanocomposite PZT film on the electroless-plated Pt showed the peaks at 2θ = 22˚, 31˚, 38˚ and 45˚, which are corresponding to the perovskite phase. On the other hand, the peaks corresponding to the pyrochlore phase, which generally appear at 29˚, 34˚ and 47˚ [13] [14], were not observed. Thus, this PZT nanocomposite film had a pure perovskite phase without a pyrochlore phase, and this PZT film can be expected to show a ferroelectric property.

The PZT (111) peak, which is important for the piezoelectric property, of the nanocomposite PZT film on the electroless-plated Pt was observed. Thus, it is expected that the bimorph PZT actuator with the nanocomposite PZT structure can be actuated by piezoelectric effect. However, the intensity of the PZT (111) peak is lower than the PZT (110). The PZT (111) peaks of both sol-gel PZT and nanocomposite PZT films on the sputtered Pt films are higher than that on the electroless-plated Pt films, which means that the as-prepared electroless-plated Pt film has limited ability to enhance (111) orientation of the PZT films. This result is confirmed from the fact that the Pt (111) peak of the sputtered film is higher than that of the electroless-plated film. In the XRD pattern of nanocomposite PZT on the sputtered Pt film, a large PZT (110) peak are shown. On the other hand, such a large peak is not seen in the sol-gel PZT film on the sputtered Pt film. It is considered that this PZT (110) peak is originated from the PZT nanopowders. It is expected that the (111) orientation of

the electroless-plated Pt film can be enhanced by post-annealing process at the high temperature such as 600 ˚C.

C. Characterization of the ferroelectric property of the nanocomposite PZT film

The ferroelectric property of the nanocomposite PZT film was investigated by measuring the polarization-electric field (P-E) hysteresis loop. In this experiment, an 8-μm-thick nanocomposite PZT film was deposited on an electroless-plated Pt/ALD-deposited Al$_2$O$_3$ films. Then, the upper Pt electrode was fabricated on the PZT film by sputtering.

As the result, the remnant polarization (P_r) and coercive field (E_c) measured 11.7 μC/cm^2 and 71.2 kV/cm, respectively, as shown in Fig. 10. It is speculated that the relative lower P_r was attributed to the small pores and gaps in the thick PZT film generated from the repetition of PZT depositions and heat treatments [6] [7]. The ferroelectric property is also expected to be improved by optimizing the process conditions such as mixture ratio of the composite sol-gel solution, the annealing temperature, and the orientation of the Pt film.

D. Composition analysis of nanocomposite PZT in Si trench

Finally, the composition difference of the nanocomposite PZT between on the top side and filled in the trench was investigated by Energy-Dispersive X-ray (EDX) analysis.

The EDX pattern shown in Fig. 11 indicates that the PZT structure filled in the trench had relatively larger amount of carbon compared to the PZT film on the top side. It was attributed to imperfect release of the organic compound in the sol-gel slurry from the PZT structure during the drying and RTA processes. It is considered that the release speed of the decomposed organic compound from the PZT structure in the trench was slower than that on the top side because the thicker film was deposited in the trench in one deposition process. This issue is also expected to be improved by extending the time of the two-step drying and RTA.

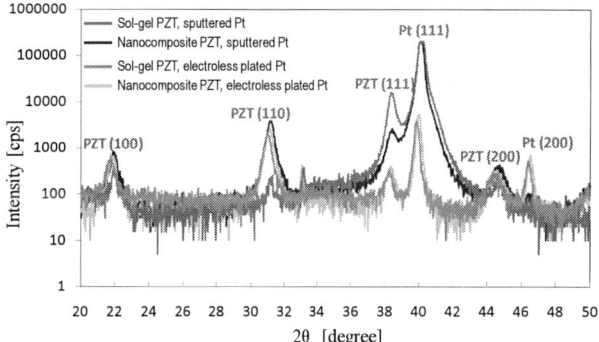

Fig. 9. Investigation of the difference of the XRD patterns of the nanocomposite PZT and sol-gel PZT.

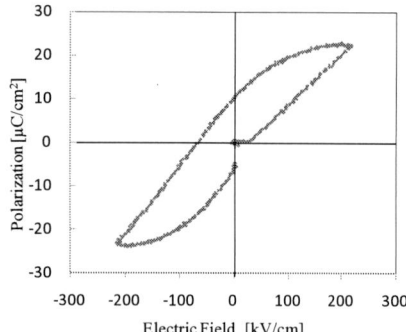

Fig. 10. Polarization-electric field (P-E) hysteresis loop of the nanocomposite PZT film.

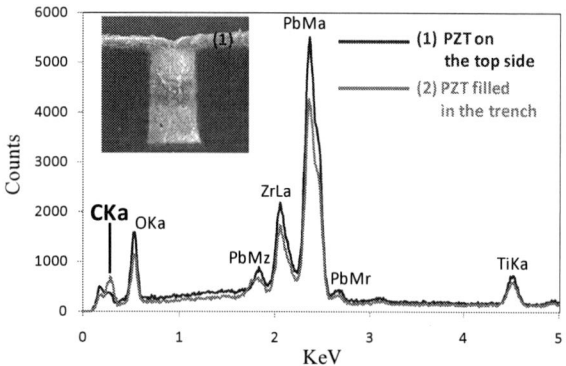

Fig. 11. EDX spectra of the nanocomposite PZT structures (1) on the top side and (2) in the trench.

V. CONCLUSION

In this study, we proposed the fabrication process of the laterally-driven piezoelectric MEMS actuator with high-AR PZT structure which is formed using nanocomposite sol-gel method. As the preliminary study, it was demonstrated that the Al_2O_3 and Pt films can be conformally formed on the deep Si trench by ALD and electroless plating, respectively. And, the availability of the FAB etching process for fabrication of the side-wall electrode was also demonstrated. In addition, the PZT structures with the AR of 4 were successfully fabricated using the Si templates and nanocomposite sol-gel processing. The XRD measurement indicated that the nanocomposite PZT had a pure perovskite phase and (111) orientation component. Its remnant polarization (P_r) and the coercive field (E_c) measured 11.7 $\mu C/cm^2$ and 71.2 kV/cm, respectively. The optimization of the process condition is expected to improve the density, the (111) orientation and ferroelectric performance of the composite film. We believe that this novel fabrication method for high-AR PZT structures can provide a novel laterally-driven PZT actuator, which results in realizing a large-force and fast-response piezoelectric MEMS switch.

ACKNOWLEDGMENT

This work was supported in part by a Grant in Aid for Scientific Research from the Japanese Ministry of Education, Culture, Sports, Science and Technology. This work was also supported by World Premier International Research Center Initiative (WPI Initiative), MEXT, Japan.

REFERENCES

[1] Yogesh B. Gianchandani, Osamu Tabta, Hans Zappe, *Comprehensive Microsystems,* Vol.3, pp. 325-326, 2008

[2] Lee H-C, Park J-Y, "Piezoelectrically actuated RF MEMS DC contact switches with low voltage operation". *IEEE MicroWirel Compon Lett,* Vol. 15(4), pp. 202-204, 2005

[3] I. Kanno, Y. Tazawa, T. Suzuki and H. Kotera, "Piezoelectric unimorph microactuators with X-shaped structure composed of PZT thin films" *Microsyst Technol,* Vol. 13, pp. 825-829, 2007

[4] Masaaki Moriyama, Yusuke Kawai, Shuji Tanaka, and Masayoshi Esashi, "Low-Voltage-Driven Thin Film PZT Stacked Actuator for RF-MEMS Switches", *The 28th Sensor Symposium on Sensors, Micromachines and Applied Systems,* Japan, pp. 9, 2011.

[5] Takashi Kaneko, Nobuyuki Ohya, Nobuaki Kawahara, "Optical Microscope Expanded Depth of Field using Dynamic Focusing Lens for Micro Parts Assembling" *T.IEE Japan,* Vol.118-E, No. 7/8, pp.364-370, 1998

[6] Q.F.Zhou, H.L.W.Chan, C.L.Choy, "PZT ceramic/ceramic 0-3 nanocomposite films for ultrasonic transducer applications" *Thin Solid Films,* Vol. 375, pp. 95-99, 2000

[7] Changlei Zhao, Zhihong Wang, Weiguang Zhu, Ooikiang Tan and Hueyhoon Hng, "Microstructure and properties of PZT 53/47 thick films derived from sols with submicron-sized PZT particle", *Ceramic International,* Vol. 30, pp. 1925-1927, 2004

[8] J. A. Aboaf, "Deposition and properties of Aluminum Oxide Obtained by Pyrolytic Decomposition of an Aluminum Alkoxide" *J. Electrochem. Soc.,* Vol.114, pp. 948-952, 1967

[9] L Zhang, H C Jiang, C Liu, JW Dong and P Chow, "Annealing of Al_2O_3 thin films prepared by atomic layer deposition" *J. Phys. D: Appl. Phys.* Vol. 40, pp. 3707–3713, 2007

[10] Q.F. Zhou, K.K. Shung, and Y. Huang, "Improvement electrical properties of sol-gel derived lead zironate titanate thick films for ultrasonic transducer application"*J Mater Sci* Vol. 42, pp.4480-4484, 2007

[11] D.A. Barrow, T.E. Petroff, R.P. Tandon, and M. Sayer. "Characterization of thick lead zirconate titanate films fabricated using a new sol gel based process", *J. Appl. Phys.* Vol. 81(2) , pp. 876-881, 1997

[12] Vinay Chikarmane, Jiyong Kim, Chandra Sudhama, Jace Lee and Al Tasch, "Annealing of Lead Zirconate Titanate (65/35) Thin Films for Ultra Large Scale Intergation Storage Dielectric Applications: Phase Transformation and Electrical Characteristics", Journal of Electronic Materials, Vol. 21, No. 5, pp. 503-512, 1992

[13] Radhouane Bel Hadj Tahar, Noureddine Bel Hadj Tahar, and Abdelhamid Ben Salah, "Low-temperature processing and characterization of single-phase PZT powders by sol-gel method", *J Mater Sci,* Vol. 42, pp. 9801-9806, 2007

[14] Zhenxing Bi, Zhisheng Zhang and Panfeng Fan, "Characterization of PZT Ferroelectric Thin Films by RF-magnetron Sputtering" *Journal of Physics: Conference Series,* Vol. 61, pp. 120-124, 2007

Silicon-Polymer Electro-thermal bimorph actuators with SiC bottom-layer for large out-of-plane motion and improved power efficiency

M. Aarts, J. Wei, P.M. Sarro

*Department of Electronic Components, Technology and Materials, TU Delft, NETHERLANDS (TU Delft, Netherlands)
Mark-aarts@hotmail.com

Abstract—**This paper presents the fabrication and characterization of a silicon-polymer electro-thermal out-of-plane bimorph actuator with SiC as bottom layer, for improved motion and better energy efficiency. The proposed concept improves on an earlier design using aluminum by making the bottom layer of SiC, a more robust material with a lower CTE. A process is developed to fabricate out-of-plane actuators with either Al or SiC as bottom layer. Both devices are characterized and their performances are compared. The new actuator with SiC as bottom layer shows a higher displacement (15.5μm) with less actuation voltage (4 V) and an average temperature increase on the actuator of 82°C.**

Keywords-component; formatting; style; styling

I. INTRODUCTION

Polymer based electro-thermal actuators have been widely explored in recent years for their large motion, low working temperature and low actuation voltage. Among them, the silicon-polymer laterally stacked electro-thermal actuators have shown to provide large motion, large force, low actuation voltage and relatively low operation temperature in both in-plane and out-of-plane actuation [1,2].

The commonly used electrothermal bimorph actuators, consisting of an actuation layer on the top of a suspended bottom layer can generate a downward out-of-plane motion (Fig 1).

The performance of the actuator depends highly on the properties of the working material and for the actuation layer they benefit from the silicon-polymer laterally stacked structure and the constraint polymer effect [3].

However, for the eventual performance of the out-of-plane bimorph actuator the suspended bottom layer is also of great importance. In the previously used aluminum bottom layer fatigue was easily observed and its large coefficient of thermal expansion (CTE) limited its performance. To overcome these problems and to improve the overall performance of the actuator a material with a lower CTE should be consider so to achieve a greater displacement before the material gets fatigued.

Silicon carbide (SiC) as material for the bottom layer is proposed in this paper. Offering both more robustness and a lower CTE, SiC should ensure an actuator capable of greater displacements.

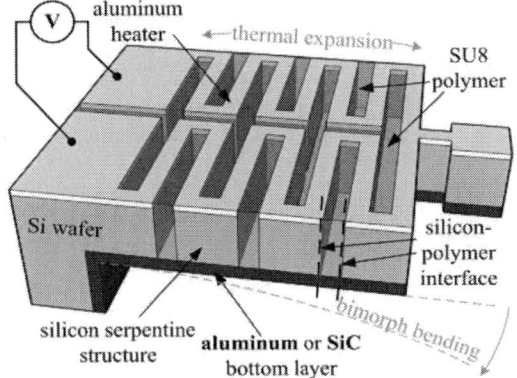

Figure 1: Schematic drawing of the silicon-polymer laterally stacked output-plane actuator based on bimorph structure.

The basic principle, fabrication process and experimental comparison with the aluminum layer will be illustrated and discussed in the following sections.

II. DESIGN & SIMULATION

The actuator consists of two parts that influence its total displacement, the top actuation layer and the bottom layer.

The combined material upper block (the top layer of the bimorph actuator) consists of polymer and silicon layers. The polymer layer is constrained at the interface and it will therefore expand perpendicular to the interface instead of expanding in all three dimensions. [4, 5].

Modeling the polymer layers as infinitely wide in parallel direction to the silicon-polymer interface, a lamellar model [5] is used to express the CTE (α_{con}) and Young's Modulus (E_{con}) of the constrained polymer as:

$$\alpha_{con} = \alpha_{polymer} + \frac{2\upsilon_{polymer}\left(\alpha_{polymer} - \alpha_{plate}\right)}{1 - \upsilon_{polymer} + \gamma\left(1 - \upsilon_{plate}\right)} \qquad (1)$$

$$E_{con} = E_{polymer} \Bigg/ \left[1 - \frac{2\upsilon_{polymer}\left(\upsilon_{polymer} - \eta\upsilon_{plate}\right)}{1 - \upsilon_{polymer} + \gamma\left(1 - \upsilon_{plate}\right)}\right] \qquad (2)$$

978-1-4673-1122-9/12 $31.00 © 2012 IEEE

in which $\eta=E_{polymer}/E_{plate}$ and $\gamma=t_{polymer}E_{polymer}/t_{plate}E_{plate}$. The symbols α, E and υ denote the CTE, Young's modulus and Poisson's ratio, respectively, while the subscripts *polymer* and *plate* denote unconstrained polymer and stiff plate respectively. The terms $t_{polymer}$ and t_{plate} represent the thickness of polymer layers and stiff plates, respectively. Using the formulas for the constrained polymer the CTE and Young's modulus of the silicon-polymer block are given by [5]:

$$\alpha_{block}=\Phi\alpha_{con}+(1-\Phi)\alpha_{plate} \qquad (3)$$

$$E_{block}=\frac{E_{plate}E_{con}}{(1-\Phi)E_{con}+\Phi E_{plate}} \qquad (4)$$

where $\Phi=t_{polymer}/(t_{plate}+t_{polymer})$ is the volume fraction of the polymer in the block.

To calculate the combined material block SU8 polymer [6], silicon and a Φ value of 0.5 are chosen while the bottom layer consists of either aluminum or SiC. The material parameters used are shown in table 1.

Table 1, material parameters used in the calculation

	Bottom layer		Top layers	
	Aluminum	SiC	SU8	Silicon
α(ppm/K)	23	2.77	150.7	2.6
E(GPa)	69	450	3.2	130
υ	-	-	0.33	0.28

The displacement of the cantilever can be estimated based on Timoshenko's calculation [7]:

$$d=\frac{3L^2\left(\alpha_{block}-\alpha_{bottom}\right)(1+1/m)^2\,\Delta T}{\left(h_{block}+h_{bottom}\right)\left[3(1+1/m)^2+(1+1/mn)\left(1/m^2+mn\right)\right]} \qquad (5)$$

where $m=h_{block}/h_{bottom}$ and $n=E_{block}/E_{bottom}$. The term h_{block} is the thickness of the combined material block. α_{bottom}, h_{bottom} and E_{bottom} are the CTE, thickness and Young's modulus of the aluminum bottom layer, respectively. The bending stiffness and force can be calculated as [8]:

$$stiffness=\frac{width\ h_{block}h_{bottom}^2E_{block}}{4L^3\left(1+mn\right)}\cdot$$
$$\left[4+6m+4m^2+nm^3+1/mn\right] \qquad (6)$$

$$force=stiffness\times d \qquad (7)$$

Using equations 5, 6 and 7 the displacement and force of the actuator with either aluminum or SiC bottom layers can be calculated. The calculations are done considering a length of 480 μm, a width of 160 μm, a thickness of the bottom layer of 3 μm and a temperature change of 80°C. Figure 2 and 3 show the results for both layers versus the thickness of the actuation layer. The effect of the bottom layer thickness can be calculated in a similar way.

From these results it is clear that the SiC bottom provides both a better displacement and applied force which can be

accredited to the lower CTE and higher Young's modulus. These two factors contribute to both a higher displacement and stiffness and thus also increase the applied force.

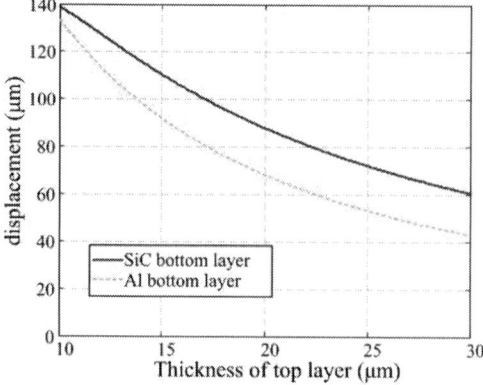

Figure 2: Calculated out-of-plane displacement versus the thickness of the top layer

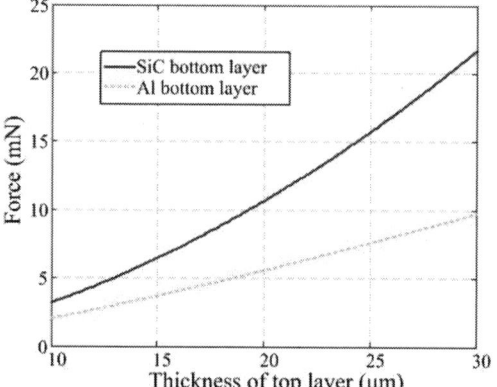

Figure 3: Calculated applied force versus the thickness of the top layer.

III.   FABRICATION

To compare the performance of the actuators with aluminum and silicon carbide bottom layer, respectively, devices with same dimensions, but different bottom layer were fabricated. The process, shown in Fig.2, was developed based on [2].

In this process, the two types of devices were fabricated on the same substrate. The aluminum heater and deep reactive-ion etching (DRIE) mask (silicon oxide) were patterned at the front side of the wafer (Fig.4a), and DRIE is used on the back side of the wafer to define the thickness of the silicon-polymer block. After a thin layer of aluminum (~100 nm) deposition inside the cavity (as an etch-stop layer for the later DRIE process), a silicon carbide layer (1.5 μm undoped) was deposited using plasma enhanced chemical vapor deposition (PECVD) with process parameters as listed in Table.2. The silicon carbide layer was patterned with a spray-coated photoresist, and was dry-etched with the recipe given in Table.3. A 3 μm thick aluminum layer was then deposited inside the cavity by sputtering. A thin layer of PECVD silicon oxide was immediately coated and patterned on the surface of aluminum layer as a mask in the later aluminum wet-etching steps (Fig.4b). After the DRIE and SU8 lithography defining the silicon-polymer block on the front side of the wafer (Fig.4c), the devices were released by wet-etching to partly remove the aluminum support layer (Fig.4d), and to define the aluminum and SiC bottom layers. Finally, the oxide mask on both sides of the wafer was etched away in a buffered HF solution.

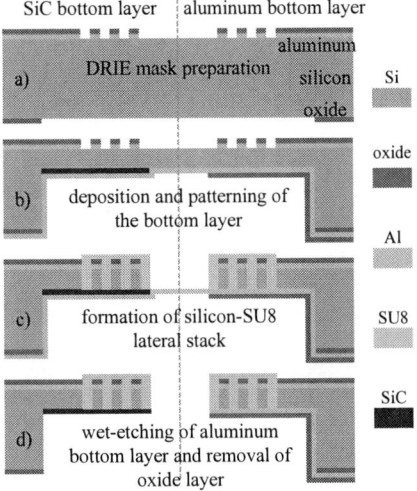

Figure 4: Schematic view of the fabrication process.

Table 2: Main process parameters used in the PECVD SiC deposition

Process parameter	Value
Plasma Power HF (13.56 MHz)	450 W
Plasma Power LF (280 kHz)1	150 W
Deposition Temperature	400 °C
Pressure	1800 mT
SiH$_4$ gas flow	250 sccm
CH$_4$ gas flow	4167 sccm

^{1} a low frequency power source is used to tune the stress of the deposited SiC layer, avoiding high tensile stress.

Table 3: Main process parameters used in the PECVD SiC etching

Process parameter	Value
ICP RF Power	500 W
Bias RF Power	50 W
Temperature	10°C
Pressure	50 mT
SF$_6$ gas flow	20 sccm
O$_2$ gas flow	20 sccm

IV. RESULTS

The fabricated device with SiC bottom layer is shown in Fig. 5. The total structure is 500 µm long and 250 µm wide. The thickness of the top layer is 30 µm, and the thickness of the bottom layer is 3 µm and 1.5 µm, respectively for the aluminum and silicon carbide. The silicon structure, together with the aluminum heater on the top, is completely embedded

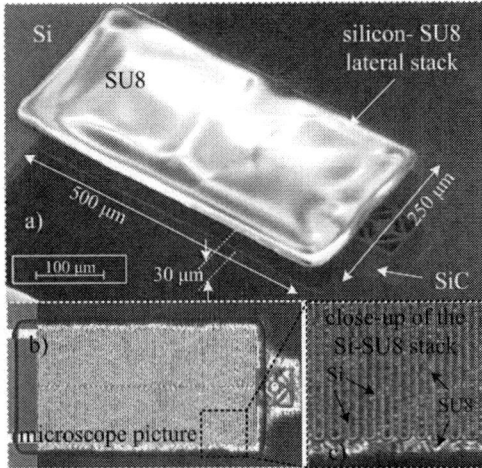

Figure 5: photos of the fabricated out-of-plane actuator: a) the SEM photo, silicon structure is buried in SU8; b) the microscope photo, showing the silicon structure and transparent SU8; c) a close-up of the Si-SU8 stack.

in the SU8 material (Fig.5a). Each silicon beam is 3 µm wide, and the gap between two adjacent silicon beams is 3 µm as well and is filled with SU8 polymer.

Fig. 7 shows the measured displacement on the structure tip versus the applied voltage The out-of-plane displacement of both types of actuators is measured by focusing the microscope on the cantilever tip before applying the voltage, reading the focus wheel, then focusing it again after the out-of-plane actuation and reading the distance the lens moved to get the actuated device in focus, as schematically shown in Fig.6.

The actuator with the SiC bottom layer achieved a maximum of 15.5 µm out-of-plane displacement with an actuation voltage of 4 V, outperforming the 10 µm displacement obtained by the actuator with the aluminum bottom layer at 6.5 V.

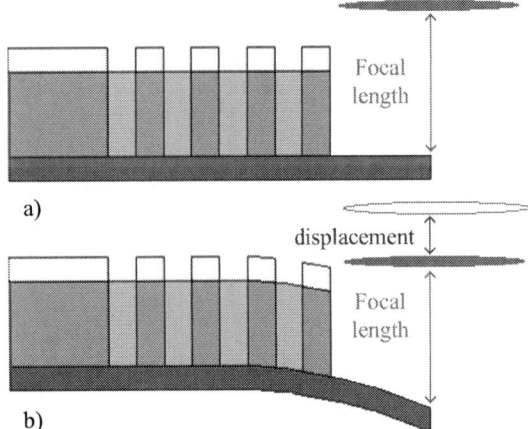

Figure 6: Schematic drawing of the measurement. (a) the cantilever tip in focus before applying voltage; (b) the tip in focus after actuation

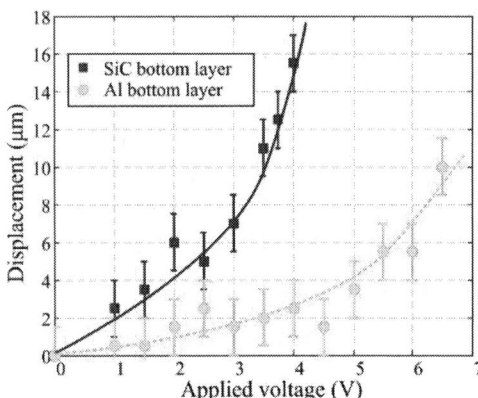

Figure 7: Measured displacement versus the applied voltage.

Apart from the absolute displacement at a certain voltage of the device another point of interest is the displacement at a certain temperature. This is shown in Figs.8&9. These are measured by monitoring the resistance variation of the aluminum heater with the voltage and then combining this with the Fig.7.

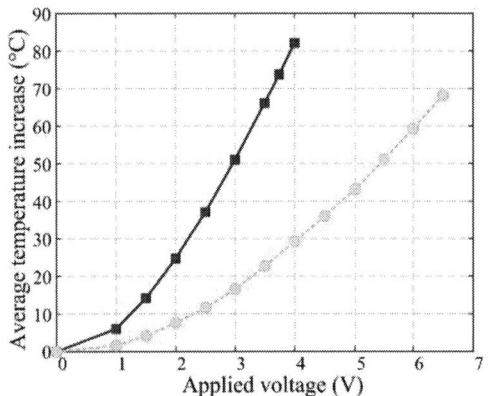

Figure 8: Measured average temperature increase (by monitoring the resistance variation of the heater) versus the applied voltage

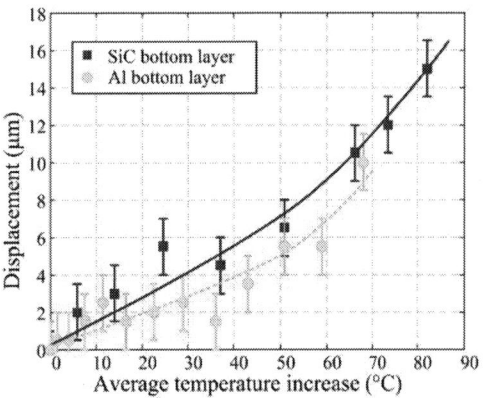

Figure 9: Displacement versus the average temperature increase.

From these two figures it can also be seen that the SiC layer behaved as expected. The devices were measured up to a certain voltage at which the device would not bend back (due to plastic deformation) once the voltage was removed and thus was not fully functional anymore, it is shown that the SiC bottom is indeed more robust then the aluminum, and can operate over a larger displacement range (Fig.7). Also the lower CTE of the material resulted in a larger motion at a given actuation temperature (Fig. 9). In addition, the new device has a lower thermal loss to the substrate, because as shown in fig.8, a higher average temperature is reached under the same applied voltage.

V. CONCLUSION

In this paper SiC as a new material for the bottom layer of an electrothermal bimorph actuator is proposed and tested. The comparison with a previously developed bimorph actuator with Al as bottom material proves the better operation of SiC. A 15.5 µm displacement at 4V for the SiC against 10 µm at 6.5V for aluminum was measured.

Next to larger displacements at lower voltages and temperatures, a reduction in thermal loss to the substrate is obtained, making the total performance of the device with the SiC bottom clearly superior to the one with aluminum.

ACKNOWLEDGMENT

The authors would like to acknowledge the DIMES ICP group for technical contributions, and the Honours Programme Bachelor from the TU Delft funding for financial support.

REFERENCES

[1] T. Chu Duc and etc., Proc. IEEE MEMS 2007, pp. 687-690.
[2] J. Wei and etc., Proc. IEEE MEMS 2008, pp 46-49.
[3] G. K. Lau and etc., Appl. Phys. Lett. 90, 214103 (2007); DOI:10.1063/1.2742599.
[4] T. Chu Duc, G. K. Lau, J. Wei, P. M. Sarro, 5th IEEE Sensor Conference, 2006.
[5] G. K. Lau, J. F. L. Goosen, and F. van Keulen, Appl. Phys. Lett. 90, 214103, 2007.
[6] R. Feng, R.J. Farris, J.Mater. Sci. 37, pp.4793-4799, 2002
[7] S. Timoshenko, J. Opt. Soc. Am. 11, pp. 233, 1925
[8] W.C. Young, R.G. Budynas, Roark's Formulas for Stress and Strain, McGraw-Hill, p. 138, 2002

Study of a Novel Bi-Stable and Easy Integrated MEMS ETBS

Yue Zhao*, Wenzhong Lou, Dongguang Li

*National Key Laboratory of Mechatronics Engineering and Control, Beijing Institute of Technology, Beijing, China
ayan_1113928@163.com

Abstract—**This paper presents a novel bi-stable and integrated single use MEMS blowout switch based on electro-thermal theory. The switch mechanically breaks a metallic line. Switching between the two stable states is accomplished by a DC signal through an integrated heating resistance underneath the electrical lines to be melted. Some key features are that switches are bi-stable, integratable and IC compatible. They are compatible with various voltages operation for different applications, operation in ambient environment may be possible; predicted lifetime is very long and therefore both switches can be used for long life systems. Batch fabrication using planar processing methods is used. The ETBS could be manufacture easily in great amount. The switch can also be applied in some areas of high-energy control such as initiation security and self-control in failure.**

Keywords-ETBS; electro-thermal; MEMS; bi-stable

I. INTRODUCTION

Due to the rapidly growing MEMS initiator market and the needs for smaller, safer and higher integration, more advanced switch are in demand [1]. To be compared with fuse, ETBS has great advantages in huge energy control, its work can be controlled by external command and there are no moveable parts which can improve security of MEMS initiator. ETBS has received much attention as an externally controlled fuse. There were some MEMS ETBS to be reported [2, 3]. But there were some shortcomings of them, such as more power consumption, low heating efficiency, complex process and high cost. In this paper, a novel MEMS ETBS is presented which overcomes those shortcomings.

In this paper, we present a novel bi-stable, easy integrated and low power consumption MEMS Electro-thermal blowout switch (ETBS). This switch is used for special initiator or fuze. In traditional design, there were movable components in switch, which increased the volume and power consumption. The ETBS works by electrifying the underlying resistance to heat the signal line and melt it. The ETBS has small size and low consumption. It also can improve the reliability of special initiator or fuze working in high dynamic environment. The ETBS could be manufacture easily in great amount. The switch can also be applied in some areas of high-energy control such as initiation security and self-control in failure.

II. THEORY AND DESIGN

A. Principle of Operation

This paper introduces the MEMS ETBS which have a structure of six levels. These levels are substrate, bottom insulation, heating resistance, middle insulation, metallic line, top insulation and metallic pads. The basic action process of the ETBS is that metallic line remains conducting in normal state. When connecting a constant current source with certain amplitude or pulse source to the heating resistance of the ETBS, the resistor is heated rapidly. Role to heat conduction, the heat is passed to the metallic line which was an aluminum line deposited on top of middle insulation (Fig.1 and Fig.2). When the temperature reaches the melting temperature of metallic line (660 °C), which is causing the ETBS rupture.

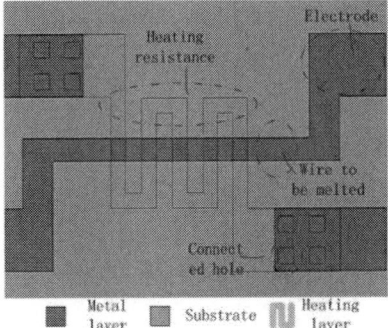

Fig.1. The structure of the MEMS ETBS

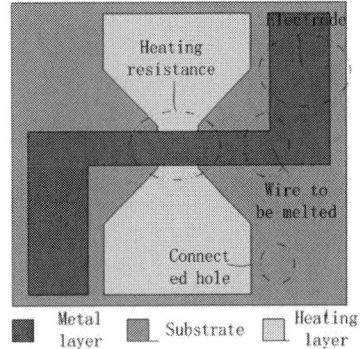

Fig.2. The other structure of the MEMS ETBS

B. Working Principle of ETBS

The ETBS working based on Electro-thermal theory. The resistance of ETBS can be expressed by equation 1.

$$R_{\blacksquare} = \frac{\rho}{t_x} \qquad (1)$$

Here : $R_{\blacksquare}$—— The sheet resistance of fuse;

ρ——Metal resistivity;

t_x——the thickness of metal band.

According to Ohm's law, the resistance of metal band can approximate as equation 2:

$$R = R_{\blacksquare} \frac{1}{w} \qquad (2)$$

Here : l——the length of metal band ;

W——the width of metal band。

Here the metal pad and the wire resistance are ignored. When the heating resistance connects with a constant voltage source, Thermal power is produced as equation 3.

$$P = \frac{U^2}{R} \qquad (3)$$

The temperature and heat have the following relationship:

$$Q = C\Delta T \qquad (4)$$

Here : C—— metal specific heat ;

Q——the heat absorb by metal bands。

Type the above 4 equations:

$$\frac{U^2 t_x w}{\rho l} \Delta t = C\Delta T \qquad (5)$$

According to equation 2-5, if the process and material are determined without considering heat radiation and heat conduction. When the phase transition of the metal band doesn't occur, the temperature of the metal strip increases with the ratio of length to bandwidth; when the phase transition of the metal band occurs, the material parameters will be changed significantly, but the regular of temperature still follows the above rules. As, in the air, the thermal conduction, convection and thermal radiation exist at the same time, the metal band follows the following two equations simultaneously.

$$Q_T = \frac{\Delta T}{t_x / \lambda w l} \qquad (6)$$

Here : QT——conduction heat between the device;

λ——Thermal conductivity for the substrate.

$$Q_c = hlw\Delta T \qquad (7)$$

Here : QC——convection heat loss of the device and the air ;

h——Heat convection coefficient for air.

If take the heat conduction and heat convection correction factors consideration, the equation 2-5 should be modified as follows:

$$\frac{U^2 t_x w}{\rho l} \Delta t - \frac{\Delta T}{t_x / \lambda w l} - hlw\Delta T = C\Delta T \qquad (8)$$

C. Size and Simulation

The ETBS has following characteristics: (1) within the range of safety margin as large as possible, (2) good high dynamic characteristic, (3) easily controlled. To combine with equation 2-8, the following conclusions can be made: within security and time allowed, devices with low resistivity have batter performance. To find reasonable design parameters for the ETBS, COMSOL multi-physics analysis software is applied.

We have done some simulations on the characteristic of the ETBS with various parameters. Materials of heating resistance layer are poly-silicon and titanium. In the same time, the heating resistor shapes of a broken line and the box-shaped were considered. Simulation model and the results are shown in Fig. 3a and Fig. 3b.

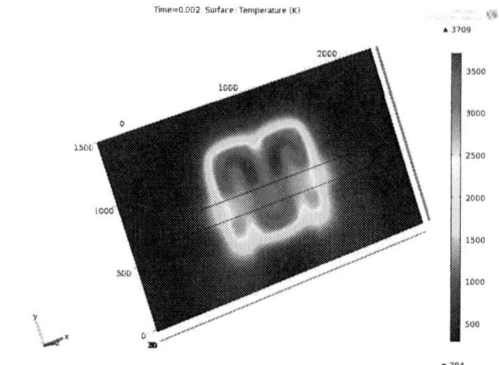

Fig.3a. simulation result of the ETBS with the heating resistor shapes of a broken line

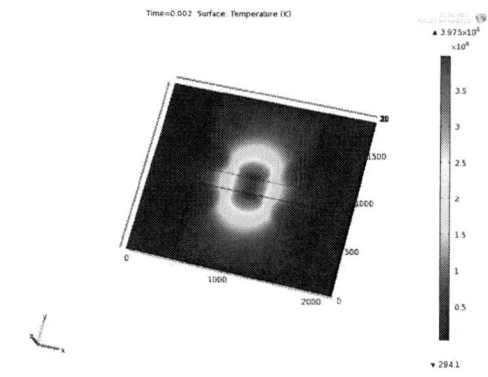

Fig.3b. simulation result of the ETBS with the heating resistor shapes of box-shaped

Through simulation, we can find that there is good performance in reducing blowout time and low power consumption.

III. FABRICATION

The MEMS ETBS is composed by a thin heating resistor layer, a metal signal layer and another layers used to insulate or connect different layers. The switch was produced on a silicon substrate through oxidation, photolithography, CVD, ion implantation, sputtering, and wet etch.

The ETBS was fabricated with MEMS Planar process. The process of fabrication is completely planar and IC compatible. The 350 um thick (1 0 0) oriented 4 in. silicon wafer is thermally oxidized to a depth of 0.4um at 1150 °C.

A. Poly-Silicon Heating Resistor

In this paper, we use two material to fabricate heating resistor. One is poly-silicon, the other is titanium. A 0.75um thick poly-silicon layer is deposited by LPCVD at 605 °C and N doped by the diffusion is illustrated in figure 4. The process for fabrication of ETBS of phosphorus. The resulting $R_\blacksquare$ is 120 Ω / $\blacksquare$. The poly-silicon layer is patterned using photolithography to define heater resistors and etched by reactive ion etching (RIE). In our process, we do ion implantation with a max concentration of $1 \times 10^{16}/mm^3$.

B. Titanium Heating Resistor

titanium was used as another material of heating resistor for low voltage application. A 0.7um thick titanium layer is then deposited by magnetron sputtering and patterned using photolithography and removed in a HF bath everywhere except on the heater resistance.

After resistance formation, a 0.7um thick oxide layer is deposited by LPCVD. The next step consists of making the metal layer. A 0.7um thick layer of aluminum is evaporated. The thin aluminum layer makes it possible to reduce the volume of aluminum to be melted.

In some applications, an unoxidized aluminum line is often used to transfer energy. An oxide layer is needed to protect aluminum unoxidized. In order to ensure the release of aluminum vapor, you need to create holes on oxide layer which cover the aluminum.

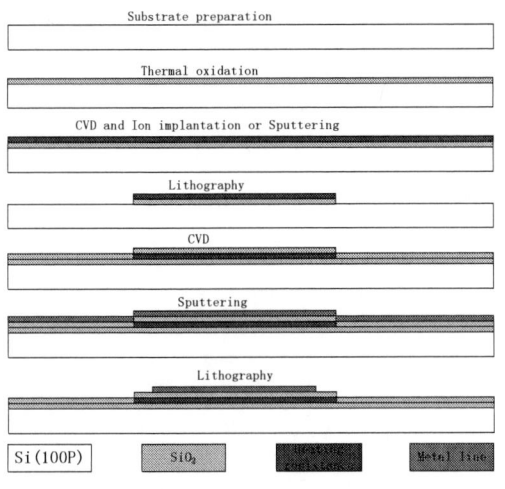

Fig.4. Process of MEMS ETBS

IV.   RESULTS

Through simulation and fabrication, some performances of ETBS can be obtained. From Fig. 2a, simulation results show that ETBS could achieve the required operating temperature in 2ms under 5V voltage and had a good high dynamic performance (where the electric conductivity of heating resistor is about 3×10^5S/m). Focus on relationship between steady-state temperature and electric conductivity of ETBS,

we do lots of simulations. Simulation result is shown in Fig. 5 as a curve. We can find that there is an excellent linear relationship between them. The ETBS is a high reliability, bi-stable one-time switch, which is often used as a one-time switch or a high reliability fuse of some circuits. Because the ETBS has good linearity, we can choose devices with different electric conductivity for various applications.

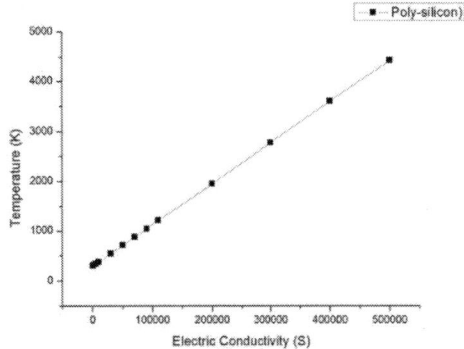

Fig.5. relationship between steady-state temperature and electric conductivity

Different heating area can bring a different heating effect. Three kinds of ETBS of different heating area were simulated. We found that for same equivalent heating resistor, ETBS of larger heating area can achieve a higher steady-state temperature (table 1).

TABLE I.   STEADY-STATE TEMPERATURE OF DIFFERENT HEATING AREA

Electric conductivity (S/m)	Area(um)		
	100×100	100×80	100×60
	Temperature (K)		
4×10^5	3602	3290	2991
3×10^5	2776	2541	2316
2×10^5	1951	1792	1642

From fig.2a and 2b, such conclusion can be got: ETBS of box-shaped resistor has better performance than broken line shaped. The ETBS of box-shaped resistor can be seen as a semiconductor bridge initiator which drives flyer. The working process of this type ETBS is equivalent to initiation. In this process, the metallic line was destroyed not only by electro-thermal effective, there was High-temperature plasma to be generated. The ETBS of box-shaped resistor has less transfer time from one stable state to the other. This kind ETBS has less power consumption than broken line shaped and the action time is down to tens microseconds.

The ETBS, with different heating resistance materials and shapes are being produced. But all process well not be completed until The end of December. So far, the preproduction of ETBS in which heating resistance material was poly-silicon (figure 5) has been completed. The size of this chip is $2 \times 3mm^2$, which was fabricated on a 4 inches silicon wafer. Hundreds of devices were produced with different ships, resistor value. Table 2 shows the manufacturing parameters of them.

TABLE II.   THE MANUFACTURING PARAMETERS OF ETBS WITH POLY-

SILICON RESISTOR

type	Resister values	Heating arm number	Heating arm area
A	2.28k	3	100*150
B	3.0k	4	100*150
C	3.72k	5	100*150
D	3.0k	5	100*150
E	3.0k	5	80*150
F	3.0k	5	60*150

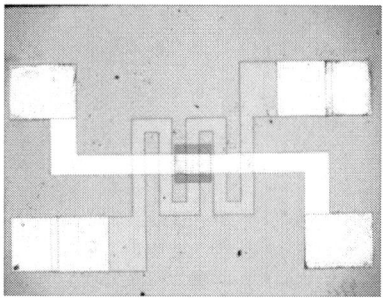

Fig.6 A sample of the ETBS

The switch has two stable states, on and off. It is mainly used to keep initiator safe as a fuse. The fuse function circuit is shown in figure 6. Usually the switch is closed as a part of exploding circuit. System is in working mode. After a DC current flow through the heating resistance, the temperature of resistance increased rapidly until it above the melt point of the metal line. The metal line will be mechanically broken. The switch turns open and the exploding circuit disconnect with MEMS initiator. Exploding system became invalid. The performances of ETBS have been tested.

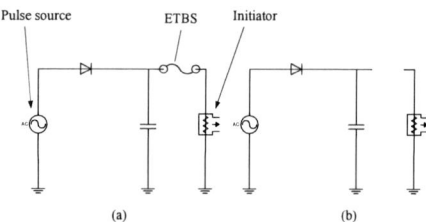

Fig.7 Fuse functions of ETBS with initiator

Through test, the ETBS with poly-silicon did not take effect under 60 V DC voltage because the resistor value was too big to act. The experiment with bigger excitation didn't to be done for safe.

V.   CONCLUSION

The detailed design, sizing and fabrication process of MEMS ETBS have been presented. Through above analysis, such conclusions can be got.

1) Based on the different shape, material and resistance of heating resistance, there are differences in the performance of ETBS.

2) The performance of ETBS is closely related to the resistivity of heating resistance. The ones which have lower resistivity have good performance in low voltage.

3) The ETBS of box-shaped resistor can destroy metallic line batter. With the same heat resistance, the ETBS of a more focused heating area have better heating effect.

The proposed ETBS have valuable advantages for redundancy applications, when electrical switching is necessary once and after a long storage time.

The other devices with low resistivity are processing, there will be prefect performances.

ACKNOWLEDGEMENTS

The authors would like to thank Junfeng Li and Guilei Wang from the institute of microelectronics of the chinese academy of sciences for their cooperation on the ETBS processing.

REFERENCES

[1] Pezous. H. , Rossi. C. , Sanchez. M. , Mathieu. F. , Dollat. X. , Charlot. S. , Salvagnac. L. , Coné de´ra. V. Integration of a MEMS Based Safe Arm and Fire Device. Sensors And Actuators: A Physical, v159(2010), n2, p157-67.

[2] Pennarun. Pierre, Rossi. Carole, Estève. Daniel, Colin. René David. Single Use, Robust, MEMS Based Electro-Thermal Micro-switches for Redundancy and System Reconfiguration. Sensors and Actuators, A: Physical, v 136(2007), n1, p273-281.

[3] W. R. Nie, Z. W. Xi, N. Gong. A MEMS Igniter Design for Fuze Safety & Arming. 3rd International Conference on Mechanical Engineering and Mechanics, Beijing, China, OCT 21-23, 2009, p44-50.

Rapid Thermal Plasma Deposition of Transparent Nanocrystalline ZnO Thin Films and the Effects of Annealing

Kwok Siong Teh[*], Joachim Pedersen, Heather Esposito
[*]School of Engineering, San Francisco State University, USA
kwok.siong@gmail.com

Abstract— **Conductive, undoped zinc oxide nanocrystalline thin film with predominant c-axis orientation is prepared on crystalline and amorphous substrates using a rapid, one-step ambient-pressure, thermal plasma chemical vapor deposition process. Nonporous and conformal zinc oxide films can be prepared at temperature as low as 160°C, with an average grain size of 25 nm. Scanning electron micrographs indicate a growth rate of 15~50 nm/min, depending on factor including source temperature, deposition temperature, and pressure. X-ray diffraction shows a predominant (002) grain orientation that is independent of the substrate's crystallinity. For films with thickness of 200 nm, the average electrical conductivity ranges from 60-910 S/m. The results demonstrate the potential of thermal plasma CVD for the rapid synthesis of conductive zinc oxide film at ambient condition.**

Keywords- zinc oxide, transparent nanocrystalline film, thermal plasma chemical vapor deposition, annealing, nanorods

I. INTRODUCTION

The increasing demand for solar cells, flat-panel displays, and touch screens has driven the impetus to develop low-cost, mass produced transparent oxide electrode (TCE) materials in recent years. Among the TCEs, which include tin oxide (SnO_2), indium-doped tin oxide (ITO), and zinc oxide (ZnO) ZnO and doped ZnO such as aluminum-doped ZnO (AZO) is a choice material owing to its high electron mobility (Hall mobility of up to $200 cm^2/V-s$), the abundance and low cost of Zn, and the ease of forming functional ZnO at relatively benign processing conditions. In spite of these desirable attributes, most current methods of synthesizing ZnO thin films – including rf or dc magnetron sputtering [1], metal organic chemical vapor deposition (MOCVD) [2], spray pyrolysis [3], pulsed laser deposition [4], thermal evaporation [5], hydrothermal [6], and sol-gel processes [7] – often require substantial vacuum, expensive consumables (e.g. diethyl zinc, dimehtyl zinc, ZnO sputter target), catalyst (e.g. gold), and lengthy synthesis time.. While solution-based methods—such as hydrothermal and sol-gel—generally require much lower processing temperature (~100°C), vapor phase methods such as sputtering, thermal evaporation, and MOCVD tend to produce films with higher crystallinity and better electrical conductivity. Nevertheless, because vapor phase methods are usually performed at much higher temperatures (80°C and above) and require high vacuum ($10^{-4} \sim 10^{-5}$ torrs), the process is not CMOS-compatible. A low-temperature vapor phase synthesis method is therefore highly desirable from a processing and manufacturing standpoint.

To address such challenges, this paper reports a rapid, ambient-pressure, one-step, thermal plasma chemical vapor deposition (CVD) process for depositing conformal, non-porous nanocrystalline ZnO thin film on various crystalline and amorphouse substrates ranging from silicon to polyamide. Thermal plasmas—high power discharges—can be produce at or near ambient pressure using high-power sources, such as RF induction plasma system [8]. Previous research has shown that inductive heating can provide a useful and efficient means to rapidly introduce large amount of heat for nanomaterial synthesis [9-11]. This is attributed to the high enthalpy of RF induction plasma and its being capable of high-frequency (13.56MHz) switching capabilities, making it well suited for applications where high-temperature and high-heating rate heat treatments are needed [12]. In particular, RF induction plasma systems have shown industry-scale utility for synthesis of high-quality nanoparticles at ambient pressures [13]. In induction plasma nanoparticle synthesis methods, concurrent introduction of complex liquid, gas, or powder precursors enables a one-step, cost-efficient, and time-efficient synthesis. During synthesis, the reagents are introduced into a plasma-entrained flow, become fully ionized, and condense as droplets as they leave the plasma region. This paper reports ZnO nanocrystalline film synthesis results from a thermal plasma CVD system developed in house. Unlike MOCVD which uses diethyl zinc as a source for ZnO nucleation, the only reagents used here are solid zinc and oxygen gas.

II. EXPERIMENTAL DETAILS

A. ZnO Nanocrystalline Film Synthesis

Fig 1 shows the schematic diagram of a 13.56 MHz RF thermal plasma CVD system used to synthesize conformal ZnO nanocrystalline thin film. The synthesis system consists of two main components: (i) source and (ii) growth substrate, both of which are contained within a sealed quartz chamber flushed with argon and oxygen at a ratio of 99.97% to 0.03% at a flow rate of 301 sccm. The source is made up of solid zinc (99.99% purity) contained within a nickel heating chamber.

The nickel heating chamber has an emission orifice—a single opening—on its upper cover. When the RF power is turned on, rapid switching of current in the induction coil heats up the argon-oxygen gases to form plasma. Rapid heat transfer from the plasma to the nickel heating chamber heats up the solid zinc contained within the chamber. Solid zinc is ionized to form Zn ions, which are ejected from the emission orifice in the form of a Zn plasma jet. As the Zn ions are transported away from the orifice toward the fringe, they react with oxygen in the synthesis chamber to form ZnO nanoparticles. These nanoparticles that are formed in-flight supersaturate in the boundary layer of the growth substrate and deposit on the growth substrate surface as ZnO nuclei, forming the foundation for subsequent deposition of ZnO nanocrystalline films. The growth substrate used ranges from silicon (100), mica (muscovite), fused quartz, c-plane and a-plane sapphire, soda lime glass, ITO, to polyamide (Kapton®). The deposition rate (15-50nm/min) is tightly controlled by a closed-loop temperature control algorithm where the output RF power is modulated by the source temperature.

B. Post-Process Film Characterization and Heat Treatment

The surface morphology, film thickness, and crystal dimensions of the synthesized ZnO nanocrystalline films are characterized by scanning electron microscopy (SEM) on a Zeiss Ultra 55 that is equipped with a Schottky field emission gun. Elemental analysis is conducted using an Oxford energy dispersive x-ray probe. Film crystallinity is investigated using an X-ray diffractometer (Bruker D8 ADVANCE) with Cu-Kα radiation ($\lambda = 1.54178$ Å) and a scanning range of 2θ between 24° and 100°. Electrical conductivity measurement is conducted using a four-point probe and transmittance of the as-deposited film is measured using a Lambda UV-Vis spectrometer (Perkin Elmer) with an integrating sphere. The spectra are collected in the 200-800 nm spectral range. Thermal annealing of samples is performed in a tube furnace (MTI GSL-1100X) at 300 sccm of argon flow at temperatures ranging from 750°C to 950°C for 1 hour.

III. RESULTS AND DISCUSSION

The morphological and dimensional properties of the ZnO nanocrystalline thin films are found to be dependent on factors including but not limited to substrate temperature, source temperature, and extent of thermal annealing. On the contrary, the types of substrate on which the ZnO films are deposited do not seem to have a substantial effect on the morphology of the ZnO films.

A. General Properties of As-Deposited ZnO

Fig. 2(a) shows ZnO film deposited using a symmetric heating profile where the rates of heating and cooling are identical. Using an RF power of 100W, the maximum source temperature attained by the nickel heating chamber is 570°C and the corresponding deposition temperature experienced by the substrate is 330°C at a distance of 15 mm away from the source. The ZnO hence deposited on p-silicon(100) is nonporous and highly conformal, where the measured grain sizes are distributed between 16 nm to 88 nm, with a mean of 40 nm ± 2 nm. The mean thickness of the film is 42 nm ± 0.5 nm. For ZnO with 200 nm nominal thickness, the electrical conductivity is measured to be between 60 – 910 S/cm, whereas the percentage optical transmittance is above 80%.

B. Influence of Substrate Temperature

Substrate temperature plays an important role in determining the mode of formation of nuclei and subsequently the grain size, crystallinity, and material properties of the film. For thermal plasma CVD synthesis process described in this paper, we have found ZnO to deposit on substrates at a temperature as low as 160°C when the growth substrate is placed 25 mm from the source. Distances of 5mm to 25mm from the plasma jet successfully produce ZnO nanocrystalline thin films. The further the growth substrate is away from the source, the lower the surface temperature of the growth substrate and hence the smaller is the grain size—a typical Volmer-Weber growth. This process allows a range of

Fig. 1. Thermal plasma chemical vapor deposition system for depositing ZnO nanocrystalline thin films. (Left) Anatomy of the synthesis setup consisting of solid zinc enclosed in a nickel heating chamber. When RF is activated, the nickel heating chamber is resistively heated by the RF magnetic flux and by the inductively coupled argon/oxygen plasma. Zinc is quickly vaporized and ejected from the orifice. Subsequently, the zinc vapor reacts with oxygen to form ZnO, which deposits on the growth substrate. (Center) Synthesis chamber showing the position of the nickel heating chamber in relation to the induction coil. (Right) During synthesis, zinc ions form a plasma jet and escapes from the emission orifice.

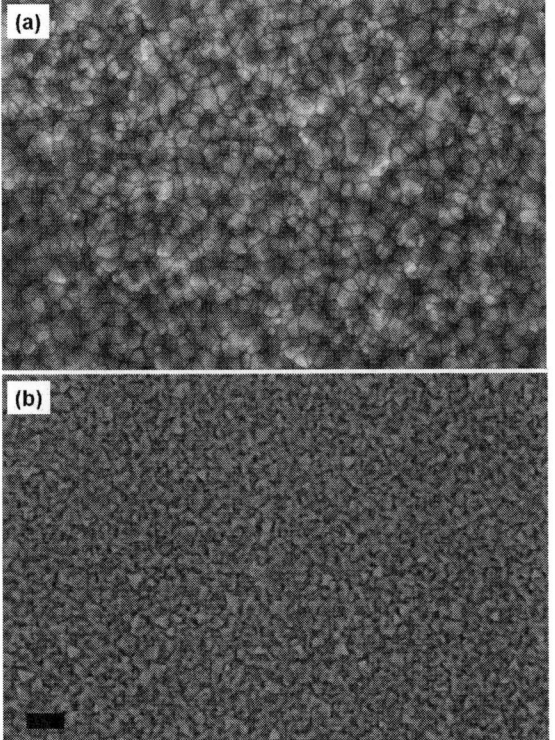

Fig. 2. (a) High substrate temperature (320°C) favors the formation of larger grains with an average diameter of 40 nm. (b) Smaller grains with an average diameter of 25 nm form at lower substrate temperature (160°C). Scale bar = 100 nm.

polycrystalline film morphologies to be synthesized. In general, for refractive materials such as ZnO, high substrate temperature favors the formation of few but large nuclei at the initial stage. Subsequent growth of the nuclei gives rise to a polycrystalline film made up of large grains with clearly defined grain boundaries. For application such as dye-sensitized solar cells and transparent conducting electrodes, the goal is to grow ZnO at low temperatures—as low as 100°C—so as to be able to deposit the films on plastic substrates such as polyethylene terepthalate (PET).

Fig. 2 (b) shows the ZnO film deposited at a substrate temperature of 160°C. Compared to the film shown in Fig. 2 (a), the average grain size is smaller (mean of 25 ± 2 nm) and the grains are less defined. Such a decrease in grain size is attributed to lower surface mobility, hence reduced surface diffusion during the formation of nuclei.

C. Influence of Source Temperature

For a high rate of deposition, the RF power and plasma intensity must be high so that Zn droplets do not form in transit, while crucible temperature must be maintained well below the boiling point of Zn. This is achieved through issuing a saw toothed temperature profile for the crucible temperature control. Controller RF output pulses to a high power to maintain the appropriate crucible temperature rate increase. As an upper temperature limit is reached RF power is automatically reduced allowing the crucible to cool to a predetermined temperature. Further pulses can be programmed until the zinc source is completely depleted.

We find clear indication in the experimental data that correlates the total durations of the Zn source heated above a threshold temperature (420°C) in the crucible to the as-deposited film thicknesses, grain structures and sizes. Growth substrates exposed to a triangular source temperature profile at temperatures above 420°C shows a positive correlation between the heating durations (above 420°C) and the film thicknesses; and to a lesser extent, the number of pulses and the film thicknesses, as shown in Figure 3 (a), (b), and (c). As the number of pulses (from 1 to 3 pulses) and total synthesis time above 420°C increase, the nominal thicknesses of the films increase proportionately from 25 nm (1 pulse), 70 nm (3 pulses), to 110 nm (5 pulses). Another factor that may seem to influence film thickness is the resident time the source temperature stays at the peak temperature (570°C). We compare two samples, Figure 3 (a) and (d), where each has an identical triangular heating profile. The sample in Figure 3 (a) has a 1-second resident time at 570°C while that of Figure 3 (d) has a 5-second resident time at 570°C. Results show that there is no significant difference in the thicknesses between these two samples—sample in Figure 3 (a) has a nominal thickness of 25 nm, while sample in Figure 3 (d) has a nominal thickness of 22nm. This indicates that—while the number of triangular pulses at temperature at the peak temperature of 570°C seemingly influences the thicknesses of the film—the total duration the source temperature stays above 420°C plays a more critical role in influencing the thicknesses of the films. As shown, there is no major differences in the thicknesses in spite of the fact that sample in Figure 3 (d) has 4 more seconds at peak temperature of 570°C. On the other hand, when comparing Figures 3 (d) and (e), the effect of heating duration above a critical temperature of 420°C becomes even more obvious—longer synthesis time leads to thickening of the film (to 57 nm) as shown in Figure 3 (e), the sample of which is exposed to 75 s longer than sample in Figure 6 (a) at 570°C.

It is evident that the thickness of the film is predominantly influenced by the duration of heating at and above 420°C and to a number of triangular pulses, and to a much lesser extent by the resident time at the peak temperature.

D. Influence of Thermal Annealing

To better understand the state of the as-deposited film we have annealed films at a variety of temperatures ranging from 750°C-950°C in argon (Fig. 4). Samples were annealed for 90 minutes in a tube furnace supplied with 100sccm argon under a vacuum of 130torr. SEM has shown several interesting morphological changes resulting from the annealing process. Application of moderate heat at 750°C results in a segregation into larger grains with greater definition at the grain boundaries. At 800°C surface height variation increases, grain size increases further.

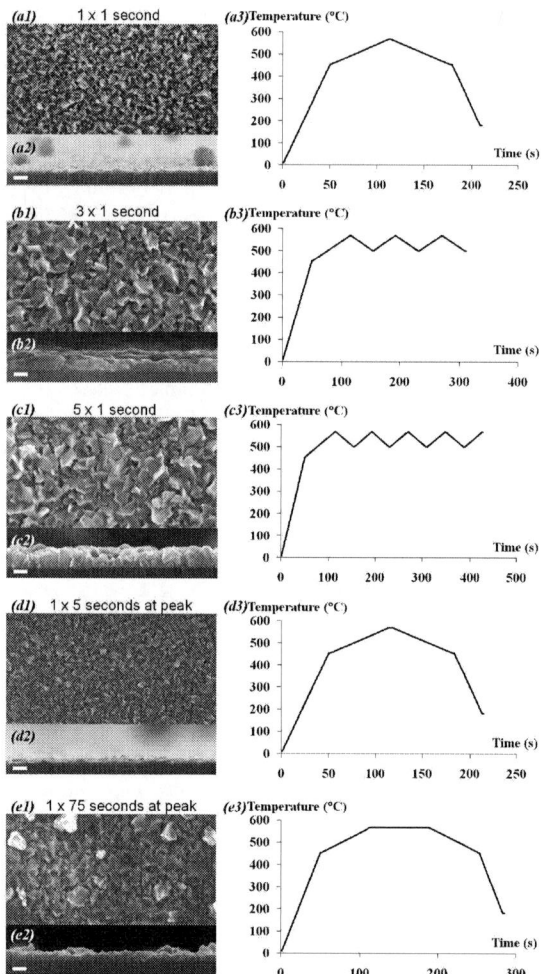

Fig. 3. SEMs and deposition temperature profiles of ZnO films deposited using (a1-a3) 1x, (b1-b3) 3x, (c1-c3) 5x triangular pulses that have a resident time of 1 second at the maximum source temperature of 570°C. The corresponding nominal thicknesses are (a2) 25 nm, (b2) 70 nm, and (c2) 110 nm. 3(d1-d3) and (e1-e3) show the SEMs of ZnO films deposited using 1x triangular pulses with the same rise and decay profiles, but with modified resident times at the peak temperature that are fixed at (d3) 5 seconds and (e3) 75 seconds. The respective thicknesses of films deposited using these profiles are (d2) 22 nm and (e2) 57 nm. Scale bar = 100 nm.

Exposed normal facets, identified to be c-planes, continue to grow and appear to be consuming adjacent ZnO. As temperature is further increased, nascent nanowire stubs form as growth continues in the c-axis. We hypothesize this behavior to be attributed to grain boundary and bulk diffusion that occurs at elevated temperature, which causes ZnO to restructure and grow in the plane with the highest surface energy (002). Such growth mechanism favors the formation of

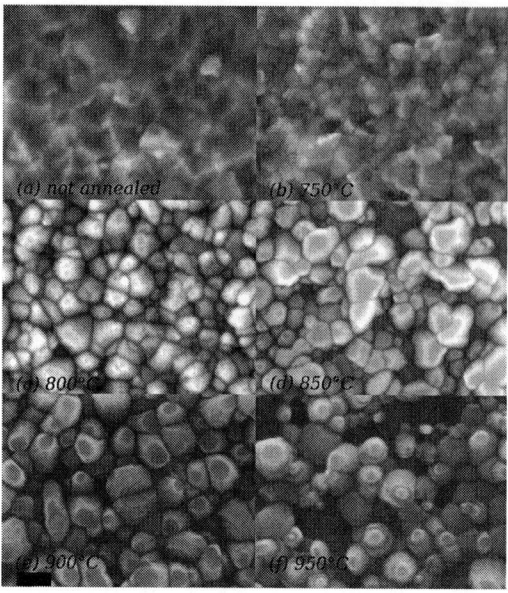

Fig. 4. Annealed samples from films of initially identical morphology and average grain sizes show increasing restructuring of film texture with higher annealing temperatures. Scale bar = 100nm

a stable, faceted large crystal with a surface texture dominated by (002).

E. Influence of Substrate Type

We have deposited ZnO nanocrystalline films on various substrates—crystalline or amorphous, polymer or ceramic and conductive or insulating. Our process appears to be largely deposition surface-independent for materials capable of withstanding the growth conditions and process temperature, which can be as low as 160°C. We have successfully grown ZnO thin films on Si (100), fused quartz, soda lime glass, muscovite, c- and a-plane sapphire (Al_2O_3), and the common polymer polyimide (Kapton®). Film coverage on these substrates is continuous where the grains are contiguous with no observed porosity under the SEM. During visual inspection of the cross section of the film under an SEM, we observe no epitaxial growth of ZnO on both Si (100) and sapphire (a- or c-plane) due to large lattice mismatches between the crystals, hence ruling out the epitaxial growth mode. Films deposited on polyimide have a slightly different nanocrystalline structure as observed using SEM. Energy dispersive x-ray spectroscopy confirms the presence of ZnO on polyimide in Figure 5. ZnO on polyimide is conductive enough that SEM imaging can be achieved without an additional conductive coating such as carbon or gold/palladium.

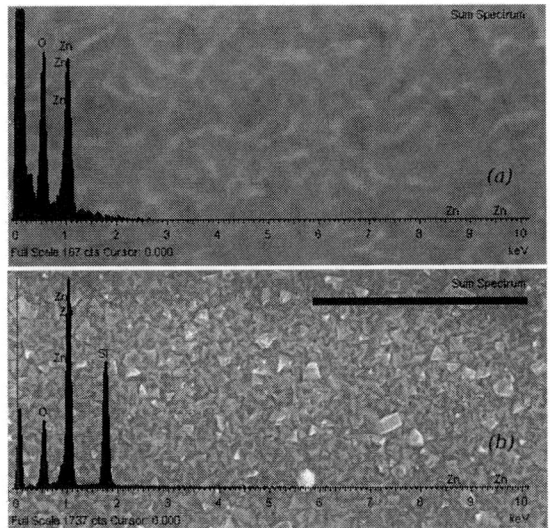

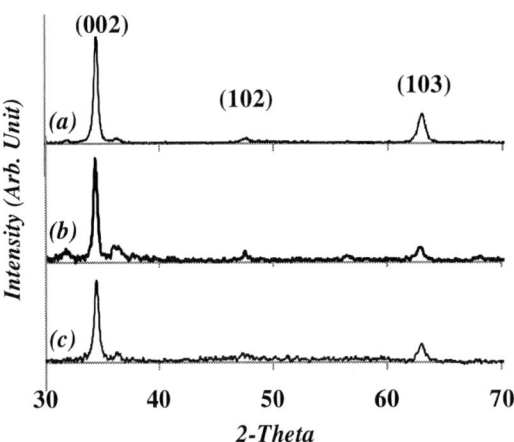

Fig. 6. XRD of as-deposited films on (a) Si(100), (b) c-plane Al$_2$O$_3$, (c) fused quartz.

Fig. 5. Energy dispersive x-ray spectroscopy of ZnO on (a) polyimide and (b) silicon substrates. Scale bar = 2μm.

Figure 6 shows the x-ray diffractographs of ZnO films deposited on crystalline materials (Si(100) and c-plane Al$_2$O$_3$) and amorphous materials (fused quartz). For each substrate type, the growth of the ZnO nanocrystals is predominantly c-axis-oriented, yielding largely (002) peaks. ratios of the intensity peaks of (002)/(102) and (002)/(103) planes are between 10~20 and 5~7, respectively. SEM shows the presence of (102) and (103) peaks to be largely attributed to the tapering of the growing faces of the crystals. The morphologies of as-deposited ZnO appear be independent of substrate crystallinity—a result of lattice mismatches between ZnO and the growth substrates used.

IV. CONCLUSIONS

We have successfully demonstrated a one-step, ambient-pressure, catalyst-free synthesis method of depositing c-axis oriented, conformal nanocrystalline ZnO film on crystalline and amorphous substrate. SEM and XRD indicates film is highly conformal with evenly distributed grain sizes and a strong presence of (002) planes. The obtained results demonstrate the high potential of the plasma-assisted inductive heating technique for the rapid synthesis of zinc oxide at ambient condition.

ACKNOWLEDGMENT

The authors thank (i) Dr. Andrew Ichimura for his assistance with x-ray diffraction, UV-vis spectrometry and fruitful discussions, (ii) Tom Franco and Richard Moore for helping to design and machine mechanical fixtures used in the experiments, (iii) Curtis Hilger for work on the implementation and tuning of the control system, (iv) Mark Brunson for a portion of the data collection work. Acknowledgment is made to the Donors of the American Chemical Society Petroleum Research Fund for partial support of this research under grant #49524-UNI 10.

REFERENCES

[1] A. Tanusevskia and V. Georgieva, "Optical and electrical properties of nanocrystal zinc oxide films prepared by dc magnetron sputtering at different sputtering pressures", Appl. Surf. Sci., vol. 256, pp. 5056-5060, 2010.

[2] S. T. Tan, B. J. Chen, X. W. Sun, and W. J. Fan , "Blueshift of optical band gap in ZnO thin films grown by metal-organic chemical-vapor deposition", J. Appl. Phys., vol 98, pp. 013505/1-013505/5, 2005.

[3] M. Subramanian, M. Tanemura, T. Hihara, V. Ganesan, T. Soga, T. Jimbo, "Magnetic anisotropy in nanocrystalline Co-doped ZnO thin films", Chem. Phys. Lett., vol. 487, pp. 97-100, 2010.

[4] Z. Xu, H. He, L. Sun, Y. Jin, B. Zhao, and Z. Ye "Localized exciton emission from ZnO nanocrystalline films", J. Appl. Phys., vol. 107, pp. 052524/1-052524/5, 2010.

[5] S.J. Chen, Y.C. Liu, J.G. Ma, D.X. Zhao, Z.Z. Zhi, Y.M. Lu, J.Y. Zhang, D.Z. Shen, X.W. Fan "High-quality ZnO thin films prepared by two-step thermal oxidation of the metallic Zn", J. Crys. Gro., vol. 240, pp. 467-472, 2002.

[6] Shuzhi Li, Shengming Zhou, Hongxia Liu, Yin Hang, Changtai Xia, Jun Xu, Shulin Gu and Rong Zhang , "Low-temperature hydrothermal growth of oriented [0001] ZnO film", Mat. Lett., vol. 61, pp. 30-33, 2006.

[7] Y. Zhang, B. Lin, X. Sun, and Z. Fu , "Temperature-dependent photoluminescence of nanocrystalline ZnO thin films grown on Si (100) substrates by the sol–gel process", Appl. Phys. Lett., vol. 86, pp. 131910/1-131910/3, 2005.

[8] M. Shigeta, A. B. Murphy, "Thermal Plasmas for Nanofabrication", J. Phys. D: Appl. Phys., vol. 44, pp. 1-16, 2011.

[9] L. Luo, B. Sosnowchik and L. Lin, "Room temperature fast synthesis of zinc oxide nanowires by inductive heating", Appl. Phys. Lett., vol 90, pp. 093101, 2007.

[10] B. Sosnowchik, L. Lin, "Rapid synthesis of carbon nanotubes via inductive heating", Appl. Phys. Lett., vol. 89, pp. 193112, 2006.

[11] J. Y. Guo, F. Gitzhofer and M. I. Boulos, "Induction plasma synthesis of ultrafine SiC powders from silicon and CH4", J. Mat. Sci., vol 30, pp. 5589-5599, 1995.

[12] M. I. Boulos, B. Fauchais, and E. Pfender, Thermal Plasmas: Fundamentals and applications, Springer, pp. 37-38, 1994.

[13] T. M. Barnes, J. Leaf, C. Fry, and C. A. Wolden, "",J. Crys. Gro. Vol. 274, pp. 412, 2005.

Various Carbon Composite Pyrolyzed Polymers and Their Electrical Characterization

Akira Akazawa, Kanji Okamoto, Atsushi Syunori and Satoshi Konishi
Department of Micro System Technology, Ritsumeikan University, JAPAN
rt000061@ed.ritsumei.ac.jp

Abstract—In this paper, we deals with various carbon composite pyrolyzed polymers. We present carbon composite pyrolyzed polymers using glassy carbon particles ($\varphi = 8$ μm), carbon microspheres ($\varphi = 260$ nm) and carbon black particles ($\varphi = 70$ nm). Carbon composite pyrolyzed polymers were prepared by pyrolysis of carbon composite photoresists at temperatures ranging from 600 to 1000°C. The carbon composite pyrolyzed polymers were characterized by a scanning electron microscopy (SEM), an atomic force microscope (AFM), four point probe measurements and cyclic voltammetry. This paper reports evaluation results of various carbon composite pyrolyzed polymers of semiconducting and electrochemical properties.

Keywords-Pyrolyzed polymer; Carbon composite; Four point probe mesurment; Cyclic voltammetr

I. INTRODUCTION

This paper deals with various carbon composite pyrolyzed polymers. Pyrolyzed polymer is carbon material that provided by pyrolysis of polymer material in an iner atomosphere. We can form various microstructures of polymer materials by microelectromechanical systems (MEMS) fabrication technology. The patterns of polymer materials are transformed into carbon patterns through the pyrolysis. Various pyrolyzed polymer microstructures are already reported [1, 2]. Pyrolyzed polymer itself has attractive characteristics, especially electrical properties [3-5]. Furthermore, composite materials of carbon and polymer have been extensively explored. So far we know, carbon nanotubes (CNTs) composite polymer films are reported [6, 7]. The CNTs composite polymer materials which have electrical, thermal conductive and mechanical properties are expected to be an application of MEMS devices [7-9]. We presented the first report of carbon composite pyrolyzed polymer using glassy carbon particles elsewhere [10]. In this study, we mixed glassy carbon particles, carbon microspheres and carbon black particles with positive photoresist. Introducing different carbon materials such as glassy carbon to pyrolyzed polymer modifies and improves characteristics of pyrolyzed polymer (Fig. 1). Composite materials of positive photoresist and carbon particle were formed by MEMS fabrication technology and pyrolyzed under an inert atmosphere. This paper reports further evaluation results of various composites in addition to composites with glassy carbon particles. We consider that carbon composite pyrolyzed polymer has attractive material property of carbon particles and good formability of pyrolyzed polymer [10].

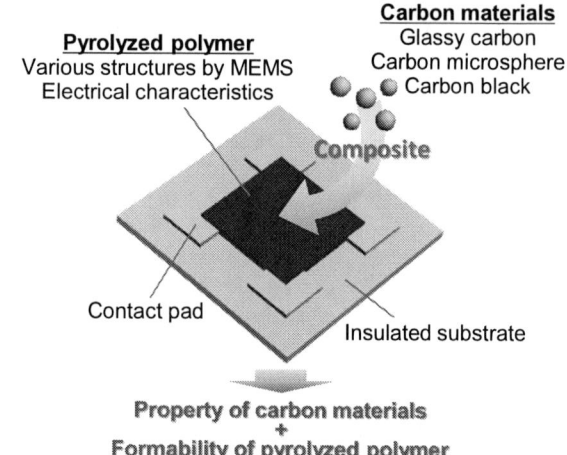

Fig. 1. *Introducing different carbon materials to modifies and improves the characteristics of pyrolyzed polymer. In this study, we used glassy carbon particles, carbon microspheres and carbon black particles as composite carbon materials.*

II. MATERIALS AND FABRICATION

A. Carbon materials

In this study, we prepared glassy carbon particles ($\varphi = 8$ μm), carbon microspheres ($\varphi = 260$ nm) and carbon black ($\varphi = 70$ nm) (Tokai carbon Co., Ltd., Tokyo, Japan) as composite materials. Glassy carbon is the most popular carbon material as working electrode. It is amorphous structure and impermeable to liquids and gases [11]. The structure of glassy carbon is interwoven ribbons of the graphite structure, so it is much harder. Carbon microspheres are an amorphous structure similar to glassy carbon. It is a single particle, and it is not agglutinate as such. Carbon black is widely used as filler for modifying the mechanical and electrical characteristics [12, 13].

978-1-4673-1122-9/12 $31.00 © 2012 IEEE

TABLE I. MIX RATIO OF PHOTORESIST AND CARBON MATERIALS

Carbon material	Ratio by weight (Photoresist : Carbon material)
Glassy carbon	5:1
Carbon microsphere	100:1
Carbon black	100:1

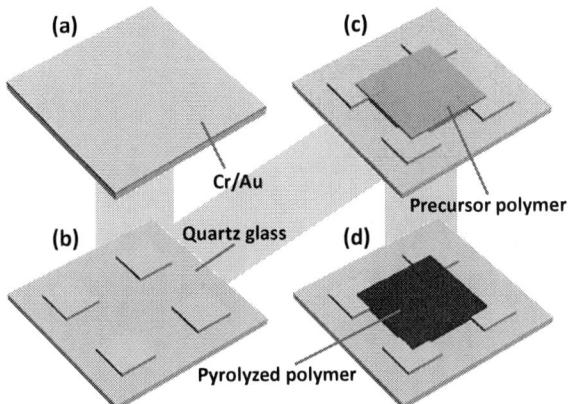

Fig. 2. *Fabrication process of the four probe measurements device. (a) Cr/Au layer forming on the quartz glass substrate, (b) Cr/Au contact electrodes pattering, (c) AZ P4620 or carbon particles composite AZ P4620 pattering and (d) at 600°C, 800°C and 1000°C for two hours at a maximum temperature in N₂ atmosphere.*

Each carbon particles was mixed with positive photoresist (AZ P4620). The mixed ratio of photoresist and each carbon particles is shown in Table 1. If there are a lot of mixed quantities of carbon particles, the thin film of pyrolyzed polymer peeled off from the substrate. Table 1 is the greatest mixed ratio that carbon composite pyrolyzed polymer did not peel off from the substrate. The mixed ratio decreased with carbon particle size becomes small.

B. Fabrication

Fabrication process of the four point probe measurements device is illustrated in Fig. 2. Four chromium/gold (Cr/Au) contact electrodes were formed on the quartz glass substrate. The size of Cr/Au contact electrodes are 2.0×2.0 mm². Next, each carbon particle was mixed with AZ P4620 (see Table. 1) and spincoated on the quartz glass substrate. The precursor polymer structure was patterned by photolithography and pyrolyzed in nitrogen (N₂) atmosphere. The pyrolysis process was executed at 600°C, 800°C and 1000°C for two hours at a maximum temperature in N₂ atmosphere. The flow rate and the pressure of N₂ were 2.0 L/min and 0.2 MPa, respectively. The precursor polymer was heated at 5.0 °C/min, and cooled to room temperature at 2.0 °C/min immediately after the desired temperature was reached. The size of the fabricated carbon pattern is 5.0×5.0 mm².

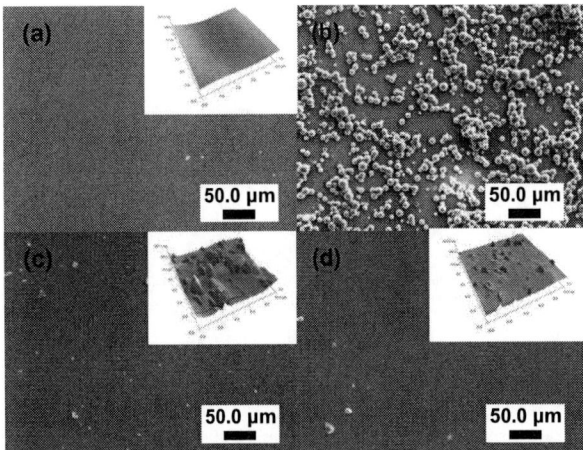

Fig. 3. *The SEM and the AFM images of fabricated pyrolyzed polymers. (a) Pyrolyzed polymer derived from AZ P4620, (b) glassy carbon composite pyrolyzed polymer, (c) carbon microspheres composite pyrolyzed polymer and (d) carbon black composite pyrolyzed polymer.*

III. EXPERIMENTS

A. Surface observations

The SEM and the AFM images of fabricated carbon surfaces are shown in Fig. 3. Pyrolyzed polymer film thickness was approximately 1 μm. Glassy carbon particles were observable on the pyrolyzed polymer surface. Surface asperity of glassy carbon composite pyrolyzed polymer was approximately 10 μm. On the other hand, carbon microsphere and carbon black hid in the thin film of pyrolyzed polymer. Surface asperity of carbon microspheres composite pyrolyzed polymer and carbon black composite pyrolyzed polymer were approximately 200 and 30 nm, respectively.

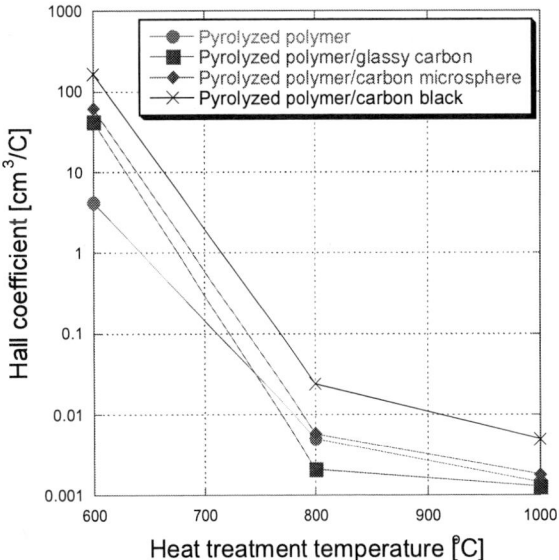

Fig. 4. *Measurement results of the hall coefficient at fabricated pyrolyzed polymers by four point probe measurements.*

B. Semiconductor Characteristics

Measurement results of the hall coefficient at fabricated carbons by four point probe measurements are shown in Fig. 4. Fig. 4 tells larger heat treatment temperature at lower hall coefficients. Especially, carbon black composite pyrolyzed polymer has the highest hall coefficient in fabricated carbons. At the heat treatment temperature of 600°C, hall coefficient of carbon composite pyrolyzed polymers is higher than that of pyrolyzed polymer. According to heat treatment temperature rise, the hall coefficient of carbon composite pyrolyzed polymer and pyrolyzed polymer shows similar characteristics. The fabricated carbons have positive sign.

C. Electrochemical Characteristics

Electrochemical measurements were conducted using standard three-electrode configuration. Pyrolyzed precursor polymers prepared at 1000°C pyrolysis were used as a working electrode. The fabricated carbons were activated by air plasma treatment before experiments. A Pt mesh was used as a counter electrode. A silver/silver chloride (Ag/AgCl) was used as a reference electrode.

Measurement results of the electrochemical potential window by fabricated carbons in 0.5 M KCl are shown in Fig. 5. Sweep speed was 0.1 V/s. Fig. 5 tells that carbon black composite pyrolyzed polymer has the widest electrochemical potential window and the lowest residual current in fabricated carbons.

Fig. 6 shows cyclic voltammograms by the fabricated carbons in 0.5 mM $K_3Fe(CN)_6$ with 0.5 M KCl. Sweep speed was 0.5 V/s. The potential was stepped from -0.1 to 0.6 V vs Ag/AgCl. Peak to peak separations, ΔE_p, of fabricated carbons are shown in Table 2. ΔE_p of glassy carbon composite pyrolyzed polymer (= 76 mV) and carbon microspheres composite pyrolyzed polymer (= 77 mV) were lower than ΔE_p of pyrolyzed polymer (= 89 mV). On the other hand, ΔE_p of carbon black composite pyrolyzed polymer (= 127 mV) was the highest in ΔE_p of fabricated carbons. These measurement results tells that glassy carbon composite pyrolyzed polymer and carbon microspheres composite pyrolyzed polymer are enough sensitive to the electron transfer in oxidation-reduction of electrolyte.

IV. DISCUSSIONS

First of all, relations between heat treatment temperatures of precursor materials and hall coefficients are discussed. We already presented to semiconductor characteristics of pyrolyzed polymer derive from photoresist [3]. Fig. 4 tells that each carbon composite pyrolyzed polymer change from semiconductor to conductor as heat treatment temperatures rise. Carbon composite pyrolyzed polymers are P-type semiconductor because they showed positive sign.

Next, we evaluated electrochemical properties of carbon composite pyrolyzed polymer by cyclic voltammetry. Fig. 7 shows cyclic voltammograms by planar glassy carbon in 0.5 mM $K_3Fe(CN)_6$ with 0.5 M KCl. ΔE_p of planar glassy carbon (= 78 mV) is similar to ΔE_p of glassy carbon composite pyrolyzed polymer and carbon microspheres composite

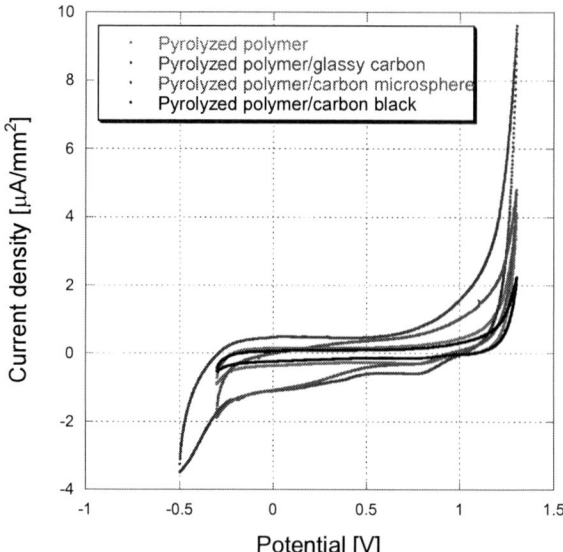

Fig. 5. *Measurement results of the potential window by fabricated carbon electrodes in 0.5 M KCl. Scan rate was 0.1 V/s. Pt and Ag/AgCl were used as a counter electrode and a reference electrode.*

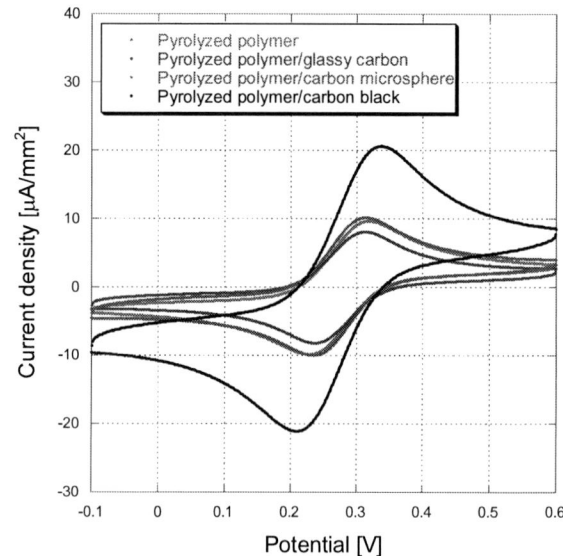

Fig. 6. *Cyclic voltammograms by fabricated carbon electrodes in 0.5 mM $K_3Fe(CN)_6$ with 0.5 M KCl. Scan rate was 0.5 V/s. Pt and Ag/AgCl were used as a counter electrode and a reference electrode.*

TABLE II.   PEAK TO PEAK SEPARATIONS OF CARBON ELECTRODES

Carbon electrode	⊿Ep [mV]
Pyrolyzed polymer	89
Pyrolyzed polymer/gassy carbon	76
Pyrolyzed polymer/carbon microsphere	77
Pyrolyzed polymer/carbon black	127

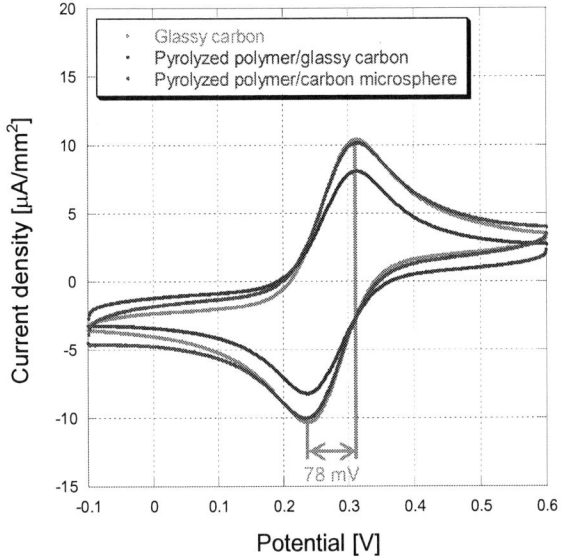

Fig. 7. *Cyclic voltammograms by planar glassy carbon in 0.5 mM K₃Fe(CN)₆ with 0.5 M KCl. Scan rate was 0.5 V/s. Peak to peak separations (ΔEₚ) of planar glassy carbon was compared with ΔEₚ of glassy carbon composite pyrolyzed polymer and carbon microsphere composite pyrolyzed polymer.*

pyrolyzed polymer (see Table. 2). Carbon black composite pyrolyzed polymer showed the highest ΔE_p value. Carbon black exhibited a higher ΔE_p (approximately 100 mV) for $Fe(CN)_6^{3-/4-}$ redox couple in other report [14]. However, carbon black composite pyrolyzed polymer showed a large peak current. This result suggests that carbon black composite pyrolyzed polymer has large surface area. Each carbon particles modified electrochemical characteristics of pyrolyzed polymer.

V. CONCLUSIONS

This paper reports electrical properties of carbon composite pyrolyzed polymers. Pyrolyzed polymer and carbon composite pyrolyzed polymers showed similar heat treatment temperature dependence of the hall coefficient. Especially, carbon black (φ = 70 nm) composite pyrolyzed polymer has the highest hall coefficient. We analyzed carbon composite pyrolyzed polymers in terms of electrochemical properties. Glassy carbon (φ = 8 μm) composite pyrolyzed polymer, carbon microsphere (φ = 260 nm) composite pyrolyzed polymer and planar glassy carbon showed similar peak to peak separations (ΔE_p) (see Fig. 7). Carbon particles affect electrochemical characteristics of pyrolyzed polymer as working electrode. Furthermore, carbon black composite pyrolyzed polymer has the widest electrochemical potential window and the lowest residual current. The results suggest that potential possibility of carbon black is attractive as composite material of pyrolyzed polymer for electrochemical sensor. As we can see, various carbon composite pyrolyzed polymers would contribute to attractive materials in MEMS systems and μTAS.

REFERENCES

[1] K. Naka, H. Hayashi, M. Senda, H. Shiraishi, and S. Konishi, "Effect of Nano Stripe Carbonized-Polymer Electrode on High S/N Ratio in Electrochemical Detector," *MEMS 2007 conference*, Kobe, Japan, January 2007, pp. 195-198.

[2] K. Yamamoto, K. Naka, Y. Nagaura, S. Shoji, and S. Konishi, "Pyrolyzed Polymer Mesh Electrode Integrated into Fluidic Channel for Gate type Sensor," *MEMS 2007 conference*, Kobe, Japan, January 2007, pp. 271-274.

[3] S. Konishi, H. Nagae, T. Sagawa, K. Naka and K. Yoshioka, "Semiconductivity of Pyrolyzed Polymer for MEMS Application," *Transduces 2005 conference*, Seoul, Korea, Jun 2005, pp. 271-274.

[4] S. Konishi, M. Ligeu, T.A. Hardel and Y-C. Tai, "PARYLENE-PYROLYZED CARBON FOR MEMS APPLICATIONS," *MEMS 2004 conference*, Maastricht, Netherlands, January 2004, pp. 161-164

[5] J. Kim, X. Song, K, Kinoshita, M. Madou, and R. White, "Electrochemical Studies of Carbon Films from Pyrolyzed Photoresist," *Journal of Electrochemical Society*, Vol. 145, No. 7, pp. 2314-2319, 1998.

[6] Y. Show, K. Takahashi, "Stainless steel bipolar plate coated with carbon nanotube (CNT)/polytetrafluoroethylene (PTFE) composite film for proton exchange membrane fuel cell (PEMFC)," *Journal of Power Sources*, Vol. 190, Issue 2, pp. 322-325, 2009.

[7] K.S. Teh and L. Lin, "MEMS sensor material based on polypyrrole-carbon nanotubenanocomposite: film deposition and characterization," *J. Micromech. Microeng 15*, pp. 2019-2027, 2005.

[8] Y. Xu, G. Ray, B. Abdel-Magid, "Thermal behavior of single-walled carbon nanotubepolymer–matrix composites," *Composites Part A: Applied Science and Manufacturing*, Vol. 37, Issue 1, pp. 114-121, 2006.

[9] B. Ashrafi, P. Hubert and S. Vengallatore, "Carbon nanotube-reinforced composites asstructural materials for microactuators in microelectromechanical systems," *Nanotechnology 17*, pp. 4895-4903, 2006.

[10] K. Okamoto, A. Syunori, R. Iwata, H. Saiki, H. Nakanishi and S. Konishi, "PYROLYZED POLYMER AND ITS COMPOSITE FOR ¹⁸F⁻ CONCENTRATION ELECTRODE TOWARD PET APPLICATION", *Transducers 2009 conference*, Denver, Colorado, USA, June 2009, pp. 1904-1907.

[11] C.E. Banks and R.G. Compton, "New electrodes for old: from carbon nanotubes to edge plane pyrolytic graphite," *Analyst*, 131, pp. 15-21, 2006.

[12] T. Ding, L. Wang and P. Wang, "Changes in Electrical Resistance of Carbon-Black-FilledSilicone Rubber Composite During Compression," *Journal of Polymer Science Part B: Polymer Physics*, Volume 45, Issue 19, pp. 2700–2706, 2007.

[13] J. Hwang, J. Muth and T. Ghosh, "Electrical and Mechanical Properties of Carbon-Black-Filled, Electrospun Nanocomposite Fiber Webs", *Journal of Applied Polymer Science*, Volume 104, Issue 4, pp. 2410-2417, 2007.

[14] S. Goeringer, N.R de Tacconi, C.R Chenthamarakshan, K. Rajeshwar and W.A Wampler, "Redox characterization of furnace carbon black surfaces," *Carbon* 39, pp. 515-522, 2001.

978-1-4673-1122-9/12 $31.00 © 2012 IEEE

Reliability Prediction of 3C-SiC Cantilever Beams using Dynamic Raman Spectroscopy

Raden Dewanto[*], Tao Chen, Rebecca Cheung, Zhongxu Hu, Barry Gallacher, John Hedley

[*]Mechanical and Systems Engineering, Newcastle University, Newcastle upon Tyne, UK

r.s.dewanto@ncl.ac.uk

Abstract—**We propose an extension and improvement to reliability predictions in epitaxially grown 3C-SiC cantilever beam MEMS by utilizing dynamic Raman spectroscopy to allow the gathering of Weibull fracture test data to be done directly on devices thereby taking account of actual geometrical tolerances, dynamic load conditions and effects from the microfabrication process due to high lattice and thermal mismatch between 3C-SiC and Si. In this work, 3C-SiC devices were fabricated, modeled and actuated to determine both theoretical and experimentally measured strain levels within the device during operation. Initial results indicate both characteristic Raman peaks of 3C-SiC are suitable for this characterization and measurement resolution of 0.02cm^{-1} is demonstrated. As the technique is performed directly on devices, it simplifies the frequently found time consuming methodology of preparations of micron-sized specimen fracture test pieces and gives a mechanism for feedback to optimize the fabrication process.**

Keywords-silicon carbide, Raman spectroscopy, MEMS, charatcerization, reliability

I. INTRODUCTION

Micro-electro-mechanical system (MEMS) technology has evolved rapidly in recent decades. MEMS sensors such as gyroscopes and accelerometers have matured and seen wide application in industry and commercial applications. Up to now, most MEMS devices are based on a silicon substrate due to easy incorporation into the fabrication process of integrated circuits, thus reducing the cost. However, as applications are being extended to more extreme conditions, a more robust material is required. Silicon carbide (SiC) features high Young's modulus, wide bandgap, and chemical inertness. Although the physical properties of various SiC polytypes differ from each other, the average Young's modulus is more than two times that of silicon. Moreover, the bandgap of SiC is beyond 2.2eV, almost double that of silicon. The SiC MOSFET has been reported to work at temperatures as high as 923K [1]. These robust properties of SiC make it extremely suitable for harsh environment applications such as aerospace and deep space technology. For MEMS applications, 3C-SiC has been widely used because it is relatively simple to grow on Si substrates [2,3,4].

Microfabrication processes of 3C-SiC MEMS devices always induce either residual stresses or undesirable structure deformation due to the large lattice mismatches and thermal expansion coefficient differences between 3C-SiC and silicon [5, 6]. This condition can produce unstable structures that are vulnerable to cracking and thereby increase the risk of device failure [7]. Improvement of work on reliability prediction of single-crystal silicon MEMS reported by Fitzgerald et al [8] is implemented here in 3C-SiC single crystal by combining it with our previous work in dynamic Raman MEMS characterization [9]. The dynamic Raman characterization we use in this work is a continuous beam Raman spectroscopy that measures the maximum volumetric strain of the structure by taking account of Raman profile broadening during structure vibrations. The main differences of our method compared to the work of Fitzgerald are that our method is employed in a dynamic system rather than static and the Weibull parameters being in use for the reliability prediction are taken in micron-sized specimens rather than in the macro scale.

II. DEVICE FABRICATION

The process steps are schematically shown in Fig. 1. The fabrication of 3C-SiC cantilevers and bridges starts from a 3C-SiC/Si substrate (Fig. 1a). The 3C-SiC layer is 1.8µm thick, epitaxially grown on top of the <100> surface of 4" silicon wafer. A 3.5µm layer SiO$_2$ was then grown on top of 3C-SiC layer by plasma enhanced chemical vapour deposition (PECVD) (Fig. 1b). The oxide layer is patterned by reactive ion etching (RIE) (Fig. 1c, d). Next, the 3C-SiC is etched by inductively coupled plasma (ICP) with a mixture of SF$_6$ and O$_2$ (Fig. 1 e). The 3C-SiC structures are finally released by the etching of the Si with XeF2, resulting in the presence of an undercut at the anchors of the 3C-SiC cantilever beams (Fig. 1f). RIE has been used again to clean the residual oxide on the 3C-SiC (Fig. 1g).

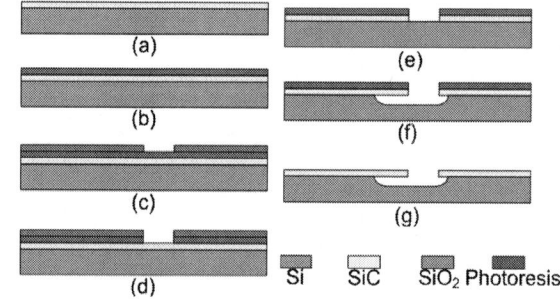

Fig. 1. Schematic of the process flow for fabricating the devices. (a) SiC/Si substrate. (b) SiO$_2$ grown by PECVD. (c) Photolithography. (d) SiO$_2$ etching by RIE. (e) SiC etching by ICP. (f) Etching of Si by XeF$_2$ to release the cantilever. (g) Removal of SiO$_2$ by RIE.

(a)

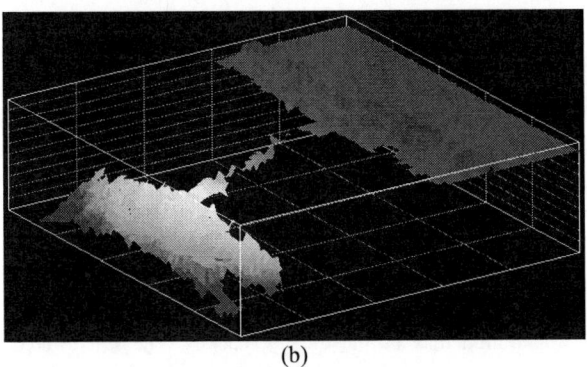

(b)

Fig. 2. (a) Scanning electron microscope image of some of the fabricated devices. (b) A surface profiler measurement indicates a downwards deflection of 2.5μm due to strain induced during the fabrication process.

The fabricated devices are composed of three different surfaces to examine for the Weibull fracture parameters, these being a CVD grown top layer surface, a XeF$_2$ etched Si release substrate-film interface and ICP etched surfaces, each surface having a characteristic finish. The series of structures were designed with fundamental resonant frequencies between 10 KHz to 1 MHz so that complimentary characterization of device dynamics is possible with conventional optical metrology techniques [10]. Fig. 2 shows a scanning electron microscope (SEM) image of one of the fabricated devices. The large mass towards the cantilever free end is to induce a higher strain in the structure when it is driven into resonance. For the same purpose, some cantilevers are of triangular shape with a narrow anchor connected to the bulk. The released cantilevers are deflected downwards due to the tensile strain near the 3C-SiC/Si interface caused by lattice mismatch, as shown by Fang [11].

III. MODELLING

The finite element analysis package ANSYS was used to create a basic model of the test structures. Devices were modelled with a 3D structural element with the material

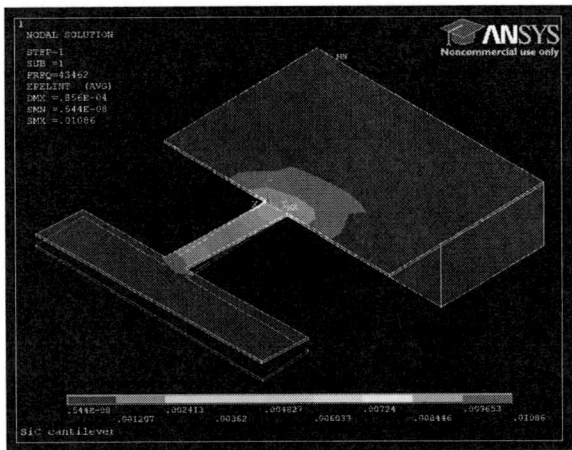

Fig. 3. A finite element package is used model device resonant frequencies and the resulting induced strain during operation.

properties for 3C-SiC being 3100kgm^{-3}, 410GPa and 0.14 for density, Young's modulus and Poisson's ratio respectively. The test device examined in this work had cantilever dimensions of 100μm × 20μm with an end paddle dimension of 45μm × 220μm. SiC thickness was 1.8μm. To model the inherent strain, a tensile force was applied to the top half of the end of the structure whilst a compressive force was applied to the lower half. A force value that resulted in the measured out of plane deflection of 2.5μm was used. Model geometry was updated accordingly and then a modal analysis and harmonic analysis performed to determine resonant frequencies and induced dynamic strain respectively. The driving force used in the harmonic analysis is adjustable so that the experimentally measured deflection may be obtained. Results from these analyses are shown in Fig. 3.

IV. EXPERIMENTAL DETAILS

Devices were mounted onto a piezoelectric disk for actuation and mounted in a vacuum chamber. The pressure was reduced to <1mbar to reduce the effects of squeeze film damping. A laser vibrometer was used to record the resonant frequencies of the devices. The devices, still contained within the vacuum chamber, were then mounted onto a FHR1000 Horiba Jobin Yvon Raman system. The 50X objective used gave a probe beam diameter of 3μm. The device was driven into resonance in its fundamental out-of-plane mode at 20.5KHz and the Raman spectrum measured at its clamped end. A schematic of the setup is shown in Fig. 4.

V. RESULTS

The Raman spectrum of SiC shows two characteristic peaks at 796.7cm^{-1} and 974.7cm^{-1}. A Voigt profile was fitted to each profile using a Matlab non linear least squares fitting routine with fitting parameters of Gaussian width (ΔG), Lorentzian width (ΔL), peak position, peak intensity and background

978-1-4673-1122-9/12 $31.00 © 2012 IEEE

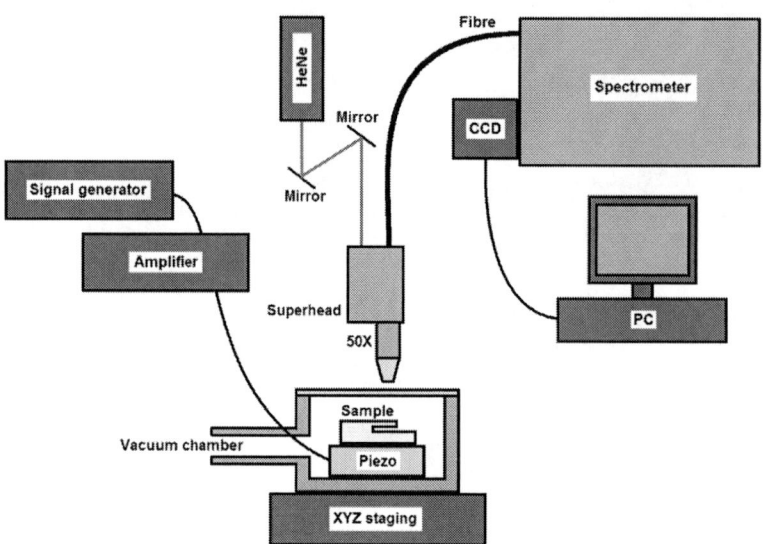

Fig. 4. Schematic of the experimental setup. The sample is mounted on a piezo disk to allow for easy actuation. The optical head incorporates a long working distance 50X objective giving a probe beam spot diameter of 3μm. The Raman signal is analyzed via a FHR1000 Horiba Jobin Yvon Raman system with liquid nitrogen cooled CCD camera.

level. From this, the Voigt width (ΔV) was calculated using the Whiting modified expression from Olivero [12]

$$\Delta V = 0.5346 \Delta L + (0.2166 \Delta L^{2} + \Delta G^{2}). \qquad (1)$$

7 spectra from the actuated device were compared with 6 spectra obtained with no actuation, examples are shown in Fig. 5, broadening of the peaks during device actuation is clearly visible. The Voigt widths for the 796.7cm⁻¹ and 974.7cm⁻¹ peaks were measured to be 2.27±0.02cm⁻¹ and 3.23±0.03cm⁻¹ during actuation respectively compared with 2.03±0.01cm⁻¹ and 3.09±0.02cm⁻¹ when stationary. Experimentally

quantifying the amplitude of vibration proved problematic due to poor reflectivity of the microcantilever surface. However, an upper limit may be put on the analysis. The silicon etch depth was 85μm and therefore the maximum volumetric strain induced in the structure at the point of measurement is modelled as 4130μstrain. As the Raman probe beam would penetrate the depth of the microcantilever, the profile would be an integral of volumetric strains through this depth.

Fig. 6 shows the fits obtained for both the 796.7cm⁻¹ and 974.7cm⁻¹ peaks. The residuals are amplified to visually indicate the goodness of fit. The Voigt gives a good approximation to the profile however for the 796.7cm⁻¹ peak,

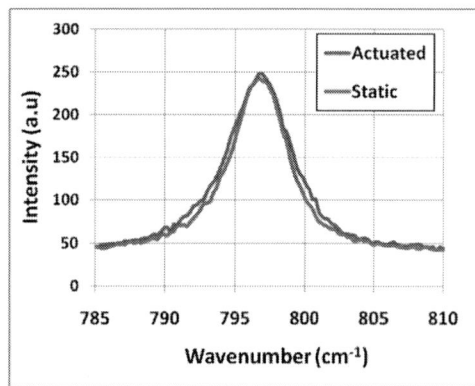

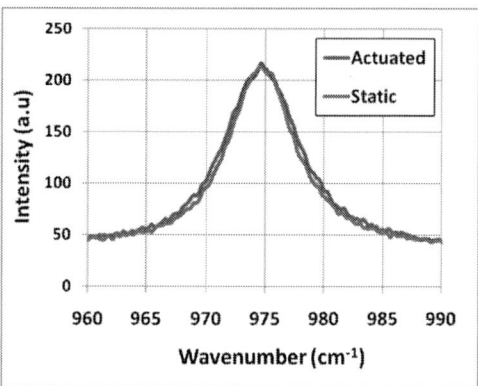

Fig. 5. The two profiles shows the variation in Raman profiles between a static measurement and one taken during device actuation. Both the 796.7cm⁻¹ and 974.7cm⁻¹ Raman peaks of SiC show broadening during actuation.

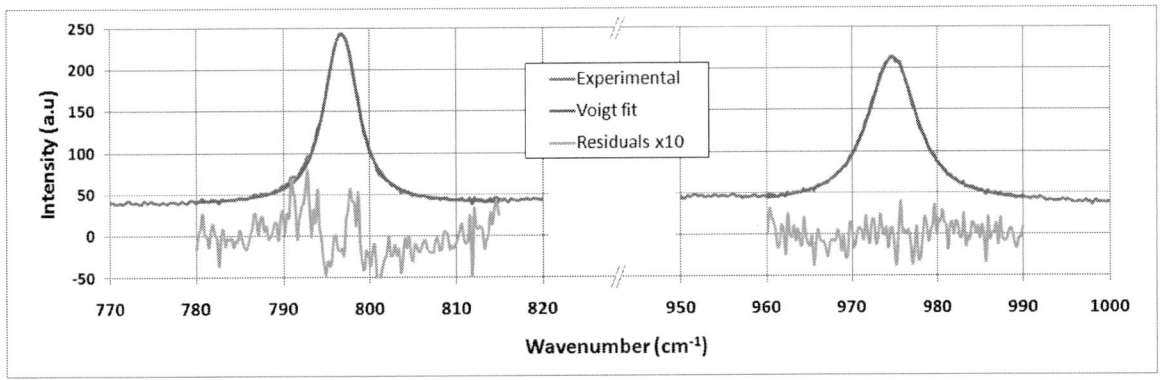

Fig. 6. A Voigt profile is used to fit each peak. The structure seen in the residuals of the 796.7cm⁻¹ fit indicate an asymmetry term is required in the profile descriptor.

an asymmetry is seen in the residuals with more intensity being seen in the lower wavenumber inner shoulder. For the 974.7cm^{-1} peak, despite there being a prominent outer shoulder towards lower wavenumbers, limiting the fit to within 15 cm^{-1} of the to the peak position allows the Voigt profile to be a suitable descriptor, as indicated by the randomly scattered residuals.

VI. CONCLUSIONS

3C-SiC cantilevers have been designed to manifest the induced stress during vibration. The 3C-SiC layer on a Si substrate have been patterned by ICP and released in gaseous phase by XeF$_2$. Raman spectroscopy is demonstrated to be a suitable approach to measuring dynamic strains induced in SiC MEMS as devices are driven into resonance. Both the 796.7cm^{-1} and 974.7cm^{-1} Raman peaks characteristic of SiC show broadening during operation. The work acts as a first step towards investigating reliability and predicting failure in SiC MEMS.

ACKNOWLEDGMENT

R.S.D. acknowledges the Directorate General of Higher Education, Ministry of National Education, Indonesia for the funded studentship. T.C. acknowledges Chinese Scholarship Council and The University of Edinburgh for providing a joint scholarship for PhD study. J.H. thanks EPSRC for the research grant EP/C015045/1 thereby enabling the Raman spectroscopy aspects of this work to be performed.

REFERENCES

[1] J.W. Palmour, H.S. Kong, and R.F. Davis, "High-temperature depletion-mode metal-oxide-semiconductor field-effect transistors in beta-SiC thin films,"*Appl. Phys. Lett.* vol. 51, pp. 2028-2030, December 1987.

[2] C. Locke et al, "3C-SiC Films on Si for MEMS application: Mechanical Properties," *Materials Science Forum*, vols. 615-617, pp. 633-636, March 2009.

[3] E. Mastropaolo, I. Gual, G. Wood, A. Bunting, and R. Cheung, " Piezoelectrically driven silicon carbide resonators," *J. Vac. Sci. Technol. B*, vol. 28, pp. 1071-1023, November/December 2010.

[4] Ch. Forster, V. Cimalla, K. Brucknerr, M. Hein, J. Pezoldt, an dO. Ambacher, "Micro-electromechanical systems based on 3C-SiC/Si heterostructures," Materials Science and Engineering C, vol. 25, pp. 804-808, December 2005.

[5] M. Bosi et al, "Wafer curvature analysis in 3C-SiC layers grown on (0 0 1) and (1 1 1) Si substrates," *Journal of Crystal Growth,* vol. 318, pp. 401-405, 2011.

[6] G.S. Wood, I. Gual, P. Parmiter, and R. Cheung, "Temperature stability of electro-thermally and piezoelectrically actuated silicon carbide MEMS resonators", *Microelectronics Reliability*, vol. 50, pp. 1977-1983, 2010.

[7] W.L. Zhu, J.L. Zhu, S. Nishino, G. Pezzotti, "Spatially resolved Raman spectroscopy evaluation of residual stresses in 3C-SiC layer deposited on Si substrates with different crystallographic orientations," *Applied Surface Science*, vol. 252, pp. 2346-2354, 2006.

[8] A. M. Fitzgerald et al, *Journal of Microelectromechanical Systems*, 14 (2009), pp. 962-970.

[9] Z. X. Hu, J. Hedley, B.J. Gallacher and I. Arce-Garcia, "Dynamic characterization of MEMS using Raman spectroscopy," *Journal of Micromechanics and Microengineering*, vol. 18, 095019, 2008.

[10] J. Hedley, A. Harris, J. Burdess, and M. McNie, "The development of a workstation for optical testing and modification of IMEMS on a wafer," *Proceedings of SPIE*, vo. 4408, pp. 402-408, April 2001.

[11] W. Fang and J.A. Wickert, "Determining mean and gradient residual stresses in thin films using micromachined cantilevers," *J. Micromech. Microeng.*, vol. 6, pp. 301–309, 1996.

[12] J.J. Olivero and R.L. Longbothum, "Empirical fits to the Voigt line width: A brief review," *Journal of Quantitative Spectroscopy and Radiative Transfer*, vol. 17, pp. 233–236, 1977.

Characterization of Wafer-level XeF$_2$ Gas-phase Isotropic Etching For MEMS Processing

Dehui Xu[1], Bin Xiong[1], Guoqiang Wu[1,2], Yinglei Ma[1], Yuelin Wang[1] and Errong Jing[1,3]

[1]State Key Laboratory of Transducer Technology, Science and Technology on Micro-system Laboratory, Shanghai Institute of Microsystem and Information Technology, Chinese Academy of Sciences, Shanghai, China
[2] Graduate School of Chinese Academy of Sciences, Beijing, China
[3] CSMC Technologies Corporation, Wuxi, China
dehuixu@mail.sim.ac.cn

Abstract—**This paper reports the characteristics of wafer-level XeF$_2$ gas-phase etching. Compared with chip-level XeF$_2$ etching, the silicon etch rate for wafer-level XeF$_2$ process is much smaller, which is mainly caused by the large exposed silicon area in wafer-level process. Additionally, the silicon etch rate drops off as etching time increased. The aperture size effect is apparent in wafer-level XeF$_2$ processing. However, for etching window with large size, the aperture size effect will be minimized. The vertical aperture size effect is direct proportion to the number of etch cycle, while the lateral aperture size effect is first increase then decrease with the number of etch cycle increasing. Slight anisotropy of wafer-level XeF$_2$ etching is also observed. Based on the characteristics of XeF$_2$ etching, layout design rule for MEMS device with XeF$_2$ releasing is developed and demonstrated.**

Keywords-XeF$_2$ etching;wafer-level; MEMS; silicon etching; dry isotropic etching;

I. INTRODUCTION

Silicon etching is a key process for microelectromechanical systems (MEMS) fabrication [1]. For sensing and actuating purpose, the microstructure should be free from the underlying support. Generally speaking, silicon micromachining technology for MEMS application can be divided into wet chemical etching and dry etching. Potassium hydroxide (KOH), tetramethylammonium hydroxide (TMAH), ethylenediamine pyrocatechol (EDP), and hydrofluoric acid-nitric acid-acetic acid (HNA) are usually used for wet chemical etching of silicon. Stiction problem in solution is a bottleneck that limits the yield. Stiction problem can be avoided by using a dry etching process. Current state of art silicon dry etching includes deep reactive ion etching (DRIE) and vapor-phase chemical etching. Because complicated and expensive equipment is needed to start and maintain plasma, the overall cost of DRIE process is high. Since materials directly etch silicon in their gaseous forms, no external energy sources or ion bombardment are required for vapor-phase chemical etching. Thus, vapor-phase chemical etching is an enabling technology for MEMS device fabrication [2].

Gas etchants for silicon etching include xenon difluoride (XeF$_2$), bromine trifluoride (BrF$_3$), chlorine trifluoride (ClF$_3$) and fluorine (F$_2$) [2]. As a silicon etchant, XeF$_2$ has unique advantages such as high selectivity for silicon, gas-phase isotropic etching and ease of operation [3]-[5].

Due to its unique capabilities, much attention has been focused on XeF$_2$ isotropic etching [6]-[9]. However, previous reported work was mainly performed for small samples cut from silicon wafer. Since both top surface and cross-sectional surface are exposed during chip-level XeF$_2$ etching, XeF$_2$ gas will be wasted on etching the cross-sectional silicon surface. Additionally, for large-volume processing, wafer-level XeF$_2$ etching is required. However, there are few reports for wafer-level XeF$_2$ etching process.

In this paper, we present the study of wafer-level XeF$_2$ bulk silicon etching. The silicon etch rate, aperture effect and etching anisotropy are studied to provide a guide for micromachining applications. Moreover, design rules are developed to help designers during the layout of microstructure with XeF$_2$ releasing. The etching windows layout design methodology for a micromachined thermopile radiation sensor with XeF$_2$ etching is also presented.

II. EXPERIMENTS

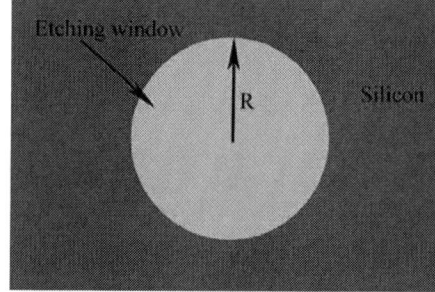

Fig. 1. Circle etch window for XeF2 etching. The lateral undercut illustrates the isotropic etching of XeF2 gas.

A 300 nm thermal oxide was grown on the 4-inch silicon wafer as etching mask. Etching windows were patterned on the oxide layer by RIE process. The total exposed area of silicon is about 5.8% of the 4-inch wafer. The gas-phase etching was performed with a commercially available XeF$_2$ pulse-etching system (Xactix Xetch e1 Series).

As shown in Fig. 1, circle etch window is used for XeF$_2$ etching. The radius of circle etching windows is among 2.5 μm, 5 μm, 7.5μm and 10 μm. The parameters used for the pulse-etching are 4 Torr XeF$_2$ pressure, 60 seconds etching

978-1-4673-1122-9/12 $31.00 © 2012 IEEE

time of each cycle. The number of etch cycle is set among 15, 35, 60, 90, 120.

III. RESULTS AND DISCUSSION

Fig. 2 shows the etching results for circle etching windows. The wafer-level XeF$_2$ processing exhibits much lower silicon etch rate than previous reports [3]–[5]. This is because that the amount of etched silicon is determined by the interaction between XeF$_2$ gas and silicon. The exposed silicon area for wafer-level XeF2 processing is much larger than that of chip-level XeF2 processing. Moreover, due to the fixed expansion chamber size of the etching system, the XeF$_2$ gas amount used per cycle is the same for both wafer-level and chip-level XeF$_2$ etching. If the wafer-level XeF$_2$ processing is performed with a larger expansion chamber to hold more XeF2 gas, the silicon etch rate can be increased.

From Fig. 2(a), it can be seen that the lateral etch rate will drop as etch cycle increasing. The same phenomena can also be observed in the vertical etch rate results (Fig. 2(b)). Silicon etch rate is much higher in the initial of etching than in the rest of etching. Because XeF$_2$ etching is performed without external energy source, the adsorption of XeF$_2$ molecule is regulated by XeF$_2$ gas diffusion. After bulk silicon was removed by initial XeF$_2$ etching, XeF$_2$ gas molecule should enter the small etching cavity under etching mask through the small etching window to react with the bulk silicon. Since both the lateral etch length and vertical etch depth are proportional to the number of etch cycle, the difficulty for XeF$_2$ gas diffusion into the bulk silicon surface will increase as etch cycle increase. In other words, the adsorption of XeF$_2$ molecule will decrease as etch cycle increase. Hence, the etch rate is diminished for a long time etching. Furthermore, as etch cycle increasing, the lateral etch length and vertical etch depth incremental step will slow down due to the decreased etch rate. Thus, as etch time increasing, the impact of lateral etch length and vertical etch depth on adsorption of XeF$_2$ molecule will decrease, and both the lateral and vertical etch rate decreasing step will slow down.

The dependence of etch rate on etching window size can also be seen in Figs. 2(a) and 2(b). The aperture size effect is because the etching window size also influences XeF$_2$ gas diffusion. For a large size etching window, there will be more XeF$_2$ adsorption events, which results in large silicon etch rate. In order to express the aperture size effect, a ratio of etch rate was calculated by dividing the etch rate at an aperture radius of x μm by the etch rate at an aperture radius of 2.5 μm. Therefore, the lateral etch rate ratio (R$_{LR}$(x)) and vertical etch rate ratio (R$_{VR}$(x)) can be defined as:

$$R_{LR}(x) = \frac{LR(x)}{LR(2.5)}; R_{VR}(x) = \frac{VR(x)}{VR(2.5)} \qquad (4)$$

where LR(x) and LR(2.5) are the lateral etch rates for circle etching window with radius of x μm and 2.5 μm, VR(x) and VR(2.5) are the vertical etch rates for circle etching window with radius of x μm and 2.5 μm. Figs. 2(c) and 2(d) present the etch rate ratio as a function of the number of etch cycle. The

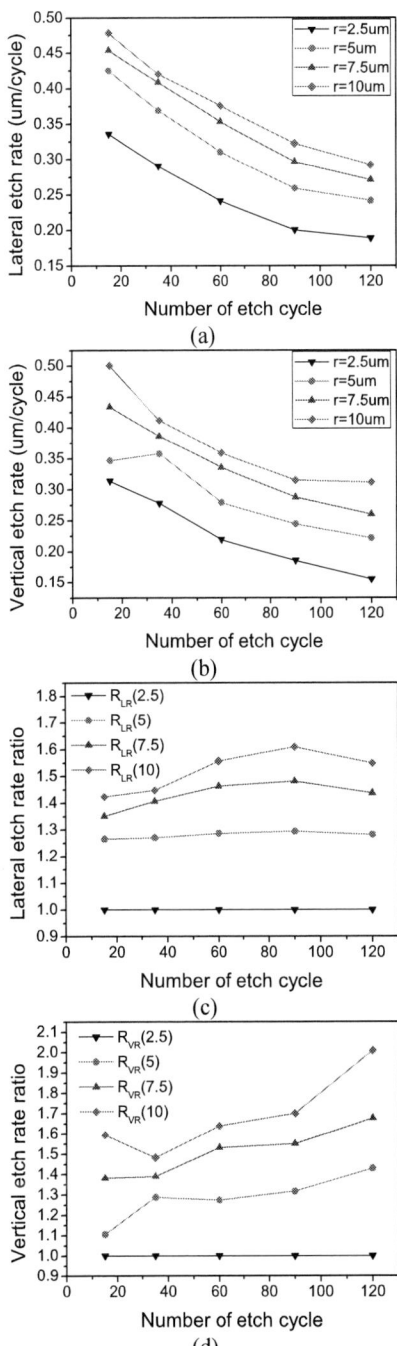

Fig. 2. Dependence of (a) lateral etch rate, (b) vertical etch rate, (c) lateral etch rate ratio and (d) vertical etch rate ratio on etching window size and number of etch cycle for circle etching window. Etch rates drop as number of etch cycle increase, and the aperture size effect can also be seen in the etch rate results. Aperture size effect is enhanced for etching window with small radius. Aperture size effect is also function of number of etch cycle.

aperture size effect is enhanced for small etching windows and appears to minimize for large etching windows. For silicon etching in lateral direction, $R_{LR}(x)$ is first rise and then drop as number of etch cycle increase. But for vertical direction silicon etching, $R_{VR}(x)$ is direct proportional to number of etch cycle.

The essence of the aperture size effect is the influence of etching window size on XeF_2 adsorption. Due to the undercut around the edges of mask, the XeF2 molecule may stick to the floating mask layer during lateral diffusion. For XeF_2 etching with small number of etch cycle, the undercut is small and the stiction event is few occurred. Thus, the lateral etching is mainly influenced by lateral gas diffusion. The lateral aperture size effect is direct proportional to number of etch cycle. But for XeF_2 etching with large number of etch cycle, the undercut is large and the XeF_2 molecule adsorption on the floating mask layer is apparent. XeF_2 stiction on floating mask also plays a role in lateral silicon etching. The impact of lateral diffusion on lateral silicon etch rate is reduced. Therefore, the lateral aperture size effect will be inverse proportional to number of etch cycle. However, the gas diffusion in vertical direction is only limited by the etching window size and no XeF_2 stiction will happen in the vertical direction. Hence, a longer etch time will lead to a higher vertical aperture size effect.

IV. MEMS APPLICATION

Knowing the characteristics of wafer-level XeF_2 etching, the undercut of XeF_2 etching will be predictable, and XeF_2 etching process will be better control. Thus, layout design rule can be developed. To release microstructure, the etching window layout should be designed to ensure that the undercut of XeF_2 etching cover the area for release. Moreover, since the undercut of XeF_2 etching is predictable, the XeF_2 release time will also be predictable. This will contribute to more freedom for MEMS device design.

Fig. 3(a) illustrates the application of the design rule to a micromachined thermopile radiation sensor. The width of rectangle with circle end (RCE) etching window is 10 μm, the radius of cicrle etching window is 5 μm. The undercut of each etching window in the membrane was predicted, and the etching windows were designed to make sure that the membrane was almost covered by the undercut. The undercuts in Fig. 8 were predicted as 30 μm and 25 μm for RCE and 232 circle etching windows, respectively. This undercut prediction corresponds to about 90 cycles XeF_2 etching. Most of the membrane is covered by the predicted undercut, which implies that most of the membrane structure will be released after 90 cycles of XeF_2 etching. The area that not covered by undercut prediction in the membrane will enhance silicon etching efficiency.

Fig. 3(b)~3(d) presents the membrane release revolution by wafer-level XeF_2 etching. The silicon around etching window is isotropic etched by XeF_2 gas in the lateral direction as design. Because of the undercut overlap design, the silicon

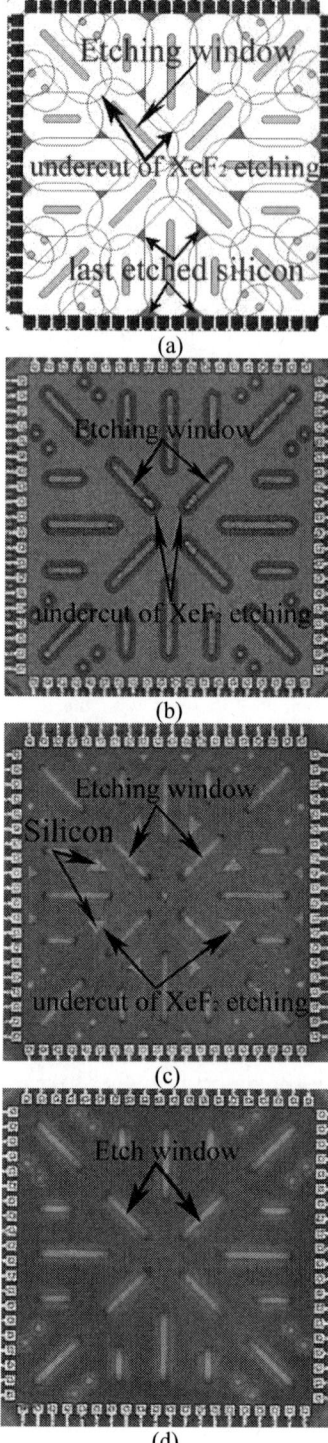

Fig. 3. (a) Etch windows layout design for membrane structure release by XeF2 etching, where the predicted undercut of XeF_2 etching is designed to cover the whole area for release. Membrane structure after (b) 40 cyelces, (c) 90 cycles and (d) 120 cycles XeF_2 etching. The last etched silicon area in the microstructure is predicted and the time etching evolution demonstrates that silicon is etched as the design. This result also verifies the feasibility of the design rule for microstructure design with XeF_2 release.

design. The last etched silicon area prediction is also verified by Fig. 3(c). After 90 cycles XeF2 etching, most of the bulk silicon under membrane was removed. The membrane is completely released after 120 cycles XeF2 etching (Fig. 3(d)).

V. CONCLUSION

We have reported on the wafer-level XeF_2 pulse etching with 4 Torr pressure and each cycle time of 60 seconds. The silicon etch rate and aperture size effect for etching windows with different sizes and different shapes have been characterized. The silicon etch rate is only about 0.5~0.2 µm/cycle due to the large silicon exposure area in wafer-level processing. Both the lateral and vertical silicon etch rate are inverse proportion to number of etch cycle. The aperture size effect is extremely high for small etching windows and appears to minimize for large etching windows. By increasing the number of etch cycle, the lateral aperture size effect will be increased at first and then be reduced. However, the vertical aperture size effect will be enhanced as number of etch cycle increase. The wafer-level XeF_2 etching also exhibits slight anisotropy. For small etching windows, the vertical etch rate is larger than lateral etch rate. The design rule for XeF_2 etching was proposed to help designers during microstructure layout. The application of the design rule to a micromachined thermopile sensor layout was also demonstrated, and experimental result also validates the design rule.

ACKNOWLEDGMENT

This work was supported in part by National S&T Major Project of China under Grant 2011ZX02507-003, in part by a grant from the Major State Basic Research Development Program of China (973 Program) (No. 2011CB309501).

REFERENCES

[1] G. T. A. Kovacs, N. I. Maluf and K. E. Petersen, "Bulk Micromachining of Silicon," Proceedings of the IEEE, vol. 86, no. 8, pp. 1536-1551, Aug. 1998.

[2] L. R. Arana, N. de Mas, R. Schmidt, A. J. Franz, M. A. Schmidt and K. F. Jensen, "Isotropic etching of silicon in fluorine gas for MEMS micromachining," J. Micromech. Microeng., vol. 17, no. 2, pp. 384-392, Jan. 2007.

[3] P. B. Chu, J. T. Chen, R. Yeh, G. Lin, J. C. P. Huang, B. A. Warneke, and S. J. Pister, "Controlled pulse-etching with xenon difluroide,"in Proc. Int. Conf. Solid State Sens. Actuators, TRANSDUCERS, vol. 1, pp.665-668, 1997.

[4] K. Sugano, and O. Tabata, "Reduction of surface roughness and aperture size effect for etching of si with XeF2," J. Micromech. Microeng., vol. 12, no. 6, pp.911-916, Nov. 2002.

[5] C. Easter, and C. B. O'Neal, "Characterization of high-pressure XeF2 vapor-phase silicon etching for MEMS processing," J. Microelectromech. Syst., vol. 18, no. 5, pp.1054-1061, Oct.2009.

[6] B. Warneke and K. S. J. Pister, "In situ characterization of CMOS post-process micromachining," Sens. Actuators A, vol. 89, no. 1, pp. 142-151, Mar. 2001.

[7] B. Bahreyni and C. Shafai, "Investigation and simulation of XeF2 isotropic etching of silicon," J. Vac. Sci. Technol. A, vol. 20, no. 6, pp. 1850-1854, Dec. 2002.

[8] R. C. Hefty, J. R. Holt, M. R. Tate, and S. T. Ceyer, "Mechanism and dynamics of the reaction of XeF2 with fluorinated Si(100): Possible role of gas phase dissociation of a surface reaction product in plasmaless etching," J. Chem. Phys., vol. 130, no. 16, 164714, Apr. 2009.

[9] M. J. M. Vugts, G. L. J. Verschueren, M. F. A. Eurlings, L. J. F. Hermans and H. C. W. Beijerinck, "Si/XeF2 etching: Temperature dependence," J. Vac. Sci. Technol. A, vol. 14, no. 5, pp. 2766-2774, Sep/Oct 1996.

Precise Width Control of Single Crystalline Silicon Nano-Wall Structure Based on Wet Etching Process on (111) Wafer

Xiao Yu, Qinhua Jin, Tie Li, Yuelin Wang

State Key Laboratories of Transducer Technology & Science and Technology on Microsystem Laboratory, Shanghai
Institute of Microsystem and Information Technology, Chinese Academy of Sciences, Shanghai, CHINA
tli@mail.sim.ac.cn

Abstract—**This paper reports a novel method for precise width control of single crystalline silicon nano-wall structures using conventional top-down micro-fabrication techniques on (111) wafers. Nano-scaled walls with perfect silicon lattices on the surface were fabricated by wet etching process. The width can be controlled at the highest resolution of 80 nm when rotating the wafer by each step of 0.5 degree in alignment, achieving to fabricate silicon walls of the width as low as 134 nm by a micron level lithography mask. These nano-wall structures can be further used to fabricate high-quality silicon-nano-wires (SiNWs) with self-limiting oxidation process.**

Keywords-top-down, precise width control, nano wall.

I. INTRODUCTION

Being the most important semiconductor used in micro/nano-electromechanical system (M/NEMS), Single Crystalline Silicon (SCS) has attracted wide attentions [refs]. In recent years, researchers have found silicon nano-structures (such as nanobeams and nanowires) can exhibit unique electrical, mechanical, thermal, and optical properties[1-4], potentially applied in field-effect transistors (FETs), fieldemission devices, chemical sensors, nanoresonators, photonics, and so forth.

In general, most nanofabrication techniques for single crystalline silicon nano structure can be specified as bottom-up and top-down type approaches. The bottom-up type, usually working for silicon nanowires, is in principle simple but requires complicated fabrication procedures and techniques for positioning and aligning. On the other hand, the top-down type mainly adopts nanopatterning, deposition, etching processes on planar substrates, which is very suitable for the fabrication of the nano structure in a large quantity at the wafer scale.

Conventional top-down method using precision engineering of dry etching and lithography, provides high spatial resolution fabrication, while is extremely expensive and accessible only to large-scale integrated circuit manufacturers. Furthermore, a challenge in this method is to synthesize high aspect ratio at the level of sub-micron or nanometer. Patterning like optical and electron-beam lithographies, without concurrent sidewall damage, is some

kind of compromise for nano-scale device operation. To reduce the cost and overcome sidewall damage, wet etching for top-down silicon nanostructure fabrication were developed recently[5~7], with challenges including precision, lateral uniformity, and procedure simplification.

In this study, we present a low-cost top-down fabrication method for volume-producible SCS nanostructure using wet etching method. We developed a unique precise size control approach and fabricated SCS nanostructure with high uniformity and simple procedures on a (111) silicon wafer. By anisotropic wet etching, a straight silicon wall structure along <110> orientation with high uniform {111} sidewalls was formed by two hexagon-shaped etching cavities. The width of the wall structure was preset precisely by employing a wafer rotation angle during the wet etching mask window patterning. A resolution of 80~200 nm was achieved when the angle changed every 0.5 degree. We reduced the width of SCS wall from ~1.5 μm to ~100 nm, and fabricated suspended silicon nano wire (SiNW) of 30nm in width from the wall structure by self-limiting oxidation. The design, processing, formation and size reduction mechanism are presented and discussed subsequently.

II. FABRICATION

A. Overall fabrication process.

The proposed SCS nanostructure fabrication process is schematically shown in Fig.1.

The SCS nanosturcture were fabricated on a (111) silicon wafer covered with a 200 nm thick low pressure chemical-vapor-deposited (LPCVD) silicon nitride (Si_3N_4) layer (Fig.1(a)). Then anisotropic wet etching windows were patterned on the Si_3N_4 layer, which was then removed by reactive ion etching (RIE) (Fig.1.(b). Silicon underneath was subsequently etched by Deep RIE, for depth of t, as the height of the nano wall structure, as shown in Fig.1(c). Anisotropic wet etching process was employed to fabricate a straight <110> oriented silicon wall structure with high uniform {111} sidewalls via the crystalline wet etching of KOH solution (40%, 50℃) for 2 hour (Fig.1(d)). SiNW could be then formed by a self-limiting oxidation (Fig.1(e)), and it would suspend in free space by removing the Si_3N_4 layer and oxide layer, as shown in Fig.1(f).

978-1-4673-1122-9/12 $31.00 © 2012 IEEE

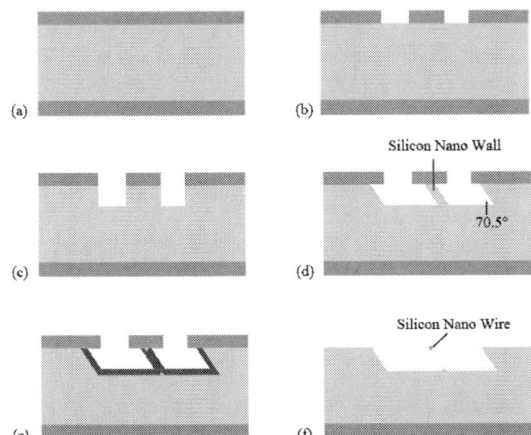

Fig.1.Fabrication processes of the sub-micro line structure.(a) oxidizing the (111) wafer, (b) pattens developing on the oxide, using conventional lithography and RIE, (c) RIE etching for silicon underneath, (d) KOH anisotropic wet etching, (e) self-limiting oxidizing, (f) Si₃N₄, SiO₂ layers removed.

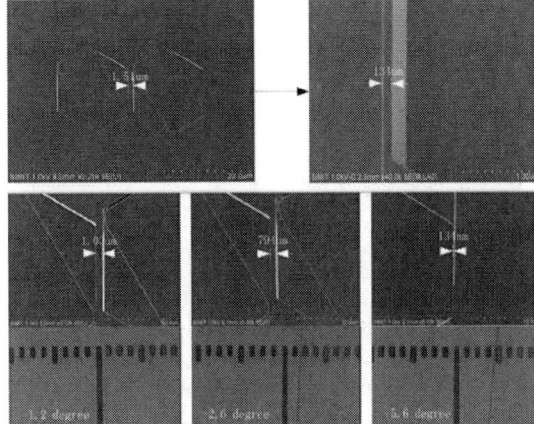

Fig.3. Width reduced for wall structure from the original (a)w_0=1.51um down to (b)w=134nm for a l=15um type cell.The rotation angles vs widths in this process were showed in (c):1.2 degree vs 1.03um，2.6 degree vs 794nm，and 5.6 degree vs 134nm.The length of silicon wall was also decreased as the rotation angle increased. The photos below the SEM views in (c) were alignment keys on the mask used to set rotation angles precisly, and the thin black lines were the <110> crystal orientation lines on the wafer.

B. Anisotropic wet etching windows designing

Fig. 2(a) is the top view of the etch cavity shape for the designed mask windows (white area), which are two tilted rectangle-shaped trenches positioning side by side. The edges of each trench are aligned along the two <110> directions. After anisotropic wet etching, two hexagon-shaped etching cavities would form as shown by the gray area, with each side along the <110> orientation. The cross-section between the two cavities was a parallelogram with a obliquity of 70.5° (Fig.2(b)). The length and width of the silicon line were determined by the scale and position of the original trenches, and the height by the etching depth t in Fig. 1(c).

This method enables us to fabricate sub-micron scaled silicon wall sturcture easily by standard MEMS processes. The lateral distance between the two trenches is designed to be more than 2 um, meeting the requirement of conventional lithography technique.

C. Size reduction process for silicon wall structure

The designed silicon wall width is determined by the distance between the projections of point A and B on the x axis set as shown in Fig.2(a). However, if rotating the wafer clockwise during wet etching (Fig.1(b)) by a special angle between <110> crystalline and the original orientation (y axis in Fig.1), the width would decrease as the new-formed etching cavities deformed (Fig.3). The black hexagon-shaped frame indicates the cavities formed after anisotropic wet etching with no wafer rotation, while

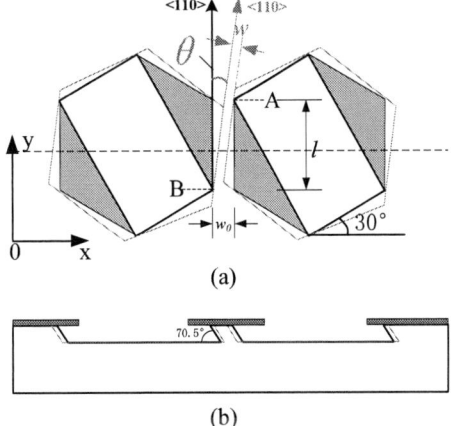

(a)

(b)

Fig.2.(a)Mask window of the titled rectangle trenches(white area), two hexagon-shaped cavities (gray area) formed after anisotropic process. If rotating the wafer in the alignment process, with the same mask window, etching cavities would be changed as red frames showed. (b)The cross-section view of the anisotropic wet etching cavities.

the red frame indicates the new-formed cavities formed with a special angle employed. As a result, the lateral dimension of the wall structure is changed, and the value can be defined as

$$w = w_0 \cdot \cos \theta - l \cdot \sin \theta - \rho \cdot t \qquad (1)$$

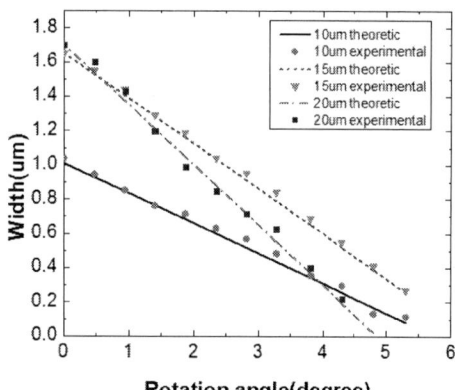

Fig.4.Width of the wall structure vs rotation angle. The theoretic values are given by formula (1). The experiments were proceed with angle resolution of 0.5 degree, and the values of l in those experiments were 10um, 15um and 20um. Wet etching rate to {111} plane in this case was ~7.4nm/h, whlie the etching time was about 2 hours.

where w_0 is the designed width, l the distance between projections of A and B on y axis, which is the designed length of the original wall structure, θ the rotation angle of the wafer meeting $\theta < arctg(w_0/l)$, ρ the wet etching rate for {111} planes, and t the etching time in KOH solution.

At fixed etching rate and time, the size reduction accuracy can achieve as high as ~80nm, for angle resolution of 0.5 degree, mainly restricted by the accuracy of lithography and RIE etching process. However, a higher accuracy could be acquired with low-rate anisotropic wet etching solution. In this case, the KOH solution had a etching rate of ~7.4 nm/h for {111} plane. To obtain ultra-thin wall structure, we calculated the rotation angle using (1) to reduce the width down to ~150nm for 1 hour anisotropic wet etching, forming the two hexagon-shaped cavities, and extended the etching time for more precise width control. In this way, we finally reduced the width down to ~70 nm.

III. RESULTS & DISCUSSION

We repeated the experiments and found that the width reduction with fixed anisotropic wet etching time had different resolution of 80~200 nm when the rotating angle changed every 0.5 degree for different designed length and width of the wall structure. Those results were showed in Fig.4, and the experimental data was well fitted to the theoretic value calculated by (1), with deviation no more than 20%. Obviously, the longer the designed length is, the lower size control precision can be achieved.

The silicon wall made by this method had a high uniformity in {111} plane without any damages to the lattices, enabling applications for optical and biochemical sensors. The height of the wall and wet etching cavities was determined by DRIE etching depth with minimal influence by the anisotropic wet etching time. Therefore, the fabrication only requires a thin top layer of the (111) wafer compatible with CMOS and other MEMS fabrication processes.

Moreover, by employing self-limiting oxidation process, high straight suspended SiNWs of width from 20nm~70nm with perfect lattices on surface were fabricated in silicon walls when we reduced the width to less than 300 nm as shown in Fig.5. The lateral accuracy could be influenced by many factors, such as the oxidizing time, the width and the height of the wall structure. Correlations of those parameters is under working.

IV. CONCLUSION

In this report, we presented a novel volume-producible SiNWs fabrication method that requires only conventional MEMS processes and commercial (111) wafers.

By adopting a wafer-rotating aligning method, we are able to control the width of a nano-scaled silicon wall structure with a highest resolution of ~80 nm in a large range after anisotropic wet etching process in KOH solution. The sidewalls had a high uniformity in {111} plane without any damages to the lattices. Moreover, by employing self-limiting oxidation process, high straight suspended SiNWs with perfect lattices on surface were fabricated in l=1.5 um silicon walls when we reduced the width to less than 300 nm.

This simple method is thus compatible for the production of high-uniform silicon nano structures on the structures of various functions. It can be predicted fabrication of silicon nano structure with this method can be potentially applied on nanoresonators, chemical sensors, biosensors, and etc.

ACKNOWLEDGMENT

This work was supported by the National Basic Research Program of China granted No. 2011CB300501, , No. 2012CB934102, the Fund for Creative Research of NSFC granted No. 60721004 and 61021064, and the key project of NSFC granted No. 60936001.

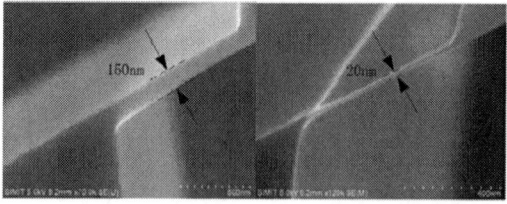
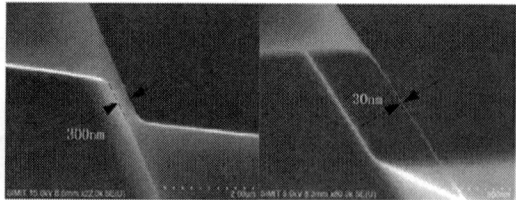

(a)   (b)

Fig.5.SEM views of (left) silicon walls and (right) SiNWs fabricated after self-limiting oxidation process from silicon walls. The width of the silicon walls and the lateral dimension of the SiNWs were (a)150nm and 20nm,(b)300nm and 30nm.

978-1-4673-1122-9/12 $31.00 © 2012 IEEE

REFERENCES

[1] K. K. Lew, L. Pan, T. E. Bogart, et al., "Structural and electrical properties of trimethylboron-doped silicon nanowires", *Appl. Phys. Lett.*, vol. 85, no. 15, pp. 3101-3103 , 2004.

[2] A. Gaidarzhy, G. Zolfagharkhani, R. L. Badzey, et al., "Evidence for quantized displacement in macroscopic nanomechanical oscillators", *Phys. Rev. Lett.*, vol. 94, Article ID 030402, 2005.

[3] D. Li, Y. Wu, P. Kim, et al., "Thermal conductivity of individual silicon nanowires", *Appl. Phys. Lett.*, vol. 83, no. 14, pp. 2934-2936 , 2003.

[4] C. T. Huang, C. L. Hsin, K. W. Huang, et al., "Er-doped silicon nanowires with 1.54 μm light-emitting and enhanced electrical and field emission properties", *Appl. Phys, Lett.*, vol. 91, Article ID 093133, 2007.

[5] Liu Wenping, Li Tie,Yang Heng,Jiao Jiwei,Li Xinxin,and Wang Yuelin, "Silicon Nanowires Fabricated by MEMS Technology and Their Electronic Performance", *Chinese Journal of Semiconductors*, 2006-09

[6] Patrick Ginet, Sho Akiyama, Nobuyuki Takama, Hiroyuki Fujita and Beomjoon Kim, "CMOS-compatible fabrication of top-gated field-effect transistor silicon nanowire-based biosensors", *J. Micromech. Microeng*, 21 (2011) 065008 (7pp).

[7] Sung-Sik Yun, Sung- Keun Yoo, Sung Yang and Jong-Hyun Lee, "Volume-producible fabrication of a silicon nanowire via crystalline wet etching of (110) silicon", *J.Micromech.Microeng*, 18(2008) 095017 (7pp).

Conductive Micro Silver Wires via Aerosol Deposition

Bulei Xu[1], Wenlong Lv[1,2], Xiang Wang[1], Tingping Lei[1], Gaofeng Zheng[1], Yang Zhao[1]*, Daoheng Sun[1]*

[1]Department of Mechanical and Electrical Engineering, Xiamen University, CHINA

[2]Pen-Tung Sah Micro-Nano Technology Institute, Xiamen University, CHINA

zhaoy@xmu.edu.cn, sundh@xmu.edu.cn

Abstract—A novel non-contact micro manufacturing technology of aerosol deposition is presented to fabricate conductive micro silver wires with width of several micrometers from silver inks. The ceramic nozzle with inner diameter of 250 μm is utilized to print the micro silver wires, the width of wires ranges from 15 μm to 70 μm and the thickness ranges from 200 nm to 2 μm. The printed silver wires are cured to decrease the resistance. High curing temperature and long curing time result in better conductivity of deposited wires due to the sufficient melt and combine of sliver particles. The minimum resistivity of 3.32 μΩ·cm is demonstrated for a continuous track cured at 300 ℃.

Keywords-aerosol deposition; silver wires; resistivity

I. INTRODUCTION

Fabrication technologies with the advantage of faster, lower cost and better process compatible on the flexible substrate have become the key for the rapid development of flexible electronics develops rapidly, such as flexible flat panel displays [1], polymer solar cells [2][3], and flexible printed circuits [4][5]. Traditional lithography which used in electronic industry requires masks, is a wasteful and time-consuming technology to fabricate micro conductive structure. As a subtractive process, lithography has low ratio of material utilization and pollution to the environment. Furthermore, traditional lithography is main used for the silicon based electronics, but isn't compatible with the flexible substrate process.

Aerosol Deposition (AD) method is a non-contact and maskless patterning technology, which provides an alternative method for the micro wires production on various substrates, such as PI, PET, glass, and so on. This deposition process can be done at atmosphere pressure and room temperature. Beside the wire from metical nano particle solution, dielectric components can be also printed from polymer particle solution. At present, aerosol deposition has been used to fabricate wires and dielectric parts in touch screen [6][7]. For the various advantages, aerosol deposition has great potential application in the electronics, such as flexible displays, touch screen, RFID tag, and so on.

The aerosol deposition system of the company Optomec is showed in Fig. 1, which consists of a pneumatic atomization

component, a virtual impactor, a nozzle, and a XY stage. The pneumatic atomization component consists of a reservoir and an atomizer nozzle. The atomizer nozzle uses a high-velocity gas stream to shear a liquid stream into droplets. Large droplets impact the sidewalls of the reservoir and drain back into it while small droplets go out with the gas flow. The virtual impactor is used to purge excess atomization gas away and stable the gas flow with nano metal particle. The remaining aerosol particles come to the nozzle and focused in one line by the sheath gas. Fine jet is ejected from the nozzle and deposited on the substrate to form micro wire. Substrate attached to the XY stage, printed wire can be direct-written to gain pre-design pattern. The XY stage also has the function of heating. There is a shutter used to stop and collect aerosol particles between nozzle and substrate to fabricate discontinuous wires, which can be used to cut off the gas jet from the nozzle.

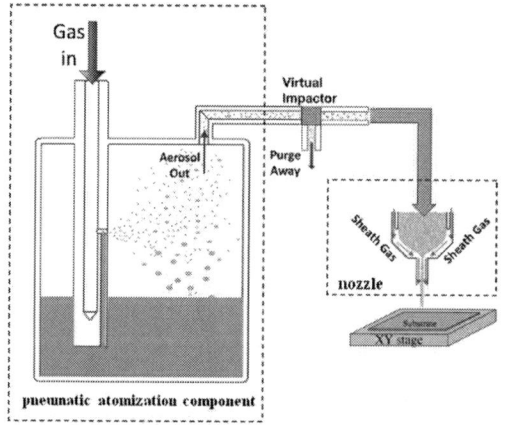

Fig. 1. Aerosol deposition system: pneumatic atomization; nozzle; XY stage.

II. EXPERIMENT

A. Surface Treatment

The surface energy is a key role that defines the

morphology of wires. Improper surface energies may have a deleterious effect on overall print quality that metal particle inks printing process must take surface energies of glasses into account. The type of ink we used is the ink of CSD-32 gained from Cabot Corporation is used in our work. For this kind of ink, the surface treatment includes five steps: (1) wipe surface of glasses; (2) put glasses in heated acetone; (3) place glasses in IPA for a minimum of 30 minutes, (4) dry glasses; (5) wait for 60 minutes before printing.

B. Printing

Fig. 2 shows the processing flow chart of aerosol deposition. Pressure test must be taken before print. And then, set the sheath gas flow with appropriate rate. When the sheath gas is stable, turn on the intake valve to input the gas and atomize the sliver ink. In the meantime, open the virtual impactor to purge away excessive aerosol particles. By adjusting the flow rate of gas in, sheath flow, purge velocity and the speed of XY stage, desired lines can be acquired. During the processing the substrate should be heated to 80 ℃ by XY stage as presintering.

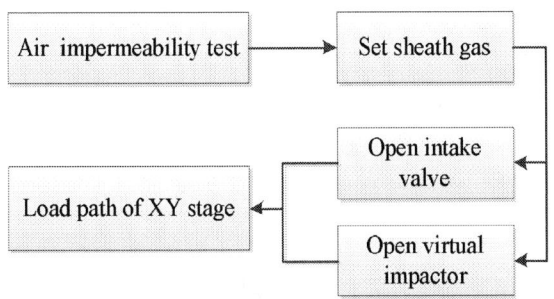

Fig. 2. Processing flow chart of aerosol deposition. The processing includes air impermeability, set sheath gas, open intake valve, open virtual impactor and load path of XY stage.

C. Thermal Treatment and Resistivity measurements

In order to know the resistivity varies with the change of time and temperature, wire samples were heated at five different temperatures and measured resistance every 5 minutes.

The width and thickness of wires were measured by Stylus Profiler and microscope. And then, the printed samples were measured by the SEM.

III. RESULTS and DISCUSSION

A. Printing Characteristics

Fig. 3 shows printed circuits. The wires are 20 μm in width and fabricated with the speed of 5mm/s in liner and 1mm/s in corner.

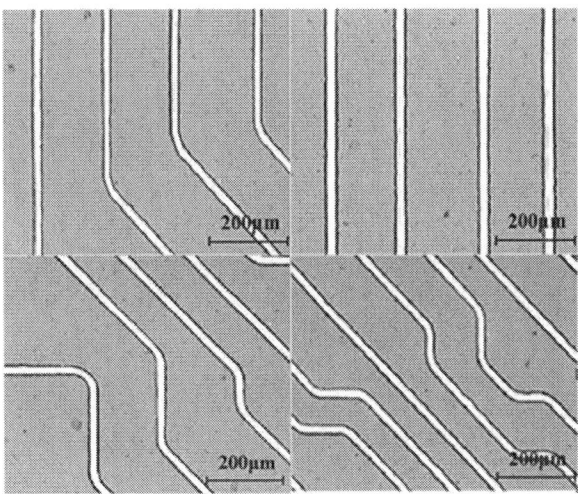

Fig. 3. AD printed circuit wires with corners and straight wires. The width of printed wires is 20 μm.

The quality of wires mainly controlled by the atomization gas flow, rate of purge away, sheath flow and speed of XY stage, so the four parameters must be mutual coordination, each factor has influence on quality of wires. Fig. 4 shows defects that may occur in processing. Experimental result shows slow corner speed can cause accumulated flow (Fig. 4a); unstable atomization condition is the main reason of nonuniform thickness (Fig. 4b). If the sheath flow is too small, the aerosol line cannot be focused and may lead to many spots at the edge of wires (Fig. 4c). To the contrary, large flow may result in discontinuous wires (Fig. 4d).

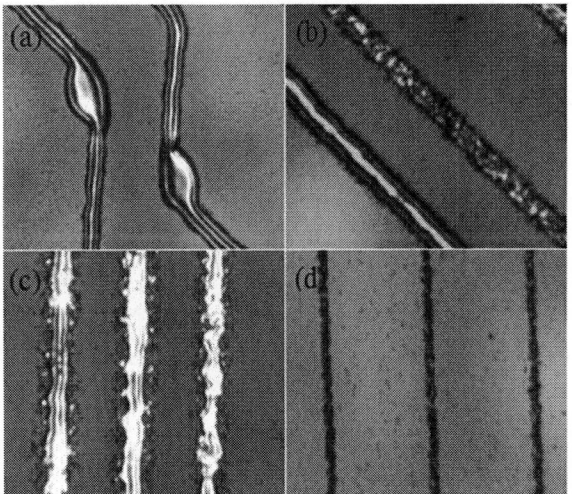

Fig. 4. Defected printed wire: (a) accumulated corner flow; (b) nonuniform thickness; (c) spots drop at the edge of wires; (d) imperfect wires.

978-1-4673-1122-9/12 $31.00 © 2012 IEEE

The velocity of atomization gas flow and the flow of purge away should match each other. Too large velocity of atomization gas flow with too small flow of purge away will lead to excessive aerosol come to the printing nozzle and block it. One the other hand, too small velocity of atomization gas flow with too large flow of purge away may result in inadequate ink and discontinuous printed wires.

B. Thermal Treatment

Thermal treatment was used to study the resistivity changes of wires. Wires were heated at five different temperatures and the resistance is measured every 5 minutes. Width and thickness of printed wires were measured by Stylus Profiler and microscope. The length of wire samples is 20mm, the width is 50 μm and the thickness is about 5 μm. Equation (1) was used to calculate the resistivity.

$$\rho = r \cdot s / l \qquad (1)$$

Where ρ is the mean resistivity, r is resistance, s is sectional area, and l is length of wires.

Five samples were fabricated, and each sample includes ten wires with the length of 20 mm.

Fig.5 shows the resistivity versus curing time under various curing temperatures. From the curves we can find out that the higher curing temperature and longer curing time result in better conductivity. The minimum resistivity of 3.32 μΩ·cm is demonstrated for a continuous track cured at 300 ℃.

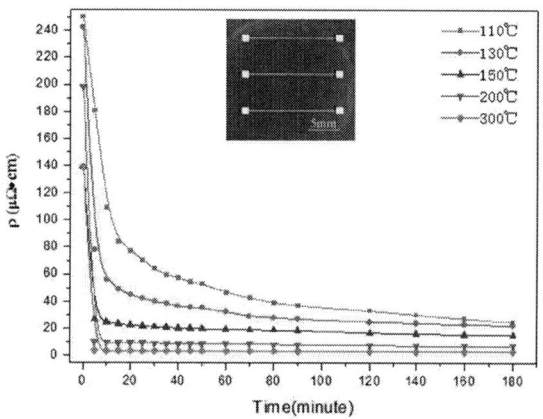

Fig. 5.   Resistivity versus with curing time under various temperatures. The top inset is a testing sample.

In order to study the resistance characters of AD printing wires further, the printing samples were measured by the SEM. Fig. 6 shows SEM images of sliver wires by aerosol deposition at different temperatures. The density of sliver particles increases with the temperature increasing. Higher curing temperature and longer curing time result in better conductivity of deposited wires, due to the sufficient

combustion and combine of sliver particles.

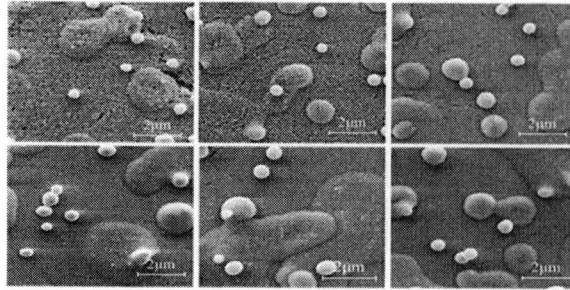

Fig. 6: SEM images of sliver wires by aerosol deposition cured at different temperatures. The curing temperature is (a) 90 ℃; (b) 110 ℃; (c) 130 ℃; (d) 150 ℃; (e) 200 ℃; (f) 300 ℃.

C. Printed Components

With the aerosol deposition system, basic electronic components can be fabricated. Fig. 7a shows a RFID tag antennae with 100 μm in wire width and 15 μm in wire thickness. The RFID tag antennae has 8 circles, the distance between each circle is 1 mm, the shape dimension is 2 cm×2 cm. Fig. 7b shows a solenoid circle printed with wire width of 20μm, the gap between two circles is 20μm. A touch screen unit is showed in Fig. 7c, which consisted of two layers. The lower layer was printed on a glass and the top layer was printed on PET. There are two printed structures in the shape of rhombuses are deposited on each layer. The rhombuses are 5 mm in width and 5 μm in thickness, which are fabricated with contiguous wires. The two layers are assembled to each other by alignment marks. The capacitance value is about 15 pf.

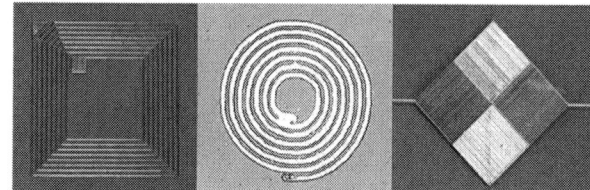

Fig. 7.   Printed components. (a) RFID tag antennae; (b) solenoid circles; (c) touch screen unit.

IV. CONCLUTION

This paper studied the fabrication of conductive micro silver wires via aerosol deposition. Influence factors of atomization gas flow, rate of purge away, sheath flow and move speed of XY stage to wire quality was discussed.

Furthermore, thermal treatment was used to study the conductive properties of sliver wires. The resistivity of sliver wires decreases with both curing temperature and time. High curing temperature and long curing time can improve the interconnection of silver particles to reduce contact resistance.

In this study, the minimum resistivity of 3.32 μΩ·cm is demonstrated for a continuous track cured at 300 ℃.

In summary, aerosol deposition is used to fabricate RFID tag antennae, solenoid circles and touch screen units. The experimental results demonstrate the potential of aerosol deposition in micro electronic manufacturing.

ACKNOWLEDGEMENT

We would like to thank Ru Xue in the Nanotechnology Center of Xiamen University for providing SEM observation. This work is supported by the National Natural Science Foundation of China (No. 51035002, No. 51105320), Natural Science Foundation of Fujian Province (No. 2011121045) and Fundamental Research Funds for the Central Universities (No. 2010121039, No. 2011121045).

References

[1] J. Jang, "Displays develop a new flexibility," *Mater. Today*, vol. 9 (2006), pp. 46-52.

[2] F. C. Krebs, "Fabrication and processing of polymer solar cells: A review of printing and coating techniques," *Sol. Energy. Mater. Sol. Cells*, vol. 93 (2009), pp. 394-412.

[3] A. C. Mayer, S. R. Scully, B. E. Hardin, "Polymer-based solar cells," *Proc. Mater. Today*, vol. 10 (2007), pp. 28-33.

[4] M. Singh, H. M. Haverinen, P. Dhagat, G. E. Jabbour, "Inkjet printing—process and its applications," *Adv. Mater.*, vol. 22 (2010), pp. 673-685.

[5] S. Bidoki, J. Nouri, A. Heidari, "Inkjet deposited circuit components," *J. Micromech. Microeng.*, vol. 20 (2010), pp. 055023.

[6] W. Hegge, D. Bohling, J. Chou, M. McAllister, "Direct dielectric line printing for touch panel display jumpers using transparent dielectric inks and aerosol Jet deposition methods," *SID Symposium Digest of Technical Papers*, vol. 42(2011), pp. 837-840.

[7] M. Essien, W. Hegge, J. Chou, P. Schottland, "Direct Aerosol Jet Printing of Fine Conductor and Transparent Dielectric Jumpers for Touch Screen Display Substrates," *LOPE-C 2011*, pp. 209-212.

978-1-4673-1122-9/12 $31.00 © 2012 IEEE

Dry Mechanical Liftoff Technology for Metallization on Parylene-C Using SU-8

Jay Han-Chieh Chang[*], Dongyang Kang, Yu-Chong Tai

[*]Department of Electrical Engineering, California Institute of Technology, Pasadena, USA

jaychang@caltech.edu

Abstract —**This paper presents a new fabrication method to employ dry mechanical liftoff of SU-8 mask to pattern metals on parylene-C film. Soaking tests were done to examine the adhesion between SU-8 and parylene-C film with different interface treatments. SU-8 masks with different thicknesses were tested in order to find out the best trade-off between the feature sizes and the easiness of removal from parelene-C film after metal deposition. Features from 10μm to 300μm in diameter and width were successfully patterned on parylene-C film by this method. The 15μm thick SU-8 liftoff mask is highly flexible and can be peeled off mechanically from parylene-C film by tweezers in seconds without any visible residue. Metal lines with testing pads from 10μm to 100μm wide and 1cm long were also fabricated and tested to evaluate their resistances. This new method provides an alternative way of metallization on parylene-C film which benefits the application in MEMS area.**

Keywords- parylene-C; liftoff; metallization; SU-8

I. INTRODUCTION

Parylene-C, a biocompatible material, is selected as the main structure and packaging material in neural interface devices because of its many favorable properties [1]. The parylene-metal-parylene sandwich skin structures which consist of one metal layer between two parylene layers were also largely utilized in our retinal implant application [2]. Due to the limit of the size of the cut on eyeball, line width of less than 10μm was often desirable in our application. The standard metal line wet etching process, however, is greatly affected by the isotropic etching time and the metal etching solution, which means thin metal lines are very easily to be etched away and the achievable highest resolution becomes difficult to control. On the other hand, liftoff process is an alternative to metal etching; unlike metal etching process where the metal is deposited before the lithography step, liftoff process requires the lithography steps to be taken before the metal deposition so the metals can be deposited in the "trenches" to achieve very high resolution metal lines. In addition to the regular photoresist used in the standard lithography process, a special photoresist, Lift-off Resist (LOR), has also been developed and widely used to create the dual-layer photoresist patterns with undercut to facilitate the liftoff process [3].

However, it was found that patterning metal features by classical metal liftoff process on parylene-C layer is still a time-consuming and difficult process. Besides, some residues of metal particles, caused by particles left on substrates during photoresist spinning, can be easily observed after liftoff process which may cause unwanted short circuits between metal lines. In addition, scratch and cracks, caused by cotton rods used to help remove metal residues during liftoff process, also appeared on metals which may affect the electrical properties of the entire devices, as shown in Figure 1.

In this work, we propose a new method to pattern metals directly on parylene-C film by a dry mechanical liftoff using SU-8 sacrificial masks. The estimated process time of this method is less than half of the time compared to classical metal liftoff process. Most importantly, it is a dry process without soaking parylene-C film in acetone or high temperature photoresist stripper (ST-22) for a long time (Table I), which will prevent delamination and increase lifetime of parylene devices as well [4]. Features from 10μm to 300μm in diameter and width have been successfully patterned on parylene-C film by this method. Smaller feature sizes can also be achieved by high-resolution masks. Metal lines with testing pads from 10μm to 100μm wide and 1cm long have also been fabricated and tested to evaluate their resistances. The results show that there is almost no observable difference in resistance among theoretical values, metals patterned by classical liftoff, and metals patterned by our proposed dry mechanical liftoff process, which demonstrates the feasibility of this method to be applied to pattern metals on parylene-C films.

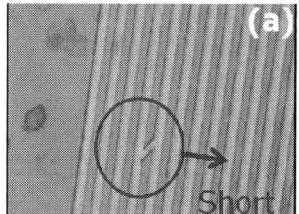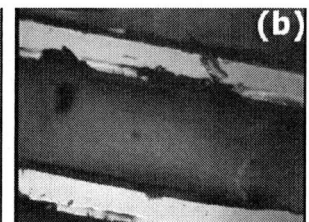

Figure 1: (a) Particles left on parylene-C substrate during liftoff photoresist spinning will cause the unwanted short circuits between two metal lines. (b) Cotton rods were often used during classical metal liftoff process to help remove the metal residues which will easily damage the metal surfaces to affect the electrical properties.

TABLE I. The estimated time of the wet process

Solution	Acetone	ST-22	Total
Time	~2hr	~2hr	>4hr

II. IDEA

During our parylene-C device process, patterning metals on parylene-C film was found to take the most time in the whole process. Besides, many defects on metal patterns were also found after this long process. These problems can probably be solved if we can pre-pattern a mask on parylene-C film before metal deposition and peel off this sacrificial mask afterward. However, this mask should satisfy several requirements: 1. to have desired adhesion to parylene-C film, (e.g. not too strong or too weak), 2. easy to be patterned by photolithography process, 3. and the features of less than 10μm can be achieved.

Therefore, several tested samples were designed and tested in high temperature (90°C) saline solution to examine their adhesion to parylene-C films by daily undercut observation to find out the best materials to be sacrificial masks.

A. Design of soaking samples

Figure 2 shows the basic structure of the soaking samples. Uniform trenches on the masks were developed away by developing solution or etched away by O_2 plasma. The width of trenches which serves as reference was carefully measured by P15 surface profiler to calculate the undercut rates. Different interface treatments were done on the interfaces between parylene-C films and the coated masks to examine the adhesion affected by the treatments.

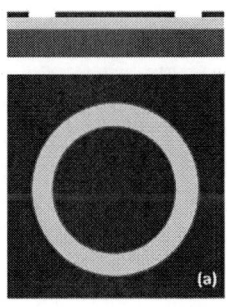 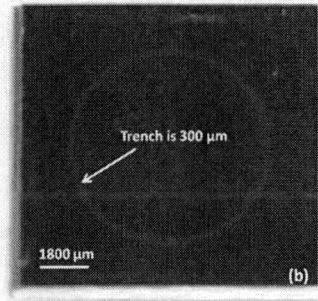

Figure 2: (a) Sample layout for soaking test. (b) Real fabricated device with 300μm trench.

B. Examination of undercut rates

Figure 3 shows the soaking results. The saline solution can attack interface between parylene-C films and the coated masks by delaminating parylene to form the so called "undercut". Undercut was observed under microscope and images were analyzed for precise calculation.

SU-8, a negative photoresist, has strong adhesion to parylene-C film after hard bake at 120°C [5]. However, soaking tests, (Table II), show that the adhesion between SU-8 and parylene-C film decreased a lot without the final hard bake step. Besides, SU-8 can be easily patterned by photolithography on parylene-C film with feature size less than 10μm. Therefore, with its desirable adhesion, SU-8 without hard baking satisfies all the requirements of the dry mask and hence has been selected as sacrificial masks for metallization on parylene-C film.

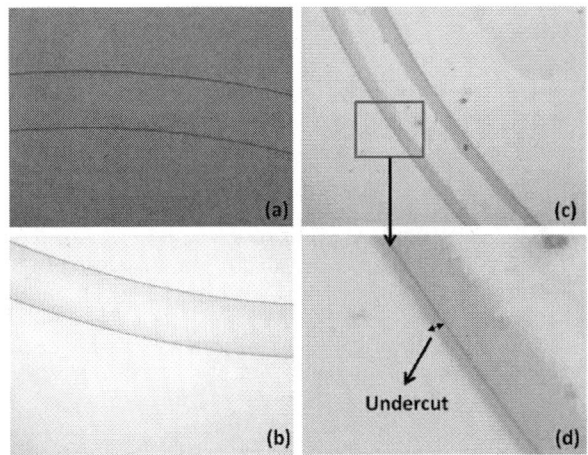

Figure 3: (a) Trench of the initial soaking device (0 days). (b) Trench of the soaking device with hard-bake after 7 days. (c) Trench of the soaking device with micro-90 detergent treatment after 7 days. (d) Close-up view of (c); undercut can be easily observed.

TABLE II. Undercut rate of the samples soaked in high temperature saline by different interface treatments (μm/day)

Treatment	Hard-bake	No treatment	Micro-90 detergent
Undercut	~0	12.85	12.48

III. FABRICATION PROCESS

As shown in Figure 4, the fabrication process started with the SU-8 spin coating on parylene-C film, followed by pre-bake at 65°C on a hotplate with controlled ramps. The samples were then exposed by UV light to pattern the desired features. Next, samples were post-bake at 95°C before developing the uncross-linked SU-8. Ultrasonic bath was also used to accelerate the developing process. After metal deposition (Ti/Au) by E-beam, the sacrificial SU-8 masks were peeled off from parylene-C films to complete the whole dry mechanical liftoff process.

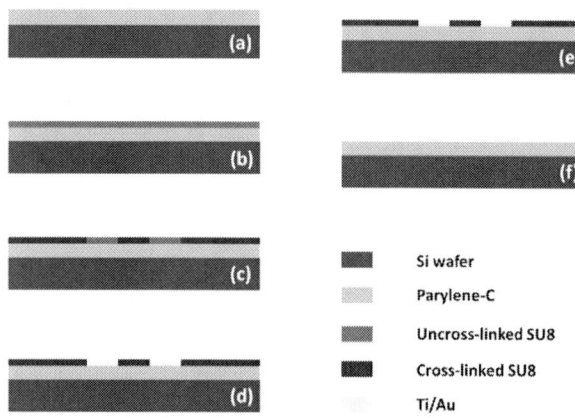

Figure 4: Fabrication process of the direct metallization on parylene-C film by sacrificial SU-8 masks. (a) Deposit parylene-C film on silicon wafer, (b) spin coat SU-8 with desired thickness on parylene-C film, followed by pre-soft bake, (c) expose by UV light and post bake, (d) develop the uncross-linked SU-8, (e) metal (Ti/Au)deposition by E-beam, and (f) peel off SU-8 mask.

As shown in Figure 5, experiments show that thinner SU-8 layer (~15μm) can be easily peeled off from parylene-C substrates mechanically by tweezers in seconds before or after metal deposition because of its flexibility. In addition, almost no visible residues or damages were observed on parylene-C films after dry peeling process.

Features as small as 10μm in diameter and width have been demonstrated on parylene-C films, as shown in Figure 6. Smaller feature sizes can be achieved by high-resolution masks. It was noted that the sacrificial SU-8 masks became more brittle after metal deposition which might be easily broken and hence the peeling process needs great care.

In addition, metal lines with different widths were also patterned on parylene-C film, as shown in Figure 7, to compare to the resistances of the metal lines formed by different methods.

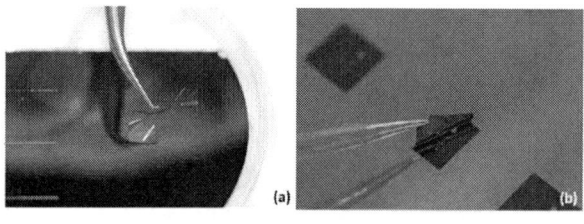

Figure 5: (a) 15μm thick SU-8 is highly flexible. (b) After

metal deposition, sacrificial SU-8 mask can be easily peeled off by tweezers in seconds, and no visible residues were observed.

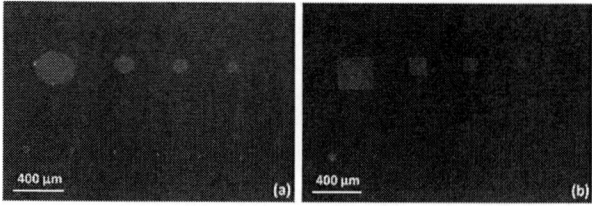

Figure 6: (a) Circles from 10μm to 300μm in diameter can be fabricated on parylene-C film by sacrificial SU-8 masks. (b) Squares from 10μm to 300μm in side length can also be fabricated on parylene-C film

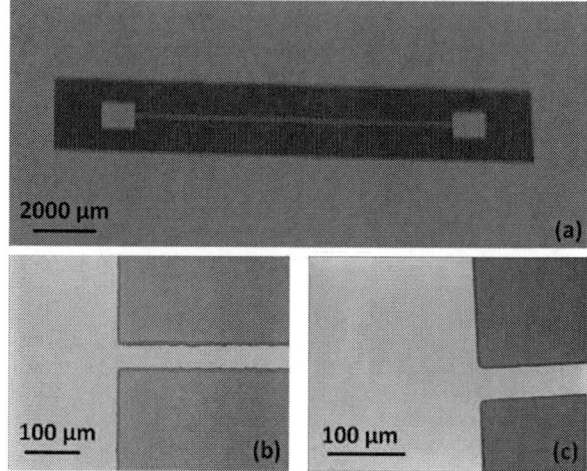

Figure 7: (a) Image of 1cm long and 40μm wide metal line along with two metal pads. (b) Microscopic image of the pattern fabricated by liftoff process. (c) Microscopic image of the pattern fabricated by sacrificial SU-8 masks.

IV. DISCUSSION

For the different metal pattern techniques, in addition to the time, the most important considerations are the resolution and the resistances of the patterned metals.

A. SU-8 thickness

SU-8 masks with different thicknesses were tried on parylene-C films to compare the easiness of removal from parylene-C films after metal deposition. It was found that thicker SU-8 masks were much stronger after metal deposition compared to the thinner ones. However, there were observable residues and even damages left on parylene-C films for these thicker SU-8 masks after peeling process which was undesirable. Therefore, despite of its more difficult handling, thinner SU-8 masks of around 15μm were still selected in this dry mechanical liftoff process.

B. Smallest feature

For our dry mechanical liftoff process, the smallest feature size only depends on the smallest feature size that SU-8 can be achieved. Theoretically, features as small as less than 10μm, which are good enough for MEMS application, can be achieved by high-resolution masks.

C. Comparison of resistance

Metal lines with different widths from 10μm to 100 μm and 1cm in length were patterned on parylene-C substrate to compare to the resistances of the metal lines formed by different methods. The theoretical electrical resistance of a wire is calculated by:

$$R = \frac{\rho L}{A} = \frac{L}{\sigma A} \qquad (1)$$

where ρ is resistivity, σ is conductivity, L is the length of the wire, and A is the cross sectional area of the wire.

In our case, the metal structures contain Ti (20nm) and Au (230nm). When estimating the resistance, we simplified the calculation by only considering the resistivity of Au (22.14 nΩ/m@20°C [6]) to represent metals to simplify the calculation. As shown in Figure 9, there are almost no observable differences in resistance among the theoretical value of metals, metals patterned by classical wet chemical liftoff process, and metals patterned by our dry mechanical liftoff process.

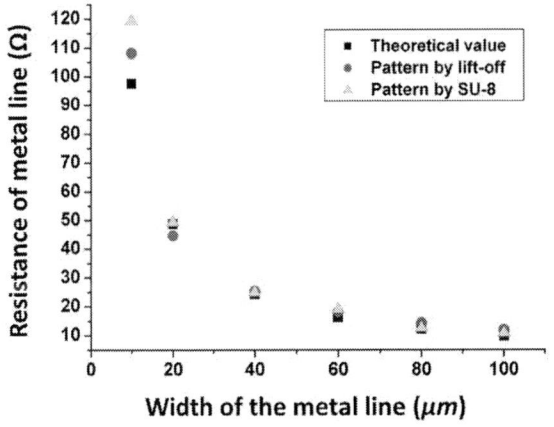

Figure 9: Comparison of the resistance of the metal lines patterned by different methods.

CONCLUSION

In this paper, we present a new alternative fabrication method to classical wet chemical liftoff by employing dry mechanical liftoff of SU-8 mask to pattern metals on parylene-C film. This kind of SU-8 without hard bake process has proper adhesion to parylene-C films, and is easily to be patterned with small feature sizes by photolithography. Sacrificial SU-8 masks with different thicknesses were tested in order to find out the best trade-off between the feature sizes and the easiness of removal from paryelene-C film after metal deposition. The 15μm thick SU-8 liftoff mask has been found to be highly flexible and can be peeled off mechanically from parylene-C film by tweezers in seconds without any visible residues. Features from 10μm to 300μm in diameter and width were successfully patterned on parylene-C film by this method. Metal lines with testing pads from 10μm to 100μm wide and 1cm long were also fabricated and tested to evaluate their resistances. The results show that metals of small feature sizes (< 10μm) and similar resistances can be patterned on parylene-C directly by our proposed SU-8 mechanical liftoff method.

This dry liftoff technology possesses several advantages, especially for metallization on parylene substrate. The totally dry process can greatly prevent delamination and increase lifetime of parylene device. This new technique provides an alternative way to pattern metals on parylene-C film which benefits the application in MEMS area.

ACKNOWLEDGMENT

This work is supported by Biomimetic MicroElectronic Systems (BMES). The authors would also like to thank Dr. Wentai Liu from UCSC for providing 268 channel retinal stimulation chips and Mr. Trevor Roper and other members of the Caltech Micromachining Laboratory for assistance with fabrication and fruitful discussion.

REFERENCES

[1] S. Nancy, " Literature Review: Biological safety of parylene C," Medical Plastics and Biomaterials, 3 (1996) pp. 30-35.

[2] J. H.C. Chang, R. Huang, and Y.C. Tai, "High-Density 256-Channel Chip Integration with Flexible Parylene Pocket," *Proc. TRANSDUCERS 2011*, pp. 378-381.

[3] MicroChem: http://www.microchem.com/pmgi-lor_faq.htm

[4] J. H.C. Chang, B. Lu, and Y.C. Tai, "Adhesion-Enhancement Surface Treatment for Parylene Deposition," *Proc. TRANSDUCERS 2011*, pp. 390-393.

[5] MicroChem: http://www.microchem.com/pdf/SU8_2-25.pdf

[6] Wikipedia: http://en.wikipedia.org/wiki/Gold

Fabrication of Nanogap Electrode Using Electromigration Method During Metal Deposition

Tatsuhiko Ohata[1], Yasuhisa Naitoh[2,*], Masayo Horikawa[2], Dong F Wang[1,*], and Ryutaro Maeda[3]

[1] Micro Engineering & Micro Systems Laboratory, Ibaraki University (College of Eng.), Hitachi, Ibaraki 316-8511, JAPAN
[2] Nanosystem Research Institute, AIST, Tsukuba, Ibaraki 305-8564, JAPAN
[3] Research Center for Ubiquitous MEMS and Micro Engineering (UMEMSME), AIST, Tsukuba, Ibaraki 305-8564, JAPAN
([2,*] E-mail: ys-naitou@aist.go.jp; [1,*] E-mail: dfwang@mx.ibaraki.ac.jp)

Abstract—**In this paper, we suggest a new electormigration method to improve the previous electromigration method. The method is that electrical break-down is carried out during metal deposition for the fabrication of electrodes. As a result, the nanogap structure can be confirmed and break-down current showed lower magnitudes than the previous method. These result indicated possibility of applications to mass production of nanogap electrodes.**

Keywords- nanogap; electromigration

I. INTRODUCTION

Recently, there has been well studied into nanogap electrodes for applications of electronic devices [1], optical devices [2] and bio-devices [3]. Nanogap electrodes comprise two electrodes facing each other across nanometer-separation. Characteristics of such devices with nanogap electrodes are closely dependent on the gap width. Especially, under 10 nm gap width is necessary for applications of a resistance switch [1]. Therefore, nanogap fabrication with well-defined gap widths is very important. Because the gap width is extremely smaller than the current fabrication accuracy (approximately 35 nm in 2011), an additional fabrication process is necessary for application of the nanogap devices. Currently, various methods have been studied about the process; electron beam lithography [4], electroplating [5], molecular lithography [6], and electromigration [7]. Especially, the electromigration method is one of the most popular methods for the fabrication of nanogap electrodes. In this method, nanogap structures are fabricated using electrical break-down by applying overcurrent through conductive nanowires. However, when this method is adapted on very thin nanowires, this method has poor reproduction. It has been reported that process currents are dependent on the critical current density, the magnitude of which is $0.7\text{-}3.1 \times 10^{12}$ A/m^2 for gold nanowires [8]. Recently, feedback controlled electromigration (FC-EM) method [9], which is an improved electromigration method, was investigated. This method is highly reproducible production of nanogap electrodes. However it is not suitable for mass production for two problems. One is that processing currents are too large. For example, around 1 mA, which is estimated by the above critical current density, is necessary for gold nanowire, the cross-section area of which is 45 nm × 45 nm. Another is that this method is very time-consuming since

applied voltages are controlled with checking process current carefully during the FC-EM method. In order to solve these problems, we demonstrated a new method which is simultaneous process of applying a voltage for electromigration and metal deposition for the fabrication of electrodes. Figure 1 shows a schematic diagram of the new methods. The voltage is applied between the facing first electrodes which are fabricated as pre-structure. During the metal deposition, metal islands are initially fabricated (Fig. 1(a)). Then, as soon as deposited metal layer are electrically connected the first electrodes, current flows are concentrated through the metal bridged parts and the bridges are removed by electromigration. During the process, metal deposition and electrical break-down of metal bridges are also caused at every moment (Fig. 1(b) and 1(c)). When electromigration processes are enough to remove the all bridges, it is expected that cross-section areas of the removed bridges during the electromigration can be kept small area and nanogap structures are formed with low process currents compared with the previous electromigration methods. In this study, the electrical properties during the new electromigration method were investigated. These properties and fabricated nanogap structures are discussed herein.

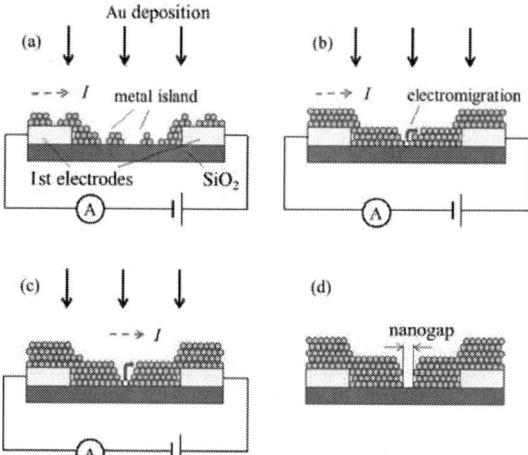

Fig. 1. Diagrammatic illustration of electromigtration during metal deposition (a)-(c) and after fabrication (d).

II. EXPERIMENTAL

Nanogap electrodes were prepared by two cycles of photolithography and Au evaporation on a Si substrate covered with 300-nm-thick thermally oxidized layer. Figure 2 shows a schematic diagram of the fabrication process. In the first cycle, pre-structures which have 100-nm-thick Au electrodes and 5-μm-gap structures were fabricated (Fig.2 (a)). Then, 1.5-μm-width slit was bridged across the 5-μm-gap by the second photolithography (Fig.2 (b)), and the second Au deposition was carried out on the substrate (Fig.2 (c)). The deposition rate and film thickness was 0.02 nm/s and 20 nm, respectively. During the second deposition, a voltage of 10 V is applied for electromigration with measuring process currents using a Keithley 4200 Semiconductor Characterization System at room temperature. These Au films have 2-nm-thick Cr layer for adhesion layer. Electrical properties measurements were carried out without exposing to air after the depositions. Fig. 2(d) are illustrated the fabricated electrodes after lift-off.

during the process. It is suggested that the bridged Au atoms were removed by electromigration at each moment, as predicted in Fig. 1(b) and 1(c). In the region D, the currents merely change. This indicates that fabricated structures are stable with applying of 10 V after the Au deposition. In spite of requiring several ten mA which is estimated by the above critical current density for break-down of an Au wire with 1.5 μm × 20 nm in the previous methods, we achieved that process current was suppressed under one thousandth. Figure 4(b) shows a field-emission scanning electron microscopy (FE-SEM) image of fabricated sample just after the process as shown in Fig. 2(c). Fig. 4 (a) shows a FE-SEM image of the sample without applying any voltages for a reference. The specific structure is not shown in Fig. 4(a). However Au electrical break-down structure is shown on the second Au layer between electrodes of pre-structures in Fig. 4(b). This clearly indicates that electromigration method could be combined with metal deposition process.

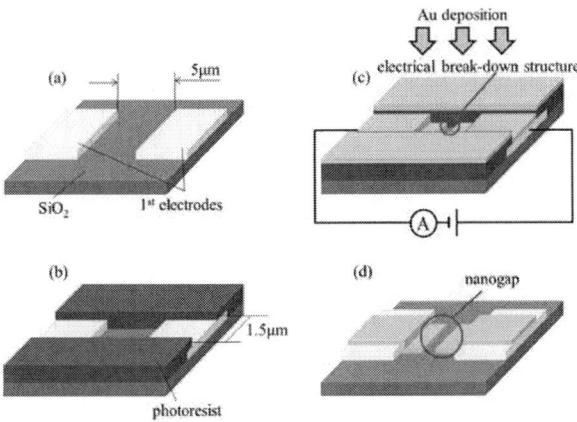

Fig. 2. Fabrication procedure of nanogap electrodes, where the electrical breakdown structure in Fig. 2 (c) corresponds to that observed in Fig. 4 (b).

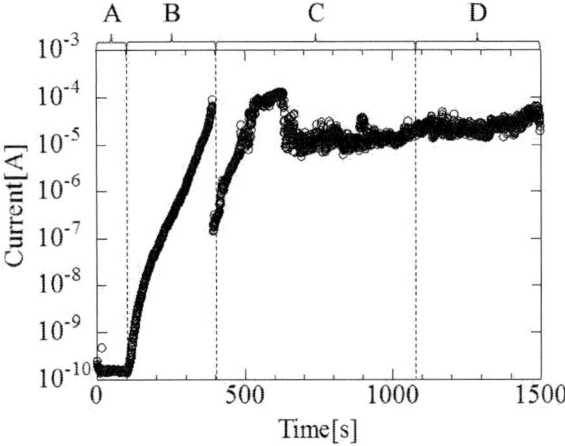

Fig. 3. The time dependence of current and resistance during Au deposition with applying 10V.

III. RESULTS AND DISCUSSION

Fig. 3 shows the time dependence of current during the second Au deposition with applying 10 V. Au deposition was carried out from 100 to 1080 s (range B and C). Before the Au deposition, the observed currents were merely changed in region A. In the region B, currents were exponentially increased. This implies that the exponential change is explained by tunneling current change. It is considered that the tunnel resistance decreases exponentially by the metal island spacing decreases with the growth of the island structure as shown in Fig. 1(a). In the region C, some current peaks are observed, however the current becomes approximately constant. The maximum current during the process is about 130 μA, the conductance of which is smaller than quantization conductance (77.5 μS: the conductance of single Au atom) [10]. This indicates that metal bridges could not be maintained

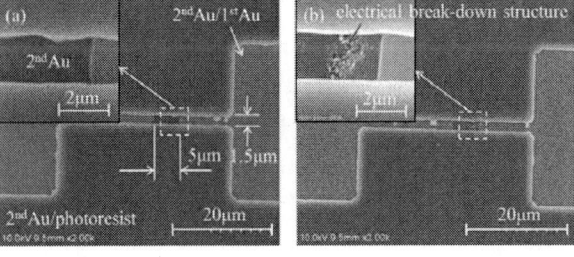

Fig. 4. The field emission scanning electron microscope (FE-SEM) image of fabricated sample just after the process: (a) fabricated sample with non-voltage and (b) fabricated sample with 10 V.

Fig. 5 shows the current-voltage characteristics of the fabricated sample in the range from -1V to 1V. Open circles and a solid line are indicated the observed currents and a fitting curve, respectively. The observed currents are quite below the quantized conductance. This shows that the electrodes do not contain metal bridges. The fitting curve was estimated from the tunneling equation [11]. In this equation, the tunneling current is described as

$$I = \frac{k_1 A}{s^2}[x^2 \exp(-k_2 sx) - y^2 \exp(-k_2 sy)] \qquad (1)$$

where $x = \sqrt{\phi - V/2}$, $y = \sqrt{\phi + V/2}$, $k_1 = 6.32 \times 10^{10}$ Vs^{-1}, and $k_2 = 1.025$ J$^{-1/2}$. The s, A, and ϕ stand for the gap length, the tunneling-emission-area, and the barrier height, respectively. From the results of fitting curve, the parameters were obtained as shown in table I. The gap width is estimated to be 0.55 $\pm$ 0.0086 nm.

NDR in current-voltage curve is characteristic phenomena in nanogap electrodes which has under 10 nm of gap width [1]. These results clearly indicate that the fabricated sample has nanogap structures. In Fig. 6(b), the resistance exponentially decreases at 4V, and resistance gradually increases according to be the high voltage. The resistance gradually decreases according to be lower the voltage from 10 V. It stops at around 4V and resistance gradually increases from 4 to 0 V. It shows clear resistance hysteresis. The on/off resistance ratio measured at voltages lower than 3V is about 10^2. It is conjectured that the change in the nanogap structure causes a large resistance change like the change in the previous study [1]. This resistance hysteresis could be observed over a few ten times. Thus, the fabricated sample also exhibited a potential reversible resistance memory effect.

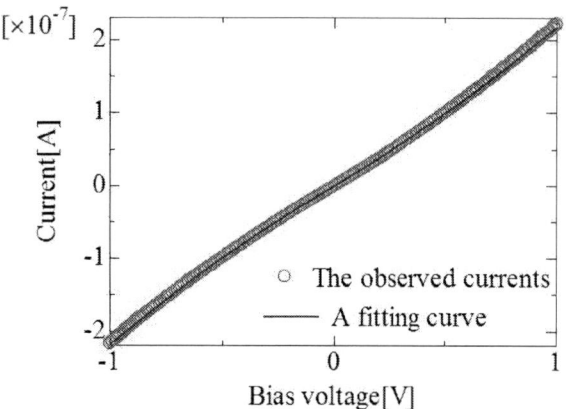

Fig. 5. The current-voltage characteristics of fabricated nanogap electrode when applying a voltage swept from -1 to 1V and a fitting curve estimated from tunneling equation.

TABLE I.   THE PARAMETERS OF A FITTING CURVE

	Coefficient values $\pm$ standard deviation
s (nm)	0.54629 $\pm$ 0.00858
A (nm^2)	0.82502 $\pm$ 0.0203
ϕ (eV)	0.75341 $\pm$ 0.0279

Fig. 6(a) and 6(b) shows the current-voltage and resistance-voltage characteristics of the fabricated sample. The blue circles are the results obtained when the voltage varied from -1 to 10 V, and the red circles are those obtained from 10 to -1V. Before the measurements, the sample was initially applied the voltage of 10V. These red circles clearly show the negative differential resistance (NDR) with a peak voltage of 4V. The

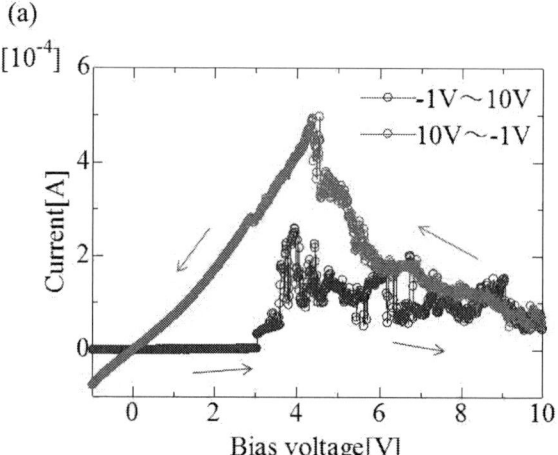

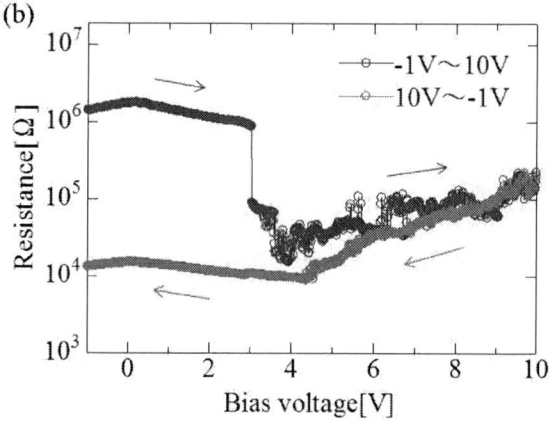

Fig. 6. The current-voltage (a) and resistance-voltage (b) characteristics of the fabricated sample when applying a voltage swept from -1 to 10V and from 10 to -1V.

IV. CONCLUSIONS

In this paper, we have proposed a new method of fabricating nanogap electrode using electromigration method during a metal deposition. As a result, compared with the previous electromigration methods, reduction of process current could be achieved. The resistance of the fabricated sample is over quantized resistance. Clear negative differential resistance and the resistance hysteresis were also observed. These imply that nanogap structure was formed in the sample. This technique achieved simultaneously fabrication of nanogap structure and metal electrodes. We expect that this method is adaptable to mass production of nanogap devices.

ACKNOWLEDGMENTS

This work was performed in the Nanosystem Research Institute of National Institute of Advanced Industrial Science & Technology (AIST). Part of this work was also supported by MEMS Inter University Network via Research Center for Ubiquitous MEMS & Micro Engineering (UMEMSME) of National Institute of Advanced Industrial Science & Technology (AIST).

REFERENCES

[1] Y. Naitoh, M. Horikawa, H. Abe and T. Shimizu, "Resistance switch employing a simple metal nanogap junction," Nanotechnology 17, p. 5669, 2006.

[2] K. Ueno, S. Takabatake, Y. Nishijima, V. Mizeikis, Y. Yokota and H. Misawa, "Nanogap-Assisted Surface Plasmon Nanolithography," J. Phys. Chem. Lett., 1 (3), pp. 657–662, 2010.

[3] T. L. Changa, , C. Y. Tsaib, C. C. Sunb, C. C. Chenc, L. S. Kuob, P. H. Chenb, "Ultrasensitive electrical detection of protein using nanogap electrodes and nanoparticle-based DNA amplification," Biosensors and Bioelectronics 22 pp. 3139–3145, 2007.

[4] M. S. M. Saifullah1, T. Ondarçuhu, D. K. Koltsov, C. Joachim and M. E. Welland, "A reliable scheme for fabricating sub-5 nm co-planar junctions for single-molecule electronics," Nanotechnology 13, p. 659, 2002.

[5] A. F. Morpurgo, C. M. Marcus, and D. B. Robinson, "Controlled fabrication of metallic electrodes with atomic separation," Appl. Phys. Lett. 74, p. 2084, 1999.

[6] H. Tanaka, M. E. Anderson, M. W. Horn1 and P. S. Weiss, "Position-Selected Molecular Ruler," Jpn. J. Appl. Phys. 43, L950, 2004.

[7] H. Park, A. K. L. Lim, A. P. Alivisatos, J. Park and P. L. McEuen, "Fabrication of metallic electrodes with nanometer separation by electromigration," Appl. Phys. Lett. 75, p. 301, 1999.

[8] C. Durkan, M. A. Schneider, and M. E. Welland, "Analysis of failure mechanisms in electrically stressed Au nanowires," J. Appl. Phys. 86, p. 1280, 1999.

[9] D. R. Strachan, D. E. Smith D. E. Johnston, T. –H. Park, Michael J. Therien, D. A. Bonnell, A. T. Johnson, "Controlled fabrication of nanogaps in ambient environment for molecular electronics," Appl. Phys. Lett. 86, p. 043109, 2005.

[10] J. L. Costa-Kra¨mer, N. Garcı´a and H. Olin, "Conductance quantization histograms of gold nanowires at 4 K," Phys. Rev. B 55, p. 19, 1997.

[11] G. Simmons, "Generalized Formula for the Electric Tunnel Effect between Similar Electrodes Separated by a Thin Insulating Film," J. Appl. Phys. 34, p. 1793, 1963.

Pseudo-spin filter in metallic single-walled carbon nanotubes

Dario Bercioux[1†], Leonhard Mayrhofer[2]

[1]Freiburg Institute for Advanced Studies, Albert-Ludwigs-Universität, D-79104 Freiburg, Germany
[2]Fraunhofer IWM, Wöhlerstraße 11, D-79108 Freiburg, Germany
[†]Email: dario.bercioux@frias.uni-freiburg.de

Abstract—We analyze the scattering properties of low-energy electrons in metallic single-walled carbon nanotubes in the presence of structural defects. In particular, we focus on vacancy defects characterized by the lack of two carbon atoms. We show how the geometry of the defect with respect to the tube symmetry establishes the available scattering channels. Furthermore, we show how the reconstruction of the chemical bond among the carbon atoms at the place of the defect can induce a selection rule based on the pseudo-spin of the scattered electrons. This selection rule results in an effective filtering of the scattered electrons with respect to the pseudo-spin.

I. INTRODUCTION

A single-walled carbon nanotube (SWNT) can be regarded as a strip of graphene rolled into a hollow cylinder with diameters typically in the order of one nanometer and lengths ranging from a few microns to a few centimeters [1]. Because of the very small ratio between diameter and length, a SWNT is generally considered as a quasi-one-dimensional system. SWNTs can be either metallic or semiconducting with energy band-gaps varying from zero to above 1 eV, depending on their geometrical structure [1]. This unique electronic properties of SWNTs have attracted great attention in fundamental and applied research due to the possibility of exploring distinctive phenomena of one-dimensional systems [1], [2], [3]. This interest is mainly a consequence of the unusual band structure of SWNTs that in the metallic case results in an effective one-dimensional Dirac-Weyl theory [2]. From the theoretical and experimental point of view, SWNTs without defects are well characterized — however, the presence of defects or tube endings can be employed for the observation of quantum mechanical coherence phenomena such as quantum interference [4], [5], [6].

In this work we focus on the low-energy electrons in metallic SWNTs — these that can be described with an effective one-dimensional Dirac-Weyl theory. In this regime, electrons are characterized by a linear energy dispersion and by a quantum number σ named *pseudo-spin* that is related to the effective electron motion direction. This quantum number is the one-dimensional projection of the pseudo-spin of electrons in the underlying graphene lattice in the low-energy approximation [2]. In particular, we investigate the electron scattering properties in the presence of structural defects. We consider a defect constituted by the lack of two carbon atoms — a *di*-vacancy — in the SWNT structure. We show that the interplay between the geometry of the defect and

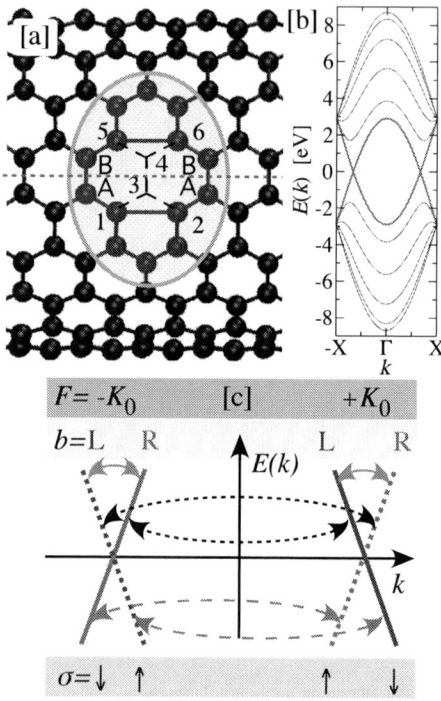

Fig. 1. Panel [a]: a double-vacancy defect that is symmetric with respect to the pseudo-spin symmetry axes parallel to the tube axes (red dotted line). Panel [b]: the energy spectrum of an armchair single-walled carbon nanotube. Panel [c]: linearization of the complete energy spectrum around the charge neutrality point — scheme of the various scattering processes. On the bottom, the actual value of the pseudo-spin is given. Solid arrows indicate intra-valley backscattering, dotted arrows represent inter-valley backscattering and finally dashed arrows denote inter-valley forward scattering processes.

the fundamental symmetry of the tube determines the open scattering channels, whereas the reconstruction of chemical bonds between the carbon atoms at the defect site induces an asymmetry in the scattering property of electrons with different pseudo-spins.

II. MODEL AND SYMMETRY ANALYSIS

We consider an armchair SWNT in the presence of a *di*-vacancy defect (DVD).[1] Several simulations [7], [8], [9] and

---

[1]The approach and the results we present are quite general and we have also extended our investigation to other structural defects, *e.g.* Stone-Wales defects [13].

experiments [10], [11], [12] have indicated that DVDs are likely the prevailing defect-type in SWNTs and graphene.

In proximity of the charge neutrality point, the energy spectrum of armchair SWNTs can be approximated by two linearly dispersing branches $\mathcal{E} = \pm\hbar v_F |k - K_0|$ centered around the two valleys $K_0 = \pm 4\pi/3\sqrt{3}a_{CC}$, where $a_{CC} \approx 0.142$ nm is the carbon-carbon bond length and $v_F \sim 10^6$ m/s is the Fermi velocity. In this approximation the slowly oscillating part of the electronic wave function can be expressed as [14]

$$|\psi_\sigma\rangle = \frac{1}{\sqrt{2}} \left(|A\rangle + \sigma|B\rangle \right). \quad (1)$$

Here, $|A\rangle$ and $|B\rangle$ refer to the wave functions delocalized along the two sub-lattices of A and B of the underlying graphene structure. The pseudo-spin σ is defined as

$$\sigma = \text{sgn}(bF), \quad (2)$$

where $b = \pm$ = Right/Left is the electron motion direction and $F = \pm K_0$ is the valley index (c.f. Fig. 1c).

The structure in Fig. 1a exhibits a useful mirror symmetry at the DVD. A simultaneous exchange of the two carbon atoms — A and B — of the honeycomb SWNT lattice leaves the system invariant. In the low-energy approximation this operation is described by the pseudo-spin symmetry operator $\mathcal{S}$ [13]. It commutes with the SWNT translational operator parallel to the tube axes. On the subspace of the two sub-lattices it can be represented by $\mathcal{S} = \sigma_x$, where σ_x is the corresponding Pauli matrix. The symmetric [antisymmetric] eigenstates of the pseudo-spin operator are the pseudo-spin + [-] eigenstates of the SWNT from Eq. (1). Hence no scattering between states with different pseudo-spin occurs at the DVD shown in Fig. 1a. Additionally, for any defect in a SWNT that does not mix p_z electrons from the same sub-lattice, the particle-hole symmetry around the charge neutrality point that is inherent to the Dirac-Weyl description is conserved [13]. Hence, disregarding a reconstruction of chemical bonds at the defect site, the DVD shown in Fig. 1a has a particle-hole symmetric electronic structure. Thus, in this approximation, the transmission probability across the DVD for pseudo-spin up ($\sigma = +$) and pseudo-spin down ($\sigma = -$) states can be expected to be identical at the charge neutrality point,

$$T_{\sigma=-} = T_{\sigma=+}. \quad (3)$$

Note, that Eq. (3) holds for any defect that is symmetric on the two sub-lattices and avoids mixing of p_z states from the same sub-lattice. So in general for this kind of defects no pseudo-spin filtering effect can be expected. However, theoretical studies [7], [8], [9] have clearly shown that a recombination of dangling bonds at the DVD site is energetically most favorable, leading to the bonding of atoms A_5-A_6 and B_1-B_2 in Fig. 1a. Therefore, in the next section we present a more quantitative description of the electronic scattering properties of the DVD, that allows to examine the effect of the actual atomistic structure of the DVD.

III. PERTURBATIVE APPROACH

Here we present a more quantitative understanding of a possible pseudo-spin filtering effect of DVDs — based on Fermi's Golden Rule (FGR) approximation [15]. The Hamiltonian of the defected SWNT can be split into three parts containing the defect-less tube $\mathcal{H}_0$, the DVD term containing the two missing carbon atoms $\mathcal{H}_{DVD}$, and the reconstruction term containing the possible reconstruction of chemical bonds between the carbon atoms at the DVD site $\mathcal{H}_R$. For the defect-less tube we have the standard tight-binding description

$$\mathcal{H}_0 = -t \sum_{\langle i,j \rangle} c_i^\dagger c_j,$$

where the sum runs over all pairs of nearest-neighbor carbon atoms, t is the tight-binding hopping parameter and $c_i^\dagger$ [c_i] is the creation [annihilation] operator for a p_z electron at atom i. In the basis of the localized p_z electrons, the DVD Hamiltonian is expressed as

$$\mathcal{H}_{DVD} = \begin{pmatrix} \ddots & \vdots & \vdots & \vdots & \vdots & \vdots & \vdots \\ & B_1 & B_2 & A_3 & B_4 & A_5 & A_6 & \\ \cdots B_1 & 0 & 0 & t & 0 & 0 & 0 & \cdots \\ \cdots B_2 & 0 & 0 & t & 0 & 0 & 0 & \cdots \\ \cdots A_3 & t & t & 0 & 0 & 0 & 0 & \cdots \\ \cdots B_4 & 0 & 0 & 0 & 0 & t & t & \cdots \\ \cdots A_5 & 0 & 0 & 0 & t & 0 & 0 & \cdots \\ \cdots A_6 & 0 & 0 & 0 & t & 0 & 0 & \cdots \\ & \vdots & \vdots & \vdots & \vdots & \vdots & \vdots & \ddots \end{pmatrix} \quad (4)$$

where the tags A_i and B_j in red represent the carbon atoms i, j on sub-lattices A and B as shown in Fig. 1a. Finally, the Hamiltonian representing the reconstruction of the chemical bonds at the DVD site is given by the following matrix,

$$\mathcal{H}_R = \begin{pmatrix} \ddots & \vdots & \vdots & \vdots & \vdots & \vdots & \vdots \\ & A_1 & B_2 & A_3 & B_4 & B_5 & A_6 & \\ \cdots B_1 & 0 & -t_R & 0 & 0 & 0 & 0 & \cdots \\ \cdots B_2 & -t_R & 0 & 0 & 0 & 0 & 0 & \cdots \\ \cdots A_3 & 0 & 0 & 0 & 0 & 0 & 0 & \cdots \\ \cdots B_4 & 0 & 0 & 0 & 0 & 0 & 0 & \cdots \\ \cdots A_5 & 0 & 0 & 0 & 0 & 0 & -t_R & \cdots \\ \cdots A_6 & 0 & 0 & 0 & 0 & -t_R & 0 & \cdots \\ & \vdots & \vdots & \vdots & \vdots & \vdots & \vdots & \ddots \end{pmatrix}, \quad (5)$$

where we have used the same notation for the carbon atoms as in Hamiltonian (4). The actual value of the hopping element t_R associated to the lattice reconstruction of the chemical bond between the carbon atoms at the DVD site is determined via DFT calculation [16]. For the DVDs we are presenting here, we have determined a value of $t_R/t \sim 0.5$.

In the real space representation the wave function (1) is expressed as $[\psi_\sigma(\mathbf{r}) \to \psi_{\sigma F \kappa}(\mathbf{r})]$

$$\psi_{\sigma F \kappa}(\mathbf{r}) = \langle \mathbf{r} | \sigma F \kappa \rangle = \sum_{p=A,B} f_{p\sigma} \varphi_{pF\kappa}(\mathbf{r}). \quad (6)$$

Here κ is the momentum relative to $F = \pm K_0$, and σ is the pseudo-spin (2). The coefficients $f_{p\sigma}$ account for the two inequivalent carbon atoms ($p = $ A,B) in the SWNT lattice structure. In accordance with Eq. (1) they are given by

$$f_{p\sigma} = \frac{1}{\sqrt{2}} \begin{cases} 1 & p = \text{A} \\ \sigma & p = \text{B} \end{cases}. \tag{7}$$

The function $\varphi_{pF\kappa}(\boldsymbol{r})$ describes the armchair SWNT wave function component on sub-lattice p. The functions $\varphi_{pF\kappa}(\boldsymbol{r})$ in Eq. (1) are defined as

$$\varphi_{pF\kappa}(\boldsymbol{r}) = \frac{1}{\sqrt{N_L}} \sum_{\boldsymbol{r}_p} e^{i(F+\kappa)R_x} \chi(\boldsymbol{r} - \boldsymbol{r}_p),$$

where the sum runs over the SWNT lattice sites, $2N_L$ is the total number of carbon atoms, and $\chi(\boldsymbol{r} - \boldsymbol{r}_p)$ are the localized p_z orbitals at the positions $\boldsymbol{r}_p = \boldsymbol{R} + \boldsymbol{\tau}_p$ on sub-lattice p, and $\boldsymbol{\tau}_p$ is the displacement vector within the graphene lattice unit cell [14]. Finally, R_x is the projection of the lattice vector onto the tube axis.

With the use of the previous definitions, we can evaluate the transition probability between generic eigenstates $|\psi_{\sigma F\kappa}\rangle$ and $|\psi_{\sigma' F'\kappa'}\rangle$. By using the FGR approximation, the transition probability between these states is given by

$$\mathcal{P}_\sigma^{(\sigma')} \propto |_f\langle \sigma' F' \kappa' | \mathcal{H}_{\text{DVD}} + \mathcal{H}_{\text{R}} | \sigma F \kappa \rangle_i|^2 \, \delta(E_f - E_i)$$

where $E_{i/f}$ are the energies of the initial and final states, respectively. The previous expression explicitly accounts for the microscopic details of the defect.

All the Hamiltonians we have introduced so far commute with the pseudo-spin operator $\mathcal{S}$. Therefore the pseudo-spin is a good quantum number, as a consequence it is conserved during a scattering process [13]. This means that a DVD — with the same spatial orientation as the one in Fig. 1a — can only induce scattering processes between states characterized by the same initial and final pseudo-spin state, $\sigma = \sigma'$. If we refer to all the possible scattering processes among states on the two valleys $\pm K_0$ — Fig. 1c — we hence observe that only inter-valley backscattering processes have a finite probability. On the contrary, intra-valley backward and inter-valley forward processes are characterized by a zero scattering probability [13].

Let's start by analyzing the results in the case where the lattice reconstruction is neglected $t_{\text{R}} = 0$. Here we obtain

$$\mathcal{P}_\sigma^{(\sigma')}(\mathcal{E}) \propto \delta_{\sigma\sigma'} \left[2\sigma t \cos\left(\frac{2\pi}{3} + \sigma \frac{\mathcal{E}R_x}{\gamma} \right) \right]^2 \tag{8}$$

where $\mathcal{E} = \gamma\kappa$ with $\gamma = 3a_{\text{CC}}t/2$ is the energy dispersion around the charge neutrality point. These results are in agreement with the one proposed by Eq. (3); at the charge neutrality point ($\mathcal{E} = 0$) the scattering and therefore also the transmission probability does not depend on the pseudo-spin σ of the considered state. Moreover, particle-hole symmetry is conserved, i.e., $\mathcal{P}_\sigma^{(\sigma)}(\mathcal{E}) = \mathcal{P}_{-\sigma}^{(-\sigma)}(-\mathcal{E})$ as shown in Fig. 2. When lattice reconstruction is considered $t_{\text{R}} \neq 0$, the physics of the scattering processes is enriched with new physical phenomena.

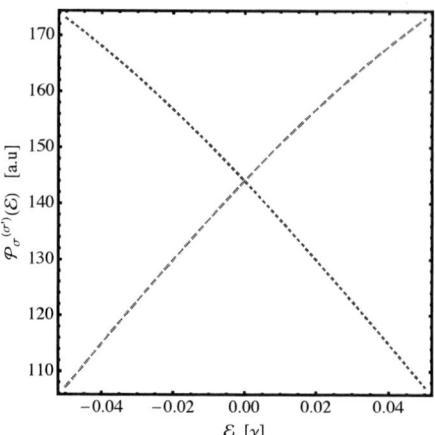

Fig. 2. Not normalized transition probability $\mathcal{P}_\sigma^{(\sigma')}(\mathcal{E})$ as a function of the energy states for electrons in a metallic SWNT in the case of a DVD as in Fig. 1a. The red-dashed refers to $\mathcal{P}_+^{(+)}$ probability, whereas the dotted-blue to $\mathcal{P}_-^{(-)}$ one.

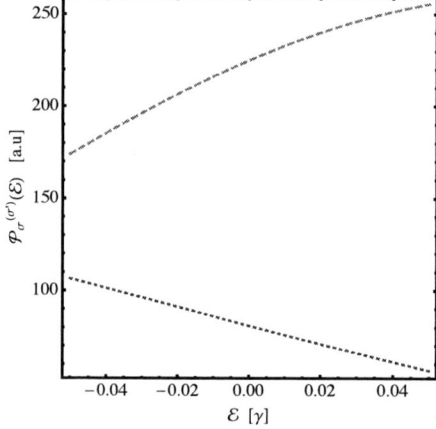

Fig. 3. Not normalized transition probability $\mathcal{P}_\sigma^{(\sigma')}(\mathcal{E})$ as a function of the energy states for electrons in a metallic SWNT in the case of a DVD with $t_{\text{R}} = 0.5t$ as in Fig. 1a. The red-dashed refers to $\mathcal{P}_+^{(+)}$ probability, whereas the dotted-blue to $\mathcal{P}_-^{(-)}$ one.

As it can be seen from the structure of Hamiltonian (5), the pairs of carbon atoms B_1–B_2 and A_5–A_6 are now connected by the hopping element t_{R}. Clearly these carbon atoms are part of the same sub-lattice — as emphasized by the letters A and B, respectively. This leads to the breaking of the particle-hole symmetry [13]. A first important consequence of the lack of this symmetry becomes observable in form of an additional term in the scattering probability Eq. (8):

$$\mathcal{P}_\sigma^{(\sigma')}(\mathcal{E}) \propto \delta_{\sigma\sigma'} \left[-t_{\text{R}} + 2\sigma t \cos\left(\frac{2\pi}{3} + \sigma \frac{\mathcal{E}R_x}{\gamma} \right) \right]^2. \tag{9}$$

Now we clearly observe that for a finite value of $t_{\text{R}}/t > 0$, the probability for the processes $\mathcal{P}_-^{(-)}$ differs from the one for the processes $\mathcal{P}_+^{(+)}$ at the charge neutrality point as can be clearly deduced from Fig. 3. It is now clear that the breaking of particle-hole symmetry due to a localized defect is leading to a drastically higher scattering probability at a DVD for pseudo-spin + electrons compared to pseudo-spin − ones.

IV. DISCUSSION

We have shown the possibility of creating a filtering effect for the charge carriers in metallic SWNTs by simply analyzing the interplay between the tube symmetry and the symmetry of a DVD. Our findings are well explaining experimental results obtained by analyzing the Friedel oscillations of a defected metallic SWNT [17].

Choosing the symmetry of the defect accordingly, it is possible to modify the ratio of the transmission probabilities of the two pseudo-spin species [13]. Therefore, the prospect of engineering definite defects in SWNTs, would permit to filter electrons with a specific degree-of-freedom — the pseudo-spin — in analogy to a similar effect on valley-spin induced by line defects in graphene [18]. In this last case, the two-dimensional nature of the graphene flake permits to use the injection angle upon the line defect [19] as a degree of freedom for choosing the valley-spin to be selected. However, in the case we have considered, the quasi-one-dimensional nature of the metallic SWNT does not permit this freedom. Therefore the only way to modify the ratio between pseudo-spin + *versus* pseudo-spin − is simply by increasing the degree of reconstruction — but this is not easily achievable because it is fixed by the chemistry of the considered problem. In this specific case, a sequence of defects — as DVDs with different orientations — could result in a sort of diode system for a specific pseudo-spin species. Moreover, a possibility to affect the defect chemistry in a controlled way might be reversible adsorption of atoms or molecules. Hereby, interesting applications in the field of gas sensing might arise.

ACKNOWLEDGMENT

We thank G. Buchs, H. Grabert, L. Lenz, C. Lieber, O. Gröning, and M. Moseler for useful discussions. The work of DB is supported by the DFG grant BE 4564/1-1 and by the Excellence Initiative of the German Federal and State Governments.

REFERENCES

[1] J.C. Charlier, X. Blase and S. Roche, Rev. Mod. Phys. **79**, 677 (2007).
[2] T. Ando, J. Phys. Soc. Jpn. **74**, 777 (2005).
[3] H. W. C. Postma, T. Teepen, Z. Yao, M. Grifoni, and C. Dekker, Science **293**, 76 (2001).
[4] L. C. Vanema, J. W. G. Wildöer, J. W. Janssen, S. J. Tans, H. L. J. T. Tuinstra, L. P. Kouwenhoven, and C. Dekker, Science **283**, 52 (1999).
[5] M. Bockrath, W. Liang, D. Bozovic, J. H. Hafner, C. M. Lieber, M. Tinkham, and H. Park, Science **291**, 283 (2001).
[6] S. G. Lemay, J. W. Janssen, M. van den Hout, M. Mooij, M. J. Bronikowski, P. A. Willis, R. E. Smalley, L. P. Kouwenhoven, and C. Dekker, Nature **412**, 617 (2001).

[7] G. D. Lee, C. Z. Wang, E. Yoon, N.-M. Hwang, D.-Y. Kim, and K. M. Ho, Phys. Rev. Lett. **95**, 205501 (2005).
[8] R. G. Amorim, A. Fazzio, A. Antonelli, F. D. Novaes, and A. J. R. da Silva, Nano Lett. **7**, 2459 (2007).
[9] S. Berber and A. Oshiyama, Phys. Rev. B **77**, 165405 (2008).
[10] A. Hashimoto, K. Suenaga, A. Gloter, K. Urita1, and S. Iijima, Nature **430**, 870 (2004).
[11] G. Buchs, D. Bercioux, P. Ruffieux, P. Gröning, H. Grabert, and O. Gröning, Phys. Rev. Lett. **102**, 245505 (2009).
[12] D. Bercioux, G. Buchs, H. Grabert, and O. Gröning, Phys. Rev. B **83**, 165439 (2011).
[13] L. Mayrhofer and D. Bercioux, Phys. Rev. B **84**, 115126 (2011).
[14] L. Mayrhofer and M. Grifoni, Eur. Phys. J. B **63**, 43 (2008).
[15] H. Bruus, and K. Flensberg, *Many-Body Quantum Theory in Condensed Matter Physics: An Introduction* (Oxford University Press, USA , 2004).
[16] L. Mayrhofer and D. Bercioux, *in preparation.*
[17] M. Ouyang, J.-L- Huang, and C. M. Lieber, Phys. Rev. Lett. **88**, 066804 (2002).
[18] D. Gunlycke and C. T. White, Phys. Rev. Lett. **106**, 136806 (2011).
[19] J. Lahiri, Y. Lin, P. Bozkurt, I. I. Oleynik, and M. Batzill, Nature Nanotech. **5**, 326 (2010).

Top-down vs. bottom-up coarse-graining of graphene and CNTs for nanodevice simulation

David Kauzlarić[*], Orly Liba[†], Yael Hanein[†], Pep Español[‡],
Andreas Greiner[§], Sauro Succi[*,¶] and Jan G. Korvink[*,§]

[*]Freiburg Institute for Advanced Studies (FRIAS), University of Freiburg, Albertstr. 19, 79104 Freiburg, GERMANY,
Email: david.kauzlaric@frias.uni-freiburg.de
[†]Department of Physical Electronics, Tel-Aviv University, P.O.B. 39040, Ramat Aviv, Tel-Aviv 69978, ISRAEL
[‡]Departamento Física Fundamental, Universidad Nacional de Educación a Distancia, Aptdo. 60141 E-28080, Madrid, SPAIN
[¶]Istituto Applicazioni Calcolo, CNR, via dei Taurini 9, 00185, Rome, ITALY
[§]Department of Microsystems Engineering, University of Freiburg, Georges-Köhler-Allee 103, 79110 Freiburg, GERMANY

Abstract—**We present and compare two approaches for the coarse-graining (CG) of models for graphene and carbon nanotubes (CNTs). Such models are required to enable mechanical device simulation on mesoscopic time and length scales hardly reachable by the molecular dynamics method. The first is a heuristic top-down approach while the second performs a rigorous bottom-up CG based upon an atomistic description. Both models belong to the family of dissipative particle dynamics. The top-down model already allows to analyze CNT self assembly and the temperature dependent resonance behavior of resonators. Correct relaxation time-scales required, e.g., for the Q-factor of resonator-devices are hard to adjust in this model. Therefore, a statistical projection-operator based bottom-up approach was investigated. This model allows to reproduce the correct time scales of autocorrelation functions on a CG-level. For correct cross-correlations and hence the correct decay of eigenmodes, further improvements are necessary.**

Keywords-carbon nanotubes; graphene; coarse-graining;

I. INTRODUCTION

Due to their unique mechanical properties such as very high in-plane elastic modulus and thermal conductivity the interest in Graphene [1] and carbon nanotubes (CNTs) [2] as building blocks for NEMS-devices is constantly growing. CNT based resonators, e.g., have the potential to become excellent sensors for tension, strain [3], or small masses [4], [5]. This work addresses the development of advanced simulations techniques allowing for an advanced design of these devices.

At the largest scale, mechanical simulation methods (e.g. finite element methods) completely discard the atomic structure of the CNTs and rely only on their macroscopic elastic parameters. To account for the discrete nature of the tubes and to facilitate simulations concerning CNT-interactions, atomistic methods such as molecular dynamics (MD) are often used.

There can be found numerous MD simulations of graphene and CNTs (see, e.g., [5]–[8]). However, the length and time scales relevant to device simulations, require substantial computational resources and, consequently, coarse-grained methods emerge as an attractive alternative [9]–[13]. For the case of atomistic systems, the main task of coarse-graining (CG) is to link the molecular specificity of the system to its macroscopic or mesoscopic description. A well-defined theory of CG is Mori-Zwanzig theory [14]–[16] whose basic idea is to select those variables with the longest time scales of evolution, and derive closed equations for their dynamics. Besides this systematic approach, empirical approaches containing adjustable parameters are possible as well. These parameters may for example be those of the well-known dissipative particle dynamics (DPD) model [17], [18]. The aim of this work is to compare these two approaches for the specific example of coarse-graining graphene based materials. We will focus on the practical utility for the device engineer on the one hand and on the rigorous representation of intrinsic dissipation mechanisms in these materials on the other hand. In section II we will first introduce the DPD model in a convenient form as the common basis for an empirical top-down model which will be presented in section III and a more rigorous bottom-up model which will follow in section IV. Then, we will present simulation results from both models in section V and discuss their advantages, disadvantages, and challenges in section VI.

II. THE DPD-MODEL

For both approaches the coarse-graining (CG) is based on the center of mass (COM) variables $\{\mathbf{R}_\mu, \mathbf{P}_\mu\}$ of "blobs". Following the indexing of, e.g., Ref. [13], each blob is uniquely identified by a Greek index μ. Then, we can define the center of mass variables as follows:

$$\mathbf{R}_\mu = \frac{1}{M_\mu} \sum_i m_i \mathbf{r}_i \delta_\mu(i), \qquad \mathbf{P}_\mu = \sum_i \mathbf{p}_i \delta_\mu(i) \qquad (1)$$

where the symbol $\delta_\mu(i)$ takes the value 1 if atom i with position $\mathbf{r}_i$, momentum $\mathbf{p}_i$, and mass m_i is in blob μ and zero otherwise, and $M_\mu = \sum_i m_i \delta_\mu(i)$ is the total mass of blob μ. It has been shown rigorously [13], [19] that the resulting evolution equations for the COM variables are those of the dissipative particle dynamics (DPD) model [17], [18]. The DPD equations of motion for the position $\mathbf{R}_\mu$ and momentum $\mathbf{P}_\mu$ of a blob μ are

$$\dot{\mathbf{R}}_\mu(t) = \frac{\mathbf{P}_\mu(t)}{M_\mu}, \qquad \dot{\mathbf{P}}_\mu(t) = \mathbf{F}_\mu^C(t) + \mathbf{F}_\mu^D(t) + \mathbf{F}_\mu^R(t), \quad (2)$$

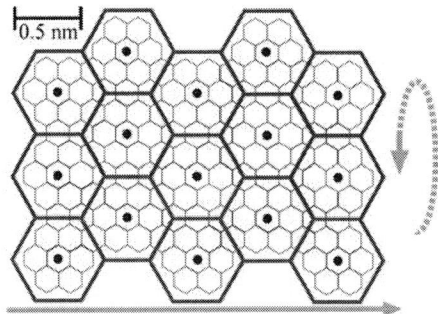

Fig. 1. The coarse graining of a CNT by grouping 24 carbon atoms. Solid dots represent the blobs. The solid arrow represents the axis of the tube. The dashed arrow represents the wrapping of the graphene sheet to form a tube.

where $\mathbf{F}_\mu^C$, $\mathbf{F}_\mu^D$, $\mathbf{F}_\mu^R$ are conservative, dissipative, and stochastic forces, respectively, the latter two being related via a fluctuation-dissipation theorem. These forces read as follows:

$$\mathbf{F}_\mu^C(t) = -\frac{\partial V^{cg}}{\partial \mathbf{R}_\mu}(R), \qquad \mathbf{F}_\mu^D(t) = -\sum_\nu \gamma_{\mu\nu}(R)\frac{\mathbf{P}_\nu(t)}{M_\nu} \quad (3)$$

where $V^{cg}(R)$ is a CG potential depending on the positions of the blobs $R = \{\mathbf{R}_\mu\}$ and $\gamma_{\mu\nu}(R) = \gamma_{\nu\mu}(R)$ is a friction tensor that satisfies momentum conservation by $\sum_\nu \gamma_{\mu\nu}(R) = 0$. Eqs. (2) are an approximation to the *exact* equations for $\mathbf{R}_\mu, \mathbf{P}_\mu$, derived via a projection operator technique and including a memory kernel given by the correlation of the *projected forces* $\mathbf{F}_\mu^R$. The main approximation leading to Eq. (2) is the Markovian assumption [16] which states that, whenever the correlation time of the microscopically defined projected force $\mathbf{F}_\mu^R(t)$ is much smaller than the correlation time of the momentum $\mathbf{P}_\mu(t)$, then the memory kernel is

$$\langle \mathbf{F}_\mu^R(t')\mathbf{F}_\nu^R(t)\rangle \approx 2k_B T \gamma_{\mu\nu}(R)\delta(t'-t) \quad (4)$$

where $\gamma_{\mu\nu}(R)$ is given by a Green-Kubo formula. Eq. (4) implies that the projected force may be modeled as a white noise, and memory effects can be disregarded. The Markovian approximation was shown in [20] to hold approximately for the microscopic model described later in section IV-A.

In summary, our task is to determine $V^{cg}(R)$ and $\gamma_{\mu\nu}$ in Eqs. (3). We first present a heuristic top-down approach where the explicit connection between COM-variables and atomistic variables Eq. (1) is not directly used. Instead, matching procedures are performed, e.g., to macroscopic measured quantities such as the elastic modulus. On the other hand, the second approach directly applies Eq. (1) finding a relation between the equations' left-hand-side and models for $V^{cg}(R)$ and $\gamma_{\mu\nu}$.

III. TOP-DOWN MODEL

We perform CG of a single walled CNT with a diameter of 1 nm. As in [21] we group 24 C-atoms into one DPD-particle, i.e., the coarse-graining number is $N_{cg} = 24$. The DPD model consists of rings of blobs which obey the equation of motion (2) and retain the tubular structure of the CNT (see Fig. 1).

The elastic properties of the simulated tube are determined by a model for conservative bonding forces between blobs. This model derives from the following coarse-grained potential

$$\begin{aligned} V^{cg}(R) &= \frac{k_s^{cg}}{2}\sum_{P_{\mu\nu}}(R_{\mu\nu}-r_0^{cg})^2 \\ &+ \frac{k_a^{cg}}{2}\sum_{T_{\mu\nu\lambda}}(\cos\alpha_{\mu\nu\lambda}-\cos\alpha_0^{cg})^2 \quad (5) \end{aligned}$$

where the first term is a Hookian spring bonding neighbouring blobs $P_{\mu\nu}$, $R_{\mu\nu} \equiv |\mathbf{R}_{\mu\nu}| \equiv |\mathbf{R}_\mu - \mathbf{R}_\nu|$, and r_0^{cg} is the equilibrium distance between blobs. The angular force applies to triplets of neighbouring blobs $T_{\mu\nu\lambda}$, which are aligned along the axis of the tube (cf. Fig. 1), hence $\alpha_0^{cg} = \pi$. The model-parameters k_s^{cg} and k_a^{cg} have been calibrated by simulation experiments stretching and bending the CG tube to reproduce the elastic properties of CNTs. As a result, the CG-CNT behaves like an Euler-Bernoulli beam for small deformations and non-linearly for large deformations [21]. Additionally, we let all blobs interact by a Lennard-Jones interaction to be able to model tube-tube Van der Waals interactions. The determination of the parameters σ and ϵ can be found in [21].

For the top-down approach we have made no attempt to quantitatively match the dissipation matrix $\gamma_{\mu\nu}$. This could have been done in principle by considering measurements of the Q-factor of CNT resonators either by a real experiment or by MD-simulation. Both approaches are quite unsatisfactory due to the following reasons: i) Q-factors from experimental measurements are distributed over a wide range (see, e.g., [22]) and are never purely intrinsic Q-factors as required for the DPD-model but include external damping mechanisms. ii) Measuring the Q-factor of CNTs from MD requires long simulation runs, ideally, for the tube-length, we are aiming to model with DPD. But due to the computational cost there exist only MD-simulations for short tubes over short times [23], [24], which, in fact, is exactly the motivation for developing a CG DPD-model. Hence, only the measurement of observables from short MD-runs is useful for the construction of a CG-model. This approach is adopted for the bottom-up model described in the next section.

For the top-down model, the dissipation constants are therefore only adjusted according to the requirements of giving stable simulations at reasonable integration time-steps. Since the chosen dissipation rates are always much smaller than the resonance frequencies of the CNT-resonators investigated below, the resonance behaviour does not depend on this choice. The following bottom-up approach performs a step towards the goal to systematically reproduce a resonator's Q-factor.

IV. BOTTOM-UP MODEL

A. Microscopic model

In contrast to the top-down approach we require an explicit microscopic Hamiltonian as a starting point for the bottom-up approach. For the sake of simplicity, we use a simple two dimensional force field based on pairwise and angular springs for the in-plane motion of a periodic graphene sheet instead of more realistic potentials such as the Tersoff-Brenner potential [25]. The interaction potential between C-atoms is assumed to have the form $U = U_s + U_a$ The pair potential U_s models a harmonic spring and is defined as

$$U_s = \sum_{P_{ij}} U_{ij}^s, \qquad U_{ij}^s = \frac{1}{2}k_s(r_{ij} - r_0)^2 \qquad (6)$$

where $k_s > 0$ is a stiffness constant, r_0 is the equilibrium bond-length, $r_{ij} = |\mathbf{r}_{ij}|$, $\mathbf{e}_{ij} = \mathbf{r}_{ij}/r_{ij}$, $\mathbf{r}_{ij} = \mathbf{r}_i - \mathbf{r}_j$, and P_{ij} denotes all and only those pairs of neighbouring C-atoms i and j that are bonded and assumed to interact by the given potential. The angular potential $U(\alpha_{ijk})$ is defined as

$$U_a = \sum_{T_{ijk}} U_{ijk}^a, \qquad U_{ijk}^a = \frac{1}{2}k_a\left(\cos\alpha_{ijk} - \cos\alpha_0\right)^2 \qquad (7)$$

where T_{ijk} denotes all those and only those triplets of neighbouring C-atoms $\{i, j, k\}$ forming an angle α_{ijk}. The equilibrium angle is $\alpha_0 = 2\pi/3$ and k_a is a second angular stiffness constant.

The unit of length l^* is fixed by setting $l^* \equiv r_0 \approx 0.142$ nm in graphene. The unit of mass is the mass of a carbon atom, i.e., $m^* \equiv m_C \approx 2.0 \times 10^{-26}$ kg. The unit of time $t^* \approx 9.56$ fs and temperature $T^* \approx 3.2 \times 10^5$ K result from choosing $k_a \equiv 1$ in these units and from adjusting the speed of sound as obtained from the MD-model, to experimental values [26].

B. Coarse-grained model

Equation (4) is to be understood as the Fluctuation-Dissipation theorem for the present model giving the amplitude of the random forces in terms of the transport coefficients of the system. The random forces need to be specified as linear combinations of the Wiener process $d\mathbf{W}_\mu(t)$ as, for example, $\mathbf{F}_\mu^R dt = d\tilde{\mathbf{P}}_\mu = \sum_\nu \mathbf{B}_{\mu\nu} d\mathbf{W}_\nu(t)$. The explicit form of the noise terms is given in [20].

For the dissipative force in Eq. (3) we use the form

$$\mathbf{F}_\mu^D = -\sum_{P_{\mu\nu}}\left[\gamma_{||}\mathbf{e}_{\mu\nu}\mathbf{e}_{\mu\nu} + \gamma_\perp(1 - \mathbf{e}_{\mu\nu}\mathbf{e}_{\mu\nu})\right] \cdot \mathbf{V}_{\mu\nu} \quad (8)$$

for neighbouring blobs μ, ν and zero otherwise, where $\mathbf{V}_{\mu\nu} = \mathbf{V}_\mu - \mathbf{V}_\nu$. We determine the value of the friction coefficients $\gamma_{||}, \gamma_\perp$ by requiring that a prediction of the CG DPD model, which is the value of the slope of the blob velocity correlation function (BVCF) near $t \sim 0$, reproduces the MD results for this same quantity. We know from the formal derivation of the DPD equations from projection operators that the DPD equations are not valid *until* the memory kernel has decayed, and the Markovian assumption applies. The memory decays in a time scale given by $\tau_\mathbf{F}$. Therefore, we will estimate the friction coefficients from

$$\frac{d}{dt}\langle \mathbf{V}_\mu \cdot \mathbf{V}_\mu(t)\rangle^{\text{MD}}\Big|_{t \sim \tau_\mathbf{F}} = -\frac{6k_BT}{M_\mu^2}\left(\gamma_{||} + \gamma_\perp\right)$$

$$\frac{d}{dt}\langle \mathbf{V}_\mu \cdot \mathbf{e}_{\mu\nu}^0 \mathbf{V}_\nu(t) \cdot \mathbf{e}_{\mu\nu}^0\rangle^{\text{MD}}\Big|_{t \sim \tau_\mathbf{F}} = \frac{k_BT}{M_\mu M_\nu}\gamma_{||}. \quad (9)$$

In this way the left hand sides of Eqs. (9) have well-defined non-zero values from MD simulations [20], [27]. The vectors $\mathbf{e}_{\mu\nu}^0$ are the equilibrium analogues of $\mathbf{e}_{\mu\nu}$.

The model for the conservative forces that we propose derives again from the coarse-grained potential (5) except that now, $\alpha_0^{\text{cg}} \equiv \pi/3$ is assumed to be the angle between triplets of blobs $T_{\mu\nu\lambda}$, where all three pairs $P_{\mu\nu}$, $P_{\mu\lambda}$, $P_{\nu\lambda}$ of the blobs are nearest neighbors in the hexagonal blob lattice (cf. Fig. 1). The two parameters k_s^{cg} and k_a^{cg} of the conservative DPD-forces are determined by adjusting them to produce the correct value for two target observables computed from MD, which are the second moments of fluctuations of bond distances $\mu_2^R \equiv \langle (R_{\mu\nu} - r_0^{\text{cg}})^2\rangle$ and angles $\mu_2^\alpha \equiv \langle (\cos\alpha_{\mu\nu} - \cos\alpha_0^{\text{cg}})^2\rangle$. As shown previously [20], Shell's relative entropy method [28] leads to an inverse problem for k_s^{cg} and k_a^{cg} containing exactly these two observables. This non-linear problem is then solved by the Newton-Raphson method and we obtain the following values of the spring constants: $k_s^{\text{cg}} = 0.78, k_a^{\text{cg}} = 10.0$ for the blob size $N_{cg} = 24$ and $k_s^{\text{cg}} = 1.13, k_a^{\text{cg}} = 25.5$ for $N_{cg} = 96$. Note that even if the model for the conservative forces were completely identical for the top-down and for the bottom-up approach, we can not expect the obtained force constants to be identical. The reason is that different target observables have been used for the optimisation (macroscopic elastic constants vs. bond length and angle fluctuations) and it is known that CG models do not necessarily reproduce properties they haven't directly been designed for.

V. RESULTS

A. Top-down model

First, we show an example of the ability of the top-down model to reproduce experimental observations, such as the binding of CNTs in the CVD (chemical vapour deposition) process. Two tubes are initially positioned as shown in Fig. 2 (top). The tubes are anchored at one end and positioned in such a way to form an initial angle of $2\theta = 5°$. The distance between their fixed ends is $2\Delta = 11$ nm and the distance at the crossing point is 5 nm thus, quite far from the cutoff distance $r_{VdW} = 2.5\sigma \approx 1.11$ nm of the Van der Waals (VdW) force. The temperature in the simulation is 900 °C, similar to the temperature in the CVD chamber. Due to the large thermal vibration, the CNTs partially enter each other's attractive range of the VdW force and start binding. The average equilibrium configuration is a Y-junction as shown in Fig. 2 (bottom) and as observed in experiments [29]. According to continuum mechanics theory the binding energy per unit-length ϵ_b of two partially zipped CNTs is [30]

$$\epsilon_b = \frac{4YI}{L^2}\left(\theta - 3\frac{\Delta}{L}\right)^2. \qquad (10)$$

As illustrated in Fig. 2 2θ is the opening angle between the tubes, and L is the distance from the fixed ends of the tube to the Y-junction. Y and I are Young's modulus and the second moment of inertia, respectively. In our simulation we obtain a binding energy per unit-length of $\epsilon_b = (0.93 \pm 0.12) \times 10^{-10}$J/m. The theoretical binding energy per unit-length is

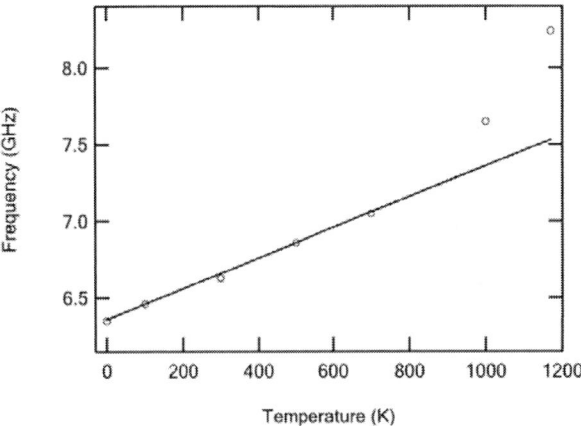

Fig. 2. Initial (top) and equilibrium average configuration of the simulation of two anchored CNTs performing a zipper effect. Symbols are explained in the text.

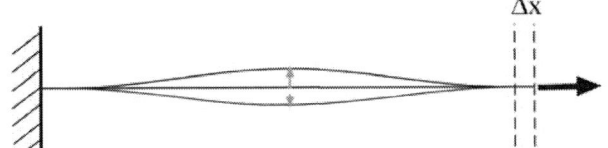

Fig. 3. Illustration of a nanometer scale displacement sensor based on a single CNT used as a doubly clamped cantilever beam. The arrow shows the direction of the displacement, the left edge of the tube remains fixed.

Fig. 4. The first resonance frequency of a 57 nm tube versus temperature. Empty circles are the simulation results and the solid line is calculated based on the Duffing theory.

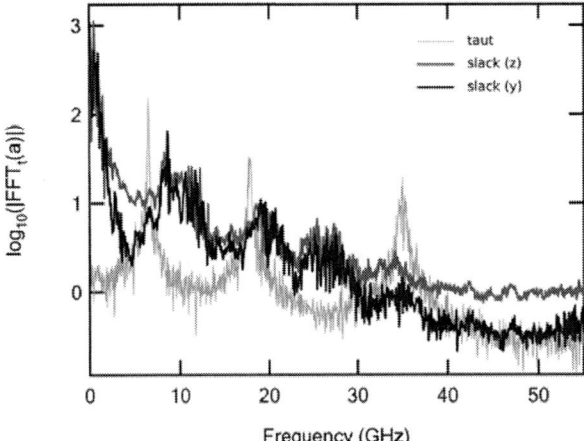

Fig. 5. Spectral analysis of the first three vibration modes of a slack (black and dark grey lines) and a taut (light grey line) tube. The natural length of both tubes is 57 nm and the temperature of the simulation is 300 K. The slack tube is produced by compressing the edges to 99% of the natural length. The taut tube is neither strained nor compressed.

$\epsilon_b = 1.191 \times 10^{-10}$ J/m according to [31] which is very close to our result. Note that the investigated DPD-CNTs are single-walled and posses the same helicity as in [31].

The main purpose of the top-down model is to analyse the dynamic behaviour of CNT resonators [32]. We consider double-clamped resonators as sketched in Fig. 3. We can investigate now, e.g., the dependence of the resonance frequencies on the temperature. For the first mode we see in Fig. 4 that the resonance frequency is increasing. The linear dependency can be predicted by the so called Duffing oscillator [33]. The thermal vibrations induce an additional tension in the beam which leads to a frequency shift. In the Duffing theory the shift is proportional to the temperature. This can be observed in the simulation up to a temperature of 700 K. For higher temperatures we observe circular vibration modes as the lowest order modes in the simulation which cannot be described by a Euler-Bernoulli theory and therefore we see a deviation from the linear behaviour.

It is also interesting to compare taut and slack tubes. Slack tubes may either result from the production process or because the resonator is used as a displacement sensor with one movable clamping side as indicated in Fig. 3. Fig. 5 shows the spectral analysis for the taut and the slack case. We see that for the spectrum of the slack tube, the resonance peaks have almost completely disappeared, which means that the quality factor has decreased. In Fig. 6 the slack tube was prestressed by a force in the y-direction and we can make two observations: i) We obtain sharper resonance peaks. ii) The peaks in the y-direction are shifted considerably relative to those in the z-direction, showing that the resonator differentiates between the vibration spectra in the two directions and hence, detects the direction of the load.

B. Bottom-up model

Once we have computed the parameters of the DPD model from the microscopic information generated from MD, we may compare how well the DPD model reproduces the MD results for the dynamics of the blobs. In Fig. 7 we plot the auto-correlation function of the blob velocities (BVACF) as computed from MD and from the DPD model for $N_{cg} = 96$. For $N_{cg} = 24$, these and all following results are qualitatively similar. We observe a very good agreement between both simulations including the anti-correlation effect in the blob velocity. Note that the input information from MD is obtained from the decay at $t \sim 0$ and, nevertheless, the correct description of the auto-correlation at later times is predicted by the DPD model. The NDPD model (for Non-Dissipative Particle Dynamics) is identical to the DPD model, except for the dissipative and random forces that are set to zero. It is obvious from the plot that the NDPD model cannot reproduce the proper CG dynamics, despite the fact that it features the correct CG *conservative* forces. This result shows that the blob

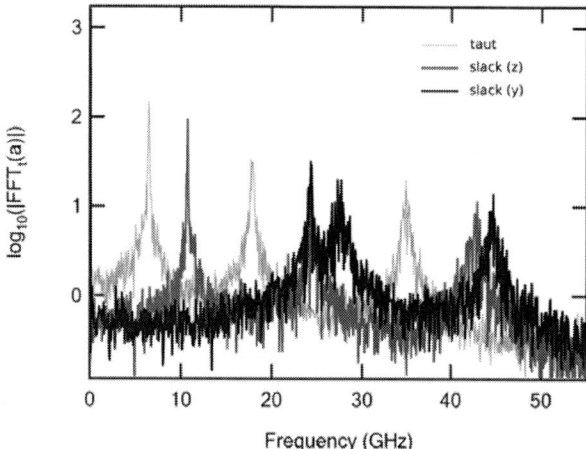

Fig. 6. Similar to Fig. 6, however the slack tube is pulled by a constant uniform force perpendicular to the tubes axis. The force applied to the slack tube is $F_y \approx 7$ pN. The dark grey and black lines are the spectra of the vibrations in the z and y directions respectively.

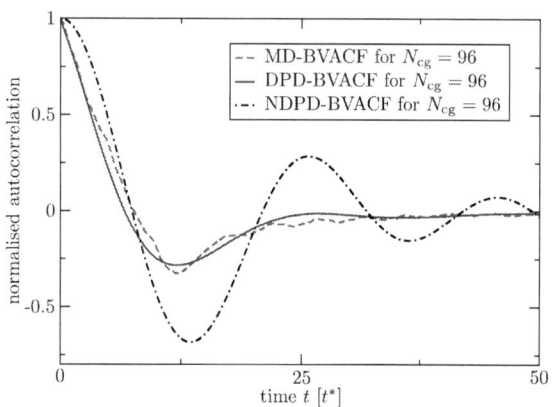

Fig. 7. Velocity auto-correlation functions of blobs (BVACFs) for $N_{cg} = 96$ normalized with $2k_BT/M$. MD-BVACF stands for the BVACF as computed from MD. DPD-BVACF is the BVACF as computed from DPD. NDPD-BVACF denotes the result obtained by removing friction and noise from DPD.

CG dynamics of graphene requires the introduction of non-trivial dissipative and random forces. The cross-correlations between the velocities of neighbouring blobs, as presented in Fig 8, although qualitatively correct, show some quantitative discrepancies. The amplitude of the cross-correlation function (CCF) as described by the DPD model, is smaller than the one obtained from MD, suggesting that the very same friction effects that are crucial to get the right ACF, exceed the values required to correctly recover the CCF. Interestingly, if no friction is considered at all, the results are very poor, as illustrated in Fig. 8, where results of the NDPD model are shown as dashed-dotted lines. Again, these results show that dissipation and thermal noise play an important role for the dynamics of the blobs. The blob velocity CCF reflects in its structure the basic interaction mechanism between blobs, which is the propagation of sound. Therefore, we may expect

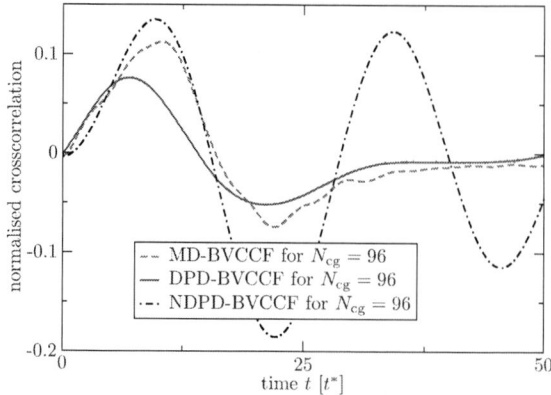

Fig. 8. Velocity cross-correlation functions $\langle \mathbf{V}_\mu \cdot \mathbf{e}^0_{\mu\nu} \mathbf{V}_\nu(t) \cdot \mathbf{e}^0_{\mu\nu} \rangle$ of neighboring blobs (BVCCFs) for $N_{cg} = 96$, normalized with $2k_BT/M$. MD-BVCCF stands for the BVCCF as computed from MD. DPD-BVCCF is the BVCCF as computed from DPD. NDPD-BVCCF denotes the result obtained by removing friction and noise from DPD.

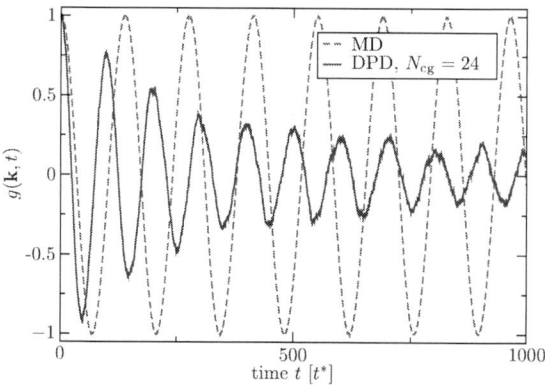

Fig. 9. Comparison of the decay of a shear wave of wavelength equal to the size of the system as a function of time, for MD and DPD models. While the MD simulation decays on a time scale not appreciable in the plot (0.1% per oscillation), the DPD simulation displays significant dissipation.

that a shear or a pressure wave in the DPD model would display also a larger dissipation than the corresponding MD simulation. In order to check this we add to every particle from an equilibrated MD configuration a velocity perturbation of the form $\Delta \mathbf{v}_i = \mathbf{v}_0 \cos(\mathbf{k} \cdot \mathbf{r}_i)$ and measure the Fourier transform of the momentum density field, which is defined as

$$
\begin{aligned}
\mathbf{g}(\mathbf{k}, t) &= \sum_i m_i \mathbf{v}_i(t) \exp\{i\mathbf{k} \cdot \mathbf{r}_i(t)\} \\
&\approx \sum_\mu M_\mu \mathbf{V}_\mu(t) \exp\{i\mathbf{k} \cdot \mathbf{R}_\mu(t)\}
\end{aligned}
\tag{11}
$$

where the approximation is valid for wavelengths much larger than the blob separation. Fig. 9 shows the comparison between the MD and DPD simulations, and, indeed the DPD results are subject to a higher damping than their MD counterpart.

VI. DISCUSSION AND CONCLUSION

It was shown that the bottom-up DPD-model can be used as a design-tool to study CNT-self-assembly and the temperature and strain-dependent resonance behaviour of CNT-resonators. The relaxation times of the system remain an independent adjustable parameter in this model. As discussed earlier, relaxation times are of minor interest when analysing the resonance frequencies in the limit of low damping. Also, in the case of CNT-self-assembly, we might only be interested in the assembled structure representing the energy minimum and whether it can be reached at the given temperature. But as soon as we are asking for the resonator's Q-factor or for the time scale on which the self-assembly can be observed, we must be sure about the magnitude of the dissipation matrix $\gamma_{\mu\nu}$. As discussed in section III this can only be achieved with great computational effort or experimental uncertainty. Clearly, we can adjust the dissipation of the top-down model to a desired value consistent with a specific experimental observation and hence mimic the overall dissipation and Q-factor. But this would not allow to systematically identify and split the intrinsic dissipation mechanisms from all external sources of damping. Additionally, we require a systematic means to assess the validity of this operating-point approximation and, especially, how good the Markovian approximation really is. This is provided by bottom-up coarse-graining.

Even though, the bottom-up approach can not yet fully replace the top-down approach, it is the only way to systematically derive CG-expressions. Here, the dissipation matrix $\gamma_{\mu\nu}$ represents the intrinsic phonon-mediated dissipation of the underlying MD-model in a well defined way. As shown, the diffusive behaviour of the blobs is already very well reproduced, since the velocity autocorrelation functions of the coarse-grained model are correct. The remaining challenge is a better reproduction of the cross-correlations and hence of the dissipation of sound waves.

ACKNOWLEDGMENT

The authors acknowledge funding by the Deutsche Forschungsgemeinschaft (DFG, via SFB 499), the University of Freiburg (via the German excellence initiative), and the Israel Science Foundation (via grant 1138/04).

REFERENCES

[1] A. K. Geim and K. S. Novoselov, "The rise of graphene," *Nature Materials*, vol. 6, pp. 183–191, 2007.

[2] S. Iijima, "Helical microtubules of graphitic carbon," *Nature*, vol. 354, pp. 56–58, 1991.

[3] V. Sazonova, Y. Yaish, H. Ustunel, D. Roundy, T. A. Arias, and P. L. McEuen, "A tunable carbon nanotube electromechanical oscillator," *Nature*, vol. 431, pp. 284–287, 2004.

[4] P. Poncharal, Z. L. Wang, D. Ugarte, and W. A. de Heer, "Electrostatic deflections and electromechanical resonances of carbon nanotubes," *Science*, vol. 283, pp. 1513–1516, 1999.

[5] C. Y. Li and T. W. Chou, "Vibrational behaviors of multiwalled-carbon-nanotube-based nanomechanical resonators," *Appl. Phys. Lett.*, vol. 84, pp. 121–123, 2004.

[6] Q. Lu, M. Arroyo, and R. Huang, "Elastic bending modulus of monolayer graphene," *J. Phys. D*, vol. 42, p. 102002, 2009.

[7] S. Iijima, C. Brabec, A. Maiti, and J. Bernholc, "Structural flexibility of carbon nanotubes," *J. Chem. Phys.*, vol. 104, pp. 2089–2092, 1996.

[8] C. L. Zhang and H. S. Shen, "Self-healing in defective carbon nanotubes: a molecular dynamics study," *J. Phys.: Condens. Matter*, vol. 19, p. 386212, 2007.

[9] J. Q. Broughton, F. F. Abraham, N. Bernstein, and E. Kaxiras, "Concurrent coupling of length scales: Methodology and application," *Phys. Rev. B*, vol. 60, no. 4, pp. 2391–2403, 1999.

[10] S. Succi, O. Filippova, G. Smith, and E. Kaxiras, "Applying the lattice boltzmann equation to multiscale fluid problems," *Computing in Science and Engineering*, vol. 3(6), pp. 26–37, 2001.

[11] W. E and B. Engquist, "Heterogeneous multiscale method: a general methodology for multiscale modeling," *Phys. Rev. B*, vol. 67, p. 092101, 2003.

[12] S. Succi, "Lattice boltzmann across scales: from turbulence to DNA translocation," *Eur. Phys. J. B*, vol. 64, pp. 471–479, 2008.

[13] C. Hijón, P. Español, E. Vanden-Eijnden, and R. Delgado-Buscalioni, "Mori-zwanzig formalism as a practical computational tool," *Faraday Discuss.*, vol. 144, pp. 301–322, 2010.

[14] H. Mori, "Transport, collective motion, and brownian motion," *Prog. Theor. Phys.*, vol. 33, pp. 423–455, 1965.

[15] R. Zwanzig, "Memory effects in irreversible thermodynamics," *Phys. Rev.*, vol. 124, pp. 983–992, 1961.

[16] H. Grabert, *Projection Operator Techniques in Nonequilibrium Statistical Mechanics*, ser. Springer Tracts in Modern Physics. Berlin; Heidelberg: Springer, 1982.

[17] P. J. Hoogerbrugge and J. M. V. A. Koelman, "Simulating microscopic hydrodynamic phenomena with dissipative particle dynamics," *Europhys. Lett.*, vol. 19, no. 3, pp. 155–160, 1992.

[18] P. Español and P. Warren, "Statistical mechanics of dissipative particle dynamics," *Europhys. Lett.*, vol. 30, no. 4, pp. 191–196, 1995.

[19] T. Kinjo and S. Hyodo, "Equation of motion for coarse-grained simulation based on microscopic description," *Phys. Rev. E*, vol. 75, no. 5, p. 051109, 2007.

[20] D. Kauzlarić, J. T. Meier, P. Español, S. Succi, A. Greiner, and J. G. Korvink, "Bottom-up coarse-graining of a simple graphene model: the blob picture," *J. Chem. Phys.*, vol. 134, p. 064106, 2011.

[21] O. Liba, D. Kauzlarić, Z. R. Abrams, Y. Hanein, A. Greiner, and J. G. Korvink, "A dissipative particle dynamics model of carbon nanotubes," *Molecular Simulation*, vol. 34, pp. 737–748, 2008.

[22] B. Witkamp, M. Poot, and H. S. J. van der Zant, "Bending-mode vibration of a suspended nanotube resonator," *Nano Lett.*, vol. 6, pp. 2904–2908, 2006.

[23] H. Jiang, M. F. Yu, B. Liu, and Y. Huang, "Intrinsic energy loss mechanisms in a cantilevered carbon nanotube beam oscillator," *Phys. Rev. Lett.*, vol. 93, p. 185501, 2004.

[24] P. A. Greaney, G. Lani, G. Cicero, and J. C. Grossman, "Anomalous dissipation in single-walled carbon nanotube resonators," *Nano Lett.*, vol. 9, pp. 3699–3703, 2009.

[25] D. W. Brenner, O. A. Shenderova, J. A. Harrison, S. J. Stuart, B. Ni, and S. B. Sinnott, "A second-generation reactive empirical bond order (REBO) potential energy expression for hydrocarbons," *J. Phys.: Condens. Matter*, vol. 14, pp. 783–802, 2002.

[26] C. Lee, X. Wei, J. W. Kysar, and J. Hone, "Measurement of the elastic properties and intrinsic strength of monolayer graphene," *Science*, vol. 321, pp. 385–388, 2008.

[27] D. Kauzlarić, P. Español, A. Greiner, and S. Succi, "Three routes to the friction matrix and their application to the coarse-graining of atomic lattices," *Macromolecular Theory and Simulations*, vol. 20, no. 7, pp. 526–540, 2011.

[28] M. S. Shell, "The relative entropy is fundamental to multiscale and inverse thermodynamic problems," *J. Chem. Phys.*, vol. 129, no. 14, p. 144108, 2008.

[29] Z. R. Abrams and Y. Hanein, "Tube-tube and tube-surface interactions in straight suspended carbon nanotube structures," *J. Phys. Chem. B*, vol. 110, pp. 21 419–21 423, 2006.

[30] B. Chen, M. Gao, J. M. Zuo, S. Qu, B. Liu, and Y. Huang, "Binding energy of parallel carbon nanotubes," *Appl. Phys. Lett.*, vol. 83, pp. 3570–3571, 2003.

[31] L. A. Girifalco, M. Hodak, and R. S. Lee, "Carbon nanotubes, buckyballs, ropes, and a universal graphitic potential," *Phys. Rev. B*, vol. 62, pp. 13 104–13 110, 2000.

[32] O. Liba, Y. Hanein, D. Kauzlarić, A. Greiner, and J. G. Korvink, "Investigation of the mechanical properties of bridged nanotube resonators by dissipative particle dynamics simulation," *Int. J. Multiscale Comp. Eng.*, vol. 6, pp. 549–562, 2008.

[33] H. W. C. Postma, I. Kozinsky, A. Husain, and M. L. Roukes, "Dynamic range of nanotube-and nanowirebased electromechanical systems," *Appl. Phys. Lett.*, vol. 86, p. 223105, 2005.

Electrowetting droplets investigated with smoothed particle hydrodynamics and moving least squares

Dennis Weiß*, Andreas Greiner*, Jan Lienemann‡ and Jan G. Korvink*†

*Department of Microsystems Engineering, University of Freiburg, Germany
{dennis.weiss, andreas.greiner}@imtek.uni-freiburg.de
†Freiburg Institute for Advanced Studies, University of Freiburg, Germany
jan.korvink@frias.uni-freiburg.de
‡Schmidt&Partner Engineering AG, Zuerich, Switzerland
lienemann@speag.com

Abstract—The numerical analysis of oscillating electrowetting droplets is challenging predominantly due to the frequent change in topology of the phases involved and the coupling of electrostatic forces into the system. Damped systems require an even larger number of oscillations or simulation time steps, respectively, until resonance becomes apparent, and an efficient model is required.

We perform an investigation on the transient regime of an electrowetting setup with a sessile droplet in a three-phase system by meshless particle methods. The droplet shape and the electric field distribution are modelled self-consistently with a coupled smoothed particle hydrodynamics and moving least squares approach. We expose the surface energy between liquid and solid phase to harmonic excitation patterns and exhibit the droplet's shape evolution in terms of Legendre polynomials in time and frequency domain.

Fluidics is described with a multiphase capable smoothed particle hydrodynamics model. Adjustable surface tensions allow for establishing a droplet's equilibrium contact angle according to Young's law. The electric field distribution is calculated on the Lagrangian discretisation points by solving iteratively the Laplace equation with moving least squares methodology. The Korteweg-Helmholtz body force then contributes to the momentum equation of the fluid particles. For a simplified model, a time-harmonic component is added to the liquid-solid surface energy.

We present a fully coupled electrowetting model and successfully use particle methods for a dynamic electrowetting simulation by employing an effective, simplified model.

I. INTRODUCTION

In this work, the dynamics of oscillating electrowetting droplets is modelled and characterised numerically. Applications which require tailored flow patterns or in reverse specific droplet mode shapes might benefit substantially from an insight in the transient regime.

The electrowetting effect has received increasing attention in recent years. The contact angle can be controlled electrically enabling the manipulation and transport of liquid droplets by electrocapillary forces. Fundamental research in terms of experimental work (e.g. [1]–[4]) has been smoothed the way for applications like lab-on-a-chip devices, liquid lenses, and optical displays (e.g. [5]–[8]).

The interesting characterisation of the droplet flow field and in particular the crucial time evolution of mode shapes is challenging with experimental setups, especially when focussing higher order mode shapes and a possible coupling

among them. Analytical investigations performed e.g. in [9]–[11] are justified for small amplitude oscillations only due to the non-linearity of the Navier-Stokes equations.

Smoothed particle hydrodynamics (SPH) as spatial discretisation method is ideally suited for being cheap and inherently conserving phase interfaces, even in the long run. But it is not a Galerkin method and comes with deficiencies for boundary conditions and long term interactions. For the electric domain, discretising with an adaptation of moving least squares (MLS) is beneficial: additional equations for the Laplace equation are introduced and solved on the SPH discretisation points.

We want to setup a fully coupled electrowetting model and investigate on the transient regime of oscillating droplets. An initial contact angle can be established according to the Young equation. The electrowetting effect contributes to the interfacial energy and in effect causes a substrate to be more hydrophilic. We add a time-harmonic component which changes the contact angle periodically. This can be achieved either by applying the forces induced from electric fields or by altering the solid-liquid surface energy of the phases involved.

We want firstly to investigate on the feasibility of our coupled electrowetting model. Secondly, we want to provide and characterise a simplified model. This might be highly justified or demanded, since it allows performing parameter studies on vibrating droplets and the investigation of damped systems with small characteristic times with small computational and overall costs.

II. MODEL AND DISCRETISATION

We model an electrowetting on dielectrics (EWOD) setup with a droplet on a dielectric substrate having a liquid, solid and vapor phase labelled by l, s and v, respectively and schematically shown in figure 1a. In equilibrium, the Young equation relates the droplet's contact angle ϕ to the surface energies γ between the phases:

$$\gamma_{sl} - \gamma_{sv} = -\cos(\theta)\gamma_{lv} \qquad (1)$$

An electric potential between the two buried electrodes causes a change in the contact angle. By altering the potential periodically, oscillations in the droplet can be induced.

978-1-4673-1122-9/12 $31.00 © 2012 IEEE

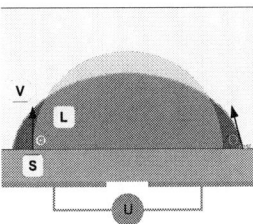

(a) Schematic: electric field changes contact angle

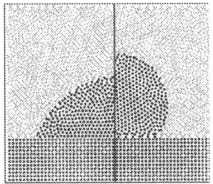

(b) Simulation: droplet at steady-state contact angles $60°$ and $120°$

(c) Experimental EWOD devices with aluminium electrodes on ceramics substrate

Fig. 1. Electrowetting on dielectrics

A. Governing equations

We consider isothermal, incompressible flow with constant density ρ. Mass is inherently conserved by our Lagrangian, particle based approach. The momentum equation for the electrowetting fluid phases reads then

$$\frac{d\mathbf{v}}{dt} = -\frac{1}{\rho}\nabla p + \mathbf{g} + \frac{\eta}{\rho}\nabla^2\mathbf{v} + \mathbf{f}^{(sf)} + \mathbf{f}^{(ew)} \quad (2)$$

and contains contributions of the pressure gradient ∇p, the gravitational acceleration $\mathbf{g}$, a viscous force with the dynamic viscosity η, and surface tension and electrowetting force densities $\mathbf{f}^{(sf)}$ or $\mathbf{f}^{(ew)}$, respectively.

Surface tension force $\mathbf{F}^{(sf)}$ is calculated via Brackbill's continuum surface force (CSF) model [12]

$$\mathbf{F}^{(sf)} = -(\alpha\kappa\mathbf{n} + \nabla_s\alpha)\delta_{sf} \quad (3)$$

with the surface tension coefficient α, the interface curvature κ, the surface normal $\mathbf{n}$ and the surface delta function δ_{sf} which generates a body force from Cauchy stress.

For calculating the electrowetting forces $\mathbf{F}^{(ew)}$, we consider the effects of an electric field $\mathbf{E}$ on ponderable, dielectric matter being incompressible, homogeneous and electrically linear. The electric field distribution is calculated via the Laplace equation with piecewise constant coefficients ϵ in the domain Ω

$$\nabla \cdot (\epsilon(\mathbf{x})\nabla u(\mathbf{x})) = 0 \quad \text{in } \Omega \quad (4)$$
$$u = g_D \quad \text{on } \Gamma_D \quad (5)$$
$$\frac{\partial u}{\partial \mathbf{n}} = g_N \quad \text{on } \Gamma_N \quad (6)$$

with the dielectricity number $\epsilon = \epsilon_0\epsilon_r$, and Neumann and Dirichlet Boundary conditions in (5, 6) on the boundary Γ. The effects of the electric field on phase interfaces is given by the simplified Korteweg-Helmholtz body force density $\mathbf{f}_{KH}$

[13], [14], where we can write

$$\mathbf{f}_{KH} = -\nabla\epsilon E^2 \quad (7)$$

B. Discretisation of the electric domain

We apply an adaptation [15] of the MLS method [16] with special regards to the Laplace equation in (4),

$$\sum_{j\in J}(m(\mathbf{x}) - b_j)^2\theta(\|\mathbf{x} - \mathbf{x}_j\|) = 0 \quad (8)$$

with the pivot particle at $\mathbf{x}$, the set of neighbouring particles J, a weighting function θ, a continuous function $m(\mathbf{x})$ and discrete function values $b_j|_{\mathbf{x}_j}$ defined at each computational node j. For the sake of simplicity, we employ our SPH smoothing kernel for θ. As continuous function, we expand a Taylor series around $\mathbf{x}_j$.

Via the weighted least squares measure in (8), the error $\mathbf{e} = M\mathbf{a} - \mathbf{b}$ is minimized particle by particle, where M contains the expansion coefficients of the Taylor series, $\mathbf{b}$ the given function values and $\mathbf{a}$ stands for the calculated electric potential and its derivatives. The set of equations in (8) is extended with further equations for the Laplace equation and boundary conditions in eqs. (4)-(6) by setting the M entries accordingly. The solution is obtained iteratively where the Laplace equation is fulfilled particle by particle and boundary conditions are propagated through the whole domain. Iterations are performed until the local error $u_e = |u^{(t+1)}/u^{(t)}| \leq z$ is under a given, positive threshold z. $\mathbf{a}$ finally holds the solutions for the potential u and its derivatives.

C. Discretisation of the fluid phases, static and dynamic contact angle for the droplet

We employ a multiphase capable SPH model from Hu and Adams [17], which we instantiated first in [18]. The SPH density equation then reads

$$\rho_i = m_i\sum_j W_{ij} = m_i\sigma_i \quad (9)$$

so that the density of a particle i does not depend on the mass m_j of neighbouring particles but only on its specific volume $\mathcal{V}_i = \sigma_i^{-1}$. For discretising the momentum equation in (2), we limit onto the contributions being in particular relevant for a contact angle to arise.

Surface tension force is not calculated directly by discretising (3). Instead, calculating the surface stress between two phases k and l

$$\Pi_{kl}^{(sf)} = \alpha^{kl}\frac{1}{\nabla C^{kl}}\left(\frac{1}{d}\mathbf{I}|\nabla C^{kl}|^2 - \nabla C^{kl}\nabla C^{kl}\right) \quad (10)$$

is more accurate [19]. Here, $\mathbf{I}$ stands for the unit tensor, d for the dimensionality and C describes a color field which indicates to which phase a particle belongs to. By setting the surface energy coefficients α^{kl}, an equilibrium contact angle can be established as shown in figure 1b and corresponding

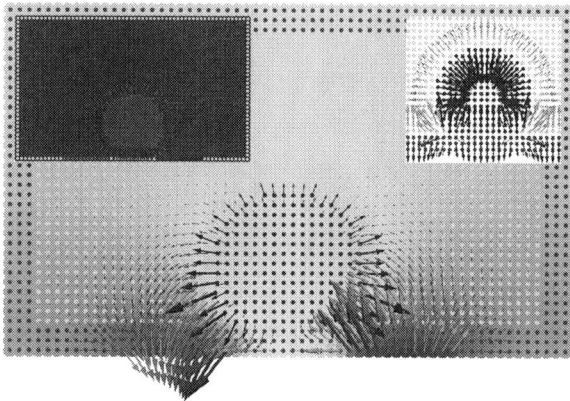

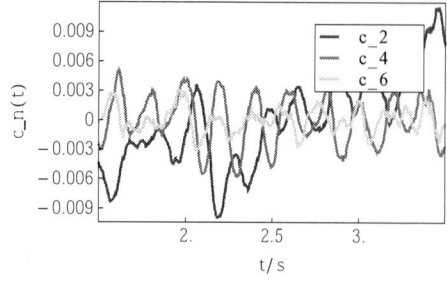

(a) Amplitudes c_n of the first Legendre polynomials in time domain

Fig. 2. MLS/SPH simulation setup: (bottom) droplet with electric potential as scalar quantity and ∇u as vectorial quantity; blue: calculated Korteweg-Helmholtz body force (left) scalar quantity for the dielectricity number and boundary conditions; vectorial quantity fot $\nabla \epsilon$ (right) droplet phases as scalar quantity; surface tension forces as vectorial quantity

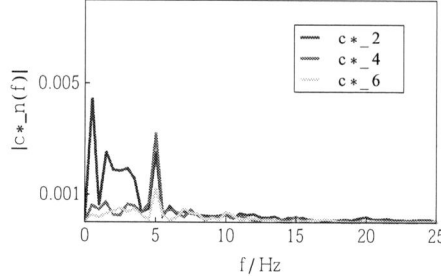

(b) Fourier transformation of normalised amplitudes c_n^* in frequency domain

Fig. 3. Simplified setup: analysis of the P_2, P_4 and P_6 mode

to (1). From the global surface stress $\Pi^{(\mathrm{sf})} = \sum_{k,l}^{k \neq l} \Pi_{kl}^{(\mathrm{sf})}$, the surface tension force is calculated:

$$\mathbf{F}^{(\mathrm{sf})} = -\frac{1}{\rho} \nabla \Pi^{(\mathrm{sf})} \qquad (11)$$

.

III. RESULTS

A. Coupled SPH/MLS electrowetting droplet model

A droplet resides in a rectangular domain $[-0.5\mathrm{m}; 0.5\mathrm{m}] \times [-0.3\mathrm{m}; 0.3\mathrm{m}]$ filled with 2160 particles. A quintic smoothing kernel with a smoothing length of $h = 0.01\bar{6}\mathrm{m}$ and a support of $3h$ is used. The fluid phase is assigned a dielectricity number of 80, the surrounding media's dielectricity number is set to 1. Corresponding gradients $\nabla \epsilon$ are calculated as seen in figure 2 in the left inset. Dirichlet boundary conditions of $u = \pm 1\mathrm{V}$ are adjusted for the electrodes, Neumann boundary conditions $\partial u / \partial \mathbf{n} = 0$ for the rest of the boundary.

The initial surface energies α_{sl}, α_{lv} and α_{sv} are set to 1 for a steady state contact angle of $90°$. Corresponding surface tension forces on liquid and vapor domain nearby the triple phase contact line are equal as shown in figure 2.

With the MLS solution to converge as described in section II-B, the distribution of the potential and the electric field forms out, and the Korteweg Helmholtz body force on the fluid phase can be calculated indicated by blue vectors in the main part of figure 2.

This configuration was very sensitive to changes in topology. For particles nearby the triple phase contact line, the number of neighbouring particles within the same phase degraded for sharp contact angles and thus the rank of the MLS interpolant. Although the set of equations in (8) is not underestimated with the setup described here, the simulation may become unstable and makes transient simulations difficult to run.

B. Transient analysis with a simplified model

The next electrowetting droplet experiment is performed in a rectangular domain $[-0.5\mathrm{m}; 0.5\mathrm{m}] \times [-0.4\mathrm{m}; 0.4\mathrm{m}]$. The complete domain consists of 2800 particles, which have initially positions on a regular lattice. The identical smoothing kernel and smoothing length were employed like in III-A. Viscosities were set to $\eta = 0.05\,\mathrm{Pa\,s}$ and densities to $\rho = 1\mathrm{kg/m}^3$. Static surface tension coefficients where chosen to $\alpha_{sl} = \alpha_{lv} = \alpha_{sv} = 1$ corresponding to a steady state contact angle of $90°$. The simulation runs were driven with a Reynolds number of $\mathrm{Re} = 10$ and a Ohnesorg number of $\mathrm{Oh} = 0.1$, so that inertia and surface tension dominate viscous forces.

The solid-liquid surface energy α_{sl} is excited with a harmonic pattern: The time harmonic component $\hat{\alpha}_{sl}$ is given by

$$\hat{\alpha}_{sl}(t) = \alpha_{sl,A} \,\mathrm{sgn}\left[\cos(2\pi f t)\right] \qquad (12)$$

where the amplitude $\alpha_{sl,A}$ is chosen to correspond to maximum contact angles of $60°$ and $120°$. The excitation frequency was set to $f = 5\mathrm{Hz}$, which can be seen in the frequency domain of the mode shape amplitudes in figure 3b. A loose coupling for low frequencies is furthermore visible. In the time domain in figure 3a, the P_2 mode starts to oscillate.

IV. CONCLUSION

A fully coupled model for electrowetting droplets has been presented. The SPH/MLS basing model is only suited when there are sufficiently particles available to guarantee a good quality for the MLS interpolants. Transient electrowetting simulations with changing contact angles and in general single particles repulsing a neighbouring phase may lead to an unstable simulation. A way out might be a larger smoothing length for the SPH kernel, which makes additional effort necessary to fight the well known phenomenon of particle clustering and tensile instability. Artificial pressure and adaptive smoothing length kernels could be used to address this issue.

Since dipoles in fluids polarize on a much smaller timescale than inertial forces take effect, it might be justified to employ a simplified model with a purely surface tension based approach without the need to calculate the costly Laplace equation. A verification that dynamics is not influenced and the accuracy is comparable is still needed.

In the simplified setup, we observe resonance behaviour to start. However, for a higher efficiency, the resonance frequency should be matched. We suggest to analyse a step response of an excited droplet. The eigenfrequencies identified thereby might be helpful for estimating resonance frequencies.

The simplified, efficient approach can be used for characterising electrowetting based droplet vibrations with different transport parameters within a whole frequency range. This way, flow patterns can be determined or tailored with respect to different Reynolds and Ohnesorge-numbers. In the perspective, we want to be able to simulate on small, possibly mesoscopic length-scales to predict droplet dynamics for MEMS relevant regimes.

ACKNOWLEDGMENT

The authors acknowledge complete funding by the DFG via the project *Electrowetting-Simulation mit Partikelmethoden* (grant no LI 1831/1-2).

APPENDIX

A. Droplet mode shape

Surface particles with position $\mathbf{r} = (r, \vartheta)$ form the shape of the droplet.

Fig. 4. Surface extraction

By sorting the particles in equally sized ϑ slots and extracting the respective, furthermost surface particle as seen in figure 4, the droplet shape can be expressed in terms of Legendre polynomials

$$r(\theta) = \sum_{n=0}^{\infty} c_n P_n(x) \qquad (13)$$

in the interval $[-1, 1]$ for $x = \cos\theta$ with the individual Legendre Polynomial P_n of order n and the attributed amplitude c_n. These are calculated with

$$c_n = \frac{2n+1}{2} \int_{-1}^{1} \hat{r}(x) P_n(x)\, \delta x \qquad (14)$$

where $\hat{r}$ is the radius of a surface particle within a δx slot.

B. Characteristic numbers

Flow is described by different dimensionless numbers. The Reynolds number is given by

$$\mathrm{Re} = \frac{\rho v L}{\eta} \qquad (15)$$

where L stands for a characteristic length. The Ohnesorge number is given as

$$\mathrm{Oh} = \frac{\eta}{\sqrt{\rho \sigma L}} \qquad (16)$$

and relates the viscous forces to inertial and surface tension forces. The Weber number gives a measure of fluid's inertia compared to its surface tension:

$$We = \frac{\rho v^2 L}{\sigma} \qquad (17)$$

REFERENCES

[1] F. Mugele and J.-C. Baret, "Electrowetting: from basics to applications," *Journal of Physics: Condensed Matter*, vol. 17, pp. R705–R774, July 2005.

[2] J.-C. Baret, M. M. J. Decré, and F. Mugele, "Self-Excited Drop Oscillations in Electrowetting," *Langmuir*, vol. 23, pp. 5173–5179, Apr. 2007.

[3] J. M. Oh, S. H. Ko, and K. H. Kang, "Shape Oscillation of a Drop in ac Electrowetting," *Langmuir*, vol. 24, no. 15, pp. 8379–8386, 2008.

[4] M. G. Pollack, R. B. Fair, and A. D. Shenderov, "Electrowetting-based actuation of liquid droplets for microfluidic applications," *Applied Physics Letters*, vol. 77, no. 11, p. 1725, 2000.

[5] G. Beni, "Electrowetting displays," *Applied Physics Letters*, vol. 38, no. 4, p. 207, 1981.

[6] V. Srinivasan, V. K. Pamula, and R. B. Fair, "An integrated digital microfluidic lab-on-a-chip for clinical diagnostics on human physiological fluids.," *Lab on a chip*, vol. 4, pp. 310–5, Aug. 2004.

[7] V. Srinivasan, "Droplet-based microfluidic lab-on-a-chip for glucose detection," *Analytica Chimica Acta*, vol. 507, pp. 145–150, Apr. 2004.

[8] S. Kuiper and B. H. W. Hendriks, "Variable-focus liquid lens for miniature cameras," *Applied Physics Letters*, vol. 85, no. 7, p. 1128, 2004.

[9] A. Prosperetti, "Free oscillations of drops and bubbles: the initial-value problem," *journal of fluid mechanics*, vol. 100, pp. 333–347, 1980.

[10] A. Prosperetti, "Normal-mode analysis for the oscillations of a viscous liquid drop in an immiscible liquid," *J Mécanique*, vol. 19, no. 1, pp. 149–182, 1980.

[11] A. Prosperetti, "Viscous effects on perturbed spherical flows," *Q. Appl. Maths*, vol. 35, pp. 339–352, 1977.

[12] J. U. Brackbill, D. B. Kothe, and C. Zemach, "A Continuum Method for Modeling Surface Tension," *J Comput Phys*, vol. 100, pp. 335–354, June 1992.

[13] J. R. Melcher, *Continuum Electromechanics*. Cambridge, MA: MIT Press, 1981.

[14] E. M. L. S. C. Landau, *Electrodynamics of Continuous Media*. Pergamon Press, 1960.

[15] O. Iliev and S. Tiwari, "A Generalized (Meshfree) Finite Difference Discretization for Elliptic Interface Problems," pp. 488–497, 2003.

[16] P. Lancaster and K. Salkauskas, "Surfaces Generated by Moving Least Squares Methods," *Mathematics of Computation*, vol. 37, no. 155, pp. pp. 141–158, 1981.

[17] X. Y. Hu and N. A. Adams, "A multi-phase SPH method for macroscopic and mesoscopic flows," *J. Comput. Phys.*, vol. 213, pp. 844–861, Apr. 2006.

[18] D. Weiss, J. Lienemann, A. Greiner, D. Kauzlarić, and J. G. Korvink, "Smoothed particle hydrodynamics-based numerical investigation on sessile, oscillating droplets.," *Philosophical transactions A.*, vol. 369, pp. 2565–73, June 2011.

[19] J. P. Morris, "Simulating surface tension with smoothed particle hydrodynamics," *International Journal for Numerical Methods in Fluids*, vol. 33, pp. 333–353, 2000.

Processing of 3D multilevel SU-8 fluidic network assisted by PerMX dry-photoresist lamination

R. Ch. Meier[*,1], V. Badilita[2], U. Wallrabe[2], J. G. Korvink[1,3]

[1]Department of Microsystems Engineering, Laboratory for Simulation, University of Freiburg, GERMANY
[2]Department of Microsystems Engineering, Laboratory for Microactuators, University of Freiburg, GERMANY
[3]Freiburg Institute for Advanced Studies - FRIAS, University of Freiburg, GERMANY
meierr@imtek.de

Abstract— **In this work we present a new manufacturing method for MEMS based multi-level microfluidic devices. The combination of spin-coating SU-8 (MicroChem) and lamination of PerMX dry-photoresist (DuPont) enables the fabrication of a complex 3-level microfluidic channel network with channel aspect ratios ranging from 0.3 up to 3. On 13 mm² fluidic chip area we realized a complex 3D fluidic network by interconnecting three individual fluidic levels. The unique use of UV-photolithography tools and high-precision UV-lasering for the fluidic chip manufacturing results in 25 μm alignment accuracy between the fluidic levels. We further report on the high material compatibility of SU-8 and PerMX which results in high substrate adhesion of the fluidic devices (26.5 MPa).**

Keywords - multi-level fluidic network; dry-resist lamination;

I. INTRODUCTION

Recent fluidic applications led to the development of Lab-on-a-chip devices with 3D multi-level fluidic networks. This trend towards dense and complex fluidic systems requires a straightforward and versatile technology for the reliable fabrication of highly integrated microfluidic chips.

A widely used fabrication method for the manufacturing of such 3D fluidic channel networks nowadays is bonding. An open channel network is patterned on a substrate and then subsequently bonded onto a second substrate to form sealed cavities. This technique has been transferred to different materials, such as poly(dimethyl-siloxane) (PDMS) [1], poly-methyl-methacrylate (PMMA) [2,3], cyclic olefin copolymer (COC) [4] and SU-8 photoresist [5-8]. Complex multilevel fluidic devices have also been achieved in SU-8 [8] and PDMS [9-11] with the stacking and bonding of multiple fluidic layers. These step-by-step assembly techniques are time-consuming and further difficult to control with high accuracy. The proposed method for SU-8 microfluidic chips [8] includes delicate fabrication steps for the manual release of thin wafer-size SU-8 layers from Kapton films. Multi-layer PDMS fluidic networks require inaccurate manual alignment and bonding steps. Activation of PDMS in an O₂ plasma, the following alignment and pressing of the PDMS slabs should be carried out within a few minutes [9] to achieve best bonding results. As a consequence, time consuming alignment using conventional equipment is often replaced by manual alignment under an optical microscope. In recent reports on PDMS multi-level microfluidics, this drawback remains unsolved [10,11]. Fluidic networks are also fabricated in a sacrificial layer technique. Fluidic channels are the result of the selective removal of material from its surrounding medium [12-14]. However, the complete removal of the sacrificial material from large-scale channel circuits is time-consuming and the fluidic device can be damaged due to the exposure to removal chemicals or to high temperatures required for the diffusion of sacrificial material [13,14].

A new approach for the fabrication of complex fluidic networks is the use of dry SU-8 photoresist film [15], where a thin film of soft baked SU-8 is laminated onto a substrate to enclose fluidic channels. This technique is efficient, however, fluidic geometries with an high-aspect ratio are not achievable in a simple way. The required total photoresist thickness has to be built up from multiple dry-resist layers which is not practical. It was one of our objectives to develop a dry-resist lamination process that overcomes this drawback [16].

In this report we combined PerMX 3050 dry-resist with thick SU-8 photoresist processing and developed a full wafer fabrication technique which enables a versatile and straightforward route to the manufacturing of multi-level 3D fluidic channel networks. We present a microfluidic chip with a stack of three individual fluidic levels accurately aligned by standard UV-photolithography and automatically interconnected by automatic high-precision UV-lasering. The process integration of standard thick SU-8 lithography allows for a high degree of freedom in terms of microfluidic design, as the lateral and vertical channel dimensions can be adjusted over a wide range. PerMX and SU-8 show good compatibility which results in high interlayer adhesion.

II. EXPERIMENTAL

A. Process compatibility

With the method explained here for the manufacturing of 3D fluidic networks, PerMX is combined with conventional high viscosity SU-8 2150 photoresist. As we combine two different photoresist materials, process compatibility between them is a major concern for this fabrication strategy to succeed. Processing steps and the corresponding parameters are similar for both SU-8 and PerMX. The process compatibility is quantified in a shear force experiment, where the adhesive force present at the SU-8 and PerMX material boundary is evaluated. For this purpose, we fabricated SU-8 posts on predefined PerMX plateaus (Fig. 1) in a process with modified parameters, as explained in the next section (Fabrication). Post

978-1-4673-1122-9/12 $31.00 © 2012 IEEE

exposure baking temperatures are set to 55 °C (15 h) for PerMX and 65 °C (5 h) for SU-8, instead of the conventional 95 °C to minimize thermal stress. Details of the test configuration are shown in Fig 1.

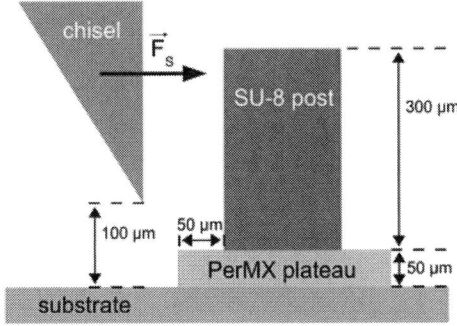

Fig. 1. DAGE 4000 Shear-Tester setup for the measurement of the failure force of SU-8 posts (SU-8 2150) formed on top of PerMX plateaus. The chisel width was 1 mm, the velocity of the chisel was 50 µm/s.

The chisel of the shear tester is moved in front of the post. After a touchdown to the substrate, the chisel moves to the desired shear height h_S of 100 µm above the substrate and pushes with increasing force F_S against the post. The force that is required to shear off the post is defined as the failure force F_{fail} and used to calculate the substrate adhesion. The result of this test is that for SU-8 posts on a PerMX plateau the failure force thus the substrate adhesion is increased by more than 200 % for all post diameters.

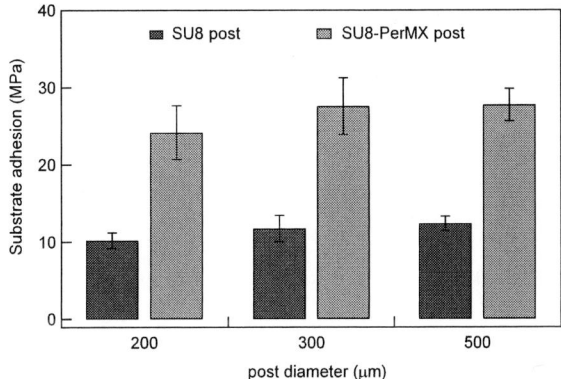

Fig. 2. Result of shear test experiment on Si substrates. SU8 post adhesion is 11.4±0.92 MPa and SU8-PerMX post adhesion is 26.5±1.64 MPa.

The posts without PerMX plateau have a mean substrate adhesion of (11.4±0.92) MPa. The posts with PerMX plateau have a mean substrate adhesion of (26.5±1.64) MPa. This result is a proof of the compatibility of both PerMX and SU-8 photoresists.

B. Fabrication

In this study, we have used 50 µm thick PerMX 3050 (DuPont) dry-photoresist which is processable with standard UV-photolithography equipment. The dry film is based on a negative acting photo-imageable epoxy polymer that defines permanent and durable structures and good side wall shape. In terms of microfluidic fabrication, PerMX 3050 is hot-roll laminated onto a pre-structured substrate to seal channel networks (Fig. 3).

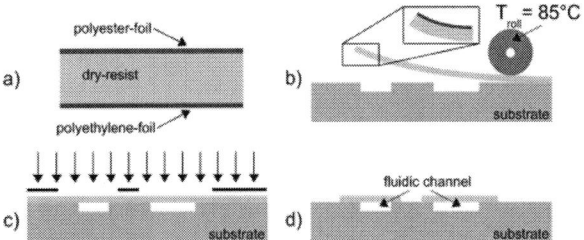

Fig. 3. Lamination of PerMX to seal fluidic channels a) Three layer package of PerMX. Dry-resist is covered by polyester and polyethylene protective foil. b) Polyethylene foil is removed and the dry-resist is hot-roll laminated onto the substrate. c) UV-photolithography. d) Post exposure bake and development to reveal structures.

Besides the photolithography of PerMX and SU-8, we include high-precision UV-laser-machining in the process. in order to gain flexibility regarding the topology of the fluidic network. On this fabrication basis, we demonstrate a microfluidic channel network with three individually addressable fluidic levels.

On a 4 inch quartz glass substrate, the first microfluidic layer was structured using UV-micro laser-maching (Fig. 4a). We use a Trumpf TruMark Station 5000 (355 nm). The lateral geometry of the channels was defined in CAD software, the vertical dimensions are set by the parameters of the laser system. After lasering the glass wafer was cleaned in acetone, isopropanol followed by a DI water rinse and baking at 120 °C (1 h). In step 4b, one layer of PerMX 3050 dry resist was laminated onto the glass substrate to enclose the lasered glass channels using a DuPont laminator (Riston HRL). The temperature of the lamination roll was set to 85 °C, the lamination speed was set to 1 cm/s and the applied roll pressure approximately 0.1 MPa. The dry resist is then UV-patterned with a dosis of 1125 mJ/cm^2 (i-line). For all photolithography steps, the exposure dose recommended in the data sheet was increased by a factor of 2.25 to take into account the absorption of transparency masks as well as the use of quartz glass substrates. The post exposure bake was carried out at 55 °C (15 h). We lowered the temperature for the PerMX post exposure bake in order to reduce deformations of the dry-resist due to the expanding air volume inside the enclosed channels. The protective polyester foil of the PerMX was removed at the end of the post exposure bake when the substrate was at 55 °C. This first PerMX layer was not developed to leave a planar surface for the following spin coating step.

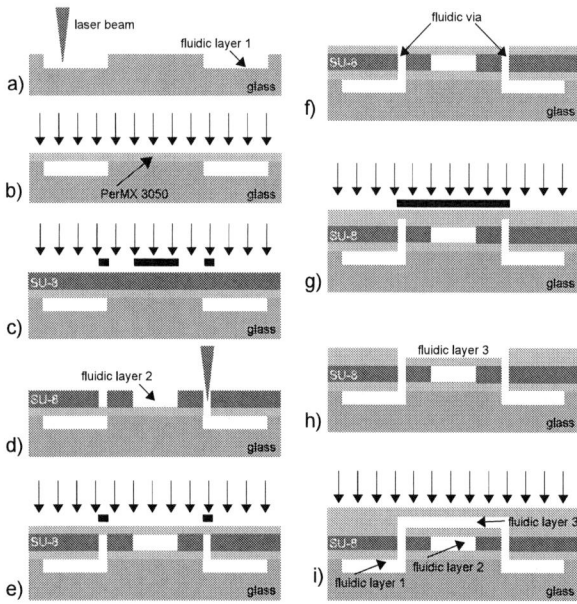

Fig. 4. Process flow: a) UV-laser machining of quartz glass. b) PerMX photolithography. c) SU-8 photolithography. d) PerMX opening for fluidic via between fluidic layer 1 and 2 with UV-laser. e) PerMX photolithography. f) Development of fluidic vias g) Lamination of one or multiple PerMX films to adjust the channel height of fluidic layer 3. h) Fluidic layer 3 is defined. i) Photolithography of final PerMX.

For the second fluidic level, SU-8 2150 photoresist was spin coated on the wafer at 500 rpm (10 s) and 2000 rpm (30 s) to define a resist layer of 300 μm thickness. The soft bake was carried out at 65 °C (15 min) and 95 °C (1 h) on a leveled hotplate (Fig. 4c). This soft bake has no visible negative effect on the PerMX layer underneath the SU-8 resist. The applied exposure dose for the SU-8 layer was set to 1125 mJ/cm² (i-line). The post exposure bake was carried out at a temperature of 65 °C to minimize thermally induced stress. To compensate for the low curing temperature, the time of the post exposure bake was extended to 5 hours. Both the SU-8 and the first PerMX layer were then developed simultaneously in PGMEA developer (15 min). The top layer for fluid handling was realized by the lamination of multiple PerMX layers (Fig. 4e-i) using the same parameters as previously.

Initially, the fluidic layers are not connected with each other. The fluidic interconnections (vias) are defined at positions where the PerMX separating two levels is removed. Between fluidic level 2 and 3, vias are realized using photolithography (Fig. 4f). However, level 1 and 2 cannot be connected with photolithographically removed dry-resist. The first PerMX layer will melt at positions where it was not cross-linked when heated up above its glass transition temperature of approximately 60 °C. This has to be taken into account for the soft bake of the next SU-8 resist layer. At a

required soft bake temperature of 95 °C, sealing of the fluidic channel is not reliable in places where the PerMX layer was not crosslinked. To make sure that the viscous SU-8 does not penetrate into the already sealed glass channels during the soft bake step, the firstly laminated dry-resist was flood exposed (Fig. 4b). Vias between layer 1 and 2 are then achieved with the local removal of PerMX caused by a focused UV-laser beam (Fig. 4d). Fully automatic lasering of the vias is achieved with a computer controlled xy-table. The accuracy of the positioning system is better than 25 μm.

III. DISCUSSION

The capability of the reported technique is demonstrated in this section. A 3D helical channel configuration is chosen in order to realize a topologically complex fluidic network. Two individual channels are arranged on a fluidic chip such that both form the shape of a spiral and continuously wind around a third channel. Thus, the straight channel is passed over and under at the same x-y-position. A stack of three separate channels all passing each other at the same lateral position is realized. In total, the fluidic chip has 16 crossing points in an area of 4.8 mm x 2.7 mm (Fig. 5).

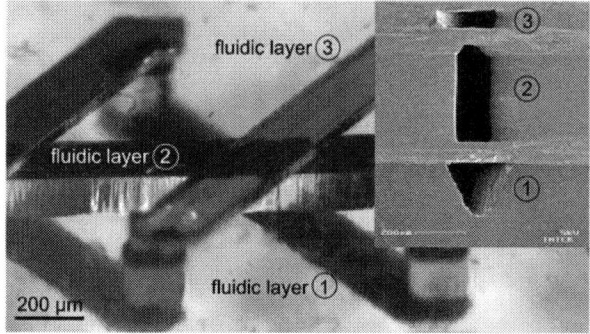

Fig. 5. The fluidic double helix geometry in a close-up view. The centre channel (blue) is passed over and under alternately at the same lateral position by the green and red channel. The smaller image shows the cross section of the fluidic layers where they cross each other.

Compared to similar geometries in PDMS, which required several time-consuming bonding steps for each fluidic device, the method developed by us provides full-wafer manufacturing of compact multi-level fluidic devices. The combination of SU-8 and PerMX processing further offers high flexibility for the dimensioning of the fluidic device. The high substrate and interlayer adhesion of PerMX and SU-8 allows also for expanded chip sizes in the centimeter region without an increased risk of delamination failure. For each fluidic layer the channel geometry, including the height, can be adjusted separately in order to meet specific design requirements. In particular the SU-8 and the glass channels allow for a considerable freedom as they can cover channel heights over a wide range.

Further advantages arise from the properties of the materials used in this process. With one of the fluidic layers

978-1-4673-1122-9/12 $31.00 © 2012 IEEE

manufactured in SU-8, it is possible to integrate SU-8 designs into the fluidic chip and benefit from high-resolution and high aspect-ratio UV-lithography. PerMX dry-resist is suitable for microfluidic applications with highly constant and accurately defined layer thicknesses. PerMX and SU-8 are transparent for visible light, thus the fluidic chip is suitable for applications where optical access is required. Bio-compatibility, chemical and thermal stability are further demands that are met.

The presented technique utilizes UV-lasering and standard UV-lithography procedures that reduce manual interventions to a minimum. Compared to bonding methods that achieve the same network complexity, our approach has the advantage of high alignment accuracy between the fluidic levels. Sacrificial layer techniques require complex UV-lithography methods or high-temperature steps that increase the process complexity significantly compared to our strategy.

IV. CONCLUSION

We presented a new fabrication method to implement complex 3D multi-level microfluidic networks. Based on the combined use of dry-photoresist lamination and SU-8 UV-photolithography, the demonstrated approach for the manufacturing of microfluidic networks facilitates an increase in achievable topological complexity. Both types of permanent photoresists used in this study are suitable for microfluidic applications and further show good process compatibility and high interlayer adhesion. Those facts will lead to the ability to manufacture complex 3D fluidic circuits removing many of the constraints currently associated with the 2D manufacturing process for SU-8 photoresist.

ACKNOWLEDGMENT

This work is supported by European NEST project 028533 and by an operating grant of the University of Freiburg. We gratefully acknowledge the help of DuPont, the IMTEK INUMAC group as well as the IMTEK RSC.

References

[1] M. A. Unger, H.-P. Chou, T. Thorsen, A. Scherer and S. R. Quake, „Monolithic microfabricated valves and pumps by multilayer soft lithography", *Science*, 2000, 288, 113-116.

[2] Y. Sun, Y. C. Kwok and N.-T. Nguyen, „Low-pressure, high-temperature thermal bonding of polymeric microfluidic devices and their applications for electrophoretic separation", *J. Micromech. Microeng.*, 2006, 16, 1681-1688.

[3] B. Bilenberg, T. Nielsen, B. Clausen and A. Kristensen, „PMMA to SU-8 bonding for polymer based lab-on-a-chip systems with integrated optics", *J.Micromech. Microeng.*, 2004, 14, 814-818.

[4] C. W. Tsao, L. Hromada, J. Liu, P. Kumar and D. L. DeVoe, „Low temperature bonding of PMMA and COC microfluidic substrates using UV/ozone surface treatment", *Lab Chip*, 2007, 7, 499-505.

[5] F. J. Blanco, M. Agirregabiria, J. Garcia, J. Berganzo, M. Tijero M. T. Arroyo, J. M. Ruano, I. A. and K. Mayora, „Novel three-dimensional embedded SU-8 microchannels fabricated using a low temperature full wafer adhesive bonding", *J. Micromech. Microeng.*, 2004, 14, 1047-1056.

[6] P. Svasek, E. Svasek, B. Lendl and M. Vellekoop, „Fabrication of miniaturized fluidic devices using SU-8 based lithography and low temperature wafer bonding", *Sensors and Actuators A*, 2004, 115, 591-599.

[7] S. Tuomikoski and S. Franssila, „Free-standing SU-8 microfluidic chips by adhesive bonding and release etching", *Sensors and Actuators A*, 2005, 120, 408-415.

[8] M. Agirregabiria, F. J. Blanco, J. Berganzo, M. T. Arroyo, A. Fullaondo, K. Mayora and J. M. Ruano-López, „Fabrication of SU-8 multilayer microstructures based on successive CMOS compatible adhesive bonding and releasing steps", *Lab Chip*, 2005, 5, 545-552.

[9] J. R. Anderson, D. T. Chiu, R. J. Jackman, O. Cherniavskaya, J. C. McDonald, H. Wu, S. H. Whitesides and G. M. Whitesides, „Fabrication of Topologically Complex Three-Dimensional Microfluidic Systems in PDMS by Rapid Prototyping" *Anal. Chem.*, 2000, 72, 3158-3164.

[10] M. Zhang, J. Wu, L. Wang, K. Xiao and W. Wen, „A simple method for fabricating multi-layer PDMS structures for 3D microfluidic chips", *Lab Chip*, 2010, 10, 1199-1203.

[11] B. Mosadegh, M. Agarwal, Y.-S. Torisawa and S. Takayama, „Simultaneous fabrication of PDMS through-holes for three-dimensional microfluidic applications", *Lab Chip*, 2010, 10, 1983-1986.

[12] K.-S. Yun and E. Yoon, „Fabrication of complex multilevel microchannels in PDMS by using three-dimensional photoresist masters", *Lab Chip*, 2008, 8, 245-250.

[13] H. A. Reed, C. E. White, V. Rao, S. A. B. Allen, C. L. Henderson and P.A. Kohl, „Fabrication of microchannels using polycarbonates as sacrificial materials", *J. Micromech. Microeng.*, 2001, 11, 733-737.

[14] S. Metz, S. Jiguet, A. Bertsch and Ph. Renaud, „Polyimide and SU-8 microfluidic devices manufactured by heat-depolymerizable sacrificial material technique", *Lab Chip*, 2004, 4, 114-120.

[15] P. Abgrall, Ch. Lattes, V. Conédéra, X. Dollat, S. Colin and A. M. Gué, „A novel fabrication method of flexible and monolithic 3D microfluidic structures using lamination of SU-8 films", *J. Micromech. Microeng.*, 2006, 16, 113-121.

[16] R. Ch. Meier, V. Badilita, J. Brunne, U. Wallrabe, Jan G. Korvink „Complex three-dimensional high aspect ratio network manufactured in combined PerMX dry-resist and SU-8 technology", *Biomicrofluidics*, 2011, 5, 034111.

978-1-4673-1122-9/12 $31.00 © 2012 IEEE

Conductive and Transparent Gel Microstructures Fabricated by Inkjet Printing of Ionic Liquid Based Fluids

Ute Löffelmann[1], Dario Mager[1], Jan G. Korvink[1,2]

[1]Department of Microsystems Engineering (IMTEK), Albert-Ludwigs-Universität Freiburg, GERMANY
[2]Freiburg Institute for Advanced Studies (FRIAS), Albert-Ludwigs-Universität Freiburg, GERMANY
ute.loeffelmann@imtek.uni-freiburg.de

Ionic liquid-polymer gel microstructures with high spatial resolution (line widths of 35 - 40 µm), good electrical conductivity (5 - 30 mS/cm), optical transparency and mechanical flexibility are fabricated using a new and solvent-free inkjet printing process. Homogeneously gelled structures are achieved by carrying out the printing and polymerization process in nitrogen atmosphere eliminating the inhibiting influence of oxygen. Printable low viscosities are achieved without adding extra solvents by the careful selection and combination of ionic liquids (ILs) and unsaturated monomers. The final properties of the printed IL-polymer gels can be controlled in terms of ionic conductivity, optical transmission and mechanical flexibility by using different types and amounts of ILs and monomers

Ionic liquids, inkjet printing, conductive and transparent gel, microstructures, photopolymerization

I. INTRODUCTION

Transparent conductive materials are important constituents of electronic devices like sensors, displays and solar cells. Currently, the materials most often used are transparent conductive oxides (TCOs), mainly indium tin oxide (ITO). TCOs are brittle and are becoming increasingly expensive due to decreasing availability [1].

To meet the increasing technological demand for the application of transparent conductive materials in flexible electronic devices, considerable research is being done in order to find suitable flexible materials with good availability. That is why organic materials gain increasing interest. One class of organic conductive materials includes mechanically flexible polymer-electrolyte gels. In these gels, high conductivities are combined with interesting mechanical properties. Ionic liquid-polymer gels represent a quite new and interesting subclass of polymer-electrolyte gels. In ionic liquid-polymer gels, polymer chains or networks surround the ionic liquids (ILs) and provide good stability.

ILs are salts that are liquid below 100°C or even at room temperature [2, 3]. Beside their good solvent properties that offer the possibility for the combination with different reagents, specific properties such as high ionic conductivity, transparency, low vapor pressure and good chemical and thermal stability make them of great interest for use in flexible electronic devices [4].

Unstructured layers of IL-polymer gels can be achieved by casting. But there is also great interest in precise micro-structured geometries. This led to the target of the work reported here: the development of a reliable fabrication process for creating ion-conductive, flexible and transparent IL-polymer gel microstructures by inkjet printing.

Inkjet printing has gained considerable interest as an attractive fabrication technique. Picoliter-sized droplets can be precisely deposited at predetermined positions on a substrate. No masks are needed and the number of process steps is reduced. This enables cost savings, highly efficient use of materials and the elimination of waste.

Polymer matrices containing ILs can be achieved in two ways: either by swelling of existing polymers in ILs or by in situ polymerization, carried out in an IL.

Ref. [5] describes the swelling of polymers in ILs by printing ILs on a substrate covered with a polymeric gelling agent. This process has several obvious disadvantages. Covering the substrate with the gelling agent implies an additional upstream process step. The gelling agent in the unprinted areas is not swelled and remains as excess on the substrate or has to be removed after printing. Furthermore, the dimension accuracy of the structures is limited due to the spreading of the ILs into the gelling agent.

The second approach of in situ polymerization was used by [6]. UV-initiated radical polymerization of monomers in ILs was carried out after inkjet printing and led to IL-polymer gel structures. The described ink comprised a telechelically methacrylate-functionalized low molecular mass polyethylenoxide (PEO) as reactive macromonomer, IL and organic solvent. Printing on paper and subsequent polymerization was performed in normal atmosphere. 50 % of organic solvent was added to the ink to reduce the viscosity to a printable value. However, organic solvents in inkjet inks provide several disadvantages. Solvents like ethanol with a higher vapor pressure can already evaporate at the nozzle orifice and cause instable jetting or nozzle clogging. Solvent residues require an additional process step of solvent removal during the fabrication and account for emissions of volatile organic compounds (VOCs) into the environment. Solvents with a low vapor pressure tend to remain in the gel, lower the amount of ILs and might reduce the electrical conductivity. Deformation of the printed structures due to controlled or uncontrolled evaporation of solvents after polymerization, especially when multiple layers are printed, is another negative effect of solvents in the printing fluid.

ILs are ideal reagents for inkjet printing because they are non-volatile. Inks containing exclusively ILs and no volatile organic solvent would minimize the risk of instabilities or nozzle clogging during printing [7]. In our solvent-free approach the given disadvantages and risks caused by volatile or non-conducting solvents should be avoided.

The target of the presented research was hence to develop a solvent-free inkjet printing process for the fabrication of in situ polymerized IL-polymer gels with good conductivity, transparency and flexibility [8].

Inkjet printing requires low ink viscosities (< 30 mPa·s). Monomers with low viscosities seemed to be the most practicable and promising way to achieve printable fluids without adding further solvents. A method using low-viscosity methyl methacrylate monomers (MMA) in ILs without additional organic solvents has been presented by [9]. But the process described can be carried out only by casting in covered systems as MMA is a relatively high volatile monomer and is, hence, not applicable to inkjet printing.

For our purpose we chose methacrylate or acrylate monomers with low viscosities and negligibly low vapour pressures. Out of the huge variety of available ILs, those with high conductivities, low viscosities and good compatibility with the monomers were identified. Furthermore, we used the initiation by UV irradiation enabling the possibility to further lower the ink's viscosity by increasing the print head temperature without the risk of polymerization in the print head. Printing the most promising chemical formulations on non-porous polymer foil revealed significant requirements for the surrounding atmosphere during printing and polymerization and will be discussed in detail in the results part. Ionic conductivity values of IL-polymer gels were obtained from printed and successfully polymerized lines but also from cast samples. Investigation of mechanical flexibility was conducted by recording compressive stress-strain curves of flat cylindrical gel disks and by measuring conductivity on printed lines in unbowed, bowed and kinked shape. The transparency of the gels was evaluated by transmission measurements of cast films.

II. EXPERIMENTAL

A. Materials

ILs were purchased from IoLiTec (Heilbronn, Germany). Acrylate and methacrylate monomers were free samples from Sartomer Europe (Paris, France). The UV initiator Irgacure 1700 was donated by BASF (Ludwigshafen, Germany).

B. Methods

1) Fluid properties: Viscosity measurements were performed using a Haake RheoStress 300 Rheometer (Thermo Fisher Scientific, Karlsruhe, Germany). Volumetric weights were determined using a pycnometer (Carl Roth, Karlsruhe, Germany).

2) Printing tests: The diameter of a single printed droplet on a 220 µm thick PET foil was approximately 40 µm. Overlapping droplets with a dot spacing of 25 µm were printed on the same flat PET foil in order to achieve homogeneous lines. For conductivity measurements lines with a width of 25 pixels (approx. 625 µm) and a height of 2 layers (medial height of one layer approx. 10 µm) were printed. Printing tests were carried out in normal and nitrogen atmosphere with a piezoelectric driven Dimatix Material Printer DMP-2800 (Fujifilm Dimatix, Santa Clara, USA) equipped with a 10 picoliters printhead. Printing in nitrogen atmosphere was performed in an MBraun Glovebox System MB-20 G (MBraun, Garching, Germany) conditioned to less than 0.1 ppm oxygen and less than 1 ppm moisture. After printing and still in the nitrogen-filled glovebox, the radical polymerization was UV-initiated using an UV-lamp (3UV-38, 8 Watt, UVP, Upland, USA) emitting at 254 nm. Digital photographs of the printed lines were taken with a Keyence VHX-100K Digital Microscope.

3) Analysis of polymerization: Characterization of the polymerization process was done using ATR-IR spectroscopy (Varian Excalibur FTS 3000 Spectrometer). Spectra of cast and printed IL-monomer compositions were recorded before and after UV irradiation. Pure ILs and monomers were measured in order to complete the analysis.

4) Ionic conductivity: Ionic conductivity was determined by complex impedance measurements using a Solartron SI 1260 impedance gain-phase analyzer and a Solartron SI 1287 electrochemical interface (Solartron analytical, Hampshire, UK) over the frequency range from 1 Hz to 100 kHz at an AC amplitude of 10 mV.

For the measurement of ionic conductivity, two different kinds of setups were used. The ionic conductivity of the bulk material was measured on cast gel cylinders that were prepared in UV-transparent PMMA pipe sections fixed between two stainless steel electrodes. UV curing was carried out for 5 minutes by irradiating the sample through the UV-transparent sidewall.

For the conductivity measurement of printed lines, they were directly contacted with flattened crocodile clamps. In order to determine the flexibility of the printed lines, conductivity was measured in original unbowed as well as in bowed and kinked shape (Fig. 1). All conductivity measurements were done in normal atmosphere.

5) Optical transparency: For transmission measurements 170 µm thick cast films were fabricated between two quartz glass slides. UV curing was carried out for 5 minutes. For the subsequent transmission measurement a Cary 50 UV-Vis Spectrometer (Agilent, Santa Clara, USA) was used.

6) Mechanical properties: Stress-strain curves during compression were recorded using a Haake RheoStress 300 Rheometer (Thermo Fisher Scientific, Karlsruhe, Germany) in a modified way. During downward movement of the upper metal plate with a constant velocity of 0.15 mm/min, the integrated sensor recorded the force for the linear compression of flat cylindrical gel disks lying on the lower metal plate.

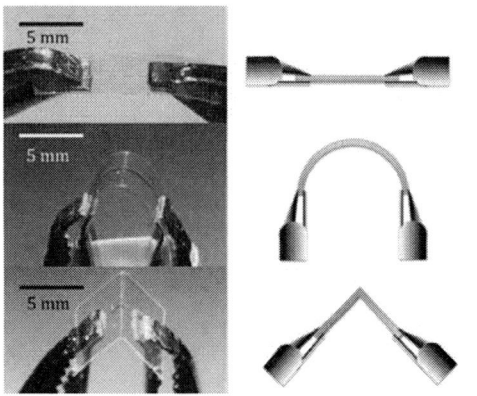

Fig. 1. Photos and schematics of the conductivity measurement setup for printed lines in unbowed, bowed and kinked shape. The length of the measured line section was 5 mm for the unbowed and the kinked shape, and 10 mm for the bowed shape.

The disks were fabricated in the nitrogen-filled glovebox by casting IL-monomer solutions into plastic rings. UV curing was carried out for 5 minutes.

III. RESULTS AND DISCUSSION

A. Preliminary tests

Mixing tests of low viscous mono- or bifunctional monomers with low viscous and highly conductive ILs in different ratios resulted in compatible monomer-IL systems. A clear solution without phase separation indicated compatibility. The maximum fluid viscosity ejectable with the Dimatix materials printer (DMP 2800) is limited to 30 mPa·s. This further narrowed down suitable monomer-IL systems. Advantageously, low viscosities of ILs are accompanied by good electrical conductivities as both properties depend on the ion mobility of the IL [10]. Table I shows viscosity values of favored ILs, monomers and mixtures. Composition C shows that adding the monomer to the IL can mean a significant decrease in viscosity. Consequently, even ILs with higher viscosities can be used if the addition of the monomer leads to a printable viscosity. First impressions of the gel properties of various compositions were obtained from droplets, applied with a pipette and UV-cured on a glass slide.

B. Casting

The appropriate method to achieve good casting results in normal atmosphere is to cover the surface during UV irradiation with a UV-transparent slide. Without slide the gel films tend to show a liquid or at least a sticky surface after UV irradiation. No or incomplete polymerization takes place at the surface in an oxygen-containing atmosphere. By comparing the printing and casting process in the next text section a more detailed description of the observed oxygen-caused inhibition concerning radical polymerization in ILs is given.

C. Printing

Printing was done on transparent and flexible PET foil that did not soak up the ink. Tests whether the printed lines could be converted into gelatinous form or whether they remained liquid have been carried out. First of all printing and subsequent UV-radiation were tested in normal atmosphere. However, the printed lines stayed completely liquid which indicates no or minor polymerization. It is known that oxygen has an inhibiting influence on radical polymerization. It scavenges very efficiently both the initiating and the polymer radicals [11]. Therefore, oxygen-caused inhibition of free radical polymerization leads to slower reaction rates and lower degree of cure.

To overcome this inhibiting influence of oxygen, it was tested to run the polymerization in the oxygen-free nitrogen atmosphere of a glovebox system. The upstream application of a thin film onto a glass slide was done with a pipette in normal oxygen-containing atmosphere. The glass slide was then transferred into the nitrogen filled glovebox system and exposed to UV-light. Successful polymerization and no liquid or sticky film on the surface could be achieved. By reason of this observation the printing process was done in the same way. Samples printed in normal atmosphere were transferred for polymerization into the oxygen-free atmosphere of the glovebox. However, in contrast to the samples prepared with a pipette, the inkjet printed lines did not polymerize and stayed liquid.

These observations are in accordance with the description of [12] about air entrapment at the nozzle of piezo-driven inkjet printheads. It is comprehensible that the steps of formation, ejection, flight and landing of very small droplets during inkjet printing in normal atmosphere promote diffusion and entrapment of oxygen to a high saturation level. Attempts to avoid curing defects by purging the ink with nitrogen before printing were not successful. Even traces of oxygen in the ink seem to have an inhibiting effect and prevent proper curing. The only way to achieve UV-cured inkjet printed IL-polymer gels was to perform the printing as well as the polymerization in an oxygen-free atmosphere provided in the nitrogen-filled glovebox.

This was performed with the three compositions listed in Table I. Free of particles and having low vapor pressures, the compositions showed excellent ink and printing properties. No interruptions caused by nozzle clogging or evaporation of solvent could occur.

Successfully polymerized lines are shown in Fig. 2. Using a dot spacing of 25 μm, line widths between 35 and 45 μm could be achieved.

D. Analysis of polymerisation

ATR-IR spectra of cast and inkjet printed composition C (see Table I) before and after polymerization are presented in Fig. 3. For comparison, the spectra of the pure IL, monomer and polymer were added. Non-polymerized samples clearly show the distinct peak of the conjugated carbon-carbon double bond of the monomer at 1640 cm^{-1} in their spectra. This peak confirms that polymerization has not yet started. After successful polymerization this peak is not visible any more in the spectra of the polymerized samples. This is the case in the spectrum of the pure polymer as well as in the spectrum of the polymerized printed layer of composition C and proves that the applied process of inkjet printing and subsequent

TABLE I. VISCOSITY AND CONDUCTIVITY VALUES.

	Chemical composition (% w/w)	Viscosity @ 30 °C (mPa·s) before polymerization	Conductivity of printed lines @ 22 °C (mS/cm) after polymerization	Conductivity of cast specimens @ 22 °C (mS/cm) after polymerization
A	79 % EMIM DCA 20 % PEG(200)DMA 1 % Irgacure 1700	16.7 Pure IL: 13.1 Pure monomer: 13.1	Unbowed: 29.5 Bowed: 24.9 Kinked: 11.8	15.9 (Pure IL: 27.4)
B	49 % EMIM DCA 50 % 2-(2-EOEO)EA 1 % Irgacure 1700	11.2 Pure IL: 13.1 Pure monomer: 3.3	Unbowed: 15.9 Bowed: 11.8 Kinked: 3.2	9.1 (Pure IL: 27.4)
C	49 % BMIM DCA 50 % 2-(2-EOEO)EA 1 % Irgacure 1700	13.6 Pure IL: 24.8 Pure monomer: 3.3	Unbowed: 5.8 Bowed: 5.6 Kinked: 3.0	4.9 (Pure IL: 11.1)

UV irradiation in nitrogen atmosphere assures successful and complete polymerization. Band shifts were observed in other sections of the IR spectra. This may be caused by interaction between polymer and IL.

E. Mechanical properties

Recording stress-strain curves of cast IL-polymer gel disks during compression provided information on the mechanical properties of the IL-polymer gels. Fig. 4 shows a stress-strain diagram of gels comprising 30 % of either a mono- or a bifunctional monomer. To obtain the Young's modulus E of the gels the stress σ was devided by the strain ε in the elastic initial linear portion of the stress-strain curves.

$$E = \sigma / \varepsilon \qquad (1)$$

The Young's modulus of the gel based on the bifunctional monomer was approximately 12 times higher than the Young's modulus of the gel based on the monofunctional monomer. This means a much higher dimensional stability and a much lower flexibility. As bifunctional monomers do not form only polymer chains but also polymer networks they provide significantly higher dimensional stability and cohesion than monofunctional monomers. Generally, there is a direct correlation between the percentage and the functionality of the applied monomer in the gel and the measured Young's modulus. A higher amount of monomer or a monomer with a higher functionality in the gel causes an increase in dimensional stability and a decrease in flexibility. Consequently there is always a trade-off between these two properties.

F. Conductivity

The formulations listed in Table I were used for printing tests. They are showing comparably good flexibility after polymerization by using different concentrations of the monofunctional monomer (2-(2-EOEO)EA) and of the bifunctional monomer (PEG(200)DMA). Less of the bifunctional monomer is required to achieve the same gel strength as of the monofunctional monomer. High conductivity values between 5 and 30 mS/cm were measured on inkjet printed lines in the unbowed shape. In the bowed shape 74 - 97 % and in the kinked shape 40 - 52 % of the original conductivity values could still be achieved. This proves the good flexibility of the printed IL-polymer gels, as no complete interruptions caused by cracks occurred. The conductivity of the incorporated ILs preset the order of magnitude of the conductivity of the synthesized IL-polymer gels. Within this order of magnitude the IL-monomer ratio strongly affects the conductivity of the gels [9]. The higher the amount of non-conducting monomer in the gel, the lower the conductivity. This explains why using a bifunctional monomer instead of a monofunctional monomer makes it possible to achieve higher conductivity by lowering the amount of monomer, without loss in mechanical strength.

Printed lines showed significantly higher conductivity values than cast specimens.

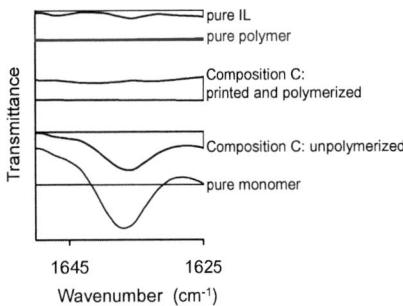

Fig. 3. Section of the ATR-IR spectra of composition C (see Table I) before and after polymerization. For comparison the spectra of the pure IL, monomer and polymer are added.

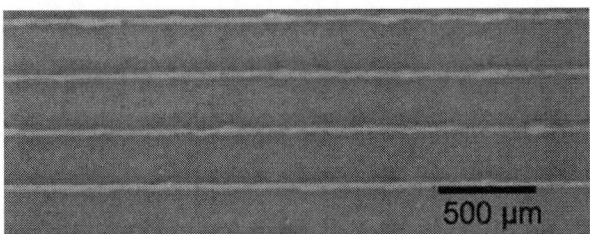

Fig. 2. About 34 – 45 μm wide printed and polymerized lines of composition C (see Table I).

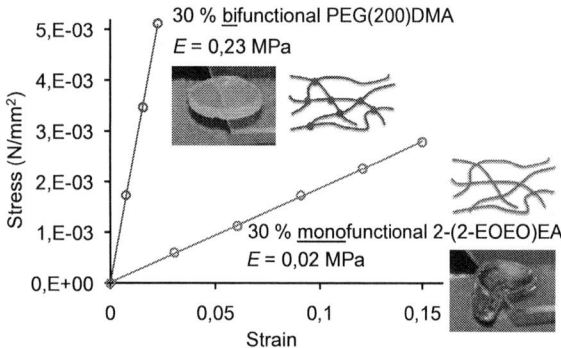

Fig. 4. Compression stress–strain diagram of IL-polymer gel disks (height 2 mm, diameter 13 mm) polymerized with 30 % of monofunctional or bifunctional monomer in 1-ethyl-3-methyl-imidazolium dicyanamide (EMIM DCA). The Young's modulus (E) was deduced from the slope of the graphs.

For conductivity measurements the samples were taken out of the glovebox into normal atmosphere that enabled moisture absorption by the samples. Regarding the cast samples the measuring cell protected their surface against quick moisture uptake. In contrast, diffusion of water into printed lines could take place easily because of the high surface-to-mass ratio and the lack of any shielding effect. Similar behavior has been reported by [13], where an increase in conductivity of ILs by adding water was observed.

G. Transparency

Transmission spectra of the polymerized compositions of Table I are presented in Fig. 5. 170 µm thick films of all three compositions exhibited transmission values of approximately 90 %. In the range of lower wavelength higher absorption caused by the yellowish inherent color of the incorporated ILs was observed. As BMIM DCA is an almost colorless IL, composition C showed the highest transmission values. As the film of composition A looked a little bit hazy, comparable lower transmission values were measured. The inset in Fig. 5 shows a photo of a 2 mm thick film and provides a good impression of the transparency of the gels.

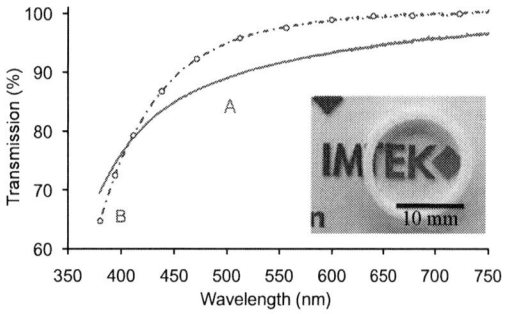

Fig. 5. Transmission spectra of compositions A, B, and C (see Table I) in the visible range of cast 170 µm thick IL-polymer films, measured after 14 days in normal atmosphere. The inset shows a photo of a 2 mm thick film of composition A cast in the ring.

IV. CONCLUSIONS

Concerning the fabrication of microstructures, ILs combined with unsaturated monomers are very suitable fluids for inkjet printing of ion-conductive, flexible and transparent structures with high spatial resolution. Solvent-free printing can be performed by choosing low-viscosity and compatible IL-monomer systems. Following this, UV initiated in situ polymerization leads to gelatinous IL-polymer electrolyte structures. To avoid high loss in conductivity by adding non-conductive monomers to the ILs, bifunctional monomers can be used that provide sufficient dimensional stability at low concentrations. IL-polymer gels with good conductivity and transparency are achieved by using ILs with high conductivity and low inherent color in combination with suitable monomers leading to compatible gel systems. Against the background of the inhibition of the free radical polymerization by oxygen, the required atmosphere during application and polymerization depends on the manufacturing process. For the case of casting in a closed mould, almost no part of the bulk of the IL-monomer solution is in contact with oxygen. Therefore, application and polymerization can be carried out in a normal oxygen-containing atmosphere. This does not apply to inkjet printing because there is a lot of interaction at the interface between the liquid phase of the solution and the surrounding gas phase. The movement of the meniscus at the printing nozzle outlet and the processes of ejection, flight and landing of the droplets support the diffusion of oxygen into the ink. For complete polymerization, an oxygen-free atmosphere as provided in a nitrogen-filled glovebox is of crucial importance during inkjet printing and UV irradiation. The whole process is fast, stable and robust and no nozzle clogging or interruption during printing was observed while 10 millions of droplets were ejected. Even after a jetting break of several days, ejection of droplets could be started again without problems.

ACKNOWLEDGMENT

We wish to express our gratitude to the University of Freiburg and the Excellence Initiative of the German Federal and State Governments for financial support. Furthermore, we would kindly like to thank our cooperation partner IoLiTec (Heilbronn, Germany) for expert advice in the field of ILs. We also thank Sartomer Europe (Paris, France) for the kind donation of numerous monomer samples. Our colleagues from the Department of Microsystems Engineering (IMTEK) of the University of Freiburg working at the Chair for Biomedical Microtechnology, the Laboratory for Chemistry and Physics of Interfaces and the Laboratory for Micro-optics, we thank for providing access to specialized equipment used for sample preparation and characterization.

REFERENCES

[1] Z. Chen, B. Cotterell, and W. Wang, "The fracture of brittle thin films on compliant substrates in flexible displays," *Eng. Fract. Mech.*, 2002, vol. 69, pp. 597- 603.

[2] P. Kubisa, "Ionic liquids in the synthesis and modification of polymers," *J. Polym. Sci., Part A: Polym. Chem.*, 2005, vol. 43, pp. 4675-4683.

[3] I. Krossing, J. M. Slattery, C. Daguenet, P. J. Dyson, A. Oleinikova, and H. Weingärtner, "Why are ionic liquids liquid? A simple explanation based on lattice and solvation energies," *J. Am. Chem. Soc.*, 2006, vol. 128, pp. 13427-13434.

[4] M. Deetlefs, K. R. Seddon, "Ionic liquids: fact and fiction," *Chem. Today*, 2006, vol. 24, pp. 16-23.

[5] T. Kugler, U.S. Patent 20090042346.

[6] J. T. Delaney, A. R. Liberski, J. Perelaer, and U. S. Schubert, "A practical approach to the development of inkjet printable functional ionogels-bendable, foldable, transparent, and conductive electrode materials," *Macromol. Rapid Comm.*, 2010, vol. 31, pp. 1970–1976.

[7] T. Welton, "Room-temperature ionic liquids. Solvents for Synthesis and Catalysis," *Chem. Rev.*,1999, vol. 99, pp. 2071-2083.

[8] U. Löffelmann, N. Wang, D. Mager, P. J. Smith, and J. G. Korvink, "Solvent-free inkjet printing process for the fabrication of conductive, transparent, and flexible ionic liquid-polymer gel structures," *J. Polym. Sci. Pol. Phys.*, 2012, vol. 50, pp. 38-46.

[9] M. Susan, T. Kaneko, A. Noda, and M. Watanabe, "Ion gels prepared by in situ radical polymerization of vinyl monomers in an ionic liquid and their characterization as polymer electrolytes," *Am. Chem. Soc.*, 2005, vol. 127, pp. 4976-4983.

[10] J. M. Slattery, C. Daguenet, T. Schubert and I. Krossing, "How to predict the physical properties of ionic liquids: a volume-based approach," *Angew. Chem. Int. Ed.*, 2007, vol. 46, pp. 5384–5388.

[11] C. S. B. Ruiz, L. D. B. Machado, J. E. Volponi, and E. S. Pino, "Oxygen inhibition and coating thickness effects on UV radiation curing of weatherfast clearcoats studied by photo-DSC," *J. Therm. Anal. Calorim.* 2004, vol. 75, pp. 507–512.

[12] J. de Jong, H. Reinten, M. van den Berg, H. Wijshoff, G. J. de Bruin, M. Versluis, and D. Lohse, "Air entrapment in piezo-driven inkjet printheads," *J. Acoust. Soc. Am.*, 2006, 120, 1257-1265.

[13] A. Jarosik, S. R. Krajewski, A. Lewandowski, and P. Radzimski, "Conductivity of ionic liquids in mixtures," *J. Mol. Liq.*, 2006, vol.123, pp. 43-50.

Nanomechanical DNA Origami Devices as Versatile Molecular Sensors

Akinori Kuzuya[*], Takahiro Yamazaki, Kohei Yasuda, Yusuke Sakai, Yusei Yamanaka, Yan Xu, Yuichiro Aiba,
Yuichi Ohya, Makoto Komiyama
[*]Department of Chemistry and Materials Engineering, Kansai University, JAPAN
kuzuya@kansai-u.ac.jp

Abstract—**We propose versatile sensing systems for a variety of chemical and biological targets at molecular resolution. We have designed functional nanomechanical DNA origami devices that can be used as "single-molecule beacons", which consist of two levers approximately 170 nm long connected at a fulcrum. Various single-molecule inorganic/organic targets ranging from metal ions to proteins as well as a unique binding event of a nucleic acid analogue can be visually detected on AFM by a shape transition of the origami devices.**

DNA; Nanotechnology; DNA Origami; Nanomechanical Device

I. INTRODUCTION

Rapid development of nanotechnology has enabled precise manipulation of nanomaterial. However, typical analytical methods for chemical or biochemical targets are still based on spectroscopic principles, which treat average behavior of vast molecules. To analyze individual molecule's behavior, nanomechanical devices that can work with target molecules in single-molecule manner are required. Structural DNA nanotechnology [1]–[3], which is based on programmed assembly of branched DNA helices, is one of a key technology that provides functional nanomechanical devices. Various fine nanomechanical DNA devices have been constructed to date. However, individual molecules of most of them are still too small to be analyzed one by one. DNA origami [4]–[7], in which long single-stranded DNA is folded into designed nanostructure with the aid of many short staple strands, is a latest powerful tool in structural DNA nanotechnology that provides quite robust and precise nanostructures both in 2D and 3D, which are all visible with atomic force microscopy (AFM) or electron microscopy (EM). Despite such advantages, DNA origami itself has scarcely been studied as a building material for nanomechanical DNA devices [8], [9]. Here we present unique application of nanomechanical DNA origami devices that can be used as versatile and visible "single-molecule" beacons (DNA origami beacons) [10]. By using "DNA origami pliers", which consist of two ca. 170-nm-long levers connected at a fulcrum, co-existence of various inorganic/organic targets from metal ions to proteins in a solution as well as an invasion event of a triplex-forming peptide nucleic acid (PNA) into a short DNA duplex attached to DNA origami beacons were visually detected on AFM in

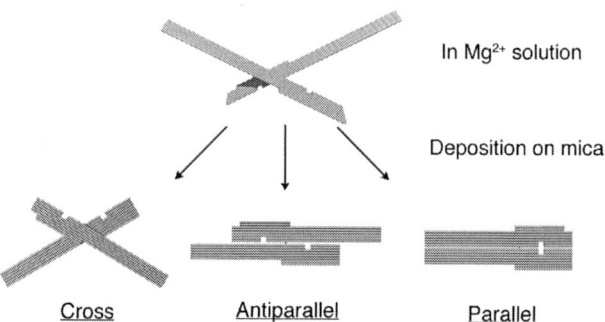

Fig. 1. The structure and shape change of DNA origami pliers used in this study.

single-molecule manner as a shape transition of DNA origami beacons.

II. EXPERIMENTAL PROCEDURES

Staple DNA strands and dye-labeled staples were purchased from Sigma Genosys (Japan) and used without further purification. Biotin-TEG and FAM modified anchor strands were chemically synthesized using appropriate CPG columns (Glen Research, VA), and purified by reverse-phase HPLC. Triplex-forming bis-PNA was synthesized by Boc-based approach [11]. Formation of nanomechanical DNA origami was performed with M13mp18 ssDNA (4 nM, Takara, Japan), staples and anchor strands (16 nM for each strand) in a solution containing Tris (40 mM), acetic acid (20 mM), EDTA (10 mM), and magnesium acetate (12.5 mM, 1 X TAE/Mg buffer, 50 μL). This mixture was cooled from 90°C to 25°C at a rate of -1.0°C/min using a PCR thermal cycler to anneal the strands. After excess staples and anchors were removed from the mixture by using an ultrafiltration microtube, 2 eq. of SA to DNA origami beacons (final concentration, 4 nM) was added to the solution and immediately subjected to AFM measurement. Selective release of SA via strand displacement was done with 10 eq. of unset strands at room temperature for 2 h. Unzipping of pre-closed DNA pliers with triplex invasion by triplex-forming bis-PNA was done with 100 or 1000 eq. to the closing elements of bis-PNA in 5 mM HEPES (pH 7) in the presence of 12.5 mM Mg^{2+} at 16°C for 3 days. AFM imaging of DNA origami was performed on a SPA-300HV system (SII, Japan). The mixture (3 μL) was deposited on a

978-1-4673-1122-9/12 $31.00 © 2012 IEEE

Scheme 1.

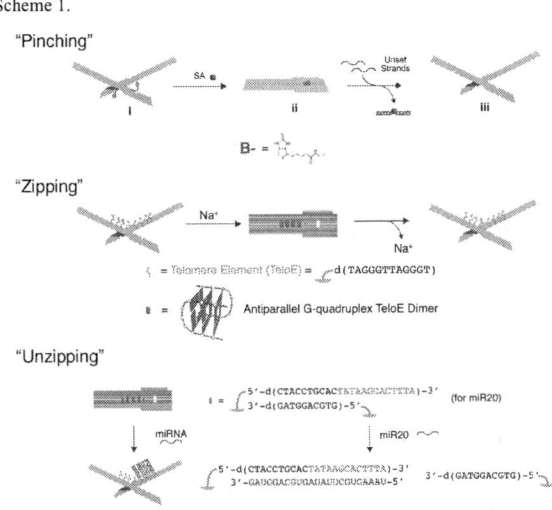

freshly cleaved mica, additional 1X TAE/Mg buffer (or 1 X TAE/Mg buffer with 200 mM NaCl for pre-closed DNA pliers, 5 mM HEPES buffer with 12.5mM Mg^{2+} for triplex invasion system, 200 µL) was added, and the imaging was done using fluid DFM scanning mode with a BL-AC40TS tip (Olympus, Japan).

III. RESULTS AND DISCUSSION

Fig. 1 shows the structure of "DNA origami pliers" used in this study. It consists of two ca. 170-nm long lever domains, which are made of six parallel DNA helices of M13 scaffold and 117 staple strands. These levers are joined together at a fulcrum via two phosphodiester linkages in the M13 scaffold so that an immobile Holliday junction is formed between them. In a solution containing Mg^{2+} ion, which is a typical condition for DNA origami preparation, DNA four-way junction is known to be in right-handed, antiparallel stacked X-structure with a small angle of 60° [12]–[14]. DNA origami pliers in solution are thus expected to be twisted as illustrated in Fig. 1. AFM measurements of DNA origami beacons require deposition of them on mica surface. There are three possible

forms of origami beacons to adhere on a 2D surface. The most feasible one is a "cross", which directly represents the expected twisted structure of origami pliers. The second feasible structure is an "antiparallel form", in which two levers are aligned in parallel on a plane but point opposite directions. Configuration of the M13 scaffold at the fulcrum in this form is in "antiparallel" mode, which is rather relaxed and a typical configuration of DNA strands in DNA origami design. The third feasible form of origami beacons is a "parallel closed form", in which both of the levers are aligned in parallel pointing the same direction. This configuration of DNA four-way junction at the fulcrum is considered to be unfavorable compared to the antiparallel mode [12]–[14]. Thus, origami beacons in the parallel closed form are expected to be the least popular species in the AFM images unless any additional interaction spontaneously rotates the fulcrum. Each of the levers has a small concavity of two helical turns long and two helices wide (ca. 7 nm x 7 nm). These concavities serve as the jaws of pliers to pinch target molecule in a single-molecule manner. When DNA pliers are in parallel closed form, these concavities are placed next to each other to form larger cavity, which accommodates pinched target quite stable throughout AFM scanning, as shown in our previous studies [15], [16].

DNA origami beacons used in this study sense various targets in a single-molecule manner by three independent mechanisms; "pinching", "zipping", and "unzipping" (Scheme 1). Pinching is a process to detect single target molecule that binds multiple ligands on a molecule. For this purpose, two ligands are attached to each of the staple strands placed in the concavities (anchor strands) and co-operatively capture single target molecule between the jaws. This intramolecular process triggers shape transition of origami beacons from cross or antiparallel form to parallel closed form, which can be visualized on AFM imaging or can be monitored in solution with spectroscopic analyses.

Fig. 2 shows typical AFM images of selective and single-molecular pinching of a streptavidin tetramer (SA) by DNA origami pliers. Here, DNA pliers were modified with a biotin group in each of the jaws. The dominant species initially observed in 1xTAE/ Mg^{2+} buffer solution without SA (i) was DNA pliers in cross (58% yield), which is consistent with the X-structure of DNA four-way junction. The population of

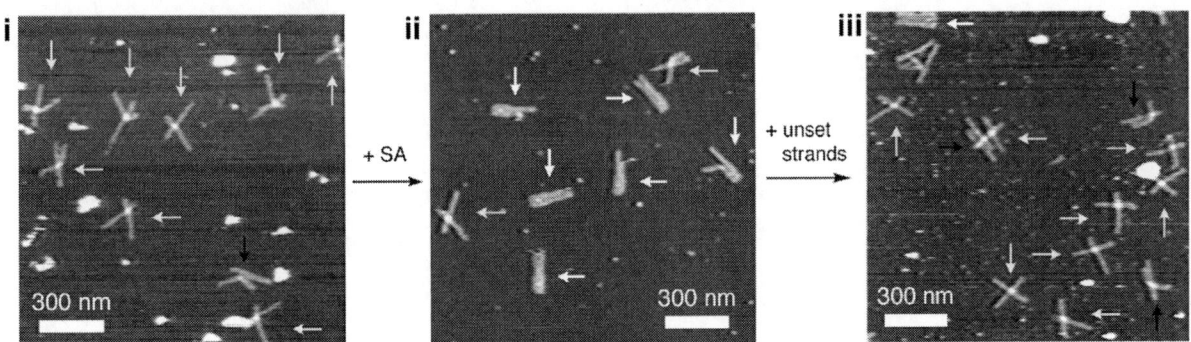

Fig. 2. AFM images of selective pinching of SA.

DNA pliers in antiparallel form was 16%, and only 5% of DNA pliers were in parallel closed form. Preference to antiparallel over parallel form is in good accordance with previous observations. When SA was added to the solution, on the other hand, the population of each forms drastically altered. Almost 58% of DNA pliers were found in parallel closed form in the presence of SA (ii). In addition, a bright spot of 5-nm height that corresponds to the expected diameter of the pinched SA molecule was found in the jaws of most of the parallel pliers. The yields of antiparallel pliers and cross were 5% and 23%, respectively. The only reasonable explanation for such drastic shape change is that single SA tetramer bridged the two levers by binding to each of the biotins attached to them. The potential of nanomechanical DNA origami as a single-molecular detector for biochemical non-covalent and multidentate interactions has been shown. To further prove that the shape transition to parallel pliers was solely achieved by biotin–SA–biotin bidentate binding, the biotinylated anchor strands were selectively detached from DNA pliers after SA pinching by using DNA strand displacement technique with the aid of extra 8-nucleotide toehold sequence on the anchor strands (iii). After 2 h incubation at room temperature in the presence of 10 eq. of unset strands, which is complementary to the anchor strands, the population of parallel pliers significantly decreased down to 10%, and the pliers in cross became the dominant species (53%) again as expected.

Although selective pinching and detection of a target molecule by utilizing protein-ligand bindings, which are among the strongest in biological interactions, is successful, single-molecular pinching of other targets that do not show such a strong binding is usually difficult. Zipping mechanism, which involves multiple binding events, is the second detection mechanism of origami beacons appropriate for such targets (Scheme 1). Here, multiple elements that bind together in the presence of the target are introduced to each of the levers, and cooperatively trigger selective closing of origami beacons. For example, Na^+- and K^+-ion sensing are possible utilizing G-quadruplex formation [10].

Unzipping, which is a reverse process to the zipping mechanism, is the third detection mechanism of origami beacons. The zipper elements for this mechanism are designed to bind together at the initial stage, and selectively unbind in the presence of target molecules. As shown in The presence of human micro RNA (miRNA) can be clearly detected with origami beacons by employing unzipping mechanism [10].

The present DNA origami beacons are also applicable to observe and confirm unique binding events of PNA, a nucleic acid analogue with polypeptide backbone [11], [17]. A PNA strand with poly-pyrimidine sequence (bis-PNA) is known to invade and denature a DNA duplex with matching poly-pyrimidine–poly-purine sequence by forming a DNA-PNA-PNA hetero-triplex. However, direct observation of such invasion event in a single-molecule manner has not yet been accomplished. Unzipping of DNA origami beacons provides a

simple mean for the purpose (Fig. 3). A pre-closed DNA origami pliers bearing four pairs of pre-closing elements with a 10-bp sequence of 5'-CTCTCTTTCC-3'/5'-GGAAAGAGAG-3' on each lever was prepared (Fig. 3a). Simple incubation of these pre-closed DNA origami pliers for 3 days at 16°C in pH 7 HEPES buffer containing Mg^{2+} ion did not change the population of closed DNA origami pliers (ca. 80%). Addition of complementary bis-PNA to the solution, on the other hand, significantly altered the population (Fig.3b). Almost 40% of DNA origami pliers were in closed form in the presence of 100 eq. bis-PNA to the closing elements, whereas the fraction was only 28% in the presence of 1000 eq. bis-PNA. The opening was completely sequence-selective, and observed overall shape of DNA origami pliers on AFM was not damaged by bis-PNA addition (data not shown). These results clearly show that what opened the DNA origami pliers was indeed co-existing bis-PNA.

IV. CONCLUSIONS

We have successfully applied nanomechanical DNA

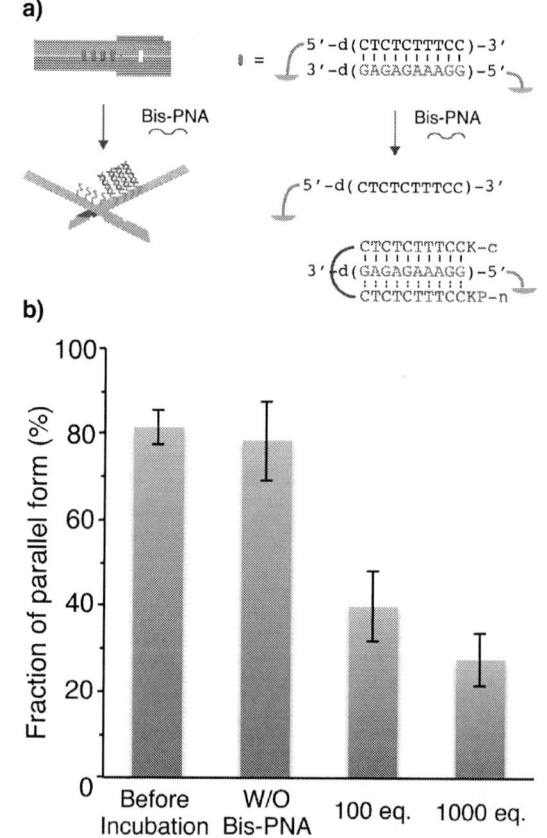

Fig. 3. Observation of triplex invasion by bis-PNA into 10-bp DNA duplex attached to DNA origami pliers. a) Schematic illustration of the system. b) Yield of opened DNA origami pliers after 3 days incubation at 16°C.

origami devices to construct visual "single-molecule" beacons that can detect various targets of a quite wide range of molecular weights from metal ions (a few tens of Da) to proteins (hundreds of kDa) as well as unique biding events of nucleic acid analogues in a single-molecule manner on the same platform. Observation of origami beacons on AFM is totally on single-molecular basis, thus theoretical detection limit of the systems should be enormously small if the reaction volume is further downsized with the aid of MEMS technology. Not only detection systems, development of allosteric metaenzymes by attaching another functional nanomaterial such as enzymes at the other end of the levers and switching their activity by mechanical movement is another interesting application of DNA pliers. The present system may be the first step toward strong tools in future biochemical studies on various interactions.

ACKNOWLEDGMENT

This work was partially supported by a Grant-in-Aid for Specially Promoted Scientific Research (22000007), Grant-in-Aid for Scientific Research (S) (22220001), and Grant-in-Aid for Young Scientists (B) (22750144) from the Ministry of Education, Science, Sports, Culture and Technology, Japan. Supports from the Global COE Program for Chemistry Innovation and from the Association for the Progress of New Chemistry are also acknowledged.

REFERENCES

[1] N.C. Seeman, "Nucleic-Acid Junctions and Lattices." *J. Theor. Biol.*, vol. 99, pp. 237-247, 1982.

[2] H. Li, J.D. Carter, and T. H. LaBean, "Nanofabrication by DNA Self-Assembly." *Mater. Today,* vol. 12, pp. 24-32, 2009.

[3] J. Bath and A.J. Turberfield, "DNA Nanomachines." *Nature Nanotech.* vol. 2, pp. 275-284, 2007.

[4] P.W.K. Rothemund, "Folding DNA to Create Nanoscale Shapes and Patterns." *Nature,* vol. 440, pp. 297-302, 2006.

[5] A. Kuzuya and M. Komiyama, "DNA Origami: Fold, Stick, and Beyond." *Nanoscale,* vol. 2, pp. 310–322, 2010.

[6] W.M. Shih and C. Lin, "Knitting Complex Weaves with DNA Origami." *Curr. Opin. Struct. Biol.*, vol. 20, pp. 276-282, 2010.

[7] J. Nangreave, D. Han, Y. Liu, and H. Yan, "DNA Origami: a History and Current Perspective." *Curr. Opin. Chem. Biol.*, vol. 14, pp. 608-615, 2010.

[8] E.S. Andersen, et al. "Self-Assembly of a Nanoscale DNA Box With a Controllable Lid." *Nature,* vol. 459, pp. 73-76, 2009.

[9] A. Kuzuya and M. Komiyama, "Design and Construction of a Box-Shaped 3D-DNA Origami." *Chem. Commun.*, pp. 4182-4184, 2009.

[10] A. Kuzuya, Y. Sakai, T. Yamazaki, Y. Xu, and M. Komiyama, "Nanomechanical DNA Origami 'Single-Molecule Beacons' Directly Imaged by Atomic Force Microscopy," *Nature Commun.*, vol. 2, article # 449, 2011.

[11] M. Komiyama, Y. Aiba, T. Ishizuka, and J. Sumaoka, "Solid-Phase Synthesis of Pseudo-Complementary Peptide Nucleic Acids." *Nature Protoc.*, vol. 3, pp. 646-654, 2008.

[12] A.I.H. Murchie, et al. "Fluorescence Energy Transfer Shows that the Four-Way DNA Junction is a Right-Handed Cross of Antiparallel Molecules." *Nature,* vol. 341, pp. 763-766, 1989.

[13] C.D. Mao, W.Q. Sun, and N.C. Seeman, "Designed Two-Dimensional DNA Holliday Junction Arrays Visualized by Atomic Force Microscopy." *J. Am. Chem. Soc.*, vol. 121, pp. 5437-5443, 1999.

[14] P.A. Khuu, A.R. Voth, F.A. Hays, and P.S. Ho, "The Stacked-X DNA Holliday Junction and Protein Recognition." *J. Mol. Recognit.*, vol. 19, pp. 234-242, 2006.

[15] A. Kuzuya, K. Numajiri, and M. Komiyama, "Accommodation of a Single Protein Guest in Nanometer-Scale Wells Embedded in a "DNA Nanotape"." *Angew. Chem. Int. Ed.* vol. 47, pp. 3400-3402, 2008.

[16] A. Kuzuya, et al. "Precisely Programmed and Robust 2D Streptavidin Nanoarrays by Using Periodical Nanometer-Scale Wells Embedded in DNA Origami Assembly." *ChemBioChem,* vol. 10, pp. 1811-1815, 2009.

[17] P.S. Shirude, V.A. Kumar, and K.N. Ganesh, "BisPNA Targeting to DNA: Effect of Neutral Loop on DNA Duplex Strand Invasion by aepPNA-N7G/aepPNA-C Substituted Peptide Nucleic Acids." *Eur. J. Org. Chem.*, pp. 5207-5215, 2005.

Gold-Coated Polystyrene Bead Array and the Investigation of Their Plasmon Coupling Abilities

[1]Hsin-Yi Hsieh, [2]Tsu-Wei Huang, [2,3]Chung-Shi Yang, [2]Pen-Cheng Wang, [4,5]Jian-Long Xiao, [4,5]Chau-Hwang Lee, and [1,2,4*]Fan-Gang Tseng

[1]Institute of NanoEngineering and MicroSystems, [2]Department of Engineering and System Science, National Tsing Hua University, Hsinchu 300, Taiwan R.O.C.
[3]Center for Nanomedicine Research, National Health Research Institutes, Miaoli 350, Taiwan R.O.C.
[4]Research Center for Applied Sciences, Academia Sinica, Taipei 115, Taiwan R.O.C.
[5]Institute of Biophotonics, National Yang-Ming University, Taipei 112, Taiwan, R.O.C.

e-mail : fangang@ess.nthu.edu.tw

Abstract—**Many of SERS nanoparticles took advantage of the surface roughness for the significant improvement of their Raman sensing ability. Nevertheless, few papers analyzed the characteristics of surface roughness nanostructures that contribute to the SERS. Thus, this paper investigates the characteristics of the corrugated polystyrene bead (PSB) array etched by a series of oxygen plasma etching time for giving a criterion to fabricate appropriate SERS-active nanoparticles. Three factors were considered in this paper: (1) the effect of plasma coupling among neighboring particles, (2) the vertical surface roughness of nanocorrugations, and (3) the pitch size, the lateral surface roughness, of nanocorrugations. By the analysis of SEM and AFM images, those factors were quantifiable. The correlation coefficient between each factor and SERS Raman enhancement was also investigated to verify that the pitch size of nanocorrugations (ranging from ~6 nm to ~12 nm on the surface of PSBs) dominates the SERS enhancement. Therefore, the maximum improvement of Raman intensity that derives from surface roughness treatment is 12 times compared to smooth surface. Moreover, it has a high enhancement factor of ~10^6.**

Keywords-SERS; polystyrene; roughness; plasma coupling

I. INTRODUCTION

Recently, there have been various researches that demonstrate the chemical [1] or physical [2] methods to obtain nanoparticles with surface roughness for the surface enhanced Raman scattering (SERS) enhancement. One popular method among these studies focuses on the polystyrene nanoparticles, such as the fabrication of PSBs with corrugated surface by RIE etching [2-3]. Besides of the enhancement from surface roughness, Masson *et. al.* reported the gap-to-diameter ratio of metal film over etched PSB array will affect the plasma coupling effect [4-5]. They conclude that the maximum Raman enhancement falls in the gap-to-diameter ratio of ~0.2. Although previous researches suggest that surface roughness and plasmonic coupling effect of two neighboring metal nanoparticles can fairly contribute to the SERS enhancement ability, those factors have not yet been quantitatively investigated. Therefore, we selected five sizes of PSBs to be densely-packed on a glass substrate and then used oxygen plasma to etch the PSB array in a series of

etching time, 0-120 sec. By the analysis of SEM and AFM images, we can compare the correlation coefficient of Raman intensity with the characteristics of metal coated corrugated polystyrene beads.

II. EXPERIMENTS

A. Materials

Four sizes, 870 nm (PS03N/5714), 640 nm (PS03N/4136), 420 nm (PS02N/2141), and 300 nm (PS02N/5378) of PSBs were purchased from Bangs Laboratories Inc. The other size of the 250-nm PSB (G250) was purchased from Duke Scientific Corp. Rhodamine 6G (R6G) was purchased from Sigma-Aldrich, R4127. AFM tips with 8-nm tip radius of curvature (NCH-50, PointProbe Plus) were purchased from NANOSENSORS.

B. Fabrication process

To obtain PSB array, cleaned and oxygen plasma treated substrates was firstly covered with a PDMS wall to form a liquid container. The diameter of the container was 2.86 mm and the depth of that was 225 μm. Then 0.1~0.5 % in solids of PSB solution was dropped into the PDMS container. When the water dried up, densely packed PSB array formed from the center to the wall. After the formation of PSB array and the detachment of PDMS wall, the PSB array was etched under oxygen plasma condition, 60W/60mtorr, for 20-120 sec (Fig. 1). Finally, we coated Au/Ti 20nm/5nm or 60nm/5nm on the corrugated PSB array to give SERS ability.

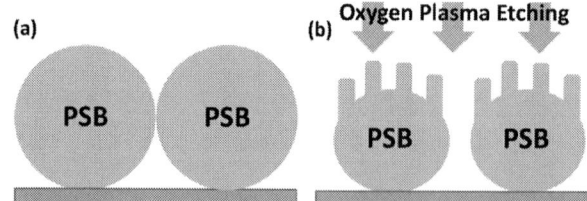

Fig. 1. *Schematic illustration of the effects on a polystyrene bead array that is etched by oxygen plasma: (1) shrinking bead diameter and (2) producing corrugated surface.*

C. Definition and the analysis of nanocorrugations

The surface roughness of corrugated PSB is defined as two characteristics, the lateral roughness of pitch size (**p**) and the vertical roughness of average roughness (**R**$_{average}$) and peak-to-valley roughness (**R**$_{peak-to-valley}$ or **R**$_v$). Fig. 2(a) is a schematic illustration of corrugated PSB, in which **D** is the diameter of corrugated PSB and **p** is the minimum pitch size between two nanocorrugations. The definition of **g** is the difference value before and after plasma etching in Fig. 2(b). To measure the diameter (**D**) of the polystyrene beads with or without oxygen plasma etching, the SEM images that captured by JEOL 6330F were analyzed by the software of Image-Pro Plus 6.0. Besides, the pitch size (**p**) was also defined by the enlargement of the SEM images by averaging 20 data through the software. Because of the non-uniformed distribution of the nanocorrugations, only the minimum distance to the surrounding pillars was selected as **p** of the randomly chosen nanocorrugation. The gap-to-diameter ratio of PSB array was calculated through the division of **g** by **D** for each PSB array before or after oxygen plasma etching. Another analysis method for the vertical roughness was by the AFM images. Each value of the **R**$_{average}$ and **R**$_v$ was averaged by randomly selecting 6 PSBs in AFM heigh images and analyzing from the center 0.4**D** and 0.2**D** of the PSBs, respectively, through JPK software (Fig. 2(c)).

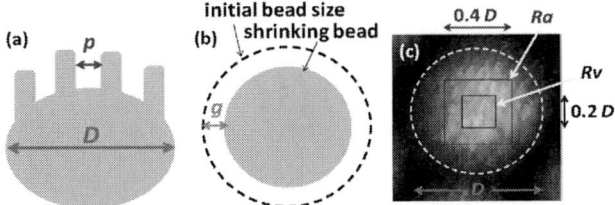

Fig. 2. The definition of corrugated polystyrene beads. (a) **p** is the pitch size between two nanopillars and **D** is the bead diameter. (b) **g** represents the gap size. (c) In a top view of the corrugated polystyrene bead, the measurement areas for the average roughness (**R**$_{average}$) and the peak-to-valley roughness (**R**$_{peak-to-valley}$) are 0.4**D** and 0.2**D**, respectively.

D. The plasmonic coupling effect

The plasmonic coupling effect causes the locally enhanced electric field between two closed metal spheres with a nano-scale gap [5]. One of the common methods to observe this effect is by measuring the absorption or transmission spectrum. When the absorption peak shows a larger red shift, it implies the plasmonic coupling effect is higher. The plasma ruler equation (1) shows that the fractional plasmon shift can be evaluated by the **s/D** ratio, equivalent of twice the gap-to-diameter (**g/D**) ratio in this paper. The equation also suggests that the smaller gap between two nanoparticles, the larger red shift of the absorption peak, as well as the higher increscent of the local electric field between two particles. However, it becomes different when the case of two closed metal nanoparticles is replaced to a non-surface roughness treated nanoparticle array. The maximum red shift, also the maximum electric field and the best SERS Raman enhancement ability, will fall in a specific value of the

gap-to-diameter (**g/D**) ratio, ~0.2 [4].

$$\frac{\Delta\lambda}{\lambda_0} \cong 0.18\exp(\frac{-(s/D)}{0.23}) \qquad (1)^{[5]}$$

,where $\Delta\lambda/\lambda_0$ is the fractional plasmon shift and **D** is the particle diameter. **s**, which is equivalent to 2**g**, is the interparticle edge-to-edge separation.

III. RESULTS and DISCUSSION

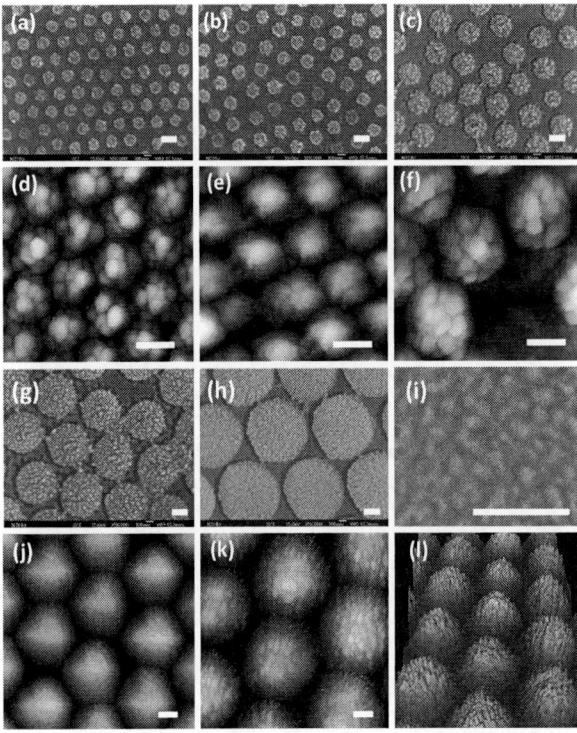

Fig. 3. The SEM and AFM images of oxygen plasma etched PSB array: (a) SEM / 250 nm / 100 sec, (b) SEM / 300 nm / 100 sec, (c) SEM / 420 nm / 120 sec, (d) AFM / 250 nm / 100 sec, (e) AFM / 300 nm / 100 sec, (f) AFM / 420nm / 120s, (g) SEM / 640 nm / 120 sec, (h) SEM / 870 nm / 120 sec, (i) the enlargement of the SEM image in (h) and (j) AFM / 870 nm / 120 sec, (k) AFM / 870 nm / 120 sec, and (l) the 3D AFM image of the 870 nm / 120 sec PSB array. (Scale bar: 200 nm)

Different oxygen plasma etching time will cause different surface morphologies and different interparticle distances. In Fig. 2, we define the characteristics of these corrugated PSBs and the roughness measurement area. After the plasma treatment, the SEM and AFM images show the sizes and surface morphologies of those five initial diameters of PSBs in Fig. 3(a)(b)(c)(g)(h) and Fig. 3(d)(e)(f)(j)(k), respectively. Moreover, Fig. 3(i) shows the enlargement SEM image of the 870-nm PSB after 120 sec oxygen plasma etching. Fig. 3(l) shows the 3D reconstruction of the 870-nm PSB after 120 sec oxygen plasma etching. All SEM images of those five sizes of PSB array can refer to Hsieh et al.'s work [3]. From SEM images [3], the longer plasma etching time contributes the larger gap-to-diameter ratio. In the LSPR spectra of those

978-1-4673-1122-9/12 $31.00 © 2012 IEEE

metal coated oxygen plasma treated PSB arrays in Fig. 4, the longer oxygen plasma treatment time presents larger red shift (Fig. 4(a)(b) are more distinguishable). However, this result violates the plasma ruler equation (1) [5]. Since the red-shift represents larger plasmonic coupling [4-5], the better Raman intensity is supposed to appear at smaller gap-to-diameter ratio that represents shorter plasma etching time. Thus, besides of the gap-to-diameter ratio, it implies that the surface roughness on corrugated PSBs also contributes to the plasma coupling effect.

smoothen the surface roughness that results in the reduction of surface roughness contributed plasma coupling effect. It verifies that the roughness factors, p, $R_{average}$, and $R_{peak-to-valley}$ might be further important than the factor of gap-to-diameter ratio.

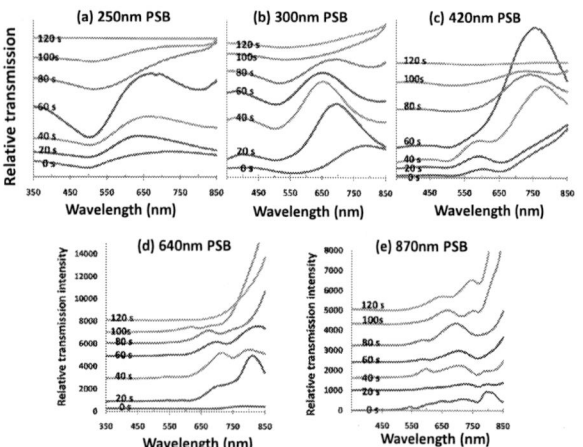

Fig. 4. The relative transmission spectra of (a) 250nm, (b) 300nm, (c) 420nm, (d) 640nm, and (e) 870nm PSB array after 0-120s oxygen plasma etching and Au/Ti deposition.

Table I. Summary of the relative Raman enhancement (%), Raman Intensity (cnt), gap-to-diameter ratio (%), pitch size (nm), and surface roughness ($R_{average}$ and $R_{peak-to-valley}$) at the maximum relative Raman intensity of each 20-nm gold coated polystyrene bead array.

Bead size etching time	250nm 100s	300nm 100s	420nm 120s	640nm 120s	870nm 120s
enhancement (%)	215	328	631	163	1224
Raman intensity (cnt)	3905	1791	2607	775	2803
gap-to-diameter (%)	63.3	85.2	44.5	24.1	15.7
pitch size (nm)	9.71	8.99	7.28	11.55	6.14
$R_{average}$ (nm)	6.89	4.78	13.73	10.72	9.65
$R_{peak-to-valley}$ (nm)	13.53	13.16	55.35	41.10	32.66

Table I summarizes the results on the 20-nm gold-coated polystyrene bead array. Only the characteristics of the maximum Raman intensity in each initial bead diameter were listed in Table 1 and discussed in this paper. Fig. 5(a) shows the relative Raman intensity of R6G on the 20-nm Au-coated PSB array. According to the previous finding [4], the maximum Raman intensities should be in the gap-to-diameter of ~0.2. However, the highest Raman intensities were not in the pink region, 0.05~0.3, of the recommended gap-to-diameter ratio, ~0.2. Besides, when the gold deposition increases to 60 nm that further smoothens the surface roughness, the highest relative Raman intensities then fall in the gap-to-diameter ratio of 0.05~0.3 (Fig. 5(b)). This result also suggests that the surface roughness can further promote the plasma coupling ability because thicker metal film

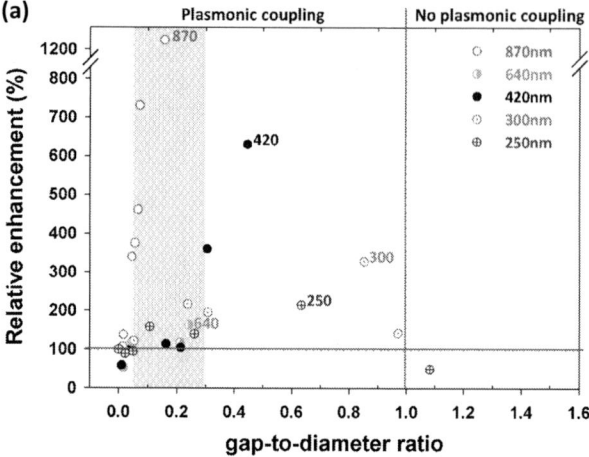

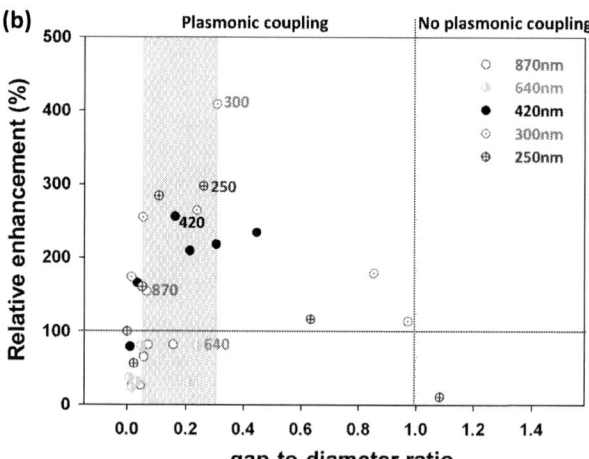

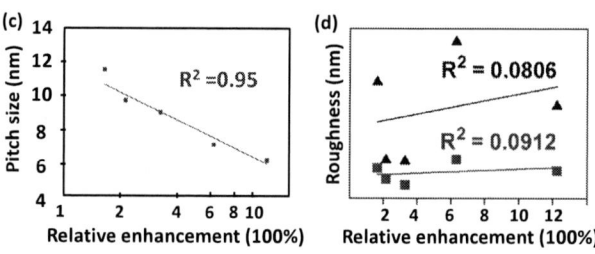

Fig. 5. Relative Raman intensity of 100 μM rhodamine 6G solution on a (a) 20-nm gold coated PSB array and (b) 60-nm gold coated PSB array. The bead diameters are labeled near the maximum relative intensity of each series of initial bead diameter with different plasma etching time. (c) The correlation coefficient, 0.95, of the relative enhancement and the factor of pitch size (**p**). (d) The correlation coefficient, 0.0806 and 0.0912, respectively, of the relative enhancement and the factor of $R_{average}$, and $R_{peak-to-valley}$.

When we correlate the Raman enhancement percentage with the factor of pitch size (Fig. 5(c)) and the factor of $R_{average}$ and $R_{peak-to-valley}$ (Fig. 5(d)), it shows that the factor of pitch size (p) (smaller pitch size (<12 nm)) dominates the other factors, the factor of $R_{average}$ and $R_{peak-to-valley}$ and the factor of the gap-to-diameter ratio. The reason is that the correlation coefficient with the factor of pitch size is 0.95 that extremely larger than the correlation coefficient with the factor of $R_{average}$ and $R_{peak-to-valley}$, 0.08 and 0.09, respectively.

For the 60-nm Au deposition PSB arrays, most of the surface roughness on the oxygen plasma etched PSBs was diminished. Moreover, the pitch, which is the hot-spot area that dominates the surface Raman enhancement scattering intensity, between two nanopillars was filled with the metal deposition. Therefore, the main factor that contributes to SERS Raman intensity becomes the gap-to-diameter ratio for the 60-nm gold coated oxygen plasma etched PSBs array. This result is in agreement with previous finding [4].

IV. CONCLUSION

Nanocorrugations possessing large surface-to-volume ratios of hot-spot areas have been considered potential tools to detect single molecule for biological applications. In this paper, we use oxygen plasma to treat polystyrene beads and obtain surface roughness, through the nanomasking effect of oxygen plasma induced carboxyl groups on the PSB surface [3]. Based on this fabrication platform, we defined and investigated the surface roughness morphologies. The characteristics of the pitch size (p), lateral roughness, and the degree of the vertical surface roughness ($R_{average}$, and $R_{peak-to-valley}$) and their contributions to the SERS enhancement. For initially densely-packed polystyrene beads on the substrate surface, plasma etching will cause the gap-to-diameter ratio to increase and generate interparticle plasma coupling. Therefore, we also considered the plasmonic coupling effect generated by the plasma resonance of neighboring particles while the gap-to-diameter ratio (g/D) is smaller than 1. From the experimental results measured by SEM images (p and g/D), AFM images ($R_{average}$, and $R_{peak-to-valley}$), and LSPR, we conclude the effect of pitch size (ranging from ~6 nm to ~12 nm on the surface of polystyrene beads) dominates the SERS enhancement, rather than the degree of $R_{average}$, and $R_{peak-to-valley}$, and the effect of plasmonic coupling among neighboring particles. Based on the minimum pitch size of ~6 nm on the surface of 870 nm/120s oxygen plasma etched PSB array, the Raman intensity is promoted by 12.24-folds, compared to nontreated 870-nm PSB (while the Au/Ti coating is 20 nm/5 nm). This promotion could still be optimized while considering the laser wavelength, the thickness of metal deposition, and the metal type.

ACKNOWLEDGMENT

This work was financially supported by the National Science Council of the R.O.C., Taiwan, under Contract No. NSC-98-2120-M-007-001, NSC-99-2120-M-007-009, and NSC-99-2113-M-007-013 and the Academia Sinica Research Program on Nanoscience and Nanotechnology. In addition, the authors thank the Center for Nanoscience and Technology, National Chiao Tung University, Taiwan, R.O.C., for the supplement of a High-Resolution Confocal Raman Microscope and the Xingtian Temple Foundation talented student scholarship for the financial supporting during Hsin-Yi Hsieh's Ph. D program.

REFERENCES

[1] J. Xie, Q. Zhang, J. Y. Lee, and D. I. C. Wang, "The Synthesis of SERS-Active Gold Nanoflower Tags for In Vivo Applications," *ACS Nano*, 2 (12), 2473-2480.

[2] Wu, L. Y.; Ross, B. M.; Hong, S.; Lee, L. P., "Bioinspired Nanocorals with Decoupled Cellular Targeting and Sensing Functionality," Small 2010, 6 (4), 503-507.

[3] Hsieh, H.-Y.; Xiao, J.-L.; Lee, C.-H.; Huang, T.-W.; Yang, C.-S.; Wang, P.-C.; Tseng, F.-G., "Au-Coated Polystyrene Nanoparticles with High-Aspect-Ratio Nanocorrugations via Surface-Carboxylation-Shielded Anisotropic Etching for Significant SERS Signal Enhancement," The Journal of Physical Chemistry C 2011, 115 (33), 16258-16267.

[4] Masson, J.-F.; Gibson, K. F.; Provencher-Girard, A., "Surface-Enhanced Raman Spectroscopy Amplification with Film over Etched Nanospheres," The Journal of Physical Chemistry C 2010, 114 (51), 22406-22412.

[5] Jain, P. K.; Huang, W.; El-Sayed, M. A., "On the Universal Scaling Behavior of the Distance Decay of Plasmon Coupling in Metal Nanoparticle Pairs: A Plasmon Ruler Equation," Nano Letters 2007, 7 (7), 2080-2088.

Development of Microfabricated Phononic Crystal Resonators Based on Two-dimensional Silicon Slab

Nan Wang, Fu-Li Hsiao, Moorthi Palaniapan, and Chengkuo Lee[*]

*Department of Electrical and Computer Engineering, National University of Singapore, Singapore
elelc@nus.edu.sg

Abstract— This paper shows the design, fabrication and characterization of novel micromechanical resonators with Bloch-mode resonance by creating defects on a two-dimensional (2-D) silicon phononic crystal (PnC) slab. The PnC slab was made by etching a square array of cylindrical air holes in a 10μm thick free-standing silicon plate, while the defects are created by replacing periodically arranged three rows of air holes with one row of air holes at the centre of the PnC region. The radius of the central air holes (*r'*) is also varied to study the effect of *r'* on the performance of the PnC resonators. Piezoelectric aluminium nitride (AlN) film is employed as the inter-digital transducers (IDT) to transmit and detect acoustic waves, thus making the whole microfabrication process CMOS-compatible. We also fabricate a pure PnC structure with a stopband of 140MHz < *f* <195MHz which agrees quite well with the simulation results. The characterized resonant frequency of microfabricated PnC resonators reaches its maximum value (174.67MHz) when central-hole radius (*r'*) reaches 8μm, while Q factor reaches maximum (893) at *r'*=6μm. The Bloch-mode PnC resonators based on square lattice PnC structure show promising acoustic resonance characteristics and may be further optimized for applications such as microfluidics, biomedical devices and RF communications in GHz range.

Keywords- Microelectromechanical Systems (MEMS); CMOS-compatible; phononic crystal; resonator

I. INTRODUCTION

Due to their unique acoustic characteristics, phononic crystals (PnCs) have become a popular research topic [1-8]. PnCs are the acoustic wave equivalent of the well-known photonic crystals (PhCs), which consist of periodically arranged scattering centres embedded in a homogeneous background matrix. From the aspect of elastic properties, PnCs are inhomogeneous materials with periodic variations. Thus, the dispersion characteristics of the PnCs lead to the existence of phononic band gaps, in which the propagation of elastic waves within a certain frequency range is prohibited in any direction. PnCs with properly engineered band gaps can be the basis of realizing a variety of functionalities such as acoustic waveguides, cavities and filters. We can obtain such functionalities by modifying portions of the PnC structure for various applications in wireless communication and sensors.

So far, various types of acoustic and elastic wave propagation have been studied in different classes of PnC structures. Surface acoustic waves in semi-infinite PnCs with cylindrical air holes etched in a single material or a solid/solid composition whereby the air holes are filled with other kinds of materials, have been studied [2, 3]. Recently, guided waves in 2-D PnC slabs have attracted more attention due to the fact that the 2-D nature of PnC slabs can provide better confinement of elastic energy. Various configurations of PnC slabs, such as cylindrical air holes etched in freestanding membrane [5-7], cylindrical rods inserted into the air holes [3, 9], cylindrical rods deposited on the top of the membrane [10], as well as inverse acoustic band gap (IABG) structure which is formed based on the use of a high acoustic velocity centre media (AlN and Pt) held by four thin tethers in a low acoustic velocity matrix (air)[11], have been proved to have phononic band gaps for elastic waves travelling in any direction. Researchers have also demonstrated the PnC slabs which can be operated in GHz frequencies [12, 13], making PnCs a big step closer to applications in RF communications.

By adding defects to PnC structure, devices of various functionalities like waveguides and resonators have been reported [9, 10, 14-21]. For example, waveguide is formed by adding a line defect (e.g., removing one row of air holes) to a PnC structure [18], while resonator can be formed by adding either a point defect or a line defect in the form of a Fabry–Perot structure [22]. Also, a waveguide is also demonstrated in an inverse acoustic band gap (IABG) structure [23] to achieve the function of ultrasonic focusing.

This paper shows the design, fabrication and characterization of novel micromechanical resonators based on Bloch-mode resonance on a 2-D PnC slab. For the purpose of easy comparison, the thickness of the PnC structures is fixed at 10μm and this is also the thickness of the device layer of the silicon-on-insulator (SOI) substrate used in later microfabrication. The transmission spectra and the mode profiles of displacement of the designed Bloch-mode PnC resonators are analyzed using FEM, while experimental results of the microfabricated resonators using 2-D silicon-based PnC slab with high-Q resonant peaks in the hundred MHz range are presented.

II. MODELING AND DESIGN

In phononic structures, phononic band gap can be formed when scattering inclusions arranged periodically in a homogeneous host material which causes waves in certain frequencies to be completely reflected by the structure. When defects are introduced to the PnC structure by partially or completely removing lines of air holes, a resonant cavity can be formed within the PnC structure. As such, a resonant peak can then appear within the stopband to form a phononic resonator.

The studies of the propagation of bulk waves in two- or three-dimensional phononic crystals include the analysis of the band gap effect in finite structures as well as the confinement and guiding of acoustic energy through the use of defect inclusions in the PnC structures. COMSOL Multiphysics software is employed for the Finite Element Modelling (FEM) of the band structure of a 2-D PnC slab. The main principle behind the FEM of the band structure follows closely from the first-principles physics, i.e., the combination of the Newton's Second Law of motion and a constitutive relation. The structural mechanics formulation is shown in (1).

$$\rho \frac{\partial^2 u}{\partial t^2} - \nabla \cdot c \nabla u = F = 0 \qquad (1)$$

In (1), ρ is the mass density, c is the stiffness tensor or the elasticity matrix, and u is the solution vector which consists of displacements u_x, u_y and u_z. In our study, we set $F = 0$ because the solutions we are looking for are the eigenvalues. Upon Fourier Transform of (1), the resulting harmonic time dependence term, $\exp(i\omega t)$, can be factored out. The time derivative can then be replaced by $-\omega^2$, as shown in (2).

$$\frac{\nabla \cdot c \nabla}{\rho} u = \omega^2 u; \qquad (2)$$

From (2), the mode frequencies can be derived from the eigenvalues, ω^2, which could be solved by FEM. The details of FEM was described in [24] and will not be repeated here for brevity.

In this work, the lattice constant (a) of the PnC is 18.18μm, the thickness of the silicon slab (d) is 10μm and the radius of the air holes (r) is 8.18μm. The PnC resonators are formed by removing three rows of air holes (cavity width, $w=3a$) and replacing them with one row of air holes at the centre [Fig.1 (a)]. Structures with various radii of the central holes (r'), including 2μm, 4μm, 6μm and 8μm, are designed but only structure with $r'=4$μm is shown here for the purpose of illustration.

Using the model built in Fig.1(a), we analysed the transmission spectra of the designed PnC resonator structures. A flat silicon plate without any PnC structure and with the same length of wave propagation is also constructed for the purpose of normalization. Periodic boundary conditions are applied along y direction. Frequency response solver is used and the transmission spectra for various r' are simulated and the respective resonant frequencies are obtained. In this paper, we only show the case of $r'=6$μm is as an example in Fig.1(b), which reveals that the resonant frequency is 160.95MHz and the estimated Q factor is 4.5×10^4.

We also analysed the mode profiles of displacement in x, y and z directions of all designed PnC resonators under their respective resonant frequencies using FEM. Again, periodic boundary conditions are applied along y direction. As elastic

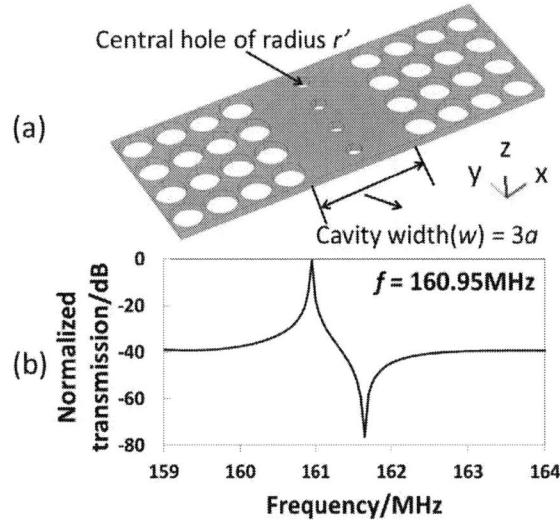

Fig. 1. (a)Schematic drawing of the PnC resonator formed by removing three rows of air holes (cavity width, $w=3a$) and replacing them with one row of air holes with radius $r'=4$μm at the centre; (b) Simulated transmission spectrum of the PnC resonator with $w=3a$ and $r'=6$μm.

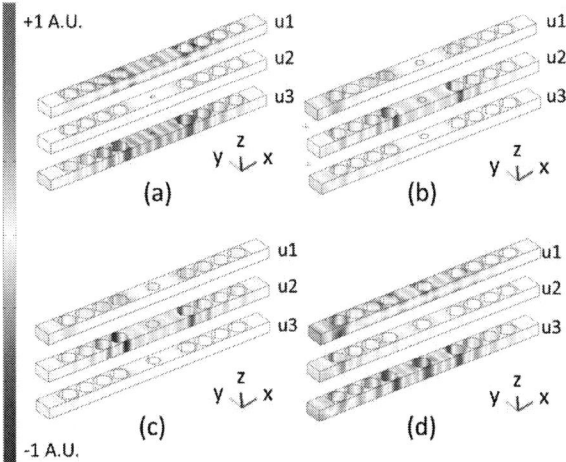

Fig. 2. Simulated mode profiles of displacement of the PnC resonator with (a) $r'=2$μm, (b) $r'=4$μm, (c) $r'=6$μm, and (d) $r'=8$μm. The $u1$, $u2$ and $u3$ represent the displacement vector components in x, y, and z directions, respectively. The colour bar indicates the amplitude of displacements in an arbitrary unit.

waves in silicon plate propagate by the interactions among the silicon atoms when they are displaced from their equilibrium positions, the energy stored in any solid structure is then associated with the displacements of the silicon atoms within the silicon plate. Thus, by analysing the displacements of all the silicon atoms within the silicon plate (mode profiles of displacement), we can get information about the energy distribution along the structure. Fig. 2 shows the mode profiles of displacement of the PnC Bloch-mode resonator with one row of central holes and (a) $r'=2$μm, (b) $r'=4$μm, (c) $r'=6$μm, and (d) $r'=8$μm, The $u1$, $u2$ and $u3$ represent the displacement vector components in x, y, and z directions, respectively. The colour bar indicates the amplitude of displacements in an arbitrary unit. For the cases whereby the displacement vector components in all three directions are concentrated at the central defect region [e.g., Fig. 2(c)], which means the energy of the structure are concentrated at the central defect region, we can expect a higher Q factor as a result of the better confinement of the energy by the phononic structure surrounding the central defect region. On the other hand, lower Q factors are expected for the cases whereby the displacement vector components are distributed evenly along the structure or concentrated at one end of the structure [e.g., Fig. 2(d)]. For these cases, energy is confined poorly in the central defect region, resulting in lower Q factors expected.

III. MICROFABRICATION

To realize the 2-D PnC slab structure, a silicon-on-insulator (SOI) wafer with a device layer thickness of 10μm was deposited with a 1μm AlN layer and a 0.5μm top Al electrode layer firstly. The deposited AlN layer and top Al electrode were patterned subsequently using reactive ion etching (RIE) to form the Inter-Digital Transducer (IDT) as the input and output of the acoustic waves. Cylindrical air holes of the square lattice were created on silicon device layer by deep reactive ion etching (DRIE). Lastly, the 2-D silicon PnC slab becomes a free-standing plate which was released from the SOI substrate by using back-side DRIE for silicon handle wafer portion and RIE for silicon insulator layer, i.e., buried oxide layer (BOX). Fig. 3(a) shows a microfabricated pure PnC structure of 10 periods (rows) of air holes, with IDT formed by Al on the two sides of the PnC structure. The close-up view of the PnC structure is shown in Fig. 3(b). Fig. 3(c) shows a PnC Bloch-mode resonator with $w=3a$ and $r'=4$μm. Four periods (rows) of air holes are patterned at each side of the defect region for all the PnC resonators. Fig 3(d) shows the close-up view of defect region of the resonator. Four types of r', including 2μm, 4μm, 6μm and 8μm, are prepared for the PnC resonators in current study.

IV. DEVICE CHARACTERIZATION

To experimentally characterize the microfabricated PnC resonators, a network analyser (Model: Agilent E8364B, Short-Open-Load-Through calibrated) was used to measure the transmission spectrum against frequency via IDT on the two sides of the phononic structure. Due to the piezoelectric properties of AlN film, acoustic waves were launched toward the PnC structure when an AC signal was applied on the IDT of the input port. As the piezoelectric AlN film can convert the displacements caused by transmitted acoustic waves into electrical signals, the transmission through the PnC structure could be picked up by the IDT of the output port on the other side of the PnC structure. We then extracted the S_{21} parameter which is essentially the transmission of acoustic waves after the interaction with PnC structure. A free-standing silicon slab without any PnC structure and with the same length is

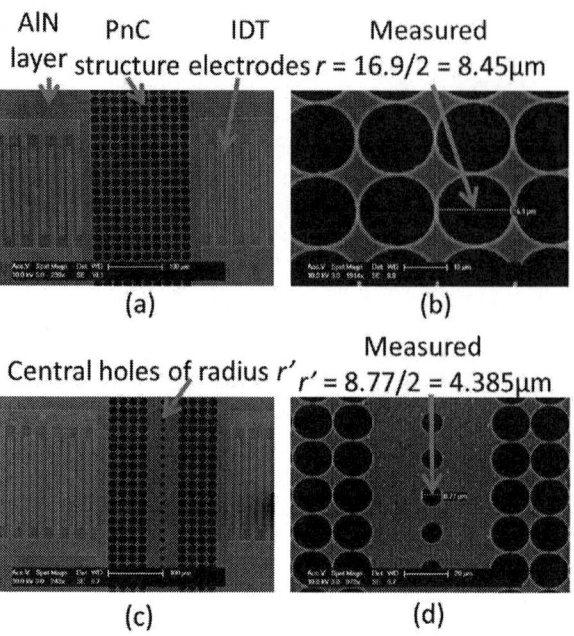

Fig. 3. SEM images of (a) the pure PnC structure (b) Close-up view of the pure PnC structure (c) the PnC resonator with $r'=4$μm (d) Close-up view of the defect region of the PnC resonator with $r'=4$μm.

characterized by the same measurement setup such that the wave propagation of the tested PnC devices is normalized with respect to the transmission of the free-standing silicon slab in the same frequency range.

Fig. 4 shows the measured transmission spectra of the PnC resonator with $w=3a$ and with (a) $r'=2$μm, (b) $r'=4$μm, (c) $r'=6$μm, and (d) $r'=8$μm. The measured data are in excellent agreement with the modelling results [e.g., 160.77MHz in Fig. 4(c) versus 160.95MHz in Fig. 1(b)]. We believe the discrepancies are due to the drift of critical dimension control in microfabrication process, in which the radii of the air holes in microfabricated devices are larger than designed. As r' increases from 2μm to 6μm, the resonant frequency remains relatively constant, which then increases sharply as r' further increases to 8μm. Besides, as r' increases from 2μm to 4μm, Q factor drops from 847 to 766, while Q factor rises to 893 when r' further increases to 6μm. Upon further increment of r' to 8μm, Q factor drops again to only 700. The tested Q factor deviates from the simulated value by an order of magnitude. This could be due to two reasons. First, as periodic boundary conditions were applied in the simulation, scattering loss, which is a major source of energy loss, is not considered. Second, the loss due to electro-acoustic coupling is not considered, when loads are directly applied to the structure in our simulation while acoustic waves are launched towards the phononic structure through the piezoelectric effect of IDT in actual experiment. From the tested results, we can see that the optimized condition for Q factor is $r'=6$μm, which leads to a conclusion that the resonance conditions of Bloch-mode

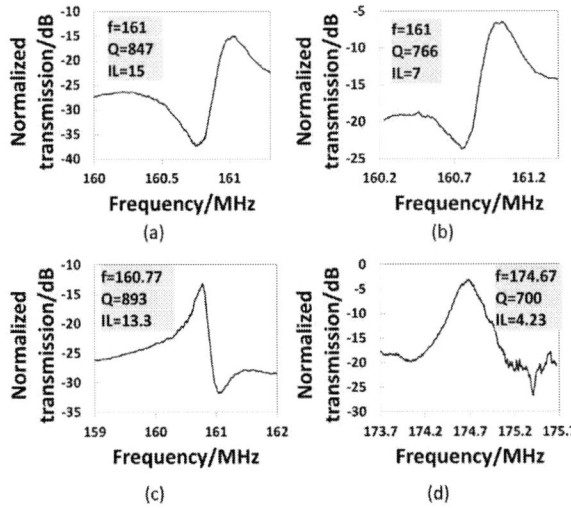

Fig. 4. Measured transmission spectrum of PnC resonator with $w=3a$ and with (a) $r'=2\mu m$, (b) $r'=4\mu m$, (c) $r'=6\mu m$, and (d) $r'=8\mu m$.

resonators with central-hole radii of 4μm and 8μm are slightly deviated from the optimized condition as achieved in the case of $r'=6\mu m$. Also, the measured data confirm the simulated mode profiles of displacement shown in Fig. 2. The displacement vector components for the case of $r'=6\mu m$ [Fig. 2(c)] are more concentrated at the central defect region than other designs, which means the best confinement of the energy by the surrounding phononic structure, rendering the highest Q factor obtained among all the four cases. On the other hand, for the cases of $r'=4\mu m$ [Fig. 2(b)] and $r'=8\mu m$ [Fig. 2(d)], the displacements vectors are large at one end of the structures and then distributed almost evenly along the structures, which yield poor confinement of energy by the surrounding phononic structure and thus lower Q factors.

S. Mohammadi *et al*, have reported a cavity mode based PnC resonator using 2-D silicon PnC slab of hexagonal lattice [22]. Two kinds of resonant peaks were reported. One type of resonant peak with Q factor of 680 and insertion loss (IL) of 4.6 dB was obtained at 149.5MHz, and the other type of resonant peak with Q factor of 6300 and insertion loss (IL) of 21 dB was obtained at 126.52MHz. Due to the nature of better confinement of acoustic waves and 4 rows of air holes in one period, the cavity-mode resonator of hexagonal lattice using three periods provides resonance of higher Q factor with compromise of higher IL. In the case of the cavity-mode resonator using two periods, i.e., 8 rows of air holes, resonant peak of lower Q factor and lower IL was obtained. In our current study, four periods, i.e., 4 rows of air holes, is employed for all Bloch-mode PnC resonators of square lattice. Although the acoustic wave confinement of PnC structures of square lattice is not as good as the one of hexagonal lattice, the measured Q factor of characterized Bloch-mode PnC resonators of square lattice with 4 rows of air holes is slightly better than the Q factor of cavity-mode PnC resonators with 8 rows of air holes. One possible mechanism to explain this

scenario could be as the propagation of acoustic waves is forbidden within a certain frequency band gap, the Bloch waves must be evanescent so that the phenomena of exponential decay of waves can be explained [25]. For the case of cavity-mode PnC resonator, acoustic waves are of cavity-mode in the defect region (cavity region) while acoustic waves are of Bloch-mode in the PnC structure surrounding the defect region. The conversion between the two modes comes with substantial amount of energy loss. However, in the case of Bloch-mode resonator, the acoustic waves in both the PnC structure and the defect region are of Bloch-mode. This means that no mode conversion is involved when acoustic waves are confined within the defect region, thus less energy is lost and higher Q factor can be achieved. As a result, for similar design of Bloch-mode PnC resonators of square lattice using more periods, we can expect to get resonant peak of higher Q factor with slightly increased IL. Another advantage of Bloch-mode PnC resonators of square lattice refers to the narrower width of one period used to form the square lattice of the PnC structures such that the device area of PnC square-lattice-based resonator is smaller than the other cases.

V.　Concluding Remarks

In this paper, we reported the experimental data of Bloch-mode PnC resonators of square lattice, which was realized from a micrfabricated silicon freestanding plate. We also characterized resonators with different central-hole radius (r') to explore the effects of r' on resonant frequency and Q factor. We found that the resonant frequency reaches its maximum value (174.67MHz) when $r'=8\mu m$ and Q factor reaches its maximum (893) when $r'=6\mu m$. As the feasibility of PnCs operating in GHz frequencies have been demonstrated [12, 13], our proposed Bloch-mode PnC resonators of square lattice are very promising to be further optimized for applications such as sensing and RF communications in the GHz range.

Acknowledgment

This work was supported in research grants of Academic Research Committee (ARC) Fund MOE2009-T2-2-011 (R-263000598112) at the National University of Singapore.

References

[1] J. O. Vasseur, P. A. Deymier, B. Chenni *et al.*, "Experimental and theoretical evidence for the existence of absolute acoustic band gaps in two-dimensional solid phononic crystals," *Physical Review Letters*, vol. 86, no. 14, pp. 3012-3015, Apr, 2001.

[2] S. Benchabane, A. Khelif, J. Y. Rauch *et al.*, "Evidence for complete surface wave band gap in a piezoelectric phononic crystal," *Physical Review E*, vol. 73, no. 6, Jun, 2006.

[3] I. El-Kady, R. H. Olsson, and J. G. Fleming, "Phononic band-gap crystals for radio frequency communications," *Applied Physics Letters*, vol. 92, no. 23, pp. 233504, Jun, 2008.

[4] A. Khelif, F. L. Hsiao, A. Choujaa *et al.*, "Octave Omnidirectional Band Gap in a Three-Dimensional Phononic Crystal," *IEEE Transactions on Ultrasonics Ferroelectrics and Frequency Control*, vol. 57, no. 7, pp. 1621-1625, Jul, 2010.

[5] S. Mohammadi, A. A. Eftekhar, A. Khelif *et al.*, "Evidence of large high frequency complete phononic band gaps in silicon phononic crystal plates," *Applied Physics Letters*, vol. 92, no. 22, pp. 221905, Jun, 2008.

[6] S. Mohammadi, A. A. Eftekhar, A. Khelif *et al.*, "Complete phononic bandgaps and bandgap maps in two-dimensional silicon phononic crystal plates," *Electronics Letters*, vol. 43, no. 16, pp. 898-899, August, 2007.

[7] T. T. Wu, L. C. Wu, and Z. G. Huang, "Frequency band-gap measurement of two-dimensional air/silicon phononic crystals using layered slanted finger interdigital transducers," *Journal of Applied Physics*, vol. 97, no. 9, pp. 094916, May, 2005.

[8] Y. Pennec, B. Djafari-Rouhani, H. Larabi *et al.*, "Phononic crystals and manipulation of sound," *Physica Status Solidi C - Current Topics in Solid State Physics, Vol 6, No 9*, Physica Status Solidi C-Current Topics in Solid State Physics 9, E. M. Sheregii, W. Gebicki and Y. M. Azhniuk, eds., pp. 2080-2085, Weinheim: Wiley-V C H Verlag Gmbh, 2009.

[9] F.-L. Hsiao, A. Khelif, H. Moubchir *et al.*, "Waveguiding inside the complete band gap of a phononic crystal slab," *Physical Review E*, vol. 76, no. 5, pp. 056601, Nov, 2007.

[10] R. H. Olsson, I. F. El-Kady, M. F. Su *et al.*, "Microfabricated VHF acoustic crystals and waveguides," *Sensors and Actuators A: Physical*, vol. 145, pp. 87-93, Jul-Aug, 2008.

[11] N. K. Kuo, C. J. Zuo, and G. Piazza, "Microscale inverse acoustic band gap structure in aluminum nitride," *Applied Physics Letters*, vol. 95, no. 9, pp. 093501, Aug, 2009.

[12] M. F. Su, R. H. Olsson, Z. C. Leseman *et al.*, "Realization of a phononic crystal operating at gigahertz frequencies," *Applied Physics Letters*, vol. 96, no. 5, Feb, 2010.

[13] Y. M. Soliman, M. F. Su, Z. C. Leseman *et al.*, "Phononic crystals operating in the gigahertz range with extremely wide band gaps," *Applied Physics Letters*, vol. 97, no. 19, pp. 193502, Nov, 2010.

[14] J. O. Vasseur, A. C. Hladky-Hennion, B. Djafari-Rouhani *et al.*, "Waveguiding in two-dimensional piezoelectric phononic crystal plates," *Journal of Applied Physics*, vol. 101, no. 11, pp. 114904, Jun, 2007.

[15] S. Benchabane, A. Khelif, A. Choujaa *et al.*, "Interaction of waveguide and localized modes in a phononic crystal," *Europhysics Letters*, vol. 71, no. 4, pp. 570-575, Aug, 2005.

[16] A. Khelif, P. A. Deymier, B. Djafari-Rouhani *et al.*, "Two-dimensional phononic crystal with tunable narrow pass band: Application to a waveguide with selective frequency," *Journal of Applied Physics*, vol. 94, no. 3, pp. 1308-1311, Aug, 2003.

[17] A. Khelif, A. Choujaa, B. Djafari-Rouhani *et al.*, "Trapping and guiding of acoustic waves by defect modes in a full-band-gap ultrasonic crystal," *Physical Review B*, vol. 68, no. 21, pp. 214301, Dec, 2003.

[18] A. Khelif, S. Mohammadi, A. A. Eftekhar *et al.*, "Acoustic confinement and waveguiding with a line-defect structure in phononic crystal slabs," *Journal of Applied Physics*, vol. 108, no. 8, pp. 084515, Oct, 2010.

[19] S. Mohammadi, A. A. Eftekhar, and A. Adibi, "Support loss-free micro/nano-mechanical resonators using phononic crystal slab waveguides," in *2010 IEEE International Frequency Control Symposium (FCS)*, 2010, pp. 521-523.

[20] M. Oudich, M. B. Assouar, and Z. Hou, "Propagation of acoustic waves and waveguiding in a two-dimensional locally resonant phononic crystal plate," *Applied Physics Letters*, vol. 97, no. 19, pp. 193503, Nov, 2010.

[21] Y. W. Gu, X. D. Luo, and H. R. Ma, "Low frequency elastic wave propagation in two dimensional locally resonant phononic crystal with asymmetric resonator," *Journal of Applied Physics*, vol. 105, no. 4, Feb, 2009.

[22] S. Mohammadi, A. A. Eftekhar, W. D. Hunt *et al.*, "High-Q micromechanical resonators in a two-dimensional phononic crystal slab," *Applied Physics Letters*, vol. 94, no. 5, pp. 051906, Feb, 2009.

[23] K. Nai-Kuei, and G. Piazza, "Evidence of acoustic wave focusing in a microscale 630 MHz Aluminum Nitride phononic crystal waveguide," in *2010 IEEE International Frequency Control Symposium (FCS)*, 2010, pp. 530-533.

[24] N. Wang, J. M. Tsai, F. L. Hsiao *et al.*, "Experimental Investigation of a Cavity-Mode Resonator Using a Micromachined Two-Dimensional Silicon Phononic Crystal in a Square Lattice," *Electron Device Letters, IEEE*, vol. 32, no. 6, pp. 821-823, Jun, 2011.

[25] V. Laude, Y. Achaoui, S. Benchabane *et al.*, "Evanescent Bloch waves and the complex band structure of phononic crystals," *Physical Review B*, vol. 80, no. 9, pp. 092301, Sep, 2009.

Electric Field Design of Metallic Sub-Wavelength Hole Arrays for Optical Permittivity Sensing

Etsuo Maeda, Takanori Matsuki, Ichiro Yamada, and Jean-Jacques Delaunay[*]

[*]Department of Mechanical Engineering, School of Engineering, The University of Tokyo, Tokyo, JAPAN
jean@mech.t.u-tokyo.ac.jp

Abstract—The transmission spectrum of metallic sub-wavelength hole arrays strongly depends on the shape, size, and material properties of structures and therefore enables the arrays to be used for sensing applications. We investigated hydrogen detection to prove the concept of permittivity sensing. Palladium (Pd) was used as the metallic material because of the large change in Pd permittivity upon hydrogen absorption. The fabricated Pd sub-wavelength hole arrays work as permittivity sensors having a strong main transmittance peak that show a redshift upon hydrogen exposure. The electric field distribution within the sub-wavelength holes was studied as a function of nanoplasmonic-based waveguide shape and length. The lower order optical modes within rectangular sub-wavelength holes in a thin metallic layer array were found to enhance the sensor response.

Keywords-plasmonic waveguide; permittivity sensor; hydrogen sensor; propagation constant

I. INTRODUCTION

The surface plasmon resonance (SPR) technique has been applied to biological, biomedical, and chemical sensors for over 40 years because of the high sensitivity of surface plasmon to changes in the optical properties of the metallic surface [1]. Recent advances in nano-micro-scale fabrication processes have made it possible to achieve nano-scale sub-wavelength metallic structures that induce unique optical responses at light wavelengths with localized surface plasmon resonances (LSPR). LSPR is determined by the size, shape, and optical properties of nano-scale sub-wavelength scaled noble metals [2].

The extraordinary optical transmission (EOT) phenomenon with metallic sub-wavelength hole arrays was discovered by Ebbesen and his co-workers [3]. An EOT spectrum can be designed by the periodicity and optical property of metallic sub-wavelength hole arrays. EOT is enhanced by the surface plasmon polariton (SPP) on a periodic sub-wavelength structure. The transmission spectrum has been applied in fabricating new optical sensors, as seen in the optical SPR and LSPR sensors [4].

EOT shows strong zero-order transmission, whereas the general zero-order of the transmitted light wave is diffracted through sub-wavelength scaled apertures. The amplitude of the transmittance is beyond Bethe's theory of transmission through small apertures [3, 5]. The wavelength for SPP, λ_{sp}, with two-dimensional cubic sub-wavelength hole arrays in metal film is described as follows

$$\lambda_{sp} (i^2 + j^2)^{0.5} = P \{ (\varepsilon_m \varepsilon_d) / (\varepsilon_m + \varepsilon_d) \}^{0.5}, \quad (1)$$

where ε_m and ε_d are the permittivity of the metal and dielectric medium, respectively, and (i, j) is associated with the periodicity.

In the last decade, researchers have designed optical properties and periods of sub-wavelength metallic holes for EOT-based optical permittivity sensors following (1). However, the main transmittance peak shows a larger response to permittivity changes than the wavelength described by (1) [4, 7]. Many questions still remain about the EOT spectrum response to permittivity changes [6]. We propose an electric field design of metallic sub-wavelength hole arrays to achieve large optical response of EOT upon changes in the permittivity of the sensing medium.

Previous studies have mainly used round sub-wavelength holes for optical permittivity sensors. Compared to these isotropic sub-wavelength holes, anisotropic sub-wavelength holes show unique responses [8, 9, 10]. Therefore, we applied polarized illumination and rectangular sub-wavelength holes for our optical permittivity sensor.

II. TRANSMISSION SPECTRA WITH DIFFERENTLY SHAPED METALLIC SUB-WAVELENGTH HOLE ARRAYS

Palladium (Pd) was used as the metallic material because the absolute value of its permittivity decreases upon hydrogen (H_2) absorption. The use of Pd also makes it easy to control the permittivity of the metallic sub-wavelength hole array. Additionally, we can fabricate not only an optical permittivity sensor but also an optical H_2 sensor with a Pd sub-wavelength

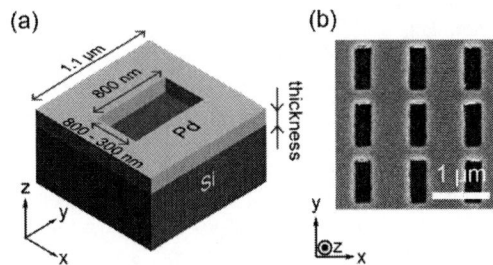

Fig. 1. (a) Schematic of Pd sub-wavelength hole in array on Si substrate for RCWA simulation. (b) SEM image of (long side in nm, short side in nm) = (800, 300) hole array. Black areas are Si, and gray area is Pd.

hole array. Pd thin film shows a long propagation length of surface plasmon in the infrared region on the surface with the dielectric medium. Silicon (Si) was selected as a substrate because it is transparent in infrared light. The hole period was fixed at 1.1 μm. The long sides of the rectangular holes were fixed at 800 nm, and the short sides were varied from 300 to 800 nm. The simulated and measured samples had hole dimensions (long side, short side, both in nm) of (800, 800), (800, 700), (800, 600), (800, 500), (800, 400), and (800, 300). A schematic of the Pd sub-wavelength hole array on a Si substrate is shown in Fig. 1.

The transmission spectra with rectangular sub-wavelength hole arrays were simulated and measured. Rigorous coupled-wave analysis (RCWA) was performed for the EOT simulation [14]. DiffractMOD (Rsoft Design Group, Ossining, NY, USA) was used to calculate the transmission spectra of Pd sub-wavelength hole arrays (Fig. 1(a)). The measured samples were fabricated by electron beam (EB) direct writing (F5112, Advantest, Tokyo, Japan) followed by a lift-off process (Fig. 2(a)). The far-field transmission spectra with differently shaped hole arrays were measured by Fourier transform infrared spectroscopy (Infinity FT-IR, Mattson Technology, Fremont, CA, USA). The illumination was x-polarized, and the optical axis was in the z-direction (Fig. 2(b)).

Simulated and measured transmission spectra as a function of the hole shape are shown in Fig. 3. All spectra were normalized by the transmittance of the Si substrate and their respective hole areas. Here, λ_{sp} is the wavelength related to the SPP in (1), and ε_m and ε_d correspond respectively to Pd and Si dielectrics. The position of the (i, j) integer is (1, 0). As a result of the decrease in the short side length, the transmittivity at λ_{sp} is attenuated; in other words, the amplitude of SPP resonance gets stronger. Furthermore, the main transmittance peak shifts toward the longer wavelength region.

The optical responses to permittivity change with 2% H_2 exposure as a function of the short side length of the sub-wavelength holes is shown in Fig. 4. The concept with a rectangular H_2 sensor is shown in our previous reports [10, 11]. In our RCWA simulation for EOT response upon 2% H_2 exposure, the absolute value of Pd permittivity was taken as a 20% decrease [12]. The change in the dimensions of the holes was simulated by a 3.5% expansion of the Pd lattice [13]. In our experimental study on EOT response upon 2% H_2 exposure, dry air and H_2 gas cylinders were prepared. A stainless steel gas flow cell with two barium fluoride windows was prepared for gas exposure. Mass flow controllers (SEC series, HORIBA, Kyoto, Japan) were prepared for the gas mixture. As a result of the Pd hydrogenation with 2% H_2 exposure, the wavelength of the main transmittance peak moved toward the longer wavelength region. When the short side length of the Pd sub-wavelength hole was reduced, the amount of the peak shift value increased and was saturated around 200 nm.

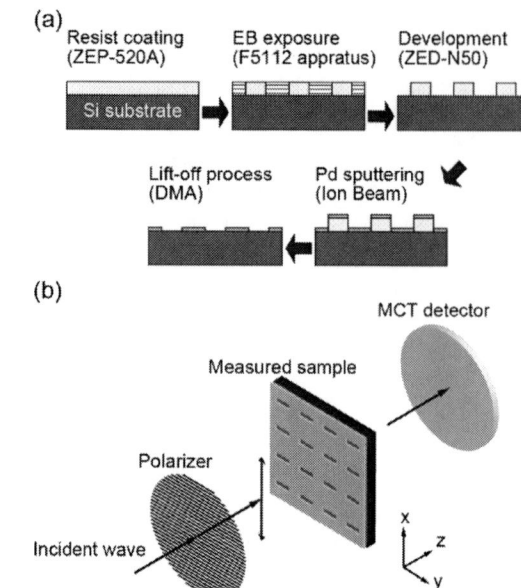

Fig. 2. (a) Fabrication sequence for Pd sub-wavelength hole array on Si substrate. EB lithography was carried out with a 400-nm spin-coated resist film (ZEP-520A, Zeon, Japan) and subsequent resist development (ZED-N50). Pd film was deposited by sputtering, and finally, the resist was removed by a lift-off process in DMA. (b) The dimensions for the EOT measurement with Pd sub-wavelength hole array on Si substrate using FT-IR. The illumination was polarized. The transmitted light wave was measured with an MCT (HdCdTe) detector.

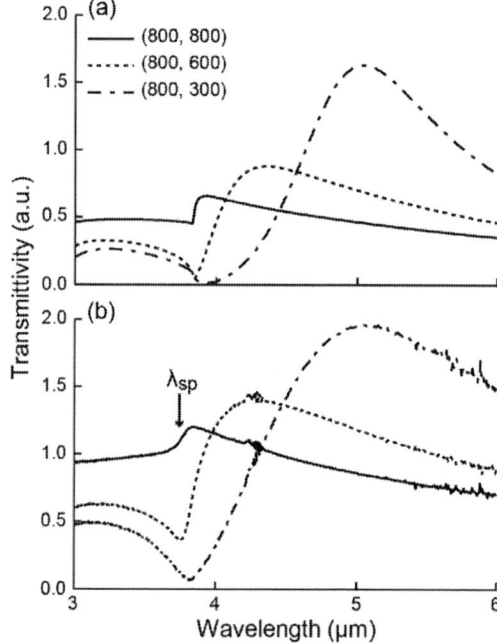

Fig. 3. Transmission spectra as a function of the hole shape of three sub-wavelength hole arrays, (length in nm, width in nm) = (800, 800), (800, 600), and (800, 300), (a) simulated and (b) measured. The Pd layer thickness was fixed at 100 nm. The amplitudes of all spectra were normalized by the transmittance of the Si substrate and their respective hole areas. λ_{sp} represents the wavelength in (1).

From these results (Fig. 3 and Fig 4), it was revealed that the short side length of the rectangular shaped hole is one of the factors that can be designed to increase the optical response of a permittivity sensor with a Pd sub-wavelength hole array. The amplitude of SPP resonance becomes stronger, and the response of the optical permittivity sensor gets larger when the length of the short sides of rectangular holes in an array is reduced, and the illumination is polarized to the short side of the rectangular sub-wavelength holes.

III. TRANSMISSION SPECTRA WITH VARIOUS THICKNESS OF METALLIC SUB-WAVELENGTH HOLE ARRAYS

The mechanism of EOT with metallic sub-wavelength hole arrays has been studied over the last decade, but there remain some mismatches between theoretical and experimental studies [6]. Here, our Pd sub-wavelength hole arrays were regarded as one kind of nanoplasmonic waveguide because the wavelength that can propagate the sub-wavelength holes is determined by the hole shape, not unlike general waveguides (Fig. 3).

To improve the optical response of the rectangular-shaped Pd sub-wavelength hole array upon H_2 exposure, the electric field distributions in sub-wavelength holes as a function of Pd layer thickness were investigated. Our rectangular Pd sub-wavelength hole array on a Si substrate has two interfaces, Pd/Air and Pd/Si along the optical axis in the z-direction. The measured far-field transmittance was believed to be affected by the interaction between the two interfaces.

The Pd layer thickness corresponds to the length of the nanoplasmonic waveguide that has two different interfaces. The normalized $|E_x|$ field distribution of the (800, 300) hole arrays as a function of Pd layer thickness is shown in Fig. 5. It can be seen in the holes that the strong $|E_x|$ field is enhanced near the edge of the rectangular holes in both Pd layer thicknesses. The difference between the 200-nm-thick layer and the 50-nm-thick layer is the amount of $|E_x|$ field dispersion. As a result of reducing the Pd layer thickness, the $|E_x|$ field dispersion within the sub-wavelength holes is attenuated. The dispersion of the waveguide mode within the holes as seen from the Ex field distribution was expected to disturb the peak shift of the transmission spectrum upon H_2 exposure.

The transmission spectra with various Pd layer thicknesses of sub-wavelength hole arrays were examined in Fig. 6. As a result of the reduced Pd layer thickness, the peak shift value was enhanced with the (800, 300) hole array. On the contrary, the peak shift value was not enhanced with the (800, 800) hole array. This difference between the two hole arrays can be explained by the amplitude of SPP resonance that is related to the wavelength, λ_{sp} in Fig. 3. With the (800, 300) hole array, the amplitude of SPP resonance is stronger than that with the (800, 800) hole array. This is evidence for the difference in the observed peak shift values as a function of the Pd layer thickness. From these results, we can conclude that the attenuation of $|E_x|$ field dispersion with a thin metallic layer

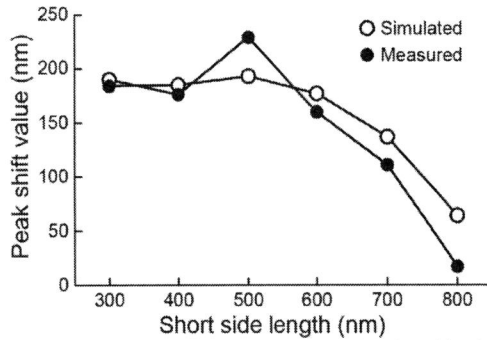

Fig. 4. Peak shift values upon 2% H2 exposure as a function of the short side length of the holes. The variation in the optical properties of the Pd layer was simulated by a decrease in the absolute permittivity value of Pd taken as 20%. The change in the dimensions of the hole array was simulated by a 3.5% expansion of the Pd lattice.

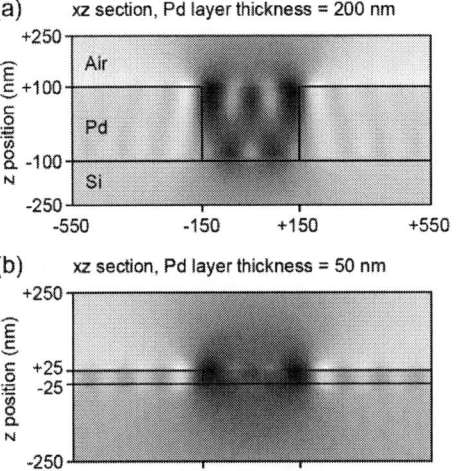

Fig. 5. Normalized $|E_x|$ field distribution of the (long side in nm, short side in nm) = (800, 300) hole arrays at the wavelength corresponding to the peak of the transmission spectrum, 5.04 μm. The Pd layer thickness is 200 nm in (a) and 50 nm in (b). Both $|E_x|$ field distributions are plotted in the xz section.

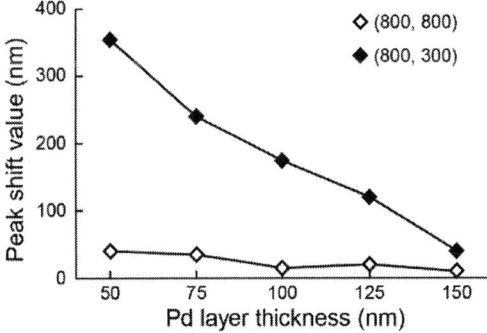

Fig. 6. Observed peak shift values upon 2% H2 exposure as a function of Pd layer thickness for (long side in nm, short side in nm) = (800, 800) and (800, 300) hole arrays. All peak shift values were measured by our gas flow system and FT-IR.

makes it possible to improve the response of our optical permittivity sensor.

Meanwhile, the thinner, less than 50-nm-thick Pd layer is semi-transparent in the infrared wavelength region. The sharpness of the transmission spectrum is broadened by the increase in the background transmittance amplitude. There is thus a limit to the metallic layer thickness that still allows the measurement of EOT and the optical response to permittivity changes.

IV. CONCLUSION

This report presented the electric field design of an optical permittivity sensor based on EOT with a metallic sub-wavelength hole array. This hole array showed strong zero-order transmission with SPP resonance beyond Bethe's theory for transmittance with small apertures. The EOT phenomenon was applied for this new optical permittivity sensor because of the sensitivity of SPP resonance to changes in optical properties. We applied Pd hydrogenation in our optical permittivity sensor in order to control the permittivity. The Pd sub-wavelength hole arrays were regarded as nanoplasmonic waveguides. The shape and length of the waveguide were examined in the infrared region.

The rectangular Pd sub-wavelength hole array with polarized illumination applied parallel to the short side of the rectangular holes showed stronger EOT as the length of the short sides decreased. The simulated and measured transmission spectra showed a large optical response with the reduced short side length of the rectangular holes upon the permittivity change in the Pd layer with the hydrogenation.

Our electric field distribution analysis within the sub-wavelength holes as a function of Pd layer thickness showed $|E_x|$ field dispersion along the incident wave axis. The Pd layer thickness is related to the propagation length of the nanoplasmonic waveguide. We expected that the attenuation of the $|E_x|$ field dispersion within the sub-wavelength hole would induce a large optical response to the permittivity change in the Pd layer upon H_2 exposure. The experimental study on the optical responses of the rectangular sub-wavelength hole array with varied Pd layer thicknesses showed that the thinner Pd layers induced a larger optical response to Pd hydrogenation.

These results lead to the conclusion that the short side length and the metallic layer thickness can be designed to achieve a large optical response to a permittivity sensor with EOT of a metallic sub-wavelength hole array. The strong SPP is enhanced by the reduction in the short side length of the sub-wavelength rectangular holes, and the $|E_x|$ field dispersion within the hole is attenuated by the reduction in the metallic layer thickness.

ACKNOWLEDGMENT

This work was supported by the Grant-in-Aid for Challenging Exploratory Research 22656036 and Grant-in-Aid for JSPS Fellows by Japan Society for the Promotion of Science. The EB direct writings were carried out using EB lithography apparatus F5112+CD01 of the VL01 of the VLSI Design and Education Center (VDEC) of the University of Tokyo donated by ADVANTEST Corporation. Part of this work was conducted at the Center for Nano-lithography & Analysis of the University of Tokyo and was supported by the Ministry of Education Culture, Sports, Science and Technology (MEXT), Japan.

REFERENCES

[1] E. Kretshmann, "The angular dependence and the polarisation of light emitted by surface plasmons on metals due to roughness," *Optics Communications*, vol. 5, pp. 331-336, August 1972.

[2] J.N. Anker, W.P. Hall, O. Lyandres, N.C. Shah, J. Zhao, and R.P. van Duyne, "Biosensing with plasmonic nanosensors," *Nature Materials*, vol. 7, pp. 442-453, June 2008.

[3] T.W. Ebbesen, H.J. Lezec, H.F. Ghaemi, T. Thio, and P.A. Wolf, "Extraordinary optical transmission throught sub-wavelength hole arrays," *Nature*, vol. 391, pp. 667-669, February 1998.

[4] C. Genet, T.W. Ebbesen, "Light in tiny holes," *Nature*, vol. 445, pp. 39-46, January 2007.

[5] H.A. Bethe, "Theory of diffraction by small holes," *Physical Review*, vol. 66, pp. 163-182, October 1944.

[6] H. Liu and P. Lalanne, "Microscopic theory of the extraordinary optical transmission," *Nature*, vol. 452, pp. 728-731, April 2008.

[7] A.G. Brolo, R. Gordon, B. Leathem, and K.L. Kavanagh, "Surface plasmon sensor based on the enhanced light transmission through arrays of nanoholes in gold films," *Langmuir*, vol. 20, pp. 4813-4815, June 2004.

[8] K.L. van der Molen, K.J. Klein Koerkamp, S. Enoch, F.B. Segerink, N.F. van Hulst, and L. Kuipers, "Role of shape and localized resonances in extraordinary transmission through periodic arrays of subwavelength holes: Experiment and theory," *Physical Review B*, vol. 72, pp. 045421, July 2005.

[9] F.J. Garcia-Vidal, E. Moreno, J.A. Porto, and L. Martin-Moreno, "Transmission of light through a single rectangular hole," *Physical Review Letters*, vol. 95, pp. 103901, September 2005.

[10] E. Maeda, S. Mikuriya, K. Endo, I. Yamada, A. Suda, and J.-J. Delaunay, "Optical hydrogen detection with periodic subwavelength palladium hole arrays," *Applied Physics Letters*, vol. 95, pp. 133504, September 2009.

[11] E. Maeda, S. Mikuriya, I. Yamada, and J.-J. Delaunay, "Hydrogen absorption effects on the transmittance of sub-wavelength palladium hole arrays with different thicknesses," *Proceedings of SPIE*, vol. 7941, pp. 79410O, February 2011.

[12] J. Isidorsson, I.A.E.M. Giebels, H. Arwin, and R. Griessen, "Optical properties of MgH2 measured in situ by ellipsometry and spectrophotometry," *Physical Review B*, vol. 68, pp. 115112, September 2003.

[13] J.A. Eastman, L.J. Thompson, and B.J. Kestel, "Narrowing of the palladium-hydrogen miscibility gap in nanocrystallinepalladium," *Physical Review B*, vol. 48, pp. 84-92, July 1993.

[14] M.G. Moharam, D.A. Pommet, E.B. Grann, and T.K. Gaylord, "Stable implementation of the rigorous coupled-wave analysis for surface-relief gratings - enhanced transmittance matrix approach," *Journal of the Optical Society of America A-Optics Image Science and Vision*, vol. 12, pp. 1077-1086, May 1995.

A Stable Frequency References Platform based on Packaged Microsphere-Taper Coupling System

Ji-Jun Xiong*[1,3], Ying-Zhan Yan[1], Chang-Ling Zou[2], Fang-Wen Sun[2], Shu-Bin Yan[3],
Chen-Yang Xue[1], Jun Liu[3], and Wen-Dong Zhang[1]

[1]*Key Laboratory of Instrumentation Science & Dynamic Measuremen, North university of China,
Taiyuan, China*
[2]*Key Lab of Quantum Information,University of Science and Technology of China, Hefei, China*
[3]*Science and Technology on Electronic Test & Measurement Laboratory, North university of China,
Taiyuan, China*
(xiongjijun@nuc.edu.cn)

Abstract—**"Whispering gallery modes" (WGM) optical micro- and nano resonators are important category in nanophotonics researches. The ultra-high quality factor (Q) makes them promising ranging from sensors, filters to low threshold sources. In the above supplications, the frequency stability of the WGM is critical, which is determined by several factors. In this paper, using the microsphere resonator we construct a robust packaged microsphere-taper coupling system (MTCS) to stabilize the resonant WGMs. Furthermore, a stable frequency references platform based on the packaged microcavity unit is constructed. This stable frequency references platform can eliminate the noise of the WGM, and promote the microresonator researches.**

Keywords-Frequency References; Packaged structure; Stable

I. INTRODUCTION

The Stable Frequency References is important for the optical metrology. It not only acts as the basis of the metrology but also limit the measuring accuracy.

High quality factors (Q) Whispering gallery mode (WGM) optical microcavities [1] have attracted more and more attention in applied researches ranging from sensors, filters to low threshold sources, due to the characteristics of small size, high transparency windows and narrow resonances [2-9]. In the above supplications, a critical point is the frequency stability of the resonator mode. For example, the stable resonant frequency makes resonators be natural candidates for use in laser stabilization. The stability of the resonator mode is also very important for microcavity-based sensing applications. Because in sensing, the resonant spectra shift functions as the sensing signal to detect the nearby substances, and an unstable resonant spectra can induce large sensing errors [10-15].

For a resonant dip in a microresonator coupling regime, the resonant location is determined by the following factors: (i) Structure parameters of the coupling system, including the size of the microresonator and the taper, the coupling gap between them. (ii) Physical parameters of the coupling system, including the RI of the coupling system as well as the purity, the RI of the surroundings. (iii) The Temperature of the coupling system and its surroundings. (iiii) The noise of the laser. In our former research, we have realized a wholly-

packaged microsphere-taper coupling structure, in which the coupling system was encapsulated and solidified into an entire body. In such a packaged structure, the coupling gap is fixed and the coupling system is isolated form the surroundings. Thus the first two items (i and ii) of the factors mentioned above are ensured in the packaged structure. Consequently, the rest of the two items (iii and iiii) are dominant. Actually, when considering the frequency stability of the resonator mode, it refers to two aspects which are supplemented each other: (i) How to stabilize the WGM. (ii) How to construct a system to provide a frequency references to eliminate the limiting factors.

In this paper, we optimize the packaged structure and realize a much more robust, modularized and integrated microresonator unit, which also can be moved freely. Then using this unit, we attempt to stabilize the resonant WGMs. Furthermore, a stable frequency references platform based on the robust packaged microcavity unit is constructed. Using this references platform, we improve the microcavity based thermal sensing application.

II. EXPERIMENTS

In experiments, the microspheres are used to illustrate the stable frequency references platform, and individual microspheres are fabricated by melting the end of the optical fiber. The microsphere WGMs are excited by a tunable laser (1550nm wavelength band, linewidth< 300kHz) through a fiber taper. And the gap between microsphere and taper is controlled by electromechanical 3D X-Y-Z stages with 20nm resolution.

A. The WGMs of the traditional microsphere-taper coupling system

In the traditional microsphere-taper coupling system, the construction of the coupling structure bases on controlling the relative positions of the microsphere and taper, to adjust the fine coupling. And the maintenance of the coupling state depends on the fixing of the position by using two respective 3D stages. At the meantime, the relative positions and the gap between the two parts are strict necessary condition for the resonance, which directly determines a certain WGM. This means that the strict coupling conditions for the microstructure are ensured by the two discrete bulky stages. In such a regime, it is a great challenge to maintain a certain coupling (or a certain WGM) for a long time, because a tiny vibration or shake can change the relative positions and the gap of the microcavity and the taper, which subsequently interfere the coupling and the frequency of the WGMs. Fig. 1 shows the irregular drift of one certain WGM with time passing by about 135 seconds. It reveals that the WGM marked by the blue arrow fluctuates in a range about 1.5GHZ irregularly. We attribute the WGM drift to the slight shake of the taper or the microsphere, which results from the inevitable air flow. Even we can estimate that the coupling can be broken by strenuous vibration.

Fig. 1. Typical drifts of two certainWGMs during 135 seconds for traditional MTCS.

In addition, thermal noise can also result in the WGM shift, because the light traveling in the microresonator is always confined within the dielectric material. The resonant light also functions as a heater to rise the temperature of the microcavities, and a small temperature change can result in a large frequency shift of the WGMs through the thermo refractive and thermal expansion fluctuations. This phenomenon leads to thermal instability in high-QWGM resonators. In order to avoid the thermal effect as far as possible, the laser power is controlled to be lower than 5mW. Actually, although the laser power is low, there is still a little thermal fluctuations superimposing on the WGMs. In Fig. 1 the ruleless fluctuations imply that, the thermal noise is much smaller and is annihilated by the vibration noise mentioned above.

In short, eliminating the vibration noise to achieve the stable coupling is the precondition for the WGM reference. Additionally, thermal noise also needs to be suppressed and eliminated. In the following section, we take some measures to improve the traditional coupling system.

B. The WGMs of the hermetically-sealed and packaged microsphere-taper coupling system

In our former researches, we have fabricated robust packaged microsphere-taper structures. Here, In order to eliminate the vibration noise of the discrete microcavity-taper coupling system, we further improve the package technology to realize a hermetically-sealed and packaged coupling structure, in which the fragile taper and the microsphere are all encapsulated together to fabricate a functional module. Simply speaking, the hermetically-sealed and packaged structure is fabricated as following. First, a robust packaged microsphere-taper system is obtained according to the fabrication procedure as reported in the reference, as shown in Fig. 2(a). Then, similar to the potted circuit module we further coat the package material (low RI polymer) on the structure by using a designed slot as the mould to package the fragile taper totally. In this way, the microcavity coupling system is not suspended, moreover, for the first time, an integrated microcavity-taper-clamp functional unit is obtained. Fig. 2(b) shows the microcavity in the preliminary stage of the potting. Fig. 2(c) shows the microcavity which is almost entirely submerged in the package. Fig. 2(d) shows a microsphere submerged totally in the package. Afterwards, the hermetically-sealed and package is solidified through 30 minutes exposure under the UV lamp. Fig. 2(e) shows the semi-finished products of the hermetically-sealed and packaged microcavity unit. Fig. 2(f) shows a typical hermetically-sealed and packaged module. In this module the microcavity, its coupling system and the fragile taper are all solidified into an entire body which in integrated with the fixture. Thus, the microsphere-taper coupling system is embedded deeply in the hermetically-sealed and packaged structure. This hermetically-sealed and packaged structure possesses obvious merits, including the robustness, portability, which make the structure much more practical, especially in the stable frequency reference application.

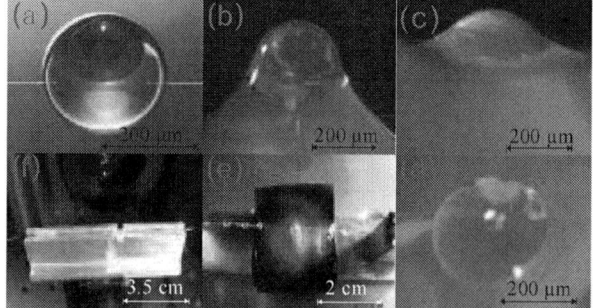

Fig. 2. The fabrication Workflow of the packaged microspheres.

In order to stabilize the temperature, a heat insulating bucket filled with ice-water mixture is used to contain the microcavity. The combination of the heat insulating bucket and the hermetically-sealed and packaged microcavity structure endows the microcavity regime with excellent thermal and mechanical performance.

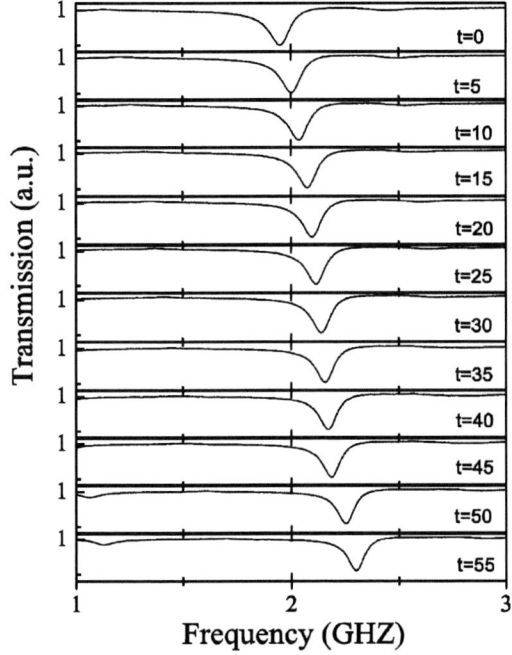

Fig. 2. A series of resonant dips of one selected WGM within 45min for the packaged structure embedded in the ice-water mixture.

We first measure the resonant spectra of the hermetically-sealed and packaged microspherecoupling structure. Fig. 3 shows a series of resonant dips of a selected WGM within 55min. As clearly shown, the selected WGM is much more stable than that for a microsphere in the open air. The WGM no longer fluctuate erratically and randomly, moreover, shows a stable resonant line shape (linewidth or the Q). This is mainly because of the robustness of the hermetically-sealed and packaged microcavity structure and the stability of the temperature in the heat insulating bucket. However, the WGM can also shift after a long time (e.g. half an hour). In Fig. 3(a) the selected WGM shifts about 355 MHZ during 55 min with an average shift speed about 6.45MHZ/min, corresponding to a shift of 0.0516pm per min. Sometimes, the shift appear in the form of "hopping", in which the resonant frequency can abruptly move to another location. As shown in Fig. 3(b), in the first 1000 second the WGM slowly shift about 107 MHZ. At the moment of 1005 second, the WGM suddenly move to a far position with a shift about 2.575 GHZ. Both of the slow shift and the hopping are all independent of the microcavity coupling system, and they result from the laser. Generally, the resonant WGMs are measured by repetitively sweeping the wavelength of the probe lase by using external scan voltage.

Thus, the conversion of the scan voltage to the output wavelength provides a significant source of error. Previous research has estimated that such error can be even greater than the resonance linewidth when the Q is larger than 107. Actually, the long-term frequency shift (slow shift and hopping) can exceed the linewidth of high-Q resonators such as microtoroids as well as microspheres. Moreover, this shift is intrinsic with the laser and is inevitable in many commercial tunable lasers. Consequently, the above shift is hard to eliminate and the resonant frequency shift can introduce errors in microcavity-based application. However, this noise resulted from the laser has a characteristic that the uncertain fluctuations contribute equally to the WGMs. This provides us a feasible way to suppress even eliminate the laser noise, to construct a stable frequency references.

As shown in Fig. 4, we test the resonant spectra for two individual hermetically-sealed and packaged microspheres which are placed in the ice-water. A 1550-nm tunable laser (Newport Velocity 6328) is driven by a voltage triangular wave, and the laser output is split into two branches by a directional coupler. In each branch the laser is used to monitor the resonance of the two microcavity structures respectively. Fiber taper is used to couple the microspheres, and the coupled power is kept to be lower than 5mW. The two hermetically-sealed and packaged microsphere structures are in the ice-water to ensure the thermal and mechanical isolation for several hours before use. The two packaged microspheres each has a Q about 4.6×10^6 and 4.8×10^6, respectively. The dual outputs are detected using two photoreceptors with the same product model.

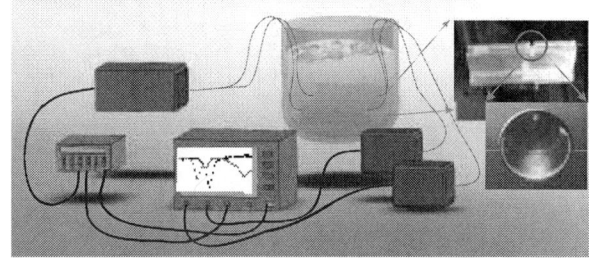

Fig. 3. The schematic diagram of the testing system for the packaged MTCS-based frequency reference platform.

In experiments, we respectively select one WGM from the resonant dips of the two packaged microspheres. As shown in the inset of the Fig. 5, the black and the red lines denote the selected resonant dips for NO. 1 and No. 2 packaged structures respectively. Due to the laser noise the two selected resonant dips all shift to the same direction. We measured the frequency difference (FD) between the two dips over a long period of time about 400 min. As shown in Fig. 5, the FD exhibits fine stability, which has an average value about 330.1132 MHZ with the standard deviation about 1.4 MHZ. In the insets, we show three groups of data at different times (t=0 min, 160 min and 320 min).

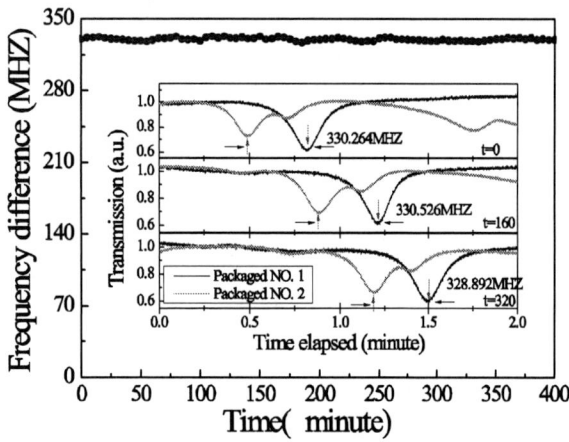

Fig. 4. The tested FD between the two dips over a long period of time about 400 min. Inset: three groups of data at different times (t=0 min, 160 min and 320 min).

III. CONCLUSION

In this paper, using the microsphere resonator we construct a robust packaged microsphere-taper coupling system (MTCS) to stabilize the resonant WGMs. Furthermore, a stable frequency references platform based on the packaged microcavity unit is constructed. This stable frequency references platform can eliminate the noise of the WGM, and promote the microresonator researches.

ACKNOWLEDGMENT

This work was supported by the NSFC under Grant 91123036 and Grant 61178058, by the National Basic Research Program of China under Grant 2009CB326206, and by the Innovation Project under Grant 7130907 and Grant 9140C1204040706.

REFERENCES

[1] K. J. Vahala, "Optical microcavities," Nature (London) 424, 839–846 (2003).

[2] A. Chiasera, Y. Dumeige, P. Fron, M. Ferrari, Y. Jestin, G. Nunzi Conti, S. Pelli, S. Soria, and G. C. Righini, "Spherical whispering-gallery-mode microresonators," Laser Photon. Rev. 4, 457–482 (2010).

[3] A. B. Matsko and V. S. Ilchenko, "Optical Resonators With Whispering-Gallery Modes-Part I: Basics," IEEE J. Quantum Electron. 12, 3–14 (2006).

[4] F. Vollmer and S. Arnold, "Whispering-gallery-mode biosensing: label-free detection down to single molecules," Nat. Methods 5, 591–596 (2008).

[5] Y. Sun and X. Fan, "Analysis of ring resonators for chemical vapor sensor development," Opt. Express 16, 10254–10268 (2008).

[6] M. Sumetsky, R. S. Windeler, Y. Dulashko, and X. Fan, "Optical liquid ring resonator sensor," Opt. Express 15, 14376–14381 (2007).

[7] F. Xu, V. Pruneri, V. Finazzi, and G. Brambilla, "An embedded optical nanowire loop resonator refractometric sensor," Opt. Express 16, 1062–1067 (2008).

[8] F. Xu and G. Brambilla, "Demonstration of a refractometric sensor based on optical microfiber coil resonator," Appl. Phys. Lett. 92, 101126 (2008).

[9] I. M. White and X. Fan, "On the performance quantification of resonant refractive index sensors," Opt. Express, 16, 1020–1028 (2008).

[10] M. Cai, O. Painter, and K. J. Vahala, "Observation of Critical Coupling in a Fiber Taper to a Silica-Microsphere Whispering-Gallery Mode System," Phys. Rev. Lett. 85, 74–77 (2000).

[11] J. C. Knight, G. Cheung, F. Jacques, and T. A. Birks, "Phase-matched excitation of whispering-gallery-mode resonances by a fiber taper," Opt. Lett. 22, 1129–1131 (1997).

[12] M. Hossein-Zadeh and K. J. Vahala, "Fiber-taper coupling to Whispering-Gallery modes of fluidic resonators embedded in a liquid medium," Opt. Express 14, 10800–10810 (2006).

[13] M. L. Gorodetsky, A. A. Savchenkov, and V. S. Ilchenko, "Ultimate Q of optical microsphere resonators," Opt. Lett. 21, 453–455 (1996).

[14] D. W. Vernooy, V. S. Ilchenko, H. Mabuchi, E. W. Streed, and H. J. Kimble, "High-Q measurements of fused silica microspheres in the near infrared," Opt. Lett. 23, 247–249 (1998).

[15] A. M. Armani, D. K. Armani, B. Min, K. J. Vahala, and S. M. Spillane, "Ultra-high-Q microcavity operation in H_2O and D_2O," Appl. Phys. Lett. 87, 151118 (2005).

Arrayed Metallic Micro/Nano Particles for Localized Surface Plasmon Resonance Based on Metal Contact Transfer Lithography

H. Y. Chung, C. Y. Wu, C. H. Chen, Y. C. Lee[*]

[*] Department of Mechanical Engineering, National Cheng-Kung University, Tainan, R.O.C

yunglee@mail.ncku.edu.tw

Abstract—In this study, we demonstrate a rapidly, low cost, and mass production process to fabricate arrayed metallic nanoparticles on a variety of substrates based on contact transfer and metal mask embedded lithography (CMEL). A hexagonal arrayed metallic nanoparticles deployed on ITO/glass substrate with sub-micron periodicity is achieved. It is observed in optical transmittance measurements that noble metallic arrayed nanoparticles deployed on ITO/glass substrate result in a spectrally narrowband of extinction in visible range, and is in good agreement with the simulated results using finite-element method (FEM). It is found that the narrowband extinction spectrum is associated with electromagnetic field coupling between the arrayed metallic nanostructures and the ITO layer. This electromagnetic field coupling induces significant plasmon resonance in the ITO layer underneath the arrayed metallic nanostructures. Based on this observed phenomenon and our innovative large-area nano-fabrication processes, optoelectronic devices with arrayed metallic nanostructures can be easily designed and developed.

Keywords- localized surface plasmon resonance; guided mode resonance; nanoimprinting lithography

I. INTRODUCTION

Arrayed metallic nanoparticles have gained lots of attentions in both scientific researches as well as engineering application during last few decades. Metallic nanostructures exhibit a rich variety of intriguing optical properties due to the interaction of the electromagnetic field with the free electrons of the metal. Such an excitation can occur at a metal-dielectric interface and is called surface-plasmon polariton or at a metallic nanoparticle, and in this case it is termed as particle-plasmon polariton [1-3].

There are several ways to achieve metallic nanoparticles, such as laser ablation method, chemical reduction method and pyrolysis method [4-5]. These methods can produce nanoparticles over large area but have limitations to efficiently deploy nanoparticles in specified arrangements. Electron beam lithography is an excellent method to fabricate arrayed metallic nanoparticles. The size of nanoparticle can be well controlled to about several tens of nanometers and arranged into square or triangular lattice [1]. However, the costs of equipment and time-consuming issues limit the capability to mass produce large-area devices.

In this study, we demonstrate a rapidly, low cost, and mass productive process to fabricate arrayed metallic nanoparticles on a variety of substrates based on contact transfer and metal mask embedded lithography. A hexagonal arrayed metallic nanoparticles deployed on ITO/glass substrate with sub-micron periodicity is obtained. Moreover, the optical transmittance spectrum of the sample is measured via spectrophotometer experimentally and a numerical simulation using finite-element method (FEM) is carried out to identify the mechanisms of resonance characteristics observed in the spectrum measurements.

II. EXPERIMENTAL DETAILS AND RESULTS

This section describes the experimental details to fabricate arrayed metallic nanoparticles on a variety of substrates. A nanoimprinting process presented in our previous study [6], contact transfer and metal mask embedded lithography (CMEL), is applied to define and pattern arrayed metallic structures which have feature sizes in sub-micron or nanometer scales. First of all, a hexagonal arrayed structured silicon mold is prepared using electron beam lithography. The diameter of each holed structure is 200 nm and the periodicity of hexagonal array is 400 nm. A flexible h-PDMS mold replicated from the silicon mold is then obtained to act as the imprinting mold used in CMEL process. This h-PDMS mold is inexpensive compared to the primary silicon mold thus could be disposable after being contaminated during the imprinting process. Furthermore, this flexible mold has benefit to minimize the contact issue in imprinting process and utilize large-area pattern transfer successfully. Figure 1 illustrates the procedures to fabricate arrayed metallic nanostructures. An h-PDMS concave mold deposited with metal film is contacted to the top surface of polymer film coated on substrate and then heating to the glass transition temperature (Tg) of the polymer film. After cooling down to room temperature, the patterned metal film is transferred from the h-PDMS mold to the polymer film owing to the good adhesion between metal-polymer interfaces. This patterned metal film is then acted as etching mask in dry etching process to obtain a patterned polymer nanostructure. Combining with lift-off and thermal annealing processes, it is possible to fabricate arrayed metallic nanoparticles on a variety of substrates. Figure 2 shows the

978-1-4673-1122-9/12 $31.00 © 2012 IEEE

SEM images of the hexagonal arrayed Au nanoparticles (AuNPs) and Au nanorods (AuNRs) deployed on a 230nm thick ITO film deposited on a soda-lime glass substrate. The periodicity of hexagonal array is 400 nm and the radius of AuNPs is 85 nm; the dimensions of AuNRs are 200 nm in diameter and 40 nm in thickness. According to the experimental results, arrayed metallic nanoparticles deployed on a variety of substrates can be achieved. The dimensions, shapes, and arrangements of these arrayed metallic nanostructures can be easily adjusted by using different type of imprinting molds. The size and material combination of the obtained metal particles can be controlled by the thicknesses and varieties of metallic films deposited during the lift-off process.

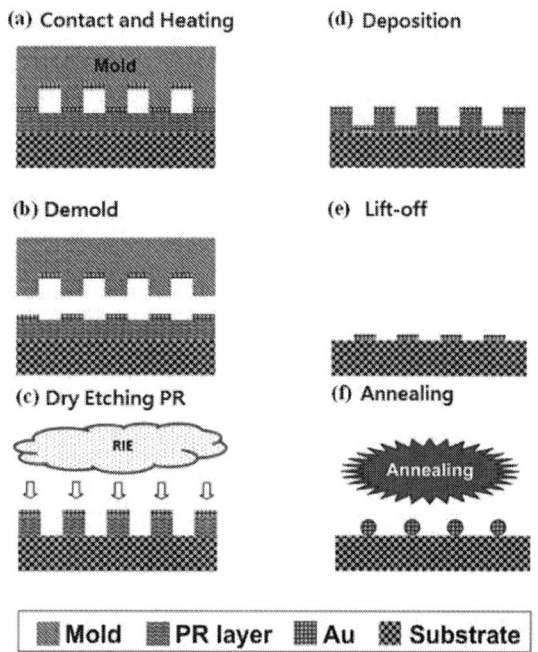

Figure 1. The procedures of using (a-c) CMEL process, (d-e) lift-off and (f) thermal annealing processes to achieve an arrayed metallic nanoparticle.

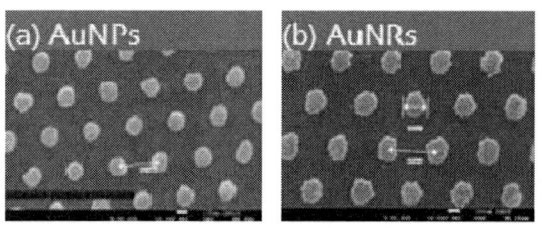

Figure 2. SEM images of (a) the hexagonal arrayed AuNPs deployed on 230nm-thick ITO film/glass substrates. The periodicity of hexagonal array is 400nm; the radius of AuNPs is 85nm, (b) the hexagonal arrayed AuNRs deployed on 230nm-thick ITO film/glass substrates. The periodicity of hexagonal array is 400nm; the dimensions of AuNRs are 200nm in diameter and 40nm in thickness.

III. OPTICAL MEASUREMENTS AND SIMULATIONS

The optical transmittance spectrum measurement of these arrayed metallic nanoparticles is obtained using a Hitachi U-3010 spectrophotometer. An unpolarized light with wavelength λ_0 ranging from 400 nm to 1000 nm is normally incident onto the metal side of these samples and the transmitted power is collected at the substrate side. Figure 3 shows the transmittance spectrum of arrayed AuNRs on glass substrate and on a 230 nm thick ITO/glass substrate. It is observed that the transmittance spectrum of both samples exhibits a spectrally wideband extinction in near-IR range due to localized surface plasmon resonance. Moreover, significant narrowband extinction in visible range is observed in case of arrayed AuNRs deployed on ITO/glass which means a strong electromagnetic field coupling between the arrayed metallic nanostructures and the ITO layer.

A numerical simulation using finite-element method is carried out to identify the resonance phenomenon. Both the transmittance spectrum and the electromagnetic field distribution are obtained to clarify the different mechanisms of resonances in visible and near-IR spectrum. Figure 4 illustrates the geometric model of hexagonal arrayed AuNRs on ITO/glass using in the simulation. The height of each AuNRs is 40 nm and the diameter is 200 nm. The periodicity of hexagonal array is 400 nm and the thickness of ITO layer is 230 nm. The dielectric functions of Au and ITO are described as Lorentz-Drude model [7-8]. The dielectric constants of air and glass substrate are $\varepsilon_{air} = 1$ and $\varepsilon_{glass} = 2.31$, respectively. The incident electromagnetic wave is Ex-polarized and normally incident onto the AuNRs. Figure 5 shows the variation of transmittance spectrum with the periodicities of hexagonal array ranging from 300 nm to 450 nm. The transmittance spectrum exhibits spectrally red-shift while the periodicity of hexagonal array increasing from 300 nm to 450 nm.

Figure 6 demonstrates the comparison between the calculated transmittance spectrum and the experimental result. It is shown that the simulated result is in good agreement with the experimental result. The arrayed AuNRs deployed on ITO/glass substrate result in a spectrally narrowband extinction in visible and a wideband extinction in near-IR both in experiments and in simulations. Regarding to the simulated electric field distribution while the incident wavelength $\lambda_0 = 870$ nm, as shown in Fig. 7(a), an enhanced electric field around the metallic nanorod due to localized surface plasmon resonance is observed. Furthermore, Fig. 7(b) shows the significant guided mode resonance in the ITO layer underneath the arrayed metallic nanostructures when $\lambda_0 = 579$ nm.

A more detail discussion about the guided mode resonance is started from a simple empty-lattice approximation [9]. As a first step of this approximation, the energy dispersions of the lowest transverse electric (TE$_0$) and transverse magnetic (TM$_0$) guided modes of a homogeneous ITO layer could be found from the solution of the transcendent equations for the asymmetric waveguide slab [10]. Figure 8 demonstrates the calculated dispersions of TE$_0$ and TM$_0$ guided mode of a 230 nm thick ITO slab on the top of a soda-lime glass substrate. When assuming a surface corrugation with periodicity d of hexagonal array in a second step (in this case, d = 400 nm), the propagation constant is normalized to the reciprocal lattice of the 2D photonic crystal slab which equals to $4\pi/d\sqrt{3}$. While the propagation constants of TE$_0$ and TM$_0$ equal to the reciprocal lattice, it is shown that the photon energies of TE$_0$ and TM$_0$ guided modes equal to 2.034 eV and 2.110 eV, respectively. These calculated photon energies of TE$_0$ and TM$_0$ guided modes are in good agreement with the simulated results using FEM. The shifts of center wavelength might be caused by the neglecting of material and geometry of metallic deployed on the homogeneous ITO waveguide slab.

According to the FEM simulation and empty-lattice approximation discussed above, it suggests that the spectrally narrowband in visible range owing to the guided mode resonance has quite different resonance mechanism with respect to the wideband extinction in near-IR range due to the localized surface plasmon resonance. When an electromagnetic wave normally incident onto arrayed metallic nanostructures, a Bragg's diffraction phenomenon occurs and then couples the electromagnetic wave into the waveguide slab. At certain photon energies of the propagating guided waves which the propagation constant equal to the reciprocal lattice of arrayed nanostructures, a guided mode resonance happens and enhanced electric field distributes within waveguide slab can be observed.

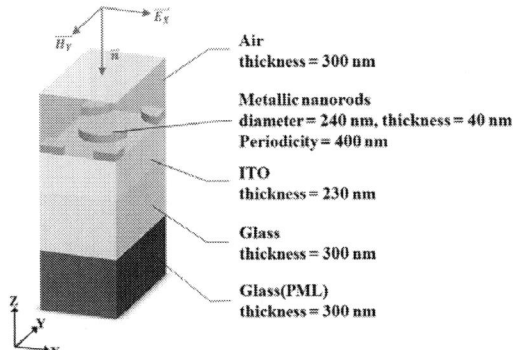

Figure 4. The geometric model of hexagonal arrayed AuNRs on ITO/glass using in the FEM simulation. The height of each AuNRs is 40 nm and the diameter is 200 nm; the periodicity of hexagonal array is 400 nm and the thickness of ITO layer is 230 nm. The incident electromagnetic wave is Ex-polarized and normally incident onto the AuNRs.

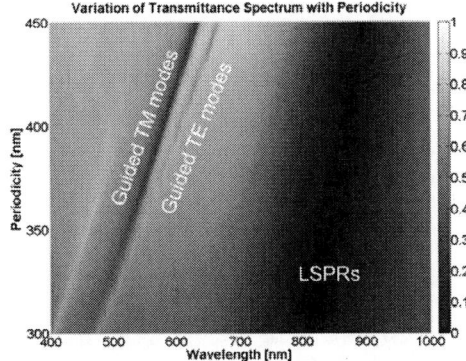

Figure 5. The variation of transmittance spectrum with the periodicities of hexagonal arrayed AuNRs.

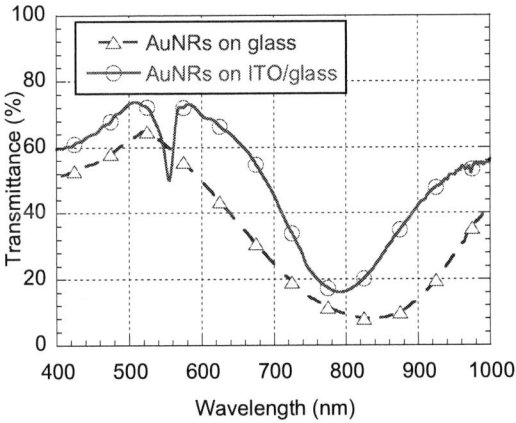

Figure 3. The normal incidence transmittance spectrum measurements of hexagonal arrayed AuNRs on glass substrate and on 230nm-thick ITO film/glass substrate.

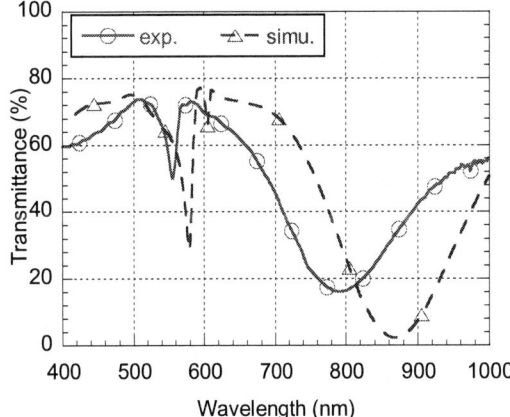

Figure 6. The normal incidence transmittance spectrum measurements and the simulated results using FEM. It is observed that these hexagonal arrayed AuNRs deployed on the surface of ITO/glass substrate result a spectrally narrowband extinction both in experiments and in simulations.

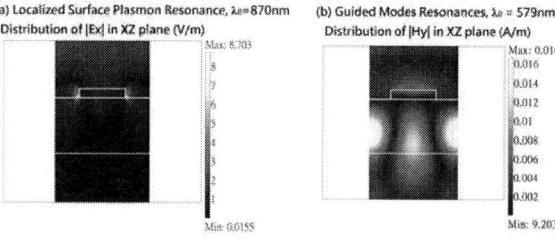

(a) Localized Surface Plasmon Resonance, λ_0=870nm
Distribution of |Ex| in XZ plane (V/m)

(b) Guided Modes Resonances, λ_0 = 579nm
Distribution of |Hy| in XZ plane (A/m)

Figure 7. The electromagnetic field distribution of arrayed AuNRs deployed on ITO layer. (a) When incident wavelength λ_0=870nm, only the particle-plasmon polariton is excited around the AuNRs. (b) When incident wavelength λ_0=579nm, there exhibits a strong electric field enhancement within ITO thick film due to guided modes resonance.

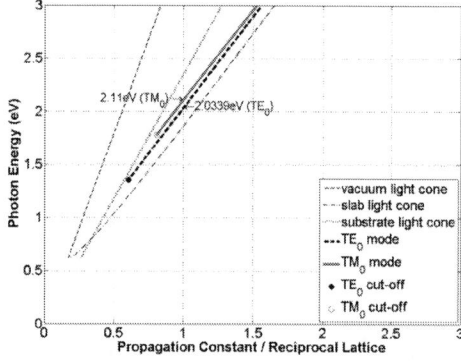

Figure 8. The lowest TE and TM polarized guided modes of a homogeneous and 230 nm thick ITO waveguide. The propagation constant is normalized to the reciprocal lattice to figure out the wavelength of guided mode resonance.

IV. CONCLUSIONS

In this study, we demonstrate a rapidly, low-cost, large-area, and mass productive fabrication process to obtain arrayed metallic nanostructures on a variety of substrates. The key element in this fabrication method is to combine an innovative metal contact printing lithography with conventional lifting-off and thermal annealing processes. Hexagonal arrays of metallic nanoparticles with sub-micron periodicity are successfully deployed on an ITO/glass substrate. The dimensions, shapes, and arrangements of these arrayed metallic nanostructures and nanoparticles can be easily adjusted by using different pattern designs in the imprinting molds. The sizes and material compositions of the obtained metal nanoparticles can be easily controlled by the deposition thicknesses and material varieties of films deposited during the sample preparation process.

Optical transmittance measurements show that certain kinds of noble metallic arrayed nanoparticles deployed on an ITO/glass substrate can result in a phenomenal narrowband of extinction in spectral range of visible light. Theoretical analysis indicates this narrowband extinction spectrum is associated with electromagnetic field coupling between the arrayed metallic nanostructures and the underlying ITO layer. Numerical simulation based on finite element method is carried out to demonstrate the electromagnetic field distributions of the localized surface plasmon resonance of arrayed metallic nanostructures and the excited waveguide modes within the ITO layer. This electromagnetic field coupling induces significant plasmon resonance in the ITO layer underneath the arrayed metallic nanostructures. A further evidence is attained by comparing the measured transmittance spectrum of a similar noble metallic arrayed nanoparticles deployed on a glass substrate. Experimental results show that the narrowband extinction in visible spectrum is vanished since there is no ITO layer to support guided modes resonance. Based on this observed phenomenon and our innovative large-area nano-fabrication processes, optoelectronic devices with arrayed metallic nanostructures can be easily designed and implemented in the future.

ACKNOWLEDGMENT

This work is supported by the National Science Council (NSC) of Taiwan through projects NSC 098-2811-M-006-012 and NSC 100-2120-M-006 -007 -CC1.

REFERENCES

[1] S. Linden, J. Kuhl, H. Giessen, "Controlling the interaction between light and gold nanoparticles: Selective suppression of extinction," *Phys. Rev. Lett.*, vol. 86, pp. 4688-4691, 2001.

[2] E. Hutter, J. H. Fendler, "Exploitation of localized surface plasmon resonance," *Adv. Mater.*, vol. 16, pp. 1685-1706, 2004.

[3] V. Yannopapas, N. Stefanou, "Optical excitation of coupled waveguide-particle plasmon modes: A theoretical analysis," *Phys. Rev. B*, vol. 69, 012408, 2004.

[4] F. Mafune, J. Y. Kohno, Y. Takeda, T. Kondow, "Dissociation and aggregation of gold nanoparticles under laser irradiation," *J. Phys. Chem.*, vol. 105, pp.9050-9056, 2001.

[5] S. Pillai, K. R. Catchpole, T. Trupke, M. A. Green, "Surface plasmon enhanced silicon solar cells," *J. Appl. Phys.*, vol. 1019, 093105, 2007.

[6] Y. C. Lee, C. Y. Chiu, "Micro-/nano-lithography based on the contact transfer of thin film and mask embedded etching," *J. Micromech. Microeng.*, vol. 18, 075013, 2008.

[7] A. D. Rakic, A. B. Djurisic, J. M. Elazar, M. L. Majewski, "Optical properties of metallic films for vertical-cavity optoelectronic devices," *Appl. Opt.*, vol. 37, pp.5271-5283, 1998.

[8] C. W. Chen, Y. C. Lin, C. H. Chang, P. C. Yu, J. M. Shieh, C. L. Pan, "Frequency-dependent complex conductivities and dielectric responses of indium tin oxide thin films from the visible to the far-infrared," *IEEE J. Quant. Electron.*, vol. 46, pp.1746-1754, 2010.

[9] A. Christ, T. Zentgraf, J. Kuhl, S. G. Tikhodeev, N. A. Gippius, H. Giessen, "Optical properties of planar metallic photonic crystal structures: Experiment and theory," *Phys. Rev B*, vol. 70, 125113, 2004.

[10] M. K. Barnoski, *Introduction to Integrated Optics*, Univ. of Calif., Santa Barbara, 1973, pp.53-72.

Ligand-Exchange of TOPO-capped CdSe Quantum Dots with Quinuclidines

Judy M. Obliosca,[1,2] Fan-Gang Tseng,[1] Ching-Mao Huang,[3] Leu-Wei Lo[3] and Pen-Cheng Wang[1*]

[1]Department of Engineering and System Science, National Tsing Hua University, 101, Section 2, Kuang-Fu Road, Hsinchu 30013, Taiwan

[2]Nano Science and Technology, Taiwan International Graduate Program, Institute of Physics, Academia Sinica, 128, Academia Road, Section 2, Nankang, Taipei, 11529, Taiwan

[3]Division of Medical Engineering Research, National Health Research Institutes, Zhunan 35053, Taiwan

[*]E-mail: wangpc@ess.nthu.edu.tw

Abstract— A series of tertiary amine quinuclidine-capped cadmium selenide (Q-CdSe) quantum dots (QDs) of ~4 nm in diameter was successfully synthesized via ligand exchange process in which the original hydrophobic trioctylphosphine oxide (TOPO) ligand bound to QDs was replaced with stronger quinuclidine derivatives, quinuclidine (Q1), 3-quinuclidinone (Q2) and 3-quinuclidinol (Q3). The ligand exchange of TOPO by Q probed using the combined fluorescence and absorption spectroscopy was achieved in just only a few minutes. Moreover, disappearance of prominent C-H aliphatic stretching (~2900 cm^{-1}) and phosphate signal (35 ppm) of TOPO-capped CdSe after replacement with Q as revealed in FT-IR and solid state ^{31}P-NMR spectra was observed indicating efficient fast ligand exchange.

Keywords-ligand-exchange; CdSe quantum dots; quinuclidines; tertiary amines, ^{31}PNMR spectroscopy

I. INTRODUCTION

The use of semiconductor quantum dots (QDs) as fluorescence probes plays a significant role in the areas of medicine, chemistry and engineering as they possess remarkable advantages over conventional organic fluorophores in terms of narrow luminescence emission profiles, resistance to photobleaching, and continuous absorption above the bandgap [1]. In the case of CdSe, typical preparation of monodispersed QDs involves thermal decomposition of organometallic precursors in organic solvents using hydrophobic ligand layers that usually consist of original phosphine surface ligand, trioctylphosphine oxide (TOPO) [2]. However, this route affords hydrophobic QDs that are not soluble in aqueous and protic solvents and therefore, not useful in many biological applications such as cellular imaging [3], DNA hybridization [4], immunoassays [5], photodynamic therapy [6] and optical barcoding [7].

Several strategies have been developed to modify the surface of QDs aiming for biocompatibility, bioconjugation, nonspecific adsorption to cells, efficient fluorescence and active targeting. These include ligand exchange, one step in situ functionalization, silica coatings, addition of a second ligand layer and incorporation of QDs into polymeric particles and gels. Ligand exchange is the most facile process in which the original phosphine ligands bound to QDs are replaced with stronger ligand (e. g., amines, thiols and carboxylic acids) highly capable of binding to the QD surfaces. Many studies utilized nitrogen-containing ligands since amine moieties are ideal as they are present in peptides, proteins and nucleic acids allowing studies on their interaction and behavior with QDs in biological environments [8]. Recent reports have demonstrated the use of pyridine [1], 4-amino-2,2,6,6-tetramethylpiperidine oxide (4-amino-TEMPO) [9], 1,4-diazabicyclo[2.2.2]octane (DABCO), triethylamine (TEA) [8], poly(amidoamine) (PAMAM) dendrimers [10], N,N-dimethylaminoethyl methacrylate (PDMA) [11] as ligands. However; longer ligand exchange was reported taking place in a couple of hours to days [1,8,11].

Here, we present an effective ligand exchange of TOPO-capped CdSe with tertiary amine quinuclidines, achieved in just only a few minutes producing products of improved solubility readily dispersed in ethanol. We have examined ligand exchange of TOPO on QD with Qs in detail using the combined techniques of fluorescence and absorption profile monitoring and the FTIR, NMR, solid state ^{31}PNMR spectroscopy and TEM microscopy.

II. EXPERIMENTAL SECTION

A. Materials and Instrumentation

Reagents, quinuclidine (Q1), potassium carbonate (K$_2$CO$_3$), 1,2-dimethoxyethane (DME), sodium borohydride (NaBH$_4$), deuterium oxide (D$_2$O), 1-octadecene (ODE), tributylphosphine (TBP), cadmium oxide (CdO), oleic acid (OA), selenium (Se), octadecylamine (ODA) and pentane were purchased from Sigma-Aldrich while quinuclidinone hydrochloride (Q2 HCl) was acquired from Alfa Aesar, dichloromethane (DCM) from Fluka and ethanol from Shimakyu's Pure Chemicals. Deionized water (DI) was used for all solution preparations. Samples were centrifuged using

KUBUTA 3740. UV–vis absorption spectra were recorded on a Jasco V-670 spectrophotometer. FT-IR spectra were obtained using a Jasco FT/IR-4200 spectrometer. Fluorescence spectra were measured using Jasco FP-6500 spectrofluorometer. ^{1}H- and ^{13}C-NMR spectra were recorded using a Varian-Unity INOVA 500 NMR. Samples for ^{31}P-NMR analysis were measured using a BRUKER DSX 400 solid state NMR.

B. Synthesis of 3-Quinuclidinone (Q2)

The synthesis of Q2 was carried out following the procedure of Horenstein, et al. [12]. 10 mL of 2 M Q2 HCl (3.2 g in 10 mL DI) was added with 2 M K_2CO_3 (2.8 g in 10 mL DI). Q2 was extracted with 30 mL DME using a separatory funnel. The upper layer containing Q2 was separated from the lower layer containing KCl salt. The extraction of the remaining Q2 in the lower layer was repeated twice. The solvent in the upper layer was then removed using a rotary evaporator at 40-60 °C resulting to the waxy white powder, neutralized Q2. The purity of the product was analyzed under UV-vis, FT-IR and NMR spectroscopy.

^{1}H-NMR (500 MHz, D_2O): δ 1.852 (2H, m), 1.995 (2H, m), 2.388 (3H, qnt), 2.749 (2H, m), 2.857 (2H, m); 13C-NMR (500 MHz, D_2O): δ 24.310, 39.356, 45.490, 225.305; IR (neat) v_{max}: 1150, 1470, 1550, 1760, 2860, 2900 cm^{-1}.

C. Synthesis of 3-Quinuclidinol (Q3)

In a mixture containing 0.40 g (~0.1 mol) $NaBH_4$ in 3 mL of DI water and 3 mL of ethanol at 4 °C, 2.3 g of previously neutralized Q2 (~0.2 mol) was added while stirring [13]. Thereafter, formation of bubbles was observed. The solution was further stirred for 2 hrs at room temperature. Q3 product was then extracted with DME thrice and the extract was dried with enough amount of K_2CO_3. The drying agent was then removed by decanting the product in the translucent solution into another flask. The solvent was removed by rotary evaporator producing fine white powder Q3. Purity of samples was analyzed under UV-vis, FT-IR and NMR spectroscopy.

^{1}H-NMR (500 MHz, D_2O): δ 1.360 (1H, m), 1.425 (1H, m), 1.620 (1H, m), 1.717 (1H, m), 1.767 (1H, m), 2.380 (2H, m), 2.493 (1H, m), 2.601 (1H, m), 2.663 (2H, m), 2.995 (1H, m), 3.860 (1H, s); 13C-NMR (500 MHz, D_2O): δ 17.909, 23.578, 27.048, 45.202, 46.227, 55.586, 67.445; IR (neat) v_{max}: 1065, 1150, 1470, 1550, 2860, 2900, 3300 cm^{-1}.

D. Preparation of TOPO-capped CdSe

A mixture of 0.6 mmol CdO, 0.8mL OA and 8mL ODE was heated to 300°C under argon flow to generate a homogeneous transparent solution. Then, the mixture was cooled to 60°C. 5g ODA and 1g TOPO were added into the Cd/OA/ODE mixture and heated to 280~320°C. 3.6 mmol Se and 4mL TBP were mixed and heated to 50°C under argon flow to remove O_2 and H_2O. The temperature of the Cd solution was reduced to 280~320°C. A solution of 4 mL TBP/Se at 40°C was rapidly injected into the Cd solution. The resulting mixture was stirred for few minutes at 280°C and finally quenched with liquid N_2.

E. Ligand-Exchange of TOPO-capped CdSe with Quinuclidines (Qs)

Freshly prepared CdSe were then dispersed in DCM. 1.8 mL of this quantum dot solution was added with 0.2 mL of 10 mg/mL of Q in DME. The mixture was sonicated for every 5-min interval over a 25-min period. For each time interval, UV-vis spectra were recorded subsequent to the fluorescence measurement.

F. Purification of Q-CdSe Products for ^{31}P-NMR Analysis

1.8 mL of 75μL/mL of TOPO-capped CdSe in DCM was added with 0.2 mL of 500 mg/mL of Q2 (or Q1) in DME. The mixture was sonicated for a minute in the case of Q1 while 5 min in Q2 and Q3. The products were then precipitated with 6 mL of pentane and centrifuged at 4000 rpm at 4 °C for 20 min. All solvents were discarded while precipitates (Q-CdSe) were washed with ethyl acetate twice by centrifugation at 4000 rpm at 4 °C for 20 min. The products were oven dried at 50 °C overnight. Samples were then added with NaCl, ground into fine powders and analyzed under ^{31}P-NMR.

III. RESULTS AND DISCUSSION

A. Synthesis of Qs

A series of quinuclidine derivatives, Q1, Q2 and Q3, was utilized to assess the influence of the amine nature and basicity on the kinetics of the exchange reaction. Two compounds were synthesized in which Q2 was prepared by the base neutralization of commercially available hydrochloride salt of 3-quinuclidinone while Q3 was synthesized by the reduction of the ketone Q2 with sodium borohydride.

Results of ^{1}H-NMR and ^{13}C-NMR spectra indicated high purity of these products without unexpected signals from unreacted starting materials. Careful examinations of the FT-IR and absorption profiles showed all functional groups expected for the desired Qs.

B. Ligand-exchange of TOPO-capped CdSe with Qs

The ligand exchange of TOPO-capped CdSe with tertiary amines was achieved by equilibrating three different Qs in DME with the TOPO-capped CdSe in DCM under mild sonication for a few minutes to afford the products illustrated in Fig. 1. The exchange kinetics was monitored by measuring the fluorescence and absorption profiles for every 5-min

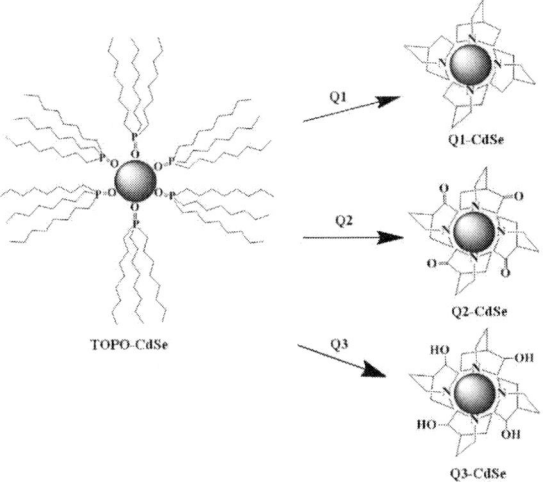

Fig. 1. Schematic diagram on the ligand exchange of TOPO-capped CdSe with quinuclidines.

interval over a 25-min period. Figs. 2a-c show the summary of the fluorescence and absorption profiles of TOPO-CdSe during the exchange reaction. For both measurements, a blue shift behavior (Figs. 2a,c) was observed attributed to the effective replacement of longer alkyl chain TOPO with shorter ligand Q. Addition of Q for the first few minutes results in an abrupt fluorescence quenching attributed to the strong amine binding to QD surface (Fig. 2b). Longer sonication results in an increase in fluorescence signal that may be attributed to the amine binding of QD to lower energy trap sites [5]. Further sonication leads to the removal of ligands from the surface, creating defect sites on QD and giving a complete quenched behavior.

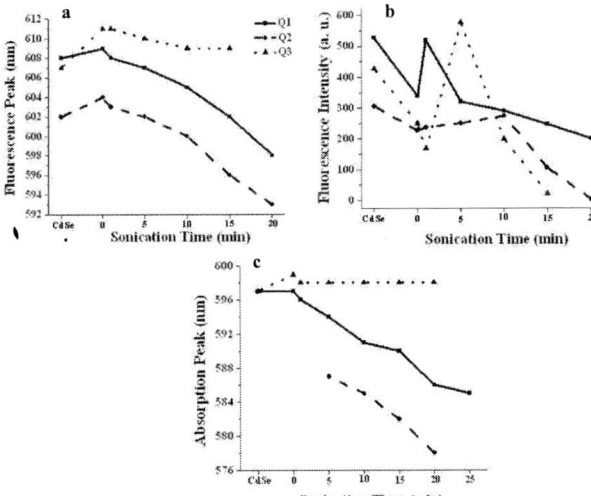

Fig. 2. Summary of the a) fluorescence peak, b) fluorescence intensity and c) absorbance peak during the ligand exchange of TOPO-CdSe with Qs.

C. UV-vis absorption, FTIR and and ^{31}P-NMR Spectra of Quinuclidine-capped CdSe

Absorption profile of the quantum dot exhibits similar trend after the ligand exchange of TOPO with Q3 (Fig. 3a). TEM analysis shows TOPO-CdSe (~4 nm in diameter) remains its size after replacement with Q3. However, some become bigger (> 5nm) due to purification process after the exchange by precipitation and centrifugation (Figs. 3b-c).

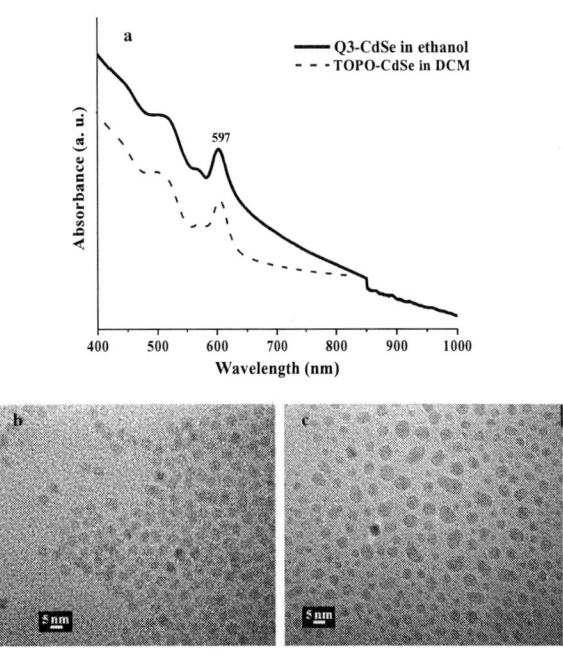

Fig. 3. a) UV-vis absorption and TEM images of b) TOPO-CdSe in DCM and c) Q3-CdSe in ethanol after 5-min ligand exchange.

Subsequent to the ligand-exchange, the quantum dot products were further analyzed using FT-IR spectrometer. It was observed that C-H stretching of TOPO's dominant aliphatic chain at ~2900 cm^{-1} disappears after exchange with Q, giving rise to another signal at ~1150 cm^{-1} showing C-N stretching of tertiary amine. Presence of O-H stretch for all Q-CdSe products was also found that is due to the hygroscopic property of the samples.

Moreover, analysis on solid state ^{31}P-NMR shows disappearance of phosphine signal at 35 ppm after replacement reaction. The most effective ligand exchange of TOPO by Qs happened at 1-min sonication for Q1 sample, followed by Q3 and Q2 samples for sonication time (ligand-exchange time) of at least 5 or10 min.

978-1-4673-1122-9/12 $31.00 © 2012 IEEE

IV. CONCLUSION

In summary, we have successfully prepared the quinuclidine derivatives and utilized these as ligands to replace the hydrophobic TOPO on quantum dot CdSe surface. Based from the fluorescence and absorption profiles, ^{31}PNMR and TEM results, the exchange gives efficient fast replacement resulting products of improved solubility and nanoparticle stability. The replacing ligands, consisting of amine functional group that binds to metal surfaces and an exposed hydrophilic carbonyl or hydroxyl group at the other side, can offer bifunctionality for improved water solubility, biocompatibility and bioconjugation.

ACKNOWLEDGMENT

We acknowledge the financial support from National Science Council of Taiwan (grant numbers: 98-2113-M-007-017, 99-2113-M-007-013 and NSC-99-2113-M-007-010). We also thank Miss Su-Yun Fang for the assistance of the solid state ^{31}PNMR analysis.

REFERENCES

[1] T. Emrick, and H. Skaff, "The use of 4-substituted pyridines to afford amphiphilic, pegylated cadmium selenide nanoparticles," *Chem. Commun.*, pp. 52-53, 2003.

[2] J. K. Oh, "Surface modification of colloidal CdX-based quantum dots for biomedical applications," *J. Mater. Chem.*, vol. 20, pp. 8433-8445, 2010.

[3] X. Y. Wu, H. J. Liu, J. Q. Liu, K. N. Haley, J. A. Treadway, J. P. Larson, N. F. Ge, F. Peale, and M. P. Bruchez, "Immunofluorescent labeling of cancer marker Her2 and other cellular targets with semiconductor quantum dots (vol 21, pg 41, 2003)," *Nat. Biotechnol.*, vol. 21, pp. 452-452, April 2003.

[4] T. H. Wang, Y. P. Ho, M. C. Kung, and S. Yang, "Multiplexed hybridization detection with multicolor colocalization of quantum dot nanoprobes," *Nano Lett.*, vol. 5, pp. 1693-1697, September 2005.

[5] H. S. Kim, E. Oh, M. Y. Hong, D. Lee, S. H. Nam, and H. C. Yoon, "Inhibition assay of biomolecules based on fluorescence resonance energy transfer (FRET) between quantum dots and gold nanoparticles," *J. Am. Chem. Soc.*, vol. 127, pp. 3270-3271, March 16 2005.

[6] A. J. MacRobert, E. Yaghini, and A. M. Seifalian, "Quantum dots and their potential biomedical applications in photosensitization for photodynamic therapy," *Nanomedicine-Uk*, vol. 4, pp. 353-363, April 2009.

[7] D. Y. Wang, M. Kuang, H. B. Bao, M. Y. Gao, H. Mohwald, and M. Jiang, "Fabrication of multicolor-encoded microspheres by tagging semiconductor nanocrystals to hydrogel spheres," *Adv. Mater.*, vol. 17, pp. 267-270, February 2005.

[8] J. C. Scaiano, and R. E. Galian, "Fluorescence quenching of CdSe quantum dots by tertiary amines and their surface binding effect," *Photoch. Photobio. Sci.*, vol. 8, pp. 70-74, 2009.

[9] J. C. Scaiano, V. Maurel, M. Laferriere, P. Billone, and R. Godin, "Free radical sensor based on CdSe quantum dots with added 4-amino-2,2,6,6-tetramethylpiperidine oxide functionality," *J. Phys. Chem. B*, vol. 110, pp. 16353-16358, August 2006.

[10] L. Balogh, C. X. Zhang, and S. O'Brien, "Comparison and stability of CdSe nanocrystals covered with amphiphilic poly(amidoamine) dendrimers," *J. Phys. Chem. B*, vol. 106, pp. 10316-10321, October 2002.

[11] M. A. Winnik, M. F. Wang, J. K. Oh, T. E. Dykstra, X. D. Lou, and G. D. Scholes, "Surface modification of CdSe and CdSe/ZnS semiconductor nanocrystals with Poly(N,N-dimethylaminoethyl methacrylate)," *Macromolecules*, vol. 39, pp. 3664-3672, May 2006.

[12] N. A. Horenstein, F. M. Leonik, and R. L. Papke, "Multiple Pharmacophores for the Selective Activation of Nicotinic alpha 7-Type Acetylcholine Receptors," *Mol. Pharmacol.*, vol. 74, pp. 1496-1511, December 2008.

[13] M. R. Johnson, and B. Rickborn, "Sodium Borohydride Reduction of Conjugated Aldehydes and Ketones," *J. Org. Chem.*, vol. 35, pp. 1041-1045, 1970.

A Study of Nano-structured Manganese Dioxides and Their Composites as Electrode Materials for Micro Supercapacitors

Siwei Li[1,2], Xiaohong Wang[1,2], Caiwei Shen[1,2], Jian'gan Wang[1,3], and Feiyu Kang[1,3]

[1]Tsinghua National Laboratory for Information Science and Technology
[2]Institute of Microelectronics, Tsinghua University, China
[3]Department of Materials Science and Engineering, Tsinghua University, China
wxh-ime@tisnghua.edu.cn

Abstract—**This paper presents the design, synthesis and characterization of nano-structured manganese dioxides (MnO₂) and their composites as electrode materials for micro supercapacitors. An optimized composite containing nano-MnO₂ powder exhibits a specific capacitance as high as 160F/g in 0.2M K₂SO₄ aqueous solution, and is compatible with micro fabrication process. Then a micro supercapacitor with 3D interdigital structure using the composite material as electrodes is achieved by MEMS fabrication technology. The micro supercapacitor has well-performed capacitive behavior under various scanning rates and has a large specific capacitance of 30mF/cm², which indicates that the composite is a promising material for energy storage on chips.**

Keywords- nano-MnO₂; liquid-phase redox; micro supercapacitor

I. INTRODUCTION

Supercapacitors, i.e. electrochemical capacitors, are a kind of energy storage devices based on non-faradic electric double layer capacitance (EDLC) and faradic pseudocapacitance [1]. Supercapacitors have attracted intensive research interests in recent years due to their high charge/discharge rate, long cycle life and high energy efficiency compared with traditional rechargeable batteries. The energy density of supercapacitors is generally lower than batteries, but with the development of advanced materials, reasonable energy density close to that of batteries can be achieved while high power density is still retained [2].

Micro supercapacitors are orienting on-chip power system applications which can be expected in areas such as energy harvesting systems, energy storage units and power suppliers for micro sensors, electric devices, etc.. Notable work has been developed on both electrode structures and advanced electrode materials of micro supercapacitors to improve the performance in a limited area. Compared with planar structures [3,4], 3D electrodes with good extensibility enhance the performance greatly [5-8]. To take full advantages of 3D electrode structures, several qualities of the electrode materials are required, including mechanical stability, active material loading mass and utilization rate of the materials. In a word,

how to apply advanced materials to achieve efficient electrode structures is a challenge for micro supercapacitors.

Generally, pseudocapacitance materials provide much higher specific capacitance, which means higher energy density, than EDLC, Typical materials with pseudocapacitive behavior are various kinds of transition metal oxides and conducting polymers. MnO₂ is a kind of promising pseudocapacitive materials. It has been intensively studied in recent years since the first report published in 1999 [9] due to its high theoretical specific capacitance. MnO₂ is also abundant, environment friendly and can be easily derived by many simple synthesize routines [10]. With these merits, it's an important material with great potential in energy storage use. However, Seldom works on applying MnO₂ in micro supercapacitors has been reported probably because of the issues including the lack of mechanical stability and the low loading mass.

This paper studies the nano-MnO₂ and its composites as electrode materials for micro supercapacitors. Different nano-MnO₂ samples were synthesized by liquid-phase redox reactions. Characterizations including XRD, SEM and electrochemical test were carried out. One composite was demonstrated as the optimized material due to its overall properties. A micro supercapacitor device was also fabricated with the composite as electrodes and then tested.

II. EXPERIMENTAL

A. Synthesis of nano-MnO₂

Three samples, we called sample A, B and C, were synthesized by co-precipitation liquid-phase redox reactions.

The sample A was conducted from a reaction between KMnO₄ and MnSO₄ as follows. 0.474g KMnO₄ was dissolved in 60mL DI water and 0.65g ammonia (25% wt.) was droped into the solution. Then a solution containing 0.760g MnSO₄·4H₂O in 90mL DI water was added to the previous one. Thus, the mole ratio of Mn⁷⁺ to Mn²⁺ in the reactant is 2:3. After a 40min-reaction under ultrasound agitation, the pH value of the final solution was about 5.5, i.e. a weak acid medium. The precipitate was then filtered, washed carefully, and dried at 60℃ finally.

In order to increase the specific capacitance and conductivity, we tried to add CNT during the process to synthesize sample B. 0.474g $KMnO_4$ was dissolved in 60mL DI water, 0.48g KOH and 0.165g carbon nanotube (CNT) were added to the $KMnO_4$ solution. 0.760g $MnSO_4 \cdot 4H_2O$ in 90mL DI water was then mixed with the solution above. Reaction under vigorous mechanical stirring lasted for 6 hours. The precipitate was seperated from the solution by filtration, washed and dried to attain MnO_2-CNT composite.

To make better dispersal of nano-particles, the sample C was gained by a reaction between $KMnO_4$ and polyethylene glycol (PEG 400). PEG was expected to serve as both reductant and dispersant in the process. Thus 4.00g $KMnO_4$ was dissolved in 400mL DI water and PEG was dropped to the solution while stirring. After the fuchsia color of the solution was faded, several more drops were added to ensure a fully transformation from $KMnO_4$ to MnO_2. Then the precipitate was filtered, washed with both ethanol and DI water and dried finally.

B. Characterization

To evaluate the properties of the materials synthesized, XRD and SEM were used to characterize their crystalline and nano structure. Electrodes for electrochemical tests were prepared as follows. A composite mixing 75% MnO_2 sample, 20% acetylene black and 5% polyvinylidene fluoride (PVDF) was dispersed in N-methyl pyrrolidone (NMP) under mechanical stirring and ultrasonic agitation to form slurry. Ni foam (about $1cm^2$) served as a current collector and the slurry was coated on it to make the test electrode. The electrode was finally dried to evaporate the NMP solvent.

The electrochemical tests were carried out by cyclic voltammetry (CV) method in a three-electrode configuration with the test electrode as the working electrode, a Pt foil as the counter electrode and Ag/AgCl electrode as the reference electrode. 0.2M K_2SO_4 solution was used as electrolyte.

C. Frabrication of the device

The micro supercapacitor illustrated in figure 1 was fabricated. The fabrication process to achieve such a device with self-supported interdigital electrodes is briefly described as follows [8]. Silicon substrate was etched by ICP to form 3D interdigital channels. An insulation layer and then a metal layer were deposited and partially etched, so that the metal served as current collector at the bottom. The self-supported electrode material consisting of one of the MnO_2 samples as active material, acetylene black as conductive agent and PVDF as binder in a weight proportion of 80:10:10 was filled into the channels. The silicon wall that separated the interdigital electrodes was finally etched and the space was left for the electrolyte. The prototype was then sealed by a PDMS cap, with electrolyte (0.2M K_2SO_4) injected into the effective area. The thickness of stacked electrodes is

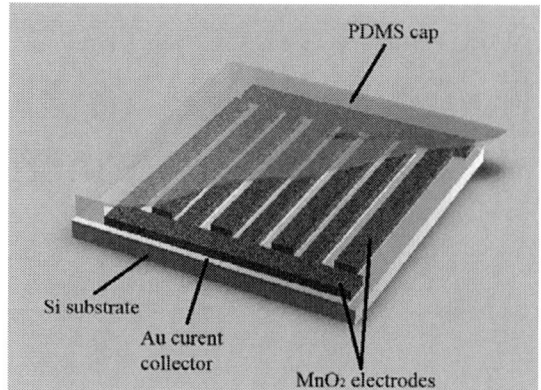

Fig. 1. Schematic to illustrate the 3D structure of a symmetric micro supercapacitor with MnO_2 electrode material.

dependent on the fabrication process, and in this case, it was about 5-7 micrometers.

III. RESULTS AND DISCUSSIONS

A. Characteristics of synthesized smaples

Figure 2 shows the XRD patterns of the three samples above. The board peaks in each pattern indicate that all the three samples are mainly amorphous MnO_2. Generally, samples synthesized through liquid-phase redox under room temperature are amorphous. The peak at about 26.5 degree in figure 2(b) is due to the CNT additive.

SEM images of the as-prepared samples are shown in figure 3 (a)~(c).

Figure 3(a) shows the image of sample A. Aggregation can

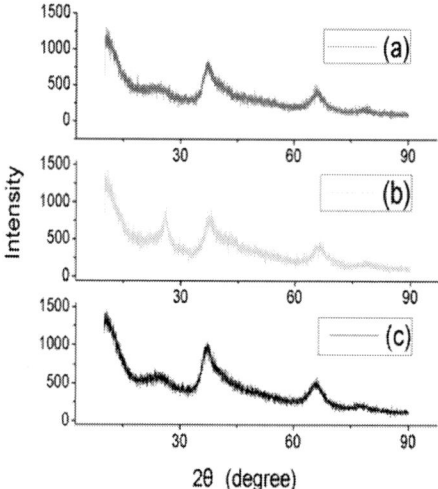

Fig. 2. XRD patterns of MnO_2 powder synthesized by (a) reacting $KMnO_4$ with $MnSO_4$ (sample A), (b) reacting $KMnO_4$ with $MnSO_4$ adding CNT (sample B) and (c) reacting $KMnO_4$ with PEG (sample C).

Fig. 3. SEM images of MnO_2 synthesized by (a) reacting $KMnO_4$ with $MnSO_4$ (sample A), (b) reacting $KMnO_4$ with $MnSO_4$ and adding CNT (sample B), (c) reacting $KMnO_4$ with PEG (sample C), and (d) composite of sample C with acetylene black and PVDF added.

be observed and the nano-particles are not fully dispersed. Figure 3(b) shows the image of sample B. Some MnO_2 nano-particles adhered to CNT, but the CNT were not fully dispersed. The entanglement and the stress introduced by CNT during drying process leaded to mechanical damage to the electrodes. Desquamation of the composite can be seen on both foam Ni current collector and Si/Au substrate. In figure 3(c), the last sample was shown. The nano-particles were dispersed and the average diameter was about 50~80nm.

Figure 3(d) shows the SEM image of the self-supported electrode material. The mixture of MnO_2 nano-particles and acetylene black can be seen in the image.

CV tests were carried out at a scan rate of 5mV/s in a potential range of 0.0~1.0V vs. Ag/AgCl. For sample A, we got only a low specific capacitance of about 90F/g, which may be a result of the aggregation. The capacitance of sample B is about 140F/g, which is higher than sample A due to the enhanced surface area and conductivity brought by CNT. For sample C, the capacitance was a little higher than B..The CV curve of sample C is shown in figure 4. In the range of 0.0~0.9V, the nearly rectangular shape of the curve verifies a nearly ideal capacitive behavior. At about 1.0V, a peak appears due to the voltage window of water [1]. From the CV

curve, a specific capacitance of 160F/g for the composite and about 200F/g for pure MnO_2 are calculated.

In summary, sample A provided low specific capacitance due to a large agglomeration of nano-particles. For sample B, mechanical stability was a big challenge when used as electrodes. Sample C exhibited reasonable specific capacitance, and the composite containing these dispersed nano-particles can be adhered on the substrate with enough

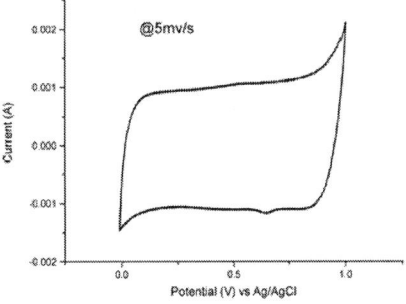

Fig. 4. Cyclic voltammetric curve of the composite containing MnO2 sample C. The test is based on a three-electrode configuration and the working voltage range is 0.0~0.9V vs. Ag/AgCl.

mechanical strength. Therefore, it was the most suitable active material for the self-supported electrode material and it was selected to develop the device.

B. Characterizations of the device

The device was tested by cyclic voltammetry method and impedance spectrum.

Figure 5 shows the CV curves at different scan rates of 20mV/s, 50mV/s and 100mV/s. At a scan rate of 20mV/s, the shape of the curve is nearly rectangle, which indicats good capacitive behavior. At higher scan rates, the capacitive behavior is still preserved. A specific capacitance of 30mF/cm^2 is calculated at the scan rate of 20mV/s and a little decrease in the value can be observed at higher scan rates.

Nyquist plot based on impedance test is shown in Figure 6. The nearly vertical line at low frequency is also good demonstration of capacitive nature.

The test results above including capacitive behavior and high capacitance indicate fine capability of the device, and the feasibility of the material developed in this work.

IV. CONCLUTIONS

To develop a well-performed MnO$_2$ material for energy storage use, nano-MnO$_2$ was synthesized by liquid –phase redox. Production by reacting KMnO$_4$ with MnSO$_4$ in ammonia showed relative low specific capacitance. By adding CNT into the reactant to increase the capacitance, electrochemical performance improved but the poor mechanical stability made it unsuitable for self-supported electrodes. MnO$_2$ derived by reacting KMnO$_4$ with PEG showed a specific capacitance of 200F/g, high capacitance, reliable stability and good compatibility with micro fabrication made it an optimized material for on-chip use.

The optimized sample was then used as the active material in the self-supported electrode of a micro supercapacitor prototype. The device showed good capacitive behavior and a large specific capacitance of 30mF/cm^2. As such, the material

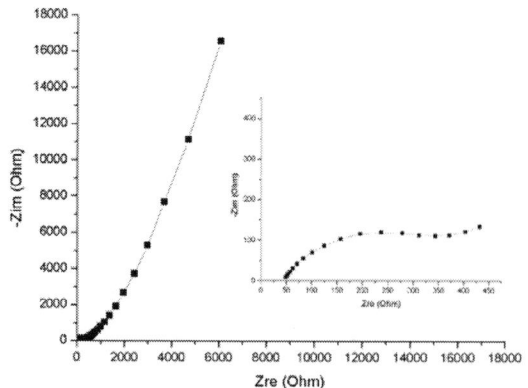

Fig. 6. Nyquist plot of the two-electrode device.

in this work is applicable to on-chip energy storage and the micro supercapacitor is promising in various energy systems.

ACKNOWLEDGMENT

This work is supported by the National Natural Science Foundation of China (No. 60936003), 973 program (No. 2009CB320304), and 863 program (No. 2009AA04Z319) of China.

REFERENCES

[1] B.E Conway, *Electrochemical Supercapacitors: Scientific Fundamentals and Technological Applications,* Kluwer Academic/Plenum Publishers, 1999.

[2] P. Simon and Y. Gogotsi, "Materials for electrochemical capacitors", *Nat. Mater.,* vol. 7, pp. 845~854,2008.

[3] H. J. In, S. Kumar, S.H. Yang, G. Barbastathis, "Origami fabrication of nanostructured, three-dimensional devices: Electrochemical capacitors with carbon electrodes", *Appl. Phys. Lett.,* vol. 88, 083104, 2006.

[4] H. X. Ji, Y. F. Mei, O. G. Schmidt, "Swiss roll nanomembranes with controlled proton diffusion as redox micro-supercapacitors", *Chem. Commun.,* vol. 46, pp. 3881-3883, 2010.

[5] W. Sun, X. Chen, "Fabrication and tests of a novel three dimensional micro supercapacitor", *Microelectron. Eng.,* vol. 86, pp. 1307-1310, 2009.

[6] D. Pech, et al., "Elaboration of a microstructured inkjet-printed carbon electrochemical capacitor", *J. Power Sources,* vol. 195, pp. 1266-1269, 2010.

[7] Y. Q. Jiang, Q. Zhou, L. Lin, "Planar MEMS supercapacitor using carbon nanotube forests", *Proc. MEMS 2009,* pp. 587-590, 2009

[8] C. Shen, et al, "A high-performance three-dimensional micro supercapacitor based on self-supporting composite materials", *J. Power Sources,* vol. 196, pp. 10465-10471, 2011.

[9] H. Y. Lee and J. B. Goodenough, "Supercapacitor behavior with KCL electrolyte", *J. Solid State Chem.,* vol. 144, pp. 220-223, 1999.

[10] W. Wei, et al., "Manganese oxide-based materials as electrochemical supercapacitor electrodes", *Chem. Soc. Rev.,* vol. 40, pp. 1697-1721, 2011.

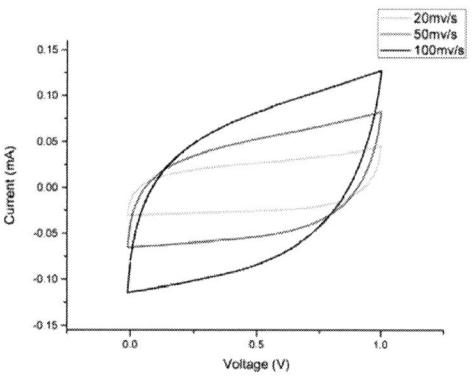

Fig. 5. Cyclic voltammetric curves of the two-electrode device at different scan rates.

CoNiMnP-AAO Hard Magnetic Nanocomposite Film for MEMS Applications

Tzu-Yuan Chao, Jian-Ruei Lin, and Y. T. Cheng*, *Senior Member, IEEE*

*Microsystems Integration Laboratory, Department of Electronics Engineering & Institute of Electronics,
National Chiao Tung University, HsinChu, Taiwan R.O.C.
ytcheng@g2.nctu.edu.tw

Abstract—This paper presents a novel permanent magnetic CoNiMnP-AAO nanocomposite film comprising of a CoNiMnP nanowire array with a high aspect ratio (>150). The nanocomposite synthesized by a low temperature electroplating process (at 30°C) with AAO template can exhibit a unique magnetic characteristic of vertical shape anisotropy which results in a coercivity (H_c) of 2512Oe, a remanence (B_r) up to 3616Gauss and a maximum energy product, $(BH)_{max}$, of 12.02kJ/m³. Such a simple fabrication process and excellent magnetic property reveal that permanent magnetic CoNiMnP-AAO nanocomposite is with great potential for magnetic N/MEMS applications.

Keywords-AAO; CoNiMnP; hard magnet; nanocomposite

I. INTRODUCTION

Hard magnets, or permanent magnets, are critical components to provide an external magnetic field in MEMS device for microactuation [1-3], magnetic field sensing [4] and microscale magnetic power generation [5]. For magnetic MEMS applications, such a hard magnet should exhibit the following characteristics: (1) high coercivity, H_c, to prevent a reversal of the magnetic moment, (2) high remanence, B_r, for provide a magnetic field without external power, (3) high maximum energy product, $(BH)_{max}$, for supplying most magnetic energy for actuation and (4) process integration capability. Since the device performance of permanent magnet-based MEMS device is usually limited by the magnet capability owing to its minute size, how to realize a micromagnet with high magnetic properties using conventional wafer-level fabrication processes to make itself fully integrated with N/MEMS has become a critical research topic in recent years [1].

Conventional deposition methods for micromagnet fabrication include physical vapor deposition (sputtering and pulsed laser deposition) and electrochemical deposition (electroplating). Although PVD can provide various material choices like high performance rare-earth transition metal alloy, SmCo and NdFeB [6,7], in comparison with the electroplating technique, its application in N/MEMS society is still limited because the thickness of the magnets is only several micrometers resulted by low deposition rate (~3 nm/sec). In addition, process difficulty in etching and patterning of these magnetic film and the necessity of high temperature thermal

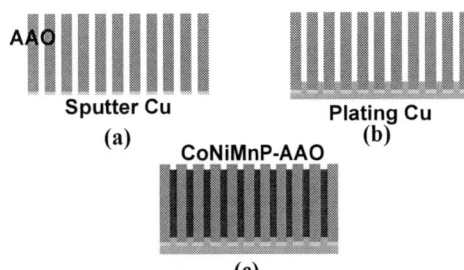

Fig. 1. Illustration of fabrication process. (a) A Cr/Cu layer sputtering onto one side of AAO template. (b) Electroplating 5μm Cu to cover nanopores. (c) CoNiMnP electrodeposition in low pressure chamber.

TABLE I. COMPOSITION OF CoNiMnP ELECTROPLATING BATH AND EXPERIMENT PARAMETERS.

$CoCl_2 \cdot 6H_2O$	24 g/l	$C_{12}H_{25}O_4NaS$	3 g/l
$NiCl_2 \cdot 6H_2O$	24 g/l	Saccharin	0.9 g/l
$MnSO_4 \cdot H_2O$	3.4 g/l	pH	3.6
NaH_2PO_2	4.4 g/l	Current Density	4mA/cm²
H_3BO_3	25 g/l	Temperature	30°C
NaCl	24 g/l	Agitation	200 rpm

anneal (>400°C) for better magnetic characteristics would make the PVD process not practical for N/MEMS fabrication. Therefore, several metal-alloy magnets, FePt [8], CoPt [9], and CoNiMnP [10-12], have been widely studied by electrodeposition owing to the characteristics of low fabrication cost, low process temperature, wide thickness range and simplicity process.

Among these magnetic materials, CoNiMnP can provide various material and process advantages including strong magnetic properties in the as-deposition state and relative low cost in comparison with FePt and CoPt. Meanwhile, previous study has shown patterned CoNiMnP film can have a larger H_c resulted by the reduction of demagnetization effect [10]. The result also indicated that the magnetic properties of CoNiMnP could be further improved by continuously increasing the aspect ratio of the film. Therefore, the electrodeposition of nano-size hard magnet array using template substrate is studied in this work.

Anodic alumina oxide (AAO) is a common template material for synthesis of nanowire array owing to its self-ordering behavior and wide range of the interpore distance, pore diameter and length. It was found that the H_c and B_r of

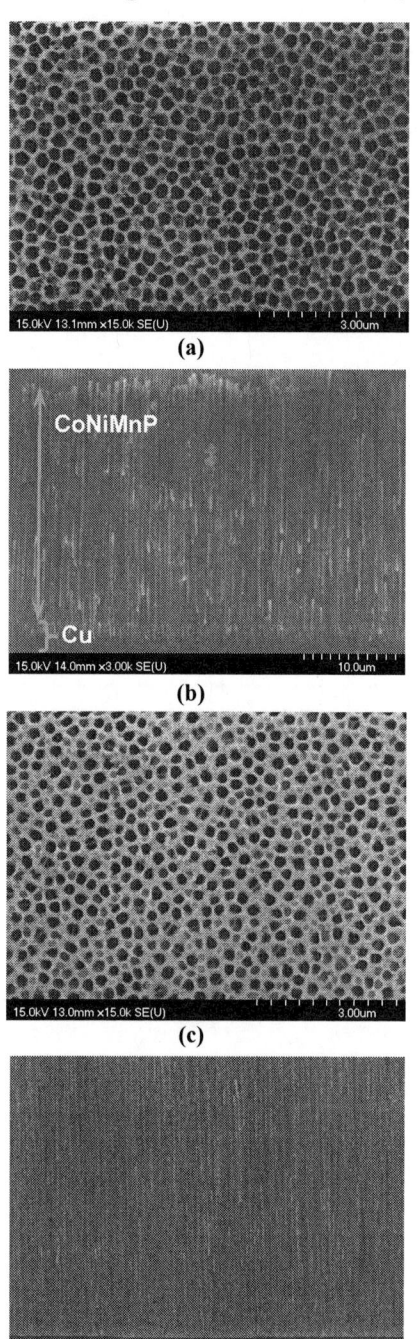

Fig. 2. The top view and cross section view SEM micrographs of as-deposited CoNiMnP-AAO magnetic nanocomposite film with (a), (b) 200nm and (c), (d) 20nm nominal pore diameter, respectively.

nickel nanowire arrays formed in an anodic alumina oxide template could increase with the diameter reduction of the nanowires [13], i.e. reducing pore diameter of AAO template. For the case where the magnetic easy axis is along the wire axis, the dependence of H_c on radius of nanowire is derived as follows [14]:

$$H_c = \frac{2\pi k A}{M_s} \frac{1}{r^2} + \frac{2K_u}{M_s} \tag{1}$$

where k, A, r, M_s, and K_u are material shape constant (1.08 for an infinite cylinder), exchange stiffness, radius of the nanowire, saturation magnetization, and uniaxial first anisotropy constant, respectively. Thus, the H_c of magnetic nanowire array could be adjusted by controlling the diameter of nanowire. On the other hand, the remanence ratio (M_r/M_s) could also approach to 1 when decreasing the diameter due to larger magnetic anisotropy which also results in a larger B_r simultaneously. Since two important factors of hard magnet, H_c and B_r, could be improved at the same time while patterning thin film into nanowire array, it is our belief that such a hard magnetic nanocomposite would provide promising magnetic behavior for magnetic MEMS devices. Therefore, in this work, a hard magnetic nanocomposite is firstly synthesized by electroplating CoNiMnP into a commercial AAO template and characterized using vibrating sample magnetometer (VSM) for MEMS applications.

II. NANOCOMPOSITE FILM SYNTHESIS

In this experiment, CoNiMnP-AAO nanocomposite films are synthesized by electroplating CoNiMnP alloy into the nanopores of commercial AAO templates (Whatman, Anodisc 47, 60μm thickness) with 20nm (Sample A) and 200nm (Sample B) nominal pore diameter, respectively. Fig. 1 shows the process flow of nanocomposite film synthesis. The film synthesis begins with a sputtered Cr (100nm)/Cu (150nm) layer on one side of AAO template which is attached to a 4" Si handle wafer, followed by the electrodeposition of a 5μm thick Cu on the template to form a continuous Cu seeding layer. The AAO template is then electroplated with CoNiMnP in the bath with the composition and plating parameters listed in

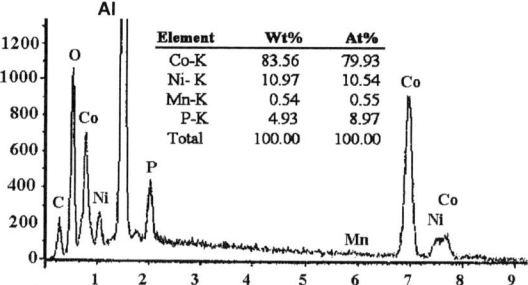

Fig. 3. The EDS analysis of CoNiMnP-AAO nanocomposite. The composition of CoNiMnP is 83.56 wt% of Co, 10.97 wt% of Ni, 4.93 wt% of P and 0.54 wt% of Mn.

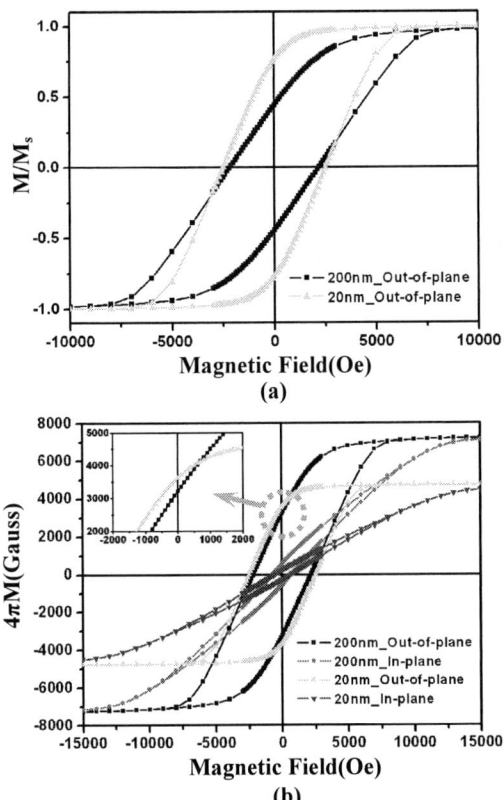

Fig. 4. The M-H hysteresis loops of CoNiMnP-AAO nanocomposite with different pore diameter (20nm and 200nm) and different direction of applied magnetic field (in-plane and out-of-plane). The M-H loops are normalized with (a) the saturation magnetization and with (b) the volume of nanocomposite film, respectively.

Table I to form a CoNiMnP nanocomposite film. Prior to CoNiMnP electroplating, it is noted that the AAO template is dipped into de-ionized water tank in a low-pressure chamber for preventing the existence of trapped bubbles in the high aspect ratio nanopores. The wetted AAO template was then immersion into the CoNiMnP bath for the plating process.

Fig. 2 shows the top and cross sectional SEM views of as-

deposited CoNiMnP-AAO nanocomposite films. Since the CoNiMnP nanowires are uniformly grown within both the 200nm and 20nm pore diameter AAO template, respectively, it reveals that the plating composition as listed in Table I and condition are good for the nanocomposite synthesis. The actual pore diameters of 200nm and 20nm AAO templates are measured by SEM. In fact, the pore diameters are about 235 ± 45nm and 185 ± 15nm, respectively. The composition of $Co_{83.5}Ni_{11}Mn_{0.5}P_5$ is verified by energy dispersive spectrometer analysis as shown in Fig. 3.

III. RESULTS AND DISCUSSIONS

The hysteresis loops of the CoNiMnP-AAO nanocomposite films are measured by VSM with out of plane and in plane applied magnetic field at room temperature as shown in Fig. 4. The measured magnetic momentums are firstly normalized with M_s in order to compare the difference of magnetic behavior between sample A and sample B as shown in Fig. 4(a). The H_c and M_r/M_s ratio of sample A is obviously larger than that of sample B. It means the smaller the AAO nanopores diameter is, the higher the vertical anisotropy can be achieved. Similar trend was also observed in the previous work about Ni nanowire arrays [13]. In the work, the measured magnetic momentums are normalized with the volume of measured composite film instead of the volume of CoNiMnP nanowires to extract magnetic properties as shown in Fig. 4(b). Table II summarizes the perpendicular magnetic performance of the CoNiMnP-AAO nanocomposite films. Although sample B has larger saturated magnetization than that of sample A due to larger porosity, i.e. more weight percent of the magnetic material in the composite, sample A has higher B_r than that of sample B, as shown in the enlarged inset figure of Fig. 4(b), because of high squareness. Furthermore, high H_c and $(BH)_{max}$ also make sample A more suitable for MEMS applications.

In comparison with the state-of-the-art CoNiMnP films as listed in Table III, the composite film can provide the best ever performance for MEMS applications. Moreover, it could be expected that the magnetic properties of CoNiMnP-AAO can be further optimized. Instead of using a commercial AAO

TABLE II. SUMMARY OF PERPENDICULAR MAGNETIC PROPERTIES OF CoNiMnP-AAO NANOCOMPOSITES.

	Diameter(nm)	Thickness(μm)	H_c(Oe)	$4\pi M_s$(Gauss)	B_r(Gauss)	$(BH)_{Max}$(kJ/m^3)
Sample A	185±15	35	2512	4730	3616	12.02
Sample B	235±45	33	2171	7132	3218	8.42

TABLE III. COMPARISONS BETWEEN PREVIOUS ARTS AND THIS WORK [10-12].

	Material	Thickness(μm)	H_c(Oe)	B_r(Gauss)	$(BH)_{Max}$(kJ/m^3)
Ref. 10	CoNiMnP	10~45	900~1300	2000~3000	14[*]
Ref. 11	CoNiMnP	2~40	1200~2000[**]	700~1100	1.8~2.6
Ref. 12	CoNiMnP	2	2262	2200	5.3
This Work (Sample A)	CoNiMnP	35	2512	3616	12.02

*Calculated by internal magnetic field (H_{int}) instead of applied magnetic field (H_{appl}).

**Measured from *B-H* loop, not *M-H* loop.

template, the pore diameter and porosity of the AAO template can be well controlled through chosen of electrolyte and process conditions, such as applied voltage and temperature, in the formation of AAO using an Al foil or thick evaporated Al film on Si substrate. Increasing the porosity of AAO can allow more CoNiMnP incorporation in the composite film to have higher M_s and, meanwhile, decreasing pore diameter to tens nanometer could expect to provide larger H_c and M_r/M_s ratio. On the other hand, the further investigation is required regarding the integration of CoNiMnP-AAO nanocomposite film with MEMS device fabrication process and the variation in magnetic properties with different thickness of CoNiMnP.

IV. CONCLUSION

CoNiMnP-AAO magnetic nanocomposite film has been successfully synthesized by electrodeposition of CoNiMnP alloy into a commercial AAO template. Such a magnetic nanocomposite reveals its superior performance for MEMS microactuator fabrication in terms of excellent magnetic properties, low manufacture cost and low process temperature.

ACKNOWLEDGMENT

This work is supported by the National Science Council of Taiwan, R. O. C. under Grant NSC 100-2220-E-009-007 and in part by the Ministry of Education in Taiwan under ATU Program. The authors would like to thank Prof. Tseng with Department of Materials Science and Engineering at National Chiao Tung University for the support of vibrating sample magnetometer measurement, and Nano Facility Center of National Chiao Tung University for the support of manufacturing facilities.

REFERENCES

[1] D. P. Arnold and N. Wang, "Permanent magnets for MEMS," *J. Microelectromech. Syst.*, vol. 18, no. 6, pp. 1255-1266, 2009.

[2] H. J. Cho and C. H. Ahn, "A bidirectional magnetic microactuator using electroplated permanent magnet array," *J. Microelectromech. Syst.*, vol. 11, no. 1, pp. 78-84, 2002.

[3] J. Sutanto, P. J. Hesketh, and Y. H. Berthelot, "Design, microfabrication and testing of a CMOS compatible bistable electromagnetic microvalve with latching/unlatching mechanism on a single wafer," *J. Micromech. Microeng.*, vol. 16, pp. 266, 2006.

[4] D. Ettelt, P. Rey, M. Savoye, C. Coutier, M. Cartier, O. Redon, M. Audoin, A. Walther, P. Robert, Y. Zhang, F. Dumas-Bouchiat, N.M. Dempsey, and J. Delamare, "A new low consumption 3D compass using integrated magnets and piezoresistive nano-gauges," in *TRANSDUCERS 2011*, 5-9 June 2011, pp. 40-43.

[5] D. P. Arnold, "Review of microscale magnetic power generation," *IEEE Trans. Magn.*, vol. 43, pp. 3940, 2007.

[6] A. Walther, D. Givord, N. M. Dempsey, K. Khlopkov, and O. Gutfleisch, "Structural, magnetic, and mechanical properties of 5 μm thick SmCo films suitable for use in microelectromechanical systems," *J. Appl. Phys.*, vol. 103, no. 4, pp. 043911, 2008.

[7] N. M. Dempsey, A. Walther, F. May, D. Givord, K. Khlopkov, and O. Gutfleisch, "High performance hard magnetic Nd–Fe–B thick films for integration into micro-electro-mechanical systems," *Appl. Phys. Lett.*, vol. 90, no. 9, pp. 092509, 2007.

[8] M. Nakano, S. Shibata, T. Yanai, and H. Fukunaga, "Anisotropic properties in Fe–Pt thick film magnets," *J. Appl. Phys.*, vol. 105, no. 7, pp. 07A732, 2009.

[9] N. Wang and D. P. Arnold, "Thick electroplated Co-rich Co–Pt micromagnet arrays for magnetic MEMS," *IEEE Trans. Magn.*, vol. 44, no. 11, pp. 3969-3972, 2008.

[10] T. M. Liakopoulos, W. Zhang, and C. H. Ahn, "Micromachined thick permanent magnet arrays on silicon wafers," *IEEE Trans. Magn.*, vol. 32, no. 5, pp. 5154-5156, 1996.

[11] S. Guan and B. J. Nelson, "Electrodeposition of low residual stress CoNiMnP hard magnetic thin films for magnetic MEMS actuators," *J. Magn. Magn. Mater.*, vol. 292, pp. 49-58, 2005.

[12] M. D. Grapes and C. J. Morris, "Improving the CoNiMnP electrodeposition process using taguchi design of Experiments," *J. Electrochem. Soc.*, vol. 157, no. 12, pp. D642-D647, 2010.

[13] K. Nielsch, R. B. Wehrspohn, J. Barthel, J. Kirschner, U. Gösele, S. F. Fischer, and H. Kronmüller, "Hexagonally ordered 100nm period nickel nanowire arrays," *Appl. Phys. Lett.*, vol. 79, no. 9, pp. 1360-1362, 2001.

[14] J. Mallet , K. Y. Zhang , C. L. Chien , T. S. Eagleton, and P. C. Searson, "Fabrication and magnetic properties of fcc Co_xPt_{1-x} nanowires," *Appl. Phys. Lett.*, vol. 84, pp. 3900, 2004.

An Adhesion Strength Measurement Method for Particle Transfer and Assembly in Dry Environment

Sho-Fu Wang, Yen-Ting Lin, Yan-Bo Lin, Kerwin Wang

Department of Mechatronics Engineering, National Changhua University of Education, Changhua, Taiwan

kerwin@cc.ncue.edu.tw

Abstract—**This research presents a new and noncontact method to measure the adhesion strength between micro-particles and substrates in parallel. It is cost effective and time saving. In this study, the adhesion strength between micro solder ball and PDMS (polydimethylsiloxane) and ReproRubber®️ (polymethylhylhydrogen) are investigated. The setup is capable of evaluating more than 8 particles separately each run. The experiment results indicated that the adhesion strength is highly dependent on surface conditions, parts geometry and material properties. The testing results also have been applied to successive parts transferring and assembling.**

Keywords- PDMS; adhesive strength; adhesion; assembly; flip-chip

I. INTRODUCTION

Dry micro-assembly usually involves multistep parts transferring between different (rigid or flexible) substrates. In order to transfer the parts from donor substrate to receiver substrate successfully, the adhesive forces between the parts and the receivers have to be stronger than that between the parts and the donors. Micro-assembly in dry environment is one of the most convenient and straightforward method for flip-chip bonding and deterministic micro-assembly [1-6]. The adhesive force arrangement among thee parts, donor and the receiver substrates are critical for multi-batch parts transferring. However, there is a lack of efficient way to measure the adhesion strength between pats and substrates for static analysis. Most of the published studies use contact methods to measure the forces by elastic micro-testing-structures [6-11].

This paper presents a noninvasive tool for interfacial adhesion gauging between two solid surfaces. It provides a simplified platform in which more realistic testing can be readily developed to estimate the particle adhesion strength and its detachment conditions. The distribution of detachment condition is a straightforward mechanics, which is calculated from centrifuge force of the tested particles. The following sections are going to introduce the platform design, testing results. The testing results have been verified by solder ball transferring process in dry environment.

II. TEST PLATFORM DESIGN

The configuration of the test platform is shown in Fig. 1. The noninvasive examination tool includes a rotating wheel on a motor, a glass cover, a laser, an optical window, a CMOS camera, a light-to-voltage optical converter and a strobe. It is driven by a DC motor which has controllable angular speed from 33 rpm to 196 rpm. The centripetal acceleration is equal to 1.03 G. Since the velocity vectors of solder balls is constantly changing that moves in a circular path. The solder ball has a centripetal acceleration force. When the centripetal acceleration force or the torque generated by the centripetal force is over the limit of the adhesion force or the torque generated by the adhesion force, the micro-ball will slip or rotate away from the center, respectively. A glass cover is place on the top of the test platform to prevent the air flow or turbulent interference (Fig. 2).

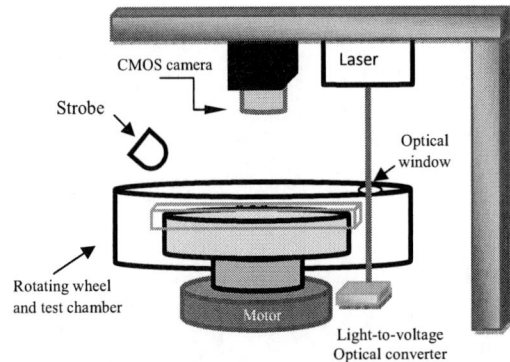

Fig. 1. The test setup, it includes an optical window near the edge. The chamber is 1.5 mm height (radius = 12.7mm). The platform is capable of independent measurement at multiple testing spots in parallel.

To assess the adhesion forces of solder ball in parallel. A 5mW, 650nm laser diode is placed on the top of the platform. We focused the laser through the optical window (shutter) onto a high speed light-to-voltage converter (TAOS TSL 254). This converter is designed to trigger an ultra bright LED (460 nm). Because the frame rate (30Hz) of the camera is slower than the chamber rotating speed 33~196Hz, it is also possible to use visual staying phenomenon to inspect the parts slip-sliding movement on the substrate.

978-1-4673-1122-9/12 $31.00 © 2012 IEEE

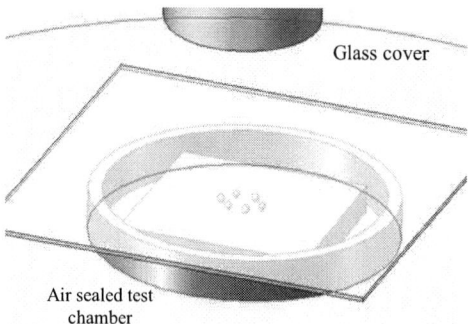

Fig. 2. Using air sealed chamber to minimize air flow disturbing

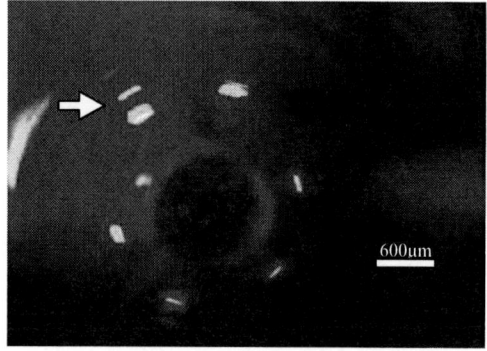

(a) A 500μm bead is detaching from the surface

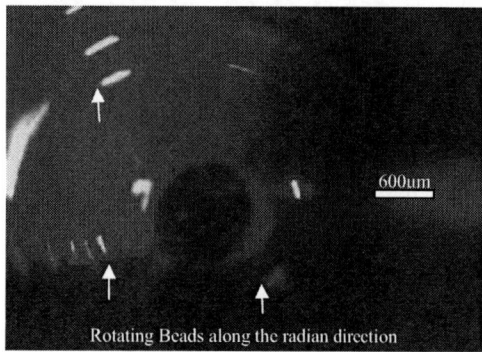

(b) One can estimate the detaching rates of a lot of microbeads with a syncronized strobe ilumination

Fig. 3. The strobe is syncronized the rotating wheel. It illuminates only at a specific phase angle of the periodic motion and produces a quasi-static image. The visual staying phenomenon has been used for dynamic analysis.

The LED acts as a strobe light to generate visual staying images or videos, one of the visual staying images is shown Fig. 3 in dark surrounding. These images and videos present rich information of a group of particles during the adhesion strength testing.

The frequency of the strobe light is synchronized to the rotating wheel and is monitored by an oscilloscope to calculate the angular speed. The circuit of the auto-triggered synchronized strobe test is shown in Fig 4; it is capable of time-resolved testing. The test chamber located at the center of the rotating wheel is capable of holding a 10mm×10mm×1mm square sample substrate. After placing solder balls on the top of the sample substrate and sealed the chamber with a glass cover. The platform is capable of independent measurement at multiple-spot testing in parallel. It is convenient and efficient to record data for static analysis.

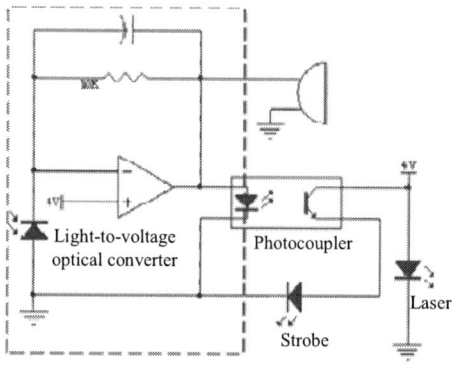

Fig. 4. Focused laser passes through the optical window and illuminates the light-to-voltage converter when the window is open which generates a pulse output to trigger the strobe.

III. EXPERIMENTS AND RESULTS

For the surface adhesion test, we test three different surfaces in experiment, which include two PDMS surface with different mixing ratio of cure agent and base material and ReproRubber®. Both two materials have been used in micro-assembly for making flexible electronic devices [12-13]. The flexible substrates have the advantages of cost effective; it would also compatible to roll to roll fabrication process. Both two PDMS and ReproRubber® can be spinning coated on glass substrates for testing. One can adjust driving voltage of the DC motor to control the rotating speed of the wheel. We gradually increase the rotating speed and inspect visual staying phenomenon to monitor solder ball slip-sliding in the test chamber on the rotating wheel.

In this experiment, we test two sizes (300μm and 500μm in diameter) of micro solder balls for transfer adhesion strength assessment. When the solder balls are placed onto the flexible substrate, they stick to each other and form a temporal contact area. In general, if there is no external force, the size of contact area is the strength of contacts. Fig. 5 is taken from an inverted setup for contact force characterization. A series of test conditions is listed in Table 1. The results are shown in Fig. 6.

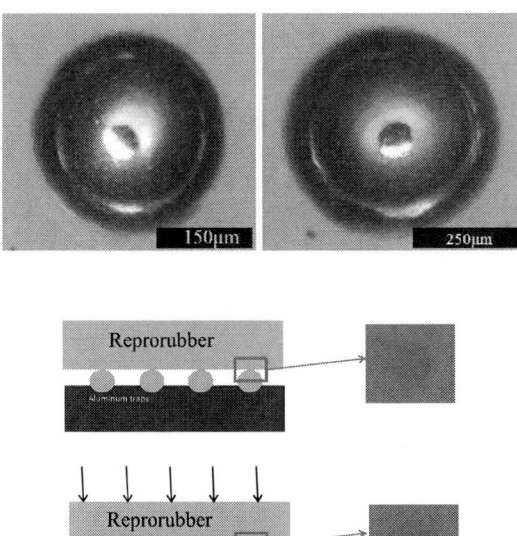

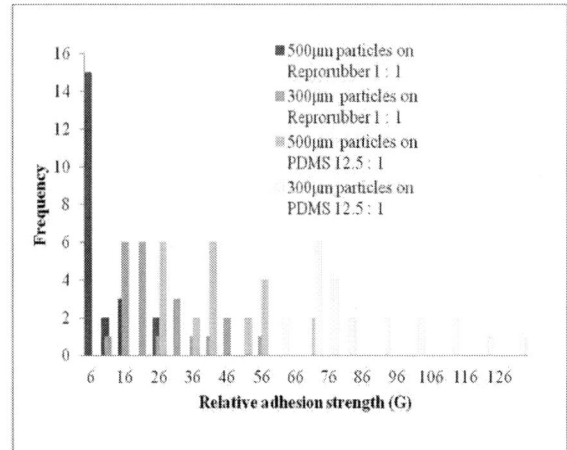

Fig. 6. The static results from different particles /substrates combinations. The result reveals that contact force has strong dependence on parts scale and material properties.

The experiment results indicated that the contact forces are highly dependent on parts scale, material properties, parts geometry and surface conditions.

Based on the adhesion strength testing results, shows in Fig. 6, one can design a multistep parts transferring system. 500 micron solder balls could be successfully transferred from Reprorubber® to PDMS with almost 100% yield. To verify the conclusion, a series of parts transferring experiments are delivered, shown in Fig. 7.

Fig. 5. Size of the solder balls, contact areas and interface conditions will dominate the adhesion strength of s. These pictures are taken from an inverted setup for contact force characterization.

Table 1: Adhesion strength competition between acceptor and donor *vs.* article transfer successful rate [2]

		REPRORUBBER®	PDMS(10:1)	PDMS(12.5:1)
REPRORUBBER®	YIELD RATE	50%	100%	100%
	N*	1	1.67	3.33
PDMS(10:1)	YIELD RATE	0%	50%	100%
	N*	0.6	1	1
PDMS(12.5:1)	YIELD RATE	0%	20%	45%
	N*	0.3	0.5	1

N* denotes the ratio of adhesion strength of balls between donor and acceptor substrates.

According to the test results, the contact force between PDMS (base agent: cure agent = 12.5:1) and solder ball dominates the contact force between Reprorubber® (base agent: cure agent = 1:1) and solder ball. Larger solder balls would provide more reproducible results than the smaller solder balls.

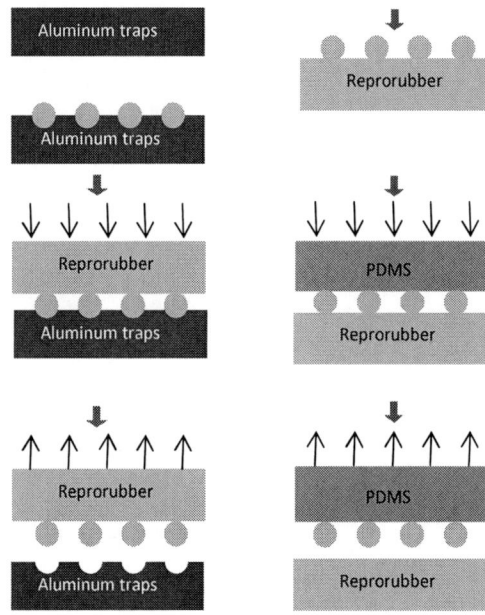

Fig. 7. The experiment is designed to demonstrate a multistep parts transferring process. 300μm solder balls are first placed in Aluminum traps, than transfer to ReproRubber®. In the second step, the solder balls are transferring from ReproRubber® to PDMS.

Because the adhesion strength between micro solder balls and substrates are highly sensitive to the surface condition. The surface cleanliness is extremely important for multistep parts transferring process; however, atmospheric N_2 plasma cleaning is not recommended for this purpose. It will completely disable both PDMS and ReproRubber® surfaces adhesion. The results of multistep parts transferring test are shown in Fig. 8. The results show that the yield ratio of 300μm solder ball transferring (1) from Aluminum mold to ReproRubber® and (2) from ReproRubber® to PDMS can achieve yield ratio up to 95.9% and 100% respectively.

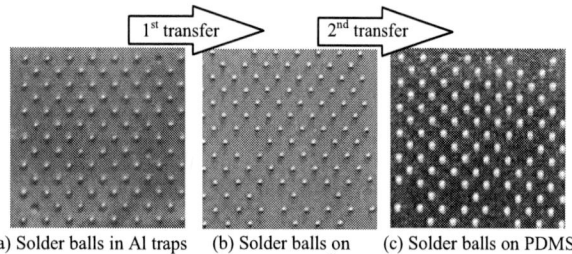

(a) Solder balls in Al traps   (b) Solder balls on   (c) Solder balls on PDMS
ReproRubber®

Fig. 8. From the test results, one can conclude that 300 micron solder ball could be transferred from Reprorubber® to PDMS with almost 100% yield., The multi-batch transferring (a)-(c) show that the yield ratio of 300 m solder ball transferring (1) from Aluminum mold to ReproRubber® and (2) from ReproRubber® to PDMS can achieve yield ratio up to 95.9% and 100% respectively.

The yield ratio for parts transferring is also dependent the adhesive strength contrast. The adhesive force between solder balls and acceptor substrates should be 1.67 times greater than the adhesive force between solder balls and donor substrates to ensure good parts transfer rate. The number would be varied, it is depend on the properties of different particles and substrate system, shows in Fig. 9.

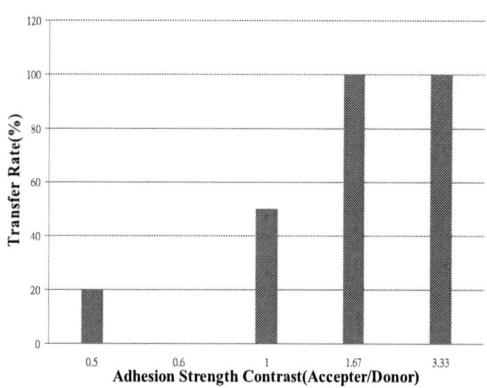

Fig. 9. The adhesive strength arrangements are critical for successful parts transferring.

IV. CONCLUSIONS AND DISCUSSIONS

This paper presents a non-contact and a noninvasive contact force test platform. In this study, the interfacial properties of the contact force are investigated with varying parts and substrates. Gauging the adhesive force is critical for dry assembly process. In order to transfer the parts from donor substrate to receiver substrate, the adhesive force between the parts and the receiver have to be stronger than that between the parts and the donor for good yield ratio. The adhesive force arrangement among the parts, donor and the receiver substrates are critical for successive multi-batch parts transferring in dry environment.

The novel adhesion strength measurement method can measure more than one particle at once in parallel for statics analysis.

Acknowledgment

This work is supported by National Science Council, NSC 98-2221-E-018-012-MY2 and NSC 100-2221-E-018-015.

REFERENCES

[1] K. Wang, R. Baskaran, and K.F. Böhringer, "Template Based High Packing Density Assembly For Microchip Solid State Cooling Application," *FNANO 2006*, April 23-27, 2006.

[2] H. C. Lin, C. C. Wang, K. Wang,"*Deterministic Three-Dimensional Micro-Assembly in Parallel*," IEEE International Conference on Nano/Micro Engineered and Molecular Systems 2011, pp.1128-1131, 2011.

[3] E. Menard, R.G. Nuzzo, and J.A. Rogers,"*Bendable Single Crystal Silicon Thin Film Transistors Formed by Printing on Plastic Substrates*," IEEE 2005, v86.pp 093507-039507-3, 2005

[4] Z. Bao, W. Zhongand, W. Xiangjun, "*Technology of Stacking Vertical Interconnecting Micro-Assembly*," IEEE Proceedings of HDP, 1-4244-1253-6, 2007.

[5] Y. R. Chion, S. Y. Huang, K. Wang,"*Multi-Step Dimensional Micro Assembly for a Flexible LED Display*," IEEE MEMS 2010, pp.59-62, 2010.

[6] H. O. Jacobs, A. R. Tao, A. Schwartz, D. H. Gracias and G. M. Whitesides,"*Fabrication of a Cylindrical Display by Pattered Assembly,"* Science 2003, vol296. pp.323-325, 2002.

[7] M. A. Meitl, Z-T. Zhu, V. Kumar, K. J. Lee, X. Feng, Y. Y. Huang, I. Adesida, R. G. Nuzzo & J. A. Rogers, et al,"*Transfer Printing by Kinetic Control of Adhesion to An Elastomeric Stamp*," Nature Materials January 2006.

[8] B. L. Pruitt, W. T. Park, and T. W. Kenny, "Measurement System for Low Force and Small Displacement Contacts," *J. Microelectromech. Syst.*, v.13, no. 2, pp.220-229, 2004.

[9] E. Iwase, H. Onoe, K. Matsumoto, and I. Shimoyama, "Hidden vertical comb-drive actuator on PDMS fabricated by parts-transfer," *IEEE MEMS 2008*, pp.116-119, 2008.

[10] H. Onoe, E. Iwase, K. Matsumoto, and I. Shimoyama,"*Three-Dimensional Integration of Heterogeneous Silicon Micro-Structures by Liftoff and Stamping Transfer,*" J. Micromech. Microeng 2007, v.17,pp1818-1827, 2007.

[11] H. Lee, J. Chung, S. Chang, and E. Yoon, "Normal and Shear Force Measurement Using a Flexible Polymer Tactile Sensor With Embedded Multiple Capacitors," *J. Microelectromech. Syst.*, v.17, no. 4, pp.934-942, 2008.

[12] C. Yang, W. Wang, and Z. LI, "Optimization of Corona-triggered PDMS-PDMS Bonding Method," *IEEE International Conference on Nano/Micro Engineered and Molecular Systems 2009*, pp319-322, 2009.

Automatic Somatic Cell Operating Process for Nuclear Transplantation

Wang Yiliao, Zhao Xin*, Zhao Qili, Sun Mingzhu, Lu Guizhang

Institute of Robotics and Automatic Information System, Nankai University, Tianjin, China

Abstract—**Somatic cell operating, a key procedure of nuclear transplantation, is a process of picking up an appropriate donor cell. Improving its degree of automation and reducing operating time is very crucial for enhancing success rate of cloning. This paper presents an automatic somatic cell operating process for nuclear transplantation. Firstly, somatic cells are localized employing texture image segmentation and then the most active cell is selected for operation by cell morpha. Secondly, injection pipette is positioned utilizing template matching. Afterwards path planning for injection pipette to pick the selected cell is conducted considering obstacle avoidance, preventing other cells from sticking the pipette. Finally the pipette moves to the target cell along the path automatically under the guidance of visual servo control. Validity and efficiency of this operating process are proved by a dozen time of experiment.**

Keywords-Nuclear Transplantation;Automatic Operating Process;Texture Image Segmentation; Visual servo;

I. INTRODUCTION

Benefiting from high transfection efficiency and stable character of offspring, somatic cell cloning [1-2] becomes one of significant means of transgenic engineering. General procedures of somatic cell cloning contain remove of hereditary substance from oocyte (denucleation), implantation donor somatic cell (nuclear injection), fusion of oocyte and donor cell, activation and cultivation of reconstructed embryo. Denucleation and nuclear injection, joined together to be nuclear transplantation, are key procedure of somatic cell cloning. Somatic cell operating in nuclear transplantation is a process of picking up an appropriate donor cell for subsequent nuclear injection, including 2 steps: 1) Find the most active somatic cell in microscopic view; 2) Move the injection pipette to the target cell while avoiding other cells' adhesion. Traditional somatic cell operating relied on professional operator to accomplish numerous times of such tasks manually, which is very time and effort consuming. Our team reported a batch-operation process of nuclear transplantation based on global field of view [3], optimizing the overall process; however the somatic cell operating part remains manual, proved unfavorable to efficient and stable nuclear transplantation. From motives of further improvement of nuclear transplantation, an automatic somatic cell operating process, including 4 procedures, is put forward in this paper. Firstly, somatic cells are localized employing texture image segmentation and then the most active cell is selected for operation by cell morpha. Secondly, injection pipette is positioned utilizing template matching. Afterwards path

planning for injection pipette to pick the selected cell is conducted considering obstacle avoidance, preventing other cells from sticking the pipette. Finally the pipette moves to the target cell along the path automatically under the guidance of visual servo control. Validity and efficiency of this operating process are proved by a dozen time of experiment.

II. CELLRECOGNITION ALGORITHM

A. Texture Image Segmentation

Texture image [4] presents irregularity in local region but regularity in the whole image. Images of somatic cells in microscopic view are shown in Fig.1 in comparison with ordinary texture images in Fig.2, demonstrating that though its grayscale differentiation with background is not obvious, the somatic cells exhibit texture. Therefore texture image segmentation methods are suitable for the localization of somatic cells.

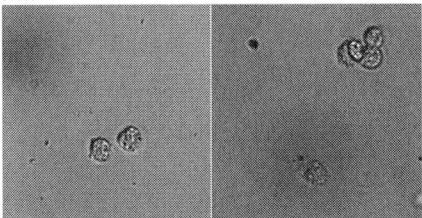

Fig. 1: Images of somatic cells in microscopic view

Fig. 2: ordinary texture images

On the consideration of real-time requirements, statistical method based on local illumination histograms to quantify the texture. This method, usually considered as a simplification of conventional vein analysis means such as Gray Level Co-occurrence Matrix (GLCM) [5-6], is appropriate for segmentation of target with simple texture and enjoys low computation complexity, coinciding with the application demand of this paper.

It can be found in Fig.1 that the inner image of somatic cell demonstrates dark and bright alternately with high regional contrast while the background and interfering shadow are smooth with slow changing gray scale. Therefore the standard deviation $\sigma = \sqrt{\mu_2(z)} = \sqrt{\sigma^2}$ (1) is employed to measure the regional contrast of image. Where $\mu_n = \sum_{i=0}^{L-1}(z_i - m)^n p(z_i)$ (2)

is the moment of nth order of mean $m = \sum_{i=0}^{L-1} z_i p(z_i)$ (3) and

$p(z_i)$ is the histogram of local image.

To realize the binarization and segmentation of somatic cell image, we calculated the standard deviation of histogram on local image for every pixel by choosing a proper neighborhood and then settled a threshold according to the deviation, which equals to apply a texture filtering to an image.

The segmentation result of somatic cells image in real operation (Fig.3 and Fig.4) after employed the foregoing segmentation method based on texture analysis are in comparison with traditional OTSU [7] method(Fig.5),proving the texture image segmentation method are capable of dividing the somatic cells out of the background.

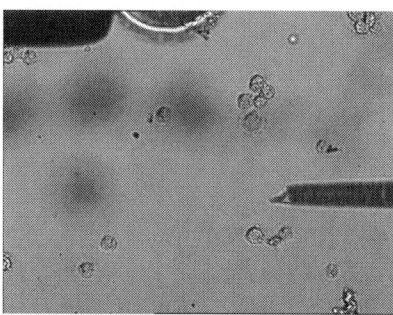

Fig. 3: Somatic cells in operation

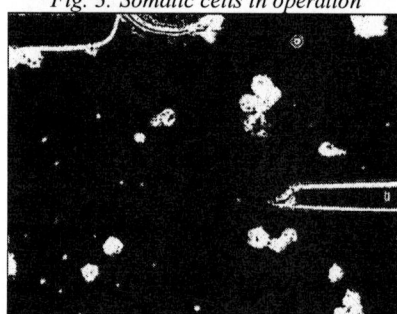

Fig. 4: Texture segmentation result of somatic cell image

Fig. 5: OTSU segmentation result of somatic cell image

B. Target Cell Selection

To further localize the somatic cells and select the target cell for operation after segmentation, their own characters such as shape and size need to be considered.

According to the experience from nuclear transplantation experiments, single somatic cells with most round shape and smooth surface often have high level of activity. After combination of analysis of size, shape and contour, somatic cells can be localized as well as the target cell selected (Fig.6).

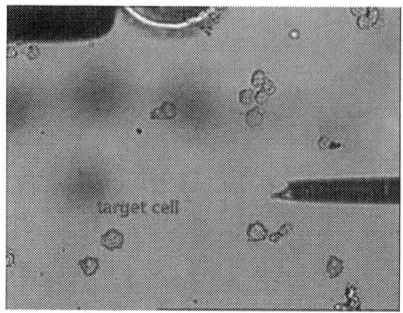

Fig. 6: Localization of somatic cells and the target cell

III. PATH PLANNING

A. Injection Pipette Localization

To locate the injection pipette precisely and stably in complex background, we utilize template matching method to trace the moving injection pipette in operation. The specific procedures are as follows：

1) At the very first stage of nuclear transplantation, the microscopic view contains only holding pin, injection pipette and oocyte, lacking of interference from somatic cells and other impurities. The injection pipette tip is positioned by analyzing grayscale change on vertical scanning line and then the template of pipette tip is acquired automatically. Localization result and template of injection pipette tip are shown in Fig.7.The Region of Interest (ROI) is set to be the left/right half part according to the injection pipette is installed in left/right arm.

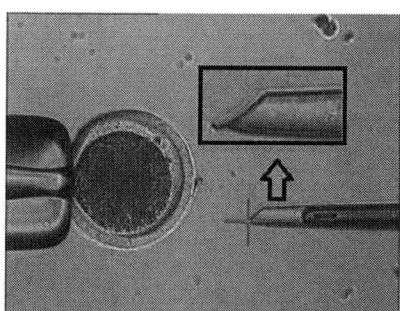

Fig. 7: Automatic template acquisition at the first stage of nuclear transplantation

2) In somatic cell operation when various kind of interference exist, template matching [8-9] is utilized in ROI to find out the current position of injection pipette tip (Fig.8),

978-1-4673-1122-9/12 $31.00 © 2012 IEEE

showing the reliability of localization method both in focusing and defocusing situation.

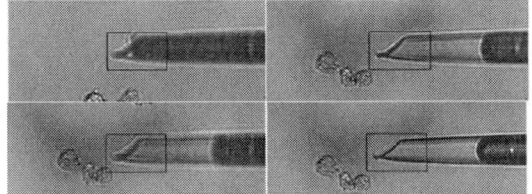

Fig.8: Micropipette position result using template matching from defocusing to focusing situation

3) The ROI is updated according to the latest localization result (Fig.9).

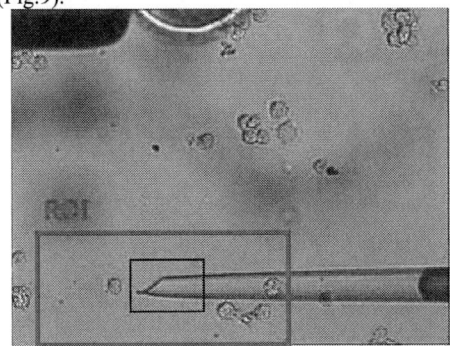

Fig.9: Updated ROI for template matching

B. Obstacle Avoidance Strategy

To realize automatic picking up of the target somatic cell, the injection pipette's moving path needs to be planned, enabling the injection pipette to move to the target cell as quickly as possible and at the same time avoiding conglutinating other somatic cells. Therefore the major task of path planning is obstacle avoidance analysis.

Fig.10 shows the Position relationship of target cell and injection pipette and cell C is the selected most active cell after contour analysis. Then the region the injection pipette needs to move across to pick up the target cell can be determined as the region enclosed by horizontal line of injection pipette and link between injection pipette and target cell which is noted in Fig.10. Whether this area contains other somatic cells decides the necessary of obstacle avoidance. For example, if the initial position of injection pipette is 1 in Fig.10, obstacle avoidance is necessary since cell B is located in the area the pipette needs to pass. Otherwise, when injection pipette lie in position 2, direct motion to cell C can be conducted without obstacle avoidance.

At the beginning of somatic cell operation, the somatic cells and injection pipette are adjusted to focus plane, therefore their positions can be settled using above-mentioned cell recognition algorithm and template matching method. Then the decision of whether obstacle avoidance is necessary can be made according to the existence of non-target cells in the region the injection need to pass across. The detailed obstacle avoidance approach is designed as below.

Firstly, the microinjection pipette raise up 20 μ m along axis Z to avoid from touching and conglutinating non-target cells since the maximum diameter of somatic cell is 10 μ m. Secondly, the microinjection pipette move along axis Y upon X-Y plane to position $Ob_y + \Delta$, given the coordinate of non-target cell on X-Y plane is (Ob_x, Ob_y) and Δ is the additional distance to avoiding contact, usually set to be 10 μ m. In case of more than one cell $(Ob_x1, Ob_y1)(Ob_x2, Ob_y2)\ldots(Ob_xn, Ob_yn)$ needed to be avoided, the microinjection pipette will travel along axis Y to position $\max\{Ob_y1, Ob_y2, \ldots, Ob_yn\} + \Delta$. Lastly, the injection pipette descend 20 μ m to accomplish obstacle avoidance. Experimentally the cells need to be avoided will not more than 3 since the arrangement of somatic cell will not be overcrowded. After obstacle avoidance, the injection pipette can move straightly towards the target cell under guidance of visual servo control since it will appear in focusing plane again. The overall process of obstacle avoidance is shown in Fig.10. On the other hand, if obstacle avoidance is not necessary, the injection pipette will move directly to the position of target cell using with visual servo control, which will be introduced in section IV.

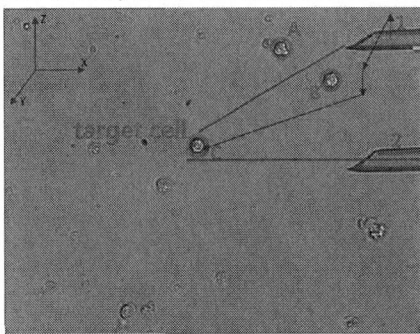

Fig. 10: Path planning for the micropipette to pick the target cell C considering obstacle avoidance

IV. VISUAL SERVO CONTROL

The visual servo control system will be positioning the microinjection pipette from current position to the target position to pick up the target cell.

Two requirements should be considered in designed the controller. First of all, the control system should be very precise and at the meantime be capable of approach the target as fast as possible. Therefore, minimizing the overshoot as well as enhancing the rapidity of the control system are highly desirable. To achieve this purpose, a PID based controller with visual feedback of injection pipette is designed. The visual servo control scheme is shown in Fig.11.

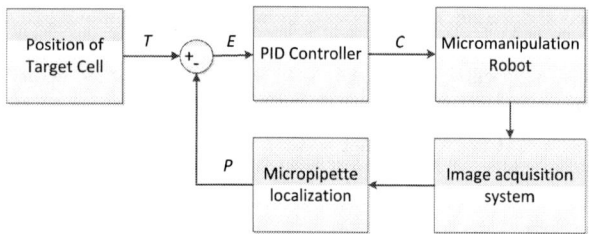

Fig. 11: Visual Servo Control scheme based on PID controller

The control signal C in Fig.11 is calculated as follows:

$$CS(t) = J_v^+ \cdot [CS_p(t) + CS_i(t) + CS_d(t)] \quad (4)$$

where J_v^+ is the Image Jacobian of micromanipulation robot[10].The micromanipulation robot NKMR601 this paper supported by have three orthogonal arms and move without coupling motion. Considering the visual servo control in this paper include solely motion in X-Y plane, the analysis can be

simplified, giving result $J_v^+ = \begin{bmatrix} \dfrac{1}{\lambda} & 0 \\ 0 & -\dfrac{1}{\lambda} \end{bmatrix}$ (5)where λ is

the pixel density of the microscopic view of the micromanipulation robot which can be obtained by dividing the magnification of microscope's object lens by interval between pixels of CCD.

$CS_p(t)$, $CS_i(t)$ and $CS_d(t)$ are proportional, integral and derivative control signals respectively. The ultimate goal of the controller is to make $error(t)$, defined as target position minus current position at time t, decrease to zero. Denote by:

$$CS_p(t) = K_p \cdot error(t) \quad (6)$$

$$CS_i(t) = CS_i(t-1) + K_i \cdot error(t) \cdot \Delta t \quad (7)$$

$$CS_d(t) = K_d \left(\frac{error(t) - error(t-1)}{\Delta t} \right) \quad (8)$$

, where K_p, K_i and K_d are proportional, integral and derivative gains respectively and Δt is the time between two times of visual feedback.

V. Experiment Results

Employing the proposed image processing method, 80 somatic cells were 93% correctly recognized. In nuclear transfer experiment, to guarantee the effect of tracking microinjection pipette, the moving speed of the pipette should be limited from being too fast, so that the injection pipette tip will lie in updated ROI at next time of image acquisition. Therefore, in experiments, the speed of injection pipette is restricted below 100μ m/s and the success rate of micropipette positioning under this limitation is 99.3% on 796 images within 10ms for one frame of picture. Since the operation is conducted in ten-fold objective lens thus the microscopic view is relatively small, speed limitation of 100μ m/s is capable of

meeting the experiment demand of picking up target somatic cell quickly. The Image Jacobian J_v^+ can be gained after calibration of the microscopic view of NKMR601.The controller gains were well tuned experimentally to following values:

$$K_p = 0.5$$
$$K_i = 0.001$$
$$K_d = 0.02$$

Figure.12 shows the performance of the designed controller, illustrating the system is capable of moving the micropipette accurately to target cell within 30 control cycles and make the error completely from two different X and Y coordinate without overshoot.

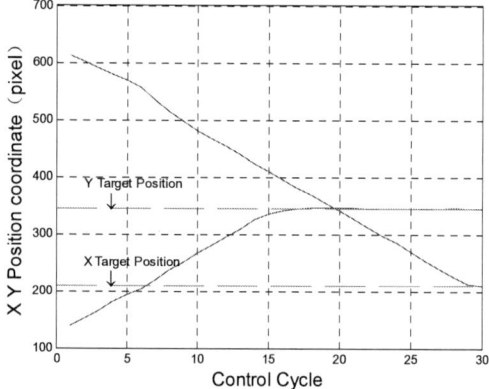

Fig. 12: Control effect of the visual servo control system

Applying proposed automatic operating process to experiments, a dozen time of somatic cell operation all received success. Fig. 13 shows the final result of automatic operating.

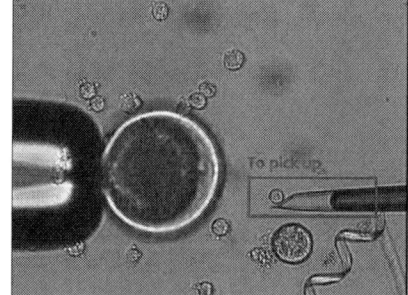

Fig. 13: Final result of automatic somatic cell operating

VI. CONCLUSION

Somatic cell operating in nuclear transplantation is a process of picking up an appropriate donor cell for subsequent nuclear injection, including 2 steps: 1) Find the most active somatic cell in microscopic view; 2) Move the injection pipette to the target cell while avoiding other cells' adhesion. From motives of further improvement of nuclear transplantation, an automatic somatic cell operating process is put forward in this paper.

Corresponding to step 1 of somatic cell operating, we propose an image processing algorithm based on texture image segmentation to identify somatic cells in microscopic field since the cell presents texture while background and interfering shadow are smooth. Then cell with most smooth contour and round shape is selected to be the target cell. An obstacle avoidance strategy and visual servo control design are presented to realize step 2's automation. Giving pipette's location acquired by template matching, the path for the micropipette to pick the target cell can be planned, deciding if the micropipette needs to move in Z axis to avoid unexpected cells. A PID based servo control system with real-time position feedback from vision is designed to precisely approach the target as fast as possible. A dozen time of experiment prove the validity and efficiency of the automatic operation process.

ACKNOWLEDGMENT

The authors would like to thank the Institute of Robot and automatic information system of Nankai University. This research was supported by National High Technology Research and Development Program ("863"Program) of China (2009AA043703), National Natural Science Foundation of China(NSFC: 60875059, 91023045, 61105107), the Fundamental Research Funds for the Central Universities(65012081).

REFERENCES

[1] I.Wilmut, A. E. Schnieke, J. McWhir, A. J.Kind, K. H. S. Campbell. "iable Offspring Derived From Fetal and Adult Mammalian Cells.", *Letters to Nature*, 1997, pp. 810-813.

[2] Campbell, K.H.S., McWhir, J., Ritchie, W.A. & Wilmut, I. Sheep Cloned by Nuclear Transfer from ACultured Cell Line", *Nature* , vol.380, 1996, pp.64–66.

[3] ZHAO Qili, CUI Maosheng, ZHAO Xin, SUN Mingzhu, Wang Yiliao, Feng Jianzhong."Batch-operation Process of Nuclear Transplantation Based on Global Field of View",*CCC2011*

[4] T. Randen and J. H. Husoy, "Texture segmentation using filters with optimized energy separation", *IEEE Trans. Image Process.*, vol. 8, no. 4, 1999, pp.571 - 582.

[5] ROBERT M. HAWLICK. "Statistical and Structural Approaches to Texture". *Proceedings of the IEEE*, vol. 67, no.5, 1979, pp.786-804.

[6] ROBERT M. HARALICK, K. SHANMUGAM, AND ITS'HAK DINSTEIN. "Textural Features for Image Classification", *IEEE Transactions on Systems, Man and Cybernetics*, vol. SMC-3(6), 1973, pp.610-621.

[7] NOBUYUKI OTSU, ".A threshold selection method from gray-level histograms"*IEEE Transactions on Systems, Man and Cybernetics*,vol. 9-1,1979, pp. 62-66

[8] "Coarse-Fine Template Matching", *IEEE Transactions on Systems, Man and Cybernetics*, vol. 7, no. 2,1977, pp. 104 - 107

[9] A. Goshtasby, S.H. Gage, and J.F. Bartholic, "ATwo-Stage Cross Correlation Approach to TemplateMatching", *IEEE Trans. Pattern Analysis and Ma-chine Intelligence*, vol. 6, no. 3, 1984, pp.374-378.

[10] Hashimoto, K.; Kimoto, T. Ebine, T., Kimura, H. "Manipulator control with image-based visual servo", *IEEE International Conference onRobotics and Automation Proceedings*,1991

High Speed Enucleation of Oocyte
Using Magnetically Actuated Microrobot on a Chip

Masaya Hagiwara[*], Akihiko Ichikawa, Tomohiro Kawahara, Fumihito Arai
[*]Department of Micro-Nano System Engineering, Nagoya University, JAPAN
hagiwara@biorobotics.mech.nagoya-u.ac.jp

Abstract— We have developed novel microfluidic chip for enucleation of oocytes by magnetically actuated microrobots. Si and Ni composite microrobot driven by permanent magnets was developed and used as high power, high speed and high accuracy microarm to transport, rotate and cut swine oocyte in a microfluidic chip. The novelties of developed microrobot are as follows. 1) The drive frequency capability of microrobot improved by 10 times comparing to previously developed microrobot. 2) Si-Ni composite microrobot has no risk of biocompatibility. 3) A force of order of millinewton can be outputted from permanent magnets. The microfluidic chip was specially designed by fluid analysis of FEM for stable supply of the oocytes to the MMT manipulation area from the inlet. The integration of this high speed microrobot and microfluidic chip enables high throughput enucleation process in a microfluidic chip.

Keywords-component; Microrobot, micro TAS, Magnetically driven microtool, Enucleation,

I. INTRODUCTION

Cloning has been actively investigated for use in food production, organ transplantation, and the development of genetically similar laboratory animals and so on. The enucleation of oocyte is one of the cloning processes, which is time consuming and requires operation skill. In current operation, manually operated micromanipulators with glass capillaries are used to remove a nucleus under a microscope [1-3]. However, the manipulation is conducted in a open environment to the air due to the huge size of the manipulator and it leads to the cell contamination issues. In addition, the manipulation requires a high skill to an operator because the manipulator has to be controlled in 6 degrees of freedom. Therefore, automation of the cloning process can be breakthrough for the development of this fields owing to high-throughput production of processed cells with high quality and homogeneity. Ichikawa et al. proposed to cut oocyte for enucleation by controlling syringe pump to generate shear stress on an oocyte [4]. But the speed of cutting an oocyte was slow due to difficulty to control flow and it was difficult to remove nucleus precisely because there was no oocyte posture control.

A magnetically driven microtool (MMT) actuated by a permanent magnet can output a force of the order of millinewtons, which is in keeping with the small size of the drive unit. In fact, a permanent magnet possesses a magnetic field that is more than ten times the strength of a magnetic field of an electromagnetic coil of the same size [5-7] and therefore, it can be applied to wide range of cell manipulations such as sorting [8] cell cutting [7], arranging, and rotating [9].

We have continuously improved positioning accuracy of the MMT by a horizontally arranged permanent magnets drive in order to supply greater magnetic force into driving direction on an MMT [10] as well as employing ultrasonic vibration to glass substrate to reduce effective friction on the MMT. As a result, we previously achieved 1.1 μm positioning accuracy and several mN output force [9]. However, the friction reduction ratio by the ultrasonic vibration depends on the MMT velocity and it decreases considerably when the velocity of the MMT increases. In fact, the error between the magnet and the MMT position became more than 100 μm when the drive speed exceeded 5 Hz.

Furthermore, the MMT used to be made by nickel, which is not bio-compatible material in general. The fabrication of MMT was based on nickel electroplating, which is difficult to achieve precise fabrication accuracy and complex three dimensional shape.

Here, we propose new type of MMT composed of hybrid structure of Ni and Si. Taking advantage of flexibility of Si

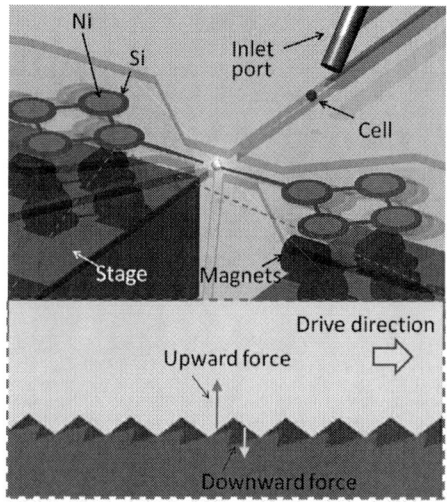

Fig. 1. Concept of high speed magnetically driven microtool (MMT) composed of Ni and Si structure. The MMT surface has riblet shape in order to reduce fluid friction on the MMT and it enables MMT to actuate high speed in a microfluidic chip.

978-1-4673-1122-9/12 $31.00 © 2012 IEEE

fabrication, three dimensionally patterned surface is produced on the MMT in order to reduce fluid friction on the MMT. Figure 1 shows the concept of Si-Ni composite MMT. The actuation part is composed of Ni in order to obtain strong magnetic force from the permanent magnet, but Si is surrounding Ni to have bio-compatibility when the MMT manipulates biological cells. In addition, regularly arrayed V groove, which is called "riblet", is patterned on the MMT and it reduces fluid friction force on the MMT. As a result, MMT can achieve precise accuracy at high speed region as well as high power output from the permanent magnet. It leads to high speed enucleation of oocyte process in a microfluidic chip.

II. FABRICATION AND EVALUATION OF SI-NI COMPOSITE STRUCTURE OF MMT WITH RIBLET

A. Fabrication of Si – Ni composite MMT with riblet

The riblet shape, which is regularly arrayed V groove and also known as shark skin, is practically used for airplane and ship to reduce fluid friction force [11-13]. The basic principle to reduce fluid friction force on MMT by riblet is that squeezing liquid by the V groove generates upward fluid force on the riblet surface and it makes the lubricant film thickness increase. As a result, the shear stress from the wall reduces and the fluid friction force decreases.

The precise three-dimensional fabrication is required in order to produce the riblet shape. However, it is quite difficult to produce micron meter size riblet on the MMT by Ni electroplating. Therefore, the composite structure of Si and Ni is employed.

Figure 2 shows the fabrication process of Ni and Si composite fabrication. First, SiO_2 was sputtered on the 200 μm thickness Si wafer with (100) crystal orientation (thickness = 300 nm). Then, OAP (Tokyo Ohka Kogyo Co. LTD.) was spin-coated on SiO_2 to bond OFPR and SiO_2 properly. Then,

OFPR-800 200cp (Tokyo Ohka Kogyo Co. LTD.) was spin-coated on the OAP (thickness = 4 μm). The OFPR was exposed and stripes were patterned. Acid treatment was then conducted by 10:1 BHF buffered hydrogen fluoride (Stella Chemifa Co.) to etch SiO_2. After removing the OFPR by acetone, Si was etched by 50 % KOH. Then, undercut phenomenon occurs and riblet shape was fabricated on Si wafer (process 5 in figure 2). Figure 3 (a) shows the SEM image of Si after wet etching. The designed riblet shape was fabricated precisely.

Then, Cr/Au was sputtered on the riblet surface (thickness = 300 nm) and Ni was formed on the riblet surface by electroplating as a support layer. After the SU-8 was coated on the opposite surface (thickness = 15 μm), it was exposed to pattern MMT shape. Then, deep reactive ion etching (DRIE) was conducted to penetrate Si to form MMT.

After the Cr film was removed by Cr etchant, Ni was formed by electroplating. Finally, Si-Ni composite structure of MMT was fabricated after Cr / Au and Ni support sheet were removed by Cr and Au etchant. Figure 2 (b) shows SEM image of fabricated MMT with riblet. The riblet shape was patterned on Si and Ni surface and Ni was covered by Si. Therefore, the MMT becomes bio-compatible tools.

B. Experimental evaluations

The experiment was conducted in order to evaluate the effect of the riblet surface. Figure 4 shows the experimental setup. The linear stage was actuated in 1 DOF sine wave with 0.5 mm stroke from 0.1 Hz to 100 Hz. The corresponding MMT positions were measured by high speed camera (1000 frame / sec) through the microscope and stage position was recorded through the encoder. The four of ϕ 1.0 mm × 1.0 mm columnar neodymium magnets were used as the drive unit in Horizontal polar drive conditions [10]. 150 Vp-p were applied to piezoelectric ceramics to induce the ultrasonic vibrations [9]. The culture solution (Medium 199 with follicular fluid) was used for the liquid.

The riblet has orientation and thus the MMT actuation must be different by the drive direction. Therefore, the frequency characteristic was measured for Si MMT in two different directions as well as Ni MMT and Si-Ni hybrid MMT without

1. Sputter SiO₂ SiO₂	6. Sputter Cr/Au	11. Remove Cr
Si		
2. OAP / OFPR coating OAP OFPR	7. Ni electroplating	12. Ni electroplating
3. Exposure	8. SU-8 coating	13. Cr, Au development
4. Acid treatment	9. Exposure	
5. Undercut wet Etching	10. DRIE	

Fig. 2. Fabrication process of Si and Ni composite structure with riblet surface. Si anisotropic wet etching was employed to produce V groove and dry etching was employed to fabricate MMT frame. Ni electroplating made ferromagnetic material to obtain high power from permanent magnet.

 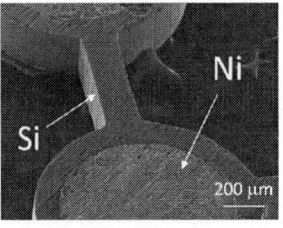

(a) SEM image of after wet etching   (b) SEM image of Si-Ni hybrid MMT

Fig. 3. SEM image of Si-Ni composite MMT. (a) after wet etching of Si wafer (b) complete form of Si – Ni hybrid MMT with riblet surface.

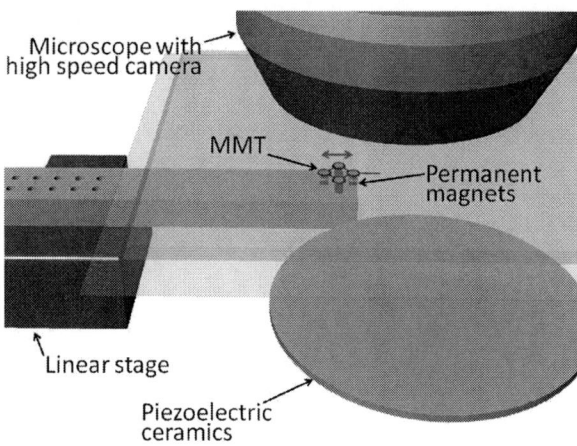

Fig. 4. Experimental setup for MMT drive frequency characteristics. Piezoelectric ceramics were attached to the glass substrate to induce the ultrasonic vibrations. Fabricated MMT with riblet surface were evaluated as well as MMT without riblet.

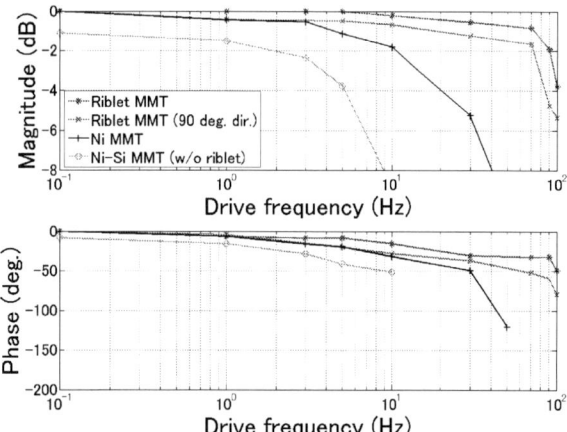

Fig. 5. Frequency response characteristics of MMT from 0.1 Hz to 100 Hz when the drive stage was actuated with 0.5 mm stroke sine wave.

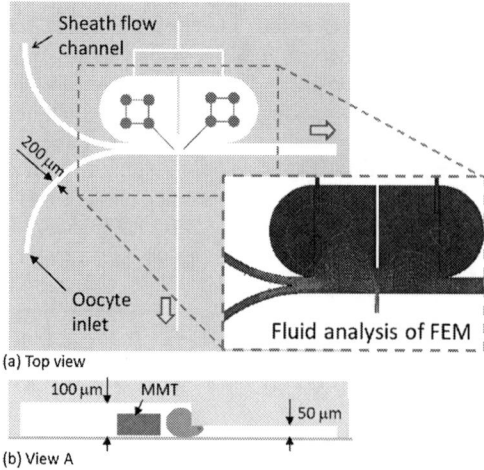

(a) Top view

(b) View A

Fig. 6. Design of the microfluidic chip for the enucleation of oocyte.

riblet. Figure 5 shows the experimental result of frequency response characteristics.

The result shows that the MMT with riblet surface can follow the stage up to 90 Hz when the drive direction was forward direction, while the conventional Ni based MMT does not follow properly after 10 Hz. Also, Si-Ni hybrid MMT without riblet delayed at 3 Hz and does not follow properly after 5 Hz. This result shows that the riblet surface successfully improved the MMT capability of the drive speed by 10 times.

On the other hand, the MMT driven in parallel to the riblet grooves can be actuated properly up to 70 Hz. Comparing to the result of MMT with riblet moved in forward direction, the drive capability of the MMT in parallel to the riblet was deteriorated, however, the following response improved considerably from the MMT without riblet. This is because of the smaller contact area to the substrate as well as the orifice effects.

III. ENUCLEATION OF OOCYTE

A. Experimental setup

The MMT can be now driven with high speed (more than 70 Hz) in x-y directions owing to the riblet surface, as well as high precision (minimum 1.1 μm), and high power (several mN order) [9], thus, the application for the cell manipulation is broaden. Here we conducted enucleation of oocyte by MMT in a microfluidic chip.

In advance of oocyte enucleation process, the swine oocyte has to be prepared with hylauronidase (0.1% of TCM 199) for 10 min in order to remove cumulus cells surrounding the oocytes and pronase (0.5% of PBS) for 10 min to remove zona pellucida. Then, Hoechst 34580 is applied for staining the nucleus of the oocyte.

Figure 6 shows the design of the microfluidic chip for the enucleation of oocytes. The oocytes are inserted at the inlet port and flow in a channel of 200 μm height and 200 μm width. The height of the channel to the nucleus collection port is set lowered (50 μm) so that the area containing nucleus cannot escape from the MMT to z direction. The dual-arm MMT then rotates the oocyte and area containing the nucleus is withdrawn into the channel. The dual-arm MMT cuts the oocyte by the tip of the blade and the nucleus part flows to the nucleus collection port. The remainder of the oocyte flows to oocyte collection port to be used in subsequent processes.

The sheath flow channel is set for the oocyte to flow along the PDMS wall and reaches to the manipulation area properly. The FEM result for fluid analysis in Figure 6 shows the sheath flow restricts for the fluid from oocyte inlet port to flow to the area where MMT is placed, and thus oocyte can keep flowing along the wall. Also, the center PDMS wall between the dual-

978-1-4673-1122-9/12 $31.00 © 2012 IEEE   366

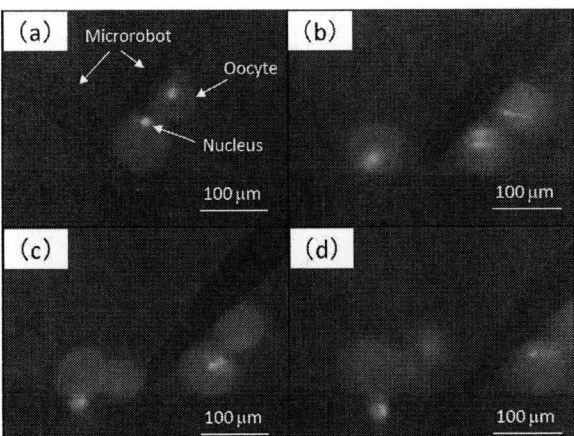

Fig. 5. Experimental setup for MMT drive frequency characteristics. Piezoelectric ceramics were attached to the glass substrate to induce the ultrasonic vibrations. Fabricated MMT with riblet surface were evaluated as well as MMT without riblet.

arm MMT prohibit fluid flow from occurring in MMT actuation area.

B. Experimenatal results

Figure 7 shows the experimental result of swine oocyte enucleation process. The oocyte inserted from the inlet flowed along the PDMS wall due to the sheath flow and reached to suction area (Fig. 7 (a)). After the nucleus position was confirmed, the oocyte was rolled by the MMT (Fig. 7 (b)) and pushed to the narrow channel so that the nucleus containing area was trapped in the channel (Fig. 7 (c)). Then the tip of the MMT pressed oocyte and cut the nucleus part of the oocyte (Fig. 7 (d)). It can be seen that the nucleus was successfully removed. The processing time, which is the duration from the time when oocyte reached the narrow channel to the time when oocyte was cut, was less than 10 s and the amount of oocyte removed was about 20% of the original size. The flow control in the microfluidic chip supply oocyte continuously and it enables continuous process of the enucleation which leads to high throughput.

IV. CONCLUSIONS AND FUTURE WORK

In this paper, we conducted high speed enucleation of oocyte in a microfluidic chip. The high speed magnetically driven microtool was developed as an enucleation tool. The microfluidic chip was designed to control oocyte flow and gave stable conditions for the enucleation.

For the fabrication of MMT with riblet, Si-Ni composite fabrication process was employed to obtain precise three dimensional shape on the MMT as well as taking advantage of bio-compatibility of Si material. Anisotropic wet etching of Si wafer was conducted to obtain riblet shape and DRIE and electroplating gave Si-Ni composite structure of the MMT.

The evaluation experiment successfully proved that the advantage of the riblet was getting large in high speed region. The drive capability of the MMT with riblet surface improved by 10 times comparing to the MMT without riblet.

The enucleation process in the designed microfluidic chip was successfully achieved by dual-arm MMT. Owing to the microfluidic chip, the oocyte can be delivered to the MMT manipulation area continuously and MMT can be conducted one process after another.

The final goal of this research is fully automated enucleation in a microfluidic chip and dispensing system for non-nucleus oocyte to culture well. Therefore, the detection of the nucleus position by image processing followed by feedback MMT position and trajectory control is our future work.

ACKNOWLEDGMENT

This work is partially supported by SENTAN, JST, and the Research Fellowship for Young Scientists, Japan Society for the Promotion of Science

REFERENCES

[1] F. Barnes, M. Endebrock, C. Looney, R. Powell, M. Westhusin and K. Bondioli, "Embryo cloning in cattle: the use of in vitro matured oocytes", Journal of Reproduction and Fertility, vol. 91, pp. 317-320, 1993.

[2] T.T. Peura, I.M. Lewis, and A.O. Trounson, "The Effect of Recipient Oocyte Volume on Nuclear Transfer in Cattle", Molecular Reproduction and development, vol. 50, pp.185 – 191, 1998.

[3] A. Baguisi, E. Behboodi1, D. T. Melican1, J. S. Pollock, et al., Production of goats by somatic cell nuclear transfer, Nature Biotechnology, vol. 17, pp. 456-461, 1999.

[4] A. Ichikawa, T. Tanikawa, S. Akagi, and K. Ohba, "Automatic Cell Cutting by High-Precision Microfluidic Control", Journal of Robotics and Mechatronics, vol. 23, pp. 13-18, 2011.

[5] J. J. Abott, K. E. Peyer, M. C. Lagomarsino, L. Zhang, L. Dong, I. K. Kaliakatsos, and B. J. Nelson, "How should microrobots swim?", The international journal of Robotics Research, vol. 28, pp. 1434-1448, 2009.

[6] M. S. Sakar, E. B. Steager, D. H. Kim, M. J. Kim, G. J. Pappas, and V. Kumar, "Single cell manipulation using ferromagnetic composite microtransporters", Applied physics letters, vol. 96, issue. 4, pp.96-98, 2010.

[7] Y. Yamanishi, S. Sakuma, K. Onda, and F. Arai, "Powerful Actuation of Magnetized Microtools by Focused Magnetic Field for Particle Sorting in a Chip", Biomed. Microdevices, vol. 12, pp. 745-752, 2010.

[8] N. Inomata, T. Mizunuma, Y. Yamanishi, and F. Arai, "Omnidirectional Actuation of Magnetically Driven Microtool for Cutting of Oocyte in a Chip", Journal of Microelectromechanical Systems, vol. 20, pp. 383-388, 2011.

[9] M. Hagiwara, T. Kawahara, L. Feng, Y. Yamanishi, T. Masuda, F. Arai, "On-chip magnetically actuated robot with ultrasonic vibration for single cell manipu-lations", Lab on a Chip, vol. 11, No.12, pp.2049-2054, 2011.

[10] M. Hagiwara, T. Kawahara, Y. Yamanishi, and F. Arai, "Driving Method of Microtool by Horizontally-arranged Permanent Magnets for Single Cell Manipulation", Applied Physics Letters, vol. 97, pp. 013701-1 -013701-3, 2010.

[11] Walsh,M. J., "Drag Characteristics of V-Groove and Transverse Curvature Riblets", AIAA, 72, pp. 168-184, 1980.

[12] H. Choi, P. Moin,. and J. Kim, "Direct numerical simulation of turbulent flow over riblets",J.Fluid Mech.,255, pp. 503-539, 1993.

[13] Dean B, Bhushan B, "Shark-skin surfaces for fluid-drag reduction in turbulent flow", a review. Philosophical Transactions of the Royal Society A: Mathematical, Physical and Engineering Sciences, 368, pp.4775-4806, 2010.

978-1-4673-1122-9/12 $31.00 © 2012 IEEE

Catalytic Nano-mobile Robot with Finely Designed Geometry

Jingjing Bao[1]* and Masahiro Nakajima [2], Zhan Yang [1], Masaru Kojima[1], Toshio Fukuda [1,2]

1 Department of Micro-Nano Systems Engineering, Nagoya University, Furo-cho, Chikusa-ku, Nagoya 464-8603, Japan.
2 Center For Micro-nano Mechatronics, Nagoya University, Furo-cho, Chikusa-ku, Nagoya 464-8603, Japan.
E-mail: bao@robo.mein.nagoya-u.ac.jp

Abstract— In this paper, we develop the Catalytic Nano-mobile Robot (CNMR) with finely designed geometry swimming in hydrogen peroxide solution as a novel type of nano machine. First we use Focus Ion Beam (FIB) etching to fabricate the nano- mobile robot into the geometry or shape as we desire freely with little constrain. And then an effective and stable manipulation method in sub micrometer scale and liquid environment is shown to operate the fabricated nano-mobile robot. Following up the manipulation method, we get the motion trajectory and carry out the kinetic analysis of the nano-mobile robot with sophisticated geometry. To understand its unique phenomenon, we try to explain the mechanism by showing the relationship between nano-mobile robot's motion and time duration, concentration of hydrogen peroxide.

Keywords- catalytic nano-mobile robot; Focus Ion Beam; sub micro manipulation method in liquid.

I. INTRODUCTION

To compete with the biological nanomotor, researchers have done marvelous works on its counterpart-artificial nanomotor, which can be used to delivery cargo to the destination or actuate micro-nano mechanical system.

Among all of the candidates for promising man made nanomotor, the spotlight is always focused on fuel driven nanomotor. As related previous work, Pt and Au bimetallic nanowire motor was synthesized by Aluminum membrane template (TDEP) [1]-[5], where hydrogen peroxide is used as fuel. And the Pt and Au hybrid nanorod [1][4]or nanorotor[2] demonstrated translation or rotation in nanometer scale.

Recently, they introduced a new type of nanomotor based on Cu and Pt segmented nanobattery in dilute aqueous Br_2 or I_2 solutions [6]. The motion of the two types of nanomotor mentioned above are both powered by self-electrophoresis caused by redox reactions occurring on the two different metal segments. Beside this, Pt (or Ag) and Si nanomotor grown by dynamic shadow growth (DSG) method is also introduced, which even can grow nanospring structure [7]. In this case, the nanomotor is actuated by the bubble generated from hydrogen peroxide decomposition into water and oxygen. And Pt or Ag works as an engine because it is strong catalyst of this chemical reaction; while Si serves as the body of the nanomotor without reaction with the solution. Moreover, not only metal material can be used for development of nanomotor, even polymer can realize the actuation in micro and nanometer scale. Researchers devise the motor powered by a polymerization reaction outside biological systems. The motor is powered by ring-opening metathesis polymerization

(ROMP) of norbornene in solutions of the monomer [8]. In fact, any design of nano scale object, which can create the gradient of concentration, has promise to generate the desired phenomenon of actuation. But there are some drawbacks in the previous works: the fabrication method is complex and time-consuming, and the shape or geometry is hard to control, which is critical for mobile nano machine.

In this paper, we propose our novel design of catalytic nano-mobile robot. Our nano-mobile robot with finely designed geometry can swim in the hydrogen peroxide solution in certain orbit that is the reason why we name it nano-mobile robot instead of just nanomotor. We use Focus Ion Beam (FIB) etching, which allows us to fabricate nearly any shape we want. Then, we demonstrated a stable method to manipulate the fabricated nano-mobile robot, which can be introduced widely to other cases with the common features: in liquid environment and sub micrometer scale.

II. EXPERIMENT PROCEDURE

A. Fabrication process of catalytic nano-mobile robot

We use Focus Ion Beam (FIB) to fabricate finely designed nano-mobile robot. Compare with previous work [1]-[5], our process is simple and effective, most significant advantage is that we can work out nearly any shape as we need. That is important, because the shape and dimension of the mobile object in micro and nanometer scale has great influence on its movement [9].

The fabrication process is shown in Fig. 1. The gallium primary ion beam, which is accelerated with energy of 30 Kev, hits the sample and sputters atoms to form the surface. This etching technique allows precision milling of a specimen down to the nanometer scale.

Firstly, the tip of Pt probe is etched to a narrow strip with the width of 0.8um (Fig. 1(a)). And then turn around the probe with 90 degree and continue etching the plate to the desired shape, here we take the 'T' letter shape nano- mobile robot as an example (Fig. 1(b) and (c)). And finally we get the sophisticatedly designed nano-mobile robot with controllable dimension (Fig. 1(d)). Following the same process, we can get various shapes of nano-mobile robot. It is fabricated from a commercial Platinum probe (STM Pt probe P-100PtIr

978-1-4673-1122-9/12 $31.00 © 2012 IEEE

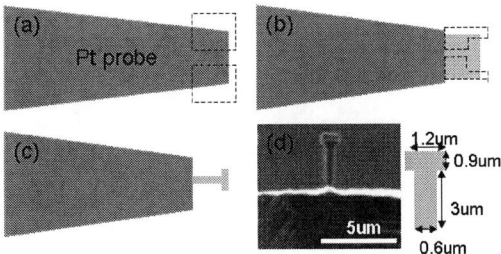

Fig. 1. Nano-mobile robot fabrication process and the FIB images. (a) The tip of Pt probe is etched to a narrow strip, shown by the black dotted line. (b) Turn over Pt probe 90° and etch the plate. (c) Fabricate the nano-mobile robot into 'T' letter shape. (d) FIB picture and design drawing with dimension.

（S），UNISOKU.) by FIB instrument (SMI2050, Seiko Instrument Inc.). In this paper, we show two types: 'T' letter shape and 'V' letter shape like an arrow which will be shown later in this paper.

B. Ultrasonic cleaning of substrate

Before the experiment, we have to do some prepare work like cleaning the contaminant on the glass substrate. Our nano-mobile robot is in sub micrometer scale and the driven force is very weak [1]-[2], so even a tiny contaminant or impurity can prevent the nano-mobile robot from swimming smoothly. We put the substrate into ethanol solution, and then place it into ultrasonic cleaning machine to remove the contaminant. And the cleaning time lasts 15 mins.

C. Manipulation method of geometric Platinum nano-mobile robot

Because the nano-mobile robot is extremely small, it is important how to manipulate it. At the beginning, we proposed a process in which we used Tungsten probe as tools to cut off it from the main body and then released it onto the glass substrate by adhesion force, and finally injected solutions [10]. But this former method is time consuming for preparing, because the Tungsten probe the tip of which needs to be etched sharply, furthermore the former method is not effective with low success rate, especially in the steps of release and injection of solution.

So we make big improvements for the manipulation method and make it evolve better. And the experiment setup and schematic of the method is shown Fig. 2 and Fig. 3.

First, we use gold coated glass capillary (GD-1, NARISHIGE JAPAN) to take place of Tungsten probe. The tip of glass capillary can be made by heat puller (PC-10, NARISHIGE JAPAN), which just takes about twenty seconds.

And in order to make it easier to observe under optical microscope, we coat gold on the transparent surface with sputtering machine (E-200S Canon ANELVA) (Fig. 2(c)).

And after cutting off the nano-mobile robot from the main body of Pt probe with the above mentioned glass capillary connected with manipulator1 (Fig. 3(a)), we remove Pt probe

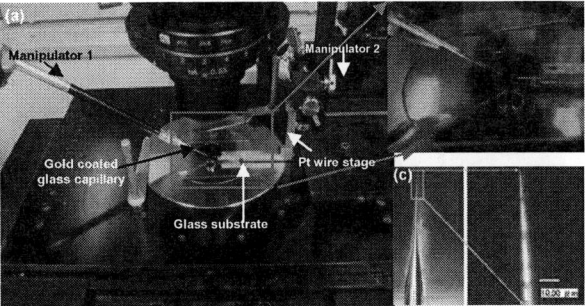

Fig. 2. Experiment setup photo. (a) The overall view of the setup containing two manipulators, glass capillary, Pt stage and so on. (b) Inset picture of glass substrate. (c) Gold coated glass capillary with 1um shape tip.

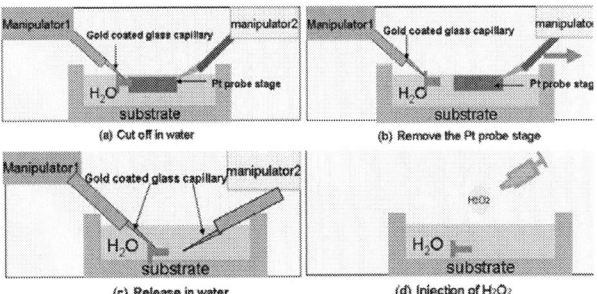

Fig. 3. Schematic of manipulation method for nano-mobile robot. (a) Cut off the 'T' letter shape nano-mobile robot. (b)Remove the Pt probe stage. (c)Release nano-mobile robot into water with chopsticks-like manipulators. (d) Injection of hydrogen peroxide.

stage and bring in the second glass capillary controlled by manipulator2 (Fig. 3(b)(c)). Two glass capillaries are operated together as a pair of chopsticks to release it into solutions. The process goes like this: use one glass capillary press onto the second one adhered with nano-mobile robot, it will be bent because of the pressure. And we release the pressure rapidly, the bend glass capillary will vibrate. Due to the vibration, the nano-mobile robot on it can be shaken off from it. Moreover, we do not have to worry about which place the nano-mobile robot will fly to, in fact in air we have to, but in liquid environment for micro nano size object, the surface viscosity drag force is dominant over inertia force [6]. As a result, the vibrated nano-mobile robot will be decelerated quickly and constrained within the observation area. The new and effective manipulation method increases success rate dramatically and can be used widely in other similar experimental cases.

III. EXPERIMENT RESULT

A. Trajectory of nano-mobile robot

As the catalyst of hydrogen peroxide's decomposition into water and oxygen, Platinum can increase the reaction rate and make the activity more intensive. As the result, there will be plenty of oxygen bubbles generated at the interface between Pt nano-mobile robot and solutions. Then the bubbles keep growing in size until big enough to detach from the it.

978-1-4673-1122-9/12 $31.00 © 2012 IEEE

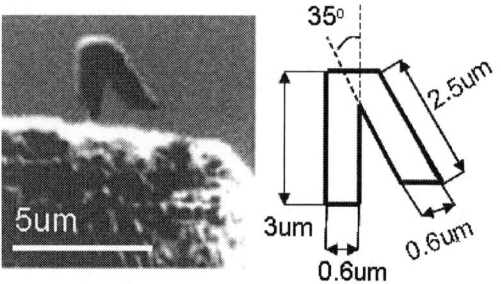

Fig. 4. FIB picture of 'V' letter shape nano-mobile robot with thickness of 0.8um.

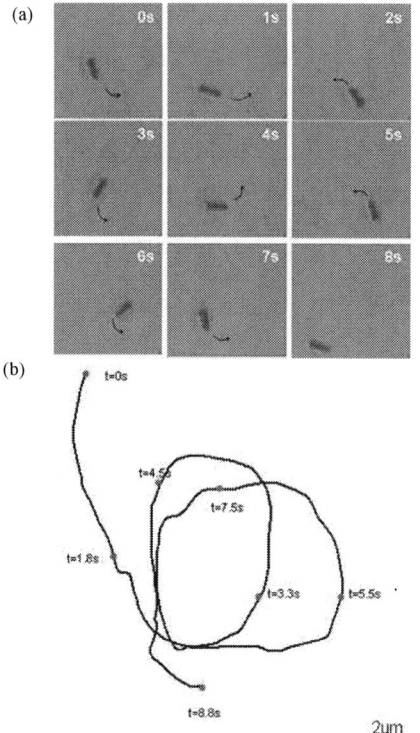

Fig. 5. Continuous photos and trajectory of 'T' letter shape nano-mobile robot (shown in Fig. 1). (a) The continuous photos taken in 8 seconds. (b) The trajectory of 'T' letter shape nano-mobile robot.

The momentum change of bubbles and nano-mobile robot's system leads to Platinum nano-mobile robot's movement in the opposite direction by recoiling [10].

But if the structure of nano-mobile robot is symmetric, for example like a cantilever beam, the movement will be hard to control and is kind of random movement. Because the oxygen bubbles' propulsion take place on the whole symmetric surface, the propulsion is offset with each other. That is the reason why we concentrate on the geometric design and also is the advantage of our method.

So we design and fabricate Platinum nano-mobile robot with special geometry like the above mentioned 'T' letter shape one (Fig. 1.). Follow the same fabrication process, we

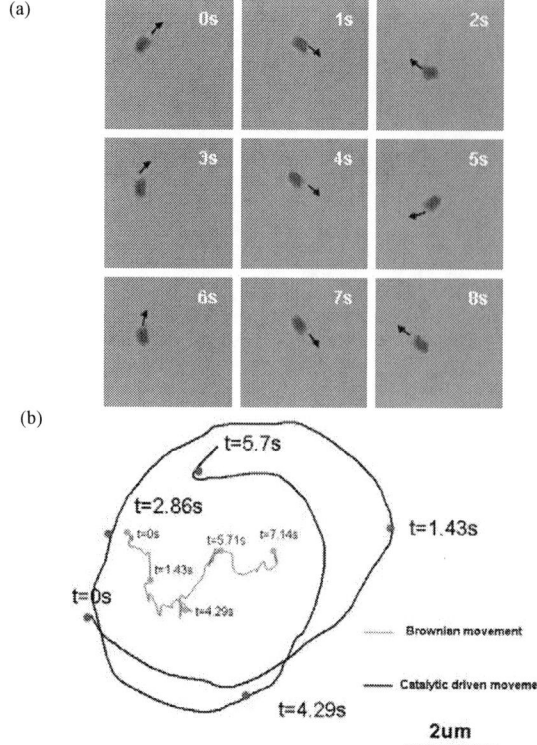

Fig. 6. Continuous photos and trajectory of 'V' letter shape nano-mobile robot (shown in Fig. 4). The pink line is Brownian movement, while catalytic driven movement trajectory is in dark blue. (a) The continuous photos taken in 8 seconds. (b) The trajectory of 'V' letter shape nano-mobile robot.

also work out the second type with 'V' letter shape shown in Fig. 4.

Follow the manipulation method (Fig. 3.), we can get the trajectory of nano-mobile robots with 'T' letter shape and 'V' letter shape respectively, which are shown in Fig. 5 and 6. We carry out the image processing of our experiment video with ImageJ (An open source software). Also the dimensions of the fabricated nano-mobile robot marked in this paper are also determined with ImageJ.

Fig. 5 demonstrates the trajectory of 'T' letter shape nano-mobile robot (shown in Fig. 1) with continuous photos during 8 seconds. In fact, it is keep swimming in rotary orbit and the trajectory and photos are just a small part of the whole process. Furthermore it is obvious that the trajectory is not a circle and the curvature is change every time, and the reason is the fluctuation of Brownian movement.

In order to compare oxygen bubble propulsion movement with random Brownian movement, we depict their trajectories in Fig. 6 with different colors respectively.

B. Analysis on the motion of nano-mobile robot.

In our experiment, we find an interesting result, as time goes nano-mobile robot's speed is decreasing slowly, if we do not add more solutions. The reason is considered to be, as we

978-1-4673-1122-9/12 $31.00 © 2012 IEEE

Table 1 Nano-mobile robot's rotation changes over time from high speed to low speed.

	Time (min)	Speed (um/s)
High speed	2	7.46
Middle speed	10	6.35
Low speed	23	5.56

Table 2 The relationship between concentration and average speed of nano-mobile robot.

Con. (wt.%)	Average Speed (um/s)	SD (um/s)
1.25	2.35	0.078
3.4	2.39	0.055
4.3	2.91	0.132
7.4	3.15	0.152
13.2	3.59	0.561
16.6	4.82	0.402
19.6	6.84	0.799
21.3	7.28	0.480
23.6	11.62	1.344

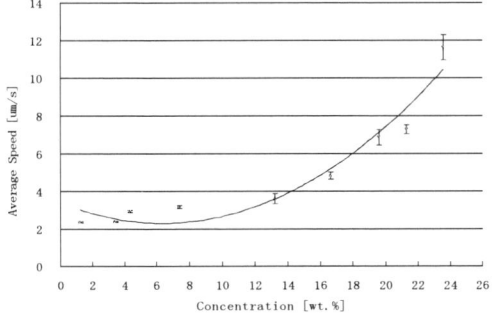

Fig. 7. The graph shows the speed of nano-mobile robot increases as the concentration of hydrogen peroxide changes from 1.25% to 23.6%, the trend line equation is $y=0.027x^2-0.348x+3.39$

mentioned above, our nano-mobile robot is propelled by the oxygen bubbles produced in the decomposition of hydrogen peroxide. So the concentration of the solution is changing from time to time. The longer time lasts, the lower the concentration is. Gradually the chemical reaction is becoming less intensive, which leads to the "fatigue" of our nano-mobile robot, just like a Marathon runner, it is hard to keep fast pace over all the distance. The relationship is shown in Table 1.

Inspired by this phenomenon, we use pipette to add more solutions to change the concentration shown in Fig. 3(d). And try to unveil how the concentration influences nano-mobile robot motion.

We use 'V' letter shape nano-mobile robot and work out two groups of data, one is concentration and corresponding speed, the other one is concentration and the mean diameter of nano-mobile robot's rotary trajectory. The relationships are depicted in Fig. 7 and Fig. 8. And the data is shown in Table 2 and Table 3, we calculate the average speed and mean diameter based on image processing with ImageJ.

As the concentration increases from 1.3% to 23.6%, the 'V' letter shape nano-mobile robot's corresponding speed increases from the original 2.76um/s by about 5 times to 11.67um/s.

Meanwhile, its rotation orbit is gradually enlarging with mean diameter changing from 32um to 52.2um. In fact, during

Table 3 The relationship between concentration and mean diameter of nano-mobile robot's rotary trajectory.

Con. (wt.%)	Mean diameter (um)	SD (um)
1.25	32.69	9.71
3.4	38.00	8.18
7.4	40.49	5.20
13.2	43.31	19.36
19.6	45.28	3.57
21.3	51.59	31.03

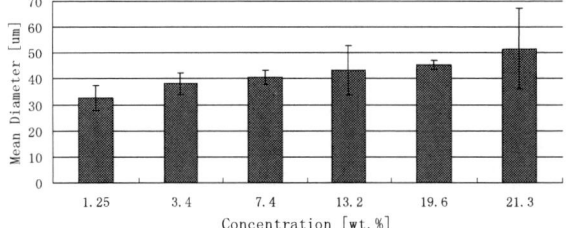

Fig. 8. The graph shows the mean diameter of nano-mobile robot's trajectory becomes longer as the concentration of hydrogen peroxide increases.

our experiment as the concentration increases, we have to adjust the optical microscope from high magnification to low level, or else the nano-mobile robot will move out of the observation area. So the analysis result keeps harmonious with our first hand experiment experience.

IV. CONCLUSION

We developed catalytic nano-mobile robot as a novel type of nano machine, which can be fabricated and manipulated with the simple and effective method introduced in the paper. The evaluation of its movement was presented from the aspects of solution concentration and its geometry. From its trajectory, averaged speed change was calculated to understand towards its mechanism.

As next step we will try to find some cases where our nano-mobile robot can be applied, especially in biological system.

References

[1] W. F. Paxton, K. C. Kistler, C. C. Olmeda, A. Sen, S. K. St. Angelo, Y. Y. Cao, T. E. Mallouk, P. E. Lammert, and V. H. Crespi, Catalytic Nanomotors: Autonomous Movement of Striped Nanorods. J. Am. Chem. Soc., 2004, 126 (41), 13424–13431.
[2] W. F. Paxton, S. Sundararajan, T. E. Mallouk, A. Sen, Chemical Locomotion. Angew. Chem., Int. Ed. 2006, 45, 5420–5429.
[3] G. A. Ozin, I. Manners, S. Fournier-Bidoz, A. Arsenault, A. Dream Nanomachines. Adv. Mater. 2005, 17, 3011–3018.
[4] W. F. Paxton, A.Sen, T.E. Mallouk, Motility of Catalytic Nanoparticles Through Self-Generated Forces. Chem. OEur. J. 2005, 11, 6462–6470.
[5] S. Fournier-Bidoz, A. Arsenault, I. Manners, G. A. Ozin, Synthetic Self-Propelled Nanorotors. Chem. Commun. 2005, 4, 441–443.
[6] R. Liu, A. Sen, Autonomous Nanomotor Based on Copper–Platinum Segmented Nanobattery, J. Am. Chem. Soc., Article ASAP.
[7] Y. He, J. Wu, and Y. P. Zhao, Designing Catalytic Nanomotors by Dynamic Shadowing Growth, Nano Lett., 2007, 7 (5), pp 1369–1375.
[8] R. A. Pavlick, S. Sengupta, T. McFadden, H. Zhang, and A. Sen, A Polymerization-Powered Motor, Angew. Chem. Int. Ed. 2011, 50, 9374–9377
[9] E. M. Purcell, life at low Reynolds number Am. J. Phys. 1977, 45, 3.
[10] J. J. Bao, M. Nakajima, Z. Yang, T.Fukuda, Fabrication and evaluation of Pt andAu hybrid and Geometric Pt nano vehicle, Proc. of MHS 2011 222-227.

Design and Fabrication of Diffractive Phase Element for Minimizing the Focusing Spot Size beyond Diffraction Limit

N. Atthi[1,*], S. Boonruang[2], W. Mohammed[3], W. Jeamsaksiri[1], C. Hruanun[1], and A. Poyai[1]

[1]Thai Microelectronics Center, National Electronics and Computer Technology Center, National Science and Technology Development, Chachoengsao, Thailand

[2]Photonics Technology Laboratory, National Electronics and Computer Technology Center, National Science and Technology Development, Pathumthani, Thailand

[3]School of engineering, Bangkok University, Pathumthani, Thailand

Corresponding author e-mail: nithi.atthi@nectec.or.th

Abstract—**This paper proposes a fabrication apparatus of high numerical aperture (NA) diffractive lens (Concentric Chirped Grating, CCG). The fabrication scheme is based on photolithography incorporating with Double Patterning (DP) technique and Litho-Etch-Litho-Etch (LELE) process. The CCG element having NA up to 1.4 in a glass substrate (n=1.5) at 940 nm wavelength and feature size down to 320 nm is successfully fabricated. The fabricated element can be very useful in integrated Surface Plasmon sensors and beam shaping applications. When controlling the shift between the odd and even rings during the second exposure, the novel phase element that can minimize the optical beam spot size beyond the diffraction limit is generated.**

Keywords-Diffractive Optical Element, Double Patterning Techinique, Concentric Chirped Grating

I. INTRODUCTION

Diffractive Optical Elements (DOEs) are commonly used in integrated optical systems as they are compact and integration of multiple elements on both sides of silica substrates can minimize the mechanical alignments [1]. The structure can be designed to enhance the system efficiency through controlling the structure parameters such as periodicity, fill factor and depth, for instance. Gratings, micro lenses, vortex elements and phase elements are common examples. Diffractive lens, such as concentric chirped grating (CCG) and Fresnel lens, is a common approach of realizing micro lenses. Achieving high numerical aperture (NA) micro-lenses is not trivial especially when implementing refractive optics. Using CCG element, high NA can be achieved through the proper design of the grating period rendering them more suitable for system integration. In this paper, CCG is implemented as a coupling device for integrated Surface Plasmon resonance (SPR) sensors proposed in reference [2.] The coupling scheme is to use a focusing beam at normal incidence for a compact and polarization independence system. Hence, the element should have high NA to ensure that the beam convergence covers the desired SPR angle and to enhance the dynamic range of the sensor. Achieving high NA requires small feature size and large element dimensions which can be in terms of few hundred nanometers feature size and > 5-mm in diameter.

Higher resolution structures with optical lithography requires a Reticle enhancement technology (RETs) such as Phase shifted mask (PSM), Optical proximity correction (OPC), Off-axis illumination, or Hyper NA immersion scanning system [3-4]. In contrary, the other non-optical lithography techniques have also been developed such as e-beam lithography (EBL) [5], and X-ray projection lithography (XPL) [6]. However, the limitation of these techniques is low throughput, small area fabrication, and high process cost. The complicity and cost of the process further increase when large element or a three-dimensional structure is desired. For low cost photo-lithography schemes utilizing i-line UV light (λ = 365 nm), the feature size can typically goes down to around 500 nm. To reduce the feature size without extra investment, it requires engineering the process such as double patterning (DP) technique using two consequent photolithographic steps [7, 8].

In this paper, DP technique with Litho-Etch-Litho-Etch (LELE) is the major fabrication scheme to fabricate CCG on silicone substrate. The design approach of CCG will be discussed in section II. The periodical structure such as the CCG structure is divided into two patterns: even and odd rings. Each pattern is exposed separately and aligned center to center to generate the micro-lens. The fabrication process and the experimental results will be included in section III and IV, respectively. In addition to the micro-lens realization, DP technique provides a flexibility of controlling the shift between the odd and even rings. With small shift, the element can be considered as an overlapping pattern of two CCGs that cause two focal points with close proximity. The interference of these two beams can modify the focal spot profile. With proper control of the shift, spot size reduction can be achieved below the diffraction limit. This novel element would be helpful for high resolution imaging and photolithography applications as reported using super lens, radial polarization, and 2-dimensional (2-D) plasmonic lenses [9-11], for example. The optical properties of the novel phase element will be calculated and discussed in section V.

II. CONCENTRIC CHIRPED GRATING

The CCG structure and focusing scheme are demonstrated in Fig. 1. The element can focus the beam into a spot via the first-order diffraction. At any radial location, r, on the CCG surface, the first order diffracted light converges into a spot at the focal point with an angle $\psi(r)$. Thereby, the period of the grating, $p(r)$, changes with radial location as defined in Eqn. (1).

$$p(r) = \frac{\lambda}{n_0 \sin \psi(r)} = \frac{\lambda}{n_0} \sqrt{1 + \left(\frac{f}{r}\right)^2} \quad (1)$$

Where λ is wavelength of operation. n_0 is refractive index of the transmission region. f is CCG's focal length.

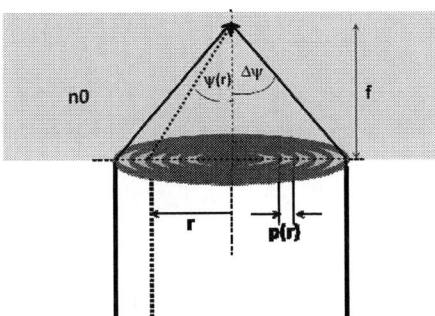

Fig. 1. CCG structure and focusing scheme.

For high NA, the sinusoidal term in Eqn. (1) approaches unity and the minimum period is approximately λ/n_0. The radius of the diffractive element has to be large enough to guarantee the desired diffraction angles. The minimum radius (r_{min}) of CCG that guarantee the desired NA can be defined as

$$r_{min} > \frac{f}{\sqrt{\frac{n_0^2}{NA^2} - 1}} \quad (2)$$

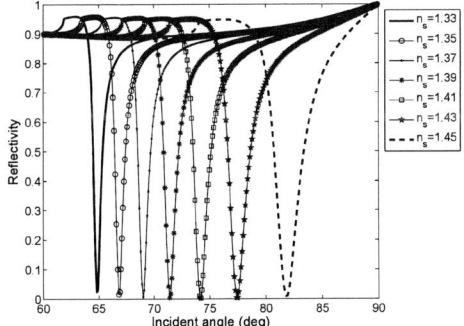

Fig. 2. SPR spectrum when using 940 nm wavelength light source.

For CCG to be used in integrated SPR sensor, NA should be high enough to ensure that the angular spectrum of the focused light covers the desired range of operation as defined in Eqn. (3)

$$NA = n_0 \sin(\Delta \psi) > \left(\Re \left\{ \sqrt{\frac{\varepsilon_m(\lambda)\varepsilon_{s,max}}{\varepsilon_m(\lambda) + \varepsilon_{s,max}}} \right\} \right) \quad (3)$$

In Eqn. (3), $\Delta\psi$ is a focusing beam's convergence angle as defined in Fig. 1. ε_m is permittivity of gold film. $\varepsilon_{s,max}$ is maximum value of substance's permittivity that can be detected.

SPR angle typically varies with exciting spectrum as well as range of detecting substance's refractive index (n_s). For detecting solution-based substance, SPR angle is at ~70 degree for red spectrum and at ~65 degree for near infrared light source. Realizing the fabrication limitation, CCG is designed here for near infrared light source with 940 nm wavelength. The SPR spectrum is plotted in Fig. 2.

As demonstrated, selecting an operation range $n_{s,max}$ =1.35 (SPR angle is at 67 degree), the CCG has to have a NA greater than 1.38 according to Eqn. (3). The minimum period and element size from Eqns. (1) and (2) are then 627 nm and $2.6f$, respectively. For a standard 1 mm thick glass slide (n_0=1.5, f=1 mm.), the element diameter is approximately 5 mm.

III. FABRICATION TECHNIQUE

Typically, the high density Sub-R pattern size can be fabricated by using DP technique. In DP technique, the design is split over two lithography layers in a way that the minimum pitch is relaxed (and preferably doubled) with respect to the target pitch [7]. The sequential step of litho-etch-litho-etch (LELE) process that used to produces a double trench patterns as shown in Fig. 3. [8].

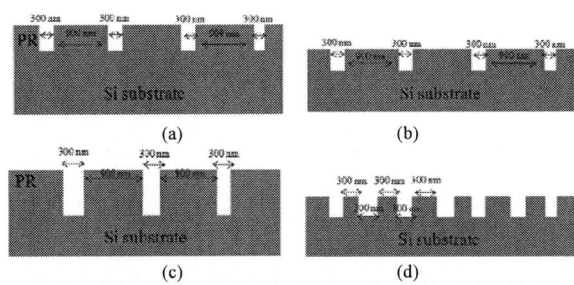

Fig. 3. The schematic of the typical double patterning (Litho-Etch-Litho-Etch, LELE) technique without shifting the pattern position.

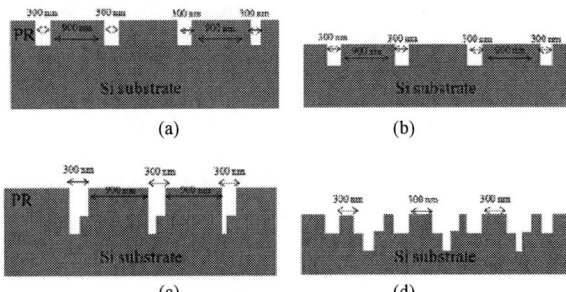

Fig. 4. The schematic of the double patterning (Litho-Etch-Litho-Etch, LELE) technique with shifting the pattern position.

978-1-4673-1122-9/12 $31.00 © 2012 IEEE   373

However, when shifting the wafer position during the second exposure step, the overlay between layer 1 (*L1*) and layer 2 (*L2*) is shifted and can be generated the 3-D sub-wavelength CCG with difference etch depth as the concept shown in Fig.4. To fabricate this structure, the designed chip diameter is 1 cm and the CCG pattern sizes (CD) are varied from 6.35 μm at the center point to 250 nm at the outer ring. The adjacent rings are separated into two mask layers as a green region (*L1*) and blue region (*L2*) shown in Fig. 5. First, the *L1* pattern was transferred to Si wafer coated with 1.09 μm of photoresist (PR) film. The photolithography process is carried out in Stepper with λ = 365 nm, and R = 0.50 μm. After developed and hard baked, the PR pattern is transferred to Si substrate by using reactive ion etching (RIE) technique with CF_4 plasma. The target etch through the Si substrate is 0.5 μm. After stripped the remained PR by using a mixture of H_2SO_4 and H_2O_2, the wafer is then coated with 1.09 μm of PR film again. During the second exposure, the *L2* pattern is shifted from *L1* etched pattern position to 0, 4, and 8 μm when compared to the origin. Then, the samples were characterized by optical microscope (OM), Scanning electron microscope (SEM), and the optical characterization.

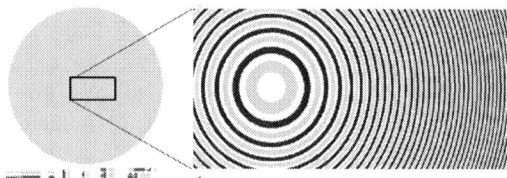

Fig. 5. The CCG layouts for LELE double patterning process; green region (*L1*) and blue region (*L2*).

IV. EXPERIMENTAL RESULT

After LELE process, the actual CD can be scaled down to 231 nm with 403 nm depth in the case of unshifted DP as depicted in Fig. 6a. and 6b. When the shifting distance is 0.4μm, there were some overlapped patterns between *L1* and *L2*, then the bottom CD was the combination between first and second exposure which totally CD was 555 nm with 417 nm depth as shown in Fig. 6c. and 6d. When the shifting distance has increased to 0.8 μm, the bottom CD is 126 nm with 820 nm depth as shown in Fig. 6e. and 6f. This means, the shifting distance during second exposure has an effects to the CCG pattern shape. The results in Fig. 7. show obvious mechanical interference pattern due to the shift. The interference between the two focusing spots can be simulated using Fresnel propagation of a phase pattern representing the designed Diffractive Optical Phase Element (DOPE). Controlling this shift can as well be used in trimming the intensity distribution at the focus plane as will be explained in the next section.

V. CONTROLLING THE SPOT SIZE

In most of the proposed schemes to reduce the point spread function of any optical element (OE), destructive interference between different waves or modes is commonly used. Following this concept, the proposed OE here utilizes destructive interference between two spots generated by shifting two high numerical CCG. Spot size is controlled by the shift. When applying the shift of the odd and even rings in the exposure process, an overlapping pattern is introduced as shown in experimental results Fig. 6 and 7. This pattern can be considered as an overlapping of two CCGs giving two focusing spots with close proximity.

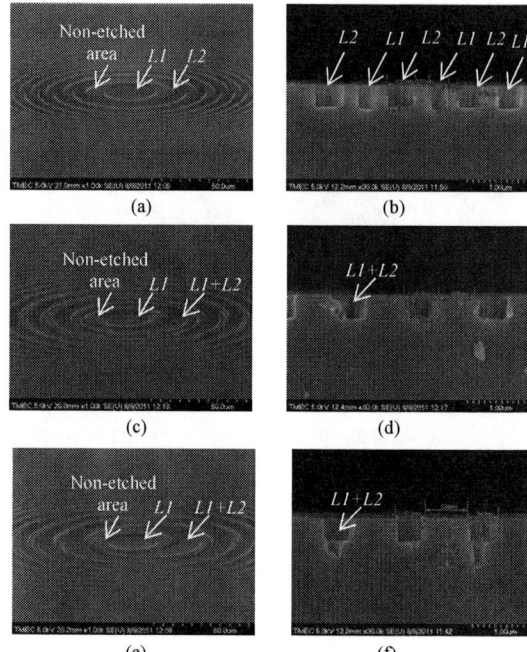

Fig. 6. 30° side view and cross-sectional view SEM images of the CCG structure on Si wafer (a-b) without shifted, (b) 4μm shifted, (c) 8 μm shifted.

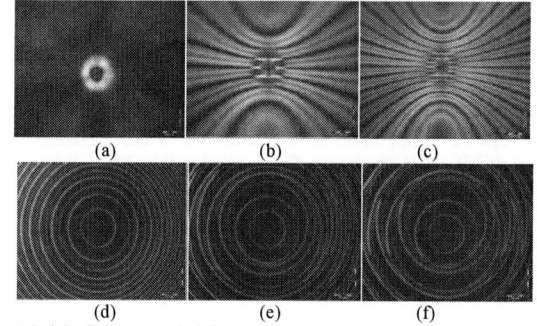

Fig. 7. Bright field and dark field top view optical microscope images of CCG pattern (a-b) without shifted, (c-d) 4μm shifted, (e-f) 8 μm shifted.

When optimally designed, the interference of two shifted focus spots can result in trimming the side slopes and hence reducing its extension in one direction. Using Fresnel propagation, the field profile at the back focal point of the element is calculated. In the simulations, the DOE is defined as a phase element as shown in Fig. 8a. The results in Fig. 8 are when applying the shift by 0.9 μm in y-direction. As seen in Fig. 8b, the modification of the sport size is clearly shown where the beam is trimmed along the y direction due to the interference effect

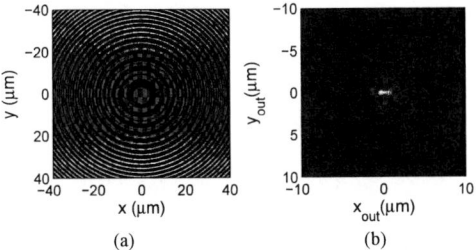

(a)   (b)

Fig. 8. The simulation aerial images when shifting the even and odd patterns by 0.9 μm (a) the phase pattern (b) the light intensity at the focal plane.

Changing the shift between the odd and even rings centers, cross section of the intensity profiles along the y-axis are plot in Fig. 9. The graphs show clearly an obvious reduction for shift values between 0.75 to 1.25 μm

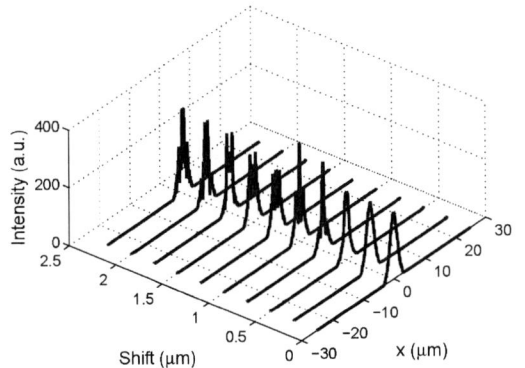

Fig. 9. Intensity profile of the focal point when varied the pattern shifting distance from 0.5 to 1.5 μm.

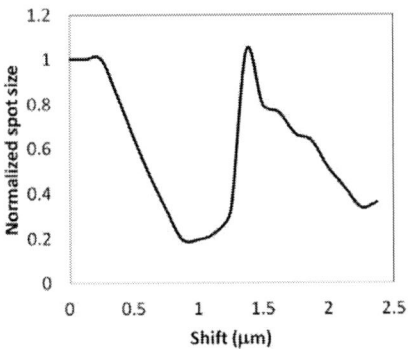

Fig. 10. The plots between the focal spot size and the pattern shifting distance.

The spot size is calculated as the full width half maximum of the profiles as depicted in Fig. 10. The plot shows spot size reduction up to 0.2 of the non-shifted case at a shift of 0.9 μm. The results clearly indicate the feasibility of achieving a spot size smaller than the case of no shift. The reduction is a result of destructive interference due to the production of two patterns: even and odd rings.

VI. SUMMARY

The paper presented a fabrication scheme to realize small feature sizes (below 400 nm) using double patterning technique and photo-lithography. Using this scheme, a dense high numerical aperture concentric chirped grating is realized in silicon substrate. The DP scheme divides the total pattern into even and odd rings where the total pattern is produced when aligning the centers of these rings. Controlling the separation between the two rings set, spot size reduction is achieved due to the destructive interference at the edges. In the calculations, a reduction of 0.2 is produced for a shift of 0.9 mm when using 940 nm light source.

REFERENCES

[1] N. Bokor1 and N. Davidson, "Ideal collimation, concentration, and imaging with curved diffractive optical elements," *Rev. Sci. Instrum.* vol. 76, pp. 111101, November 2005.

[2] S. Boonruang, M. Pitchumani, and W. S. Mohammed, "Polarization-independent on-axis light coupler for surface plasmon resonance using a concentric chirped grating," *Opt. Lett.,* vol. 36, pp. 3524-3526, September 2011.

[3] B.J. Lin, "Optical lithography— present and future challenges," *C. R. Physique,* vol. 7, pp. 858–874, November 2006.

[4] K. Ronse, "Optical lithography—a historical perspective," *C. R. Physique,* vol. 7, pp. 844–857, November 2006.

[5] R.F. Pease and S.Y. Chou, "Lithography and other patterning techniques for future electronics," *Proc. of the IEEE,* vol. 96, no. 2, pp. 248-270, February 2008.

[6] J.H. Bruning, "Optical lithography 40 years and holding," *Proc. of SPIE,* vol. 6520, pp. 1-13, Febuary 2007.

[7] K. Ronse, Ph. Jansen, R. Gronheid, E. Hendrickx, M. Maenhoudt, M. Goethals, G. Vandenberghe, "Lithography options for the 32nm half pitch node and beyond," *Proc. of IEEE Custom Intergrated Circuits Conference (CICC)* , pp. 15-1-1 – 15-1-8, 2008.

[8] R. S. Ghaida, G. Torres, and P. Gupta, "Single-Mask Double-Patterning Lithography," *Proc. of SPIE,* vol. 7488, pp. 74882J1-74882J-11, 2009.

[9] K. W. Kho, S. Xiang, and O. Malini, "Hyper-spectral confocal nano-imaging with a 2D super-lens ," *Opt. Exp.* vol. 19, pp 2502-2518 , January 2011.

[10] K. Kitamura, K. Sakai, and S. Noda, "Sub-wavelength focal spot with long depth of focus generated by radially polarized, narrow-width annular beam," *Opt. Exp.* , vol. 18, pp 4518-4525, Febuary 2010.

[11] X. Chen, V. Sandoghdar, and M. Agio, "Nanofocusing radially-polarized beams for high-throughput funneling of optical energy to the near field ," *Opt. Exp.,* vol. 18, pp 10878-10887, May 2010.

Enhancing Light Output of GaN-based Light-emitting Diodes With Nanoparticle-assembled On-top Layers

Cheng Zheng[1], Ling Sun[1], Xi Chen[1], Yan Shen[2], Peng Mao[1], Min Han[1]

[1]National Laboratory of Solid State Microstructures & Department of Materials Science and Engineering, Nanjing University,
Nanjing University, Nanjing 210093, China
[2]State Key Laboratory of Crystal Materials, Shandong University, Jinan 250100, China
sjhanmin@nju.edu.cn

Abstract—**We present a systemic study on the tailoring of light emission properties of the GaN-based light-emitting diodes (LEDs) with nanoparticle-assembled on-top layers. A layer of silver nanoparticles is deposited on the top of the InGaN/GaN quantum wells(QWs) in gas-phase. The coupling of spontaneous emission from InGaN QWs into the surface plasmon modes of silver nanoparticle layer is demonstrated. The effect of the silver nanoparticle arrays on the extraction efficiency of the emitted light is also investigated. We show that the nanoparticles on-top layers can be used to enhance the light output of LEDs either by increasing the internal quantum efficiency through surface plasmon coupling or by increasing the light extraction efficiency through near field interaction with the evanescent field induced by the total internal reflection of the light.**

Keywords: light-emitting diodes; surface plasmons; internal quantum efficiency; light extraction efficiency

I. Introduction

Light-emitting diodes (LEDs) are the subject of considerable interest for a wide range of applications from displays to optical communications. Recently, high-efficiency GaN-based LEDs have gained much attention because of their importance in the white LED lighting which is expected to replace the fluorescent lamps in the near future. The efficiency of LEDs has important consequences on such applications and thus has to be further improved. The internal quantum efficiency (IQE) is strongly influenced by nonradiative recombination processes. The light extraction efficiency (LEE) is limited by the total internal reflection (TIR) of generated light and successive re-absorption due to the high refractive index difference between LED structures and air [1]. Several techniques are being investigated to enhance the LED lighting efficiency, including internal quantum efficiency and light extraction efficiency. A promising approach that is attracting a lot of attentions is the use of metal/dielectric nanostructures [2-5]. It is believed that the coupling of spontaneous emission from InGaN/GaN quantum wells (QWs) with the surface plasmon (SP) modes of metal/dielectric nanostructures can be an effective way to enhance the IQE of GaN-based LEDs [6-8]. Nanoscale texturing on the surface of a light emitter is confirmed to be useful by enhancing the LEE from the dialect medium to the air [9-11].

In this work, we present a systemic study on the tailoring of light emission properties of the GaN-based LEDs with nanoparticle-assembled on-top layers. We have demonstrated that light output enhancement can be achieved by both increasing IQE and LEE of the light emitter. The two important factors, IQE and LEE of the light emitters, are investigated respectively by fabricating a dielectric spacer of different thickness between the QWs and the nanoparticle layer.

II. Experimental Details

The LED sample used in our experiment is consisted of InGaN/GaN multiple quantum wells (MQWs) grown onto the sapphire substrates by metal-organic chemical vapor deposition (MOCVD) method. Dielectric Si_xN spacer-layers either with a thickness of 10nm or with a thickness of 500nm were grown on the surface of the MQWs by plasma-enhanced chemical vapor deposition (PECVD). A layer of silver nanoparticles was then fabricated on top of the spacer-layers by means of gas-phase cluster beam deposition [12]. Figure 1 interprets the configuration of the fabricated GaN-based LEDs. With a thin spacer layer, the quenching of the luminescence due to the direct contact of the light emitter with the metal layer should be eliminated [13]. Taking into account the penetration depth of the SP evanescent field of Ag metal [14], a spacer layer of 500nm is thick enough to exclude QW-SP coupling and thus the exclusive influence of the nanoparticles on the light extraction can be investigated.

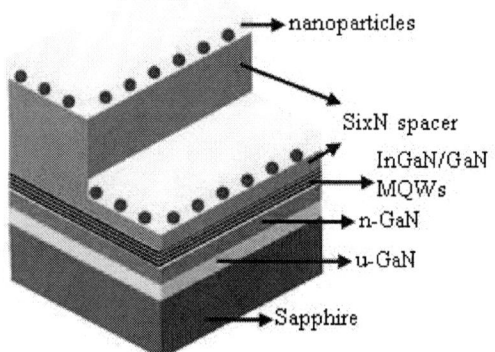

Fig. 1. 3D configuration of a InGaN/GaN light emitting diode(LED) with Ag nanoparticles assembled on-top layers. Si_xN spacer layers with different thickness are grown between the QWs and the nanoparticle layers.

978-1-4673-1122-9/12 $31.00 © 2012 IEEE

The scheme of the nanoparticle depositing process is show in Figure 2. Silver clusters were generated by a gas aggregation cluster source. A collimated beam was formed through a set of skimmers and collimators which separate the differential pumping stages. The nanoparticles could be deposited on the selected areas of the QWs samples by moving the selected area of the substrates into the cluster beam exposition spot by means of a micro-positioner. In the depositing process, the equivalent deposition rate was carefully stabilized at 0.2Å/s. A series of nanoparticle stripes with different coverage were fabricated on the top layer by controlling the deposition time. After deposition, the GaN-based LED sample is annealed at 200°C for 30 minutes in the vacuum.

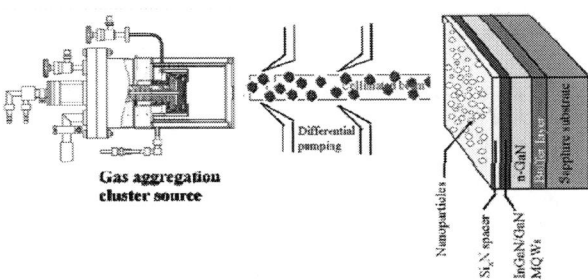

Fig. 2. The scheme of the depositing process by gas-phase cluster beam deposition.

Photoluminescence(PL) measurements were carried out by exciting the fabricated LED sample from the bottom side of the transparent sapphire substrate with a 405nm laser diode and collecting PL signals either from the same side of laser irradiation or from the obverse side which has the nanoparticle layers. The PL intensity was measured with a monochromator (Zolix Omni-λ 300) equipped with a Hamamatsu CR2658 photomultuplier-tube.

III. RESULTS AND DISCUSSION

Figure 3(a) shows photoluminescence (PL) spectra measured from the InGaN/GaN light emitting diode(LED) sample coated with silver nanoparticles on top of the 10nm-thick spacer layer. A systemic change of the PL intensity with the nanoparticle deposition mass can be directly observed. The PL intensity of the uncoated emitter peaks at 450nm, and no obvious shift of the PL peak has been observed after Ag deposition. With a low coverage of silver nanoparticles (deposition time of 50s and 100s), little changes on the PL intensity are shown as comparing to the uncoated emitter. However, the PL intensity presents an increase with the growing coverage over 300s deposition time. At high silver nanoparticle coverage with a deposition time of 2100s, a 4-fold enhancement in peak PL intensity is achieved. It indicates that, with a thin dielectric spacer layer of 10nm, the coupling of the spontaneous emission from InGaN/GaN quantum wells (QWs) into the surface plasmon (SP) modes of the silver

nanoparticle arrays is convincible. The electron oscillations of the silver nanoparticle layer generate dense intensity of SP modes within the InGaN/GaN active layer, which accelerates the electron–hole pair recombination process. On the other hand, when the energy of electron–hole recombination in InGaN/GaN QWs is coupled into the silver SPs, it can also make contribution to improve the spontaneous recombination rate of the InGaN/GaN emitter, and ulteriorly the internal quantum efficiency (IQE) [2,15,16].

The extinction spectra of silver nanoparticle arrays provide further evidences for the existence of QW-SP coupling in the InGaN/GaN light emitting diode (LED) samples. As shown in Figure 3(b), a red-shift of the surface plasmon resonance (SPR) wavelength of silver nanoparticles occurs with the increase of the deposition mass. The component between 405nm (PL excitation wavelength) and 450nm (PL emission wavelength) increases significantly at higher nanoparticle coverage. The coverage dependence of the PL enhancement can therefore be interpreted in terms of the matching of the SP modes of the silver nanoparticle arrays with the wavelength of the PL excitation and emission.

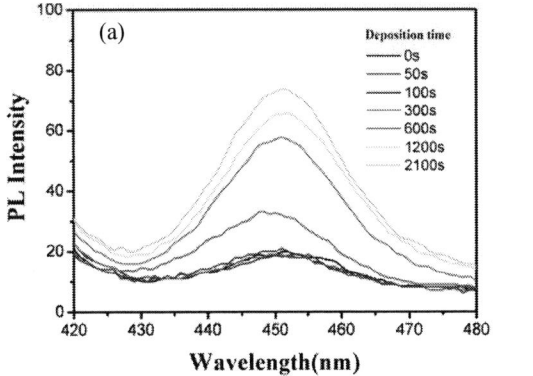

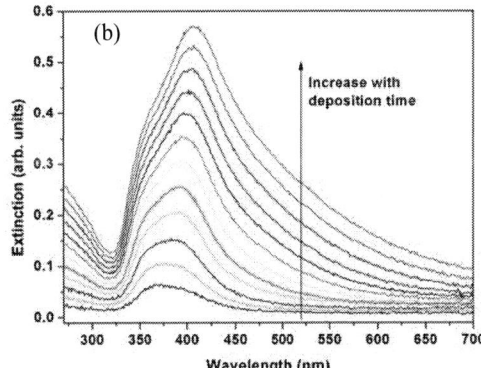

Fig. 3. (a) Photoluminenscence(PL) spectra of the InGaN/GaN QWs coated with Ag nanoparticle on-top layers. The legend shows the deposition time of the nanoparticles. (b) Extinction spectra of Ag nanoparticle layer with different deposition time.

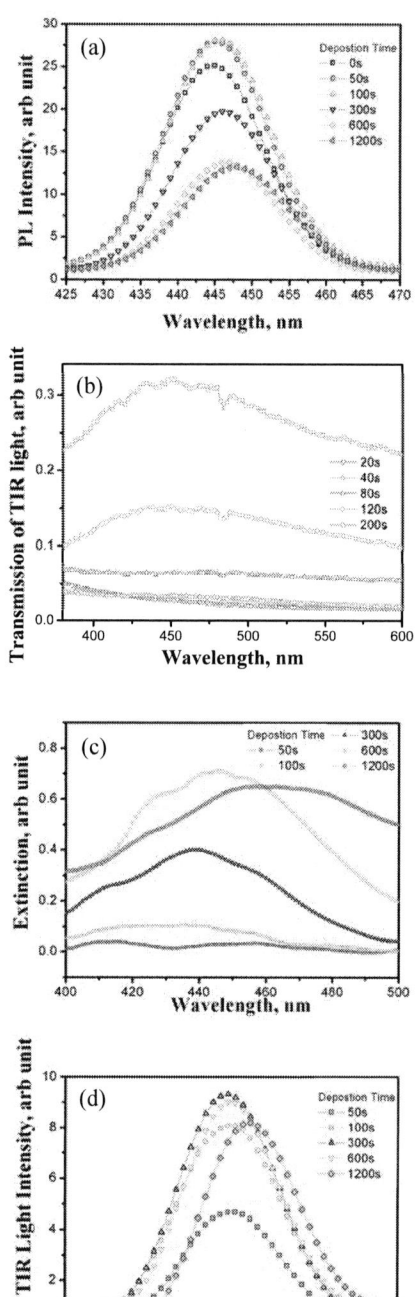

Figure 4(a) shows the PL spectra of the InGaN/GaN QWs sample coated with silver nanoparticles on top of the 500nm-thick spacer layer. At low nanoparticle coverage e.g. with a deposition time of 50s and 100s as shown in Fig. 4(a), a 10% enhancement of peak PL intensity is observed. It indicates a positive impact on the optimizing of light extraction efficiency of the LEDs, for the nanoparticle layers are so far away from the QWs layer that the influence of QW-SP coupling on the internal quantum efficiency can be eliminated.

To interpret the light extraction enhancement, we propose a mechanism of scattering of evanescent field by nanoparticle arrays. There is a distribution of evanescent field near the dielectric medium/air interface induced by the total internal reflection (TIR) of the light emission inside the GaN active layer. The SPs of the silver nanoparticles can interact to the evanescent field, which scatter a significant part of the evanescent wave into the radiative far field. We performed a quantitative measurement of such process on an analogous system by fabricating silver nanoparticle arrays on a spectro-prism surface. The results presented in Figure 4(b) show that the total internal reflection (TIR) of the light can be effectively extracted out. Keeping a relatively low silver nanoparticle coverage, the transmission of TIR is found to increase with the deposition time. With a 200s silver nanoparticle deposition, about 30% TIR light can be extracted at the peak wavelength of QWs light emission (450nm).

In Figure 4(a), no direct enhancement of the PL intensity is found on the area of the QWs surface with various heavier silver nanoparticle coverages (deposition time larger than 100s). However, if we take into account the extinction of the fabricated nanoparticle layers, as shown in Figure 4(c), the decrease of the PL intensity with the silver nanoparticle deposition time can be attribute to the light absorption of the metal layers, which may induce significant loss of the PL light that directly emit outside the QWs and counteract the benefit from TIR extraction. In Figure 4(d), the net PL intensity that comes from the extracted TIR is shown. A systemic change of TIR light extracted intensity with the nanoparticle coverage is revealed. At low (50s and 100s) and medium (300s) nanoparticle coverage, the extracted TIR light intensity increases with the deposition mass. But when it comes to high nanoparticle coverage (600s and 1200s), the extracted TIR light decreases with the increase of the deposition mass.

Furthermore, the peak of the spectrum also shows red shift, which corresponds to the red shift of the surface plasmon resonance (SPR) peak of the silver nanoparticle arrays, as shown in Fig. 4(c). There is a good consistency between the extracted TIR light intensity and the SPR intensity of the silver nanoparticle arrays, which demonstrate that the extraction of the TIR light can be attributed to the interaction of the SPs of the silver nanoparticles with the evanescent field accompany with the total internal reflection (TIR).

Fig. 4. (a) Photoluminenscence(PL) spectra of the GaN-base LEDs coated with Ag nanoparticle on-top layers. (b) Extraction efficiency of TIR light from glass with nanoparticle layers measured with a spectro-prism. (c) Extinction spectra of deposited Ag nanoparticles on LEDs. (d) Net PL intensity spectra of the GaN-base LEDs excluding the distribution of extinction.

IV. CONCLUSION

In summary, we have achieved enhancing light output of GaN-based light emitting diodes (LEDs) with silver nanoparticles assembled on-top layers. We demonstrate that the coupling of spontaneous emission from InGaN QWs into the surface plasmon modes of silver nanoparticle layer can significantly enhance the light output of LEDs by increasing the internal quantum efficiency. We also show that the nanoparticle on-top layers can be used to increasing the light extraction efficiency through near field interaction with the evanescent field that induced by the total internal reflection of the light. Our results indicate that to generate an exact enhancement of the light output of GaN-based LED with nanoparticle-assembled on-top layers, the coverage of the nanoparticles needs to be carefully controlled.

ACKNOWLEDGMENT

This work is supported by the National Basic Research Program of China (973 Program, Grant No. 2009CB930501), NSFC (Grant Nos. 10974092 and 51171077) and the Fundamental Research Funds for the Central Universities (Grant No. 1114021303).

REFERENCES

[1] Min-Ki Kwon, Ja-Yeon Kim, Baek-Hyun Kim, Il-Kyu Park, Chu-Young Cho, Clare Chisu Byeon, and Seong-Ju Park, Adv. Mater. 20, 1253–1257 (2008).

[2] K. Okamoto, I. Niki, A. Shvartser, Y. Narukawa, T. Mukai, A. Scherer, Nature Materials 3, 601 (2004).

[3] D. M.Yeh, C. F. Huang, C. Y. Chen, Y. C. Lu, C. C. Yang, Appl. Phys. Lett. 91, 171103(2007).

[4] R. Paiella, Appl. Phys. Lett. 87,111104(2005).

[5] J. J. Wierer, Jr, A. David, M. M. Megens, Nature Photonics 3, 163 (2009)

[6] C. Y. Chen, D. M. Yeh, Y .C. Lu, C. C. Yang, Appl. Phys. Lett. 89, 203113 (2006).

[7] Kwon M.K., Kim J.Y., Kim B.H., Park I.K., Cho C.Y., Byeon C.C., Park S.J., Adv. Mater. 20, 1253–1257 (2008).

[8] Chu-Young Cho, Min-Ki Kwon, Sang-Jun Lee, Jang-Won Kang, Se-Eun Kang, Sang-Heon Han, Seong-Ju Park, Dong-Yul Lee, Nanotechnology, 21, 205201(2010).

[9] Chul Huh, Kug-Seung Lee, Eun-Jeong Kang, and Seong-Ju Park, J. Appl. Phys. 93, 9383 (2003).

[10] J. K. Kim, A. N. Noemaun, F. W. Mont, D. Meyaard, E. F. Schubert, D. J. Poxson, H. Kim, C. Sone, and Y. Park, Appl. Phys. Lett. 93, 221111 (2008).

[11] F. Ishida, K. Yoshimura, K. Hoshino, and K. Tadatomo, Phys. Status Solidi C 5, 2083 (2008).

[12] M. Han, C. Xu, D. Zhu, L. Yang, J. Zhang, Y. Chen, K. Ding, G. Wang, Adv. Mater. 19, 2979 (2007).

[13] Ya-Ping Hsieh, Chi-Te Liang, Yang-Fang Chen, Chih-WeiLai and Pi-Tai Chou, Nanotechnology 18, 415707(2007).

[14] I. Gontijo M. Broditsky, E. Yablonvitch, S. Keller, U. K. Mishra, and S. P. DenBaars, Phys. Rev. B 60, 11564 (1999).

[15] J. Vuckovic, M. Loncar, and A. Scherer, IEEE J. Quantum Electron. 36, 1131 (2000).

[16] A. Neogi, C. W. Lee, H. O. Everit, T. Kuroda, A. Tackeuchi, and E. Yablonvitch, Phys. Rev. B 66, 153305 (2002).

The Chromatic Dispersion Module with Large Chromatic Focal Shift

M. C. Wei, K. Y, Hung[*], Y. J. Chuang, S. H. Huang
[*]Institute of Mechanical and Electrical Engineering, Ming-Chi University of Technology, Taiwan
kuoyung@mail.mcut.edu.tw

Abstract—**This paper integrates micro-lenses to develop a miniature chromatic dispersion module with 1.89 mm (400-800 nm) chromatic focal shift for the application of the profile and roughness detection of roll to roll process. The dimension of the designed and fabricated lens module is about 6 mm (long) X 2 mm (diameter) which is far smaller to the commercial system with 136 mm (long) X 25.4 mm (diameter). The module includes two lenses and self-alignment lens holder for assembling and alignment. The system design is based on the principle of afocal, transmission, stacked and refractive type lenses. This paper also compares the commercial lenses (3mm) and self-fabricated micro lenses (2mm) system for spot size and focal shift length. White light source was used to test the module could have similar spot sizes after light through the 10X/0.25 objective lenses at individual wavelength (500-650 nm). And the polymer convex lenses have 95 % high transparence in 400-800 nm. As a result of its long chromatic dispersion distance, causes the system could linearity operating at the region not to need the mechanical vertical scanning.**

Keywords— Chromatic dispersion, Afocal, Roll-to-Roll.

I. INTRODUCTION

Chromatic dispersion is a variation in the velocity of light according to wavelength. The chromatic dispersion technology has been widely used in conventional laser scanning or scanning confocal optical microscopes for biological tissues or living cells imaging and sample profile detection [1-4]. Popularly applications is due to the ability to reject light from out of focus planes and provide a clear in focus image of a specific cross section within the sample with high resolution. For focus of light onto the sample, the use of diffractive lenses [5] was presented. The dispersion resulting from diffraction phenomena is normally stronger than that from refraction, providing more sensitive wavelength selectivity. But diffractive lenses do not possess the same aberration as they are designed and fabricated with the exact parabolic structure.

Eric Altendorf (Mitutoyo Corp., Japan) [6] have filed a patent about the pen type chromatic dispersion module in 2007. The modules including the housing, doublet lens, positive power lens and aperture. Positive power lens includes the biconvex lens and meniscus lens. This patent has described the lenses design of the chromatic dispersion module but it didn't declare the optical specifications that this module could reach. Micro Epsilon Inc. (2006) [7] also announces several types about confocal chromatic optical pens for different applications,

such as: profilometry, microtopography, roughness, autofocus vibrometry, in-line inspection quality control and thickness measurements. The customers could choice based on the requirements about depth of field (130 µm to 27 mm), spot size, focal lengths (3.3 mm to 29 mm), working distance, object slope and photometric efficiency. The system price is about 15,000 US dollars.

The conventional chromatic dispersion modules are bulky and expensively not appropriate for many new applications where size reduction and portability are required. Actually, the miniaturized imaging or chromatic dispersion systems offering both high resolution and imaging capability are not commercially available. Size reduction and portability are desired for applications such as in vivo imaging, R2R sample profile, and handheld micro fluidic chips.

So the purpose of this paper is to provide a new design concept of a miniature chromatic dispersion module with 2 mm focal shift, the major components are integrated within 10 mm long by use of Micro-opto-electro-mechanical systems (MOEMS). And the module (lenses and lens-holder) could be batch fabricated, inexpensive and that can be arrayed for various parallel imaging applications. We present a method of achieving these purposes by stacking microfabricated optical components in the vertical direction. And high optically transparent polymer lens material NOA 63 (95% at 405nm) was applied to fabricate the refractive plano-convex lenses. It has the advantages in alignment and integration, since the optical path is passed through multiple substrates.

II. DESIGN AND SIMULATION

Confocal techniques in a number of related literatures, there are discussions to the lens, wavelength, refractive index and the relationship between the curvature radius [8-10]. As (1) shows:

$$F(\lambda) = \frac{1}{(n(\lambda)-1)(\frac{1}{R_1} - \frac{1}{R_2})} \qquad (1)$$

where R_1, R_2 are the curvature radius of the lenses, $n(\lambda)$ is the refractive index of the lens material. Differential (1) could have the change of the focal length due to chromatic dispersion of the material:

$$\delta F = -\frac{\delta n}{(n-1)}F \tag{2}$$

So the relative position δz, at which different spectral component of the light is focused, is given by the following (3)

$$\delta z = -\frac{\delta n}{(n-1)}(F_1 + F_2)\frac{F_3^2}{n_o F_2^2} \tag{3}$$

where n_o is the refractive index of the objective lens material. F_n is the focal length of the objective lens in the system.

The intensity in the z direction can be written approximately as (4)

$$I(z) = \left\{\frac{\sin[kz(1-\cos\alpha)]}{kz(1-\cos\alpha)}\right\}^2 \tag{4}$$

where k is the wave number $k = \frac{2\pi}{\lambda}$, $\sin(\alpha)$ is the numerical aperture of the image-forming system. In the confocal systems, if we consider a planar reflector axially scanned through the focus, the variation in the detected intensity as a function of its position is given by (5) [11]

$$I(u) = \left(\frac{\sin(u/2)}{u/2}\right)^2, u = \frac{8\pi}{\lambda}z\sin^2(\frac{1}{2}\alpha) \tag{5}$$

where u is a normalized axial coordinate related to the real axial distance z from the planar reflector position to the focal plane. In (4) the intensity drops to its half-maximum, i.e., to the 3-dB points. For the FWHM we find (6)

$$FWHM = \frac{0.44\lambda}{1-\cos\alpha} \tag{6}$$

α is the half angle of incident light. In this paper, we are interested in **longitudinal aberration (focal shift)** which the position of the focus image varies with the wavelength (F (λ_1) and F (λ_2)). If λ_1 and λ_2 are two different wavelength of the light source, then their focal distances $f(\lambda_1)$ and $f(\lambda_2)$ are formed on different positions along the longitudinal optical axis. The principle is come true by a diffractive lens in much paper [12–14]. The chromatic dispersion properties can be characterized for order +1 by

$$f(\lambda) = \frac{r_1^2}{2\lambda} = f(\lambda_d)\frac{\lambda_d}{\lambda} \tag{7}$$

where f is the focal position, λ is the operating wavelength, r_1 is the inner most radius of the Fresnel lens, and λ_d and $f(\lambda_d)$ are the design wavelength and the corresponding design focal length, respectively. Expressing (7) as a Taylor expansion around the design wavelength and eliminating the high-order terms, the chromatic dispersion can be expressed as

$$f(\lambda) \approx f(\lambda_d) - 2f(\lambda_d)(\frac{\lambda}{\lambda_d}) + 6f(\lambda_d)(\frac{\lambda}{\lambda_d})^2$$
$$-3f(\lambda_d)(\frac{\lambda}{\lambda_d})^3 + f(\lambda_d)(\frac{\lambda}{\lambda_d})^4 \tag{8}$$

If $f(\lambda_1)$ and $f(\lambda_2)$ are calculated for λ_1 and λ_2 respectively, then a segment of wavelengths $\Delta\lambda$ is created, and it is defined as $\Delta\lambda = f(\lambda_1) - f(\lambda_2)$.

According to design theory and reference literatures above. This paper initially intended to use ZEMAX software to design the specifications of optical components, such that the overall size of the dispersion module can be less than 10 mm.

Fig. 1a shows the design concept of the miniature chromatic dispersion module. This module includes two plano-convex micro-lenses and lens holder. Fig. 1b shows the diagram of application system. After simulating by ZEMAX software, we have the optimize lenses specific, dimension and its optics characteristics shown in Table 1. Figure 2 shows the optimize simulations results about the spot size at different wavelength (Fig. 2a) and chromatic focal shift (Fig. 2b). The specification of the commercial lenses module (Diameter is 3 mm) was also designed to compare with our micro lenses module (Diameter is 2 mm). Based on the Fig. 2 and Table 1, we could know the design in this paper have smaller diameter, shorter module length and larger focal shift that compare to commercial lenses module.

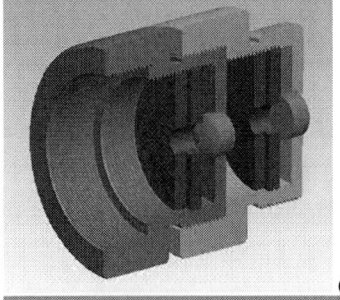

(a)

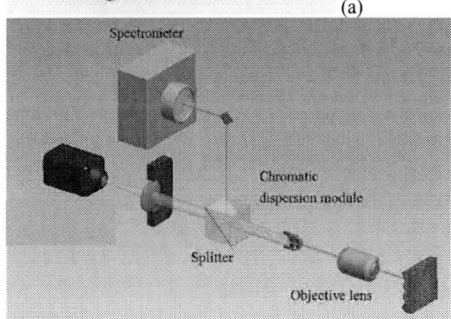

(b)

Fig. 1. (a) It shows the design concept of the miniature chromatic dispersion module. This module includes two convex micro-lenses. (b) Application system diagram.

The plano-convex microlens were designed as the hybrid structures, is composed of Norland Optical Adhesive 63

(NOA63) (refractive index: 1.5802 at 405nm) and BK-7 glass (refractive index: 1.5302). The lens profile is control by balancing **the gravity force and surface tension force**. Because of the surface tension, the lenses with smooth surface and spherical shape could be obtained.

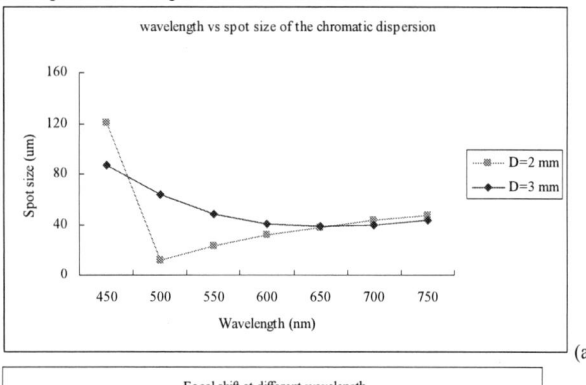

(a)

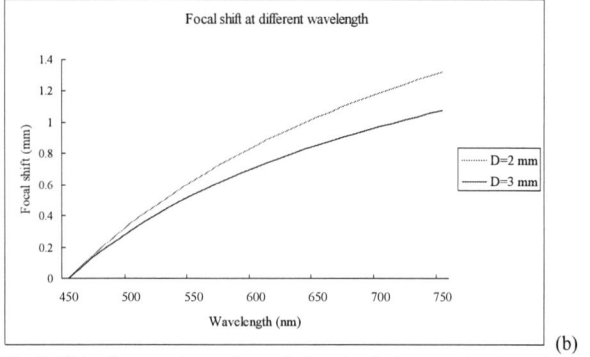

(b)

Fig. 2. This diagram shows the optimize simulations results about (a) spot size at each wavelength. (b) focal shift at different wavelength.

III. FABRICATE METHODS AND DISCUSSIONS

Figure 3 shows the fabricated process of the plano-convex lenses. The plano-convex microlens were designed as the hybrid structures, is composed of Norland Optical Adhesive 63 (NOA 63) (refractive index: 1.5802 at 405nm) and BK-7 glass (refractive index: 1.5302). The lens is fabricated by dispense technique to control the volume of the NOA 63 and to form the hydrophobic/hydrophilic area. The hydrophobic/hydrophilic area was forming by lift-off process. Because of the surface tension, the lenses with smooth surface and spherical shape could be obtained. Besides, the SU-8 post and hole were fabricated on BK-7 glass for precision alignment and for combine the two lenses. The lens holder was machined by us for alignment and fixing the two lenses.

Figure 4 shows the pictures of fabricate convex lenses. Figure 5a shows the test system diagram of chromatic module. Fig. 5b shows the picture of the real test system. It includes a white light source was coupling into the fiber, chromatic dispersion module, objective lens and Spectrometer. Fig. 5c-5d shows pictures of chromatic module after assembling. Fig. 5e-5f shows the spot diagram of the beam-profiler. It shows that the

commercial module and micro lenses module have similar results. It proves the micro lenses have the skill to replace the commercial lenses. Besides, the alignment problem was also resolved in this research. Fig. 6 shows the test results of chromatic modules in Fig. 1b's application system. The focal shift is about 2.2 mm (Fig. 6a) and spot size is about 50um (500-650 nm wavelength) (Fig. 6b) for our module.

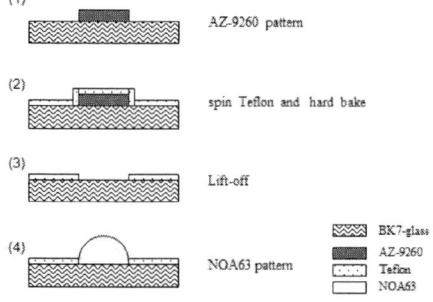

Fig. 3: The fabricated process of the plano-concave process.

(a)

(b)

Fig. 4. Convex lens (a) single lens (b)1 X 5 lens array

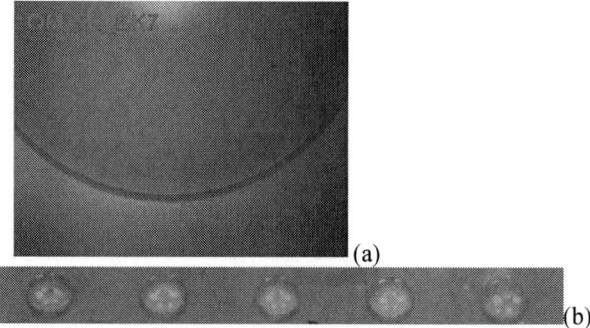

(a)

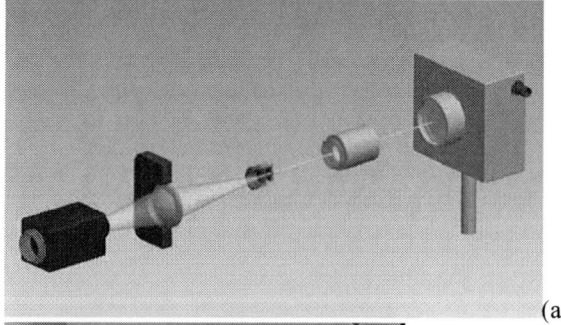

(b)

978-1-4673-1122-9/12 $31.00 © 2012 IEEE

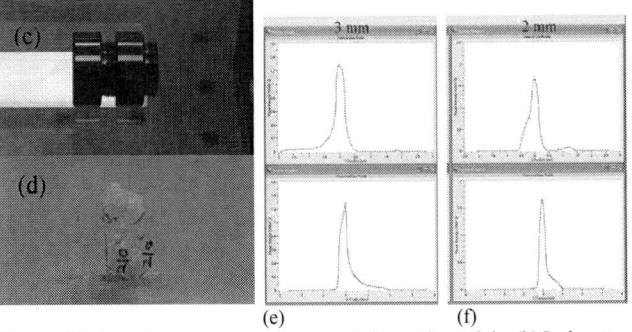

Fig. 5. (a) It shows the test system diagram of chromatic module. (b) It shows the picture of the real test system. The picture of chromatic module after assembling (c) commercial lenses (d) our micro lenses. The spot diagram of the beam-profiler for (e) commercial lenses (f) our micro lenses.

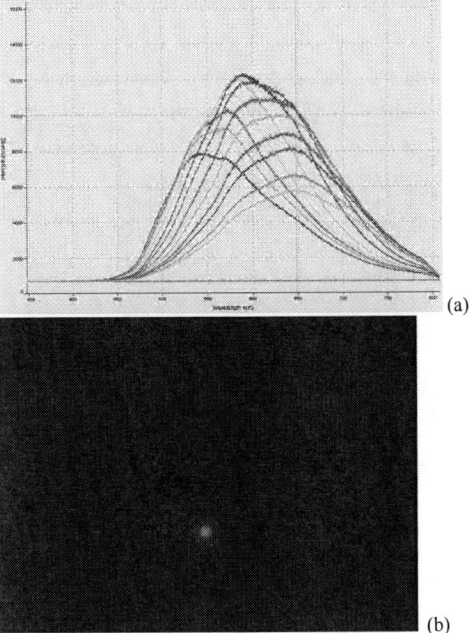

Fig. 6. Test results of the chromatic module in the application system of Fig. 1b (a) Spectrometer (focal shift~2.2 mm) (b) Spot size of our lenses module ~50um (500-650 nm)

CONCLUSION

In summary, we have demonstrated the design of the miniature chromatic dispersion module by use of Micro-opto-electro-mechanical systems (MOEMS). The module including two plano-convex micro-lenses and the dimension is smaller than 6 mm (long) X 2 mm (diameter). We demonstrated stacked two-dimensional microlens that achieved two-dimensional transmissive optical path that passed through multiple substrates. This system could realize the chromatic dispersion with 1.89 mm long chromatic focal shift, causes the system could linearity operating at the region not to need the mechanical vertical scanning.

REFERENCES

[1] K. Shi, P. Li, S. Yin, and Z. Liu, "Chromatic confocal microscopy using supercontinuum light," *Optics express*, vol. 12, pp. 2096, 2004.
[2] K. Shi, S. H. Nam, P. Li, S. Yin, Z. Liu, "Wavelength division multiplexed confocal microscopy using supercontinuum," *Optics Communications*, vol. 263, pp. 156–162, 2006.
[3] S. Bargiel, L. Nieradko, M. Jozwik, and C. Gorecki, J. A. Dziuban, "New generation of fully integrated optical microscopes on-chip:application to confocal microscope," *MEMS, MOEMS, and Micromachining II, Proc. of SPIE* 6186, 618602, pp. 1-9, 2006
[4] J. Garzon, T. Gharbi and J. Meneses, "Real time determination of the optical thickness and topography of tissues by chromatic confocal microscopy," *Journal of Optics A: Pure and Applied Optics*, vol. 10, no. 10, 104028, 2008.
[5] E. Papastathopoulos, K. Korner, and W. Osten, "Chromatic confocal spectral interferometry," *Applied Optics*, vol. 45, pp. 8244-8252, 2006.
[6] Eric Altendorf, "Chromatic Sensor Lens Configuration," *US patent* 7626705, December 2009.
[7] Micro epsilon Inc. datasheet, "Confocal chromatic displacement sensors," pp 1-12.
[8] J. Garzon, T. Gharbi and J. Meneses, "Real time determination of the optical thickness and topography of tissues by chromatic confocal microscopy," *J. Opt. A: Pure Appl. Opt.*, vol. 10, 104028, 2008.
[9] H. J. Tiziani and H.-M. Uhde, "Three-dimensional image sensing by chromatic confocal microscopy," *Applied Optics*, vol. 33, no. 10, pp. 1838-1843, 1994.
[10] K. Shi, P. Li, S. h. Yin, and Z. W. Liu, "Chromatic confocal microscopy using supercontinuum light," *Optics Express*, vol. 12, no. 10, pp. 2096-2101, 2004.
[11] T .Wilson, "*Confocal Microscopy,*" (London: Academic), 1990.
[12] D. Faklis and G. M. Morris, "Spectral properties of multiorder diffractive lenses," *Appl. Opt., vol.* 34, pp. 2462–2468, 1995.
[13] J. Jahns and S. J. Walker, "Two-dimensional array of diffractive microlenses fabricated by thin film deposition," *Appl. Opt., vol.* 29, pp. 931–936, 1990.
[14] S. L. Dobson, P. C. Sun and Y Fainman, "Diffractive lenses for chromatic confocal imaging," *Appl. Opt., vol.* 36, pp. 4744–4748, 1997.

Table 1. After simulating by ZEMAX software, we have the optimize lenses specific, dimension and its optics characteristics (our lens diameter is 2 mm).

	Material	Curvature radius	Thickness (mm)	Refractive index	module length
Plano-Convex lens (2 mm)	BK-7+ NOA 63	1.98	0.27	Nd:1.580, Vd:42.40 (NOA 63)	5.74 mm
Chromatic focal shift: 1.89 mm (400-800 nm)					
Commercial lenses (3 mm)	BK-7	3.1	1.8	1.5302 (at λ =404 nm)	9 mm
Chromatic focal shift: 1.19 mm (400-800 nm)					

A Multi-view Reflective Three-dimensional Display

Jun-Fu Chuang, Kerwin Wang

Department of Mechatronics Engineering, National Changhua University of Education
Changhua, Taiwan
kerwin@cc.ncue.edu.tw

Abstract—**This research presents a novel volumetric 3D display which takes advantages on the self-assembly technology to prepare a multi-view reflective screen. The screen consists of a simple cubic metallic microsphere array. Each microsphere is well aligned to 69 independently-projected-image-pixels. This process is called pixel-mapping. The purpose of pixel-mapping process is to assign a image pixel to its own position on the reflective microsphere. It can reflect the incident lights to a predetermined orientation because of the spherical reflective surface. That can reduce the perspective phenomenon in traditional volumetric 3D display images. The testing results prove that the self-organized metallic microsphere array works well corresponded to carefully aligned and projected pixels and frames. The novel design successfully exhibit different volumetric 3D images from different viewing angles.**

Keywords–volumetric display; self-organize; multi-view.

I. INTRODUCTION

3D display technologies attract many researchers' attention in recent years. Traditional 3D display technologies, such as optical shutter glasses [1] and parallax barrier 3D LCD [2], provide good experiences of 3D vision; however, users must wear specific glasses or situate in a particular position to watch the screen. Volumetric display technologies have been grabbing a significant of attention lately [3-8], such as Felix's spiral surface system [3]; viewers can watch stereoscopic images directly without wearing specific glasses or sitting in a specific position.

In this paper, we propose a novel volumetric 3D display, which consists of an LCoS projector with a focusing lens set, an optic stop, a coil motor and a reflective screen. The sweeping reflective screen, shown in Fig. 1, integrates a 760μm microsphere array by self-assembly technique to improve the 3D image rendering. "Self-assembly" is a terminology to describe the metallic microsphere self-organization process. With the assistance from intrinsic potential trap, one can assemble randomly dispensed microspheres to form a simple-cubic-packed microsphere array. The reflective screen is driven by a coil motor for voxel sweeping. In what follows, the design and fabrication of the reflective screen are introduced and that are followed by the discussions and the experiment results.

II. DESIGN AND FABRICATION

The configuration of this volumetric 3D display is integrated with LCoS projector with a focusing lens set, an optic stop, a coil motor and a reflective screen. Fig. 1 shows

the self-organized metallic microsphere array and the principle of this reflective screen. The screen is made of a self-organized metallic microsphere array. Each microsphere is well aligned to projected image pixels which are produced through pixel-mapping process. In this process, we create the projecting image pixels by using MATLAB program to assign individual image pixel to a specific location on the spherical surface. The incident lights will reflect separately to a predetermined direction because of its spherical surface, as shown in Fig. 2. That can reduce the perspective phenomenon in traditional volumetric three-dimensional display.

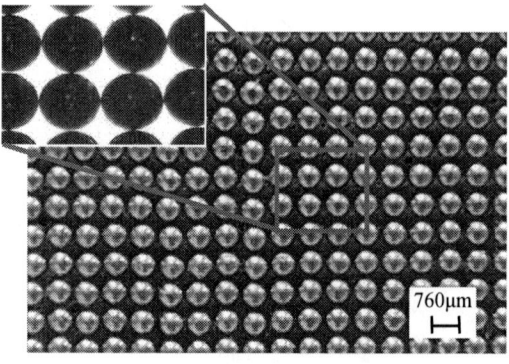

Fig. 1. A self-organized metallic microsphere array, which formed by a simple cubic lattice structure.

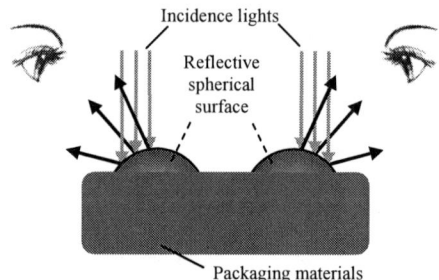

Fig. 2. The incident lights will reflect separately to a predetermined direction because of its spherical surface.

Fig. 3 shows the process flow of self-organized metallic microsphere array. First, we prepared an $18\times12\times0.6\text{mm}^3$ screen template. We tilted the template ((α, γ) = (5°, 45°)) to keep the corner of template at the lowest potential energy as shown in Fig. 4. It can help us to form a simple-cubic-packed array. Second, we dripped IPA (isopropyl alcohol) with

978-1-4673-1122-9/12 $31.00 © 2012 IEEE

760μm metallic microspheres into the screen template. IPA can help microspheres rolling smoothly during the assembly process. With the assistance from gravity; randomly dispensed microsphere formed a single layer simple cubic lattice array (Fig. 5). After transferring by blue tape with glass slice, we use epoxy glue to immobilize microsphere array. Because of epoxy glue is contained ferromagnetic ingredients, placing a magnet under glass slice can help epoxy glue pouring more quickly and smoothly than regular epoxy. It is also good for bubble removing and contrast enhancing. After epoxy glue curing, we remove the blue tape and clean the microsphere array with acetone to wash and to remove the residual from blue tape. The textured reflective screen is completed and shown in Fig. 6.

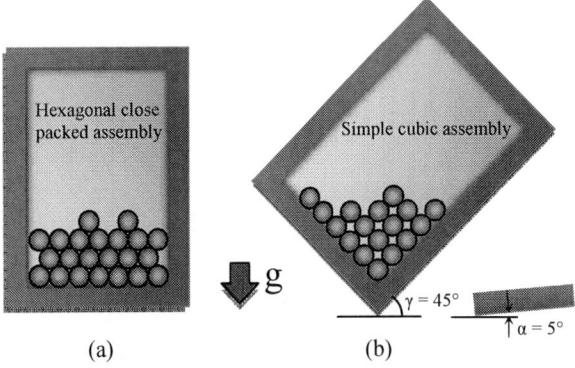

Fig. 4. Square template tilting for potential traps manipulation. (a) HCP lattice. (b) Simple cubic packing.

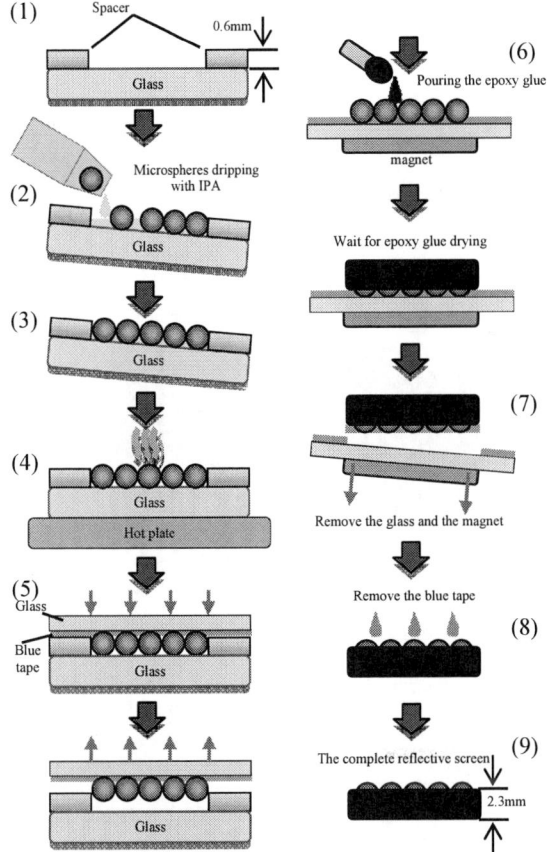

Fig. 3. The fabrication process for the reflective screen. (1) preparing a screen template, (2) metallic microspheres spreading, (3) metallic microsphere self-organization, (4) baking and drying, (5) microspheres transferring by blue tape, (6) microspheres immobilization, (7) screen peeling, (8) blue tap cleaning and removinge, (9) a fabricated reflective screen. Its size is 18×12 (mm²), thickness is 2.3mm, and the diameter of microsphere is 760μm.

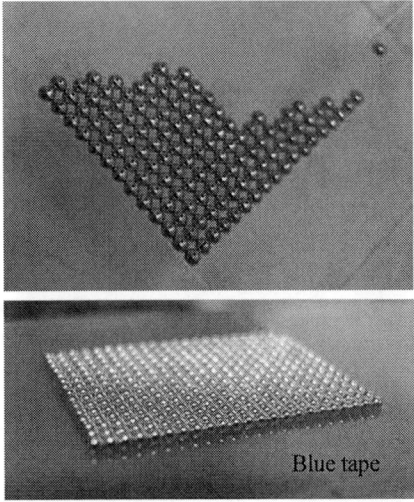

Fig. 5. With the assistance from gravity and screen template; randomly dispensed microsphere formed a single layer simple cubic lattice array.

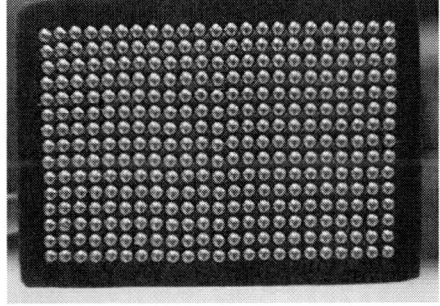

Fig. 6. The completed reflective microsphere array, it immobilized by the epoxy glue.

The reflective screen, consists of 345 (23×15) microspheres, uses an embedded microsphere array as an anisotropic diffuser for multi-view rendering to reconstruct 3-D scenes.

Fig. 7 shows the optical simulation diagram of reflective microsphere array which is generated by LightTools® software. When the light grid (parallel lights in this simulation) incidents to spherical surface, the reflected light beams will be isotropically dispersed. If the light beams are limited to specifically designed regions on spherical surface, it will reflect to predetermined direction only, as shown in Fig. 8.

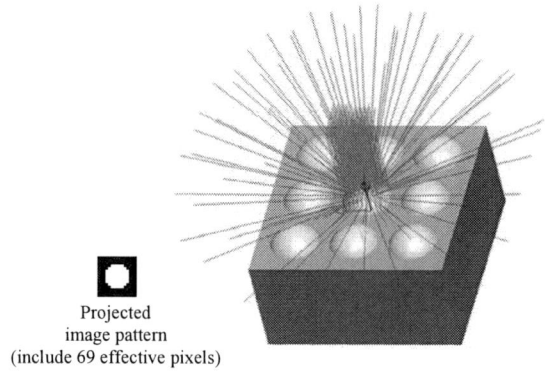

Projected
image pattern
(include 69 effective pixels)

Fig. 7. The optical simulation diagram of reflective microsphere array which simulated by LightTools® software. When the light grid incidents to spherical surface, the reflected light beams will be dispersed isotropically.

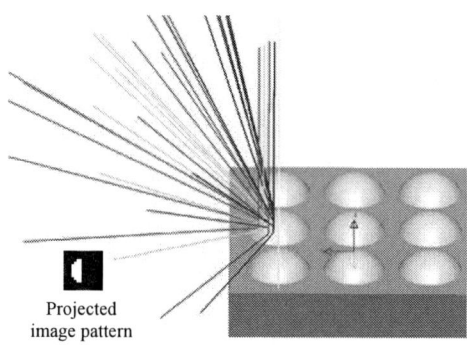

Projected
image pattern

Fig. 8. If the light beams are limited to specific region on spherical surface, it will reflect to a predetermined direction.

III. EXPERIMENTS AND RESULTS

A. Experiment Setup and 3D Imaging

Fig. 9 shows the experiment setup of the volumetric display. The sweeping microsphere array can be regarded as a continuous moving reflective screen. The configurations of projected image pixels are created by a homemade MATLAB program (Fig. 10). By a well aligned pixel-mapping process, the image pixel unit is consists of two kinds of pixels, bright and dark. The bright pixels, called effective pixels, can incident to spherical region on the screen. The dark pixels are the beams which incident to packaging materials (black cured epoxy). Because of the epoxy glue can't reflect the image pixels, the dark pixels, defined as ineffective, pixels will not be illuminated. Each spherical reflective imaging unit has 225 pixels, which includes 69 effective pixels. The fillfactor of projecting images is defined as (1):

$$Fillfactor = \frac{Effective\ pixels}{Total\ pixels} \qquad (1)$$

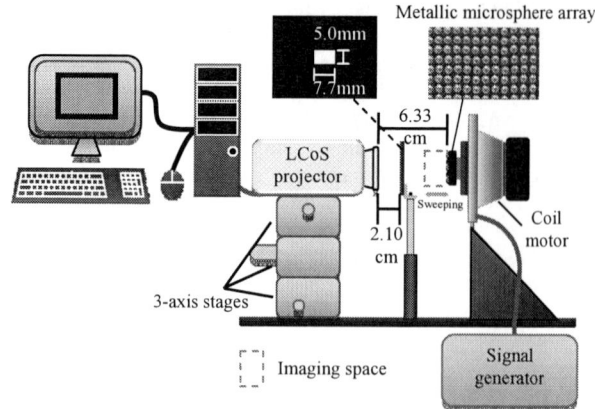

Fig. 9. The experiment setup of multi-view reflective 3D display, it includes a LCoS projector with a focusing lens set, an optic stop, a screen and a coil motor.

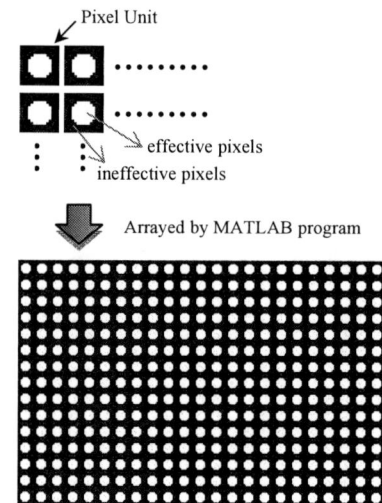

Fig. 10. Projecting images created by a homemade MATLAB program.

According to (1), the fillfactor of designed projected images is 0.307. The effective pixels are precisely aligned to the center of each single spherical surface. It projects a serial of two-dimensional image frames on reflective screen (Fig. 11(a)). When the reflective screen sweeps by coil motor, one can generate stereo images in the space. The reflected image frames can be viewed in all directions. The three-dimensional image produced by volumetric 3D display shows in Fig. 11(b).

B. Image Resolution Estimating

In previous research [9], we estimated the space resolution of 3D images of our volumetric display. The Z resolution, R_Z, is defined as the number of 3D voxel levels of 3D images in Z-direction. In bidirectional sweeping projection, the number of 3D voxel levels is an integer number derive by image output frequency divided by the image screen sweeping frequency. The space resolution R_{sp} is the product of projecting 3D images' resolution (800×600 in this case) and Z resolution.

Table I shows the results of image resolution testing.

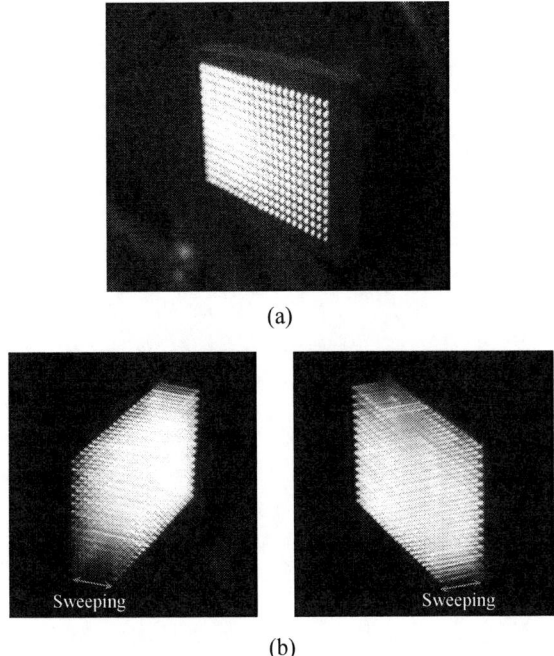

(a)

(b)

Fig. 11. (a) The two-dimensional image frames which is projected on reflective screen. (b) The three-dimensional image produced by volumetric 3D display.

TABLE I.   RESULTS OF IMAGE RESOLUTION TESTING

	Test I	Test II
Screen sweeping frequency (Hz)	60	30
Image output frequency (Hz)	60	60
The number of 3D voxel levels of 3D images in Z-direction. (R_Z) (estimate)	1	2
Space resolution (voxels) (estimate)	480000	960000
Effective space resolution (voxels) (estimate)	23805	47610

C. Multi-view 3D Imaging and Testing Results

In order to produce multi-view 3D images, we assign fillfactor designed image pixels to their own position on the reflective microsphere. Fig. 12(a) shows the testing image pattern which produced by pixel-mapping process. The image pattern is projected to reflective screen. The incident lights can be reflected separately to a predetermined orientation (Fig. 2). In this way, users can watch different image patterns from different viewing directions (Fig. 12(b, c)). That can reduce the perspective phenomenon in traditional volumetric 3D display images by this imaging method.

IV. CONCLUSIONS AND DISCUSSIONS

According to the experiment results, a self-organized microsphere array can be used to generate multi-view three-dimensional image. The projected image pixels need to have

precision alignment to the spherical reflective region. A screen made of simple cubic microsphere lattice has lower fill factor than hcp lattice, however, it is easy to use and easy to align. It can yield improved contrast.

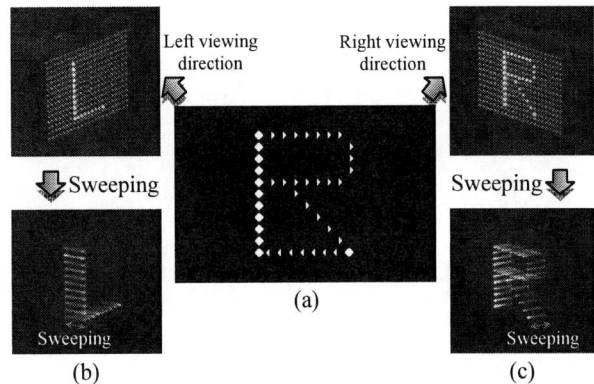

(a)

(b)   (c)

Fig. 12. (a) The testing image pattern which produced by pixel-mapping process. (b) The image pattern which displayed in left viewing direction. (c) The image pattern which displayed in right viewing direction.

ACKNOWLEDGMENT

The authors thank National Science Council. This work is supported by National Science Council, NSC 98-2221-E-018-012-MY2 and NSC 100-2221-E-018-015. The authors also thank Optical Research Associates to provide LightTools® educational software for concept verification.

REFERENCES

[1] R. Okoshi, "Three-Dimensional Displays," *proceeding of IEEE*, vol.68, pp. 548-564, 1980.

[2] I. Sexton, "Parallax Barrier Display Systems," *IEE Colloquium on Stereoscopic Television*, pp. 5/1 - 5/5, 1992.

[3] K. Langhans, D. Bahra, D.Bezecnya, D.Homanna, K. Oltmanna, K. Oltmanna, C. Guilla, E. Riepera, and G. Ardeyb, "Felix 3D Display: An Interactive Tool for Volumetric Imaging," *Proceedings of SPIE*, "Stereoscopic Displays and Virtual Reality Systems IX," *IS&T/SPIE's 14th International Symposium at Photonics West 2002 On "Electronic Imaging: Science And Technology"*, 20-25 January 2002.

[4] S. Noguchi,T. Inaba, H. Igarashi, "Semi-Three-Dimensional Visualization of Electromagnetic Field Analysis Result With Volumetric Display," *IEEE Transactions on Magnetics*, vol. 47, no. 5, May 2011.

[5] B. Mora, R. Maciejewski, M. Chen, and D. S. Ebert, "Visualization and Computer Graphics on Isotropically Emissive Volumetric Displays," *IEEE transactions on visualization and computer graphics*, vol. 15, no. 2, March/April 2009.

[6] D. Miyazaki, T. Honda, K. Ohno, and T. Mukai, "Volumetric Display System Using A Digital Micromirror Device Based On Inclined-Plane Scanning", *journal of display technology*, vol. 6, no. 10, October 2010

[7] Q. Fan, C. Shen, and Li Li, "Research of A High-Resolution Volumetric 3D Display System," *2009 Second International Conference on Intelligent Networks and Intelligent Systems*, pp. 397-400, 2009.

[8] J. Xing, H. Gong, Li Li, and W. Pan, "A Highly Parallel Beam-Addressed True Three-Dimensional Volumetric Display System," *2010 Symposium on Photonics and Optoelectronics*, pp. 1-5, 2010.

[9] J.F. Chuang and K. Wang, "Using a Self-Organized Metallic Microsphere Array to Enhance Image Brightness and Contrast for Volumetric 3D Displays," *2011 World Congress on Intelligent Control and Automation (WCICA 2011)*, p.p. 904-907.

Use Bionic Microlens Array and CMOS Image Sensor for Three-Dimensional Motion Detection

Chung-You Liu, Jun-Fu Chuang, Ting-Chieh Yu, Kerwin Wang

Department of Mechatronics Engineering, National Changhua University of Education, Taiwan
kerwin@cc.ncue.edu.tw

Abstract—**This paper proposed a novel three dimensional motion detection design. This design consists of a bionic microlens array, which integrates an aberration limited four-lens system and a CMOS image sensor to mimic insect compound eye vision. The microlens array, made from time multiplexed SF_6/O_2 plasma and PDMS (polydimethylsiloxane)-molding, is directly placed on the top of CMOS image sensor. A LED testing setup is prepared to project referent rays through the artificial ommatidium to the image sensor for 3D motion analysis and characterization. A computational method also has been established to convert sensor image into displacement. Both analysis and experiment results demonstrate that the design can predict the three-dimensional positions with average error less than 2.41%.**

Keywords-bionic; microlens; motion detection; 3D positioning; insect vision; compound eyes.

I. Introduction

Three-dimensional motion detection (TDMD) is one of the most important base stones for advanced autonomous robots. These robots can perform independent movement without continuous operator guidance with accurate position sensing or object tracking. A few three-dimensional motion detection technologies have been proposed in the past decay, these methods include gyroscopic sensing, magnetic sensing and video motion detection methods. Video motion detection [1-4] is an ideal approach because it can provide significant information without reengineering the environment. However, the computational cost and response speed would be a major concern when detecting a fast moving object.

In the natural of insect motion vision, a compound eye, consists of a significant amount of individual photoreceptor units (eyes), can detect fast movement by combining a number of optical information from each eye. This study presents the design, process, configuration, experiment setup, and testing and analysis results of a bionic compound system for 3D motion positioning [4]. Comparing with vision-recognition based detection method, bionic microlens array possesses simplified motion strategies [5].

II. System Design

The novel 3D motion detection architecture and image recognition methodology are developed by biologically inspired artificial compound eyes without involving complex lens configuration or expensive camera array.

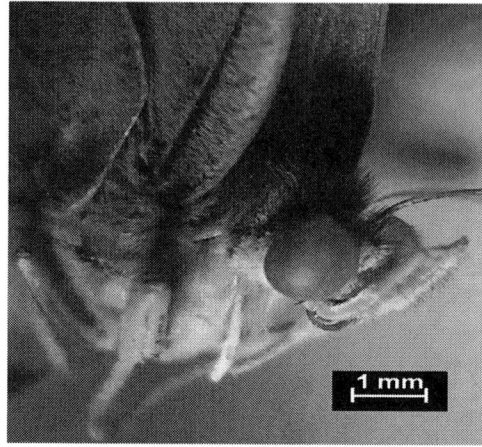

(a) Moth compound eyes

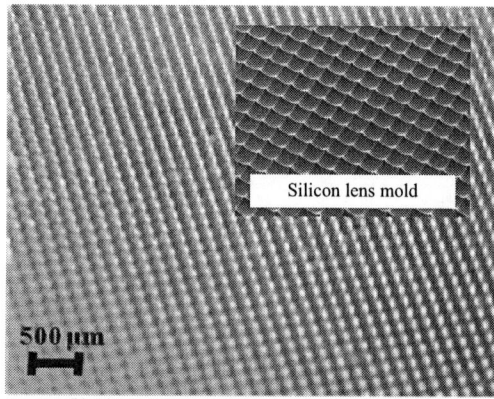

(b) Bionic microlens array (artificial ommatidia)

Fig. 1. (a) moth compound eyes, (b) a PDMS microlens array and a silicon lens mold (microlens pitch:125µm).

The artificial compound eye system consists of a microlens array, a CMOS image sensor and an aberration limited four-lens system (Fig. 1). The micro-lens molds are made from time-multiplexed plasma etching [6] and PDMS micromolding. After integrated with the image sensor, the bionic microlens array can project reference light onto CMOS image sensor for image processing. The configuration of the three-dimensional motion detector is shown in Fig. 2. The parameters of the microlens array are listed in Table 1.

978-1-4673-1122-9/12 $31.00 © 2012 IEEE

TABLE 1.THE PARAMETERS OF THE MICROLENS ARRAY AND THE IMAGE SENSOR

Microlens array	
Material	PDMS
Optical index	1.5
Pitch	200 μm
Focal length	83.3μm
Image sensor	
Image Resolution	VGA 640×480
Size	1/4 inch VGA Sensor
Frame rate	30 FPS

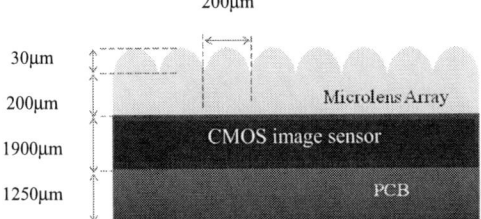

(a) A schematic diagram of the cross-section view of the artificial ommatidium, it consists of a bionic microlens array and a CMOS image sensor

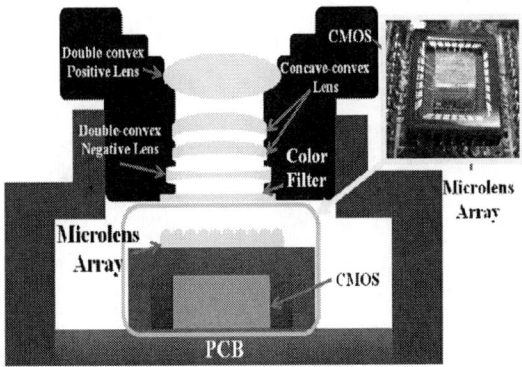

(b) The configuration of the 3D motion detector

Fig. 2. The 3D motion detector consists of a microlens arrays, an aberration limited four-lens system, and a 1.3 million-pixel CMOS image sensor.

Without using pinhole array [7] or gradient-index microlens, this design integrates the flat artificial compound eyes with an aberration limited four-lens system. It has adjustable focus, which can project referent light spot onto the artificial ommatidium.

III. EXPERIMENT SETUP

Fig. 3 shows the experiment set up of the bionic detector. The detection area of microlens array is exposed to a reference light spot. Each CMOS pixel receives rays from reference light sources; the point light sources are located at predetermined positions. A set of LEDs project referent rays

to a CMOS image sensor. The center of the optical referent center and the optic axis of the lens system is aligned together. The bionic 3D motion detector is mounted to a fine-tuning stage which can move along a rail toward or backward to the reference light for detector characterization. Three major modes, include one single-reference-mode and two dual reference modes, are tested in the experiment. In the single-reference-mode, only one LED is illuminated. In the dual reference mode, two sets of LEDs are arrange along x and y axes, respectively (Fig. 4). Over 64 pictures have been taken and processed in series for image analysis. One can move the bionic detector around the referent LED. We only test one mode each times.

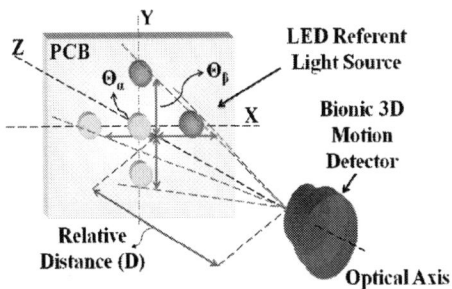

Fig. 3. The experiment set up of the bionic 3D motion and the referent light sources. One can move the bionic detector around the referent LED. We only charge one or two LED for each test mode.

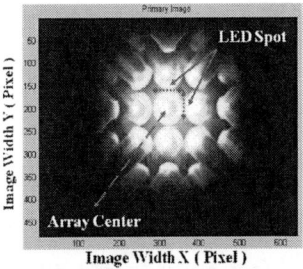

(a) Single reference mode

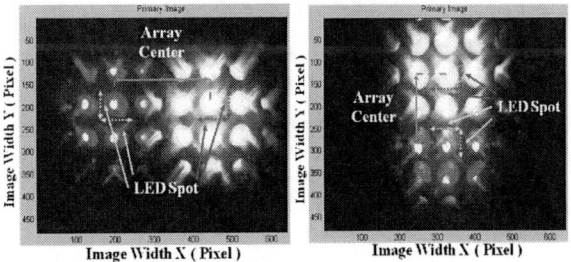

(b) Dual reference mode in X- direction (c) Dual reference mode in Y-direction

Fig. 4. The testing images grabbed from three major test modes. LED modules are divided into a single LED light source as shown in (a). Two dual light sources are set along x and y axes respectively, as shown in (b), (c). Set the direction and CCD camera lens was level with the vertical array.

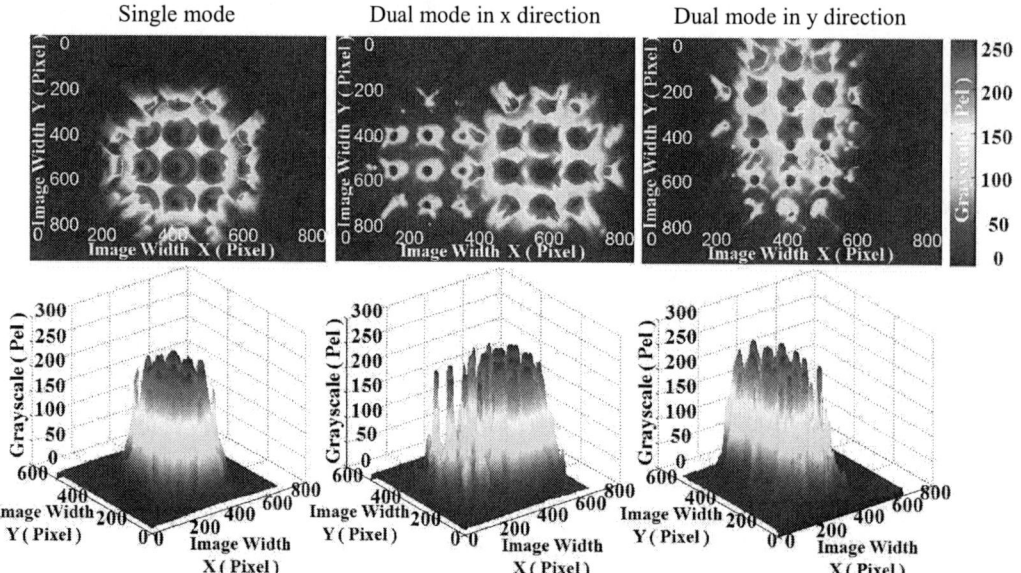

Fig. 5. Using MATLAB software analysis light intensity from gray scale images and convert image information into 2D and 3D coordinate systems.

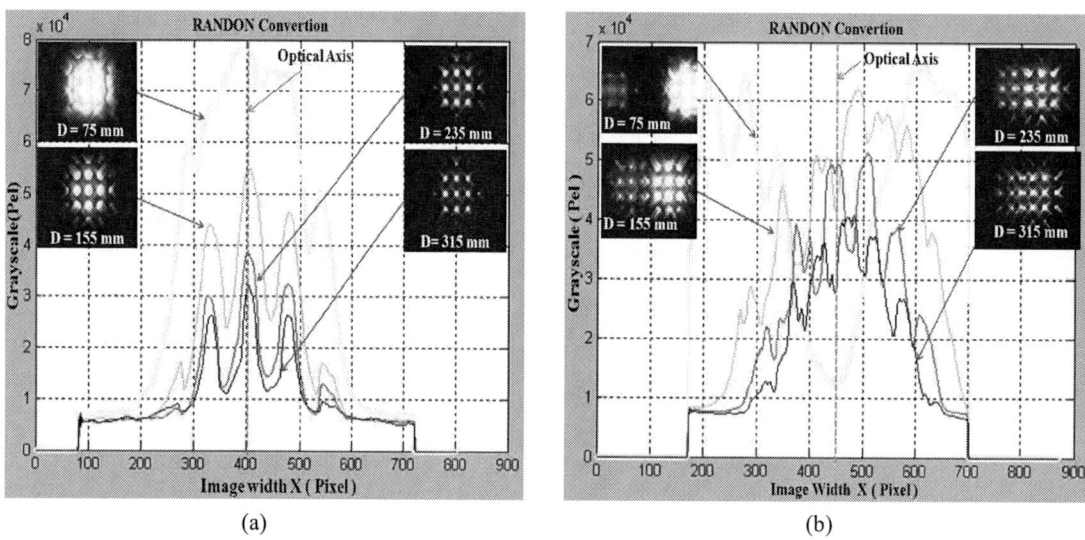

Fig. 6. After MATLAB post image processing, optical illumination distribution is shown above: (a) Single source and 200μm microlens array; (b) Double source and 200μm microlens array (vertical).

In the testing setup, a fine-tuning knob controls compound eye module movement toward or backward to the reference light spot for three-dimensional motion detection experiments. Two types of light source, including single source and double sources, are tested with bionic microlens array of photographic image tests. The frame rate is 30FPS. The VGA resolution is 640 x 480 pixels.

IV. EXPERIMENT RESULTS

The full-filled densely packaged bionic microlens can help the sensor gather enough light (Fig. 4). To determine the

relative orientation and distance of the referent ray, three dimensional relative positions are extracted by image analysis (Fig. 5). After the analysis, the three dimensional position sensing is achieved by evaluating the threshold value and weight distribution of the ray intensity spread, distributed along x, y and z directions in the images (Fig. 6(a, b)).

V. ANALYSIS AND 3D MOTION DETECTION

The projected images show angular and distance sensitivity to a given test. One can use a still image or consequence moving pictures to determine the location or movement of the referent light respectively. The angular acceptance function of an artificial compound eye is shown in Fig. 7.

The aberration limited four-lens system is place at front of the microlens array, it also determine the sensitivity of the measurement, as (1).

$$Seneitivity = \frac{pixel\ shifs}{displacement} \qquad (1)$$

However, the sensitivity and coupling efficiency reveal nonlinear dependency to the distance and incident angle between the referent rays and motion detector. The results of pixel shifting versus light source displacement are shown in Fig. 8. The distance between reference light source and detector is 155mm to 355mm. The optical characterizations of artificial compound eyes, sensor images and the real relative displacements are investigated with both analysis and experiment method.

A post process method also has been established to convert sensor image into space geometry. After post image processing and evaluation by computational method, the simulated position of light source can be obtained. Fig. 9 shows the comparisons between simulated positions and actual positions.

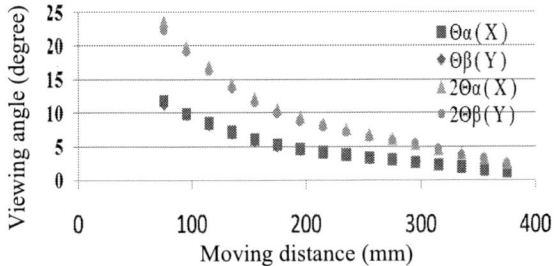

Fig. 7. The maximum viewing angle of the reference light spot versus relative distance (Z), the maximum viewing angle in this system is 23.6°

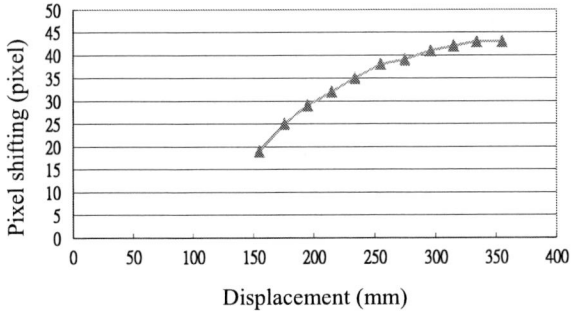

Fig. 8. The pixel shifts versus light source displacement.

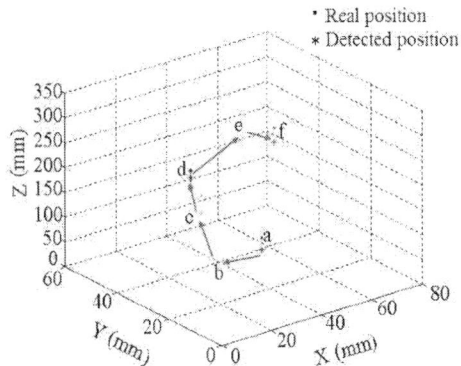

Fig. 9. Post image processing, optical weight (intensity) distribution function is obtained after conversion to estimate the real position and detected position of the referent target.

VI. CONCLUSIONS AND DISSCUSSIONS

Defer to artificial compound eyes made by 3D microfabrication method, which locates thousands of artificial ommatidium on a convex surface to increase the view angle; this work put significant effort to integrate flat artificial compound eyes with an aberration limited four-lens system and a CMOS image sensor without using pinhole array, gradient-index microlens or 3D microlens array. This research may provide significant impact in fast 3D-motion detecting, self-guided robot development, insect vision and insect path finding behavior studies. Both theoretic analysis and experiment results demonstrate that the design can achieve sensitivity to 0.286 (pixel/mm) along the optic axis.

ACKNOWLEDGMENT

The authors gratefully acknowledge Prof. C.-C. Lin for valuable discussions.

REFERENCES

[1] K. M. Lee and S. Foong, "Lateral optical sensor with slip detection for locating live products on moving conveyor," *Automation Science and Engineering, IEEE Transactions*, Volume: 7, Issue: 1, pp.123-132, January 2010.

[2] P.-L. Chang, F.-H. Hsieh, W.-L. Hsu, and H.-L. Shieh, "An Efficient Approach for Motion Detecting and Tracking," *Computer Communication Control and Automation (3CA), 2010 International Symposium*, Volume: 2, pp.330-333, 2010.

[3] A. Bak, S. Bouchafa, and D. Aubert, "Detection of Independently Moving Objects Through Stereo Vision and Ego-Motion Extraction," *Intelligent Vehicles Symposium (IV), 2010 IEEE*, pp.863-870, 2010.

[4] C. L. Nelson, "3-Dimensional Video Motion Detection and Assessment," *38th Annual 2004 International Carnahan Conference on Security Technology*, pp.295-302, October, 2004.

[5] K.H. Jeong, J. Kim, L.P. Lee "Polymeric Synthesis of Biomimetic Artificial Compound Eyes", *The 13th International Conference on Solid-State Sensors, Actuators and Microsystems, 2005 IEEE*. 2005.

[6] K. Wang and K.F. Böhringer, "Time-Multiplexed Plasma-Etching of High Numerical Aperture Paraboloidal Micromirror Arrays," *IQEC/CLEO-PR*, July 11-15, 2005.

[7] M. Sekine and K. Umeda, "Thin compound eye camera," *Applied Optics*, Vol. 44, Issue 15, pp. 2949-2956, 2005.

978-1-4673-1122-9/12 $31.00 © 2012 IEEE

A High Sensitivity and Low-Cost Polycarbonate (PC)-based Biosensor

Yu-Shan Chen and Gou-Jen Wang*
*Graduate Institute of Biomedical Engineering, National Chung-Hsing University, Taiwan
gjwang@dragon.nchu.edu.tw

Abstract—**This study integrates the techniques of nano electroforming, hot embossing, and electrochemical deposition to develop a disposable, low cost, and high sensitivity nanostructure biosensor. A modified anodic aluminum oxide (AAO) barrier-layer surface is used as the template for nickel thin film deposition. After etching the AAO template off, a 3D mold of the concave nano structure array is created. The fabricated 3D nickel mold is further used for replica molding of a nano-structure polycarbonate (PC) substrate by hot embossing. An Au thin film is than sputtered on the PC substrate to form the electrode followed by the deposition of an orderly and uniform gold nanoparticle (GNP) layer on the 3D Au electrode using electrochemical deposition. Finally, silver nanoparticles (SNP) are deposited on the uniformly deposited GNPs to enhance the conductivity of the sensor. Electrochemical impedance spectroscopy (EIS) analysis is then used to detect the concentration of the target element. The sensitivity of the proposed scheme on the detection of the dust mite antigen Der p2 can reach 0.1pg/ml.**

Keywords- nano electroforming, nano-structure polycarbonate (PC) substrate, gold nanoparticles, silver nanoparticles, EIS analysis

I. INTRODUCTION

The current trend for point of care device development requires that a device be low cost, sensitive, specific, easy to use, rapid and robust, and disposable, as well as having a small sample requirement. This is especially the case for people who live in developing countries where even common tests are not affordable and the fundamental infrastructure is often not available [1-3]. A variety of microfluidic devices such as the micro total analysis system (μTAS) have been developed which partially meet these demands [4-7]. However, the requirement of lithographic equipment and the relatively high cost of such devices can prove to place limitations for mass production and general applications.

Microfluidic paper-based analytical devices (μPADs) as initially proposed by Whitesides group can be good alternatives to satisfy the low cost and ease of use requirements of modern point-of-care diagnostic devices [8-10]. A μPAD generally consists of a paper-based microfluidic testing platform based on patterned channels on hydrophilic paper delimited by partitions of hydrophobic material and a light reflectance detector for quantitative colorimetric detection of target analytes at the test zone. Methods for paper patterning include photolithography [8,11], plotter printing of PDMS patterns in paper [12], inkjet printing [13,14], and wax printing [15,16]. A simple reflectance detection device such as a desktop scanner or a digital camera is quite adequate for use with a μPAD. Other detection methods such as quantitative fluorescence and absorbance measurements obtained using a microplate reader [17], electrochemistry [18-20], or electro-chemiluminescence [21] can also be implemented for the quantitative detection of the target analytes. The μPEDs, although still in their infancy, have proven to be a promising solution for simple and low-cost quantitative detection of biological and inorganic species in aqueous solutions. However, their relatively low sensitivity may limit their applications for the detection of rare molecules. A low-cost diagnostic device, with high sensitivity, is still desired for precise detection of low levels of species in biological samples, including bodily fluids.

Nanomaterials, which can provide a substantially larger surface area than that of bulk material or thin films, have been used to either simplify the readout or magnify the detection signal of a diagnostic device [22-24]. Several nanomaterial-based devices have also been approved by the FDA for in vitro diagnosis [25-28].

This study demonstrates the integration of nano electroforming, hot embossing, electrochemical deposition techniques for the fabrication of a polycarbonate (PC) based biosensor platform. The key component in this PC-based platform is a GNP-deposited nano-structured PC substrate which serves as the transducer for the detection of the target analytes by binding to it. The nano-structure PC substrates can be replicated using a nickel mold by means of nano hot embossing or nano injection molding. This makes the platform disposable and low cost. In addition, the high surface to volume ratio of the nano-structure PC substrate and the uniformly deposited GNPs allows the attachment of many more analytes to the transducer which in turn means considerable enhancement of the sensitivity. In this study, the detection of the dust mite antigen Der p2 is carried out using electrochemical impedance spectroscopy (EIS) analysis to demonstrate the performance of the proposed PC-based biosensor scheme.

II. EXPERIMENTAL SECTION

A. PC-based Transducer Fabrication

Figure 1 shows a schematic illustration of the proposed PC-based biosensor design. It consists of a PC substrate with a nano-hemisphere array, a gold thin film sputtered on the PC substrate to form the electrode, gold nanoparticles uniformly deposited on the film, and smaller silver nanoparticles deposited on the GNPs. The sequential fabrication processes include: fabrication of a nano-structured PC substrate; deposition of an Au thin film; annealing; electrochemical deposition of Au nanoparticles; and electrochemical deposition of Ag nanoparticles.

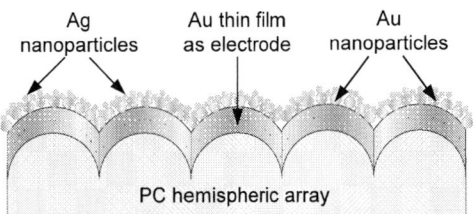

Fig. 1. Schematic illustration of the proposed PC-based biosensor.

(1) Fabrication of a nano-structured PC substrate

The nano-structured PC substrate is hot-embossed using a nickel replica mold. The fabrication procedures of the nano replica mold are briefly described as follows. A modified anodic aluminum oxide (AAO) barrier-layer surface is used as the master mold for nickel thin film deposition. After etching off the AAO template, a 3D replica of the concave nanostructure array can be obtained.

The nano-structured PC substrates were produced by Leadoptical Co., Ltd., Taichung, Taiwan. Leadoptical specializes in microlens array fabrication. Since the glass transition temperature (T_g) of the PC used is 120°C, the embossing temperature is set at 150°C. After holding for 50 sec, a cooling process at 100°C for 180 sec is carried out.

(2) Deposition of Au thin film

An Au thin film electrode is deposited on the hemispheric array of the PC substrate using radio frequency (RF) magnetron sputtering. The deposition is conducted under a 4.0×10^{-3} torr pressure at room temperature with a 20 sccm argon injection and an 80 W power lasting for 1.5 min. This creates a 30 nm thick Au thin film.

(3) Annealing

To further refine the surface structure of the Au thin film electrode and make it homogeneous, an annealing process is employed. The sequential annealing procedures include: heating the sample to 120 °C at a rate of 5 °C/min, maintaining that temperature for 60 min, followed by cooling the sample in open air to room temperature.

(4) Packaging

To ensure the uniformity of the sensing area and prevent other solutions from getting in touch with the sample, precise packaging before electrochemical depositing of Au nanoparticles is needed. The packaging procedures are described as follows:

(*i*) Prepare a 2.5×2.5 cm² square parafilm.

(*ii*) Make a ϕ=6 mm hole in the center of the parafilm.

(*iii*) Daub a thin layer of epoxy on the bottom surface of the parafilm square, followed by bonding of the parafilm to the Au film deposited PC substrate to enclose the sensing area of the device.

(*iv*) Package the device using silica gel to ensure air tightness.

(4) Deposition of Au nanoparticles

The electrochemical deposition is processed by an SP-150 an electrochemical analyzer (EC-Lab, USA). The packaged sample is placed at the working electrode (WE), with the Au thin film serving as the electrode. The counter electrode and the reference electrode (RE) are a Pt film and an Ag/AgCl electrode, respectively. The deposition processes are described below.

(*i*) Electrolyte preparation:

The electrolyte is prepared by dissolving 1 mL of 0.02 M $HAuCl_4$ (Aldrich Inc.) solution in 39 mL of deionized water.

(*ii*) Reducing potential measurement:

The cyclic voltammetric method is conducted using the SP-150 to examine the reduction potential of the $HAuCl_4$ solution. The scanning range is set to be between -0.8 V and +0.8V. Three reducing peaks, at 0.8V, -0.2V, and -0.8V, respectively, are detected, where -0.8V denotes the reducing peak of a monovalence Au ion. Several experiments with reducing potentials set around -0.8 V are conducted. The results indicate that -0.7 V is a more suitable reducing potential for the $HAuCl_4$ solution used in this study.

(*iii*) Electrochemical deposition:

The electrochemical deposition is processed under a DC - 0.7V electric potential at room temperature, lasting for 180 sec. Several experiments with different deposition durations are carried out. It is found that 180 sec can perform better deposition of GNP.

(5) Deposition of Ag nanoparticles

Due to its extra high electric and thermal conductivity, silver has been widely applied to micro/nano systems to enhance conductivity. In this study, silver nanoparticles are produced using the Meisel rodex method [29]. Ag nanoparticles are then electrochemically deposited on the Au nanoparticles to further increase the device's conductivity. The deposition processes are described below.

(*i*) Electrolyte preparation:

The electrolyte for producing Ag nanoparticles is a mixture of 1mM water bathed $NaBH_4$ and 1mM $AgNO_3$. Canary yellow Ag nanoparticles can be synthesized by stirring the electrolyte at 40 °C for 2 min. Continuing stirring for 1 hr is essential to prevent aggregation of the Ag nanoparticles.

(*ii*) Reducing potential measurement:

The reducing potential for better reduction of Ag ions is detected to be -0.4 V through the cyclic voltammetric procedures same as the measurement of the reducing potential for the $HAuCl_4$ solution.

(*iii*) Electrochemical deposition:

Electrochemical deposition of Ag nanoparticles was conducted by the application of a DC -0.4V electric potential for 150 sec at room temperature.

B. Electrochemical Characterization [30]

The SP-150 electrochemical analyzer used for nanoparticle deposition is also employed for cyclic voltammetry (CV) testing. A Pt wire and an Ag/AgCl electrode are used as the counter and the reference electrodes, respectively. The active surface area is estimated with steady-state voltammetry using a 0.5M H_2SO_4 solution as the supporting electrolyte.

The cyclic voltammogram trace can be used to calculate the actual area of the sensing device which is most effective for the adhesion of analytes.

C. Testing Sample Preparation

In this study, the dust mite allergen Der p2 is used for the performance evaluation of the proposed PC-based biosensor. The Der p2 is prepared using the procedures proposed by Tsai et al.[31]. The self-assembled monolayer (SAM) method is adopted for the immobilization of Der p2 on the sensor. The procedures for SAM are briefly described below.

First, the surface of the sensor is cleaned by sequentially soaking it in ethanol, acetone and deionized (DI) water, and then vibrating it using ultrasonic waves for 5 min. 10μL of a 10mM 11-MUA (11-mercaptoundecanoic acid) solution is then dispensed onto the sensor surface to construct a 11-MUA SAM anchor membrane. A 20μL sample of the mixed NHS (N-hydroxysuccinimide) and EDC (1-Ethyl-3-(3-dimethyl-aminopropyl)-carbodiimide) solution (molar ratio, 1:2) is further allotted to the sensor for 10 min to activate the carboxylic groups on the 11-MUA membrane. After two washings with a PBS buffer solution, the sensor is bathed in a 20μL (10 μg/mL) dust mite monoclonal antibody (IgG) solution and then incubated for 10 min. The IgG is supplied by Bethyl Laboratories Inc. (Texas, USA). The NH_2^+ group in an IgG is substituted for the NHS functional group and chemically binds with the $COOH^-$ group on the MUA membrane. The sensor is again rinsed twice with a PBS buffer solution (pH 7.4). To ensure binding uniqueness, a 20μL 1% BSA solution is used as the blocking layer to seal off the binding sites of those IgGs which do not bond with the MUA membrane. Finally the sensing samples are acquired by dipping the sensor into dust mite allergen Der p2 solutions of different concentrations (10 ng/ml, 1 ng/ml, 0.1 ng/ml, 0.01 ng/ml, 1 pg/ml, and 0.1 pg/ml) at 4 °C for 30 min.

D. Electrochemical impedance spectroscopy (EIS) analysis

EIS analysis is a method commonly used for determining information about an interface. The impedance information of the system can be evaluated by applying a periodic AC signal of small amplitude to the interface and detecting the actual system response. In this study, this method of analysis is implemented for the detection of the Der p2 concentration in different solutions.

The PC-based biosensor, Pt film, and Ag/AgCl serve as the working electrode, counter electrode and reference electrode, respectively, for the EIS. The buffer solution is a mixture of 5 mM $Fe(CN)_6^{4-}$ and 5 mM $Fe(CN)_6^{3-}$ in 100 mM 2-(N-morpholino) ethanesulfonic acid (MES) (pH=6.0). The applied DC power and AC power are 0 V and 5mV, respectively. The AC frequency ranges from 50mHz to 50kHz.

III. RESULTS AND DISCUSSION

A. Transducer Fabrication Results

Figure 2 shows images of the hot-embossed PC-based substrates. From the SEM image, it is observed that high order nano-hemispheres with a diameter of around 80 nm can be transferred from the Ni replica mold to the PC material.

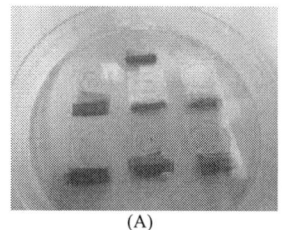

(A)   (B)

Fig. 2. Images of a PC nano-hemisphere array created using hot embossing; (A) real products; (B) SEM image.

Figure 3 shows an SEM image of a GNP deposited electrode. The uniformly and compactly deposited GNPs have an average diameter of around 20 nm. The uniformly propagated electric flux perpendicular to the hemispheric Au thin film electrode attracts the positive charges of the Au nanoparticles in the electrolyte. It is revealed that the GNPs can be densely deposited onto the surface of the Au thin film electrode without adding a reducing agent or stabilizer.

Chemical analysis electron spectroscopy (PHI 5000, ULVAC-PHI Inc., Kanagawa, Japan) is used to detect the content of the SNP deposition. It is found that the SNP content is 0.3%.

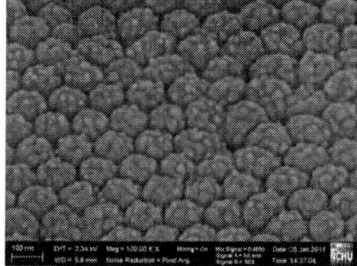

Fig. 3. SEM image of GNPs synthesized by electrochemical deposition.

B. Electrochemical Characterization Results

Figure 4 shows the steady-state CVs and the I-t curves transformed from the CVs for the nanostructure Au thin film before GNP deposition and the GNP deposited electrodes in 0.5M H_2SO_4 at 100 mV/s versus that of the Ag/AgCl reference electrode. The total amount of electric charge in each electrode can be estimated by integrating the area of the reducing peak in Figure 4(B). The total amount of electric charges of the film before GNP depositing and the

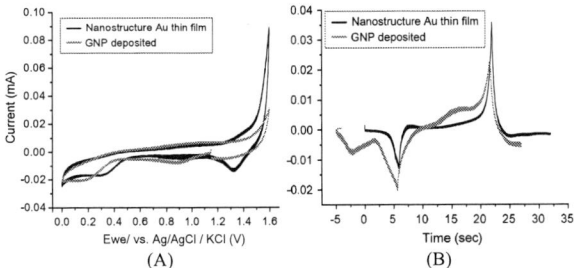

(A)　　　　　　　　(B)

Fig.4. CVs and I-t curves transformed from the CVs for the film before deposition of the nanostructure Au thin film and the GNP deposited electrodes.

PC/Au/GNP electrode are estimated to be 16.51 and 53.90 μC, respectively. Since the charges required to form AuO per 1 cm^2 of Au electrode is 386 μC, the effective areas of the Au thin film before GNP nanostructure deposition and the PC/Au/GNP electrode are calculated to be 0.043 (16.51 μC/386 μC) and 0.14 (53.90 μC/386 μC) cm^2, respectively. The effective area is enhanced 3.26-fold due to the deposition of GNPs.

C. Electrochemical Impedance Spectroscopy Analysis Results

The EIS analysis results obtained using an SP-150 electrochemical analyzer are presented in Figure 5. Figure 5(A) shows the Nyquist plots for a GNP deposited electrode, an electrode after the anti-dust mite monoclonal IgG is bound to the NHS-EDC, and electrodes after the Der p2 (with various concentrations) are immobilized for the anti-dust mite monoclonal IgG. The diameter of each individual semicircular EIS curve in Figure 5(A) represents the charge transfer resistance of the electrode with respect to this specific EIS analysis. Figure 5(A) reveals that the charge transfer resistance for each individual electrode increases with an increase in the Der p2 concentration. In other words, the impedance of the Der p2 binding device increases with an increase in the Der p2 concentration. The EIS analysis results can be modeled by the Randles's equivalent circuit shown in the inset to Figure 5(A).

The total impedance in the Randles's equivalent circuit is composed of the electrolyte resistance (R_s), the charge transfer resistance (R_{et}), and the double layer capacitance (C_{dl}). Since R_s represents the bulk properties of the electrolyte solution, it is unrelated to the chemical reactions coming about at the electrode interface. The values of R_{et} and C_{dl} change with the substance bonded onto the electrode surface. Table 1 presents the fitting parameters for the equivalent circuit model of the PC-based biosensor corresponding to the Nyquist plots in Figure 5(A), in which, ΔR_{et} denotes the different levels of resistance between the electrode in the individual SAM stage and the IgG bonding electrode. As tabulated in Table 1, the value of each R_s is much smaller, and can be neglected when compared with its corresponding R_{et} value. Hence, the Randles's equivalent circuit can be simplified as follows:

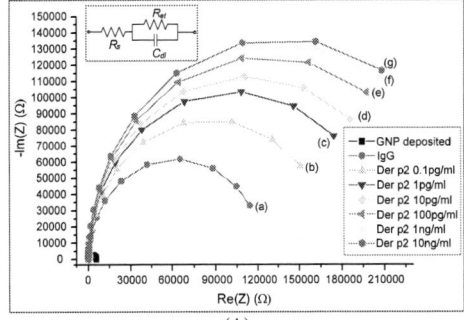

(A)

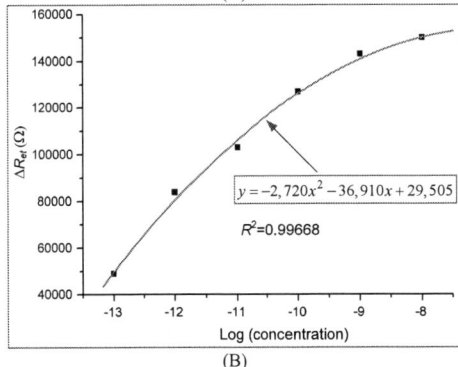

(B)

Fig.5. EIS analysis results: (A) Impedance plots for Der p2 at various concentrations: (a) IgG, (b) 10^{-13}g/ml, (c) 10^{-12}g/ml, (d) 10^{-11}g/ml, (e) 10^{-10}g/ml, (f) 10^{-9}g/ml, (g) 10^{-8}g/ml; (B) ΔR_{et} as a function of the logarithmic concentration of Der p2 for the proposed PC-based bionsensor.

TABLE I.　　FITTING PARAMETERS FOR THE RANDLES'S EQUIVALENT CIRCUIT

	GNP	IgG	0.1 pg	1 pg	10 pg	100 pg	1 ng	10 ng
$R_{et}(10^5\Omega)$	0.053	1.24	1.73	2.07	2.26	2.51	2.67	2.73
$R_s(\Omega)$	185	181	178	179	181	181	181	180
C_{dl} $(10^{-6}F)$	3.95	3.73	3.51	3.35	3.38	3.31	3.24	3.26
ΔR_{et} $(10^5\Omega)$			0.49	0.84	1.03	1.27	1.43	1.50

$$Z(\omega) = \frac{R_{et}}{1+\omega^2 R_{et}^2 C_{dl}^2} - j\frac{\omega R_{et}^2 C_{dl}}{1+\omega^2 R_{et}^2 C_{dl}^2}. \quad (1)$$
$$= R + jX$$

As expressed in Eq. (2), both R_{et} and C_{dl} affect the impedance plot. However, Table 1 shows that the difference in R_{et} between the electrode for the individual SAM stage and the GNP deposited electrode is more substantial when compared with the change of C_{dl}. Hence, ΔR_{et} can be used as the parameter to relate the sensor response with different analyte concentrations.

Figure 5(B) shows ΔR_{et} as a function of the logarithmic concentration of Der p2 for the proposed PC-based nanobiosensor. The sensor response is well fitted by a conic

978-1-4673-1122-9/12 $31.00 © 2012 IEEE

section with R^2=0.99668. The detection limit is around 0.1pg/ml as indicated by curve (a) in Figure 5(A). A dynamic range of up to 10 ng/ml (curve (g) in Figure 5(A)) is also presented. When the ΔR_{et} of a desired Der p2 solution is detected, the concentration of Der p2 can be determined by this standard regression curve.

A test sample of 1pg/ml is used for further verification. The ΔR_{et} of the test sample is measured to be $8.27 \times 10^4 \Omega$. The logarithmic concentration of the test sample is calculated to be -11.8 using the standard regression curve. It is only 1.7% off the real value of -12.

It is presumed that the uniformly distributed Au nanoparticles on the hemispheric array enable the MUA molecules to reach individual gold nanoparticles; therefore, the sequential bindings of EDC/NHS molecules and IgG molecules are thus increased. Hence, the effective binding between Der p2 and IgG can be enhanced. As a result, the sensitivity of the sensor is considerably enhanced. The proposed low-cost and high sensitivity PC-based nanobiosensor can be useful for the rapid detection of rare molecules in an analyte.

IV. CONCLUIONS

In this study, we have designed a low-cost and highly sensitive PC-based nanostructured biosensor. The sequential synthesis processes of the biosensor include: fabrication of a nano-structured PC substrate; deposition of an Au thin film; annealing; electrochemical deposition of Au nanoparticles, and electrochemical deposition of Ag nanoparticles. The sensing limit and dynamic range of the proposed design for Der p2 detection are investigated using EIS analysis and found to be 0.1 pg/ml and 10 ng/ml, respectively. It is presumed that the uniformly distributed Au nanoparticles on the hemispheric array enable the MUA molecules to reach individual gold nanoparticles; thereby increasing the sequential bindings of EDC/NHS molecules and IgG molecules. Hence the effective binding between Der p2 and IgG can be enhanced, resulting in the sensitivity of the sensor being enhanced considerably. Fluorescence analysis results obtained using EGFP further indicate that the high sensitivity of the proposed PC-based nanobiosensor can be attributed to the intensity and uniformity of the Au nanoparticles on the sensor. The proposed low-cost and high sensitivity sensor can be useful for the rapid detection of rare molecules in an analyte.

In future work, the PC-based biosensor will be integrated into the conventional printed circuit board (PCB) fabrication process to produce a final product. Mass production tasks will be conducted. A portable EIS device is currently being designed and fabricated for fast and low-cost detection.

ACKNOWLEDGMENT

The authors would like to offer their thanks to the Department of Health of Taiwan for their financial support of this research under grant number DOH100-TD-N-111-006.

REFERENCES

[1] S. K. Sia and L. J. Kricka, "Microfluidics and point-of-care testing," *Lab Chip*, vol. 8, pp. 1982-1983, 2008.

[2] C. D. Chin, V. Linder, and S. K. Sia, "Lab-on-a-chip devices for global health: past studies and future opportunities," *Lab Chip*, vol. 7, pp. 41–57, 2007.

[3] P. Yager, T. Edwards, E. Fu, K. Helton, K. Nelson, M. R. Tam, and B. H. Weigl, "Microfluidic diagnostic technologies for global public health," *Nature*, vol. 442, pp. 412-418, 2006.

[4] Y. C. Lim, A. Z. Kouzani, W. Duan, "Lab-on-a-chip: a component view," *Microsyst. Technol.*, vol. 16, pp.1995-2015, 2010.

[5] S. Haeberle and R. Zengerle, "Microfluidic platforms for lab-on-a chip applications," *Lab Chip*, vol. 7, pp. 1094-1110, 2007.

[6] A. G. Crevillen, M. Pumera, M. C. Gonzalez, A. Escarpa, "Towards lab-on-a-chip approaches in real analytical domains based on microfluidic chips/electrochemical multi-walled carbon nanotube platforms," *Lab Chip*, vol. 9, pp. 346-353, 2009.

[7] Y. F. Lee, K. Y. Lien, H. Y. Lei, and G. B. Lee, "An integrated microfluidic system for rapid diagnosis of dengue virus infection," *Biosens. Bioelectron.* vol. 25, pp. 745-752, 2009.

[8] A. W. Martinez, S. T. Phillips, M. J. Butte, and G. M. Whitesides, "Patterned paper as a platform for inexpensive, low-volume, portable bioassays," *Angew. Chem., Int. Ed.*, vol. 46, pp.1318-1320, 2007.

[9] A. W. Martinez, S. T. Phillips, E. Carrilho, S. W. Thomas III, H. Sindi, and G. M. Whitesides, "Simple telemedicine for developing regions: camera phones and paper-based microfluidic devices for real-time, off-site diagnosis," *Anal. Chem.*, vol. 80, pp. 3699-3707, 2008.

[10] A. W. Martinez, S. T. Phillips, and G. M. Whitesides, "Diagnostics for the developing world: microfluidic paper-based analytical devices," *Anal. Chem.*, vol. 82, pp. 3-10, 2010.

[11] A. W. Martinez, S. T. Phillips, B. J. Wiley, M. Gupta, and G. M. Whitesides, "FLASH: A rapid method for prototyping paper-based microfluidic devices," *Lab Chip*, vol. 8, pp. 2146-2150, 2008.

[12] D. A. Bruzewicz, M. Reches, and G. M. Whitesides, "Low-cost printing of poly (dimethylsiloxane) barriers to define microchannels in paper," *Anal. Chem.*, vol. 80, pp. 3387-3392, 2008.

[13] K. Abe, K. Suzuki, and D. Citterio, "Inkjet-printing microfluidic multianalyte chemical sensing paper," *Anal. Chem.*, vol. 80, pp. 6928-6934, 2008.

[14] X. Li, J. Tian, G. Garnier, and W. Shen, "Fabrication of paper-based microfluidic sensors by printing," *Colloids and Surfaces B: Biointerfaces*, vol. 76, pp. 564-570, 2010.

[15] Y. Lu, W. Shi, J. Qin, and B. Lin, "Fabrication and characterization of paper-based microfluidics prepared in nitrocellulose membrane by wax printing," *Anal. Chem.*, vol. 82, pp. 329-335, 2010.

[16] E. Carrilho, A. W. Martinez, and G. M. Whitesides, "Understanding wax printing: a simple micropatterning process for paper-based microfluidics," *Anal. Chem.*, vol. 81, pp. 7091-7095, 2009.

[17] E. Carrilho, S. T. Phillips, S. J. Vella, A. W. Martinez, G. M. Whitesides, "Diagnostics for the developing world: microfluidic paper-based analytical devices," *Anal. Chem.*, vol. 81, pp. 5990-5998, 2009.

[18] W. Dungchai, O. Chailapakul, and C. S. Henry, "Electrochemical detection for paper-based microfluidics," *Anal. Chem.*, vol.81, pp. 5821-5826, 2009.

[19] Z. Nie, C. A. Nijhuis, J. Gong, X. Chen, A. Kumachev, A. W. Martinez, M. Narovlyansky, and G. M. Whitesides, "Electrochemical sensing in paper-based microfluidic devices," *Lab Chip*, vol. 10, pp. 477-483, 2010.

[20] Z. Nie, F. Deiss, X. Liu, O. Akbulut, and G. M. Whitesides, "Integration of paper-based microfluidic devices with commercial electrochemical readers," *Lab Chip*, vol. 10, pp. 3163-3169, 2010.

[21] J. L. Delaney, C. F. Hogan, J. Tian, and W. Shen, "Electrogenerated chemiluminescence detection in paper-based microfluidic sensors," *Anal. Chem.*, vol. 83, pp. 1300-1306, 2011.

[22] Y. H. Yun, A. Bange, W. R. Heineman, H. B. Halsall, V. N. Shanov, Z. Dong, S. Pixley, M. Behbehani, A. Jazieh, Y. Tu, D. K. Y. Wong, A. Bhattacharya, and M. J. Schulz, "A nanotube array immunosensor for direct electrochemical detection of antigen-antibody binding," *Sensors and Actuators B*, vol. 123, pp. 177-182, 2007.

[23] J. Huang, G. Yang, W. Meng, L. Wu, A. Zhu, X. Jiao, "An electrochemical impedimetric immunosensor for label-free detection of

campylobacter jejuni in diarrhea patients' stool based on O-carboxymethylchitosan surface modified Fe_3O_4 nanoparticles," *Biosens. Bioelectron.* vol. 40, pp. 893-896, 2009.

[24] J. J. Tsai, I. J. Bau, H. T. Chen, Y. T. Lin, and G. J. Wang, "A novel nanostructured biosensor for the detection of the dust mite antigen Der p2," *Inter. J. Nanomedicine*, vol. 6, pp 1-8, 2011.

[25] G. A. Posthuma-Trumpie, J. Korf J, and A. V. van Amerongen, "Lateral flow immunoassay: its strengths, weaknesses, opportunities and threats: a literature survey," *Anal. Bioanal. Chem.* Vol. 393, pp. 569-82, 2009.

[26] J. M. Nam, C. S. Thaxton, C. A. Mirkin, "Nanoparticle-based bio-bar codes for the ultrasensitive detection of proteins," *Science*, vol. 301, pp. 1884-1886, 2003.

[27] Lefferts JA, Jannetto P, Tsongalis GJ. Evaluation of the Nanosphere Verigene System and the Verigene F5/F2/MTHFR Nucleic Acid Tests. Exp Mol Pathol 2009;87:105-8.

[28] C. S. Thaxton, R. Elghanian, and A. D. Thomas, "Nanoparticle-based bio-barcode assay redefines undetectable PSA and biochemical recurrence after radical prostatectomy," *Proc Natl Acad Sci USA*, vol. 106, pp.18437-18442, 2009.

[29] J. Gehl, T. H. Sorensen, K. Nielsen, P. Raskmark, S. L. Nielsen, T. Skovsgaard, and L. M. Mir, "In vivo electroporation of skeletal muscle: Threshold, efficacy and relation to electric field distribution," *Biochimica et Biophysica Acta*, vol. 1428, pp. 233-240, 1999.

[30] S. Hrapovic, Y. Liu, K. B. Male, and J. H. T. Luong, "Electrochemical biosensing platforms using platinum nanoparticles and carbon nanotubes," *Anal. Chem.*, vol. 76, pp. 1083-1088, 2004.

An electro-enzymatic flexible molecular lactate sensor

Nicole Thomas, Ilkka Lähdesmäki, and Babak A. Parviz*
*Department of Electrical Engineering, University of Washington, USA
parviz@uw.edu

Abstract—We present the fabrication and electrochemical characterization of an electronic molecular lactate sensor comprised of an optimized three-electrode setup on a transparent, flexible polymer substrate. The sensors are functionalized with the flavoenzyme lactate oxidase as a selective, nanoscale recognition element. They exhibit an exceptionally high sensitivity in their linear range of ~ 370 $\mu AmM^{-1}cm^{-2}$, a very low detection limit of ~ 5 μM, and good thermal stability, with a working electrode area of only ~ 0.19 mm^2. We envision the sensors being utilized for one-time measurements of lactate levels in food or agricultural products for monitoring of their freshness, and show results of lactate level measurements for various dairy products.

Amperometric sensing; lactate oxidase; flexible substrate; enzyme-based sensors

I. INTRODUCTION

Enzymes are nanometer sized molecules that catalyze the reaction of a respective enzyme substrate. Enzymes are therefore considered highly suited as selective recognition elements for certain molecules: the presence of the desired molecule may be determined by the detection of the products of the catalytic reaction.

The α-hydroxy-acid flavoenzyme lactate oxidase (LOx), for example, catalyzes the oxidation of L-lactate to hydrogen peroxide and pyruvate by a reduction of the cofactor flavin mononucleotide (FMN) by 2 electrons [1,2]:

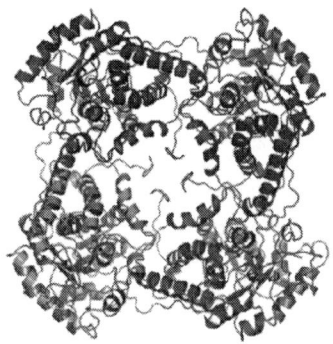

Fig. 1. Single tetrameric unit of LOx. Each of the four monomers contains the cofactor FMN. Reprinted with permission from [2].

$$L-lactate + O_2 \xrightarrow{LOx} pyruvate + H_2O_2. \qquad (1)$$

LOx appears as two tightly-packed tetramers with fourfold symmetry, respectively [2]. Fig. 1 shows one of the tetrameric units. The dimensions of each tetramer are reported to be ~ 5 nm by 10 nm by 10 nm [2]. An active FMN cofactor was found in the center of each of the monomers.

Under an appropriate potential, hydrogen peroxide can be oxidized at an electrode surface:

$$H_2O_2 \rightarrow 2H^+ + O_2 + 2e^- \qquad (2)$$

In the presence of a second electrode, a current flow arises that is proportional to the amount of hydrogen peroxide produced by the catalytic action of LOx. This transduction can be used to determine the lactate concentration in solution. Different amperometric sensors for the detection of lactate have been devised [3]: typical sensing structures are screen-printed and microelectrodes as well as needle-type sensors. Numerous electrode materials have been utilized, e.g. carbon paste, metals like Pt and Au and metal oxides. Immobilization of LOx on the sensors was achieved by very different means, such as entrapment or containment, inclusion into the electrode material or covalent binding, usually by cross-linking with glutaraldehyde.

However, the fabrication process is often very complex and costly, and results may not be reproducible [3]. In addition, none of the previous developed sensors are completely flexible and small enough to be employed in a wide range of applications, for example in small-sized test tubes for food or agricultural products.

We present the development of lactate sensors that are built directly onto a flexible, transparent polymer substrate. We used widely available fabrication technologies, which are also suitable for the mass production of the devices. The sensors utilize LOx, covalently immobilized on the substrate, as nanoscale detection element for the presence of L-lactate. We employ a three-electrode setup in which a constant reference voltage V_{ref} is applied between a reference and a working electrode (RE and WE, respectively) and the current is drained by an auxiliary counter electrode (CE). We determined basic sensor characteristics such as sensitivity, detection limit and the sensor response to high substrate concentrations, and utilized the sensors to determine the concentration of lactate in several dairy products.

978-1-4673-1122-9/12 $31.00 © 2012 IEEE

II. SENSOR FABRICATION AND TESTING

A. Fabrication on Flexible Substrate

We designed the amperometric sensors with working and counter electrode as concentric rings surrounding the reference electrode. Pt was chosen as an appropriate metal for the WE due to its low H_2O_2 oxidation potential. To simplify the fabrication process, all electrodes are made out of the same materials. The WE and CE are 75 μm and 100 μm wide, respectively, and the radius of the RE is 200 μm. The area of the WE is ~ 0.19 mm^2, and the total footprint of the sensing area is ~ 1.43 mm^2. The electrodes were connected to larger electrode pads for wiring to an external potentiostat via Pt lines.

100 μm thick polyethylene terepthtalate (PET, *Policrom*, USA) was laser cut into 100 mm diameter wafers for further processing. The substrates were cleaned in acetone, IPA and DI water and spun dry, followed by an oxygen plasma treatment. The substrates were then photolithographically patterned with AZ4620 positive resist (*Microchem*, USA). Due to the thermal limitations of the PET substrate, the softbake was carried out for 5 min at a temperature of 80 °C. The sensing structures were formed by e-beam evaporation of a 10 nm Ti/20 nm Pd/100 nm Pt stack onto the substrate, and subsequent lift-off in acetone. Finally, the substrates were rinsed in acetone, IPA and DI water, and dried. Fig. 2(a) shows a schematic of the fabrication sequence for the sensor structures. The individual sensors were released from the substrate by CO_2 laser cutting. Fig. 2(b) displays a close-up of a single sensor in the three-electrode setup. The flexible substrates allow for shaping (or molding) of the sensors into the desired form (Fig. 2(c)).

B. Enzyme Immobilization

We utilized the flavoenzyme LOx for the selective detection of lactate by sensors on the flexible, transparent PET substrate. Reliable immobilization of LOx onto the substrate surface was

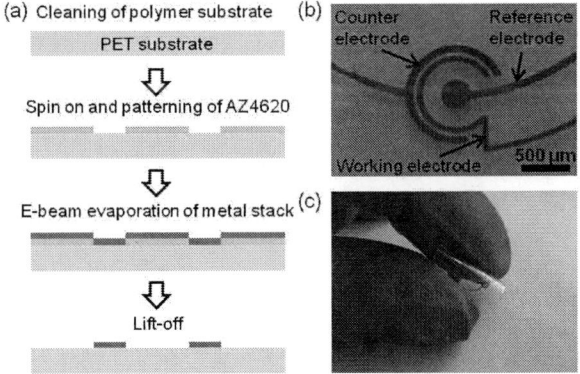

Fig. 2. (a) Schematic of the microfabrication process. (b), (c) Optical micrographs of the fabricated sensing structure on the flexible, transparent PET substrate.

achieved by the covalent cross-linkage with glutaraldehyde (GTA) and bovine serum albumin (BSA), a well-established procedure [3-5]. Earlier reports indicate the active groups of LOx being enclosed by the GTA-BSA matrix, thereby keeping the enzyme functional [5,6].

A phosphate buffered saline solution (PBS) with a pH of ~ 7.4 was prepared from 130 mM NaCl (*EMD*, USA), 10 mM monobasic sodium phosphate (*J.T. Baker*, USA) and 10 mM dibasic sodium phosphate (*J.T. Baker*, USA) in DI water. All chemicals were reagent grade and used as obtained by the supplier. The enzyme was immobilized from a PBS solution containing 2% LOx, 3% BSA and 0.3% GTA [4]. 93.8 μl PBS was added to 3 mg of ≥ 98% BSA (lyophilized powder, *Sigma Aldrich*, USA) and vortexed until the powder was dissolved. 1.2 μl of 25% GTA (in H_2O, grade I, *Sigma Aldrich*, USA) was added to the solution, which was again thoroughly mixed. 2 mg LOx from *Aerococcus viridans* (> 20 U/mg lyophilized powder, *A.G Scientific*, USA) were suspended in the GTA-BSA-PBS solution, and the mixture was vortexed until all solids were dissolved. The final solution was left to rest for about 10 min to equilibrate concentration differences.

The sensors were functionalized by pipetting 10 μl of the enzyme solution onto the electrode area, and drying the membrane for 90 min at room temperature. The sensing structures were then rinsed with DI water and PBS to remove excess, unbound LOx from the substrate and blown dry with N2. The devices were stored in PBS at 4 °C.

Remaining enzyme stocks were stored at 4 °C for up to 24 hours.

C. Test Setup

The sensors were tested in a beaker (V=100 ml) with stirred PBS (pH ~ 7.4). The lactate concentration in the solution was increased stepwise by addition of appropriate amounts of a 100 mM ≥ 99.0% sodium L-lactate (reagent grade, *Sigma Aldrich*, USA) stock to the solution.

An EC epsilon potentiostat (*BASi*, USA) was used to apply a reference voltage between the working and reference electrode and measure the current response of the sensor between the working and counter electrode. The current was recorded every 1 s. Measurements for basic sensor characteristics were carried out at room temperature.

III. SENSOR PERFORMANCE

A. Basic Sensor Characteristics

Firstly, we determined the sensor response in dependence of the applied reference voltage V_{ref} between the working and the reference voltage to find the optimum operating conditions. Fig. 3 shows time traces of the current response of a single sensor for lactate concentrations between 0 and 1 mM with V_{ref} ranging from 50 mV to 500 mV. The sensors show a fast response time of ~ 40 s. We define the response time as the time passed between the addition of lactate to the solution and the sensors reaching 90% of the final current level. The current response for each concentration increases with

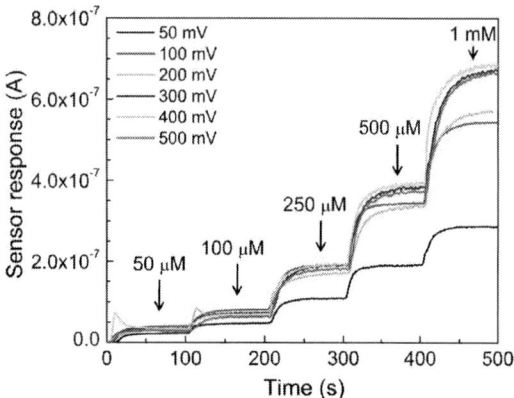

Fig. 3. Measured sensor response as function of the applied reference voltage V_{ref} and lactate concentration. The current for the respective concentrations increases for increasing V_{ref} up to 300 mV, and stabilizes for higher voltages.

increasing V_{ref} up to 300 mV. We chose a reference voltage of 400 mV for the following measurements to allow both for a stable sensing behavior with sufficient sensitivity as well as low power sensor operation.

Fig. 4 shows the calibration curve for the linear range of our devices. We measured the current response of a sensor for 200 s after each increase in lactate concentration, and averaged the data between 100 s and 200 s. The data points in Fig. 4 are the mean values obtained from measurements with six different sensors. The dependency between the sensor response and lactate concentration can be approximated by a linear function with a correlation factor R^2 of 0.997. We determined an average sensitivity of the sensors of ~370 $\mu AmM^{-1}cm^{-2}$. The high sensitivity can be explained by the small area of the working electrode of only ~ 0.19 mm^2, the improved sensor design with a minimum spatial separation

between WE and CE for an optimum utilization of the redox reactions at both electrodes, and efficient immobilization of LOx in the GTA-BSA matrix. The error bars in this and the following graphs represent the standard errors. The sensors show a very low variation in current response between different devices, which is promising for the potential mass production.

We further studied the sensor characteristics for very low and high concentrations of lactate in solution. Fig. 5 displays the sensor response for lactate concentrations between 0 and 100 μM. We measured an average sensor response of ~ 4 nA for a lactate concentration of only 5 μM. Fig. 6 shows the typical saturation of the current response for high substrate concentrations in enzyme based amperometric sensors. This behavior is likely caused by a lack of oxygen at the reaction site [7,8], which limits the amount of lactate converted into pyruvate and hydrogen peroxide. We fitted the data according to Michaelis-Menten enzyme kinetics:

$$I = \frac{I_{max}c}{c + K_M} \qquad (3)$$

with I the respective sensor current measured for a lactate concentration c, I_{max} the maximum observed current for a single sensor and K_M the Michaelis-Menten constant. We determined an average K_M of ~ 0.7 mM for our sensors.

B. Temperature stability

We further studied the sensor response to different lactate concentrations in dependence of the temperature of the solution to determine the range of appropriate operating temperatures. We heated the PBS solution to temperatures between 20 °C (room temperature) and 50 °C and then added the lactate stock. The temperature was constantly measured during the experiment. Fig. 7 displays the results of temperature stability experiments. The sensor shows a stable

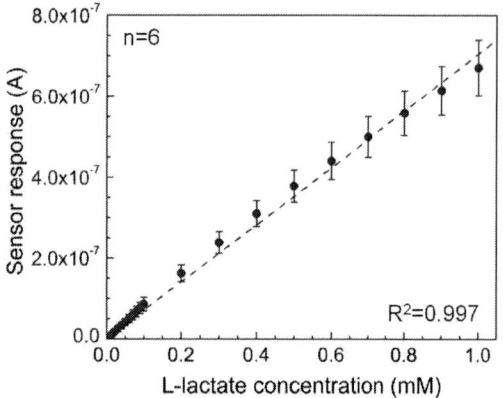

Fig. 4. Calibration curve for the approximated linear range of the sensor response between 0 and 1 mM lactate. The data points are averaged from measurements with six different sensors. The error bars in this and the following graphs represent the standard error.

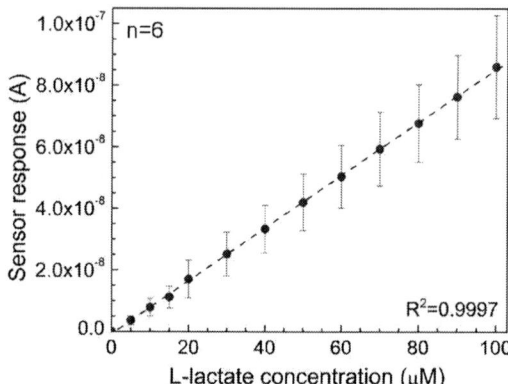

Fig. 5. Measured sensor response to small concentrations of lactate in solution. The data was averaged from 6 sensor measurements. We can reliable detect lactate concentrations as low as 5 μM.

978-1-4673-1122-9/12 $31.00 © 2012 IEEE

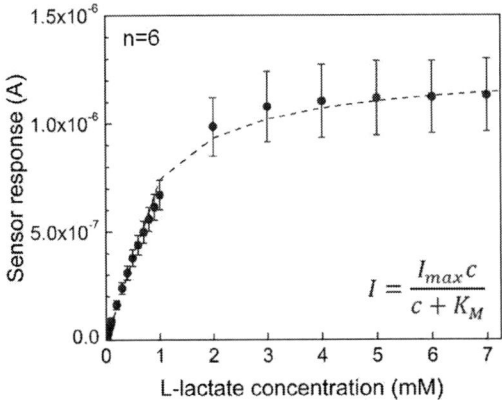

Fig. 6. Sensor response to large concentrations of lactate in solution. The data was averaged from measurements with 6 different sensors. The measured current follows Michaelis-Menten kinetics with a mean K_M of ~ 0.7 mM.

response between room temperature and 35 °C. For higher temperatures, the current response decreases significantly. LOx is known to be a rather unstable enzyme that denatures, and thereby loses its functional structure, at temperatures above ~ 35 °C [9]. Efforts are underway to engineer the enzyme towards a more stabile chemical structure [10]. Even at this stage, however, the sensors are suitable for applications which only require testing at room temperature or slightly above.

IV. LACTATE MEASUREMENTS IN DAIRY PRODUCTS

To show the applicability of the sensor design for practical use, e.g. the monitoring of food products, we measured the lactate content in several dairy products such as cottage cheese, cultured buttermilk, and yogurt.

The samples were diluted in PBS (1:100 dilution) to allow for a similar measurement routine as applied for the basic

sensor characteristics. We measured the current response of three different sensors in the stirred solution, and then compared the measured current levels to previously obtained calibration curves for the respective sensors. We further recorded the current response of bare Pt electrodes, i.e. sensors not functionalized with LOx, exposed to the different dairy samples to test for any interference signals arising due to other component present in the samples. In all cases bare sensors only gave rise to baseline level signals.

Table 1 shows the estimated lactate concentrations in diluted samples of non-fat cottage cheese, low-fat cultured buttermilk, and non-fat blueberry yogurt. The data is averaged from measurements with three different sensors. The concentrations are well within the linear range of the sensors and are comparable to those measured by other groups [11-13].

We further utilized the sensor to study the development of lactate levels in milk over an extended period of time. The milk was stored at room temperature and exposed to light over a period of several days. Under these conditions, milk sugar (lactose) is fermented into lactic acid by bacteria in the milk. Measurements were carried out in fresh milk (0 h), undiluted milk (24 h), in the liquid part of the fermented milk (72 h and 120 h), and the liquid part diluted in PBS (168 h). We could indeed observe a significant increase in lactate levels (Fig. 8).

TABLE I. ESTIMATED LACTATE CONCENTRATIONS FROM MEASUREMENTS IN DIFFERENT DAIRY SAMPLES.

Sample	Lactate concentration in diluted sample (mM)
Non-fat cottage cheese	0.21±0.02
Low-fat cultured buttermilk	0.58±0.03
Non-fat blueberry yogurt	0.57±0.15

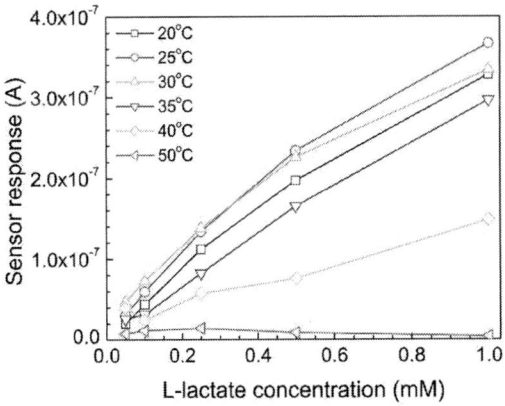

Fig. 7. Temperature dependent sensor measurements for a single lactate sensor. The sensor response is stable up to 35 °C, and the sensors are functional to at least 40 °C.

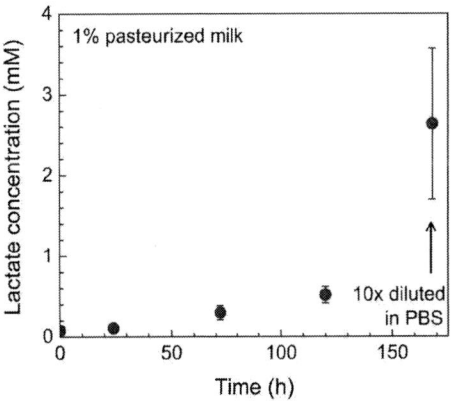

Fig. 8. Development of lactate levels in low-fat, pasteurized milk stored at room temperature and exposed to light. The lactate concentrations rose significantly due to fermentation of lactose into lactic acid by bacteria in the milk. The data is averaged from measurements with three different sensors prepared shortly before the respective tests.

V. CONCLUSION

We demonstrated the fabrication of amperometric sensors on transparent and flexible PET substrates. Functionalization of the sensing structures with the enzyme LOx via cross-linkage with GTA and BSA allows for a selective recognition of L-lactate. The sensors show a high sensitivity within their linear range, a low detection limit and satisfactory thermal stability for applications close to room temperature. We further tested them for practical applications, i.e. the estimation of lactate concentrations in dairy samples. Due to their small size and the flexibility of the substrate, they can be formed or molded into the desired shape, and may therefore be suitable for utilization in test tubes or capillaries. With appropriate biocompatible enclosure the sensors may also be applied for measurements of lactate as a product of the human metabolism. Efforts are underway to develop sensors on contact lenses for the detection of lactate in tear fluid.

ACKNOWLEDGMENT

The sensor fabrication was carried out in the University of Washington Microfabrication Facility. Financial support for this work was provided by the National Science Foundation of the United States.

REFERENCES

[1] K. Yorita et al., "On the reaction mechanism of L-lactate oxidase: quantitative structure-activity analysis of the reaction with *para*-substituted L-mandelates", Proc. Natl. Acad. Sc., vol. 94. 18, pp. 9590-9595, September 1997.

[2] I. Leiros et al., "The 2.1 Å structure of *Aerococcus viridans* l-lactate oxidase (LOX)", Acta Crystallogr. Sect. F Struct. Biol Cryst. Commun., vol. 62, Pt 12, pp. 1185-1190, December 2006.

[3] N. Nikolaus and B. Strehlitz, "Amperometric lactate biosensors and their application in (sports) medicine, for life quality and wellbeing", Microchim. Acta vol. 160, 1-2, pp. 15-55, January 2008.

[4] O. M. Schuvailo, O. O. Soldatkin, A. Lefebvre, R. Cespuglio, and A.P. Soldatkin, "Highly selective microbiosensors for in vivo measurement of glucose, lactate and glutamate", Anal. Chim. Acta vol. 573-574, pp. 110-116, July 2006.

[5] S. A.M. Marzouk, V. V. Cosofret, R. P. Buck, W. E. Cascio, and S. S. M. Hassan, "Amperometric monitoring of lactate accumulation ischemic myocardium", Talanta vol. 44, 9, pp. 1527-1541, September 1997.

[6] S. S. Wong, and L.-J. C. Wong, "Chemical crosslinking and the stabilization of proteins and enzymes", Enzyme Microb. Techn. vol 14, 11, pp. 866-873, November 1992.

[7] R. Zimei, E. Leitao, J. Popplewell, B. Alp, and P. Vadgama, "Needle Enzyme Electrode for Lactate Measurement In Vivo", IEEE Sens. J. vol. 8, 1, pp.113-120, January 2008.

[8] J. J. Burmeister, M. Palmer, and G. A. Gerhardt, "L-lactate measures in brain tissue with ceramic-based multisite microelectrodes", Biosens. Bioelectron. vol. 20, 9 pp. 1772-1779, March 2005.

[9] H. Misaki, Y. Horiuchi, K. Matsuura, and S. Harada, "Lactate oxidase process for the manufacture thereof and analytical method and kit for the use of the same", US Patent # 4237222, 1980.

[10] H. Minagawa, J. Shimada, and H. Kaneko, "Effect of the mutations at Glu160 and Val198 on the thermostability of lactate oxidase", Eur. J. Biochem. vol. 270, 17, pp. 3628-3633, September 2003.

[11] W. A. Collier, P. Lovejoy, and A. L. Hart, "Estimation of soluble L-lactate in dairy products using screen-printed sensors in a flow injection analyser", Biosens. Bioelectron. vol. 13, 2, pp. 219-225, February 1998.

[12] W. J. Mullin and D. B. Emmons, "Determination of organic acids and sugras in cheese, milk and whey by high performance liquid chromatography", Food Res. Internat. vol. 30, 2, pp. 147-151, March 1997.

[13] F. Palmisano, M. Quinto, R. Rizzi, and P. G. Zambonin, "Flow injection analysis of L-lactate in milk and yoghurt by on-line microdialysis and amperometric detection at a disposable biosensor", Analyst vol. 126, 6, pp. 866-870, June 2001.

Spontaneous Motion of a Water Droplet on Hydrophilic and Curvature Gradient Conical-Shaped Surfaces

Y. C. Chuang[1], H.Y. Hsieh[2], Quanshui Zheng[3], and F. G. Tseng[1,2,4*]

[1] Department of Engineering and System Science, National Tsing Hua University, Hsinchu 30013, Taiwan R.O.C.
[2] Institute of NanoEngineeringand MicroSystems, National Tsing Hua University, Hsinchu 30013, Taiwan R.O.C.
[3] Department of Engineering Mechanics, Tsinghua University, Beijing 100084, China.
[4] Research Center for Applied Sciences, Academia Sinica, Taipei 11529, Taiwan R.O.C.
*F. G. Tseng, E-mail: fangang@ess.nthu.edu.tw

Abstract—**This paper reports the study on spontaneous and fast motion for a microliter water droplet on nanotextured glass capillary surfaces with a wide range of curvature gradients. The surface is highly related to the surface tension gradient that is mainly formed by three distinct driving forces, including surface hydrophilicity gradients, chemically patterned nanotexture, and curvature gradient capillaries. In the experiments, the droplet velocity shows a dependency to the droplet position on the conical capillary curvature surface and moves toward the more wettable part of the gradient. The speed of the droplet on the oxygen plasma treated nanotextured glass capillary is up to 238.5 mm/s with more than two times of that, 101.7 mm/s, on the untreated surface. Therefore, we can conclude that a gradual variation of wettability property governs the droplet motion.**

Keywords- Droplet motion; surface tension; curvature gradient

I. INTRODUCTION

Self-propelled drops is important due to the disuse of pumps and valves, increasing mixing velocity, minimization of cross-contamination, and dispersionless transport [1]. This technology could become a valuable microfluidic tool in drug screening and DNA analysis [2], and even extended life science application such as quick drainage of condensed liquid drops for fuel cell and semiconductor devices and directional water collection on wetted spider silk, as shown in Fig. 1 [3]. However, one major challenge in moving discrete liquid droplets is to overcome the hysteresis force and viscous shear force. These two forces will lead to deceleration and stop of the droplet, unless other external forces are applied. Thus, conventionally, spontaneous motion efforts toward the use of surface tension gradient for liquid transport. The most rapid moving speed of the microliter droplet on the plate has been reported up to 500 mm/s in the literature [1]. To further increase this moving tendency, this research incorporates the third driving force: surface curvature gradients [4] for generating ultrahigh speed and multi-directed droplet movement. The experiments were performed by dispensing a small water droplet (~1 µL) on a conical glass capillary surface with three different surface conditions, including original glass surface, oxygen plasma treated surface, and

oxygen plasma treated nanotextured surface. Droplet movement behavior on these three different surfaces were also compared in this study.

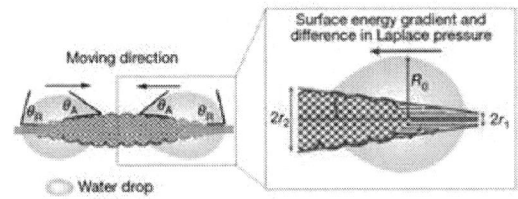

Fig. 1. Schematic the mechanism of directional water collection on wet-rebuilt spider [3].

II. THREORY

Considering a droplet moving on a unbalanced interfacial forces between their front and rear end imposed by the unique surface (Fig. 2 (a-c)), conservation of momentum is employed to analyze the dynamics of the droplet moving on the outer of the spherical surface (Fig. 2(b)). It can be written as $F = -\partial E / \partial s$, and the developing equation is as follows:

$$-2kE_0 r_0 \left(1 - \frac{4}{3} \cdot \frac{1}{sin^2\theta}\right) \cdot \frac{\partial H}{\partial s} = \rho V_0 \ddot{s} \quad ,$$

where $E = E_0 \left[1 + 2kr_0 \left(1 - \frac{4}{3} \cdot \frac{1}{sin^2\theta}\right)H\right], k = -\frac{4}{3} \cdot \frac{sin^4\theta}{2 - 3cos\theta + cos^3\theta}$ (1)

Here, the droplet with radius r_0, volume V_0, and contact angle θ on a flat surface (Fig. 2(a)), ρ is the density of the droplet, E_0 is the surface tension of the droplet such as $E_0 = \gamma(A_{LV} - A_{SL} \cdot cos\theta)$, γ is surface energy, and A_{LV} and A_{SL} are the liquid –vapor and solid-liquid interfacial surface. Besides, $H = 1/|R|$ is the mean curvature of the spherical surface, R is the radius of the contact area. When a droplet is moving on the inner of the spherical surface (Fig. 2(c)), we can also get the same expression of the Eq(1) and $H = -1/|R|$.

978-1-4673-1122-9/12 $31.00 © 2012 IEEE

In the other words, whether H is positive or negative depends on the shape of the contact area. Then we can calculate and integrate the above equation to get the following equation:

$$\dot{s}^2 = \frac{6\eta\gamma_{LV}}{\rho}\left(\frac{1}{R_0}-\frac{1}{R}\right), \quad \eta = \frac{(1+cos\theta)(1+3cos^2\theta)}{2(1-cos\theta)(2+cos\theta)} \quad (2)$$

Her, R_0 and R are the initial and the stop local radiuses of the contact area. Moreover, we also need to consider the droplet affected by the hysteresis force and viscous shear force during the movement:

$$\dot{s}^2 = \frac{6\eta\gamma_{LV}}{\rho}\left(\frac{1}{R_0}-\frac{1}{R}\right) - \kappa\cdot\Delta\theta\cdot\frac{\gamma_{LV}}{\rho}s\cdot V_0^{-2/3} - \xi\cdot\frac{\tau}{\rho}\cdot s\cdot V_0^{-1/3}$$

Where

$$\kappa = 4\left[\frac{3}{\pi(2-3cos\theta+cos^3\theta)}\right]^{1/3}\cdot sin\theta \text{ ,}$$

$$\xi = 2\pi\left[\frac{3}{\pi(2-3cos\theta+cos^3\theta)}\right]^{2/3}\cdot sin^2\theta \quad (3)$$

Here, $\Delta\theta$ is the contact angle hysteresis, and $\tau = \mu\frac{\partial u}{\partial z}$ is the viscous shear force.

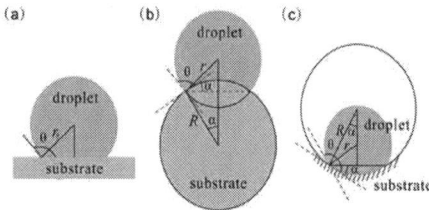

Fig. 2. Schematic illustration of a droplet on (a) flat, (b) outer, (c) inner of curved surfaces[4]

III.   EXPERIMENTAL

Fig. 3 shows the fabrication process to form three different statue surfaces and the contact angles on each surface. The static contact angles (CA) of the original glass and oxygen plasma treated surface are around 28° and 5°, respectively. The chemical patterned nanotextured surface was treated by phase separation of methyltrichlorosilane (CH_3SiCl_3:MTS) [2]. The 3D nanostructures were synthesized by immersing glass capillaries in anhydrous toluene with MTS solution at a concentration of 0.014 M for 75 mins under 23 ℃ and 75 %RH environmental condition. After the treatment, the nanotexture provides the superhydrophobic surface with a static contact angle (CA) and hysteresis of about 147° and 10°, respectively. The surface was then treated by oxygen plasma to decrease the contact angle to about 0° on the nanotextured surface. Three different surface conditions, including original surface, O_2 plasma treated one, and O_2 plasma treated MTS-modified glass capillaries will be compared their droplet moving velocity herein.

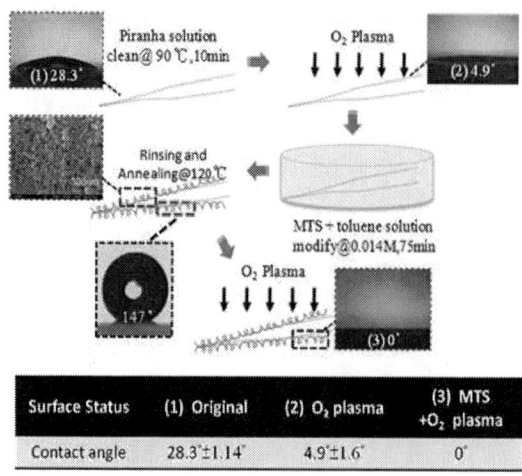

Surface Status	(1) Original	(2) O_2 plasma	(3) MTS $+O_2$ plasma
Contact angle	28.3°±1.14°	4.9°±1.6°	0°

Fig. 3. Fabrication process to form three different statuses on glass capillary surfaces.(Values represent the mean ±maximum/minimum of at least five samples).

IV.   RESULTS AND DISCUSSION

In all cases, water droplets can be self-directed and transported toward the capillary's larger cross-section region (with lower curvature). The fastest velocity always occurs on the surface with the largest curvature gradient. The evolution of droplets movement are shown from Fig. 4(a-g) representing droplet moving process on original (CA= 28°), O_2 plasma treated (CA=5°), and MTS with O_2 plasma treated (CA= 0°) capillary surfaces, respectively. Furthermore, droplets deposit on these three different kinds of curvature gradients were compared in this study, and the diameter of curvature gradients are selected to 0.3–1 mm, 0.3–1.5 mm, and 0.1–1 mm. The 1-μL droplet cannot move on the two former sizes on the original and oxygen plasma treated surfaces, the droplet just wetting the surfaces. However, the droplet moving speed on the smallest size capillary (0.1–1mm) can approach a maximum of 101.7 mm/s, 174.2 mm/s and 238.6 mm/s for original, O_2 plasma treated ,and MTS treated capillary surfaces, respectively, as shown in Fig 5-6 In the same figure, droplet occupies larger curvature gradients also carry out a higher moving velocity. The droplet always move starting from the region with the smallest radius (the largest curvature) and stop at the area without curvature gradient. In other words, the droplet velocity along the trajectory presents an accelerating speed at first and then gradually reduces the speed to the zero. Fig. 7 shows calculated results from theory formulation that has same trend with the previous experimental results and difference less than one order.

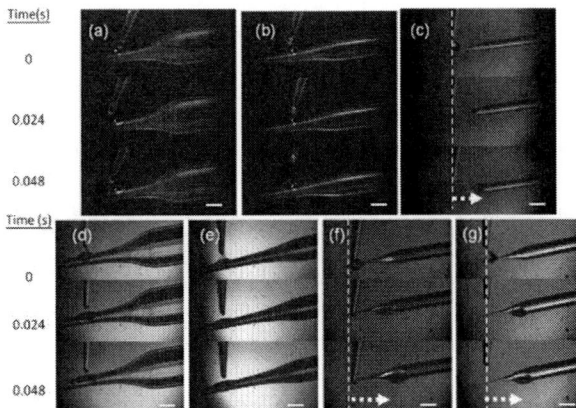

capillary.

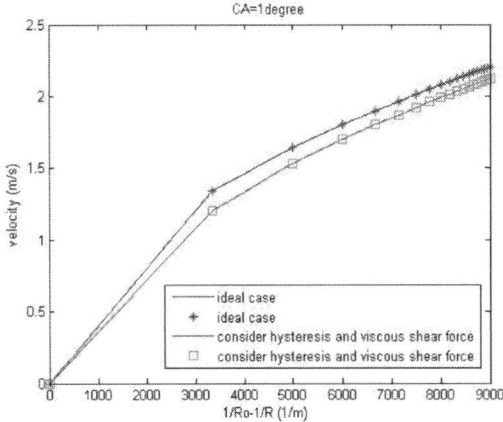

Fig. 7 Calculated results from theory formulation. The relationship between the velocity and curvature gradient of $1\mu L$ *moving* on contact angle around 1 degree surface.

Fig. 4. Self-motion behavior of a 1 µL water droplet occupying different curvature ranges (a, d) 0.3–1 mm,(b, e) 0.3–1.5 mm, (c, f, g) 0.1-1 mm. (where (a-c) are original surface, (d-f) are O_2 plasma treated surface, (g) is MTS nanotexture and O_2 plasma modified surface) (scale bar: 2 mm),

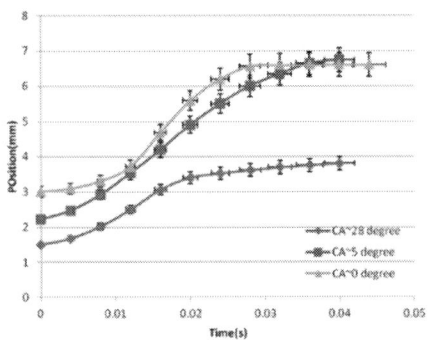

Fig. 5. Position evolution of 1 µL droplet motion on hydrophilic glass capillary surfaces with 0.1-1 mm of diameter gradients

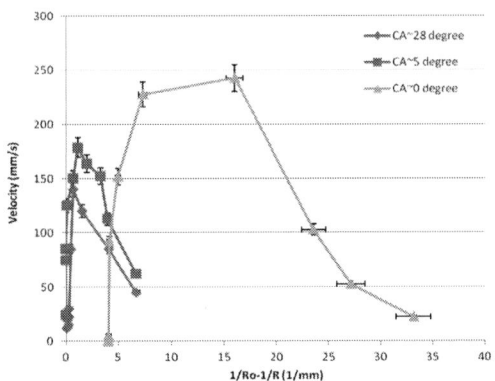

Fig. 6. The relationship of curvature difference and velocity of a droplet moving on 0.1-1 mm diameter gradient glass

V. CONCLUSION

In summary, we have applied the curvature gradient conical capillary surface to facilitate the microliter droplet spontaneous fast motion on the hydrophilic surfaces. The droplet profile and velocity are related to the surface hydrophilicity (CA) and the capillary size (curvature). For a water droplet, the fastest speed is established on the superhydrophilic and the smallest size surface (largest curvature). The intrinsic behavior of the droplet could be concluded and predicted by using submicroliter droplet moving on a capillary with smaller than 0.1 mm diameter size to obtain ultrahigh speed droplet movement. The finding is potentially useful for manipulating water droplet of micro-nano-liter and could be a valuable microfluidics tool for the applications in drug discovery, even for life science such as fuel cell and semiconductor devices.

ACKNOWLEDGMENT

National Nanoscience and Nanotechnology Program, National Science Council, Taiwan through NSC-99-2120-M-007-009.

REFERENCES

[1] Yongmei Zheng, Hao Bai, Zhongbing Huang, Xuelin Tian, Fu-Qiang Nie, Yong Zhao, Jin Zhai, and Lei Jiang, "Directional water collection on wetted spider silk", Nature 08729,640-643(2010)

[2] H. S. Khoo, F. G. Tesng, "Spontaneous high-speed transport of subnanoliter water droplet on gradient nanotextured surfaces," *Appl. Phys. Lett.*, **95**, 063108, 2009.

[3] Yongmei Zheng, Hao Bai, Zhongbing Huang, Xuelin Tian, Fu-Qiang Nie, Yong Zhao, Jin Zhai, and Lei Jiang, "Directional water collection on wetted spider silk", Nature 08729,640-643(2010)

[4] H. S. Khoo, F. G. Tesng, "Engineering the 3D architecture and hydrophobicity of methyltrichlorosilane nanostructures," Nanotechnology, **19**, 345603, 2008.

Electric Manipulations of Hydrogel on a Digital Microfluidic Platform

Min-Yu Chiang[1], Shih-Kang Fan[1,2]*

[1]Department of Materials Science and Engineering, National Chiao Tung University, Hsinchu, Taiwan
[2]Department of Mechanical Engineering, National Chiao Tung University, Hsinchu, Taiwan
skfan@mail.nctu.edu.tw

Abstract—For the first time, we demonstrate the electric manipulations of (1) crosslinkable polymer solution droplets by electrowetting and dielectrophoresis, (2) crosslinked polymer and hydrogel blocks by dielectrophoresis, and (3) pre-polymer solution and crosslinked hydrogel containing cells. Different hydrogel blocks cured from individually-driven pre-polymer droplets provide 3D microenvironments for in vitro cell culture. With the ability to actively drive the pre-polymer droplets, assemble the hydrogel blocks, and arrange the cells inside the hydrogel, this work provides an essential approach to tissue engineering, cell-cell interaction, stem cell differentiation studies on a highly programmable digital microfluidic device.

Keywords-electrowetting on dielectric; dielectrophoresis; hydrogel; tissue engineering

I. INTRODUCTION

Most natural tissues are composed of repeated microscale functional units with specific cell-cell and cell-extracellular matrix (ECM) arrangements. In order to generate bio-mimic microscale tissue modules, the regeneration of structural features is important in enabling the resulting function. These tissue modules could be made to generate macroscale tissue from microgel units made of cell-seeded or cell-laden hydrogels. Hydrogels are the hydrophilic materials widely used in biomedical applications including drug delivery, microfluidics-based immunoassays and tissue engineering due to its excellent properties such as its porous network structure, good biocompativity and highly absorbent and tunable physical and chemical properties. Typically, cells are encapsulated in hydrogels by mixing a cell suspension with hydrogel pre-polymer solution followed by crosslinking of the network. The crosslinking reaction may be controlled by variety of methods such as changing temperature, changing pH value, the addition of chelating ions, and exposing ultraviolet light. Moreover, the microscale functional units were fabricated by micro-molding or photo-patterning. Fabrication of microscale hydrogel such as agarose [1], gelatin [2] and poly(ethylene glycol) diacrylate (PEG-DA) [3] has been demonstrated on continuous-flow microfluidic devices. There are few investigators who proposed several techniques for the combination of hydrogels with digital microfluidics (DMF) [4, 5]. In this study, we utilized two electric manipulations which include electrowetting-on-dielectric (EWOD) and dielectrophoresis (DEP) to manipulate pre-polymer solution, pre-polymer hydrogel/cells mixture, and crosslinked cell-laden hydrogel on a digital microfluidic platform. In an EWOD/DEP-based digital microfluidic device,

the crosslinkable droplets containing cells can be manipulated for the fabrication of building blocks of tissue engineering. Photo-crosslinkable polymer, PEG-DA, was driven on-chip by EWOD/DEP with adjustable shapes and sizes. By utilizing EWOD [6], basic microfluidic functions including creation, transportation, cutting, and merging pre-polymer droplets were implemented. In addition, multiple polymer droplets containing different mechanical and chemical cues for cell culture will be simultaneously created on a single device. Moreover, DEP force was employed to draw dielectric liquids to the region of stronger electric field on chip.

II. PRINCIPLE

A. Electrowetting-on-Dielectric (EWOD)

In 1875, the electrocapillary phenomenon was first recognized by Lippmann [7]. When adding an extra electrostatic charge, the capillary forces at an interface may change significantly. This phenomenon was not noticed for many years. However, with the development of micro electro mechanical systems (MEMS) technology, the EWOD phenomenon had drawn much attention. Recently, many studies showed that the wettability of liquids on a dielectric surface could be controlled with a high reversibility. When an electric voltage V was applied between the aqueous droplet and the electrode, the droplet spontaneously spreads out on the hydrophobic surface (as shown in Fig. 1(a)). The contact angle θ_v is modulated by the applied voltage according to the Young-Lippmann equation:

$$\cos\theta_v = \cos\theta_0 + \frac{\varepsilon_0\varepsilon_r}{2\gamma_{LG}t}V^2, \tag{1}$$

where ε_0 denotes the permittivity of the vacuum, ε_r the dielectric constant of the dielectric layer, γ_{LG} the gas-to-liquid interfacial tension and t the thickness of the dielectric layer. When the electric potential V was applied, the contact angle of liquids would be changed. When the electric potential V is removed, the changed contact angle returns to the initial contact angle θ_0. The EWOD driving force in the parallel-plate device is described as:

$$F_{EWOD} = \frac{\varepsilon_0\varepsilon_D W}{2t}V_D^2, \tag{2}$$

where ε_0 is the permittivity of vacuum, ε_D is the relative permittivity of the dielectric layer, W is the width of EWOD driving electrodes, t is the thickness of the dielectric layer, and

V_D is the applied voltage across the dielectric layer. The method of EWOD is an excellent tool to manipulate the fluid interfaces.

B. Dielectriphoresis (DEP)

Another driving force of liquid droplet, DEP, is extensively studied for many years. DEP has been known for the abilities of particles [8] and liquid [9] manipulations by non-uniform electric fields. Dielectric liquids are driven by DEP with none or limited contact angle change when applying a signal on the electrodes. By DEP actuation, the dielectric droplet is manipulated between two parallel plates when a sufficient electric field generated by a signal (AC or DC) is perpendicular to the liquid/medium (e.g., liquid/air boundary) of normal vector boundary [10]. The DEP force drives a liquid with higher permittivity towards high electric field regions (as shown in Fig. 1(b)), which is explained by Maxwell stress tensor and expressed as:

$$F_{DEP} = \frac{\varepsilon_0 \varepsilon_D W}{2t} V^2 \left(\frac{\varepsilon_{liquid}}{\varepsilon_{liquid} t + \varepsilon_D d} + \frac{\varepsilon_{air}}{\varepsilon_{air} t + \varepsilon_D d} \right), \qquad (3)$$

where V is the applied voltage, ε_{liquid} and ε_{air} are the permittivities of hydrogel and air, respectively, and d is the gap between the parallel plates.

III. MATERIALS AND METHODS

A. Fabrication of Digital Microfluidic Device

The tested devices were fabricated on glass substrates for ideal electric isolation between the patterned driving electrodes and for easier observation of the droplets. For the same observation purposes, the reference electrode on the top plate was made of transparent and conductive ITO (indium tin oxide). A layer of ITO was first deposited on the top glass substrate and subsequently covered by a low surface energy material, Teflon. Teflon was dissolved in FC-77 (3M) at a concentration of 0.5 w/v% and spun on the ITO-coated top plate at 3000 rpm for 30 s. Meanwhile, a layer of ITO was deposited on the bottom glass substrate. An array of driving electrodes, measuring 1 mm × 1 mm, was patterned by wet chemical etching. Teflon layer was spun on the bottom plate as the top plate. Before testing, a dielectric droplet was dispensed onto the bottom plate, where proper spacers were attached. Device assembly was completed by placing the top plate on the spacers that were used to define the droplet thickness and the gap height between plates. Finally, wires were connected between the device, the control circuits, and a computer. The device structure was showed in Fig. 1.

B. Cell culture

The NIH-3T3 fibroblast cells were kept at 37 ℃ in a 5 % CO_2 and 95 % air incubator and maintained in Dulbecco's Modified Eagle's Medium (DMEM, BioWest) with 10 % Fetal bovine serum (FBS, BioWest) and 1 % antibiotic (BioWest). Confluent cells were subcultured every 2-3 days by trypsonization with 0.25 w/v% trypsin and 0.13 w/v% EDTA.

C. Manipulation of Hydrogel on Microfluidic Platform

We utilized two different concentrations of poly (ethylene glycol) diacrylate (PEG-DA, Mw = 575 Da) solution. The concentrations of PEG-DA we used were 20% and 100% (pure), respectively. The cell/hydrogel pre-polymer solution was prepared by dissolving 20 v/v% PEG-DA, in DMEM. Immediately before UV polymerization, 1 % photoinitiator and 2×10^7 cell/ml NIH-3T3 fibroblast cell suspension were added to pre-polymer solution. After the preparation of cell/hydrogel pre-polymer solution, a 0.7 μl droplet was dropped to the reservoir on the bottom plate. Meanwhile, the top plate covered to bottom plate.

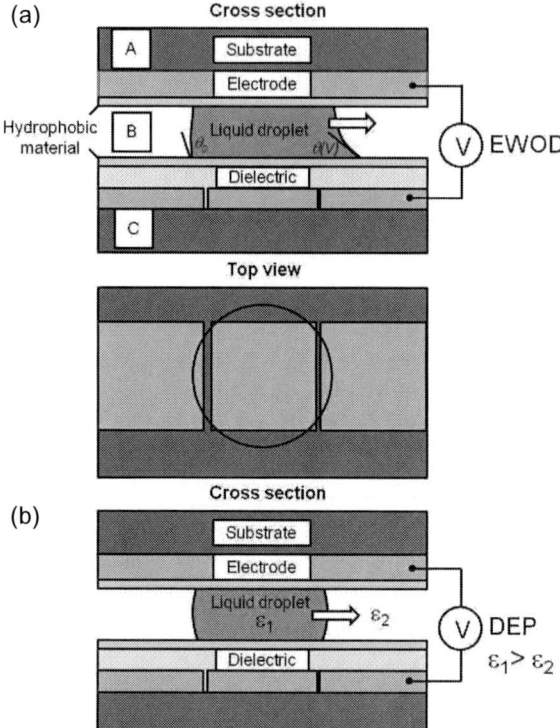

Fig. 1: A parallel plate device manipulating droplets by EWOD and DEP on an array of dielectric-covered driving electrodes. (a) Conductive pre-polymer droplet driven by EWOD within a device containing A: top plate, B: liquid droplet, and C: bottom plate. (b) Dielectric pre-polymer droplet with a higher permittivity actuated by DEP.

IV. RESULTS AND DISCUSSION

In this study, we successfully demonstrate the actuations of different concentrations hydrogel with different electric manipulations on digital microfluidic platform. As shown in Fig. 2, the EWOD study of pre-polymer solutions with different PEG-DA concentrations in water. From the contact angle versus applied voltage curves, we realized that EWOD occurred obviously at low PEG-DA concentrations (Fig. 2(b)), while contact angle barely changed for pure PEG-DA (Fig. 2(a)). Therefore, we presumed the pure PEG-DA should be

978-1-4673-1122-9/12 $31.00 © 2012 IEEE

controlled by DEP force because the conductivity of pure PEG-DA is relatively low (0.11 μS/cm) resulting the most of voltage drop across PEG-DA. And the diluting PEG-DA should be controlled by EWOD force because the higher water content of diluting PEG-DA solution.

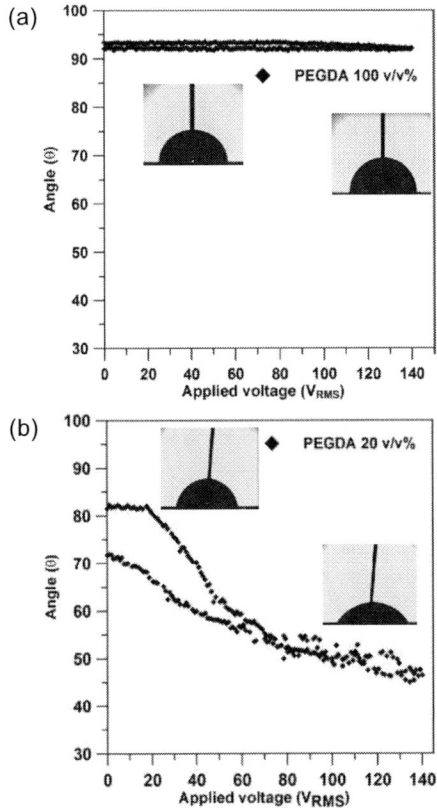

Fig. 2: The contact angle (θ) and applied voltage (V_RMS) curve of different pre-polymer droplets. (a) Pure PEG-DA. (b) 20 v/v% PEG-DA.

Fig. 3 shows the non-conductive pure PEG-DA solution has a higher permittivity and was driven by DEP in our experiment. When applying 1 kHz AC signal, the pure PEG-DA pre-polymer solution with higher permittivity could be driven to the region of high electric field. After crosslinking, the crosslinked PEG-DA blocks was similarly moved, stacked, and separated by DEP force in water medium (Fig. 4). When applying electric potential V between the top plate and bottom plate, the crosslinked PEG-DA blocks would be moved to the region of weak electric field. The phenomenon was resulted from that the water moved to the energized electrode and then pushed the crosslinked PEG-DA block to the adjacent electrode. According to the DEP phenomenon of crosslinked PEG-DA blocks, these hydrogel blocks could be arranged, separated and stacked by different electrode design. Simultaneously, the microgels production and manipulations by this technique might provide a stable platform for immobilizing entities including proteins and cells.

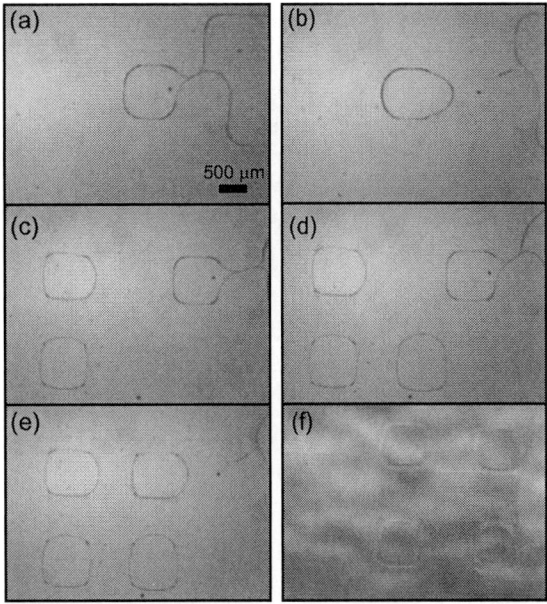

Fig. 3: Manipulation of pre-polymer droplets by DEP. (a)-(e) Creating and moving droplets by applying 72 - 92 VRMS and 1 kHz AC signals. (f) PEG-DA blocks after UV light crosslinking.

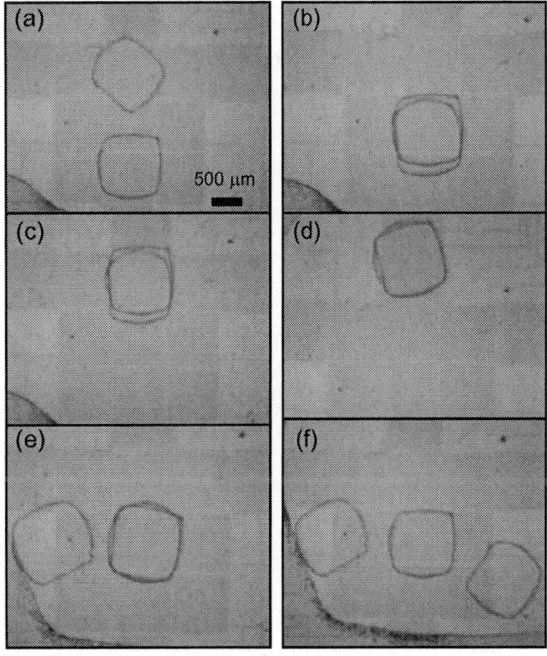

Fig. 4: Manipulation of crosslinked hydrogel blocks by DEP. (a)-(b) Crosslinked hydrogel blocks were stacked. (c)-(d) The stacked hydrogel blocks were moved together. (e)-(f) The stacked hydrogel blocks were separated. All the manipulations were completed by applying a 48 V_RMS and 100 kHz AC signal.

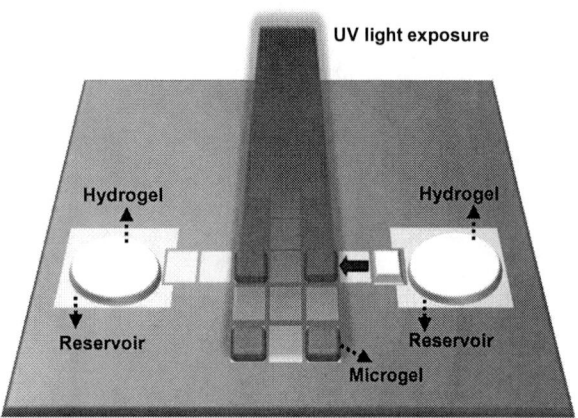

Fig. 5: Preparation of microgels by UV crosslinking on the digital microfluidic platform. The hydrogel were created from reservoir and cured on electrodes.

In order to achieve bio-mimic tissue-like construct in vitro, we presented the production and manipulations of cell-laden hydrogel blocks on our digital microfluidic platform. We demonstrated cell-laden hydrogel blocks prepared as shown in Fig. 5 by UV crosslinking. At the first step, we drop the NIH-3T3 cells/ PEG-DA pre-polymer solution to reservoirs, following we apply an AC signal to the actuated electrodes to manipulate the microgels. The NIH-3T3/PEG-DA pre-polymer solution was manipulated by EWOD (Fig. 6(a)-(e)).

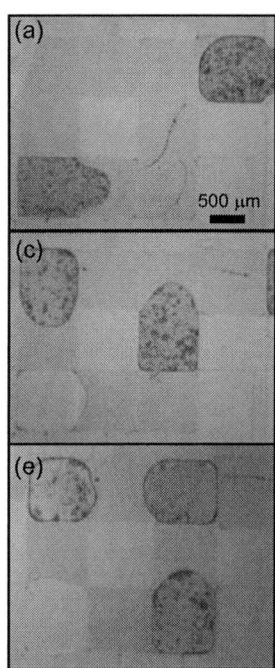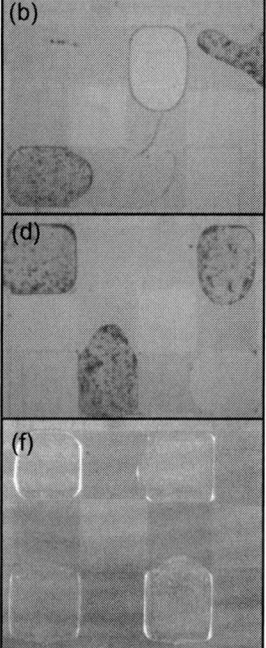

Fig. 6: Manipulation of NIH-3T3 fibroblasts/pre-polymer droplets by EWOD. (a)-(e) Creating and moving droplets by applying 72 - 92 VRMS and 1 kHz AC signals. (f) Cells/PEG-DA blocks after UV light crosslinking.

The PEG-DA pre-polymer droplet containing cells could be transported, moved, cut, and merged at 1 kHz AC signal by EWOD force. After UV crosslinking (Fig. 6(f)), cell-laden PEG-DA hydrogel blocks were obtained.

In this study, we utilized PEG-DA hydrogel as a model and successfully manipulated pure PEG-DA hydrogel and cells encapsulated PEG-DA hydrogel on a digital microfluidic platform. However, cells encapsulated in PEG-DA would lose their viability because the nutrient in culture medium is difficult to transport to the inside of cell-laden microgels (Data was not showed). From our preliminary results, PEG-DA (575 Da) is not suitable for the application of tissue engineering. Therefore, the hydrogel materials with good nutrient transportation, biocompatibility, suitable for cell growing, and good mechanical properties are still finding out. In brief, these results reveal the possibility of assembling hydrogel blocks to form macroscale tissue-like structure features, and provide a suitable model on stem cell differentiation, cell-cell interaction, and tissue engineering.

V. CONCLUSION

We successfully demonstrated the hydrogel manipulations by two different electric manipulations. With the ability to actively drive the hydrogel polymer droplets, assemble the hydrogel blocks, and arrange the cells inside the hydrogel, this work provides an essential approach to tissue engineering, cell-cell interaction, stem cell differentiation studies on a highly programmable digital microfluidic device.

ACKNOWLEDGMENT

This work was partially supported by National Science Council, Taiwan, under grants NSC 100-2120-M-009-009.

REFERENCES

[1]   Y. Ling, J. Rubin, Y. Deng, C. Huang, U. Demirci, J. M. Karp and A. Khademhosseini, *Lab Chip*, 7 (2007), pp. 756-762.

[2]   A. P. Golden, J. Tien, *Lab Chip*, 7 (2007), pp. 720-725.

[3]   M. P. Cuchiara, A. C. B. Allen, T. M. Chen, J. S. Miller, J. L. West, *Biomaterials*, 31 (2010), pp. 5491-5497.

[4]   R Fair, A. Khlystov, T. D. Tailor, V. Ivanov, R. D. Evans, V. Srinivasan, V. K. Pamula, M. G. Pollack, P. B. Griffin, J. Zhou, *Design & Test of Computers, IEEE*, 24 (2007), pp. 10-24.

[5]   V. N. Luk, L. K. Fiddes, A. H. C. Ng, E. Kumacheva, A. R. Wheeler, *μTAS* (2011), Seattle, pp. 1373-1375.

[6]   S.-K. Cho, H. Moon, and C.-J. Kim, *J. Microelectromech. Syst*, 12 (2003), pp. 70-80.

[7]   M. G. Lippmann, *Ann. Chim. Phys.* 5 (1875), pp. 494–549.

[8]   S.-K. Fan, C.-P. Chiu, P.-W. Haung, *Biomicrofluidics*, 4 (2010), 043011.

[9]   T. B. Jones, J. D. Foler, Y.S. Chang, C.-J. Kim, *Langmuir*, 19 (2003), pp. 7646-7651.

[10] S.-K. Fan, T.-H. Hsieh, and D.-Y. Lin, *Lab Chip*, 9(2009), pp.1236-1242.

978-1-4673-1122-9/12 $31.00 © 2012 IEEE

Microbubble and Microplasma Manipulations for Gas Analyses

Yan-Ting Shen, Ling-Pin Tsai, and Shih-Kang Fan[*]

[*] Department of Materials Science and Engineering, National Chiao Tung University, Hsinchu, Taiwan
skfan@mail.nctu.edu.tw

Abstract—**For the first time, we demonstrate the manipulations of microbubble and excitation of microplasma within a 200 nL microbubble in an inert oil environment using dielectrophoresis (DEP). Microbubble generation, transportation, and mixing were examined to prepare various microbubbles with tunable gas compositions further analyzed by optical emission spectrum (OES) of the excited and DEP-driven microplasma from the prepared 200 nL microbubble. Here, microbubbles containing helium, argon and their mixtures were ignited to the plasma state on one of the electrodes where a breakdown voltage applied. Characteristic peaks of helium, argon, and the mixtures were found and analyzed.**

Keywords: dielectrophoresis; microplasma; microbubble

I. INTRODUCTION

To look back to the development history of gas detection, as early as the mid-19th century Bunsen and Kirchhoff had developed the using of gas characteristic emission spectrum as a basis of analyses [1]. They introduced elements into the flame to emit radiation and got atomic emission spectroscopy. At the same time they also observed that the spectral intensity of a particular element would change with the element amount. Therefore, the qualitative and quantitative analysis of atomic emission spectrum had started.

Since the 1990s, the concept of micro total analysis systems (μ-TAS) or Lab-on-a-chip (LOC) was proposed and started to develop [2,3]. This technique integrates laboratory functions such as sample processing, transmission and testing into one tiny chip. μ-TAS has a good perspective in the future due to the breakthroughs of development in miniaturization technology. Miniaturization has the advantages of size reduction, economical amount of samples, efficiency, and automatic processing of samples and thus can reduce costs, etc., which can overcome the shortcomings of traditional test methods which take longer time and more expensive.

In all of the techniques of μ-TAS, "digital microfluidics" is an important technology of sample transmission , which is a non-continuous control over the behavior of fluidics by definition. Electrowetting-on-dielectric (EWOD) [4] and dielectrophoresis (DEP) [5] are the two of the most important mechanism for the current digital microfluidics control. In this research, a one-dimensional array chip is designed. We utilize silicone oil as a bubble medium to drive bubbles on the chip by the mechanism of DEP [6]. This platform has the functions of bubbles creating, mixing, splitting, expelling and transportation [7] for the sampling of test gas and carrier gas.

Plasma is wildly used in industrial application. Recently, micoplasma on the chip also attracts lots of interest. Atmospheric-pressure plasma draws lots of attention due to its useful applications such as biomedical and plasma reactors [8,9]. In previous literatures, however, high carrier gas consumption was necessary when plasma was lighted up especially for inductive coupled plasma optical emission spectrum (ICP-OES) [10]. Therefore, a plasma system with high efficiency and low sampling gas and carrier gas consumption is urgently required. In this study, a closed and micro-scale bubble plasma has been demonstrated, which can overcome the disadvantages those mentioned above.

We present a bubble plasma detection chip without the conventional apertures of inflow and outflow carrier gas. Different from the conventional open system of microplasma [11], the microbubble can be regarded as a closed microenvironment with no carrier gas flow. We can manipulate the microbubble, ignite the microbubble into a plasma state, and further manipulates of bubble plasma is achieved. The ability of manipulating microbubble and microplasma presents potentials to quantitative measurements of gas compositions and concentrations through sequential dilutions of the detecting and reference microbubbles with tunable ratios by mixing.

II. PRINCIPLE

A. Dieletrophoresis

The manipulation of the bubble is based on the DEP principle. The control of the bubbles is trough the manipulation of the surrounding medium, such as silicone oil. According to the DEP principle, we know that the dielectric liquids will be attracted to the region of high electric field strength and driven by the Maxwell stress tensor [12]. DEP force can be rewritten as follows:

$$\Delta P = T_{gas} - T_{liguid} = \frac{(\varepsilon - \varepsilon_O)}{2d} wV^2 \qquad (1)$$

where ΔP is the pressure difference, T_{gas} is the gas tensor at the liquid-gas interface, T_{liquid} is the liquid tensor at the liquid-gas tensor, ε is the dielectric constant of the liquid, ε_o is the dielectric constant of the air, d is the distance between two parallel electrode plates, w is the width of the electrodes, and V is the applied voltage. Through the equation we can find that due to the dielectric constant of silicone oil is about 2.5, which is larger than the dielectric constant of air whose value is about 1, that is, matters which will trend toward high density electric field are silicone oils.

978-1-4673-1122-9/12 $31.00 © 2012 IEEE

B. Plasma Ignition

To benefit observation and detection purpose, we adopt the discharge type of dielectric barrier discharge (DBD)[13], This discharge type has sufficient area to produce a uniform discharge and to facilitate the optical signal receivingr. The main factors which affecting plasma ignition involve gas species, gas pressure, applied voltage and the spacer height between electrodes. Different types of gas under the same environmental conditions have different breakdown voltage. Gas pressure mainly affects the mean free path (MPF) of gas molecules. Applied voltage and spacer height are the energy factors that influence the transformation from gas to plasma state. According to the Paschen curve from Paschen's law [14], breakdown voltage and products of pressure and electrode spacing can be plotted to get a minimum of the curve with a breakdown voltage. The following equation can be used to express the breakdown voltage:

$$V_b = \frac{B(P \cdot d)}{\ln[A(P \cdot d) - \ln[\ln(1 + \frac{1}{\gamma_{se}})]}$$ (2)

where V_b is the breakdown voltage, P is the gas pressure, A and B are the parameters from experiments, γ_{se} is the cathode secondary electron emission coefficient.In this study, the test gas currently utilize argon and helium due to both of them are monoatomic molecule, which is simpler for energy level transition. The characteristic radiation lines are also relatively uncomplex to facilitate the spectrum study. The bubbles are at atmospheric pressure in this system. Therefore, the distance between the parallel plates was 50 μm to minimize the required microbubble breakdown voltage derived from the Paschen curve. A voltage of 400 to 800 V_{RMS} at 2.4 kHz was applied onto the corresponding electrode underneath the microbubble to ignite the gas to the plasma state.

III. EXPERIMENTAL SETUP

We demonstrated a gas detection chip by optical emission spectroscopy (OES) measurement of microbubble plasma in silicone oil medium, as shown in Fig. 1. We used a function generator as a power supply, connected with an amplifier to amplify the signal in the range of 400 to 800 V_{RMS}, and the setup was shown in Fig. 1(a). We controlled the bubbles by switching the applied voltage signals with the LabVIEW program from National Instrument Corporation. The electrodes were patterned on Indium-Tin-Oxide (ITO) glass with lithography method which is commercially available. Then negative photoresist (SU8-2002) was used to be dielectric layer by spin coating and positive photoresist (AZ-P4620) was applied as a physical wall structure between two parallel electrodes on the device. The spacer height was 50 μm from the result of the Paschen curve prediction where the breakdown voltage is close to the minimum value. The bubble plasma emission spectrum was available from spectrometer measurement after bubble plasma ignition, as shown in Fig. 1(a). The setup of the whole instrument of the experiment is also shown in Fig. 1(a).

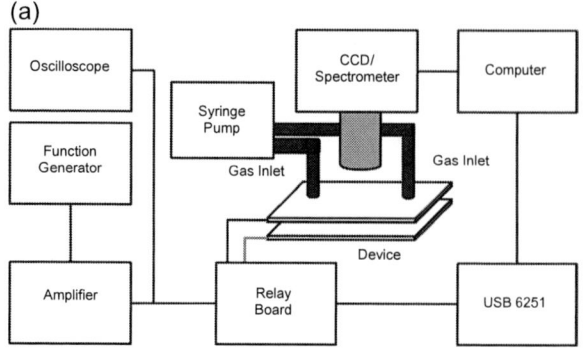

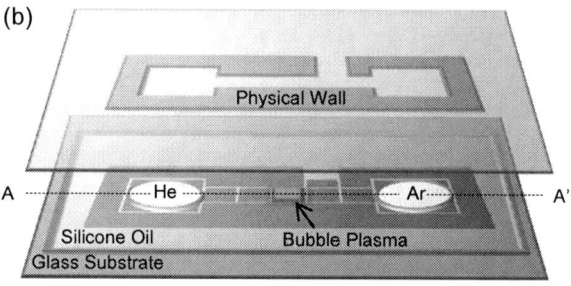

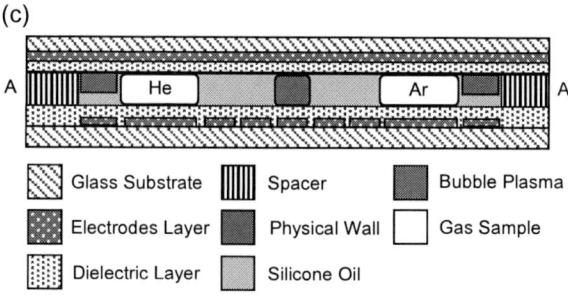

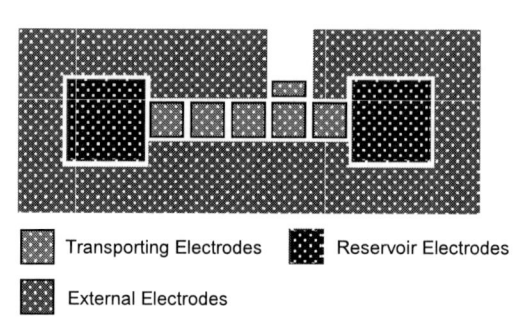

Fig. 1. Experimental setup and device design. (a) Experimental setups. (b) The angled view of the device having a top plate with physical walls, a bottom plate with patterned electrodes, and filled silicone oil between the plates. (c) Cross-sectional view of the device. (d) The electrodes design on the bottom plate.

The helium, argon and helium/argon mixing bubble with the theoretical emission peaks and the emission peaks was successfully detected on the spectra. From these spectra, the gas detection chip was proved the ability for gas detection.

As shown in Fig. 1(b), 1(c), and 1(d), the gas reservoir was fabricated by lithography to generate bubbles by DEP force. In the middle part of the device are five transporting electrode (area of 2 mm × 2 mm, and the gap between each electrodes is 30μm). These electrodes play a role of plasma ignition and bubble manipulation. The surrounding electrode was fabricated to trap bubbles on the transporting electrodes. In addition, it is important to note that the medium was 20 cSt silicone oil. The silicone oil has a significant advantage of low volatile, resulting in little gas formed from oil volatile in the bubble.

The silicone oil was first pumped to the spacing between the two plates by a syringe pump and trapped in the chip due to capillary. The silicone oil would expel the air in the pumping tube and the air in the device, and the following pumping of the detection gas was also operated by the syringe pump. When the gas was pumped to the chip, the operation of bubble creating and cutting could be operated, and therefore a bubble in the chip was formed. The function generator supply a voltage signal which would be amplified by the amplifier on the electrodes of the chip. And the switch of each electrodes was controlled by the computer with the LabVIEW program. Therefore the manipulation and the ignition of bubbles were operated by computers.

IV. RESULT AND DISSCUSSION

Fig. 2 shows the continuous manipulation images of bubbles on the chip, showing the digital microfluidics behavior such as creating, merging, splitting and transportation. This platform is a one-dimensional array electrodes designed with gas reservoirs at left and right side. There are five transportation electrodes which have the functions of bubbles manipulation and plasma ignition at the central part of the chip. The reservoirs image is not shown in Fig. 2. In this study, bubbles can be generated from each reservoir by DEP, and two bubbles cab be mixed and then split and expelled. Therefore, if the composition of the reference gas is known, we can mix the reference and test bubble trough the bubble manipulations on this platform and ignite plasma. We eliminate the characteristic peaks of reference gas from the plasma spectra of the mixture to analyze the test gas composition. Fig. 3 are the images of plasma ignition and manipulations. When we mean to move the bubble plasma to the adjacent electrode, as shown in Fig. 3, we have to switch on the proper electrode first and then switch off the original electrode. When we switch on the neighboring electrode, it is found that the edge of the bubble will spread toward to this electrode, as seen in Fig. 3(c). Then we switch off the original plasma maintaining electrode, the part of the plasma of the original electrode will extinguish immediately and the bubble will move to the neighboring electrode quickly. This driving technique of bubble plasma expands the applications afterward.

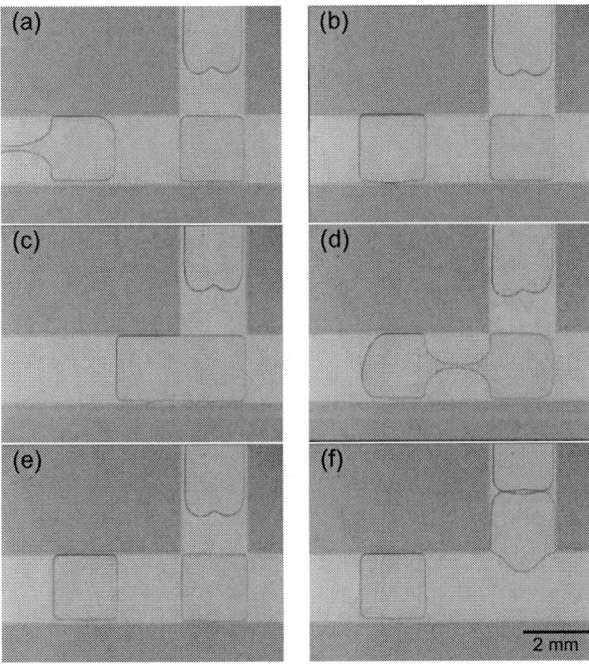

Fig. 2. Manipulations of bubbles. (a) and (b) Generation of a helium bubble from the left reservoir (not shown) and an argon bubble from the right reservoir. (c) Helium and argon mixture preparation by merging bubbles. (d) and (e) Splitting bubbles. (f) Expelling excess bubble.

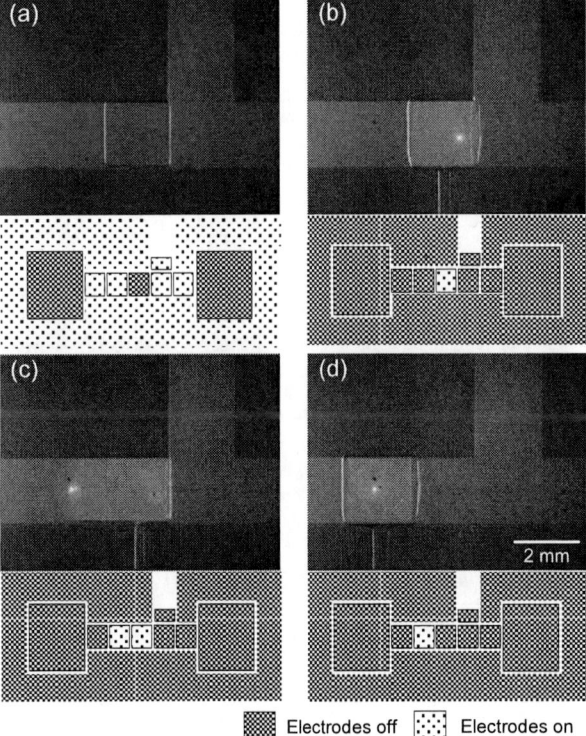

Fig. 3. Argon plasma ignition and transportation. (a) Microplasma ignited in the microbubble. (b)-(d) Microplasma and microbubble moving by applying appropriate voltage on the neighbor electrode.

The first step we use this chip system to analyze the spectrum of single type gas bubble. With regard to the emission spectrum of argon, the emission spectrum at the wavelength of 484.8 nm, 696.5 nm, 706.7 nm, 750.4 nm, and 811.5 nm has the characteristic peaks[15], as shown in Fig. 4(a). We can see from the spectrum at 750 nm with a very strong peak. Compared with this strong peak, others characteristic peaks are relatively weak but still observable. Regarding the emission spectrum of helium, the characteristic peaks are respectively at 388.8 nm, 492.1 nm, 501.5 nm, 587.5 nm, 667.8 nm, 706.5 nm and 728.1 nm [15], as seen in Fig. 4(b). Fig. 4(c) shows the spectrum of the argon-helium mixture bubble in plasma state. From this mixed gas spectrum we can observe the characteristic peaks of both argon and helium simultaneously.

(a)

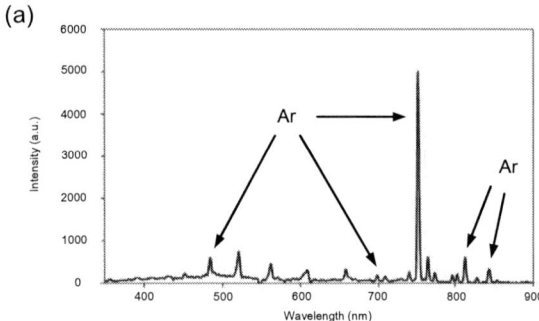

(b)

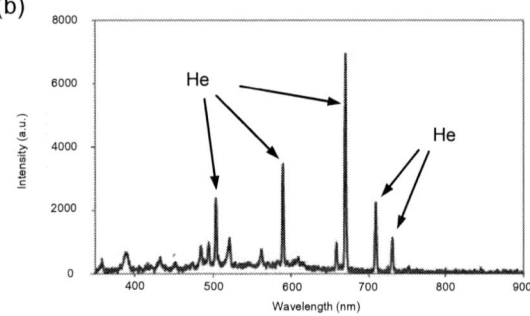

(c)

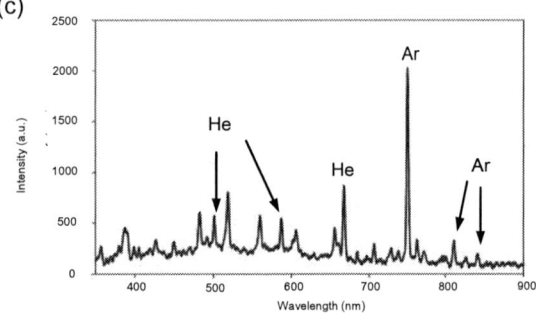

Fig. 4. The spectra of microplasma containing different gas compositions. (a) Argon. (b) Helium. (c) Mixture of argon and helium.

V. CONCLUSION

A gas detection chip based on bubble plasma by optical emission spectroscopy is demonstrated in this study. The functions of creating, merging, splitting, and expelling of bubble for preparing detecting gas bubbles is utilized in this system. After the bubble plasma ignition, the manipulation of bubble plasmas is achieved. By analyzing the spectra of bubbles in the discharging state, two species of test gas and its mixture bubble is recognizable in this system.

ACKNOWLEDGMENT

This work was partially supported by National Science Council, Taiwan, under grants NSC98-2221-E-009-129-MY3.

REFERENCES

[1] G. Kirchhoffg and R. Bunsen, "Chemical Analysis by Spectrum Observations," *Philosophical. Magazine and Journal of Science*, S. 4, Vol. 22, pp. 329-349, 1861

[2] A. Berg and T. S. J. Lammerink, "Micro Total Analysis Systems: microfluidic aspects, integration concept and applications," *Topics in Current Chemistry*, Vol. 194, pp. 22-49, 1998.

[3] A. Manz, N. Graber, and H. M. Widmer, "Miniturized total chemical analysis system: a novel concept for chemical sensing," *Sensor and Actuators B*, Vol. 1, pp. 244-248, 1990.

[4] G. Beni and M. A. Tenan, "Dynamics of electrowetting displays," *Journal of Applied Physics*, Vol. 52, pp. 6011-6015, 1981.

[5] T. B. Jones, "Multipole corrections to dielectrophoretic force," *Transacttions on Industry Applications*, Vol. IA-21, pp.930-934, 1985.

[6] S.-K. Fan, T.-H. Hsieh and D.-Y. Lin, "General digital microfluidic platform manipulating dielectric and conductive droplets by dielectrophoresis and electrowetting", *Lab on a Chip*, vol. 9, pp. 1236-1242. 2009.

[7] S. K. Cho, H. Moon, and C.-J. Kim, "Creating, transporting, cutting, and merging liquid droplets by electrowetting-based actuation for digital microfluidic circuits," *Journal of Microelectromechanical Systems*, Vol. 12, pp. 70-80, 2003.

[8] M. Laroussi and X. Lu, "Room-temperature atmospheric pressure plasma plume for biomedical applications ," *Journal of Physics D: Applied Physics.* Vol. 42, pp. 085204–085212, 2009

[9] E. El Ahmar, C. Met, O. Aubry, A. Khacef and J. M. Cormier, "Hydrogen enrichment of a methane–air mixture by atmospheric pressure plasma for vehicle applications, " *Journal of Chemical Engineering*, Vol. 116, pp. 13–18, 2006

[10] A. J. Beck, R. D. Short and A. Matthewa, "Deposition of functional coatings from acrylic acid and octamethylcyclotetrasiloxone onto steel using an atmospheric pressure dielectric barrier discharge," *Surface and Coatings Technology*, Vol. 203, pp. 822-825, 2008

[11] J. C. T. Eijkel, H. Stoeri, A. Manz, "A dc Microplasma on a Chip Employed as an Optical Emission Detector for Gas Chromatography", *Analytical Chemistry*, Vol. 72, pp. 2547-2552, 2000.

[12] J. R. Melcher, "Dielectrophoresis Liquid Explusion", *Journal of Spacecraft and Rockets*, Vol. 6, pp. 961-967, 1969.

[13] B. Eliasson and U. Kogelschatz, "Nonequilibrium volume plasma chemicalprocessing," *IEEE Trans. Plasma Sci.,* vol. 19, pp. 1063–1077, Dec. 1991.

[14] M. A. Lieberman and A. J. Lichtenberg, "Principles of Plasma Discharges and Materials Processing" , John Wiley, New York, 1994.

[15] R. W. B. Pearse, A. G. Gaydon, "The identification of molecular spectra " Fourth Edition, John Wiley & Sons, Inc., New York, 1976.

Three-dimensional Digital Microfluidics and Applications

Gary Wang[*], Daniel Teng, Shih-Kang Fan

[*]Department of Electrical and Computer Engineering, University of Saskatchewan, Saskatoon, Canada
gary.wang@usask.ca

Abstract—**Digital microfluidics, by the electrowetting-on-dielectric (EWOD), is a promising way to manipulate biological targets like DNA, proteins, or cells in very small liquid volumes. Conventionally, digital microfluidic operations are performed on a two-dimensional electrode array. This research expands the two-dimensional conventional digital microfluidic architecture into a three-dimensional architecture. The concept relies on the possibility that back and forth motions between open and covered regions are possible under electrowetting actuation. Two face-to-face plates based on "microelectrode dot array" (MEDA) architecture form the foundation of the two-layer microfluidic operations and a dual open/covered hybrid design adds the inter-layer microfluidic connection to complete the three-dimensional system. By combining the advantages of the open/covered systems, a three-dimensional digital microfluidics may achieve better droplet routing capabilities and fit more functions into a given footprint.**

Keywords-EWOD; three-dimensional; microfluidics; MEDA

I. INTRODUCTION

A three-dimensional digital microfluidic system that can perform microfluidic operations on stacked layers has many advantages over the two-dimensional systems. This research leverages the flexibility of Microelectrode Dot Array (MEDA) architecture [1], [2] and the droplet motion between covered and open regions of EWOD microsystems [3], [4] by extending the two-dimensional conventional digital microfluidic architecture into a three-dimensional architecture. In covered EWOD microsystems, droplets are confined between two plates. In open systems, the sessile droplet is sitting freely on a horizontal solid substrate. Each one of these systems has his own advantages. Drop dispense, motion and slitting are easier in covered EWOD systems whereas mixing and evaporation are preferably performed in the open configuration [5], [6]. The three-dimensional architecture combines the advantages of the open and covered EWOD systems, a dual open/covered system.

A three-dimensional digital microfluidic system is fundamentally constructed by two face-to-face plates based on MEDA architecture. Then, different physical arrangements of the two plats provide inter-plate connecting solutions for different applications. Two fundamental three-dimensional microfluidic systems are studied in this research. The first one is "Dual-Layer System" which enables microfluidic operations on dual layers. As the complexity of digital microfluidics-based "lab-on-a-chip" (LOC) systems is increasing, the droplet routing capability of the system has significant impact on its performance. This dual-layer system has better capabilities to route droplets to blocked locations or to avoid unwanted path contaminations. Naturally, more functions also can fit into a given footprint in a three-dimensional system.

The second three-dimensional microfluidic system is "Inter-Chip Bridge" which enables a wide range of applications in modular LOC system design by the inter-chip droplet transport capabilities. LOC system can be partitioned into discrete scalable, reusable modules consisting of isolated, self-contained elements to have advantages in cost, flexibility, reliability and development time. Moreover, components with incompatible manufacturing could be combined in a single LOC system for heterogeneous integration. This means that components can be optimized to a much greater degree than if they were built together on a single fabrication technology.

The rest of the paper is organized as follows. Section II covers the analytical model and the experiment results of the Dual-Layer System. The Inter-Chip Bridge is discussed in section III. Then section IV concludes the study.

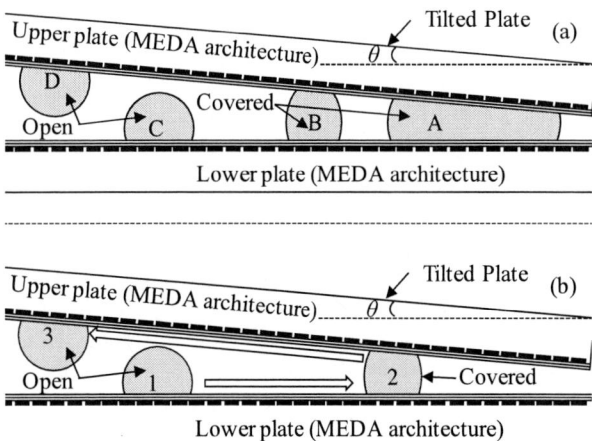

Fig. 1. Sketch of a dual-layer digital microfluidic system constructed by two face-to-face MEDA plates: (a) droplet B is created from reservoir A in covered region and sessile droplets C & D (open) are formed by moving droplet B from covered region to open region; (b) inter-layer connection is illustrated as the droplet motion from position 1 through position 2 into position 3.

II. DUAL-LAYER SYSTEM

A. System Structure and Microfluidic Operations

As illustrated in Fig. 1, a coplanar microelectrode array based on MEDA architecture is designed as the bottom plate and another MEDA coplanar microelectrode array is used as the top plate. The coplanar structure of the microelectrode array plus the tilted top plate that provides the flexible gap adjustment capability forms a three-dimensional microfluidic system.

The concept relies on the possibility that back and forth motions between open and covered regions are possible under electrowetting actuation. Microfluidic operations of the dual-layer system leverage the best of a dual open/covered hybrid design that droplet transporting and mixing are performed on the open region of the stacked dual-layer system and droplet creation and splitting are performed at the closed region. As illustrated in Fig. 1(a), droplet B is created from reservoir A under covered configuration. At the closed region, MEDA architecture provides the flexibility of switching microfluidic actuations between the top and bottom plates. While the droplet actuation is on the top plate then the bottom plate is configured into a zero potential plate and vice versa. A top-plate actuation of droplet B to the left eventually will result a sessile droplet D breaking up from the bottom plate when the gap is wide enough as indicated in Fig. 1(a). Similarly, a bottom-plate actuation will result a sessile droplet C on the bottom plate. A coplanar actuation of the sessile droplet is then provided to move the sessile droplet around. The inter-layer connection is illustrated as the droplet motion from position 1 (open) through position 2 (closed) into position 3 (open) in Fig. 1(b). MEDA architecture allows dynamic configurations and activations of identical basic microfluidic unit called "micro-electrode cells" to actuate droplets. This architecture has flexibilities in performing open (coplanar) and closed (top or bottom) actuations that have to consider the droplet sizes and overall system integration variables.

The tilted angle (θ) of the top plate can be one dimensional that gap differences are only along one axis (x- or y-axis) or two dimensional that both x-axis and y-axis have different gaps. To simply the analysis, one dimensional tilting was used in our studies.

B. Analytical Model of Dual-Layer System

For the dual-layer system, there are two directions of the droplet motion between the open region and the closed region. As illustrated in Fig. 2, the first motion is from the covered region to the open region. Suppose the droplet starts from the closed region by actuating electrodes toward the left side as shown in Fig. 2(a). When the droplet has reached the point where the droplet internal pressure of Fig. 2(a) is greater than Fig. 2(b), the droplet breaks up from the top plate and becomes a sessile droplet in the open region.

To simplify our analysis, the covered droplet pressure is approximated by removing the tilted angle. Due to a small Bond number and the flat top and bottom surfaces, the profile of the droplet in a vertical cross section is nearly a circular arc. Taking advantage of this observation, we can derive a formulation for the calculation of the volume of the droplet. Droplet pressure in each region can be calculated by using the Laplace's law. First, for a drop of volume V, confined between two horizontal plates separated by a distance h, internal pressure is given by [4]

$$P_c = \gamma \left(\frac{-\cos\theta_t - \cos\theta_b}{h} + \sqrt{\frac{\pi h}{V}} \right), \quad (1)$$

where θ_t and θ_b are the contact angles with the top and bottom plates and γ the surface tension in covered and open regions. In (1), the first term of the right hand side corresponds to the vertical curvature, and the second term to the horizontal curvature. For a sessile drop of same volume, using the Laplace's law, the droplet pressure is given by [4]

$$P_o = 2\gamma \left(\frac{3V}{\pi(2 - 3\cos\theta_o + \cos^3\theta_o)} \right), \quad (2)$$

where θ_o is the contact angle with the bottom plates. The condition for the motion from covered to open region is

$$P_{\text{covered,actuated}} > P_{\text{open,actuated}}. \quad (3)$$

Let θ_a be the actuated contact angle and θ_n the open (non-actuated contact angle), so that

$$P_c(\theta_t = \theta_n, \theta_b = \theta_a) - P_o(\theta_o = \theta_a) \geq 0, \quad (4)$$

$$\gamma \left(\frac{-\cos\theta_n - \cos\theta_a}{h} + \sqrt{\frac{\pi h}{V}} \right) - 2\gamma \left(\frac{3V}{\pi(2 - 3\cos\theta_a + \cos^3\theta_a)} \right) \geq 0. \quad (5)$$

The second motion is from the open region to the covered region. When the motion to the right as shown in Fig. 3(a) continues, the droplet touches the top plate and the region transition from open into covered can be easily completed. Also, as shown in Fig. 3(b), the de-actuation of the droplet could speed up the transition due to increased height of the non-wetting sessile droplet. In a non-wetting case, the height of a sessile droplet is

$$h = R + R\cos(\pi - \theta_n) = R(1 - \cos\theta_n). \quad (6)$$

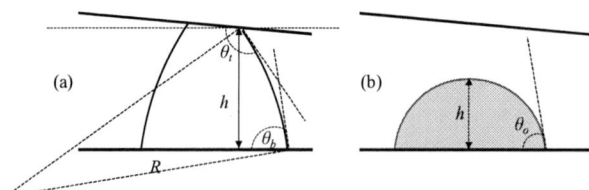

Fig. 2. Droplet motion from the closed region (a) to the open region (b).

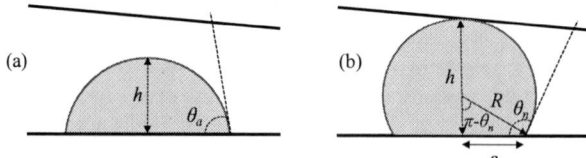

Fig. 3. From open region (a) to closed region (b).

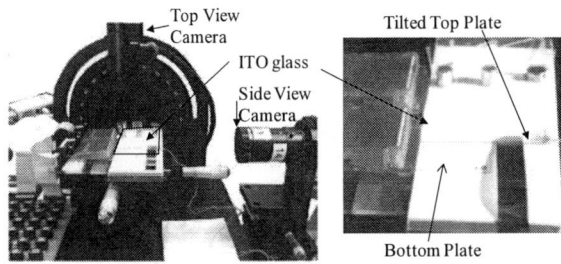

Fig. 4. The experiment setup for the dual-layer system.

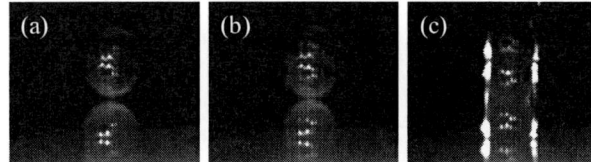

Fig. 6. From left to right, side experimental views show a droplet motion from open sessile droplet into covered droplet for the tested dual-layer system

C. Excperimental Results of Dual-Layer System

A number of prototypes are fabricated on ITO glasses to conduct experiments of the dual-layer system. The test system is shown in Fig. 4. Deionized (DI) water is used to form the droplet and the air is the medium. The non-actuated contact angle is close to 115°, and the electrode activating voltage was chosen so that the actuated contact angle is about 80°.

Fig. 5 shows that the motion from covered to open region is possible. The lighter circle in Fig. 5(a) clearly shows the cross section on the top plate and the dark outline indicates the cross section of the bottom plate. Fig. 5(b) and (e) show the top view and the side view at moment before the breaking-up of the droplet from the top plate. Fig. 5(c) and (f) illustrate the cross section and the side view of the sessile droplet.

From left to right as shown in Fig. 6, the side experimental views illustrate a droplet motion from open sessile droplet into covered droplet of the dual-layer system. Fig. 6(b) shows an actuated droplet moved into the position where the top of the droplet almost touched the top plate. Then, a de-actuation made the droplet touch the top plate as shown in Fig. 6(c) and the droplet transited into covered region. Gap height is an important parameter of the motion. The gap height needed for the covered-to-open transition (h in Fig. 2(a)) is greater than the required height of the open-to-covered transition (h in Fig. 3(b)). Different droplet volumes also change the transition heights. To assure the smooth back-and-forth transitions between the covered and open regions, MEDA architecture is leveraged to provide the flexibility of the droplet manipulations for varied transition positions between covered and open regions.

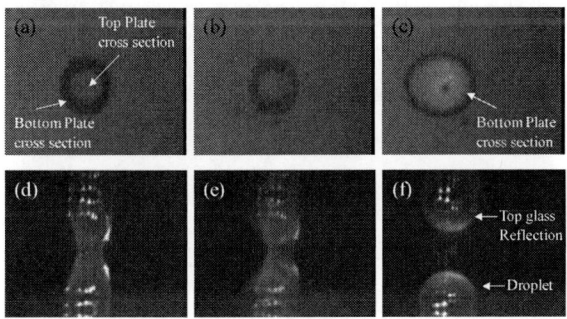

Fig. 5. From left to right, top and side experimental views show a droplet motion from a covered droplet into an open sessile droplet for the tested dual-layer system.

III. INTER-CHIP BRIDGE

A. System Structure and Microfluidic Operations

Another design of the three-dimensional structure is the inter-chip bridge. This design is based on the concept of the dual open/covered EWOD microsystem developed by Berthier and et al. [3]. It has been shown that back and forth motions between open and covered regions are possible under electrowetting actuation motion, even though it is closely related to the difference of drop internal pressure between the departing and arriving regions. The motion from open to covered region is not possible if the vertical gap is too small related to the liquid droplet volume and the motion in the reverse direction is not possible if the gap is too large [3], [4].

In an inter-chip bridge system, droplet A moves from LOC #1 (open region) into a covered region formed by the bridge plate and LOC #1 which is illustrated as droplet B in Fig. 7. Then droplet B keeps moving to the left into the open region on the bridge illustrated as droplet C. Again, droplet C moves into the covered region between the bridge plate and LOC #2 as droplet D. The droplet motion from position A on LOC #1 to position D on LOC #2 completes the inter-chip bridging function of a droplet. The most critical element of the inter-chip bridge design is the range of the gap height that both motions are possible and this range is sufficiently large to leave room to build the system. To ensure the smooth inter-chip bridging operations, electrodes at the covered/open region boundaries must be adequately designed in function of the liquid droplet volume. The flexibility of the field-programmability of electrodes in MEDA architecture [2] is a great help in designing reliable and effective inter-chip bridges.

B. Analytical Model of Inter-Chip Bridge

The motion from the open region to the covered region starts from de-actuating electrodes in the open region and actuating electrodes in the covered region. The droplet then moves towards the covered region. When it has crossed the boundary, and is located in the covered region, the actuation is

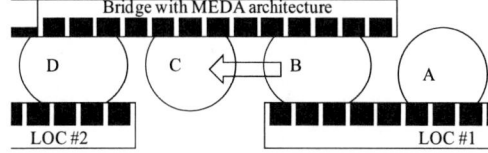

Fig. 7. Sketch of an inter-chip bridging based on the three-dimensional structure. The back and forth motion between open and covered region make it possible to transfer droplets from one LOC to another.

978-1-4673-1122-9/12 $31.00 © 2012 IEEE

switched off. The actuation in the open region is then switched on and the droplet moves back to open region. The condition for the motion from open to covered region is

$$P_o(\theta_o = \theta_n) - P_c(\theta_t = \theta_n, \theta_b = \theta_a) \geq 0, \quad (7)$$

and conversely, for the motion from covered to open region

$$P_c(\theta_t = \theta_n, \theta_b = \theta_n) - P_o(\theta_o = \theta_a) \geq 0. \quad (8)$$

The motion from the open to covered region derived from (1), (2) and (7) is

$$2\gamma \left(\frac{3V}{\pi(2 - 3\cos\theta_n + \cos^3\theta_n)} \right) - \gamma \left(\frac{-\cos\theta_n - \cos\theta_a}{h} + \sqrt{\frac{\pi h}{V}} \right) \geq 0. \quad (9)$$

Also, the motion from the open to covered region derived from (1), (2) and (8) is

$$\gamma \left(\frac{-\cos\theta_n - \cos\theta_n}{h} + \sqrt{\frac{\pi h}{V}} \right) - 2\gamma \left(\frac{3V}{\pi(2 - 3\cos\theta_a + \cos^3\theta_a)} \right) \geq 0. \quad (10)$$

C. Experimental Results of Inter-Chip Bridge

Experiments have been performed on prototypes fabricated from ITO glasses. For a droplet of DI-water in air, motion from open to covered region is possible provided that the vertical gap in the covered region is not too small, and motion from covered to open region is possible if the vertical gap is not too large as shown in Fig. 8. The actuated contact angle θ_a is about 80°. Contact angle with the top plate θ_t (115°) is an important parameter of the motion. Below 90° the droplet will have difficulties exiting the covered region due to hydrophilic grip on the top plate and above 120° the hydrophobic repulsion will make motion towards the covered region difficult [4]. A vertical distance of about 200 microns seems to work well for liquid drops of 0.03 to 0.05 μl on Teflon hydrophobic layer with electrodes of 600 μm × 600 μm. Our experiment data showed no droplet motion from open to covered region when the gap height is below 120 μm which is in line with Berthier's work. Also, when the gap is wider than 220 μm the drop started to have difficulty to move from covered to open region. In comparison, Berthier's work showed the gap can be 500 μm and the motion from covered to open region should be still good at θ_t = 115° [3]. We believe the difference is caused by the actuation of the open region. A catena is used in the droplet motion study by Berthier as a zero

potential contact for open EWOD systems. From a system integration point of view, a catena is difficult to implement in real inter-chip bridge applications, so MEDA architecture is used to provide the dual open/covered droplet actuation capability for the system. The coplanar actuation is achieved by configuring half of the electrode on the bottom plate into the zero potential contact and half of the electrode as the actuating electrode. The reduced size of the actuating electrode would have less pulling force for the droplet.

IV. CONLUSIONS

The mechanism of this three-dimensional architecture is a combination of two face-to-face plates and careful arranged gaps and electrodes. Two different three-dimensional digital microfluidic systems are discussed in this research. The inter-chip bridge structure is useful in a wide range of applications in modular LOC system design and the system integration of biochips. Nonetheless, it also has constrains in the gap height and the size of the droplet. MEDA architecture can be leveraged to smooth the microfluidic operations and to minimize constrains. Dual-layer system with the tilted top plate provides the flexibility of adjusting the gap height and therefore has less constrain than the inter-chip bridge model. This dual-layer system has better capabilities to route droplets and has higher functional density in a given footprint.

Combinations of the two basic structures are possible and could be more powerful in constructing future LOC system structures and advanced multi-chip system integrations.

ACKNOWLEDGMENT

The authors are grateful for the support of Department of Materials Science and Engineering, National Chiao Tung University (NCTU), Taiwan. All experiments and fabrications for proof of the concept presented in this paper were performed at the FAN-TASY Lab at NCTU.

REFERENCES

[1] G. Wang, D. Teng, and S.-K. Fan, "Digital Microfluidic Operations on Micro-Electrode Array Architecture", *in Proceedings of the IEEE Int. Conf. on Nano/Micro Engineered and Molecular Systems (NEMS), Kaohsiung, Taiwan*, pp. 1227-1230, 2011.

[2] G. Wang, D. Teng, S.-K. Fan, "Digital Microfluidic Operations on Micro-Electrode Array Architecture," *IET Nanobiotechnology*, 2011, in press.

[3] J. Berthier, Ph. Clementz, J.-M. Roux, Y. Fouillet and C. Peponnet, "Modeling microdrop motion between covered and open regions of EWOD microsystems," *Proceedings of the 2006 Nanotech Conference, Boston, USA*, pp. 685-688, 2006.

[4] J. Berthier, "Microdrops and Digital Microfluidics," *William Andrew Publishing*, Norwich, NY, 2008.

[5] J. Berthier, Ph. Clementz, O. Raccurt, D. Jary, P. Claustre, C. Peponnet and Y. Fouillet, "Computer aided design of an EWOD microdevice," *Sensors and Actuators A: Physical*, Vol. 127, pp. 283-294, 2006.

[6] P. Clementz, J. Berthier, J.M. Roux, R. Blanc, G. Castellan, C. Chabrol, P. Claustre, O. Constantin, D. Jary, D. Lauro, O. Raccurt, Y. Fouillet and C. Peponnet, "EWOD contribution to sample preconcentration: droplet geometrical transformation from drop formation to evaporation step," *NanoBio Europe Conference, Münster*, pp. 22–24, September 2005.

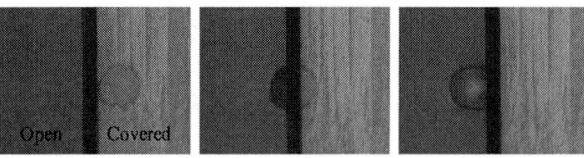

Fig. 8. Experimental view of a droplet crossing the frontier between the covered region and the open region. The droplet can move on both directions.

Specific Design and Implementation of a Piezoelectric Droplet Actuator for Evaporative Cooling of Free Space

Hsun-You Wang, Cheng Huang, Chin-Tai Chen[*]

[*]Department of Mechanical Engineering, National Kaohsiung University of Applied Sciences, TAIWAN, ROC
chintai@kuas.edu.tw

Abstract—We report the experimental results of evaporative cooling effect of a novel piezoelectric droplet actuator designed for a free space utilizing dynamic infrared thermographs as their space-time courses of temperatures recorded. These droplets of pure water were jetted as working liquid of an evaporating cooling system comprising a PZT micro-actuator, controller and liquid delivery for the circulation of air with the cooling space. Cyclic cooling tests demonstrated that these micro droplets well performed the evaporative cooling effectively. The space cooling for such an evaporative cooling system was studied by a high-performance infrared imager that enabled all the temperature changes of the space to be recorded and analyzed per three seconds, yielding uniform water droplets (< 100 μm) with cooling rate ~ -1.1 °C/min.

Keywords- droplet generation; evaporative cooling; spray

I. INTRODUCTION

Ventilated water droplets have demonstrated great potential in many applications of heat and mass transfer as cooling agents [1]. One of very recent attentions is paid to the evaporative cooling efficiency using those micro droplets on macro- to micro-scale objects like human farming environments and electronic devices for more energy conservation. However, few studies focused on the droplets used in cooling space of human daily life. This novel application at least requires low-cost micro components and electromechanical systems, user-friendly robust operations for delivery of working liquid, and a low-noise power unit with the programmable on/off switch controls, in order to establish such high-efficiency evaporative cooling capability. Meanwhile the jetted droplets have been previously demonstrated by a MEMS-based spray technology for the medical-care use of human life [2]. In this study, we further propose a novel micro piezoelectric actuator for the application, with its experimental results for the evaporative cooling effect of water droplets on a regular open space [3]. The cooling measurement and configuration for the evaporative cooling system described herein are compared with those of the other spray systems. The infrared thermographs, in this study, with the space-time temperature data acquired offer the real-time high-resolution observation and analysis for multiple locations of space, demonstrating the proven accuracy, non-contact and long-distance measurement

capacities [4]. The PZT actuator implemented here for such an evaporative cooling effect has been not optimized, which should be taken under the considerations about the higher throughput and uniformity of droplets in the future application [5, 6]. Finally, we used a function generator (SFG-2120) and amplifier (SVR 150-3) to control the PZT actuator, in which the preliminary results show that maximum of jet velocity appear at the first-order natural frequency of vibration; a higher driving voltage can improve the jet velocity as well [7].

II. FABRICATION OF NOZZLE PLATES

A nozzle plate with desirable diameters of nozzles is crucial for determining the jetted droplet size using the evaporating cooling actuator. In the study, the nozzle plate was micro-manufactured by photolithography techniques as shown in Fig. 1. First, at the step (a), the acetone and dilute hydrochloric acid were used to clean a stainless steel sheet for removing dust particles, followed by the spin-coating of the photoresist (SU-8) at 500 rpm for 60 sec and 3000 rpm for 60 sec. As shown in the steps (b)~(c), after hot-plate pre-baking at 90°C for 60 sec, the coated photoresist was further exposed (exposure intensity with 27s yielding 14 mJ/cm^2), developed, and post-baked (at 90°C for 30sec) to obtain the corresponding nozzle barrier areas, as shown in Fig. 1(c).

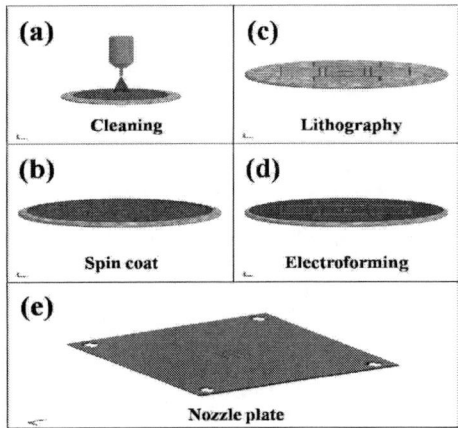

Fig. 1. The manufacturing processes for a nozzle plate: (a) cleaning stainless steel sheet (b) spin-coating thin photo-resists (c) exposing and the structure of the photo-resists (d) electroforming the nickel nozzles of the plate (e) the formation of the nozzle plate (Nickel).

In the final electroforming process steps (d)~(e), as shown in Fig. 1(d), the necessary electroforming material (nickel) was connected to the anode, in the electroforming solution composed of $Ni(SO_3NH_2)_2$ (340c.c.), $NiCl_2 \cdot 6H_2O$ (3.3g), and H_3BO_3 (40g). Thus, with the electroforming base of a stainless steel plate connected on the cathode, the electroforming process was performed under the electrical current 0.2 A , followed by adjusting the voltage up to 20V, and took the electroforming time of 7 hrs (420 min) to yield the nozzle plate thickness of ~52 μm, as shown in Fig. 1(e). As a result, each nozzle (holes of 5×5 array) was obtained with a diameter of ~30 μm, as demonstrated in Fig. 2.

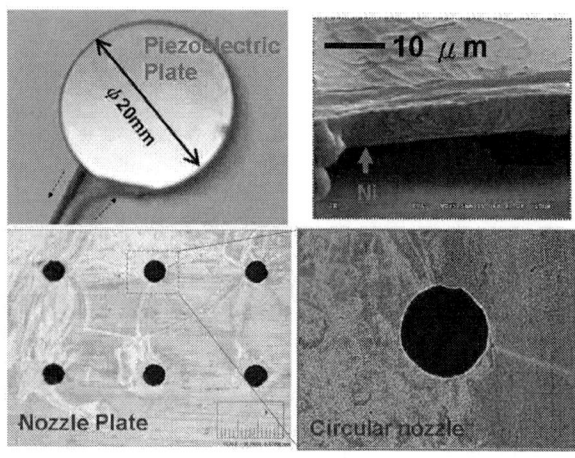

Fig. 2. The PZT plate and nozzle plate electroformed were bonded as a micro piezoelectric droplet actuator.

III. EXPERIMENTAL

A novel micro-machined piezoelectric (PZT) actuator for droplet generation was proposed and assembled as shown in Fig. 3.

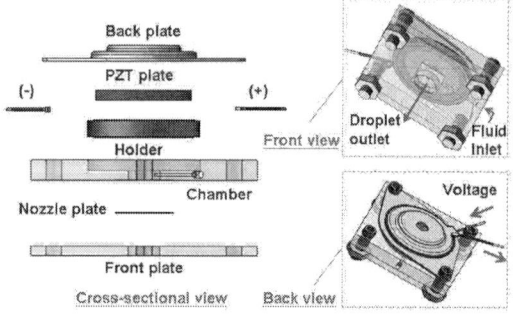

Fig. 3. Schematic configuration of a piezoelectric (PZT) actuator together with the fluidic and electric connection as a droplet microsystem.

Therefore, a stream of droplets could be jetted out of the 5×5 nozzles (diameter of 30μm), as connected to water reservoir, and underwent an evaporation process to completely dry out for cooling the surrounding space. This microsystem implemented to cool a small space of ~14×12×9.3 cm³ for an evaporative cooling test as shown in Fig.4, with the driving controller for power and signal connection. Such a cooling process of a space for evaporative cooling test was carried out as shown in Fig. 5. The infrared thermographs would be captured and adopted for measuring the temperature of the cooled space, providing further data analysis in the next section.

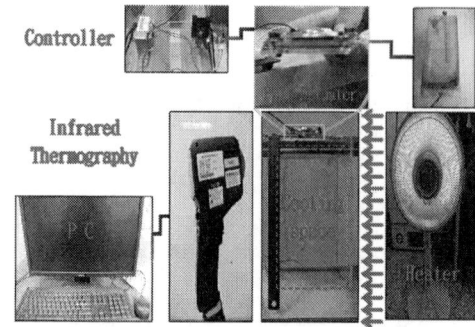

Fig. 4. Experiment setup of the evaporative cooling system in the study.

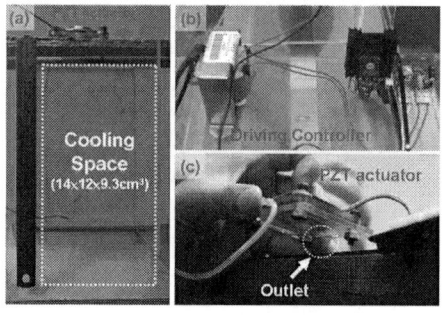

Fig. 5. Implementation of the droplet microsystem: (a) the PZT actuator set up above the cooling free space (b) driving controller (c) outlet of fluid droplet from the PZT actuator.

Note that the driving controller was mainly composed of the AC transformer and piezoelectric driver, in order to provide the appropriate power (voltage ~ 72 V) and the operating frequency (1.7 MHz) as shown in Fig. 6. Hence, a typical jetting situation for evaporative-cooling tests under normal test environments was completed and demonstrated in Fig. 7.

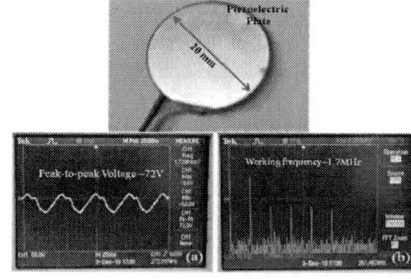

Fig. 6. Signal controlling for driving the piezoelectric microsystem with 72 V voltage and 1.7 MHz into the PZT plate.

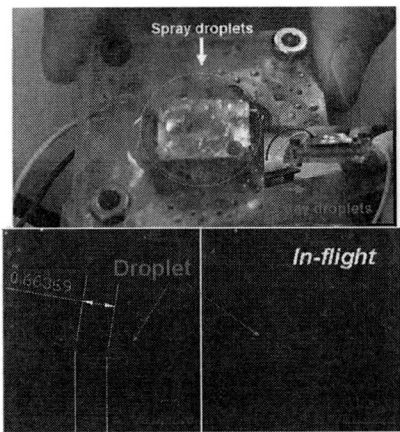

Fig. 7. Images of spray droplets generated from the outlet of the nozzle plate in the piezoelectric microsystem.

To explore the droplet cooling effect, the space was first measured for the change of the temperature using the infrared imager for 600 frames (3 seconds per frame) to evaluate the natural cooling rate without the jetted droplets. Under the same initial conditions, this space was measured for temperature again with 300 frames (3 seconds per frame) during a spray cooling cycle of 3min in spray situation. Finally, using the infrared thermal image analysis software (TAS14), we analyzed those measurement data in detail as described in the next section. Also a Doppler instrument was applied to measure the flight speed and size of the droplets. Each image was captured every 0.25 s, and thus the speed of ~2.7 mm/s for droplets was estimated in the tests.

IV.   ANALYSIS AND DISCUSSION

We evaluated the cooling effect of a spray on / off switch on the temperature variations of the five specified locations (P1 ~ P5) within the cooling space during a cyclic time (~ 3min) as shown in Fig. 8(a), while as a temperature distribution without spray-on as shown in Fig.8 (b) was compared to identify the cooling performance within the occupied space only because of droplet evaporation and heat transfer.

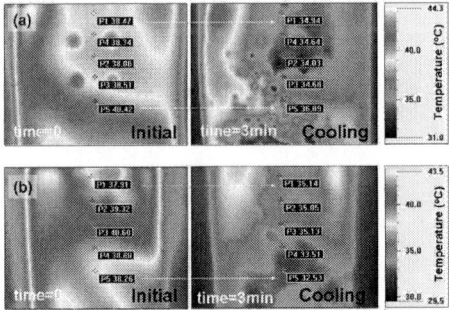

Fig. 8. Infrared thermographs: (a) spray cooling of free space for a periodic of 3 minutes with constant background heating powered on (b) spray cooling of free space for 15 minutes after the background heater powered off.

It was found that the temperatures of natural cooling decreased from 40°C to 32°C during 900 seconds, yielding a temperature rate of -0.53 °C/ min as shown in Fig. 9 (b). However, in the spray cooling cycles, as shown in Fig 9 (b), the spray-off during the second cycle exhibited obvious cooling effect with the temperature of 43°C down to 37°C during 180 seconds, i.e. cooling rate up to 1.1 °C/min. Obviously, jetting the water droplets within ~5min exhibited the decreasing rate of temperature (-1.1°C/min) larger than that of the space without spray cooling (-0.5°C/min) within 15 min.

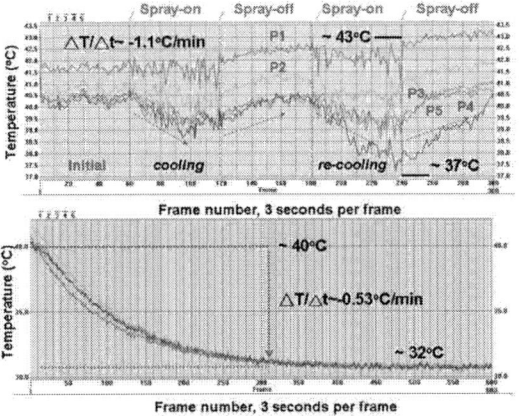

Fig. 9. The measured temperature evolution of the five locations within the cooling free space: (a) spray cooling during 3-minute cycles with the constant background heating powered on (b) the background cooling with heater powered off.

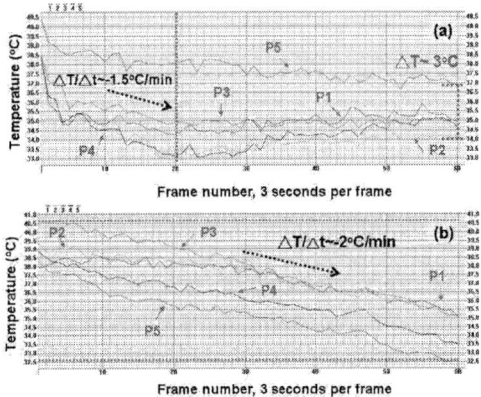

Fig. 10. The measured temperature evolution of the five locations within the cooling free space: (a) spray cooling during 3-minute cycles with the constant background heating powered on (b) the background cooling with heater powered off.

In order to show the spray cooling system having a good cooling effect, a separate heating fan was initially used to heat the same cooling space up to 40°C, and then followed by spray cooling for 3 minutes (with continuous fan heating) down to the lowest temperature; thus the cooling rate was calculated. In the natural cooling experiments, as shown in Fig. 9, that the

978-1-4673-1122-9/12 $31.00 © 2012 IEEE   421

natural cooling down to 32°C with the rate of -0.53°C /min. Compared to the result of natural cooling, the spry cooling rate in this case was found with about -1.5°C /min as shown in Fig. 10(a), showing that the cooling effect was much improved by the present spray system. In addition, as shown in Fig. 10(b), this cooling effect (rate ~ -2.0°C /min) was enhanced by four times when compared to that in natural process without spray cooling (rate ~ -0.53°C /min) as seen in Fig. 9(b), demonstrating the higher cooling efficiency by the present spray system.

V. CONCLUSION

In the study, the experimental results showed that the present droplet-based cooling system performed well for cooling the small space. According to the thermal image data, the jetted water droplets within ~5min exhibited the decreasing rate of temperature about ~ -1.1°C/min. It is expected that since smaller droplets than current ones can evaporate quickly, the cooling system using the smaller nozzles will further improve the cooling performance. Hence, in the future, we might enhance the cooling mechanism by generating the droplet size smaller and more uniform to yield a high-performance evaporative-cooling micro actuator.

ACKNOWLEDGMENT

The authors thank research grants by the National Science Council (NSC) under NSC-98-2622-E-151-018-CC3 and NSC-100-2221-E-151-042, Taiwan, ROC. The Research Center for MEMS and Precision Machines in National Kaohsiung University of Applied Sciences (KUAS), Taiwan, ROC, is also appreciated for the access of the surface processing and analysis equipments.

REFERENCES

[1] J. Smolik, L. Dzumbova, J. Schwarz, and M. Kulmala, "Evaporation of ventilated water droplet: connection between heat and aass transfer," *Aerosol Science*, vol.32, pp. 739-748, 2001.

[2] S. C. Shen, Y. J. Wang, Y. Y. Chen, "Design and fabrication of medical micro-nebulizer," *Sensors Actuators A*, vol.144, pp. 135-143, 2008.

[3] H. Y. Wang, C. Huang, and C. T. Chen, "Design, fabrication and verification of piezoelectric droplet actuator for space-cooling application," *The 15th Nano and Microsystem Technology Conference*, Taipei, Taiwan, Sep. 6-7, 2011. (Chinese)

[4] P. Tartarini, M. A. Corticelli, L. Tarozzi, "Dropwise cooling: experimental tests by infrared thermograph and numerical simulations," *Applied Thermal Engineering*, vol. 29, pp.1391-1397, 2009.

[5] A. A. Pavlova, K. Otani, M. Amitay, "Active performance enhancement of spray cooling", *J. Heat Fluid Flow*, vol. 29, pp.985-1000, 2008.

[6] Y. L. Lin, I. D. Yang, C. C. Chieng, F. G. Tseng, "High throughput micro droplet generator array controlled by two-dimensional dynamic virtual walls", *Microfluid Nanofluid*, vol. 9, pp.681-693, 2010.

[7] J. Deng, W. Yuan, J. Luo, D. Shen, and B. Ma, "Design and fabrication of a piezoelectric micro synthetic jet actuator," *Proc. of 2011 IEEE International Conference on Nano/Micro Engineered and molecular Systems (NEMS)*, pp. 301-304, Kaohsiung, Taiwan, Feb. 20-23, 2011.

Micro-droplet Formation with Non-Newtonian Solutions in Microfluidic T-junctions with Different Inlet Angles

Zhipeng Gu*, Jong-Leng Liow
*School of Engineering and Information Technology, UNSW@ADFA, AUSTRALIA
Zhipeng.Gu@student.adfa.edu.au

Abstract—**T-junctions with various inlet angles (θ=30°, 60°, 90°, 120° and 150°) were employed to produce microdroplets. Experimental results showed that the inlet angles play an important role in controlling the droplet size and the T-junction with an inlet angle of 90° always generating the largest droplets for a given continuous phase flow rate. The results for the droplet diameters were separated into two groups, based on the range of the inlet angles (acute angles of 30°, 60° and 90°, and obtuse angles of 90°, 120° and 150°). For the acute inlet angles, the droplet size was larger for an inlet angle of θ=30° than for an inlet angle of θ=60°. For the obtuse inlet angles, the diameter of droplets produced by T-junctions with inlet angles of 120° and 150° were similar in size but smaller than that for 90°. The effect of flow rates of both phases on the droplet formation mechanism was discussed. Water and non-Newtonian xanthan gum and PEO (Polyethylene oxide) solutions were used as the dispersed phase and the differences in droplet sizes were studied.**

Keywords-T-junction; microchannel; droplet formation; inlet angle; non-Newtonian fluids

I. INTRODUCTION

Micro-droplets are used extensively in the promotion of chemical reactions, bioassays, food processing, material production and health care products [1, 2]. Compared to traditional methods, the advantages of using microfluidic techniques for droplet formation is the monodispersity and well-controlled morphology of the droplets. Currently, a number of methods are used to generate microfluidic droplets, including T-junctions [3], flow-focusing devices [4], co-flow devices [5] and microchannel terraces [6]. The T-junction is the most frequently used and widely researched microfluidic geometry. It produces monodispersed droplets by shearing an immiscible fluid with a cross-flow. The droplets generated are highly monodispersed and equally spaced in the outlet stream.

Three major parameter groups govern the droplet formation mechanism and breakup dynamics; the channel geometry, structure and properties (such as channel type, dimensions and hydrophobicity), fluid properties (such as density, viscosity, interfacial tension and contact angle), and operating parameters (such as pressure, flow rate ratio, temperature, *etc.*). A substantial number of studies involving the effect of these parameters on droplet generation have been reported in the literature [7-9].Previous studies [7] have provided an understanding the role played by the design of the

microchannel in determining the formation mechanism. In spite of the numerous investigations into the effects of flow rates, fluid properties and other controlling parameters on droplet production, it is only very recently that the role of the inlet geometry has been studied, particularly that of the effect of the inlet angle. Although the results from different literature sources do not agree, they do indicate that the inlet design affects the droplet formation process significantly.

Recent investigation of the inlet angle of a T-junction by Fries and von Rohr [10] and Tan *et al.* [11] provided some experimental data but the effect of inlet angles was not fully elucidated in these papers. They only studied the bubble formation process whereas the formation processes for droplets and bubbles are substantially different due to differences in density and compressibility properties of gases and liquid [12].

Hitherto, only Newtonian fluids have been employed in the study of droplet formation while non-Newtonian fluids, despite their widespread applications, have been ignored due to their high viscosity, elastic behaviour and large molecular weight which severely complicate the analysis. Investigations into the behaviour of multiphase non-Newtonian-Newtonian systems in microchannels are of importance for microfluidic applications in the medical research and biological fields that employ non-Newtonian biological and pharmaceutical fluids.

Xanthan gum, a typical shear thinning fluid with properties similar to blood plasma, is used extensively in biological studies and food processing [13]. Xanthan gum exists in aqueous solutions as a double-stranded helix conformation, although the single-stranded structure has also been found due to variation in the bacteria origin and solvent conditions used in the extraction process. Rheological studies have shown that it is a shear thinning fluid with viscosity decreasing with increasing shear rate. Two critical concentrations (c^* and c^{**}) exist; the concentration at which individual polymer molecules begin to physically interact is defined as c^* [14] and the concentration at which anisotropic aggregation occurs as c^{**} [15]. Each critical concentration is accompanied by an increase in the viscosity hence, the rheological characteristics of xanthan gum solutions are a strong function of the xanthan gum concentration. The values of c^* have been reported with values from 0.02 wt% to 0.12 wt% and for c^{**} from 0.07 wt%

978-1-4673-1122-9/12 $31.00 © 2012 IEEE

to 0.78 wt%, which suggests that both the thermal history and xanthan source have a significant influence on rheological behaviour. Polyethylene oxide (PEO) is a polyether compound used in industrial manufacturing and the medical field [16]. A number of studies have been carried out to examine the flow behaviour of PEO solutions in microdevices in terms of their viscoelastic properties [17, 18]. For this study, a PEO with a relatively small molecular weight of 200,000 was chosen as a weak non-Newtonian dispersed phase.

This paper reports on an experimental study of droplet formation process in T-junctions where the inlet angles were varied (θ=30°, 60°, 90°, 120° and 150°), and water and two non-Newtonian fluids (xanthan gum and PEO solutions) were used as the dispersed phase. Droplet formation pictures taken under 3000 frame/second were provided to discuss the effect of the inlet angles.

II. EXPERIMENTAL

A. The T-junction microdevice

The T-junction microfluidic device (dimensions shown in Fig. 1) used in this study was fabricated on a 50.0 (L) × 30.0 (W) × 3.0 (T) mm poly(methyl methacrylate) (PMMA) wafer by micro-end-milling [19]. It was sealed with a PMMA cover plate which was solvent-bonded with acetone at room temperature. Two syringe pumps (Cole-Parmer 7600) were used to control the flow rates of the continuous and dispersed phases into the microfluidic device.

B. Materials

Canola oil (Coles brand) was used as the continuous phase. Different concentrations of deionised water, xanthan gum (Sigma-Aldrich) solutions and PEO (Sigma-Aldrich, with a molecular weight of 200,000) in water-glycerol (Sigma-Aldrich) mixtures were used as the dispersed phases, all of which contained 0.5 wt% Tween 80 (Sigma-Aldrich). The viscosity data for the xanthan gum solutions, measured with a Bohlin Visco88, were fitted with a power-law model $\mu = k\gamma^{n-1}$, where μ is the viscosity (cP), k the consistency index (cP·s^{n-1}), γ the shear rate (s^{-1}), n the power-law index and R^2 is the coefficient of determination [20]. Fluids physical properties are listed in Table 1. Note that only the 0.25% xanthan gum showed a strong variation with shear rate, while the lower concentration solutions of xanthan gum and PEO do not.

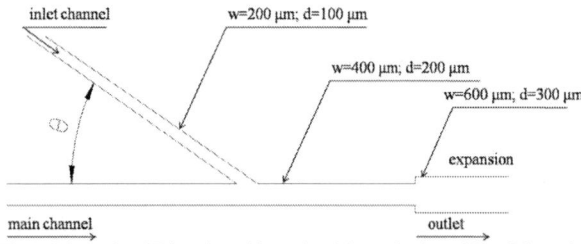

Fig 1. Schematic of T-junction with varying inlet angles (w -width of channel; d -depth of channel). (As expansion downstream of the main channel allows droplets to relax into spherical shapes, diameters can be measured.)

TABLE I.   PHYSCIAL PROPERTIES OF DISPERSED PHASES

Solution	Viscosity/cP			Interfacial tension with canola oil (mN/m)	Density (g/ml)
	k	n	R^2		
Water	1.04			9.46	0.9982
0.2% PEO in 30% glycerol-water	6.67[a]			11.12	1.0714
0.025% xanthan gum	2.02[a]			11.49	0.9983
0.25% xanthan gum	824	0.3594	0.9987	8.26	0.9989

a: low concentrations of the xanthan gum and PEO solutions show Newtonian-like behaviour for the measurable shear rate.

C. Apparatus and analysis

The surface and interfacial tensions were measured by a KSV CAM 200 pendent -drop tensiometer and the fluid density by a density bottle. All the physical property measurements and the experiments were carried out at room temperature (20°C). A Nikon SMZ1000 stereomicroscope, set at magnifications from 8 to 80×, was used to observe droplet formation which was filmed by a Redlake high-speed CCD video camera. The images for droplet formation were recorded at a frequency of 3000 frames/second (see Fig. 8). The diameters of the droplets were measured from the video images using an imaging software (Image tool 3.0) in the expansion region of the microchannel where the drops took on a spherical shape. At least 50 droplets were used to determine the average droplet diameter and polydispersity index. For each droplet, the geometric mean diameter was obtained [$(D_h \times D_v)^{1/2}$] from the horizontal (D_h) and vertical (D_v) diameters. The percentage size distribution (λ) was defined by the equation $\lambda = \delta/D_{av} \times 100\%$, where δ is the standard derivation and D_{av} is the average diameter.

III. EXPERIMENTAL ANALYSIS

A. Effect of inlet angles on droplet size under different flow rates- water as dispersed phase

In order to examine the effect of the inlet angles, various T-junctions (with θ=30°, 60°, 90°, 120° and 150°) were employed in this batch of experiments. Water was chosen as the dispersed phase and canola oil as the continuous phase. The continuous phase flow rates (Q_c) used were 1.0, 5.0 and 10.0 ml/hr and for each flow rate the dispersed phase flow rates (Q_d) was varied. The results are shown in Figs. 2 to 5.

Figs. 2 to 5 show that the inlet angles play an important role in controlling the droplet size. The T-junction with an inlet angle of 90° (black symbols in all figures) always generates the largest droplets for a given continuous phase flow rate. The results for the droplet diameters have been separated into two groups, based on the range of the inlet angles (acute angles of 30°, 60° and 90° in Figs. 2 and 4, and obtuse angles of 90°, 120° and 150° in Figs. 3 and 5).

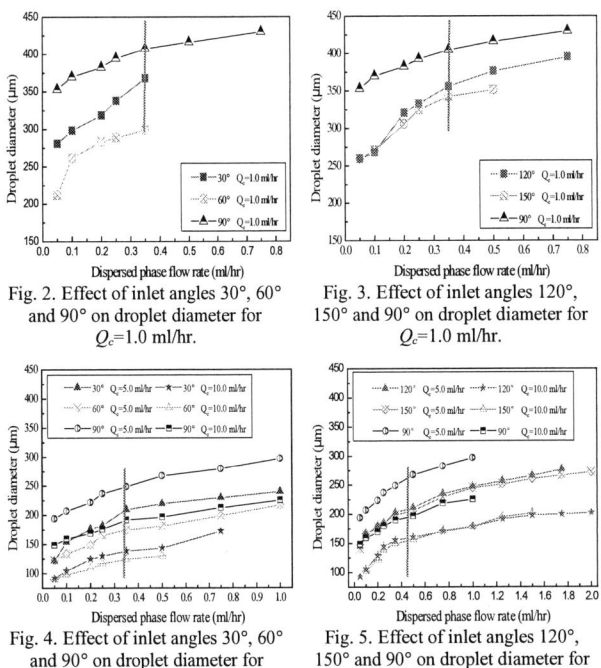

Fig. 2. Effect of inlet angles 30°, 60° and 90° on droplet diameter for Q_c=1.0 ml/hr.

Fig. 3. Effect of inlet angles 120°, 150° and 90° on droplet diameter for Q_c=1.0 ml/hr.

Fig. 4. Effect of inlet angles 30°, 60° and 90° on droplet diameter for Q_c=5.0 and 10.0 ml/hr.

Fig. 5. Effect of inlet angles 120°, 150° and 90° on droplet diameter for Q_c=5.0 and 10.0 ml/hr.

For the acute inlet angles, the droplet size is larger for an inlet angle of θ=30° than for an inlet angle of θ=60°. This occurs for all the continuous phase flow rates studied. This suggests that there may be a minimum in droplet size in the acute inlet angle range. Thus, the droplet size goes through a minimum as the angle increases in the acute angle range.

For the obtuse inlet angles, the droplet sizes are not significantly different for the inlet angles of 120° and 150° but are smaller than that for 90°. The results do show that the droplet sizes are slightly larger for the 120° inlet angle which becomes more evident as the dispersed phase flow rate is increased for a continuous phase flow rate of 1.0 ml/hr. Thus, the droplet size decreases with increasing angle for the obtuse inlet angle range. The results also suggest that the droplet size may be independent of large inlet angles for the obtuse inlet angle range.

The results for acute and obtuse inlet angles are quite different and it is possible that the effect of the inlet angles may be different for these two inlet angle groups.

The droplet size is observed to be dependent on the continuous phase flow rate. A higher Q_c results in the production of smaller droplets which is consistent with our previous investigation [21]. The differences in diameter for the acute and obtuse inlet angle sets are reduced when higher continuous phase flow rates are employed; for example, for the T-junctions with an inlet angle of θ=30° and 60°, the variations in droplet size are smaller for Q_c=5.0 and 10.0 ml/hr (Fig. 4) than for Q_c=1.0 ml/hr (Fig. 2).

The same phenomena are observed in Figs. 3 and 5. The effect of varying inlet angles on droplet diameter diminishes as the continuous phase flow rate is increased. At low continuous phase flow rates, the contribution of the dispersed phase momentum is significant and this leads to the observed variation in droplet diameters. At higher continuous phase flow rate, the shear force generated in the continuous phase flow begins to dominate over the formation resulting in the variations in droplet size becoming less dependent on the inlet angles.

Although the droplet diameter increase with increasing Q_d, there is a critical value (0.35 ml/hr, the red dashed line in Figs. 2 to 5) where a change in the rate at which the droplet diameter increase is observed, especially for the higher Q_c values (5.0 and 10.0 ml/hr). The rate of change of the droplet diameter with an increasing Q_d is faster below 0.35 ml/hr and slower above it. It is noted that the continuous phase flow rate is constant so the shear force and interfacial tension values do not vary across the critical Q_d value. It is proposed that this change is brought about by the change in the droplet formation mechanism.

The mechanisms controlling the droplet formation in a micro-channel give rise to three different regimes, namely squeezing, dripping and jetting. For a fixed liquid-liquid system in a confined T-junction, the three regimes can be identified on a 2D plot of the continuous versus the dispersed phase flow rates. This variation of droplet formation at a T-junction for an inlet angle of 90° is shown in Table 2. For experiments carried out under the same Q_c, the controlling mechanism for droplet formation can change with Q_d. For example, the droplet formation mechanism shifts from dripping to squeezing as Q_d increases for a Q_c of 1.0 ml/hr. It is possible that the critical dispersed phase flow rate is a result of the mechanism change.

For a large Q_c (e.g. 10.0 ml/hr), the shear force plays a key role in determining the droplet size due to its high magnitude relative to the other forces present. When Q_d is higher than 0.35 ml/hr, the jetting mechanism dominates the formation process and breakup dynamics due to the higher dispersed phase momentum. The instability of a jet results in drop breakup which is determined by the diameter and length of the jet. As such, the dispersed phase flow rate only has a marginal effect on the droplet diameter. However, as Q_d decreases below 0.35 ml/hr, the dripping mechanism controls the droplet formation process which occurs when the formation process is governed by the balance between the interfacial tension and shear force. In this case, the droplet size is dependent on the dispersed phase flow rate.

For the case of a low Q_c of 1.0 ml/hr, the interfacial tension (σ) and pressure drop (ΔP) dominates the force balance as the shear force is small. The low shear force results in the continuous phase flow behind the droplet being blocked by the growing droplet and this pressure drop scales as Q_c/D^4, where Q_c is the volumetric flow rate, D the channel diameter and ΔP the pressure drop over the droplet. For Q_d below 0.35 ml/hr, the dripping mechanism still governs droplet breakup and the droplet size is a function of Q_d.

Q_c (ml/hr)	Q_d=0.1 ml/hr	Q_d=0.35 ml/hr	Q_d=0.5 ml/hr
1.0			
5.0			
10.0			

TABLE II.   THE FORMATION PICTURE FOR T-JUNCTION WITH 90° INLET ANGLE.

Fig. 2 shows that no droplets could be produced for Q_d greater than 0.35 ml/hr for acute angles but rather a two-phase flow regime occurred. For the obtuse angles, Fig. 3 provides evidence of the existence of the critical dispersed phase flow rate. As the formation process is controlled by a squeezing mechanism, the droplets produced are comparable to the main channel width and block it causing a surge in the pressure drop. This accelerates the breakup process and the droplet detaches when it reaches a size comparable to the main channel width. This is reflected in the slow rate of change in droplet size as the dispersed phase flow rate increases past 0.35 ml/hr.

B. Effect of inlet angles on droplet diameter for various polymeric dispersed phases (Newtonian-like and non-Newtonian)

In this series of experiments, solutions of polymeric material with Newtonian-like and non-Newtonian solutions (0.025% xanthan gum and 0.2% PEO in 30% glycerol-water solutions show a Newtonian-like behavior while the 0.25% xanthan gum shows a highly Non-Newtonian property (Table 1)) were used for the dispersed phase. As the previous experiments with water showed that the effect of the inlet angles on droplet size was more pronounced for small Q_c's, experiments were carried out at a lower Q_c of 1.0 ml/hr with varying Q_d values. Canola oil is the continuous phase fluid.

Although the rheological properties of the non-Newtonian fluid are different, the results (Figs 6 and 7) are not too different from that for water. Fig. 6 shows the result for the acute angles where θ=30° (blue square) always give rise to larger droplets than that for θ=60° (orange hollow circle) regardless of the dispersed phases employed. The droplet diameter increases with increasing dispersed phase flow rates but the largest droplets are not necessarily from the right angle T-junction. Generally, in the lower dispersed phase flow rate regime, the right angle T-junction does produce the largest droplet size but for two particular cases, the 0.2%PEO in 30% glycerol-water and 0.025% xanthan gum, the θ=30° T-junction produced the largest droplet size. For the θ=30° and 60° cases, a two-phase parallel flow was obtained at higher dispersed phase flow rates and this limited the results available for these two angles.

Fig. 7 shows that an obtuse angle of θ=120° (green square) results in a larger droplet size than for θ=150° (magenta hollow circle) over the range of dispersed phase flow rate

investigated. Generally the T-junction with θ=90° produces the largest droplet diameters although some of the droplets produced by the T-junction with θ=120° are of comparable size. The rate at which the droplet size increases with the dispersed phase flow rate is steeper for the 0.25% xanthan gum case which has a higher viscosity than the other two fluids, although the θ=30° case for the 0.025% xanthan gum does shows a steep gradient.

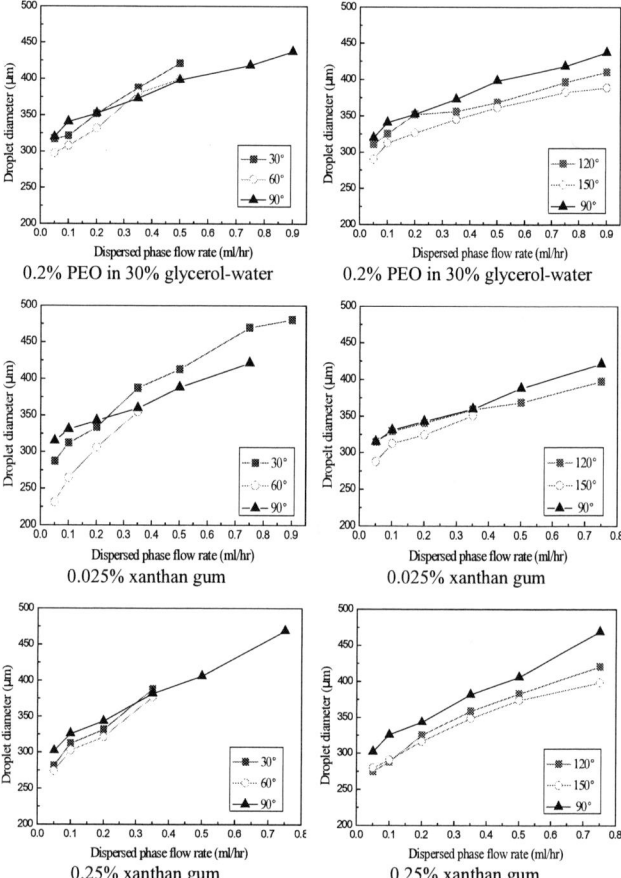

Fig. 6. Effect of inlet angles 30°, 60° and 90° on droplet diameter for different dispersed phases.

Fig. 7. Effect of inlet angles 120°, 150° and 90° on droplet diameter for different dispersed phases.

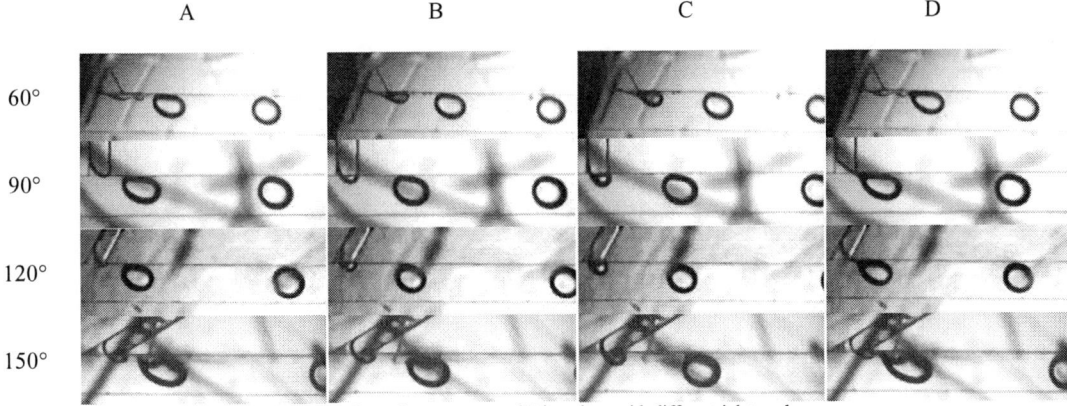

Fig. 8. Droplet formation processes of T-junctions with different inlet angles.

The droplet formation processes were recorded at 3000 frames/second with a Redlake (MotionPro SI-4) camera. The formation process is fairly similar between the different polymeric solutions used and the 0.2% PEO solution was chosen to illustrate the droplet formation process. The continuous phase and dispersed phase flow rates were 1.0 and 0.2 ml/hr respectively with canola oil as the continuous phase.

The droplet formation processes for T-junctions with inlet angles of 60° (first row), 90° (second row), 120° (third row) and 150° (fourth row) are shown in Fig.8. For each T-junction angle, four images were chosen to represent the droplet formation process that was observed. Frame A locates the time frame when the droplet starts to grow. Frame B shows the growing droplet. Frame C shows the neck connecting the droplet to the dispersed phase in the feed channel starting to narrow. Frame D locates the time frame when the droplet is just fully detached.

For the T-junction with a 30° inlet angle, there is a long filament present that cannot be seen within the viewing area so that the growing and detached droplets cannot be accommodated within the same picture frame. Hence no images are provided in Fig 8.

From Fig. 8 and Table 3, we find that, for T-junctions with obtuse inlet angles there is a progressive narrowing of the neck connecting the dispersed phase to the growing droplet which does not appear in those with θ=30° and 60°. This narrowing is caused by the shear stresses acting on the droplet starting to overcome the surface tension force and the droplet is moved downstream. The dispersed phase flow rate is unable

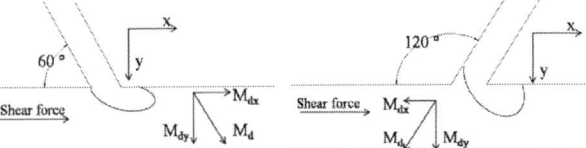

Fig. 9. Schematic of the formation process at a T-junction with inlet angles of 60° and 120°. (M_d is total dispersed phase momentum, and M_{dx} and M_{dy} are the x and y coordinate components of dispersed phase momentum respectively.)

to supply enough fluid to ensure a connection exists and the neck narrows resulting in droplet breakup. For a T-junction with a 60° inlet angle, the droplet begins with an ellipsoid shape that is longer in the streamwise direction and for this shape, the droplet will become unstable at a lower critical Capillary number which results in it elongating indefinitely downstream with the flow resulting in droplet breakup. For this case, the narrowing of the neck is not seen.

For the obtuse inlet angles, the growth stage is longer, up to twice, than that for acute angles. For example, the formation time for a T-junction with a 120° inlet angle is 420.6 ms but is only 212.0 ms for a T-junction with a 60° inlet angle. This occurs for two reasons. The first is that the droplet must initially build up the pressure to overcome the Laplace pressure for a hemispherical cap. The droplet growth rate at the early stage is slow as the pressure rises in the dispersed phase flow channel. The second is that the disperse phase momentum is directed differently for the obtuse inlet angles compared to the acute inlet angles. Other than a right angled T-junction, the momentum in the direction of the continuous phase channel (taken as the x-direction), labeled as M_{dx} in Fig. 9 cannot be neglected. M_{dx} is in the same direction of the continuous phase shear force for acute angles and hence assists the drop in deforming to an ellipsoid in the direction of the flow. This has the effect of causing the drop to be unstable at a lower critical Capillary number (which in turn, occurs at a lower shear rate). In contrast, for an obtuse inlet angle, the disperse phase M_{dx} acts against the direction of the continuous phase shear force which has a stabilizing effect on the drop that is forming. Consequently, the growing droplet is stable for

TABLE III.   TIME INTERVALS OF THE DROPLET FORMATION PROCESSES IN FIG. 8 FOR T-JUNCTIONS WITH VARIOUS INLET ANGLES

Angles	I - A to C (ms)	II - C to D (ms)	III - A to D (ms)
30°	$-^a$	187.7	187.7
60°	$-^a$	212.0	212.0
90°	141.7	183.7	325.3
120°	210.3	212.3	420.6
150°	150.7	293.0	443.7

a: the time interval was not measurable for this T-junction. I is the time from picture A to C shown in Fig.8. II is the time from the narrowing process to the droplet breakup point. III is the total time of the formation process.

a longer time period and when the shear force overcomes the surface tension force, the droplet first begins to move downstream with the formation of a neck which subsequently narrows prior to droplet detachment.

IV. CONCLUSION

The effect of inlet angles (θ=30°, 60°, 90°, 120° and 150°) of T-junctions was investigated using Newtonian and non-Newtonian solutions for the dispersed phase. The experiments were carried out with continuous phase flow rates, Q_c, of 1.0, 5.0 and 10.0 ml/hr for varying Q_d for water. The result showed that the inlet angles play an important role in controlling droplet size where the T-junction with a right inlet angle usually generating the largest droplets for a given continuous phase flow rate. The results for the droplet diameters were separated into two groups, based on the range of the inlet angles (acute angles of 30°, 60° and 90°, and obtuse angles of 90°, 120° and 150°). For the acute inlet angles, the T-junction with an inlet angle of θ=30° resulted in larger droplets than that for θ=60°. For the obtuse inlet angles, the diameter of droplets produced by T-junctions with inlet angles of 120° and 150° were similar in size but smaller than that for 90°. It was also found that the diameter difference due to varying inlet angles was reduced at higher continuous phase flow rates. A critical dispersed phase flow rate (0.35 ml/hr), at which there is a change in the rate the droplet diameter increased with Q_d, was identified and the mechanism behind this change was discussed.

Non-Newtonian xanthan gum and PEO solutions, which are promising for application in biological and medical areas, were also used as dispersed phases to study the effect of the inlet angles. The results suggested that the effect of the non-Newtonian fluids were not as pronounced as expected although differences particularly with θ=30° was observed. There was a tendency to form two-phase parallel flows which was not found with the water results. Droplet formation was imaged at 3000 frames/second to highlight the effect of acute and obtuse inlet angles on the droplet forming mechanism.

ACKNOWLEDGMENT

Z.P. Gu gratefully acknowledges the financial support provided by the Chinese Scholarship Council. The authors would like to thank AP Vincent Craig (ANU) for the use of the tensiometer and his insightful suggestions.

REFERENCES

1. Engl, W., R. Backov, and P. Panizza, *Controlled production of emulsions and particles by milli- and microfluidic techniques.* Current Opinion in Colloid & Interface Science, 2008. **13**(4): p. 206-216.
2. Christopher, G.F. and S.L. Anna, *Microfluidic methods for generating continuous droplet streams.* Journal of Physics D-Applied Physics, 2007. **40**(19): p. R319-R336.
3. Thorsen, T., et al., *Dynamic pattern formation in a vesicle-generating microfluidic device.* Physical Review Letters, 2001. **86**(18): p. 4163-4166.
4. Anna, S.L., N. Bontoux, and H.A. Stone, *Formation of dispersions using "flow focusing" in microchannels.* Applied Physics Letters, 2003. **82**(3): p. 364-366.

5. Cramer, C., P. Fischer, and E.J. Windhab, *Drop formation in a co-flowing ambient fluid.* Chemical Engineering Science, 2004. **59**(15): p. 3045-3058.
6. Kobayashi, I., et al., *Generation of Geometrically Confined Droplets Using Microchannel Arrays: Effects of Channel and Step Structure.* Industrial & Engineering Chemistry Research, 2009. **48**(19): p. 8848-8855.
7. Abate, A.R., et al., *Impact of inlet channel geometry on microfluidic drop formation.* Physical Review E, 2009. **80**(2).
8. Tice, J.D., A.D. Lyon, and R.F. Ismagilov, *Effects of viscosity on droplet formation and mixing in microfluidic channels.* Analytica Chimica Acta, 2004. **507**(1): p. 73-77.
9. Tostado, C.P., J.H. Xu, and G.S. Luo, *The effects of hydrophilic surfactant concentration and flow ratio on dynamic wetting in a T-junction microfluidic device.* Chemical Engineering Journal, 2011. **171**(3): p. 1340-1347.
10. Fries, D. and P.R. von Rohr, *Impact of inlet design on mass transfer in gas-liquid rectangular microchannels.* Microfluidics and Nanofluidics, 2009. **6**(1): p. 27-35.
11. Tan, J., et al., *Gas-liquid flow in T-junction microfluidic devices with a new perpendicular rupturing flow route.* Chemical Engineering Journal, 2009. **146**(3): p. 428-433.
12. Wang, K., et al., *Generation of Micromonodispersed Droplets and Bubbles in the Capillary Embedded T-Junction Microfluidic Devices.* Aiche Journal, 2011. **57**(2): p. 299-306.
13. Krstonosic, V., et al., *Effects of xanthan gum on physicochemical properties and stability of corn oil-in-water emulsions stabilized by polyoxyethylene (20) sorbitan monooleate.* Food Hydrocolloids, 2009. **23**(8): p. 2212-2218.
14. Rodd, A.B., D.E. Dunstan, and D.V. Boger, *Characterisation of xanthan gum solutions using dynamic light scattering and rheology.* Carbohydrate Polymers, 2000. **42**(2): p. 159-174.
15. Rodd, A.B., D.E. Dunstan, and D.V. Boger, *Dynamic light scattering of dilute and semi-dilute Xanthan solutions and comparison with Rheological Characteristics.* Hydrocolloids, Pt 1 - Physical Chemistry and Industrial Application of Gels, Polysaccharides, and Proteins, ed. K. Nishinari. 2000, Amsterdam: Elsevier Science Bv. 151-157.
16. Lomax, G.R., *Breathable polyurethane membranes for textile and related industries.* Journal of Materials Chemistry, 2007. **17**(27): p. 2775-2784.
17. Tirtaatmadja, V., G.H. McKinley, and J.J. Cooper-White, *Drop formation and breakup of low viscosity elastic fluids: Effects of molecular weight and concentration.* Physics of Fluids, 2006. **18**(4).
18. Husny, J. and J.J. Cooper-White, *The effect of elasticity on drop creation in T-shaped microchannels.* Journal of Non-Newtonian Fluid Mechanics, 2006. **137**(1-3): p. 121-136.
19. Liow, J.L., *Mechanical micromachining: a sustainable micro-device manufacturing approach?* Journal of Cleaner Production, 2009. **17**(7): p. 662-667.
20. Macosko, C.W., *Rheology: Principles, Measurements and Applications.* Wiley-VCH, 1994.
21. Gu, Z.P. and J.L. Liow, *Microdroplet formation in a T-junction with xanthan gum solutions.* CHEMECA. Sydney, 2011, p.1-10.
22. Niu, H., et al., *Flow Pattern, Pressure Drop, and Mass Transfer in a Gas−Liquid Concurrent Two-Phase Flow Microchannel Reactor.* Industrial & Engineering Chemistry Research, 2008. **48**(3): p. 1621-1628.
23. Cubaud, T. and C.M. Ho, *Transport of bubbles in square microchannels.* Physics of Fluids, 2004. **16**(12): p. 4575-4585.
24. Berthier, J., et al., *Highly viscous fluids in pressure actuated flow focusing devices.* Sensors and Actuators A: Physical, 2010. **158**(1): p. 140-148.

Investigation of Electrical Properties of DNA-Attached Carbon Nano-Particles for Biological Applications

Mengxing Ouyang[1], Wen J. Li[2,*], *Fellow, IEEE*, Ka Wai Wong[3], and Wing Keung Liu[4]

[1]Centre for Micro and Nano Systems, The Chinese University of Hong Kong, Shatin, N. T., Hong Kong SAR

[2]Dept. of Mechanical and Biomedical Engineering, City University of Hong Kong, Kowloon, Hong Kong SAR

[3]Chengdu Green Energy and Green Manufacturing Technology R&D Center, Chengdu, Sichuan, China

[4]School of Biomedical Sciences, The Chinese University of Hong Kong, Shatin, N. T., Hong Kong SAR

[*]wenjli@cityu.edu.hk

Abstract—**Carbon nano-particles are nano-sized crystalline with predominantly graphitic structure. Our recent work showed that DNA-attached Carbon Nano-Particles (DAC) can be positioned between microelectrodes in a microfluidic system to investigate its electrical properties. Dielectrophoretic based "deposition" led to a robust adhesion of various DACs with the chip substrate even after repeatedly DI water flushing. The IV characteristics and stability of three types of DACs under different conditions (i.e., towards open environment, sealed in dry microchannel, or immerged in DI water) were compared and analyzed. In addition, experiments were conducted to determine the temperature and humidity dependency of the DACs.**

Keywords—*Carbon Nano-Particles; DNA attached Carbon Nano-Particles; Dielectrophoresis Manipulation.*

I. INTRODUCTION

DNA, which carries the genetic code of living species, is essential to the maintenance of homeostasis of life [1]. A wide range of biological process requires the understanding of the electronic properties of DNA, for example, repair of damaged bonds in cells, detection of mutations, as well as protein-DNA interactions [2]. Both temperature and humidity have been known to affect the electrical conductivity of DNA. On the other hand, nanomaterials such as Carbon Nano-Particles are suitable for various biomedical applications, for instance, they can serve as biomolecular carriers during cell delivery due to their simple geometry (e.g., spherical), freedom from entanglement [3], and low toxicity compared to quantum dots. Therefore, by attaching DNAs onto the surface of CNPs, it is possible to transport DNAs into cells along with the CNPs for further manipulation and analysis. Some preliminary study has been published recently discussing the interaction between DNA and carbon nanoparticles [4].

In this paper, the electrical properties of DNA-attached CNPs in liquid environment (i.e., DI water) were investigated to provide reference for further biological applications such as cellular uptake.

II. PREPARATION OF SENSOR CHIP

Carbon Nano-Particles with diameter of 3-6 nm were prepared by oxidative chemical treatment and different types of DNAs (e.g., single-strand DNA and double-strand DNA)

were attached onto the CNPs, as illustrated in Fig. 1 (a). Then, **D**NA-attached **C**NPs (DAC) were deposited between microelectrodes and encapsulated in a micro-fluidic platform.

The fabrication process is described as following: Au microelectrodes were firstly deposited on Silicon substrate using sputtering (with Cr as adhesion layer) and patterned with photolithography. Then, the 'deposition' of DAC between a pair of Au microelectrodes was accomplished by dielectrophoretic manipulation. We have demonstrated the DEP manipulation for CNPs assembly before in [5]. Based on our previous work, a very similar process is adopted in this paper. Then, the top cover of this microfluidic system was prepared as follows: A PDMS microchannel was fabricated using cast molding method and irreversibly bonded on top of the Si chip using oxygen plasma. Inlet and outlet were created and connected by tubes for later injection of solutions into the microchannel. A photo of the fabricated microfluidic system is illustrated in Fig. 1 (b).

During experiments, aqueous flow was introduced into the microchannel and the resistance changes of the DAC sensors were recorded by source meter. We have noticed that DEP manipulation was not only able to deposit and position DACs between microelectrodes with 2 µm gap successfully and efficiently, but also able to form a strong adhesion between DACs with both Silicon substrate and Au microelectrodes. In fact, the adhesion force is so strong that even after the sensors were flushed with aqueous flow repeatedly, the linkage still remained and was not easily washed off.

III. ELECTRICAL PROPERTY OF DACs

A. Current-Voltage Characterization

The IV characteristics of three types of DACs were collected, i.e., ds- (double-strand), ss- (single strand) and n- (neutral) DNA. For each type of DAC, its behavior in the microchannel towards ambient air or DI water was compared. In addition, for experiments carried under each condition, 2 consecutive cycles of measurement were conducted to verify the repeatability of the results. As shown in Fig. 2, solid lines and dash lines represent DI water condition and ambient air condition, respectively. By comparing all six groups of I-V

978-1-4673-1122-9/12 $31.00 © 2012 IEEE

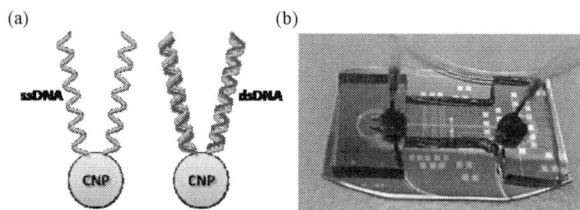

Fig. 1. (a) Conceptual illustration of DNA-attached CNPs; (b) A photo of the microfluidic chip used in this project.

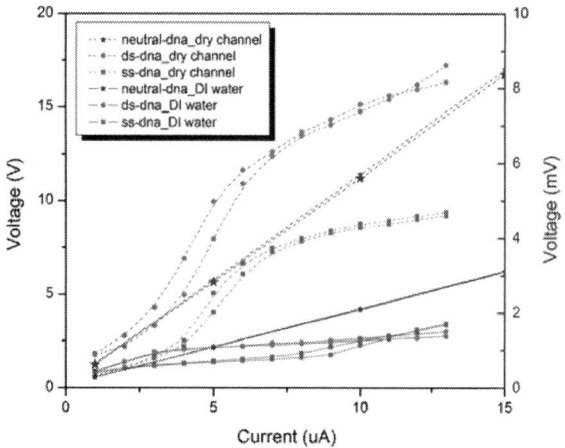

Fig. 2. IV Characteristics of DNA-attached CNPs.

curves, we found that the sensor resistance for all three types of the DACs decreased when they are immersed in DI water instead of exposed to ambient air. Also, by comparing the three groups of experimental data conducted under the same condition, i.e., either ambient air or DI water, it is clearly indicated that the linearity of both ds-DAC and ss-DAC , which showed a similar I-V trend, were not as good as neutral-DAC. In addition, all three groups of I-V curves carried out in DI water showed better linearity compared to the corresponding results when they were exposed to ambient air. Above all, by immersing the DACs into DI water, the sensor resistance would decrease and the linearity of I-V characteristics would increase.

B. Stability

We have analyzed the stability of three types of DACs under three conditions, i.e., sensor surface open to ambient air (pre-bonding), sensor sealed in a microchannel exposed to ambient air, and sensor sealed in the microchannel immersed in DI water. In Fig. 3, these three conditions are denoted as open (square), channel (dot), and DI-water (triangular), respectively. For each of these nine groups, a stability test with duration of 12 minutes was carried out, and three different activating current were used for further comparison. For ds-DACs and ss-DACs, the activating current applied was 5ua, 10ua and 15ua; for neutral-DACs, current of 1ua, 5ua and 10ua were chosen. This difference of current range was based on the resistance difference between normal-DACs (i.e., ds-,

ss-) and neutral-DACs, as the resistance of the former type is in Mega-Ohm level and the latter is typically only in the order of hundreds of Ohms. If normal-DAC was applied with a current of 15ua, the resistance would have exceeded the maximum detection range of the source meter; on the other hand, if we activated the neutral-DACs with only 1ua current, the voltage recorded would be only a few mV and too small to stay in the optimal range of the source meter.

For analysis purpose, we estimated the fluctuation of each of the 27 cycles of measurement using the following equation (1):

$$\text{Fluctuation} = \max\left(\frac{R_i - R_{average}}{R_{average}}\right) \times 100\% \quad (1)$$

Therefore, smaller fluctuation value is desired and indicates better stability. The fluctuation under different conditions and DACs types is summarized in Fig. 3 in terms of both activating current and activating power, where DAC types were illustrated by ss-DACs (solid line), ds-DACs (dash line) and neutral-DAC (dot line), respectively; measurement conditions were denoted as open (square), channel (dot), and DI-water (triangular), respectively.

A few observations were made from this figure. (1) While comparing different DACs types, neutral-DACs in general showed smaller fluctuation compared to that of normal-DACs. (2) While comparing different sensor conditions, the sensors tended to exhibit smaller fluctuation when they were exposed to ambient environment without microchannel, i.e., category open. This may be due to the fact that heat accumulated around the sensor surface in the microchannel cannot be dispersed in time and this would in turn cause the fluctuation of the sensor resistance. This phenomenon was especially severe for normal-DACs (i.e., ds-, ss-). (3) While comparing fluctuation change versus increasing current, we observed the typical trend of higher fluctuation under higher activating current for 7 out of the 9 groups. The only exceptions were 2 groups of data from neutral-DACs (i.e., denoted in the figure as neutral-open and neutral-DI-water, respectively), as the fluctuation value did not rise with increasing applied current. However, given that for both groups, the overall change of the fluctuation was less than 0.5%, we treated these two groups as they showed fairly constant fluctuation when the current was increasing. (4) While comparing the fluctuation change versus activating power of the sensor for each cycle of measurement, a clear trend is shown in Fig. 3 (b), that is, higher power level would result in higher fluctuation, i.e., poorer stability of the sensor. Therefore, for future sensor applications, the sensor is more desirable to perform under reasonably lower power level to minimize extra noise that would affect the sensor performance.

C. Temperature Dependency

It has been reported previously that the electrical conductivity of DNA is temperature dependent [6]. Our experiments confirmed that the conductivity of DACs also

978-1-4673-1122-9/12 $31.00 © 2012 IEEE

(a)

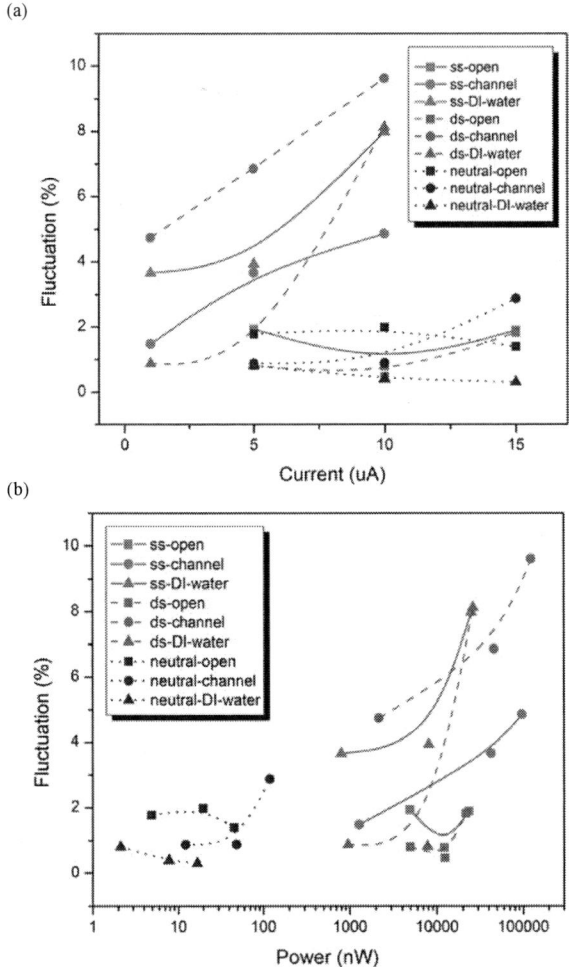

(b)

Fig. 3: Stability comparison of DNA-attached CNPs under (a) different current and (b) different power.

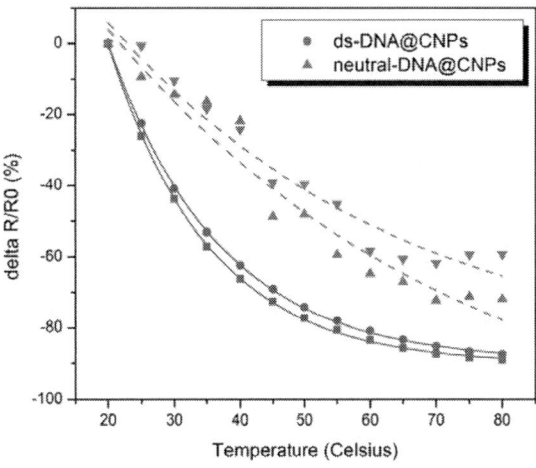

Fig. 4. Temperature dependency of DNA-attached CNPs.

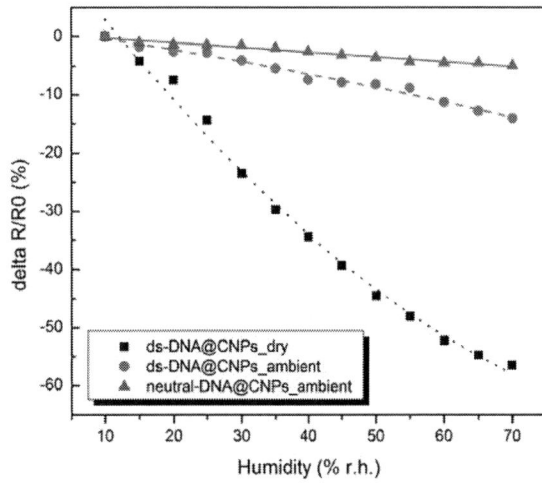

Fig. 5. Humidity dependency of DNA-attached CNPs.

have this temperature dependency property. The sensors were put into an oven where both humidity and temperature were controllable. As shown in Fig. 4, both ds- and neutral-DACs experienced a dramatic resistance drop (> 60%) when the temperature was increased by 60 °C. For both types of DACs, the temperature dependency curves were not linear, especially when the temperature was higher than 55 °C. By comparison, ds-DACs exhibited larger resistance decrease (88.34%) than that of neutral-DACs (65.64%) as well as a larger curvature of their fitting line. During the test, the humidity was kept constant at 25% r. h.

D. Humidity Dependency

Other than temperature, humidity also has a significant influence on the electrical conductivity of DNA. Researchers have reported that the resistance of a DNA film would decrease exponentially with increasing relative humidity [7]. The proposed theory is that water molecules are easily adsorbed and form hydrogen bonds between the phosphoric

acid and the water molecule, because of the hydrophilic phosphoric acid around the base pairs of DNA.

During our experiments, the humidity dependency of DACs has been investigated and the results demonstrated similar trend. While the temperature was kept constant (see Fig. 5), the resistance of DACs decreased (< 15%) when the ambient humidity increased. Additionally, we found that when the DACs were kept in extremely low humidity (e.g. ~7% r. h.) environment before measurements, the resistance drop increased to ~55%. It implies that the pre-exposure status of DACs is crucial as it may largely improve the sensor response towards humidity.

To further investigate the humidity dependence of DACs, we conducted a series of flow rate tests using DACs sensors. As shown in Fig. 6, the resistance change of ds-DACs, ss-DACs and neutral-DACs towards flow rate in the microchannel were normalized and compared. While exposing to the aqueous flow with a certain flow velocity, the resistance

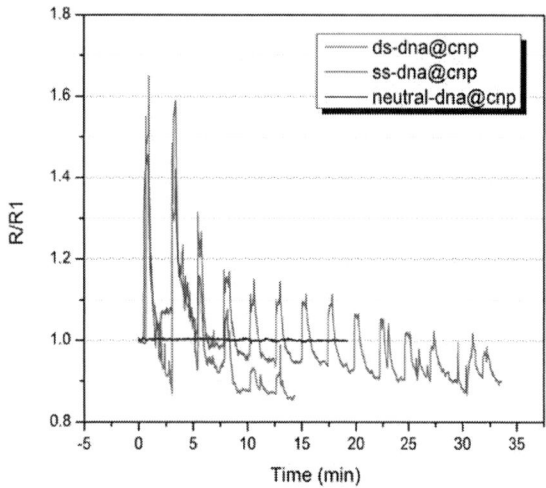

Fig. 6. Response of DNA attached CNPs towards flow rate.

of the DACs sensors exhibited a sudden resistance increase, and when the flow was stopped, the resistance would decrease immediately. For each type of DACs (i.e., ds-, ss-, and neutral-), aqueous flow was introduced into the microchannel for a period of 30 seconds followed by a resting period. During experiments, a syringe pump was used to provide controllable flow rate. As shown in Fig. 6, both normal-DACs (i.e., ds-, ss-) demonstrated a clear trend of decreasing of the sensor response (R/R_1) towards aqueous flow rate after several cycles of experiment. This insensitivity towards flow rate occured even faster for ss-DACs than ds-DACs.

Besides, we would like to stress that the flow rate for each peak was neither constant nor continuously slower, but rather random. Take the curve of ds-DACs for instance, the flow rate was firstly increased gradually from 50μm/s to 250μm/s from cycle 1 to 5 with a step of 50μm/s; after that, the flow rate was decreased to 50μm/s from cycle 6 to 10; then, the velocity was raised towards 250μm/s again till cycle 14 with a step of 50μm/s. In spite of the flow rate change, as shown in the figure, the sensor response kept decreasing from cycle 1 to cycle 14. Similar experiments were carried out for ss-DACs and neutral-DACs. However, unlike normal-DACs, which showed gradually decreased sensitivity towards flow rate, neutral-DACs exhibited much smaller response towards flow rate change, i.e., the resistance almost kept constant when the sensor was repeatedly exposed to aqueous flow. Proposed explanation is that since the decrease in resistance is affected by water molecules adsorbed by DNAs, when DNAs have adsorbed the largest amount of water molecules, the resistance change would gradually stop.

To summarize, the response of DACs sensor does not depend on the magnitude of flow rate. The sensitivity towards flow rate could be either minimized after repeatedly exposure to aqueous flow (for normal-DACs) or considered insensitive towards flow rate (for neutral-DACs). This implies that DAC

sensors are not suitable for flow rate measurement. But they may be applicable to serve as humidity sensors if a pre-designed process is available to eliminate the water molecules attached onto the DNAs. Another potential application of DACs is for chemical or biological detection in the microfluidic systems with flowing solution without worrying about its sensitivity towards aqueous flow, i.e., after proper treatment of the sensor before detection, the resistance change of the sensor during chemical or biological sensing in the flow is more likely to be caused by the target instead of the flow velocity.

IV. CONCLUSION

Three types of DNA attached Carbon Nano-Particles were deposited on a Silicon chip using DEP manipulation and sealed in a microfluidic channel to study their electrical properties. The linearity of the IV characteristics was improved when DAC sensors were immerged in DI water. However, less fluctuation were observed when the sensors were exposed to open environment rather than in a microchannel, or when the sensors were activated with smaller current/power level. Moreover, we have proved that increasing temperature and humidity would cause the electrical conductivity of DACs to increase accordingly. Also, the response of DACs sensors do not depend on the magnitude of flow rate. By repeated exposure to aqueous flow, sensors would gradually become insensitive to flow rate. In addition, compared to normal ds- and ss- DACs, neutral-DACs demonstrated better linearity and stability, and they are almost not responsive to flow rate. We will investigate DAC sensors as humidity sensors and chemical/biological detection sensors in micro flow systems in the future.

REFERENCES

[1] M. Zhang, and T. J. Tarn, "DNA electrical properties and potential nano-applications", *Conf. Proced. Of IEEE-NANO 2003*, vol. 2, pp. 512-515, 2003.

[2] V. Hodzic, V. Hodzic, and R. W. Newcomb, "Modeling of the electrical conductivities", *IEEE Trans. on Circuit and Systems*, vol. 54, No. 11, pp. 2360-2636, 2007.

[3] A. Yan, B. W. Lau, B. S. Weissman, I. Kulaots, N. Y. C. Yang, and A. B. Kane, and R. H. Hurt, "Biocompatible, hydrophilic, supramolecular carbon nanoparticles for cell delivery", *Adv. Mater.*, Vol. 18, pp. 2373-2378, 2006.

[4] H. An, Q. Liu, Q. Ji, and B. Jin, "DNA binding and aggregation by carbon nanoparticles", Biochemical and Biophysical Research Communications, vol. 393, pp. 571-576, 2010.

[5] M. Ouyang, Y. M. Ho, Wen J. Li, and Ka Wai Wong, "Fabrication and Manipulation of Fluorescent Carbon Nanoparticles for Biosensing Applications", *IEEE Int. Conf. on Nano/Micro Engineered and Molecular Systems (IEEE-NEMS 2011)*, pp. 893-896, Feb. 20-23, 2011.

[6] P. Tran, B. ALavi, and G. Gruner, "Charge transport along the x-DNA double helix", *Phys. Rev. Lett*, vol. 85, no.7, pp. 1564-1967, 2000.

[7] Y. Otsuka, H. Lee, J. Gu, J. Lee, K. Yoo and H. Tanaka et al, "Influence of humidity on the electrical conductivity of synthesized DNA film on nanogap electrode", *Jpn. J. Appl. Phys.*, Vol. 41, pp. 891-894, 2002.

Electrical Performance of Micro-assembled Beads under Different Temperatures and Loadings

Yen Lin Tzeng, Kerwin Wang

Department of Mechatronics Engineering, National Changhua University of Education, Changhua, Taiwan
kerwin@cc.ncue.edu.tw

Abstract—**Micro-assembly is an efficient tool to build electrical connections with metallic micro-beads. This process uses patterned photoresist AZ1512 as an adhesion for micro-bead arrangement. The assembled beads is immobilized with underfill embedment (ZYMET 2821). This method allows arbitrary geometric pattern designs. All of these processes can be completed below 150°C. This paper characterizes the electrical performance of these densely-arranged anisotropic conductive tunnels under different temperatures and stresses loading Experiment results suggest that using photoresist to assemble micro conductive beads with underfill immobilization can yield stable performance.**

Keywords-assembly; micro-assembly; micro beads; vertical conductive channel; underfill.

I. INTRODUCTION

In recent years, 3D flip-chip packaging becomes an important technology to integrate miniaturized systems. Overall, the packaging scheme can involve high density electrical interconnections between chips and integral circuits or between vertical conductive channels and through silicon vias (TSV) [1, 2]. Through the vertical conductive path, the transmission and processing speed of the microelectronic device can be improved [3]. Anisotropic conductive film (ACF) bonding is one of the well knowing technologies for building vertical conductive paths. It is widely used in flat panel displays, chip packaging and ACF wafer-level packaging [4]. However, it is well known that the thermo-mechanical reliability of these vertical conductive paths is depended on the thermo-mechanical properties of ACF. Many researchers have conducted some useful research [5-6] in this area. Instead of using ACF, we presented a photolithography based micro-conductive-bead assembly method our previous research [7]. It can provide miniaturized and densely packed conductive paths. The assembled micro-conductive-beads have to be permanently immobilized and electrically connected to the binding sites for practical applications. In this paper, we arrange the micro-scaled conductive beads on the lithographic patterns and inject the unferfill into the gap between chips; it can immobilize the beads and help the conductive chip to against the mechanical stress and moisture. The effect of the mechanical stress and temperature variation of electrical conductivity was also investigated. In the following sections, we first discuss the materials, experimental procedures and methods. Then, we present the test results and conclusions.

II. MATERIALS AND EXPERIMENT PROCEDURES

A. Matrials

3 μm micro-composite conductive beads are selected for building anisotropic conductive electrical channels. The beads consist of a 0.025 μm gold-nickel metal layer and an encapsulated benzoguanamine resin core.

AZ1512 is used to define the require patterns for micro-beads assembly by photolithography process. It is a widely used positive photoresist for micro-fabrication.

ZYMET X2821 is an encapsulation underfill materials, it can be quickly cured under low temperature. In this paper, it is used to immobilize the conductive beads and to hold the assembled chip. It can also protect the connections from damages.

TABLE I.   WL-CSP AND BGA UNDERFILL ENCAPSULANT

Type	ZYMET X2821(underfill)
Viscosity, 25°C	500 (cps)
Cure conditions	150°C, 1minutes
CTE	38ppm/°C
Tg, (TMA)	135°C
Shear storage modulus	3GPa

TABLE II.   BEADS COEFFICIENT OF THERMAL EXPANSION

Layer	Coefficient of Thermal Expansion
Au	14.2 ppm /°C
Ni	13 ppm /°C
Micro conductive beads	45 ppm /°C

B. Micro-assembly Process

To achieve densely packed electrical connections between chip and substrate, the procedures for micro conductive beads assembly and underfill process are listed in Fig. 1. A similar process of micro conductive beads assembly has been published in our previous research [7]. This process is start from an oxidized and aluminum deposited 4" wafer. AZ1512 is spin coated on Si wafer to pattern the aluminum. Depends on the applications, after lithography, developing and etching process, one can create desired patterns on the photoresist and aluminum for micro conductive beads assembly. The beads are spread densely and evenly on the photoresist patterns.

978-1-4673-1122-9/12 $31.00 © 2012 IEEE

The Hotplate heating makes the photoresist soften at 120°C. It can increase sticking force between beads and patterns. Micro-conductive beads are selectively immobilized on the photoresist patterns. After gently washed redundant beads away with deionized water, we injected underfill into the chip's gap and heated it up to 150°C for 1 minute to bond and to immobilize the combinations.

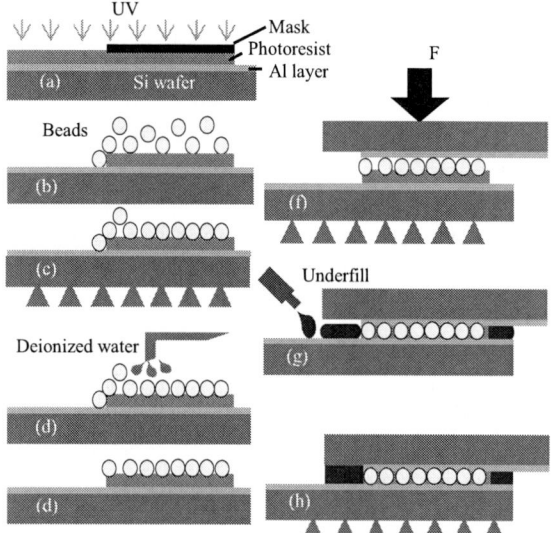

Fig. 1: Process steps: (a) photolithography; (b) conductive-bead spreading; (c) photoresist soften and bead immobilization; (d) redundant bead cleaning; (e) patterned beads; (f) photoresist reflow and heating impression; (g) underfill injection; (h) underfill curing.

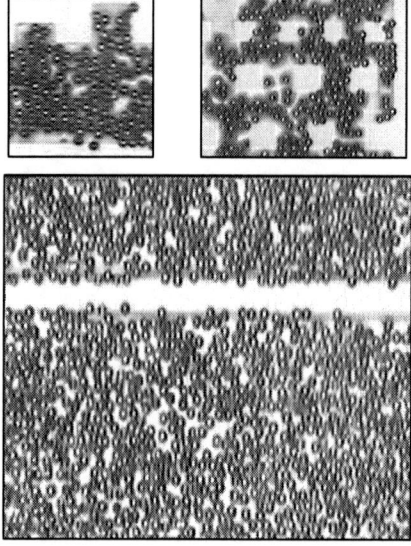

Fig. 2: The testing result shows that this technique can assemble a single layer of microbeads on well-defined photoresist (AZ1512) patterns.

Fig. 2 shows the immobilized micro-conductive bead on the photoresist. The diameter of the micro conductive beads is around 3μm. The micro conductive beads are selectively immobilized on the patterned region. This immobilization process does not require external force fields. It is a cost and time efficient process.

The bead distributions on the patterns are random in nature; however, they are confined to predetermined area. To ensure a good encapsulation process, the designed pattern should have topologically opened cavity geometry to spread the underfill.

This study is focus on the electrical performance of micro-assembled beads under different temperatures and loadings. The simplified bead assembly region is prepared to ensure the applicable array arrangement, which provides improved filling results on the curing, velocity and encapsulant distributions for capillary-driven by dispensing. Fig. 3 shows an assembled combination of top and bottom chip with sandwiched conductive beads, underfill and photoresist.

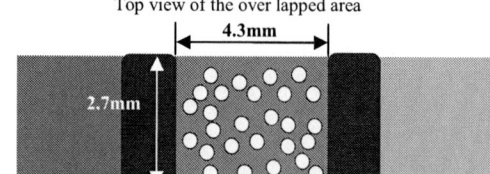

Fig. 3: The schematic top view and cross-section diagram shows the configuration of thickness and size of photoresist and micro conductive bead assembly. The overlap area of electrodes is 11.6mm².

III. EXPERIMENTAL SETUP AND MEASUREMENT

Unlike traditional soldering processes, the electrical connections are not built by chemical bonding or phase transition. There are only physical contacts between electric pads and micro-conductive beads. That means contact sliding gap deformation and stress variation may cause the resistance change. Measurements have been made of the electrical conductive performance of the assembled conductive beads, namely its resistance at different temperature, compressive stress and shear stress. The experiment setups of these tests are shown in Fig. 4.

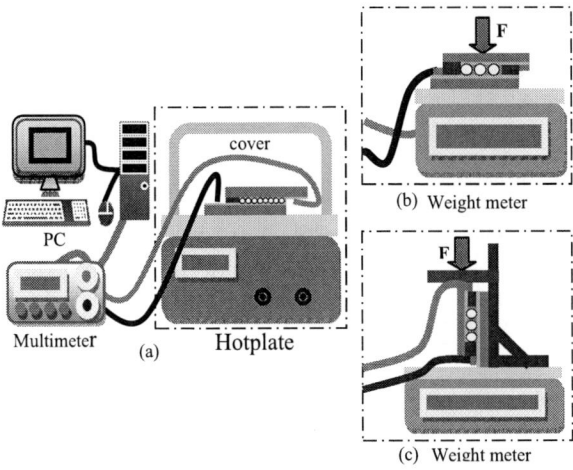

Fig. 4: Three kinds of experiment setups : That are measure and transmit data by mutimeter and computer (a) Use hotplate to Heat the products in the close cover. (b) Give the products downward stress on the weight meter. (c) Give the upright products shear stress on the weight meter and .

The test conditions and results of the resistances at different temperatures and loading stress are listed below:

A. Electrical resistance of this connection at different temperatures

The assembled chip is placed on a hot plate; the electrodes are connected to a digital multimeter (Agilent 34410A) to measure the resistance. The setup is installed in a sealed chamber which provides a closed environment and good temperature uniformity. The heating rate of this setup is around 29°C/min and the cooling rate is around 4°C/min. As shown in Fig. 5, the resistance measurement range is setup from 25 to 150 ^{O}C, the data are automatically collected by a computer.

Contact resistance measurement results corresponding to the thermal cycle is shown in Fig. 6, the contact resistance between 3µm Ni/Au coated Benzoguanamine resin beads and bonding pads are increased after temperature increased from 27 to 148 ^{O}C. Photoresist above reflow temperature was thermally and mechanically less stable above 110°C; however, been embedded in the underfill, photoresist reflow did not show any significant impact to resistance.

There are various explanations on what causes the resistance variation (Fig. 6) during the thermal cycle, the most possible explanation being Matthiessen's rule and the coefficient of thermal expansion differences among the photoresist, micro metallic beads and the underfill.

B. Electrical resistance of this connection under different compressive loadings

The sample was tested under compressive loadings (Fig. 7). The loadings are gradually increased from 0 to 800kN/m^2. Then it is gradually decreased back to 0kN/m^2 to form an enclosed loop.

The assembled connection had low contact resistance at the initial stage. When the pressure force is gradually applied to the assembly, compressive loading pressure would promote contact stress redistribution which increases the resistance of the electric assembly. However, the data measured within the linear elastic deformation loop are steady and repeatable.

C. Electrical resistance of this connection at different shear stresses

To precede the resistance investigation under shear stress loading, an assembled chip is installed onto a testing platform with a force gauge. The shear stress was continuously loaded to the assembled chips until the catastrophic damage occurred.

The value of contact resistance, measured correspond to the shear loading, is shown in Fig. 8. The experiment results suggest that shear stress may cause significant degradation on the contact resistance. It can cause physical detaching and delaminating which leads to mechanical and electrical failure.

The composite metallic beads aging caused by high temperature can lead further degradation of the contacts. High bonding and reflow temperatures above 200 ^{O}C are not recommended for reliability issue.

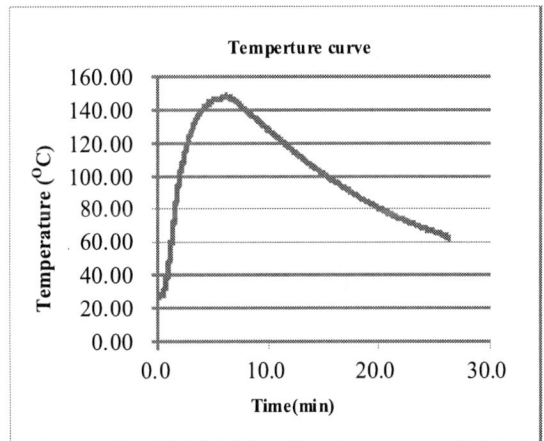

Fig. 5: The temperature curve of the platform during heating and cooling process.

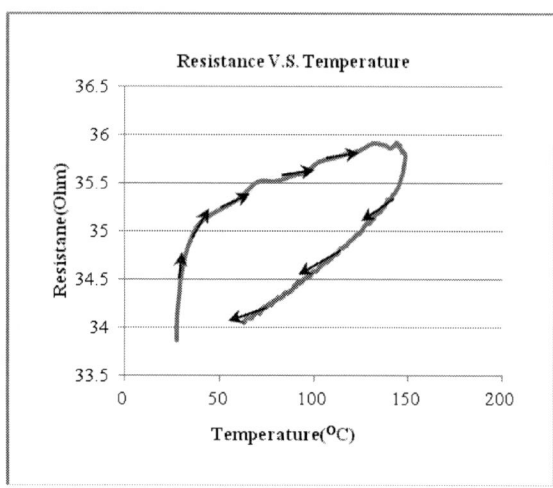

Fig. 6: The resistances versus temperature characteristics of the bonded chip.

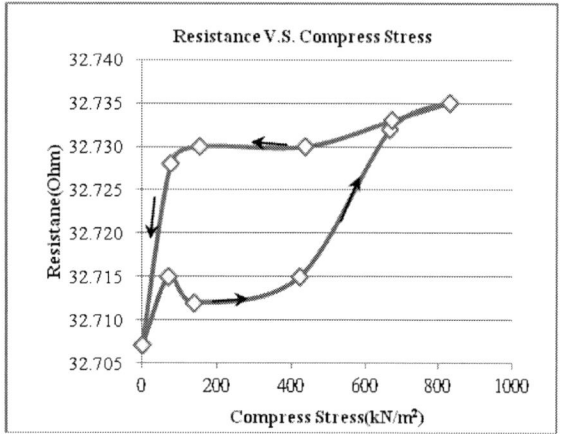

Fig. 7: The resistances versus compress stress characteristics of the bonded chip.

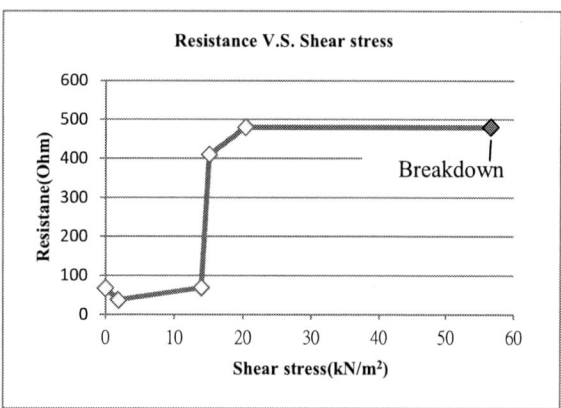

Fig. 8: The resistances versus shear stress characteristics of the bonded chip.

IV. CONCLUSION AND SUMMARY

Micro-assembly is an efficient tool to build electrical connections with metallic microbeads. This process uses patterned photoresist AZ1512 as an adhesion for micro-bead adhesion. This method allows arbitrary geometric pattern designs. All of these processes can be completed below 150°C.

This paper characterizes the electrical performance of this densely-arranged anisotropic conductive tunnels under different temperatures and stresses. Experiment results suggest that using photoresist to assemble micro conductive beads with underfill embedment (XYMET 2821) can yield stable performance with appropriate shear stress protection.

The results indicate that the micro-bead assembly with the underfill encapsulation method developed in this study is adequate for micro conductive beads assembly in flip chip bonding process.

ACKNOWLEDGMENT

The authors would also like to thank National Chiao Tung University, and the Nano Facility Center at National Changhua University of Education, for providing the fabrication facility. This work is supported by National Science Council, NSC 100-2221-E-018-015 and NSC-98-2221-E-018-012-MY2.

REFERENCES

[1] S. F. Al-Sarawi, D. Abbott, and P. Franzon, "A review of 3-D-packaging technology," *IEEE Trans. Compon. Packag. Manuf. Technol*, vol. 21,no. 1, pp. 2–14, 1998.

[2] Y. Yano, T. Sugiyama, S. Ishihara, Y. Fukui, H. Juso, K. Miyata,Y. Sota, and K. Fujita, "Three-dimensional thin stacked packaging technology for SiP," *in Proc. Electron. Compon. Technol. Conf*, pp. 1329–1334, 2002.

[3] M, Motoyoshi, "Through-Silicon Via (TSV)," Proceedings of the IEEE, Vol. 97, No.2, 2009, p43-48H.-Y. Son, C.-K. Chung, "Wafer-Level Flip Chip Packages Using Preapplied Anisotropic Conductive Films (ACFs)," *IEEE Trans. on Electron. Packag. Manuf*, 30, pp. 221-227, 2007.

[4] K.-W Jang, and K.-W. Paik, "Effects of Heating Rate on Material Properties of Anisotropic Conductive Film (ACF) and Thermal Cycling Reliability of ACF Flip Chip Assembly," *IEEE Trans. on Components and Packaging Technologies*, 32, pp. 339-346, 2009.

[5] S-T Lu, W-H Chen, "Reliability and Flexibility of Ultra-Thin Chip-on-Flex (UTCOF) Interconnects With Anisotropic Conductive Adhesive (ACA) Joints," *IEEE Trans. on Advanced Packaging*, 33, pp.702-712, 2010.

[6] T. Kubono, S. Teraoka, "Relationship between the contact resistance and the temperature near electrical contacts switching continually," *Electronics and Communications in Japan (Part II: Electronics)*, 72(5), pp. 89-96, 1989.

[7] K.-W Wang, C.-H Chen, "Binding Site Design For The Micro Conductive Bead Assembly In Flip Chip Bonding" *in The 4th Asia Pacific Conference on Transducers and Micro/Nano Technologies*, Tainan, Taiwan, June 22-25, 2008.

Vertical Deposition of Nanospheres on the Open Sidewalls of Silicon Pillars

Y.F. Wang, Y. Tian, K.J. Feng, C. Li, D.D. She and W.G. Wu*

*National Key Laboratory of Science and Technology on Micro/Nano Fabrication, Institute of Microelectronics, Peking University, Beijing 100871, P. R. CHINA
Email: wuwg@pku.edu.cn

Abstract—**In this paper, we report a simple method, via colloidal suspension evaporation, to deposit monolayer nanospheres vertically onto the open sidewalls of manifold silicon pillar arrays. In experiments we utilize hydrophobic polystyrene (PS) nanospheres and hydrophilic silica nanospheres, and find that the former require more hydrophilic pillar surfaces than the latter do. In addition, proper quantity of nanosphere suspension depends in larger part upon intervals of silicon pillar arrays than upon diameters, and large intervals require more initial nanosphere suspension. More importantly, sidewall angels influence nanosphere distribution significantly and 'evaporation angles effect' should be avoided if homogeneous distribution is expected.**

Keywords-vertical deposition; self-assembly; nanosphere; geometric characteristics

I. INTRODUCTION

Self-assembly of nanoparticles is of great significance in the realm of nano-fabrication. With the aid of nanoparticle self-assembly, we could obtain large scale of nanostructure without traditional method such as lithography, which is always of a high price. Since large area of nanostructure array or crystal is extremely useful in researches of photonic crystal, superhydrophobicity and nano-lithography, the convenient and economical way of nanoparticle self-assembly has drawn a lot of attention.

During the past few years, several approaches of nanoparticle self-assembly have been developed, such as Gravitational Sedimentation [1-4], Emulsion Crystallization [5], self-assembly under physical confinement [6-7] and Solvent evaporation combined with Vertical Deposition [8-9]. Among these methods, the first one is always extremely slow (takes several days or weeks), or we must rely on some equipment like centrifuge. And the crystallization way requires an extremely strict physical environment, which leads to great difficulty in the fabrication. As for the third way, it requires a pretty complicated equipment set. Finally, the method of Vertical Deposition (VD) proves to be an easy but efficient way for colloidal assembly and it has already been used to obtain controllable large-scale multilayer of close-packed nanospheres.

However, all these methods so far focus on one simple case, which is self-assembly of nanoparticles on a flat substrate; there have never been any research about the influence of the geometric characteristics of the substrate. Various substrate patterns provide us with a lot of probability in the fabrication of nanoarrays; with nanospheres deposited on microstructures, we'll get more complicated hierarchical structure. According to the work of several groups, hierarchical structure is able to enhance kinds of physical phenomena, such as the superhydrophobicity. Obviously, to improve the deposition result on various substrates, it is necessary to find out the relation between the geometric feature of substrate and the distribution of nanospheres.

In this paper, we utilize evaporation-induced self-assembly of nanospheres to explore this kind of relationship. The substrates are of silicon pillars with different conformation and length sizes. In experiments, via controlling significant parameters of evaporation condition, we succeeded in making nanospheres adsorbed onto the open sidewalls of such silicon pillars, and in finding out how open sidewalls influence the distribution effect of nanosphere vertical deposition.

II. DESIGN AND FABRICATION

Fig. 1 depicts the schematic diagrams of the vertical deposition fabrication process. The silicon pillar arrays are fabricated by regular lithography and reactive ion etching.

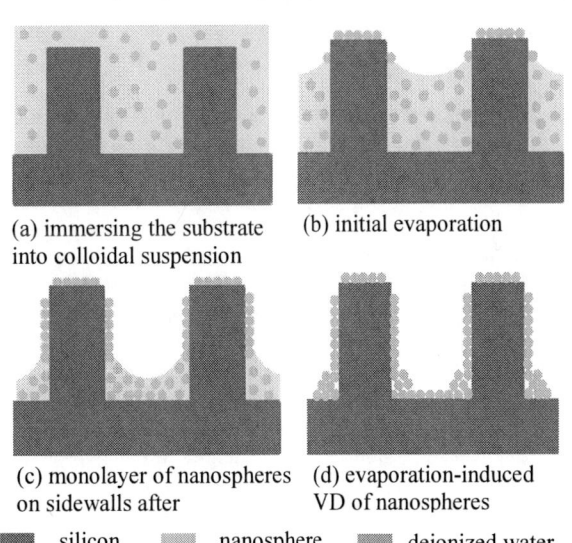

(a) immersing the substrate into colloidal suspension

(b) initial evaporation

(c) monolayer of nanospheres on sidewalls after

(d) evaporation-induced VD of nanospheres

▓ silicon   ░ nanosphere   ▒ deionized water

Fig. 1. Schematic fabrication process of vertical deposition of nanospheres on the open sidewalls of silicon pillars.

978-1-4673-1122-9/12 $31.00 © 2012 IEEE

Oxygen plasma etching is then utilized to remove the polymer resist layer and to make the pillar surfaces hydrophilic, for the purpose of facilitating the nanosphere adsorption. Afterwards we immerse the substrates into colloidal suspension with a proper total quantity of nanospheres (Fig. 1(a)). Intrinsic hydrophilic silica nanospheres and intrinsic hydrophobic polystyrene (PS) nanospheres are considered to be used, in order to observe the influence of nanosphere properties to distribution effect.

During the process of deionized water evaporation, crescentic liquid level pushes those nanospheres attached to sidewalls of silicon pillars (Fig. 1(b-c)) [10]. Meanwhile, some special chemical groups on these two kinds of nanospheres would hydrolyze to silanols, while dangling bonds on the pillar surfaces would adsorb hydroxyls in deionized water. These silanols and hydroxyls would further combine together to become stable Si-O-Si bonds or analogues. Other forces like electrostatic force, intermolecular interactions and hydrogen bonds being considered, these nanospheres would finally stick tightly onto the silicon pillar surfaces even after water having evaporated completely (Fig. 1(d)).

III. Experiments and Results

PS nanospheres in the experiments were purchased from Bangs Laboratories Inc. (Fishers, IN, USA) with a diameter of 700 nm and silica nanospheres were purchased from Microspheres-Nanospheres (Cold Spring, NY, USA) with a diameter of 800 nm. The silicon pillars we fabricated were from 5 μm high to 40 μm high. In the experiments we found that too high pillars (for example 40 μm) led to many difficulties of vertical deposition, especially on the high part sidewalls of those pillars; on the other hand, too low pillars (for example 5 μm) made crescentic liquid level flatter and resulted in insufficient surface tension to push those nanospheres onto sidewalls. In the experiments, we finally chose the pillar arrays with height of 20 μm as major research objects

Then we immersed those substrates with pillar arrays into nanosphere colloidal suspension, with a constant evaporation condition of temperature 23°C and humidity 20%. This evaporation condition could lead to a proper evaporation rate and then efficient distribution result of nanospheres. Higher evaporation rate could make more self-assembly defects because there were not enough time for nanospheres to arrange freely; however, lower evaporation rate not only impeded the fabrication efficiency but also led to coagulation of colloidal solid, so the defects became more as well.

The typical vertical deposition distribution of self-assembly nanospheres onto silicon sidewalls are depicted in Fig. 2(a-b), silica nanospheres and PS nanospheres respectively. In experiments we found that PS nanospheres required more hydrophilic pillar surfaces than silica nanospheres did, possibly because those chemical groups on silica nanospheres could hydrolyze more easily. It might also result from the

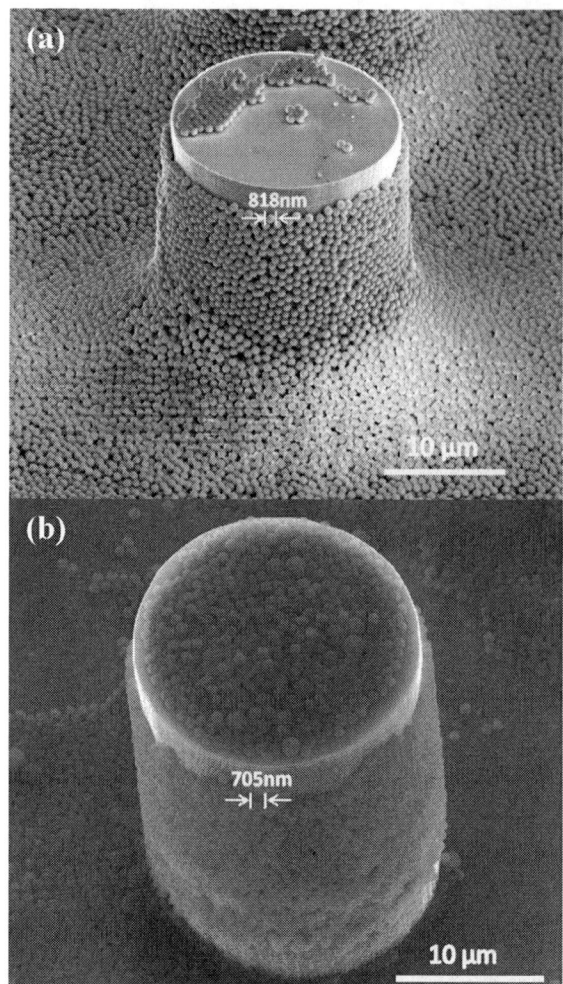

Fig. 2. SEM of close-up pictures of self-assembly distribution. (a) hydrophilic silica nanospheres. (b) hydrophobic PS nanospheres.

nanosphere intrinsic properties of hydrophobicity or hydrophilicity.

As we observed in our experiments, different from horizontal evaporation, the result of vertical sidewall evaporation had little to do with the concentration of colloidal suspension. Accurately, the denser the colloidal suspension was, the better the result was. It is because of the crescentic liquid level: only very thin liquid layer which was parallel close to silicon sidewalls contributed to nanosphere distribution, while in such thin liquid layer there were almost monolayer nanospheres even with a very high concentration. In our experiments we ignored concentration parameter but cared total quantity of nanospheres in colloidal suspension, on the premise that suspension concentration was enough.

Another significant point different from horizontal evaporation was to keep balance of evaporation dishes, especially at the end of evaporation. It was the time that nanospheres could be detached from silicon pillars easily even

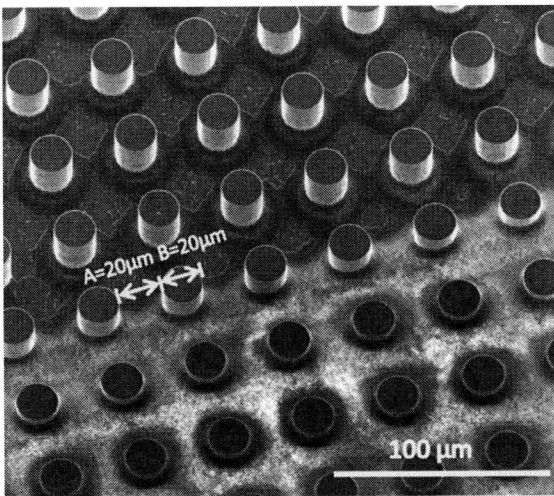

Fig. 3. Inhomogeneous distribution of nanospheres with silicon pillar arrays.

when there was a slightest vibration. Additionally, if the liquid level was not exactly horizontal, even with a tiny angel, the distribution effect might be influenced significantly. Those nanospheres might be no longer distributed homogeneously and form like the structure shown in Fig.3. We doubt that the level of nanosphere distribution indicated the level of liquid evaporation. During evaporation process, with sloping evaporation dishes, the pillar bottoms could hold on some nanospheres but the sidewalls could not. Another explanation for the structure in Fig. 3 is droplet agglomeration at the end of evaporation. The droplet agglomeration happened because of the liquid metastable state. At the end of evaporation the liquid level was so low that real liquid contact angle was in fact smaller than what it should be. In that case any tiny vibration could lead liquid to recover its contact angel, which would contribute to the nanosphere re-distribution. Therefore, nanospheres might be detached from silicon pillars and distribute like the structure shown in Fig.3.

In experiments we tried silicon cylinder arrays with different ratios of interval to diameter. Theoretical calculation indicated that larger diameters or smaller intervals required larger total quantity of nanospheres in colloidal suspension because there was more surface area to adsorb nanospheres. However, experimental results didn't always agree with that. When the intervals were the same and diameters were

different, experimental results corresponded with the theory; but when diameters were the same and intervals were smaller, the cylinder arrays actually required less total quantity of nanospheres. TABLE I. shows initial suspension quantity of PS nanospheres (not the diluted suspension) based on different silicon arrays, which led to pretty distribution effect. Fig. 4 depicts related SEM photos.

The fact contradicting to theory may be explained by the reason that intervals actually played a more important role in distribution effect. Larger interval led to weaker capillary force, so more nanospheres were necessary to increase vertical deposition probability. It may also explain the lack of uniformity of vertical deposition on cylinders with large intervals. Actually, the scale of homogeneous distribution could be only up to centimeters. It is better to utilize overdose of initial suspension quantity to increase the probability of homogeneous distribution.

In the further research, we tried noncircular pillars with

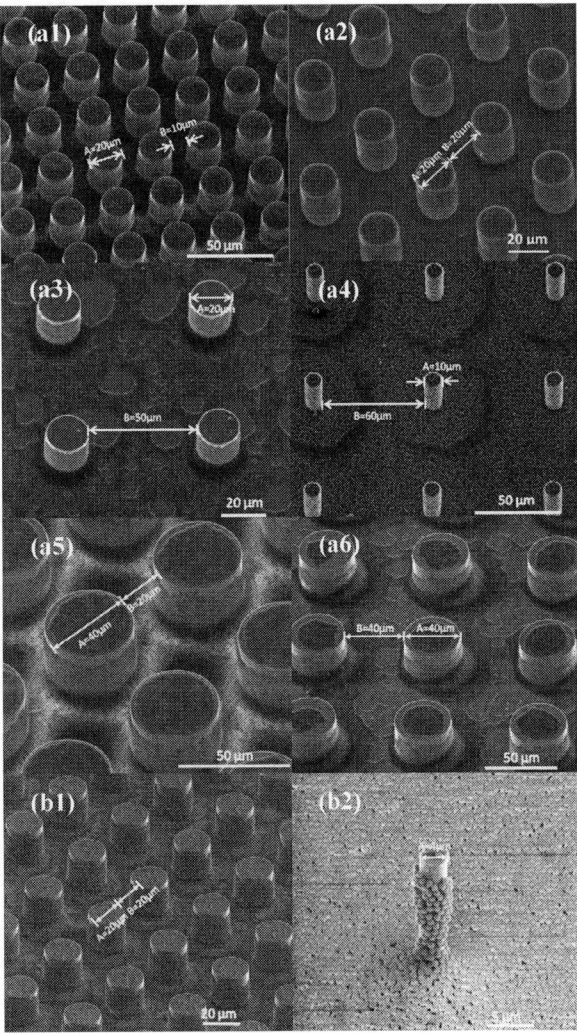

Fig. 4. SEM of self-assembly of nanospheres on different cylinder arrays. (a1)-(a6) PS nanospheres. (b1)-(b2) silica nanospheres.

TABLE I. INITIAL SUSPENSION QUANTITY OF PS NANOSPHERES BASED ON DIFFERENT SILICON ARRAYS

Diameter/μm	Interval/μm	Initial suspension quantity/μL
20	10	40
20	20	60
20	50	80
40	20	80
40	40	100
10	60	120

different morphology: circular sector, rectangle, trapezia and so on. Different from cylinder arrays, based on which we cared about the relationship between feature sizes and total quantity of colloidal suspension, we here paid attention to how sidewall angels influenced the nanosphere distribution of vertical deposition. Then we found 'evaporation angles effect' (EAE), as shown in Fig. 5. The sector pillars with 270° angle would result in nanosphere accumulation, while obtuse angles could only adsorb a few nanospheres, let alone right angles and acute angles. Actually EAE reflected the liquid distribution at the end of evaporation: 270° angles could hold on liquid while there was no liquid accumulated with acute angles, which caused particular nanosphere distribution. We can see EAE obviously from Fig. 5(b1-2), in which right angles influence the distribution of nanospheres to nearly half-oval. Therefore, if we want to get homogeneous distribution of vertical deposition, sidewall angels, even obtuse angles, should be avoided. In order to obtain homogeneous distribution on all surfaces of silicon pillars, half-ellipsoid-like pillars have been considered to be fabricated. In this case, novel micor-/nano-hierarchical structures might be achieved and new substrates for application of superhydrophobicity or optical crystal may come out.

IV. CONCLUSION

Via evaporation-induced self-assembly of nanospheres, we have succeeded in obtaining monolayer nanospheres deposited

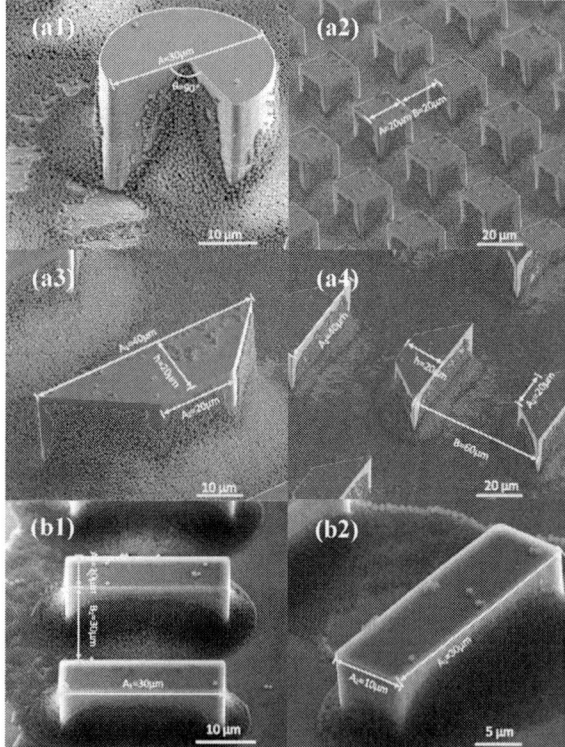

Fig. 5. SEM of EAE. (a1)-(a4) silica nanospheres. (b1)-(b2) PS nanospheres.

vertically onto sidewalls of different silicon cylinder arrays or non-cylinder pillar arrays, and we find out the relationship between the result of evaporation-induced self-assembly and such geometric factors. Besides some evaporation conditions that should be noticed, array periods of substrates and morphology of pillar sidewalls could influence the self-assembly distribution significantly as well. All above are to be an inspiration for homogeneous distribution of nanospheres, for the purpose of fabrication of micor-/nano-hierarchical structures and their multifarious future applications such as superhydrophobicity, chemical and bio-sensors and photonic crystal.

ACKNOWLEDGMENT

This work is supported by the National Basic Research Program of China (973 Program, Grant No.2009CB320300) and Peking University Presidential Research Fund.

REFERENCES

[1] A. van Blaaderen, R. Ruel, and P. Wiltzius, "Template-Directed Colloidal Crystallization," *Nature*, vol. 385, 1997, pp. 321-324.

[2] K.E. Davis, W.B. Russel, and W.J. Glantschnig, "Disorder-to-Order Transition in Settling Suspensions of Colloidal Silica-X-Ray Measurements", *Science*, vol. 245, pp. 507-510, August 1989

[3] H. Miguez, et al., "Evidence of FCC Crystallization of SiO2 Nanospheres," *Langmuir*, vol. 13, 1997, pp. 6009-6011.

[4] P.N. Pusey, and W. Vanmegen, "Phase-Behavior of Concentrated Suspensions of Nearly Hard Colloidal Spheres," *Nature*, vol. 320, pp. 340-342, March 1986.

[5] S. Dosho, et al., "Recent Study of Polymer Latex Dispersions," *Langmuir*, vol. 9, 1993, pp. 394-411.

[6] S.H. Park, and Y.N. Xia, "Assembly of Mesoscale Particles over Large Areas and its Application in Fabricating Tunable Optical Filters," *Langmuir*, vol. 15, 1999, pp. 266-273.

[7] S.H. Park, D. Qin, and Y. Xia, "Crystallization of Mesoscale Particles over Large Areas," *Advanced Materials*, vol. 10, 1998, pp. 1028-1032.

[8] S. CHANG, Self-assembly of Polystyrene Nanospheres and its Applications as Templates for Plasmonic Structures, TX: University of Texas at Arlington, 2010.

[9] Z.C. Zhou, et al., "Flow-Controlled Vertical Deposition Method for Fabrication of Photonic Crystals," *Langmuir*, vol. 20, 2004, pp. 1524-1526.

[10] M. Abasaki, et al., "Selective Self-assembly of Nanoparticles on Trench Sidewalls and its Relationship with Scallop Nanostructure," *IEEE MEMS*, Mexico, pp. 41-44, January 2011.

Nanoscale Laser Writing of Indium-Tin-Oxide Nanowires

M.Afshar[1,*], D. Feili[1], H. Voellm[1], M. Straub[2], K. Koenig[2], and H. Seidel[1]

1 :Institute of Micromechanics, Microfluidics, and Microactuators, Faculty of Physics and Mechatronics, Saarland University, Saarbrücken, Germany

2: Institute of Biophotonics and Laser Technology, Faculty of Physics and Mechatronics, Saarland University, Saarbrücken, Germany

m.afshar@lmm.uni-saarland.de

Abstract— **In this study we report on sub-wavelength nanostructuring of sputtered Indium-Tin-Oxide (ITO) films using a high-repetition rate near-infrared Ti:Sapphire laser system based on a 85 MHz, sub-10 fs resonator.**

Our experiments demonstrate that cuts as small as 20 nm in width can be generated by ablation. ITO nanowires ranging in size down to 50 nm were produced by laser writing at radiant exposure below the ablation threshold followed by etching in hydrochloric acid. The dependence of the minimum structure size on irradiation and material parameters as well as the electrical properties of the nanowires were investigated.

Keywords: sub-15 fs laser pulses, sub-100 nm structures, indium tin oxide, self-organization, surface ripples, thin films, nanowires.

I. INTRODUCTION

Because of its enormous application potential the fabrication of micro- and nanostructures by pulsed laser machining has become the subject of intense current research activities. ITO is a well known transparent conducting oxide (TCO) layer material, which facilitates transparent electrical contacts in many electro-optical devices. Thin films of this n-type wide band gap semiconductor (3.5-4.06 eV) have electrical and optical properties which strongly depend on sputter parameters and on the resulting morphology on different substrates [1-4]. Recent studies on laser structuring of ITO thin films revealed a significant reduction of the threshold fluence by shifting of laser wavelength toward the UV range [5-7]. As shown in fig. 1a, due to its high reflectivity at infrared and transparency at visible wavelengths there is almost no single-photon absorption of ITO in the spectral range of our laser (730-850 nm). However, ultrashort laser pulses induce multiphoton absorption and, hence, initiate material change [8]. Exploiting this nonlinear effect we readily obtained nanoscale structural dimensions.

II. EXPERIMENTAL

A. Sample preperation

We deposited polycrystalline ITO films with a thickness of approximately 150 nm on 170 μm thick coverslips by magnetron reactive DC-sputtering in a *Von Ardenne*

machine (type LS 730S) using a compound indium-tin (90:10) target. The deposition was optimized in an Ar/O_2 plasma regarding the crystallographic structure and the optoelectrical properties of the films, which strongly depend on their oxygen content. Controlling the oxygen to argon ratio in the gas flow (total flow 60 sccm) at a defined chamber pressure of 0.003 mbar we deposited three sorts of samples with 2 (ITO-2), 4 (ITO-4), and 6 (ITO-6) sccm oxygen fraction corresponding to 3.3, 6.6, and 10 percent oxygen flow ratio.

B. Sample characterization

The crystallinity of films was analyzed by grazing incidence X-ray diffractometry (GIXD), which showed a strong peak from (222) crystal planes (fig. 1b). The surface morphology was examined by atomic force microscopy (AFM). On increase of the oxygen flow it revealed a slight increase in

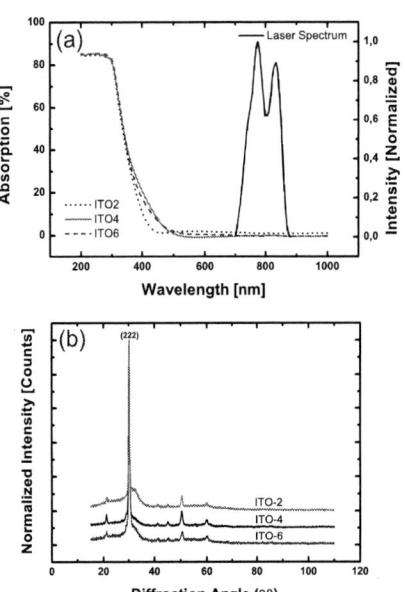

Fig. 1. (a) Absorption spectrum of ITO samples compared with the laser spectrum; (b) Normalized XRD spectra reveal a strong peak

RMS-roughness (tab. 1), whereas the XRD-spectrum remained almost unchanged (fig. 1b).

The optical characterization of ITO films was performed by transmission/reflection UV-VIS-spectroscopy (Cary 5000 spectrometer, VARIAN). Absorption curves of three samples are shown in fig. 1a in addition to the normalized laser spectrum. Varying the oxygen flow resulted in minor changes of the absorption edge, but absorption at the laser wavelengths was always negligible (fig. 1a). Thus, two photon (TPA, $\lambda \approx$ 395 nm) or even three photon absorption ($\lambda \approx$ 265 nm) prevailed.

The film conductivity was examined by four-point measurements and, as expected [9], exhibited a tendency towards higher electrical resistivity (ρ) with increasing oxygen content (tab. 1). In tin-doped indium oxide tin atoms replace the In^{3+} ions and establish bonds to the oxygen-atoms as Sn^{2+} or Sn^{4+}. The latter serve as electron donors and, in combination with oxygen vacancies, augment the conductivity [10]. By optimizing the composition electrical conductivities of the ITO films in the range of metals were obtained. The temperature coefficient of resistivity (TCR) as measured at low temperatures gave evidence of their metallic character with positive metal-like TCR values (α).

O_2(%)	R_{rms}(nm)	$\rho(10^{-4}\Omega cm)$	$\alpha(10^{-3}K^{-1})$	d (nm)
3.3	0.516	4.37	1.3	150
6.6	0.531	14	1	120
10	0.584	200	5	130

Tab. 1. Properties of three samples with different oxygen content. Flow ratios of 3.3, 6.6, and 10 percent of 60 sccm total flow correspond to ITO-2, ITO-4, and ITO-6. The surface roughness R_{rms} was obtained from an AFM measurements. The electrical resistivity (ρ) and temperature coefficient (α) result from TCR measurements. The film thickness (d) was determined by SEM images of focused ion beam sections.

C. Laser processing

Nanostructures were generated using the compact optical setup shown in fig. 2. The laser beam was launched into an inverse microscope (*Zeiss AxioObserver.D1*), which focused the beam onto the sample by a *Zeiss EC Plan Neofluar 40x NA 1.3* oil immersion objective mounted on a piezoactuator to adjust the focus vertically with nanometer accuracy. The experiments were carried out with a Ti:Sapphire laser (*Integral Pro 400, FemtoLasers*) operating in the wavelength range 730 - 850 nm (fig. 1a), which generated sub-10 fs laser pulses with a maximum pulse energy (E_p) of 7 nJ at 85 MHz repetition rate. A dispersion precompensation unit was used to chirp the laser pulses and, in combination with a glass pulse stretcher, yielded a pulse length of approximately 12 fs at the sample position. By galvanometric scan mirrors the ITO films were illuminated through the coverslip with linearly polarized light using different scan parameters. A *GSI Lumonics SC2000* scan controller allowed for manufacturing of complex

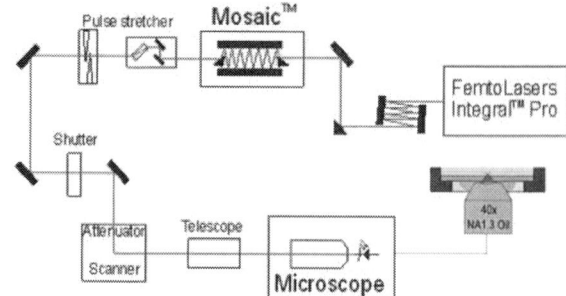

Fig. 2. Laser setup based on a sub-10 femtosecond laser mounted on a inverse microscope. The precompensation unit and pulse stretcher are between laser and attenuator, afterwards the laser beam goes through a telescope over galvano scanner in the microscope to be focused on the sample.

structures with a positioning accuracy of 9 nm over a scan range of 550 µm. A mechanical shutter controlled the exposure of the samples with millisecond response time. The entire scan procedure was controlled by a home-made LabView program.

High-repetition rate near-infrared ultrafast lasers provide a high ablation efficiency compared to excimer lasers or higher harmonics of Nd:YAG, Nd:YLF lasers at a minimum of collateral sample or substrate damage [11-13]. In our experiments we studied the influence of the conductivity of the layers, which is based on a combination of carrier concentration and carrier mobility, on nanostructure formation by laser machining.

D. Wet etching

Subsequent to laser processing, the degree of crystallization of illuminated regions was examined by room temperature etching in 37% hydrochloric acid (HCl), which is the standard etchant of ITO [14]. It appeared that the laser-modified regions were more resistive against HCl and persisted, whereas pristine film areas were removed. This phenomenon was exploited to generate nanostructures from ITO by direct laser writing in combination with an etching step.

III. RESULTS AND DISCUSSION

The average focal power varied between 13.5 and 25.5 mW, and at each intensity the scan speed ranged from 11.5 to 185 µm/s. Thus, a broad range of exposure conditions was available for laser processing. Scanning electron microscope (SEM) images showed three different regimes: total ablation of grooves, periodic self-assembly of ablated cuts, and structural modification (fig. 3). In figure 3b, three different morphologies can be identified. Region 1 corresponds to the pristine ITO surface. Region 2 corresponds to a modified region about the most intensely illuminated region 3, which consists of periodic or totally ablated structures depending on the exposure as well as on material parameters.

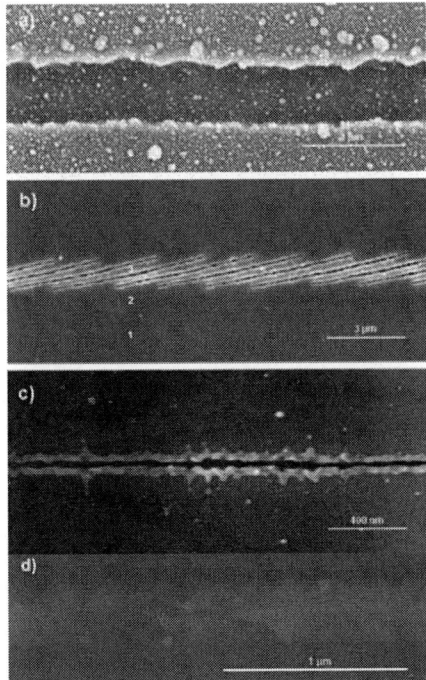

Fig. 3. SEM images of (a) a groove with 500 nm width; (b) nano-ripples perpendicular to the polarization direction of laser beam; (c) a single cut with lateral dimension of 20 nm in width; (d) a modified zone in ITO film by a laser line scan.

In the ablation regime multiphoton excitation of electrons seeds avalanche processes which generate an electron-hole plasma that ultimately heats up the ITO lattice beyond its melting point. The average width of cuts reduces from 500 nm down to 200 nm on reduction of the radiant exposure using faster scan speeds, while the cut width varies periodically by 10-20%. At near-threshold exposure periodic cuts were generated at sub-20 nm lateral dimensions (fig. 3b). Rotation of the polarization showed that cuts were oriented perpendicular to the laser polarization. Single cuts were observed on reduction of the irradiation to values extremely close to the ablation threshold (fig. 3c) with the polarization perpendicular to the scan direction. However, their length was limited to a few micrometers.

Another aspect of our study deals with the dependence of the ablation threshold on the resistivity of the ITO films. As already mentioned, our layers were almost identical in their crystal structure, but their resistivity varied with oxygen content. The conductivity of ITO depends on the carrier density and mobility [10]. The grain size has a strong influence on the mobility due to the density of traps and scattering of electrons at grain boundaries. In ITO-2 total ablation at the slowest scan speed required an average power of 25.5 mW, whereas 22 mW in ITO-4 and 18 mW in ITO-6 were sufficient. An increase of the electrical resistivity of

about 0.02 Ωcm from ITO-4 to ITO-6 yielded a decrease of 30 mJ/cm^2 in the threshold exposure.

Below the ablation threshold SEM images of illuminated traces revealed modified zones of different texture (fig. 3d). In these locations laser illumination yielded a more resistive phase in ITO (fig. 4a). The modification was largely of crystallographic nature as illustrated in fig. 3b. On etching with HCl we observed that the laser-annealed regions of the polycrystalline film were more resistive against etching. The annealing effect may be attributed to heat accumulation caused by the short time of merely 12 ns between laser pulses, which prevents heat from diffusing out of the focal region completely [15]. Moreover, the correlation between electrical and thermal conductivity (Wiedemann-Franz law [16]) may explain the higher ablation threshold of samples of lower resistivity.

The size of the modified region can reach very small dimensions by optimizing the effective absorption volume via the focal intensity. In figure 4 we present representative structures, which were generated by sub-ablation threshold laser writing and subsequent etching. The logo of the conference ("NEMS IN JAPAN") was written with lines of 500 nm in width (fig. 4a). Repeating the scan at the same speed but at a slightly reduced focal power resulted in a line width of 250 nm (fig. 4b). Figure 4c shows the logo of our group ("LMM") written under the same conditions as fig. 4a. Figure 4d illustrates a meander ITO structure of an average lateral size of 170 nm with a magnified section shown in fig. 4e. The highlight of structures produced by our method is an ITO nanowire of as little as 50 nm in width (fig. 4f).

The crystallographic analysis of annealed ITO compared to the original layer shows an improvement of crystallinity corresponding to a change in diffraction patterns. We also

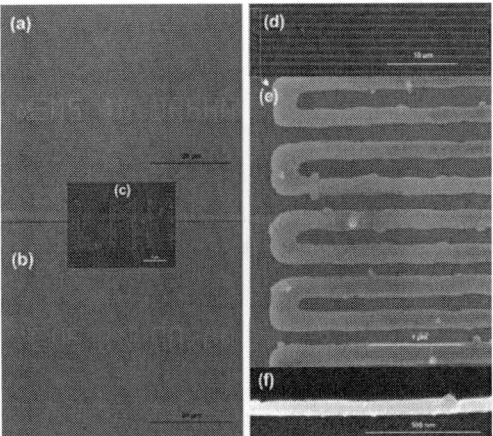

Fig. 4. SEM images of (a) "NEMS IN JAPAN" with 500 nm width and (b) 250 nm width; (c) logo of our group in 500 nm thick lines; (d) a meander structure of ITO width 170 nm average width and 30 µm length; (e) the same meander with higher magnification; and (f) a single nanowire with lateral dimension of 50 nm in width.

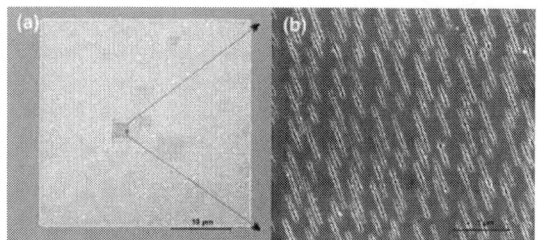

Fig. 5. (a) SEM image of ITO layer with laser-induced periodic nanocuts; (b) magnified section. The cuts of sub-20 nm lateral dimension are perpendicular to the laser polarization all over the irradiated square of 35 µm length.

performed first experiments to characterize the electrical properties of such nanowires. Our measurement chips were designed for four-point measurements of single nanowires with contacts. An increase of approximately 10% in conductivity of ITO was observed after laser processing.

Other representative structures are with nanocuts decorated ITO thin films such as those shown in fig. 3b over a larger area (fig. 5). These periodic cuts penetrate the whole layer and regarding their dimensions may be used as filters for nanoparticles or components of microoptical devices. A change in cuts density can be achieved by different scan parameters.

The laser-induced ITO nanostructures have a great potential to be applied in various optoelectrical elements. The piezoresistive character of ITO in addition to its conductivity and relatively high Young's modulus may open a new field in nanoscale processing of electromechanical components. We are planning to realize demonstrators for such applications in the near future.

ACKNOWLEDGMENT

We want to acknowledge German Research Foundation (DFG) for the financial support of this project within Priority Program 1327 "Sub-100 nm structures for biomedical and technical applications". We are also grateful to Prof. K. Jacobs (Saarland University, Soft Matter Research Group) and Dr. K. Jilavi (Leibniz Institute for New Materials) for their encouragement.

REFERENCES

[1] D.C. Paine, H.-Y. Yeom, and B. Yaglioglu, "Transparent Conducting Oxide Materials and Technology, in Flexible Flat Panel Displays," (ed G. P. Crawford), John Wiley & Sons, Ltd, Chichester, UK (2005).

[2] I. Hamberg, C. Granqvist, "Evaporated Sn-doped In_2O_3 films: Basic optical properties and applications to energy-efficient windows," J. Appl. Phys. 60, (11) 123 (1986).

[3] S. Ray, R. Banerjee, N. Basu, A. K. Batabyal, A. k. Barua, "Properties of tin doped indium oxide thin films prepared by magnetron sputtering," J. Appl. Phys. 54, 3497 (1983).

[4] H. Hartnagel, A. Dawar, A. Jain, C. Jagadish, "Semiconducting transparent thin films," Institute of physics Publishing 1995.

[5] O. Yavas, M. Takai, "Effect of substrate absorption on the efficiency of laser patterning of indium tin oxide thin films ," J. Appl. Phys. 85, 4207 (1999).

[6] J. G. Lunny, R.R O'neil, K. Schulmeister, "Excimer laser etching of transparent conducting oxides," Appl. Phys. Lett. 59, 647 (1991).

[7] M.Y. Xu, J. Li, L.D. Lilge, P.R. Herman, "F2-laser patterning of indium tin oxide (ITO) thin film on glass substrate," Appl. Phys. A 85, 7 (2006).

[8] K. Koenig, "Femtosecond laser nanoprocessing," In: P. So and B.R. Masters (eds.): Handbook of Biological Nonlinear Optical Microscopy, Oxford University Press, 689 (2008).

[9] T. Ashida, A. Miyamura, N. Oka, Y. Sato, T. Yagi, N. Taketoshi, T. Baba, and Y. Shigesato, "Thermal transport properties of polycrystalline tin-doped indiumoxide films," J. Appl. Phys. 105, 073709 (2009).

[10] H. Kim, C. M. Gilmore, A. Pique´, J.S. Horwitz, H. Mattoussi, H. Murata, Z.H. Kafafi,and D.B. Chrisey, "Electrical, optical, and structural properties of indium–tin–oxide thin films for organic light-emitting devices," J. Appl. Phys. 86, 11 (1999).

[11] M.-F. Chen, Y.-P. Chen, W.-T. Hsiao, and Z.-P. Gu, "Laser direct writie patterning technique of indium tin oxide film," Thin solid Films 515, 8515 (2007).

[12] R. Tanaka, T. Tokaoka, H. Mizukami, T. Arai, Y. Iwai, "Laser etching of indium tin oxide thin films by ultra-short pulsed laser," SPIE Proceedings 5063, 370 (2003).

[13] M. Park, B.H. Chon, H.S. Kim, S.C. Jeoung, D. Kim, J. Lee, H.Y. Chu, H.R. Kim, "Ultrafast laser ablation of indium tin oxide thin films for organic light-emitting diode application," Optics and Lasers in Engineering 44, 138 (2006).

[14] C. J. Huang, Y. K. Su, and S. L. Wu, "The effect of solvent on the etching of ITO electrodes," Materials Chemisttry and Physics 84, 146-150 (2004).

[15] C. B. Schaffer, J. F. Garcia, and E. Mazur, "Bulk heating of transparent materials using a high-repetition-rate femtosecond laser," Appl. Phys. A 76, 351-354 (2003).

[16] G. Wiedemann, and R. Franz, " Ueber die Wrme- Leitungsfhigkeit der Metalle," Annalen der Physik 165, (8) 497 (1853).

Formation and Interaction of Micro/Nano Multi-scale Structures

Xiao-Sheng Zhang, Fu-Yun Zhu, Hai-Xia Zhang[*]

[*] Science and Technology on Micro/Nano Fabrication Lab, Institute of Microelectronics, Peking University, Beijing, 100871, CHINA

zhang-alice@pku.edu.cn

Abstract—**This paper presents the fabrication of squama-shape micro/nano multi-scale structures and the analysis of the interaction among different scale structures during fabrication processes. Well-designed microstructures made of inverted pyramids and V-shape grooves are fabricated by KOH wet etching. High-dense high-aspect-ratio nanostructures are realized atop microstructures by an improved maskless deep reactive ion etching (DRIE) process to form micro/nano dual-scale structures (MNDS). Due to the impact of the profile of microstructure on the shape of nanostructure, dissymmetrical (i.e. squama-shape) nanopillars have been formed atop inclined surfaces of microstructures, while symmetrical nanopillars are formed atop horizontal surfaces with different formation velocity. Furthermore, the optical properties of MNDS are not sensitive to structural parameters of microstructures, which makes it more tolerant to process errors and overcomes the lithography limitation of conventional processes for photo-devices. Additionally, three-level structures are fabricated by sputtering gold particles on MNDS, and the profile of MNDS is selective in the deposition of gold.**

Keywords- DRIE; multi-scale structures; squama-shape nanopillars

I. INTRODUCTION

Micro/nano dual-scale structures (MNDS) show attractive properties, such as anti-reflectance [1,2] and superhydrophobicity [3,4]. The conventional processes to fabricate nanostructures (NS) atop microstructures (MS), such as reactive ion etching [5-7], metal-catalyzed chemical etching [8], nanosphere/nanoimprinting lithography [9-12] and femtosecond laser [13], are uncontrollable, low-production-speed or low-aspect-ratio. Notably, some of them need masks and contain several steps. In order to overcome these drawbacks, we present an improved deep reactive ion etching (DRIE) process, which is maskless, single-step, controllable and large-size [14]. By optimizing the process parameters of DRIE, the polymer deposited during passivation step cannot be removed completely. Therefore, the residual nanoparticles serve as self-masks to protect the substrate during the next etching step, which results in high-dense high-aspect-ration nanopillars. Fig. 1(a) shows the principle of formation of nanostructures atop microstructures.

This improved DRIE process is suitable for realize MNDS resulting from maskless and simple feature. However, due to the combination of multi-scale structures, more factors have to

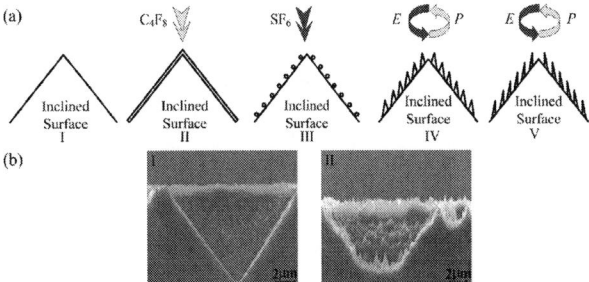

Fig. 1. Schematic illustration and SEM images of producing nanostructures atop microstructures.

be considered for the fabrication procedure, and the chemical-physical energy transfer becomes more complicated. In one word, the combination of multi-scale structures brings some new challenges and phenomena.

In order to investigate the interaction of different scale structures, the microstructures are specifically designed to inverted pyramids and V-shape grooves for photo-device application. We further study the formation of nanostructures atop different microstructures (e.g., horizontal and inclined surface) by this improved DRIE process, and the relationship between the optical property of MNDS structure and its morphology. High-compact dissymmetrical (i.e. squama-shape) and symmetrical nanostructures have been realized atop inclined and horizontal surfaces of microstructures, respectively. Additionally, the formation velocities of nanostructures atop inclined and horizontal surfaces are different. These new observed fabrication and morphology properties may extend the application of micro/nano dual-scale structures and induce some new devices.

II. FABRICATION

An N-type (100) silicon substrate was used. Firstly, a Si_3N_4 layer of ~2000Å was deposited on the silicon substrate by low pressure chemical vapor deposition process. Sequentially, by combining photolithography and reactive ion etching (RIE) process, some windows were defined and opened. After KOH etching and removal of Si_3N_4 layer by RIE, microstructures were formed. And then, high-dense high-aspect-ratio nano-pillars were formed atop microstructures using the improved

DRIE process. The inductively coupled plasma (ICP) etcher (Surfacing Technology Systems plc, Multiplex ICP 48443) was used. The optimized recipe is shown in Table I.

The schematic process flow of MNDS is shown in Fig. 2. After fabrication of MNDS, a thin gold membrane of ~500Å was sputtered on MNDS, which realized micro/nano multi-scale structures (i.e. MNDS with gold membrane). The fabrication results are shown in Fig. 3.

TABLE I.   THE OPTIMIZED RECIPE OF DRIE PROCESS

Process Parameters	Gas Flow		Platen Power	Time Ratio of Etching to Passivation (E/P)
Value	SF₆	C₄F₈	10.5 W	6s : 6s
	30 sccm	50 sccm		

Process Parameters	RF Power	Total Cycle of E/P
Value	825 W	80

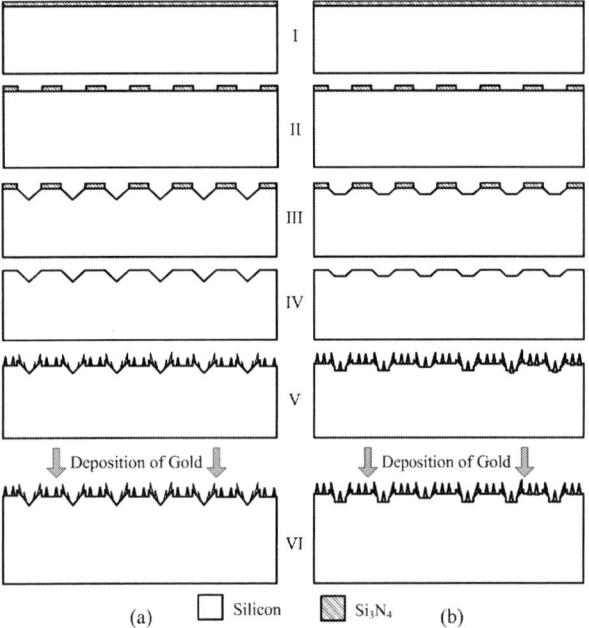

(a)   □ Silicon   ▨ Si₃N₄   (b)

Fig. 2. Schematic fabrication processes of micro/nano dual-scale structures. (I) Depositions of ~2000Å Si₃N₄; (II) Photolithography and RIE to form pattern windows; (III) KOH-etching to form microstructures with different cross-section, triangle (a) and trapezoid (b) respectively; (IV) Removal of the Si₃N₄ layer by RIE; (V) Improved DRIE to form high-compact high-aspect-ratio nanotips atop the microstructures; (VI) Deposition of gold.

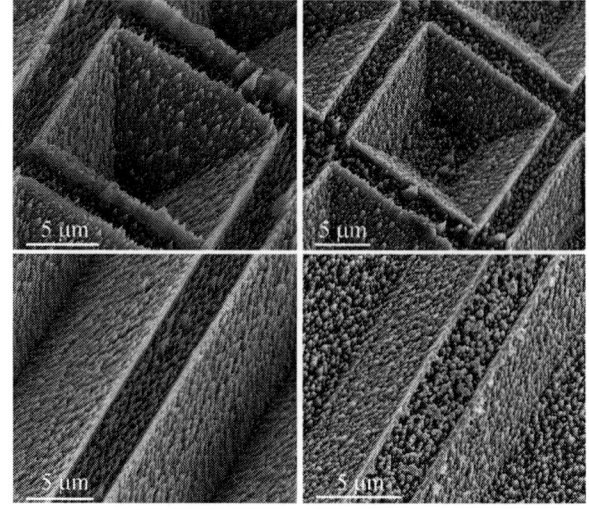

Fig. 3. The SEM images of micro/nano dual-scale structures with gold membrane.

III.   RESULTS AND DISCUSSION

A.   Formation of Dual-scale Structures

By investigating the fabrication results, we find that the profile of microstructure will affect the shape of nanostructure. The nanostructures formed on a horizontal microstructure surface are symmetrical, while those formed on an inclined microstructure surface are dissymmetrical (i.e. squama-shape), which is schematically illustrated in Fig. 1 and Fig. 4.

The energy flux vector field of the ion beam $\vec{F}$ with the direction normal to the horizontal direction can be decomposed into two orthogonal components, i.e., $\vec{F_1}$ and $\vec{F_2}$, which are parallel and normal to the inclined surface, respectively (Fig. 4 (a)).

$$\vec{F_1} = \vec{F} \times \sin \theta \qquad (1)$$

$$\vec{F_2} = \vec{F} \times \cos \theta \qquad (2)$$

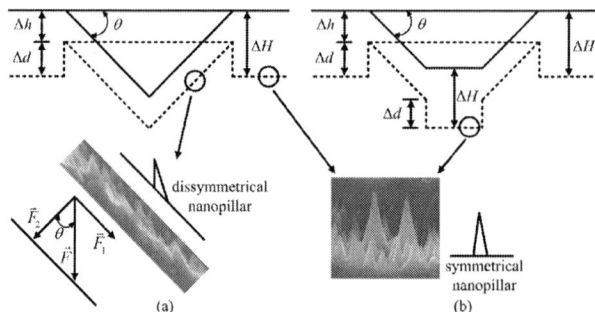

Fig. 4. Illustration of micro/nano dual-scale structures formation (solid line: the profile of microstructure; dash line: the profile of micro/nano dual-scale structure): (a) triangle cross-section (SEM of fig. 3a and fig. 3c); (b) trapezoid cross-section (SEM of fig. 3b and fig. 3d).

Therefore, because of the parallel component $\vec{F_l}$, the nanostructure is etched faster on one side than on the other side, resulting in a dissymmetrical profile. Moreover, the incline angle of microstructure affects the etching velocity of DRIE process. In Fig. 4, the incline angle θ is approximately constant during the formation of MNDS structure, but the depth change of the surfaces in the horizontal direction (ΔH) is different from that in the inclined direction (Δh). In fact, there is a relation between ΔH and Δh:

$$\Delta H\ (=\Delta h + \Delta d) > \Delta h \tag{3}$$

By defining the corresponding average etching velocity $V_h = \Delta H / t$ and $V_g = \Delta h / t$, we can conclude the relation of $V_h > V_g$ with the same etching time. Therefore, the relation between V_h and V_g can be determined when the incline angle θ is fixed, i.e.,

$$V_g = f(V_h,\ \theta) \tag{4}$$

B. Formation of Multi-scale Structures

Because of altitude difference (i.e., $\Delta H > \Delta h$) and the effect of electron accumulation, the profile of MNDS is selective in deposition of gold membrane. The accumulated electron forms an opposite electrical field which decreases the gold membrane deposition at the bottom of MNDS. According to SEM images of micro/nano multi-scale structures shown in Fig. 5, gold particles can only be sputtered on inclined surfaces. Meanwhile, there are no gold particles on horizontal surfaces and at the bottom of microstructures (i.e., A and B zones in Fig. 5).

However, more experiments and investigations need to be done to verify this analysis. But, this unique property is very useful when we fabricate micro/nano fluidic surface-enhanced Raman scattering (SERS) sensor [15]. MNDS with high-dense high-aspect-ratio will be an outstanding potential material for SERS sensor. Meanwhile, the deposition of noble metal, which can rapidly enhance the sensitivity of SERS sensor, will be controllable on MNDS.

C. Optical Property of Dual-scale Structures

The optical properties of pure microstructures, which only consist of inverted pyramids or V-shape grooves, are sensitive to their structural parameters [16,17]. Apparently, the denser the pure microstructures are, the more efficient the light trap is. In other words, the reflectance of pure microstructures decreases as the surface area in the horizontal direction decreases. However, the conventional fabrication process (i.e. lithography) limits the minimal space of patterns. MNDS structures can overcome the drawback above due to combining wideband antireflective nanostructures with microstructures. In order to facilitate the description, the ratio of surface area in the horizontal direction to the total surface area of a sample is defined as RSA. By assuming that there are n inverted pyramids/grooves on the whole sample, and that the area of a single inverted pyramid or groove is S_g (as indicated in Fig. 3), we can obtain:

$$RSA = (S_{total} - n * S_g)/S_{total} \tag{5}$$

here S_{total} is 2 cm x 2cm.

Optical reflectances of MNDS were then measured by a UV-3600 spectrophotometer (SHIMADZU Corporation). According to Fig. 6, the hemispherical reflectance of microstructures increases from 13.2% to 24% as the RSA increases from 41.6% to 61.7%. In contrast, the variation of hemispherical reflectance of MNDS structures is limited to less than 2%, although the RSA changes during a wide range from 13.4% to 79.9%. Therefore, the optical property of MNDS structure is not sensitive to structural parameters, which makes it more tolerant to process errors and solves the limitation of lithography.

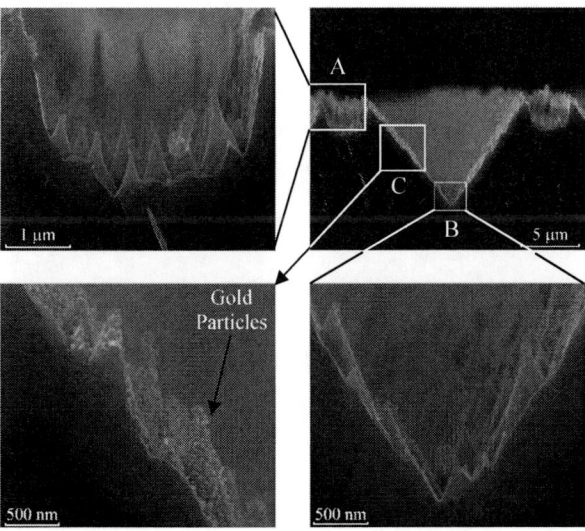

Fig. 5. The SEM images of micro/nano multi-scale structures (i.e., MNDS with gold particles).

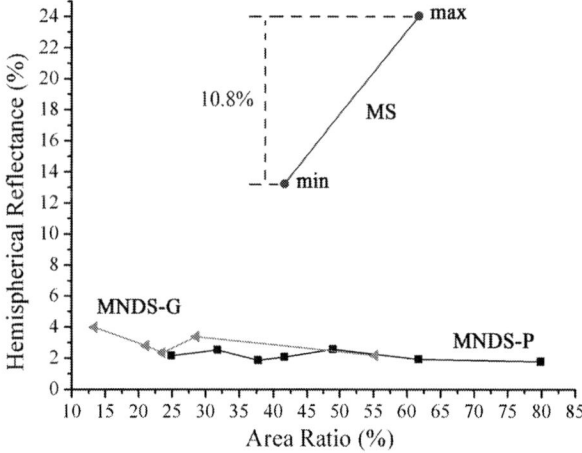

Fig. 6. Relation between hemispherical reflectance and surface area ratio (RSA) of horizontal surface to the total surface under an incidence wavelength of 700 nm. MS: microstructure; MNDS-P: MNDS inverted pyramids; MNDS-G: MNDS grooves.

IV. CONCLUSION

The fabrication of micro/nano multi-scale structures mainly contains three steps, (i.e. KOH wet etching, improved DRIE process and gold sputtering), which realize microstructures, nanostructures and sub-nano structures, respectively. The new phenomena were present, and the interaction of different scale structures was investigated. The effect of microstructure's profile to nanostructure's formation results in squama-shape micro/nano dual-scale structures. Moreover, the observed relationship between the morphology and formation velocity of nanostructures (i.e. symmetrical or dissymmetrical nanotips) and the profile of microstructures (i.e. horizontal or inclined surfaces) may extend the application of micro/nano dual-scale structures and induce some new devices. Due to the accumulated electron, the MNDS show selectivity to the gold deposition. Meanwhile, the optical property of MNDS structure is not sensitive to structural parameters, which makes it more tolerant to process errors and solves the limitation of lithography.

ACKNOWLEDGMENT

The authors would like to thank the State Key Laboratory on Integrated Optoelectronics at Jilin University for their cooperation on the measurements of optical properties. This work is supported by the National Natural Science Foundation of China (Grand No. 91023045 and No. 61176103), and Key Laboratory Fund (No. 9140C790103110C7903).

REFERENCES

[1] J. Yoo, G. Yu, and J. Yi, "Large-area multicrystalline silicon solar cell fabrication using reactive ion etching (RIE)," *Sol. Energy Mater. Sol. Cells*, vol. 95, pp.2-6, 2011.

[2] X.S. Zhang, Q.L. Di, F.Y. Zhu, G.Y. Sun, and H.X. Zhang, "Wideband anti-reflective micro/nano dual-scale structures: fabrication and optical properties," *Micro & Nano Letters*, vol. 6, pp.947-950, 2011.

[3] C. Lee and C.J. Kim, "Maximizing the giant liquid slip on superhydrophobic microstructures by nanostructuring their sidewalls," *Langmuir*, vol. 25, pp.12812-12818, 2009.

[4] X.S. Zhang, Q.L. Di, F.Y. Zhu, G.Y. Sun, and H.X. Zhang, "Superhydrophobic micro/nano dual-Scale structures," *Journal of Nanoscience and Nanotechnology*, in press.

[5] H. Jansen, M. de Boer, R. Legtenberg, and M. Elwenspoek, "The black silicon method: a universal method for determining the parameter setting of a fluorine-based reactive ion etcher in deep silicon trench etching with profile control," *J. Micromech. Microeng.*, vol. 5, pp.115-120, 1995.

[6] Y. Inomata, K. Fukui, and K. Shirasawa, "Surface texturing of large area multicrystalline silicon solar cells using reactive ion etching method," *Sol. Energy Mater. Sol. Cells*, vol. 48, pp.237-242, 1997.

[7] D. S. Ruby, S. H. Zaidi, S. Narayanan, B. M. Damiani, and A. Rohatgi, "Rie-texturing of multicrystalline silicon solar cells," *Sol. Energy Mater. Sol. Cells*, vol. 74, pp.133-137, 2002.

[8] X. Li, B. K. Tay, P. Miele, A. Brioude, and D. Cornu, "Fabrication of silicon pyramid/nanowire binary structure with superhydrophobicity," *Appl. Surf. Sci.*, vol. 255, pp.7147-7152, 2009.

[9] I. Maximov, E.L. Sarwe, M. Beck, K. Deppert, M. Graczyk, M.H. Magnusson, and L. Montelius, "Fabrication of Si-based nanoimprint stamps with sub-20 nm features," *Microelectron. Eng.*, vol. 61-62, pp.449-454, 2002.

[10] J. Zhu, Z. Yu, G.F. Burkhard, C.M. Hsu, S.T. Connor, Y. Xu, Q. Wang, M. McGehee, S. Fan, and Y. Cui, "Optical absorption enhancement in amorphous silicon nanowire and nanocone arrays," *Nano Lett.*, vol. 9, pp. 279-282, 2009.

[11] L.J. Guo, "Nanoimprint lithography: methods and material requirements," *Adv. Mater.*, vol. 19, pp. 495-513, 2007.

[12] S. Krishnamoorthy, Y. Gerbig, C. Hibert, R. Pugin, C. Hinderling, J. Brugger, and H. Heinzelmann, "Tunable, high aspect ratio pillars on diverse substrates using copolymer micelle lithography: an interesting platform for applications," *Nanotechnology*, vol. 19, pp.285301, 2008.

[13] T.H. Her, R.J. Finlay, C. Wu, S. Deliwala, and E. Mazur, "Microstructuring of silicon with femtosecond laser pulses," *Appl. Phys. Lett.*, vol. 73, pp.1673-1675, 1998.

[14] G.Y. Sun, T.L. Gao, X. Zhao, and H.X. Zhang, "Fabrication of micro/nano dual-scale structures by improved deep reactive ion etching," *J. Micromech. Microeng.*, vol. 20, pp.075028, 2010.

[15] H.Y. Mao, P.P. Lv, and W.G. Wu, "Microfluidic surface-enhanced raman scattering sensors for online monitoring trace chemical mixing and reaction," *Transducers'11*, Beijing, June 5-9, 2011, pp. 1970-1973.

[16] P. Campbell and M. A. Green, "Light trapping properties of pyramidally textured surfaces," *J. Appl. Phys.*, vol. 62, pp.243, 1987.

[17] P. Campbell and M.A. Green, "High performance light trapping textures for monocrystalline silicon solar cells," *Sol. Energy Mater. Sol. Cells*, vol. 65, pp.369-375, 2001.

A High Efficient POM Micro-Methanol Reformer

Hsueh-Sheng Wang[1], Kuo-Yang Huang[1], Hsien-Chih Peng[1], Yuh-Jeen Huang[2] and <u>Fan-Gang Tseng</u>[1*]

[1] Department of Engineering and System Science, National Tsing Hua University, Hsinchu 300, Taiwan, ROC

[2]Department of Biomedical Engineering and Environmental Sciences, National Tsing Hua University, Hsinchu 300, Taiwan, ROC

E-mail: fangang@ess.nthu.edu.tw

Abstract—In the present study, a novel micro-channel methanol reformer with a finger-shaped groove structure was successfully demonstrated to enhance the methanol conversion rate and the hydrogen yield. By introducing a centrifugal technique, a porous and gradient distribution of the catalyst layer thickness can be obtained inside the micro-channels so as to force the methanol steam to react sufficiently with high surface area catalysts. As the ratio of binder to catalysts varied from 60 to 0, the methanol conversion rate, hydrogen selectivity and hydrogen yield of the micro-methanol reformer at 250°C can approach ~100%, 92% and 1.56×10^{-5} mole min^{-1}, respectively. Moreover, a high performance output can still be obtained even at 200°C, which is superior to our previous studies .

Keywords: *PEMFC*; *Hydrogen*; *micro-methanol reformer*; *centrifugal process.*

I. INTRODUCTION

In recent years, the global warming and energy crisis have threatened the existence of creatures around the world. For this reason, it's necessary to develop green, clean and renewable energies to solve such big issues. In varieties of methods to create green energies, fuel cells have emerged strongly as alternative power sources owing to their high overall system efficiency and eco-friendly nature, furthermore, one of the promising and attractive applications is its use as portable making them striking alternatives to conversional lithium ion batteries. There are mainly two types of small fuel cells can be used: direct methanol fuel cell (DMFC) and proton exchange membrane fuel cell (PEMFC). DMFC can be operated at lower temperature as its advantages but supplies low power density in virtue of methanol crossover through the membrane and the low reaction rate of methanol oxidation at anode. On the contrary, PEMFCs have higher power density because of few phenomenon of fuel crossover at membrane area and it requires gaseous H_2 as a fuel. Hence, a system capable of providing a stable and economical supply of pure H_2 gas is a prerequisite before any other challenges in the development of PEMFCs. Conventionally, hydrogen generated by electrolyzing water or coal gasification is believed to be a feasible solution for vehicle applications (>50kilowatt), but it is not suitable for portable system (<100watt) [1]. Recently, it's reported that micro-PEMFC may cover the application range with the power output form 1watt (W) to 50 watt (W) under a considerable small size system through using MEMs technique and equipping with a micro-channel methanol reformer to generate hydrogen [2]. Generally the micro methanol reformer produces hydrogen via catalytic action of the Cu/Mn/Zn catalyst coated in the micro-channel of reformer, so far there are several reactions that can transform steam methanol into hydrogen, such as steam reforming of methanol (SRM), deposition of methanol (DM), oxidative steam reforming of methanol (OSRM) and partial oxidant of methanol (POM), etc., and in this article we adopted the POM reaction thanks to its characteristic of quick start-up and exothermic process.

The POM reaction of methanol over the Cu/Mn/Zn catalyst can be expressed as the following reactions [3-5]:

$$CH_3OH\ (g) + 0.5O_2 \rightarrow CO2 + 2H_2, \Delta H = -192\ kJ/mol \qquad (1)$$

$$CO + 0.5O_2 \rightarrow CO_2, \Delta H = -283\ kJ/mol \qquad (2)$$

$$CO + H_2O\ (g) \rightarrow CO_2 + H_2, \Delta H = -41\ kJ/mol \qquad (3)$$

Eq. (1) shows POM is a highly exothermic reaction, which can which can be used to construct highly dynamic and fast reforming systems; the preferential oxidation reaction(PROX) (Eq. (2)) and the water gas shift reaction (Eq.(3)) are clean-up steps to reduce the CO concentration in the productive gas. However, the influence on the performance of micro-methanol reformer includes not only the efficiency of catalyst but the structure, minimization and channel design of micro channel. Casio presented that the utilizing of MEMs technology can integrate micro channel on the silicon wafer to minimize the volume of micro reformer and concluded an appropriate length of micro channel can effectively improve the methanol conversion and precisely control the variation of temperature and pressure [6]. S. S. Hsieh et al. found out the mesh-shaped have the smallest variation of pressure in the channel and interdigitated channel obtained the best performance in suitable operating temperature among the different kinds of channel design involving interdigitated, mesh-shaped and serpentine-shaped channel [7]. Hung-Ming Yan et al. used Cu/ZnO-CNTs with size of Cu particles ranging from 8 nm to 10 nm as the catalyst for methanol reforming reaction after preprocess of nitric acid and sulfuric acid, and considered that CNTs as catalyst supporter can offer high reactive area so that it can modify the methanol conversion and experimental result showed that the 83% and 100% of methanol conversion could be obtained at operating temperature 280 ℃ and 320 ℃ , respectively [8].

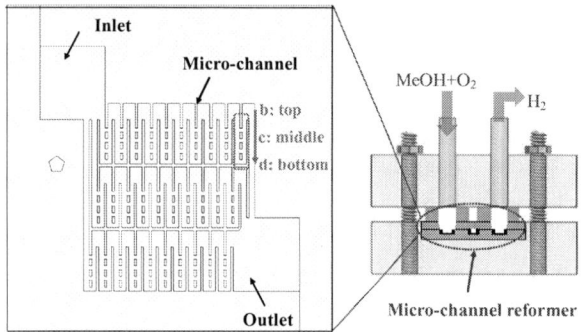

Fig 1. The schematic diagram of a micro-channel methanol reformer. The b, c and d indicate the direction of the micro-channels from top to bottom.

The main objective of this study was the development of new methods of coating catalyst into micro channels, and tried to decrease the ratio of binder in the water-based suspension of catalyst particles to improve the performance of micro-channel methanol reformer by increasing the catalyst content in micro channels.

II. EXPERIMENTAL METHOD

II.1 Preparation of micro channel

The fabrication step is shown in Fig. 1. Finger-shaped grooves were first fabricated on silicon (100) wafer by photolithography and deep silicon reactive ion etching (DRIE) process. Then, hydrophilic groove-walls were obtained by oxygen plasma manner. Pyrex glasses with inlet/outlet holes were machined by laser dripping process and then bonded on the chips to obtain micro-channels by using anodic bonding process.

II.2 Catalyst preparation with different ratio of binder to catalyst

We prepared different catalyst solutions containing different binder concentrations for the experiment. First, 10ml DI water was mixed with a constant content of catalyst and a specific amount of binder, and well mingled catalyst solutions were then mixed by stirring and sonicating for one hour; the mixed solutions were injected into the fabricated micro channel, and then put in a centrifugal apparatus for packing self-assembled catalysts into the micro channels. The solvents in the catalyst slurries were dried out at $105^{\circ}C$ for 30 min. The catalyst preparation was repeated ten times by the above-mentioned steps and then finally obtained after annealing at $400^{\circ}C$ for 2 hr.

II.3 The analytic and testing system of micro-channel methanol reformer

The schematic diagram of whole testing system has been set up successfully. At the beginning, argon, oxygen and steam methanol were well mingled in the mixing chamber as reactive gaseous in the ratio of 81.7 (ml/min), 6.1(ml/min) and 12.2(ml/min), respectively. Then, the fabricated micro-channel methanol reformer was operated on a stainless stage and put on the hotplate to be heated to the required temperature for testing. Afterward, reactive gaseous were sent in the inlet hole of micro-channel methanol reformer in the flow rate of 2

(ml/min) via mass flow controller to transform methanol into hydrogen accompanying with some of carbon dioxide, carbon monoxide and water, and then these productive gas would be analyzed through gas chromatography (GC) to indentify the performance of micro-channel methanol reformer. The catalyst content and cross-section image of the tested micro-channel methanol reformer will be confirmed via Inductively Coupled Plasma-Mass Spectrometer (ICP-MS) and field emission gun scanning electron microscopy (FEGSEM, JEOL-6330F, Japan).

III. RESULTS AND DISCUSSION

III.1 The influence of the centrifugal technique for catalyst deposition

To increase the reactive surface area of the catalyst layer inside the micro-channels for the partial oxidation of methanol (POM), the catalyst deposition was coated on the sidewalls of the micro-channels by the centrifugal technique and non-centrifugal technique as shown in Fig. 2 and It was clearly observed that the deposition of the catalyst layer became thicker from top to bottom of the channels, as shown in Figs. 2(b)-(d), and catalyst layer without spin technique was uniform. The methanol conversion rate, hydrogen selectivity were calculated via the following Eq.(4) and Eq. (5), respectively :

$$cov = \frac{n1 - n2}{n1} \times 100\%$$

n1 : nMeOH$_{in}$; n2 : nMeOH$_{out}$ $\qquad$ (4)

$$S_{H2} = \frac{nH_2}{nH_2 + nH_2O} \times 100\% \qquad (5)$$

where n1 is the content of steam methanol sent first into the micro reformer, n2 is the remained content of steam methanol that be used, nH$_2$ and nH$_2$O are the content of the productive hydrogen and water after POM reaction, respectively.

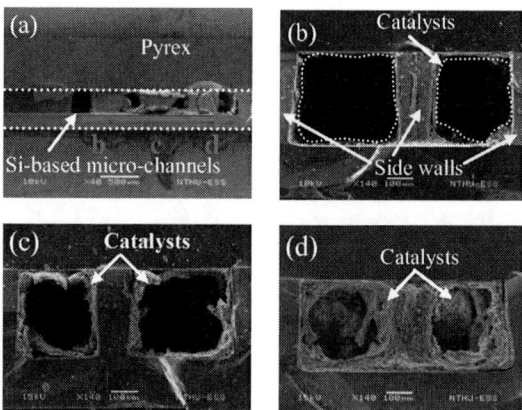

Fig. 2. SEM images of the Si-based micro-channel chip. (a) The cross-sectional view of the channels from top to bottom indicated along the direction of the red arrow. The figures (b), (c) and (d) are the high-magnification images of the top, middle and bottom of the channel.

Experimental results showed the methanol conversion rate, hydrogen selectivity and hydrogen yield of the micromethanol reformer with the centrifugal technique, as shown in Fig. 6, were improved about 10%, 16% and 56% at 250 °C, respectively. The highest methanol conversion rate reached up to 80% meanwhile, the hydrogen yield of this type micro-methanol reformer was improved from 1.70×10^{-6} to 2.77×10^{-6} (mole/min) at 250℃ for binder/catalyst ratio at 60.

III.2 *The influence of the ratio of binder to catalysts*

To reduce the coverage of binder on active sites of the catalysts, the weight ratio of binder to solid catalysts (B) was adjusted in this study, as shown in Table 1, indicating that the composition percentage of the catalysts was fixed (Cu/Mn/Zn: 20/30/50%). The catalyst content in the micro-channel methanol reformer was raised from 2.73 microgram (μ g) to 15.7 microgram (μ g) when B was reduced from 60 to 0. Fig.3 shows that the performance of micro-channel methanol with centrifugal process was improved by reducing the contents of binder from 60 to 0, demonstrating that the methanol conversion, hydrogen selectivity and hydrogen yield of the micro-methanol reformer without any binder incorporation at 250 °C approaches approximately 100%, 92% and 1.56×10^{-5} mole min^{-1}, respectively. Moreover, this novel micro-channel reformer can maintain a high performance output at low temperature (200℃, B:0), which was superior to our previous work [9-11]. Because the existence of Al_2O_3-based binder in the micro reformer may provide extra capture of hydrogen to carry out oxidative reaction with remained oxygen so that it would deteriorate the hydrogen selectivity and hydrogen yield. Moreover, the methanol conversion rate and hydrogen selectivity is close to that of the fixed bed from 200 to 250 ℃ in comparison with the fixed bed result.

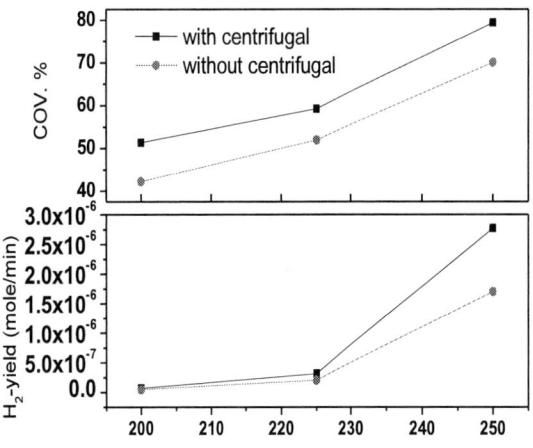

Fig. 3: The influence of centrifugal process on conversion and hydrogen yield of micro-methanol reformer.

IV. CONLUSIONS

A low-temperature and high performance micro-channel methanol reformer was successfully demonstrated to enhance the methanol conversion rate, hydrogen selectivity and hydrogen yield at 200 ℃ which can reach 91%, 94% and 1.44×10^{-5} mole min^{-1} without any binder addition to catalysts by a centrifugal technique. As the ratio of binder to catalysts varied from 60 to 0 and applying centrifugal process, the methanol conversion rate, hydrogen selectivity and hydrogen yield of the micro-methanol reformer at 250°C can approach ~100%, 92% and 1.56×10^{-5} mole min^{-1}, respectively.

Table 1: the catalyst content in micro channels with different ratio of binder

The ratio of binder	The content of catalyst and ingredients			
	Cu	Mn	Zn	The Catalyst content (mg)
B60	18.9%	30.6%	50.4%	2.73
B35	18.8%	30.1%	50.9%	8.21
B10	18.7%	30.1%	51.2%	15.7

*B: the weight ratio of binder to catalysts, $M_{Cu+Mn+Zn}$: mass density of the catalysts.

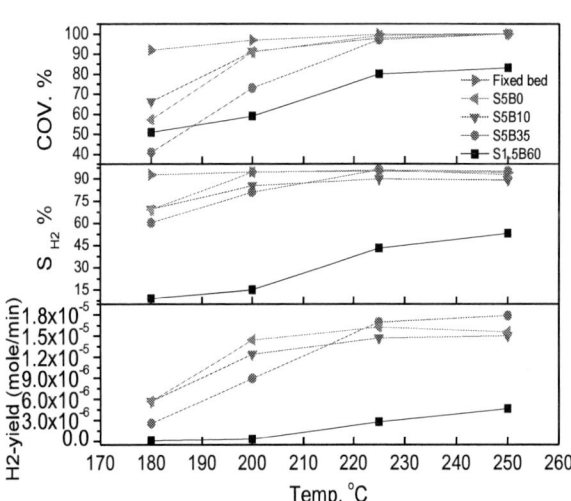

Fig. 4: The influence of different ratio of binder in the catalyst solution on the performance of micro-methanol reformer.

ACKNOWLEDGMENT

The work was financially supported by National Science Council (NSC), Taiwan under grant number NSC-100-3113-E-007-001.

REFERENCES

[1] J.H.WEE "A comparison of sodium borohydride as a fuel for proton exchange membrane fuel cells and for direct borohydride fuel cells" J. Power Sources, vol. 155 , pp. 329–339,2006

[2] F. Joensen, and J. R. Rostrup-Nielsen, "Conversion of hydrocarbons and alcohols for fuel cells", J. Power Sources, vol.105, pp. 195,2002

[3] R.F.Horng, H.M. Chou, C.H.Lee, H.T.Tsai, "Characteristics of hydrogen produced by partial oxidation and auto-thermal reforming in a small methanol reformer" ,J. Power Sources, vol.161, pp.1225-1233,2006

[4] B. Lindstro¨m, L. J. Pettersson,"Development of a methanol fuelled reformer for fuel cell applications", J. Power Source, vol.118, pp.71-78, 2003

[5] G.G. Park, D.J. Seo, S.H. Park, Y.G. Yoon, C.S. Kim, W.L.Yoon, " Development of microchannel methanol steam reformer " , Chemical Engineering Journal , vol.101,pp. 87–92,2004

[6] Y. Kawamura, K. Yamamoto, N. Ogura, T. Katsumata, A. Igarashi, "Preparation of Cu/ZnO/Al$_2$O$_3$", *J. Power Sources* , vol.150, pp. 20-26. 2005

[7] S. S. Hsieh, S. H. Yang, J. K. Kuo, C. F. Huang, H. H. Tsai, "Study of operational parameters on the performance of micro PEMFCs with different flow fields", Energy Conversion and Management, vol. 47, pp. 1868-1878,2006

[8] H. M. Yang and P. H Liao, "Preparation and activity of Cu/Zno-CNTs nano-catalyst on steam reforming of methanol", *App. Catalysis, General, vol.* 317, pp. 226-233, 2007.

[9] Kuo-Yang Huang, Shu-Ping Lai, Hsien-Chih Peng, Fan-Gang Tseng, Yuh-Jeen Huang, Fuel Cell Science & Technology conference, Zaragoza, Spain, 2010

[10] Shu-Ping Lai, Kuo-Yang Huang, Yuh-Jeen Huang, and Fan-Gang Tseng, Power MEMS, 2010

[11] H.S. Wang, K.Y. Huang, H.C. Peng, Y.J. Huang, F.G. Tseng, ISMM, 2011

Proton Exchange Membranes Based on Aryl Epoxy Resin for Fuel Cells Operated at Elevated Temperatures

[1] Tung-Yuan Lee, [1] Tsung-Cheng Ho, [2] Chia-Jung Chang, [1] Pen-Cheng Wang* and [1,3] Fan-Gang Tseng*

[1] Department of Engineering and System Science, National Tsing Hua University, Hsinchu 300, Taiwan

[2] Industrial Technology Research Institute ITRI South, Microsystem Technology Center, Taiwan

[3] Division of Mechanics, Research Center for Applied Sciences, Academia Sinica, Taiwan

E-mail: wangpc@ess.nthu.edu.tw, Tel: 886-3-5715131 ext. 42372, Fax: 886-3-5720724

E-mail: fangang@ess.nthu.edu.tw, Tel: 886-3-5715131 ext. 34270, Fax: 886-3-5720724

Abstract—**This paper reports the characterization of a new type of low-cost proton exchange membrane (PEM) based on photo-patternable nano porous aryl epoxy resin (npAER) sulfonated by sulfanilic acid. The npAER PEM fabrication process involves solvent-casting nanoporous structure formation combined with standard photolithography steps for microstructure fabrication. The PEM was placed in the cathode of a half-fuel cell for testing in 0.5M H_2SO_4 at different temperatures with constant oxygen flow. When compared to commercial PEM Nafion®, this npAER PEM exhibits increased current density by about 170% as temperature increased from 60°C to 90°C, while the current density of Nafion® dropped by about 70%. The new npAER PEM demonstrates decent thermal stability, mechanical strength and proton transport ability at a higher temperature (90°C).**

Keywords – Nanoporous structure; Proton Exchange Membrane (PEM); Fuel Cell.

I. INTRODUCTION

Fuel cells are applicable to rechargeable batteries, vehicles, and portable electronic devices [1-2]. For proton exchange membrane fuel cells (PEMFC) or direct methanol fuel cells (DMFC), Nafion® is commonly used as the PEM and can maintain high moisture content, high proton transportation and low fuel crossover at room temperature. However, the performance of Nafion® will drop rapidly when a fuel cell is operated at an elevated temperature above 80ºC [3]. Many researchers have been focusing on this issue recently, including filling Nafion® in mesoporous microstructures [4] to reduce the fuel cross-over problem. However, this method still cannot solve moisture maintenance problem at an elevated temperature for Nafion®.

This paper describes a new npAER PEM material sulfonated by sulfanilic acid with controllable nanopore structure to reduce fuel cross-over and maintain proton transport stability.

II. EXPERIMENTAL

We prepared the proton exchange membrane (PEM) material by mixing aryl epoxy resin and sulfanilic acid. The mixed material was filled in PEM-making mold. Ultraviolet light was then used to polymerize the mixed material. The temperature was maintained at 75^0C for hours until the solvent was eliminated. Finally, the npAER PEM was removed from PEM-making mold.

The obtained npAER PEM was characterized by FEGSEM, FTIR, BET, Thermal stability test by cyclic voltammetry, and half-fuel cell test.

A. FEGSEM characterization

The npAER thin film with a thickness of ~50 nm was prepared photochemcially on a bare silicon wafer. After evaporation of the solvent, a thin Au film was then sputtered on the obtained material. Images were taken by FEGSEM (JEOL-6330F, Japan) at 15 kV for acquiring the morphologies of npAER.

B. FTIR characterization

A Fourier transform infrared (FTIR) spectrum of the npAER PEM was recorded using a Fourier transform infrared spectrometer (Jasco FT/IR-4200, Japan) in the range of 4000-400cm^{-1}.

978-1-4673-1122-9/12 $31.00 © 2012 IEEE

C. BET characterization

The sample was cut into 1cm x 1cm pieces. The cut sample was loaded into a sample cell and kept at 120^0C for more than 12 hours to remove solvent and water from the sample. Then, nitrogen gas was passed into the sample cell to perform physical gas absorption and desorption using Autosorb-1-C gas sorption analyzer (Quantachrome, USA). Finally, the absorption and desorption curves were analyzed by DFT/Monte-Carlo Pore Volume Histogram software to determine the pore size distribution.

D. Thermal stability characterization

The sample was made on carbon cloth electrode directly, and just exposed 1 cm x 1 cm reaction area. Using the electrochemical test(CHI 1140B, U.S.A.) method to process the thermal stability analysis, the sample was immersed in 0.5 M H_2SO_4 reaction solution, and was performed at different reaction solution temperatures, 60℃ and 90℃, shown by Fig.1.

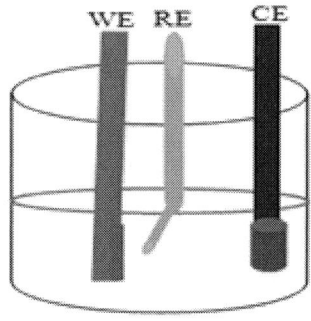

Fig. 1 Schematic of thermal stability analysis: WE: Working electrode; CE: Counter electrode; RE: Reference electrode; ▬▬▬: PEM.

E. Cathode preparation for the half-cell test

The PEM sample (1cm x 1cm) was sandwiched in the half-fuel cell test apparatus and characterized by Linear Sweep Voltammetry (LSV) three-electrode electrochemical test. The half-cell test module was immersed in 0.5 M H_2SO_4 reaction solution with a constant oxygen flow. The test was performed for several samples at different reaction solution temperatures, fig.2.

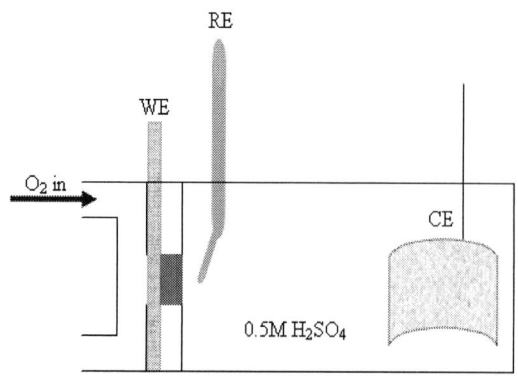

Fig. 2 The half-fuel cell test set-up. WE:Working electrode, CE:Counter electrode, RE: Reference electrode, ▬▬▬ working electrode, ▬▬▬ PEM, ▬▬▬ cathodic electrode.

III.   RESULTS AND DISCUSSION

Figure 3 shows the schematics of proton transportation mechanisms of npAER/Sulfanilic acid PEM. Due to better moisture maintaining for npAER/Sulfanilic acid PEM at higher temperature that, it has better transport capability at elevated temperature.

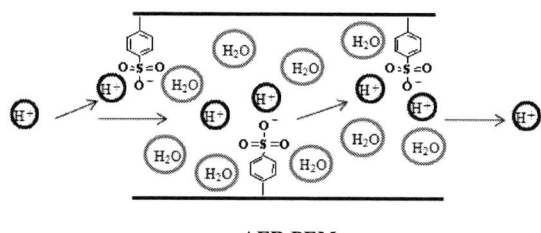

npAER PEM

Fig. 3 Schematic of proton transport mechanisms of npAER/Sulfanilic acid PEM.

Figure 4(a) shows the FEGSEM image of the microstructured in npAER and Figure 4(b) shows controllable nano pore size on different npAER soild ratio by solvent-casting fabrication process.

Figure 5 shows the Fourier transform infrared (FTIR) spectrum of the npAER PEM, and the result shows that the material contains -SO₃H functional group.

978-1-4673-1122-9/12 $31.00 © 2012 IEEE   454

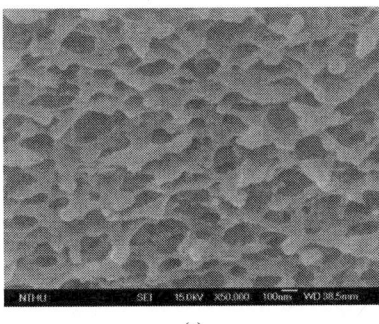

(a)

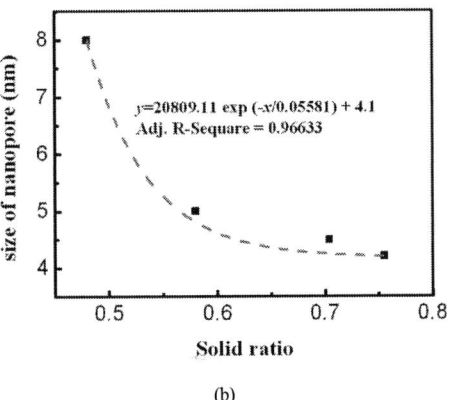

$y = 20809.11 \exp(-x/0.05581) + 4.1$
Adj. R-Sequare = 0.96633

(b)

Fig. 4 (a) FEGSEM image of the micro/nano structures of sulfonated nano porous aryl epoxy resin (npAER/Sulfanilic acid) and (b) size-of-nanopore vs. npAER-soild-ratio curve.

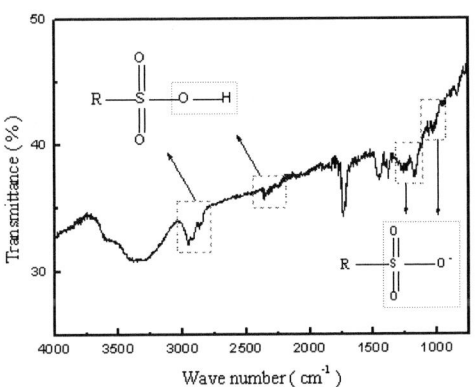

Fig. 5 Fourier transform infrared (FTIR) spectrum of the npAER PEM.

The nanoporous size distribution for different PEM materials are compared in Table 1. It shows that the major pore distribution of npAER/Sulfanilic acid PEM is very similar to that of Nafion® 115. And the large pore distribution also shows that npAER appears to have less pores with pore width >6 nm than Nafion 115, suggesting less fuel cross-over for the npAER sample.

Material	Major pore distribution (ratio)	large pore distribution (ratio)
Nafion® 115	3 ~ 6 nm (73%)	7 ~ 9 nm (20%)
npAER/sulfanilic acid PEM	3 ~ 6 nm (80.9%)	7.5 ~8.5 nm (4.1%)

Table. 1. The comparisons of nanopore distributions for different nanoporous materials.

Figure 6 and Figure 7 show the half-cell testing result of Nafion®-115 and npAER/sulfanilic acid PEM as the cathode at constant oxygen flow, reaction solution 0.5M H_2SO_4 and different temperatures, respectively. Table 2 shows that at $60^{\circ}C$ in 0.5M H_2SO_4, Nafion®-115 has high moisture content. Therefore, sulfonic groups can transport proton efficiently and the current density is -78.3 (mA/cm^2). At $90^{\circ}C$ in 0.5M H_2SO_4, Nafion®-115 moisture content decreases significantly and the current density is decreased to -23.4 (mA/cm^2). However, for npAER/sulfanilic acid PEM, it shows higher current density at $90^{\circ}C$ (-0.384 mA/cm^2) than at $60^{\circ}C$ (-0.142 mA/cm^2). When compared to PEM Nafion®, this npAER/sulfanilic PEM exhibits increased current density by about 170% as temperature increased from $60^{\circ}C$ to $90^{\circ}C$, while the current density of Nafion® dropped by about 70%.

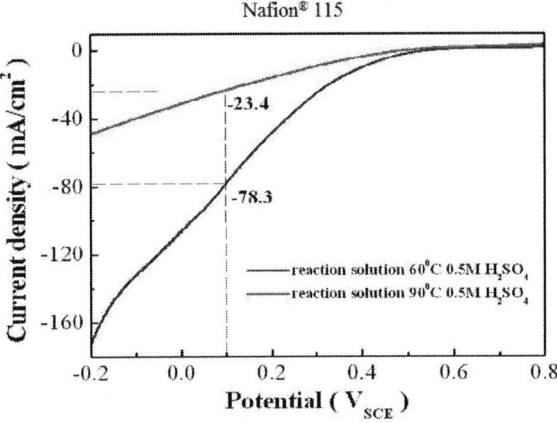

Fig. 6 Cathode proton transport test for Nafion® 115 in a half-cell at $60^{0}C$ and $90^{0}C$, respectively

978-1-4673-1122-9/12 $31.00 © 2012 IEEE

455

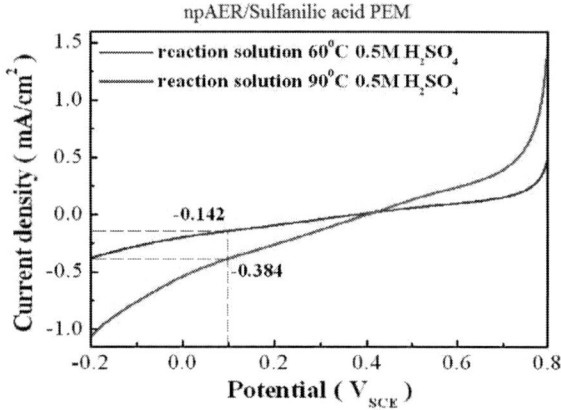

Fig. 7 Cathode proton transport test for nano porous aryl epoxy resin in half-cell filled with 0.5M H₂SO₄ at 60⁰C and 90⁰C, respectively.

Material	60°C 0.5M H₂SO₄	90°C 0.5M H₂SO₄	Change
Nafion® 115	-78.3 (mA/cm²)	-23.4 (mA/cm²)	Decrease 70%
npAER/sulfanil ic acid PEM	-0.142 (mA/cm²)	-0.348 (mA/cm²)	Increase 140%

Table. 2. The comparison of the half-cell testing result of Nafion®-115 and npAER/Sulfanilic acid PEM.

Figure 8 shows the npAER/Sulfanilic acid PEM thermal stability at 60°C and 90°C reaction solutions in 0.5M H₂SO₄ by cyclic voltammetry.

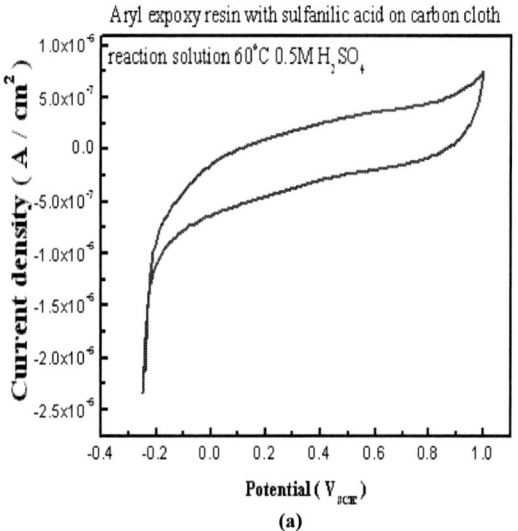

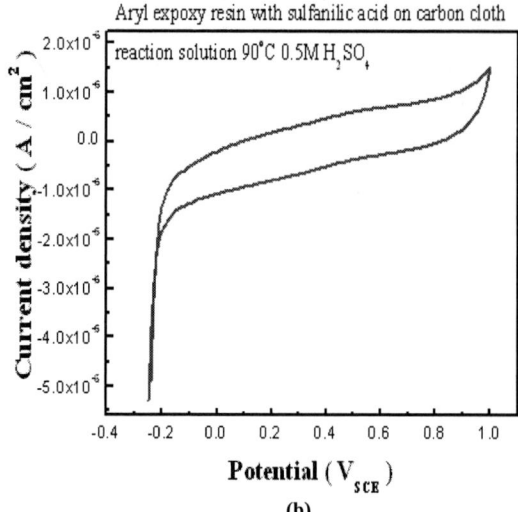

Figure 8: Test of npAER/Sulfanilic acid PEM thermal stability at 60°C (a) and 90°C (b) in 0.5M H₂SO₄ reaction solutions by cyclic voltammetry.

IV. CONCLUSION

This paper proposes a new proton exchange membrane (PEM) based on photochemically synthesized nano porous aryl epoxy resin (npAER) sulfonated by sulfanilic acid. From BET characterization, the npAER PEM pore size distribution is similar to Nafion 115. On the other hand, the result also shows that npAER has less pores with pore width >6 nm, indicating that npAER is more effective in decreasing fuel cross-over in fuel cells. In the half-fuel cell test, as temperature increased from 60⁰C to 90⁰C, Nafion® demonstrates current density drop by about 70%. However, the current density of npAER PEM increases about 170%. The new npAER PEM demonstrates decent thermal stability, mechanical strength and proton transport ability at an elevated temperature (90⁰C).

ACKNOWLEDGMENT

The authors greatly appreciate the financial support from National Science Council of Taiwan ROC through National Energy Program under Contract NSC98-3114-E-007-009.

REFERENCES

[1] S. J. Peighambardoust, S. Rowshanzamir and M. Amjadi, "Review of the proton exchange membranes for fuel cell applications," *Journal of Hydrogen Energy*, vol.35, pp. 9349-9384, 2010.
[2] S. Bose, T. Kuila, T. X. H. Nguyen, N. H. Kim, K.-t. Lau, and J. H. Lee, "Polymer membranes for high temperature proton exchange membrane fuel cell: Recent advances and challenges", Progress in Polymer Science, vol. 36, pp. 813-843, 2011.
[3] D-A. Dupuis, "Proton exchange membranes for fuel cells operated at medium temperatures: Materials and experimental techniques," *Progress in Materials Science*, vol. 56, pp.289-327, 2011.
[4] B. A. Holmberg, X. Wang and Y. Yan, "Nanocomposite fuel cell membranes based on Nafion and acid functionalized zeolite beta nanocrystals," *Journal of Membrane Science*, vol. 320, pp.86-92, 2008.

Material Nonlinearity Limits on a Lamé-Mode Single Crystal Bulk Resonator

Haoshen Zhu[*], Cheng Tu, Joshua E.-Y. Lee

[*]Department of Electronic Engineering, City University of Hong Kong, Hong Kong

haoshezhu2@student.cityu.edu.hk

Abstract—**In this paper, the material nonlinearity in single crystal silicon (SCS) square-plate bulk-mode resonators is described as a strain-dependent shear modulus and incorporated into a lumped element model. The equivalent circuit of the fully differential setup is developed for the nonlinear oscillation cases. According to measurement results, the spring hardening effect in a bulk resonator is reported for the first time. Opposite nonlinear behaviors (spring softening and spring hardening) are detected in devices with same dimensions and driving mechanism but different crystalline orientations ([110] and [100]) indicating the material nonlinearity dominance. The experimental results are compared against the circuit simulation results in various conditions which validate the proposed model.**

Keywords-Material nonlinearity; MEMS; bulk resonator; modeling

I. INTRODUCTION

Micromechanical resonators (µresonators) have attracted much attention over the last decade. They have shown excellent potential for use in timing and frequency control applications due to their ultra-high quality factor and frequency stability. Moreover, the manufacturing processes of these devices are largely compatible with modern IC fabrication technologies. These advantages plus its small form factor make the µresonator a promising alternative for fully-integrated oscillators over embedded LC. Discretely packaged quartz crystal resonators can be eliminated by integration of high performance µresonators with ICs. This makes products more reliable to shock and reduces the overall cost.

Nonlinearity is a critical issue in µresonator design. The performance of the resonator-based devices is largely constrained by nonlinear effects in terms of quality factor (Q factor), power handling capability and phase noise. Nonlinearity in SCS µresonators mainly originate from mechanical and electrical nonlinear effects [1]. For flexural-mode resonators, the most dominant nonlinearity arises from the geometric deformation that is always associated with a spring hardening type of nonlinearity [2]. In bulk mode resonators, nonlinearity associated with the nonlinear material properties. The property of SCS becomes dominant. These two effects are both classified as mechanical nonlinearity. Electrical nonlinearity is induced by capacitive transduction since the generated electrostatic driving force is nonlinear and displacement-dependent.

Bulk mode µresonators are capable of achieving high quality factors (over 1 million) and excellent power handling capability compared to their flexural mode counterparts. These in turn lead to better phase noise performance. For instance, a Square-Extensional (SE) mode bulk resonator that satisfies the GSM specification in terms of phase noise was demonstrated by [3]. Since SCS is an anisotropic material, different nonlinear material property has been found in devices oriented in different directions, like longitudinal beam resonator [4]. Thus, it is worthwhile to study the Lamé-mode resonators with crystalline orientations. In this paper, we launch an experimental-based study on the nonlinear behavior of the bulk mode µresonator in different crystal orientations and aim to understand the source of nonlinearity in the Lamé-mode [5] which has not been fully explored. The spring hardening effect previously only observed in flexural mode beam resonators is also detected in this bulk resonator aligned with [100] orientation. This is opposite to the spring softening behavior reported in [110] devices according to [6]. This paper is organized as follows: the bulk resonator under forced vibration is modeled with nonlinear effects in Section II and the material nonlinearity is highlighted. In Section III, an equivalent circuit is employed to compute nonlinear oscillations. In Section IV, the measured nonlinear dynamic behavior of the bulk mode devices is compared with the circuit simulation results. This paper is concluded in Section V.

II. MODELING NONLINEAR OSCILLATION WITH MATERIAL NONLINEARITY

To model the nonlinear dynamic behavior of the continuum mechanical system, the lumped element model approach remains highly attractive for its simplicity without greatly sacrificing accuracy. Here, the forced vibration of the µresonator can be transformed into a single degree of freedom (DOF) mass-damper-spring system and the general equation of motion is expressed as follows:

$$m\ddot{X} + \gamma\dot{X} + k_m X = F \qquad (1)$$

where X is the mechanical displacement, m is the effective mass, γ is the damping coefficient, k_m is the effective mechanical spring constant and F stands for the driving force.

Material nonlinearity is known to be the major source of nonlinearity in bulk mode resonators. SCS is elastically anisotropic whose linear elastic characteristics can be described in a stiffness matrix comprising second-order

978-1-4673-1122-9/12 $31.00 © 2012 IEEE

stiffness tensors. The third-order stiffness tensors of SCS was reported for the first time in 1960's [7], indicating its nonlinear strain-stress relationship. Since the Q factor reaches as high as several million in bulk mode μresonators, even small material nonlinearity of SCS becomes significant.

In order to quantitatively study the material nonlinearity and illustrate its effect in bulk mode μresonators through lumped element modeling, Kaajakari et al. [1] expressed the mechanical nonlinearity in the form of a nonlinear strain-dependent Young's modulus. Derived using large deformation theory involving the third-order stiffness tensor, this nonlinear Young's modulus incorporates both nonlinear geometric deformation and elastic properties. By applying this method, they were able to differentiate the material from geometric nonlinearity. It was pointed out that material nonlinearity is significant in both longitudinal beam and SE-mode plate bulk resonators [1, 3]. While these two resonators are always measured in one-port configuration which suffers from heavy feedthrough, the Lamé-mode bulk resonator is particularly of interest due to its reduced feedthrough interference achieved through a differential setup as well as high Q factor.

The 2D layout of the square plate bulk resonator studied in this paper is shown in Fig. 1(a). The Lamé-mode shape is illustrated in Fig. 2, obtained by finite element (FE) simulation. The stresses exerted in the x and y directions are equal and opposite directions. By assuming plane stress, the structure can be considered to be in pure shear within xy plane. Similar to [3], to incorporate the material nonlinearity, we define a nonlinear engineering strain-dependent shear modulus (G_γ) as follows:

$$G_\gamma = \frac{\tau}{\gamma} = G_0\left(1 + G_1\gamma + G_2\gamma^2\right) \quad (2)$$

where τ is the shear stress, γ is the corresponding shear strain, G_0 is the linear shear modulus, G_1 and G_2 are first and second order correction terms relative to G_0.

The lumped parameters m and k_m can be obtained based on Raleigh's method [8] by equating the maximum kinetic energy to the maximum potential energy at resonance. For linear Lamé-mode resonators, the mode shape in can be approximated as below:

$$U_x(x,y) = X\cos\left(\pi\frac{y}{L}\right)\sin\left(\pi\frac{x}{L}\right)$$
$$U_y(x,y) = Y\cos\left(\pi\frac{x}{L}\right)\sin\left(\pi\frac{y}{L}\right) \quad (3)$$

where L is the edge length, and X and Y are the amplitude of displacement (at the middle point of each edge) in the x and y directions respectively. In the case of symmetric motion in the Lamé-mode, the amplitude of displacements are equal on each side, giving $X = Y$. The normalized x-displacement along five parallel lines (Fig. 1(a)) is obtained by FE simulation. From Fig. 1(b), these indicate that the largest displacement occurs at the middle points and verifies (3). The mode shape under nonlinear conditions can still be approximately described by (3) due to the weak nonlinear material property in SCS.

In this case, the potential energy per unit volume is given as $G_\gamma\gamma^2/2$, where γ is the shear strain. Considering the structure to be in pure shear within the xy plane, the planar shear strain is given by:

$$\gamma = \frac{\partial U_x}{\partial y} + \frac{\partial U_y}{\partial x} \quad (4)$$

According to [9], the potential energy can be determined by integrating the unit potential energy over the mode shape:

$$E_{potential} = \frac{1}{2}k_m X^2 = \iiint_V \frac{G_\gamma\gamma^2}{2}dV \quad (5)$$

By substituting (2), (3) and (4) into (5), k_m for the nonlinear case is obtained as follows:

$$k_m = k_{1m} + k_{2m}X + k_{3m}X^2 \quad (6)$$

where h is the thickness of the device, $k_{1m} = \pi^2 G_0 h$, $k_{2m} = 0$, and $k_{3m} = 9\pi^4 G_0 G_2 h/4L^2$ are the linear, quadratic and cubic mechanical spring constants, respectively. It is interesting that the material nonlinearity will not result in an asymmetric restoring force ($k_{2m} = 0$) for Lamé-mode resonators in contrast to the longitudinal and SE modes. The effective mass m can be determined by calculating total kinetic energy [9]:

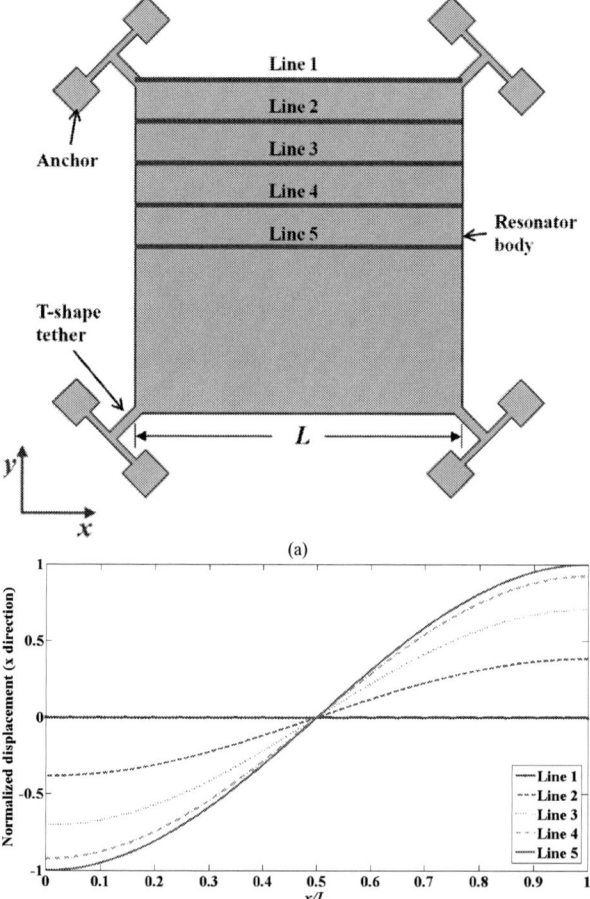

Fig. 1. (a) 2D layout of the resonator and (b) the normalized x-direction displacement along different lines.

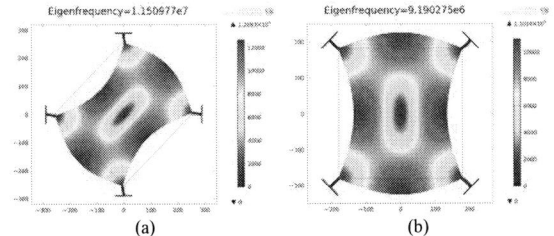

Fig. 2. FE-simulated vibration modes in (a) [110] and (b) [100] orientations using COMSOL

$$m = \frac{\rho h L^2}{2} \qquad (7)$$

where ρ is the material density of silicon. Then the resonant frequency can be roughly estimated by:

$$f_0 = \frac{1}{2\pi}\sqrt{\frac{k_{1m}}{m}} = \frac{1}{\sqrt{2}L}\sqrt{\frac{G_0}{\rho}} \qquad (8)$$

where G_0 relates with Young's modulus (Y) and Poisson's ratio (v) by $G_0 = Y/2(1+v)$ in linear case.

According to the setup, the resonator is surrounded by four capacitive transducers for driving and sensing. The electrostatic driving force generated by a capacitive transducer is given by:

$$
\begin{aligned}
F_e &= \frac{U_c^2}{2}\frac{\partial C}{\partial X} = \frac{\varepsilon_0 L_e h}{2(g-X)^2}\left(U_b - U_{ac}\cos\omega t\right)^2 \\
&= \frac{\varepsilon_0 L_e h}{2g^2}\left(1 + 2\frac{X}{g} + 3\left(\frac{X}{g}\right)^2 + 4\left(\frac{X}{g}\right)^3 + \ldots\right)\left(U_b - U_{ac}\cos\omega t\right)^2
\end{aligned}
\qquad (9)
$$

where g is the transducer gap distance, U_c is the voltage over the transducer gap which is the difference between DC bias (U_b) and AC signal ($U_{ac}\cos\omega t$) in this case, ε_0 is the permittivity of free space, and L_e is the length of electrode.

Finally, the governing equation of an electrostatic-driven Lamé-mode SCS bulk resonator can be expressed as below:

$$m\ddot{X} + \gamma\dot{X} + k_{1m}X + k_{2m}X^2 + k_{3m}X^3 = F_{e,tot}(X) \qquad (10)$$

where $F_{e,tot}$ stands for the total force generated by the transducers. As indicated by the Taylor's expansion in (9), the electrostatic force is intrinsically nonlinear and involves higher harmonics that cause parametric excitation [10].

III. EQUIVALENT CIRCUIT FORMATION

The Harmonic Balance (HB) method has been proved as a computationally efficient approach to approximately solve the nonlinear differential equations such as (10) [11]. The mechanical resonator and its coupled interface circuits are mapped into a new equivalent circuit formulation. Nonlinear effects are accounted for using customized circuit blocks such as a charge controlled capacitor and a gap modulated variable capacitor [12]. These circuit blocks are based on the lumped parameters derived in Sec. II and are applicable to nonlinear resonators of various configurations. The complete schematic of the equivalent circuit for nonlinear resonator using a fully differential setup is given in Fig. 3.

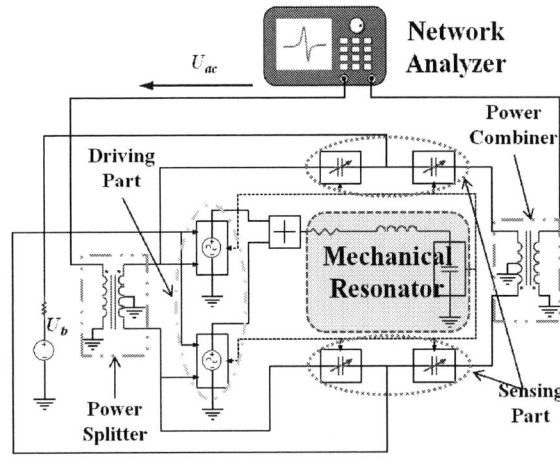

Fig. 3. The schematic of equivalent circuit for the nonlinear Lamé-mode μresonator with fully differential setup.

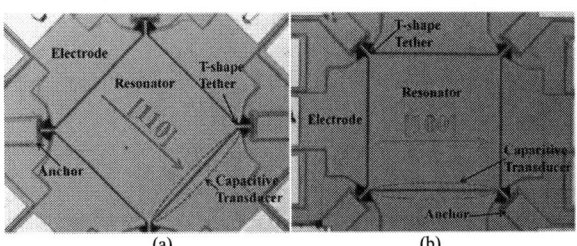

Fig. 4. The micrographs of the fabricated resonators. (a) [110] (b) [100]

IV. EXPERIMENTAL RESULTS VS SIMULATION

Square plate resonators of the same dimensions but aligned against two different orientations (namely [110] and [100]) were fabricated in a foundry MEMS process. Both devices were fabricated on the same die. The length of each square-plate is 360μm, yielding a designed resonant frequency of 11.5MHz and 9.2MHz in the [110] and [100] devices respectively according to (8). The design parameters are tabulated in Table I.

TABLE I. DESIGN PARAMETERS OF THE RESONATORS

Parameters (Symbol)	Orientations		Unit
	[110]	[100]	
Edge length (L)	360		μm
Device thickness (h)	25		μm
Electrode length (L_e)	340		μm
Gap distance (g)	2		μm
Silicon density (ρ)	2330		kg/m³
Linear shear modulus (G_0)	79.56	50.90	GPa

The micrographs of the fabricated devices in both orientations are demonstrated in Fig. 4.

The fully differential setup for measurement is illustrated in Fig. 5 [13]. Both devices were measured in a Janis Research cryogenic probe station with GSG probes and an Agilent E5061A vector network analyzer. The pressure in the chamber was kept around 60μTorr under room temperature. For the

[110] devices, the transmission S21 is measured first with fixed source power (10dBm), but ramping the DC bias voltage from 20V to 50V. Fig. 6 shows that the transmission increases while the resonant peak tilts toward a lower frequency to indicate spring softening. In Fig. 7, the source power is ramped up while the DC bias was kept constant at 50V. These observations are within expectations since increasing AC and/or DC levels result in larger mechanical displacement X. This in turn results in a more prominent nonlinear oscillation defined by (10). The observed features in [110] devices are well agreed with reported results in [6].

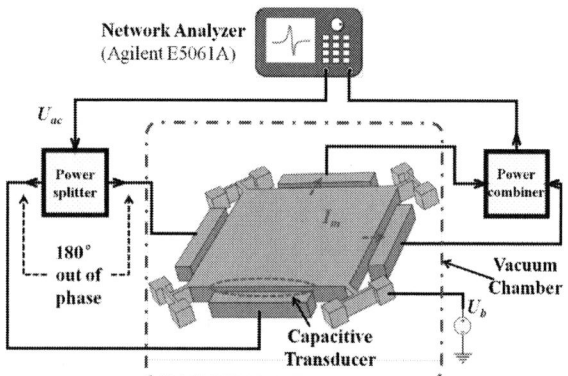

Fig. 5. Fully differential characterization setup for Lamé-mode μresonators.

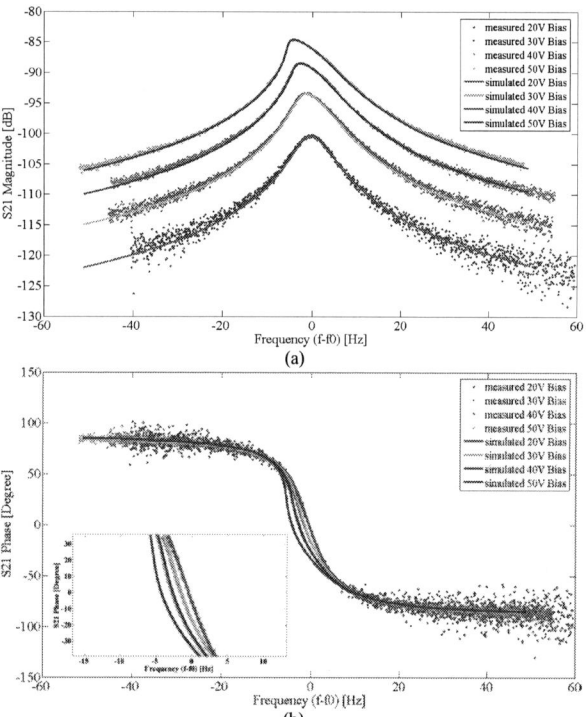

Fig. 6. Measured and simulated transmission (a) magnitude and (b) phase for increasing DC bias with fixed source power (10dBm) in [110] resonator.

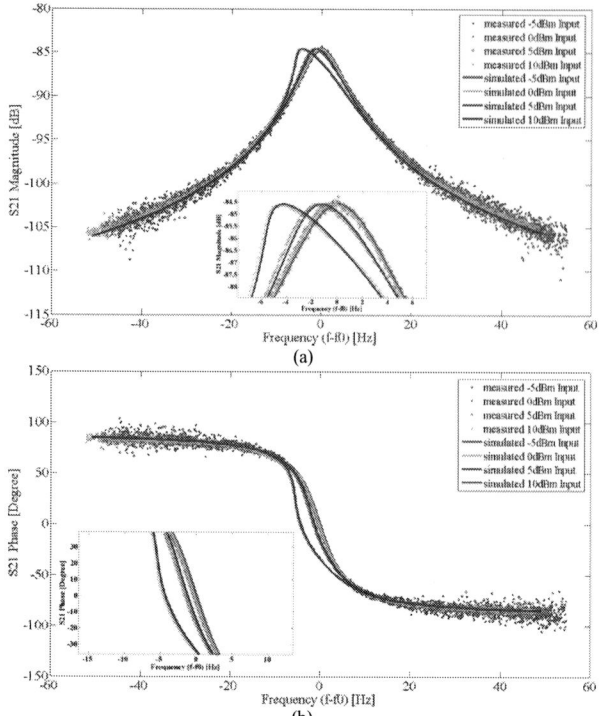

Fig. 7. Measured and simulated transmission (a) magnitude and (b) phase for increasing source power with fixed DC bias (50V) in [110] resonator.

With the [100] devices, spring hardening was remarkably detected as the DC bias and AC power were each independently increased (Fig. 8 and Fig. 9). To date, the spring hardening effect has only been observed in flexural beam μresonators due to large geometric deformation. No such observations have been reported for bulk mode resonators. Moreover, it can be seen that the effect of nonlinearity feature is much more prominently than in the [110] devices under the same amount of driving power. For instance, the resonant frequency shift in the [100] device is around 10Hz as opposed to against 5Hz in the [110] case under 50V DC bias and 10dBm AC input.

According to FE simulation in Fig. 2, the vibration modes is unaffected by the orientation. The highest transmission level in both cases are around -85dB under same driving condition indicating similar degrees of geometric deformation. Besides, the capacitive nonlinearity, which only results in spring softening, is negligible due to the large transducer gap. Thus, since geometric nonlinearity is theoretically the same in both cases, the reversal in the nature of nonlinearity behaviors is likely to originate in the material's elastic characteristics.

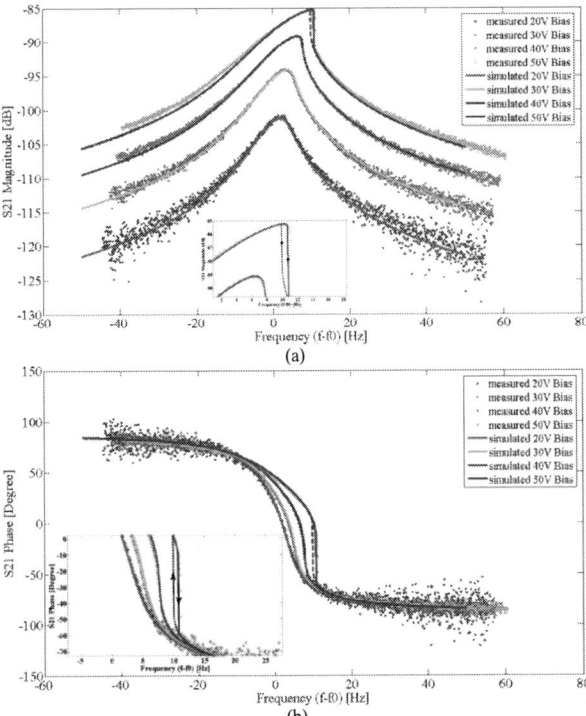

Fig. 8. Measured and simulated transmission (a) magnitude and (b) phase for increasing DC bias with fixed source power (10dBm) in [100] resonator.

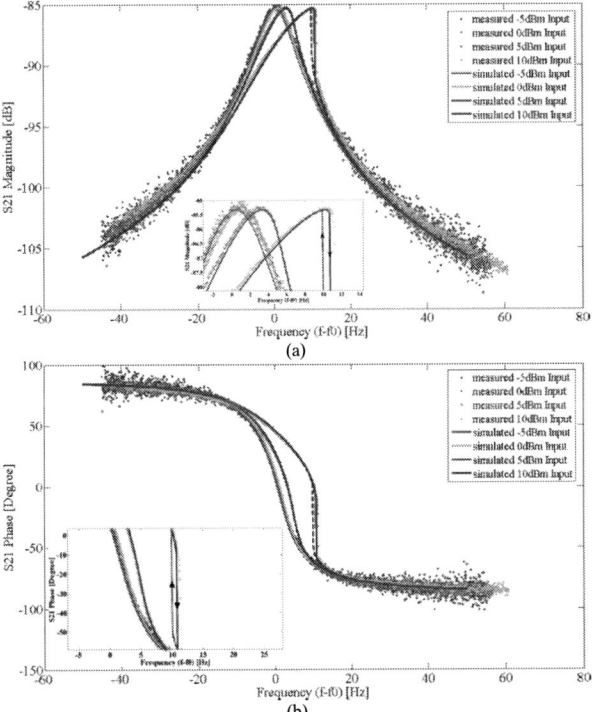

Fig. 9. Measured and simulated transmission (a) magnitude and (b) phase for increasing source power with fixed DC bias (50V) in [100] resonator.

To quantitatively verify the lumped parameter model, the simulated transmission data using the equivalent circuit are fitted to the measured data. However, since γ and k_{3m} are difficult to be determined in the first place (as G_2 is unknown), experimentally extracted values are applied based on two relationships listed below:

$$\gamma = \frac{\sqrt{k_{1m}m}}{Q} \qquad (11)$$

$$\Delta f = f_0{'} - f_0 \approx \frac{3k_{3m}}{8k_{1m}}X^2 \qquad (12)$$

where Δf is the resonant frequency shift from linear to nonlinear cases. The nonlinear electrostatic driving force can be assumed to be pure sinusoidal for large gap distance while k_{3m} is roughly proportional to Δf according to Landau [14]. The lumped parameters used in the circuit simulation are summarized in Table II. The calculated (using (8)), FE simulated and experimentally measured resonant frequencies are also listed for comparison.

TABLE II.   CHARACTERIZATION PARAMETERS

Parameters (Symbol)		Orientations		Unit
		[110]	[100]	
Effective mass (m)		3.77×10^{-9}		kg
Linear spring constant (k_{1m})		1.96×10^7	1.26×10^7	N/m
Cubic spring constant (k_{3m})		-8.9×10^{15}	12.06×10^{15}	N/m³
Quality factor (Q)		1.32×10^6	9.5×10^5	
Damping coefficient (γ)		2.05×10^{-7}	2.29×10^{-7}	N•s/m
Resonant frequency (in linear case)	Calculated	11.4776	9.1804	MHz
	Simulated	11.5098	9.1903	MHz
	Measured	11.5115	9.1846	MHz

Simulation results are embedded with the measurement results in Fig. 6, 7, 8 and 9. In all the cases, both magnitude and phase frequency responses are well matched. Once the lumped parameters have been correctly determined, the established model can well predict the nonlinear behavior under various driving conditions. It is noted that for the [100] device cases, the maximum resonant frequency shift measured was 10Hz, which is beyond the theoretically estimated frequency shift at the critical vibration point ($\approx$7.3Hz) [14]. The discrepancy is due to hysteresis loop, which was verified by the circuit simulation (as dash line is the response when sweeping from higher frequency to lower and arrows indicate the sweeping directions).

V.   CONCLUSIONS

In this paper, the crystal orientation dependence of material nonlinearity in SCS square-plate bulk-mode resonators is highlighted. A lumped element model that accounts for this effect is established for the Lamé-mode. This model is transformed into an equivalent circuit and solved by the HB method. Observation of obvious spring hardening in a bulk mode resonator due to material nonlinearity is reported for the first time when aligned along the [100] direction. We have also observed that devices aligned in the [100] direction exhibit stronger nonlinearity compared with their [110] counterparts under the same driving conditions. The overall measurement results are verified against the derived model through circuit simulations, showing a good match between analytical prediction and the measured nonlinear behavior of the resonators. These results could provide useful insight in understanding the material nonlinear limits in bulk mode resonators, which could pave the way for potential methods for nonlinearity cancellation in order to improve the performance of MEMS devices fabricated in SCS.

ACKNOWLEDGMENT

This work was partially supported by a grant from the Research Grant Council of Hong Kong, China (Project No. CityU 111510), and partially by a grant from the City University of Hong Kong (Project No. 7008088).

REFERENCES

[1] V. Kaajakari, T. Mattila, A. Oja and H. Seppa, "Nonlinear limits for single-crystal silicon microresonators," *Microelectromechanical Systems, Journal of*, vol. 13, pp. 715-724, 2004.

[2] L. C. Shao, C. L. Wong and M. Palaniapan "Study of the nonlinearities in micromechanical clamped–clamped beam resonators using stroboscopic SEM," *J Micromech Microengineering*, vol. 18, pp. 085019, 2008.

[3] V. Kaajakari, T. Mattila, A. Oja, J. Kiihamäki, and H. Seppä, "Square extensional mode single-crystal silicon micromechanical resonator for lowphase noise oscillator applications," *IEEE Electron Device Letters*, vol. 25, pp. 173-175, 2004.

[4] V. Kaajakari, T. Mattila, A. Lipsanen and A. Oja, "Nonlinear mechanical effects in silicon longitudinal mode beam resonators," *Sensors and Actuators A: Physical*, vol. 120, pp. 64-70, 4/29, 2005.

[5] H. Majjad, J.-R. Coudevylle, S. Basrour and M. de Labachelerie, "Modeling and characterization of Lamé-mode microresonators realized by UV-LIGA," in TRANSDUCERS, 12th International Conference on Solid-State Sensors, Actuators and Microsystems, 2003, pp. 300-303 vol.1.

[6] L. C. Shao, M. Palaniapan, L. Khine and W. W. Tan, "Nonlinear behavior of lamé-mode SOI bulk resonator," in *IEEE International Frequency Control Symposium*, 2008, pp. 646-650.

[7] H. McSkimin and P. Andreatch Jr., "Measurement of third-order moduli of silicon and germanium," J. Appl. Phys., vol. 35, no. 11, pp. 3312–3319, Nov. 1964.

[8] W. Weaver, S. Timoshenko and D. H. Young, *Vibration Problems in Engineering*. New York: Wiley, 1990.

[9] J. E.-Y. Lee, "Silicon Micromechanical Resonators for Measurements of Mass and Charge," Ph.D. dissertation, Department of Engineering, University of Cambridge, Cambridge, UK, 2008.

[10] W. Zhang and G. Meng, "Nonlinear dynamical system of micro-cantilever under combined parametric and forcing excitations in MEMS," *Sensors and Actuators A: Physical*, vol. 119, pp. 291-299, 4/13, 2005.

[11] T. Veijola and T. Mattila, "Modeling of nonlinear micromechanical resonators and their simulation with the harmonic-balance method," *International Journal of RF and Microwave Computer-Aided Engineering*, vol. 11, pp. 310-321, 2001.

[12] H. Zhu, J. Lee, "Simulating Nonlinearity in MEMS Resonators by a Charge Controlled Capacitor," *Eurosensors*, 2011.

[13] S. A. Bhave, Di Gao, R. Maboudian and R. T. Howe, "Fully-differential poly-SiC lame mode resonator and checkerboard filter," in *18th IEEE International Conference on Micro Electro Mechanical Systems*, 2005, pp. 223-226.

[14] L. D. Landau and E. M. Lifshits, *Mechanics*. Oxford: Pergamon, 1976.

Evidence on the Impact of T-shaped Tether Variations on Q Factor of Bulk-mode Square-plate Resonators

Yuanjie Xu[*] and Joshua E.-Y. Lee

[*]Department of Electronic Engineering, City University of Hong Kong, Kowloon, Hong Kong SAR
charleshsu1986@hotmail.com

Abstract— **This paper reports evidence from measurements that anchor design has no obvious impact on the quality factor (Q) of a square-plate resonator vibrating in the Lamé mode. This observation is attributed to uniquely locating the anchors at the nodes positioned conveniently at the corners. The measured Q and eigenfrequency show no dependence on variations in the length of the T-shaped tethers. In contrast, the same variations were shown to significantly affect both Q and eigenfrequency when the resonator was excited in the square-extensional (SE) mode. These trends were consistently observed between measurements on two different fabricated SOI dies. For further validation, the measured correlation between the eigenfrequency and tether length for both eigenmodes agree well with those predicted by finite-element (FE) simulations in COMSOL.**

Keywords-component; T-shaped tether; resonator; Lamé mode; square-extensional mode

I. INTRODUCTION

MEMS resonators have gained increasing popularity in the last decade as replacements to quartz resonators for timing applications. MEMS resonators benefit from their small form factor and possibility for monolithic integration with CMOS [1]. On this note, a high Q is desirable for lower phase noise in MEMS oscillators. For filtering applications, a high Q yields lower insertion loss and greater selectivity, which are desirable for frequency control applications [2]. Anchor designs can prove to be a limiting factor on Q [3]. On this note, the overall Q of a resonator is governed by the net sum of losses due to different sources of energy dissipation:

$$\frac{1}{Q_{overall}} = \frac{1}{Q_{anchor}} + \frac{1}{Q_{viscous}} + \frac{1}{Q_{TED}} + \frac{1}{Q_{AKE}} \quad (1)$$

where Q_{anchor}, Q_{air}, and Q_{TED} each respectively correspond to energy losses due to clamping loss at the anchors, viscous damping, and phonon-phonon interaction induced scattering resulting in thermoelastic damping (TED) and Akhieser effect (AKE) loss. By operating the device under moderate vacuum , energy loss due to viscous damping in air can be significantly reduced. In practical implementations, such a clean vacuum environment can be realized through wafer-level vacuum encapsulation [4] techniques. Also, as indicated in [5], due to large surface to volume ratio for bulk mode resonator, the surface loss can be considered as negligible. For bulk modes,

it has been shown that Q_{TED} is typically much larger than Q_{anchor} in the VHF and UHF range. Q_{AKE} sets the ultimate upper bound on the realizable Q in the limit where all other sources for dissipation can be neglected. Hence from the view point of design, anchor loss can be the dominating energy loss mechanism limiting the overall Q in bulk mode resonators. The basis of anchor loss stems from the irreversible flow of energy from the resonator body to the substrate via the anchors [6]. This is illustrated shown in Fig. 1.

T-shaped anchors have been previously employed in square-plate resonators for Q's >10^6 [7-8]. While comparisons between T-shaped and straight tethers have been reported previously in [3], the impact of T-shaped tether length (the section physically connected to the anchors) on Q for square-plate resonators is still unknown. Hence we examine the impact of varying the length of the tether on Q for the Lamé and square-extensional (SE) modes. The mode shape of the Lamé mode is illustrated by Fig. 2 a), which shows the nodes as located at the corners where the resonator is clamped. Fig. 2 b) describes the mode shape of the SE mode.

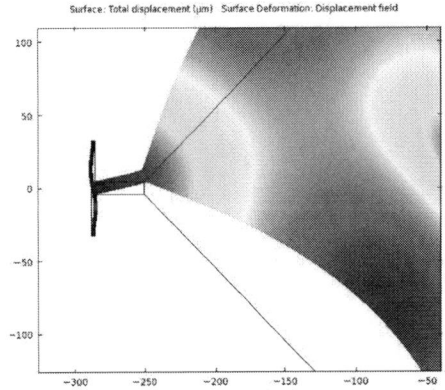

Fig. 1. COMSOL simulation showing an elastic wave propagating through the anchor.

978-1-4673-1122-9/12 $31.00 © 2012 IEEE

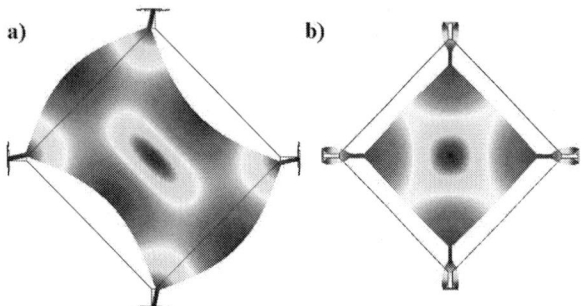

Fig. 2 2D finite-element simulation using COMSOL of the a) Lamé mode and b) SE mode in a square-plate resonator.

II. DEVICE AND MEASUREMENT SETUP

A. Device Description

A total of 5 square-plate resonators with T-shaped tethers were fabricated in a foundry SOI MEMS process, with length of 360μm and 25μm thickness. The tether lengths, as defined in Fig. 3, were varied with reference to a standard length of 65μm by factors of 0.5, 0.75, 1.0, 1.5, 2.0 for the 5 resonators fabricated on the same die. Fig. 4 shows a micrograph of the resonator with a standard tether length. For the resonator with tether length 1.5 times that of the standard, the tether at one corner was unreleased to observe its impact on both the SE and Lamé modes. All the resonators were aligned along the [110] direction on the (100) silicon wafer.

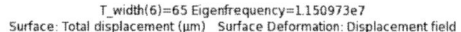

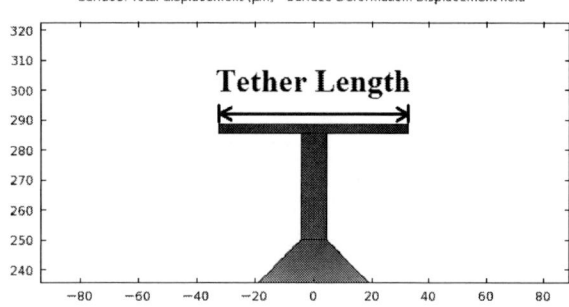

Fig. 3. Definition of the T-shaped tether length.

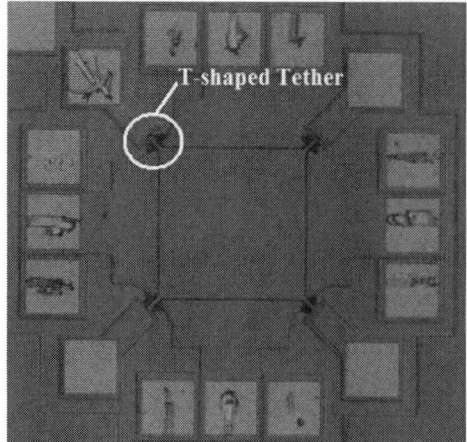

Fig. 4. Micrograph showing the device with standard tether length

The resonant frequency of a square plate resonator vibrating in Lamé mode can be analytically calculated by [9]:

$$f_{0,\text{Lame}} = \frac{1}{\sqrt{2}L}\sqrt{\frac{G}{\rho}} \qquad (2)$$

Here L refers to the square plate's length, G is the shear modulus, and ρ is the density of silicon. Based on a length of 360μm, shear modulus G of 79.4 GPa in the [100] direction from [10], and a density of 2330 kg/m^3 for silicon the calculated resonant frequency based on (2) was found to 11.48 MHz.

For the SE mode, the resonant frequency can also be calculated with the analytical formula below based on [11]:

$$f_{0,SE} = \frac{1}{2L}\sqrt{\frac{E}{\rho(1-\nu)}\left[1+\left(1-\frac{8}{\pi^2}\right)\left(\frac{\nu}{\nu-1}\right)\right]} \qquad (3)$$

E and ν are respectively the Young's modulus and Poisson's ratio of silicon in the [110] direction. Based on the elastic properties reported in [10] ($E = 169$GPa and $\nu = 0.064$), the resonant frequency according to (3) is about 12.14MHz.

B. Measurement Setup

Resonators from two dies were electrically characterized under vacuum at 5×10^{-4} Torr in a Janis Research cryogenic probe station using GSG probes. The electrical transmissions of the resonators were measured using an Agilent E5061A network analyzer. A circuit schematic for the Lamé mode measurement setup is described in Fig. 5, which shows a fully differential configuration to suppress feedthrough. The schematic for the measurement setup used for the SE mode is shown in Fig. 6.

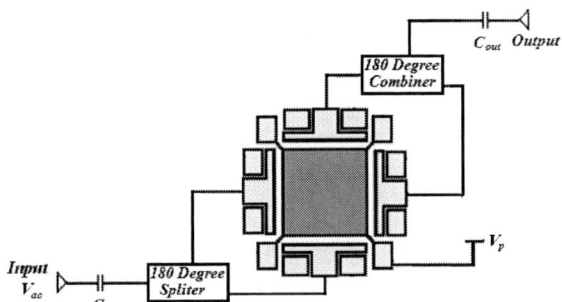

Fig. 5. Schematic showing the setup for Lamé mode measurement

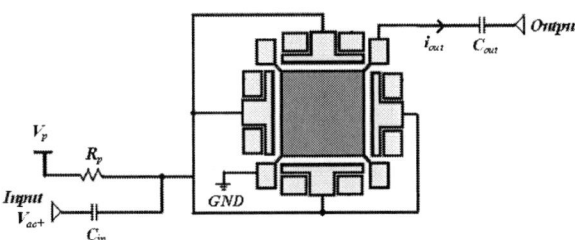

Fig. 6. Schematic showing the setup for SE mode measurement

III. MEASUREMENT RESULTS

A. Lamé Mode Measurement Results

Figs. 6-7 show the electrical transmission of the Lamé mode measured for each respective tether length variant from two dies. It should be noted that the resonators fabricated on other die exhibit the same trends. From Figs. 8-9, it can be seen that the tether length variations have no observable impact on the eigenfrequency. The variation between the 5 devices is about 120 ppm, normalized to the center frequency. The variation is more like due to fabrication tolerances than the effect caused by differences in the anchor geometry. The measured eigenfrequencies are all around 11.51MHz, which coincide well with both the analytical calculation and FE simulations in COMSOL. In the FE simulations, the tether length was parametrically varied according to the fabricated designs. The resonant frequencies predicted by FE simulation are plotted against the measured resonant frequenices to illustrate the close agreement between the two sets of values. Figs. 10-11 show the measured Q as a function of tether length, indicating no significant and consistent relationship between the two parameters.

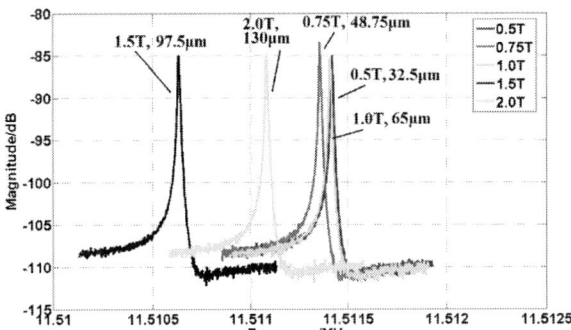

Fig. 6. Measured transmission of Lamé mode resonators from Die 1

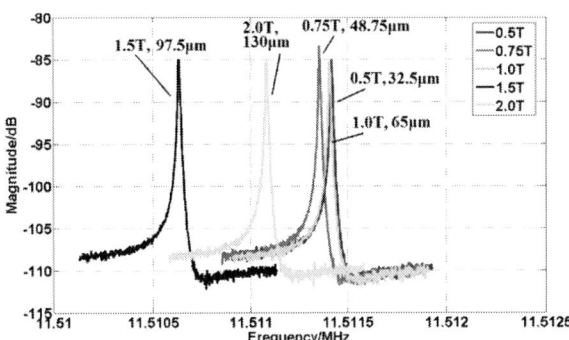

Fig. 7. Measured transmission of Lamé mode resonators from Die 2

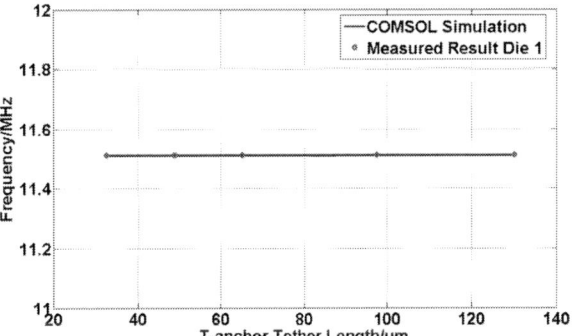

Fig. 8. Measured and FE simulated Lamé mode eigen- frequencies as a function of T-shaped tether length from Die #1.

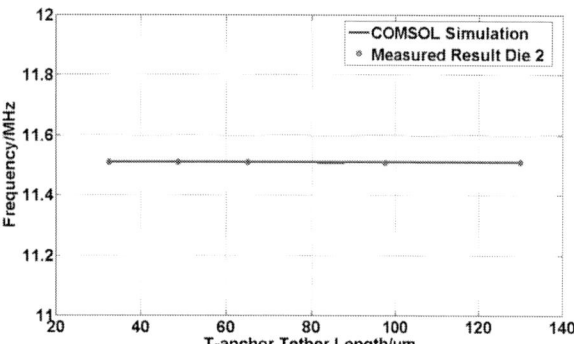

Fig. 9. Measured and FE simulated Lamé mode eigen- frequencies as a function of T-shaped tether length from Die #2.

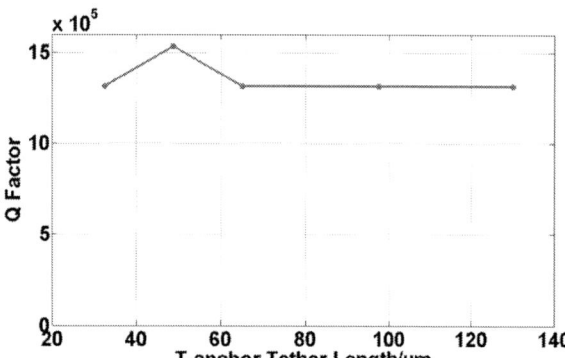

Fig. 10. Measured Lamé mode Q-factors as a function of T-shaped tether length from Die #1.

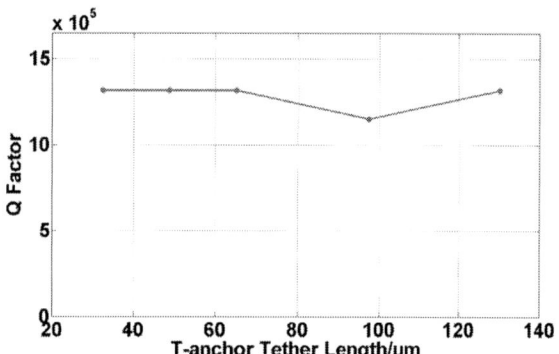

Fig. 11. Measured Lamé mode Q-factors as a function of T-shaped tether length from Die #2.

The reason that the variations on the tether length have no impact on the eigenfrequency and Q factor is likely due to the observation from FE simulations that the displacement on the four corners of the square is insignificant relative to the maximum displacement. This can be seen previously from Fig. 1 a. As the displacement on the four corners of the square is

small, the stress imposed on the clamped section of the tethers will also be correspondingly small. According to (2) and (3), the respective eigenfrequency is determined by the dimensions of the square-plate and acoustic velocity. As such, the eigenfrequency is unaffected by the variation in tether length since it will neither alter the size of the square nor its mode shape. With regards to Q factor, as previously mentioned, the stress imposed on tethers is negligible since the resonator is clamped at the nodes. Hence regardless of the length of the clamped section in the T-shaped tether, strain energy stored on the tethers remains insignificantly small in comparison to the strain energy stored in the square-plate. Furthermore, the fQ (frequency-Q) product is about 1.5×10^{13}, which is close to the Akhieser limit for single crystal silicon in the Akhieser regime [12]. This thereby implies that AKE loss is the dominant loss mechanism rather than the anchors. As such, it is expected that variations in the tether geometry has no observable effect on Q.

B. SE Mode Measurement Results

Figs. 12-13 show the electrical transmission measured for the same 2 dies but now exciting the same devices in the SE mode. It can be seen in this case that variation in tether length results in notable variations in the SE mode resonant frequency which are significantly larger compared to the Lamé mode. Interestingly, the variations in Q and resonant frequency as a function of tether length exhibit opposing trends when comparing Figs. 14 and 16, the SE mode resonant frequency decreases first as the tether length is increased from 0.5T to 1.0T before increasing again between 1.0T to 2.0T. Inversely, it can be seen that Q increases first as the tether length is increased from 0.5T to 1.0T and then drops when the tether length increases to 2.0T. Figs. 15 and 17 from Die #2 also show the same trends. As such, an increase in the tether length does not always bring about an increase in Q. In contrast, the measurement results imply that there exists an optimal tether length for maximizing Q.

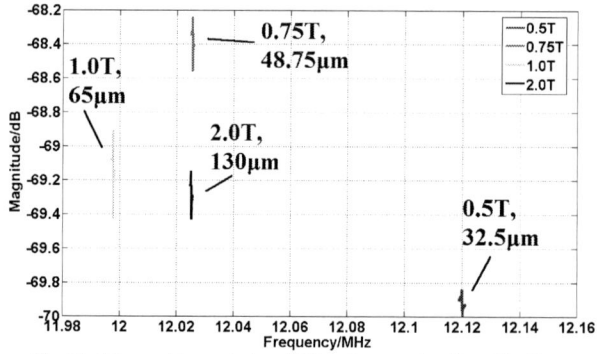

Fig. 12. Measured transmission of SE mode resonators from Die #1; no observable peak for the 1.5T tether length design (unreleased at one corner).

978-1-4673-1122-9/12 $31.00 © 2012 IEEE

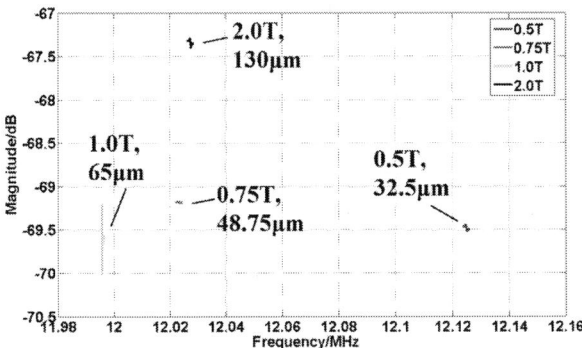

Fig. 13. Measured transmission of SE mode resonators from Die #2; no observable peak for the 1.5T tether length design (unreleased at one corner).

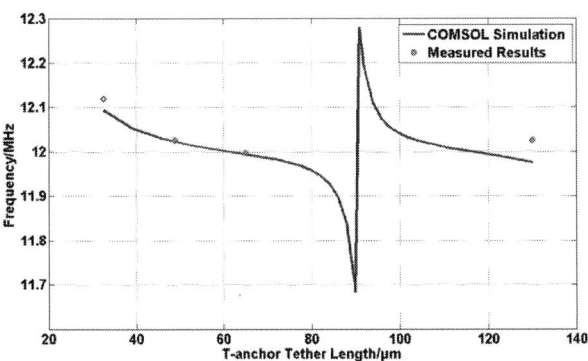

Fig. 14. Measured & FE simulated SE mode eigenfrequencies as a function of T-shaped tether length from Die #1.

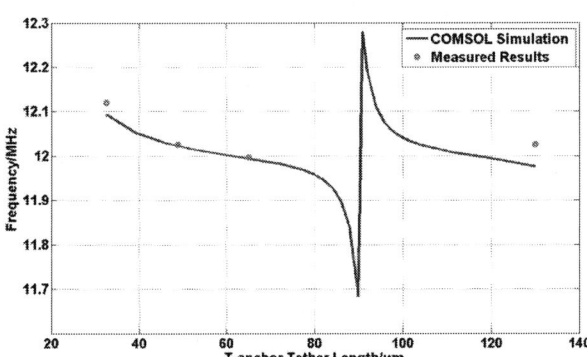

Fig. 15. Measured & FE simulated SE mode eigenfrequencies as a function of T-shaped tether length from Die #2.

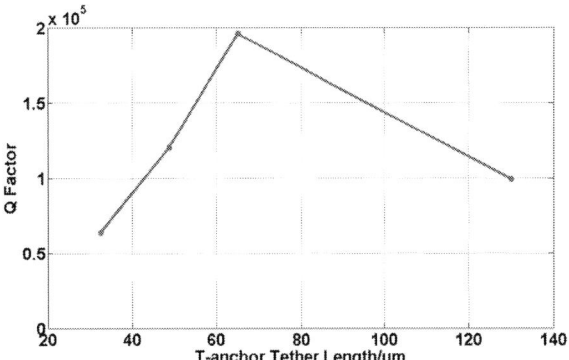

Fig. 16. Extracted SE mode Q-factors as a function of T-shaped tether length measured from Die #1.

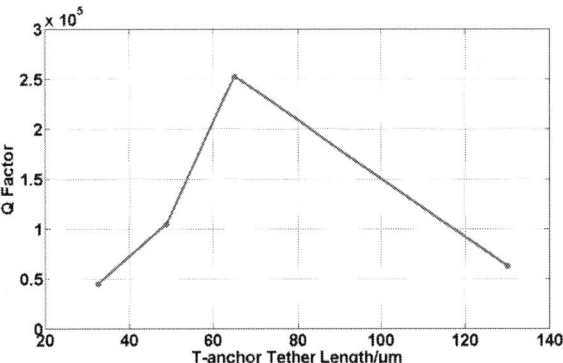

Fig. 17. Extracted SE mode Q-factors as a function of T-shaped tether length measured from Die #2.

The reason that the variations of the tether length have an impact on both the resonant frequency and the Q factor can be explained in Fig. 18. Fig. 18 shows a COMSOL simulation at the frequency of SE mode, but clearly as shown in Fig. 12, the tether begin to resonant and the mode shape has been altered due to the resonating tether. So the frequency will change and the Q factor will drop due to large stress imposed on the tether causing the energy flow to the substrate with elastic wave propagation.

By comparing the ratio of distributed strain energies in the tether and the entire device, it was previously found that the tether design had little impact on Q for the Lamé mode but notable for the SE mode [12]. These measurement results for both the Lamé and SE mode reported in this paper have now further verified these predictions based on FE simulations in [12].

978-1-4673-1122-9/12 $31.00 © 2012 IEEE   467

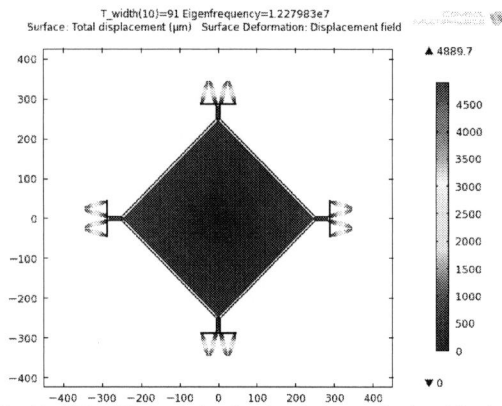

Fig. 18. COMSOL simulation showing the large tether length begins to alter the mode shape.

Fig. 19 further shows the variation of the mode shape as a function of changes in the T-shape tether length from 65μm to 86μm, to 94μm and finally to 104μm.

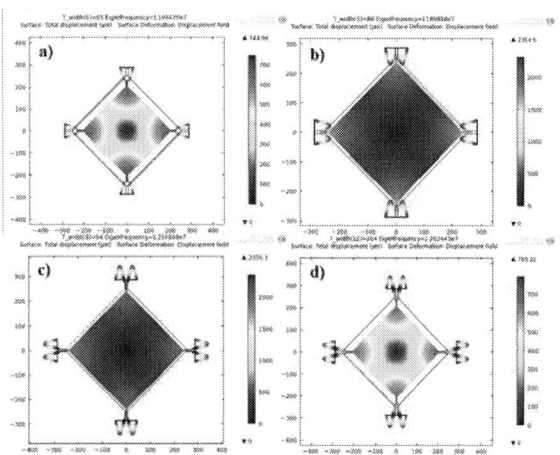

Fig. 19. COMSOL simulation showing the mode shape change due to the tether length change: a) 65μm, b) 86μm, c) 94μm and d) 104μm

IV. CONCLUSION

We have examined the impact of varying the length of the clamped section in a T-shaped tether on the resonant frequency and Q factor for both the Lamé and SE modes in square plate resonators. For the Lamé mode, neither the resonant frequency nor the Q factor shows any observable change resulting from variations in the clamped section length of the T-shaped tether length. This is largely attributed to the location of the nodes at the corners of the square-plate where the resonator can be conveniently clamped. However, for the SE mode resonators, variation of the clamped section length in the T-shaped tethers demonstrates significant changes in both Q and the resonant frequency. These observations are consistent across repeated measurements carried out on two different dies. These trends observed from measurements of the resonant frequency agree well with those obtained by FE simulations for both the Lamé and SE modes. These results verify a previously reported hypothesis that anchor loss plays a prominent role in determined the over Q for an SE mode but not the Lamé mode. This thereby points to the need for additional care to be taken when designing for the SE mode.

V. ACKNOWLEDGMENT

The work described in this paper was fully supported by a grant from the City University of Hong Kong (Project No. 7002690).

REFERENCES

[1] J. Verd, A. Uranga, G. Abadal, J.L. Teva, F. Torres, J.L. Lopez, E. Perez-Murano, J. Esteve, N. Barniol, "Monolithic CMOS MEMS Oscillator Circuit for Sensing in the Attogram Range," *IEEE Electron Device Letters*, vol.29, no.2, pp.146-148, Feb. 2008

[2] C. T.-C. Nguyen, "MEMS technology for timing and frequency control," *IEEE Trans. On Ultrasonics, Ferroelectrics, and Frequency Control*, vol. 54, no. 2, pp. 251-270, Feb. 2007.

[3] L.Khine, M.Palaniapan, and W.K.Wong, "12.9MHz Lamé-Mode Differential SOI Bulk Resonators," *Transducers'07*, pp.1753-1756, June 2007.

[4] K. S. Lebouitz, A. Mazaheri, R.T. Howe, A. P. Pisano, "Vacuum encapsulation of resonant devices using permeable polysilicon," Twelfth IEEE International Conference on Micro Electro Mechanical Systems, 1999, vol., no., pp.470-475, 17-21 Jan 1999.

[5] Z. Hao and F. Ayazi, "Support loss in micromechanical disk resonators", IEEE MEMS Conf., 2005, pp. 137-141.

[6] P. G. Steeneken, J. Ruigork, S. Kang, J. van Beek, J. Bontemps, and J. J. Koning, "Parameter Extraction and Support-Loss in MEMS Resonators," in COMSOL Users Conference 2007 Grenoble, 2007.

[7] J. E.-Y. Lee, J. Yan, and A. A. Seshia, "Low loss HF band SOI wine glass bulk mode capacitive square-plate resonator," *Journal of Micromechanics and Microengineering*, vol. 19, 074003, Jul 2009.

[8] J. E.-Y. Lee, Y. Zhu, and A. A. Seshia, "Bulk acoustic mode single-crystal silicon microresonator with high quality factor," *Journal of Micromechanics and Microengineering*, vol. 18, 064001, Jun 2008.

[9] S. Basrour, H. Majjad, J. R. Coudevylle, and M de Labachelerie, 2001 "Simulation and characterization of high Q microresonators fabricated by UV-LIGA," in Proc. 2001 Int. Conf. on Modeling and Simulation of Microsystems (MSM'01), pp. 294–297

[10] W. A. Brantley, "Calculated elastic constants for stress problems associated with semiconductor devices," Journal of Applied Physics, vol. 44, no. 1, pp.534 - 535, 1973

[11] D. Berlincourt and H. Jaffe, "Elastic and piezoelectric coefficients of single-crystal Barium Titanate", *Physical Review*, vol. 111, no. 1, pp. 143-148

[12] J. E.-Y. Lee, J. Yan, and A. A. Seshia, "Study of lateral mode SOI-MEMS resonators for reduced anchor loss," *Journal of Micromechanics and Microengineering*, vol. 21, 045010, Apr. 2011

978-1-4673-1122-9/12 $31.00 © 2012 IEEE

Analysis of Air Damping in Micromachined Resonators

Guoqiang Wu[1,2], Dehui Xu[1], Bin Xiong[1], Yinglei Ma[1], Yuelin Wang[1] and Errong Jing[1,3]

[1]State Key Laboratory of Transducer Technology, Science and Technology on Micro-system Laboratory, Shanghai Institute of Microsystem and Information Technology, Chinese Academy of Sciences, Shanghai, China;
[2]Graduate School of Chinese Academy of Sciences, Beijing, China
[3] CSMC Technologies Corporation, Wuxi, China
dehuixu@mail.sim.ac.cn

Abstract—**An approach to the modeling and simulating the air damping effect on quality factor (Q) of a square plate Lamé mode microresonator is presented in this paper. Both the squeeze film damping and slide film damping are considered in the analysis procedure. The Reynolds equation and the Stoke-flow model are used to investigate the reactions of the resonant plate with the air flows in the transduction gaps and around the resonator plate surfaces, respectively. An electrical equivalent model has been derived for a microresonator operating in air. The model is realized with resistors equivalent for the slide film damping force and frequency-dependent resistors and capacitances connected in series equivalent for the squeeze film damping force. The simulated transmission characteristics are in good agreement with the experimental results for a 4.13 MHz Lamé mode microresonator.**

Keywords-air damping; bulk mode microresonator; transmission characteristic; electrical equivalent model

I. INTRODUCTION

Silicon micromechanical resonators are very promising devices for RF application in recent years as they take advantage of small size, low power consumption and the potential for integration with mainstream integrated circuit [1-3]. However, the dynamic performances of the microresonators are strongly affected by several energy dissipation mechanisms. The energy dissipation influences the behavior of MEMS devices in various ways, depending on their design criteria and operating conditions. For microresonators operating in ambient conditions, it has been verified that air damping is the dominant energy loss mechanism [4, 5]. Although the air damping effect could be eliminated by encapsulating the micromechanical devices in vacuum, the packaging method is unacceptable in some areas, especially in mass-sensing and biochemical detections. So it is necessary to analyze and estimate the effect of air damping on the dynamic performance of the micromechanical structures. In the implementation of micromechanical resonators, low damping is required in order to obtain a high quality factor (Q), which is desirable for realizing low-phase-noise micromechanical resonators as well as high-resolution resonant sensors.

In this paper, we present a model for extracting the air damping effect on Q of the microresonators. The Reynolds

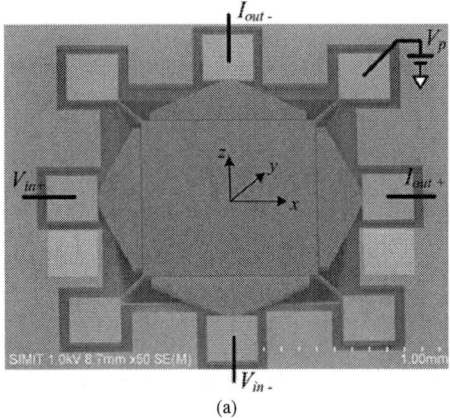

(a)

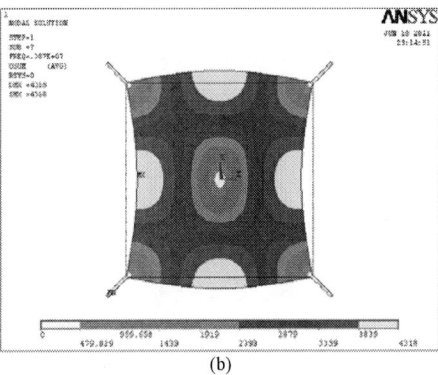

(b)

Fig. 1. Schematic of the square plate Lamé mode resonator. (a) Scanning electron microscope of the resonator, along with the driving and sensing setups. (b) A contour plot of the relative displacement of each point of the resonator obtained by ANSYS simulation.

equation is used to present the squeeze film damping effect and the Stoke-flow model is employed to analysis the slide film damping effect. The transmission characteristics of the microresonators operating in air are predicted and simulated. An electric equivalent circuit model for the forces created by squeeze film damping and slide film damping is proposed. The simulation results are shown to accurately reproduce the

978-1-4673-1122-9/12 $31.00 © 2012 IEEE

experimentally measured resonator response using our approach.

II. THEORY

Fig. 1(a) shows the schematic diagram of the square plate microresonator fabricated for our damping test. The square plate resonator is vibrating in Lamé mode excited by the electrostatic actuation, which is achieved by applying a superposition of dc component V_{dc} and an ac component V_{ac} to the transduction gaps. In this mode, the square plate resonator may be described as a thin square plate that is contracting along one axis, while simultaneously extending along the orthogonal axis in the plate, as shown in Fig. 1(b).

The basic mechanisms of air damping are squeeze film damping and slide film damping [6, 7]. As the square plate resonator vibrating in Lamé mode, the vibration of the square plate will lead to the compression of the air fluid in the transduction gaps, causing squeeze film air damping. The slide film air damping is induced by the lateral movement of both the top and bottom side of the resonator plate. Both of them will affect the transmission characterization of the resonator device.

In order to simplify the analysis procedure, the Lamé mode shape is approximated as a superposition of two orthogonal sound waves with the displacements given as $u_x = U\sin(\pi x/L)$ and $u_y = -U\sin(\pi y/L)$ [8], where L is the plate size, U is the maximum vibration amplitude along x- and y- axis, as shown in Fig. 1. Thus the vibration problem of the square plate resonator could be simplified as one dimensional nature of a beam resonator operating in the longitudinal mode.

The squeeze film damping force consists of two main components: the viscous damping force due to viscous and the elastic damping force due to the compressibility of the fluid. The normalized squeeze film damping force for a pair of parallel plates can be obtained as:

$$F_{squ} = -k_e u - c_d \tilde{u} \qquad (1)$$

where F_{squ} is the squeeze film damping force, u denotes the displacement of the plate with reference to its balanced position, $\tilde{u}$ is vibrating velocity of the plate, k_e and c_d present the coefficient of the elastic damping force and coefficient of the viscous damping force, respectively.

According to [10], k_e and c_d can be obtained as:

$$k_e = \frac{64\sigma^2 p_a A_e}{\pi^8 h_0} \sum_{m,n \; odd} \frac{1}{(mn)^2\left\{\left[m^2+(\frac{n}{\beta})^2\right]^2+\frac{\sigma^2}{\pi^4}\right\}} \qquad (2)$$

$$c_d = \frac{64\sigma p_a A_e}{\pi^6 h_0 \omega} \sum_{m,n \; odd} \frac{m^2+(\frac{n}{\beta})^2}{(mn)^2\left\{\left[m^2+(\frac{n}{\beta})^2\right]^2+\frac{\sigma^2}{\pi^4}\right\}} \qquad (3)$$

where m and n are odd integers 1, 3, 5,..., β is transduction area aspect ratio, ω, h_0, A_e, M and p_a are radial oscillation frequency, gap thickness, transduction area, mass of the

resonator and the ambient pressure, respectively. The squeeze number σ is expressed as $\sigma = 12\mu\omega L_e^2/p_a h_0^2$, where μ denotes

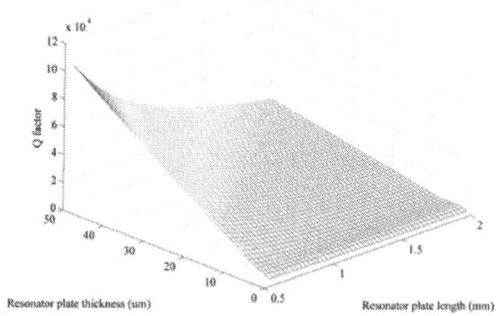

Fig. 2. The influence of the resonator dimension on Q.

TABLE I
COMPARISON THE CONTRIBUTIONS OF DIFFERENT
ENERGY DISSIPATION MECHANISMS TO THE OVERALL ESTIMATED Q

	Q_{squ}	Q_{sli}	Q_{other}	Q_{all}
Value	15172	1.26×10^5	1.02×10^6	13359
Contribution to total Q	88.1%	10.6%	1.3%	—

TABLE II
MEASURED Q FOR SIX DEVICES TESTED IN ATMOSPHERIC PRESSURE

	1#	2#	3#	4#	5#	6#	Average
Q_{mea}	11152	12386	12260	11746	13100	12360	12167

the coefficient of viscosity of the fluid, L_e is the length of the electrode.

The quality factor, Q_{squ}, due to the squeeze film damping of the air fluid is expressed as [9, 10]:

$$Q_{squ} = \frac{\pi^6 h_0 M \omega^2}{64\sigma P_a A_e} \sum_{m,n \; odd} \frac{(mn)^2\left\{\left[m^2+(\frac{n}{\beta})^2\right]^2+\frac{\sigma^2}{\pi^4}\right\}}{m^2+(\frac{n}{\beta})^2} \qquad (4)$$

As the resonator vibrating in ambient air, the air film behaves as a slide film damper to both top and bottom sides of the resonator. The Stoke-flow model is used to assume the slide film damping [6]. The moving velocity for different point of the square resonator vibrating in Lamé mode is not uniform. However, we could calculate the total slide film damping force on the resonator plate through integrating over the plate area. Then the energy dissipated by the slide film damping could be deduced. The quality factor due to the stokes-type damper underneath the plate can be obtained as [11]:

$$Q_{sli\_bottom} = \frac{\pi M \omega \delta}{\mu A_s}\left(\frac{\cosh 2d^* - \cos 2d^*}{\sinh 2d^* + \sin 2d^*}\right) \qquad (5)$$

where A_s is area of the top surface of the resonator plate. The normalized variables $d^* = d/\delta$, where d is the gap distance between the resonator plate and the substrate, δ is the "effective decay distance", which is described as: $\delta = \sqrt{2\mu/\rho\omega}$, in which ρ denotes the density of the fluid.

Similarly, the quality factor caused by the stokes-type damping above the plate ($d=\infty$) is expressed as [11]:

$$Q_{sli\_top} = \frac{\pi M \omega \delta}{\mu A_s} \qquad (6)$$

When the gap distance is large enough ($d \gg \delta$), Q_{sli_bottom} equals Q_{sli_top}. The quality factor, Q_{sli}, due to the slide film damping of the air fluid on the top and bottom side of the resonator is expressed as

$$\frac{1}{Q_{sli}} = \frac{1}{Q_{sli\_bottom}} + \frac{1}{Q_{sli\_top}} \qquad (7)$$

For a microresonator vibrating in air fluid, the overall Q factor can be estimated as:

$$\frac{1}{Q_{all}} = \frac{1}{Q_{squ}} + \frac{1}{Q_{sli}} + \frac{1}{Q_{others}} \qquad (8)$$

where Q_{other} denotes contribution of energy losses mechanisms other than air damping. The air damping effect could be reduced in vacuum, so Q_{other} can be obtained through testing the resonator in low pressure.

In order to study the influence of the resonator dimension on air damping effect, a parametric study is performed to evaluate the influence of geometric parameters such as the length of square resonator plate L, and the thickness of the resonator plate t on Q value. Fig. 2 illustrates the influence of the resonator dimension on Q caused by air damping. For a resonator vibrating in air, Q rises as the plate thickness increases, which is due to the fact that the thicker the resonator plate is, the more energy in the vibrating system stores. While Q decreases as the plate side length increases. Since the larger the plate size is, the more energy dissipation caused by air will occur.

III. Experiment Results

The resonator size is 1000 μm×1000 μm×46 μm (Length × Width × Thickness). The transduction gap is measured as 2.4 μm. The length of the electrode is 900 μm. The density and Young's modules for single crystal silicon are 2330 kg/m^3 and 180 GPa, respectively. The coefficient of viscosity for air is 18.1×10^{-6} Pa-sec.

A. Comparison of the estimated Q with the measured ones

Table 1 compares the contribution of squeeze film damping (Q_{squ}), slide film damping (Q_{sli}) and other energy loss mechanisms (Q_{other}) with the total Q (Q_{all}). Q_{sli} and Q_{sli} are calculated according to (4) and (7). Q_{other} denotes the contribution of energy losses mechanisms other than air damping, which is obtained through testing the resonator in low pressure. The measured Q values for six devices tested in atmospheric pressure are summarized in Table 2. The average value of the measured Q is 8.9% lower than the estimated one. The discrepancy may be caused by the approximated vibration mode for the square plate, which takes the Lamé mode shape approximated as a superposition of two orthogonal acoustic waves, ignoring the interactions between the two motions of x- and y- axis.

B. Equivalent circuit for the air damping force

The electrical equivalent circuit for a microresonator can be model as a serial R-L-C circuit. The equivalent inductance L_m, resistance R_m, and capacitance C_m are given as [12]:

$$L_m = \frac{M}{\eta^2} \qquad R_m = \frac{c_0}{\eta^2} = \frac{\sqrt{k_0 M}}{\eta^2 Q} \qquad C_m = \frac{\eta^2}{k_0} \qquad (9)$$

where k_0 is the spring constant, the damping constant c_0 is expressed as: $c_0 = \sqrt{k_0 M}/Q$. The electromechanical coupling coefficient of the resonator system η is defined as: $\eta = V_{dc}\varepsilon A_e/g^2$, where V_{dc} is the bias voltage, ε is the permittivity of air, g is the transduction gap width. Considering the effects of the squeeze film damping and slide film damping, the electrical equivalent model for the resonator operating in air are obtained using following approach.

The electrical equivalent model for the forces caused by squeeze film damping, can be presented by resistors and capacitances connected in series form. The coefficient of the viscous damping force ($c_{d\,m,n}$) and the coefficient of the elastic damping force ($k_{e\,m,n}$) could be obtained according to (2) and (3). The component values $R_{m,n}$ and $C_{m,n}$ in the series expansion are expressed as:

$$R_{m,n} = \frac{c_{d\,m,n}}{\eta^2} = \frac{64\sigma p_a A_e}{\pi^6 h_0 \omega \eta^2} \sum_{m,n\,odd} \frac{m^2 + (\frac{n}{\beta})^2}{(mn)^2 \left\{ \left[m^2 + (\frac{n}{\beta})^2 \right]^2 + \frac{\sigma^2}{\pi^4} \right\}} \qquad (10)$$

$$C_{m,n} = \frac{\eta^2}{k_{e\,m,n}} = \frac{\pi^8 h_0 \eta^2}{64\sigma^2 p_a A_e} \sum_{m,n\,odd} (mn)^2 \left\{ \left[m^2 + (\frac{n}{\beta})^2 \right]^2 + \frac{\sigma^2}{\pi^4} \right\} \qquad (11)$$

The slide film damping effect can be equivalent as a resistive element in the electrical equivalent circuit of the microresonator. The equivalent coefficient of the slide film damping force (c_{sli_bottom} and c_{sli_top}) can be obtained using the approach illustrated in [6]. Thus, the equivalent resistors for damping forces caused by the fluid underneath the plate and above the plate can be written as:

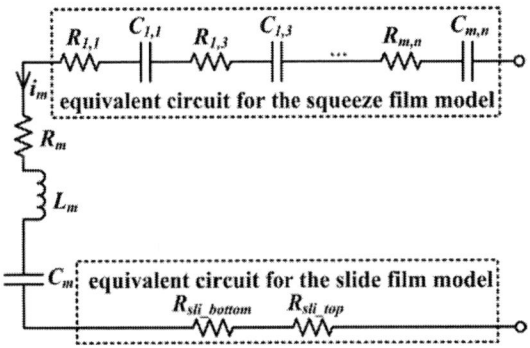

Fig. 3. Equivalent circuit for the resonator with the effect of air damping. The microresonator can be modeled as a serial R-L-C circuit. R_m, L_m and C_m denote the equivalent resistance, inductance and capacitance, respectively. The squeeze films damping force is modeled using series-connected R-C sections and the slide film damping forces are modeled using the resistive elements. R_{sli_bottom} and R_{sli_top} present the slide damping underneath the resonator plate and above the plate, respectively.

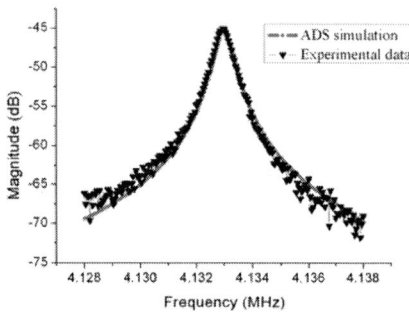

Fig. 4. Measured and ADS simulated transmission curves of the square plate resonator in terms of magnitude operating in ambient pressure, using 60V dc bias.

$$R_{sli\_bottom} = \frac{c_{sli\_bottom}}{\eta^2} = \frac{\mu A_s}{\pi \delta \eta^2}(\frac{\sinh 2d^* + \sin 2d^*}{\cosh 2d^* - \cos 2d^*}) \qquad (12)$$

$$R_{sli\_top} = \frac{c_{sli\_top}}{\eta^2} = \frac{\mu A_s}{\pi \delta \eta^2} \qquad (13)$$

The complete equivalent circuit for the resonator with squeeze film damping and slide film damping forces is shown in Fig. 3.

The differential drive and sense measurement setup is used in this work. The electrical transmission characteristic of the resonator is measured using an HP 4395A network analyzer. The circuit design program ADS is composed to obtain the resonator response. In our simulations nine RC sections are used to model the squeeze film damping. The component values of the remaining sections are insignificant and can therefore be ignored. Fig. 4 shows the measured and simulated transmission curves of the resonator. The simulated curve agrees well with the measured one, which confirms the rationality of our method.

IV. CONLUSIONS

The effect of air damping on the transmission characteristics of a bulk mode microresonator is reported. Considering the slide film damping and squeeze film damping, an equivalent circuit model for a bulk mode microresonator operating in air is proposed. We conclude that the effect of air damping on the dynamic performance of a mircoresonator can be equivalent as spring and damper elements in series form in an electrical equivalent circuit. The simulation results accord well with the measured ones. The results of this study give a very useful way to estimate and simulate the dynamic performances of micromechanical structure operating in ambient pressure.

ACKNOWLEDGMENT

This work was supported in part by National S&T Major Project of China under Grant 2011ZX02507-003, in part by a grant from the Major State Basic Research Development Program of China (973 Program) (No. 2011CB309501). The authors would like to thank Yan Li for her help in preparing this paper.

REFERENCES

[1] C. T.-C. Nguyen, "MEMS technology for timing and frequency control," *IEEE Trans. Ultrason. Ferroelectr. Freq. Contr*, vol. 54 (2), pp. 251-270, February 2007.

[2] J. E.-Y. Lee, B. Bahreyni, Y. Zhu, and A. A.Seshia, "A single-crystal-silicon bulk-acoustic-mode microresonator oscillato," *IEEE Electr. Dev. Lett.* vol. 29(7), pp. 701-703, July 2008.

[3] T. H. Metcalf, B. B. Pate, D. M. Photiadis, and B. H. Houston, "Thermoelastic damping in micromechanical resonators," *Appl. Phys. Lett.* vol. 95, 061903, August 2009.

[4] J. E. Y. Lee, J. Yan, and A. A. Seshia, "Low loss HF band SOI wine glass bulk mode capacitive square-plate resonator," *J. Micromech. Microeng.* vol. 19(7), 074003, July 2009.

[5] B. L. Bircumshaw, "High-Q COMS-compatible poly-silicon germanium electrostatic RF MEMS resonators," *Ph.D. dissertation, California Univ. Berkeley USA*, 2005.

[6] M. H. Bao, *Analysis and Design Principles of MEMS Devices, Elsevier, Amsterdam*, 2005.

[7] T. Veijola, "Quality factor and resonance frequency shift due to air in RF MEMS radial disk resonators, " *Transducers '07 & Eurosensors Xxi, Digest of Technical Papers*, 1&2, pp. 643-646, 2007.

[8] J. E. Y. Lee and A. A. Seshia, "5.4-MHz single-crystal silicon wine glass mode disk resonator with quality factor of 2 million," *Sens. Actuators A* , vol. 156(1) pp. 28-35, February 2009.

[9] Sarne M. Hutcherson, "Theoretical and Numerical Studies of the Air Damping of Micro-Resonators in the Non-Continuum Regime," *M.S. thesis, Georgia Institute of Technology, USA*, 2004.

[10] M. H. Bao and H. Yang, "Squeeze film air damping in MEMS," *Sens. Actuators A*, vol. 136(1), pp. 3-27, January 2007.

[11] Y. H. Cho, A. P. Pisano, and R. T. Howe, "Viscous damping model for laterally oscillating microstructures," *J. Microelectromech. Syst.* vol. 3(2), pp. 81-87, June 1994.

[12] T. Mattila, J. Kiihamäki, T. Lamminmäki, O. Jaakkola, P. Rantakari, A. Oja, H. Seppä, H. Kattelus, and I. Tittonen, "12 MHz micromechanical bulk acoustic mode oscillator," *Sens. Actuators A Phy* vol. 1(1), pp. 1–9, September 2002.

978-1-4673-1122-9/12 $31.00 © 2012 IEEE

Benchmarking the Passive Differential Input Technique to Shielded GSG Probes

Yuanjie Xu, Haoshen Zhu[*] and Joshua E.-Y. Lee

[*]Department of Electronic Engineering, City University of Hong Kong, Kowloon, Hong Kong SAR
haoshezhu2@student.cityu.edu.hk

Abstract— **We have previously showed that feedthrough in MEMS resonators of generic mode shapes can be suppressed by as much as close to 40dB by applying a passive differential input technique. This paper compares the degree of feedthrough suppression achieved via the passive differential input technique against electrical measurements carried out using shielded GSG microwave probes. The shielded GSG microwave probes in themselves provide for excellent isolation between the input and output ports. The device under test was a square-extensional (SE) mode micromechanical resonator probed at 4 electrodes for actuation. Our results show, that despite as simple an implementation as it presents, the passive differential input technique offers additional 20dB suppression relative to the GSG probes.**

Keywords- resonator; square-extensional mode; passive differential input; shielded GSG probes

I. INTRODUCTION

MEMS resonators possess small form factors and possibility for integration with CMOS interface electronic circuitry [1]. Their small form factor provides MEMS devices with tremendous portability and capability for many MEMS devices to be integrated together to form a system [2]. The electro-mechanical-electro manner of transduction for MEMS resonators enables MEMS devices to electrically interface with the input/output circuit while allowing their mechanical resonance to provide them with a very high Q [3] (over a million), which is desirable for oscillator circuits to reduce phase noise. These advantages have motivated research on MEMS resonators for timing applications as an attractive replacement over quartz crystals [4].

Apart from timing applications, MEMS resonators are also used for sensing applications.

The detection of a desired input stimulus is achieved through monitoring shifts in the resonant frequency. A major advantage over of resonant sensing over static sensing is that it allows implementation of a digital output. Typical examples include resonant accelerometers [5] and mass detection for small particles such single virus strand [6].

Detection of the resonant structure's motion can be done using a variety of means that include optical detection [6] and magnetic induction electrical readout [7]. The drawback of these detection methods lies in their cumbersome setup that is unsuitable for integrated systems. For practical applications, full electrical characterization (whereby the input and output are both electrical signals) remains the most desirable. But in

moving to full electrical characterization, electrical crosstalk (feedthrough) between input and output ports presents a well-known challenge to characterization.

The problem of feedthrough grows in significance as the size of the MEMS device is scaled down to achieve higher resonant frequencies. Due to the small form factor of the device, the feedthrough current through a parasitic capacitor appearing between the input and output ports can dominate the overall transmission particularly since the motional impedance scales inversely with the size of the device. For timing applications and resonant microsensors where the operating frequencies can range from 100 kHz to a few GHz, electrical feedthrough could hijack the output of the device such that its resonance cannot be uncovered. This can be shown in Fig. 1, where a parallel capacitance C_f models the feedthrough and a series resonant circuit is used to represent the resonator. Here, R_m, L_m, and C_m respectively denote the motional resistance, inductance and capacitance when a DC bias voltage (V_P) is applied to the MEMS device.

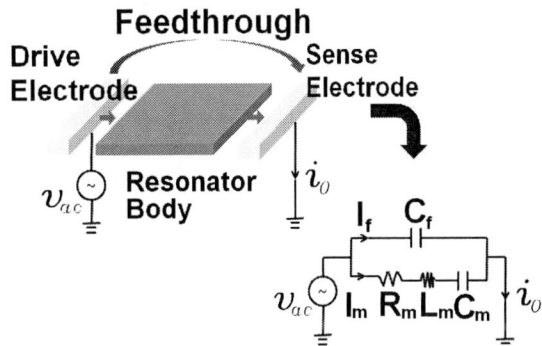

Fig. 1. Electrical characterization of a MEMS resonator and its equivalent circuit showing the feedthrough capacitor across the input and output ports

The parameters in the resonant circuit can be calculated by:

$$L_m = \frac{M}{\xi^2 V_P{}^2}, R_m = \frac{\sqrt{KM}}{\xi^2 V_P{}^2}, C_m = \frac{\xi^2 V_P{}^2}{K} \qquad (1)$$

K and M are the resonator's effective spring constant and mass respectively; V_P is the applied DC bias voltage; $\xi =$

978-1-4673-1122-9/12 $31.00 © 2012 IEEE

dC/dx, which is the normalized transduction coefficient [8]. Thus the overall admittance can be deduced as:

$$Y(\omega) = j\omega C_f + \frac{j\omega C_m}{\left[1 - (\omega/\omega_0)^2 + j(\omega/\omega_0)/Q\right]} \quad (2)$$

ω_0 is the mechanical resonant frequency of the resonator, and Q is the quality factor. From (2), the impact of increasing feedthrough level on the transmission can be illustrated in Fig. 2.

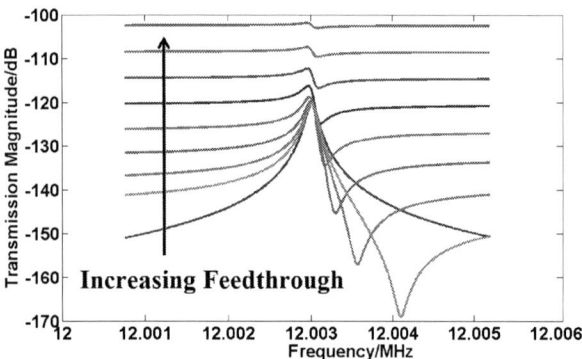

Fig. 2. Impact of increasing feedthrough level on the transmission

As such, the problem of feedthrough presents a challenge to full electrical characterization of MEMS resonators since the motional current may be fully buried under feedthrough to render the resonant peak undetectable [9].

One way of suppressing electrical feedthrough is to employ a fully-differential transduction configuration [10]. However, the requirement of a differential mode of resonance narrows the scope of application for this method. Furthermore, for a fully-differential transduction configuration, the transduction area is halved compared to a one-port configuration as in [11]. While all of the electrodes in [11] can be used for actuation, only half of the electrodes can be used for actuation in a fully-differential configuration while the other half is used for sensing. Another disadvantage of a fully-differential configuration is that alternative mechanisms of sensing the motional current cannot be realized in a differential mode. One such example is piezoresistive sensing, which has been shown to provide superior transduction and scales better with frequency compared with a capacitive sensing configuration [12].

We have previously demonstrated feedthrough suppression by 50dB in a symmetric mode by using a novel passive differential input technique [13]. In this paper, the aim to compare the effectiveness of this method for suppressing electrical feedthrough against measurements obtaining using GSG probes. GSG microwave probes are commonly used in RF electrical characterization as these are designed to reduce probe to probe crosstalk for unpackaged die measurements.

As such, this provides an excellent reference to compare our package-level feedthrough cancellation method against.

II. PASSIVE DIFFERENTIAL INPUT VS GSG PROBES

A. Passive differential input

The basic idea for passive differential input technique is to construct a matched negative feedthrough capacitance through applying a differential input signal to a non-actuating pad that is adjacent to the actuating pad. As shown in Fig.3, the actuating pad together with its active signal carrying wire forms a feedthrough capacitance C_f with the output port. Meanwhile, the non-actuating pad (i.e. the passive pad) together with the passive signal carrying wire forms another feedthrough capacitance C_f' with the output port. As the passive pad is lies adjacent to the actuating pad, C_f' will be approximately equal to C_f (thus matching C_f and C_f' with each other). By applying a differential input signals between the active and passive pads, the matched C_f' can be used to suppress C_f. The parasitic feedthrough capacitances C_f and C_f' are further illustrated in the micrograph of Fig. 4.

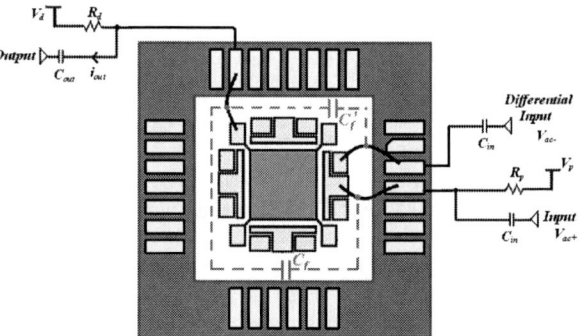

Fig. 3. Schematic showing the negative capacitance produced by differential input to non-actuating pad adjacent to actuating pad

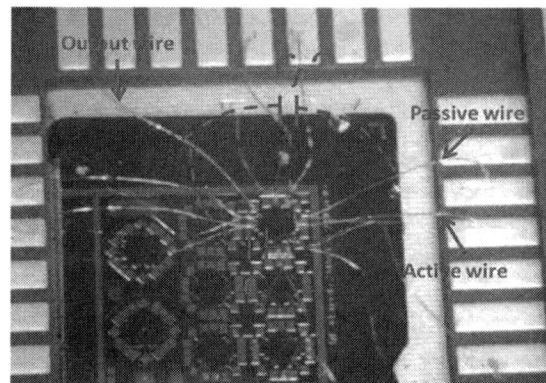

Fig. 4. Micrograph of the feedthrough capacitances C_f between the active and the output wire bonds, and C_f' between the passive and output wire bonds

978-1-4673-1122-9/12 $31.00 © 2012 IEEE

B. Shielded GSG probes

Shielded GSG microwave probes help to suppress electrical feedthrough by trapping fringing fields from the RF signal to the adjacent ground pads. Fig. 5 shows a side profile of an RF GSG probe. The signal probe is in the center surrounded by two ground probes on each side. When probing the metal pads, RF radiation from the signal probe in the center will couple to the two ground probes, so that the RF fringing field from the center signal probe will be trapped locally by the ground probes. The net effect is that the electrical feedthrough seen by the output is greatly reduced.

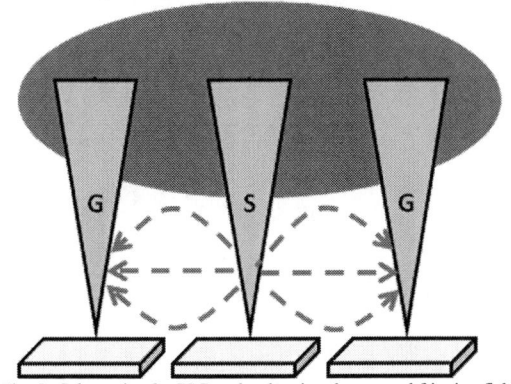

Fig. 5. Schematic of a GSG probe showing the trapped fringing field

Here, we compare the our novel passive differential input technique against measurements using shielded GSG probes in terms of its effectiveness for suppressing feedthrough.

III. DEVICE AND MEASUREMENT SETUP

A. Device Description

Square-plate resonators with a length of 360μm and 25μm thickness were fabricated in an SOI MEMS foundry process. The resonators were excited in the square-extensional (SE) mode, as described by the FE simulation in Fig. 6. The resonators were aligned along the [110] direction on a (100) silicon wafer.

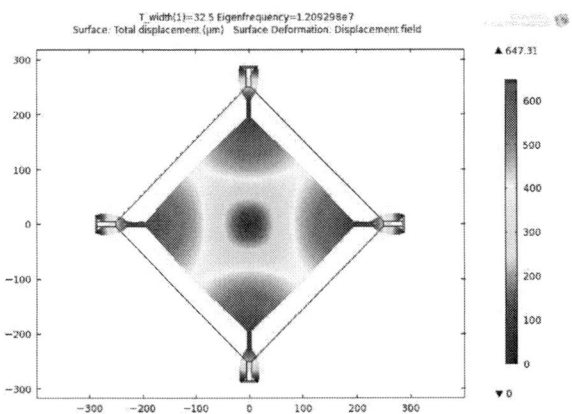

Fig. 6. COMSOL simulation of an SE mode resonator

The resonant frequency of a square plate resonator vibrating in SE mode can be analytically calculated by [14]:

$$f_{0,SE} = \frac{1}{2L} \sqrt{\frac{E}{\rho(1-\nu)}\left[1 + \left(1 - \frac{8}{\pi^2}\right)\left(\frac{\nu}{\nu-1}\right)\right]} \quad (3)$$

Here L refers to the square plate's length, ρ is the density of silicon, E and υ are respectively the Young's modulus and Poisson's ratio of silicon in the [110] direction. Based on a length of 360μm, the elastic properties reported in [15] (E = 169GPa and ν = 0.064), and a density of 2330 kg/m³ for silicon, the calculated resonant frequency based on (3) is about 12.14MHz.

B. Measurement Setup

1) Packaged die

One of the devices was wire bonded in a 28-DIL package according to the layout shown in the optical micrograph of Fig. 7, and characterized in a custom-built vacuum chamber at a pressure of 10mTorr. With reference to the schematic in Fig. 8, the electrical transmission of the packaged device was measured with and without differential inputs applied. The electrical transmissions of the resonators were measured using an Agilent E5061A network analyzer.

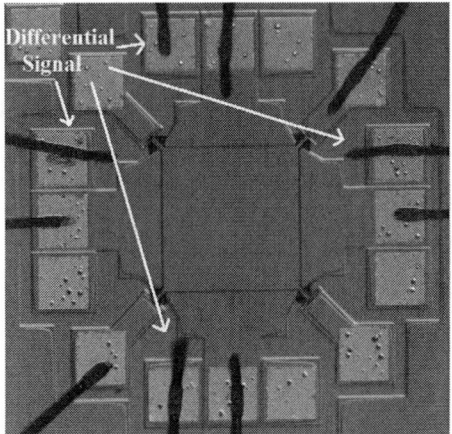

Fig. 7. Optical micrograph for passive differential input pads

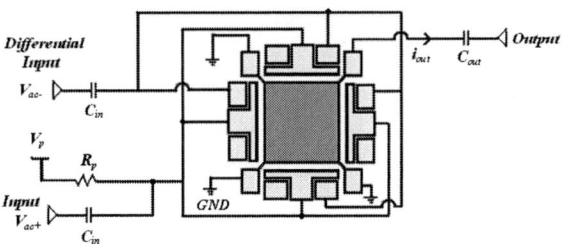

Fig. 8. Circuit schematic for passive differential input

978-1-4673-1122-9/12 $31.00 © 2012 IEEE

2) Unpackaged bare die

Another device on a bare die was electrically characterized under vacuum at 0.5 mTorr in a Janis Research cryogenic probe station using GSG RF probes (Fig. 9). The electrical transmission of the resonator was measured using an Agilent E5061A network analyzer. A circuit schematic for the SE mode measurement setup with GSG probes is described in Fig. 10.

Fig. 9. Micrograph of the device electrically interfaced with GSG probes

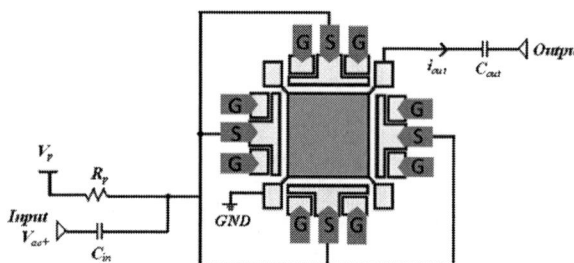

Fig. 10. Circuit schematic for measurement with GSG probes

IV. MEASUREMENT RESULTS

A. Passive Differential Input Measurement Results

Fig. 11 shows the electrical transmission of the packaged device with an AC and DC drive voltages applied to the active wire while the passive wire is grounded. In this configuration, the resonant peak height was 0.04dB above a floor of -54dB. Fig. 12 shows the electrical transmission of the same device when an AC input out of phase with active wire is applied to passive wire. Using differential inputs, the peak height was as high as 2dB above a floor of -90dB. Since the measured floor is set by parasitic feedthrough, we see that a 37dB reduction in feedthrough was achieved.

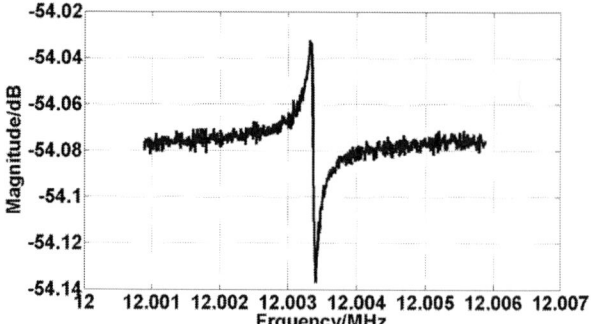

Fig. 11. Measured transmission magnitude for packaged device without differential input applied (grounded passive wires)

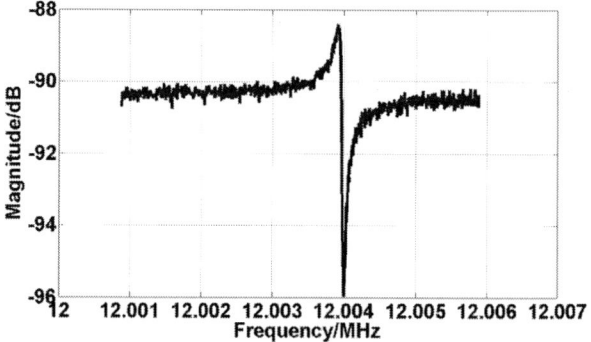

Fig. 12. Measured transmission magnitude for packaged device with differential input applied.

B. GSG Probes Measurement Results

Fig. 13 shows the measured transmission with the GSG probes. The resonant peak height is 0.2dB above a -69.1dB floor.

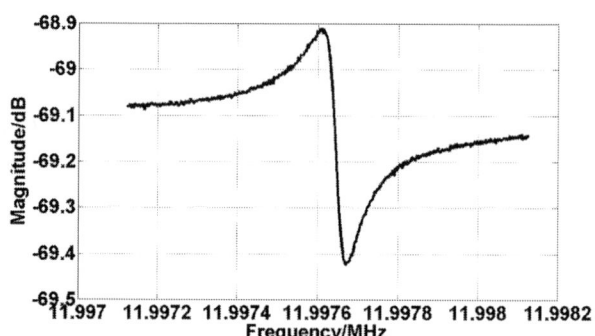

Fig. 13. Measured transmission magnitude for device (unpackaged) characterized with GSG probes

Fig. 14 compares the three measurements. We can see that use of GSG probes reduces feedthrough by 15dB relative to the packaged device when the passive wire is simply grounded. When differential inputs are applied to the packaged device, (an AC input out of phase with the active wire is applied to the passive wire) the measured feedthrough is lower than with GSG probes used by 20dB.

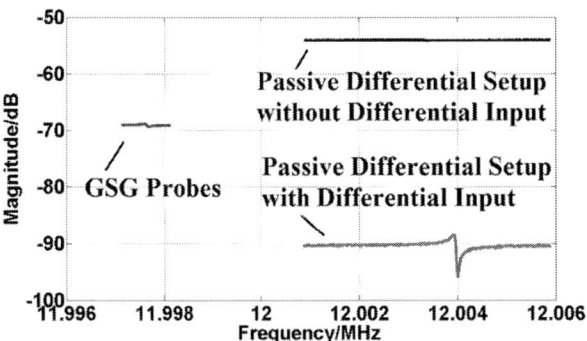

Fig. 14. Comparison between measurements of the bare and packaged devices showing a 20dB suppression using the passive differential technique

Grounding the passive wire in the packaged device mimics the function of a GS probe, which we have found to provide an additional isolation of 10 dB compared to leaving this point to float. GSG probes enhance this effect, hence providing even further isolation. The observed additional 20dB suppression in the packaged device employing differential inputs shows the effectiveness the feedthrough cancellation scheme. Since the design of GSG probes minimizes probe to probe coupling, the 20dB enhancement indicates that the observed feedthrough stems from coupling through the substrate. In short, the results here show that the passive differential input technique, though simple is still nonetheless highly effective for providing feedthrough cancellation.

V. CONCLUSION

We have compared the degree of feedthrough suppression achieved via the passive differential input technique against measurements carried out using shielded GSG RF microwave probes. The device under test was a square-extensional (SE) mode micromechanical resonator. Electrical characterization using shielded GSG microwave probes provides for excellent isolation between the input and output ports, as can be seen from the -69.1dB feedthrough level. Our results show, that despite as simple an implementation as it presents, the passive differential input technique offers additional 20dB suppression relative to the GSG probes. These results further promote the effectiveness of the novel passive differential input technique for reducing electrical feedthrough at the package-level.

VI. ACKNOWLEDGEMENT

The work described in this paper was supported by a grant from the City University of Hong Kong (Project No. 7008088).

REFERENCES

[1] J. Verd, A. Uranga, G. Abadal, J.L. Teva, F. Torres, J.L. Lopez, E. Perez-Murano, J. Esteve, N. Barniol, "Monolithic CMOS MEMS Oscillator Circuit for Sensing in the Attogram Range," *IEEE Electron Device Letters*, vol.29, no.2, pp.146-148, Feb. 2008.

[2] C. T. C. Nguyen, "Method and apparatus for selecting at least one desired channel utilizing a bank of vibrating micromechanical apparatus," U.S. Patent 6,680,660. Jan. 2004.

[3] Y. Xu, J. E.-Y. Lee, "Feedthrough cancellation in micromechanical square resonators via differential transduction," IEEE International Conference on Electron Devices and Solid-State Circuits, Dec 2010.

[4] C. T.-C. Nguyen, "MEMS technology for timing and frequency control," IEEE Trans. On Ultrasonics, Ferroelectrics, and Frequency Control, vol. 54, no. 2, pp. 251-270, Feb. 2007.

[5] A. A. Seshia, M. Palaniapan, T. A. Roessig, R. T. Howe, R. W. Gooch, T. R. Schimert, S. Montague, "A vacuum packaged surface micromachined resonant accelerometer", JMEMS, vol. 11, no. 6 pp. 784 – 793, Dec 2002.

[6] A. Gupta, D. Akin, & R. Bashir, "Single virus particle mass detection using microresonators with nanoscale thickness", Appl. Phys. Lett. 84, pp.1976–1978, 2004.

[7] A. N. Cleland and M. L. Roukes, "A nanometre-scale mechanical electrometer," *Nature*, vol. 392 pp. 160-162, 1998

[8] J. E.-Y. Lee, A. A. Seshia, "Parasitic feedthrough cancellation techniques for enhanced electrical characterization of electrostatic microresonators", Sensors and Actuators A, Vol. 156, pp.36–42, 2009.

[9] J. E.-Y. Lee and A. A. Seshia, "Direct parameter extraction in feedthrough-embedded capacitive MEMS resonators," Sensors and Actuators A: Physical, vol. 167, no. 2, pp. 237-244, Jun 2011

[10] S. A. Bhave, D. Gao, R. Maboudian and R. T. Howe, "Fully-differential poly-SiC Lame-mode resonator and checkerboard filter," 18th IEEE Int Conf on Micro Electro Mechanical Systems (MEMS 2005), Miami, Florida, Jan 30 - Feb 3, 2005, pp. 223-226.

[11] J. E.-Y. Lee, Y. Zhu, and A. A. Seshia, "Bulk acoustic mode single-crystal silicon microresonator with high quality factor," Journal of Micromechanics and Microengineering, vol. 18, 064001, Jun 2008.

[12] J. T. M. van Beek, G. J. A. M. Verheijden, G. E. J. Koops, K. L. Phan, C. van der Avoort, J. van Wingerden, D. Ernur Badaroglul, J. J. M. Bontemps, "Scalable 1.1 GHz fundamental mode piezo-resistive silicon MEMS resonator", IEEE International Electron Device Meeting, pp. 41-414, 2007.

[13] Y. Xu and J. E.-Y. Lee, " Single-device and On-chip Feedthrough Cancellation for Hybrid MEMS Resonators," IEEE Transactions on Industrial Electronics, 2011 (in press).

[14] D. Berlincourt and H. Jaffe, "Elastic and piezoelectric coefficients of single-crystal Barium Titanate", Physical Review, vol. 111, no. 1, pp. 143-148

[15] W. A. Brantley, "Calculated elastic constants for stress problems associated with semiconductor devices," Journal of Applied Physics, vol. 44, no. 1, pp.534 - 535, 1973

Study on Thermoelastic Dissipation in Bulk Mode Resonators with Etch Holes

Cheng Tu[*], Joshua E-Y Lee

[*]Department of Electronic Engineering, City University of Hong Kong, Kowloon, Hong Kong
chengtu3@cityu.edu.hk

Abstract—**This paper aims to investigate the primary cause for marked drops in quality factor (by over 98%) of a bulk mode resonator as a result of introducing etch holes into the structure. We show that thermoelastic damping (TED) appears to be the dominant cause of energy dissipation. The resonator is fabricated in single-crystal silicon (SCS). According to finite-element (FE) analysis, the uniform isochoric property of the Lamé mode is no longer preserved when etch holes are introduced. This results in marked drop in quality factor (Q) by 98.5% for the Lamé mode and 75.7% for the extensional from measurements. FE analysis based on coupled thermoelastic equations is used to compute the the thermal gradients that give rise to TED when etch holes are added to the resonator.**

Keywords-Etch holes; Lamé mode resonator; quality factor drop; thermoelastic dissipation

I. INTRODUCTION

Micromechanical resonators have shown great potential for use in time referencing and wireless applications. Their advantages over traditionally favored quartz crystals include their compact size, low power consumption and easier path towards integration with IC electronics [1]. Among the possible structural topologies realized from silicon micromachining, flexural and bulk mode micromechanical resonators are most common. Compared with flexural mode resonators, bulk mode resonators have been shown to possess a higher frequency-Q product [2] and better power handling capability [3]. In the fabrication process of bulk mode resonators (e.g. using bonded SOI technology), etch holes are often employed. The etch holes are introduced to facilitate the timed HF etch of the buried oxide (BOX) layer underneath the bonded silicon device layer that defines the resonator structure. Undercutting the structure by etching the BOX releases the resonator from the handling substrate [4-6]. In addition, it has also been demonstrated that etch holes can be used to help realize a temperature compensation scheme for microresonators [7-8]. In all these works, it was found that the resonant frequencies of the resonators with etch holes were consistently lower than those without etch holes. The drop in frequency is attributed to a reduction in the stiffness of the resonator to a larger degree over the mass through the introducing etch holes [8]. In fact, the use of etch holes for accurate frequency trimming has been proposed by [9]. Rabinovich [10] and Sharpe [11] have studied the effect of etch holes on the mechanical properties of micro-devices, showing that the strength of the material drops significantly

with the presence of the etch holes. The focus of these studies have been on the effect of etch holes on the elastic properties. In contrast, there remains little information on the effect of etch holes on the Q factor. Recently, it was pointed out by L. Shao et al. [12] that introducing square etch holes into a Lamé mode resonator degraded the Q by more than an order of magnitude. The extent of Q drop varied according to the location and distribution of the etch holes. The drop in Q was therein attributed to changes in the mode shape. However, no further quantitative study on particular energy loss mechanisms was reported. The motivation for this paper arises from previous report [13] which demonstrates that the slots (similar to etch holes in a bulk mode structure) in flexural beam resonators affect the thermomechanical coupling drastically; this in turn impacts the quality factor. Although that work was based on beam resonators, it is likely that the same physics applies to bulk mode resonators.

For a silicon-based microresonator, the net quality factor Q_{net} can be expressed as

$$\frac{1}{Q_{net}} = \frac{1}{Q_{viscous}} + \frac{1}{Q_{surface}} + \frac{1}{Q_{anchor}} + \frac{1}{Q_{TED}} + \frac{1}{Q_{AKE}} \quad (1)$$

where $Q_{viscous}$, $Q_{surface}$, Q_{anchor}, Q_{TED} and Q_{AKE} relates to the energy loss mechanism from air damping, surface loss, anchor loss, thermoelastic damping (TED) and Akhieser effect (AKE) loss respectively. By designing a resonator with resonant frequency at several MHz and operating it in vacuum, the effect of surface loss and viscous damping can be suppressed to a level where other sources of energy loss will dominate. This leaves only Q_{anchor}, Q_{TED} and Q_{AKE} from equation (1) to consider. Since the Lamé mode of the square-plate resonator has a uniformly isochoric strain pattern, the driving mechanism for TED is absent. Hence Q_{TED} is expected to be very high [14]. Furthermore, the commonly adopted T-type corner anchoring method [2] greatly minimizes the energy leakage to the substrate, resulting in a high Q_{anchor} as well. As such, it is believed that the overall observable Q, Q_{net} of the Lamé mode square resonator is set by Q_{AKE} [15]. In this paper, we demonstrate that when etch holes is uniformly distributed over a square resonator, the TED is increased to a level whereby Q_{TED} sets the lower bound limit for Q_{net}.

978-1-4673-1122-9/12 $31.00 © 2012 IEEE

II. SIMULATION AND ANALYSIS

To investigate the primary source of energy dissipation stemming from the introduction of etch holes into the structure, we first examined for changes in the mode shape. Such changes in the mode shape change could shift the node positions in the Lamé mode away from the corners (where the plate is clamped by tethers) to within the square-plate. This in turn would lead to additional energy loss at the anchors. This comparison was made by computing the longitudinal displacements along 3 selected line sections across the square-plate using finite-element (FE) simulations of the eigenmode as illustrated in Fig. 1. This was done for a plain resonator without etch holes and another with the same dimensions but with distributed etch holes. The FE computed longitudinal displacements along each line section (normalized against the maximum displacement) are given in Fig. 2. As Fig. 2 shows, the Lamé mode shape remains mostly unchanged with the addition of etch holes. As such, the significant drop in Q is unlikely to stem from additional anchor loss, indicating that the primary source of dissipation arises from the resonator itself.

Fig. 3 shows the FE computed strain components in both in-plane orthogonal axes of line section 2 for both resonators. With the addition of etch holes into the structure, we see that the distribution of strain in either in-plane axes is periodic about a mean standing waveform. In contrast, the strain distribution is a smooth standing wave for the plain resonator without etch-holes. Superposition of both in-plane strain components along the chosen line section yields the distributed volumetric change. Considering the symmetry of the Lamé mode, this observation would apply to the rest of the structure. Hence for the plain resonator, the distributed volumetric change is zero everywhere for the Lamé mode, thus indicating the absence of thermal gradients. This in turn implies that the driving mechanism for TED is absent. But when etch holes are introduced, as illustrated by the periodic variation in the distributed volumetric change, this condition is broken. The consequence is the insertion of thermal gradients that give rise to TED.

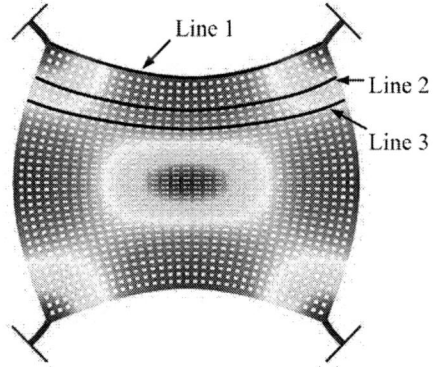

Fig. 1. The Lamé mode shape for a square-plate resonator with etch holes. Three line sections on the resonator along which the displacement and strain distributions were extracted and plotted in Figs. 2 and 3 respectively.

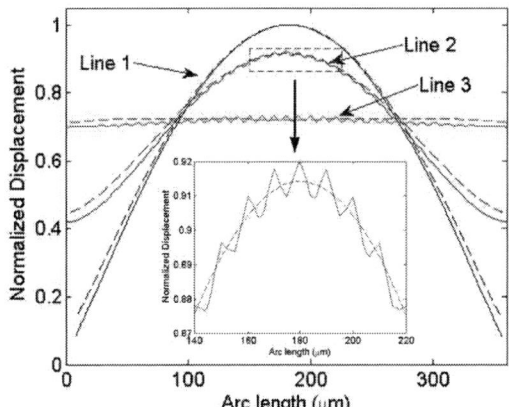

Fig. 2. Normalized displacement distribution along the 3 line sections for the Lamé mode with [solid lines] and without etch holes [broken lines].

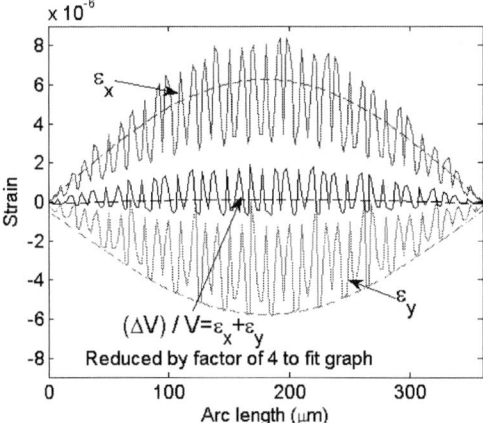

Fig. 3. FE simulated orthogonal in-plane strains along line 2 for the Lamé mode with [solid lines] and without etch holes [broken lines]. ε_x: x-direction strain; ε_y: y-direction strain component].

To further substantiate this hypothesis, the Q factor set by TED can be computed using FE analysis by considering the 3D thermal dynamics in a solid given by [16]

$$\kappa\nabla^2 T - C_v \frac{\partial T}{\partial t} = \alpha(3\lambda + 2\mu)T_o(\frac{\partial}{\partial t}\frac{\partial u}{\partial x} + \frac{\partial}{\partial t}\frac{\partial v}{\partial y} + \frac{\partial}{\partial t}\frac{\partial w}{\partial z}) \quad (2)$$

where κ represents the thermal conductivity, T denotes the absolute temperature, t denotes time, α represents the coefficient of thermal expansion. λ, μ denote the elastic Lamé parameters and T_o is the reference temperature. Here, u, v, w denote displacement in the x, y, and z directions respectively. The right hand side of equation (2) represents the heat source. By applying the equation to points on the resonator where the etch holes are located, and neglecting the out-of-plane z-displacements, the right hand side of the equation was found to be non-zero. This result thereby indicates generation of heat. Since the superposed orthogonal strain components vary in magnitude at different locations on the resonator, this establishes a temperature gradient around the etch holes as shown in Fig. 4. This leads to heat flow, which results in

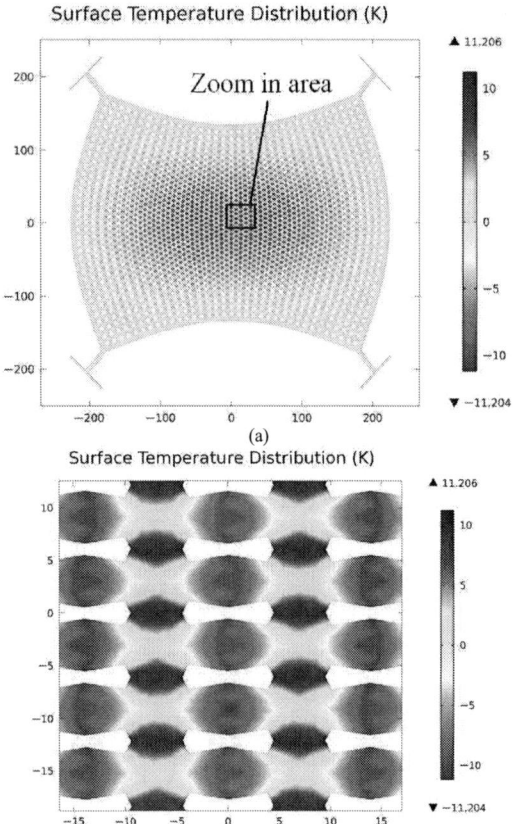

Fig. 4. (a) FE simulated temperature distribution in the resonator with uniformly distributed square etch-holes (The displayed temperature scale is only relative); (b) Zoomed in area.

irreversible thermal dissipation. Equation (2) describes the coupling of mechanical strain to the thermal domain, while the coupling of temperature to the mechanical domain along the x direction can be expressed as

$$
\rho \frac{\partial^2 u}{\partial t^2} = \mu \left(\frac{\partial^2 u}{\partial x^2} + \frac{\partial^2 u}{\partial y^2} + \frac{\partial^2 u}{\partial z^2} \right) + (\lambda + \mu)\left(\frac{\partial^2 u}{\partial x^2} + \right.
$$
$$
\left. \frac{\partial^2 v}{\partial x \partial y} + \frac{\partial^2 w}{\partial x \partial z} \right) - \alpha(3\lambda + 2\mu)\frac{\partial T}{\partial x}
$$

(3)

where ρ denotes density.

III. MEASUREMENT AND DISCUSSION

Two single-crystal silicon (SCS) square-plate resonators of the same length (360μm) were fabricated on the same die with a device layer thickness of 25μm. One of the two resonators contained uniformly distributed square etch holes with 5μm in size and an edge-to-edge pitch of 5μm as shown in Fig. 5, while the other was free of etch holes. Each pair of devices were fabricated on the same die to avoid possible process-related discrepancies between different dies when making comparisons. To mitigate the effects of parasitic feedthrough, a fully-differential setup [18] was adopted, which is illustrated

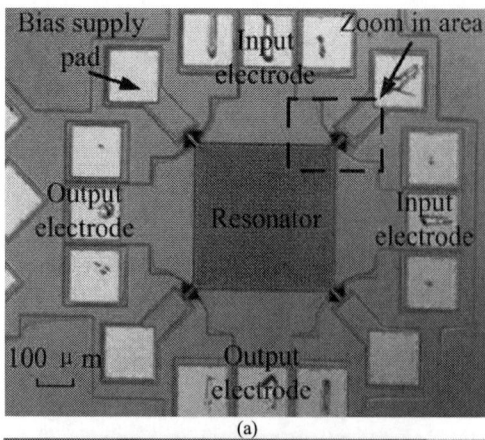

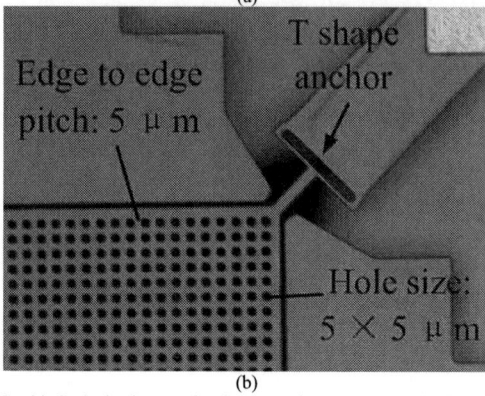

Fig. 5. (a) Optical micrograph of square-plate resonator with etch-holes; (b) Zoomed in area.

in Fig. 6. The resonators were both characterized at pressures below 0.4 mTorr in a Janis Research cryogenic probe station using GSG probes. The electrical transmissions of the resonators were measured using an Agilent network analyzer (E5061A). The measured electrical transmission plot is given in Fig. 7, which shows an approximately 98.5% drop in Q (in other words only 1.5% of the original value). For reference, the earlier work by Shao on etch holes induced effects on Q [12] showed a reduction of the same order (93.1%).

For the etch-hole free plain resonator, at a resonant frequency of 11.5MHz and Q factor of 1.3 million, this yields a frequency-Q (fQ) product of 1.51×10^{13}. This is close to the limit set by AKE loss [15], suggesting that the measured Q is set by AKE loss. In the absence of TED, AKE sets the bound on Q for the bulk mode resonator. On the other hand, the measured Q for the square-plate resonator with etch-holes drops to just about 1.5% that of the etch-hole free device. The associated fQ product is just 1.83×10^{11}, which is significantly lower than the AKE limit by almost 2 orders of magnitude. Hence it strongly appears that the perforation of etch holes in the square-plate causes the resonator degrades Q through a significant increase in TED. These are summarized in Table I.

TABLE I. COMPARISON OF MEASURED Q FOR RESONATORS WITH AND WITHOUT ETCH HOLES

Square-plate Bulk Mode Resonator		Lamé mode		SE mode	
		Plain	Etch holes	Plain	Etch holes
Resonator Frequency	Measured	11.51 MHz	9.67 MHz	11.99 MHz	10.19 MHz
	Simulated	11.51 MHz	9.78 MHz	11.99 MHz	10.27 MHz
Measured Q		1.31×10^{6}	18970	2.5×10^{5}	61332
Measured fQ product		1.51×10^{13}	1.83×10^{11}	3.02×10^{12}	6.25×10^{11}
Primary loss mechanism		AKE	TED	Anchor Loss	Unknown

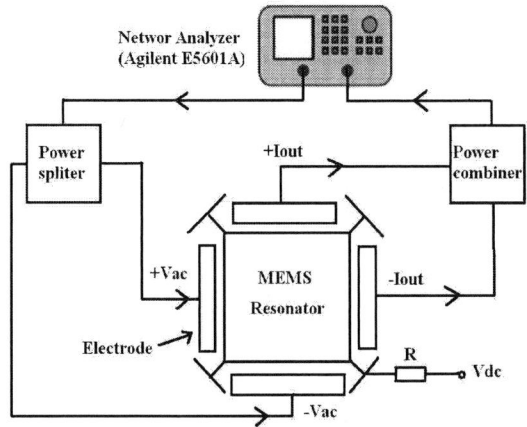

Fig. 6. Schematic showing measurement setup for characterizing the Lamé mode.

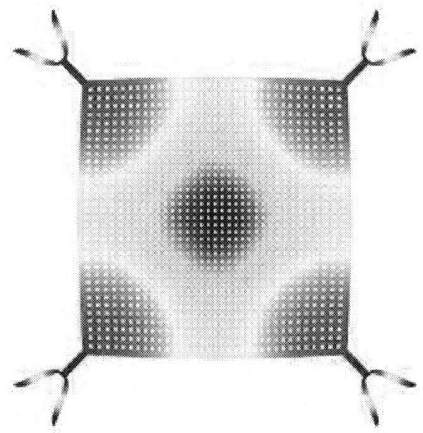

Fig. 8. Simulated SE mode shape for resonator with uniformly distributed etch holes.

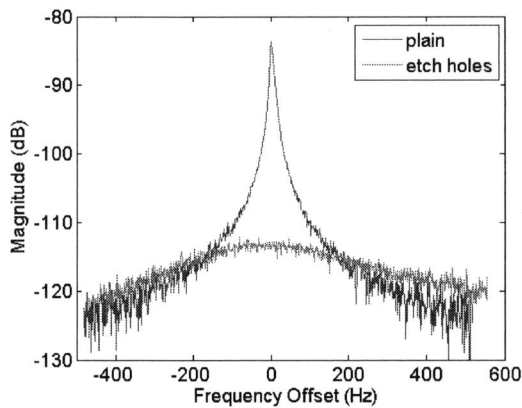

Fig. 7. Comparison between the measured transmission of the Lamé mode resonators with and without etch holes.

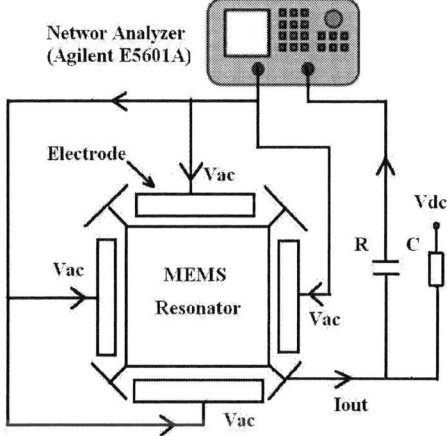

Fig. 9. Schematic showing measurement setup for characterizing the SE mode.

For reference, we have also excited the same set of resonators to vibrate in the extensional (SE) mode. The simulated SE mode shape for the resonator with uniformly distributed etch holes is shown in Fig. 8. The SE mode was excited by applying AC drive signals symmetrically at all four electrodes as depicted in Fig. 9. Fig. 10 shows the measured electrical transmission for each of the resonators when excited in the SE mode. The measured Q with and without etch holes have also been included in Table I. It can be seen from Table I that the introduction of etch holes leads to a Q drop of 75.7% in the SE mode resonator. In comparison, this drop in Q for the SE mode is less than that observed in the Lamé mode

(98.5%). In contrast to the Lamé mode where the nodes are located at the corners, in the SE mode the corners of the square-plate now correspond to the antinodes. Clamped at these corners, substantial anchor loss is likely to occur when exciting the SE mode [19]. Hence there is a possibility that adding etch holes increases anchor loss for the SE mode, which in turn degrades the measured Q. This hypothesis awaits further quantitative validation, which is a direction that will be explored in future works.

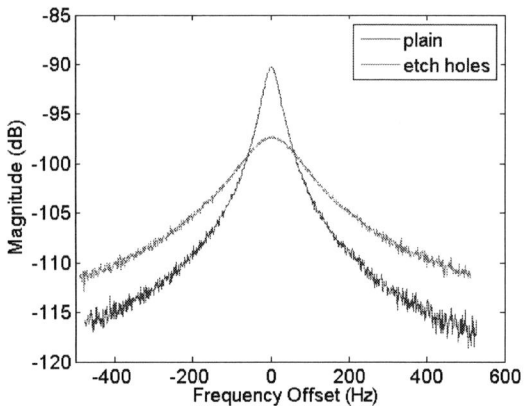

Fig. 10. Comparison between the measured transmission of the square extensional mode resonators with and without etch holes.

IV. CONCLUSION

In this paper, we have characterized two square-plate resonators in both the Lamé and SE modes. One of these is designed with uniformly distributed etch holes. A drop in Q by as much as over 98% is observed in the Lamé mode. Using finite-element (FE) computation, we have shown that the isochoric characteristic of the Lamé mode is no longer preserved when etch holes are uniformly distributed over the resonator. This results in thermoelastic dissipation (TED) which causes the drop in Q. By solving coupled thermoelastic equations using FE analysis, we can observe the thermal gradients that accompany the volumetric change that result from adding etch holes. In the SE mode, the drop is less drastic (75.7%). From FE analysis, it appears that TED is not dominant in either device in the SE mode.

V. ACKNOWELDGEMENT

The work described in this paper was supported by a grant from the City University of Hong Kong (Project No. 9667049).

REFERENCES

[1] C. T. C. Nguyen, "MEMS technology for timing and frequency control," IEEE Trans. Ultrason., Ferroelectr., Freq. Control, vol. 54, no. 2, pp. 251–270, Feb. 2007.

[2] J. E.-Y. Lee, J. Yan, and A. A. Seshia, "Low loss HF band SOI wine glass bulk mode capacitive square-plate resonator," J. Micromech. Microeng., vol. 19, no. 7, 074003, July 2009.

[3] V. Kaajakari, T. Mattila, A. Oja, and H. Seppä, "Nonlinear limits for single-crystal silicon microresonators," J. Microelectromech. Syst., vol. 13, no. 5, pp. 715–724, Oct. 2004.

[4] O. Holmgren, K. Kokkonen, T. Veijola, T. Mattila, V. Kaajakari, A. Oja, J. V. Knuuttila, and M. Kaivola, "Analysis of vibration modes in a micromechanical square-plate resonator", J. Micromech. Microeng., vol. 19, pp. 015028, 2009.

[5] V. Kaajakari, T. Mattila, A. Oja, J. Kiihamäki, H. Kattelus, M. Koskenvuori, P. Rantakari, I. Tittonen, and H. Seppä, "Square-extensionalmode single-crystal silicon micromechanical RF-resonator," in Proc. Transducers'03, The 12th International Conference on Solid-State Sensors and Actuators, Boston, MA, June 8–12, 2003, pp. 425–432.

[6] V. Kaajakari, T. Mattila, A. Oja, J. Kiihamaki, and H. Seppä, "Square extensional mode single-crystal silicon micromechanical resonator for low-phase noise oscillator applications," IEEE Electron Device Lett., vol. 25, no. 4, pp. 173–175, Apr. 2004.

[7] D. Grogg, H. Tekin, N. Ciressan-Badila, D. Tsamados, M. Mazza, and A. Ionescu, "Bulk Lateral MEM Resonator on Thin SOI With High Q-Factor," J. Microelectromech. Syst. vol. 18, no. 2, pp. 466–479, 2009.

[8] F. Casset et al, "MEMS Resonator temperature compensation," IEEE 10th International Conference EuroSimE, 2009.

[9] Y. Civet, S. Skandar; F. Casset, B. Icard, D. Mercier, J.-F. Carpentier, J. Bustos, F. Leverd, "Holed MEM resonators with high aspect ratio, for high accuracy frequency trimming Solid-State Device Research Conference," Proceedings of the European 123 – 126, 2011.

[10] V. L. Rabinovich, R. K. Gupta, and S. D. Senturia, "The effect of release-etch holes on the electromechanical behavior of MEMS structures," Dig. Tech. Papers, Transducers'97, Chicago, USA, pp. 1125–1128, 1997.

[11] W. N. Sharpe, R. Vaidyanathan, B. Yuan, G. Bao, and R. L. Edwards, "Effect of etch holes on the mechanical properties of polysilicon," J. Vac. Sci. Technol. B, vol.15, pp. 1599–1603, 1997.

[12] L. Shao, and M. Palaniapan, "Effect of etch holes on quality factor of bulk-mode micromechanical resonators," Electronics Letters, vol. 44, no. 15, pp. 938–939, 2008.

[13] R. N. Candler, A. Duwel, M. Varghese, S. A. Chandorkar,M. A. Hopcroft, P. Woo-Tae, K. Bongsang, G. Yama, A. Partridge, M. Lutz, and T. W. Kenny, "Impact of geometry on thermoelastic dissipation in micromechanical resonant beams," J. Microelectromech. Syst., vol. 15, no. 4, pp. 927–934, Aug. 2006.

[14] S. A. Chandorkar, R. N. Candler, A. Duwel, R. Melamud, M. Agarwal, K. E. Goodson, and T. W. Kenny, "Multimode thermoelastic dissipation," J. Appl. Phys. vol. 105, pp. 043505, 2009.

[15] S. A. Chandorkar, M. Agarwal, R. Melamud, R. N. Candler, K. E. Goodson and K. E. Kenny, "Limits of quality factor in bulk-mode micromechanical resonators," Proc. 21st IEEE Micro Electro Mechanical Systems Conference, pp. 74–77, 2008.

[16] A. Duwel, R. N. Candler, T. W. Kenny, and M. A. V. M. Varghese, "Engineering MEMS resonators with low thermoelastic damping," J. Microelectromech. Syst., vol. 15, pp. 1437–1445, 2006.

[17] COMSOL Multiphysics, v4.1.0.88, 2010.

[18] S. A. Bhave, G. Di, R Maboudian, and R. T. Howe, "Fully-differential poly-SiC Lame mode resonator and checkerboard filter," in Proc. Int. Conf. Micro Electro Mech. Syst., (MEMS 2005), 2005, pp. 223–226.

[19] V. Taş, S. Olcum, M. D. Aksoy, and A. Atalar, "Reducing anchor loss in micromechanical extensional mode resonators", IEEE Transactions on Ultrasonics, Ferroelectrics, and Frequency Control, vol. 57, No.2, pp. 448-454, 2010.

A Piezoresistive Normal and Shear Force Sensor Using Liquid Metal Alloy as Gauge Material

Xiaomei Shi[1], Ching-Hsiang Cheng[1*], Chen Chao[2], Like Wang[2], Yongping Zheng[2]

[1*] Department of Industrial and Systems Engineering, The Hong Kong Polytechnic University, Hong Kong
[2] Department of Health Technology and Informatics, The Hong Kong Polytechnic University, Hong Kong
mfcheng@inet.polyu.edu.hk

Abstract—We present a novel normal and shear force sensor by using liquid metal alloy (Ga-In-Sn) as piezoresistive gauge material encapsulated in a polydimethylsiloxane (PDMS) substrate. By using liquid metal alloy as gauge material, it can detect large forces without breaking the sensor wires. Since the liquid-metal piezoresistors deform with the elastomeric substrate, shear and normal forces can be detected with resistance changes of the piezoresistors. Each force sensor comprises a pair of symmetric piezoresistors, which is screen-printed on the cavity of PDMS substrate with tilt angle around 30° to be sensitive to both normal and shear forces. Normal force will compress both piezoresistors as common mode while shear force will shorten one piezoresistor but elongate the other as differential mode. The testing results demonstrate the sensitivity of the force sensor in both normal and shear directions. The hysteresis of the force sensor was also measured.

Keywords-normal and shear force sensor; PDMS; liquid metal

I. INTRODUCTION

It is important to identify the human body 3D reaction force produced in the study of human-machine interaction and biomechanical analysis [1] [2].Some sensor systems have been developed for applications in human dynamics analysis. In the market, most of the force sensors only measure the magnitude of the resultant force. Separating the effects from normal and shear force is essential for biomechanics research including monitoring of diabetic foot and ergonomic studies on foot wears, finger and hand rehabilitation, dental occlusal and joint prosthetic analysis, lip interface for dysphagia, car seats, bed mattresses, wheelchairs, sports wears and equipments, etc. Although the shear force can be extracted indirectly from the distribution of resultant forces, however, it requires complicated calculation formulated by a computer program. Some silicon sensors are developed to measure both normal and shear force at the skin object interface [3] [4], but the force levels of these sensors are limited in the measure of small forces. It means these sensors can get broken when a large force is applied. And the silicon substrates usually are too thick and lack of flexibility to fit into a curved interface.

Identifying both normal and shear forces produced in the human-machine interface is essential to understand their separate effects. Capacitive normal and shear force sensors with polymer substrate have been reported [5] [6]. They both

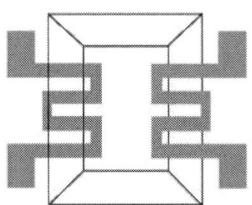

Fig. 1. The top view of the interface shear stress sensor

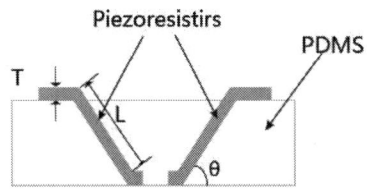

Fig. 2. Two serpentine force sensors are laid on the slopes symmetrically with opposite tilted angle.

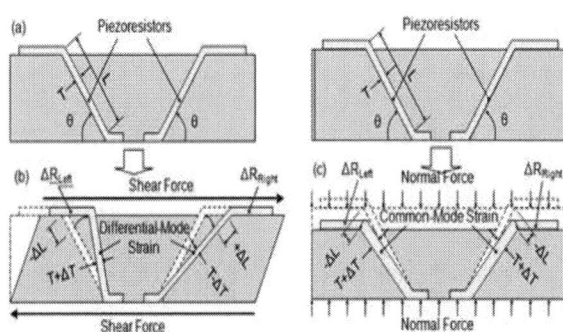

Fig. 3. An interfacial normal and shear force sensor has two piezoresistors aligned in mirror-image directions with tilt angles of θ to the measured interface. (b) The shear force (c) The normal force.

use parallel plate capacitors with four electrodes to differentiate normal and shear forces. A normal and shear piezoresistive tactile sensor that uses inclined micro-cantilevers covered with elastomer has also been developed with crystalline silicon as piezoresistive material [7]. All of them use PDMS as flexible substrate material to encapsulate sensing materials. However, solid conductors were still used as sensing materials that can be broken when a large force is applied.

As discussed above, it's necessary to develop an effective way for fabricating the flexible force sensor. The current project was started to fabricate a flexible normal and shear force sensor by using new materials since we believe there is a great potential to improve the flexibility and durability without reducing the sensitivity.

II. BASIC PRINCIPLE

An interface shear and normal force sensor can measure the interfacial shear and normal strain by inserting the sensor in the interface between adjacent mated bodies. Two serpentine strain gauges are laid on the slopes symmetrically with opposite tilted angle. The serpentine design is to increase the resistance of the metal gauge wire to reduce the self-heating effect from joule heating when applying an electrical current (Figure1, Figure 2).

$$\Delta R_{Left}=\Delta R_{Right} \quad normal\ force\ is\ applied \quad (1)$$

$$\Delta R_{Left}=-\Delta R_{Right} \quad shear\ force\ is\ applied \quad (2)$$

The force can be measured with the strain-induced resistance change that depends on the gauge factor (GF) when a metallic piezoresistive material is employed. By using ΔR_{Left}-ΔR_{Right}, the common-mode (normal) strain induced resistance change can be cancelled, which can determine the shear force (Figure 3(b)). By using $\Delta R_{Left}+\Delta R_{Right}$, the differential-mode (shear) strain induced resistance change can be cancelled, which can determine the normal force (Figure 3(c)).

III. FABRICATION

A. Material—Coollaboratory Liquid Pro

Liquid metal alloy means alloy of several very low melting point metal at room temperature. As a replacement for toxic mercury, liquid metal alloy is used in various applications. These liquid metal alloys have a high degree of thermal conductivity far superior to ordinary nonmetallic liquids. Another advantage of these liquid metals is lower toxic, lower vapor pressure, high densities and electrical conductivities. Liquid metal alloy will wet and adhere to most metallic surface. It forms a thin looking oxide layer on the surface of metallic.

In this design, we choose a kind of liquid metal called Coollaboratory Liquid Pro. Alloy of the metal components are gallium, indium, rhodium, silver, zinc and stannous. It has high performance hear conducting and good electrical conducting that consists of 100% liquid metal alloy. It is liquid at room temperature (freezing temperature 8 ℃), but it is absolutely nontoxic and has a high moistening ability for several materials. The Coollaboratory Liquid Pro is especially recommended for nickel-plated copper. Coollaboratory Liquid

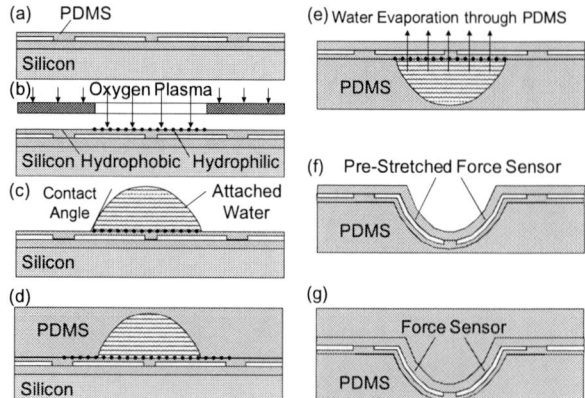
Fig. 4. Fabrication process of the shear and normal force sensor

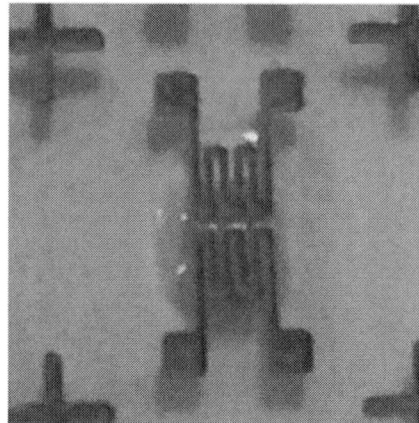
Fig. 5. The photo of the fabricated force sensor

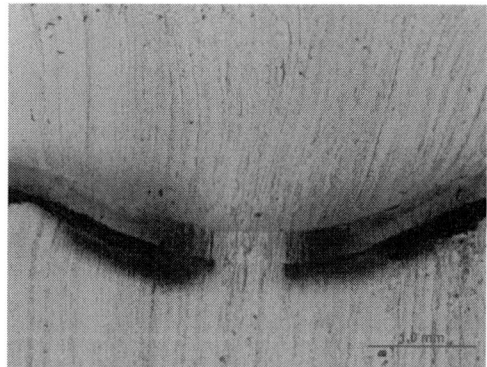
Fig. 6. The cross section of the force sensor. The gray metallic area is the liquid-metal piezoresistor.

Pro contain no non-metallic additives (like silicone, oxides etc.) at all. It also does not contain any solid particles. Due to these properties, Coollaboratory Liquid Pro surpasses the best high performance heat conducting pastes by a multiple.

B. Fabrication of Force Sensor

The fabrication process of a single force sensor is shown in Figure 4. A thin layer of PDMS film was made by coating the

Dow Corning Sylgard 184 PDMS pre-polymer and its curing agent mixture on a glass substrate then cured at elevated temperature of 70°C in a convection oven (Figure 4(a)). The cured PDMS film was then covered with the shadow mask for either sputtering with a thin layer of gold around 20nm or treated by oxygen plasma around 1min and print liquid metal pad. After last step, another thin layer of PDMS was needed by coating for protecting the metal piezoresistors and the bond pads then cured at room temperature (Figure 4(b)). The cured PDMS film was then covered with another shadow mask, treated by oxygen plasma around 1min, printed water ball pad (Figure 4(c)) and covered with a thick layer of PDMS (Figure 4(d)). After the PDMS cured and water evaporated (Figure 4(e)), the fabrication process was completed (Figure 4(g)).

C. Fabrication of the hollow on the surface

The step of produced the hollow is the key of the whole fabrication. A thin layer of PDMS film was made by coating the PDMS pre-polymer and its curing agent mixture on a silicon substrate. After curing at the room temperature, the cured PDMS film was then covered with a shadow mask, treated by oxygen plasma around 1min, printed water ball pad and covered with a thick layer of PDMS. After the PDMS cured and water evaporated, two layers of PDMS will connected with each other by van der Waals force. The contact angle depends on the timing of oxygen plasma treat.

D. Dimensions of force sensor

We successfully fabricated several shear and normal force sensor arrays. The photographs of the fabricated sensor were shown in Figure 5. The area of the piezoresistor pair is 2mm x 1mm, the substrate thickness is 2mm and the width of gauge is 100μm. The serpentine design is to reduce the self-heating effect by increasing the resistance of the metallic gauge wire. To further investigate the structure, we cleaved one cell and used an optical microscope to image the photo of the part of liquid metal force sensor. As shown in Figure 6, the fabricated sensors have a tilt angle around 30°. The initial resistance of the force sensor was 10 ohm in average. The force can be measured with the strain-induced resistance change that depends on the gauge factor (GF).

IV. MEASUREMENT AND DISCUSSION

Figure 7 and Figure 8 are the output of the schematics of the experimental setup for measuring normal force and shear force, respectively. A force gauge, whose maximum force is 50N, is used to measure the applied force. The force gauge is fixed on vertical translational stage whose displacement resolution is 1μm. The resistances of the sensing element are measured by LabView with a control DC circuit. For normal force measurement, a pair of force sensor is connected to the force gauge. As the z-axis stage table moves down, the sensing element is pushed by the force gauge. For shear force measurement, an I-shaped PMMA rod is connected to the force gauge. As the z stage table moves laterally, the sensing element is pushed by the I-shaped rod.

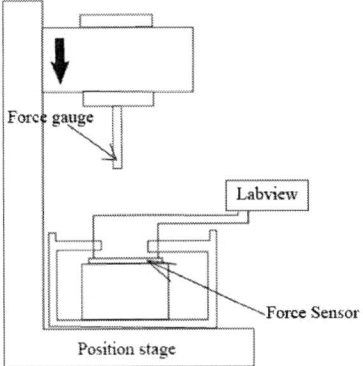

Fig. 7. Experimental setup for measuring the resistance of force sensor elements with applied normal force.

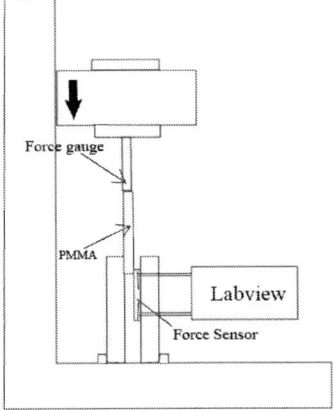

Fig. 8. Experimental setup for measuring the resistance of force sensor elements with applied shear force.

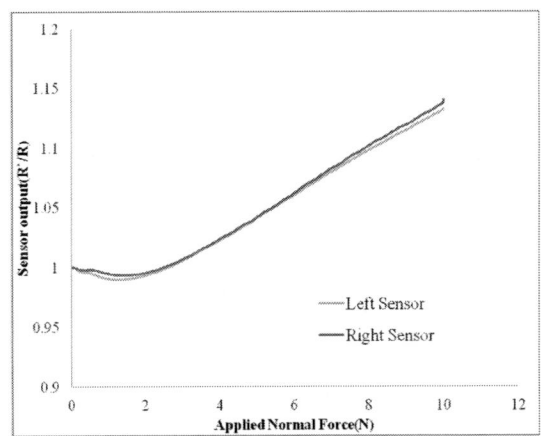

Fig. 9. The measured relationships of resistance vs. applied normal force (N) for a force sensor. The applied force compressed both piezoresistors, which reduced their cross-sectional area to increase the resistance.

Figure 9 shows the testing results when a normal force is applied in the Z-direction. For each measured point, the displacement of a stage slowly increases until the force (measured by the force gauge) applied on the sensing element

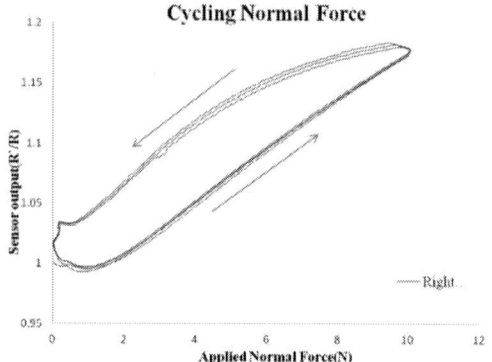

Fig. 10. The hysteresis of the force sensor when a cycling normal force is applied

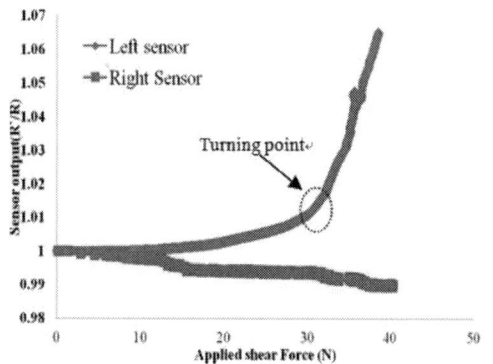

Fig. 11. The measured relationships of resistance vs. applied shear force (N) for a force sensor. The force is applied in the right direction, which causes the right piezoresistor shortened (resistance decreased) while the left elongated (resistance increased)

reaches to certain predefined value, and then the resistance is measured by the LabView analyzer system. The y-axis of Figure 9 represents the ratio of the measured resistance to its initial resistance (the normalized resistance).

Figure 10 shows the hysteresis of the force sensor when applying a normal force. Each data point in figure 10 is the result by measuring a sensing element 6 Cycles. The response time of the force sensor depends on the speed of the applying force (2mm/min) due to the PDMS elastomeric substrate. It shows the curve of increasing the applied force is slightly lower than that of decreasing applied force. The phenomenon can also be considered as the working range of the force sensor. This phenomenon indicates that the liquid metal electrode of the PDMS structure layer almost completely contacts the force gauge.

Figure 11 shows the testing results when a shear force is applied in the Y-direction. In this configuration, a torque was induced on the sensing element, so the resistance of the left piezoresistor was increased, while that of the right was decreased. Compared with the curve of right sensor, the curve of left sensor has a turning point. Before this point, the applied force need to do work by overcome the static friction force.

The figure also presents the average sensitivities of the devices, which are defined as the slopes in the linear ranges of these curves. The phenomenon can also be considered as the working range of the force sensor. With several working cycles, the force sensor is still working well.

V. CONCLUSIONS

The development of a liquid metal piezoresistive shear stress sensing array was presented in this paper. The proposed sensing array, which employs a novel resistive sensing mechanism, consists of a PDMS substrate and a liquid metal circuit. Each shear sensing element consists of two sensing liquid metal force sensor. The design can effectively reduce the complexity of device structure and thus makes the device highly manufacturable. It can also improve the flexibility and durability of the force sensor without reduced the sensitivity. The characteristics of the devices were measured and discussed. The measured resistance distributions induced by applying normal and shear forces are also successfully detected.

ACKNOWLEDGMENT

The work described in this paper was partially supported by a grant from the RGC General Research Fund (RGC project reference number PolyU 513208), The Hong Kong Polytechnic University, Hong Kong; and partially supported by a grant from the Departmental General Research Fund, Department of Industrial and Systems Engineering, The Hong Kong Polytechnic University, Hong Kong (Project Account Code G-UA02).

REFERENCES

[1] Gerald F.Harris, Peter A. Smith, "Human Motion Analysis," Reading, IEEE, Inc., pp12-30, 1996.

[2] Mehmet Engin, Alparslan Demirel, Erkan Zeki Engin, Musa Fedakar, "Recent developments and trends in biomedical sensors," Measurement, vol.37, pp. 173-188, 2005.

[3] Beccai L, Roccella S, Arena A, Valvo F, Valdastry P, Menciassi A, Valdasri P, Menciassi A, Carrozza M.C, Dario P, "Design and fabrication of a hybrid silicon three-axial force for biomechanical applications," Sensors and Actuators A: Physical, Vol.120, pp.370-382, 2005.

[4] Valdastri P, Roccella S, Beccai L, Cattin E, Menciassi A, Carrozza M.C, Dario P, "Characterization of a novel hybrid silicon three-axial force sensor," Sensors and Actuators A: Physical, Vol.123-124, pp.249-257, 2005.

[5] Ming-Yuan Cheng, Chun-Liang Lin, Yu-Tse Lai and Yao-Joe Yang, "A Polymer-Based Capacitive Sensing Array for Normal and Shear Force Measurement," Sensors 2010, pp. 10211-10225.

[6] Hyung-Kew Lee, Jaehoon Chung, Sun-Il Chang, and Euisik Yoon, "Normal and Shear Force Measurement Using a Flexible Polymer Tactile Sensor With Embedded Multiple Capacitors," Journal of Microelectromechanical Systems, Vol. 17, No. 4, August 2008, pp. 934-942.

[7] Masayuki Sohgawa, Yu-Ming Huang, Minoru Noda, Takeshi Kanashima, Kaoru Yamashita, Masanori Okuyama, Masaaki Ikeda, and Haruo Noma, "Fabrication and Characterization of Normal and Shear Stresses Sensitive Tactile Sensors by Using Inclined Micro-cantilevers Covered with Elastomer," MRS Proceedings (2007), 1052.

Development of A Novel Force Sensor System Built with an Industrial Multilayer Ceramic Capacitor (MLCC)

Keng-Ren Lin[1], Cheng-Hung Chiang[2], Chih-Han Chang[1], Che-Hsin Lin[2*]

[1]Department of Biomedical Engineering, National Cheng Kung University, Tainan, TAIWAN
[2]Department of Mechanical and Electro-Mechanical Engineering, National Sun Yat-sen University, Kaohsiung, TAIWAN
linkanshin@mail.bme.ncku.edu.tw

Abstract— This paper presents the development and normalization of a novel sensor system built with low-cost industrial-grade multilayer ceramic capacitors (MLCC). MLCCs play both the roles of force sensing elements and force sustaining elements. With adopting this low-cost industrial component of MLCC (less than 1 cent USD for a sensing component), the time and the cost for producing a high performance force sensor system can be achieved. Practically, industrial-grade MLCCs are produced for capacitor applications but not for a sensing element for force measurement. However, the sensitivity for each independent MLCC presents a large variation for force sensing (coefficient of variation around 0.6). This study successfully overcome this drawback, a simple poling process is demonstrated to both increase the sensitivity and to reduce the variation coefficient of MLCC force sensors. Results indicate that the variation for the MLCCs after re-poling treatment is reduced to 0.02. The effects on the poling voltage and poling time are experimentally investigated. Results show that the MLCC treated with a high poling field can increase sensitivity and also reduce the variation for force sensing. This re-poling treatment is essential for developing a simple and low cost MLCC-based force sensor system that can be used in a wide variety of applications.

Keywords- piezoelectric; multilayer ceramic capacitors; poling

I. Introduction

In general, multilayer ceramic capacitors (MLCCs) as capacitors are widely used in 3C products. The MLCC element is composed of a stack of piezoelectric ceramic materials, $BaTiO_3$, as the dielectric layer, and nickel metal is used as the conducting layer [1]. The present study demonstrated that the MLCC has excellent repeatability in a cyclic loading test (a variation of less than 2%). It also showed a nice performance with regard to the developed sensor array, confirming the feasibility of the sensor for measuring forces in both static and dynamic conditions. In addition, the composed materials provided effective strength in the case of a small size MLCC to sustain large forces. These features have demonstrated that MLCCs have high performance as large force piezoelectric sensing elements [2]. Moreover, industrial-grade MLCC can be easily assembled to electrode substance using a semiconductor-grade automation machine which simplifies and accelerates the sensor manufacture process. After research has demonstrated the MLCC can perform with

good sensibility, it is a prerequisite to make an overall inquiry elucidating further details concerning the MLCC sensor system (Fig. 1). Note the inclusion of an MLCC sensor, a home-built charge amplifier, and a 16-channel data logger (INET-100 InstruNet, USA).

The major piezoelectric material in MLCCs is $BaTiO_3$, which preforms cubic structure with paraelectric behavior above the Curie temperature and relative has a tetragonal structure as a ferroelectric material with a spontaneous polarization below the Curie temperature. In order to reduce the free energy, all the dipoles align in the same direction within a ferroelectric domain. In the neighboring ferroelectric domains, the polarization may orientation in different direction (antiparallel or perpendicular). A well-known process to reverse the domain direction to pointing the same direction is called "poling". In the typical poling process, an appropriate direct current (DC) voltage field is applied to piezoelectric material to align the anisotropic ferroelectric domains in the direction of the applied field [3]. This is a very important process for piezoelectric material. Because before the poling process, where the ferroelectric domains are random arrangement. There is no piezoelectric contribution in this situation. Therefore, the anisotropic ferroelectric domain structure in $BaTiO_3$ crystals with very low polarization level can be treated using the poling process. Generally, the poling processes for piezoelectric material, such as PZT and a $BaTiO_3$, require a high voltage (several kV/cm^2) and a high temperature environment (around the Curie temperature) [4]. In order to reduce the poling environment requirements for an MLCC sensor, this study presented a simple poling method with only a DC voltage supply designed to simplify the poling process.

This study focuses on measurement system development especially through the use of MLCC sensing element normalization. In a simple MLCC sensor poling process, the effects of poling time and voltage is evaluated. Further piezoelectric behaviors after poling will be also discussed in the present study. Instead of using an expensive single-channel charge amplifier system, a home-built charge amplifier circuit with a switchable converting rate was applied to the sensing system. Systematic investigations into the force sensing performance of the MLCC device are experimentally measured and analyzed.

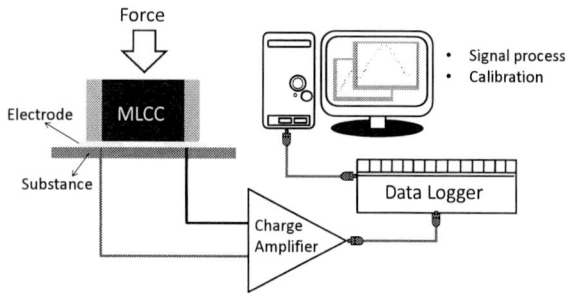

Figure 1: Schematic diagram of the system, which uses a low-cost industrial-grade MLCC as a high strength force sensor. The charge signal from the sensor is converted to a voltage signal by the home-built charge amplifier and collected by a data logger.

II. MATERIALS AND METHODS

A. Sensor system development

In order to develop a sensor system suitable for measuring a large force on an irregular surface, the MLCC force sensor was mounted on a flexible polyimide film. A 1×10^{-6}F, 25V MLCC (1206F105Z500CT, Walsin Technology Corporation, Taiwan) in 3.2mm length, 2.6mm weight, and 0.95mm thickness was employed in this study. Fig. 2 shows the simplified fabrication process for the electrode and the MLCC sensing element. The flexible PI-based electrode was fabricated using a simple and standard MEMS process. A thin layer of positive photoresist layer (AZ1642) was coated on flexible copper clad polyimide tape (Fig. 2A, 2B). After coating the photoresist, the electrode was then patterned using a standard photo-lithography process (Fig. 2C). Then, the developing process was performed using developer solution (Fig. 2D. Finally, the electrode was fabricated by etching the undesired copper with standard ferro-chloride solution (Fig. 2E). After the electrode patterning process, the MLCC component was aligned under a microscope and then soldered onto the electrode plate (Fig. 2F).

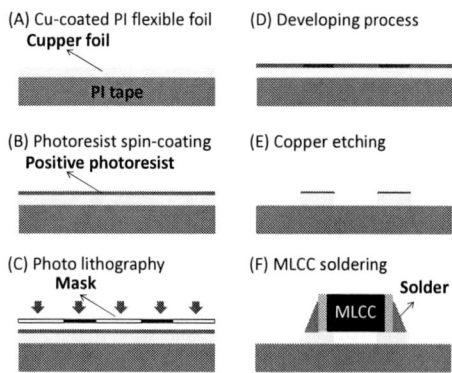

Figure 2: Illustration of the process for fabricating the flexible MLCC force sensor. (A) ~ (E) shows the simplified MEMS fabrication process for fabricating the electrode, including a positive photoresist spin coating, photo-lithography, and etching. (F) The MLCC was soldered onto the electrode.

Fig. 3 shows the photographs of both the fabricated electrode and the MLCC sensing element. The tested MLCC sample was connected to a self-fabricated charge amplifier to convert the charge variation into output voltage signal (Fig. 4). The piezoelectric charge amplifier was based on an OP amplifier, OPA129 (OPA129, Burr-Brown Corporation, U.S.A.) with a parallel capacitor (C_F). The output voltage was converted by the ratio of the released charge (ΔQ), and (C_F) [5] as (1).

$$V_{out} = -\Delta Q / C_F \qquad (1)$$

It can be determined that the converting rate of the amplifier can be adjusted by swapping the C_F for the measuring various output range of MLCC. Note that this study employed the C_F in 47nF. The converting rate was 47000pC/V.

B. Sensor normalization

Generally, MLCCs are used as capacitors. In order to utilize MLCCs as force sensors, they should be further normalized for better performance. This study presented a simple process to normalize the sensitivity of MLCC sensors using a simple poling process. The MLCC sensors were treated using the following process: at first, the specimens were heated with a 500°C solder for 1 minute to exceed the Curie temperature of the BaTiO$_3$ [6]. After heating, and after waiting for a few minutes, the sensors were applied to various DC electrical fields (1.5V to 60V) for different lengths of treatment times (0.5 to 300 minutes) at room temperature. Then, short two electrodes for 5 minutes to eliminate the remained charge of the capacitor. It is because the released charge from the self-discharge process of the capacitor would influence the force measurement signal. After poling, the poled MLCC sensors were tested using a mechanical test machine from 0N to 100N. This study evaluated the effect of poling conditions, including poling voltage and poling time. In the case of the poling voltage, five MLCCs were poled using 1.5V, 3V, 12V, 24V, 36V, 48V and 60V for 30 seconds at room temperature. For comparison, 7 MLCC sensors were poled at room temperature for 0.5, 5, 10, 30, 60, 90, 120, 150, 180, 210, and 240 with 12V and 60V, respectively.

Figure 3: Photo image for a MLCC force sensor after fabrication and an inset close-up view of the assembled MLCC element.

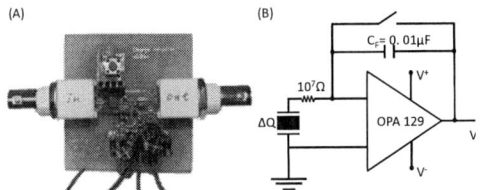

Figure 4: (A) Photograph of a home-built charge amplifier. The major component is an OPA129 amplifier. (B) The equivalent circuit of the charge amplifier.

III.   Results and Discussion

A. *Effects of the poling voltages*

Like general piezoelectric materials, industrial-grade MLCCs can also generate great piezoelectric properties. Fig. 5 shows that the sensor response corresponded to the various applied poling voltages from 1.5V to 60V when the poling time was 30 seconds at room temperature. This demonstrated that the sensor responses increased with increases in poling voltages. The response of 60V applied (952.35pC/N) was over 100 times higher than 1.5V poling (9.42pC/N).

This also demonstrated that the coefficient of variation decreased from 0.69 to 0.02 with poling fields from 1.5V to 60V. This means that the variation of each MLCC specimen could be reduced by applying higher poling voltages because the applied electric field provided a force for ferroelectric domain reversal in the piezoelectric material poling process [7], and the higher poling voltages led to a reversal in the large polarization degree of ferroelectric domain.

B. *Effects of the poling time*

To evaluate the effects of the poling time, 7 MLCCs were used in this study as a group to process the simple poling treatment. Fig. 6 shows the results of the 7 MLCCs after a simple poling process under the same conditions (12V DC field and poling at room temperature) for various poling times, from 0.5 minutes to 4 hours, respectively. The results indicated that the sensitivity increased along with increases in poling time and converged beyond 180 minutes (only increasing by 1.7% after another hour of poling).

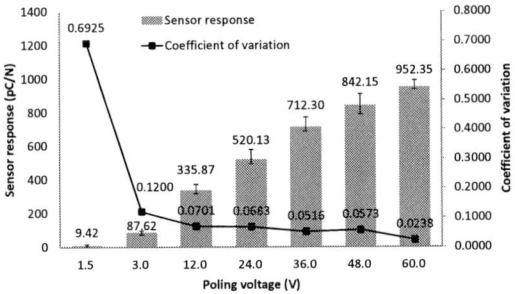

Figure 5: The sensor response of the various poling fields and variation coefficients for the MLCCs group. The histogram indicates the sensor response; the solid line indicates the coefficients of variation of the MLCC sensors.

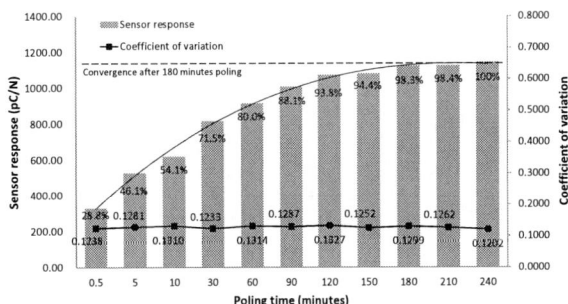

Figure 6: The sensor response of the various poling times and the coefficients of variation for the MLCC group. The histogram indicates the sensor response; the solid line indicates the coefficients of variation of the MLCC sensors.

The poling level is therefore directly affected by the poling time, and the efficiency slows down with longer poling time. This may be because the some ferroelectric domain wall is hard to reverse, and a longer period of time is required to treat the domain in order to ingratiate the electric field direction [7]. When the poling time reached the best poling degree that the applied voltage could provide, the sensitivity resulted in a saturation situation.

However, it can be observed that the variation of each MLCC did not decline as a result of longer poling time. This may be due to the fact that only 12V was used in this poling process to reduce the effect form poling voltage. There was therefore not enough poling voltage to reverse a part of the ferroelectric domain which required higher poling energy.

Fig. 7 shows the results for the same group after the simple poling process under the 60V applied field to 300 minutes and after the sensors were set aside for 24 hours. After testing the sensors that were aged for 24 hours, the sensors were subjected to 60V for another 300 minutes. It can be seen that the sensitivity was raised to 2200pC/N after a 60V poling process, indicating excellent sensitivity. This also confirms that the sensitivity of the MLCC sensors was almost converged after 180 minutes of poling, shown by the fact that the response value of the MLCC sensors was reduced by 27% after 24 hours of aging. The ferroelectric behavioral properties decayed over time, which is in agreement with prior studies on piezoelectric materials [8]. Moreover, after the sensor aging, the sensors were poled with 60V for another 300 minutes. The results indicated that the response is reverted to its original behavior, with only a 0.6% error. This indicated that the MLCC sensors can be re-poled to the same response behavior if the sensor has been poled to the saturation situation. This feature also indicates that an MLCC sensor can be easily and precisely used by poling the sensor to saturation. Note that the mean coefficient of variation of this group was 0.053, which is lower than the 12V poling process (0.1273). This confirms that the higher poling voltage can provide better normalized level sensitivity in the MLCC group.

978-1-4673-1122-9/12 $31.00 © 2012 IEEE   489

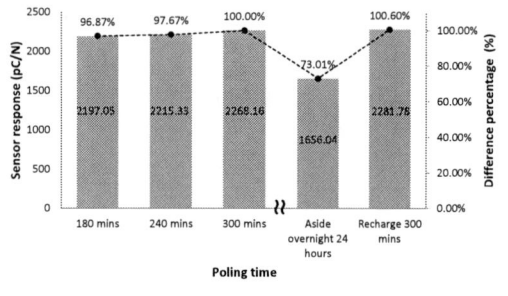

Figure 7: The sensor responses at 180, 240, and 300 minutes and after the sensors are set aside overnight and then recharged for another 300 minutes. The histogram indicates the sensor response; the dashed line indicates the difference ratio with the response after poling for 300 minutes.

Fig. 8 shows the force input and response voltage output results for the 7 MLCC sensors before and after the poling process. The charge amplifier was connected to the sensor to convert and amplify the charge signal to the voltage signal output. Both data are fitted using a linear polynomial. The R-square values were 0.4625 and 0.9804, respectively. These results demonstrated that the MLCC sensors can be efficiency normalized using a simple poling process and can keep good linearity.

The results showed that the MLCC sensor can be easily poled using a low DC voltage field. This may be because the multilayer structure of an MLCC provides a very thin BaTiO₃ layer. The thickness of each layer is only a few micrometers [9]. The multiple thin layer structure allows the material to be poled without a specific poling system. On the other hand, in contrast to prior research, the MLCC sensors in this study were poled at room temperature without a high temperature oil bath. We believe that the higher poling temperature of an MLCC sensor can provide increased piezoelectric properties. However, even though the MLCC sensors in this study were poled only at room temperature, the results still showed that they could demonstrate good poling behavior. By simplifying the process, we predict that the application of MLCC sensors will result in higher efficiency and lower costs.

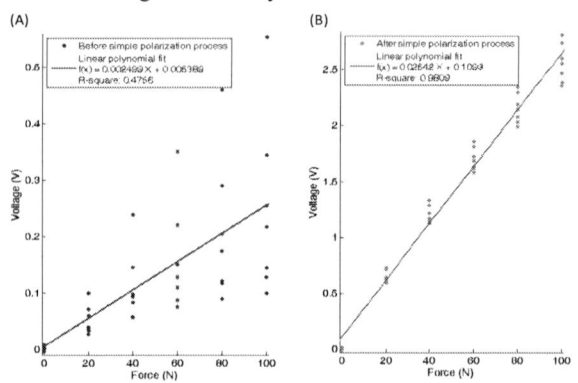

Figure 8: The sensitivity of 7 MLCC sensors before and after a simple polarization process by applying a DC field for the same time. The R-square values are 0.4625 and 0.9804, respectively.

IV. CONCLUSIONS

A simple re-poling method for MLCC sensor normalization was successfully demonstrated. Both the poling voltage and poling time were shown to directly influence the sensitivity of MLCC sensors. Results showed that a high poling voltage can increase the sensitivity and also reduce the variation for force sensing. Results show that the treated MLCC elements exhibited a low variation of 0.02 which was 30 times smaller than the MLCCs without re-poling treatment. The sensitivity was also increase for 200 times. Additionally, MLCC sensors poling reached to saturation after a poling time longer than 180 minutes. Moreover, once the MLCC sensor has been poled to saturation, the aged MLCC sensor can recover response behavior to the saturation condition. Therefore, the developed simple poling method significantly enhances the treated MLCC force sensor with both more precise and more accurate. Accompanied with the home-built piezoelectric charge amplifier, the established MLCC sensor system has shown its excellent potential for a wide variety of applications with a simple, low cost approach.

ACKNOWLEDGMENTS

Financial support from National Science Council of Taiwan is greatly acknowledged.

REFERENCES

[1] H. Kishi, Y. Mizuno, and H. Chazono, "Base-metal electrode-multilayer ceramic capacitors: Past, present and future perspectives," *JPN J APPL PHYS*, vol. 42, no. 1, pp. 1-15, Jan, 2003.

[2] K.-R. Lin, C.-H. Chang, T.-H. Liu et al., "Experimental and numerical estimations into the force distribution on an occlusal surface utilizing a flexible force sensor array," *J Biomech*, vol. 44, no. 10, pp. 1879-1884, 2011.

[3] M. McQuarrie, "Role of Domain Processes in Polycrystalline Barium Titanate," *J. Am. Ceram. Soc.*, vol. 39, no. 2, pp. 54-59, 1956.

[4] S. Wada, K. Yako, T. Muraishi et al., "Enhanced Piezoelectric Property of Barium Titanate Single Crystals by Domain Wall Engineering Using Patterning Electrode," *Ferroelectr.*, vol. 340, no. 1, pp. 17-24, 2006.

[5] T. I. Incorporated, "Burr Brown OPA129 Ultra-Low Bias Current Difet® OPERATIONAL AMPLIFIER," OP129, B.-B. corporation, ed., 2005.

[6] M. H. Garrett, J. Y. Chang, H. P. Jenssen et al., "A method for poling barium titanate, BaTiO3," *Ferroelectr.*, vol. 120, no. 1, pp. 167-173, 1991.

[7] H. Shifeng, C. Jun, L. Futian et al., "Poling process and piezoelectric properties of lead zirconate titanate / sulphoaluminate cement composites," *J. Mater. Sci.*, vol. 39, no. 23, pp. 6975-6979, 2004.

[8] D. Damjanovic, "Ferroelectric, dielectric and piezoelectric properties of ferroelectric thin films and ceramics," *Rep. Prog. Phys.*, vol. 61, no. 9, pp. 1267, 1998.

[9] X. Wang, B. I. Lee, M. Z. Hu et al., "Mechanism of nanocrystalline BaTiO3; particle formation by hydrothermal refluxing synthesis," *J. Mater. Sci. - Mater. Electron.*, vol. 14, no. 8, pp. 495-500, 2003.

Implementation of a Subwavelength Bragg Reflector for Terahertz Applications

Vikas Singal, Sami Smaili, and Yehia Massoud
Electrical and Computer Engineering Department
University of Alabama at Birmingham, Birmingham, AL-35205

Abstract—In recent years, the usage of the Terahertz (THz) portion of the electromagnetic spectrum has been investigated in the field of communications, medicine and biology. The evaluation of the THz wave performance in the applications requires supportive THz based devices such as emitter, detectors and filters. The efficient integration of the THz devices with the state of the art silicon based devices is limited by its typical diffraction limit (millimeters). In this abstract, a subwavelength Surface Plasmon Polaritons (SPPs) based Bragg reflector is using Indium antimonide (InSb), Silicon-dioxide (SiO2) and Porous-SiO2, is proposed. The SPPs based wave propagation at THz allows the nanoscale realization of the THz based devices. The proposed plasmonic Bragg reflector utilizes the periodic changes of the dielectric material in the core layer, while InSb is used in the cladding layer. Finite Element Method (FEM) based simulations are used to demonstrate the working of the proposed subwavelength Bragg reflector.

I. INTRODUCTION

THz radiation and waves have recently become under increased attention due to the wide range of application of THz radiations, such as security scanners, biomedical imaging and sensing [1]. Achieving on-chip integration of THz devices promise to mitigate many of the limitations faced by current VLSI systems [2]–[46]. However, the THz range occupies what is called the THz gap, resulting from the lack of devices to manipulate THz radiations, such as the case with the lower frequency RF and microwave radiations [38], [47]–[71] and the higher frequency optical radiations [72]–[78]. Such a gap remains the main bottleneck that faces applications involving THz radiations.

Numerous efforts are currently undergoing in order to close the THz gap, such as the development of small and efficient THz sensors, detectors, sources, mixers, waveguides, and so on [79], [80]. Many of such devices are based on plasmons, which are the oscillations of two-dimensional electron density for example at the interface between a dielectric and a metal or a doped semiconductor. Plasmonic devices, in turn, have been of high interest due to the high levels of miniaturization that such devices allow, and because of their use in sub-wavelength light manipulation. Many of these plasmonic devices studied utilize surface plasmons at a metal-dielectric interface and, thus, operate in the optical domain. However, it is possible to utilize the excellent properties of plasmons in order to realize THz devices by using plasmons at the interface of a dielectric and a doped semiconductor.

In this paper we propose a THz plasmon-based Bragg filter that consists of Indium antimonide (InSb), Silicon-dioxide (SiO2) and Porous-SiO2 to realize the structure. Plasmonic Bragg filters in the optical domain utilize the band-gap arising from the alternating dielectric material in a metal-insulator-metal structure to achieve the filtering objective, and are characterized by high levels of light confinement. For THz filtering, we use a semiconductor-dielectric-semiconductor (SIS) structure consisting of $InSb$ as the semiconductor material and $PSiO_2$ and SiO_2 as alternating dielectric material. Using the finite element method (FEM), we simulate the structure and study its transmission spectrum. We use the Drude's model for the wavelength-dependent dielectric function of the $InSb$ from [80].

The rest of the paper is organized as follows. Section II presents the THz semiconductor-insulator-semiconductor (SIS) structure which is the basis for the Bragg reflector presented in this paper. Section III presents the THz Bragg reflector based on the SIS structure and the main results of this paper. Section IV concludes the paper.

II. THz PLASMONS ON A SEMICONDUCTOR-INSULATOR INTERFACE

Surface plasmons refer to the oscillation of a two-dimensional charge density, such as the electron gas at

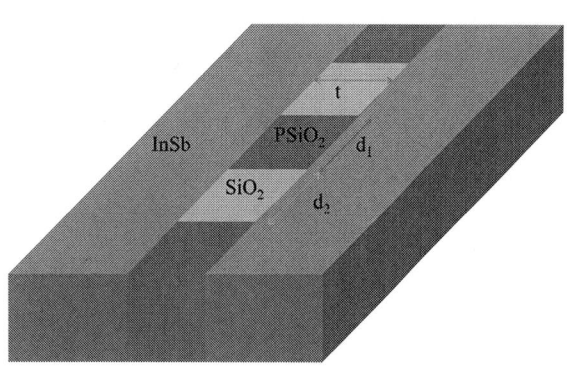

Fig. 1. Schematic of a THz Bragg reflector

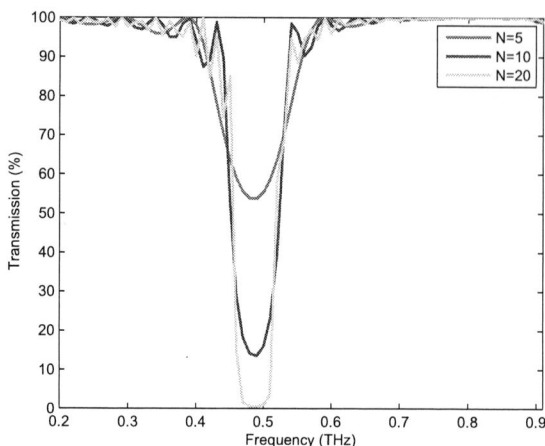

Fig. 2. Transmission percentage for a Bragg filter with various N values. The filtering is effective for $N \geq 20$.

a metal-dielectric interface. Metal-insulator-metal structures are popular because of the high levels of field confinement in the device and the various functionalities that can be realized based on such structure, among other features of the MIM structure. Also, plasmonic devices allow sub-wavelength manipulation of electromagnetic radiation and thus, a plasmonic device typically can achieve high levels of miniaturization as compared to non-plasmonic devices.

Plasmon oscillations can also occur at the interface between a semiconductor and a dielectric, resulting in THz surface plasmons, which can be utilized to build THz plasmon devices. Similar to an MIM structure, an SIS structure, consisting of two layers of a semiconductor sandwiching an insulator, can be used to realize a wide range of THz plasmon devices that have similar properties to its optical counterparts. Because the plasma frequency of semiconductors is in the THz range, semiconductor plasmonic devices operate at THz frequencies. The various optical plasmonic devices can be realized for THz applications by replacing the metal by a semiconductor. In order to model the semiconductor, we use a Drude's dielectric function, which is given by [80]

$$\varepsilon_s(\omega) = \varepsilon_\infty(1 - \frac{\omega_p^2}{\omega^2 + \tau^{-2}} + i\frac{\omega_p^2 \tau^{-1}}{\omega(\omega^2 + \tau^{-2})})) \quad (1)$$

where ε_∞ is the background dielectric constant (high frequency dielectric function), τ is the average mean free time, and ω_p is the plasma frequency given by

$$\omega_p = \sqrt{\frac{ne^2}{\varepsilon_{infty}\varepsilon_0 m^*}} \quad (2)$$

with n being the charge carrier concentration of the semiconductor and m^* is the effective mass of the charge carriers. In this paper, we use the Drude's model given for the $InSb$ dielectric function in [80], where the electron mobility $\mu = 7.7 \times 10^4 cm^2 V^{-1} s^{-1}$ and the $\tau = \mu m^*/e$. The electron effective mass is $m^* = 0.014 m_0$, where m_0 is the electron mass, the room temperature intrinsic carrier concentration is $10^{16} cm^{-3}$, and $\varepsilon_\infty = 15.7$. The fundamental TM mode for plasmon oscillations in an SIS structure satisfies the following relations, which are used to derive the dispersion relation of the oscillations,

$$\epsilon_s k_{z1} coth(\frac{-ik_{z1}t}{2}) + \epsilon_d k_{z2} = 0$$
$$k_{z1,2}^2 = \epsilon_{s,d}\frac{\omega}{c} - k_x^2 \quad (3)$$

where ϵ_s and ϵ_d are the dielectric functions of the semiconductor and the dielectric layer respectively, k_{z1} and k_{z2} are the wavevectors of surface plasmons in the z direction in the two semiconductor layers, respectively, and k_x is the wave vector of the plasmon wave. Typically, wave propagation is described using the dimensionless effective index, $n_{eff} = k_x/k_o$, where k_o is the free space wave vector, which is related to the free space wavelength, λ_o as $k_o = 2\pi/\lambda_o$. k_{z1} and k_{z2} reflect

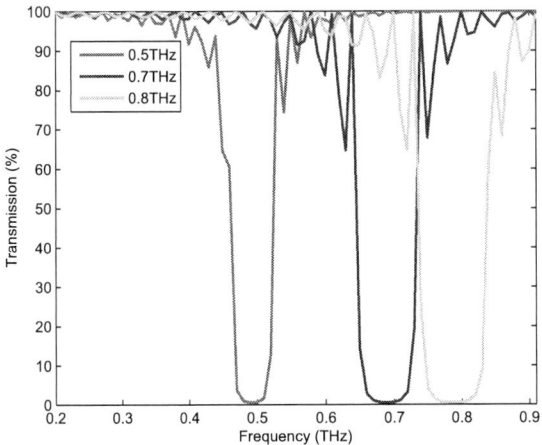

Fig. 3. Transmission percentage for Bragg reflectors with various center frequencies. The Bragg filter is chaacterized by $d_1 Real(n_{eff,1}) = d_2 Real(n_{eff,2}) = \frac{\lambda_c}{4}$.

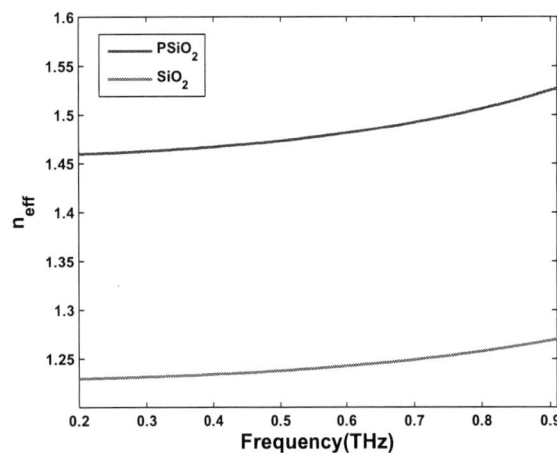

Fig. 4. Effective index of refraction for the SiO_2 and $PSiO_2$ versus frequency.

the confinement of the electromagnetic fields in the dielectric layer, as the decay factor of the fields in the two semiconductor layers is $e^{-ik_{z1,2}z}$.

III. A

SEMICONDUCTOR-INSULATOR-SEMICONDUCTOR BASED THZ BRAGG REFLECTOR

A THz plasmonic Bragg reflector consists of two layers of doped semiconductor sandwiching a periodic alternation of two dielectric media, as shown in figure 1. The alternation between two dielectrics in the dielectric layer creates a band gap in the spectrum of surface plasmons. The Bragg filter structure can be studied using the transfer matrix theory, as was shown in [81], where the effective refractive indices of the alternating media are used as the refractive indices. This is because the plasmon oscillations on the surface of each medium should satisfy the dispersion relations in 3. The first band gap in the Bragg reflector is given by

$$\Delta\omega_g = \frac{4}{\pi}\omega_c sin^{-1}\left(\frac{n_{eff,2} - n_{eff,1}}{n_{eff,2} + n_{eff,1}}\right) \quad (4)$$

where $n_{eff,1}$ and $n_{eff,2}$ are the effective indices of refraction in the two dielectric materials. If the thickness of the dielectric materials are denoted by d_1 and d_2, then the Bragg condition is given by

$$d_1 Real(n_{eff,1}) + d_2 Real(n_{eff,2}) = \frac{\lambda_c}{2} \quad (5)$$

where $\lambda_c = 2\pi/\omega_c$. Even though there exist infinitely many combinations of d_1 and d_2 that satisfy the Bragg condition 5, it is desirable to have $d_1 Real(n_{eff,1}) = d_2 Real(n_{eff,2}) = \frac{\lambda_c}{4}$ because this combination results in an equivalent wave concentration in each dielectric medium.

The results presented in this section are for an infinite Bragg reflector, in which the insulator materials alternate infinitely. In practice, a finite number of periods is present; however, when the number of periods is large, the results are valid. Also, the quality of the Bragg filtering increases as the number of periods increase. That is, the transmission coefficient in the stop zone goes to zero as the number of periods increases. Figure 2 shows the transmission percentage in a Bragg filter for various N, where it is shown that the filtering is effective for $N = 20$ while it is not effective for $N = 5$ and $N = 10$. In the rest of the paper, we use $N = 20$.

Figure 3 shows the transmission percentage for a Bragg filter with various center frequencies. In this figure, the Bragg filter is characterized by $d_1 Real(n_{eff,1}) = d_2 Real(n_{eff,2}) = \frac{\lambda_c}{4}$. The center frequency, which in turn decides the stop band of the Bragg filter, can be controlled through the design parameters d_1 and d_2. In particular, d_1 and d_2 are chosen so that $d_1 Real(n_{eff,1}(\lambda_c)) = d_2 Real(n_{eff,2}(\lambda_c)) = \frac{\lambda_c}{4}$, where $n_{eff,1}(\lambda_c)$ and $n_{eff,2}(\lambda_c)$ are the effective refractive indices of the insulating materials at the desired center frequency. However, $n_{eff,1}$ and $n_{eff,2}$ change

978-1-4673-1122-9/12 $31.00 © 2012 IEEE

with wavelength, so that the relation $d_1 Real(n_{eff,1}) = d_2 Real(n_{eff,2}) = \frac{\lambda_c}{4}$ might not hold for wavelengths different from λ_c. For this reason, the effective indices should not vary drastically with wavelength, but rather, it is desired to have as small as possible a variation in $n_{eff,1}$ and $n_{eff,2}$ with wavelength. This is indeed that case in our case as shown in figure 4.

The bandwidth of the Bragg reflector relative to the center frequency, which satisfies 4, does not vary with the center frequency, since $(n_{eff1} - n_{eff1})/(n_{eff1} + n_{eff1})$ does not vary drastically with frequency. This means that there is no trade off between the center frequency and the relative bandwidth of the Bragg filter, so that increasing or decreasing the center frequency does not result in decreasing the relative bandwidth of the Bragg filter. Additionally, a non varying relative bandwidth means that the absolute bandwidth can be increased by increasing the center frequency of the filter.

IV. CONCLUSION

The wide range of applications of THz radiation makes it more important to develop efficient devices to manipulate THz radiation and narrow what is called the THz gap. THz plasmonic devices, in which the metal in regular plasmonic devices is replaced by a semiconductor, promise to play a key role in this regard because such devices allow for sub-wavelength processing of radiation and can be used to build a wide range of THz devices. In this paper, we presented a THz Bragg reflector based on the SIS structure, which consists of two semiconductor layers sandwiching a layer of two alternating materials. Our simulation results show that a THz Bragg filter can be realized using SnIb as the semiconductor and silicon dioxide and porous silicon dioxide as the alternating insulator materials.

REFERENCES

[1] W. Stillman and M. Shur, "Closing the gap: Plasma wave electronic terahertz detectors," *J. of Nanoelectronics and Optoelectronics*, vol. 2, no. 3, 2007.
[2] Y. Massoud, S. Majors, T. Bustami, and J. Whites, " techniques for minimizing on-chip interconnect self-inductance," *IEEE DAC*, 1998.
[3] D. M. Y. Massoud, J. Kawa and J. White, "Modeling and Analysis of Differential Signaling for Minimizing Inductive Cross-Talk," *IEEE DAC*, 2001.
[4] A. Nieuwoudt and Y. Massoud, "Evaluating the Impact of Resistance in Carbon Nanotube Bundles for VLSI Interconnect using Diameter-Dependent Modeling Techniques," *IEEE TED*, October 2006.
[5] "Modeling and Design Challenges and Solutions for Carbon Nanotube-Based Interconnect in Future High Performance Integrated Circuis," *ACM JETC*, July 2006.
[6] Y. Massoud and J. White, "Managing On-Chip Inductive Effects," *IEEE TVLSI*, Dec. 2002.

[7] A. Nieuwoudt and Y. Massoud, "Understanding the impact of inductance in carbon nanotube bundles for VLSI interconnect using scalable modeling techniques," *IEEE TNANO)*, Nov. 2006.
[8] Y. Massoud and Y. Ismail, "Grasping the Impact of On-Chip Inductance in High Speed ICs," *IEEE Circuits and Devices Magazine*, July 2001.
[9] A. Nieuwoudt and Y. Massoud, "On the impact of process variations for carbon nanotube bundles for VLSI interconnect," *IEEE TED*, March 2007.
[10] ——, "Assessing the implications of process variations on future carbon nanotube bundle interconnect solutions," *IEEE ISQED*, 2007.
[11] M. Mondal, A. Ricketts, S. Kirolos, T. Ragheb, V. N. G. Link, and Y. Massoud, "Thermally robust clocking schemes for 3D integrated circuits," *IEEE DATE*, 2006.
[12] M. Mondal and Y. Massoud, "Reducing pessimism in RLC delay estimation using an accurate analytical frequency dependent model for inductance," *IEEE ICCAD*, November 2005.
[13] S. Kim, Y. Massoud, , and S. Wong, "On the Accuracy of Return Path Assumption for Loop Inductance Extraction for 0.1 um Technology and Beyond," *IEEE ISQED*, 2003.
[14] S. Eachempati, A. Nieuwoudt, A. Gayasen, V. Narayanan, and Y. Massoud, "Assessing carbon nanotube bundle interconnect for future FPGA architectures," *IEEE DATE*, 2007.
[15] Y. Massoud and J. White, "Simulation and Modeling of the Effect Substrate Conductivity on Coupling Inductance and Circuit Crosstalk," *IEEE TVLSI*, June 2002.
[16] A. Nieuwoudt and Y. Massoud, "On the optimal design, performance, and reliability of future carbon nanotube-based interconnect solutions," *IEEE TED*, August 2008.
[17] ——, "Accurate resistance modeling for carbon nanotube bundles in VLSI interconnect," *IEEE Nano*, 2006.
[18] A. Nieuwoudt, M. Mondal, and Y. Massoud, "Predicting the Performance and Reliability of Carbon Nanotube Bundles for On-Chip Interconnect," *IEEE ASP-DAC*, 2007.
[19] A. Nieuwoudt and Y. Massoud, "Performance implications of inductive effects for carbon-nanotube bundle interconnect," *IEEE EDL*, April 2007.
[20] A.Nieuwoudt and Y.Massoud, "Predicting the performance of low-loss on-chip inductors realized using carbon nanotube bundles," *IEEE TED*, January 2008.
[21] M. Mondal, T. Ragheb, X. Wu, A. Aziz, and Y. Massoud, "Provisioning on-chip networks under buffered rc interconnect delay variations," *ISQED*, 2007.
[22] A. Hosseini, T. Ragheb, and Y. Massoud, "A fault-aware dynamic routing algorithm for on-chip networks," *IEEE ISCAS*, 2008.
[23] Y. Massoud and J. White, "Improving the generality of the fictitious magnetic charge approach to computing inductances in the presence of permeable materials," *IEEE DAC*, 2002.
[24] A. Nieuwoudt and Y. Massoud, "RC circuit model for multi-walled carbon nanotubes," *IEEE Nano*, 2007.
[25] Q. Su, J. Kawa, C. Chiang, and Y. Massoud, "Accurate modeling of substrate resistive coupling for floating substrates," *ACM TODAES*, January 2006.
[26] M. Mondal, A. Ricketts, S. Kirolos, T. Ragheb, G. Link, V. Narayanan, and Y. Massoud, "Mitigating thermal effects on clock skew with dynamically adaptive drivers," *IEEE ISQED*, 2007.
[27] M. Alam, A. Nieuwoudt, and Y. Massoud, "Wavelet-based passivity preserving model order reduction for wideband interconnect characterization," *IEEE ISQED*, 2007.
[28] M. Mondal, K. Mohanram, and Y. Massoud, "Parameter-Variation-Aware Analysis for Noise Robustness," *IEEE ISQED*, 2007.
[29] Y. Massoud, J. Wang, and J. White, "Accurate inductance extraction with permeable materials using qualocation," *International Conf. on Modeling and Simulation of Microsystems*, 1999.
[30] S. Kirolos and Y. Massoud, "Adaptive Ratio-Size Gates for Minimum-Energy Operation," *IEEE TCAS II*, November 2006.
[31] S. Kirolos, Y. Massoud, and Y. Ismail, "Power-supply-variation-aware timing analysis of synchronous systems," *IEEE ISCAS*, 2008.
[32] ——, "Accurate analytical delay modeling of cmos clock buffers considering power supply variations," *IEEE ISCAS*, 2008.
[33] S. Kirolos and Y. Massoud, "Adaptive SRAM Design for Dynamic Voltage Scaling VLSI Systems," *IEEE MWSCAS*, 2007.

[34] X. Wu, T. Ragheb, A. Aziz, and Y. Massoud, "Implementing DSP algorithms with on-chip networks," *IEEE NOC*, 2007.

[35] M. Alam, A. Nieuwoudt, and Y. Massoud, "Model order reduction using spline-based dynamic multi-point rational interpolation for passive circuits," *Analog Integrated Circuits and Signal Processing*, March 2007.

[36] M. Mondal and Y. Massoud, "Analytical Modeling of Loop Self Inductance Bound for Inductance-Aware Physical Synthesis," *IEEE ISCAS*, 2005.

[37] Y. Massoud and J. White, "Fast Inductance Extraction of 3-D Structures with Non-ConstantPermeabilities," *International Conference on Modeling and Simulation of Microsystems*, pp. 190–193, April 1998.

[38] M. A. Y. Massoud and A. Nieuwoudt, "On the Selection of Spectral Zeros for Generating Passive Reduced Order Models," *IEEE IWSOC*, 2006.

[39] M. Mondal and Y. Massoud, "Estimation of capacitive crosstalk-induced short-circuit energy," *JCSC*, June 2007.

[40] Y. Massoud and J. White, "FastMag: a 3-D Fast Inductance Extraction Program for Structures with Permeable materials," *IEEE ICCAD*, 2002.

[41] S. Kirolos and Y. Massoud, "Supply Voltage Adaptive Low-Power Circuit Design," *IEEE DCAS*, 2006.

[42] . M. Modal and Y. Massoud, "Accurate loop self inductance bound for efficient inductance screening," *IEEE TVLSI*, December 2006.

[43] S. Kirolos and Y. Massoud, "Robust Wide Range of Supply-Voltage Operation using Continuous Adaptive Size-Ratio Gates," *IEEE ISCAS*, 2008.

[44] Y. Massoud, S. Kirolos, and K. Mohanram, "Analytical model-based technique for efficient evaluation of noise robustness considering parameter variations," *Analog Integrated Circuits and Signal Processing*, August 2009.

[45] S. Kirolos, M. Mondal, K. Mohanram, and Y. Massoud, "A Model-Based Technique for Efficient Evaluation of Noise Robustness," *IEEE ISCAS*, 2008.

[46] S. Kirolos, Y. Massoud, and Y. Ismail, "Mitigating Power-Supply Induced Delay Variations using Self Adjusting Clock Buffers," *IEEE MWSCAS*, 2008.

[47] A. Nieuwoudt and Y. Massoud, "Variability-aware multilevel integrated spiral inductor synthesis," *IEEE TCAD*, Dec. 2006.

[48] ——, "Multi-level Approach for Integrated Spiral Inductor Optimization," *IEEE DAC*, 2005.

[49] A. Nieuwoudt, T. Ragheb, and Y. Massoud, "SOC-NLNA: synthesis and optimization for fully integrated narrow-band CMOS low noise amplifiers," *IEEE DAC*, 2006.

[50] A. Nieuwoudt and Y. Massoud, "Robust automated synthesis methodology for integrated spiral inductors with variability," *IEEE ICCAD*, 2005.

[51] A. Nieuwoudt, M. McCorquodale, R. Borno, and Y. Massoud, "Efficient analytical modeling techniques for rapid integrated spiral inductor prototyping," *IEEE CICC*, 2005.

[52] T. Ragheb, S. Kirolos, J. Laska, M. Strauss, A. Gilbert, R. Baraniuk, and Y. Massoud, "Implementation models for analog-to-information conversion via random sampling," *IEEE MWSCAS*, 2007.

[53] A. Nieuwoudt, M. McCorquodale, R. Borno, and Y. Massoud, "Accurate analytical spiral inductor modeling techniques for efficient design space exploration," *IEEE EDL*, December 2006.

[54] A. Nieuwoudt and Y. Massoud, "Analytical wide-band modeling of high frequency resistance in integrated spiral inductors," *Analog Integrated Circuits and Signal Processing*, February 2007.

[55] S. Kirolos, T. Ragheb, J. Laska, M. Duarte, Y. Massoud, and R. Baraniuk, "Practical Issues in Implementing Analog-to-Information Converters," *IEEE IWSOC*, 2006.

[56] A. Nieuwoudt and Y. Massoud, "Efficient modeling of substrate eddy currents for integrated spiral inductor design automation," *IEEE MWSCAS*, 2005.

[57] T. Ragheb, A. Nieuwoudt, and Y. Massoud, "Efficient modeling of integrated narrow-band low noise amplifiers for design space exploration," *IEEE GLSVLSI*, 2006.

[58] A. Nieuwoudt, T. Ragheb, and Y. Massoud, "Hierarchical optimization methodology for wideband low noise amplifiers," *IEEE ASP-DAC*, 2007.

[59] T. Ragheb, J. N. Laska, H. Nejati, S. Kirolos, R. G. Baraniuk, and Y. Massoud, "A prototype hardware for random demodulation based compressive analog-to-digital conversion," *IEEE MWSCAS*, 2008.

[60] A. Nieuwoudt, T. Ragheb, H. Nejati, and Y. Massoud, "Increasing manufacturing yield for wideband RF CMOS LNAs in the presence of process variations," *IEEE ISQED*, 2007.

[61] ——, "Numerical design optimization methodology for wideband and multi-band inductively degenerated cascode CMOS low noise amplifiers," *IEEE TCAS I*, June 2009.

[62] T. Ragheb, A. Nieuwoudt, and Y. Massoud, "Modeling of 3.1-10.6 GHz CMOS Filter-Based Low Noise Amplifier for Ultra-Wideband Receivers," *IEEE WAMICON*, 2006.

[63] H. Najati, T. Ragheb, A. Nieuwoudt, and Y. Massoud, "Modeling and design of ultrawideband low noise amplifiers with generalized impedance matching networks," *IEEE ISCAS*, 2007.

[64] A. Nieuwoudt and Y. Massoud, "Optimizing the Design of Tunable Spiral Inductors for On-Chip Wireless Applications," *IEEE WAMICON*, 2006.

[65] H. Nejati, T. Ragheb, A. Nieuwoudt, and Y. Massoud, "Analytical modeling methodology for ultrawideband low noise amplifiers with generalized filter-based impedance matching," *Analog Integrated Circuits and Signal Processing*, May 2007.

[66] A. Nieuwoudt, T. Ragheb, and Y. Massoud, "Systematic Design Optimization Methodology for Multi-Band CMOS Low Noise Amplifiers," *IEEE ISVLSI*, 2007.

[67] H. Nejati, T. Ragheb, and Y. Massoud, "On the design of customizable low-voltage common-gate LNA-mixer pair using current and charge reusing techniques," *IEEE GLSVLSI*, 2008.

[68] S. Pfetsch, T. Ragheb, J. Laska, H. Nejati, A. Gilbert, M. Strauss, R. Baraniuk, and Y. Massoud, "On the Feasibility of Hardware Implementation of Sub-Nyquist Random-Sampling Based Analog-to-Information Conversion," *IEEE ISCAS*, 2008.

[69] Y. Massoud, A. Nieuwoudt, and T. Ragheb, "Automated design solutions for fully integrated narrow-band low noise amplifiers," *IEEE IWSOC*, 2006.

[70] A. Nieuwoudt and Y. Massoud, "Design optimization of switchable multi-port spiral inductors," *Springer ALOG*, June 2007.

[71] H. Nejati, T. Ragheb, and Y. Massoud, "Analytical Modeling of Common-Gate Low Noise Amplifiers," *IEEE MWSCAS*, 2008.

[72] A. Hossieni, A. Nieuwoudt, and Y. Massoud, "Efficient simulation of subwavelength plasmonic waveguides using implicitly restarted Arnoldi," *Optics Express*, August 2006.

[73] M. Alam and Y. Massoud, "A closed-form analytical model for single nanoshells," *IEEE TNANO*, May 2006.

[74] ——, "RLC ladder model for scattering in single metallic nanoparticles," *IEEE MTT-S*, 2006.

[75] A. Hossieni, H. Nejati, and Y. Massoud, "Design of a maximally flat optical low pass filter using plasmonic nanostrip waveguides," *Optics Express*, November 2007.

[76] M. Alam and Y. Massoud, "An accurate closed-form analytical model of single nanoshells for cancer treatment," *IEEE MWSCAS*, 2005.

[77] A. Hossieni and Y. Massoud, "Optical range microcavities and filters using multiple dielectric layers in metal-insulator-metal structures," *Journal of the Optical Society of America A*, January 2007.

[78] M. Alam and Y. Massoud, "An RLC ladder model for the equivalent impedance of single metal nanoparticles in electromagnetic field," *IEEE Transactions on Nanotechnology (TNANO)*, September 2006.

[79] M. Dyakonov and M. Shur, "Detection, mixing, and frequency multiplication of terahertz radiation by two-dimensional electronic fluid," *IEEE TED*, vol. 43, no. 3, 1996.

[80] J. Rivas, C. Janke, P. Bolivar, and H. Kurz, " Transmission of THz radiation through InSb gratings of subwavelength apertures ," *Opt. Express*, vol. 13, no. 2, Apr. 2005.

[81] A. Hosseini, H. Nejati, and Y. Massoud, "Modeling and design methodology for metal-insulator-metal plasmonic Bragg reflectors," *Optics Express*, Feb. 2008.

On Sustaining Robustness of Molecular Pathway Circuits of the HSR Network of *E. coli* under Spatial Configuration

Jian-Qin Liu[*+], Teruya Yamanishi[†], Haruhiko Nishimura[‡], Sou Nobukawa[‡], Hiroaki Umehara[+]

[+]Brain ICT Laboratory, Advanced ICT Research Institute, National Institute of Information and Communications Technology,
588-2 Iwaoka, Iwaoka-cho, Nishiku, Kobe, Hyogo, 651-2492 JAPAN *Corresponding Author
[†]Department of Management Information Science, Fukui University of Technology, Gakuen, Fukui, 910-8505 JAPAN
[‡]Graduate Scholl of Applied Informatics, University of Hyogo, Chuo-ku, Kobe, 650-0044 JAPAN
*liu@nict.go.jp, yamanisi@fukui-ut.ac.jp, haru@ai.u-hyogo.ac.jp, ab10y406@ai.u-hyogo.ac.jp, ume@nict.go.jp

Abstract—In systems biology, the robustness of molecular circuits has been intensively studied. Especially, the temporal dynamics mechanism of robust molecular circuits has been verified by experimental evidences. But its spatial dynamics mechanism is still an open problem yet. To solve this problem, it is necessary to introduce the communication channel into the dynamics model based on the spatial configuration of molecular circuits. In this paper, we study how the robustness of molecular circuits constructed by signaling pathways is sustained. The information-theoretic model called network coding is presented to formalize the molecular pathway circuits to introduce the channel unit into the controller of the signaling pathway, which is a networked control system (NCS). By using the Heat Shock Response (HSR) network of the model organism *E. coli* as an example of robust molecular pathway circuits, we quantitatively analyze its robustness behavior under channel delay and show the robustness is sustained under the stochastic channel of intra-cellular communication. The result on the robustness of molecular circuits under the spatial configuration is significant for designing and controlling engineered molecular circuits *in vivo* by synthetic biological technology. It is concluded that the robustness mechanism of signaling pathway networks may contribute to the application of molecular circuits in the molecular biomedical engineering systems built by the means of synthetic biology.

Keywords-robustness; molecular circuits; networked control systems; neytwork coding

I. INTRODUCTION

As one of the applications of robustness theory in bioinformatics [1~11], the robustness models of molecular circuits [5~8] in cells are significant for systematically understanding the signaling mechanism of cells. It is imperative to quantitatively investigate the robust circuits constructed by signaling pathways of the cell – molecular pathway circuits. Because of the importance of the cellular communication mechanism of molecular pathway circuits determined by the dynamics of signal transduction of the cell, the temporal dynamics aspect of the signaling pathway networks has been clarified in bioinformatics [1~11]. However, it is still a challenging task to quantitatively analyze the spatial dynamics mechanism of robust molecular pathway circuits, which is the objective of our research.

In this paper *robustness* is defined as *a mechanism that guarantees the stability under the condition of the variation of the parameters and the network structure of the underlying system*. With respect to the network dynamics including robustness, the dynamics of a state is given in the following form

$$\frac{dX(t)}{dt} = f(X(t)), \qquad (1)$$

where $X(t) = (x_1(t),..., x_n(t))$ is the vector of the state variables in the network, $\frac{dX(t)}{dt}$ denotes the derivative of $X(t)$, corresponding to $x_i(t)$ ($i = 0,1,...,n$-1, $n \in N$), with respect to time, and $f(X(t))$ is a function defined in a real-valued space R.

Then,

$$\frac{dX(t)}{dt} = 0 \qquad (2)$$

refers to the steady state for stability, and

$$\lim_{t \to t_s} \frac{dX(t)}{dt} = 0 \qquad (3)$$

refers to the asymptotically steady state that is the common case in the robustness of molecular circuits when $t \geq t_s$ (t_s is the time of the beginning at the steady state).

978-1-4673-1122-9/12 $31.00 © 2012 IEEE

Considering the parameter set Φ ($\varphi \in \Phi$) in the underlying network, the robustness of molecular circuits is given in the following form

$$\lim_{t \to t_s} \frac{dX(\varphi, t)}{dt} = 0 . \qquad (4)$$

When the variation of the parameters is considered, the robustness, i.e., robust stability, is given by

$$\lim_{t \to t_s} \frac{\partial X(\varphi, t)}{\partial t} = 0 , \qquad (5)$$

when $\Delta \varphi \neq 0$. This quantifies the invariance of the steady state of $X(\varphi, t)$ independently on the parametric variation $\Delta \varphi$.

In this paper, the issue on sustaining the robustness in spatially configured molecular circuits is presented in which we formulate the spatially configured molecular pathway circuits by the integration of NC (network coding) and NCS (networked control system) models and quantitatively analyze the robustness of the molecular pathway circuits under the channel delay. The remainder of the paper is organized as follows. Section II presents the network coding method for the spatial representation of molecular circuits and discusses the robustness of molecular circuits. Section III presents the research on the NCS (networked control systems) model for spatially configured molecular circuits by using the example of the HSR (heat shock response) network. The simulation result shows that the robustness of the HSR network is sustained under the spatial configuration. Section IV concludes the paper by briefly summarizing the main points of our work.

II. ROBUST MOLECULAR CIRCUITS FORMULATED BY THE INFORMATION THEORETIC MODEL OF NETWORK CODING AS A ROBUST NETORKED CONTROL SYSTEM

A. Network Coding for the Representation of Moelcular Circuits

Because information theory is the theoretical foundation for synthesizing and analyzing signal transmission processes of communication systems, formulating the biochemical mechanism of cellular functions by information theory can help us to understand the communication process of cellular signaling. As a promising paradigm of information theory, network coding [12~14] shows the advantages in analyzing communication networks. Based on the abstract structure of networks which rely on the store-and-forward principle [14], an equivalence between the classic network structure of linear network coding illustrated in Fig. 1(b) of [12] and a formalized network structure of signaling pathway networks is established [15, 16] as the logical form of the network structure to formalize molecular pathway circuits.

B. Reversible Molecular Switches for Molecular Patwhay Circuits

In cells, there exist the molecular circuits constructed by the reversible switches of molecular signals, i.e., the switch of phosphorylation and dephosphorylation and the switch of the GTP-bound state and the GDP-bound state of GTPases. The phosphorylation and the GTP-bound state are defined as 1; the dephosphorylation and the GDP-bound state are defined as 0. The molecule with the state of phosphorylation or dephosphorylation and the GTPase with the GTP-bound state or the GDP-bound state can be used to represent the variable in a molecular circuit. The two states provide the binary values of molecular switches, which is a kind of building blocks for constructing molecular circuits. These switches are reversible. A schematic diagram of a molecular switch is given in Fig. 1. Kinase and phosphatase activate the phosphorylation and dephosphorylation processes, respectively. GEF and GAP activate the processes of the GTP-bound state and the GDP-bound state of GTPases, respectively.

C. Formalzing a Binary Moelcular Pathway Circuits by Linear Network Coding

The binary form of the molecular pathway circuits is the basis of network analysis of real-valued molecular pathway circuits where the dynamics of the real-valued states is assigned. An example of binary pathway circuits is constructed in terms of linear network coding [12] (LNC) as shown in Fig. 2.

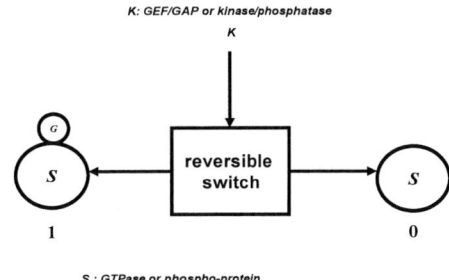

Fig. 1. Reversible switch of pathways

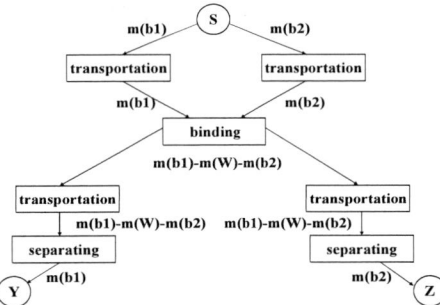

Fig. 2 An example of the binary pathway circuit by LNC

Let *path (A, E)* be a predicate (i.e., the logical form of the reasoning operator) that denotes a pathway whose input is *A* and output is *E*. The binary values of data bits *b1* and *b2* (1 and 0) are defined by the phosphorylation and dephosphorylation states, respectively. The term "data bits" in LNC refers to the variables of molecular circuits assigned with two states. The two states defined by phosphorylation or dephosphorylation can also be mapped to the GTP-bound or GDP-bound states of GTPases. With the predicate form for molecular circuits, we may describe a binary molecular pathway circuits. The nodes *S, T, W, U, X, Y,* and *Z* in Fig. 1(b) of [12] are used as the reference. Let the molecules at *S* node be denoted as *m(b1)* and *m(b2)*. We have *path (m(b1), m(b1))* for the channel from *S* to *T* denoted as *channel (S,T)* and the channel from *T* to *W* denoted as *channel (T, W)*. The channel from *S* to *U* and then from *U* to *W* are defined in a similar way and is denoted as *path (m(b2), m(b2))*. We use *site(m(b1)* to represent the channel from *T* to *Y* and *site(m(b2))* to represent the channel from *U* to *Z*. The information flow of the linear network coding model obeys the max-flow min-cut rule. So it is the optimized configuration of the molecular pathway circuit. Source *S* is set as the starting point of the intra-cell communication process of molecular pathway circuits and is mapped to the output of a switch of phosphorylation and dephosphorylation processes. The channel from *S* to *T* and the channel from *T* to *W* are simplified as one process, that is, they are mapped to one (active or passive) transportation process of the cell and won't change its state of a signaling molecule, which is used to represent *b1*. *b1* and *b2* meet at node *W*, which construct a block of data. The molecules *m(b1)* and *m(b2)* will be bound to the molecule which represents the node *W* and is denoted as *m(W)* in different sites of *m(W)*. The binding operation for *b1* and *b2* corresponds to the encoding operation in linear network coding. The notation of *b1+b2* in linear network coding is transformed into the compound of *m(b1)-m(W)-m(b2)*. The channels from *W* to *X* and from *X* to *Y* are unified and defined as a transportation process denoted as *path(m(b1)-m(W)-m(b2), m(b1)-m(W)-m(b2))*. The channels from *W* to *X* and from *X* to *Z* are unified and defined in the same way. At *Y* and *Z*, a pathway is used to separate the bound *b1* and *b2* into the individual state of *b1* and *b2*, respectively. Since the sites of the compound are the references to separate *m(b1)* and *m(b2)* from the compound *m(b1)-m(W)-m(b2)*, the channel from *T* to *Y* and the channel from *U* to *Z* are equivalently reflected here, i.e., the information of the sites for *Y* equals to the reference of the channel from *T* to *Y* and the information of the sites for *Z* equals to the reference of the channel from *X* to *Z*. Consequently, the structure of linear network coding for temporal dynamics is mapped into an information network of signaling pathways. In general, a configuration of signal pathways by the linear network coding provides us a way to formalize molecular pathway circuits where we can introduce the dynamics mechanism.

D. Network Coding for the Dynamics of Molecular Patwhay Circuits

In order to formulate the spatial configuration of robust molecular pathway circuits, the network coding scheme is presented to describe the intra-cellular communication process between two signaling molecules *x* and *y*. As shown in Fig. 3, the node *s* denotes the signal that activate *x* and *y*, respectively. The routes for *x* and *y* are given by the follows:

$$s \rightarrow activate\ (x) \rightarrow Tx(x)$$
$$\rightarrow channel\ (x,y) \rightarrow x+y$$
$$\rightarrow biochemical\text{-}reaction\ (x,y)$$
$$\rightarrow x+y \rightarrow channel\ (x,y)$$
$$\rightarrow Rx(x)$$

(6)

and

$$s \rightarrow activate\ (y) \rightarrow Tx(y)$$
$$\rightarrow channel\ (y,x) \rightarrow x+y$$
$$\rightarrow biochemical\text{-}reaction\ (y,x)$$
$$\rightarrow x+y \rightarrow channel\ (y,x)$$
$$\rightarrow Rx(y).$$

(7)

Then, according to the representation scheme of LNC by Li et al in [12], the dynamics of *x* and *y* is denoted as *f(x,y)* and *g(x,y)*, respectively. In order to describe the spatially configured molecular pathway circuits, it is necessary to integrate the representation forms of LNC for temporal and spatial dynamics into a layered LNC model. As shown in Fig. 4, a three-layered LNC model is proposed to describe the joint temporal and spatial dynamics of a molecular pathway circuits with two variables *x* and *y*. The red and blue colored parts represent the feedback and feedforward, respectively. The network coding scheme can be also applied for the formulation of the HSR network (where *x* and *y* are defined as *DnaK* and σ^{32}, respectively) by which the robustness of real-valued circuits will be discussed.

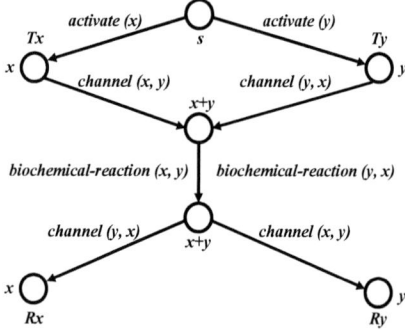

Fig. 3 Network coding for the spatial configuration of signaling pathways

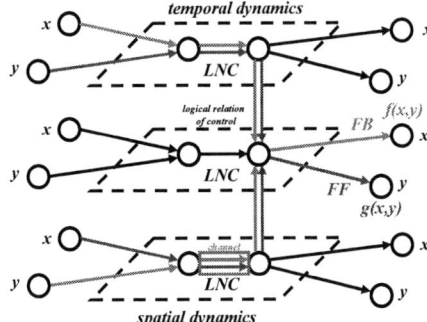

Fig. 4 Three layered representation for the joint spatial and temporal dynamics of molecular pathway circuits

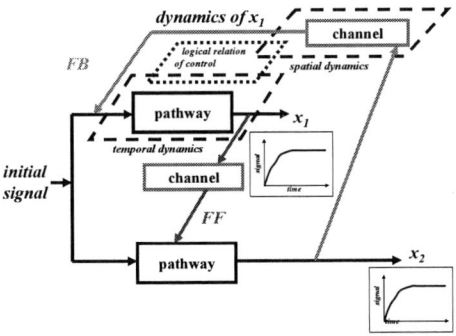

Fig. 5 A NCS model for a molecular pathway circuit with two variables

E. The Equivalent Model of Networked Control Systems for Molecular Patwhay Circuits

Since the FF and FB units are inferred by the integrated structure of LNCs, a pathway circuit is formulated as a networked controller shown in Fig. 5 in order to study the network dynamics. Based on the state representation for a linear form of a controller, a molecular pathway circuit that is constructed based on the LNC form is modeled as follows:

$$\frac{dX}{dt} = AX + U,$$ (8)

$$\frac{dY}{dt} = BX.$$ (9)

Here,

X is the state of the system, defined as a vector,
U is the input of the system, defined as a vector,
A is the matrix of the system with feedback,
B is the matrix of the system with feedforward, and
Y is the output of the system, defined as a vector.

With respect to the general form of network coding (NC), we denote the nonlinear form of a molecular pathway circuit by the following equations in general:

$$\frac{dX(t)}{dt} = f(t, X(t), U(t)),$$ (10)

$$\frac{dY(t)}{dt} = g(t, X(t)),$$ (11)

where
$U(t)$ is the input vector,
$X(t)$ is the state vector,
$Y(t)$ is the output vector, and
$f(.)$ and $g(.)$ are functions dependent on time.

Thus, the FF and FB units in the networked controller are coupled with the channel units that demonstrate the complex behavior of nonlinear pathway circuits.

III. ROBUSTNESS OF MOLECULAR PATWHAY CIRCUITS

Considering a molecular pathway circuit with two variables X_1 and X_2, we denote the nonlinear form of the molecular pathway circuit with two variables by the following equations:

$$\frac{dX(t)}{dt} = f(t, X_1(t), X_2(t)),$$ (12)

where $X_1(t)$ and $X_2(t)$ are the two variables which refer to two distant nodes of a molecular pathway circuit. By introducing the parameter θ that denotes the effect of the communication process (e.g., channel loss), (10) for temporal dynamics becomes the following formulae for the joint spatial and temporal dynamics:

$$\frac{dX(t)}{dt} = f(t, X_1(t), X_2(t + \tau), \theta).$$ (13)

Then, we have that

$$\frac{\partial X(t)}{\partial \theta} = f(t, X_1(t), X_2(t + \tau), \theta),$$ (14)

$$\frac{\partial X(t)}{\partial \theta} = 0.$$ (15)

where parameter τ refers to the channel delay. The parameter τ is a measure for the quantitative analysis of the signaling mechanism that sustains the stochastic robustness under the channel constraint (Cf. Fig. 6 where $p(.)$ refers to the transition probability between different states). It is obvious that the delay in the networked control can be considered as a filtering factor for the network dynamics that will be simulated in next section.

IV. ROBUSTNESS OF SPATIALLY CONFIGURED HSR CIRCUITS: THE DELAY CASE

E. coli, a bacterium with a genome size of 4,600,000 bps

(base pairs), is commonly used as a model organism in molecular biology research. The most obvious merit of this model organism is the ease and efficiency of its generation. Owing to the conservation of genomes among different species, it can be used for a wide range of applications to study biological functions at the genome level. The heat shock response of *E. coli* refers to a kind of chemical reaction that sustains the biological function of protein folding in the bacteria when the temperature is higher than a certain threshold. The HSR network of *E. coli* has been investigated for more than twenty years in biology. In recent years, models in control theory are applied in the HSR network analysis. El Samad *et al.* [9] suggested three derivative types of PID (Proportional, Integral, Derivative) controllers to model the heat shock response (HSR) process. Kurata *et al.* [10] proposed a module-based method to analyze the robustness of the HSR networks. The dynamic model described by Chen *et al.* [11] is a good example to design models for the HSR network analysis. Here we select the HSR network of *E. coli* as an example to study the robustness of molecular pathway circuits. The reasons are briefly given as follows:

(1) The HSR network of *E. coli* shows strong robustness in cellular signaling in the presence of disturbances.

(2) The structure of HSR network is formulated in terms of control theory and has been quantitatively studied by simulation in bioinformatics [9~11].

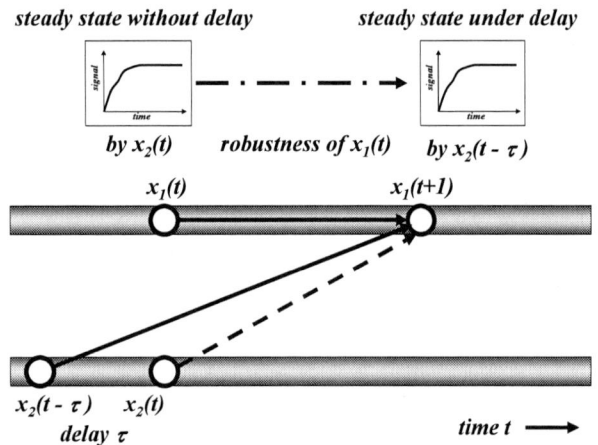

Fig. 6 Robustness of the molecular pathway circuit with respect to channel delay

As a networked control system that responds to heat shock signal, the molecular pathway circuit of the HSR network consists of a feedforward that corresponds to the signaling process centered on σ^{32} and a feedback loop to actuate the HSR signal negatively according to the output of the signaling process centered on *DnaK*. σ^{32} is generated when the heat shock signal is activated. *RNAP* is bound with σ^{32}. This binding leads to the generation of *DnaK* and *FtsH*. The binding state of *DnaK* and σ^{32} causes a negative feedback within the HSR network owing to the fact that it activates the

pathway for protein degradation by *FtsH*. High temperature is a kind of stress on the cell and activates the translation σ^{32} first by changing the folding state of *mRNA* for σ^{32}. σ^{32} then binds with *RNAP* to generate *FtsH* and *DnaK*. *DnaK* promotes the activity of protein folding, and *FtsH* exerts an inhibitory effect on the compound formed by σ^{32} and *DnaK*.

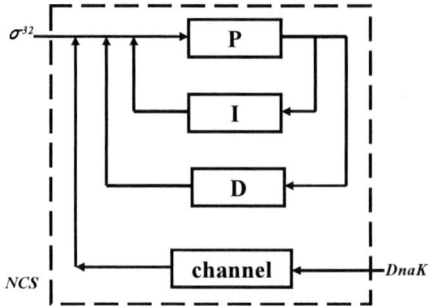

Fig. 7 The NCS representation of the HSR network

Shown in Fig. 7 (where P, I, and D refer to Proportional, Integral, Derivative, respectively), a networked control model for a pathway circuit constructed by the HSR network is presented under the conditions that

(1) the process on *mRNA(σ^{32})* is a constant process of feeding to σ^{32} that activates the HSR network;

(2) the binding states of *DNAP* and σ^{32} are approximated by the linear function of σ^{32};

(3) the partial differential equation of σ^{32} is approximated by a differential equation of σ^{32} dependent on time;

(4) the integration of σ^{32} is approximated by the linear function of σ^{32}.

The robustness of the pathway circuit is represented as

$$\frac{d\sigma^{32}(t)}{dt} = f(t, \sigma^{32}(t), DnaK(t+\tau), \theta) = 0.$$

(16)

where $DnaK(t+\tau)$ is a function of σ^{32} dependent on time t and delay τ. Based on the group of differential equations of the HSR network presented by Kurata *et al.* in [10], the direct relation between σ^{32} and *DnaK* is inferred as follows

$$\frac{d}{dt}\sigma^{32} = b_0 \exp(-t) - b_1 \sigma^{32} - b_2 \sigma^{32}(\exp(-a_1 t)) + \exp(-a_2 t + a_3 \sigma^{32})$$
$$- b_3 \sigma^{32}[Dnak](\exp(-a_1 t) + \exp(-a_2 t) + \sigma^{32}).$$

(17)

when the steady state is reached (e.g., $f(x,y) = x$).

Based on the LNC model and (17), we can quantitatively analyze the delay effect of the HSR network on its spatial robustness. In the simulation, the channel delay is set as a uniform distribution whose mean is 5 ms and variance is 1.2 ms. Fig. 8 and Fig. 9 give the result on the case without delay and under delay where the unit of concentration is nM (10⁻

[9]Mole per unit liter). The channel capacity is set as a uniform distribution where the mean is 5 nM/ms and the variance is 1.2nM/ms. It is observed that the robustness is sustained in the pathway circuits under the spatial configuration.

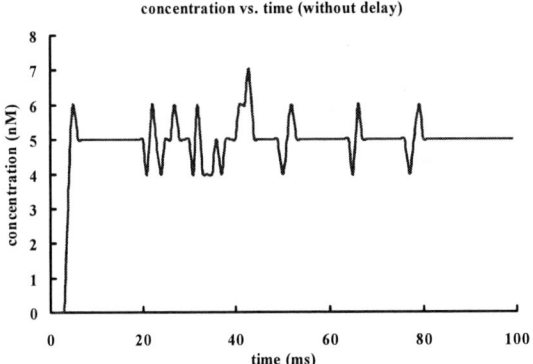

Fig. 8 Concentration versus time curve of σ^{32} without delay

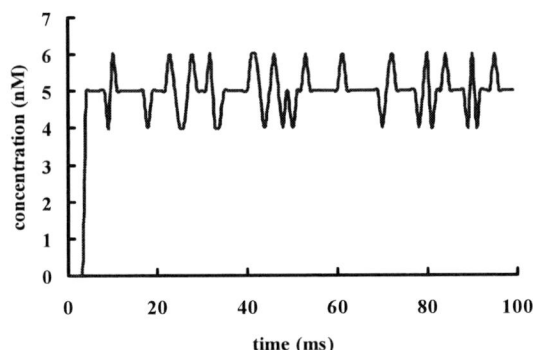

Fig. 9 Concentration versus time curve of σ^{32} under delay

V. CONCLUSION

In this paper, the robustness of spatially configured molecular pathway circuits is discussed, where the linear network coding is applied for formulating the molecular pathway circuits into a networked control system. By using the HSR network of *E. coli* as an example, it is observed from the numerical simulation that the robustness of the molecular pathway circuits is sustained when the channel delay is embedded into the network dynamics. Although our current study on the effect of the communication channel [1,2] on the robustness of molecular pathway [3,4] circuits with a spatially configured network structure is carried out in the case of the channel delay, in the next step of our research we will consider multiple parameters such as channel loss, width, and delay under different distributions, in order to establish a unified framework for the robustness analysis of molecular pathway circuits.

ACKNOWLEDGMENT

The authors sincerely thank Hiroyuki Kurata for his important help on the heat shock response network, Hiroaki Kojima and Kazuhiro Oiwa on molecular biology, and Kazuyuki Aihara on nonlinear dynamics.

REFERENCES

[1] J. Q. Liu, A robust controller of dynamic networks and its verification by the simulation of the heat shock response network with reliable signal transmisison, *Proc. 1st IEEE International Worshop on Molecular and Nano Scale Communciation (MoNaCom2011,* Shanghai, China, April, 2011), pp. 495-500.

[2] J. Q. Liu, "Molecular informatics of nano-communication based on cells: a brief survey", Journal of Nano Communication Networks, vol.1, no.2, pp. 118-125, 2010.

[3] J. Q. Liu, T. Nakano, "Principles and methods for nanomechatronics: signaling, structure, and functions towards nanorobots," IEEE Transactions on Systems, Man, and Cybernetics, Part C, early access at IEEE Xplore® April 2011, in publication.

[4] J. Q. Liu, K. Shimohara, *Biomolecular Computation for Bionanotechnology*, Boston|London: Artech House, 2007.

[5] B. S. Chen, W. S. Wu, Y. C. Wang, and W. H. Li, "On the robust circuit design schemes of biochemical networks: steady-state approach," *IEEE Transactions on Biomedical Circuits and Systems,* vol.1,no.2, pp.91-104, April 2007.

[6] B. S. Chen and P. W. Chen, "Robust engineered circuit design principles for stochastic biochemical network with parameter uncertainties and disturbance," *IEEE Transactions on Biomedical Circuits and Systems*, vol. 2, no. 2, pp.114-132, April 2008.

[7] A. Tamsir, J. J. Tabor and C. A. Voigt, "Robust multicellular computing using genetically encoded NOR gates and chemical 'wires", *Nature*, vol.469, pp.212-215, January 2011.

[8] J. E. Ferrell, Jr. and E. M. Machleder, "The biochemical basis of an all-or-none cell fate switch in *Xenopus* oocytes," *Science*, vol.280, pp.895-898, 1998.

[9] H. El-Samad, et al., "Surviving heat shock: control strategies for robustness and performance," *Proceedings of National Academy of Sceicne of United States of America*, vol.102, no.8,, pp.2736-2741, 2005.

[10] H. Kurata, H. El-Samad, R. Iwasaki, H. Ohtake, J. C Doyle, I. Grigorova, C. A. Gross, and M. Khammash, "Module-based analysis of robustness tradeoffs in the heat shock response system," *PLoS Computational Biology*, vol.2, no.7, e59, 2006.

[11] H. C. Chen, H. C. Lee, T. Y. Lin and B. S. Chen, "Stabilization control of small heat shock proteins in cellular chaperone protection systems," *Journal of the Chinese Institute of Engineers*, vol.29,no.2, pp.364-373, 2006.

[12] S.-Y. R. Li, R.W. Yeung, N. Cai, "Linear network coding," *IEEE Transactions on Information Theory*, Vol.49, No.2, February 2003, pp.371-381.

[13] S.-Y. R. Li, N. Cai, R. W. Yeung, On theory of linear network coding, Proceedings of the 2005 IEEE International Symposium on Information Theory (ISIT 2005), Sep. 4-9, 2005, pp. 273-277.

[14] S.-Y. R. Li, Network coding theory via commutative algebra, 2009, http://www.tims.ntu.edu.tw/ntumath2009/pdf/S.-Y.%20R.%20Li%20.pdf.

[15] J.-Q. Liu, On information-theoretical formalization of intracellular communications based on linear network coding, SICE Annual Conference 2007 (Sept.17-20, 2007, Takamatsu, Japan), pp. 909-912.

[16] J.-Q. Liu, On computational complexity of pathway-Inspired networking, The 24th Annual Conference of the Japanese Society for Artificial Intelligence (June, 2010, Nagasaki, Japan), 2010.

Numerical Simulation of CZTS Thin Film Solar Cell

Wenhao Zhao[1], Wenli Zhou[*1, 2], Xiangshui Miao[1, 2]

[*1]Department of Electronic Science and Technology, Huazhong University of Science and Technology, China
[2]Wuhan National Laboratory for Optoelectronics, Huazhong University of Science and Technology, China
wlzhou@mail.hust.edu.cn

Abstract—**The performance of CZTS (Cu₂ZnSnS₄) thin film solar cell is numerically simulated in this paper. The influences of structural and physical parameters are studied, including thickness, carrier density and defect density of CZTS absorber, thickness of CdS buffer layer. It can be found in the simulation results that to reach high conversion efficiency, the cell should have a thin buffer layer and a thick absorber layer. It is necessary to control the defect density in the absorber one order lower than carrier density. The effect of operating temperature on the cell performance shows that increased temperature will strongly affect the efficiency. The presence of low band gap interfacial layer between CZTS absorber and Mo back contact may lower the open circuit voltage.**

Keywords- CZTS solar cell; numerical simulation; AMPS-1D.

I. INTRODUCTION

Over recent years, the study of the kesterite compound, Cu_2ZnSnS_4 (CZTS), is one of the most attractive research projects in the photovoltaic field. The CZTS thin film is a promising candidate absorber material for low cost and sustainable solar cells because of its excellent material properties, such as suitable band gap energy of 1.4-1.5 eV and absorption coefficient over 10^4 cm^{-1} [1, 2]. In addition, all constituents of this CZTS thin film are naturally abundant and nontoxic.

A variety of physical and chemical techniques have been employed to fabricate CZTS-based thin film solar cells, including co-sputtering [3], thermal evaporation [4], pulse laser deposition [5], sol-gel method [6] and electrodeposition [7], etc. The conversion efficiency of CZTS-based solar cells has been improved from 0.66% in 1996 to 6.8% in 2010 [3, 4].

Despite the development of different techniques and the improvement of conversion efficiency, our understanding of this type of solar cell is relatively limited compared to other cells, like silicon-based solar cells and CIGS solar cells, a kind of cell based on chalcopyrite compounds, $CuIn_xGa_{1-x}Se$, which have already been explored extensively [8].

In order to fully understand the performance of CZTS thin film solar cell and to fabricate cells with higher efficiency, a complete investigation of the operation mechanism of CZTS solar cell is needed. In this paper, we evaluated the performance of CZTS thin film solar cells using the AMPS-1D simulator. We have studied the performance dependency on a set of geometric and physical parameters such as absorber layer / buffer layer thickness, carrier density, defect density, temperature, etc. In addition, we investigated the effect of low

band gap interfacial layer between CZTS absorber and Mo back contact on the output of CZTS solar cells.

II. METHOD AND RESULT

The AMPS-1D simulation software used in this work is developed by a group of Pennsylvania State University [9]. The simulations are carried out by specifying the device structure and electrical and optical parameters of materials as input arguments of the simulator. Then calculated the band diagram, recombination profile and carrier transport properties are calculated on the basis of the Possion equation as well as hole and electron continuity equations. Recombination in deep bulk levels and their occupation are described by Shockley-Read-Hall (SRH) model.

In this study, we consider the structure of CZTS thin film solar cell which is consisted of the following material layers: ZnO window layer, CdS buffer and CZTS absorber. An interfacial layer is added under the absorber in the last stage of simulation to study its influence. The schematic structure of the cell is shown in Fig. 1.

The basic input parameters used in the simulation are adopted from literatures, theories, or reasonable estimates in some cases [1, 4, 10], as shown in TABLE. I. The front surface reflectivity R_f and the back surface reflectivity R_b are set to 0.1 and 0.9, respectively. The default illumination spectrum is set to the global AM1.5 standard and the default operation temperature is set to 300K.

TABLE. I. Parameters used in simulation

Parameters	ZnO	CdS	CZTS
Thickness ,W(nm)	200	50	2000
Relative permittivity, $\varepsilon/\varepsilon_0$	9.0	9.0	10.0
Electron mobility, μ_n (cm²/Vs)	100	100	100
Hole mobility, μ_p (cm²/Vs)	25	25	25
Acceptor concentration, N_A(cm⁻³)	0	0	2.00e14
Donor concentration, N_D(cm⁻³)	1.00e18	1.00e17	0
Band gap, E_g (eV)	3.30	2.40	1.50
Effective density of state, N_C(cm⁻³)	2.20e18	2.20e18	2.20e18
Effective density of state, N_V (cm⁻³)	1.80e19	1.80e19	1.80e19
Electron affinity , χ(eV)	4.40	4.20	4.10
Gaussian defect density density ,N_G (cm⁻³)	1.00e18	1.00e18	1.00e12

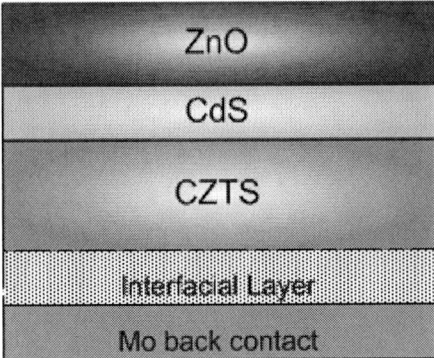

Fig. 1. Structure of CZTS solar cell

The resulting J-V and quantum efficiency (QE) output are shown in Fig. 2. Compared to CIGS solar cell with a similar structure [10], CZTS solar cell has a higher open circuit voltage (Voc) and a lower short circuit current density (Jsc) due to its larger band gap ($E_{g\text{-CIGS}}$ = 1.10eV). The spectrum response, QE curves of these two cells are almost identical in 300-520 nm region. This is because of absorption of high energy photons and recombination of electron-hole pairs in the ZnO window layer and CdS buffer layer. In 520-820 nm region, the two cells show high QEs close to the maximum possible QE, $1-R_f$. The cutoff of QE of CZTS model at 820 nm corresponds to the CZTS band gap, while the cutoff of QE of CIGS model is at the wavelength of 1100 because of its smaller band gap. The efficiency of this CZTS model is 16.38%, which outperforms most experimental devices by a large margin, indicating that there may possibly be huge potential for the improvement for conversion efficiency of CZTS solar cell.

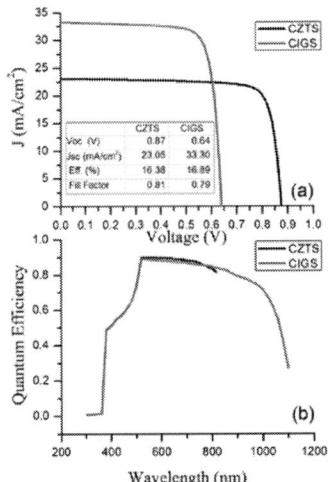

Fig. 2. J-V and Q.E. of CZTS cell and CIGS cell

The dependency of CZTS cell performance on the absorber layer thickness is studied by varying the thickness of CZTS layer varies from 1000 nm to 3000 nm while other input parameters are kept unchanged. Fig. 3 shows the detailed effects of the layer thickness on the cell performance parameters such as J_{SC}, V_{OC}, efficiency and fill factor. J_{SC} and V_{OC} both increase with the CZTS layer thickness. This may mainly because that more photons with longer wavelength can be absorbed by the absorber if the absorber layer is thicker, and thus more electron-hole pairs are generated. Hence, J_{SC} and V_{OC} will increase and finally efficiency will be improved. However, considering the cost of material and fabrication process, it would not be cost effective to produce CZTS solar cells with very large absorber thickness. Thus a tradeoff between cell efficiency and cost is needed for mass production.

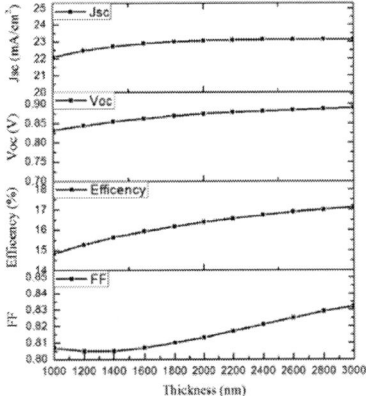

Fig. 3. The dependency of cell performance on the CZTS layer thickness

Fig. 4 illustrates the light J-V characteristics of cells with various carrier densities (N_A) in CZTS absorber. It shows that the V_{OC} increases with the carrier concentration while the J_{SC} decreases with it. The changes caused by the N_A can be explained with the simple PN junction model:

$$V_{OC} = \frac{kT}{q}\ln(\frac{I_L}{I_0}+1), I_0 = Aqn_i^2(\frac{D_e}{L_e N_A}+\frac{D_h}{L_h N_D}) \qquad (1)$$

The saturation current I_0 will be reduced if N_A increases. Therefore V_{OC} will increase. On the other hand, J_{SC} will decrease when N_A increases. It is probably because the recombination process is enhanced by the rising N_A. This will lower the possibility of collection of the electron that generated by incident light, and thus lower the short circuit current. This will also result in a drop of the quantum efficiency of long wavelength photons, which are absorbed deeper in the CZTS material so that the electrons created there are more dependent on diffusion effect to be collected effectively.

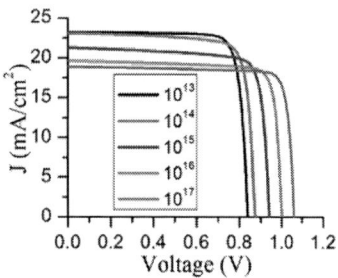

Fig. 4. J-V of cells with variation of CZTS carrier density

The light J-V characteristics of the CZTS solar cells with different defect densities in the absorber layers are shown in Fig. 5. It can be seen that the performance remains nearly unchanged when the defect density is less than 10^{14} cm^{-3}. When the defect density is beyond 10^{14} cm^{-3}, however, the output parameters are strongly affected. The increase of the number of defect states in the absorber layer introduces additional recombination centers, which will cause a stronger recombination of photogenerated carriers in the absorber layer and lead to the decrease of J_{SC}, V_{OC}, conversion efficiency and fill factor.

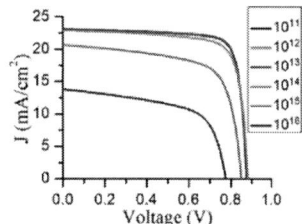

Fig. 5. J-V of cells with variation in CZTS defect density

Fig. 6 shows quantum efficiency of CZTS cell with variable CdS buffer layer thickness. Changes of CdS layer thickness affect the QE, since there exists considerable absorption of photons whose energies are between the band gap energy of ZnO (3.3 eV) and CdS(2.4 eV). Increasing of CdS layer thickness causes stronger absorption of these photons and therefore decreases the QE of the corresponding spectral range. Thus in order to obtain high efficiency, it is necessary to make CdS buffer layer thin.

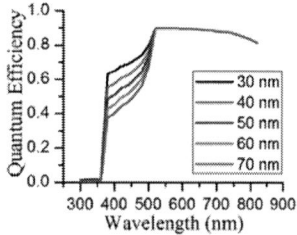

Fig. 6. Quantum efficiency of CZTS solar cell with variable CdS layer thickness

Since the cell can be heated up under the sun, there is a need to investigate the stability of the cell at different operating temperature. The influences of operating temperature on the output parameters have been studied with the temperature ranged from 280 K to 360 K. As shown in Fig.7, V_{OC} and efficiency are strongly affected by the variation of operating temperature while J_{SC} and fill factor are nearly unchanged. When the operating temperature rises, electrons in the cell gain additional energies. Then the electrons become unstable and more likely to recombine with holes before they reach the depletion region and being collected. This will lead to the decrease of V_{OC} and fill factor.

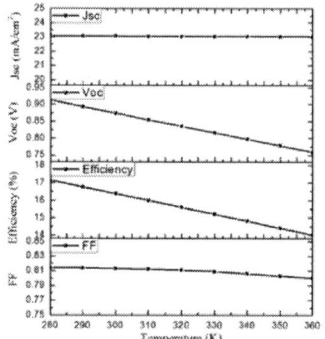

Fig. 7. The dependency of cell performance on the operation temperature

It is reported that the experimental CZTS thin film solar cells exhibit low open circuit voltage (~ 0.6V) which are lower than expected, considering the band gap of CZTS [4]. It is estimated that the existence of interfacial layer between the absorber and the back contact, which is consist of lower band gap phases such as Cu_2SnS_3 (~ 0.95 V) , may contribute to the loss of V_{OC} [11]. In order to simulate this phenomenon, an interfacial layer is added to the model. The interfacial layer is set as a p-type semiconductor with a band gap of 0.95 eV and a thickness of 50 nm. The simulation result is shown in Fig. 8. It can clearly be seen that the presence of secondary phase with lower band gap has a significant effect on V_{OC}, which is in agreement with the speculation in [11].

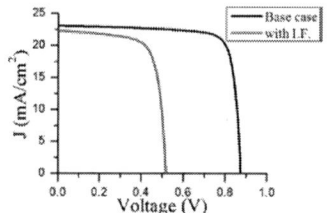

Fig. 8 Influence of interfacial layer with low band gap

III. CONCLUSION

In conclusion, a comprehensive study numerical simulation of CZTS thin film solar cell is performed in this study. The simulation results demonstrate that the cell parameters such as CZTS absorber layer thickness, carrier density, defect density, CdS layer thickness have different effects on cell performance. The impact of interfacial layer is also studied. From the simulation, it shows that there is great potential to improve the conversion efficiency of CZTS thin film solar cell by reducing the thickness of CdS buffer layer and optimizing the quality of CZTS abosorber layer. Efforts to improve the fabrication process are also needed to prevent the formation of low band gap phases near the back contact.

ACKNOWLEDGMENT

This research is supported by Frontier Exploration Project of Independent Innovation Foundation of Huazhong University of Science & Technology (No. 2011TS108).

REFERENCES

[1] J. S. Seol, S. Y. Lee, J. C. Lee, H. D. Nam, and K. H. Kim, "Electrical and optical properties of Cu_2ZnSnS_4 thin films prepared by RF magnetron sputtering process," *Solar Energy Materials and Solar Cells,* vol. 75, no. 1-2, pp. 155-162, 2003.

[2] F. Y. Liu, K. Zhang, Y. Q. Lai, J. Li, Z. A. Zhang *et al.*, "Growth and characterization of Cu_2ZnSnS_4 thin films by DC reactive magnetron sputtering for photovoltaic applications," *Electrochemical and Solid State Letters,* vol. 13, no. 11, pp. H379-H381, 2010.

[3] H. Katagiri, K. Jimbo, W. S. Maw, K. Oishi, M. Yamazaki *et al.*, "Development of CZTS-based thin film solar cells," *Thin Solid Films,* vol. 517, no. 7, pp. 2455-2460, Feb, 2009.

[4] K. Wang, O. Gunawan, T. Todorov, B. Shin, S. J. Chey *et al.*, "Thermally evaporated Cu_2ZnSnS_4 solar cells," *Applied Physics Letters,* vol. 97, no. 14, pp. 143508-3, 2010.

[5] A. V. Moholkar, S. S. Shinde, A. R. Babar, K. U. Sim, Y. b. Kwon *et al.*, "Development of CZTS thin films solar cells by pulsed laser deposition: influence of pulse repetition rate," *Solar Energy,* vol. 85, no. 7, pp. 1354-1363, 2011.

[6] N. Moritake, Y. Fukui, M. Oonuki, K. Tanaka, and H. Uchiki, "Preparation of Cu_2ZnSnS_4 thin film solar cells under non-vacuum condition," *physica status solidi (c),* vol. 6, no. 5, pp. 1233-1236, 2009.

[7] A. Ennaoui, M. Lux-Steiner, A. Weber, D. Abou-Ras, I. Kötschau *et al.*, "Cu_2ZnSnS_4 thin film solar cells from electroplated precursors: Novel Low-cost Perspective," *Thin Solid Films,* vol. 517, no. 7, pp. 2511-2514, 2009.

[8] I. Repins, M. A. Contreras, B. Egaas, C. DeHart, J. Scharf *et al.*, "19.9%-efficient ZnO/CdS/CuInGaSe2 solar cell with 81.2% fill factor," *Progress in Photovoltaics,* vol. 16, no. 3, pp. 235-239, May, 2008.

[9] H. Zhu, A. K. Kalkan, J. Y. Hou, and S. J. Fonash, "Applications of AMPS-1D for solar cell simulation," *Aip Conference Proceedings.* pp. 309-314.

[10] M. Gloeckler, A. L. Fahrenbruch, and J. R. Sites, "Numerical modeling of CIGS and CdTe solar cells: setting the baseline," Proceedings of 3rd World Conference on Photovoltaic Energy Conversion, Vols a-C, Tokyo, 2003, pp. 491-494.

[11] K. Wang, B. Shin, K. B. Reuter, T. Todorov, D. B. Mitzi *et al.*, "Structural and elemental characterization of high Efficiency Cu_2ZnSnS_4 solar cells," *Applied Physics Letters,* vol. 98, no. 5, pp. 051912-3, 2011.

On the Design of Subwavelength Waveguiding Structures for Terahertz Applications

Vikas Singal, Sami Smaili, and Yehia Massoud
University of Alabama at Birmingham, Birmingham, AL-35205
massoud@uab.edu

Abstract—**Closing the THz gap would lead to a tremendous of advancement in a wide range of applications such as biomedical imaging, security, and material inspection. The gap refers to the lack of devices for the manipulation of THz radiation as compared to its microwave and optical counterparts. Plasmonic devices based on semiconductors rather than metals allow the realization of efficient and small scale THz devices by utilizing the unique properties of plasmon oscillations. In this paper, we investigate the performance of an $InSb - SiO_2 - InSb$ structure for THz waveguiding. We study the propagation length and the skin depth of the symmetric and antisymmetric transverse magnetic modes of these waveguides. We use numerical techniques to solve for the dispersion relation and derive the propagation length and the skin depth as a function of frequency.**

Index Terms—**Plasmonics, Strip Waveguide, Finite Element Method**

I. INTRODUCTION

Recent advances in THz technology has shed the light on the various applications possible with THz radiation, from communications and biomedical imaging, to security and material inspection [1]. Additionally, THz or sub-THz radiation can play an important role in mitigating the limitations faced by current VLSI systems [2]–[46]. . The Thz region lies between the microwave spectrum and the optical (infrared and visible) spectrum, and it is this unique position between two different regimes that lends THz technology its importance [1]. However, unlike its microwave [38], [47]–[71] and optical counterparts [72]–[78], the THz regime lacks the wide range of devices required to utilize the potential of THz radiations. This THz gap is the main bottleneck facing the development of THz technology [1].

Current efforts are dedicated to develop various THz devices such as THz mixers and receivers [79]. Many of these devices use high electron mobility transistor (HEMT), which utilize surface plasmon oscillations in order to achieve the desired functionality. Plasmon oscillations refer to the oscillations of electron charge density at, for example, a metal-insulator interface or, in the case of THz plasmons, at semiconductor-insulator interface. The field of plasmonics has been developing rapidly and promise to provide solutions for many problems in a wide range of applications, such as future VLSI systems, subwavelength waveguiding of light, and nanoscale light manipulation [80]. Replacing the metal in typical plasmonic devices by a semiconductor results in THz plasmonic devices, that can realize many highly desired functionalities such as waveguiding and filtering.

In this paper we investigate a semiconductor-insulator-semiconductor structure for THz waveguiding, which is similar to the metal-insulator-metal structure for optical waveguiding. We consider $InSb$ for the semiconductor and SiO_2 for the insulator and use a Drude's model for the $INSb$ dielectric function. We study the effect of the waveguide thickness on the propagation length of the plasmon wave and the confinement of the wave within the waveguide.

The rest of the paper is organized as follows. Section II describes the SIS structure and the associated dispersion relations for the symmetric and asymmetric and modes. Section III presents the main results concerning the propagation length and skin depth of the SIS structure and the tradeoff between the two. Section IV concludes the paper.

II. SEMICONDUCTOR-INSULATOR-SEMICONDUCTOR STRUCTURES FOR THZ WAVEGUIDING

Plasmons allow the waveguiding of electromagnetic radiation at sub-wavelength scales, beyond the diffraction limit. Among the most promising plasmon waveguiding structure are those consisting of an insulator core with a cladding made of materials with free charge carriers, such as metals or semiconductors. A metal-insulator-metal (MIM) structure made of a metal cladding and an

978-1-4673-1122-9/12 $31.00 © 2012 IEEE

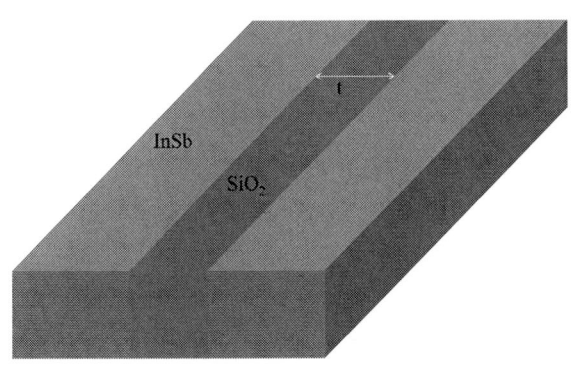

Fig. 1.   Schematic of a THz Bragg reflector

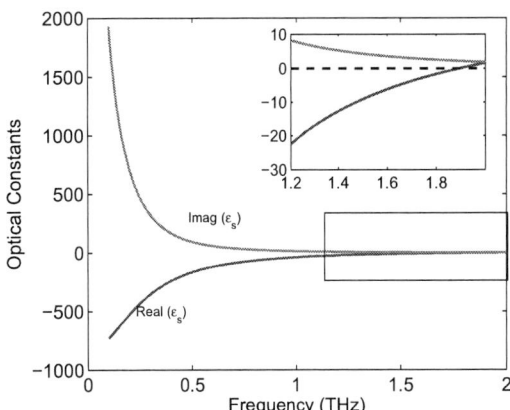

Fig. 2.   InSb dielectric function versus frequency.

insulator core provides high levels of wave confinement because of the skin depth effect experienced by optical or infra-red radiation inside metals, and can be used to achieve nanoscale optical waveguides. Since the plasma frequency of a typical semiconductor is in the THz regime, and since the dielectric function of semiconductors as a function of wavelength behaves similarly to that of a metal at optical wavelengths (for example both can be approximated by a Drude's dielectric), it is natural to expect a semiconductor-insulator-semiconductor (SIS) structure to have favorable properties in the THZ regime similar to those of MIM structures in the optical regime. In this paper, we consider silicon dioxide (SiO_2) as the insulator and Indium antimonide (InSb) for the semiconductor material, for which a schematic is shown in figure 1. The dielectric function of InSb is modeled as a Drude's function

$$\varepsilon_s(\omega) = \varepsilon_\infty(1 - \frac{\omega_p^2}{\omega^2 + \tau^{-2}} + i\frac{\omega_p^2\tau^{-1}}{\omega(\omega^2 + \tau^{-2})})) \quad (1)$$

where ε_∞ is the background dielectric constant (high frequency dielectric function), τ is the average mean free time, and ω_p is the plasma frequency given by

$$\omega_p = \sqrt{\frac{ne^2}{\varepsilon_{infty}\varepsilon_0 m^*}} \quad (2)$$

with n being the charge carrier concentration of the semiconductor and m^* is the effective mass of the charge carriers. In this paper, we use the Drude's model given for the $InSb$ dielectric function in [81], where the electron

mobility $\mu = 7.7 \times 10^4 cm^2 V^{-1} s^{-1}$ and the $\tau = \mu m^*/e$. The electron effective mass is $m^* = 0.014 m_0$, where m_0 is the electron mass, the room temperature intrinsic carrier concentration is $10^{16} cm^{-3}$, and $\varepsilon_\infty = 15.7$. The dielectric function of InSb is shown in figure 2.

The SIS structure supports two transverse magnetic modes of plasmon oscillations, symmetric ($L-$) and antisymmetric ($L+$) modes, with dispersion relations

$$L- \quad : \quad \epsilon_s k_{z1} coth(\frac{-ik_{z1}t}{2}) + \epsilon_d k_{z2} = 0 \quad (3)$$

$$L+ \quad : \quad \epsilon_s k_{z1} tanh(\frac{-ik_{z1}t}{2}) + \epsilon_d k_{z2} = 0 \quad (4)$$

$$k_{z1,2}^2 = \epsilon_{s,d}(\frac{\omega}{c})^2 - k_x^2 \quad (5)$$

A. The SIS Dispersion Relations

Figure 3 shows the dispersion relation of the SIS structure in the range [0.1 THz, 2 THz] for the symmetric and antisymmetric modes, respectively. Typically, the dispersion relation shows three modes of propagation: the surface plasmon polariton (SPP) mode or the bound mode, the Quasi-bound mode, and the radiative mode. However, the analysis we present in this paper pertains to the SPP mode only. For instance, figure 3 does not show the radiative mode for all thicknesses, but quasi-bound state can be seen in the figure (the part of the dispersion relation where the wave vector decreases as the frequency increases). The antisymmmetric radiative state in an SIS structure with a thickness of $5\mu m$ can be seen in figure 3. The dispersion curve for the symmetric mode for small thicknesses moves to the left as the thickness increases. However, this trend stops and

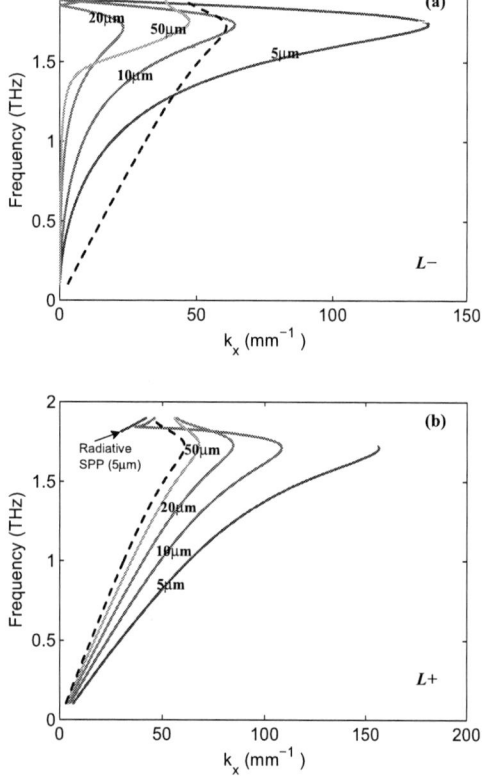

Fig. 3. Dispersion relation of the (a) symmetric and (b) antisymmetric mode of an SIS structure with various thicknesses. The dashed lines in both panels correspond to an SIS structure with infinite thickness (a semiconductor-insulator structure).

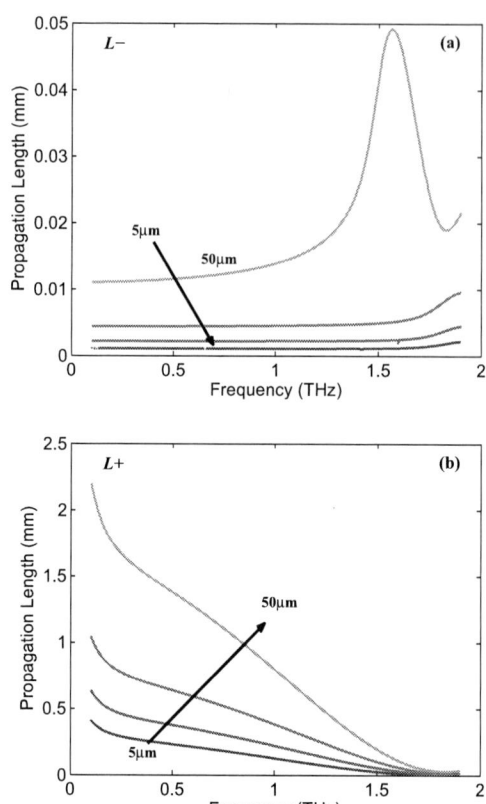

Fig. 4. Propagation length of the (a) symmetric and (b) antisymmetric mode of an SIS structure with various thicknesses.

the dispersion curve starts moving to the right again. However, for the antisymmetric mode, the dispersion curve moves to the left as the thickness increases. The dispersion curve lies to the right of the infinite thickness SIS structure dispersion curve.

III. SKIN DEPTH AND PROPAGATION LENGTH

Among the most important parameters for assessing the performance of a waveguide, in general, are the propagation length and the skin depth. The propagation length is the distance that the wave travels before decaying to half its original value. The skin depth measures the penetration into the semiconductor media and is defined as the depth into the semiconductor material after which the wave amplitude is reduced to $\ln e$ of its value at the semiconductor-insulator interface. The propagation length and the skin depth can be deduced from the imaginary part of the wavevector in the propagation direction

and the imaginary part of the wave vector perpendicular to the semiconductor-insulator interface, and are shown in figures 4 and 5, respectively. The Propagation length increases as the waveguide thickness increases for both, the symmetric mode and the antisymmetric mode. However, the propagation length of the antisymmetric mode decreases rapidly as the frequency increases. The skin depth at frequencies below 1THz increases as the thickness increases for the symmetric mode and decreases as the thickness increases for the antisymmetric mode. It is noted though that the thickness does not have a large effect on skin depth below 1THz. For frequencies above 1THz, the skin depth does not follow a monotonic trend with thickness for the symmetric mode, but rather increases as the thickness increases from $5\mu m$ to $20\mu m$ and starts decreasing again as the thickness is increased to $50\mu m$. The skin depth of the antisymmetric mode increases as the thickness increases for frequencies above 1THz. Thus, there is a tradeoff between propagation length and skin depth or confinement, where a smaller

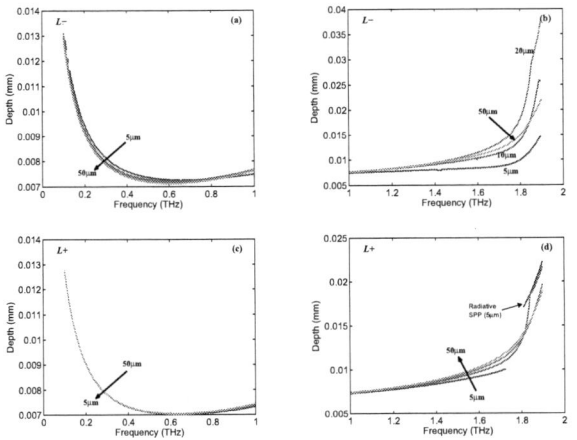

Fig. 5. Skin depth of the (a and b) symmetric and (c and d) antisymmetric mode of an SIS structure with various thicknesses.

thickness, in general, results in a lower propagation length but higher confinement (lower skin depth).

IV. CONCLUSION

Plasmonic devices are being increasingly used in a wide range of applications involving the nanoscale manipulation of light. Plasmonic devices using semiconductors can be used similarly to realize miniaturized THz devices, which in turn promise to reduce the THz gap. In this paper, we study the SIS THz waveguide and study its propagation length and skin depth, using $InSb$ for the semiconductor and SiO_2 for the insulator.

REFERENCES

[1] W. Stillman and M. Shur, "Closing the gap: Plasma wave electronic terahertz detectors," *J. of Nanoelectronics and Optoelectronics*, vol. 2, no. 3, 2007.

[2] Y. Massoud, S. Majors, T. Bustami, and J. Whites, " techniques for minimizing on-chip interconnect self-inductance," *IEEE DAC*, 1998.

[3] D. M. Y. Massoud, J. Kawa and J. White, "Modeling and Analysis of Differential Signaling for Minimizing Inductive Cross-Talk," *IEEE DAC*, 2001.

[4] A. Nieuwoudt and Y. Massoud, "Evaluating the Impact of Resistance in Carbon Nanotube Bundles for VLSI Interconnect using Diameter-Dependent Modeling Techniques," *IEEE TED*, October 2006.

[5] "Modeling and Design Challenges and Solutions for Carbon Nanotube-Based Interconnect in Future High Performance Integrated Circuis," *ACM JETC*, July 2006.

[6] Y. Massoud and J. White, "Managing On-Chip Inductive Effects," *IEEE TVLSI*, Dec. 2002.

[7] A. Nieuwoudt and Y. Massoud, "Understanding the impact of inductance in carbon nanotube bundles for VLSI interconnect using scalable modeling techniques," *IEEE TNANO)*, Nov. 2006.

[8] Y. Massoud and Y. Ismail, "Grasping the Impact of On-Chip Inductance in High Speed ICs," *IEEE Circuits and Devices Magazine*, July 2001.

[9] A. Nieuwoudt and Y. Massoud, "On the impact of process variations for carbon nanotube bundles for VLSI interconnect," *IEEE TED*, March 2007.

[10] ——, "Assessing the implications of process variations on future carbon nanotube bundle interconnect solutions," *IEEE ISQED*, 2007.

[11] M. Mondal, A. Ricketts, S. Kirolos, T. Ragheb, V. N. G. Link, and Y. Massoud, "Thermally robust clocking schemes for 3D integrated circuits," *IEEE DATE*, 2006.

[12] M. Mondal and Y. Massoud, "Reducing pessimism in RLC delay estimation using an accurate analytical frequency dependent model for inductance," *IEEE ICCAD*, November 2005.

[13] S. Kim, Y. Massoud, , and S. Wong, "On the Accuracy of Return Path Assumption for Loop Inductance Extraction for 0.1 um Technology and Beyond," *IEEE ISQED*, 2003.

[14] S. Eachempati, A. Nieuwoudt, A. Gayasen, V. Narayanan, and Y. Massoud, "Assessing carbon nanotube bundle interconnect for future FPGA architectures," *IEEE DATE*, 2007.

[15] Y. Massoud and J. White, "Simulation and Modeling of the Effect Substrate Conductivity on Coupling Inductance and Circuit Crosstalk," *IEEE TVLSI*, June 2002.

[16] A. Nieuwoudt and Y. Massoud, "On the optimal design, performance, and reliability of future carbon nanotube-based interconnect solutions," *IEEE TED*, August 2008.

[17] ——, "Accurate resistance modeling for carbon nanotube bundles in VLSI interconnect," *IEEE Nano*, 2006.

[18] A. Nieuwoudt, M. Mondal, and Y. Massoud, "Predicting the Performance and Reliability of Carbon Nanotube Bundles for On-Chip Interconnect," *IEEE ASP-DAC*, 2007.

[19] A. Nieuwoudt and Y. Massoud, "Performance implications of inductive effects for carbon-nanotube bundle interconnect," *IEEE EDL*, April 2007.

[20] A.Nieuwoudt and Y.Massoud, "Predicting the performance of low-loss on-chip inductors realized using carbon nanotube bundles," *IEEE TED*, January 2008.

[21] M. Mondal, T. Ragheb, X. Wu, A. Aziz, and Y. Massoud, "Provisioning on-chip networks under buffered rc interconnect delay variations," *ISQED*, 2007.

[22] A. Hosseini, T. Ragheb, and Y. Massoud, "A fault-aware dynamic routing algorithm for on-chip networks," *IEEE ISCAS*, 2008.

[23] Y. Massoud and J. White, "Improving the generality of the fictitious magnetic charge approach to computing inductances in the presence of permeable materials," *IEEE DAC*, 2002.

[24] A. Nieuwoudt and Y. Massoud, "RC circuit model for multi-walled carbon nanotubes," *IEEE Nano*, 2007.

[25] Q. Su, J. Kawa, C. Chiang, and Y. Massoud, "Accurate modeling of substrate resistive coupling for floating substrates," *ACM TODAES*, January 2006.

[26] M. Mondal, A. Ricketts, S. Kirolos, T. Ragheb, G. Link, V. Narayanan, and Y. Massoud, "Mitigating thermal effects on clock skew with dynamically adaptive drivers," *IEEE ISQED*, 2007.

[27] M. Alam, A. Nieuwoudt, and Y. Massoud, "Wavelet-based passivity preserving model order reduction for wideband interconnect characterization," *IEEE ISQED*, 2007.

[28] M. Mondal, K. Mohanram, and Y. Massoud, "Parameter-Variation-Aware Analysis for Noise Robustness," *IEEE ISQED*, 2007.

[29] Y. Massoud, J. Wang, and J. White, "Accurate inductance extraction with permeable materials using qualocation," *International Conf. on Modeling and Simulation of Microsystems*, 1999.

[30] S. Kirolos and Y. Massoud, "Adaptive Ratio-Size Gates for Minimum-Energy Operation," *IEEE TCAS II*, November 2006.

[31] S. Kirolos, Y. Massoud, and Y. Ismail, "Power-supply-variation-aware timing analysis of synchronous systems," *IEEE ISCAS*, 2008.

[32] ——, "Accurate analytical delay modeling of cmos clock buffers considering power supply variations," *IEEE ISCAS*, 2008.

[33] S. Kirolos and Y. Massoud, "Adaptive SRAM Design for Dynamic Voltage Scaling VLSI Systems," *IEEE MWSCAS*, 2007.

[34] X. Wu, T. Ragheb, A. Aziz, and Y. Massoud, "Implementing DSP algorithms with on-chip networks," *IEEE NOC*, 2007.

[35] M. Alam, A. Nieuwoudt, and Y. Massoud, "Model order reduction using spline-based dynamic multi-point rational interpolation for passive circuits," *Analog Integrated Circuits and Signal Processing*, March 2007.

[36] M. Mondal and Y. Massoud, "Analytical Modeling of Loop Self Inductance Bound for Inductance-Aware Physical Synthesis," *IEEE ISCAS*, 2005.

[37] Y. Massoud and J. White, "Fast Inductance Extraction of 3-D Structures with Non-ConstantPermeabilities," *International Conference on Modeling and Simulation of Microsystems*, pp. 190–193, April 1998.

[38] M. A. Y. Massoud and A. Nieuwoudt, "On the Selection of Spectral Zeros for Generating Passive Reduced Order Models," *IEEE IWSOC*, 2006.

[39] M. Mondal and Y. Massoud, "Estimation of capacitive crosstalk-induced short-circuit energy," *JCSC*, June 2007.

[40] Y. Massoud and J. White, "FastMag: a 3-D Fast Inductance Extraction Program for Structures with Permeable materials," *IEEE ICCAD*, 2002.

[41] S. Kirolos and Y. Massoud, "Supply Voltage Adaptive Low-Power Circuit Design," *IEEE DCAS*, 2006.

[42] . M. Modal and Y. Massoud, "Accurate loop self inductance bound for efficient inductance screening," *IEEE TVLSI*, December 2006.

[43] S. Kirolos and Y. Massoud, "Robust Wide Range of Supply-Voltage Operation using Continuous Adaptive Size-Ratio Gates," *IEEE ISCAS*, 2008.

[44] Y. Massoud, S. Kirolos, and K. Mohanram, "Analytical model-based technique for efficient evaluation of noise robustness considering parameter variations," *Analog Integrated Circuits and Signal Processing*, August 2009.

[45] S. Kirolos, M. Mondal, K. Mohanram, and Y. Massoud, "A Model-Based Technique for Efficient Evaluation of Noise Robustness," *IEEE ISCAS*, 2008.

[46] S. Kirolos, Y. Massoud, and Y. Ismail, "Mitigating Power-Supply Induced Delay Variations using Self Adjusting Clock Buffers," *IEEE MWSCAS*, 2008.

[47] A. Nieuwoudt and Y. Massoud, "Variability-aware multilevel integrated spiral inductor synthesis," *IEEE TCAD*, Dec. 2006.

[48] ——, "Multi-level Approach for Integrated Spiral Inductor Optimization," *IEEE DAC*, 2005.

[49] A. Nieuwoudt, T. Ragheb, and Y. Massoud, "SOC-NLNA: synthesis and optimization for fully integrated narrow-band CMOS low noise amplifiers," *IEEE DAC*, 2006.

[50] A. Nieuwoudt and Y. Massoud, "Robust automated synthesis methodology for integrated spiral inductors with variability," *IEEE ICCAD*, 2005.

[51] A. Nieuwoudt, M. McCorquodale, R. Borno, and Y. Massoud, "Efficient analytical modeling techniques for rapid integrated spiral inductor prototyping," *IEEE CICC*, 2005.

[52] T. Ragheb, S. Kirolos, J. Laska, M. Strauss, A. Gilbert, R. Baraniuk, and Y. Massoud, "Implementation models for analog-to-information conversion via random sampling," *IEEE MWSCAS*, 2007.

[53] A. Nieuwoudt, M. McCorquodale, R. Borno, and Y. Massoud, "Accurate analytical spiral inductor modeling techniques for efficient design space exploration," *IEEE EDL*, December 2006.

[54] A. Nieuwoudt and Y. Massoud, "Analytical wide-band modeling of high frequency resistance in integrated spiral inductors," *Analog Integrated Circuits and Signal Processing*, February 2007.

[55] S. Kirolos, T. Ragheb, J. Laska, M. Duarte, Y. Massoud, and R. Baraniuk, "Practical Issues in Implementing Analog-to-Information Converters," *IEEE IWSOC*, 2006.

[56] A. Nieuwoudt and Y. Massoud, "Efficient modeling of substrate eddy currents for integrated spiral inductor design automation," *IEEE MWSCAS*, 2005.

[57] T. Ragheb, A. Nieuwoudt, and Y. Massoud, "Efficient modeling of integrated narrow-band low noise amplifiers for design space exploration," *IEEE GLSVLSI*, 2006.

[58] A. Nieuwoudt, T. Ragheb, and Y. Massoud, "Hierarchical optimization methodology for wideband low noise amplifiers," *IEEE ASP-DAC*, 2007.

[59] T. Ragheb, J. N. Laska, H. Nejati, S. Kirolos, R. G. Baraniuk, and Y. Massoud, "A prototype hardware for random demodulation based compressive analog-to-digital conversion," *IEEE MWSCAS*, 2008.

[60] A. Nieuwoudt, T. Ragheb, H. Nejati, and Y. Massoud, "Increasing manufacturing yield for wideband RF CMOS LNAs in the presence of process variations," *IEEE ISQED*, 2007.

[61] ——, "Numerical design optimization methodology for wideband and multi-band inductively degenerated cascode CMOS low noise amplifiers," *IEEE TCAS I*, June 2009.

[62] T. Ragheb, A. Nieuwoudt, and Y. Massoud, "Modeling of 3.1-10.6 GHz CMOS Filter-Based Low Noise Amplifier for Ultra-Wideband Receivers," *IEEE WAMICON*, 2006.

[63] H. Najati, T. Ragheb, A. Nieuwoudt, and Y. Massoud, "Modeling and design of ultrawideband low noise amplifiers with generalized impedance matching networks," *IEEE ISCAS*, 2007.

[64] A. Nieuwoudt and Y. Massoud, "Optimizing the Design of Tunable Spiral Inductors for On-Chip Wireless Applications," *IEEE WAMICON*, 2006.

[65] H. Nejati, T. Ragheb, A. Nieuwoudt, and Y. Massoud, "Analytical modeling methodology for ultrawideband low noise amplifiers with generalized filter-based impedance matching," *Analog Integrated Circuits and Signal Processing*, May 2007.

[66] A. Nieuwoudt, T. Ragheb, and Y. Massoud, "Systematic Design Optimization Methodology for Multi-Band CMOS Low Noise Amplifiers," *IEEE ISVLSI*, 2006.

[67] H. Nejati, T. Ragheb, and Y. Massoud, "On the design of customizable low-voltage common-gate LNA-mixer pair using current and charge reusing techniques," *IEEE GLSVLSI*, 2008.

[68] S. Pfetsch, T. Ragheb, J. Laska, H. Nejati, A. Gilbert, M. Strauss, R. Baraniuk, and Y. Massoud, "On the Feasibility of Hardware Implementation of Sub-Nyquist Random-Sampling Based Analog-to-Information Conversion," *IEEE ISCAS*, 2008.

[69] Y. Massoud, A. Nieuwoudt, and T. Ragheb, "Automated design solutions for fully integrated narrow-band low noise amplifiers," *IEEE IWSOC*, 2006.

[70] A. Nieuwoudt and Y. Massoud, "Design optimization of switchable multi-port spiral inductors," *Springer ALOG*, June 2007.

[71] H. Nejati, T. Ragheb, and Y. Massoud, "Analytical Modeling of Common-Gate Low Noise Amplifiers," *IEEE MWSCAS*, 2008.

[72] A. Hossieni, A. Nieuwoudt, and Y. Massoud, "Efficient simulation of subwavelength plasmonic waveguides using implicitly restarted Arnoldi," *Optics Express*, August 2006.

[73] M. Alam and Y. Massoud, "A closed-form analytical model for single nanoshells," *IEEE TNANO*, May 2006.

[74] ——, "RLC ladder model for scattering in single metallic nanoparticles," *IEEE MTT-S*, 2006.

[75] A. Hossieni, H. Nejati, and Y. Massoud, "Design of a maximally flat optical low pass filter using plasmonic nanostrip waveguides," *Optics Express*, November 2007.

[76] M. Alam and Y. Massoud, "An accurate closed-form analytical model of single nanoshells for cancer treatment," *IEEE MWSCAS*, 2005.

[77] A. Hossieni and Y. Massoud, "Optical range microcavities and filters using multiple dielectric layers in metal-insulator-metal structures," *Journal of the Optical Society of America A*, January 2007.

[78] M. Alam and Y. Massoud, "An RLC ladder model for the equivalent impedance of single metal nanoparticles in electromagnetic field," *IEEE Transactions on Nanotechnology (TNANO)*, September 2006.

[79] M. Dyakonov and M. Shur, "Detection, mixing, and frequency multiplication of terahertz radiation by two-dimensional electronic fluid," *IEEE TED*, vol. 43, no. 3, 1996.

[80] J. A. Dionne, L. A. Sweatlock, and H. A. Atwater, "Planar Metal Plasmon Waveguides Frequency dependent Dispersion Propagation Localization and Loss Beyond the Free Electron Model," *Physical Review B*, vol. 72, no. 075405, 2005.

[81] J. Rivas, C. Janke, P. Bolivar, and H. Kurz, " Transmission of THz radiation through InSb gratings of subwavelength apertures ," *Opt. Express*, vol. 13, no. 3, Apr. 2005.

Polypyrrole (PPy) nanowire arrays entrapped with glucose oxidase biosensor for glucose detection

G.Q. Xu*, J. Lv, Z.X. Zheng, Y.C. Wu*

*Laboratory of Functional Nanomaterials and Devices, School of Material Science and Engineering, Hefei University of Technology, Hefei 230009, China (gqxu1979@126.com)

Abstract—**Well aligned PPy nanowire arrays with diameter about 20 nm have been accomplished by electrochemical polymerization within anodized aluminum oxide (AAO) template. Glucose oxidase (GOx) was galvanostatically entrapped within the PPy nanoarrays. Scanning electron microscopy (SEM) confirmed the presence of PPy nanowire arrays. The electrochemical properties and response of the PPy-GOx nanowire arrays biosensor to glucose were characterized by cyclic voltammetry and amperometry. Evident current response can be observed in the cyclic voltammogram of the PPy-GOx nanowire arrays biosensor when exposed to glucose in a phosphate buffer solution. Amperometric detection of glucose was successfully achieved and the established optimum conditions with the PPy(GOx) biosensor were 0.4 M pyrrole, 200 U/ml GOx, applied current density of 0.1 mA/cm^2, polymerization time of 600 s. The biosensor achieved a detection limit of 50 μM, a linear concentration range of 0.1 mM to 8 mM, and sensitivity of 9.97 μA cm^{-2} mM^{-1}.**

Keywords- PPy nanowire arrays; glucose oxidase; amperometry; biosensor

I. INTRODUCTION

Hyperglycemia is a serious worldwide public health problem, which leads to numerous complications, such as higher risks of heart disease, kidney failure and blindness. These complications may be greatly reduced through stringent control of blood sugar level by glucose sensors. Electrochemical biosensors based on glucose oxidase (GOx) have played a leading role due to its simplicity, reliability, cheapness, and easy to use [1, 2]. In most cases, the glucose detection was performed by measuring the current response originated from the electrochemical oxidation of hydrogen peroxide, a product during the course of the enzymatic reaction.

Conducting polymers such as polypyrrole (PPy), polyaniline (PAn) and nafion are often used for immobilization of bioactive enzymes because of their electron conductivity and biocompatibility. Many researches have been performed on biosensors based on conducting polymer films entrapped with enzymes (such as glucose oxidase, nitrate reductase, sulfite oxidase, etc.)[3-5]. However, the performances of these film-based biosensors are limited by the achievable effective surface area and enzyme concentration. The application of nanotechnology in this field is hopeful to resolve this problem.

There has been a considerable interest in recent years on the use of nanomaterials for fabrication of electrochemical sensors and biosensors because of their unique electronic, chemical,

and physical properties. Many nanoparticles[6,7] and nanotubes[8] have been employed in the conductive polymers for electrode modification to achieve enhancement of conductivity and other electrochemical performances of film-based biosensors. Also, well aligned one-dimensional nanoarray, such as TiO$_2$ nantubes array obtained by anodic oxidation [9] and noble metal nanoarray obtained by electrochemical deposition in AAO templates [10] have been used for fabrication of biosensors.

In this study, well aligned PPy nanowire entrapped with glucose oxidase (GOx) were formed by electrochemical polymerization in AAO templates, for amperometric detection of glucose. The concentration of GOx, current density and time of polymerization are investigated for the optimizing of glucose detection.

II. EXPERIMENTAL

2.1 Chemicals

High purity aluminium foil (99.999%) was purchased from Beijing Cuibolin Non-Ferrous Technology Developing Co., Ltd, China. Glucose oxidase (type II) extracted from Aspergillus niger was purchased from Sigma-Aldrich Company and used as received. Pyrrole, potassium chloride, glucose, oxalic acid, phosphoric acid, chromic acid, and other chemicals of analytical reagent grade were purchased from Sigma-Aldrich Company. The pyrrole was distilled at about 130℃ prior to use, and was stored in a bottle covered with aluminum foil in the freezer to prevent UV degradation.

The buffer solution was 0.05 M phosphate solution (pH = 7) obtained by adjusting ratio of 0.05 M Na$_2$HPO$_4$ and 0.05 M NaH$_2$PO$_4$. The enzyme solution was prepared by dissolving GOx in 0.05 M phosphate buffer to give 1000 U/mL solution and was kept at 4 ℃ in fridge.

2.2 Fabrication of PPy (GOx) nanowire arrays within AAO template

PPy nanowire arrays were obtained by electrochemical polymerization of pyrrole monomer within AAO template. AAO template was fabricated by a two-step anodization on an aluminum foil (99.999%, 0.2 mm thick) in a 0.3 M oxalic acid electrolyte with the application of a potential of 45 V. After the 2nd anodization, the aluminum on the back side was removed by immersing sample in 0.5 M copper chloride solution to get the AAO template. The template was immersed in 0.5 M phosphoric acid solution at 40 ℃ for 20 min to remove the barrier layer and enlarge the holes. A gold layer was deposited on one side of the template by vacuum deposition to make it conductive.

A gold-disk electrode was cleaned thoroughly with Milli-Q water and dried with fibre-free tissue paper. The template was fixed on the gold-disk electrode with a small cap and used as the working electrode. The PPy nanowire arrays without and with GOx were formed by galvanostatic polymerization in a pyrrole monomer solution which contained 0.4 M pyrrole, 0.1 M KCl and different concentration of GOx at variable applied current density with a three-electrode system using an Ag/AgCl electrode as reference electrode and a Pt/Ti porous electrode as auxiliary electrode.

After the polymerization, the electrode was washed several times with Milli-Q water to remove remaining monomer solution. Then the electrode was placed into 0.5 M NaOH solution to partly remove AAO template, then the electrode was subsequently washed thoroughly with water.

2.3 Detection of glucose

The amperometric detection of glucose was performed in a three-electrode cell with the PPy(GOx) nanowire arrays as the working electrode in 2 mL buffer solution. The applied potential was determined by the cyclic voltammograms. 20 μL glucose standard solution of different concentration was continuously added when a steady-state current was reached.

III. RESULTS AND DISCUSSION

PPy nanowire arrays entrapped with GOx were fabricated by electrochemical polymerization using AAO template fixed on gold-plate electrode as working electrode. Fig.1 gives the Chronopotentiogram obtained for growth of PPy nanowire arrays in AAO templates with the current density of 0.1 mA/cm². The electrode potential goes up and fluctuates dramatically at the commencement of the polymerization due to the irregular surface of gold layer and the difficulty of pyrrole diffusion in nano-holes of template. The potential goes down after 70 s and stabilizes at 538 eV. When adding different concentration of enzyme (100 - 700 U/ml) in pyrrole monomer solution to fabricate PPy entrapped with GOx nanowire arrays, the Chronopotentiograms are almost the same. Only the stabilized potentials change in the range from 535 mV to 550 mV, but no evident regularity can be obtained.

Inset in fig.1 is the morphology of PPy nanowire arrays with amplification factor of 16000. PPy nanowires are uniformly grown from the holes of AAO template with the diameter about 20 nm. PPy (GOx) nanowires have the same morphology with PPy nanowires.

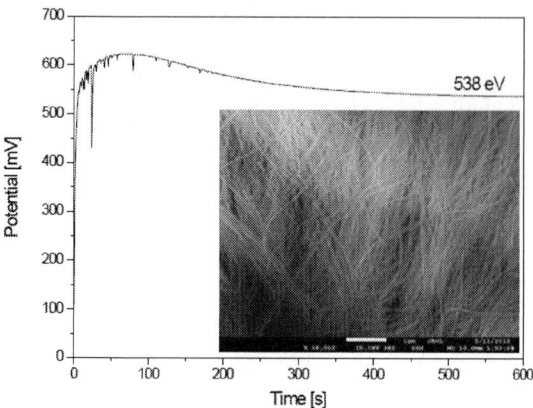

Fig.1 Chronopotentiogram obtained for growth of PPy nanowire arrays, the inset is the SEM view of PPy nanowire arrays. [Py] = 0.4 M, [KCl] = 0.1 M, current density is 0.1 mA/cm² and time is 600 s.

Fig.2 compares the cyclic voltammograms of PPy and PPy (GOx) nanowire arrays biosensor in 0.05 M buffer solution in absence and presence of 10 mM glucose. The characteristic reduction/oxidation couple of PPy appeared at -0.6 V and -0.26 V (vs Ag/AgCl) for both samples in phosphate buffer solution, which is similar to those obtained for PPy film on Pt-disc electrode [3]. The cathodic current is much higher than the anodic current, which means negative potential is inappropriate to be used for glucose detection because of the high background current. When adding 10 mM glucose in buffer solution, PPy nanowire arrays show very small amperometric response when applied potential is higher than 0.7 V due to the electrochemical oxidation of glucose.

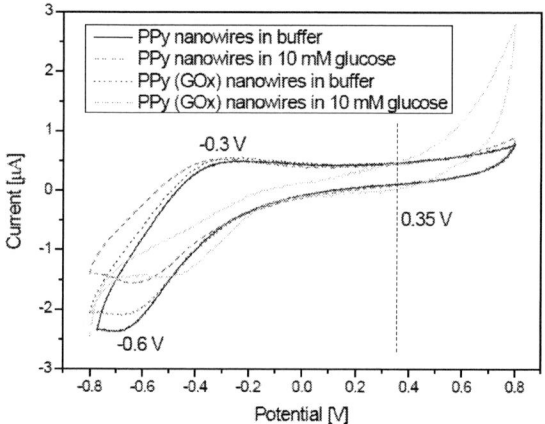

Fig.2 Cyclic voltammograms of PPy and PPy(GOx) nanowire arrays biosensor in 0.05 M phosphate buffer solution without and with 10 mM glucose.

The comparison of CVs of PPy (GOx) in buffer solution without and with 10 mM glucose shows that PPy (GOx) electrode gives evident amperometric response to glucose when applied potential higher than 0.35 V, accompanied by a little decrease in the cathodic current. The resulting current response is caused by the electrochemical oxidation of hydrogen peroxide (H_2O_2) generated during the enzymatic reaction. The increment of the current density should have linear relationship with the concentration of H_2O_2, which indicates the linear relationship between current density and glucose concentration.

The optimizations of (a) current density of polymerization, (b) polymerization time, and (c) GOx concentration performed for amperometric detection of 1 mM glucose are showed in fig.3. Fig.3 (a) shows that the use of a current density of 0.1 mA/cm^2 led to optimum amperometric response for glucose. At a low current density (less than 0.1 mA/cm^2) short length of PPy (GOx) nanowires gives low current response. Especially when the current density of polymerization is 0.025 mA/cm^2, the length is too short to expose from the AAO nanohole after the template is partly removed. The lower sensitivities are observed when current density was higher than 0.1 mA/cm^2. The increase of length and thickness of nanowires causes high diffusion and electron transmission barriers of hydrogen peroxide. When current density is higher than 0.15 mA/cm^2, no current response can be detected. Fig.3 (b) shows the optimum polymerization time of 600 s to get the best amperometric response, which maybe arises from the same reason with current density of polymerization.

Fig.3 (c) shows that the optimum condition of the GOx concentration is 200 U/ml. The concentration of GOx in pyrrole monomer solution determines the concentration of GOx in PPy nanowires which affects the enzymatic reaction sensitivity and the conductivity of nanowires. High GOx concentration gives the high sensitivity and low conductivity. When the GOx concentration is too high, the conductivity of nanowire is too low to give high current response. In the same way, when GOx concentration is too low, the low enzymatic reaction sensitivity gives poor current response.

Galvanostatic method is used for the polymerization of PPy nanowires. So, the amount of PPy is determined by the current density and time of polymerization process. Concentration of pyrrole has no effect on the polymerization process of PPy.

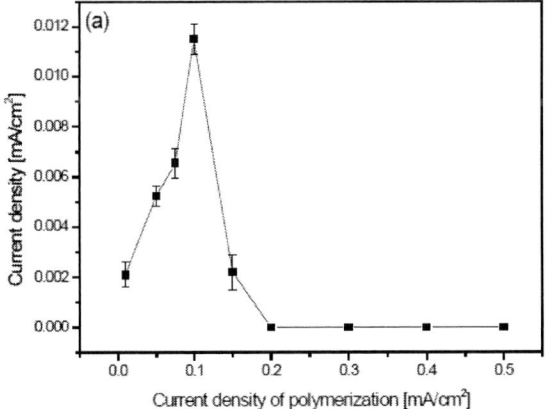

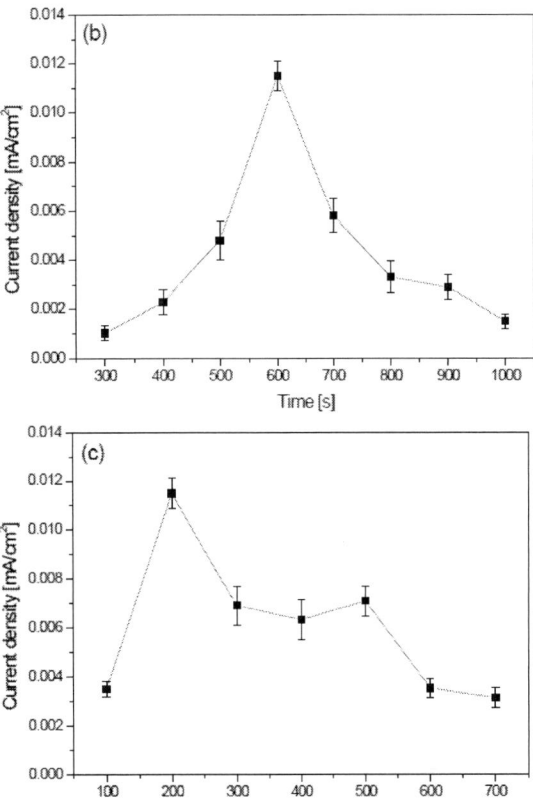

Fig.3 Effect of (a) current density of polymerization, (b) polymerization time, and (c) concentration of GOx on the amperometric detection of glucose. The applied potential is 700 mV for the amperometric detection, and the increment of glucose is 1 mM. The optimizing fabrication parameters: current density of polymerization = 0.1mA/cm^2, time = 600 s, [GOx] 200 U/ml.

The optimum conditions achieved in this study for obtaining sensitive PPy (GOx) biosensor are 0.4 M pyrrole, 200 U/ml GOx, applied current density of 0.1 mA/cm^2, polymerization time of 600 s. Fig.4 gives the typical amperometric response at applied potential of 0.7 V and the calibration curve of PPy (GOx) nanowire biosensor to successive additions of glucose in 0.05 M phosphate buffer solution. The anodic current increases stepwise with successive injection of glucose. The rapid response time (7s) to the changes in glucose concentrations indicates excellent electrocatalytic behavior of the nanowire biosensor. The calibration curve of PPy (GOx) nanowire biosensor shows that the amperometric response increases linearly with glucose concentration from 0.1 mM to 8 mM (I[mA/cm^2] = 0.0019+0.0099 C[M], R^2 = 0.99316). The sensitivity of the PPy (GOx) nanowire arrays biosensor obtained with an applied potential of 700 mV is 9.9 μA cm^{-2} mM^{-1}. The detection limit (dl) of the PPy-Pt-GOx nanowire arrays biosensor can be calculated as dl = 3σ/m, where σ is the standard deviation of the background current (0.32μA cm^{-2}) of the potentiostat, and m the slope of the linear part of the calibration curve (9.9 μA cm^{-2} mM^{-1}). A detection limit of 96 μM can be obtained, which is in the typical range of detection limits of glucose sensors [10-12].

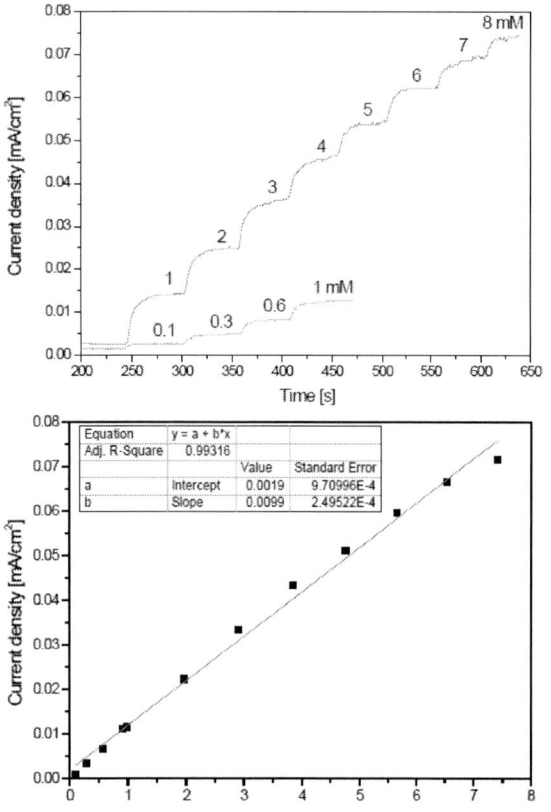

Fig.4 Amperometric response of (a) PPy-GOx nanowire biosensor to successive additions of glucose at 700 mV in 0.05 M phosphate buffer solution and (b) the calibration curve of PPy-GOx nanowire toward concentration of glucose.

IV. CONCLUSIONS

Well aligned PPy nanowire arrays entrapped with GOx with a diameter of about 20 nm within AAO template has been successfully fabricated. After optimization, good analytical performances have been obtained for amperometric detection of glucose. The optimizing fabrication parameters are 0.4 M pyrrole, 0.1 M KCl, 200 U/ml GOx, current density of 0.1mA/cm^2, and polymerization time of 600 s. The amperometric detection gives a sensitivity of 9.9μA cm^{-2} mM^{-1} and a linear concentration range from 0.1 to 8 mM with a correlation coefficient of 0.99316.

ACKNOWLEDGMENT

This work was supported by Nature Science Foundation of China (51102071, 51072044) and China Postdoctoral Science Foundation (20100480678).

REFERENCES

[1] M. Przybyt, E. Miller, T. Szreder. "Thermostability of glucose oxidase in silica gel obtained by sol-gel method and in solution studied by fluorimetric method", Journal of Photochemistry and Photobiology B: Biology. vol. 103, pp. 22-28, 2011.

[2] D. Olea, O. Viratelle, C. Faure, et al. "Polypyrrole glucose oxidase biosensor Effect of enzyme encapsulation in multilamellar vesicles on analytical properties", Biosensors and Bioelectronics. vol. 23, pp. 788-794, 2008.

[3] S.B. Adeloju, A.N. Moline. "Fabrication of ultra-thin polypyrrole-glucose oxidase film from supporting electrolyte-free monomer solution for potentiometric biosensing of glucose", Biosensors & Bioelectronics. vol.16, pp. 133-139, 2001.

[4] A. Ameer, S.B. Adeloju. "Galvanostatic Entrapment of Sulfite Oxidase into Ultrathin Polypyrrole Films for Improved Ampeometric Biosensing of Sulfite", Electroanalysis, vol. 20, pp. 2549-2556, 2008.

[5] M. Sohail, S.B. Adeloju. "Fabrication of Redox-Mediator Supported Potentiometric Nitrate Biosensors with Nitrate Reductase" Electroanalysis, vol. 21, pp. 1411-1418, 2009.

[6] X.Y. Jiang, Y.H. Wu, X.Y. Mao, et al. "Amperometric glucose biosensor based on integration of glucose oxidase with platinum nanoparticles/ordered mesoporous carbon nanocomposite", Sensors and Actuators B: Chemical. vol. 153, pp. 158-163, 2011.

[7] Z.H. Wen, S.Q. Ci, J.H. Li. "Pt Nanoparticles Inserting in Carbon Nanotube Arrayss: Nanocomposites for Glucose Biosensors", J. Phys. Chem. C, vol. 113, pp. 13482-13487, 2009.

[8] L.N. Cella, W. Chen, N.V. Myung, et al. "Single-Walled Carbon Nanotube-Based Chemiresistive Affinity Biosensors for Small Molecules: Ultrasensitive Glucose Detection", J. Am. Chem. Soc., vol. 132, pp. 5024-5026, 2010.

[9] X. Xu, J.Q. Zhao, D.C. Jiang, et al. "TiO$_2$ sol-gel derived amperometric biosensor for H$_2$O$_2$ on the electropolymerized phenazine methosulfate modified electrode", Anal. Bioana. Chem., vol. 374, pp. 1261-1266, 2002.

[10] X.Y. Zhang, D. Li, L. Bourgeois, et al. "Direct electrodeposition of Porous Gold Nanowire Arrayss for Biosensing Applications", ChemPhysChem, vol. 10, pp. 436-441, 2009.

[11] J. Li, X.Q. Lin. "Glucose biosensor based on immobilization of glucose oxidase in poly(o-aminophenol) film on polypyrrole-Pt nanocomposite modified glassy carbon electrode", Biosensors & Bioelectronics, vol. 22, pp. 2898-2905, 2007

[12] E.M.I. M EKanayake, D.M.G. Preethichandra K. Kaneto. "Polypyrrole nanotube arrays sensor for enhanced adsorption of glucose oxidase in glucose biosensors", Biosensors & Bioelectronics, vol. 23, pp. 107-113, 2007.

The Electromigration Investigation of Cu-Ni Nanocomposites

Y. C. Chen, Chia Wei Chu, Tzu-Yuan Chao, Y. T. Cheng*, *Senior Member, IEEE*, and Chih Chen**

*Microsystems Integration Laboratory, Department of Electronics Engineering & Institute of Electronics,
**Department of Materials Science and Engineering,
National Chiao Tung University, HsinChu, Taiwan R.O.C.
ytcheng@mail.nctu.edu.tw

Abstract—The paper presents the electromigration behavior study of Cu-Ni nanocomposite for low-power electromagnetic microactuator fabrication. The nanocomposite is characterized based the striped with 50μm width and synthesized in the Cu plating bath with the 2g/L concentration of 50nm Ni nanopowders. About 2.03% Ni weight percentage in the Cu-Ni nanocomposite stripe is characterized using inductively coupled plasma mass spectrometer (ICP-MS). The drift velocity, critical length, critical product, and activation energy of the Cu-Ni nanocomposite are 565nm/hr, 14μm, 1714 A/cm, and 0.39eV respectively. In comparison with the values of Cu which are 88nm/hr, 20μm, 2365 A/cm, and 1.09eV, respectively, the poor electromigration behavior of the nanocomposite is dominated by the surface diffusion mechanism of Cu atoms due to the void formation in the interface between itself and the passivation oxide.

Keywords-Cu-Ni; electromigration; nanocomposite

I. INTRODUCTION

Previously, we had reported a Cu-Ni nanocomposite for low power magnetic microactuation [1,2]. Via the incorporation of Ni ferromagnetic nanoparticles into Cu matrix, the magnetic characteristic of the Cu can be modified from diamagnetism to ferromagnetism. While ~50nm Ni nanopowders were added in the alkaline noncyanide copper plating solution to form a colloidal bath for the synthesis of Cu-Ni nanocomposite film. We have shown that a 200μm wide Cu-Ni composite coil can realize ~50% power saving of the EM-driven microspeaker as compared with the Cu coil [3].

Electromigration is the most common failure mechanism of electric-related failure modes in MEMS devices. For example, electromigration in the micro four-point probe metal layer [4], in the heater of microthermal inkjet printhead [5], in the high power RF MEMS switch [6] and in the electro-thermal micro gripper [7]...etc. For the case of previously designed magnetic microactuator, the inductive metal coil with the dimensions of 200μm in width and 5μm in thickness carrying the operating current of 250mA would have a current density of 2.5×10^4 A/cm^2, void and extrusion formation due to the electromigration effect can cause the inductive coil failure, an serious reliability problem for magnetic microactuation.

Therefore, in this work, a reliability study of the composite material in terms of the electromigration behavior is

performed for future applications of the material in magnetic microactuation.

II. TESTING SAMPLE DESIGN AND FABRICATION

In this experiment, two kinds of Blech specimens [8] are made of Cu-Ni nanocomposite and pure Cu, respectively. Each specimen constitutes of a 600nm-thick SiO2 insulating layer on which 1μm-thick electroplated Cu-Ni nanocomposite or pure Cu stripes are deposited using a sandwich seed layer of Ti (2μm) /Cu (120nm) /Ti (5nm). The preparation of the Cu-Ni nanocomposite stripes is listed as follows with an optimal composite concentration referred to [3] based on the power saving ratio which is determined by the trade-off between the electrical and magnetic properties of Cu-Ni nanocomposite. In this work, the tested composite samples are chosen with the design of 50μm line width and synthesized in the Cu plating bath with the 2g/L concentration of 50nm Ni nanopowders.

A 600nm-thick SiO$_2$ insulating layer was thermal grown on a 4" silicon wafer and then a sandwich layer of Ti (2μm)/Cu (120nm)/Ti (5nm) was deposited by e-gun evaporation. The 2μm-thick Ti layer is utilized as the conductive baseline of Blech structure and the Cu layer is the seed layer for Cu-Ni nanocomposite electroplating. It is noted that the 5nm thick Ti layer is required in the nanocomposite case for improving the adhesion of photoresist AZ10XT which is photo-patterned for following test stripe fabrication electroplated in the alkaline based copper plating solution. After patterning the test stripe regions photolithographically, two kinds of Blech specimens which are ~1μm-thick Cu-Ni nanocomposite stripes and pure Cu stripes respectively were electroplated. 5μm-thick Cu pads were then patterned and electroplated on the top of the stripes for probing with electrical current input. Once the stripes were fabricated, the Ti(2μm)/Cu(120nm)/Ti(5nm) layer besides the one underneath the stripes were etched away to complete the fabrication Cu-Ni nanocomposite or pure Cu stripes along the Ti base line, i.e. the Blech structure. At final, a 1μm-thick SiO$_2$ film was deposited by plasma-enhanced CVD on the stripes as a passivation layer which is photolithographically and chemically etched to open the Cu pad surfaces for probing.

Fig. 1 shows the top-view SEM image of the as-fabricated Cu-Ni nanocomposite Blech specimen. Electromigration test

978-1-4673-1122-9/12 $31.00 © 2012 IEEE

was done on an integral hot plate with the desired temperature and conducting the stress current by probing. Most of the current drifted into the test stripes because of the much larger resistance of the Ti base line than the test stripes. Percentage of the stress current detoured into the test stripes was estimated to decide the stress current density in the stripes. The surface profiler, ET-4000, was utilized to measure the thickness of the test stripes for well controlling stress current density for following electromigration test. About 2.03% Ni weight percentage in the Cu-Ni nanocomposite stripes was characterized using inductively coupled plasma mass spectrometer (ICP-MS). In addition, optical microscope, scanning electron microscope (SEM) and focused ion beam (FIB) were utilized to observe the electromigration phenomena as well as measure the average depletion length from which average drift velocity of the metal atoms can be derived.

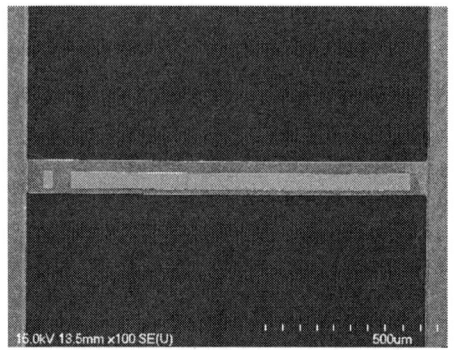

Fig. 1. Top-view SEM image of the as-fabricated Cu-Ni nanocomposite Blech specimen.

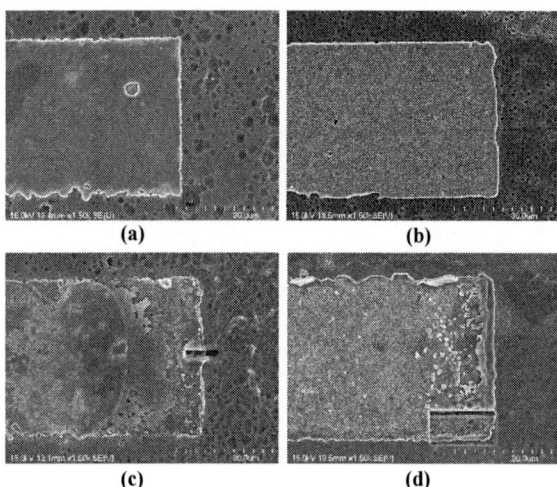

(a) **(b)**

(c) **(d)**

Fig. 2. (a) and (b) are the top-view SEM images on the cathode side of the 1000µm-long Blech specimens of the pure Cu and the Cu-Ni nanocomposite, respectively, before current stressing. (c) and (d) are the SEM images after current stressing by 1.19×10^6 A/cm² at 275°C for 312 hrs corresponding with (a) and (b). The SEM images were taken after removing the top oxide passivation.

III. RESULTS AND DISCUSSION

Electromigration is the phenomenon happening in an electrical conductive stripe loaded with high density electrical current. Under the influence of electric field, the atomic displacement and accumulation based on mass transport would take place in the conductive stripes. The force applied on atoms (ions) for moving can be calculated as follows [8]:

$$F_{em} = (Z_{el}^{*} + Z_{wd}^{*})eE = Z^{*}eE \qquad (1)$$

where e is the charge of an electron, E is the applied electric field along the line and Z_{el}^* is the nominal valence of the ion moving in the metal while the dynamic screening effect is ignored. It only stands for the field effect and Z_{el}^*eE is called the direct force. Z_{wd}^* is the charge number responsible for the momentum exchange effect and Z_{wd}^*eE is called the electron wind force. For a good conductor, the electron wind force is much greater than the direct force. Z^* is the effective charge number while the electromigration occurs. Previously, Blech and Herring proposed that electromigration in a short segment tends to induce back stress [9, 10]. The back stress is the driving force on atoms to move in an opposite direction to the atomic flow induced by electromigration and it is resulted by the vacancy concentration gradient in a stressed stripe caused by electromigration. The stress can, therefore, be taken as the gradient of chemical potential (μ) as follows:

$$F_{me} = -\nabla \mu = -\frac{d\sigma\Omega}{dx} \qquad (2)$$

where σ is hydrostatic stress in the metal and Ω is atomic volume. Thus, combining the influences of electrical and the back stress, the next atomic flux is derived as follow:

$$J_{em} = -C\frac{D}{kT}d\sigma\Omega/dx + C\frac{D}{kT}Z^*eE \qquad (3)$$

where J_{em} is atomic flux in units of atoms/m²-sec, C is the number of atoms per unit volume and D/kT is atomic mobility.

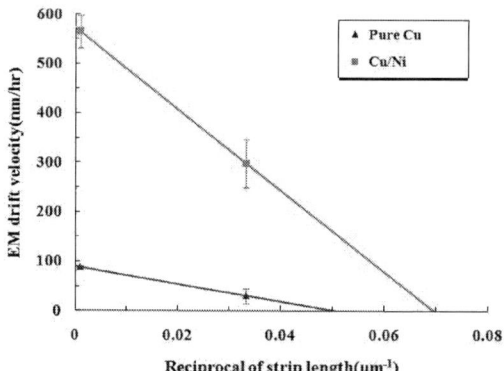

Fig. 3. Average EM drift velocity of the pure Cu and Cu-Ni nanocomposite stripes, respectively, as a function of reciprocal stripe length. The critical lengths were obtained by extrapolating the fitted lines to zero drift velocity.

978-1-4673-1122-9/12 $31.00 © 2012 IEEE

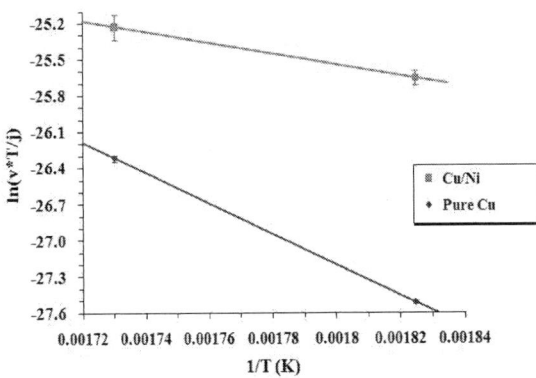

Fig. 4. Plot of the $\ln(vT/1.19\times106\ (A/cm^2))$ as a function of reciprocal temperature. Here, the stressed temperature of 275°C and 305°C are chosen in the work. The activation energies of pure Cu and Cu-Ni nanocomposite are 1.09 and 0.39 eV respectively, which are derived from the slopes of the fitted lines.

The first term of Eq. (3) indicates the atomic flux caused by the back stress (the sign "−" represents the direction of the flux opposite to the one by electromigration) depends on the length of the stripe; the shorter the strip, the greater the back flow stress. At a certain length defined as the "critical length", the gradient in the first term would be large enough to balance the electromigration term, so no electromigration damage occurs. In Eq. (3), if we set $J_{em} = 0$, we obtain the expression for the critical length as:

$$\Delta x = \Delta\sigma\Omega/(Z^*eE) \qquad (4)$$

Since the resistivity of the metal can be regarded as a temperature-dependent constant, the $j\Delta x$ value, recognized as the "critical product", can be derived by moving the current density j from the right-hand to the left-hand side of the equation ($E = \rho j$, ρ is resistivity and j is current density),

$$j\Delta x = \Delta\sigma\Omega/(Z^*e\rho) \qquad (5)$$

The critical product is negative-correlated to the operating temperature as indicated in Eq. (5). Meanwhile, another constant acquired by dividing the critical product by the line length is called the "threshold current density", which means the maximum current density can be conducted in the metal line without having electromigration effect. In other words, if a device structure, i.e. the Belch stripe in this work, can be designed to have a product of $j\Delta x$ less than the critical product the structure can then be immune from the electromigration damage.

Fig. 2 shows the SEM images of the pure Cu (Fig. 2(a)&(c)) and Cu-Ni nanocomposite (Fig. 2(b)&(d)) 1000µm-long stripes respectively at the cathode end before and after current stressing by 1.19×10^6 A/cm² at 275°C for 312 hrs, respectively. In general, the average drift velocity of the metal atoms due to electromigration is derived by Huntington and Grone as follows [8]:

$$v=J/C=\frac{D}{kT}Z^*ej\rho=\frac{D_0}{kT}Z^*ej\rho\exp(-E_a/kT) \qquad (6)$$

where J is the atom flux, C is the density of metal ions, D is the diffusivity, D/kT is the mobility, k is Boltzmann's constant, T is the absolute temperature, Z^*e is the effective charge of the ions, j is the electrical current density, ρ is the metal resistivity, D_0 is the prefactor of diffusivity and E_a is the activation energy of diffusion. The equation can be rewritten as:

$$\ln\frac{vT}{j}=-(E_a/kT)+\ln\frac{D_0Z^*e\rho}{k} \qquad (7)$$

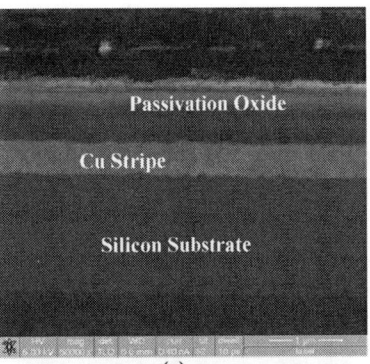

(a)

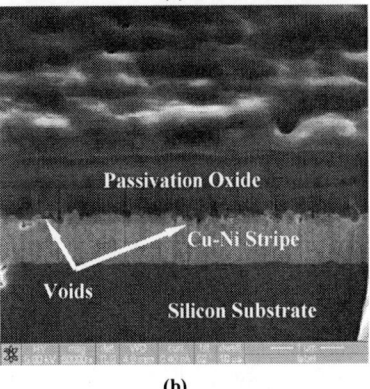

(b)

Fig. 5. (a) and (b) Cross-sectional FIB images of the pure Cu and the Cu-Ni nanocomposite Blech specimens respectively. The specimens are as-fabricated samples before electromigration testing.

Therefore, by measuring the drift velocity as a function of reciprocal temperature, the activation energy E_a of the diffusion process in electromigration can be determined. Fig. 3 shows the average atomic drift velocity, v, of the Cu-Ni nanocomposite and pure Cu stripes, respectively, as a function of the reciprocal of stripe length. The critical lengths are obtained to be ~20 and 14µm for pure Cu and Cu-Ni nanocomposite stripes, respectively. The shorter critical length of the Cu-Ni nanocomposite stripe means the poor electromigration resistance. By multiplying the acquired

978-1-4673-1122-9/12 $31.00 © 2012 IEEE

critical lengths with the stressed current density, the critical product of ~2365 and 1714 A/cm and the threshold current density of 2.37×10^4 and 1.71×10^4 A/cm^2 are derived for the pure Cu and nanocomposite stripes respectively while the length of line segment is designed to be 1000μm at temperature condition 275°C. These values represent the maximum current densities that the pure Cu and Cu-Ni nanocomposite can carry without electromigration damage at the above conditions.

Fig. 4 shows the plot of $\ln(vT/1.19 \times 10^6$ (A/cm^2)) as a function of the reciprocal temperature. The activation energies (E_a) of ~1.09 and $0.39 eV$ are then derived for pure Cu and Cu-Ni nanocomposite respectively in the temperature range of 275~305°C. The lower activation energy of diffusion is found for the Cu-Ni nanocomposite. The $0.39 eV$ activation energy in the nanocomposite system suggests that surface diffusion of Cu atoms could be the primary mechanism resulting in electromigration in the Cu-Ni nanocomposite stripe. Liu et al. have proposed the embedded atom method (EAM) for calculating the activation energy of surface diffusion on {100} surface of copper to be $0.38 eV$ [11]. Experimental study investigated by M. Breeman and D. O. Boerma using low-energy ion scattering (LEIS) gave the value of $0.39 \pm 0.06 eV$ for the diffusion mechanism on same plane [12]. Here the measured activation energy of diffusion of the Cu-Ni nanocomposite stripe capped by oxide passivation is ~$0.39 eV$ which is quite close to the values obtained theoretically and experimentally.

Meanwhile, B. H. Jo and R. W. Vook have reported that the copper films exhibited surface diffusion mechanism in electromigration test under free surface and ultra-high vacuum conditions [13]. In comparison with the measured activation energy of pure Cu which is ~$1.09 eV$, indicating grain-boundary diffusion mechanism under electromigration, the experimental results suggest that surface diffusion is the primary deterministic mechanism for the electromigration behavior in the Cu-Ni nanocomposite stripe. According to the voids at the interface between the Cu-Ni nanocomposite stripes and the oxide layer could make the composite film exhibit a similar situation.

Fig. 5(a) and (b) show the cross sectional FIB images of the Cu and Cu-Ni nanocomposite stripes, respectively. It can be found that the interface of pure Cu film and its upper SiO$_2$ capping layer is smooth and has no noticeable defects. On the contrary, there are voids at the interface between the Cu-Ni nanocomposite film and the oxide layer. Therefore, the source to the diffusion can be attributed to the voids formed in the interface between the Cu-Ni nanocomposite and capped oxide film.

IV. CONCLUSION

In summary, the electromigration behavior of Cu-Ni nanocomposite film has been investigated. The reliability degradation of the composite in comparison with that of Cu is mainly caused by the imperfect interface structure between the composite film and its passivation oxide.

ACKNOWLEDGMENT

This work is supported by the National Science Council of Taiwan, R. O. C. under Grant NSC 100-2220-E-009-007 and in part by the Ministry of Education in Taiwan under ATU Program. The authors would like to thank the Instrument Center at National Chiao Tung University (NCTU) for the support of fabrication facility and SQUID measurement.

REFERENCES

[1] Y. W. Huang, T. Y. Chao, C. C. Chen, and Y. T. Cheng, "Power consumption reduction scheme of magnetic microactuation using electroplated Cu-Ni nanocomposite," *Applied Physics Letters*, vol. 90, pp. 244105-244107, 2007.

[2] Yu Wen Huang, Tzu-Yuan Chao, and Y.T. Cheng, "Synthesis and device fabrication of Cu-Ni nanocomposite for low power magnetic microactuation," in *Proc. IEEE-NANO 2007*, Hong Kong, China, pp. 899-902, Aug. 2-5, 2007.

[3] Y. C. Chen, Wei-Ting Liu, Tzu-Yuan Chao, and Y. T. Cheng, "An optimized Cu-Ni nanocomposite coil for low-power electromagnetic microspeaker fabrication," in *Proc. Transducers 2009*, Denver, Colorado, USA, Jun. 21-25, 2009.

[4] T. Ansbaek, D. H. Petersen, O. Hansen, J. B. Larsen, T. M. Hansen, and P. Boggild, "Fundamental size limitations of micro four-point probes," *Microelectron. Eng.*, vol. 86, pp. 987-990, 2009.

[5] J. H. Lim, K. Kuk, S. J. Shin, S. S. Baek, Y. J. Kim, J. W. Shin, and Y. S. Oh, "Failure mechanisms in thermal inkjet printhead analyzed by experiments and numerical simulation," *Microelectron. Reliab.*, vol. 45, pp.473-478, 2005.

[6] K. Grenier, D. Dubuc, B. Ducarouge, V. Conedera, D. Bourrier, E. Ongareau, P. Derderian, and R. Plana, "High power handling RF MEMS design and technology," in *Proc. 18th IEEE Int. Conf. MEMS*, pp. 155-158, 2005.

[7] A. Deutschinger, U. Schmid, M. Schneider, W. Brenner, H. Wanzenböck, B. Volland, T. Ivanov, and I. W. Rangelow, "Characterization of an electro-thermal micro gripper and tipsharpening using FIB technique," *Microsyst. Technol.*, vol. 16, pp. 1901-1908, 2010.

[8] H. B. Huntington and A. R. Grone, "Current-induced marker motion in gold wires," *J. Phys. Chem. Solids*, vol. 20, pp.76-87, 1961.

[9] I. A. Blech, "Electromigration in Thin aluminum films on titanium nitride," *J. Appl. Phys.*, vol. 47, pp. 1203-1208, 1976.

[10] I. A. Blech and C. Herring, "Stress generation by electromigration," *Appl. Phys. Lett.*, vol. 29, pp. 131-133, 1976.

[11] C. L. Liu, J. M. Cohen, J. B. Adams, and A. F. Voter, "EAM study of surface self-diffusion of single adatoms of fcc metals Ni, Cu, Al, Ag, Au, Pd, and Pt," *Surface Science*, vol. 253, pp. 334-344, 1991.

[12] M. Breeman and D. O. Boerma, "Migration of Cu adatoms on a Cu(100) surface, studied with low-energy ion scattering (LEIS)," *Surface Science*, vol. 269/270, pp. 224-228, 1992.

[13] B. H. Jo and R.W. Vook, "In-situ ultra-high vacuum studies of electromigration in copper films," *Thin Solid Films*, vol. 262, no. 1\2, pp.129-134, 1995.

A SWNTs Thin Film Solar Microcell Prepared by Simple Solution-Evaporation Method

C.C. Chen, Y.Y. Chang and J. Zhang*

*National Key Lab of Micro/ Nanometer Fabrication Technology, Institute of Microelectronics, Peking University, China
zhangjinwen@pku.edu.cn

Abstract—This paper reports a solar microcell based on single-walled carbon nanotubes (SWNTs) thin film-silicon heterojunction prepared by very simple and low cost solution-evaporation (SE) method. The nano-material of SWNTs, which has a one-dimensional structure and direct band gap with unique electric properties, is applied and plays the role of the energy conversion in the solar microcells, including exciton generation, carrier collection and transportation. The intrinsic p-SWNTs film was deposited conformally on the n-type silicon substrate to form the p-n heterojunction by SE method. Under 100mA/cm^2 illumination, the SWNTs thin film microcell shows the open voltage (V_{oc}) of 230mV, short circuit current density (J_{sc}) of 73.7μA/cm^2, and fill factor (FF) of about 19%, proving that SE method is promising for achieving SWNTs thin film for application in microdevices.

Keywords-solar microcell; SWNTs thin film; heterojunction; soloution-evaporation

I. INTRODUCTION

Solar energy, as a type of clean and renewable power, has already been increasingly concerned. It is an important form of power supply in the micro-electro-mechanical system (MEMS), which has been deeply developed in the last few decades, with different kinds of devices micromachined into the size of millimeter to micrometer and even nanometer. However, the study on power supply of MEMS meets the bottleneck when it comes to the conflict between size limit and power density.

Thin-film solar microcell meets the same situation. It is a big challenge to improve the performance of the traditional silicon-based solar cells in energy conversion efficiency, power density, and costs. In spite of these difficulties, thin-film solar microcell still gains a lot of popularity and attention on account of its wide range of choices of materials including ZnO and CdTe [1-2] or substrate including GaN and SiC [3-4] to design a variety of structures and of its immense potentials in improving efficiency and reducing costs.

Currently, solar cells based on nano-materials such as quantum dots [5] and semiconducting nanostructures [6] are expected to become the next generation batteries. They attract tremendous interests, because nano-materials have potentials to break the upper limit of theoretical efficiency, which is about 31%, of a traditional monocrystalline silicon solar cell, and even increase this percentage to 66% theoretically [7]. Carbon nanotube probably is a perfect material for solar microcells owing to the one-dimensional structure, defect-free

to greatly decrease carrier recombination, high carrier mobility and low carrier scattering [8-9]. Previous reports on Si-carbon nanotube (Si-CNT) heterojunction solar cells showed a good performance [10]. However, their fabrication process of the carbon nanotube (CNT) films, such as chemical vapor deposition method (CVD), is very complicated and expensive.

In this paper a very simple SWNTs thin-film preparation method, such as solution-evaporation (SE), is proposed. It is applied to prepare p-SWNTs thin film on n-silicon substrate forming p-n heterojunction which is functioned to separate and transport electron hole pairs (EHPs) excited by solar illuminating to outside circuit forming the current. The design, fabrication and measurement of this SWNTs thin film solar microcell are presented.

II. THEORY AND DESIGN

Because SWNTs is intrinsic p-type, our device mainly has a SWNTs film prepared on an n-type silicon wafer (2-4Ωcm) forming p-n heterojunction as shown in Fig. 1. The front and back electrodes are used to transport carriers to external circuit. The gold (Au) is chosen as electrode materials for effectively reducing the contacting resistance. The front electrode has network structure with many Au strips orthogonal across with each other. The width of Au strip is 20μm. The SiO$_2$ layer having same structure as Au strips is used to insulate the front electrode from the underlying silicon substrate. In fact, the effective device area is composing with those square arrays surrounded by Au strips, where deposited SWNTs thin film can contact with silicon directly. There have 2784 (80μm×80μm) square array and so the total effective area is 17.8mm^2.

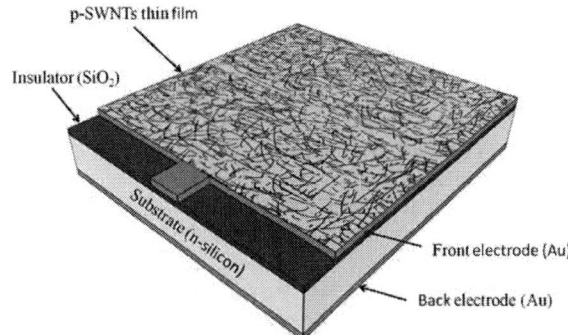

Fig. 1. A three-dimension schematic view of the SWNTs thin film-Si heterojunction solar microcell.

At last, SWNTs thin film is deposited on the substrate except pad area which will be wire bonded to outside circuit.

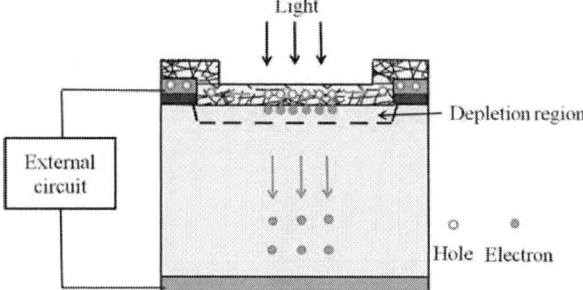

Fig. 2. The principle of the SWNTs thin film-Si heterojunction solar microcell.

When the SWNTs thin film and silicon heterojunction is illuminated, EHPs are excited both in p-type SWNTs thin film and n-type monocrystalline silicon substrate. Only those generated in the depletion region can be separated and swept across the junction by the build-in electric field into external circuit. Holes are collected by the p-type SWNTs, at the same time electrons are collected by the n-type monocrystalline silicon, which leads to the production of the electrical current as shown in Fig. 2. The equivalent electrical circuit of our microcell is as shown in Fig. 3. Current source (I_g) reveals the current due to EHPs generated by solar illumination. Resistor (R_{sh}) models the shunt leakage in the SWNTs thin film and silicon p-n heterojunction resulting from carrier recombination generation in the depletion region. It should be significantly larger than the load impedance R_l to decrease current losses as much as possible. Resistor R_s is the series resistance due to the diode neutral regions and contacts.

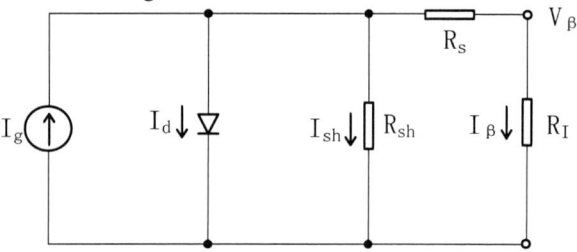

Fig. 3. The equivalent circuit of the SWNTs thin film solar microcell.

III. FABRICATION

The fabrication process (as shown in Fig. 4) started with the n-type (100) monocrystalline silicon substrate (Fig. 4(a)). On the beginning, 300nm SiO_2 was grown on the substrate by thermal oxidation. The SiO_2 on the backside of the substrate was corroded by buffered hydrofluoric acid (BHF). The next step was ion implantation to the rear surface of the substrate by phosphorus followed by annealing at 1000°C so as to form a good ohmic contact with back Au electrode. Then, the SiO_2 on the front surface was patterned by wet-etching, so that square array was formed on the substrate. The following process was sputtering 30nm/150nm Cr/Au onto the front

surface of the substrate and came up with patterning the metal layer into the same shape with the SiO_2 layer by wet-etching Cr and Au respectively. After that, 30nm/150nm Cr/Au were also sputtered onto the rear surface. And, Cr/Au layer of back side was annealed at 320°C to form a good ohmic contact with silicon (Fig. 4(b)). At last, the wafer was dicing to 6.6mm×6.6mm as pieces on which the SWNTs thin film will be deposited.

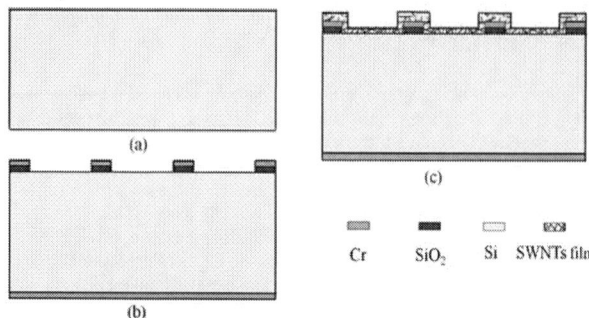

Fig. 4. The fabrication process flow. (a) the n-type monocrystalline Si substrate; (b) the front surface with patterned SiO_2 and Au electrode and the rear surface with back Au electrode; (c) the front surface deposited by SWNTs.

SWNTs powder which was produced by Chengdu Organic Chemicals Company with high purity (>90%). It was ultrasonically dispersed in dimethylformamide (DMF) with 10ug/ml concentration for 2hours with the water bath heating. Thenceforth, the SWNTs solution could be used to make thin films. Before forming the thin film, the silicon substrate was treated by BHF to wipe off the native oxide layer in order to attain a better contact between SWNTs and the silicon substrate. Then drops of solution were continuously introduced in turn onto the substrate which was heated to a constant temperature (150°C) by a thermostat. It made the DMF in each drop of the suspension evaporating in a few seconds, and the SWNTs remained and deposited onto the substrate achieving SWNTs thin film (Fig. 4(c)).

Fig. 5 showed the SEM images of the SWNTs film on silicon substrate. It could see that SWNTs randomly distributed on the substrate forming a porous structure (Fig. 5(a)). The cross section of SWNTs thin film was shown in Fig. 5(b), and the thickness of the film was about 1.25μm.

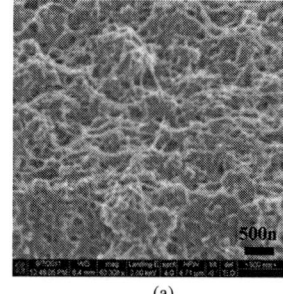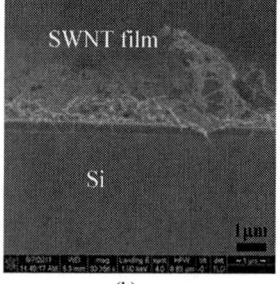

Fig. 5. The SEM images of SWNTs thin film-Si heterojuction: (a) the network structure of SWNTs thin film; (b) the cross section of SWNTs thin film.

IV. TEST AND RESULTS

Our testing system is as shown in Fig. 6. Before testing, the device was metallic packaged at first. The solar microcell was adhered onto a PCB board. The front electrode of the device was wire bonded to a pad of the PCB and the back electrode was electrically connected to another pad by conductive tape on the PCB board. Then, the solar microcell was laid into the illumination chamber and tested in dark and under 100mW/cm^2 illumination supplied by Keithley 2400. The I-V charactersistics both in dark and under illumination were also measured by Keithley 2400 .

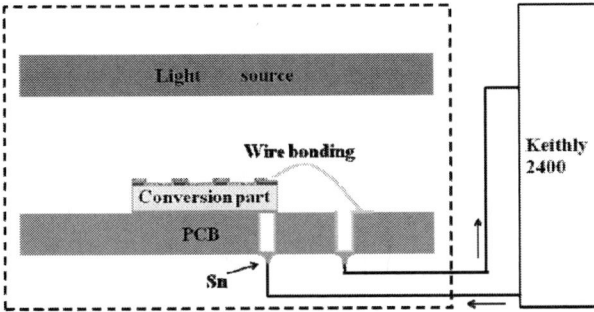

Fig. 6. Testing system.

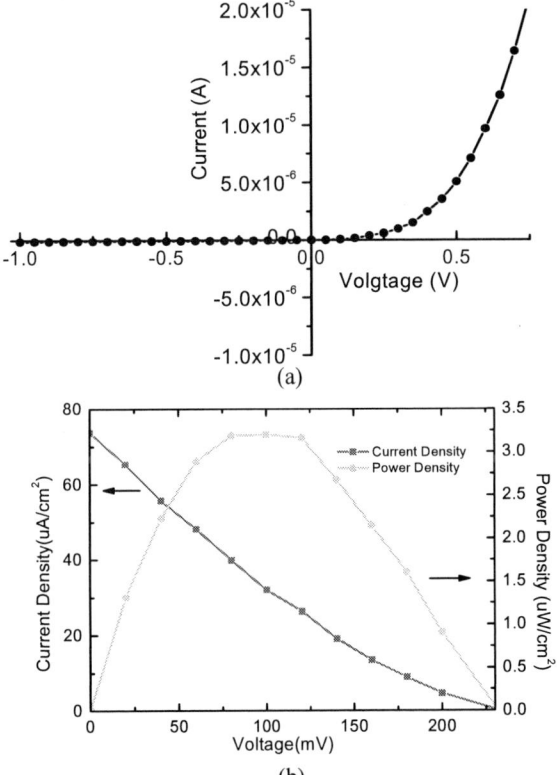

(b)

Fig. 7. (a) the I-V curve of the microbattery in the dark, (b) Current density-voltage (J-V) and power density-voltage (P-V) curves under illumination.

The measurement results were shown in Fig. 7. The I-V curve in the dark (Fig. 7(a)) exhibits that SWNTs thin film-Si heterojunction realizing a good p-n junction behavior. The reverse current (1.26×10^{-7}A when the bias voltage is -1V) was about 516 times lower than the forward current (6.50×10^{-5}A at 1V) when the bias voltage swept from -1V to 1V. The fourth quadrants of the J-V and P-V curve were mapped into the first quadrant, so that the generated current density and power density can be displayed as positive numbers (Fig. 7(b)). J-V and P-V curves under solar illumination showed V_{oc} of 230mV and J_{sc} of 73.7μA/cm^2. The fill factor (FF) of the solar microcell can be calculated by the equations as following:

$$FF = \frac{V_m J_m}{V_{OC} J_{SC}}$$

where J_m and V_m are the current density and voltage at the peak power point, J_{sc} and V_{oc} are short current density and open voltage. According to Fig. 6(b), J_m is about 32.0μA/cm^2, and V_m is about 100mV. Therefore, the FF is 19%.

It is proved that SWNTs has a considerable potential for application in solar microcells. And the SE method is effective and attainable, though the performance of our device is not as good as that of other solar microcell based on carbon nanotubes film prepared by those relatively complicated and expensive methods. It is believed that the properties of the solar microcell could be enhanced if the process of SE method is meliorated so that the quality of SWNTs thin film is improved, and the heterojunction structure is optimized further.

V. CONCLUSION

In summary, a very simple and low temperature SWNTs thin-film preparation method, such as solution-evaporation (SE), is proposed. It was applied to prepare p-SWNTs thin film on n-silicon substrate forming p-n heterojunction which is functioned to separate and transport electron hole pairs (EHPs) excited by solar illuminating to outside circuit forming the current. The whole fabrication is complicated with micromachining technology. I-V curves in the dark and under illumination were measured and the performance of the solar microcell was extracted. It is indicated that p-SWNTs thin film-Si heterojunction has a good p-n junction behavior. Under 100mW/cm^2 illumination, the solar microcell showed the V_{oc} of 230mV, J_{sc} of 73.7μA/cm^2, and the FF of 19% respectively. The performance of our device could be enhanced by improving the quality of SWNTs thin film and optimizing the heterojunction structure in further.

ACKNOWLEDGMENT

This work is supported in part by a grant from the National Basic Research Program of China (973 Program) (No. 2009CB320300).

REFERENCES

[1] O. Kluth, A. Loffl, S. Wieder, C. Beneking, W. Appenzeller, L. Houben, B. Rech, H. Wagner, S. Hoffman, R. Waser, J.A.A. Selvan, H. Keppner, "Texture etched Al-doped ZnO: a new material for enhanced

light trapping in thin film solar cells," *26th IEEE PVSC*, pp. 715-718, Oct 1997.

[2] S.H. Demtsu, J.R. Sites, "Quantification of losses in thin-film CdS/CdTe solar cells," *31st IEEE PVSC*, pp. 347-350, Jan 2005.

[3] Y. Jung, S. Chun, D. Kim, J. Kim, "Growth of p-CdTe thinfilms on n-GaN/sapphire," *Journal of Crystal Growth*, vol. 326, Issue 1, pp. 69-72, July 2011.

[4] M. Konagai, S. Miyajima, Y. Yashiki, T. Watahiki, K.L. Narayanan, A. Yamada, "Progress in the development of microcrystalline 3C-SiC, SiGeC and GeC thin films for solar cell applications," *31st IEEE PVSC*, pp. 1424-1427, Jan 2005.

[5] P. Martyniuk, A. Rogalski, "Quantum-dot infrared photodetectors: Status and outlook," *Quantum Electronics*, vol.32, Issues 3-4, pp. 89-120, Aug 2008.

[6] K.R. Catchpole, K.L. Lin, M.A. Green, A.G. Aberle, R. Corkish, J. Zhao, A. Wang, "Thin semiconducting layers and nanostructures as active and passive emitters for thermophotonics and thermophotovoltaics," *Physica E: Low-dimensional Systmes and Nanostructures*, vol. 14, Issues 1-2, pp. 91-95, April 2002.

[7] W.A. Tisdale, K. J. Williams and B.A. Timp, D.J. Norris, E.S. Aydil, X.Y. Zhu, "Hot-electron transfer from semiconductor nanocrystals," *Science*, vol. 328, pp. 1543–1547, Jun 2010.

[8] M.S. Fuhrer, B.M. Kim, T. Durkop and T. Brintlinger, "High-mobility nanotube transistor memory," *Nano Lett*, vol.2, Issue 7, pp. 755-759, May 2002.

[9] C. Chen, L. Yang, K. Eric, Y. Zhang and S. Lee, "Nanowelded carbon-nanotube-based solar microcells," *Small*, vol.4, Issue 9, pp. 1313-1318, Sep 2008.

[10] Y. Jia, A. Cao, X. Bai, Z. Li, L. Zhang and D. Wu, "Achieving high efficiency silicon-carbon nanotube heterojunction solar cells by acid doping," *Nano Lett*, vol.11, Issue 5, pp. 1901-1905, Mar 2011.

Investigation of Particle Dispersion and Deposition in a Channel with Elliptic Obstructions using Lattice Boltzmann Method

A. Tehrani[*], A. Moosavi

[*]Department of Mechanical Engineering, Sharif University of Technology, IRAN

moosavi@sharif.edu

Abstract—**Particle transport and deposition in a channel flow with elliptic obstruction is studied. Numerical simulation of fluid flow is performed using two-dimensional lattice Boltzmann method, while one-way coupling Lagrangian method for particle tracking is used. Standard particles are injected in the inlet of the channel. Gravity, Drag force, Brownian forces, and the Saffman lift are considered in equation of particle motion. The influence of geometrical parameter, ellipse aspect ratio, is studied on dispersion and deposition of particles as well as the flow parameters, such as Reynolds number. In addition, the effect of particles size -particles of 0.01-10μm in diameter- on dispersion and deposition patterns is studied. Results reveal more inertial and gravitational trapping on obstacle surface for particles of larger diameter and obstacles of higher aspect ratio. In addition, the Brownian diffusion and the vortexes formed behind the obstacle, which occurs in high Reynolds numbers, dominantly affect finer particles trajectories.**

Keywords-lattice Boltzmann method; one-way coupling Lagrangian particle tracking; Brownian diffusion

I. INTRODUCTION

Investigation of transport and deposition of aerosol particles have attracted interests in many fields of sciences such as therapeutic drug delivery, aerosol particles filtration, clean rooms, and micro-contamination control [1]. Patterns of particles deposition and dispersion as well as their deposition efficiency and trajectories are of great importance and can be used in order to optimize designs in such areas of work [1].

Numerous experimental and numerical studies have been performed on transport and deposition of aerosol particles in channel flows with obstructions of different geometries such as cylinder, sphere, and square [2-4]. In these studies, Brownian force has been rarely taken into account [3, 4]. In addition, the effect of vortex shedding downstream of the obstacle on particles deposition and dispersion has recently been studied in a channel flow with square obstruction by [4].

Accuracy of particles trajectories can be affected considerably by results obtained from fluid flow numerical simulation. In the present study, numerical approach used for fluid flow simulation is two-dimensional lattice Boltzmann method (LBM) which is now widely employed for various types of flows [5] including laminar and turbulent flows [5, 6], non-Newtonian fluids [7, 8], and micro- and nanoscale flows [9]. This method provides an accurate fluid flow simulation which is needed for particle tracking. The inlet and outlet boundary conditions implementation has been developed so that numerical simulations are facilitated while the accuracy of the solution and its consistency are enhanced [9]. In addition, curved wall boundary conditions implementations are improved [11, 12].

In the present work, the influence of fluid flow properties, i.e. Reynolds number, as well as the particle size and the obstruction geometry (ellipse aspect ratio) is studied on the dispersion and deposition of aerosol particles in a channel flow with elliptic obstructions of various aspect ratios. The effect of vortex shedding downstream of the obstacle which occurs in high Reynolds numbers and Brownian motion which is important for ultrafine particles are investigated.

II. METHODOLOGY

A. Fluid Flow Simulation by LBM

The two-dimensional lattice Boltzmann method was used for fluid flow simulations. In LBM, the fluid flow is interpreted in terms of $f_i(x, t)$ which is particle density distribution as a function of position, x, and time, t:

$$f_i = f_i(x,t) \quad , i = 0,1,...,M , \qquad (1)$$

where M is the number of directions of nonzero discrete velocities in a lattice. In the present study, D2Q9 model of LBM is used, representing two-dimensional model with nine discrete velocities, i.e. $M=8$ (as shown in Fig. 1).

Lattice Boltzmann equation (LBE) is the discretized form of Boltzmann equation in which the collision term can be approximated by several methods. Among these methods, the linear collision operator proposed by the Bhatnagar-Gross-Krook (BGK) is now widely-used and reported to be efficient [6]:

$$f(x + c\Delta t, t + \Delta t) = f(x,t) - \left[f(x,t) - f^{eq}(x,t)\right]/\tau , \qquad (2)$$

where τ is the dimensionless relaxation time which controls the rate of approaching local equilibrium state (f_i), in other words, τ is a representation of the viscosity of the fluid. The relation between τ and viscosity and the equilibrium distribution function in D2Q9 model are given as:

978-1-4673-1122-9/12 $31.00 © 2012 IEEE

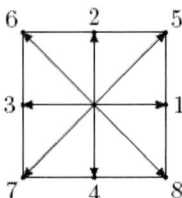

Fig. 1. A D2Q9 lattice, discrete velocities are shown as vectors.

$$v = (2\tau - 1)/6, \tag{3}$$

$$f_i^{eq} = \rho \omega_i \left[1 + 3e_i.u + 4.5(e_i.u)^2 - 1.5u^2 \right], \tag{4}$$

with:

$$\omega_0 = 4/9, \omega_1 = \omega_2 = \omega_3 = \omega_4 = 1/9, \omega_5 = \omega_6 = \omega_7 = \omega_8 = 1/36, \tag{5}$$

and e_i are local particle velocities defined as:

$$e_i = \begin{cases} 0 & \text{for } i = 0 \\ (\cos((i-1)\pi/4), \sin((i-1)\pi/4)) & \text{for } i = 1,2,3,4 \\ \sqrt{2}(\cos((i-1)\pi/4), \sin((i-1)\pi/4)) & \text{for } i = 5,6,7,8 \end{cases} \tag{6}$$

Starting by discrete LBE, LBGK equation (2), and applying Chapman–Enskog procedure, the incompressible Navier–Stokes equations can be obtained with the kinematic viscosity defined by (3).

B. Boundary Conditions

Inlet and outlet boundary conditions are implemented using a method proposed by [9]. Velocity inlet with parabolic profile is used for inlet of the channel. Pressure outlet is applied for outlet of the channel. Regular bounce-back boundary condition for channel walls are used which is reported as a consistent and accurate method. Boundary condition on the surface of the obstacle is implemented by method developed by [10, 11] for curved walls.

C. Particle Equation of Motion

Aerosol particles injected to the fluid domain are standard particles, namely, particles of spherical shape with liquid water density (998 kg/m³), which expeience these four forces: drag force, gravity, Brownian diffusion and Saffman lift. The most important of which is drag force causing particles to follow fluid flow streamlines. Gravity is important for particles of larger diameter but its effect is negligible for finer particles. On the other hand, Brownian force considerably affects ultrafine particles motion and has negligible influence on larger particles trajectories. Thus, particles equation of motion can be written as,

$$du_i/dt = F_d/m + g_i + F_{Saffmann}/m + n_i \tag{7}$$

where u is the velocity of the particle, t is the time, m is particle mass, g_i is gravity acceleration in i-th direction, F_d is drag force, $F_{Saffman}$ is Saffman lift, and n_i is Brownian acceleration. Drag force, and Saffman lift are given as,

$$F_d = (u^f - u^p)/\tau, \tag{8}$$

and

$$F_{L(Saffman)} = 1.615\rho v^{1/2} d_p^2 (u^f - u^p) |du^f/dy|^{1/2} \operatorname{sgn}(du^f/dy), \tag{9}$$

where u^f is the fluid velocity, u^p is the particle velocity, ρ and v are fluid density and viscosity, and τ is particle relaxation time defined as follows:

$$\tau = C_c \rho_p d_p^2 /(18\mu), \tag{10}$$

and

$$C_c = 1 + (2\lambda/d)(1.257 + 0.4e^{-1.1d/2\lambda}) \tag{11}$$

where d_p is the particle diameter, λ is molecular mean free path of the gas, and C_c is Stokes-Cunningham correction factor [1].

Brownian force simulation is performed using Gaussian white noise random process. The more complete formulation is presented in [13, 14].

D. Simulation Parameters

Simulations are performed for Re=1, 10, 50, 100, 200. The fluid is taken to be air at T=20°C:

$$\mu = 1.81 \times 10^{-5} \, kg/(m.s)$$
$$\rho = 1.225 \, kg/m^3 \tag{12}$$

In calculating Reynolds number, vertical diameter of the ellipse is taken to be the characteristic length because it represents the area obstructed in the channel, and the maximum velocity in the inlet of the channel is considered as characteristic velocity.

In each Reynolds number, 10 aspect ratios for ellipses have been studied. In each case particles of 0.01-10 µm in diameter are injected in channel inlet. For each particle diameter, 100 particles of same diameter are injected in various initial positions in channel inlet. Due to the large amount of data and plots, selected results are presented in order to demonstrate effective parameters.

III. RESULTS

It has been reported that for channel flow with circular cylinder obstruction, vortex shedding starts to appear in approximately Re=90 [15]. In present study, steady particle tracking simulations are performed for Re=1, 10, 50, and simultaneous particle tracking is performed for Re=100, 200 to consider the effect of unsteady vortices shaped downstream of the obstacle. Figs. 2-8 show the velocity magnitude contours and streamlines of the solution obtained by LBM.

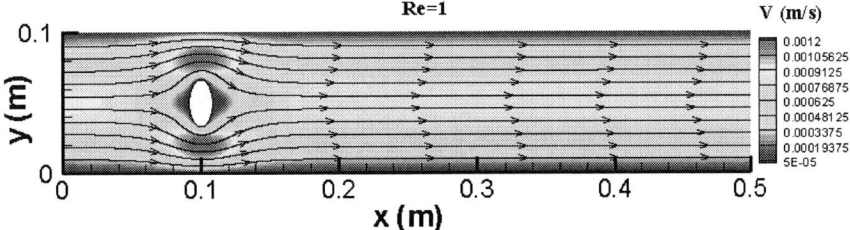

Fig. 2. Velocity magnitude and streamlines for *Re*=1, *a/b*=2.

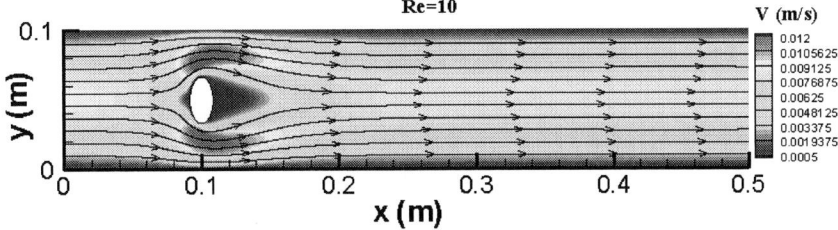

Fig. 3. Velocity magnitude and streamlines for *Re*=10, a/b=2.

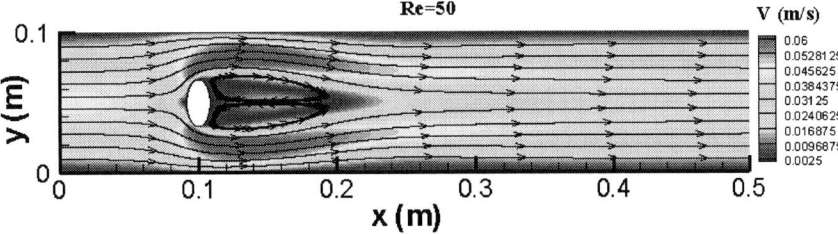

Fig. 4. Velocity magnitude and streamlines for *Re*=50, *a/b*=2. Steady vortices formed behind the obstacle.

Results obtained for particle trajectories indicate that in lower Reynolds numbers, particles follow fluid flow streamlines so that they do not deposit on obstacle surface due to inertial trapping. However, gravitational trapping is the dominant trapping process because of lower velocities. In other words, due to low velocities of the fluid, particles are suspended in the channel for considerable time before they leave the channel, thus, gravity effect is observable for particles of larger diameter (d_p=10 µm). As the aspect ratio of the ellipse increases, i.e. the obstruction area of the channel, particle deposition on the obstacle surface increases because the higher the aspect ratio is, the more accelerates the fluid in throats of the channel, and hence in upstream of the obstacle (Fig. 10). By decreasing the aspect ratio, the streamlines are less deviated and particles of larger diameter drop on the bottom wall of the channel without facing the obstacle. As far as the gravity and inertial effects are negligible for finer particles (d_p= 1, 0.1, 0.01 µm) they perfectly follow fluid flow streamlines and leave the channel with no deposition, showing no interesting result (Fig.9). The plots of 10nm and 100nm particles trajectories are not presented for this reason.

On the other hand, in higher Reynolds numbers it is expected that inertial trapping increases, especially for larger particles, but it is not always for sure. Results for *Re*=50 (Fig. 12, 13) indicate that particles deposition on the obstacle surface increases by decreasing the ellipse aspect ratio. It can

be justified for two reasons: first, as the horizontal area of the obstacle increases gravitational deposition on its surface increases and, as a result, total deposition increases. Second, in lower aspect ratios for the ellipse streamlines are less deviated and particles are more likely to pass near the obstacle so the interception and diffusion trapping mechanisms occur more frequently (Fig. 11).

In higher Reynolds numbers vortices formed behind the obstacle affect finer particles trajectories. This may result in suspension of the particles in wakes formed behind the obstacle (Fig. 14). In addition to vortices, Brownian diffusion has a great influence on ultrafine particles motion for which gravity force is negligible.

It has also been observed that obstacle geometrical characteristics, ellipse aspect ratio, influence greatly the particles deposition trends. For high aspect ratios, the projection surface obstructed in the channel increases which may result in more inertial trapping but this phenomenon does not take place, as mentioned. Wakes formed behind the obstacle grows as the aspect ratio increases: as a result finer particles are trapped in these wakes. In contrary, lower aspect ratios may cause the vortices to grow less or even disappear (Fig. 6). However, gravitational trapping increases when horizontal surface of the obstacle increases, i.e. in lower aspect ratios.

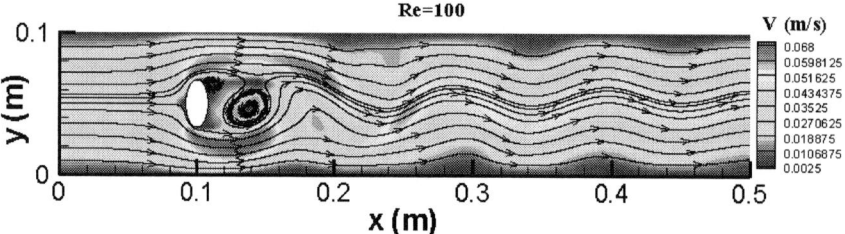

Fig. 5. Velocity magnitude and streamlines for *Re*=100, *a/b*=2. Unsteady vortices formed behind the obstacle.

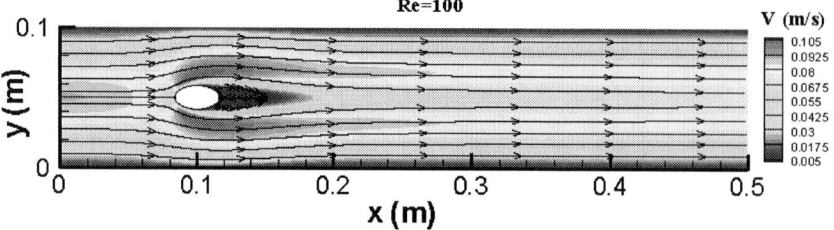

Fig. 6. Velocity magnitude and streamlines for *Re*=100, *a/b*=0.5. Vortices formed behind the obstacle are steady because the aspect ratio of the ellipse is low, i.e. the ellipse cross less streamlines in comparison with Fig. 6, hence perturbation made downstream is not large enough to make vortices unsteady.

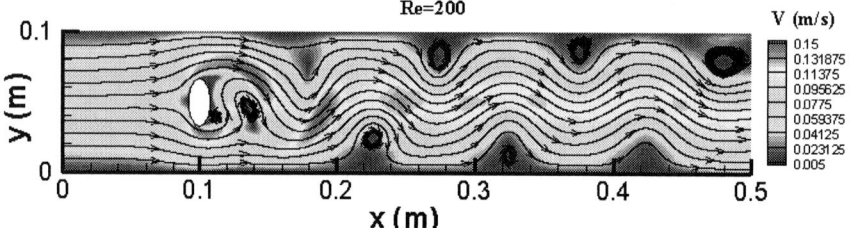

Fig. 7. Velocity magnitude and streamlines for *Re*=200, *a/b*=2.

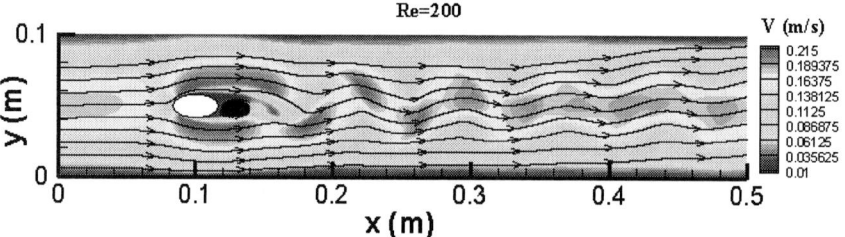

Fig. 8. Velocity magnitude and streamlines for *Re*=200, *a/b*=0.5. Although the aspect ratio is low, Reynolds number is so high that perturbations lead in unsteady vortices.

IV. CONCLUSION

LBM is used for fluid flow simulations coupled with Lagrangian point of view for particle tracking. Based on the results reported, following conclusions can be pointed out:

Particles of large diameters are highly influenced by drag force and gravity. Impaction, interception, and gravitational trapping mechanism result in deposition of particles on the obstacle. Ultrafine particles are so mobile that follow fluid streamlines perfectly and are affected mainly by drag and Brownian forces. For nano-particles, gravity and inertial effects are negligible and diffusion trapping is the dominant mechanism. As well, they may trap in the wakes formed behind the obstacle.

In low Reynolds numbers, fine particles follow fluid streamlines and rarely deposit on obstacle surface, while larger particles motion are so slow that they are greatly affected by gravity.

Higher aspect ratios of ellipse, which represents more obstruction area, induces more deviations in streamlines and hence more inertial trapping on obstacle surface particles of 10 μm in diameter. Lower aspect ratios, when the obstacle is more flat, result in increase in trapping by gravitational and interception mechanisms.

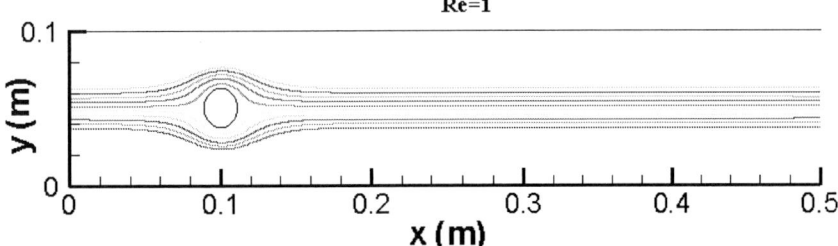

Fig. 9. *Re*=1, *a/b*=1.125, number of particles=10, d_p=1 micron. Particles trajectories are perfectly lie on streamlines. Trajectories of 10 nm and 100 nm diameter particles are totally same as trajectories above and there is no observable difference among them.

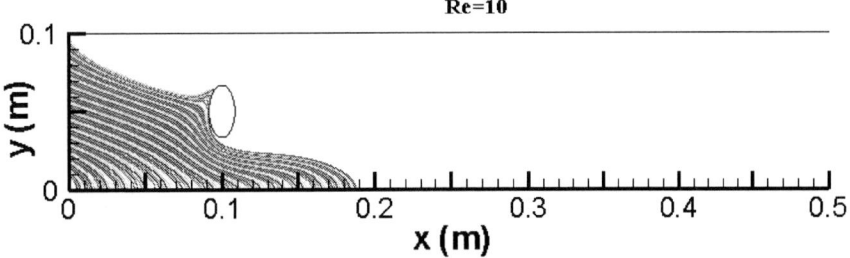

Fig. 10. *Re*=10, *a/b*=2, number of particles=100, d_p=10 micron. Deposition percentage on the obstacle is 9%. The deposition percentage reduces to zero for a/b=0.5.

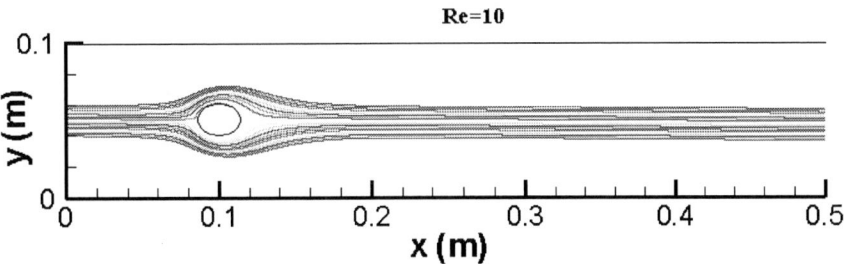

Fig. 11. *Re*=10, *a/b*=0.77, number of particles=21, d_p=1 micron. Particles are injected in the projected area of obstruction in the channel inlet. Deposition percentage on the obstacle is approximately 5%. It is noteworthy that 1 micron particles are mobile enough to not trap on the obstacle except the one which is trapped by interception mechanism.

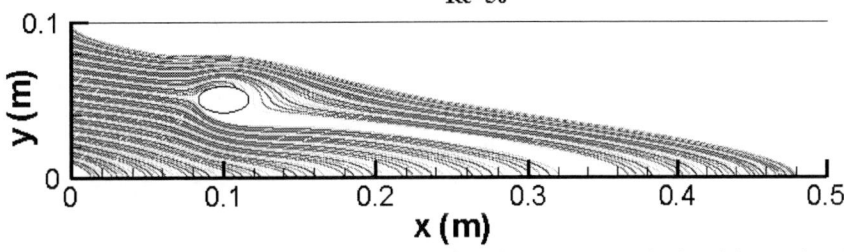

Fig. 12. *Re*=50, *a/b*=0.5, number of particles=100, d_p=10 micron. Deposition percentage on the obstacle is approximately 6%. Particles of 10 micron are deposited by gravitational trapping mechanism. As the horizontal area increases, the number of particles trapped on the obstacle due to gravitational mechanism increases.

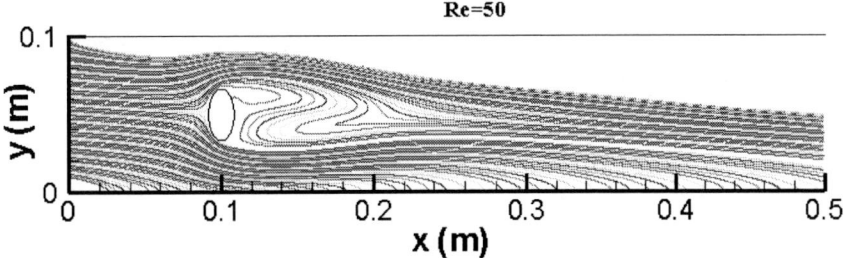

Fig. 13. *Re*=50, *a/b*=2, number of particles=100, d_p=10 micron. Deposition percentage on the obstacle is approximately 2%. It is unexpected that increase of obstruction area results in increase in number of particles escaping in the channel outlet. Another surprising result is that impaction deposition which is expected to be increased in higher aspect ratios is reduced.

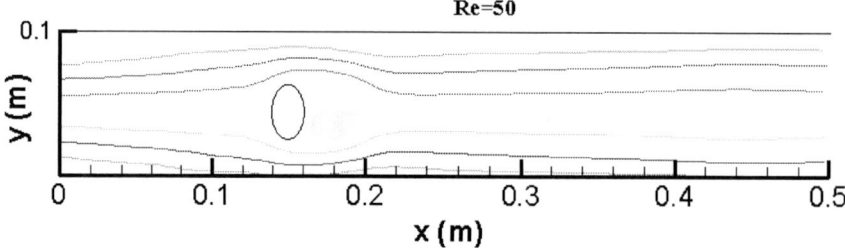

Fig. 14. *Re*=50, *a/b*=2, number of particles=10, d_p=100 nm. Particle injected near the midline of the channel is trapped in the wake formed behind the obstacle. It can be deduced that its trajectory deviation toward the wake is due to Brownian motion.

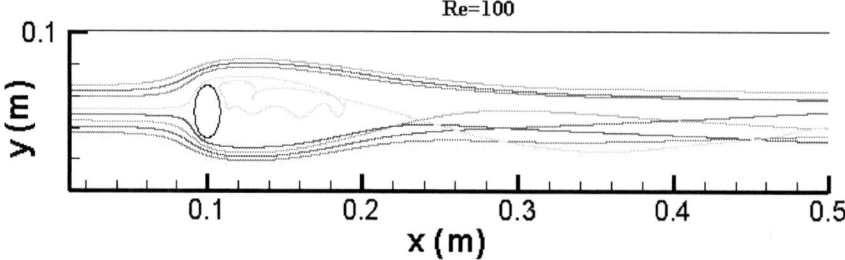

Fig. 15. *Re*=50, *a/b*=2, number of particles=100, d_p=1 micron. Particles trajectories are deviated by unsteady vortices downstream of the obstacle and Brownian diffusion.

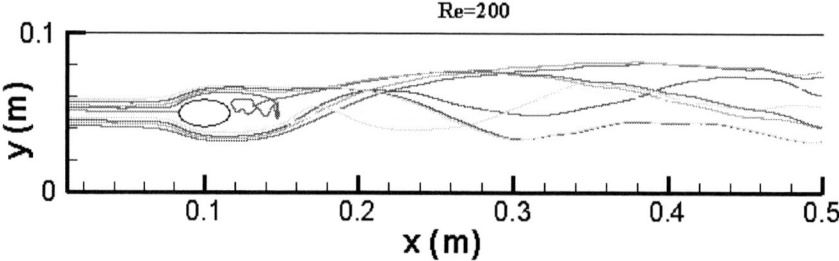

Fig. 16. *Re*=200, *a/b*=0.5, number of particles=10, d_p=10 nm. Particles are injected in the projected area of obstruction in the channel inlet. Particles injected near the midline of the channel can be trapped. Brownian motion near the obstacle resulted in the particle deposition in this case. High Reynolds number and flat obstacle which causes the streamlines deviate smoothly are not affectless. One particle is trapped in the wake formed behind the obstacle. Brownian motion is clearly observable for this particle.

REFERENCES

[1] W.C. Hinds, *Aerosol Technology: Properties, Behavior, and Measurement of Airborne Particles*. 2nd ed., John Wiley & Sons, 1999.

[2] K. R. May, R. Clifford, "The impaction of aerosol particle on cylinder, sphere, ribbons and disks," *Ann. Occup. Hyg.*, vol. 10, pp. 83-95, 1967.

[3] D.J. Brandon, S.K. Aggarwal, "A numerical investigation of particle deposition on a square cylinder placed in a channel flow," *Aerosol. Sci. Tech.*, vol. 34, pp. 340-352, 2001.

[4] S. Jafari, M. Salmanzadeh, M. Rahnama, G. Ahmadi, "Investigation of particle dispersion and deposition in a channel with a square cylinder obstruction using the lattice Boltzmann method," *J. Aerosol. Sci.*, vol. 41, pp. 198-206, 2010.

[5] S. Chen, G.D. Doolen, "Lattice Boltzmann method for fluid flows," *Ann. Rev. Fluid. Mech.*, vol. 30, pp. 329-364, 1998.

[6] S. Succi, *The Lattice Boltzmann Equation for Fluid Dynamics and Beyond*. Oxford: Oxford University, 2001.

[7] S. Gabbanelli, G. Drazer, J. Koplik, "Lattice Boltzmann method for non-Newtonian (power-law) fluids," *Phys. Rev. E.*, vol. 72, 2005.

[8] A. Nejat, V. Abdollahi, K. Vahidkhah, "Lattice Boltzmann simulation of non-Newtonian flows past confined cylinders," *J. non-Newton. Fluid*, vol. 166, pp. 689-697, 2011.

[9] Y. Zhou, R. Zhang, I. Staroselsky, H. Chen, W.T. Kim, M.S. Jhon, "Simulation of micro- and nano-scaled flows via lattice Boltzmann method," *Physica A*, vol. 362, pp. 68-77, 2006.

[10] Q. Zou, X. He, "On pressure and velocity boundary conditions and bounce-back for the lattice Boltamann BGK model," *Phys. Fluids. 9*, vol. 21, pp. 1591-1598, 1997.

[11] O. Filippova, D. Hanel, "Grid refinement for lattice-BGK models," *J. Compt. Phys.*, vol. 147, pp. 219-228, 1998.

[12] R. Mei, L.S. Luo, W. Shyy, "An accurate curved boundary treatment in the lattice Boltzmann method," *J. Comput. Phys.*, vol. 155, pp. 307-329, 1999.

[13] A. Li, G. Ahmadi, "Dispersion and deposition of sphere particles from point source in a turbulent channel flow," *Aerosol. Sci. Tech.*, vol. 16, pp. 209-226.

[14] A. Li, G. Ahmadi, "Deposition of aerosols on surfaces in a turbulent channel flow," *Int. J. Eng. Sci.*, vol. 31, pp. 435-451.

[15] D.J. Tritton, "Experiments on the fllow past a circular cylinder at low Reynolds numbers," *J. Fluid. Mech.*, vol. 6, 1959.

A Study of Hydrogen Peroxide Microfluidic Fuel Cells

J. C. Shyu[1]*, C. L. Huang[1], T. S. Sheu[2], H. Ay[3], J. W. Huang[4]

[1]Department of Mechanical Engineering/National Kaohsiung University of Applied Sciences, Kaohsiung, Taiwan
[2]Department of Mechanical Engineering/R. O. C. Military Academy, Kaohsiung, Taiwan
[3]Department of Mold and Die Engineering/National Kaohsiung University of Applied Sciences, Kaohsiung, Taiwan
[4]Department of Physics/R .O. C Military Academy, Kaohsiung, Taiwan

Abstract—This study investigated various effects, including reactant concentrations, volumetric flow rates and microchannel width, as well as the electrode distance, on the performance of microfluidic fuel cells employing hydrogen peroxide dissolved in alkaline and acid electrolytes as both fuel and oxidant, respectively. Three concentrations ranging from 0.1 M to 0.6 M and five volumetric flow rates ranging from 0.01 mL/min to 1.0 mL/min were tested in the present study for cell performance measurement and discussion. Three microfluidic fuel cells were tested here. Two of them have rectangular microchannel of 0.5 mm and 1.0 mm in width with electrode distance of 0.4 mm. An additional 0.5-mm-wide microchannel fuel cell was also tested with a shorter electrode distance of 0.2 mm. Results show that cell performed at either larger volumetric flow or with smaller microchannel width usually had higher current output at a given cell voltage. The highest cell output at 0.1 V and 0.1 M among the present cells was approximately 100 mA/cm^2 produced by the cell whose microchannel width and electrode distance are 0.5 mm and 0.2 mm, respectively. However, with a higher reactant concentration of 0.6 M, the highest cell output at 0.1 V and 0.1 M among the present cells was 2.5 times higher than the abovementioned value, namely 250 mA/cm^2, produced by the cell with microchannel width and electrode distance of 0.5 mm and 0.4 mm, respectively.

Keywords-microfluidic; fuel cell; hydrogen peroxide; bubble

I. INTRODUCTION

Fuel cells are considered as one of the promising power sources for numerous applications such as microelectronics device, several kW backup power units and automotive industry. One of the main concerns hindering fuel cells from popularizing can be its durability that mainly arises from the degradation of polymer electrolyte membrane. Fortunately, due to the advance of the MEMS technology, fabricating miniature fuel cells which are capable of generating electricity without polymer electrolyte membrane becomes feasible. In recent years, numerous types of such microfluidic fuel cells employing various combinations of fuel and oxidant with novel microchannel designs have been presented [1]. Among those microfluidic fuel cells, methanol and formic acid are two commonly used aqueous fuels that emit carbon dioxide at the anode during electricity generation process. Instead of CO_2 emission [2-4], one particular kind of microfluidic fuel cells that emits oxygen gas at anode was proposed by Hasegawa et al. [5]. This cell, which is briefly called as direct hydrogen

peroxide microfluidic fuel cell (DHPFC), generates electricity by pumping two hydrogen peroxide streams mixed with alkaline and acid electrolytes as fuel and oxidant, respectively, into a microchannel deposited with platinum catalyst. A fairly high cell output, i.e. 78 mA/cm^2 and 23 mW/cm^2 at 0.3 V, with $[H_2O_2]$ = 0.75 M was reported. Since no hydrocarbon was used as reactant, no CO_2 was produced during the electric generation process. However, the so-called direct H_2O_2 fuel cell was tested only at an extremely large flow rate (~1.5 mL/min) compared with similar cells published in the literature for a unique cell having a 1-mm-wide microchannel with H_2O_2 concentrations ranging from 0.1 M to 0.75 M.

Therefore, the objective of this study is to investigate various effects on the performance of such microfluidic fuel cells. Microfluidic fuel cells having rectangular microchannel of 0.5 mm and 1.0 mm in width were tested in this study at various reactant concentrations (0.1 M ~ 0.6 M) with volumetric flow rates ranging from 0.01 mL/min to 1.0 mL/min to measure and analyze the cell performance.

II. EXPERIMENTS

A schematic of the present experimental setup is shown in Fig. 1. It consists of four major components, including a microfluidic fuel cell with a T-shaped micro-channel, a fluid delivery system, and an electronic load for cell performance measurement.

Microfluidic fuel cells consisting of a T-shaped rectangular microchannel and electrode were fabricated and tested in the present study. The fabrication process of the microfluidic fuel cells is schematically shown in Fig. 2. Firstly, a polydimethylsiloxane (PDMS) based T-shaped microchannel was fabricated using standard soft lithography with photoresist of SU-8 and mold replica technique. Followed by the PDMS microchannel fabrication, lift-off process consisted of patterning photoresist (AZ-4620), and depositing two metal layers (titanium and platinum) by sputtering was implemented to form catalyst and electrodes of the cells on prescribed region of the glass slide. Subsequently, PDMS microchannel activated using oxygen plasma was bonded to the Pt-patterned glass slide with the aid of the alignment mark on both PDMS and glass slide to form the microfluidic fuel cell.

978-1-4673-1122-9/12 $31.00 © 2012 IEEE

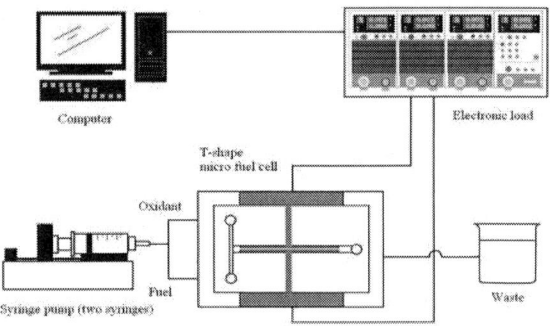

Fig. 1. Schematic of the present experiment setup

Finally, the PDMS microchannel chip of the microfluidic fuel cell was then drilled with three holes to form the aqueous reactant inlets and outlet.

During the experiments, both electrodes were connected to an electronic load (KIKUSUI PLZ-70UA) to measure the voltage-current curve at room temperature and ambient pressure under stepwise potentiostatic control. The electronic load enabled a maximum voltage and current of 150 V and 15 A, respectively. The resolution of voltage and current for the present electronic load are 0.1 mV and 0.01 mA, respectively.

For the fluid delivery system, a syringe pump (KDS 210, KD Scientific Inc.) with two syringes was employed. Polyethylene tubing was used to deliver liquid into the microfluidic fuel cell and to guide the waste stream out of the fuel cell. Both aqueous fuel and oxidant having the same volumetric flow rate were simultaneously pumped into the microfluidic fuel cell via individual inlets. Note that the volumetric flow rate in this study represents the inlet volumetric flow of one single stream, and the total volumetric flow rate in the micro-channel is twice that value.

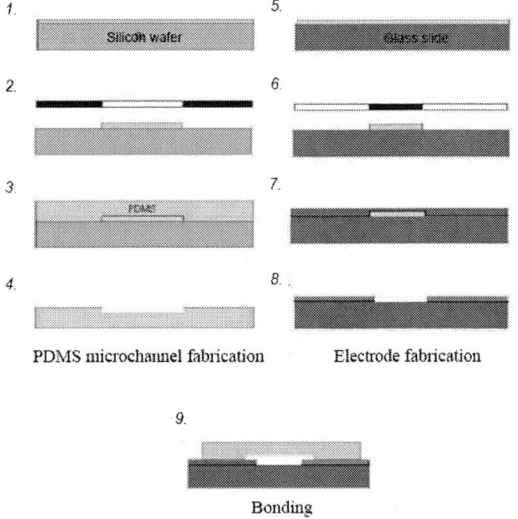

PDMS microchannel fabrication

Electrode fabrication

Bonding

Fig. 2. Fabrication process of the present microfluidic fuel cells

TABLE 1. EXPERIMENTAL CONDITIONS

Reactant	Fuel	Mixture of H_2O_2 and NaOH, $[H_2O_2]:[H_2O_2] = 1:1$
	Oxidant	Mixture of H_2O_2 and H_2SO_4, $[H_2O_2]:[H_2SO_4] = 2:1$
	H_2O_2 concentration	0.1 M, 0.3 M, 0.6 M
Microchannel	Width	0.5 mm, 1.0 mm
	Length	20 mm
	Depth	50 μm
Catalyst / Electrode		Platinum
Electrode distance		0.2 mm, 0.4 mm
Flow rate		0.01, 0.05, 0.1, 0.5, 1.0 mL/min
Reynolds No.		1.0 mm-wide microchannel: 0.74 ~ 73.83 0.5 mm-wide microchannel: 1.41 ~ 140.94

The fuel of the present cell was a mixture of NaOH (SHOWA 96%) solution and H_2O_2 (Merck KGaA 30-35%) solution in the ratio of 1:1, while the oxidant used here was a mixture of H_2SO_4 (Merck KGaA 95-97%) solution and H_2O_2 solution in the ratio of 1:2. The H_2O_2 concentration in both anode and cathode for each test is kept identical. Millipore water (18.3 MΩ-cm, Millipore Milli-Q) was used for the preparation of all aqueous solutions.

Three microfluidic fuel cells having microchannel width of 0.5 mm and 1.0 mm were tested at volumetric flow rates ranging from 0.01 mL/min to 1.0 mL/min in the present study. The distance between two platinum electrodes in a cell is 0.4 mm except one 0.5-mm-wide microfluidic fuel cell, whose electrode distance is 0.2 mm. All experimental conditions are shown in Table 1 in detail.

III. RESULTS AND DISCUSSION

Figures 3 and 4 show both the *V-I* (green) and *P-I* (blue) curves of the present cells having width of 1.0 mm and 0.5 mm, respectively, operated with $[H_2O_2]$ = 0.1 M and volumetric flow rate ranging from 0.01 mL/min to 1.0 mL/min. Note that both cells have identical electrode distance, 0.4 mm. It can be found that the cell performance usually increases

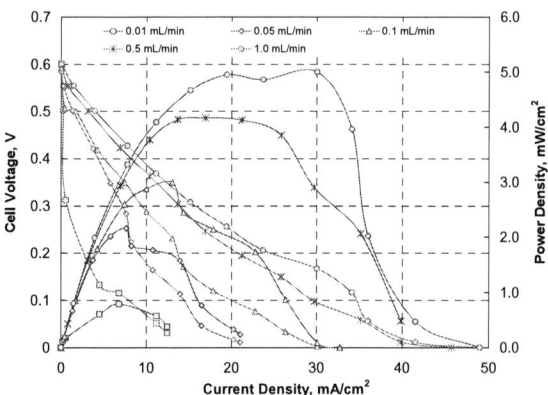

Fig. 3 *V-I* (green) and *P-I* (blue) curves of the present cell having width and electrode distance of 1.0 mm and 0.4 mm, respectively, operated with $[H_2O_2]$ = 0.1 M with volumetric flow rate ranging from 0.01 mL/min to 1.0 mL/min

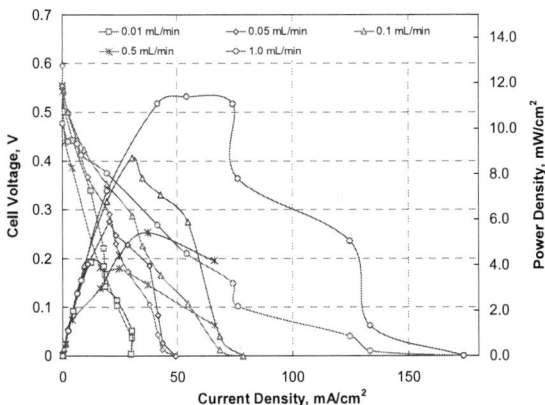

Fig. 4 *V-I* (green) and *P-I* (blue) curves of the present cell having width and electrode distance of 0.5 mm and 0.4 mm, respectively, operated with $[H_2O_2] = 0.1$ M with volumetric flow rate ranging from 0.01 mL/min to 1.0 mL/min

with the increase of the flow rate for each cell. Besides, the cell having smaller width in Fig. 4 outperforms that having larger width in Fig. 3 operated at a given flow rate and H_2O_2 concentration.

One of the reasons attributed to the abovementioned results is because either higher volumetric flow rate at a given microchannel or smaller microchannel at a given flow rate establishes not only a thinner depletion boundary layer over the electrode surface but also a narrower inter-diffusion zone between the interface of the two streams [6] so that the replenishment of the depletion zone over the electrodes can be accelerated and thus generate a higher cell current and power density. The maximum power densities are 5 mW/cm^2 occurring at 1.0 mL/min and 11 mW/cm^2 occurring also at 1.0 mL/min as shown in Figs. 3 and 4, respectively.

Both the *V-I* (green) and *P-I* (blue) curves of the present cell having width and electrode distance of 0.5 mm and 0.2 mm, respectively, operated with $[H_2O_2] = 0.1$ M and volumetric

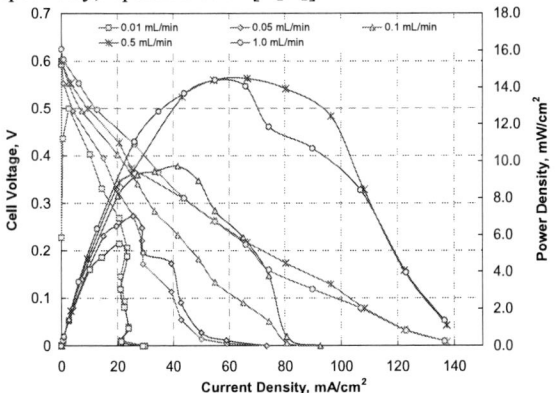

Fig. 5 *V-I* (green) and *P-I* (blue) curves of the present cell having width and electrode distance of 0.5 mm and 0.2 mm, respectively, operated with $[H_2O_2] = 0.1$ M with volumetric flow rate ranging from 0.01 mL/min to 1.0 mL/min

flow rate ranging from 0.01 mL/min to 1.0 mL/min is shown in Fig. 5.

Compared with Fig. 4, it can be observed that the cell having less electrode distance showed a little higher power density output than the cell shown in Fig. 4, since cell having shorter electrode distance could reduce its ohmic overpotential arising from the ion exchange process between both electrodes at a given microchannel width and electrolyte concentration.

Besides, based on the present measurement, cell performance is usually less stable at less flow rate like the curve shown in Fig. 5 at volumetric flow rate of 0.1 mL/min. It could be because the reactant streams in the microchannel were likely to be disturbed by the oxygen bubbles generated during the electricity generation. Such bubble influence would be more pronounced at less flow rate and thus resulted in an unstable cell output. However, such bubble effect on the *V-I* curve is less at larger microchannel width with less volumetric flow rate like the result demonstrated in Fig. 4. It is due to less current density output under such condition generates less oxygen gas [7].

Figure 6 shows the *V-I* curves of the three cells presently tested with reactant concentration of 0.1 M at volumetric flow rates of 0.1 mL/min and 1.0 mL/min. The highest current density output shown in Fig. 6 at 0.2 V and 1.0 mL/min is about 70 mA/cm^2 generated by the cell having channel width of 0.5 mm and electrode distance of 0.2 mm, and the lowest current density output at 0.2 V and 1.0 mL/min is about 25 mA/cm^2 generated by the cell having channel width of 1.0 mm. However, with the increase of the reactant concentration shown in Fig. 7, those cells revealed different cell outputs. Figure 7 shows the *V-I* curves of the three cells tested with reactant concentration of 0.6 M at volumetric flow rates of 0.1 mL/min and 1.0 mL/min. The cell having 1.0-mm-wide microchannel still performed the worst. However, the cell having microchannel width of 0.5 mm and electrode distance of 0.4 mm shows the best performance.

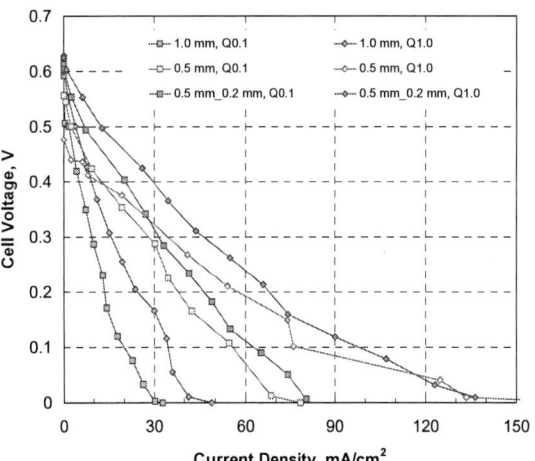

Fig. 6 *V-I* curves of the present cells operated with $[H_2O_2] = 0.1$ M with volumetric flow rates of 0.1 mL/min and 1.0 mL/min

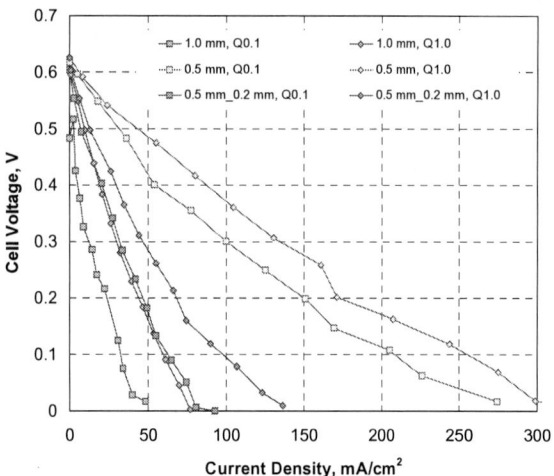

Fig. 7 *V-I* curves of the present cells operated with [H₂O₂] = 0.6 M with volumetric flow rates of 0.1 mL/min and 1.0 mL/min

IV. CONCLUSIONS

This study investigated various effects on the performance of microfluidic fuel cells employing hydrogen peroxide dissolved in alkaline and acid electrolytes as both fuel and oxidant, respectively. Two microfluidic fuel cells having rectangular microchannel of 0.5 mm and 1.0 mm in width with electrode distance of 0.4 mm were tested in this study at various reactant concentrations, 0.1 M, 0.3 M and 0.6 M, with volumetric flow rates ranging from 0.01 mL/min to 1.0 mL/min for cell performance measurement and discussion. Besides, an additional 0.5-mm-wide microchannel fuel cell was also tested with a shorter electrode distance, 0.2 mm.

Results show that cell performed at either larger volumetric flow or with smaller microchannel width usually had higher current output at a given cell voltage. Besides, for cells having 0.5-mm-wide microchannel but with different distance between two electrodes, it can be observed that the cell having shorter electrode distance had higher current density output with reactant concentration of 0.1 M. However, as demonstrated in Fig. 7, with the increase of the reactant concentration, cell having larger electrode distance yielded better cell performance than the one having shorter electrode distance. The highest cell output at 0.1 V and 0.1 M among the present cells was approximately 100 mA/cm² by the cell with microchannel width and electrode distance of 0.5 mm and 0.2 mm, respectively. With a higher concentration of 0.6 M, The highest cell output at 0.1 V and 0.1 M among the present cells was 2.5 times higher than the abovementioned value, namely 250 mA/cm², by the cell with microchannel width and electrode distance of 0.5 mm and 0.4 mm, respectively.

ACKNOWLEDGEMENT

The authors are indebted to National Science Council of Taiwan for financial support under the contracts of NSC 99-2221-E-151-028 and NSC 100-2221-E-151 -066.

REFERENCES

[1] E. Kjeang, N. Djilali, and D. Sinton, "Microfluidic Fuel Cells: A Review", *J. Power Sources*, vol. 186, pp. 353–369, 2009.

[2] H. B. Park, D. H. Ahmed, K. H. Lee, H. J. Sung, "An H-shaped design for membraneless micro fuel cells." *Electrochimica Acta*, vol. 54, pp. 4416–4425, 2009.

[3] R. S. Jayashree, S. K. Yoon, F. R. Brushett, P. O. Lopez-Montesinos, D. Natarajan, L. J. Markoski, P. J. A. Kenis, "On the performance of membraneless laminar flow-based fuel cells." *J. Power Sources*, vol. 195, pp. 3569-3578, 2010.

[4] E. Kjeang, A. G. Brolo, D. A. Harrington, N. Djilali, and D. Sintona, "Hydrogen Peroxide as an Oxidant for Microfluidic Fuel Cells", *J. Electrochem. Soc.*, vol. 154, pp. B1220–B1226, 2007.

[5] S. Hasegawa, K. Shimotani, K. Kishi, and H. Watanabe, "Electricity Generation from Decomposition of Hydrogen Peroxide", *Electrochem. Solid-State Lett.*, vol. 8, pp. A119–A121, 2005. [6] A. Li, S. H. Chan and N. Nguyen, "A laser-micromachined polymeric membraneless fuel cell", *J. Micromech. Microeng.*, vol. 17, pp. 1107–1113, 2007.

[6] S. A. M. Shaegh, N.-T. Nguyen, S. H. Chan, "A review on membraneless laminar flow-based fuel cells", *Int. J. Hydrogen Energy*, vol. 36, pp. 5675-5694, 2011.

[7] J.-C. Shyu and C.-L. Huang, "Characterization of bubble formation in microfluidic fuel cells employing hydrogen peroxide", *J. Power Sources*, vol. 196, pp. 3233–3238, 2011.

Microfluidic Circulatory System for the Raise of Liver Urea Assay

Yu-Shih Chen[1*], Tzu-Hsuan Dai[2], Ling-Yi Ke[1], and Cheng-Hsien Liu[1,2]

[1] Institute of NanoEngineering and MicroSystems, National Tsing Hua University, Hsinchu, Taiwan.
[2] Department of Power Mechanical Engineering, National Tsing Hua University, Hsinchu, Taiwan.
*ericys2004@gmail.com

Abstract— In this research, we propose an in vitro liver circulatory system via the concept of cell pattern technology and microfluidic system. Liver lobule structure in our body consisted of several blood vessel and hepatic tissue. The fluid direction of blood would get in from hepatic portal vein, liver lobule tissue to the hepatic central vein. Hence, in order to mimic the liver circular system in vitro, a connected microchannel system is proposed in this research. This connected system consisted of three parts of channel design, blood vessel chamber, Lobule-flow-mimic channel and liver tissue chamber. Liver circulatory system chip is consisted of blood vessel chamber and liver tissue chamber via dielectrophoresis pattern method and microfluidic channel design.

Keywords- Microfluidics Circulatory System; cell pattern; Liver Urea Assay

I. INTRODUCTION

The liver plays a major role in metabolism and detoxification. This significant organ consists of many sorts of cells proportionally. The proliferation and migration of hepatocytes would be experienced by culturing with hepatic stellate cell was proposed earlier [1, 2]. Different cell ratio in co-culture environment would influence the ability of each cell growth. The secretion from the cell would enhance the growth of neighbor cells.

Researchers desire to find a way to construct bio-tissue in vitro to mimic the real tissue in vivo and utilize the engineered tissue in the biomedical field. For example, reducing the need for organ replacement, accelerating the development of new drugs, or even replacing the need of organ transplants altogether by artificial organ [3]. For mimicking the real tissue in our body, many researchers in the field of tissue engineering focus on how to construct a three dimension tissue structure. They must find a solution to support cell growth in vertical direction that cannot use the traditional culture method. One of the challenges in realizing tissue engineering is the reasonable design of scaffolds that mimic the autologous extracellular matrix (ECM). Those scaffolds provided a temporary mechanical support and suitable architecture to reconstruct three dimensional tissue structures [4]. The ability to arrange heterogeneous cells onto ordered tissue-mimetic patterns not only benefits the studies of how cells interacted with other cells or its microenvironment [5-7] but also promotes the developments of engineered complex tissue reconstruction, which highly rely on the cell patterning techniques.

This microchip mimic cross talk between cells in different culture chambers connected only by flow, such that cell–cell interaction is mediated by soluble ligands as occurs in the body. [8] Use this concept to design the circulatory system for drug studying. Hence, the interaction of cell and cell could be studied via connected chamber instead of using cell co-culture. Those biotechnological tools has been developed using biomedical engineering concepts based on biomimetics, which is composed of cell culture chambers, connected channel and in parallel to form a closed loop biological system in vitro.

II. METHODS

A. Microfluidic Circulatory System

This connected system consisted of three parts of channel design, blood vessel chamber, Lobule-flow-mimic channel and liver tissue chamber as shown in Figure 1.

Hence, the volumetric flow rate in all six microchannels is similar if the branching microchannels are of the same cross sectional area and length. Figure 2 shows on-chip experimental setting process of urea assay. Step I: HMEC-1 was patterned as the first pattern and C2C12 was patterned as the second pattern. Then, at step II, change the outlet/inlet setting of chip, the HepG2 and 3T3 was patterned into the liver-lobule-mimetic chamber. Step III, change the outlet/inlet again and pump fresh medium that flow in from the first inlet and flow out from the central outlet of liver-lobule-mimetic chamber for overnight. By this developed Labchip, the fluid direction and volumetric flow rate could be controlled well in all the branching lobule-flow-mimic channel.

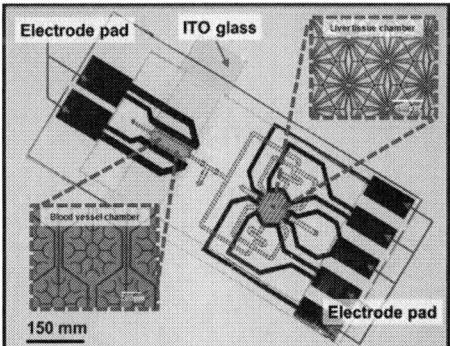

Figure 1: Liver circulatory system chip. It consisted of blood vessel chamber and liver tissue chamber via dielectrophoresis pattern method and microfluidic channel design.

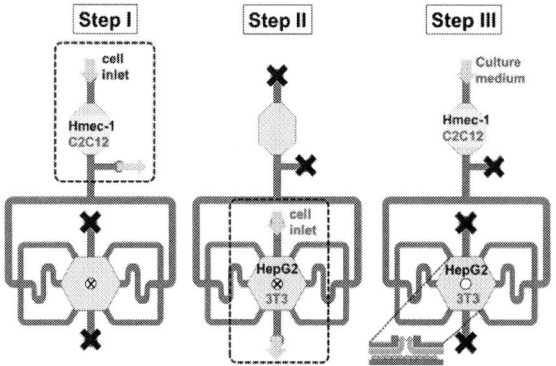

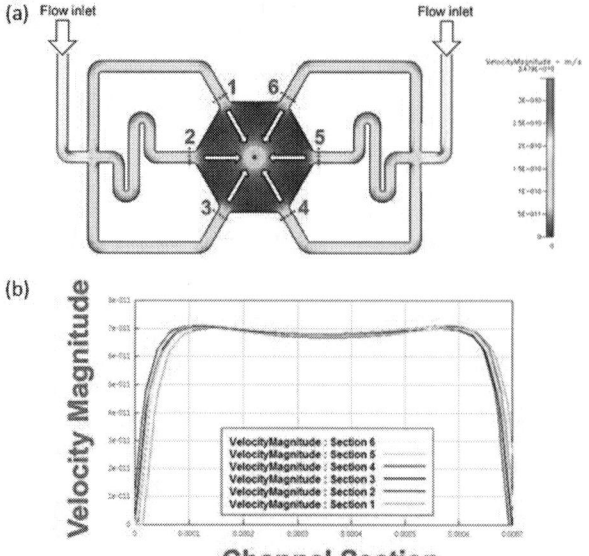

Figure 2: On-chip experimental setting process of urea assay. Step I: HMEC-1 was patterned as the first pattern and C2C12 was patterned as the second pattern. Then, at step II, change the outlet/inlet setting of chip, the HepG2 and 3T3 was patterned into the liver-lobule-mimetic chamber. Step III, change the outlet/inlet again and pump fresh medium that flow in from the first inlet and flow out from the central outlet of liver-lobule-mimetic chamber for overnight.

Figure 4: (a) CFD-ACE simulation results of flow velocity distribution for a microfluidic system. (b) The simulation results for the flow velocity profile of branching channel across the red dotted line 1, 2, 3, 4, 5, 6 which indicated the correspond positions in Figure 4(a).

B. Simulation

Sequence image of circulatory system simulation result in figure 3, medium pumped from the upstream inlet, through the first chamber, lobule-flow-mimic channel, and flow out from the central outlet of liver-lobule-mimetic chamber.

By the way, CFD-ACE simulation results of flow velocity distribution for a microfluidic system with a flow rate of 0.4 mm/sec at the two upstream inlets in figure 4(a). Figure 4(b) shows the simulation results for the flow velocity profile of branching channel across the red dotted line 1, 2, 3, 4, 5, 6 which indicated the correspond positions in Figure 4(a). For the step I, on-chip cell-patterning result of HMEC-1 by green dye and C2C12 by red dye cells in Figure 5. After pattern cells, C2C12 was patterned on the snow-flake-like electrode via vertical negative DEP manipulation. Then DEP buffer replaced with the fresh culture medium. For the step II, we patterned cells in liver tissue chamber by DEP force.

And the fluid direction from branching channel to the central outlet is simulated as shown in Figure 3, which shows the fluid direction directing towards the center.

III. RESULTS AND DISCUSSION

As illustrated in Figure 7, the cell ratio could be defined via the size of electrode. The number of trapped cell was fixed according to the electrode area. For confirming the cell ratio, we separate the electrode to eight equal shape and count the number of each cell. Because the chamber was full of snow-flake-like electrode, we can easily get the cell ratio of endothelial cell to muscle cell in the whole chamber by counting the cell number of any equal shapes. In this case, there are 18 endothelial cells and 75 muscle cells in one equal shape. It means that the ratio of endothelial cell to muscle cell is 1 to 4 via this electrode design.

After patterning the HMEC-1 cell, C2C12, muscle cells with the size of about 10~15 μm in diameter, were patterned second in the ratio control chip. C2C12 were drawn into the Labchip with the DEP buffer (the cell concentration is 1×10^7

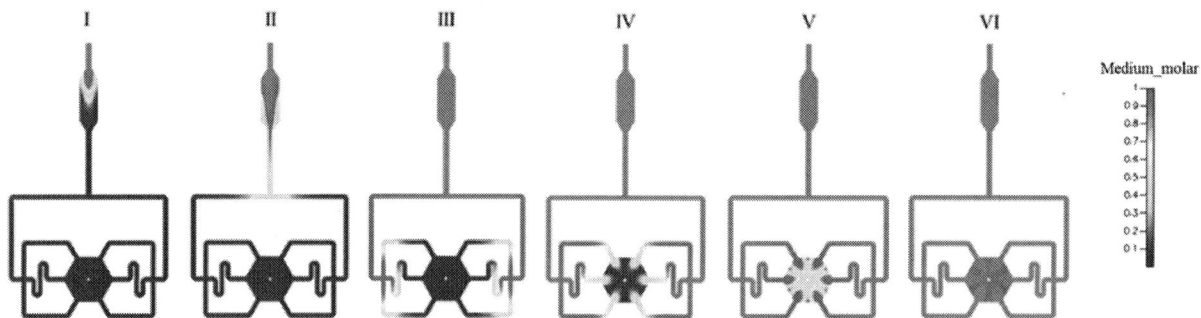

Figure 3: Sequence image of circulatory system simulation result, medium pumped from the upstream inlet, through the first chamber, lobule-flow-mimic channel, and flow out from the central outlet of liver-lobule-mimetic chamber.

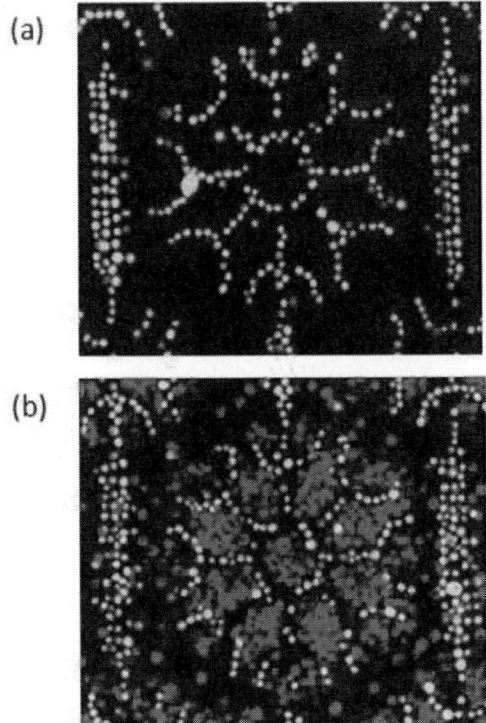

Figure 5: In step I, on-chip cell-patterning result of HMEC-1 (green) and C2C12 (red) cells. After pattern the HMEC-1, C2C12 was then patterned on the snow-flake-like electrode via vertical negative DEP manipulation. Then DEP buffer replaced with the fresh culture medium.

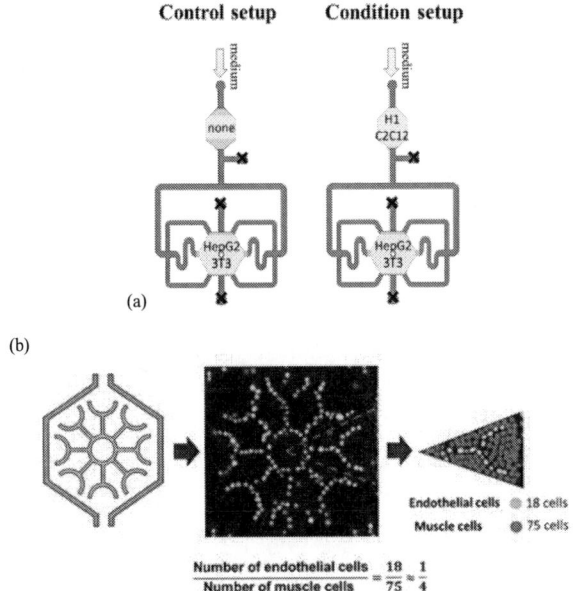

Figure 7: Concept of ratio control via snow-flake-like electrode. We design the electrode for the ratio of endothelial to muscle cell is 1 to 4 with the different size of octagon electrode.

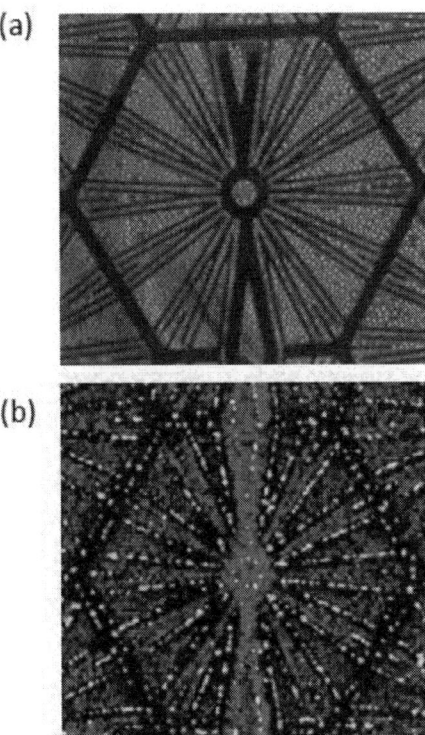

Figure 6: In step II, patterning in liver tissue chamber. (a) On-chip cell-patterning demonstrations of HepG2 cells. (b)Lobule-mimetic-stellate patterns of HepG2 (red) and 3T3 (green) cells mimic the pattern of the classic hepatic lobule.

cells/ml). When the cell was distributed uniformly over the chamber, the ac voltage of 7 Vpk-pk at 1 kHz was turned on. Then the DEP buffer in the micro chamber replaced with fresh culture medium. Figure 5 shows the pattern result of the C2C12 and HMEC-1. The micro-environment of blood vessel was accomplished in the first upstream chamber.

The positive parallel DEP operation attracted the 3T3 to form the patterns in-between each previous patterned cells. Finally, we are able to obtain the lobule-mimetic cell pattern as shown in Figure 6. Then the DEP buffer in the micro chamber replaced with fresh culture medium.

We finished the pattern of liver-lobule-mimetic chamber without patterning the blood vessel chamber. For the condition experiment setup as shown in Figure 7(a), we have achieved the both cell micro-environment in the liver-lobule-mimetic chamber and blood vessel chamber. While pumping the medium flow, the medium will pass through the blood vessel chamber and transport the cell secretion of blood vessel to the downstream liver-lobule-mimetic chamber. Via the lobule-flow-mimicchannel, the hexagonal chamber which mimics liver tissue could be supplied the medium flow like real fluid field of liver lobule.   As illustrated in Figure 7(b), the cell ratio could be defined via the size of electrode. The number of trapped cell was fixed according to the electrode area. For

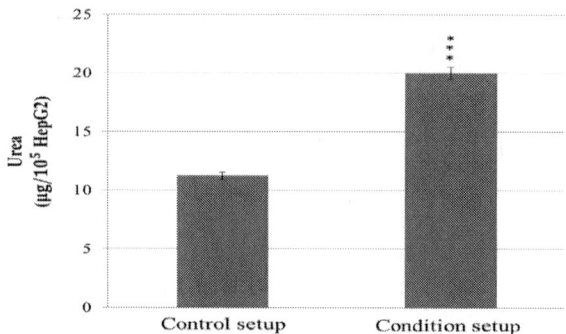

Figure 8: On-chip urea assay result, urea secretion of liver chip contain blood vessel environment (condition setup) was 78% higher than the urea secretion of liver chip without blood vessel environment (control setup).

confirming the cell ratio, we separate the electrode to eight equal shape and count the number of each cell. Because the chamber was full of snow-flake-like electrode, we can easily get the cell ratio of endothelial cell to muscle cell in the whole chamber by counting the cell number of any equal shapes. In this case, there are 18 endothelial cells and 75 muscle cells in one equal shape. It means that the ratio of endothelial cell to muscle cell is 1 to 4 via this electrode design.

The urea assay result for two conditions, are as shown in Figure 8, urea secretion of liver chip contain blood vessel environment was 78% higher than the urea secretion of liver chip without blood vessel environment. We find the value of condition medium is higher than the condition of control medium. It proves that the medium which used to culture blood vessel environment will enhance the ability of urea production by HepG2. The secretion is from the environment of blood vessel influence the urea produce ability of hepatocytes, HepG2 significantly.

IV. CONCLUSION

In the field of in vitro liver tissue reconstruct, the research combine liver-lobule-mimetic tissue, blood vessel environment and lobule fluid field is unusual. In this thesis, we successfully propose the liver circulatory system via microfluidics and cell pattern technology. To mimic the real liver circulatory system, this Lab chip consisted of three major parts, cell ratio controllable blood-vessel-environment-mimetic micro chamber, liver-lobule-mimetic micro chamber, lobule-flow-mimic micro channel. In this chip, C2C12, HMEC-1 and HepG2, 3T3 were used to mimic blood vessel condition and liver lobule, separately. The cell-cell interaction can be studied via this connected chip. The experiment result shows the liver function, urea producing ability, of HepG2 improved in this developed in vitro micro environment.

Nevertheless, this result is just an initiative work of tissue engineering, the next step or challenge is finding the optimal micro environment in this proposed system, like the optimal cell ratio of blood vessel, suitable cell source. For the ultimate

purpose of developing disease model studied at liver cancer, hepatitis B, hepatitis C and serving as a platform to screen drugs. These initiative achievements lead us to find a closer platform between in vivo and in vitro studies.

ACKNOWLEDGMENT

This work was supported partially by Cheng-Hsien Liu of Department of Power Mechanical Engineering, National Tsing Hua University of R.O.C. under the grant NSC 99-2l20-M-007-00 1.

REFERENCES

[1] R. Langer and J. P. Vacanti, "Tissue engineering," *Science*, VOL 260, 920-926, 14 May 1993.

[2] L. V. McIntire, H. P. Greisler, L. Griffith, P. C. Johnson, D. J. Mooney, M. Mrksich, N. L. Parenteau, D. Smith, "WTEC Panel Report on Tissue Engineering Research," *International Technology Research Institute*, January 2002.

[3] L. Griffith and G. Naughton, "Tissue Engineering - Current Challenges and Expanding Opportunities," *Science*, VOL 295, 1009-1014, 8 Feb 2002.

[4] S. J. Hollister, "Porous scaffold design for tissue engineering," *Nature materials*, VOL 4, 518-524, July 2005.

[5] A. Khademhosseini, R. Langer, J. Borenstein, and J. P. Vacanti, "Microscale technologies for tissue engineering and biology," *Proc Natl Acad Sci USA*, VOL 103, 2480-2487, 21 February 2006.

[6] H. Andersson, and A. V. D. Berg, "Microfabrication and microfluidics for tissue engineering: state of the art and future opportunities," *Lab Chip*, VOL 4, 98-103, 2004.

[7] A. Kikuchi and T. Okano, "Nanostrutured designs of biomedical materials: applications of cell sheet engineering to functional regenerative tissues and organs," *Journal of Control Release*, VOL 101(1-3), 69-84, January 2005.

[8] J. H. Sung and M. L. Shuler, "A micro cell culture analog (μCCA) with 3-D hydrogel culture of multiple cell lines to assess metabolism-dependent cytotoxicity of anti-cancer drugs," *Lab Chip*, VOL 9, 1385–1394, 2009.

Preparation and Physicochemical Study of Liposomes Containing Nicotinamide

N. Langlah[*], S. Pinsuwan[*], T. Amnuaikit[**]

[*]Department of Pharmaceutical Technology, Prince of Songkla University, Thailand
[**]Nanotec Center of Excellence, Prince of Songkla University, Songkhla, Thailand
Nare_mon@windowslive.com

Abstract— **This study was undertaken to evaluated preparation and physicochemical properties of liposomes formulations containing nicotinamide (NCM), sixteen different liposomal formulations were prepared using phosphatidylcholine from soybean (SPC), cholesterol (CHO) and Tween80 by modified ethanol injection method. Total lipid ratio of all formulations were varied in rang 20-80 (μmole/ml). All liposomal formulations prepared showed mean particle size of about 430 nm or less. Among the sixteen liposomal formulations prepared, liposomal formulation number 12 showed the best result in the physicochemical properties such as entrapment efficiency, particle size, polydispersity index (PI), and zeta potential. In additional this formulation has been stable when stability studies for 2 month. SPC:CHOL:Tween80 (80 μmole/ml) were composed of liposomal formulation number 12, it has percent entrapment efficiency equal 73.86±0.09 while that in others formulations less than.**

Keywords-component; nicotinamide; liposomes; entrapment efficiency

I. INTRODUCTION

Nicotinamide, or vitamin B_3, or niacinamide, is a derivative of niacin obtained through diet from meat, fish, milk, egg, and nuts. Its deficiency is one of the causing factors of pellagra. Nicotinamide is part of the coenzymes nicotinamide adenine dinucleotide (NAD), NAD phosphate (NADP). These molecules are important in many cellular metabolic enzyme reactions [1, 2].

In cosmetic, nicotinamide has been reported due to their potential for anti-inflammatory, anti-acne actions. It is also believed that may its anti-inflammatory effect may improve skin appearance by reducing leucocyte peroxidase systems that may lead to localized tissue damage as well as by ameliorating the cutaneous barrier [3]. In a comparative study, the anti-inflammatory effect of 4% nicotinamide gel in the management of acne vulgaris was as good as the benefits of 1% clindamycin gel [4]. This anti-inflammatory effect is also useful to reduce cutaneous erythema in various disorders [5]. Normally, the most drugs for acne vulgaris treatment was

antibiotic drug such as benzoyl peroxide, erythromycin, clidamycin, etc. these drug has side effect with the skin [6,7]. Then the nicotinamide has been interested because it is nonirritating to facial skin, easily formulated, chemically stable, and compatible with other formulation components [8]. Because the barrier property of the skin which is stratumcorneum and the hydrophilicity of nicotinamide make it difficult to permeate through the skin and reach to its site of action. Topical delivery of drugs by lipid vesicles is good candidate for considerable interest. Then, we want prepared liposomes containing nicotinamide and developed formulation of it for receives the best physicochemical formulation.

Since of discovery of liposomes or lipid vesicles derived from self-forming enclosed lipid bilayers upon hydration, liposomes drug delivery systems have play a significant role in formulation of potent drugs to improve therapeutics. Currently, most of liposomal formulations are designed to reduce toxicity, inhibition of rapid clearance of liposomes controlling size, charge, and surface hydration. Enhanced safety and heightened efficacy have been achieved for a wide range of drug classes [9]. However, the skin permeation of liposomes vesicles significantly varies depending on lipid composition. Different lipid composition that constitute the physicochemical property of liposomes (entrapment efficiency, particle size, and zeta potential) were differently too. Though, in this study have purpose for development and evaluate of liposomes containing nicotinamide formulations for find out the formulation which it has best physicochemical properties. In addition, these best formulations are taken to skin permeation study in the future.

In the present study, we have prepare sixteen of nicotinamide-loaded liposomal formulations consisting of phosphatidylcholine from soybean (SPC), cholesterol (CHO) and Tween80 which four group was divided; SPC:CHO, SPC:Tween80, SPC:CHO:Tween80, and only SPC, after that evaluated their physicochemical characteristics in term of mean percent entrapment efficiency, particle size, polydispersity index (PI), zeta potential. Percent entrapment efficiency and size evolution were evaluated during storage period of 2 months at 4°C.

II. MATERIALS AND METHODS

A. Materials

Phosphatidylcholine from soybean (SPC), type IV-S, ≥ 30% (TLC) (Sigma- Aldrich®, USA). Cholesterol from lanolin (CHO) (Fluka®, Buchs, Japan). Polyoxyethylene (20) sorbitan monooleate, Tween80® (Srichand Co., Ltd., BKK, Thailand). Nicotinamide (NCM) and standard purchase from P.C. Drug Center Co., Ltd., BKK,Thailand and Sigma-Aldrich®, USA,respectively. Absolute ethanol, AR grade, ethanol 95%, potassium dihydrogen phosphate (KH2PO4) from Merck®, Darmstadt, Germany. Triton®X-100 (Baker analyzed®, USA) and other chemicals, reagents used were of analytical grade.

B. Methods

Formulation and preparation of nicotinamide liposomes

Liposomes were prepared by modified ethanol injection method. Compositions for various formulations of liposomes are listed in TABLE I. Ingredient for use was lipid, phosphatidylcholine from soybean (SPC) in formulation. Additive use Tween80 and cholestertol (CHO) and concentration of total lipids were starting at 20-80 µmole/ml. For each of the formulation, a lipid phase was prepared in a round-bottom flask by dissolving appropriate quantities of SPC and additive (CHO) in an absolute ethanol solution, most of water phase (NCM and Tween80) was dissolving in an distilled water after that take both phase to sonicated at 30 ºC until homogeneity. Both oil and aqueous phases were heated at 60 ºC and mixed by swirl continuously for 1 minute in round bottom flask followed by removal of organic solvent by rotary evaporator at 60 ºC (Rotary evaporator, N-1000 series, Eyela, Tokyo) to obtain the liposomes suspension.

Physicochemical characteristics of liposomes

Entrapment efficiency was determined with centrifuge by ultracentrifuge (Backman Coulter ultracentrifuge, USA) at 60,000 rpm, temperature 4 ºC at time 2 hour, separation supernatant from lipid part followed bring part of supernatant was transferred to vials and the amount of drug entrapped was measured by HPLC after destruction of NCM-liposomes formulation with 20% triton®X-100 as ratio 1:1 (v/v). Percent entrapment efficiency was calculated by the following equation:

$$\text{Entrapment efficiency (\%)} = \left[1 - \left(\frac{NCM_F}{NCM_T} \right) \right] \quad (1)$$

TABLE I. COMPOSITION OF NICOTINAMIDE LIPOSOMES FORMULATIONS

Formulary	Ratio of lipid composition	Total lipid (µmole/ml)
1.SPC:CHO		20
2.SPC:CHO		40
3.SPC:CHO	84:16*	60
4.SPC:CHO		80
5.SPC:Tween80		20
6.SPC:Tween80		40
7.SPC:Tween80	4:1**	60
8.SPC:Tween80		80
9.SPC:Tween80:CHO		20
10. SPC:Tween80:CHO		40
11. SPC:Tween80:CHO	4:1:1**	60
12. SPC:Tween80:CHO		80
13. SPC		20
14. SPC		40
15. SPC		60
16. SPC	4**	80

SPC= Phosphatidylcholine from soybean; CHO= Cholesterol; *Weight ratio; **Molar ratio. All formulation was added 5% nicotinamide (w/v)

NCM_F and NCM_T are free nicotinamide which non-entraped and total amount of nicotinamide in formulations, respectively.

Particle size and zeta potential of liposomes were measured by zeta potential analyzer (ZetaPALS, Brookhaven, USA) at room temperature. Liposomes suspension was diluted with water before the measurement to adjust the intensity. Polydispersity index (PI) was also determined as a mearsurement of particle size homogeneity of the prepared liposomes. A small value of PI (<0.2) indicates homogeneous vesicle population.

Stability study of liposomes formulations were evaluated for their entrapment efficiency and physicochemical properties such as color of it, particle size, and zeta potential after storage at 4 ºC for 2 months, these formulations were compared the value with freshly prepared liposomes formulation. Samples were measurement at 0, 4, 8 weeks (2 months). Amount of nicotinamide retained in the vesicles was determined by HPLC.

High Performance Liquid Chromatography (HPLC) assay

Concentration of nicotinamide was assayed by HPLC system equipped with model Agilant 1100 series, pump, UV detector, Degasser, Autosampler (Palo Alto, USA).UV-VIS detector at 254 nm. The column used was reverse phase BDS HYPERSIL C18 column (150× 4.6 mm, particle size 5 µm) and mobile phase consisted of a mixture of methanol and buffer (75:925) of 0.05 M potassium dihydrogen phosphate having pH 3.6±0.1, adjusted with orthophosphoric (85%). Injection volume was 20 µl and flow rate was 0.6 ml/min. The validation of the HPLC assay was performed by repeating three times a day for three continual days using the same condition in the range of 1-40 µg/ml.

Data analysis

All the data were analyzed using SPSS 18.0 for windows program and significance level of $p < 0.05$ was used to indicate the statically significant difference between data sets.

III. RESULTS

Sixteen different liposomes formulations with varied total lipid of SPC:CHO, SPC:Tween80, SPC:Tween80:CHO, and only SPC were prepared by modified ethanol injection method. The total lipids were in range of 20-80 μmole/ml as shown in TABLE I. The result of physicochemicals studies of liposomal formulations shown as TABLE II. The mean particle size of liposome was about 173-430 nm, SPC:CHO formulations (1-3) have about 173-213nm, SPC:Tween80(5-8) and SPC:Tween80:CHO (12) have about 331-431 and 185 nm, respectively. Formulation 12 was significant difference of mean particle size when compared with formulation 6. PI of formulation was showed in TABLE II no greater than 1, indicating homogeneous size distribution of the liposome vesicles and zeta potential was in range -16 to -40 mV. For percent entrapment efficiency of satisfied formulations were indicated in TABLE II, from find out the formulation 12 has maximum of percent entrapment efficiency which composition were SPC:Tween80:CHO (80μmole/ml) in formulation. From observation find out the percent entrapment was increased when increased total lipid to 60μmole/ml but up to 80μmole/ml percent entrapment was decreased for SPC:CHO and SPC:Tween80 formulations except SPC:Tween80:CHO formulation.

Stability studies of eight formulations which showed in TABLE II. shows in Fig. 1(A). In term percent entrapment of all formulation was decreased after 2 months, liposomal formulation 12 was smallest decreased about 22 % after 2 months in the same time other formulations were decreased more than 30 %, in the same condition the particle size of all formulations have little or not changed, some formulations-

TABLE II. PHYSICOCHEMICAL PROPERTIES OF NICOTINAMIDE-ENTRAPPED LIPOSOMES

Formulary	Entrapment efficiency (%)	Size (nm)	PI	Zeta potential (mV)
1.	50.06 ± 0.02	172.96 ± 1.79	0.12±0.02	-16.47±4.65
2.	65.61 ± 1.15	206.28 ± 2.95	0.15±0.01	-19.57±5.44
3.	60.21 ± 0.18	212.89 ± 3.04	0.13±0.09	-21.56±4.71
5.	58.59 ± 0.10	331.20 ± 3.18	0.23±0.02	-31.75±4.95
6.	64.85 ± 0.58	430.59 ± 6.86	0.16±0.01	-30.22±2.50
7.	69.33 ± 0.03	380.29 ± 1.27	0.13±0.02	-37.9±2.45
8.	66.43 ± 0.90	355.1 ± 9.9	0.05±0.02	-40.3±1.40
12.	73.86 ±0.09	185.4 ± 1.25	0.08±0.02	-32.20±1.87

Each value represents the mean ± S.D. (n = 3). PI, Polydispersity-index. $p < 0.05$ of formulary 12 compared with formulary 6.

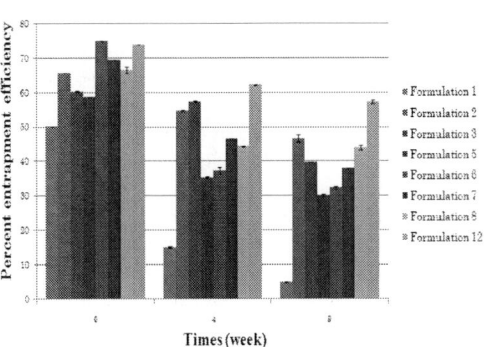

A

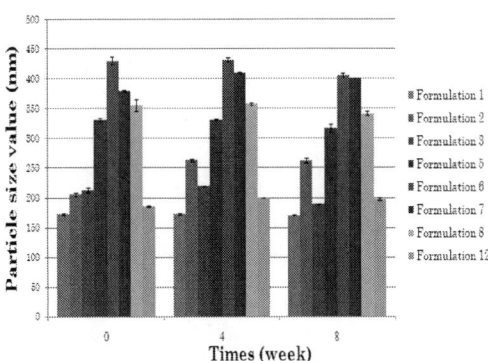

B

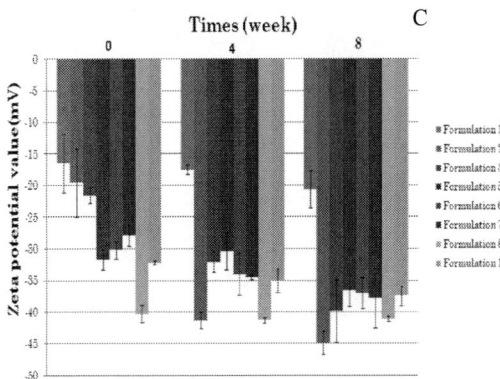

C

Fig. 1. Shows percent entrapment efficiency (A), particle size (B), and zeta potential (C) of liposomal formulations which stability studies with storage at 4°C at 0, 4, 8 weeks (for 2 months).

were increased or decreased after storage period because liposome particles were accumulated or peeled of liposome bilayers. The term of zeta potential of all formulations were increased less than 26 % but excepted formulation 2 and 3 increased much to 56 and 45 % after 2 months, respectively. Especially, the formulation 2 has maximum to -45 mV.

978-1-4673-1122-9/12 $31.00 © 2012 IEEE   539

IV. DISCUSSION AND CONCLUSION

Most of the liposomes formulations approved for human use contain phosphatidylcholine (neutral charge), with fatty akyl chains of verying lengths and degree of saturation. A fraction of cholesterol (~30 mol %) is often included in the lipid formulation to modulate rigidity and to reduce serum-induced instability caused by the binding of serum protein to the liposomes membrane. Cellular and physiological mechanisms explain the variations of liposomes size, charge, surface hydration, membrane fluidity, and clearance of lipid associated drug. Physical characteristics that determine liposome stability in storage are some of the most important parameters for parenteral preparations of liposome-based therapeutics [9].

Based on the head group composition of the lipid and pH, liposomes may bear a negative, neutral, or positive charge on their surface. The nature of density of charge on the surface of the liposomes influences stability, kinetics, and extend of biodistributiion, as well as interaction with and uptake of liposomes by target cells. Liposomes with a negative charged have a reduce aggregation and have increased stability in suspension [9]. Which in this study surface charge of all formulations were negative charged because composition was CHO and Tween80 and plus nicotinamide in formulation too. Each formulation group have increased tendency of zeta potential when the total lipid was increased.

Phosphatidylcholine from soybean have mainly structure in liposomal formulations, and CHO and Tween80 was additived. Tween80, a nonionic surfactant with a large head group, is miscible water [10]. It produces flexibility favoring penetration and further increases membrane hydrophilicity and diffusability [11]. However, a balance concentration of CHO is essential to from stable liposomes to prevent leakage, while excess or deficit CHO would result in destabilization of lipid bilayer membrane leading to decreased drug entrapment and increased leakage. This phenomenon is clearly evident in formulation 3 which shows decreased percent entrapment and stability when compared with formulation 12, as same as Tween80 have similar tendency with CHO which shows in formulation 8, but both CHO and Tween80 were added together in liposomal formulation make to increased percent entrapment when total lipid was increased up to 80μmole/ml which plays in formulation 12. This benefit is obtain together working of CHO and Tween80, which can be improved formulations for suitable.

In summary, among the eight liposomes formulations tested, formulation 12 showed the best result in terms of physicochemical study such as percent entrapment efficiency, particle size, and zeta potential. Additionally, this formulation had stable when stability tested at 4 °C for 2 months, which have percent entrapment particle size, and zeta potential value to be equal to 73.86±0.09, 184.4±1.25, -32.50±1.87, respectively. And appearance of nicotinamide was entrapped in liposomes with determined by SEM show as Fig. 2.

FUTURE WORK

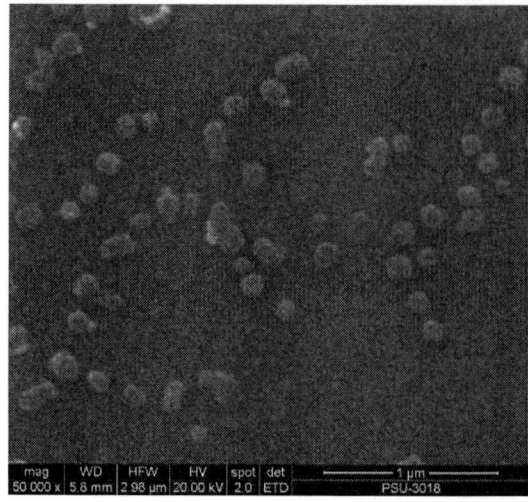

Fig. 2. Scanning electron micrograph showing nicotinamide liposomes vesicles formulation 12 (× 50,000 magnification).

After suitable formulations was selected, I would like to conduct their liposomes to cream formulations preparation, and skin permeation study of nicotinamide liposomes versus nicotinamide solution, and nicotinamide liposomes cream versus nicotinamide solution cream. In addition, I would like to study commercial product together.

ACKNOWLEDGMENT

We also thank the Faculty of Pharmaceutical Sciences for providing laboratory facilities. And also thank my advisor and co-advisor, Assistant Professor Dr. Thanaporn Amnuaikit and Assistant Professor Dr. Sirirat Pinsuwan, for suggestions and support throughout this work.

REFERENCES

[1] A. Dahiaya and J.F. Romano. "Cosmeceuticals: a review of their use for aging and photo aged skin," *Cosmet Dermatol*, vol. 19, pp. 84-479, 2006.

[2] J.E. Oblong and D.L. Bisset, "Retinoids," In: Z.D. Draelos, editor. *Cosmeceuticals*. 1st ed, Philadelphia:Elsevier Saunders, 2005, p. 35-45.

[3] D.S. Berson, D.K. Chalker, J.C. Harper and *et al*. "Current concepts in the treatment of acne: report from a clinical roundtable," Cutis vol.72, p.5-19, 2003.

[4] A.R. Shalita, J.G. Smith, L.C. Parish and *et al*. "Topical nicotinamide compared with clindamycin gel in the treatment of inflammatory acne vulgaris," *Int J Dermatol*, vol. 34, pp. 7-434, 1995.

[5] D.L. Bisset, J.E. Oblong and C.A. Berge. "Niacinamide: a B vitamin that improves aging facial skin appearance," *Dermatol Surg*, vol. 31, pp. 6-860, 2005.

[6] N. Rebecca and S. John, "Treatment of acne vulgaris," Prediatrics and child health 21:3, *Symposium: Dermatol*, pp. 119-125, 2010.

[7] J.J. Leyden, MD, Review article, "Therapy for acne vulgaris," J.J. Alastair, Wood, M.D., editer. *The New Eng. J. of med*, pp. 1156-1162, April 17, 1997.

[8] M. Manela-Azulay and E. Bagatin, "Cosmeceuticals vitamins," *Clinics in Dermatol,* vol. 27, pp. 469-474, 2009.

[9] L. Tianshun and H. J.Y. Rodney, "Minireview: Trends and developments in liposome drug delivey systems," *J.of Pharm.Sci*, vol. 90, No. 6, pp. 667-680, June 2001.

[10] G.M.M El Maghraby, A.C. Williams and B.W. Barry, "Interaction of surfactants (edge activators) and skinpenetration enhancers with liposomes," *Int. J. of Pharm*. Vol. 276, pp.143-161, 2004.

[11] M. Pichandy, M. Mishra, S. Kanaiyan and *et al.* " Formulation and psychopharmacological evaluation of surfactant modified liposome for parkinsonism disease," Research Article-ISSN 0974-2441, *Ass. J. of pharm. and Clinical Research*, vol. 3, Issue 1, January-March 2010.

Fabrication of Controllable Profile Microlens Array by Nanoimprinting Process

M. C. Cheng[*], L. K. Chen, C. K. Sung

[*]Department of Power Mechanical Engineering, National Tsing Hua University, Taiwan

d913790@gmail.com

Abstract—**A thermal imprinting process with a cylindrical mold is proposed to fabricate the microlens array using glass as raw material. In this study, the function of the mold is to decide the position of the microlens array instead of define the profile of the microlens, which is different from that of the conventional imprinting process. Microlens arrays are shaped because of the cohesion and surface tension of the glass material. Because of the surface of microlens didn't contact the mold, no adhesion phenomena occurred when implementing the procedure, which can eliminate the adhesive issue. By regulating the imprinting force, various focal lengths of microlens are produced. Both the molecular dynamics simulation data and experimental results show the same trend and also proof the feasibility of this method.**

Keywords-nanoimprinting; microlens array; molecular dyanmics simulation; nanofabrication

I. INTRODUCTION

The optical performance of measuring, imaging, and display systems are heavily influenced by the applied microlens arrays (MA) which enable a better illumination, focusing and collimation. Historically, various techniques include photoresist reflow [1], deep lithography [2], laser beam shaping [3], LIGA [4], microjet printing [5], laser ablation [6] and direct laser writing [7] were developed to suit specific requirements in manufacturing microlens array. Materials such as glass, polymers and semiconductors are reported to be suitable candidates and show different levels of complexities in the MA production. Polymers are frequently used due to its low glass transition temperature (T_g) which makes the reflow process possible. Glass microlens arrays also show better transparency and chemical stability than polymers. However, the complexities and the cost of the manufacturing process are greatly enhanced when glass is used in a mass production of making microlens arrays [8-10].

NIL technique shows a great potential in producing smaller size microstructure (<10 micro) [11] and a wider range in material selections [11-14] compared to the current inkjet molding. A vast number of research groups have already focused on the integration of NIL and photolithography processes, namely thermal NIL [11], UV-NIL [12], Roll-to-Roll NIL [13] and the hybrid processes [14]. The general procedure of NIL manufacturing process is to fill the partially melted material into the cavity of the mold where the profile of the structure is defined. External forces are applied on the mold to further achieve a solid structure before the demolding process takes place. The quality of the final product usually depends on the following factors: (1) the amount of material impregnated into the cavity during the imprinting. (2) the adhesive problem between the mold and the imprinted microstructure of the final product during the demolding. The former issue dominates the completion of the structure during imprinting while the latter one dictates final configuration of pattern after removing the mold. Numerous projects have been proposed to investigate issues mentioned above and concluded that factors that affect the filling ratio of materials in a confined volume are temperature, pressure from external forces, the substrate effect and the vacuum level. During the NIL manufacturing process, the ability of material to flow into a cavity increases with increasing temperature. The temperature is kept slightly above the glass transaction temperature under high vacuum and the external force on the mold is set to be constant. The melted material gradually diffuse into the cavity and the resulting filling ratio is increased. Furthermore, an anti-adhesion layer coated is applied on the mold before the imprinting to avoid the adhesive problem.

Up to date, there is no study reported of making microlens array using glass as raw material. The current NIL technique may provide a good chance to produce glass MA. However, the drawbacks are the initial cost of the mold, the cleaning process after the mold is used as well as keeping the pattern of the mold undamaged.

In this study, a thermal imprinting process is adopted to fabricate the MA using glass as raw material. The imprinting mold employed in this research is simple and easy to make, and the function is also different from that of the conventional imprinting process. In addition, by regulating the imprinting force; various focal lengths of microlens array can be generated. The entire process is investigated by molecular dynamics simulation and experiments. The advantages of the proposed process are discussed, and the formation mechanism of various focal length microlens is explored. Meanwhile, comparisons with other existing imprinting processes are reported.

II. SIMULATION METHODOLOGY

Over the past few decades, molecular dynamics (MD) simulation is applied in predicting or interpreting experimental results of the change in physical properties of materials. In this study, a full scale three dimensional MD simulation is employed to explore the mechanism in the formation of MA using glass as material.

A. Physics Model

A single cell simulation model demonstrates a thermal nanoimprinting process of a silicon mold on an amorphous glass (SiO$_2$) substrate (Fig. 1). The mold is made in a cylindrical shape with a diameter in 15 nm. The main function of the mold is to define the dimensions and the position of the microlens arrays whereas the profile of microlens is based on the force control of the thermal imprinting process. This manufacturing concept is quite different from the traditional thermal imprinting. The simulation system is divided into three stages. First, the model is heated over the glass transition temperature (T$_g$), at 345 °C under the Microcanonical ensemble (NVE) condition, second, the amorphous glass material is steered upward to implement the thermal imprinting process with a constant loading rate 17.37 Nt/sec under the Canonical ensemble (NVT) environment, and third, the removal of the mold after the model is cooling down to the room temperature. The atom positions and corresponding steering forces (imprinting force) are monitored and recorded during the entire simulation; furthermore, the radius of curvature can be determined through the atom positions.

B. Potential Function

In the molecular dynamics simulation, the mathematical model is based on the classical Newton's law in the phase space. Equations of motion for Newtonian atoms are derived based on the potential function and the numerical algorithm. In the current work, a fifth-order Gear's predictor-corrector algorithm was used in the simulation and the modified

Born-Mayer-Huggins (BMH) potential [15] is employed to be the interactive energy which has been demonstrated that the properties of amorphous silica glass [15-19]. The potential function is given as [15]

$$\phi_{ij} = A_{ij} \exp\left(-\frac{r_{ij}}{\rho}\right) - \frac{z_i z_j e^2}{r_{ij}} \xi\left(\frac{r_{ij}}{\beta_{ij}}\right) \quad (1)$$

where A_{ij} depends upon ion sizes and is used to obtain the correct interatomic bond lengths, r_{ij} is the distance between atoms (ion) i and j; ρ is a constant (= 0.29x10^{-8} cm), z_i and z_j represents electronic charge, +4 for Si, -2 for O, β_{ij} are constants for particular pairs of species. The ξ function is a complementary error function and employed as a convergence term in the Coulomb portion of the potential which is both size-independent and species-independent. The parameters used in the BMH potential are listed in Table I [17].

III. EXPERIMENT

A. Material Preparation

The selected materials of microlens and the mold were K-PG375 and silica glass, according to the single cell simulation model. The entire imprinting process was in-situ observed and recorded because both substrate and the mold were transparent. The physical properties were altered by regulating the system temperatures due to different T$_g$ of selected materials. Both the substrate and the mold were sliced in circular plate (20mm in diameter and 5mm in thickness). The mold was machined with 5 × 5 cylindrical holes array with 400µm in diameter by using femtosecond laser ablation process. The SEM image of the mold is shown in Fig. 2. The preparation of the mold was not as complicate as the conventional imprinting process. Furthermore, the requirement of machining precision is not high. The most important concerns are the accuracy of the position and the diameter of the cylindrical holes; as for the portion inside the hole in this manufacturing method, it didn't affect the imprinting results significantly.

B. Thermal Imprinting Process

Thermal imprint process can affect the quality of microlens arrays. Therefore, to obtain a better quality of microlens arrays, the parameters in each stage of thermal imprint process is required to be precisely controlled. The experiment was conducted using a lab-made thermal imprinting machine.

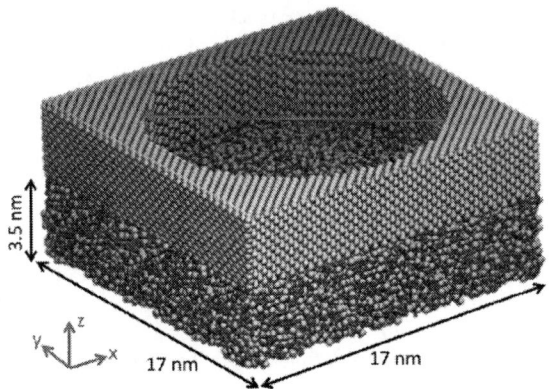

Fig. 1. Physical model of simulation.

TABLE I. THE PARAMETERS USED IN THE BMH POTENTIAL FOR SILICA INTERACTIONS

	A_{ij} ($\times10^{-8}$) (ergs)	B_{ij} ($\times10^{-7}$) (cm)	Z_i	Z_j	ρ_{ij} ($\times10^{-8}$) (cm)	σ_{ij} ($\times10^{-8}$) (cm)
Si-Si	0.1877	0.260	4	4	0.29	1.148
Si-O	0.2962	0.255	4	-2	0.29	1.284
O-O	0.0725	0.253	-2	-2	0.29	1.420

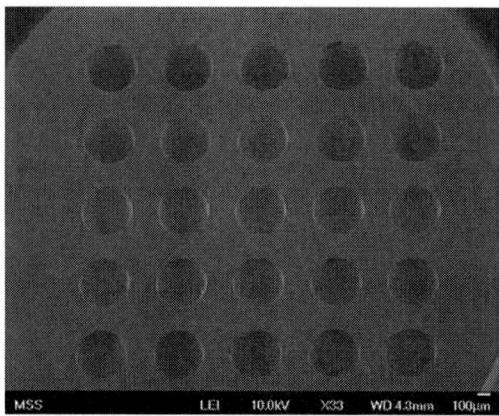

Fig. 2. A 5 x 5 SEM image of mold.

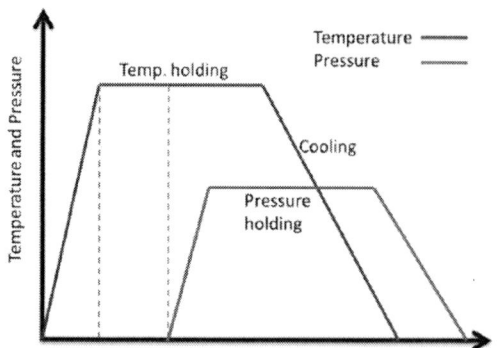

Fig. 3. The temperature and pressure procedure and sequence w.r.t. time.

The thermal imprint processes was divided into three stages, including (1) heating the system to glass transition temperature (2) executing imprinting process (3) cooling the system to room temperature and removing the mold. The procedure and sequence are shown in Fig. 3.

IV. RESULTS AND DISCUSSTIONS

A. Formation Mechanism and Advantage

In this study, a mold with cylindrical shape is utilized in NIL to define the dimensions and the positions of the microlens array, which means the boundary is confined by the cylindrical. When the external force exert on the mold, the melted glass starts to extrude into the cave, because of the confine of the boundary, the material extruded in the cave formed a velocity gradient inside the boundary, research [20] shows that the velocity in the middle is higher than that at the border, therefore, it forms the model of lens during imprinting (Fig. 4 (a)). Additionally, the surface tension drives the surface energy to the minimum value which forms the most stable status and least surface roughness.

As the imprinting process continues, more and more material are extruded into the mold, under the effect of velocity gradient, the material flowed at the middle of the lens is larger than that at the border, which means various profiles of lens are generated at different steps (Fig. 4 (b), (c)). Each step corresponds to an external force that provides us a parameter to manipulate the focal length. The variation of the surface profile stops until the velocity gradient disappears which represents the material compassed by the boundary starts to raise; and eventually it become a flat-convex lens (Fig. 4 (d)).

For the general thermal imprinting process, the function of the mold is to determine the profile of the microstructure. The issues that will encounter in the conventional thermal imprinting process have been described in section I. In this study, when the microlens arrays are generated, the surface is

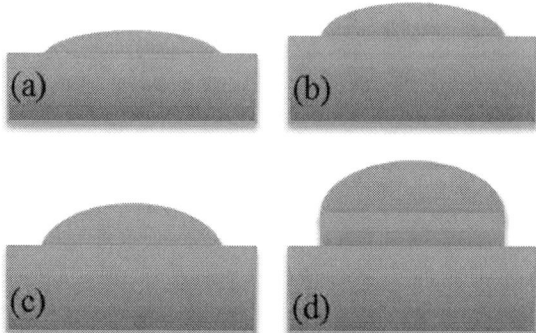

Fig. 4. The schematics of the formation of microlens.

a free surface, from the previous description, the minimum energy equilibrium to the most stable status which also forms a microlens array, no filling issue happens in this process. Sequentially, because of the free surface of microlens didn't contact the mold, no adhesion phenomena occurred when implementing the procedure, which eliminate the adhesive problem and also remove the issue caused from adhesive issue, such as the mold life, the fidelity of duplicating the pattern.

B. Simulation Results

The thermal imprinting process includes the heating process which may change the material property of silica. The material property of amorphous silica needs to be checked to decrease the uncertainty of the results. In MD simulation, radius distribution function (RDF) provides criteria to judge the status of crystalline. Fig. 5 showed the RDF of the utilized material in the simulation, the curves indicated that the materials were in amorphous status before and after imprinting.

Fig. 6 exhibited the microlens array after imprinting process, and proved the concept of utilizing cylindrical mold for thermal imprinting is feasible. From the simulation resulting data, the relationship between the exerted force and the positions of atoms were obtained which implied the radius of curvature of microlens is inversely proportional to the applied force (Fig. 7).

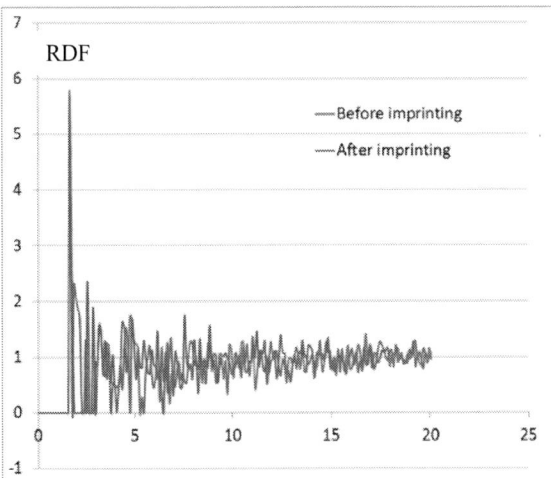

Fig. 5. The radius distribution function (RDF) before and after imprinting

Fig. 6. Microlens array after imprinting process.

The focal lengths of the microlens can be derived and the values are also determinable through the lens-maker's formula which is shown as

$$\frac{1}{f} = (n-1)\left(\frac{1}{R_1} - \frac{1}{R_2}\right) \qquad (2)$$

where f is the focal length, n represents the index of the glass, R_1 and R_2 represent the radius of the curvature of microlens, in this study, R_2 is infinite.

C. Experimental Results

The experiment was implemented by a lab-made thermal imprinting machine and followed by the sequence in Fig. 3. The microlens array generated by imprinting process by utilizing the cylindrical shows a good result (Fig. 8) which proved that the mold with cylindrical shape had high feasibility to generate microlens array by thermal imprinting process. From the simulation results, the applied force played the most significant role in determining the radius of curvature of microlens. The SEM images in Fig. 9 also verified the

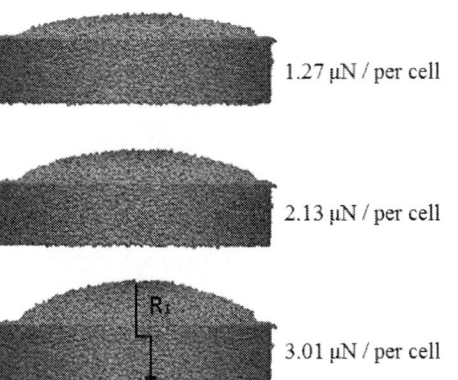

Fig. 7. The relationship between curvature radius (r) of microlens and force.

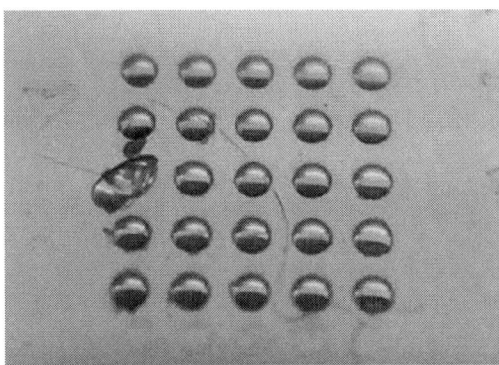

Fig. 8. A 5 × 5 microlens array spacemen.

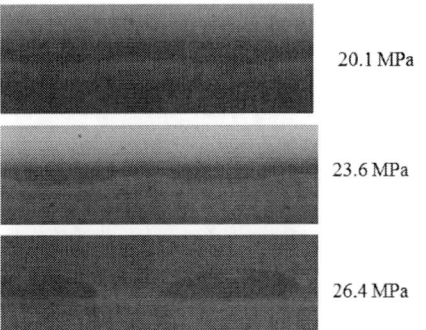

Fig. 9. The SEM image of microlens for different imprinting force.

prediction of the simulation and obtained the same trend compared to the simulation results.

V. CONCLUSIONS

In this study, the proposed thermal imprinting process has been proved to successfully fabricate 5 × 5 glass microlens arrays with a diameter of 400 µm by using the cylindrical mold. From the manufacturing point of view, the mold is easy to make and is apt to restore, in addition, the adhesive problem can be eliminated, which increases the mold life and low down the cost. Utilizing a cylindrical mold and regulating the imprinting force can generate the various focal lengths of microlens. Both simulation and experimental results show good agreement that the radius of curvature of microlens is inversely proportional to the applied force.

ACKNOWLEDGMENT

This study was supported in part by the National Science Council of Taiwan under Grant No. NSC-97-2221-E-007-048-MY3. The authors are thankful for this support.

REFERENCES

[1] Z.D. Popovic, R.A. Sprague, and G.A. Neville Connell, "Technique for monolithic fabrication of microlens arrays," *Appl. Opt.* vol. 27, pp. 1281–1284, 1988.

[2] H. Ottevaere, B. Volckaerts, J. Lamprecht, J. Schwider, A. Hermanne, I. Veretennicoff, and H. Thienpont, "2D plastic microlens arrays by deep lithography with protons: fabrication and characterization," *J. Opt. A: Pure Appl. Opt.*vol. 4, pp. S22–8, 2002.

[3] T.R. Jay, and M.B. Stem, "Preshaping photoresist for refractive microlens fabrication," *Opt. Eng.* vol.33, 3552–3555, 1994.

[4] C. Ossmann, J. G¨ottert, M. Ilie, J. Mohr, and P. Ruther, *Proc. Microlens Arrays EOS Topical Mtg*, pp. 54–58, 1997.

[5] D.L. MacFarlane, V. Narayan, W.R. Cox, T. Chen, and D.J. Hayes, "Microjet fabrication of microlens arrays," *IEEE PTL* vol. 6, pp. 1112–1114, 1994.

[6] K. Naessens, H. Ottevaere, R. Baets, P. Van Daele, and H. Thienpont, "Direct writing of microlenses in polycarbonate with excimer laser ablation," *Appl. Opt.* vol. 42 pp. 6349–6359, 2003.

[7] M.T. Gale, "Direct writing of continuous relief micro-optics," *Micro-optics: Elements, Systems and Applications*, chapter 4, London: Taylor and Francis, 1997.

[8] M. Wakaki, Y. Komachi, and G. Kanai, "Microlenses and microlens arrays formed on a glass plate by use of a CO_2 laser," *Appl. Opt,* vol. 37, pp. 627-631, 1998.

[9] M. Oikawa, K. Iga, T. Sanad, N. Yamamoto, and K. Nishizawa, "Array of distributed-index planar microlens prepared from ion-exchange technique," *Jap. J. Appl. Phys*, vol. 20, pp. L296-298, 1981.

[10] N.F. Borrelli, D.L. Morse, R.H. Bellman, and W.L. Morgan, "Photolytic technique for producing microlenses in photosensitive glass," *Appl. Opt.* vol. 24, pp. 2520-2525, 1985.

[11] S.Y. Chou, P.R. Krauss, and P.J. Renstrom, "Imprint lithography with 25-nanometer resolution," *Science*, vol. 272, pp. 85-87, April, 1996.

[12] M. Colburn, S. Johnson, M. Stewart, S. Damle, T. Bailey, B. Choi, M. Wedlake, T. Michaslson, S. V. Sreenivasan, J. Ekerdt, and C. G. Willson, "Step and Flash Imprint Lithography: An Alternative Approach to High Resolution Patterning," *Pro. of SPIE*, vol. 3676, pp. 379-389, 1999.

[13] H. Tan, A. Gilbertson and S.Y. Chou, "Roller nanoimprint lithography," *J. Vac. Sci. Technol. B*, vol.16, pp. 3926-3928,1998.

[14] Y. Xia, and G. M. Whitesides, "Soft Lithography," *Angewandte Chemie Int. Ed.*, vol. 37, pp. 550-575, 1998.

[15] S.H. Garopalini, "Molecular dynamics simulation of the frequency spectrum of amorphous silica," J. Chem. Phys., vol. 76, pp. 3189-3192, March, 1982.

[16] S.H. Garopalini, "A molecular dynamics simulation of the vitreous silica surface," J. Chem. Phys., vol. 78, pp. 2069-2072, February, 1983.

[17] A.B. Rosenthal, and S.H. Garofalini, "Molecular dynamics study of amorphous titanium silicate, " *J. of Non-Cryst. Solids*, vol. 107, pp. 65-72, 1988.

[18] R. Ochoa, and J.H. Simmons, "High strain rate effects on the structure of a simulated silica glass," *J. of Non-Cryst. Solids*, vol. 75, pp. 413-418, 1985.

[19] S.H. Garopalini, "Molecular dynamics computer simulations of silica surface structure and adsorption of water moluculars," *J. of Non-Cryst. Solids*, vol. 120, pp. 1-12, 1990.

[20] Y.M. Hung, "The Study of Glass Micro/Nanostructure Formation by Imprinting Process," *Doctorial Dissertation*, 2009.

Analysis on 3-Dimensional spatial electric field of AFM based anodic oxidation

Zenglei Liu, Niandong Jiao, Zhidong Wang, Zaili Dong
Shenyang Institute of Automation Chinese Academy of Sciences, Shenyang, China
liuzenglei@sia.cn

Abstract—Atomic force microscope (AFM) based anodic oxidation is an important method to fabricate nano-structures and nano-devices. To realize precise fabrication, electric field between AFM tip and substrate should be under precise control. For precise control of the electric field, a necessary topic is to find out the distribution of the spatial electric field and the relationship between the electric field and parameters. By theoretical analysis we simulated the spatial distribution of the tip/substrate electric field and analyzed the relationship between the electric field and parameters, which were verified by experiments. Our work can provide theoretic support for electric field assisted nanofabrication.

Keywords-AFM; electrical fabrication; anodic oxidation

I. INTRODUCTION

In 1993, H. C. Day first reported anodic oxidation with STM[1]. Since then researchers have investigated AFM based anodic oxidation. Just one year later, E. S. Snow presented a method for fabricating Si nano-structures with an air-operated AFM[2]. As regards mechanism, P. Avouris examined kinetics and mechanism of AFM based anodic oxidation in 1997[3]. To improve anodic oxidation method B. Legrand proposed an original way which could both increase the reliability and improve the nanolithography resolution in 1999[4]; X.-D. Hu analyzed current of dynamic electric field induced oxidation nanofabrication with AFM in 2010[5].

To improve the reliability and resolution of anodic oxidation, we focused on the tip/substrate 3-dimensional spatial electric field in this paper. First we constructed the simulation model for the spatial electric field and attained its distribution. The vertical section of the spatial electric field showed that the electric field was mostly confined within the water meniscus at the tip-substrate junction. J. A. Dagata indicated that product ion concentrations associated with the water meniscus[6] and R. Garcia realized to confine the oxidation of silicon surfaces with a non-contact AFM by controlling the nanometer-size water bridges[7]. Our simulation coincided with Dagata and Garcia. The horizontal section showed that the electric field wears off from inside out. Next the electric field variation under the existence of nano-oxide line was simulated. After that the relationship between the electric field and parameters was analyzed. Ten simulations were carried out with tip-substrate voltage varying from 1V to 10V and illustrated a linear relationship with the voltage. Another ten simulations with tip/substrate separation varying from 1nm to 10nm confirmed an exponential

relationship with the tip/substrate separation. Then the impact of the water meniscus on the electric field was discussed. However the heights of the water films on the substrate and the tip surface had a slight impact on the electric field deducing from our simulation. Next two groups of nano-dots were fabricated on Si substrate to illustrate our simulations. One group was made under the same experimental conditions except different tip-substrate separations and the other group was fabricated under the same conditions except different voltages. The variation trends of the nano-dots coincided with our simulations. At last a group of nano-lines was fabricated on Si substrate under instructions of our simulation. The mean value of the heights of these lines was 1.64nm and the variance was $0.13nm^2$. The data showed that the lines were of good repetition. In brief, our works can help to plan procedure for fabricating nano-transistors or nano-masks by anodic oxidation.

II. SIMULATION MODEL

Gomer R set up a model for STM tip electric field in 1986 [8]. STM Tip was equivalent to a sphere in the model, as shown in Fig. 1(a). Gomer R also deduced an equation by using image potentials shown as (1) to calculate the electric field on substrate surface. E(r) represented Electric field intensity at point A in fig. 1(a). V, D, R and r were as shown in Fig. 1(a).

$$E(r) = \frac{V}{D}\left[1+\left(\frac{r}{R+D}\right)^2\right]^{-\frac{3}{2}} \tag{1}$$

AFM has similar structure with STM. However AFM works in air and the tip is connected with the substrate by a meniscus [9], as shown in fig. 1(b). The existence of the meniscus

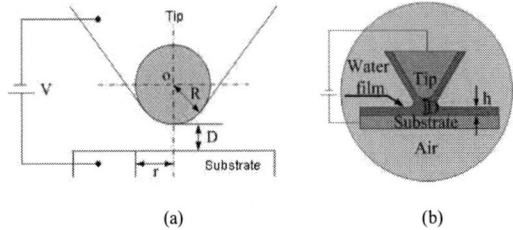

(a)   (b)

Fig. 1. (a) STM model for electric field distribution. (b) AFM model for electric field distribution.

This research is supported by National Natural Science Foundation of China (Grant No. 61106109), and the CAS/SAFEA International Partnership Program for Creative Research Teams.

978-1-4673-1122-9/12 $31.00 © 2012 IEEE

makes it hard to build up a formula to calculate the electric field on the substrate.

We adopted simulation method to study the spatial distribution of the electric field between AFM tip and substrate. Electric field in space rather than just that on the substrate surface could be studied through simulation method. Further more electric field variation when parameters changed could be discussed by simulation. According to Poisson's equation, available mathematic model for space electric field was shown as (2):

$$-\nabla \cdot [(\varepsilon/T + \sigma)\nabla V - (J^e + P/T)] = \rho/T \qquad (2)$$

Among (2), ∇ is the Hamiltonian operator, ε is the dielectric constant, σ is the conductivity, J^e is the external current density, ρ is the space charge density, P is the polarization vector, and T is the time step.

Equation (2) was broken down into two parts in the simulation software we used in order to reduce difficulty of solving. The first part was shown as (3). This part was used to calculate electric field within the thin water film, as shown in Fig. 1(b). The other part was shown as (4) and was used to calculate electric field in the air sub-domain, as shown in fig. 1 (b).

$$-\nabla \cdot (\sigma \nabla V - J^e) = Q_j \qquad (3)$$

$$-\nabla \cdot \varepsilon_0 \varepsilon_r \nabla V = \rho \qquad (4)$$

The geometric model of AFM tip was based on NSC15/Ti-Pt, which was a conductive tip from MikroMasch Company coated with 15nm Ti first and then 10nm Pt. The tip was equivalent to a cone with a sphere on the end point, as shown in Fig. 1(b). The height of the water film h was a variable in the simulation. Substrate was simplified to a round panel with a radius as long as 500nm. A bigger radius was proved to be useless to improve simulation results because electric filed gathered closely around the peak point of AFM tip. The height of the panel was ignorable because electric field inside the panel was zero.

III. SIMULATION ANALYSIS

A. Simulation Results

The spatial distribution of the electric field was shown in Fig. 2. Fig. 2 was simulation results under conditions that h equaled 5nm, tip-sample separation D equaled 5nm, and tip bias V equaled 10V. Only electric field distribution in the water film was presented and that in the air was blocked because electric field in the air was relatively less important.

The coordinate system of Fig. 2 was set as below. Y-axis was along the central axis of the tip. XZ plane paralleled to the substrate surface. The origin was located at the end point of the tip. Cut the Fig. 2(a) along Y-axis, and the vertical section of the electric field could be derived, shown as Fig. 2(b). As we could see from Fig. 2(b), the electric field was symmetrical about Y-axis, and the farer from Y-axis, the lower the electric

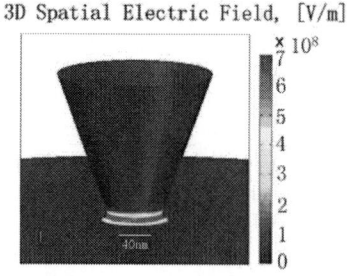

(a)

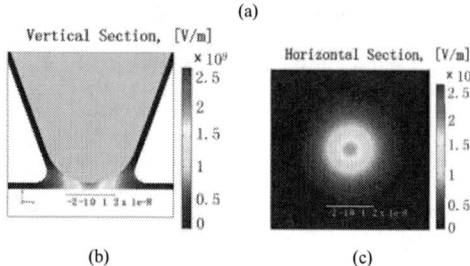

(b)   (c)

Fig. 2. Distribution of the 3-dimensional spatial electric field. (a) Outlook of the electric field in the water film. (b) Vertical section of the electric field. The electric field is mostly confined within the meniscus. (c) Horizontal section of the electric field. The electric field wears off from inside out.

was. In the meniscus the electric field displayed a funnel-shape and decreased from top to down. Fig. 2(c) was the horizontal section of Fig. 2(a). The electric field spread in round-shape and decreased outward from the center as shown in Fig. 2(c).

Electric field on a line across the center of substrate surface could represent that on the whole surface as the electric field on the substrate surface was symmetrical about the center. Every one of the four curves in Fig. 3 represented electric field on the substrate surface. X-axis was distance from the center and Y-axis was normal value of electric field in Fig. 3. Curve 1 was calculated by (1). Curve 2 was simulation result assuming the tip was a sphere and without considering the water film. Curve 1 and Curve 2 were obtained under same geometry models and physical conditions but using different methods, so they were supposed to match with each other. Actually they were not, as shown in Fig. 3. On possible reason was that in the derivation of (1) with image potential

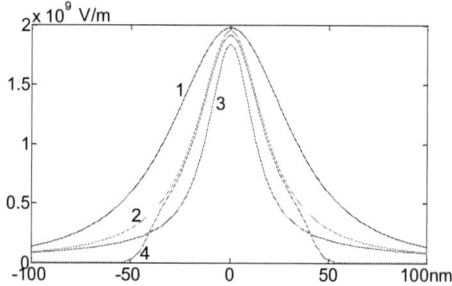

Fig. 3. Electric field distribution on the substrate surface under different geometric model.

method the tip was equivalent to a sphere first and then assumed as a point charge which was located in the center of the sphere. Actually the position of the point charge was a little below the center. The assumption was made to avoid solving high-order equation. Curve 3 was simulation result on conditions that AFM tip was equivalent to a cone and ignore water films over tip and substrate. Curve 2 and curve 3 were derived under same conditions except the shape of the tip. A cone-shape was closer to the real shape of AFM tip. The difference between curve 2 and curve 3 reflected the effect of the tip-shapes on the electric field. Curve 4 was derived considering the water film and the tip as a cone. The difference between curve 3 and curve 4 reflected the effect of the water film on the electric field. Under consideration of water film, the electric field further concentrated to center.

Among anodic oxidation processing, already generated oxide might disturb the existed electric field. Fig. 4 showed the electric field distribution in the presence of line-shape oxide as high as 2nm. Comparing Fig. 4 with Fig. 3(b), the electric field in Fig. 4 was bigger.

B. Electric field under different tip bias

The electric field between the tip and substrate is aroused by tip bias voltage, so the bias has a significant impact on the electric field. When the water film was as thick as 5nm and tip-substrate distance was 5nm, tip bias varied from 1V to 10V, thus ten corresponding distributions of the electric field were obtained, as shown in Fig. 5(a). Image 1 was electric field when tip bias was 1V, and so on. The bigger the bias voltage was, the stronger the electric field was, as we can see from Fig. 5(a). When the bias voltage was under 6V, the magnitude of the electric field was less than 10^9V/m, which was the threshold voltage for anodic oxidation. For precious investigation of the electric field on substrate surface, image data of Fig. 5(a) were exported and data on a line across center

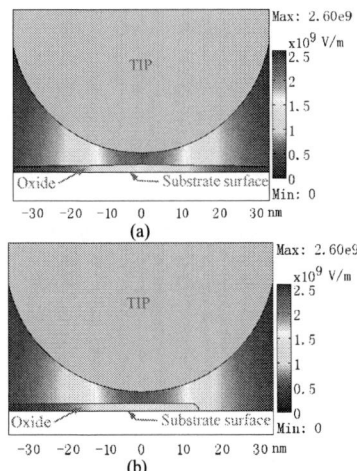

Fig. 4. Electric field distribution during fabricating nano-oxide line. (a) Front view. Tip moving from back to front. (b) Side view. Tip moving from left to right. Tip-Substrate separation is 5nm. The height of the nano-line is 2nm and the width is 200nm. The line has a curved section.

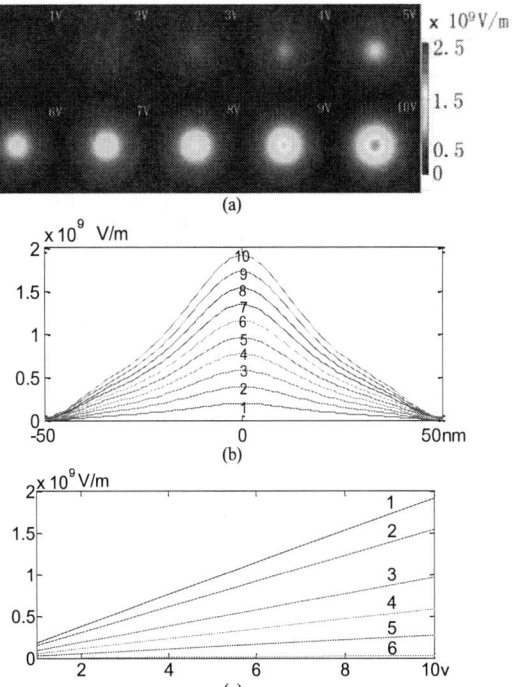

Fig. 5. (a) Two-dimensional electric field distributions on substrate surface when tip bias varies from 1V to 10V. Image 1 represents electric field distribution when tip bias is 1V, and so on. (b) One-dimensional electric field distributions on substrate surface when tip bias varies from 1V to 10V. Curve 1 represents electric field distribution when tip bias is 1nm. (c) Electric filed variation on a certain point. Curve 1 represents electric variation on the substrate surface center. Curve 2 represents electric field variation on the points 10nm far away from the center, and so on.

of substrate surface were picked up to form ten curves as shown in Fig. 5 (b). Curve 1 in Fig. 5(b) corresponded to image 1 in Fig. 5(a). Curve 2 in Fig. 5(b) corresponded to image 2 in Fig. 5(a), and so on. From Fig. 5(b) we could conclude that the peak value of the substrate surface electric field located at the center, and the electric field was symmetrical about the center. The electric field on a certain point of the substrate surface was linear with the tip bias, as shown in Fig. 5(c). Six points on the substrate surface were chosen. Point 1 corresponded to curve 1 in Fig. 5(c) and located on the center. Point 2 was 10nm away from the center, and so on.

C. Electric field under different tip-substrate seperation

Tip-substrate separation is another important factor to affect the electric field. When tip bias was kept as 10V and water-film height was kept as 5nm, tip-substrate separation varied form 1nm to 10 nm, thus a collection of ten images was obtained as shown in Fig. 6(a). Image 1 was electric field distribution when tip-substrate distance was 1nm, and so on. The smaller the separation was, the stronger the electric field was, as we could see from Fig. 6(a). When the separation was over 9nm, the magnitude of the electric field was less than

978-1-4673-1122-9/12 $31.00 © 2012 IEEE

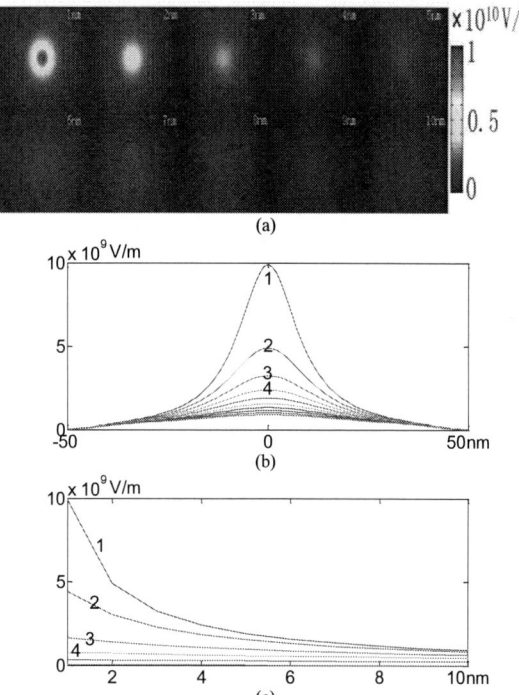

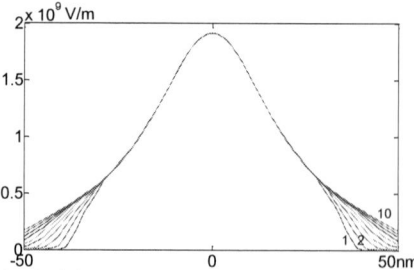

Fig. 7. Electric field variation when water-film heights change from 1nm to 10nm. Curve 1 represents electric field distribution on the substrate surface when water film height is 1nm, and so on.

Fig. 6. (a) Two-dimensional electric field distributions on substrate surface when tip-substrate distance varies from 1nm to 10nm. Image 1 represents electric field distribution when tip-substrate distance is 1nm, and so on. (b) One-dimensional electric field distributions on substrate surface when tip-substrate distance varies from 1nm to 10nm. Curve 1 represents electric field distribution when tip-substrate distance is 1nm. (c) Electric filed variation on a certain point. Curve 1 represents electric variation on the substrate surface center. Curve 2 represents electric field variation on the points 10nm far away from the center, and so on.

10^9V/m, which was the threshold voltage for anodic oxidation. For precious investigation of the electric field on substrate surface, image data of Fig. 6(a) were exported and data on a line across center of substrate surface were picked up to form ten curves as shown in Fig. 6(b). Curve 1 in Fig. 6(b) corresponded to image 1 in Fig. 6(a). Curve 2 in Fig. 6(b) corresponded to image 2 in Fig. 6(a), and so on. From Fig. 6(b) we could also conclude that the peak value of the substrate surface electric field located at the center, and the electric field is symmetrical about the center. The electric field of a certain point on the substrate surface was non-linear with the separation, as shown in Fig. 6(c). Also six points on the substrate surface were chosen. Point 1 corresponded to curve 1 in Fig. 6(c) and located on the center. Point 2 was 10nm away from the center, and so on.

D. Electric field under different heights of water film

Both AFM tip and substrate was cover by water films. The height of the water film was related to the humidity of the air. We simulated variation of the electric field under different water-film heights. In Fig. 7 the difference between electric fields with and without water film had been discussed. The

impact of the water-film height on the electric field was going to be discussed in this section. Kept the tip bias 10V constantly and tip-substrate separation 5nm and varied the height of the water film from 1nm to 10nm, thus a cluster of curves was obtained as shown in Fig. 7. Curve 1 represented the electric field on the substrate surface when the water film height was 1nm, and so on. The water film height only played a role in the area approximately 25nm far away from the substrate surface center. Within 25nm, the electric field kept constant no matter what was the height. One possible reason was the water film over the tip and that over the substrate connected with each other to form a water bridge. When the water film varied the width of the bridge changed. The electric field was strong mostly in the bridge. Outside the bridge the electric field decreased rapidly to zero. The boundary between the bridge and the air constrained the electric field inside the bridge. So when the water bridge grew broader, the electric field reached out.

IV. EXPERIMENTAL RESULTS

Experiments fabricating nano-dots and nono-lines on silicon substrate have been conducted for analyzing relations between results of electric field simulation and nano-structure fabrication. Fig. 8(a) shows results of six nano-dots fabricated with two dfferent conditions, and Fig. 9 shows results of nine nano-lines fabricated under the same condition. The six nano-dots in Fig. 8(a) are divided into two groups. Group 1 included Dots D1, D2 and D3, which were fabricated under same tip bias but different tip-substrate distance. Dots D4, D5 and D6 were group 2, which were fabricated under different tip bias while tip-substrate distance was kept constant. Take D1 for example, the mark on the upper left corner of D1 (10V, 5nm) means D1 was fabricated under 10V tip bias and 5nm tip-substrate distance. The time period of tip bias was 1 second for all the six nano-dots.

Fig. 8(b) showed the widths and heights of dots D1 to D6. The higher curve in Fig. 8(b) was height values on the line across D1, D2 and D3 in Fig. 8(a). The lower curve in Fig. 8(b) was height values on the line across D4, D5 and D5 in Fig. 8(a). Dot D1 was about 3.7nm high. The heights of dots D2 and D3 were very close. Fig. 8(b) approximately represented that the heights of D1, D2 and D3 had an inverse relationship

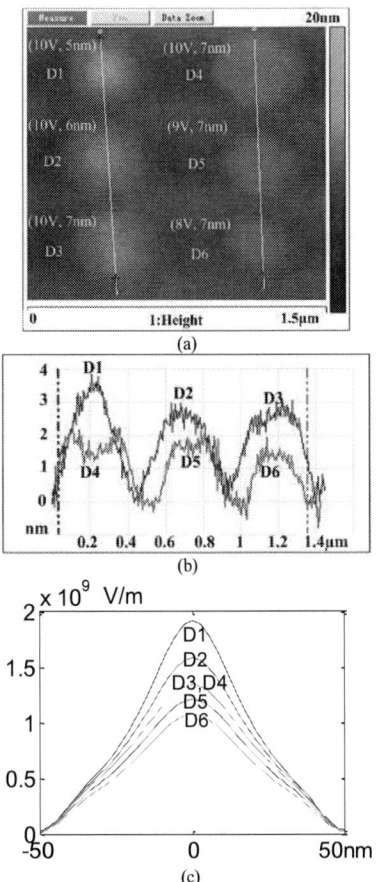

(a)

(b)

(c)

Fig. 8. Experimental results. (a) Two groups of nano-dots were fabricated by anodic oxidation on Si substrate under different parameters. (10V, 5nm) indicates the dot was fabricated under 10V with the tip 5nm away from substrate. (b) Measurement of dots D1 to D6. (c) Simulation results of initial electric fields of fabrications of D1, to D6.

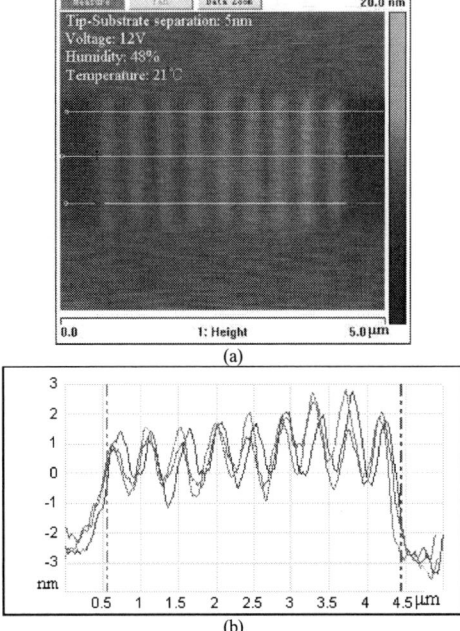

(a)

(b)

Fig. 9. Experimental results. (a) A group of nano-lines of good repetition which were fabricated under instructions of our simulation results. The lines are as long as 2μm. (b) Heights of the lines.

R1 was bigger, the dot was wider. This was true inside group 1 and group 2 respectively, but not so between the two groups. One possible explanation was the substrate environment of group 2 differed from that of group 1. Furthermore R1 was corresponding to R2. Take D1 for example, R1 of D1 was smallest in group1, correspondingly R2 was the smallest one in group 1.

Fig. 9 showed nine nano-lines fabricated on silicon substrate under same parameters. The nine nano-lines were made in one

with tip-substrate distance and the relationship was nonlinear, while the heights of D4, D5 and D6 were proportional to the tip bias.

However the widths grew boarder from D1 to D3 while the widths grew narrower from D4 to D6. The widths of these nano-dots were not simply proportional to the amplitude of the electric field. The simulation results of electric field when fabricating D1 to D6 was put on in Fig. 8(c) to discuss the relationship between the shapes of D1 to D6 and the electric field. The curve D1 in Fig. 8(c) represented the electric field under which dot D1 was fabricated, and so on. D3 and D4 were fabricated under same parameters. However the experimental results were different. One possible reason was the substrate surface was covered by a natural generated silicon dioxide film and the film might be thicker on D4. Table 1 recorded the heights and widths of the experimental nano-dots in Fig. 8(a), the amplitude of electric field simulation results and the width within which the simulation result was over 10^9V/m. From table 1 we could see that when

TABLE 1 Size of experimental dots and shapes of electric field of simulation

	Group1			Group2			
	D1	D2	D3	D4	D5	D6	Unit
H	3.7	2.8	2.9	2	1.9	1.8	nm
W1	295	324	330	371	312	277	nm
R1= W1/H	80	116	114	186	164	154	
E	1.92	1.58	1.34	1.34	1.21	1.08	V/nm
W2	39	35	29	29	23	13	nm
R2= W2/E	20.3	22.2	21.6	21.6	19	12	
R1/R2	3.9	5.2	5.3	8.6	8.6	12.8	

H: Heights of experimental dots
W1: Widths of experimental dots
E: Amplitude of electric field
W2: Width of electric field over 1 V/nm

fabrication. The AFM tip was 5nm high from substrate and moved from up to down by 200nm/s after a 12V bias loaded to AFM tip. After a nano-line was fabricated, the above process was repeated automatically until the last line was finished. The mean value of the heights of these lines is 1.64nm and the variance is $0.13nm^2$ from three scanning measurements crossing the lines. The data and scanned image show that the lines are of good repetition.

V. CONCLUSION

We studied the electric field caused by the tip bias of AFM by simulation and analyzed the relationship between the electric field and parameters such as tip bias, tip-substrate distance and water-film height. Variation of electric field under different tip bias and tip-substrate distance was discussed. The impact of water films over tip and substrate was studied. Then some simple nano-structures were fabricated to verify simulation result. Experimental results approximately meet simulation results. So our simulation can provide helpful instruction for anodic oxidation processing.

ACKNOWLEDGMENT

Thanks to Wenfeng Liang for his help about simulation software.

REFERENCES

[1] H. C. Day and D. R. Allee, "SELECTIVE AREA OXIDATION OF SILICON WITH A SCANNING FORCE MICROSCOPE," Applied Physics Letters, vol. 62, pp. 2691-2693, May 1993.

[2] E. S. Snow and P. M. Campbell, "FABRICATION OF SI NANOSTRUCTURES WITH AN ATOMIC-FORCE MICROSCOPE," Applied Physics Letters, vol. 64, pp. 1932-1934, Apr 1994.

[3] P. Avouris, et al., "Atomic force microscope tip-induced local oxidation of silicon: Kinetics, mechanism, and nanofabrication," Applied Physics Letters, vol. 71, pp. 285-287, Jul 1997.

[4] B. Legrand and D. Stievenard, "Nanooxidation of silicon with an atomic force microscope: A pulsed voltage technique," Applied Physics Letters, vol. 74, pp. 4049-4051, Jun 1999.

[5] X.-D. Hu and M.-M. Huang, "Current analysis of dynamic electric field induced oxidation nanofabrication with AFM," Nami Jishu yu Jingmi Gongcheng/Nanotechnology and Precision Engineering, vol. 8, pp. 352-355, 2010.

[6] J. A. Dagata, et al., "Role of space charge in scanned probe oxidation," Journal of Applied Physics, vol. 84, pp. 6891-6900, Dec 1998.

[7] R. Garcia, et al., "Patterning of silicon surfaces with noncontact atomic force microscopy: Field-induced formation of nanometer-size water bridges," Journal of Applied Physics, vol. 86, pp. 1898-1903, Aug 1999.

[8] K. Akiyama, et al., "Functional Probes for Scanning Probe Microscopy," in Proceedings of the International Conference on Nanoscience and Technology. vol. 61, E. H. M. G. C. G. H. J. Meyer, Ed., ed, 2007, pp. 22-25.

[9] M. Bartosik, et al., "Role of humidity in local anodic oxidation: A study of water condensation and electric field distribution," Physical Review B, vol. 79, May 2009.

Scanning electron beam induced deposition for conductive tip modification

P. L. Chen[1*], James Su[1], M. H. Shiao[1], M. N. Chang[2], C. H. Lee[3] and C. W. Liu[3]

[1, *] Instrument Technology Research Center, National Applied Research Laboratories, Hsinchu, 300, Taiwan
[2] Department of Physics, National Chung Hsing University, Taichung, 402, Taiwan
[3] Department of Electrical Engineering, National Taiwan University, Taipei, 106, Taiwan
poli@itrc.narl.org.tw

Abstract— In this study, we have developed conductive tips with high aspect ratio and good sharpness by scanning electron beam induced deposition (SEBID) method. The structure and morphology of modified conductive tips were performed by analytical transmission electron microscopy (TEM) methods. Atomic force microscope (AFM), scanning Kelvin probe microscopy (SKPM) and electrostatic force microscopy (EFM) analysis were done by using these modified tips and compared to standard commercially scanning probe microscopy (SPM) probes. The SEBID modified tips demonstrate signals with smaller full-width at half-maximum (FWHM) value in EFM measurements. The prepared tips could provide high spatial resolution and high stability to avoid a perturbation of the sensor characterization due to the sample's stray field.

Keywords- scanning electron beam induced deposition; conductive tip; modification

I. INTRODUCTION

The scanning probe microscopies (SPMs) based on two-pass technique including electrostatic force microscopy (EFM) [1], magnetic force microscopy (MFM) [2], and scanning Kelvin probe microscopy (SKPM) [3] are important analysis tools for electrical properties of nanoscopic systems, magnetic field and potential distribution measurement on material surfaces, respectively. However, SPMs have suffered from spatial resolution problems due to the urgent need in nano-metrology. In general, the sharpness and shape of the tip apex of a SPM probe directly affects the SPM measurements on surface properties including electric field, magnetic field and morphologic roughness. A standard commercial SPM probe with PtIr conductive coating have tips of pyramid shape, which is insensitive to the change of a tiny area and may result in an image with a poor spatial resolution, leading to detail information lost. Furthermore, typical probes in electric field or magnetic field distribution resulting in resolution degradation show stray field effect [4]. An ideal SPM probe should have properties of high spatial resolution and high stability to avoid a perturbation of the sensor characterization due to the sample's stray field. For this reason, a tip apex with high aspect ratio and good sharpness is needed to provide field images without stray field effect.

Various SPM probes with high aspect ratio tips have been fabricated by mechanical and electrochemical etching methods. For instance, carbon contamination tips and metallic tips are created via electron beam induced deposition (EBID) on SPM tips [5]. Carbon tips are prepared by catalyst patterning through EBID based technique [6]. Sharp tips fabricated with focused ion beam (FIB) techniques [7]. Atomic force microscope (AFM) tips attached with carbon nanotubes (CNTs) [8]. Conductive tips fabricated by electrochemical etching method [9]. In addition, AFM tips that works by adding nanoparticles on their surface with an ion cluster source [10].

However, these fabrication methods have their own drawbacks. The EBID based techniques need additional coating process to perform conductive tips. Besides, tips created by EBID method need instruments equipped with a computer-controlled procedure which laterally shifts the electron beam in spot mode by a fixed step. The FIB and CNTs attachment methods require a serial process which is time-consuming. Moreover, the attachment angle and the number of CNTs attached are not always controllable. The electrochemical etching methods produce pollution during fabrication. Although the method by adding nanoparticles lead to AFM images of higher aspect ratios and spatial resolution, it is hard to generate a tip with an aspect ratio of 1:10.

For these reasons, we have developed high aspect ratio conductive platinum (Pt) tips by a novel technique called scanning electron beam induced deposition (SEBID) method in a dual beam system. In this paper, the structure and morphology of conductive Pt tips were performed by analytical transmission electron microscopy (TEM) methods. We also studied the electrical characterizations of the modified tips and demonstrate that these tips could be useful for electrical measurements. Comparisons between modified tips and standard commercially SPM probes were done by AFM, SKPM and EFM measurements. The prepared tips with high aspect ratio and good sharpness could provide high spatial resolution and high stability. With all of the experimental results, we exhibit that tip apex by this technique can not only improve the tip apex but also recycle the used probes, reuse the probes as well as reduce the consumption of probes.

978-1-4673-1122-9/12 $31.00 © 2012 IEEE

II. EXPERIMENTS

For modifying a conductive tip apex, we employed a dual beam FIB system (FEI NOVA 200) equipped with a gas injection system (GIS) which allows either the deposition of finely shaped metallic and insulating nanostructures or a locally selective etching of various materials. FIB induced material deposition provides the ability to deposit materials on solid surface with very high spatial precision.

The SEBID process for AFM tip apex modification is shown in Figure 1. The precursor (methylcyclopentadienyl platinum trimethyl; $(CH_3)_3CH_3C_5H_4Pt$) was initially vaporized by heating to about 41 °C and injected into the path of the electron beam. In order to control the growth rate and avoid anisotropic growth effects, we adopted the gas precursor without needle insertion [11] while freestanding Pt nano-rod growth was performed. The electron beam was then focused on the apex of an AFM probe and scanning area which was about 30 x 100 nanometers at 1280 kX magnification. The electron beam shifted along the lateral direction of AFM probe and decomposed the gas creating a deposition to fabricate a freestanding Pt nano-rod on the tip apex. The reaction byproducts were finally removed by the instruments vacuum system. The nominal electron beam spot size was 5 nm, probe current was 0.6 nA and the accelerating voltage was 20 kV; the chamber pressure was 5×10^{-4} Pa during deposition, which was performed at room temperature. The probes used in this work were standard commercial SPM probes with PtIr coating produced by NANOSENSORS.

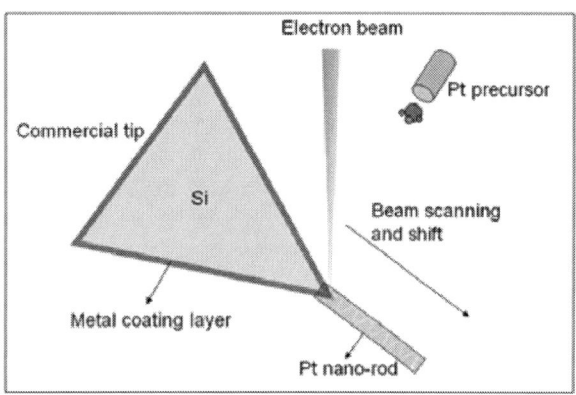

Fig.1 The SEBID process for AFM tip apex modification

The SEBID Pt nano-rod samples deposited directly onto TEM grids were analyzed by a JEOL TEM system (JEM-2010F) with an accelerating voltage of 200 kV and a point resolution of 0.23 nm. All the AFM and local surface potential distribution analysis were performed in an environment with well-controlled temperature and humidity, using a Veeco di Dimension 3100 scanning probe microscope equipped with the Extender Electronics Module (R) to acquire the related phase information. Whereas the Extender Electronics Module

provides phase linearity range nominally up to ±30 degrees. Surface potential microscopy detects the effective surface voltage of the sample by adjusting the voltage on the tip to compensate that of the surface, therefore minimizes the electric force between the tip and the sample. Samples may include both conducting and nonconducting materials. Samples with regions of different materials will show image contrast due to contact potential differences (CPD).

III. RESULTS AND DISCUSSION

Figure 2 shows the front-view SEM images of the tip apex (a) before and (b, c) after Pt nano-rod modification. The nano-rod is about 270 nm long and 20 nm in diameter with an aspect ratio of 1:13.5. The aspect ratio of modified tip is controllable because of the adjustability of electron beam for growth site and growth rate. Due to the distribution of the interaction volume, the nano-rod size in the side-view SEM image is slightly wider than that in the front-view SEM image, as shown in figures 2(b) and 2(c).

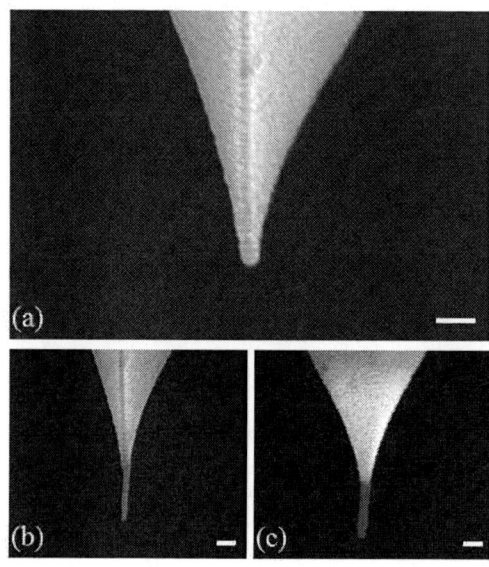

Fig.2 Front-view SEM images of the tip apex (a) before and (b) after Pt nano-rod modification. The side-view SEM image after tip apex modification is shown in (c). Each image has the same scale bar of 100 nm.

Figure 3(a) represents a low-magnification TEM image of a SEBID Pt nano-rod and a magnified image of the central part of the nano-rod is shown in Fig.3(b). The images exhibit the structure and the morphology of the deposited Pt nano-rod. It is composed of an amorphous carbon thin layer with low electro-optical density (bright contrast) and Pt grains with high electro-optical density (dark contrast) [11]. The selected area electron diffraction (SAED) pattern is displayed in Fig. 3(c). From conductive atomic force microscopy (C-AFM) measurements, the average probe resistances for the modified tip and a commercial tip are 1.16×106 and 1.59×106 ohms, respectively. This result implies that the modified tips can be used for electrical measurements.

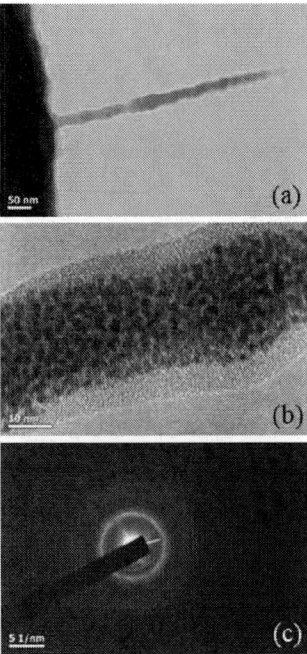

Fig.3 (a) TEM image of a SEBID Pt nano-rod. (b) Magnified image of the central part of the nano-rod. (c) SAED of the central part of the nano-rod.

The samples used in this study were SiGe quantum rings (QRs) on a silicon substrate. Figures 4(a) and 4(b) are the AFM and SKPM images of a QRs sample measured by a standard commercial SPM probes, respectively. As shown in Figures 4(c) and 4(d), a Pt nano-rod modified probe with a sharp tip apex demonstrates a small island structure at the central area of a SiGe quantum ring. From the stactistics analysis of figures 4(a) and 4(c), the diameter range is (42.85~186.54) nm and (34.36~192.19) nm, respectively. As shown in Fig.5, using SKPM with a SEBID-modified conductive tip, we have further observed that there is a Ge-like area at the central region of the quantum ring. Figure 6 displays AFM (a, c) and EFM (b, d) images of SiGe quantum rings on a semiconductor substrate obtained by using a standard commercial SPM probe and a Pt nano-rod modified tip, respectively.

From figures 4 and 6, the SKPM results were consistent with the EFM results. The quality of prepared conductive tip has been tested through a range of practical applications in electrical testing. These tips demonstrate signals with smaller full-width at half-maximum (FWHM) value in EFM measurements, implying that images with a better spatial resolution can be obtained.

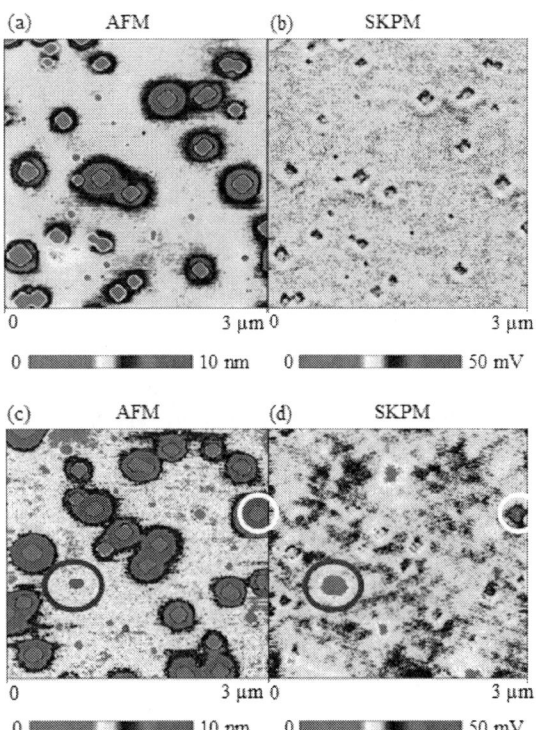

Fig.4 AFM (a and c) and SKPM (b and d) images of SiGe quantum rings on a semiconductor substrate obtained by using a standard commercial SPM probe and a Pt nano-rod modified tip, respectively.

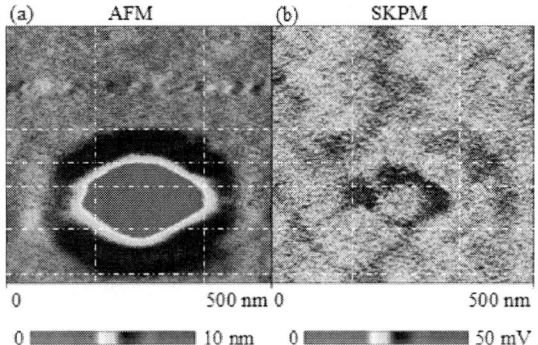

Fig.5 (a) AFM image and (b) SKPM image of a SiGe quantum ring scanned by a Pt nano-rod modified tip.

IV. CONCLUSIONS

Using one-step SEBID process, one can easily modify a single conductive tip on the commercial SPM probes. The quality of SEBID-modified conductive tips has been tested through a range of practical applications in electrical testing. The AFM image shows a small island structure at the central area of a SiGe quantum ring with a sharp tip apex. Using SKPM with a SEBID-modified conductive tip, we have further observed that there is a SiGe-like area at the central

978-1-4673-1122-9/12 $31.00 © 2012 IEEE

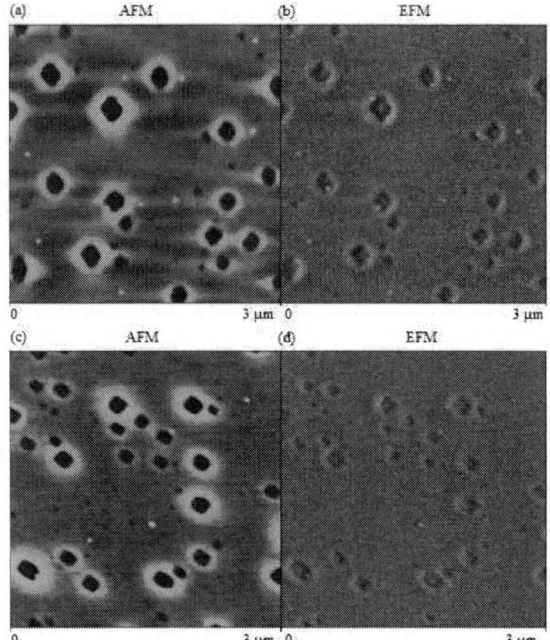

Fig.6 AFM (a, c) and EFM (b, d) images of SiGe quantum rings on a semiconductor substrate obtained by using a standard commercial SPM probe and a Pt nano-rod modified tip, respectively.

region of the quantum ring. The prepared conductive tips demonstrate signals with smaller FWHM value in EFM measurements, implying that images with a better spatial resolution can be obtained. Moreover, the experimental results indicate that tip apex by this one-step technique can not only improve the tip apex but also recycle the used probes, reuse the probes as well as reduce the consumption of probes.

ACKNOWLEDGMENT

We would like to acknowledge the assistance of the technical staff at National Nano Device Laboratories at National Applied Research Laboratories in Taiwan where the TEM analysis was performed.

REFERENCES

[1] G. M. Sacha, A. Verdaguer, J. Martínez, J. J. Sáenz, D. F. Ogletree, and M. Salmeron, "Effective tip radius in electrostatic force microscopy," *Appl. Phys. Lett.*, vol.86, pp.123101, 2005.

[2] L. Folks, M. E. Best, P. M. Rice, B. D. Terris, D. Weller and J. N. Chapman, "Perforated tips for high-resolution in-plane magnetic force microscopy," *Appl. Phys. Lett.*, vol.76, pp.909, 2000.

[3] T. Arai, S. Gritschneder, L. Troger and M. Reichling, "Carbon tips as sensitive detectors for nanoscale surface and sub-surface charge," *Nanotechnology*, vol.15, pp.1302–1306, 2004.

[4] A. Wadas and H. J. Hug, "Models for the stray field from magnetic tips used in magnetic force microscopy," *J. Appl. Phys.*, vol.72 (1), pp.203-206, 1992.

[5] J. D. Beard and S. N. Gordeev, "Fabrication and buckling dynamics of nanoneedle AFM probes," *Nanotechnology*, vol.22, pp.175303 (8pp), 2011.

[6] I. C. Chen, L. H. Chen, C. Orme, A. Quist, R. Lal and S. Jin, "Fabrication of high-aspect-ratio carbon nanocone probes by electron beam induced deposition patterning," *Nanotechnology*, vol.17, pp.4322–4326, 2006.

[7] M. J. Vasile, C. Biddick and H. Huggins, "Formation of probe microscope tips in silicon by focused ion beams," *Appl. Phys. Lett.*, vol.64 (5), pp.575–576, 1994.

[8] H. W. Lee, S. H. Kim, Y. K. Kwak, and C. S. Han, "Nanoscale fabrication of a single multiwalled carbon nanotube attached atomic force microscope tip using an electric field," *Rev. Sci. Instrum.*, vol.76, pp.046108, 2005.

[9] P. Kim J. H. Kim, M. S. Jeong, D. K. Ko, J. Lee, and S. Jeong, "Efficient electrochemical etching method to fabricate sharp metallic tips for scanning probe microscopes," *Rev. Sci. Instrum.*, vol.77, pp. 103706, 2006.

[10] L. Martínez, M. Tello, M. Díaz, E. Román, R. Garcia, and Y. Huttel, "Aspect-ratio and lateral-resolution enhancement in force microscopy by attaching nanoclusters generated by an ion cluster source at the end of a silicon tip," *Rev. Sci. Instrum.*, vol.82, pp.023710, 2011.

[11] S. Frabboni, G. C. Gazzadi, A. Spessot, "TEM study of annealed Pt nanostructures grown by EBID," *Physica E*, vol.37, pp.265–269, 2007.

In situ study of thermal deformation of metal resistive heater on silicon nitride membrane by digital holographic microscopy

Yiu Wai Lai and Joshua E.-Y. Lee*

*Department of Electronic Engineering, City University of Hong Kong, Kowloon, Hong Kong

joshua.lee@cityu.edu.hk

Abstract—**Metal resistive heater on dielectric membrane structures are common in MEMS. In this paper, the evolution of the surface topography of this type of structure during operation is studied by *in situ* digital holographic microscopy with nanometer-scale resolution. Devices of a typical design with platinum resistive heater lying on 200 nm silicon nitride membrane were fabricated by standard MEMS processes. A permanent out-of-plane surface deformation up to 200 nm could be detected when applying heating cycles via real-time *in situ* images of the device surface profile. Such deformation bears the risk of failure in the thin membrane device.**

Keywords-membrane, microheater, digital holograpic microscopy

I. INTRODUCTION

Metal resistive heater on dielectric membrane structures have been employed in a wide variety of applications, e.g. micro-hotplates [1]. The membrane structure, whose thickness leis in the order of 100 nm, possesses the advantage of rapid heating and cooling rates owing to their small thermal mass. Applications of micro-hotplates include localized CNT growth [2], which can be further configured to realize CMOS based gas sensors [3]. For such a thin released structure, study of the surface topography during device operation is either slow (e.g. by atomic force microscopy) or invasive (e.g. by stylus profilometry). In this work, a customized lab-built digital holographic microscope (DHM) is used to monitor the surface topography of devices with a typical heater-membrane design. DHM is an optical interferometry-based technique that allows fast and non-invasive capture of topographical information of the device surface with nanometer resolution [1, 4, 5].

Digital holographic microscopy analyses the interference patterns of a light beam scattered by an object, with a coherent reference beam. It has recently been used previously to obtain three-dimensional surface information of MEMS devices [1, 4, 5]. Compared to other optical surface analysis techniques (e.g. white light interferometry – WLI for short), DHM possesses similar vertical and lateral resolutions. Although the maximum measurable step height in DHM is limited by the wavelength of the light source, the absence of any high-precision moving

component (e.g. the vertical scanning unit in WLI) leads to a simpler, cheaper and more robust measurement setup in DHM.

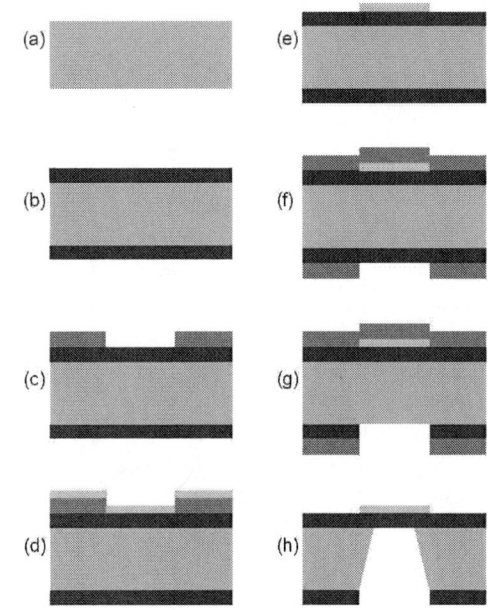

Fig. 1. Fabrication scheme of the heater-membrane structure. See text for detailed description.

II. EXPERIMENTS

A. Fabrication of MEMS heater

The heater-membrane structure was fabricated by standard MEMS processes. Fig. 1 shows the fabrication sequence. A double-side polished Si wafer was used as the substrate of the device (fig. 1a). After removing the native oxide layer by buffered HF solution, 200 nm silicon nitride layers were grown on both sides of the wafer by either low-pressure chemical vapor deposition (LPCVD) or plasma enhanced chemical vapor deposition (PECVD) (fig. 1b). The properties of the nitride films prepared by these methods are different in several aspects [6]. Depending on the required material properties and fabrication conditions, both technologies are widely used for silicon nitride deposition in industry. The top side of the nitride-coated wafer was then patterned by standard photolithography for defining the structure of the metal heater (fig. 1c). 190 nm Pt was deposited by sputtering with a layer of 10 nm Ti as an adhesion layer between Pt and silicon nitride (fig. 1d). This was followed by a lift-off process such that only the desired metal pattern was left on the wafer (fig. 1e). Photolithography was then performed again to pattern the bottom side to define the opening for wet etching of the bottom side trench (fig. 1f). At the same time, the top layer was protected by a thick photo resist layer. The nitride at the bottom opening was removed by reactive-ion-etching (fig. 1g). Finally the membrane was released by performing single side KOH etching of the silicon wafer from the bottom side (fig. 1h). Fig. 2 shows a schematic diagram of the as-fabricated structure and the corresponding optical micrograph.

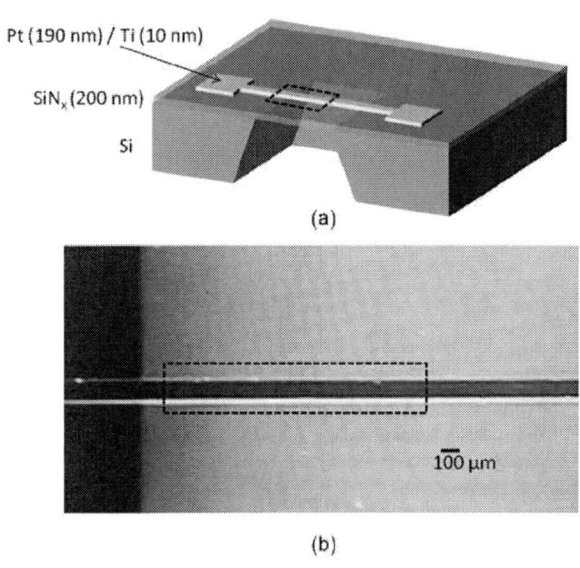

Fig. 2. (a) A schematic diagram of the as-fabricated heater-membrane structure. The squared area corresponds to the area imaged in (b). (b) An optical micrograph showing the metal structure lying on the silicon nitride membrane. The squared area corresponds to the area imaged by DHM.

To operate the microheater, an electrical current of up to 30 mA was applied to the metal beam by an external DC source meter through the square bond pads at the ends of the structure. Heat is generated due to the electrical resistance of the metal. The surface deformation of the metal beam due to thermal expansion as current is passed was monitored by DHM. The technique of DHM is described below.

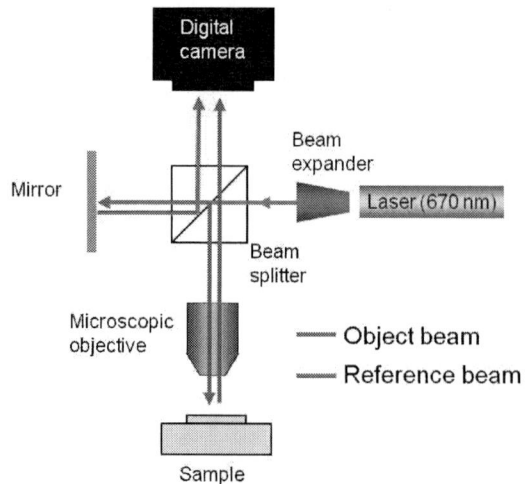

Fig. 3. Scheme of the experimental setup of the DHM.

B. Digital holographic microscopy (DHM)

To study the surface topography of the heater-membrane structure, a DHM setup (fig. 3) was developed for the *in situ* observation of the device during operation. A diode laser beam (670 nm) is spatially expanded by a beam expander before being split into two coherent beams by a beam splitter. The reference beam is reflected by a mirror. The object beam shines on the sample through a microscope objective. The scattered light from the sample returns through the beam splitters and interferes with the reference beam. The interference pattern, which is called a hologram, is recorded by a digital camera (2592 x 1728 pixels with pixel size of 8.6 x 8.6 μm^2). As the formation of the hologram relies on the light reflected from the sample, a sufficiently reflective sample surface is required for use of DHM. A numerical reconstruction method, based on Fresnel transformation, is used to obtain the intensity and phase information of the object beam from the recorded hologram [7]. An example of the reconstruction procedure can be found in [5]. While the reconstructed intensity shows the 2D picture of the sample, the phase contains the 3D surface profile information. However, due to the aberrations caused in the optical components, the phase image is distorted, which consequently leads to a distorted reconstructed surface profile. A quasi-double-exposure scheme is employed to solve this problem [8]. This is done by recording a reference hologram from a flat reference surface on the sample close to the area of interest followed by subtracting its phase from the original phase. Once the reference hologram is obtained at the beginning of the experiment, only a single exposure is needed for visualizing the topography at any stage of the experiment. Fig. 4 shows the reconstructed DHM images for a section of the heater-membrane structure in the as-fabricated state. Note that the vertical and lateral axes are in nm and μm respectively. Therefore the scale of the vertical topography is exaggerated relative to the lateral features for better visualization. The lateral resolution of the DHM depends on the magnification of the microscope objective. The vertical resolution of the DHM has been shown to be in the order of 10 nm [1].

III. RESULTS AND DISCUSSION

Fig. 4 shows the DHM images of a section of the metal beam on a LPCVD nitride membrane in the as-fabricated state corresponding to the marked area in fig 2 (b). The step height of the beam is 200 nm, which agrees well with the designed thickness of the metal layers. By passing electrical current along the metal, heat is generated locally due to the electrical

resistance of the material. Current was applied stepwise up to 30 mA and DHM images were taken for every step. Fig. 5 shows images of the same section after passing 30 mA current (image taken after switching off the current). A comparison between fig. 4 and fig. 5 reveals that the topography of the membrane is dramatically changed. As a result of the large difference in thermal expansion coefficients between Pt/Ti and silicon nitride, stress is developed during the heat induced expansion. After switching off the current, instead of returning to the original state, the stressed structure deforms to accommodate the stress. Fig. 6 shows images of the back side of the membrane corresponding to the same section depicting in figs. 4 and 5. It can be seen that the originally flat surface has deformed in the opposite sense with respect to the front

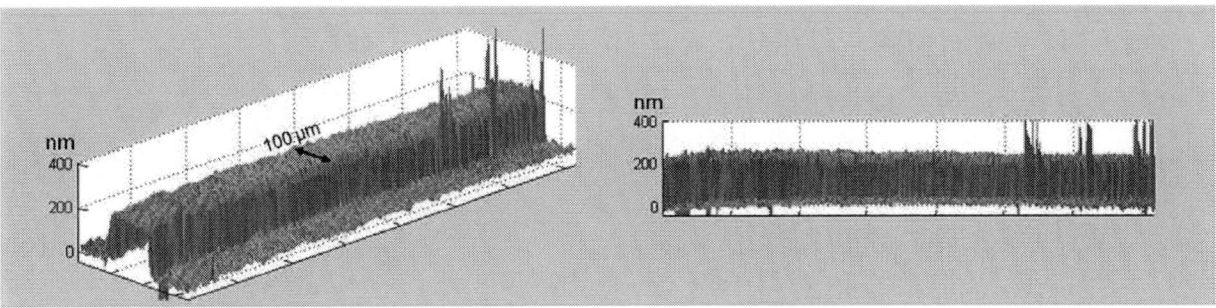

Fig. 4. DHM images show the structure after the releasing the membrane by wet etching in perspective (left) and transverse (right) views.

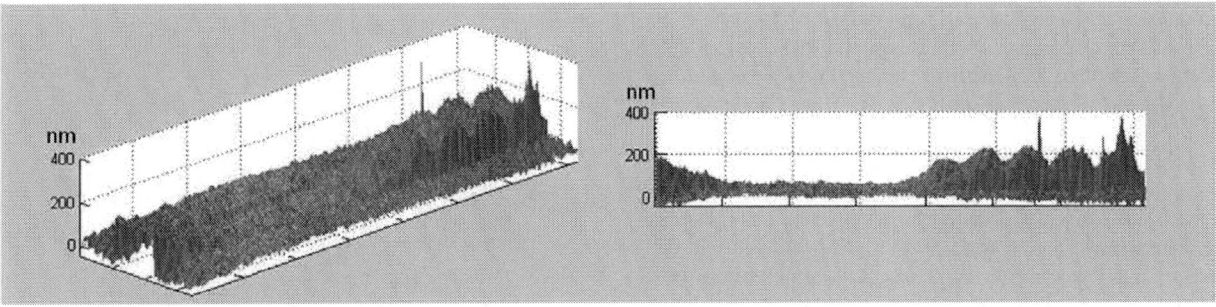

Fig. 5. DHM images show the structure after switching off the 30 mA electrical current in perspective (left) and transverse (right) views.

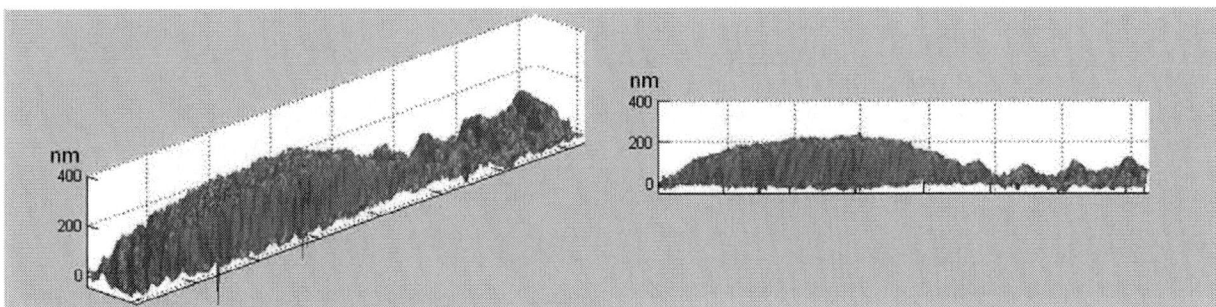

Fig. 6. DHM images show the back side of the membrane after switching off the 30 mA electrical current in perspective (left) and transverse (right) views.

978-1-4673-1122-9/12 $31.00 © 2012 IEEE

side topology DHM image. It should also be noted that the inward bending of the metal beam on the front side results in the outward bending on the bottom side with the same shape and magnitude. This shows that the deformation is mainly a mechanical one without a significant redistribution of materials during heating.

Fig. 7 contains a series of successive *in situ* DHM images taken during the heating/cooling process as the applied current through the metal is ramped up. The images correspond to the same area as shown in fig. 4, but only the deformation of the metal beam is shown for better visualization. Fig. 7(a) was taken at 5 mA. No significant change in surface topography is observed compared to the as-fabricated state shown in fig. 4. Fig. 7(b) was taken when the current was 19 mA. It can be seen that at this stage, the metal beam has started to deform out-of-plane to form a rippled deformation pattern across the underlying membrane. Fig. 7(c) was taken when the current reached 21 mA. The metal beam has now deformed further with a sinusoidal profile. The largest vertical deformation is around 400 nm, which is comparable to the total thickness of the metal-membrane structure. The current was increased further up to 30 mA. At this point, the deformation became too large for an accurate measurement of the surface profile by DHM due to the step height limitation in the technique. The current was then reduced stepwise. Fig. 7(d) was taken at 19 mA after reaching the maximum applied heater current levels of 30 mA. As can be seen from Fig. 7(d), the profile was changed significantly in that half of the observed metal track section has bent downward to form a shallow U-shape. Fig. 7(e) was taken when the current was further reduced to 5 mA. It can be seen that the U-shape bending still remains, but the scale of the sinusoidal deformation has been largely reduced.

The *in situ* observation of the surface deformation of the metal beam shows how the structure accommodates the induced stress developed during the temperature ramp. Due to the difference in thermal expansion coefficients between the dielectric and the metal layers, the heat generated by the electrical current induces different degrees of volumetric change in the respective layers of the heater-membrane structure. Because of the strong adhesion between layers (nitride and metal), the individual layers are not able to expand freely. Mechanical stress is developed across each of the layers as a result. Eventually the whole structure deforms in such a way in order to balance the thermal stress and adhesion force between layers. The observed nano-scale out-of-plane deformation bears the risk of leading to integrity problems of the device. Compared to the membrane thickness itself (only 200 nm thick), the 400 nm out-of-plane deformation during heating would be larger than the membrane thickness. Furthermore, the permanent deformation after heating/cooling cycles may lead to functional inaccuracy since the geometry of the device is distorted.

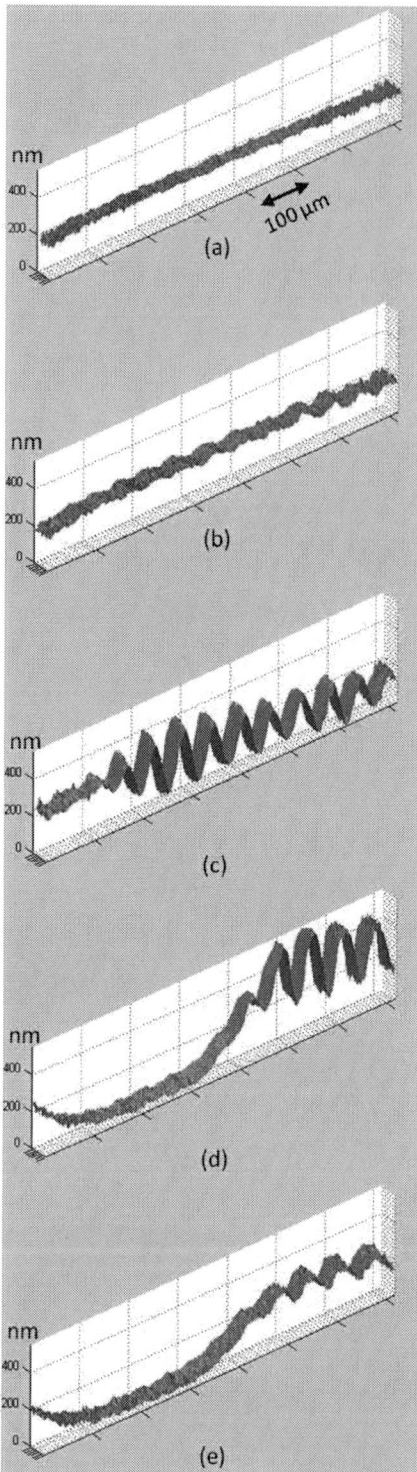

Fig. 7. In-situ DHM images during the heat/cooling process of the metal beam corresponding to the same area as in fig. 4. See text for detailed description

IV. SUMMARY

The use of DHM, as shown in this work, reveals otherwise not easily available information on out-of-plane deformations in MEMS structures due to thermal expansion. The usefulness of this technique is highlighted by the importance of such information for investigating the general integrity of MEMS devices with nanometer resolution *in situ*.

The nano-scale out-of-plane sinusoidal deformations that have been observed for the microheater during operation stem from differences in the thermal expansion coefficients between the structural layers. Test carried out on nitride membranes separately prepared by LPCVD and PECVD each show very similar results. This therefore suggests that the observed deformation using DHM during the heating/cooling operation is fabrication process independent. Further study by using finite element simulation will be beneficial to quantify the stress evolution.

V. ACKNOWLEDGMENT

The work described in this paper was supported by a grant from the City University of Hong Kong (Project No. 7002690).

REFERENCES

[1] Y. W. Lai, N. Koukourakis, N. C. Gerhardt, M. R. Hofmann, R. Meyer, S. Hamann, M. Ehmann, K. Hackl, E. Darakis, and A. Ludwig, "Integrity of micro-hotplates during high-temperature operation monitored bu digital holographic microscopy" *J. Microelectromech. Syst.*, vol. 19, pp. 1175-1179, 2010.

[2] M. S. Haque, K. B. Teo, N. L. Rupensinghe, S. Z. Ali, I. Haneef, S. Maeng, J. Park, F. Udrea, and W. I. Milne, "On-chip deposition of carbon nanotubes using CMOS microhotplates," *Nanotechnology*, vol. 19. 025607, 2007.

[3] J. W. Gardner, P. K. Guha, F. Udrea, and J. A. Covington, "CMOS interfacing for integrated gas sensors: A review," *IEEE Sens. J.*, vol. 10, no. 12, pp. 1833-1848, 2010.

[4] Y. W. Lai, M. Krause, A. Savan, S. Thienhaus, N. Koukourakis, M. R. Hofmann, and A. Ludwig, "High-throughput characterization of stresses in thin film materials libraries using Si cantilever array wafers and digital holographic microscopy" *Rev. Sci. Inst.*, vol. 82, 063903, 2011.

[5] Y.W. Lai, M. Krause, A. Savan, S. Thienhaus, N. Koukourakis, M.R. Hofmann, and A. Ludwig, "High-throughput characterization of film thickness in thin film materials libraries by digital holographic microscopy" *Sci. Technol. Adv. Mater.*, vol. 12, 054201, 2011.

[6] A. Stoffel, A. Kovacs, W. Kronast, and B. Mueller, "LPCVD against PECVD for micromechanical applications," *J. Micromech. Microeng*, vol. 6, pp. 1-13, 1996.

[7] U. Schnars and W. Jueptner, "Digital Holography," Berlin, Germany: Springer-Verlag, pp. 44–51, 2005.

[8] G. Coppola, P. Ferraro, M. Iodice, S. De Nicola, A. Finizio, and S. Grilli, "A digital holographic microscope for complete characterization of microelectromechanical systems," *Meas. Sci. Technol.*, vol. 15, no. 3, pp. 529-539, 2004.

Comparison of Glass Etching Properties Between HCl and HNO₃ Solution

Wei Tao[1], Wenlong lv[1,2], Zhan Zhan[1], Wenjia Zuo[1], Xiaochun Qiu[1], Linyun Wang[1*], Daoheng Sun[1*]

[1] Department of Mechanical & Electrical Engineering, Xiamen University, CHINA
[2] Pen-Tung Sah Micro-Nano Technology Institute, Xiamen University, CHINA
rabitwangly@yahoo.com.cn, sundh@xmu.edu.cn

Abstract—A comparison of glass etching properties between HCl and HNO₃ solution is presented in this paper, which allows us to predict the etched product's shape under a variety of etching conditions, mask compensation and multiple processing steps. Four conclusions could be draw from the experiments. First, the best concentration ratio of the etching solution to protect the mask from damage and get a channel with depth of 40 μm is HF:HCl:NH₄F=5.5mol/L:4mol/L:2.5mol/L. Second, as the temperature increases, the longitudinal etching rate increases. However, the temperature has little influence on the lateral erosion ratio when the temperature gets high. Third, HCl has a better surface morphology against HNO₃ as an addition to solution. Last, the mask will introduce strain because of sputtering, which is harmful to the glass etching..

Keywords-glass etching; masking layer; HCl; HNO₃

I. INTRODUCTION

With the rapid development of MEMS, glass is gradually playing an important part in the fabrication of MEMS devices since glass has lots of advantages such as good optical properties, good mechanical properties, high electrical insulation and it can be easily bonded to silicon substrates at temperatures lower than for fusion bonding.

High precision bulk micromachining of glass is a key process step to shape spatial structures for fabricating different types of microsensors and microactuators. In the fabricating process of silicon micro-gyroscope, it is required that the etching depth of glass is 40 μm and the non-etching area is free from defect. There are mainly three methods used for glass etching: mechanical, dry and wet. Mechanical methods include ultrasonic drilling, electrochemical discharge as well as powder blasting. Ultrasonic drilling does not thermally damage the workpiece or appear to introduce significant levels of residual stress, which is important for the survival of brittle materials in service [1]. Electrochemical discharge is suitable for the machining of a large class of materials such as glass, quartz, various ceramics and others. Not only simple structures such as holes but as well as very complex structures like threads can be machined [2]. Powder blasting has a higher etching rate and a high machining efficiency. In addition to etching the channels by powder blasting, the in- and outlet holes can also be etched by the same process technique [3]. However, it's not easy to get smooth surface with these mechanical methods. Dry etching technique of glass in ICP

reactors had previously been reported in [4] using SF₆ and electroplated Ni as the mask. In [5], a new method of dry etching of glass using bulk silicon mask and C₄F₈ as etching gas is reported. However, the etching rate is relatively low and the materials for erosion are refrained to certain materials, which is not beneficial for the enlargement of the technique scale. As wet etching has a lot of advantages such as the etching rate is fast and a large quantity of glass wafers can be processed simultaneously, it remains the most effective method currently.

For glass etching, the most important thing is to obtain high etching rates and high aspect ratio structures. Researchers mainly focus on the choice of etching solutions and masking layer. On the one hand, researchers are trying to find the most suitable etching solutions. For example, highly concentrated HF (49%) was used as the etching solution in [6], in order to get through holes etched from both sides of a 500μm thick Pyrex glass wafer. In [7], HF/HCl was chosen as the etching solution for the achievement of smooth surface. In addition, the classical HNA (HF/HNO₃/H₂O) was also introduced as the etching solution in [8], to get the influence of the composition of the etch solution on glass etch velocity, undercut phenomenon and the quality of the structures. On the other hand, researchers are doing their best to search for the best masking layer. For instance, photoresist was used as the masking layer in [9], but unfortunately the photoresist in HF solution would experience a delamination with a relative long etching time, which limits its applications. In [10], Cr/Au is also used as a masking layer, where Cr layer is used to improve the adhesion of gold to glass. Another commonly used masking layer is silicon which is deposited by different method such as PECVD (amorphous silicon) [11], LPCVD (polysilicon) [12] as well as bulk silicon [13]. However, all of the methods illustrated on the above have its merits and demerits, which are not suitable for all the circumstances.

Here, in this work, an etching solution which is the composition of HF/HCl(HNO₃)/NH₄F and a masking layer consisting of two multilayers of Cr/Au and thick photoresist(BP212), was investigated to get a channel with depth of 40 μm and to get good surface morphology. A comparison was made between two group etching solutions with HCl and HNO₃ as an addition to the etching solution.

978-1-4673-1122-9/12 $31.00 © 2012 IEEE

II. EXPERIMENTS

In our experiments, we choose the Pyrex glass of Corning 7740 with a content of Al_2O_3 for wet etching process. There are mainly two reasons for our choice. For the first one, the composition of this glass is: SiO_2 81%, B_2O_3 13%, Na_2O 4%, Al_2O_3 2%. From the composition, we could discover that the Pyrex glass of Corning 7740 has a low content Al_2O_3 which is beneficial for the uniformity of the etching process. As we all know that, Al_2O_3 will give insoluble products after reacting with HF as AlF_3 which will be deposited on the generated surface and can act as a masking layer, as a result, the generated surface can be rough or, in the worst case, the etch rate can be drastically reduced. For the other one, Corning 7740 is easy to be integrated with microstructures fabricated in/on silicon and its thermal coefficient of expansion is similar to silicon which can result in low residual stress induced in the MEMS structure.

A schematic diagram of the glass wet etching process is shown in Fig.1. The experiment was performed as the follows. The wafers were first cleaned in acetone and alcohol for five minutes respectively, followed by being rinsed in potassium dichromate for 24h, then rinsed in DI water and spun-dried. Before sputtering the wafers were baked in an oven at 135℃ for 30 min. For patterning of Cr/Au mask, positive photoresist BP212 (from Clariant) and wet Cr and Au etchant were used. The thickness of photoresist layer was 1.5 μm. A hard baking process on a hot plate at 140℃ for 10 min was performed.

A schematic diagram of the glass wet etching device is shown in Fig.2. The device consists of a magnetic stirrer, a water bath heating beaker, a Teflon cup, a magnetic particle and a Teflon shelf. The magnetic stirrer is used to keep the water temperature at a constant magnitude and provide the power to drive the magnetic particle. The magnetic particle is used to stir the etching solution to make the solution uniform and the wafer to be fully etched. The Teflon shelf is used to support the wafer.

There are mainly two groups of experiments. The first group is to discover the effect of etching temperature on etching process and compare the difference between HCl and HNO_3 as an addition to the solution. The composition of the solution with HCl as an addition in the first group is $HF/NH_4F/HCl/H_2O$=32.6ml/19.27g/42.4ml/65ml. While the composition of the solution with HNO_3 as an addition in the first group is $HF/NH_4F/HNO_3/H_2O$ =32.6ml/19.27g/34.3ml/65 ml. The etching time is 20min. In the experiment, the composition of the solution and the etching time are constant, while the temperature varies from 50℃ to 70℃, as illustrated in table 1.

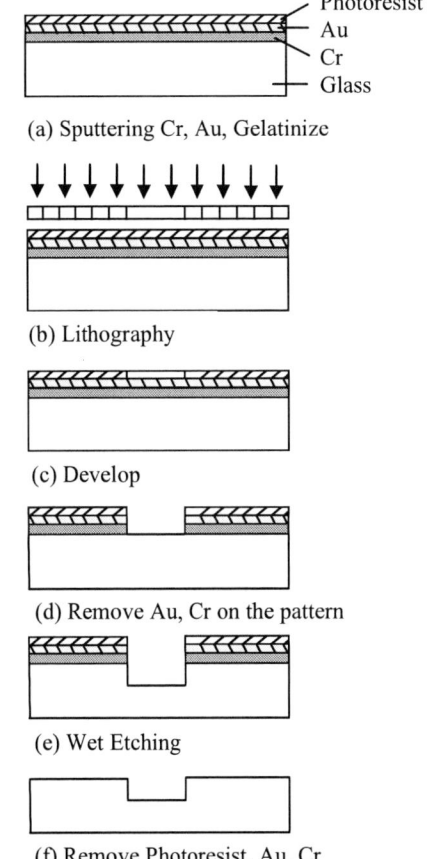

(a) Sputtering Cr, Au, Gelatinize

(b) Lithography

(c) Develop

(d) Remove Au, Cr on the pattern

(e) Wet Etching

(f) Remove Photoresist, Au, Cr

Figure1. Glass wet etching process

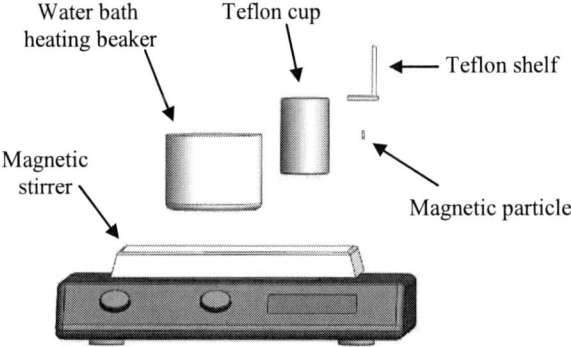

Figure2. Glass wet etching device

TABLE 1

Temperature （℃）	50	55	60	65	70
HCl Etching depth （μm）	13.48	29.74	30.51	36.70	47.77
HNO₃ Etching depth （μm）	29.61	39.89	46.32	51.77	65.00

The second group is to explore the effect of concentration on etching process and compare the difference between HCl and HNO_3 as an addition to the solution. The composition of the solution with HCl as an addition in the second group is $HF/NH_4F/HCl/H_2O$. The solution has a total volume of 200ml and the amount of substance concentration of HF and NH_4F are both 2mol/L. In the experiment, the amount of substance concentration of HCl is chosen as a variable from 1mol/L to 5mol/L, as depicted in table 2. While the composition of the solution with HNO_3 as an addition in the second group is $HF/NH_4F/HNO_3/H_2O$. The solution has a total volume of 200ml and the amount of substance concentration of HF and NH_4F are both 4mol/L, while the amount of substance concentration of HNO_3 varies from 2mol/L to 10mol/L as suggested in table 3. On the other hand, the etching time is 5min and the etching temperature is 60℃.

TABLE2

Concentration of HCl (mol/l)	1.00	2.00	3.00	4.00	5.00
Etching depth (μm)	5.21	6.68	6.89	7.36	11.86

TABLE 3

Concentration of HNO_3 (mol/l)	2.00	4.00	6.00	8.00	10.00
Etching depth (μm)	8.49	13.33	18.54	24.90	30.12

III. RESULTS AND DISCUSSIONS

The longitudinal etching rate versus temperature is shown in Fig.3. As the curve shows, we could draw a conclusion that the temperature has a great influence on the etching rate, in addition, the etching rate increases as the temperature increases. This is probably due to the fast movement of the molecule in the etching solution, which accelerates the reaction between the glass and the solvent with the increase of the temperature. It could also be drawn from Fig.3 that HNO_3 has a higher etching rate than HCl in the same situation. As a result, a higher temperature is preferred to get a higher etching rate.

Fig.4 shows the lateral erosion ratio versus temperature. We could draw a conclusion that the aspect ratio is slightly decreased but hasn't changed too much as the temperature increases. This is likely attribute to the simultaneous increase of transverse etching rate and longitudinal etching rate. Fig.4 also provides us a proper etching temperature since when the temperature exceed 60℃，the lateral erosion ratio decreases. However, when the temperature is higher than 65℃, the mask

has a tendency of delamination. therefore, 60℃ is the proper etching temperature.

From Fig.3 and Fig.4, we could see that HNO_3 has a faster longitudinal etching rate than HCl. This is perhaps due to the strong oxidability of HNO_3 which facilitates the reaction between HF and glass. However, HCl is preferred since HCl also has a fast longitudinal etching rate and has little influence on the mask layer. On the other hand, there are no much differences between HNO_3 and HCl in the lateral erosion ratio. So the temperature has little influence on the lateral erosion ratio when the temperature gets high.

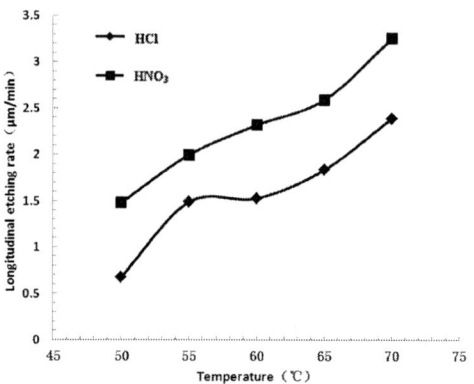

Figure3. Etching rate versus temperature

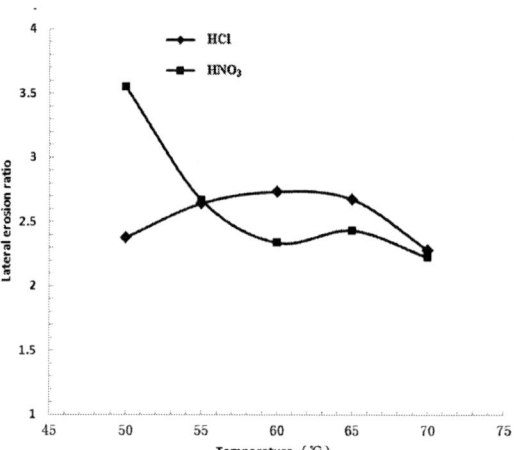

Figure4. Lateral erosion ratio versus temperature

Fig.5 presents the effect of the amount of substance concentration of HCl on etching process. The etching rate increases with the increase of the amount of substance concentration of HCl. It's likely that the increase of the amount of substance concentration of HCl will provide more and more H^+ ions which bind the F^- ions in the solution, causing the increase of the amount of substance of HF which will intensify the reaction with glass. Fig.5 also suggests that when the amount of substance concentration of HCl increases, the aspect ratio has a decrease tendency. This phenomenon can be explained as that when the amount of substance concentration of HCl increases, transverse etching rate is smaller than longitudinal etching rate.

Fig.6 presents the effect of the amount of substance concentration of HNO_3 on etching process. The etching rate increases with the increase of the amount of substance concentration of HNO_3, which can be drawn as a conclusion from Fig.6. It's possible that the increase of the amount of substance concentration of HNO_3 will provide more and more H^+ ions which bind the F^- ions in the solution, causing the increase of the amount of substance of HF which will intensify the reaction with glass. This process is almost the same with HCl since they all play a role in providing H^+ ions. Fig.6 also suggests that when the amount of substance concentration of HNO_3 increases, the aspect ratio has a decrease tendency. This phenomenon can also be explained as that when the amount of substance concentration of HNO_3 increases, transverse etching rate is smaller than longitudinal etching rate.

From Fig.5 and Fig.6, we could observe that HCl and HNO_3 have similar change tendency in longitudinal etching rate and lateral erosion ratio versus concentration. The longitudinal etching rate increases as the concentration increases, while the lateral erosion ratio decreases as the concentration increases. So we choose HCl concentration of 4mol/L because the longitudinal etching rate is high and the lateral erosion ratio is low. When the concentration exceeds 4mol/L, the longitudinal etching rate increases fast, which will result in a rough surface because the etching remains could not remove promptly. Although HNO_3 has a higher longitudinal concentration rate and a lower lateral erosion ratio than HCl (Fig6), HCl is preferred than HNO_3 as an addition to solution since HCl has a better surface morphology than HNO_3 (Fig7), which is due to HNO_3 has a strong oxidability so as to do damage to the mask.

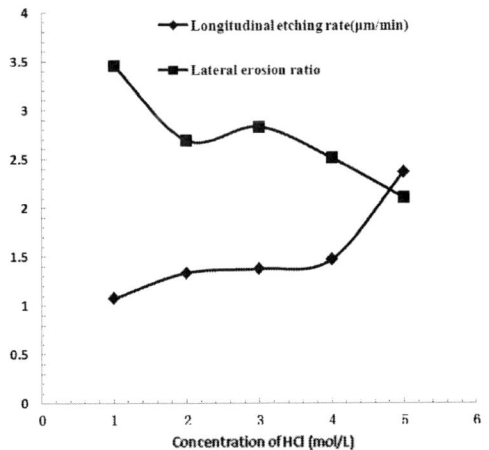

Figure5. Etching rate/Lateral erosion ratio versus concentration of HCl

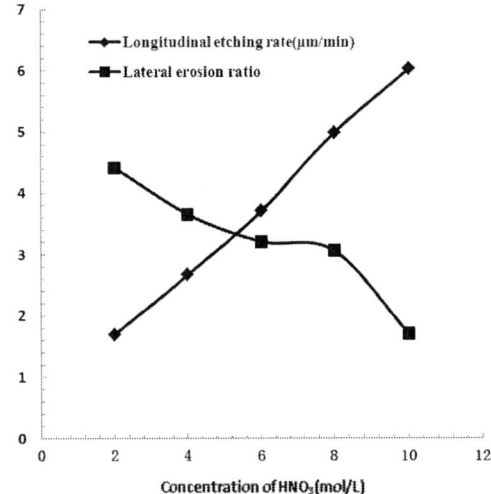

Figure6. Etching rate/Lateral erosion ratio versus concentration of HNO_3

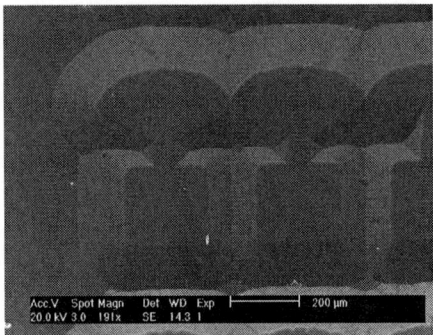

(a) 4mol/L HCl

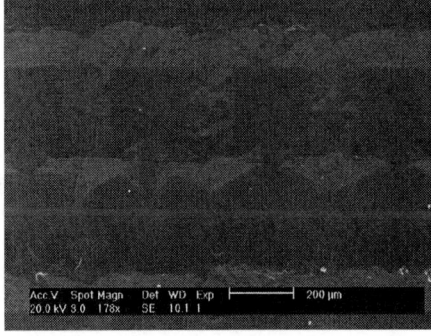

(b) 4mol/L HNO_3

Figure7. Surface morphology

From Fig.8, some boundary defects could be found in the etching area but in the non-etching area the surface morphology is perfect. So the Au/Cr mask is powerful other than the strain introduced by sputtering which results in defects. However, the strain could be removed by annealing after sputtering.

978-1-4673-1122-9/12 $31.00 © 2012 IEEE

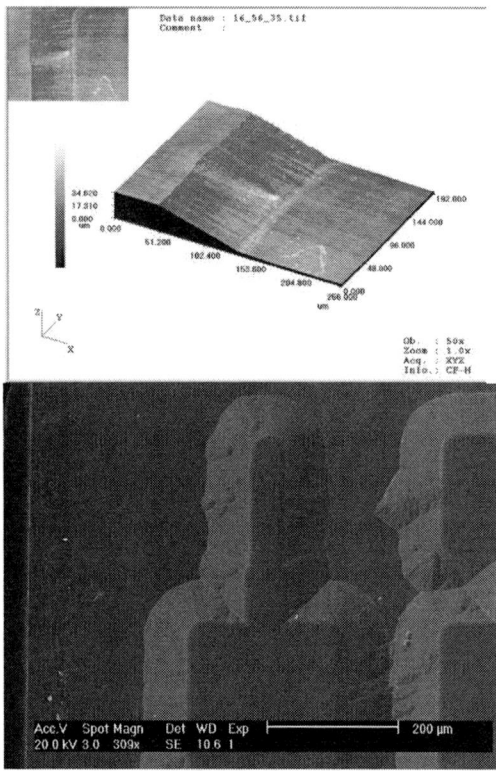

Figure8. Boundary defects

IV. CONCLUSION

A comprehensive comparison is made from temperature, concentration and surface morphology between HCl and HNO_3 as an addition to solution which is beneficial for us to predict the etched product's shape under a variety of etching conditions, mask compensation and multiple processing steps. Four conclusions could be drawn from the experiments. First, the best concentration ratio of the etching solution to protect the mask from damage and get a channel with depth of 40 μm is $HF:HCl:NH_4F=5.5mol/L:4mol/L:2.5mol/L$. Second, as the temperature increases, the longitudinal etching rate increases. However, the temperature has little influence on the lateral erosion ratio when the temperature gets high. Third, HCl has a better surface morphology against HNO_3 as an addition to solution. Last, the mask will introduce strain because of sputtering, which is harmful to the glass etching.

ACKNOWLEDGMENT

This work is supported by the National Natural Science Foundation of China (No. 51035002, No. 51105320), Natural Science Foundation of Fujian Province (No. 2011121045) and Fundamental Research Funds for the Central Universities (No. 2010121039, No. 2011121045).

REFERENCES

[1] T.B. Thoe, D.K. Aspinwall and M.L.H. Wise, "Review on ultrasonic machining," *International Journal of Machine Tools & Manufacture*,vol. 38 (1998), pp. 239–255.

[2] R.Wüthrich and V. Fascio, "Machining of non-conductive materials using electrochemical discharge phenomenon - An overview," *International Journal of Machine Tools and Manufacture*,vol.45 (2005), pp. 1095–1108.

[3] S. Schlautmann, H. Wensink, R. Schasfoort, M.C. Elwenspoek and A. van den Berg, "Powder-blasting technology as an alternative tool for microfabrication of capillary electrophoresis chips with integrated conductivity sensors," *J. Micromech. Microeng*,vol. 11 (2001), pp. 386–389.

[4] X. Li, T. Abe and M. Esashi, "Fabrication of high-density electrical feed-throughs by deep-reactive-ion etching of Pyrex glass," *J. Microelectromech. Syst*,vol.11 (2002), pp. 625–630.

[5] C. Iliescu, K.L. Tan, F.E.H. Tay and J.M. Miao, "Deep wet and dry etching of Pyrex glass: a review," *Proceedings of the ICMAT (Symposium F) Singapore*, July (2005), pp. 75–78.

[6] M. Bu, T. Melvin, G.J. Ensell, J.S. Wilkinson and A.G.R. Evans, "A new masking technology for deep glass etching and its microfluidic application," *Sens. Actuators A*,vol.115 (2004), pp. 476–482.

[7] C. Iliescu, J. Jing, F.E.H. Tay, J. Miao and T. Sun, "Characterization of masking layers for deep wet etching of glass in an improved HF/HCl solution," *Surf. Coat. Technol*.vol.198 (2005), pp. 314–318.

[8] Y. Mourzina, A. Steffen and A. Offenhäusser, "The evaporated metal masks for chemical glass etching for BioMEMS," *Microsyst. Technol.* vol.11 (2005), pp. 135–140.

[9] M. Stjernström and J. Roeraade, "Method for fabrication of microfluidic systems in glass," *J. Micromech. Microeng*,vol.8 (1998), pp. 33–38.

[10] F.E.H. Tay, C. Iliescu, J. Jing, J. Miao, "Defect-free wet etching through Pyrex glass using Cr/Au mask," *Microsyst. Technol*, vol.12 (2006), pp. 935–939.

[11] C. Iliescu, B. Chen, J. Miao, "Deep wet etching-through of 1mm Pyrex glass wafer for microfluidics applications," in: *Proceedings of the IEEE MEMS 2007*, January 21–25, Kobe, Japan, 2007.

[12] M.A. Grettilat, F. Paoletti, P. Thiebaud, S. Roth, M. Kondelka-Hep, N.F.de Rooij, "A new fabrication method for borosilicate glass capillary tubes with lateral inlets and outlets," *Sens. Actuator A: Phys*,Vol.60 (1997),pp. 219–222.

[13] T. Corman, P. Enokson, G. Stemme, "Deep wet etching of borosilicate glass using anodically bonded silicon substrate as mask," *J. Micromech. Microeng,*vol.8 (1998), pp. 84–87.

Surface Analysis and Process Optimization of Black Silicon

Fu-Yun Zhu, Qian-Li Di, Xing-Juan Zeng, Xiao-Sheng Zhang, Xin Zhao and Hai-Xia Zhang*

Science and Technology on Micro/Nano Fabrication Lab, Institute of Microelectronics, Peking University, Beijing, 100871, China

zhang-alice@pku.edu.cn

Abstract—This paper puts forward a description method for surface topography of black silicon using SF_6/C_4F_8 in a cyclic etching–passivation DRIE process. Three main parameters, i.e. density, height and width, are defined and used to describe black silicon and can be extended to several other parameters, such as aspect ratio, duty ratio and so on. By means of these parameters we can also establish a standard modal to provide the very basic data for other kind of research. So a program is developed to achieve these parameters expediently and accurately. Then we discuss the influence of the process parameters to surface topography and finally obtain a group of optimum parameters to fabricate black silicon. Through these results we are expecting to get better cognition of black silicon and form more controllable surface structures for mass production of black silicon.

Keywords- Black silicon, surface topography, DRIE, process parameter, classification analysis

I. INTRODUCTION

Black silicon is a new type of silicon and attracts extensive attention worldwide because of its excellent physical properties such as ultra-low reflectivity, super hydrophobicity, and high sensitization with light and so on. Black silicon was produced for the first time during the formation of Si trenches by reactive ion etching (RIE) in fluorine, bromine and chlorine plasmas [1]. It was just a by-product which was supposed to be avoided until its special properties were noticed.

Semiconductor processes widely use plasma etching to produce micromorphology through a lithographic etching mask. So the combination of lithography and mask-assisted etching is a typical technique to fabricate black silicon [2]. For example, fabrication of black silicon using conventional processes, such as nanosphere lithography [3] and silicon oxide nanopatterns [4], usually consist of several steps, and more notably, an etching mask is required. In general, the feature size is determined by the mask size. The wavelength of the exposure light source, high facility cost and low process throughput limit the achievable smallest feature size and thus represent a significant challenge. Unlike those other fabrications, we implement a single-step maskless process to fabricate large area of black silicon surfaces using an improved DRIE by optimizing its process parameters, which is low-cost, controllable, compatible with IC process and high-rate.

However, the detailed mechanism of the black silicon formation and the origin of its properties are yet to be

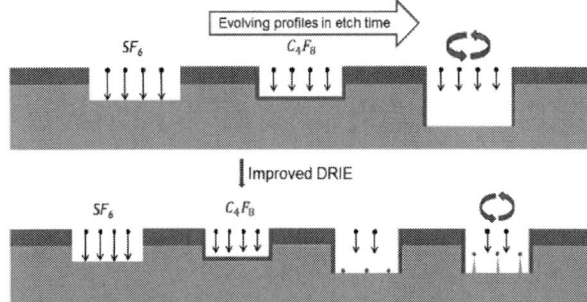

Fig. 1. Top: Conventional DRIE; Bottom: Improved DRIE

understood. In consideration of the tight staunch and strong relation between the characteristics and surface topography of black silicon, it is of vital importance to get a better cognition of black silicon surface topography, and then we can form more controllable black silicon and have further study.

This paper puts forward a description method for surface topography of black silicon and discusses the influence of the process parameters to surface topography. Through these results we are expecting to get better cognition of black silicon and form more controllable surface structures for mass production of black silicon.

II. EXPERIMENT

In this study, 4 inch polished N-type highly doped (100) silicon wafers with a resistivity of 0.05-0.2 Ω •cm were used. Experiments were carried out in fluorinated (SF_6 and C_4F_8) plasma in the cyclic etching/passivation mode which is called improved DRIE.

The top picture in Fig.1 shows the process of conventional

Fig. 2. SEM pictures of black silicon with different magnification.

978-1-4673-1122-9/12 $31.00 © 2012 IEEE

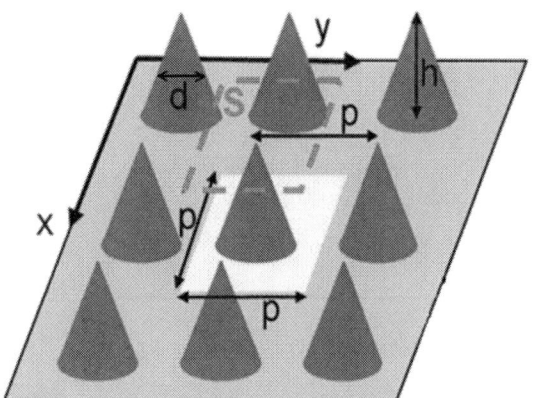

Fig. 3. Standard model of black silicon.

DRIE that can be divided into two steps: etching and passivation [5]. It needs to etch Si substrate using SF_6 in the first place and then deposit substrate using C_4F_8. Those two steps are repeated in turn and finally the structure with highly vertical sidewalls and high aspect ratio can be obtained. However, during the improved DRIE process which is obtained by optimizing the process parameters, there appear nanoparticals after a few etching/passivation cycles. Although we are not sure about the formation mechanism of nanoparticals, they will work as the nanomasks during the following DRIE process, thus black silicon is formed. Fig.2 shows the SEM pictures of black silicon with different magnification which represent good uniformity and high aspect ratio.

III. DESCRIPTION OF BLACK SILICON SURFACE TOPOGRAPHY

A. Defination

Black silicon has been demonstrated for numerous times by many researchers. However, because of its less irregular roughness about the surface, there are no uniform standards used when describing it. This obviously increases the difficulty to learn from each other.

In this paper, three main parameters, i.e. density, height and width, are defined as follows.

Density: Average number of silicon cones within $1 um^2$ area;

Height (h): Distance from the top to the bottom of the silicon cone;

Width (d): Diameter of the silicon cone at half of the height [6].

They are defined to describe black silicon, and will be used as the only uniform standard in the whole research group.

B. Extension

The three main parameters above can be extended to several other parameters, such as aspect ratio, occupied ratio,

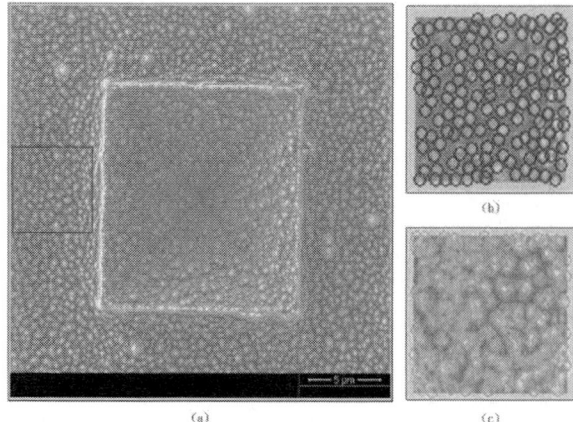

Fig. 4. Density extraction method of black silicon with a top view SEM picture.

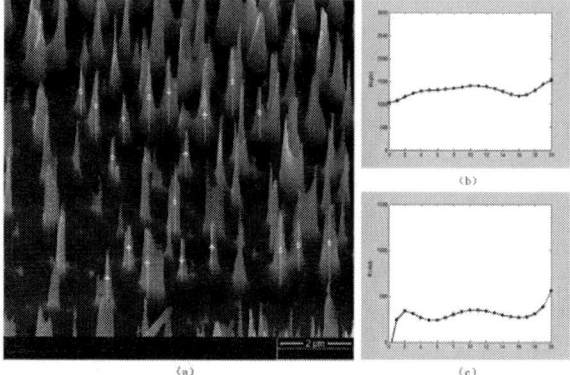

Fig. 5. Height and width extraction method of black silicon with a side view SEM picture

periodicity (p), elementary area (s) and so on. It can be achieved easily after a few simple calculations or conversion.

By means of these parameters we can also establish a standard model [7] to provide the very basic data for other kind of research, as is shown in Fig.3. To simplify the model, the average values of the parameters are often used.

C. Parameter Extraction

Given the huge workloads, a program is developed to achieve these parameters expediently and accurately. Fig.4 shows the density extraction method with a top view SEM picture while Fig.5 shows the height and width extraction method with a side view SEM picture. At the same time, the program can also export the excel forms with all of these data and draw the curves of them in order to offer more convenience to deal with them, for example, to get the average values of them.

By the above method we can get all the information we want about black silicon and continue further research such as optical simulation.

TABLE I. PROCESS PARAMETERS OF IMPROVED DRIE

SF_6 flow (sccm)	C_4F_8 flow (sccm)	Flat plate power(w)	Substrate power(w)	Gas pressure(mT)	Time ratio(s)	Cycles	Black Silicon
130	85	12	825	23	6/6	80	N
30	50	12	825	23	6/6	80	Y
30	50	12	600	23	6/6	80	Y
30	50	12	825	30	6/6	80	Y
30	50	12	825	23	8/5	80	N
30	50	12.5	825	23	6/6	80	N

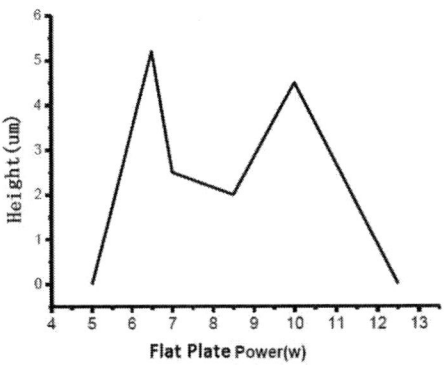

(a)

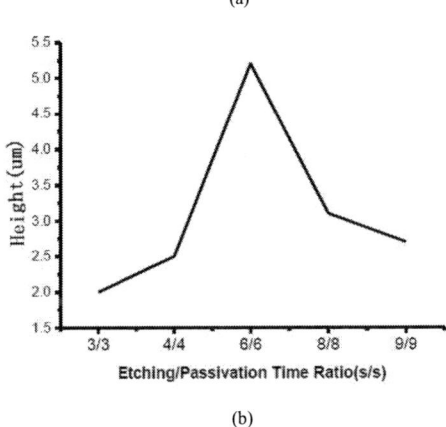

(b)

Fig. 6. (a) Relationship between Height and Flat Plate Power;
(b) Relationship between Height and Etching/Passivation Time Ratio.

IV. PROCESS OPTIMIZATION OF BLACK SILICON

In this study, the black silicon was formed by an improved DRIE by optimizing its process parameters including SF_6 flow, C_4F_8 flow, flat plate power, substrate power, gas pressure, etching/passivation time ratio and cycles. As is partly shown in Table. I, for every experiment we changed only one of the process parameters to see if black silicon can be formed. And then SEM pictures of all the samples with black silicon are taken. With the program introduced above, we can get all the information including density, height and width of every black silicon sample very easily and accurately.

In consideration of the strong relation between the characteristics and surface topography of black silicon, so we'd like to discuss the influence of the process parameters to surface topography as illustrated in Fig.6. The curves are drawn with the data obtained by the program and their corresponding process parameters. (a) shows the relationship between height and flat plate power. We can see that at 6.5w and 10.5w, the average height is the highest and 6.5w is better. (b) shows the relationship between height and etching/passivation time ratio. In other experiment we have already proved that it is best for forming black silicon when etching time and passivation time are the same. So after this curve we can determine 6s/6s is the best time ratio to fabricate black silicon. Also we have got several other curves between density, width and all the process parameters with which we can find all the best process parameters.

Finally we get three key process parameters which have the greatest impact on the result, including SF_6 flow, Flat plate power and etching/passivation time ratio. After all of these analyses, we can obtain a group of optimum parameters to fabricate black silicon, as shown in Table. II, and it has been verified and applied for many times during experiments.

TABLE II. OPTIMUM PARAMETERS OF BLACK SILICON.

SF_6 flow (sccm)	C_4F_8 flow (sccm)	Flat plate power(w)	Substrate power(w)	Gas pressure(mT)	Time ratio(s)	Cycles
30	50	6.5	825	23	6/6	80

V. CONCLUSION

This paper put forward a description method for surface topography of black silicon which is fabricated by an improved DRIE process. We defined density, height and width as three main parameters to describe black silicon. They could be extended to several other parameters and further established a standard model to provide the very basic data for other kind of research. A program was developed to achieve these parameters expediently and accurately. Then we discussed the influence of the process parameters to surface topography and also gave a group of optimum parameters to fabricate black silicon. Our further work is concentrated on achieving more controllable surface structures for mass production of black silicon and trying to research the generating mechanism of black silicon.

ACKNOWLEDGMENT

The authors would like to thank the Institute of Robotics and Automatic Information System of Nankai University for their cooperation on developing the parameter extraction program. This work is supported by the National Natural Science Foundation of China (Grand No. 91023045 and No. 61176103), and Key Laboratory Fund (No. 9140C790103110C7903).

REFERENCES

[1] S. Kalem, P. Werner, O. Arthursson, V. Talalaev, B. Nilsson, M. Hagberg, H. Frederiksen and U. S odervall, "Black silicon with high density and high aspect ratio nanowhiskers", Nanotechnology(2011) , doi:10.1088/0957-4484/22/23/235307

[2] Jiann Shieh, Srikanth Ravipati, Fu-Hsiang Ko and Kostya (Ken) Ostrikov. "Plasma-made silicon nanograss and related nanostructures", J. Phys. D: Appl. Phys. 44 (2011) 174010 (6pp).

[3] Yunfeng Li, Junhu Zhang, Shoujun Zhu, etc. "Bioinspired silicon hollow-tip arrays for high performance broadband anti-reflective and water-repellent coatings ". Advanced Materials (2009), pp. 4731-4734.

[4] H.Y. Mao, D. Wu, W.G. Wu, etc. "The fabrication of diversiform nanostructure forests based on residue nanomasks synthesized by oxygen plasma removal of photoresist ". MEMS2009, pp. 677-680.

[5] H V Jansen, M J de Boer, S Unnikrishnan, M C Louwerse and M C Elwenspoek. "Black silicon method X: a review on high speed and selective plasma etching of silicon with profile control: an in-depth comparison between Bosch and cryostat DRIE processes as a roadmap to next generation equipment", J. Micromech. Microeng. 19 (2009) 033001 (41pp).

[6] M. Mehran, Z. Sanaee, M. Abdolahad and S. Mohajerzadeh, "Controllable silicon nano-grass formation using a hydrogenation assisted deep reactive ion etching", Materials Science in Semiconductor Processing (2011), doi:10.1016/j.mssp.2011.02.014.

[7] K.N. Nguyen, P. Basset, F. Marty and E. Richalot, "On the Use of Black Silicon Obtained by Reactive Ion Etching as the Hot Spot of a Thermoelectric Generator Heated by Electromagnetic Radiation", Ph.D. Research in Microelectronics and Electronics (2010), pp.1-4.

Application of Nonlinear Driving in Frequency Matching of Tunneling Gyroscope

Lingyun Wang[a]*, Xiaohui Du[b], Yuanzhe Su[c], Zhan Zhan[d], Wenjia Zuo[e] and Daoheng Sun[g]*

*Department of Mechanical and Electrical Engineering, Xiamen University, Xiamen, CHINA

[a]rabitwangly@yahoo.com.cn, [g]sundh@xmu.edu.cn

Abstract—The nonlinear vibration behavior of driving mode of a tunneling gyroscope is measured and simulated. Based upon these testing results, a new method for frequency matching is presented. Compared to traditional harmonic vibration, large amplitude and wide shift resonance frequency could be obtained in stable region of nonlinearity by using this approach, especially, in this region gyroscope could spontaneously vibrate at a certain driven frequency without needing sweep-up frequency excitation. In our tunneling gyroscope, the vibration amplitude changes from 1μm to 5μm at jumping-up point J' with the amplitude of the excitation electrostatic force increasing. At the same time, the resonant frequency of driving mode can be shifted from nature frequency 1.81 kHz to 2.2 kHz, which is much closed to the resonant frequency of sensing mode (2.23 kHz). This effect can be used as an alternative approach to trim the frequency and eliminate the frequency mismatching for the post-fabrication tunneling gyroscope, eventually, higher sensitivity could be gotten than using linear driving mode.

Keywords-tunneling gyroscope ;nonlinear effect; electrostatic force ;frequency mismatching

I. INTRODUCTION

To achieve high sensitivity in microgyroscopes based on harmonic oscillators, the drive and the sense resonant frequencies are typically designed and tuned to match. Resonant frequency of the fabricated mcirogyroscope, however, tends to deviate from designed values due to the errors in the microstructure dimensions. Although solution to overcome frequency mismatching have been pursued, many of them involve adding complexity to the system by including additional controllers[1], additional degrees of freedom[2], utilizing multiple drive mode oscillators[3] or needing special driving signal[4]. This paper will show a new means for the post-fabrication frequency-matching of tunneling gyroscope by using stable region of the nonlinear driving response. This method utilizes the inherent characteristic of resonance frequency shift with driving force variation as the matching criterion.

II. THEORY OF NONLINEAR VIBRATION

A. Principle of Tunneling Gyroscope

A simple schematic diagram and SEM images of the vibratory tunneling microgyroscope, which senses x-axis

rotation, is shown in Fig.1. The microgyrsocope is based on silicon-on-glass compound structure through silicon-glass anodic bonding technique. The anchor bonded with the glass substrate supports the silicon frame and proof mass by sensing beam, which are floating 4μm above the substrate and free for sensing and driving vibration. The proof mass connected with silicon frame through four driving beams can vibrate along Y direction because of driving comb finger. Two driving comb anchors are also bonded with the glass substrate. The deflection electrode under the proof mass could deflect the proof mass and silicon frame into the tunneling position. The silicon tip at the end of the silicon frame is used to sense the vibration of the silicon frame at Z direction.

A schematic diagram of the parallel comb is shown in Fig.2 and the structure parameters of the parallel comb in TABLE I. During the operation, AC and DC driving voltage are respectively applied to stator and rotor. This induces alternating electrostatic forces on the rotor driving fingers along Y direction in a "push-pull" mode. The proof mass will vibrate along Y direction. If there is an angular velocity along the X direction, the proof mass will experience an alternating Coriolis' force along Z direction. This in turn will excite the proof mass and silicon frame to vibrate along Z direction. The tunneling gap between silicon tip and detected electrode will be changed due to the sensing vibration along Z direction. By measuring the tunneling current change, we can get the value of the angular velocity [5].

The sensitivity of tunneling gyroscope depends strongly on the amplitude of the drive mode and the matching of the resonance frequencies of driving and sensing mode. For this

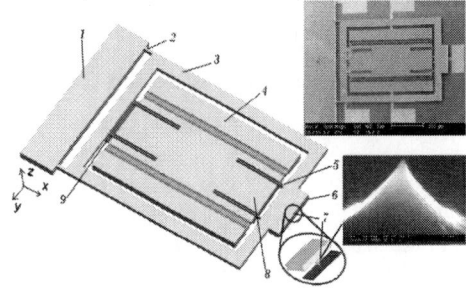

Fig. 1. Schematic diagram and SEM images of bulk- micromachined tunneling gyroscope, in it: 1.anchor 2.sensing beams 3.silicon frame 4.driving comb anchors 5.driving beams 6.detected electrode 7.silicon tip 8.proof mass 9. deflection electrode.

reason, gyroscope are generally driven in resonance by forcing both frequencies as close as possible, on the one hand to get the best coupling between the two modes and on the other hand to obtain large driving amplitudes. But large amplitudes will result in mechanical nonlinearities which would change the coupling character of two modes.

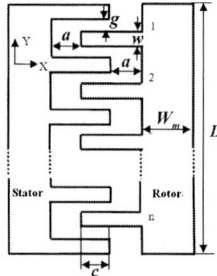

Fig. 2. Schematic diagram of the parallel comb and its structure parameters.

TABLE I.   STRUCTURE PARAMETERS OF PARALLEL COMB

Parameter	Symbol	Numerical
Comb insert spacing	g	4.6μm
Comb tenoning width	w	4.4μm
Overlap length	c	20μm
Distance between stator and rotor	a	25μm
Number of movable fingers	N	73
Thickness of rotor	b	36μm

B.   Nonlinear-force vibration with damping

For a MEMS vibratory microgyroscope, the mathematical model of driven mode can be represented by the following differential equations when taking account of nonlinear phenomena [8]:

$$m_d \ddot{x} + c_d \dot{x} + k_d x + k_3 x^3 = F_e \cos \omega t \quad (1)$$

Where C_d is the damping ratio of the driving mode, k_d is the linear stiffness, k_3 is the cubic stiffness, F_e is the amplitude of the driving force, ω is the frequency of the driving force. This equation of motion is known as the "Duffing Equation".

A nonlinear-force model with damping is used to describe the vibration of parallel comb at atmosphere, and the damping is ζ, the natural frequency is ω_n, then the driving amplitude X under different frequency ω could be analytically given as follows[6~7]:

$$\frac{3\mu X^3}{4} = \left(\frac{\omega^2}{\omega_n^2} - 1\right) X + \frac{F_e}{\omega_n^2} \sqrt{1 - \frac{(2\zeta\omega_n \omega X)^2}{F_e^2}} \quad (2)$$

Where μ ($0<\mu<1$) is the coefficient of cubic term (nonlinear stiffness) in the elastic expression. The amplitude frequency response characteristics can be obtained after damping ζ and natural frequency ω_n are measured.

A method of free vibration attenuation is applied to measure the damping ζ and natural frequency ω_n.

Theoretically, the ratio of logarithmic attenuation δ can be obtained by[9]

$$\delta = \zeta\omega_n T_d = \frac{2\pi\zeta}{\sqrt{1-\zeta^2}} \quad (3)$$

Where T_d is the attenuation vibration period. During measurement, the vibration curve is recorded after the rotor is impelled by step excitation, and then the damping ζ and natural frequency ω_n are calculated by measuring the ratio of logarithmic attenuation δ and attenuation vibration period T_d.

III.   MEASUREMENT AND RESULTS DISCUSSION

A.   Extraction of Device Dynamic Parameters

How to extract the mechanical stiffness parameters (linear and nonlinear) and the damping is the research base of nonlinear vibration. Firstly, the linear and cubic stiffness is extracted by fitting the measurement results of electrostatic force versus the rotor displacement with polynomial equation at different input DC voltage. The rotor displacement is measured by optical microscopy and the force from an interdigitated comb driver is given by

$$F_{comb} = \frac{N\varepsilon_0 b}{g} V^2 \quad (4)$$

Where N is the number of movable fingers, b is the thickness of the rotor, g is the gap between the movable and fixed electrodes, ε_0 is the dielectric constant of the material between the comb fingers, and V is the potential difference between the electrodes. The test result of distance-electrostatic force relationship is shown in Fig.3. The device obviously exhibits a nonlinear effect at large deflect distance. This is due to the cubic nonlinearity of the springs as the motion exceeds the linear range. From the fitting equation, we can get that the linear and cubic spring constants of the device is 8.1287 and 0.2751 respectively.

Secondly, the damping coefficient and inherent frequency for the rotor vibration of tunneling gyroscope can be extracted from a step-down stimulation response. A Schematic diagram of test set-up for the detection of the proof mass motion is shown in Fig.4. A low-noise current-to-voltage converter circuit, which has 5pA resolution, was used to test the rotor displacement. A step down voltage (Vi=40V) is applied on one side of the comb fingers and the other side of the comb

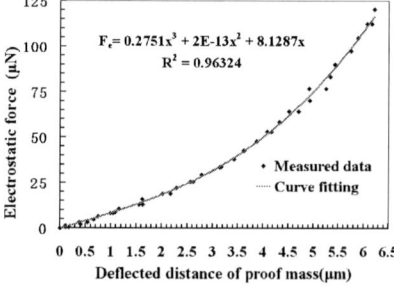

Fig. 3. The characteristic of electrostatic force against deflection of proof mass and the extraction of the linear and cubic elastic constants.

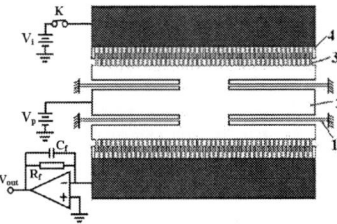

Fig. 4. Schematic diagram of test set-up for the detection of the proof mass motion.

fingers simultaneously senses the vibration of the proof mass by recording the output voltage of *I-V* transresistance amplifier.

Response of the drive mode to a step-down excitation voltage at atmospheric pressure is shown in Fig.5. The true motion, as revealed by the measurement, shows behavior typical of an underdamped oscillator. By observing the motion of the devices as it settles to a new position, the damping coefficient and the undamped resonant frequency in drive direction can be determined by counting the cycles and calculated from the amplitude envelope of proof mass motion respectively. As a result, the natural resonance frequency of driving mode is 1.81kHz and the damping coefficient is 0.0094.

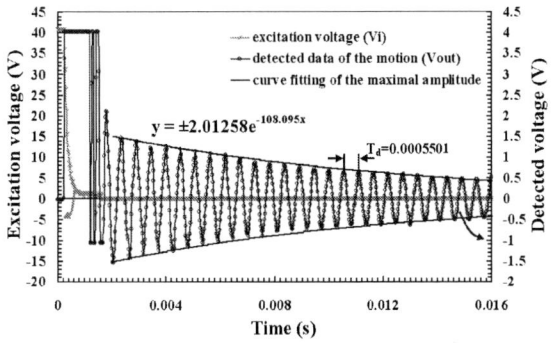

Fig. 5. Response of the drive mode to a step-down excitation voltage at atmospheric pressure.

B.   Approach of Frequency Matching

During the operation, nonlinearity effect will occur in case at large amplitude of driving mode because of the cubic stiffness. Using the extracted dynamic characterizations and optical microscopy described above, the frequency response of the gyroscope at driven mode can be measured and simulated. As seen in equation (2), the resonant frequency increases with amplitude. Fig.6 shows experimental and simulation frequency-response curves for the driven mode of tunneling gyroscope. The curve is obtained by sweeping the frequency for varying amplitudes at fixed exciting voltage (30V DC voltage and 3V AC voltage). The response curve shape is distorted and clearly shows hysteresis. The resonance peak is deflected towards high frequencies. Significantly, the

curve could be repeatable in stable region, regardless of frequency sweeping up or sweeping down, that is why we called this curve range as stable region.

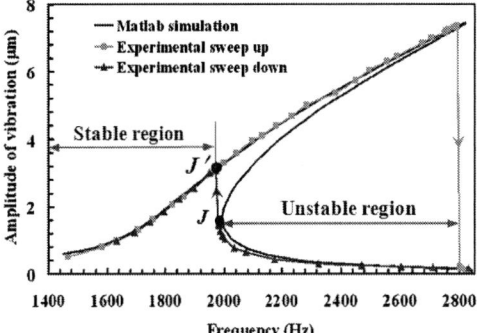

Fig. 6. The frequency response of driving mode at atmospheric pressure with a driving V_d=30V and V_a=3V.

When changing the driving voltage (Using different DC base voltage and AC sinusoidal voltage), the curves, as shown in Fig.7, obviously show that the resonant frequency increase with increasing driving amplitude and is characteristic of this device with a hard non-linear stiffness due to the cubic nonlinearity. At the same time, we also find that the higher the excitation forces, the wider the stable region could be obtained.

Although large amplitude could be obtained in unstable region, a sweep-up frequency excitation must be used due to the bistable effect in this region. Compared to harmonic vibration, large amplitude and shift resonance frequency could also be observed without sweep-up frequency excitation in stable region. In addition, the amplitude and frequency of jumping up point can be changed with driving force. When the resonance frequency of the sensing mode is higher than the one of the driving mode in gyroscope, the jump-up frequency can be shifted to higher frequency by increasing the amplitude of the driving voltage until the driving oscillation is close to the resonance frequency of the sensing mode. Therefore, some Hz less than the jumping-up point J'

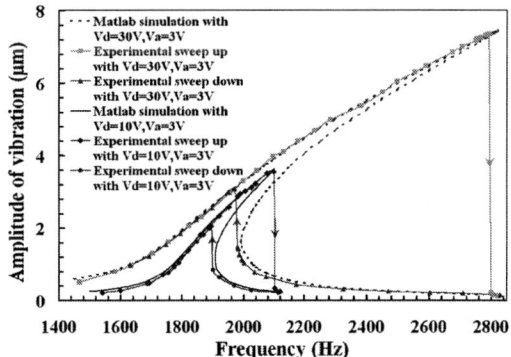

Fig. 7. The vibration amplitude response of driving mode under different actuation voltages.

978-1-4673-1122-9/12 $31.00 © 2012 IEEE   573

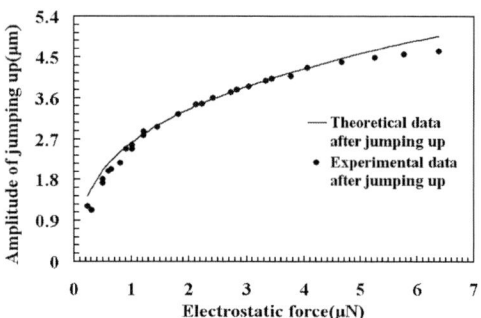

Fig. 8. Vibration amplitude of the driving mode at jumping-up point against amplitude of excitation electrostatic force.

frequency could be selected as working frequency of driving mode for achieving the maximum of sensitivity. For a theoretical form, the amplitude X_j of jump-up point J' at different driven force can be calculated from (2) by Shengjin's Distinguishing Means

$$X_{j'} = -\frac{8\zeta^2\omega_{j'}^2}{3\mu F} + \left(\frac{16F}{3\mu\omega_n^2}\right)^{\frac{1}{3}} \quad (5)$$

Where ω_j, X_j are vibration frequency at point J' and homologous amplitude , respectively. Fig.8 shows the vibration amplitude at jumping-up point J' against the excitation electrostatic force amplitude, which changes from 1μm to 5μm. Thus, the theoretical expression given by equation (5) is sufficient to describe the observed tendencies, since it is in good agreement with the experimental results.

Fig.9 shows the resonance frequency of the driving mode as a function of the electrostatic force and frequency matching could be achieved by increasing the amplitude of the driving electrostatic force. Therefore, not only could the matching frequency be obtained but also the amplitude of driving vibration could be increased with the amplitude of driving electrostatic force increasing. Calculated from above measurement results, about 9 times amplitude of tunneling tip could be obtained under 6.37μN amplitude of driving electrostatic force, which tunes the driving frequency at 2.2kHz (nearly matching with the sensing frequency (2.23kHz)) and gets 5μm amplitude of vibration, compared

with 0.24μN amplitude of driving electrostatic force, which tunes the driving frequency at 1.84k Hz (nearly closed to the nature driving frequency) and gets 1.2μm amplitude of vibration. In contrast to standard approaches where nonlinearities are being avoided, nonlinear driving can be used for an easier matching of frequencies.

IV. CONCLUSION

The nonlinear behavior of driving mode of a tunneling gyroscope is measured and simulated. In nonlinear driving response, different amplitude could be gotten over a wide range of stable region. The experimental results are compared to a well-established theoretical model. Good qualitative agreement is obtained. Thus, this effect can provide a new approach to eliminate the frequency mismatching for the post-fabrication frequency-matching of tunneling gyroscope to increase the sensitivity of gyroscope. In contrast to standard approaches where nonlinearities are being avoided, the characteristics of nonlinear driving, where the shift of resonance frequency with driving force variation, can be used for an easier matching of frequencies.

ACKNOWLEDGMENT

The authors would like to thank Pen-Tung Sah Micro-Electro-Mechanical Systems Research Center of Xiamen University for the fabrication of gyroscopes. This work is supported by the Fundamental Research Funds for the Central Universities (No.2010121039) and the National Natural Science Foundation for Youth of China (No. 51105320).

REFERENCES

[1] S.Park and R. Horowitz, "Adaptive control for the conventional mode of operation of MEMS gyroscopes," *Journal of microelectromechanical systems*, Vol.12, pp.101-108, Feb 2003.

[2] C.Acar and A. Shkel, "Nonresonant micromachined gyroscopes with structural mode-decoupling," *IEEE Sensors Journal*, Vol.3, No.4, August 2003.

[3] C. Acar and AM Shkel, "An approach for increasing drive-mode bandwidth of MEMS vibratory gyroscopes," *Journal of microelectromechanical systems*, Vol.14, June 2005.

[4] Laura A. Orapeza-Ramos, Christopher B.Burgnar, and Kimberly L. Turner, "Inherently robust micro gyroscope actuated by parametric resonance," *MEMS 2008*, pp.872-875.

[5] Lingyun Wang, Yuanzhe Su, Yi Lei, Wenwang Li, Yifang Liu, Daoheng Sun, "A novel bulk micromachined tunneling gyroscope," *proceedings of the Nano/Micro Engineered and Molecular Systems (NEMS)*, 2011. 873-876 [2011 IEEE International Conference on, 20-23 Feb. 2011].

[6] Timoshenko S,Yonug DH and Weaver W JR. *Vibration problems in engineering (Fourth-edition)*, New York: John Wiley & Sons,1974:177-186.

[7] O. Schwarzelbach, G.Fakas, W.Nienkirchen. "New approach for frequency matching of tuning fork gyroscopes by using a nonlinear driving concept," *TRANSDUCERS 2001*, pp.464-467, 2001.

[8] K. Maenaka, Y. Konishi, T. Fujita, M.Maeda, "Analysis and Design Concept of Highly Sensitive Silicon Gyroscope", *Transducers'95*, Vol.2, June 25-29, 1995, pp.612-615 [The 8th international conference on Solid-State Sensors and Actuators, 1995 and Eurosensors IX].

[9] Li Xiaolei, Yu Defu, Sun Fengchun. *The basis of mechanical vibration*, BeiJing : Beijing University of Technology Press,1996:21-24.

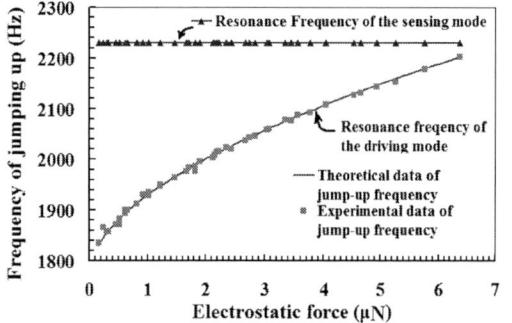

Fig.9. Resonance frequency of the driving mode as a function of the electrostatic force.

Integrated Flexible Micro Pressure, Temperature and Flow Sensors for Use in PEMFC

Chi-Yuan Lee*, Tachung Yang, Yu-Ming Lee, Tzu-Hao Chien, Yen-Tin Cheng

Department of Mechanical Engineering, Yuan Ze Fuel Cell Center, Yuan Ze University, Taoyuan, Taiwan, R.O.C.

cylee@saturn.yzu.edu.tw

Abstract — **Temperature, flow and pressure are critical parameters that influence the performance of fuel cells, in terms of potential, current and power density, for example. Hence, the monitoring of non-uniform temperature/flow rate/pressure within a proton exchange membrane fuel cell (PEMFC) is crucial. To prevent degradation of the performance of a PEMFC, the size of sensors is reduced herein to a μm scale using micro-electro-mechanical systems (MEMS). Integrated flexible micro sensors were fabricated and embedded in a PEMFC to measure local pressure, temperature, and flow rate. The temperatures upstream and downstream of the PEMFC were respectively 64.8°C and 64.7°C at RH50% and 0.1 A/cm^2, and 62.7°C and 64.3°C at RH100% and 0.1 A/cm^2.**

Keywords — ***PEMFC; MEMS; micro pressure sensor; micro flow sensor***

I. INTRODUCTION

A proton exchange membrane fuel cell (PEMFC) is a complex system that is characterized by several mutually coupled operating parameters, such as temperature, humidity, pressure drop and flow rate. Hence, optimizing operating conditions is very important. Lee [1] integrated micro temperature and humidity sensors using micro-electro-mechanical systems (MEMS) to measure local values in a membrane electrode assembly (MEA). They found a temperature difference between the MEA and a bipolar plate of 5.7°C. Inman [2] measured *in-situ* the reaction temperature in an operating fuel cell by placing five fiber temperature sensors in it. Wang [3] employed an infrared temperature device to measure external temperature distribution under various operating conditions. When flow humidity is too low, dehydration occurs inside the MEA. However, excessively high flow humidity condenses as water droplets, flooding either the MEA or the flow channel. Dehydration and flooding worsen the performance and the durability of the fuel cell. Flooding also makes cell performance unpredictable [4, 5].

Methods for diagnosing water flooding can be classified into imaging approaches and physical indicator measurement approaches. Imaging schemes include direct visualization, neutron imaging, magnetic resonance imaging, and X-ray imaging [6-8]. However, they have several limitations. Martins [9] developed a reliable simulation model that accounts for power losses that are associated with a drop in pressure in the gas channels. This model can be adopted to examine the relationship between pressure drop and cell performance. For example, when liquid water condenses in a porous gas diffusion layer (GDL) or flow field, the flow resistance rises, increasing the pressure drop between the inlet

and the outlet of the PEMFC [10, 11]. Additionally, water flooding can be diagnosed by determining the pressure drop in the GDL or flow field.

Properly controlled temperature, humidity, pressure drop, and flow rate are critical to ensure favorable PEMFC performance at low temperature. However, relevant studies have not utilized micro pressure sensors to determine the internal pressure drop or identify water flooding in a PEMFC. Therefore, this work involves the design and fabrication of flexible integrated micro pressure, thermal and flow sensors to measure internal pressure, temperature and flow in a PEMFC.

II. METHODOLOGY

A. Theory of Pressure Sensor

Conventional pressure sensors can be classified as piezoresistive, capacitive, and resonant pressure sensors [11-13]. A piezo-resistive pressure sensor is extremely sensitive to temperature; therefore, it is unsuited to high-temperature environments. A capacitive pressure sensor is less sensitive to temperature, but its sensitivity is higher under lower pressure [14]. The fabrication of resonant pressure sensors is highly complex. For all these reasons, the capacitive pressure sensor is most suitable for fuel cell applications.

Fig. 1 schematically depicts a capacitive pressure sensor (800μm×800μm). When pressure P is applied to the electrode surface, the gap between the two parallel electrodes changes as the dielectric layer deforms, and so the capacitance changes simultaneously. Equation (1) describes the operation of a capacitive pressure sensor;

$$\Delta C = \varepsilon_r \varepsilon_0 \frac{A}{\Delta d} \qquad (1)$$

where ΔC is the change in capacitance; ε_r represents the dielectric constant; ε_0 is the electric constant; A denotes the area of overlap of the two parallel electrodes, and Δd is the change in the distance between the two parallel electrodes under an applied pressure, which is the change in thickness of the dielectric layer.

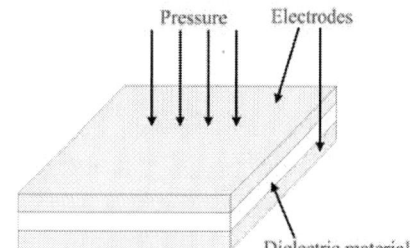

Fig. 1. Schematic diagram of a capacitive pressure sensor.

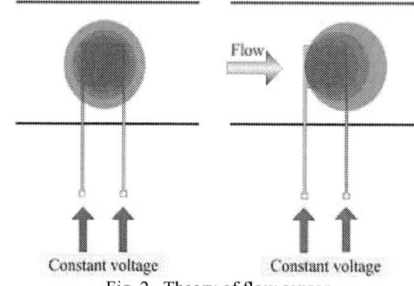

Fig. 2. Theory of flow sensor.

B. Theory of Thermal Sensor

In this work, a resistance thermal detector (RTD) is used as the thermal sensor. General resistance is given by (2)

$$R = \rho \frac{L}{A} \qquad (2)$$

where ρ is resistivity (Ωm); L denotes the length of component (m); and A represents the cross-sectional area of the component (m^2). In the linear range, (2) can be rewritten as (3).

$$R_t = R_r \left(1 + \alpha \Delta T\right) \qquad (3)$$

where R_t is the resistance at t°C; R_r is the resistance at r°C, and α is sensitivity. The temperature can be determined from the resistance using a calibration curve.

C. Theory of Flow Sensor

Flow sensors are grouped into thermal flow sensors and non-thermal flow sensors. The structure of the flow sensor makes the fabrication and the design simple. A hot-wire flow sensor is adopted herein. The main structure of hot-wire flow sensor is that of a micro serpentine heater. The operation of a hot-wire flow sensor involves the supply of a constant voltage to it, that it heats the environment, establishing a stable temperature environment. If the flow sensor is placed in a constant flow field, then the temperature of flow sensor changes, as shown in Fig. 2. In a hot-wire, Joule heating converts electrical energy to thermal energy. The increase in wire temperature changes the resistance to an extent determined by the temperature coefficient of resistance of the wire material, according to [15], the relationship between the thermal energy dissipation rate and the flow speed is given by (4).

$$Q = I^2 \times R = I \times V = \left(A + B \times U^n\right)\left(T_s - T_o\right) \qquad (4)$$

where Q represents the power that is supplied by the external power supply to a hot wire; U denotes the flow speed; T_s represents the temperature of the hot wire; T_o is the flow temperature; and n is the coefficient that relates U to Q. A involves a constant coefficient of transfer of energy by the heater at a flow speed of 0, whereas B involves a constant coefficient of transfer of energy by the heater at a non-zero flow speed.

A micro temperature sensor can be also use as a micro flow sensor [16].

III. FABRICATION OF FLEXIBLE MICRO SENSORS

Fig. 3 shows the fabrication processes of flexible micro sensors. The processes are as follows: (a) Particles on the stainless steel surface are removed by acetone and methanol; stainless steel oxide is then cleaned using solvent (H_2SO_4:H_2O_2 = 3:1). (b) Sputtering aluminum nitride (AlN) as the bottom insulation layer by RF sputter. (c) The Cr/Au as the bottom electrodes is evaporated by e-beam evaporator. Micro sensors are patterned by lithography process and wet etching process; (d) Copper is evaporated on the micro sensors and pad part as a hard mask; (e) Sylgard 184 Polydimethylsiloxane (PDMS) of DOW CORNING®, was mixed with base and curing agent in an appropriate ratio, followed by degassing in a vacuum chamber. Next, PDMS mixture solvent was spin coated onto stainless steel foil, followed by curing in an oven. PDMS surface present a hydrophobic surface, and plasma oxidation treatment can render a PDMS hydrophilic surface; (f) Second time copper deposition as a hard mask is performed to prevent PDMS, which was etched away by a reactive ion etching (RIE); (g) PDMS was patterned by RIE; (h) Copper was removed by $FeCl_3$ etching solvent and plasma treatment was performed on the PDMS surface again; (i) Copper was patterned as the upper electrode of a capacitive micro pressure sensor; (j) Double side photoresist (PR) coating was performed, followed by AlN and SS foil etching by phosphoric acid and aqua regia, respectively. (k) Upper dielectric layer was patterned by PR. Fig. 4 presents the finally obtained flexible micro sensors.

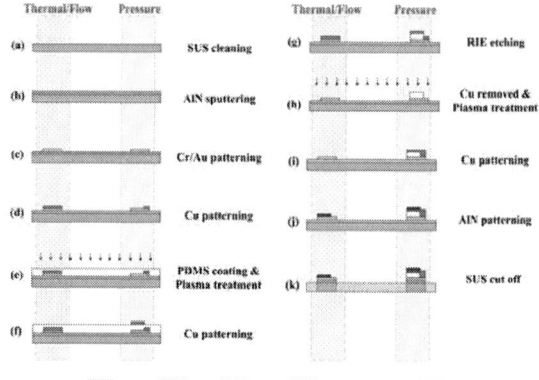

Fig. 3. Procedure for fabricating flexible micro sensors on stainless-steel foil substrate.

Fig. 4. Final flexible micro sensors.

IV. RESULTS AND DISCUSSION

A. Calibration of Micro Pressure Sensor

The variation of capacitance of the micro pressure sensor was measured using an LCR meter. Pressure range was from abs. 1 to 1.3 atm gauge pressure. Fig. 5 plots the calibration curves of the micro pressure sensor. Calibration was conducted seven times to confirm the repeatability of the measurements made using the micro pressure sensors.

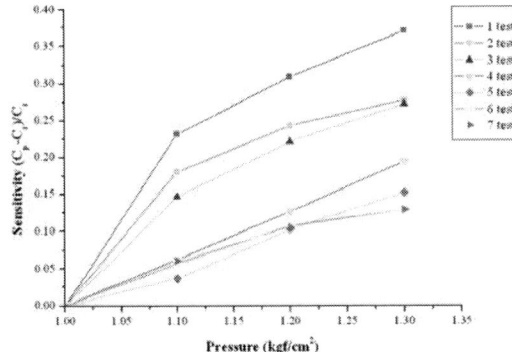

Fig. 5. Calibration curve of micro pressure sensor.

B. Calibration of Micro Thermal Sensor

The variation of the resistance of the micro thermal sensor was measured using an oven and a multimeter. Fig. 6 plots the calibration curve of the micro thermal sensors. The calibration results demonstrate that the resistance of the micro thermal sensors varies highly linearly with temperature.

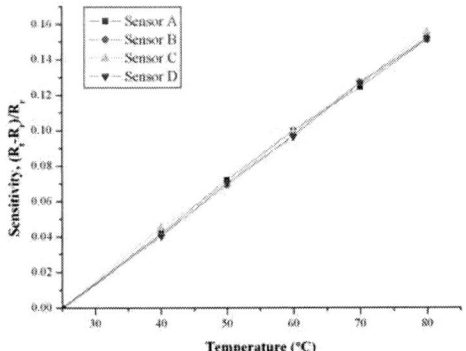

Fig. 6. Calibration curve of micro thermal sensors.

C. Calibration of Micro Flow Sensor

A micro flow sensor that was embedded in a PEMFC was calibrated. Micro flow sensors were placed upstream and downstream of the flow channel of the PEMFC. Fig. 7 plots the calibration curve of the micro flow sensor. The calibration range was from 0 to 250 SCCM. I_f is the resistance at flow f, I_r is the resistance at point of reference.

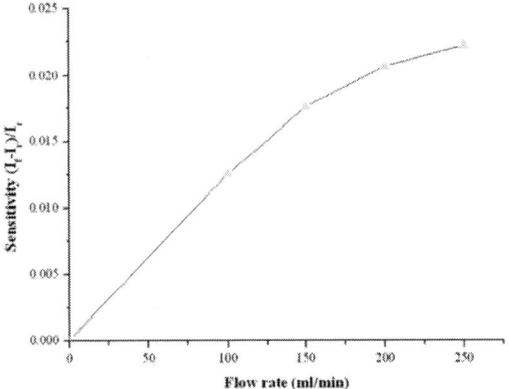

Fig. 7. Calibration curve of micro flow sensor.

D. Micro Sensors are embedded in a PEMFC

Measurements were made using a 100W/A fuel cell and an integrated fuel cell test station (850C). The fuel inlet and outlet were connected to the fuel control system in the test station. Its main function was to control the flow rates of hydrogen and air, humidity, temperature and other parameters.

To monitor the temperature of the fuel cell, two multimeters were required to measure the resistance of the micro sensors upstream (fuel inlet) and downstream (fuel outlet). Flexible micro sensors were embedded into an MEA in a flow channel at 31.5mm and 189mm from the fuel inlet.

E. Test with Constant Current of 0.1 A/cm² at 65℃

Fig. 8 displays the temperature of the micro sensors that was embedded in a PEMFC at a constant current of 0.1 A/cm², 65°C, and various relative humidities. When the relative humidity was 50% RH, the temperature of the thermocouple was 64.9°C, the temperatures measured by the upstream and downstream micro sensors were 64.8°C and 64.7°C, respectively. In the experiment, the difference between the upstream and downstream temperatures was 0.1°C. The temperature measured using the thermocouple was not very close to that measured using the micro sensor. This result may be attributable to the incomplete electrical reaction.

When the relative humidity was 100% RH, the temperature of the thermocouple was 63.5°C, and the temperatures measured by the upstream and downstream micro sensors were 62.7°C and 64.3°C, respectively. The difference between the upstream and downstream temperatures was 1.6°C. At a constant current of 0.1 A/cm², the fuel that entered the fuel cell reacted incompletely. The reaction proceeded to completion downstream. The generation of thermal

978-1-4673-1122-9/12 $31.00 © 2012 IEEE   577

downstream increase the temperature. The upstream temperature was unstable, because of the water flooding that occurred when the fuel entered through the fuel inlet.

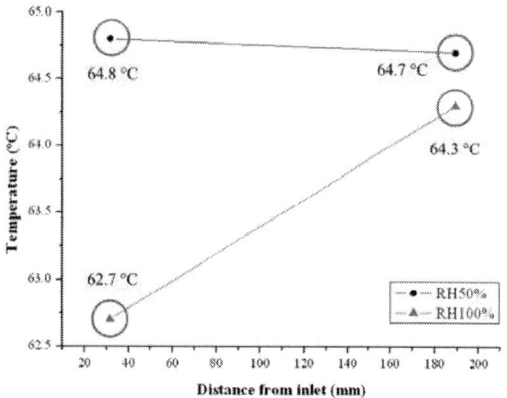

Fig. 8. *Temperatures upstream and downstream of PEMFC.*

V. CONCLUSIONS

This work demonstrates integrated micro pressure, temperature and flow three-in-one flexible sensors on a 40μm-thick stainless steel foil substrate. When micro sensors are embedded in PEMFC for *in-situ* diagnosis, they successfully measured local data. This method is more precise than other invasive or *ex-situ* measurement methods and provides more detailed information about the interior of fuel cells. This information can enable a user to reduce make measurements with less of an impact on fuel cell performance.

ACKNOWLEDGMENT

This work was accomplished with much needed support and the authors would like to thank for the financial support by National Science Council of R. O. C. through the grant NSC 100-2628-E-155-002-MY2. The authors also like to thank Professors Lung-Jieh Yang, Shuo-Jen Lee, Shih-Hung Chan, Ay Su, Fangbor Weng and Guo-Bin Jung for their valuable advice and assistance in experiment. In addition, we would like to thank the TKU MEMS Lab, YZU Fuel Cell Center and NEMS Common Lab for providing access to their research facilities.

REFERENCES

[1] C. Y. Lee, W. J. Hsieh, and G. W. Wu, "Embedded flexible micro-sensors in MEA for measuring temperature and humidity in a micro-fuel cell," *Journal of Power Sources*, vol. 181, pp. 237-243, 2008.

[2] K. Inman, X. Wang, and B. Sangeorzan, "Design of an optical thermal sensor for proton exchange membrane fuel cell temperature measurement using phosphor thermometry," *Journal of Power Sources*, vol. 195, pp. 4753-4757, 2010.

[3] M. Wang, H. Guo, and C. Ma, "Temperature distribution on the MEA surface of a PEMFC with serpentine channel flow bed," *Journal of Power Sources*, vol. 157, pp. 181-187, 2006.

[4] J. St-Pierre, "PEMFC contamination model: foreign cation exchange with ionomer protons," *Journal of Power Sources*, vol. 196, pp. 6274-6283, 2011.

[5] A. Mughal, and X. Li, "Experimental diagnostics of PEM fuel cells," *International Journal of Environmental Studies*, vol. 63, pp. 377-389, 2006.

[6] X. G. Yang, F. Y. Zhang, A. L. Lubawy, and C. Y. Wang, "Visualization of liquid water transport in a PEFC," *Electrochemical and Solid-State Letters*, vol. 7, pp. A408-A411, 2004.

[7] K. Tüber, D. Pócza, and C. Hebling, "Visualization of water buildup in the cathode of a transparent PEM fuel cell," *Journal of Power Sources*, vol. 124, pp. 403-414, 2003.

[8] R. Satija, D. L. Jacobson, M. Arif, and S. A. Werner, "In situ neutron imaging technique for evaluation of water management systems in operating PEM fuel cells," *Journal of Power Sources*, vol. 129, pp. 238-245, 2004.

[9] L. S. Martins, J. E. F. C. Gardolinski, J. V. C. Vargas, J. C. Ordonez, S. C. Amico, and M. M. C. Forte, "The experimental validation of a simplified PEMFC simulation model for design and optimization purposes," *Applied Thermal Engineering*, vol. 29, pp. 3036-3048, 2009.

[10] R. Anderson, L. F. Zhang, Y. L. Ding, M. Blanco, X. T. Bi, and D. P. Wilkinson, "A critical review of two-phase flow in gas flow channels of proton exchange membrane fuel cells," *Journal of Power Sources*, vol. 195, pp. 4531-4553, 2010.

[11] C. C. Chiang, C. K. Lin, and M. S. Ju, "An implantable capacitive pressure sensor for biomedical applications," *Sensors and Actuators A*, vol. 134, pp. 382-388, 2007.

[12] D. W. Lee, and Y. S. Choi, "A novel pressure sensor with a PDMS diaphragm," *Microelectronic Engineering*, vol. 85, pp. 1054-1058, 2008.

[13] G. Koley, J. Liu, M. W. Nomani, M. Yim, X. Wen, and T. Y. Hsia, "Miniaturized implantable pressure and oxygen sensors based on polydimethylsiloxane thin films," *Materials Science and Engineering C*, vol. 29, pp. 685-690, 2009.

[14] M. Elwenspoek, and R. Wiegerink, "*Mechanical Microsensors*," Springer, 2001.

[15] L. V. King, "On the convection of heat from cylinders in a stream of fluid: Determination of the convection constants of small platinum wires with application to hot wire anemometry," *Proc. Roy. Soc. London A*, vol. 90, pp. 563-570, 1914.

[16] C. Y. Lee, P. C. Chan, and C. J. Lee, "Use of multi-functional flexible micro-sensors for *in situ* measurement of temperature, voltage and fuel flow in a proton exchange membrane fuel cell," *Sensors*, vol. 10, pp. 11605-11617, 2010.

Sensitivity Enhancement in SGOI Nanowire Biosensor Fabricated by Top Surface Passivation

Kow-Ming Chang[a,b], Chu-Feng Chen[a], Chiung-Hui Lai[c], Cheng-Ting Hsieh[a], Chin-Ning Wu[a], Yu-Bin Wang[a], and Chung-Hsien Liu[a],

[a] Dept. of Elec. Engineering and Inst. of Elec., National Chiao-Tung University
No. 1001, University Rd., Hsinchu 300, Taiwan, R.O.C.
[b] Department of Electronic Engineering, I-Shou University
No.1, Sec. 1, Syuecheng Rd., Kaohsiung 840, Taiwan, R.O.C.
[c] Dept. of Electronics. Engineering, Chung Hua University
No. 707, Sec.2, WuFu Rd., Hsinchu, 300, Taiwan, R.O.C.

kmchang@faculty.nctu.edu.tw; kmchang@isu.edu.tw

Abstract—Increasing the fraction of Ge in SiGe-on-Insulator (SGOI) using Ge condensation by oxidation significantly increases hole mobility. This effect can be exploited to improve the sensitivity of SGOI nanowire. However, our previous studies found that the sensitivity of an SGOI nanowire is degraded as the Ge fraction increases over 20%, because of the surface state of SiGe is unstable when the Ge fraction is high. In this work, a top surface passtivation SiO2 layer was deposited on an $Si_{0.8}Ge_{0.2}$ nanowire and successfully improve its sensitivity around 1.3 times that of the nanowire sample without top a passivation layer.

Keywords-component; SiGe-on-Insulator, bio-sensor, passivation

I. INTRODUCTION

The biological, chemical and optical applications of nano-wires, have recently become an important topic of research. [1,2]. Nano-wires have attracted substantial interest as essential building blocks for functional electronics devices especially owing to their use in nano-electronics [3-5] and highly sensitive biosensors [6-7]. The formation of SiGe-on-Insulator (SGOI) nanowire by Ge-condensation process can enhance the mobility of hole carriers in nanowire's conductance. Furthermore, the sensitivity of the SGOI nanowire can be enhanced by increasing its surface to volume ratio to condense carriers in the very thin conductivity layer. In our previous studies, the surface-to-volume ratio of SGOI nanowire was increased by utilizing SiGe/ α -Si stacking structure [8]. Furthermore, increasing the fraction of Ge in $Si_{1-x}Ge_x$ nanowire improves the sensitivity of a nanowire biosensor by increasing the mobility [9]. An ultrathin SiGe-on-Insulator with a high Ge fraction was fabricatedby Ge condensation and the Ge piled-up at the $SiO_2/SiGe$ interface upon Ge condensation process. However, oxidation increased the trap density at the $SiO_2/SiGe$ interfaces about 10^{12} cm^{-2} [10]. Many dangling bonds at the free surface make the surface state of the semiconductor unstable and a high Ge fraction in $Si_{1-x}Ge_x$ is associated with high surface state and therefore faster oxidation [11]. Fast oxidation rate can make the create even more unstable [12]. To reduce the instability of the surface state, a SiO_2 passivation layer is formed to reduce interface state of the surface of SGOI nanowire to less than 10^{11} $cm^{-2}eV^{-1}$. The interface state of the free surface is around 10^{15} $cm^{-2}eV^{-1}$. In this work, a PECVD SiO_2 layer is used as a passivation layer to improve the interface state and an O_2 gas buffer layer is adopted to reduce the oxidation rate for 20% Ge in SiGe nanowire sensors.

II. EXPERIMENTAL PROCEDURE

α -Si/ $Si_{1-x}Ge_x$ was deposited on patterned 300 nm-hight bottom oxide. The deposition thickness of α -Si was 200Å, and Ge fraction of $Si_{1-x}Ge_x$ splits were 7%, 14% and 20%,. To clarify the influence of the passivation layer on the sensitivity of the nanowire, two kinds of nanowire sensors were fabricated with and without passivation an SiO_2 passivation layer. The SiO_2 layer was one of split in two thicknesses- 100 Å and 200 Å. The poly-Si nanowire was also fabricated as control group to verify the rate of oxidation. After the formation of the α -Si /SiGe layer formation, the samples were annealed at 950ºC for 180 sec in a mixture of O_2 gas and

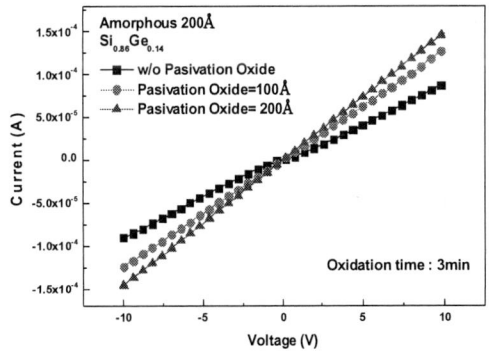

Figure 1 The characterization of I-V curve of a-Si=200 Å /$Si_{0.86}Ge_{0.14}$ nanowires

978-1-4673-1122-9/12 $31.00 © 2012 IEEE

13% N$_2$ gas. After annealing, the Al films were thermally evaporated and the electrodes were defined by the mask process. The poly-Si and SiGe nanowires were implanted in the p-type nanowire. To functionalize α-Si / Si$_{1-x}$Ge$_x$ nanowires, the wires were adopted the 3-aminopropyltri-ethoxysilane (APTMS) was adopted to modify the surface of the silicon oxide around the nanowires. A hydroxyl functional group on the surface of the oxide was replaced by methoxy groups of APTMS modules, and simultaneously, the surface of the nano-wire was terminated by amine groups. From authors' earlier investigations, the amine groups are prone to deplete positive carriers, reducing the conductivity of the p-type nanowire. Next, bis (3-sulfo-N-hydroxysuccinimide ester) sodium salt (BS3) was utilized as a linker between APTMS and IgG antibody. BS3 was prone to becoming negatively charged, increasing the conductivity of the p-type nanowire, because of the holes were accumulated by BS3 on the surface of the nano-wire.

A Hewlett Packard HP 4156A instrument was used in this study to measure the electric characteristics of the nanowire sensor. The drain voltage (VD) was varied from -10 to 10V in steps of 500 mV, and the back gate voltage was 0 V. The electric characteristics were measured at every stage of the surface modification, and the average conductance was then extracted from the ID-VD characteristics with VD = 3~6 V. The sensitivity (S) of a nanowire-based sensor is defined as the ratio of the magnitude of the change in conductance the baseline conductance :

$$S = \frac{|G - G_0|}{G_0} = \frac{\Delta G}{G_0} \quad (1)$$

, where G_0 denotes the conductance before capture of a molecule; G is the conductance after molecule capture, and ΔG is the difference between G and G_0.

III. RESULTS & DISCUSSION

Figure 1 presents the characterization of I-V curve of α-Si=200Å/Si$_{0.86}$Ge$_{0.14}$ nanowires with and without a passivation SiO$_2$ layer. The nanowire with a 200 Å-thick SiO$_2$ passivation layer has a higher conductance than that with a 100 Å-thick passivation layer or without no passivation layer. The sensitivity of the nanowire can be improved by increasing the conductance because more mobile carriers close to the surface of the nanowire are then more carriers were sensitized without being trapped by the interface state. Figure 2 plots the sensitivity of SiGe nanowires structure with different fractions of Ge condition. The higher sensitivity was obtained when it was capped with a passivation layer. It was obvious that the increasing in sensitivity of Si$_{0.8}$Ge$_{0.2}$ is higher than the other SiGe nanowire samples. The sensitivity of the Si$_{0.8}$Ge$_{0.2}$ nanowire sample was poorer than that of other nanowire samples with a lower Ge fraction when no passivation layer was used to cap the SiGe nanowire. This finding is consistent with our previous work, in which the sensitivity of SiGe nanowire was degraded as the fraction of Ge increased above

14%. Therefore, the sensitivity of the SiGe nanowire increased with the fraction of Ge when a suitable capping layer was used.

Figure 3 compares the sensitivities of samples with 20% Ge nanowire following treatment with various ambient. Treatment in pure N$_2$ ambient did not increase sensitivity because Ge condensation did not occur without O$_2$ gas. The sensitivity of the pure O$_2$-treated nanowire sample was lower than that of the sample that was treated in O$_2$ gas mixed with some N$_2$ gas. Although in pure O$_2$ gas, more Ge piled up at the surface of the nanowire upon the Ge condensation process, the number

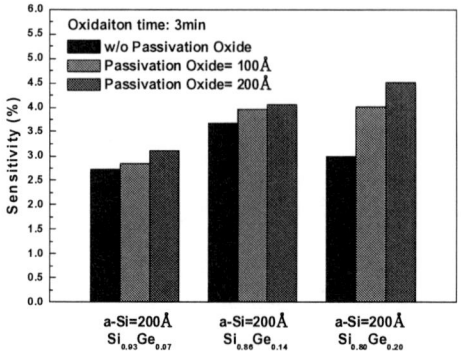

Figure 2 The sensitivity of different Ge fraction nanowires with and without PECVD SiO$_2$ passivation

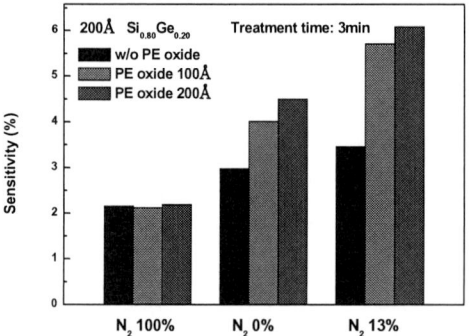

Figure 3 The sensitivity of Si$_{0.8}$Ge$_{0.2}$ nanowires with and without PECVD SiO$_2$ passivation in different treatment ambient.

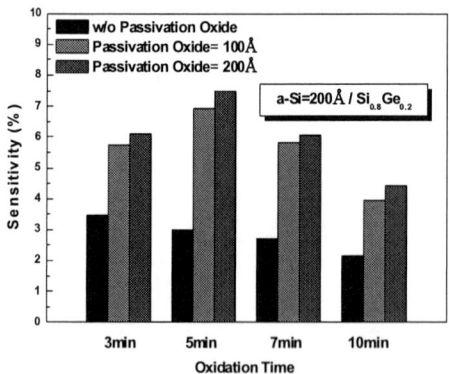

Figure 4 The sensitivity of Si$_{0.8}$Ge$_{0.2}$ nanowire structure with different oxidation time.

of surface defects simultaneously increased. These surface defects degraded the sensitivity of the SiGe nanowire. Hence, the oxidation rate can be reduced using O_2 gas that is diluted with N_2 gas to prevent the formation of interface states during Ge condensation. As shown in Fig. 3, the sensitivity of the nanowire in the 13% N_2 and O_2 gas mixture is higher than that of in pure N_2 or O_2 gas even though the nanowire samples has no paasivation layer. Figure 4 displays the effect of oxidation time under different passivation conditions of the $Si_{0.8}Ge_{0.2}$ nanowire. A peak value of sensitivity was obtained at an oxidation time of 5 min from the nanowire that was capped with PECVD oxide. The sensitivity was improved approximately 1.3 times higher than that of the nanowire sample in the same oxidation time without a top passivation layer. There is a degradation trend of sensitivity in a long period of oxidation on nanowire with and without capping layer, because the Ge did not only pile up at the surface in the Ge condensation process but also diffused into the buried oxide layer. The O atoms did not pass through the thicker top oxide layer more easily as the oxidation time increased. Hence, the Ge rejection rate and Ge pile-up rate were reduced and the diffusion rate started to dominate the Ge distribution. Although the passivation layer suppresses oxidation like a capping layer, the suppression of the unstable surface state suppressed in oxidation is the major mechanism of the improvement of sensitivity. The Ge condensation still proceeds if a suitable oxidation time.

To clarify the influence of the oxidation rate on the sensitivity of the naowire with the passivation layer, poly-silicon nanowire was fabricatied. Figure 5 displays the effect of oxidation time under different passivation conditions of the poly-silicon nanowire. The sensitivity of the poly-silicon nanowire without the SiO_2 passivation layer exceed that of the nanowire with the layer, because the former was oxidized at a higher rate, yielding a high surface-to-volume ratio of poly-silicon. The oxidation rate decreased as the thickness of the passivation layer increased, reducing the surface-to-volume ratio and the sensitivity. Hence, this passivation oxide layer effectively reduced the rate of oxidation and suppressed the instability of the surface state. Finally, Fig. 6 shows the sensitivity results for the SiGe nanowire samples with different Ge fractions after 5 min of oxidation. In this work, the maximum improvement of sensitivity was achieved for the nanowire with a Ge fraction of 20%. Accordingly, the rate of oxidation and the instability of surface state was reduced using a suitable capping layer, which therefore increased the sensitivity of high fraction Ge .

CONCLUSION

In this study, an SiO_2 passivation layer was used to reduce surface state result from rate of oxidation is reduced to suppress the formation of an interface at the surface of SiGe. Hence, the sensitivity of a nanowire with a high fraction of Ge can be improved by Ge condensation. The maximum improvement of the sensitivity of the $Si_{0.8}Ge_{0.2}$ nanowire was

achieved using a 5 min oxidation time and a 200 Å-thick passivation oxide layer. The maximum improvement in the sensitivity of the passivated nanowire is around 1.3 times higher than that of the nanowire without.

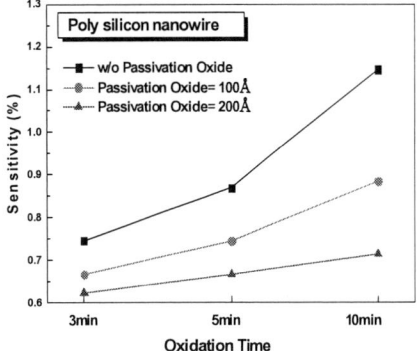

Figure 5 The sensitivity of $Si_{0.8}Ge_{0.2}$ nanowire structure with different oxidation time.

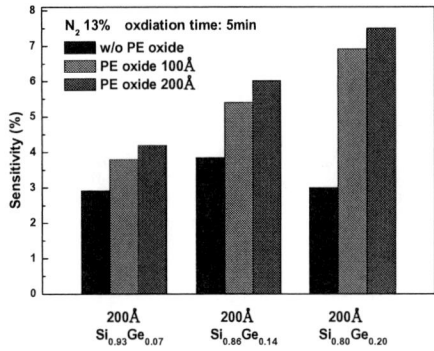

Figure 6 The sensitivity result of SiGe nanowire in 5 min oxidation time with N_2 13% diluted O_2 gas treatment.

ACKNOWLEDGMENT

The authors would like to thank the staff of the National Nano Device Laboratory for their technical help. They also acknowledge the financial support of the National Science Council (NSC) under Contract Nos. NSC 98-2221-E-009-174 MY3. Ted Knoy is appreciated for his editorial assistance.

REFERENCES

[1]   Y. Cui, Q. Wei, H. Park and Charles M. Lieber , Science, Vol. 293,No. 5533,pp. 1289-1292, August 17, 2001.

[2]   G.Brambilla, F. Xu, and X. Feng, Electronic Letters, Vol 42, 8, 2006.

[3]   P. R. Nair, and M. A. Alam, Applied Physics Letters, Vol. 88, No. 23, 2006.

[4]   Duan, C. Niu, V. Sahi, J.Chen, J. Parce, and S. Empedocles, Nature, 425, pp. 274-278, 2003

[5]   A.M. Morales and C.M. Lieber, Science 279, 208,1998.

[6] Zhaohui Zhong, Deli Wang, Yi Cui, Marc W. Bockrath, Charles M. Lieber, Science Vol. 302. no. 5649, pp. 1377 – 1379, 2003.

[7] G. Zheng, F. Patolsky, Y. Cui, W. U. Wang, C. M. Lieber, Nature Biotech., 2005

[8] K.M Chang, et al. 4nd IEEE International Nanoelectronics Conference, 2011 IEEE INEC

[9] K.M Chang, et al., 2nd IEEE International Nanoelectronics Conference, INEC, pp. 315-319,2008.

[10] F.K LeGoues, R. Rosenberg, T. Nguyen, F. Himpsel, and B.S. Meyerson, J.Appl.Phys, 65(4), 15 February, 1989

[11] Masanori Tanaka, Tatsuo Ohka, Taizoh Sadoh, and Masanobu Miyao, J.Appl.Phys, 103, 054909,2008

[12] Haigui Yang, et al. Applied Physics Letters, 93, 072104, 2008

A Silicon-on-Glass Z-axis Accelerometer with Vertical Sensing Comb Capacitors

Jiankun Wang, Zhenchuan Yang*, Guizhen Yan
National Key Laboratory of Science and Technology on Micro/Nano Fabrication
Institute of Microelectronics, Peking Univ., Beijing, 100871, CHINA
z.yang@ pku.edu.cn

Abstract—A Z-axis accelerometer with asymmetrical vertical sensing comb capacitors and high aspect ratio single crystal mechanical structures is presented. A 5-mask silicon-on-glass (SOG) process combined with silicon/glass anodic bonding and multiple deep ion reactive etching (DRIE) is used to fabricate the accelerometer. With the specially designed out-of-plane springs, the accelerometer shows a cross-axis sensitivity of 0.46%. The scale factor and the non-linearity of the accelerometer are measured to be 172.5 mV/g and 0.47% with the input range of ±1 g (gravity), respectively. The noise floor of the accelerometer is 0.22mg/Hz$^{1/2}$ at 100Hz and for a bandwidth of 400Hz. The short term (10 min) zero bias stability of the fabricated accelerometer is also evaluated to be about 0.47mg.

Keywords-MEMS; Accelerometer; SOG; Asymmetrical Vertical Comb;

I. INTRODUCTION

For the last decades, the demands on MEMS inertial sensors have grown rapidly for the application on consumer electronics, automotive electronics and so on [1-3]. Among the reported inertial sensors, differential sensing accelerometers especially with differential sensing comb attracted widely attentions with the advantages of high sensitivity and high reliability [4-5] while many differential sensing mechanisms of Z-axis accelerometers had been developed. References [2-3] reported gap closing differential sensing electrodes, and reference [6] realized the differential sensing by rotational displacement detection, while asymmetrical vertical comb sensing were reported by [5,7].

Miniature inertial measurement unit (MIMU) realized with MEMS inertial sensors are especially attractive for their compact size and low cost. Monolithic integration of multiple inertial sensors on a single chip is an efficient approach to implement the low cost MIMU with moderate performance. In our previous work, fully asymmetrical vertical comb capacitors have been successfully demonstrated with a 5-mask Silicon-on-Glass (SOG) process for lateral-axis gyroscopes [8], which make it feasible to co-fabricate both lateral-axis and z-axis on a single chip. To extend the previous work, a Z-axis accelerometer with asymmetrical vertical sensing comb capacitors is demonstrated with the same process, leading to the potential for monolithic integrated MIMU.

In this paper, a novel Z-axis SOG accelerometer which is originally designed for a vertical electrostatic actuator and here used to demonstrate the Z-sensing with fully asymmetrical vertical comb capacitors as shown in Fig. 1 is presented. Four sets of springs and fully asymmetrical vertical comb capacitors are distributed symmetrically to realize movement in the Z direction and the differential capacitive sensing. The fully asymmetrical comb shows high linearity with the position of the proof mass which ensures the high linearity performance of the

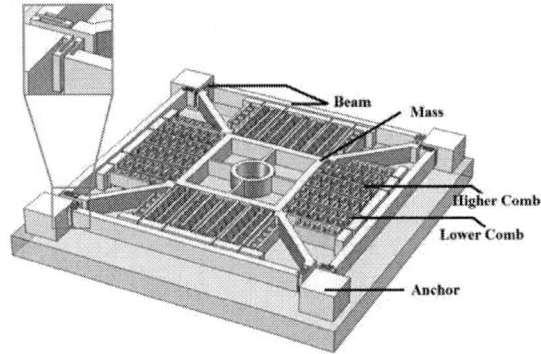

Fig. 1. The schematic of the Z-axis accelerometer.

accelerometer.

II. STRUCTURE DESIGN AND ANALIZE

A. Structure Design

As shown in Fig. 1, the accelerometer is suspended by four sets of vertical spring which is composed of folded torsion beams which can realize rotational movement and a seesaw beam which can convert the rotational movement to vertically translational movement of the proof mass. The details of the vertical spring is also presented in Fig. 1, which shows that the spring has a long seesaw beam to realize small Z axis stiffness and relatively large lateral stiffness to suppress cross-axis sensitivity and realize large sensitivity in Z axis.

The asymmetrical comb are also distributed symmetrically to suppress the effect of fabrication imperfection and also to reduce the influence of higher vibration modes of the structure with the rotational movement around the lateral axis or laterally translational movement. All the comb can function as sensing capacitors for Z direction detection as a accelerometer, or a part of them realize the Z position sensing with a voltage

978-1-4673-1122-9/12 $31.00 © 2012 IEEE

applied on the other to produce the vertically electrostatic force to function as an vertical electrostatic actuator.

B. FEM Simulation

A FEM analyze was performed to analyze the influence of vibration modes of the structure, with the mode shape of the first two modes shown in Fig. 2. The first mode is a piston movement in the Z direction with a resonant frequency of 564 Hz, which is the working mode with the input of a static or dynamic acceleration. The second mode and third mode both are rotational modes around the lateral axis with the same resonant frequency of 2233 Hz, which will deteriorate the linearity performance of the device. But the symmetrically distributed comb and mass of the structure will suppress these two modes, which will not be excited by the input acceleration.

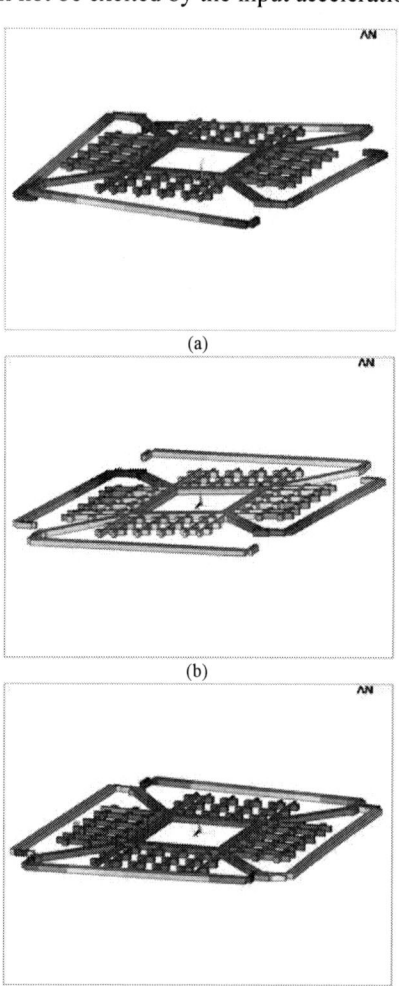

Fig. 2. (a) The First Mode of the Accelerometer. (b) The Second and Third Mode of the Accelerometer. (c) The Fourth and Fifth Mode of the Accelerometer.

The fourth and fifth mode have rotational movement in plane around the Z axis as shown in Fig. 2(c), which will rarely affect the performance with the symmetrically distributed comb which can cancel the effect of the movement. It also has a rather high resonant frequency of 8832 Hz.

III. DEVICE FABRICATION

The fabrication of fully asymmetrical vertical comb capacitors based on SOG process has been successfully demonstrated for lateral-axis gyroscopes [8], which is

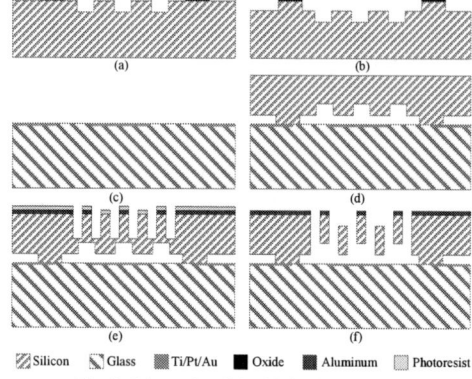

Fig. 3. Schematic of the fabrication process.

schematically depicted in Fig. 3.

The process begins with a double polished 4 inch bulk silicon wafer and a 4 inch glass wafer. The anchor of the structure is patterned by the first mask and then the second mask defines the step between the lower comb and higher comb on one side of the silicon wafer which defines the higher comb. Then a two-step deep reactive ion etching (DRIE) which defines the step between lower and higher comb before the accomplishment of the anchor etching with the composite etching mask formed by thermal silicon oxide and photoresist. The pad and interconnect are patterned on the glass wafer by the third mask which will increase the design flexibility of the structure. Then the silicon wafer will be bonded with the glass wafer by a wafer bonding process with the processed side toward the glass side with pad and interconnect on it. The silicon layer is then thinned by KOH. After that, a composite etching mask on the top side of the bonded wafer which is formed by PECVD (plasma enhanced chemical vapor deposition) silicon oxide and sputtered Al is performed by the fourth and fifth mask. The structure is released by a DRIE process and then, the Al mask will be removed by wet etching. Then the higher comb fingers have PECVD silicon oxide on them while the lower comb fingers haven't. The second DRIE is then performed to realize the step between the higher and lower comb on the top side of the bonded wafer which defines the lower comb. The device is then accomplished after the silicon oxide removed after the DRIE. The fabrication process realizing the lower comb finger is the key step in this process, while too much over-etching will cause serious footing effect and then damage the structure.

The SEM photo of the fabricated device with details of the asymmetrical comb and suspend spring is presented in Fig. 4. The step between the higher comb finger and the lower comb finger is 8 μm on the top side of the device, while the step on the other side of the silicon wafer is 10 μm which is precisely defined by the first DRIE before the anchor etching. The whole device has a dimension of 4.5 mm by 4.5 mm, with silicon layer thickness of 80 μm.

The fabrication process is capable with the fabrication of symmetrical comb capacitors and then it is feasible to co-fabricate both lateral-axis and z-axis gyroscopes and accelerometers on a single chip to be a MIMU.

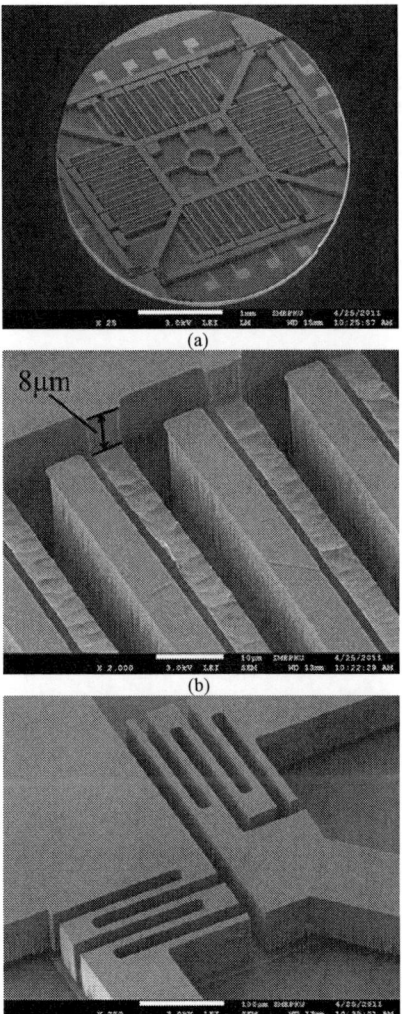

Fig. 4. SEM pictures of the fabricated accelerometer. (a) Overview of the accelerometer. (b) Details of the asymmetrical comb. (c) Details of the suspend springs

IV. TEST RESULTS

The characteristics of the fabricated accelerometer are tested with a commercial ASIC chip that schematically shown in Fig.5. The minimum capacitance can be distinguished is 40aF/Hz$^{1/2}$ for the ASIC. The higher and lower combs are connected to carrier waves with anti-phase while the mass is connected to the amplifier. The output voltage is recorded as the output of the accelerometer both for static test or noise test and bias drift test.

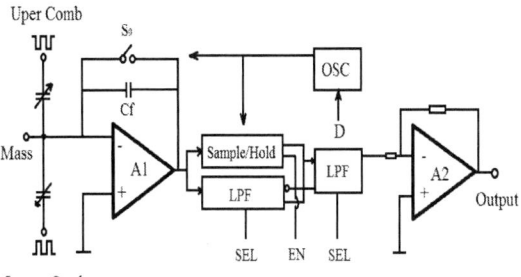

Fig. 5. Schematic of the test circuit for the accelerometer.

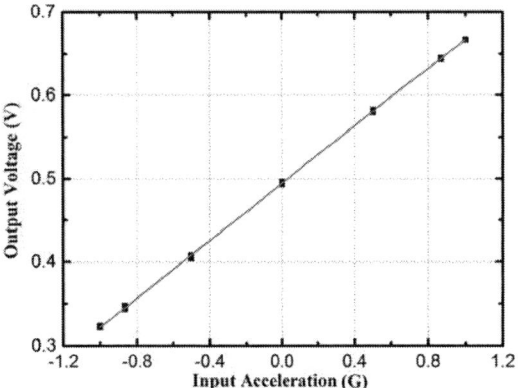

Fig. 6. Static test result of the accelerometer.

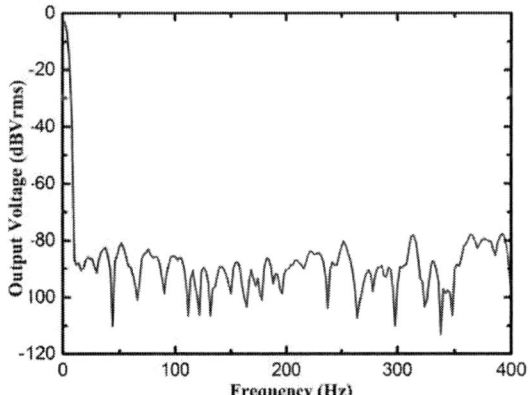

Fig. 7. FFT result with 1G applied on the accelerometer.

978-1-4673-1122-9/12 $31.00 © 2012 IEEE

The static test result of the presented accelerometer is shown in Fig.6. The open loop sensitivity of the Z-axis accelerometer is 172.5mV/g with a non-linearity of 0.47% for the measurement range of ±1 g. The cross-axis sensitivity which is mainly introduced by the lateral movement of the mass is evaluated to be of 0.46%, as the springs are specially designed for a small Z direction stiffness and relatively large lateral stiffness to suppress the cross-axis sensitivity.

The noise characteristic was also studied by HP35670A with 1g applied on the accelerometer by the method of FFT analyze. The noise test result is shown in Fig.7 with a noise floor of 0.22mg/Hz$^{1/2}$ at 100Hz for a bandwidth of 400 Hz. Short term zero bias stability of the represented accelerometer was also studied, as shown in Fig.8; the zero bias stability is 0.47mg for equivalent input acceleration over a test period of 10 minute in normal pressure.

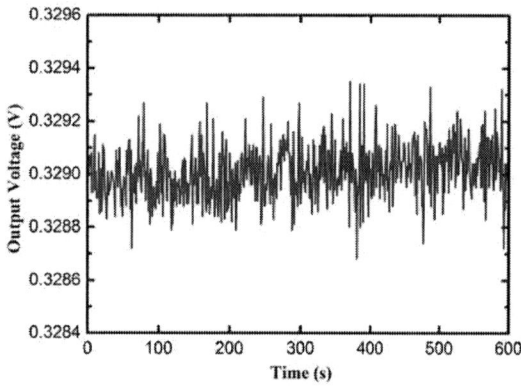

Fig. 8. Zero bias stability of the accelerometer.

V. CONCLUSION

A silicon-on-glass Z-axis accelerometer with vertical sensing comb capacitors is designed and analyzed. Static test and noise test as well as bias drift test are carried out to verify the accelerometer. The tested sensitivity of the Z-axis accelerometer is 172.5mV/g with a non-linearity of 0.47% for the measurement range of ±1 g. The cross-axis sensitivity is evaluated to be of 0.46%. The noise floor of the accelerometer is 0.22mg/Hz$^{1/2}$ at 100Hz while the short term zero bias stability of the represented accelerometer is 0.47mg for a test period of 10 minute. The cross-axis sensitivity is well suppressed by the specially designed springs for small Z direction stiffness and relatively large lateral stiffness.

ACKNOWLEDGMENT

The authors are grateful to the technical staffs of National Key Laboratory of Science and Technology on Micro/Nano

Fabrication for their supports on devices fabrication. The work is partially supported by the National Natural Science Foundation of China under Grant No. 90923037.

REFERENCES

[1] H. Hamaguchi, K. Sugano, et al., "A Differential Capacitive Three-Axis SOI Accelerometer using Vertical Comb Electrodes," *transducers'2007*, pp.1483-1486.

[2] C.P.Hsu, M.C.Yip, et al., "A novel soi Z-axis accelerometer with gap closing differential sensing electrodes," *transducers'2009*, pp.1154-1157.

[3] M. Fujiyoshi , Y. Nonomura, et al., "An SOI 3-axis accelerometer with a zigzag-shaped Z-electrode for differential detection," *transducers'2011*, pp. 1010-1013.

[4] G. Li, et al., "Design and fabrication of a highly symmetrical capacitive triaxial accelerometer," Journal of Micromechanics and Microengineering, vol. 11, pp. 48-54, 2001.

[5] T. Tsuchiya, H. Funabashi, "A Z-axis differential capacitive SOI accelerometer with vertical comb electrodes," *MEMS'2004*, pp. 524-527.

[6] Y. Hirata, N. Konno, et al., "A new z-axis capacitive accelerometer with high impact durability," *transducers'2009*, pp. 1158-1161.

[7] Z.C. Yang, G.Z. Yan, et al., "Design and fabrication of a torsional z-axis capacitive accelerometer with novel comb capacitor," *ICSICT'2004*, pp. 1792- 1795.

[8] Z.Y. Guo, L.T. Lin, et al., "A Lateral-Axis Microelectromechanical Tuning-Fork Gyroscope With Decoupled Comb Drive Operating at Atmospheric Pressure," J. Microelectromech. Syst, vol. 19, pp. 458-468, 2010

The Vertical MSM Diamond X-ray Detector

S. Cheirsirikul[1], S. Jesen[1], C. hruanun[2]

[1]*Department of Electronics Engineering, Faculty of Engineering, KMITL, Bangkok, Thailand*
[2] *Thai Microelectronics Center, National Electronics and Computer Center, Chachoengsao, Thailand*

Abstract— **Development synthesizing diamond film on (100) silicon substrate was processed by HFCVD (Hot Filament Chemical Vapor Deposition). The processes consisted of using methane (CH_4) to produce intrinsic diamond and built MSM device on the diamond film. The top and the bottom of diamond is Schottky junctions were produced by aluminum which caused the carriers were able to move in the vertical direction to reduce the effect of grain boundaries. After that, the result of detecting an X-ray of MSM diamond was satisfactory because it could respond along with increasing of intensity of an X-ray and X-ray expose time**

Keyword—*diamond-film fabrication, vertical MSM diamond, X-ray sensor*

I. INTRODUCTION

Nowadays diamond film is well known as a semiconductor material having good physical, optical and electrical properties [1]. These properties have attracted researchers to develop diamond film synthesis by using CVD technologies to achieve growth of diamond films over relatively large areas. Among the electronic applications of such material is its use as a practical detector. A conspicuous property of special interest is the large band gap energy (5.5eV, 225nm) of diamond film, making it a suitable material for photo detection of short wavelength [2]. After that photoconductive and photodiode were fabricated by diamond film, the carrier generate from the light will change electric and light characteristic [3, 4]. This paper presents a MSM diamond X-ray detector fabricated on free-standing diamond film, which operates in a vertical direction. This structure, the energy are absorb in depletion region and similar to boundary grain. So, it reduces the result of carrier move pass boundary grain.

II. METHOD

Diamond film was synthesized on silicon substrate by HFCVD (Hot Filament Chemical Vapor Deposition). The processes consisted of using methane (CH_4) to produce intrinsic diamond. It was cleaned in a saturated solution by CrO_3 in H_2SO_4 at 170°C and then cleaned with a 1:1 boiling solution of H_2O_2 and NH_4OH at 80°C to eliminate surface graphitic material, and DI water [5]. A SEM photograph of the diamond film is show in Fig.1 and a Raman analysis of the diamond film show in Fig.2 reveals no other structure than a sharp peak at 1332 cm^{-1}. The Vertical MSM diamond was fabrication on free standing diamond film by etching the silicon substrate with EPD solution to produce diaphragm of diamond film[6]. The thickness of diamond films is 7 μm and the device area is 4 mm^2. We designed it to work in verticals

and parallel with the boundary grain, that it's reduce the effect of boundary grain for direction of current.

The photolithography technique was use to from the planar array electrode contact at top surface and Aluminum metal was used as a schottky contact. Fig.3 shows the device structure. The structure of diamond is polycrystalline.

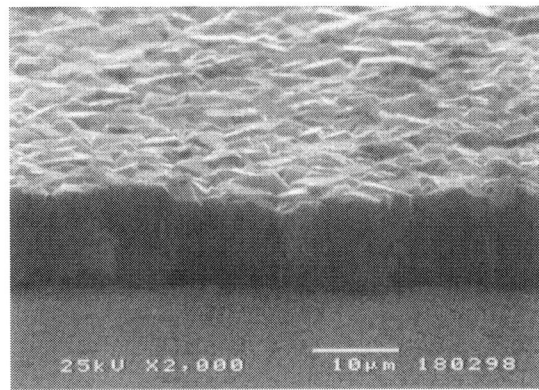

Fig.1 SEM photograph of diamond film.

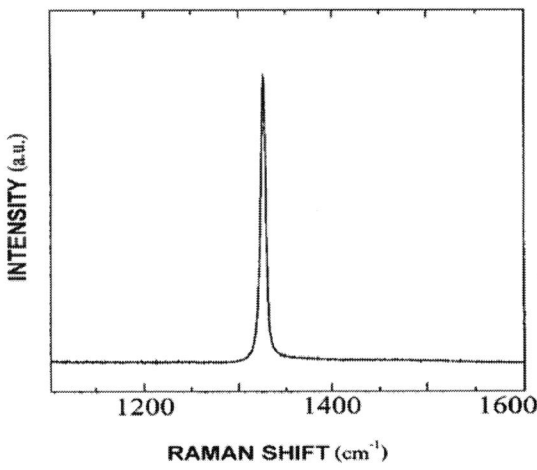

Fig.2 Raman spectrum of diamond film.

It has many crystals and many boundary grains. Therefore, affect to the characteristics of device. While the device was reversed bias and cause the depletion region. When the light against to the device, it's generate electron-hole pair and move to the pole by electric field in depletion region. However, the device work in horizontal on diamond film face a problem with electron-hole pair move pass boundary grain and effect to reduce the amount of carrier. So, we designed to work in verticals and parallel with the boundary grain, that it's reduce the effect of boundary grain for direction of current.

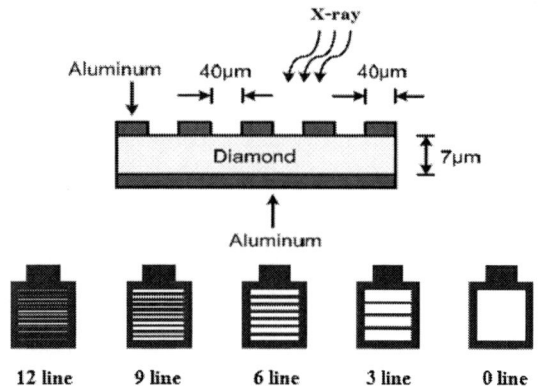

Fig.3 Structure of the vertical MSM diamond X-ray detector.

III. EXPERIMENTAL RESULTS

A. characteristic of Voltage and Current of MSM diamond device

I-V Characterization of the vertical MSM diamond was determined by Hp4061-A semiconductor parameter analysis at the temperature of 25°C, using input voltage from -20 V to +20 V.

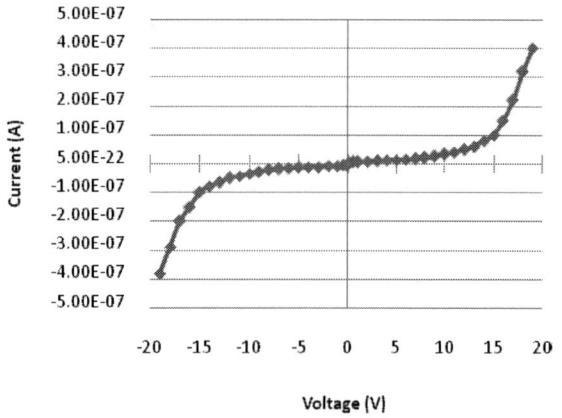

Fig.4 The I-V Characteristic of vertical MSM diamond device.

The result from Fig.4 shows the amount of the dark current that leaks from the MSM diamond device will increase as the voltage bias increase. From the result, the dark current will be around 35 nA at -10 V, 33 nA at 10 V and less than 10 pA at -3 V.

B. The response of X-ray

The of MSM diamond X-ray detector structure is show in Fig.3. The condition of metal line on the active surface area was 12, 9, 6, 3 and none line

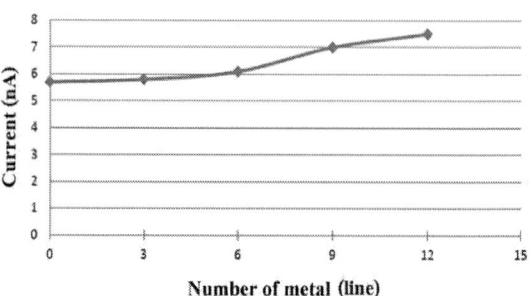

Fig.5 Relationship of metal line and signal current of MSM diamond detector

MSM diamond were expose X-ray by Varian model S-785 at 150 µSv at room temperature (25°C) as show in Fig.5. It found that, structure of MSM diamond detector 12 metal line is the best response detector. Since, MSM diamond detector is highest active area and depletion region. So, it cause free carrier more and more, and height current output. For structure none metal line, free carrier is small, and output current is small. It found that number of metal line was increased. Thus, output current of MSM diamond x-ray detector was increased.

C. Relationship of x-ray energy, current and rise time of MSM diamond detector

MSM diamond detector 12 metal line was expose X-ray by Varian model S-785 using high voltage bias x-ray source from 50 kV to 100 kV at 6 mA, expose time 0.5 sec at room temperature (25°C).

We expose direct to detector. In order to investigated the response of MSM diamond detector. The result show in Fig.6. The increasing of X-ray energy, cause increasing of output current. Since, increasing of X-ray energy caused increasing of free carrier, so output current was increased.

The result show in Fig.7. It found that, for direct expose the rise time was decrease when we increased X-ray energy 50 kV to 100 kV. at 100 kV the rise time is 140 µs. Since increasing of X-ray energy cause free carrier move to electrode in shot time so the rise time was decreased.

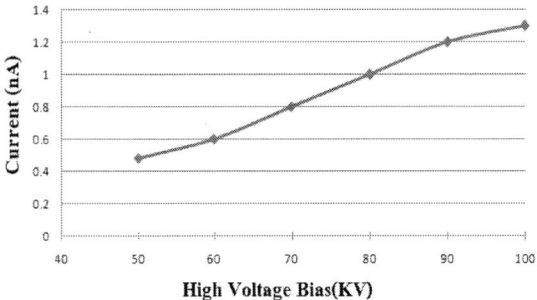

Fig.6 Relationship between X-ray energy and current
of MSM diamond detector.

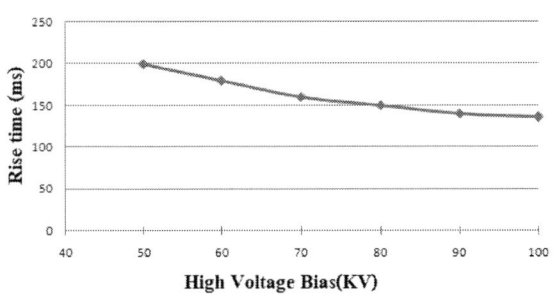

Fig.7 Relationship between of X-ray energy and rise time
of MSM diamond

D. Relationship of expose time of X-ray, current, and rise time of MSM diamond detector.

MSM diamond detector 12 metal line were expose X-ray by Varian model S-785 at 26 µSv, expose time 0.1 sec -0.5 sec at room temperature (25°C). We exposed direct to detector. In order to investigated the response of MSM diamond detector.

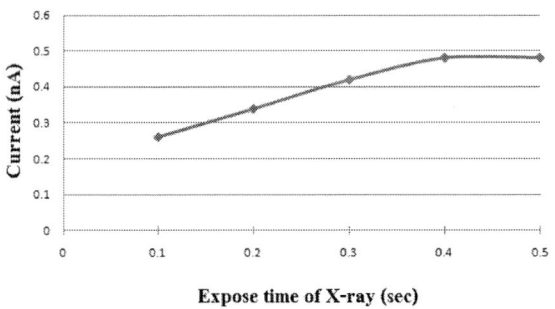

Fig.8 Relationship between expose time of X-ray and current
of MSM diamond detector.

The result show in Fig.8. Since, for direct expose from 1 sec to 3 sec, increasing of expose time caused increasing of free carrier, so output current was increased. And for 0.4 sec-

0.5 sec, amount of free carrier was not increased, so output current was stable.

The result show in Fig.9. It found that, the direct expose time, the current and rise time was increased when we increased the expose time from 0.1 sec to 0.3 sec and the expose time at 0.1 sec, the rise time value was 55 msec. And, the rise time value was stabled from 4 sec to 5 sec about 200 msec.

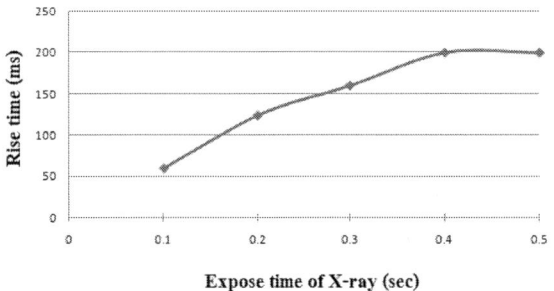

Fig. 9 Relationship between expose time of X-ray and
rise time of MSM diamond detector.

Since, the direct expose time, from 0.1 sec to 0.3 sec, the increasing of expose time caused increasing of the free carrier in depletion region. So, output current was increased. And the expose time from 0.4 sec – 0.5 sec, the X-ray intensity was absorbed to the depletion region. But the electron-hole pair was unchanged and amount of the free carrier and output current was stabled.

IV. CONCLUSION

The vertical MSM diamond X-ray detector was fabricated on the free standing diamond film obtained by etching the silicon substrate. The operation of the devices in a vertical direction implies that X-ray is absorbed directly and higher output current than planar device that is fabricated on the same material. The X-ray absorption within the depletion region is expected to be the main mechanism for activating the devices. Therefore, the carrier generated in the depletion region are transported to the both of contacts in the same direction as grain boundaries and the width of active region is the same order as the grain size of the diamond film. Hence, the effect of grain boundaries was reduced and the device also has response to the X-ray, UV light and low response to the visible light. The dark current of the device is less than 10 pA. We conclude that the MSM diamond structure allowing activation in the vertical direction is suitable for crystallite boundaries on the generation and transport of carriers is reduced. Finally, the structure of the vertical MSM diamond should be increase number of fine metal line at active area for high output current which can detect the low energy X-ray, expose time below 0.1 sec.

ACKNOWLEDGMENT

The author would like to thank National Science and Technology Development Agency (NSTDA) and National Electronics and Computer Technology Center (NECTEC) support in research.

Thank you for Assoc.Prof Manas Sangworasil and Assoc. Prof Chuchart Pintuviruj who assist to use X-ray tool. Finally, this work might not have been possible the Thailand Graduate Institute of Science and Technology (TGIST) support in Scholarship.

REFERENCES

[1] J.E. Field. The properties of diamonds, Academic Press, London, 1979.

[2] D.R. Kania, M.I. Landstrass, M.A. Plano, L.S. Pan, S. Han, Diamond radiation detectors, Diamond Relat. Mater 2 (1993) 1012-1019.

[3] S.S., Chan, R.D. McKeag, M.D. Whitfield, R.B. Jackman, UV Photo detectors from thin film diamond, Phys. Stat. Sol. (a) 154 (1996) 445-454, January.

[4] R.D. McKeag, M.D. Whitfield, S.Sm Chan, L. Ys Pang, R.B.Jackman, A high performance UV Photodetector from thin film diamond, Mater. Res, Soc. 416(1996)419-424.

[5] L. Thaiyotin, E. Ratanaudompisut, T. Phetchakul, S. Cheirsirikul, S. Supadech, UV Photodetector from Schottky diode diamond film, Diamond Relat. Mater 11(2002)442-445.

[6] W. Ebert, A Vescan, T.H. Borst, E. Kohn, High current p/p+ diamond Schottky diode, IEEE Electron Device Lett. 15 (2) (1994) 289-291, August.

978-1-4673-1122-9/12 $31.00 © 2012 IEEE   590

Effect of Geometrical Design of Support on Frequency Shift and Energy Loss of Piezoelectric Ring Resonator Applicable to Liquid Circumstance

Takahisa Sagawa[1], Dong F Wang[1,*], Jian Lu[2], and Ryutaro Maeda[2]

[1] Micro Engineering & Micro Systems Laboratory, Ibaraki University (College of Eng.), Hitachi, Ibaraki 316-8511, JAPAN
[2] Research Center for Ubiquitous MEMS and Micro Engineering (UMEMSME), AIST, Tsukuba, Ibaraki 305-8564, JAPAN
([*] Tel: +81-294-38-5024; Fax: +81-294-38-5047; E-mail: dfwang@mx.ibaraki.ac.jp)

Abstract— **A ring-shaped piezoelectric resonator was proposed for health monitoring. The proposed resonator is believed to be experienced several energy loss mechanisms, when used in air or liquid circumstance. Actually, the quality factor of film bulk acoustic resonator in liquid might be decreased to 5% of that in air. So, it is necessary to accomplish suitable structural design and select proper vibration mode to reduce the energy loss of the resonator especially in liquid application. For the pursuit of high sensitivity, we studied the effect of support geometry on frequency shift and energy loss. It was found that low-amplitude vibration area is smaller at narrow support, and frequency shift of trapezoid support is higher than that of rectangle support.**

Keywords- Ring-shaped resonator; Piezoelectric thin film; Support loss; Frequency shift; Vibration mode shape; Health monitoring

I. HEALTH MONITORING AND ITS EXAMPLE

Aging causes increasing demand of health monitoring. Most important point of medical care is to discover disease at early stage. To monitor health condition is the main research purpose of the proposed resonator.

For example, CRP is one of the proteins in blood. It has 187 amino acids. CPR molecular weight is about 21500 [1]. The weight of one molecular is 0.00003571 fg. Size of CRP molecular ranges from 48.6 Å to 49.0 Å. When someone becomes disease, quantity of the CRP in blood will increase. So, CRP detection enables to monitor health condition and detect disease at early stage. Quantity of CRP in blood of health fitness human is 580 ng/ml. When some health problems happen in human body, the quantity of CRP is usually changed to be over 20000 ng/ml.

Current major method for CRP monitoring is Latex agglutination test. CRP and Latex are combined by antibody response in liquid, and far-red light is exposed to combined CRP. The quantity of CRP is measured from change of absorbance, but dedicated equipment and higher cost are required.

In contrast, proposed sensing resonator does not require dedicated equipment but all-purpose one, and can be fabricated by MEMS fabrication technology in mass. So, the total cost is lower, and the monitoring is real time. However, the proposed resonator is believed to be experienced several energy loss mechanisms, when used in air or liquid circumstance. Usually, energy loss is very large in liquid, compared to that in air. According to recent studies [2-4], quality factor of film bulk acoustic resonator in liquid decreased to 5% of that in air. So, it is necessary to accomplish suitable structural design and select proper vibration mode to reduce the energy loss of the resonator especially in liquid application. In order to reduce the energy loss, geometrical design of support was therefore investigated from view points of frequency shift and vibration mode shape.

II. PROPOSED DEVICE IN PRESENT STUDY

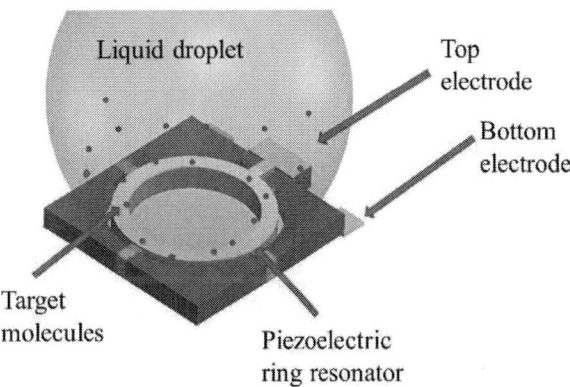

Fig. 1. Schematic of ring-shaped piezoelectric resonator for bio-sensing and health-care applications in liquid environment.

Fig. 1 shows the proposed ring-resonator for detecting target molecules. The resonator is comprised of SiO_2/ Pt / Ti / PZT / Ti / Pt / SiO_2 / Si layers. SiO_2, Pt, Ti and PZT are respectively insulation layer, electrode layer, adhesion layer and piezoelectric layer. Especially, Ti layer is used to improve the adhesion between PZT and SiO_2. PZT is used as

978-1-4673-1122-9/12 $31.00 © 2012 IEEE

piezoelectric layer [5] in present study. The ring is suspended by 4 supports which are arranged at nodal point of boundary of ring to reduce support loss. Antigen for combining target molecules is arranged on the surface of resonator. Adsorption of target molecules causes frequency shift due to mass change of resonator. And target molecules are thus detected by the frequency shift.

The ring resonator has some problems for achieving high sensitivity of detection of target molecules in liquid. These problems are energy losses which mainly include support loss, liquid damping and thermo elastic dissipation. Support loss is happened due to propagation of elastic wave from ring [6-9]. Liquid damping is caused by friction between ring resonator and liquid [10-11]. Vibration mode of the ring resonator is in-plane vibration which contributes to low liquid damping. Amplitude of in-plane vibration mode of the ring resonator is generally low than that of beam resonator. Therefore, liquid damping of ring resonator is smaller than beam resoantor devices.

This paper focuses on support loss, because at bulk mode resonator, support loss is most important [12]. Support loss depends on geometry and alignment of support. Support loss can be reduced by putting support on nodal point of ring.[13-14]

However, to achieve expected sensitivity in the proposed ring resonator, support loss must be further reduced. Support loss is caused by shear stress of support. This stress depends on amplitude and geometry of support. Therefore, geometry of support is important for reducing support loss. The effect of length of support was investigated in our recent paper [15]. The shorter support becomes, the higher frequency shift of ring resonator shows. Therefore, the effect of geometry of support, especially the width and shape, was investigated based on frequency shift and mode shape,

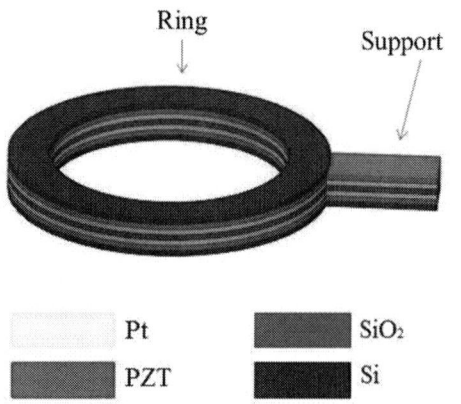

Fig. 2. Simulation model constructed in present study, comprised of Pt, PZT, SiO₂, and Si substrate.

Fig. 2 is the simulation model for modal analysis by finite element method. This model was comprised of Si / SiO$_2$ / Pt / PZT / Pt / SiO$_2$ / Si layers. Upper Si layer is tentative mass layer. Ti layer is cut to simplify the structure of model. Ti layer is much thinner than other layers except for tentative mass layer. So, the effect of Ti layer is small. Internal radius of ring is 30 μm, outer radius is 70 μm. Simulation was done by using ANSYS. When calculating frequency shift, mass of upper Si layer is changed from 10 ng to 20 ng. And the end section of the support where connecting with substrate is fixed. Mode shape and frequency shift were calculated by modal analysis.

III. STRUCTURAL DESIGN OF RING RESONATOR

A. Effect of support on frequency shift

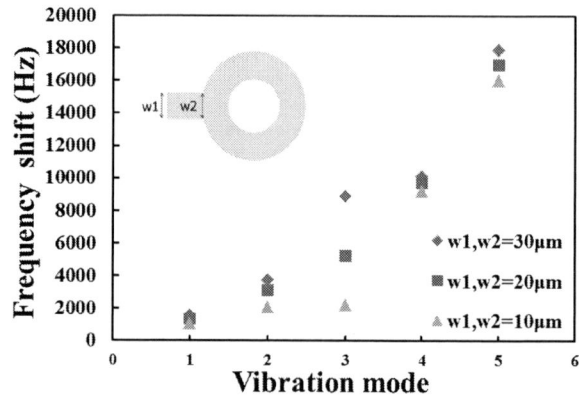

Fig. 3. Frequency shift as a function of vibration mode with a respect to support geometry.

Fig. 3 shows frequency shift when changing width of rectangle support. w 1 and w 2 are parameters of support size, which are defined by the upper left model. Fig. 3 shows that frequency shift changing depends on width of support and mode of vibration. Higher mode leads to high frequency shift and wider support results in high frequency. It is caused by changing stiffness of support. Higher width support translates to higher stiffness support. Frequency shift of mode 5 is higher than other mode. Comparing to mode 1, frequency shift of mode 5 is ten times as higher as that of mode 1. So, vibration mode is very important factor from view point of frequency shift. The highest frequency shift was obtained at mode 5 of 30 μm width support. The shift was measured as 17.9 kHz when changing the mass perturbation from 10 to 20 ng with the eigen frequency of 1176.8 kHz. Thus, from the point of view of frequency shift, mode 5 and 30 μm are better than those of other for rectangle support. Amount of change of frequency shift of mode 3 is bigger than other modes. The phenomenon suggests that effect of support width of is much larger than other modes.

978-1-4673-1122-9/12 $31.00 © 2012 IEEE

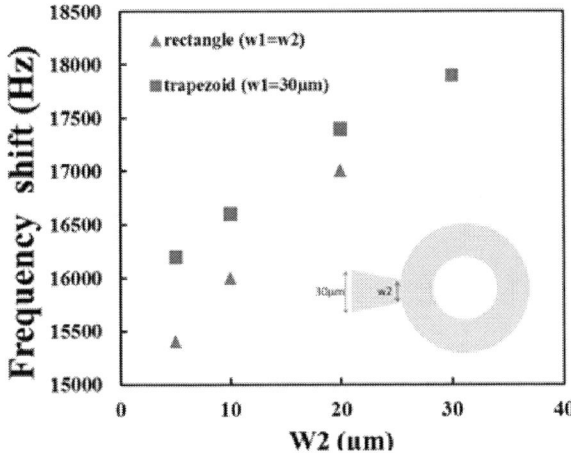

Fig. 4. Frequency shift as a function of boundary width with respect to support geometry.

Fig. 4 is frequency shift of rectangle support and trapezoid support when changing boundary width of support. w 1 of trapezoid support is fixed at 30 μm. w 1 of rectangle support is equivalent to w 2.

It is noted that frequency shift of trapezoid is higher than that of rectangle support. To sum up, the smaller w 2 becomes, the bigger frequency shift differences of trapezoid support and rectangle support show. Highest frequency was seemed at 30 μm width rectangle support. Frequency shift of 30 μm rectangle support was 17.9 kHz. So, to get higher frequency shift, 30 μm width rectangle support is the best of those 8 models that plotted in Fig. 4.

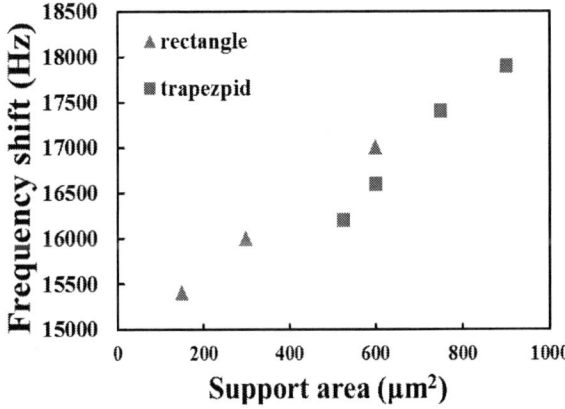

Fig. 5. Frequency shift as a function of support area with respect to support geometry.

Fig. 5 shows the interaction of frequency shift and support area of trapezoid support and rectangle support. Frequency shift becomes higher when support area becomes larger. It is

resulted by changing stiffness of support. Stiffness of large support is higher than that of small support. So, frequency shift of large support is higher.

B. *Effect of support on energy loss*

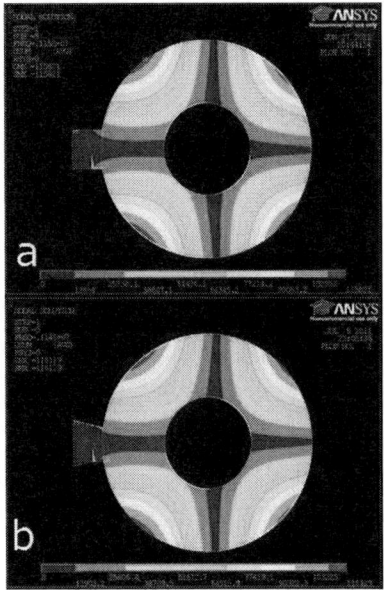

Fig. 6. Comparison of different vibration mode shapes (mode 5) at different support geometry, where a: rectangle support; b: trapezoid support.

Fig. 6 shows mode shape of ring resonator. Mode shape only near the support is somewhat different. However, most of the mode shape is similar. Support loss can be varied by changing support geometry without potential influence on most of the mode shape.

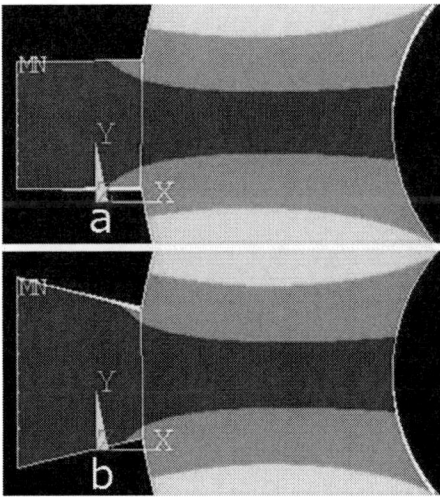

Fig. 7 Comparison of low-amplitude vibration area with a respect to support geometry, where (a) for rectangle support, and (b) for trapezoid support.

Fig. 7 shows the vibration mode shapes of trapezoid support and rectangle support when w 2 is 20 µm. Fig.7 (a) is the enlarged image of Fig. 6 (a), and Fig.7 (b) is the enlarged image of Fig.6 (b). The blue area is sandwiched by the aqua blue area. The amplitude of the aqua blue area is larger than that of the blue one. The aqua blue area of rectangle support is larger than that of trapezoid support. In the present case, it can be concluded that the amplitude of rectangle support is larger than that of trapezoid support, based on the above analytical observation. As a result, the support loss of rectangle support is higher than that of trapezoid support when w 2 is fixed at 20 µm.

IV. CONCLUSIONS

The frequency shift and amplitude of some support shapes in the proposed ring-shaped resonator were analytically studied. It was found that low-amplitude vibration area is larger at narrow support, and frequency shift of trapezoid support is higher than that of rectangle support.

ACKNOWLEDGMENTS

Part of this work was supported by MEMS Inter University Network and performed in Research Center for Ubiquitous MEMS & Micro Engineering (UMEMSME) of National Institute of Advanced Industrial Science & Technology (AIST).

REFERENCES

[1] Eduardo B. Oliveira, Emil C. Gotschlich, and The-yung E. Legua, "Primary Structure of Human C-reactive Proteins" The Journal of biological chemistry vol. 254 No.2 Issue of January 25, pp489-502(1979)

[2] Wencheng Xu, Seokheun Choi, and Junseok Chaea Arizona state University "A contour-mode film bulk acoustic resonator of high quality factor in a liquid environment for biosensing applications" American Institute of Physics, (2010)

[3] J Weber, W Albers, J Tuppurainen, M Link, R Gabl, W Wersing, M schreiter, Sens Actuators "Multilayered shear wave resonator consisting of c-axis tilted ZnO films" 128 (2006)

[4] H Zhang, M Marma, E kim, C Mckenna, M Thompdom Micromech Microeng "A film bulk acoustic resonator in liquid environments" 15 1911(2005)

[5] Jian LU, Takeshi Kobatashi, Yi Zhang, Ryutaro Maeda, Takashi Mihara" Wafer scale lead zirconate titanate film preparation by sol-gel method using stress balance layer" Thin solid films 515 1506-1510 (2006)

[6] Zhili Hao, Ahmet Erbil, Farrokh Ayazi "An analytical model for support loss in micromachined beam resonators with in-plane flexural vibrations" Sensor and Actuators A 109 (2003)

[7] T Ikehara, J Lu, R Maeda T Mihara "A high quality-cactor silicon cantilever for a low detection limit resonant mass sensor operated in air" J. Micromech Microeng 17(2007) 2491-2494

[8] Yong-Hwa, Park and K C Park, "High-Fidelity Modeling of MEMS Resonators Part 1 Anchor Loss Mechanisms Through Substrate" Journal of Microelectromechanical systems vol 13 no2 (2004)

[9] Yong-Hwa, Park and K C Park, "High-Fidelity Modeling of MEMS Resonators Pat 2 Coupled Beam Substrate Dynamics and Validation" ournal of Microelectromechanical systems vol 13 no2 (2004)

[10] Zhili Hao and Farrokh Ayazi"Support loss in micromechanical disk resonator" (2005) IEEE 137-141

[11] Joydeep Basu Tarun Knti Bhattacharyya "Microelectromechanical resonators for radio frequency communication applications" Microsystem Technologies volume17 (2011)

[12] Wang J, Ren Z Nguyen CTC, "1.156-GHz self-aligned vibrating micromechanical disk resonator" IEEE Trans Ultra son Ferroelectric Freq Control 1607-1628

[13] M Konno, T Ikehara S Murakami R Maeda, M Kimura, T Fukawa, T Mihara "Novel MEMS oscillator using in-plane disk resonator with sensing platform and its mass sensing characteristics"Transducers'11 518-521(2011)

[14] Jian Lu, Tadatomo Suga, Yi Zhang, Toshihiro Itoh, Ryutaro Maeda Takashi Mihara "Micromachined Silicon Disk resonator transduced by piezoelectric lead zirconate titanate thin films" Japanese jornal of applied physics 49 (2010)

[15] Dong F Wang, Xiaoqiang Li, Jian Lu, Takahisa Sagawa, Ryutaro Maeda, "Ring-shaped PZT Film Resonator for Bio-sensing Applications in liquid environment" Eurosensors 2011

Effective Force Generation for ECLIA Composed of Si Bone Structure and Conductive Polymer Flexible Slider

Tuan Anh Nguyen* and Satoshi Konishi
*Graduate school of Science and Engineering, Ritsumeikan University, Shiga, JAPAN
gr0081pp@ed.ritsumei.ac.jp

Abstract -In this paper, we present the design and fabrication of a new type of slider for ECLIA (Electrostatic Control Linear Actuator). The slider is composed of a flexible conductive polymer and Silicon bone structure, hereafter called Polymer–Si bone slider, in order to generate electrostatic and suction force for the clutch mechanism and pushing force of the slider. The Polymer–Si bone slider can improve the stability of slider movement compared to a polymer slider. The flexible conductive polymer film, which consists of a Aramid base thin film and coated conductive PEDOT:PSS layer, can be deformed, and fitted to the opposite electrode surface at low applied voltages. In addition, the Si bone structure is rigid for stability movement and large pushing force of the slider. The characteristics of this new slider were compared to conventional sliders.

Keywords- ECLIA; Si bone structure; flexible polymer slider

I. INTRODUCTION

Microactuators are widely used MEMS devices, taking advantage of a scaling effect in the micro domain. They are actuated by using different principles such as electrostatic [1], thermal [2] and piezoelectric actuation [4]. A piezoelectric actuator can provide precise motion with nanometer resolution and large force. Moreover, electrostatic controlled stepwise motion can make a large stroke. Actuators based on the stepwise actuation principle present an interesting combination of advantages for such micro system applications. The stepwise concept is to convert short range actuation mechanism into arbitrarily long range translation via the coordinated sequencing of three actuators: two clamping actuators and one translational actuator. Several stepwise embodiments have been proposed [3], and are based mostly on piezoelectric actuators. ECLIA with precise steps, parallel motion and large stroke of sliders has been reported in [4]. The sliders require straight movement and low applied voltages. To satisfy these requirements, guide structures and flexible thin film sliders are proposed.

ECLIA has been applied in WDM (Wavelength Division Multiplexing) spectral attenuation [5] and micro syringe pumps for picoliter scale liquid manipulation [6]. In these applications, the sliders made of Si cause a clearance (air gap) between contacting electrodes due to surface conditions. In addition, slider movement generates dynamic clearance because of undesired out-of-plane motion. The clearance decreases electrostatic clamping forces between the slider and

electrodes, and thus decreases the pushing force generated by the slider. Reports of studies on an electrostatic microactuator using polymer materials such as SU-8 and polydimethylsiloxane (PDMS) can be found [7, 8]. In general, polymer materials are flexible enough to generate large deformation due to their low stiffness. Moreover, polymers can be deposited by simple and low cost process. To improve the electrostatic force, a flexible polymer slider with a probe dipping into µPool connection (PDP-connection) was proposed in [9]. The flexible slider can be deformed and fitted to the opposite electrode when applying an electrostatic force. The applied voltage can be reduced while electrostatic force is improved.

In this paper, we propose a new design of slider composed of Si bone structure and flexible polymer thin film. The combination of advantages of both Si and polymer thin film can improve stability of movement and the pushing force of the sliders. The fabrication, characteristics and application of these new sliders are also described in this paper.

II. DESIGN AND FABRICATION OF POLYMER - SI BONE STRUCTURE SLIDER FOR ECLIA

Fig. 1 shows a schematic of an ECLIA developed in our laboratory. ECLIA is composed of a base frame, a guide structure, holding, driving electrodes, a piezoactuator, and sliders. The guide structure employs linear movement of sliders. ECLIA can provide precise motion due to a piezoelectric component of the oscillator and a large stroke through electrostatic controlled stepwise motions. The electrostatic clutch method makes it possible to control parallel motion of multi sliders.

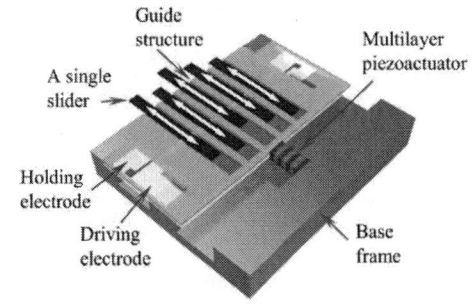

Fig. 1. Composition of ECLIA (Electrostatic Control Linear Actuator).

Sliders require linear movement and low applied voltages for their operation. A Si slider has an insulating parylene layer with thickness of 1 μm is easy in fabrication, but requires high applied voltages for the electrostatic clutch mechanism. In this research, the flexible polymer slider has been studied.

A polymer slider is composed of Aramid base film (Mictron$_{TM}$, Toray, thickness: 12, 4.4 and 3.6 μm, permittivity: 3.27×10^{-11} F/m, Young's modulus: 10 GPa) and PEDOT:PSS conductive layer (thickness: 160 nm, fabricated condition: spin-coated on a base polymer film at spinning speed of 1500 rpm and dried at the temperature of 75^0C for 15 minutes).

Surface roughness (Ra) of the flexible polymer slider is 12.152 nm by AFM measurement. The Si wafer used in this work has a surface roughness of several nanometers (Ra = 3.536 nm). The parylene coated Si slider has a surface roughness of 6.614 nm. However, there might be unexpected particles existing between contacting surfaces. This causes undesired contacting conditions due to surface rigidity; and thus forms a small air gap between the Si slider and the electrode. We also used an Al electrode that had larger surface roughness (Ra = 42.82 nm) than the Si electrode in order to compare frictional force of the flexible polymer slider as discussed below.

The frictional force between two contacting surfaces is linear with the friction coefficient and normal load. The friction coefficient depends on parameters, such as surface roughness and material. If the surface roughness is large, both of the friction coefficient and the air gap between contacting surface will be large. Increasing the air gap causes the decreasing electrostatic force (normal load). Hence, the frictional force decreases. Consequently, an increase of the friction coefficient results in an increase of electrostatic force. The previous discussion of the roughness of different surfaces found that the layer, which consists of the air gap and the insulating layer between the Si slider and Si electrode, and the friction coefficient (μ = 0.16) is smallest. With the same flexible polymer slider (12-μm insulating layer), the air gap between the polymer slider and Si electrode is smaller than that of Al electrode. The friction coefficients between the polymer slider and Si electrode; and the polymer slider and Al electrode are 0.25 and 0.3, respectively.

When the frictional force of the polymer slider with the Si electrode was evaluated, it was found that the frictional force was much larger than that of the Al electrode (Fig. 3). The flexible polymer slider with the same nanometer level roughness can be deformed and better fitted to the Si electrode

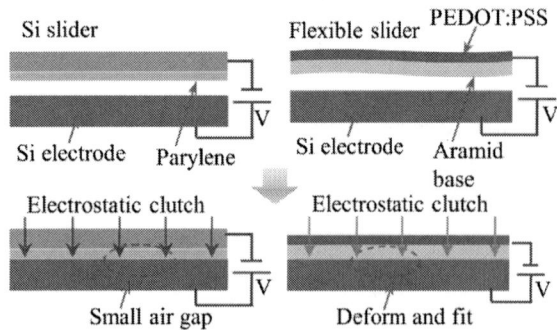

Fig. 2. Schematic of contacting sliders to Si electrodes.

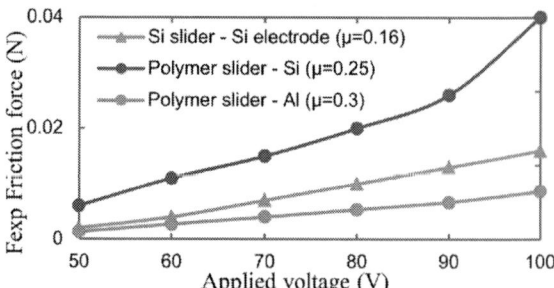

Fig. 3. Experimental value of frictional force for different sliders.

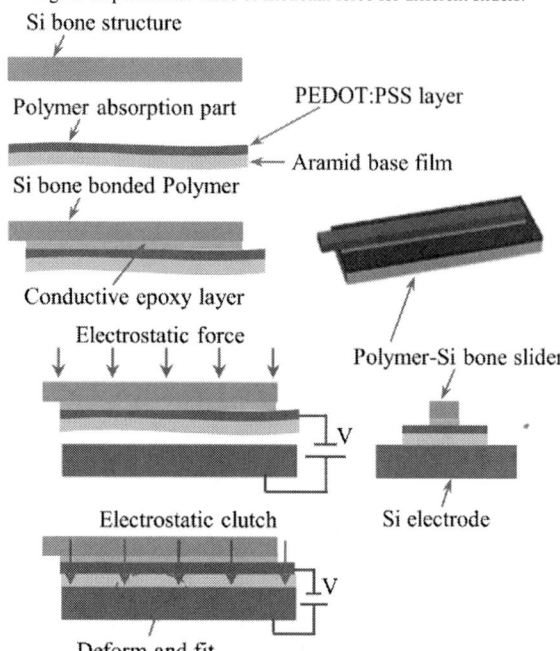

Fig. 4. Fabrication process of a Polymer-Si bone slider and schematic of the contacting slider to a Si electrode.

to enhance the attractive force (Fig. 2). Therefore, the polymer slider and the Si electrode generate the largest frictional force and require the lowest applied voltages (Fig. 3).

However, the flexible slider has no stable motion. It is difficult to employ guide structures for straight movement. The combination of the Si bone structure and the conductive polymer flexible slider gains the advantages of Si and polymer. Si bone structure is intended to enhance stability and the pushing force of the slider motion. In addition, the flexible polymer is an absorption part for low applied voltages.

In this study, we fabricated two Polymer-Si bone sliders with the same size of the polymer absorption part but with a different size of Si bone structure. The polymer absorption part was cut into a rectangular shape of 20-mm length, and 4.5-mm width. The Si bone structures were cut by a dicing machine into dimensions of 20 mm × 1.5 mm (150-μm-thick Si wafer), and 20 mm × 2 mm (525-μm-thick Si wafer). The Si bone was bonded to the polymer absorption part by conductive epoxy resin and dried at a temperature of 80^0C for 120 minutes. The Si slider (150-μm-thick Si, 1-μm-thick Parylene) was also fabricated with dimension of 20 mm × 4.5 mm. Fig. 4 shows the fabrication process of a Polymer-Si bone slider and schematic of a Polymer-Si bone slider contacts

978-1-4673-1122-9/12 $31.00 © 2012 IEEE

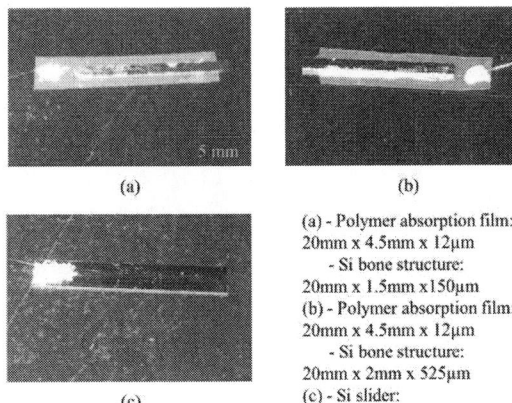

(a)

(b)

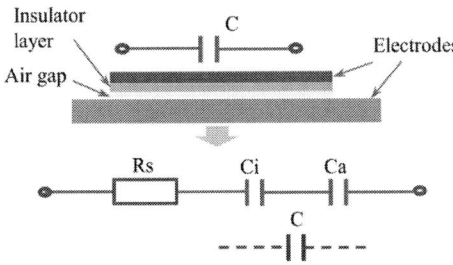

(c)

(a) - Polymer absorption film:
20mm x 4.5mm x 12µm
- Si bone structure:
20mm x 1.5mm x150µm
(b) - Polymer absorption film:
20mm x 4.5mm x 12µm
- Si bone structure:
20mm x 2mm x 525µm
(c) - Si slider:
20mm x 4.5mm x 150µm

Fig. 5. Image of Polymer-Si bone sliders (a), (b) and Si slider (c).

Fig. 6. Schematic of an equivalent capacitor.

TABLE I.   THE FEATURES OF DESIGNED SLIDERS

Material	Slider without bone structure	Slider with bone structure
Si – Parylene 1 µm	Si 150 µm (Fig. 5c) - Hard - High voltage	Si 525 µm + Polymer film 12 µm (Fig. 5b) - Hard guide structure - Flexible absorbed part - Medium voltage
Polymer (PEDOT-Aramid 12 µm)	- Flexible - Low voltage. - Not stability motion	Si 150 µm + Polymer film 12 µm (Fig. 5a) - Higher performance that of case in Fig. 5b

on to a Si electrode. Images of the fabricated Polymer-Si bone sliders and Si slider are shown in Fig. 5. A summary of features of sliders are listed in Table I.

III.   CHARACRTERISTICS OF THE POLYMER - SI BONE SLIDER

A. Capacitance

The slider and Si electrode with a dielectric layer form a capacitor. There is an air gap between the slider and the electrode due to the surface roughness of electrodes and slider curvature. The clearance is changed by changing the applied voltage. It can be assumed the slider and the electrode is equivalent to a resistor R_S and series capacitors C_i (insulating layer) and C_a (air gap) as shown in Fig. 6. The equivalent capacitance C is calculated by

$$C = \frac{C_i . C_a}{C_i + C_a} \qquad (1)$$

TABLE II.   EXPERIMENTAL RESULTS OF DIFFERENT SLIDERS

Sliders / Parameter	Si slider (Fig. 5c)	Polymer slider (20x4.5mm x 12 µm)	Polymer-Si bone slider (Fig. 5b)	Polymer-Si bone slider (Fig. 5a)
C [pF]	77	95	42	50
g [µm]	8.8	5.14	15.7	12.7
Fr [mN]	16	40	26	30

C: Capacitance
g: Estimated air gap of capacitor
Fr: Frictional force at 100 V applied

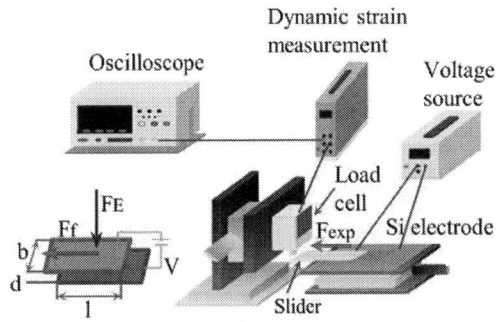

Fig. 7. Apparatus set up for measuring friction force (F_f).

Measured results of the capacitance, C for the various sliders and Si electrode without an applied voltage are shown in Table II. The results show that the capacitance of the polymer slider is largest. The Polymer-Si bone slider has a bonding layer between the Si bone and the polymer slider which causes the curvature of the absorption part. Hence, the capacitance of the Polymer-Si bone slider decreases. The estimated air gaps are also shown in Table II.

B. Electrostatic force for clutch

The step motion of sliders in ECLIA is generated by the PZT actuator and an electrostatic clutch mechanism between the slider and driving electrode. The electrostatic force F_E is illustrated in Fig. 7 and calculated by

$$F_E = -\frac{A\varepsilon}{2d^2}V^2 \qquad (2)$$

where, $A = lb$; ε is the permittivity of the dielectric layer between electrodes; d is the thickness of the dielectric layer. Electrostatic force F_E creates frictional force F_f for the clutch of the slider and the electrode

$$F_f = \mu F_n \qquad (3)$$

where, μ is the friction coefficient of the slider and the electrode. Fig. 7 shows the experimental set up used to determine the frictional forces. Driving voltage was applied to the slider and the Si electrode by DC voltage source.

Frictional force was measured by the load cell using an oscilloscope. Fig. 8 compares experimental frictional forces (F_{exp}) of different sliders at different voltages (V). Fig. 8 shows that the frictional force of the 150-µm Si bone structure slider is larger than that of the 525-µm Si bone structure slider and much larger than that of the Si slider. The frictional force generated by polymer sliders is largest and differed from the theoretical value at 100 V.

978-1-4673-1122-9/12 $31.00 © 2012 IEEE

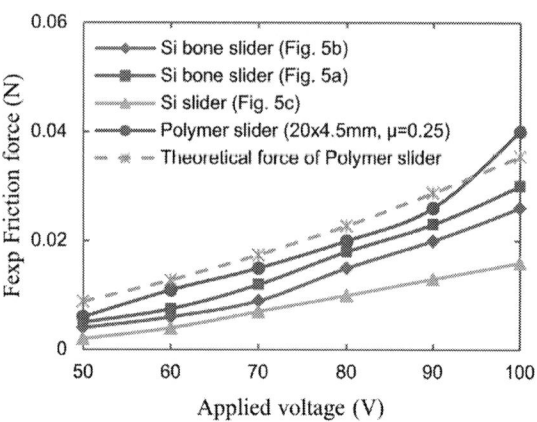

Fig. 8. Experimental values of frictional force (Fexp) of sliders.

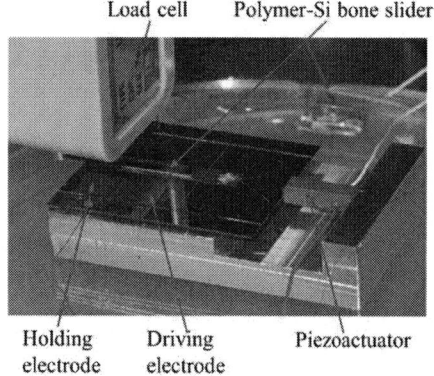

Fig. 9. Image of experimental pushing force measurement of the Polymer-Si bone slider in ECLIA.

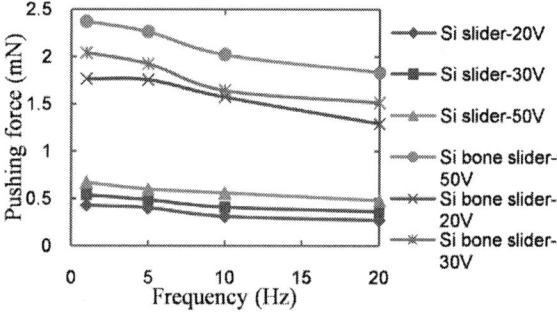

Fig. 10. Experimental values of pushing force of sliders in ECLIA.

IV. ELECTROSTATIC STEPPING MOTOR "ECLIA"

An improved method of the new slider is discussed. The slider is composed of the Si bone structure and the flexible polymer slider to provide highly efficient force generation in ECLIA. Fig. 9 shows the image of the experimental pushing force measurement of Polymer–Si bone slider in ECLIA. The slider generates a pushing force in response to electrostatic clutch mechanism and piezoactuator. Fig. 10 shows variation of pushing force of the Si slider and the Polymer-Si bone slider with respect to the frequency of PZT. PZT is applied at various voltages, and 100 V is applied to the electrostatic clutch. Fig.10 shows that the generated pushing force of the

Polymer–Si bone slider is much larger than that of the Si slider. Decreasing the frequency and increasing the applied voltage results in an increase in the pushing force. The results suggest that the Polymer–Si bone slider considerably improves the pushing force.

V. CONCLUSION

A Polymer–Si bone slider has been successfully demonstrated to improve electrostatic clutch mechanism, stable movement and a generated pushing force. We used the Si bone structure as a rigid part, and the flexible polymer as an absorptive part. Low stress and the deformable polymer leads to a high electrostatic force. Moreover, the Si bone achieves stable movement and large generated pushing force. The Polymer–Si bone slider is more effective than the conventional Si slider in terms of frictional force generation. The Polymer–Si bone slider showed good performance for ECLIA (Fig. 10). However, the frictional force generated by the polymer slider is in disagreement with expected theoretical value (Fig. 8). It is likely that friction coefficient depends on applied voltages. Further work is currently investigating the advantages of polymer sliders.

ACKNOWLEDGMENT

The authors would like to thank Mr. Mori for his help on using the AFM for measuring surface roughness. We wish to thank Mr. Kuwamura, Mr. Ishi, and other members in our Lab for their advice and apparatus setup.

REFERENCES

[1] M. A. Erismis, H. P. Neves, R. Puers and C. V. Hoof, "A low-voltage, large-displacement, large-force inchworm actuator," *J.Microelectromech. Syst.*, vol. 17, pp. 1294-1301, 2008.

[2] Geisberger A A, Sarkar N, Ellis M and Skidmore G, "Electrothermal properties and modeling of polysiliconmicrothermal actuators," *J. Microelectromech. Syst,* vol. 12, pp. 513–523, 2003.

[3] P. Cusin, T. Sawai, S. Konishi, "Compact and precise positioner based on the Inchworm principle," J. Micromech. Microeng. 10 (4) (2000) 516–521.

[4] S. Konishi, K. Ohno and M. Munechika, "Parallel Linear Actuator System with High Accuracy and Large Stroke," *Sensors and Actuators A: Physical*, Vol. 97-98, pp. 610-619, 2002.

[5] S. Konishi, A. Oshima, N. Kinoshita, I. Kumagaya, T. Kishi, T. Shimazu and M. Katayama, "Batch-fabricated High Dense Multi Slider for WDM Spectral Attenunation," *Proc. of TRANSDUCERS'05 Conference*, Seoul, Korea, June. 5-9, 2005, pp. 1242-1245.

[6] Ryuji Yokokawa, Tomohiko Saika, Testuya Nakayama, Hiroyuki Fujita and Satoshi Konishi, "On-chip syringe pumps for picoliter-scale liquid manipulation," Lab Chip, 2006, 6, 1062–1066.

[7] T. Fujita, K. Maenaka and Y. Takayama, "Dual-axis MEMS mirror for large deflection-angle using SU-8 soft torsion beam," *Sensors and Actuators A: Physical*, Vol. 121, pp. 16-21, 2005.

[8] S. K. Lee, M. G. Kim, J. Y. An, M. H. Jun, S. Yang and J. H. Lee, "Polymeric (SU-8) Optical Microscanner Driven by Electrostatic Actuator," *Proc. of MEMS 2009 Conference*, Sorrent, Jan. 25-29, 2009, pp. 988-958.

[9] M. Ito, T. Kuwamura, J. Komoda, and S. Konishi, "Conductive polymer coated flexible electrode for highly efficient force generation of electrostatic actuator," *Proc. of MEMS 2011 Conference*, Cancun, Mexico, January 23-27, 2011, pp. 392-395.

The Study of Forward and Reverse Schottky Junction for Dual Magnetodiode

T. Phetchakul [1], W. Luanatikomkul[1], W. Yamwong[2], A. Poyai[2]

[1]Department of Electronics, Faculty of Engineering, King Mongkut's Institute of Technology Ladkrabang, Bangkok, Thailand.
[2]Thai Microelectronics Center, National Electronics and Computer Technology Center, Chachoengsao, Thailand.

Abstract— **This paper presents schottky diode for sensing magnetic field. The structure is the same as dual magnetodiode that had been reported by the same group. The device can operate both in forward and reverse biasing for magnetic field device. The sensitivity ($\Delta V_o/\Delta B$) by simulation of forward bias is 0.8 mV/T at the current 1.93 μA and reverse bias is 6 nV/T at the current 13.49 pA . Two modes of operation show linearity which electron is the only type of carrier. The mechanism of forward and reverse modes of operation are described.**

Keywords- Sentaurus TCAD; Magnetodiode; Schottky junction

I. INTRODUCTION

There are many kinds of magnetic sensor devices which most of them use physical phenomena depend on the measured magnetic field strength [1, 2]. They can be classified according to whether they measure the total magnetic field or the vector components of the magnetic field. First type is the devices that insensitive with the rotational vibrations such as optically pumped magnetometer, nuclear-precession magnetometer and overhauser magnetometer. The another type is, for example, search-coil magnetometer, fluxgate magnetometer, superconductor magnetometer, Hall effect sensor, magnetoresistor, magnetodiode, magnetotransistor, magnetostrictive magnetometer, magnetooptical sensor and NEMS based magnetometer [3-6]. The measured magnetic field varied in the order from pico-tesla to mili-tesla depends on magnetic sensing technique.

The classical magnetodiode is one of magnetic sensors that is a two-terminal semiconductor device whose operation is based on the superposition of high carriers injection by one or two(a p$^+$-n and an n$^+$-n) junctions, magnetoconcentration and surface recombination [7, 8]. It operates by the difference in recombination rate between two interfaces depending on the direction of the magnetic induction. It is the device that can detect the magnetic in the technical field order mili-tesla as Hall device and magnetotransistor. The magnetodiode is not popular as Hall Effect device and magnetotransistors because its mechanism is based on the recombination rate of interface. In recent year, dual magnetodiode was proposed which the mechanism is difference from the classical one [9]. Its three-terminal structure composes of two symmetrical diodes with one common anode and two separated cathodes. The mechanism is mainly explained by carrier deflection from Lorentz's force acting on injected carriers from anode to

separated cathode [10]. It detects the magnetic field by the difference of cathodes current.

The Dual schottky magnetodiode has the same structure as dual magnetodiode which schottky junction is in place of p-n junction. The carrier is only electron differ from dual magnetodiode that is hole. In this work, the forward and reverse bias modes were studies for detecting magnetic field. The device was compared between measurement and simulation by TCAD sentaurus program. The mechanism of electron injection from anode to cathode, forward bias, and from cathode to anode by thermionic emission and reverse bias was described.

II. DEVICE STRUCTURE AND OPERATION PRINCIPLE

The structure of schottky magnetodiode is show in Fig. 1(a), the common anode terminal is aluminum represent to p-type in dual magnetodiode (p^+). The symmetrical separated cathodes are n-type substrate with the n^+ diffusion window for ohmic contact with aluminum.

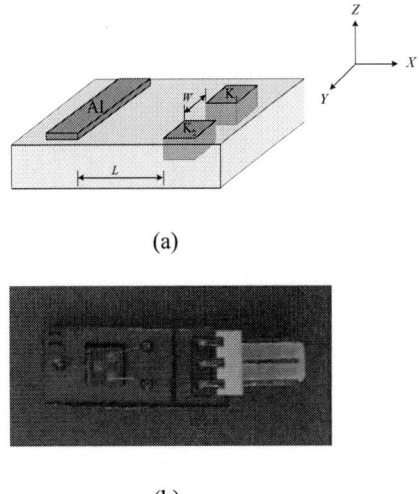

(a)

(b)

Fig.1. The dual schottky magnetodiode structure. (a) Top view and (b) Packaged device.

The width of anode terminal as wide as the combined width of two cathodes is 500 μm. Two cathode terminals are equal in size and symmetrical to each other. The cathode terminals are separated by 10 μm(W) gap and the spacing between anode and cathode terminal is 15 μm(L). The device is fabricated on n-type (100) substrate of resistivity 15 Ω.cm and thickness 400 μm. The cathodes were fabricated by diffusion of phosphorus, $N_A = 10^{15}$ cm^{-3}. The aluminum was evaporated for anode schottky junction and cathode ohmic contact. Full packaging of the device has been done as shown in Fig. 1(b). It contains a single chip of device which already has been wire bonded on PCB for practical and convenient operation.

The operating principle of the dual schottky diode is based on the carrier deflection in the presence of magnetic field and can be analyzed by applying the principle of Hall Effect. This can be analyzed in Fig. 2, it shows the displacement of electron current, J_{nx}, at an angle of θ_H, because of the Lorentz force due to magnetic field (B) perpendicular to the direction of electron current. The anode current is a result of combination between cathode 1 and cathode 2 currents as shown in (1). The anode current can also be shown in term of current density, J_{nx}, during the state without magnetic field and can be express by

$$I_D = I_{D1} + I_{D2} \qquad (1)$$

where I_{D1} and I_{D2} is cathode current.

$$I_D = J_{nx}.d.W_A \qquad (2)$$

where J_{nx} is current density along x-axis, d is effective depth of the current and W_A is emitter width.

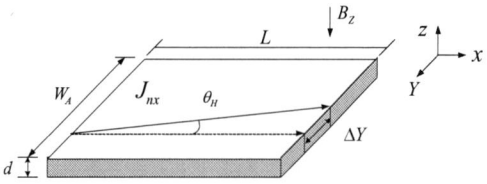

Fig. 2 Deflection carrier area of electron.

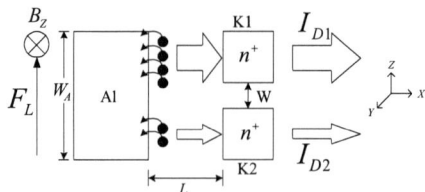

Fig. 3 Electron deflection in Al-Substrate junction.

Figure 3 shows the electron deflection responding to the magnetic field in vertical direction of the dual schottky diode. In an anode terminal (Al-Substrate junction), conduction electrons move from dual cathode to the anode and the cathode 1 and cathode 2 current are equal in absent magnetic field. The respond magnetic field perpendicular to the chip surface (B_Z) and electron current in x direction. Therefore, the resulting Lorentz force and electron deflection will be in -y direction. As a result, both cathode 1 and cathode 2 current will be changed. The electron deflection of ΔI_D can cause the output voltage to magnetic field will be higher. The increases in magnetic field density can cause larger electron deflection, and therefore, larger difference between the two currents. The difference of cathode 1 and cathode 2 current (ΔI_D) during application of magnetic field in vertical direction will cause charge carriers to travel in y-axis and the change of $\theta_{H(y)}$ as show in (3). $\theta_{H(y)}$ can be found by $tan\,\theta_{H(y)} = \mu.B_Z$.

$$\Delta I_D = J_{nx}.d.\Delta Y \qquad (3)$$

where

$$\Delta Y = \mu.L.B_Z \qquad (4)$$

Using (3) and (4) obtain for ΔI_D

$$\Delta I_D = J_{nx}.d.L.\mu.B_Z \qquad (5)$$

where μ is electron mobility, d is effective depth of the current and B_Z is magnetic field density in z axis.

III. SIMULATION AND DISCUSSION

Figure 4 shows simulation results of virtual structure replicate the device. The structure was generated and simulated by three dimension hall sensors simulation package [11]. Figure 4 (a) shows current density flow in the device was shown as a color contour plot in absence magnetic field. The color shade same as between cathode 1 and cathode 2 represent current density in dual cathode area are equal. As can be seen when magnetic field has been applied to the device in Fig. 4(b), the deflection current has been added to the terminals which causes current density especially on the cathode 2 terminal more intensified compared to the result with no magnetic field applied in Fig. 4(a).

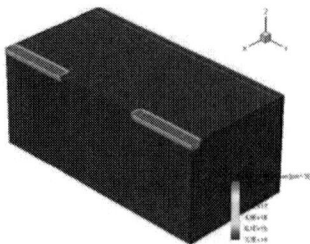

Fig. 4(a). The current density by 3-D modeling with zero magnetic field.

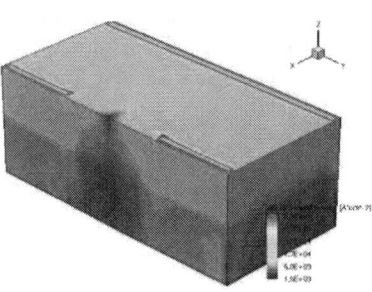

Fig. 4(b) The current density by 3-D modeling with vertical magnetic field.

A. The sensor Interface Circuit

The interface circuit used in this experiment is shown in Fig. 5 the dual schottky diode is biased by forcing a constant current into anode. The cathode 1 (D1) and the cathode 2 (D2) are connecting to the variable resistor R_{D1} and R_{D2}, respectively. An ideal dual schottky diode structure will have a current gain of one, since the device relies on D1 and D2 current to be identical during zero input state. The imbalance of these two currents will introduce an offset voltage that will determine the effectiveness of this dual schottky diode. This offset voltage can be adjusted through an electrical circuit, as shown in Fig. 5, by adjusting the resistor. The differential current will be converted to voltage can be express as

$$V_o = A \, (\Delta I_{D1}.R_{D1} - \Delta I_{D2}.R_{D2}) \qquad (6)$$

where ΔI_{D1} and ΔI_{D2} are the change of D1 and D2 currents, respectively.

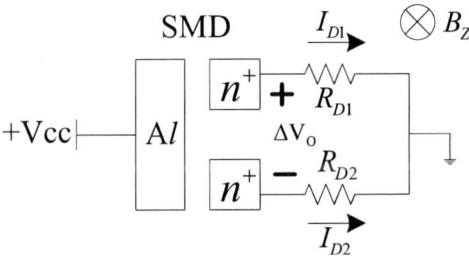

Fig. 5. The sensor interface circuit for magnetic field measurement in simulation.

Figure 6 shows output response of forward bias for dual schottky diode at difference diode current I_D during 1.93-30.20 μA. The magnetic field density $(+B_z)$ is varied in the range of 0 to 1 tesla. The result from Fig. 6 shows simulation curve output voltage in forward bias of dual schottky magnetodiode.

The output voltage of cathode1 higher than cathode 2 can cause electron deflection into cathode 1 and accumulation of electron deflection and movement in θ_H into cathode 1 as in (2).

Figure 7 shows the output response of reverse bias for dual schottky diode at difference diode current I_D during 13.49 - 253.74 pA The magnetic field density $(+B_z)$ is varied in the range of 0 to 1 tesla. The result from Fig. 7 shows simulation curve output voltage in reverse bias of dual schottky magnetodiode. The simulation of reverse bias curve shows cathode 2 current higher than cathode 1 current, therefore output voltages are negative curve.

The result in Fig. 8 shows that measurement curve at magnetic field density were in the range of 0 to 1T in vertical magnetic field direction. The measurement was comparison between forward and reverse bias at forward current 0.2 nA. From the result, it shows monotonic increase of output response as a function of field density. At the same field density, the absolute output voltage of forward bias and reverse bias are equal. It means that the mechanism of injection current in forward bias and thermionic emission in reverse bias do not affect the responses, only the amount electron carrier deflection relates to output voltage.

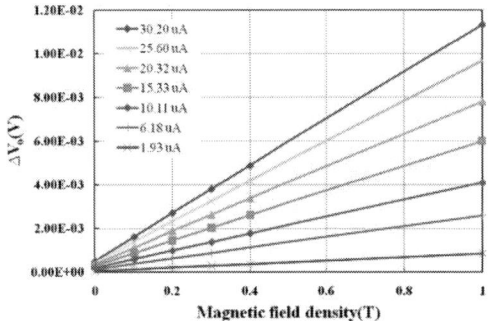

Fig. 6. The vertical magnetic field responses of forward bias by simulation of dual schottky magnetodiode.

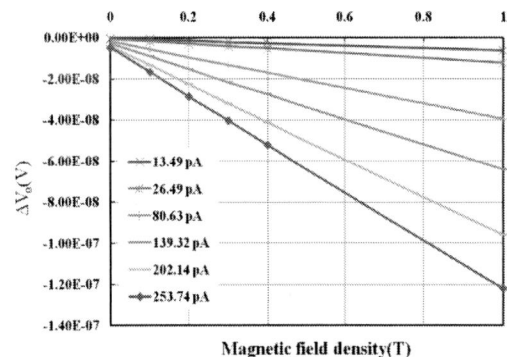

Fig. 7. The vertical magnetic field responses of reverse bias by simulation of dual schottky magnetodiode.

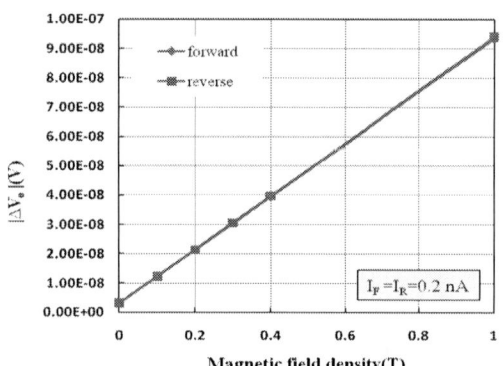

Fig. 8. Comparison between forward bias and reveres bias measured at vertical magnetic field response.

IV. CONCLUSION

The dual schottky diodes have been fabricated with standard bipolar technology and can be easily integrated into a single chip. They have the ability to detect magnetic field in vertical direction. The devices rely on the carrier deflection mechanism causing the difference between cathode 1 and cathode 2. The comparison between Forward bias and revere bias are similarly so dual schottky diode can be operate in both of them. The simulated results are possibility to apply these devices as magnetic sensors using both vertical and lateral axes for magnetic field detection. The advantages the dual schottky diode are its convenient use as a discrete device, high sensitivity to magnetic field and wide range of applications.

ACKNOWLEDGMENT

The authors would like to thank to National Research Council of Thailand for financial support.

REFERENCES

[1] H.P. Baktes. R.S. Popovic, "*Integrated semiconductor magnetic field sensors*", Proc, IEEE 74(8) (1986) 1107-1132.

[2] R.S. Popovic, H.P. Balts and F.Rudolf, "An Integrated Silicon Magnetic Field Sensor Using the Magnetodiode Principle", IEEE Trans. Electron Devices, vol. 31, pp.286-291, 1984.

[3] Lj. Ristic, M.T. Doan, M. Paranjape, "*2-D integrated magnetic Field sensor in CMOS technology*", Circuit Syst (1989) 701-704.

[4] A. Nagy, Hector Trujillo, "*3D magnetic-field sensor using only a pair of terminals*" Sensor and ActuatorsA, 58(1997)137-140.

[5] James Lenz and Alan S. Edelstein, "Magnetic Sensors and Their Application" IEEE Sensors Journal, vol. 6, No. 3, pp. 631-649, 2006.

[6] T. Nakamula and K. Maenaka, "Integrated Magnetic Sensor", Sensors and Actuators, vol. 35, pp. 21-23, 1990.

[7] H. Pfleiderer, "Magnetodiode model", Solid-State Electron., vol. 15, pp. 335, 1972.

[8] O.S. Lutes, P.S. Nussbaum and O.S. Adland, "Sensitivity limit in SOS magnetodiode", IEEE Trans., Electron Devices, vol. Ed-27, p. 68, 1979.

[9] T. Phetchakul and S. Jankamkaw, "Dual magneticdiode" The 12th International Symposium on Integrated Circuits, ISIC' 09, Singapore 2009.

[10] C. Leepattarapongpan, T. Phetchakul, N. Penpondee, P. Pengpad, E. Chaowicharat, C. Hruanun and A. Poyai, "Magnetotransistor Based on the Carrier Recombination-Deflection Effect", IEEE Sensors Journal, Vol. 10, no.2, pp. 294-299, 2010.

[11] TCAD Sentaurus Manual, sysnopsis®, version A-2007.12.

Electrode Design Optimization of a CMOS Fringing-Field Capacitive Sensor

Yu-Ting Li, Yen-Lin Tzeng, Chih-Ming Chao, Kerwin Wang
Department of Mechatronics Engineering, National Changhua University of Education,
Changhua, Taiwan
kerwin@cc.ncue.edu.tw

Abstract—**The capacitance of a micro-capacitive-sensor is mainly depends on its electrode structures, electrode arrangement and the dielectric strength of its surrounding. This paper presents a novel electrode design to enhance the fringing-field and to increase the capacitance of a CMOS MEMS capacitive sensor. The design optimization process is assisted by COMSOL, a multi-physics finite-element simulation tool. The design goal is to maximize the fringing-field of capacitive sensor. The capacitor has been built with a VLSI Schmitt trigger, and temperature compensator to probe the capacitance. The experiment results show that proposed capacitor can achieve the capacitance to 75.2 pF.**

Keywords- fringing field, capacitance ; mesh-electrode

I. INTRODUCTION

Capacitive sensors, built with conductive electrodes and dielectric materials can sense a variety of physic parameters, such as displacement, electric fields, fluid level, humidity and fluid composition. They have many useful applications [1-4], including touch sensor, motion control, and so on. Most of the capacitive sensors have fringing-field. In the some cases, it draws unwanted parasitic capacitance and nonlinearity into systems; however, fringing effect can be used in micro-capacitance sensor for practical applications [5-14]. The electrode geometry arrangement will affect the electrostatic field distribution and its strength.

This study presents a refined fringe-field capacitive sensor configuration. To enhance the fringing-field effect of the capacitive sensor, various micro-mesh-electrode arrangements are explored. The sensor performance is characterized and discussed by both simulation and experiment results. The capacitive sensor design is realized by TSMC 2P4M CMOS-MEMS process. The advantages of developing a CMOS compatible fringe capacitance sensor are small chip size, fast response, good sensitivity, high reliability, good accuracy, low driving voltage and cost effective process. In following sections, the electrode geometry and sensor design, simulation and measurement results are presented. The goal is to maximize the fringing-field for sensing.

II. ELECTRODE DESIGN

Electrode design is focused on the optimal arrangement of geometry and configurations in a constrained design space, such as chip size and mask resolution. Due to the rapid evolution of fabrication technology, CMOS circuits are becoming more and more complex; however, the electrode design problem is a constrained optimization problem.

To enhance fringing-field, a CMOS base capacitive sensor various electrode designs are introduced and discussed. The capacitive sensor design is realized by TSMC 2P4M CMOS-MEMS process, as shown in Fig. 1. We use 4-layer metal conductors (M1~M4) as the electrodes, six different electrode arrangements have been evaluated by simulation, as shown in Table 1. The electrodes are made of thin film aluminum; the capacitors comprise aluminum electrodes and silicon dioxide insulators. The capacitive sensor has staggered arrangement of electrodes in three-dimensions.

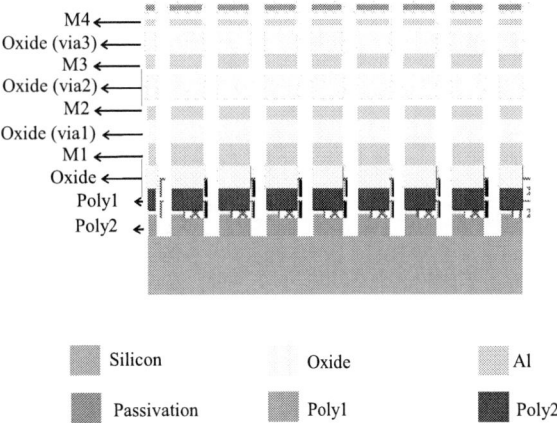

Fig. 1. Cross section diagram of the electrode configuration of a CMOS fringing field capacitive sensor

TABLE 1. THE ARRANGEMENT OF SIX DIFFERENT ELECTRODE POLARIZATIONS AND CONFIGURATIONS

	Type 1	Type 2	Type 3	Type 4	Type 5	Type 6
Metal 4	+		+		+	
Via3		+		+		+
Metal 3			-			
Via2	-	-			-	
Metal 2			+			
Via1		+		-		-
Metal 1	+		-		+	

In the electrode arrangement phase, all the possible configurations are placed on a blueprint to facilitate the design. A good arrangement is required to maximize the fringing-field for sensing. Fringing-field is the external electrostatic field pass around the surrounding. The area of the capacitive electrode has less contribution to fringing-field than the effective edge length of electrodes [1]. The goal of this design should be focused on the maximization of fringing-field. To exam the strength of fringing-field, it is subtracted from the total capacitance by a parallel plate capacitance (1). In general, the total capacitance can be estimated by finite element simulation. Using this method, the fringe capacitance can be estimated more precisely. The thickness of each layer is illustrated in Table 2.

TABLE 2. THE THICKNESS OF EACH LAYER

Layer	Thickness
Metal 1 (M1)	6650Å
Metal 2 (M2)	6400Å
Metal 3 (M3)	6400Å
Metal 4 (M4)	9250Å
Contact	11000Å
Via1	10000Å
Via2	10000Å
Via3	10000Å
Electrode width	2μm

The CMOS capacitance block is presented as four capacitive cells and the clearance of each cell is rectangular. The polarity of their configurations can be found in Table 1. Increasing fringing-field strength contributes to the increasing of sensor sensitivity too. The optimized designs are also selected as a turnkey for CMOS capacitive sensor design conceptualization. The design criterion is considered in maximizing fringing-field. We explored the constrained area and the design problems. Our previous work [1] shows that fringing-field can be applied to novel sensor development.

To enhance the fringing-field, we use the mesh-type capacitive sensing structures, it has high-density mesh hole to generate fringing-field for capacitive sensing. The influence of the dimension of mesh pattern is discussed in our previous research [1]. Because the charge and electric field is gathering easily near the edges of electrodes, the porous mesh-type capacitance can increase the effective fringe electrode length of capacitance sensor.

The strength of parallel capacitance of the electrode is propositional to the area of the electrode; however, in this case, the strength of parasitic capacitance is mainly depends on the effective electrode edge length. Two types of mesh designs with six types of electrode polarity arrangements are simulated and analyzed by a multi-physics finite-element simulation, COMSOL. To minimize the computational cost of simulation, the periodical mesh electrodes are simplified to single unit cells to check and to evaluate the configuration of the electrodes. Width of electrode structures is shown in Table 2. To evaluate the capacitances of six different electrode arrangements, 3D multi-physics simulation tool used to analysis and to estimate the total capacitance of these electrode structures.

A mesh electrode design will reduce the capacitor plate area and parallel capacitance, however, it can increase the effective electrode edge length, which is contribute to the increasing of fringing-field. Base on the fundamental physics [16], fringing-field can be estimated from equation (1).

$$C_{fringe\ capacitance} = C_{total} - C_{parallel\ capacitance} \quad (1)$$

$$C_{parallel\ capacitance} = \varepsilon_r \varepsilon_0 \frac{A}{d}$$

Here, A is the electrode area, d is the distance between the electrode plates, ε_0 is the vacuum permittivity, ε_r is the relative permittivity, $C_{fringing\ capacitor}$ is the fringe capacitance value. One can calculate the fringing capacitance by subtracting the parallel capacitance from total capacitance. Table 1 shows the polarity arrangement design of electrodes. The electrode polarity arrangements results the differences in the fringe-field strength. Table 3 shows the capacitance analysis results from COMSOL, the simulations show that the arrangement of Type 3 has the greatest capacitive fringing-field. Its positive and negative electrodes are staggered from top to bottom. The electrode M1 and M3 are connected to each other as the base electrodes, electrode M2 and M4 are connected to each other as top electrodes. Type 3 is selected for prototyping. An oscillator and a second order operational amplifier are placed against the capacitors. Its oscillating frequency can be changed according to the capacitance of the sensor. The capacitor structures are placed on the top of a heater and a thermistor. The heater structure is made of polysilicon, which has low resistivity and operation voltage. Thermal compensation can improve the stability of the heater.

TABLE 3. FRINGE CAPACITANCE OF SIX DIFFERENT ELECTRODE CONFIGURATIONS

	Type 1	Type 2	**Type 3**	Type 4	Type 5	Type 6
Total capacitance	5.21 fF	7.40 fF	7.19 fF	1.78 fF	3.38 fF	1.78 fF
Parallel plate capacitance	3.57 fF	5.58 fF	5.35 fF	0.78 fF	2.18 fF	3.23 fF
Fringe capacitance	1.64 fF	1.83 fF	**1.84 fF**	0.99 fF	1.21 fF	1.45 fF
Plate-fringe capacitance ratio	0.46	0.33	0.34	1.28	0.55	0.45

When the sensing structure is experienced the surrounding dielectric constant changes; the changing of the capacitance creates an alternating oscillation of the electrical signal. The Sensing circuit consists of a Schmitt trigger oscillating circuit and an operational amplifier (OPA). One can estimate the capacitance according to the value of output frequency. Table 5 shows the specifications and analysis results of the sensor with second order operational amplifier after taking RC parasitic effect into consideration. We use HSPICE to estimate the output frequency from the sensor.

The capacitor structures are placed on the top of a polysilicon heater and a polysilicon thermistor (TSR) for temperature variation compensation. The temperature compensator has low resistivity and low operation voltage, can improve stability of the heater. The thermistor structure uses polysilicon (Poly1) as a temperature sensing layer. The heater structure is made of polysilicon (Poly2). We remove a part material away from the base silicon by plasma etching to make efficient heat convection and conductive heat isolation. Figure 3 shows the fabricated chip structure, it contains 4 groups of electric capacitive sensing regions.

III. ANALYSIS AND TEST RESULTS

To enhance fringing-field, the CMOS base capacitive sensor various electrode designs are tested and discussed. The capacitive sensor design is realized by TSMC 2P4M CMOS-MEMS process. It use 4-layer metal conductor layers (M1~M4) as the sensing electrodes.

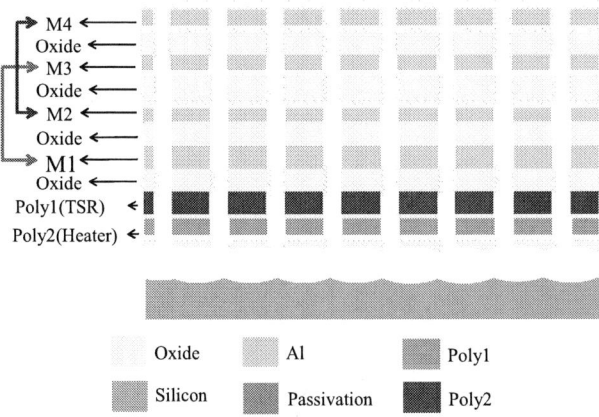

Fig. 2. Cross-section diagram of the electrode configuration of sensor Type 3

Each sensor cell takes up 480X480μm², which includes four bridges, composed of marginal sensing capacitor block, heater and thermistor. The pads on the sensor chips are connected with probes to measure the total capacitance. They are driven by a dual power supply (± 3.3V). The non-steady-state oscillating frequency is corresponded to the changes of capacitance. The measurement and simulation values of the capacitance of 4 capacitive sensing regions are slightly different. The sensor has good ability to against the temperature variation (Fig. 4).

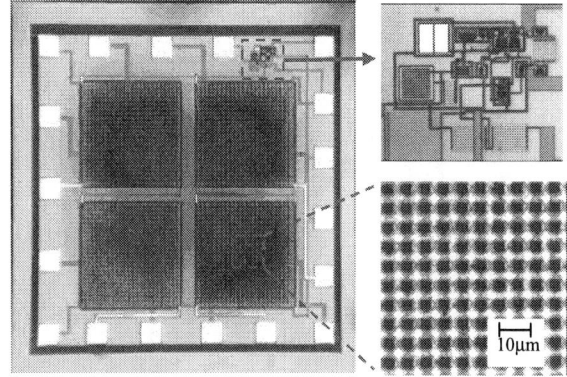

Fig. 3: The capacitive sensing regions and electrode of device

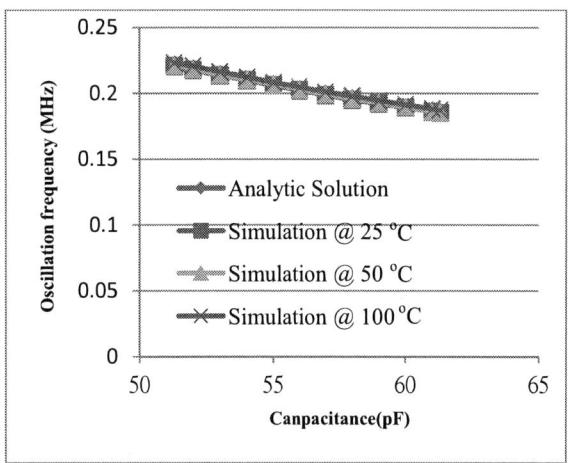

Fig. 4: Capacitance stability of the sensor with thermal compensation.

The capacitance of each sensing structure is 75.2pF (shows in Table 4). The capacitances at different temperatures are presented in Fig. 4. According to analyses and simulation results, the capacitance values between different temperatures are stabilized after thermal compensation.

TABLE. 4: CAPACITANCE OF THE CMOS CAPACITIVE SENSOR

	Simulation results without substrate (pF)	Simulation results with substrate (pF)	Experiment value (pF)
Capacitance	51.3~61.5	56.4~66.6	75.2

TABLE. 5: SEPCTICATION OF OPERATIONAL AMPLIFIERS

Process	TSMC 0.35μm mixed signal process
Specification	Post-simulation
Vdd	3.3V
C_L / RL	10pF/10kΩ
Small Signal Gain	78.9732
Phase Margin	59°
Gain-Bandwidth	120 MHz
f-3db	26.7KHz
SR (rise/fall) V/μs	119 / 175
ICMR	0.171V~3.3V
CMRR	80.6dB
PSRR	112db
Offset Voltage	0.167V
Settling Time	Up：31ns, Down：380ns
Output Swing	0 .088V ~ 3.2 V
Power Dissipation	5.5916 mW

IV. CONCLUTION

This paper presents a refined configuration of fringing-field capacitive sensor and staggered mesh electrodes, which can take the advantages of fringing effect. This paper solved an electrode configuration optimization problem for fringing-field enhancement. The multi-layer meshes sensing chip simulation results show an improved configuration of the electrode structure, based on the enhancement of fringing-field. It also can improve the sensor sensitivity. Associated with the fringing effect, it would be able to indirectly applied to an electrostatic self-assembly to gauge the yield of a micro-assembly process. Enhancing the fringing effect can contribute to the performance of capacitive sensors.

ACKNOWLEDGMENT

The authors would like to thank Prof. C. S. Shen for helpful discussion. This work was supported by the National Science Council, NSC-98-2221-E-018-012-MY2 and National Chip Implementation Center (CIC).

REFERENCES

[1] Y.-T. Li, C.-M. Chao, K. Wang, "A Capacitance Level Sensor Design and Sensor Signal Enhancement,"*the 6th Annual IEEE Int. Conf.on NEMS*, Feb. 20-23, Kaohsiung, pp.888-891, 2011.

[2] Z.-H. Zhang, Y.-H. Zhang, L.-T. Liu, and T.-L. Ren, "A novel MEMS pressure sensor with MOSFET on chip," *in IEEE Sensor, Lecce*, Italy, pp.1564-1567, 2008.

[3] P. E. Allen and D. R. Holberg, *CMOS Analog Circuit Design, Second Edition*: Oxford University Press, Inc., 2002.

[4] C.-L. Dai, J.-H. Chiou, and M. S.-C. Lu, "A maskless post-CMOS bulk micromachining process and its application," *Journal of micromechanics and microengineering* vol. 15, 2005

[5] J.-H. Kim., S.-M. Hong, J.-S. Lee, B.-M Moon, K. Kim, "High sensitivity capacitive humidity sensor with a novel polyimide design fabricated by MEMS technology," *Nano/Micro Engineered and Molecular Systems, NEMS 2009. 4th IEEE International Conference*, 2009.

[6] L. Cu, Q.-A. Huang, and M. Qin, "A novel capacitive-type humidity sensor using CMOS fabrication technology," *Sensors and Actuators B: Chemical*, 99(2-3): p. 491-498, 2004.

[7] C.-C. Lu, Y.-S. Huang, J.-W Huang, "A Macroporous TiO2 Oxygen Sensor Fabricated Using Anodic Aluminium Oxide as an Etching Mask," *Sensors*. 10(1): p. 670-683, 2010.

[8] J. Huang and Q. Wan, "Gas Sensors Based on Semiconducting Metal Oxide One-Dimensional Nanostructures," *Sensors*, 9(12): p. 9903-9924, 2009.

[9] T. Nitta and S. Hayakawa, "Ceramic Humidity Sensors. Components, Hybrids, and Manufacturing Technology," *IEEE Transactions*, 3(2): p. 237-243, 1980.

[10] N. Lazarus, S.S. Bedair, C.-C Lo, G.K. Fedder, "CMOS-MEMS Capacitive Humidity Sensor. in Micro Electro Mechanical Systems," *MEMS 2009. IEEE 22nd International Conference*, 2009.

[11] A. Kawalec and M. Pasternak, "Microwave Saw Humidity Sensor. in Microwaves," *Radar & Wireless Communications, 2006. MIKON 2006. International Conference*, 2006.

[12] B.E. Noltingk, "A novel proximity gauge," *Journal of Physics E: Scientific Instruments*, 2(4): p. 356-360. 1969

[13] R.C. Luo and Z. Chen, "Modeling and implementation of an innovative micro proximity sensor using micromachining technology," *Intelligent Robots and Systems '93, IROS '93. Proceedings of the 1993 IEEE/RSJ International Conference*, 1993.

[14] Z. Chen ; R.C. Luo,, "Design and implementation of capacitive proximity sensor using microelectromechanical systems technology," *IEEE Transactions on Industrial Electronics*, 45(6): p. 886 – 894, 1998.

[15] M.C. Hegg and A.V. Mamishev, "Influence of variable plate separation on fringing electric fields in parallel-plate capacitors," *Electrical Insulation, 2004. Conference Record of the 2004 IEEE International Symposium*, 2004.

[16] V. Leus and D. Elata, "Fringing field effect in electrostatic actuators," Technical report ETR-2004-02, Israel Institute of Technology.

[17] P. E. Allen and D. R. Holberg, *CMOS Analog Circuit Design, Second Edition*: Oxford University Press, Inc., 2002.

Dimensions and Capillary Effects of Microfluidic Channel for Blood Plasma Separation

Yu-Hui Zhan and Ju-Nan Kuo[*]

[*]Department of Automation Engineering, National Formosa University, Yunlin, Taiwan
junan@nfu.edu.tw

Abstract—**This paper presents a cross-flow microchannel network for separation blood plasma on a microfluidic chip. The whole blood could be introduced into the microchannel via the effects of capillary forces alone, i.e., there was no need for any external driving force. A microfluidic chip for plasma separation including a cross-flow microchannel consisting of a straight main microchannel and a filtration microchannel has been proposed. The microfluidic chip used for plasma separation is fabricated from a single polydimethylsiloxane (PDMS) mold and sealed with an etched glass substrate. Since filtration microchannel depth is less than 2 μm, plasma can be extracted from whole blood. The merits of this design are its simple structure, high separation efficiency and without any external driving force. The experimental results show that this design can extract at least 0.02 μl of plasma from whole blood within 25 s and with a residual cell concentration of less than 0.07 % in the purified plasma.**

Keywords- Blood plasma; Capillary; Cross-flow microchannel; Whole blood.

I.   INTRODUCTION

Blood plasma separation is the first required step for subsequent blood tests in clinical diagnosis and basic research assays. In practice, blood cells must be separated by centrifugation from whole blood collected from a patient before the plasma can be analyzed for electrolyte concentration, glucose, cholesterol, [1] etc. However, standard clinical methods for separating whole blood into blood cells and cell-free plasma require large facilities and complex equipment. The sequential separation and extraction procedures take several hours to complete and require the intervention of qualified technical personnel. Therefore, there is a need for systems capable of performing the blood separation process in an automated and more rapid manner.

Recently developed microfluidic devices based on microelectromechanical system (MEMS) technologies include micrototal analysis systems (μTAS), biochemical micro-sensors, and microvalves. Chemical and biological experiments can be performed simultaneously, which reduces the reagent volume, the sample volume, and the time required to perform the assay [2]. Microfluidic chip blood separation can be performed with built-in porous filters [3], microchannel structures [4], and cross-flow filters [5, 6]. The MEMS microfluidic devices are simultaneously used for cell separation and collection in many processes, including microfabricated fluorescence-activated cell sorting [7], magnetic-activated cell separation [8, 9], and dielectrophoresis

force separation [10, 11]. However, these reported cell separation microchips have significant disadvantages. In addition to pumping, the inherent centrifugal forces can also be used for blood separation based on the variance of mass density among different components [12, 13]. But it requires the device rotating to provide centrifugal pumping, and the centrifugal force induced by microchannel design. The objective of this work is to design a simple microchannel network that can spontaneously separate blood plasma from whole blood into different reservoirs.

II.   DESIGN AND FABRICATION

In this study, a plasma separation microfluidic chip including a cross-flow microchannel consisting of a straight main microchannel and a filtration microchannel was tested. The photomask pattern of the chip was generated using AutoCAD software and transferred to an inexpensive plastic mask. The cross-flow microchannel is schematically illustrated in Fig. 1.

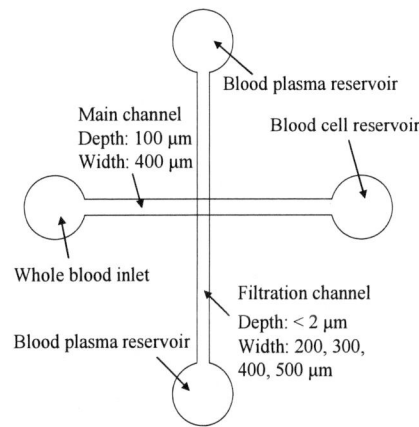

Fig. 1. Schematic diagram of cross-flow microchannel configuration on microfluidic chip for separating blood plasma.

Figure 2 schematically depicts the cross-flow microchannel fabrication steps. The filtration channels were fabricated on a glass substrate. In the present study, photoresist (S1818) was used as the etching mask. The S1818 photoresist was spin coated onto the glass substrate and soft baked at 95 °C for 3 min. The photoresist was then UV-exposed, developed in a

MP351 solution, and hard baked at 120 °C for 15 min. The glass substrate was immersed in a buffered oxide etchant (BOE 6:1) at room temperature. A precise microchannel depth was achieved by carefully adjusting the etching time. Following the etching process, an acetone solution was used to strip the remaining photoresist from the substrate surface. The main channels were fabricated using conventional soft lithography techniques. A conventional photolithography technique was used to fabricate an SU-8 master of the main channels. The master was then used to transfer the main channels to a polydimethylsiloxane (PDMS) layer. For whole blood perfusion, a blunt needle was used to drill a 5 mm diameter hole into the whole blood inlet region of the PDMS layer. After positioning the PDMS layer on the glass substrate surface of the filtration channels, it was bonded to the surface using an oxygen plasma adhesion process then cleaned using a plasma cleaning process (PDC-001, Harrick). Figures 3 presents photograph of the completed blood plasma separation microfluidic chip.

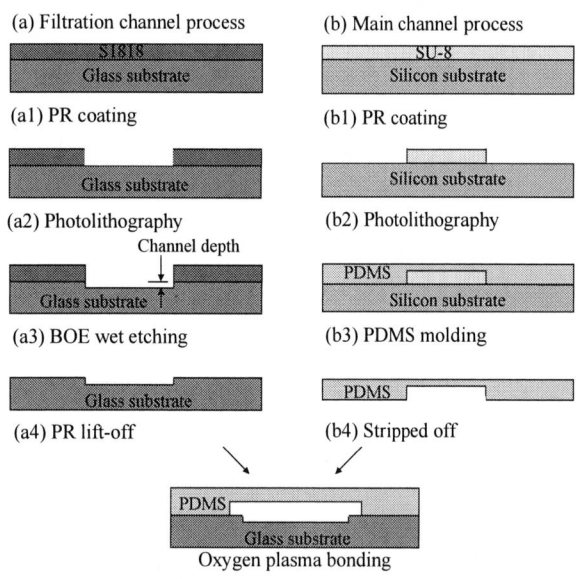

Fig. 2. Schematic overview of microfluidic chip fabrication process.

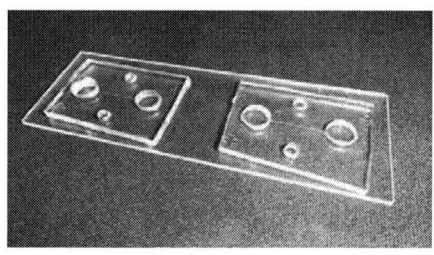

Fig. 3. Photograph of plasma separation microfluidic chip.

III. RESULTS AND DISCUSSION

The glass substrate was immersed in a buffered oxide etchant (BOE 6:1) without agitation. Figure 4 shows the relationship between filtration channel depth and etching time. It is shown that BOE etches glass very fast and the etch rate reaches a stable speed of 0.65 µm/min over 10 min. This is most likely because borosilicate glass is a multicomponent mixture of SiO_2 and other metal oxides. When etching glass in a BOE solution, the insoluble products will be present during etching and the formation of crystalline precipitates will hinder the etching process by preventing etchants from contacting the glass substrate [14]. The surface profiles of the etched filtration channels were measured with a surface profilometer (ET3000, Kosaka Lab Ltd.). Figure 5 presents the typical results obtained for a filtration channel with a depth of 1.6 µm after 2.5 min etching.

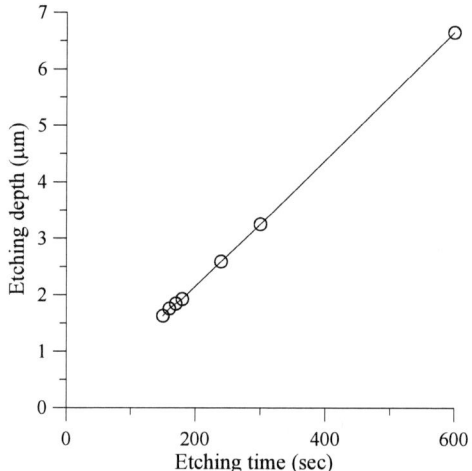

Fig. 4. Filtration channel depth as a function of etching time of glass in BOE (6:1) etchant. The average etching rate is 0.65 µm/min.

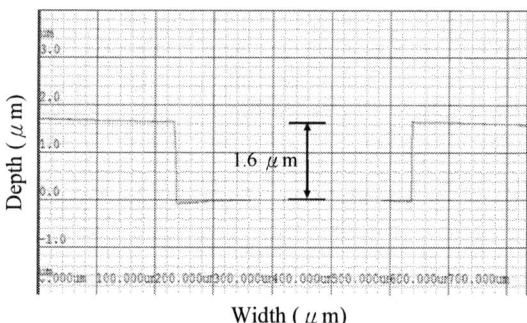

Fig. 5. Surface profile of 1.6-µm-deep filtration channel.

During oxygen plasma treatment, PDMS/glass bonds converts the hydrocarbon groups of PDMS on the surface to silanol groups (-OH) or to dangling bonds that can form strong (Si-O-Si) bonds with silanol groups on the glass surface [15]. Filtration microchannels of varying widths and depths were fabricated to test the limit of this PDMS/glass bonding technique under normal bonding conditions. Due to the depth of the filtration microchannel less than 2 μm, it may cause the channels to collapse after PDMS/glass bonding. Figures 6(a) and 6(b) are optical micrographs of sealed filtration channels with depths of 1.0 μm and 1.6 μm, respectively. It is seen that the channel with a depth of 1.6 μm remains well-defined and unobstructed along its entire length. However, the channel with a depth of 1.0 μm collapses at a certain point along its length.

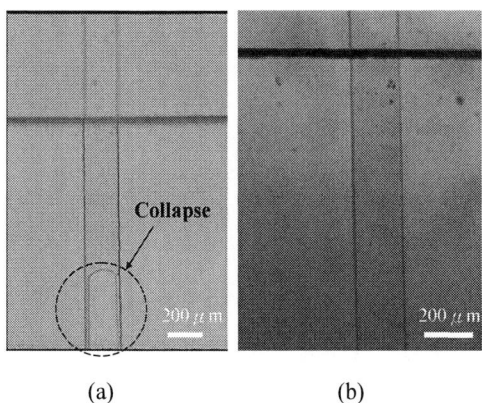

(a)   (b)

Fig. 6. Optical micrographs of 200 μm wide filtration channels showing (a) collapse of 1.0 μm deep microchannels but (b) survival of 1.6 μm deep microchannels after bonding.

Plasma was separated from whole blood in the chip by dimension exclusion. Whole blood is a suspension of RBCs, WBCs, and platelets in plasma. The RBCs are discoid, anuclear cells ~8 μm in diameter and ~2.5 μm in thickness; WBCs are spherical cells 8-12 μm in diameter; and platelets are discoid particles 2-3 μm in diameter. Because the filtration microchannels in this study had depths less than 2 μm, they could be used to extract plasma from whole blood. In the experimental procedure, a volume of 2 μL whole blood sample was introduced into the whole blood inlet using a pipette. Plasma flow velocity in the microchannels was also measured. Figures 7 presents extract images of filtration channels with a depth of 1.6 μm, and widths of 200 μm, and 400 μm, respectively. Experimental observations showed that whole blood could be introduced into the microchannel via the effects of capillary forces alone, i.e., there was no need for any external driving force. The plasma could then be extracted from the whole blood and separated into the filtration channel. The experimental results show that this method provide more than 0.02 μl of plasma was extracted from whole blood within 25 s and the residual cell concentration in the purified plasma amounts to less than 0.07 % has been demonstrated.

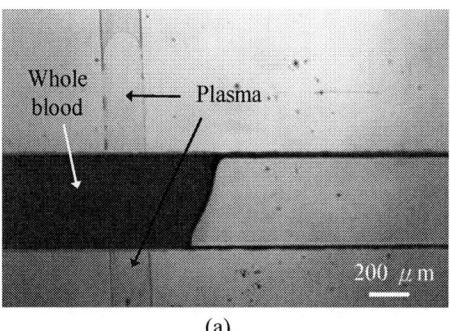

(a)

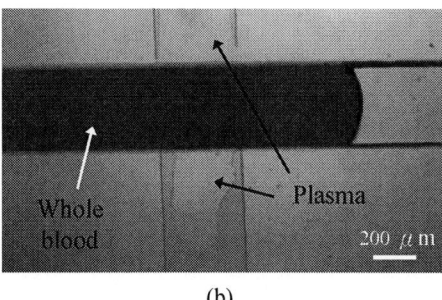

(b)

Fig. 7. Image of cross-flow microchannel filled with whole blood. Plasma extracted from whole blood was separated into filtration channels with widths of (a) 200 μm and (b) 400 μm.

IV. CONCLUSION

In this study, we design of microfluidic chip with cross-flow microchannel structures utilizing the effects of capillary forces and dimensions of the filtration microchannel, cell-free plasma was effectively extracted from whole blood samples. The merits of this microfluidic chip based on cross-flow filtration microchannel are its simple structure, it high plasma separation efficiency, and its capability of operating without external driving force and with only a low volume of sample and reagents. The microfluidic chip is disposable because it is simply and inexpensively fabricated from a single PDMS mold sealed by a glass substrate. The simple method of fabrication of the cross-flow separation microchannel also facilitates its integration with on-chip biochemical plasma assay elements. The integrated chips have potential applications in point-of-care clinical diagnostics required for rapid biochemical tests.

ACKNOWLEDGMENT

The authors gratefully acknowledge the financial support provided to this study by the National Science Council of Taiwan under Grant No. NSC 100-2221-E-150-019. In addition, the access provided to fabrication equipment by the Common Lab for Micro/Nano Science and Technology, Research and Services Headquarters at National Formosa University is greatly appreciated.

REFERENCES

[1] S. Yang, A. Undar, and J. D. Zahn, "A microfluidic device for continuous, real time blood plasma separation," *Lab Chip*, vol. 6, pp. 871-880, 2006.

[2] D. J. Harrison, K. Fluri, K. Seiler, Z. Fan, C. S. Effenhauser, and A. Manz, "Micromachining a miniaturized capillary electrophoresis-based chemical analysis system on a chip," *Science*, vol. 261, pp. 895-897, 1993.

[3] J. Moorthy and D. J. Beebe, "In situ fabricated porous filters for microsystems," *Lab Chip*, vol. 3, pp. 62-66, 2003.

[4] C. T. Huang, P. N. Li, C. Y. Pai, T. S. Leu, and C. P. Jen, "Design and simulation of a microfluidic blood-plasma separation chip using microchannel structures,"*Sep. Sci. Technol.*, vol. 45, pp. 42-49, 2010.

[5] V. Van Delinder and A. Groisman, "Separation of plasma from whole human blood in a continuous cross-flow in a molded microfluidic device," *Anal. Chem.*, vol. 78, pp. 3765-3771, 2006.

[6] X. Chen, D. F. Cui, C. C. Liu, and H. Li, "Microfluidic chip for blood cell separation and collection based on crossflow filtration," *Sens. Actuator B-Chem.*, vol. 130, pp. 216-221, 2007.

[7] A. Wolff, I. R. Perch-Nielsen, U. D. Larsen, P. Friis, G. Goranovic, C. R. Poulsen, J. P. Kutter, and P. Telleman, "Integrating advanced functionality in a microfabricated high-throughput fluorescent-activated cell sorter," *Lab Chip*, vol. 3, pp. 22-27, 2003.

[8] D. W. Inglis, R. Riehn, R. H. Austin, and J. C. Sturm, "Continuous microfluidic immunomagnetic cell separation," *Appl. Phys. Lett.*, vol. 85, pp. 5093-5095, 2004.

[9] K. Smistrup, O. Hansen, H. Bruus, M. F. Hansen, "Magnetic separation in microfluidic systems using microfabricated electromagnets-experiments and simulations," *J. Magn. Magn. Mater.*, vol. 293, pp. 597-604, 2005.

[10] Y. Huang, S. Joo, M. Duhon, M. Heller, B. Wallace, and X. Xu, "Dielectrophoretic cell separation and gene expression profiling on microelectronic chip arrays," *Anal. Chem.*, vol. 74, pp. 3362-3371, 2002.

[11] J. Voldman, M. L. Gray, M. Toner, and M. A. chmidt, "A microfabrication-based dynamic array cytometer," *Anal. Chem.*, vol. 74, pp. 3984-3990, 2002.

[12] S. Haeberle, T. Brenner, R. Zengerle, and J. Ducr'ee, "Centrifugal extraction of plasma from whole blood on a rotating disk," *Lab Chip*, vol. 6, pp. 776-781, 2006.

[13] J. Zhang, Q. Guo, M. Liu, and J. Yang, "A lab-on-CD prototype for high-speed blood separation," *J. Micromech. Microeng.*, vol. 18, pp. 125025, 2008.

[14] C. H. Lin, G. B. Lee, Y. H. Lin, and G. L. Chang, "A fast prototyping process for fabrication of microfluidic systems on soda-lime glass," *J. Micromech. Microeng.*, vol. 11, pp. 726-732, 2001.

[15] S. Bhattacharya, A. Datta, J. M. Berg, and S. Gangopadhyay, "Studies on surface wettability of poly(dimethyl) siloxane (pdms) and glass under oxygen-plasma treatment and correlation with bond strength," *J. Microelectromech. Syst.*, vol. 14, pp. 590-597, 2005.

Fabrication and Analysis of Integrated MEMS Pyramidal Horn Antenna for Terahertz Applications

C. Li[1], L. Guo[2], W.G. Wu[1,*], X.S. Tang[2] and F.Y. Huang[2]

[1]National Key Laboratory of Science and Technology on Micro/Nano Fabrication,
Institute of Microelectronics, Peking University, Beijing 100871, CHINA
[2]Institute of RF- & OE-ICs, National Mobile Communications research laboratory,
Southeast University, No.2 Sipailou, Nanjing 210096, CHINA
wuwg@pku.edu.cn

Abstract— **This paper reports the fabrication and analysis of a back-to-back pyramidal horn antenna based on Micro-electromechanical Systems (MEMS) technology, which is used in terahertz (THz) applications. This method is able to manufacture antenna with various geometrical configuration embedded in Si substrates. The dimensional tolerance of this technique is determined by the lithography process, which could be controlled down to tens nm. Therefore, this technique is controllable and of high precision and quality. The performance of fabricated antenna device was verified by Ansoft High Frequency Structure Simulator (HFSS) simulation. The result shows that the antenna gain ranges from 10-15dB, and voltage standing wave ratio (VSWR) is less than 2 in the frequency band of 3.25-3.55THz.**

Keywords-THz; Horn Antenna; MEMS; KOH etching

I. INTRODUCTION

The terahertz (THz) technology presently focuses on the frequencies range from 0.1 THz to 10THz. Its spectrum locates in the transition area from macrophysics to microcosmic quantum theory which covers the frequency of protein molecular surge [1]. Using THz technology, wide range of applications therefore can be implemented with natural advantages in areas such as spectroscopy [2], imaging [3, 4], plasma diagnostics [5] and communication systems [6-8], etc.

While major progresses continue to be made in submillimeter astronomy and remote sensing, the past few years have seen an unprecedented expansion of THz applications, components and instruments. THz region is one of the least explored regions of the spectrum. Because THz field has multidisciplinary characters, researching the THz technique requires a deep knowledge of microwave engineering and semiconductor physics, optics and photonics.

The size of THz antenna is expected to closely relate to wavelength, which is about 30 μm (10 THz) to 3 mm (0.1 THz), thus Micro-electromechanical Systems (MEMS) technology is highly suitable for manufacturing THz devices since nm-level precision can be achieved using this technology.

The THz applications rely on the antenna elements with compact size, wide bandwidth, and electrically large radiation aperture. The size of the radiation aperture is an important parameter to enhance directional gain that is beneficial to the high directivity imaging or light gathering applications. Planar antennas [9] are used extensively in THz applications, because they can be integrated with other devices easily, such as diodes. The planar half-wavelength and full-wavelength

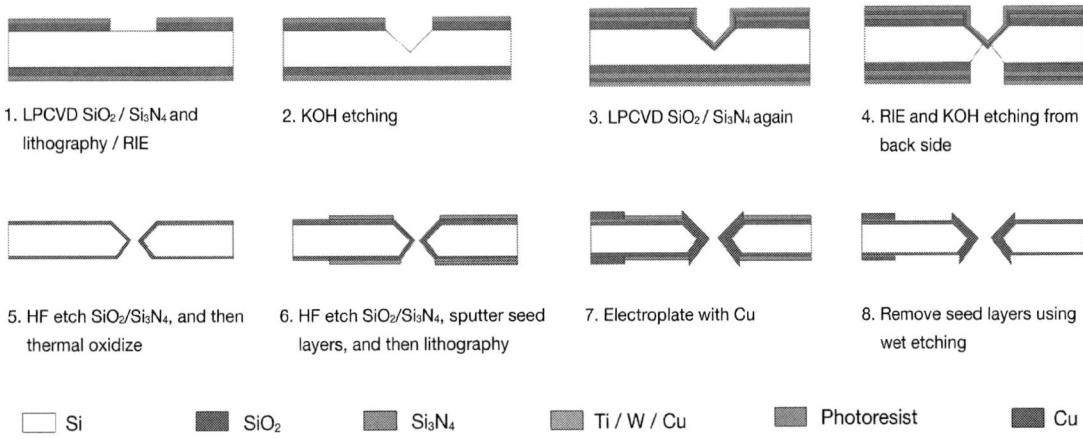

1. LPCVD SiO₂/ Si₃N₄ and lithography / RIE

2. KOH etching

3. LPCVD SiO₂/ Si₃N₄ again

4. RIE and KOH etching from back side

5. HF etch SiO₂/Si₃N₄, and then thermal oxidize

6. HF etch SiO₂/Si₃N₄, sputter seed layers, and then lithography

7. Electroplate with Cu

8. Remove seed layers using wet etching

☐ Si   ▨ SiO₂   ▨ Si₃N₄   ▨ Ti / W / Cu   ▨ Photoresist   ▨ Cu

Figure 1: Schematic fabricating process of proposed pyramidal horn antenna.

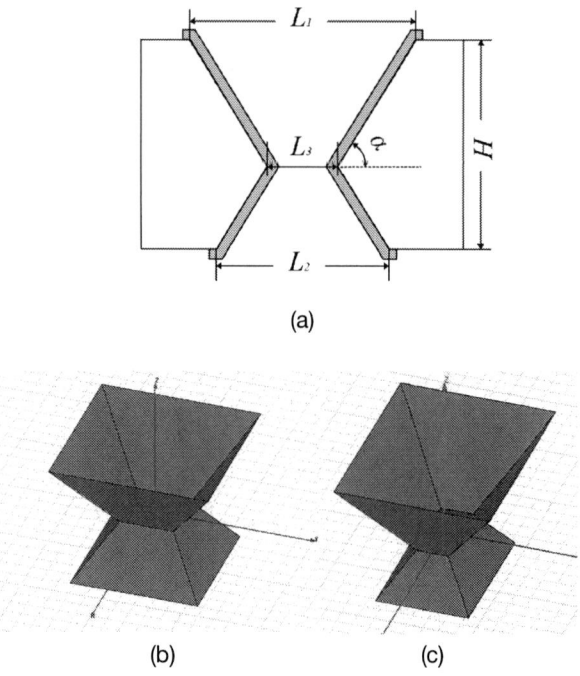

(a)

(b)   (c)

Figure 2: (a) Geometry and configuration of fabricated antenna; 3D model with (b) square and (c) rectangle window which is used in HFSS simulation

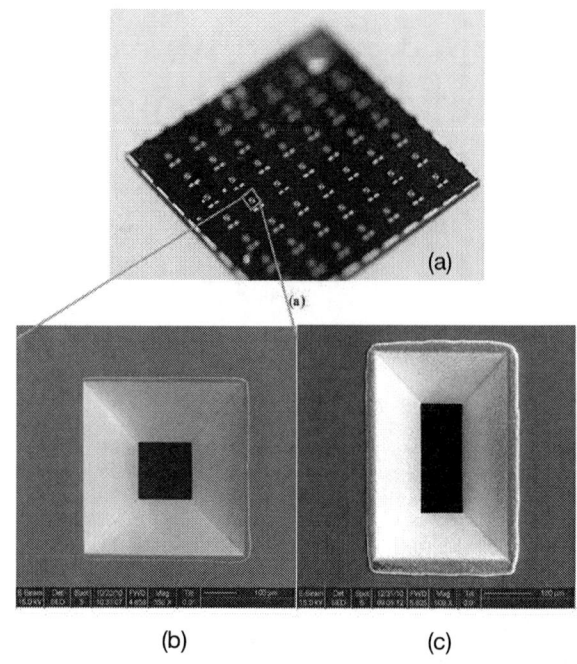

(a)

(b)   (c)

Figure 3: (a) Photographs of integrated chip with proposed antennas array; (b) SEM top view image of proposed antenna with square and (c) rectangle window.

dipoles are the most commonly used antennas in the present terahertz antenna design. However, the impedance bandwidths of the dipoles are inherently relative narrow. Several broadband antennas have been devoted in the terahertz system including the spiral antenna and log-periodic antenna. However, the gains of these broadband antennas are relatively low and their geometries results in a difficulty to make them as a planar array. Several approaches to fabricate THz antenna employing MEMS technology has also been reported in recent years, in which the fabrication process, however, is highly complicated and antennas with different size are difficult to integrate [9, 10]. In this paper we report the back-to-back pyramidal horn antenna, which could overcome these drawbacks.

In this paper we report the back-to-back pyramidal horn antenna, which could overcome these drawbacks. This method employs KOH etching to fabricate antenna structures on one wafer, which is capable of parallel fabrication of multiple antennas with different geometrical configuration. The dimensional tolerance can be controlled down to tens nm determined by lithography. Electro-magnetic simulation was conducted by High Frequency Structure Simulator (HFSS) to explore and optimize the proposed antenna.

II. DESIGN & FABRICATION

The proposed antennas were fabricated in single-crystal silicon substrates using conventional bulk micromachining, and the detailed process was illustrated in Figure 1. A Si_3N_4 layer was firstly deposited by low pressure chemical vapor deposition (LPCVD) on a prior thermal SiO_2 layer, which is used as KOH etching mask. (Figure 1.1) The cavity etching was divided into two steps: (1) After KOH etch stop from top side, SiO_2 and Si_3N_4 layers were again deposited so that back side etch is able to stop at this layer. (Figure 1.2-1.3) (2) After lithography at back side, which defined back-side window, KOH etch was again employed to fabricate back cavity. (Figure 1.4) The isolation of between substrate and metallic antenna body was achieved by thermal oxidized layer for 400 nm, followed by removal of previous deposited SiO_2 and Si_3N_4 layers. (Figure 1.5)

Since the inclination of <111> and <100> plane is fixed at α=57.4 degree, the geometric parameter can be adjusted by lithography on both sides. Figure 2 shows geometrical configuration of proposed antenna, which can be determined by the configuration of etched cavity. Figure 2(a) shows the two-dimensional cross-section of the antenna, which clearly indicates that the geometrical parameter can be determined by following equation.

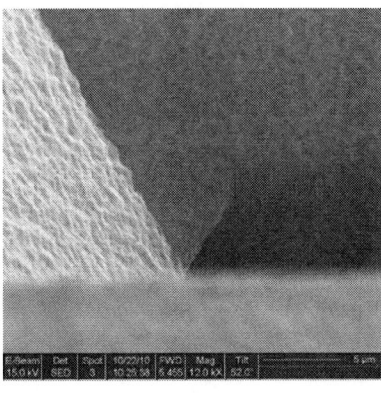

(a)

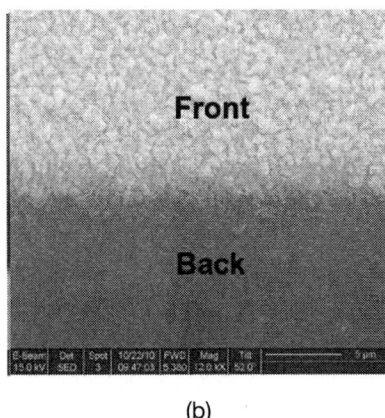

(b)

Fig. 4: Magnified SEM image at 52 degree tilt shows surface roughness less than 50 nm.

$$L_3 = \frac{L_1 + L_2 - H / \tan\alpha \times 2}{2}$$

$$H_1 = \frac{L_1 - L_3}{2} \times \tan\alpha \tag{1}$$

The geometrical parameter along x-axis and y-axis can be determined by equation 1 independently. Therefore both antennas with both square and rectangle are able to be fabricated. Figure 2(b, c) shows three dimensional antenna models used in HFSS simulation with respectively square and rectangle window.

After the formation of the cavities, thin layers of Ti/W/Cu were deposited by sputtering as adhesion, resistance, and seed layer, respectively, which is used for the following electroplating. After that, lithography with photoresist spraying were firstly employed to provide patterns (Figure 1.6), and then 5μm thick metallic antenna body and possible transmission line such as coplanar waveguide (CPW) were electroplated following the photoresist template. (Figure 1.7)

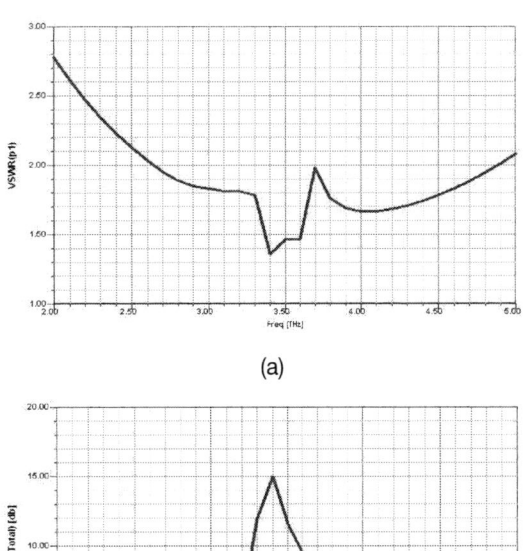

(a)

(b)

Figure 5: Simulation of (a) voltage standing wave ratio (VSWR) and (b) gain of fabricated antenna.

TABLE I.   GEOMETRIC CHARACTERIZATION OF ANTENNA USED IN HFSS SIMULATION

Parameter	L_{1x}	L_{1y}	L_{2x}	L_{2y}	H
Value (μm)	420	290	610	480	400

After removal of photoresist and Ti/W/Cu layers, the antennas were isolated from each other. (Figure 1.8)

Figure 3(a) shows the photograph of fabricated antenna arrays on integrated chip. Figure 3(b) and figure 3(c) show the SEM image of respectively square and rectangle window of fabricated antenna. The surface roughness and coverage of electroplating can be characterized by magnified SEM image with different tilt and view, which is shown in figure 4. From the images, we can see the surface roughness is less than 50 nm, and corners can be perfectly covered.

III.   SIMULATION & ANALYSIS

We use HFSS to analyze the performance of one of fabricated antennas. The antenna has rectangle window as detailed parameter shown in Table 1, where L_{1x} and L_{1y} refer to L_1 in x axis and y axis, and the same in L_2, H is the thickness of wafer. The three dimensional model is shown in Figure 2(c).

978-1-4673-1122-9/12 $31.00 © 2012 IEEE   613

Figure 5 shows simulated voltage standing wave ratio (VSWR) and gain, from which we can see that the antenna gain ranges from 10 dB to 15 dB and VSWR is less than 2 in the frequency band of 3.25-3.55THz.

IV. CONCLUSION

In summary, detailed fabrication process and results as well as characterization by HFSS simulation of integrated MEMS pyramidal horn antenna have been reported. The fabrication method is highly cost effective and of high fabrication precision and quality. HFSS simulation results show the fabricated antenna has good performance, which indicates this method is able to fabricate high performance antenna works in THz spectrum.

ACKNOWLEDGMENT

This work is supported by the Research Fund of National Key Laboratory of Science and Technology on Micro/Nano Fabrication and the National High-Technology Research and Development Program of China (863 Program, Grant No. 2009AA01Z228 and Grant No. 2009AA01Z261).

REFERENCES

[1] D. Dragoman and M. Dragoman, "Terahertz fields and applications," *Progress in Quantum Electronics,* vol. 28, pp. 1-66, 2004.

[2] M. C. Beard, *et al.*, "Terahertz Spectroscopy," *The Journal of Physical Chemistry B,* vol. 106, pp. 7146-7159, 2002/07/01 2002.

[3] H.-T. Chen, *et al.*, "Terahertz imaging with nanometer resolution," *Applied Physics Letters,* vol. 83, pp. 3009-3011, 2003.

[4] B. B. Hu and M. C. Nuss, "Imaging with terahertz waves," *Opt. Lett.,* vol. 20, pp. 1716-1718, 1995.

[5] S. P. Jamison, *et al.*, "Plasma characterization with terahertz time-domain measurements," *Journal of Applied Physics,* vol. 93, pp. 4334-4336, 2003.

[6] R. Piesiewicz, *et al.*, "Short-Range Ultra-Broadband Terahertz Communications: Concepts and Perspectives," *Antennas and Propagation Magazine, IEEE,* vol. 49, pp. 24-39, 2007.

[7] T. Kleine-Ostmann, *et al.*, "Audio signal transmission over THz communication channel using semiconductor modulator," *Electronics Letters,* vol. 40, pp. 124-126, 2004.

[8] R. Piesiewicz, *et al.*, "Scattering Analysis for the Modeling of THz Communication Systems," *Antennas and Propagation, IEEE Transactions on,* vol. 55, pp. 3002-3009, 2007.

[9] G. M. Rebeiz, "Millimeter-wave and terahertz integrated circuit antennas," *Proceedings of the Ieee,* vol. 80, pp. 1748-1770, 1992.

[10] S. Biber, *et al.*, "Design and Measurement of a 600 GHz Micromachined Horn Antenna Manufactured by Combined DRIE and KOH-Etching of Silicon," in *Sixteenth International Symposium on Space Terahertz Technology,* 2005, pp. 507-512.

Characterization of Super-harmonic Effect Using Piezoelectric Film Cantilever with A Proof Mass in the Point

Haruki Ishinabe [1], Takeshi Kobayashi [2,*], Dong F. Wang [1,*], Toshihiro Itoh [2], and Ryutaro Maeda [2]

[1] Micro Engineering & Micro Systems Laboratory, Ibaraki University (College of Eng.), Hitachi, Ibaraki 316-8511, JAPAN
[2] Research Center for Ubiquitous MEMS and Micro Engineering (UMEMSME), AIST, Tsukuba, Ibaraki 305-8564, JAPAN
([2,*] E-mail: takeshi-kobayashi@aist.go.jp; [1,*] E-mail: dfwang@mx.ibaraki.ac.jp)

Abstract— The super-harmonic effect has been studied using a piezoelectric micro-cantilever with a proof mass. The fabricated prototype device, comprised with a multilayer of Pt/Ti/PZT/Pt/Ti/SiO$_2$ deposited on a silicon-on-insulator (SOI) wafer, has been measured to have an eigenstate frequency of around 816 Hz. The driven frequencies of 204, 272, 409 and 816 Hz have been then converted into the resonant one via the super-harmonic vibration, corresponding to a sudden increase of output voltage, based on our preliminary characterizations.

Keywords - Super-harmonic effect; Micro-cantilever; PZT thin film; Proof mass; Eigenstate frequency

I. INTRODUCTION

Recently, there has been a growing interest in research and development of the wireless sensor network. Wireless sensor network put a restriction on install of wireless sensor nods. Therefore, it is possible to apply to a condition that the sensing object move [1-2].

The chicken health monitoring system by wireless sensor that can be applied to earlier detection system of avian influenza in the chicken farm was development. It is constructed wireless sensor terminal attached temperature sensor and acceleration sensor, and the system of reception and analysis from wireless sensor nodes. Event driven switch was developed for "ultra-low" power of wireless sensor nodes. This is used piezoelectric thin film. And the serpentine cantilever structural have proof mass at the end. Output voltage is attributable to frequency by chicken moves. The projecting output voltage is obtained in both a characteristic frequency and super-harmonic resonance. Therefore, it is an acceleration sensor selectively detected by setting the threshold from this to the output voltage [3-6]. However, there is no reported on influence of the super-harmonic resonance in a MEMS device, especially the output voltage characteristic to the input vibration has many still indefinite points.

Fig. 1 shows the device for evaluation of characteristics. The micro cantilever used for the study has proof mass in the point, and use the PZT thin film. The present study aimed to clarify vibration property and the characteristic of output voltage by the super-harmonic resonance.

Fig. 1. Design of proposed super harmonic oscillators.

II. SUPER-HARMONIC RESONANCE

Fig. 2 explains the principle of the super-harmonic resonance. The resonance is caused when the excitation vibration (ω) is close to the eigenstate frequency (ω_0) in a nonlinear vibration system. In other excitation vibrations, the sub-resonance can be caused. When the n is not so large integer, the resonance in Equation 1 is harmonics resonance of n order.

$$\omega = \omega_0 / n \qquad (1)$$

The oscillatory waveform like (c) by getting in tune with harmonics vibration of n degree of excited vibrational component by vibration (a) and element that vibrates by $1/n$ of characteristic frequency (b) is obtained [7-8].

III. MICRO-FABRICATION OF OSCILATORS

We have fabricated the oscillator using the PZT thin film. Fig. 3 shows the layers of oscillator. The fabrication process began from Pt/Ti/PZT/Pt/Ti/SiO$_2$ multilayers deposition on silicon-on-insulator (SOI) wafers with a 5 µm thick structural Si layer. At first, the SOI wafers were oxidized at 1100 ℃ to form thermal oxide (1 µm). Next, Pt (0.2 µm)/Ti (0.05 µm) thin films as a bottom electrode were deposited by sputtering. After, 2.5 µm thick PZT thin film was formed by sol-gel deposition. Finally, Pt (0.2 µm)/Ti (0.05 µm) thin films as upper electrode were deposited by sputtering.

The multilayers were etched by dry and wet etching to form the oscillator. Pt/Ti thin films as a top electrode were etched used Ar ions to determine the top electrode. PZT thin films were wet-etched used mixture of HF, HNO_3 and HCl. Pt/Ti thin films as a bottom electrode were etched used Ar ions. Thermal SiO_2 thin films were etched by reactive ion etching (RIE) with CHF_3 gas to pattern the bottom electrodes. Structural Si and buried oxide (BOX) layers were etched by RIE with SF_6 (for Si) and CHF_3 (for BOX) to determine the cantilever. Finally, Si substrates and BOX were etched from their back side.

The micro cantilever used for the study has proof mass in the point. The resonance frequency of this cantilever is 808 Hz by numerical result based on theoretical formula [9].

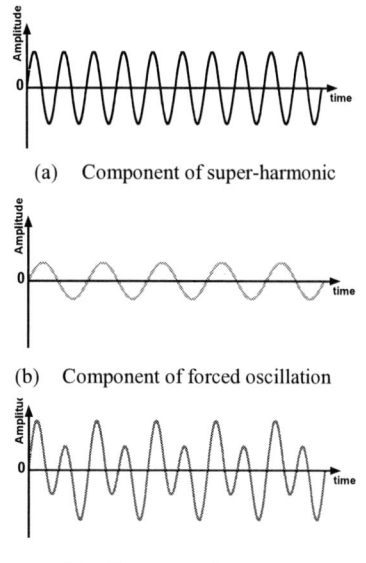

(a) Component of super-harmonic

(b) Component of forced oscillation

(c) Output waveform

Fig. 2. Schematic of response curves of super-harmonic component, forced oscillation component, and output waveform.

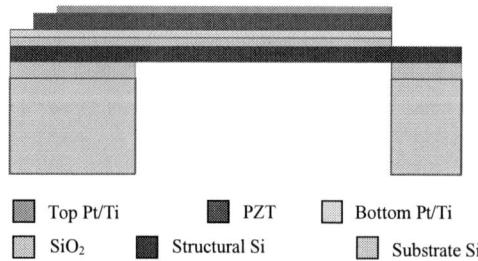

| Top Pt/Ti | PZT | Bottom Pt/Ti |
| SiO_2 | Structural Si | Substrate Si |

Fig. 3. The cross sectional structure of PZT cantilever.

IV. MEASURING SYSTEM

Fig. 4 shows composition of experiment system. The oscillator is mounted on the 16-pin ceramic package by the die bonding and wired by the wire bonding. This is installed on the breadboard fixed on a shaker (Showa Sokki Model-8100, Japan) at the time of experiment. The excitation frequency can be set with the function generator (Tektronix AFG 3022, USA), and the vibration acceleration can be set arbitrarily with the shaker. A shape of waves and the maximum value of the output voltage obtained by vibration were measured by a digital oscilloscope (Tektronix TDS 2004B, USA).

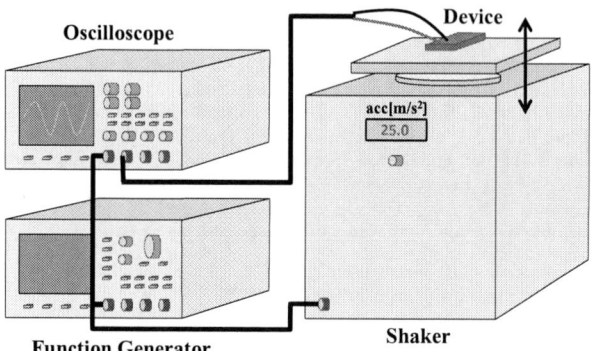

Fig. 4. Design of proposed super harmonic oscillators.

V. EXPERIMENTAL RESULTS AND DISCUSSION

A. Output voltage as a function of excited frequency

Resonance frequency of the fabricated oscillator resulted in 816.5 Hz when this oscillator was excited vibration changing excitation frequency. It is a little bit larger than the numerical result value. In this report, the excitation vibration experiment was done that the acceleration was changed to 5.0, 10.0, and 25.0 m/s^2 as search the relevance of vibration acceleration and output voltage by super-harmonic resonances. And the influence by the value of the excited frequency when vibrational frequency sets to 409 Hz which super-harmonic frequency of order two was investigated by changing acceleration into 5 to 25 m/s^2.

Fig. 5 shows outputting corrugate with the oscilloscope of the output voltage that excitation frequency is (a) resonance frequency: 812 Hz, (b) the half of resonance frequency: 402 Hz, (c) the one third of resonance frequency: 272 Hz and (d) the frequency that is not divisor of resonance frequency: 400 Hz obtained with excitation acceleration 25.0 m/s^2.

The shape of waves shown in (b) and (c) is similar to the shape of wave shown in (c) of Fig. 2. Meanwhile, the shapes of waves don't appear like (d) that the frequency that is not divisor of resonance frequency.

As a result of doing the excitation experiment that is changed the frequency and the acceleration, Fig. 6 shows the relation between the excitation frequency and the output voltage. The output voltage indicates the peak to peak value of the oscilloscope output corrugate of the voltage caused from the oscillator. The maximum output voltage in the resonance frequency 816 Hz in each acceleration was obtained. And, the output voltage has increased in the frequency of the divisor of the resonance frequency in excitation acceleration 25.0 m/s². On the other hand, the output voltage doesn't increase in the frequency since 1/3 of the resonance frequencies excitation acceleration 5.0 m/s² and since 1/4 of the resonance frequencies excitation acceleration 10.0 m/s². Therefore, to enlarge the output voltage by the super-harmonic resonance, it has been understood that the threshold exists that depend on a structure of the oscillator by size of the input excitation vibration.

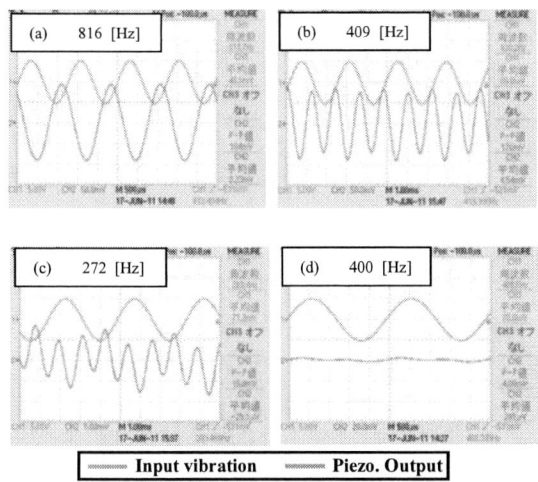

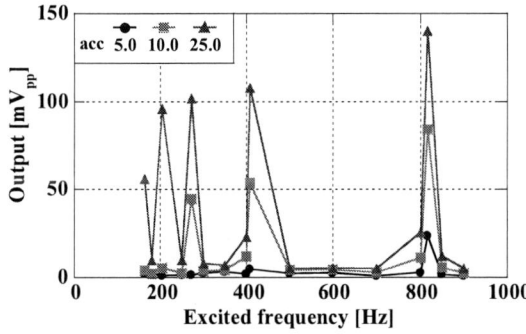

Fig. 5. Waveforms of the output voltage from the PZT cantilever excited at acceleration of 25.0 m/s².

Fig. 6. Output voltage of PZT cantilever at acceleration of 5.0, 10.0, 25.0 m/s².

B. Output voltage as a function of excitation acceleration

Next, excited frequency was set to 409 Hz which super-harmonic frequency of order two, and acceleration was changed into 5 to 25 m/s². Fig. 7 shows outputting corrugate with the oscilloscope of the amplitude of PZT cantilever in

the point that accelerations (a) 5.0 m/s², (b) 8.0 m/s², (c) 9.0 m/s², (d) 10.0 m/s², (e) 15.0 m/s², (f) 25.0 m/s² obtained with excitation frequency 409 Hz. And the output voltage waveform of (a), (b), and (c) is used as the 10 times as many range as (d), (e), and (f) in order to clarify a waveform. Moreover, the displacement the cantilever in the point was investigated by using a laser Doppler vibrometer. It can check that the waveform by super-harmonic frequency of order two is obtained in the output voltage of all the acceleration. Meanwhile, compared with 5 to 9 m/s², in displacement at the cantilever in the point have appeared the super-harmonic frequencies of order two as excitation acceleration the way of 10 to 25 m/s² notably. It seems that the difference of the oscillatory wave type at output voltage and cantilever in the point is based on the measurement method. As for output voltage, modification of the whole cantilever appears as output waveform. However, only the irradiation part of laser can measure a laser Doppler vibrometer as an output waveform. Therefore, we surmised that displacement of the whole cantilever was performing modification like super-harmonic resonance of order two but the cantilever in the point has not carried out modification like super-harmonic resonance of order two.

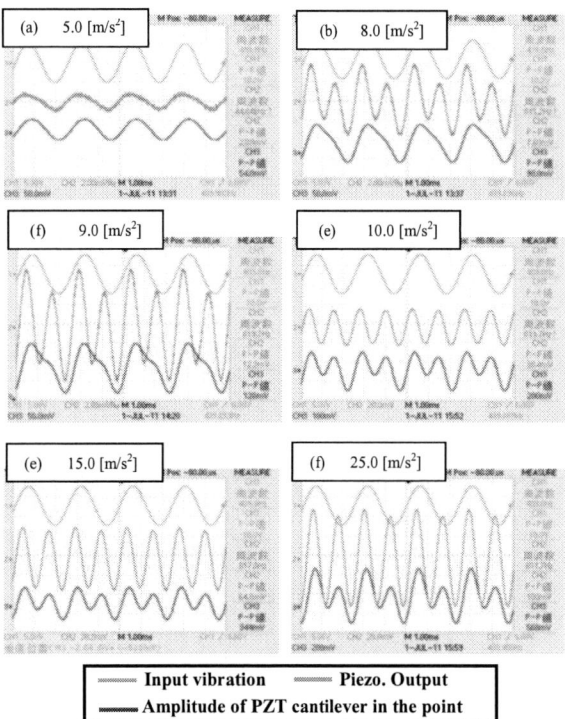

Fig. 7. Waveforms of the output voltage from the PZT cantilever excited at frequency of 409 Hz.

Fig. 8 shows excitation vibration experiment in excited frequency 409 Hz in second super-harmonic frequency changed acceleration 5 to 25 m/s². Output voltage is increase when excited frequency carries out changing 5 to 9 m/s². But output voltage increased rapidly at excited frequency changed 9 to 10 m/s². Output voltage is increase when excited frequency carries out changing 10 to 25 m/s².

Fig. 9 shows amplitude of PZT cantilever in the point as a function of acceleration varying from 5 to 25 m/s² at 409 Hz. When excitation acceleration was increased, the amplitude at the cantilever in the point increased. Moreover, it has checked that amplitude of PZT cantilever in the point increased like the output voltage to the increase in excitation acceleration.

Then, the relation of the output voltage and amplitude of PZT cantilever in the point were obtained in Fig. 10. It was checked that amplitude and output voltage are proportionality relation.

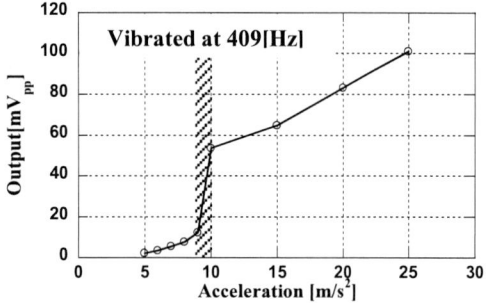

Fig. 8. Output voltage of PZT cantilever as a function of acceleration varying from 5 to 25 m/s² at 409 Hz, showing a sudden transition from 9 to 10 m/s².

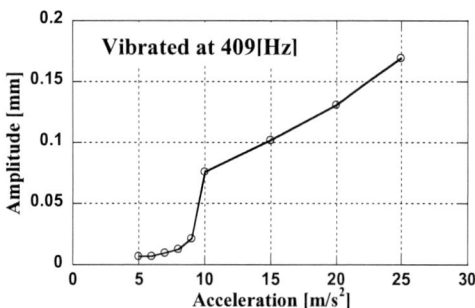

Fig. 9. Amplitude of PZT cantilever in the point as a function of acceleration varying from 5 to 25 m/s² at 409 Hz.

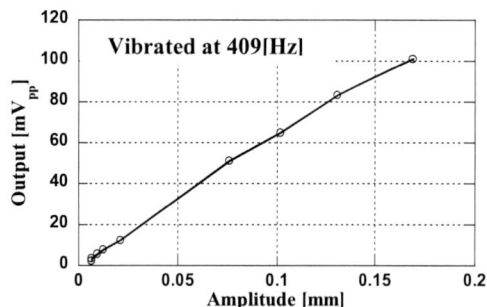

Fig. 10. Output voltage as a function of amplitude of PZT cantilever in the point, showing an approximately linear relationship.

VI. CONCLUSIONS

Using the micro-fabricated oscillators, the super-harmonic resonances were observed and studied from the view points of the corresponding output voltages and amplitudes.

As a result, the corresponding output voltages were found to be greatly increased by the super-harmonic oscillation, and to be approximately proportional to the amplitude changes.

ACKNOWLEDGEMMENTS

Part of this research is granted by the Japan Society for the Promotion of Science (JSPS) through the "Funding Program for World-Leading Innovative R&D on Science and Technology (FIRST Program)," initiated by the Council for Science and Technology Policy (CSTP).

Part of this work was also supported by MEMS Inter University Network and performed in Research Center for Ubiquitous MEMS & Micro Engineering (UMEMSME) of National Institute of Advanced Industrial Science & Technology (AIST).

REFERENCES

[1] K. U. Roscher, W. J. Fischer, J. Landgraf, G. Pfeifer, and E. Starke, Sensor Networks for Integration intoTextile-reinforced Composite", in *Digest Tech. Papers Transducers '07 Conference*, Lyon, June 10-14, 2007, pp. U803-U803.

[2] E. Yoon and K. S. Yun, "Development of a Wireless Environmental Sensor System and MEMS-based RF Circuit Components", in *Digest Tech. PapersTransducers '05 Conference*, Seoul, June 05-09, 1016 2005, pp. 1981-1985.

[3] T. Kobayashi, H. Okada, T. Masuda, and T. Itoh, "A Digital Output Piezoelectric Accelerometer Using Patterned Pb(Zr,Ti)O3 Thin Films Electrically Connected in Series", in *Tech. Digest of MEMS 2009 Conference*, Sorento, Jan. 25-29, 2009, pp. 801-804.

[4] T. Itoh, T. Masuda, and K. Tsukamoto, "Development of a sensor system for animal watching to keep human health and food safety" *Synthesiology*, Vol.3. No.3 pp.231-240 (Aug. 2010).

[5] H. Okada, T. Kobayashi, T. Masuda, and T. Itoh, "Ultra-Low Power Event-Driven Wireless Sensor Node Using Piezoelectric Accelerometer for Health Monitoring", in *JPN. J. APPL. PHYS.*, 48 (2009), PP. 070222 (3PP).

[6] T. Kobayashii, H. Okada, V. Z. Gang, R. Maeda, T. Masuda, and T. Itoh, "Piezoelectric Switch to Active Event-Driven Wireless Sensor Node by Several Hz of Vibration" *IEEE Transducers 2011*, pp.1014-1017 (2011).

[7] A. H. Nayfeh "Quenching of primary resonance by a superhamonic resonance" *Journal of Sound and Vibration* (1984) 92, 363-377.

[8] A. H. Nayfeh and M. I. Younis, "Dynamics of MEMS resonators under superharmonic and subharmonic excitations", *J. Micromech. Microeng.*, 15 (2005), pp. 1840-1847.

[9] D. Shen, J-H. Park, J. Ajitsaria, S-Y. Choe, H. C. Wikle III, and D-J. Kim, "The design, fabrication and evaluation of a MEMS PZT cantilever with an integrated Si proof mass for vibration energy harvesting" *J. Micromech. Microeng*, 18 (2008), pp 055017 (7pp).

High-Q Maintenance of Microcavity by using a Sealed and Packaged Structure

Shu-Bin Yan*[1], Ying-Zhan Yan[2], Yu-Guang Zhang[2], Li-Wang[1], Chen-Yang Xue[1],
Jun Liu[2], Wen-Dong Zhang[1], and Ji-Jun Xiong[1,2],

[1]Key Laboratory of Instrumentation Science & Dynamic Measuremen, North university of China,
Taiyuan, China
[2]Science and Technology on Electronic Test & Measurement Laboratory, North university of China,
Taiyuan, China
(shubin_yan@nuc.edu.cn)

Abstract— **The high-Q maintenance of microcavities greatly challenges the microresonator-based practical application. In this paper, the Q spoiling factors are demonstrated experimentally to show the Q spoiling which originated from the water and the particulate in the surroundings. Then we propose and realize the Q maintenance through constructing a sealed and packaged microcavity regime. In the packaged structure the Q decreases a little but a high Q larger than 10^6 can be achieved continuously. Moreover, the sealed structure has good performance to maintain the high Q for a long time with the standard deviation about 10^4, because the Q spoiling factors are isolated by the package layer. Additionally, the package also enhances the robustness. These merits can promote the practical application of the microcavities.**

Keywords-Microcavity; High-Q maintenance;Package;

I. INTRODUCTION

Whispering gallery mode(WGM) optical microresonators [1] are widely researched for a number of passive and active devices as lters, lasers, sensors and modulators [2] due to their ultra-high quality factor (Q) values, very small volume and good compatibility with traditional fiber optics. In these applications the WGMs, which behave as resonant dips, are every important. On the one hand, the dips reveal the Q through the resonant linewidth. On the other hand, the resonant spectra are the basis of the above applications. For example, in sensing applications [3, 4, 5, 6, 7, 8, 9] the resonant frequency shift is used as the sensing signal normally, and the detecting resolution depends on the resonant linewidth (or the Q) greatly [10]. However, the Q (or the resonant linewidth) can be disturbed easily in practical application, because the actual environment where the microcavity lies is complex and mutable, which can induce extra losses and impact the Q inevitably. This can subsequently cause sensing errors. Therefore, a stable Q with a steady resonant linewidth is the prerequisite for the practical sensors and other devices, indicating the necessity and importance of the high-Q maintenance.

Generally speaking, there are two kinds of the Q spoiling mechanism, the absorption loss caused by the water molecules and the scattering loss induced by the micro-dust adhering to the microcavity [11]. The two Q spoiling factors are great obstacles to the development of the microcavity-based practical devices. Consequently, in practical application how to eliminate these spoiling factors to obtain the stable Q is of great significance for the microcavity-based practical device investigation. In former researches, a hermetic box is used to protect the coupling system from being polluted by the micro-dust in the air [11]. Although the contaminations can be excluded by the hermetic box effectively, the box is bulky and occupies large space, making this manner only appropriate in laboratory experiments but not applicable in practical applications.

In this paper, we propose a novel method to realize the Q-maintenance in the open air. In this method a sealed and packaged microcavity structure is realized, in which the microcavity and the fragile taper are encapsulated wholly to isolate the microcavity-taper coupling system from the surroundings by using low refractive index (RI) ultraviolet (UV) polymer as the encapsulating material. Thus, the Q spoiling factors are eliminated ultimately. This processing method is of great significance to promote the development of the microcavity based practical devices as filters, lasers and sensors.

II. FABRICATION

According to the theoretical model introduced above, the Qtot is unstable and can be spoiled easily by the water and the micro-dust in the open environment. In this paper, we eliminate the Q spoiling factors by encapsulating the microcavity-taper coupling system wholly to fabricate a sealed and packaged structure to isolate the microcavity-taper coupling system from the surroundings. Here, the microspheres are used to illustrate the package process. And the fiber taper coupler is used to excite and probe the microsphere WGMs evanescently. The method introduced here is not just for the microsphere. All other kinds of WGM microresonators (microdisks, microtoioids, and bottle microcavity et.al.) are still applicable.

A. Construction of the coupling system

In this paper, we have fabricated microspheres with D ranging from 180μm to 620μm. The microspheres are fabricated by using thermal melting method. Depending on the surface tension, it is easy to obtain spheres with smooth surface. The tapered fiber is fabricated by using the thermal stretching technique with low insertion loss (less than 5%), in which the fiber is stretched while being heated by oxy-hydrogen flame. A tunable laser (1550nm wavelength band, linewidth < 300kHz) is used to explore the WGMs. The transmission spectra are collected by a photoreceptor and displayed on a digital oscilloscope. The diagrammatic sketch of the experimental setup is shown in Fig.1 (a). 3D X-Y-Z stages with 20nm resolution are used for controlling the air gap between the two parts to adjust the coupling strength. Fig. 1(b) is an experimental photo of the testing system. Fig. 1(c) shows a typical microsphere resonant dip. The dip reveals two pieces of important information, the coupling efficiency about 11db and the high Q about 4.8×10^7. It is worth noting that, during experiments a protective covering is used to protect the coupling system. The following section introduces a novel method to maintain the Q by isolating the coupling system in an encapsulating manner. The encapsulated structure is independent and irrelevant to the surroundings.

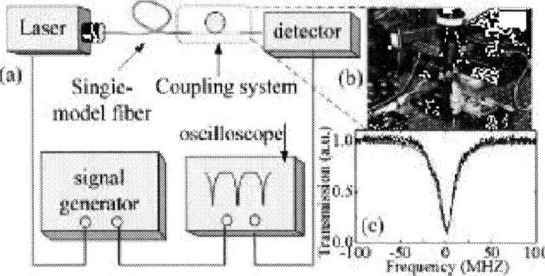

Fig. 1 (a) The diagrammatic sketch of the testing equipment. (b) The photo of the testing system. (c) A typical resonant dip of the microsphere with D about 280μm.

B. Packaging experiment

It is a feasible way to maintain the Q in practical applications by isolating the microsphere-taper coupling system from the outside. A sealed and packaged structure in which the microsphere-taper coupling system is embedded deeply and isolated wholly from the surroundings can be realized in a capsulation manner by using the low RI (1.35) solidifiable UV polymer. Two important factors should be considered when selecting the polymer. One is that the RI must be lower than that of the microsphere, because low RI makes the optical energy confined in microresonators. The other is that the absorption coefficient of the polymer at the working wavelength should be as low as possible, because the absorption loss induced by the packaged body can decrease the loaded Q.

The packaged structure has been introduced for the first time in our former researches [15]. Here we improve the packaging technology to realize a sealed and packaged structure, in which the microsphere and the taper are wholly encapsulated to realize an integrated bulk, as shown in Fig.

2(a). Briefly, there are five steps to perform the package. First, we need to obtain an effective coupling between the microsphere and the taper in the air before the package, as shown in Fig. 2(b1). The effective coupling is confirmed through the resonant dips with a large coupling strength and a narrow linewidth. Then, as illustrated in Fig. 2(b2), the UV polymer is coated on the coupling system in a dropping manner. Afterward, as shown in Fig. 2(b3) an ultraviolet lamp is used to irradiate the capsulated structure to solidify the UV polymer. The microsphere-taper coupling system here is capsulated by the solidified UV polymer. In the fourth step, the microsphere stem, mounted on the 3D stages, is truncated by using a heat burning manner. Then the microsphere coupling system is independent of the 3D stages and can be moved freely, as shown in Fig. 2(b4) and Fig. 2(c1). Finally, similar to the potted circuit module we further package the structure by using a designed slot as the mould to package the fragile taper totally. Fig. 2(c2) shows the semi-finished products of the sealed and packaged microcavity unit. Fig. 2(c3) shows a typical sealed and packaged module. In this module the microcavity, its coupling system and the fragile taper are all solidified into an entire body. Moreover, the microsphere-taper coupling system is embedded deeply in the sealed packaged structure. And the thinnest layer of the package is tested to be 2500μm.

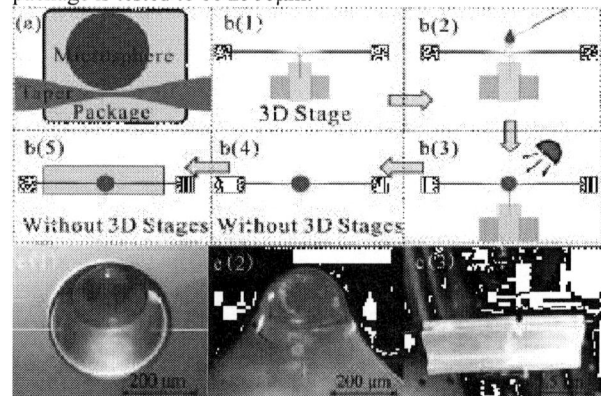

Figure 2. (a) The schematic diagram of the sealed and packaged structure. b(1)-b(5) Illustration of the package process. c(1) The micrograph of the free microsphere coupling system after the forth step of the package. c(2) The micrograph of the semi-finished products of the packaged microcavity unit. c(3) A typical sealed and packaged microsphere-taper coupling system.

III. TEST AND RESULTS

It is particularly important to maintain the Q because it can be spoiled easily by the water and the dust. Here, different mal-conditions have been established to illustrate this spoiling effect. Firstly, we test the Q in the clean air with different relative humidity (RH: 90%, 60%, 40%, 20%) at the room-temperature. Fig. 3(a) shows a typical time-dependence of Q, with the time = 0 point corresponding to the fabrication of the microsphere (its removal from the flame after the sphere formation by surface tension forces). The Q-versus-t plot indicates quick decay of the record Q within the first 13 mins toward 35% of the first Q and slower saturation toward Q≈4 × 10^7. The results also demonstrate that the Q shows a sharper decline in a more humid condition. This phenomenon is mainly due to a faster formation of the adsorbed water monolayer on the microsphere surface in a condition with high RH. In this experiment, one microsphere is used to test its Q in the four conditions, and a bakeout method is used to restore the Q.

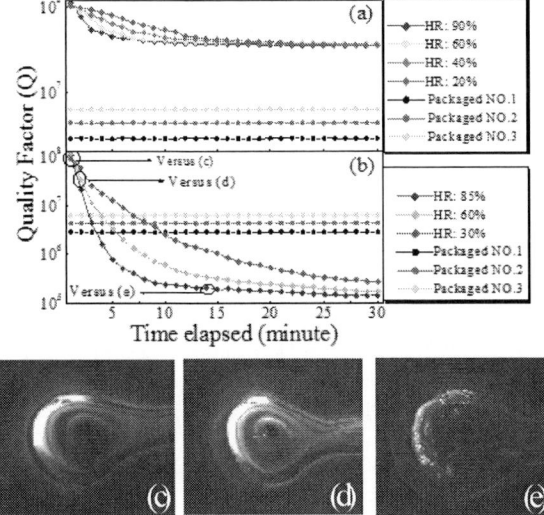

Figure 3. Tested and fitted Q versus D, for untreated and packaged microspheres, respectively. (c), (d) and (e) are the microsphere surface at different time.

We further simulate a more complex environment by adding the particulate into the humid air. The particulate we use is aerogel with diameter ranging from 200nm to 1200nm with the quantitative density about 800 per cubic centimeter. As shown in Fig. 3(b), the recorded Q shows a sharp decline with the time elapsed. Especially in the first 15 minutes, the recorded Q is reduced by about 2 ∼ 3 orders of magnitude. And the descending depends on the HR. Experiment results indicate that the microsphere in the higher HR condition shows a faster descending speed. This is mainly because the water molecules can help the particulate to adhere to the microsphere surface in the more humid condition. We also estimate that the Q will behave as a much more drastic reduction in a more formidable environment with higher concentration dust and higher RH. The further reduction of the Q here results from the anabatic microsphere surface roughness. Fig. 3(c), (d) and (e) show the microsphere surface in different time (as marked in Fig. 3). As is clearly shown, the surface becomes dirtier and dirtier. It can be concluded that the microsphere surface can be broken easily by the complex environment which subsequently spoils the Q drastically.

Under the same experimental conditions as mentioned above, we also test the Q of the sealed and packaged

microspheres. As shown in Fig. 3, the round spot and their lines show the tested Q of three samples. The green, red and black dots are for the microspheres with D about 485μm, 350μm and 270μm, respectively. The results reveal that, the Q of the sealed and packaged microsphere structures are independent from the outside surroundings, and shows great stability. The three packaged structure respectively have an average Q of about 6.25 × 10^6, 4.35 × 10^6, 2.86 × 10^6, with standard deviation (SD) about 1.5811 × 10^4, 1.3663 × 10^4, 1.4944 × 10^4, respectively. In the above experiment, the SD for the testing in humid condition is about 1.5199 × 10^7, while in the particulate condition is about 1.9723 × 10^7. The S D of the tested Q is improved about 3 orders of magnitudes. The Q maintenance is mainly because the package functions as a protecting layer to eliminate the Q spoiling factors introduced above. In fact, we have maintained the Q with the low S D for a few months.

IV. Conclusion

This paper propose a novel method to maintain the Q of the microspheres by constructing a sealed and packaged microcavity structure to eliminate the Q spoiling factors in the open air. Although there is a reduction of the Q due to the radiation and the absorption loss, the packaged structure has excellent performance to maintain the high Q greater than 10^6 with the SD about 10^4. In other words, we realize the Q−maintenance at the expense of a portion of the Q value through the sealed and packaged structure. Moreover, the package also enhances the robustness. These merits can promote the practical applications of the microcavities.

Acknowledgment

The authors greatly appreciate the helpful discussion with Chang-Ling Zou in University of Science and Technology of China. The work is supported by the National Basic Research Program of China under Grant No. 2009CB326206 and the Innovation Project under Grant Nos. 7130907 and 9140C1204040706.

References

[1] K. J. Vahala, Nature. 424 (2003) 839.

[2] V. S. Ilchenko and A. B. Matsko, IEEE J. Sel. Top. Quantum Electron. 12 (2006) 15.

[3] F. Vollmer, S. Arnold, Nat. Methods. 5 (2008) 591.

[4] F. Vollmer, D. Braun, A. Libchaber, M. Khoshsima, I. Teraoka, and S. Arnold, Appl. Phys. Lett. 80 (2002) 4057.

[5] H. C. Ren, F. Vollmer, S. Arnold and A. Libchaber, Opt. Express. 15 (2007) 17410.

[6] J. T. Gohringa, P. S. Daleb, X. Fan, Sens. Actuators, B. 146 (2010) 226.

[7] M. Sumetsky, R. S. Windeler, Y. Dulashko, and X. Fan, Opt. Express. 15 (2007) 14376.

[8] J. Zhu, S. K. Ozdemir, Y. F. Xiao, L. Li, L. He, D. R Chen, L. Yang, Nat. Photonics. 4 (2010) 46.

[9] T. Lu, H. Leea, T. Chena, S. Herchakb, J. H. Kima, S. E. Frasera, R. C. Flaganb, and K. Vahalaa, PNAS. 108 (2011) 1017962.

[10] I. M. White and X. Fan, Opt. Express. 16 (2008) 1020.

[11] M. L. Gorodetsky, A. A. Savchenkov, and V. S. Ilchenko, Opt. Lett. 21 (1996) 453.

[12] M. Cai, O. Painter, and K. J. Vahala, Phys. Rev. lett. 85 (2000) 74.

[13] D. W. Vernooy, V. S. Ilchenko, H. Mabuchi, E. W. Streed, and H. J. Kimble, Opt. Lett. 23 (1998) 247.

[14] X. Zhang, H. S. Choi, and A. M. Armani, Appl. Phys. Lett. 96 (2010) 153304.

[15] Y.-Z. Yan, Z. Ji, S.-B. Yan, J. Liu, C.-Y. Xue, W.-D. Zhang, J.-J. Xiong, Chin. Phys. Lett. 28 (2011) 034208.

[16] A. M. Armani, D. K. Armani, B. Min, K. J. Vahala, and S. M. Spillane, Appl. Phys. Lett. 87, (2005) 151118.

[17] Y.-Z. Yan, C.-L. Zou, S.-B. Yan, F.-W. Sun, Z. Ji, J. Liu, Y.-G. Zhang, L. Wang, C.-Y. Xue, W.-D. Zhang, Z.-F. Han, and J.-J. Xiong, Opt.Express. 19 (2011) 5753.

On the Effect of Width of Metallic Armchair Graphene Nanoribbons in Plasmonic Waveguide Applications

Sami Smaili, Vikas Singal, and Yehia Massoud

Electrical and Computer Engineering Department,University of Alabama at Birmingham, Birmingham, Al 35294

Abstract—Graphene demonstrates superior electronic properties that make it a potential candidate for future electronic systems. Graphene, additionally, support surface plasmon oscillations, which in turn makes graphene attractive for optoelectronics because of its planar structure and its conductivity properties. When a graphene layer is confined in one dimension, a graphene nanoribbon arises, with proerties differing from the original two dimensional graphene. In this paper we study the main properties of plasmon oscillations on metallic armchair graphene nanoribbons using the dielectric function obtained through the random phase approximation. We mainly study the effect of the graphene nanoribbon width on the plasmon propagation length using numerical techniques to extract the dispersion relation of graphene nanoribbons and the propagation properties of palsmons on graphene nanoribbons.

I. INTRODUCTION

The excellent electrical properties of graphene allows this material to play an important role in a wide range of applications such as high density interconnects for VLSI systems, new field effect transistors, and THz devices [1]. Graphene based electronics are suggested for applications in analog and RF systems to enhance the performance of such systems and to mitigate the limitations that faces current technologies [2]–[27] . The planar nature of graphene makes graphene based devices extremely small. Thus, while allowing high levels of miniaturization, graphene has excellent electronic properties that allow the realization of high performance electronic and optoelectronic devices.

Plasmonic devices, in general, are characterized mainly by their ability to manipulate light at sub-wavelength scales [28]–[34]. Combining the excellent electronic properties of graphene with the power of plasmonics promise to provide novel solutions for a wide range of applications. For instance, graphene plasmonic devices promise to play an important role in future VLSI systems to mitigate the effects of their down scaling using current technologiessystems [26], [35]–[78].

Graphene consists of a sheet of honeycomb lattice of carbon atoms in which the electrons can be treated as a two dimensional electron gas (2DEG). Graphene support plasmon oscillations (the oscillations of charge density) at THz frequencies, and thus can be used to build nanoscale THz waveguides. Plasmonic devices that utilize the plasmon oscillation on metal-dielectric interfaces have been used for subwavelength waveguiding of optical waves [1].

Graphene Nanoribbons (GNRs) consist of graphene sheets with finite width. The confined dimension makes the properties of GNRs different from those of inifinite graphene layers [79]. The type of GNR depends on the way the graphene sheet is cut, and two types are mostly famous: armchair GNRs and zigzag GNRs. Zigzag GNRs do not support plasmon oscillations because those are damped by surface states. Armchair GNRs can be either metallic or insulating depending on their width. We consider in this paper metallic armchair GNRs.

978-1-4673-1122-9/12 $31.00 © 2012 IEEE

II. GRAPHENE NANORIBBON DIELECTRIC FUNCTION

We assume in this paper that the GNR is confined in the x direction and charge density oscillations occur in the y direction. The confinement of the two dimensional graphene in one direction results in the quantization of energy states, with energies [79]

$$E_{ns}(k_y) = sa_0\gamma\sqrt{k_n^2 + k_y^2} \quad (1)$$

where a_0 is the graphene lattice constant, $s = -1$ for a valence band and $s = 1$ for a conduction band. The index n labels the subband and k_n is given by

$$k_n = \frac{2\pi n}{2W + a_0} + \frac{2\pi}{3a_0} \quad (2)$$

where W is the GNR width. The dielectric function can be described by means of a tensor whose entries are given as [79]

$$\epsilon_{ijmn}(q,\omega) = \delta_{im}\delta_{jn} - v_{ijmn}(q)\Pi_{m,n}(q,\omega) \quad (3)$$

The matrix elements of the coulomb interaction are

$$v_{ijmn} = \int_0^1 du \int_0^1 du' cos[\pi(i-j)u]$$
$$cos[\pi(m-n)u']\frac{2e^2}{\epsilon_o}K_o(qW|u-u'|) \quad (4)$$

where e is the electronic charge, ϵ_o is the average of the dielectric constants of the media surrounding the GNR, and K_o is the zeroth-order modified Bessel function of the second kind. The dielectric function of GNRs can be approximated by

$$\epsilon = 1 - v_{0,0}\sum_n \chi_n \quad (5)$$

where

$$\chi_n = \frac{g_s}{L_y}\sum_{k_y} \frac{2\Delta_n(k_y,q)}{(\hbar\omega + i\delta)^2 - (\Delta_n(k_y,q))^2}$$
$$F_{n^-,n^+}(k_y, k_y + q) \quad (6)$$

where g_s is the spin degeneracy (2), $\Delta_n(k_y,q) = |E_n(k_y + q)| + |E_n(k_y)|$ and $F_{n^-,n^+}(k_y, k_y + q) = \frac{1}{2}(1 - cos\theta)$. We note here that the sum over wavevectors is in fact an integral, and the δ is added for the integration to convergence. The integration is evaluated as $\delta \to 0$ using the Sokhatsky-Weierstrass

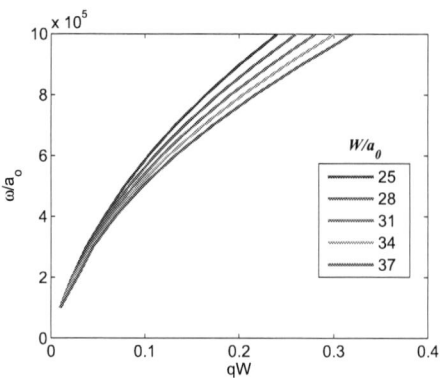

Fig. 1. Dispersion relation for an armchair GNR with various widths.

theorem to get the imaginary part and the Kramers-Kronig relations to obtain the real part of the integral.

We used complex frequencies to account for scattering by two main mechanisms, scattering of electrons in graphene and scattering due to edge roughness. For the electron scattering, we used a scattering time of $0.5ps$ while for the edge roughness we used [80]

$$\tau(E) = \frac{\hbar W^2}{\pi E_n^2 H^2}\frac{1 + 4k_y^2\Lambda^2}{\Lambda}\frac{1}{\rho(E)(1 + cos\theta_{kk'})} \quad (7)$$

where $\rho(E)$ is the density of states, H is the amplitude of edge roughness, and Λ is the correlation length of roughness. The dispersion relation is obtained by setting the determinent of the tensor in 5, and is shown in figure 1 for an armchair GNR with various widths. The dispersion curve moves to the right as the GNR width increases. This means that the plasmon wavevector at a given frequency increases as the width of the GNR increases. Figure 2 shows the propagation length of an armchair GNR for various GNR widths. The propagation length decreases as the width of the GNR increases. Typically, there is a tradeoff between propagation length and confinement in the waveguide. This behavior is shown in figure 3, which shows the imaginary part of the wavevector perpendiculat to the graphene sheet versus frequency for a GNR with various widths. This quantity measures for the decay of fields away from the GNR (z direction); the higher $Imag(k_z)$, the more confined is the wave on the GNR surface. Figure 3 shows that as the GNR

978-1-4673-1122-9/12 $31.00 © 2012 IEEE

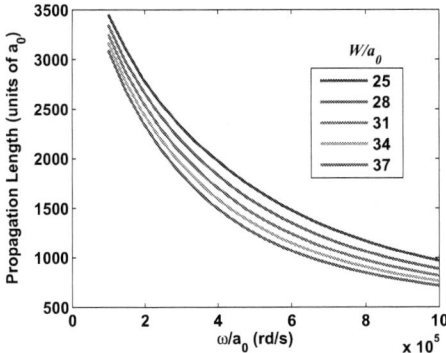

Fig. 2. Propagation length of an armchair GNR vs. frequency for various GNR widths

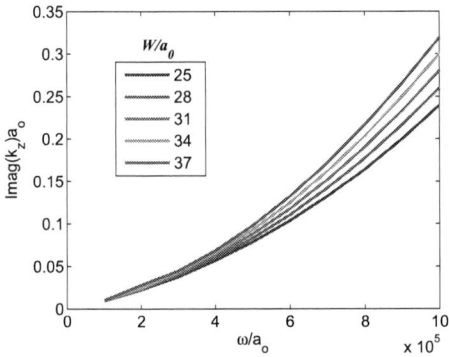

Fig. 3. Imaginary part of the wavevector perpendiculat to the graphene sheet versus frequency.

width increases, the confinement of the wave on the GNR surface increases. Thus, as the GNR width increases, the propagation length decreases but the confinement increases.

III. CONCLUSION

Graphene plasmonics is an emerging field that combines the power of plasmon oscillations with the excellent properties of graphene. In this paper we investigated the propagation length of plasmons on graphene nanoribbons, which is one of the most important properties of waveguiding structures. The results show that GNRs support propagation lengths much alrger than the dimensions of the GNR.

REFERENCES

[1] A. Geim and K. Novoselov, "The Rise of Graphene," *Nature Materials*, vol. 6, 2007.
[2] A. Nieuwoudt and Y. Massoud, "Variability-aware multilevel integrated spiral inductor synthesis," *IEEE TCAD*, Dec. 2006.
[3] ——, "Multi-level Approach for Integrated Spiral Inductor Optimization," *IEEE DAC*, 2005.
[4] A. Nieuwoudt, T. Ragheb, and Y. Massoud, "SOC-NLNA: synthesis and optimization for fully integrated narrow-band CMOS low noise amplifiers," *IEEE DAC*, 2006.
[5] A. Nieuwoudt and Y. Massoud, "Robust automated synthesis methodology for integrated spiral inductors with variability," *IEEE ICCAD*, 2005.
[6] A. Nieuwoudt, M. McCorquodale, R. Borno, and Y. Massoud, "Efficient analytical modeling techniques for rapid integrated spiral inductor prototyping," *IEEE CICC*, 2005.
[7] T. Ragheb, S. Kirolos, J. Laska, M. Strauss, A. Gilbert, R. Baraniuk, and Y. Massoud, "Implementation models for analog-to-information conversion via random sampling," *IEEE MWSCAS*, 2007.
[8] A. Nieuwoudt, M. McCorquodale, R. Borno, and Y. Massoud, "Accurate analytical spiral inductor modeling techniques for efficient design space exploration," *IEEE EDL*, December 2006.
[9] A. Nieuwoudt and Y. Massoud, "Analytical wide-band modeling of high frequency resistance in integrated spiral inductors," *Analog Integrated Circuits and Signal Processing*, February 2007.
[10] S. Kirolos, T. Ragheb, J. Laska, M. Duarte, Y. Massoud, and R. Baraniuk, "Practical Issues in Implementing Analog-to-Information Converters," *IEEE IWSOC*, 2006.
[11] A. Nieuwoudt and Y. Massoud, "Efficient modeling of substrate eddy currents for integrated spiral inductor design automation," *IEEE MWSCAS*, 2005.
[12] T. Ragheb, A. Nieuwoudt, and Y. Massoud, "Efficient modeling of integrated narrow-band low noise amplifiers for design space exploration," *IEEE GLSVLSI*, 2006.
[13] A. Nieuwoudt, T. Ragheb, and Y. Massoud, "Hierarchical optimization methodology for wideband low noise amplifiers," *IEEE ASP-DAC*, 2007.
[14] T. Ragheb, J. N. Laska, H. Nejati, S. Kirolos, R. G. Baraniuk, and Y. Massoud, "A prototype hardware for random demodulation based compressive analog-to-digital conversion," *IEEE MWSCAS*, 2008.
[15] A. Nieuwoudt, T. Ragheb, H. Nejati, and Y. Massoud, "Increasing manufacturing yield for wideband RF CMOS LNAs in the presence of process variations," *IEEE ISQED*, 2007.
[16] ——, "Numerical design optimization methodology for wideband and multi-band inductively degenerated cascode CMOS low noise amplifiers," *IEEE TCAS I*, June 2009.
[17] T. Ragheb, A. Nieuwoudt, and Y. Massoud, "Modeling of 3.1-10.6 GHz CMOS Filter-Based Low Noise Amplifier for Ultra-Wideband Receivers," *IEEE WAMICON*, 2006.
[18] H. Najati, T. Ragheb, A. Nieuwoudt, and Y. Massoud, "Modeling and design of ultrawideband low noise amplifiers with generalized impedance matching networks," *IEEE ISCAS*, 2007.
[19] A. Nieuwoudt and Y. Massoud, "Optimizing the Design of Tunable Spiral Inductors for On-Chip Wireless Applications," *IEEE WAMICON*, 2006.
[20] H. Nejati, T. Ragheb, A. Nieuwoudt, and Y. Massoud, "Analytical modeling methodology for ultrawideband low noise amplifiers with generalized filter-based impedance matching," *Analog Integrated Circuits and Signal Processing*, May 2007.
[21] A. Nieuwoudt, T. Ragheb, and Y. Massoud, "Systematic Design Optimization Methodology for Multi-Band CMOS Low Noise Amplifiers," *IEEE ISVLSI*, 2007.
[22] H. Nejati, T. Ragheb, and Y. Massoud, "On the design of customizable low-voltage common-gate LNA-mixer pair using current and charge reusing techniques," *IEEE GLSVLSI*, 2008.
[23] S. Pfetsch, T. Ragheb, J. Laska, H. Nejati, A. Gilbert, M. Strauss, R. Baraniuk, and Y. Massoud, "On the Feasibility of Hardware Implementation of Sub-Nyquist Random-Sampling Based Analog-to-Information Conversion," *IEEE ISCAS*, 2008.
[24] Y. Massoud, A. Nieuwoudt, and T. Ragheb, "Automated design solutions for fully integrated narrow-band low noise amplifiers," *IEEE IWSOC*, 2006.
[25] A. Nieuwoudt and Y. Massoud, "Design optimization of switchable multi-port spiral inductors," *Springer ALOG*, June 2007.
[26] M. A. Y. Massoud and A. Nieuwoudt, "On the Selection of Spectral Zeros for Generating Passive Reduced Order Models," *IEEE IWSOC*, 2006.

[27] H. Nejati, T. Ragheb, and Y. Massoud, "Analytical Modeling of Common-Gate Low Noise Amplifiers," *IEEE MWSCAS*, 2008.

[28] A. Hossieni, A. Nieuwoudt, and Y. Massoud, "Efficient simulation of subwavelength plasmonic waveguides using implicitly restarted Arnoldi," *Optics Express*, August 2006.

[29] M. Alam and Y. Massoud, "A closed-form analytical model for single nanoshells," *IEEE TNANO*, May 2006.

[30] ——, "RLC ladder model for scattering in single metallic nanoparticles," *IEEE MTT-S*, 2006.

[31] A. Hossieni, H. Nejati, and Y. Massoud, "Design of a maximally flat optical low pass filter using plasmonic nanostrip waveguides," *Optics Express*, November 2007.

[32] M. Alam and Y. Massoud, "An accurate closed-form analytical model of single nanoshells for cancer treatment," *IEEE MWSCAS*, 2005.

[33] A. Hossieni and Y. Massoud, "Optical range microcavities and filters using multiple dielectric layers in metal-insulator-metal structures," *Journal of the Optical Society of America A*, January 2007.

[34] M. Alam and Y. Massoud, "An RLC ladder model for the equivalent impedance of single metal nanoparticles in electromagnetic field," *IEEE Transactions on Nanotechnology (TNANO)*, September 2006.

[35] Y. Massoud, S. Majors, T. Bustami, and J. Whites, " techniques for minimizing on-chip interconnect self-inductance," *IEEE DAC*, 1998.

[36] D. M. Y. Massoud, J. Kawa and J. White, "Modeling and Analysis of Differential Signaling for Minimizing Inductive Cross-Talk," *IEEE DAC*, 2001.

[37] A. Nieuwoudt and Y. Massoud, "Evaluating the Impact of Resistance in Carbon Nanotube Bundles for VLSI Interconnect using Diameter-Dependent Modeling Techniques," *IEEE TED*, October 2006.

[38] "Modeling and Design Challenges and Solutions for Carbon Nanotube-Based Interconnect in Future High Performance Integrated Circuis," *ACM JETC*, July 2006.

[39] Y. Massoud and J. White, "Managing On-Chip Inductive Effects," *IEEE TVLSI*, Dec. 2002.

[40] A. Nieuwoudt and Y. Massoud, "Understanding the impact of inductance in carbon nanotube bundles for VLSI interconnect using scalable modeling techniques," *IEEE TNANO)*, Nov. 2006.

[41] Y. Massoud and Y. Ismail, "Grasping the Impact of On-Chip Inductance in High Speed ICs," *IEEE Circuits and Devices Magazine*, July 2001.

[42] A. Nieuwoudt and Y. Massoud, "On the impact of process variations for carbon nanotube bundles for VLSI interconnect," *IEEE TED*, March 2007.

[43] ——, "Assessing the implications of process variations on future carbon nanotube bundle interconnect solutions," *IEEE ISQED*, 2007.

[44] M. Mondal, A. Ricketts, S. Kirolos, T. Ragheb, V. N. G. Link, and Y. Massoud, "Thermally robust clocking schemes for 3D integrated circuits," *IEEE DATE*, 2006.

[45] M. Mondal and Y. Massoud, "Reducing pessimism in RLC delay estimation using an accurate analytical frequency dependent model for inductance," *IEEE ICCAD*, November 2005.

[46] S. Kim, Y. Massoud, , and S. Wong, "On the Accuracy of Return Path Assumption for Loop Inductance Extraction for 0.1 um Technology and Beyond," *IEEE ISQED*, 2003.

[47] S. Eachempati, A. Nieuwoudt, A. Gayasen, V. Narayanan, and Y. Massoud, "Assessing carbon nanotube bundle interconnect for future FPGA architectures," *IEEE DATE*, 2007.

[48] Y. Massoud and J. White, "Simulation and Modeling of the Effect Substrate Conductivity on Coupling Inductance and Circuit Crosstalk," *IEEE TVLSI*, June 2002.

[49] A. Nieuwoudt and Y. Massoud, "On the optimal design, performance, and reliability of future carbon nanotube-based interconnect solutions," *IEEE TED*, August 2008.

[50] ——, "Accurate resistance modeling for carbon nanotube bundles in VLSI interconnect," *IEEE Nano*, 2006.

[51] A. Nieuwoudt, M. Mondal, and Y. Massoud, "Predicting the Performance and Reliability of Carbon Nanotube Bundles for On-Chip Interconnect," *IEEE ASP-DAC*, 2007.

[52] A. Nieuwoudt and Y. Massoud, "Performance implications of inductive effects for carbon-nanotube bundle interconnect," *IEEE EDL*, April 2007.

[53] A.Nieuwoudt and Y.Massoud, "Predicting the performance of low-loss on-chip inductors realized using carbon nanotube bundles," *IEEE TED*, January 2008.

[54] M. Mondal, T. Ragheb, X. Wu, A. Aziz, and Y. Massoud, "Provisioning on-chip networks under buffered rc interconnect delay variations," *ISQED*, 2007.

[55] A. Hosseini, T. Ragheb, and Y. Massoud, "A fault-aware dynamic routing algorithm for on-chip networks," *IEEE ISCAS*, 2008.

[56] Y. Massoud and J. White, "Improving the generality of the fictitious magnetic charge approach to computing inductances in the presence of permeable materials," *IEEE DAC*, 2002.

[57] A. Nieuwoudt and Y. Massoud, "RC circuit model for multi-walled carbon nanotubes," *IEEE Nano*, 2007.

[58] Q. Su, J. Kawa, C. Chiang, and Y. Massoud, "Accurate modeling of substrate resistive coupling for floating substrates," *ACM TODAES*, January 2006.

[59] M. Mondal, A. Ricketts, S. Kirolos, T. Ragheb, G. Link, V. Narayanan, and Y. Massoud, "Mitigating thermal effects on clock skew with dynamically adaptive drivers," *IEEE ISQED*, 2007.

[60] M. Alam, A. Nieuwoudt, and Y. Massoud, "Wavelet-based passivity preserving model order reduction for wideband interconnect characterization," *IEEE ISQED*, 2007.

[61] M. Mondal, K. Mohanram, and Y. Massoud, "Parameter-Variation-Aware Analysis for Noise Robustness," *IEEE ISQED*, 2007.

[62] Y. Massoud, J. Wang, and J. White, "Accurate inductance extraction with permeable materials using qualocation," *International Conf. on Modeling and Simulation of Microsystems*, 1999.

[63] S. Kirolos and Y. Massoud, "Adaptive Ratio-Size Gates for Minimum-Energy Operation," *IEEE TCAS II*, November 2006.

[64] S. Kirolos, Y. Massoud, and Y. Ismail, "Power-supply-variation-aware timing analysis of synchronous systems," *IEEE ISCAS*, 2008.

[65] ——, "Accurate analytical delay modeling of cmos clock buffers considering power supply variations," *IEEE ISCAS*, 2008.

[66] S. Kirolos and Y. Massoud, "Adaptive SRAM Design for Dynamic Voltage Scaling VLSI Systems," *IEEE MWSCAS*, 2007.

[67] X. Wu, T. Ragheb, A. Aziz, and Y. Massoud, "Implementing DSP algorithms with on-chip networks," *IEEE NOC*, 2007.

[68] M. Alam, A. Nieuwoudt, and Y. Massoud, "Model order reduction using spline-based dynamic multi-point rational interpolation for passive circuits," *Analog Integrated Circuits and Signal Processing*, March 2007.

[69] M. Mondal and Y. Massoud, "Analytical Modeling of Loop Self Inductance Bound for Inductance-Aware Physical Synthesis," *IEEE ISCAS*, 2005.

[70] Y. Massoud and J. White, "Fast Inductance Extraction of 3-D Structures with Non-ConstantPermeabilities," *International Conference on Modeling and Simulation of Microsystems*, pp. 190–193, April 1998.

[71] M. Mondal and Y. Massoud, "Estimation of capacitive crosstalk-induced short-circuit energy," *JCSC*, June 2007.

[72] Y. Massoud and J. White, "FastMag: a 3-D Fast Inductance Extraction Program for Structures with Permeable materials," *IEEE ICCAD*, 2002.

[73] S. Kirolos and Y. Massoud, "Supply Voltage Adaptive Low-Power Circuit Design," *IEEE DCAS*, 2006.

[74] . M. Modal and Y. Massoud, "Accurate loop self inductance bound for efficient inductance screening," *IEEE TVLSI*, December 2006.

[75] S. Kirolos and Y. Massoud, "Robust Wide Range of Supply-Voltage Operation using Continuous Adaptive Size-Ratio Gates," *IEEE ISCAS*, 2008.

[76] Y. Massoud, S. Kirolos, and K. Mohanram, "Analytical model-based technique for efficient evaluation of noise robustness considering parameter variations," *Analog Integrated Circuits and Signal Processing*, August 2009.

[77] S. Kirolos, M. Mondal, K. Mohanram, and Y. Massoud, "A Model-Based Technique for Efficient Evaluation of Noise Robustness," *IEEE ISCAS*, 2008.

[78] S. Kirolos, Y. Massoud, and Y. Ismail, "Mitigating Power-Supply Induced Delay Variations using Self Adjusting Clock Buffers," *IEEE MWSCAS*, 2008.

[79] L. Brey and H. Fertig, "Elementary electronic excitations in graphene nanoribbons," *Phys. Rev. B*, vol. 75, 2007.

[80] T. Fang, A. Konar, H. Xing, and D. Jena, "Mobility in semiconducting graphene nanoribbons: Phonon, impurity, and edge roughness scattering," *Phys. Rev. B*, vol. 78, 2008.

978-1-4673-1122-9/12 $31.00 © 2012 IEEE

The Facile Transferral of Graphene onto Interdigitated Electrodes for Sensing Applications

C. Dale*, S. Rana, R. H. Page, J. Hedley, N. Keegan

* *School of Mechanical and Systems Engineering, Newcastle University, UK*
* carl.dale@newcastle.ac.uk

Abstract- **This research uses the facile transferral of graphene onto gold interdigitated electrodes (IDEs) such that the adjacent digits of the electrode array form contacts to graphene flakes. Successful transferral is measured using environmental scanning electron microscopy (ESEM) and Raman spectroscopy. Following successful transferral the graphene has been employed as a sensing material in regard to its electrical characteristics. The presence of pyrene butyric acid (PBA) between 5 µM and 50 nM can be measured as a function of graphene electrical resistance. The initial results presented indicate a resistance change of 770 mΩ at saturation. Measurement variability is attributed to the reproducibility of the graphene which is currently being assessed.**

Keywords-Graphene transferral; interdigitated electrodes; biosensor

I. INTRODUCTION

The material graphene is a 2-dimensional allotrope of carbon that exhibits extraordinary physical and electrical properties [1]. The one atom thick wonder material has quickly become a star attraction in condensed matter physics and material science, but its unique properties (strength, dimensions, electron transfer) have also catapulted it to the forefront of research across multiple disciplines and fields [2]. One disparate field is the construction of novel biological sensors. The vast majority of publications have focused on utilizing graphene as a composite material to enhance the electrical properties of existing electrodes [3,4].

Graphene is usually available in the form of flakes, which are, in most cases, less than 20 µm in diameter and produced by mechanical exfoliation. Excellent research is being carried out to grow graphene over extended areas on numerous substrates [5,6] but until this reaches maturity the most readily available material remains graphene flakes and graphene grown on copper foil that has to undergo transferral on to an appropriate substrate. One of the current challenges in graphene research is creating electrical contacts using the existing material. In this work, we utilize an existing IDE geometry to construct a simple graphene based sensor that modulates the electrical resistance of graphene as material is immobilized at the interface. The IDEs were fabricated on silica substrates with 2 µm digit spacing and large contact pads for wire bonding; the IDE digits and bond pads are displayed in Fig. 1. The transfer of epitaxially grown graphene

from copper foil onto the IDEs was performed using the thermal release tape method [7]. To confirm that graphene had been transferred onto the IDEs Raman spectroscopy was utilized to assess the quality of the graphene and environmental scanning electron microscopy (ESEM) was employed to observe graphene placement. The electrical property of the transferred graphene was measured on the IDEs and gave an average resistance value of 18 ±1 Ω. Finally the graphene bridges were functionalized with PBA which contains a four carbon ring structure that interacts strongly with graphene via non-covalent pi stacking. The PBA molecule also includes a terminal carboxyl functional group that can be employed to chemically crosslink to biological molecules. The functionalization of graphene with PBA was monitored in real time using the change in the resistance at the graphene bridges.

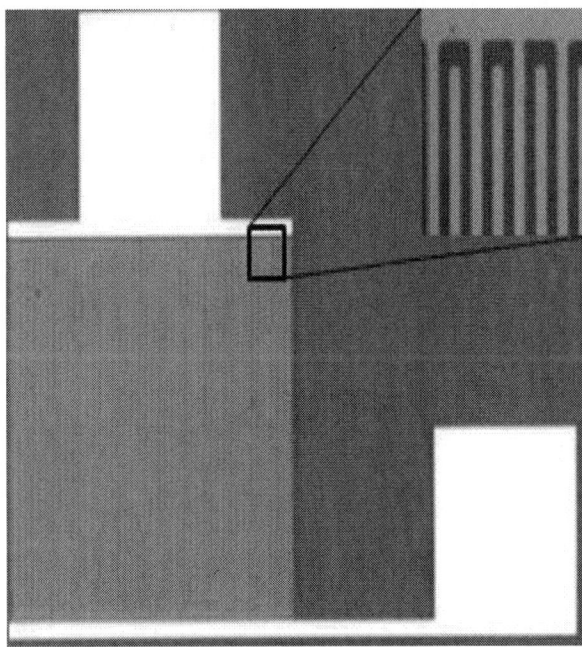

Fig. 1. A light microscope image of a set of interdigitated electrodes. The upper inset shows a high magnification image of interdigitated digits.

978-1-4673-1122-9/12 $31.00 © 2012 IEEE

II. MATERIALS AND METHODS

A. IDE Fabrication

The interdigitated electrodes were fabricated at Lionix, Netherlands on fused silica substrates. Each set of the interdigitated electrode array occupied an area of 1 mm². Each digit was 2 μm wide with a 2 μm inter-digit space. The fabrication process flow is shown in Fig. 2.

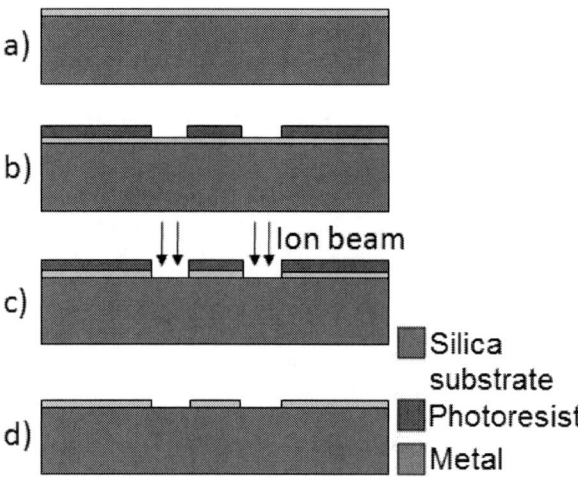

Fig. 2. IDE fabrication process flow: a) metal deposition - Cr/Au (15/200 nm) b) photoresist spin-on and pattern, c) ion beam etch of the exposed regions, d) photoresist removal and clean.

B. Graphene Tranferral

The IDEs were first cleaned by immersing in piranha solution (70 % H_2SO_4: 30 % H_2O_2) for 20 minutes to remove any organic contamination. The IDEs were dried in a stream of nitrogen to make the substrate suitable for graphene transferal. Graphene produced by chemical vapor deposition (CVD) onto a copper substrate was purchased from Graphene Laboratories (Calverton, NY, 11933, USA). The graphene was placed firmly onto thermal release tape and the copper/graphene/tape was placed into a copper etching solution of 40 % Fe_3Cl_2 for 60 minutes, etching the 20 μm of copper and leaving graphene deposited on the tape. The tape was left to dry at room temperature for 20 minutes until the adhesive properties were restored. The IDEs were placed on to the tape so that the graphene would be deposited onto the digits. The samples were placed in an oven at 95 °C for 15 minutes so that the graphene would be successfully transferred. After 15 minutes the tape lost its adhesive properties leaving graphene alone on the IDEs. Raman spectroscopy and environmental scanning electron microscopy (ESEM) were utilized to confirm the presence of graphene.

C. The confirmation of graphene transfer to IDEs

Raman spectroscopy is a non-contact optical technique that can confirm the presence of graphene using its unique spectral characteristics. The instrument consists of a 632 nm laser beam with a spot size of 3 μm. The IDEs were exposed to this laser to obtain characteristic intensity peaks for graphene. Graphene has three main peaks known as the D peak (1320 cm⁻¹), the G peak (1600 cm⁻¹) and the 2D peak (2650 cm⁻¹) [8]. The peaks give an indication of the quality of the graphene layer, the D peak gives an indication of the defects in the graphene C-C bonds, the G peak intensity gives an indication of the graphene layers and the 2D peak wave number is used to determine the layers of graphene present. The Raman spectroscopy was performed immediately after graphene transferral to confirm successful deposition before further imaging and electrical measurements.

ESEM was performed instead of standard SEM so that imaging could be performed outside of a vauum. In standard vacuum SEM charging can occur on non-conducting substrates such as silica, causing a buildup of electrons on the sample, which can be an impediment to high quality images. Most commonly it is overcome by coating the sample with a thin conducting layer such as gold, but the dimensions of graphene make this an unsuitable method. The imaging was performed at the environmental imaging suite at Newcastle University.

D. PBA immobilization

First the IDEs were stuck onto a PCB dipstick and wirebonding was used to make electrical contacts from the die to the PCB. This allowed dipping of individual IDEs into a deposition solution for real time monitoring of the graphene's electrical characteristics. The PBA was dissolved in methanol (5 μM, 50 nM) and effective dipping of the IDE active surface

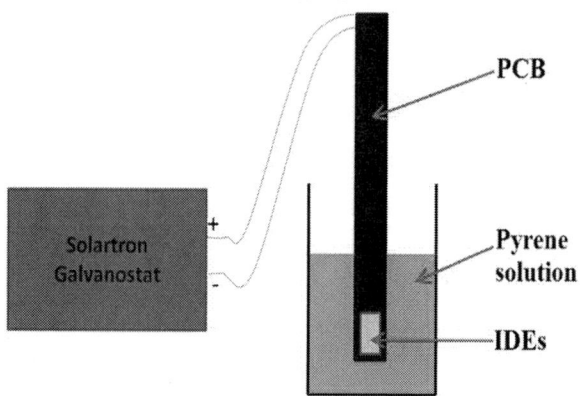

Fig. 3. An illustration of the electrically connected graphene functionalised IDEs during electrical resistance measurements. The IDE die stuck to a PCB allowed the electrical connection to be made to the Solartron galvanostat to measure the electrical resistance of graphene.

was achieved by monitoring the IDE surface with a camera; see Fig. 3 for an illustration of the basic setup. Once the IDE was dipped in the deposition solution the change of resistance was monitored. The resistance measurement scans on the IDEs were carried out in real time using the galvanostat unit on a commercially available electrical impedance spectroscopy instrument (Solartron Analytical, Ametek, Inc.). The Galvanostat provided a constant current supply of 1 mA to the setup and monitored the changes in voltage during the graphene functionalization with PBA. This in turn provided the resistance change in the graphene layer.

III. RESULTS AND DISCUSSION

A. Raman spectroscopy

The Raman spectroscopy analysis indicated that monolayer graphene was present on the IDEs. This was determined by the 2D peak characteristics as displayed in Fig. 4 which are in agreement with the literature [8]. A comparison of the D and G peak intensity confirms that graphene was transferred onto the IDEs without producing large defects in the crystal lattice. The defect peak observed was attributed to the dangling sp[3] carbon bonds at the edge sites of graphene flakes, which is expected due to the decreasing lateral graphene size in the current samples.

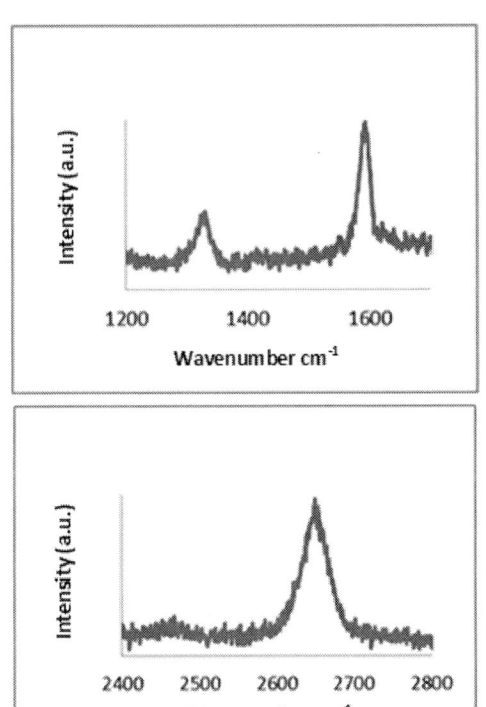

Fig. 4. Raman spectroscopy confirms that graphene was present on the IDE support. The upper chart shows the D peak and G peak and the lower chart shows the 2D peak which indicates monolayer graphene.

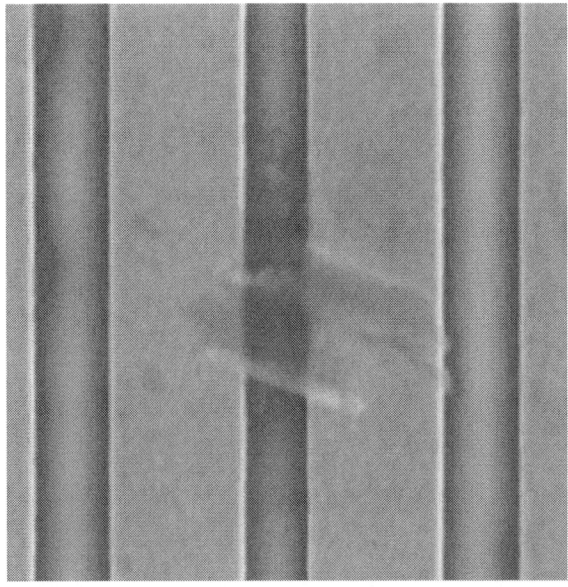

Fig. 5. An ESEM image showing a single flake graphene spanned between interdigitated electrodes.

B. Graphene functionalised IDEs and ESEM imaging

To ensure that the graphene present on the IDEs was spanned between the digits effectively to form a bridge, ESEM was used to image the devices. The ESEM imaging confirmed that graphene bridges were positioned between the digits to construct an electrically conducting path. The average bridge size observed was approx. 5 μm in diameter that easily spanned the digits, see Fig. 5.

C. Real time electrical resistance measurements

The immersion of the graphene functionalized IDEs in the PBA solution produced an increase in the electrical resistance measured. As represented in Fig. 6, the placement of the graphene IDEs in a 5 μM and a 50 nM PBA solution caused a rapid increase of the electrical resistance. It is shown that using 5 μM PBA solution caused a resistance plateau within a 30 minute deposition time. The 50 nM PBA showed a slower rate of resistance change, reaching a plateau after 4 hours. The control sample was performed by placing the IDEs in methanol alone, showing a very small decrease in resistance, which may be attributable to temperature drift with time. The resistance plateau that was observed for all PBA concentrations employed (5 μM, and 50 nM) indicates that the surface becomes saturated at all concentrations, although the kinetics of saturation become slower. PBA is known to associate with carbon ring structures via a pi stacking interaction, thus disrupting the flow of electrons at the graphene interface and disturbing the electrical properties of graphene. The PBA baselayer will allow additional layers to be formed by chemically crosslinking to the carboxyl functional group. If the additional layers also modulate the

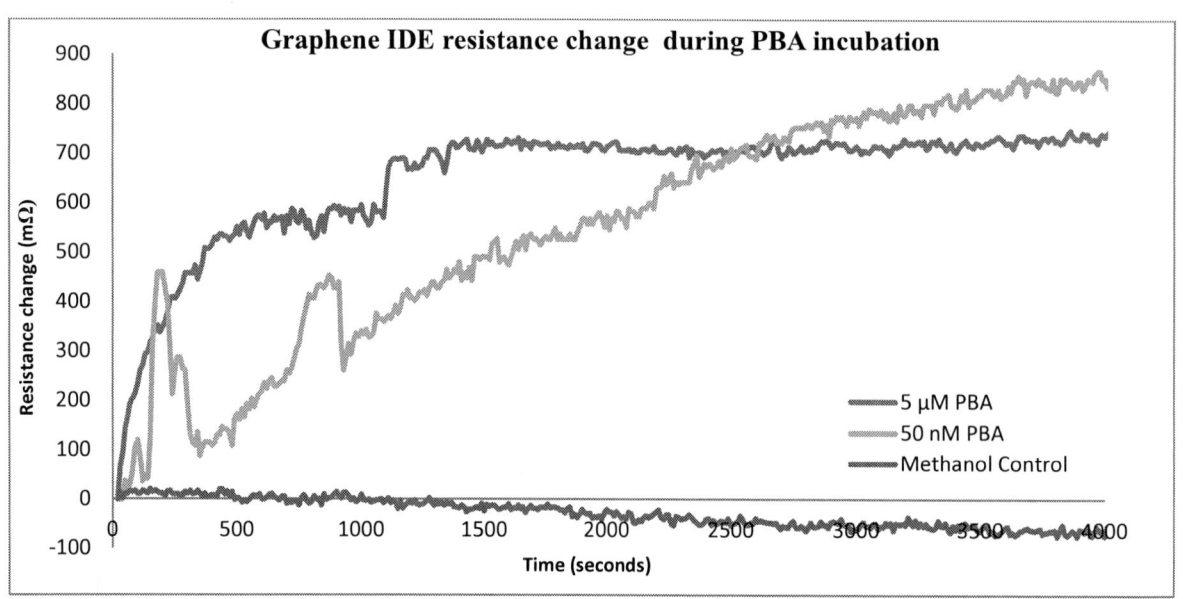

Fig. 6. A representative real time electrical resistance measurement of graphene placed on an IDE in a 5 μM and 50 nM PBA deposition solution and methanol control. The data was normalised for ease of observation from the initial values prior to PBA; 5μM (+18761 mΩ), 50 nM (+17764 mΩ) and methanol (+19243 mΩ).

resistance in the underlying graphene, it would be reasonable to envisage a very simple and potentially sensitive biosensor for all manner of applications.

IV. CONCLUSION

The initial research demonstrates that the facile transfer of graphene from a copper substrate can be performed to produce electrical contacts in the form of graphene bridges. The graphene functionalized IDEs were successfully employed to measure the resistance changes that occurred with the immobilization of a range of PBA concentrations. However the transferral method was uncontrollable and the repeatability to produce identical coverage of graphene between IDEs was not achieved. The variability of the electrical resistance measurements may be determined by the graphene coverage between the IDE sets. Research is currently underway to determine the effect of chemically crosslinking additional layers to the surface. The maximum distance from the graphene interface, that still modulates the electrical properties, will define whether this approach has utility as a biosensor.

V. ACKNOWLEDGMENT

The authors would like to thank the EPSRC for the feasibility research grant on graphene sensors EP/I015930/1 and the previous research grant EP/C015045/1 thereby enabling the Raman spectroscopy aspects of this work to be performed. In addition, we would like to thank the Commission of the European Communities, Framework 7

programme "Coeliac Disease Management Monitoring and Diagnosis using Biosensors and an Integrated Chip System, CD MEDICS [FP7-2007- ICT-1 IST-2007-2-216031] for the provision of IDEs.

VI. REFERENCES

[1] K.S. Novoselov, A. K. Geim, S.V. Morozov, D. Jiang, Y. Zhang, S.V. Dubonos, I.V. Grigorieva, A.A Firsov, "Electric Field Effect in Atomically Thin Carbon Films." *Science,* Vol. 306, pp. 666-669, 2003.

[2] A.K. Geim and K. S. Novoselov, "The rise of graphene."*Nature Materials,* Vol. 9, pp. 183-191, 2007.

[3] Y.J. Shao, J.W. Wang, J.Wu, H.Liu, J. Aksay, I. A. Lin, "Graphene based electrochemical sensors and biosensors: A review." *Electroanalysis, Vol.* 22, pp. 1027-1036, 2010.

[4] W. Yang, K.R. Ratinac, S.P Ringer, P. Thordarson, J.J. Gooding, F. Braet, "Carbon Nanomaterials in Biosensors: Should You Use Nanotubes or Graphene?" *Angewandte Chemie International Edition*, Vol. 49, pp. 2114–2138, 2010

[5] K.V. Emtsev, A. Bostwick., H. Karsten., J. Johannes, G. L. Kellogg, L. Lothar., J.L. McChesney, T. Reshanov, A. Sergey A, R. Jonas., E. Rotenberg, A.K. Schmid, D. Waldmann, "Towards wafer-size graphene layers by atmospheric pressure graphitization of silicon carbide." *Nature Materials, Vol.* 8, pp. 203-207, *2009.*

[6] X. Li, and W. Cai, "Large-Area Synthesis of High-Quality and Uniform Graphene Films on Copper Foils." *Science*, Vol. 324, pp.1312-1314, 2009.

[7] S. Bae, H. Kim, Y. Lee, X. Xu, J.S. Park, Y. Zheng, J. Balakrishnan, T. Lei, H. Ri Kim, Y. Song, Y.J. Kim, K.S. Kim, B. Ozyilmaz, J.H. Ahn, B.H. Hong, S. Iijima, "Roll-to-roll production of 30-inch graphene films for transparent electrodes." *Nature Nanotechnology, Vol.* 5, pp. 574-578, 2010.

[8] A.C. Ferrari, J.C. Meyer, V. Scardaci, C. Casiraghi, M. Lazzeri, F. Mauri, S. Piscanec, D. Jiang, K.S. Novoselov, S. Roth, A.K. Geim, "Raman Spectrum of Graphene and Graphene Layers," *Physical Review Letters,* Vol. 97, pp. 187401-187405, 2006.

Excitation of mechanical oscillations in double-carbon-nanotube system by terahertz radiation

Vyacheslav Semenenko[1], Vladimir Leiman[1], Aleksey Arsenin[1], Yury Stebunov[1], and Victor Ryzhii[2]

[1]Laboratory of Nanooptics and Femtosecond Electronics,
Moscow Institute of Physics and Technology, Dolgoprudny 141700, Russia
[2]Computational Nanoelectronics Laboratory, University of Aizu, Aizuwakamatsu, Fukushima 965-8580, Japan

Abstract— **The system of the two same, placed side-by-side and double clamped single-walled carbon nanotubes with metallic conductivity in the electromagnetic field of modulated and non-modulated terahertz radiation is considered. Forced oscillations of the carbon nanotubes electron plasma are calculated. The lumped parameters of the mechanical resonators that the nanotubes represent by themselves are determined. It is shown that the considered system of the nanotubes can serve as a detector of modulated terahertz radiation. The responsivity of the detector is estimated. The threshold value of the electric field amplitude of the incoming monochromatic terahertz radiation, above which the self-excitation of the nanotube mechanical resonators occurs is estimated.**

Keywords-carbon nanotubes; microelectromechanical systems; terahertz radiation; sensors; actuators

I. INTRODUCTION

Terahertz radiation has a great potential for different applications [1]. Recently, a number of devices for detection of modulated terahertz radiation, based on the concept of the field-effect transistor with a mechanically floating gate was proposed [2,3-5]. Such devices have a small size (~ 1 um) and their operating frequency can be tuned in quite wide range, from one up to several THz. A high responsivity value in the devices in question can be achieved if single-walled carbon nanotubes (SWCNTs) would be used as mechanical cantilevers because of their unique mechanical properties [4,5]. However, the excitation of electron plasma oscillations in the nanotubes is suppressed due to the high contact resistance ($R_c \geq 6.5$ kOhm). To achieve the higher quality factor of the electron plasma oscillations we propose a new scheme with the excitation realized without electrical contacts. In this scheme CNTs are influenced directly by the electric field of the incoming terahertz wave (see Fig. 1). In the system in question the electric field of the THz wave causes plasma oscillations in the nanotubes. Redistribution of the electric charge results in the repulsive mechanical force acting from one nanotube to another, that is quadratic on the incoming signal. Because of this non-linearity the spectrum of the mechanical force has the low-frequency harmonics of the modulation signal. Thus, if the modulation signal gets into resonance with the mechanical oscillations in the nanotubes, their excitation is possible. Also, it is known, that in such systems, even in the absence of the modulation of the THz

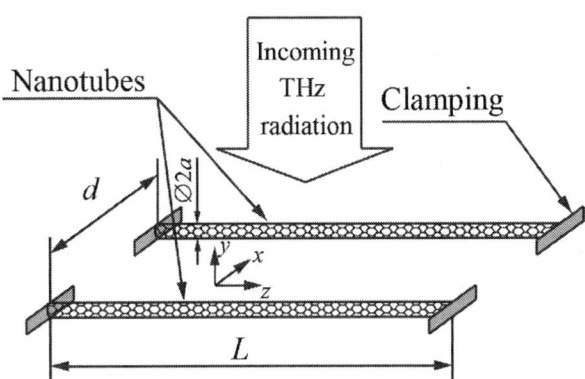

Fig. 1. Double-SWCNT system in the electromagnetic field of THz radiation.

carrier signal, the self-excitation of the mechanical low-frequency resonator can occur [6,7]. In this work we develop a model describing both plasma and mechanical oscillations in the considered system and their interference. In the case of the modulated THz radiation incident on the system we calculate the amplitude of the nanotubes mechanical oscillations and propose a scheme for its measurement (detection). Also we estimate the threshold value of the electric field in the incident monochromatic THz wave above which the self-excitation of the mechanical oscillations in the nanotubes is induced.

II. THEORETICAL TREATMENT

A. Forced oscillations of the electron plasma in the nanotubes

To describe electron transport in the SWCNTs we use the hydrodynamic formalism that is valid for the nanotubes with metallic conductivity [8]. According to it, linearized equations for the surface electron charge σ and the velocity of the electron fluid u in the nanotubes are the following:

$$\begin{cases} \dfrac{\partial \sigma}{\partial t} - e n_s \dfrac{\partial u}{\partial z} = 0 \\[2mm] \dot{u} + \gamma_e u = \dfrac{v_F^2}{e n_s} \dfrac{\partial \sigma}{\partial z} - \dfrac{e}{m^*} E_z, \end{cases} \quad (1)$$

978-1-4673-1122-9/12 $31.00 © 2012 IEEE

where n_s is the sheet density of free electrons in the nanotubes, γ_e is the electrons collision frequency, v_F - is the Fermi velocity, m^* is the electron effective mass, E_z is the component of the macroscopic electric field directed along the nanotubes, and e is the electron charge (absolute value).

Combining equations (1) with the Maxwell's equations and using the boundary conditions on the surface of the nanotubes

$$\nabla_{\bar{n}} E_n = 4\pi\sigma, \quad \nabla_{\bar{n}} E_z = 0, \quad (2)$$

where E_n is the component of the electric field vector directed along the normal vector $\bar{n}$ to the nanotubes surface, one can find the dispersion relation for the plasma waves traveling along the considered system of the two nanotubes:

$$\omega(\omega - i\gamma) = v_\phi^2 k^2, \quad v_\phi^2 = v_F^2 + v_{ED}^2 \cdot \theta,$$
$$v_{ED}^2 = \frac{4\pi e^2 n_S a}{m^*}, \quad \theta = K_0(ka) \pm K_0(kd), \quad (3)$$

where a is the radius of the nanotubes, d is the spacing between their axes, ω and k are the wave frequency, and its wave vector correspondingly. Sign "+" in the expression for θ corresponds to the symmetric wave mode and sign "-" corresponds to the anti-symmetric one. The latter is common in radio engineering as it propagates in any kind of transmission lines. The relation (3) was obtained in the assumption of the axially symmetric distribution of charge density in the both nanotube, that is valid for our case when $a \ll d$. Also the following assumptions was used: $\omega/c \ll k$ and $kd \ll 1$. The first one corresponds to the case when the electrons are affected only by the Coulomb forces, that is valid when ω is of THz frequency range. And the second one was invoked for simplifying our calculations using the approximation formulas for the Bessel functions, e.g. $K_1(x) \approx 1/x$, $x \ll 1$. Besides the dispersion relation, from the equations mentioned above we obtain equation for forced oscillations of the symmetric plasma modes in Fourier space:

$$\begin{cases} i\omega\sigma_{\omega k} + ik \cdot en_S \cdot u_{\omega k} = 0, \\ u_{\omega k}(i\omega + \gamma) = -ik\dfrac{v_\phi^2}{en_S}\sigma_{\omega k} - \dfrac{e}{m^*}E_{z,\omega k}, \end{cases} \quad (4)$$

here $\sigma_{\omega k}$, $u_{\omega k}$ and $E_{z,\omega k}$ are the amplitudes in ωk-space of Fourier harmonics of the corresponding functions $\sigma(z,t)$, $u(z,t)$ and $E_z(z,t)$, in the initial tz-space. Taking into account that the phase velocity of plasma waves v_ϕ depends slowly on the wave vector k, one can rewrite eq. (4) in the tz-space:

$$\begin{cases} \dfrac{\partial\rho}{\partial t} + \dfrac{\partial I}{\partial z} = 0 \\ \dfrac{\partial I}{\partial t} + \gamma I + v_\phi^2\dfrac{\partial\rho}{\partial z} = v_{ED}^2 E_z, \end{cases} \quad (5)$$

here $\rho = 4\pi a \times \sigma$ is the linear charge of the considered system of two nanotubes, and $I = -4\pi a \cdot en_S \times u$ is the electric current in it. Note, that eqs. (5) were obtained for the system

of the nanotubes of infinite length. But neglecting the boundary effects, one can consider that they are valid also when the lengths of our two-nanotube system is finite. In our case, when the lengths of the system is L, eqs. (5) should be supplemented with the following boundary conditions:

$$I(z,t)\big|_{z=-L/2} = 0, \quad I(z,t)\big|_{z=L/2} = 0. \quad (6)$$

Because of the small size of the system ($L \sim 1$ um), the electric field of the THz wave incident on it, can be considered as spatially uniform, so we can put: $E_z(z,t) = E_z(t)$. This corresponds to the incident linear polarized wave with the electric field vector directed along the nanotubes. Solving eq. (5) with the boundary conditions (6) we obtain:

$$\rho_\omega = \frac{1}{k}\frac{v_{ED}^2}{v_\phi^2}E_{z,\omega}\frac{\sin kz}{\cos k\dfrac{L}{2}}, \quad I_\omega = \frac{i\omega}{k^2}\frac{v_{ED}^2}{v_\phi^2}E_{z,\omega}\left(\frac{\cos kz}{\cos k\dfrac{L}{2}} - 1\right), (7)$$

where $E_{z,\omega}$ is the Fourier component of the $E_z(t)$ function, $\rho_\omega(z)$ and $I_\omega(z)$ are the Fourier components in ωz-space of the real functions $\rho(z,t)$ and $I(z,t)$.

Expressions for ρ_ω and I_ω have poles that are given as follows:

$$\omega_n = \frac{\pi v_\phi}{L}(1+2n) + \frac{i \cdot \gamma_e}{2}; \quad (8)$$

real parts of ω_n are the resonant frequencies of the considered plasma resonator based on the two-carbon-nanotube system, and $Q_n = \operatorname{Re}\omega_n / 2\operatorname{Im}\omega_n$ are the quality factors of the corresponding resonance.

B. Mechanical oscillations of the nanotubes

Repulsive force acting on the nanotubes due to the presence of distributed charge $\rho(z,t)/2$ on both of them, in the case when $d \ll L$, can be expressed as $F_l = \rho^2(z,t)/2d$. If the quality factor of the general resonance is quite big ($Q_0 \gg 1$) and the Fourier spectrum of the external signal $E_z(t)$ is concentrated in the area of some single resonance ω_n, the linear force F_l in tz-space can be represented as follows:

$$F_l(z,t) \approx f_l(t)g(z), \quad f_l(t) = \frac{\rho_0^2(t)}{2d}, \quad g(z) = \sin^2(k_n z),$$
$$k_n = \frac{\pi}{L}(1+2n), \quad \rho_{0,\omega} \approx \frac{1}{k_n}\frac{v_{ED}^2}{v_0^2}\frac{E_{z,\omega}}{\cos k\dfrac{L}{2}}, \quad (9)$$

where $\rho_0(t)$ is the linear charge density at the right end of the two-nanotube system.

Given the linear force acting on a nanotube, we can write the equation for its mechanical oscillations:

$$\frac{\partial^2 x}{\partial t^2} + \gamma_m\frac{\partial x}{\partial t} + \frac{J_b}{M_l}\frac{\partial^4 x}{\partial z^4} = \frac{F_l}{M_l}, \quad x\big|_{z=\pm L/2} = 0, \quad \frac{\partial x}{\partial z}\bigg|_{z=\pm L/2} = 0, (10)$$

where x is the displacement along $0X$-direction of the axis of the nanotube from its non-deformed position, γ_m is the oscillations damping constant, M_l is the linear mass of the nanotube, J_b is its flexural stiffness. One can find the solution of this equation in the following form:

$$x(z,t) = \sum_{j=1}^{\infty} x_j(t)\psi_j(z), \qquad (11)$$

where $\psi_j(z)$ are normalized to 1 spatial parts of the solutions of the of homogeneous equation (Eq. (10) when its right hand side is zero):

$$\Omega(\Omega - i\gamma_m)\psi_j(z) + \frac{J_b}{M_l}\frac{\partial^4\psi_j}{\partial z^4}, \qquad (12)$$

where Ω is the complex frequency of the nanotube free oscillation modes. The dispersion equation for Ω is the following [9]:

$$\cos\xi = \frac{1}{\cosh\xi}, \quad \xi = \sqrt[4]{\frac{\Omega(\Omega - i\gamma_m)}{J_b / M_l}} \times L. \qquad (13)$$

Substituting the expression (11) into the Eq. (10) and using the series expansion of $g(z)$ in functions ψ_j

$$g(z) = \sum_{j=1}^{\infty}\eta_j\psi_j(z), \quad \eta_j = \frac{1}{L}\int_{-L/2}^{L/2} g(z)\psi_j(z)dz, \qquad (14)$$

one obtains a set of separate equations for the amplitudes of the free oscillation modes:

$$\ddot{x}_j + \gamma_m\dot{x}_j + \Omega_j^2 x = \frac{\eta_j f_l(t)}{M_l}, \qquad (15)$$

where Ω_j is the resonant frequency corresponding to a given function ψ_j and is expressed as follows:

$$\Omega_j = \sqrt{\frac{J_b}{M_l}}\frac{\xi_j^2}{L^2}, \quad \cos\xi_j = \frac{1}{\cosh\xi_j}. \qquad (16)$$

Here, the roots ξ_j of the dispersion equation are numerated from 0, thus $\xi_0 = 0$ and $\psi_0(z) \equiv 0$.

C. Coupled modes equations for the mechanical and plasma oscillations

From now we will consider such an incoming signal $E_z(t)$, whose Fourier spectrum is concentrated near the general plasma resonance in the nanotubes $\omega_e = \mathrm{Re}\,\omega_0 = \pi v_\phi / L$ (see Eq. (8)) and that has a width about nanotubes general mechanical resonance $\omega_m = \Omega_1$ (see Eq. (16)). Under such conditions, one can neglect the higher harmonics of plasma and mechanical oscillations, and consider the excitation only of the general ones. According to the Eqs. (16) and (9) we have:

$$\ddot{x} + \gamma_m\dot{x} + \omega_m^2 x = \frac{\eta\rho_0^2}{2M_l d}, \quad \rho_{0,\omega} = \frac{L}{\pi}\frac{v_{ED}^2}{v_0^2}\frac{E_{z,\omega}}{\cos k\frac{L}{2}}. \qquad (17)$$

Here we put $x = x_1$ and $\eta = \eta_1$. Now relation between ρ_0 and E_z are written in ω-space, but if we use the approximation for $1/\cos kL/2$, that is valid for $\omega \sim \omega_e$,

$$\frac{1}{\cos kL/2} \approx \frac{4}{\pi}\frac{\omega_e^2}{\omega_e^2 - \omega^2 + i\omega\gamma_e}, \quad \omega_e - \gamma_e \lesssim \omega \lesssim \omega_e + \gamma_e, (18)$$

we can rewrite it in the t-space as follows:

$$\ddot{\rho}_0 + \gamma_e\dot{\rho}_0 + \omega_e^2\rho_0 = \frac{4L}{\pi^2}\frac{v_{ED}^2}{v_\phi^2}\omega_e^2 \cdot E_z(t). \qquad (19)$$

Next, take into account that the resonant frequency ω_e is varied as the nanotubes deform. To calculate the frequency shift for a small deformation of the nanotubes ($x \ll d$) one should solve the Eqs. (5) with the boundary conditions (6) for the spatially dependent phase velocity of plasma waves:

$$\left(v_\phi'(z)\right)^2 \approx v_\phi^2 - \frac{2x}{d}\psi_1(z), \quad x \ll d, \qquad (20)$$

where v_ϕ is calculated using Eqs. (3) for the non-deformed nanotubes. It's easy to show that in comparison with the non-deformed case, when the resonant condition is $kL/2 = \pi/2$, for the deformed nanotubes it is transformed as follows:

$$\int_0^{L/2} k'(z)dz = \frac{\pi}{2}, \quad k'(z) = \sqrt{\frac{\omega(\omega - i\gamma_e)}{v_\phi'^2(z)}}. \qquad (21)$$

Performing the calculation, for the shifted resonant frequency ω_e' we obtain:

$$\frac{\omega_e'}{\omega_e} = 1 - \zeta\frac{v_{ED}^2}{v_0^2}\frac{x}{d}, \quad \zeta = \frac{1}{L}\int_{-L/2}^{L/2}\psi_1(z)dz. \qquad (22)$$

Substituting ω_e' into the Eq. (19) and write it together with the Eq. (17) for the mechanical resonator, we obtain the following system of equations:

$$\begin{cases} \ddot{x} + \gamma_m\dot{x} + \omega_m^2 x = \dfrac{\eta\rho^2}{2M_l d} \\[2mm] \ddot{\rho} + \gamma_e\dot{\rho} + \omega_e^2\rho\left(1 - \zeta\dfrac{v_{ED}^2}{v_0^2}\dfrac{x}{d}\right) = \dfrac{4L}{\pi^2}\dfrac{v_{ED}^2}{v_0^2}\omega_e^2 \cdot E_z(t), \end{cases} \qquad (23)$$

here for the sake of simplicity we put $\rho = \rho_0$.

Thus, given the incoming signal $E_z(t)$, by solving the Eqs. (23), we can find dependencies $x(t)$ and $\rho(t)$, and express in terms of them any electrical or mechanical parameters of the state of the system.

III. RESULTS

A. Excitation of the mechanical oscillations in the nanotubes by modulated THz radiation

In this paragraph we consider the modulated signal of the incoming radiation:

$$E_z(t) = E_0(1 + m\cdot\cos\Omega t)\times\cos\omega t, \qquad (24)$$

where E_0 is the amplitude of the carrier wave, and m is the modulation depth. Solving Eqs. (23) for such $E_z(t)$, we can neglect the dependence of the resonant frequency of the plasma resonance in the nanotubes on the mechanical deformation. It is valid for this problem because of the small frequency shift even for mechanical oscillations of the nanotubes with the maximum amplitude under which them touch one another. In this case $2x \cdot \psi_1(0) = d$ or $x/d \approx 0.3$ and the relative frequency shift is $\left| \omega_e' / \omega_e - 1 \right| \approx 0.035$, that is less then the relative resonance width $\gamma_e / \omega_e \sim 0,1$. Thus, we can consider that the mechanical motion of the nanotubes doesn't affect the plasma oscillations in them, and for $\rho(t)$ find the following formula:

$$\rho(t) = \frac{4E_0 L}{\pi^2} \frac{v_{ED}^2}{v_0^2} \times 2\,\mathrm{Re}\left[G_e(\omega) f(t) \right],$$

$$G_e(\omega) = \omega_e^2 / (\omega_e^2 - \omega^2 + i\omega\gamma_e), \qquad (25)$$

$$f(t) = 0.5 \times \left(e^{i\omega t} + m/2 \cdot e^{i(\omega - \Omega)t} + m/2 \cdot e^{i(\omega + \Omega)t} \right).$$

Here we used approximation $G_e(\omega \pm \Omega) \approx G_e(\omega)$.

Substituting (25) to the equation for the mechanical oscillations (23) one obtains the following dependence for $x(t)$:

$$x(t) = \frac{4m\eta}{\pi^4} \frac{E_0^2 L^2}{M_l \omega_m^2 d} \left(\frac{v_{ED}}{v_\phi} \right)^4 \left| G_e(\omega) \right|^2 2\,\mathrm{Re}\left[G_m(\Omega) e^{i\Omega t} \right],$$
$$(26)$$

$$G_m(\Omega) = \omega_m^2 / (\omega_m^2 - \omega^2 + i\omega\gamma_m).$$

In the expression for $x(t)$ we omitted constant term and high frequency harmonics due to their smallness.

To demonstrate the efficiency of the excitation of the mechanical oscillations in the nanotubes, we propose a scheme for their detection shown in the Fig. 2. In this device the two considered nanotubes are charged with a DC voltage V_0. The electric charge on the nanotubes is locked on them because of high value of the resistance R in the circuit of the DC voltage. The value of the resistance is estimated as $R > 1/\omega_m C_l L$, or $R > 160$ MOhm, where $C_l = 1/4 \ln d/a$ is the linear electrostatic capacitance between the nanotubes. In such conditions, amplitude of the AC voltage at the detector output is given as $\delta V \approx 4 C_l V_0 \times \langle \delta d \rangle / d$, where $\langle \delta d \rangle$ is the average change in distance between the nanotubes that can be estimated as $\langle \delta d \rangle \approx 2x \cdot \zeta$.

The responsivity of the detection R_V is the ratio of the detector output signal δV to the total power P_Σ incoming to it. The value P_Σ consists of the power P_e dissipating in the nanotubes plasma due to the electron collisions, the power P_m dissipating under the mechanical motion of the nanotubes, and the reradiation power P_{rad}. In the conditions of combined plasma and mechanical resonances, when $\omega = \omega_e$ and $\Omega = \omega_m$, the ratio of P_m to P_e is given by the following formula:

$$\frac{P_m}{P_e} = \frac{4m^2\eta^2}{\pi^4} \frac{E_0^2 L^2}{M_l \omega_e^2 d^2} \left(\frac{v_{ED}}{v_\phi} \right)^6 Q_e^3 Q_m n, \qquad (27)$$

where $Q_e = \omega_e / \gamma_e$ and $Q_m = \omega_m / \gamma_m$ are the quality factors of the plasma and mechanical resonances correspondingly, $n = \omega_e / \omega_m$ is the ratio of the resonant frequency of the plasma resonator to the mechanical one. For the nanotubes length $L = 0.5$ um, their radius $a = 1$ nm and the spacing between their axes $d = 20$ nm, $Q_m \sim 10^3$, $\eta \approx 0.438$, $\zeta \approx 0.831$, $m = 0.1$ and $E_0 \approx 0.87$ V/cm (that corresponds to the intensity of the incoming radiation 1 mW/cm^2) we obtain $v_{ED} \approx 2.2 \times 10^8$ cm/s, $v_\phi \approx 6.1 \times 10^8$ cm/s, $\omega_e \approx 6.1$ THz, $Q_e \approx 74$, $\omega_m \approx 0.2$ GHz, and the ratio $P_m / P_e \sim 10^{-10}$. Also, it is known (see for example Refs. [10,11]), that the efficiency (i.e. ratio P_{rad} / P_e) of the metallic carbon nanotube antennas in THz region is about 10^{-6}. Thus, one can put $P_\Sigma \approx P_e$ and find the peak responsivity of the detection $R_V = \delta V / P_\Sigma$ as follows:

$$R_V = \frac{8m\eta\zeta}{\pi} \frac{C_l V_0}{M_l \omega_m^2 d^2 v_\phi} \left(\frac{v_{ED}}{v_\phi} \right)^2 Q_e Q_m. \qquad (28)$$

Performing calculation, for the value of the DC voltage $V_0 = 1$ V, one obtains $R_V \sim 10^6$ V/W.

B. Parametric excitation of the mechanical oscillations in the nanotubes

Structure of the equations (23) allows excitation of the mechanical oscillations in the considered system with the monochromatic high-frequency incoming signal:

$$E_z(t) = E_0 \cos \omega t, \quad \omega \sim \omega_e. \qquad (29)$$

Because of the oscillations in plasma resonator are set non-immediately under the change of the geometry of the system, the term in right-hand side of the Eqs. (23) for the mechanical

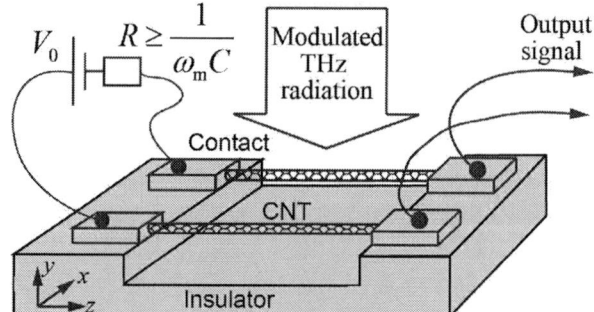

Fig. 2. Scheme for detection of the mechanical oscillations of the nanotubes induced by modulated THz radiation.

oscillator can be considered as time-retarded:

$$\ddot{x} + \gamma_m \dot{x} + \omega_m^2 x = F(t - \tau), \qquad (30)$$

where $\tau \sim \gamma_e^{-1}$ is the relaxation time of the plasma oscillations. The oscillations damping constant γ_m in (30) can get a negative additive depending on the amplitude of the incoming signal. For the values E_0 above some threshold value E_{th} the decrement the oscillations turns to be negative that corresponds generation of the mechanical oscillations. The detail procedure of calculation of the threshold value of the incoming signal for the equations system like (23) is described in Refs. [6,7,12]. Using it, one can estimate E_{th} as follows:

$$E_{th}^2 = \frac{\pi^4}{\sqrt{3}\eta\zeta} \frac{M_l \omega_e^2}{n Q_e^4 Q_m} \left(\frac{v_\phi}{v_{ED}}\right)^6 \frac{d^2}{L^2}. \tag{31}$$

For the parameters of the stated in the previous paragraph we obtain $E_{th} \sim 1$ kV/cm or 50 mV per the length of the system L. According to antenna theory (see, for example, Ref. 13), such a value of the electric field can be obtained in the feeder area of an antenna (e.g. dipole type) under the incident radiation of intensity about 1 W/cm^2.

IV. CONCLUSION

In conclusion, we proposed a model for describing the excitation of mechanical and plasma oscillations in the double-SWCNT system by the incident THz radiation. The detector of modulated terahertz radiation based on this system has a rather high responsivity, about 10^6 V/W. Monochromatic THz radiation can cause self-excitation of the mechanical oscillations in the CNTs, but for this one requires the radiation of very high intensity and some amplifying system (antenna or lens) which magnifies the electric field in the wave at least several dozen times.

ACKNOWLEDGMENT

This work was supported in part by the Russian Foundation for Basic Research (grants no. 11-07-12072-ofi-m, 10-07-00618 and 11-07-00505), by the Ministry of Education and Science of the Russian Federation (grants no. 16.513.11.3129, 07.514.11.4086, 14.740.11.0888 and 14.740.11.1135) and by the grant MK-334.2011.9 of the President of the Russian Federation.

REFERENCES

[1] M. Tonouchi, *Nature Photon*, 1, pp. 97-105, 2008

[2] H.C. Nathanson, W.E. Newell, R.A. Wickstrom, J.R. Davis, " The resonant gate transistor", *IEEE trans. on Electron Devices*, vol. 14, pp. 117-133, March 1967.

[3] V. Ryzhii, M. Ryzhii, Y. Hu, I. Hagiwara, and M.S. Shur, "Resonant detection of modulated terahertz radiation in micromachined high-electron-mobility transistor," *Appl. Phys. Lett.*, vol. 90, 203503, May 2007.

[4] V.G. Leiman, M. Ryzhii, A. Satou, N. Ryabova, V. Ryzhii, T. Otsuji, and M.S. Shur, "Analysis of resonant detection of terahertz radiation in high-electron mobility transistor with a nanostring/carbon nanotube as the mechanically floating gate," *J. Appl. Phys.*, vol. 104, 024514, 2008.

[5] Yu. Stebunov, V. Leiman, A. Arsenin, A. Gladun, V. Semenenko, V. Ryzhii, "Detection of Modulated Terahertz Radiation Using Combined Plasma and Mechanical Resonances in Double-Carbon-Nanotube Device," *Appl. Phys. Express*, vol. 4, 075101, 2011.

[6] A.V. Arsenin, A.D. Gladun, V.G. Leiman, V.L. Semenenko and V.I. Ryzhii, "Parametric instability in a nanoelectromechanical detector of modulated terahertz radiation on the basis of a high electron mobility transistor with a mobile elastic gate," J. Commun. Technol. Electron., vol. 54, pp. 1322-1330, 2009.

[7] A.V. Arsenin, A.D. Gladun, V.G. Leiman, V.L. Semenenko and V.I. Ryzhii, "Parametric Instability in the Resonance Detector of Terahertz Radiation Based on FET with Cylindrical Gate Electrode," *J. Commun. Technol. Electron.*, vol. 56 (10), pp. 1242-1248, 2011.

[8] A. Maffucci, G. Miano, and F. Villone, "A transmission line model for metallic carbon nanotube interconnects," *Int. J. Circ. Theor. Appl.*, 1, No. 36, pp. 31-51, 2008.

[9] L.D. Landau and E.M. Lifshitz, *Theory of Elasticity*, Pergamon, New York, 1986.

[10] G.V. Hanson, "Fundamental Transmitting Properties of Carbon Nanotube Antennas," *IEEE Trans. Antennas and Prop.*, vol. 53 (11), Nov 2005.

[11] M. V. Shuba, G. Ya. Slepyan, S. A. Maksimenko, C. Thomsen, A. Lakhtakia, "Theory of multiwall carbon nanotubes as waveguides and antennas in the infrared and the visible regimes," *Phys. Rev. B*, vol. 79 (15), 155403, 2009.

[12] V.B. Braginsky, A.B. Manukin, *Measurement of Weak Forces in Physics Experiments*, 1st ed. (Univ of Chicago Pr, 1977), pp. 29-35.

[13] R. S. Elliott, *Antenna Theory and Design.* Englewood Cliffs, NJ: Prentice-Hall, 1981.

Capillary Kinetics of Water in Hydrophilic Microscope Coverslip Nanochannels

Ju-Nan Kuo[*] and Yi-Kai Lin

[*]Department of Automation Engineering, National Formosa University, Yunlin, Taiwan
junan@nfu.edu.tw

Abstract—**This study analyzed the capillary filling speed of water in hydrophilic microscope coverslip nanofluidic channels with depths ranging from 40 to 575 nm. Nanofluidic channels were fabricated on a substrate of borosilicate glass (thickness of 160 μm) using the buffered oxide wet etching and glass-glass fusion bonding technique. The capillary filling speed was measured and compared with theoretical values. Comparison of capillary filling speed with theoretical values showed that filling speed inside the coverslip nanochannel was lower than the theoretical speed. A finite-element model was established to analyze the capillary filling speed of water in nanochannels. Finite-element analysis and experimental results show that the conventional theoretical formula for predicting the capillary filling speed is inaccurate without adjustable parameters. Experiments show that the capillary filling speed decreases with a decreasing depth of nanochannels. It need fill 8 mm long, 200 μm wide and 40 nm deep nanochannels around 90 s has been demonstrated.**

Keywords- Capillary filling; Coverslip; Hydrophilic; Nanochannels.

I. INTRODUCTION

Nanochannels are defined as channels in which at least one cross-sectional dimension is less than 100 nm. Fluid behavior in nanoscale is a new field of study. Capillary action is very prominent in nanochannels, due to the extremely large surface-to-volume ratio. For example, when nanochannels are filled by capillary action, the wetting properties of channel walls play a crucial role. The physical origin and shape of the meniscus of a wetting liquid in contact with a surface have been a subject of specific interest in capillarity since a long time [1-3]. The capillary phenomenon was first studied in the early period of the 19th century. However, very little progress has been made since the 1920s due to constraints in experimental facilities and techniques. The capillary phenomenon can be defined quantitatively in terms of surface tension, which causes the surface of a liquid act to behave like an elastic sheath. By minimizing the surface area of liquid, surface tension also minimizes energy in fluidic systems.

In nanochannels, the domination of bulk phenomena by physical surface phenomenon is even greater than that in microchannels [4]. Therefore, capillary kinetics is of fundamental importance for to investigate the behavior of liquid in nanometer-scale [5-7]. Silicon-based microfluidic devices are frequently used because the composition of the channel walls can be hydrophilic for a wide range of biotechnology applications [8-10]. Bubble formation during capillary filling has also been studied [11]. Microscope coverslip nanochannels with predictable and carefully measured depths ranging from 40 and 575 nm were successfully fabricated and filled with de-ionized water. In this study, the capillary filling speed of water in nanochannels was analyzed using theoretical, simulation and experimental methods.

II. NANOCHANNEL FABRICATION

Each of the test patterns produced in this study had an array of parallel lines 8 mm long and of varying widths. For 1D nanochannels, photolithography was used to pattern the channel area with a nano-scale depth. The depth depends on the etching time of the glass substrate.

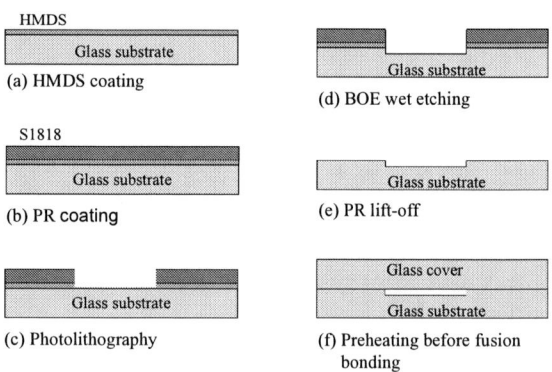

Fig. 1. Schematic diagram of nanochannel fabrication process.

The nanochannels fabrication steps are schematically shown in Fig. 1. The nanofluidic channels were fabricated on a 160-μm-thick borosilicate glass coverslip. The surface patterning of glass substrates is generally performed using an evaporated metal film as the etching mask. However, in the present study, photoresist (S1818) was used as the etching mask since the intended channel depth was in nanometer scale. To improve the adhesion of the photoresist, the borosilicate coverslip was vapor primed with hexamethyldisilazane (HMDS). The glass substrate was immersed in a 0.5 % diluted buffered oxide etchant (BOE 6:1) at room temperature. The depth of the resulting nanochannel was carefully controlled by tuning the etching time. Following the etching process, the remaining photoresist was stripped from the substrate surface in an acetone solution, and then the substrate was dipped in a 1 M HCL solution for 2 min. The etched glass substrate and an

978-1-4673-1122-9/12 $31.00 © 2012 IEEE

unpatterned borosilicate coverslip were preheated in a furnace at 400 °C for 8 hours. Finally, a programmable furnace was used to obtain a permanent bond between the two coverslips by means of thermal fusion bonding process (JH-1, Kingtech Scientific Co. Ltd.) at a temperature of 580 °C and with a bonding time of 8 hours.

Figure 2 shows the cross-sectional SEM images of nanochannels of varying depths. The etched channels are completely open and have retained their original shape with good uniformity, there is no significant deformation or sagging of the nanochannels even in the channel with a depth of just 20 nm. In addition, no obvious interface is visible between the upper and lower coverslips. In other words, the effectiveness of the glass-glass bonding process is confirmed.

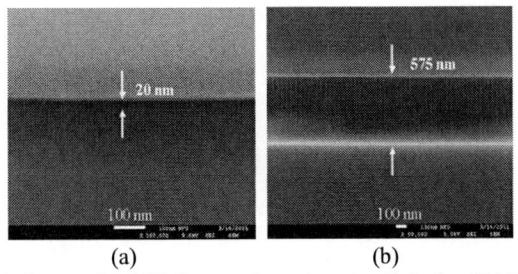

(a)   (b)

Fig. 2. Cross-sectional SEM images of nanochannels with depths of (a) 20 nm and (b) 575 nm fabricated on a glass substrate bonded with a glass cover.

III. THEORY AND SIMULATION

The liquid flow rate in a rectangular channel can be obtained by solving the Navier-Stokes equation for incompressible fluids. The Laplace-Young equation expresses the pressure drop across the free surface (Δp) as described in [12]:

$$\Delta p = \frac{2\gamma \cos\theta}{h}, \qquad (1)$$

where γ is the surface tension of the liquid in air, θ is the contact angle of the liquid to the channel walls, and h is the channel height.

Theoretically, the filling speed can be described by the classical Washburn model [13], modified for the rectangular cross section of the channels. The position of the moving meniscus x as a function of time t [14] is

$$x = \sqrt{\frac{\gamma \cos(\theta)ht}{3\mu}}, \qquad (2)$$

where μ is the viscosity of the liquid.

Therefore, the relationship between the position of the moving meniscus x and the flow time t for viscous flow between parallel plates driven by capillary action was achieved.

A finite-element model was established to analyze the capillary filling speed of water in nanochannels. The model was used to determine the flow between two parallel flat plates, one stationary and one moving at a uniform velocity. If the two parallel plates are both stationary, the fully developed

flow between the plates requires the boundary conditions as shows in Fig. 3. The velocity profile of the fully developed laminar flow between two parallel plates is parabolic. This type of flow with parabolic velocity distribution is generally referred to as a plane poiseuille flow.

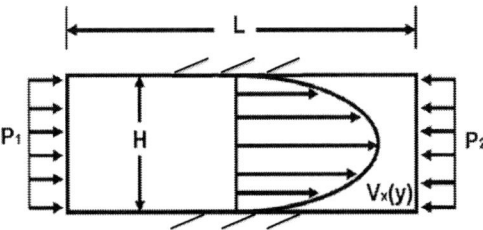

Fig. 3. Schematic illustration of plane poiseuille flow.

In this study, a commercial software tool (ANSYS) was used to clarify the experimental observations. In order to analyze the capillary filling speed of water in nanochannels, a two-dimensional finite-element model was established. As an approximation, the pressure drop across the free surface (Δp) can be modeled as a input data from Eq. (1). To enter the pressure drop into the simulation program, finite-element simulation was performed to estimate the two-dimensional state of velocity and velocity profile across channel. Therefore, the capillary filling speed of water in nanochannels can be numerically simulated. The physical properties of de-ionized water are listed in Table I. Figure 4 presents the ANSYS results obtained for the velocity distribution of channels with a depth (or height) of 40 nm.

Table I. Physical properties of water at a temperature of 30 °C used in this study.

Density (kg/m^3)	Viscosity $(N \cdot S/m^2)$	Surface Tension (N/m)
995.7	7.975×10^{-4}	7.118×10^{-2}

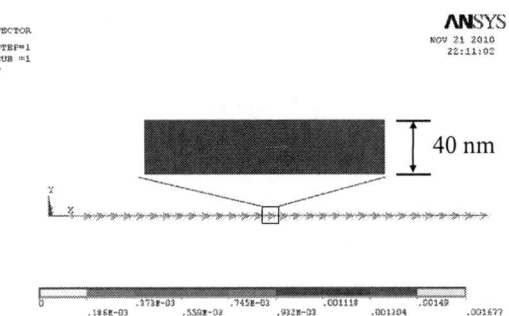

Fig. 4. The ANSYS simulation results for flow rate of nanofluidic channels.

IV. RESULTS AND DISCUSSION

Since a hydrophilic surface improves the bonding strength of the glass substrate and liquid moving in a channel, the glass substrates were dipped in a 1 M HCL solution for 2 min to remove precipitated particles and to increase surface flatness. The substrates were then cleaned with piranha solution. To evaluate the hydrophilicity and flow rate in the glass channels, the advancing contact angle was measured with a contact angle analyzer (FTA 188, First Ten Angstroms) on a flat glass surface, the results are shown in Fig. 5. Note that the real contact angle of a moving liquid in a channel is not constant and depends on the wetting-line velocity [15]. The contact angle in this analysis is assumed to be constant and independent of flow rate. The measurement results of the advancing contact angles on clean and unclean glass substrates were approximately 23.2° and 59.0°, respectively and did not appreciably change over time.

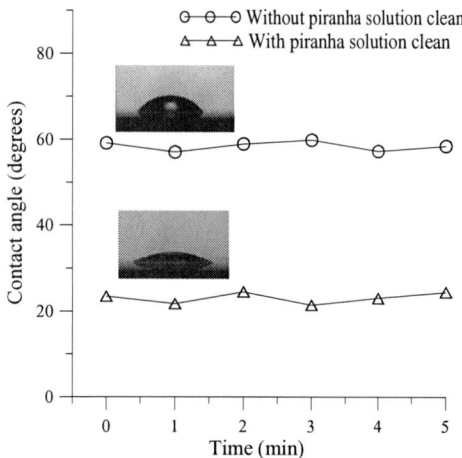

Fig. 5. Advancing contact angles on glass surfaces showing no appreciable changes over time.

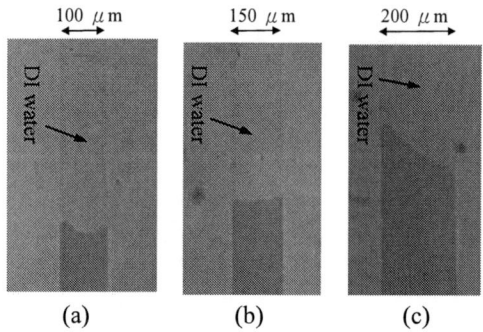

Fig. 6. Images of the capillary filling in 40-nm-deep nanochannels. The widths of channels are (a) 100 μm, (b) 150 μm, and (c) 200 μm, respectively.

Figure 6 shows the fabricated in 40-nm-deep with three different widths of nanofluidic channels being filled with DI

water. Due to the strong capillary force, DI water was introduced into the 40-nm-deep nanofluidic channel without any external driving force. The filling process was video recorded at a frame rate of 30 images/s. The videos were then analyzed frame by frame to measure the position of the meniscus as a function of time.

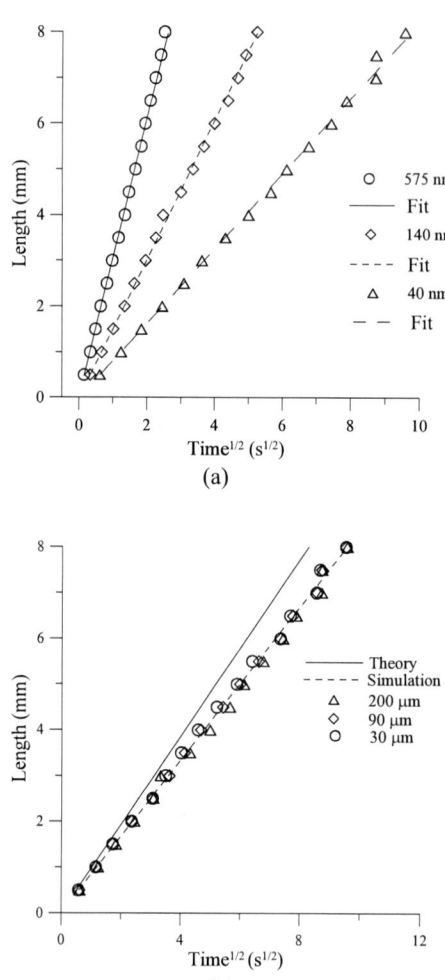

Fig. 7. Position measurement as a function of the square root of filling time. (a) Width of 200 μm with three channel depths and (b) depth of 40 nm with three channel widths, the theoretical and simulation values were also plotted for comparison.

Figure 7(a) shows the measured position of the moving meniscus as a function of the square root of time, for three channel depths (40 nm, 140 nm, 575 nm) with a width of 200 μm. A near perfect linear relation was noted between the position and the square root of time. It is observed that the flow rates decreased dramatically with decreasing depths. The average flow rates were approximately 1.27, 0.30, and 0.09 mm/s for depths of 575, 140, and 40 nm, respectively, over a total channel length of 8 mm. Figure 7(b) shows the measured position of the moving meniscus as a function of the square root of time, for three channel widths (30 μm, 90 μm, 200 μm) with a depth of 40 nm. Experiments show that the flow rates are independent of micrometer-scale widths. The experimental data agree well with the simulation results. The theoretical predictions are notably higher than the experimental values. Some discrepancies might be attributed to the unsteady flow profiles. Since the channel flow could be hindered by surface conduction and double-layer overlap in nanoscale channels [16], it would be very difficult to accurately predict the flow rate without adjustable parameters.

V. CONCLUSION

This study has proposed the capillary filling speed of water inside homogenous coverslip nanochannels. The capillary filling speed was measured and compared with theoretical and simulation values. Although filling speeds inside coverslip nanochannels were lower than the theoretical values, they could be accurately described by the simulation. Simulation and experimental results indicate that the commonly used theory formula for predicting the capillary filling speed is not accurately without adjustable parameters. Experiments confirm that capillary filling speed decreases as nanochannel depth decreases but is independent of micrometer-scale width. The flow rate inside the coverslip nanochannels was sufficiently high even for nanochannels with a depth of 40 nm. The coverslip nanochannel proposed in this study has potential applications in pumpless high speed analytical devices and in studies of flow kinetics in nanoscale regimes.

ACKNOWLEDGMENT

The authors gratefully acknowledge the financial support provided to this study by the National Science Council of Taiwan under Grant No. NSC 100-2221-E-150-019. In addition, the access provided to fabrication equipment by the Common Lab for Micro/Nano Science and Technology, Research and Services Headquarters at National Formosa University is greatly appreciated.

REFERENCES

[1] N. R. Tas, P. Mela,T. Kramer, J. W. Berenschot, and A. V. D. Berg, "Capillarity induced negative pressure of water plugs in nanochannels," *Nano Letters*, vol. 3, no. 11, pp. 1537-1540, 2003.

[2] L. J. Yang, T. J. Yao, and Y. C. Tai, "The marching velocity of the capillary meniscus in a microchannel," *J. Micromech. Microeng.*, vol. 14, pp. 220-225, 2004.

[3] J. W. V. Honschoten, M. Escalante, N. R. Tas, and M. Elwenspoek, "Formation of liquid menisci in flexible nanochannels," *J. Colloid Interface Sci.*, vol. 329, no. 1, pp. 133-139, 2009.

[4] S. W. P. Turner, M. Cabodi, and H. G. Craighead, "Confinement-Induced Entropic Recoil of Single DNA Molecules in a Nanofluidic Structure," *Phys. Rev. Lett.*, vol. 88, no. 12, pp. 128103, 2002.

[5] H. E. Jeong, P. Kim, M. K. Kwak, C. H. Seo, and K. Y. Suh, " Capillary kinetics of water in homogeneous, hydrophilic polymeric micro- to nanochannels," *Small*, vol. 3, no. 5, pp. 778-782, 2007.

[6] A. Hibara, T. Saito, H.-B. Kim, M. Tokeshi, T. Ooi, M. Nakao, and T. Kitamori, "Nanochannels on a Fused-Silica Microchip and Liquid Properties Investigation by Time-Resolved Fluorescence Measurements," *Anal. Chem.*, vol. 74, no. 24, pp. 6170-6176, 2002.

[7] J. Haneveld, N. R. Tas, N. Brunets, H. V. Jansen, and M. Elwenspoek, "Capillary filling of sub-10 nm nanochannels," *J. Appl. Phys.,* vol. 104, pp. 014309, 2008.

[8] J. Haneveld, H. Jansen, E. Berenschot, N. Tas, and M. Elwenspoek, "Wet anisotropic etching for fluidic 1D nanochannels," *J. Micromech. Microeng.*, vol. 13, pp. S62-66, 2003.

[9] J. Han, and H. G. Craighead, "Separation of long DNA molecules in a microfabricated entropic trap array," *Science*, vol. 288, pp. 1026-1029, 2000.

[10] H. T. Hoang, I. M. S.Nolten, J. W. Berenschot, M. J. D. Boer, N. R. Tas, J. Haneveld, and M. C. Elwenspoek, "Fabrication and interfacing of nanochannel devices for single-molecule studies," *J. Micromech. Microeng.*, vol. 19, pp. 065017, 2009.

[11] L. H. Thamdrup, F. Persson, H. Bruus, A. Kristensen, and H. Flyvbjerg, "Experimental investigation of bubble formation during capillary filling of SiO$_2$ nanoslits," *Appl. Phys. Lett.*, vol. 91, pp. 163505, 2007.

[12] M. K. Schwiebert and W. H. Leong, "Underfill flow as viscous flow between parallel plates driven by capillary action," *IEEE Trans. Compon. Packaging Technol. Part C*, vol. 19, no. 2, pp. 133-137, 1996.

[13] E. W. Washburn, "The dynamics of capillary flow," *Phys. Rev.*, vol. 17, pp. 273-283, 1921.

[14] N. R. Tas, J. Haneveld, H. V. Jansen, M. Elwenspoek, and A. van den Berg, "Capillary filling speed of water in nanochannels," *Appl. Phys. Lett.*, vol. 85, pp. 3274-3276, 2004.

[15] P. G. de Gennes, "Wetting: statics and dynamics," *Rev. Mod. Phys.*, vol. 57, pp. 827-863, 1985.

[16] J. C. T. Eijkel, J. Bomer, N. R. Tas, and A. van den Berg, "1-D nanochannels fabricated in polyimide," *Lab Chip*, vol. 4, pp. 161-163, 2004.

A Visualization Study of Venting Gas via Hydrophobic Nanoporous Membrane

Jin-Cherng Shyu[1,*], Sin-Min Dai[1], Kai-Shing Yang[2] and Chi-Chuan Wang[3]

[1]Department of Mechanical Engineering/National Kaohsiung University of Applied Sciences, Kaohsiung, Taiwan
[2]Green Energy and Environment Research Lab./Industrial Technology and Research Institute, Hsinchu, Taiwan
[3] Department of Mechanical Engineering/National Chiao Tung University, Hsinchu, Taiwan

jcshyu@cc.kuas.edu.tw

Abstract—This study examines the venting gas characteristics in a microchannel through a hydrophobic nanoporous membrane with a pore size of 0.22 µm and a porosity of 70%. A total of three microchannels were made and tested, including a serpentine, a parallel and a contraction-expansion configuration. The mass flux (G) is 5, 7.5 and 10 kg/m^{2}s with the quality (x) ranging from 0 to 0.1. Flow visualization is also performed to observe the bubbles movement under different experimental conditions. The tested results show that the amount of venting gas is strongly related to the contact time and configuration of gas slug and the development of flow pattern in the channel.

Keywords - two phase flow; microchannel; hydrophobic; nanoporous membrane; venting gas

I. INTRODUCTION

Portable electronic devices are now almost everywhere around our daily life. In addition to their basic functions, f smaller power source, low cost, environmentally benign, highly efficient and longevity [1] are regarded as the most important features. The DMFCs have unique advantages such as high energy-conversion efficiency, easy storage of liquid fuel, ambient-temperature operation, and simple construction [2] are therefore considered as one of the most potential candidates of portable power source. Typically, the DMFC is operated at a temperature lower than 100 °C and consists of a proton exchange membrane with an anode and a cathode catalyst layer on each side [3]. The methanol is electrochemically oxidized to become carbon dioxide, while oxygen is reduced to water at cathode. According the overall electrochemical reaction of DMFC, the Carbon dioxide gas bubbles constantly generated in anodic side. The generated gas in the microchannels would increase flow resistance and even block the flow. Moreover, gas bubble can block the reactant from the surfaces of the catalyst and electrodes, negating the inherent benefit of microreactors [4]. Therefore, gas venting is an important issue in the development of DMFC.

Venting gas in microchannels has received much attention due to its importance in practical applications of direct methanol fuel cells (DMFCs) which show potential applications as a portable power source. The gas venting in a microchannel had been studied by several investigators and various methods such as dynamic bubble traps and diffusion/capillarity based devices [5]. The gas/liquid separation using porous membrane has been successful demonstrated of remove bubbles in micochannels by diffusion/capillarity methods [6-7]. However, until now, there still lacks of experimental results for the gas venting performance subject to the configurations of microchannel. In this regard, the objective of this study is to provide some detailed experimental data to explore the associated influence.

II. EXPERIMENTAL SETUP

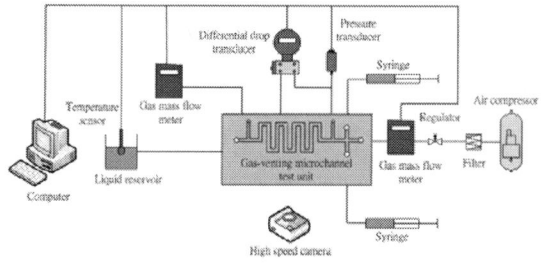

Fig. 1. Experimental set up.

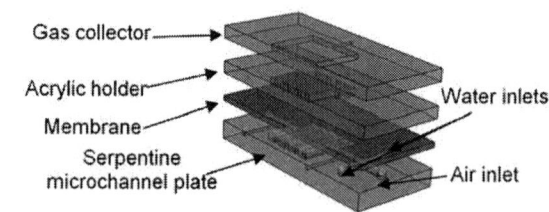

Fig. 2. Schematic of the test section.

978-1-4673-1122-9/12 $31.00 © 2012 IEEE

The schematic of the experimental apparatus and test section is shown in Fig. 1 and Fig. 2. The microchannels are with a serpentine winding channel as the test section, two liquid inlets and one air inlet, as well as an outlet which was fabricated in an acrylic plate. A Fluoropore membrane (Millipore), with an average pore size and a porosity being 0.22 μm and 70%, respectively, was attached to the test section. An acrylic plate and the microchannel were carefully joined together with several screws to prevent leakage. Besides, an acrylic holder was fixed on the nanoporous membrane to collect the air discharged from the microchannel via the nanoporous membrane. A differential pressure transmitter (YOKOGAWA EJA110A) having an adjustable span of 1300 to 13000 Pa was used to measure the differential pressure of the air/water two-phase flow across the test section of the present microchannel. The ambient air, served as the gas phase in the present study, was compressed to a higher pressure with an air compressor and then stored in a reservoir. Between the air reservoir and the microchannel, a filter and a regulator were set up to adjust air to a specified flow rate before went into the microchannel.

Two gas flow meters (NEW-FLOW, TLF-04-A-1-W-1-1-2) were employed in the present study. One was used to measure the inlet air flow rate, and the other was connected to an acrylic holder fixed on the nanoporous membrane to measure the flow rate of discharged air from the microchannel through the porous membrane. The accuracy of the flow meter for air is within ±1% of the test spans. The reading difference between these two gas flow meters represents the air flow rate that is not discharged through the membrane and is residual in the microchannel.

In addition, a dual-syringe pump (KD Scientific, KDS-210) was used to pump the deionized water into the microchannel. The contact angle between the water drop and the present membrane was 118.4°. From the measurements of the flow rate of the air and water, mass flux (G) and quality (x) can thus be estimated in terms of cross-sectional area of the microchannel (A), volumetric flow rate ($\dot{Q}$) and density (ρ) as follows,

$$G = G_L + G_G = \frac{\dot{Q}_L \rho_L}{A} + \frac{\dot{Q}_G \rho_G}{A} \quad (1)$$

$$x = \frac{G_G}{G} \quad (2)$$

Where the subscripts G and L denote gas phase and liquid phase. The mass flux tested in the present study was 5, 7.5, and 10 kg/m²s, respectively. In addition, the quality tested in the present study ranged from 0 to 0.1.

All measured data, including pressure drop and flow rate, were transmitted to a data acquisition unit (Yokogawa MX100) for further analysis.

A visualization instrument including a lens (Nikon AF-S) mounted on a high-speed video camera (Fastec imaging, TroubleShooter 500), whose maximum shutter speed is 1/2000 second, and an illuminator (Dolan Jenner, Fiber-Life) was simultaneously used for the air/water two-phase flow visualization in the microchannel.

In this study, three different type microchannels were made of acrylic via precise machining, the corresponding channels patterns are of (a) serpentine, (b) parallel; and (c) contraction-expansion configurations. Detailed geometries and dimensions of tested samples are shown in Fig. 3. The depth of corresponding microchannels of serpentine, parallel and contraction-expansion are 530 μm, 500 μm, and 500 μm, respectively.

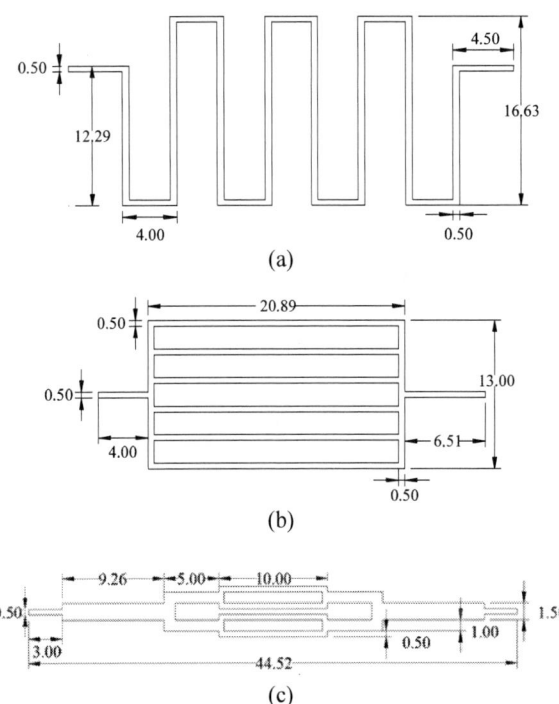

Fig. 3. All tested microchannels in the present investigation (a). serpentine, (b). parallel and (c). contraction-expansion. (Unit : mm)

III. Results and Discussion

In order to examine the effects of operating conditions on tested samples for gas/liquid separation, various mass fluxes (G) of 5, 7.5 and 10 kg/m²s with the quality (x) ranging from 0 to 0.1 was tested. Typical gas-venting processes at G = 7.5 kg/m²s, x = 0.02 of serpentine microchannels are shown in Fig. 4. This results show that the gas slug can be completely vented via the hydrophobic nanoporous membrane. During the gas-venting process, there is no liquid water presented in the acrylic holder, indicating no liquid water can penetrate through the hydrophobic nanoporous membrane.

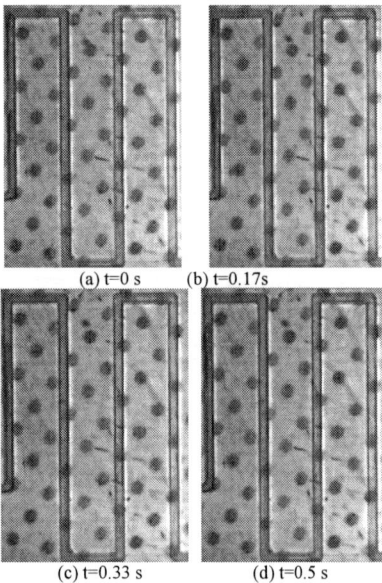

Fig. 4. Exemplary venting gas process vs. time of serpentine microchannel at G = 7.5 kg/m²s, x = 0.02.

The high speed flow visualization of gas-venting processes at $G = 10$ kg/m²s, $x = 0.09$ of parallel microchannels and $G = 7.5$ kg/m²s, $x = 0.05$ of contraction-expansion microchannels are shown in Fig. 5. and Fig. 6. The pictures show that the gas/liquid flow in the parallel and contraction-expansion microchannels is non-uniform. However, the results show that the gas slug can be completely vented via the hydrophobic nanoporous membrane of the tested condition. As show in Fig. 6, the region of contraction-expansion can enhance the performance of gas venting.

For further comparison the performance of gas venting, the experimental results of quality vs. farthest distance and gas permeation flux for the serpentine microchannel are shown in Fig. 7 and Fig. 8. The tested results show that the amount of venting gas is strongly related to the contact time, configuration of gas slug and the development of flow pattern in the channel. At a lower quality the gas permeation flux rise rapidly, but the gas permeation flux tends to be saturated with a further increase of quality. At a higher quality and mass flux, we find that a long bubble may be stretched and break into two parts as shown in Fig. 9. This is because the swirled motion caused by the turning. In fact, three

stages of the marching distance vs. gas quality are identified in serpentine microchannel in this study. Namely an initially linear relationship, followed by an unchanged marching distance and finally a considerable increase at a higher gas quality, this phenomenon is particularly conspicuous at a higher mass flux. The serpentine microchannel shows higher permeation flux values than the other microchannels in this study. The gas and liquid flow paths is in a random in parallel and contraction-expansion microchannels, resulting in an unstable permeation flux. The corners effect of venting gas becomes even more significant in contraction-expansion microchannel.

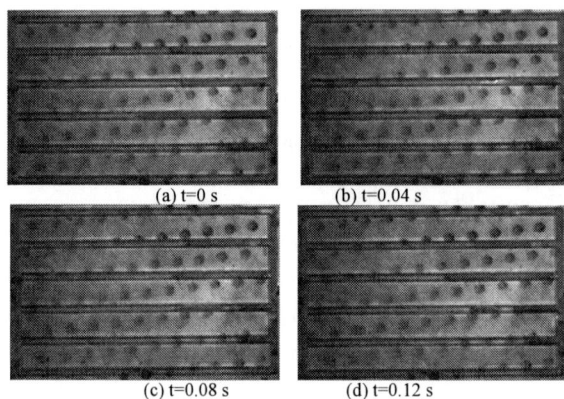

Fig. 5. Exemplary venting gas process vs. time of serpentine microchannel at G = 10 kg/m²s, x = 0.09.

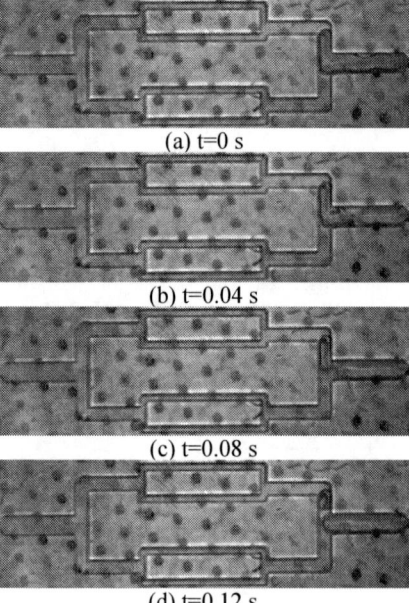

Fig. 6. Exemplary venting gas process vs. time of contraction-expansion microchannel at G = 7.5 kg/m²s, x = 0.05.

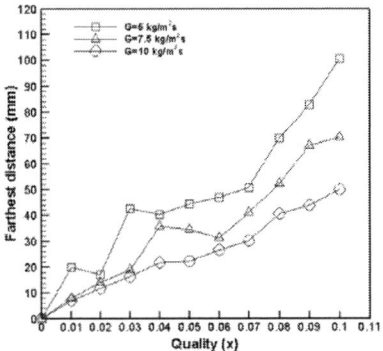

Fig. 7: Quality vs. farthest distance for the serpentine microchannel.

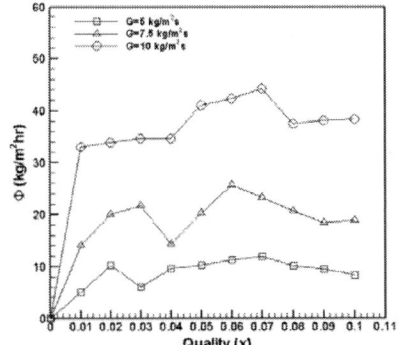

Fig. 8: Quality vs. gas permeation flux for the serpentine microchannel.

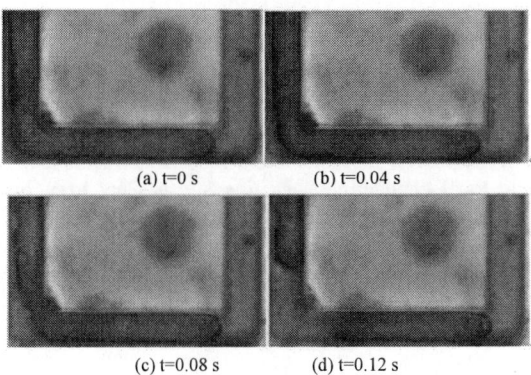

(a) t=0 s   (b) t=0.04 s

(c) t=0.08 s   (d) t=0.12 s

Fig. 9. High speed flow visualization of break-up process.

IV. Conclusions

This study examines the venting gas characteristics in microchannel by a hydrophobic nanoporous membrane with a pore size of 0.22 μm and a porosity of 70%. A total of three microchannels were made and tested, including serpentine, parallel and contraction-expansion. The mass flux (G) tested in the present study is 5, 7.5 and 10 kg/m^{2}s with the quality (x) ranging from 0 to 0.1.

Flow visualization is also performed to observe the bubbles movement under different experimental conditions. The tested results show that the amount of venting gas is strongly related to the contact time and configuration of gas slug and the development of flow pattern in the channel. At a lower quality, the gas permeation flux rise rapidly, but the gas permeation flux tends to be saturated with a further increase of quality. At a higher quality and mass flux, we find that a long bubble may be separated into two at the corner. Three stages of the marching distance vs. gas quality are identified in serpentine microchannel in this study. Namely an initially linear relationship, followed by an unchanged marching distance and finally a considerable increase at a higher gas quality, this phenomenon is particularly conspicuous at a higher mass flux. The serpentine microchannel shows higher permeation flux values than the other microchannels. The corners effect on venting gas becomes even more significant in the contraction-expansion microchannel.

ACKNOWLEDGMENT

The authors are indebted to the financial support from the Bureau of Energy of the Ministry of Economic Affairs, Taiwan.

REFERENCES

[1] Q. Liao, X. Zhu, X. Zheng, and Y. Ding, "Visualization study on the dynamics of CO2 bubbles in anode channels and performance of a DMFC," *Journal of Power Sources*, vol. 171, pp. 644–651, 2007.

[2] C. W., Wong, T. S. Zhao, Q. Ye, and J. G. Liu, "Transient Capillary Blocking in the Flow Field of a Micro-DMFC and Its Effect on Cell Performance," *Journal of the Electrochemical Society*, vol. 152, pp. A1600-A1605, 2005.

[3] H., Yang, T.S. Zhao, and Q. Ye, "In situ visualization study of CO$_2$ gas bubble behavior in DMFC anode flow fields," *J. Power Sources*, vol. 139, pp. 79-90, 2005.

[4] D. D. Meng, T. Cubaud, C. M. Ho, snd C. J. Kim, "A Methanol-Tolerant Gas-Venting Microchannel for a Microdirect Methanol Fuel Cell", *J. Microelectromech. Syst*, vol. 16, pp.1403-1410, 2007.

[5] J. Xu, R. Vaillant, and D. Attinger, "Use of a porous membrane for gas bubble removal in microfluidic channels: physical mechanisms and design criteria," *Microfluidics and Nanofluidics*, vol. 9, pp. 765-772, 2010.

[6] D. D. Meng, J. Kim, and C. J. Kim, "A degassing plate with hydrophobic bubble capture and distributed venting for microfluidic devices," *J Micromech Microeng*, vol. 16, pp. 419–424, 2006.

[7] A Kamitani, S. Morishita, H. Kotaki, and S. Arscott, "Improved fuel use efficiency in microchannel direct methanol fuel cells using a hydrophilic macroporous layer," *J Power Sources*, vol. 187, pp. 148–155, 2009.

[8] X. Zhu, "Micro/nanoporous membrane based gas–water separation in microchannel," *Microsyst Technol*, vol. 15, pp. 1459–1465, 2009.

Switching Characteristic Model and Biochemical Application Analysis for Electrolyte-oxide-semiconductor Structure Diodes

G. C. Sun[1], X. Y. Ma[1], A. S. Tang[2], Y. F. Chen[1], W. G. Wu[1]

[1]National Key Laboratory of Science and Technology on Micro/Nano Fabrication, Institute of Microelectronics, Peking University, Beijing, P. R. China;

[2]Department of Materials Science and Engineering, Massachusetts Institute of Technology, Cambridge, MA, USA

E-mail: wuwg@pku.edu.cn

Abstract—We present a model and applications analysis of the switching characteristic of electrolyte-oxide-semiconductor (EOS) structure diodes. The EOS structure consists of a heavily-doped silicon layer, an SiO_2 layer, and an electrolyte-solution layer, and exhibits diode characteristics when a sweeping voltage is applied. A conduction model of the switching characteristic of EOS diodes is suggested, and implements aspects of ion diffusion theory. The application potential of the EOS structure is also analyzed in accordance with the suggested model. EOS structures provide a simple means to fabricate half-fluidic diodes without photolithography and other sophisticated structure-fabrication processes; hence, they can be utilized for many biochemical applications, such as ion detection and current control in microfluidic devices.

Keywords— *electrolyte-oxide-semiconductor structure; switching characteristic model; ion detection*

I. INTRODUCTION

Lab-on-a-chip technology, which is based on micromachining techniques that shrink the size of fluid-handling systems, can reduce the size, cost, and speed of chemical and biological experimental analyses [1]. With the miniaturization of devices, interfacial effects become more and more important due to increasing surface area-to-volume ratios [2]. During the past few decades, the developments in micro- and nanotechnology have made it possible to fabricate channels with at least one dimension ranging from 1 to 100 nm [3]; as a result, interfacial effects are of great importance in these devices. Techniques such as electrophoresis, electroosmotic flow, and electrofluidic gating are employed to control the ion flow in micro- and nanochannels. When electrolytes are placed onto the surface of a SiO_2 layer, an electrical double layer (EDL), a shielding layer that is naturally created within the liquid near a charged surface [4], will form at the interface. Due to the different electrochemical properties of the insulator and the electrolyte, the EDL's properties can vary widely and be strongly influenced by the applied voltage, suggesting possible applications in sensors and other devices. Therefore, the electrolyte-insulator-semiconductor (EIS) structure, which first appeared in the ion-sensitive field-effect transistors proposed by Bergveld [5], has

been widely studied, using SiO_2 as an insulator layer. Based on this structure, C. D. Fung et al proposed devices called electrolyte-insulator-semiconductor field-effect transistors (EISFETs) [6]; these devices were broadly studied to determine their ability to measure pH values and detect ion concentrations in electrolytes [7].

Theoretically, SiO_2 is perfect insulator that either cuts off leakage current or stops charged species diffusion. However, the insulator layer was found to only be partially insulated under specific conditions. For example, hydrogen-related currents induced by vacuum ultraviolet radiation in ultrathin SiO_2/Si structures have been reported by V.V.Afanas'ev and A.Stesmans [8]. The mechanism for the degeneration of the thin SiO_2 layer has been researched to improve the properties of MOSFETs [9] [10].

In our experiment, when the thickness of the SiO_2 layer exceeded 100nm, we observed a significant leakage current through the electrolyte-oxide-semiconductor structure under a specific voltage range that resulted in the EOS structure displaying an obvious one-way electrical conductivity [11]. Thus far, however, the conducting mechanism for this one-way electrical conductivity characteristic has not been fully understood [1]. In this paper, we build a conduction model describing the switching characteristic of EOS structure diodes. The switching characteristic model, which is based on the transportation of positive ions in the SiO_2 layer, successfully fits the experimental phenomenon. EOS structure diode samples with SiO_2 layers of different thicknesses were fabricated. The electrical characteristics of EOS structure diodes were tested using different electrolyte-solutions, and subsequently analyzed. The results show that the I-V relationship of this structure correlates with the thickness of the SiO_2 layer and the properties of the electrolytes, including the diffusion coefficients and the concentrations.

In accordance with the model and analysis, biochemical applications influenced by the critical factors, such as ion detection and electricity control in microfluidic devices, have been analyzed. In addition, the theory can provide suggestions for insulating lab-on-a-chip systems.

978-1-4673-1122-9/12 $31.00 © 2012 IEEE

II. FABRICATION

The fabrication process of EOS structure diodes is schematically depicted in Fig.1. First, phosphorus ions were diffused into the <100> silicon substrates at a temperature of 900℃ for 30 min to produce n-type doping. This was followed by phosphorus ion implantation on the backside of the substrates (Dose: $6 \times 10^{15}/cm^2$, Energy: 80 KeV) to make the substrates heavily n-type doped. Then, the substrates were annealed at 1000℃ for 10s to further the ohmic contact to the aluminum electrode. SiO_2 layers of different thicknesses, including 1000Å, 3000Å, and 1um, were subsequently fabricated on these heavily-doped surfaces using low-pressure-chemical-vapor-deposition (LPCVD), as shown in Fig.1(b). 500nm layers of aluminum were finally sputtered on the backside of the substrates as electrodes, and the structures were annealed at a temperature of 430℃ to await further electrical investigations (Fig.1(c)).

Following this, polydimethylsiloxane (PDMS) prepolymer and a curing agent were mixed in a 10:1 weight ratio. The mixture was cured in an oven at a temperature of 60 ℃ for 60 minutes. 3 millimeter-diameter holes were then punched on the dried PDMS slice at specific positions; these would serve as electrolyte reservoirs. After a 10-second oxygen-plasma treatment on the front sides of the SiO_2-covered substrate and PDMS slice, the two parts were bonded in an oven at a temperature of 120 ℃ for 2 hours (Fig.1(d)). The final EOS structure was completed after filling the reservoirs with electrolytes, as shown in Fig.1(e).

III. MODELING

It is widely believed that SiO_2 is a perfect insulator in integrated circuits. However, it shows low-resistivity under specific conditions. In EOS structures, when the electrolyte is in contact with the SiO_2 layer, some positive ions in the solution will diffuse into the SiO_2 layer. The diffusion will be affected by the applied voltage. As a result of some ions'

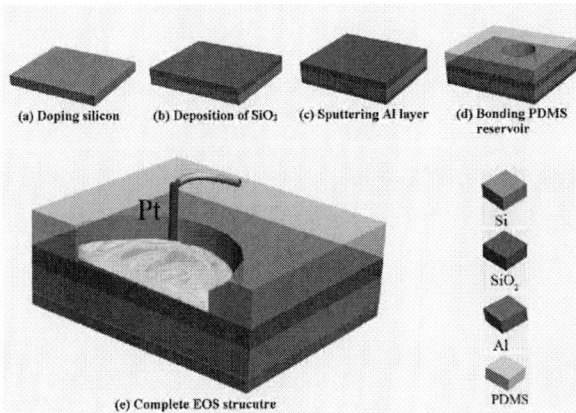

Fig. 1: The fabrication process of an EOS structure diode.

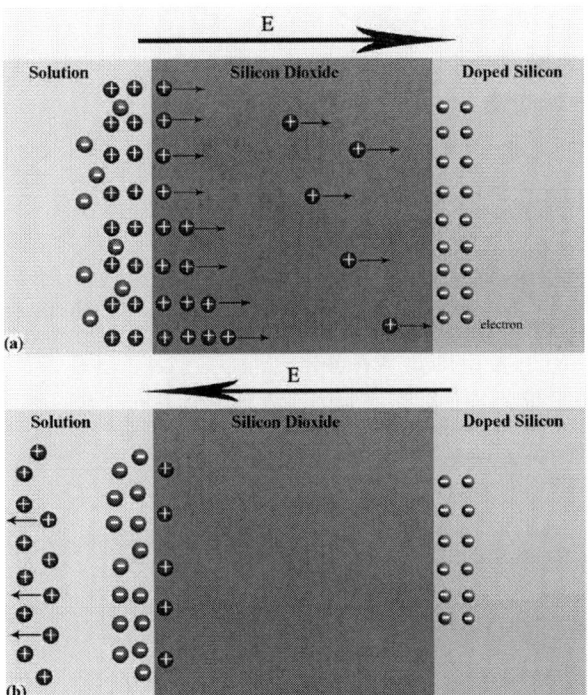

Fig. 2: Schematic model of the switching characteristic of an EOS structure diode: (a) on-state of the EOS structure under forward bias; (b) off-state of the EOS structure under reverse bias.

relatively high mobility in SiO_2 [12], positively-charged ions such as hydrogen ions and alkali ions act as current carriers in the EOS structure. When applying a forward bias (from electrolyte to semiconductor), positive ions in the electrolyte solution will diffuse into the SiO_2 layer due to the concentration gradient and the applied electric field, turning the EOS diode to the on-state. In contrast, when applying a reverse bias, the positive ions will diffuse out of the SiO_2 layer into the electrolyte, and will move away from the electrolyte-oxide interface, switching the EOS diode to the off-state (Fig.2).

In the on-state, after a short time to allow the distribution of ions in the SiO_2 layer to stabilize, the SiO_2 layer can be divided into two regions based on the boundary layer approximation [13]: a thin boundary layer near the electrolyte-oxide interface that has a relatively large ion concentration, in which the ion transportation is dominated by diffusion; the remainder of the SiO_2 layer, in which the ion transportation is dominated by drift (Fig.2(a)). Positive ions will move to the oxide-silicon interface and recombine with electrons, creating the forward on-state current.

According to this approximation, the I-V relationship can be stated in terms of the Poisson-Boltzmann description, as follows:

Within the boundary layer:

$$\frac{\partial C}{\partial t} = D\frac{\partial^2 C}{\partial x^2} \qquad 0 < x < x_1 \quad t > 0$$

$$J_1 = -eD\frac{dC}{dx}$$

$$\frac{d^2 V}{dx^2} = -\frac{eC}{\varepsilon_0 \varepsilon_r}$$

C is the carrier ion (positive-ion) concentration, D is the carrier ion diffusion coefficient in SiO_2, e is the electron charge, ε_0 is the vacuum permittivity, and ε_r is the relative permittivity of SiO_2. J_1 refers to the current density, while x_1 refers to the location of the two SiO_2 regions' boundary.

Since the ion distribution in the SiO_2 layer is supposed to be stable, the ion concentration will not vary with the time, which means

$$\frac{\partial C}{\partial t} = 0.$$

The boundary conditions for the equation are

$$C(0) = C_0 \qquad C(x_1) = C_1$$

Here, C_0 is the initial carrier ion concentration at the electrolyte-oxide interface, and C_1 is the carrier ion concentration at the boundary, below which we can ignore the diffusion part of the current.

Outside the boundary layer:

The carrier ion concentration is taken as a constant, and the current consists only of the drift current induced by the electric field:

$$J_2 = C_1 e\mu E \qquad x_1 < x < L$$

The carrier ion concentration in this region is the same C_1 as above. μ is the ion mobility in SiO_2, E is the applied electrical field strength, and L is the thickness of the SiO_2 layer.

Consider the boundary conditions of the electrolyte and the semiconductor, and the continuity of the current density at the boundary of the two regions in the SiO_2 layer:

$$V(0) = V_0 \qquad V(L) = 0$$

$$J_1(x_1) = J_2(x_1)$$

V_0 refers to the voltage applied between the electrolyte and the semiconductor.

Hence, the I-V relationship of an EOS structure diode can be derived from the model above. The theoretical I-V curves are shown in Fig.3, illustrating the EOS structure diode's switching characteristic and distinct threshold voltage.

IV. CHARACTERIZATION AND DISCUSSION

In order to verify the switching characteristic model of EOS structure diodes, an Auger Electron Spectroscopy test was employed to examine the ion distribution in the SiO_2 layer after applying an external forward voltage.

An EOS structure sample with potassium chloride (KCl) solution as the electrolyte was galvanized under a 6V forward bias (from electrolyte to semiconductor) for 100 seconds. The ion distribution in the SiO_2 layer of this sample was then determined with an Auger Electron Spectroscopy test; the result is shown in Fig.4. It is clear that K^+, the carrier ion, has a relatively high concentration in the very thin layer near the electrolyte-oxide interface. Outside the boundary layer, the K^+ concentration is low enough to be treated as a constant, which indicates that the diffusion current can be ignored in this region. This result confirms that the model presumed above is reasonable.

The practical I-V relationships of EOS structure diodes possessing different structural parameters were investigated using a HP 4156b semiconductor parameter analyzer. A Pt probe was dipped into the electrolyte solution as an electrode; this is schematically depicted in Fig.1(e).

In order to test the influence of the SiO_2 layer's thickness on the I-V relationship of the EOS structure diodes, diodes with SiO_2 thicknesses of 100nm, 300nm and 1000nm were tested, using deionized water (DI water) as the electrolyte. Fig.5 shows that as the thickness of the SiO_2 layer increases, the on-state current magnitudes for a given voltage correspondingly decrease. This is because the electric field in the SiO_2 layers change due to the different thicknesses of the SiO_2 layers.

I-V relationships of the EOS structure diodes with different electrolytes, indicating different carrier ions in the SiO_2 layers, were also measured. Sodium chloride solution (1 M, in which Na^+ serves as carrier ion) and DI water (in which H^+ serves as carrier ion) were tested. The result, shown in Fig.6, illustrates the significant influence on EOS structure diode properties that different carrier ions have. Due to the different mobilities of the carrier ions in the SiO_2 layer, the on-state current of the

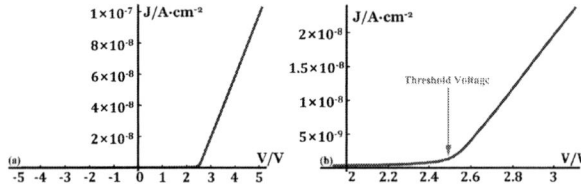

Fig. 3: (a) Theoretical I-V relationship of an EOS structure diode, (b) theoretical threshold voltage of an EOS structure diode, derived via the switching characteristic model.

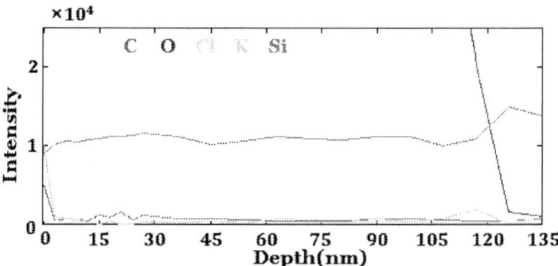

Fig. 4: Auger Electron Spectroscopy test of a 100nm SiO_2 layer EOS structure diode with potassium chloride solution as the electrolyte solution.

978-1-4673-1122-9/12 $31.00 © 2012 IEEE

EOS structure diode is greater when the carrier ion has a relatively large mobility in SiO_2, as is the case with Na^+ (Fig.6(a)). It is also worth noting that the threshold voltage varies when using different electrolyte-solutions in the electrolyte layer; this is probably due to the different diffusion activation energies of the various carrier ions in SiO_2.

From the measurements above, we find that our switching characteristic model for EOS structure diodes fits the practical electrical characteristics well. According to the model and the experimental analysis, both the carrier ion species and the initial carrier ion concentration at the electrolyte-oxide interface strongly affect the I-V relationship of the EOS structure diode. In order to analyze the significant influence of carrier ions on the switching characteristic, the I-V relationships of the EOS structure diodes with different carrier ions (different diffusion coefficient) and different initial carrier ion concentrations at the electrolyte-oxide interface were simulated, as shown in Fig. 7. The results indicate that increasing diffusion coefficient and initial carrier ion concentration can increase the on-state current dramatically while leaving the switching characteristic unchanged. Thus,

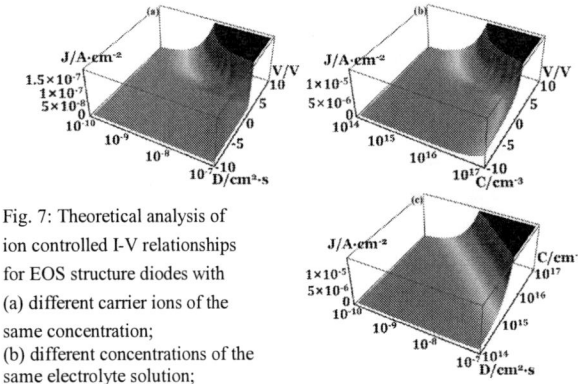

Fig. 7: Theoretical analysis of ion controlled I-V relationships for EOS structure diodes with (a) different carrier ions of the same concentration; (b) different concentrations of the same electrolyte solution; (c) different carrier ions of different concentrations under a fixed bias.

J refers to current density; D refers to diffusion coefficient of the carrier ions in SiO_2; C refers to the initial carrier ion concentration at the electrolyte-oxide interface.

EOS structure diodes can be utilized for ion detection in liquid environments, and has a simple fabrication process and high sensitivity.

V. CONCLUSIONS

An analysis of the switching characteristic of electrolyte-oxide-semiconductor (EOS) structure diodes is presented in this paper. The switching characteristic model is built from the Poisson-Boltzmann description and boundary layer approximation, which successfully explain the conduction mechanism and key features of EOS structure diodes. We believe that, in the EOS structure, positive ions in the electrolyte will diffuse into the SiO_2 layer due to the concentration gradient and applied electric field, resulting in the one-way conductivity observed. An Auger Electron Spectroscopy test was employed to confirm the ion distribution in the SiO_2 layer that was predicted by the model. The practical electrical characteristics of EOS structure diodes with different SiO_2 layer thicknesses and different electrolyte layers (different carrier ions) were also measured. Finally, the model is used to analyze the effect of the carrier ion, a key factor, on the EOS diode characteristic. It shows that EOS structure diodes have a high sensitivity to different electrolyte solution carrier ions. Hence, EOS structures provide a simple approach to constructing half-fluidic diodes without sophisticated fabrication processes, and can be utilized for chemical and biochemical applications, including ion detection and current control in micro/nanofludic devices.

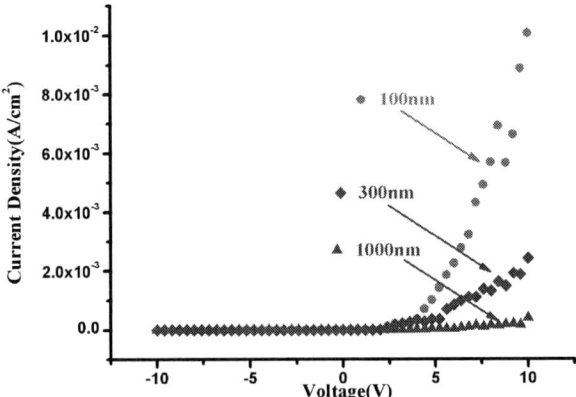

Fig. 5: I-V relationship of EOS structure diodes with different SiO_2 layer thicknesses.

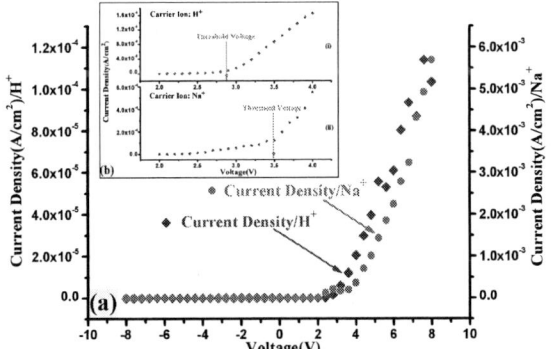

Fig. 6: (a) I-V relationship of EOS structure diodes with sodium chloride solution and DI water as the electrolyte layers; (b) threshold voltage comparison of EOS structure diodes with different electrolyte layers.

ACKNOWLEDGMENT

This work is supported by the National Basic Research Program of China (973 Program, Grant No.2009CB320300) and the Peking University Presidential Research Fund. We thank Yulong Zhang for initiating this work. G.C. Sun would also like to thank Tiancong Zhu for helpful discussions regarding the model.

REFERENCES

[1] Zhijun Jiang and Derek Stein, "Electrofluidic Gating of a Chemically Reactive Surface", *Langmuir* 2010, vol.26, No.11, pp.8616-8173

[2] Adrien Plecis, Reto B. Schoch, and Philippe Renaud, "Ionic Transport Phenomena in Nanofluidics: Experimental and Theoretical Study of the Exclusion-Enrichment Effect on a Chip", *NANO LETTERS*, vol.5, No.6, pp.1147-1155, 2005

[3] Patrick Abgrall and Nam Trung Nguyer, "Nanofluidic Devices and Their Applications", *Analytical Chemistry*, vol.80, No.7, pp.2326-2341, 2008

[4] Reto B. Schoch, Jongyoon Han and Philippe Renaud, "Transport Phenomena in Nanofluidics", *Reviews of Modern Physics*, vol.80, No.3, pp.839-883, 2008

[5] P.Bergveld, "Development of an Ion-Sensitive Solid-State Device for Neurophysiological Measurements", *IEEE Transactions on Bio-medical Engineering*, vol. BME-17, pp.70-71, 1970

[6] Clifford D. Fung, Peter W. Cheung and Wen H. Ko, "A Generalized Theory of an Electrolyte-Insulator-Semiconductor Field-Effect Transistor," *IEEE Transactions on Electron Devices*, vol.ED-33, pp.8-18, 1986

[7] Michael J.Schoning, Anette Simonis, Christian Ruge, Holger Ecken, Mattea Muller-Veggian, and Hans Luth, "A (Bio-)Chemical Field-Effect Sensor with Macroporous Si as Substrate material and a SiO_2/LPCVD-Si_3N_4 Double Layer as pH Transducer", *Sensors*, vol.2, pp.11-22, 2002

[8] V.V.Afanas'ev and A.Stesmans, "Hydrogen-Related Leakage Currents Induced in Ultrathin SiO_2/Si Structures by Vacuum Ultraviolet Radiation", *Journal of The Electrochemical Society*, vol.146, No.9, pp.3409-3414, 1999

[9] Peter E.Blochl and James H. Stathis, "Hydrogen Eletrochemistry and Stress-Induced Leakage Current in Silica", *Physical Review Letters*, vol.83, No.2, pp.372-375, 1999

[10] D.J.DiMaria and E.Cartier, "Mechanism for stress-induced leakage currents in thin silicon dioxide films", *Journal of Applied Physics*, vol.78, No.6, pp.3883-3894, 1995

[11] Y. L. Zhang, G. C. Sun, W. G. Wu, "Diode Characteristic of Electrolyte-Oxide-Semiconductor for Potential Chemical and Biological Applications", *Transducers 2011*, pp.330-333

[12] S. M. Sze, *Semiconductor Devices, Physics and Technology*, 2nd ed.; Wiley: New York, 2002.

[13] E. H. Snow, A. S. Grove, B. E. Deal, and C. T. Sah, "Ion Transport Phenomena in Insulating Films", *Journal of Applied Physics*, vol.36, No.5, pp.1664-1673, 1965

CONTACT

W.G. Wu, Tel: 86-10-62757163; Fax: 86-10-62751789;
E-mail: wuwg@pku.edu.cn

A 3D Micro-channel Cooling System Embedded in LTCC Packaging Substrate

Songtao Jia[1], Min Miao[2,3], Runiu Fang[1,2], Shichao Guo[1,2], Duwei Hu[1,2], Yufeng Jin[1,2]
[1] Information Engineering Institute of Peking University, Shenzhen, China
[2] National Key Laboratory of Science and Technology on Micro/Nano Fabrication, Peking University, China
[3] Institute of Information Microsystem, Beijing Information Science & Technology University, China
jiasongtaohao@163.com

Abstract--**Micro fluidic channel embedded in LTCC packaging substrate is reported in IEEE-NEMS 2010 by our team , which showed that the micro channel can promise a great effect on heat-dissipating of the heat resource packaged in the LTCC substrate. Yet we find there is more room to intensify this effect. So we design a 3-D micro channel embedded in LTCC substrate which is not fabricated in one single layer but fabricated through multilayer substrate and constitute a channel circle surrounding the heat resource packaged inside. We accomplished some simulation on heat distribution, flow pressure and flow velocity field with commercial software FLUENT. This new structure can promise a much better effect on heat redistribution compared to single layer micro channel embedded in LTCC substrate. With this fluid channel the temperature of the outside package is tremendously cut down.**

Keywords: system-in-package (SiP), LTCC(low temperature co-fired ceramic), micro fluid channel

I. INTRODUCTION

From 1960s, the industry of semiconductor has developed by an amazing speed, forming a very significant field in the whole world's economy. And with the technology improves generation by generation, the volume of the chip is decreasing and the density of the transistor is increasing. Just as the prediction of Moore's Law, which indicates that the quantity of the transistors in one chip is doubled every 18 months. As a result of this trend, the thermal density increased greatly to a degree today that we need to set a fan on the CPU to help it work normally despite of the high temperature. So the dissipation of heat generated by the chip itself is becoming more and more important.

Engineers have tried many ways to solve this problem from the packaging process, either by cutting the power of chip, redistributing the circuit design or by adding some sort of heat dissipation part in the package. For the power cutting always leads to performance discount and the redistribution of circuit helps little in heat decreasing, so the third one is desired to be the most hopeful way.

LTCC(Low Temperature Co-fired Ceramic) has been used as a substrate for system-in-package(SiP) and system-on-package(SoP) because of its multilayer fabrication process and capability to embed passives and functional structures. [1] In recent years , by ways of simply stacking up multiple layers of micro-machined LTCC green-tape, the capacity of flexibly fabricating 3-D structures with much lower cost and complexity than Si micro-machining have attracted great interests. Furthermore devices made of LTCC promises wider operational temperature range.

So the idea of fabricating a pipe array embedded in the package of LTCC in which some kinds of big thermal capacity fluid is passing by to take away the heat appears. Some researchers has made some experimental attempt in this direction. Reference [2-3] proved that the micro-channels embedded into LTCC substrate may decrease the temperature rise obviously. Our former work on this field[4] showed that as for the case of LTCC substrate with vacuum micro system packaged inside, the substrate embedded with cooling micro channel cut over 70% of rising temperature compared with the one without embedded cooling micro channel. But LTCC substrate integrated with micro channel and cavity inside has a problem of defects and impurities caused by the machining process, which will affect the reliability of the whole system[5].

Now we know the fluid channel in the package really works. We can go further. The single layer micro channel limited the effect of taking heat power away from the heat resource. So based on former work, we improved the structure of the micro fluid channel ,to change it from single layer to 3-D structure, which surrounds the chip cavity as a square circle. The fluid is introduced from inlet on the top side of the shell, and then divided into two direction, and each direct to the same outlet at the bottom side. With this design, the heat can spread not only from the bottom of the chip, but also from the lateral side and the top side, which will surely promise a much better heat dissipating effect.

II. PHYSICAL PRINCIPLE

We use water as the material of fluid, which can be considered to be uncompressed, so the inlet boundary condition is set to velocity-inlet. Consequently, we can use the given velocity and the density of the water to calculate mass flow rate, momentum flow rate and energy flux. Equation for mass flow rate is as (1):

$$\dot{m} = \int \rho \vec{v} \bullet d\vec{A} \qquad (1)$$

where ρ indicate the density, A indicate the area of cross section.

The wall between fluid channel and the LTCC package is set to be a coupled bilateral wall, and the wall between heater cavity and LTCC package is set to a fixed-temperature wall.

This is for convenience of calculation and well imitation of the actual condition. In this case we can calculate the heat flux on the temperature-fixed wall, the equation is as (2):

$$q = \frac{k_s}{\Delta n}(T_w - T_s) + q_{rad}$$

(2) where s refer to solid body, and w refer to Wall, q refer to heat flux, and q_{rad} refer to radiation flow flux, Δn refer to the distance between wall and the center of solid body.

III. STRUCTURE INTRODUCTION

The main structure is designed as a symmetric channel array consists of 8 parallel channels, as Fig. 1 and Fig. 2 shows. The whole package is made of 10 layers of Dupont 9k7 LTCC green tape, with each layer thickness about 130um. The horizontal channel is 100um in width and 130um in height, and the vertical via channel is cylinder with a diameter of 100um.

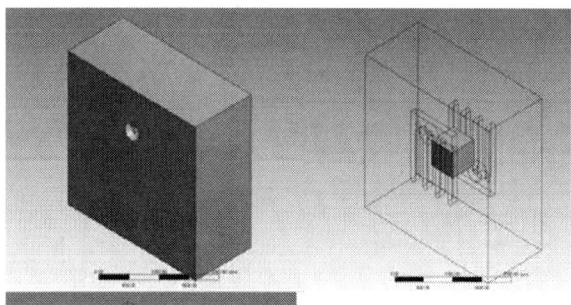

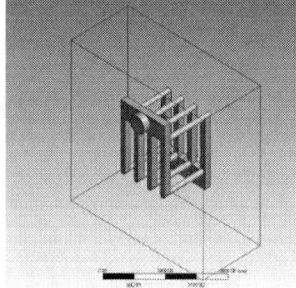

Fig. 1 the main structure of the micro fluid channel system, from left to right: the LTCC package, the cavity as a heat resource, the micro fluid channel

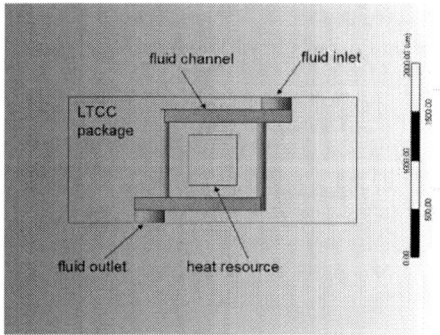

Fig. 2 the lateral view of the fluid channel

The heat cavity is surrounded by the cage of channels from 2 dimensions. We introduce the fluid from the inlet and then it will be divided into 8 ways of stream, four flowing horizontally and the others vertically. With ceaseless supply of cold water, the heat generated by the cavity is taken away efficiently.

with this kind of structure, the heat generated by the chip in the cavity will be taken away from 4 directions, which can promise a better dissipation performance compared to single layer fluid channel. We can even intensify the dissipation by increasing numbers of parallel fluid channels. For fabrication concern, the channel will collapse during co-firing process if the density and the width of the channel is too big. The only way to solve this problem is to use sacrifice process, which is proved to be a efficient way.[6]

IV. SIMULATION AND RESULT

The whole simulation consists of two sections. In first section, we only simulate the fluid channel and try to find out the flow characteristics. While in the second section, we embed the channel into the package, and added a heating cavity in the middle to simulate the whole structure.

A. Section 1

We consider the fluid is water and the initial temperature is set to 300k. While the temperature of channel wall is set to 500k. With this temperature drop we want to see how does the water take energy from the outside environment. The inlet fluid velocity is set respectively to 1um/s, 10um/s,100um/s,1000um/s and even 10mm/s which we know won't be true in reality. As a result, we get corresponding simulation data.

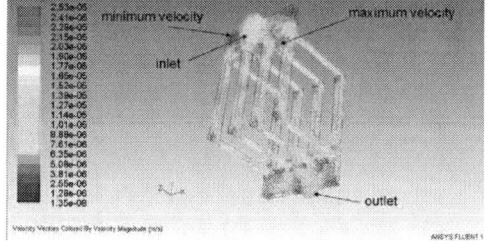

Fig.3 Velocity Vectors(inlet velocity:10um/s)

Fig. 4 Temperature Vectors(inlet velocity:10um/s)

From Fig. 3 we find that the maximum velocity spot is at the vertical via beneath the inlet, which is nearly 25um/s, about 3 times of the initial velocity. While at the corner of the channel the velocity is nearly zero. Where water flow quicker the temperature possibly remain lower. On the contrary, the temperature remain higher where water flow not so quick.

978-1-4673-1122-9/12 $31.00 © 2012 IEEE

This can be certified by comparing Fig. 3 and Fig. 4, where we can find the minimum velocity spot has a much higher temperature than the spots nearby. So we can assume that with a bigger inlet velocity, the dissipation effect will be better. And after simulation of different velocity, we can intensify the conclusion(Fig. 5). As the fluid velocity increases, the temperature in the channel is decreasing. Applying a 1000um/s velocity, we can find from Fig.5 that the channel is rarely affected by the heat. Also pressure applied to the internal wall of the fluid channel increases with the fluid velocity introduced from the inlet(Fig. 6). When the velocity reaches too big, the pressure will exceed the limit of the LTCC material. For a concern of the reliability, the velocity should stay in a proper range.

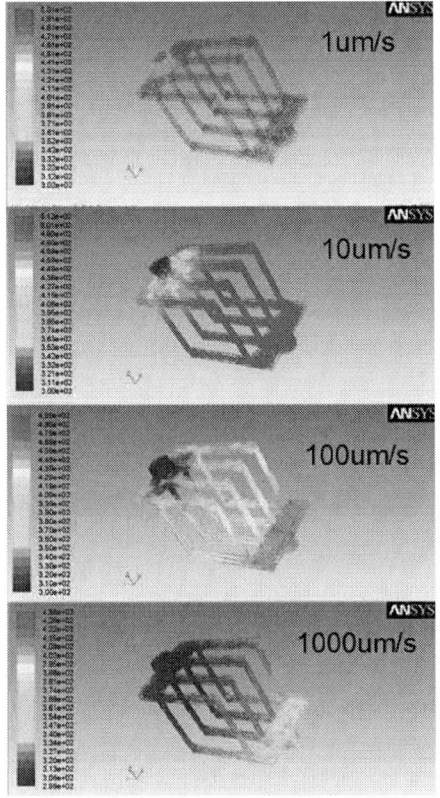

Fig. 5 Temperature Vector vs. Inlet Velocity

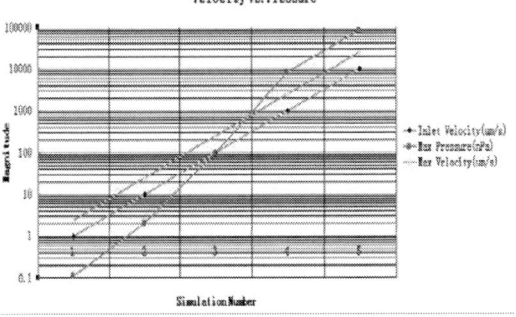

Fig.6 Max Surface Pressure vs Velocity

From this section the dissipation capability of fluid channel is proved. With a proper velocity the heat which spreads to the channel wall is quickly taken away. The channel arrays take a role of heat insulation ,to prevent the heat spreading outwards.

B. Section 2

In section two, for better understanding the effect, we reconstructed a 2-d model of the whole package system from lateral side. The fluid is introduced from the inlet and flow through the circular channel to the outlet. The temperature of the cavity is set to 500k, and the other part is set to 300k as an initial condition.

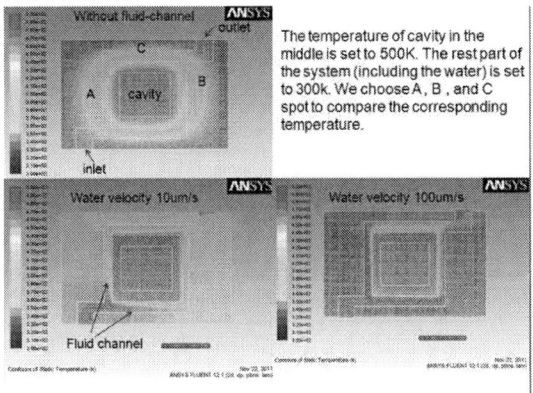

Fig. 7 Temperature simulation of lateral cross section vs different fluid velocity

The solid material is Dupont 9k7 , with a parameter of density (2719kg/m3),Cp(871j/kg-k),and thermal conductivity (2.62w/m-k). Fig. 7 shows the simulation result, in which we choose A, B, and C spot as a comparation. Directly judging from the contour , we can obviously observe that the heat is spreading further in the body of package in the case without fluid-channel, while the heat is efficiently limited in the cavity area in the case with fluid dissipation. Refer to Fig. 8 the result is clearer, where we can find out the difference between temperature of A, B, and C spot. The average temperature is 365k, 337k, and 308k for three kinds of conditions respectively. The case with 100um/s fluid has a temperature cut of 57k compared to the one without fluid.

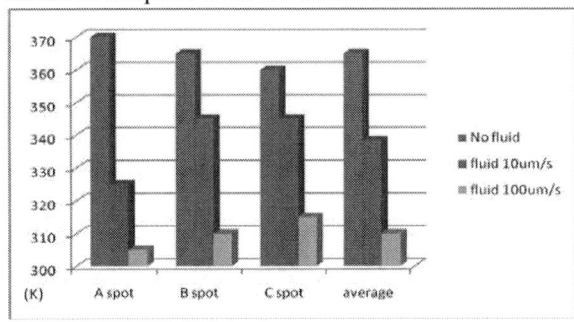

Fig. 8 Temperature of three spot vs. velocity

Fig. 9 visually describe how the fluid channel take effect. The figure shows out how the temperature distribute from the center of heat cavity to the outside part of package along X axis. When there is no fluid channel the temperature is gradually decreasing from the center to the edge with a tiny degree. While with the channel packaged in, the temperature is descending greatly at the channel field. And even more, outside the channel cage ,the temperature has been cut down tremendously by 170k(compared to the center temperature 500k). The simulation result support the conclusion we get in section 1 that the fluid channel plays a role of heat insulator, and certify our prediction of the dissipation effect of the fluid channel.

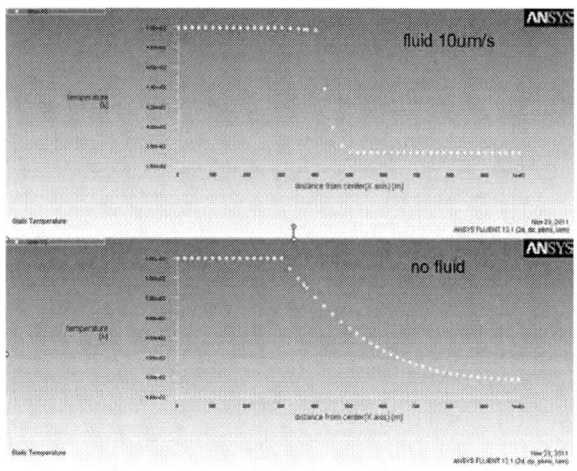

Fig. 9 Temperature distribution (x-axis)

V. CONCLUSION

In this paper we present a new structure of micro fluid channel packaged in LTCC substrate, and construct geometry model with the commercial software Gambit. The model contains 3 main part: the package itself, the cavity in which an imaged chip is packaged, and the fluid channel circle. We choose water of 300k as the coolant, and LTCC as the solid package material.

Then we simulate the model with software FLUENT, and get a series of result. We calculate the flow velocity, wall pressure, temperature distribution of the structure, and come out with the conclusion that with the help of the 3d fluid channel, the heat generated by the packaged chip is taken away efficiently, and the temperature of the outside of package is decreasing greatly with the help of water flowing. With a fluid velocity of 10um/s and the center temperature of 500k, the temperature outside the fluid channel cage is cut by 170k compared to all solid dissipation. And with higher fluid speed the temperature will be cut even more.

ACKNOWLEDGMENT

The work presented here has been funded by National Natural Science Foundation of China (Project No. 60976083 and 60501007), Beijing Natural Science Foundation (Project No. 3102014), China, and Chinese National High-tech R&D Program (863 Program, Project No. 2007AA04Z352). In addition, research funding from National S&T Major Project with the contract No. 2009ZX02038 is greatly acknowledged.

REFERENCES

[1] R. Tummala, "Emerging trends in 3-D systems packaging technologies(invited talk)", *2009 international Conference on Electronic Packaging Tech.& High Density Packaging.* Aug 11th~ 13th ,2009, Beijing, China

[2] LJ Golonka, T. Zawada, J.Radojewski et al. "LTCC micro fluidic system", *International Journal of Applied Ceramic Technology*, 2006, 3,pp.150-156

[3] K. Malecha, L. J. Golonka; "Micro-channel fabrication process in LTCC ceramics", Microelectronics Reliability 48 (2008), pp. 866-871

[4] Min Miao , Jing Zhang; "A LTCC microsystem vacuum package substrate with embedded cooling microchannel and Pirani Gauge", *5th IEEE International Conference on Nano/Micro Engineered and Molecular Systems(2010)*, January 20~23, Xiamen, China

[5] Yang-Fei Zhang, Jia-Qi Chen,"Nanoscale mechanical properties and microstructure of 3D LTCC substrate", *International Conference on Electronic Packaging Technology & High Density Packaging(ICEPT-HDP 2009)*, Aug 10th-13th, Beijing, China

[6] Fangqing Mu, Yangfei Zhang,"Fabrication of micro fluid channel system", *Journal of Institute of Electronic Science of China, 2011 January*, pp 20-23,China

A Flexible Evaporation Micropump with Precision Flow Rate Control for Micro-Fluidic Systems

Kuan-Yu Chen, Kuang-En Chen, Kerwin Wang

Department of Mechatronics Engineering, National Changhua University of Education, Changhua, Taiwan

kerwin@cc.ncue.edu.tw

Abstract—This paper proposes an active-evaporation micropump for micro-fluidic system. The flexible micropump featured power, cost and space efficient structure, capable of stable and adjustable flow-rate control during long working periods. Both passive and active evaporation pumping modes have been studied. Without using external power, passive mode allows the pump to work under well-controlled temperature and humidity (e.g. in incubator). Active evaporation pumping allows the pump to work under ambient which makes this design less sensitive to temperature and relative humidity fluctuation. This paper also presents a novel method for the flexible micropump fabrication.

Keywords- active-evaporation micropump; micro-fluidic; flow-rate control.

I. INTRODUCTION

In the past few years, a significant amount of microfluidic devices and systems have been developed for biomedical detection systems. Micropumps, which are capable of handling the microliquid transportation, play important roles in the system. Most of previous published micropumps used mechanical force (thermal bubble, PZT, magnetic coil, or other mechanical forces) as the active energy source to drive micro-fluid. These pumps may have difficult to maintain smooth driving force for steady flow [1-2]. The moving parts of a mechanical pump could also cause unwanted stress or impact to living cells. Inspired from the evaporation and vascular tissue capillary effects in plants (Fig.1), this paper introduces a new method base an active capillary mechanism for micro fluidic pumping. The capillary evaporation pumping mechanism should be able to keep living cells away from pressure impact. It also deliveries reasonable good flow rate with precision velocity control.

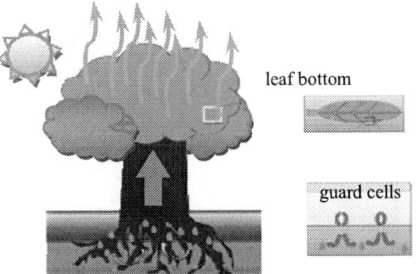

Fig. 1. Schematic diagram of plant's evaporation network

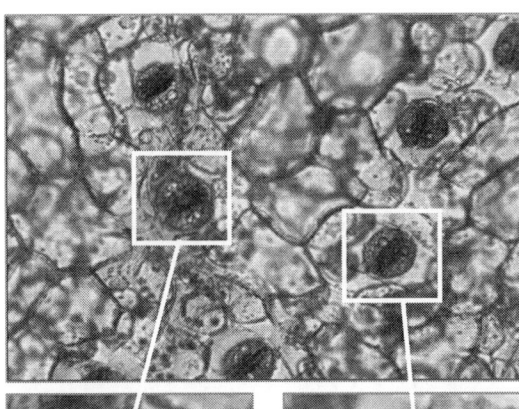

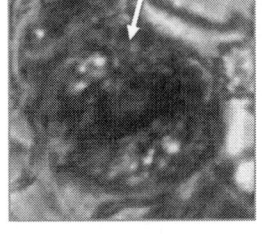 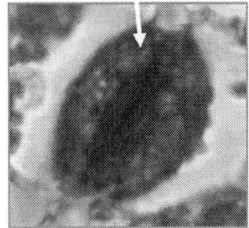

Fig. 2. Auto-graphic images of guard cells of a zamioculcas zamiifolia's leaf

In general, the capillary flow rates of a plat are mainly depend on the geometric opening of guard cell, the environment humidity, temperature and air flow. Fig. 2 shows the micrographic images for the guard cells on the bottom of a zamioculcas zamiifolia's leaf. A few of researchers have proposed several useful passive capillary pump designs [3, 4]. The methodology of how to design passive capillary pumps with good controlling of flow rate is also discussed. A good filling behavior is important for capillary pump is also important to avoid air bubble trapping; however, working under similar space constrains, the flow rates of passive capillary micro-pumps are still lower than regular mechanical micro-pumps.

This paper is going to present an active capillary pump. In the following sections, we are going to present the design idea, process steps, and testing results of a flexible active capillary micro-pump, accordingly.

II. DESIGN AND FABRICATION

The proposed pump consists of two parts, active evaporator and capillary reservoir, shown in Fig. 3. The active evaporator has sandwiched structure which composites a micro nickel-alloy-heater between two porous filters. One can apply voltage to heat the wires thus manipulates the temperature of evaporator to control the evaporation rate and flow rate of micro fluid.

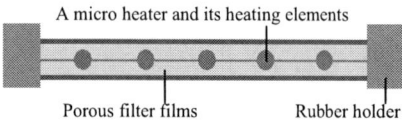

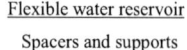

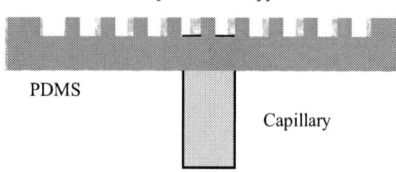

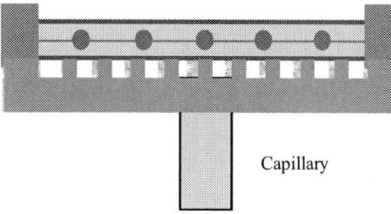

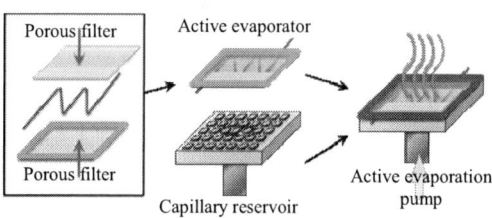

Fig. 3 The design and configurations of the active evaporation micropump is shown in the figure. Each pump consists of two parts, active evaporator and capillary reservoir.

The reservoir, made of PDMS (polydimethylsiloxane) and ReproRubber® (vinyl dimethyl polydimethylsiloxane) by micromolding, is shown in Fig. 4. The master mold is prepared by photolithography and aluminum patterning. It creates a mesh-like fluidic network in the reservoir. The flexible capillary reservoir can be easily assembled to an active evaporator (Fig 5). The capillary reservoir is designed to have a small flow resistance and is preceded by a wide opened microfluidic network. Evaporation pumping mechanism is similar to the guard cell and vascular tissue systems in a plant. They can move water and nutrition against earth's gravity by capillary force.

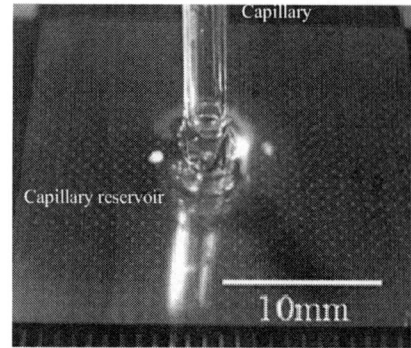

Fig. 4 Use patterned silicon wafer to mold PDMS fluidic network

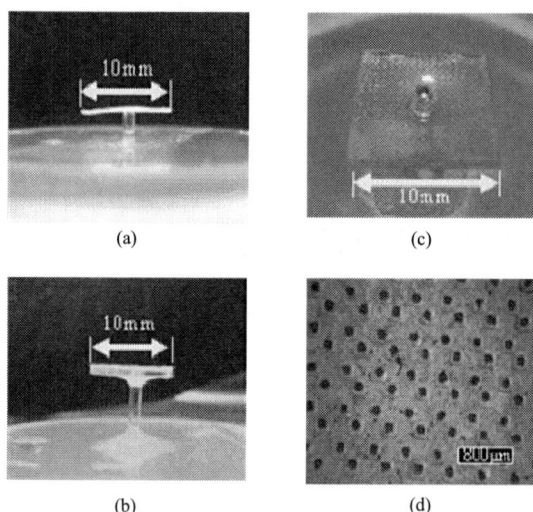

Fig. 5 Experiment setups: (a) a double sides setup; (b) a single side evaporation setup; (c) the capillary reservoir is below the active evaporator (the evaporator is removed); (d) The mesh-like fluidic network;

To prepare the active evaporator, a micro nickel-alloy-heater is sandwiched between two porous filters. Three different heat designs, including 3, 5, and 9 element heaters, have been prepared or testing.

Fig. 6 presents the process flow of the active evaporator. The process includes heater holder placement, bottom porous filter placement, nickel-alloy-heater arrangement, top porous filter placement and sandwiched components edge sealing. After ReproRubber® edge sealing process, the rods are removed. The active evaporator is assembled to capillary reservoir for final assembly and integration. The completed evaporation pump is shown in Fig. 7.

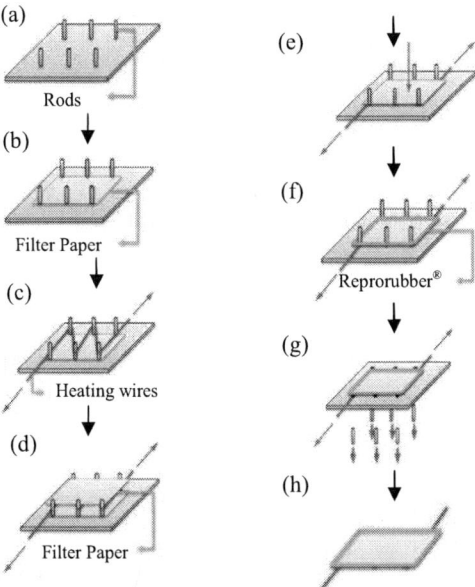

Fig. 6 (a)-(h) are the process flow of an active evaporator: (a) a rod array is used to hold the heater; (b) put a piece of porous filter between rods; (c) place, align and hold the Ni heater in place, keep the heater compact; (d) put another porous filter on top; (e) give pressure on top; (f) seal the porous filters and sandwich the heater between them; (g) remove the rods after curing and remove the active evaporator from the frame; (h) prepare the active evaporator for final assembly and integration

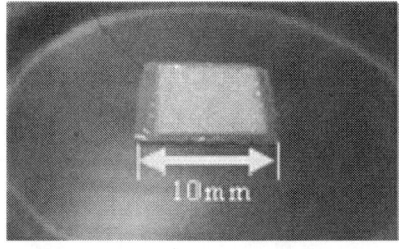

Fig. 7. A completed active evaporator, microheater are sandwiched between two porous filters.

III. EXPERIMENT AND RESULTS

Five different evaporation conditions have been tested under different working modes (Table 1). Under passive evaporation mode, the pump can pump water without external power supply. The pumping rate is almost proportional to the passive evaporation area of the evaporator. Under active evaporation mode, one can use Ohmic heating to increase the evaporation rate on the evaporator thus increase the flow rate in the micro-channel.

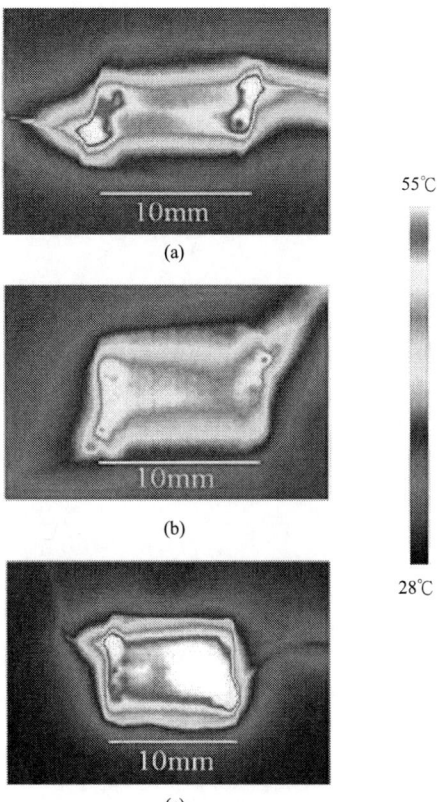

Fig. 8 The infrared images of working evaporators for or evaluation and analysis. (a) 3 heating elements (b) 5 heating elements (c) 9 heating elements

TABLE 1.   A EVAPORATION RATE COMPARISON OF DIFFERENT EXPERIMENT SETUPS

Setup	Flow rate (μl/min)	Power (W)
Passive mode		
Analyze (single side)	0.53	0
Single side(with mesh)	0.42	0
Double sides	0.69	0
Active mode		
3 heating elements	6.67	2.16
5 heating elements	10.00	1.58
9 heating elements	13.33	0.87

Working area=100 mm², temperature 25 ℃, humidity 47%

IV. CONCLUSIONS AND DISCUSSIONS

A comparison of different pumping modes is listed in Fig. 9 and Table 2. The flow rate of evaporation pump does look smaller than some other published pumping technologies, however, it is sufficient for the needs (circulating culture medium 0~30μl in a day) of a microcell culture system. One can simply increase the evaporation area to increase the pumping rate. Infrared images of working evaporators are shown in Fig. 8 for evaluations and analysis.

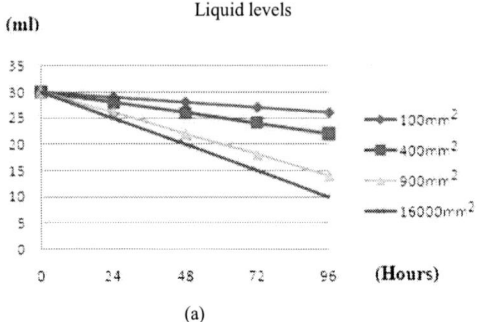

(a)

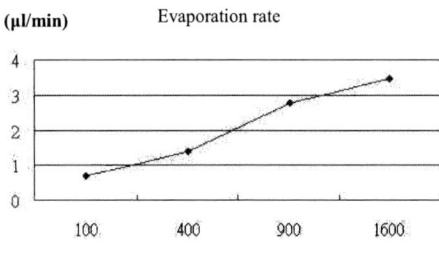

(b)

Fig. 9 Evaporation velocity versus evaporation panel area and

TABLE 2.   A COMPARISON OF DIFFERENT PUMPING MODES

	Power	Pressure	Flow rate	Flow rate control
Mechanical active microvalve [5]	medium	large	Fast (190+µl/min)	pumping frequency
Thermal-bubble-actuator pump [6]	high	low	Slow (1~6µl/min)	pumping frequency
PZT pump[7]	high	low	Fast (300+µl/min)	pumping frequency
Passive evaporation pump [3]	needless	low	very slow (0.41µl/min) (per 100mm²)	pressure
Passive evaporation pump [this work]	needless	medium	very slow (0.42µl/min) (per 100mm²)	temperature & humidity
Active evaporation pump [this work]	low(*1)	medium	Slow-medium (13.33µl/min) (per 100mm²)	active heating elements

*1 0.87W, the power consumption of active evaporator.

ACKNOWLEDGMENT

Part of this work is supported by National Science Council, NSC 98-2221-E-018-012-MY

REFERENCES

[1] Linfen Yu, et al., *Electrophoresis*, 2008, pp. 5055-5060, 2008

[2] Balasubramania G. S., et al., *Journal of Micromechanics and Microengineering*, 2008.

[3] Yan-Xia Guan, et al., *Talanta 68*, 2006, pp.1384-1389, 2006

[4] Martin Zimmermann, Heinz Schmid, Patrick Hunziker and Emmanuel Delamarche, "Capillary pumps for autonomous capillary systems," Lab Chip, 7, pp. 119–125 2007.

[5] Kwang W. Oh, et al., *Journal of Micromechanics and Microengineering*, 2006.

[6] Jr-Hung Tsai, et al., *Journal of Microelectromechanical systems*, v.11, no. 6, 2002.

[7] Amos Ullmann, et al., *Journal of Microelectromechanical systems*, v.11, no. 6, 2002.

Nanofluidic Device with Self-assembled Nafion Membrane Utilizing Capillary Valve

Shuo Wang, Huaiqiang Yu, Wei Wang and Zhihong Li
National Key Laboratory of Science and Technology on Micro/Nano Fabrication,
Institute of Microelectronics, Peking University, Beijing, China
zhhli@pku.edu.cn

Abstract—**In this paper, a PDMS-based nanofluidic device for ICP (ion concentration polarization) was designed, fabricated and tested. In this device, a Nafion membrane, which acted as a network of nanochannels and provided paths for particular ions to migrate between main channel and buffer channel, was formed in 20μm-wide microchannel using capillary valve structure. Ion concentration polarization effect was observed when a voltage of 80 volts was applied across the Nafion junction. The fabrication process was simple and inexpensive. The manipulation of cells movement was successfully demonstrated using our PDMS micro-chip.**

Keywords-nanofluidcs; ion concentration polarization; nanochannel; Nafion;

I. INTRODUCTION

In the recent two decades, much attention has been drawn to nanofluidics in the field of lab-on-chip, for its unusual phenomena, which are impossible in microfuidics [1] [2]. Ion enrichment and ion depletion in the nanochannel structure provides a novel way to control the charged or polarized solute or particles in the solution, as reported by Pu [3]. Many nanochannel devices utilizing ICP (ion concentration polarization) effect were widely used in protein preconcentration and seawater desalination [4].

Recently, Nafion, a kind of material with numerous nanometer-scale pores, which are originally used in fuel cells, [5] has been applied to form 2-D nanochannel structure because of its simple fabrication process and low cost without the requirement for photolithography of nano precision [6]. Liu [7] further facilitates the nafion membrane formation process by using the capillary burst valve in nafion filling procedure to form nano structures.

In this paper, nanofluidic device with self-assembled nafion membrane utilizing capillary valve was designed and tested. By applying one-valve structure and modifying nafion self-assembling method, we successfully fabricated devices with high yield rate which can be used for cell movement manipulation.

II. CAPILLARY-VALVE-BASED FABRIACTION

Although capillary-valve-based fabrication method of nafion membrane is simple with only one mask, experiment conditions are hard to control. Nafion solution is filled through nafion filling channel to create interconnection between main channel and buffer channel. Nafion is supposed to pass the first capillary burst valve and stop at the second valve. After the filling procedure, nafion is heated at 80℃ for an hour. The width of first valve is 20 μm and the depth of all channels is 50 μm,. However, due to the hydrophobic property of PDMS (Polydimethylsiloxane), nafion solution of 5wt % stopped at the first valve in our experiment even if it was funnel-type. To alleviate this problem, we tried to fill nafion solution for 0, 1, 2, 5, 12, 24, and 48 hours after oxygen plasma bonding, respectively. If the time was short, nafion passed the second valve easily because PDMS after oxygen plasma treatment is hydrophilic; if the time was long, nafion solution still stopped at the first valve when the property of PDMS changed back to hydrophobic. No optimized time was found and the repeatability was bad.

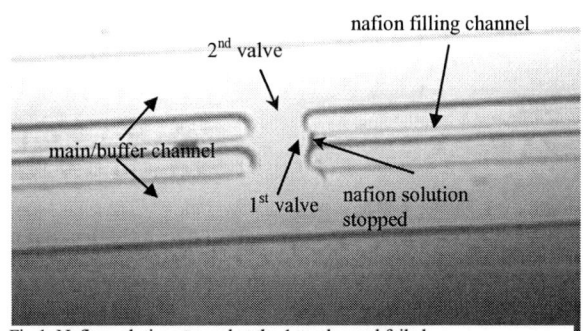

Fig.1. Nafion solution stopped at the 1st valve and failed.

To solve the contradiction discussed above, the width of nafion filling channel was enlarged to 100 μm. We found nafion was not retarded at the first capillary valve even if the filling procedure was conducted one week after oxygen plasma bonding and PDMS was hydrophobic. However, nafion membrane could not fill up the space between main channel and buffer channel after thermal treatment or drying up at room temperature, as shown in Fig. 2.

978-1-4673-1122-9/12 $31.00 © 2012 IEEE

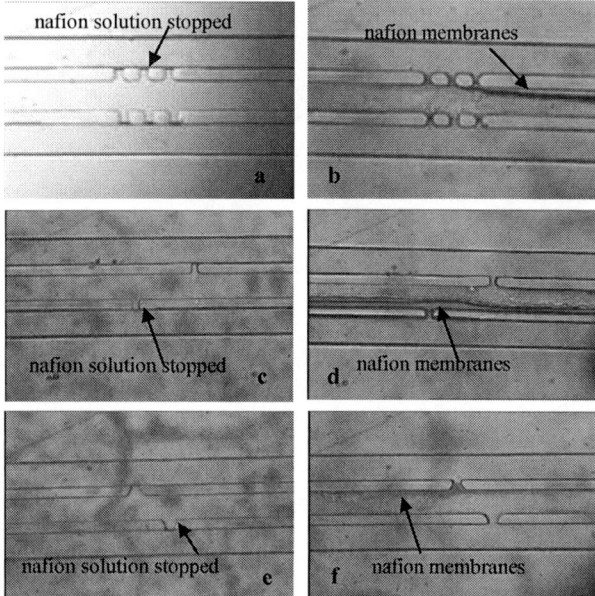

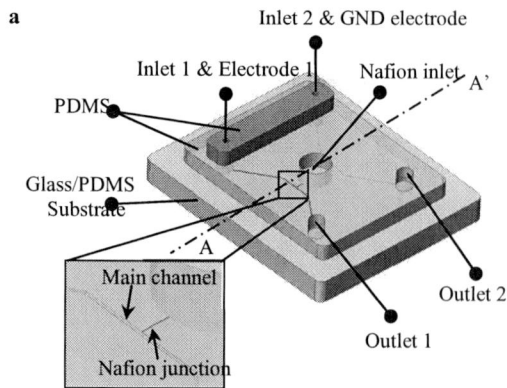

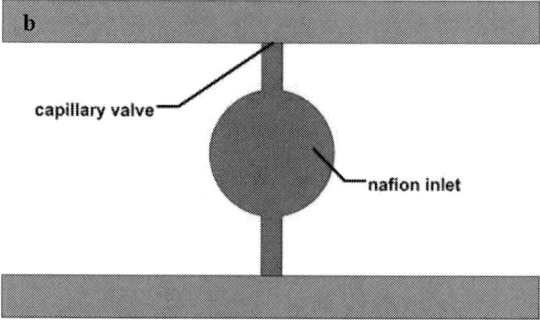

Fig. 2. a) & c) & e) nafion stopped at the second capillary burst valve after filling procedure; b) & d) & f) nafion membrane was damaged after solution stopped at thermal treatment.

One possible key factor leading to this fabrication failure was owing to the weight percentage of nafion solution. We changed the weight percentage from 5 wt% to 10 wt% and 20 wt%. Nafion membrane formed by nafion solution of 10 wt% could not fill up the space between the main channel and buffer channel either. Nafion membrane made by nafion solution of 20 wt% was too viscous to fill in the channel and stopped before arriving at the first capillary burst valve. Even filling procedure was done immediately after bonding, nafion solution of 20 wt% could not form compete and satisfactory membrane after thermal treatment. We cannot get complete nafion membrane by simply increasing the weight percentage of nafion solution.

The width of nafion filling channel is a critical parameter to not only filling procedure, but also drying up and membrane formation process. A larger width leads to smaller capillary force, which makes nafion solution pass the first valve easily, and may also cause incompleteness of nafion membrane. However, a smaller width helps to form satisfactory nafion membrane, but it would retard nafion solution to pass the first valve at the same time.

To ravel this knot, we designed and fabricated devices of a one-valve structure.

III. DESIGN AND FABRICATION

A. Device design

Fig. 3 presents the design of the single junction device. The PDMS micro-chip consisted of a 100 μm-wide main channel, a 100μm-wide buffer channel and a 20 μm-wide nafion junction connecting these two channels. All channels were 50

Fig. 3. a) The 3-D structure b) Vertical view of PDMS micro-chip for ion concentration polarization.

μm deep. A designed inlet of 4mm diameter was the entrance of nafion solution droplet to form nafion junction. The nafion solution was supposed to stop at the capillary valve. PDMS or glass was adopted as substrate material.

B. Fabrication

A single mask fabrication process flow was developed to fabricate the device, as shown in Fig. 4.

First, the PDMS micro-chip was made out of silicon mould (Fig. 4.1). Si mould was fabricated using DRIE. Next, all inlets and outlets were punched to get reservoirs and the chip was bonded with a flat PDMS or glass substrate after oxygen plasma treatment for 10s (Fig. 4.2). Then a droplet of 10 uL nafion solution (5 wt% ~ 15 wt%) was injected into nafion junction through the middle inlet using pipette, and nafion stopped at the capillary valve as expected.

Due to the hydrophobic property of PDMS, it might take a long period of time for nafion solution to fill in the junction. Using the tip of pipette to tickle the bottom of nafion inlet could accelerate the filling procedure without breaking the capillary valve. Subsequently, a PDMS block was placed on the nafion inlet to slow down the drying up process and get a better nafion membrane. Finally, the device was accomplished after being heated on a hotplate at 80℃ for an hour (Fig. 4.3).

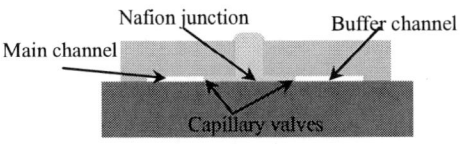

(1) PDMS fabrication

(2) Bonding with glass/PDMS substrate

(3) Nafion junction fabrication using the capillary valve structure.

Fig. 4. Sectional view of fabrication process flow of single junction device.

As shown in Fig. 5, after the filling procedure, nafion solution was stopped in front of main/buffer channel. After the thermal procedure, nafion membrane remained complete and the fabrication of the device was finished.

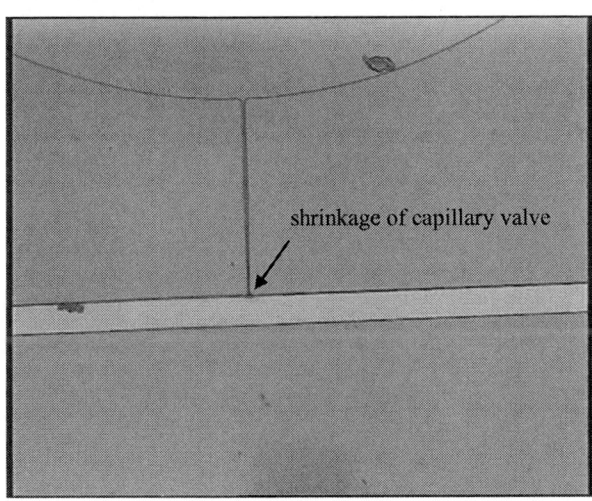

Fig. 5. Device after nafion filling procedure and thermal treatment.

We observed that the capillary valve shrank after thermal treatment. This might be caused by expansion of PDMS when heated and could be alleviated by using a glass substrate instead of a PDMS substrate. However, the shrinkage did not damage the device, but, on the contrary, it prevented incomplete filling of nafion membrane happened in Fig. 2.

IV. IN-VITRO EXPERIMENT RESULT

As shown in Fig. 6, the *in-vitro* test of the device was performed using HEK-293 cells which were of great significance in gene expressions. A signal generator (Agilent 33120A) and a power amplifier (NF BA4825) were used to provide high voltage. Microliter syringes samplers (Shanghai Gaoge Industry Trade Co.,Ltd) were used to provide stable rate of flow (500 nL/min).

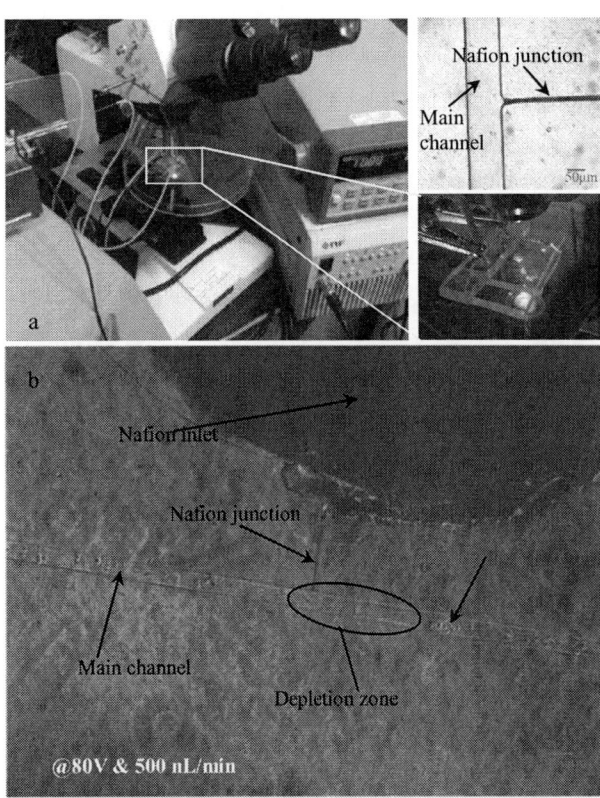

Fig. 6. The test of manipulating cells' movement using the PDMS microchip. a) Experiment equipment and conditions; b) Cells were trapped.

A depletion zone of 500 μm long was set up when 80V voltage was applied between the main channel and buffer channel. The picture in Fig. 6 was taken five seconds after the high voltage was applied. Cells flowing down the main channel were found rebounding back and concentrating at the boundary of depletion zone. A disturbance to the flow rate caused cells to move forward and backward, as shown in Fig. 7. Once the electrical power was cut off, these cells immediately flew down the main channel smoothly.

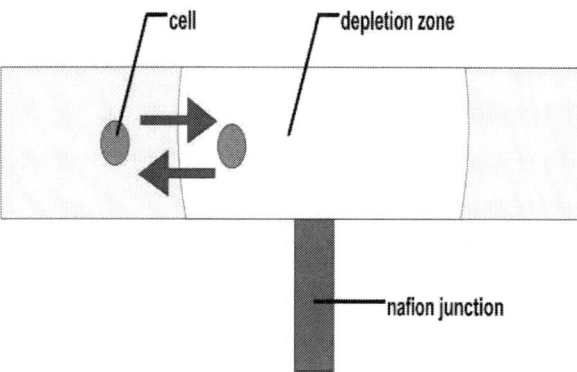

Fig. 7. Cells moved forward and backward when flow rate was not stable.

Cell wreckage and debris were found concentrating at the interface between nafion junction and main channel. Fig. 8 demonstrates this phenomenon. The mechanism was stilled under investigation. As the time went on, the main channel was clogged and the device failed to work. To solve this problem, the width of main/buffer channel should be designed as about 1 mm.

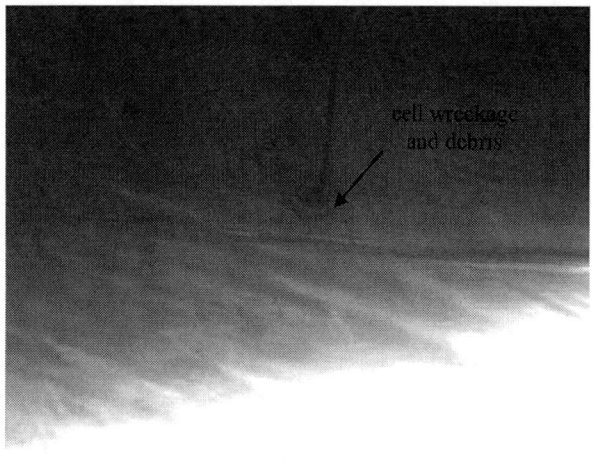

Fig. 8. Cell wreckage and debris concentrated at the interface between nafion junction and main channel .

V. CONCLUSION

This study demonstrated a modified nanofluidic device with self-assembled nafion membrane utilizing capillary valve. By reducing two capillary burst valves to only one, we solved the contradiction between nafion filling procedure and nafion membrane formation process. The width of nafion filling

channel was designed as 20 μm because device of this size could provide enough capillary force to stop the nafion solution at nafion junctions and at the same time result in better membrane formation. Due to the simple structure of the device, fabrication process was simple and the nafion filling procedure could be done long after oxygen plasma bonding. The *in vitro* test of manipulating movement of HEK-293 cells was performed to validate the performance of this device. Cells were successfully trapped at the boundary of the depletion zone when voltage was applied across the nafion junction. This device can be introduced to large and complex microfluidic systems.

ACKNOWLEDGMENT

This work is supported by the National Natural Science Foundation of China (Grant No. 61176111). The authors would like to thank the stuff in the National Key Laboratory of Science and Technology on Micro/Nano Fabrication in Peking University for their valuable assistance.

REFERENCES

[1] JonasO.Tegenfeldt, et al, "Micro-and nanofluidics for DNA analysis," *Analytical and bioanalytical chemistry*, Volume 378, Number 7, 1678-1692.

[2] RB Schoch, et al,"Transport phenomena in nanofluidics," *Rev. Mod. Phys.* 80, 839–883 (2008).

[3] Qiaosheng Pu, Jongsin Yun, Henryk Temkin, and Shaorong Liu, "Ion-Enrichment and Ion-Depletion Effect of Nanochannel Structures," *Nano Letters*, 2004, 4 (6), pp 1099 –1103.

[4] Sung Jae Kim, Sung Hee Ko, Kwan Hyoung Kang, and Jongyoon Han, "Direct seawater desalination by ion concentration polarization," *Nature Nanotechnology* ,2010,5, pp 297 - 301.

[5] Supramaniam Srinivasan, David J. Manko, Herman Koch, Mohammad A. Enayetullah, A.John Appleby, "Recent advances in solid polymer electrolyte fuel cell technology with low platinum loading electrodes," *Journal of Power Sources*, Volume 29, Issues 3-4, February 1990, pp 367-387.

[6] Jeong Hoon Lee, Yong-Ak Song and Jongyoon Han, "Multiplexed proteomic sample preconcentration device using surface-patterned ion-selective membrane," *Lab Chip*, 2008, 8, 596-601.

[7] Vincent Liu, Yong-Ak Song and Jongyoon Han, "Capillary-valve-based fabrication of ion-selective membrane junction for electrokinetic sample preconcentration in PDMS chip," *Lab Chip*, 2010, 10, 1485-1490.

Effect of Coating Organic Film on Dropwise Condensation in Microgrooves with Nanostructure Surface

T. S. Sheu[*], J. C. Shyu, J. W. Hsiao, Y. C. Pan

[*]Department of Mechanical Engineering, R. O. C. Military Academy, TAIWAN

tssheu@mail.cma.edu.tw

Abstract—The Standard photolithography and wet etching technology were applied to prepare the hydrophobic of microgrooves surface on copper plate. The organic STA film were coating on microgrooves surface to promote the dropwise condensation. The experimental data show that the STA coated microgrooves with nanostructure surface show a better performance of hydrophobicity and a larger apparent contact angle. The STA coated microgrooves with nanostructure surface shows a better the heat transfer performance than the microgrooves surface with nanostructures. An increase of the droplet departure diameter and a lower departure frequency, for a larger contact angle hystersis on the microgrooves surface during the DWC process, also results in a decrease in the heat transfer performance of microgrooves nanostructure surface without STA.

Keywords-organic film; dropwise condensation; microgroovrs

I. INTRODUCTION

Dropwise condensation (DWC) has shown the ability to increase condensation heat transfer coefficients by an order of magnitude over filmwise condensation. Rapid removal of large condensate drops is essential to continuous DWC, for which hydrophobic surfaces appear to be an ideal solution. The low surface energy or the hydrophobic property of solid surface plays an important role in the DWC process. Over the past few decades, considerable attention has been paid toward the development of suitable DWC promote film or geometric structure. The surface wettability of solid surface was modified to realize the dropwise condensation [1].Organic and polymer materials with low surface energy were used to promote DWC for their hydrophobic properties [2–7]. Das et al. [8] applied an organic self-assembled monolayers coating to enhance the dropwise condensation heat-transfer by a factor of 4. The lower the surface energy is, the higher the heat-transfer coefficient the dropwise condensation exhibits [9]. Neumann et al. [10] investigated the effect of the contact angle hysteresis on the heat-transfer characteristics during DWC. It was found that the heat-transfer during DWC is related to the contact angle hysteresis of the surface. The larger the contact angle hysteresis is, the lower the heat-transfer coefficient is. Superhydrophobic surfaces exhibit high contact angle (between 150∘ and 180∘), low water contact angle hysteresis and low water roll-off angle [11–13]

A variety of elegant techniques have been rapidly proposed for constructing superhydrophobic materials. Guo et al. [14] employed a wet chemical reaction to fabricate the rough structure on a polished copper substrate and obtained a stable superhydrophobic surface after chemical modification with poly(dimethysiloxane) vinyl terminated (PDMSVT). Yang et al. [15] fabricated stable, transparent and superhydrophobic carbon nanotube (CNT) nanocomposite films. Song et al. [16] prepared reproducibly smooth aminopropyltrimethoxysilane (APTMS) surface through chemical vapor deposition (CVD) on a silicon surface and obtained dual self-assembled and superhydrophobic films. Xiu et al. [17] described amethod for the preparation of inorganic superhydrophobic silica coatings using sol–gel processing with tetramethoxysilane and isobutyltrimethoxysilane as precursors. Baldacchini et al. [18] presented an approach for fabricating superhydrophobic silicon surfaces which consisted of femtosecond laser pulses and then coating the surfaces with a layer of fluoroalkylsilane molecules.

However, most of those methods are subject to certain limitations, such as severe conditions, expensive materials. Chemical etching and coating organic film are a simple and cheap technique for fabricating rough structures and hydrophobic on the surface of solid materials. In this work, we used a photo-lithography and chemical etching method for fabricating hydrophobic microgrooves surface on copper plate. The steam condensation experiments on a vertical plate were conducted to determine the heat transfer characteristics. The effect of microgrooves with nanostructure surface and coating organic film on DWC was investigated.

II. EXPERIMENTAL

A. Surface frabication

Copper plate was hand polished with emery papers of grades 240 to 1200. All the polished samples were ultrasonicated in ethyl acetate, isopropanol, and de-ionied water for 10 min each and then blown dry with oxygen gas. Standard photolithographic practices were employed in preparing the plates for etching in Fig. 1. First, we then spin coated photo-resist AZ4620 with thickness of 10 μm and the copper plate was soft baked to prevent mask sticking. After an

978-1-4673-1122-9/12 $31.00 © 2012 IEEE

OAI-200IR mask aligner was then used to align the mask over the copper substrate and exposure the photoresist to UV light and development, a etch mask of photo-resist AZ4620 was formed. The above pretreated copper plate was immersed in hydrochloric acid (HCl), hydrogen peroxide, and water in a mixture of approximately 5:1:10 by volume at room temperature. A trapezoidal shape of microgrooves is showed that is the result of the wet etching process used. The width of the traces at the bottom is about 50 μm and the width at the top of the trace is about 38 μm. The wet etch process used caused undercutting of approximately 6 μm from each side of the mask was shown in Fig. 2. Therefore, the microgrooves surface with nanostructures is produced. Then, the etched copper plates were immersed in a solution of 0.002 mol/L ethanol solution of STA for 24 h and then dried in air.

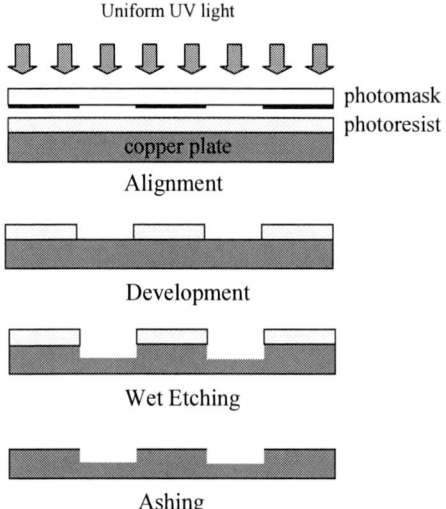

Fig. 1. Standard photolithography and wet etching process.

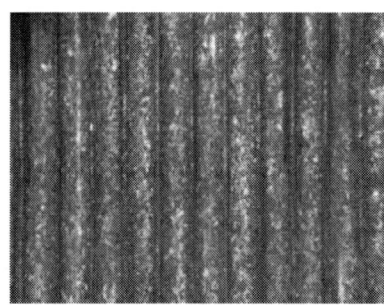

Fig. 2. The top view of the etched Copper plate.

B. Contact Angle Measurement

The static contact angles, a measure of surface hydrophobicity, were measured using an optical contact angle meter with an accuracy of ±5 and water droplets of deionized water. For the measurement of static contact angle of patterned surfaces, the droplet size should be small but larger than dimension of the structures present on the surfaces. Droplets of about 5 mL in volume were gently deposited on the substrate using a microsyringe for the static contact angle. The contact angle values were obtained by averaging five measurement results on different areas of the copper surface.

C. Condensation measurement

The condensation experimental apparatus is shown schematically in Fig. 3, comprised of four parts: stream generator, cooling water, condensing chamber, and data acquisition and control system. Stream at about 100℃ was supplied from the stream generator. After stream flows through stainless steel tubing before into the upper entrance of a vertical rectangular condensing chamber and condensed on the condensing heat transfer surface (the copper plate). The rectangular condensing chamber was 60mm by 60mm. The stream pressure was maintained at the approximately atmospheric pressure. Water was used to cool the test surface. The water was kept at a constant temperature in a chiller and was delivered to the backside of the condensing heat transfer plate by means of a pump. Varying the cooling water temperature changed the heat flux of dropwise condensation.

A Schematic of the measurement system is shown in Fig. 4. The heat-transfer coefficient can be determined from the following relationships:

$$h = \frac{q''}{T_{steam} - T_{surface}} \qquad (1)$$

and

$$q'' = -k_1 \cdot \frac{dT}{dx} \qquad (2)$$

where the heat flux q'' was obtained by averaging the through the condensing heat transfer plate

$$q'' = \frac{1}{3}\left(k_2 \frac{T_1 - T_2}{d_1} + k_3 \frac{T_2 - T_3}{d_2} + k_4 \frac{T_3 - T_4}{d_3} \right) \qquad (3)$$

A mean temperature of the condensing surface $T_{surface}$ of the condensing surface can be determined an extrapolation of tthe temperatures measured inside the plate to the condensing surface:

$$T_{surface} = T_1 + \frac{q''}{k_1} \cdot d_0 \qquad (4)$$

III. RESULTS AND DISCUSSION

The condensing heat transfer plate was set vertically and the microgrooves were arranged in vertical position. Every experimental run was kept operated at the approximately atmospheric pressure. Both microgrooves and STA coated microgrooves surfaces have promoted the DWC. Fig. 5 shows the measured Condensation heat transfer coefficients in

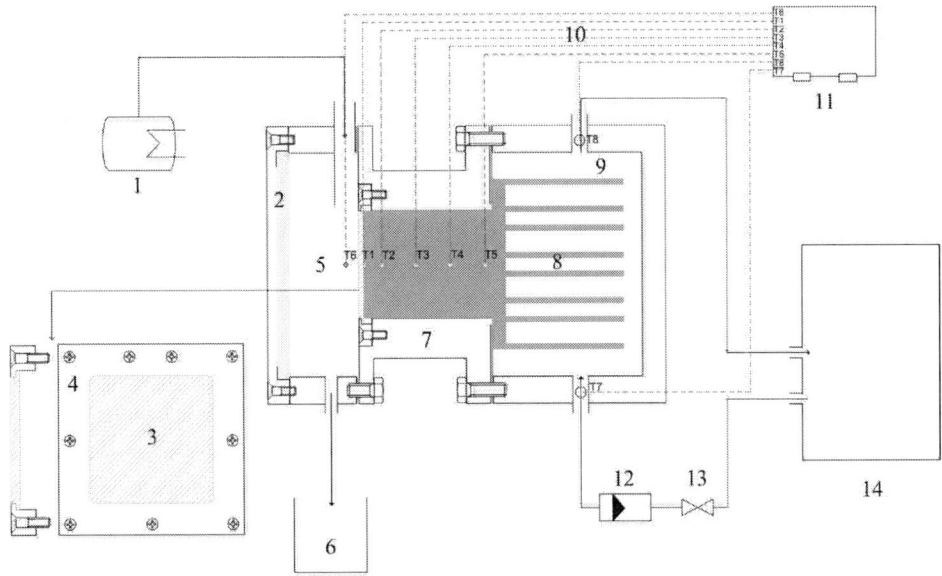

Fig. 3. Experimental apparatus for heat transfer measurements: (1) steam generator; (2) glass window; (3) condensing plate; (4) insulator; (5) condensing chamber; (6) cooling water tank; (7) insulator; (8) condensing block; (9) cooling chamber; (10) thermocouples; (11) data acquisition system; (12) flow meter; (13) Valve; (14) Chiller.

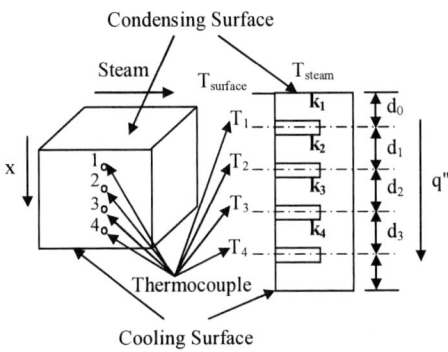

Fig. 4. Schematic diagram of the condensing block.

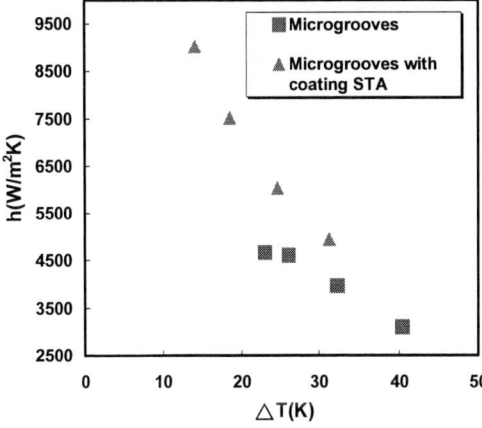

Fig. 5. Condensation heat transfer coefficients.

dependence on the surface subcooling temperature. The heat transfer coefficient decreased with the increase in the surface subcooling temperature. The STA coated microgrooves surface produced heat transfer coefficients that were higher than microgrooves surface, a significant improvement. According to modified Cassie–Baxter model, the microgrooves surface would show a better performance of hydrophobicity and a larger apparent contact angle in the air-steam environment. Fig. 6 is the contact angle measurements for microgrooves surface and STA coated microgrooves surface before experimentation. The contact angle of microgrooves surface and STA coated microgrooves surface

were 136° and 147°, respectively. These values were the increase of 87° and 104° from the contact angle (43°) of the polished only copper surface. The hydrophobicity of microgrooves surface and STA coated microgrooves surface, micropillarstructured surface was ascribed to the air trapped between micropillar. The trapped air can prevent the intrusion of water into the microstructures, resulting in increase of the contact angle on the microgrooves surface.

However, in steam condensation process, the condensate filled the cavities of the microgrooves surface without air.

This implies that the DWC takes place at a composite surface, comprising the microgrooves surface and condensate, so it is in a Wenzel state. Therefore, the STA coated microgrooves surface indicating that air pockets exit, so that the droplet is in Cassie state. The heat transfer process associated with dropwise condensation is an extremely short cyclic process of formation of microscopic droplets, droplet growth, coalescence/growth, and downflow. The period of the condensation cycle is an important determinant for the heat flux associated with dropwise condensation. The condensation cycle on the vertically microgrooves plate was faster. Consequently, the DWC heat-transfer coefficients for the microgrooves surface are lower than those of STA coated microgrooves surface. As the contact angle hystersis is larger for microgrooves surface, the droplet departure diameter on the microgrooves surface during DWC is larger than that on STA coated microgrooves surface, due to the adhesion effect of Wenzel or Cassie state. The departure diameter and the growth cycle of the droplets on the microgrooves surface both are larger than those on STA coated microgrooves surface. These also lead to an increase in the heat transfer performance for STA coated microgrooves with nanostructure surface.

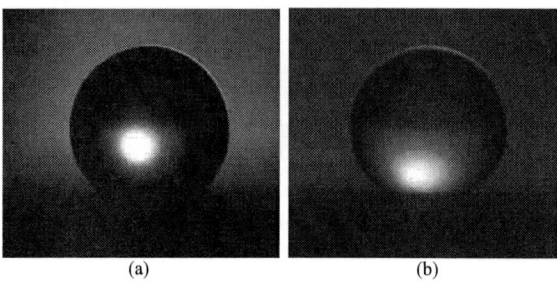

(a)   (b)

Fig. 6. Contact angle measurement for (a) microgrooves surface, (b) coated STA microgrooves surface.

IV. CONCLUSIONS

The experimental data show that the STA coated microgrooves with nanostructure surface show a better performance of hydrophobicity and a larger apparent contact angle. The STA coated microgrooves with nanostructure surface shows a better the heat transfer performance than the microgrooves surface with nanostructures. An increase of the droplet departure diameter and a lower departure frequency, for a larger contact angle hystersis on the nanostructure surface during the DWC process, also results in a decrease in the heat-transfer performance of microgrooves nanostructure surface without STA.

V. ACKNOWLEDGMENT

The work was supported by National Science Council of the Republic of China under contracts NSC 98-2221-E-145 -002.

REFERENCES

[1] Y.L. Lee, T.H. Fang, Y.M. Yang, J.R. Maa, "The enhancement of dropwise condensation by wettability modification of solid surface," *Int. Commun. Heat Mass Transfer*, vol. 25, pp. 1095–1103, 1998.

[2] P.J. Marto, D.J. Looney, J.W. Rose, A.Wanniarachchi, "Evaluation of organic coatings for the promotion of dropwise condensation of steam," *Int. J. Heat Mass Transfer*, vol. 29, pp. 1109–1117, 1986.

[3] K.M. Holden, A. Wanniarachchi, P.J. Marto, D.H. Boone, J.W. Rose, "The use of organic coatings to promote dropwise condensation of steam," *ASME J. Heat Transfer*, vol.109, pp. 768–774, 1987.

[4] T. Haraguchi, R. Shimada, S. Kumagai, T. Takeyama, "The effect of polyvinylidene chloride coating thickness on promotion of dropwise steam condensation," *Int. J. Heat Mass Transfer*, vol. 34, pp. 3047–3054, 1991.

[5] X.H. Ma, D.Q. Xu, J.F. Lin, "Dropwise condensation on super thin polymer surface," *J. Chem. Ind. Eng*, Vol. 44, pp. 165–170, 1993.

[6] Q. Zhao, D. Zhang, J.F. Lin, "Surface material with dropwise condensation made by ion implantation technology," *Int. J. Heat Mass Transfer*, vol. 34, pp. 2833–2835, 1991.

[7] X.H. Ma, J.B. Chen, D.Q. Xu, J.F. Lin, C.S. Ren, Z.H. Long, 'Influence of processing conditions of polymer film on dropwise condensation heat transfer," *Int. J. Heat Mass Transfer*, pp. 45, pp. 3405–3411, 2002.

[8] A.K. Das, H.P. Kilty, P.J. Marto, B.G. Andeen, A. Kumar, "The use of an organic selfassembled monolayer coating to promote dropwise condensation of steam on horizontal tubes," *ASME J. Heat Transfer*, vol. 122, pp. 278–286, 2000.

[9] M. Izumi, S. Kumagai, R. Shimada, N. Yamakawa, "Heat transfer enhancement of dropwise condensation on a vertical surface with round shaped grooves," *Exp. Therm. Fluid Sci.*, vol. 28 243–248, 2004.

[10] A.W. Neumann, A.H. Abdelmessih, A. Hameed, "The role of contact angles and contact angle hysteresis in dropwise condensation heat transfer," *Int. J. Heat Mass Transfer*, vol. 21, pp. 947–953, 1978.

[11] C.W. Extrand, "Model for contact angle and hysteresis on rough and ultraphobic surfaces," *Langmuir*, vol. 18, pp. 7991–7999, 2002.

[12] J. Kijlstra, K. Reihs, A. Klami, "Roughness and topology of ultrahydrophobic surfaces," *Colloid Surf. A*, vol. 206, pp. 521–529, 2002.

[13] Y.C. Jung, B. Bhushan, "Contact angle, adhesion and friction properties of microand nanopatterned polymers for superhydrophobicity," *Nanotechnology*, vol. 17, pp. 4970–4980, 2006.

[14] Z.G. Guo, J. Fang, L.B. Wang, and W.M. Liu, "Fabrication of superhydrophobic copper by wet chemical reaction," *Thin Solid Films*, vol. 515, pp.7190-7194, 2007.

[15] coolingJ. Yang, Z.Z. Zhang, X.H. Men, and X.H. Xu, "Fabrication of stable, transparent and superhydrophobic nanocomposite films with polystyrene functionalized carbon nanotubes," *Appl. Surf. Sci.*, Vol. 255 pp. 9244-9247, 2009.

[16] X.Y. Song, J. Zhai, Y.L. Wang, and L. Jiang, "Self-assembly of amino-functionalized monolayers on silicon surfaces and preparation of superhydrophobic surfaces based on alkanoic acid dual layers and surface roughening," *J. of Colloid Interface Sci.*, pp. 298, pp. 267-273, 2006.

[17] Y.H. Xiu, D.W. Hess, and C.P. Wong, "UV and thermally stable superhydrophobic coatings from sol-gel processing," *J. of Colloid Interface Sci.*, vol. 326, pp. 465-470, 2008.

[18] T. Baldacchini, J.E. Carey, M. Zhou, and E. Mazur, "Superhydrophobic surfaces prepared by microstructuring of silicon using a femtosecond laser," *Langmuir*, vol. 22, pp.4917-4919, 2006.

978-1-4673-1122-9/12 $31.00 © 2012 IEEE

Novel Core Etching Technique on Synthesized Gold Nanoparticles for Colorimetric Detection of Dopamine Biosample

Ho-Cheng Lee[1], Tzu-Heng Chen[2], Wei-Lung Tseng[2], Che-Hsin Lin[1, *]

[1]Department of Mechanical and Electro-Mechanical Engineering, National Sun Yat-sen University, TAIWAN
[2]Department of Chemistry, National Sun Yat-sen University, TAIWAN
chehsin@mail.nsysu.edu.tw

Abstract— **This study develops a novel and high performance colorimetric probe for dopamine (DA) detection. Aqueous-phase gold nanoparticles (AuNPs) extracted with 4-(dimethylamino) pyridine (DMAP) from toluene solvent is used as the reaction probes. The conjugated AuNPs of the diameter around 13 nm disperse into the size of 2-5 nm while adding dopamine (DA), resulting in the color change of the AuNPs solution from red to green. Therefore, the extracted AuNP solution can be a good indicator for detecting DA biosample. Transmission electron microscope (TEM) observations and dynamic light scattering (DLS) tests show that the AuNPs turn into smaller size right after adding DA. Results confirm that the DMAP capped AuNPs are etched by the DA molecules due to the stronger affinity between DA and AuNPs and thus cause a blue shift in the absorption spectrum. The concentration of dopamine is quantitatively monitoring using an UV-vis spectrometer with a limit of detection (LOD) of lower than 5 nM. The developed AuNPs etching protocol for dopamine detection provides a novel and versatile approach for rapid biosensing applications.**

Keywords-gold nanoparticles; dopamine; colorimetric probe

I. INTRODUCTION

During the last decade, dopamine (DA) has been well known to be one of the most important neurotransmitters that plays a significant role in the central nervous, renal, and hormonal systems [1]. DA molecules in human brain is for transmitting neuro-signals regarding usual movements, feedings and neuro-cognitions in daily life and the concentration of DA should be maintained at a stable level. A number of diseases such as Parkinson's disease and schizophrenia [2] are highly correlating to DA concentration in human brain. Therefore, many researchers have been focused on finding analytical methods for detecting DA with high sensitivity and good selectivity. Typical methods for detecting DA have been established for years such as electrochemistry [3], high performance liquid chromatography (HPLC) [4] and chemiluminescence [5] have been reported.

However, most of these methods suffered from the drawbacks that these detection techniques required sophisticate equipment and time consuming analytical processes for DA detection. Furthermore, there is a problem happens when detecting bio-sample consist with ascorbic acid (AA) while taking the voltammetric detection for DA analysis,

especially under the typical concentration of 10^{-9}~10^{-5} M. This is due to which the oxidized potential of ascorbic acid is almost overlapped with the redox potential of DA under the standard electrodes configuration. This also caused the electrodes suffer from fouling effect that is due to the accumulation of oxidized products on the electrode surface. Although several approaches have been proposed for solving these problems by using the polymer-modified electrodes, self-assembled monolayers of the electrodes [6], but these processes also brought the problem that need to preconcentrate the sample solutions before the measurement. In this regards, it is essential to develop a simple and low cost method for high performance DA detection.

Gold nanoparticles (AuNPs) have unique electrical and optical properties which have been investigated and applied for molecular catalysis and biosensors during the last decades. AuNPs exhibit the special absorption band at 520 nm. The apparent color of AuNP solution would change from red to blue while aggregating of AuNPs which is easily to observe by the naked eye [7]. Therefore, most of the AuNPs based sensors adopt the colorimetric assay as the detection principle. In order to conquer the difficulties of the above proposed methods as using for the detection of DA, many researches have been proposed that based on the cross-linking of AuNPs aggregation for provides a sensitivity detection of DA [8-10]. The lowest limit of detection (LOD) out of the literature of using cross-linking of AuNPs was reported to be 36 nM [10].

Alternatively, this study presents a different approach that uses AuNPs core-etching induced colorimetric assay for DA detection. Differ from typical nanoparticle-based colorimetric measurements of increasing the NP size due to NP conjugation. The size of the AuNPs reduces after adding the DA such that a blue shift on the absorption spectra can be observed. The sensitivity of the developed DMAP-AuNPs core etching technique for DA detection is experimentally investigated using the UV-Vis spectrometry and the limit of detection can be low as 5 nM. TEM microscopy is also used to confirm the morphology of the capped AuNPs before and after the addition of the DA analyte. With the simple and efficient method of using DMAP-AuNPs for the DA probe, a rapid and high sensitive DA detection can be achieved.

978-1-4673-1122-9/12 $31.00 © 2012 IEEE

II. PRINCIPLE

Scheme 1 presents the brief mechanism for the developed AuNP core etching technique for dopamine detection. The fundamental of the developed method is the preparation of functionalized DMAP-AuNPs with ligand exchange and phase transfer (toluene/water). Prior to the core etching process, AuNPs were first synthesized in a solvent of toluene then extracted into an aqueous solution utilizing DMAP. DMAP molecules spontaneously attach onto AuNP surface due to the positive charge of DMAP. The DMPA capped AuNPs will move into the water phase due to the hydrogen bond between water molecule and the $-NH_2$ end of DMAP. The adding of DA into the aqueous AuNP solution will cause the core etching effect of AuNPs such that the size of AuNPs decreases. The corresponding chemical schematics and the size for the AuNPs at different stages are labeled in the scheme below.

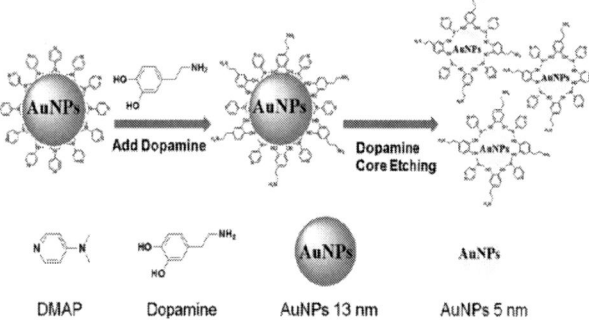

Scheme 1: DA detection based on the core etching of DMAP-AuNPs.

III. EXPERIMENTAL

A. Reagents and apparatus

The section presents the details for the preparation of the functionalized DMAP-AuNPs for dopamine detection. In this study, deionized and distilled water was used throughout the experiments. Hydrogen tetrachloroaurate ($HAuCl_4 \cdot 3H_2O$) (99.99%), tetra-n-octylammonium bromide (98%), 4-(Dimethylamino) pyridine (DMAP) were purchased from Alfa Aesar® and sodium borohydride ($NaBH_4$) was supplied by Acros®. Dopamine hydrochloride (DA) was purchased from Sigma-Aldrich®. All experiment solutions were fresh prepared within 1 h at ambient temperature (25 ± 2 °C) prior to the experimental procedure. The stock solutions of 1 M of gold salt ($HAuCl_4$), and 1 M of dopamine were prepared with the volume of 1 mL with DI water.

UV-Visible absorption spectrum was recorded by the SP-880 (Metertech) spectrophotometer with a 3-mL of plastic cuvette. The size of AuNPs was characterized by transmission electron microscopy (TEM JEM-3010 JEOL Corp., USA). The sample for TEM characterization was prepared by dropping 5 μL of colloidal AuNPs solution on the carbon-coated copper grid then dried at 100 °C for 24 h. Dynamic light scattering (DLS; N5 submicrometerparticle size analyzer, Beckman Coulter Inc., U.S.) was used for measuring the size distribution of AuNPs at different stages. Electrochemical analyzer (Model 611C, CH Instruments, Inc., U.S.A) was used for confirming the DA bonding on AuNPs surface.

B. Preparation of the gold nanoparticles probe

Based on the Brust two-phase method [11], a 1 M aqueous gold-salt solution with the volume of 1.0 mL in a centrifugal tube was added with a 30.0 mL toluene dissolved with a 20 mM of tetraoctylammonium bromide (TOAB). Since the water and toluene is immiscible to each other, aqueous solution will locate at the bottom and the toluene solvent will stay at top. The centrifugal tube was then stirred for 30s. As the salt ions transferred to the toluene phase, the color of the toluene layer changed from yellow to brown. The brown toluene phase was moved to another centrifugal tube. The reduction agent of a 0.4 M $NaBH_4$ solution (10 mL) was then added into the toluene bath and stirred for 1 min. The reduction of AuNPs with the size 13 nm occurred and got the color of the toluene layer turned into ruby red after the reduction reaction. The toluene layer was again move into a new tube for AuNPs extraction.

The procedure for extracting the synthesized AuNPs back into water phase from solvent phase was referred to the work reported by Gittins *et. Al.*[12] A 1.0 mL of prepared toluene with TOAB-AuNPs was added with a 1.0 mL of 0.1 M 4-dimethyl-aminopyridine (DMAP) solution for AuNPs extraction. The hydrophilic DMAP molecules replaced the hydrophobic TOAB molecules on AuNP surface such that DMAP-AuNPs move into the water phase. The color of the water phase changed to ruby red and the toluene layer became colorless. The synthesis of DMAP stabilized gold nanoparticles (DMAP-AuNPs) in water phase was prepared for colormetric detection of DA samples.

C. Colorimetric detection of DA

The proposed DMAP-AuNPs for colorimetric measurement of DA biosample was realized by following steps. Fist, diluted the prepared of stock DA solution with DI water and added 0.5 mL into the above prepared DMAP-AuNPs solution. Then mixture the solution and reacted for 5 min at room temperature. Absorption spectrum of the reacted solutions was recorded with 1 cm path-length of plastic cuvette shown in figure 1. The concentration of DA was also quantified by the absorption ratio (A_{415}/A_{520}). Also, it is taken by the TEM images and DLS for observed the morphology and the distribution.

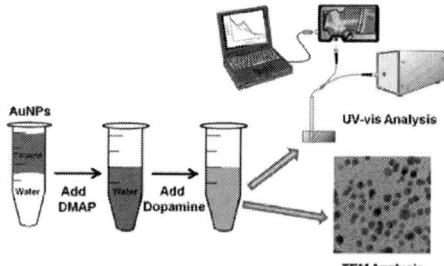

Fig. 1: A simplified experimental process for detection of DA by DMAP-AuNPs in aqueous solution.

IV. RESULTS AND DISCUSSION

According to the typical AuNPs-based of colorimetric measurements for DA, the NP is conjugate as the addition of DA in AuNPs colloid. The proposed core etching method tends to split the original DMAP-AuNPs into smaller ones. Scheme 1, have been illustrated the mechanism of the proposed DMAP-AuNPs for colorimetric detection of DA. The below of the given experiment results, also shows that DA molecular would caused the removal of Au atoms from the outer surface layer and to form the smaller AuNPs size.

In order to analyze the sensitivity of proposed DMAP-AuNPs for the detection of DA. It is diluted the prepared 1 M DA solution into $1 \sim 1 \times 10^{-9}$ M concentration orderly and with the equal volume. Then, took out 0.5 mL from DA solution and added it into the prepared of 1 mL DMAP-AuNPs colloid. As figure 2 shows, the visual color has changed from ruby red to brown green. Which the DA concentration changed from $1 \sim 10^{-4}$ M can also easily be observed by the naked eye. To confirm the property of the proposed detection concept, a series of UV-Vis spectrum were measured under different experimental conditions.

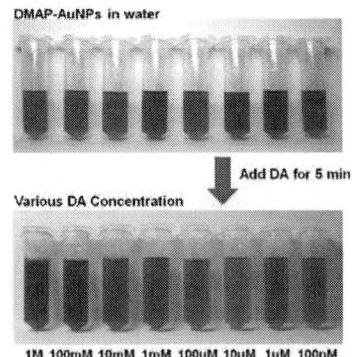

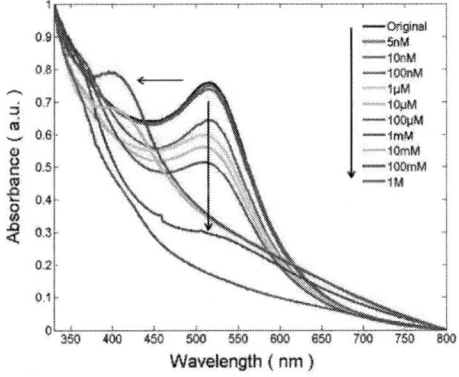

Fig. 2: Photographs showing the color change of DMAP-AuNPs reacted with various concentrations of DA solution.

![Absorption spectra graph]

Fig. 3: Absorption spectra of AuNPs in the presence of different concentration of DA; incubation time: 5 min; room temperature.

As shown in figure 3, the surface plasmon resonance (SPR) absorption band had a blue shift from 520 nm to 415 nm as presence in different concentration of DA. Under the low concentration (from 10^{-5} to 10^{-8} M) of the addition DA, there

only can be observed with one absorption peak at 520 nm. That shows the spectrum only gives the changed of the absorbance intensity under this range. This result indicates that exist of DMAP-AuNPs with original size of 13 nm have been decreased by the higher concentration of added DA. As the concentration of DA is getting about 10^{-3} M, there shows the absorption peak at 415 nm. It means that the number of original AuNPs with size in 13 nm has decreased and transformed into the smaller sizes. Note that the spectrum has no specific absorption peak when the concentration of DA is up to 1 M. This shows that all the original DMAP-AuNPs have been etched into many smaller sizes and caused the lack of a significant plasmon resonance so that it can't be monitor by the UV-Vis spectrometer.

To give the further confirmation of the change in AuNPs morphology, both TEM and DLS are utilized to analyze for the detail. Figure 4 (a), presents the morphology of DMAP-AuNPs transferred into the water phase from toluene. It obviously shows the original DMAP-AuNPs in the uniform average size around 13 nm. While added with 10^{-1} M DA into the DMAP-AuNPs colloid, the relatively small size of the particles is obtain as shown in Fig 4 (b). Furthermore, by utilizing DLS analysis, it is obviously indicates that the average diameter of the AuNPs tends to decrease after adding with DA (figure 5). The above results strongly gave the proof that DA molecular plays an important role for the core etching of the AuNPs. It is considered that based on the proposed method can provide a novel and simple process for detection of DA biosample.

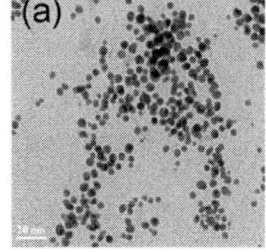 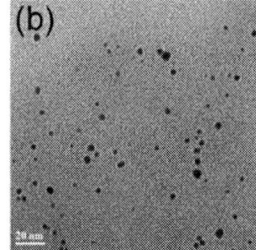

Fig. 4: TEM images of (a) DMAP stabilized AuNPs in water, and (b) after addition of 100 mM DA caused the decreased of AuNPs size.

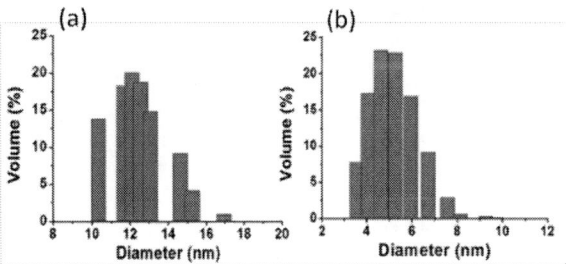

Fig. 5: Size distribution of (a) DMAP-AuNPs and (b) after added 100 mM DA to the AuNPs obtained by DLS. With mean diameter changed from 12.5 nm to 5.2 nm.

Figure 6 also shows the cyclic voltammogram responses of 50 mM DA in the water and then after added with DMAP-AuNPs. Because of DA has been bonded to the gold surface, which decreased the concentration of original DA in the solution. According to these results, it is certainly determined that DA plays an important rule for the size variation of the AuNPs. DA would cause the smaller size of DMAP-AuNPs due to the core etching effect.

With this approach, the change in DA concentration can be easily monitored by the naked eye and also capable of detecting a nano-molar concentration by the spectrometer. As shown in figure 7, it is found that the absorption ratio (A_{415}/A_{520}) would be raised as DA concentration increased. Also, the proposed assay exhibited a linear correlation to DA concentration in the range of $10^{-5} \sim 10^{-9}$ M with correlation coefficients of 0.9706. Most of all, the presence limit of detection for DA can be as low as 5 nM, it benefits the sensitivity for detecting DA comparing to other proposed colorimetric methods.

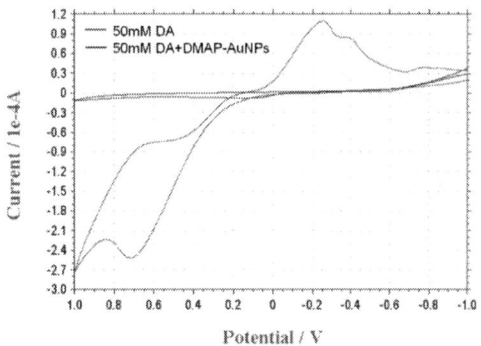

Fig. 6: Cyclic voltammogram responses of 50 mM DA and after added DMAP-AuNPs, shows that the concentration of DA has been reduced by the AuNPs.

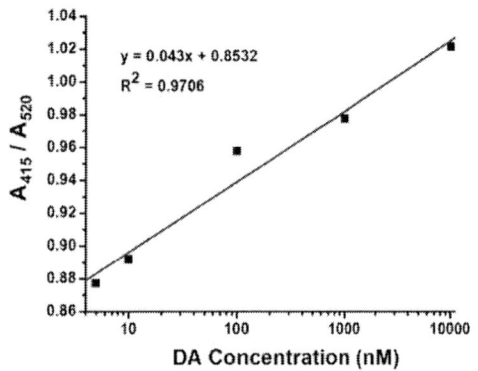

Fig. 7: Calculated absorption ratio (A_{415}/A_{520}) versus DA concentration. Experimental conditions: AuNPs, 6 nM 1 mL; DA, (10^{-5}-10^{-9} M) 0.5 mL; incubation time 5 min; room temperature.

V. CONCLUSION

This research found that DA has provided the DMAP-AuNPs for core etching effect and made the particles size changed into the smaller ones. Under the proposed DMAP-AuNPs as the colorimetric probe has presence the limit of detection for DA lower than 5 nM. Also, the proposed assay exhibited a linear correlation to DA concentration in the range

of $10^{-5} \sim 10^{-9}$ M with correlation coefficients of 0.9706. These kinds of results have provided AuNPs for a simple and sensitive detection for DA biosamples. The proposed of method also consider as a novel and versatile approach for rapid biosensing applications.

ACKNOWLEDGMENT

The authors would like to thank the financial supports from National Science Council of Taiwan.

REFERENCES

[1] Heien MLAV, Phillips PEM, Stuber GD, Seipel AT, and Wightman RM, "Overoxidation of Carbon-Fiber Microelectrodes Enhances Dopamine Adsorption and Increases Sensitivity," *Analyst,* vol. 128, pp. 1413-1419, 2003.

[2] Jiang D, Li X, Williams R, Patel S, Men L, Wang Y, and Zhou F, "Ternary Complexes of Iron, Amyloid-B, and Nitrilotriacetic Acid: Binding Affinities, Redox Properties, and Relevance to Iron-Induced Oxidative Stress in Alzheimer's Disease," *Biochemistry,* vol. 48, pp. 7939-7947, 2009.

[3] Kim YR, Bong S, Kang YJ, Yang Y, Mahajan RK, Kim JS, and Kim H, "Electrochemical Detection of Dopamine in the Presence of Ascorbic Acid Using Graphene Modified Electrodes," *Biosensors and Bioelectronics,* vol. 25, pp. 2366-2369, 2010.

[4] Li N, Guo J, Liu B, Yu Y, Cui H, Mao L, and Lin Y, "Determination of Monoamine Neurotransmitters and Their Metabolites in a Mouse Brain Microdialysate by Coupling High-Performance Liquid Chromatography with Gold Nanoparticle-Initiated Chemiluminescence," *Analytica Chimica Acta,* vol. 645, pp. 48-55, 2009.

[5] Zhao S, Huang Y, Shi M, Liu R, and Liu YM, "Chemiluminescence Resonance Energy Transfer-Based Detection for Microchip Electrophoresis," *Analytical Chemistry,* vol. 82, pp. 2036-2041, 2010.

[6] Mo JW and Ogorevc B, "Simultaneous Measurement of Dopamine and Ascorbate at Their Physiological Levels Using Voltammetric Microprobe Based on Overoxidized Poly (1, 2-Phenylenediamine)-Coated Carbon Fiber," *Analytical Chemistry,* vol. 73, pp. 1196-1202, 2001.

[7] Zhao W, Brook MA and Li Y, "Design of Gold Nanoparticle‐Based Colorimetric Biosensing Assays," *ChemBioChem,* vol. 9, pp. 2363-2371, 2008.

[8] Baron R, Zayats M and Willner I, "Dopamine-, L-Dopa-, Adrenaline-, and Noradrenaline-Induced Growth of Au Nanoparticles: Assays for the Detection of Neurotransmitters and of Tyrosinase Activity," *Analytical Chemistry,* vol. 77, pp. 1566-1571, 2005.

[9] Zhang Y, Li B and Chen X, "Simple and Sensitive Detection of Dopamine in the Presence of High Concentration of Ascorbic Acid Using Gold Nanoparticles as Colorimetric Probes," *Microchimica Acta,* vol. 168, pp. 107-113, 2010.

[10] Zheng Y, Wang Y and Yang X, "Aptamer-Based Colorimetric Biosensing of Dopamine Using Unmodified Gold Nanoparticles," *Sensors and Actuators B: Chemical,* 2011.

[11] Brust M, Walker M, Bethell D, Schiffrin DJ, and Whyman R, "Synthesis of Thiol-Derivatised Gold Nanoparticles in a Two-Phase Liquid–Liquid System," *J. Chem. Soc., Chem. Commun.,* pp. 801-802, 1994.

[12] Gittins DI and Caruso F, "Spontaneous Phase Transfer of Nanoparticulate Metals from Organic to Aqueous Media," *Angewandte Chemie International Edition,* vol. 40, pp. 3001-3004, 2001.

Fast Self-resonant Startup Procedure for Digital MEMS Gyroscope System

Fei Ge, Dachuan Liu, Longtao Lin, Zhenchuan Yang, Guizhen Yan[*]
[*]Institute of Microelectronics, Peking University, CHINA
gzyan@pku.edu.cn

Abstract—This paper describes a control procedure for the startup of a digital MEMS gyroscope system. Unlike the normal operating mode in which the gyroscope is actuated by synthesized sinusoidal drive signal, self-resonance of the gyroscope is stimulated by feeding back the drive-sense signal and converting it to a square-wave drive signal. During this procedure the resonant frequency that is near the natural frequency of the gyroscope can be calculated, and then provides a key parameter for the sinusoidal wave synthesizer, which will generate high-quality reference signals for driving and demodulation. Experiment shows a rapid rising of the magnitude of vibration, resulting a faster startup procedure for the system.

Keywords-MEMS gyroscope; self-resonace; Direct Digital Synthesis; FPGA

I. Introduction

With rapid advances in MEMS (Micro Electro-Mechanical Systems) technology, micromachined inertial sensors have found tremendous applications in various fields [1]. MEMS vibratory gyroscope is a tiny and cost-effective angular rate sensor, and can be easily integrated into devices where rotation detection is required. The working principle of vibratory gyroscopes is based on the Coriolis effect, which transfers the energy from the primary vibration mode (drive mode) to the secondary mode (sense mode).

Since the angular rate readout is through amplitude demodulation of the displacement of proof mass in the sense mode, it is fundamental to maintain a stable vibration with constant magnitude in the drive mode [2-4]. Traditionally, the gyroscope is part of the analog oscillator and it is stimulated into self-resonance and stabilized using automatic gain control (AGC) technique [4].

Digital control and signal processing of MEMS gyroscope systems is favored nowadays, because it is precise, robust against environment changes, and flexible to implement. In literature, digital gyroscopes have reached very high accuracy [5]. In a typical FPGA-based digital gyroscope system [6] shown in Fig. 1, the actuating (drive) circuit is connected to the D/A converter after the digital sinusoidal wave synthesizer. The A/D converters measure the voltages that reflect the displacement of proof mass in both drive and sense mode. The sinusoidal wave synthesizer uses direct digital synthesis (DDS) algorithm to generate high quality signals that are also used in demodulation, so it governs the working frequency of the gyroscope system. The frequency parameter was obtained by automatic scanning at powering up in [6], but obviously this approach is time-consuming.

To shorten the startup time while performing automatic frequency calibration, we borrow the idea from analog oscillator and propose to stimulate the gyroscope with a signal derived from the drive-sense signal. In this procedure, the D/A converter outputs a square wave having the maximum voltage swing, whose phase is controlled by a monitor module reading the gyroscope's response. In this way the gyroscope would self-resonate near its natural frequency. Meanwhile, using certain counting method, this resonant frequency can be calculated and finally the core parameter of the wave synthesizer is obtained. Once the synthesizer is initialized, the startup process completes and the gyroscope can be driven by a sinusoidal signal as in normal mode.

II. Princple of Operation

A. Gyroscope Model

The MEMS vibratory gyroscope can be viewed as a two-dimensional mass-spring system, in which the proof mass is suspended on a silicon framework. The drive and sense vibration modes are orthogonal and mechanically decoupled. A simplified model is illustrated in Fig. 2 and can be described by the following equations:

$$m\frac{d^2 u_x(t)}{dt} + c_x\frac{du_x(t)}{dt} + k_x u_x(t) = F_{el}(t), \quad (1)$$

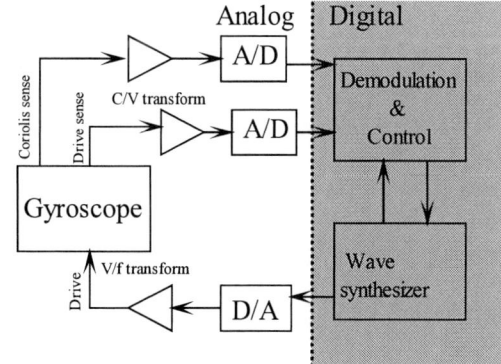

Fig. 1. An FPGA based digital gyroscope system

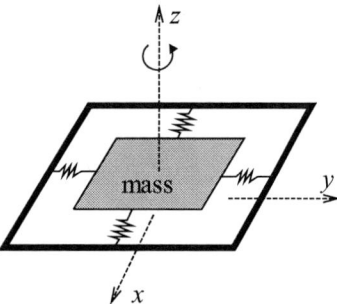

Fig. 2. A simplified model of MEMS vibratory gyroscope.

$$m\frac{d^2 u_y(t)}{dt} + c_y\frac{du_y(t)}{dt} + k_y u_y(t) = F_c(t), \quad (2)$$

$$F_c(t) = -2m\Omega \times \frac{du_x(t)}{dt}. \quad (3)$$

In the above equations, u_x, u_y are the displacement in x (drive) and y (sense) direction, k_x, k_y are stiffness coefficients, and c_x, c_y are damping coefficients, respectively. F_{el} is the electrostatic force actuating the drive mode, and F_c is the Coriolis force that is proportional to the angular rate Ω along z axis.

In this paper we only discuss the drive mode, so the subscript x will be omitted in the following paragraphs. Using Laplace transform we can derive the transfer function of this mode to be

$$G(s) = \frac{U(s)}{F(s)} = \frac{1}{m} \cdot \frac{1}{s^2 + 2\zeta\omega_n s + \omega_n^2}, \quad (4)$$

where $\zeta = c/(2\sqrt{mk})$, and $\omega_n = \sqrt{k/m}$ is the natural frequency of the drive mode. Clearly this is a second-order low-pass system. If F_{el} is a sinusoidal $F\sin(\omega t + \varphi)$, the gyroscope's response is also sinusoidal when the oscillation is stable. Especially if $\omega = \omega_n$, the amplitude is maximum when the amplitude of actuation is specified.

B. Self-resonance

In order to get sufficient signal-to-noise ratio, the gyroscope is designed to operate close to its natural frequency. With displacement sensing of the drive mode by capacitance variance detection, an analog oscillator can be constructed easily. Because at frequency ω_n, the phase of displacement is 90 degrees behind the phase of actuating force, the gyroscope will run into self-resonance at ω_n if the drive-sense signal is fed back to the driving circuit with 90 degrees phase shift and enough gain [4].

In such digital system as Fig. 1, however, all signals processed by FPGA is discrete, i.e., sample-by-sample and quantized. There is a constant time interval between any two consecutive samples, delays in A/D and D/A converters. Due to these restrictions we cannot form the same feedback path as that in the analog circuit. Fortunately, if the phase difference

of drive signal and drive-sense signal is not far from 90 degrees, the gyroscope can still resonate at a frequency close to ω_n.

Instead of amplifying the noisy drive-sense signal, we propose to directly form a binary signal to actuate the gyroscope. Because of the high quality factor, the vibration of gyroscope will still be sinusoidal. The binary signal is much simpler, as we only need to determine when to change its sign, which corresponds to the direction of driving force. Since a phase shift of 90 degree is needed, the drive signal should change from $+D$ to $-D$, when the drive-sense signal is at positive peak value, and conversely, from $-D$ to $+D$ at negative peak value. To detect the peaks, current sample of drive-sense signal must be compared with two preceding samples, as follows:

Algorithm: *Peak detector of* $u(t)$

a) If $u(t) > u(t-1)$, $d(t) = 1$.
   If $u(t) < u(t-1)$, $d(t) = -1$.
   If $u(t) = u(t-1)$, $d(t) = -d(t-1)$.

b) If $d(t) = 1$ and $d(t-1) = -1$, a "negative peak."
   If $d(t) = -1$ and $d(t-1) = 1$, a "positive peak."

To get the fastest startup, the D/A converter outputs the highest voltage, if the drive signal is $+D$, otherwise it outputs the lowest voltage. Fig. 3 sketches the drive signal and drive-sense signal, using this approach. Actually, the sign change of drive signal is delayed by one sample interval due to discrete signal processing.

C. Start of Oscillation

One question remains that if such digital resonator could start to work, when initially the gyroscope is not vibrating. Similar question applies to analog oscillator, in which the noise or fluctuations act as the trigger. Usually, an A/D converter's output has small disturbance, even if its input is at constant DC value, thus the peak detector will always give

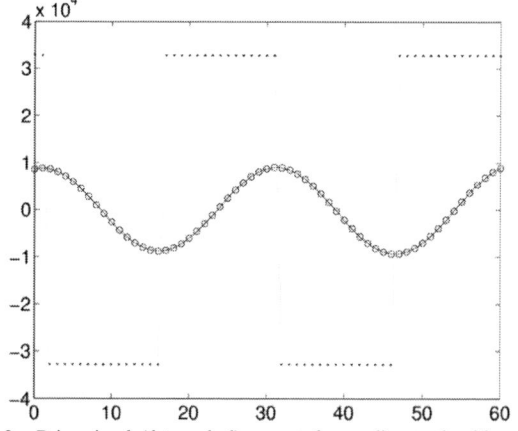

Fig. 3. Drive signal (dot marked) generated according to the drive-sense signal (circle marked) in self-resonant mode.

some instructions to change the sign of drive signal.

A square wave can be viewed as the sum of a series of step signals. The step response of a damped system (4) is

$$u(t) = 1 - \frac{\exp(-\zeta\omega_n t)}{\sqrt{1-\zeta^2}}\sin(\omega_n\sqrt{1-\zeta^2}\,t+\varphi), \qquad (5)$$

where $\varphi = \operatorname{atan}(\sqrt{1-\zeta^2}/\zeta)$, and $\omega_d = \omega_n\sqrt{1-\zeta^2}$. Fig. 4 shows one such signal. As a MEMS gyroscope has high quality value, the damped oscillation is significant. As long as the response of the gyroscope given a strong step signal can be detected by the A/D converter, the oscillation will be reinforced by the following step signals.

D. Frequency Calculation

If the peaks of a modulated sinusoidal signal can all be detected, it is easy to calculate the frequency. For example, if there are n sample intervals between $N+1$ detected positive or negative peaks, the frequency will be

$$f = \frac{f_s N}{2n}, \qquad (6)$$

with f_s being the sampling rate. However, since the actual peak may arrive at any time between two samples, the equivalent n is actually not an integer and thus introduces an error. This absolute error is no larger than 2, so for longer time or higher sampling rate, the error will be smaller.

If the wave synthesizer has a phase resolution of $2\pi/M$, the frequency resolution would be f_s/M. At each sample time, if the phase counter is added by p, the synthesized frequency will be $f_s p/M$. Together with (6), we can find that it is just needed to obtain an integer

$$p \approx \frac{MN}{2n} \qquad (7)$$

as parameter for the wave synthesizer. To avoid division, we can solve p iteratively, as follows:

Algorithm: *frequency parameter calculation*
If $pn > MN/2$, $p_{next} = p-1$;
Otherwise, if $pn < MN/2$, $p_{next} = p+1$.

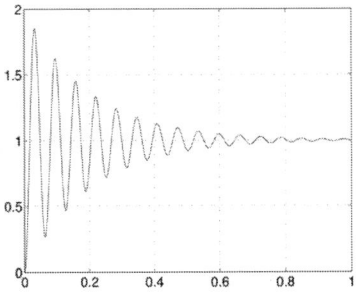

Fig. 4. Typical step response of a second-order damped system.

Note that the counters for n and N are only updated when a peak is detected, but p value is updated more often for quick adaptation.

E. Amplitude Detection and Control

A direct amplitude detection method is subtracting the most recent positive peak value with the most recent negative peak value of sampled signal. This approach has larger error than common demodulation methods, but it is enough to monitor the strength of vibration.

As soon as the amplitude of drive-sense signal has reached a proper level, which is usually covering most of the A/D converter's input range, automatic gain control must be applied to limit the amplitude of oscillation. A proportional or proportional-integral (PI) controller can be applied to adjust the level of square wave.

Switching from square wave mode to sinusoidal wave mode takes place when the frequency parameter of the wave synthesizer is obtained with desired accuracy.

III. Experiment and Results

Fig. 5 shows the experiment gyroscope system board. The gyroscope, the analog and digital components are all mounted on the PCB, which is powered by ±5V dual DC supply. The z-axis gyroscope used in the test is vacuum packaged and has a quality factor about 3000.

Fig. 6 sketches the waveform of drive-sense signal at startup, when AGC is not used. The plot data were collected by built-in memory card interface circuit. We can observe that the self-resonance started very quickly, and the amplitude of oscillation rose to almost the full input range in less than 0.06 second.

The sampling rate of our test system is 97.656kHz, while the gyroscope resonates at about 3.2kHz at room temperature. Since frequency calculation using data within less than 0.06 second may introduce an error of more than 1Hz, it is better to switch to sinusoidal wave driving later.

Fig. 7 is captured by a digital oscilloscope measuring the DAC output, showing the driving wave switching from square wave mode to sinusoidal wave mode. The transition is made with minimum phase change. Fig. 8 and 9 respectively show the DAC output in these two modes, with contrast to the

Fig. 5. Photo of the PCB board of experiment digital gyroscope system.

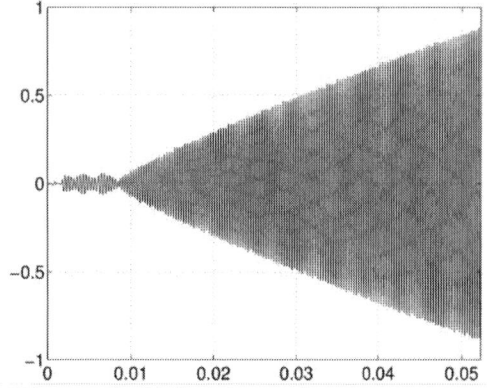

Fig. 6. Drive-sense signal in the startup procedure.

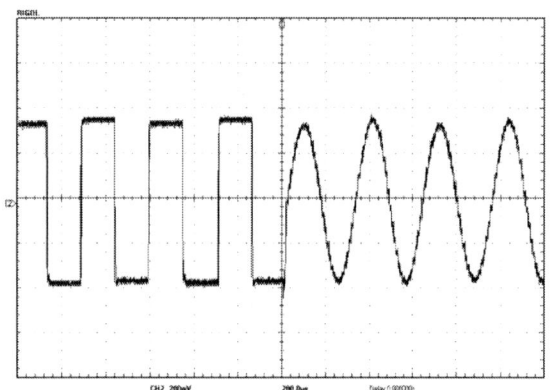

Fig. 7. DAC output when the driving switches from square wave mode to sinusoidal wave mode.

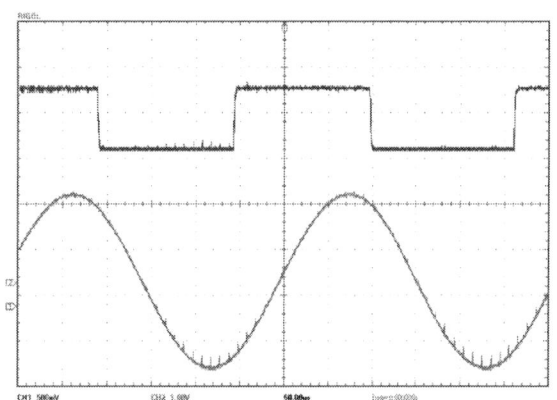

Fig. 8. Drive signal (upper) and drive-sense signal (lower), in the self-resonant square wave driving mode.

analog drive-sense signal. The startup procedure successfully found a working frequency for the gyroscope in the experiment.

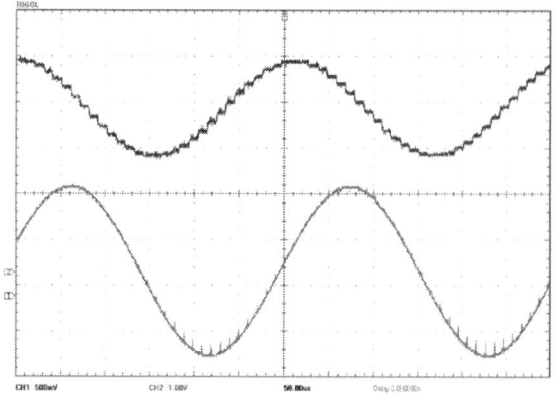

Fig. 9. Drive signal (upper) and drive-sense signal (lower), in sinusoidal wave driving mode. The frequency parameter was determined in square wave driving mode.

IV. CONCLUSION

In an FPGA based digital gyroscope system, the frequency at which the gyroscope resonates is governed by the digital wave synthesizer. A procedure to start the oscillation and initialize the synthesizer is proposed, which is based on feedback technique adopted by analog oscillators. Experiment shows that the rising of oscillation is very fast, and a working frequency can be determined in a short time. This startup procedure has significant advantages over frequency scan approach.

ACKNOWLEDGMENT

The authors are grateful to the technical staffs of National Key Laboratory of Science and Technology on Micro/Nano Fabrication for their supports on device fabrication.

REFERENCES

[1] N. Yazdi, F. Ayazi, and K. Najafi, "Micromachined inertial sensors," *Proceedings of IEEE*, vol. 86, no. 8, 1998, pp. 1640-1659.

[2] A. Shaban, M. El-Badry, and A. El-Sayed, "Analysis and design of gyro-drive mode loop with amptitude control," *Proc. Int. Design and Test Workshop*, 2009, pp. 1-4.

[3] R. Oboe, R. Antonello, E. Lasalandra, G. S. Durante, and L. Prandi, "Control of a z-axis MEMS vibrational gyroscope," *IEEE. Trans. Mechatronics*, vol. 10, no. 4, 2005, pp. 364-370.

[4] J. Cui, et al., "Transient response and stability of the AGC-PI closed-loop controlled MEMS vibratory gyroscopes," *J. Micromechan. Microeng.*, vol. 19, 2009, pp. 125015.

[5] H. Rodjegard, et al., "A digitally controlled MEMS gyroscope with 3.2 deg/hr stability," *Digest of Technical Papers. IEEE Transducers'05, 2005*, pp. 535-538.

[6] L. Lin, et al., "Digital closed-loop controller design of a micromachined gyroscope based on auto frequency swept," *Proc. 6th IEEE NEMS*, 2011, pp. 654-657.

The Influence of Experimental Parameters on the Assembly of SWNTs by AC Dielectrophoresis

Zhihui Wang, Fangzhou Yu, Wei Li, Jinwen Zhang*

*National Key Laboratory of Science and Technology on Micro/Nano Fabrication,
Institute of Microelectronics, Peking University, CHINA
zhangjinwen@pku.edu.cn

Abstract—In this paper, we investigated the influence of experimental parameters on the assembly of single-walled carbon nanotubes (SWNTs) by dielectrophoresis (DEP) process. After W/Au electrodes with designed structure parameters were prepared, designed experimental parameters were adopted to assemble SWNTs by DEP. I-V characteristics were measured and the assembled number of electrodes was enumerated by scanning electron microscopy (SEM). The results show that there was an ohmic contact formed between the assembled metallic SWNTs and the electrode. The variation of experimental parameters such as the SWNT solution concentration, deposition time, voltage magnitude and frequency can have significant impacts on the assembly productivity. Optimum for high assembly productivity of single-bundle of SWNTs was discussed and proposed at last.

Keywords- Experimental Parameters; SWNTs; Productivity; DEP Assembly

I. INTRODUCTION

Carbon nanotubes (CNTs), due to their unique one-dimensional structure and outstanding physical properties, are considered as promising building blocks for next generation nanoelectronic devices. A major technological challenge is the selective and controllable patterning of CNTs. DEP[1], which refers to the polarization effects in a nonuniform electric field, has shown tremendous promise to be an efficient technique to control the alignment and location of the SWNTs in the recent years. M. Senthil Kumar et al.[2] used this technique to align CNTs between microelectrodes. And Krupke et al.[3] achieved large-scale precise assembly of SWNTs. However, there is lack of systematic research on optimum experimental conditions for DEP process.

In this paper, we investigated to assemble single or bundles of SWNT by DEP. The influence of experimental parameters such as the SWNTs solution concentration, deposition time, voltage magnitude and frequency were studied experimentally for improving the productivity of SWNTs DEP assembly.

II. THEORY AND DESIGN

Using the concept of "effective dipole moment"[1], Jingqi Li et al.[4] analyzed the induced force on the SWNTs in the DEP process. The SWNTs bundle can be modeled as a prolate spheroid. When the electric field is nonuniformly distributed, the dipole will experience a net force：

$$F_{DEP} = \frac{\pi r^2 l}{2} \varepsilon_1 \operatorname{Re}(K) \nabla E^2 \qquad (1)$$

where $K = \dfrac{\varepsilon_2^* - \varepsilon_1^*}{\varepsilon_1^* + (\varepsilon_2^* - \varepsilon_1^*)L_{//}}$ and $\varepsilon_{1,2}^* = \varepsilon_{1,2} + \dfrac{\sigma_{1,2}}{i\omega}$.

where r and l are the radius and length of the SWNTs bundle respectively, ε_1 and ε_2 are the permittivities of the media and the ellipsoid respectively, σ_1 and σ_2 are the conductivities of the media and the ellipsoid respectively. $L_{//} \approx 4 r^2/l^2 [\ln(l/r)-1]$ is the depolarization factor. If $L_{//}$ is very small, we can derive that

$$\operatorname{Re}(K) = \frac{\sigma_1(\sigma_2 - \sigma_1) + \omega^2 \varepsilon_1(\varepsilon_2 - \varepsilon_1)}{\sigma_1^2 + \omega^2 \varepsilon_1^2} \qquad (2)$$

At high frequency, equation (2) can be simplified as

$$\operatorname{Re}(K) = (\varepsilon_2 - \varepsilon_1)/\varepsilon_1 \qquad (3)$$

So the direction of F_{DEP} is determined by the permittivity. As the permittivity of metallic SWNT is much larger than that of solution, the metallic SWNT will travel in the direction of the gradient of $\mathbf{E}^2$. In other words, the metallic SWNTs will move towards the electrodes where the electric field is the strongest.

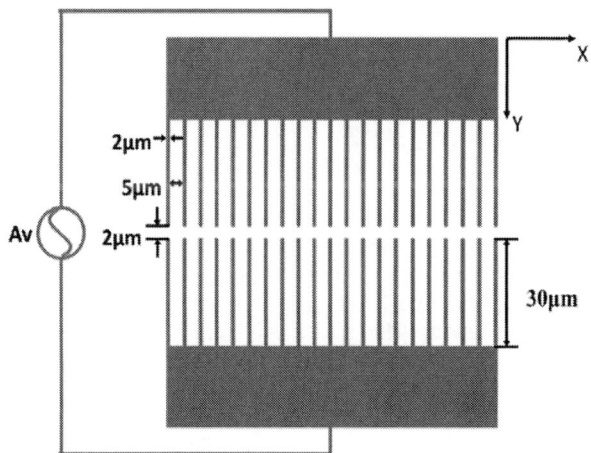

Fig. 1. Schematic graph of electrode structure.

The electrode was designed with optimal dimensions based on our previous work[5]. It consists of 20 opposed finger pairs 30μm long and 2μm wide and the gap of opposed and adjacent fingers is 2μm and 5μm respectively (Figure 1). With DEP process, the high-frequency ac bias will be applied to the ends of comb electrodes and the SWNTs are expected to be bridged between opposite fingers along Y orientation.

The experimental parameters, such as the SWNTs solution concentration(C), deposition time(T), peak-to-peak voltage magnitude(V_{PP}) and frequency(F), were chosen as our research objects. They are varied as follows:(1)C: 0.02, 0.1, 0.5, 2, 10μg/ml; (2)T: 0, 10, 30, 60, 120, 300s; (3)V_{PP}: 1, 3, 5, 7, 10V; (4)F: 10, 100, 1K, 10K, 100K, 1M, 5M, 10M, 15MHz.

III. PREPARATION AND ASSEMBLY

The process of preparation and assembly is shown in figure 2. N-type Si <100> wafer with 300nm thermal oxide was used as the substrate. W/Au (20nm/150nm) was sputtered and patterned by lithography and wet-etching.

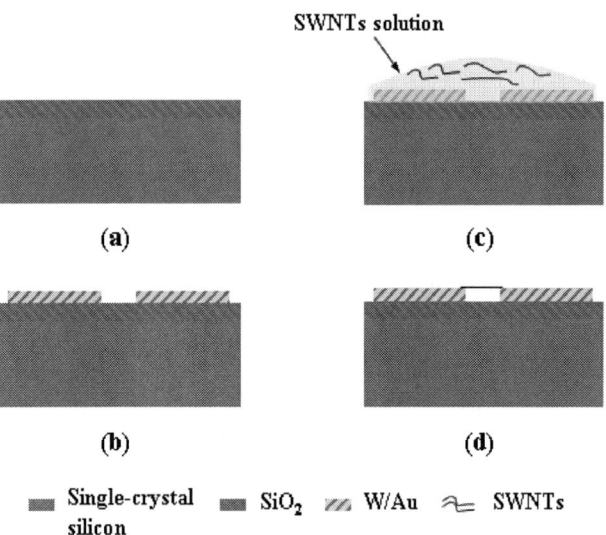

(a) (c)

(b) (d)

▬ Single-crystal silicon ▬ SiO₂ ▨ W/Au ⌁ SWNTs

Fig. 2. Process of preparation and assembly.

SWNTs with high purity (>90 %) were produced by Chengdu Organic Chemicals Company. The diameter of SWNTs was 0.8-1.6nm. The SWNTs were ultrasonically dispersed in N, N-dimethylformamide (DMF) for 2 hours and the solutions with the different designed concentrations were prepared respectively. Then a drop of the solution was introduced onto the chip after an AC bias having different V_{pp} and F was applied to the electrode. After a delay of T, the drop was blown off using nitrogen gas. Figure 3 shows SEM image of parts of assembled SWNTs bundles and there were 8 pairs of electrodes assembled by single-bundle of SWNTs in 20 pairs.

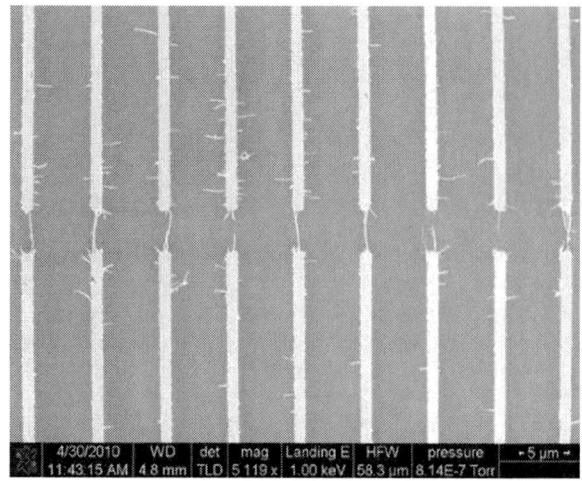

Fig. 3. SEM image of assembled SWNT bundles.

IV. RESULTS AND DISCUSSION

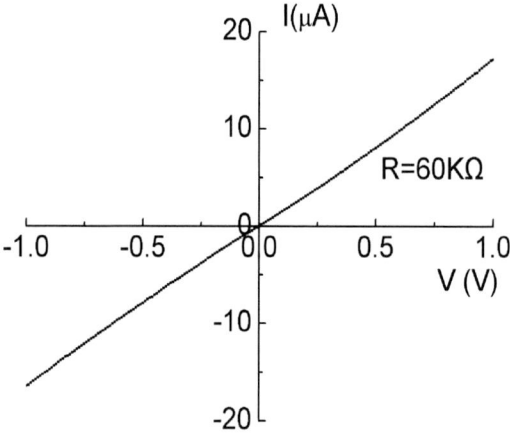

Fig. 4. I-V curve of the SWNT bundles.

The linear I-V characteristic (Figure 4) measured by HP4145B implied there was a good ohmic contact formed between the assembled metallic SWNT and the electrode. The resistances of assembled SWNTs (containing tube and contact resistances) were approximate to 60KΩ.

The assembled number of electrodes was enumerated by scanning electron microscopy (SEM) and the assembly productivity was calculated by P=(assembled number of electrodes)/20. P_M, P_S, P are defined as productivities of single-bundle, multi-bundles and both kinds of SWNTs respectively, i.e. $P=P_S+P_M$.

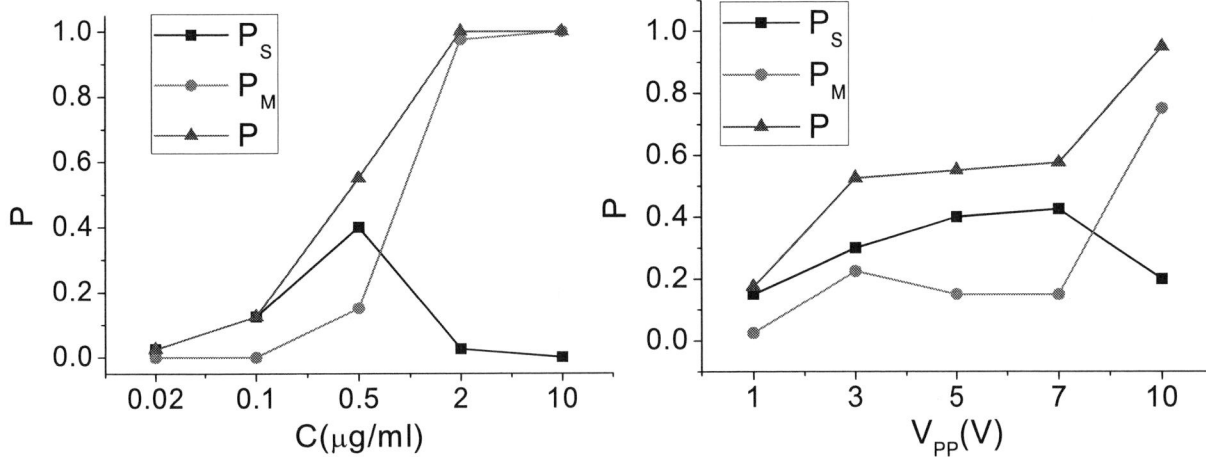

Fig.5 Productivity vs. C(solution concentration) with T=60s, V_{PP}=5V, F=5MHz.

Fig.7 Productivity vs. V_{PP}(peak-to-peak voltage magnitude) with C=0.5μg/ml, T=60s, F=5MHz.

P and P_M increase with the solution concentration and are higher to 100% when C=10μg/ml (Figure 5). However, the relationship of P_S and C is near parabola. And the maximum of P_S is 40% when C=0.5μg/ml. With the increase of C, the amount of SWNTs around the electrodes increases. This may be the reason why P increases with C. The phenomenon that P_M increases with C may indicates that the assembled SWNTs may have less impact on the assembly of other SWNTs between the same electrode pair. So the concentration of SWNTs should be controlled carefully to get high P_S. And C= 0.5μg/ml may be an appropriate value from the picture.

With the rise of V_{PP}, P and P_M increase substantially. According to theoretical research, the DEP force induced on SWNTs is proportional to $\bigtriangledown E^{2[4]}$. Therefore, the productivity should increase with the voltage magnitude. While the relationship of P_S and V_{PP} is near parabola (Figure 7). And V_{PP}=7V is the turning point. The reason may be that the voltage magnitude is too large to avoid multi-bundle of SWNTs assembled between same pair of comb, which needs to be verified by further study. Besides, P_M is almost 0 with V_{PP}=7V. So, when C=0.5μg/ml, T=60s, F=5MHz, V_{PP} has better to be 7V to get high P_S and low P_M simultaneously.

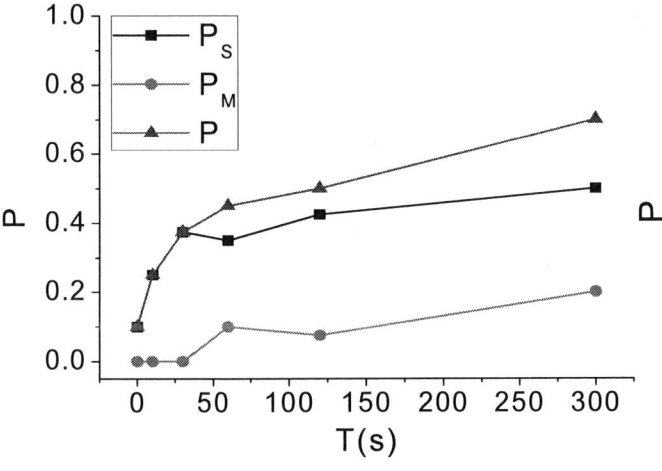

Fig.6 Productivity vs. T(deposition time) with C=0.5μg/ml, V_{PP}=5V, F=5MHz.

Fig.8 Productivity vs. F(voltage frequency) with C=0.5μg/ml, V_{PP}=5V, T=60s.

All three productivities rise basically with deposition time (Figure 6). The reason may be that the more time used for SWNTs assembly, the more SWNTs will move to the electrodes and be assembled. However, the productivity of multi-bundle SWNTs is also be enhanced with T. So it is impossible to get high P_S and low P_M at the same time by simply prolonging the deposition time.

All three kinds of productivity fluctuate with the frequency （Figure 8）. The maximum of P and P_M are both higher to 100% when F=10kHz. While the maximum of P_S is 35% when F=5MHz. Meanwhile, when F=5MHz, P_M is less than 10%. So frequency has better to be 5MHz to get high productivity of single-bundle SWNTs with few multi-bundles of SWNTs assembled.

978-1-4673-1122-9/12 $31.00 © 2012 IEEE

In sum, the variation of experimental parameters such as the SWNTs solution concentration, deposition time, voltage magnitude and frequency can have significant impacts on the assembly productivity. To get high assembly productivity of single-bundle of SWNTs, T should be lower than 120s with C=0.5μg/ml, V_{PP}=7V, F=5MHz.

V. CONCLUSION

In this paper, the influence of experimental parameters on the assembly of SWNTs by the DEP process is discussed in detail, which is helpful to DEP assembly in further study. Using proper W/Au electrodes based on our previous work, we assembled SWNTs in designed experimental conditions. I-V characteristics of the assembled SWNTs were measured and the assembled number of electrodes was enumerated by SEM.

The results turned out that the assembled metallic SWNTs formed an ohmic contact with the electrode. The variation of experimental parameters such as C, T, V_{PP} and F can have significant impacts on the assembly productivity. The relationship of P_S and C is near parabola, so is P_S and V_{PP}. All three productivities increase basically with T and fluctuate with the frequency. Optimum for high assembly productivity of single bundle of SWNTs was T<120s, C=0.5μg/ml, V_{PP}=7V, F=5MHz.

ACKNOWLEDGMENT

This work is supported in part by a grant from the National Basic Research Program of China (973 Program) (No. 2009CB320300).

REFERENCES

[1] T. B. Jones, *Electromechanics of Particles*, Cambridge University Press, Cambridge, 1995.

[2] M. Senthil Kumar, S. H. Lee, T. Y. Kim, T. H. Kim, S. M. Song, J. W. Yang, et al, "DC electric field assisted alignment of carbon nanotubes on metal electrodes," *Solid-State Electronics*, vol.47, pp.2075-2080, 2003.

[3] A. Vijayaraghavan, S. Blatt, D. Weissenberger, M. Oron-Carl, F. Hennrich, D. Gerthsen, et al, "Ultra-Large-Scale Directed Assembly of Single-Walled Carbon Nanotube Devices," *Nano Lett.*, vol.7, no.6, pp. 1556-1560, 2007.

[4] N. Peng, Q. Zhang, J. Li and N. Liu, "Influences of ac electric field on the spatial distribution of carbon nanotubes formed between electrodes," *Journal of Applied Physics*, 100, 024309, 2006.

[5] Z. Wang, W. Li and J. Zhang, "The Effect of Electrode Physical Design Parameters on the Assembly of Single-walled Carbon Nanotubes by AC Dielectrophoresis," *Proceedings of 2011 IEEE NEMS*, pp.160-163, 2011.

The manufacture of micropillars with high depth-to-width ratio, and the comparison between two typical materials

Zhe Wang [*], Xian Qin,

[*] Singapore-MIT Alliance for Research and Technology (SMART) Center, Biosym, Singapore
Department of Micro-electronics, Tsinghua University, China
wangzhe@innovation-china.org

Abstract—This paper demonstrates a new method in manufacturing micropillars with a relative high depth-to-width ratio, compared with former works. As is known that micropillars enjoy a wide range of applications in the field of biology and chemistry, such as cancer drug screening, the auto-formation of artificial vessels, and so forth, the strict demand for high depth-to-width ratio haven't effectively settled in conventional MEMS technology. Also considering other means, for instance, through the stereolithographic printing process is of high cost, we propose a rapid way that combined the laser drilling as well as molding to realize the whole procedure. And we will discuss the diverseness of two materials, PMMA (Polymethylmethacrylate) and PDMS (polydimethylsiloxane), in fabricating the micropillars. Some simulations contribute to our deeper comprehension of the pillar structure, which mainly lie in strain behavior.

Keywords-micropillar, molding micro-fabrication, PDMS, PMMA, CO_2 laser drilling technique.

I. INTRODUCTION

Studying three-dimensional (3D) in vitro tissues has revolutionized many aspects of cellular function such as embryonic development, wound healing, and malignancy. This 3D model avoids two-dimensional (2D) monolayer culture, which mimics a better in vivo tissue-like situation in terms of morphology, cell-cell interaction, signalling transduction, and mechanical stimulation thus providing a physiological and pathological relevant environment. Multi-cellular 3D vitro is one of the models that have been widely applied for embryonic stem cell formed embryo bodies directing specific cell lineage. How can we utilize the 3D culture technique to reliably reproduce the vivo tissues, such as artificial blood vessel, is still a pending issue and act as a promising point due to be efficiently settled. Some former research propose to assist in micro-structure to realize the attachment that artificial blood vessel could grow along the surface of these micro-structures, with necessary medium on them leading the direction of different types of cells' activities.

To address the need of control-size micro-structure with high throughput capability, various approaches have been proposed using micro electro mechanical system (MEMs) technology. Generally, a cell unattachable surface with microstructure is used to isolate particular number of cells in the region of interest. Parallel culture control-size microstructure with high-throughput nature is feasible. Observation and change of medium are also made easy. However, based on replication of these fabrication methods mentioned so far are not only relatively complex and laborious but also expensive compared to laser fabrication. The principle objective of this paper was to develop a rapid fabrication of micropillar for generation of multi-cellular 3D aggregates with controllable size [1] [2].

In this work, we present a rapid prototyping technique using CO2 laser, accompanied by molding approaches, for generating controllable micropillars. CO2 laser directly writes structure on substrate, allowing the fabrication of even a complex microstructure w/without high aspect ratio in a few seconds. We implement a CO2 laser ablation to create size- and depth-controllable microwell on two types of conventionally used substrates, polymethylmethacrylate (PMMA), polydimethylsiloxane (PDMS), to characterize the size and depth and geometry of micropillar configuration. Similar size and morphology of the micropillars would be fabricated under different combination of prototyping recipe. Results serve as a guide to optimize the fabrication of desire micropillars [3].

II. METHOD OF MANUFACTURING ARRAY OF MICROPILLARS

A. Fabrication of PDMS as well as PMMA mould

We use 10.6 um CO2 laser operated in TEM00 mode, where the spatial energy distribution appears to be a Gaussian profile (Fig. 1 (a)). High energy density of laser beam is capable of affecting surface of the substrate in short time. Substrate absorbing laser energy reaches to melting point or gasification temperature of the materials PDMS or PMMA, causing these two materials to be vapor or melted substance. High pressure of vaporized gas is capable of removing out melted particles on surface, where leave out a space for docking cells.

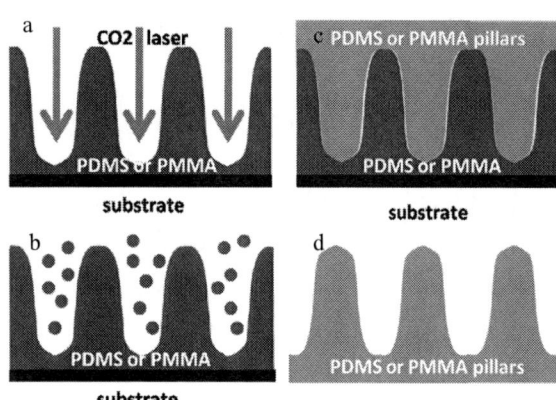

Fig. 1. (a) CO2 laser abalate the substrates manufactured by PDMS or PMMA to form the mould. (b) Coat a thin layer of silane to prevent the material bonding with each other. (c) Transfuse some liquid PDMS or PMMA, and wait for coagulation. (d) Strip out the two parts, we can get the array of micropillars.

Laser prototyping was implemented by a commercial air-cooled CO2 laser engraving/cutting system (VLS-2.30, Universal Laser System Inc., U.S.A.) with a high power density focusing optic (HPDFO) and maximum output power of 10 W. Pattern drawing of well structure, that serves as the die, was designed using commercial software (AutoCAD 2007, Autodesk Inc.) and transferred to laser system. Laser scanning was done by a XY mobile Focus Carriage (FC) travelling to the location of interest. Characterization of micropillar for desired laser output was characterized with laser power and FC scanning speed on PMMA (1mm thick,), and PDMS (Sylgard 184, Dow Corning) [4].

B. Following steps of manufacturing micropillars

We coated a layer of silane (Dynasylan®) preserving for 3 hours, though the relative strong Van der Waals force between the same materials existed, it can still prevent the two part from tightly bonding with each other (Fig. 1 (b)). The coating process should be conducted in ventilation environment, for silane is very toxic, as well as volatile. Next, we transfused PDMS (1mm thick,), in liquid state, and PMMA, onto the surface of the microwell-structured substrate (Fig. 1 (c)), and placed them in flat smooth plane (70° C) until solidification completed. In the meanwhile, we should also sometimes locate it into the air exhaust machine, in order to let the viscous materials effectively inflow respectively into the wells without remaining air bubble [5]. Detaching the two parts by slightly and slowly riving them apart (Fig. 1 (d)), we finally obtained the array of micropillars with high depth-to-width ratio. The dimension of the micropillars can be precisely controlled by adjusting the power of CO_2 laser, however there exists a threshold value of maximum imposing laser power, and if the power exceed beyond the value, the PDMS or PMMA substrate might become scorched condition that cannot be used to mold micropillars.

C. CHARACTERIZATION OF MICROPILLARS

To understand how the power and scanning speed of CO2 laser could affect the material, we first to determine the range of power output that could generate proper micropillar formation by using PDMS substrate. For the criteria of complete micropillar formation, geometry should be symmetric in shape and with a relative high depth-to-width ratio. Results showed that the size of microwell was enlarged along with increasing laser power from 1W-5W at scanning speed of 25 and mm/s. However, the micropillar became irregular and obscure or absent in shape after scanning speed above 50 mm/s. This is due to that the increasing speed diminished the duration of laser energy acting on the substrate, affecting the laser-generated heat incapable of properly melting the substrate. Besides, we found that there is a similarity among the shape of microwell in scanning speed at 25 and 50 mm/s, whereas the energy shifted 1W at 50 mm/s. For instance, the microwell at 2W and 50 mm/s is alike to parameter at 1W and 25mm/s and so forth. Under this circumstance, we believe similar shape of well-structure could be generated by varying the combinations of power and scanning speed, which faster speed should coordinate with higher power to compensate the energy loss from the speed for the hole completion vice versa. Therefore, scanning speed at 25 mm/s was considered as representative parameters throughout the study.

We further investigated microwell formation using PMMA, PDMS materials to understand how well-structure formation, as the mould for manufacturing respective micropillars, over both materials. Representative side view of well features show marked differences both between materials and as a function of substrate material. PMMA and PDMS micropillars revealed extent conical shape along with increasing laser power. Diameter, depth, and aspect ratio of two materials of pillar-structure were quantified, which reflect the corresponding conditions of micropillars after molding procedure. It is worth mentioning that the surface patterns of micropillars primely reveal the many features of the array of microwells, and in fact, it can also be regarded as qualitative method of analyzing semblable micro-structures [6].

In order to more closely examine micropillars structure, isometric-view SEM images were obtained for both materials. PMMA micropillars exhibited an uneven surface at the periphery, with micro-size pored distributed over much of the internal surface. PDMS micropillars, in contrast, possessed a smooth surface, but optical resolution was impeded by the sharp geometry of the bottom. We can also quantify the relevance between the powers of the CO_2 laser and the diameter as well as the height of micropillars (Fig. 2 (b)). Fig. 2 (3) demonstrates the array of micropillars with 10W CO2 laser, and presents a higher depth-to-width ratio. There existed a disproportionate relevance that the depth-to-width ratio illustrates an roughly ascending trend, with the increasing powers of the CO_2 laser. As we attempt to adopt 15W CO_2

laser, the PDMS or PMMA substrates were scorching, and the molding result is unacceptable.

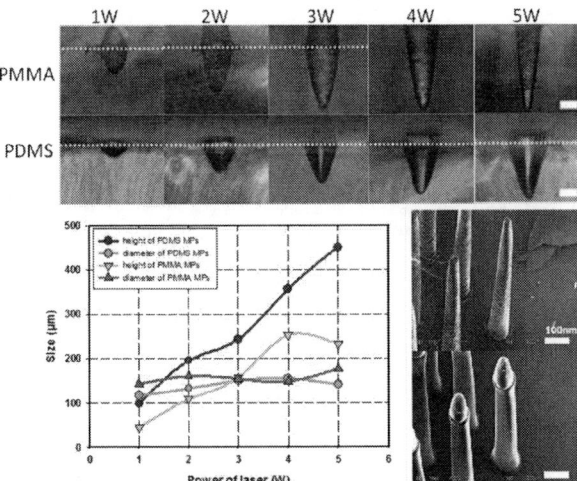

Fig. 2. (a) Demonstrate the profile of holes with different power of laser ranging from 1W to 5W. (b) Present the relation between the power of laser and the height, depth of micropillars. (c) Present the appearance of PMMA and PDMS micropillars.

D. Discussion on the freatures of micropillars

We have developed a simple and economical high-throughput method for rapid micropillar prototyping. The fabricated micropillars have been proved to be good outcomes suited for forming artificial blood vessel in separate part, which is in progress now, and preliminary results account for the array of micropillars can efficaciously shape the Dulbecco's modified Eagle medium (Gibco/Invitrogen 12100) supplemeted with 10% fetal bovine serum (Gibco/Invitrogen), 1% penicillin-streptomycin. The cells we utilized is A549 human lung adenocarcinoma epithelial cells that transfected with histone H2B-mCheery cDNA in nucleus were cultured on T75 flask (Nunclon™Δ Surface) [7].

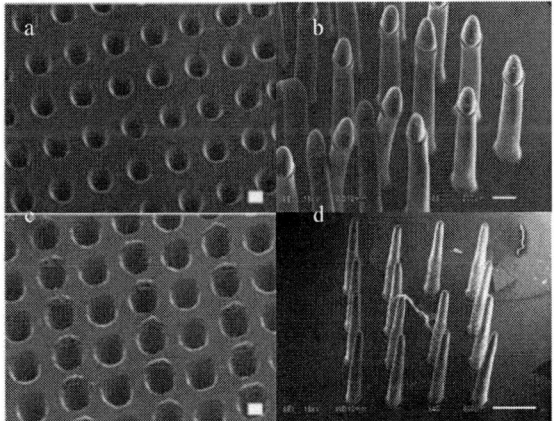

Fig. 3. (a) The mould array manufactured by PDMS. (b) The array of micropillars manufactured by PDMS. (c) The mould array manufactured by PMMA (d) the array of micropillars manufactured by PMMA.

Observing from top view, the array of micropillars as well as its mould (Fig. 3 (a)-(d)), we can apparently figure out that owing to the prototyping technique of fabricating mould would manufacture higher depth-to-width ratio wells than common etching, and the micropillars not only exhibit excellent character in uniform, but could also realize abundant number. Therefore, this method efficaciously expresses outstanding results ranging from time, and energy saving, to low cost.

From the SEM picture, we can draw the approximate depth-to-width ratio of PDMS micropillars are around 10:1, and PMMA micropillars are around 15:1. Both are remarkable consequences as we expected.

The surface structures of both micropillars were investigated using scanning electron microscope (SEM) (JEOL Ltd., JSM-6510LV). Prior to SEM imaging of aggregates, medium was aspirated and rinsed with PBS. Primary fixation was done using 0.1 M Na-Cacodylate buffer, 2.0 % paraformaldehyde, and 2.5% glutaraldehyde experienced 4 hours in fridge. Then we rinsed sample with 0.1 M Na-Cacodylate buffer on shaker table for 10mins. Post-fixation was done using 1.0% aqueous osmium tetroxide in 0.1 M Na-Cacodylate buffer in dark fume hood for 90mins, and followed by rinsing with 0.1 M Na-Cacodylate buffer on shaker table for another 10min. Finally, samples were serially dehydrated by different concentration of ethanol (37%, 67%, 95%, and 100%) on shaker table for 10min each, following by critical point drying (Balzers, BALSERS CPD 030) and platinum sputtered in vacuum (JEOL Ltd., JFC-1600).

III. SIMULATION RESULTS

We also run some simulation works (Comsol v7.0) in order to reveal the intensive property of micropillars, and the model's dimension, features of material, and surroundings are all resemble with the actual micropillars. We separately applied 10, 20, and 30 pressures on the upper surface of micropillars. Solving the model with proper external effect after meshing, we can uncover the carrying capability of micropillars, and we can also approximately estimate the maximum bearing capability of the structure (Fig. 4 (a) (b)).

The maximum force tension occurred in the medium position that is the most fragile point of micropillars. From lateral comparison, the situation of material PDMS also differs from PMMA that the stress conditions present contrast to some extent, and vulnerable areas of PDMS micropillars are also narrower than PMMA micropillars. In some special cases, the bearing force is not centrally symmetry distributed that demand further persuasive explanations [8] [9].

The substrates weren't considered in the model when running simulation; while actually, the part may exert an impact on the stress state of the micropillars, and the joint section would also testify to be an intensive region of external force. The simulation models have some aspects need to be ulteriorly optimized

978-1-4673-1122-9/12 $31.00 © 2012 IEEE

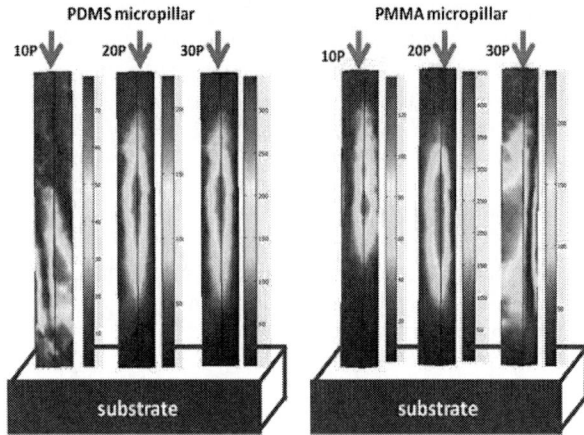

Fig. 4. (a) The simulation work of separately applying 10, 20, and 30 pressures on the upper surface of PDMS micropillars. (b) The simulation work of separately applying 10, 20, and 30 pressures on the upper surface of PMMA micropillars.

IV. CONCLUSIONS

A primary purpose of this study was to develop a rapid micropillar prototyping technique using CO2 laser as well as molding for generation of artificial blood vessel. And the whole article presents an efficient method in manufacturing plentiful of micropillars array with high depth-to-width ratio within very limiting time. We also discussed the differences between two suitable materials PDMS as well as PMMA. Respectively, the PDMS micropillars possess a highly smooth profile, and the PMMA micropillars have an even higher depth-to-width ratio. Adopting these fabricated micropillars, we can shape the configuration of artificial blood vessel that made of specially compounded medium mentioned in the above section. Cells might have the ability of interacting with each other in the surface of pillar-structure, and form the artificial vascular wall of certain diameter, length as well as other controllable factors. This greatly promising point is under study.

The subsequent simulation works are conducive to optimizing parameters that varies with sundry of situations. However, some designed flaws still exist in the simulation model. The precision and the accuracy of simulations are due to be enhanced.

ACKNOWLEDGMENT

This work was supported by the Singapore-MIT Alliance for Research and Technology (SMART) Biosym lab. The authors would like to thank Dr. Chang Yu Chen, Institute of Applied Mechanics, National Taiwan University for his advice in laser fabrication, thank to Ting-Yuan Tu, Mechanobiology Institute, National Singapore University, for his instruction in cell culture.

REFERENCES

[1] I. G. Wilson, "Inhibition And Facilitation of Nucleic acid Amplification" Appl. Environ. Microbiol., vol. 63, pp. 3741-3751, 1997

[2] J. Wen, L. A. Legendre, J. M. Bienvenue, J. P. Landers, "Purification of Nucleic Acids in Microfluidic Devices" Anal. Chem., vol. 80, pp. 6472-6279, 2008

[3] L. Chen, A. Manz, P. J. R. Day, "Total Nucleic Acid Analysis Integrated on Microfluidic Devices" Lab Chip, vol. 7, pp. 1413-1423, 2007

[4] O. Carp, C. L. Huisman, A. Reller, "Photoinduced Reactivity of Titanium Dioxide" Prog. Solid State Chem., vol. 32, pp. 33-177, 2004

[5] K. Y. Hwang, H. K. Lim, S. Y. Jung, K. Namkoong, J. H. Kim, N. Huh, C. Ko, and J. C. Park, "Bacterial DNA Sample Preparation from Whole Blood Using Surface-modified Si Pillar Arrays" Anal. Chem., vol. 80, pp. 7786-7791, 2008

[6] S. Bhattacharya, A. Datta, J. M. Berg, and S. Gangopadhyay, "Studies on Surface Wettability of Poly(dimethyl)siloxane (PDMS) And Glass under Oxygen-plasma Treatment And Correlation with Bond Strength" J. Microelectromech. Syst., vol. 14, pp. 590-597, 2005

[7] Stephen Y. Chou and Lei Zhuang, Lithographically induced self-assembly of periodic polymer micropillar arrays, Volume 17/Issue 6/NANOTECHNOLOGY: DEVICES AND FABRICATION.

[8] Daniele Bajoni, Polariton Laser Using Single Micropillar GaAs-GaAlAs Semiconductor Cavities Phys. Rev. Lett. 100, 047401 (2008)

[9] R. Idrissi Kaitouni, O. El DaÄ³f, A. Baas, M. Richard, T. Paraiso, P. Lugan, T. Guil-let, F. Morier-Genoud, J.D. Ganiµere, J.L. Staehli, V. Savona, B. Deveaud, "Zero dimensional exciton-polaritons" Phys.Rev. B 74, 155311 (2006).

Surface-modified diamond embedded in nickel matrix composite for intrinsic polishing application

Ching-Jui Shih[*], Wei-Chih Lin, Chao-Sung Lin, Yung-Ning Pan

[*]Department of Mechanical Engineering, National Taiwan University, Taipei, Taiwan

d97522024@ntu.edu.tw, wcl27@cam.ac.uk, csclin@ntu.edu.tw, panyn@ntu.edu.tw

Abstract—An approach for fabricating a diamond conditioner by using a micro patterning and nickel electroforming method is demonstrated to improve the material removal rate (*MRR*) and the diamond worn ratio. By compared with the currently commercial conditioner, our developed diamond conditioner performs approximate 35% of the *MRR* and 18% of diamond worn ratio. The contact surface between the diamond and the counterpart is increased by the well-controlled of the grit protrusion which is within the range of 1.36 micrometer. This protrusion control not only results in the *MRR* and frictional coefficient increase, but also the load force reduction on each diamond grit during a wearing test. Furthermore, a chemical treatment is utilized to functionalize the surface of the diamond grits for the adhesion enhancement which can eliminate the matrix defects at the interface between the grits and nickel matrix.

Keywords - diamond conditioner; electroforming; Micro-patterning

I. INTRODUCTION

The surface uniformity of the wafer is obtained by the chemical mechanical polishing process (CMP). Due to the debris and byproducts produced during the CMP process, a glazing effect gradually occurs on the pad which is a crucial component in the CMP process for achieving efficient material removal. Therefore, a diamond conditioner is used to regenerate the surface morphology of the pad in order to maintain a consistent material removal. A traditional diamond conditioner is fabricated by brazing and electroplating. Although the diamond grits are regularly distributed, the height level of diamond tips is not uniform [1-2]. Even though an interfacial layer with a sufficient bonging strength forms between the diamond grits and the matrix, the active elements in the brazing alloy can act as a catalyst causing the degradation of diamond to graphite once an interfacial layer is formed. Meanwhile, in nickel electroplating, the composite using surface-modified diamond grits as reinforcement particles has been fabricated by nickel electroplating during which the surface characteristics of a diamond are retained [3].

In general, micro electro mechanical systems (MEMS) include lithography, micro machining, thin film and etching process have been used for micro-fabrication processes to develop miniaturization of features or to improve the capacity and performance of component. In particular, lithography has been widely used in many applications, such as hard mask, polymer adhesive bonding [4], etc. However, a few reports have been mentioned about using the MEMS procedures and surface modified diamond grits as abrasive particles to fabricate lapping tools for polishing application. The present study introduces a new approach to reserve the properties of diamond by employing surface-modified diamond as abrasive particles for the fabrication of diamond conditioners, and at the same time, to achieve the uniformity of the distribution and height level of diamond cutting tips on stainless steel using micro patterning and nickel electroforming.

II. EXPERIMENTAL PROCEDURES

A. The modification of the diamond surface

Commercial synthetic diamond powders (Diamond innovation MBG EP3 170/200; average diameter 100μm; sharp type) were used as the abrasive particles for the diamond conditioner in this study. The substrate for the diamond conditioner is stainless steel of 25mm (length), 25mm (width), and 1mm (height). The substrate was first polished down to 1μm by using Al_2O_3 slurry. Before photoresist coating, the substrate was subsequently cleansed in acetone, isopropanol, methanol, and distilled (DI) water. To obtain the hydrophilic property of diamond grits, oxygenated diamond was prepared following the procedures described in the literatures [3, 5]. In brief, the diamond grits were cleansed by 38% HNO_3 at 80℃ for 4hs. The diamond grits were subsequently immersed in a $HF + HNO_3$ mixed acid solution at room temperature for 4hs and then rinsed with DI water. After drying, the diamond grits were then immersed in $H_2SO_4 + HNO_3$ solution at 80℃ for 4hs, followed by rinsing with DI water and then dried in an oven. After the modification process, the functional groups on the diamond grits were identified by diffuse reflectance infrared Fourier-transform (DRIFT) spectroscopy (Varian 640-IR FT-IR Spectrometer) and Raman spectroscopy (He-Ne laser type: λ=633nm). The surface-modified diamond grits were then ready for the fabrication of diamond conditioners.

B. Preparation of the diamond conditioner

Two types of diamond conditioner were fabricated via lithography using photoresist as the mask and electrodeposited nickel as the matrix, as schematically shown in Fig. 1. On a 25mm (length) x 25mm (width) x 1mm (height) stainless steel substrate, one half carat of diamond grits were added to use an abrasive particle. In order to design specific morphologies of diamond composites, two types of photoresists were used; a columnar type designated as N (JSR-151N) and a hemispherical type designated as E (EPG-510).

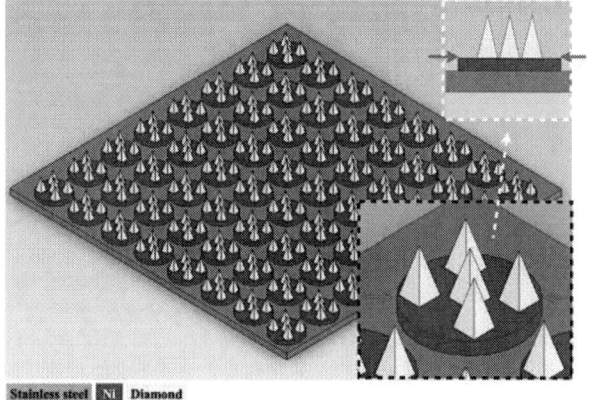

Fig. 1. A schematic view of the cross section of the developed diamond conditioner.

Both positive and negative photoresist (EPG-510 and JSR-151N) were prepared by spin coating a thin photoresist layer of 15μm and 75μm, respectively, onto the stainless steel substrate using an SSP-02A spin coater. After pattern definition, nickel matrix was electrodeposited to fill up the cavities, as surface-modified diamond grits was placed, in the stainless steel substrate. The parameters for Ni electroplating are presented in Table. 1. An entrapment layer was first deposited at 0.5ASD for 30 minutes in the electrolyte at 50℃.

The residual photoresist layer was then removed by a stripper. The electroforming time and the dimensions of the pattern can determine the morphology of the electroforming mold. In addition, the length of the mold and the height level of diamond tips can be controlled by varying electroforming parameters. To further improve the bonding strength of each column to the stainless steel substrate, an enhanced layer was deposited in the electrolyte at 0.5ASD for 5hs. A commercial diamond conditioner manufactured by 3M (BD-A188J) was

Table 1. Experimental parameters for nickel electroplating

Bath compositions	Parameter
Nickel sulfamate	90 g/l
Boric acid	40 g/l
Nickel chloride	3 g/l
Wetting agent	1-1.8 ml/l
Stress reducing agent	0-3 g/l
Diamond	0.5 carats
Operating conditions	
Temperature	50 ℃
Current density	0.5-2 ASD

chosen for comparison, in which the diamond grits are regularly arranged and brazed onto the stainless steel substrate by using Ni-based alloy. The multi designedly diamond composites of type E, type N and type BD are shown in Fig. 2.

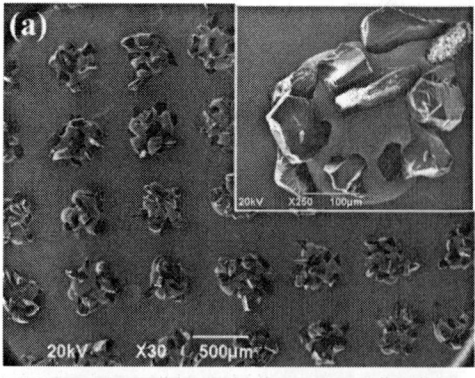

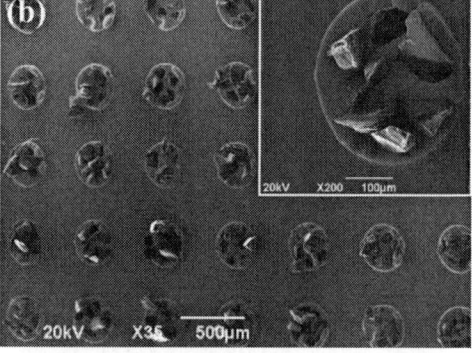

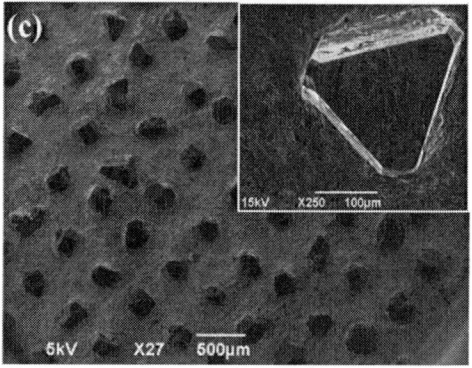

Fig. 2. SEM micrographs of conditioners: (a) E, corner picture at 250x; (b) N, corner picture at 200x and (c) BD, corner picture at 200x.

C. Wear resistance experiments

The wear resistance experiments were performed using a block-on-ring test in a solution close to the CMP slurry at room temperature. The test solution is an H_2O_2 solution that contains 6 wt% Al_2O_3 abrasives with size ranging from 50 to 300nm. Prior to the wear test, the ceramic ring was polished to a roughness of Ra 0.5μm and cleansed ultrasonically in DI water. All experimental parameters for block-on-ring test are presented in Table 2. The sampling rate for frictional coefficient was 30 counts per second. After the wear test, the conditioner and Al₂O₃ ring were washed with DI water and dried with a hot air stream. Finally, the weight loss was measured using a precision balance. Each datum point of the wear test was the average of three measurements.

Table 2. Experimental parameters for block-on-ring test

Sintered Al₂O₃ ceramic ring		Operating conditions		
Outer diameter (mm)	Inner diameter (mm)	Applied Load (g)	Sliding speed (rpm)	Periodic time (h)
22	13	300	100	1

III. RESULTS & DISCUSSION

To characterize the surface-modified diamond grits, the functional groups of the diamond surface and the nickel matrix in the vicinity of diamond grits were analyzed. The FTIR measurement results indicate that three basic functional groups, namely C-O-C, C=O and -OH are present in the oxygenated diamond grits (Fig. 3a), whereas C-O-C, -CH and -OH, appear in the unmodified diamond grits (Fig. 3c). However, it has to be noted that the peak -OH in the unmodified case and the peak C=O in the oxygenated case are very minute. In addition,

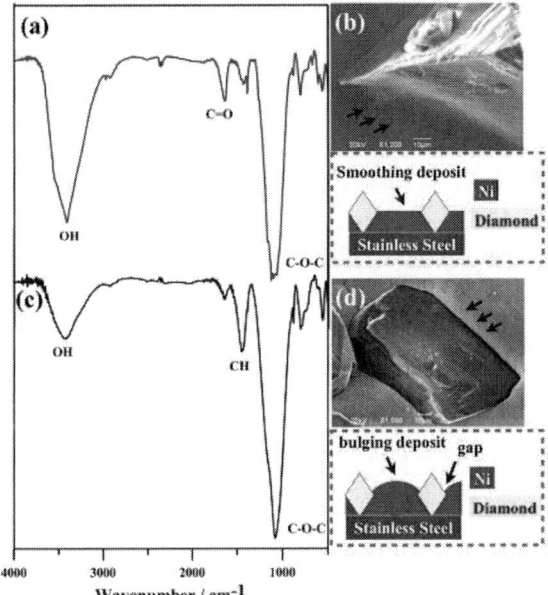

Fig. 3. IR spectra of the diamond grits and SEM of the nickel matrix deposit in the vicinity of diamond grit, (a) and (b) oxygenated diamond; (c) and (d) unmodified diamond.

the interface between the modified diamond grits and the nickel matrix is relatively smooth, indicating a good bonding, as shown in Fig. 3b. On the other hand, the adhesion between

diamond and the nickel matrix in the unmodified case is poor, as can be observed in which a significant bulging deposit is evident, even though a small peak of -OH is present (Fig. 3d).

The present results are in agreement with those of Tsubota et. al., who pointed out that the C=O and -OH groups that are present on the surface of the oxygenated-diamond exhibit a hydrophilic property. This hydrophilic property favors the incorporation of the surface-modified diamond grits in the composite coating by a conventional electroplating technique [3].

To evaluate the conditioning characteristics of the conditioner, block-on-ring tests were conducted in CMP slurry. The comparisons in weight losses of both conditioner and the Al₂O₃ ring among the three diamond conditioners investigated

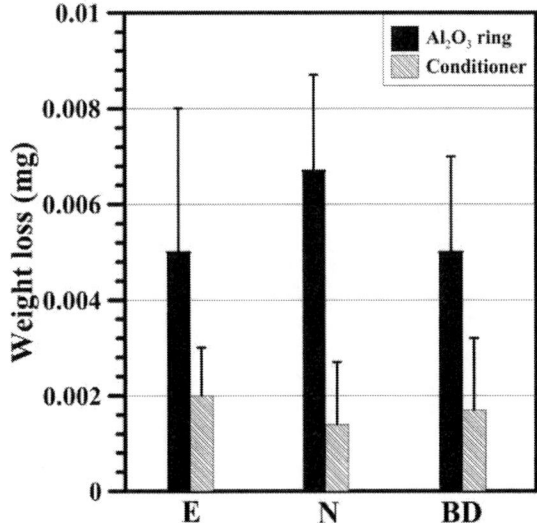

Fig. 4. Comparisons of weight losses of different conditioners and Al₂O₃ rings in the wear test.

herein are shown in Fig. 4.

Conditioner N shows the highest weight loss of Al₂O₃ rings, but the lowest weight loss of itself among the three conditioners studied, implying that conditioner N exhibits the best conditioning ability of the three conditioners investigated. The columnar shape of the diamond composites in conditioner N promotes the uniformity in height level of the diamond grits, which provides a higher amount of working diamond grits than either type E or type BD for sustaining the impact force. In addition, although the diamond grits in commercial conditioner BD are distributed in a discrete array and more uniformly than type E, the weight loss of the Al₂O₃ counterpart is similar to that of conditioner E. This is because a smaller amount of working diamond grits in conditioner BD, and hence, the contact surface area between diamond grits and Al₂O₃ ring is also smaller, leading to less ploughing action of diamond grits or Al₂O₃ ring. On the other hand, the shape of the diamond composite in conditioner E is hemispherical and that causes variations in height level of the diamond grits and results in less working diamond grits at the interface.

The average frictional coefficients of conditioner E, N, and BD are 1.1×10^{-4}, 1.4×10^{-4} and 1.2×10^{-4}, respectively, as indicated in Fig. 5. The results of both wear test (Fig. 4) and frictional coefficient (Fig. 5) are in agreement with those of Tan [6], who reported that the material removal rate is positively related with the frictional coefficient. Based upon the results, we can conclude that conditioner N exhibits the highest weight loss of Al_2O_3 ring among the three conditioners studied, this is due to the highest frictional coefficient and the most uniformity of height level of the conditioner N among the three conditioners.

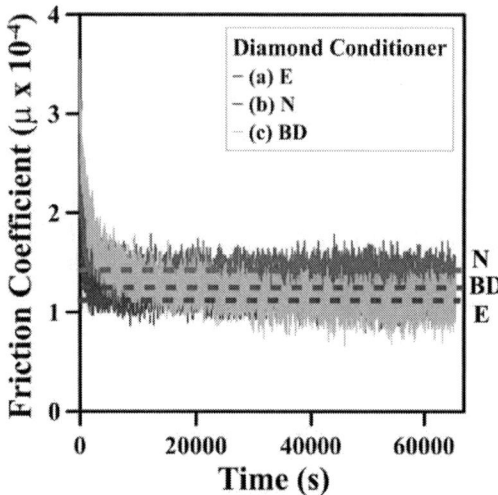

Fig. 5. Frictional coefficient of conditioners as a function of time: (a) E; (b) N and (c) BD.

Based upon the above discussion, the presently developed conditioner N was by means of micro patterning and nickel electroforming and by using surface-modified diamond grits as abrasive particles to form a columnar type of diamond composite exhibits the best overall conditioning ability in the CMP process among the three conditioners studied, and possesses the advantage of cost saving in manufacturing.

IV. CONCLUSIONS

A patterned diamond nickel composites array of the conditioner based on the process of micro patterning and electroforming is presented for fabricating N and E type of conditioners which are in form of columnar shape and hemispherical shape, respectively. In the manufacture of the conventional conditioner, brazing is commonly utilized as an adhesive material to connect distributed diamond grits with the flat substrate. By compared with conventional method, the nickel composite is utilized for fabricating the columnar shape of the nickel microstructure to instead of the brazing

connective approach. Single diamond grit is, then, placed and fixed by the micro patterning method and the protrusion size of the diamond grits can be well-controlled by the proposed method. Furthermore, the diamond grits are vertical fixed into the nickel matrix placed on the stainless steel substrate. Due to the surface modification treatment, the adhesion and defects between the diamond grits and the nickel deposit are enhanced and declaimed, respectively.

Two developed conditioners are characterized and the experimental results show that N type has the lowest height level of the diamond grits. The variation rate of height level in N type is approximately 1.36μm which is smaller than the 6.98μm measured from the conditioner BD.

In addition, conditioner N exhibited some 35% increase in the removal rate of the Al_2O_3 counterpart, but still maintained a lower weight loss, as compared to the commercial conditioner BD. As a result, the developed conditioner shows higher material removal rates, but less worn, than the currently available conditioners.

ACKNOWLEDGMENT

The authors would like to thank Nanoscience centre, University of Cambridge and the National Science Council of Taiwan, for financial support under Contract No. NSC-100-3113-E-002-002.

REFERENCES

[1] James C. Sung, Michael Sung, "The brazing of diamond," *Int. Journal of Refractory Metals & Hard Materials*, 27 (2009), pp. 382-393.

[2] Ming-Yi Tsai, "Polycrystalline diamond shaving conditioner for CMP pad conditioning," *J. Materi, Process. Techn.*, 210 (2010), pp. 1095-1102.

[3] Toshiki Tsubota, Shunsuke Tanii, Toshihito Ishida, Masanori Nagata, Yasumichi Matsumoto, "Composite electroplating of Ni and surface-modified diamond particles with silane coupling regent," *Diamond & Related Materials*, 14 (2005), pp. 608-612.

[4] C-T Pan, H Yang, S-C Shen, M-C chou and H-P Chou, "A low-temperature wafer bonding technique using patternable materials," *J. Micromech. Microeng.* 12 (2002) 611-615.

[5] Shintaro Ida, Toshiki Tsubota, Osamu Hirabayashi, Masanori Nagata, Yasumichi Matsumoto, Akira Fujishima, "Chemical reaction of hydrogenated diamond surface with peroxide radical initiators," *Diamond & Related Materials*, 12 (2003), pp. 601-605.

[6] An-Hung Tan, "Wear-corrosion properties of diamond conditioners in CMP slurry," *Wear*, 262(2007), pp. 693-698.

Fabrication and Performance Optimization of the Microplasma Reactor with Composite Dielectrics

Zhen Yuan, Li Wen[*], Leili Cheng, Jianqiang Ma and Jiaru Chu

[*]Department of Precision Machinery and Precision Instrumentation, University of Science and Technology of China,
Hefei, Anhui, China.
lilywen@ ustc.edu.cn

Abstract—A microplasma reactor with composite dielectric layers for maskless micro/nano plasma etching is presented. The composite dielectric layers include the Si_3N_4, SiO_2 and polyimide film. In order to obtain optimum dielectric properties of dielectric films, the process parameters for depositing the Si_3N_4 and SiO_2 film were obtained through a large number of experimental data. Then the microplasma reactor having inverted pyramidal hollow cathode and metal-composite dielectric layers-metal sandwich structure, is successfully fabricated. The experiment results show that the lifetimes, stability and microdischarge characteristics of the devices are superior to those of earlier fabricated with a single polymer dielectric layer. These microplasma devices are expected to operate in the further maskless scanning plasma etching.

Keywords- composite dielectric layers; microplasma reactor; maskless etching; V-I characteristics

I. INTRODUCTION

Due to many advantages such as small size, low power consumption, high electron density, high-pressure operation and convenient portability, microplasma reactors fabricated by MEMS technology have gained more and more attention in recent years. There is a rapid growth of microplasma applications, both in terms of the diversity of applications and in the breadth and depth in a given area of application[1-2]. Now the main applications of the microplasma devices include the light source display[3-5], the high-precision etching and the material surface treatment[6-7], the micro-thruster in microspacecraft projects[8] and so on. In our previous work, a novel maskless microplasma etching method based on scanning probe microscope with advantages of submicro/nano resolution and high etching rate was proposed[9]. The microdischarge reactor integrated on the cantilever probe with inverted pyramidal hollow cathode and Ni/ polyimide(PI) /Ni sandwich structure is designed and successfully fabricated[10]. However during the microdischarge experiment, we found that these devices with the single dielectric layer only have lifetimes of five to ten minutes, and some devices were unstable after ignition. We consider that the dielectric properties of the polyimide are greatly affected in the microfabrication process. In order to reinforce the dielectric strength and improve the discharge performance of the microplasma reactor, we use composite dielectric layers of SiO_2-Si_3N_4-PI in place of the single dielectric layer of PI as shown in Fig. 1.

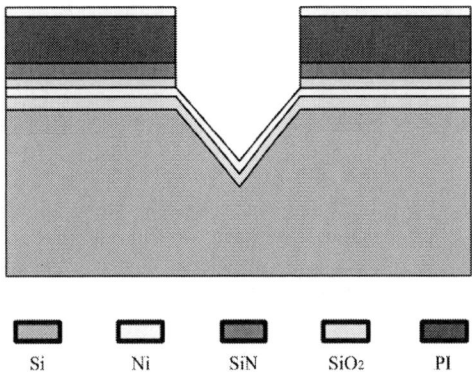

Fig 1. Schematic diagram of microplasma reactor with composite dielectric layers.

There are many methods to fabricate the dielectric material such as sputtering, chemical vapor deposition, evaporation and so on. The thin dielectric film deposited by plasma enhanced chemical vapor deposition (PECVD) system has many advantages such as low deposition temperature, good uniformity and well step-coverage[11]. When we use PECVD to deposit Si_3N_4 and SiO_2, the film quality and deposition rate are greatly influenced by the fabrication processing parameters[12]. In order to obtain the Si_3N_4 and SiO_2 film with good dielectric properties, series of experiments were done under the same conditions of the laboratory, and then we got the processing parameters which are able to obtain the optimal dielectric properties of the Si_3N_4 and SiO_2 films. Subsequently, the microplasma discharge test system is used to get the V-I characteristics in different proportions of the CHF_3 and Ar gas mixture about these devices with composite dielectric layers.

II. EXPERIMENT RESULTS.

A. Determination of the Process Parameters

There are many research directions about properties of the dielectric material such as the internal stress, insulation properties, refractive index, thickness uniformity and so on. We try to obtain the good dielectric properties film which can have a high deposition rate and a high breakdown voltage after applied with the voltage across the thin film. To this end, we made series of experiments that focused on the relation about the RF power, pressure and the reactive gas flow ratio with the dielectric properties and deposition rate of the film.

In these experiments, the p-type Si (100) wafers were used as the substrate. The thin dielectric film was deposited by the plasma enhanced chemical vapor deposition system (SAMCO PD-220 made in Japan). The frequency of the RF plasma source was 13.56MHz. The sample was heated at the temperature of 300℃. We prepared the Si_3N_4 and SiO_2 film which were used the different process parameters such as different RF power, pressure and the reactive gas flow ratio, under the same conditions of the laboratory. Then we measured the thickness and breakdown strength of the films respectively.

As shown in Fig. 2(a) and (b), when the RF power and reactive gas flow ratio increases, the deposition rate of the Si_3N_4 film shows a linear increasing trend, and the breakdown strength does not change significantly which is about 450V/μm when the RF power increases. However when the gas flow ratio SiH_4/NH_3 increases, the breakdown strength has the trend that firstly increases to a maximum and then decreases gradually. The reason is that when the gas flow ratio of SiH_4/NH_3 increases, the ratio of Si/N in the film is gradually approaching to the standard ratio of the Si_3N_4 film. At this time the breakdown strength comes to a maximum. With the continuant increase of it, the element N can not combine more Si, then the breakdown strength decreases when the excess Si form Si-Si bond by their own. At the same time as the increasing of SiH_4 gas, the film will be introduced

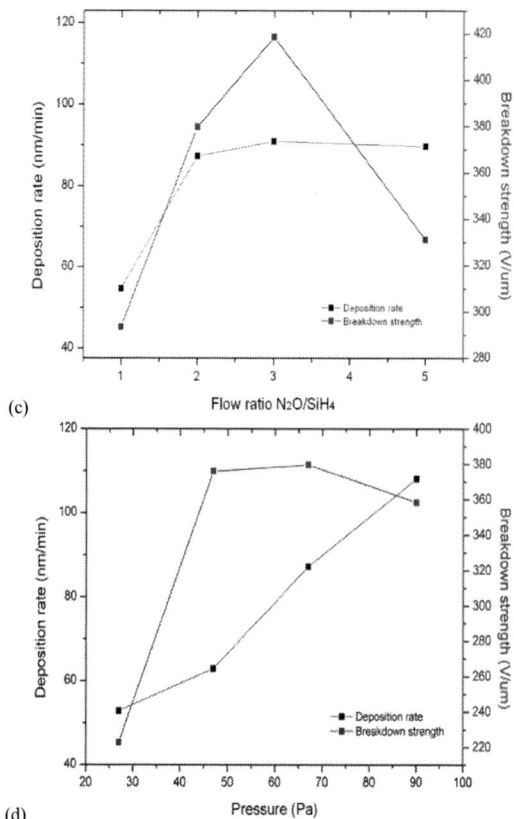

(c)

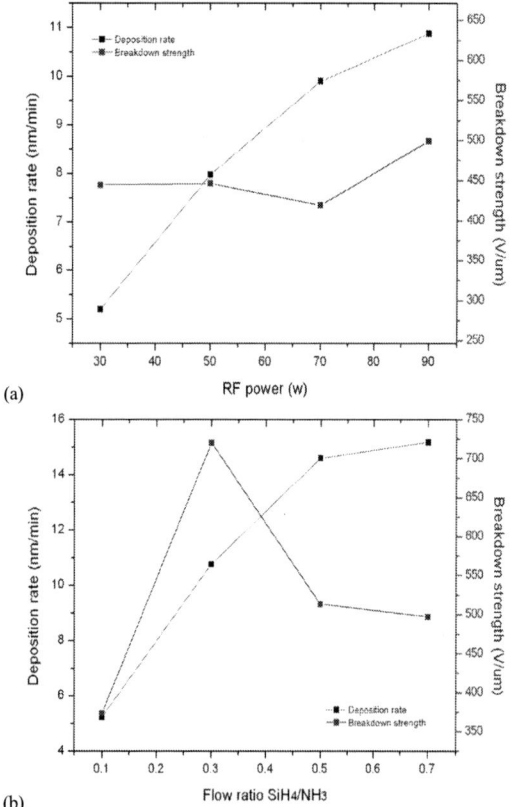

(a)

(b)

Fig.2: (a) The relation among the RF power, the deposition rate and the breakdown strength of the Si_3N_4 film. (b) the relation among the flow ratio SiH_4/NH_3, the deposition rate and the breakdown strength of the Si_3N_4 film. (c) the relation among the flow ratioN_2O/SiH_4, the deposition rate and the breakdown strength of the SiO_2 film. (d) the relation among the pressure, the deposition rate and the breakdown strength of the SiO_2 film.

more element H. Then, the quality of the film and the breakdown strength will decline. From the same analysis of Fig. 2 (c) and (d), we can get the relation about the pressure and reactive gas flow ratio with the breakdown strength and deposition rate of the SiO_2 film. The deposition rate shows a linearly increasing trend as the pressure increases, and when the gas flow ratio SiH_4/N_2O increases, it becomes increasing firstly and then gradually turns to be a constant value about 90nm/min. The breakdown strength also becomes increasing firstly to a maximum and then decreases gradually, but as the pressure increases, the decreasing trend is not very apparent. After the comparison of the experiment data, we get the optimal processing parameters for depositing the Si_3N_4 and SiO_2 film. When we use these processing parameters as shown in Tab 1 which include the temperature of the substrate, the pressure of reactive gases, the flow ratio, RF power and the flow of N_2, the deposition rate of Si_3N_4 and SiO_2 film can be about 11nm/min and 91nm/min, and the breakdown strength can reach to 720V/μm and 420V/μm. These thin films can be used as the dielectric layer in the microplasma reactor to obtain the higher breakdown voltage than only with a single dielectric layer of PI.

TABLE 1: THE PROCESSING PARAMETERS WHICH OBTAIN THE OPTIMAL DIELECTRIC PROPERTIES OF THE Si_3N_4 AND SiO_2 FILM.

Material	Temperature (℃)	Pressure (Pa)	Flow ratio	N_2 (sccm)	RF power (w)	Deposition rate (nm/min)	Breakdown strength (V/μm)
Si_3N_4	300	80	$SiH_4/NH_3=3:10$	400	75	11	720
SiO_2	300	67	$SiH_4/N_2O=1:3$	0	100	91	420

B. Devices Fabrication

The inverted pyramidal microplasma reactors have metal-composite dielectric layers-metal sandwich structures with inverted square pyramidal hollow cathodes, which are 50 or 100μm square at the base and 35 or 70μm in depth, respectively. The dielectric materials choose the polyimide film 8um (ZKPI-305IIG), the Si_3N_4 film 500nm and the SiO_2 film 1um. The fabrication process of microplasma reactors with composite dielectric layers is similar to the single polyimide dielectric layer structure which we have designed before.

The devices are fabricated in p-type Si (100) wafers having a thickness of typically 380μm. Pyramidal microcavities are firstly produced by anisotropic wet etching in a 30% KOH etchant at 70℃. Then, 1.4μm SiO_2 films are formed on both sides of the substrate by wet oxidation. The cathode film Ni is sputtered and patterned by lift-off process using negative resist. Comparing with the previous fabrication process, only to add the deposition and patterning Si_3N_4 and SiO_2 inorganic dielectric films. The processing parameters of the Si_3N_4 and SiO_2 film by PECVD which we used just as Table. 1 shows. Subsequently, the dielectric is formed by spin coating an 8μm-thick dry etchable polyimide, followed by RF sputtering and lift-off process using image reversal resist to pattern the anode Ni film. The desired discharges cavities in dielectric film are defined and etched by reactive ion etched process. After that, the anode film Ni is patterned by $FeCl_3$ wet etching. Fig. 3 is the optical microscope image of the microplasma reactor with composite dielectric layers. From the images we can see that the microplasma reactor has good cavity and sandwich structure. The entire device surface is clean and without any apparent cracks after fabricated.

C. V-I Characteristics

The microdischarge devices which have not been released with composite dielectric layers are used to do some electrical performance tests. Devices are fixed in a turbo molecular pumped vacuum station. The anode and cathode are contacted to the test circuit through mire with stainless steel probes. The chamber is pumped down to 10^{-2}Pa by the pump and backfilled with CHF_3/Ar gas mixture to the desired pressure. The DC source voltage applied between the cathode and anode is gradually increased until breakdown occurs. Then we can get the V-I characteristics of the device. In this certain case, external ballast is used during device operation to stabilize discharge performance and protect the device. The V-I characteristics of the 50μm device related to the gas pressure, source voltage and ballast resistance in CHF_3 /Ar

gas mixture are shown in Fig. 4. We can judge that when the content of CHF_3 gas and the ballast resistance increase, the ignition voltage will become higher gradually, because of the ignition voltage of molecular gas is higher than the rare gas. The discharge status shows an abnormal discharge mode because of the positive slope of V-I characteristics. That is different from V-I characteristics we have tested before which is the negative slope. The composite dielectric layers structure of the microplasma reactor can have a higher breakdown voltage about 900V. This can lay a good foundation to increase the life of the reactor. The experiment results show that the microplasma reactors with composite dielectric layers have longer lifetime and more discharges stability than those

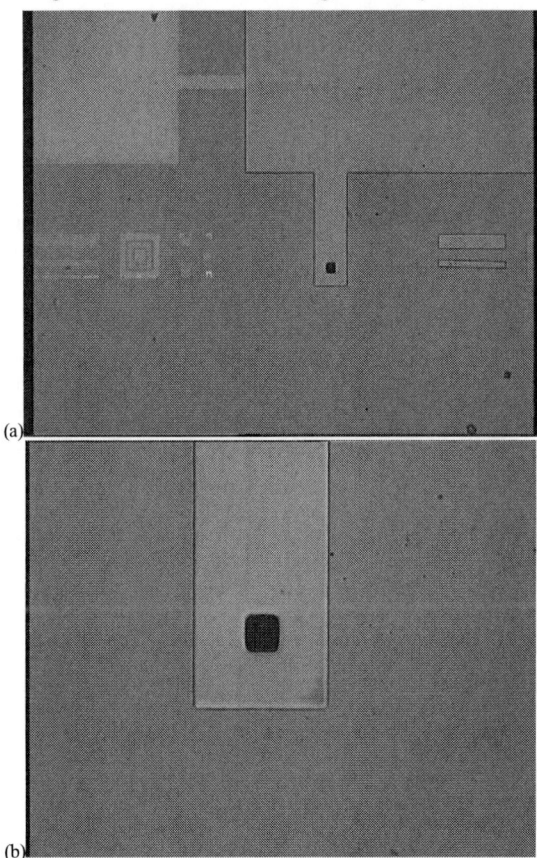

Fig. 3 The optical microscope images of the 50μm microplasma reactor after microfabricated. (a) 5X; (b) 20X.

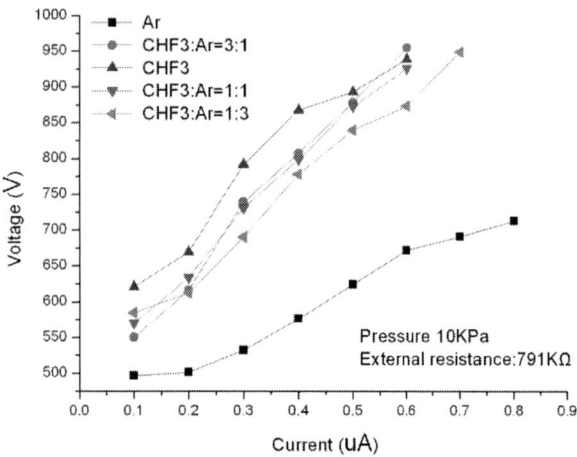

Fig. 4: The V-I characteristics of microplasma reactor operating in different proportions of the CHF₃ and Ar gas mixture.

of the single dielectric. It is able to meet the requirements of maskless micro/nano plasma etching system.

III. CONCLUSION

In order to improve the performance of microplasma reactors, its structure was changed with the single dielectric layer to composite dielectric layers. From a large number of experiment data we get the optimal processing parameters for depositing the dielectric films just on the plasma enhanced chemical vapor deposition (PECVD) system in our laboratory. Then microplasma reactors with inverted pyramidal hollow cathodes and SiO_2-Si_3N_4-polyimide composite dielectric layers for maskless micro/nano plasma etching are fabricated and characterized in CHF₃/Ar gas mixture. The lifetimes, stability and discharge characteristics of the devices are superior to those of earlier fabricated with a single polymer dielectric layer from the electrical performance experiment results. These microplasma devices are expected to operate in the further maskless scanning etching.

ACKNOWLEDGMENTS

This work is supported by the National Natural Science Foundation of China (No. 50605061) and the Fundamental Research Funds for the Central Universities (2090090001).

REFERENCES

[1] K.H Becher, K.H Schoenbach and J. G. Eden, "Microplasmas and applications", *J. Phys. D: Appl. Phys.* 39 (2006), pp. 55-70.

[2] K. Tachibana, "Current Status of Microplasma Research", *IEEJ Trans 2006*, pp. 145–155.

[3] J. G. Eden, Sung-Jin Park, C.M Herring and J.M Bulson, "Microplasma light tiles: thin sheet lamps for general illumination", *J. Phys. D: Appl. Phys.* 44 (2011) 224011 (6pp).

[4] S.J. Park, P. A. Tchertchian and J. G. Eden, "Arrays of Addressable Microcavity Plasma Devices", *IEEE Transactions on Plasma Science*, Vol. 35, No. 2, pp. 215-222 April 2007.

[5] J. Chen, S.J. Park and C. Liu, "Development and Characterization of Micromachined Hollow Cathode Plasma Display Devices", *Journal of Microelectromechanical Systems*, 2002, pp. 536-543.

[6] L. Wen, Z. Yuan, L.L. Cheng, and J.R Chu, "Characteristics of Inverted Pyramidal Microdischarge Devices Operating in CHF3 and CHF3/Ar for Maskless Microplasma Etching", *Transducers'11*, pp.454-457.

[7] T. Shao, C. Zhang, K.H. Long and Y.X. Zhou, "Surface modification of polyimide films using unipolar nanosecond-pulse DBD in atmospheric air", *Applied Surface Science* 256 (2010) pp.3888–3894.

[8] M. Mihailovic, T.V. Mathew, Zandbergen and P.M. Sarro, "MEMS Silicon-based Resistojet Micro-thruster for Attitude Control of Nano-satellites", *Transducers'11*, pp. 261-265.

[9] Q.P. Zhang, L. Wen, W.W. Xiang and J.R. Chu, "Fabrication and Performance of Microplasma Reactor for Maskless Scanning Plasma Etching", *IEEE-NEMS2010*, pp. 211-214.

[10] L. Wen, H. Wang, L.W. He and J.R. Chu, "Design and fabrication of microcantilever probe integrated with microplasma reactor for maskless scanning plasma etching Comprehensive Microsystems", *Sensors and Actuators*, A 169 (2011) 362–366.

[11] X.D. Xu, L.C. Zhang, L. Huang and Z. Yang, "A comparison of structures and properties of SiNx and SiOx films prepared by PECVD", *Journal of Non-Crystalline Solids* 358 (2012), pp.99–106.

[12] J.C. Yang , Y.T. Wu and X.H. Xu, "Experimental Study of the Internal Stress of Silicon Nitride Films Deposited by PECVD", *New Technology & New Process*,2008, pp. 77-80.

Fabrication of Deep Lateral Single-Crystal-Silicon Blaze Micro-grating by Inductively-Coupled-Plasma Reactive Ion Etch

Y. H. Lin[1,2], C. J. Weng[2], C. Y. Su[2], W. Hsu[1]

[1] Department of Mechanical Engineering, National Chiao Tung University , Hsinchu, Taiwan, R.O.C.
[2] Instrument Technology Research Center, National Applied Research Laboratories, Hsinchu, Taiwan, ROC.
yhlin@itrc.narl.org.tw

Abstract—This paper presents a method by using a compensative structure assisted to fabricate deep lateral single-crystal-silicon (SCS) blaze micro-grating at Inductively-Coupled-Plasma Reactive Ion Etch (ICP-RIE). Due to the high resolution of blaze micro-grating, it's hard to maintain the teeth structure of blaze micro-grating under deep silicon etch in ICP-RIE process. Here, the independent rectangular structure and symmetrical structure to micro-grating is designed to obstruct the non-vertical plasma ion to etch the sidewall of micro-grating structure and to get better the profile control at deep micro-grating structure. The lateral silicon blaze micro-grating with 100 μm thickness by compensative structure assisted etch process have been successfully demonstrated this method.

Keywords-blaze graing; Inductively-Coupled-Plasma Reactive Ion Etch (ICP-RIE).

I. INTRODUCTION

Due to excellent material properties of the single-crystal-silicon (SCS), the lateral silicon blaze micro-grating have potential to integrate the MEMS sensors and actuators [1]. The micro-grating is an important optical device, especially in applications of optical coherence tomography (OCT), micro-spectrometer, tunable laser, dense wavelength division multiplexer (DWDM) etc. Profile control of teeth structures is important issue in manufacturing lateral blaze micro-grating. In anisotropic silicon etching, under the Bosch patent, sequentially alternating etch and passivation cycles can easily achieve high-aspect-ratio silicon structures [2,3]. Many fabrication parameters such as uniformity, etch lag and geometrical effect in ICP-RIE have been investigated to obtain high-aspect-ratio structures [4-8]. There is rare research to discuss the corner structure like teeth of micro-grating under deep ICP-RIE etching. In fabricating lateral SCS blaze micro-grating by ICP-RIE process, the teeth structure of micro-grating can't still maintain under deep etching. In this paper, we propose a method using a compensative structure assisted etch process to fabricate better profile of grating at the deep ICP-RIE.

II. DESIGN

The experimental study is performed in the STS Multiplex ICP-RIE here. This study generates the source plasma of an

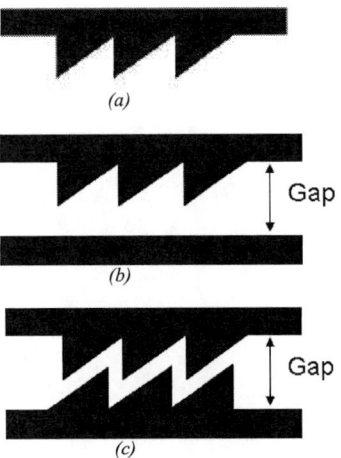

Fig. 1. Micro-grating (a) without compensative structure, (b) with rectangular compensative structure and (c) with symmetrical compensative structure to assisted profile control in ICP-RIE etching.

inductively coupled coil by a 1kW 13.56 MHz R.F generator, and uses another 13.56MHz generator as a platen power to independently control the bias potential of the wafer relative to the source plasma. The fabrication process maintains at low temperature using helium as cooling gas supplied to the backside of the wafer. Sulfur hexafluoride (SF_6) and octafluorocyclobutane (C_4F_8) are used as the main etch and passivation gases, respectively.

A. Concept Design

In the proposed method, a compensative structure near silicon blaze micro-grating structure is designed to assist etching process in ICP-RIE. Here, the lateral blaze micro-grating structure with 8.9 degrees blaze angle, 10 μm periodic grating structures as the critical device is used to investigate the profile control of teeth structures after ICP-RIE deep etch, as shown in figure 1(a). The rectangular compensative structure and symmetrical compensative structure near micro-grating structure that assisted ICP-RIE etching is design, as shown in figure 1(b) and figure 1(c). The gap between teeth structures and compensative structure is 5 μm .

978-1-4673-1122-9/12 $31.00 © 2012 IEEE

B. Fabrication Process

Figure 2 schematically shows the fabrication flowchart using the compensative structure assisted ICP-RIE etch in fabricating lateral silicon blaze micro-grating. First, the negative photo-resist (SU-8 2002) is spun and patterned as an etch mask on a SOI wafer, as shown in figure 2(a). The SU-8 is baked by 95°C 20minutes to enhance the resist the plasma ion etching. The lateral silicon blaze micro-grating structure is etched by Bosch silicon cyclic anisotropic etch, as shown in figure 2(b). Finally, the compensative structure is removed by hydrofluoric acid (HF) wet etch.

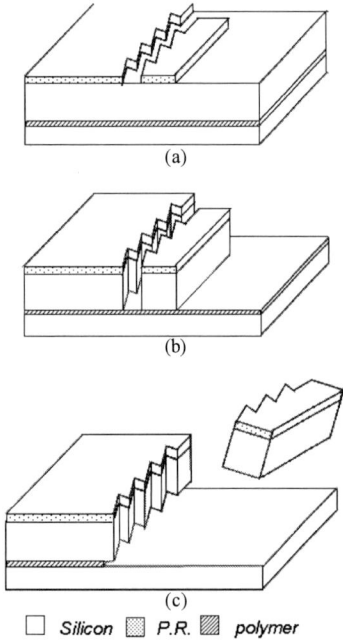

Fig. 2. Fabrication flowchart of lateral silicon blaze micro-grating using compensative structure assisted RIE etch.

III. EXPERIMENTAL RESULTS AND DISSCUSSIONS

A. Process results

This paper uses lateral silicon blaze micro-grating without/with rectangular and symmetrical compensative structure to discuss the profile control at the teeth of micro-grating structure at deep ICP-RIE. The compensative structure is a temporary structure at fabrication process. So, the compensative structure is designed as independent structure that easily to remove after anisotropic etching. The SOI wafer with 100 μm device thickness is used here. The silicon cyclic anisotropic etching recipe used in this study, which is based on Bosch anisotropic etching method, is 800/12 W source/bias power, 130/13 sccm SF_6/O_2 flow rate, 12 seconds in etching step, and 800/0 W source/bias power, 85

sccm C_4F_8 flow rate, 8 seconds in passivation step. Etch rate is about 2.3 μm /minutes.

Since the Bosch patent, etch and passivation cyclic mechanism developed, the high-aspect-ratio anisotropic silicon structures can be easily achieved. The pattern fom photo-resist to silicon is fine transferred at the beginning etch, even in corner structures. But the structures at the corner can't be maintained as original designed under deep etch. This is seriously issue influenced efficiency at some devices like optical grating that need accurate structure's profile.

Figure 3(a) shows the scanning electron microscope (SEM) photograph of the lateral silicon blaze micro-grating under ICP-RIE etching. The teeth structures of micro-grating are measured by Atomic Force Microscopy (AFM) and profile photograph, as shown in figure 3(b) and figure 3(c). The profile of the teeth is approach flat under 50 μm depth. The efficiency of dispersion of light at grating structures is low caused by not bad teeth structures.

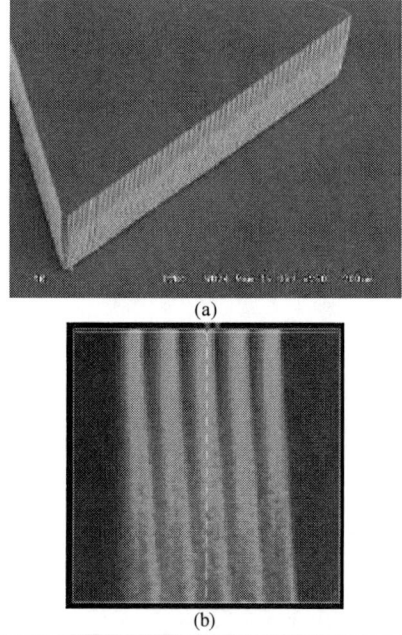

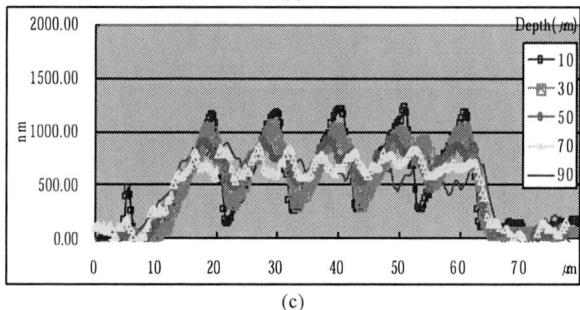

Fig. 3. The lateral silicon blaze micro-grating (a)SEM, (b)AFM, (c)profile photography under ICP-RIE etching without any compensative structure.

In ICP-RIE etching, the structures at the corner is hard be maintained after deep etch. Most plasma ion has been moved vertically by bias voltage attraction and performed spontaneous etch to the base silicon. But, some plasma applied non-vertical movement, due to the collision between plasma. The non-vertical movement plasma has spontaneous etching reactions with silicon material at the sidewall. The etching of silicon at the sidewall becomes serious at deep position which can explain the damage of the grating structure at deep position in figure 3(a). Enlarged bias voltage and lowered process chamber pressure can slightly reduce the non-vertical plasma ion. But, it is not effective method to solve this problem.

In this paper, an independent and compensative structure near the blaze micro-grating is design to assist the corner profile control. This compensative structure not only can obstruct the most non-vertical plasma to etch the silicon at the sidewall of the grating, but also can reduce the effect of etch

lag around the teeth structures.

A rectangular compensative structure near blaze micro-grating structure is design, as shown in figure 1(b). After anisotropic silicon etch and rectangular compensative structure removal, the teeth structures at the deep position obviously have been improved, as shown in figure 4(a). The AFM and profile photograph of teeth are shown in figure 4(b). and figure 4(c). The profile of teeth structures at 70 μm deep position is similar to at 10 μm deep. The teeth structures still have maintained even in 90 μm deep.

The SEM photography of grating with symmetrical compensative structure assisted etch is shown in figure 5(a). The AFM and profile photograph of teeth are shown in figure 5(b). and figure 5(c). The profile of the teeth structures with symmetrical compensative structure is also obviously better than without compensative structure.

Although, the compensative structure can improve the profile of teeth of the grating structures at deep etch, but some extra defects have be found at sidewall silicon at 30~50 μm

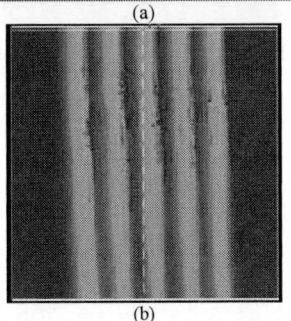

(a)

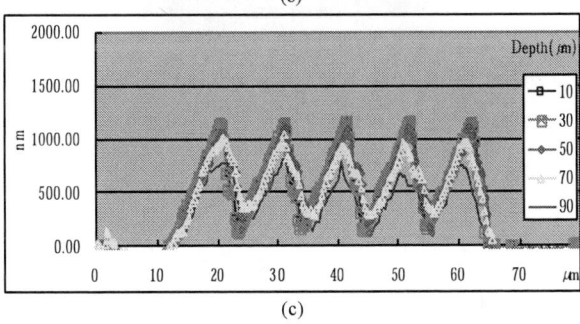

(b)

(c)

Fig. 4. The lateral silicon blaze micro-grating (a)SEM, (b)AFM, (c)profile photography under ICP-RIE etching with rectangular compensative structure assisted etch.

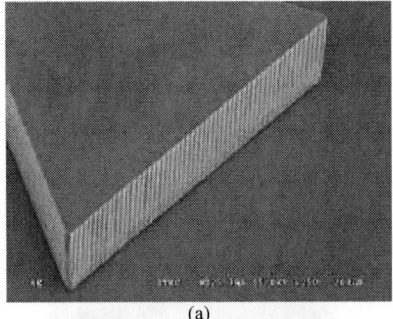

(a)

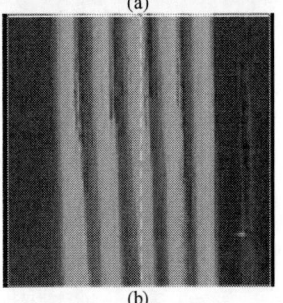

(b)

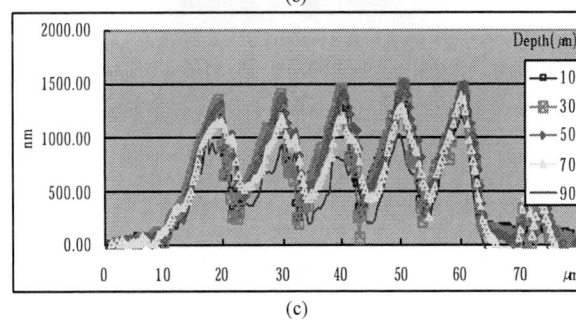

(c)

Fig. 5 The lateral silicon blaze micro-grating (a)SEM, (b)AFM, (c)profile photography under ICP-RIE etching with symmetrical compensative structure assisted etch.

deep as shown in AFM photography, as shown in figure 4(b) and figure 5(b). These defects didn't happen at grating without compensative structure, as shown in figure 3(b). These defects of the sidewall caused by reflected ion from the mask side on the compensative structure [4].

B. Optical measurement results

The schematic diagram of measurement setup is shown in figure 6. The measurement equipments include 6W power 450~2000nm wavelength supercontinnum light source, cylindrical lens for focusing output spectrum, CMOS linear sensor arrays for optical amplitude measurement.

Figure 7 shows the optical measurement results of lateral blaze micro-grating without compensative structure assisted etch and with rectangular/symmetrical compensative structure

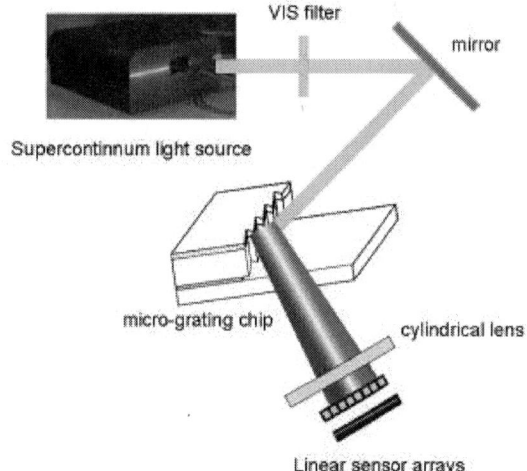

Fig. 6 The schematic diagram of optical measurement.

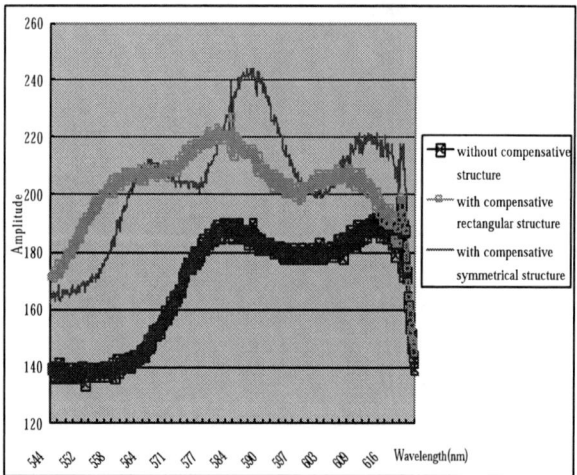

Fig. 7 Optical amplitude of grating without, with rectangular compensative structure and symmetrical compensative structure.

assisted etch. The amplitude of background is 120 measured by the same CMOS linear sensor arrays. The amplitude of micro-grating with compensative structure assisted profile control is obviously higher than without compensative structure. The amplitude of grating with symmetrical compensative structure is higher than with rectangular compensative structure. The grating with symmetrical compensative structure let the opening gap be the same around the teeth of the grating. So, the symmetrical compensative structure with grating not only can block non- vertical plasma to etch the silicon material at the depth position, but avoiding etch lag effect around the teeth structures.

IV. SUMMARY

A method to fabricate the deep lateral SCS blaze micro-grating by add a compensative structure near the micro-grating to assist ICP-RIE etch has been proposed. The rectangular and symmetrical compensative structure is design to obstruct the non-vertical plasma ion etch and reduce the etch lag effect. The better profile of the grating structure with 100 μm has been successfully fabricated using the symmetrical compensative structure method.

ACKNOWLEDGMENT

This work is supported by Instrument Technology Research Center of National Applied Research Laboratories for providing fabrication facilities and technical supports.

REFERENCES

[1] K. E. Peterson, "Silicon as a mechanical material", *Proc. IEEE*, vol.70, pp. 420-457, 1982.

[2] F. Larmer, and A. Schilp, " Method of anisotropically etching silicon ", German Patent DE4241045C1, USA patents 4855017 and 478472.

[3] A.M. Hynes, H. Ashraf, J.K. Bhardwaj, J. Hopkins, I. Johnston, J.N. Shepherd, "Recent advances in silicon etching for MEMS using the ASE process", *Sensors and Actuators,* vol. 74, pp. 13–17, 1999.

[4] G. Marcos, A. Rhallabi, and P. Ranson, "Monte Carlo simulation method for etching of deep trenches in Si by a SF6/2 plasma mixture", *J. Vac. Sci. Technol.* A 21(1), 2003.

[5] C. K. Chung, "Geometrical pattern effect on silicon deep etching by an inductively coupled plasma system", *J. Micromech. Microeng.* vol. 14, pp. 656–662, 2004.

[6] C. K. Chung, H. C. Lu, and T. H. Jaw, "High aspect ratio silicon trench fabrication by inductively coupled plasma", *Microsystem Technologies*, vol. 6, pp. 105-108, 2000.

[7] J. Yeom, Y. Wu, J. C. Selby, and M. A. Shannon, "Maximum achievable aspect ratio in deep reactive ion etching of silicon due to aspect ratio dependent transport and the microloading effect", *J. Vac. Sci. Technol. B*. 23(6), pp. 2319-2329, 2005.

[8] B. Wu, A. Kumar, and S Pamarthy, "High aspect ratio silicon etch: A review", *JOURNAL OF APPLIED PHYSICS*, 108, 051101 _2010.

A Study of Tin Oxide Thin Film Gas Sensors with High Oxygen Vacancies

Chen Lin, Dacheng Zhang[*], Xiaodi Liu
[*]National Key Laboratory of Science and Technology on Micro/Nano Fabrication,
Institute of Microelectronics, Peking University, Beijing, China
dchzhang@ime.pku.edu.cn

Abstract—In this paper, gas sensors employing nanoscale tin oxide films of high oxygen vacancies were successfully made and tested in order to determine which factors affected the gas sensitivity of them. Firstly electrothermal simulations were carried out to predict the steady-state temperature distributions of the tin oxide thin film and the overall gas sensor using ANSYS[TM]. Then the composition of the tin oxide calcined twice in O_2 was analyzed by X-ray photoelectron spectroscopy (XPS). The result of XPS showed that the tin oxide under this condition had high oxygen vacancies. Finally, the ethanol sensing and humidity sensing properties of the gas sensors at various operating voltages and calcined in different conditions were tested. The experimental results showed that high gas sensitivity of the gas sensors could be achieved by changing the internal and external factors.

Keywords-tin oxide; gas sensors; high oxygen vacancies; gas sensitivity

I. Introduction

Because of society's concern with toxic gases leading to environmental pollution and inflammable gases resulting in combustion accidents, gas sensors employing semiconducting metal oxides have been widely used to detect various kinds of gases [1-4]. Today the most widely used commercial gas sensors are tin oxide based gas sensors [5-7]. Gas sensors employing tin oxide as the gas sensing film have many advantages, such as low power consumption, high sensitivity, simple design, fast response and low cost. Tin oxide (SnO_{2-x}) which has been well studied is known as a non-stoichiometric material. Because the composition and morphology of SnO_{2-x} have much effect on the gas sensitivity of tin oxide, controlling the composition of SnO_{2-x} is very important. For a long time the characteristics of tin oxide thick films and powders have been studied [8-11]. Because IC fabrication is similar to micromachining, micromechanics and microelectronics may be integrated on a single chip, which results in the term microelectromechanical systems (MEMS) to describe this technology. A lot of advantages can be achieved by the use of mircomachining in the gas sensor field, such as reducing power consumption and improving the sensitivity of the gas sensors [12].

In this paper, gas sensors employing nanoscale tin oxide films of high oxygen vacancies were successfully made by choosing high Ar/O_2 ratio in the gas mixture during sputtering and the factors which affected the gas sensitivity of them will be discussed.

II. Design and Simulation of the Gas Sensor

A. Design of the Tin Oxide Thin Film Gas Sensor

The package structure and the schematic diagram of the tin oxide thin film gas sensor are shown in Fig. 1 respectively.

The gas sensor chip mainly consisted of the heating electrode, the measuring electrode, the tin oxide film which was 90 nm thick and the quartz glass substrate. Quartz glass was chosen as the material of the substrate due to its low thermal conductivity, which resulted in low power consumption. The comb-shaped electrode which was located beneath the tin oxide film was composed of a heating electrode and a measuring electrode. The heating electrode acted as a heater to heat the tin oxide film to a proper temperature and the measuring electrode was used to measure the resistance variations of the tin oxide film. Moreover, in order to achieve high sensitivity to the resistance variations, the comb-shaped electrode was selected because the shape of the electrode could measure relatively greater resistance variations.

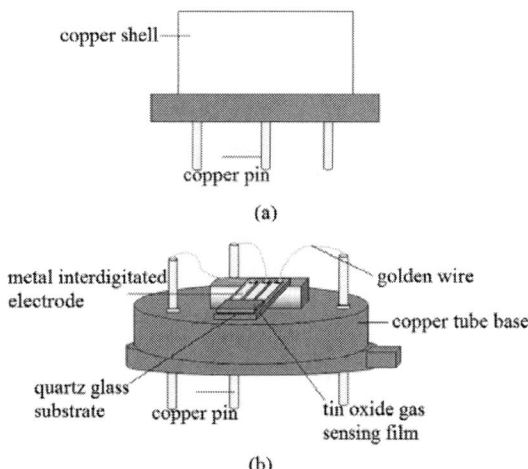

Fig. 1. (a) The package structure and (b) the schematic diagram of the tin oxide thin film gas sensor.

B. Simulation of the Tin Oxide Thin Film Gas Sensor

In order to obtain the steady-state temperature distribution of the tin oxide thin film gas sensor, a model of the device was simulated by ANSYS™. Various boundary conditions of thermal conduction, radioactive loss and convection were taken into account.

The theoretical model of the overall tin oxide thin film gas sensor is shown in Fig. 2. The steady-state temperature distributions of the overall gas sensor and the tin oxide thin film are shown in Fig. 3. The operating voltage and the ambient temperature were taken as 4.75 V and 70 ℃ respectively. According to the simulation results, the temperature distribution of the film ranged from 350.2 ℃ to 414.5 ℃ which was appropriate to detect ethanol.

III. EXPERIMENT

A. Fabrication of the Tin Oxide Thin Film Gas Sensor

The fabricated tin oxide thin film gas sensor was cantilever beam structure. Quartz glass with a thickness of 300 μm was chosen as the material of the substrate. Platinum with a thickness of 2000 Å was used as the material of the comb-shaped electrode. The comb-shaped electrode was made by sputtering.

The nanoscale tin oxide films with a thickness of 90 nm were fabricated using the techniques of reactive direct current magnetron sputtering and calcination. In order to obtain tin oxide films with high oxygen vacancies, high Ar/O_2 ratio in the gas mixture during sputtering was chosen. After the films were sputtered, they were calcined under two different conditions. One was calcined in O_2 at 500 ℃ for 2 h. The other was firstly calcined in O_2 at 500 ℃ for 2 h, then calcined in O_2 at 600 ℃ for 2 h. The XPS result of the tin oxide material calcined twice in O_2 is shown in Fig. 4. It can be concluded from Fig. 4 that the tin oxide under this condition has high oxygen vacancies.

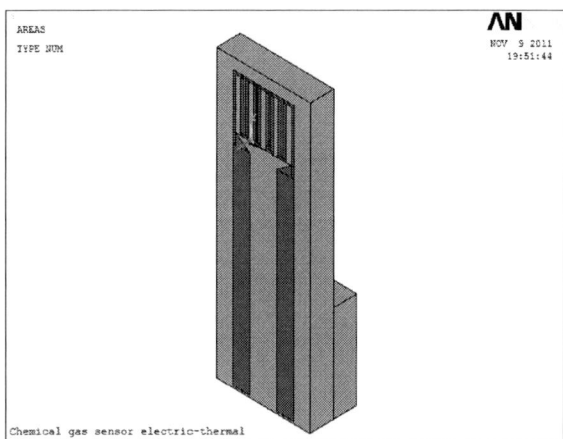

Fig. 2. The theoretical model of the overall tin oxide thin film gas sensor.

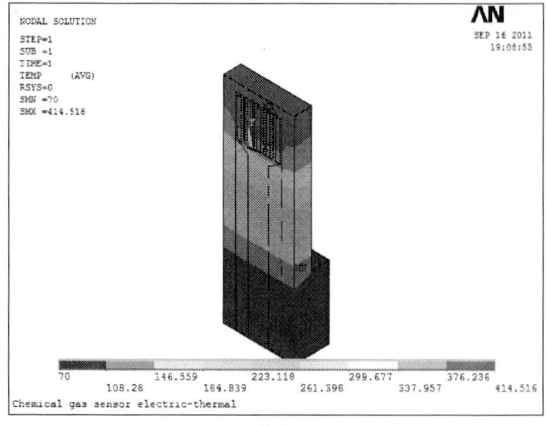

(a)

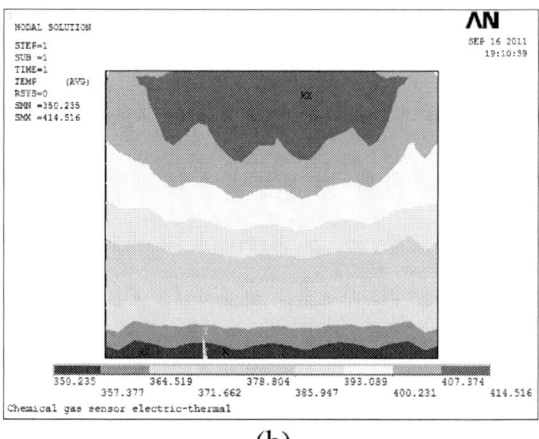

(b)

Fig. 3. The temperature distributions of (a) the overall gas sensor and (b) the tin oxide thin film.

B. Test the Gas Sensing Properties of the Devices

The measurement circuit used for the gas sensor system is shown in Fig. 5. As shown in Fig. 5, V_{DD} is the heating voltage, V_{OUT} is the output voltage, R is the load resistance which has a resistance value of 100 Ω. By changing the heating voltage, the operating temperature of the tin oxide gas sensor could be well adjusted.

In the first place, the ethanol sensing properties of the tin oxide thin film gas sensors at various operating voltages and calcined in different conditions were tested. It is widely accepted that the mechanism for the tin oxide thin film gas sensor to respond to ethanol can be explained in the following process. Firstly, oxygen species such as O^{2-}, O^- and O_2^- are absorbed by the surface of tin oxide thin film. Secondly, when the tin oxide thin film is exposed to ethanol, ethanol reacts with the absorbed oxygen species, which contributes to the release of electrons to the conduction band [13]. As a result of the reactions between ethanol and the tin oxide thin film, the resistance of the n-type materials decreases while the resistance of the p-type materials increases [14-17]. During the

978-1-4673-1122-9/12 $31.00 © 2012 IEEE

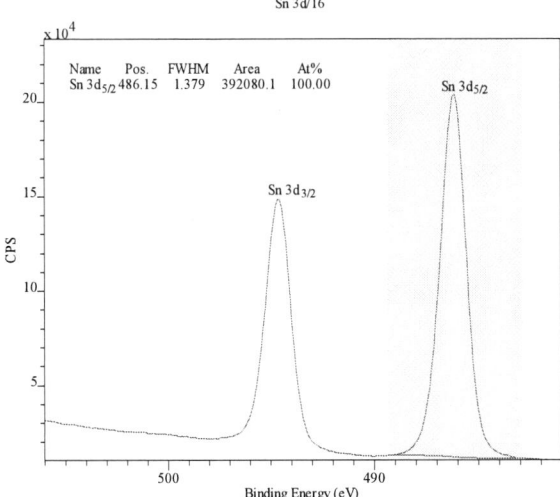

Fig. 4. Sn3d level of the tin oxide material calcined twice in O_2 analyzed by XPS.

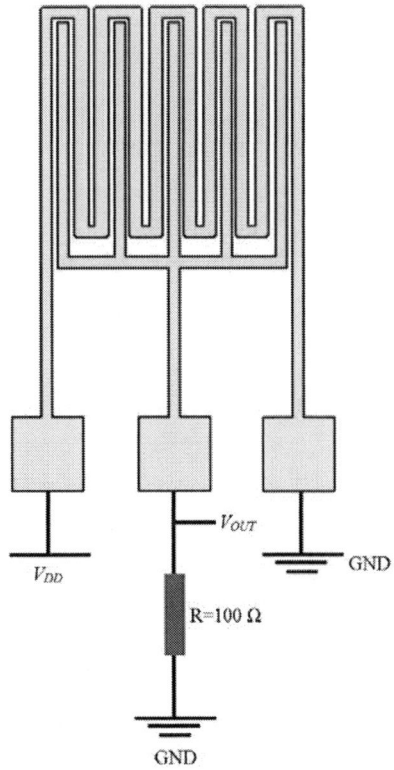

Fig. 5. The measurement circuit used for the gas sensor system.

process of testing the ethanol sensing properties of the gas sensors, dry nitrogen was chosen as both a diluting gas and a reference gas. In addition, a syringe was used to inject saturated ethanol vapor into the test chamber through a rubber plug. After ethanol was adequately mixed with the diluting gas, the tin oxide thin film gas sensor was put into the test chamber. The sensor should not be taken out of the test chamber until the response of the tin oxide thin film gas sensor reached a constant value.

In the second place, the humidity sensing properties of the tin oxide thin film gas sensors at various operating voltages and calcined in different conditions were also tested. During the process of testing the humidity sensing properties of the gas sensors, the sensors were placed near the surface of the saturated aqueous solution. The relative humidity above the $MgCl_2$ saturated aqueous solution at 25 ℃ is 33.2% and the relative humidity above the KCl saturated aqueous solution at 25 ℃ is 84.3%.

IV. RESULTS AND DISCUSSION

A. Results of the Gas Sensing Properties

The gas sensing properties of the tin oxide thin film gas sensors based on experimental results are shown in Table I. The gas sensitivity S is defined as $1000 \times \Delta V_{out}/V_{out}$, where ΔV_{out} is the change in the output voltage upon exposure to the test gas and V_{out} is the output voltage in N_2. The gas sensitivity to ethanol and water vapor can be increased by increasing the concentration of the gases and choosing suitable calcination conditions based on Table I. For example, S can be increased from 22.16 to 57.62 if the concentration of ethanol is increased from 100 ppm to 1000 ppm in Sample A according to Table I. Moreover, the ethanol gas sensitivity can be increased by increasing the operating voltage in the appropriate range according to Table I. For instance, the gas

sensitivity to 1000 ppm ethanol can be increased from 57.62 to 67.04 if the operating voltage of the gas sensor which is calcined in O_2 at 500 ℃ for 2h is increased from 4.500 V to 5.000 V based on Table I.

B. Discussion of the Gas Sensing Properties

As demonstrated in Table I, the output voltage of the measurement circuit can be increased by increasing the concentration of ethanol and water vapor. As a result of it, the measurement circuit can be used for measuring the concentration of ethanol and water vapor.

Furthermore, the ethanol sensitivity of the tin oxide thin film gas sensor increases with the working temperature of the device in the appropriate range based on Table I, which can be also explained with the mechanism of gas adsorption and desorption on SnO_2 films [18]. Oxygen from the atmosphere both in the O^- and O_2^- species can be adsorbed by an n-type metal oxide. O^-, the more reactive oxygen ion, contributes to the higher sensitivity to reducing gases, in our case ethanol. With the preferential adsorption of O_2^- at relatively low temperature, the sensitivity of tin oxide is very small accordingly. With the increase of the temperature in the appropriate range, the dominant process comes to be the adsorption of O^-. As a result of it, the sensitivity of tin oxide increases [19].

TABLE I. THE GAS SENSING PROPERTIES OF THE TIN OXIDE THIN FILM GAS SENSORS BASED ON EXPERIMENTAL RESULTS

Gas Sample		N_2	*$MgCl_2$	*KCl	1000 ppm ethanol	500 ppm ethanol	100 ppm ethanol
*A	V_{out} [V]	1.128	1.130	1.134	1.193	1.169	1.153
	S	-	1.773	5.319	57.62	36.35	22.16
*B	V_{out} [V]	1.198	1.202	1.205	1.272	1.256	1.234
	S	-	3.339	5.843	61.77	48.41	30.05
*C	V_{out} [V]	1.238	1.241	1.243	1.321	1.305	1.282
	S	-	2.423	4.039	67.04	54.12	35.54
*D	V_{out} [V]	1.128	1.129	1.131	1.138	1.137	1.136
	S	-	0.8865	2.660	8.865	7.979	7.092

*In the table above, Sample A: V_{DD}=4.500 V, calcined in O_2 at 500 ℃ for 2 h; Sample B: V_{DD}=4.750 V, calcined in O_2 at 500 ℃ for 2 h; Sample C: V_{DD}=5.000 V, calcined in O_2 at 500 ℃ for 2 h; Sample D: V_{DD}=5.000 V, firstly calcined in O_2 at 500 ℃ for 2 h, then calcined in O_2 at 600 ℃ for 2 h; $MgCl_2$: above the $MgCl_2$ saturated aqueous solution at 25 ℃ (RH=33.2%); KCl: above the KCl saturated aqueous solution at 25 ℃ (RH=84.3%).

V. CONCLUSION

In this paper, gas sensors employing nanoscale tin oxide films of high oxygen vacancies were successfully fabricated and the gas sensitivity to ethanol and water vapor of the devices was tested.

Firstly, an electrothermal finite element model of the tin oxide thin film gas sensor was simulated using ANSYS™ for the purpose of predicting the steady-state temperature distributions of the tin oxide thin film and the overall gas sensor. The temperature distribution of the tin oxide thin film which ranged from 350.2 ℃ to 414.5 ℃ was appropriate to detect ethanol based on the simulation results.

Secondly, the techniques of reactive direct current magnetron sputtering and calcinations were used in order to prepare the nanoscale tin oxide films. In addition, high Ar/O_2 ratio in the gas mixture during reactive direct current magnetron sputtering was selected in order to produce a great number of oxygen vacancies in the tin oxide films. Platinum and quartz glass were chosen as the materials of the comb-shaped electrode and the substrate respectively.

At last, the ethanol sensing and humidity sensing properties of the tin oxide thin film gas sensors were tested. Moreover, the factors which affected the gas sensitivity to ethanol and water vapor of the devices were discussed. The test results showed that by choosing proper calcination conditions and increasing the concentration of the gases the gas sensitivity to ethanol and water vapor could be enhanced. Besides, another way to enhance the ethanol sensitivity of the tin oxide thin film gas sensor was to increase the operating voltage in the appropriate range.

ACKNOWLEDGMENT

This research was supported by National Key Laboratory of Science and Technology on Micro/Nano Fabrication, Institute of Microelectronics, Peking University. It was financed by the National High Technology Research and Development Program of China (863 Program, Project No. 2011AA040401) and the National Basic Research Program (also called 973 Program, Project No. 2011CB309502).

REFERENCES

[1] T. Maekawa, K. Suzuki, T. Takada, T. Kobayashi, and M. Egashira, "Odor identification using a SnO_2-based sensor array," Sens. Actuators B, vol. 80, pp. 51-58, November 2001.

[2] S.C. Ray, M.K. Karanjai, and D. DasGupta, "Tin dioxide based transparent semiconducting films deposited by the dip-coating technique," Surf. Coat. Tech., vol. 102, pp. 73-80, April 1998.

[3] R.S. Niranjan and I.S. Mulla, "Spin coated tin oxide: a highly sensitive hydrocarbon sensor," Mater. Eng. B, vol. 103, pp. 103-107, October 2003.

[4] A.D. Brailsford and E.M. Logothetis, "Selected aspects of gas sensing," Sens. Actuators B, vol. 52, pp. 195-203, September 1998.

[5] G. Gaggiotti, A. Galdikas, S. Kac̆iulis, G. Mattogno and A. S̆etkus, "Temperature dependencies of sensitivity and surface chemical composition of SnO_x gas sensors," Sens. Actuators B, vol. 24-25, pp. 516-519, April 1995.

[6] J. Watson, K. Ihokura and G.S.V. Coles, "The tin dioxide gas sensor," Meas. Sci. Technol., vol. 4, pp. 711-719, 1993.

[7] R. Rella, A. Serra, P. Siciliano, L. Vasanelli, G. De, A. Licciulli and A. Quirini, "Tin oxide-based gas sensors prepared by the sol-gel process," Sens. Actuators B, vol. 44, pp. 462-467, October 1997.

[8] K.S. Yoo, N.W. Cho, H.S. Song and H.J. Jung, "Surface morphology and gas sensing characteristics of SnO_{2-x} thin films oxidized from Sn films," Sensors and Actuators B, vol. 24-25, pp. 474-477, April 1995.

[9] X. Vilanova, E. Llobet, R. Alcubilla, J.E. Sueiras and X. Correig, "Analysis of the conductance transient in thick-film tin oxide gas sensors," Sensors and Actuators B, vol. 31, pp. 175-180, March 1996.

[10] H. Teterycz, J. Kita, R. Bauer, L.J. Golonka, B.W. Licznerski, K. Nitsch and K. Wiśniewski, "New design of an SnO_2 gas sensor on low temperature cofiring ceramics," Sensors and Actuators B, vol. 47, pp. 100-103, April 1998.

[11] J.P. Coleman, A.T. Lynch, P. Madhukar and J.H. Wagenknecht, "Antimony-doped tin oxide powders:: Electrochromic materials for printed displays," Solar Energy Materials and Solar Cells, vol. 56, pp. 375-394, January 1999.

[12] T. Demirci, D. Guney, A. Bozkurt and Y. Gurbuz, "Electro-thermal simulations and modelling of micromachined gas sensor," Proceedings of the 14th Microelectromechanical Systems Conference (MEMS 2001), pp. 99-102, 2001.

[13] Sung-Soon Park and J.D. Mackenzie, "Thickness and microstructure effects on alcohol sensing of tin oxide thin films," Thin Solid Films, vol. 274, pp. 154-159, March 1996.

[14] Shanxing Huang, Hongwei Qin, Peng Song, Xing Liu, Lun Li, Rui Zhang, Jifan Hu, Hongdan Yan and Minhua Jiang, "The formaldehyde sensitivity of $LaFe_{1-x}Zn_xO_3$-based gas sensor," Journal of Materials Science, vol. 42, pp, 9973-9977,2007.

[15] K. Galatsis, L. Cukrov, W. Wlodarski, P. McCormick, K. Kalantar-zadeh, E. Comini and G. Sberveglieri, "p- and n-type Fe-doped SnO_2 gas

sensors fabricated by the mechanochemical processing technique," *Sensors and Actuators B*, vol. 93, pp. 562-565, August 2003.

[16] McGeehin, P., "Gas sensors for improved air quality in transportation," *Sensor Review*, vol. 20, pp. 106-112, 2000.

[17] Z. Liu, T. Yamazaki, Y. Shen, T. Kikuta, N. Nakatani and T. Kawabata, "Room temperature gas sensing of p-type TeO_2 nanowires," *Appl. Phys. Lett.*, vol. 90, pp. 173119-1-3, 2007.

[18] N. Yamazoe, J. Fuchigama, M. Kishikawa and T. Seiyama, "Interactions of tin oxide surface with O_2, H_2O and H_2," *Surf. Sci.*, vol. 86, pp. 335-344, July 1979.

[19] D. Manno, G. Micocci, R. Rella, A. Serra, A. Taurino and A. Tepore, "Titanium oxide thin films for NH_3 monitoring: structural and physical characterizations," *J. Appl. Phys.*, vol. 82, pp. 54-59, 1997.

In situ Monitoring of Temperature using Flexible Micro Temperature Sensors inside Polymer Lithium-ion Battery

Chi-Yuan Lee*, Shuo-Jen Lee, Yu-Ming Lee, Ming-Shao Tang, Pei-Chi Chen, Yu-Ming Chang
Department of Mechanical Engineering, Yuan Ze Fuel Cell Center, Yuan Ze University, Taoyuan, Taiwan, R.O.C.
cylee@saturn.yzu.edu.tw

Abstract — **Polymer lithium-ion batteries must rapidly charge and discharge, rapidly increasing their inner temperature. This heating raises a safety issue. Therefore, a flexible micro temperature sensor for the *in-situ* monitoring of temperature within a polymer lithium-ion battery must be developed. Traditional thermocouples are too large to be used to measure temperature inside polymer lithium-ion batteries. In this study, a novel method for the *in situ* monitoring of temperature inside a polymer lithium-ion battery involves flexible micro temperature sensors. Temperature is successfully measured *in situ* using flexible micro temperature sensor.**

Keywords — *MEMS; flexible micro temperature sensors; polymer lithium-ion battery; in- situ monitoring*

I. INTRODUCTION

In recent years, lithium-ion batteries have been widely used in commercial products, including smart phones, notebooks and electric vehicles. Therefore, the safety and efficiency of lithium-ion batteries are important. The safety of a polymer lithium-ion battery depends on the electrolyte, separator, anode and cathode [1-3]. Metallic lithium can separate out as dendrites and acicular crystals and cause several problems during rapid charging and discharging [4], reducing the efficiency of the battery, and raising a safety concern [5, 6]. The internal temperature of a polymer lithium-ion battery is typically measured using a thermocouple. Traditional thermocouples are too large for measuring temperature in batteries without damaging them [7]. Thermal modeling has also been adopted to simulate the temperature distribution inside a lithium-ion battery [8].

This work proposes a flexible micro temperature sensor that is manufactured using micro-electro-mechanical system (MEMS) technology. It is fabricated on a flexible substrate and inside a polymer lithium-ion battery. The micro temperature sensor is calibrated to read out the interior temperature of the lithium-ion battery.

II. METHODOLOGY

The four major classes of temperature sensor are resistance temperature detection sensors (RTDs); thermally sensitive resistors (thermistor); thermocouples and mercury-in-glass thermometers.

Thermocouples and RTDs, whose voltage and resistance varies with temperature, are both in common use. The internal temperature of a lithium battery is normally measured using a conventionally adopted thermocouple. However, conventional thermocouples are large, making it impossible to determine the optimal measurement position accurately. An RTD micro sensor is characterized by their small volume, high accuracy, short response time, fabrication simplicity, appropriateness of mass production, and ability to measure the temperature more effectively than traditional thermocouples. Therefore, in this work, RTDs are utilized to measure temperature.

Commonly used sensing materials in temperature sensors include Pt and Au; however, Pt is expensive and Au has greater conductivity and flexibility.

The resistance of a general metal is expressed as

$$R = \rho\, L/A \qquad (1)$$

where R is resistance (Ω); ρ is resistivity (Ω m); L is wire length (m), and A is cross-sectional area (m^2). In the range of linear variation of resistance of an RTD with temperature, the relationship between the measured resistance and the change in temperature satisfies

$$R_t = R_r\,(1 + \alpha_T\,\Delta T) \qquad (2)$$
$$\Delta T = t - t_r \qquad (3)$$

where R_t and R_r are the resistance of the RTD at t°C and r°C, respectively; α_T is the positive temperature coefficient of the RTD; ΔT is the variation in temperature from a reference temperature and t and t_r are the temperatures of the RTD at t°C and r°C, respectively. Hence, (2) can be rewritten as

$$\alpha_T = (R_t - R_r) / (R_r\,\Delta T) \qquad (4)$$

where α_T is the temperature coefficient of resistance (TCR) of the sensor [9].

The micro temperature sensor that is fabricated herein is a gold RTD, and is shown in Fig. 1.

Fig. 1. Schematic diagram of flexible micro temperature sensor.

III. Fabrication of Flexible Micro Temperature Sensors

In this work, parylene thin film is anti-erosive, and resists both high temperature and stress corrosion. Parylene thin film is utilized as a protected layer and a substrate. Pareylen is a thin film that contains a high number of molecules. It was deposited by physical vapor deposition (PVD) at a low temperature, enabling the thickness of the thin-film to be controlled accurately. The process of fabricating flexible micro temperature sensor is described below. Fig. 2 displays the fabrication of the flexible micro temperature sensor.

A layer of copper (Cu, 300Å) was deposited as a sacrificial layer on a silicon wafer substrate, and then a 5μm-thick parylene thin film was deposited onto the layer of copper, as displayed in Fig. 2(a) and (b).

(c) The parylene thin film acted as a protective layer, an isolation layer and a substrate.

(d) The first lithographic process was carried out to define the pattern of the flexible micro temperature sensor.

(e) Both Cr (250Å) and Au (2000Å) are then deposited in that order on the parylene substrate using an e-beam evaporator. The structure of the flexible micro temperature sensors was determined using the lift-off process.

(f) Another parylene layer was then deposited to protect the structure.

(g)(h) The second lithographic process, reactive ion etching (RIE), defined the pattern on the contact pads and the sensing region.

Fig. 3 shows an optical micrograph of the flexible micro temperature sensor.

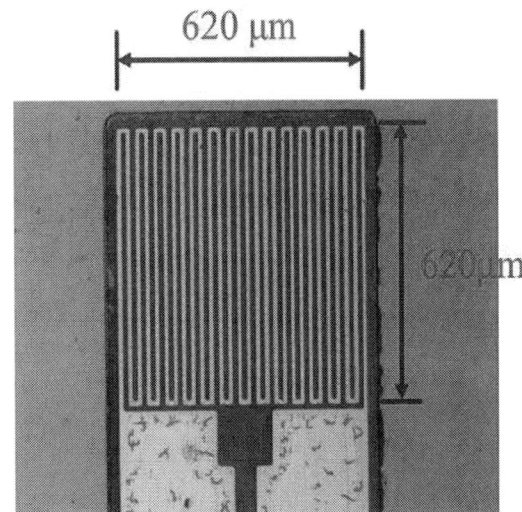

Fig. 3. Optical microscopic photograph of flexible micro temperature sensor.

IV. Results and Discussion

The flexible micro temperature sensor was placed in an oven (DENG YNG DS-45). The resistance signal was picked up by a Data Acquisition system. The temperature of the oven was increased from 20°C to 70°C three times.

Fig. 4 plots the calibration curve of the flexible micro temperature sensor. The calibration curve reveals high repeatability and linearity of the relationship between temperature and resistance.

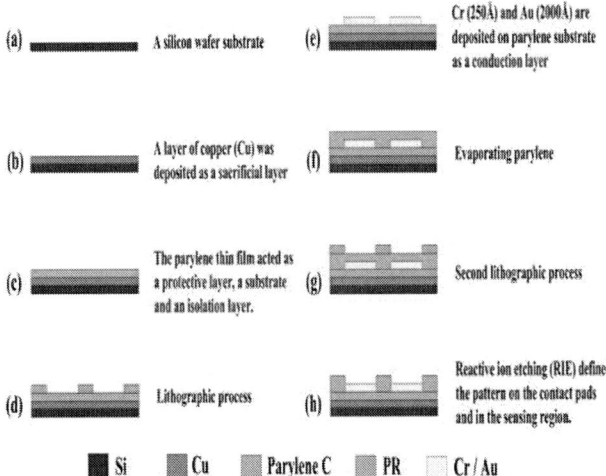

Fig. 2. Procedure for fabricating flexible micro temperature sensors on silicon substrate.

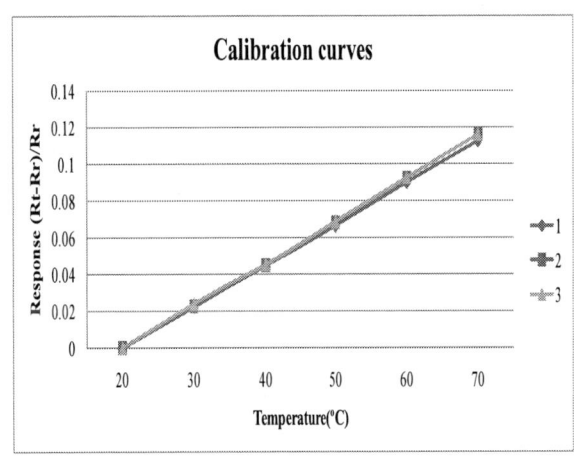

Fig. 4. Calibration curve of a flexible micro temperature sensor.

The flexible micro temperature sensors were inserted into a polymer lithium-ion battery, as shown in Fig. 5.

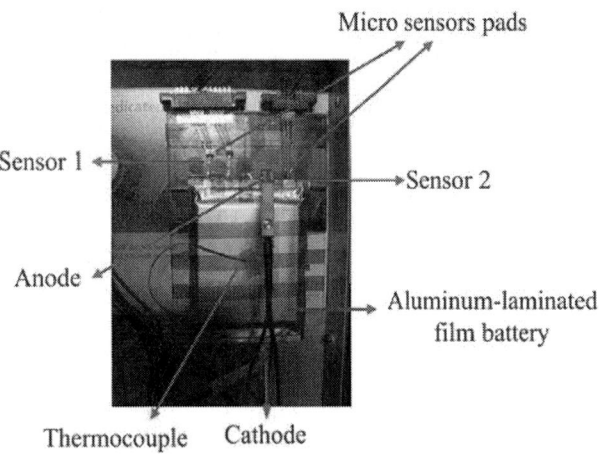

Fig. 5. Flexible micro temperature sensors were inserted into a polymer lithium-ion battery.

NI PXI 1033 was used to pick up the signal from flexible micro temperature sensor, and the polymer lithium-ion battery was charged and discharged using GBT 2211, as displayed in Fig. 6.

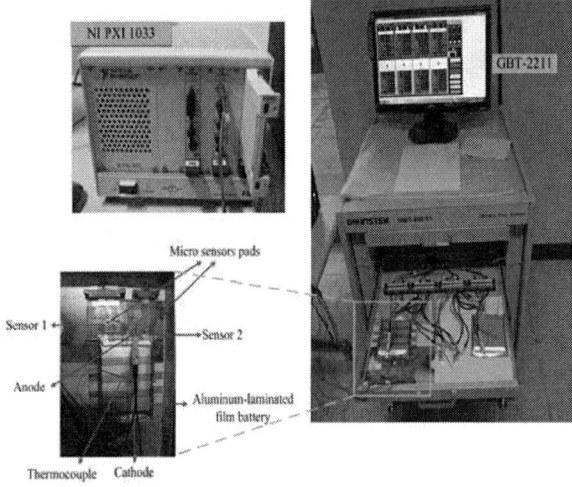

Fig. 6. Testing system that comprises micro temperature sensors in lithium battery.

In situ measurements of temperature were made successfully. The results demonstrate that when charging and discharging reactions occurred in the lithium battery, the internal temperature was 3°C higher than the external temperature, as shown in Fig. 7.

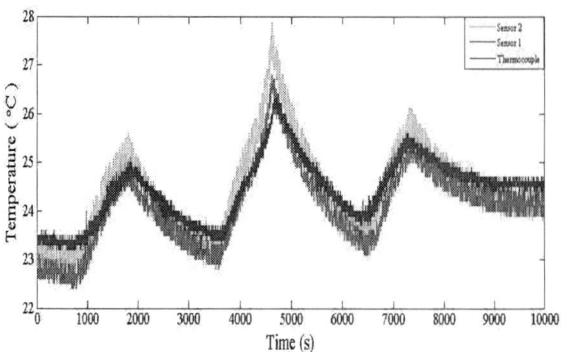

Fig. 7. Temperature curve during 1C charging and discharging.

V. CONCLUSIONS

In this investigation, the flexible micro temperature sensor that is fabricated herein is a gold thermally sensitive resistor (RTD). Parylene thin film is utilized as a protected layer and a substrate. Parylene thin film is anti-erosive, and resists both high temperature and stress corrosion. The *in situ* measurements of temperature are picked up successfully. The results demonstrate that when the lithium battery 1C charging and discharging reactions occur in the battery, the inner temperature is 3°C higher than the outer.

ACKNOWLEDGMENT

This work was accomplished with much needed support and the authors would like to thank the National Science Council of R.O.C. through the grant NSC 99-2632-E-155-001-MY3. The authors also like to thank EXA Energy Technology Company and Professors Ming-Yu Chung and Kuo-Chang Han of the Yuan Ze University for their valuable advice and assistance in experiment. In addition, we would like to thank the YZU NENS Common Lab, TKU MEMS Lab for providing access to their research facilities.

REFERENCES

[1] B. Scrosati, and J. Garche, "Lithium batteries: status, prospects and future," *Journal of Power Sources*, vol. 195, pp. 2419-2430, 2010.

[2] Y. Guo, "Safety," *Encyclopedia of Electrochemical Power Sources*, vol. 1, pp. 241-253, 2009.

[3] S. Tobishima, "Secondary batteries – lithium rechargeable systems – lithium-ion," *Encyclopedia of Electrochemical Power Sources*, vol. 1, pp. 409-417, 2009.

[4] Z. Chen, Y. Qin, and K. Amine, "Redox shuttles for safer lithium-ion batteries," *Electro. Acta*, vol. 54, pp. 5605-5613, 2009.

[5] C. Delmas, M. M. Trier, L. Croguennec, S. Levasseur, J. P. Pe´re`s, C. Pouillerie, G. Prado, L. Fourne`s, and F. Weill, "Lithium batteries: a new tool in solid state chemistry," *International Journal of Inorganic Materials*, vol. 1, pp. 11-19,1999.

[6] A. Fernicola, F. Croce, B. Scrosati, T. Watanabe, and H. Ohno, "LiTFSI-BEPyTFSI as an improved ionic liquid electrolyte for rechargeable lithium batteries," *Journal of Power Sources*, vol. 174, pp. 342-348, 2007.

[7] C. Forgez, D. V. Do, G. Friedrich, M. Morcrette, and C. Delacourt, "Thermal modeling of a cylindrical LiFePO4/graphite lithium-ion battery," *Journal of Power Sources*, vol. 195, pp. 2961-2968, 2010.

[8] G. Guo, B. Long, B. Cheng, S. Zhou, P. Xu, and B. Cao, "Three-dimensional thermal finite element modeling of lithium-ion battery in thermal abuse application," *Journal of Power Sources*, vol. 195, pp. 2393-2398, 2010.

[9] C. Y. Lee, W. Y. Fan, and C. P. Chang, "A novel method for in-situ monitoring of local voltage, temperature and humidity distributions in fuel cells using flexible multi-functional micro sensors," *Sensors*, vol. 11, pp. 1418-1432, 2011.

Design and Simulation of Fully-symmetrical Resonant Pressure Sensor

Yiwen Jiang[a*], Xiaohui Du[b], Zhan Zhan[c], Bulei Xu[d], Wenlong Lv[e], Lingyun Wang[f*], Daoheng Sun[g]

Department of Mechanical and Electrical Engineering, Xiamen University, CHINA

[f]rabitwangly@yahoo.com.cn, [g]sundh@xmu.edu.cn

Abstract — **A fully-symmetrical resonant pressure sensor based upon lateral drive is presented, which can avoid stress concentration that arises from temperature or vacuum packaging. Unlike conventional driving method, resonant structure with laterally driven comb capacitance allows the linear characteristic of driving force and also obtains high quality factor for its slide-film air damping. Furthermore, the detection sensitivity of the device can be improved by using differential capacitance, which will reduce shared-frequency interference phenomenon at the same time. According to the FEM analysis, the structural parameters of resonant pressure sensor are optimized. Meanwhile, the pressure sensitivity of the sensor has designed to be 22.602Hz/kPa for a 18μm thick diaphragm over a pressure range of 550kPa. From temperature simulation, the temperature coefficient of sensor is -1.8233Hz/°C in the range of -20°C~60°C without any temperature compensation. Finally, the frequency domain characteristics have been confirmed and the quality factors of sensor under different damping ratios are identified, it offers reliable reference for the choice of vacuum in resonant pressure sensor packaging.**

Keywords- resonant; pressure sensor; MEMS; sensitivity; quality factor

I. INTRODUCTION

Miniature silicon resonant pressure sensor based upon MEMS (Microelectromechanical system), which has the highest precision for frequency output, measures pressure by detecting the natural frequency of the resonator. Because of its strong anti-jamming capability, excellent stability, quick response, wide frequency band, compact conformation, and low consumption, the resonant pressure sensor has been used in many fields such as aerospace, automatic system and military affairs, attracting much attention in many countries [1].

Up to date, resonant pressure sensor has been well developed and its measurement accuracy has reached up to 0.01%F.S./year. Harada [2] utilized a self-aligning selective epitaxial method and a hybrid selective etching method to fabricate a pressure sensor based on electro-magnetic excitation and detection. Two resonators were encapsulated into the micro-vacuum cavities to measure applied pressure from the difference of two resonant frequencies, but the sensor was large in size and difficult to integrate with magnets. Greenwood [3] manufactured an electrostatically driven and capacitively sensed sensor which utilizes deep boron diffusion etch-stop layer silicon wafer process. However, the upper and lower plates driving method could cause mechanical coupling

between resonator and diaphragm, which made it difficult to obtain a wider measuring range.

In order to overcome these defects, a fully-symmetrical resonant pressure sensor based upon lateral drive was presented in this paper, as shown in Fig. 1. Due to electrostatic excitation and capacitance detection adopted in this sensor, a finite-element simulation was performed with ANSYS software to obtain the natural frequency of the resonator. The related parameters such as applied pressure and temperature to frequency were analyzed and the frequency response was then given.

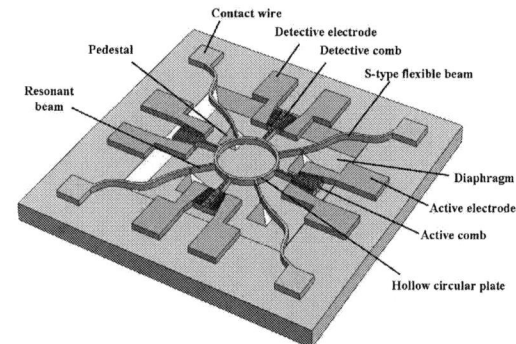

Fig. 1. Schematic of a novel silicon resonant pressure sensor

II. THEORY

The frequency of the resonator is determined by structural dimensions and material parameter, each vibration mode corresponding to a certain natural frequency. Since structures like fixed comb drive and detective electrode, contact wire and the frame have no causal effect on natural frequency and sensitivity, the sensor model was simplified in ANSYS software for an improvement in simulation speed.

Figure 2 shows the vibration modes of the resonator. A Fan-shaped comb was employed in fully-symmetrical resonant pressure sensor to realize lateral drive, which will improve quality factors of the sensor. As can be seen in Fig. 2, both the first and the sixth modes of resonator are parallel to diaphragm, which can reduce damping and meet the requirements of lateral drive. However, the resonator in the other vibration modes produces a certain displacement in the vertical direction of diaphragm, and it is impossible to apply lateral drive when ambient pressure reaches a certain value, its

978-1-4673-1122-9/12 $31.00 © 2012 IEEE

vibration mode tends to cause coupling effect with diaphragm. Compared with the sixth mode, the first mode has a lower natural frequency, less energy required, and easy to be excited [4], meanwhile, the process of electrostatic drive and the detection for differential capacitance can be realized. Thus, the first mode was selected as the working mode for the resonator.

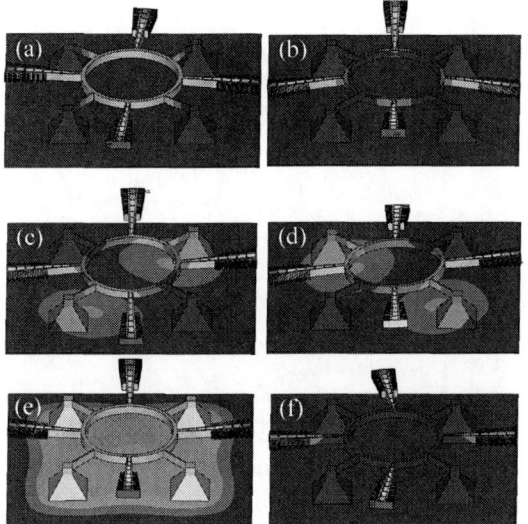

Fig. 2. Vibrate mode of the resonator. (a) mode 1, (b) mode 2, (c) mode 3, (d) mode 4, (e) mode 5, (f) mode 6

With the aim to verify the exactness of ANSYS simulation analysis, structural parameters of the sensor based on the theoretical analysis were chosen as follows: the length, width and thickness of supporting beam were 375, 18 and 100μm, respectively; the external and internal radius of hollow round plate were 600 and 550μm, respectively; thickness of the diaphragm and pedestal was 20 and 300μm, respectively.

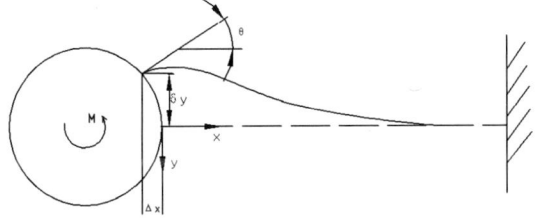

Fig. 3. The geometric relationships between support beam

In the theoretical analysis, suppose that the process by which the supporting beam rotates with the hollow round plate, as shown in Fig.3, the right end of beam was fixed, and the external radius of round plate is R. Rayleigh-Ritz method [5] was used to analyze the relational expression of the rotational stiffness. When the supporting beam is rotating, suppose that the rotation angle of round plate is θ (small angle) by the action of external torque, the deflection curve of beam is y_x, and the axial displacement caused by the axial force is

Δx. The boundary conditions are as follows:

Left end: $y\big|_{x=0} = R\sin\theta \approx R\theta$, $\quad y'\big|_{x=0} = \tan\theta \approx \theta$.

Right end: $y\big|_{x=L} = 0$, $\quad y'\big|_{x=L} = 0$.

Axial constraint: Δx refers to the axial displacement of beam that caused by the rotation of round plate, and the geometrical relationship shows:

$$-\Delta x = -(R - R\cos\theta) \approx -\frac{1}{2}R\theta^2 \qquad (1)$$

A polynomial of three terms is adopted to approximate y_x, thus assume $y_x = ax^3 + bx^2 + cx + d$, the boundary conditions are substituted for it, and the following can be obtained:

$$y_x = \frac{(L+2R)\theta}{L^3}x^3 - \frac{(2L+3R)\theta}{L^2}x^2 + \theta x + R\theta \qquad (2)$$

Ultimately, according to the theorem of stationary value of total potential energy:

$$\frac{dI}{d\theta} = \frac{dE_1}{d\theta} + \frac{dE_2}{d\theta} + \frac{dE_p}{d\theta} = 0 \qquad (3)$$

So the rotational stiffness of single supporting beam is:

$$k = \frac{M}{\theta} = \frac{4EJ(L^2 + 3LR + 3R^2)}{L^3} \qquad (4)$$

Since the frequency formula of resonator can be expressed as $2\pi f = \sqrt{\dfrac{4k}{I_p}}$, and then can be written as:

$$f = \frac{2}{\pi}\sqrt{\frac{EJ(L^2 + 3LR + 3R^2)}{I_p L^3}} \qquad (5)$$

Where E is the elastic modulus of the material; J is moment of inertia; L is the length of supporting beam; R is the external radius of round plate; I_p is the rotational inertia of the resonator.

According to the theoretical formula, the natural frequency of resonator can be calculated and the natural frequency by ANSYS analysis is $f=29013$(Hz), thus their relative deviation is about 1.89%. This similar result in simulation and in theoretical calculation is good for optimizing design of the sensor.

III. SIMULATION

When designing the structure of senor, it is necessary to pay close attention to the following three key points: (1) The work described in this paper [6] is the natural frequency of the sensor under its working mode should be more than 20% when the full range is 300kPa, thus it can ensure that the value of Q is greater than 10000, and it is possible to keep low phase noise in the process of resonance. Therefore, with the permission of the process conditions, the sensitivity of senor is designed can be over 20Hz/kPa; (2) In order to avoid the influence of ambient noise on the senor, the natural frequency of resonator must be greater than sound frequency (20kHz); (3) To avoid the cross interference between the working mode and other modes, it is necessary to separate the working mode from other modes within the measuring range. By grasping the above three important points, finite element simulation analysis is conducted for those parameters such as the length, width and thickness of supporting beam, the thickness of diaphragm as well as the thickness of pedestal, to get a knowledge about the impact of the structural parameters on the performance of resonator, and ultimately to determine the physical dimensions of the sensor.

A. Length of Support Beam

The natural frequency of the first mode shows linearly to the ambient pressure (see Fig.4), which can be attributed to an increase in ambient pressure resulting in supporting beam of resonator working under the action of tension stress. The natural frequency of resonator decreases with increasing the supporting beam length, which improves sensitivity of the sensor even through the maximum value was obtained at the length of 425μm in supporting beam, consistent with the results of formula (5). However, supporting beam of 425μm or even longer will generate cross interference between the mode of working and other modes within the measuring range for greater interval of frequency (see Table I). For this reason, the length of 400μm was determined for the supporting beam, because sensitivity of 27.438Hz/kPa and the natural frequency of over 20kHz can be satisfied at this length.

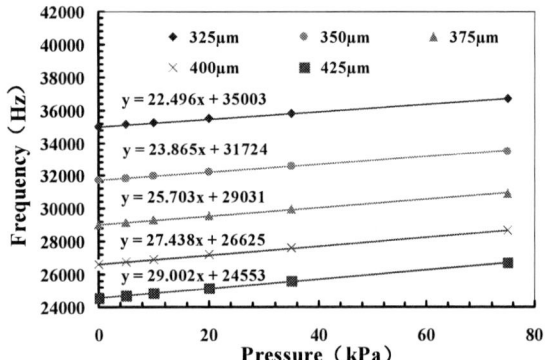

Fig.4 Relationship between length of support beam and resonator performance parameters

TABLE I. RELATIONS BETWEEN LENGTH OF SUPPORT BEAM AND NATURAL FREQUENCY OF THE RESONATOR

Length of beam (μm)	The natural frequency of first mode ((Hz))	The natural frequency of second mode ((Hz))
325	34911	38224
350	31708	37845
375	29013	37648
400	26603	37370
425	24528	37345

B. Width of Support Beam

The natural frequency of resonator will increase with the increase of supporting beam width, while the sensitivity of senor will decrease with the increase of supporting beam

width (Fig.5), it shows a more significant impact on the variation tendency of its sensitivity than the length, consistent with the results of theoretical analysis. Because the natural frequency of resonator should be more than 20 kHz and the sensitivity of senor should exceed 20Hz/kPa, so the qualified widths of beam are 18μm and 15μm. (see Fig.5). Compared with the width of 15μm, the beam of 18μm has a greater natural frequency of resonator, which it is much easier to avoid the influence of ambient noise. Moreover, the beam of 15μm always tends to cause cross interference between the working mode and other modes within the measuring range, thus the width of supporting beam is set to 18μm.

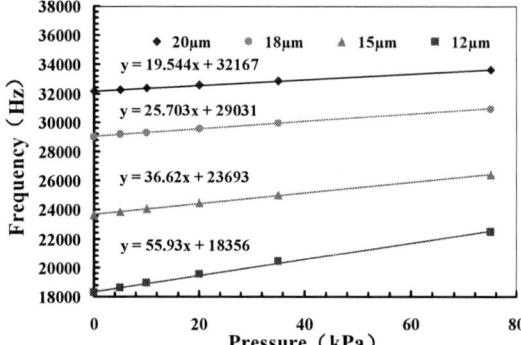

Fig.5 Relationship between width of support beam and resonator performance parameters

C. Thickness of Support Beam

The variation of supporting beam thickness has little influence on the natural frequency of resonator (Fig.6), which is in line with the computational formula theoretically. As is shown in Fig.6, The sensitivity will decrease with the increase of supporting beam thickness, however, it is necessary to bonding the resonator with the pedestal when making the senor, and then the resonator must use the process of lapping to lower thickness, so the supporting beam thickness of 100μm is easier to be obtained than other thickness. Therefore, when the thickness of supporting beam is 100μm, the sensitivity has reached up to 25.703Hz/kPa, which completely complies with the design requirements above.

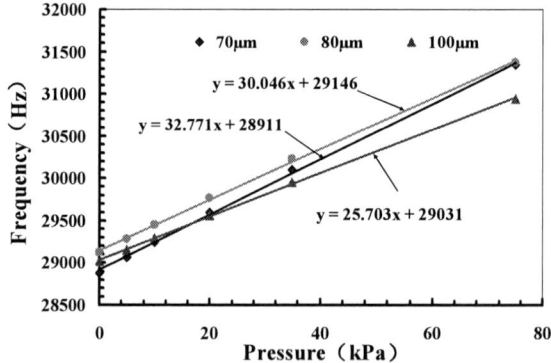

Fig.6 Relationship between thickness of support beam and resonator performance parameters

D. Thickness of Diaphragm

The thickness of diaphragm does not significantly influence the natural frequency of resonator (see Fig.7), because the natural frequency of resonator mainly depends on the structural parameters of resonator, and it has little relationship with the diaphragm and pedestal. The thickness of diaphragm mostly affects the sensitivity and measuring range of senor. In the Fig.7, it is shown that the sensitivity of senor will decrease with the increase of diaphragm thickness, although the sensitivity of senor reaches its maximum value when the thickness of diaphragm is 15μm, but the reduction of diaphragm thickness will reduce the measuring range of senor, thus the thickness of diaphragm is set as 18μm in the end.

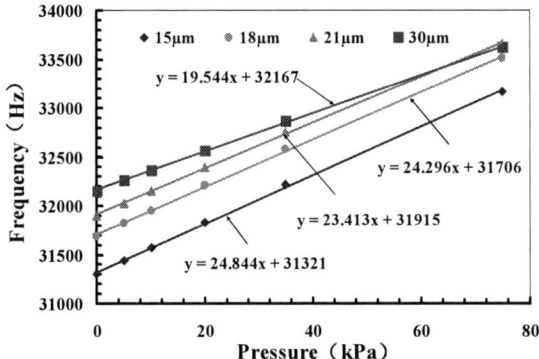

Fig.7 Relationship between thickness of diaphragm and resonator performance parameters

E. Thickness of Pedestal

Like the diaphragm, the variation of pedestal thickness puts little impact on the natural frequency of resonator, and the sensitivity of senor will go up with the increase of pedestal thickness (see Fig.8). Because pedestal which the shape is a type of prismatic table, is obtained by using wet etching, and the increase of thickness will result in the enlargement of contact area between pedestal and diaphragm, and then causing that the stress which is transferred to the supporting beam will decrease, so that the sensitivity of resonator goes down. To sum up, the thickness of pedestal is set as 250μm at last.

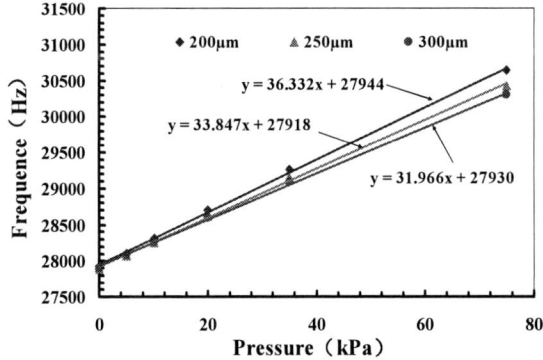

Fig.8 Relationship between thickness of pedestals and resonator performance parameters

According to the above analysis, structural parameters of the resonant pressure sensor have been confirmed: the length, width and thickness of supporting beam are 400, 18 and100μm, respectively; the thickness of diaphragm and the thickness of pedestal are 18 and 250μm, respectively.

IV. ANALYSIS

A. Sensitivity

The ultimate purpose of resonant pressure sensor is to get the relationship between pressure and natural frequency, and make every effort to keep the linear relationship between pressure and natural frequency within measuring range. The relationship between pressure and vibration frequency is obtained through the prestressed modal analysis on the resonant pressure sensor.

The results of simulation analysis show that the natural frequency of resonator is 27885Hz without any applied pressure. In the process of ANSYS simulation analysis, first of all, pressure is put on the bottom of diaphragm, and static analysis is conducted to get the deformation result of resonator; and then the prestressed modal analysis is made for resonator to get the natural frequency of resonator. The relationship between the natural frequency of resonator and pressure is shown in Fig.9, the supporting beam is under the action of tension stress in the whole range of pressure variation, thus the natural frequency will increase with the increase of pressure. When pressure is within the full range of 0kPa~ 550kPa, the sensitivity of sensor has reached up to 22.602Hz/kPa, which is very close to the performance of overseas mature products [7].

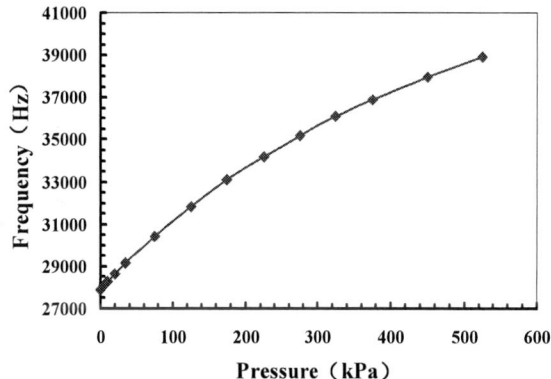

Fig. 9. Relationship between natural frequency and applied pressure

B. Temperature

The performance of sensor is easy to be influenced by temperature, thus it is necessary to take the impact of temperature on the natural frequency of resonator when analyzing the vibration characteristics of resonator. In the process of simulation analysis, first of all, the thermal stress analysis is made for resonator, to get its deformation result and then conduct the prestressed modal analysis. Based on the simulation analysis, the relationship between temperature and natural frequency is shown in Fig. 10.

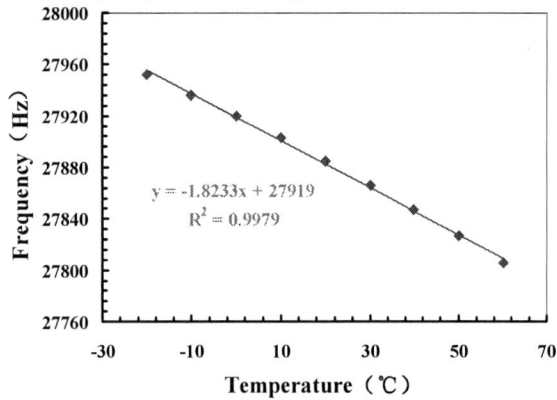

Fig. 10.   Relationship between natural frequency and applied temperature

According to the analysis results in Fig.10, The temperature drift is lower than the laterally driven sensor without using fully-symmetrical structure [8]. The natural frequency of resonator will decrease with the increase of temperature, which shows negative temperature characteristic, and the temperature coefficient of thermal sensitivity is -1.8233Hz/℃. The paper [9] shows that the Young modulus and shear modulus can be easily influenced by temperature, the increase of temperature can result in the reduction of supporting beam rotational stiffness, and then the natural frequency of senor will decrease, therefore the natural frequency of resonator shows negative temperature characteristic. However, the external temperature will influence the accuracy of senor to some extent, so it is necessary to take the temperature compensation into consideration when designing of resonant pressure sensor, to eliminate the impact of temperature on the performance of the senor as much as possible.

C.   Frequency Domain Response Characteristic

Quality factor is one of the main performance indexes of resonator, which can directly influence the frequency selection capability and anti-jamming capability of resonator, the harmonic response analysis technique of ANSYS is adopted to analyze the frequency domain response characteristics of resonator.

The natural frequency of resonator is 27885Hz, sinusoidal excitation force at an amplitude value of 1μN is exerted at the excitation point around the natural frequency. The system damping in the structural design is mainly slide-film damping, thus according to the analysis in this paper [10], the damping ratio is set as 1×10^{-5}. The relationship between the amplitude of vibration pickup point and frequency is shown in Fig. 11.

The resonator shows higher amplitude of vibration around the natural frequency, but the amplitude of vibration will go down as the frequency deviates from the centre frequency. Therefore, the quality factor can be obtained by the curve of frequency domain characteristics, different damping ratios are correspondent with different quality factors (Fig.12).

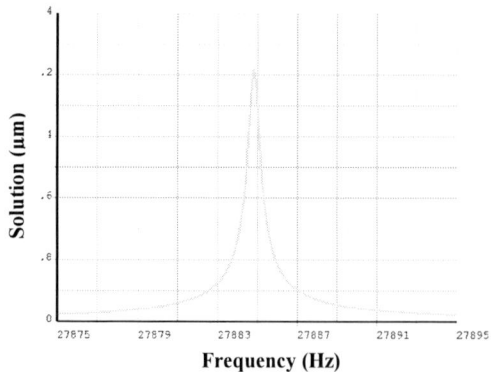

Fig. 11.   Frequency domain response characteristic of the resonator

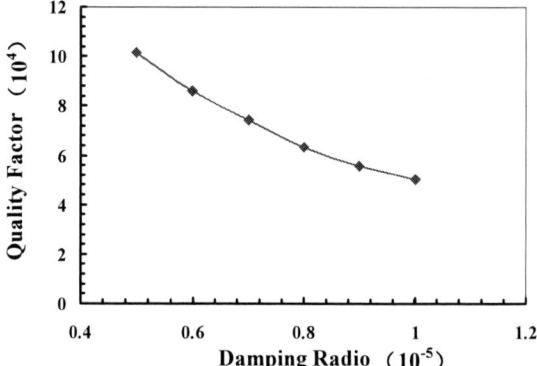

Fig. 12.   The relations between damping radio and quality factor

According to the analysis on the curve in Fig.12, it can be found that fully-symmetrical resonant pressure sensor can get better quality factor, it is mainly because that its structure is lateral drive, and the damping of system is mostly slide-film damping. High quality factor means that the resonator can only vibrate under the natural frequency, and its vibration is extremely weak under other frequencies in spite of suffering vibration interference from the external, so this kind of resonator has a good property of frequency-selecting and anti-vibration.

V.   CONCLUSION

With the help of ANSYS simulation analysis, the impact of senor physical dimensions on the performance of resonator has been discussed, and ultimately the structural parameters of resonator are identified: the length, width and thickness of supporting beam are 400, 18 and 100μm, respectively; the thickness of diaphragm and pedestal are 18 and 250μm, respectively. Based on the analysis on the vibration characteristics of resonator, it can be concluded that the sensitivity of senor can come up to 22.602Hz/kPa under the pressure of 0kPa~550kPa; when the external temperature is 20℃~60℃, the temperature coefficient of thermal sensitivity is -1.8233Hz/℃. Finally, by the harmonic response analysis on resonator, not only could the curve of frequency domain characteristics be obtained, but also the relationship between

978-1-4673-1122-9/12 $31.00 © 2012 IEEE

damping ratio and quality factor could be confirmed. The harmonic response analysis offers reliable reference for the choice of vacuum in resonant pressure sensor packaging.

ACKNOWLEDGMENT

The authors would like to thank Department of Mechanical and Electrical Engineering of Xiamen University for the research of resonant pressure sensor. This work is supported by the Fundamental Research Funds for the Central Universities (No.2010121039) and the National Natural Science Foundation for Youth of China (No. 51105320).

REFERENCES

[1] R M Landon, "Resonant sensors-a view," *J. Phys. E: Sci. Instrum.*, Vol.18, pp. 103-115, 1985.

[2] K. Harada, K. Ikeda, H. Kuwayama, et al. "Various applications of resonant pressure sensor chip based on 3-D micromachining," *Sensors and Actuators*, 73(1999), pp. 261-266.

[3] J.C. Greenwood, and D.W. Satchell, "Miniature silicon resonant pressure sensor," *IEEE*, pp. 369-372, 1998.

[4] Luo Zhigao, Li Ju, Wang Xiang, et al. "The Reacher of Reducing Vibration of Circular Saw Based on Experimental Modal Analysis," *Journal of Vibration and Shock,* vol.28, pp. 124-127, 2009.

[5] Chen Zhiyong, Gao Zhongyu, and Zhou Bin, "Nonlinear mechanical characteristics in micromachine vibratory sensor," *Chinese Journal of Mechanical Engineering*, vol.37, pp. 78-81, April 2001.

[6] Christopher J. Welham, John Greenwood, and Michael M. Bertioli. "A high accuracy resonant pressure sensor by fusion bonding and trench etching," *Sensors and Actuators*, pp. 298-304, 1999.

[7] Jessica Melin, Peter Enoksson, et al. "A low-pressure encapsulated deep reactive ion etched resonant pressure electrically excited and detected using 'burst' technology," *Journal of Micromechanics and Microengineering*, vol.10, pp. 209–217, 2000.

[8] Deyong Chen, Yuxin Li, Meng Liu, et al. "Design and experiment of a laterally driven micromachined resonant pressure sensor for barometers," Procedia Engineering, vol.5, pp. 1490-1493, 2010.

[9] R. A Buser, and N. F. de Rooij. "Resonant silicon structures," *Sensors and Actuators,* vol.17, pp. 145-154, 1989.

[10] S. Bianco, M. Cocuzza, et al. "Silicon resonant microcantilevers for absolute pressure measurement," *Journal of Vacuum Science & Technology B,* vol.24, pp. 1803-1809, 2006.

Effect of Oxidation on SGOI Nanowire Biosensor Fabrication Using Ge Condensation

Kow-Ming Chang[a,b], Chu-Feng Chen[a], Chiung-Hui Lai[c], Chin-Ning Wu[a], Cheng-Ting Hsieh[a], Yu-Bin Wang[a], and Chung-Hsien Liu[a],

[a] Dept. of Elec. Engineering and Inst. of Elec., National Chiao-Tung University
No. 1001, University Rd., Hsinchu 300, Taiwan, R.O.C.
[b] Department of Electronic Engineering, I-Shou University
No.1, Sec. 1, Syuecheng Rd., Kaohsiung 840, Taiwan, R.O.C.
[c] Dept. of Electronics. Engineering, Chung Hua University
No. 707, Sec.2, WuFu Rd., Hsinchu, 300, Taiwan, R.O.C.

kmchang@faculty.nctu.edu.tw; kmchang@isu.edu.tw

Abstract—Ge condensation offers an attractive way to increase Ge the fraction of Ge in SGOI. From authors' previous investigations, increasing the fraction of Ge increases the sensitivity of the SiGe nanowire sensor. To understand how Ge condensation on an SGOI nanowire sensor helps to optimize the conditions of oxidation and improve the sensitivity of the sensor, the effect of oxidation gas and SiGe/α-Si stacked structure on the movement of Ge is examined. The results reveals that SiGe nanowire has a maximum sensitivity when it includes a 14% Ge containing $Si_{1-x}Ge_x$ layer that is stacked on a 200 Å-thick α-Si layer and is treated for 3 min with O_2 gas to which is added 13% N_2 gas.

Keywords-component; SiGe-on-Insulator, bio-sensor,oxidation

I. INTRODUCTION

The biological, chemical and optical applications of nano-wires, have recently been an signification topic of research [1,2]. Nano-wires have attracted substantial interest as essential building blocks for functional electronics and especially in nano-electronics [3-5] and highly sensitive biosensors [6-7]. In our earlier studies, the Ge in $Si_{1-x}Ge_x$ improved the sensinanowire biosensor's sensitivity due to higher carrier mobility [8]. Ge condensation process is an attractive way to increase the concentration of Ge at the SiO_2/SiGe interface [9]. Since surface charges on the surface of a nanowire alter its conductivity, nano-wire sensors can be used in chemical or biological detection. Although Ge condensation enhances the fraction of Ge at the surface, it simultaneously produces the interface traps [10]. Ge not only accumulates at the SiO_2/SiGe interface, but also diffuses from the surface into burrier oxide layer [11]. Reducing the Ge concentration at the SiGe surface degraded the nanowire sensitivity, as revealed our previous studies [12]. Accordingly, a better understanding of the characteristics induced by Ge condensation will help to optimize and improve biosensor sensitivity. In this work, the oxidation ambient, oxidation time and nanowire structure were varied. Oxidation temperature was fixed at 950°C to prevent any undesirable consequences of an inappropriate lower or higher of temperature during Ge condensation.

II. EXPERIMENTAL PROCEDURE

Figure 1 displays the structure of SiGe nanowire. α-Si/ $Si_{1-x}Ge_x$ was deposited on a patterned 300 nm-high bottom oxide. The deposition thickness of α-Si was 200Å or 400Å and the Ge fraction of $Si_{1-x}Ge_x$ splits in this experiment was 7%, 14% or 20%. To determine the influence of Ge condensation properties of interest, the oxidation ambient (pure N_2, pure O_2, and diluted O_2 gas), oxidation time (3min, 5min , and 10min) and nanowire structure (layers with different fractions of Ge stacked on layers of α-Si of various thicknesses) were varied . The poly-Si nanowire was also fabricated as a control group to confirm the oxidation rate under each condition. After the α-Si /SiGe layer was formed and implanted with P-type species, the samples were annealed at 950°C for 180 sec in N_2 gas to activate the implanted species and cause re-crystallization. After annealing, the Al films were thermally evaporated and the electrodes were defined by the mask process. The poly-Si and SiGe nanowires were implanted in p-type nanowire. To functionalize α -Si / $Si_{1-x}Ge_x$ nanowires, the wires were adopted the 3-aminopropyltri-ethoxysilane (APTMS) was adopted to modify the surface of the silicon oxide around the nanowires. A hydroxyl functional group on the surface of the oxide was replaced by methoxy groups of APTMS modules, and simultaneously, the surface of the nano-wire was simultaneously terminated by amine groups. From our earlier investigations, the amine groups are prone to deplete positive

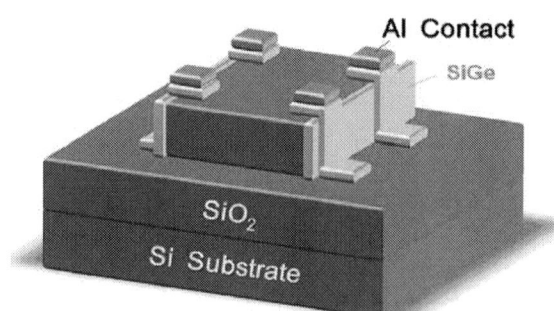

Figure 1 Schematic representation of SiGe nanowire structure by the side-wall spacer technique.

carriers, reducing the conductivity of the p-type nano-wire. Next, bis (3-sulfo-N-hydroxysuccinimide ester) sodium salt (BS3) was used as a linker between APTMS and IgG antibody. BS3 was prone to becoming negatively charged, increasing the conductivity of the p-type nanowire. This is because of the accumulation of holes on the surface of the nano-wire.

A Hewlett Packard HP 4156A instrument was used to measure the electric characteristics of the nanowire sensor. Drain voltage (VD) was varied from -10 to 10V in step of 500 mV, and the back gate voltage was 0 V. The electric characteristics were measured in every stage of the surface modification, and the average conductance was then determined from the ID-VD characteristics with VD = 3~6 V. The sensitivity (S) of a nanowire-based sensor is defined as the ratio of the magnitude of the change in conductance to the baseline conductance value：

$$ S = \frac{|G - G_0|}{G_0} = \frac{\Delta G}{G_0} \qquad (1) $$

, where G_0 is the conductance before capture of molecule; G is the conductance after molecule capture, and ΔG is the difference between G and G_0.

III. RESULTS & DISCUSSION

Figure 2 presents the sensitivities of nanowires in different oxidation gases condition. Using N_2 gas ratio in the fabrication of the poly-silicon nanowire samples degrades their sensitivity, by reducing the surface-to-volume ratio of the nanowire below that obtained using pure O_2 gas (N =0%). Therefore, the sensitivity was degraded when the proportion of O_2 gas ratio was reduced and that of N_2 increased. SiGe nanowire samples yield a peak value of sensitivity when 13% of N_2 is added to O_2 gas, because doing so increased the density of traps at the SiO_2/SiGe interfaces by approximately about 10^{12} cm^{-2} after oxidation. Accordingly, diluting the O_2 gas can help to reduce the formation of interfaces on the surface of nanowire in the Ge condensation process. However, increasing the proportion of N_2 ratio in all instances degraded the sensitivity as it did for poly-silicon nanowire samples. Figure 3 plots the sensitivities of the nanowires that were oxidized fir various times. α-Si=200Å and stacked nanowires with various fraction of Ge were utilized. The oxidation atmosphere was 13% gaseous N_2 with the O_2 gas. The maximum sensitivity value was obtained at 3 min. of oxidation; beyond that period, the sensitivity degraded. We suggested that this phenomenon was caused by competition between the piling-up of Ge with the diffusion of Ge in the Ge condensation process, because Ge did not pile up at the surface but also diffused into the buried oxide layer. The O atoms did not easily pass through the thick_top oxide layer when oxidation time increased. Hence, the Ge rejection rate and pile-up rate were reduced and the diffusion rate started to dominate the Ge distribution at particular oxidation time. The sensitivity of SiGe increased when the fraction of Ge without oxidation. This result is consistent with our previous study [8]. However, the sensitivity did not increase with the Ge fraction

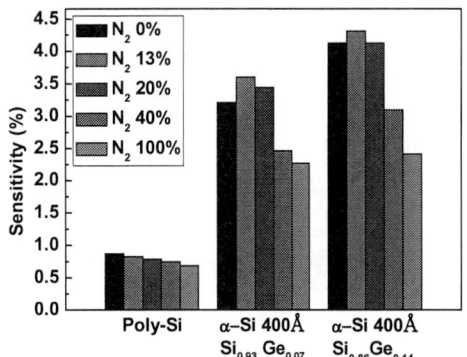

Figure 2 The sensitivity of nanowire samples in different oxidation gas condition

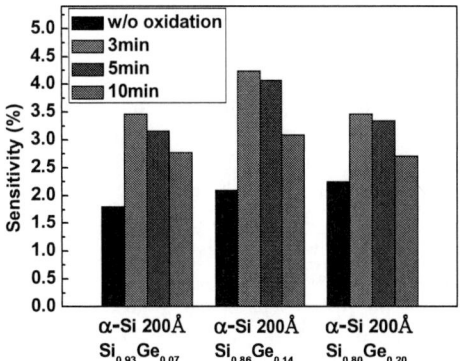

Figure 3 The sensitivity of nanowire in different oxidation time.

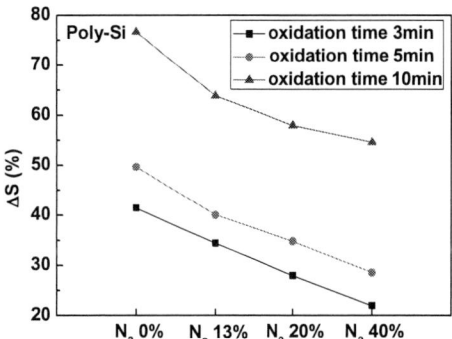

Figure 4 The sensitivity of poly-silicon nanowire structure with different oxidation time and ambient.

following oxidation. This finding may result from the increasing of the interface state with the fraction of Ge in the nanowire in the oxidation process. To verify the oxidation rate in these process split. The poly-Si nanowire was also fabricated as a control. Figure 4 displays the sensitivities of the poly-silicon nanowire structures that were oxidized for various times under various ambient conditions. The ΔS term as shown in Eq. (2) is the percentage of change on the sensitivity where S_0 sensitivity without treatment.

$$\Delta S = \frac{S - S_0}{S_0} \times 100 \% \qquad (2)$$

An oxidation time of 10 min yields a higher the percentage of change in the sensitivity of poly-siliocon nanowire than oxidation time of 3min or 5min, because of the former was oxidized at a longer oxidation time, yielding a high surface-to-volume ratio of poly-silicon. The percentage of change in the sensitivity is also degraded as the proportion of N_2 gas into the O_2 gas reduces the rate of oxidation to cause a lowe surface-to-volume ratio of poly-silicon. Reducing the rate of oxidation in Ge condensation suppresses the formation of surface defects.

Figure 5 plots the percentage of change in the sensitivity of $Si_{0.93}Ge_{0.07}$ obtained after three oxidation process in different oxidation gases. The peak of the percentage of change in the sensitivity y was obtained in 13% N_2 condition for all periods of oxidation. There is a degradation of the percentage of change in the sensitivity of the SiGe nanowire was observed that differs from that of poly-silicon nanowire. Extending the oxidation time reduces the sensitivity of the SiGe nanoweire. This trend is consistent with the results in Fig. 3. Increasing the proportion of N_2 above 20% reduces the sensitivity of SiGe nanowire. At this time, the Ge diffusion effect dominated the Ge distribution and Ge did not pile-up at the surface of the SiGe nanowires. Figure 6 plots the percentage of change in the sensitivity of $Si_{0.86}Ge_{0.14}$ in different oxidation gases for three oxidation times. The peak of the percentage of change in the sensitivity was obtained when 13% N_2 was used to dilute O_2 gas and a longer oxidation time was associated with lower sensitivity. Comparing Figs. 5 and 6, indicate that nanowire with 14% Ge was more sensitive than the nanowire with 7% Ge following a short period of oxidation. A large gap is observed between the sensitivity at 5 min and that at 10 min in figure 6. We suggested that there the sensitivity does not differ between the 7% Ge and the 14% Ge nanowire after 10 min of oxidation because Ge diffusion dominated. The condensation process on the sample that contained a large fraction of Ge clearly caused the oiling-up of Ge at the surface in a short oxidation period. A better understanding of Ge condensation will help to optimize the process and improve the sensitivity of the biosensor.

CONCLUSION

This investigation examines the effect of oxidation gas and the SiGe/α-Si stacked structure on the movement of Ge in the structure. An SiGe nanowire has maximum sensitivity when an $Si_{1-x}Ge_x$ layer that contains 14% Ge is stacked on 200 Å-thick layer of α-Si and when the nanowire is treated for 3 min with O_2 gas that is diluted by adding 13% N_2 gas. Therefore, this study improves our understanding of the mechanisms by which oxidation time, oxidation ambient and

the Ge fraction of $Si_{1-x}Ge_x$ affect the sensitivity of SiGe based nanowire.

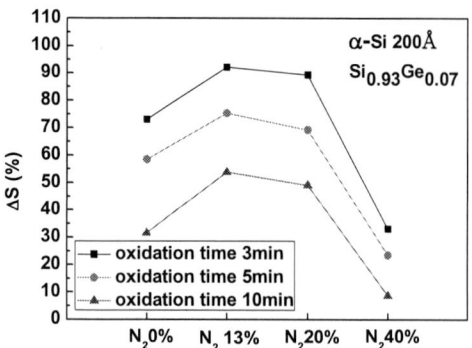

Figure 5 The percentage of change in the sensitivity of $Si_{0.93}Ge_{0.07}$ in different oxidation gas with three oxidation times.

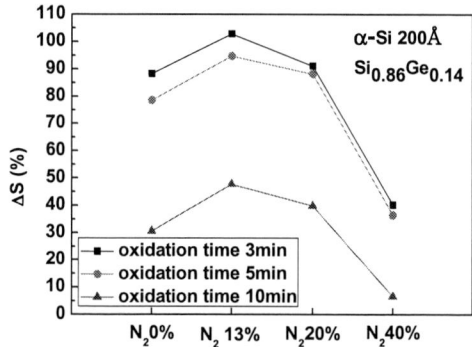

Figure 6 The percentage of change in the sensitivity of $Si_{0.86}Ge_{0.14}$ in different oxidation gas with three oxidation times.

ACKNOWLEDGMENT

The authors would like to thank the staff of the National Nano Device Laboratory for their technical help. They also acknowledge the financial support of the National Science Council (NSC) under Contract Nos. NSC 98-2221-E-009-174 MY3.

REFERENCES

[1] Y. Cui, Q. Wei, H. Park and Charles M. Lieber , Science, Vol. 293,No. 5533,pp. 1289-1292, August 17, 2001.

[2] G.Brambilla, F. Xu, and X. Feng, Electronic Letters, Vol 42, 8, 2006.

[3] P. R. Nair, and M. A. Alam, Applied Physics Letters, Vol. 88, No. 23, 2006.

[4] Duan, C. Niu, V. Sahi, J.Chen, J. Parce, and S. Empedocles, Nature, 425, pp. 274-278, 2003

[5] A.M. Morales and C.M. Lieber, Science 279, 208,1998.

[6] Zhaohui Zhong, Deli Wang, Yi Cui, Marc W. Bockrath, Charles M. Lieber, Science Vol. 302. no. 5649, pp. 1377 – 1379, 2003.

[7] G. Zheng, F. Patolsky, Y. Cui, W. U. Wang, C. M. Lieber, Nature Biotech., 2005

[8] K.M Chang, et al., 2nd IEEE International Nanoelectronics Conference,INEC., pp. 315-319,2008

[9] S.Balakumar, et al., Applied Physics Letters, 90, 03211, 2007

[10] F.K LeGoues, R. Rosenberg, T. Nguyen, F. Himpsel, and B.S. Meyerson, J.Appl.Phys, 65(4), 15 February, 1989

[11] N.Sugiyama, T.Tezuka, T. Mizuno, and M. Suzuki, et al. J.Appl.Phys, vol. 95, No.8, 2004, pp. 4007-4011

[12] K.M Chang, et al., International Conference on Biomedical Electronics and Devices,BIODEVICES, 2011

[13] S.Balakumar, et al., Proceeding of 13th IPFA,2006, pp. 301-305

A Capacitive Readout Circuit with DC Sensing Method for Micromachined Gyroscopes

Kailong Zhou, Lina Sun, Fei Ge, Zhenchuan Yang* and Guizhen Yan

National Key Laboratory of Science and Technology on Micro/Nano Fabrication

Institute of Microelectronics, Peking University, Beijing 100871 China

Email: z.yang@pku.edu.cn

Abstract— **In this work, a capacitive readout circuit for micromachined capacitive gyroscope is presented. Compared with the commonly used AC sensing which need a high frequency carrier to modulate the signal into high frequencies, the DC sensing only need a high DC voltage applied on the proof mass of the gyroscope, therefore simplify the circuit structure. In addition, the proposed circuit can partially suppress the mechanical coupling from the drive mode to the sense mode of the gyroscope, automatically cancel the DC offset due to capacitance mismatch, lower the input referred noise by two optimized stages of amplifier, and maximize the circuit signal-to-noise (SNR). The circuit is designed and fabricated by Charter 0.35um mixed-signal CMOS process, and the chip size is 1.2mm×1.2mm. The circuit was successfully used to test a tuning fork gyroscope, and a 10 dB reduction of the mechanical coupling has been demonstrated.**

Keywords-capacitive;DC sensing; mechanical coupling

I. INTRODUCTION

Owing to the advantages of small size, low cost, low-power consumption, etc, capacitive accelerometers and gyroscopes have drawn much attention [1, 2]. The signal processing circuits for capacitive MEMS accelerometers and gyroscopes have made great progress for the last decades. Most sensing circuits use AC sensing methods, such as AC-bridge with voltage amplifier, transimpedance amplifier and switched-capacitor circuit. In these AC sensing methods, the chopper stabilization (CHS) or correlated double sampling (CDS) techniques are generally used to reduce the 1/f noise and offset, which need complicated circuit modules such as synchronous demodulators and low pass filters. However, modulating the signal to a much higher frequency is not the only way to suppress 1/f noise. By carefully partition and optimization of the signal amplification path, the circuit can be low noise, low offset with a simple configuration. In this work, a DC sensing method is proposed.

II. WORKING PRINCIPLE

The basic schematic of the DC sensing is shown in Fig. 1. A high DC voltage set to 10 V is directly applied on the proof mass of the gyroscope, and the variation of the two differential sensing capacitances is converted to the voltage by a charge sensitive amplifier (CSA).

The schematic of the CV converting circuit is depicted in Fig. 2. It is mainly composed of an input buffer, a two stage

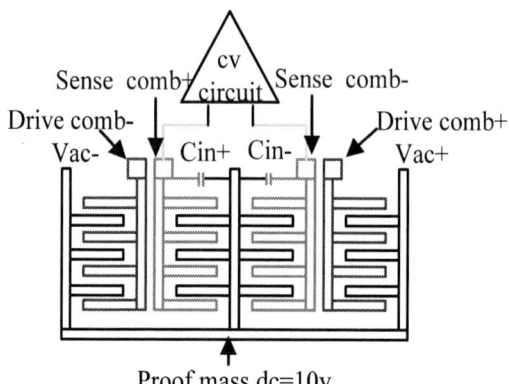

Fig. 1. Schematic diagram of DC sensing scheme.

amplifier, an out buffer, a feedback capacitor and a coupling cancellation capacitor. The input buffer is optimized to minimize the input referred noise by matching the capacitance between the gyroscope and the parasitic capacitor of the input PMOS transistors[1]. The capacitance matching equation is:

$$C_{gs} + C_{gd} = \eta(2C_s + C_p) \quad (1/3 < \eta < 1) \qquad (1)$$

where C_s is the sensing capacitance, C_p is the interconnect parasitic capacitance, C_{gs} and C_{gd} are the gate-to-source and gate-to-drain capacitances of the input MOS transistor.

According to capacitance matching equation, the gate

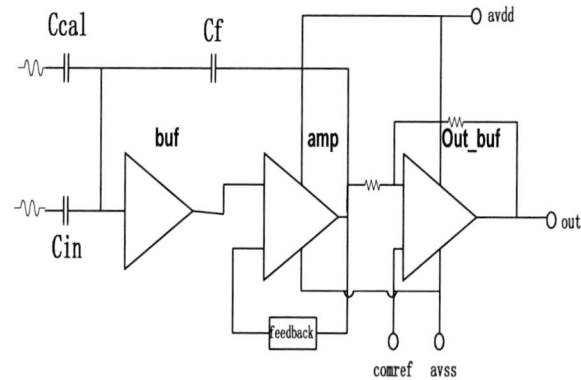

Fig. 2. Schematic of the circuit module.

capacitance must be equal or smaller than the combined sensing and interconnect capacitance to achieve maximum SNR and minimum input-referred noise floor. So we set small size of the input transistors to meet the equation, and a source follower that provides only a small gain is used. The second stage is a two stage high gain amplifier which offers the main gain.

A feedback loop is used to minimize the output offset due to MEMS fabrication mismatch and previous buffer mismatch [3]. Due to DC offset, the output may reach 0V or 5V. Even if the output is 3V, the output swing is restricted to only 2V. So the output signal is forced to be 2.5V through the feedback loop and the output swing is maximized [4].

A mechanical coupling cancellation technique is also proposed and depicted in Fig. 3. The mechanical coupling or

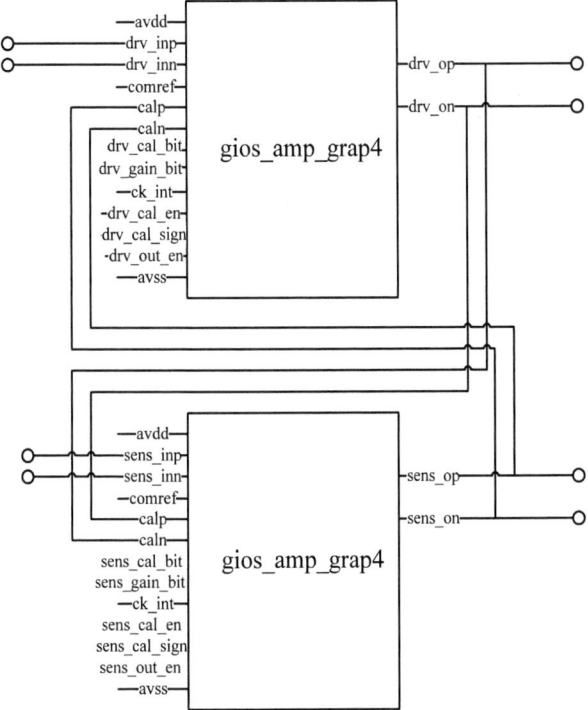

Fig. 3. Circuit schematic for the mechanical coupling cancellation.

the quadrature error from the drive mode to the sense mode of the gyroscope is a major source to degrade the performance of a gyroscope. Generally the coupling is larger than the sense signal, so it is necessary to suppress the coupling. As the quadrature error is always proportional to the amplitude of the driving mode, a portion of the driving output is introduced to the sensing circuit of the sensing mode, which is used to cancel out the sensing output caused by the quadrature error and the portion can be controlled through programmable capacitors.

Either positive driving output signal or negative driving output signal used depends on which one has contrast phase

with the coupling signal. Through changing the value of sen_cal_sign signal(as shown in Fig. 3), we can choose the coupling signal plus positive driving output signal or negative driving output signal. If positive driving output signal has 180 phase difference with the coupling signal, it will be chosen, or negative driving output signal is chosen.

The coupling cancellation capacitor consists of a binary bank of capacitors. In this work, a 7-bit capacitor bank is selected. The most significant bit capacitor is composed of 64 unit capacitors that is 12.5fF. The last significant capacitor is equal to the unit capacitor. There are 7 paths and every path is controlled by a transmission gate to decide whether the corresponding capacitor is connected in the circuit or not. Every transmission gate is controlled by a digital input bit of sense_cal_bit signal (as shown in Fig.3). So we could change the value of the coupling cancellation capacitor from 12.5fF to 1.6pF through setting the sense_cal_bit signal. With the change of the coupling cancellation capacitor, the proportion of driving output signal that is used to cancel the coupling signal is changed.

The layout of the designed circuit is shown in Fig. 4. As shown in Fig. 3, there are two paths and one is used to process the driving input signal and the other is used to process the sensing input signal. Both of them are designed to be symmetrical completely in the layout.

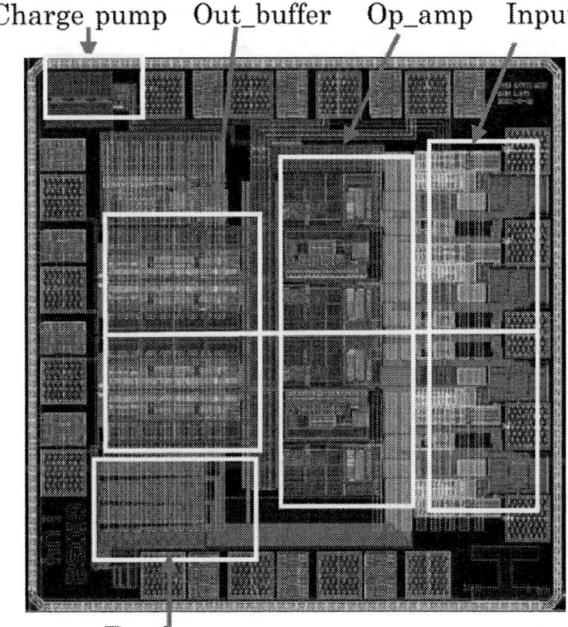

Fig. 4. Layout of the interface chip.

978-1-4673-1122-9/12 $31.00 © 2012 IEEE

III. Function Test

After tape out, the chip is connected with a tuning fork gyroscope on a PCB [5]. The PCB (as shown in Fig.5) is a two-layer board. The gyroscope is fixed in front of the board and the chip is fixed in back of the board. The red wire is used to connect the ports of the gyroscope. The blue wire is used to connect the ports of the chip to the pads of the PCB. The

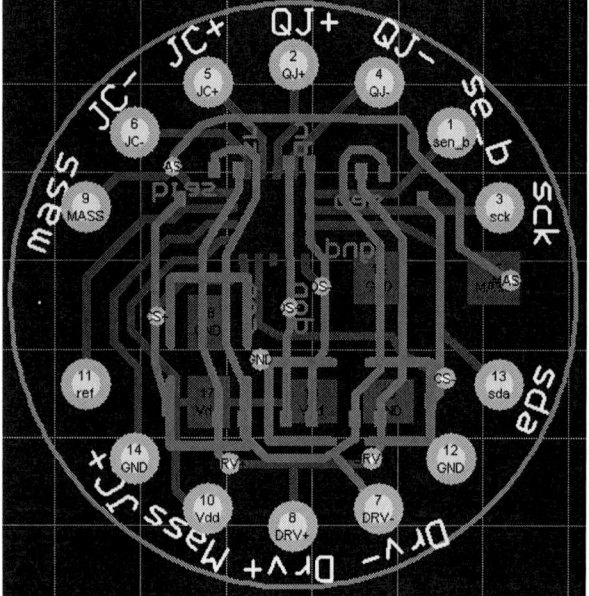

Fig. 5. The PCB for the chip and the gyroscope.

excitation voltage signal of the drive mode is added through DRV+ and DRV- pads. The driving output voltage signal is measured through QJ+ and QJ- pads, and the sensing output voltage signal is measured through JC+ and JC- pads. The DC voltage is supplied by a charge pump and applied to the mass. Through the mass port we could measure whether the DC voltage meets our requirement or not. The feature of the board is that the driving input paths, the driving output paths and the sensing output paths are designed as symmetrical as possible, also these paths are isolated by the ground. So the electrical coupling from the driving input signal to the driving output signal or the sensing output signal are largely suppressed.

The amplitude frequency characteristic of the drive mode of the gyroscope was measured and shown in Fig. 6, which indicates the proposed DC sensing circuit can be used as readout circuit for capacitive gyroscopes. The mechanical coupling cancellation is also verified. With the value of cancelling coupling capacitor increasing, the sense coupling signal is reduced, which is -39.8 dB when the capacitor is 75fF. While the capacitor gets larger than 75fF, the sense coupling signal begins to rise as shown in Fig.7. So the measured coupling signal is reduced from -29.7 dB to -39.8 dB, about 10 dB suppression on the mechanical coupling achieved in Fig.8.

IV. Conclusion

In conclusion, it has been shown that DC sensing circuit can simplify the overall complexity of the signal processing circuits. What is more, making use of a portion of the driving output to cancel out the sensing coupling signal could suppress the mechanical coupling and improve the performance of a gyroscope greatly.

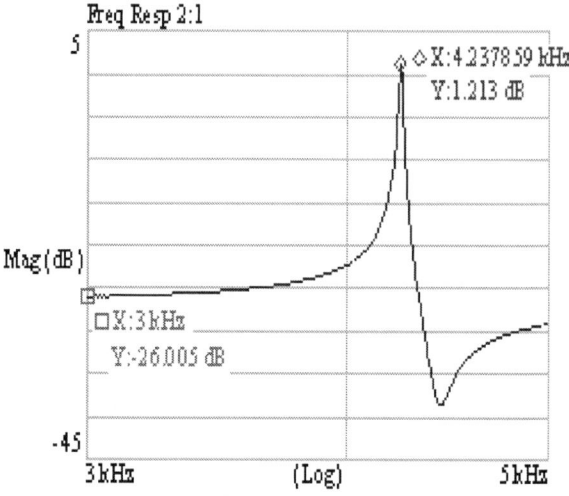

Fig. 6. Amplitude-frequency response of the drive mode of the gyroscope.

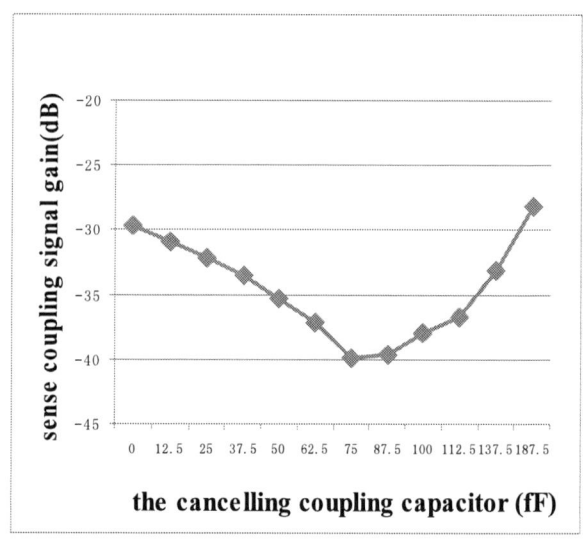

Fig. 7. Sense coupling signal gain depending on the value of the cancelling coupling capacitor.

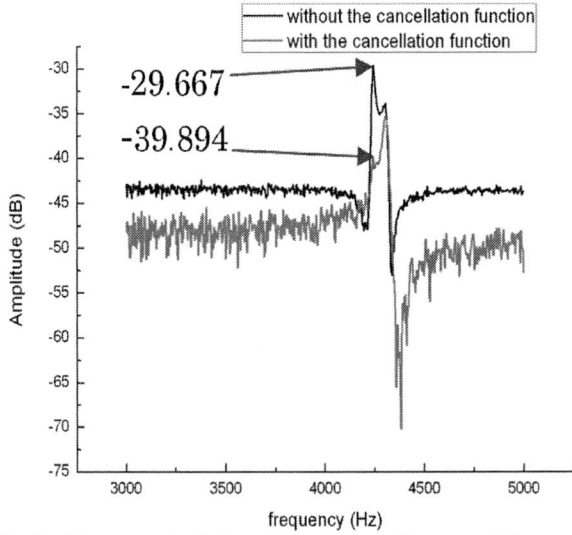

Fig. 8. Measure mechanical coupling signal at drive resonant frequency (4.2379kHz) with and without the cancellation function applied.

ACKNOWLEDGEMENT

The authors are grateful to the technical staffs of National Key Laboratory of Science and Technology on Micro/Nano Fabrication for their supports on chip test.

REFERENCES

[1] J. Wu, G.K. Fedder, L.R. Carley, "A low-noise low-offset capacitive sensing amplifier for a 50-μg/Hz$^{1/2}$ monolithic CMOS MEMS accelerometer", IEEE J. of Solid-State Circuits, vol. 39, May 2004, pp. 722-730.

[2] Yazdi, N. Kulah, H. Najafi, K., "Precision readout circuits for capacitive microaccelerometers", IEEE Sensors, 2004, pp. 28 - 31 vol.1

[3] H. Cheung, K. Cheung, J. Lau, "A Low Power Monolithic AGC with Automatic DC Offset Cancellation for Direct Conversion Hybrid CDMA Transceiver Used in Telemetering," IEEE ISCAS Conference, pp.390-393, Oct. 2001.

[4] M. Elmala, B. Carlton, R. Bishop, K. Soumyanath, "A 1.4V, 13.5mW, 10/100MHz 6th Order Elliptic Filter/VGA with DC-Offset Correction in 90nm CMOS," RFIC Symp. Dig. Papers, pp. 189-192, June 2005.

[5] J. Chae, H. Kulah, and K. Najafi, "An in-plane high-sensitivity, Lownoise micro-G silicon accelerometer," in Proc. 16th IEEE Int. Conf.Micro ElectroMechanical Systems (MEMS 2003), Kyoto, Japan, 2003, pp. 466–469.

The Optimal Vibrational Shear Stress for Bovine Endothelial Cell Proliferation

Ching-Wen Li[1], Jau-Liang Chen[1], Chia-Ching Wu[3], Gou-Jen Wang[1, 2*]

[1]Department of Mechanical Engineering, National Chung-Hsing University, Taichung, Taiwan
[2]Graduate Institute of Biomedical Engineering, National Chung-Hsing University, Taichung, Taiwan
[3]Department of Cell Biology and Anatomy, National Cheng Kung University, Taiwan
gjwang@dragon.nchu.edu.tw

Abstract—In this study, a novel method for quantitative analysis of the optimal shear stress for better cell growth using a micro-positioning piezoelectric lead zirconate titanate (PZT) stage is proposed. Compared with the conventional fluidic sheer stress, a micro- positioning PZT stage not only provides cultured cells with laminar flow, but also is able to generate large shear stress gradient by precise reciprocating motions. Bovine endothelial cells (BEC) are than cultured on different scaffolds for investigation of cell proliferation under different vibrational excitations by a PZT stage. The fluorescence labeling is adopted to estimate the adhesive area of a cultured cell. It is observed that cells cultured on different artificial scaffolds adjusted their adhesion area to respond to the shear stress induced stimulus. Optimal growth curves for BECs on different scaffolds are drawn. The optimal shear stress for the proliferation of BECs on different scaffolds is found to be closely identical. It is suggested that a micro-positioning PZT stage may be a more cost and time effective solution than the nanostructured scaffold approach for the enhancement of cell growth.

Keywords-optimal proliferation shear stress; micro vibration; bovine endothelial cells

I. Introduction

Recent advances in biomaterials, stem cells, growth and differentiation factors, and biomimetic environments have enabled the fabrication of artificial tissues in the laboratory [1]. The process involves the implantation or culture of cells in a biocompatible scaffold that is capable of supporting three-dimensional tissue formation. This scaffold serves a number of purposes, such as the foundation for cell attachment and migration, for the exchange of nutrients, and for delivering and retaining the cells and biochemical factors. The proliferation and characteristics of cells in the scaffold are influenced by the material of the scaffold, the growth factors, and the culture environment. To facilitate effective growth of cells on a scaffold, several approaches have been studied and reported. The existing approaches can be briefly categorized into biochemical and physical treatments.

For those approaches of biochemical treatment, modification of surface functional group, coating of growth factor, and using highly biocompatible materials such as collagen or chitonsan are the general methods to enhance the adhesion and growth of cells on a scaffold [2-4]. However, the probability of cell variation will also be increased when more biological components and/or chemical substances are involved in a scaffold. Hence, physical approaches such as

morphology modification of scaffold surface and stimulus of mechanical force are increasingly being studied.

A mechanical force applied to a cell can be classified into normal force, shear stress, tension, and compression force. To investigate the influence of shear stress on cell growth, usually a stimulus of applied force is continuingly applied so the biochemical response, the distribution and delivery of the applied force in the cell skeleton, and the deformation of cell can be observed. Micropipette aspiration, optical tweezers/laser traps, magnetic twisting cytometry, AFM indentation, cytointer, and fluid shear flow are the commonly used approaches for shear stress implementation [5-8].

Since the purpose of a stimulus by an applied force is the enhancement of cell adhesion and proliferation, advancement of the mass transfer efficiency of cell's nutrient supply is the main stream of interesting. Hence shear stress is applied on a cell's surface by a fluid shear flow to enhance its metabolism. In general, the shear stress applied on a cell's surface due to fluid shear flow ranges from 10 to 80 dyne/cm^2 [9, 10]. Davies et al. [11] applied a 1-15 dyne/cm^2 steady shear stress and a 3-13 dyne/cm^2 intermittent shear stress that circulated every 15 min to BECs to provide a condition of shear stress approximating to the condition of regular pulse circulation. It was observed that the pinocytosis of BEC was significantly influenced by the alternating shear stress. Albuquerque et al. reported that human umbilicalvein endothelial cells (HUVEC) can effectively differentiate thus heal a wound when stimulated by a 12 dyn/cm^2 shear stress for 18 hours followed by either a 3 dyn/cm^2 or 12 dyn/cm^2 shear stress [12]. Weber et al. used a 15 dyne/cm^2 unidirectional shear flow to investigate the influence of shear stress induced MiR-21 on cell apoptosis and eNOS activity [13]. Chen et al. applied a consecutively fluid shear flow to bone cells for 60 min and found that a 12 dyne/cm^2 shear stress could provide the best condition for cell proliferation. However, the proliferation of cell sharply degraded when the shear stress reached 14dyne/cm^2 [14]. Except the fluid shear flow, ultrasound and motor were employed to directly impose stimulus on cells. Pre et al. used eccentric motor to applied stimulus with frequency ranging 1-120Hz and amplitude of 11mm to SAOS-2 cells. Under this condition, cell differentiation was enhanced while cell proliferation was downgraded [15]. Rosenberg et al. applied horizontal vibrations with amplitude being 25±5μm and frequencies ranging from 20Hz to 60Hz to osteoblast. It was observed that osteoblast proliferated the most under a horizontal vibration having frequency of 20Hz and acceleration of 0.5 ± 0.1 m/sec^2. While a horizontal vibration with frequency being 20Hz and acceleration of 1.3 ± 0.1 m/sec^2 could provide the best efficiency of metabolism [16].

When a cell is subject to a shear stress induced stimulus, the receptors for shear stress perception on the surface of a cell membrane transform the stimulus into inner cell signal through cell skeleton. Hence the cell functions such as proliferation, apoptosis, migration, permeablility, and remodeling are influenced, especially the cell proliferation [17]. Early research reported that fluid shear flow induced shear stress could downgrade the proliferation rate of endothelia cell. The DNA synthesis rate of endothelia cells under large shear stress of a laminar flow is less than that of the static culture cells [18]. Recent reports indicated that a laminar flow can reduce the cell numbers going into the cell cycle because that most of the cells stay in G0 and G1 periods [19, 20]. On the contrary, a turbulent flow can create a lower shear stress but a higher gradient of shear stress at the boundary layer of a cell. Those DNAs close to the back and forth region of flow have higher synthesis rate. Hence cells possess a higher proliferation rate [17].

Under normal physiological conditions, different shear stresses will induce expressions of certain proteins such that cell apoptosis can be restrained. In responding to the stimulus due to shear stress, cells self-regulate their apoptosis by means of different pathways. Dimmeler et al. reported that the shear stress induced protein expressions such as TNF-α, exogenous oxygen radicals, and oxidized LDL can self-regulate apoptosis of endothelia cell by means of restraining interleukin 1β-converting enzyme and caspase-3. It was also discovered that shear stress can activate the survival pathway of PI3K-Akt, hence NO and the mechanism of phosphorylation of Akt can be restrained and cell apoptosis can be suppressed [21-23]. In addition, shear stress can also be used to enhance the amount of eNOS and Cu/Zn SOD, thus cell apoptosis is repressed [24].

Compared with those researches on molecule mechanism of cell growth and cell survival, relatively rarer reports are on molecule mechanism of cell migration. The existing reports basically are based on static culture. It is thus desired to investigate the molecule mechanism of cell migration. Sprague et al. and Wu et al. used microchannels and blood vessel implants respectively to investigate the influence of fluid field on cell migration both *in vitro* and *in vivo*. Experimental results indicated that laminar flow induced shear stress can enhance wound-healing capability of cell migration [25, 26]. Else researches revealed that cell use Rho family small GTPase to moderate its pseudopod, focal adhesion, and actin cytoskeleton such that cell can migrate and align along the direction of shear stress [9, 27, 28].

As mentioned above, an appropriate shear stress induced stimulus can enhance proliferation and adhesion of cells on a biomaterial. How to precisely control the applied shear stress such that the proliferation and adhesion of cells can be achieved becomes an imperative issue in tissue engineering. In this study, a novel method for quantitative analysis of the optimal shear stress for better cell growth using a micro-positioning piezoelectric lead zirconate titanate (PZT) stage is proposed. Compared with the conventional fluidic sheer stress, a micro- positioning PZT stage not only provides cultured cells with laminar flow, but also is able to generate large shear stress gradient by precise reciprocating motions. For a start, we cultured bovine endothelial cells (BEC) on different scaffolds for further investigations of cell proliferation under different vibrational excitations by a PZT stage. The fluorescence labeling is than adopted to estimate the adhesive area of a cultured cell. Optimal growth curves for a certain cell cultured on different scaffolds are then drawn. Resultantly optimal shear stress for the proliferation of a certain cell can be estimated. It is hoped that a micro-positioning PZT stage may be a more cost and time effective solution than the nanostructured scaffold approach for the enhancement of cell growth.

II. Material and Method

A. Fabrication of PLGA and PLA Scaffolds

The 85/15 Poly (lactide-co-glycolide) (PLGA) (IV：1.6-1.99 dl/g, Bio Invigor Corp., Taiwan) solution was prepared by dissolving in acetone with 1:4 w/w ratio, then stirring the mixture with a magneto agitator at room temperature for 1-2 hr. The Poly (DL-lactide) (PLA) (IV : 0.63 dl/g,)solution was prepared in a similar process with a w/w ratio of 1:6. Next, an ultrasonic cleaner was employed to clean a ϕ15mm circle cover glass using ethanol, acetone and DI water in turn for 10 min. 1ml of the polymer solution was dropped onto the clean cover glass. The sample was then placed at room overnight to enable complete evaporation of the acetone. Finally, the sample was put into an oven at 40°C for further use.

B. Culturing of Bocine Endothelial Cells(BEC)

BECs were maintained in DMEM supplemented with 10% (v/v) FBS, 100 U/ml penicillin. The cells were maintained in T-75 flasks in an incubator at 37°C in the presence of 5% CO_2. The medium was renewed every 2 days. The cells were expanded to 80–90% confluence prior to trypsinization for passage or scaffold seeding.

C. Micro Vibrational Shear Stress Stimulus

The piezoelectric micro vibrational stage used for providing micro vibrational stimuli of different frequencies is shown in Figure 1. The horizontal vibration is driven by a piezoelectric actuator with a maximum frequency of up to 100 kHz and maximum amplitude of 135.83 μm.

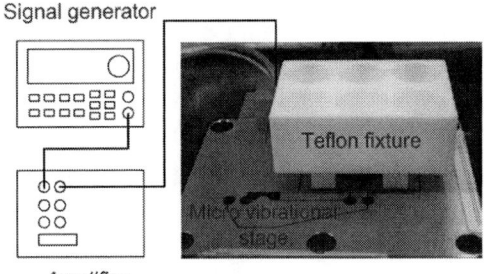

Fig. 1. Piezoelectric micro vibrational stage

In experiment process, the PLA and PLGA scaffolds were sterilized with 70% ethanol followed by washing with sterilized PBS, Then, scaffolds were placed into 24-well plate for static cultured. BECs were trypsinized and diluted to obtain a concentration of 2×10^4 cells/ml. The suspensions of dilute cell were seeded on scaffolds and cultured in static for 24H. A TCPS 24-well plate was used as the control. After 24H static cultured, scaffold was placed into a Teflon fixture, and then incubated in the presence of mechanical vibration at various frequencies of 0.5Hz, 1Hz, 1.5Hz, 2Hz and 2.5Hz, and amplitude of 20 μm. The stage was vibrated for 30 min, followed by a resting period of 30 min, for a total of 3 times. The numbers of cells attached onto the scaffolds or TCPS plates were counted by the trypan blue exclusion method after 48H incubation.

D. Measurement of BECs Thrown off from the Scaffold

After 24H static cultured, scaffold was placed upside down onto the 15ml centrifugal tube top and rotating at various speeds of 1300rpm, 1600rpm and 1900rpm. Next, the cells which thrown off from scaffold through centrifugal force were collected and counted by the trypan blue exclusion method.

E. Hydrophilicity Measurement of the Scafold

The surface hydrophilicity was the important factor for cell attached material [10]. The hydrophilicity of TCPS, PLA scaffolds and PLGA scaffolds were measured in relation to the contact angle using Contact Angle Measurement System (FTA2000, First Ten Angstro).

F. Statisitical Analysis

All the data obtained were expressed as the means with standard deviation (mean ± SD) and analyzed using two tailed t-test (SAS Institute Inc.). $p < 0.05$ denotes significant difference.

III. RESULTS AND DISCUSSIONS

A. The Influence of Scaffolds Hydrophilicity on BECs Growth

Cells adhesion and future proliferation are influenced by the hydrophility of a scaffold [3]. Figure 2 shows the hydrophilicity of different scaffolds in terms of the contact angle. The PLGA scaffolds possess a notable hydrophobic property, compared with commercial TCPS plates and PLA scaffolds. The results lead to a poor proliferation of BECs on

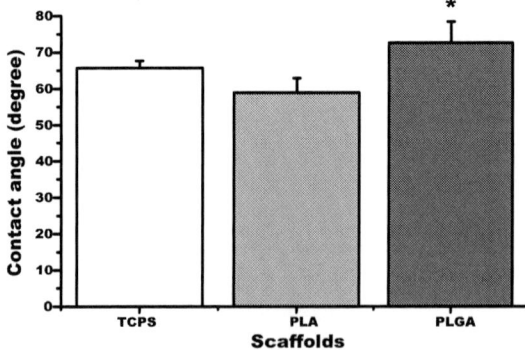

Fig. 2. The contact angle measurement results.
* :P<0.05.

PLGA scaffolds (Figure 3).

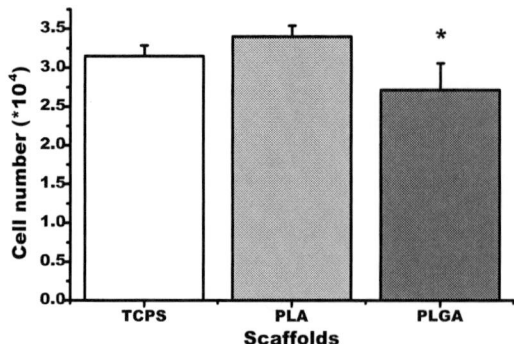

Fig. 3. The proliferation of BECs culture on different materials for 24H.
* :P<0.05.

To further demonstrate that the cells have a lower affinity on PLGA scaffolds, the amount of BECs thrown off from different scaffold under various rotating speeds are counted. The results are shown in Figure 4. The results indicate that a higher number of BECs were thrown off from PLGA scaffolds regardless of the rotating speed.

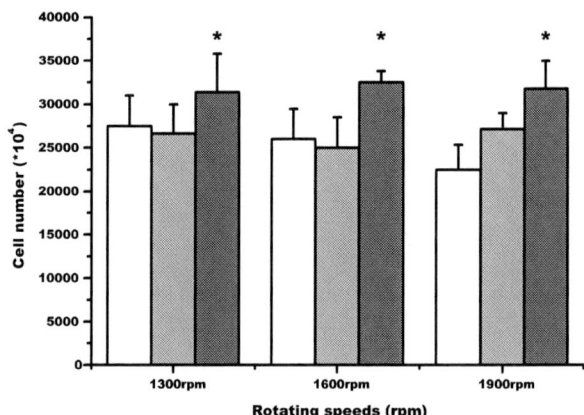

Fig .4.The amounts of BECs thrown off from different scaffolds under various rotating speeds. * : P<0.05.

These measurements confirm that hydrophilic scaffolds

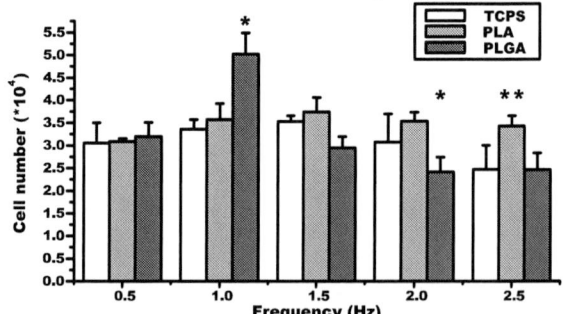

Fig. 5.The cultural results of BEC on different materials under various of vibrational frequencies.
* : PLGA vs.TCPS and PLA P<0.05.
** : PLA vs.TCPS and PLGA P<0.05

have a better affinity to BECs. The BECs cultured on hydrophilic scaffolds have better adhesion force than those cultured on hydrophobic scaffolds.

B. The Reciprocating Shear Stress Effects for BECs Growth

Figure 5 illustrates the cultural results of BEC on different materials under various of vibrational frequencies. The optimal frequencies for BEC proliferation on TCPS and PLA are about at 1.5Hz and are almost identical, while the best growth frequency of PLGA is at 1Hz. It can be found that after optimal proliferation frequency, the proliferation of BECs on PLGA scaffolds sharply decreased at 1.5Hz while the proliferation of cells on TCPSs gradually reduced. However, the proliferations of BECs on PLA scaffolds were almost identical. This phenomenon can be attributed to the hydrophilicity of scaffolds. As PLGA scaffolds have a poor affinity to BECs, the vibration at 1Hz is enough to generate a critical stimulus for cell proliferation. Since TCPSs have a better hydrophilic property than PLGAs, the frequency of the critical stimulus is shifted to 1.5 Hz. For those BECs cultured on PLA scaffolds, they are subject to a relatively better adhesion force than those cultured on the other two scaffolds; therefore, the frequency of the critical stimulus is also at 1.5 Hz but with better cell proliferation than TCPSs.

As mentioned above, the shear stress plays an important role on cell proliferation. It is interesting to investigate the shear stress acting on the cultured cells by the scaffold. Since the cells have various sizes, it is more feasible to replace the shear stress with a normalized shear stress. The normalized shear stress acting on a cell can be calculated according to Equation (1).

$$\tau = a/A \tag{1}$$

where τ is the shear stress; a is the horizontal reciprocating acceleration of the vibration; A is the adhesion area of a cell on the scaffold. Figure 6 shows the normalized shear stresses regarding to various vibrational frequencies. The normalized shear stress provided by PLGA scaffolds sharply rises at 1.5Hz while that provided by the TCPS raises at 2Hz. It is also observed that the normalized shear stress on PLA scaffolds have a larger but smoother increase at 2 Hz.

Figure 7 displays the optimal growth curve for BECs on difference types of scaffold. The optimal shear stress for the proliferation of BECs on different scaffolds is found to be closely identical. The results also reveal that the optimal proliferation shear stress for BECs ranges from $0.85(\text{um} \times \text{s}^2)^{-1}$ to $13.355(\text{um} \times \text{s}^2)^{-1}$ and is irrelevant to the material of the scaffold.

IV. CONCLUSIONS

Both the surface characteristics of the scaffold and applied stimuli can influence the growth of cultured cells. In this study, a simple method based on the vibrational stimuli for the enhancement of cell proliferation and differentiation on scaffolds is proposed. A piezoelectric micro vibrational stage is used to provide micro vibrational stimuli of different

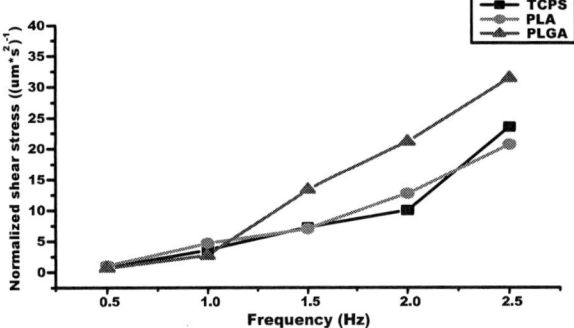

Fig. 6. Normalized shear stresses with respect to various vibrational frequencies..

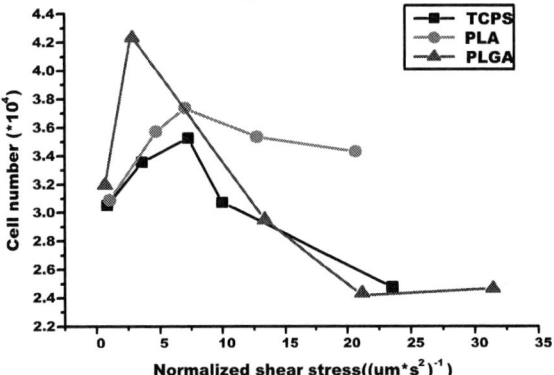

Fig. 7. The optimal growth curve for BECs

frequencies to generate various sheer stresses. TCPS（tissue culture polystyrene）, biodegradable poly lactide-co-glycolide (PLGA) scaffolds, and poly lactide acid (PLA) scaffolds are used for the cell culture. It is observed that cells cultured on different materials adjusted their adhesion area to respond to the shear stress induced stimulus. It can also be inferred that the vibration induced sheer stress can effectively enhance BEC growth as long as the corresponding normalized sheer force is within a specific window. It is also suggested that micro vibration stimulus may be a more cost and time effective solution than the nanostructured scaffold approach for the enhancement of BEC growth.

References

[1] R. Langer, and J.P. Vacanti, "Tissue engineering," *Science*, vol. **260**, pp. 920-926, 1993.

[2] E.S. Thian, Z. Ahmad, J. Huang, M.J. Edirisinghe, S.N. Jayasinghe, D.C. Ireland, R.A. Brooks, N. Rushton, W. Bonfield and S.M. Best, "The role of surface wettability and surface charge of electrosprayed nanoapatites on the behaviour of osteoblasts," *Acta Biomater*, vol. **6**(3), pp. 750-755, 2010.

[3] M.S. Ravenscroft-Chang, J.M. Stohlman, P. Molnar, A. Natarajan, H.E. Canavan, M. Teliska, M. Stancescu, V. Krauthamer and J. J. Hickman, "Altered calcium dynamics in cardiac cells grown on silane-modified surfaces," *Biomaterials*, vol. **31**(4), pp. 602-607, 2010.

[4] A. Phadke, C. Zhang, Y. Hwang, K. Vecchio, and S. Varghese, "Templated mineralization of synthetic hydrogels for bone-like composite materials: role of matrix hydrophobicity," *Biomacromolecules,* vol. **11**(8), pp. 2060-2068, 2010.

[5] C.T. Lim, E.H. Zhou, and S.T. Quek, "Mechanical models for living cells--a review," *J Biomech*, vol. **39**(2), pp. 195-216, 2006.

[6] D. Stamenovic, and D.E. Ingber, "Models of cytoskeletal mechanics of adherent cells," *Biomech Model Mechanobiol,* vol. **1**(1), pp. 95-108, 2002.

[7] D. Stamenović, B. Suki, B. Fabry, N. Wang, J. J. Fredberg and J.E. Buy, "Rheology of airway smooth muscle cells is associated with cytoskeletal contractile stress," *J Appl Physiol,* vol. **96**(5), pp. 1600-1605, 2004.

[8] G.N. Maksym, B. Fabry, J.P. Butler, D. Navajas, D.J. Tschumperlin, J.D. Laporte and J.J. Fredberg, "Mechanical properties of cultured human airway smooth muscle cells from 0.05 to 0.4 Hz." *J Appl Physiol,* vol. **89**(4), pp. 1619-32, 2000.

[9] P.F. Davies, and S.C. Tripathi, "Mechanical stress mechanisms and the cell. An endothelial paradigm," *Circ Res,* vol. **72**(2), pp. 239-245, 1993.

[10] C.J. Jen, S.J. Jhiang, and H.I. Chen, "Invited review: effects of flow on vascular endothelial intracellular calcium signaling of rat aortas ex vivo," *J Appl Physiol,* vol. **89**(4), pp. 1657-1662; discussion 1656, 2000.

[11] P.F. Davies, C.F. Dewey, S.R. Bussolari, E.J. Gordon, and M.A. Gimbrone, "Influence of hemodynamic forces on vascular endothelial function. In vitro studies of shear stress and pinocytosis in bovine aortic cells," *J Clin Invest,* vol. **73**(4), pp. 1121-1129, 1984.

[12] M.L. Albuquerque, C.M. Waters, U. Savla, H.W. Schnaper, A.S. Flozak, "Shear stress enhances human endothelial cell wound closure in vitro," *Am J Physiol Heart Circ Physiol,* vol. **279**(1), pp. H293-302, 2000.

[13] M. Weber, M.B. Baker, J.P. Moore and C.D. Searles, "MiR-21 is induced in endothelial cells by shear stress and modulates apoptosis and eNOS activity," *Biochem Biophys Res Commun,* vol. **393**(4), pp. 643-648, 2010.

[14] W. Chen, J. Qiao, S. Luo, S. Li, W. Tian and P. Li, "The effect of fluid shear stress on the proliferation of rat primary osteoblast-like cells in vitro," *Hua Xi Yi Ke Da Xue Xue Bao,* vol. **32**(2), pp. 232-234, 239, 2001.

[15] D. Pre, G. Ceccarelli, L. Benedetti, G. Magenes and M.G. De Angelis, "Effects of low-amplitude, high-frequency vibrations on proliferation and differentiation of SAOS-2 human osteogenic cell line," *Tissue Eng Part C Methods,* vol. **15**(4), pp. 669-679, 2009.

[16] N. Rosenberg, M. Levy, and M. Francis, "Experimental model for stimulation of cultured human osteoblast-like cells by high frequency vibration," *Cytotechnology,* vol. **39**(3), pp. 125-130, 2002.

[17] Y.S. Li, J.H. Haga, and S. Chien, "Molecular basis of the effects of shear stress on vascular endothelial cells," J *Biomech,* vol. **38**(10), pp. 1949-1971, 2005.

[18] M.J. Levesque, R.M. Nerem, and E.A. Sprague, "Vascular endothelial cell proliferation in culture and the influence of flow," *Biomaterials,* vol. **11**(9), pp. 702-707,1990.

[19] S. Akimoto, M. Mitsumata, T. Sasaguri, Y. Yoshida, "Laminar shear stress inhibits vascular endothelial cell proliferation by inducing cyclin-dependent kinase inhibitor p21(Sdi1/Cip1/Waf1)," *Circ Res,* vol. **86**(2), pp. 185-190, 2000.

[20] K. Lin, P.P. Hsu, B.P. Chen, S. Yuan, S. Usami, J.Y. Shyy, Y.S Li and S. Chien, "Molecular mechanism of endothelial growth arrest by laminar shear stress," *Proc Natl Acad Sci U S A,* vol. **97**(17), pp. 9385-9389, 2000.

[21] S. Dimmeler, J. Haendeler, M. Nehls, A.M. Zeiher, "Suppression of apoptosis by nitric oxide via inhibition of interleukin-1beta-converting enzyme (ICE)-like and cysteine protease protein (CPP)-32-like proteases," *J Exp Med,* vol. **185**(4), pp. 601-607, 1997.

[22] S. Dimmeler, B. Assmus, C. Hermann, J. Haendeler, A.M. Zeiher, "Fluid shear stress stimulates phosphorylation of Akt in human endothelial cells: involvement in suppression of apoptosis," *Circ Res,* vol. **83**(3), pp. 334- 341, 1998.

[23] S. Dimmeler, J. Haendeler, A. Sause, A.M. Zeiher, "Nitric oxide inhibits APO-1/Fas-mediated cell death," *Cell Growth Differ,* vol. **9**(5), pp. 415-422, 1998.

[24] S. Dimmeler and A.M. Zeiher, "Nitric oxide-an endothelial cell survival factor," *Cell Death Differ,* vol. **6**(10), pp. 964-968, 1999.

[25] E.A. Sprague, J. Luo and J.C. Palmaz, "Human aortic endothelial cell migration onto stent surfaces under static and flow conditions," *J Vasc Interv Radiol,* vol. **8**(1 Pt 1), pp. 83-92, 1997.

[26] M.H. Wu, Y. Kouchi, Y. Onuki, Q. Shi, H. Yoshida, S. Kaplan, R.F. Viggers, R. Ghali and L.R. Sauvage, "Effect of differential shear stress on platelet aggregation, surface thrombosis, and

endothelialization of bilateral carotid-femoral grafts in the dog," *J Vasc Surg,* vol. **22**(4), pp. 382-90; discussion 390-2, 1995.

[27] P.P. Hsu, S. Li, Y.S. Li, S. Usami, A. Ratcliffe, X. Wang, and S. Chien, "Effects of flow patterns on endothelial cell migration into a zone of mechanical denudation," *Biochem Biophys Res Commun,* vol. **285**(3), pp. 751-759, 2001.

[28] Y.L. Hu, S. Li, H. Miao, T.C. Tsou, M.A. del Pozo, S. Chien, "Roles of microtubule dynamics and small GTPase Rac in endothelial cell migration and lamellipodium formation under flow," *J Vasc Res,* vol. **39**(6), pp. 465-476, 2002.

An Optimized Fabrication of High Yield CMOS-compatible Silicon Carbide Capacitive Pressure Sensors

B. Meng, W. Tang, Z.R. Wang, H.X. Zhang[*]

[*]Institute of Microelectronics, Peking University, CHINA

hxzhang@pku.edu.cn

Abstract—Using anodic bonding, we fabricated a silicon carbide absolute capacitive pressure sensor. Low process temperature below 430℃ made the whole fabrication process CMOS compatible. Choosing gold as electrodes, good bonding between gold bottom electrode and SiC layer was available, which made the testing results agree well with the finite element method (FEM) simulations, i.e. the sensor with a square sensing membrane of 200×200 μm² shows a sensitivity of 0.09494 pF/bar over a pressure range of 5 bars, while the simulation result is 0.1035pF/bar. The use of gold increased the yield of devices, for its lower strain, compared to tungsten. Additionally, owing to PECVD carbon silicon and gold's excellent corrosion resistance, this device could be used in harsh environment.

Keywords-Silicon Carbide; Pressure Sensor; anodic bonding; CMOS-compatible

I. INTRODUCTION

Silicon carbide was well-known as a promising material for Micro-Electro-Mechanical Systems [1], especially when applied in harsh environments [2,3]. A low-temperature PECVD process for Low-stress SiC films which is IC-compatible was developed and demonstrated excellent properties [4]. Using surface micromachining, Sarro et.al [5] fabricated a CMOS compatible absolute pressure sensor employing PECVD SiC as its structure layer. However, this sensor's performance is limited by the restricted vertical and horizontal fabrication dimensions of surface micromachining. Our previous work presented a bulk micro-machined absolute capacitive pressure sensor which used silicon carbide film as its mechanical membrane [6]. The use of anodic bonding process makes the sensor's geometric size immune to the limitation of surface micromachining. However, the large compressive strain of tungsten hurt the yield.

In this work, tungsten was replaced by gold. Thus, not only the yield increased a lot, but also good bonding between gold bottom electrode and SiC layer was available. The fabricated sensor with a square sensing membrane of 200×200 μm² shows a sensitivity of 0.09494 pF/bar over a pressure change ranging from 1 bar to 6 bar. Thanks to the good bonding between gold electrode and SiC layer, this measurement result quite corresponds with the finite element method (FEM) simulation results. Additionally, since both PECVD carbon silicon and gold process excellent corrosion resistance, this pressure sensor is promising to be used in harsh environment.

II. DESIGN AND THEORETICAL ANALYSIS

A. Sensors design

As the device's schematic is shown in Fig. 1. The sensing membrane consists of multiple layers of SiC/Au/SiC. A thin film of chromium was deposited between the gold layer and the SiC layer serving as the adhesion layer. The thicknesses of the five layers are 1μm, 20nm, 150nm, 20nm and 0.5μm respectively from top to bottom. Vertically under the sensing membrane there is a sealed cavity on the glass, which is designed to be 2-μm deep. The capacitance increases when the external pressure bends the suspended sensing membrane towards the bottom of the cavity [6].

B. Theoretical analysis

When the applied external pressure is low, the sensor's sensitivity S is nearly a constant and could be approximately figured out, as reported in Ref. [6].

$$S = 12 \, \varepsilon \, a^6 \, (1 - v_c^2) / (E_c \, h^3 \, G^2 \, \pi^6). \qquad (1)$$

In (1), ε is the vacuum permittivity. a is the side length of the suspended square membrane, and G stands for the deepness of the sealed cavity. E_c, v_c and h represent the Young's modulus, Poisson's ratio and thickness of the multiple layers of SiC/Au/SiC, respectively.

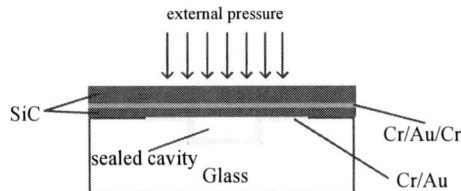

Fig. 1. Schematic of silicon carbide capacitive pressure sensor. The sensing membrane consists of multiple layers of SiC/Au/SiC. A thin film of chromium between the gold layer and the SiC layer serves as the adhesion layer. When the applied external pressure bends the suspended sensing membrane towards the bottom of the cavity, the capacitance of the sensor increases.

According to Ref. [7], the multiple layers could be considered as a composite layer, and E_c, v_c could be figured out from these equations:

$$\frac{E_c h}{1-v_c^2} = \frac{E_{SiC} h_{SiC}}{1-v_{SiC}^2} + \frac{E_{Au} h_{Au}}{1-v_{Au}^2} + \frac{E_{Cr} h_{Cr}}{1-v_{Cr}^2}.$$ (2)

$$v_c h = v_{SiC} h_{SiC} + v_{Au} h_{Au} + v_{Cr} h_{Cr}.$$ (3)

h_{SiC} (E_{SiC}, v_{SiC}), h_{Au} (E_{Au}, v_{Au}) and h_{Cr} (E_{Cr}, v_{Cr}) represent the thickness (the Young's modulus, Poisson's ratio) of SiC, gold and chromium layer respectively.

Since the sensor's sensitivity depends much on side length of the suspended membrane, sensors with variety sensitivity are available by changing the membranes' side length. In this work, we chose 100μm and 200μm as the membranes' side length.

Meanwhile, FEM simulations by ANSYS were made to estimate the sensors' sensitivity. Table I shows the calculation results of the sensors' sensitivity, as well as the FEM simulation results.

III. FABRICATION

The fabrication of pressure sensors in this work is based on bulk micromachining. The whole fabrication process is illustrated in Fig. 2.

On glass, a square cavity of 2μm in depth was first patterned by DRIE. And then a gold bottom electrode was formed by a lift-off process with Cr as the adhesive layer.

Multiple layers of SiC/Au/SiC membranes were deposited in sequence on silicon substrate. Between the gold layer and the SiC layers, thin films of chromium were deposited serving as the adhesion layers. The top SiC layer and the gold Cr/Au/Cr layer were then pre-patterned to release a window for the bottom electrode.

Next, the sealed cavity was formed by a silicon-glass anodic bonding process. SiC and gold served as the media in the anodic bonding process. And then silicon was removed by KOH etching. Finally, electrode pads got released after SiC was patterned by DRIE.

In this work, silicon carbide was deposited by PECVD at 300°C and then annealed at 430°C for 50 minutes. Some of the PECVD parameters are listed in Table II. Since the process temperatures of both PECVD and the following annealing process were below 430°C, the whole fabrication process was CMOS compatible.

TABLE I. THE THEORETICAL CALCULATION AND FEM SIMULTION RESULTS OF THE SENSORS' SENSITIVITY

Side length (μm)	Theoretical calculation sensitivity (pF/bar)	FEM simulation sensitivity (pF/bar)
100	0.0033	0.0018
200	0.1962	0.1035

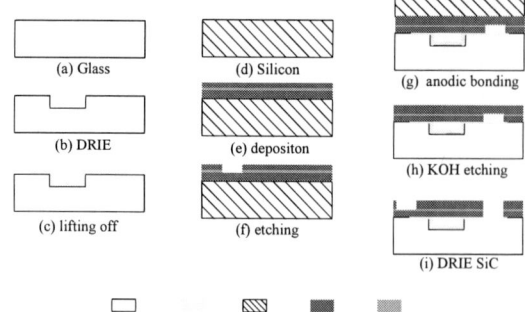

Fig. 2. Fabrication process of the silicon carbide capacitive pressure sensor. (a) to (c) a square cavity of 2μm in depth was patterned by DRIE on glass, and a bottom electrode was formed by a lift-off process. (d) to (e) multiple layers of SiC/Au/SiC membranes were deposited in sequence on silicon substrate, and the top SiC/Au layers were pre-patterned to release a window for the bottom electrode. (g) to (h) The sealed cavity was formed by a silicon-glass anodic bonding process, and then silicon was removed by KOH etching. Electrode pads got released fter SiC was patterned by DRIE.

Sensors with square membranes of different size were fabricated. In Fig. 3, The SEM picture shows the top view of a fabricated pressure sensor with a square membrane of 200×200 μm².

IV. RESULTS AND DISCUSSION

A traditional PCB testing circuit based on an integrated circuit BD031 was used as the read-out circuit of the pressure sensors, as reported in Ref. [8].

$$V_{out} = K_1(C_{sen} - C_{Ref}) + K_2.$$ (4)

In (4), K_1 and K_2 are constants. By choosing an suitable reference capacitor C_{Ref} which has a capacitance of around 3pF, an output voltage V_{out} changing linearly with the capacitance of the sensor could be measured. Thus, the sensors capacitance is measured indirectly.

The measurement was carried out in a pressure chamber at a room temperature. Fabricated sensors with a square membrane

TABLE II. PARAMETERS OF PECVD SIC

Parameters of PECVD SiC	Values
Pressure (mTorr)	1000
Flow rate of CH₄ (sccm)	400
Flow rate of Ar (sccm)	400
Flow rate of NH₃ (sccm)	5
Flow rate of SiH₄ (sccm)	20
Power (W)	300
Temperature (°C)	300

Fig. 3. The SEM picture of a fabricated pressure sensor with a square membrane of 200×200 μm².

of 200×200 μm² or 100×100 μm² have been measured.

The fabricated sensor with a square membrane of 200×200 μm² has a linear response to the pressure which increasing from 1 bar to 6 bar and then decreasing down to 1 bar, as showed in Fig. 4. This pressure sensor shows a sensitivity of 0.09494pF/bar over a pressure change of 5 bars, which quite corresponds with the FEM simulation result of 0.1035pF/bar. And the capacitances measured in the pressure increasing process and the decreasing process are almost the same.

The sensor with a square membrane of 100×100 μm², though shows a sensitivity of nearly 0.003pF/bar which agrees well with the theoretical calculation result, has an evident capacitance difference between the pressure increasing process and the decreasing process, as showed in Fig. 5. Since

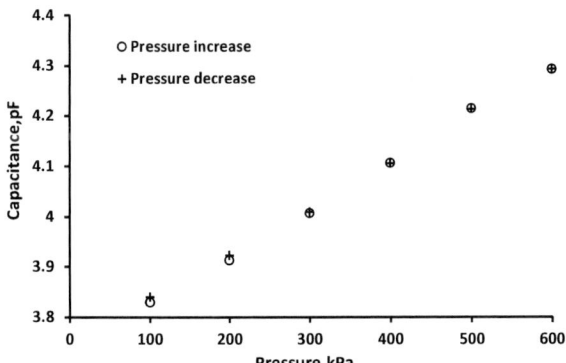

Fig. 4. The absolute capacitance of a fabricated sensor with a square membrane of 200×200 μm² shows a linear response to a pressure increasing from 1 bar to 6 bar and then decreasing down to 1 bar.

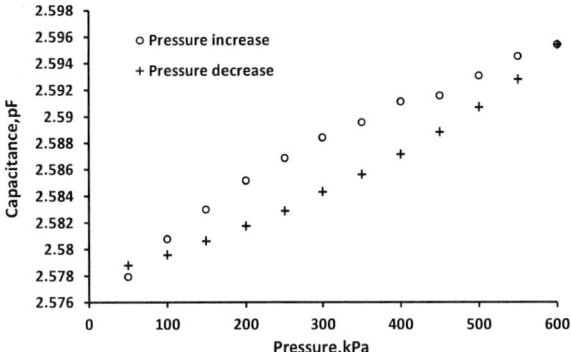

Fig. 5. The absolute capacitance of a fabricated sensor with a square membrane of 100×100 μm² shows an evident difference between the pressure increasing process and the decreasing process.

this sensor's capacitance change over the whole pressure range is very small, it could be impacted by the reference capacitor. More detailed measurements are needed to find out the cause of this difference.

V. CONCLUTION

Silicon carbide absolute capacitive pressure sensors were designed and fabricated based on bulk micromachining, which made the sensor's geometric size immune to the limitation of surface micromachining. Silicon carbide films were deposited at a low temperature by PECVD. Since the process temperatures of both PECVD and the following annealing process were below 430°C, the whole fabrication process was CMOS compatible. Gold was chosen as electrode material instead of tungsten, which not only made good bonding between gold bottom electrode and SiC film available but also increased the yield of sensors. Owing to the good bonding between gold bottom electrode and SiC layer, the measurement results of the sensors' sensitivity agrees well with the FEM simulations. Additionally, excellent corrosion resistance of PECVD carbon silicon and gold makes this sensor promising to be applied in harsh environment.

ACKNOWLEDGMENT

This work was supported by the National Natural Science Foundation of China (Grand No. 61176103). The authors appreciate the first MEMS Company in Peking University for assist in testing work.

REFERENCES

[1] P.M. Sarro, "Silicon carbide as a new MEMS technology," *Sensors and Actuators A: Physical*, vol. 82, pp. 210-218, May 2000.

[2] M. Mehregany, C.A. Zorman, N. Rajan, and H.W. Chien, "Silicon Carbide MEMS for Harsh Environments," *Proceedings of the IEEE*, vol. 86, pp. 1594-1609, August 1988.

[3] H.X. Zhang, H. Guo, Y. Wang, G.B. Zhang, and Z.H. Li, "Study on a PECVD SiC-coated pressure sensor," *J. Micromech. Microeng.*, vol. 17, pp. 426-431, March 2007.

978-1-4673-1122-9/12 $31.00 © 2012 IEEE

[4] P.M. Sarro, C.R. deBoer, E. Korkmaz, and J.M.W. Laros, "Low-stress PECVD Sic thin films for IC-compatible microstructures," *Sensors and Actuators A: Physical*, vol. 67, pp. 175-180, May 1998.

[5] L.S. Pakula, H. Yang, H.T.M. Pham, P.J. French, and P.M. Sarro, "Fabrication of a CMOS compatible pressure sensor for harsh environments," *J. Micromech. Microeng.*, vol. 14, pp. 1478-1483, November 2004.

[6] W. Tang, B.X. Zheng, L. Liu, Z. Chen, and H.X. Zhang, "Complementary metal-oxide semiconductor-compatible silicon carbide pressure sensors based on bulk micromachining," *Micro & Nano Letters*, vol. 6, pp. 265-268, April 2011.

[7] J. Zhou, Y.P. Huang, and S.C. Mantell, "Measurement of thin film Young's modulus from multilayer membranes," *J. Electron Dev.*, vol. 24, pp. 7-13, March 2001.

[8] X.W. Ni, B.X. Mo, B. Xiang , H. Zhu, B.J. Ning, Z.F. Wang, et al. "The design and test of MEMS signal processing circuit," *J. Test Meas. Technol.*, vol. 18 Supp, pp. 63-68, June 2004.

Study of Thin Film Adhesion Properties of Multi-layer Flexible Electronics Composites

C.C. Li[1], Z.H. Liu[1], <u>C.T. Pan</u>[1], J.K. Tseng[2], H.L. Huang[2], S.W. Mao[2], S.C. Shen[3], S.J. Chang[4]

[1]Department of Mechanical and Electro-Mechanical Engineering, National Sun Yat-Sun University, Kaohsiung 804, Taiwan
[2]Department of Mechanical Engineering, R.O.C Military Academy, Kaohsiung 804, Taiwan
[3]Department of Systems and Naval Mechatronic Engineering, National Cheng Kung University, Tainan 701, Taiwan
[4]Department of Mechanical Engineering, National Yunlin University of Science and Technology, Yunlin 64002, Taiwan
panct@faculty.nsysu.edu.tw

Abstract—The deformation between interface and adhesion mechanism of multi-layer flexible electronics composite were discussed. First, ITO (Indium tin oxide), Al (Aluminum) and ZnO (Zinc oxide) were directly deposited on a PET (Polyethylene terephthalate) substrate by RF (Radiofrequency) magnetron sputtering in room temperature to form the flexible multi-layer structures (ZnO/ITO/PET and ZnO/Al/PET) for piezoelectric transducers. ZnO thin film reveals a high (002) c-axis preferred orientation at $2\theta=34.45°$ with an excellent piezoelectric property. To analyze adhesion following a periodic mechanical stress by vibrating flexible composite plates, nano-scratching test (Nanoindenter XP system) was used for scratch process to learn the relationship between normal force and penetration depth. The results show that the plastic deformation can be observed from SEM and OM observation between Al film and PET substrate. This behavior means that the deposited ZnO film has excellent adhesion with Al/PET conductive substrate compared with ITO/PET. Through nano-indentation test, ZnO film deposited on Al/PET substrate decreased the elastic modulus and hardness compared with ITO/PET substrate.

Keywords- Adhesion; Flexible electronics composite; PET; ZnO; ITO; Al; Nano-scratching

I. INTRODUCTION

In flexible electronics lifetime, it would be subjected to external mechanical forces, such as normal force, shear or a periodic stress by vibrating due to their setting functions or environments. The flexible substrates deposited with functional and conductive materials were applied to self-powered piezoelectric energy harvesting system [1]. Piezoelectric functional materials can be manufactured on PET-based substrate such as ZnO [2-4]. For flexible composite plates under ambient vibrations, the delamination phenomenon obviously induced among the functional films, conductive films and polymer-based flexible substrates would influence conductive, mechanical and even optical properties. Therefore, it is worthy to demonstrate the interfacial property of coating adhesion by using nano-indentation and nano-scratching measurement system (NanoIndenter XP System, MTS co.). In this study, to analyze the adhesion performance among the ZnO piezoelectric film, conductive films (Al and ITO) and PET substrate, nano-indentation and nano-scratching test were discussed. After nano-indentation test, ZnO film deposited on Al/PET substrate decreased the elastic modulus

compared with ITO/PET substrate. Nano-scratching test shows better adhesion between Al film and PET substrate; and Al film has a good adhesion and lattice constant matching with ZnO film. After vibration test, between ZnO and Al/PET substrate didn't any delamination phenomenon. It demonstrates that the conspicuous adhesion exists by using Al film as electrode between ZnO film and PET substrate.

II. EXPERIMENTAL DETAILS

The conductive films of Al and ITO with piezoelectric film of ZnO were directly deposited on a PET substrate (Elastic modulus of 3.5×10^9 N/m^2 and Poisson ratio of 0.36) by radio-frequency (RF) magnetron sputtering in room temperature to form flexible multi-layer structures (Al/PET, ITO/PET, ZnO/Al/PET and ZnO/ITO/PET). The fabrication procedure of multi-layer flexible electronics composites are described below:

A. Flexible composite plate fabrication

(a) Cleaning of PET substrate:

The 188 μm thick PET substrates was ultrasonically cleaning in toluene, isopropyl alcohol and methanol for 15 min each, respectively.

(b) Sputtering Al and ITO thin film:

Tab. 1 The parameters of Al sputtering during the deposition

Target	Substrate temperature	Ar concentration (Sccm)	Work pressure (mTorr)	DC power (W)
Al (99.9%)	R. T.	7.6	3	300

Tab. 2 The parameters of ITO sputtering during the deposition

Alloy target	Substrate temperature (℃)	Ar concentration (Sccm)	Work pressure (mTorr)	RF power (W)
In$_2$O$_3$: SnO$_3$ (90 : 10)	30	6.5	15	30

Al thin film with 200 nm thick was selected as the bottom electrode by DC reactive magnetron sputtering deposition (see Tab. 1) due to its low sheet resistance, good adhesion and lattice constant matching with ZnO film. ITO film with 200 nm thick was also deposited onto the PET substrate as the conductive electrode. ITO films were deposited by RF reactive magnetron sputtering. We adjusted process conditions, such as RF power, work pressure and substrate temperature to get good electrical properties (see Tab. 2).

(c) Sputtering piezoelectric ZnO thin film:

Tab. 3 The parameters of ZnO sputtering during the deposition

Target	Substrate temperature	O$_2$/Ar+ O$_2$ (%)	Work pressure (mTorr)	RF power (W)
Zn (99.995%)	R. T.	60	15	75

RF reactive magnetron sputtering method was employed to deposit ZnO piezoelectric thin films. To achieve the required deposition conditions, the sputtering system was evacuated to a base pressure of 2×10^{-6} Torr. Then ZnO thin film with 1.5 μm thick of (002) c-axis preferred orientation was deposited on the Al/PET and ITO/PET flexible conductive substrate. The parameters of sputtering ZnO film are listed in Tab. 3.

The surface morphologies of low-temperature deposited thin films (Al, ITO, and ZnO) on PET substrate were also investigated using SEM observation. The SEM images of films are shown in Fig. 1 (a-d), respectively. It can be observed that the crystallite size of Al films deposited on PET substrate is uniform and clean on PET substrate. The uniform grain size distribution of ZnO film is quite homogeneous with smooth surface.

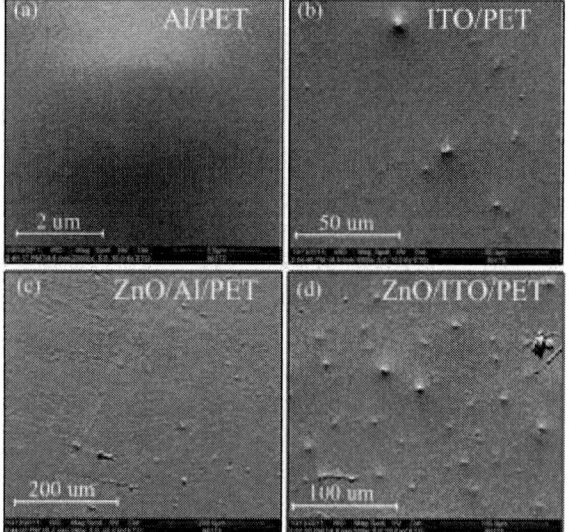

Fig. 1: The prototype of the flexible composite plates and the observations of the surfaces of the coating layers deposited on PET substrate in low-temperature: (a) Al/PET, (b) ITO/PET, (c) ZnO/Al/PET, and (d) ZnO/ITO/PET.

B. Testing the flexible composite plates under vibration

To realize the adhesion performance of flexible composite plates, they were tested under a periodic mechanical force. Fig. 2 (a) shows that ITO, Al and ZnO were directly deposited on a PET substrate to form flexible multi-layer structures (ZnO/ITO/PET and ZnO/Al/PET). Photo of the measurement apparatuses are shown in Fig. 2 (b). The amplitude and frequency were controlled to supply sufficient external force to deflect flexible composite plates as shown in Fig. 3.

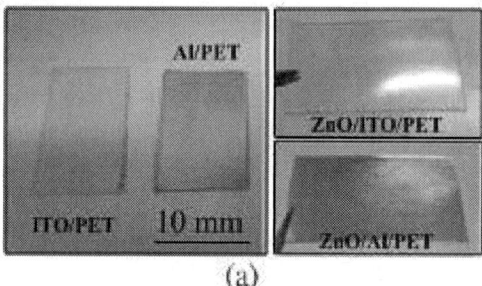

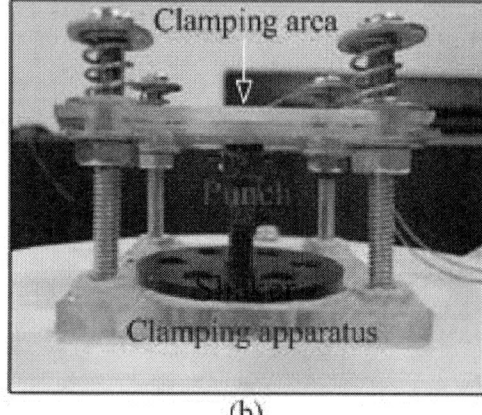

Fig. 2: (a) The flexible composite plates with a (b) homemade clamping apparatus fixed on the shaker.

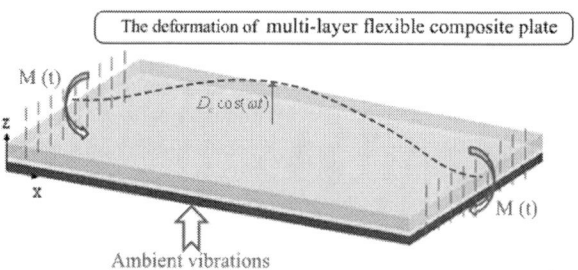

Fig. 3: Schematic layout of the deflected flexible composite plate under a periodic mechanical force.

III. RESULTS AND DISCUSSION

A. Nano-indentation test

The critical information on mechanical property of elastic modulus was measured by nano-indentation technique with CSM (Continuous stiffness measurement) option. The comparison of modulus of PET substrate and PET-based composite plates is shown in Fig. 4. As result of ZnO/ITO/PET sample, it reveals that the initial modulus is much higher than the others. Then the modulus decreases quickly doe to flexible substrate effect. After displacement into surface about 1000 nm, the modulus is close to ITO/PET substrate. The resistance between the probe and ZnO/ITO coating is gradually decreased until the displacement into surface 2000 nm. At this time, the tip fully pierces ITO film and the modulus is close to PET substrate. Especially the substrate effect of Al/PET and ZnO/Al/PET samples is really conspicuous due to the ductility of Al film compared with ITO film.

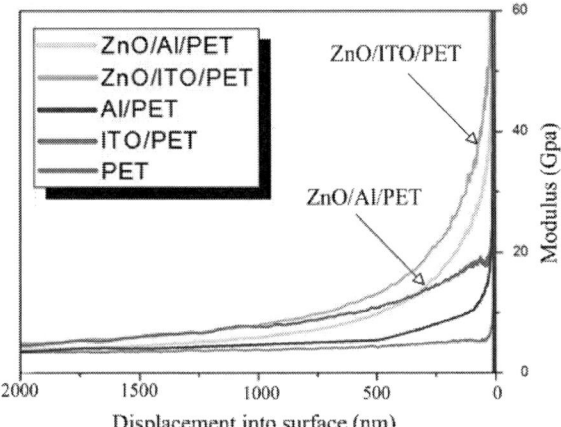

Fig. 4: modulus as a function of depth; (b) hardness as a function of depth.

B. . Nano-scratching test using Berkovich probe

Nano-scratching test (XP standard scratch with cross profile) was used to analyze the interfacial adhesion performance among the ZnO piezoelectric film, conductive films (Al, ITO) and PET substrate. Fig. 5 (a) shows the load and penetration curve diagram of ITO/PET and Al/PET. The residual depth and width of the track groove were scanned by probe, and the phenomena of crack, interfacial crack propagation and serious delamination on the surface can accurately be measured. Since the hardness of ITO film is higher than Al film, the penetration curve of Al/PET is deeper than ITO/PET at the same normal force of 5 mN. When the penetration curve of ITO/PET is reaching at 400 nm, the area of scratching increased and induced a big crack due to augmenting abrasive wear resistance. Compared with ITO/PET, the curve of Al/PET is smoother without any fluctuation. SEM observation of Al/PET shows abrasive wear phenomena and slight plastic deformation as show in Fig. 5 (b).

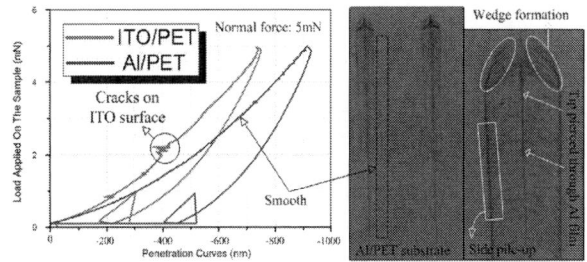

Fig. 5: (a) the load and penetration curve diagram of ITO/PET and Al/PET; (b) OM observation of Al/PET after scratching test.

Fig. 6 (a) shows the load and penetration curve diagram of ZnO/ITO/PET and ZnO/Al/PET. The curve of ZnO/Al/PET is smooth without any distinctly fluctuation on the surface. But the observation of ZnO/ITO/PET has two clearly beats at 1500 nm and 2000 nm depths. It reveals that the interfacial adhesion strength between ZnO and ITO films has been destroyed. It suggested that the interfacial crack propagation and a little delamination phenomenon would be induced. Fig. 6 (b) shows slight fishbone pattern within scratching groove, it reveals that the deposited ZnO thin film exhibits good adhesion with Al film. Fig. 6 (c) shows some chips and crack propagation problem on the surface of ZnO/ITO coating along scratching direction.

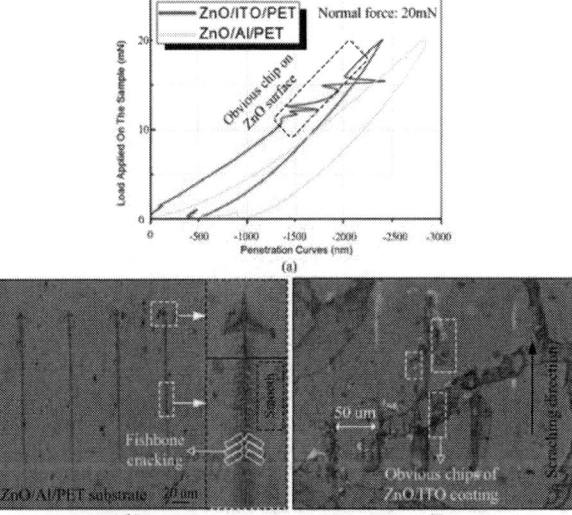

Fig. 6: (a) the load and penetration curve diagram of ZnO/ITO/PET and ZnO/Al/PET; (b) OM observation of ZnO/Al/PETand (c) ZnO/ITO/PET plate after scratching test.

To realize the difference of adhesion after vibration test, the flexible composite plates were placed on shaker and subjected to a periodic mechanical force using a punch. Compared with the difference of ITO/PET and Al/PET after vibration test, adhesion performance of Al film on PET substrate is much better than ITO film as show in Fig. 7 (a). In the comparison of ZnO/ITO/PET and ZnO/Al/PET after vibration test as shown in Fig. 7 (b), adhesion of Al film between ZnO film and PET substrate is much better than ITO film.

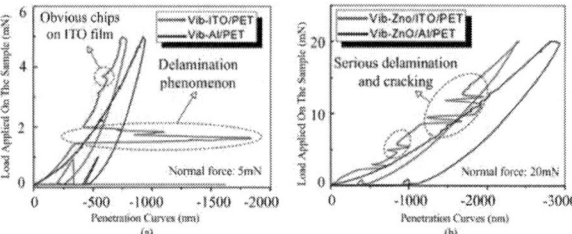

Fig. 7: The load and penetration curves of four composite plates following vibration test (a) ITO/PET and Al/PET; (b) ZnO/ITO/PET andZnO/Al/PET

After vibration test, SEM observation of ZnO/ITO/PET demonstrates that the phenomena of crack, interfacial crack propagation and serious delamination on the surface of ZnO/ITO coating can be obviously measured. When the area of scratching is increased gradually, abrasive wear appears between scratch face of probe and ZnO film surface lead to a large area crack and serious delamination as shown in Fig. 8 (a). Fig. 8 (b) shows that slight fishbone pattern and fracture appear in the scratching groove, but there are no delamination phenomenon generated between ZnO and Al/PET substrate.

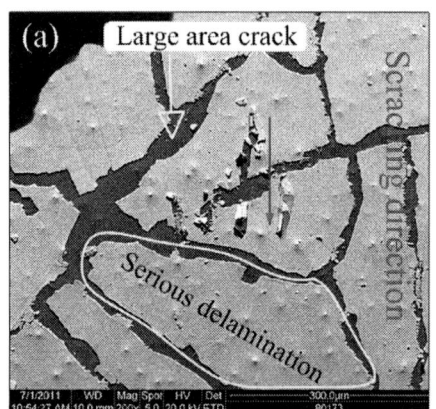

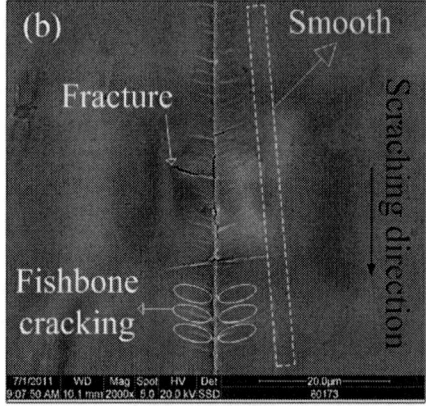

Fig. 8: SEM observations of (a) ZnO/ITO/PET and (b) ZnO/Al/PET following vibration test

IV. CONCLUSIONS

Interfacial adhesion mechanisms of multi-layer flexible electronics composite were investigated in this study. Fist, the sputtering Al/PET and ITO/PET flexible conductive substrates were subjected to periodic mechanical force. Compared with ITO film, Al film has a good adhesion on PET substrate. Then ZnO piezoelectric film of high (002) c-axis preferred orientation was successfully deposited on ITO/PET and Al/PET conductive substrates in room temperature. However, the phenomena of crack, interfacial crack propagation and delamination on the surface would be induced after vibration test to influence the mechanical properties of flexible electronics composite. After nano-indentation test, the elastic modulus of ZnO film deposited on Al/PET substrate decreased compared with ITO/PET substrate due to substrate effect. As the results of nano-scratching test, there is no delamination phenomenon between ZnO and Al/PET substrate observed after vibration test. This behavior demonstrated obviously that the outstanding adhesion exists by using ductile Al film as electrode between ZnO film and PET substrate due to the lattice constant of Al film matching with ZnO compared with ITO film.

ACKNOWLEDGMENT

The authors would like to thank National Science Council Core Facilities Laboratory for Nano-Science and Nano-Technology in National Sun Yat-Sen University, Kaohsiung-Pingtung area, Taiwan for supports.

REFERENCES

[1] C.T. Pan, Z.H. Liu, Y.C. Chen, W.T. Chang, and Y.J. Chen, "Study of Vibration-induced Broadband Flexible Piezoelectric ZnO Micro-harvester with Storage System", Solid-State Sensors, Actuators and Microsystems Conference (TRANSDUCERS), 16th International, pp. 1669-1672, 2011.

[2] W.T. Chang, Y.C. Chen, R.C. Lin, C.C. Cheng, K.S. Kao, Y.C. Huang, B.R. Wu, Y.C. Huang, "Design and Fabrication of a Piezoelectric Transducer for Wind-Power Generator", Thin Solid Films, Vol. 519, pp. 4687-4693, 2011

[3] C.T. Pan, Z.H. Liu, Y.C. Chen, C.F. Liu, "Design and Fabrication of Flexible Piezo-microgenerator by Depositing ZnO Thin Films on PET substrates", Sensors and Actuators A-Physcial, Vol. 159, pp. 96-104, 2010.

[4] C.T. Pan, Z.H. Liu, Y.C. Chen, "Study of broad bandwidth vibrational energy harvesting system with optimum thickness of PET substrate", Current Applied Physics, 2011

Using a Canny-Edge-Detection Based Method to Characterize In-Plane Micro-Actuators

Chao-Yu Cheng, Yan-Bo Lin, Kerwin Wang
*Department of Mechatronics Engineering. National Changhua University of Education
Changhua, Taiwan
kerwin@cc.ncue.edu.tw

Abstract— This paper developed an automatic optical inspection (AOI) technology for characterizing the motion amplitude and the vibration mode shape of in-plane micro-actuators. The technology involves a multi-stage image process method. It includes an image contrast enhancement step and a Canny edge detection method. To evaluate this approach, a rotational electrostatic actuator has been prepared for testing. This actuator performed low contrast optical micrographs under first vibration mode. The testing result shows that the technology can handle low contrast microscopic images quite well. The suggested method successfully provides quantitative information to identify the vibration amplitudes and mode shape of the tested actuator with a space resolution of 0.46μm/pixel.

Keywords-automatic optical inspection; AOI;Canny edge detection; actuator, in-plane motion.

I. INTRODUCTION

Microstructures excited at resonance frequency will perform blurred images. Blurry image analysis is a proven and a cost-effective technique for MEMS motion and resonance frequency measurements [1, 2]. In previous study, in order to calculate the displacements of a vibrating microstructure with image blur, a good microscopic photography with adequate contrast is important [1]. Blurry image pixel-intensity-scanning has often been used for parameter characterizations [1, 2]. However, a high-contrast image is not always available. The brightness profile of a line of pixels is sensitive to illumination conditions [1, 3].

This paper introduces a multi-stage image processing algorithm to measure 2D in-plane motion without using high-contrast images. The multi-stage image processing algorithm is developed based on Canny edge detection method [3]. In general, an edge detection method can effectively reduce the data amount of an image without losing important structural properties. Canny edge detection method developed by John F. Canny in 1986 [3] has become one of the standard edge detection methods. In this paper, a multi-stage image preprocessing method has been investigated to ensure the successfulness of the MATLAB® implementation of the algorithm. The goal is to maximize the probability of detecting real edge of an in-plane micro-actuator from a low contrast micrograph.

To initiate the multi-stage image process design for practical use, we first check in Section II the low contrast blur image given by the comb-drive rotational actuator. The actuator vibrating at fundamental resonance mode can be useful for determining the image preprocessing strategies and their threshold values. Then we describe a DRIE-based process for the actuator etching on silicone on oxide (SOI) wafers.

In Section III, we review and introduce the image preprocessing methods. We also discuss the different in detail by comparing the edge estimated with different approaches and with a real edge. As a consequence, in Section IV, we establish a multi-stage Canny-edge-detection based method. In Section V, we describe testing results. This method is used to measure the fundamental resonance frequency of the in-plan micro actuator.

II. TEST PLATFORM OF MICROACTUATOR MOVEMENT

To develop the automatic optical inspection (AOI) technology, a rotational electrostatic actuator (Fig. 1) with 15 μm thick comb-drive fingers has been prepared by deep reactive ion etching. Fig. 2 presents the actuator design, fabrication steps and geometric dimension of the electrostatic actuator. The actuator is fabricated by deep reactive-ion etching (DRIE, Oxford PlasmaLab 100) in Washington Technology Center at the University of Washington. To achieve a good aspect ratio, Bosch process was selected for the vertical comb etching. The etching process does not stop when it reaches the bottom of the oxide layer. Plasma reflected from silicon dioxide releases narrow silicon cantilevers (Fig 2). After DRIE process, the micro-structures which are larger than $75x75 \ \mu m^2$ will be anchored to box the oxide on substrate, the structures which are smaller than $15x15\mu m^2$ will be released. The rotational micro-actuator excited at resonance frequency performs blurred images (Fig. 3). The actuators are tested on a probe station (Fig. 4). The probe station is equipped with an Optem-zoom adjustable microscope (2000X magnification) and an external CCD camera. Driving signals are sent from a function generator to an amplifier and contacted to the device under test (DUT) through microprobes. The blurry images have low contrast back ground. The blurry image is hard to analysis because the vague and gradually weakened material boundaries in micrographic images. It is even more challenge to fully capture the structure in vibration.

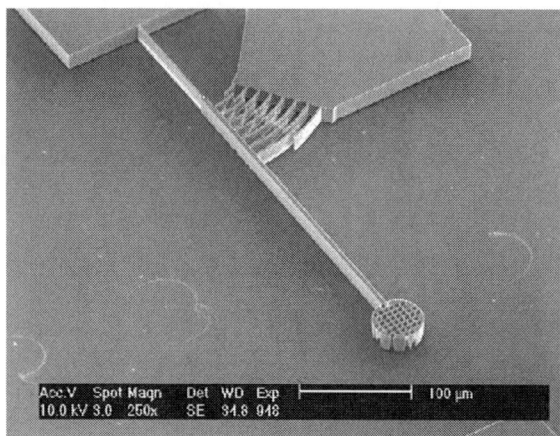

Fig. 1. SEM micrographs of comb-drive rotational actuator, perspective view. The actuator was originally designed by Michael Sinclair at Microsoft Research.

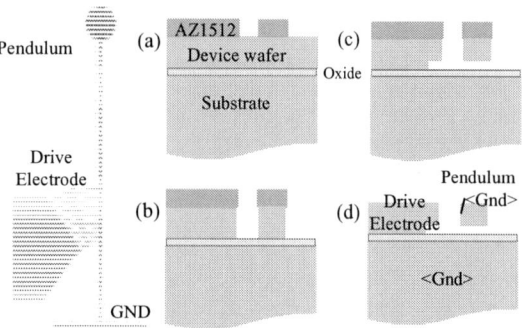

Fig. 2. CAD layout diagram of the actuator and the process flow chart for the fabrication of the rotational electrostatic actuator: (a) photolithography; (b) deep reactive ion etching; (c) silicon-cantilever release-etching; (d) Oxygen plasma photoresist cleaning. The process was completed at Washington Technology Center, Seattle, USA.

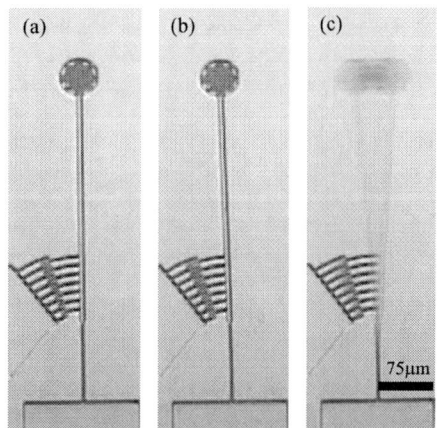

Fig. 3: Optical micrographs (a)–(c) show, respectively, images of a comb-drive actuator at rest, driven by 90Volts, and driven at resonance. 4053Hz.

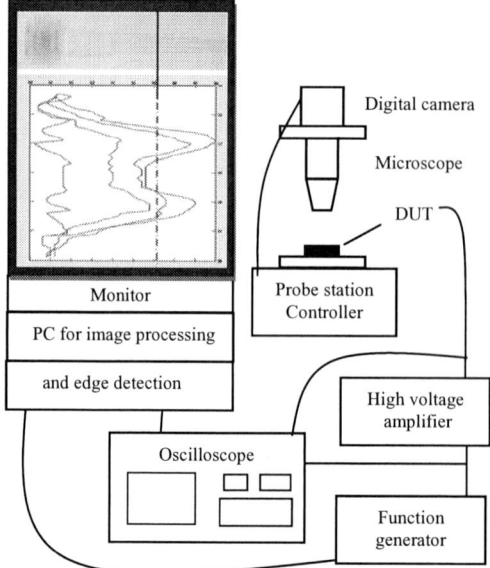

Fig. 4. Test setup with optical microscope, function generator, high voltage amplifier oscilloscope and device under test (DUT). The brightness profile of a line of pixels is shown in the upper-left corner. In the figures, the cursor is positioned to select pixel brightness scanning line; however, each line only offers limited geometric information of a moving device.

III. IMAGE PREPROCESS METHOD

To estimate the deflection and the frequency response of the in-plane micro-actuator, the cursor is positioned to select pixel brightness scanning line in Figure 2; however, each line only offers limited geometric information of a moving device. A two-dimensional edge detection method would provide more information of a vibrating actuator, including the mode shape by detecting image features, such as edges, gradient direction base on a measure of the strength of the edge response. A Canny edge detection based method should have accurate and low noise gradient information to specifically measure the edge strength in detail.

When detecting the edge of grayscale image data, the information that can be extracted may be strongly dependent on the contrast and the range of the histogram distribution of the data. An investigation is presented for selection of scale levels when adjusting the contrast of the image by mapping the values of the input image to new one. An important consequence of this stage is that it allows the scale levels to enlarge cross or along the edges.

A contrast-limited adaptive-histogram-equalization process [4] is adopted to further enhance the image contrast in specific data regions without amplifying noise which might be revealed in the image. The parameters were set according to the grayscale level of the noise in the image.

978-1-4673-1122-9/12 $31.00 © 2012 IEEE

All of these image preprocessing strategies can work together no against with each other to optimize the performance of Canny edge detection method; however, one real edge should not result in multiple detected edges.

To determine the best strategist would be time consuming, this paper suggests a methodology to find the short path. Several image refining processes have been investigated to validate this approach. It should be able to catch the shape of the structure of the actuator when it vibrates or moves to the endpoints.

IV. MULTISTAGE EDGE DETECTION

It is necessary to establish a methodology to determine the parameters in order to optimize the use of image preprocessing algorithm. A methodology for these criteria was presented to find optimal operators for edge detections. The procedure consists of determining the sequence and parameters for the image contrast adjusting processes, the contrast-limited adaptive-histogram-equalization and a Canny edge detection algorithm.

We use a tree analysis diagram to break this complex-engineering-problem down. The tree diagram is shown in Fig 5. The analysis results are shown in Fig 5. It was shown that edge detection results are sensitive to the sequences of multi-stage image preprocessing implantations. Approach II and Approach III are suggested by the analysis results.

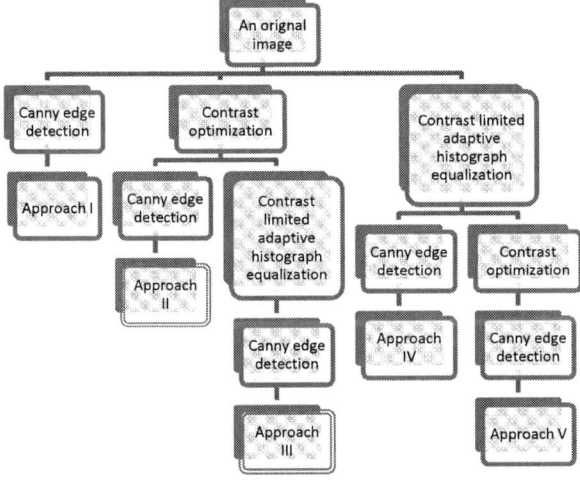

Fig. 5. We use tree analysis diagram to find the optimized path to adjust the image. The image preprocess methods match with the best image Canny edge detection. The input low and high threshold values for image intensity adjustment are set from 0.3 to 0.98 after contrast optimization. The optimization process can minimize the noise without losing any important information of the image. The clip limit and the specified histogram shape of the contrast-limited-adaptive-histograph-equalization are set to 0.001and bell-shape, respectively.

Both approaches can catch the deflection, the frequency response and the mode shape of the actuator; however, only Approach III can catch the part of the moving structure when it moves to the death points (the endpoints of the maximum sweeping amplitude). Detailed quantitative information of rotational comb-drive actuator will be described in the next Section.

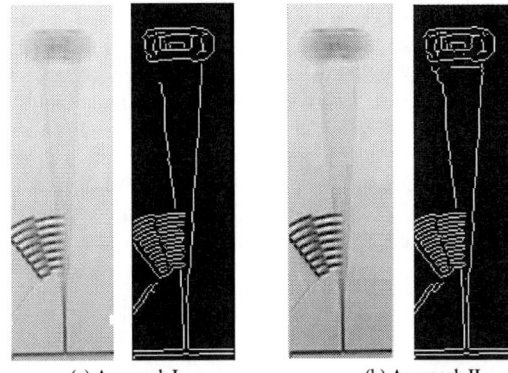

(a) Approach I   (b) Approach II

Fig. 6. Image adjustment and edge detection results from (a) Approach I and (b) Approach II.

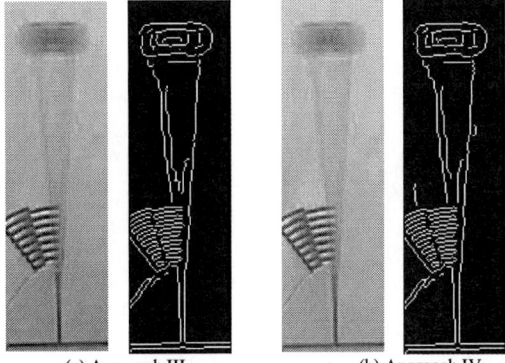

(a) Approach III   (b) Approach IV

Fig. 7. Image adjustment and edge detection results from (a) Approach III and (b) Approach IV

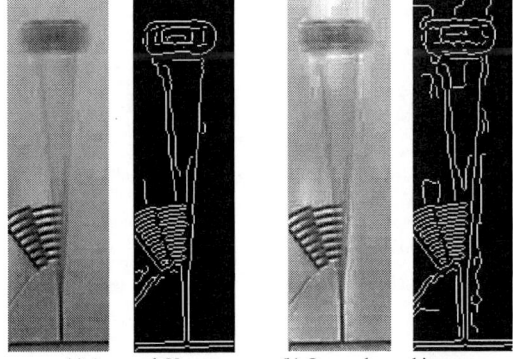

(a) Approach V   (b) Over enhanced image

Fig. 8. Image adjustment and edge detection results from (a) Approach V and (b) over enhanced image will cause ill defined edges

V. DETECTION RESULT

The experiment setup of this optical inspection system for micro-actuator is shown Fig. 4. Blurry images of a vibrating comb drive at rest and driven at resonance are shown in Fig. 3. The images are recorded and preprocessed for the multi-stage Canny edge detection. A graph comparing the processed images of swiping in-plane actuator at resonance is shown in Fig. 6~8. Over enhanced images can induce unwanted noise into the edge detection process. One can calculate the resonance amplitude by evaluating sweep rang of the actuator from the edge detector defined area, which is enclosed by bright pixels. Fig. 6(b) and Fig. 7(a) are shows the displacement versus driving voltage and dynamic response of this micro-actuator, respectively. With suggested edge-detection method, one can easily gather important clues to identify the vibration amplitudes (Fig. 9) and the mode shape (Fig. 10) of the tested actuator with a space resolution of 0.46μm/pixel. The algorithm is capable of handling low contrast images.

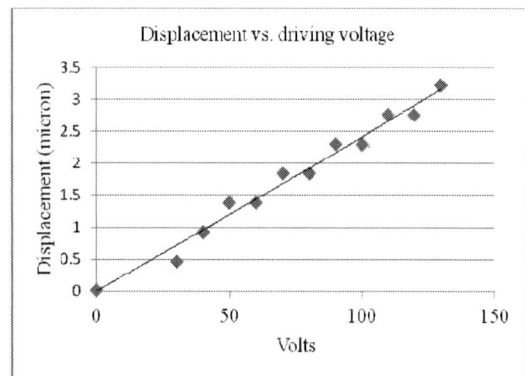

Fig. 9. Displacement versus driving efforts, analysis from the blurry images.

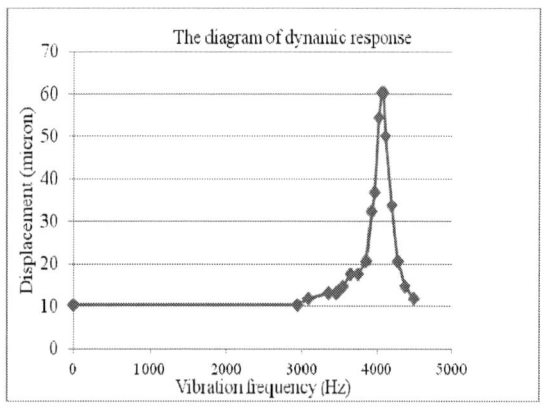

Fig. 10. Dynamic response diagram.

VI. CONCLUSIONS

This paper describes a procedure for the implementation of a Canny-edge-detection based method to characterize in-plane micro-actuators. The multi-stage scheme is combined with the contrast enhancement, the adaptive-histogram-equalization, and the Canny edge detection algorithms into a single description. It can catch the shape of the structure of the actuator when it vibrates or even moves to the death points (endpoint with the maximum deflection). It is shown how the methodology can be implemented in experiments on real-world data.

ACKNOWLEDGMENT

The author would like to thank Prof. M. F. Chen for his influence on the course of this work. Part of this work is supported by National Science Council, NSC 98-2221-E-018-012-MY2 and NSC 100-2221-E-018-005. The authors gratefully acknowledge M. Sinclair, and K. F. Böhringer for valuable discussions and suggestions. The authors would like to thank K. Nishimura, E. Miller, A. Tombros, G. Holman, and K. Ritala of the Washington Technology Center; D. Qin and G. Golden of the Center for Nanotechnology at the University of Washington; and W. Walker of Oxford Instruments for their technical training and assistance.

REFERENCES

[1] D. J. Burns and H. F. Helbig, "A System for Automatic Electrical and Optical Characterization of Microelectromechanical Devices,"*J. Microelectromech. Syst*, v8, n4, pp. 473-482. 1999.

[2] M. Yang, M. Castellani, R. Landot, F. Beyeler, Z. Wang, D. T. Pham, "Automated optical inspection method for MEMS fabrication," *Proc. IEEE, Int. Conf. on Mechatronics and Automation,* (2010), pp. 1923-1931.

[3] J. Canny, "A Computational Approach to Edge Detection on Pattern Analysis and Machine Intelligence," *IEEE Trans*, 8(6), pp.679–698. 1986.

[4] K. Zuiderveld, "Contrast Limited Adaptive Histograph Equalization," *Graphic Gems IV*. San Diego: Academic Press Professional, 474–485. 1994.

[5] T. Lindeberg, "Edge detection and ridge detection with automatic scale selection," *International Journal of Computer Vision*, 30, 2, pp 117-154, 1998.

Thermal switch and variable capacitance designed for micro electrostatic converter by using CMOS MEMS process

Jin-Chern Chiou[1], Lei-Chun Chou[1], You-Liang Lai[2], Sheng-Chieh Huang[1]

1. Dept. of Electrical Control Engineering, National Chiao Tung University, Hsinchu, Taiwan 300, R.O.C
2. National Chip Implementation Center (CIC), Hsinchu, Taiwan 300, R.O.C

E-mail: ypray123@gmail.com

Abstract- **This paper focuses on implementing a novel thermal switch and variable capacitance design by using commercially available CMOS MEMS process which can approach in a micro electrostatic converter system. In this system, there are two major parts. First is the variable capacitance, and the second is the thermal switch. In the variable capacitance, it implement by UMC 0.18μm one-poly seven-metal (1P7M) CMOS MEMS process. In the post-process, the silicon-oxidation have been released and the gap between two metal layers filled with PDMS (Polydimethylsiloxane). Filling with PDMS is to significantly increase C_{max}.**

In the thermal switch design, there are two novel designs in this switch: first, the soft contact structure and post-processing fabrication; second, using residual stress to achieve large structural deformation and a new design of thermal switch. To create the soft contact structure, residual stress effect has been utilized to make different bending curvatures. According to the experiments, the layer Metal1 has the largest residual stress [1] effect that can achieve the largest deflection in z-axis. Because the residual stress of the layer metal-1 is negative, the structure will bend down after release, the largest contact area which has been set up to get the lowest contact miss ability. In the post-processing fabrication, 0.3μm thickness gold will be patterned on the contact tips. Due to gold, rather than Aluminum, has no oxidation issue, it has more reliability on preventing the problem of oxidation than Aluminum. In the new thermal actuator design, we design a novel folded-flexure [2, 3] with the electro-thermal excitation to turn the switch on or off. In the prototype, the device size is 500 μm * 400 μm and the gap between two contact pads is 9 μm in off-state.

Depending on the experimental results, the C_{max} is 19.22 nF, and the C_{min} is 10.65 nF in variable capacitance. The switch can work stably at 3 volts, and the working temperature and operating bandwidth are individually 20-200 °C. The natural frequency of the switch is 42.9k Hz and the response time is 14.28Hz.

I. INTRODUCTION & SYSTEM

Electrostatic capacitive energy conversion utilizes a variable capacitor to convert vibration energy into electric energy. The electrical energy W stored in a capacitor with capacitance C and voltage V is $W = \frac{1}{2}CV^2 = \frac{1}{2}\frac{Q^2}{C}$. If the capacitance of a pre-charged capacitor with constant charge Q is decreased due to vibration, the stored electrical energy in the capacitor will increase, thus converting the kinetic energy into electrical energy. The main concern of the capacitive energy conversion

is how to extract the stored electrical energy in a properly controlled timing scheme. Capacitive energy conversion is more suitable for the steady frequency moderate amplitude vibration source mentioned before.

Fig. 1 is the micro electrostatic energy converter system. The converter is composed of an auxiliary battery supply V_{in}, a vibration driven variable capacitor C_v and an output storage capacitor C_{stor}, which is connected to the load R_L. Two switches, SW1 and SW2, are used to connect these components and control the charge-discharge conversion timing. The variable capacitor serves as the conversion transducer and the auxiliary battery supply is used to pre-charge the capacitor. A basic operation cycle begins when the variable capacitor C_v is charged by the auxiliary voltage supply V_{in} through SW1 at its maximum Cmax. After C_v is charged to V_{in}, SW1 is opened and the capacitance changes from C_{max} to C_{min} due to vibration driven displacement. In this process, the charge Q on the capacitor remains constant (SW1 and SW2 both open). Therefore, the terminal voltage on the capacitor is increased, converting the kinetic energy of vibration into electrical energy stored in C_v. When the capacitance reaches C_{min} and terminal voltage reaches V_{max}, SW2 closes and allows C_{stor} to be charged by C_v through charge redistribution, transfer the energy to the output port. SW2 is then opened and C_v varies back to C_{max}, preparing for the next conversion cycle.

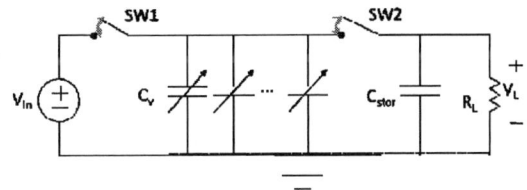

Figure 1. Micro electrostatic energy converter system.

II. FABRICATION

CMOS MEMS utilizes the CMOS stacked layers to form the micro-sensors and the micro-actuators. It has great potential for commercial production. However, stacked CMOS layers are composed of compound materials like metal, via, poly-silicon, and oxide layers. There is stress in and between these layers. The extraction for individual layer in CMOS spends large chip area. And the simulation time for microstructure with the

978-1-4673-1122-9/12 $31.00 © 2012 IEEE

complicated multi-layers is much longer by using this method. The efficient method is to extract the effective mechanical properties of some basic metal/oxide combination structures. It not only costs less testing area and simulation time but also includes the stress of interlays. This paper used the effective mechanical properties extracted from eight combinations of CMOS metal/oxide stacked structures to simulate and predict the static and dynamic behavior of MEMS device. The problem of the lateral contact switch has been solved in this study by depositing gold on the Aluminum layer.

The CMOS MEMS multi-project wafer (MPW) was fabricated by TSMC 0.35 µm two-poly four-metal (2P4M) CMOS process and CIC micromachining post process. The major materials of metallization and dielectric layer in this CMOS process are respectively aluminum and silicon oxide. Fig. 2 shows the cross-section views of the MPW with post-CMOS micromachining steps. By the way, in this study we use Bio CMOS MEMS process, a gold layer is deposited on top layer.

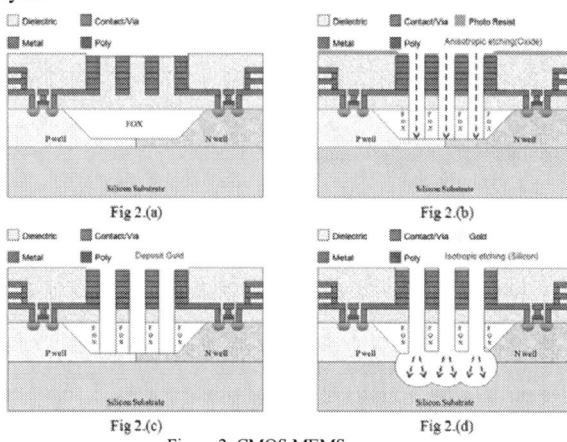

Figure 2. CMOS MEMS process.

For alleviating the dry etching process on dielectric layers, as Fig. 2(a) shows, the CMOS passivation layers on etching selective regions are removed during the CMOS process. Before the anisotropic dry etching, an additional photo-resist layer is spun and lithographed on the wafer for the post-process. This lithography enables the definition of the protective region of photo-resist layer over the metal of the inductor. The thickness of the photo-resist requests at least 5µm to avoid the RIE etching directly on the top meal. The anisotropic RIE etching with CF4, CHF3 and O2 is subsequently used to remove the exposed dielectric oxide in Fig. 2(b). In this step, most photo-resist layers would exhaust with the RIE gas. When all metal layers are done, a gold layer is deposited on the top layer, as seen in Fig. 5. As Fig. 2(d) indicates, the following isotropic dry etching with SF6 and O2 is then included to remove the underlying silicon substrate and release the microstructures.

III. DEVICE DESIGN

A. Variable capacitance design

In the variable capacitance design, a out-plane gap-closing comb structure is used for the variable capacitor. It implement by UMC 0.18 μm 1P7M CMOS process. In the post process, the silicon dioxide layer removed by HF vapor, thus there are 7 metal layers can be used in variable capacitor. It means that there are up to 6 available variable capacitors in one block. In order to increase the capacitance value, the gap of the metal layer is filled with PDMS (Polydimethylsiloxane), as shown in Fig 3. The relative permittivity of PDMS is 10.

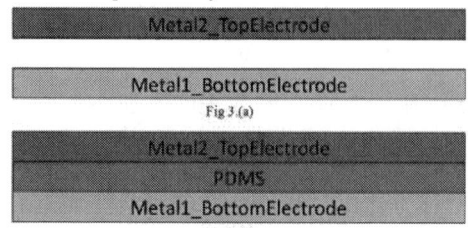

Figure 3(a). Remove SiO_2 by HF vapor. (b) The gap filled with PDMS.

B. Mechanical switch design

The switches SW1 and SW2 are realized as lateral contact mechanical switches. Conventional design of the charge-discharge timing control switches utilize diodes or clocked active switches. In order to prevent charge leakage out of C_v and C_{stor}, the switches must have a reverse leakage current lower than a few nA. This is not common in commercially available diodes and other switching circuitry. Capacitive coupling is another problem, in which the capacitance of the switch contributes to parasitic capacitance. Our design of SW1 and SW2 has barely zero charge leakage and very low capacitance coupling effect. Other advantages are the low energy consumption, the synchronous operation to the variable capacitor, and the monolithic integration with the whole device structure.

SW1 is designed to contact simultaneously when or merely after the metal layer have touched. SW2 is designed by a thermal type. SW2 is turn on at initial time, and there are current through pass the switch. The temperature of the switch will increase caused by the current. And it will pull-back, thus the SW2 will turn off. To design the thermal switch, there are two parts should discuss. First is the soft contact structure, and second is new design of thermal actuator.

I. The soft contact structure

In a normal switch design, the contact area is too small to make a poor contact. To solve this problem, a new design of the contact profile which used negative residual stress effect and the different configurations of the contact tip has proposed in this study. Every metal used in the TSMC 0.35 µm 2P4M CMOS process has its own effective residual stress. Combine different metal layer also can get different residual stress [1].

In the Fig. 4, only the layer Metal 1 has the negative residual stress. Thus the contact profile will bend down after release.

978-1-4673-1122-9/12 $31.00 © 2012 IEEE   734

With this profile, it provides lager contact area than the normal design.

This study also designs different configurations of the t‑ with different contact profile which can get more contact ar (Fig. 5a, 5b).

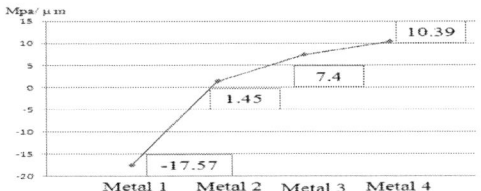

Figure 4. Effective gradient stress vs. metal layer.

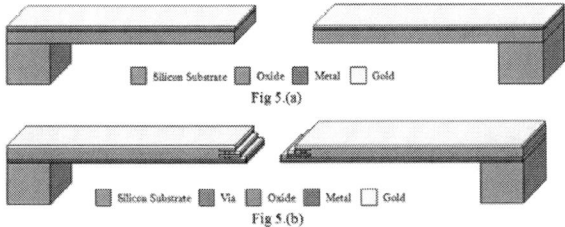

Figure 5(a). Cross-section view of normal contact tip. (b). Cross-section view of proposed contact tip.

In the post-processing, 0.3μm thickness gold will be deposited on the surface of the element. Since gold is non-active, it has more reliability on avoiding the problem of oxidation and on increasing the conductivity at contact area.

II. A novel design of thermal switch

In the novel thermal switch design, this paper proposed a novel folded-flexure with the electro-thermal excitation to control the switch on or off, as seen in Fig. 6. In the normal design [2, 3], the numbers of the layers used are metal 1, metal 2 and metal 3. In this proposed design, the different numbers of layers are used.

In this design, effective gradient stress and thermal coefficient are used finite-element-method software in simulation.The stress increases with the numbers of metal layers, as shown in Fig. 7. It indicated that oxide layer in CMOS process has compressive stress and the metal layer has tensile stress before post-processing release, as illustrated in Fig. 6(a). Due to the isotropic dry silicon etch and structural release. The oxide layer has tensile stress and the metal layer has compressive stress after release, as shown in Fig. 6(b).

When the switch turn on, there are current pass by the switch and let the temperature of the switch rise up. Because oxide layer (8.5 μm/m-℃) and metal layer (27.4 μm/m-℃) [4] have different thermal coefficient. The folded-flexure of the switch will pull back and the switch will turn off, as shown in Fig. 6(c). So we can use this folded-flexure structure to control the switch turn on or turn off.

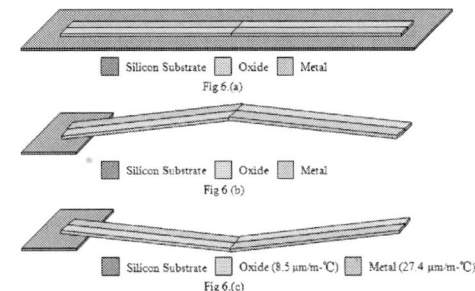

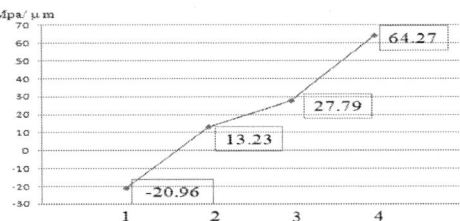

Figure 6(a) is the top view of micro-cantilever beam before release. (b) is the structure after release at 293 K (20 ℃). (c). Due to different thermal Coefficient, the structure will pull back with temperature increase to 473 K (200 ℃).

Figure 7. Effective residual stress vs. numbers of metal layer.

IV. SIMULATION AND EXPERIMENTAL RESULT

The simulation software (Coventor-Ware) also used to verify the switch and capacitance design. By the simulation result, the initial gap is 11 μm at 293 K (20 ℃). Due to temperature increase, the gap pulling back and the displacement of the contact-tip is 6 μm at 473 K (200 ℃).

Fig. 8 is the layout of the variable capacitance, and it has implemented by UMC 0.18 μm CMOS MEMS process, as shown in Fig. 9(a). Fig. 9(b) shown the SiO$_2$ layers of the capacitance have been etched by HF and filled with the PDMS. The size of the capacitor is 1100 μm * 1100 μm. Fig. 10 is the thermal switch implement by TSMC 0.35 μm 2P4M CMOS process and CIC micromachining post process. The width of the device is 500 μm, and length is 400 μm.

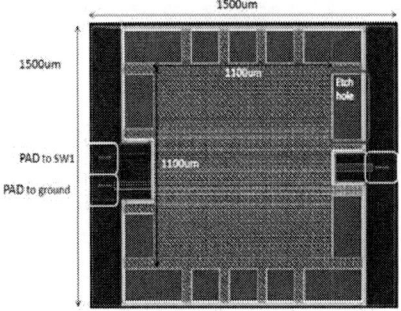

Figure 8. Layout of the variable capacitor.

978-1-4673-1122-9/12 $31.00 © 2012 IEEE   735

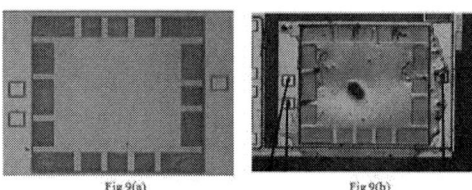

Fig 9(a)　　　　　　　　Fig 9(b)

Figure 9(a). Device implement by UMC 0.18 μm process and remove the SiO₂ layer. (b) The gap filled with PDMS.

Figure 10. device implement by TSMC 0.35 μm process.

The switch and variable capacitor were first tested by a probe station to see if they were properly released. The switch was measured by the MEMS Motion Analyzer (MMA) system at the National Chip Implementation Center (CIC). The MMA system utilizes image processing technique and measures the periodic relative motion between the movable and the still structures of the device. The photograph of the MMA measurement setup provided by CIC is shown in Fig. 11. The MMA measurement results are plotted in Fig. 12(a). The resonant frequency is about 42.9 K Hz with a corresponding −90° phase shift at resonant. From the measured amplitude response curve, it is seen that the 3dB bandwidth is about $\Delta f_{3dB} = 250Hz$ in Fig. 12(b). The response time of the switch is 14.28Hz, as shown in Fig. 13.The system is used to bio-medical environment. The operation frequency is about 1 ~ 2 Hz, so the response time of the switch is suitable for the system.

Figure 11. MEMS Motion Analyzer (MMA)

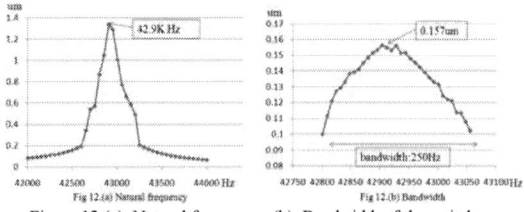

Figure 12.(a). Natural frequency. (b). Bandwidth of the switch.

In the static state of the switch, we use probes to control the voltage given and signal analysis. Fig. 14(a) is the photo by Optical Microscope (OM) and Fig. 14(b) is the measurement result by optical profiler. The gap is 19.863μm at initial time (0 V and 300 K). Fig. 15(a) is the switch in 20℃, and Fig. 15(b) is the switch in 200℃ by using MMA.

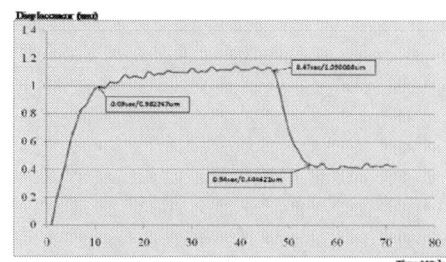

Figure 13. Response time of the switch.

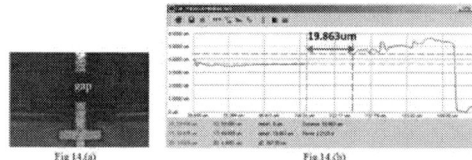

Fig 14.(a)　　　　　　　Fig 14.(b)

Figure 14.(a). One side of the switch. (b). Measurement result by optical profiler.

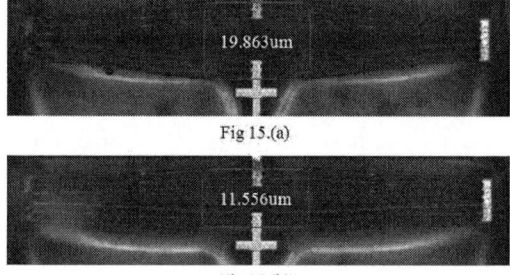

Figure 15.(a). Thermal switch at 20℃. (b). Thermal switch at 200℃

V. CONCLUSION

In this study, a MEMS-based variable capacitor and a new thermal switch was proposed and implemented by using CMOS MEMS process and CIC micromachining post process has been developed. In the prototype, the device size of the variable capacitor is 1100μm * 1100 μm, and the thermal switch size is 500 μm * 400 μm. The gap between two contact pads of the switch is 9 μm in off-state. According to the measurement result and the photo by OM, the switch can work stably at 3 volts, and the working temperature and operating bandwidth are individually 20-200 °C. The natural frequency of the switch is 42.9K Hz and the response time is 14.28Hz.

978-1-4673-1122-9/12 $31.00 © 2012 IEEE

ACKNOWLEDGMENT

This work was supported in part by National Science Council, Taiwan, under Contract Number 99-2218-E-039-001, 99-2220-E-009- 019, 99-2220-E-009-002, 99-2220-E-009-072-, 100-2220-E-009-031-, 100-2220-E-009-032-, 100-2220-E-009-019-, 100-2220-E-009-018- and in part by Taiwan Department of Health Clinical Trial and Research Center of Excellence under contract Number DOH99-TD-B-111-004 and Number DOH99-TD-C-111- 005. This work was also supported in part by the UST-UCSD International Center of Excellence in Advanced Bio-engineering sponsored by the Taiwan National Science Council I-RiCE Program under Grant Number NSC-99-2911-I-009-101. The authors would also like to thank National Chip Implementation Center (CIC) for chip fabrication.

REFERENCES

[1] Y. L. LAI, Y. Z. JUANG, S. C. HUANG, S. H. TSENG, H. H. TSAI,"Effective Mechanical Properties Measurements in Standard CMOS Process," *Asia-Pacific Conference on Transducers and Micro-Nano Technology (APCOT2010)* Perth, WA, June 2010, pp. 126.

[2] A.Oz, G. K. Fedder, "RF CMOS-MEMS capacitor having large tuning range," in *Proc. 2003 IEEE Int. Conference on Solid-State Sensors and Actuators (Transducers '03)* Boston, MA, June 2003, pp. 851-854.

[3] A.Oz, G. K. Fedder, "CMOS/BiCMOS Self Assembling and Electrothermal Microactutors for Tunable Capacitors, Gap-closing Structures and Latch Mechanisms," *2004 Solid-State Sensor, Actuator and Microsystems Workshop,* Hilton Head Is., SC, pp. 212-215.

[4] Callister, W.D., *Materials Science and Engineering, An Introduction*, 4th ed. New York: John Wiley and Sons, 1997.

A Research of the Bandwidth of a Mode-Matching MEMS Vibratory Gyroscope

Chunhua He, Qiancheng Zhao, Jian Cui, Zhenchuan Yang*, Guizhen Yan

National Key Laboratory of Science and Technology on Micro/Nano Fabrication,
Institute of Microelectronics, Peking University, Beijing 100871, P. R. China
z.yang@pku.edu.cn

Abstract—A novel electrostatic force feedback approach is presented for extending the bandwidth of a mode-matching vibratory microgyroscope. A 0.02 Hz mode-mismatch gyroscope is achieved by applying a DC voltage of 26 V to the squeeze-film combs to adjust the stiffness of the sense mode. Sweep-frequency tests demonstrate that the open loop frequency response is close to a one-order system with a bandwidth of 7.9 Hz, which agrees well with the theoretical simulation. Moreover, both experiment and simulation results show that the bandwidth can be extended to about 94Hz from 7.9Hz after adopting an electrostatic force feedback control.

Keywords- mode-matching; MEMS vibratory gyroscope; force feedback control; bandwidth

I. INTRODUCTION

Recently, mode-matching and high quality factor MEMS gyroscopes have been widely reported for improving the dynamic and static performances [1-5]. Many automatic mode-matching methods including one-time adjustment or on-line adjustment techniques have been proposed and proved effective at the same time [6, 7]. Thanks to the high signal-noise-ratio brought by mode-matching and high quality factor, the bias drift instability can reach sub-1 degree per hour, but it is at the cost of its bandwidth [1-5]. The higher of the quality factor, the smaller of the bandwidth. The bandwidth is sometimes smaller than 1Hz so that it cannot satisfy many application demands. Therefore, the contradiction between high performances and moderate bandwidth should be solved. Confronted with these problems, it is very necessary to investigate the bandwidth of the mode-matching gyroscope and extend it to the demands.

This paper extracts an equivalent linear model of the sense mode for the open loop control system by theoretical analyses, firstly. Then a novel generalized PID controller is designed for the electrostatic force feedback control to extend the bandwidth of the mode-matching gyroscope. Afterwards, simulations and experiments are carried out and the results are compared, followed by a conclusion in the end.

II. THEORETICAL ANALYSES

The block diagram of the electrostatic force rebalance control for the sense mode is shown in Fig.1, in which F_c and F_f are Coriolis force and feedback force, respectively. k_{cv} is the conversion gain from differential capacitances to voltage. Due

to the mode-matching, the gyroscope brings about a phase lag of ($\pi/2$) for the signals whose frequencies are equal to the drive resonant frequency ω_d, hence a signal of $2\cos(\omega_d t)$ and a two-order low pass filter $K_1(s)$ are used for demodulation to obtain the maximum in-phase component. After the control of the generalized PID controller $K_2(s)$, a remodulation process is required to construct the in-phase feedback voltage. Here, a signal of $2\sin\omega_d t$ is used for remodulation, and $y_i(t)$ is then used for the final output of the closed loop. After the function of a V/F converter, the final feedback forces F_f is generated to balance the Coriolis force of the sense mode.

However, the control loop is a nonlinear system due to the introduction of demodulation and remodulation, which makes it impossible to use classical linear control theory to design the controller. Fortunately, since the remodulation operation can retain the identical phase information between the feedback force and the external force, the Coriolis force rebalance control loop can be simplified to an equivalent linear model, as shown in Fig.2. H is the coefficient from angular rate to Coriolis force. Linear model $G(s)$ represents the response from the external force input to the output of the low pass filter.

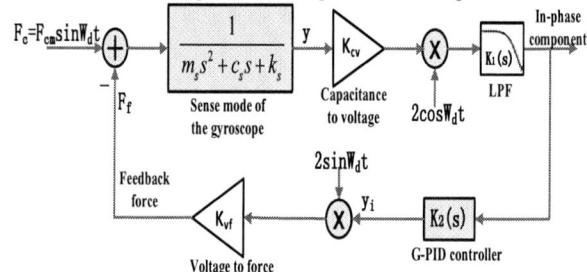

Fig.1. Block diagram of the sense force rebalance control.

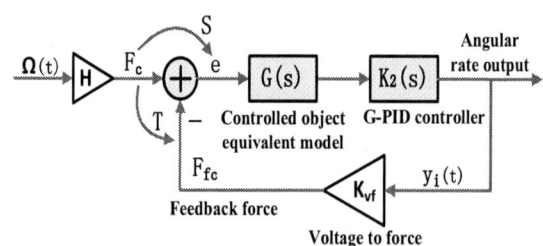

Fig.2. Equivalent linear model of the sense force rebalance control for in-phase component.

In order to maintain the MEMS gyroscope in an equilibrium position which means the sense displacement is close to zero, a negative force feedback control is utilized to counteract the external forces exerted on the proof mass in the sense direction. Hence, the equilibrium condition is described as,

$$F_f = F_C \tag{1}$$

Assume that the angular rate and the velocity of the drive mode are,

$$\Omega = \Omega_r \cos(\omega_r t) \tag{2}$$

$$v(t) = A_v \sin \omega_d t \tag{3}$$

Where ω_d represents the resonant frequency of the drive mode, and A_v is the stable amplitude controlled by velocity AGC control loop. The dynamical equation of the sense mode can be expressed,

$$m_s \ddot{y} + c_s \dot{y} + k_s y = -2m_p \Omega v \tag{4}$$

Where m_s and m_p are the sense mass and proof mass of the z-axis double decoupled tuning fork gyroscope, and c_s, k_s and y are the damping coefficient, stiffness coefficient and the displacement of the sense mode, respectively. Now, it is crucial to identify the precise linearization model of $G(s)$. The Coriolis force can be written as,

$$F_C = 2m_p A_v \Omega_r \cos(\omega_r t) \sin(\omega_d t)$$
$$= F_{cm} \cos(\omega_r t) \sin(\omega_d t) \tag{5}$$

Where $F_{cm} = 2m_p A_v \Omega_r$ is the amplitude of the Coriolis force. Then, the output of the capacitance-to-voltage component is expressed as,

$$y_{out} = \frac{F_{cm} k_{cv}}{2m_s} \{ -y_1 \sin[(\omega_d + \omega_r)t + \varphi_1]$$
$$+ y_2 \sin[(\omega_d - \omega_r)t + \varphi_2] \} \tag{6}$$

Where y_1, y_2, φ_1 and φ_2 are the functions of ω_r, as shown below,

$$y_1 = \frac{1}{\sqrt{[\omega_s^2 - (\omega_d + \omega_r)^2]^2 + \omega_s^2(\omega_d + \omega_r)^2 / Q_s^2}} \tag{7}$$

$$y_2 = \frac{1}{\sqrt{[\omega_s^2 - (\omega_d - \omega_r)^2]^2 + \omega_s^2(\omega_d - \omega_r)^2 / Q_s^2}} \tag{8}$$

$$\varphi_1 = \frac{\omega_s(\omega_d + \omega_r)}{Q_s[(\omega_d + \omega_r)^2 - \omega_s^2]} \tag{9}$$

$$\varphi_2 = \frac{\omega_s(\omega_d - \omega_r)}{Q_s[\omega_s^2 - (\omega_d - \omega_r)^2]} \tag{10}$$

Where ω_s and Q_s represent the resonant frequency and quality factor of the sense mode, respectively. Y_{out} is demodulated by a signal of $2\cos(\omega_d t)$ and then passes through a LPF to remove the high-frequency components. The in-phase component is thus described as,

$$V_{out} = \frac{F_{cm} k_{cv}}{2m_s} [y_1 \sin(\omega_r t + \varphi_1) + y_2 \sin(\omega_r t - \varphi_2)]$$
$$= A \cos(\omega_r t + \theta) \tag{11}$$

Where

$$A = \frac{F_{cm} k_{cv}}{2m_s} \sqrt{y_1^2 + y_2^2 + 2y_1 y_2 \cos(\varphi_1 - \varphi_2)} \tag{12}$$

$$\theta = \pi + arc \cot \frac{y_1 \sin \varphi_1 + y_2 \sin \varphi_2}{y_1 \cos \varphi_1 + y_2 \cos \varphi_2} \tag{13}$$

Therefore, the sense open loop normalized scale factor is written as,

$$SF_{open} = \frac{A \cos(\omega_r t + \theta)}{F_{cm} \cos(\omega_r t)} \tag{14}$$

When it comes to frequency-domain, and taking the LPF into account, the equivalent linear model of $G(s)$ can be derived as,

$$G(s) = SF_{open}(s) \times K_1(s) = SF_{open}(s) \times \frac{\omega_0^2}{s^2 + 2\xi_0 \omega_0 s + \omega_0^2} \tag{15}$$

Where ω_0 and ξ_0 are the cut-off frequency and damping ratio of the two-order low pass filter $K_1(s)$, and $s=j\omega_r$. $SF_{open}(s)$ is the original equivalent frequency response of the gyroscope, which can be obtained with (12)-(14). .

After the identification of $G(s)$, a generalized proportional-integration-derivative (GPID) controller can be designed for making a force feedback control to adjust the bandwidth. The transfer function of GPID controller is expressed as,

$$K_2(s) = \frac{k(s + \omega_2)(s + \omega_3)}{(s + \omega_1)(s + \omega_4)} \tag{16}$$

Where k is the overall gain. $(s+\omega_2)/(s+\omega_1)$ is a classical PI controller or can be regarded as a phase-lag compensator. $(s+\omega_3)/(s+\omega_4)$ is a special derivative controller or can be regarded as a phase-lead compensator to obtain better phase margin and gain margin.

When a DC voltage V_p is exerted to the squeeze-film combs in the sense mode, a corresponding electrostatic force is generated as,

$$F_e = \frac{N\varepsilon h l V_p^2}{2d^2} \tag{17}$$

Where h, l, d, ε, and N are the comb thickness, overlapped length, gap, permittivity and the number of the comb fingers, respectively. Then the electrostatic stiffness resulted from the derivative of F_e can be described as,

$$k_e = -\frac{\partial F_e}{\partial d} = \frac{N\varepsilon h l V_p^2}{d^3} \tag{18}$$

Therefore, the stiffness coefficient of the sense mode can be adjusted to achieve mode-matching based on this principle.

III. EXPERIMENTAL RESULTS

In this paper, we use a Z-axis double decoupled tuning fork gyroscope for experiments, as shown in Fig.3. There are two groups of combs in the drive mode, namely slide-film driving combs and slide-film drive-sensing combs. Simultaneously, there are three groups of combs in the sense mode, namely squeeze-film sensing combs, slide-film force feedback control combs, and squeeze-film stiffness adjustment combs. Fig.4 shows the experiment circuits which include the drive closed loop, mode-matching control loop and sense closed loop. Making use of the electrostatic stiffness modulation effect shown in (18), a DC voltage V_p of 26 V is exerted to the squeeze-film combs to adjust the stiffness coefficient of the sense mode. Sweep-frequency test results figure out that the drive resonant frequency and quality factor are 8965.56 Hz (f_d) and 4200 (Q_d), while those of sense are 8965.54 Hz (f_s) and 570 (Q_s), respectively. As shown in Fig.5, the mismatch frequency is only 0.02 Hz.

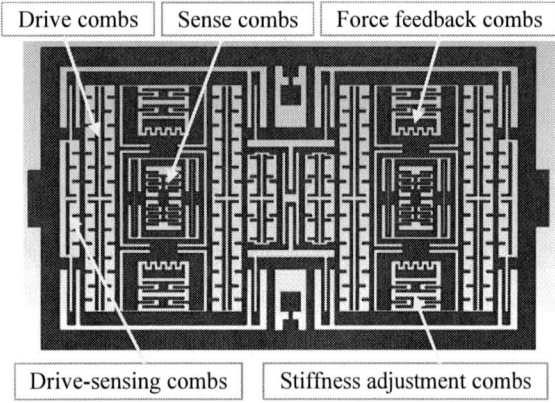

Fig.3. The schematic of the Z-axis double decoupled tuning fork gyroscope.

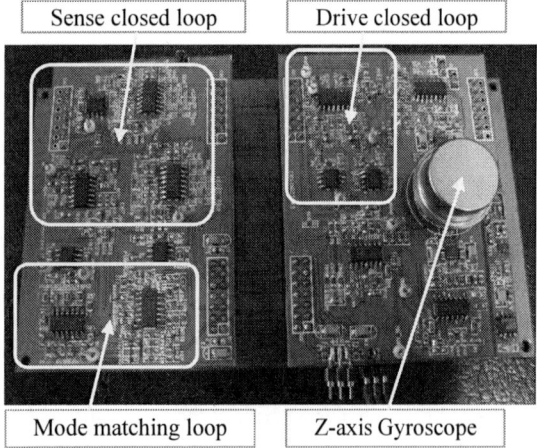

Fig.4. Test circuits for the Z-axis microgyroscope.

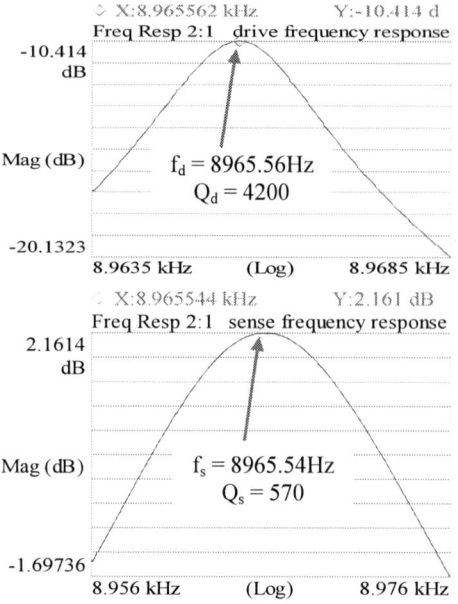

Fig.5. Sweep-frequency tests of the drive and sense modes.

The experimental result of the frequency response of $SF_{\text{open}}(s)$ is obtained with sweep-frequency test, and the simulation result is made with (12)-(14), as shown in Fig.6. From the amplitude-frequency responses and phase-frequency responses, we can see the experiment result agrees with the theoretical simulation result, and they both approach to an ideal one-order system. The bandwidth is 7.9 Hz which is equal to the theoretic value of $f_s/(2Q_s)$ [8]. Therefore, in the sense open loop, the mode-matching gyroscope system is approximately a one-order system with a very narrow bandwidth. If the quality factor of the sense mode is higher, the bandwidth will be smaller, which affects the application of the gyroscope.

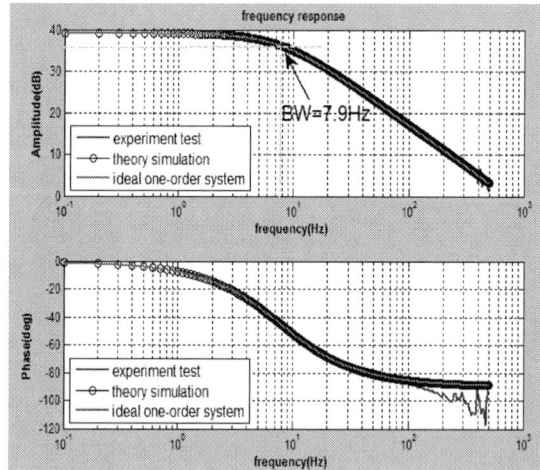

Fig.6. Open loop frequency responses of experiment test, theory simulation and ideal one-order system.

978-1-4673-1122-9/12 $31.00 © 2012 IEEE

After the sense closed loop control with a GPID controller, the bandwidth is extended to about 94 Hz which can satisfy more application demands, as shown in Fig.7. The experiment and simulation results coincide with each other and prove the effectiveness of the force rebalance control methodology. Besides, the phase margin and gain margin are 51° and 21dB, respectively, demonstrating that the sense closed loop is stable and robust.

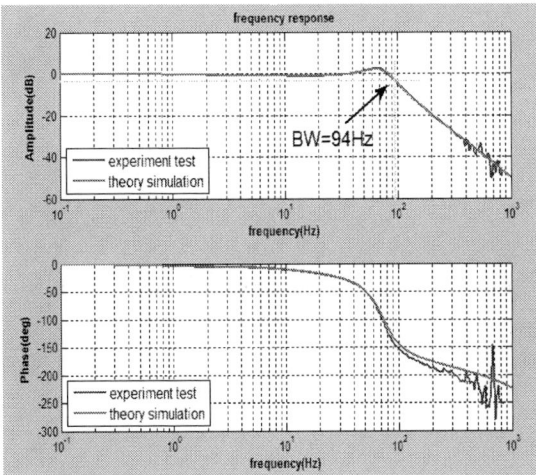

Fig.7. Frequency responses of experiment test and theory simulation for the sense closed loop system.

IV. CONCLUSION

In this paper, a 0.02 Hz mode-mismatch gyroscope is achieved by exerting a DC voltage of 26 V to the squeeze-film combs to adjust the stiffness coefficient of the sense mode. Then an equivalent linear model for the sense open loop is extracted by theoretical analyses. Both experiment and simulation results demonstrate the sense open loop system is a one-order system with the cut-off frequency of 7.9 Hz. Afterwards, a novel GPID controller is designed for the force feedback control in the sense mode. Experiment results show that the bandwidth is extended to about 94Hz in the sense closed loop system, which can satisfy more application demands.

ACKNOWLEDGMENT

The authors are grateful to the technical staffs of National Key Laboratory of Science and Technology on Micro/Nano Fabrication for their supports on MEMS gyroscopes fabrication.

REFERENCES

[1] M.F. Zaman, A. Sharma, Z. L. Hao, F. Ayazi, "A mode-matched silicon-yaw tuning-fork gyroscope with subdegree-per-hour Allan deviation bias instability", *J. Microelectromech. Syst.* Vol.17, pp.1526–1536, 2008.

[2] A. Sharma, M. F. Zaman, M. Zucher, and F. Ayazi, "A 0.1 deg/HR bias drift electronically matched tuning fork microgyroscope," in *21st Int. Conf. on. Micro Electro Mechanical System (MEMS2008)*, Tucson, AZ, pp. 6–9, 2008.

[3] A. Sharma, M. F. Zaman, and F. Ayazi, "A sub-0.2 °/hr bias drift micromechanical silicon gyroscope with automatic CMOS mode-matching", *IEEE J. Solid-State Circuits*, vol. 44, pp. 1593-1608, 2009.

[4] I.P. Prikhodko, S.A. Zotov, A.A. Trusov, and A.M. Shkel, "Sub-degree-per-hour silicon MEMS rate sensor with 1 million Q-factor", in *16th Int. Conf. on. Solid-state Sensors, Actuators and Microsystems (Transducers'11)*, Beijing, China, pp.2809 - 2812, 2011.

[5] W. K. Sung, M. Dalal, and F. Ayazi, "A mode-matched 0.9 MHz single proof-mass dual-axis gyroscope", in *16th Int. Conf. on. Solid-state Sensors, Actuators and Microsystems (Transducers'11)*, Beijing, China, pp.2821-2824, 2011.

[6] B.S. Chang, W.T. Sung, J. G. Lee, K.Y. Lee, and S.K. Sung, "Automatic mode matching control loop design and its application to the mode matched MEMS gyroscope", in *2007 ICVES*, pp. 1-6, 2007.

[7] R. Antonello, R. Oboe, L. Prandi, F. Biganzoli, "Automatic mode matching in MEMS vibrating gyroscopes using extremum-seeking control", *IEEE Trans. Ind. Electron.*, vol.56, pp. 3880-3891, 2009.

[8] M. H. Bao, *Micro mechanical transducers: pressure sensors, accelerometers, and gyroscopes*. 1st ed. Amsterdam, The Netherlands: Elsevier, 2000.

Graphene Nanoribbon Based AM Demodulator of Terahertz Radiation

Yury Stebunov[1], Aleksey Arsenin[1], Vladimir Leiman[1], Vyacheslav Semenenko[1], and Victor Ryzhii[2]

[1]Laboratory of Nanooptics and Femtosecond Electronics,
Moscow Institute of Physics and Technology, Dolgoprudny 141700, Russia
[2]Computational Nanoelectronics Laboratory, University of Aizu, Aizuwakamatsu, Fukushima 965-8580, Japan
ystebunov@gmail.com

Abstract—**We proposed a novel scheme of AM demodulator of terahertz radiation modulated by megahertz or gigahertz signals. Demodulator model is based on the nano-electromechanical system with the sufficiently narrow graphene nanoribbon as a mechanically moving part. Motion of a nanoribbon is supported by the electromagnetic ponderomotive force appearing due to the plasmon resonance exited in the system. Using developed model we obtained that the proposed demodulator has a number of advantages such as higher output signal and higher modulation frequency of incoming radiation in comparison with the previously proposed demodulators of terahertz radiation.**

Keywords-graphene nanoribbon, AM demodulator, terahertz radiation, nano-electromechanical system

I. INTRODUCTION

In the last years technology in the terahertz (THz) frequency band attracts a lot of attention. The terahertz radiation shows great potential application in a large variety of areas [1]. In spite of this the use of terahertz radiation is complicated by the absence of convenient, small-scaled and reliable sources and detectors, and also other devices required for operation with the radiation [2]. Nevertheless one of the hopeful applications of the THZ radiation is communication and sensing, where it could be more profitable than the radiation in the megahertz range [3]. The key components of the communication process are detectors and demodulators of radiation. Recently there were proposed highly efficient demodulators of terahertz radiation based on the concept of the transistor with a mechanically floating gate [4-9]. This idea initially appeared in the work of Nathanson [10]. The operation of such devices is based on a principle of influence on the current in a channel through the mechanical movement of a gate. In the previous works the role of channel was played by the two-dimensional gas, and authors studied the influence by the floating gate on plasma waves in channel [4-8]. In recent work [9] we considered device model of terahertz AM demodulator, where carbon nanotube was used simultaneously as a channel for plasmon excitations and a floating gate of the device. Output characteristics of such device exceed those for demodulators based on the two-dimensional gas.

Now we propose device model of terahertz demodulator with plasmon oscillations excited in the sufficiently narrow graphene

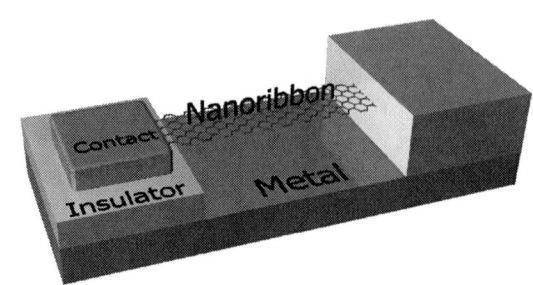

Fig. 1. General view of the device under consideration

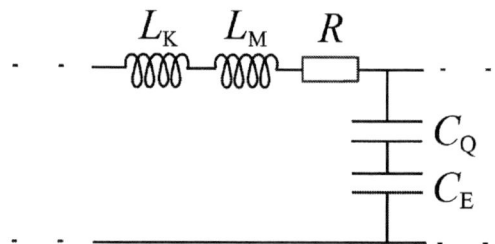

Fig. 2. Transmission line model of AM demodulator

nanoribbon (GNR) (Fig. 1). The possibility of plasmon excitation in GNR was observed experimentally [11]. It was shown that narrow graphene nanoribbons could behave as quantum wires [12]. One of possible ways to describe plasmons in quantum wires is usage of the model based on the theory of Tomonaga-Luttinger liquid (TLL) [13]. Convenient description of this approach is based on the use of the model of transmission line with additional elements, that describe quantum properties of the electron behavior in such systems (Fig. 2) [14-15]. It is worth noting another advantage of graphene nanoribbons use in the model of demodulator as their unique mechanical properties [16]. GNRs could be used as mechanical resonators in schemes of nano-electromechanical system exploiting quite high frequencies of mechanical resonance in the range of megahertz and gigahertz [17-25].

978-1-4673-1122-9/12 $31.00 © 2012 IEEE

II. DEVICE MODEL

A. Principles of operation

Device model of proposed AM demodulator of terahertz radiation consists of double clamped narrow graphene nanoribbon placed over surface of metal (Fig. 1). Under influence of incoming terahertz radiation plasmon sare exited in GNR. Graphene nanoribbon and underlying metal form a capacitor, so between them the ponderomotive force exists. This mechanical force leads to excitation of mechanical resonator associated with the vibrations of GNR at a modulation frequency of incoming THz radiation. So mechanical and electromagnetic resonators are coupled through the nonlinear ponderomotive force.

Bias voltage is applied between GNR and metal surface. Therefore alternating current appears due to mechanical oscillations of GNR and amplitude of this current is proportional to the squared amplitude of plasmon wave in GNR, modulation depth of incoming radiation and biased voltage. So this ac current constitutes the output signal of the demodulator.

B. TLL model for the electron transport in a narrow GNR

Electron oscillations in a narrow graphene nanoribbon could be described in the framework of the theory of Tomonaga-Luttinger liquid applicable to the description of plasmons in quantum wires [13]. It is possible to represent this theory as an equivalent circuit model of transmission line (TL) (Fig. 2) [14-15]. In the case of GNR we introduce the quantum capacitance $C_Q = 8e^2 / h v_F$, the kinetic inductance $L_K = h / 8e^2 v_F$ and the quantum resistance in the equivalent circuit in addition to the ordinary electric capacitance C_E and magnetic inductance L_M (h is the Plank's constant, e is the charge of an electron) Equations describing this TL could be represented in the following form:

$$-\frac{\partial v}{\partial x} = L_{eff} \frac{\partial i}{\partial t} + R_{eff} i; \quad -\frac{\partial i}{\partial x} = C_E \frac{\partial v}{\partial t}. \quad (1)$$

Here $v = v(x,t)$ and $i = i(x,t)$ are responsively the distributions of the voltage and current along TL,

$$L_{eff} = \frac{L_M + L_K}{1 + C_E / C_Q},$$

$$R_{eff} = \frac{R_{pul}}{1 + C_E / C_Q},$$

where $R_{pul} = R_Q / 4 l_{mfp}$, l_{mfp} is the electron mean free path, x is directed along GNR.

We have chosen the length of GNR to be equal to an odd number of one-quarter of plasmon wavelength. It is known that such open-circuited TL is equivalent to a series resonant circuit [26]. For definiteness we consider one quarter wavelength transmission line. So we obtain the solution for the voltage amplitude of standing plasmon wave represented in the form of TL: $v_0 = v_\omega H(\omega)$, where v_ω is an amplitude on incoming terahertz radiation,

$$H(\omega) = \frac{Q}{\sqrt{Q^2 (1 - \omega^2 / \Omega^2)^2 + \omega^2 / \Omega^2}} \quad (2)$$

is the frequency response function. Here Q is the quality factor of a plasmon resonator, $\Omega = \pi v_F / 2l \sqrt{C_Q / C_E}$ is the resonant frequency of a plasmon resonator. For the typical parameters of considered system (distance between GNR and underlying metal $W = 15$ nm, length $l = 500$ nm, GNR width $d = 10$ nm, $l_{mfp} = 500$ nm) $L_K = 3.2 \cdot 10^{-3}$ H/m, $L_K \gg L_M$, $C_E = 2.3 \cdot 10^{-11}$ F/m, $C_E / C_Q \approx 0.1$ and taking into account the contact resistance we obtain $\Omega / 2\pi \approx 1.8$ THz and $Q \approx 4$.

C. Mechanical oscillation of the doubly clamped GNR

Graphene and different graphene structures are the exceptional candidates for using in nano-electromechanical systems due to their unique mechanical properties and lightness [16]. Different mechanical resonators based on graphene were studied both in theory and experiment[17-25]. One of the theoretical model which describes mechanical properties of graphene mechanics is those based on the structural mechanics, where Young's module E, Poisson's ratio σ, density ρ and thickness of graphene monolayer t were computed [22]. In our estimations we will use following mechanical properties: $E = 1$ TPa, $\sigma = 0.16$, $\rho = 2250$ kg/m^3 and $t = 0.335$ nm [27]. Also one of the benefits of GNR is their possibility to perform transverse [17-20, 22-25] and torsional vibrations [21]. Theory of elasticity gives us the solutions for vibration eigenmodes [28]. So frequencies for transverse eigenmodes are $\Omega_{tr,i} = k_i \sqrt{E / 12 \rho h^2}$ (k_i are the solutions of $\cosh kl \cos kl = 1$) and for torsional eigenmodes are $\Omega_{tor,i} = i\pi / l \sqrt{4h^2 / d^2 \cdot E / 2\rho(1 + \sigma)}$ (i is an integer). It is worth noting a fact that for size of GNRs considered in this work frequencies of first modes of torsional vibrations are one-two orders of magnitude greater than of transverse vibrations. So for typical parameters we received for transverse vibrations: $\Omega_{tr,1} = 29$ MHz, $\Omega_{tr,1} = 80$ MHz, and for torsional vibrations: $\Omega_{tor,1} = 0.93$ GHz, $\Omega_{tor,2} = 1,85$ GHz .

978-1-4673-1122-9/12 $31.00 © 2012 IEEE

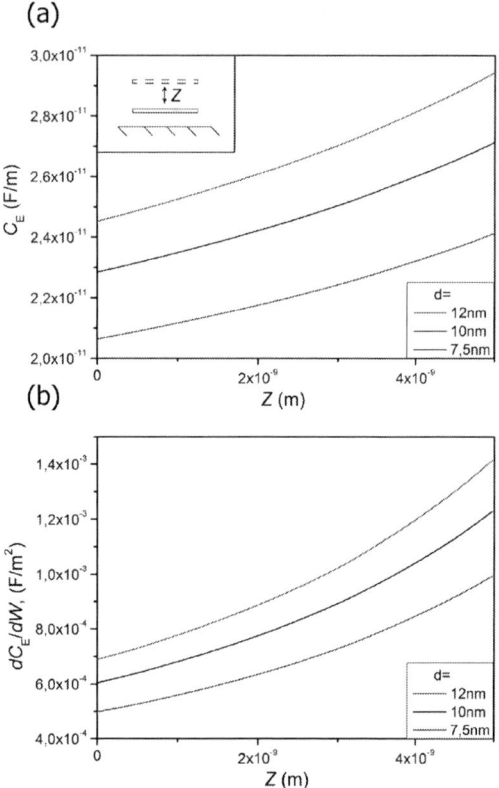

Fig. 3. (a) Capacitance and (b) derivative of the capacitance of a system under consideration as a function of GNR spacing

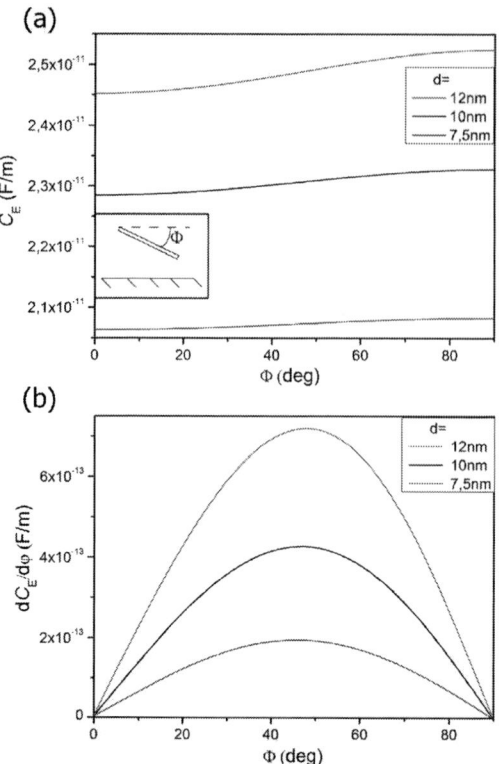

Fig. 4. (a) Capacitance and (b) derivative of the capacitance of a system under consideration as a function of GNR rotating

D. Ponderomotive forces in the demodulator and equations of GNR motion

As kinetic inductance is much higher than magnetic one we could neglect the magnetic constitute of electromagnetic energy. Therefore the force per unit length acted on the graphene nanoribbon in the case of transverse vibrations is

$$f \approx -\frac{d}{dW}\left(\frac{C_E v^2}{2} + \frac{L_M i^2}{2}\right) \approx -\frac{v^2}{2}\frac{dC_E}{dW} \qquad (3)$$

Here $v = v_0(1 + \cos\omega_m t)\cos\omega t$, where ω_m is the modulation frequency of incoming radiation.

And moment of force in the case of torsional vibrations is

$$M_f \approx -\frac{d}{d\varphi}\left(\frac{C_E v^2}{2} + \frac{L_M i^2}{2}\right) \approx -\frac{v^2}{2}\frac{dC_E}{d\varphi} \qquad (4)$$

So ponderomotive force depends on the change of systems capacitance due to motion of GNR. It is possible to estimate electric capacitance in the framework of finite element method. We consider two-dimensional system of the equipotential plate over the ground with the help of commercial program COMSOL.

We computed capacitance of such system as the function of plate position defined in the case of transversal vibrations by the distance between plate and ground (Fig. 3), and in the case of torsional vibrations by the angle of rotational displacement (Fig. 4). It is worth noting that for the case of torsional oscillations the capacitance derivative is in maximum when GNRs are inclined, so to rise the device responsivity GNRs should be positioned at an angle to metal plane.

In our model the equation of GNR movement for transverse vibrations is

$$\ddot{Z} + 2\gamma_{tr}\dot{Z} + \Omega_{tr}^2 Z = \frac{1}{M}\int_0^l f\,dx, \qquad (5)$$

where Z is characteristic variation of the spacing between GNR and metal (so that $W = W_0 + Z$, W_0 is the spacing in equilibrium), γ_{tr} is pertinent damping coefficient of transverse oscillations.

And the equation of GNR movement for torsional vibrations is

$$\ddot{\Phi} + 2\gamma_{tor}\dot{\Phi} + \Omega_{tor}^2 \Phi = \frac{1}{I}\int_0^l M_f\,dx, \qquad (6)$$

where Φ is characteristic variation of the rotating of oscillating GNR, γ_{tor} is the pertinent damping coefficient of torsional oscillations, $I = d^3 h l \rho / 12$ is the moment of inertia.

Solutions of these equations are respectively:

$$Z = H(\omega)^2 H_{tr}(\omega_m) \frac{\alpha_m v_m^2 l}{4M\Omega_m^2} \frac{\partial C_E}{\partial W} \cos(\omega_m t), \qquad (7)$$

$$\Phi = H(\omega)^2 H_{tor}(\omega_m) \frac{\alpha_m v_m^2 l}{4I\Omega_m^2} \frac{\partial C_E}{\partial \varphi} \cos(\omega_m t), \qquad (8)$$

where

$$H(\omega) = \frac{Q}{\sqrt{Q^2(1 - \omega^2/\Omega^2)^2 + \omega^2/\Omega^2}} \qquad (9)$$

is the frequency response function of mechanical resonator.

III. DEVICE RESPONSIVITY

One of major parameters that defines the quality of demodulation by proposed device is its responsivity. It defines as the ration of alternating current amplitude in output circuit to the energy of terahertz radiation consumed by the system:

$$R_{tr,tor} = \frac{J_{tr,tor}}{P_\omega}. \qquad (10)$$

Here $J_{tr,tor}$ are the current amplitudes appearing in the output circuit in the case of transverse and torsional oscillations. Current amplitudes depend on the amplitudes of mechanical oscillations:

$$J_{tr} = 2\omega_m V_0 l \frac{\partial C_E}{\partial W} Z \text{ and } J_{tor} = 2\omega_m V_0 l \frac{\partial C_E}{\partial \varphi} \Phi, \qquad (11)$$

where V_0 is a bias voltage.

Incoming radiation received by the antenna is:

$$P_\omega = \frac{2Gc\varepsilon_0}{\pi} v_\omega^2, \qquad (12)$$

where G is the antenna gain, c is the speed of light and ε_0 is the vacuum dielectric constant.

So we could define demodulator responsivity:

$$R_{tr} = V_0 \frac{\pi \alpha_m l^2 \omega_m}{GM\Omega_{tr}^2 c_i \varepsilon_0} \left(\frac{\partial C_E}{\partial W} \right)^2 H(\omega)^2 H_{tr}(\omega_m). \qquad (13)$$

Analogous expression we could obtain for the responsivity of demodulator based on torsionally oscillating GNR.

For typical parameters of the system and $Q = 4$, $\alpha_m = 0.1$, $Q_M = 10^3$ we obtain values:

$$R_{tr} = 2 \text{ [A/W]}, \ R_{tor} = 2 \cdot 10^{-3} \text{ [A/W]} \qquad (14)$$

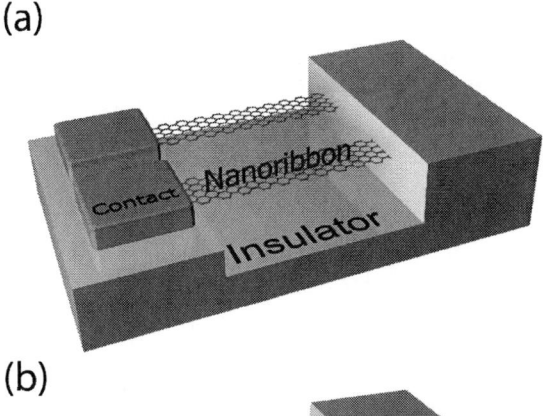

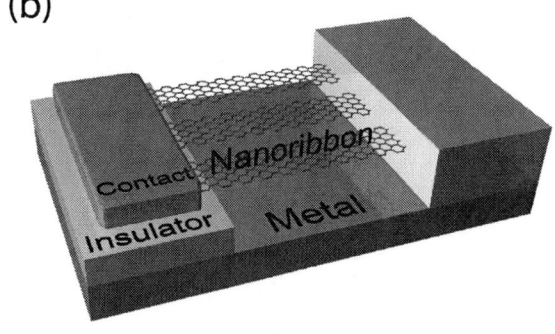

Fig. 5. Schematic views of AM demodulators based on (a) double GNR, (b) GNR array.

Obtained value of responsivity is much higher than for demodulators based on 2DEG [1-8], and it is approximately equal to the value of responsivity of demodulators based on CNTs [9].

Besides the considered scheme of demodulator based on the GNR it is possible to propose another schemes based on the double GNR over insulating surface (Fig. 5a) and on the GNR array over metal (Fig. 5b). In the case of the GNR array the value of responsivity is proportional to the number of graphene nanoribbons.

Also there is another approach to rise the responsivity by varying a number of quantum channels in the GNR through the change in its width. Therefore we could rise the quality factor of plasmon resonance.

IV. CONCLUSION

In conclusion we proposed schemes of AM demodulator of terahertz radiation based on GNRs. We developed the model that describes electrodynamics and mechanics of proposed device. Different mechanic vibration modes of GNR were considered. Also the responsivity of the device was calculated as a function of carrier and modulation frequencies. Responsivity exhibits sharp resonance at combined plasma-mechanical resonance and in the peak of this resonance it exceeds analogous values for previously proposed devices.

ACKNOWLEDGMENT

This work was supported in part by the Russian Foundation for Basic Research (grants no. 11-07-12072-ofi-m, 10-07-00618 and 11-07-00505) and by the Ministry of Education and Science of the Russian Federation (grants no. 16.513.11.3129, 07.514.11.4086, 14.740.11.1412 and 14.740.11.1135).

REFERENCES

[1] P.H. Siegel, "Terahertz Technology", *IEEE Trans. on Microwave Theory and Techniques*, vol. 50, pp. 910-927, March 2002.

[2] M. Tonouchi, "Cutting-edge terahertz technology", *Nat. Photonics,* vol. 1, pp. 97-105, February 2007.

[3] J. Federici and L. Moeller, "Review of terahertz and subterahertz wireless communications", *J. Appl. Phys.*, vol. 107, 111101, 2010.

[4] V. Ryzhii, M. Ryzhii, Y. Hu, I. Hagiwara, and M.S. Shur, "Resonant detection of modulated terahertz radiation in micromachined high-electron-mobility transistor", *Appl. Phys. Lett.*, vol. 90, 203503, May 2007.

[5] Y. Hu, M. Ryzhii, I. Hagiwara, M.S. Shur and V. Ryzhii, "Combined resonance and resonant detection of modulated terahertz radiation in a micromachined high-electron mobility transistor", *Phys. Stat. Sol. C.*, vol. 5, pp. 277-281, 2008.

[6] V.G. Leiman, M. Ryzhii, A. Satou, N. Ryabova, V. Ryzhii, T. Otsuji, and M.S. Shur, "Analysis of resonant detection of terahertz radiation in high-electron mobility transistor with a nanostring/carbon nanotube as the mechanically floating gate", *J. Appl. Phys.*, vol. 104, 024514, 2008.

[7] A.V. Arsenin, A.D. Gladun, V.G. Leiman, V.L. Semenenko and V.I. Ryzhii, "Parametric instability in a nanoelectromechanical detector of modulated terahertz radiation on the basis of a high electron mobility transistor with a mobile elastic gate", J. Commun. Technol. Electron., vol. 54, pp. 1322-1330, 2009.

[8] A.V. Arsenin, A.D. Gladun, V.G. Leiman, V.L. Semenenko and V.I. Ryzhii, " Plasma oscillations of the two-dimensional electron gas in the field-effect transistor with a cylindrical gate electrode", J. Commun. Technol. Electron., vol. 55, pp. 1285-1294, 2010.

[9] Yu. Stebunov, V. Leiman, A. Arsenin, A. Gladun, V. Semenenko, V. Ryzhii, *"Detection of Modulated Terahertz Radiation Using Combined Plasma and Mechanical Resonances in Double-Carbon-Nanotube Device"*, *Appl. Phys. Express*, vol. 4, 075101, 2011.

[10] H.C. Nathanson, W.E. Newell, R.A. Wickstrom, J.R. Davis, " The resonant gate transistor", *IEEE trans. on Electron Devices*, vol. 14, pp. 117-133, March 1967.

[11] L. Ju, B. Geng, J. Horng, C. Girit, M. Martin, Z. Hao, H.A. Bechtel, X. Liang, A. Zettl, Y.R. Shen, F. Wang, "Graphene plasmonics for tunable terahertz metamaterials", *Nat. Nanotechnology,* vol. 6, pp. 630-634, 2011.

[12] X. Wang, Y. Ouyang, L. Jiao, H. Wang, L. Xie, J Wu, J. Guo, H Dai, "Graphene nanoribbons with smooth edges behave as quantum wires", *Nat. Nanotechnology,* vol. 6, pp. 563-567, 2011.

[13] T. Giamarchi, *Quantum Physics in One Dimension.* Oxford: Oxford University Press, 2003.

[14] P.J. Burke, "Luttinger Liquid Theory as a Model of the GHz Electrical Properties of Carbon Nanotubes", *IEEE Transactions on Nanotechnology*, vol. 1, pp. 129-144, 2002.

[15] C. Rutherglen, P.J. Burke, "Nano-Electromagnetics: Circuit and Electromagnetic Properties of Carbon Nanotubes", *Small*, vol. 5, pp. 884-906, 2009.

[16] K. Eom, H.S. Park, D.S. Yoon, T. Kwon, "Nanomechanical resonators and their applications in biological/chemical detection: Nanomechanics principles", *Phys. Rep.,* vol. 503, pp. 115-163, 2011.

[17] J.S. Bunch, A.M. van der Zande, S.S. Verbridge, I.W. Frank, D.M. Tanenbaum, J.M. Parpia, H.G. Craighead and P.L. McEuen, " Electromechanical Resonators from Graphene Sheets", *Science*, vol. 26, pp. 490-493, January 2007.

[18] D. Garcia-Sanchez, A.M. van der Zande, A. San Paulo, B. Lassagne, P.L. McEuen, and A. Bachtold, "Imaging Mechanical Vibrations in Suspended Graphene Sheets", *Nano Lett.*, vol. 8, pp. 1399-1403, 2008.

[19] J.T. Robinson, M. Zalalutdinov, J.W. Baldwin, E.S. Snow, Z. Wei, P. Sheehan, and B.H. Houston, "Wafer-scale Reduced Graphene Oxide Films for Nanomechanical Devices", *Nano Lett.*, vol. 8, pp. 3441-3445, 2008

[20] Y. Xu, C. Chen, V.V. Deshpande, F.A. DiRenno, A. Gondarenko, D.B. Heinz, S. Liu, P. Kim, and J. Hone, "Radio frequency electrical transduction of graphene mechanical resonators", *Appl. Phys. Lett.*, vol. 97, 243111, 2010.

[21] B.D. Urso, M. V. Gurudev Dutt, S. Dhingra, and N.M. Nusran, "Quantum measurements between a single spin and a torsional nanomechanical resonator", *New J. of Phys.*, vol. 13, 045002, 2011.

[22] S. Kitipornchai, X.Q. He, and K.M. Liew, "Continuum model for the vibration of multilayered graphene sheets", *Phys. Rev. B*, vol. 72, 075443, 2005.

[23] A. Sakhaee-Pour, M.T. Ahmadian, and R Naghdabadi, "Vibrational analysis of single-layered graphene sheets", *Nanotech.*, vol. 19, 085702, 2008.

[24] S.Y. Kim and H.S. Park, "The Importance of Edge Effects on the Intrinsic Loss Mechanisms of Graphene Nanoresonators", *Nano Lett.*, vol. 9, pp. 969-974, 2009.

[25] R. Chowdhury, S. Adhikari, F. Scarpa, and M.I. Friswell, "Transverse vibration of single-layer graphene sheets", *J. Phys. D: Appl. Phys.*, vol. 44, 205401, 2011.

[26] R. E. Collin: Foundations for Microwave Engineering, Wiley, New York, 2nd ed, 2000.

[27] C. Lee, X. Wei, J.W. Kysar, J. Hone, "Measurement of the Elastic Properties and Intrinsic Strength of Monolayer Graphene", *Science*, vol. 321, pp. 385-388, 2008.

[28] L.D. Landau and E.M. Lifshitz, *Theory of Elasticity*, Pergamon, New York, 1986.

Atomic layer deposited protective coatings for integrated MEMS flow sensor

Dan Li[1*], Aziz Abdulagatov[2], Fang Yang[1], D. C. Zhang[1]

[1] Institute of Microelectronics, Peking University, Beijing, 100871, China

beidalidan@126.com

[2] Department of Mechanical Engineering, University of Colorado, Boulder, CO, 80309, USA

Abstract— **This paper investigates the Atomic layer deposited (ALD) coatings on an integrated Microelectromechanical (MEMS) flow sensor to resist ironic contamination of on-chip CMOS circuits. 25nm Al_2O_3/TiO_2 multilatyer thin film was deposited on the integrated flow sensor. To assess the ability of ALD coatings to resist K^+ and Na^+ contamination, the flow sensor was immersed into NaCl and KCl solutions for 1 week, respectively. The CMOS transistors fabricated on the same chip were tested after long time immersion and results showed that the performance of the CMOS circuits were not degraded, which implied this integrated flow sensor with ALD protective coatings could be used under K^+ and Na^+ environment.**

Keywords-ALD protective coatings; integrated MEMS flow sensor; on-chip CMOS circuits ;

I. INTRODUCTION

As is well known, the most important failure mechanism of semiconductor devices is associated with ionic contamination (mainly Na^+ and K^+). The phenomenon is particularly serious in n-channel MOS, where the presence of a positive charge in the gate oxide induces silicon surface inversion and therefore the current may flow from source to drain out of the control of the gate [1]. For integrated MEMS devices with on-chip CMOS circuits, it is sometimes inevitable for the device to be exposed to K^+ and Na^+ environment. Thus, it is urgent to develop an effective way to protect the on-chip CMOS circuits from ironic contamination and at the same time not affect the performance of the MEMS device. Atomic layer deposition (ALD) has become an attractive thin film deposition technology during the last decades and semiconductor processing has been one of the main motivations for the recent development of ALD, including high dielectric constant gate oxides in the MOSFET structure, copper diffusion barriers in backend interconnects and high quality dielectrics to fabricate trench capacitors for DRAM [2]. Recently, ALD is applied to MEMS devices to improve the reliability. For example, moving parts can be protected by such conformal coatings from wear, stiction in suspended MEMS devices can be eliminated by such thin films, adhesion caused by the build-up of static charge could be alleviated using conducting ALD coatings and ALD coating using biocompatible materials such as TiO_2 can be used in biocompatible devices [3, 4].

ALD shows a number of unique characteristics comparing with other thin film technique, such as chemical vapor deposition (CVD) and self-assembling monolayer (SAM) techniques. First, ALD offers conformal deposition on high aspect structures and precise thickness control at the Ångstrom or monolayer level using sequential, self-limiting surface reactions, which remarkably meets the requirement of MEMS protective coatings. ALD films will cover all sides of a released MEMS device as long as the gaps between the different components are larger than the ALD precursor molecules. Secondly, ALD can be performed at very low temperature (as low as 35℃ for Al_2O_3), which makes it possible to be applied to the CMOS integrated circuit without damaging the device. In addition, ALD films tend to be very continuous and pinhole-free because the reactions are driven to completion during every reaction cycle and no surface sites are left behind during film growth. Ceramic corrosion protection coatings have typical thickness of at least one micro. However, such continuous and pinhole-free ALD coating is promising to protect the underlying objects with very thin films. Another advantage of ALD worth mentioning is the large-area/large-batch processing capability of ALD, which makes it possible to achieve competitive productivities.

This paper investigates the Al_2O_3/TiO_2 multilayer ALD coatings on an integrated MEMS flow sensor to protect the on-chip CMOS circuits from K^+ and Na^+ contamination. In this study, released MEMS flow sensor with on-chip CMOS circuits was coated by Al_2O_3/TiO_2 ALD with film thickness of 25 nm. The effect of the ALD layers on the corrosion protection was evaluated by testing the characteristics of on-chip CMOS transistors and an amplifier before and after long-term immersion into 3.5 wt. % NaCl and 5 wt. % KCl solutions.

II. DEPOSITION PROCESS

The integrated flow sensor used in this study was fabricated through the 3.0 μm single poly single metal process in National Key Laboratory of Micro/Nano Fabrication Technology in Peking University based on an inter - CMOS MEMS process. Detailed design, fabrication process and testing results are described in our earlier publications [5]. The sensing element of this flow sensor contains four released diaphragms with the thickness of 30μm and the on-chip CMOS signal conditioning circuit. The thin film deposition was carried out in a viscous-flow, hot-wall type ALD reactor shown in figure 1 [6]. The ALD process is demonstrated in figure 2. The notches in the starting substrate for reaction A represent discrete reactive surface sites. Exposing this surface to reactant A results in the self-terminating adsorption of a monolayer of A species. The resulting surface becomes the starting substrate for reaction B. Subsequent exposure to molecule B will cover the surface with a monolayer of B species. Consequently, one AB cycle deposits one monolayer of the compound AB and regenerates the initial substrate. By repeating the binary reaction sequence ABAB... fashion, a film of any thickness can be deposited [3]. Such thin-film growth technique relies on a binary reaction sequence of self-limiting chemical reactions between gas phase precursor molecules and a solid surface, which allows atomic scale thickness control. Deposition was performed in static mode in order to conformably coat the high aspect diaphragms. Prior to TiO_2 deposition, about 60 Å thick ALD Al_2O_3 (alumina) was pre-deposited as a seed layer. This layer is needed to insure proper TiO_2 nucleation. The sequential dosing of trimethylaluminum (TMA) and water were used for ALD Al_2O_3 where titanium tetrachloride (TiCl4) and water were used for ALD TiO_2. The surface reactions to form this Al_2O_3/TiO_2 film are demonstrated in equation (1) and (2). The chemistry of ALD Al_2O_3 and TiO_2 were well studied and understood and has been shown to result in high quality film [7, 8]. The coating process was performed at a reaction temperature of 120 °C. At this deposition temperature, both Al_2O_3 and TiO_2 are in amorphous state [4, 10].

$$2Al(CH_3)_3 + 3H_2O \rightarrow Al_2O_3 + 3CH_4 \qquad (1)$$

$$TiCl_4 + 2H_2O \rightarrow TiO_2 + 4HCl \qquad (2)$$

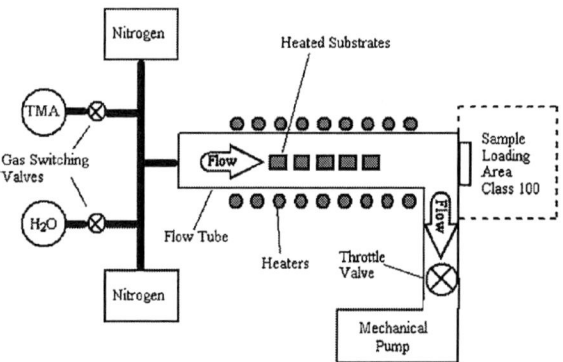

Fig. 1. Schematic of viscous flow reactor for Al_2O_3/TiO_2 [6].

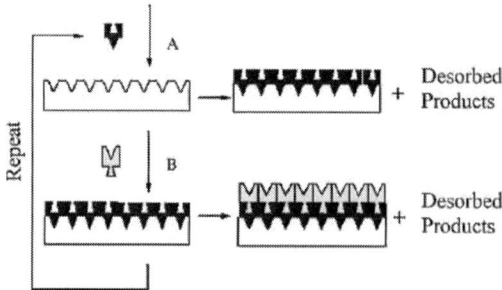

Fig. 2. Schematic of ALD process using a binary reaction sequence [3].

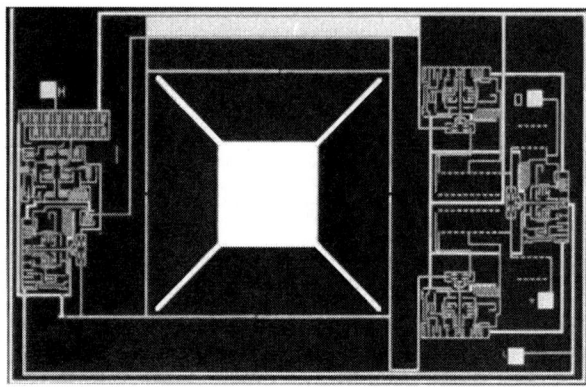

Fig. 3. Integrated MEMS flow sensor with ALD coating

III. TESTS AND RESULTS

The Al_2O_3/TiO_2 films grown in this study were amorphous, uniform and well adhered to the substrate. To illustrate the ability of this thin film to protect ironic contamination, the ALD coated flow sensors were immersed into 3.5 wt. % NaCl and 5 wt. % KCl solutions for 1 week at room temperature, respectively. The transfer characteristics of on-chip CMOS transistors before and after immersion were both tested. HP 4156 precision semiconductor parameter analyzer was used in this test and both NMOS and PMOS were tested. Figure 4 (a) and (b) show the transfer characteristics of NMOS and PMOS transistors before and after 1 week immersion in 3.5 wt. % NaCl, respectively. Figure 5 (a) and (b) show the transfer characteristics of NMOS and PMOS transistors before and after 1 week immersion in 5 wt. % KCl, respectively. No visible deviations of the transfer curves have been detected after immersion in both NaCl and KCl solutions. The threshold voltages for the transistors before and after corrosion in different solutions with different concentrations are listed in table 1 and table 2 and the differences are within acceptable range. A closed-loop amplifier fabricated on the same integration chip was also tested. Agilent 93000 was used to test the characteristics of the closed loop amplifier and the dc curves of both the amplifiers before and after immersion into 3.5 wt. % NaCl and 5 wt. % KCl solutions are shown in figure 6 and figure 7, respectively. It is observed that the performance of the amplifier has little change after long term immersion. This revealed the integrated flow sensor with ALD protective coatings could be used under K^+ and Na^+ environment.

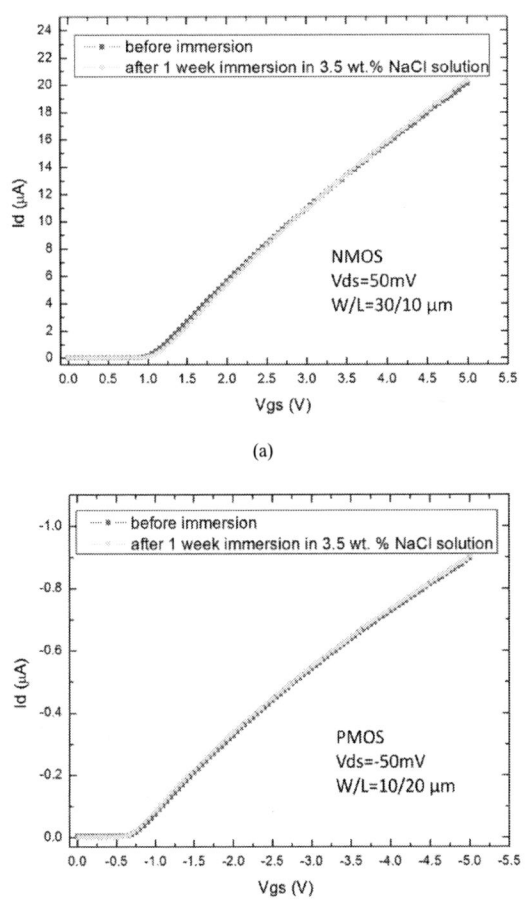

Fig. 4. Transfer characteristics of CMOS transistors before and after 1 week immersion in 3.5 wt. % NaCl

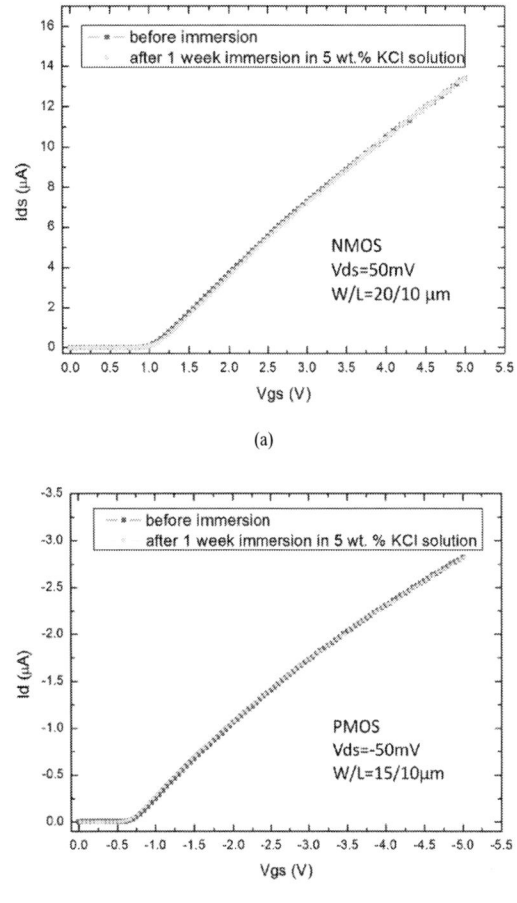

Fig. 5. Transfer characteristics of CMOS transistors before and after 1 week immersion in 5 wt. % KCl

TABLE I.   THRESHOLD VOLTAGES OF CMOS TRANSISTORS BEFORE AND AFTER CORROSION

Solution	Transistor type	V_{th} before corrosion (V)	V_{th} after corrosion (V)
3.5 wt. % NaCl	NMOS	1.056	1.053
	PMOS	-0.906	-0.905
5 wt. % KCl	NMOS	1.062	1.060
	PMOS	-8.928	-8.93

TABLE II.   THRESHOLD VOLTAGES OF CMOS TRANSISTORS BEFORE AND AFTER CORROSION

Solution	Transistor type	V_{th} before corrosion (V)	V_{th} after corrosion (V)
10.5 wt. % NaCl	NMOS	1.056	1.052
	PMOS	-0.906	-0.903
15 wt. % KCl	NMOS	1.062	1.059
	PMOS	-8.928	-8.933

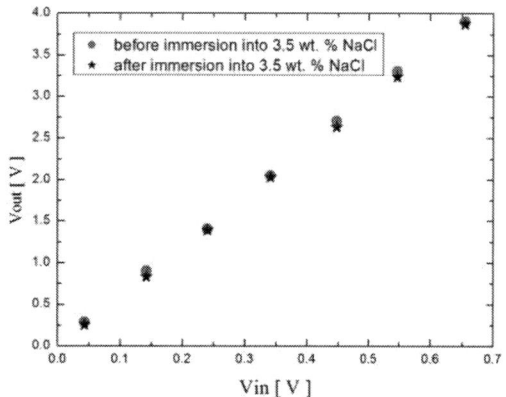

Fig. 6. Closed-loop dc characteristic of an amplifier. The pink circle is the testing data of the amplifier before immersion into 3.5 wt. % NaCl. The black star is the testing results of the dc characteristic of the amplifier after immersion into 3.5 wt. % NaCl.

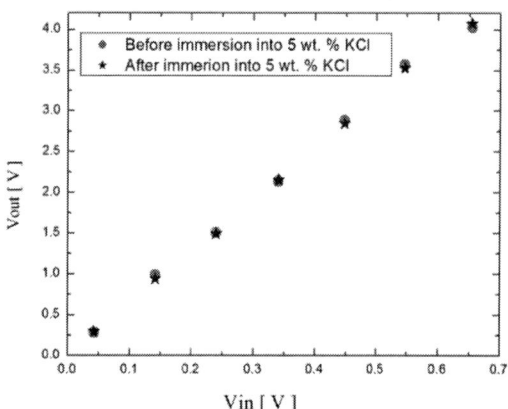

Fig. 7. Closed-loop dc characteristic of an amplifier. The pink circle is the testing data of the amplifier before immersion into 5 wt. % KCl. The black star is the testing results of the dc characteristic of the amplifier after immersion into 5 wt. % KCl.

IV. CONCLUSION

This paper investigates the atomic layer deposited protective coatings for an integrated MEMS flow sensor. 25nm Al_2O_3/TiO_2 multilayer ALD were coated on an integrated MEMS flow sensor to protect the on-chip CMOS circuits from K^+ and Na^+ contamination. The thin film was deposited in a viscous-flow, hot-wall type ALD reactor and the process was performed at the temperature of 120℃, which would not damage the CMOS circuits. To evaluate the protective ability of Al_2O_3/TiO_2 against the K^+ and Na^+ contamination, the flow sensor was immersed into 3.5 wt. % NaCl and 5 wt. % KCl solutions for 1 week at room temperature, respectively. The characteristics of both NMOS and PMOS transistors were tested before and after long time immersion. The dc characteristic of a closed loop amplifier was also tested. The testing results indicated that the performance of the CMOS circuits were not degraded due to the long time immersion, which suggests that this integrated MEMS flow sensor with ALD protective coatings can be used under K^+ and Na^+ environment. The influence of this ALD coating on the sensor diaphragm stiffness will be the future work.

ACKNOWLEDGMENT

The authors would like to thank the staff of National Key Laboratory of Nano/Micro Fabrication Technology for cooperation and help in the fabrications. The work is financially supported by National Basic Research Program of China (863 Program) under Grant No. 2011AA040401 and National Basic Research Program of China (973 Program) under Grant No. 2011CB309502.

REFERENCES

[1] F. Fantini, "Reliability Problems with VLSI," *Microelectron. Reliab.*, vol. 24, pp. 529-551, 1984.

[2] S. M. George, "Atomic layer deposition: An overview," *Chem. Rev.*, vol. 110, pp. 111–131, 2010.

[3] Nils D. Hoivik, Jeffrey W. Elam, Ryan J. Linderman, Victor M. Bright, Steven M.George, Y.C. Lee, "Atomic layer deposted protective coatings for micro-electromechanical systems," *Sensors and Actuators A*, vol. 103, pp. 100-108, 2003.

[4] J. Aarik, A. Aidla, T. Uustare, V. Sammelselg, J. Crystal Growth, "Morphology and structure of TiO_2 thin films grown by atomic layer deposition," *J. Crystal Growth*, vol. 148, pp. 268-275, 1995.

[5] D. Li, T. Zhao, Z. C. Yang and D. C. Zhang, "monolithic integrated micromachined piezoresistive flow sensor," *J. Micromech. Microeng.*, vol. 20, pp. 035024.A, 2010.

[6] Elam, J. W., M. D. Groner, and S. M. George, "Viscous flow reactor with quartz crystal microbalance for thin film growth by atomic layer deposition" *Review of Scientific Instruments*, vol. 73, pp. 2981-2987, 2002.

[7] Puurunen, R. L., "Surface chemistry of atomic layer deposition: A case study for the trimethylaluminum/water process," *Journal of Applied Physics*, vol 97, pp. 121301, 2005.

[8] Ritala, M., et al., "Growth of Titanium-Dioxide Thin-Films by Atomic Layer Epitaxy," *Thin Solid Films*, vol. 225, pp.288-295, 1993.

[9] Jakschik, S., Schroeder, U., Hecht, T., Gutsche, M., Seidl, H., Bartha, J. W., "Crystallization behavior of thin ALD-Al_2O_3 films," *Thin Solid Films*, vol. 245, pp. 216-220, 2003.

NEMS Based Tools for Nanoscience and Atomic Clocks

N.F. de Rooij[*,1,2], S.Gautsch[1], T. Akiyama[1], F. Loizeau[1], G. Mileti[3], Yves Pétremand[1], U. Staufer[4],
Rahel Straessle[1], G. Yoshikawa[5]

[1] Ecole Polytechnique Fédérale de Lausanne (EPFL), The Sensors Actuators and Microsystems Laboratory (SAMLAB)
[2] Centre Suisse, d'Electronique et de Microtechnique SA, both at Rue Jaquet-Droz 1, CH-2002 Neuchâtel, Switzerland.
[3] Université de Neuchâtel, Laboratoire Temps-Fréquence (LTF), Neuchâtel, Switzerland
[4] Technical University of Delft, Dpt. of Precision and Microsystems Engineering, Delft, The Netherlands
[5] World Premier International (WPI) Research Center, International Center for Materials Nanoarchitectonics (MANA),
National Institute for Materials Science (NIMS), 1-1 Namiki, Tsukuba, Ibaraki 305-0044, Japan
Nico.derooij@epfl.ch

Abstract—Nanoscience is a thriving multi-disciplinary activity, which aims at understanding the properties and the interaction of very small objects on the nanometer scale. In this endeavor, tools for the observation, analysis and modification of individual objects like macromolecules, clusters or even single atoms are required. The development of dedicated microfabricated instruments to measure physical and chemical interactions at this scale is therefore required. This talk will give an overview of microfabrication techniques employed to shape such NEMS based tools and introduce the audience to several probing techniques. In a second part, we focus on the principles and fabrication techniques of atomic clocks

Keywords; NEMS, AFM, SNOM, Mars, Arthoscope, MSS, Insulated conductive probes, atomic clocks

I. TOOLS FOR NANOSCIENCE

As tweezers are common tools used as an extension of the human hand to manipulate smaller objects, specific NEMS devices allow for the manipulation of objects in the nanometer range and even below.

The most common tool known today to perform such operation is the Atomic Force Microscope (AFM) [1], which allows to measure the interaction forces between a sharp tip and the surface to be assessed. Designed initially to image the topography of surfaces down to the atomic scale, this technique has evolved to give birth to a whole range of AFM related diagnostic tools, where additional physical and chemical properties can be inferred. For instance, an insulated conductive apex, shown in figure 1, will allow to immerse the tool into electrolytic solutions and to assess transports properties of single molecules under applied external electric

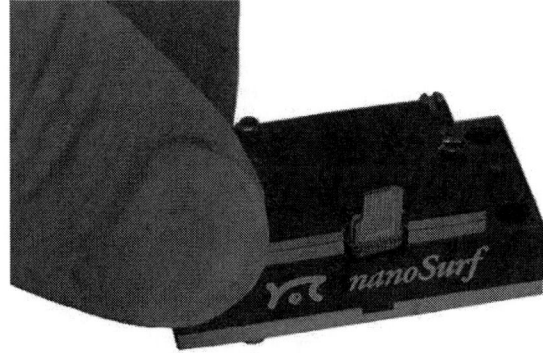

Fig. 2. AFM setup designed for the Phoenix mission to Mars in 2008. The picture shows the scanning head (black box) with mounted AFM chip equipped with 8 support beams and cantilevers.

fields, and simultaneously record its mechanical contraction [2]. An other example of such tip functionalization is the scanning nearfield optical microscope (SNOM) [3], which guides light through the apex of a modified AFM tip, and allows for simultaneous assessment of optical and topographical information.

In general, apart from the cantilever and the tip, AFM based instruments are quite bulky in order to guarantee as sufficient level of noise isolation from the outside. For some specific applications, the whole measurement setup has to be miniaturized to an extreme level in order to complete its duty. This was the case when designing an AFM instrument for a Mars mission (figure 2) to measure the properties of dust on

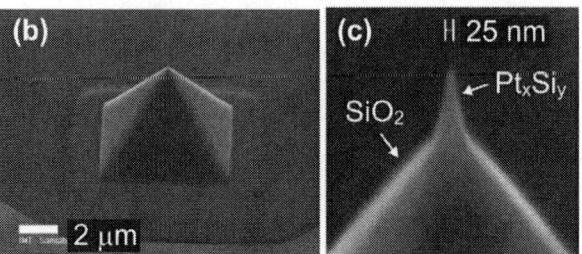

Fig. 1. SEM images of an AFM cantilever with integrated conductive path to the tip apex. The whole cantilever is insulated, except for the very apex of the tip.

Fig. 3. Schematic illustration of the membrane type surface stress sensor with p-type piezoresistors on n-type single crystal Si(100)

978-1-4673-1122-9/12 $31.00 © 2012 IEEE

the surface of the planet [4], or for developing a scanning force arthroscope to investigate the properties of cartilage inside the human knee joint [5].

Other tools have been developed to detect the presence of nanosized objects in gases or liquids. By transduceing the surface stress of a thin polymeric layer to 4 piezoresistive suspension hinges (figure 3), the membrane type surface stress sensor (MSS) presented below is capable of detecting species bound to the polymeric receptor layer with extremely high sensitivity [6].

The major drawback of most probe-based technique is the low throughput and limited area that can be investigated in a single measurement. Efforts are thus often focusing on parallelization. In the following example, the measurement of nanomechanical properties of cells as well as cell-cell interactions are of particular interest for cancer research. For this method to acquire statistical information, a probe array system is suggested that allows measuring multiple cells in parallel [7]. In contrast to the sharp tips of AFM imaging, well defined round tips are required for cell investigations (figure 4).

II. ATOMIC CLOCKS

Atomic frequency standards provide stable and accurate

Fig. 5. Micro-fabricated Rubidium vapour cell (courtesy of C. Schori and Y. Petremand). Dimensions: 10x10x1.5 mm.

Fig. 4. Cantilever array for the mechanical assessment of cell properties. Spring constants of the cantilever and tip radius of the tip apex can be controlled through the molding process used to fabricate the array.

timing references in all those applications that require better performances than quartz oscillators.

During the last forty years, a number of developments concerned atomic clocks for space applications (GPS, GLONASS, GALILEO, etc.) and fundamental science. More recently, there has been a strong effort to drastically reduce the size and power consumption of vapour cell atomic clocks, (figure 5) [8]. One very promising development in this field concerns the use of microresonators [9].

A low-temperature Indium hermetic bonding technique on wafer level without using flux, active atmosphere has been developed very recently [10]. The simplicity and low temperatures compared to other indium sealing techniques [11, 12] allow encapsulation of sensitive MEMS devices. The technique applied to fabricate Rb vapor-cells for chip-scale atomic clocks (CSAC) at low temperature, which is a major step towards wall-coated alkali vapor-cells for CSACs [[1]13].

III. ACKNOWLEDGEMENTS

We would like to acknowledge the Swiss National Science Foundation (SNF) for the support on the insulated conductive probe activities under project Sinergia CRSII2_126969. The AFM on Mars mission was financially supported by the Wolfermann-Naegeli Foundation, SPP Minast, the Space Center at EPFL and IMT. The MSS activities run under direct collaboration between IMT-EPFL and MANA-NIMS. The cantilever array activities are supported by the Swiss federal initiative Nanotera through the project PATLiSci. The activities on miniature atomic clocks were funded by the Conférence Universitaire Suisse, the Swiss National Science Foundation and the European Space Agency. We gratefully thank the clean room staff of CSEM's Microsystems Technology Division for their technical assistance in microfabrication.

REFERENCES

[1] G. Binnig, et al., Physical Review Letters, vol. 56, p. 930, 1986

[2] T. Akiyama et al., Japanese Journal of Applied Physics, Vol. 43, pp. 3865-3867, 2003

[3] L. Aeschimann, et al., Journal of Microscopy, Vol. 209, pp. 182-187, 2003,

[4] S. Gautsch et al., Surface and Interface Analysis, Vol. 33, pp. 163-167, 2002,

[5] R. Imer et al., Journal of Physics: Conference Series, Vol. 61, pp. 467-471, 2006

[6] G. Yoshikawa et al. Nano Lett., 2011, 11 (3), pp 1044–1048

[7] F. Loizeau et al., to be published in Micro and Nanoletters

[8] S. Knappe et al., Appl. Phys. Lett., vol. 85, pp. 1460-1462, (2004)

[9] P. Del'Haye et al., , Nature 450, 1214 (2007)

[10] R. Straessle et al., Proceedings of MEMS 2012, to be published

[11] M. Volpert et al., Proc. ETCT 2009, pp. 224-231

[12] T. F. Marinis and J. W. Soucy, Proc. ECTC 2011, pp. 1552-1561

[13] J. Kitching et al., Metrologia 42 (2005), pp. 100-104

978-1-4673-1122-9/12 $31.00 © 2012 IEEE

Beyond Watson and Crick: Programming DNA Self-assembly for Nanofabrication

Paul W.K. Rothemund

*Department of Bioengineering, California Institute of Technology, USA

pwkr@dna.caltech.edu

Abstract—**The specificity of Watson-Crick base-pairing allows great control over the design and synthesis of artificial DNA nanostructures. Periodic one-dimensional (1D) tubes, 2D lattices, and 3D crystals are easily created from "DNA tiles" with spacings of 10–20 nanometers. "DNA origami" allows the folding of long DNA strands into arbitrary shapes and patterns up to about 100 nanometers in size with five nanometer features. This suggests that DNA self-assembly might be used for sublithographic fabrication of devices or even circuits. There are, however, a number of challenges. For example, compared to the inorganic materials used as semiconductors, DNA has poor electronic properties and must be augmented by other materials. Such challenges must be solved before DNA self-assembly can be used in a complete process for nanofabrication. We will present partial solutions to a few of them.**

Keywords-DNA; self-assembly; nanofabrication

I. MOTIVATION AND RESULTS

DNA origami provides a simple solution-phase synthesis for arbitrary two- [1] and three-dimensional [2] shapes and patterns (Fig. 1). Each origami is typically comprised of ~200 short "staple strands" of DNA, which fold a ~7000 base long "scaffold strand" of DNA in an annealing process which takes approximately two hours. Each staple strand can be chemically modified as desired. Thus an origami shape has ~200 features, which can be used as binding sites for other, more functional, materials. For example (Fig. 2), we used a rectangular DNA origami to organize DNA-functionalized carbon nanotubes into cross junctions, and verified that the crosses could serve as field-effect transistors [3].

Two difficulties arise when trying to organize such functional nanodevices using DNA origami. First, if the DNA origami are coupled to the nanodevices in solution, nanodevices which carry multiple DNA linkers can crosslink the origami into undesired aggregates. Second, when origami-device constructs are deposited on a substrate they do so at random positions with random orientations. This means that the devices must be located by ultramicroscopy, and "wired-up" in order to be tested—typically using a custom e-beam pattern. To better integrate DNA origami with microfabrication, we have developed a process [4] by which an e-beam generated lithographic pre-pattern is used to define origami-shaped sticky patches, to which DNA origami bind and orient from solution (Fig. 3). Once down on the surface, origami may be functionalized by nanodevices, such as plasmonic gold nanoparticles, without the possibility of aggregation. Further, such origami-organized devices are pre-registered for downstream processing.

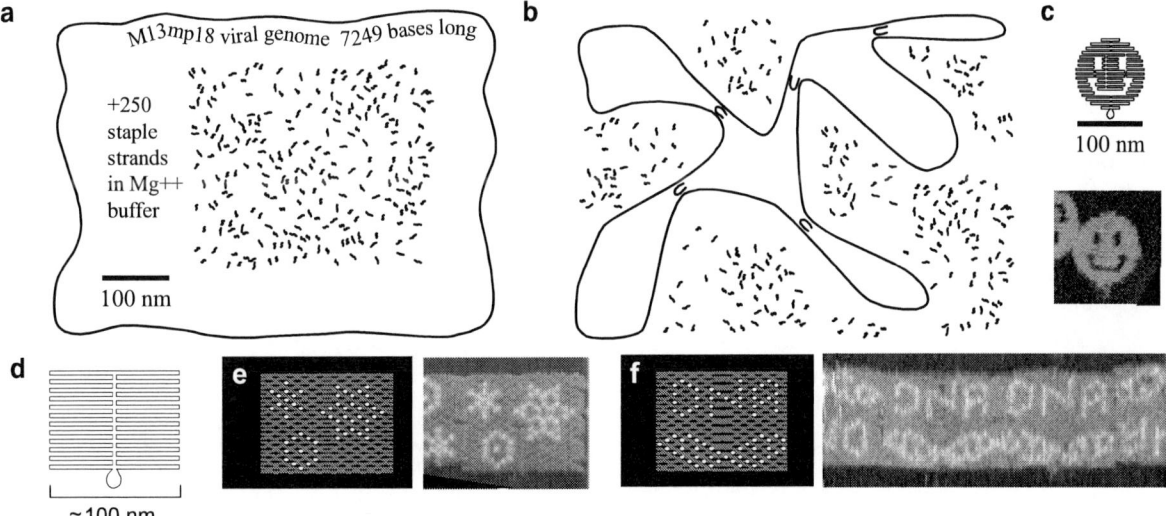

Fig. 1 The technique of DNA origami. (a) unfolded scaffold at 90°C. (b) cooling allows staples to bind and bring distant scaffold segments together. (c) schematic folding and AFM image of a DNA origami "smiley face" at 20°C. (d) schematic folding for a rectangle. (e-f) two different patterns of DNA hairpins demonstrating 200 pixel patterns.

978-1-4673-1122-9/12 $31.00 © 2012 IEEE

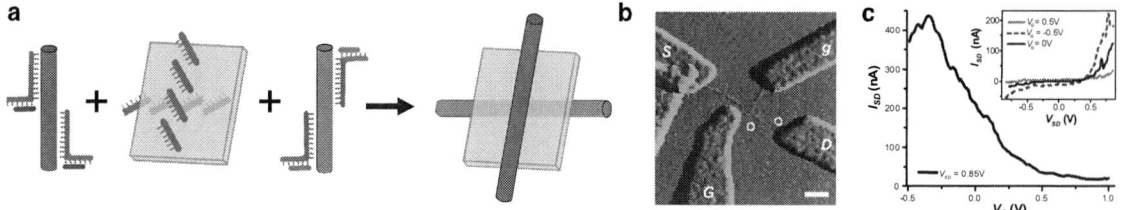

Fig. 2: DNA origami used to template a carbon nanotube field-effect transistor. (a) assembly scheme for a carbon nanotube cross junction. (b) cross junction between e-beam written electrodes. (c) switching behavior of the device.

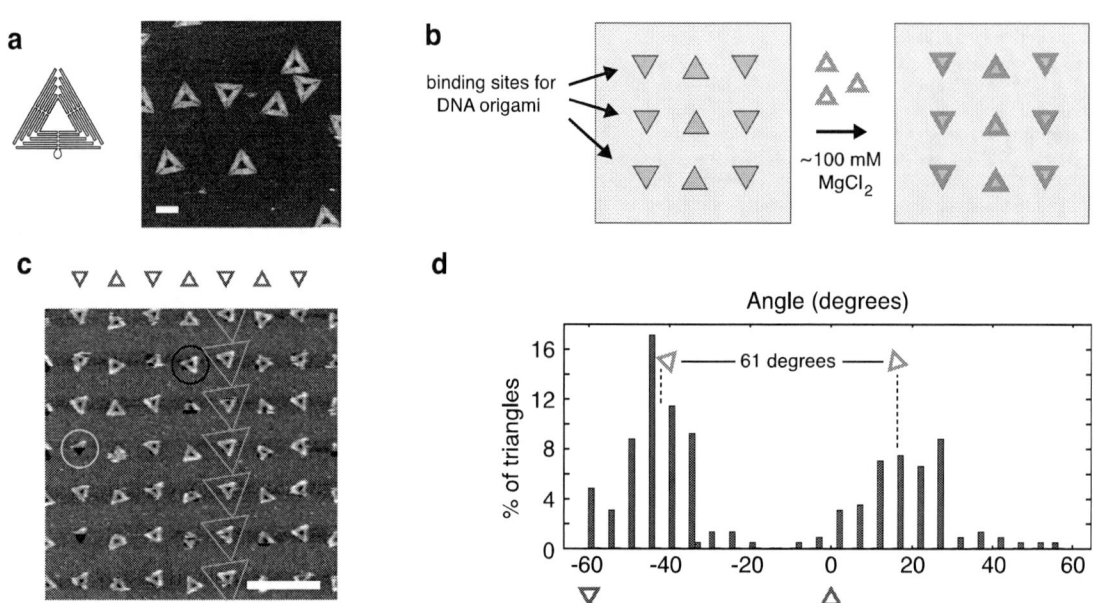

Fig. 3: Directed self-assembly of DNA origami on lithographic substrates. (a) DNA origami triangles randomly deposited on mica. (b) scheme for assembly on e-beam written sticky patches. Yellow is a non-sticky diamond-like carbon (DLC) background. Gray indicates oxygen plasma-etched triangles, which bind magnesium ions, which in turn bind DNA origami. (c) AFM data of triangles oriented on patterned DLC. (d) histogram of triangle orientations, demonstrating fidelity of alignment to two different orientations of sticky patch.

Finally, just as real origami are limited by the size of the paper from which they are made, so too are DNA origami limited by the length of the scaffold strand. To make larger structures with more features, we explored a method of combining DNA origami that does not depend on Watson-Crick base pairing: we created "stacking bonds" [5] based on the geometric arrangement of blunt-end stacking interactions. We showed that both binary codes and shape complementarity can serve as a basis for stacking bonds, and explored their specificity, thermodynamics and binding rules. Orthogonal stacking bonds were used to connect five distinct DNA origami to achieve ~1000 features.

REFERENCES

[1] P.W.K. Rothemund, "Folding DNA to create nanoscale shapes and patterns," *Nature*, vol. 440, pp. 297-302, March 2006.

[2] S.M. Douglas et al., "Self-assembly of DNA into nanoscale three-dimensional shapes," *Nature*, vol. 459, pp. 414-418, May 2009.

[3] H.T. Maune et al., "Self-assembly of carbon nanotubes into two-dimensional geometries using DNA origami templates," *Nature Nanotechnology*, vol. 5, pp. 61-66, November 2010.

[4] R.J. Kershner et al., "Placement and orientation of individual DNA shapes on lithographically patterned surfaces," *Nature Nanotechnology*, vol. 4, pp. 557-561, August 2009.

[5] S. Woo and P.W.K. Rothemund, "Programmable molecular recognition based on the geometry of DNA nanostructures," *Nature Chemistry*, vol. 3, pp. 620-627, August 2011.

Recent Advances on Nano-materials for Advanced Packaging Applications

C.P. Wong

Dean of the Faculty of Engineering
The Chinese University of Hong Kong
Shatin, NT, Hong Kong
Email: cpwong@cuhk.edu.hk

On leave from
School of Materials Science and Engineering
Georgia Institute of Technology
Atlanta, GA 30332-0245
Phone: 404-894-6631; Fax: 404-894-9140
Email: cp.wong@mse.gatech.edu

The advance of semiconductor technology is mainly due to the advances of materials, especially polymeric materials. These include the use of polymers as: resists (for deep submicron lithography), adhesives (both conductive and non conductive for die attach and assembly interconnects), interlayer dielectrics (low k, low loss dielectrics for high speed and low loss signal transmission), encapsulants (discrete and wafer level packages for device protection), embedded passives (high K capacitors, high Q inductors for high density PWB substrates), superhydrophobic self-cleaning lotus effect surfaces, etc. In this presentation, I will review some of the recent advances on polymeric materials and polymer nanocomposites that are currently being investigated for these types of applications, such as : lead-free electrically conductive adhesives (ECAs) with self assembly monolayer molecular wires for fine pitch and high current density interconnects, flip chip and wafer level underfills, nano lead-free alloys for low temperature interconnects, nanometal particle composites for high k embedded passives, well-aligned carbon nanotubes and graphenes for high current and high thermal interface materials(TIMs), and superhydrophobic self-clean lotus surface coatings for high efficiency solar cell applications.

References

[1] C. P. Wong, K. Moon, and Y. Li Ed., *"Nano-Bio-Electronic, Photonic and MEMS Packaging"*, Springer, (2010)

[2] Y. Li, D. Lu, C. P. Wong, *"Electrically Conductive Adhesives with Nanotechnologies"*, Springer, published (2010)

[3] Y. Li, K. Moon and C. P. Wong, *"Nano Conductive Materials For Electronic Packaging"*, Chapter 2, in "Nano-Bio- Electronic, Photonic and MEMS Packaging", C. P. Wong, K. Moon and Y. Li, Ed., Springer (2010)

[4] H. Jiang, K. Moon and C. P. Wong, "Nano Lead-free Solder Pastes for Low Processing Temperature Interconnect Applications in Microelectronic Packaging", Chapter 8, in "Nano-Bio- Electronic, Photonic and MEMS Packaging", C. P. Wong, K. Moon and Y. Li, Ed., Springer (2010)

[5] W. Lin and C. P. Wong, "Applications of Carbon Nano-materials as Electrical Interconnects and Thermal Interface Materials", Chapter 17, in "Nano-Bio- Electronic, Photonic and MEMS Packaging", C. P. Wong, K. Moon and Y. Li, Ed., Springer (2010)

Printed Carbon Nanotube Devices and Their Applications

Zheng Cui, Jianwen Zhao, Ting Zhang

Printable Electronics Research Center, Suzhou Institute of Nanotech,
Chinese Academy of Science, Suzhou 215123, China
zcui2009@sinano.ac.cn

Printed electronics represents a paradigm shift in electronics manufacturing after 50 years of development in silicon microelectronics. Manufacturing of electronic devices by printing can significantly reduce the equipment investment which is nowadays astronomical in the latest generation of very large scale integrated circuits manufacturing. Printing is an additive process where electronic materials are printed layer by layer to form the final device structure, which is independent of the properties of substrate materials. Such a feature enables electronic devices to be made on flexible substrate, such as plastic or paper, and of much larger size than a silicon wafer.

The advances in nanomaterials have led to the rapid development of printed electronics in recent years [1]. Caron nanotube is one of the most promising nanomaterials which is of high electric charge mobility and chemical stability, and can be formulated into ink form and printed as conducting or semiconducting layers in a device structure. However, carbon nanotubes are generally in a mixture of metallic and semiconducting nanotubes. To use them as either conducting or semiconducting ink, they have to be well separated. To form a printable ink, various surfactants have to be used in order to thoroughly disperse the nanotubes in a solvent. The presence of surfactants in a device layer is not desirable and has to be removed. In addition, the printing methods and process conditions can make big differences in the performance of printed devices. In general, making electronic devices by printing nanomaterials is not a straightforward process and many issues have to be solved.

In this paper, development in formulation of single-wall carbon nanotubes ink and printing of field effect transistors are reported. A simple and scalable method to print high performance thin-film transistors (TFTs) with high yield by an aerosol jet printer has been developed, using water-based and chemically functionalized single-wall carbon nanotube inks as the semiconducting layer. Printed TFTs with high yield were obtained by tuning the concentrations of single-wall carbon nanotubes and surfactants in the inks. The printed side-gate and bottom-gate TFTs exhibited the effective mobility of ~1 cm^2/Vs and on/off ratio over 10^3. In addition, these printed TFTs were hysteresis-free and able to operate at low-voltage when using printable high-capacitance ion gel as the dielectric layer. Figure 1 shows the all printed carbon nanotube based thin film transistor and its characteristics. Applications of printed carbon nanotube transistors as nanosensors are presented. Figure 2(a) shows a simple printed carbon nanotube sensor. By suitable functionalization, it can be used as a gas sensor. Figure 2(b) shows the response curves of the printed sensor for detection of ammonia gas.

References

[1] Dan Rogers, "Nanomaterials are becoming synonymous with printed electronics", +Plastic Electronics, Vol.3(6), p.35, 2011

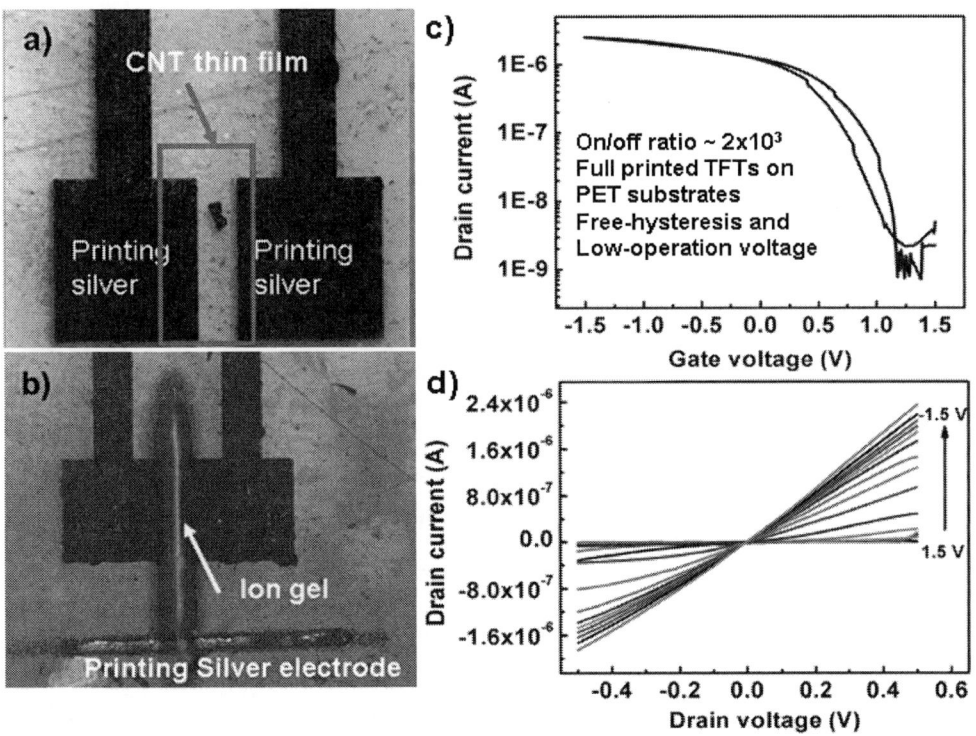

Fig. 1 All printed thin film transistor (a)after printing CNT layer, (b) after printing gate electrode and dielectric layer, (c) and (d)measured transfer and I-V characteristics

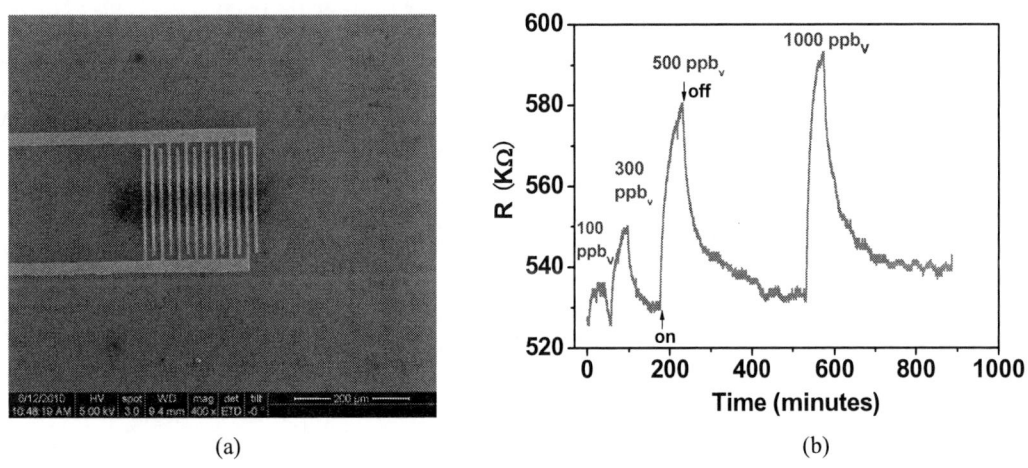

Fig.2 Printed thin film sensor based carbon nanotube inks (a) photo of printed sensor device, (b)measurement of ammonia gas using the printed CNT sensor

Sensing and Noise Characteristics of Si-Nanowire Ion-Sensitive-Field-Effect-Transistors for Future Biosensor Applications

Jeong-Soo Lee

Division of IT-Convergence Engineering POSTECH, Pohang, 790-784, Korea

Si-nanowire biologically-active field effect transistors (Si-NW BioFETs) have been considered recently for biosensing applications because the nanowire structures with the high surface-to-volume ratio are capable of detecting extremely low concentration of biomolecules in solution. Charges from specific binding of biomolecules on the surface of nanowires can affect the channel conductance of the entire volume of nanowires. In order to achieve high performance with low power consumption of the device, understanding of the sensing and noise mechanisms is very important. This work will present the pH sensing characteristics and the low-frequency noise (1/f noise) characteristics of the Si-NW BioFETs. The Si-NW BioFETs have been fabricated using well-established microfabrication technology which enables large wafer processing of sensor devices with high regularity to reduce the production costs. From the carcinoembryonic antigen (CEA) sensing performances, the sensitivity and the resolution of devices have been also discussed. This work is supported by the WCU program through the Korea Science and Engineering Foundation funded by the Ministry of Education, Science, and Technology.

Atomic Resolution Nanofabrication and Dynamic Characterization

Litao Sun

SEU-FEI Nano-Pico Center, Key Laboratory of MEMS of Ministry of Education, Southeast University,
Nanjing 210096, China
E-mail: slt@seu.edu.cn

Experiments on individual nanoparticles are generally difficult but can be carried out by the techniques of modern *in-situ* electron microscopy. With the continuous improvement of *in situ* techniques inside transmission electron microscope (TEM), the capabilities of TEM extend beyond structurual characterization to high-precision nanofabrication and property measurement, which not only enriches the experimental methods of nanoresearch, but also provides new opportunities for the development in nanoscience and nanotechnology. Based on the idea of "setting up a nanolab inside a TEM", we review our recent progress in atomic resolution nanofabrication and dynamic characterization of structure and properties of nanomaterials.

The electron beam can be used as a tool to induce nanofabrication on an atomic scale. Selective irradiation of MWNTs with a focused electron beam can induce changes of their shape such as cutting, bending, welding and drilling a hole on them. In addition, metal crystals can be encapsulated inside graphitic nanocontainers that were designed for *in-situ* electron irradiation experiments. Under irradiation, the carbon shells contract and lead to compressive forces. By measuring the lattice spacings in HRTEM images of metal crystals inside graphitic shells, pressures on the order 10-20 GPa were determined. Non-hydrostatic pressure, e.g., inside collapsing nanotubes, may deform the crystals considerably. This can be used for studying the deformation behaviour of individual nanometer-sized crystals. Fig. 1 shows how a Co wire inside a carbon nanotube is deformed and extruded when the the nanotube collapses locally under electron irradiation. Another deformation cell, allowing the detailed study of crystal deformation, was designed by electron-beam structuring, as shown in Fig. 2. A 'carbon onion', encapsulating a Au crystal, was punctured by a fully focused electron beam and subsequently exposed to uniform irradiation so that the Au crystal was slowly extruded through the hole. The interaction between metal and carbon was also studied on an atomic scale by introducing Au and Pt atoms into graphene layers. Individual Au or Pt atoms were observed by HRTEM at different temperatures and their migration was monitored. Fig. 3 shows drilling a 5nm-diameter hole by electron beam on graphene for third-generation gene sequencing. The mechanism of electron-beam induced high-resolution nanofabrication was also discussed.

References

[1] L. T. Sun, F. Banhart, A.V. Krasheninnikov, J. A. Rodriguez-Manzo, M. Terrones and P.M. Ajayan. *Science*, 312 (2006), pp. 1199-1202.

[2] J. A. Rodríguez-Manzo, M. Terrones, H. Terrones, L. T. Sun, F. Banhart, H. W. Kroto. *Nature Nanotechnology*, 2 (2007), pp. 307-311.

[3] L. T. Sun, A. V. Krasheninnikov, T. Ahlgren, K. Nordlund and F. Banhart. *Phys. Rev. Lett.*, 101 (2008) pp. 156101.

[4] S. Suresh, J. Li, *Nature*, 456 (2008), pp. 716-717.

[5] O. Cretu, A. V. Krasheninnikov, J. A. Rodríguez-Manzo, L. T. Sun, R. M. Nieminen and F. Banhart. *Phys. Rev. Lett.*, 105 (2010), pp. 196102.

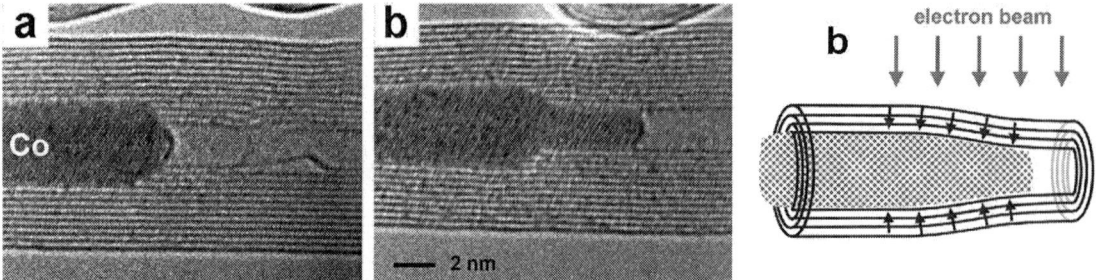

Fig. 1: Electron irradiation of a carbon nanotube encapsulating a Co crystal. The collapse of the tube leads to the extrusion of the metal.

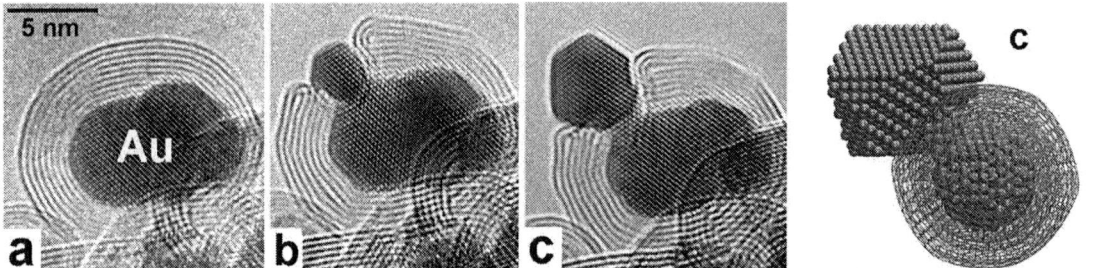

Fig. 2: Extrusion of a Au crystal through a beam-induced hole (top left) from a spherical carbon shell under electron irradiation.

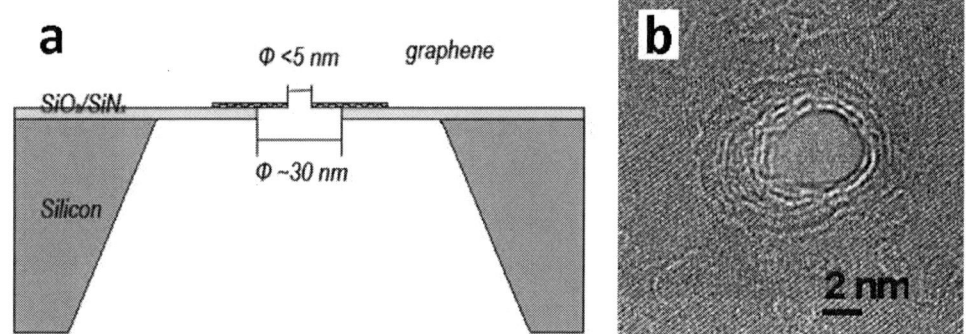

Fig. 3: Drilling a 5nm-diameter hole by electron beam on graphene for third-generation gene sequencing.

Nanopatterning by Phase Change Nanolithography

X. S. Miao, B. J. Zeng, Z. Li, W. L. Zhou
Huazhong University of Science and Technology, Wuhan 430074, CHINA
miaoxs@mail.hust.edu.cn

Several techniques such as the electron beam technology and ion beam technology are employed to obtain the high resolution nanopatterns. However, these techniques cause some problems, for example, the vacuum installation, high voltage power supply and low throughput, which consequently make these techniques more expensive. On the other hand, the organic resists used for lithography are important to control the shape and size of the patterns. The reactivity of the resist is dominated by the total irradiation amounts of a beam, that is, the number of electrons or ions or photons absorbed by the resist. As a result, the precision control of the shape and size for nanopatterns become difficult due to their intrinsic accumulation effect of the absorption in the resists.

In this paper, a heat-mode lithography technology is presented, which calls phase change lithography (PCL). An inorganic phase change material is used as the resist that changes the phase from as-deposited amorphous to crystalline when exposed to laser. The laser beam will induce the structural change by thermal effect. The irradiated and non-irradiated regions can be selectively removed by an etching process. No photon-accumulation effect is observed. The PCL technology is cost-effective, reliable and stable with loose environment requirement.

The structure diagram and photos of PCL system are shown in Fig.1 and Fig.2. The GeSbTe chalcogenide phase change material and metallic glass materials are used as the active materials for PCL. The etching depth of GeSbTe material for PCL is shown in Fig.3. More detailed results including the metallic glass materials will be presented in the meeting.

This work was supported by the grants from the Ph.D Programs Foundation of Ministry of Education of China (No. 20090142110003).

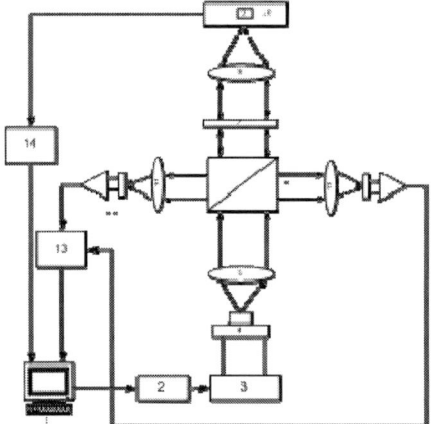

1.Computer. 2, Single-chip processor. 3, Pulse generator. 4, Laser. 5, Collimator 6, polarization beam splitter. 7, quarter wave plate. 8, Focusing object lens. 9,   sample. 10, Step-motor. 11,lens. 12, Laser detector . 13, Oscilloscope. 14, Step-motor controlling box

Fig.1: Structure diagram of phase change lithography system.

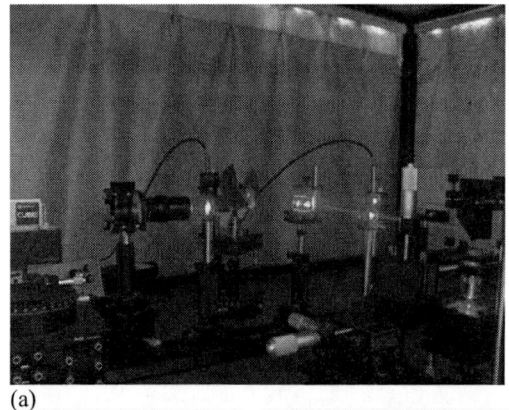

(a)

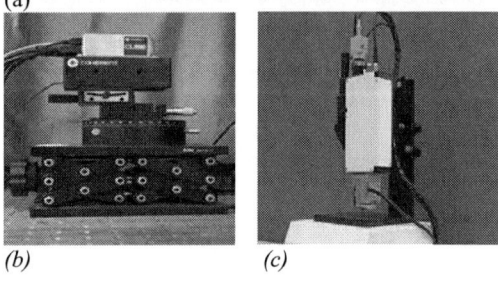

(b)   *(c)*

Fig. 2: Photos of (a)phase change lithography system (b) nano-platform (c)auto focusing system.

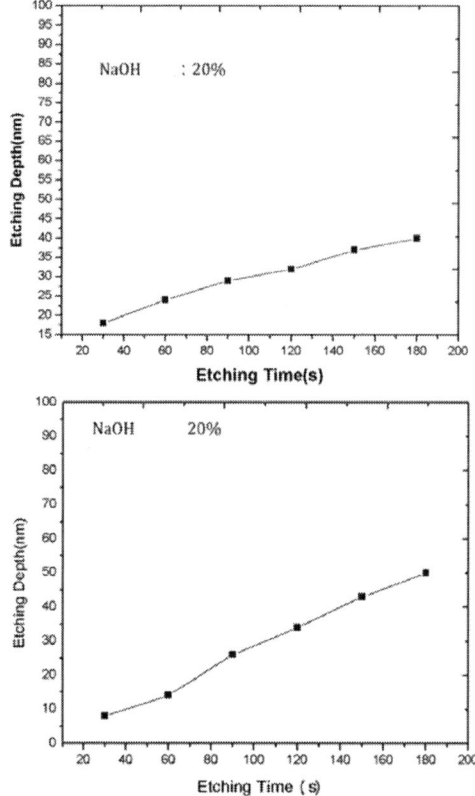

Fig. 3: Etching depth of GeSbTe material for phase change lithography (a) GeSbTe/ZnS-SiO$_2$/Si substrate (b) GeSbTe/Ag/Si substrate

Author Index

A

Aarts, M.	253
Abdulagatov, A.	747
Abelmann, L.	86
Afshar, M.	441
Aiba, Y.	318
Akazawa, A.	266
A iyama, .	751
Amnuaikit, T.	537
Arai, F.	53, 129, 364
Arsenin, A.	631, 742
Atthi, N.	372
Ay, H.	529
Azeta, T.	213

B

Badilita, V.	308
Bao, J.	368
Bercioux, D.	294
Bi, X.-P.	222
Boonruang, S.	372
Burouni, N.	86
Busch, T.	90

C

Chang, C.-H.	487
Chang, C.-J.	453
Chang, D. C.	150
Chang, J. H.-C.	226, 286
Chang, K.-M.	579, 708
Chang, K.-W.	133, 162, 218
Chang, M. N.	553
Chang, S. J.	725
Chang, W.-H.	204
Chang, Y. J.	45
Chang, Y. Y.	519
Chang, Y.-M.	698
Chao, C.	483
Chao, C.-M.	603
Chao, T.-Y.	351, 515
Chatani, K.	69
Chau, L.-H.	195
Cheirsirikul, S.	587
Chen , Y.-S.	191
Chen, B.-H.	175
Chen, C.	515
Chen, C. C.	519
Chen, C.-A.	183
Chen, C.-F.	579, 708
Chen, C.-H.	183, 339
Chen, C.-T.	179, 419
Chen, H.-P.	158, 218
Chen, J.-L.	716
Chen, K.-E.	653

Chen, K.-Y.	653
Chen, L. K.	542
Chen, P. L.	553
Chen, P.-C.	698
Chen, P.-J.	137
Chen, T.	270
Chen, T.-H.	665
Chen, T.-Y.	204
Chen, X.	376
Chen, Y.	230
Chen, Y. C.	515
Chen, Y. F.	644
Chen, Y. J.	125
Chen, Y.-S.	154, 166, 392, 533
Cheng, C.-H.	104, 483
Cheng, C.-Y.	729
Cheng, F.-Y.	200
Cheng, L.	685
Cheng, M. C.	542
Cheng, Y. T.	515
Cheng, Y.-T.	351, 575
Cheung, R.	270
Chiang, C.-H.	487
Chiang, M.-Y.	407
Chien, T.-H.	575
Chiou, J.-C.	733
Chomnawang, N.	112
Chou, L.-C.	733
Chu, C. W.	515
Chu, H.-T.	141
Chu, J.	685
Chu, Y.-H.	120
Chuang, J.-F.	384, 388
Chuang, Y. C.	403
Chuang, Y. J.	380
Chuang, Y.-Y.	141
Chung, H.-Y.	339
Chuo, T.-M.	187
Cole, M. T.	11
Covington, E.	57
Crum, B. P.	222
Cui, J.	738
C i, .	75

D

Dai, S.-M.	640
Dai, T.-H.	533
Dale, C.	15, 627
Deekla, P.	112
Delaunay, J.-J.	331
Dewanto, R.	15, 270
Di, Q.-L.	567
Dong, Z.	547
Du, X.	571, 702

E

Esashi, M.	247
Español, P.	298
Esposito, H.	261

F

Fan, S.-K.	183, 187, 407, 411
	415
Fang, R.	649
Fang, W.	116
Feili, D.	441
Feng, K. J.	437
Feng, T.	7
Fu, C.-C.	230
Fu, Y.	230
Fukuda, T.	368

G

Gallacher, B.	15, 270
Gallacher, B. J.	73
Gao, X.	65
a tsch S.	751
Ge, F.	669, 712
Ghaemmaghami, A. M.	183
Greiner, A.	298, 304
Gu, Z.	423
Guan, X.	108
Guo, L.	611
Guo, S.	649

H

Hagiwara, M.	364
Han, M.	376
Hanein, Y.	298
He, C.	738
Hedley, J.	15, 73, 270, 627
Hibbert, O.	90
Ho, T.-C.	453
Horikawa, M.	290
Hoshino, T.	82
Hruanun, C.	372, 587
Hsiao, F.-L.	326
Hsiao, J. W.	661
Hsieh, C.-T.	579, 708
Hsieh, H. Y.	403
Hsieh, H.-Y.	322
Hsieh, Y.-T.	40
Hsu, L.	162
Hsu, W.	187, 689
Hu , D.	649
Hu, Z.	15, 270
Hu, Z. X.	73
Huang, C.	419
Huang, C. L.	529
Huang, C.-C.	200
Huang, C.-M.	343

Huang, F. Y.	611
Huang, G.-W.	179
Huang, H. L.	725
Huang, J. W.	529
Huang, K.-F.	31
Huang, K.-Y.	449
Huang, S. H.	380
Huang, S.-C.	733
Huang, S.-H.	141
Huang, T.-W.	322
Huang, Y.	230
Huang, Y.-J.	449
Hung, K. Y.	380
Hung, K.-Y.	120
Hung, L.-Y.	200
Hung, T.-Y.	179

I

Ichikawa, A.	364
Ikehara, T.	69, 95
Ishii, K.	27
Ishinabe, H.	615
Ishizawa, M.	213
Itoh, T.	95, 615

J

Jan, F.-J.	204
Jeamsaksiri, W.	372
Jesen, S.	587
Jia, S.	649
Jiang, Y.	702
Jiao, N.	547
Jin, Q.	278
Jin, Y.	649
Jing, E.	274, 469

K

Kang, D.	226, 286
Kang, F.	347
Kauzlarić, D.	298
Kawahara, T.	364
Kawai, Y.	247
Ke, L.-Y.	154, 166, 191, 533
Keegan, N.	15, 73, 627
Kerdcharoen, T.	1
Kim, J. Y.-H.	226
Kobayashi, M.	78
Kobayashi, T.	615
Koenig, K.	441
Kojima, M.	368
Komiyama, M.	318
Konishi, S.	266, 595
Korvink, J. G.	298, 304, 308, 312
Kumagai, S.	78
Kumano, M.	247

Kuo, C. L.	45
Kuo, J.-N.	607, 636
Kuo, Z.-K.	166, 191
Kurdak, C.	57
Kuriki, H.	129
Kuzuya, A.	318
Kwong, D.-L.	99

L

Lähdesmäki, I.	398
Lai, C.-H.	579, 708
Lai, H. W.	125
Lai, Y. W.	557
Lai, Y.-L.	733
Langlah, N.	537
Lee, C.	99, 326
Lee, C. H.	553
Lee, C.-F.	120
Lee, C.-H.	322
Lee, C.-T.	218
Lee, C.-Y.	575, 698
Lee, G.-B.	195, 200, 204
Lee, H.-C.	665
Lee, J. E.-Y.	457, 463, 473, 478
	557
ee, . S.	75
Lee, S.-J.	698
Lee, T.-Y.	453
Lee, Y.-C.	31, 36, 40, 339
Lee, Y.-K.	150, 170
Lee, Y.-M.	575, 698
Lei, H.-Y.	200
Lei, K. F.	208
Lei, T.	282
Leiman, V.	631, 742
Li , P.-C.	204
Li, C.	437, 611
Li, C. C.	725
Li, C.-W.	716
Li, D.	257, 747
Li, L.	27
Li, S.	347
Li, T.	23, 65, 278
Li, W.	57, 222, 673
Li, W. J.	195, 429
Li, X.	7, 238
Li, Y.-T.	603
Li, Z.	7 1, 49, 238, 657
Liang, W.	195
Liba, O.	298
Lienemann, J.	304
Lim, L.	99
Lin, C.	693
Lin, C.-H.	487, 665
Lin, C.-S.	681
Lin, C.-W.	191
Lin, H.-L.	133, 158

Lin, J. C.-H.	137
Lin, J.-R.	351
Lin, K.-R.	487
Lin, L.	669
Lin, L. W.	125
Lin, R.	150
Lin, W.-C.	681
Lin, Y.-B.	355, 729
Lin, Y.-H.	689
Lin, Y.-K.	636
Lin, Y.-T.	355
Lindvall, N.	11, 19
Liow, J.-L.	423
Liu , C.-H.	166
Liu, C. W.	553
Liu, C.-H.	133, 154, 158, 162
	191, 218, 533, 579
	708
Liu, C.-Y.	388
Liu, D.	669
Liu, D. Y.	45
Liu, J.	154, 335, 619
Liu, J.-Q.	496
Liu, W. K.	195, 429
Liu, X.	693
Liu, Y.	49
Liu, Z.	547
Liu, Z. H.	125, 725
Lo, H.	146
Lo, L.-W.	343
Löffelmann, U.	312
oizea , F.	751
Lorwongtragool, P.	1
Lou, L.	99
Lou, W.	257
Lu, G.	359
Lu, J.	591
Luanatikomkul, W.	599
Lv, J.	511
Lv, W.	282, 562, 702

M

Ma, J.	685
Ma, X. Y.	644
Ma, Y.	274, 469
Maeda, E.	331
Maeda, R.	69, 95, 290, 591
	615
Mager, D.	312
Mao, P.	376
Mao, S. W.	725
Mason, A. J.	57
Massoud, Y.	491, 506, 623
Matsuki, T.	331
Mayrhofer, L.	294
McNeil, C. J.	73
Meier, R. C.	308

Meng, B. 721
Miao, M. 649
Miao, X. 7 1, 502
ileti, . 751
Mizuno, J. 27
Mohammed, W. 372
Moosavi, A. 523
Morishima, K. 82
Mu, X. 57
Muševič, I. 61

N

Naitoh, Y. 290
Nakajima, M. 95, 368
Nandra, M. 234
Nguyen, T. A. 595
Nishimura, H. 496
Nobukawa, S. 496
Nose, H. 213

O

Obliosca, J. M. 343
Ohata, T. 290
Ohya, Y. 318
Okamoto, K. 266
Okamoto, N. 78
Onda, K. 129
Ou, Z. Y. 125
Ouyang, M. 195, 429

P

Page, R. H. 627
Palaniapan, M. 326
Pan , C. T. 125, 725
Pan, Y. C. 661
Pan, Y.-N. 681
Pantong, C. 112
Park, W.-T. 99
Parviz, B. A. 398
Pedersen, J. 261
Peng, H.-C. 449
Peng, J. 104
Peng, P. 7
Petremand, . 751
Phatthanakun, R. 112
Phetchakul, T. 599
Pinsuwan, S. 537
Poyai, A. 372, 599
Pummara, W. 112

Q

Qin, X. 677
Qiu, X. 562

R

Rana, S. 627

Ren, T. 7
ooi N.F. 751
othem nd, P. 75
Ryzhii, V. 631, 742

S

Sagawa, T. 591
Sakai, Y. 318
Sakuma, S. 53, 129
Sarajlic, E. 86
Sarro, P. M. 253
Satsanarukkit, P. 146
Seesaard, T. 1
Seidel, H. 441
Semenenko, V. 631, 742
Serrano, G. 57
She, D. D. 437
Shen, C. 347
Shen, S. C. 725
Shen, Y. 376
Shen, Y.-T. 411
Sheu, T. S. 529, 661
Shi, X. 104, 483
Shiao, M. H. 553
Shih, C.-J. 681
Shoji, S. 27
Shyu, J. C. 529, 661, 640
Siekman, M. 86
Singal, V. 491, 506, 623
Sivashankar, S. 133, 158, 162, 218
Smaili, S. 491, 506, 623
Spoors, J. A. 73
Sriphung, C. 112
Sta fer, . 751
Stebunov, Y. 631, 742
Štefane, B. 61
Straessle, . 751
Straub, M. 441
Strle, D. 61
Su, C. Y. 689
Su, J. 553
Su, Y. 571
Succi, S. 298
Sugita, M. 53
Sun, D. 282, 562, 571, 702
Sun, F.-W. 335
Sun, G. C. 644
Sun, J. 11, 19
Sun, L. 75 , 376, 712
Sun, M. 359
Sun, X. 238
Sung, C. K. 542
Syunori, A. 266

T

Tai, Y.-C. 137, 146, 226, 234
286
Tang, A. S. 644
Tang, M.-S. 698
Tang, S. 230

Tang, W.	721
Tang, X. S.	611
Tao, W.	562
Tas, N.	86
Teh , K. S.	261
Tehrani, A.	523
Teng, D.	415
Teo, K. B. K.	11
Thomas, N.	398
Tian, H.	7
Tian, Y.	437
Tongta, C.	1
Tsai, L.-P.	411
Tsai, S.-W.	31
Tsai, Y.-C.	200
Tsai, Y.-W.	120
Tseng, F.-G.	322, 343, 403, 449
	453
Tseng, H.-W.	166, 191
Tseng, J. K.	725
Tseng, W.-L.	665
Tsutsui, Y.	27
Tu, C.	457, 478
Tung, S.	90
Tushar Harishchandra, P.	218
Tzeng, Y.-L.	433, 603

U

Ukita, Y.	213
Umehara, H.	496
Utsumi, Y.	213

V

Valagerahally Puttaswamy, S.	
	133, 158, 162, 218
Voellm, H.	441

W

Wallrabe, U.	308
Wang, C.-C.	640
Wang, C.-H.	204
Wang, D. F.	69, 95, 290, 591
	615
Wang, G.	415
Wang, G.-J.	392, 716
Wang, H.-S.	449
Wang, H.-Y.	419
Wang, J.	347, 583
Wang, K.	355, 384, 388, 433
	603, 653, 729
Wang, L.	483, 562, 571, 619
	702
Wang, N.	247, 326
Wang, P.-C.	322, 343, 453
Wang, S.	657

Wang, S.-F.	355
Wang, T.-H.	243
Wang, W.	23, 49, 170, 657
Wang, X.	282, 347
Wang, Y.	23, 65, 274, 278
	359, 469
Wang, Y. F.	437
Wang, Y.-B.	579, 708
Wang, Z.	547, 673, 677
Wang, Z. R.	721
Ward, N.	57
Ward, N. L.	222
Waugh, W.	73
Wei, J.	253
Wei, M. C.	380
Weiß, D.	304
Wen, J.-H.	116
Wen, L.	685
Weng, C. J.	689
Wong, C.P.	755
Wong, K. W.	429
Wu, C.-C.	716
Wu, C.-N.	579, 708
Wu, C.-W.	133, 141
Wu, C.-Y.	339
Wu, G.	274, 469
Wu, W.	49
Wu, W. G.	437, 611, 644
Wu, Y. C.	511
Wu, Z.	108

X

Xiao, J.-L.	322
Xie, D.	7
Xiong , J.-J.	335
Xiong, B.	274, 469
Xiong, J.-J.	619
Xu, B.	282, 702
Xu, D.	274, 469
Xu, G. Q.	511
Xu, L.	65
Xu, Y.	318, 463, 473
Xue, C.-Y.	335, 619

Y

Yamada, I.	331
Yamanaka, Y.	318
Yamanishi, T.	496
Yamanishi, Y.	129
Yamashita, I.	78
Yamazaki, T.	318
Yamwong, W.	599
Yan , Y.-Z.	335, 619
Yan, G.	583, 669, 712, 738
Yan, J.-N.	36
Yan, S.-B.	335, 619
Yang, C.-S.	322

Yang, F.	747
Yang, K.-S.	640
Yang, L.-J.	175, 243
Yang, S.-M.	158, 162
Yang, S.-Y.	204
Yang, T.	575
Yang, W.	230
Yang, Y.	7
Yang, Z.	368, 583, 669, 712
	738
Yasuda, K.	318
Yeh, C.-S.	200
Yeh, C.-T.	133, 158
Yoshida, S.	247
oshi awa, .	751
Yu, C.-C.	234
Yu, F.	673
Yu, H.	657
Yu, M.	238
Yu, T.-C.	388
Yu, X.	278
Yuan, Q.	238
Yuan, Z.	685
Yueh, T.-R.	218
Yurgens, A.	11, 19

Z

Zellers, E. T.	57
eng, . .	7 1
Zeng, X.-J.	567
hao, .	75
Zhan, Y.-H.	607
Zhan, Z.	562, 571, 702
Zhang, D.	693
Zhang, D. C.	747
Zhang, H.-X.	49, 238, 445, 567
	721
Zhang, J.	108, 519, 673
Zhang, S.	99
hang, .	75
Zhang, W.-D.	335, 619
Zhang, X.-S.	445, 567
Zhang, Y.-G.	619
Zhao, Q.	359, 738
Zhao, W.	7 1, 502
Zhao, X.	359, 567
Zhao, Y.	137, 234, 257, 282
Zhao, Z.	108
Zheng, C.	376
Zheng, G.	282
Zheng, Q.	403
Zheng, Y.	238, 483
Zheng, Z. X.	511
Zhou, C.	7
Zhou, K.	712
Zhou, W.	502
Zhu, F.-Y.	445, 567
Zhu, H.	7, 457, 473
Zou, C.-L.	335
Zuo, W.	562, 571